LANGE'S
HANDBOOK
OF
CHEMISTRY

LANGE'S HANDBOOK OF CHEMISTRY

John A. Dean
Professor Emeritus of Chemistry
University of Tennessee, Knoxville

Fourteenth Edition

McGRAW-HILL, INC.
New York   St. Louis   San Francisco   Auckland   Bogotá
Caracas   Lisbon   London   Madrid   Mexico   Milan
Montreal   New Delhi   Paris   San Juan   São Paulo
Singapore   Sydney   Tokyo   Toronto

Library of Congress Catalog Card Number 84-643191

ISSN 0748-4585

4 5 6 7 8 9 0   DOC/DOC   9 8 7 6 5 4 3

ISBN 0-07-016194-1

The sponsoring editor for this book was Gail F. Nalven, the editing supervisor was Stephen M. Smith, and the production supervisor was Pamela A. Pelton. It was set in Times Roman by Progressive Typographers.

Printed and bound by R. R. Donnelley & Sons Company.

CONTENTS

For the detailed contents of any section, consult the title page of that section. See also the alphabetical index in the back of this handbook.

PREFACE TO FOURTEENTH EDITION

Perhaps it would be simplest to begin by stating the ways in which this new edition, the fourth under the aegis of the present editor, has *not* been changed. It remains the one-volume source of factual information for chemists, both professionals and students—the first place in which to "look it up" on the spot. The aim is to provide sufficient data to satisfy all one's general needs without recourse to other reference sources. Even the worker with the facilities of a comprehensive library will find this volume of value as a time-saver because of the many tables of numerical data which have been especially compiled.

The changes, however, are both numerous and significant. First of all, there is a change in the organization of the subject matter. For example, material formerly contained in the section entitled Analytical Chemistry is now grouped by operational categories: spectroscopy; electrolytes, electromotive force, and chemical equilibrium; and practical laboratory information. Polymers, rubbers, fats, oils, and waxes constitute a large independent section.

Descriptive properties for a basic group of approximately 4000 organic compounds are compiled in Section 1. These follow a concise introduction to organic nomenclature, including the topic of stereochemistry. Nomenclature is consistent with the 1979 rules of the Commission on Nomenclature, International Union of Pure and Applied Chemistry (IUPAC). All entries are listed alphabetically according to the senior prefix of the name. The data for each organic compound include (where available) name, structural formula, formula weight, Beilstein reference, density, refractive index, melting point, boiling point, flash point, and solubility (citing numerical values if known) in water and various common organic solvents. Structural formulas either too complex or too ambiguous to be rendered as line formulas are grouped at the bottom of the page on which the entries appear. Alternative names, as well as trivial names of long-standing usage, are listed in their respective alphabetical order at the bottom of each page in the regular alphabetical sequence. Another feature that assists the user in locating a desired entry is the empirical formula index.

Section 2 combines the former separate section on Mathematics with the material involving General Information and Conversion Tables. The fundamental physical constants reflect values recommended in 1986. Physical and chemical symbols and definitions have undergone extensive revision and expansion. Presented in 14 categories, the entries follow recommendations published in 1988 by the IUPAC. The table of abbreviations and standard letter symbols provides, in a sense, an alphabetical index to the foregoing tables. The table of conversion factors has been modified in view of recent data and inclusion of SI units; cross-entries for "archaic" or unusual entries have been curtailed.

Descriptive properties for a basic group of approximately 1400 inorganic compounds are compiled in Section 3. These follow a concise, revised introduction to inorganic nomenclature that follows the recommendations of the IUPAC published in 1990. In this section are given the exact atomic (or formula) weight of the elements accompanied, when available, by the uncertainty in the final figure given in parentheses.

In Section 4 the data on bond lengths and strengths have been vastly increased so as to include not only the atomic and effective ionic radii of elements and the covalent radii for atoms, but also the bond lengths between carbon and other elements and between elements other than carbon. All lengths are given in picometers (SI unit). Effective ionic radii are tabulated as a function of ion charge and coordination number. Bond dissociation energies are given in kilojoules per mole with the uncertainty of the final figure(s) given in parentheses when known. New tables include bond dipole moments, group dipole moments, work functions of the elements, and relative abundances of the naturally occurring elements. The table of nuclides has been shortened and includes only the more commonly encountered nuclides; tabulations list half-life, natural abundance, cross-section to thermal neutrons, and radiation emitted upon disintegration. Entries have been updated.

Revised material in Section 5 includes an extensive tabulation of binary and ternary azeotropes comprising approximately 850 entries. Over 975 compounds have values listed for viscosity, dielectric constant, dipole moment, and surface tension. Whenever possible, data for viscosity and dielectric constant are provided at two temperatures to permit interpolation for intermediate temperatures and also to permit limited extrapolation of the data. The dipole moments are often listed for different physical states. Values for surface tension can be calculated over a range of temperatures from two constants that can be fitted into a linear equation. Also extensively revised and expanded are the properties of combustible mixtures in air. A table of triple points has been added.

The tables in Section 6 contain values of the enthalpy and Gibbs energy of formation, entropy, and heat capacity at five temperatures for approximately 2000 organic compounds and 1500 inorganic compounds, many in more than one physical state. Separate tabulations have enthalpies of melting, vaporization, transition, and sublimation for organic and inorganic compounds. All values are given in SI units (joule) and have been extracted from the latest sources such as *JANAF Thermochemical Tables,* 3d ed. (1986); *Thermochemical Data of Organic Compounds,* 2d ed. (1986); and *Enthalpies of Vaporization of Organic Compounds,* published under the auspices of the IUPAC (1985). Also updated is the material on critical properties of elements and compounds.

The section on Spectroscopy has been expanded to include ultraviolet-visible spectroscopy, fluorescence, Raman spectroscopy, and mass spectrometry. Retained sections have been thoroughly revised: in particular, the tables on electronic emission and atomic absorption spectroscopy, nuclear magnetic resonance, and infrared spectroscopy. Detection limits are listed for the elements when using flame emission, flame atomic absorption, electrothermal atomic absorption, argon ICP, and flame atomic fluorescence. Nuclear magnetic resonance embraces tables for the nuclear properties of the elements, proton chemical shifts and coupling constants, and similar material for carbon-13, boron-11, nitrogen-15, fluorine-19, silicon-29, and phosphorus-31.

Section 8 now combines all the material on electrolytes, electromotive force, and chemical equilibrium, some of which had formerly been included in the old "Analytical Chemistry" section of earlier editions. Material on the half-wave potentials of inorganic and organic materials has been thoroughly revised. The tabulation of the potentials of the elements and their compounds reflects recent IUPAC (1985) recommendations.

An extensive new Section 10 is devoted to polymers, rubbers, fats, oils, and waxes. A discussion of polymers and rubbers is followed by the formulas and key properties of plastic materials. For each member and type of the plastic families there is a tabulation of their physical, electrical, mechanical, and thermal properties and characteristics. A similar treatment is accorded the various types of rubber materials. Chemical resistance and gas permeability constants are also given for rubbers and plastics. The section concludes with various constants of fats, oils, and waxes.

The practical laboratory information contained in Section 11 has been gathered from many of the previous sections of earlier editions. This material has been supplemented with new material under separation methods, gravimetric and volumetric analysis, and laboratory solutions. Significant new tables under separation methods include: properties of solvents for chromatography, solvents having the same refractive index and the same density, McReynolds' constants for stationary phases in gas chromatography, characteristics of selected supercritical fluids, and typical performances in HPLC for various operating conditions. Under gravimetric and volumetric analysis, gravimetric factors, equations and equivalents for volumetric analysis, and titrimetric factors have been retained along with the formation constants of EDTA metal complexes. In this age of awareness of chemical dangers, tables have been added for some common reactive and incompatible chemicals, chemicals recommended for refrigerated storage, and chemicals which polymerize or decompose on extended storage at low temperature. Updated is the information about the U.S. Standard Sieve Series. Thermometry data have been revised to bring them into agreement with the new International Temperature Scale – 1990, and data for type N thermocouples are included.

Every effort has been made to select the most useful and most reliable information and to record it with accuracy. However, the editor's many years of involvement with handbooks bring a realization of the opportunities for gremlins to exert their inevitable mischief. It is hoped that users of this handbook will offer suggestions of material that might be included in, or even excluded from, future editions and call attention to errors. These communications should be directed to the editor at his home address (or by telephone).

201 Mayflower Drive   *John A. Dean*
Knoxville, TN 37920-5871
(615) 573-1602

PREFACE TO
THIRTEENTH EDITION

In this edition, the third under the aegis of the present editor, the large section devoted to the general description of 7600 organic compounds has been thoroughly revised. Nomenclature is now consistent with the 1979 rules of the Commission on Nomenclature, International Union of Pure and Applied Chemistry. A synopsis of the extensive nomenclature rules precedes the tabulation. All entries are listed alphabetically according to the senior prefix of the name rather than by indexing according to the Chemical Abstracts system. With the latter system there may be a bewildering array of subordinate entries listed under a key index name. The data for each organic compound include: name, structural formula, formula weight, Beilstein reference, density, refractive index, melting point, boiling point, flash point (introduced for the first time), and solubility in water and various organic solvents. Structural formulas are drawn for compounds either too complex or ambiguous to render by line formulas; these are grouped at the bottom of the same page on which the entry appears rather than being gathered together in a remote and separate listing. Many compounds will possess more than one approved name. These alternative names, as well as trivial names in long-standing usage, are listed in their respective alphabetical order at the bottom of each main page in the regular alphabetical sequence. Another aid to assist the user in locating a desired entry is the empirical formula index.

Expanded coverage is given to the areas of:

pK_a values of organic acids

Temperature dependence of selected values of pK_a and pK_{sp} in water

Properties of combustible mixtures; in particular, the autoignition temperature and the flammable limits in percent by volume, upper and lower limits

The section on thermodynamic properties has been revised to reflect the latest recommended values for heats of formation and Gibbs energies of formation, entropies, and heat capacities for the members of the alkali family and the compounds of uranium, protactinium, thorium, and actinium. These data, plus heats of melting, vaporization, and sublimation, are gathered into two sets of two tables each, one for 2400 inorganic compounds and the other for 1500 organic compounds. The editor feels that related properties are thus more readily available to the user than if they were scattered over several separate tabulations.

Offered for the first time are carbon-13 NMR data involving chemical shifts and spin-spin coupling constants. This addition recognizes the increased role played by carbon-13 NMR in the elucidation of chemical structures.

In response to user requests, gravimetric conversion factors, equations and equivalents for volumetric analyses, standard volumetric solutions, and volumetric factors have been restored and updated in this edition.

The mathematical section has been restructured to exclude the tables for logarithms and trigonometric functions, data now easily obtained with the ubiquitous hand calculator.

This Handbook still remains the only one-volume source for extensive entries involving solubility products, the estimation of vapor pressure at various temperatures for inorganic and organic compounds—Antoine equation data, formation constants of metal complexes both inorganic and organic, equivalent conductance, critical volumes among the other critical data of temperature and pressure, and Hammett and Taft substituent constants.

Grateful acknowledgment is extended to Mr. L. P. Buseth (Norway) who consented to revise the table of conversion factors in Section 2.

It is hoped that users of this edition will continue to offer friendly criticism and suggestions and call attention to errors. These communications should be directed to the editor at his home address.

201 Mayflower Drive   *John A. Dean*
Knoxville, TN 37920

PREFACE TO
FIRST EDITION

This book is the result of a number of years' experience in the compiling and editing of data useful to chemists. In it an effort has been made to select material to meet the needs of chemists who cannot command the unlimited time available to the research specialist, or who lack the facilities of a large technical library which so often is not conveniently located at many manufacturing centers. If the information contained herein serves this purpose, the compiler will feel that he has accomplished a worthy task. Even the worker with the facilities of a comprehensive library may find this volume of value as a time-saver because of the many tables of numerical data which have been especially computed for this purpose.

Every effort has been made to select the most reliable information and to record it with accuracy. Many years of occupation with this type of work bring a realization of the opportunities for the occurrence of errors, and while every endeavor has been made to prevent them, yet it would be remarkable if the attempts towards this end had always been successful. In this connection it is desired to express appreciation to those who in the past have called attention to errors, and it will be appreciated if this be done again with the present compilation for the publishers have given their assurance that no expense will be spared in making the necessary changes in subsequent printings.

It has been aimed to produce a compilation complete within the limits set by the economy of available space. One difficulty always at hand to the compiler of such a book is that he must decide what data are to be excluded in order to keep the volume from becoming unwieldy because of its size. He can hardly be expected to have an expert's knowledge of all branches of the science nor the intuition necessary to decide in all cases which particular value to record, especially when many differing values are given in the literature for the same constant. If the expert in a particular field will judge the usefulness of this book by the data which it supplies to him from fields other than his specialty and not by the lack of highly specialized information in which only he and his co-workers are interested (and with which he is familiar and for which he would never have occasion to consult this compilation), then an estimate of its value to him will be apparent. However, if such specialists will call attention to missing data with which they are familiar and which they believe others less specialized will also need, then works of this type can be improved in succeeding editions.

Many of the gaps in this volume are caused by the lack of such information in the literature. It is hoped that to one of the most important classes of workers in chemistry, namely the teachers, the book will be of value not only as an aid in answering the most varied questions with which they are confronted by interested students, but also as an inspiration through what it suggests by the gaps and inconsistencies, challenging as they do the incentive to engage in the creative and experimental work necessary to supply the missing information.

While the principal value of the book is for the professional chemist or student of chemistry, it should also be of value to many people not especially educated as chemists. Workers in the natural sciences—physicists, mineralogists, biologists, pharmacists, engineers, patent attorneys, and librarians—are often called upon to solve problems dealing with the properties of chemical products or materials of construction. For such needs this compilation supplies helpful information and will serve not only as an economical substitute for the costly accumulation of a large library of monographs on specialized subjects, but also as a means of conserving the time required to search for information so widely scattered throughout the literature. For this reason especial care has been taken in compiling a comprehensive index and in furnishing cross references with many of the tables.

It is hoped that this book will be of the same usefulness to the worker in science as is the dictionary to the worker in literature, and that its resting place will be on the desk rather than on the bookshelf.

Cleveland, Ohio   *N. A. Lange*
May 2, 1934

ACKNOWLEDGMENTS

Grateful acknowledgment is hereby made of an indebtedness to those who have contributed to previous editions and whose compilations continue in use in this edition. In particular, acknowledgment is made of the contribution of L. P. Buseth, who prepared the conversion tables and offered revisions for this edition and who prepared the table on the U.S. Standard Sieve Series.

LANGE'S
HANDBOOK
OF
CHEMISTRY

SECTION 1
ORGANIC COMPOUNDS

1.1 NOMENCLATURE OF ORGANIC COMPOUNDS

The following synopsis of rules for naming organic compounds and the examples given in explanation are not intended to cover all the possible cases. For a more comprehensive and detailed description, see J. Rigaudy and S. P. Klesney, *Nomenclature of Organic Chemistry,* Sections A, B, C, D, E, F, and H, Pergamon Press, Oxford, 1979. This publication contains the recommendations of the Commission on Nomenclature of Organic Chemistry and was prepared under the auspices of the International Union of Pure and Applied Chemistry (IUPAC).

1.1.1 Nonfunctional Compounds

1.1.1.1 Alkanes. The saturated open-chain (acyclic) hydrocarbons (C_nH_{2n+2}) have names ending in -ane. The first four members have the trivial names *methane* (CH_4), *ethane* (CH_3CH_3 or C_2H_6), *propane* (C_3H_8), and *butane* (C_4H_{10}). For the remainder of the alkanes, the first portion of the name is derived from the Greek prefix (see Table 2.4) that cites the number of carbons in the alkane followed by -ane with elision of the terminal -a from the prefix, as shown in Table 1.1.

TABLE 1.1   Names of Straight-Chain Alkanes

n^*	Name	n^*	Name	n^*	Name	n^*	Name
1	Methane	11	Undecane‡	21	Henicosane	60	Hexacontane
2	Ethane	12	Dodecane	22	Docosane	70	Heptacontane
3	Propane	13	Tridecane	23	Tricosane	80	Octacontane
4	Butane	14	Tetradecane			90	Nonacontane
5	Pentane	15	Pentadecane	30	Triacontane	100	Hectane
6	Hexane	16	Hexadecane	31	Hentriacontane	110	Decahectane
7	Heptane	17	Heptadecane	32	Dotriacontane	120	Icosahectane
8	Octane	18	Octadecane			121	Henicosahectane
9	Nonane†	19	Nonadecane	40	Tetracontane		
10	Decane	20	Icosane§	50	Pentacontane		

* n = total number of carbon atoms.
† Formerly called enneane.
‡ Formerly called hendecane.
§ Formerly called eicosane.

For branching compounds, the parent structure is the longest continuous chain present in the compound. Consider the compound to have been derived from this structure by replacement of hydrogen by various alkyl groups. Arabic number prefixes indicate the carbon to which the alkyl group is attached. Start numbering at whichever end of the parent structure that results in the lowest-numbered locants. The arabic prefixes are listed in numerical sequence, separated from each other by commas and from the remainder of the name by a hyphen.

If the same alkyl group occurs more than once as a side chain, this is indicated by the prefixes di-, tri-, tetra-, etc. Side chains are cited in alphabetical order (before insertion of any multiplying prefix). The name of a complex radical (side chain) is considered to begin with the first letter of its complete name. Where names of complex radicals are composed of identical words, priority for citation is given to that radical which contains the lowest-numbered locant at the first cited point of difference in the radical. If two or more side chains are in equivalent positions, the one to be assigned the lowest-numbered locant is that cited first in the name. The complete expression for the side chain may be enclosed in parentheses for clarity or the carbon atoms in side chains may be indicated by primed locants.

If hydrocarbon chains of equal length are competing for selection as the parent, the choice goes in descending order to (1) the chain that has the greatest number of side chains, (2) the chain whose side chains have the lowest-numbered locants, (3) the chain having the greatest number of carbon atoms in the smaller side chains, or (4) the chain having the least-branched side chains.

These trivial names may be used for the unsubstituted hydrocarbon only:

Isobutane	$(CH_3)_2CHCH_3$	Neopentane	$(CH_3)_4C$
Isopentane	$(CH_3)_2CHCH_2CH_3$	Isohexane	$(CH_3)_2CHCH_2CH_2CH_3$

Univalent radicals derived from saturated unbranched alkanes by removal of hydrogen from a terminal carbon atom are named by adding -yl in place of -ane to the stem name. Thus the alkane *ethane* becomes the radical *ethyl*. These exceptions are permitted for unsubstituted radicals only:

Isopropyl	$(CH_3)_2CH—$	Isopentyl	$(CH_3)_2CHCH_2CH_2—$
Isobutyl	$(CH_3)_2CHCH_2—$	Neopentyl	$(CH_3)_3CCH_2—$
sec-Butyl	$CH_3CH_2CH(CH_3)—$	*tert*-Pentyl	$CH_3CH_2C(CH_3)_2—$
tert-Butyl	$(CH_3)_3C—$	Isohexyl	$(CH_3)_2CHCH_2CH_2CH_2—$

Note the usage of the prefixes iso-, neo-, *sec*-, and *tert*-, and note when italics are employed. Italicized prefixes are never involved in alphabetization, except among themselves; thus *sec*-butyl would precede isobutyl, isohexyl would precede isopropyl, and *sec*-butyl would precede *tert*-butyl.

Examples of alkane nomenclature are

$$\overset{4}{C}H_3—\overset{3}{C}H_2—\overset{2}{C}H—\overset{1}{C}H_3 \quad \text{2-Methylbutane (or the trivial name, isopentane)}$$
$$\underset{\underset{CH_3}{|}}{}$$

$$\overset{5}{C}H_3—\overset{4}{C}H_2—\overset{3}{C}H—CH_3 \quad \text{3-Methylpentane (not 2-ethylbutane)}$$
$$\underset{\underset{2}{C}H_2—\underset{1}{C}H_3}{|}$$

5-Ethyl-2,2-dimethyloctane (note cited order)

3-Ethyl-6-methyloctane (note locants reversed)

4,4-Bis(1,1-dimethylethyl)-2-methyloctane
4,4-Bis-1′,1′-dimethylethyl-2-methyloctane
4,4-Bis(*tert*-butyl)-2-methyloctane

Bivalent radicals derived from saturated unbranched alkanes by removal of two hydrogen atoms are named as follows: (1) If both free bonds are on the same carbon atom, the ending -ane of the hydrocarbon is replaced with -ylidene. However, for the first member of the alkanes it is methylene rather than methylidene. Isopropylidene, *sec*-butylidene, and neopentylidene may be used for the unsubstituted group only. (2) If the two free bonds are on different carbon atoms, the straight-chain group terminating in these two carbon atoms is named by citing the number of methylene groups comprising the chain. Other carbon groups are named as substituents. Ethylene is used rather than dimethylene for the first member of the series, and propylene is retained for $CH_3—CH—CH_2—$ (but trimethylene is $—CH_2—CH_2—CH_2—$).

Trivalent groups derived by the removal of three hydrogen atoms from the same carbon are named by replacing the ending -ane of the parent hydrocarbon with -ylidyne.

1.1.1.2 Alkenes and Alkynes.

Each name of the corresponding saturated hydrocarbon is converted to the corresponding alkene by changing the ending -ane to -ene. For alkynes the ending is -yne. With more than one double (or triple) bond, the endings are -adiene, -atriene, etc. (or -adiyne, -atriyne, etc.). The position of the double (or triple) bond in the parent chain is indicated by a locant obtained by numbering from the end of the chain nearest the double (or triple) bond; thus $CH_3CH_2CH=CH_2$ is 1-butene and $CH_3C≡CCH_3$ is 2-butyne.

For multiple unsaturated bonds, the chain is so numbered as to give the lowest possible locants to the unsaturated bonds. When there is a choice in numbering, the double bonds are given the lowest locants, and the alkene is cited before the alkyne where both occur in the name. Examples:

$CH_3CH_2CH_2CH_2CH=CH—CH=CH_2$ 1,3-Octadiene

$CH_2=CHC≡CCH=CH_2$ 1,5-Hexadiene-3-yne

$CH_3CH=CHCH_2C≡CH$ 4-Hexen-1-yne

$CH≡CCH_2CH=CH_2$ 1-Penten-4-yne

Unsaturated branched acyclic hydrocarbons are named as derivatives of the chain that contains the maximum number of double and/or triple bonds. When a choice exists, priority goes in sequence to (1) the chain with the greatest number of carbon atoms and (2) the chain containing the maximum number of double bonds.

These nonsystematic names are retained:

Ethylene $CH_2=CH_2$

Allene $CH_2=C=CH_2$

Acetylene $HC≡CH$

An example of nomenclature for alkenes and alkynes is

$$HC≡\overset{5}{C}—\underset{4}{\overset{CH_2—CH_2—CH_3}{\underset{|}{\overset{|}{C}}}}=\underset{|}{\overset{3}{C}}—\overset{2}{CH}=\overset{1}{CH_2}}$$

$$\underset{CH=CH_2}{}$$

4-Propyl-3-vinyl-1,3-hexadien-5-yne

Univalent radicals have the endings -enyl, -ynyl, -dienyl, -diynyl, etc. When necessary, the positions of the double and triple bonds are indicated by locants, with the carbon atom with the free valence numbered as 1. Examples:

$CH_2=CH—CH_2—$ 2-Propenyl

$CH_3—C≡C—$ 1-Propynyl

$CH_3—C≡C—CH_2CH=CH_2—$ 1-Hexen-4-ynyl

These names are retained:

Vinyl (for ethenyl) CH_2=CH—

Allyl (for 2-propenyl) CH_2=CH—CH_2—

Isopropenyl (for 1-methylvinyl but for unsubstituted radical only) CH_2=$C(CH_3)$—

Should there be a choice for the fundamental straight chain of a radical, that chain is selected which contains (1) the maximum number of double and triple bonds, (2) the largest number of carbon atoms, and (3) the largest number of double bonds. These are in descending priority.

Bivalent radicals derived from unbranched alkenes, alkadienes, and alkynes by removing a hydrogen atom from each of the terminal carbon atoms are named by replacing the endings -ene, -diene, and -yne by -enylene, -dienylene, and -ynylene, respectively. Positions of double and triple bonds are indicated by numbers when necessary. The name *vinylene* instead of ethenylene is retained for —CH=CH—.

1.1.1.3 *Monocyclic Aliphatic Hydrocarbons.*
Monocyclic aliphatic hydrocarbons (with no side chains) are named by prefixing cyclo- to the name of the corresponding open-chain hydrocarbon having the same number of carbon atoms as the ring. Radicals are formed as with the alkanes, alkenes, and alkynes. Examples:

Cyclohexane   Cyclohexyl- (for the radical)

Cyclohexene   1-Cyclohexenyl- (for the radical with the free valence at carbon 1)

1,3-Cyclohexandiene   Cyclohexadienyl- (the unsaturated carbons are given numbers as low as possible, numbering from the carbon atom with the free valence given the number 1)

For convenience, aliphatic rings are often represented by simple geometric figures: a triangle for cyclopropane, a square for cyclobutane, a pentagon for cyclopentane, a hexagon (as illustrated) for cyclohexane, etc. It is understood that two hydrogen atoms are located at each corner of the figure unless some other group is indicated for one or both.

1.1.1.4 *Monocyclic Aromatic Compounds.*
Except for six retained names, all monocyclic substituted aromatic hydrocarbons are named systematically as derivatives of benzene. Moreover, if the substituent introduced into a compound with a retained trivial name is identical with one already present in that compound, the compound is named as a derivative of benzene. These names are retained:

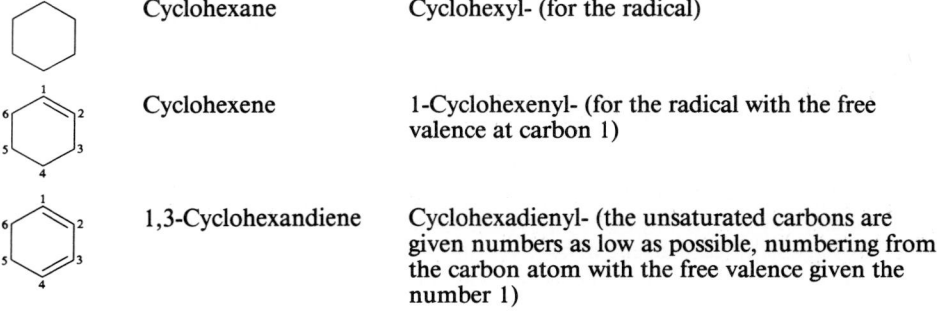

Cumene   Cymene (all three   Mesitylene
                 forms; *para-* shown)

Styrene   Toluene   Xylene (all three
                                  forms; *meta-* shown)

The position of substituents is indicated by numbers, with the lowest locant possible given to substituents. When a name is based on a recognized trivial name, priority for lowest-numbered locants is given to substituents implied by the trivial name. When only two substituents are present on a benzene ring, their position may be indicated by *o-* (*ortho-*), *m-* (*meta-*), and *p-* (*para-*) (and alphabetized in the order given) used in place of 1,2-, 1,3-, and 1,4-, respectively.

Radicals derived from monocyclic substituted aromatic hydrocarbons and having the free valence at a ring atom (numbered 1) are named phenyl (for benzene as parent, since benzyl is used for the radical $C_6H_5CH_2$—), cumenyl, mesityl, tolyl, and xylyl. All other radicals are named as substitued phenyl radicals. For radicals having a single free valence in the side chain, these trivial names are retained:

Benzyl	$C_6H_5CH_2$—	Phenethyl	$C_6H_5CH_2CH_2$—
Benzhydryl (alternative to		Styryl	$C_6H_5CH=CH$—
diphenylmethyl)	$(C_6H_5)_2CH$—	Trityl	$(C_6H_5)_3C$—
Cinnamyl	$C_6H_5CH=CH—CH_2$—		

Otherwise, radicals having the free valence(s) in the side chain are named in accordance with the rules for alkanes, alkenes, or alkynes.

The name *phenylene* (*o-, m-,* or *p-*) is retained for the radical —C_6H_4—. Bivalent radicals formed from substituted benzene derivatives and having the free valences at ring atoms are named as substituted phenylene radicals, with the carbon atoms having the free valences being numbered 1,2-, 1,3-, or 1,4-, as appropriate.

Radicals having three or more free valences are named by adding the suffixes -triyl, -tetrayl, etc. to the systematic name of the corresponding hydrocarbon.

1.1.1.5   *Fused Polycyclic Hydrocarbons.*

The names of polycyclic hydrocarbons containing the maximum number of conjugated double bonds end in -ene. Here the ending does not denote one double bond. Names of hydi   irbons containing five or more fixed benzene rings in a linear arrangement are formed from a numerical prefix (see Table 2.4) followed by -acene. A partial list of the names of polycyclic hydrocarbons is given in Table 1.2. Many names are trivial.

Numbering of each ring system is fixed, as shown in Table 1.2, but it follows a systematic pattern. The individual rings of each system are oriented so that the greatest number of rings are (1) in a horizontal row and (2) the maximum number of rings are above and to the right (upper-right quadrant) of the horizontal row. When two orientations meet these requirements, the one is chosen that has the fewest rings in the lower-left quadrant. Numbering proceeds in a clockwise direction, commencing with the carbon atom not engaged in ring fusion that lies in the most counterclockwise position of the uppermost ring (upper-right quadrant); omit atoms common to two or more rings. Atoms common to two or more rings are designated by adding lowercase roman letters to the number of the position immediately preceding. Interior atoms follow the highest number, taking a clockwise sequence wherever there is a choice. Anthracene and phenanthrene are two exceptions to the rule on numbering. Two examples of numbering follow:

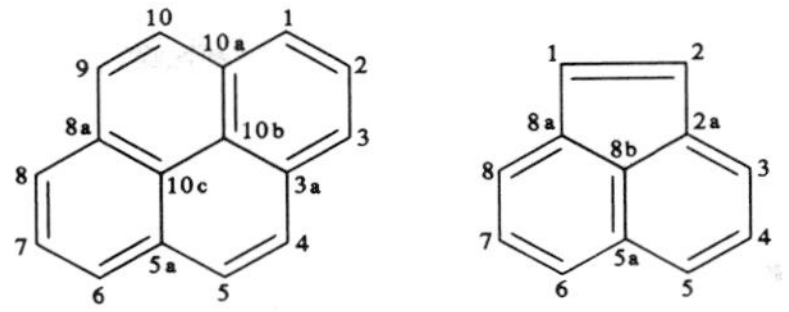

When a ring system with the maximum number of conjugated double bonds can exist in two or more forms differing only in the position of an "extra" hydrogen atom, the name can be made specific by indicating the position of the extra hydrogen(s). The compound name is modified with a locant followed by an italic capital H for each of these hydrogen atoms. Carbon atoms that carry an indicated hydrogen atom are numbered as low as possible. For example, $1H$-indene is illustrated in Table 1.2; $2H$-indene would be

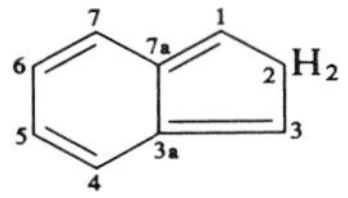

Names of polycyclic hydrocarbons with less than the maximum number of noncumulative double bonds are formed from a prefix dihydro-, tetrahydro-, etc., followed by the name of the corresponding unreduced hydrocarbon. The prefix perhydro- signifies full hydrogenation. For example, 1,2-dihydronaphthalene is

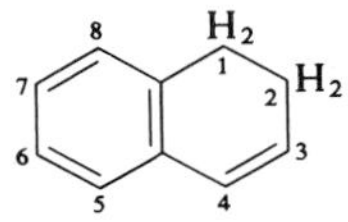

Examples of retained names and their structures are as follows:

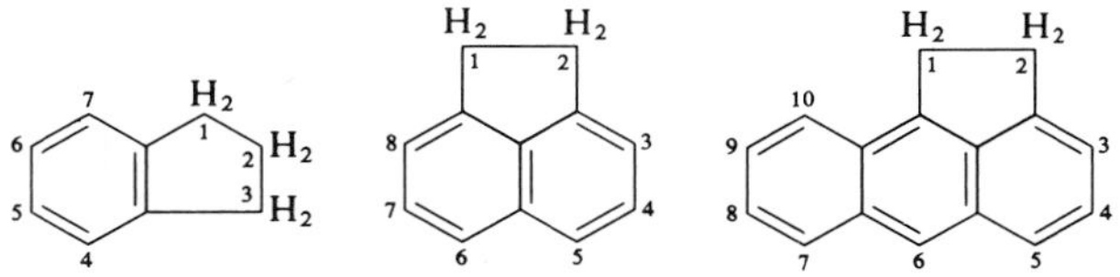

Indan   Acenaphthene   Aceanthrene

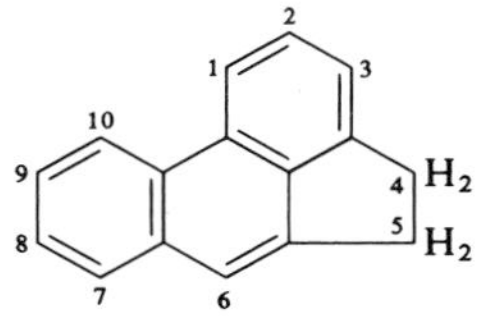

Acephenanthrene

Polycyclic compounds in which two rings have two atoms in common or in which one ring contains two atoms in common with each of two or more rings of a contiguous series of rings and which contain at least two rings of five or more members with the maximum number of

TABLE 1.2   Fused Polycyclic Hydrocarbons
Listed in order of increasing priority for selection as parent compound.

1. Pentalene

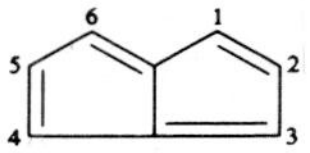

9. Acenaphthylene

2. Indene

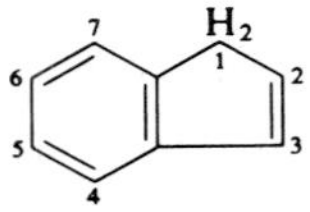

10. Fluorene

3. Naphthalene

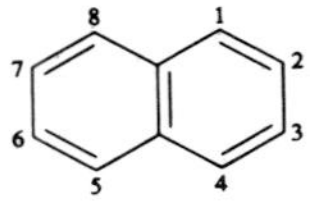

11. Phenalene

4. Azulene

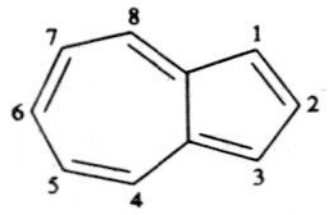

12. Phenanthrene*

5. Heptalene

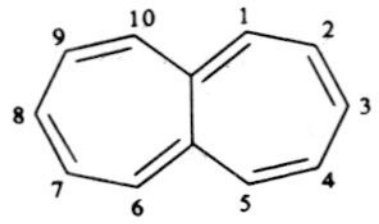

13. Anthracene*

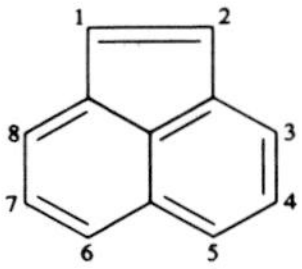

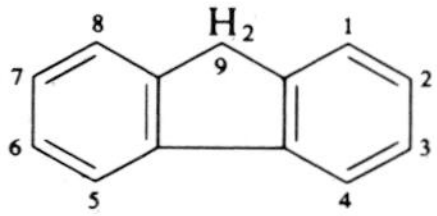

6. Biphenylene

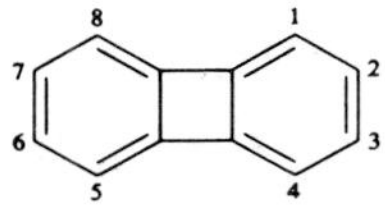

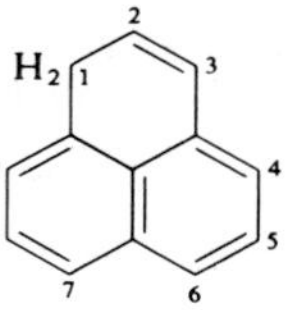

14. Fluoranthene

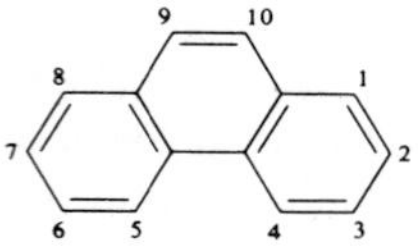

7. *asym*-Indacene

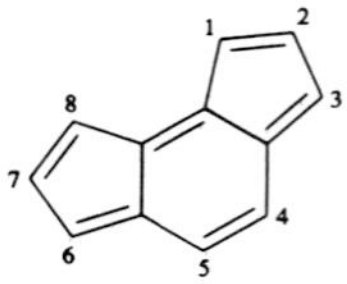

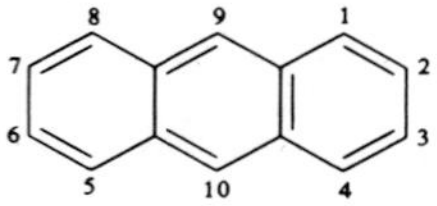

15. Acephenanthrylene

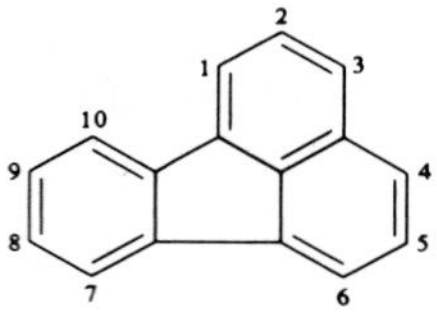

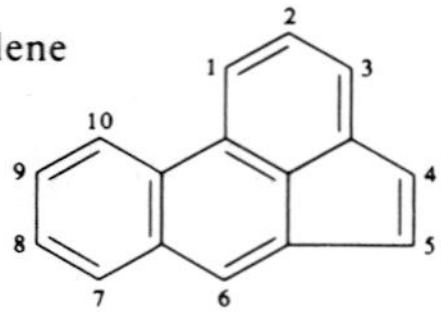

8. *sym*-Indacene

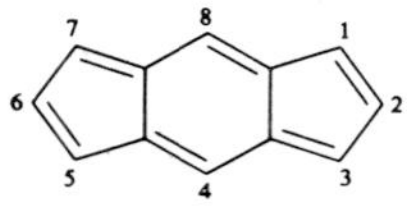

16. Aceanthrylene

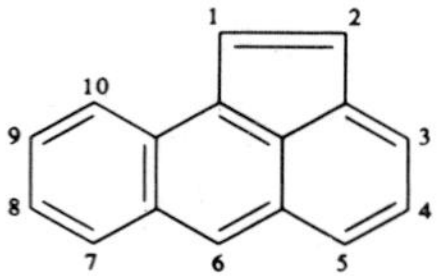

* Asterisk after a compound denotes exception to systematic numbering.

TABLE 1.2 Fused Polycyclic Hydrocarbons (*Continued*)

17. Triphenylene

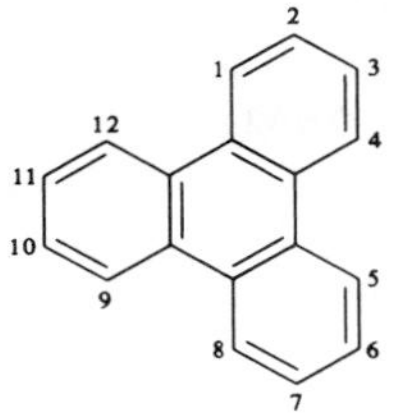

18. Pyrene

19. Chrysene

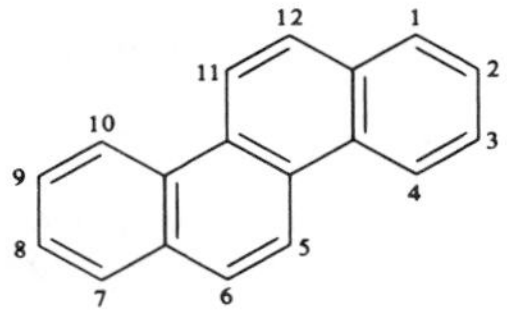

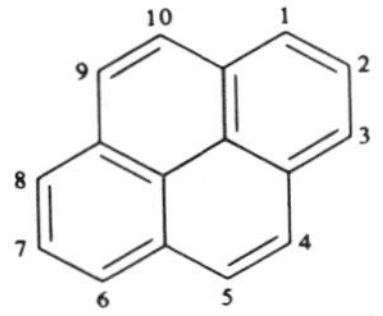

20. Naphthacene

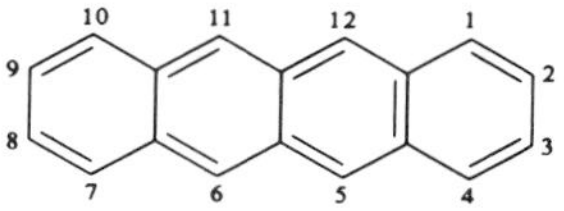

noncumulative double bonds and which have no accepted trivial name (Table 1.2) are named by prefixing to the name of the parent ring or ring system designations of the other components. The parent name should contain as many rings as possible (provided it has a trivial name) and should occur as far as possible from the beginning of the list in Table 1.2. Furthermore, the attached component(s) should be as simple as possible. For example, one writes dibenzophenanthrene and not naphthophenanthrene because the attached component benzo- is simpler than naphtho-. Prefixes designating attached components are formed by changing the ending -ene into -eno-; for example, indeno- from indene. Multiple prefixes are arranged in alphabetical order. Several abbreviated prefixes are recognized; the parent is given in parentheses:

Acenaphtho-	(acenaphthylene)	Naphtho-	(naphthalene)
Anthra-	(anthracene)	Perylo-	(perylene)
Benzo-	(benzene)	Phenanthro-	(phenanthrene)

For monocyclic prefixes other than benzo-, the following names are recognized, each to represent the form with the maximum number of noncumulative double bonds: cyclopenta-, cyclohepta-, cycloocta-, etc.

Isomers are distinguished by lettering the peripheral sides of the parent beginning with *a* for the side 1,2, and so on, lettering every side around the periphery. If necessary for clarity, the numbers of the attached position (1,2, for example) of the substituent ring are also denoted. The prefixes are cited in alphabetical order. The numbers and letters are enclosed in square brackets and placed immediately after the designation of the attached component. Examples are

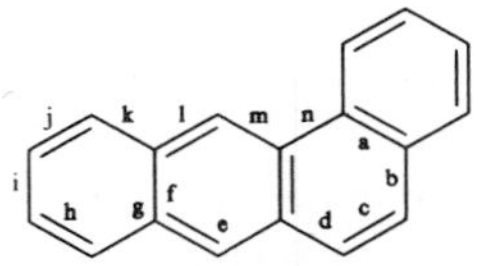

Benz[α]anthracene

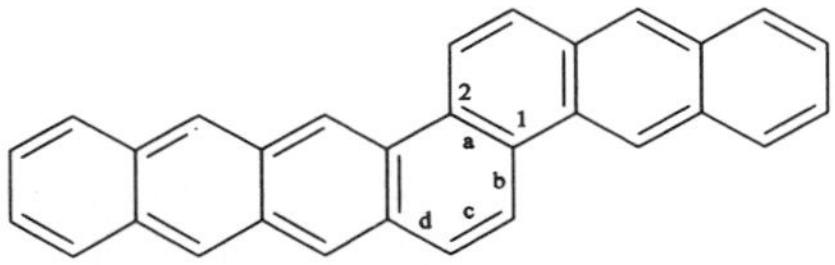

Anthra[2,1-α]naphthacene

1.1.1.6 Bridged Hydrocarbons. Saturated alicyclic hydrocarbon systems consisting of two rings that have two or more atoms in common take the name of the open-chain hydrocarbon containing the same total number of carbon atoms and are preceded by the prefix bicyclo-. The system is numbered commencing with one of the bridgeheads, numbering proceeding by the longest possible path to the second bridgehead. Numbering is then continued from this atom by the longer remaining unnumbered path back to the first bridgehead and is completed by the shortest path from the atom next to the first bridgehead. When a choice in numbering exists, unsaturation is given the lowest numbers. The number of carbon atoms in each of the bridges connecting the bridgeheads is indicated in brackets in descending order. Examples are

$$\overset{7}{C}H_2-\overset{1}{C}H-\overset{2}{C}H_2$$

Bicyclo[3.2.1]octane   Bicyclo[5.2.0]nonane

1.1.1.7 Hydrocarbon Ring Assemblies. Assemblies are two or more cyclic systems, either single rings or fused systems, that are joined directly to each other by double or single bonds. For identical systems naming may proceed (1) by placing the prefix bi- before the name of the corresponding radical or (2), for systems joined through a single bond, by placing the prefix bi- before the name of the corresponding hydrocarbon. In each case, the numbering of the assembly is that of the corresponding radical or hydrocarbon, one system being assigned unprimed numbers and the other primed numbers. The points of attachment are indicated by placing the appropriate locants before the name; an unprimed number is considered lower than the same number primed. The name *biphenyl* is used for the assembly consisting of two benzene rings. Examples are

1,1′-Bicyclopropyl or 1,1′-bicyclopropane   2-Ethyl-2′-propylbiphenyl

For nonidentical ring systems, one ring system is selected as the parent and the other systems are considered as substituents and are arranged in alphabetical order. The parent ring system is assigned unprimed numbers. The parent is chosen by considering the following characteristics in turn until a decision is reached: (1) the system containing the larger number of rings, (2) the system containing the larger ring, (3) the system in the lowest state of hydrogenation, and (4) the highest-order number of ring systems set forth in Table 1.2. Examples are given, with the deciding priority given in parentheses preceding the name:

(1) 2-Phenylnaphthalene

(2) and (4) 2-(2′-Naphthyl)azulene

(3) Cyclohexylbenzene

1.1.1.8 Radicals from Ring Systems. Univalent substituent groups derived from polycyclic hydrocarbons are named by changing the final *e* of the hydrocarbon name to -yl. The carbon atoms having free valences are given locants as low as possible consistent with the fixed numbering of the hydrocarbon. Exceptions are naphthyl (instead of naphthalenyl), anthryl (for anthracenyl), and phenanthryl (for phenanthrenyl). However, these abbreviated forms are used only for the simple ring systems. Substituting groups derived from fused derivatives of these ring systems are named systematically. Substituting groups having two or more free bonds are named as described in Monocyclic Aliphatic Hydrocarbons on p. 1.5.

1.1.1.9 Cyclic Hydrocarbons with Side Chains. Hydrocarbons composed of cyclic and aliphatic chains are named in a manner that is the simplest permissible or the most appropriate for the chemical intent. Hydrocarbons containing several chains attached to one cyclic nucleus are generally named as derivatives of the cyclic compound, and compounds containing several side chains and/or cyclic radicals attached to one chain are named as derivatives of the acyclic compound. Examples are

2-Ethyl-1-methylnaphthalene   Diphenylmethane

1,5-Diphenylpentane   2,3-Dimethyl-1-phenyl-1-hexene

Recognized trivial names for composite radicals are used if they lead to simplifications in naming. Examples are

1-Benzylnaphthalene   1,2,4-Tris(3-p-tolylpropyl)benzene

Fulvene, for methylenecyclopentadiene, and stilbene, for 1,2-diphenylethylene, are trivial names that are retained.

1.1.1.10 Heterocyclic Systems. Heterocyclic compounds can be named by relating them to the corresponding carbocyclic ring systems by using replacement nomenclature. Heteroatoms are denoted by prefixes ending in *a*, as shown in Table 1.3. If two or more replacement prefixes are required in a single name, they are cited in the order of their listing in the table. The lowest possible numbers consistent with the numbering of the corresponding carbocyclic system are assigned to the heteroatoms and then to carbon atoms bearing double or triple bonds. Locants are cited immediately preceding the prefixes or suffixes to which they refer. Multiplicity of the same heteroatom is indicated by the appropriate prefix in the series: di-, tri-, tetra-, penta-, hexa-, etc.

TABLE 1.3 Specialist Nomenclature for Heterocyclic Systems

Heterocyclic atoms are listed in decreasing order of priority.

Element	Valence	Prefix	Element	Valence	Prefix
Oxygen	2	Oxa-	Antimony	3	Stiba-*
Sulfur	2	Thia-	Bismuth	3	Bisma-
Selenium	2	Selena-	Silicon	4	Sila-
Tellurium	2	Tellura-	Germanium	4	Germa-
Nitrogen	3	Aza-	Tin	4	Stanna-
Phosphorus	3	Phospha-*	Lead	4	Plumba-
Arsenic	3	Arsa-*	Boron	3	Bora-
			Mercury	2	Mercura-

* When immediately followed by -in or -ine, phospha- should be replaced by phosphor-, arsa- by arsen-, and stiba- by antimon-. The saturated six-membered rings corresponding to phosphorin and arsenin are named *phosphorinane* and *arsenane*. A further exception is the replacement of borin by borinane.

TABLE 1.4 Suffixes for Specialist Nomenclature of Heterocyclic Systems

Number of ring members	Rings containing nitrogen		Rings containing no nitrogen	
	Unsaturation*	Saturation	Unsaturation*	Saturation
3	-irine	-iridine	-irene	-irane
4	-ete	-etidine	-ete	-etane
5	-ole	-olidine	-ole	-olane
6	-ine†	‡	-in	-ane§
7	-epine	‡	-epin	-epane
8	-ocine	‡	-ocin	-ocane
9	-onine	‡	-onin	-onane
10	-ecine	‡	-ecin	-ecane

* Unsaturation corresponding to the maximum number of noncumulative double bonds. Heteroatoms have the normal valences given in Table 1.3.

† For phosphorus, arsenic, antimony, and boron, see the special provisions in Table 1.3.

‡ Expressed by prefixing perhydro- to the name of the corresponding unsaturated compound.

§ Not applicable to silicon, germanium, tin, and lead; perhydro- is prefixed to the name of the corresponding unsaturated compound.

If the corresponding carbocyclic system is partially or completely hydrogenated, the additional hydrogen is cited using the appropriate *H*- or hydro- prefixes. A trivial name from Tables 1.5 and 1.6, if available, along with the state of hydrogenation may be used. In the specialist nomenclature for heterocyclic systems, the prefix or prefixes from Table 1.3 are combined with the appropriate stem from Table 1.4, eliding an *a* where necessary. Examples of acceptable usage, including (1) replacement and (2) specialist nomenclature, are

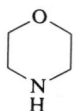

(1) 1-Oxa-4-azacyclo-   (1) 1,3-Diazacyclo-   (1) Thiacyclopropane
     hexane   hex-5-ene
(2) 1,4-Oxazoline   (2) 1,2,3,4-Tetra-   (2) Thiirane
     Morpholine   hydro-1,3-diazine   Ethylene sulfide

Radicals derived from heterocyclic compounds by removal of hydrogen from a ring are named by adding -yl to the names of the parent compounds (with elision of the final *e*, if present). These exceptions are retained:

Furyl (from furan)   Furfuryl (for 2-furylmethyl)

Pyridyl (from pyridine)   Furfurylidene (for 2-furylmethylene)

Piperidyl (from piperidine)   Thienyl (from thiophene)

Quinolyl (from quinoline)   Thenylidyne (for thienylmethylidyne)

Isoquinolyl   Furfurylidyne (for 2-furylmethylidyne)

Thenylidene (for thienylmethylene)   Thenyl (for thienylmethyl)

Also, piperidino- and morpholino- are preferred to 1-piperidyl- and 4-morpholinyl-, respectively.

TABLE 1.5 Trivial Names of Heterocyclic Systems Suitable for Use in Fusion Names

Listed in order of increasing priority as senior ring system.

Structure	Parent name	Radical name	Structure	Parent name	Radical name
	Thiophene	Thienyl		2H-Pyrrole	2H-Pyrrolyl
	Thianthrene	Thianthrenyl		Pyrrole	Pyrrolyl
	Furan	Furyl		Imidazole	Imidazolyl
	Pyran (2H-shown)	Pyranyl		Pyrazole	Pyrazolyl
	Isobenzofuran	Isobenzo-furanyl		Isothiazole	Isothiazolyl
	Chromene (2H-shown)	Chromenyl		Isoxazole	Isoxazolyl
	Xanthene*	Xanthenyl		Pyridine	Pyridyl
				Pyrazine	Pyrazinyl
	Phenoxathiin	Phenoxa-thiinyl		Pyrimidine	Pyrimidinyl
				Pyridazine	Pyridazinyl

* Asterisk after a compound denotes exception to systematic numbering.

TABLE 1.5   Trivial Names of Heterocyclic Systems Suitable for Use in Fusion Names
(*Continued*)

Structure	Parent name	Radical name	Structure	Parent name	Radical name
	Indolizine	Indolizinyl		Phthalazine	Phthalazinyl
	Isoindole	Isoindolyl		Naphthyri-dine (1,8-shown)	Naphthyri-dinyl
	3*H*-Indole	3*H*-Indolyl		Quinoxaline	Quinoxalinyl
	Indole	Indolyl		Quinazoline	Quinazolinyl
	1*H*-Indazole	1*H*-Indazolyl		Cinnoline	Cinnolinyl
	Purine*	Purinyl		Pteridine	Pteridinyl
	4*H*-Quin-olizine	4*H*-Quin-olizinyl		4α*H*-Carbazole*	4α*H*-Carbazolyl
	Isoquinoline	Isoquinolyl			
	Quinoline	Quinolyl		Carbazole*	Carbazolyl

* Asterisk after a compound denotes exception to systematic numbering.

TABLE 1.5 Trivial Names of Heterocyclic Systems Suitable for Use in Fusion Names (*Continued*)

Structure	Parent name	Radical name	Structure	Parent name	Radical name
	β-Carboline	β-Carbolinyl		Phenazine	Phenazinyl
	Phenanthri-dine	Phenanthri-dinyl		Phenarsazine	Phenarsazinyl
	Acridine*	Acridinyl		Phenothiazine	Phenothiazinyl
	Perimidine	Perimidinyl		Furazan	Furazanyl
	Phenanthroline (1,10-shown)	Phenanthrolinyl		Phenoxazine	Penoxazinyl

* Asterisk after a compound denotes exception to systematic numbering.

If there is a choice among heterocyclic systems, the parent compound is decided in the following order of preference:

1. A nitrogen-containing component
2. A component containing a heteroatom, in the absence of nitrogen, as high as possible in Table 1.3
3. A component containing the greater number of rings

TABLE 1.6 Trivial Names for Heterocyclic Systems That Are Not Recommended for Use in Fusion Names

Listed in order of increasing priority.

Structure	Parent name	Radical name	Structure	Parent name	Radical name
	Isochroman	Isochromanyl		Pyrazoline (3-shown*)	Pyrazolinyl
	Chroman	Chromanyl		Piperidine	Piperidyl†
	Pyrrolidine	Pyrrolinyl		Piperazine	Piperazinyl
	Pyrroline (2-shown*)	Pyrrolinyl		Indoline	Indolinyl
	Imidazolidine	Imidazolidinyl		Isoindoline	Isoindolinyl
	Imidazoline (2-shown*)	Imidazolinyl		Quinuclidine	Quinuclidinyl
	Pyrazolidine	Pyrazolidinyl		Morpholine	Morpholiny‡

* Denotes position of double bond.
† For 1-piperidyl, use piperidino.
‡ For 4-morpholinyl, use morpholino.

4. A component containing the largest possible individual ring

5. A component containing the greatest number of heteroatoms of any kind

6. A component containing the greatest variety of heteroatoms

7. A component containing the greatest number of heteroatoms first listed in Table 1.3

If there is a choice between components of the same size containing the same number and kind of heteroatoms, choose as the base component that one with the lower numbers for the heteroatoms before fusion. When a fusion position is occupied by a heteroatom, the names of the component rings to be fused are selected to contain the heteroatom.

1.1.2 Functional Compounds

There are several types of nomenclature systems that are recognized. Which type to use is sometimes obvious from the nature of the compound. Substitutive nomenclature, in general, is preferred because of its broad applicability, but radicofunctional, additive, and replacement nomenclature systems are convenient in certain situations.

1.1.2.1 Substitutive Nomenclature. The first step is to determine the kind of characteristic (functional) group for use as the principal group of the parent compound. A characteristic group is a recognized combination of atoms that confers characteristic chemical properties on the molecule in which it occurs. Carbon-to-carbon unsaturation and heteroatoms in rings are considered nonfunctional for nomenclature purposes.

Substitution means the replacement of one or more hydrogen atoms in a given compound by some other kind of atom or group of atoms, functional or nonfunctional. In substitutive nomenclature, each substituent is cited as either a prefix or a suffix to the name of the parent (or substituting radical) to which it is attached; the latter is denoted the parent compound (or parent group if a radical).

In Table 1.7 are listed the general classes of compounds in descending order of preference for citation as suffixes, that is, as the parent or characteristic compound. When oxygen is replaced by sulfur, selenium, or tellurium, the priority for these elements is in the descending order listed. The higher valence states of each element are listed before considering the successive lower valence states. Derivative groups have priority for citation as principal group after the respective parents of their general class.

In Table 1.8 are listed characteristic groups that are cited only as prefixes (never as suffixes) in substitutive nomenclature. The order of listing has no significance for nomenclature purposes.

Systematic names formed by applying the principles of substitutive nomenclature are single words except for compounds named as acids. First one selects the parent compound, and thus the suffix, from the characteristic group listed earliest in Table 1.7. All remaining functional groups are handled as prefixes that precede, in alphabetical order, the parent name. Two examples may be helpful:

Structure I   Structure II

Structure I contains an ester group and an ether group. Since the ester group has higher priority, the name is ethyl 2-methoxy-6-methyl-3-cyclohexene-1-carboxylate. Structure II contains a carbonyl group, a hydroxy group, and a bromo group. The latter is never a suffix. Between the other two, the carbonyl group has higher priority, the parent has -one as suffix, and the name is 4-bromo-1-hydroxy-2-butanone.

Selection of the principal alicyclic chain or ring system is governed by these selection rules:

1. For purely alicyclic compounds, the selection process proceeds successively until a decision is reached: (a) the maximum number of substituents corresponding to the characteristic group cited earliest in Table 1.7, (b) the maximum number of double and triple bonds considered together, (c) the maximum length of the chain, and (d) the maximum number of double bonds. Additional criteria, if needed for complicated compounds, are given in the IUPAC nomenclature rules.

2. If the characteristic group occurs only in a chain that carries a cyclic substituent, the compound is named as an aliphatic compound into which the cyclic component is substituted; a radical prefix is used to denote the cyclic component. This chain need not be the longest chain.

3. If the characteristic group occurs in more than one carbon chain and the chains are not

TABLE 1.7 Characteristic Groups for Substitutive Nomenclature

Listed in order of decreasing priority for citation as principal group or parent name.

Class	Formula*	Prefix	Suffix
1. Cations:		-onio-	-onium
	H_4N^+	Ammonio-	-ammonium
	H_3O^+	Oxonio-	-oxonium
	H_3S^+	Sulfonio-	-sulfonium
	H_3Se^+	Selenonio-	-selenonium
	H_2Cl^+	Chloronio-	-chloronium
	H_2Br^+	Bromonio-	-bromonium
	H_2I^+	Iodonio-	-iodonium
2. Acids:			
Carboxylic	—COOH	Carboxy-	-carboxylic acid
	—(C)OOH		-oic acid
	—C(=O)OOH		-peroxy···carboxylic acid
	—(C=O)OOH		-peroxy···oic acid
Sulfonic	—SO₃H	Sulfo-	-sulfonic acid
Sulfinic	—SO₂H	Sulfino-	-sulfinic acid
Sulfenic	—SOH	Sulfeno-	-sulfenic acid
Salts	—COOM		Metal···carboxylate
	—(C)OOM		Metal···oate
	—SO₃M		Metal···sulfonate
	—SO₂M		Metal···sulfinate
	—SOM		Metal···sulfenate
3. Derivatives of acids:			
Anhydrides	—C(=O)OC(=O)—		-carboxylic anhydride
	—(C=O)O(C=O)—		-oic anhydride
Esters	—COOR	R-oxycarbonyl-	R···carboxylate
	—C(OOR)		R···oate
Acid halides	—CO—halogen	Haloformyl	-carbonyl halide
Amides	—CO—NH₂	Carbamoyl-	-carboxamide
	(C)O—NH₂		-amide

TABLE 1.7 Characteristic Groups for Substitutive Nomenclature (*Continued*)

Class	Formula*	Prefix	Suffix
Hydrazides	$-CO-NHNH_2$	Carbonyl- hydrazino-	-carbohydrazide
	$-(CO)-NHNH_2$		-ohydrazide
Imides	$-CO-NH-CO-$	R-imido-	-carboximide
Amidines	$-C(=NH)-NH_2$	Amidino-	-carboxamidine
	$-(C=NH)-NH_2$		-amidine
4. Nitrile (cyanide)	$-CN$	Cyano-	-carbonitrile
	$-(C)N$		-nitrile
5. Aldehydes	$-CHO$	Formyl-	-carbaldehyde
	$-(C=O)H$	Oxo-	-al
	(then their analogs and derivatives)		
6. Ketones	$>(C=O)$	Oxo-	-one
	(then their analogs and derivatives)		
7. Alcohols (and phenols)	$-OH$	Hydroxy-	-ol
Thiols	$-SH$	Mercapto-	-thiol
8. Hydroperoxides	$-O-OH$	Hydroperoxy-	
9. Amines	$-NH_2$	Amino-	-amine
Imines	$>NH$	Imino-	-imine
Hydrazines	$-NHNH_2$	Hydrazino-	-hydrazine
10. Ethers	$-OR$	R-oxy-	
Sulfides	$-SR$	R-thio-	
11. Peroxides	$-O-OR$	R-dioxy-	

* Carbon atoms enclosed in parentheses are included in the name of the parent compound and not in the suffix or prefix.

TABLE 1.8 Characteristic Groups Cited Only as Prefixes in Substitutive Nomenclature

Characteristic group	Prefix	Characteristic group	Prefix
$-Br$	Bromo-	$-IX_2$	X may be halogen or a radical; dihalogenoiodo- or diacetoxyiodo-, e.g., $-ICl_2$ is dichloroido-
$-Cl$	Chloro-		
$-ClO$	Chlorosyl-		
$-ClO_2$	Chloryl-	$>N_2$	Diazo-
$-ClO_3$	Perchloryl-	$-N_3$	Azido-
$-F$	Fluoro-	$-NO$	Nitroso-
$-I$	Iodo-	$-NO_2$	Nitro-
$-IO$	Iodosyl-	$>N(=O)OH$	*aci*-Nitro-
$-IO_2$	Iodyl*	$-OR$	R-oxy-
$-I(OH)_2$	Dihydroxyiodo-	$-SR$	R-thio-
		$-SeR\ (-TeR)$	R-seleno- (R-telluro-)

* Formerly iodoxy.

directly attached to one another, then the chain chosen as parent should carry the largest number of the characteristic group. If necessary, the selection is continued as in rule 1.

4. If the characteristic group occurs only in one cyclic system, that system is chosen as the parent.

5. If the characteristic group occurs in more than one cyclic system, that system is chosen as parent which (a) carries the largest number of the principal group or, failing to reach a decision, (b) is the senior ring system.

6. If the characteristic group occurs both in a chain and in a cyclic system, the parent is that portion in which the principal group occurs in largest number. If the numbers are the same, that portion is chosen which is considered to be the most important or is the senior ring system.

7. When a substituent is itself substituted, all the subsidiary substituents are named as prefixes and the entire assembly is regarded as a parent radical.

8. The seniority of ring systems is ascertained by applying the following rules successively until a decision is reached: (a) all heterocycles are senior to all carbocycles, (b) for heterocycles, the preference follows the decision process described under Heterocyclic Systems, p. 1.11, (c) the largest number of rings, (d) the largest individual ring at the first point of difference, (e) the largest number of atoms in common among rings, (f) the lowest letters in the expression for ring functions, (g) the lowest numbers at the first point of difference in the expression for ring junctions, (h) the lowest state of hydrogenation, (i) the lowest-numbered locant for indicated hydrogen, (j) the lowest-numbered locant for point of attachment (if a radical), (k) the lowest-numbered locant for an attached group expressed as a suffix, (l) the maximum number of substituents cited as prefixes, (m) the lowest-numbered locant for substituents named as prefixes, hydro prefixes, -ene, and -yne, all considered together in one series in ascending numerical order independent of their nature, and (n) the lowest-numbered locant for the substituent named as prefix which is cited first in the name.

Numbering of Compounds. If the rules for aliphatic chains and ring systems leave a choice, the starting point and direction of numbering of a compound are chosen so as to give lowest-numbered locants to these structural factors, if present, considered successively in the order listed below until a decision is reached. Characteristic groups take precedence over multiple bonds.

1. Indicated hydrogen, whether cited in the name or omitted as being conventional

2. Characteristic groups named as suffix following the ranking order of Table 1.7

3. Multiple bonds in acyclic compounds; in bicycloalkanes, tricycloalkanes, and polycycloalkanes, double bonds having priority over triple bonds; and in heterocyclic systems whose names end in -etine, -oline, or -olene

4. The lowest-numbered locant for substituents named as prefixes, hydro prefixes, -ene, and -yne, all considered together in one series in ascending numerical order

5. The lowest locant for that substituent named as prefix which is cited first in the name

For cyclic radicals, indicated hydrogen and thereafter the point of attachment (free valency) have priority for the lowest available number.

Prefixes and Affixes. Prefixes are arranged alphabetically and placed before the parent name; multiplying affixes, if necessary, are inserted and *do not* alter the alphabetical order already attained. The parent name includes any syllables denoting a change of ring member or relating to the structure of a carbon chain. Nondetachable parts of parent names include

1. Forming rings: cyclo-, bicyclo-, spiro-

2. Fusing two or more rings: benzo-, naphtho-, imidazo-

3. Substituting one ring or chain member atom for another: oxa-, aza-, thia-

4. Changing positions of ring or chain members: iso-, *sec-, tert-,* neo-

5. Showing indicated hydrogen

6. Forming bridges: ethano-, epoxy-

7. Hydro-

Prefixes that represent complete terminal characteristic groups are preferred to those representing only a portion of a given group. For example, for the prefix —C(=O)CH$_3$, the name (formylmethyl) is preferred to (oxoethyl).

The multiplying affixes di-, tri-, tetra-, penta-, hexa-, hepta-, octa-, nona-, deca-, undeca-, and so on are used to indicate a set of *identical* unsubstituted radicals or parent compounds. The forms bis-, tris-, tetrakis-, pentakis-, and so on are used to indicate a set of identical radicals or parent compounds *each substituted in the same way.* The affixes bi-, ter-, quater-, quinque-, sexi, septi-, octi-, novi-, deci-, and so on are used to indicate the number of identical rings joined together by a single or double bond.

Although multiplying affixes may be omitted for very common compounds when no ambiguity is caused thereby, such affixes are generally included throughout this handbook in alphabetical listings. An example would be ethyl ether for diethyl ether.

1.1.2.2 *Conjunctive Nomenclature.* Conjunctive nomenclature may be applied when a principal group is attached to an acyclic component that is directly attached by a carbon-carbon bond to a cyclic component. The name of the cyclic component is attached directly in front of the name of the acyclic component carrying the principal group. This nomenclature is not used when an unsaturated side chain is named systematically. When necessary, the position of the side chain is indicated by a locant placed before the name of the cyclic component. For substituents on the acyclic chain, carbon atoms of the side chain are indicated by Greek letters proceeding from the principal group to the cyclic component. The terminal carbon atom of acids, aldehydes, and nitriles is omitted when allocating Greek positional letters. Conjunctive nomenclature is not used when the side chain carries more than one of the principal group, except in the case of malonic and succinic acids.

The side chain is considered to extend only from the principal group to the cyclic component. Any other chain members are named as substituents, with appropriate prefixes placed before the name of the cyclic component.

When a cyclic component carries more than one identical side chain, the name of the cyclic component is followed by di-, tri-, etc., and then by the name of the acyclic component, and it is preceded by the locants for the side chains. Examples are

4-Methyl-1-cyclohexaneethanol

α-Ethyl-β,β-dimethylcyclohexaneethanol

When side chains of two or more different kinds are attached to a cyclic component, only the senior side chain is named by the conjunctive method. The remaining side chains are named as prefixes. Likewise, when there is a choice of cyclic component, the senior is chosen. Benzene derivatives may be named by the conjunctive method only when two or more identical side chains are present. Trivial names for oxo carboxylic acids may be used for the acyclic component. If the cyclic and acyclic components are joined by a double bond, the locants of this bond are placed as superscripts to a Greek capital delta that is inserted between the two names. The locant for the cyclic component precedes that for the acyclic component, e.g., indene-$\Delta^{1,\alpha}$-acetic acid.

1.1.2.3   Radicofunctional Nomenclature.   The procedures of radicofunctional nomenclature are identical with those of substitutive nomenclature except that suffixes are never used. Instead, the functional class name (Table 1.9) of the compound is expressed as one word and the remainder of the molecule as another that precedes the class name. When the functional class name refers to a characteristic group that is bivalent, the two radicals attached to it are each named, and when different, they are written as separate words arranged in alphabetical order. When a compound contains more than one kind of group listed in Table 1.9, that kind is cited as the functional group or class name that occurs higher in the table, all others being expressed as prefixes.

Radicofunctional nomenclature finds some use in naming ethers, sulfides, sulfoxides, sulfones, selenium analogs of the preceding three sulfur compounds, and azides.

TABLE 1.9   Functional Class Names Used in Radicofunctional Nomenclature
Groups are listed in order of decreasing priority.

Group	Functional class names
X in acid derivatives	Name of X (in priority order: fluoride, chloride, bromide, iodide; cyanide, azide; then the sulfur and selenium analogs)
$-CN$, $-NC$	Cyanide, isocyanide
$\geq CO$	Ketone; then S and Se analogs
$-OH$	Alcohol; then S and Se analogs
$-O-OH$	Hydroperoxide
$\geq O$	Ether or oxide
$\geq S$, $\geq SO$, $\geq SO_2$	Sulfide, sulfoxide, sulfone
$\geq Se$, $\geq SeO$, $\geq SeO_2$	Selenide, selenoxide, selenone
$-F$, $-Cl$, $-Br$, $-I$	Fluoride, chloride, bromide, iodide
$-N_3$	Azide

1.1.2.4   Replacement Nomenclature.   Replacement nomenclature is intended for use only when other nomenclature systems are difficult to apply in the naming of chains containing heteroatoms. When no group is present that can be named as a principal group, the longest chain of carbon and heteroatoms terminating with carbon is chosen and named as though the entire chain were that of an acyclic hydrocarbon. The heteroatoms within this chain are identified by means of prefixes aza-, oxa-, thia-, etc., in the order of priority stated in Table 1.3. Locants indicate the positions of the heteroatoms in the chain. Lowest-numbered locants are assigned to the principal group when such is present. Otherwise, lowest-numbered locants are assigned to the heteroatoms considered together and, if there is a choice, to the heteroatoms cited earliest in Table 1.3. An example is

$$HO-\overset{13}{C}H_2-\overset{12}{O}-\overset{11}{C}H_2-\overset{10}{C}H_2-\overset{9}{O}-\overset{8}{C}H_2-\overset{7}{C}H_2-\overset{6}{N}-\overset{5}{C}H_2-\overset{4}{C}H_2-\overset{3}{N}-\overset{2}{C}H_2-\overset{1}{C}OOH$$

$$\qquad\qquad\qquad\qquad\qquad\qquad\qquad\qquad\qquad H \qquad\qquad\qquad H$$

13-Hydroxy-9,12-dioxa-3,6-diazatridecanoic acid

1.1.3 Specific Functional Groups

Characteristic groups will now be treated briefly in order to expand the terse outline of substitutive nomenclature presented in Table 1.7. Alternative nomenclature will be indicated whenever desirable.

1.1.3.1 Acetals and Acylals. Acetals, which contain the group $>C(OR)_2$, where R may be different, are named (1) as dialkoxy compounds or (2) by the name of the corresponding aldehyde or ketone followed by the name of the hydrocarbon radical(s) followed by the word *acetal.* For example, $CH_3-CH(OCH_3)_2$ is named either (1) 1,1-dimethoxyethane or (2) acetaldehyde dimethyl acetal.

A cyclic acetal in which the two acetal oxygen atoms form part of a ring may be named (1) as a heterocyclic compound or (2) by use of the prefix methylenedioxy for the group $-O-CH_2-O-$ as a substituent in the remainder of the molecule. For example,

(1) 1,3-Benzo[*d*]dioxole-5-carboxylic acid

(2) 3,4-Methylenedioxybenzoic acid

Acylals, $R^1R^2C(OCOR^3)_2$, are named as acid esters;

Butylidene acetate propionate

α-Hydroxy ketones, formerly called acyloins, had been named by changing the ending -ic acid or -oic acid of the corresponding acid to -oin. They are preferably named by substitutive nomenclature; thus

$$CH_3-CH(OH)-CO-CH_3 \qquad \text{3-Hydroxy-2-butanone (formerly acetoin)}$$

1.1.3.2 Acid Anhydrides. Symmetrical anhydrides of monocarboxylic acids, when unsubstituted, are named by replacing the word *acid* by *anhydride.* Anhydrides of substituted monocarboxylic acids, if symmetrically substituted, are named by prefixing bis- to the name of the acid and replacing the word *acid* by *anhydride.* Mixed anhydrides are named by giving in alphabetical order the first part of the names of the two acids followed by the word *anhydride,* e.g., acetic propionic anhydride or acetic propanoic anhydride. Cyclic anhydrides of polycarboxylic acids, although possessing a heterocyclic structure, are preferably named as acid anhydrides. For example,

1,8;4,5-Napthalenetetracarboxylic dianhydride (note the use of a semicolon to distinguish the pairs of locants)

1.1.3.3 Acyl Halides. Acyl halides, in which the hydroxyl portion of a carboxyl group is replaced by a halogen, are named by placing the name of the corresponding halide after that of the acyl radical. When another group is present that has priority for citation as principal group or when the acyl halide is attached to a side chain, the prefix haloformyl- is used as, for example, in fluoroformyl-.

1.1.3.4 Alcohols and Phenols. The hydroxyl group is indicated by a suffix -ol when it is the principal group attached to the parent compound and by the prefix hydroxy- when another group with higher priority for citation is present or when the hydroxy group is present in a side chain. When confusion may arise in employing the suffix -ol, the hydroxy group is indicated as a prefix; this terminology is also used when the hydroxyl group is attached to a heterocycle, as, for example, in the name 3-hydroxythiophene to avoid confusion with thiophenol (C_6H_5SH). Designations such as isopropanol, *sec*-butanol, and *tert*-butanol are incorrect because no hydrocarbon exists to which the suffix can be added. Many trivial names are retained. These structures are shown in Table 1.10. The radicals (RO—) are named by adding -oxy as a suffix to the name of the R radical, e.g., pentyloxy for $CH_3CH_2CH_2CH_2CH_2O—$. These contractions are exceptions: methoxy ($CH_3O—$), ethoxy ($C_2H_5O—$), propoxy ($C_3H_7O—$), butoxy ($C_4H_9O—$), and phenoxy ($C_6H_5O—$). For unsubstituted radicals only, one may use isopropoxy [$(CH_3)_2CH—O—$], isobutoxy [$(CH_3)_2CH_2CH—O—$], *sec*-butoxy [$CH_3CH_2CH(CH_3)—O—$], and *tert*-butoxy [$(CH_3)_3C—O—$].

TABLE 1.10 Retained Trivial Names of Alcohols and Phenols with Structures

Ally alcohol	$CH_2=CHCH_2OH$
tert-Butyl alcohol	$(CH_3)_3COH$
Benzyl alcohol	$C_6H_5CH_2OH$
Phenethyl alcohol	$C_6H_5CH_2CH_2OH$
Ethylene glycol	$HOCH_2CH_2OH$
1,2-Propylene glycol	$CH_3CHOHCH_2OH$
Glycerol	$HOCH_2CHOHCH_2OH$
Pentaerythritol	$C(CH_2OH)_4$
Pinacol	$(CH_3)_2COHCOH(CH_3)_2$
Phenol	C_6H_5OH

Xylitol

$$HOCH_2CH\underset{\underset{OH}{|}}{-}\overset{\overset{OH}{|}}{CH}-\underset{\underset{OH}{|}}{CH}-CH_2OH$$

Geraniol

$$(CH_3)_2C=CHCH_2CH_2\underset{\underset{CH_3}{|}}{C}=CHCH_2OH$$

TABLE 1.10 Retained Trivial Names of Alcohols and Phenols with Structures (*Continued*)

Phytol

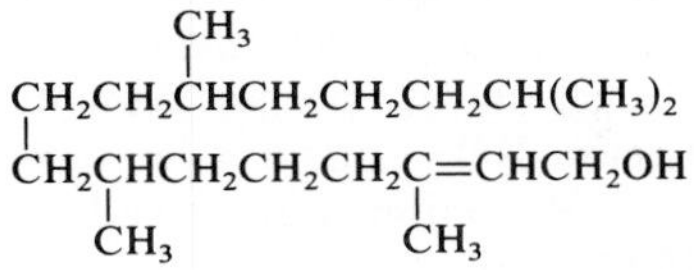

$$CH_3$$
$$CH_2CH_2CHCHCH_2CH_2CH_2CH(CH_3)_2$$
$$CH_2CHCH_2CH_2CH_2C=CHCH_2OH$$
$$CH_3 \qquad CH_3$$

Menthol

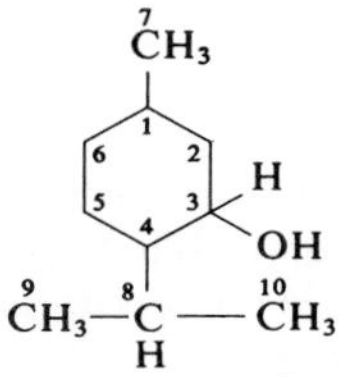

Borneol

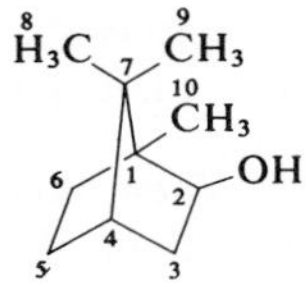

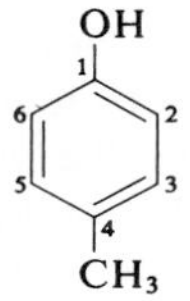

Cresol (1,4-isomer shown)

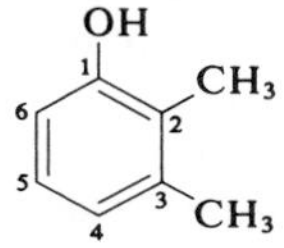

Xylenol (2,3-isomer shown)

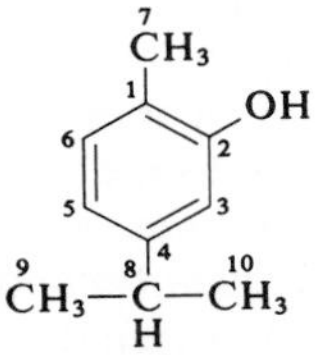

Carvacrol

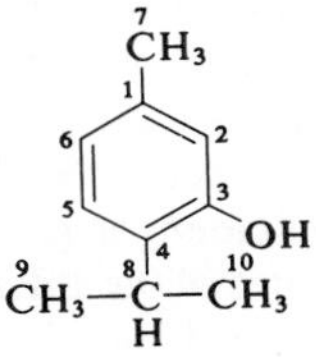

Thymol

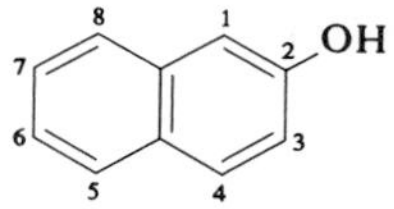

Naphthol (2-isomer shown)
2-Hydroxynaphthalene

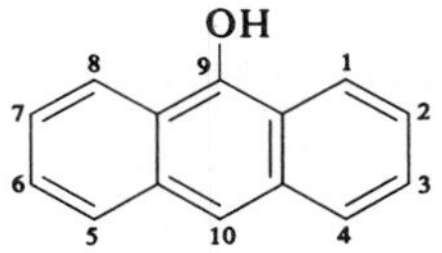

Anthrol (9-isomer shown)
9-Hydroxyanthracene

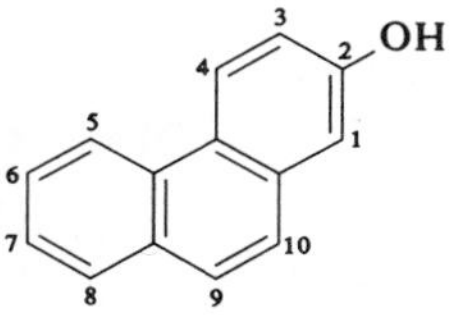

Phenanthrol (2-isomer shown)
2-Hydroxyphenanthrene

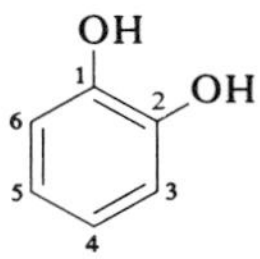

Pyrocatechol
1,2-Dihydroxybenzene

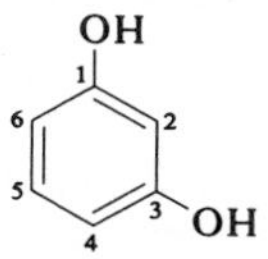

Resorcinol
1,3-Dihydroxybenzene

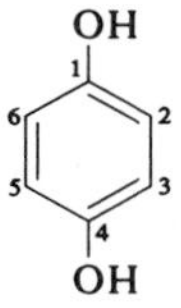

Hydroquinone
1,4-Dihydroxybenzene

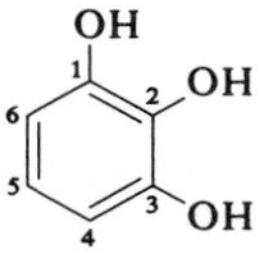

Pyrogallol
1,2,3-Trihydroxybenzene

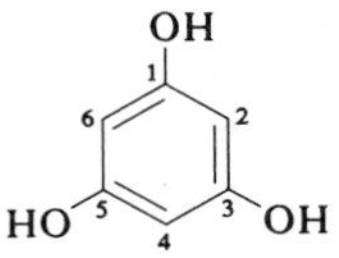

Phloroglucinol
1,3,5-Trihydroxybenzene

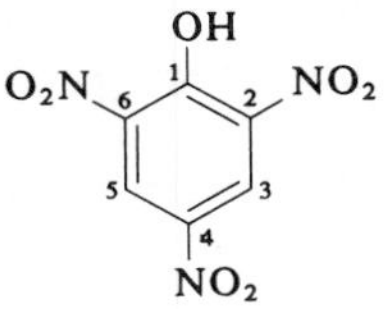

Picric acid
2,4,6-Trinitrophenol

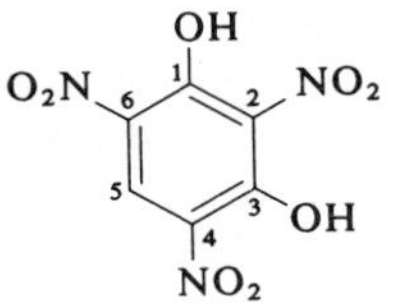

Styphnic acid
1,3-Dihydroxy-2,4,6-trinitroben-
zene

Bivalent radicals of the form O—Y—O are named by adding -dioxy to the name of the bivalent radicals except when forming part of a ring system. Examples are —O—CH$_2$—O— (methylenedioxy), —O—CO—O— (carbonyldioxy), and —O—SO$_2$—O— (sulfonyl-dioxy). Anions derived from alcohols or phenols are named by changing the final -ol to -olate.

Salts composed of an anion, RO—, and a cation, usually a metal, can be named by citing first the cation and then the RO anion (with its ending changed to -yl oxide), e.g., sodium benzyl oxide for C$_6$H$_5$CH$_2$ONa. However, when the radical has an abbreviated name, such as methoxy, the ending -oxy is changed to -oxide. For example, CH$_3$ONa is named sodium methoxide (not sodium methylate).

1.1.3.5 Aldehydes. When the group —C(=O)H, usually written —CHO, is attached to carbon at one (or both) end(s) of a linear acyclic chain the name is formed by adding the suffix -al (or -dial) to the name of the hydrocarbon containing the same number of carbon atoms. Examples are butanal for CH$_3$CH$_2$CH$_2$CHO and propanedial for, OHCCH$_2$CHO.

Naming an acyclic polyaldehyde can be handled in two ways. First, when more than two aldehyde groups are attached to an unbranched chain, the proper affix is added to -carbalde-hyde, which becomes the suffix to the name of the longest chain carrying the maximum number of aldehyde groups. The name and numbering of the main chain do not include the carbon atoms of the aldehyde groups. Second, the name is formed by adding the prefix formyl- to the name of the -dial that incorporates the principal chain. Any other chains carrying aldehyde groups are named by the use of formylalkyl- prefixes. Examples are

$$\underset{\text{OHC}-\text{CH}_2-\text{CH}_2-\text{CH}_2-\overset{\displaystyle \overset{\text{CHO}}{|}}{\text{CH}}-\text{CH}_2-\text{CHO}}{}$$

> (1) 1,2,5-Pentanetricarbaldehyde
> (2) 3-Formylheptanedial

$$\text{OHC}-\text{CH}_2-\text{CH}_2-\text{CH}_2 \diagdown \quad \overset{\text{CHO}}{\underset{\underset{\text{CH}_2-\text{CHO}}{|}}{|}}$$
$$\text{OHC}-\text{CH}_2-\text{CH}_2 \diagup \text{CH}-\text{CH}-\text{CH}-\text{CH}_2-\text{CHO}$$

> (1) 4-(2-Formylethyl)-3-(formylmethyl)-1,2,7-heptanetricarbaldehyde
> (2) 3-Formyl-5-(2-formylethyl)-4-(formylmethyl)nonanedial

When the aldehyde group is directly attached to a carbon atom of a ring system, the suffix -carbaldehyde is added to the name of the ring system, e.g., 2-naphthalenecarbaldehyde. When the aldehyde group is separated from the ring by a chain of carbon atoms, the compound is named (1) as a derivative of the acyclic system or (2) by conjunctive nomenclature, for example, (1) (2-naphthyl)propionaldehyde or (2) 2-naphthalenepropionaldehyde.

An aldehyde group is denoted by the prefix formyl- when it is attached to a nitrogen atom in a ring system or when a group having priority for citation as principal group is present and part of a cyclic system.

When the corresponding monobasic acid has a trivial name, the name of the aldehyde may be formed by changing the ending -ic acid or -oic acid to -aldehyde. Examples are

Formaldehyde	Acrylaldehyde (not acrolein)
Acetaldehyde	Benzaldehyde
Propionaldehyde	Cinnamaldehyde
Butyraldehyde	2-Furaldehyde (not furfural)

The same is true for polybasic acids, with the proviso that all the carboxyl groups must be changed to aldehyde; then it is not necessary to introduce affixes. Examples are

Glyceraldehyde

Glycolaldehyde

Malonaldehyde

Succinaldehyde

Phthalaldehyde (*o-*, *m-*, *p-*)

These trivial names may be retained: citral (3,7-dimethyl-2,6-octadienal), vanillin (4-hydroxy-3-methoxybenzaldehyde), and piperonal (3,4-methylenedioxybenzaldehyde).

1.1.3.6 Amides. For primary amides the suffix -amide is added to the systematic name of the parent acid. For example, $CH_3—CO—NH_2$ is acetamide. Oxamide is retained for $H_2N—CO—CO—NH_2$. The name -carboxylic acid is replaced by -carboxamide.

For amino acids having trivial names ending in -ine, the suffix -amide is added after the name of the acid (with elision of *e* for monoamides). For example, $H_2N—CH_2—CO—NH_2$ is glycinamide.

In naming the radical $R—CO—NH—$, either (1) the -yl ending of RCO— is changed to -amido or (2) the radicals are named as acylamino radicals. For example,

$CH_3—CO—NH—$⟨benzene ring⟩$—COOH$

(1) 4-Acetamidobenzoic acid
(2) 4-Acetylaminobenzoic acid

The latter nomenclature is always used for amino acids with trivial names.

N-substituted primary amides are named either (1) by citing the substituents as *N* prefixes or (2) by naming the acyl group as an *N* substituent of the parent compound. For example,

⟨benzene ring⟩$—CO—NH—CH_3$

(1) *N*-Methylbenzamide
(2) Benzoylaminomethane

1.1.3.7 Amines. Amines are preferably named by adding the suffix -amine (and any multiplying affix) to the name of the parent radical. Examples are

$CH_3CH_2CH_2CH_2CH_2NH_2$ Pentylamine

$H_2NCH_2CH_2CH_2CH_2CH_2NH_2$ 1,5-Pentyldiamine or pentamethylenediamine

Locants of substituents of symmetrically substituted derivatives of symmetrical amines are distinguished by primes or else the names of the complete substituted radicals are enclosed in parentheses. Unsymmetrically substituted derivatives are named similarly or as *N*-substituted products of a primary amine (after choosing the most senior of the radicals to be the parent amine). For example,

HN⟨$CH_2CH_2CH_2F$ / $CHF—CH_2CH_3$⟩

(1) 1,3'-Difluorodipropylamine
(2) 1-Fluoro-*N*-(3-fluoropropyl)propylamine
(3) (1-Fluoropropyl)(3-fluoropropyl)amine

Complex cyclic compounds may be named by adding the suffix -amine or the prefix amino- (or aminoalkyl-) to the name of the parent compound. Thus three names are permissible for

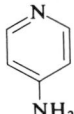

(1) 4-Pyridylamine
(2) 4-Pyridinamine
(3) 4-Aminopyridine

Complex linear polyamines are best designated by replacement nomenclature. These trivial names are retained: aniline, benzidene, phenetidine, toluidine, and xylidine.

The bivalent radical —NH— linked to two identical radicals can be denoted by the prefix imino-, as well as when it forms a bridge between two carbon ring atoms. A trivalent nitrogen atom linked to three identical radicals is denoted by the prefix nitrilo-. Thus ethylenediamine-tetraacetic acid (an allowed exception) should be named ethylenedinitrilotetraacetic acid.

1.1.3.8   Ammonium Compounds. Salts and hydroxides containing quadricovalent nitrogen are named as a substituted ammonium salt or hydroxide. The names of the substituting radicals precede the word *ammonium,* and then the name of the anion is added as a separate word. For example, $(CH_3)_4N^+I^-$ is tetramethylammonium iodide.

When the compound can be considered as derived from a base whose name does not end in -amine, its quaternary nature is denoted by adding ium to the name of that base (with elision of *e*), substituent groups are cited as prefixes, and the name of the anion is added separately at the end. Examples are

$C_6H_5NH_3^+HSO_4^-$ Anilinium hydrogen sulfate

$[(C_6H_5NH_3)^+]_2PtCl_6^{2-}$ Dianilinium hexachloroplatinate

The names *choline* and *betaine* are retained for unsubstituted compounds.

In complex cases, the prefixes amino- and imino- may be changed to ammonio- and iminio- and are followed by the name of the molecule representing the most complex group attached to this nitrogen atom and are preceded by the names of the other radicals attached to this nitrogen. Finally the name of the anion is added separately. For example, the name might be 1-trimethylammonioacridine chloride or 1-acridinyltrimethylammonium chloride.

When the preceding rules lead to inconvenient names, then (1) the unaltered name of the base may be used followed by the name of the anion or (2) for salts of hydrohalogen acids only the unaltered name of the base is used followed by the name of the hydrohalide. An example of the latter would be 2-ethyl-*p*-phenylenediamine monohydrochloride.

1.1.3.9   Azo Compounds. When the azo group (—N=N—) connects radicals derived from identical unsubstituted molecules, the name is formed by adding the prefix azo- to the name of the parent unsubstituted molecules. Substituents are denoted by prefixes and suffixes. The azo group has priority for lowest-numbered locant. Examples are azobenzene for C_6H_5—N=N—C_6H_5, azobenzene-4-sulfonic acid for C_6H_5—N=N—$C_6H_5SO_3H$, and 2′,4-dichloroazobenzene-4′-sulfonic acid for ClC_6H_4—N=N—$C_6H_3ClSO_3H$.

When the parent molecules connected by the azo group are different, azo is placed between the complete names of the parent molecules, substituted or unsubstituted. Locants are placed between the affix azo and the names of the molecules to which each refers. Preference is given to the more complex parent molecule for citation as the first component, e.g., 2-aminonaph-thalene-1-azo-(4′-chloro-2′-methylbenzene).

In an alternative method, the senior component is regarded as substituted by RN=N—, this group R being named as a radical. Thus 2-(7-phenylazo-2-naphthylazo)anthracene is the name by this alternative method for the compound named anthracene-2-azo-2′-naphthalene-7′-azobenzene.

1.1.3.10   Azoxy Compounds. Where the position of the azoxy oxygen atom is unknown or immaterial, the compound is named in accordance with azo rules, with the affix azo replaced by azoxy. When the position of the azoxy oxygen atom in an unsymmetrical compound is designated, a prefix *NNO-* or *ONN-* is used. When both the groups attached to the azoxy radical are cited in the name of the compound, the prefix *NNO-* specifies that the second of these two groups is attached directly to —N(O)—; the prefix *ONN*- specifies that the first of these two groups is attached directly to —N(O)—. When only one parent compound is cited in the name, the prefixed *ONN*- and *NNO*- specify that the group carrying the primed and unprimed substituents is connected, respectively, to the —N(O)— group. The prefix *NON*- signifies that the position of the oxygen atom is unknown; the azoxy group is then written as —N_2O—. For example,

2,2′,4-Trichloro-*NNO*-azoxybenzene

1.1.3.11 Boron Compounds.

1.1.3.11 *Boron Compounds.* Molecular hydrides of boron are called boranes. They are named by using a multiplying affix to designate the number of boron atoms and adding an Arabic numeral within parentheses as a suffix to denote the number of hydrogen atoms present. Examples are pentaborane(9) for B_5H_9 and pentaborane(11) for B_5H_{11}.

Organic ring systems are named by replacement nomenclature. Three- to ten-membered monocyclic ring systems containing uncharged boron atoms may be named by the specialist nomenclature for heterocyclic systems. Organic derivatives are named as outlined for substitutive nomenclature. The complexity of boron nomenclature precludes additional details; the text by Rigaudy and Klesney should be consulted.

1.1.3.12 *Carboxylic Acids.* Carboxylic acids may be named in several ways. First, —COOH groups replacing CH_3— at the end of the main chain of an acyclic hydrocarbon are denoted by adding -oic acid to the name of the hydrocarbon. Second, when the —COOH group is the principal group, the suffix -carboxylic acid can be added to the name of the parent chain whose name and chain numbering *does not include* the carbon atom of the —COOH group. The former nomenclature is preferred unless use of the ending -carboxylic acid leads to citation of a larger number of carboxyl groups as suffix. Third, carboxyl groups are designated by the prefix carboxy- when attached to a group named as a substituent or when another group is present that has higher priority for citation as principal group. In all cases, the principal chain should be linked to as many carboxyl groups as possible even though it might not be the longest chain present. Examples are

$CH_3CH_2CH_2CH_2CH_2CH_2COOH$ (1) Heptanoic acid
                                                      (2) 1-Hexanecarboxylic acid

$C_6H_{11}COOH$ (2) Cyclohexanecarboxylic acid

(3) 2-(Carboxymethyl)-1,4-hexanedicarboxylic acid

Removal of the OH from the —COOH group to form the acyl radical results in changing the ending -oic acid to -oyl or the ending -carboxylic acid to -carbonyl. Thus the radical $CH_3CH_2CH_2CH_2CO$— is named either pentanoyl or butanecarbonyl. When the hydroxyl has not been removed from all carboxyl groups present in an acid, the remaining carboxyl groups are denoted by the prefix carboxy-. For example, $HOOCCH_2CH_2CH_2CH_2CH_2CO$— is named 6-carboxyhexanoyl.

TABLE 1.11   Names of Some Carboxylic Acids

Systematic name	Trivial name	Systematic name	Trivial name
Methanoic	Formic	*trans*-Methylbutenedioic	Mesaconic*
Ethanoic	Acetic		
Propanoic	Propionic	1,2,2-Trimethyl-1,3-	Camphoric
Butanoic	Butyric	cyclopentanedicarboxylic	
2-Methylpropanoic	Isobutyric*	acid	
Pentanoic	Valeric	Benzenecarboxylic	Benzoic
3-Methylbutanoic	Isovaleric*	1,2-Benzenedicarboxylic	Phthalic
2,2-Dimethylpropanoic	Pivalic*	1,3-Benzenedicarboxylic	Isophthalic
Hexanoic	(Caproic)	1,4-Benzenedicarboxylic	Terephthalic
Heptanoic	(Enanthic)	Naphthalenecarboxylic	Naphthoic
Octanoic	(Caprylic)	Methylbenzenecarboxylic	Toluic
Decanoic	(Capric)	2-Phenylpropanoic	Hydratropic
Dodecanoic	Lauric*	2-Phenylpropenoic	Atropic
Tetradecanoic	Myristic*	*trans*-3-Phenylpropenoic	Cinnamic
Hexadecanoic	Palmitic*	Furancarboxylic	Furoic
Octadecanoic	Stearic*	Thiophenecarboxylic	Thenoic
		3-Pyridinecarboxylic	Nicotinic
Ethanedioic	Oxalic	4-Pyridinecarboxylic	Isonicotinic
Propanedioic	Malonic		
Butanedioic	Succinic	Hydroxyethanoic	Glycolic
Pentanedioic	Glutaric	2-Hydroxypropanoic	Lactic
Hexanedioic	Adipic	2,3-Dihydroxypropanoic	Glyceric
Heptanedioic	Pimelic*	Hydroxypropanedioic	Tartronic
Octanedioic	Suberic*	Hydroxybutanedioic	Malic
Nonanedioic	Azelaic*	2,3-Dihydroxybutanedioic	Tartaric
Decanedioic	Sebacic*	3-Hydroxy-2-phenylpropanoic	Tropic
Propenoic	Acrylic	2-Hydroxy-2,2-	Benzilic
Propynoic	Propiolic	diphenylethanoic	
2-Methylpropenoic	Methacrylic	2-Hydroxybenzoic	Salicylic
trans-2-Butenoic	Crotonic	Methoxybenzoic	Anisic
cis-2-Butenoic	Isocrotonic	4-Hydroxy-3-methoxybenzoic	Vanillic
cis-9-Octadecenoic	Oleic		
trans-9-Octadecenoic	Elaidic	3,4-Dimethoxybenzoic	Veratric
cis-Butenedioic	Maleic	3,4-Methylenedioxybenzoic	Piperonylic
trans-Butenedioic	Fumaric	3,4-Dihydroxybenzoic	Protocatechuic
cis-Methylbutenedioic	Citraconic*	3,4,5-Trihydroxybenzoic	Gallic

* Systematic names should be used in derivatives formed by substitution on a carbon atom.
Note: The names in parentheses are abandoned but are listed for reference to older literature.

Many trivial names exist for acids; these are listed in Table 1.11. Generally, radicals are formed by replacing -ic acid by -oyl.* When a trivial name is given to an acyclic monoacid or diacid, the numeral 1 is always given as locant to the carbon atom of a carboxyl group in the acid or to the carbon atom with a free valence in the radical RCO—.

---

* Exceptions: formyl, acetyl, propionyl, butyryl, isobutyryl, valeryl, isovaleryl, oxalyl, malonyl, succinyl, glutaryl, furoyl, and thenoyl.

1.1.3.13 Ethers (R^1—O—R^2). In substitutive nomenclature, one of the possible radicals, R—O—, is stated as the prefix to the parent compound that is senior from among R^1 or R^2. Examples are methoxyethane for $CH_3OCH_2CH_3$ and butoxyethanol for $C_4H_9OCH_2CH_2OH$.

When another principal group has precedence and oxygen is linking two identical parent compounds, the prefix oxy- may be used, as with 2,2'-oxydiethanol for $HOCH_2CH_2OCH_2CH_2OH$.

Compounds of the type RO—Y—OR, where the two parent compounds are identical and contain a group having priority over ethers for citation as suffix, are named as assemblies of identical units. For example, $HOOC$—CH_2—O—CH_2CH_2—O—CH_2—$COOH$ is named 2,2'-(ethylenedioxy)diacetic acid.

Linear polyethers derived from three or more molecules of aliphatic dihydroxy compounds, particularly when the chain length exceeds ten units, are most conveniently named by open-chain replacement nomenclature. For example, CH_3CH_2—O—CH_2CH_2—O—CH_2CH_3 could be 3,6-dioxaoctane or (2-ethoxy)ethoxy-ethane.

An oxygen atom directly attached to two carbon atoms already forming part of a ring system or to two carbon atoms of a chain may be indicated by the prefix epoxy-. For example, CH_2—CH—CH_2Cl is named 1-chloro-2,3-epoxypropane.

$\diagdown O \diagup$

Symmetrical linear polyethers may be named (1) in terms of the central oxygen atom when there is an odd number of ether oxygen atoms or (2) in terms of the central hydrocarbon group when there is an even number of ether oxygen atoms. For example, C_2H_5—O—C_4H_8—O—C_4H_8—O—C_2H_5 is bis-(4-ethoxybutyl)ether, and 3,6-dioxaoctane (earlier example) could be named 1,2-bis(ethoxy)ethane.

Partial ethers of polyhydroxy compounds may be named (1) by substitutive nomenclature or (2) by stating the name of the polyhydroxy compound followed by the name of the etherifying radical(s) followed by the word *ether.* For example,

CH_2O—C_4H_9 (1) 3-Butoxy-1,2-propanediol
| (2) Glycerol 1-butyl ether; also, 1-*O*-butylglycerol
HCOH
|
CH_2OH

Cyclic ethers are named either as heterocyclic compounds or by specialist rules of heterocyclic nomenclature. Radicofunctional names are formed by citing the names of the radicals R^1 and R^2 followed by the word *ether.* Thus methoxyethane becomes ethyl methyl ether and ethoxyethane becomes diethyl ether.

1.1.3.14 Halogen Derivatives. Using substitutive nomenclature, names are formed by adding prefixes listed in Table 1.8 to the name of the parent compound. The prefix perhalo- implies the replacement of all hydrogen atoms by the particular halogen atoms.

Cations of the type $R^1R^2X^+$ are given names derived from the halonium ion, H_2X^+, by substitution, e.g., diethyliodonium chloride for $(C_2H_5)_2I^+Cl^-$.

Retained are these trivial names; bromoform ($CHBr_3$), chloroform ($CHCl_3$), fluoroform (CHF_3), iodoform (CHI_3), phosgene ($COCl_2$), thiophosgene ($CSCl_2$), and dichlorocarbene radical ($\geq CCl_2$). Inorganic nomenclature leads to such names as carbonyl and thiocarbonyl halides (COX_2 and CSX_2) and carbon tetrahalides (CX_4).

1.1.3.15 Hydroxylamines and Oximes. For RNH—OH compounds, prefix the name of the radical R to hydroxylamine. If another substituent has priority as principal group, attach the prefix hydroxyamino- to the parent name. For example, C_6H_5NHOH would be named *N*-phenylhydroxylamine, but HOC_6H_4NHOH would be (hydroxyamino)phenol, with the point of attachment indicated by a locant preceding the parentheses.

Compounds of the type R^1NH—OR^2 are named (1) as alkoxyamino derivatives of compound R^1H, (2) as *N,O*-substituted hydroxylamines, (3) as alkoxyamines (even if R^1 is hydrogen), or (4) by the prefix aminooxy- when another substituent has priority for parent name. Examples of each type are

1. 2-(Methoxyamino)-8-naphthalenecarboxylic acid for CH_3ONH—$C_{10}H_6COOH$
2. *O*-Phenylhydroxylamine   for H_2N—O—C_6H_5 or   *N*-phenylhydroxylamine   for C_6H_5NH—OH
3. Phenoxyamine for H_2N—O—C_6H_5 (not preferred to *O*-phenylhydroxylamine)
4. Ethyl (aminooxy)acetate for H_2N—O—CH_2CO—OC_2H_5

Acyl derivatives, RCO—NH—OH and H_2N—O—CO—R, are named as *N*-hydroxy derivatives of amides and as *O*-acylhydroxylamines, respectively. The former may also be named as hydroxamic acids. Examples are *N*-hydroxyacetamide for CH_3CO—NH—OH and *O*-acetylhydroxylamine for H_2N—O—CO—CH_3. Further substituents are denoted by prefixes with *O*- and/or *N*-locants. For example, C_6H_5NH—O—C_2H_5 would be *O*-ethyl-*N*-phenylhydroxylamine or *N*-ethoxyaniline.

For oximes, the word *oxime* is placed after the name of the aldehyde or ketone. If the carbonyl group is not the principal group, use the prefix hydroxyimino-. Compounds with the group $>N$—OR are named by a prefix alkyloxyimino- as oxime *O*-ethers or as *O*-substituted oximes. Compounds with the group $>C=N(O)R$ are named be adding *N*-oxide after the name of the alkylideneaminc compound. For amine oxides, add the word *oxide* after the name of the base, with locants. For example, C_5H_5N—O is named pyridine *N*-oxide or pyridine 1-oxide.

1.1.3.16   Imines.   The group $>C=NH$ is named either by the suffix -imine or by citing the name of the bivalent radical $R^1R^2C<$ as a prefix to amine. For example, $CH_3CH_2CH_2CH=NH$ could be named 1-butanimine or butylideneamine. When the nitrogen is substituted, as in $CH_2=N$—CH_2CH_3, the name is *N*-(methylidene)ethylamine.

Quinones are exceptions. When one or more atoms of quinonoid oxygen have been replaced by $>NH$ or $>NR$, they are named by using the name of the quinone followed by the word *imine* (and preceded by proper affixes). Substituents on the nitrogen atom are named as prefixes. Examples are

O=⟨benzene ring⟩=NH   *p*-Benzoquinone monoimine

HN=⟨benzene ring⟩=NH   *p*-Benzoquinone diimine

1.1.3.17   Ketenes.   Derivatives of the compound ketene, $CH_2=C=O$, are named by substitutive nomenclature. For example, $C_4H_9CH=C=O$ is butyl ketene. An acyl derivative, such as CH_3CH_2—CO—$CH_2CH=C=O$, may be named as a polyketone, 1-hexene-1,4-dione. Bisketene is used for two to avoid ambiguity with diketene (dimeric ketene).

1.1.3.18   Ketones.   Acyclic ketones are named (1) by adding the suffix -one to the name of the hydrocarbon forming the principal chain or (2) by citing the names of the radicals R^1 and R^2

followed by the word *ketone*. In addition to the preceding nomenclature, acyclic monoacyl derivatives of cyclic compounds may be named (3) by prefixing the name of the acyl group to the name of the cyclic compound. For example, the three possible names of

(1) 1-(2-Furyl)-1-propanone
(2) Ethyl 2-furyl ketone
(3) 2-Propionylfuran

When the cyclic component is benzene or naphthalene, the -ic acid or -oic acid of the acid corresponding to the acyl group is changed to -ophenone or -onaphthone, respectively. For example, C_6H_5—CO—$CH_2CH_2CH_3$ can be named either butyrophenone (or butanophenone) or phenyl propyl ketone.

Radicofunctional nomenclature can be used when a carbonyl group is attached directly to carbon atoms in two ring systems and no other substituent is present having priority for citation.

When the methylene group in polycarbocyclic and heterocyclic ketones is replaced by a keto group, the change may be denoted by attaching the suffix -one to the name of the ring system. However, when ≥CH in an unsaturated or aromatic system is replaced by a keto group, two alternative names become possible. First, the maximum number of noncumulative double bonds is added after introduction of the carbonyl group(s), and any hydrogen that remains to be added is denoted as indicated hydrogen with the carbonyl group having priority over the indicated hydrogen for lower-numbered locant. Second, the prefix oxo- is used, with the hydrogenation indicated by hydro prefixes; hydrogenation is considered to have occurred before the introduction of the carbonyl group. For example,

(1) 1(2*H*)-Naphthalenone
(2) 1-Oxo-1,2-dihydronaphthalene

When another group having higher priority for citation as principal group is also present, the ketonic oxygen may be expressed by the prefix oxo-, or one can use the name of the carbonyl-containing radical, as, for example, acyl radicals and oxo-substituted radicals. Examples are

4-(4'-Oxohexyl)-1-benzoic acid

1,2,4-Triacetylbenzene

Diketones and tetraketones derived from aromatic compounds by conversion of two or four ≥CH groups into keto groups, with any necessary rearrangement of double bonds to a quinonoid structure, are named by adding the suffix -quinone and any necessary affixes.

Polyketones in which two or more contiguous carbonyl groups have rings attached at each end may be named (1) by the radicofunctional method or (2) by substitutive nomenclature.

For example,

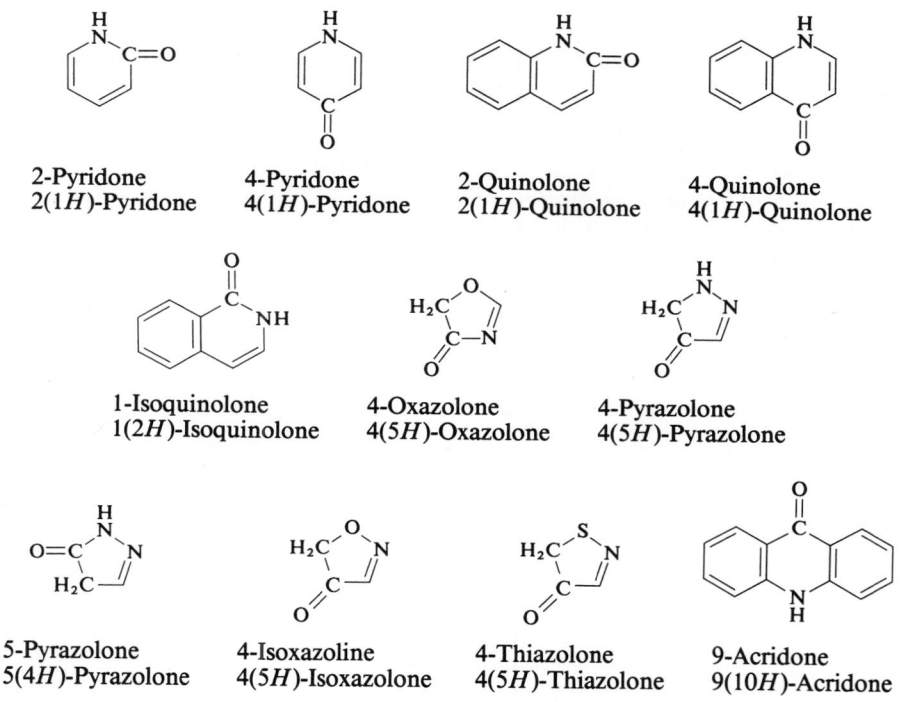

(1) 2-Naphthyl 2-pyridyl diketone
(2) 1-(2-Naphthyl)-2-(2-pyridyl)ethanedione

Some trivial names are retained: acetone (2-propanone), biacetyl (2,3-butanedione), propiophenone (C_6H_5—CO—CH_2CH_3), chalcone (C_6H_5—CH=CH—CO—C_6H_5), and deoxybenzoin (C_6H_5—CH_2—CO—C_6H_5).

These contracted names of heterocyclic nitrogen compounds are retained as alternatives for systematic names, sometimes with indicated hydrogen. In addition, names of oxo derivatives of fully saturated nitrogen heterocycles that systematically end in -idinone are often contracted to end in -idone when no ambiguity might result. For example,

2-Pyridone
2(1*H*)-Pyridone

4-Pyridone
4(1*H*)-Pyridone

2-Quinolone
2(1*H*)-Quinolone

4-Quinolone
4(1*H*)-Quinolone

1-Isoquinolone
1(2*H*)-Isoquinolone

4-Oxazolone
4(5*H*)-Oxazolone

4-Pyrazolone
4(5*H*)-Pyrazolone

5-Pyrazolone
5(4*H*)-Pyrazolone

4-Isoxazoline
4(5*H*)-Isoxazolone

4-Thiazolone
4(5*H*)-Thiazolone

9-Acridone
9(10*H*)-Acridone

1.1.3.19 *Lactones, Lactides, Lactams, and Lactims.* When the hydroxy acid from which water may be considered to have been eliminated has a trivial name, the lactone is designated by substituting -olactone for -ic acid. Locants for a carbonyl group are numbered as low as possible, even before that of a hydroxyl group.

Lactones formed from aliphatic acids are named by adding -olide to the name of the nonhydroxylated hydrocarbon with the same number of carbon atoms. The suffix -olide signifies the change of >CH···CH_3 into >C···C=O.

Structures in which one or more (but not all) rings of an aggregate are lactone rings are named by placing -carbolactone (denoting the —O—CO— bridge) after the names of the structures that remain when each bridge is replaced by two hydrogen atoms. The locant for —CO— is cited before that for the ester oxygen atom. An additional carbon atom is incorporated into this structure as compared to the -olide.

These trivial names are permitted: γ-butyrolactone, γ-valerolactone, and δ-valerolactone. Names based on heterocycles may be used for all lactones. Thus, γ-butyrolactone is also tetrahedro-2-furanone or dihydro-2(3H)-furanone.

Lactides, intermolecular cyclic esters, are named as heterocycles. *Lactams* and *lactims,* containing a —CO—NH— and —C(OH)=N— group, respectively, are named as heterocycles, but they may also be named with -lactam or -lactim in place of -olide. For example,

(1) 2-Pyrrolidinone
(2) 4-Butanelactam

1.1.3.20   Nitriles and Related Compounds.   For acids whose systematic names end in -carboxylic acid, nitriles are named by adding the suffix -carbonitrile when the —CN group replaces the —COOH group. The carbon atom of the —CN group is excluded from the numbering of a chain to which it is attached. However, when the triple-bonded nitrogen atom is considered to replace three hydrogen atoms at the end of the main chain of an acyclic hydrocarbon, the suffix -nitrile is added to the name of the hydrocarbon. Numbering begins with the carbon attached to the nitrogen. For example, $CH_3CH_2CH_2CH_2CH_2CN$ is named (1) pentanecarbonitrile or (2) hexanenitrile.

Trivial acid names are formed by changing the endings -oic acid or -ic acid to -onitrile. For example, CH_3CN is acetonitrile. When the —CN group is not the highest priority group, the —CN group is denoted by the prefix cyano-.

In order of decreasing priority for citation of a functional class name, and the prefix for substitutive nomenclature, are the following related compounds:

Functional group	Prefix	Radicofunctional ending
—NC	Isocyano-	Isocyanide
—OCN	Cyanato-	Cyanate
—NCO	Isocyanato-	Isocyanate
—ONC	—	Fulminate
—SCN	Thiocyanato-	Thiocyanate
—NCS	Isothiocyanato-	Isothiocyanate
—SeCN	Selenocyanato-	Selenocyanate
—NCSe	Isoselenocyanato-	Isoselenocyanate

1.1.3.21   Peroxides.   Compounds of the type R—O—OH are named (1) by placing the name of the radical R before the word *hydroperoxide* or (2) by use of the prefix hydroperoxy- when another parent name has higher priority. For example, C_2H_5OOH is ethyl hydroperoxide.

Compounds of the type $R^1O—OR^2$ are named (1) by placing the names of the radicals in alphabetical order before the word *peroxide* when the group —O—O— links two chains, two rings, or a ring and a chain, (2) by use of the affix dioxy to denote the bivalent group —O—O— for naming assemblies of identical units or to form part of a prefix, or (3) by use of the prefix epidioxy- when the peroxide group forms a bridge between two carbon atoms, a ring, or a ring system. Examples are methyl propyl peroxide for CH_3—O—O—C_3H_7 and 2,2'-dioxydiacetic acid for HOOC—CH_2—O—O—CH_2—COOH.

1.1.3.22 Phosphorus Compounds. Acyclic phosphorus compounds containing only one phosphorus atom, as well as compounds in which only a single phosphorus atom is in each of several functional groups, are named as derivatives of the parent structures listed in Table 1.12.

TABLE 1.12 Parent Structures of Phosphorus-Containing Compounds

Formula	Parent name	Substitutive prefix		Radicofunctional ending
H_3P	Phosphine	H_2P—	Phosphino-	Phosphide
H_5P	Phosphorane	H_4P—	Phosphoranyl-	
		$H_3P\subset$	Phosphoroanediyl-	
		$H_2P\leqq$	Phosphoranetriyl-	
H_3PO	Phosphine oxide			
H_3PS	Phosphine sulfide			
H_3PNH	Phosphine imide			
$P(OH)_3$	Phosphorous acid			Phosphite
$HP(OH)_2$	Phosphonous acid			Phosphonite
H_2POH	Phosphinous acid			Phosphinite
$P(O)(OH)_3$	Phosphoric acid	$P(O)\leqq$	Phosphoryl-	Phosphate(V)
$HP(O)(OH)_2$	Phosphonic acid	$HP(O)\subset$	Phosphonoyl-	Phosphonate
		—$P(O)OH_2$	Phosphono-	
$H_2P(O)OH$	Phosphinic acid	$H_2P(O)$—	Phosphinoyl-	Phosphinate
		$\gt P(O)OH$	Phosphinoco-	
			Phosphinato-	

Often these are purely hypothetical parent structures. When hydrogen attached to phosphorus is replaced by a hydrocarbon group, the derivative is named by substitution nomenclature. When hydrogen of an —OH group is replaced, the derivative is named by radicofunctional nomenclature. For example, $C_2H_5PH_2$ is ethylphosphine; $(C_2H_5)_2PH$, diethylphosphine; $CH_3P(OH)_2$, dihydroxy-methyl-phosphine or methylphosphonous acid; C_2H_5—$PO(Cl)(OH)$, ethylchlorophosphonic acid or ethylphosphonochloridic acid or hydrogen chlorodioxoethylphosphate(V); $CH_3CH(PH_2)COOH$, 2-phosphinopropionic acid; $HP(CH_2COOH)_2$, phosphinediyldiacetic acid; $(CH_3)HP(O)OH$, methylphosphinic acid or hydrogen hydridomethyldioxophosphate(V); $(CH_3O)_3PO$, trimethyl phosphate; and $(CH_3O)_3P$, trimethyl phosphite.

1.1.3.23 Salts and Esters of Acids. Neutral salts of acids are named by citing the cation(s) and then the anion, whose ending is changed from -oic to -oate or from -ic to -ate. When different acidic residues are present in one structure, prefixes are formed by changing the anion ending -ate to -ato- or -ide to -ido-. The prefix carboxylato- denotes the ionic group —COO^-. The phrase (metal) salt of (the acid) is permissible when the carboxyl groups are not all named as affixes.

Acid salts include the word *hydrogen* (with affixes, if appropriate) inserted between the name of the cation and the name of the anion (or word *salt*).

Esters are named similarly, with the name of the alkyl or aryl radical replacing the name of the cation. Acid esters of acids and their salts are named as neutral esters, but the components are cited in the order: cation, alkyl or aryl radical, hydrogen, and anion. Locants are added if necessary. For example,

$$CH_2-CO-OC_2H_5$$
$$HOC-COO^- \quad K^+ \quad H^+ \quad \text{Potassium 1-ethyl hydrogen citrate}$$
$$CH_2-COO^-$$

Ester groups in $R^1-CO-OR^2$ compounds are named (1) by the prefix alkoxycarbonyl- or aryloxycarbonyl- for $-CO-OR^2$ when the radical R^1 contains a substituent with priority for citation as principal group or (2) by the prefix acyloxy- for $R^1-CO-O-$ when the radical R^2 contains a substituent with priority for citation as principal group. Examples are

$$CH_2CH_2CH_2CO-OCH_3$$

Methyl 3-methoxycarbonyl-2-naphthalenebutyrate

$$CO-OCH_3$$

$$[CH_3O-CO-CH_2CH_2\overset{+}{N}(CH_3)_3]Cl^-$$

[(2-Methoxycarbonyl)ethyl]trimethylammonium chloride

$$C_6H_5-CO-OCH_2CH_2COOH \quad \text{3-Benzoyloxypropionic acid}$$

The trivial name *acetoxy* is retained for the $CH_3-CO-O-$ group. Compounds of the type $R^2C(OR^2)_3$ are named as R^2 esters of the hypothetical ortho acids. For example, $CH_3C(OCH_3)_3$ is trimethyl orthoacetate.

1.1.3.24 Silicon Compounds. SiH_4 is called silane; its acyclic homologs are called disilane, trisilane, and so on, according to the number of silicon atoms present. The chain is numbered from one end to the other so as to give the lowest-numbered locant in radicals to the free valence or to substituents on a chain. The abbreviated form silyl is used for the radical SiH_3-. Numbering and citation of side chains proceed according to the principles set forth for hydrocarbon chains. Cyclic nonaromatic structures are designated by the prefix cyclo-.

When a chain or ring system is composed entirely of alternating silicon and oxygen atoms, the parent name *siloxane* is used with a multiplying affix to denote the number of silicon atoms present. The parent name *silazane* implies alternating silicon and nitrogen atoms; multiplying affixes denote the number of silicon atoms present.

The prefix sila-designates replacement of carbon by silicon in replacement nomenclature. Prefix names for radicals are formed analogously to those for the corresponding carbon-containing compounds. Thus silyl is used for SiH_3-, silyene for $-SiH_2-$, silylidyne for $-SiH<$, as well as trily, tetrayl, and so on for free valences(s) on ring structures.

1.1.3.25 Sulfur Compounds

Bivalent Sulfur. The prefix thio, placed before an affix that denotes the oxygen-containing group or an oxygen atom, implies the replacement of that oxygen by sulfur. Thus the suffix -thiol denotes $-SH$, -thione denotes $-(C)=S$ and implies the presence of an $=S$ at a nonterminal carbon atom, -thioic acid denotes $[(C)=S]OH \rightleftharpoons [(C)=O]SH$ (that is, the O-substituted acid and the S-substituted acid, respectively), -dithioc acid denotes $[-C(S)]SH$, and -thial denotes $-(C)HS$ (or -carbothialdehyde denotes $-CHS$). When -carboxylic acid has been used for acids, the sulfur analog is named -carbothioic acid or -carbodithioic acid.

Prefixes for the groups HS— and RS— are mercapto- and alkylthio-, respectively; this latter name may require parentheses for distinction from the use of thio- for replacement of oxygen in a trivially named acid. Examples of this problem are $4\text{-}C_2H_5\text{—}C_6H_4\text{—CSOH}$ named p-ethyl(thio)benzoic acid and $4\text{-}C_2H_5\text{—S—}C_6H_4\text{—COOH}$ named p-(ethylthio)benzoic acid. When —SH is not the principal group, the prefix mercapto- is placed before the name of the parent compound to denote an unsubstituted —SH group.

The prefix thioxo- is used for naming =S in a thioketone. Sulfur analogs of acetals are named as alkylthio- or arylthio-. For example, $CH_3CH(SCH_3)OCH_3$ is 1-methoxy-1-(methylthio)ethane. Prefix forms for -carbothioic acids are hydroxy(thiocarbonyl)- when referring to the O-substituted acid and mercapto(carbonyl)- for the S-substituted acid.

Salts are formed as with oxygen-containing compounds. For example, $C_2H_5\text{—S—Na}$ is named either sodium ethanethiolate or sodium ethyl sulfide. If mercapto- has been used as a prefix, the salt is named by use of the prefix sulfido- for —S^-.

Compounds of the type $R^1\text{—S—}R^2$ are named alkylthio- (or arylthio-) as a prefix to the name of R^1 or R^2, whichever is the senior.

Sulfonium Compounds. Sulfonium compounds of the type $R^1R^2R^3S^+X^-$ are named by citing in alphabetical order the radical names followed by -sulfonium and the name of the anion. For heterocyclic compounds, -ium is added to the name of the ring system. Replacement of >CH by sulfonium sulfur is denoted by the prefix thionia-, and the name of the anion is added at the end.

Organosulfur Halides. When sulfur is directly linked only to an organic radical and to a halogen atom, the radical name is attached to the word *sulfur* and the name(s) and number of the halide(s) are stated as a separate word. Alternatively, the name can be formed from R—SOH, a sulfenic acid whose radical prefix is sulfenyl-. For example, $CH_3CH_2\text{—S—Br}$ would be named either ethylsulfur monobromide or ethanesulfenyl bromide. When another principal group is present, a composite prefix is formed from the number and substitutive name(s) of the halogen atoms in front of the syllable thio. For example, BrS—COOH is (bromothio)formic acid.

Sulfoxides. Sulfoxides, $R^1\text{—SO—}R^2$, are named by placing the names of the radicals in alphabetical order before the word *sulfoxide.* Alternatively, the less senior radical is named followed by sulfinyl- and concluded by the name of the senior group. For example, $CH_3CH_2\text{—SO—}CH_2CH_2CH_3$ is named either ethyl propyl sulfoxide or 1-(ethylsulfinyl)propane.

When an >SO group is incorporated in a ring, the compound is named an oxide.

Sulfones. Sulfones, $R^1\text{—SO}_2\text{—}R^2$, are named in an analogous manner to sulfoxides, using the word *sulfone* in place of *sulfoxide.* In prefixes, the less senior radical is followed by -sulfonyl-. When the >SO$_2$ group is incorporated in a ring, the compound is named as a dioxide.

Sulfur Acids. Organic oxy acids of sulfur, that is, —SO$_3$H, —SO$_2$H, and —SOH, are named sulfonic acid, sulfinic acid, and sulfenic acid, respectively. In subordinate use, the respective prefixes are sulfo-, sulfino-, and sulfeno-. The grouping $\text{—SO}_2\text{—O—SO}_2\text{—}$ or —SO—O—SO is named sulfonic or sulfinic anhydride, respectively.

Inorganic nomenclature is employed in naming sulfur acids and their derivatives in which sulfur is linked only through oxygen to the organic radical. For example, $(C_2H_5O)_2SO_2$ is diethyl sulfate and $C_2H_5O\text{—SO}_2\text{—OH}$ is ethyl hydrogen sulfate. Prefixes O- and S- are used where necessary to denote attachment to oxygen and to sulfur, respectively, in sulfur replacement compounds. For example, $CH_3\text{—S—SO}_2\text{—ONa}$ is sodium S-methyl thiosulfate.

When sulfur is linked only through nitrogen, or through nitrogen and oxygen, to the organic radical, naming is as follows: (1) N-substituted amides are designated as N-substituted derivatives of the sulfur amides and (2) compounds of the type R—NH—SO$_3$H may be named as *N*-substituted sulfamic acids or by the prefix sulfoamino- to denote the group HO_3S—NH—. The groups —N=SO and —N=SO_2 are named sulfinylamines and sulfonylamines, respectively.

Sultones and Sultams. Compounds containing the group —SO_2—O— as part of the ring are called -sultone. The —SO_2— group has priority over the —O— group for lowest-numbered locant.

Similarly, the —SO_2—N= group as part of a ring is named by adding -sultam to the name of the hydrocarbon with the same number of carbon atoms. The —SO_2— has priority over —N= for lowest-numbered locant.

1.1.4 Stereochemistry

Concepts in stereochemistry, that is, chemistry in three-dimensional space, are in the process of rapid expansion. This section will deal with only the main principles. The compounds discussed will be those that have identical molecular formulas but differ in the arrangement of their atoms in space. *Stereoisomers* is the name applied to these compounds.

Stereoisomers can be grouped into three categories: (1) Conformational isomers differ from each other only in the way their atoms are oriented in space, but can be converted into one another by rotation about sigma bonds. (2) Geometric isomers are compounds in which rotation about a double bond is restricted. (3) Configurational isomers differ from one another only in configuration about a chiral center, axis, or plane. In subsequent structural representations, a broken line denotes a bond projecting behind the plane of the paper and a wedge denotes a bond projecting in front of the plane of the paper. A line of normal thickness denotes a bond lying essentially in the plane of the paper.

1.1.4.1 Conformational Isomers.
A molecule in a conformation into which its atoms return spontaneously after small displacements is termed a *conformer*. Different arrangements of atoms that can be converted into one another by rotation about single bonds are called *conformational isomers* (see Fig. 1.1). A pair of conformational isomers can be but do not have to be mirror images of each other. When they are not mirror images, they are called *diastereomers*.

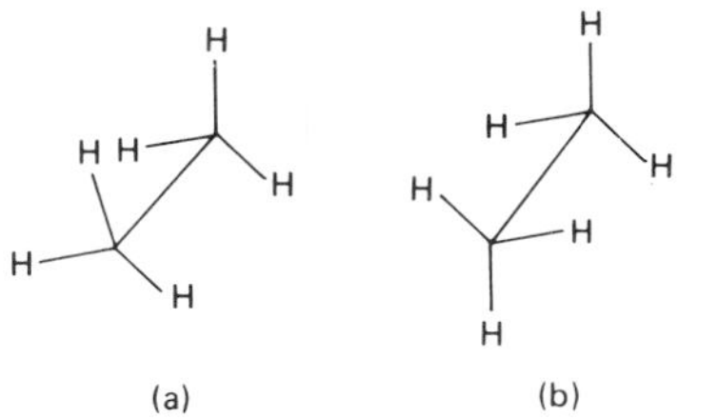

FIGURE 1.1 Conformations of ethane. (*a*) Eclipsed; (*b*) staggered.

Acyclic Compounds. Different conformations of acyclic compounds are best viewed by construction of ball-and-stick molecules or by use of Newman projections (see Fig. 1.2). Both types of representations are shown for ethane. Atoms or groups that are attached at opposite ends of a single bond should be viewed along the bond axis. If two atoms or groups attached at opposite ends of the bond appear one directly behind the other, these atoms or groups are

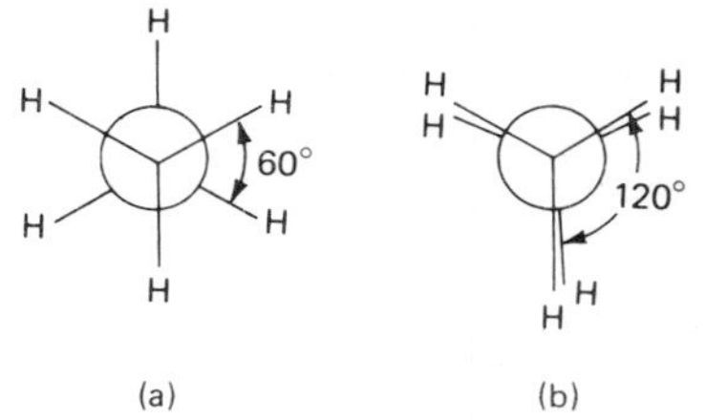

FIGURE 1.2 Newman projections for ethane.
(*a*) Staggered; (*b*) eclipsed.

described as eclipsed. That portion of the molecule is described as being in the eclipsed conformation. If not eclipsed, the atoms or groups and the conformation may be described as staggered. Newman projections show these conformations clearly.

Certain physical properties show that rotation about the single bond is not quite free. For ethane there is an energy barrier of about 3 kcal·mol^{-1} (12 kJ·mol^{-1}). The potential energy of the molecule is at a minimum for the staggered conformation, increases with rotation, and reaches a maximum at the eclipsed conformation. The energy required to rotate the atoms or groups about the carbon-carbon bond is called *torsional energy.* Torsional strain is the cause of the relative instability of the eclipsed conformation or any intermediate skew conformations.

In butane, with a methyl group replacing one hydrogen on each carbon of ethane, there are several different staggered conformations (see Fig. 1.3). There is the *anti*-conformation in which the methyl groups are as far apart as they can be (dihedral angle of 180°). There are two *gauche* conformations in which the methyl groups are only 60° apart; these are two nonsuperimposable mirror images of each other. The *anti*-conformation is more stable than the *gauche* by about 0.9 kcal·mol^{-1} (4 kJ·mol^{-1}). Both are free of torsional strain. However, in a *gauche* conformation the methyl groups are closer together than the sum of their van der Waals' radii. Under these conditions van der Waals' forces are repulsive and raise the energy of conforma-

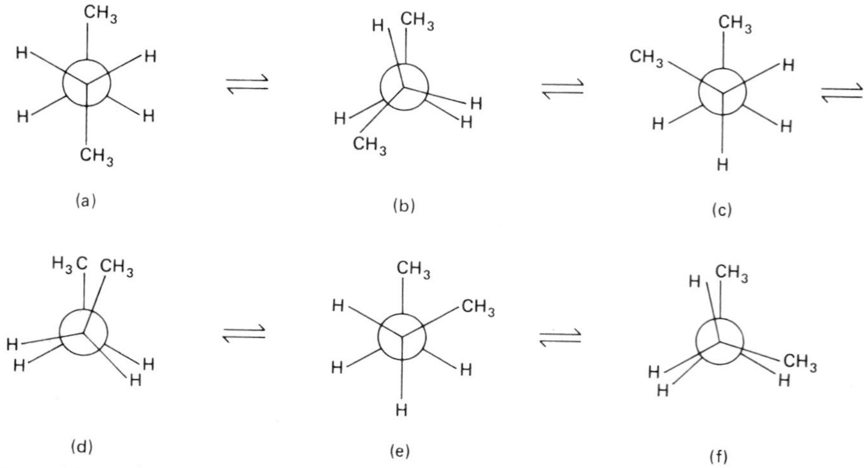

FIGURE 1.3 Conformations of butane. (*a*) *Anti*-staggered; (*b*) eclipsed; (*c*) *gauche*-staggered; (*d*) eclipsed; (*e*) *gauche*-staggered; (*f*) eclipsed. (Eclipsed conformations are slightly staggered for convenience in drawing; actually they are superimposed.)

tion. This strain can affect not only the relative stabilities of various staggered conformations but also the heights of the energy barriers between them. The energy maximum (estimated at 4.8 to 6.1 kcal·mol^{-1} or 20 to 25 kJ·mol^{-1}) is reached when two methyl groups swing past each other (the eclipsed conformation) rather than past hydrogen atoms.

Cyclic Compounds. Although cyclic aliphatic compounds are often drawn as if they were planar geometric figures (a triangle for cyclopropane, a square for cyclobutane, and so on), their structures are not that simple. Cyclopropane does possess the maximum angle strain if one considers the difference between a tetrahedral angle (109.5°) and the 60° angle of the cyclopropane structure. Nevertheless the cyclopropane structure is thermally quite stable. The highest electron density of the carbon-carbon bonds does not lie along the lines connecting the carbon atoms. Bonding electrons lie principally outside the triangular internuclear lines and result in what is known as *bent bonds* (see Fig. 1.4).

Cyclobutane has less angle strain than cyclopropane (only 19.5°). It is also believed to have some bent-bond character associated with the carbon-carbon bonds. The molecule exists in a nonplanar conformation in order to minimize hydrogen-hydrogen eclipsing strain.

Cyclopentane is nonplanar, with a structure that resembles an envelope (see Fig. 1.5). Four of the carbon atoms are in one plane, and the fifth is out of that plane. The molecule is in continual motion so that the out-of-plane carbon moves rapidly around the ring.

The 12 hydrogen atoms of cyclohexane do not occupy equivalent positions. In the chair conformation six hydrogen atoms are perpendicular to the average plane of the molecule and six are directed outward from the ring, slightly above or below the molecular plane (see Fig. 1.6). Bonds which are perpendicular to the molecular plane are known as *axial bonds,* and

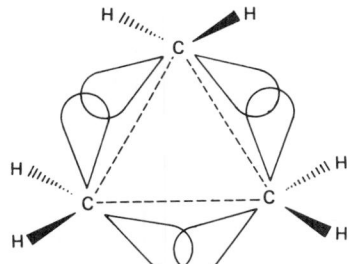

FIGURE 1.4 The bent bonds ("tear drops") of cyclopropane.

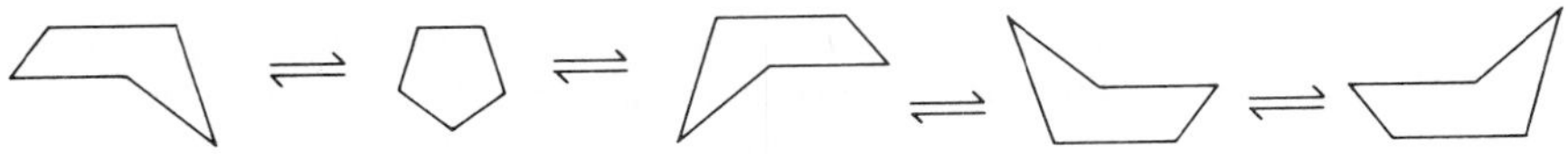

FIGURE 1.5 The conformations of cyclopentane.

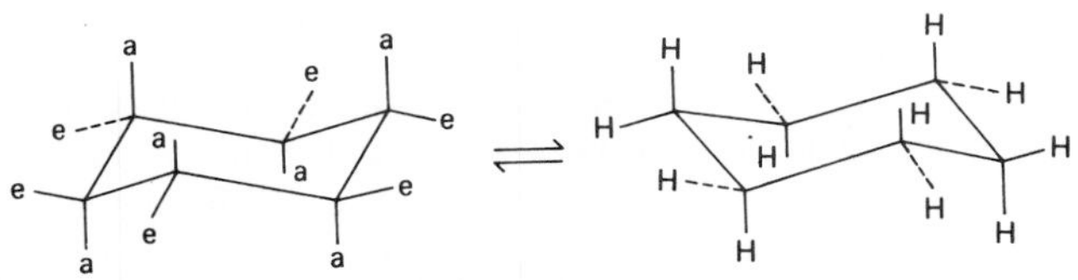

FIGURE 1.6 The two chair conformations of cyclohexane: *a* = axial hydrogen atom and *e* = equatorial hydrogen atom.

those which extend outward from the ring are known as *equatorial bonds*. The three axial bonds directed upward originate from alternate carbon atoms and are parallel with each other; a similar situation exists for the three axial bonds directed downward. Each equatorial bond is drawn so as to be parallel with the ring carbon-carbon bond once removed from the point of attachment to that equatorial bond. At room temperature, cyclohexane is interconverting rapidly between two chair conformations. As one chair form converts to the other, all the equatorial hydrogen atoms become axial and all the axial hydrogens become equatorial. The interconversion is so rapid that all hydrogen atoms on cyclohexane can be considered equivalent. Interconversion is believed to take place by movement of one side of the chair structure to produce the twist boat, and then movement of the other side of the twist boat to give the other chair form. The chair conformation is the most favored structure for cyclohexane. No angle strain is encountered since all bond angles remain tetrahedral. Torsional strain is minimal because all groups are staggered.

In the boat conformation of cyclohexane (Fig. 1.7) eclipsing torsional strain is significant, although no angle strain is encountered. Nonbonded interaction between the two hydrogen atoms across the ring from each other (the "flagpole" hydrogens) is unfavorable. The boat conformation is about 6.5 kcal · mol^{-1} (27 kJ · mol^{-1}) higher in energy than the chair form at 25°C.

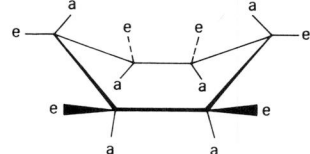

FIGURE 1.7   The boat conformation of cyclohexane. *a* = axial hydrogen atom and *e* = equatorial hydrogen atom.

FIGURE 1.8   Twist-boat conformation of cyclohexane.

A modified boat conformation of cyclohexane, known as the twist boat (Fig. 1.8), or skew boat, has been suggested to minimize torsional and nonbounded interactions. This particular conformation is estimated to be about 1.5 kcal · mol^{-1} (6 kJ · mol^{-1}) lower in energy than the boat form at room temperature.

The medium-size rings (7 to 12 ring atoms) are relatively free of angle strain and can easily take a variety of spatial arrangements. They are not large enough to avoid all nonbonded interactions between atoms.

Disubstituted cyclohexanes can exist as *cis-trans* isomers as well as axial-equatorial conformers. Two isomers are predicted for 1,4-dimethylcyclohexane (see Fig. 1.9). For the *trans* isomer the diequatorial conformer is the energetically favorable form. Only one *cis* isomer is observed, since the two conformers of the *cis* compound are identical. Interconversion takes place between the conformational (equatorial-axial) isomers but not configurational (*cis-trans*) isomers.

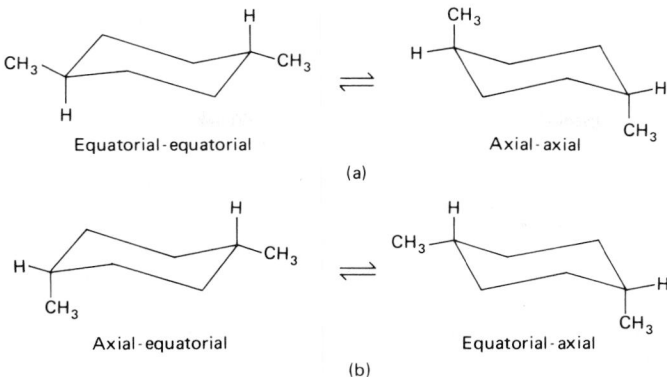

Equatorial-equatorial   Axial-axial

(a)

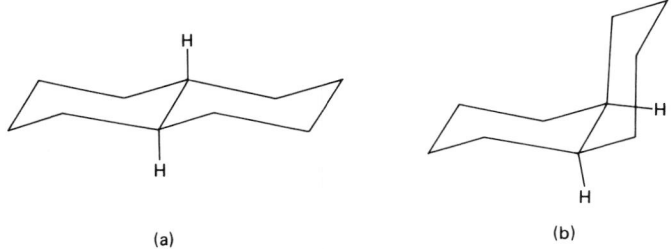

Axial-equatorial   Equatorial-axial

(b)

FIGURE 1.9   Two isomers of 1,4-dimethylcyclohexane. (*a*) *Trans* isomer;
(*b*) *cis* isomer.

FIGURE 1.10   Two isomers of decahydronaphthalene, or bicyclo[4.4.0]de-
cane. (*a*) *Trans* isomer; (*b*) *cis* isomer.

The bicyclic compound decahydronaphthalene, or bicyclo[4.4.0]decane, has two fused
six-membered rings. It exists in *cis* and *trans* forms (see Fig. 1.10), as determined by the
configurations at the bridgehead carbon atoms. Both *cis*- and *trans*-decahydronaphthalene
can be constructed with two chair conformations.

1.1.4.2   Geometrical Isomerism.   Rotation about a carbon-carbon double bond is restricted
because of interaction between the *p* orbitals which make up the pi bond. Isomerism due to
such restricted rotation about a bond is known as *geometric isomerism*. Parallel overlap of the
p orbitals of each carbon atom of the double bond forms the molecular orbital of the pi bond.
The relatively large barrier to rotation about the pi bond is estimated to be nearly
63 kcal·mol^{-1} (263 kJ·mol^{-1}).

When two different substituents are attached to each carbon atom of the double bond,
cis-trans isomers can exist. In the case of *cis*-2-butene (Fig. 1.11*a*), both methyl groups are on
the same side of the double bond. The other isomer has the methyl groups on opposite sides
and is designated as *trans*-2-butene (Fig. 1.11*b*). Their physical properties are quite different.
Geometric isomerism can also exist in ring systems; examples were cited in the previous
discussion on conformational isomers.

For compounds containing only double-bonded atoms, the reference plane contains the
double-bonded atoms and is perpendicular to the plane containing these atoms and those
directly attached to them. It is customary to draw the formulas so that the reference plane is

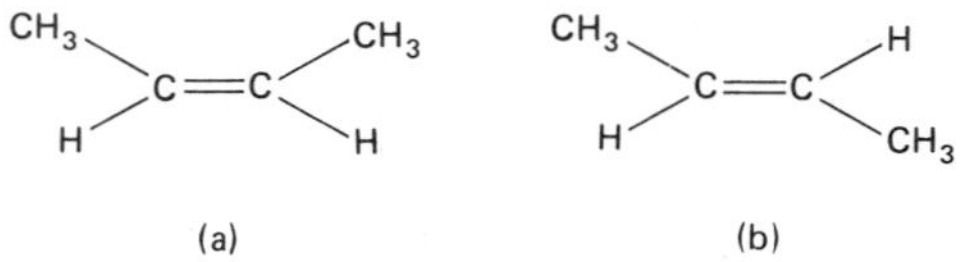

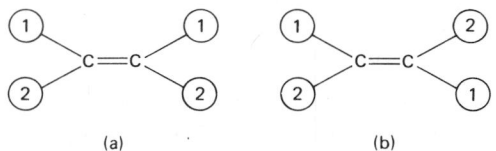

FIGURE 1.11 Two isomers of 2-butene. (*a*) *Cis* isomer, bp 3.8°C, mp −138.9°C, dipole moment 0.33 D; (*b*) *trans* isomer, bp 0.88°C, mp −105.6°C, dipole moment 0 D.

perpendicular to that of the paper. For cyclic compounds the reference plane is that in which the ring skeleton lies or to which it approximates. Cyclic structures are commonly drawn with the ring atoms in the plane of the paper.

1.1.4.3 *Sequence Rules for Geometric Isomers and Chiral Compounds.*

Although *cis* and *trans* designations have been used for many years, this approach becomes useless in complex systems. To eliminate confusion when each carbon of a double bond or a chiral center is connected to different groups, the Cahn, Ingold, and Prelog system for designating configuration about a double bond or a chiral center has been adopted by IUPAC. Groups on each carbon atom of the double bond are assigned a first (1) or second (2) priority. Priority is then compared at one carbon relative to the other. When both first priority groups are on the *same side* of the double bond, the configuration is designated as *Z* (from the German *zusammen,* "together"), which was formerly *cis*. If the first priority groups are on *opposite sides* of the double bond, the designation is *E* (from the German *entgegen,* "in opposition to"), which was formerly *trans*. (See Fig. 1.12.)

FIGURE 1.12 Configurations designated by priority groups. (*a*) *Z* (*cis*); (*b*) *E* (*trans*).

When a molecule contains more than one double bond, each *E* or *Z* prefix has associated with it the lower-numbered locant of the double bond concerned. Thus (see also the rules that follow)

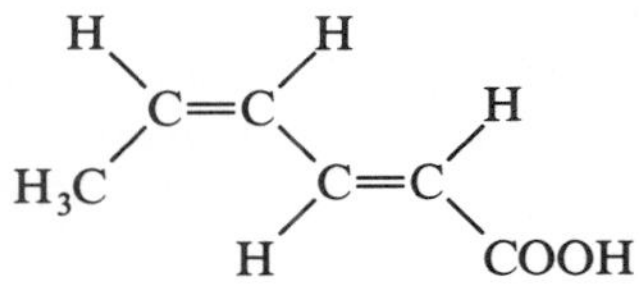

(2*E*,4*Z*)-2,4-Hexadienoic acid

When the sequence rules permit alternatives, preference for lower-numbered locants and for inclusion in the principal chain is allotted as follows in the order stated: *Z* over *E* groups and *cis* over *trans* cyclic groups. If a choice is still not attained, then the lower-numbered locant for such a preferred group at the first point of difference is the determining factor. For example,

(2Z,5E)-2,5-Heptadienedioic acid

Rule 1. Priority is assigned to atoms on the basis of atomic number. Higher priority is assigned to atoms of higher atomic number. If two atoms are isotopes of the same element, the atom of higher mass number has the higher priority. For example, in 2-butene, the carbon atom of each methyl group receives first priority over the hydrogen atom connected to the same carbon atom. Around the asymmetric carbon atom in chloroiodomethanesulfonic acid, the priority sequence is I, Cl, S, H. In 1-bromo-1-deuteroethane, the priority sequence is Cl, C, D, H.

Rule 2. When atoms attached directly to a double-bonded carbon have the same priority, the second atoms are considered and so on, if necessary, working outward once again from the double bond or chiral center. For example, in 1-chloro-2-methylbutene, in CH_3 the second atoms are H, H, H and in CH_2CH_3 they are C, H, H. Since carbon has a higher atomic number than hydrogen, the ethyl group has the next highest priority after the chlorine atom.

(Z)-1-Chloro-2-methylbutene   (E)-1-Chloro-2-methylbutene

Rule 3. When groups under consideration have double or triple bonds, the multiple-bonded atom is replaced conceptually by two or three single bonds to that same kind of atom. Thus, $=$A is considered to be equivalent to two A's, or $<^A_A$ and $\equiv$A equals $<^A_{<^A_A}$.

However, a real $<^A_A$ has priority over $=$A; likewise a real $<^A_{<^A_A}$ has priority over $\equiv$A.

Actually, both atoms of a multiple bond are duplicated, or triplicated, so that C$=$O is treated as $\overset{C-O}{\underset{O\ \ C}{|\ \ |}}$, that is $\overset{C-O}{\underset{(O)}{|}}$ and $\overset{O-C}{\underset{(C)}{|}}$, and C$\equiv$N is treated as

$\underset{(N)}{}\overset{C\text{------}N}{\underset{(N)\ \ (C)}{}}\underset{(C)}{}$. A phenyl carbon becomes $-C{\overset{CH}{\underset{CH}{\diagup\!\!\diagdown}}}C$. Only the double-bonded atoms themselves are duplicated, not the atoms or groups attached to them. The duplicated atoms (or phantom atoms) may be considered as carrying atomic number zero. For example, among the groups OH, CHO, CH_2OH, and H, the OH group has the highest priority, and the C(O, O, H) of CHO takes priority over the C(O, H, H) of CH_2OH.

1.1.4.4 *Chirality and Optical Activity.*

A compound is chiral (the term *dissymmetric* was formerly used) if it is not superimposable on its mirror image. A chiral compound does not have a plane of symmetry. Each chiral compound possesses one (or more) of three types of chiral element, namely, a chiral center, a chiral axis, or a chiral plane.

Chiral Center. The chiral center, which is the chiral element most commonly met, is exemplified by an asymmetric carbon with a tetrahedral arrangement of ligands about the carbon. The ligands comprise four different atoms or groups. One "ligand" may be a lone pair of electrons; another, a phantom atom of atomic number zero. This situation is encountered in sulfoxides or with a nitrogen atom. Lactic acid is an example of a molecule with an asymmetric (chiral) carbon. (See Fig. 1.13.)

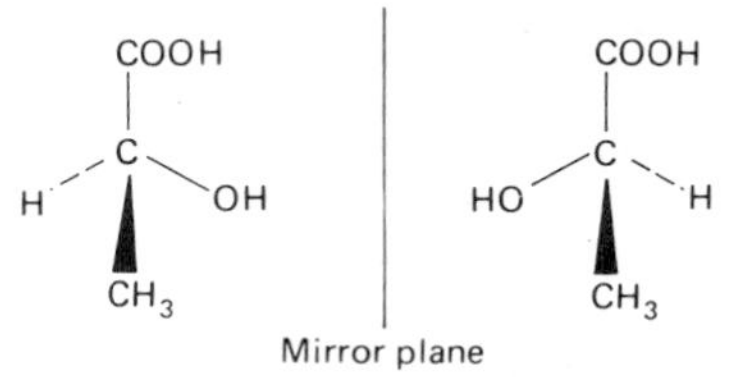

FIGURE 1.13 Asymmetric (chiral) carbon in the lactic acid molecule.

A simpler representation of molecules containing asymmetric carbon atoms is the Fischer projection, which is shown here for the same lactic acid configurations. A Fischer projection

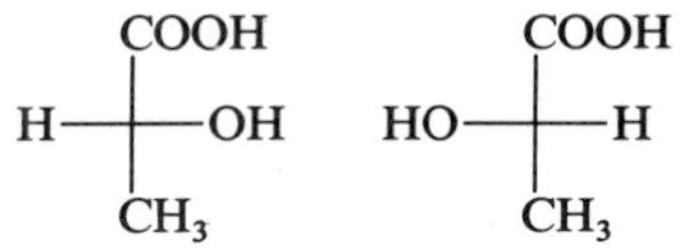

involves drawing a cross and attaching to the four ends the four groups that are attached to the asymmetric carbon atom. The asymmetric carbon atom is understood to be located where the line cross. The horizontal lines are understood to represent bonds coming toward the viewer out of the plane of the paper. The vertical lines represent bonds going away from the viewer behind the plane of the paper as if the vertical line were the side of a circle. The principal chain is depicted in the vertical direction; the lowest-numbered (locant) chain member is placed at the top position. These formulas may be moved sideways or rotated through 180° in the plane of the paper, but they may not be removed from the plane of the paper (i.e., rotated through 90°). In the latter orientation it is essential to use thickened lines (for bonds coming toward the viewer) and dashed lines (for bonds receding from the viewer) to avoid confusion.

Enantiomers. Two nonsuperimposable structures that are mirror images of each other are known as *enantiomers.* Enantiomers are related to each other in the same way that a right hand is related to a left hand. Except for the direction in which they rotate the plane of polarized light, enantiomers are identical in all physical properties. Enantiomers have identical chemical properties except in their reactivity toward optically active reagents.

Enantiomers rotate the plane of polarized light in opposite directions but with equal magnitude. If the light is rotated in a clockwise direction, the sample is said to be dextrorotatory and is designed as (+). When a sample rotates the plane of polarized light in a counterclockwise direction, it is said to be levorotatory and is designed as (−). Use of the designations *d* and *l* is discouraged.

Specific Rotation. Optical rotation is caused by individual molecules of the optically active compound. The amount of rotation depends upon how many molecules the light beam encounters in passing through the tube. When allowances are made for the length of the tube that contains the sample and the sample concentration, it is found that the amount of rotation, as well as its direction, is a characteristic of each individual optically active compound.

Specific rotation is the number of degrees of rotation observed if a 1-dm tube is used and the compound being examined is present to the extent of 1 g per 100 mL. The density for a pure liquid replaces the solution concentration.

$$\text{Specific rotation} = [\alpha] = \frac{\text{observed rotation (degrees)}}{\text{length (dm)} \times (\text{g/100 mL})}$$

The temperature of the measurement is indicated by a superscript and the wavelength of the light employed by a subscript written after the bracket; for example, $[\alpha]_{590}^{20}$ implies that the measurement was made at 20°C using 590-nm radiation.

Optically Inactive Chiral Compounds. Although chirality is a necessary prerequisite for optical activity, chiral compounds are not necessarily optically active. With an equal mixture of two enantiomers, no net optical rotation is observed. Such a mixture of enantiomers is said to be *racemic* and is designated as ($\pm$) and not as *dl*. Racemic mixtures usually have melting points higher than the melting point of either pure enantiomer.

A second type of optically inactive chiral compounds, *meso* compounds, will be discussed in the next section.

Multiple Chiral Centers. The number of stereoisomers increases rapidly with an increase in the number of chiral centers in a molecule. A molecule possessing two chiral atoms should have four optical isomers, that is, four structures consisting of two pairs of enantiomers. However, if a compound has two chiral centers but both centers have the same four substituents attached, the total number of isomers is three rather than four. One isomer of such a compound is not chiral because it is identical with its mirror image; it has an internal mirror plane. This is an example of a diastereomer. The achiral structure is denoted as a *meso* compound. Diastereomers have different physical and chemical properties from the optically active enantiomers. Recognition of a plane of symmetry is usually the easiest way to detect a *meso* compound. The stereoisomers of tartaric acid are examples of compounds with multiple chiral centers (see Fig. 1.14), and one of its isomers is a *meso* compound.

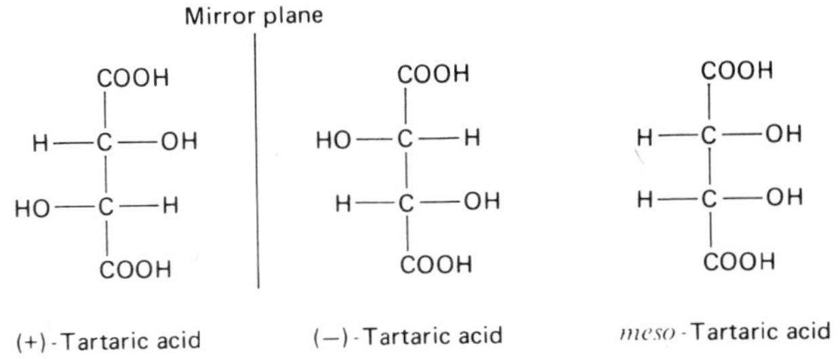

FIGURE 1.14 Isomers of tartaric acid.

When the asymmetric carbon atoms in a chiral compound are part of a ring, the isomerism is more complex than in acyclic compounds. A cyclic compound which has two different asymmetric carbons with different sets of substituent groups attached has a total of $2^2 = 4$ optical isomers: an enantiometric pair of *cis* isomers and an enantiometric pair of *trans* isomers. However, when the two asymmetric centers have the same set of substituent groups attached, the *cis* isomer is a *meso* compound and only the *trans* isomer is chiral. (See Fig. 1.15.)

Torsional Asymmetry. Rotation about single bonds of most acyclic compounds is relatively free at ordinary temperatures. There are, however, some examples of compounds in which nonbonded interactions between large substituent groups inhibit free rotation about a sigma bond. In some cases these compounds can be separated into pairs of enantiomers.

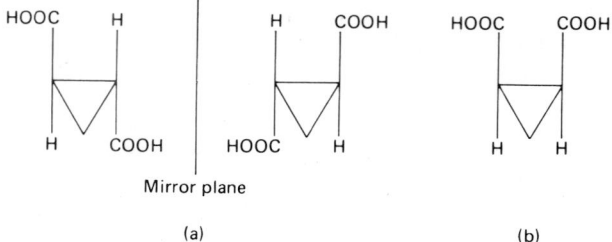

FIGURE 1.15   Isomers of cyclopropane-1,2-dicarboxylic acid. (*a*) *Trans* isomer; (*b*) *meso* isomer.

A *chiral axis* is present in chiral biaryl derivatives. When bulky groups are located at the *ortho* positions of each aromatic ring in biphenyl, free rotation about the single bond connecting the two rings is inhibited because of torsional strain associated with twisting rotation about the central single bond. Interconversion of enantiomers is prevented (see Fig. 1.16).

For compounds possessing a chiral axis, the structure can be regarded as an elongated tetrahedron to be viewed along the axis. In deciding upon the absolute configuration it does not matter from which end it is viewed; the nearer pair of ligands receives the first two positions in the order of precedence (see Fig. 1.17). For the meaning of (*S*), see the discussion under Absolute Configuration on p. 1.49.

A *chiral plane* is exemplified by the plane containing the benzene ring and the bromine and oxygen atoms in the chiral compound shown in Fig. 1.18. Rotation of the benzene ring around the oxygen-to-ring single bonds is inhibited when x is small (although no critical size can be reasonably established).

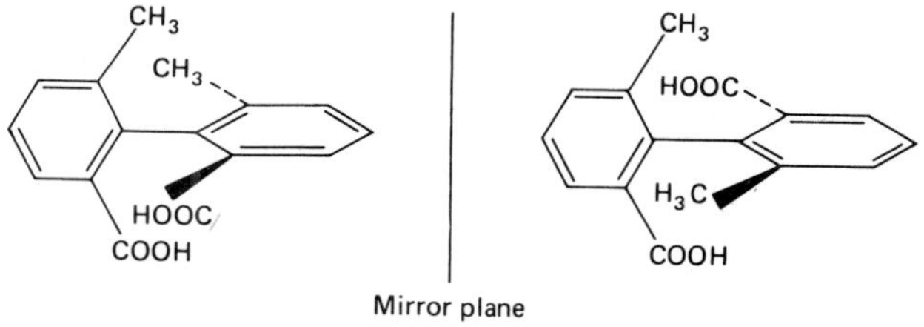

FIGURE 1.16   Isomers of biphenyl compounds with bulky groups attached at the *ortho* positions.

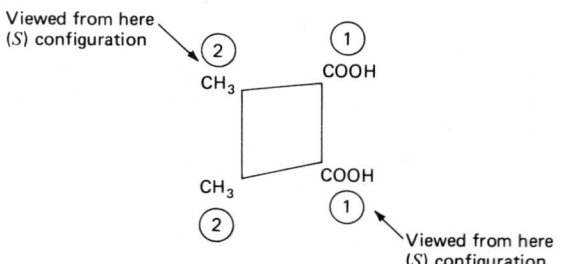

FIGURE 1.17   Example of a chiral axis.

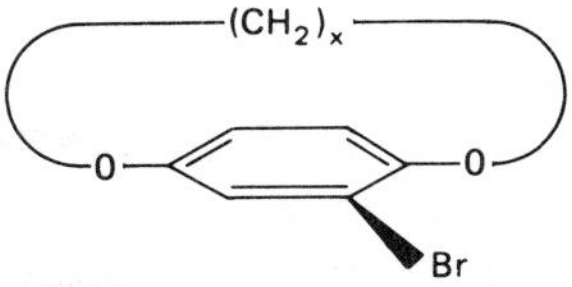

FIGURE 1.18 Example of a chiral plane.

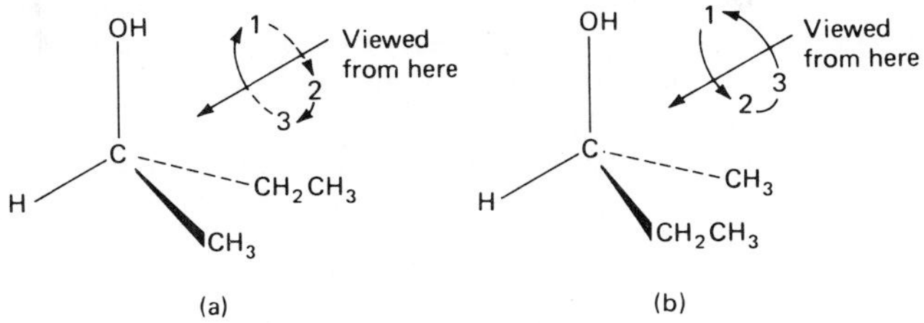

(a)   (b)

FIGURE 1.19 Viewing angle as a means of designating the absolute configuration of compounds with a chiral axis. (*a*) (*R*)-2-Butanol (sequence clockwise); (*b*) (*S*)-2-butanol (sequence counterclockwise).

Absolute Configuration. The terms absolute stereochemistry and absolute configuration are used to describe the three-dimensional arrangement of substituents around a chiral element. A general system for designating absolute configuration is based upon the priority system and sequence rules. Each group attached to a chiral center is assigned a number, with number one the highest-priority group. For example, the groups attached to the chiral center of 2-butanol (see Fig. 1.19) are assigned these priorities: 1 for OH, 2 for CH_2CH_3, 3 for CH_3, and 4 for H. The molecule is then viewed from the side opposite the group of lowest priority (the hydrogen atom), and the arrangement of the remaining groups is noted. If, in proceeding from the group of highest priority to the group of second priority and thence to the third, the eye travels in a clockwise direction, the configuration is specified *R* (from the Latin *rectus,* "right"); if the eye travels in a counterclockwise direction, the configuration is specified *S* (from the Latin *sinister,* "left"). The complete name includes both configuration and direction of optical rotation, as for example, (*S*)-(+)-2-butanol.

The relative configurations around the chiral centers of many compounds have been established. One optically active compound is converted to another by a sequence of chemical reactions which are stereospecific; that is, each reaction is known to proceed spatially in a specific way. The configuration of one chiral compound can then be related to the configuration of the next in sequence. In order to establish absolute configuration, one must carry out sufficient stereospecific reactions to relate a new compound to another of known absolute configuration. Historically the configuration of D-(+)-2,3-dihydroxypropanal has served as the standard to which all configuration has been compared. The absolute configuration assigned to this compound has been confirmed by an X-ray crystallographic technique.

1.1.5 *Chemical Abstracts* Indexing System

When compounds of complex structure are considered, the number of name possibilities grows rapidly. To avoid having index entries for all possible names, Chemical Abstracts Service has developed what might be called the principle of inversion. The indexing system employs inverted entries to bring together related compounds in an alphabetically arranged index. The *index heading parent* from the Chemical Substance Index appears in the Formula Index in lightface before the "comma of inversion." The *substituents* follow the "comma of inversion" in alphabetical order. Any *name modification* appears on a separate line. If necessary, the chemical description is completed by citation of an associated ion, a functional derivative, a "salt with" or "compound with" term, and/or a stereochemical descriptor.

Quite naturally there is a certain amount of arbitrariness in this system, although the IUPAC nomenclature is followed. The preferred *Chemical Abstracts* index names for chemical substances have been, with very few exceptions, continued unchanged (since 1972) as set forth in the *Ninth Collective Index Guide* and in a journal article.* Any revisions appear in the updated *Index Guide;* new editions appear at 18-month intervals. Appendix VI is of particular interest to chemists. Reprints of the Appendix may be purchased from Chemical Abstracts Service, Marketing Division, P.O. Box 3012, Columbus, Ohio 43210.

---

* *J. Chem. Doc.* **14**(1):3–15 (1974).

TABLE 1.13 Names and Formulas of Organic Radicals

For more comprehensive lists, see the various lists of radicals given in the subject indexes of the annual and decennial indexes of Chemical Abstracts.

Name	Formula	Name	Formula
Acenaphthenyl	$C_{12}H_9$—	Azido	N_3—
Acenaphthenylene	—$C_{12}H_8$—	Azino	$=N-N=$
Acenaphthenylidene	$C_{12}H_8=$	Azo	$-N=N-$
Acetamido	$CH_3-CO-NH-$	Azoxy	$-N(O)-N-$
Acetimidoyl	$CH_3C(=NH)-$	Azulenyl	$C_{10}H_7-$
Acetoacetyl	$CH_3-CO-CH_2-CO-$	Benzamido	$C_6H_5-CO-NH-$
Acetohydrazonoyl	$CH_3-C(=NNH_2)-$	Benzeneazo	$C_6H_5-N=N-$
Acetohydroximoyl	$CH_3-C(=NOH)-$	Benzeneazoxy	$C_6H_5-N_2O-$
Acetonyl	$CH_3-CO-CH_2-$	1,2-Benzenedicarbonyl,	
Acetonylidene	$CH_3-CO-CH=$	see Phthaloyl	
Acetoxy	$CH_3-CO-O-$	1,3-Benzenedicarbonyl	—$CO-C_6H_4-CO-$
Acetyl (*not ethanoyl*)	CH_3-CO-	(or *isophthaloyl*)	(*m*-)
Acetylamino	$CH_3-CO-NH-$	1,4-Benzenedicarbonyl	—$CO-C_6H_4-CO-$
Acetylhydrazino	$CH_3-CO-NH-NH-$	(or *terephthaloyl*)	(*p*-)
Acetylimino	$CH_3-CO-N=$	Benzenesulfinyl	C_6H_5-SO-
Acridinyl (*from acridine*)	$NC_{13}H_8-$	Benzenesulfonamido	$C_6H_5-SO_2-NH-$
Acroyloyl (*or propenoyl*)	$CH_2=CH-CO-$	Benzenesulfonyl	$C_6H_5-SO_2-$
Adipoyl (*or hexanedioyl*)	—$CO-[CH_2]_4-CO-$	Benzenesulfonylamino	$C_6H_5-SO_2-NH-$
Alanyl	$CH_3-CH(NH_2)-CO-$	Benzenetriyl	C_6H_3-
β-Alanyl	$H_2N-CH_2-CH_2-CO-$	Benzhydryl (*or diphenyl-*	$(C_6H_5)_2CH-$
Allyl (*or 2-propenyl*)	$CH_2=CH-CH_2-$	*methyl*)	
Allylidene	$CH_2=CH-CH=$	Benzidino	p-$H_2N-C_6H_4-C_6H_4-$
Allyloxy	$CH_2=CH-CH_2-O-$		$NH-$
Amidino	$H_2N-C(=NH)-$	Benziloyl (*or 2-hydroxy-*	$(C_6H_5)_2C(OH)-CO-$
Amino	H_2N-	*2,2-diphenylethanoyl*)	
Aminomethyleneamino	$H_2N-CH=N-$	Benzimidazolyl	$N_2C_7H_5-$
Aminooxy	H_2N-O-	Benzimidoyl	$C_6H_5-C(=NH)-$
Ammonio	$^+H_3N-$	Benzofuranyl	OC_8H_5-
Amyl, see Pentyl		Benzopyranyl	OC_9H_7-
Anilino	C_6H_5-NH-	Benzoquinonyl (1,2- or	$(O=)_2C_6H_3-$
Anisidino (*o-, m-, or*	$CH_3O-C_6H_4-NH-$	1,4-)	
p-)		Benzo[*b*]thienyl	SC_8H_5-
Anisoyl (*o-, m-, or*	$CH_3O-C_6H_4-CO-$	Benzoyl	C_6H_5-CO-
p-; *or methoxyben-*		Benzoylamino	$C_6H_5-CO-NH-$
zoyl)		Benzoylhydrazino	$C_6H_5-CO-NH-NH-$
Anthraniloyl	o-$NH_2-C_6H_4-CO-$	Benzoylimino	$C_6H_5-CO-N=$
Anthryl (*from anthracene*)	$C_{14}H_9-$	Benzoyloxy	$C_6H_5-CO-O-$
Anthrylene	—$C_{14}H_8-$	Benzyl	$C_6H_5-CH_2-$
Arginyl	$H_2N-C(=NH)-NH-$	Benzylidene	$C_6H_5-CH=$
	$[CH_2]_3-CH(NH)-$	Benzylidyne	$C_6H_5-C\equiv$
	$CO-$	Benzyloxy	$C_6H_5-CH_2-O-$
Asparaginyl	$H_2N-CO-CH_2-$	Benzyloxycarbonyl	$C_6H_5-CH_2-O-CO-$
	$CH(NH_2)-CO-$	Benzylthio	$C_6H_5-CH_2-S-$
Aspartoyl	—$CO-CH_2-$	Biphenylenyl	$C_{12}H_7-$
	$CH(NH_2)-CO-$	Biphenylyl	$C_6H_5-C_6H_4-$
α-Aspartyl	$HO_2C-CH_2CH(NH_2)-$	Bornenyl	$C_{10}H_{15}-$
Atropoyl (*or 2-*	$C_6H_5-C(=CH_2)-CO-$	Bornyl (*not camphyl or*	$C_{10}H_{17}-$
phenylpropenoyl)		*bornylyl*)	
Azelaoyl, *see* Nonane-		Bromo	$Br-$
dioyl		Bromoformyl	$Br-CO-$

TABLE 1.13 Names and Formulas of Organic Radicals (*Continued*)

Name	Formula	Name	Formula
Bromonio	+HBr—	Cinnamoyl (*or 3-phenylpropenoyl*)	C_6H_5—CH=CH—CO
Butadienyl (1,3- shown)	CH_2=CH—CH=CH—	Cinnamyl	C_6H_5—CH=CH—CH_2—
Butanedioyl, *see* Succinyl		Cinnamylidene	C_6H_5—CH=CH—CH=
Butanediylidene	=CH—CH_2—CH_2—CH=	Citraconoyl (*unsubstituted only*)	HC — CO —
Butanediylidyne	≡C—CH_2—CH_2—C≡		CH_3 — C — CO —
Butanoyl, *see* Butyryl		Crotonoyl	CH_3—CH=CH—CO— (*trans*)
cis-Butenedioyl, *see* Maleoyl		Crotyl, *see* 2-Butenyl	
trans-Butenedioyl, *see* Fumaroyl Butenoyl, *see* Crotonoyl and Isocrotonoyl		Cumenyl (*o-, m-, or p-*)	$(CH_3)_2$CH—C_6H_4—
		Cyanato	NCO—
		Cyano	NC—
1-Butenyl	CH_3—CH_2—CH=CH—	Cyclobutyl	C_4H_7—
2-Butenyl (*not crotyl*)	CH_3—CH=CH—CH_2—	Cycloheptyl	C_7H_{13}—
2-Butenylene	—CH_2—CH=CH—CH_2—	Cyclohexadienyl (2,4- shown)	CH = CH — CH — | CH = CH — CH_2
Butenylidene (*2- shown*)	CH_3CH=CH—CH=	Cyclohexadienylidene (2,4- shown)	CH — CH_2 — C | CH — CH = CH
Butenylidyne (*2- shown*)	CH_3—CH=CH—C≡		
Butoxy	CH_3—$[CH_2]_3$—O—	Cyclohexanecarbonyl	C_6H_{11}—CO—
sec-Butoxy (*unsubstituted, only*)	C_2H_5—CH(CH_3)—O—	Cyclohexanecarbothioyl	C_6H_{11}—CS—
		Cyclohexanecarboxamido	C_6H_{11}—CO—NH—
tert-Butoxy (*unsubstituted only*)	$(CH_3)_3$C—O—	Cyclohexanecarboximidoyl	C_6H_{11}—C(=NH)—
Butyl	CH_3—$[CH_2]_3$— or C_4H_9—	Cyclohexenyl	C_6H_9—
		2-Cyclohexenylidene	CH = CH — C | H_2C — CH_2 — CH_2
sec-Butyl (*unsubstituted only*)	C_2H_5—CH(CH_3)—	Cyclohexyl	C_6H_{11}—
tert-Butyl (*unsubstituted only*)	$(CH_3)_3$C—	Cyclohexylcarbonyl	C_6H_{11}—CO—
		Cyclohexylene	—C_6H_{10}—
Butylidene	CH_3—CH_2—CH_2—CH=	Cyclohexylidene	CH_2 — CH_2 — C | CH_2 — CH_2 — CH_2
sec-Butylidene (*unsubstituted only*)	C_2H_5C(CH_3)=		
Butylidyne	CH_3—$[CH_2]_2$—C≡	Cyclohexylthiocarbonyl	C_6H_{11}—CS—
Butyryl (*or butanoyl*)	CH_3—CH_2—CH_2—CO—	Cyclopentadienyl	C_5H_5—
Camphoroyl	$C_{10}H_{14}O_2$—	Cyclopentadienylidene	CH=CH—CH=CH—C=
Carbamoyl	H_2N—CO—	Cyclopenta[*a*]phenanthryl	$C_{17}H_{1?}$—
Carbazolyl	$NC_{12}H_8$—	1,2-Cyclopentenophenanthryl	$C_{17}H_{11}$—
Carbazoyl	H_2N—NH—CO—		
Carbonimidoyl	—C(=NH)—	Cyclopentenyl	C_5H_7—
Carbonohydrazido (*preferred to carbohydazido or carbazido*)	H_2N—NH—CO—NH—NH—	Cyclopentyl	C_5H_9—
		Cyclopentylene	—C_5H_8—
		Cyclopropyl	C_3H_5—
Carbonyl	—CO— or =C(O)	Cysteinyl	HS—CH_2—CH(NH_2)—CO—
Carbonyldioxy	—O—CO—O—		
Carboxy	HO_2C—	Cystyl	—CO—CH(NH_2)— CH_2—S—S—CH_2— CH(NH_2)—CO—
Carboxylato	—O_2C—		
Chloro	Cl—		
Chlorocarbonyl, *see* Chloroformyl		Decanedioyl	—CO—$[CH_2]_8$—CO—
Chloroformyl	Cl—C(O)—	Decanoyl	CH_3—$[CH_2]_8$—CO—
Chlorosyl	OCl—	Decyl	CH_3—$[CH_2]_9$—
Chlorothio	ClS—	Diacetoxyiodo	$(CH_3$—CO—O$)_2$I—
Chloryl	O_2Cl—	Diacetylamino	$(CH_3$—CO$)_2$N—
		Diaminomethyleneamino	$(NH_2)_2$C=N—

TABLE 1.13 Names and Formulas of Organic Radicals (*Continued*)

Name	Formula	Name	Formula
Diazo	$=N_2$	Fluorenyl	$C_{13}H_9—$
Diazoamino	$—N=N—NH—$	Fluoro	$F—$
Dibenzoylamino	$(C_6H_5—CO)_2N—$	Fluoroformyl	$F—CO—$
Dichloroiodo	$Cl_2I—$	Formamido	$OCH—NH—$
Diethylamino	$(C_2H_5)_2N—$	Formimidoyl	$CH(=NH)—$
3,4-Dihydroxybenzoyl, *see* Protocatechuoyl		Formyl (*not* methanoyl)	$OCH—$ or $—C(O)H$
2,3-Dihydroxybutane-dioyl, *see* Tartaroyl		Formylamino	$H—CO—NH—$
Dihydroxyiodo	$(HO)_2I—$	Formylimino	$H—CO—N=$
2,3-Dihydroxypropanoyl, *see* Glyceroyl		Formyloxy	$H—CO—O—$
		Fumaroyl (*or trans-butenedioyl*)	$—CO—CH=CH—CO—$ (*trans*)
3,4-Dimethoxybenzoyl, *see* Veratroyl		Furancarbonyl, *see* Furoyl	
3,4-Dimethoxyphenethyl	3,4-$(CH_3O)_2C_6H_3CH_2CH—$	Furfuryl (*2- only; preferred to 2-furylmethyl*)	
3,4-Dimethoxyphenylacetyl	3,4-$(CH_3O)_2C_6H_3CH_2CO—$	Furfurylidene (*2- only*)	
Dimethylamino	$(CH_3)_2N—$	Furoyl (*3- shown; preferred to furancarbonyl*)	
Dimethylbenzoyl	$(CH_3)_2C_6H_6H_3—CO—$		
Dioxy	$—O—O—$	Furyl	$OC_4H_3—$
Diphenylamino	$(C_6H_5)_2N—$	3-Furylmethyl	
Diphenylmethylene	$(C_6H_5)_2C=$		
Dithio	$—S—S—$		
Diethiocarboxy	$HSSC—$	Galloyl (*or 3,4,5-trihydroxybenzoyl*)	3,4,5-$(HO)_3C_6H_2—CO—$
Dithiosulfo	$HOS_2—$		
Dodecanoyl	$CH_3[CH_2]_{10}—CO—$	Geranyl (*from geraniol*)	$C_{10}H_{17}—$
Dodecyl	$CH_3[CH_2]_{11}—$	Glutaminyl	$H_2N—CO—CH_2—CH_2—CH(NH_2)—CO—$
Elaidoyl (*or trans-9-octadecenoyl*)	$CH_3[CH_2]_7CH=CH—[CH_2]_7—CO—$	Glutamoyl	$—CO—CH_2—CH_2—CH(NH_2)—CO—$
Epidioxy (*as a bridge*)	$—O—O—$	α-Glutamyl	$HOOC[CH_2]_2CH(NH_2)—CO—$
Epidiseleno (*as bridge*)	$—Se—Se—$		
Epidithio (*as a bridge*)	$—S—S—$	γ-Glutamyl	$HOOC—CH(NH_2)—[CH_2]_2—CO—$
Epimino (*as a bridge*)	$—NH—$		
Episeleno (*as a bridge*)	$—Se—$		
Epithio (*as a bridge*)	$—S—$	Glutaryl (*or pentanedioyl*)	$—CO—[CH_2]_3—CO—$
Epoxy (*as a bridge*)	$—O—$	Glyceroyl (*or 2,3-dihydroxypropanoyl*)	$HO—CH_2—CH(OH)—CO—$
Ethanesulfonamide	$C_2H_5—SO_2—NH—$	Glycoloyl (*or hydroxyethanoyl*)	$HO—CH_2—CO—$
Ethanoyl, *see* Acetyl			
Ethenyl, *see* Vinyl		Glycyl	$H_2N—CH_2—CO—$
Ethoxalyl	$C_2H_5—OOC—CO—$	Glycylamino	$H_2N—CH_2—CO—NH—$
Ethoxy	$C_2H_5—O—$	Glyoxyloyl	$OHC—CO—$
Ethoxycarbonyl	$C_2H_5—O—CO—$	Guanidino	$H_2N—C(=NH)—NH—$
Ethyl	$C_2H_5—$ or $CH_3—CH_2—$	Guanyl, *see* Amidino	
Ethylamino	$C_2H_5—NH—$	Heptanamido	$CH_3—[CH_2]_5—CO—NH—$
Ethylene	$—CH_2—CH_2—$	Heptanedioyl	$—CO—[CH_2]_5—CO—$
Ethylenedioxy	$—O—CH_2—CH_2—O—$	Heptanoyl	$CH_3—[CH_2]_5—CO—$
Ethylidene	$CH_3—CH=$	Heptyl	$CH_3—[CH_2]_5—CH_2—$
Ethylidyne	$CH_3—C≡$	Hexadecanoyl	$CH_3—[CH_2]_{14}—CO—$
Ethylsulfonylamino	$C_2H_5—SO_2—NH—$	Hexadecyl	$CH_3—[CH_2]_{14}—CH_2—$
Ethylthio	$C_2H_5—S—$	Hexamethylene	$—[CH_2]_6—$
Ethynyl	$HC≡C—$	Hexanamido	$CH_3—[CH_2]_4—CO—NH—$
Ethynylene	$—C≡C—$	Hexanedioyl (*or adipoyl*)	$—CO—[CH_2]_4—CO—$
Fluoranthenyl	$C_{16}H_9—$		

TABLE 1.13 Names and Formulas of Organic Radicals (*Continued*)

Name	Formula	Name	Formula
Hexanimidoyl	$CH_3-[CH_2]_4-$ $C(=NH)-$	Iodonio	$^+HI-$
		Iodosyl	$OI-$
Hexanoyl	$CH_3-[CH_2]_4-CO-$	Iodyl	O_2I-
Hexanoylamino	$CH_3-[CH_2]_4-CO-$ $NH-$	Isobutoxy (*unsubstituted only*)	$(CH_3)_2CH-CH_2-O-$
Hexyl	$CH_3-[CH_2]_4-CH_2-$	Isobutyl (*unsubstituted only*)	$(CH_3)_2CH-CH_2-$
Hexylidene	$CH_3-[CH_2]_4-CH=$		
Hexyloxy	$CH_3[CH_2]_5-O-$	Isobutylidene (*unsubstituted only*)	$(CH_3)_2CH-CH=$
Hippuroyl	$C_6H_5-CO-NH-CH_2-$ $CO-$	Isobutylidyne (*unsubstituted only*)	$(CH_3)_2CH-C\equiv$
Histidyl	$N_2C_3H_3-CH_2-$ $CH(NH_2)-CO-$	Isobutyryl (*unsubstituted only; or 2-methylpropanoyl*)	$(CH_3)_2CH-CO-$
Homocysteinyl	$HS-CH_2-CH_2-$ $CH(NH_2-CO-$		
Homoseryl	$HO-CH_2-CH_2-$ $CH(NH_2)-CO-$	Isocarbonohydrazido	$H_2N-N=C(OH)-$ $NH-NH-$
Hydantoyl	$H_2N-CO-NH-CH_2-$ $CO-$	Isocrotonoyl	$CH_3-CH=CH-CO-$ (*cis*)
Hydratropoyl (*or 2-phenylpropanoyl*)	$C_6H_5-CH(CH_3)-CO-$	Isocyanato	$OCN-$
		Isocyano	$CN-$
Hydrazi	$-NH-NH-$ (to single atom)	Isohexyl (*unsubstituted only*)	$(CH_3)_2CH-[CH_2]_3-$
Hydrazino	H_2N-NH-	Isoleucyl	$C_2H_5-CH(CH_3)-$ $CH(NH_2)-CO$
Hydrazo	$-NH-NH-$ (to different atoms)	Isonicotinoyl (*or 4-pyridinecarbonyl*)	NC_5H_4-CO- (4-)
Hydrazono	$H_2N-N=$		
Hydroperoxy	$HO-O-$	Isopentyl (*unsubstituted only*)	$(CH_3)_2CH-CH_2-CH_2-$
Hydroseleno	$HSe-$		
Hydroxy	$HO-$	Isophthaloyl (*or 1,3-benzenedicarbonyl*)	$-CO-C_6H_4-CO-$ (*m-*)
Hydroxyamino	$HO-NH-$		
o-Hydroxybenzoyl (*or salicyloyl*)	$o\text{-}HO-C_6H_4-CO-$	Isopropenyl (*unsubstituted only; or 1-methylvinyl*)	$CH_2=C(CH_3)-$
m-Hydroxybenzoyl	$m\text{-}HO-C_6H_4-CO-$	Isopropoxy (*unsubstituted only*)	$(CH_3)_2CH-O-$
p-Hydroxybenzoyl	$p\text{-}HO-C_6H_4-CO-$		
Hydroxybutanedioyl, *see* Maloyl		Isopropyl (*unsubstituted only*)	$(CH_3)_2CH-$
2-Hydroxy-2,2-diphenyl ethanoyl, *see* Benziloyl		*p*-Isopropylbenzoyl	$p\text{-}(CH_3)_2CH-C_6H_4-$ $CO-$
Hydroxyethanoyl, *see* Glycoloyl		Isopropylbenzyl	$(CH_3)_2CH-C_6H_4-CH_2-$
Hydroxyimino	$HO-N=$	Isopropylidene	$(CH_3)_2C=$
4-Hydroxy-3-methoxy-benzoyl (*or vanilloyl*)	$4\text{-}HO,3\text{-}CH_3O-$ C_6H_3-CO-	Isoselenocyanato	$SeCN-$
		Isosemicarbazido	H_2N-NH- $C(OH)=N-$
3-Hydroxy-2-phenylpro-panoyl (*or tropoyl*)	$C_6H_5-CH(CH_2OH)-$ $CO-$	Isothiocyanato	$SCN-$
Hydroxypropanedioyl (*or tartronoyl*)	$-CO-CH(OH)-CO-$	Isothioureido	$HN=C(SH)-NH-,$ $H_2N-C(SH)=N-$
2-Hydroxypropanoyl (*or lactoyl*)	$CH_3-CH(OH)-CO-$	Isoureido	$HN=C(OH)-NH-,$ $H_2N-C(OH)=N-$
Icosyl	$CH_3-[CH_2]_{18}-CH_2-$	Isovaleryl (*unsubstituted only; or 3-methylbutanoyl*)	$(CH_3)_2CH-CH_2-CO-$
Imino	$-NH-,$ $HN=$		
Iminomethylamino	$HN=CH-NH-$		
Iodo	$I-$	Lactoyl	$CH_3-CH(OH)-CO-$
Iodoformyl	$I-CO-$	Lauroyl (*unsubstituted only*)	$CH_3-[CH_2]_{10}-CO-$

TABLE 1.13 Names and Formulas of Organic Radicals (*Continued*)

Name	Formula	Name	Formula
Leucyl	$(CH_3)_2CH-CH_2-$ $CH(NH_2)-CO-$	5-Methylhexyl	$(CH_3)_2CH-[CH_2]_4-$
		Methylidyne	$HC\equiv$
Lysyl	$H_2N-[CH_2]_4-$ $CH(NH_2)-CO-$	Methylsulfinimidoyl	$CH_3-S(=NH)-$
		Methylsulfinohydrazonoyl	$CH_3-S(=NNH_2)-$
Maleoyl	$-CO-CH=CH-CO-$	Methylsulfinohydroxi-moyl	$CH_3-S(=N-OH)-$
Malonyl	$-CO-CH_2-CO-$		
Maloyl	$-CO-CH(OH)-CH_2-$ $CO-$	Methylsulfinyl	CH_3-SO-
		Methylsulfinylamino	$CH_3-SO-NH-$
Mercapto-	$HS-$	Methylsulfonohydrazo-noyl	$CH_3-S(O)(NNH_2)-$
Mesaconoyl (*unsubstituted only*)	$CH_3-\overset{\overset{-CO-CH}{\|}}{C}-CO-$	Methylsulfonimidoyl	$CH_3-S(O)(=NH)-$
		Methylsulfonohydroxa-moyl	$CH_3-S(O)(N-OH)-$
Mesityl	$2,4,6\text{-}(CH_3)_3C_6H_2-$	Methylsulfonyl	CH_3-SO_2-
Mesoxalo	$HOOC-CO-CO-$	Methylthio	CH_3S-
Nesoxalyl	$-CO-CO-CO-$	(Methylthio)sulfonyl	CH_3S-SO_2-
Mesyl	CH_3-SO_2-	1-Methylvinyl, *see* Iso-propenyl	
Methacryloyl (*or 2-methylpropenoyl*)	$CH_2=C(CH_3)-CO-$	Morpholino (*4- only*)	
Methaneazo	$CH_3-N=N-$	Morpholinyl (*3- shown*)	
Methaneazoxy	$(CH_3-N_2O-$		
Methanesulfinamido	$CH_3-SO-NH-$		
Methanesulfinyl	CH_3-SO-	Myristoyl (*unsubstituted only*)	$CH_3-[CH_2]_{12}-CO-$
Methanesulfonamido	CH_3-SO_2-NH-		
Methanesulfonyl, *see* Me-syl		Naphthalenazo	$C_{10}H_7-N=N-$
Methanoyl, *see* Formyl		Naphthalenecarbonyl, *see* Naphthoyl	
Methionyl	$CH_3-S-CH_2-CH_2-$ $CH(NH_2)-CO-$	Naphthoyl	$C_{10}H_7-CO-$
Methoxalyl	$CH_3OOC-CO-$	Naphthoyloxy	$C_{10}H_7-CO-O-$
Methoxy	CH_3O-	Naphthyl	$C_{10}H_7-$
Methoxybenzoyl (*o-, m-, or p-*)	$CH_3O-C_6H_4-CO-$	Naphthylazo	$C_{10}H_7-N=N-$
		Naphthylene	$-C_{10}H_6-$
Methoxycarbonyl	CH_3O-CO-	Naphthylenebisazo	$-N=N-C_{10}H_6-$ $N=N-$
Methoxyimino	$CH_3O-N=$		
Methoxyphenyl	$CH_3O-C_6H_4-$	Naphthyloxy	$C_{10}H_7-O-$
Methoxysulfinyl	CH_3O-SO-	Neopentyl (*unsubstituted only*)	$(CH_3)_3C-CH_2-$
Methoxysulfonyl	CH_3O-SO_2-		
Methoxy(thiosulfonyl)	CH_3O-S_2O-	Nicotinoyl	NC_5H_4-CO- (3-)
Methyl	CH_3-	Nitrilo	$N\equiv$
Methylallyl	$CH_2=C(CH_3)-CH_2-$	Nitro	O_2N-
Methylamino	CH_3-NH-	*aci*-Nitro	$HO-(O=)N=$
Methylazo	$CH_3-N=N-$	Nitroso	$ON-$
Methylazoxy	CH_3-N_2O-	Nonanedioyl	$-CO-[CH_2]_7-CO-$
α-Methylbenzyl	$C_6H_5-CH(CH_3)-$	Nonanoyl	$CH_3-[CH_2]_7-CO-$
Methylbenzyl	$CH_3\text{-}C_6H_4-CH_2-$	Nonyl	$CH_3-[CH_2]_7-CH_2-$
3-Methylbutanoyl	$(CH_3)_2CH-CH_2-CO-$	Norbornyl	$C_7H_{11}-$
cis-Methylbutenedioyl	$CH_3-\overset{\overset{HC-CO-}{\|}}{C}-CO-$	Norbornylyl, *see* Norbor-nyl	
trans-Methylbutenedioyl	$CH_3-\overset{\overset{-CO-CH}{\|}}{C}-CO-$	Norcamphyl, *see* Norbor-nyl	
Methyldithio	CH_3-S-S-	Norleucyl	$CH_3-[CH_2]_3-$ $CH(NH_2)-CO-$
Methylene	$-CH_2-,\ H_2C=$		
Methylenedioxy	$-O-CH_2-O-$	Norvalyl	$CH_3-CH_2-CH_2-$ $CH(NH_2)-CO-$
3,4-Methylenedioxyben-zoyl	$3,4\text{-}CH_2O_2{:}C_6H_3-$ $CO-$		
		Octadecanoyl	$CH_3-[CH_2]_{16}-CO-$

TABLE 1.13   Names and Formulas of Organic Radicals (*Continued*)

Name	Formula	Name	Formula
cis-9-Octadecenoyl	$H[CH_2]_8—CH=CH—$ $[CH_2]_7—CO—$	Phenylsulfamoyl	$C_6H_5—NH—SO_2$
		Phenylsulfinyl	$C_6H_5—SO—$
Octadecyl	$CH_3—[CH_2]_{16}—CH_2—$	Phenylsulfonyl	$C_6H_5—SO_2—$
Octanedioyl	$—CO—[CH_2]_6—CO—$	Phenylsulfonylamino	$C_6H_5—SO_2—NH—$
Octanoyl	$CH_3—[CH_2]_6—CO—$	Phenylthio	$C_6H_5—S—$
Octyl	$CH_3—[CH_2]_6—CH_2—$	3-Phenylureido	$C_6H_5—NH—CO—NH—$
Oleoyl	$H[CH_2]_8—CH=CH—$ $[CH_2]_7—CO—$	Phthalamoyl	$H_2N—CO—C_6H_4—CO—$ (*o*-)
Ornithyl	$H_2N—[CH_2]_3—$ $CH(NH_2)—CO—$	Phthalidyl	$C_6H_4—CO—O—CH—$
Oxalacetyl	$—CO—CH_2—CO—$ $CO—$	Phthalimido	$CO—C_6H_4—CO—N—$
Oxalaceto	$HOOC—CO—CH_2—$ $CO—$	Phthaloyl	$—CO—C_6H_4—CO—$ (*o*-)
Oxalo	$HOOC—CO—$	Picryl	$2,4,6-(NO_2)_3C_6H_2—$
Oxalyl	$—CO—CO—$	Pimeloyl (*unsubstituted only*)	$—CO—[CH_2]_5—CO—$
Oxamoyl	$H_2N—CO—CO—$	Piperidino (*1- only*)	$C_5H_{10}N—$
Oxido	$^-O—$ (ion)	Piperidyl (*2-, 3-, 4-*)	$NC_5H_{10}—$
Oxo	$O=$	Piperonyl	$3,4-CH_2O_2{:}C_6H_3—CH_2—$
Oxonio	$^+H_2O—$	Pivaloyl (*unsubstituted only*)	$(CH_3)_3C—CO—$
Oxy	$—O—$	Polythio	$—S_n—$
Palmitoyl (*unsubstituted only*)	$CH_3—[CH_2]_{14}—CO—$	Propanedioyl, *see* Malonyl	
Pentafluorothio	$F_5S—$	Propanoyl, *see* Propionyl	
Pentamethylene	$—CH_2—CH_2—CH_2—$ $CH_2—CH_2—$	Propargyl, *see* 2-Propynyl	
Pentanedioyl, *see* Glutaryl		Propenoyl, *see* Acryloyl	
Pentanoyl, *see* Valeryl		1-Propenyl	$CH_3—CH=CH—$
Pentenyl (*2- shown*)	$CH_3—CH_2—CH=CH—$ $CH_2—$	2-Propenyl, *see* Allyl	
		Propenylene	$—CH_2—CH=CH—$
		Propioloyl	$CH{\equiv}C—CO—$
Pentyl	$CH_3—CH_2—CH_2—$ $CH_2—CH_2—$	Propionamido	$CH_3—CH_2—CO—NH—$
		Propionyl	$CH_3—CH_2—CO—$
Pentyloxy	$CH_3—[CH_2]_4—O—$	Propionylamino	$CH_3—CH_2—CO—NH—$
Perchloryl	$O_3Cl—$	Propionyloxy	$CH_3—CH_2—CO—O—$
Phenacyl	$C_6H_5—CO—CH_2—$	Propoxy	$CH_3—CH_2—CH_2—O$
Phenacylidene	$C_6H_5—CO—CH=$	Propyl	$CH_3—CH_2—CH_2—$
Phenanthryl	$C_{14}H_9—$	Propylene	$—CH(CH_3)—CH_2—$
Phenethyl	$C_6H_5—CH_2—CH_2—$	Propylidene	$CH_3—CH_2—CH=$
Phenetidino (*o-, m-,* or *p-*)	$C_2H_5O—C_6H_4—NH—$	Propylidyne	$CH_3—CH_2—C{\equiv}$
Phenoxy	$C_6H_5—O—$	Propynoyl, *see* Propiolyl	
Phenyl	$C_6H_5—$	1-Propynyl	$CH_3—C{\equiv}C—$
Phenylacetyl	$C_6H_5—CH_2—CO—$	2-Propynyl	$HC{\equiv}C—CH_2—$
Phenylazo	$C_6H_5—N=N—$	Protocatechuoyl	$3,4-(HO)_2C_6H_3—CO—$
Phenylazoxy	$C_6H_5—N_2O—$	3-Pyridinecarbonyl	$NC_5H_4—CO—$ (3-)
Phenylcarbamoyl	$C_6H_5—NH—CO$	4-Pyridinecarbonyl	$NC_5H_4—CO—$ (4-)
Phenylene	$—C_6H_4—$	Pyridinio	$^+NC_5H_5—$ (ion)
Phenylenebisazo	$—N=N—C_6H_4—$ $N=N—$	Pyridyl	$NC_5H_4—$
		2-Pyridylcarbonyl	$NC_5H_4—CO—$ (2-)
Phenylimino	$C_6H_5—N=$	Pyridyloxy	$NC_5H_4—O—$
2-Phenylpropanoyl	$C_6H_5—CH(CH_3)—CO—$	Pyruvoyl	$CH_3—CO—CO—$
3-Phenylpropenoyl, *see* Cinnamoyl		Salicyl	$o-HO—C_6H_4—CH_2—$
		Salicylidene	$o-HO—C_6H_4—CH=$
3-Phenylpropyl	$C_6H_5—CH_2—CH_2—$ $CH_2—$	Salicyloyl	$o-HO—C_6H_4—CO—$
		Sarcosyl	$CH_3—NH—CH_2—CO—$

TABLE 1.13 Names and Formulas of Organic Radicals (*Continued*)

Name	Formula	Name	Formula
Sebacoyl (*unsubstituted only*)	$-CO-[CH_2]_8-CO-$	(Terthiophen)yl	$SC_4H_3-SC_4H_2-SC_4H_2-$
Seleneno	HOSe-	Tetradecanoyl	$CH_3-[CH_2]_{12}-CO-$
Selenino	HO_2Se-	Tetradecyl	$CH_3-[CH_2]_{12}-CH_2-$
Seleninyl	$OSe=$	Tetramethylene	$-CH_2-CH_2-CH_2-$
Seleno	$-Se-$		CH_2-
Selenocyanato	$NC-Se-$		
Selenoformyl	$HSeC-$	Thenoyl (2- *shown*)	
Selenonio	$^+H_2Se-$ (ion)		
Selenono	HO_3Se-	Thenyl	$SC_4H_3-CH_2-$
Selenonyl	O_2Se-	Thienyl	SC_4H_3-
Selenoureido	$H_2N-CSe-NH-$	Thio	$-S-$
Selenoxo	$(C)=Se$	Thioacetyl	CH_3-CS-
Semicarbazido	$H_2N-CO-NH-NH-$	Thiobenzoyl	C_6H_5-CS-
Semicarbazono	$H_2N-CO-NH-N=$	Thiocarbamoyl	H_2N-CS-
Seryl	$HO-CH_2-CH(NH_2)-$ $CO-$	Thiocarbazono	$HN=N-CS-NH-$ $NH-$
Stearoyl (*unsubstituted only*)	$CH_3-[CH_2]_{16}-CO-$	Thiocarbodiazono	$HN=N-CS-N=N-$
		Thiocarbonohydrazido	$H_2N-NH-CS-NH-$ $NH-$
Styryl	$C_6H_5-CH=CH-$	Thiocarbonyl	$-CS-$, $SC=$
Suberoyl (*unsubstituted only*)	$-CO-[CH_2]_6-CO-$	Thiocarboxy	HSOC-, $HS-CO-$
		Thiocyanato	NCS-
Succinamoyl	$H_2N-CO-CH_2-CH_2-$ $CO-$	Thioformyl	SHC-, HCS-
Succinimido		Thiophenecarbonyl, *see* Thenoyl	
		Thiosemicarbazido	$H_2N-CS-NH-NH-$
		Thiosulfino	HOS_2-
		Thiosulfo	HO_2S_2-
Succinimidoyl	$-C(=NH)-CH_2-$ $CH_2C(=NH)-$	Thioreido	$H_2N-CS-NH-$
		Thioxo	$S=$
Succinyl	$-CO-CH_2-CH_2-CO-$	Threonyl	$CH_3-CH(OH)-$ $CH(NH_2)-CO-$
Sulfamoyl	H_2N-SO_2-		
Sulfanilamido	$p-H_2N-C_6H_4-SO_2-$ $NH-$	Toluenesulfonyl (*o-, m-*)	$CH_3-C_6H_4-SO_2-$
		Toluidino(*o-, m-,* or *p-*)	$CH_3-C_6H_4-NH-$
Sulfanilyl	$p-H_2N-C_6H_4-SO_2-$	Toluoyl (*o-, m-,* or *p-*)	$CH_3-C_6H_4-CO-$
Sulfenamoyl	H_2N-S-	Tolyl (*o-, m-,* or *p-*)	$CH_3-C_6H_4-$
Sulfeno	$HO-S-$	Tolylsulfonyl	$CH_3-C_6H_4-SO_2-$
Sulfido	$^-S-$ (ion)	Tosyl (*p- only*)	$p-CH_3-C_6H_4-SO_2-$
Sulfinamoyl	H_2N-SO-	Triazano	$H_2N-NH-NH-$
Sulfino	HO_2S-	Triazeno	$H_2N-N=N-$
Sulfinyl	$-SO-$	Trichlorothio	Cl_3S-
Sulfo	$HO-SO_2-$	Tridecanoyl	$CH_3-[CH_2]_{11}-CO-$
Sulfoamino	HO_2S-NH-	Tridecyl	$CH_3-[CH_2]_{12}-$
Sulfonato	$^-O_3S-$ (ion)	Trifluorothio	F_3S-
Sulfonio	$^+H_2S-$ (ion)	3,4,5-Trihydroxybenzoyl	$3,4,5-(HO)_3C_6H_2-CO-$
Sulfonyl	$-SO_2-$	Trimethylammonio	$(CH_3)_3N^{\pm}-$ (ion)
Sulfonyldioxy	$-O-SO_2-O-$	Trimethylanilino (*all isomers*)	$(CH_3)_3C_6H_2-NH-$
Tartaroyl	$-CO-CH(OH)-$ $CH(OH)-CO-$		
		Trimethylene	$-CH_2-CH_2-CH_2-$
Tartronoyl	$-CO-CH(OH)-CO-$	Trimethylenedioxy	$-O-CH_2-CH_2-$ CH_2-O-
Tauryl	$H_2N-CH_2-CH_2-SO_2-$		
Telluro	Te replacing O	Triphenylmethyl	$(C_6H_5)_3C-$
Terephthaloyl	$-CO-C_6H_4-CO-$ (*p-*)	Trithio	$-S_3-$
Terphenylyl	$C_6H_5-C_6H_4-C_6H_4-$	Trithiosulfo	$HS-S_3-$

TABLE 1.13 Names and Formulas of Organic Radicals (*Continued*)

Name	Formula	Name	Formula
Trityl	$(C_6H_5)_3C-$	Vanilloyl	$3,4\text{-}CH_3O(HO)C_6H_3-$
Tropoyl	$C_6H_5-CH(CH_2OH)-$		$CO-$
	$CO-$	Vanillyl	$3,4\text{-}CH_3O(HO)C_6H_3-$
Tyrosyl	$p\text{-}HO-C_6H_4-CH_2-$		CH_2-
	$CH(NH_2)-CO-$	Veratroyl	$3,4\text{-}(CH_3O)_2C_6H_3-$
Undecanoyl	$CH_3-[CH_2]_9-CO-$		$CO-$
Undecyl	$CH_3-[CH_2]_9-CH_2-$	Veratryl	$3,4\text{-}(CH_3O)_2C_6H_2-$
Ureido	$H_2N-CO-NH-$		CH_2-
Ureylene	$-NH-CO-NH-$	Vinyl	$CH_2=CH-$
Valeryl	$CH_3-[CH_2]_3-CO-$	Vinylene	$-CH=CH-$
Valyl	$(CH_3)_2CH-CH(NH_2)-$	Xylidino (*all isomers*)	$(CH_3)_2C_6H_3-NH-$
	$CO-$	Xylyl (*all isomers*)	$(CH_3)_2C_6H_3-$

1.2 PHYSICAL PROPERTIES OF PURE SUBSTANCES

TABLE 1.14 Empirical Formula Index of Organic Compounds

The alphanumeric designations are keyed to Table 1.15.

ClH_4NO: h128
Cl_2H_2Si: d226
Cl_3HSi: t251

C_1

CBr_4: c13
$CBrClF_2$: b235
$CBrCl_3$: b361
$CBrF_3$: b363
CBr_2F_2: d75
$CClF_3$: c254
$CClNO_3S$: c241
CCl_2F_2: d170
CCl_3D: c128
CCl_3F: t236
CCl_3NO_2: t243
CCl_4: c14
CCl_4O_2S: t240
CCl_4S: t239
CD_4O: m36
CF_4: c15
$CHBrCl_2$: b268
$CHBr_2Cl$: d71
$CHBr_3$: t210
$CHClF_2$: c85
$CHCl_2F$: d183
$CHCl_3$: c127
CHF_3: t300

CHF_3O_3S: t301
CHI_3: i36
CHN_3O_6: t392
CH_2BrCl: b258
CH_2Br_2: d88
CH_2Cl_2: d190
CH_2I_2: d405
CH_2N_2: c286, d47
CH_2N_4: t138
CH_2O: f27
$(CH_2O)_x$: p1
CH_2O_2: f32
CH_2S_3: t441
CH_3Br: b303
CH_3Br_3Ge: m255
CH_3Cl: c138
CH_3ClHg: m298
CH_3ClO_2S: m32
CH_3Cl_3Si: t242
CH_3DO: m35
CH_3F: f18
CH_3I: i40
CH_3NO: f28
CH_3NO: m56, m317
CH_3NO_3: m316
CH_3N_5: a294
CH_4: m29
CH_4Cl_2Si: d199
CH_4N_2O: f34, u12

$CH_4N_2O_2S$: f30
CH_4N_2S: t166
$CH_4N_4O_2$: n54
CH_4O: m34
CH_4O_2: m276
CH_4O_3S: m30
CH_4S: m33
CH_5AsO_3: m125
CH_5N: m115
CH_5NO_3S: a208
CH_5N_3: g29
CH_5N_3O: s3
CH_5N_3S: t165
CH_6N_2: m271
CH_6N_4: a180, a181
CH_6N_4O: c11
CI_4: c15a
CN_4O_8: t126
CS_2: c12

C_2

$C_2Br_2ClF_3$: d72
$C_2Br_2Cl_4$: d99
$C_2Br_2F_4$: d100
$C_2Br_2O_2$: o50
$C_2Cl_2F_3$: c253
$C_2Cl_2F_4$: d227
$C_2Cl_2N_2S$: d227a

TABLE 1.14 Empirical Formula Index of Organic Compounds (*Continued*)

The alphanumeric designations are keyed to Table 1.15.

$C_2Cl_2O_2$: o51
$C_2Cl_3F_3$: t255
C_2Cl_3N: t221
C_2Cl_4: t30
$C_2Cl_4F_2$: d347, d348, t27
C_2Cl_4O: t222
C_2Cl_6: h29
C_2D_3H: a30
$C_2D_4O_2$: a21
C_2D_6OS: d617
C_2F_4: t65
C_2F_6: h44
$C_2F_6O_5S_2$: t302
$C_2HBrClF_3$: b250
$C_2HBr_2F_3$: d103
C_2HBr_2N: d63
C_2HBr_3: t209
C_2HBr_3O: t205
$C_2HBr_3O_2$: t206
$C_2HClF_2O_2$: c83
$C_2HCl_2F_3$: d232
C_2HCl_3: t234
C_2HCl_3O: d141, t219e
$C_2HCl_3O_2$: t220
C_2HCl_5: p9
$C_2HF_3O_2$: t292
C_2H_2: a41
C_2H_2BrClO: b225
$C_2H_2Br_2$: d80, d81
$C_2H_2Br_2F_2$: d74
$C_2H_2Br_2O$: b224
$C_2H_2Br_2O_2$: d62
$C_2H_2Br_4$: t10
$C_2H_2ClF_3$: c252
C_2H_2ClN: c27
$C_2H_2Cl_2$: d178, d179, d180
$C_2H_2Cl_2O$: c31
$C_2H_2Cl_2O_2$: d138
$C_2H_2Cl_3NO$: t219d
$C_2H_2Cl_4$: t29
$C_2H_2F_3NO$: t291
$C_2H_2N_2S_3$: d426a
C_2H_2O: k1
$C_2H_2O_2$: g27
$C_2H_2O_3$: g28
$C_2H_2O_4$: o48, o49
C_2H_3Br: b286
C_2H_3BrO: a35
$C_2H_3BrO_2$: b22
$C_2H_3Br_2Cl_3Si$: d82
$C_2H_3Br_3O$: t208
C_2H_3Cl: c110
$C_2H_3ClF_2$: c84
C_2H_3ClO: a37

$C_2H_3ClO_2$: c24, m188
$C_2H_3Cl_3$: t230, t231
$C_2H_3Cl_3O$: t232
$C_2H_3Cl_3Si$: t256
$C_2H_3DO_2$: a20
C_2H_3FO: a43
$C_2H_3FO_2$: f6
$C_2H_3F_3$: t296
$C_2H_3F_3O$: t297
C_2H_3IO: a48
$C_2H_3IO_2$: i25
C_2H_3N: a29
C_2H_3NO: m288
C_2H_3NS: m290, m429
$C_2H_3N_3$: t203
$C_2H_3N_3S_2$: a295
C_2H_4: e124b
C_2H_4BrCl: b256
C_2H_4BrNO: b219
$C_2H_4Br_2$: d77, d78
C_2H_4ClNO: c22
$C_2H_4Cl_2$: d176, d177
$C_2H_4Cl_2O$: d197
$C_2H_4Cl_6Si_2$: b205
C_2H_4FNO: f5
$C_2H_4F_2$: d346
C_2H_4INO: i24
$C_2H_4I_2$: d404
$C_2H_4N_2$: a106
$C_2H_4N_2O_2$: o54
$C_2H_4N_2S_2$: d712
$C_2H_4N_4$: a300, d235
$C_2H_4N_4O_2$: a329
C_2H_4O: e132
C_2H_4OS: t144
$C_2H_4O_2$: a19, h86, m251
$C_2H_4O_2S$: m14
$C_2H_4O_3$: h87, p59
$C_2H_4O_5S$: s25
C_2H_4S: e133
$C_2H_5AlCl_2$: e58
C_2H_5Br: b279
$C_2H_5BrNaO_2S$: b280
C_2H_5BrO: b281, b311
C_2H_5Cl: c103
C_2H_5ClHg: e168
C_2H_5ClO: c104, c156
$C_2H_5ClO_2S$: d19
C_2H_5ClS: c157
$C_2H_5Cl_2OPS$: e118
$C_2H_5Cl_2O_2P$: e117
$C_2H_5Cl_3Si$: c151, e229a, t235
C_2H_5DO: e22
C_2H_5F: d17

$C_2H_5FO_3S$: e137
C_2H_5I: i34
C_2H_5IO: i35
C_2H_5N: e133
C_2H_5NO: a5, a6, m249
$C_2H_5NO_2$: e189, g25, m182, n53
$C_2H_5NO_3$: e188
C_2H_5NS: t143
$C_2H_5N_3O_2$: b216, o53
C_2H_6: e14
C_2H_6AlCl: d461b
C_2H_6BrN: b283
C_2H_6Cd: d503
C_2H_6ClN: c107
$C_2H_6ClNO_2S$: d611
$C_2H_6ClO_2PS$: d506
$C_2H_6Cl_2Si$: d174
C_2H_6Hg: d548
$C_2H_6N_2$: a8
$C_2H_6N_2O$: a25, m447, n79
$C_2H_6N_2O_2$: m272
$C_2H_6N_2S$: m434
$C_2H_6N_4O_2$: o52
C_2H_6O: d520, e31
C_2H_6OS: d616, m18
$C_2H_6O_2$: e16, e131
$C_2H_6O_2S$: d615
$C_2H_6O_3S$: d614, m300
$C_2H_6O_4S$: d612
$C_2H_6O_5S_2$: m31
C_2H_6S: d613, e20
$C_2H_6S_2$: d518, e18
C_2H_6Te: d619
C_2H_6Zn: d626
$C_2H_7AsO_2$: d486
C_2H_7ClSi: c93
C_2H_7N: d463, e59
C_2H_7NO: a163, a164
$C_2H_7NO_3S$: a161
$C_2H_7NO_4S$: a169
C_2H_7NS: a162
$C_2H_7N_5$: b134
$C_2H_8N_2$: d541, d542, e15
$C_2H_8N_2O$: h120

C_3

$C_3Br_2F_6$: d85
$C_3Cl_3NO_2$: t223
$C_3Cl_3N_3$: t254
$C_3Cl_3N_3O_3$: t238
C_3Cl_6: h32

TABLE 1.14   Empirical Formula Index of Organic Compounds (*Continued*)

The alphanumeric designations are keyed to Table 1.15.

C_3Cl_6O: h23
C_3D_6O: a27
C_3F_6: h45a
C_3HCl_5O: p7
C_3H_2ClN: c32
$C_3H_2Cl_2O_2$: m6
$C_3H_2Cl_4$: t34
$C_3H_2Cl_4O$: t22
$C_3H_2Cl_4O_2$: t233
$C_3H_2F_6O$: h45
$C_3H_2N_2$: m5
$C_3H_2O_2$: p244
C_3H_3Br: b347
C_3H_3Cl: c233
C_3H_3ClO: a65
$C_3H_3Cl_3O$: e13
$C_3H_3Cl_3O_2$: m439
$C_3H_3F_3O$: t294
$C_3H_3F_3O_2$: m440
C_3H_3N: a64
$C_3H_3NOS_2$: r3
$C_3H_3NO_2$: c288
$C_3H_3NO_3O_2S$: a252
C_3H_3NS: t142
$C_3H_3N_3O_3$: a300
C_3H_4: a78, p243
C_3H_4BrClO: b343, b344
C_3H_4BrN: b342
$C_3H_4Br_2$: d95
$C_3H_4Br_2O_2$: d96
C_3H_4ClN: c221
$C_3H_4Cl_2$: d221, d222
$C_3H_4Cl_2O$: c222, c223, d139
$C_3H_4Cl_2O_2$: m220
$C_3H_4F_4O$: t66
$C_3H_4N_2$: i4, m232, p248
$C_3H_4N_2O$: c287
$C_3H_4N_2OS$: t155
$C_3H_4N_2O_2$: h84
$C_3H_4N_2S$: a296
C_3H_4O: p207, p245
$C_3H_4O_2$: a63, o59, p213
$C_3H_4O_3$: e125, o60
$C_3H_4O_4$: m3
C_3H_5Br: b266, b338, b339
C_3H_5BrO: b278
$C_3H_5BrO_2$: b340, b341, m143
$C_3H_5Br_3$: t212
C_3H_5Cl: c217
C_3H_5ClO: c102, c216, p219
C_3H_5ClOS: e102
$C_3H_5ClO_2$: c219, c220, e99, m183

$C_3H_5Cl_3$: t248
$C_3H_5Cl_3O$: t249
$C_3H_5Cl_3Si$: a102
$C_3H_5F_3O_3S$: m441
C_3H_5I: a92, i50
C_3H_5N: p218
C_3H_5NO: c291, h169, h170
$C_3H_5NO_2$: o55
C_3H_5NS: e164, m424
$C_3H_5N_3O$: c289
$C_3H_5N_3O_9$: g21
$C_3H_5N_3S$: c293
C_3H_6: c365, p208
C_3H_6BrCl: b259
C_3H_6ClNO: d504
$C_3H_6Cl_2$: d218, d219
$C_3H_6Cl_2N_2O_2$: d173
$C_3H_6Cl_2O$: d182, d220
$C_3H_6Cl_2Si$: d200
$C_3H_6Cl_4Si$: c230
$C_3H_6I_2$: d406
C_3H_6NO: a62
$C_3H_6N_2$: a279, d507
$C_3H_6N_2O$: i7
$C_3H_6N_2OS$: a58
$C_3H_6N_2O_2$: m4, m270
$C_3H_6N_2S$: a297, i5
$C_3H_6N_6$: t202
C_3H_6O: a26, a81, m449, p214, p230, t350
C_3H_6OS: m423, t164
$C_3H_6O_2$: d649, e11, e138, h89, m111, p216
$C_3H_6O_2S$: m21, m21a, m294
$C_3H_6O_3$: d398, d399, d505, L1, L2, m38, m260, t395
$C_3H_6O_3S$: p201
C_3H_6S: p209, p231, t350a
$C_3H_6S_3$: t440
C_3H_7Br: b335, b336
C_3H_7BrO: b337
C_3H_7Cl: c211, c212
C_3H_7ClO: c112, c153, c214, c215
C_3H_7ClOS: c137
$C_3H_7ClO_2$: c213
$C_3H_7ClO_2S$: p200
$C_3H_7Cl_2OP$: p239
$C_3H_7Cl_3Si$: d194, p240
C_3H_7I: i48, i49
C_3H_7N: a82, p229
C_3H_7NO: a28, d524, m110, p215

$C_3H_7NO_2$: a74, a75, a76, e92, i105a, m259, n73, n74
$C_3H_7NO_2S$: c371
$C_3H_7NO_3$: i105, n75, p236, s4
$C_3H_7NO_5S$: a293
C_3H_7NS: d622
$C_3H_7O_5P$: c17
C_3H_8: p194
C_3H_8ClN: c225
$C_3H_8Cl_2Si$: c75
C_3H_8IN: d551
$C_3H_8N_2O$: d625, e233
$C_3H_8N_2O_2$: e93, f29
$C_3H_8N_2S$: d623
C_3H_8O: e10, e174, p205, p206
$C_3H_8OS_2$: m308
$C_3H_8O_2$: d442, m65, p197, p198
$C_3H_8O_2S$: m20
$C_3H_8O_3$: g16
C_3H_8S: e185, p202, p203
$C_3H_8S_2$: p199
C_3H_9Al: t332
$C_3H_9BO_3$: t324
$C_3H_9B_3O_6$: t325
C_3H_9BrGe: b366
C_3H_9BrSi: b367
C_3H_9ClGe: c255
C_3H_9ClSi: c256
C_3H_9IOS: t385
C_3H_9IS: t394
C_3H_9ISi: i55
C_3H_9N: i88, p223, t333
C_3H_9NO: a274, a275, a276, a277, m69, m119, t334
$C_3H_9NO_2$: a273
$C_3H_9N_3Si$: a324
$C_3H_9O_3P$: d370, d553
$C_3H_9O_4P$: t369
$C_3H_{10}N_2$: m248, p195, p196
$C_3H_{10}N_2O$: d43
$C_3H_{10}O_3Si$: t327a
$C_3H_{11}Br_2N_3S$: a171

C_4

$C_4Cl_2F_6$: d185
$C_4Cl_3F_7$: h3
$C_4Cl_4N_2$: t35
C_4Cl_6: h25
$C_4Cl_6O_3$: t220a
$C_4D_6O_3$: a23

TABLE 1.14 Empirical Formula Index of Organic Compounds (*Continued*)

The alphanumeric designations are keyed to Table 1.15.

$C_4F_6O_3$: t293
C_4HBrO_3: b302
$C_4HCl_3N_2$: t250
$C_4HF_7O_2$: h2
C_4H_2: b379
$C_4H_2Br_2S$: d101
$C_4H_2Cl_2N_2$: d223
$C_4H_2Cl_2O_2$: f38
$C_4H_2Cl_2O_3$: d208
$C_4H_2Cl_2S$: d228
$C_4H_2F_6O_2$: t299
$C_4H_2O_3$: m2
$C_4H_2O_4$: q42
C_4H_3BrS: b356
C_4H_3ClS: c243
$C_4H_3Cl_2N_3O$: d193
C_4H_3IS: i52
$C_4H_3N_3O_4$: n89a
C_4H_4: b410
$C_4H_4BrNO_2$: b354
$C_4H_4Br_2O_2$: d69
$C_4H_4Br_2O_4$: d98
$C_4H_4ClNO_2$: c240
$C_4H_4Cl_2$: d168
$C_4H_4Cl_2O_2$: s20
$C_4H_4Cl_2O_3$: c25
$C_4H_4N_2$: b383, p247, p251, p271, s19
$C_4H_4N_2O_2$: d401, p272
$C_4H_4N_2O_2S$: d389
$C_4H_4N_2O_3$: b1
$C_4H_4N_2O_5$: a79
$C_4H_4N_4$: d40
C_4H_4O: f40
$C_4H_4O_2$: d423
$C_4H_4O_3$: s17
$C_4H_4O_4$: f37, m1
C_4H_4S: t157
$C_4H_5BrO_4$: b353
C_4H_5Cl: c63, c70
C_4H_5ClO: c284, c367, m28
$C_4H_5ClO_2$: a87
$C_4H_5ClO_3$: e194
$C_4H_5Cl_3O_2$: e229
$C_4H_5F_3O_2$: e230
C_4H_5N: b403, c366, m27, p273
C_4H_5NO: m291
$C_4H_5NO_2$: e107, m194, s18
$C_4H_5NO_2S$: e33
$C_4H_5NO_3$: h183
C_4H_5NS: a93, m420a
$C_4H_5N_3$: a289, i11
$C_4H_5N_3O$: a200

$C_4H_5N_3OS$: a193
$C_4H_5N_3O_2$: a154, a155, c290, m325
C_4H_6: b376, b377
$C_4H_6Br_2O$: b314a
$C_4H_6Br_2O_2$: d70
C_4H_6ClN: c73
$C_4H_6Cl_2$: c76, d166, d167
$C_4H_6Cl_2O$: c74
$C_4H_6Cl_2O_2$: m222
$C_4H_6Cl_3NSi$: c295
$C_4H_6Cl_4$: t26a
$C_4H_6N_2$: a151, m281, m282, m283
$C_4H_6N_2O_2$: e115
$C_4H_6N_2S$: a232
$C_4H_6N_2S_2$: d507a
$C_4H_6N_4O$: d39
$C_4H_6N_4O_3$: a77
C_4H_6O: c283, m24, m399
$C_4H_6O_2$: b389, b404, b405, b406, b495, b500, b501, c368, m26, m114, v2
$C_4H_6O_2S$: d369
$C_4H_6O_3$: a22, a24, m337, o56, p228
$C_4H_6O_4$: d568, s15
$C_4H_6O_4S$: m23, t151
$C_4H_6O_5$: d609a, h181, h182, o61
$C_4H_6O_6$: t1, t2, t2a
C_4H_7Br: b242, b243, b244
$C_4H_7BrO_2$: b246, b282, b310, e76, m146
C_4H_7Cl: c68, c69, c164, c165
C_4H_7ClO: b504, c67, c116, i78
$C_4H_7ClO_2$: c71, c72, e95, p226b
$C_4H_7Cl_2NSi$: c292
$C_4H_7Cl_3O$: t241
$C_4H_7Cl_3O_2Si$: c13
$C_4H_7FO_2$: e136
C_4H_7N: b502, i76
C_4H_7NO: h146, i98, m25, m99a, m335, p234, p279
$C_4H_7NO_2$: h137, m334
$C_4H_7NO_3$: a46, e195, s13
$C_4H_7NO_4$: a319, i10
C_4H_7NS: m422
$C_4H_7N_3O$: c279
C_4H_8: b398, b399, b400, c301, m386
C_4H_8BrCl: b253, b258a
$C_4H_8Br_2$: d67, d68, d68a

$C_4H_8Br_2O$: b150
$C_4H_8Cl_2$: d162, d163, d164
$C_4H_8Cl_2O$: b159, d181
$C_4H_8N_2$: d463a
$C_4H_8N_2O$: a105, a150
$C_4H_8N_2O_2$: d528, s14
$C_4H_8N_2O_3$: a318, g26
$C_4H_8N_2S$: a101, t81
C_4H_8O: b396, b407, b408, b496, e2, e235, i73, m96, m377, m388, t68
C_4H_8OS: e223, t109, t167
$C_4H_8O_2$: b401, b402, b498, d648, e52, h105a, h106, i75, m64, m392, p232
$C_4H_8O_2S$: e167, m297, t108
$C_4H_8O_3$: e24, e153, h116, h127, m292, m301
$C_4H_8O_3S$: d387
C_4H_8S: a95, t83
$C_4H_8S_2$: d709
C_4H_9Br: b240, b241, b313, b314
C_4H_9BrO: b287
C_4H_9Cl: c64, c65, c162, c163
C_4H_9ClO: a66, c111
$C_4H_9ClO_2$: c89, c105, m67
C_4H_9ClSi: c94
$C_4H_9Cl_3Si$: b486, c226
$C_4H_9Cl_3Sn$: b484
C_4H_9F: f20
C_4H_9I: i30, i31, i43, i44
C_4H_9N: e53, p274
C_4H_9NO: a326, b497, d459, i74, m391, m451
$C_4H_9NO_2$: a136, a137, a225, b467, b468, h115, i71, n50
$C_4H_9NO_2S$: a207
$C_4H_9NO_3$: a189, a190, a191, a192, i70, m327a, n51
$C_4H_9NO_5$: t435a
C_4H_9NSi: c299, t377a
$C_4H_9N_3O_2$: c278
C_4H_{10}: b381, m378
$C_4H_{10}AlCl$: d266a
$C_4H_{10}ClN$: d469
$C_4H_{10}ClO_2S$: d292
$C_4H_{10}ClO_3P$: d291
$C_4H_{10}Cl_2Si$: b161
$C_4H_{10}N_2$: p182
$C_4H_{10}N_2O$: a234
$C_4H_{10}N_2O_4S$: a8
$C_4H_{10}N_4O_2$: s16

TABLE 1.14 Empirical Formula Index of Organic Compounds (*Continued*)

The alphanumeric designations are keyed to Table 1.15.

$C_4H_{10}O$: b394, b395, e13a, e300, m384, m385, m396
$C_4H_{10}OS$: e156
$C_4H_{10}OS_2$: b187
$C_4H_{10}O_2$: b384, b385, b386, b387, b456, d439, d440, e35, m95
$C_4H_{10}O_2S$: t152
$C_4H_{10}O_2S_2$: d425
$C_4H_{10}O_3$: b182, b393, m94a, o62, t359
$C_4H_{10}O_3S$: d338
$C_4H_{10}O_4S$: d336
$C_4H_{10}S$: b391, b392, d337, i104, m381, m382, m383, m398
$C_4H_{10}S_2$: b390, d294, h118
$C_4H_{10}S_3$: b188
$C_4H_{10}Zn$: d344
$C_4H_{11}ClSi$: c167
$C_4H_{11}N$: b380, b420, b420a, b421, d267, d268, d522, i63
$C_4H_{11}NO$: a138, a165, a224, d315, d467, e39, e68
$C_4H_{11}NO_2$: a223, d245, d441
$C_4H_{11}NO_3$: t430
$C_4H_{11}O_2PS_2$: d296
$C_4H_{11}O_3P$: d314
$C_4H_{12}BrN$: t95
$C_4H_{12}ClN$: t96
$C_4H_{12}Ge$: t111
$C_4H_{12}IN$: t97
$C_4H_{12}N_2$: b382, b455, d523, m379, m380
$C_4H_{12}N_2O$: a166
$C_4H_{12}N_2S_2$: c370
$C_4H_{12}OSi$: m108
$C_4H_{12}O_3Si$: m442
$C_4H_{12}O_4Si$: t94, t116
$C_4H_{12}Pb$: t114
$C_4H_{12}Si$: t122
$C_4H_{12}Sn$: t125
$C_4H_{13}N_3$: d298
$C_4H_{14}OSi_2$: t107
$C_4H_{16}O_4Si_4$: t105

C_5

C_5Cl_5N: p12
C_5Cl_6: h27
C_5D_5N: p253
$C_5H_3Br_2N$: d97

$C_5H_3ClO_2$: f48
$C_5H_3Cl_2N$: d224
$C_5H_3F_7O_2$: m261
C_5H_4BrN: b348, b349
C_5H_4ClN: c234
C_5H_4FN: f23
$C_5H_4F_8O$: o18
$C_5H_4N_2O_3$: n76
$C_5H_4N_4O_3$: u13
C_5H_4OS: t159
$C_5H_4O_2$: f39
$C_5H_4O_2S$: t160
$C_5H_4O_3$: c272, f42
$C_5H_5ClN_2$: a149
$C_5H_5ClN_2O_2$: c168
$C_5H_5F_3O_2$: t298
C_5H_5N: p252
C_5H_5NO: h174, h175, h176, p266
$C_5H_5NO_2$: d402, h178
$C_5H_5NO_3S$: p267
$C_5H_5N_3O_2$: a251
$C_5H_5N_3O_4$: a160
$C_5H_5N_5$: a69
C_5H_6: m166
$C_5H_6Br_2N_2O_2$: d76
$C_5H_6Cl_2O_2$: d195, g15
$C_5H_6N_2$: a286, a287, a288, g14, m400, v7
$C_5H_6N_2O$: a47, a199
$C_5H_6N_2OS$: h129
$C_5H_6N_2O_2$: d392
$C_5H_6N_2O_3$: d108
C_5H_6O: m59, m253
C_5H_6OS: f44
$C_5H_6O_2$: f46
$C_5H_6O_3$: g12
$C_5H_6O_4$: c271
$C_5H_6O_4S_3$: b156
C_5H_6S: m429a, m430
$C_5H_7BrO_2$: m145
$C_5H_7BrO_3$: i80
$C_5H_7ClO_3$: m184, m184a, m189
C_5H_7N: m407, p50, p51
C_5H_7NO: f47
$C_5H_7NO_2$: e106
C_5H_7NS: t161
$C_5H_7N_3$: a231, d44
$C_5H_7N_3O$: a194
C_5H_8: c359, m147, m148, m172, p16, p17, p18, p19, p57

$C_5H_8Br_2O_2$: e116
$C_5H_8F_4O$: m417
$C_5H_8N_2$: d544, d605, e162, p276
$C_5H_8N_2O$: m452
$C_5H_8N_2O_2$: d540
$C_5H_8N_4O_{12}$: p22
C_5H_8O: c357, c369, d364, e5, m157a, m173
$C_5H_8O_2$: a80, e57, g13, i84, m58, m161, m162, m163, m193, m217a, m217b, m218, m299, p31, p211
$C_5H_8O_3$: e199, m112, o58
$C_5H_8O_4$: d547, g11, m275, m415
C_5H_9Br: b265
$C_5H_9BrO_2$: e82, e83, m144
C_5H_9Cl: c79
C_5H_9ClO: c193, d602, m180, p44
C_5H_9ClOS: c229
$C_5H_9ClO_2$: b439, c224a, e100, e101, i65, m187
$C_5H_9F_3O_2Si$: t383
C_5H_9N: d604, m179, p32, t80
C_5H_9NO: b458, b459, c358, d461a, e24a, e196, m409, m448a
$C_5H_9NO_2$: f34a, m118
$C_5H_9NO_4$: g9
$C_5H_9N_3$: i8
C_5H_{10}: c353, m158, m159, m160, p47, p48, p49
$C_5H_{10}Br_2$: d91
$C_5H_{10}ClNO$: d288
$C_5H_{10}Cl_2$: d209
$C_5H_{10}Cl_2Si$: c352
$C_5H_{10}NO_3P$: d293a
$C_5H_{10}N_2$: d293
$C_5H_{10}N_2O$: d545, p183
$C_5H_{10}N_2O_3$: g10
$C_5H_{10}O$: a90, c356, d569a, d598, i108, m164, m164a, m165, m174, m175, m418, p27, p41, p42, t78
$C_5H_{10}OS$: m428
$C_5H_{10}O_2$: b454a, b455, d543a, d600, e211, h140, h157, h158, i66, i87, m75, m102, m176, m177, m178, m287, p36, p222, t70

TABLE 1.14 Empirical Formula Index of Organic Compounds (*Continued*)

The alphanumeric designations are keyed to Table 1.15.

$C_5H_{10}O_2S$: e167a, e186, m309, m420

$C_5H_{10}O_3$: d289, d454, e165, m68, m279

$C_5H_{10}O_4$: b185

$C_5H_{10}O_5$: a315, r5, x8

$C_5H_{11}Br$: b308, b325, b326

$C_5H_{11}BrO$: b272a

$C_5H_{11}BrO_2$: b271

$C_5H_{11}BrO_2Si$: t377

$C_5H_{11}Cl$: c92, c150, c192

$C_5H_{11}ClO$: c92a

$C_5H_{11}Cl_2N$: b160

$C_5H_{11}I$: i42, i47

$C_5H_{11}N$: a90, m408, p186

$C_5H_{11}NO$: d304, d599, m310, t71

$C_5H_{11}NO_2$: a256, a257, b129, e234, i81, m310a, v1

$C_5H_{11}NO_2S$: m37

$C_5H_{11}NO_3$: n58

$C_5H_{11}NS_2$: d295

$C_5H_{11}O_5P$: t371

C_5H_{12}: d594, m149, p28

$C_5H_{12}Cl_2O_2Si$: b158

$C_5H_{12}N_2$: a272, m368, m369

$C_5H_{12}N_2O$: b490, t126

$C_5H_{12}N_2O_2$: b434, o46

$C_5H_{12}N_2S$: t124

$C_5H_{12}N_2S_2$: p275

$C_5H_{12}O$: b463, d453, d597, e212, m153, m154, m155, m156, p37, p38, p39

$C_5H_{12}O_2$: d255, d452a, d596, m57, p29a, p30

$C_5H_{12}O_2S$: e224

$C_5H_{12}O_2Si$: t375

$C_5H_{12}O_3$: e46a, h142, m66, p34, t326a, t358, t432

$C_5H_{12}O_4$: p20

$C_5H_{12}O_5$: x7

$C_5H_{12}S$: b466, e213, m150, m151, m152, p35

$C_5H_{12}Si$: t387

$C_5H_{13}N$: a254, a255, d319, d603, m168, m169, m170, p53

$C_5H_{13}NO$: a216, a258, d474, d475, e48, i89, p224

$C_5H_{13}NOSi$: t374

$C_5H_{13}NO_2$: a176, d443, d473, d525, m224

$C_5H_{13}N_3$: t112

$C_5H_{14}N_2$: d477, d595, p29, t106, t115

$C_5H_{14}N_2O$: a167

$C_5H_{14}OSi$: e51, t379

$C_5H_{15}NSi$: d624a

$C_5H_{15}N_3$: a175

C_6

C_6BrD_5: b231

C_6BrF_5: b324

$C_6Cl_4O_2$: t25, t26

$C_6Cl_5NO_2$: p10

C_6Cl_6: h24

C_6D_6: b10

C_6D_{12}: c314

C_6F_6: h43

C_6HBr_5O: p6

$C_6HCl_4NO_2$: t31

C_6HCl_5: p8

C_6HCl_5O: p11

$C_6H_2BrFN_2O_4$: b274

$C_6H_2Cl_2O_4$: d172

$C_6H_2Cl_3NO_2$: t242

$C_6H_2Cl_4$: t23, t24

$C_6H_3Br_2F$: d84

$C_6H_3Br_2NO_2$: d90

$C_6H_3Br_3O$: t211

$C_6H_3ClFNO_2$: c122

$C_6H_3ClN_2O_4$: c95, c96

$C_6H_3ClN_2O_4S$: d629

$C_6H_3Cl_2NO_2$: d203, d204, d205

$C_6H_3Cl_3$: t226, t227, t228

$C_6H_3Cl_3O$: t244, t245

$C_6H_3Cl_3O_2S$: d155

$C_6H_3FN_2O_4$: d635

$C_6H_3N_3O_6$: t389, t390

$C_6H_3N_3O_7$: p176

C_6H_4BrCl: b249, b250, b251

$C_6H_4BrClO_2S$: b232

C_6H_4BrF: b290, b291, b292

$C_6H_4BrNO_2$: b317

$C_6H_4BrN_3O_4$: b273

$C_6H_4Br_2$: d65, d112

$C_6H_4Br_2N_2O_2$: d89

$C_6H_4Br_2O$: d91a

$C_6H_4Br_3N$: t207

C_6H_4ClF: c117, c118

C_6H_4ClFO: c123

C_6H_4ClI: c136

$C_6H_4ClNO_2$: c176, c177, c178, c235, c236

$C_6H_4ClNO_3$: c187

$C_6H_4ClNO_4S$: n35

$C_6H_4ClO_2P$: p112

$C_6H_4Cl_2$: d152, d153, d154

$C_6H_4Cl_2N_2O_2$: d202

$C_6H_4Cl_2O$: d210, d211, d212, d213

$C_6H_4Cl_2O_2$: d171

$C_6H_4Cl_2O_2S$: c43

$C_6H_4Cl_3N$: t224, t225

$C_6H_4FNO_2$: f21

$C_6H_4F_2$: d345

$C_6H_4INO_2$: i45, i45a

$C_6H_4I_2$: d403

$C_6H_4N_2$: a278, c296, c297, c298

$C_6H_4N_2O_2$: b43

$C_6H_4N_2O_4$: d628

$C_6H_4N_2O_5$: d637

$C_6H_4N_4O_6$: t388

$C_6H_4O_2$: b59

$C_6H_5BO_2$: c21

C_6H_5Br: b230

C_6H_5BrO: b328, b329

C_6H_5BrS: b357

C_6H_5Cl: c41

C_6H_5ClHg: p128

$C_6H_5ClN_2O_2$: c173, c173a, c174, c175

C_6H_5ClO: c195, c196, c197

$C_6H_5ClO_2$: c86, c87

$C_6H_5ClO_2S$: b23

C_6H_5ClS: c244

C_6H_5ClSe: p155

$C_6H_5Cl_2N$: d142, d143, d144, d145, d146, d147

$C_6H_5Cl_2OP$: p139

$C_6H_5Cl_2O_2P$: p107

$C_6H_5Cl_2P$: d216

$C_6H_5Cl_2PS$: p140

$C_6H_5Cl_3Si$: p158

C_6H_5D: b9

C_6H_5F: f11

$C_6H_5FN_2O_2$: f20a

C_6H_5FO: f22

$C_6H_5FO_2S$: b24

$C_6H_5F_7O_2$: e140

C_6H_5I: b27

C_6H_5NO: p255, p256, p257

C_6H_5NOS: t155

$C_6H_5NO_2$: n30, n83, p259, p260, p261

$C_6H_5NO_3$: h177, n60, n61

$C_6H_5NO_4$: c273

TABLE 1.14 Empirical Formula Index of Organic Compounds (*Continued*)

The alphanumeric designations are keyed to Table 1.15.

$C_6H_5N_3$: b62
$C_6H_5N_3O$: h103
$C_6H_5N_3O_4$: d627
C_6H_6: b8
$C_6H_6AsNO_6$: h154
C_6H_6BrN: b226, b227, b228
C_6H_6ClN: c33, c34, c35
C_6H_6ClNO: a148, c131a, c142
$C_6H_6ClNO_2S$: c42
$C_6H_6Cl_2N_2$: d215
$C_6H_6Cl_6$: h26
C_6H_6FN: f9
$C_6H_6F_9O_3P$: t439
C_6H_6HgO: p129
C_6H_6IN: i26
$C_6H_6N_2O$: e44, p254, p254a
$C_6H_6N_2O_2$: n24, n25, n26
$C_6H_6N_2O_3$: a245, a246, a248, m84
$C_6H_6N_4O_4$: d639
C_6H_6O: p64
C_6H_6OS: a57, m431
$C_6H_6O_2$: a44, b32a, d378, d379, d380, m252
$C_6H_6O_2S$: b20, t158
$C_6H_6O_3$: h147, m254, t309, t310
$C_6H_6O_3S$: b22
$C_6H_6O_4$: d461
$C_6H_6O_4S$: h98a
$C_6H_6O_5S$: d383
$C_6H_6O_6$: p210
$C_6H_6O_8$: d382
$C_6H_6O_8S_2$: d724
C_6H_6S: t162
$C_6H_7AsO_3$: b11
$C_6H_7BO_2$: b12
$C_6H_7ClN_2$: c203, c204, c205
C_6H_7N: a303, a304, m401, m402, m403
C_6H_7NO: a260, a261, a262, m101, p268, p269
$C_6H_7NO_2$: m406
$C_6H_7NO_2S$: b21
$C_6H_7NO_3S$: a118, a119, a120
C_6H_7NS: a298
$C_6H_7N_3O$: p262
$C_6H_7N_3O_2$: n67, n68, n69
$C_6H_7O_2P$: p137
$C_6H_7O_3P$: p138
$C_6H_8AsNO_3$: a115
$C_6H_8ClN_3O_4S_2$: a139
$C_6H_8Cl_2O_2$: h62, m221

$C_6H_8N_2$: a226, a227, a228, a229, a230, d238, m258, p109, p110, p111, p120
$C_6H_8N_2O$: a211, o63
$C_6H_8N_2O_2S$: b25, s23
$C_6H_8N_2O_3$: d486a
$C_6H_8N_4$: p184
C_6H_8O: c332, d527, h40, m217
$C_6H_8O_2$: b378, c323, d365, h42, m215, v4
$C_6H_8O_3$: a36, e11a, f43
$C_6H_8O_4$: d515a, d526, d546
$C_6H_8O_6$: a317, g8, i59
$C_6H_8O_7$: c274
C_6H_9Br: b264
C_6H_9ClO: c78
$C_6H_9ClO_3$: e96, e97
$C_6H_9F_3O_2$: b487
C_6H_9NO: o57a, v11
C_6H_9NOS: m421
$C_6H_9NO_2$: b441
$C_6H_9NO_6$: n21
$C_6H_9N_3$: a158
$C_6H_9N_3O_2$: a159, c285, h83
C_6H_{10}: c331, d490, h41, h82, m338a
$C_6H_{10}N_2$: e175, p187
$C_6H_{10}N_2O_2$: c324
$C_6H_{10}N_2O_4$: d279
$C_6H_{10}N_2O_5$: a14
$C_6H_{10}N_4$: p26
$C_6H_{10}O$: c329, c331b, d26, d362, e9b, h78, m216, m351, m353, m355
$C_6H_{10}O_2$: a96, c354, d360, e41, e105, e113, e169, h61, m352
$C_6H_{10}O_3$: d437, e54, h121, m393, p217
$C_6H_{10}O_4$: d325, d610, e17, h57, m273, m301a
$C_6H_{10}O_4S$: t154
$C_6H_{10}O_4S_2$: d711, e129
$C_6H_{10}O_5$: d326
$C_6H_{10}O_6$: d618
$C_6H_{10}O_8$: t86
$C_6H_{10}S$: d27
$C_6H_{11}Br$: b263
$C_6H_{11}BrO_2$: b297, e77, e78, e79
$C_6H_{11}Cl$: c77
$C_6H_{11}ClO$: h73
$C_6H_{11}ClO_2$: b436, e98
$C_6H_{11}I$: i32

$C_6H_{11}N$: d25, h63, m342
$C_6H_{11}NO$: c330, e220, f35, m376, o57, t360
$C_6H_{11}NO_2$: e62
$C_6H_{11}O_3$: t69
C_6H_{12}: c313, d500, d501, e85, h75, m214
$C_6H_{12}Br_2$: d86
$C_6H_{12}ClNO$: c113
$C_6H_{12}Cl_2$: d172a, d187
$C_6H_{12}Cl_2O$: b162
$C_6H_{12}Cl_2O_2$: b157, d169
$C_6H_{12}Cl_3O_3P$: t424
$C_6H_{12}Cl_3O_4P$: t423
$C_6H_{12}F_3NOSi$: m443
$C_6H_{12}NO_3P$: d294
$C_6H_{12}N_2$: d45
$C_6H_{12}N_2O$: d620a
$C_6H_{12}N_2O_4S_2$: c372
$C_6H_{12}N_2S$: b175
$C_6H_{12}N_2Si$: t380
$C_6H_{12}N_4$: h52
$C_6H_{12}O$: a100, b491, c328, d499, d620, e9a, e88, h54, h72, h72a, h77, i72, m349, o47
$C_6H_{12}O_2$: b415, b416, b417, c322a, d501a, d502, e11b, e50, e89, e90, e163, h66, h143, i62, m228, m305, m343, m344, m418a, m419, m442a, p241a, t79
$C_6H_{12}O_3$: d436, d458, e38, e155, e157, i99, p2, p235
$C_6H_{12}O_4$: e134a
$C_6H_{12}O_4Si$: d23
$C_6H_{12}O_6$: f36, g1, g6, i23, m11, s6
$C_6H_{12}O_7$: g4
$C_6H_{12}S$: c327
$C_6H_{13}Br$: b296
$C_6H_{13}BrO_2$: b269
$C_6H_{13}Cl$: c130
$C_6H_{13}ClO$: c131
$C_6H_{13}ClO_2$: c81
$C_6H_{13}ClO_3$: c106
$C_6H_{13}Cl_3O_3Si$: t422
$C_6H_{13}I$: i39
$C_6H_{13}N$: c335, e171a, h51, m371, m372, m373, m374
$C_6H_{13}NO$: d260, d555, e187, h126a, h145, p190

TABLE 1.14 Empirical Formula Index of Organic Compounds (*Continued*)

The alphanumeric designations are keyed to Table 1.15.

$C_6H_{13}NO_2$: a185, a186, h122, i79, L4, L5, m227a, t435
$C_6H_{13}NO_4$: b183
C_6H_{14}: d491, d492, h55, m339, m340
$C_6H_{14}ClN$: d272
$C_6H_{14}Cl_4OSi$: b169
$C_6H_{14}N_2$: a222, c319, c320
$C_6H_{14}N_2O$: a172, h123
$C_6H_{14}N_2O_2$: L12
$C_6H_{14}N_4O_2$: a316
$C_6H_{14}O$: b452, d418, d494, d495, d496, d497, d498, d703, e84, h68, h69, h70, m346, m347, m348
$C_6H_{14}OSi$: a97, e34, t381
$C_6H_{14}O_2$: b413, d251, d252, d493, d493a, e182, h58, h59, h60, i86, m341, t327
$C_6H_{14}O_2S$: d706
$C_6H_{14}O_3$: b186b, b192, d253, e36, e159, h65, h173
$C_6H_{14}O_4$: t275
$C_6H_{14}O_4S$: d705
$C_6H_{14}O_6$: d740, m10, s5
$C_6H_{14}O_6S_2$: b189
$C_6H_{14}S$: b454, h64
$C_6H_{14}Si$: a104
$C_6H_{15}Al$: t268
$C_6H_{15}AlO$: d266b
$C_6H_{15}As$: t271
$C_6H_{15}B$: t273
$C_6H_{15}Bi$: t272
$C_6H_{15}ClO_3Si$: c232
$C_6H_{15}ClSi$: b449
$C_6H_{15}Ga$: t279
$C_6H_{15}In$: t281
$C_6H_{15}N$: d412, d696, e86, e87, h80, m353a, m354, t269, t372a
$C_6H_{15}NO$: a187, a219, a220, b422, b451, d254, d270, e119
$C_6H_{15}NO_3$: t264
$C_6H_{15}NO_3S$: t431
$C_6H_{15}NO_5S$: b181
$C_6H_{15}N_3$: a174
$C_6H_{15}O_3B$: t265
$C_6H_{15}O_3P$: d421, t287
$C_6H_{15}O_3PS$: t290
$C_6H_{15}O_4P$: t285
$C_6H_{15}O_5P$: e120
$C_6H_{15}P$: t286
$C_6H_{15}Sb$: t270

$C_6H_{16}Cl_2OSi_2$: b162a
$C_6H_{16}N_2$: d302, h56, m340a, t110
$C_6H_{16}OSi$: p221
$C_6H_{16}Br_2OSi_2$: b151
$C_6H_{16}O_2Si$: d249
$C_6H_{16}O_3SSi$: m22
$C_6H_{16}O_3Si$: t266a
$C_6H_{16}Si$: t289
$C_6H_{17}NO_3Si$: a285, t328
$C_6H_{17}N_3$: i9
$C_6H_{18}N_2Si$: b173
$C_6H_{18}N_3ClSi$: c90
$C_6H_{18}N_3OP$: h53
$C_6H_{18}N_4$: t277
$C_6H_{18}OSi_2$: h50
$C_6H_{18}O_3Si_3$: h48
$C_6H_{19}NOSi_2$: b211
$C_6H_{19}NSi_2$: h49
C_6N_4: t38

C_7

$C_7H_3BrClF_3$: b252
$C_7H_3BrF_3NO_2$: b318
$C_7H_3ClF_3NO_2$: c183, c184, c185
$C_7H_3ClN_2O_5$: d632
$C_7H_3ClN_2O_6$: c97
$C_7H_3Cl_2NO$: d215a
$C_7H_3Cl_3O$: d160, d161
$C_7H_4BrF_3$: b235, b236
C_7H_4ClFO: f14
$C_7H_4ClF_3$: c51, c52, c53
C_7H_4ClN: c47, c48
C_7H_4ClNO: c206, c207
$C_7H_4ClNO_3$: n41, n42
$C_7H_4ClNO_4$: c179, c180, c181
$C_7H_4Cl_2O$: c55, c56, d150
$C_7H_4Cl_2O_2$: d156, d157, d158
$C_7H_4Cl_3F$: t237
$C_7H_4Cl_4$: c50a, c50b
$C_7H_4F_3NO_2$: n88, n89
$C_7H_4I_2O_3$: h111
$C_7H_4N_2O_2$: n40
$C_7H_4N_2O_6$: d630, d631
$C_7H_4N_2O_7$: d640
$C_7H_4O_3S$: h104
$C_7H_4O_4S$: s26
C_7H_5BrO: b66, b229
$C_7H_5BrO_2$: b233
$C_7H_5BrO_3$: b351
$C_7H_5ClF_3N$: a144, a145, a146

C_7H_5ClO: b67, c38, c38a, c39
$C_7H_5ClO_2$: c45, c46, c47, c238, c239, p104
$C_7H_5ClO_3$: c194
$C_7H_5Cl_2F$: c120
$C_7H_5Cl_2N$: d196
$C_7H_5Cl_2NO$: d151
$C_7H_5Cl_3$: t252, t253, t253a
C_7H_5FO: b69, f10
$C_7H_5FO_2$: f12, f13
$C_7H_5F_3$: t305
$C_7H_5F_3N_2O_2$: a244
$C_7H_5F_3O$: t295
$C_7H_5F_4N$: a179
$C_7H_5IO_2$: i29
$C_7H_5IO_3$: i51
$C_7H_5I_2NO_2$: a156
C_7H_5N: b51
C_7H_5NO: b63, p123
$C_7H_5NO_3$: n27, n28
$C_7H_5NO_3S$: s1
$C_7H_5NO_4$: n37, n38, n39, p263, p264, p265
$C_7H_5NO_5$: h155
C_7H_5NS: b60, p124
$C_7H_5NS_2$: m17
$C_7H_5N_3O_2$: a241, n36, n55
$C_7H_5N_3O_2S$: a243
$C_7H_5N_3O_6$: t393
C_7H_6BrClO: b254
$C_7H_6BrNO_2$: n46
$C_7H_6BrNO_3$: h156
$C_7H_6Br_2$: b238, d102
C_7H_6ClF: c125, c126, f16
C_7H_6ClNO: c40
$C_7H_6ClNO_2$: a140, a141, c188, c189, c190, n47
$C_7H_6Cl_2$: c59, c60, d229, d230, d231
$C_7H_6Cl_2O$: d192
$C_7H_6F_3N$: a129, a130, a131
$C_7H_6INO_2$: a205
$C_7H_6N_2$: a124, a125, a126, b38
$C_7H_6N_2O_3$: n29
$C_7H_6N_2O_4$: a240, d641, d642, d643
$C_7H_6N_2O_5$: d633, d33a
$C_7H_6N_2S$: a128, m15
C_7H_6O: b3
C_7H_6OS: t145
$C_7H_6O_2$: b44, h94, h95, h96, m241
$C_7H_6O_2S$: m16

TABLE 1.14 Empirical Formula Index of Organic Compounds (*Continued*)

The alphanumeric designations are keyed to Table 1.15.

$C_7H_6O_3$: d376, d377, h99, h100, h101
$C_7H_6O_4$: d384, d385, d386
$C_7H_6O_5$: t311
$C_7H_6O_6S$: s30
C_7H_7Br: b85, b358, b359, b360
C_7H_7BrO: b237, b304, b305, b306
C_7H_7Cl: b89, c245, c246, c247
$C_7H_7ClN_4O_2$: c242
C_7H_7ClO: c57, c140, c159, c160
$C_7H_7ClO_2S$: t180
$C_7H_7ClO_3S$: m49
C_7H_7ClS: c249
$C_7H_7Cl_3Si$: b124
C_7H_7F: f24, f25, f26
C_7H_7FO: f15, f19
$C_7H_7FO_2S$: t181
C_7H_7I: i54, i54a
C_7H_7IO: i41
C_7H_7N: v9, v10
C_7H_7NO: a53, a54, a55, b4, f31
$C_7H_7NO_2$: a121, a122, a123, h97, h98, m404, m405, n84, n85, n86
$C_7H_7NO_3$: a291, a292, m81, m82, m326, m327, n44, n45
$C_7H_7NO_4S$: c16
$C_7H_7N_3$: a203, a204, m136
C_7H_8: b130, c311, t170
C_7H_8BrN: b307
C_7H_8ClN: c58, c143, c144, c145, c146, c147
C_7H_8ClNO: c138a, c139
$C_7H_8ClNO_2$: c18
$C_7H_8ClNO_2S$: c248
$C_7H_8Cl_2Si$: d198
$C_7H_8N_2O$: a114, b72, p168
$C_7H_8N_2O_2$: d33, h142, h166, m317a, m317b, m318, m319, m320, m321a, m321b, m322c
$C_7H_8N_2O_3$: m78, m79, m80
$C_7H_8N_2S$: p157
$C_7H_8N_4O_3$: t140
C_7H_8O: b78, m48
C_7H_8OS: m431a, m432
$C_7H_8O_2$: d390, d391, h105, m87, m88, m89, m277
$C_7H_8O_2S$: t176
$C_7H_8O_3$: e139, f45, m307
$C_7H_8O_3S$: m127, t179

C_7H_8S: m367, p130, t150
C_7H_9N: b79, d605a, d606, d607, d608, d609, e214, e215, e216, m122, t184, t185, t186
C_7H_9NO: a221, b99, h126, m42, m43, m44
$C_7H_9NO_2$: d457
$C_7H_9NO_2S$: t177
$C_7H_9NO_3S$: a212, a213, a299
C_7H_9NS: m295, m425, m426
$C_7H_9N_3O$: a133
C_7H_{10}: b131
$C_7H_{10}N_2$: a177, a178, d478, m363, t171, t172, t173, t174
$C_7H_{10}N_2O$: m93a, m94
$C_7H_{10}N_2OS$: h130
$C_7H_{10}N_2O_2$: e176, m233
$C_7H_{10}N_2O_2S$: t178
$C_7H_{10}O$: m61, m62, n107, t67
$C_7H_{10}O_2$: a40, c360
$C_7H_{10}O_3$: e12, m280, m336, t346
$C_7H_{10}O_4$: d552
$C_7H_{10}O_5$: d460
$C_7H_{11}Br$: b321
$C_7H_{11}BrO_4$: d286
$C_7H_{11}ClO$: c317
$C_7H_{11}ClO_4$: d290
$C_7H_{11}N$: h110
$C_7H_{11}NO$: c341
$C_7H_{11}NO_2$: a52, b440
$C_7H_{11}NO_3$: m338
$C_7H_{11}NO_5$: a45
$C_7H_{11}NS$: c342
C_7H_{12}: c312, h22, m208, m209
$C_7H_{12}N_2O$: m452a
$C_7H_{12}O$: c310, c316, c331a, m205, m206, m207, m269
$C_7H_{12}O_2$: a86, b419, b419a, c318, d356, e122
$C_7H_{12}O_3$: e171, e198, h173a
$C_7H_{12}O_4$: d317, d318, d554, d576, d577, d578, h8, m68a, m274, t128
$C_7H_{12}O_5$: g17, g18
$C_7H_{12}O_6Si$: m438
$C_7H_{13}Br$: b262, b309
$C_7H_{13}BrO_2$: e81
$C_7H_{13}ClO$: h18
$C_7H_{13}N$: a253, d333, q5
$C_7H_{13}NO$: a322, c340
$C_7H_{13}NO_2$: a152

C_7H_{14}: c307, h19, m195
$C_7H_{14}ClN$: c114
$C_7H_{14}N_2$: d417
$C_7H_{14}N_2O$: a283
$C_7H_{14}N_2O_2$: e204
$C_7H_{14}O$: c309, c343, d571, d580, h5, h15, h16, h17, m198, m199, m200, m201, m202, m203, m204, m268
$C_7H_{14}O_2$: b455a, b482, c308, d258, e124, e172d, e173, e201, h10, i71a, i80, i93a, m266, p52, p226a
$C_7H_{14}O_3$: i68
$C_7H_{14}O_6$: m257
$C_7H_{15}Br$: b293, b294
$C_7H_{15}Cl$: c129
$C_7H_{15}ClO_2$: c82
$C_7H_{15}I$: i37
$C_7H_{15}N$: c325, d593, e205, e206, m210, m211, m212, m213
$C_7H_{15}NO$: d471, e160, h125, m375, p188, p189
$C_7H_{15}NO_2$: p280
$C_7H_{15}NO_3$: m455
C_7H_{16}: d573, e200, h6, t342
$C_7H_{16}BrNO_2$: a38
$C_7H_{16}ClNO_2$: a39
$C_7H_{16}N_2$: a218, a282, m304, t372
$C_7H_{16}O$: d579, h12, h13, h14, m267, t343
$C_7H_{16}O_2$: b414c, d257, d331, m397
$C_7H_{16}O_2Si$: d256, e231
$C_7H_{16}O_3$: d702, t283, t326
$C_7H_{16}O_4$: t93, t276c
$C_7H_{16}S$: h9
$C_7H_{17}N$: h20, m269a
$C_7H_{17}NO$: d275, d276
$C_7H_{17}NO_2$: b423, d274
$C_7H_{17}NO_3S$: t434
$C_7H_{17}NO_5$: m256
$C_7H_{17}NO_7S$: t433
$C_7H_{18}N_2$: d330, h7, i103, t120
$C_7H_{18}N_2O$: b174
$C_7H_{18}N_2O_2$: a281
$C_7H_{18}O_2Si$: b489
$C_7H_{18}O_3Si$: b488, t266
$C_7H_{19}NOSi_2$: b210
$C_7H_{19}NSi$: d343, t378
$C_7H_{19}N_3$: d42, t426

TABLE 1.14 Empirical Formula Index of Organic Compounds (*Continued*)

The alphanumeric designations are keyed to Table 1.15.

$C_7H_{20}N_2OSi_2$: b214
$C_7H_{21}N_3Si$: t427

C_8

$C_8Br_4O_3$: t11
$C_8Cl_4O_3$: t32
C_8D_{10}: e68
$C_8HCl_4NO_2$: t33
$C_8H_3NO_5$: n72
$C_8H_4BrNO_2$: b298
$C_8H_4Cl_2O_2$: b14, b15, p174
$C_8H_4Cl_2O_4$: d217
$C_8H_4Cl_6$: b203
$C_8H_4F_3N$: t303
$C_8H_4F_6$: b207
$C_8H_4N_2$: d236, d237
$C_8H_4N_2O_2$: p111a
$C_8H_4O_3$: p171
$C_8H_5Br_5$: p5
$C_8H_5ClO_2$: c210
$C_8H_5Cl_3O$: t221a
$C_8H_5Cl_3O_3$: t246
$C_8H_5F_3O_2S$: t139
$C_8H_5F_6N$: b206
C_8H_5NO: b68
$C_8H_5NO_2$: i21, p173
$C_8H_5NO_3$: h167, i58
$C_8H_5NO_6$: n32, n34, n71a, n71b
C_8H_6: p82
C_8H_6BrClO: b248
C_8H_6BrN: b332
$C_8H_6Br_2O$: d64
$C_8H_6Br_4$: t12, t13, t14
$C_8H_6ClF_3$: b304
C_8H_6ClN: c61, c202
$C_8H_6ClNO_3$: c172
$C_8H_6Cl_2O$: c147a, d140
$C_8H_6Cl_2O_3$: d214
$C_8H_6Cl_4$: t36
$C_8H_6N_2$: q4
$C_8H_6N_2O_2$: a271, n65
$C_8H_6N_2O_6$: d638, m229
C_8H_6O: b42
$C_8H_6O_2$: b13, p172
$C_8H_6O_3$: b70, f33, m240
$C_8H_6O_4$: b16, b17, m242, p170
C_8H_6S: b61
C_8H_7Br: b352
C_8H_7BrO: b222, b223
$C_8H_7BrO_2$: b331

C_8H_7ClO: c28, c29, c30, p81, t191, t192, t193
C_8H_7ClOS: b91
$C_8H_7ClO_2$: b90, c201, m53, p69
$C_8H_7ClO_3$: c198, m185, m186
$C_8H_7ClO_4$: m190a
C_8H_7FO: f8
C_8H_7N: i18, p80, t188, t189, t190
C_8H_7NO: m9, m137, t194
$C_8H_7NO_2$: h134
$C_8H_7NO_3$: n22, n23
$C_8H_7NO_3S$: t182
$C_8H_7NO_4$: a116, m321, m322, m322a, m323, m324, n62, n63, n64
$C_8H_7NO_5$: m83
C_8H_7NS: b122, m135
$C_8H_7N_3O_2$: a153
C_8H_8: s11
C_8H_8BrNO: b220
$C_8H_8Br_2$: d79, d104, d105
C_8H_8ClNO: c23
$C_8H_8ClNO_3S$: a10
$C_8H_8Cl_2$: d233, d234
$C_8H_8Cl_2Si$: p169
$C_8H_8HgO_2$: p127
$C_8H_8N_2$: a263, m128
$C_8H_8N_2O$: a210
C_8H_8O: a31, e8, m126, p76a
C_8H_8OS: m427, p156
$C_8H_8O_2$: b41, b97, h90, h91, h92, m45, m45a, m46, m129, m130, m131, m132, p78, p79
$C_8H_8O_2S$: t163
$C_8H_8O_3$: d371, d381, h131, h132, h138, h139, h161, m8, m50, m51, m52, m243, m278, m413, p68, t76
$C_8H_8O_4$: d21, h133
$C_8H_8O_4S$: a33
$C_8H_8O_5$: m441a
C_8H_9Br: b285, b371, b372, b373, b374
C_8H_9BrO: b272, b288
$C_8H_9BrO_2$: b270
C_8H_9Cl: c108, c109, c259, c260, c261, c261a, c262
C_8H_9ClO: c91
$C_8H_9ClO_2$: c88
C_8H_9N: b101, c361, i22
C_8H_9NO: a18, a108, a109, a110, b96, m250

$C_8H_9NO_2$: a15, a16, a17, a214, a215, b88, d558, d559, d560, d561, e190, e217, e218, e219, m47, m116, m117, p117, t77
$C_8H_9NO_3$: a209, h164, h165, m85
$C_8H_9NO_4$: d445
$C_8H_9NO_4S$: m121
C_8H_{10}: e69, m245, x4, x5, x6
$C_8H_{10}N_2O$: d562
$C_8H_{10}N_2O_2$: c339a
$C_8H_{10}N_2O_3S$: m327b
$C_8H_{10}N_4O_2$: c1, d240
$C_8H_{10}O$: b132, d581, d582, d583, d584, d585, d586, e29, e201a, e201b, m105, m106, m107, m138, m139, m140, p114, p115
$C_8H_{10}O_2$: b18, d432, d433, d434, e46, m54, m74, p72, p113
$C_8H_{10}O_3$: c321, d447, h114, h136, m26a, m227b
$C_8H_{10}O_3S$: m437
$C_8H_{10}O_4$: d263
$C_8H_{10}O_8$: b390a
$C_8H_{10}S$: b106
$C_8H_{11}N$: b104, d479, d480, d481, d482, d483, d484, d485, e64, e65, e66, e183, e184, m141, m142, p116, t373
$C_8H_{11}NO$: a173, a259, a264, a265, a305, d472, e25, h117, m55, m71, m72, m73, p270
$C_8H_{11}NO_2$: d428, d429, d430
$C_8H_{11}NO_2S$: m436
$C_8H_{11}NO_3$: e135
$C_8H_{11}NO_3S$: d465
$C_8H_{11}N_5$: p94
C_8H_{12}: c345, c346, v6
$C_8H_{12}N_2$: d239, d587, d588, m121, t121, x9
$C_8H_{12}N_2O_2$: d411
$C_8H_{12}N_2O_3$: d280
$C_8H_{12}N_4$: a328
$C_8H_{12}O$: e237
$C_8H_{12}O_2$: d510, e222, h186, n111
$C_8H_{12}O_3$: e197
$C_8H_{12}O_4$: d24, d305, d316, m28a
$C_8H_{12}O_6Si$: t199

TABLE 1.14   Empirical Formula Index of Organic Compounds (*Continued*)

The alphanumeric designations are keyed to Table 1.15.

$C_8H_{13}N$: e238
C_8H_{14}: c350, o17, o44, v5
$C_8H_{14}N_2$: p191, p274a
$C_8H_{14}O$: c349, d512, d539a, m263
$C_8H_{14}O_2$: b462, c334, c364, d539, h78a, i69, i75a, m196
$C_8H_{14}O_3$: b418, b499, d714, e39a, e91
$C_8H_{14}O_4$: b180b, b388, d320, d335, d536, e130, e152, o24
$C_8H_{14}O_4S$: d621
$C_8H_{14}O_4S_2$: d710
$C_8H_{14}O_6$: d339, d340
$C_8H_{14}O_6Si$: t198
$C_8H_{15}BrO_2$: e78a
$C_8H_{15}ClO$: e148, o37
$C_8H_{15}N$: o27
$C_8H_{15}NO$: d368, p241
$C_8H_{15}NO_2$: d470, e208, e209
C_8H_{16}: c347, d508, d509, d509a, d509b, e109, o39, t365, t365a
$C_8H_{16}Br_2$: d90a
$C_8H_{16}ClN$: c228
$C_8H_{16}O$: c339b, c348, d511, d511a, d511b, d511c, d511d, e9d, e110, e111, m262b, o34, o35, o36, o40
$C_8H_{16}O_2$: b433, c322, c347a, c347b, e145, e146, h79, i67, m262, o29, p237
$C_8H_{16}O_2Si$: t382
$C_8H_{16}O_3$: b414b
$C_8H_{16}O_4$: e37, t127
$C_8H_{17}Br$: b323
$C_8H_{17}Cl$: c191
$C_8H_{17}Cl_3Si$: o43
$C_8H_{17}I$: i46
$C_8H_{17}N$: b483a, c351, d512a, d513, e112a
$C_8H_{17}NO_2$: p192
$C_8H_{17}NO_3$: e150d
$C_8H_{17}O_5P$: t288
C_8H_{18}: e143, e177, e178, o22, t102, t361, t362, t363
$C_8H_{18}AlCl$: d406a
$C_8H_{18}ClNO_2$: a49
$C_8H_{18}Cl_2Si$: d184
$C_8H_{18}Cl_2Sn$: d136
$C_8H_{18}F_3NOSi_2$: b213
$C_8H_{18}N_2$: c315
$C_8H_{18}N_2O$: m225, m370
$C_8H_{18}N_2O_4S$: h124

$C_8H_{18}O$: d115, d408, e147, m262a, o30, o31, o32, o33
$C_8H_{18}OSi_2$: d715
$C_8H_{18}OSn$: d137
$C_8H_{18}O_2$: b413a, d122, d537, e144, o25, o26, t363
$C_8H_{18}O_2S$: d135
$C_8H_{18}O_3$: b177, b414, d700, t282
$C_8H_{18}O_3S$: d134
$C_8H_{18}O_3Si$: t267
$C_8H_{18}O_4$: b190, o27a, t276b
$C_8H_{18}O_4S$: d131
$C_8H_{18}O_5$: t52
$C_8H_{18}S$: d132, d133, o28
$C_8H_{18}S_2$: b154, b155, d113, d114
$C_8H_{18}Si_2$: b209
$C_8H_{19}Al$: d406b
$C_8H_{19}N$: d107, d107a, d407, d413, d419, d538, e150, o41, t104
$C_8H_{19}NO_2$: b447, d247, d248
$C_8H_{19}NO_5$: b184
$C_8H_{19}O_3P$: d127
$C_8H_{20}BrN$: t49
$C_8H_{20}ClN$: t50
$C_8H_{20}Ge$: t58
$C_8H_{20}N_2$: o23, t103
$C_8H_{20}N_2O_2S$: t62a
$C_8H_{20}N_2OSi$: t48
$C_8H_{20}O_4Ti$: t168a
$C_8H_{20}O_5P_2$: t61
$C_8H_{20}O_7P_2$: t60
$C_8H_{20}Pb$: t59
$C_8H_{20}SSi$: m19
$C_8H_{20}Si$: t62
$C_8H_{20}Sn$: t64
$C_8H_{21}NO$: t51
$C_8H_{21}NOSi_2$: b208
$C_8H_{21}NO_2Si$: a280
$C_8H_{21}NO_3Si$: t329
$C_8H_{22}N_2O_3Si$: a167
$C_8H_{22}N_4$: b146
$C_8H_{22}O_2Si_2$: b212
$C_8H_{23}N_5$: t56
$C_8H_{24}Cl_2O_3Si_4$: d207
$C_8H_{24}O_2Si_3$: o21

C_9

$C_9H_2Cl_6O_3$: h30
$C_9H_3Cl_3O_3$: b32
$C_9H_4O_5$: b31

$C_9H_5BrClNO$: b257
$C_9H_5Br_2NO$: d87
C_9H_5ClINO: c135
$C_9H_5Cl_2N$: d225
C_9H_6BrN: b350
C_9H_6ClN: c237
C_9H_6ClNO: c135
$C_9H_6INO_4S$: h126d
$C_9H_6N_2O_2$: t175
$C_9H_6O_2$: b56, c277
$C_9H_6O_3$: h108, h109
$C_9H_6O_4$: i16
$C_9H_6O_6$: b28, b29, b30
C_9H_7BrO: b261
C_9H_7ClO: c269
$C_9H_7Cl_3O_3$: t247
C_9H_7N: i110, q3
C_9H_7NO: h179, i20
$C_9H_7NO_3$: h144, m286
$C_9H_7NO_4S$: h180
$C_9H_7N_3O_4S_2$: a250
C_9H_8: i17
$C_9H_8N_2$: m412
$C_9H_8N_2O_5$: n43
$C_9H_8N_2O_6$: e124a
C_9H_8O: c267, i15
$C_9H_8O_2$: c268, d354
$C_9H_8O_3$: h107
$C_9H_8O_4$: p126
C_9H_9BrO: b345
$C_9H_9BrO_2$: b85a
C_9H_9Cl: c218
C_9H_9ClO: c224
$C_9H_9ClO_3$: c200, c250, d435a
C_9H_9N: d488, m284
C_9H_9NO: m93
$C_9H_9NO_2$: a9
$C_9H_9NO_2S$: t195
$C_9H_9NO_3$: a11, a12, b71
$C_9H_9N_3O$: a268
C_9H_{10}: a84, i13, m414, m414a, v3
$C_9H_{10}F_3NO_2$: m123
$C_9H_{10}N$: a201, a202
$C_9H_{10}N_2$: a306, p121
$C_9H_{10}N_2O$: p150
$C_9H_{10}N_2O_2$: p83
$C_9H_{10}N_2O_3$: a132
$C_9H_{10}O$: a98, a99, c270, d361, e9e, i14, m113, p146, p147, p212, p220
$C_9H_{10}O_2$: b77, d487, e9a, e26, e27, e70, h171, h172, m39, m40, m41, m304a, m304b,

TABLE 1.14 Empirical Formula Index of Organic Compounds (*Continued*)

The alphanumeric designations are keyed to Table 1.15.

m304c, m359, p74, p148, t139a

$C_9H_{10}O_2S$: b121

$C_9H_{10}O_3$: d431, e30, e31, e40, e49, e154, e221, m91, m293, m301b, m301c, p75

$C_9H_{10}O_4$: d435, h126b, m289, m448

$C_9H_{10}O_8$: c355

$C_9H_{11}Br$: b300, b334, b364, b365

$C_9H_{11}BrO$: b346

$C_9H_{11}Cl$: c208a

$C_9H_{11}ClO_3S$: c115

$C_9H_{11}Cl_3Si$: c227

$C_9H_{11}N$: a83, c333, m285, t73, t82

$C_9H_{11}NO$: d464, d486b, m110a, m358, m435

$C_9H_{11}NO_2$: d466, e28, e60, e61, p84

$C_9H_{11}NO_3$: t444

C_9H_{12}: e161, i91, p225, t338, t339, t340, v8

$C_9H_{12}N_2O_4$: a249

$C_9H_{12}N_2O_6$: u14

$C_9H_{12}N_2S$: b107

$C_9H_{12}O$: b95, d549, d550, i106, i107, m356, p144, p145, p238, t366, t367, t368

$C_9H_{12}O_2$: b111, e32, i85, p73, p142, t354

$C_9H_{12}O_3$: d435b, m197, t320

$C_9H_{12}O_3S$: e225

$C_9H_{12}S$: p143

$C_9H_{13}N$: b483, d489, d624, e73, e172, e226, e227, i90, t335

$C_9H_{13}NO$: b80, m86, n112

$C_9H_{13}NO_2$: a266

$C_9H_{14}BrN$: p162

$C_9H_{14}Br_3N$: p165

$C_9H_{14}ClN$: p163

$C_9H_{14}IN$: p164

$C_9H_{14}N_2$: n94

$C_9H_{14}O$: d529, d531, i82, t345

$C_9H_{14}O_2Si$: d444

$C_9H_{14}O_3$: b193

$C_9H_{14}O_3Si$: p161

$C_9H_{14}O_5$: d262, d321

$C_9H_{14}O_6$: p204

$C_9H_{14}Si$: p166

$C_9H_{15}NO$: c362

$C_9H_{15}NO_2$: d570

$C_9H_{15}NO_5$: d261

$C_9H_{15}N_3$: t439

C_9H_{16}: h46

$C_9H_{16}N_2$: d46

$C_9H_{16}O$: d530, d533a

$C_9H_{16}O_2$: c326, n98a

$C_9H_{16}O_3$: b470

$C_9H_{16}O_4$: d303, d307, d322, d515, d532, n95

$C_9H_{17}ClO$: n101

$C_9H_{17}ClO_2$: e150a

$C_9H_{17}N$: a88, n97

$C_9H_{17}NO$: m181

$C_9H_{17}NO_2$: e180, e181

C_9H_{18}: i94, n102, p227

$C_9H_{18}NO$: t119

$C_9H_{18}O$: d533, n99a, n99b, n103, n104, t351a

$C_9H_{18}O_2$: e141, m332, n98

$C_9H_{18}O_3$: d111

$C_9H_{19}Br$: b320

$C_9H_{19}I$: i45b

$C_9H_{19}N$: i95, t337, t351

$C_9H_{19}NO$: d116

$C_9H_{19}NO_2$: e121

C_9H_{20}: n92, t352

$C_9H_{20}Cl_2Si$: m333

$C_9H_{20}N_2$: a301

$C_9H_{20}N_2S$: d136

$C_9H_{20}O$: d532a, n99, t353

$C_9H_{20}O_2$: b453, n96

$C_9H_{20}O_3$: d701, t284

$C_9H_{20}O_3Si$: a103

$C_9H_{20}O_4$: t415

$C_9H_{20}O_5$: t55

$C_9H_{21}BO_3$: t413

$C_9H_{21}ClO_3Si$: c231

$C_9H_{21}N$: n104, t414

$C_9H_{21}N_3$: t280

$C_9H_{21}O_3P$: t318

$C_9H_{22}N_2$: d327, n93

$C_9H_{22}O_3Si$: p241

$C_9H_{22}Si$: t318a

$C_9H_{24}N_4$: b148

$C_9H_{27}NO_3$: t314

C_{10}

$C_{10}H_2O_6$: b27

$C_{10}H_4Cl_2O_2$: d201

$C_{10}H_6Cl_2O$: d200a

$C_{10}H_6N_2$: b100

$C_{10}H_6N_2O_4$: d636

$C_{10}H_6N_2O_4S$: d48

$C_{10}H_6O_2$: n11

$C_{10}H_6O_3$: h153

$C_{10}H_6O_8$: b26

$C_{10}H_7Br$: b315

$C_{10}H_7BrO$: b316

$C_{10}H_7Cl$: c169, c170

$C_{10}H_7NO_2$: n57, n81, p125

$C_{10}H_7NO_8S_2$: n82

$C_{10}H_8$: a331, n2

$C_{10}H_8BrNO_2$: b289

$C_{10}H_8N_2$: d707

$C_{10}H_8O$: n9, n10

$C_{10}H_8O_2$: d393, d394, d395, d396, m192

$C_{10}H_8O_3$: h141

$C_{10}H_8O_3S$: n6

$C_{10}H_8O_7S_2$: h151, h152

$C_{10}H_9ClCrN_2O_3$: b144

$C_{10}H_9N$: a235, m410, m411, n17

$C_{10}H_9NO$: a51, a239

$C_{10}H_9NO_2$: i19

$C_{10}H_9NO_3S$: a237

$C_{10}H_9NO_4S$: a195, a196, a197, a198

$C_{10}H_9NO_6$: d563

$C_{10}H_9NO_6S_2$: a236

$C_{10}H_9N_3$: d708

$C_{10}H_{10}ClFO$: c121

$C_{10}H_{10}ClNO_2$: c26

$C_{10}H_{10}N_2$: a290, n4, n5

$C_{10}H_{10}N_2O$: m365

$C_{10}H_{10}O$: d363, m191, p98, p100

$C_{10}H_{10}O_2$: b64, m60, m191a, s2

$C_{10}H_{10}O_3$: b73

$C_{10}H_{10}O_4$: d590, d591, d592, p155

$C_{10}H_{11}BrO$: b299, b314b

$C_{10}H_{11}ClO_3$: c199, c251

$C_{10}H_{11}ClO_4$: t322

$C_{10}H_{11}IO_4$: i28

$C_{10}H_{11}N$: p103

$C_{10}H_{11}NO_2$: a32, d449

$C_{10}H_{11}NO_4$: c10, d478a

$C_{10}H_{11}NO_6$: m227

$C_{10}H_{12}$: d244, t75

$C_{10}H_{12}N_2$: a170, b81, b103

$C_{10}H_{12}O$: a94, b503, e56, i77, m97, m389, m390, m394, p97

$C_{10}H_{12}O_2$: e172a, e172b, e172c, e203, h163, m70, m92, m98, m99, m100,

TABLE 1.14 Empirical Formula Index of Organic Compounds (*Continued*)

The alphanumeric designations are keyed to Table 1.15.

m137a, p77, p101, p115a, p226
$C_{10}H_{12}O_3$: d427, e42, e166, m303, p71, p233
$C_{10}H_{12}O_4$: d26a, d448, m226, m226a
$C_{10}H_{12}O_5$: p242, t321
$C_{10}H_{12}O_6$: d516
$C_{10}H_{13}Br$: b301
$C_{10}H_{13}BrO$: b245
$C_{10}H_{13}BrO_2$: b330
$C_{10}H_{13}Cl$: b437
$C_{10}H_{13}NO$: p131
$C_{10}H_{13}NO_2$: m367a
$C_{10}H_{13}NO_2S$: b92
$C_{10}H_{13}N_5O_4$: a70
$C_{10}H_{14}$: b426, b427, b428, d283, d284, i64, i100, i101, i102, t99, t100, t101
$C_{10}H_{14}ClN$: c110a
$C_{10}H_{14}NO_5PS$: p3
$C_{10}H_{14}N_2$: n20, p141
$C_{10}H_{14}N_2O$: d323, d334
$C_{10}H_{14}N_4O_4$: d400
$C_{10}H_{14}O$: b472, b473, b474, b475, b476, b480, c20, i92, i102a, i102b, i102c, t118, t259
$C_{10}H_{14}O_2$: b435, b457, d451
$C_{10}H_{14}O_3$: c9
$C_{10}H_{14}O_4$: e17a, e131, m90, t319, t323
$C_{10}H_{15}BrO$: b247
$C_{10}H_{15}N$: b424, d277, d278, e236, i93, i99a, p99, t98
$C_{10}H_{15}NO$: b441, d273, e2
$C_{10}H_{15}NO_2$: d452, p108
$C_{10}H_{16}$: a67, c2, d513a, d566b, d651, L6, L7, m456, p25, p178, p179, t5, t6, t258
$C_{10}H_{16}ClN$: b127
$C_{10}H_{16}Cl_2O_2$: d11
$C_{10}H_{16}N_2$: d327a
$C_{10}H_{16}N_2O_4$: d27a
$C_{10}H_{16}N_2O_8$: e128
$C_{10}H_{16}O$: c3, c4, d353, d564, d565, L8, p180, p182, p246, t357
$C_{10}H_{16}O_2$: m328
$C_{10}H_{16}O_4$: c5, d266
$C_{10}H_{16}O_4S$: c7
$C_{10}H_{16}O_5$: d265, d301
$C_{10}H_{16}Si$: b128

$C_{10}H_{17}N$: a66, p278
$C_{10}H_{17}NO$: c344, m453
$C_{10}H_{18}$: b444, b445, d1, d2, p177
$C_{10}H_{18}N_2O_7$: h119
$C_{10}H_{18}O$: b217, c266, g2, i60, i83, i109, L9, m13, p193, t7, t9, t356
$C_{10}H_{18}O_2$: c339c, d15a, d566d, e112
$C_{10}H_{18}O_3$: d601, t72
$C_{10}H_{18}O_4$: b176, d9, d121, d332, d567
$C_{10}H_{18}O_4S$: d341
$C_{10}H_{18}O_6$: d422
$C_{10}H_{19}ClO$: d17
$C_{10}H_{19}N$: d12, t336
$C_{10}H_{19}NO_2$: e210
$C_{10}H_{20}$: c302, d18
$C_{10}H_{20}Br_2$: d73
$C_{10}H_{20}N_2S_4$: t63
$C_{10}H_{20}O$: b442, b443, c275, d6, d15b, d16, d358, d367, e6, e151, m12, m306
$C_{10}H_{20}O_2$: d14, e149, e193, m63, m171, o39a
$C_{10}H_{20}O_4$: b414a, b432a
$C_{10}H_{20}O_5$: p45
$C_{10}H_{20}O_5Si$: t331
$C_{10}H_{21}Br$: b267
$C_{10}H_{21}Cl$: c80
$C_{10}H_{21}I$: i33
$C_{10}H_{21}NO$: a233
$C_{10}H_{22}$: d7
$C_{10}H_{22}N_2$: d41
$C_{10}H_{22}N_6$: p23
$C_{10}H_{22}O$: d15, d566a, d653, t74
$C_{10}H_{22}O_2$: d10, d106
$C_{10}H_{22}O_3$: d699, h81a, t420
$C_{10}H_{22}O_4$: t419
$C_{10}H_{22}O_5$: b191, t53a
$C_{10}H_{22}O_7$: d650
$C_{10}H_{23}N$: d19, d108, d566c, d652
$C_{10}H_{23}NO_2$: d259
$C_{10}H_{24}N_2$: d8, t57, t113
$C_{10}H_{24}N_2O_2$: d647
$C_{10}H_{24}N_4$: b147, t87
$C_{10}H_{24}OSi$: m109
$C_{10}H_{30}O_3Si_4$: d5
$C_{10}H_{30}O_5Si_5$: d4

C_{11}

$C_{11}H_4F_{20}O$: i1
$C_{11}H_7N$: c294
$C_{11}H_7NO$: n18
$C_{11}H_8O$: n1
$C_{11}H_8O_2$: h148, m313, n3
$C_{11}H_8O_3$: h149, h149a, h150
$C_{11}H_9Br$: b312
$C_{11}H_9Cl$: c158
$C_{11}H_9N$: p151
$C_{11}H_{10}$: m311, m312
$C_{11}H_{10}N_2S$: n19
$C_{11}H_{10}O$: m76, m77
$C_{11}H_{12}N_2O$: a314
$C_{11}H_{12}N_2O_2$: t443
$C_{11}H_{12}O_2$: c269a, d359, e104, m104
$C_{11}H_{12}O_3$: e71
$C_{11}H_{13}ClO$: b431
$C_{11}H_{13}ClO_3$: b251
$C_{11}H_{13}NO$: b120
$C_{11}H_{13}NO_3$: a307, a308
$C_{11}H_{13}N_3O$: a113
$C_{11}H_{13}N_3O_3S$: d569
$C_{11}H_{14}O$: p43
$C_{11}H_{14}O_2$: a89, b429, b430, d456, e47, p116a
$C_{11}H_{14}O_3$: b412, b471, b479, e170
$C_{11}H_{14}O_4$: e158
$C_{11}H_{14}O_5$: m441b
$C_{11}H_{15}NO$: d269
$C_{11}H_{15}NO_2$: d468, e123
$C_{11}H_{16}$: b485, p24, p54
$C_{11}H_{16}N_2$: b115
$C_{11}H_{16}O$: b86, b464, b465, d589, p56
$C_{11}H_{16}O_2$: a68
$C_{11}H_{16}O_3$: m302
$C_{11}H_{16}O_4$: d716
$C_{11}H_{17}N$: b432
$C_{11}H_{17}NO$: e228
$C_{11}H_{17}NO_2$: b105
$C_{11}H_{17}O_3P$: b93
$C_{11}H_{18}O$: d308, m105
$C_{11}H_{18}O_5$: d264
$C_{11}H_{19}ClO$: u11
$C_{11}H_{20}O$: u7
$C_{11}H_{20}O_2$: e149a, u9
$C_{11}H_{20}O_4$: d119, d287, d309
$C_{11}H_{21}BrO_2$: b370

TABLE 1.14 Empirical Formula Index of Organic Compounds (*Continued*)

The alphanumeric designations are keyed to Table 1.15.

$C_{11}H_{21}N$: u2a
$C_{11}H_{22}$: u8
$C_{11}H_{22}N_2$: d697
$C_{11}H_{22}O$: u1, u5, u5a, u6, u10
$C_{11}H_{22}O_2$: e150b, e190a, m219, u3
$C_{11}H_{22}O_4Si$: e7
$C_{11}H_{23}I$: i55a
$C_{11}H_{24}$: u2
$C_{11}H_{24}O$: d310, u4
$C_{11}H_{24}O_3Si$: t316
$C_{11}H_{24}O_4$: t417
$C_{11}H_{24}O_6Si$: t437
$C_{11}H_{25}NO_2$: a302
$C_{11}H_{26}N_2$: d129
$C_{11}H_{26}N_2O_6$: b215

C_{12}

$C_{12}Br_{10}O$: b198
$C_{12}H_4Cl_6S_2$: b204
$C_{12}H_5ClO_3$: c171
$C_{12}H_6Br_4O_4S$: s27
$C_{12}H_6O_3$: n7
$C_{12}H_6O_{12}$: b19
$C_{12}H_7NO_2$: n8
$C_{12}H_8$: a3
$C_{12}H_8Br_2$: d66
$C_{12}H_8Cl_2OS$: b167
$C_{12}H_8Cl_2O_2S$: b166, c209
$C_{12}H_8N_2$: p63
$C_{12}H_8N_2O_4S_2$: b195, b196
$C_{12}H_8O$: d50
$C_{12}H_8S$: d52
$C_{12}H_9Br$: b239
$C_{12}H_9BrO$: b333
$C_{12}H_9ClO_2S$: c208
$C_{12}H_9N$: c8, d667, n16
$C_{12}H_9NO$: b74, b75, b76
$C_{12}H_9NO_2$: n48, n49
$C_{12}H_9NO_3$: n70, n71
$C_{12}H_9NS$: p66
$C_{12}H_9N_3O_4$: d634
$C_{12}H_{10}$: a2, b135
$C_{12}H_{10}ClN$: c62
$C_{12}H_{10}ClO_3P$: c664
$C_{12}H_{10}ClP$: c100
$C_{12}H_{10}Cl_2Si$: d175
$C_{12}H_{10}Hg$: d677
$C_{12}H_{10}N_2$: a327
$C_{12}H_{10}N_2O$: n80, p90
$C_{12}H_{10}N_2O_2$: n52
$C_{12}H_{10}N_3O_3P$: d684

$C_{12}H_{10}O$: d669, m314, m315, p133, p134
$C_{12}H_{10}OS$: d692
$C_{12}H_{10}O_2$: d388, h88, n14, n15
$C_{12}H_{10}O_2S$: d691, t153
$C_{12}H_{10}O_3$: n12
$C_{12}H_{10}O_3S$: b140
$C_{12}H_{10}O_4$: q1
$C_{12}H_{10}O_4S$: s29
$C_{12}H_{10}S$: d690
$C_{12}H_{10}S_2$: d666
$C_{12}H_{10}Se_2$: d665
$C_{12}H_{11}N$: a134, a135, b118, b119, d657
$C_{12}H_{11}NO$: h112, n13, p70
$C_{12}H_{11}N_3$: p88
$C_{12}H_{11}O_3P$: d683
$C_{12}H_{12}N_2$: b137, d675, p135
$C_{12}H_{12}N_2O$: o62
$C_{12}H_{12}N_2O_2$: b40
$C_{12}H_{12}N_2O_2S$: d36, d37
$C_{12}H_{12}O$: e45
$C_{12}H_{12}O_2Si$: d689
$C_{12}H_{12}O_3$: e71a, t200
$C_{12}H_{12}O_6$: t197, t341
$C_{12}H_{13}N_3$: d34
$C_{12}H_{14}$: d411b
$C_{12}H_{14}N_4O_2S$: s22
$C_{12}H_{14}O$: c363
$C_{12}H_{14}O_3$: e179
$C_{12}H_{14}O_4$: d329
$C_{12}H_{15}N$: d370
$C_{12}H_{15}NO$: b117
$C_{12}H_{15}N_3O_3$: t201
$C_{12}H_{16}$: c339, m213, p106
$C_{12}H_{16}O_2$: m364
$C_{12}H_{16}O_3$: d246
$C_{12}H_{17}N$: b116, c338
$C_{12}H_{17}NO$: d342
$C_{12}H_{18}$: c305, d415, d416, p119, t442
$C_{12}H_{18}Cl_2N_4OS$: t141
$C_{12}H_{18}N_2$: p101a
$C_{12}H_{18}N_2O_2$: i82a
$C_{12}H_{18}O$: b450, d420, e3
$C_{12}H_{18}O_2$: b448, b479
$C_{12}H_{18}O_4$: b388a, h60a
$C_{12}H_{19}N$: d414, h81
$C_{12}H_{20}O_2$: b186, b218, e103, L10
$C_{12}H_{20}O_3Si$: p160
$C_{12}H_{20}O_4$: d118
$C_{12}H_{20}O_7$: t273a

$C_{12}H_{21}N$: t438
$C_{12}H_{21}N_3$: t429
$C_{12}H_{22}$: c307, d241
$C_{12}H_{22}O$: c304, e4
$C_{12}H_{22}O_2$: e150c
$C_{12}H_{22}O_3$: h67
$C_{12}H_{22}O_4$: d130, d324, d514, d704
$C_{12}H_{22}O_6$: d135a
$C_{12}H_{22}O_{11}$: L3, m7, s21
$C_{12}H_{23}N$: d242
$C_{12}H_{23}NO$: a323
$C_{12}H_{24}$: d731
$C_{12}H_{24}O$: c303, d733, e7, m446, t355
$C_{12}H_{24}O_2$: d728, e114
$C_{12}H_{25}Br$: b277
$C_{12}H_{25}Cl$: c101
$C_{12}H_{25}Cl_3Si$: d738
$C_{12}H_{25}I$: i33a
$C_{12}H_{25}N$: c306a
$C_{12}H_{26}$: d721
$C_{12}H_{26}O$: d351, d729
$C_{12}H_{26}O_2$: d724, d725
$C_{12}H_{26}O_3$: b152
$C_{12}H_{26}O_4$: t418
$C_{12}H_{26}O_4S$: d737
$C_{12}H_{26}O_5$: t52a
$C_{12}H_{26}S$: d727
$C_{12}H_{27}Al$: t312
$C_{12}H_{27}B$: t214a
$C_{12}H_{27}BO_3$: t213
$C_{12}H_{27}ClSn$: t219
$C_{12}H_{27}N$: d350, d734, t214
$C_{12}H_{27}O_3P$: t218
$C_{12}H_{27}O_4P$: t216
$C_{12}H_{27}P$: t217
$C_{12}H_{28}BrN$: t137
$C_{12}H_{28}N_2$: d722
$C_{12}H_{28}O_4Si$: t90, t136
$C_{12}H_{28}O_4Ti$: t169, t169a
$C_{12}H_{28}O_8Si$: t91
$C_{12}H_{28}Sn$: t219b
$C_{12}H_{36}O_4Si_4Ti$: t92

C_{13}

$C_{13}H_5N_3O_7$: t391
$C_{13}H_8ClFO$: c119
$C_{13}H_8ClNO_3$: c182
$C_{13}H_8Cl_2O$: d159
$C_{13}H_8N_2O_7$: b194
$C_{13}H_8O$: f3

TABLE 1.14 Empirical Formula Index of Organic Compounds (*Continued*)

The alphanumeric designations are keyed to Table 1.15.

TABLE 1.14 Empirical Formula Index of Organic Compounds (*Continued*)

The alphanumeric designations are keyed to Table 1.15.

$C_{15}H_{26}O_6$: g19
$C_{15}H_{30}N_2$: t347
$C_{15}H_{30}O$: p14
$C_{15}H_{30}O_2$: m416
$C_{15}H_{32}$: p13
$C_{15}H_{32}O_3Si_4$: p167
$C_{15}H_{32}O_{10}$: t396
$C_{15}H_{33}NO_6$: t436

C_{16}

$C_{16}H_{10}$: b52, f1
$C_{16}H_{11}NO_2$: p152
$C_{16}H_{12}N_2O_5S$: a60
$C_{16}H_{12}N_4O_9S_2$: t3
$C_{16}H_{13}N$: p132, p132a
$C_{16}H_{14}$: d658, d659, e67
$C_{16}H_{14}O$: d660
$C_{16}H_{14}O_6S$: s28
$C_{16}H_{15}NO_4$: d446
$C_{16}H_{16}O_2$: b47, b113
$C_{16}H_{16}O_3$: d450
$C_{16}H_{18}ClN_3S$: m238
$C_{16}H_{19}ClSi$: b438
$C_{16}H_{20}N_2$: d59
$C_{16}H_{20}O_2Si$: d250
$C_{16}H_{22}O_4$: d128, d410
$C_{16}H_{22}O_{11}$: g7
$C_{16}H_{26}O_3$: d732
$C_{16}H_{26}O_7$: t53
$C_{16}H_{30}O_2$: d735
$C_{16}H_{30}O_4$: d120a, d297a
$C_{16}H_{32}$: h37
$C_{16}H_{32}O$: e9
$C_{16}H_{32}O_2$: h35
$C_{16}H_{33}Br$: b295
$C_{16}H_{33}Cl$: c129a
$C_{16}H_{33}NO$: d297
$C_{16}H_{34}$: h4, h32
$C_{16}H_{34}O$: d645a, h36
$C_{16}H_{34}O_2$: h33
$C_{16}H_{34}S$: d646, h34
$C_{16}H_{35}N$: d645, h38
$C_{16}H_{35}O_3P$: b179a
$C_{16}H_{36}BF_4N$: t20
$C_{16}H_{36}BrN$: t15
$C_{16}H_{36}BrP$: t20b
$C_{16}H_{36}ClN$: t16
$C_{16}H_{36}FN$: t17
$C_{16}H_{36}IN$: t19
$C_{16}H_{36}Sn$: t21
$C_{16}H_{37}NO_4S$: t18

C_{17}

$C_{17}H_{12}O_3$: p121
$C_{17}H_{13}N_3O_5S_2$: p175
$C_{17}H_{16}O_4$: d60
$C_{17}H_{18}O_3$: b481
$C_{17}H_{18}O_4$: b186a
$C_{17}H_{20}N_2O$: b172
$C_{17}H_{20}N_4O_6$: r4
$C_{17}H_{21}NO_4$: b276
$C_{17}H_{22}N_2$: m235
$C_{17}H_{23}NO_3$: a320
$C_{17}H_{34}O_2$: m264
$C_{17}H_{36}$: h1
$C_{17}H_{37}N$: m230

C_{18}

$C_{18}H_{12}$: b6, b7
$C_{18}H_{12}N_5O_6$: d685
$C_{18}H_{14}$: t3a, t4
$C_{18}H_{14}O$: d682
$C_{18}H_{14}O_8$: d55
$C_{18}H_{15}As$: t401
$C_{18}H_{15}ClSn$: c258
$C_{18}H_{15}N$: t399
$C_{18}H_{15}N_3Si$: a325
$C_{18}H_{15}OP$: t408
$C_{18}H_{15}O_3P$: t410
$C_{18}H_{15}O_4P$: t406
$C_{18}H_{15}P$: t407
$C_{18}H_{15}PS$: t409
$C_{18}H_{15}Sb$: t400
$C_{18}H_{16}N_2$: d682a
$C_{18}H_{16}O_2$: b425
$C_{18}H_{16}Si$: t411
$C_{18}H_{18}O_3$: e74
$C_{18}H_{20}O_2$: b48
$C_{18}H_{22}$: b174a
$C_{18}H_{25}NO_3$: i61
$C_{18}H_{26}O_6$: e159b
$C_{18}H_{30}O$: t215
$C_{18}H_{30}O_2$: o7
$C_{18}H_{32}O_2$: o1
$C_{18}H_{32}O_{16}$: r1
$C_{18}H_{33}ClO$: o12a
$C_{18}H_{34}O_2$: o10, o11
$C_{18}H_{34}O_4$: d110
$C_{18}H_{36}$: o8
$C_{18}H_{36}O$: e9c, o12
$C_{18}H_{36}O_2$: e142, o5

$C_{18}H_{37}Br$: b322
$C_{18}H_{37}Cl_3Si$: o15
$C_{18}H_{37}I$: i45c
$C_{18}H_{37}N$: o9
$C_{18}H_{37}NO$: o2
$C_{18}H_{38}$: o3
$C_{18}H_{38}O$: o6
$C_{18}H_{38}S$: o4
$C_{18}H_{39}ClSi$: t307
$C_{18}H_{39}N$: o13, t306
$C_{18}H_{39}O_7P$: t421
$C_{18}H_{40}Si$: t308

C_{19}

$C_{19}H_{15}Br$: b369
$C_{19}H_{15}Cl$: c257
$C_{19}H_{16}$: t404
$C_{19}H_{16}O$: t405
$C_{19}H_{18}BrP$: m445
$C_{19}H_{20}Br_4O_4$: i96
$C_{19}H_{20}O_4$: b87
$C_{19}H_{22}N_2O$: c265
$C_{19}H_{30}O_5$: m244
$C_{19}H_{32}$: p159
$C_{19}H_{34}ClN$: b123
$C_{19}H_{36}O_2$: m330
$C_{19}H_{37}NO$: o14
$C_{19}H_{38}O_2$: i95a, m329
$C_{19}H_{40}$: n90, t117

C_{20}

$C_{20}H_{10}Br_2O_5$: d83
$C_{20}H_{12}$: b57, b58, d49
$C_{20}H_{12}O_5$: f4
$C_{20}H_{14}O_4$: d684a, p65
$C_{20}H_{15}Br$: b368
$C_{20}H_{18}O_3Si$: t398
$C_{20}H_{19}N_3$: b2
$C_{20}H_{20}BrOP$: h126a
$C_{20}H_{22}O_6$: t276
$C_{20}H_{24}N_2O_2$: q2
$C_{20}H_{24}O_6$: d51
$C_{20}H_{26}O_4$: d243a
$C_{20}H_{30}O_2$: a1
$C_{20}H_{31}N$: d20
$C_{20}H_{35}N$: t46
$C_{20}H_{36}O_2$: e191
$C_{20}H_{38}O_2$: e192
$C_{20}H_{40}$: i3

TABLE 1.14 Empirical Formula Index of Organic Compounds (*Continued*)

The alphanumeric designations are keyed to Table 1.15.

$C_{20}H_{40}O$: o16		
$C_{20}H_{42}$: i2		

C_{21}

$C_{21}H_{15}NO$: b143
$C_{21}H_{15}N_3O_3$: t397
$C_{21}H_{21}N$: t204
$C_{21}H_{21}O_4P$: t441a
$C_{21}H_{22}N_2O_2$: s10
$C_{21}H_{24}O_2$: b145
$C_{21}H_{28}N_2O$: b171
$C_{21}H_{36}O$: p15
$C_{21}H_{39}N_3$: t260

C_{22}

$C_{22}H_{23}N_3O_9$: a321
$C_{22}H_{30}O_2S$: t146
$C_{22}H_{34}O_4$: b446
$C_{22}H_{42}O_4$: d312
$C_{22}H_{44}O_2$: b469, i71b
$C_{22}H_{46}$: d717
$C_{22}H_{46}O$: d719

C_{23}

$C_{23}H_{16}O_6$: m236
$C_{23}H_{26}N_2O_4$: b375

C_{24}

$C_{24}H_{16}N_2O_2$: b199
$C_{24}H_{18}$: t402
$C_{24}H_{20}BNa$: t130
$C_{24}H_{20}O_4Si$: t129
$C_{24}H_{20}Si$: t134
$C_{24}H_{20}Sn$: t135
$C_{24}H_{22}N_2O$: b141
$C_{24}H_{38}O_4$: b180, d313
$C_{24}H_{40}O_5$: c264
$C_{24}H_{46}O_4$: d643a, d730
$C_{24}H_{50}$: t37
$C_{24}H_{51}N$: t394
$C_{24}H_{51}O_3P$: d424, t428
$C_{24}H_{54}OSn_2$: b202

C_{25}

$C_{25}H_{48}O_4$: d411a

C_{26}

$C_{26}H_{20}$: t133
$C_{26}H_{26}N_2O_2S$: b153
$C_{26}H_{50}O_4$: b178, d311

C_{27}

$C_{27}H_{19}NO$: b149
$C_{27}H_{42}ClNO_2$: b33
$C_{27}H_{46}O$: c263
$C_{27}H_{50}ClN$: b94

C_{28}

$C_{28}H_{31}ClN_2O_3$: r2
$C_{28}H_{32}$: t131
$C_{28}H_{32}O_2Si_3$: t123

C_{30} to C_{40}

$C_{30}H_{50}$: s8
$C_{30}H_{62}$: s7
$C_{30}H_{63}O_3P$: t313
$C_{32}H_{66}$: d739
$C_{32}H_{68}O_4Si$: t88
$C_{36}H_{75}O_3P$: d644
$C_{39}H_{74}O_6$: g20
$C_{40}H_{82}O_6P_2$: b197

C_{45} to C_{57}

$C_{45}H_{86}O_6$: g24
$C_{48}H_{40}O_4Si_4$: o38
$C_{51}H_{98}O_6$: g23
$C_{57}H_{104}O_6$: g22

TABLE 1.15 Physical Constants of Organic Compounds

See also the special tables of polymers, rubbers, fats, oils, and waxes.

Names of the compounds in the table starting on p. 1.76 are arranged alphabetically. Usually substitutive nomenclature is employed; exceptions generally involve ethers, sulfides, sulfones, and sulfoxides. Each compound is given a number within its letter classification; thus compound c196 is 3-chlorophenol. Section 1.1, Nomenclature of Organic Compounds, should be consulted to familiarize oneself with present nomenclature systems.

Synonyms or Alternate Names are found at the bottom of each spread in their alphabetical listing; the number following the same refers to the numerical place of this compound in the table. For example, epichlorohydrin, c102, indicates that this compound is found listed under the name 1-chloro-2,3-epoxypropane.

Formulas are presented in semistructural form when no ambiguity is possible. Complicated systems are drawn in complete structural form and located at the bottom of each page and keyed to the number of the entry.

Beilstein Reference. In this column is found the reference to the volume and page numbers of the fourth edition of Beilstein, *Handbuch der Organischen Chemie* (Springer-Verlag, New York, 1918). Thus the entry 9, 202 refers to an entry in volume 9 appearing on page 202. When the volume number has a superscript attached, reference is made to the appropriate supplementary volume. For example, 12^2, 404 indicates that the compound will be found listed in the second supplement to volume 12 on page 404. The earliest Beilstein entry is listed. Supplementary information may be found in the supplements to the basic series; such coordinating references (series number, volume number, and page number of the main edition) along with the system number are found at the top of each *odd-numbered page*. Similarly, a back reference such as H93; E II 64; E III 190 in a volume of Supplementary Series IV means that previous items on this compound are found in the same volume of the Basic Series on page 93, of Supplementary Series II on page 64, and of Supplementary Series III on page 190. The absence of a back reference implies that the compound involved is described *for the first time* in the series concerned.

Formula Weights are based on the International Atomic Weights of 1988 and are computed to the nearest hundredth when justified. The actual significant figures are given in the atomic weights of the individual elements; see Table 3.2.

Density values are given at room temperature unless otherwise indicated by the superscript figure; thus 0.9711^{112} indicates a density of 0.9711 for the substance at 112°C. A density of 0.899_4^{16} indicates a density of 0.899 for the substance at 16°C relative to water at 4°C.

Refractive Index, unless otherwise specified, is given for the sodium line at 589.6 nm. The temperature at which the measurement was made is indicated by the superscript figure; otherwise it is assumed to be room temperature.

Melting Point is recorded in certain cases as 250 d and in some other cases as d 250, the distinction being made in this manner to indicate that the former is a melting point with decomposition at 250°C, while the latter decomposition occurs only at 250°C and higher temperatures. Where a value such as $-2H_2O$, 120 is given, it indicates a loss of 2 moles of water per formula weight of the compound at a temperature of 120°C.

Boiling Point is given at atmospheric pressure (760 mmHg) unless otherwise indicated; thus 82^{15mm} indicates that the boiling point is 82°C when the pressure is 15 mmHg. Also, subl 550 indicates that the compound sublimes at 550°C.

Flash Point is given in degrees Celsius, usually using a closed cup. When the method is known, the acronym appears in parentheses after the value: closed cup (CC), Cleveland closed cup (CCC), open cup (OC), Tag closed cup (TCC), and Tag open cup (TOC). Because values will vary with the specific procedure employed, and many times the method was not stated, the values listed for the flash point should be considered only as indicative. See also Table 5.22, Properties of Combustible Mixtures in Air.

Solubility is given in parts by weight (of the formula weight) per 100 parts by weight of the solvent and at room temperature. Other temperatures are indicated by the superscript. Another way in which solubility is explicitly stated is in weight (in grams) per 100 mL of the solvent. In the case of gases, the solubility is often expressed as 5 mL^{10}, which indicates that at 10°C, 5 mL of the gas is soluble in 100 g (or 100 mL, if explicitly stated) of the solvent.

Abbreviations Used in the Table

abs, absolute
acet, acetone
alc, alcohol (ethanol usually)
alk, alkali (aqueous NaOH or KOH)
anhyd, anhydrous
aq, aqueous, water
as, asymmetrical
atm, atmosphere
BuOH, 1-butanol
bz, benzene
c, cold
chl, chloroform
conc, concentrated
d, decomposes or decomposed
D, dextrorotatory
deliq, deliquescent
dil, dilute
diox, 1,4-dioxane
DL, inactive (50% D and 50% L)
DMF, dimethylformamide
E, trans (German "entgegen")
EtAc, ethyl acetate
eth, diethyl ether
EtOH, ethanol, 95%

expl, explodes
glyc, glycerol
h, hot
HOAc, acetic acid
hyd, hydrolysis
hygr, hygroscopic
i, insoluble
ign, ignites
i-PrOH, isopropanol, 2-propanol
L, levorotatory
m, meta configuration
Me, methyl
MeOH, methanol
misc, miscible; soluble in all proportions
NaOH, aqueous sodium hydroxide
o, ortho configuration
org, organic
p, para configuration
PE, petroleum ether
pyr, pyridine
s, soluble
sec, secondary

sl, slight, slightly
soln, solution
solv, solvent
subl, sublimes
s, symmetrical
sym, symmetrical
tert, tertiary
v, very
v s, very soluble
v sl s, very slightly soluble
vac, vacuo or vacuum
vols, volumes
Z, cis (German "zusamman")
>, greater than
<, less than
~, approximately
±, inactive [50% (+) and 50% (−)]
α, alpha (first) position
β, beta (second) position
γ, gamma (third) position
δ, delta (fourth) position
ω, omega position (farthest from parent functional group)

TABLE 1.15 Physical Constants of Organic Compounds (*Continued*)

No.	Name	Formula	Formula weight	Beilstein reference	Density	Refractive index	Melting point	Boiling point	Flash point	Solubility in 100 parts solvent
a1	(−)-Abietic acid		302.44	9^2, 424			172–175			i aq; s alc, bz, chl, eth, acet, dil alk
a2	Acenaphthene		154.21	5, 586	1.189		95	279		i aq; 3.2 alc; 20 bz
a3	Acenaphthylene		152.20	5, 625	0.8991^{16}		88–91	280		i aq; v s alc. eth
a4	Acetaldehyde	CH_3CHO	44.05	1, 594	0.7881^{16}	1.3316^{20}	−123.5	21	−38(CC)	misc aq, alc, eth
a5	Acetaldoxime	$CH_3CH{=}NOH$	59.07	1, 608	0.966	1.415^{20}	$46.5(\alpha)$ $12(\beta)$	114.5	40	v s aq, alc, eth
a6	Acetamide	CH_3CONH_2	59.07	2^2, 177	1.159^{20}	1.4158^{110}	81	222		70 aq; 50 alc; 16 pyr; s chl, glyc, hot bz
a7	Acetamidine HCl	$CH_3C({=}NH)NH_2 \cdot HCl$	94.54	2, 185			170–172			v s aq, alc; i acet, eth
a8	N-(2-Acetamido)-2-aminoethane-sulfonic acid	$H_2N(CO)CH_2NHCH_2\text{-}CH_2SO_3H$	182.20				>220 d			
a9	4-Acetamidobenz-aldehyde	$CH_3CONHC_6H_4CHO$	163.18	14, 38			154–156			s aq, bz; sl s alc
a10	4-Acetamidobenzene-sulfonyl chloride	$CH_3CONHC_6H_4SO_2Cl$	233.67	14, 702			148 d			d aq; v s alc, eth
a11	2-Acetamidobenzoic acid	$CH_3CONHC_6H_4COOH$	179.18	14, 337			185–187			sl s aq; v s alc, bz, eth, acet
a12	4-Acetamidobenzoic acid	$CH_3CONHC_6H_4COOH$	179.18	14, 432			262 d			i aq; s alc; sl s eth
a13	2-Acetamidofluorene		223.28	12, 1331	1.2932^{1}		192–196			i aq; s alc, glycols
a14	N-(2-Acetamido)-iminodiacetic acid	$H_2NCOCH_2N(CH_2COOH)_2$	190.16				219 d			
a15	2-Acetamidophenol	$CH_3CONHC_6H_4OH$	151.17	13, 370			207–210			
a16	3-Acetamidophenol	$CH_3CONHC_6H_4OH$	151.17	13, 415			146–149			
a17	4-Acetamidophenol	$CH_3CONHC_6H_4OH$	151.17	13, 460			170–172			
a18	Acetanilide	$CH_3CONHC_6H_5$	135.17	12, 237	1.2195^{15}		114.2	304	173 (OC)	s alc, acet
a19	Acetic acid	CH_3COOH	60.05	2, 96	1.0492^{20}	1.3716^{20}	16.6	117.9	39 (CC)	0.56 aq^{25}, 29 alc; 2 bz; 27 chl; 25 acet; 5 eth
a20	Acetic acid-d	CH_3COOD	61.06	2^3, 202	1.059	1.3715^{20}		115.5	40	misc aq, alc, eth, CCl_4
a21	Acetic-d_3 acid-d	CD_3COOD	64.08	2^3, 203	1.137	1.3687^{20}		115.5	40	misc aq, alc, eth, CCl_4
a22	Acetic anhydride	$(CH_3CO)_2O$	102.09	2, 166	1.0825^{15}	1.3904^{20}	−73.1	140.0	54 (CC)	13 aq; misc alc, eth; sl d aq, alc; s chl

No.	Name	Formula	Formula wt	Beil. ref.	Density	n_D	mp, °C	bp, °C	Flash pt, °C	Solubility
a23	Acetic anhydride-d_6	$(CD_3CO)_2O$	108.14		1.260^{20}	1.3875^{20}			54	See acetic anhydride
a24	Acetoacetanilide	$CH_3COCH_2CONHC_6H_5$	177.20	12, 518			85	65^{65mm} dec	185 (COC)	s alc, chl, eth, hot bz, acids, alkalis
a25	Acetoacetic acid	CH_3COCH_2COOH	102.09	3, 630			36–37	d viol 100		misc aq, alc, eth
a26	Acetone	CH_3COCH_3	58.08	1, 635	0.7908_4^{20}	1.3588^{20}	−94.6	56.5	−18 (CC)	misc aq, alc, chl,
a27	Acetone-d_6	CD_3COCD_3	64.13		0.872	1.3554^{20}	−93.8	55.5	−17	See acetone
a28	Acetone oxime	$(CH_3)_2C{=}NOH$	73.10	1, 649	0.9114_2^{6}		60	135		v s aq, alc, eth
a29	Acetonitrile	CH_3CN	41.05	2, 183	0.7868_2^{20}	1.3441^{20}	−45	81.6	5 (COC)	misc aq, alc, chl, eth
a30	Acetonitrile-d_3	CD_3CN	44.08	2^4, 428	0.844	1.3406^{20}		80.7	5	misc aq, alc, chl
a31	Acetophenone	$C_6H_5COCH_3$	120.15	7, 271	1.026_4^{20}	1.5322^{25}	19–20	202	82 (OC)	0.55 aq; s alc, eth
a33	4-Acetylbenzene-sulfonic acid, Na salt	$CH_3COC_6H_4SO_3^-Na^+$	222.20	11^2, 186			>300			
a34	4-Acetylbiphenyl	$C_6H_5C_6H_4COCH_3$	196.25	7^2, 337			116–118	325–327	>110	i aq; v s alc, acet
a35	Acetyl bromide	CH_3COBr	122.95	2, 174	1.6631^{16}	1.4486^{20}	−96	75–77	>110	dec viol by aq or alc; misc bz, chl, eth
a36	2-Acetylbutyrolactone		128.13		1.1846_4^{20}	1.4585^{20}		107^{5mm}		20 aq

Structures:

H_3C—COOH , $CH(CH_3)_2$, CH_3 — a1

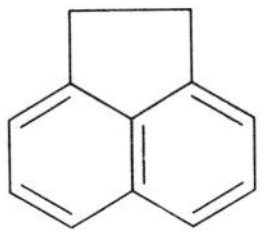

a2

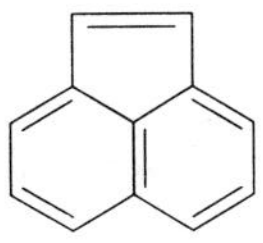

a3

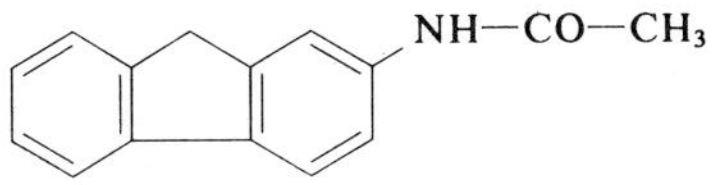

NH—CO—CH_3 — a13

$COCH_3$ (butyrolactone ring) — a36

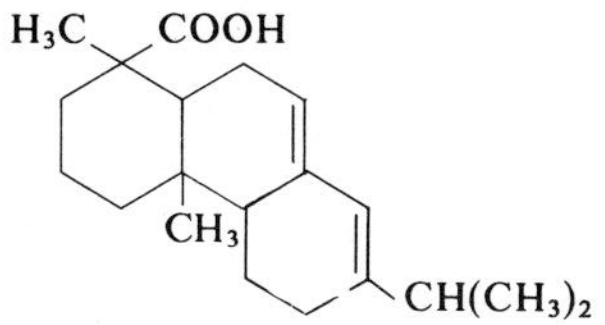

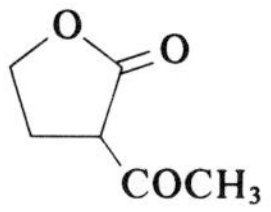

TABLE 1.15 Physical Constants of Organic Compounds (*Continued*)

No.	Name	Formula	Formula weight	Beilstein reference	Density	Refractive index	Melting point	Boiling point	Flash point	Solubility in 100 parts solvent
a37	Acetyl chloride	CH_3COCl	78.50	2, 173	1.1040^{20}_4	1.3886^{20}	−112	52	4 (CC)	dec aq, alc; misc bz, chl, eth
a38	Acetylcholine bromide	$(CH_3)_3NBrCH_2CH_2$-$OCOCH_3$	226.14	4^1, 428			144−146			v s aq (dec hot aq); s alc; i eth
a39	Acetylcholine chloride	$(CH_3)_3NClCH_2CH_2$-$OOCCH_3$	181.66	4, 281			150−152			v s aq, alc; dec hot aq; i eth
a40	2-Acetylcyclopentanone		126.16	7, 558	1.043	1.4905^{20}		$72–75^{8mm}$	72	
a41	Acetylene	$HC{\equiv}CH$	26.02	1, 228	0.90(g)		-81^{891mm}	−83.9 subl	−18	1 vol in 1 vol aq, in 6 vol HOAc or alc; s bz, eth; acet dissolves 25 vols(15)
a42	Acetylenedicarboxylic acid	$HOOCC{\equiv}CCOOH$	114.06	2, 801			180 d			v s aq, alc, eth
a43	Acetyl fluoride	CH_3COF	62.04	2, 172	1.002^{15}		<−60	20.8		5 aq(dec); sl s alc, acet, bz, eth
a44	2-Acetylfuran		110.11	17, 286	1.098	1.5065^{20}	29−30	67^{10mm}	71	
a45	N-Acetyl-L-glutamic acid	$HOOCCH_2CH_2\underset{NHCOCH_3}{CHCOOH}$	189.17	4^2, 908			200−201			
a46	N-Acetylglycine	$CH_3CONHCH_2COOH$	117.10	4, 354			207−209			$2.7\ aq^{15}$, sl s acet, chl, HOAc; i bz, eth
a47	1-Acetylimidazole		110.12				103−105			
a48	Acetyl iodide	CH_3COI	169.96	2, 174	2.0674^{20}_4	1.5491^{20}		108		dec aq, alc; s bz, eth
a49	Acetyl-2-methyl-choline chloride	$CH_3COOCH(CH_3)CH_2$-$NBr(CH_3)_3$	195.69				171−173			v s aq, alc; i eth; dec by alkalis, eth
a50	2-Acetylphenothiazine		241.31				180−185			
a51	2-Acetylphenylacetonitrile	$C_6H_5CH(CN)COCH_3$	159.19	10, 699			92−94			
a52	1-Acetyl-4-piperidone		141.17	21, 279	1.146	1.5026^{20}		218	>110	
a53	2-Acetylpyridine	$(C_5H_4N)COCH_3$	121.14	21, 279	1.080	1.5203^{20}		188−189	73	v s alc, eth
a54	3-Acetylpyridine	$(C_5H_4N)COCH_3$	131.14	21, 279	1.102	1.5336^{20}		220	150	v s acids, alc, eth; s aq
a55	4-Acetylpyridine	$(C_5H_4N)COCH_3$	121.14	21, 279	1.095	1.5350^{20}		212	>110	

No.	Name	Formula	M.W.	Ref.	Density	n	mp/bp	Solubility
a56	Acetylsalicylic acid	$HOOC_6H_4OOCCH_3$	180.16	10, 67	1.35		135	0.33 aq[25], 29 acet; 20 alc; 5.9 chl; 5 eth; sl s bz
a57	2-Acetylthiophene	$(C_4H_3S)COCH_3$	126.18	17, 287 / 3, 191	1.168^{22}_{4}	1.5564^{20}	10–11 / 166–167 / 204–206	sl s aq; misc alc, eth
a58	1-Acetyl-2-thiourea	$CH_3CONHC(S)NH_2$	118.16					s hot aq, alc; sl s eth
a59	N-Acetyl-DL-tryptophan		246.27	22^2, 469			214	s aq, alc; v s eth
a60	Acid alizarin violet N		366.33	16^2, 127 / 20, 459	1.005^{20}		106–110 / subl 100	s alc, eth, CS_2, PE
a61	Acridine		179.22				346	sl s hot aq

Acetylcyclopropane, c369
Acetylene dichloride, d179, d180
Acetylene tetrabromide, t10
Acetylene tetrachloride, t29

N-Acetylethanolamine, h115
3-Acetyl-6-methyl-2H-pyran-2,4(3H)-dione, d21
2-(Acetyloxy)benzoic acid, a56

3-Acetyl-1-propanol, j158
N-Acetylsulfanilyl chloride, a10
Aconitic acid, p210

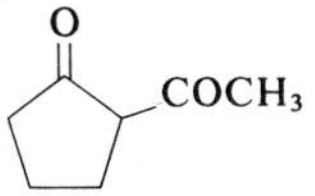

a40

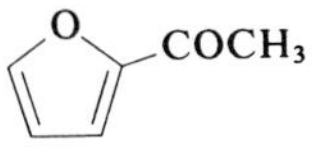

a44

a47

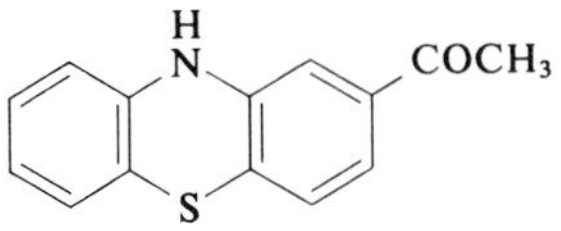

a50

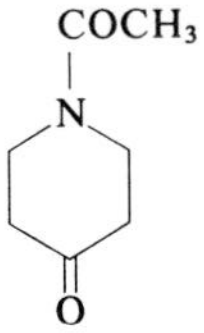

a52

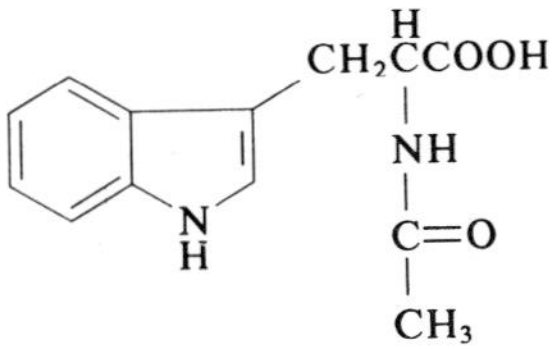

a59

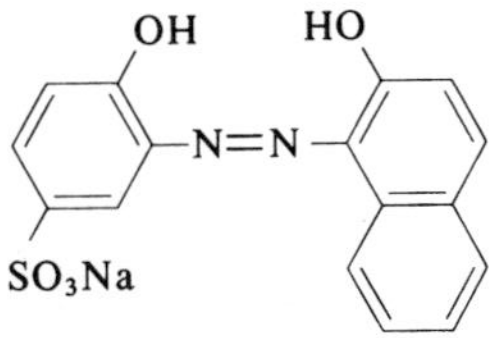

a60

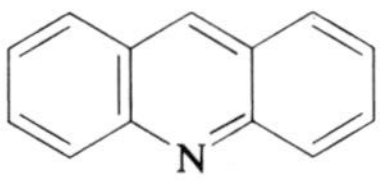

a61

TABLE 1.15 Physical Constants of Organic Compounds (*Continued*)

No.	Name	Formula	Formula weight	Beilstein reference	Density	Refractive index	Melting point	Boiling point	Flash point	Solubility in 100 parts solvent
a62	Acrylamide	$H_2C=CHCONH_2$	71.08	2, 400	1.2222^{30}_{4}		84.5	125^{25mm}		At 30°: 215 aq; 155 MeOH; 86 EtOH; 63 acet; 12.6 EtAc; 2.7 chl; 0.3 bz
a63	Acrylic acid	$H_2C=CHCOOH$	72.06	2, 397	1.0511^{20}	1.4224^{20}	14	141	68 (OC)	misc aq, alc, bz, eth, chl, acet
a64	Acrylonitrile	$H_2C=CHCN$	53.06	2, 400	0.8060^{20}	1.3911^{20}	−83.5	77.3	0 (OC)	7.3 aq; misc org solv
a65	Acryloyl chloride	$H_2C=CHCOCl$	90.51	2, 400	1.114	1.4350^{20}		72–76	16	d aq; v s chl
a66	1-Adamantanamine		151.25				206–208			sl s aq
a67	Adamantane		136.24		1.09	1.568	270	205 subl		
a68	1-Adamantane-carboxylic acid		180.25				174–175			
a69	Adenine		135.13	26, 420			>360 d	subl 220		0.05 aq; sl s alc; i chl, eth
a70	(−)-Adenosine		267.25	31, 27			234–236			s aq; i alc
a74	DL-α-Alanine	$CH_3CH(NH_2)COOH$	89.09	4, 387	1.424		289 d	subl >200		16.7 aq^{25}; 0.009 alc^{25}; i eth
a75	L-α-Alanine	$CH_3CH(NH_2)COOH$	89.09	4, 381			d 297			16.7 aq^{25}; 0.2 alc^{25}; i eth
a76	β-Alanine	$H_2NCH_2CH_2COOH$	89.09	4, 401	1.437^{-5}		197 d			v s aq; sl s alc; i eth
a77	Allantoin		158.12	25, 474			238			0.45 aq; 0.2 alc
a78	Allene	$H_2C=C=CH_2$	40.06	1, 248	1.787	1.4168	−136	−34		
a79	Alloxan monohydrate		160.09	24, 500			253 d			s alc, acet, HOAc; sl s chl, PE, Et Ac
a80	Allyl acetate	$H_2C=CHCH_2OCOCH_3$	100.12	2, 136	0.9282^{20}	1.4040^{20}		104	6	i aq; misc alc, eth
a81	Allyl alcohol	$H_2C=CHCH_2OH$	58.08	2, 436	0.8540^{20}	1.4134^{20}	−50 glass	96–97	24 (CC)	misc aq, alc, chl, eth
a82	Allylamine	$H_2C=CHCH_2NH_2$	57.10	4, 205	0.7612^{20}	1.4185^{20}	−88.2	55–58	−12 (CC)	misc aq, alc, chl, eth
a83	N-Allylaniline	$C_6H_5NHCH_2CH=CH_2$	133.19	12, 170	0.9825^{25}	1.5630^{20}		218–220	89	i aq; s alc, eth
a84	Allylbenzene	$C_6H_5CH_2CH=CH_2$	118.18	5, 484	0.8920^{20}	1.5122^{20}		156–157	33	i aq; s alc, eth
a85	Allyl bromide	$H_2C=CHCH_2Br$	120.98	1, 201	1.398^{20}_{4}	1.465^{20}	−119	71	−2	i aq; misc org solv
a86	Allyl butanoate	$CH_3CH_2CH_2CO_2CH_2CH=CH_2$	128.17	2, 272	0.902	1.4142^{20}		44^{15mm}		
a87	Allyl chloroformate	$H_2C=CHCH_2OOCCl$	120.54	3, 12	1.136	1.4223	109–110	27	31	
a88	Allylcyclohexylamine	$C_6H_{11}NHCH_2CH=CH_2$	139.24		0.962	1.4664^{20}		66^{12mm}	53	i aq; misc org solv

Acrolein, p207
Acrolein diethyl acetal, d258
Acrolein dimethyl acetal, d455
Acrylaldehyde, p207
Adipic acid, h57
Adipic acid monoethyl ester, e152

Adipolyl chloride, h62
Adiponitrile, d238
ADP, a71
Alaninol, a276, a277
Alizarin, d372
Alloocimene, d566b

Allylacetone, h78
4-Allylanisole, a94
Allyl butyrate, a86
Allyl carbamide, a105
Allyl chloride, c217
Allyl cyanide, b403

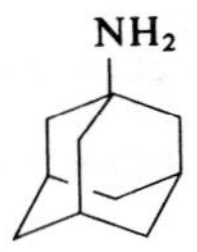

a66

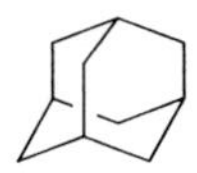

a67

a68

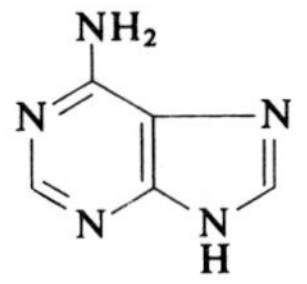

a69

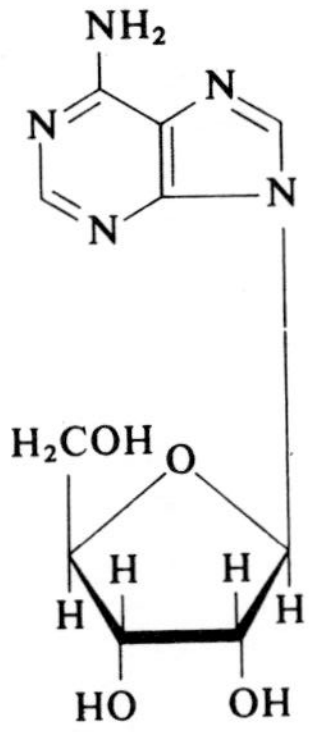

a70

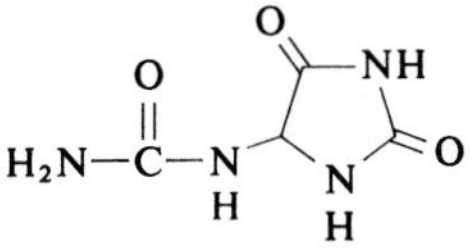

a77

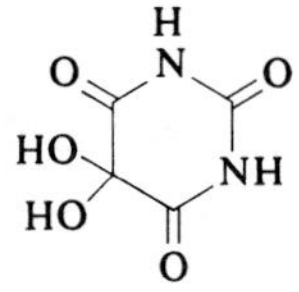

a79

1.81

TABLE 1.15 Physical Constants of Organic Compounds (*Continued*)

No.	Name	Formula	Formula weight	Beilstein reference	Density	Refractive index	Melting point	Boiling point	Flash point	Solubility in 100 parts solvent
a89	4-Allyl-1,2-dimethoxybenzene	$H_2C{=}CHCH_2C_6H_3{-}(OCH_3)_2$	178.23	6, 963	1.036	1.5344^{20}	-4	254-255		
a90	N-Allyl-N,N-dimethylamine	$H_2C{=}CHCH_2N(CH_3)_2$	85.0			1.4010^{20}		63-64		
a91	Allyl ethyl ether	$H_2C{=}CHCH_2OCH_2CH_3$	86.13	1, 438	0.7652^{20}	1.3881^{20}		64-66	-20	i aq; misc alc, eth
a92	Allyl iodide	$H_2C{=}CHCH_2I$	167.98	1, 202	1.8252^{24}			103		i aq; misc alc, eth
a93	Allyl isothiocyanate	$H_2C{=}CHCH_2NCS$	99.16	4, 214	1.0132^{25}	1.527^{20}		150	46	0.2 aq; misc org solv
a94	Allyl methacrylate	$H_2C{=}C(CH_3)CO_2CH_2{-}CH{=}CH_2$	126.16	2³, 1290	0.938	1.4360		59-61⁴³mm	33	
a95	Allyl methyl sulfide	$H_2C{=}CHCH_2SCH_3$	88.17	1, 440	0.803	1.4714^{20}		91-93	18	
a96	1-Allyloxy-2,3-epoxy-propane	$H_2C{-}CHCH_2{-}$ (epoxide) $CH{=}CH_2$	114.14		0.962	1.4332^{20}		154	57	
a97	Allyloxytrimethylsilane	$H_2C{=}CHCH_2OSi(CH_3)_3$	130.26		0.7830	1.4075^{25}		100-102	0	s alc, eth
a98	2-Allylphenol	$H_2C{=}CHCH_2C_6H_4OH$	134.18	6, 572	1.033^{5}	1.5455^{20}	10	220	88	i aq; s alc; misc eth
a99	Allyl phenyl ether	$H_2C{=}CHCH_2OC_6H_5$	134.18	6, 144	0.9831^{5}	1.5200^{20}		192	62	s alc; misc eth
a100	Allyl propyl ether	$H_2C{=}CHCH_2OC_3H_7$	100.16	1, 438	0.7670^{20}	1.3919^{20}		90-92	-5	3.3 aq; s alc; i bz; v sl s eth
a101	1-Allyl-2-thiourea	$H_2C{=}CHCH_2NHC(S)NH_2$	116.18	4, 211	1.219^{20}		77-78			sl s eth
a102	Allyltrichlorosilane	$H_2C{=}CHCH_2SiCl_3$	175.52	4³, 1909	1.2011^{20}	1.4550^{20}		117.5	31	v s aq, alc; v sl s eth
a103	Allyltriethoxysilane	$H_2C{=}CHCH_2Si(OC_2H_5)_3$	204.34	4³, 1909	0.9030^{20}	1.4062^{20}		176⁷⁴⁰mm	21	s acids, alc
a104	Allyltrimethylsilane	$H_2C{=}CHCH_2Si(CH_3)_3$	114.27		0.7193^{20}	1.4056^{20}		85-86	7	v s aq; sl s alc; i eth
a105	Allylurea	$H_2C{=}CHCH_2NHCONH_2$	100.12	4, 209			78	58¹⁵mm d		
a106	Aminoacetonitrile	H_2NCH_2CN	56.07	4, 344				d 165		
a107	Aminoacetonitrile hydrogen sulfate	$H_2NCH_2CN \cdot H_2SO_4$	154.14	4, 344			101			
a108	2'-Aminoacetophenone	$H_2NC_6H_4COCH_3$	135.17	14, 41				70³mm	>110	v sl s aq; s alc, eth
a109	3'-Aminoacetophenone	$H_2NC_6H_4COCH_3$	135.17	14, 45			98-99	289-290		
a110	4'-Aminoacetophenone	$H_2NC_6H_4COCH_3$	135.17	14, 46			106	293-295		s hot aq, alc, eth, HOAc; sl s bz
a111	1-Aminoanthraquinone		223.23	14, 177			253-255	subl		i aq; v s alc, bz, chl, eth, HOAc, HCl

a112	2-Aminoanthraquinone		223.23	14, 191			295 d	subl		i aq, eth; s alc, bz
a113	4-Aminoantipyrine		203.25	24, 273			109			s aq, alc, bz; sl s eth
a114	2-Aminobenzamide	$H_2NC_6H_4CONH_2$	136.15	14, 320			110	300 sl d		v s hot aq, alc; i bz; sl s eth
a115	4-Aminobenzenearsonic acid	$H_2NC_6H_4AsO(OH)_2$	217.06	16, 878			232			s hot aq, alk CO_3, mineral acids; i bz
a116	5-Aminobenzene-1,3-dicarboxylic acid	$H_2NC_6H_3(COOH)_2$	181.15	14^{1}, 636			>300			
a118	2-Aminobenzenesulfonic acid	$H_2NC_6H_4SO_3H$	173.19	14, 681			d 325			1.5 aq^{15}; v sl s alc, eth
a119	3-Aminobenzenesulfonic acid	$H_2NC_6H_4SO_3H$	173.19		1.69					2 aq^{15}; sl s alc
a120	4-Aminobenzenesulfonic acid	$H_2NC_6H_4SO_3H$	173.19	14, 695			d 288			1 aq^{20}; sl s hot MeOH
a121	2-Aminobenzoic acid	$H_2NC_6H_4COOH$	137.14	14, 310			144–146	subl		v s hot aq, alc, eth
a122	3-Aminobenzoic acid	$H_2NC_6H_4COOH$	137.14	14, 383	1.511^{4}		172–174			sl s aq; v s alc; s eth
a123	4-Aminobenzoic acid	$H_2NC_6H_4COOH$	137.14	14, 418	1.374		187			0.59 aq; 12 alc; 2 eth

Allyl glycidyl ether, a96
1-Allyl-2-hydroxybenzene, a98
Allyl iodide, i50
Allyl mercaptan, p209
4-Allyl-2-methoxyphenol, m99
2-Allyl-4-methylphenol, m390
2-Allyl-6-methylphenol, m389
Allyl sulfide, d27
Aluminon, a321
N-Amidinosarcosine, c278
Aminoacetaldehyde diethyl acetal, d254
Aminoacetaldehyde dimethyl acetal, d441
1-Aminoadamantane, a66
Aminoanisole, m42, m43, m44
p-Aminoazobenzene, p88
Aminobenzenethiol, a298

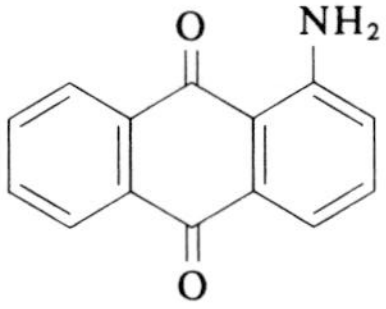

a111

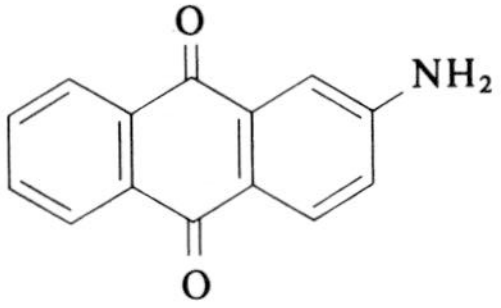

a112

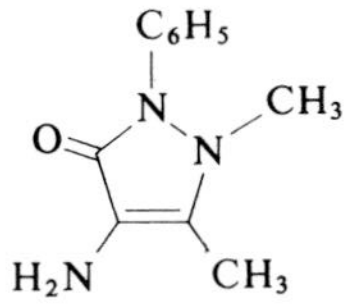

a113

TABLE 1.15 Physical Constants of Organic Compounds (*Continued*)

No.	Name	Formula	Formula weight	Beilstein reference	Density	Refractive index	Melting point	Boiling point	Flash point	Solubility in 100 parts solvent
a124	2-Aminobenzonitrile	$H_2NC_6H_4CN$	118.14	14, 322			49	268	>110	s alc, eth
a125	3-Aminobenzonitrile	$H_2NC_6H_4CN$	118.14	14, 391			53	288–290	>110	s hot aq; v s alc, eth
a126	4-Aminobenzonitrile	$H_2NC_6H_4CN$	118.14	14, 425			85	dec		v s hot aq, alc, eth
a127	2-Aminobenzophenone	$H_2NC_6H_4COC_6H_5$	197.24	14, 76			108	223–226		sl s aq; s alc, eth
a128	2-Aminobenzothiazole		150.20	27, 182			132	dec		v s alc, chl, eth
a129	2-Aminobenzotrifluoride	$H_2NC_6H_4CF_3$	161.13	12², 453	1.290²⁵	1.4785²⁵	34	175	55	
a130	3-Aminobenzotrifluoride	$H_2NC_6H_4CF_3$	161.13	12, 870	1.290	1.4800²⁰	6	187	85	
a131	4-Aminobenzotrifluoride	$H_2NC_6H_4CF_3$	161.13	12³, 2151	1.2837²⁷	1.4815²⁵	38	83¹²mm	86	
a132	N-(p-Aminobenzoyl)-glycine	$H_2NC_6H_4CONHCH_2COOH$	194.19	14², 258			198–199			i aq; s alc; bz, chl
a133	4-Aminobenzoyl hydrazide	$H_2NC_6H_4CONHNH_2$	151.17	14¹, 570			227			
a134	2-Aminobiphenyl	$H_2NC_6H_4C_6H_5$	169.23	12, 1317			53	299	>110	sl s aq; s alc
a135	4-Aminobiphenyl	$H_2NC_6H_4C_6H_5$	169.23	12, 1318			54	191¹⁵mm	>110	s hot aq, alc, eth
a136	DL-2-Aminobutanoic acid	$CH_3CH_2CH(NH_2)COOH$	103.12	4, 408			304	subl >300		21 aq: 0.2 hot alc
a137	4-Aminobutanoic acid	$H_2NCH_2CH_2CH_2COOH$	103.12	4, 413			195 d	178		v s aq; i alc, eth
a138	2-Amino-1-butanol	$CH_3CH_2CH(NH_2)CH_2OH$	89.14	4, 291	0.9442²⁰	1.4521²⁰	−2	178	74 (OC)	misc aq; s alc
a139	4-Amino-6-chloro-1,3-benzene-disulfonamide	$H_2NC_6H_2(Cl)(SO_2NH_2)_2$	285.73	14², 2810			257–261			
a140	2-Amino-4-chloro-benzoic acid	$H_2N(Cl)C_6H_3COOH$	171.58	14, 365			231–233			
a141	5-Amino-2-chloro-benzoic acid	$H_2N(Cl)C_6H_3COOH$	171.58	14, 412			188 d			
a142	2-Amino-4'-chloro-benzophenone	$H_2NC_6H_4COC_6H_4Cl$	231.68	14¹, 389			104			
a143	2-Amino-5-chloro-benzophenone	$H_2N(Cl)C_6H_3COC_6H_5$	231.68	14, 79			100			
a144	2-Amino-5-chloro-benzotrifluoride	$H_2N(Cl)C_6H_3CF_3$	195.57	12³, 1921	1.386	1.5069²⁰		66–67³mm	95	

a145	3-Amino-4-chloro-benzotrifluoride	$H_2N(Cl)C_6H_3CF_3$	195.57		1.428	1.4975[25]		82–83[9mm]	none	
a146	5-Amino-2-chloro-benzotrifluoride	$H_2N(Cl)C_6H_3CF_3$	195.37				34–36	>110		
a147	2-(3-Amino-4-chloro-benzoyl)benzoic acid	$H_2N(Cl)C_6H_3CO-C_6H_4COOH$	275.69	14, 661			171–173			
a148	2-Amino-4-chloro-phenol	$H_2N(Cl)C_6H_3OH$	143.57	13, 383			134–138			
a149	2-Amino-5-chloro-pyridine	$H_2N(Cl)C_5H_3N$	128.56	22[2], 332			135–138	128[11mm]		
a150	3-Aminocrotamide	$CH_3C(NH_2)=CHCONH_2$	100.12				102			
a151	3-Aminocrotononitrile	$CH_3C(NH_2)=CHCN$	82.11	3, 660						
a152	1-Amino-1-cyclo-hexanecarboxylic acid	$C_6H_{10}(NH_2)COOH$	143.19	14, 299			>300			
a153	5-Amino-2,3-dihydro-1,4-phthalazine-dione		177.16	25[1], 698			319–320			
a154	2-Amino-4,6-dihy-droxypyrimidine		127.10	24, 468			>300			
a155	4-Amino-2,6-dihy-droxypyrimidine		127.10	24, 469			>300			

Aminobenzoic acid hydrazide, a133
Aminobenzyl cyanide, a263
1-Aminobutane, b420
2-Aminobutane, b420a
3-Amino-2-butenamide, a150
4-Aminobutyraldehyde diethyl acetal, d247
Aminobutyric acids, a135, a137
α-Aminocaproic acid, a183
ε-Aminocaproic acid, a184
2-Amino-4-chlorophenol, c131a
4-Amino-*m*-cresol, a221
Aminocyclododecane, c306a
Aminocyclohexane, c335
Aminodecane, d19
2-Amino-5-diethylaminopentane, d327
2-Amino-1,5-dihydro-1-methyl-4*H*-imidazol-4-one, c279

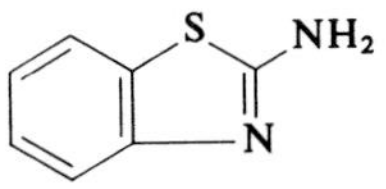
a128

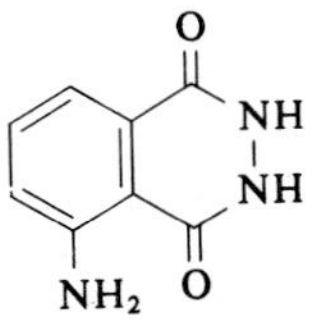
a153

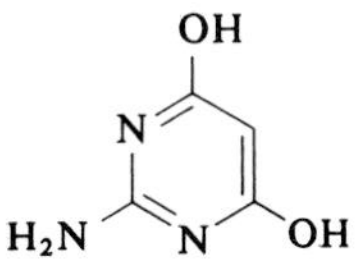
a154

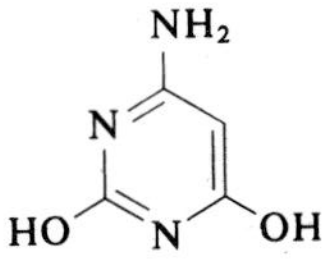
a155

TABLE 1.15 Physical Constants of Organic Compounds (*Continued*)

No.	Name	Formula	Formula weight	Beilstein reference	Density	Refractive index	Melting point	Boiling point	Flash point	Solubility in 100 parts solvent
a156	2-Amino-3,3-dimethylbutane	$(CH_3)_3CCH(NH_2)CH_3$	101.19	4, 193	0.755	1.4130^{20}	−20	102–103	1	
a157	2-Amino-4,6-dimethylpyridine	$(CH_3)_2(NH_2)(C_5H_2N)$	122.17	22, 435			64	235		
a158	4-Amino-2,6-dimethylpyrimidine		123.16	24^2, 45			184–186			156 aq; 18.9 alc
a159	6-Amino-1,3-dimethyluracil		155.16	24, 471			295 d			
a160	5-Amino-2,6-dioxo-1,2,3,6-tetrahydro-4-pyrimidine carboxylic acid		171.11	25, 264			>300			
a160a	Aminodiphenylmethane	$(C_6H_5)_2CHNH_2$	183.25	12, 1323	1.063^{20}	1.5956	12	295	>110	
a161	2-Aminoethanesulfonic acid	$H_2NCH_2CH_2SO_3H$	125.15	4, 528			d > 300			$6.45\ aq^{12}$; i abs alc
a162	2-Aminoethanethiol	$HSCH_2CH_2NH_2$	77.14	4, 286			97–99	110 d		v s aq; s alc
a163	1-Aminoethanol	$CH_3CH(OH)NH_2$	61.08				97	171		s aq; sl s eth
a164	2-Aminoethanol	$H_2NCH_2CH_2OH$	61.08	4, 274	1.0117^{25}	1.4539^{20}	10.5			
a165	2-(2-Aminoethoxy)ethanol	$H_2NCH_2CH_2OCH_2CH_2OH$	105.14	4^3, 642	1.460			218–224	93	misc aq, org solv
a166	2-(2-Aminoethylamino)ethanol	$H_2NCH_2CH_2NHCH_2CH_2OH$	104.15	4, 286	1.030	1.4861^{20}		241	110	v s aq, alc; sl s eth
a167	1-[(2-Aminoethyl)amino]-2-propanol	$CH_3CH(OH)CH_2NHCH_2CH_2NH_2$	118.18		0.9837^{25}_{4}	1.4738^{25}		112^{10mm}		
a167a	3-(2-Aminoethylamino)propyltrimethoxysilane	$H_2NCH_2CH_2NHCH_2CH_2CH_2Si(OCH_3)_3$	222.36	22^1, 642	1.01^{25}	1.4418^{25}		140^{15mm}	150	
a168	3-Amino-9-ethylcarbazole		210.28				98–100			
a169	2-Aminoethyl hydrogen sulfate	$H_2NCH_2CH_2OSO_3H$	141.15	4, 276			280 d			

No.	Name	Formula	Mol. wt.	Beil. ref.	Density	n_D	M.p.	B.p.	Flash p.	Solubility
a170	3-(2-Aminoethyl)-indole		160.22	22^1, 636			118	$137^{0.15mm}$		i aq, bz, chl, eth; s alc, acet
a171	S-2-Aminoethyl-isothiouronium bromide HBr		281.02				194–195			
a172	N-(2-Aminoethyl)-morpholine		130.19	27^3, 370	0.992	1.4755^{20}	25.6	205	175	
a173	4-(2-Aminoethyl)-phenol	$HOC_6H_4CH_2CH_2NH_2$	137.18	13, 625			164–165	175^{8mm}		1 aq^{15}, 10 boiling alc
a174	N-(2-Aminoethyl)-piperazine		129.21		0.985^{20}_{20}	1.4983^{20}	−26	222	93 (OC)	
a175	N-(2-Aminoethyl)-1,3-propanediamine	$H_2NCH_2CH_2CH_2HNCH_2CH_2NH_2$	117.20		0.928	1.4815^{20}			96	
a176	2-Amino-2-ethyl-1,3-propanediol	$HOCH_2C(NH_2)(C_2H_5)CH_2OH$	119.16		1.099^{20}_{20}	1.490^{20}	38	152^{10mm}	74	misc aq; s alc

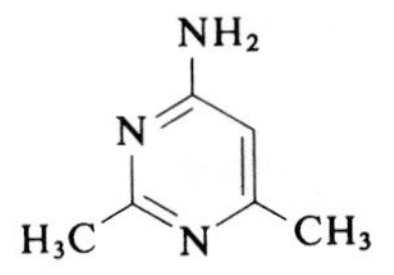

a158

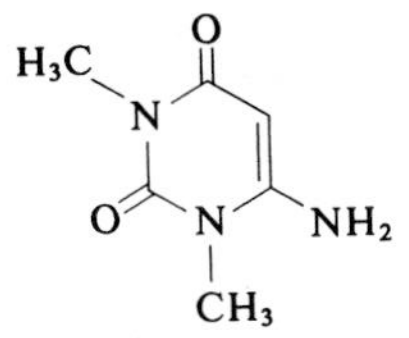

a159

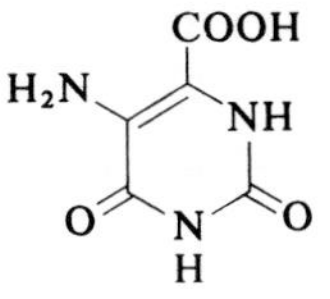

a160

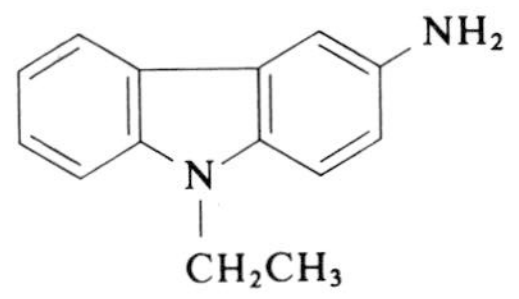

a168

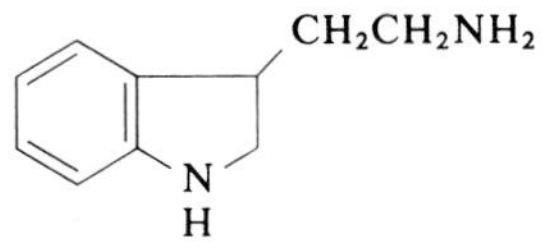

a170

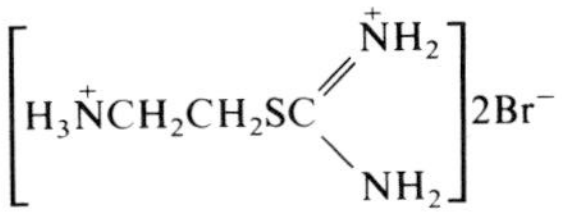

a171

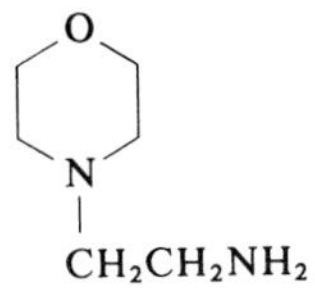

a172

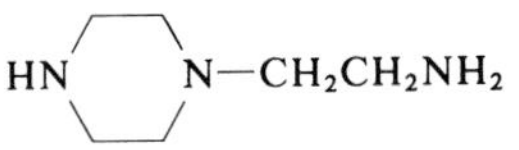

a174

$CH_2CH_2NH_2$

$CH_2CH_2NH_2$

$CH_2CH_2NH_2$

TABLE 1.15 Physical Constants of Organic Compounds (*Continued*)

No.	Name	Formula	Formula weight	Beilstein reference	Density	Refractive index	Melting point	Boiling point	Flash point	Solubility in 100 parts solvent
a177	2-(2-Aminoethyl)pyridine	$H_2NCH_2CH_2C_5H_4N$	122.17	22, 434	1.021	1.5337^{20}		93^{12mm}	100	
a178	4-(2-Aminoethyl)pyridine	$H_2NCH_2CH_2(C_5H_4N)$	122.17		1.012	1.5403^{20}		104^{9mm}		
a179	3-Amino-4-fluorobenzotrifluoride	$H_2N(FC_6H_3CF_3)$	179.0			1.4608^{20}		81^{20mm}		
a180	Aminoguanidine H_2CO_3	$H_2NNHC(=NH)NH_2 \cdot H_2CO_3$	136.11	3, 117			172 d			i aq; d hot aq
a181	Aminoguanidine nitrate	$H_2NNHC(=NH)NH_2 \cdot HNO_3$	137.11	3, 117			137			
a182	N-Aminohexamethyleneimine	$C_6H_{12}N—NH_2$	114.19		0.984	1.4850^{20}		165	56	
a185	2-Aminohexanoic acid	$CH_3(CH_2)_3CH(NH_2)COOH$	131.18	4, 433	1.172		d 327			1.15 aq^{25}; 0.42 alc
a186	6-Aminohexanoic acid	$H_2N(CH_2)_4CH_2COOH$	131.18	4, 434			204–206			v s aq; i alc
a187	6-Amino-1-hexanol	$H_2N(CH_2)_5CH_2OH$	117.19	4^2, 748			56–58	135^{30mm}		s eth
a188	1-Amino-4-hydroxyanthraquinone		239.23	14, 268			207–209			
a189	L-2-Amino-3-hydroxybutanoic acid	$CH_3CH(OH)CH(NH_2)COOH$	119.12	4, 514			d 255–257			v s aq; i alc, eth, chl
a190	DL-2-Amino-4-hydroxybutanoic acid	$HOCH_2CH_2CH(NH_2)COOH$	119.12	4, 514			188–189			s alc
a191	L-2-Amino-4-hydroxybutanoic acid	$HOCH_2CH_2CH(NH_2)COOH$	119.12	4^2, 1636			203 d			
a192	DL-4-Amino-3-hydroxybutanoic acid	$H_2NCH_2CH(OH)CH_2COOH$	119.12	4^2, 938			202 d			s aq; sl s alc, eth
a193	4-Amino-6-hydroxy-2-mercaptopyrimidine hydrate		161.18	24, 476			>300			
a194	2-Amino-4-hydroxy-6-methylpyrimidine		125.13	24, 343			>300			
a195	4-Amino-3-hydroxy-1-naphthalenesulfonic acid		239.25	14, 846			295 d			i aq, alc, bz, eth

No.	Name	Mol. wt.	Beilstein ref.	Density	Melting point, °C	Boiling point, °C	Solubility
a196	4-Amino-5-hydroxy-1-naphthalene-sulfonic acid	239.25	14, 835				sl s aq; i alc, eth
a197	5-Amino-6-hydroxy-2-naphthalene-sulfonic acid	239.25					sl s hot aq; i eth
a198	6-Amino-7-hydroxy-2-naphthalene-sulfonic acid	239.25	14, 849		>300		
a199	2-Amino-3-hydroxy-pyridine	110.12	22^2, 408	1.0384^{15}	172–174		0.77 aq; sl s alc
a200	4-Amino-2-hydroxy-pyrimidine H₂N(HO)(C₅H₃N)	111.10	24, 314		>300		
a201	1-Aminoindan	133.19	12, 1191	1.5613^{20}	1.5	97^{8mm}	sl s aq
a202	5-Aminoindan	133.19	12^1, 511		36	249^{745mm}	94	sl s aq

2-(2-Aminoethyl)-2-thiopseudourea, a171
1-Aminoheptane, h20
2-Aminoheptane, m269a
1-Aminohexane, h80
2-Aminohexane, m353a
p-Aminohippuric acid, a132
Aminohydroxybenzoic acid, a291, a292
2-Amino-2-(hydroxymethyl)-1,3-propanediol, t430
α-Amino-4-imidazolepropanoic acid, h83
Aminoiminomethanesulfinic acid, f30
N-(Aminoiminomethyl)-N-methylglycine, c278

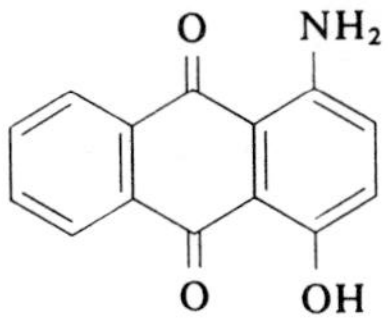

a188

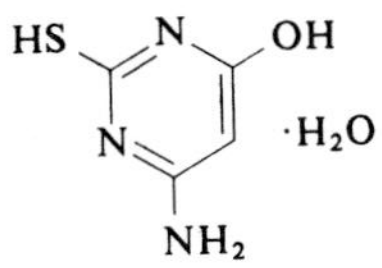

a193

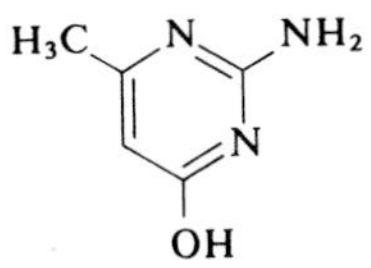

a194

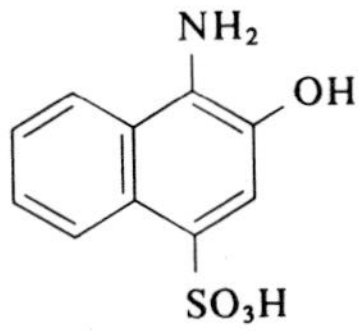

a195

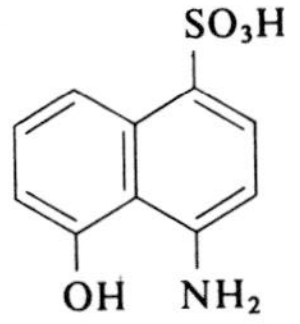

a196

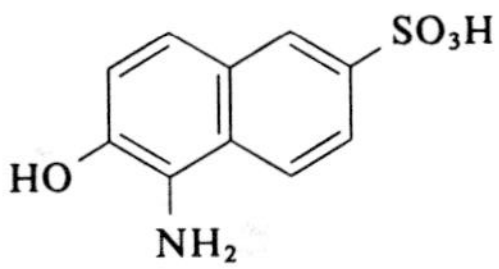

a197

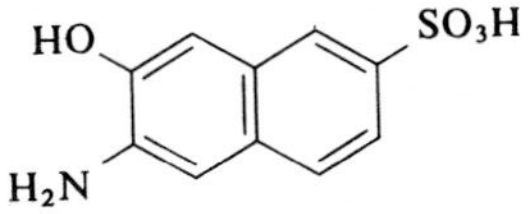

a198

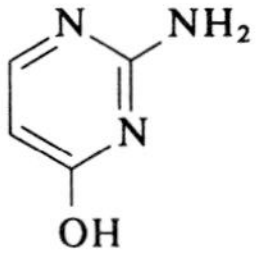

a200

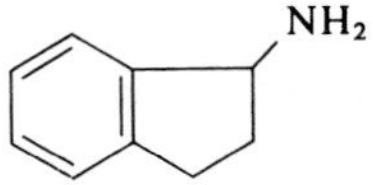

a201

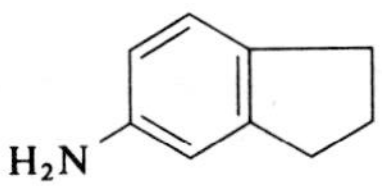

a202

TABLE 1.15 Physical Constants of Organic Compounds (*Continued*)

No.	Name	Formula	Formula weight	Beilstein reference	Density	Refractive index	Melting point	Boiling point	Flash point	Solubility in 100 parts solvent
a203	5-Aminoindazole		133.15	25[2], 308			178			sl s aq, PE; s alc
a204	6-Aminoindazole		133.15	25, 317			206 d			
a205	2-Amino-5-iodobenzoic acid	$H_2N(I)C_6H_3COOH$	263.03	14, 373			221 d			
a207	DL-2-Amino-4-mercaptobutanoic acid	$HSCH_2CH_2CH(NH_2)COOH$	135.19	4[3], 1647			232–233			v s aq
a208	Aminomethanesulfonic acid	$H_2NCH_2SO_3H$	111.12	1, 583			185 d			
a209	3-Amino-4-methoxybenzoic acid	$CH_3O(NH_2)C_6H_3COOH$	167.16	14[1], 657			241			
a210	2-Amino-6-methoxybenzothiazole		180.23	27[2], 334			165–167			
a211	5-Amino-2-methoxypyridine	$CH_3O(NH_2)—C_5H_3N$	124.14	22[2], 408		1.5745^{20}	31	90^{1mm}		
a212	2-Amino-5-methylbenzene sulfonic acid	$H_2NC_6H_3(CH_3)SO_3H$	187.22	14, 723			>300			
a213	5-Amino-2-methylbenzene sulfonic acid	$H_2NC_6H_3(CH_3)SO_3H$	187.22	14, 720			>300			
a214	2-Amino-5-methylbenzoic acid	$H_2N(CH_3)C_6H_3COOH$	151.17	14, 481			177 d			sl s aq; s alc, eth
a215	3-Amino-4-methylbenzoic acid	$H_2N(CH_3)C_6H_3COOH$	151.17	14, 487			167–169			s aq
a216	2-Amino-3-methyl-1-butanol	$(CH_3)_2CHCH(NH_2)CH_2OH$	103.17	4[3], 805	0.906	1.4543^{20}	30–32	80^{8mm}	91	
a218	2-(Aminomethyl)-1-ethylpyrrolidine		128.22		0.887	1.4665^{20}		60^{16mm}	60	
a219	2-Amino-3-methyl-1-pentanol	$CH_3CH_2CH(CH_3)CH(NH_2)CH_2OH$	117.19	4, 298	0.917	1.4589^{20}	30	97^{14mm}	100	
a220	2-Amino-4-methyl-1-pentanol	$CH_3CH(CH_3)CH_2CH(NH_2)CH_2OH$	117.19			1.4511^{20}		200	90	
a221	4-Amino-3-methylphenol	$H_2N(CH_3)C_6H_3OH$	123.16	13, 593			179			
a222	4-(Aminomethyl)piperidine		114.19			1.4900^{20}	25	200	78	

No.	Name	Formula	Formula wt.	Beilstein ref.	Density	n_D	m.p., °C	b.p., °C		Solubility
a223	2-Amino-2-methyl-1,3-propanediol	$HOCH_2C(CH_3)(NH_2)CH_2OH$	105.14	4, 303			110	151^{10mm}		250 aq^{20}; s alc
a224	2-Amino-2-methyl-1-propanol	$(CH_3)_2C(NH_2)CH_2OH$	89.14		0.9342^{20}	1.4480^{20}	30–31	165	67	misc aq; s alc, org solv
a225	2-Amino-2-methylpropionic acid	$(CH_3)_2C(NH_2)COOH$	103.12	4, 414			335 sealed tube	280 subl		v s aq
a226	2-(Aminomethyl)pyridine	$H_2NCH_2-C_5H_4N$	108.14		1.049	1.5445^{20}		85^{12mm}	90	
a227	3-(Aminomethyl)pyridine	$H_2NCH_2-C_5H_4N$	108.14		1.062	1.5510^{20}	−21	74^{1mm}	100	
a228	4-(Aminomethyl)pyridine	$H_2NCH_2-C_5H_4N$	108.14	22[3], 4181	1.065	1.5505^{20}	−8	230	108	
a229	2-Amino-4-methylpyridine	$H_2N(CH_3)-C_5H_3N$	108.14	22[2], 342			100	230		v s aq, alc, DMF
a230	2-Amino-6-methylpyridine	$H_2N(CH_3)-C_5H_3N$	108.14	22[1], 633			45	209	103	v s aq
a231	2-Amino-4-methylpyrimidine		109.13	24, 84			160	subl		s hot aq; s alc

2-Aminoisobutyric acid, a225
5-Aminoisophthalic acid, a116
2-Amino-3-mercaptopropanoic acid, c371

1-Amino-2-methoxyethane, m69
α-(Aminomethyl)benzyl alcohol, a265

3-Amino-α-methylbenzyl alcohol, a264
2-Amino-3-methylpentanoic acid, i79

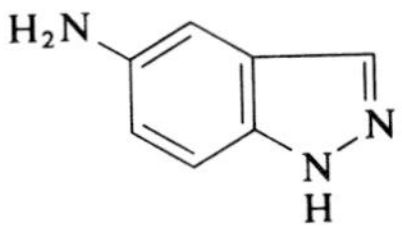

H_2N ⋯ N–N–H   a203

H_2N ⋯ N–N–H   a204

CH_3O ⋯ N=C–S, NH_2 a210

CH_2NH_2 (adamantyl)   a212

CH_2NH_2, N–CH_2CH_3 (pyrrolidine)   a218

CH_2NH_2, N–H (piperidine)   a222

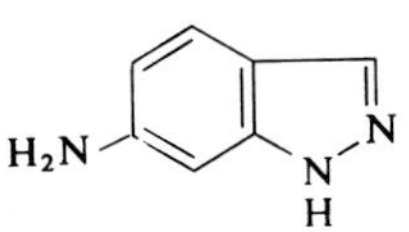

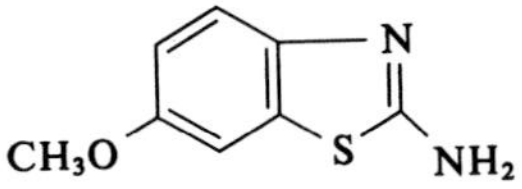

NH_2, N, N, CH_3 (pyrimidine)   a231

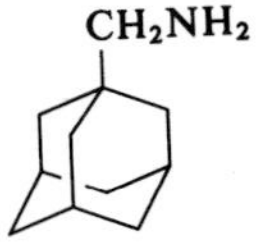

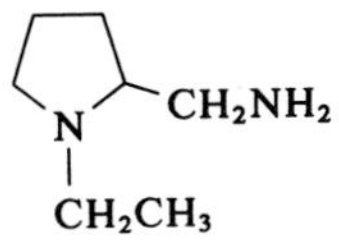

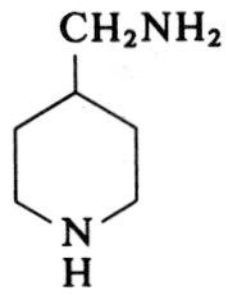

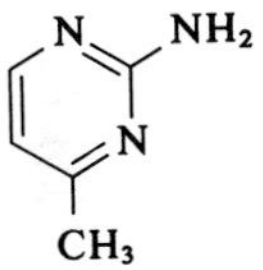

TABLE 1.15 Physical Constants of Organic Compounds (*Continued*)

No.	Name	Formula	Formula weight	Beilstein reference	Density	Refractive index	Melting point	Boiling point	Flash point	Solubility in 100 parts solvent
a232	2-Amino-4-methyl-thiazole		114.17	27, 159			45	232	>110	v s aq, alc, eth
a233	2-Aminomethyl-3,5,5-trimethylcyclohexanol		171.29		0.969	1.4904^{20}	43–48	265	>110	
a234	N-Aminomorpholine		102.14	27, 8	1.059	1.4772^{20}		168	58	
a235	1-Aminonaphthalene	$C_{10}H_7NH_2$	143.19	12, 1212	1.114		48–50	301	>110	
a236	7-Amino-1,3-naphthalenedisulfonic acid		303.14	14, 784			>300			9.2^{20} aq
a237	2-Amino-1-naphthalenesulfonic acid	$H_2N{-}C_{10}H_6SO_3H$	223.26	14, 739			dec			0.031 aq; s dil alk
a238	5-Amino-2-naphthalenesulfonic acid	$H_2N{-}C_{10}H_6SO_3H$	223.26	14, 758						
a239	8-Amino-2-naphthol	$H_2N{-}C_{10}H_6OH$	159.19	13, 685			207			
a240	2-Amino-4-nitrobenzoic acid	$H_2N(NO_2)C_6H_3COOH$	182.14	14, 374			270 d			i aq; v s alc, eth
a241	2-Amino-5-nitrobenzonitrile	$H_2N(NO_2)C_6H_3CN$	163.14	14^2, 234			200–207			
a242	2-Amino-5-nitrobenzophenone	$C_6H_5COC_6H_4(NH_2)NO_2$	242.23	14, 79			166–168			
a243	2-Amino-6-nitrobenzothiazole		195.20	27^2, 232			247–249			
a244	4-Amino-3-nitrobenzotrifluoride	$H_2N(NO_2)C_6H_3CF_3$	206.12	13^2, 192			105–106			
a245	2-Amino-4-nitrophenol	$O_2N(NH_2)C_6H_3OH$	154.13				145			
a246	2-Amino-5-nitrophenol	$O_2N(NH_2)C_6H_3OH$	154.13	13, 390			202 d			
a248	4-Amino-2-nitrophenol	$O_2N(NH_2)C_6H_3OH$	154.13	13, 520			127			
a249	D-(−)-threo-2-Amino-1-(p-nitrophenyl)-1,3-propanediol	$HOCH_2C(NH_2)C(OH)-C_6H_4NO_2$	212.21				163–165			
a250	2-Amino-5-(p-nitrophenylsulfonyl)thiazole		285.30				222–226			

a251	2-Amino-5-nitropyridine	$H_2N-C_5H_3N-NO_2$	139.11	22[1], 631			188			sl s aq, bz, eth
a252	2-Amino-5-nitrothiazole		145.14				202 d			v sl s aq; 0.7 alc; 0.4 eth
a253	exo-2-Aminonorbornane		111.19	12[3], 160	0.938	1.4807^{20}		49^{10mm}	35	s aq, alc, eth, PE
a254	2-Aminopentane	$H(CH_2)_3CH(NH_2)CH_3$	87.17	4, 177	0.739^{20}	1.4047^{20}		91–92		misc aq, alc, eth
a255	3-Aminopentane	$C_2H_5CH(NH_2)C_2H_5$	87.17	4, 178	0.749^{20}_4	1.4055^{20}		91	2	5.5 aq[18], v sl s alc, chl, eth, PE
a256	DL-2-Aminopentanoic acid	$H(CH_2)_3CH(NH_2)COOH$	117.15	4, 416			303	320 subl		v s aq; sl s alc; i eth
a257	5-Aminopentanoic acid	$H_2N(CH_2)_4COOH$	117.15	4, 418			158–161			
a258	5-Amino-1-pentanol	$H_2N(CH_2)_5OH$	103.17	4[1], 441		1.4615^{20}	37	122^{16mm}	65	
a259	2-Aminophenethyl alcohol	$H_2NC_6H_4CH_2CH_2OH$	137.18	13[3], 1679	1.045	1.5849^{20}		148^{4mm}	>112	
a260	2-Aminophenol	$H_2NC_6H_4OH$	109.13	13, 354			170–174			2 aq; 4.3 alc; v s eth; sl s bz

2-Aminomethylthiophene, t161
1-Aminonaphthalene, n17

a232

a233

a234

1-Amino-2-naphthol-4-sulfonic acid, a195
1-Amino-2-naphthol-6-sulfonic acid, a197

a235

a236

5-Aminoorotic acid, a160
1-Aminopentane, p53

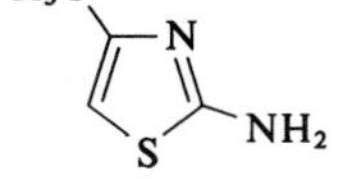

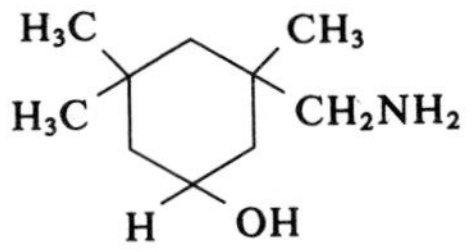

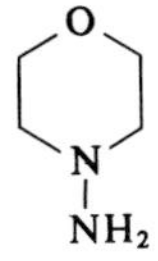

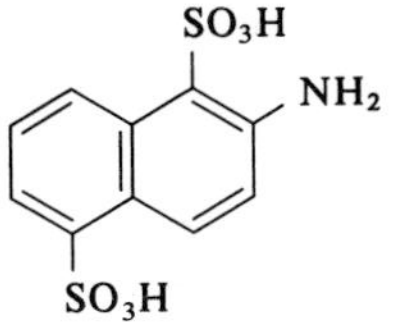

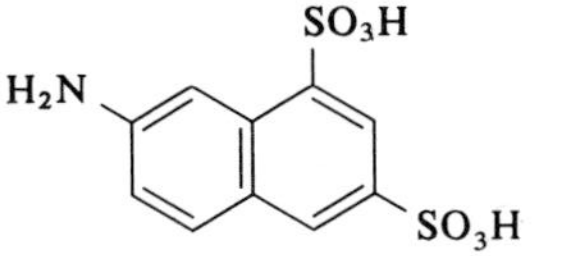

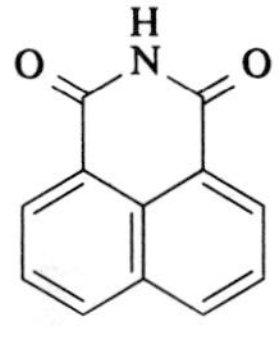

a238

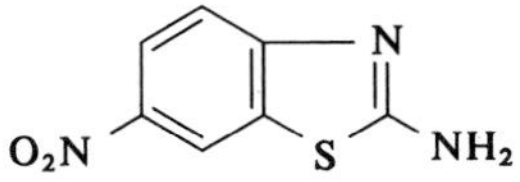

a243

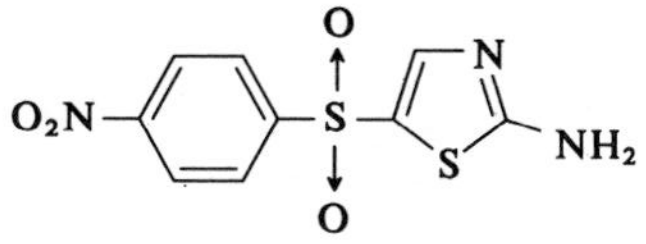

a250

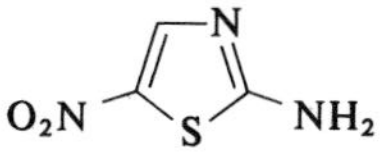

a252

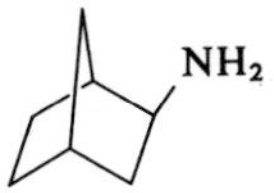

a253

1.93

TABLE 1.15 Physical Constants of Organic Compounds (*Continued*)

No.	Name	Formula	Formula weight	Beilstein reference	Density	Refractive index	Melting point	Boiling point	Flash point	Solubility in 100 parts solvent
a261	3-Aminophenol	$H_2NC_6H_4OH$	109.13	13, 401			122–123	164^{11mm}		2.5 s aq; v s alc, eth
a262	4-Aminophenol	$H_2NC_6H_4OH$	109.13	13, 427			190	284 d		0.65 aq; s alc, eth
a263	4'-Aminophenylacetonitrile	$H_2NC_6H_4CH_2CN$	132.17	14, 457			44	312	>110	sl s hot aq; s alc
a264	1-(3-Aminophenyl)-ethanol	$H_2NC_6H_4CH(CH_3)OH$	137.18	13^3, 1654			68–71			
a265	2-Amino-1-phenylethanol	$C_6H_5CH(CH_2NH_2)OH$	137.18	13^3, 361			56–57	160^{17mm}		v s aq; s alc
a266	1S,2S-(+)-2-Amino-1-phenyl-1,3-propanediol	$C_6H_5CH(OH)CH(NH_2)CH_2OH$	167.21				109–113			
a267	L-2-Amino-3-phenyl-1-propanol	$C_6H_5CH_2CH(NH_2)CH_2OH$	151.21	13^3, 1757			92–94			
a268	3-Amino-1-phenyl-2-pyrazolin-5-one		175.19				210–215			
a271	N-Aminophthalmide		162.15				200–202			
a272	N-Aminopiperidine		100.17	20, 89	0.928	1.4750^{20}		146^{730mm}	36	v s aq, alc; i eth
a273	3-Amino-1,2-propanediol	$H_2NCH_2CH(OH)CH_2OH$	91.11	4, 301	1.175	1.4920^{20}		265^{739mm}	>110	
a274	DL-1-Amino-2-propanol	$CH_3CH(OH)CH_2NH_2$	75.11	4, 289	0.973	1.4483^{20}	–2	160	73	v s aq, alc, eth
a275	DL-2-Amino-1-propanol	$CH_3CH(NH_2)CH_2OH$	75.11	4^1, 432	0.943	1.4495^{20}		173–176	83	v s aq, alc, eth
a276	S(+)-2-Amino-1-propanol	$CH_3CH(NH_2)CH_2OH$	75.11	4^1, 432	0.965	1.4495^{20}		176	62	v s aq, alc, eth
a277	3-Amino-1-propanol	$H_2NCH_2CH_2CH_2OH$	75.11	4, 288	0.982	1.4610^{20}	10–12	188	79 (TOC)	s aq, alc
a278	2-Amino-1-propene-1,1,3-tricarbonitrile	$NCC(CN){=}C(NH_2)CH_2CN$	132.13				171–173			s aq
a279	3-Aminopropionitrile	$H_2NCH_2CH_2CN$	70.09		0.9164^{20}	1.4396^{20}		185		
a280	3-Aminopropyl(diethoxy)methylsilane	$H_2N(CH_2)_3Si(CH_3)(OCH_2CH_3)_2$	191.4			1.427^{20}		88^{8mm}		
a281	N-(3-Aminopropyl)iminodiethanol	$H_2N(CH_2)_3N(CH_2CH_2OH)_2$	162.23		0.1071	1.4980^{20}		170^{2mm}	137	
a282	N-(3-Aminopropyl)morpholine	$H_2N(CH_2)_3N(CH_2CH_2)_2O$	144.22		0.9872^{20}	1.4761^{20}	–15	224	98	misc aq, alc, bz

No.	Name	Formula	Formula weight	Beilstein reference	Density	n_D	mp, °C	bp, °C	Flash point, °C	Solubility
a283	N-(3-Aminopropyl)-2-pyrrolidinone		142.20		1.014	1.5000^{20}		123^{1mm}	>110	
a284	3-Aminopropyltriethoxysilane	$H_2N(CH_2)_3Si(OC_2H_5)_3$	221.37		0.9506^{20}_4	1.4225^{20}		217	104	s aq, alc, bz, eth
a285	3-Aminopropyltrimethoxysilane	$H_2N(CH_2)_3Si(OCH_3)_3$	179.29		1.01^{25}_4	1.420^{25}	80	83^{8mm}	92	s aq, alc, bz, eth
a286	2-Aminopyridine	$(C_5H_4N)NH_2$	94.12	22, 428			58.1	210.6		s aq, alc; sl s bz, eth
a287	3-Aminopyridine	$(C_5H_4N)NH_2$	94.12	22, 431			64	248		v s aq
a288	4-Aminopyridine	$(C_5H_4N)NH_2$	94.12	22, 433			155–158	273		sl s aq; v s alc, eth acet; s hot bz
a289	2-Aminopyrimidine		95.11	24, 80			123–126	subl		0.2 aq; 4.8 alc; s dil acid, alk; sl s eth
a290	4-Aminoquinaldine		158.20	22, 453			169	333		sl s aq, alc; s acid
a291	4-Aminosalicylic acid	$H_2NC_6H_3(OH)COOH$	153.14	14, 579			147 d			v s aq
a292	5-Aminosalicylic acid	$H_2NC_6H_3(OH)COOH$	153.14	14, 579			280 d			
a293	2-Amino-3-sulfopropanoic acid	$HOOCCH(NH_2)CH_2SO_3H$	187.17	4, 533			260 d			

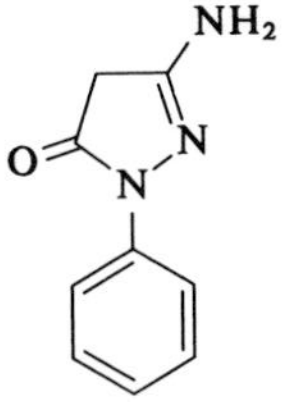

a268

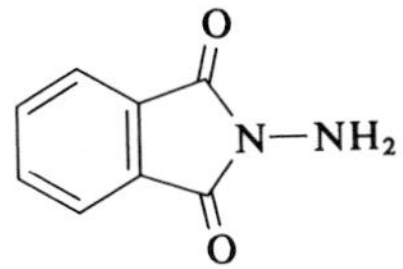

a271

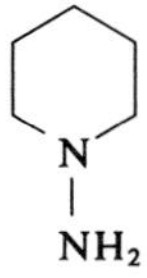

a272 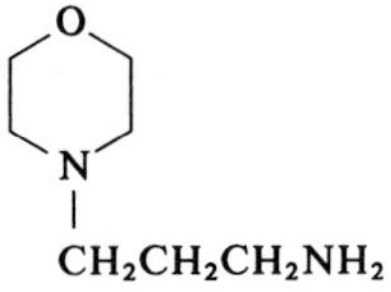   a282 $CH_2CH_2CH_2NH_2$

a283 $CH_2CH_2CH_2NH_2$

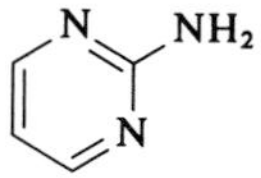

a289

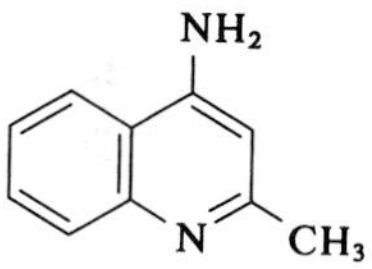

NH_2 CH_3 a290

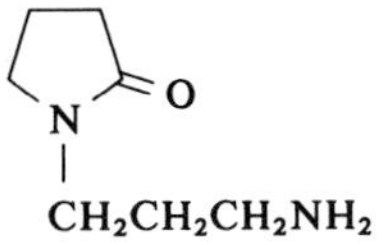

TABLE 1.15 Physical Constants of Organic Compounds (*Continued*)

No.	Name	Formula	Formula weight	Beilstein reference	Density	Refractive index	Melting point	Boiling point	Flash point	Solubility in 100 parts solvent
a294	5-Amino-1,2,3,4-tetrazole hydrate		103.08	26, 403			204 d			
a295	5-Amino-1,3,4-thiadiazole		133.20	27, 624			190–192			
a296	2-Aminothiazole		100.14	27, 155			93			sl s aq, alc, eth
a297	2-Amino-2-thiazoline		100.14	27, 136			91–93			
a298	2-Aminothiophenol	$H_2NC_6H_4SH$	125.19	13, 397		1.6405^{20}	26	234	79	l aq^{12}; v s hot aq
a299	6-Amino-3-toluenesulfonic acid	$H_2NC_6H_3(CH_3)SO_3H$	187.22	14, 723			>300			
a300	3-Amino-1,2,4-triazole		84.08	26, 137			159			
a301	5-Amino-2,2,4-trimethyl-1-cyclopentanemethylamine		156.27		0.901	1.4733^{20}		221	97	s aq, alc, chl
a302	11-Aminoundecanoic acid	$H_2N(CH_2)_{10}COOH$	201.31				190–192			
a303	Aniline	$C_6H_5NH_2$	93.13	12, 59	1.0217^{20}	1.5855^{20}	−5.98	184.40	70 (CC)	3.5 aq^{25}; s alc, CCl$_4$, eth, acids
a304	Aniline hydrochloride	$C_6H_4NH_2 \cdot HCl$	129.59	12, 182	1.222		198	245	193 (CC)	100 aq; v s alc
a305	2-Anilinoethanol	$C_6H_5NHCH_2CH_2OH$	137.18		1.085	1.5793^{20}		150–152^{10mm}	153 (OC)	sl s aq; v s alc, chl, eth
a306	3-Anilinopropionitrile	$C_6H_5NHCH_2CH_2CN$	146.19				52–53		>110	
a307	1-(o-Anisidino)-1,3-butanedione	$CH_3OC_6H_4NHCOCH_2COCH_3$	207.23	13^1, 117			84–85			
a308	1-(p-Anisidino)-1,3-butanedione	$CH_3OC_6H_4NHCOCH_2COCH_3$	207.23	13^1, 177			115–117			
a309	Anthracene		178.23	5, 657	1.24^{27}		216.3	340	121 (CC)	i aq; 1.5 alc; 1.6 bz; 1.2 chl; 3.1 CS$_2$,
a310	9,10-Anthraquinone		208.22	7, 781	1.43^{20}_4		286	377	185 (CC)	0.44 alc^{25}; 0.6 chl^{20}, 0.2 bz^{20}, 0.11 eth^{25}
a313	9,10-Anthraquinone-2-sulfonic acid Na salt		310.26							
a314	Antipyrine		188.23	24, 27	1.088^{13}		114^{174mm}	319		100 aq; 77 alc; 100 chl; 2.3 eth
a315	L-(+)-Arabinose		150.13	31, 32			160–163			100 aq
a316	L-(+)-Arginine	$H_2NC(=NH)NH(CH_2)_3$-$CH(NH_2)COOH$	174.20	4, 420			223 d			17.6 aq; sl s alc

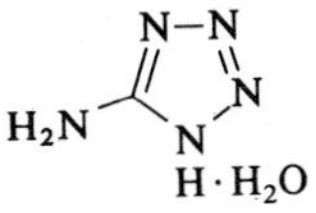

a294

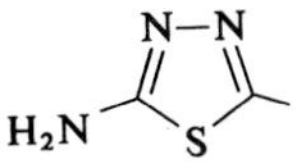

a295

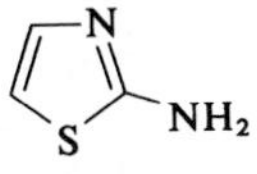

a296

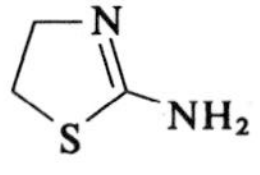

a297

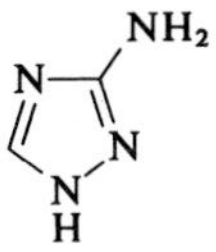

a300

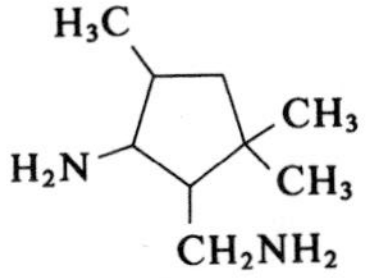

a301

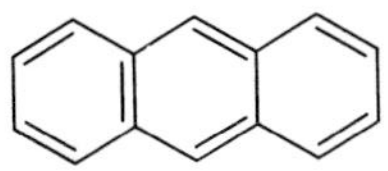

a309

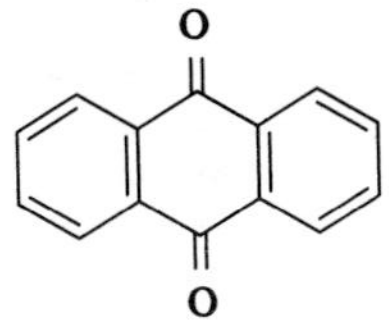

a310

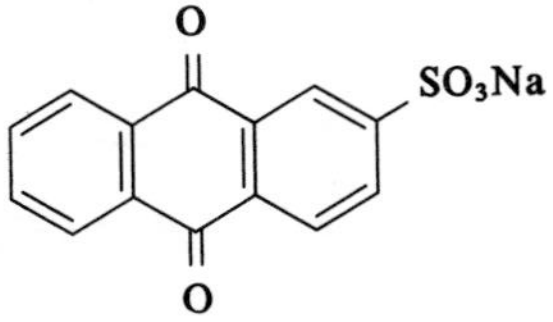

a313

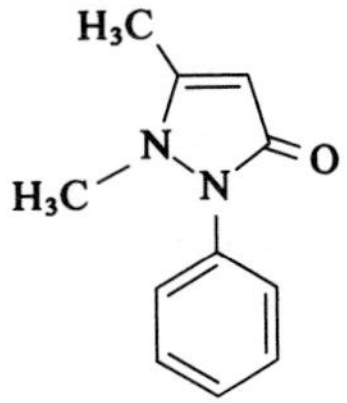

a314

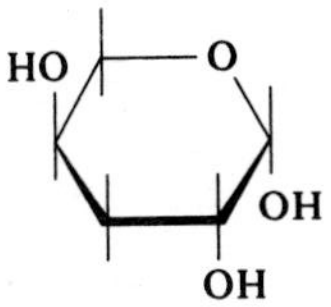

a315

TABLE 1.15 Physical Constants of Organic Compounds (*Continued*)

No.	Name	Formula	Formula weight	Beilstein reference	Density	Refractive index	Melting point	Boiling point	Flash point	Solubility in 100 parts solvent
a317	L-(+)-Ascorbic acid		176.12	18^3, 3038			192 d			100 aq; 3.3 alc
a318	L-(+)-Asparagine	$H_2NCOCH_2CH(NH_2)COOH$	132.12	4, 476			235 d			3.6 aq^{28}; s alk, acids i alc, bz, eth
a319	L-(+)-Aspartic acid	$HOOCCH_2CH(NH_2)COOH$	133.10	4, 472	1.661^{12}		270			0.45 aq; i alc, eth
a320	Atropine		289.38	21, 27			114–116			0.22 aq; s bz, dil acid
a321	Aurintricarboxylic acid, triammonium salt		473.44	10^2, 775			225 d			v s aq
a322	2-Azacyclooctanone		127.19	21, 242			35–38	148^{10mm}		
a323	2-Azacyclotridecanone		197.32				150–153			
a324	Azidotrimethylsilane	$(CH_3)_3SiN_3$	115.21		0.868	1.4142^{20}	−95	95–96	23	
a325	Azidotriphenylsilane	$(C_6H_5)_3SiN_3$	301.4				83–84	$100^{0.01mm}$		
a326	1-Aziridineethanol	$(CH_2)_2{=}NCH_2CH_2OH$	87.12		1.088	1.4560^{20}		168	67	
a327	Azobenzene	$C_6H_5N{=}NC_6H_5$	182.23	16, 8	1.203^{20}_4		68.3	293		4.2 alc^{20}, s eth, HOAc
a328	2,2′-Azobis(2-methyl-propionitrile)	$(CH_3)_2C(CN)N{=}N{-}C(CN)(CH_3)_2$	164.21	4, 563				107 d		2 EtOH; 5 MeOH; can explode in acetone
a329	Azodicarbonamide	$H_2NCON{=}NCONH_2$	116.08	3, 123			d 180–200			i aq, alc
a330	4,4′-Azoxydianisole	$CH_3OC_6H_4N{=}N{-}({\rightarrow}O)C_6H_4OCH_3$	258.28	16, 637			120			
a331	Azulene		128.17	5^2, 432			100.5	242		i aq; s org solvents
b1	Barbituric acid		128.09	24, 467			248–252 d			s hot aq, dil acid
b2	Basic fuchsin		337.86	13, 765	1.22		d 186			0.3 aq; s alc, acids
b3	Benzaldehyde	C_6H_5CHO	106.12	7, 174	1.0447^{20}	1.5455^{20}	−26	178.9	62	0.3 aq; misc alc, eth
b4	Benzamide	$C_6H_5CONH_2$	121.14	9, 195	1.341^4		127.2	288		1.3 aq; 17 alc; 30 pyr
b5	Benzanilide	$C_6H_5CONHC_6H_5$	197.24	12, 262	1.315		163.1	117^{10mm}		i aq; 1.7 alc; sl s eth
b6	1,2-Benzanthracene		228.29	5, 718			160	437.6		sl s hot alc; s most other org solvents
b7	2,3-Benzanthracene		228.29	5^2, 628	1.35		341	subl		sl s most org solvents
b8	Benzene	C_6H_6	78.11	5, 179	0.8787^{15}	1.5011^{20}	5.5	80.1	−11 (TCC)	0.17 aq; s most org solvents
b9	Benzene-*d*	C_6H_5D	79.12			1.4980^{20}		80	−11 (CC)	

b10	Benzene-d_6	C_6D_6	84.16	5^3, 519	0.950	1.4986^{20}	6.8	79.1	−11 (CC)	
b11	Benzenearsonic acid	$C_6H_5AsO(OH)_2$	202.04	16, 868	1.760^{25}		163 d			2.5 aq; 2 alc
b12	Benzeneboronic acid	$C_6H_5B(OH)_2$	121.93	16, 920			217			2.5 aq; 1.8 bz; 30 eth; 178 MeOH
b13	1,4-Benzenedicarbal-dehyde	$C_6H_4(CHO)_2$	134.13	7, 675			114	248		i aq; 6 bz; 17 acet; 2 eth; 14 diox; 46 MeOH

Arsanilic acids, a115
Aspirin, a56
Azacyclopropane, e134
Azelaic acid, n95
Azelonitrile, n94
Aziridine, e134
Azobis(isobutyronitrile), a328
4,4′-Azoxyanisole, a330
Barbitol, d280
Behenyl alcohol, d719
Benzalacetone, p98
Benzal bromide, d102
Benzalphthalide, b102
Benzanthrone, b8
Benzeneacetaldehyde, p76a
Benzeneazobenzene, a327
Benzenecarbonitrile, b51

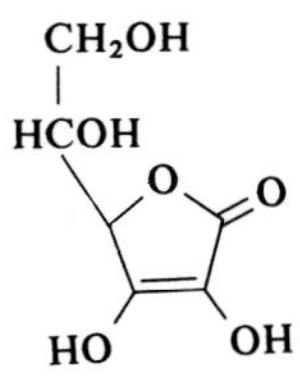

CH₂OH
HCOH
O
O
HO
OH

a317

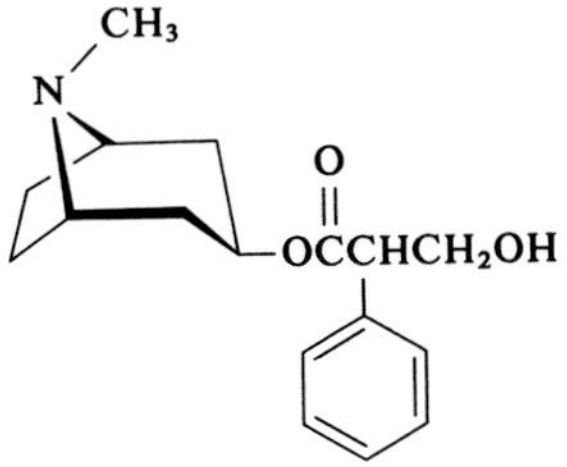

CH₃
N
O
OCCHCH₂OH

a320

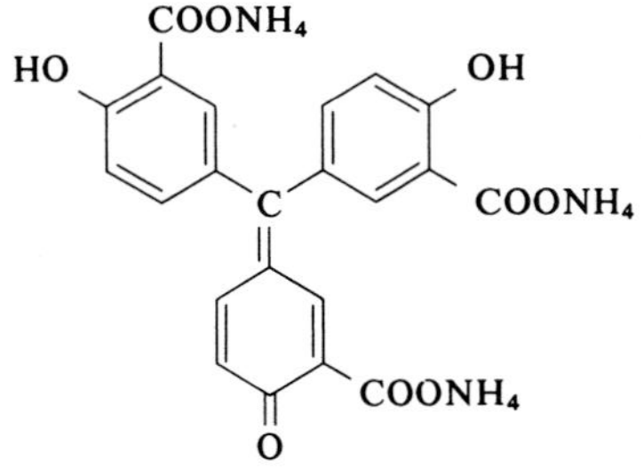

COONH₄
HO
OH
C
COONH₄
COONH₄
O

a321

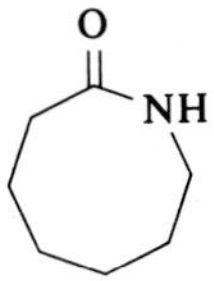

O
NH

a322

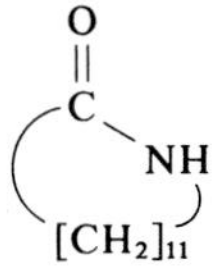

O
C
NH
[CH₂]₁₁

a323

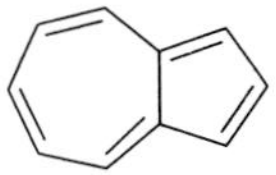

a331

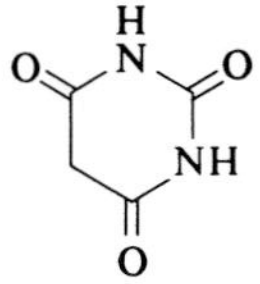

H
N
O
O
NH
O

b1

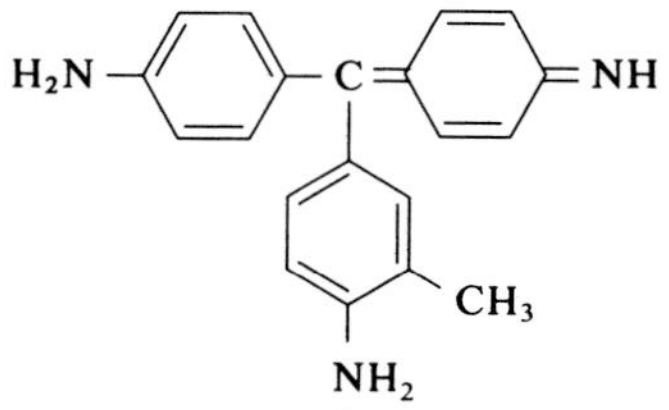

H₂N
C
NH
CH₃
NH₂

b2

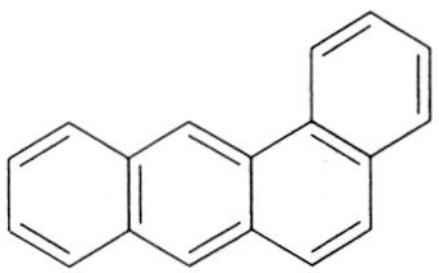

b6

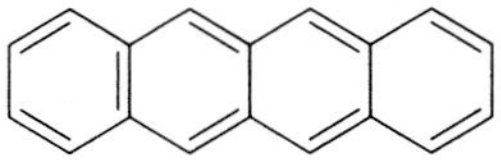

b7

TABLE 1.15 Physical Constants of Organic Compounds (*Continued*)

No.	Name	Formula	Formula weight	Beilstein reference	Density	Refractive index	Melting point	Boiling point	Flash point	Solubility in 100 parts solvent
b14	1,3-Benzenedicarbonyl dichloride	$C_6H_4(COCl)_2$	203.02	9, 834			43–44	276	180	73 bz; 62 CCl_4
b15	1,4-Benzenedicarbonyl dichloride	$C_6H_4(COCl)_2$	203.02	9, 844			81	266	180	37 bz; 9 CCl_4
b16	1,3-Benzenedicarboxylic acid	$C_6H_4(COOH)_2$	166.13	9, 832			345–348	subl		0.012 aq; v s alc, HOAc; i bz, PE
b17	1,4-Benzenedicarboxylic acid	$C_6H_4(COOH)_2$	166.13	9, 841			subl without melting			v sl s aq, chl, eth; sl s alc; s alk
b18	1,4-Benzenedimethanol	$C_6H_4(CH_2OH)_2$	138.17	6, 919	1.100^{117}		117–119	143^{1mm}	188	v s aq, alc, eth
b19	Benzenehexacarboxylic acid	$C_6(COOH)_6$	342.17	9, 1008			286 d	100 d		v s aq, alc
b20	Benzenesulfinic acid	$C_6H_5S(=O)OH$	142.16	11,2			85			sl s aq; s alc, bz, eth
b21	Benzenesulfonamide	$C_6H_5SO_2NH_2$	157.19	11, 39			152			i aq; sl s alc; s eth
b22	Benzenesulfonic acid	$C_6H_5SO_2OH$	158.18	11, 26			50–51			v s aq, alc; sl s bz
b23	Benzenesulfonyl chloride	$C_6H_5SO_2Cl$	176.62	11, 34	1.3842^{15}	1.5518^{20}	14.5	120^{10mm}	>110	i aq; s alc, eth
b24	Benzenesulfonyl fluoride	$C_6H_5SO_2F$	160.16	11^2, 23	1.3286^{20}	1.4920^{20}	−5	207–208	87	s alc, eth
b25	Benzenesulfonyl hydrazide	$C_6H_5SO_2NHNH_2$	172.21	11, 52			101–103			flammable solid
b26	1,2,4,5-Benzenetetracarboxylic acid	$C_6H_2(COOH)_4$	254.15	9, 997			276			1.5 aq; v s alc
b27	1,2,4,5-Benzenetetracarboxylic anhydride		218.12	19, 196			283–286	397–400		
b28	1,2,3-Benzenetricarboxylic acid dihydrate	$C_6H_3(COOH)_3 \cdot 2H_2O$	246.18	9, 976			192 d			sl s aq; v s eth
b29	1,2,4-Benzenetricarboxylic acid	$C_6H_3(COOH)_3$	210.14	9, 977			321 d			2.1 aq; 25.3 alc; 7.9 acet; v s eth
b30	1,3,5-Benzenetricarboxylic acid	$C_6H_3(COOH)_3$	210.14	9, 978			>330			sl s aq; v s alc; s eth

No.	Name	Formula	Formula wt	Beilstein ref.	Density	n_D	m.p., °C	b.p., °C	Solubility
b31	1,2,4-Benzenetricarboxylic anhydride		192.13	18, 468			161–164	245[14mm]	50 acet; 22 EtAc
b32	1,3,5-Benzenetricarboxylic trichloride	$C_6H_3(COCl)_3$	265.48				35–36		
b32a	1,2,4-Benzenetriol	$C_6H_3(OH)_3$	126.11	6, 1087			141		v s aq, alc, eth, EtAc
b33	Benzethonium chloride	$(CH_3)_3CH_2C(CH_3)_2C_6H_4$-$OCH_2CH_2OCH_2CH_2$-$N^+(CH_3)_2CH_2C_6H_5Cl^-$	448.10				164–166		v s aq; s alc, acet, chl
b34	Benzil	$C_6H_5COCOC_6H_5$	210.23	7, 747	1.23[15]		94.9	346	i aq; s alc, eth
b35	Benzil dioxime	$C_6H_5C(=NOH)C(=NOH)$-C_6H_5	240.26	7[3], 3816			245		i aq, eth; sl s alc; s alk
b36	Benzilic acid	$(C_6H_5)_2C(OH)COOH$	228.25	10, 342			153		sl s aq; v s alc, eth
b37	Benzil monohydrazone	$C_6H_5C(=NNH_2)COC_6H_5$	224.26	7[1], 394			150–152		
b38	Benzimidazole		118.14	23, 131			170.5	>360	sl s aq, eth; v s alc
b39	Benzo-15-crown-5		268.3				76–78		
b40	7,8-Benzo-1,3-diaza-spiro-[4,5]decane-2,4-dione		216.24				268–270		
b41	1,4-Benzodioxan		136.15	1.142		1.5485[20]		103[6mm]	87

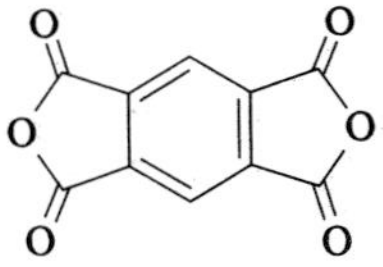

b27

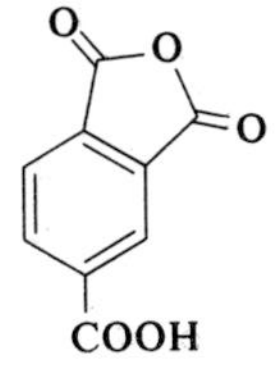

b31

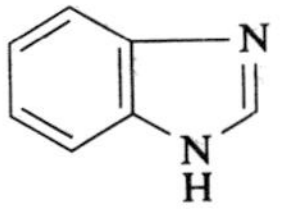

b38

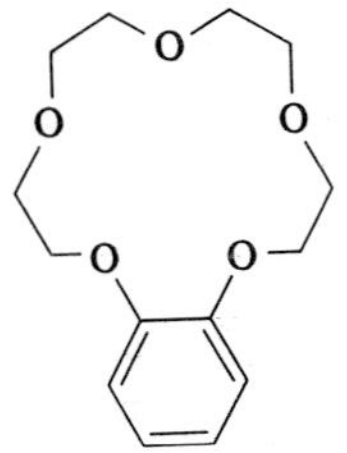

b39

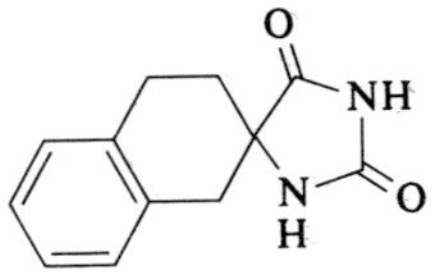

b40

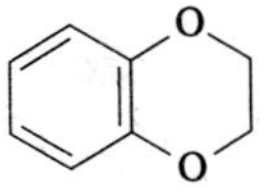

b41

TABLE 1.15 Physical Constants of Organic Compounds (*Continued*)

No.	Name	Formula	Formula weight	Beilstein reference	Density	Refractive index	Melting point	Boiling point	Flash point	Solubility in 100 parts solvent
b42	2,3-Benzofuran		118.14	17, 54	1.072	1.5660^{20}	<-18	175	56	i aq; misc bz, eth, PE
b43	Benzofurazan 1-oxide		136.11	27^1, 740			69–71			
b44	Benzoic acid	C_6H_5COOH	122.13	9, 92	1.080		122.4	249	121 (CC)	0.29 aq; 43 alc; 10 bz; 22 chl; 33 eth; 33 acet
b45	Benzoic anhydride	$(C_6H_5CO)_2O$	226.23	9, 164	1.199		39–40	360	>110	i aq; s alc, acet, chl, bz, HOAc
b46	DL-Benzoin	$C_6H_5COCHOHC_6H_5$	212.25	8, 167	1.3100^{20}	1.5727^{17}	134–136	344		s acet, 20 pyr
b47	Benzoin ethyl ether	$C_6H_5CH(OC_2H_5)COC_6H_5$	240.30	8, 174	1.1016^{17}	1.5485^{20}	61	195^{20mm}		s alc, bz, eth
b48	Benzoin isobutyl ether	$C_6H_5CH[OCH_2CH(CH_3)_2]$-$COC_6H_5$	268.36		0.985			$130^{0.5mm}$	85	
b49	Benzoin methyl ether	$C_6H_5CH(OCH_3)COC_6H_5$	226.28	8, 174	1.1278^{14}		48	189^{15mm}		v s alc, bz, eth
b50	α-Benzoinoxime	$C_6H_5CH(OH)C(=NOH)C_6H_5$	227.26	8, 175			151–152		71	sl s aq; s alc, NH_4OH
b51	Benzonitrile	C_6H_5CN	103.12	9, 275	1.2462^{20}	1.5289^{20}	−12.8	190.7		0.2 aq; misc alc, bz, chl, eth
b52	Benzo[*def*]-phenanthrene		202.26	5, 693	1.271^{23}		156	404		i aq; s alc, eth
b53	Benzophenone	$C_6H_5COC_6H_5$	182.22	7, 411	1.1108^{18}		48.5	305.4	>110	i aq; 13.3 alc; 17 eth
b54	Benzophenone hydrazone	$C_6H_5C(=NNH_2)C_6H_5$	196.25	7, 417			98	230^{55mm}		
b56	1-Benzopyran-4(4*H*)-one		146.15	17, 327			55–57			i aq; s bz; sl s alc
b57	1,2-Benzo[*a*]pyrene		252.32		1.318^{20}		179.3	495	>110	i aq
b58	4,5-Benzo[*e*]pyrene		252.32				182			i aq
b59	1,4-Benzoquinone	$O=C_6H_4=O$	108.10	7, 609			115.7			sl s aq; s alc, eth, hot bz, alk (with dec)
b60	Benzothiazole		135.19		1.2462^{20}	1.6379^{20}	2	231		sl s aq; v s alc, CS_2
b61	Benzo[*b*]thiophene		134.20	17, 59	1.1937^{40}	1.6302^{40}	31.32	221		s alc, bz, chl, eth
b62	1,2,3-Benzotriazole		119.13	26, 38	1.238	1.6420^{20}	98.5	159^{2mm}		sl s aq; s alc, bz, chl
b63	Benzoxazole		119.12	27, 42		1.5594	30	182	58	sl s aq
b64	1-Benzoylacetone	$C_6H_5COCH_2COCH_3$	162.19	7, 680	1.0906^{60}		60	260 sl d		sl s aq; v s alc, eth
b65	2-Benzoylbenzoic acid	$C_6H_5COC_6H_4COOH$	226.23	10, 747			129	265		sl s aq; v s alc, eth
b66	Benzoyl bromide	C_6H_5COBr	185.03	9, 195	1.5467^{20}	1.5883^{20}		218–219	90	d aq, alc; misc eth
b67	Benzoyl chloride	C_6H_5COCl	140.57	9, 182	1.2111^{20}	1.5525^{20}	−1.0	197.2	68 (CC)	dec aq, alc; misc bz, CS_2, eth

No.	Name	Formula	Mol. wt.	Beilstein ref.	Density	n_D	m.p., °C	b.p., °C	b.p., °C	Solubility
b68	Benzoyl cyanide	C₆H₅COCN	131.13	10, 659			32	206	48	i aq
b69	Benzoyl fluoride	C₆H₅COF	124.11	9, 181	1.140		−28	161		d hot aq; v s alc, eth
b70	Benzoylformic acid	C₆H₅COCOOH	150.13	10, 654			69			0.4 aq; 0.1 chl; 0.25 eth; sl s alc; i bz, PE
b71	N-Benzoylglycine	C₆H₅CONHCH₂COOH	179.18	9, 225		1.4960^{20}	178–179			
b72	Benzoylhydrazine	C₆H₅CONHNH₂	136.15	9, 319			117			sl s aq; s alc
b73	3-Benzoylpropanoic acid	C₆H₅COCH₂CH₂COOH	178.19	10, 696			116			
b74	2-Benzoylpyridine	C₆H₅CO—C₅H₄N	183.21	21, 330			44	317	150	s alc, bz, eth
b75	3-Benzoylpyridine	C₆H₅CO—C₅H₄N	183.21	21, 331			40	307	150	
b76	4-Benzoylpyridine	C₆H₅CO—C₅H₄N	183.21	21, 331			71	315	150	s alc, bz, eth

1,3-Benzodioxole, m241
Benzofuroxan, b43
Benzoglyoxaline, b38
Benzoic acid hydrazide, b72

o-Benzoic sulfimide, s1
Benzoresorcinol, d387
2-Benzothiazolethiol, m17
Benzotrichloride, t252

Benzotrifluoride, t305
Benzoylamide, b4
Benzoylbenzene, b53
Benzoyl peroxide, d54

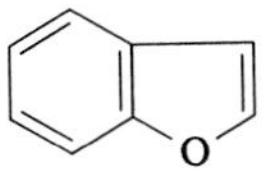

b42

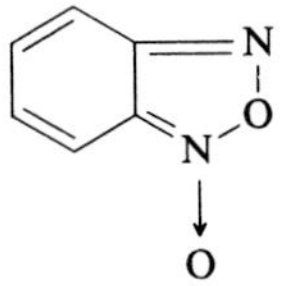

b43

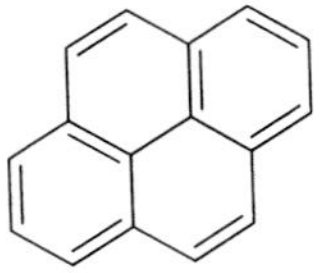

b52

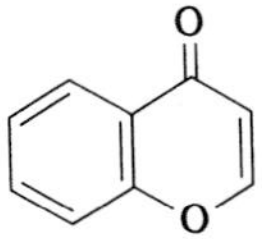

b56

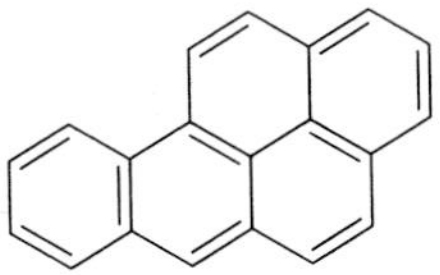

b57

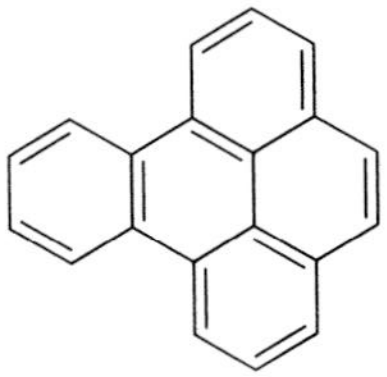

b58

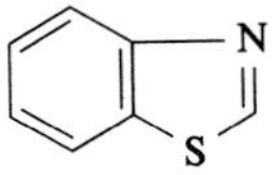

b60

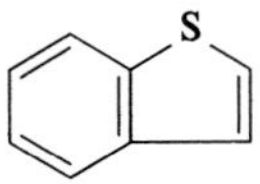

b61

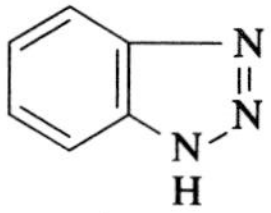

b62

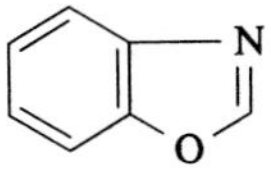

b63

TABLE 1.15 Physical Constants of Organic Compounds (*Continued*)

No.	Name	Formula	Formula weight	Beilstein reference	Density	Refractive index	Melting point	Boiling point	Flash point	Solubility in 100 parts solvent
b77	Benzyl acetate	$CH_3COOCH_2C_6H_5$	150.18	6, 435	1.0515^{25}	1.5006^{20}	-51.5	213.5	102 (CC)	sl s aq; misc alc, chl, eth
b78	Benzyl alcohol	$C_6H_5CH_2OH$	108.13	6, 428	1.0453^{20}	1.5403^{20}	-15.2	204.7	100 (CC)	0.08 aq; misc alc, eth
b79	Benzylamine	$C_6H_5CH_2NH_2$	107.16	12, 1013	0.9831^{9}	1.5401^{20}	10	185	60	misc aq, alc, eth
b80	2-Benzylaminoethanol	$C_6H_5CH_2NHCH_2CH_2OH$	151.21	12, 1040	1.065	1.5435^{20}		156^{12mm}	>110	
b81	3-(Benzylamino)-propanonitrile	$C_6H_5CH_2NHCH_2CH_2CN$	160.22		1.024	1.5308^{20}			>110	
b82	N-Benzylbenzamide	$C_6H_5CONHCH_2C_6H_5$	211.26	9, 121			106			i aq; misc alc, chl, eth
b83	Benzyl benzoate	$C_6H_5COOCH_2C_6H_5$	212.25	9², 471	1.118^{25}	1.5681^{21}	21	323–324	147	sl s aq; s alc, bz, chl, eth
b84	2-Benzylbenzoic acid	$C_6H_5CH_2C_6H_5COOH$	212.24				110–113			slowly dec aq
b85	Benzyl bromide	$C_6H_5CH_2Br$	171.04	5, 306	1.4382^{2}	1.5752^{20}	-3.9	198–199	86	
b85a	Benzyl 2-bromoacetate	$BrCH_2CO_2CH_2C_6H_5$	229.08	6¹, 220	1.446	1.5436^{20}		170^{22mm}	>110	
b86	Benzyl-*tert*-butanol	$C_6H_5CH_2C(CH_3)_2OH$	164.25	6, 548		1.5090^{20}	31–33	144^{85mm}	>110	
b87	Benzyl butyl 1,2-phthalate	$C_6H_5CH_2OOCC_6H_4COOC_4H_9$	312.37	9², 594	1.1192^{5}	1.5409^{20}			199	
b88	Benzyl carbamate	$C_6H_5CH_2OCONH_2$	151.17	6, 437			87–89	220 d		v s alc; sl s eth
b89	Benzyl chloride	$C_6H_5CH_2Cl$	126.59	5, 292	1.100^{20}	1.5381^{20}	-43	179	73	i aq; misc alc, chl, eth
b90	Benzyl chloroformate	$C_6H_5CH_2OC(O)Cl$	170.60	6, 437	1.195	1.5190^{20}		103^{20mm}	91	d aq; s eth
b91	Benzyl chlorothiol-formate	$C_6H_5CH_2S—COCl$	186.5		1.2370^{4}	1.5711^{30}		$80^{0.13mm}$	118	
b92	S-Benzyl-L-cysteine	$C_6H_5CH_2SCH_2CHCOOH$	211.28	6, 465			214 d			
b93	Benzyl diethyl phosphite	$C_6H_5CH_2P(O)(OC_2H_5)_2$	228.23		1.076	1.4930^{20}		110^{2mm}	>110	
b94	Benzyldimethylstearyl-ammonium chloride	$C_6H_5CH_2N[(CH_2)_{17}CH_3]-(CH_3)_2Cl \cdot H_2O$	442.18	12³, 2212			67–69			
b95	Benzyl ethyl ether	$C_6H_5CH_2OC_2H_5$	136.20		0.9478^{20}	1.4958^{20}		186		i aq; misc alc, eth
b96	N-Benzylformamide	$C_6H_5CH_2NHCHO$	135.17	12, 1043			60–61			
b97	Benzyl formate	$C_6H_5CH_2OOCH$	136.15	6, 440	1.0812^{4}			203		i aq; s alc; misc eth
b99	O-Benzylhydroxyl-amine	$C_6H_5CH_2ONH_2$	123.16					119^{30mm}		
b100	Benzylidenemalononi-trile	$C_6H_5CH=C(CN)_2$	154.17	9, 895			83–85			
b101	N-Benzylidenemeth-ylamine	$C_6H_5CH=NCH_3$	119.17	7, 213	0.967	1.5526^{20}		80^{18mm}	>112	

No.	Name	Formula	Formula wt	Beil. ref	Density	n_D	mp, °C	bp, °C	fp, °C	Solubility
b102	3-Benzylidenephtha-lide		222.24	17, 376			102			
b103	2-Benzyl-2-imidazo-line HCl		196.68				174			
b104	Benzylmethylamine	$C_6H_5CH_2NHCH_3$	138.23	12, 1019	0.939	1.5224^{20}		184–189	77	v s aq, alc; s chl; v sl s eth, EtAc
b105	3-(N-Benzyl-N-meth-ylamino)-1,2-pro-panediol	$C_6H_5CH_2N(CH_3)CH_2$-$CH(OH)CH_2OH$	195.26		1.084	1.5341^{20}		206^{30mm}	>110	
b106	Benzyl methyl sulfide	$C_6H_5CH_2SCH_3$	138.23	6, 453	1.015	1.5620^{20}		195–198	73	
b107	1-Benzyl-3-methyl-2-thiourea	$C_6H_5CH_2NHC(S)NHCH_3$	180.27	12, 1052			74–76			
b108	3-Benzyloxy-benzaldehyde	$C_6H_5CH_2OC_6H_4CHO$	212.25	8, 73			56–58			
b109	4-Benzyloxybenzalde-hyde	$C_6H_5CH_2OC_6H_4CHO$	212.25	8, 73			73–74			
b110	4-Benzyloxybenzyl al-cohol	$C_6H_5CH_2OC_6H_4CH_2OH$	214.26				86–87			
b111	2-Benzyloxyethanol	$C_6H_5CH_2OCH_2CH_2OH$	152.19		1.07^{20}_{20}			255.9	129	
b112	4-Benzyloxy-3-meth-oxybenzaldehyde	$C_6H_5CH_2OC_6H_4(OCH_3)CHO$	242.27				63–65			
b113	4'-Benzyloxypropio-phenone	$C_6H_5CH_2OC_6H_4COC_2H_5$	240.30				100–102			0.4 aq

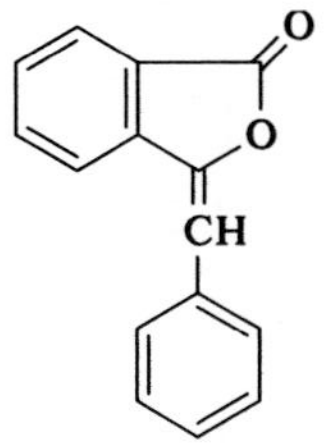

b102

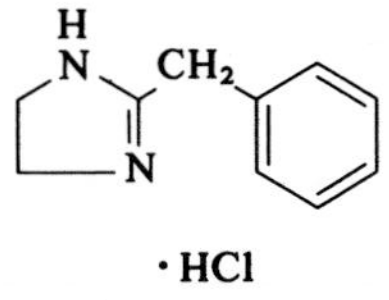

b103 · HCl

TABLE 1.15 Physical Constants of Organic Compounds (*Continued*)

No.	Name	Formula	Formula weight	Beilstein reference	Density	Refractive index	Melting point	Boiling point	Flash point	Solubility in 100 parts solvent
b114	Benzyl phenyl sulfide	$C_6H_5CH_2SC_6H_5$	200.30	6, 454			43	197^{27mm}	>110	i aq; sl s alc; s eth
b114a	DL-3-Benzylphthalide		224.26	17, 365			60–62			
b115	1-Benzylpiperazine		176.26	20, 296	1.014	1.5467^{20}		279	>110	s aq, alc, eth
b116	4-Benzylpiperidine		175.28		0.997	1.5379^{20}		134^{7mm}	>110	
b117	1-Benzyl-4-piperidone		189.26		1.021	1.5399^{20}	7		>110	
b118	2-Benzylpyridine	$C_6H_5CH_2{-}C_5H_4N$	169.23	20, 425	1.054	1.5785^{20}	10	276	125	i aq; v s alc, eth
b119	4-Benzylpyridine	$C_6H_5CH_2{-}C_5H_4N$	169.23	20, 426	1.061^{20}_{10}	1.5818^{20}		287	115	s alc; v s eth
b120	1-Benzyl-2-pyrrolidinone		175.23		1.095	1.5525^{20}			>110	
b121	(Benzylthio)acetic acid	$C_6H_5CH_2SCH_2COOH$	182.24				59–63			i aq; s alc; v s eth
b122	Benzyl thiocyanate	$C_6H_5CH_2SCN$	149.22	6, 460			43	235	>110	
b123	Benzyltributylammonium chloride	$C_6H_5CH_2N(C_4H_9)_3^+Cl^-$	312.94				155 d			
b124	Benzyltrichlorosilane	$C_6H_5CH_2SiCl_3$	225.57		1.288^4	1.526^{20}		142^{100mm}		
b125	Benzyltriethoxysilane	$C_6H_5CH_2Si(OC_2H_5)_3$	254.40		0.986^{20}_4			175^{70mm}		
b126	Benzyltriethylammonium chloride	$C_6H_5CH_2N(C_2H_5)_3^+Cl^-$	227.78				185 d			
b127	Benzyltrimethylammonium chloride	$C_6H_5CH_2N(CH_3)_3^+Cl^-$	185.70	12, 1020						
b128	Benzyltrimethylsilane	$C_6H_5CH_2Si(CH_3)_3$	164.32	4, 347	0.8933^{20}	1.4941^{20}				
b129	Betaine	$(CH_3)_3NCH_2COO^-$	117.15				dec > 310	191		160 aq; 55 MeOH; 6 alc
b130	Bicyclo[2.2.1]hepta-2,5-diene		92.14		0.909^{20}	1.4707^{20}	−20	89	−11	i aq; s PE
b131	Bicyclo[2.2.1]-2-heptene		94.16				46	96	−15	s eth
b132	Bicyclo[2.2.1]-5-heptene-2-carbaldehyde		122.16		1.018	1.4883^{20}		70^{12mm}	51	
b134	Biguanide	$H_2NC(=NH)NHC(=NH)NH_2$	101.11	3, 93			130	d 142		s aq, alc; i bz, eth
b135	Biphenyl	$C_6H_5{-}C_6H_5$	154.20	5, 578	0.9912^{75}	1.588^{77}	68.8	255.0	113 (CC)	i aq; s alc, eth
b136	4-Biphenylcarboxylic acid	$C_6H_5C_6H_4COOH$	198.22	9, 671			226	subl		i aq; s alc, eth; s bz

		Formula	M.W.	Beilstein ref.	b.p.	m.p.	Solubility
b137	(1,1'-Biphenyl)-4,4'-diamine	$H_2NC_6H_4C_6H_4NH_2$	184.23	13, 214	400^{740mm}	128	0.04 aq; s alc; 2 eth
b138	(1,1'-Biphenyl)-2,2'-dicarboxylic acid	$HOOCC_6H_4C_6H_4COOH$	242.23	9, 922		229	0.06 aq; s org solvents
b139	4-Biphenylmethanol	$C_6H_5C_6H_4CH_2OH$	184.24	6^2, 636		101	
b140	4-Biphenylsulfonic acid	$C_6H_5C_6H_4SO_3H$	234.26			138	
b141	2-(4-Biphenylyl)-5-(4-tert-butylphenyl)-1,3,4-oxadiazole		354.66			138	

Benzylphenol, h113
BES, b181
o,o-Bibenzoic acid, b138

Bibenzyl, d668
Bicine, b183
Bicyclo[4.4.0]decane, d1, d2

Bicyclo[4.3.0]nonane, h46
Biphenol, d388
Biphenylamines, a134, a135

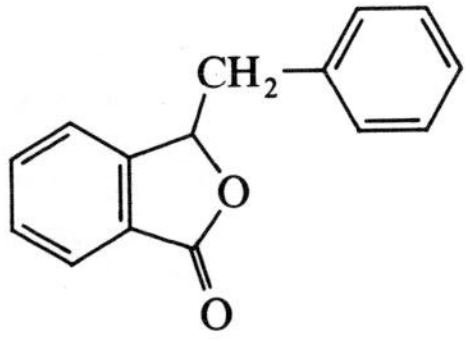

b114a

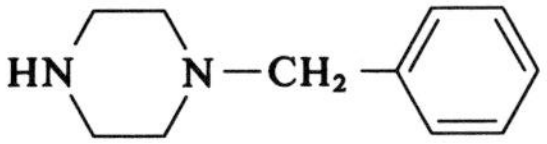

b115

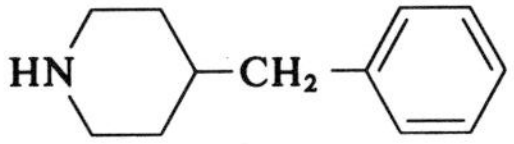

b116

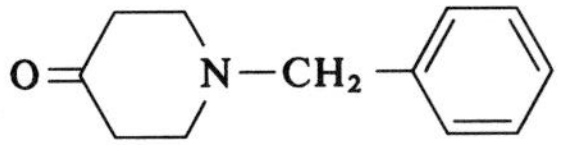

b117

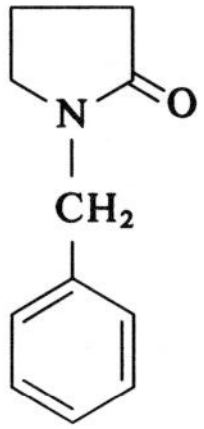

b120

b130

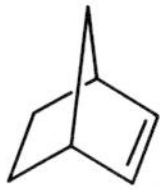

b131

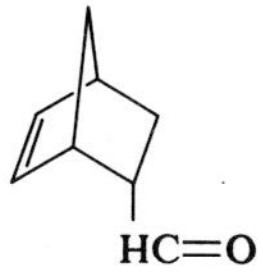

b132

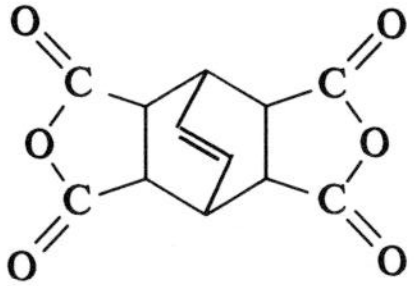

b133

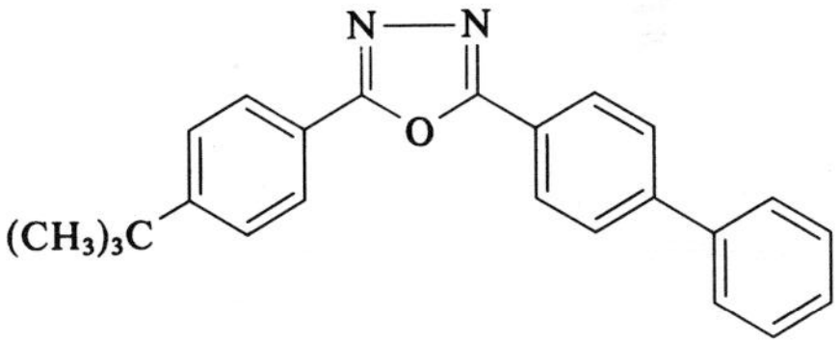

b141

TABLE 1.15 Physical Constants of Organic Compounds (*Continued*)

No.	Name	Formula	Formula weight	Beilstein reference	Density	Refractive index	Melting point	Boiling point	Flash point	Solubility in 100 parts solvent
b142	2-Biphenylyl glycidyl ether		226.28				30–32	$120^{0.1mm}$		
b143	2-(4-Biphenylyl)-5-phenyloxazole		197.36				118			
b144	2,2'-Bipyridinium chlorochromate	$C_5H_4N-C_5H_4NH^+$ $CrClO_3^-$	292.64							
b145	2,2-Bis[*p*-(allyloxy)phenyl]propane	$H_2C=CHCH_2OC_6H_4C-(CH_3)_2C_6H_4OCH_2-CH=CH_2$	308.42		1.022	1.5636^{20}			>110	
b146	*N,N'*-Bis(3-aminopropyl)ethylenediamine	$H_2N(CH_2)_3NHCH_2CH_2-NH(CH_2)_3NH_2$	174.29		0.952	1.4910^{20}		$118^{0.2mm}$	>110	
b147	*N,N'*-Bis(3-aminopropyl)piperazine		200.33	$23^3, 12$	0.973	1.5015^{20}	15	152^{2mm}	162	
b148	*N,N'*-Bis(3-aminopropyl)-1,3-propanediamine	$H_2N(CH_2)_3NHCH_2CH_2-CH_2NH(CH_2)_3NH_2$	188.32	$4^4, 1278$	0.920	1.4915^{20}		103^{1mm}		
b149	2,5-Bis(4-biphenylyl)oxazole		373.46				240			
b150	Bis(2-bromoethyl) ether	$BrCH_2CH_2OCH_2CH_2Br$	231.92					107^{20mm}		
b151	1,3-Bis(bromoethyl)-tetramethyldisiloxane	$[BrCH_2Si(CH_3)_2]_2O$	320.17		1.3918_4^{20}	1.4719^{20}		104^{15mm}		
b152	Bis(2-butoxyethyl) ether	$(C_4H_9OCH_2CH_2)_2O$	218.33		0.8853^{20}	1.4233^{20}	−60.2	254.6	47	0.3 aq; misc alc, eth, ketones, esters, CCl_4
b153	2,5-Bis(5-*tert*-butyl-2'-benzoxazolyl)-thiophene		430.57				201			
b154	Bis(*sec*-butyl) disulfide	$[CH_3CH_2CH(CH_3)]_2S_2$	178.36	$1^3, 1549$	0.957	1.4920^{20}		164^{739mm}	>112	
b155	Bis(*tert*-butyl) disulfide	$(CH_3)_3CSSC(CH_3)_3$	178.36	1, 379	0.909	1.4930^{20}		204	79	
b156	Bis(carboxymethyl) trithiocarbonate	$HOOCCH_2SC(=S)-SCH_2COOH$	226.29	3, 252			172–175			

b157	1,2-Bis(2-chloroethoxy)ethane	$(ClCH_2CH_2OCH_2{-})_2$	187.07	1[3], 2079	1.1972^{20}_4	1.4617^{20}		235	121	
b158	Bis(2-chloroethoxy)methylsilane	$H(CH_3)Si(OCH_2CH_2Cl)_2$	203.1		1.1643^{20}_4	1.4431^{20}		97^{18mm}		
b159	Bis(2-chloroethyl) ether	$ClCH_2CH_2OCH_2CH_2Cl$	143.01	1[2], 335	1.2220^{20}_{20}	1.4575^{20}	−51.7	178.5	55 (CC)	1 aq; s most org solv
b160	Bis(2-chloroethyl)-N-methylamine	$CH_3N(CH_2CH_2Cl)_2$	156.07		1.1184^{25}		−60	75^{10mm}		v sl s aq; misc most org solvents
b161	Bis(chloromethyl)dimethylsilane	$(CH_3)_2Si(CH_2Cl)_2$	157.12	4[3], 1845	1.0754^{20}_4	1.4600^{20}		160		
b162	Bis(2-chloro-1-methyl)ethyl ether	$ClCH_2CH(CH_3)OCH(CH_3)CH_2Cl$	171.07		1.1122^{20}_{20}			187.3	85	
b162a	1,3-Bis(chloromethyl)tetramethyldisiloxane	$ClCH_2Si(CH_3)_2OSi(CH_3)_2CH_2Cl$	231.3	4[3], 1864	1.050	1.4405^{20}		205	88	

3-(o-Biphenylyloxy)-1,2-epoxypropane, b142
2,2'-Bipyridine, d707

$OCH_2{-}CH{-}CH_2$ (with epoxide O)

b142

Bis(4-aminophenyl)ether, o61
1,3-Bis(aminomethyl)cyclohexane, c315

b143

1,2-Bis(benzylamino)ethane, d59
Bis(3-tert-butyl-4-hydroxy-5-methylphenyl) sulfide, t146

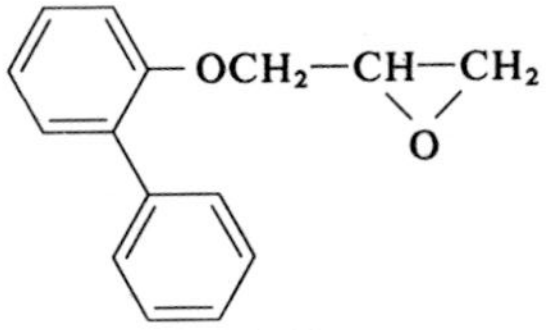

$H_2NCH_2CH_2CH_2{-}N$ (piperazine) $N{-}CH_2CH_2CH_2NH_2$

b147

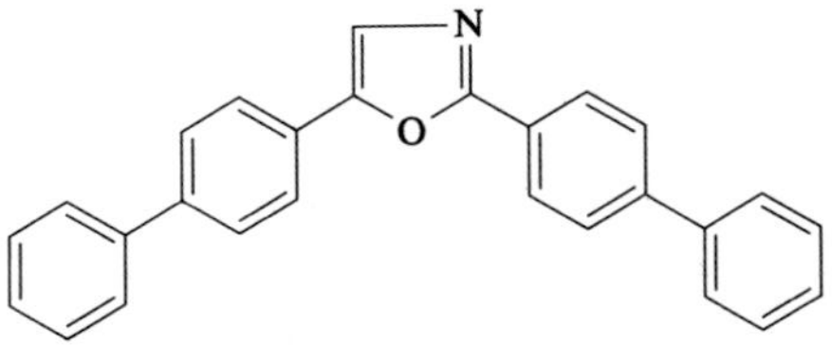

b149

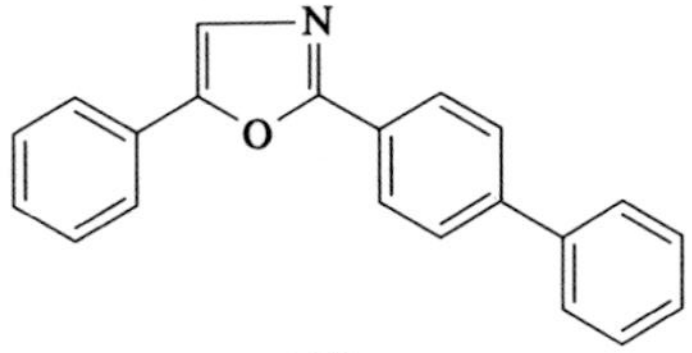

$(CH_3)_3C$ ... $C(CH_3)_3$

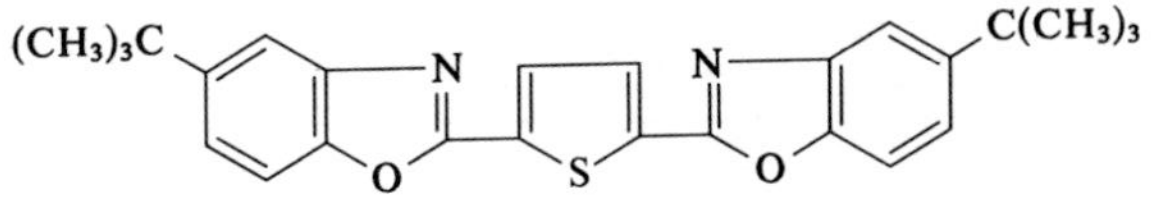

b153

TABLE 1.15 Physical Constants of Organic Compounds (*Continued*)

No.	Name	Formula	Formula weight	Beilstein reference	Density	Refractive index	Melting point	Boiling point	Flash point	Solubility in 100 parts solvent
b163	Bis(4-chlorophenoxy)acetic acid	(ClC$_6$H$_4$O)$_2$CHCOOH	313.14				142			
b164	2,2-Bis(4-chlorophenyl)-1,1-dichloroethane	(ClC$_6$H$_4$)$_2$CHCHCl$_2$	320.05	5[3], 1830			111			
b165	1,1-Bis(4'-chlorophenyl)ethanol	(ClC$_6$H$_4$)$_2$C(OH)CH$_3$	267.16	6[3], 3396			69			v sl s aq; s org solv
b166	Bis(4-chlorophenyl)sulfone	ClC$_6$H$_4$SO$_2$C$_6$H$_4$Cl	287.16	6, 327			145–148	250[10mm]		
b167	Bis(4-chlorophenyl)sulfoxide	ClC$_6$H$_4$S(O)C$_6$H$_4$Cl	271.17	6[1], 149			141–144			
b168	1,1-Bis(4-chlorophenyl)-2,2,2-trichloroethane	(ClC$_6$H$_4$)$_2$CHCCl$_3$	354.49	5[3], 1833			109			i aq; 58 acet; 78 bz; 45 CCl$_4$; v s pyr, diox
b169	1,3-Bis(dichloromethyl)tetramethyldisiloxane	[Cl$_2$CH(CH$_3$)$_2$Si]$_2$O	300.16		1.2213[20]	1.4660[20]		149[40mm]		
b170	N,N-Bis(2,2-diethoxyethyl)methylamine	[(C$_2$H$_5$O)$_2$CHCH$_2$]$_2$NCH$_3$	263.38	4, 311	0.945	1.4259[20]		222[244mm]	60	
b171	4,4'-Bis(diethylamino)benzophenone	[(C$_2$H$_5$)$_2$NC$_6$H$_4$]$_2$C=O	324.47	14, 98			95			
b172	4,4'-Bis(dimethylamino)benzophenone	[(CH$_3$)$_2$NC$_6$H$_4$]$_2$C(=O)	268.36	14, 89			172–176	d 360		i aq; s alc, warm bz
b173	Bis(dimethylamino)dimethylsilane	[(CH$_3$)$_2$N]$_2$Si(CH$_3$)$_2$	146.31		0.810[22]	1.432[22]	−98	129		
b174	1,3-Bis(dimethylamino)-2-propanol	[(CH$_3$)$_2$NCH$_2$]$_2$CHOH	146.23	4, 290	0.897	1.4422[20]			>110	
b174a	1,1-Bis(3,4-dimethylphenyl)ethane	[(CH$_3$)$_2$C$_6$H$_3$]$_2$CHCH$_3$	238.38	5[3], 1908	0.982	1.5665[20]		174[5mm]	>110	
b175	Bis(dimethylthiocarbamyl) disulfide	[(CH$_3$)$_2$NC(=S)S—]$_2$	240.43	4, 76	1.29		155–156			s alc, eth; sl s bz, acet; i aq

No.	Name	Formula	M.W.	Ref.	Density	n_D	M.P.	B.P.	Flash P.	Solubility
b176	1,4-Bis(2,3-epoxypropoxy)butane	$[H_2C{-}CHCH_2OCH_2CH_2{-}]_2$ O	202.25		1.049	1.4535^{20}		160^{11mm}	>110	v s aq, alc, org solv
b177	Bis(2-ethoxyethyl) ether	$(C_2H_5OCH_2CH_2)_2O$	162.23	1^2, 519	0.9072^4	1.4110^{20}	−44.3	188.4	54	
b177a	Bis(2-ethylhexyl)-amine	$[(CH_3(CH_2)_3CH(C_2H_5){-}CH_2]_2NH$	241.46	4^3, 388	0.805	1.4424^{20}		123^{5mm}	>110	
b178	Bis(2-ethylhexyl)-decanedioate	$CH_3(CH_2)_3CH(C_2H_5){-}CH_2OOC(CH_2)_8COOCH_2{-}CH(C_2H_5)(CH_2)_3CH_3$	426.66		0.9199^{25}	1.4496^{25}				
b179	Bis(2-ethylhexyl) hydrogen phosphate	$[CH_3(CH_2)_3CH(C_2H_5)CH_2O]_2{-}P(O)OH$	322.43	1^4, 1786	0.965	1.4450^{20}	−60	209^{10mm}	>110	
b179a	Bis(2-ethylhexyl) hydrogen phosphite	$[CH_3(CH_2)_3CH(C_2H_5){-}CH_2O]_2P(O)H$	306.43		0.916	1.4422^{20}			>110	
b180	Bis(2-ethylhexyl) o-phthalate	$[(CH_3(CH_2)_3CH(C_2H_5){-}CH_2OOC]_2C_6H_4$	390.57		0.9843^{20}	1.4859^{20}	−50	384	207	0.01 aq
b180a	Bis(4-fluorophenyl)-methane	$(FC_6H_4)_2CH_2$	204.22	5^3, 1789	1.145	1.5362^{20}	29−30	262	>110	
b180b	1,4-Bis(2-hydroxyethoxy)-2-butyne	$HOCH_2CH_2OCH_2C{\equiv}$ $CCH_2OCH_2CH_2OH$	174.20		1.144	1.4850^{20}			>110	
b181	N,N-Bis(2-hydroxyethyl)-2-amino-ethanesulfonic acid	$(HOC_2H_5)_2NCH_2CH_2SO_3H$	213.25				152−154			
b182	Bis(2-hydroxyethyl) ether	$HOCH_2CH_2OCH_2CH_2OH$	106.12	1, 468	1.1184^{20}	1.4460^{20}	−10.45	245	143	misc aq, alc, acet, eth
b183	N,N-Bis(2-hydroxyethyl)glycine	$(HOCH_2CH_2)_2NCH_2COOH$	163.17				192 sl d			sl s aq
b184	Bis(2-hydroxyethyl)-iminotris(hydroxymethyl)methane	$(HOCH_2CH_2)_2NC(CH_2OH)_3$	209.24				104			
b185	2,2-Bis(hydroxymethyl)propanoic acid	$(HOCH_2)_2C(CH_3)COOH$	134.13	3, 401			189−191			

TABLE 1.15 Physical Constants of Organic Compounds (*Continued*)

No.	Name	Formula	Formula weight	Beilstein reference	Density	Refractive index	Melting point	Boiling point	Flash point	Solubility in 100 parts solvent
b186	4,8-Bis(hydroxymethyl)tricyclo[5.2.0.0^{4,6}]decane		196.29	6^4, 5538		1.5280^{20}			>110	
b186a	4,4-Bis(4-hydroxyphenyl)pentanoic acid	$CH_3C(C_6H_4OH)_2CH_2CH_2COOH$	286.33				170			misc aq, alc
b186b	Bis(2-hydroxypropyl) ether	$HO(CH_2)_3O(CH_2)_3OH$	134.18	1^2, 537	1.0252$^{20}_{20}$	1.4407^{20}		231.8	137	
b187	Bis(2-mercaptoethyl) ether	$(HSCH_2CH_2)_2O$	138.25		1.114		−80	217		
b188	Bis(2-mercaptoethyl) sulfide	$(HSCH_2CH_2)_2S$	154.32		1.183	1.5982^{20}		136^{10mm}	90	
b189	1,4-Bis(methanesulfonoxy)butane	$(CH_3SO_2OCH_2CH_2-)_2$	246.30				115–117			sl hyd aq: 0.1 alc; 1.4 acet
b190	1,2-Bis(methoxyethoxy)ethane	$(CH_3OCH_2CH_2OCH_2-)_2$	178.23		0.990$^{20}_4$	1.4224^{20}	−45	216	110	misc aq
b191	Bis[2-(2-methoxyethoxy)ethyl] ether	$(CH_3OCH_2CH_2OCH_2CH_2-)_2O$	222.28	1^3, 2107	1.0087$^{20}_4$	1.4330^{20}	−27	275.3	140	s aq
b192	Bis(2-methoxyethyl) ether	$(CH_3OCH_2CH_2-)_2O$	134.18	1^2, 520	0.9440^{25}	1.4043^{25}	−64	162	70	misc aq
b193	Bis(2-methylallyl) carbonate	$[H_2C=C(CH_3)CH_2O]_2C(=O)$	170.21		0.943	1.4371^{20}		202	72	
b194	Bis(4-nitrophenyl) carbonate	$(O_2NC_6H_4O)_2C(=O)$	304.21	6^1, 120			141			
b195	Bis(3-nitrophenyl) disulfide	$O_2NC_6H_4SSC_6H_4NO_2$	308.33	6, 339			83			i aq; s alc, eth
b196	Bis(4-nitrophenyl) disulfide	$O_2NC_6H_4SSC_6H_4NO_2$	308.33	6, 340			181			sl s alc
b197	Bis(octadecyl)pentaerythritol diphosphite	$[C_{18}H_{37}OP(OCH_2)_2]_2$	721.01		0.925	1.457	40		261	
b198	Bis(pentabromophenyl) ether	$C_6Br_5OC_6Br_5$	969.22	6^1, 108			>300			
b199	1,4-Bis(5-phenyloxazol-2-yl)benzene		364.40				244			

No.	Name	Formula	Mol. wt.	Beilstein	Density	n_D	Melting point	Boiling point	Flash pt.	Solubility
b200	Bis(p-tolyl) disulfide	$CH_3C_6H_4SSC_6H_4CH_3$	246.39	6, 425			43–46			i aq; s alc; v s eth
b201	Bis(p-tolyl) sulfoxide	$CH_3C_6H_4S(O)C_6H_4CH_3$	230.33	6, 419			94–96			v s alc, bz, chl, eth
b202	Bis(tributyltin) oxide	$(C_4H_9)_3SnOSn(C_4H_9)_3$	596.08	5, 385	1.170	1.4864^{20}		180^{2mm}	>110	i aq; 26 acet; 38 bz;
b203	1,4-Bis(trichloromethyl)benzene	$Cl_3CC_6H_4CCl_3$	312.84				108–110			
b204	Bis(2,4,5-trichlorophenyl) disulfide	$Cl_3C_6H_2SSC_6H_2Cl_3$	425.01				140–144			
b205	1,2-Bis(trichlorosilyl)ethane	$Cl_3SiCH_2CH_2SiCl_3$	296.64		1.483^{20}_{4}	1.473^{20}	24.5	202		
b206	3,5-Bis(trifluoromethyl)aniline	$(F_3C)_2C_6H_3NH_2$	229.13		1.467	1.4335^{20}		85^{15mm}	83	
b207	1,3-Bis(trifluoromethyl)benzene	$F_3CC_6H_4CF_3$	214.0	5^3, 834	1.3790^{25}	1.3916^{25}		116	26	
b208	N,O-Bis(trimethylsilyl)acetamide	$CH_3C{=}N-Si(CH_3)_3$ $\;\;\;\;O-Si(CH_3)_3$	203.43		0.8332^{20}_{4}	1.4170^{20}		73^{35mm}	11	
b209	Bis(trimethylsilyl)acetylene	$(CH_3)_3SiC{\equiv}CSi(CH_3)_3$	170.41		0.7702^{20}_{4}	1.427^{20}		137	2	
b210	Bis(trimethylsilyl)formamide	$HC{=}NSi(CH_3)_3$ $\;\;\;\;O-Si(CH_3)_3$	189.41		0.885	1.4381^{20}		55^{13mm}		
b211	N,O-Bis(trimethylsilyl)hydroxylamine	$(CH_3)_3SiONHSi(CH_3)_3$	177.40		0.830	1.4112^{20}		80^{100mm}	28	
b212	1,2-Bis(trimethylsilyloxy)ethane	$(CH_3)_3SiOCH_2CH_2OSi(CH_3)_3$	206.43		0.842	1.4034^{20}		166	46	

Bis(4-hydroxyphenyl) sulfide, t153

4,4-Bis(4-hydroxyphenyl)valeric acid, b186a

b186

1,2-Bis-2-methoxyethoxy)ethane, t276b

Bis(phenylmethyl) disulfide, d57

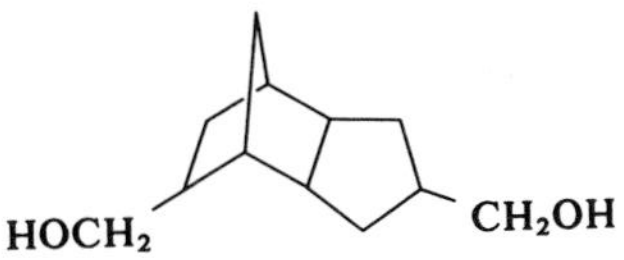

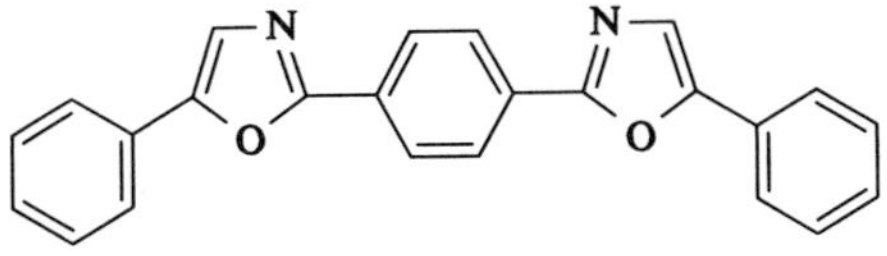

b199

TABLE 1.15 Physical Constants of Organic Compounds (*Continued*)

No.	Name	Formula	Formula weight	Beilstein reference	Density	Refractive index	Melting point	Boiling point	Flash point	Solubility in 100 parts solvent
b213	N,O-Bis(trimethylsilyl)trifluoroacetamide	$F_3C[=NSi(CH_3)_3]OSi(CH_3)_3$	257.40		0.969	1.3839^{20}	−10	50^{14mm}	23	
b214	1,3-Bis(trimethylsilyl)urea	$(CH_3)_3SiNHCONHSi(CH_3)_3$	204.4				232 d			
b215	1,3-Bis[tris(hydroxymethyl)methylamino]propane	$CH_2[CH_2NHC(CH_2OH)_3]_2$	282.34	4^3, 859			170			
b216	Biuret	$H_2NCONHCONH_2$	103.08	3, 70	1.467^{-5}_{4}		110	d 190		v s alc; 2 aq
b217	1-Borneol		154.25	6, 72	1.011^{20}_{4}		204	212	65	i aq; 176 alc; s eth bz, PE
b218	1-Bornyl acetate		196.29	6, 82	0.982	1.4626	27	224	84	sl s aq; s alc, eth
b219	N-Bromoacetamide	$CH_3CONBrH$	137.97	2, 181			102–105			sl s aq; v s eth
b220	4'-Bromoacetanilide	$BrC_6H_4NHCOCH_3$	214.07	12, 642	1.717		168			v s bz, chl, EtAc
b221	Bromoacetic acid	$BrCH_2COOH$	138.95	2, 213	1.934^{50}_{4}	1.4804^{50}	50	208	>110	v s aq, alc, eth
b222	2-Bromoacetophenone	$C_6H_5COCH_2Br$	199.05	7, 283	1.647^{20}_{4}		50	135^{18mm}	>110	i aq; v s alc, bz, chl, eth
b223	4'-Bromoacetophenone	$BrC_6H_4COCH_3$	199.05	7, 283	1.647		54	255	>110	s alc, bz, eth, HOAc
b224	Bromoacetyl bromide	$BrCH_2COBr$	201.86	2, 215	2.3172^{22}_{2}	1.5480^{20}		150	none	d aq, alc
b225	Bromoacetyl chloride	$BrCH_2COCl$	157.40	2, 215	1.908	1.4960^{20}		128	none	d aq, alc
b226	2-Bromoaniline	$BrC_6H_4NH_2$	172.03	12, 631	1.578^{20}_{4}	1.6113^{20}	31	229	>110	i aq; s alc, eth
b227	3-Bromoaniline	$BrC_6H_4NH_2$	172.03	12, 633	1.580^{20}_{4}	1.6250^{20}	16.8	251	>110	sl s aq; s alc, eth
b228	4-Bromoaniline	$BrC_6H_4NH_2$	172.03	12, 636	1.4970^{100}_{4}		66.3			i aq; v s alc, eth
b229	3-Bromobenzaldehyde	BrC_6H_4CHO	185.03	7, 238	1.587	1.5935^{20}		230	96	i aq; v s alc, eth
b230	Bromobenzene	C_6H_5Br	157.02	5, 206	1.4952^{20}_{4}	1.5602^{20}	−30.6	156.2	51	0.044 aq; 10.4 alc; misc bz, chl, PE; 71.6 eth
b231	Bromobenzene-*d₅*	C_6D_5Br	162.06					53^{23mm}		
b232	4-Bromobenzenesulfonyl chloride	$BrC_6H_4SO_2Cl$	255.52	11, 57			74.5	153^{15mm}	51	i aq; d alc; v s eth
b233	4-Bromobenzoic acid	BrC_6H_4COOH	201.02	9, 351	1.929^{25}		251–253			
b234	4-Bromobenzophenone	$BrC_6H_4COC_6H_5$	261.12	7, 422			82		350	0.18 aq^{25}; s alc, eth
b235	2-Bromobenzotrifluoride	$BrC_6H_4CF_3$	225.01		1.652^{20}	1.4817^{20}		168	51	i aq; sl s bz, eth

	Name	Formula				n_D				
b236	3-Bromobenzotrifluoride	$BrC_6H_4CF_3$	225.01		1.613	1.4749^{20}		152	43	s aq, alc, bz, eth, CS_2
b237	3-Bromobenzoyl chloride	BrC_6H_4COCl	219.47	9, 350	1.662	1.5965^{20}		$75^{0.5\text{mm}}$	107	i aq; s alc, bz, eth
b238	4-Bromobenzyl bromide	$BrC_6H_4CH_2Br$	249.94	5, 308		1.6193^{20}	61	$124^{12\text{mm}}$		i aq; s alc, bz, eth
b239	4-Bromobiphenyl	$BrC_6H_4C_6H_5$	233.11	5, 580	0.9327^{20}		87	310		<0.1 aq; vs alc, eth
b240	1-Bromobutane	$CH_3CH_2CH_2CH_2Br$	137.02	1, 119	1.2686^{25}_{4}	1.4374^{25}	−112.4	101.6	23	
b241	2-Bromobutane	$CH_3CH_2CHBrCH_3$	137.03	1, 119	1.2530^{25}_{4}	1.4360^{20}	−112.4	91	21	
b242	1-Bromo-2-butene	$CH_3CH{=}CHCH_2Br$	135.01	1, 205	1.312	1.4765^{20}		99	11	i aq; s alc, eth
b243	2-Bromo-2-butene	$CH_3CH{=}C(Br)CH_3$	135.01	1, 205	1.328	1.4613^{20}		$90^{740\text{mm}}$	1	
b244	4-Bromo-1-butene	$BrCH_2CH_2CH{=}CH_2$	135.01	1, 84	1.3230^{20}_{4}	1.4608^{20}		100	9	
b245	4-Bromobutyl phenyl ether	$C_6H_5OCH_2CH_2CH_2CH_2Br$	229.12	6^{2}, 82			41–42	$156^{18\text{mm}}$		
b246	2-Bromobutyric acid	$CH_3CH_2CH(Br)COOH$	167.01	2, 281	1.5669^{20}_{20}	1.4720^{20}	−4	$103^{10\text{mm}}$	>110	6.7 aq; s alc, eth
b247	endo-(+)-3-Bromocamphor		231.14	7, 120	1.449		76–78	244		i aq; 15 alc, 200 chl; 62 eth
b248	2-Bromo-4'-chloroacetophenone	$ClC_6H_4COCH_2Br$	233.50	7, 285			96.5			
b249	2-Bromochlorobenzene	BrC_6H_4Cl	191.46	5, 209	1.6382^{25}	1.5789^{25}		204	79	i aq; v s bz
b250	3-Bromochlorobenzene	BrC_6H_4Cl	191.46	5, 209	1.6302^{20}_{4}	1.5771^{20}	−21	196	80	i aq; v s alc, eth
b251	4-Bromochlorobenzene	BrC_6H_4Cl	191.46	5, 209	1.576^{1}	1.5531^{70}	64.5	196		0.1 aq; misc MeOH, eth

H_3C CH_3 CH_3 OH
b217

H_3C CH_3 CH_3 $OCCH_3$ O
b218

H_3C CH_3 Br H_3C O
b247

TABLE 1.15 Physical Constants of Organic Compounds (*Continued*)

No.	Name	Formula	Formula weight	Beilstein reference	Density	Refractive index	Melting point	Boiling point	Flash point	Solubility in 100 parts solvent
b252	5-Bromo-2-chlorobenzotrifluoride	$Br(Cl)C_6H_3CF_3$	259.47		1.743	1.5062^{20}			81	
b253	1-Bromo-4-chlorobutane	$ClCH_2CH_2CH_2CH_2Br$	171.47	5^3, 294	1.488	1.4875^{20}		82^{30mm}	60	i aq; s alc, chl, eth
b254	4-Bromo-6-chloro-o-cresol	$Br(Cl)C_6H_2(OH)CH_3$	221.49	6, 360			47		>110	
b255	Bromochlorodifluoromethane	$Br(Cl)CF_2$	165.4		1.83^{21}		−160.5	−4.01	none	
b256	1-Bromo-2-chloroethane	$ClCH_2CH_2Br$	143.43	1, 89	1.7392^{20}_{4}	1.4917^{20}	−18.4	106.6	none	0.7 aq; misc org solv
b257	7-Bromo-5-chloro-8-hydroxyquinoline		258.51	21^1, 222			177–179			
b258	Bromochloromethane	$ClCH_2Br$	129.39	1, 67	1.9234^{25}	1.480^{25}	−88	67.8	none	0.9 aq; misc MeOH, eth
b258a	1-Bromo-3-chloro-2-methylpropane	$ClCH_2CH(CH_3)CH_2Br$	171.47	1^3, 346	1.408	1.4836^{20}		212	95	
b259	1-Bromo-3-chloropropane	$ClCH_2CH_2CH_2Br$	157.44	1, 109	1.472	1.486^{20}	<−50	143.5	none	0.1 aq; misc org solv
b260	2-Bromo-2-chloro-1,1,1-trifluoroethane	$HC(Br)ClCF_3$	197.39	1^4, 156	1.8636^{25}	1.3738^{25}		50.2	none	
b261	2-Bromocinnamaldehyde	$C_6H_4CH=C(Br)CHO$	211.06	7, 358			66–68			
b262	Bromocycloheptane	BrC_7H_{13}	177.09	5, 29	1.2887^{22}	1.5052^{20}		72^{10mm}	68	i aq; v s chl, eth
b263	Bromocyclohexane	BrC_6H_{11}	163.06	5, 24	1.3264^{15}	1.4956^{15}		165.8	62	0.1 aq; 10 MeOH; 71 eth
b264	3-Bromocyclohexene	BrC_6H_9	161.04	5^2, 40	1.3890^{20}	1.5292^{20}		65^{15mm}	35	
b265	Bromocyclopentane	BrC_5H_9	149.04	5, 19	1.3900^{20}	1.4881^{20}		139	−6	
b266	Bromocyclopropane	BrC_3H_5	120.98		1.510	1.4605^{20}		69		
b267	1-Bromodecane	$CH_3(CH_2)_9Br$	221.19	1^2, 130	1.0658^{20}	1.4560^{20}	−30	238	94	i aq; v s chl, eth
b268	Bromodichloromethane	$HCBrCl_2$	163.83	1, 67	1.980^{20}	1.4964^{20}	−55	87	none	sl s aq; misc org solv
b269	2-Bromo-1,1-diethoxyethane	$BrCH_2CH(OC_2H_5)_2$	197.08	1, 625	1.310	1.4385^{20}		67^{18mm} 180 d	51	s hot alc
b270	4-Bromo-1,2-dimethoxybenzene	$BrC_6H_3(OCH_3)_2$	217.07	6, 784	1.702	1.5743^{20}		256	109	

No.	Name	Formula	Formula wt.	Beilstein ref.	Density	n_D	mp, °C	bp, °C	Flash pt, °C	Solubility
b271	1-Bromo-2,2-dimethoxypropane	$CH_3C(OCH_3)_2CH_2Br$	183.05		1.355	1.4475^{20}		87^{780mm}	40	
b272	4-Bromo-2,6-dimethylphenol	$BrC_6H_2(CH_3)_2OH$	201.07	6, 485			78		75	
b272a	3-Bromo-2,2-dimethyl-1-propanol	$BrCH_2C(CH_3)_2CH_2OH$	167.05	1^1, 201	1.358	1.4794^{20}		187		v s hot alc, hot acet
b273	2-Bromo-4,6-dinitroaniline	$BrC_6H_2(NO_2)_2NH_2$	262.02	12, 761			154	subl		
b274	3-Bromo-4,6-dinitrofluorobenzene	$BrC_6H_2(NO_2)_2F$	264.9				90–91			
b275	2-Bromo-2,2-diphenylacetyl bromide	$BrC(C_6H_5)_2COBr$	354.05	9^1, 283			63–65			
b276	1-Bromodiphenylmethane	$C_6H_5CH(Br)C_6H_5$	247.14	5, 592			40	184^{20mm}	>110	0.1 aq; s alc, eth
b277	1-Bromododecane	$CH_3(CH_2)_{11}Br$	249.24	1^2, 133	1.038	1.4580^{20}	−9	135^{6mm}	110	i aq; sl s alc; s eth
b278	1-Bromo-2,3-epoxypropane	$H_2C\!-\!CHCH_2Br$ (—O— epoxide)	136.98	17, 9	1.601^{20}	1.4820^{20}	−40	134–136	56	0.91 aq
b279	Bromoethane	CH_3CH_2Br	108.97	1, 88	1.451^{20}	1.4276^{15}	−118.6	38.4	−23	
b280	2-Bromoethanesulfonic acid, sodium salt	$BrCH_2CH_2SO_2^-Na^+$	211.02	4, 7			283–285 d			
b281	2-Bromoethanol	$BrCH_2CH_2OH$	124.97	1, 338	1.7629^{24}	1.4920^{20}	−13.8	150	40	misc aq; s org solv
b282	2-Bromoethyl acetate	$CH_3COOCH_2CH_2Br$	167.01	21, 57	1.5144^{20}	1.4547^{20}		159	71	v s aq; misc alc, eth
b283	2-Bromoethylamine HBr	$BrCH_2CH_2NH_2\cdot HBr$	204.90	4, 134			172–174			v s aq, alc
b285	(2-Bromoethyl)benzene	$C_6H_5CH_2CH_2Br$	185.07	5, 356	1.355	1.5563^{20}		221	89	i aq; s bz, eth

2-Bromo-p-cumene, b301
β-Bromocumene, b300

4-Bromodiphenyl ether, b333

Bromoethene, b286

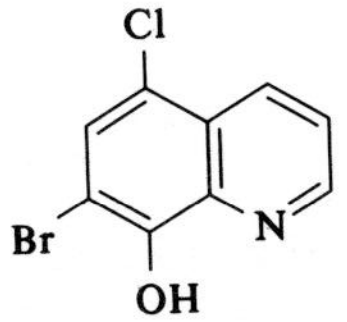

b257

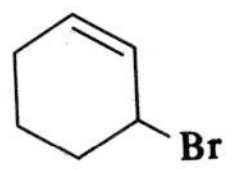

b264

TABLE 1.15 Physical Constants of Organic Compounds (*Continued*)

No.	Name	Formula	Formula weight	Beilstein reference	Density	Refractive index	Melting point	Boiling point	Flash point	Solubility in 100 parts solvent
b286	Bromoethylene	$H_2C{=}CHBr$	106.96	1, 188	1.493^{20}	1.4350^{20}	−139.5	15.8	none	i aq; misc alc, eth
b287	2-Bromoethyl ethyl ether	$BrCH_2CH_2OCH_2CH_3$	153.02	1, 338	1.3572^{20}_4	1.4450^{20}		150	21	sl s aq; misc alc, eth
b287a	1-Bromo-2-ethylhexane	$CH_3(CH_2)_3CH(C_2H_5)CH_2Br$	193.13	1^3, 475	1.086	1.4538^{20}		77^{16mm}	69	
b288	2-Bromoethyl phenyl ether	$BrCH_2CH_2OC_6H_5$	201.07	6, 142			34	144^{40mm}	65	i aq; v s alc, eth
b289	N-(2-Bromoethyl)-phthalimide		254.09	21, 461			81–84			s hot aq; v s eth
b290	1-Bromo-2-fluoro-benzene	BrC_6H_4F	175.01		1.601	1.5337^{20}		156	43	
b291	1-Bromo-3-fluoro-benzene	BrC_6H_4F	175.01		1.567	1.5257^{20}		150	38	
b292	1-Bromo-4-fluoro-benzene	BrC_6H_4F	175.01	5, 209	1.593^{15}	1.5310^{15}	−17.4	152	60	
b293	1-Bromoheptane	$H(CH_2)_7Br$	179.11	1, 155	1.1384^{20}_4	1.4505^{20}	−58	180	60	i aq; v s alc, eth
b294	2-Bromoheptane	$H(CH_2)_2CH(Br)CH_3$	179.11	1, 155	1.142	1.4470^{20}		66^{21mm}	47	
b295	1-Bromohexadecane	$H(CH_2)_{16}Br$	305.35	1^2, 138	0.9991	1.4618^{20}	17.8	336	177	i aq; misc org solv
b296	1-Bromohexane	$H(CH_2)_6Br$	165.08	1, 144	1.1763^{20}_4	1.4475	−85	154–158	57	i aq; misc alc, eth
b297	DL-2-Bromohexanoic acid	$CH_3(CH_2)_3CH(Br)COOH$	195.06	2, 325	1.370	1.4720^{20}		138^{18mm}	>110	
b298	5-Bromoisatin		226.03	21, 453			251–253			
b300	(2-Bromoisopropyl)-benzene	$C_6H_5CH(CH_3)CH_2Br$	199.10	5^1, 191	1.316	1.5480^{20}		108^{18mm}	91	
b301	2-Bromo-4-isopropyl-1-methylbenzene	$CH_3(Br)C_6H_3CH(CH_3)_2$	213.0		1.2535^{25}	1.535^{25}	−20	120		i aq; 50 MeOH; misc org solv
b302	Bromomaleic anhydride		176.96	17, 435	1.905	1.5400^{20}		215	>110	
b303	Bromomethane	CH_3Br	94.94	1, 67	1.732^{20}	1.4234^{10}	−94	3.56	none	0.1 aq; s alc, chl, eth
b304	2-Bromo-1-methoxy-benzene	$BrC_6H_4OCH_3$	187.04	6, 197	1.5018^{20}_4	1.5737^{20}	2	223	96	i aq; v s alc, eth
b305	3-Bromo-1-methoxy-benzene	$BrC_6H_4OCH_3$	187.04	6, 198	1.477	1.5635^{20}	211	93		i aq; s alc, eth
b306	4-Bromo-1-methoxy-benzene	$BrC_6H_4OCH_3$	187.04	6, 199	1.4564^{20}_4	1.5630^{20}	10	223	94	sl s aq; v s alc, eth

No.	Name	Formula	Mol wt	Beil. ref	Density	n_D	mp	bp	fp	Solubility
b307	4-Bromo-2-methyl-aniline	CH₃(Br)C₆H₃NH₂	186.06	12, 838			56	240	>110	sl s aq; v s alc
b308	1-Bromo-3-methyl-butane	(CH₃)₂CHCH₂CH₂Br	151.05	1, 136	1.2101^{15}	1.4409^{20}	−112	119.7	32	0.02 aq; misc alc, eth
b308a	(Bromomethyl)chlorodimethylsilane	BrCH₂Si(CH₃)₂Cl	187.55	44, 4024	1.375	1.4650^{20}		130^{740mm}	41	
b309	(Bromomethyl)cyclohexane	C₆H₁₁CH₂Br	177.09	5^2, 18	1.269	1.4907^{20}		77^{26mm}	57	
b310	2-Bromomethyl-1,3-dioxalane		167.01	19^2, 8	1.613	1.4817^{20}		82^{27mm}	62	
b311	Bromomethyl methyl ether	BrCH₂OCH₃	124.97	1, 582	1.531	1.4550^{20}		87	26	
b312	1-Bromo-2-methyl-naphthalene	BrC₁₀H₆CH₃	221.10	5, 568	1.418	1.6484^{20}		296	>110	0.06 aq; misc alc, eth
b313	1-Bromo-2-methyl-propane	(CH₃)₂CHCH₂Br	137.03	1, 126	1.2641^{20}	1.4362^{20}	−119	91.5	18	i aq; misc org solv
b314	2-Bromo-2-methyl-propane	(CH₃)₃CBr	137.03	1, 127	1.215^{25}_{25}	1.425^{25}	−16.2	73.1	18	
b314a	2-Bromo-2-methyl-propionyl bromide	(CH₃)₂C(Br)COBr	229.91	2, 297	1.860	1.5064^{20}		164	>110	
b314b	2-Bromo-2-methyl-propiophenone	C₆H₅COC(CH₃)₂Br	227.11	7, 316	1.350	1.5561^{20}		148^{30mm}	>112	
b315	1-Bromonaphthalene	C₁₀H₇Br	207.08	5, 547	1.4834^{20}	1.6580^{20}	−1	281.1	>110	misc alc, bz, chl, eth
b316	1-Bromo-1-naphthol	BrC₁₀H₆OH	223.07	6, 650			78	130 d		i aq; s alc, bz, eth
b317	1-Bromo-2-nitro-benzene	BrC₆H₄NO₂	202.01	5^1, 247	1.6245^{80}_4		43	261	>110	v s alc; s bz, eth

Bromoform, t210

α-Bromoisobutyrophenone, b299

2-Bromomesitylene, b365

(Bromomethyl)benzene, b85

2-Bromo-2-methylpropanoyl bromide, b314a

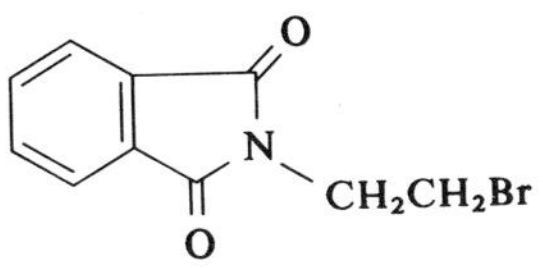

b289

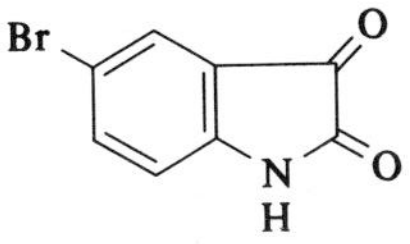

b298

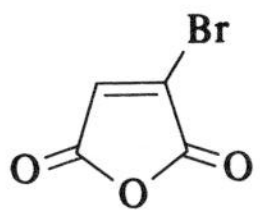

b302

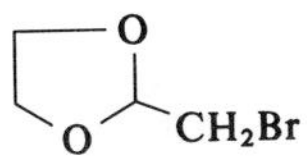

b310

TABLE 1.15 Physical Constants of Organic Compounds (*Continued*)

No.	Name	Formula	Formula weight	Beilstein reference	Density	Refractive index	Melting point	Boiling point	Flash point	Solubility in 100 parts solvent
b318	5-Bromo-2-nitro-benzotrifluoride	$O_2N(Br)C_6H_3CF_3$	270.02		1.7922^{25}	1.5180^{25}	40–44	100		
b319	2-Bromo-2-nitro-1,3-propanediol	$(HOCH_2)_2C(Br)NO_2$	199.99	1, 476			133			
b320	1-Bromononane	$H(CH_2)_9Br$	207.16	1^1, 63	1.084	1.4540^{20}		201	90	i aq; s chl, eth
b321	*exo*-2-Bromonorbor-nane		175.07		1.363	1.5148^{20}		82^{29mm}	60	
b322	1-Bromooctadecane	$H(CH_2)_{18}Br$	333.41	1^1, 69	0.976	1.4503^{25}	23	216^{12mm}	>110	i aq; s alc, eth
b323	1-Bromooctane	$H(CH_2)_8Br$	193.13	1, 160	1.1082^{25}_{4}	1.4490^{20}	−55	201	78	i aq; misc alc, eth
b324	Bromopentafluoro-benzene	BrC_6F_5	246.97		1.947^{20}		−31	137	87	
b325	1-Bromopentane	$H(CH_2)_5Br$	151.05	1, 131	1.2237^{15}	1.4444^{20}	−88	129.6	31	i aq; s alc; misc eth
b326	2-Bromopentane	$CH_3CH_2CH_2CH(Br)CH_3$	151.05	1, 131	1.2039^{20}	1.4403^{20}		117	20	
b327	9-Bromophenanthrene		257.14	5, 671	1.409^{101}		54–58	190^{2mm}	<110	i aq; s alc, eth
b328	2-Bromophenol	BrC_6H_4OH	173.01	6, 197	1.492	1.5892^{20}	6	194	42	s aq; misc chl, eth
b329	4-Bromophenol	BrC_6H_4OH	173.01	6, 198	1.5875^{80}		68	238		14 aq; v s alc, chl
b330	1-(4-Bromophenoxy)-1-ethoxyethane	$CH_3CH(OC_6H_4Br)OC_2H_5$	245.12		1.348	1.5229^{20}		125^{8mm}	106	
b331	4-Bromophenylacetic acid	$BrC_6H_4CH_2COOH$	215.05	9, 451			119			sl s aq; v s alc, eth
b332	4-Bromophenyl-acetonitrile	$BrC_6H_4CH_2CN$	196.05	9, 451			47–49		>110	i aq; sl s alc; v s bz
b333	4-Bromophenyl phenyl ether	$BrC_6H_4OC_6H_5$	249.11	6^1, 105	1.423	1.6070^{20}	18	305	>110	
b334	1-Bromo-3-phenyl-propane	$C_6H_5CH_2CH_2CH_2Br$	199.10	5, 391	1.310	1.5450^{20}		220	101	
b335	1-Bromopropane	$CH_3CH_2CH_2Br$	123.00	1, 356	1.3597^{15}	1.4370^{15}	−110.1	71.0	25	0.23 aq^{30}, misc alc
b336	2-Bromopropane	$CH_3CH(Br)CH_3$	123.00	1, 108	1.3222^{15}	1.4285^{15}	−89.0	59.5	19	0.3 aq^{18}, misc alc, bz, chl, eth
b337	3-Bromo-1-propanol	$BrCH_2CH_2CH_2OH$	139.00	1, 356	1.5374^{20}_{4}	1.4858^{20}	−116	62^{5mm}	93	s aq; misc alc, eth
b338	1-Bromo-1-propene	$CH_3CH{=}CHBr$	120.98	1, 200	1.4133^{20}_{4}	1.4538^{20}		63	4	i aq
b339	2-Bromo-1-propene	$CH_3C(Br){=}CH_2$	120.98	1, 200	1.362^{20}_{4}	1.4425^{20}	−125	49	4	
b340	2-Bromopropionic acid	$CH_3CH(Br)COOH$	152.98	2, 254	1.7000^{20}	1.4750^{20}	25.7	203	100	v s aq, alc, eth
b341	3-Bromopropionic acid	$BrCH_2CH_2COOH$	152.98	2, 256	1.480		62.5		65	s aq, alc, bz, chl, eth
b342	3-Bromopropionitrile	$BrCH_2CH_2CN$	133.98	2^2, 231	1.6152^{20}_{4}	1.4800^{20}		78^{10mm}	98	v s alc, eth

No.	Name	Formula	Formula wt	Beilstein	Density	n_D	m.p., °C	b.p., °C		Solubility
b342a	2-Bromopropionyl bromide	$CH_3CH(Br)COBr$	215.88	2, 256	2.061	1.5182^{20}		50^{10mm}	>110	d aq; s chl, eth
b343	2-Bromopropionyl chloride	$CH_3CH(Br)COCl$	171.43	2, 256	1.700^{11}	1.4800^{20}		133	51	
b344	3-Bromopropionyl chloride	$BrCH_2CH_2COCl$	171.43	2^2, 231	1.701	1.4968^{20}		57^{17mm}	79	s alc, bz, eth, acet
b345	2-Bromopropiophenone	$C_6H_5COCHBrCH_3$	213.08	7, 302	1.430^{20}	1.5715^{20}		250	>110	
b346	3-Bromopropyl phenyl ether	$C_6H_5OCH_2CH_2CH_2Br$	215.10	6, 142	1.365	1.5464^{20}	10–11	134^{14mm}	96	
b347	3-Bromopropyne	$BrCH_2C{\equiv}CH$	118.97	1, 248	1.335	1.4905^{20}		88–90	18	i aq; s org solv
b348	2-Bromopyridine	BrC_5H_4N	158.00	20, 233	1.657^{18}	1.5720^{20}		194	54	s aq; v s alc, eth
b349	4-Bromopyridine	BrC_5H_4N	158.00	20, 233	1.645^{0}	1.5695^{20}	142–143	173	51	s HOAc
b350	3-Bromoquinoline		208.06	20, 363	1.533	1.6640^{20}	15	276	>110	0.3 aq^{80}; 85 alc^{25}; 70 eth^{25}
b351	5-Bromosalicylic acid	$Br(HO)C_6H_3COOH$	217.02	10, 107			166			i aq; misc alc, eth
b352	β-Bromostyrene	$C_6H_5CH{=}CHBr$	183.05	5, 477	1.422^{20}	1.6066^{20}	7	112^{20mm}	79	18 aq; s alc
b353	Bromosuccinic acid	$HOOCCH_2CH(Br)COOH$	196.99	2, 621	2.073		172 (d)			1.5 aq; 14.4 acet; 3.1 HOAc; 0.02 CCl_4
b354	N-Bromosuccinimide		177.99	21, 380	2.098		173 sl dec			s alc; v s chl; misc bz, acet
b355	1-Bromotetradecane	$H(CH_2)_{14}Br$	277.30	1^2, 136	1.0124^{25}	1.4600^{20}	6	178^{20mm}	>110	
b356	2-Bromothiophene	$Br{-}C_4H_3S$	163.04	17, 33	1.684^{20}	1.5860^{20}		151	60	v acet, eth
b357	4-Bromothiophenol	BrC_6H_4SH	189.08	6, 330			76	239		

b321

b327

b350

b354

TABLE 1.15 Physical Constants of Organic Compounds (*Continued*)

No.	Name	Formula	Formula weight	Beilstein reference	Density	Refractive index	Melting point	Boiling point	Flash point	Solubility in 100 parts solvent
b358	2-Bromotoluene	$BrC_6H_4CH_3$	171.04	5, 304	1.4222^{25}	1.552^{25}	−26	181	78	0.1 aq; misc alc, bz, chl, eth
b359	3-Bromotoluene	$BrC_6H_4CH_3$	171.04	5, 305	1.4099^{20}	1.5517^{20}	−39.8	183.7	60	s alc, bz, eth
b360	4-Bromotoluene	$BrC_6H_4CH_3$	171.04	5, 305	1.3959^{35}	1.5490	28.5	184.5	85	s alc, bz, eth
b361	Bromotrichloromethane	$BrCCl_3$	198.28	1, 67	1.9972^{25}	1.5063	−21	103.8	none	misc org solv
b362	1-Bromotridecane	$H(CH_2)_{13}Br$	263.27	1^2, 134	1.0262^{20}	1.4592^{20}	7	$150^{10\,mm}$	>110	v s chl
b363	Bromotrifluoromethane	$BrCF_3$	148.92	1^3, 83	1.5800^{20}		−168	−57.8		v s chl
b364	5-Bromo-1,2,4-trimethylbenzene	$BrC_6H_2(CH_3)_3$	199.10	5, 403			73	235		i aq; s alc
b365	2-Bromo-1,3,5-trimethylbenzene	$BrC_6H_2(CH_3)_3$	199.10	5, 408	1.301	1.5511^{20}	2	225	96	i aq; s bz, v s eth
b366	Bromotrimethylgermane	$(CH_3)_3GeBr$	197.60		1.544^{18}	1.4705^{20}	−25	113.7	37	v s alc, eth
b367	Bromotrimethylsilane	$(CH_3)_3SiBr$	153.10		1.160	1.4145^{20}		79	32	v s alc, eth
b368	Bromotriphenylethylene	$(C_6H_5)_2C{=}C(Br)C_6H_5$	335.22	5, 722			114–115			
b369	Bromotriphenylmethane	$(C_6H_5)_3CBr$	323.24	5, 704			152–154	$230^{15\,mm}$		
b370	11-Bromoundecanoic acid	$Br(CH_2)_{10}COOH$	265.20	2^2, 315			51	$174^{2\,mm}$	>110	i aq; v s alc
b371	α-Bromo-o-xylene	$BrCH_2C_6H_4CH_3$	185.07	5, 365	1.381^{23}	1.5742^{20}	21	223–224	82	s alc, eth
b372	α-Bromo-m-xylene	$BrCH_2C_6H_4CH_3$	185.07	5, 374	1.370^{23}	1.5560^{20}		$185^{340\,mm}$	82	s alc, eth
b373	2-Bromo-p-xylene	$BrC_6H_3(CH_3)_2$	185.07	5, 385	1.340	1.5505^{20}	9–10	199–201	79	v s alc, eth
b374	4-Bromo-o-xylene	$BrC_6H_3(CH_3)_2$	185.07	5, 365	1.3701^{5}	1.5560^{20}		215	80	v s alc, eth
b375	Brucine		394.45	27^2, 797		1.4205^{1}	178			77 alc; 1 bz; 20 chl
b376	1,2-Butadiene	$CH_3CH{=}C{=}CH_2$	54.09	1, 249	0.6761^{0}		−136.2	10.9		misc alc, eth
b377	1,3-Butadiene	$H_2C{=}CHCH{=}CH_2$	54.09	1, 249	0.6504^{-6}	1.4293^{-25}	−108.9	−4.4		misc alc, eth
b378	1,3-Butadienyl acetate	$CH_3C({=}O)OCH{=}CHCH{=}CH_2$	112.13	23, 295	0.945	1.4690^{20}		$60^{40\,mm}$	33	
b379	1,3-Butadiyne	$HC{\equiv}CC{\equiv}CH$	50.06	1^3, 1056	0.7364^{0}	1.4189^{5}	−36	10.3	−19	v s eth; s bz, acet
b380	2-Butanamine	$CH_3CH_2CH(NH_2)CH_3$	73.14	4, 160	0.7308^{15}	1.3963^{15}	−104.5	66		misc aq, alc
b381	Butane	$CH_3CH_2CH_2CH_3$	58.12	1, 118	0.6011^{0}	1.3562^{-13}	−138.3	−0.50	−60 (CC)	15 mL aq; 1800 mL alc

No.	Name	Formula	Mol. wt.		Density	n_D	mp (°C)	bp (°C)		Solubility
b383	Butanedinitrile	NCCH$_2$CH$_2$CN	80.09	2, 615	0.9867^{64}_{0}	1.4173^{60}	57.9	265–267	93	11.5 aq; s acet, chl, diox; sl s bz, eth
b384	1,2-Butanediol	CH$_3$CH$_2$CH(OH)CH$_2$OH	90.12	1, 477	1.006^{18}	1.4380^{20}	<−50	207.5	121	s aq, alc, acet
b385	1,3-Butanediol	CH$_3$CH(OH)CH$_2$CH$_2$OH	90.12	1, 477	1.0053^{20}	1.441^{20}	20.9	207.5	121 (OC)	s aq, alc, acet; 9 eth
b386	1,4-Butanediol	HOCH$_2$CH$_2$CH$_2$CH$_2$OH	90.12	1, 478	1.016^{25}	1.4452^{20}		230		misc aq, alc, acet; 0.3 bz; 3.1 eth; 0.9 PE
b387	meso-2,3-Butanediol	CH$_3$CH(OH)CH(OH)CH$_3$	90.12	1, 479	0.9939^{25}	1.4324^{35}	34.4	182	85	misc aq, alc
b388	1,3-Butanediol diacetate	CH$_3$CO$_2$CH$_2$CH$_2$CH(CH$_3$)O$_2$CCH$_3$	174.20	2, 143	1.028	1.4199^{20}		99^{8mm}	85	
b388a	1,4-Butanediol dimethacrylate	[H$_2$C=C(CH$_3$)CO$_2$CH$_2$CH$_2$—]$_2$	202.25	2⁴, 1534	1.023	1.4565^{20}		134^{4mm}	110	25 aq; misc alc, eth
b389	2,3-Butanedione	CH$_3$C(O)C(O)CH$_3$	86.09	1, 769	0.990^{15}	1.3951^{20}		88	26	i aq; v s alc
b390	1,4-Butanedithiol	HSCH$_2$CH$_2$CH$_2$CH$_2$SH	122.25	1, 479	1.042	1.5290^{20}		106^{30mm}	70	
b390a	1,2,3,4-Butanetetracarboxylic acid	[—CH(CO$_2$H)CH$_2$CO$_2$H]$_2$	234.16	2, 863			196			
b391	1-Butanethiol	CH$_3$CH$_2$CH$_2$CH$_2$SH	90.19	1, 370	0.8367^{25}	1.4403^{25}	−115.7	98.5	12	0.06 aq; v s alc, eth
b392	2-Butanethiol	CH$_3$CH$_2$CH(SH)CH$_3$	90.19	1, 373	0.8246^{25}	1.4338^{25}	−165	85.0	21	sl s aq; v s alc, eth
b393	1,2,4-Butanetriol	HOCH$_2$CH$_2$CH(OH)CH$_2$OH	106.12	1, 519	1.018^{20}	1.4748^{20}		191^{18mm}	167	v s aq, alc
b394	1-Butanol	CH$_3$CH$_2$CH$_2$CH$_2$OH	74.12	1, 367	0.8097^{20}	1.3993^{20}	−88.6	117.7	35	7.4 aq; misc alc, eth
b395	2-Butanol	CH$_3$CH$_2$CH(OH)CH$_3$	74.12	1, 371	0.8069^{20}	1.3972^{20}	−114.7	99.5	31 (OC)	12.5 aq; misc alc, eth

b375

TABLE 1.15 Physical Constants of Organic Compounds (*Continued*)

No.	Name	Formula	Formula weight	Beilstein reference	Density	Refractive index	Melting point	Boiling point	Flash point	Solubility in 100 parts solvent
b396	2-Butanone	$CH_3CH_2COCH_3$	72.11	1^2, 726	0.8049_4^{20}	1.3788^{20}	-86.7	79.6	11 (CC)	24 aq; misc alc, bz, eth
b398	1-Butene	$CH_3CH_2CH{=}CH_2$	56.10	1^3, 715	0.6255^{-185}	1.3962^{20}	-185.3	-6.3	-80	i aq; v s alc, eth
b399	*cis*-2-Butene	$CH_3CH{=}CHCH_3$	56.10	1^3, 728	0.6213^{20}	1.3931^{-25}	-139.3	3.7	-73	i aq; v s alc, eth
b400	*trans*-2-Butene	$CH_3CH{=}CHCH_3$	56.10	1^3, 730	0.6041^{20}	1.3848^{-25}	-105.8	0.3	-73	i aq; v s alc, eth
b401	*cis*-2-Butene-1,4-diol	$HOCH_2CH{=}CHCH_2OH$	88.11	1^2, 567	1.0700_4^{20}	1.4779^{20}	12.5	234	128	s aq; v s alc
b402	*trans*-2-Butene-1,4-diol	$HOCH_2CH{=}CHCH_2OH$	88.11	1^3, 2252	1.070^{20}	1.4779^{20}	27.3	132		v s aq, alc
b403	3-Butenenitrile	$H_2C{=}CHCH_2CN$	67.09	2, 408	0.8341_4^{20}	1.4060^{20}	-87	119	21	sl s aq; misc alc, eth
b404	*cis*-2-Butenoic acid	$CH_3CH{=}CHCOOH$	86.09	2, 412	1.0267_4^{20}	1.4483^{14}	14	169		v s aq; s alc
b405	*trans*-2-Butenoic acid	$CH_3CH{=}CHCOOH$	86.09	2, 408	0.9648_4^{0}	1.4228^{77}	71.4	185.0	87	54.6 aq; v s EtOH, bz, acet
b406	3-Butenoic acid	$H_2C{=}CHCH_2COOH$	86.09	2, 407	1.0091_4^{20}	1.4249^{20}	-39	163	65	s aq; misc alc, eth
b407	*cis*-2-Buten-1-ol	$CH_3CH{=}CHCH_2OH$	72.11	1, 442	0.8662^{20}	1.4342^{20}		123.6	56	16.6 aq; misc alc
b408	*trans*-2-Buten-1-ol	$CH_3CH{=}CHCH_2OH$	72.11	1, 442	0.8454^{20}	1.4289^{20}		121.2	56	16.6 aq; misc alc
b409	3-Buten-2-one	$H_2C{=}CHCOCH_3$	70.09	1, 728	0.8636^{20}	1.4086^{20}	-89.4	81.4	-6	v s aq, alc, acet, eth
b410	1-Buten-3-yne	$HC{\equiv}CCH{=}CH_2$	52.07	1^3, 1032	0.7095^{1}	1.4161		5.1		
b411	4-Butoxyaniline	$CH_3(CH_2)_3OC_6H_4NH_2$	165.24	13^3, 226	0.992	1.5343^{20}		149^{13mm}	>110	
b412	4-Butoxybenzoic acid	$CH_3(CH_2)_3OC_6H_4COOH$	194.23	10^3, 93			150			
b413	2-Butoxyethanol	$CH_3(CH_2)_3OCH_2CH_2OH$	118.18	1^2, 519	0.9012_4^{20}	1.4198^{20}	-40	170.2	60	5 aq; s most org solv
b413a	1-*tert*-Butoxy-2-ethoxyethane	$(CH_3)_3COCH_2CH_2OC_2H_5$	146.23	1^3, 2085	0.834	1.4015^{20}		148	33	
b414	2-(2-Butoxyethoxy)-ethanol	$HOCH_2CH_2OCH_2CH_2OC_4H_9$	162.23	1^2, 521	0.9536_{20}^{20}	1.4306^{20}	-68.1	230.4	100	misc aq, alc, bz, acet, PE, CCl$_4$
b414a	2-(2-Butoxyethoxy)-ethyl acetate	$CH_3CO_2(CH_2CH_2O)_2{-}CH_2CH_2CH_2CH_3$	204.27	2^3, 308	0.978	1.4260^{20}		245	110	
b414b	2-Butoxyethyl acetate	$CH_3CO_2CH_2CH_2O(CH_2)_3CH_3$	160.22	2^3, 307	0.942	1.4136^{20}		192	76	
b414c	2-*tert*-Butoxy-2-methoxyethane	$(CH_3)_3COCH_2CH_2OCH_3$	132.20	1^3, 2084	0.840	1.3985^{20}		132	25	
b415	Butyl acetate	$C_4H_9OOCCH_3$	116.16	2, 130	0.8813_4^{20}	1.3941^{20}	-73.5	126.1	37 (TCC)	0.43 aq; misc alc, eth; s most org solv
b416	DL-*sec*-Butyl acetate	$CH_3COOCH(CH_3)C_2H_5$	116.16	2^2, 141	0.865^{25}	1.3840^{25}		112.3	32	0.62 aq; s alc, eth
b417	*tert*-Butyl acetate	$(CH_3)_3COOCCH_3$	116.16	2, 131	0.8665_4^{20}	1.3853^{20}		97.8	15	i aq; misc alc, eth
b417a	N-*tert*-Butylacetoacetamide	$CH_3COCH_2CONHC(CH_3)_3$	157.21				49–51	$75^{0.04mm}$	110	

No.	Name	Formula	Formula wt.	Beilstein ref.	Density	n_D	m.p., °C	b.p., °C	Flash pt.	Solubility
b418	*tert*-Butyl acetoacetate	$(CH_3)_3COC(=O)CH_2C(=O)CH_3$	158.20		0.954	1.4180^{20}			60	i aq; s alc, eth
b419	Butyl acrylate	$H_2C=CHCOOC_4H_9$	128.17	2^2, 388	0.894^{16}	1.4160^{20}		148	38	misc alc, bz, eth
b419a	*tert*-Butyl acrylate	$H_2C=CHCO_2C(CH_3)_3$	128.17	2^3, 1228	0.875	1.4108^{20}		63^{60mm}	17	misc alc, bz, eth
b420	Butylamine	$CH_3CH_2CH_2CH_2NH_2$	73.14	4, 156	0.7327^{25}	1.3992^{25}	−50.5	77.9	−12 (OC)	misc aq, alc, eth, PE
b420a	*sec*-Butylamine	$C_2H_5CH(NH_2)CH_3$	73.14	4, 160	0.7308^{15}	1.3928^{20}	−72	63	−19	misc aq, alc
b421	*tert*-Butylamine	$(CH_3)_3CNH_2$	73.14	4, 173	0.6951^{20}	1.3788^{20}	−67.5	44.4	−8	misc aq, alc
b422	2-(*tert*-Butylamino)-ethanol	$(CH_3)_3CNHCH_2CH_2OH$	117.19				42–45	92^{25mm}	68	
b423	3-(*tert*-Butylamino)-1,2-propanediol	$(CH_3)_3CNHCH_2CH(OH)CH_2OH$	147.22				70	92^{1mm}		
b424	4-Butylaniline	$CH_3CH_2CH_2CH_2C_6H_4NH_2$	149.24	12^1, 503	0.945	1.5350^{20}	100	120^{15mm}	101	
b425	2-*tert*-Butylanthraquinone		264.32							
b426	Butylbenzene	$CH_3CH_2CH_2CH_2C_6H_5$	134.22	5, 413	0.8604^{20}_4	1.4898^{20}	−88.5	183.1	71 (OC)	misc alc, bz, eth
b427	*sec*-Butylbenzene	$CH_3CH_2CH(CH_3)C_6H_5$	134.22	5, 414	0.8608^{25}	1.4902^{20}	−82.7	173.5	52 (CC)	misc alc, bz, eth
b428	*tert*-Butylbenzene	$(CH_3)_3CC_6H_5$	134.22	5, 415	0.8669^{24}	1.4927^{20}	−57.9	169.1	60 (OC)	misc alc, bz, eth
b429	Butyl benzoate	$C_6H_5COOC_4H_9$	178.23	9, 112	1.000^{20}	1.496	−22	250	106 (OC)	i aq; s alc, eth
b430	4-*tert*-Butylbenzoic acid	$(CH_3)_3CC_6H_4COOH$	178.23	9, 560	1.1142^{20}		166.3			i aq; v s alc, bz
b431	4-*tert*-Butylbenzoyl chloride	$(CH_3)_3CC_6H_4COCl$	196.68	12, 1022	1.007	1.5364^{20}		135^{20mm}	87	
b432	*N*-(*tert*-Butyl)benzylamine	$C_6H_5CH_2NHC(CH_3)_3$	163.27		0.881	1.4968^{20}		80^{5mm}	80	
b432a	Butyl 2-butoxy-2-hydroxyacetate	$CH_3(CH_2)_3OCH(OH)CO_2(CH_2)_3CH_3$	204.27	3^4, 1497	0.996	1.4291^{20}		90^{40mm}	74	

b425

C(CH₃)₃

TABLE 1.15 Physical Constants of Organic Compounds (*Continued*)

No.	Name	Formula	Formula weight	Beilstein reference	Density	Refractive index	Melting point	Boiling point	Flash point	Solubility in 100 parts solvent
b433	Butyl butyrate	$CH_3CH_2CH_2COOC_4H_9$	144.21	2, 271	0.8692^{20}	1.4064^{20}		165	53 (OC)	i aq; misc alc, eth
b434	tert-Butyl carbazate	$H_2NNHCO_2C(CH_3)_3$	132.16				39–42	$65^{0.03mm}$	91	
b435	4-tert-Butylcatechol	$(CH_3)_3CC_6H_3(OH)_2$	166.22		1.049^{25}_{25}		55	285	151	0.2 aq[80], 240 eth[25]; s alc; v s acet
b436	tert-Butyl chloroacetate	$ClCH_2COOC(CH_3)_3$	150.61	2^3, 444	1.053	1.4230^{20}		49^{11mm}	46	
b437	4-tert-Butyl-1-chloro-benzene	$(CH_3)_3CC_6H_4Cl$	168.67	5, 416	1.006	1.5108^{20}	23–25	217		
b438	tert-Butylchlorodi-phenylsilane	$(CH_3)_3CSi(C_6H_5)_2Cl$	274.87		1.057	1.5675^{20}		$90^{0.02mm}$	>110	
b439	Butyl chloroformate	$ClCOOC_4H_9$	136.58	3^2, 11	1.074^{25}	1.4114^{20}		142	25	d aq, alc; misc eth
b440	Butyl cyanoacetate	$NCCH_2COOCH_2CH_2CH_2CH_3$	141.17	2^1, 255	0.993	1.4254^{20}		115^{15mm}	87	
b441	tert-Butyl cyanoacetate	$NCCH_2COOC(CH_3)_3$	141.17		0.972	1.4200^{20}		108	91	
b442	2-tert-Butylcyclohex-anol	$(CH_3)_3C{-}C_6H_{10}OH$	156.27	6^3, 126	0.902		46		79	i aq
b443	4-tert-Butylcyclohex-anol	$(CH_3)_3C{-}C_6H_{10}OH$	156.27	6^1, 18			70	115^{15mm}	105	i aq
b444	2-tert-Butylcyclohex-anone	$(CH_3)_3C{-}C_6H_9({=}O)$	154.25	7^3, 143	0.896	1.4565^{20}		62.5^{4mm}	72	
b445	4-tert-Butylcyclohex-anone	$(CH_3)_3C{-}C_6H_9({=}O)$	154.25	7^1, 29			50	116^{20mm}	96	i aq
b446	Butyl decyl o-phthalate	$C_4H_9OOCC_6H_4CO_2C_{10}H_{21}$	362.51	4, 285	0.9942^{25}	1.4625^{20}		276	202	
b447	N-Butyldiethanol-amine	$C_4H_9N(CH_2CH_2OH)_2$	161.25		0.9862^{20}		−70		126	
b448	Butyl 3,4-dihydro-2,2-dimethyl-4-oxo-2H-pyran-6-carboxylate		226.27		1.054^{25}	1.4767^{20}		256–270	>110	misc alc, chl, eth
b449	tert-Butyldimethyl-chlorosilane	$(CH_3)_3CSi(CH_3)_2Cl$	150.7				91.5	124–126	22	
b450	6-tert-Butyl-2,4-di-methylphenol	$(CH_3)_3CC_6H_2(CH_3)_2OH$	178.28	6^3, 2020		1.5178^{20}	23	249	111	
b451	N-Butylethanolamine	$HOCH_2CH_2NHC_4H_9$	117.19		0.89^{20}	1.444^{20}	−3.5	192	77	i aq; misc alc, eth
b452	Butyl ethyl ether	$C_4H_9OC_2H_5$	102.18	1^3, 1502	0.7495^{20}	1.3818^{20}	−124	92	−5	0.8 aq
b453	2-Butyl-2-ethyl-1,3-propanediol	$HOCH_2C(C_2H_5)(C_4H_9)CH_2OH$	160.25		0.9315^{20}	1.4587^{25}	41	178^{50mm}	>110	
b454	Butyl ethyl sulfide	$C_4H_9SC_2H_5$	118.24	1^3, 1522	0.8376^{20}_{4}	1.4491^{20}	−95.1	144.2		s chl

No.	Name	Formula	Mol wt	Ref	Density	n_D	mp	bp	18 (CC)	Solubility
b454a	Butyl formate	$HCO_2CH_2CH_2CH_2CH_3$	102.13	2, 21	0.892	1.3889^{20}		107	55	
b455	Butyl glycidyl ether	$C_4H_9OCH_2CH{-}CH_2$ (O)	130.19	17^3, 988	0.910	1.4192^{20}		166		
b455a	tert-Butyl glycidyl ether	$(CH_3)_3COCH_2CH{-}CH_2$ (O)	130.19	17^3, 988	0.917	1.4166			43	s aq, alc, chl, eth
b456	tert-Butyl hydroperoxide	$(CH_3)_3C{-}O{-}OH$	90.12	1^3, 1579	0.896^{20}_4	1.4007^{20}	4–5	34^{17mm}	36	
b457	tert-Butylhydroquinone	$(CH_3)_3CC_6H_3(OH)_2$	166.22				129			
b458	Butyl isocyanate	$CH_3CH_2CH_2CH_2NCO$	99.13	4, 175	0.880	1.4061^{20}		115	26	i aq; misc alc, eth
b459	tert-Butyl isocyanate	$(CH_3)_3CNCO$	99.13	2^3, 1286	0.868	1.3865^{20}		86	26	s aq; v s alc, eth
b462	Butyl methacrylate	$H_2C{=}C(CH_3)COOC_4H_9$	142.19	1, 381	0.889^{25}	1.4220^{25}		170	49	i aq; s org solv
b463	tert-Butyl methyl ether	$(CH_3)_3COCH_3$	88.15		0.758	1.3685^{20}	−109	56	−10	
b464	2-tert-Butyl-4-methyl-phenol	$(CH_3)_3CC_6H_3(CH_3)OH$	164.25	6^2, 507	0.9247^{75}	1.4969^{75}	51.7	237	100	
b465	2-tert-Butyl-5-methyl-phenol	$(CH_3)_3CC_6H_3(CH_3)OH$	164.25		0.964	1.5192^{20}		118^{12mm}	105	
b466	tert-Butyl methyl sul-fide	$C_4H_9SCH_3$	104.21	1^3, 1591	0.826^{20}_4	1.441^{20}	−97.8	102	−3	v s alc

b448

TABLE 1.15 Physical Constants of Organic Compounds (*Continued*)

No.	Name	Formula	Formula weight	Beilstein reference	Density	Refractive index	Melting point	Boiling point	Flash point	Solubility in 100 parts solvent
b467	Butyl nitrite	C_4H_9ONO	103.12	1, 369	0.9114_4	1.3768		78	−13	misc alc, eth
b468	tert-Butyl nitrite	$(CH_3)_3CONO$	103.12	1, 382	0.8671^{20}	1.3687^{20}		63	−13	sl s aq, v s alc, chl, eth, CS_2
b469	Butyl octadecanoate	$CH_3(CH_2)_{16}COOC_4H_9$	340.60	2^2, 352	0.8551^{20}	1.4422^{25}	26.3	343	160	s alc, v s acet
b470	Butyl 4-oxopentanoate	$CH_3C(=O)CH_2CH_2COOC_4H_9$	172.22		0.9735^{20}	1.4270^{20}		107^{6mm}	91	s alc, eth, acet
b470a	4-(1-Butylpentyl)pyridine	$(C_4H_9)—CHC(CH_2)_3CH_3$ C_5H_4N	205.35	20^3, 2872	0.887	1.4877^{20}			110	
b471	tert-Butyl peroxybenzoate	$C_6H_5C(=O)O—OC(CH_3)_3$	194.23		1.021	1.4990^{20}		$76^{0.2mm}$	93	
b472	2-sec-Butylphenol	$CH_3CH_2CH(CH_3)C_6H_4OH$	150.22	6^2, 489	0.982	1.5222^{20}	12	228	112	i aq; v s alc
b473	2-tert-Butylphenol	$(CH_3)_3CC_6H_4OH$	150.22		0.9783^{20}	1.5228^{20}	−7	224	110	v s eth
b474	3-tert-Butylphenol	$(CH_3)_3CC_6H_4OH$	150.22				40–41	240	108	
b475	4-sec-Butylphenol	$CH_3CH_2CH(CH_3)C_6H_4OH$	150.22	6, 522	0.9692^{20}	1.5150	62	136^{25mm}	115	s hot aq, alc, eth
b476	4-tert-Butylphenol	$(CH_3)_3CC_6H_4OH$	150.22	6, 524	0.9084_4^{114}	1.4787^{114}	100–101	237		i aq; s alc, eth
b479	tert-Butyl phenyl carbonate	$C_6H_5OC(=O)OC(CH_3)_3$	194.23		1.047	1.4805^{20}		$79^{0.8mm}$	101	
b480	Butyl phenyl ether	$CH_3CH_2CH_2CH_2OC_6H_5$	150.22	6, 143	0.9351_4^{20}	1.4970^{20}	−19	210.3	82 (OC)	0.1 aq; 79 alc; 153 EtAc; 158 toluene
b481	4-tert-Butylphenyl salicylate	$HOC_6H_4COOC_6H_4C(CH_3)_3$	270.31				62–64			misc alc, eth
b482	Butyl propanoate	$CH_3CH_2COOC_4H_9$	130.19	2, 241	0.8818^{15}	1.3982^{25}	−89.6	145.5	38	
b483	4-tert-Butylpyridine	$(CH_3)_3C—C_5H_4N$	135.21	20, 252	0.915	1.4952^{20}		197	63	
b483a	1-Butylpyrrolidine	$(C_4H_8N)—CH_2CH_2CH_2CH_3$	127.23	20^2, 4	0.814	1.4440^{20}		93^{10mm}	36	
b484	Butyltin chloride	$C_4H_9SnCl_3$	282.17		1.693				81	
b485	4-tert-Butyltoluene	$(CH_3)_3CC_6H_4CH_3$	148.25	5, 439	0.853	1.5229^{20}		192	54	
b486	tert-Butyltrichlorosilane	$C_4H_9SiCl_3$	191.56	4^3, 1905		1.4897^{20}	100	134	40	
b487	Butyl trifluoroacetate	$CF_3COOC_4H_9$	170.1		1.0268^{22}	1.3532^{22}		100.2		
b488	Butyltrimethoxysilane	$C_4H_9Si(OCH_3)_3$	178.3		0.9312^{20}	1.3979^{20}		164–165		
b489	tert-Butyl trimethyl-silyl peroxide	$(CH_3)_3C—O—O—Si(CH_3)_3$	162.3		0.8219^{24}	1.3935^{25}	d 135	41^{41mm}		
b490	Butylurea	$C_4H_9NHCONH_2$	116.16	4^1, 371			93–95			s aq, alc, eth
b491	Butyl vinyl ether	$C_4H_9OCH=CH_2$	100.16		0.7792^{20}	1.4007^{20}	−112.7	94.2	−9	0.3 aq
b492	5-tert-Butyl-m-xylene	$(CH_3)_3CC_6H_3(CH_3)_2$	162.28	5, 447	0.867	1.4946^{20}		206	72	374 aq; 83 alc; 0.04
b495	2-Butyne-1,4-diol	$HOCH_2C≡CCH_2OH$	86.09	1^1, 261		1.450^{25}	54–58	238	152	bz; 2.6 eth; 70 acet

b496	Butyraldehyde	$CH_3CH_2CH_2CHO$	72.11	1, 662	0.8016^{20}	1.3791^{20}	-96.4	74.8	-6 (CC)	7.1 aq; misc alc, eth, acet, EtAc
b497	Butyramide	$CH_3CH_2CH_2CONH_2$	87.12	2, 275			116	216	77 (CC)	16 aq; s alc
b498	Butyric acid	$CH_3CH_2CH_2COOH$	88.11	2, 264	0.9582^{20}	1.3980^{20}	-5.3	163.3	88 (OC)	misc aq, alc, eth
b499	Butyric anhydride	$[CH_3CH_2CH_2C(O)]_2O$	158.20	2, 274	0.9668^{20}	1.4130^{20}	-65.7	199.5	60	s aq, alc(dec), eth
b500	3-Butyrolactone		86.09	17[1], 130	1.056	1.4109^{20}		73^{29mm}		misc aq, alc, acet, bz, eth, CCl_4
b501	4-Butyrolactone		86.09	17, 234	1.1244^{25}	1.4348^{25}	-43.5	204	98 (OC)	3.3 aq; misc alc, eth
b502	Butyronitrile	$CH_3CH_2CH_2CN$	69.11	2[2], 252	0.7954^{15}	1.3860^{15}	-112	117.9	29 (OC)	s aq, alc(d); misc eth
b503	Butyrophenone	$C_6H_5C(O)C_3H_7$	148.21	7, 313	1.021	1.5195^{20}	13	222	88	2.1 aq; 1.5 alc; 18 chl; 0.19 eth; 1 bz
b504	Butyryl chloride	$CH_3CH_2CH_2COCl$	106.55	2, 274	1.0263^{21}	1.4122^{20}	-89	102	21	i aq; s alc, chl, eth
c1	Caffeine		194.19	26, 461	1.23^{18}		238	subl 178		100 alc; 100 eth; 200 chl; 250 acet
c2	DL-Camphene		136.24	5, 156	0.8422^{54}	1.4551^{54}	51-52	159	36	
c3	D(+)-Camphor		152.23	7, 101	0.9920^{25}		178.8	207.4		
c4	DL-Camphor		152.24	7, 135			177	204	66 (CC)	0.8 aq; 100 alc; s chl, eth
c5	D-Camphoric acid		200.23	9, 745	1.1862^{20}		186-188			
c7	DL-10-Camphorsul-fonic acid		232.30	11, 314			194 d			deliq moist air; sl s HOAc, EtAc; i eth

b500   b501   c1   c2   c3, c4   c5   c7

TABLE 1.15 Physical Constants of Organic Compounds (*Continued*)

No.	Name	Formula	Formula weight	Beilstein reference	Density	Refractive index	Melting point	Boiling point	Flash point	Solubility in 100 parts solvent
c8	Carbazole		167.21	20, 433	1.10^{18}		245–246	355	>110	16 pyr; 11 acet; 3 eth
c9	4-Carbethoxy-3-methyl-3-cyclo-hexen-1-one		182.22	10, 631	1.078	1.4880^{20}		268–272		
c10	Carbobenzyloxyglycine	$C_6H_5CH_2OC(=O)$-$NHCH_2COOH$	209.20				122			
c11	Carbohydrazide	$H_2NNHC(=O)NHNH_2$	90.09	3, 121			d 153			v s aq; i alc, bz, eth
c12	Carbon disulfide	CS_2	76.14	3, 197	1.2632^{20}	1.6280^{20}	−111.6	46.5	−30 (CC)	0.3 aq; misc bz, chl, eth, CCl_4
c13	Carbon tetrabromide	CBr_4	331.65	1, 68	3.42		90	190	none	misc alc, bz, chl, eth, CS_2, PE
c14	Carbon tetrachloride	CCl_4	153.82	1, 64	1.589^{25}_{25}	1.4607^{20}	−23	76.7	none	
c15	Carbon tetrafluoride	CF_4	88.01	1, 59	1.89^{-183}		−183.6	−127.8		s bz, chl; d by hot alc
c15a	Carbon tetraiodide	CI_4	519.63	1, 74	4.32^{24}		171			i aq, bz, eth; v s alc
c16	4-Carboxybenzenesul-fonamide	$HOOCC_6H_4SO_2NH_2$	201.20	11, 390			d 280			
c17	2-Carboxyethylphos-phonic acid	$HOOCCH_2CH_2P(O)(OH)_2$	154.06	4^2, 976						
c18	1-(Carboxymethyl)pyr-idinium chloride		173.60				189 d			
c20	D-(+)-Carvone		150.22	7, 153	0.9652^{20}	1.4989^{20}		230	88	i aq; misc alc
c21	Catecholborane		119.92			1.5070^{20}	12	50^{50mm}	2	10 aq; 10 alc; sl s eth
c22	2-Chloroacetamide	$ClCH_2CONH_2$	93.51	2, 199			118	225 d		i aq; v s alc, eth, CS_2
c23	4'-Chloroacetanilide	$ClC_6H_4NHCOCH_3$	169.61	12, 611	1.385^{20}		179			v s aq; s alc, bz, eth
c24	Chloroacetic acid	$ClCH_2COOH$	94.50	2, 194	$1.580(c)$	1.4297^{65}	63	189	126	d aq; v s chl, eth
c25	Chloroacetic anhydride	$[ClCH_2C(O)]_2O$	170.98	2, 199	1.5494^{20}		46	203		
c26	4'-Chloroacetoacetani-lide	$CH_3COCH_2CONHC_6H_4Cl$	211.65				134	d	160 (CC)	
c27	Chloroacetonitrile	$ClCH_2CN$	75.50	2, 201	1.193	1.4225^{20}		126	47	i aq; v s alc, bz, eth
c28	2-Chloroacetophenone	$C_6H_5COCH_2Cl$	154.60	7, 282	1.324^{15}		54	245		sl s aq; s eth
c29	o-Chloroacetophenone	$ClC_6H_4COCH_3$	154.60	7, 151	1.188			228^{738mm}	88	i aq; s eth
c30	p-Chloroacetophenone	$ClC_6H_4COCH_3$	154.60	7, 281	1.192^{20}	1.5438^{20}	20–21	237	90	i aq; misc alc, eth
c31	Chloroacetyl chloride	$ClCH_2COCl$	112.94	2, 199	1.420^{20}_4	1.5549	−21.8	106	none	d by aq, MeOH
c32	2-Chloroacrylonitrile	$H_2C=C(Cl)CN$	87.51		1.096	1.4290^{20}	−65	89	6	
c33	2-Chloroaniline	$ClC_6H_4NH_2$	127.57	12, 597	1.2125^{20}_4	1.5895^{20}	−1.94	208.8	97	0.88 aq; s alc, bz, eth

No.	Name	Formula	Formula wt.	Beilstein ref.	Density	n_D	m.p., °C	b.p., °C	Solubility
c34	3-Chloroaniline	ClC$_6$H$_4$NH$_2$	127.57	12, 602	1.2150^{22}	1.5931^{20}	−10.4	123	i aq; s alc, bz, eth
c35	4-Chloroaniline	ClC$_6$H$_4$NH$_2$	127.57	12, 607	1.1697^{7}	1.5546^{85}	72.5	230.5	s hot aq; s alc, acet, eth, CS$_2$
c36	1-Chloroanthraquinone		242.66	7, 787			160	232	sl s alc; misc eth; s hot bz
c37	2-Chloroanthraquinone		242.66	7, 787			211	subl	sl s alc, bz, i eth

Capric acid, d14
Caproaldehyde, h54
Caproic acid, h66
Caproic anhydride, h67
ε-Caprolactam, o57
ε-Caprolactone, h71
Capronitrile, h63
Caproyl chloride, h73
Capryl alcohol, o30
Caprylaldehyde, o40
Caprylic acid, o29
Caprylonitrile, o27
Capryloyl chloride, o37
N-(Carbamoylmethyl)iminodiacetic acid, a14
Carbamylurea, b216

Carbanilide, d695
Carbazole, d667
Carbitol, e36
Carbitol acetate, e37
Carbobenzoxy chloride, b90
3-Carbomethoxypropionyl chloride, m189
N-Carbonylsulfamyl chloride, c241
Carboxybenzaldehyde, f33
(3-Carboxy-2-hydroxypropyl)trimethylammonium hydroxide, c18
(Carboxylmethyl)trimethylammonium hydroxide, b129
(Carboxymethylimino)bis(ethylenenitrilo)-tetraacetic acid, d299
3-Carboxypropyl disulfide, d710

Cellosolve, e35
Cellosolve acetate, e38
Cetyl alcohol, h36
Cetyl bromide, b295
Chalcone, d688
Chloramine T, c248
Chloranil, t25, t26
Chloranilic acid, d172
Chlorendic anhydride, h30
Chloroacetaldehyde diethyl acetal, c81
Chloroacetaldehyde dimethyl acetal, c89
Chloroacetone, c216
Chloroanthranilic acid, a140
5-Chloroanthranilonitrile, a141

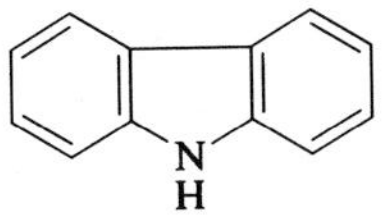

Structures:
c8 — carbazole (N–H)
c9
c18 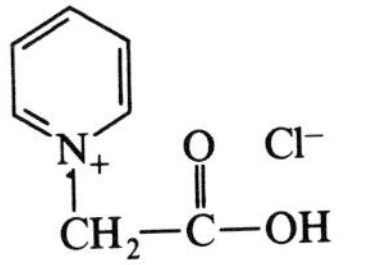
c20 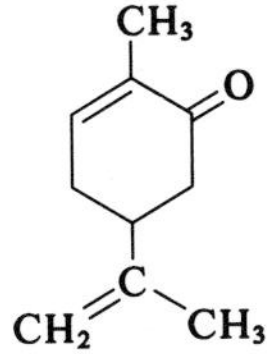
c21 — (BH) 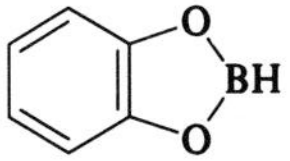
c36
c37

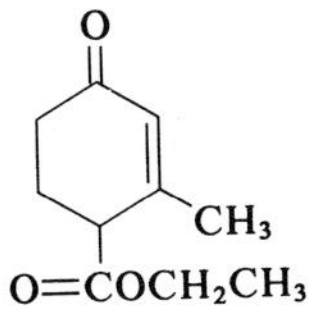

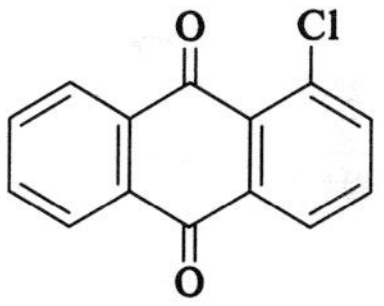

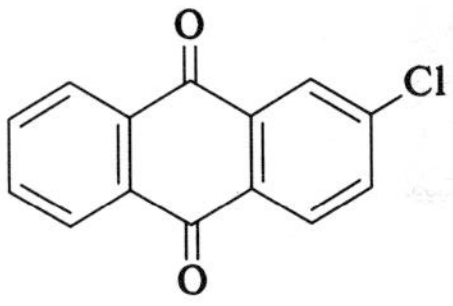

TABLE 1.15 Physical Constants of Organic Compounds (*Continued*)

No.	Name	Formula	Formula weight	Beilstein reference	Density	Refractive index	Melting point	Boiling point	Flash point	Solubility in 100 parts solvent
c38	2-Chlorobenzaldehyde	ClC_6H_4CHO	140.57	7, 233	1.2483^{20}	1.5658	11	215	87	sl s aq; s alc, bz, eth
c38a	3-Chlorobenzaldehyde	ClC_6H_4CHO	140.57	7, 234	1.241	1.5645^{20}	17–18	214	88	s aq; v s alc, bz, eth
c39	4-Chlorobenzaldehyde	ClC_6H_4CHO	140.57	7, 235	1.196_4^{61}	1.5526^1	47	214	87	s aq; v s alc, bz, eth
c40	2-Chlorobenzamide	$ClC_6H_4CONH_2$	155.58	9, 336			142–144			
c41	Chlorobenzene	C_6H_5Cl	112.56	5, 199	1.1063^{20}	1.5248^{20}	−45.3	131.7	23	0.049 aq^{30}; v s alc, bz, chl, eth
c42	4-Chlorobenzenesulfonamide	$ClC_6H_4SO_2NH_2$	191.64	11, 55			146			s hot aq, hot alc, hot eth
c43	4-Chlorobenzenesulfonyl chloride	$ClC_6H_4SO_2Cl$	211.07	11, 55			55	141^{15mm}	107	d aq, alc; v s bz, eth
c44	4-Chlorobenzhydrol	$ClC_6H_4CH(OH)C_6H_5$	218.68	6, 680			58–60			0.11 aq; v s alc, eth
c45	2-Chlorobenzoic acid	ClC_6H_4COOH	156.57	9, 334	1.544_4^{25}		142			0.04 aq; v s alc, eth
c46	3-Chlorobenzoic acid	ClC_6H_4COOH	156.57	9, 337	1.496_2^{25}		157–158			0.02 aq; v s alc, eth
c46a	4-Chlorobenzoic acid	ClC_6H_4COOH	156.57	9, 340			241–243			s alc, eth
c47	2-Chlorobenzonitrile	ClC_6H_4CN	137.57	9, 336			46	232	108	s alc, bz, chl, eth
c48	4-Chlorobenzonitrile	ClC_6H_4CN	137.57	9, 341			93	223		
c49	2-Chlorobenzophenone	$ClC_6H_4COC_6H_5$	216.67	7, 419			44–47	300	>110	
c50	4-Chlorobenzophenone	$ClC_6H_4COC_6H_5$	216.67	7, 419			77	196^{17mm}		s alc, acet, bz, eth
c50a	2-Chlorobenzotrichloride	$ClC_6H_4CCl_3$	229.92	5, 302	1.508	1.5817^{20}	29	264	98	
c50b	4-Chlorobenzotrichloride	$ClC_6H_4CCl_3$	229.92	5, 303	1.495	1.5722^{20}		245	>110	
c51	2-Chlorobenzotrifluoride	$ClC_6H_4CF_3$	180.56		1.3540^{25}	1.4513^{25}	−6.4	152.3	98	
c52	3-Chlorobenzotrifluoride	$ClC_6H_4CF_3$	180.56		1.3311^{25}	1.4438^{25}	−56.7	137.7	36	
c53	4-Chlorobenzotrifluoride	$ClC_6H_4CF_3$	180.56		1.353^{20}	1.4463	−33.2	138.7	47	
c54	2-(4-Chlorobenzoyl)-benzoic acid	$ClC_6H_4COC_6H_4COOH$	260.68	10, 750			150			s alc, bz, eth
c55	2-Chlorobenzoyl chloride	ClC_6H_4COCl	175.01	9, 336	1.382	1.5718^{20}	−3	238	110	dec aq, alc

1.132

No.	Name	Formula	Mol. wt.	Beilstein ref.	Density	n_D^{20}	m.p., °C	b.p., °C	Flash, °C	Solubility
c56	4-Chlorobenzoyl chloride	ClC_6H_4COCl	175.01	9, 341	1.377	1.5780^{20}	14	222	105	dec aq, alc
c57	4-Chlorobenzyl alcohol	$ClC_6H_4CH_2OH$	142.59	6, 444	1.164	1.5586^{20}	72	234	90	v s alc, eth
c58	4-Chlorobenzylamine	$ClC_6H_4CH_2NH_2$	141.60	12, 1074	1.274	1.5591^{20}		215	82	
c59	2-Chlorobenzyl chloride	$ClC_6H_4CH_2Cl$	161.03	5, 297			−17	214		
c60	4-Chlorobenzyl chloride	$ClC_6H_4CH_2Cl$	161.03	5, 308		1.5440^{20}	30	214	97	s alc; v s eth
c61	2-Chlorobenzyl cyanide	ClC_6H_4CN	151.60	9, 448			24		>110	
c62	4-(4-Chlorobenzyl)pyridine	$ClC_6H_4CH_2{-}C_5H_4N$	203.67		1.167	1.5900^{20}			>110	v s chl
c63	1-Chloro-1,3-butadiene	$H_2C{=}CHCH{=}CHCl$	88.54	1^3, 949	0.9601^{20}_4	1.4712^{20}		68	−20	0.11 aq; misc alc, eth
c64	1-Chlorobutane	$CH_3CH_2CH_2CH_2Cl$	92.57	1, 118	0.8864^{20}_4	1.4021^{20}	−123.1	78.44	−6	0.1 aq; misc alc, eth
c65	2-Chlorobutane	$CH_3CH_2CH(Cl)CH_3$	92.57	1, 119	0.8732^{20}	1.3971^{20}	−113.3	68.25	−15	s alc, eth
c66	4-Chloro-1-butanol	$ClCH_2CH_2CH_2CH_2OH$	108.56	1^2, 398	1.0883^{20}	1.4518^{20}		89^{20mm}	32	v s alc, eth
c67	3-Chloro-2-butanone	$CH_3CH(Cl)COCH_3$	106.55	1, 669	1.055	1.4172^{20}		117	21	s alc, acet
c68	cis-1-Chloro-2-butene	$CH_3CH{=}CHCH_2Cl$	90.55	1^2, 176	0.9426^{20}	1.4390^{20}		84.1	−15	v s acet
c69	3-Chloro-1-butene	$CH_3CH(Cl)CH{=}CH_2$	90.55	1^2, 174	0.9001^{20}	1.4155^{20}		62–65	−20	
c69a	4-Chlorobutyl acetate	$CH_3CO_2CH_2CH_2CH_2CH_2Cl$	150.61	2^2, 141	1.072	1.4338^{20}		92^{22mm}	64	
c70	3-Chloro-1-butyne	$CH_3CH(Cl)C{\equiv}CH$	88.54	1^4, 970	0.961	1.4280^{20}		68–70	1	s alc, eth
c71	3-Chlorobutyric acid	$CH_3CH(Cl)CH_2COOH$	122.55	2, 277	1.186^{20}_4	1.4442^{20}	16.3	109^{17mm}	>110	sl s aq; v s eth
c72	4-Chlorobutyric acid	$ClCH_2CH_2CH_2COOH$	122.55	2, 277	1.2236^{20}_4	1.4510^{20}	12–16	196^{22mm}	>110	s alc, eth
c73	4-Chlorobutyronitrile	$ClCH_2CH_2CH_2CN$	103.55	2, 278	1.158	1.4413^{20}		197	85	dec aq, alc; s eth
c74	4-Chlorobutyryl chloride	$ClCH_2CH_2CH_2COCl$	141.00	2, 278	1.258	1.4609^{20}		174	72	
c75	Chloro(chloromethyl)-dimethylsilane	$ClCH_2Si(CH_3)_2Cl$	143.09		1.086	1.4373^{20}	−14	114^{752mm}	21	
c76	3-Chloro-2-chloromethyl-1-propene	$H_2C{=}C(CH_2Cl)_2$	125.00	1^2, 181	1.080	1.4753^{20}		138	36	
c77	Chlorocyclohexane	ClC_6H_{11}	118.61	5, 21	1.000^{20}	1.4620^{20}	−44	142	28	i aq; s alc, eth
c78	2-Chlorocyclohexanone	$ClC_6H_9({=}O)$	132.59	7, 10	1.161	1.4835^{20}	23	83^{10mm}	82	s bz, eth, diox
c79	Chlorocyclopentane	ClC_5H_9	104.58	5, 19	1.0051^{20}	1.4512^{20}		114	15	i aq
c80	1-Chlorodecane	$CH_3(CH_2)_9Cl$	176.73	1, 168	0.868	1.4362^{20}	−34	223	83	i aq

TABLE 1.15 Physical Constants of Organic Compounds (*Continued*)

No.	Name	Formula	Formula weight	Beilstein reference	Density	Refractive index	Melting point	Boiling point	Flash point	Solubility in 100 parts solvent
c81	2-Chloro-1,1-diethoxyethane	$ClCH_2CH(OC_2H_5)_2$	152.62	1, 611	1.018	1.4157^{20}		157	29	
c82	3-Chloro-1,1-diethoxypropane	$ClCH_2CH_2CH(OC_2H_5)_2$	166.65	1, 632	0.995	1.4240^{20}		84^{25mm}	36	
c83	Chlorodifluoroacetic acid	$F_2C(Cl)COOH$	130.48	2, 201	1.118^{21}	1.3559^{20}	22.9	121.5	none	
c84	1-Chloro-1,1-difluoroethane	$CH_3C(Cl)F_2$	100.50	1^3, 138			−131	−9		0.19 aq
c85	Chlorodifluoromethane	$HCClF_2$	86.47	1^3, 41	1.209^{21}		−160	−40.8		0.30 aq
c86	1-Chloro-2,4-dihydroxybenzene	$ClC_6H_3(OH)_2$	144.56	6^2, 818			107	147^{18mm}		v s aq, alc, chl, eth
c87	2-Chloro-1,4-dihydroxybenzene	$ClC_6H_3(OH)_2$	144.56	6, 849			101–102	263		v s aq; i alc; s eth
c88	2-Chloro-1,4-dimethoxybenzene	$ClC_6H_3(OCH_3)_2$	172.61	6^3, 4432	1.211	1.5467^{20}		234	110	
c89	2-Chloro-1,1-dimethoxyethane	$ClCH_2CH(OCH_3)_2$	124.57			1.094^{20}_{20}	1.4148^{20}		130	28	
c90	Chloro-tris(dimethylamino)silane	$[(CH_3)_2N]_3SiCl$	195.8			0.9752^{20}_{4}	1.442^{20}		62^{12mm}		
c91	4-Chloro-3,5-dimethylphenol	$Cl(CH_3)_2C_6H_2OH$	156.61	6^2, 463			115.5	246		0.1 aq; l alc; s bz, eth, alk
c92	1-Chloro-2,2-dimethylpropane	$(CH_3)_3CCH_2Cl$	106.59	1, 141	0.866^{20}_{4}	1.4042^{20}	−20	84.4	32	
c92a	3-Chloro-2,2-dimethyl-1-propanol	$ClCH_2C(CH_3)_2CH_2OH$	122.60				1.4504^{20}	34–36	87^{35mm}	71	
c93	Chlorodimethylsilane	$(CH_3)_2Si(Cl)H$	94.62			0.8524^{20}	1.3827^{20}	−111	36	−28	
c94	Chlorodimethylvinylsilane	$(CH_3)_2Si(Cl)CH{=}CH_2$	120.7			0.8844^{25}	1.414^{25}		82.5		
c94a	6-Chloro-2,4-dinitroaniline	$ClC_6H_2(NO_2)_2NH_2$	217.57	12^1, 367			159			
c95	1-Chloro-2,4-dinitrobenzene	$ClC_6H_3(NO_2)_2$	202.55	5, 263	1.4982^{25}	1.5856^{60}	52–54	315	186	sl s alc; s hot alc, bz, eth
c96	1-Chloro-3,4-dinitrobenzene	$ClC_6H_3(NO_2)_2$	202.55	5, 262	1.6867^{16}	1.5870^{20}			>110	v s eth; s alc

No.	Name	Formula	M.W.	Beilstein ref.	Density	n_D	m.p., °C	b.p., °C	Flash pt, °C	Solubility
c97	2-Chloro-3,5-dinitrobenzoic acid	$ClC_6H_2(NO_2)_2COOH$	246.56	9, 415			198	241 explodes		0.3 aq
c98	Chlorodiphenylmethane	$C_6H_5CH(Cl)C_6H_5$	202.68	5^2, 500	1.1402^{20}_{4}	1.5951^{20}	17	140^{3mm}	>110	
c99	Chlorodiphenylmethylsilane	$(C_6H_5)_2Si(Cl)CH_3$	232.79	16^2, 606	1.1277^{20}_{4}	1.5742^{20}		295	>110	
c100	Chlorodiphenylphosphine	$(C_6H_5)_2PCl$	220.64	16, 763	1.229	1.6338^{20}		320	>110	
c101	1-Chlorododecane	$CH_3(CH_2)_{11}Cl$	204.79		0.8673^{20}_{4}	1.4426	−9	116	93	v s alc; s bz
c102	1-Chloro-2,3-epoxy-propane	$H_2C{-}CHCH_2Cl$ (epoxide O)	92.53	17, 6	1.1812^{20}	1.4381^{20}	−57.2	116.1	33	5.9 aq; misc alc, chl
c103	Chloroethane	CH_3CH_2Cl	64.52	1, 82	0.9214^{20}_{4}	1.3742^{10}	−136	12.3	−43	0.45 aq^0, 48 alc; misc eth
c104	2-Chloroethanol	$ClCH_2CH_2OH$	80.52	1, 337	1.197^{20}_{4}	1.4422^{20}	−67.5	128.6	60	misc aq, alc
c105	2-(2-Chloroethoxy)ethanol	$ClCH_2CH_2OCH_2CH_2OH$	124.57	1, 467	1.180	1.4529^{20}		81^{5mm}	90	
c106	2-[2-(2-Chloroethoxy)ethoxy]ethanol	$ClCH_2CH_2OCH_2CH_2OCH_2CH_2OH$	168.62	1, 468	1.160	1.4580^{20}		120^{5mm}	107	
c107	2-Chloroethylamine HCl	$ClCH_2CH_2NH_2 \cdot HCl$	115.99	4, 133			146			
c108	1-Chloro-2-ethylbenzene	$ClC_6H_4C_2H_5$	140.61		1.0552^{25}		−81	179.2	66	i aq; misc alc, eth
c109	(2-Chloroethyl)benzene	$C_6H_5CH_2CH_2Cl$	140.61	5, 354	1.069	1.5300^{20}		84^{16mm}	66	s alc, bz, eth
c110	Chloroethylene	$H_2C{=}CHCl$	62.50	1, 186	0.97^{-14}		−159.7	−13.9	>110	
c110a	N-(2-Chloroethyl)-N-ethylaniline	$C_6H_5N(C_2H_5)CH_2CH_2Cl$	183.68	12^3, 263	1.075	1.5584^{20}		164^{42mm}	>110	sl s aq; s alc
c111	2-Chloroethyl ethyl ether	$ClCH_2CH_2OCH_2CH_3$	108.57	1, 337	0.989	1.4125^{20}		107	15	
c112	2-Chloroethyl methyl ether	$ClCH_2CH_2OCH_3$	94.54	1, 337	1.035	1.4111^{20}		89–90	15	

Chlorodibromomethane, d71
2-Chloro-N,N-diethylethylamine, d272
Chlorodimethyl ether, c156

2-Chloro-N,N-dimethylethylamine, d469
4'-Chlorodiphenylmethanol, c44

2-Chloroethyl alcohol, c104
2-Chloroethyl ether, b159

TABLE 1.15 Physical Constants of Organic Compounds (*Continued*)

No.	Name	Formula	Formula weight	Beilstein reference	Density	Refractive index	Melting point	Boiling point	Flash point	Solubility in 100 parts solvent
c113	N-(2-Chloroethyl)morpholine HCl		186.08				186			
c114	N-(2-Chloroethyl)piperidine HCl		184.11	20, 17			236			
c115	2-Chloroethyl p-toluenesulfonate	$CH_3C_6H_4SO_3CH_2CH_2Cl$	234.70	11^2, 45	1.294	1.5290^{20}		$153^{0.3mm}$	>110	
c116	2-Chloroethyl vinyl ether	$H_2C{=}CHOCH_2CH_2Cl$	106.55	1^2, 473	1.0525^{15}_{15}	1.4370^{20}	−69.7	110	16	0.6 aq
c117	1-Chloro-2-fluorobenzene	ClC_6H_4F	130.55	5^1, 110	1.244	1.5010^{20}	−42.5	138.5	31	s alc, eth
c118	1-Chloro-3-fluorobenzene	ClC_6H_4F	130.55		1.219	1.4944^{20}		126	20	s alc, eth
c119	2-Chloro-4'-fluorobenzophenone	$ClC_6H_4COC_6H_4F$	234.66				60–62			
c120	2-Chloro-6-fluorobenzyl chloride	$Cl(F)C_6H_3CH_2Cl$	179.02		1.401	1.5372^{20}			93	
c121	4-Chloro-4'-fluorobutyrophenone	$FC_6H_4C({=}O)CH_2CH_2CH_2CH_2Cl$	200.64		1.220	1.5255^{20}			110	
c122	3-Chloro-4-fluoronitrobenzene	$Cl(F)C_6H_3NO_2$	175.55	5^1, 130	1.6028^{17}	1.5674^{17}	41.5	127^{17mm}		
c123	3-Chloro-4-fluorophenol	$Cl(F)C_6H_3OH$	146.55	6^4, 880			42–44	104^{11mm}	108	
c125	2-Chloro-6-fluorotoluene	$Cl(F)C_6H_3CH_3$	144.58		1.191	1.5026^{20}		156	46	
c126	4-Chloro-2-fluorotoluene	$Cl(F)C_6H_3CH_3$	144.58	5^4, 813	1.186	1.4998^{20}		158	51	
c127	Chloroform	$CHCl_3$	119.39	1, 61	1.484^{20}_{20}	1.4476^{20}	−63.5	61.7	none	0.82 aq
c128	Chloroform-d	$CDCl_3$	120.39	1^3, 63	1.50	1.4445^{20}		60.9	none	
c129	1-Chloroheptane	$CH_3(CH_2)_6Cl$	134.65	1, 154	0.8811^{16}	1.4250^{20}		159–161	41	misc alc, eth
c129a	1-Chlorohexadecane	$CH_3(CH_2)_{14}CH_2Cl$	260.89	1, 172	0.865	1.4490^{20}	−69	149^{1mm}	>110	
c130	1-Chlorohexane	$CH_3(CH_2)_5Cl$	120.62	1, 143	0.8780^{20}_{4}	1.4236^{20}	−94	134	26	i aq
c131	6-Chloro-1-hexanol	$Cl(CH_2)_6OH$	136.62		1.204	1.4557^{20}		110^{14mm}	98	sl s aq; v s alc, eth
c131a	5-Chloro-2-hydroxyaniline	$ClC_6H_3(OH)NH_2$	143.57	13, 383			138			

c132	4-Chloro-4'-hydroxy-benzophenone	$ClC_6H_4C(=O)C_6H_4OH$	232.67	8^2, 187			175–178	257^{14mm}		i alc, eth; 0.8 chl; 0.6 HOAc
c133	5-Chloro-8-hydroxy-7-iodoquinoline		305.50	21, 98			172			sl s aq HCl
c135	5-Chloro-8-hydroxy-quinoline		179.61	21, 95			130			s alc
c136	1-Chloro-4-iodoben-zene	ClC_6H_4I	238.46	5, 221	1.1865^7		53–54	226–227	108	
c137	1-Chloro-3-mercapto-2-propanol	$HSCH_2CH(OH)CH_2Cl$	126.61	1^3, 2156	1.277	1.5276^{20}		$57^{1.3mm}$	97	$0.48\ aq^{25}$, s alc; misc chl, eth, HOAc
c138	Chloromethane	CH_3Cl	50.49	1, 59	0.92^{20}	1.3712^{-24}	−97.7	−24.22	<0	
c138a	3-Chloro-4-methoxy-aniline	$ClC_6H_3(OCH_3)NH_2$	157.60	13, 511			50–55		>110	
c139	5-Chloro-2-methoxy-aniline	$ClC_6H_3(OCH_3)NH_2$	157.60	13, 383			83–85			
c140	1-Chloro-2-methoxy-benzene	$ClC_6H_4OCH_3$	142.59	6, 184	1.123	1.5445^{20}		196	76	i aq; s alc, eth
c142	2-Chloro-6-methoxy-pyridine	$CH_3O(Cl)-C_5H_3N$	143.57		1.207	1.5263^{20}		186		
c143	2-Chloro-6-methylani-line	$CH_3(Cl)C_6H_3NH_2$	141.60	12^1, 388	1.152	1.5761^{20}	2	215	98	s alc
c144	3-Chloro-2-methylani-line	$CH_3(Cl)C_6H_3NH_2$	141.60	12, 836	1.185	1.5874^{20}	2	117^{10mm}	>110	

2-Chloro-6-fluorobenzal chloride, t237
α-Chloro-4-fluorotoluene, f16

5-Chloro-2-hydroxyaniline, a148
Chlorohydroxybenzoic acids, c238, c239

1-Chloro-3-hydroxypropane, c215

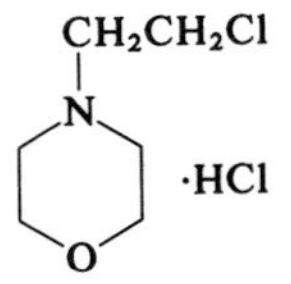

c113

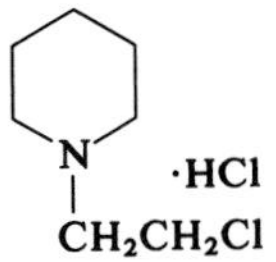

c114

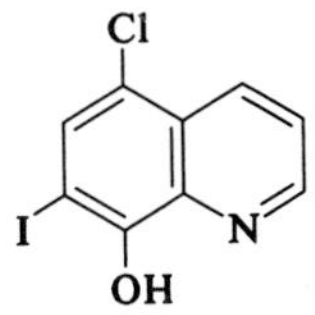

c133

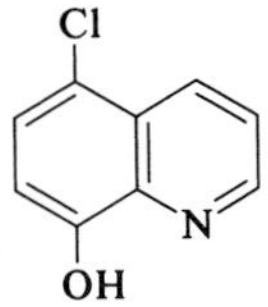

c135

TABLE 1.15 Physical Constants of Organic Compounds (*Continued*)

No.	Name	Formula	Formula weight	Beilstein reference	Density	Refractive index	Melting point	Boiling point	Flash point	Solubility in 100 parts solvent
c145	3-Chloro-4-methylaniline	$CH_3(Cl)C_6H_3NH_2$	141.60	12, 988		1.5830^{20}	25	238	100	
c146	4-Chloro-2-methylaniline	$CH_3(Cl)C_6H_3NH_2$	141.60	12, 835		1.5848^{20}	27	241	99	s hot alc
c147	5-Chloro-2-methylaniline	$CH_3(Cl)C_6H_3NH_2$	141.60	12, 835		1.5840^{20}	22	237	160	
c147a	3-Chloromethyl)benzoyl chloride	$ClCH_2C_6H_4COCl$	189.04	9^2, 325	1.330	1.5748^{20}		150^{20mm}	>110	
c148	DL-4-Chloro-2-(α-methylbenzyl)-phenol	$C_6H_5CH(CH_3)C_6H_3(Cl)OH$	232.71	6^4, 4710	1.238	1.5994^{20}		155^{2mm}	>110	i aq; s alc, eth
c150	2-Chloro-2-methylbutane	$CH_3CH_2CCl(CH_3)_2$	106.59	1, 134	0.8650^{20}_4	1.4052^{20}	−73.7	85	−9	
c151	Chloromethyldichloromethylsilane	$ClCH_2Si(Cl)_2CH_3$	163.51	4^3, 1888	1.286	1.4494^{20}		121	110	
c153	Chloromethyl ethyl ether	$ClCH_2OCH_2CH_3$	94.54	1^2, 645	1.042^{20}_4	1.4040^{20}		79–83	19	s alc; v s eth
c156	Chloromethyl methyl ether	$ClCH_2OCH_3$	80.51	1, 580	1.0703^{20}_4	1.3961^{20}	−103.5	57–59	15	dec aq; s acet, CS_2
c157	Chloromethyl methyl sulfide	$ClCH_2SCH_3$	95.48		1.153	1.4963^{20}		105	17	
c158	1-(Chloromethyl)-naphthalene	$C_{10}H_7CH_2Cl$	176.65	5, 566		1.6380^{20}	32	169^{25mm}	>110	sl s aq
c159	4-Chloro-2-methyl-phenol	$CH_3(Cl)C_6H_3OH$	142.59	6, 359			48	225	>110	
c160	4-Chloro-3-methyl-phenol	$CH_3(Cl)C_6H_3OH$	142.59	6, 381			68	235	92	i aq; s alc, bz, chl, eth, acet
c161	1-Chloro-2-methyl-2-phenylpropane	$C_6H_5C(CH_3)_2CH_2Cl$	168.67	5^2, 320	1.047	1.5240^{20}		96^{10mm}		
c162	1-Chloro-2-methylpropane	$(CH_3)_2CHCH_2Cl$	92.57	1, 124	0.8829^{15}	1.4010^{15}	−130.3	68.9	21	0.09 aq; misc alc, eth
c163	2-Chloro-2-methylpropane	$(CH_3)_3CCl$	92.57	1, 125	0.8474^{15}_4	1.3856^{20}	−25.4	50.8	18	sl s aq; misc alc, eth
c164	1-Chloro-2-methylpropene	$(CH_3)_2C{=}CHCl$	90.55	1, 209	0.9186^{20}_4	1.4225^{20}		68.1	−1	misc alc, eth

No.	Name	Formula	M.W.	Beil.	Density	n_D	m.p., °C	b.p., °C	fl.p.	Solubility
c165	3-Chloro-2-methylpropene	$ClCH_2C(CH_3)=CH_2$	90.55	1, 209	0.9210^{45}	1.4272^{20}	−80	72	−12	misc alc, eth
c167	Chloromethyltrimethylsilane	$ClCH_2Si(CH_3)_3$	122.67	4[3], 1844	0.8861^{24}	1.4180^{20}		99	−2	
c168	6-(Chloromethyl)uracil		160.56	23[1], 328			257 d			s alc, bz, PE
c169	1-Chloronaphthalene	$C_{10}H_7Cl$	162.62	5, 541	1.1938_4^{20}	1.6332^{20}	−2.3	259.3	121	s alc, bz, chl, eth
c170	2-Chloronaphthalene	$C_{10}H_7Cl$	162.62		1.1377^{71}	1.6079^{71}	59.5	256		
c171	4-Chloro-1,8-naphthalic anhydride		232.63	17, 522			210			
c172	4'-Chloro-3'-nitroacetophenone	$ClC_6H_3(NO_2)C(=O)CH_3$	199.60	7[3], 995			101			sl s aq; v s alc, eth
c173	2-Chloro-4-nitroaniline	$ClC_6H_3(NO_2)NH_2$	172.57	12, 733			109			
c173a	2-Chloro-5-nitroaniline	$ClC_6H_3(NO_2)NH_2$	172.57	12, 732			114			
c174	4-Chloro-2-nitroaniline	$ClC_6H_3(NO_2)NH_2$	172.57	12, 729			119			v s alc, eth
c175	4-Chloro-3-nitroaniline	$ClC_6H_3(NO_2)NH_2$	172.57	12, 731			101			v s alc; s eth
c176	1-Chloro-2-nitrobenzene	$ClC_6H_4NO_2$	157.56	5, 241	1.348		32–33	246	123	s alc, bz, eth
c177	1-Chloro-3-nitrobenzene	$ClC_6H_4NO_2$	157.56	5, 243	1.5344_4^{20}		46	236	103	sl s alc; v s eth, chl
c178	1-Chloro-4-nitrobenzene	$ClC_6H_4NO_2$	157.56	5, 243	1.520		82–84	242	110	sl s alc; v s eth, CS_2

Chloromethylbenzenes, c245, c246, c247

(Chloromethyl)oxirane, c102

Chloronicotinic acid, c235, c236

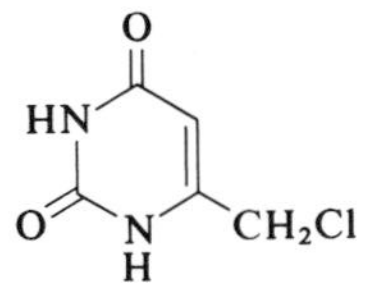

c168

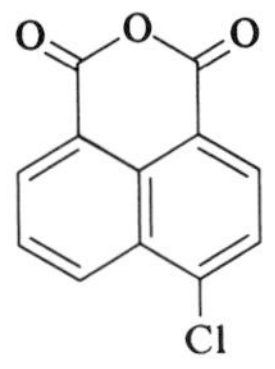

c171

TABLE 1.15 Physical Constants of Organic Compounds (*Continued*)

No.	Name	Formula	Formula weight	Beilstein reference	Density	Refractive index	Melting point	Boiling point	Flash point	Solubility in 100 parts solvent
c179	2-Chloro-4-nitrobenzoic acid	$ClC_6H_3(NO_2)COOH$	201.57	9, 404			141			s hot aq, hot bz
c180	2-Chloro-5-nitrobenzoic acid	$ClC_6H_3(NO_2)COOH$	201.57	9, 403	1.608^{18}		168			sl s aq; s alc, bz, eth
c181	4-Chloro-3-nitrobenzoic acid	$ClC_6H_3(NO_2)COOH$	201.57	9, 402	1.645^{18}		183			sl s alc; s hot aq
c182	4-Chloro-3-nitrobenzophenone	$ClC_6H_3(NO_2)C(=O)C_6H_5$	261.66	7^{I}, 230			104–106	235^{13mm}	98	
c183	2-Chloro-5-nitrobenzotrifluoride	$ClC_6H_3(NO_2)CF_3$	225.55	5, 329	1.527	1.5083^{20}		231		
c184	4-Chloro-3-nitrobenzotrifluoride	$ClC_6H_3(NO_2)CF_3$	225.55		1.511	1.4893^{20}	−2.5	222	101	
c185	5-Chloro-2-nitrobenzotrifluoride	$ClC_6H_3(NO_2)CF_3$	225.55		1.526	1.4980^{20}	21–22	224	102	
c186	o-(4-Chloro-3-nitrobenzoyl)benzoic acid	$ClC_6H_3(NO_2)COC_6H_4COOH$	305.68	10, 752			201			
c187	4-Chloro-2-nitrophenol	$ClC_6H_3(NO_2)OH$	173.56	6, 238			85–87			
c188	2-Chloro-4-nitrotoluene	$ClC_6H_3(NO_2)CH_3$	171.58	5, 329		1.5470^{70}	61	260		i aq; s alc, eth
c189	2-Chloro-6-nitrotoluene	$ClC_6H_3(NO_2)CH_3$	171.58	5, 327		1.5377^{70}	36	238	125	i aq
c190	4-Chloro-2-nitrotoluene	$ClC_6H_3(NO_2)CH_3$	171.58	5, 327			38–39	240^{718mm}	>110	i aq
c191	1-Chlorooctane	$CH_3(CH_2)_7Cl$	148.68	1, 159	0.8752^{20}	1.4298^{20}	−61	183	54	i aq; v s alc, eth
c192	1-Chloropentane	$CH_3(CH_2)_4Cl$	106.60	1, 130	0.8824^{20}_{4}	1.4118^{20}	−99.0	98.3	11	0.02 aq; misc alc, eth
c193	5-Chloro-2-pentanone	$ClCH_2CH_2CH_2COCH_3$	120.58	1^{2}, 738	1.0571^{18}_{4}	1.4375^{20}		72^{20mm}	35	s acet, eth
c194	3-Chloroperoxybenzoic acid	$ClC_6H_4C(O)OOH$	172.57	9^{4}, 972			94 d			
c195	2-Chlorophenol	ClC_6H_4OH	128.56	6, 183	1.2573^{25}	1.5579^{20}	9.3	176	63	sl s aq; v s alc, eth
c196	3-Chlorophenol	ClC_6H_4OH	128.56	6, 185	1.2455^{5}	1.5565^{40}	33.5	214	>110	sl s aq; s alc, eth
c197	4-Chlorophenol	ClC_6H_4OH	128.56	6, 186	1.2238^{8}_{4}	1.5419^{45}	43.5	220	115	sl s aq; v s alc, chl, eth
c198	4-Chlorophenoxyacetic acid	$ClC_6H_4OCH_2COOH$	186.59	6, 187			159			

No.	Name	Formula	M.W.	Beil. Ref.	Density	n	m.p., °C	b.p., °C		Solubility
c199	2-(4-Chlorophenoxy)-2-methylpropionic acid	ClC$_6$H$_4$OC(CH$_3$)$_2$COOH	214.65				122			v s aq, alc, eth; s bz
c200	DL-2-(4-Chlorophenoxy)propionic acid	ClC$_6$H$_4$OCH(CH$_3$)COOH	200.62	6^3, 695			117			
c201	4-Chlorophenylacetic acid	ClC$_6$H$_4$CH$_2$COOH	170.60	9, 448			105			
c202	(4-Chlorophenyl)acetonitrile	ClC$_6$H$_4$CH$_2$CN	151.60	9, 448			30.5	267		
c203	2-Chloro-1,4-phenylenediamine sulfate	H$_2$NC$_6$H$_3$(Cl)NH$_2$·H$_2$SO$_4$	240.67	13, 117			253			
c204	4-Chloro-1,2-phenylenediamine	ClC$_6$H$_3$(NH$_2$)$_2$	142.59	13, 25			70			
c205	4-Chloro-1,3-phenylenediamine	H$_2$N(Cl)C$_6$H$_3$NH$_2$	142.59	13, 53			90			
c206	3-Chlorophenyl isocyanate	ClC$_6$H$_4$NCO	153.57	12, 606	1.260	1.5576^{20}	−4.4	114^{43mm}	86	
c207	4-Chlorophenyl isocyanate	ClC$_6$H$_4$NCO	153.57	12, 616		1.5618^{20}	31	204	110	
c208	4-Chlorophenyl phenyl sulfone	ClC$_6$H$_4$SO$_2$C$_6$H$_5$	252.72	6^1, 149			94			74 acet; 44 bz; 5 CCl$_4$; 65 diox; 21 i-PrOH
c208a	1-Chloro-3-phenylpropane	C$_6$H$_5$(CH$_2$)$_3$Cl	154.64	5, 391	1.080	1.520^{20}		219	87	
c210	3-Chlorophthalide		168.58	17^1, 162			58	150^{10mm}		

c210

TABLE 1.15 Physical Constants of Organic Compounds (*Continued*)

No.	Name	Formula	Formula weight	Beilstein reference	Density	Refractive index	Melting point	Boiling point	Flash point	Solubility in 100 parts solvent
c211	1-Chloropropane	$CH_3CH_2CH_2Cl$	78.54	1, 104	0.8985^{15}	1.3880^{20}	-122.8	46.6	18	0.27 aq; misc alc, eth
c212	2-Chloropropane	$CH_3CHClCH_3$	78.54	1, 105	0.8563^{20}	1.3777^{20}	-117.2	35	-35	0.34 aq; misc alc, eth
c213	3-Chloro-1,2-propane-diol	$ClCH_2CH(OH)CH_2OH$	110.54		1.3218^{20}_4	1.4805^{20}		213	58	s aq, alc, eth
c214	1-Chloro-2-propanol	$CH_3CH(OH)CH_2Cl$	94.54	1, 363	1.115^{20}	1.4375^{20}		127	51	misc aq; s alc
c215	3-Chloro-1-propanol	$ClCH_2CH_2CH_2OH$	94.54	1, 356	1.1309^{20}_4	1.4460^{20}		162	73	
c216	Chloro-2-propanone	$ClCH_2COCH_3$	92.53	1, 653	1.135^{15}	1.4350^{20}	-44.5	119.7	27	10 aq; misc alc, chl
c217	3-Chloro-1-propene	$ClCH_2CH{=}CH_2$	76.53	1, 198	0.939^{20}	1.4151^{20}	-134.5	45.2	-28	0.36 aq; misc alc, chl
c218	(3-Chloropropenyl)-benzene	$C_6H_5CH{=}CHCH_2Cl$	152.62	5^5, 372		1.5845^{20}	-19	108^{12mm}	79	
c219	2-Chloropropionic acid	$CH_3CH(Cl)COOH$	108.52	2, 248	1.182	1.4345^{20}		186	101	misc aq, alc, eth
c220	3-Chloropropionic acid	$ClCH_2CH_2COOH$	108.52	2, 249			41	205	>110	v s aq, alc, chl
c221	3-Chloropropionitrile	$ClCH_2CH_2CN$	89.53	2, 250	1.1443^{18}	1.4379^{20}	-50	176	75	
c222	2-Chloropropionyl chloride	$CH_3CH(Cl)COCl$	126.97	2, 248	1.308	1.4400^{20}		111	31	dec aq, alc
c223	3-Chloropropionyl chloride	$ClCH_2CH_2COCl$	126.97	2, 250	1.3307^{13}	1.4570^{20}		145	61	i aq; d hot aq, hot alc; s alc; v s eth
c224	3'-Chloropropio-phenone	$ClC_6H_4C({=}O)CH_2CH_3$	168.62	7^3, 1028			47	124^{14mm}	>110	
c224a	3-Chloropropyl acetate	$CH_3CO_2(CH_2)_3Cl$	136.58	2^1, 58	1.111	1.4296^{20}		81^{30mm}	67	
c225	3-Chloropropylamine·HCl	$ClCH_2CH_2CH_2NH_2 \cdot HCl$	130.02	4, 148			150			
c226	3-Chloropropylmethyl-dichlorosilane	$Cl(CH_2)_3Si(CH_3)Cl_2$	191.6		1.2045^{20}_4	1.4580^{20}		70^{15}		
c227	2-Chloropropyl-(phenyl)dichlorosi-lane	$Cl(CH_2)_3SiCl_2(C_6H_5)$	253.6		1.241^{20}	1.5332^{20}		141^{10mm}		
c228	N-(3-Chloropropyl)pi-peridine HCl		198.14	20, 18			220			
c229	3-Chloropropyl thiol-acetate	$CH_3C({=}O)SCH_2CH_2CH_2Cl$	152.64	2^3, 493	1.159	1.4946^{20}		84^{10mm}	77	
c230	3-Chloropropyltri-chlorosilane	$ClCH_2CH_2CH_2SiCl_3$	212.0		1.3590^{20}_4	1.4668^{20}		181.5	66	

c231	3-Chloropropyltriethoxysilane	$Cl(CH_2)_3Si(OC_2H_5)_3$	240.8		1.0094^{20}	1.420^{20}		102^{10mm}		misc bz, alc, eth, EtAc
c232	3-Chloropropyltrimethoxysilane	$Cl(CH_2)_3Si(OCH_3)_3$	198.72		1.0772^{25}	1.4183^{25}		183	66	sl s aq; s alc, eth
c233	3-Chloropropyne	$ClCH_2C{\equiv}CH$	74.51	1, 248	1.0306^{25}	1.4349^{20}	−78	58	18	
c234	2-Chloropyridine	ClC_5H_4N	113.55	20, 230	1.205^{15}	1.5320^{20}		166^{714mm}	65	
c235	2-Chloro-3-pyridinecarboxylic acid	$C_5H_3N(Cl)COOH$	157.56	22[2], 35			d 175			
c236	6-Chloro-3-pyridinecarboxylic acid	$C_5H_3N(Cl)COOH$	157.56	22, 43			200 d			
c237	2-Chloroquinoline		163.61	20, 359	1.2464^{25}	1.6259^{25}	37	267		i aq; s alc, bz, eth
c238	4-Chlorosalicylic acid	$HO(Cl)C_6H_3COOH$	172.57	10, 101			212			
c239	5-Chlorosalicylic acid	$HO(Cl)C_6H_3COOH$	172.57	10, 102			172			1.4 aq; 0.67 alc; 2 bz; sl s chl, eth
c240	N-Chlorosuccinimide		133.53	21, 380	1.65		151			
c241	Chlorosulfonyl isocyanate	$ClSO_2NCO$	141.53		1.626	1.4467^{20}	−44	107	none	s alk
c242	8-Chlorotheophylline		214.61	26, 473			d 290			i aq; misc alc, eth
c243	2-Chlorothiophene	$Cl{-}C_4H_3S$	118.59	17, 32	1.286	1.5483^{20}	−72	129	22	
c244	4-Chlorothiophenol	ClC_6H_4SH	144.62	6, 326			51	207	>110	sl s aq; v s bz, chl, eth
c245	2-Chlorotoluene	$ClC_6H_4CH_3$	126.59	5, 290	1.0826^{20}	1.5250^{20}	−34	159.0	47	
c246	3-Chlorotoluene	$ClC_6H_4CH_3$	126.59	5, 291	1.0760^{19}	1.5218^{20}	−48.9	161.8	50	s alc, bz, chl; misc eth
c247	4-Chlorotoluene	$ClC_6H_4CH_3$	126.59	5, 292	1.0697^{20}	1.5208^{20}	7.2	162.0	49	sl s aq; s alc, bz, eth

Chloropicrin, t243
Chloroprene, c217

c228

$N{-}CH_2CH_2CH_2Cl$
·HCl

c237

c240

c242

β-Chloropropionaldehyde diethyl acetal, c82
3-Chloropropylene-1,2-oxide, c102

1-Chloro-2,5-pyrrolidinedione, c240
α-Chlorotoluene, b89

TABLE 1.15 Physical Constants of Organic Compounds (*Continued*)

No.	Name	Formula	Formula weight	Beilstein reference	Density	Refractive index	Melting point	Boiling point	Flash point	Solubility in 100 parts solvent
c248	N-Chloro-p-toluenesulfonamide, Na salt	$CH_3C_6H_4SO_2NCl^-Na^+$	227.67				167 d			s aq; i bz, chl, eth
c249	4'-Chloro-1-toluene-thiol	$ClC_6H_4CH_2SH$	158.65	6, 466	1.202	1.5893^{20}	20		76	
c250	4-Chloro-o-tolyloxy-acetic acid, Na salt	$ClC_6H_3(CH_3)OCH_2COO^-Na^+$	222.61	6³, 1265			220–225			
c251	4-(4-Chloro-o-tolyl-oxy)butyric acid	$ClC_6H_3(CH_3)O(CH_2)_3COOH$	228.68				99–100			
c251a	2-(4-Chloro-o-tolyl-oxy)propanoic acid	$ClC_6H_3(CH_3)OCH(CH_3)CO_2H$	214.65	6³, 1266			88			
c252	Chloro-2,2,2-trifluoro-ethane	CF_3CH_2Cl	118.5		1.389^0	1.3090^0	−105	6.9		
c253	Chlorotrifluoroethyl-ene	$CF_2{=}CFCl$	116.47	1³, 646	1.315		−158.2	−27.9		
c254	Chlorotrifluorometh-ane	$ClCF_3$	104.46	1³, 42			−181	−81.5		
c255	Chlorotrimethylger-mane	$(CH_3)_3GeCl$	153.16		1.2382^{22}	1.4283^{20}	−13	102	1	
c256	Chlorotrimethylsilane	$(CH_3)_3SiCl$	108.64	5, 700	0.8580^{20}_4	1.3885^{20}	−40	57	−27	v s bz, chl, eth
c257	Chlorotriphenylmeth-ane	$(C_6H_5)_3CCl$	278.78				110–112	230^{20mm}		
c258	Chlorotriphenyltin	$(C_6H_5)_3SnCl$	385.46	12, 914			108 d	240^{14mm}		
c259	α-Chloro-o-xylene	$CH_3C_6H_4CH_2Cl$	140.61	5, 364	1.063	1.5391^{20}		199	73	i aq; misc alc, eth
c260	α-Chloro-m-xylene	$CH_3C_6H_4CH_2Cl$	140.61	5, 373	1.064^{20}	1.5350^{20}		196	75	i aq; misc alc, eth
c261	α-Chloro-p-xylene	$CH_3C_6H_4CH_2Cl$	140.61	5, 384		1.5330^{20}		200	75	misc alc, bz, eth, acet
c261a	2-Chloro-p-xylene	$ClC_6H_3(CH_3)_2$	140.61	5, 384	1.049	1.5235^{20}	4.5	186	57	misc alc, bz, eth, acet
c262	4-Chloro-o-xylene	$ClC_6H_3(CH_3)_2$	140.61	5, 363	1.047	1.5283^{20}	2	223	66	1.29 alc; 35 eth; 22 chl; s bz, PE
c263	Cholesterol		386.66		1.067^{20}		148.5	360 sl d		0.028 aq; 0.06 alc; 2.8 acet; 0.036 bz; 0.5 chl
c264	Cholic acid		408.58	10³, 2162			198			
c265	Cinchonine		294.40	23³, 369			−260			1.4 alc; 0.9 chl; 0.2 eth
c266	1,8-Cineole		154.25	17, 23	0.9215^{25}	1.4572^{20}	1.5	174.4	50	misc alc, chl, eth

No.	Name	Formula								Solubility
c267	trans-Cinnamaldehyde	C₆H₅CH=CHCHO	132.16	7,348	1.0503^{25}	1.6219^{20}	−7.5	246	71	0.014 aq; misc alc, chl, eth
c268	trans-Cinnamic acid	C₆H₅CH=CHCOOH	148.16	9,573	1.2475^{4}	1.614^{43}	134	300	>110	0.05 aq; 16 alc; 8 chl
c269	trans-Cinnamoyl chloride	C₆H₅CH=CHCOCl	166.61	9^{2},390	1.1617^{25}		35–36	258	>110	s hot alc, CCl₄
c269a	Cinnamyl acetate	CH₃CO₂CH₂CH=CHC₆H₅	176.22	6^{2},527	1.057	1.5421^{20}		265	>110	s aq; v s alc, eth
c270	Cinnamyl alcohol	C₆H₅CH=CHCH₂OH	134.18	6,570	1.0397^{35}	1.5758^{33}	33	250.0	>110	v s aq, alc, eth; sl s chl; i bz, PE
c271	Citraconic acid	CH₃C(COOH)=CHCOOH	130.10	2,768	1.62		92 d			
c272	Citraconic anhydride		112.08	17,440	1.247	1.4712^{20}	8	214	101	

Chlorotoluidines, c143, c144, c145, c146, c147
2-Chlorotriethylamine, d272
Chloro-α,α,α-trifluorotoluenes, c51, c52, c53
4-Chloro-α,α,α-trifluoro-o-toluidine, a144

α'-Chloro-α,α,α-trifluoro-m-xylene, t304
Chlorotrihexylsilane, t307
Chloroxylenol, c91

Chromone, b56
Cinchophen, p152
Cinnamyl chloride, c218

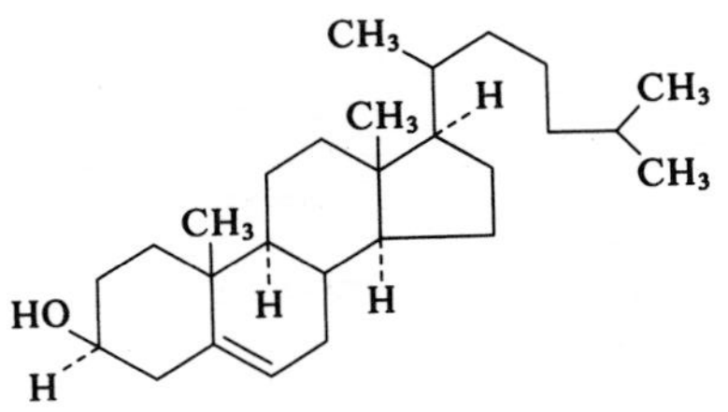

c263

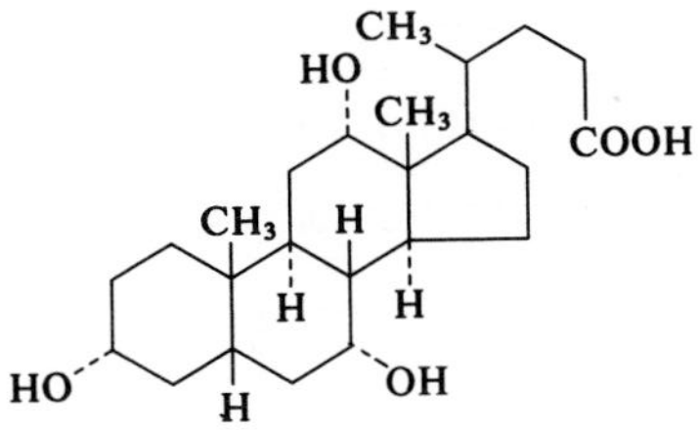

c264

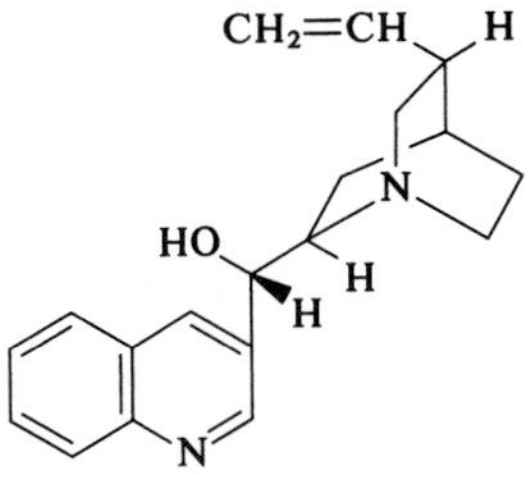

c265

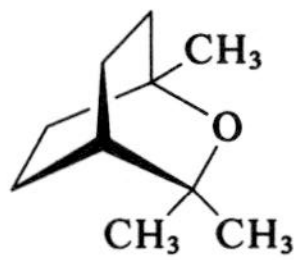

c266

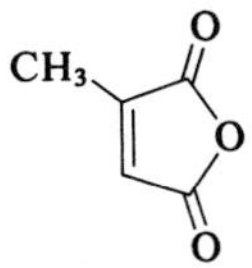

c272

TABLE 1.15 Physical Constants of Organic Compounds (*Continued*)

No.	Name	Formula	Formula weight	Beilstein reference	Density	Refractive index	Melting point	Boiling point	Flash point	Solubility in 100 parts solvent
c273	Citrazinic acid		155.11	22, 254			carbonizes without melting >300			i aq; s alk
c274	Citric acid	HOOCCH$_2$C(OH)(COOH)CH$_2$COOH	192.12	3, 556	1.665		154			59 aq
c275	β-Citronellol	(CH$_3$)$_2$C=CHCH$_2$CH$_2$CH(CH$_3$)CH$_2$CH$_2$OH	156.27	1^1, 232	0.8570^{20}_{4}	1.4556^{20}		222	79	
c276	Cocaine		303.35	22^2, 150		1.5022^{98}	98	$187^{0.1mm}$		0.17 aq; 15 alc; 140 chl; 28 eth
c277	Coumarin		146.15	17, 328	0.9354^{20}_{4}		69	298		0.25 aq; v s alc, chl, eth
c278	Creatine	HOOCCH$_2$N(CH$_3$)C(=NH)NH$_2$	131.14	4, 363			300			1.3 aq; 0.11 alc; i eth
c279	Creatinine		113.12	24, 245			255 d			8 aq; sl s alc; i eth
c280	o-Cresol	CH$_3$C$_6$H$_4$OH	108.14	6, 349	1.0273^{41}	1.5361^{41}	30.9	190.8	81	3.1 aq^{40}; misc alc, chl, eth; s alk
c281	m-Cresol	CH$_3$C$_6$H$_4$OH	108.14	6, 373	1.034^{20}_{4}	1.5438^{41}	12.2	202.7	86	2.5 aq^{40}, misc alc, chl, eth; s alk
c282	p-Cresol	CH$_3$C$_6$H$_4$OH	108.14	6, 389	1.0179^{41}	1.5312^{41}	34.8	201.9	89	2.3 aq^{40}, misc alc, chl, eth; s alk
c283	trans-Crotonaldehyde	CH$_3$CH=CHCHO	70.09	1, 728	0.8516^{20}	1.4373^{20}	−76.5	104.1	8	18.1 aq
c283a	Crotonic anhydride	(CH$_3$CH=CHCO)$_2$O	154.17	2, 411	1.040	1.4741^{20}		248	110	
c284	Crotonyl chloride	CH$_3$CH=CHCOCl	104.54	2, 411	1.091	1.4595^{20}		123	35	
c284a	Cumene hydroperoxide	C$_6$H$_5$C(CH$_3$)$_2$OOH	152.20	6^3, 1814	1.030	1.5210^{20}		101^{8mm}	56	
c285	Cupferron	C$_6$H$_5$N(NO)O$^-$NH$_4^+$	155.16	16^1, 395			163–164			v s aq, alc
c286	Cyanamide	H$_2$NCN	42.04	3^2, 63	1.282^{20}_{4}		46	83^{380mm}	>110	78 aq; 29 BuOH; 42 EtAc; s alc, eth
c287	2-Cyanoacetamide	NCCH$_2$CONH$_2$	84.08	2, 589			119.5		215	25 aq; 3.1 alc
c288	Cyanoacetic acid	NCCH$_2$COOH	85.06	2, 583			65–67	108^{15mm} dec	107	s aq, alc, eth; sl s bz
c289	Cyanoacetohydrazide	NCCH$_2$C(=O)NHNH$_2$	99.09				110			v s aq; s alc; i eth

No.	Name	Formula	M.W.	Ref.	Density	n_D	m.p., °C	b.p., °C	Solubility
c290	Cyanoacetylurea	$NCCH_2C(=O)NHC(=O)NH_2$	127.10	3, 66	1.0588^0		214 d		misc aq, alc; sl s eth
c291	2-Cyanoethanol	$NCCH_2CH_2OH$	71.08	3?, 213				$106–108^{11mm}$ 63^{4mm}	
c292	2-Cyanoethyldichloro-methylsilane	$NCCH_2CH_2Si(CH_3)Cl_2$	168.1		1.2024^{20}	1.455^{20}			
c293	1-Cyano-3-methyliso-thiourea, Na salt	$CH_3NHC(=NCN)S^-Na^+$	137.14	4, 71			290 d		i aq; v s alc, eth
c294	1-Cyanonaphthalene	$C_{10}H_7CN$	153.18	9, 649	1.1113^{25}	1.6298^{18}	38	299	s aq; v s alc, bz, eth
c295	3-Cyanopropyltrichlor-osilane	$NCCH_2CH_2CH_2SiCl_3$	202.6		1.280^{25}	1.465^{25}		$93–94^{8mm}$	
c296	2-Cyanopyridine	$NC–C_5H_4N$	104.11	22, 36	1.5288^{20}		28	215 [89]	v s aq, alc, bz, eth
c297	3-Cyanopyridine	$NC–C_5H_4N$	104.11	22, 41			52	201 [84]	s aq, alc, bz, eth
c298	4-Cyanopyridine	$NC–C_5H_4N$	104.11	22, 46			80		
c299	Cyanotrimethylsilane	$(CH_3)_3SiCN$	99.21		0.7834^{20}	1.3924^{20}	11	114–117 [1]	0.5 aq; s hot alc, pyr; i acet, bz, chl, eth
c300	Cyanuric acid		129.08	26, 239	1.768^0		dec to HOCN		i aq; v s alc, acet
c301	Cyclobutane	C_4H_8	56.10	5, 17	0.7038^0	1.3752^0	−80	13^{740mm}	

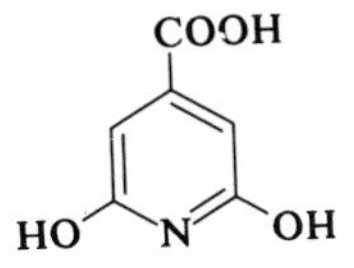

c273

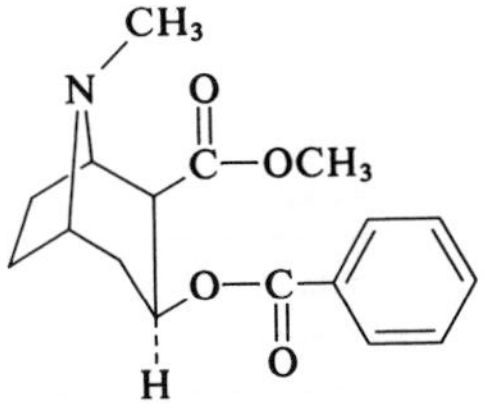

c276

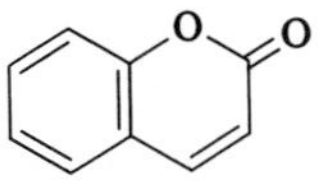

c277

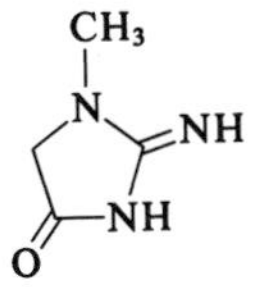

c279

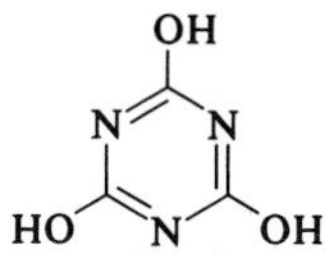

c300

TABLE 1.15 Physical Constants of Organic Compounds (*Continued*)

No.	Name	Formula	Formula weight	Beilstein reference	Density	Refractive index	Melting point	Boiling point	Flash point	Solubility in 100 parts solvent
c302	Cyclodecane	$C_{10}H_{20}$	140.27		0.871	1.4707^{20}		201	65	
c303	Cyclododecanol	$C_{12}H_{23}OH$	184.32		0.906^{62}		77	85^{1mm}		
c304	Cyclododecanone	$C_{12}H_{22}(=O)$	182.31	7^2, 48	0.9254		61	231	87	
c305	*trans,trans,cis*-1,5,9-Cyclododecatriene		162.28	5^4, 1115	0.8925^4	1.5070^{20}	−18	232–245		
c306	*trans*-Cyclododecene		166.31		0.863	1.4822^{20}		124^{7mm}	93	*v* s alc, eth
c306a	Cyclododecylamine	$(C_{12}H_{23})NH_2$	183.34				30		121	
c307	Cycloheptane	C_7H_{14}	98.18	5, 29	0.811^4	1.4455^{20}	−8.0	118.8	6	
c308	*DL-trans*-1,2-Cycloheptanediol	$C_7H_{12}(OH)_2$	130.19	6^3, 4086			61–63	139^{15mm}		
c309	Cycloheptanol	$C_7H_{13}OH$	114.19	6, 10	0.9484^{20}	1.4760^{20}	2	185	71	sl s aq; v s alc, eth
c310	Cycloheptanone	$C_7H_{12}(=O)$	112.17	7, 13	0.9490^{20}	1.4611^{20}		179–181	55	i aq; v s alc; s eth
c311	1,3,5-Cycloheptatriene		92.13	5, 280	0.888	1.5211^{20}	−75.3	115.5	26	s alc, eth; v s bz, chl
c312	Cycloheptene	C_7H_{12}	96.17	5, 65	0.8242^0	1.4585^{20}		114.7	−6	s alc, eth
c313	Cyclohexane	C_6H_{12}	84.16	5, 20	0.7786^{20}	1.4262^{20}	6.5	80.7	−18	0.01 aq; misc alc, bz, acet, eth, CCl_4
c314	Cyclohexane-d_{12}	C_6D_{12}	92.26	5^3, 36	0.89	1.4210^{20}		78	−18	
c315	1,3-Cyclohexanebis(methylamine)	$C_6H_{10}(NHCH_3)_2$	142.25						106	
c316	Cyclohexanecarbaldehyde	$C_6H_{11}CHO$	112.17	7, 19	0.926	1.4500^{20}		163	40	
c317	Cyclohexanecarbonyl chloride	$C_6H_{11}COCl$	146.62	9, 9	1.096	1.4700^{20}		184	66	
c318	Cyclohexanecarboxylic acid	$C_6H_{11}COOH$	128.17	9, 7	1.0480^{15}	1.4530^{20}	29	232.5	>110	0.21 aq; s alc, bz, eth
c319	*cis*-1,2-Cyclohexanediamine	$C_6H_{10}(NH_2)_2$	114.19	13, 1	0.931	1.4864^{20}		92^{18mm}		
c320	*trans*-1,2-Cyclohexanediamine	$C_6H_{10}(NH_2)_2$	114.19	13, 1	0.931	1.4864^{20}		92^{18mm}		
c321	*cis*-1,2-Cyclohexanedicarboxylic anhydride		154.17	17, 452		1.4893^{20}	34	158^{17mm}	>110	
c322	*cis*-1,4-Cyclohexanedimethanol	$C_6H_{10}(CH_2OH)_2$	144.21		0.978^{100}	super-cooled	43	288	74	misc aq, alc; 2.5 eth
c322a	1,4-Cyclohexanediol	$C_6H_{10}(OH)_2$	116.16	6, 741			100	150^{20mm}	65	

Code	Name	Formula								Solubility
c323	1,3-Cyclohexanedione	$C_6H_8(=O)_2$	112.13	7, 554	1.0861^{91}	1.4576^{102}	103–105		43	s aq, alc, acet, chl
c324	1,2-Cyclohexanedione dioxime	$C_6H_8(=NOH)_2$	142.16	7^2, 526			185–188	145–147	>110	s aq
c325	Cyclohexanemethylamine	$C_6H_{11}CH_2NH_2$	113.20	12, 12	0.870	1.4630^{20}				
c326	Cyclohexanepropionic acid	$C_6H_{11}CH_2CH_2COOH$	156.23	9, 82	0.912	1.4636^{20}	14–17	275.8		
c327	Cyclohexanethiol	$C_6H_{11}SH$	116.23	6, 8	0.950	1.4921^{20}		158–160	43	
c328	Cyclohexanol	$C_6H_{11}OH$	100.16	6, 5	0.9416^{30}	1.4629^{30}	25.2	161.1	67	3.8 aq^{25}, misc alc, bz
c329	Cyclohexanone	$C_6H_{10}(=O)$	98.15	7, 8	0.9478^{20}	1.4510^{20}	−45	155.7	46	15 aq^{10}, s alc, eth
c330	Cyclohexanone oxime	$C_6H_{10}(=NOH)$	113.16	7, 10			89–91	206–210		s aq, eth; sl s alc
c331	Cyclohexene	C_6H_{10}	82.15	5, 63	0.8094^{20}	1.4464^{20}	−103.5	83.0	−12	0.02 aq; misc alc, bz, acet, eth
c331a	3-Cyclohexene-1-methanol	$C_6H_9CH_2OH$	112.17	6^3, 215	0.961	1.4853^{20}		85^{18mm}	76	
c331b	Cyclohexene oxide		98.15	17, 21	0.970	1.4520^{20}	−53	130	27	v s alc
c332	2-Cyclohexen-1-one	$C_6H_8(=O)$	96.13	7^2, 55	0.993	1.4885^{20}		168	56	
c333	2,3-Cyclohexenopyridine		133.19	20^2, 176	1.025	1.5440		218	86	
c334	Cyclohexyl acetate	$CH_3CO_2C_6H_{11}$	142.20	6, 7	0.966	1.4395^{20}		173	57	sl s aq; s org solv
c335	Cyclohexylamine	$C_6H_{11}NH_2$	99.18	12, 5	0.8671^{20}	1.4593^{20}	−17.7	134.8	32	misc aq, alc, eth, chl
c338	4-Cyclohexylaniline	$C_6H_{11}C_6H_4NH_2$	175.28	12, 1209			53–56	166^{13mm}	>110	

1,5-Cyclododecadiene-9,10-epoxide, e3
Cyclododecane epoxide, e4
Cyclodecanone isooxime, a323
Cycloheptanone isooxime, a322
Cycloheptyl bromide, b262
2,5-Cyclohexadien-1,4-dione, b59
2,5-Cyclohexadiene-1,4-dione with 1,4-benzenediol (1:1), q1

Cyclohexaneacetic acid, c334
Cyclohexanecarboxylic acid chloride, c317
(E)-1,4-Cyclohexane diisocyanate, c339a
Cyclohexaneethanol, c339b
Cyclohexaneethyl acetate, c339c
Cyclohexanemethanol, c343

Cyclohexanone cyanohydrin, h110
(Z)-4-Cyclohexene-1,2-dicarboximide, t77
(Z)-4-Cyclohexene-1,2-dicarboxylic anhydride, t76
N-(1-Cyclohexen-1-yl)morpholine, m453
N-(1-Cyclohexen-1-yl)pyrrolidine, p278
Cyclohexyl alcohol, c328

c305   c306   c311   c321   c331b   c333

TABLE 1.15 Physical Constants of Organic Compounds (*Continued*)

No.	Name	Formula	Formula weight	Beilstein reference	Density	Refractive index	Melting point	Boiling point	Flash point	Solubility in 100 parts solvent
c339	Cyclohexylbenzene	$C_6H_{11}C_6H_5$	160.26	5, 503	0.9502^{20}	1.5258^{20}	5–6	240	98	i aq; v s alc, eth
c339a	trans-1,4-Cyclohexylenediisocyanate	$C_6H_{10}(NCO)_2$	166.18	13^3, 12			64–66			
c339b	2-Cyclohexylethanol	$C_6H_{11}CH_2CH_2OH$	128.22	6, 17	0.919	1.4647^{20}		207^{745mm}	86	
c339c	2-Cyclohexylethyl acetate	$CH_3CO_2CH_2CH_2C_6H_{11}$	170.25		0.949	1.4461		98^{15mm}	81	
c340	N-Cyclohexylformamide	$C_6H_{11}NHCHO$	127.18	12^2, 11			38–40	137^{10mm}	>110	
c341	Cyclohexyl isocyanate	$C_6H_{11}NCO$	125.17	12^2, 12	0.980	1.4551^{20}		168–170	48	
c342	Cyclohexyl isothiocyanate	$C_6H_{11}NCS$	141.24	12^2, 12	0.996	1.5350^{20}		219	95	
c343	Cyclohexylmethanol	$C_6H_{11}CH_2OH$	114.19	6, 14	0.9215^{25}	1.4640^{25}		181	71	s alc, eth
c343a	3-Cyclohexyl-1-propanol	$C_6H_{11}CH_2CH_2CH_2OH$	142.24	6^1, 15	0.937	1.4660^{20}		218	101	
c344	N-Cyclohexyl-2-pyrrolidinone		167.25	21^3, 3149	1.026	1.495	12	284		
c345	cis,cis-1,3-Cyclooctadiene		108.18	5^4, 401	0.869	1.4928^{20}	−53	55^{34mm}	24	
c346	1,5-Cyclooctadiene		108.18	5, 116	0.8818^{25}	1.4905^{25}	−69	150	31	s CCl$_4$
c347	Cyclooctane	C_8H_{16}	112.22	5, 35	0.834	1.4574^{20}	14.8	151.1	30	
c347a	trans-1,2-Cyclooctanediol	$C_8H_{14}(OH)_2$	144.21	6^3, 4094	1.080	1.4980^{20}	32	$94^{0.5mm}$	>110	
c347b	1,4-Cyclooctanediol	$C_8H_{14}(OH)_2$	144.21	6^4, 5228	0.9740^{20}	1.4850^{20}	83	108^{22mm}	86	
c348	Cyclooctanol	$C_8H_{15}OH$	128.22	6^2, 25	0.9584^{20}	1.6494^{20}	14–15	195–197	72	
c349	Cyclooctanone	$C_6H_{14}(=O)$	126.20	7, 21	0.846	1.4698^{20}	41–43	145–146	25	
c350	cis-Cyclooctene	C_8H_{14}	110.20	5^1, 35	0.928	1.4804^{20}	−16	190	62	
c351	Cyclooctylamine	$C_8H_{15}NH_2$	127.23		1.1558^{20}	1.4679^{20}	−48	169–170		
c352	Cyclopentamethylenedichlorosilane		169.1							
c353	Cyclopentane	C_5H_{10}	70.13	5, 19	0.7460^{20}	1.4065^{20}	−93.9	49.3	−37	i aq; misc alc, eth
c354	Cyclopentanecarboxylic acid	C_5H_9COOH	114.14	9, 6	1.0532^{20}	1.4540^{20}	4	216	93	sl s aq; s MeOH
c355	cis,cis,cis,cis-1,2,3,4-cyclopentanetetracarboxylic acid	$C_5H_6(COOH)_4$	246.17	9^2, 724			192–195 d			

c356	Cyclopentanol	C_5H_9OH	86.13	6, 5	0.9488^{20}_{4}	1.4521^{20}	−19	140.9	51	sl s aq; s alc
c357	Cyclopentanone	$C_5H_8(=O)$	84.12	7, 5	0.9509^{18}_{4}	1.4366^{20}	−58	130.6	30	sl s aq; misc alc, eth
c358	Cyclopentanone oxime	$C_5H_8(=NOH)$	99.13	7, 7			53–55	196	92	s aq, alc, bz, chl, eth
c359	Cyclopentene	C_5H_8	68.11	5, 61	0.774	1.4228^{20}	−135.1	44.2	−34	
c360	2-Cyclopentene-1-acetic acid	$C_5H_7CH_2COOH$	126.16	9, 42	1.047	1.4675^{20}	19	$94^{2.5mm}$	>110	
c361	2,3-Cyclopenteneopyridine		119.17		1.018	1.5445^{20}		88^{11mm}	67	
c362	N-(1-Cyclopenten-1-yl)morpholine		153.23		0.957	1.5105^{20}		106^{12mm}	60	
c363	Cyclopentyl phenyl ketone	$C_5H_9COC_6H_5$	174.24		1.034	1.5428^{20}			>110	
c364	3-Cyclopentylpropionic acid	$C_5H_9CH_2CH_2COOH$	142.20		0.996	1.4570^{20}		130^{12mm}	46	
c365	Cyclopropane	C_3H_6	42.08	5, 15	0.720^{-79}_{4}	1.4207^{20}	−127.4	−32.8		37 mL/100 mL aq[15]; v s alc, eth
c366	Cyclopropanecarbonitrile	C_3H_5CN	67.09	9, 4	0.911^{16}	1.4522^{20}		135	32	s eth
c367	Cyclopropanecarbonyl chloride	C_3H_5COCl	104.54	9, 4	1.152	1.4522^{20}		119	23	
c368	Cyclopropanecarboxylic acid	C_3H_5COOH	86.09	9, 4	1.008	1.4380^{20}	17–19	182–184	71	sl s hot aq; s alc, eth

Cyclohexylbenzene, p106
Cyclohexyl bromide, b263
Cyclohexyl chloride, c77
Cyclohexyl ketone, c329

Cyclohexyl mercaptan, c327
Cyclohexylmethane, m195
Cyclohexylmethyl bromide, b309
Cyclopentanepropanoic acid, c364

Cyclopentene oxide, e38
Cyclopentyl bromide, b265
Cyclopentyl chloride, c79

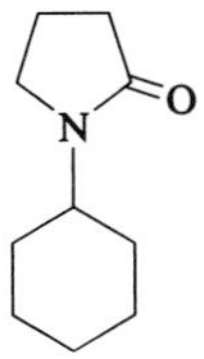

c344

c345

c346

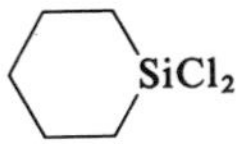

c252

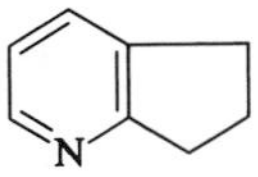

c361

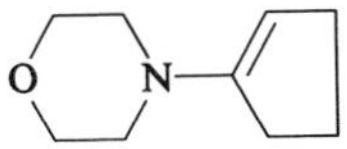

c362

TABLE 1.15 Physical Constants of Organic Compounds (*Continued*)

No.	Name	Formula	Formula weight	Beilstein reference	Density	Refractive index	Melting point	Boiling point	Flash point	Solubility in 100 parts solvent
c369	Cyclopropyl methyl ketone	$C_3H_5COCH_3$	84.12	7, 7	0.8993^{20}_4	1.4241^{20}		114	21	s aq, alc, eth
c370	Cystamine dihydrochloride	$H_2NCH_2CH_2SSCH_2CH_2NH_2 \cdot 2HCl$	225.20	4, 287			217 d			
c371	L-(+)-Cysteine	$HSCH_2CH(NH_2)COOH$	121.16	4, 506			220 d			v s aq, alc; i bz, eth
c372	L-Cystine	$HOOCCH(NH_2)SSCH_2CH(NH_2)COOH$	240.30	4, 507			d 240			0.01 aq; s acid, alk; i alc
d1	cis-Decahydronaphthalene	$C_{10}H_{18}$	138.26	5, 92	0.8963^{20}_4	1.4810^{20}	−43.0	195.8	57 (CC)	v s alc, chl, eth; misc most ketones, esters
d2	trans-Decahydronaphthalene	$C_{10}H_{18}$	138.26	5^2, 56	0.8700^{20}_4	1.4697^{20}	−30.4	187.3	52	see under cis
d4	Decamethylcyclopentasiloxane	$[-Si(CH_3)_2O-]_5$	370.8		0.9593^{20}	1.3982^{20}	−38	101^{20mm}		i aq
d5	Decamethyltetrasiloxane	$(CH_3)_3SiO[Si(CH_3)_2O]_2Si(CH_3)_3$	310.7	4^3, 1879	0.8536^{20}	1.3880^{20}	−70	194–195	62	sl s alc; s bz, PE
d6	Decanal	$H(CH_2)_9CHO$	156.27	1, 711	0.830^{15}	1.4280^{20}		207–209	85	i aq; s alc, eth
d7	Decane	$CH_3(CH_2)_8CH_3$	142.29	1, 168	0.7301^{20}_4	1.4119^{20}	−29.7	174.1	46	0.07 aq
d8	1,10-Decanediamine	$H_2N(CH_2)_{10}NH$	172.32	4, 273			62–63	140^{12mm}		
d9	Decanedioic acid	$HOOC(CH_2)_8COOH$	202.25	2, 718	1.207^{20}_4	1.422^{134}	134.5	295^{100mm}		0.1 aq; v s alc, esters, ketones
d10	1,10-Decanediol	$HO(CH_2)_{10}OH$	174.28	1^2, 560			72–75	170^{8mm}		sl s alc, eth; v s alc
d11	Decanedioyl dichloride	$ClC(O)(CH_2)_8COCl$	239.14	2, 719	1.1212^{15}	1.4678^{20}		220^{75mm}	>110	d aq, alc
d12	Decanenitrile	$CH_3(CH_2)_8CN$	153.27	2, 356	0.8295^{15}	1.4295^{20}	−15	235–237		misc alc, chl, eth
d13	1-Decanethiol	$CH_3(CH_2)_9SH$	174.35	1^2, 459	0.841	1.4536^{20}	−26	114^{13mm}	98	i aq
d14	Decanoic acid	$CH_3(CH_2)_8COOH$	172.27	2^2, 309	0.8782^{50}	1.4288^{40}	31.4	270	>110	0.015 aq; s alc, chl, bz, eth, CS_2
d15	1-Decanol	$CH_3(CH_2)_9OH$	158.29	1, 425	0.8297^{20}_4	1.4371^{20}	6.9	230.2	82 (OC)	i aq; s alc, eth
d15a	δ-Decanolactone		170.25		0.945	1.4584^{20}		$120^{0.02mm}$	>110	
d15b	3-Decanone	$CH_3(CH_2)_6COC_2H_5$	156.27	1^1, 367	0.825	1.4241^{20}	−3.8	205	26	i aq; misc alc, eth
d16	4-Decanone	$CH_3(CH_2)_6C(=O)(CH_2)_2CH_3$	156.27	1, 711	0.824^{20}_6	1.4237^{20}		207	71	d aq, alc; s eth
d17	Decanoyl chloride	$CH_3(CH_2)_8C(=O)Cl$	190.71	2, 356	0.919	1.4410^{20}	−34.5	96^{5mm}	106	i aq; misc alc, eth
d18	1-Decene	$CH_3(CH_2)_7CH=CH_2$	140.27	1^3, 858	0.7408^{20}_4	1.4215^{20}	−66.3	170.6	47	sl s aq; misc alc, bz, eth, acet
d19	Decylamine	$CH_3(CH_2)_9NH_2$	157.30	4, 199	0.787	1.4360^{20}	12–14	216–218	85	

No.	Name	Formula	Mol. wt.	Beilstein ref.	Density	n_D	mp	bp	Flash pt.	Solubility
d20	Dehydroabietylamine	(see structure d20)	285.48	12^4, 3005		1.5460^{20}	111–113	269.9	>110	22 acet; 18 bz; 5 MeOH
d21	Dehydroacetic acid	(see structure d21)	168.15	17, 559						
d22	Deoxybenzoin	$C_6H_5CH_2C(=O)C_6H_5$	196.25	7, 431	1.2010^4		55–56	320		
d23	Diacetoxydimethyl-silane	$(CH_3)_2Si(OOCCH_3)_2$	176.3		1.0544^{20}	1.4030^{20}		164–166	>110	i aq; v s alc, eth
d24	1,1-Diacetoxy-2-butene	$(CH_3CO)_2CHCH=CHCH_3$	172.18	2, 154	1.057	1.4295^{20}		106^{20mm}		
d25	Diallylamine	$(H_2C=CHCH_2)_2NH$	97.16	4, 208	0.787	1.4405^{20}	−88	111–112	15	
d26	Diallyl ether	$(H_2C=CHCH_2)_2O$	98.15	1^2, 477	0.8050^{18}	1.4240^{20}	−47	94	−6 (OC)	
d26a	Diallyl maleate	$H_2C=CHCH_2O_2CCH=CHCO_2CH_2CH=CH_2$	196.20	2^3, 1926	1.073	1.4702^{20}		116^{4mm}	>110	i aq; misc alc, eth
d26b	Diallyl 1,2-phthalate	$C_6H_4(CO_2CH_2CH=CH_2)_2$	246.27	9^3, 4120	1.121	1.5187^{20}		167^{5mm}	110	
d27	Diallyl sulfide	$(H_2C=CHCH_2)_2S$	114.21	1, 440	0.8877^{47}	1.4889^{20}	−83	138	46	sl s aq; misc alc, eth
d27a	(+)-N,N'-Diallyl tartar-diamide	$[-CH(OH)CONHCH_2CH=CH_2]_2$	228.25	4, 218			186–188			
d28	1,2-Diaminoanthra-quinone	(see structure d28)	238.25	14^1, 459			289–291			sl s alc, eth

d20 (structure): CH_3, CH_2NH_2, H_3C, $CH(CH_3)_2$

d21 (structure): CH_3, CH_3, O, O, $C=O$

d28 (structure): NH_2, NH_2, O, O

TABLE 1.15 Physical Constants of Organic Compounds (*Continued*)

No.	Name	Formula	Formula weight	Beilstein reference	Density	Refractive index	Melting point	Boiling point	Flash point	Solubility in 100 parts solvent
d29	1,4-Diaminoanthraquinone		238.25	14, 197			265–268			sl s aq, alc; v s bz
d29a	1,5-Diaminoanthraquinone		238.25	14, 203			308 d			sl s hot aq, pyr
d30	2,6-Diaminoanthraquinone		238.25	14, 215			>325			
d33	3,5-Diaminobenzoic acid	$(H_2N)_2C_6H_3COOH$	152.15	14, 453			228	$-H_2O$, 110		sl s aq; s alc, eth
d34	4,4'-Diaminodiphenylamine sulfate	$H_2NC_6H_4NHC_6H_4\text{-}NH_2 \cdot H_2SO_4$	297.33	13, 110			300			
d35	4,4'-Diaminodiphenylmethane	$H_2NC_6H_4CH_2C_6H_4NH_2$	198.27	13, 238			91–92	398	221	sl s aq; v s alc, bz, eth
d36	3,3'-Diaminodiphenyl sulfone	$H_2NC_6H_4SO_2C_6H_4NH_2$	248.30	13, 426			167–170			i aq; s alc, bz
d37	4,4'-Diaminodiphenyl sulfone	$H_2NC_6H_4SO_2C_6H_4NH_2$	248.30	13, 536			175–177			i aq; s alc, acet, HCl
d39	2,4-Diamino-6-hydroxypyrimidine		126.12	24, 469			285 d			s aq
d40	Diaminomaleonitrile	$NCC(NH_2){=}C(NH_2)CN$	108.10	4², 949			178–179			
d41	1,8-Diamino-p-menthane		170.30	13, 4	0.914	1.4805^{20}	−45	107–125^{10mm}	93	
d42	3,3'-Diamino-N-methyldipropylamine	$CH_3N[(CH_2)_3NH_2]_2$	145.25	4⁴, 1279		1.4725^{20}		112^{6mm}	102	
d43	1,3-Diamino-2-propanol	$H_2NCH_2CH(OH)CH_2NH_2$	90.13	4, 290			40–45	235	>110	
d44	2,6-Diaminopyridine	$(H_2N)_2C_5H_3N$	109.13	22¹, 647			118–120			s aq, alc
d45	1,4-Diazabicyclo[2.2.2]octane		112.18	23³, 484			158	174		45 aq; 77 EtOH; 51 bz; 13 acet; 26 MeEtKe
d46	1,8-Diazabicyclo[5.4.0]undec-7-ene		152.24		1.018	1.5219^{20}		80$^{0.6mm}$	>110	s eth, diox
d47	Diazomethane	$CH_2{=}\overset{+}{N}{=}\overset{-}{N}$	42.04	23, 25			−145	−23	VERY EX-PLO-SIVE	

1,4-Diaminobutane, b382
1,2-Diaminocyclohexane, c319, c320
1,10-Diaminodecane, d8
p-Diaminodiphenyl, b137
3,3′-Diaminodipropylamine, i9
1,12-Diaminododecane, d722
1,2-Diaminoethane, e15
1,7-Diaminoheptane, h7
1,6-Diaminohexane, h56
1,3-Diamino-2-hydroxypropane, d43
Diaminonaphthalenes, n4, n5

1,2-Diamino-4-nitrobenzene, n68
1,4-Diamino-2-nitrobenzene, n67
1,9-Diaminononane, n93
1,8-Diaminooctane, o23
1,5-Diaminopentane, p29
2,5-Diaminopentanoic acid, o46
1,2-Diaminopropane, p195
1,3-Diaminopropane, p196
4,6-Diamino-4-pyrimidinol, d39
Diaminotoluenes, t171, t172, t173, t174

1,3-Diaminourea, c11
4,5-Diamino-o-xylene, d587, d588
Diamylamine, d652
Diamyl ether, d653
Diamyl ketone, u6
1,2-Dianilinoethane, d671
Diazoacetic ester, d115
1,3-Diazole, i4
Dibenzo-18-crown-6, d51

d29   d29a   d30

d45   d46   d48   d39   d49   d41   d50

TABLE 1.15 Physical Constants of Organic Compounds (*Continued*)

No.	Name	Formula	Formula weight	Beilstein reference	Density	Refractive index	Melting point	Boiling point	Flash point	Solubility in 100 parts solvent
d51	2,3,11,12-Dibenzo-1,4,7,10,13-hexaoxacyclooctadeca-2,11-diene		360.41				162–164			
d52	Dibenzothiophene		184.26	17, 72			97–100	333		s aq; v s alc, bz
d53	Dibenzoylmethane	$C_6H_5C(=O)CH_2C(=O)C_6H_5$	224.26	7, 769			78–79	220^{18mm}		s alc; v s eth
d54	Dibenzoyl peroxide	$C_6H_5C(O)O$—$OC(O)C_6H_5$	242.23	9, 179			103–106	may explode when heated		sl s aq, alc; s bz, chl, eth
d55	(−)-Dibenzoyl-L-tartaric acid hydrate	$[(C_6H_5COOCH(COOH)$—$]_2 \cdot H_2O$	376.34	9, 170						
d56	Dibenzylamine	$C_6H_5CH_2NHCH_2C_6H_5$	197.28	12, 1035	1.026	1.5731^{20}	−26	300	143	i aq; s alc, eth
d57	Dibenzyl disulfide	$C_6H_5CH_2SSCH_2C_6H_5$	246.39	6, 465			69	d > 270		s hot alc, bz, eth
d58	Dibenzyl ether	$C_6H_5CH_2OCH_2C_6H_5$	198.27	6, 434	1.0014^{20}	1.5610^{20}	3.5	298 d	135 (CC)	misc alc, acet, chl, eth
d59	N,N'-Dibenzylethylenediamine	$(C_6H_5CH_2NHCH_2$—$)_2$	240.35	12, 1067	1.024^{20}	1.5624^{20}	26	195^{4mm}	>110	v s alc, bz, chl, eth
d60	Dibenzyl malonate	$CH_2(COOCH_2C_6H_5)_2$	284.31	6, 436	1.137	1.5447^{20}		$188^{0.2mm}$	>110	
d61	Dibenzyl phosphonate	$(C_6H_5CH_2O)_2P(O)H$	262.25	2, 218	1.187	1.5540^{20}	−5 to +5	$110^{0.01mm}$	>110	
d62	Dibromoacetic acid	$Br_2CHCOOH$	217.86	2, 219			39–41	130^{16mm}	>110	
d63	Dibromoacetonitrile	Br_2CHCN	198.86	7, 285	2.296			69^{24mm}	none	
d64	2,4'-Dibromoacetophenone	$BrC_6H_4C(=O)CH_2Br$	277.96			1.5393^{20}	108–110		none	s warm alc, eth
d65	1,4-Dibromobenzene	$C_6H_4Br_2$	235.92	5, 211	0.9641^{100}	1.5743^{100}	87.3	219	none	1.4 alc; s bz; 101 eth
d66	4,4'-Dibromobiphenyl	$BrC_6H_4C_6H_4Br$	312.00	5, 580			162–163	355–360		s bz; sl s hot alc
d67	1,3-Dibromobutane	$CH_3CH(Br)CH_2CH_2Br$	215.93	1, 120	1.800^{20}	1.5085^{20}		175	none	s chl, eth
d68	1,4-Dibromobutane	$BrCH_2CH_2CH_2CH_2Br$	215.93	1, 120	1.8080^{20}	1.5186^{20}	−20	198	>110	s chl
d68a	2,3-Dibromobutane	$CH_3CHBrCHBrCH_3$	215.93	1, 120	1.756	1.5126^{20}			>110	s chl
d69	1,4-Dibromo-2,3-butanedione	$BrCH_2C(=O)C(=O)CH_2Br$	243.89	1, 774			116–117	108^{160mm}		
d70	trans-2,3-Dibromo-2-butene-1,4-diol	$HOCH_2C(Br)=C(Br)CH_2OH$	245.91	1', 260			112–114			
d71	Dibromochloromethane	$HCClBr_2$	208.29	1, 67	2.451	1.5465^{20}	−22	120^{748mm}	none	misc alc, bz, eth

No.	Name	Formula	Mol. wt.	Beil. ref.	Density	n_D	m.p., °C	b.p., °C	Flash, °C	Solubility
d72	1,2-Dibromo-2-chloro-1,1,2-trifluoroethane	$FCCl(Br)C(Br)F_2$	276.5		2.2478^{20}	1.4275^{20}		94		sl s alc; s eth
d73	1,10-Dibromodecane	$Br(CH_2)_{10}Br$	300.09	1^1, 64	1.335^{30}	1.4912^{20}	27	160^{15mm}	>110	i aq
d74	1,2-Dibromo-1,1-difluoroethane	$CH_2BrC(Br)F_2$	223.87	1, 92	2.2238^{20}	1.4456^{20}	−61.3	93.4		
d75	Dibromodifluoromethane	Br_2CF_2	209.81	1^1, 16	2.288^{15}	1.3999^{12}	−141.6	24		0.1 aq; misc alc, bz, chl, eth
d76	1,3-Dibromo-5,5-dimethylhydantoin		285.93				197 d			
d77	1,1-Dibromoethane	CH_3CHBr_2	187.87	1, 90	2.0552^{20}	1.5379^{20}		113	none	i aq; v s alc, eth
d78	1,2-Dibromoethane	$BrCH_2CH_2Br$	187.87	1, 90	2.1802^{20}	1.5416^{15}	10.0	131.7	none	0.43 aq; misc alc, eth
d79	(1,2-Dibromoethyl)benzene	$C_6H_5CH(Br)CH_2Br$	263.97	5, 356			70–74	140^{15mm}		
d80	cis-1,2-Dibromoethylene	$BrCH=CHBr$	185.86	1, 190	2.211^7	1.5431^{18}	−53	112.5	none	s alc, bz, chl, eth
d81	trans-1,2-Dibromoethylene	$BrCH=CHBr$	185.86	1, 190	2.246	1.5505^{18}	−6.5	108	none	
d82	1,2-Dibromoethyltrichlorosilane	$BrCH_2CH(Br)SiCl_3$	321.3		2.0462^{20}	1.537^{20}		90^{11mm}		
d83	4',5'-Dibromofluorescein		490.12	19, 228			270–273			s hot alc, HOAc
d84	2,4-Dibromo-1-fluorobenzene	$Br_2C_6H_3F$	253.91		2.047^{20}	1.5840^{20}		105^{22mm}	92	

Dibenzo[b,e]pyridine, a61
Dibenzopyrrole, d667

Dibenzoyl, b34
Dibenzyl, d668

Dibenzyl ketone, d686

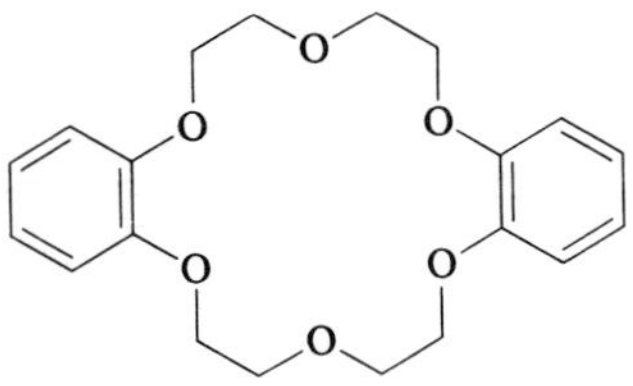

d51

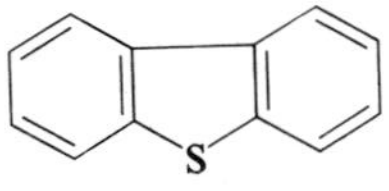

d52

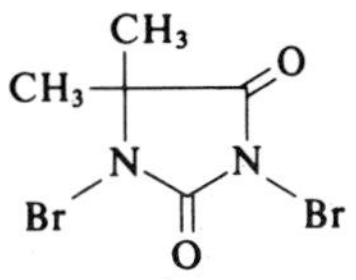

d76

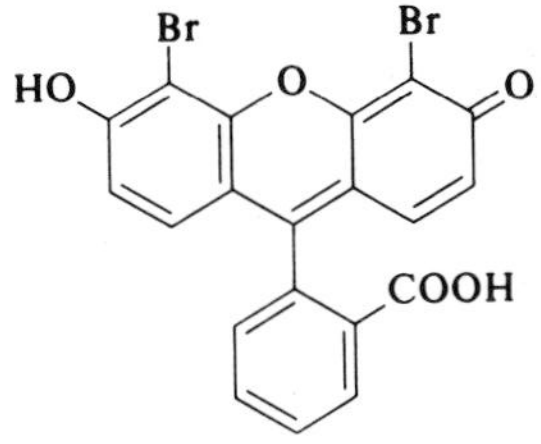

d83

TABLE 1.15 Physical Constants of Organic Compounds (*Continued*)

No.	Name	Formula	Formula weight	Beilstein reference	Density	Refractive index	Melting point	Boiling point	Flash point	Solubility in 100 parts solvent
d85	1,2-Dibromohexafluoropropane	$CF_3CF(Br)C(Br)F_2$	309.83				72.8			misc eth
d86	1,6-Dibromohexane	$Br(CH_2)_6Br$	243.98	1, 145	1.5864^8	1.5066^{20}		243	110	s alc, bz; v s eth
d87	5,7-Dibromo-8-hydroxyquinoline		302.96	21, 97			200–201	subl		
d88	Dibromomethane	CH_2Br_2	173.85	1, 67	2.4956^{24}	1.5419^{20}	−52.7	96–97	none	1.15 aq; misc alc, bz, acet, chl, eth
d89	2,6-Dibromo-4-nitroaniline	$Br_2C_6H_2(NO_2)NH_2$	295.93	12, 743			206–208			sl s aq; s HOAc
d90	2,5-Dibromonitrobenzene	$Br_2C_6H_3NO_2$	280.91	5, 250	1.9581^{111}		82–84			s bz, hot alc
d90a	1,8-Dibromooctane	$Br(CH_2)_8Br$	272.03	1, 160	1.477	1.4981^{20}	16	272	>110	
d91	1,5-Dibromopentane	$Br(CH_2)_5Br$	229.95	1, 131	1.6879^{15}	1.5092^{15}	−34	$110^{1.5mm}$	110	
d91a	2,4-Dibromophenol	$Br_2C_6H_3OH$	251.92	6, 202			42	154^{11mm}	110	
d92	1,2-Dibromopropane	$CH_3CH(Br)CH_2Br$	201.90	1, 109	1.933^{20}	1.5203^{20}	−55.5	139.6	none	0.2 aq; misc alc, bz, chl, eth
d93	1,3-Dibromopropane	$BrCH_2CH_2CH_2Br$	201.90	1, 110	1.9712^{25}	1.5233^{20}	−34	166.8	54	0.17 aq; s alc, eth
d93a	1,3-Dibromo-2-propanol	$BrCH_2CH(OH)CH_2Br$	217.90	1, 365	2.136	1.5514^{20}		837^{mm}	46	
d94	2,3-Dibromo-1-propanol	$BrCH_2CH(Br)CH_2OH$	217.90	1, 357	2.1204^{20}	1.5599^{20}		97^{10mm}	>110	sl s aq; misc alc, bz, eth, acet
d95	2,3-Dibromopropene	$BrCH_2C(Br)=CH_2$	199.88	1, 201	1.9336^4	1.5470^{20}	64–66	140–143	81	s aq, alc, bz
d96	2,3-Dibromopropionic acid	$BrCH_2CH(Br)COOH$	231.88	2, 258				160^{20mm}		
d97	2,6-Dibromopyridine	$Br_2C_5H_3N$	236.91	20², 153			118–119	255		v s aq, alc
d98	meso-2,3-Dibromosuccinic acid	$HOOCCH(Br)CH(Br)COOH$	275.89	2, 625			>275 subl			
d99	1,2-Dibromotetrachloroethane	$BrCCl_2CCl_2Br$	325.65	1, 93	2.713		220–222		none	
d100	1,2-Dibromotetrafluoroethane	$BrCF_2CF_2Br$	259.83		2.163^{25}	1.367^{25}	−110.5	47.3		
d101	2,5-Dibromothiophene	BrC_4H_2S	241.94	17, 33	2.1472^{23}	1.6289^{20}		211	99	i aq; v s alc, eth
d102	α,α-Dibromotoluene	$C_6H_5CHBr_2$	249.94	5, 308	1.510^{15}	1.6147^{20}	−6	156^{23mm}	110	i aq; misc alc, eth
d103	1,2-Dibromo-1,1,2-trifluoroethane	$HC(Br)FC(Br)F_2$	241.8	1, 92	2.274^{27}	1.4191^{24}		76.5		

No.	Name	Formula	Mol. wt.	Beil. ref.	Density	n	m.p. °C	b.p. °C	F.p. °C	Solubility
d104	α,α-Dibromo-o-xylene	$C_6H_4(CH_2Br)_2$	263.97	5, 366	1.960		92–94	245		sl s alc, chl, eth
d105	α,α-Dibromo-p-xylene	$C_6H_4(CH_2Br)_2$	263.97	5, 385	2.012^0		142–143	203.6		v s alc, chl; s eth
d106	1,2-Dibutoxyethane	$C_4H_9OCH_2CH_2OC_4H_9$	174.28		0.8374^{20}	1.4131^{20}	−69.1			0.2 aq; misc alc, acet
d107	Dibutylamine	$(C_4H_9)_2NH$	129.25	4, 157	0.7601^{20}	1.4177^{20}	−62	159.6	33	0.47 aq; s alc, acet, eth, EtAc, PE
d107a	Di-sec-butylamine	$[C_2H_5CH(CH_3)]_2NH$	129.25	4, 162	0.753	1.4100^{20}		135	20	
d108	N,N-Dibutylaminoethanol	$(C_4H_9)_2NCH_2CH_2OH$	173.29		0.860^{20}_{20}	1.444^{20}	<−70	227–230	104 (OC)	
d109	N,N-Dibutylaniline	$C_6H_5N(C_4H_9)_2$	205.34	12^2, 95	0.904^{20}	1.5197^{20}		267–275	110	i aq, MeOH; s acet, bz, EtOH, EtAc, eth
d110	Dibutyl decanedioate	$C_4H_9OOC(CH_2)_8COOC_4H_9$	314.45	2, 719	0.9366^{20}	1.4415^{20}	1.0	345	177	0.004 aq
d111	Di-tert-butyl dicarbonate	$(CH_3)_3COC(O)OC(CH_3)_3$	218.25		0.950	1.4103^{20}	23	$56^{0.5mm}$	37	
d112	2,5-Di-tert-butyl-1,4-dihydroxybenzene	$[(CH_3)_3C]_2C_6H_2(OH)_2$	222.33				217–219			
d113	Dibutyl disulfide	$C_4H_9SSC_4H_9$	178.36	1^2, 400	0.93383^{20}	1.4920^{20}	−71	231.2	93	i aq; misc alc, eth
d114	Di-tert-butyl disulfide	$(CH_3)_3CSSC(CH_3)_3$	178.36		0.935	1.4920		229–233	93	
d115	Dibutyl ether	$C_4H_9OC_4H_9$	130.22	1, 369	0.76894^{20}	1.3992^{20}	−97.9	142.4	25	0.03 aq; misc alc, eth
d116	N,N-Dibutylformamide	$HC(=O)N(C_4H_9)_2$	157.26		0.864	1.4429^{20}		120^{15mm}	100	
d116a	Dibutyl hexanedioate	$[—CH_2CH_2CO_2(CH_2)_3CH_3]_2$	258.36	2^2, 575	0.962	1.4358^{20}		305	110	
d117	3,5-Di-tert-butyl-4-hydroxybenzoic acid	$[(CH_3)_3C]_2C_6H_2(OH)COOH$	250.34				206–209			

5,7-Dibromo-8-quinolinol, d87
Dibutyl adipate, d116a
Dibutyl 1,2-benzenedicarboxylate, d128

Dibutyl butanedioate, d130
Dibutyl cellosolve, d106

Di-tert-butylcresol, d120
2,5-Di-tert-butylhydroquinone, d112

d87

TABLE 1.15 Physical Constants of Organic Compounds (*Continued*)

No.	Name	Formula	Formula weight	Beilstein reference	Density	Refractive index	Melting point	Boiling point	Flash point	Solubility in 100 parts solvent
d118	Dibutyl maleate	$C_4H_9OOCCH{=}CHCOOC_4H_9$	228.28	2^3, 1925	0.9950^{20}	1.4454^{20}	<-80	d 280	110	0.05 aq
d119	Di-*tert*-butyl malonate	$CH_2COOC(CH_3)_3\ |\ COOC(CH_3)_3$	216.27	2^3, 1621	0.8947^{5}	1.4184^{20}	-6.0	93^{10mm}	88	
d120	2,6-Di-*tert*-butyl-4-methylphenol	$[(CH_3)_3C]_2C_6H_2(CH_3)OH$	220.36	6^3, 2073		1.4859^{75}	70	265		i aq; s alc, bz, acet
d120a	Dibutyl octanedioate	$[{-}(CH_2)_3CO_2(CH_2)_3CH_3]_2$	286.41	2^3, 1767	0.948	1.4390^{20}		$176^{4.5mm}$	>110	
d121	Dibutyl oxalate	$C_4H_9OOC{-}COOC_4H_9$	202.25	2, 540	0.9862^{20}	1.4232^{20}	-30.0	239–240	108	misc alc, ketones, PE
d122	Di-*tert*-butyl peroxide	$(CH_3)_3CO{-}OC(CH_3)_3$	146.23	1^3, 1580	0.794^{20}	1.3890^{20}	-40	110	1	misc acet, octane
d123	2,4-Di-*tert*-butylphenol	$[(CH_3)_3C]_2C_6H_3OH$	206.33				56.5	263.5	115	s hot alc; i alk
d124	2,6-Di-*sec*-butylphenol	$[CH_3CH_2CH(CH_3)]_2C_6H_3OH$	206.33		0.918	1.5100^{20}	-42	255–260	127	
d125	2,6-Di-*tert*-butylphenol	$[(CH_3)_3C]_2C_6H_3OH$	206.33	6^3, 2061			35–38	253	118	s hot alc; i alk
d126	3,5-Di-*tert*-butylphenol	$[(CH_3)_3C]_2C_6H_3OH$	206.33				87–89			
d127	Dibutyl phosphonate	$(C_4H_9O)_2P(O)H$	194.21	1^1, 187	0.9954^{20}	1.4231^{20}		118^{11mm}	121	sl s (hyd) aq; misc alc, acet, eth
d128	Dibutyl 1,2-phthalate	$C_6H_4[COOC_4H_9]_2$	278.35	9^2, 586	1.0465^{30}	1.4926^{20}	-35	340	171	0.01 aq; v s alc, bz, acet, eth
d129	*N,N*-Dibutyl-1,3-propanediamine	$C_4H_9NHCH_2CH_2CH_2NHC_4H_9$	186.34		0.827	1.4463^{20}		205	103	
d130	Dibutyl succinate	$[C_4H_9OOCCH_2{-}]_2$	230.30	2^2, 551	0.9768^{20}	1.4299^{20}	-29.0	274.5		i aq; s alc, eth
d131	Dibutyl sulfate	$C_4H_9OSO_2OC_4H_9$	210.29		1.0592^{5}	1.4213^{20}		132^{11mm}		
d132	Dibutyl sulfide	$C_4H_9SC_4H_9$	146.30	1, 370	0.8396^{16}	1.4530^{20}	-75.0	188.9	76	i aq; v s alc, eth
d133	Di-*tert*-butyl sulfide	$(CH_3)_3CSC(CH_3)_3$	146.30		0.815	1.4506^{20}		151	48	
d134	Dibutyl sulfite	$(C_4H_9O)_2S(O)$	194.29	1^2, 397	0.9944^{22}	1.4310^{20}		108^{15mm}		i aq; s alc, eth
d135	Dibutyl sulfone	$(C_4H_9)_2SO_2$	178.29	1, 371			46	295	143	
d135a	Dibutyl tartrate	$[{-}CH(OH)CO_2(CH_2)_3CH_3]_2$	262.31	3, 518	1.091	1.4465^{20}	22	175^{5mm}	>110	i aq; s alc, eth
d136	*N,N'*-Dibutylthiourea	$C_4H_9NC({=}S)NHC_4H_9$	188.34				63–65			i aq; s alc; sl s eth
d136a	Dibutyltin dichloride	$(C_4H_9)_2SnCl_2$	303.83				39–41	135^{10mm}	>110	
d137	Dibutyltin oxide	$(C_4H_9)_2SnO$	248.92				>300		>110	
d138	Dichloroacetic acid	$Cl_2CHCOOH$	128.94	2, 202	1.5634^{20}	1.4642^{20}	9–11	194	>110	misc aq, alc, eth
d139	1,1-Dichloroacetone	$CH_3C(O)CHCl_2$	126.97	1, 654	1.3051^{5}	1.4455^{20}		120	24	sl s aq; s alc; misc eth
d140	2',4'-Dichloroacetophenone	$Cl_2C_6H_3COCH_3$	189.04	7^2, 219		1.5635^{20}	33–34	145^{15mm}	>110	i aq
d141	Dichloroacetyl chloride	$Cl_2CHCOCl$	147.39	2, 204	1.5315^{15}	1.4603^{20}		108	none	dec aq, alc; misc eth
d142	2,3-Dichloroaniline	$Cl_2C_6H_3NH_2$	162.02	12, 621	1.5969^{20}		23–24	252	>110	s alc; v s eth

No.	Name	Formula	MW	Beil.	d	n	mp	bp	Flash	Solubility
d143	2,4-Dichloroaniline	$Cl_2C_6H_3NH_2$	162.02	12, 621			59–62	245	>110	sl s aq; s alc, eth
d144	2,5-Dichloroaniline	$Cl_2C_6H_3NH_2$	162.02	12, 625			49–51	251	>110	s alc, bz, eth
d145	2,6-Dichloroaniline	$Cl_2C_6H_3NH_2$	162.02	12, 626			38–41			s alc, eth; sl s bz
d146	3,4-Dichloroaniline	$Cl_2C_6H_3NH_2$	162.02	12, 626			70–72	272		i aq; s alc, eth
d147	3,5-Dichloroaniline	$Cl_2C_6H_3NH_2$	162.02	12, 626			51–53	259^{741mm}	>110	sl s alc, bz, acet
d148	1,5-Dichloroanthraquinone		277.11	7, 787	1.567^{20}		245–247			
d149	1,8-Dichloroanthraquinone		277.11	7, 788			202–203			sl s alc
d150	2,4-Dichlorobenzaldehyde	$Cl_2C_6H_3CHO$	175.01	7, 236			69–73	233		i aq; s alc
d151	2,4-Dichlorobenzamide	$Cl_2C_6H_3CONH_2$	190.03	9^3, 1376			191–194			
d152	1,2-Dichlorobenzene	$C_6H_4Cl_2$	147.01	5, 201	1.30592^{20}	1.5515	−17.0	180.4	65	misc alc, bz, eth
d153	1,3-Dichlorobenzene	$C_6H_4Cl_2$	147.01	5, 202	1.2884^{20}	1.5459	−24.8	173.1	63	0.01 aq; s alc, eth
d154	1,4-Dichlorobenzene	$C_6H_4Cl_2$	147.01	5, 203	1.2417^{6}	1.5285	53	174.1	65 (CC)	s alc, bz, chl, eth
d155	2,5-Dichlorobenzene-sulfonyl chloride	$Cl_2C_6H_3SO_2Cl$	245.51	11^1, 15			36–37		>110	d hot aq, hot alc
d156	2,4-Dichlorobenzoic acid	$Cl_2C_6H_3COOH$	191.01	9, 342			157–160			s hot aq, alc, bz, chl
d157	2,5-Dichlorobenzoic acid	$Cl_2C_6H_3COOH$	191.01	9, 342			151–154	301		sl s aq; s alc, eth
d158	3,4-Dichlorobenzoic acid	$Cl_2C_6H_3COOH$	191.01	9, 343			207–209			s hot aq, eth; v s alc
d159	4,4'-Dichlorobenzo-phenone	$(ClC_6H_4)_2CO$	251.11	7, 420			144–146	353		s hot alc; v s chl, eth

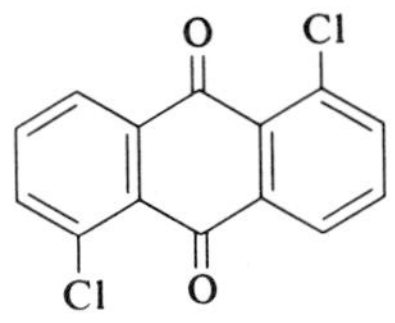

d148

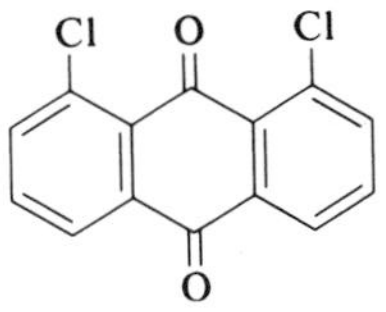

d149

TABLE 1.15 Physical Constants of Organic Compounds (*Continued*)

No.	Name	Formula	Formula weight	Beilstein reference	Density	Refractive index	Melting point	Boiling point	Flash point	Solubility in 100 parts solvent
d160	2,4-Dichlorobenzoyl chloride	$Cl_2C_6H_3COCl$	209.46	9, 342	1.494	1.5297^{20}	16–18	150^{34mm}	137	dec aq, alc
d161	3,4-Dichlorobenzoyl chloride	$Cl_2C_6H_3COCl$	209.46	9, 344			30–33	242	142	dec aq, alc
d162	1,3-Dichlorobutane	$CH_3CH(Cl)CH_2CH_2Cl$	127.01	$1^1, 38$	1.1182^{20}_{4}	1.4474^{15}		134	30	i aq; s chl, eth
d163	1,4-Dichlorobutane	$ClCH_2CH_2CH_2CH_2Cl$	127.01	1, 119	1.1598^{20}_{4}	1.4566^{20}	−38	155	40	i aq; s chl
d164	meso-2,3-Dichlorobutane	$CH_3CH(Cl)CH(Cl)CH_3$	127.01	1, 119	1.1025^{25}_{4}	1.4386^{25}	−80	115.9	18	i aq; s chl
d165	cis-1,4-Dichloro-2-butene	$ClCH_2CH=CHCH_2Cl$	125.00	$1^3, 743$	1.1882^{5}_{4}	1.4887^{25}	−48	152	55	i aq; s org solv
d166	trans-1,4-Dichloro-2-butene	$ClCH_2CH=CHCH_2Cl$	125.00	$1^3, 743$	1.1832^{5}_{4}	1.4861^{25}	1–3	76^{40mm}	53	i aq; s org solv
d167	3,4-Dichloro-1-butene	$ClCH_2CH(Cl)CH=CH_2$	125.00	$1^3, 927$	1.150	1.4658^{20}	−61	123	28	
d168	1,4-Dichloro-2-butyne	$ClCH_2C\equiv CCH_2Cl$	122.98		1.2584^{20}_{4}	1.5048^{20}		165–168	160	
d169	1,1-Dichloro-2,2-diethoxyethane	$Cl_2CHCH(OC_2H_5)_2$	187.07	1, 614	1.138	1.4360^{20}		183–184	60	
d170	Dichlorodifluoromethane	Cl_2CF_2	120.92	1, 61	1.486^{-30}		−158	−29.8		0.02 aq; 9 bz; 5.5 chl; 6 diox; s alc, eth
d171	4,6-Dichloro-1,3-dihydroxybenzene	$Cl_2C_6H_2(OH)_2$	179.00	$6^1, 403$			104–106	254		sl s aq, bz; s eth
d172	2,5-Dichloro-3,6-dihydroxy-p-benzoquinone		208.98	8, 379			283–284			
d172a	1,1-Dichloro-3,3-dimethylbutane	$(CH_3)_3CCH_2CHCl_2$	155.07	$1^3, 409$	1.027	1.4388^{20}	−56	148	36	
d173	1,3-Dichloro-3,5-dimethylhydantoin		197.02	$24^2, 158$			134–136			
d174	Dichlorodimethylsilane	$(CH_3)_2SiCl_2$	129.06		1.064^{20}_{4}	1.4038^{20}	−16	70	−16	dec aq, alc
d175	Dichlorodiphenylsilane	$(C_6H_5)_2SiCl_2$	253.20	16, 910	1.2222^{20}	1.582^{20}		308–309	157	dec aq, alc
d176	1,1-Dichloroethane	CH_3CHCl_2	98.96	1, 83	1.1757^{20}_{4}	1.4164^{20}	−97.0	57.3	−18 (OC)	0.51 aq; misc alc
d177	1,2-Dichloroethane	$ClCH_2CH_2Cl$	98.96	1, 84	1.2531^{20}_{4}	1.4448^{20}	−35.7	83.5	15	0.8 aq; misc alc, chl, eth

d178	1,1-Dichloroethylene	H2C=CCl2	96.94	1, 186	1.2129^{20}_{4}	1.4247^{20}	-122.6	31.6	-15	0.02 aq; s alc, bz, chl, eth
d179	cis-1,2-Dichloroethylene	ClCH=CHCl	96.94	1, 188	1.28182^{20}	1.4490^{20}	-80.1	60.7	6	0.7 aq; s alc, eth
d180	trans-1,2-Dichloroethylene	ClCH=CHCl	96.94	1, 188	1.2546^{20}_{4}	1.4462^{20}	-49.8	47.7	6	0.6 aq; s alc, eth
d181	2,2'-Dichloroethyl ether	ClCH2CH2OCH2CH2Cl	143.01	1[2], 335	1.2220^{20}_{20}	1.457^{20}		178.5	55	1.1 aq; s alc, bz, eth
d182	2,2'-Dichloroethyl methyl ether	Cl2CHCH2OCH3	128.99		1.226	1.4375^{20}			33	
d183	Dichlorofluoromethane	FCHCl2	102.92	1, 61	1.345^{30}		-135	8.9		69 HOAc; 108 diox; s alc, eth; i aq
d184	Dichloroheptylmethylsilane	C7H15Si(CH3)Cl2	225.2		0.9780^{20}_{4}	1.4396^{25}		207-208		
d185	1,2-Dichlorohexafluorocyclobutane	F6C4Cl2	233.0			1.3342^{25}		59-60		
d187	1,6-Dichlorohexane	Cl(CH2)6Cl	155.07	1, 144	1.068	1.4568^{20}		87^{15mm}	73	s chl
d190	Dichloromethane	CH2Cl2	84.93	1, 60	1.3255^{20}_{4}	1.4246^{20}	-96.7	40.5	none	1.3 aq; misc alc, eth
d192	3,5-Dichloro-1-methoxybenzene	Cl2C6H3OCH3	177.03	6, 190			40-42	132^{49mm}		
d193	2,4-Dichloro-6-methoxy-1,3,5-triazine		179.99				86-88			
d194	(Dichloromethyl)dimethylchlorosilane	Cl2CHSi(Cl)(CH3)2	177.5		1.237^{20}_{4}	1.461^{20}	-49	149		

2,6-Dichlorobenzyl chloride, t253
2,2'-Dichlorodiethyl ether, b159
5,5'-Dichloro-2,2'-dihydroxydiphenylmethane, m234

1,1-Dichlorodimethyl ether, d197
Dichlorohydrin, d220

Dichloroisopropyl alcohol, d220
4,4'-Dichloro-α-methylbenzhydrol, b165

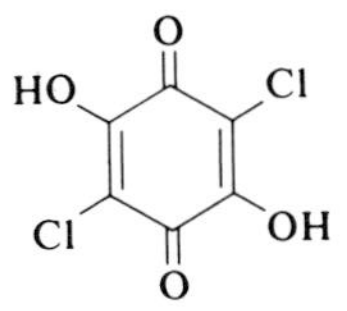

d172

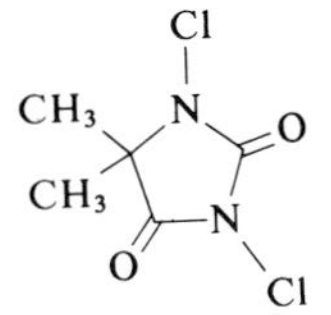

d173

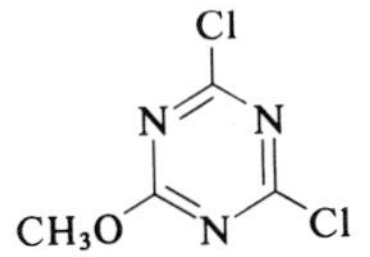

d193

TABLE 1.15 Physical Constants of Organic Compounds (*Continued*)

No.	Name	Formula	Formula weight	Beilstein reference	Density	Refractive index	Melting point	Boiling point	Flash point	Solubility in 100 parts solvent
d195	2,2-Dichloro-1-methyl-cyclopropane carboxylic acid	$Cl_2(C_3H)(CH_3)COOH$	169.01				60–65	85[8mm]		
d196	N-(Dichloromethyl-ene)aniline	$C_6H_5N=CCl_2$	174.03	12, 447	1.265	1.5710^{20}		106[30mm]	79	
d197	Dichloromethyl methyl ether	Cl_2CHOCH_3	114.96		1.271	1.4300^{20}		85	42	
d198	Dichloro(methyl)-phenylsilane	$C_6H_5Si(CH_3)Cl_2$	191.13		1.176	1.5190^{20}		205	82	
d199	Dichloro(methyl)silane	$HSi(CH_3)Cl_2$	115.04	4^1, 581	1.105		–93	41	–32	
d200	Dichloro(methyl)vinyl-silane	$H_2C=CHSi(CH_3)Cl_2$	141.07		1.0872^{20}_4	1.4300^{20}		92–93	4	
d200a	2,4-Dichloro-1-naphthol	$Cl_2C_{10}H_5OH$	213.06	6, 612			108			sl s alc, bz, eth
d201	2,3-Dichloro-1,4-naphthoquinone		227.05	7, 729			190–192			s PE
d202	2,6-Dichloro-4-ni-troaniline	$Cl_2C_6H_2(NO_2)NH_2$	207.02	12, 735			190–192			s hot alc; misc eth
d203	2,3-Dichloronitroben-zene	$Cl_2C_6H_3NO_2$	192.00	5, 245	1.721^{14}		61–62	258	123	
d204	2,4-Dichloronitroben-zene	$Cl_2C_6H_3NO_2$	192.00	5, 245	1.439^{80}		29–32	258	>110	
d205	3,4-Dichloronitroben-zene	$Cl_2C_6H_3NO_2$	192.00	5, 246	1.456^{75}		42	256	123	
d206	2,4-Dichloro-6-nitro-phenol	$Cl_2C_6H_2(NO_2)OH$	208.00	6, 241			120			
d207	1,7-Dichlorooctameth-yltetrasiloxane	$[Cl(CH_3)_2SiOSi(CH_3)_2-]_2$	351.6		1.0114^{20}	1.403^{20}	125–128	222	100	sl s aq; s alc, hot bz
d208	2,3-Dichloro-4-oxo-2-butenoic acid	$ClC(CHO)=C(Cl)COOH$	168.96	3, 727						
d209	1,5-Dichloropentane	$Cl(CH_2)_5Cl$	141.04	1, 131	1.10584^{15}	1.4553^{20}	–72	63[10mm]	26	i aq; s alc, eth
d210	2,3-Dichlorophenol	$Cl_2C_6H_3OH$	163.00	6, 102			58–60	206		s alc, eth
d211	2,4-Dichlorophenol	$Cl_2C_6H_3OH$	163.00	6, 189			42–43	210	113	v s alc, bz, chl, eth
d212	2,5-Dichlorophenol	$Cl_2C_6H_3OH$	163.00	6, 189			56–58	211		v s alc, bz, eth
d213	2,6-Dichlorophenol	$Cl_2C_6H_3OH$	163.00	6, 190			65–68	218–220		v s alc, eth

1.164

No.	Name	Formula	Formula wt	Ref.	Density	n_D	mp, °C	bp, °C	Flash pt, °C	Solubility
d214	2,4-Dichlorophenoxy-acetic acid	Cl$_2$C$_6$H$_3$OCH$_2$COOH	221.04				138	$160^{0.4\text{mm}}$		s alc, bz, chl, eth
d215	2,5-Dichloro-p-phenyl-enediamine	Cl$_2$C$_6$H$_2$(NH$_2$)$_2$	177.03	13, 118			165 d			
d215a	3,4-Dichlorophenyl isocyanate	Cl$_2$C$_6$H$_3$NCO	188.01	12[3], 1405			44	$120^{18\text{mm}}$	>110	
d216	Dichlorophenylphos-phine	C$_6$H$_5$PCl$_2$	178.99	16, 763	1.319	1.5980^{20}	−51	222	>112	s aq; v s eth
d217	4,5-Dichloro-o-phthalic acid	Cl$_2$C$_6$H$_2$(COOH)$_2$	235.02	9[1], 366			193–195			0.26 aq; misc alc, bz, chl, eth
d218	1,2-Dichloropropane	CH$_3$CH(Cl)CH$_2$Cl	112.99	1, 105	1.1558^{20}	1.4390^{20}	−100.4	96.4	4	v s alc, eth
d219	1,3-Dichloropropane	ClCH$_2$CH$_2$CH$_2$Cl	112.99	1, 105	1.1878^{20}_{4}	1.4487^{20}	−99.5	120.5	32	9.1 aq; misc alc, eth
d220	1,3-Dichloro-2-propanol	ClCH$_2$CH(OH)CH$_2$Cl	128.99	1, 364	1.3672^{20}_{4}	1.4835^{20}	−4	174.3	74	
d221	1,3-Dichloropropene	ClCH$_2$CH=CHCl	110.97	1, 199	1.217^{20}_{4}	1.470^{20}		112	27	i aq; s chl, eth
d222	2,3-Dichloro-1-propene	ClCH$_2$C(Cl)=CH$_2$	110.97	1, 199	1.2042^{25}_{4}	1.4611^{20}		94	10	misc alc; s eth
d223	3,6-Dichloropyridazine	Cl$_2$—C$_5$H$_3$N	148.98	20, 231			66–69			
d224	2,6-Dichloropyridine		147.99				86–88			
d225	4,7-Dichloroquinoline		198.05				84–86	$148^{10\text{mm}}$		
d226	Dichlorosilane	H$_2$SiCl$_2$	101.01				−122	8.3		
d227	1,2-Dichloro-1,1,2,2-tetrafluoroethane	ClCF$_2$CF$_2$Cl	170.93	1[3], 152	1.470^{20}_{4}	1.290^{20}	−94	3.6	none	s alc, eth
d227a	3,4-Dichloro-1,2,5-thiadiazole		155.01		1.648	1.5810^{20}		158		
d228	2,5-Dichlorothiophene	Cl$_2$—C$_4$H$_2$S	153.03	17, 33	1.442	1.5621^{20}	−40.5	162	59	i aq; misc alc, eth

1,1-Dichloro-2-propanone, d139

4,6-Dichlororesorcinol, d171

d201

d223

d225

d227a

TABLE 1.15 Physical Constants of Organic Compounds (*Continued*)

No.	Name	Formula	Formula weight	Beilstein reference	Density	Refractive index	Melting point	Boiling point	Flash point	Solubility in 100 parts solvent
d229	2,4-Dichlorotoluene	$Cl_2C_6H_3CH_3$	161.03	5, 295	1.2460^{20}_{20}	1.5454^{20}	−13	200.5	79	i aq
d230	2,6-Dichlorotoluene	$Cl_2C_6H_3CH_3$	161.03	5, 296	1.254	1.5507^{20}		196–203	82	i aq; s chl
d231	3,4-Dichlorotoluene	$Cl_2C_6H_3CH_3$	161.03	5, 296	1.251^{25}_{15}	1.5472^{20}	−14	201^{740mm}	85	i aq
d232	2,2-Dichloro-1,1,1-trifluoroethane	CF_3CHCl_2	152.9					28		
d233	α,α-Dichloro-*p*-xylene	$C_6H_4(CH_2Cl)_2$	175.06	5, 384			100	254		22.5 acet; 20 bz; 4.5 CCl_4; 11 eth; 18 EtAc
d234	2,5-Dichloro-*p*-xylene	$Cl_2C_6H_2(CH_3)_2$	175.06	5, 384			71	222		27 acet; 44 bz; 39 eth; 32 EtAc; 5 MeOH
d235	Dicyanodiamide	$H_2NC(=NH)NHCN$	84.08	3^2, 75	1.400^{25}		208–211			2.3 aq; 1.3 alc; i bz
d236	1,2-Dicyanobenzene	$C_6H_4(CN)_2$	128.13	9, 815			139–141			v s bz, alc; s hot eth
d237	1,3-Dicyanobenzene	$C_6H_4(CN)_2$	128.13	9, 836			158–160			s alc, bz, chl, eth
d238	1,4-Dicyanobutane	$NC(CH_2)_4CN$	108.14	2, 653	0.951	1.4380^{20}	1–3	295	>110	
d239	1,6-Dicyanohexane	$NC(CH_2)_6CN$	136.20	2, 694	0.954	1.4436^{20}	−3.5	185^{15mm}	>110	
d240	2,4-Dicyano-3-methylglutaramide	$CH_3CH[CH(CN)CONH_2]_2$	194.19	2^2, 704			159–160			
d241	Dicyclohexyl	$C_6H_{11}C_6H_{11}$	166.31	5, 108	0.864	1.4782^{20}	3–4	227	92	7 MeOH; misc bz, acet, eth
d242	Dicyclohexylamine	$(C_6H_{11})_2NH$	181.32	12, 6	0.910	1.4842^{20}	−0.1	255.8	96	misc alc, bz, chl, eth
d243	*N,N′*-Dicyclohexylcarbodiimide	$C_6H_{11}N=C=NC_6H_{11}$	206.33				34–35	124^{6mm}	>110	
d243a	Dicyclohexyl 1,2-phthalate	$C_6H_4—(CO_2C_6H_{11})_2$	330.43	9, 799			66			
d244	Dicyclopentadiene		132.21	5, 495	0.930^{25}	1.5050^{25}	−1	170	26	s alc, eth
d245	Diethanolamine	$HOCH_2CH_2NHCH_2CH_2OH$	105.14	4, 283	1.0881^{30}	1.4747^{30}	28.0	268.0	137	96 aq; 4 bz; 0.8 eth; misc MeOH, acet
d246	2,2-Diethoxyacetophenone	$C_6H_5C(=O)CH(OC_2H_5)_2$	208.26	7^1, 361	1.034	1.4995^{20}		134^{10mm}	110	
d247	4,4-Diethoxybutylamine	$H_2N(CH_2)_3CH(OC_2H_5)_2$	161.25	4, 319	0.933	1.4275^{20}		196	62	
d248	2,2-Diethoxy-*N,N*-dimethylethylamine	$(C_2H_5O)_2CHCH_2N(CH_3)_2$	161.25	4, 308	0.883	1.4129^{20}		170	45	
d249	Diethoxydimethylsilane	$(C_2H_5O)_2Si(CH_3)_2$	148.28		0.840^{20}_{4}	1.3811^{20}	−87	114	11	

No.	Name	Formula	MW	Ref.	d	n_D	mp	bp	fp	Solubility
d250	Diethoxydiphenylsilane	(C₂H₅O)₂Si(C₆H₅)₂	272.42		1.0329_4^{20}	1.5269^{20}		130^{mm}	−21 (CC)	5 aq; misc alc, eth
d251	1,1-Diethoxyethane	CH₃CH(OC₂H₅)₂	118.18	1, 603	0.8254^{20}	1.3819^{20}	2.8	102.7	27	21 aq
d252	1,2-Diethoxyethane	C₂H₅OCH₂CH₂OC₂H₅	118.18	1, 468	0.842	1.3922^{20}	−74	121.4	67	s alc, eth
d253	2,2-Diethoxyethanol	(C₂H₅O)₂CHCH₂OH	134.18	1, 818	0.8884^{24}	1.4160^{20}		167	45	
d254	2,2-Diethoxyethyl-amine	(C₂H₅O)₂CHCH₂NH₂	133.19	4, 308	0.916	1.4170		163		
d255	Diethoxymethane	(C₂H₅O)₂CH₂	104.15		0.839	1.3732^{20}		88	−5	
d256	Diethoxymethylvinyl-silane	(C₂H₅O)₂Si(CH₃)CH=CH₂	160.29	4⁴, 4183	0.8584^{20}	1.400^{20}		134	17	
d257	1,1-Diethoxypropane	CH₃CH₂CH(OC₂H₅)₂	132.20	1, 630	0.8233_4^{20}	1.3884^{20}		122.8	7	v s alc, eth
d258	3,3-Diethoxy-1-pro-pene	(C₂H₅O)₂CHCH=CH₂	130.19	1, 727	0.854	1.4000^{20}		125	4	
d259	2,2-Diethoxytriethyl-amine	(C₂H₅O)₂CHCH₂N(C₂H₅)₂	189.30	4, 309	0.850	1.4189^{20}		195	65	
d260	N,N-Diethylacetamide	CH₃C(=O)N(C₂H₅)₂	115.18	4, 110	0.925	1.4401^{20}		182–186	70	
d261	Diethyl acetamido-malonate	C₂H₅OOCCH(NHCOCH₃)-COOC₂H₅	217.22	4², 891			97–98	185^{20mm}		
d262	Diethyl 1,3-acetonedi-carboxylate	C₂H₅OOCCH₂COCH₂-COOC₂H₅	202.21	3, 791	1.113	1.4385^{20}		250	86	
d263	Diethyl acetylenedicar-boxylate	C₂H₅OOCC≡CCOOC₂H₅	170.16	2, 803	1.063	1.4426^{20}		107^{11mm}	94	
d264	Diethyl 2-acetyl-glutarate	C₂H₅OOCCH₂CH₂CH-(COCH₃)COOC₂H₅	230.26	3, 809	1.071	1.4386^{20}		154^{11mm}	>110	
d265	Diethyl acetylsuccinate	C₂H₅OOCCH₂CH(COCH₃)-COOC₂H₅	216.23	3, 801	1.081	1.4346^{20}		183^{50mm}	>110	
d266	Diethyl allylmalonate	C₂H₅OOCCH-(CH₂CH=CH₂)COOC₂H₅	200.23	2, 776	1.015	1.4304^{20}		223	92	

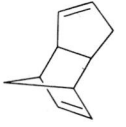

d244

TABLE 1.15 Physical Constants of Organic Compounds (*Continued*)

No.	Name	Formula	Formula weight	Beilstein reference	Density	Refractive index	Melting point	Boiling point	Flash point	Solubility in 100 parts solvent
d266a	Diethylaluminum chloride	$(C_2H_5)_2AlCl$	120.56	4³, 1972	0.961		−50	126^{50mm}	92	
d266b	Diethylaluminum ethoxide	$(C_2H_5)_2AlOC_2H_5$	130.17	4³, 1972	0.850		2.5–4.5	109^{10mm}		
d267	Diethylamine	$(C_2H_5)_2NH$	73.14	4, 95	0.7074^{20}_4	1.3864^{20}	−50.0	55.5	−28	misc aq, alc
d268	Diethylamine HCl	$(C_2H_5)_2NH \cdot HCl$	109.60	4, 95	1.0482^1		226	320–330		s aq, alc, chl; i eth
d268a	2-(Diethylamino)acetonitrile	$(C_2H_5)_2NCH_2CN$	112.18	4, 350	0.866	1.4260^{20}		170	53	
d269	4-(Diethylamino)benzaldehyde	$(C_2H_5)_2NC_6H_4CHO$	177.25	14², 25			39–41	174^{7mm}	>110	
d270	2-Diethylaminoethanol	$(C_2H_5)_2NHCH_2CH_2OH$	117.19	4, 282	0.8800^{25}	1.4389^{25}	−70	163	48	s aq, alc, bz, eth
d271	2-(Diethylaminoethyl)-4-aminobenzoate	$H_2NC_6H_4COOCH_2CH_2N(C_2H_5)_2$	236.30	14, 424			61			0.5 aq; s alc, bz, eth
d272	2-Diethylaminoethyl chloride HCl	$ClCH_2CH_2N(C_2H_5)_2 \cdot HCl$	172.10	4², 618			208–210			
d273	3-(Diethylamino)phenol	$(C_2H_5)_2NC_6H_4OH$	165.24	13, 408			65–69	170^{15mm}		s aq, alc, eth
d274	3-Diethylamino-1,2-propanediol	$(C_2H_5)_2NCH_2CH(OH)CH_2OH$	147.22	4, 302	0.9732^{20}_{20}	1.4602^{20}	233–235		107	s aq, alc, chl, eth
d275	1-Diethylamino-2-propanol	$(C_2H_5)_2NCH_2CH(OH)CH_3$	131.22	4², 737	0.889	1.4255^{20}	13.5	$55–59^{13mm}$	33	s alc
d276	3-Diethylamino-1-propanol	$(C_2H_5)_2NCH_2CH_2CH_2OH$	131.22	4, 288	0.884	1.4435		83^{15mm}	65	
d277	N,N-Diethylaniline	$C_6H_5N(C_2H_5)_2$	149.24	12, 164	0.9302^{25}	1.5394^{25}	−38	216	97	1 aq; sl s alc, eth
d278	2,6-Diethylaniline	$(C_2H_5)_2C_6H_3NH_2$	149.24		0.906	1.5452^{20}	3	243	123	
d279	Diethyl azodicarboxylate	$C_2H_5OOCN{=}NCOOC_2H_5$	174.16	3, 123	1.106	1.4280^{20}		106^{13mm}	>110	
d280	5,5-Diethylbarbituric acid		184.19	24², 279	1.220		188–192		>110	0.7 aq; 7 alc; 1.3 chl; 3.2 eth; s acet, HOAc
d281	Diethyl benzalmalonate	$C_6H_5CH{=}C(COOC_2H_5)_2$	248.28	9, 892	1.107	1.5365^{20}		215^{30mm}	>110	
d283	1,3-Diethylbenzene	$C_6H_4(C_2H_5)_2$	134.22	5, 426	0.8640^{20}_4	1.4950^{20}	−83.9	181.1	50	s alc, eth
d284	1,4-Diethylbenzene	$C_6H_4(C_2H_5)_2$	134.22	5, 426	0.8620^{20}_4	1.4940^{20}	−42.85	183.8	56	s alc, eth

No.	Name	Formula	Formula Weight	Ref	Density	n_D	mp	bp	Flash	Solubility
d285	Diethyl benzylmalonate	$C_6H_5CH_2CH(COOC_2H_5)_2$	250.29	9, 869	1.064	1.4868^{20}		162^{10mm}	>110	i aq; misc alc, eth
d286	Diethyl bromomalonate	$BrCH(COOC_2H_5)_2$	239.07	2, 594	1.4022^{25}	1.4550^{20}	−54	233–235 d		v s alc, eth
d287	Diethyl butylmalonate	$C_4H_9CH(COOC_2H_5)_2$	216.28	2^1, 282	0.983	1.4220		235–240	93	
d288	Diethylcarbamoyl chloride	$(C_2H_5)_2NCOCl$	135.59	4, 120		1.4515^{20}		187–190	75	d hot aq, hot alc
d289	Diethyl carbonate	$(C_2H_5O)_2C=O$	118.13	3, 5	0.9764^{20}	1.3843^{20}	−43.0	126.8	25 (OC)	69 aq; misc alc, bz, eth, esters
d290	Diethyl chloromalonate	$ClCH(COOC_2H_5)_2$	194.61	2^2, 537	1.2040^{20}	1.4310^{20}		223		misc alc, chl, eth
d291	Diethyl chlorophosphate	$(C_2H_5O)_2P(O)Cl$	172.55	1, 332	1.194	1.4165^{20}		60^{2mm}	61	
d292	Diethyl chlorothiophosphate	$(C_2H_5O)_2P(S)Cl$	188.61		1.200	1.4715^{20}		45^{3mm}	110	
d293	Diethylcyanamide	$(C_2H_5)_2NCN$	98.15	4, 121	0.846	1.4229^{20}		188	69	
d293a	Diethyl cyanophosphonate	$(C_2H_5O)_2P(O)CN$	163.11		1.075	1.4012^{20}		105^{19mm}	80	
d294	N,N-Diethylcyclohexylamine	$C_6H_{11}N(C_2H_5)_2$	155.29	12, 6	0.850	1.4562^{20}		195	57	
d294a	Diethyl disulfide	$C_2H_5SSC_2H_5$	122.25	1, 347	0.9984^{20}	1.5063^{20}	−101.5	154.0	40	sl s aq; misc alc, eth
d295	Diethyldithiocarbamic acid, Na salt	$(C_2H_5)_2NC(=S)S^-Na^+ \cdot 3H_2O$	225.31	4^2, 613			95–99			
d296	Diethyl dithiophosphate	$(C_2H_5O)_2P(S)SH$	186.23	1, 333	1.111	1.5120^{20}		60^{1mm}		
d297	N,N-Diethyldodecanamide	$CH_3(CH_2)_{10}C(O)N(C_2H_5)_2$	255.45		0.847	1.4545^{20}		166^{2mm}	>110	

Diethylaminoacetaldehyde diethyl acetal, d259
3-Diethylaminopropylamine, d330

Diethyl (Z)-2-butenedioate, d316

Diethyl carbitol, b177

d280

TABLE 1.15 Physical Constants of Organic Compounds (*Continued*)

No.	Name	Formula	Formula weight	Beilstein reference	Density	Refractive index	Melting point	Boiling point	Flash point	Solubility in 100 parts solvent
d297a	Diethyl dodecanedioate	$C_2H_5O_2C(CH_2)_{10}CO_2C_2H_5$	286.41	2^2, 616	0.951	1.4402^{20}	15	193^{14}	110	
d298	Diethylenetriamine	$(H_2NCH_2CH_2)_2NH$	103.17	4, 255	0.9542^{20}	1.4826^{20}	−35	207	94	misc aq, alc, bz, eth
d299	Diethylenetriamine-pentaacetic acid	$[(HOOCCH_2)_2NCH_2CH_2]_2N\text{-}CH_2COOH$	393.35	4^4, 2454			220 d			
d300	Diethyl ether	$C_2H_5OC_2H_5$	74.12	1, 314	0.7134^{20}_{4}	1.3527^{20}	−116.3	34.6	−40	6 aq; misc alc, bz, chl
d301	Diethyl ethoxymethyl-enemalonate	$(C_2H_5OOC)_2C=CHOC_2H_5$	216.23	3, 469	1.070	1.4620^{20}		279−281	155	
d302	N,N-Diethylethylene-diamine	$(C_2H_5)_2NCH_2CH_2NH_2$	116.21	4, 251	0.827	1.4360^{20}		145−147	30	
d303	Diethyl ethylmalonate	$C_2H_5CH(COOC_2H_5)_2$	188.22	2, 644	1.0042^{20}_{20}	1.4158^{20}		77^{5mm}	88	sl s aq; v s alc, eth
d304	N,N-Diethylform-amide	$(C_2H_5)_2NCHO$	101.15	4, 109	0.908	1.4340^{20}		177	60	misc aq; v s alc, eth
d305	Diethyl fumarate	$C_2H_5OOCCH=CHCOOC_2H_5$	172.18	2, 742	1.0522^{20}_{4}	1.4406^{20}	1−2	219	91	
d307	Diethyl glutarate	$C_2H_5OOCCH_2CH_2CH_2COOC_2H_5$	188.22	2, 633	1.022	1.4240^{20}	−23.8	237	96	0.9 aq; v s alc; s eth
d308	2,4-Diethyl-2,6-hepta-dienal	$H_2C=CHCH_2CH(C_2H_5)\text{-}CH=C(C_2H_5)CHO$	166.27		0.862	1.4676^{20}		91	86^{12mm}	
d309	Diethyl heptanedioate	$C_2H_5OOC(CH_2)_5COOC_2H_5$	216.28	2, 671	0.9945^{20}	1.4280^{20}	−24	192^{100mm}	>110	i aq; s alc, eth
d310	2,4-Diethyl-1-heptanol	$CH_3CH_2CH_2CH(C_2H_5)CH_2CH(C_2H_5)CH_2OH$	172.31					109^{12mm}	110	
d311	Di-(2-ethylhexyl) decanedioate	$C_4H_9CH(C_2H_5)CH_2OOC(CH_2)_8COOCH_2CH(C_2H_5)C_4H_9$	426.68		0.912^{24}_{4}	1.451^{25}		256^{5mm}	227	i aq; s alc, bz, acet
d312	Di-(2-ethylhexyl) hexanedioate	$C_4H_9CH(C_2H_5)CH_2OOC(CH_2)_4COOCH_2CH(C_2H_5)C_4H_9$	370.57		0.925^{25}_{25}	1.4474^{20}		214^{4mm}	193	s alc, eth, acet; i aq
d313	Di-(2-ethylhexyl)-o-phthalate	$C_6H_4[COOCH_2CH(C_2H_5)C_4H_9]_2$	390.56	10, 1248	0.981^{25}_{25}	1.4853^{20}	−50	384	207	
d314	Diethyl hydrogen phosphonate	$(C_2H_5O)_2P(O)H$	138.10	1, 330	1.079^{20}_{4}	1.4076^{20}		51^{2mm}	90	s aq (hyd), alc, eth
d315	N,N-Diethylhydrox-ylamine	$(C_2H_5)_2NOH$	89.14	4, 536	1.867	1.4195^{20}	−25	125−130	45	
d316	Diethyl maleate	$C_2H_5OOCCH=CHCOOC_2H_5$	172.18	2, 751	1.0687^{20}	1.4400^{20}	−8.8	225.3	93	1.4 aq; s alc, eth
d317	Diethyl malonate	$C_2H_5OOCCH_2COOC_2H_5$	160.17	2, 573	1.0550	1.4136^{20}	−48.9	199.3	100	2.7 aq; misc alc, bz, eth

No.	Name	Formula	Formula wt	Beilstein ref.	Density	n_D	M.p., °C	B.p., °C	Flash p., °C	Solubility
d318	Diethylmalonic acid	$HOOC(C_2H_5)_2COOH$	160.17	2, 686			127	d 170–180		v s aq, alc, eth
d319	N,N-Diethylmethyl-amine	$(C_2H_5)_2NCH_3$	87.17	4, 99	0.720	1.3887^{20}		63–65	−23	
d320	Diethyl methyl-malonate	$C_2H_5OOCCH(CH_3)COOC_2H_5$	174.20	2, 629	1.018^4	1.4130^{20}		198	76	sl s aq; v s alc, eth
d321	Diethyl 2-methyl-2-oxosuccinate	$C_2H_5OOCCH(CH_3)C(=O)C(=O)OC_2H_5$	202.21	3, 794	1.073	1.4313^{20}		138^{23mm}	>110	
d322	Diethyl methyl-succinate	$C_2H_5OOCCH_2CH(CH_3)COOC_2H_5$	188.22	2, 639	1.012	1.4199^{20}		218		
d323	N,N-Diethyl-4-nitrosoaniline	$C_6H_4(NO)N(C_2H_5)_2$	178.24	12, 684			82–84			i aq; s alc, eth
d324	Diethyl octanedioate	$C_2H_5OOC(CH_2)_6COOC_2H_5$	230.30	2, 693	0.9822^{20}	1.4323^{20}	5.9	282	>112	
d325	Diethyl oxalate	$C_2H_5OOCCOOC_2H_5$	146.14	2, 535	1.0785^{20}	1.4102	−40.6	185.4	75 (OC)	3.6 aq(gradual dec); misc alc, eth
d326	Diethyl oxydiformate	$[C_2H_5OC(=O)]_2O$	162.14		1.112^4	1.3980^{20}		93^{18mm}	69	s alc, esters, ketones
d327	N^1,N^1-Diethyl-1,4-pentanediamine	$CH_3CH(NH_2)(CH_2)_3N(C_2H_5)_2$	158.29		0.817	1.4429^{20}		200^{753mm}	68	s aq, alc, eth
d327a	N,N-Diethyl-1,4-phenylenediamine	$(C_2H_5)_2NC_6H_4NH_2$	164.25	13, 75	0.988			116^{5mm}	>110	
d328	Diethyl phenyl-malonate	$C_6H_5CH(COOC_2H_5)_2$	236.27	9, 854	1.0950^{20}	1.4913^{20}	16	170^{14mm}	>110	i aq; s alc
d329	Diethyl o-phthalate	$C_6H_4(COOC_2H_5)_2$	222.24	9, 798	1.2324^{14}	1.5049^{14}	−3	299	160 (OC)	i aq; misc alc, eth
d330	N,N-Diethyl-1,3-propanediamine	$(C_2H_5)_2NCH_2CH_2CH_2NH_2$	130.24		0.826	1.4416^{20}		159	58	
d331	2,2-Diethyl-1,3-propanediol	$(C_2H_5)_2C(CH_2OH)_2$	132.20		1.052^{20}	1.4574^{25}	61.3	125^{10mm}		
d332	Diethyl propyl-malonate	$C_2H_5OOCCH(C_3H_7)COOC_2H_5$	202.25	2, 657	0.987	1.4185^{20}		221–222	91	25 aq; v s alc, eth

TABLE 1.15 Physical Constants of Organic Compounds (*Continued*)

No.	Name	Formula	Formula weight	Beilstein reference	Density	Refractive index	Melting point	Boiling point	Flash point	Solubility in 100 parts solvent
d333	1,1-Diethyl-2-propynylamine	$HC{\equiv}CC(C_2H_5)_2NH_2$	111.19		0.828	1.4409^{20}		71^{90mm}	21	
d334	N,N-Diethyl-3-pyridinecarboxamide	$C_5H_4N{-}C({=}O)N(C_2H_5)_2$	178.24	22², 34	1.0602^{25}	1.5240^{20}	24–26	296–300	>110	
d335	Diethyl succinate	$C_2H_5OOCCH_2CH_2COOC_2H_5$	174.20	2, 609	1.040^{20}	1.4200^{20}	−21	217.7	110	i aq; misc alc, eth
d336	Diethyl sulfate	$(C_2H_5O)_2SO_2$	154.18	1, 327	1.1722^{25}	1.4004^{20}	−25	209 d	78	i aq; misc alc, eth
d337	Diethyl sulfide	$(C_2H_5)_2S$	90.19	1, 344	0.8367^{20}	1.4430^{20}	−103.9	92.1	−9	i aq; misc alc, eth
d338	Diethyl sulfite	$(C_2H_5O)_2S(O)$	138.19	1, 325	1.0772^{25}			157.7	53	s aq(dec), alc
d339	(+)-Diethyl L-tartrate	$[{-}CH(OH)COOC_2H_5]_2$	206.19	3, 512	1.204^{20}	1.4459^{20}	17	280	93	sl s aq; misc alc, eth
d340	(—)-Diethyl D-tartrate	$[{-}CH(OH)COOC_2H_5]_2$	206.19	3¹, 181	1.205	1.4467^{20}		162^{19mm}	93	
d341	Diethyl 3,3'-thiodipropionate	$S(CH_2CH_2COOC_2H_5)_2$	234.32		1.095	1.4655^{20}		121^{2mm}		
d342	N,N-Diethyl-m-toluamide	$CH_3C_6H_4C({=}O)N(C_2H_5)_2$	191.27	9², 325	0.9962^{20}	1.5212^{20}		111^{1mm}	>110	
d343	N,N-Diethyl-1,1,1-trimethylsilylamine	$(C_2H_5)_2NSi(CH_3)_3$	145.32		0.767	1.4081^{20}		126	10	
d344	Diethylzinc	$(C_2H_5)_2Zn$	123.49	6, 672	1.2065^{20}	1.4983^{20}	−28	118	−23	
d345	1,4-Difluorobenzene	$C_6H_4F_2$	114.09	5, 199	1.1701^{20}	1.4415^{20}	−23.7	88.9	2	
d346	1,1-Difluoroethane	CH_3CHF_2	66.05	1³, 130	0.9092^{1}		−117	−24.7		
d347	1,1-Difluorotetrachloroethane	Cl_3CCClF_2	203.83	1, 86	1.649	1.413	41	91	none	0.32 aq; sl s alc; v s eth
d348	1,2-Difluorotetrachloroethane	FCl_2CCCl_2F	203.83	1³, 365	1.6447^{25}	1.413^{25}	23.8	203.8		i aq; s alc, eth
d350	Dihexylamine	$(C_6H_{13})_2NH$	185.36	4¹, 384	0.795	1.4320^{20}		192–195	95	s alc, eth
d351	Dihexyl ether	$(C_6H_{13})_2O$	186.34	1³, 1656	0.7936^{20}	1.4204^{20}		226.2	77	i aq; s eth
d352	9,10-Dihydroanthracene		180.25	5, 641	0.880		108–110	312		i aq; s alc, bz, eth
d353	(+)-Dihydrocarvone		152.24	7³, 337	0.929^{19}	1.4718^{20}		222	81	
d354	Dihydrocoumarin		148.16	17, 315	1.169^{18}	1.5563^{20}	25	272	>110	sl s alc, eth; s chl
d355	10,11-Dihydro-5H-dibenzo-[a,d]-cyclohepten-5-one		208.26		1.156	1.6332^{20}	32–34	$148^{0.3mm}$	>110	
d356	3,4-Dihydro-2-ethoxy-2H-pyran		128.17		0.957	1.4394^{20}		42^{16mm}	24	
d357	2,3-Dihydrofuran		70.09	17³, 141	0.927	1.4239^{20}		54–55	−24	

No.	Compound	Formula	Mol. wt.	Beilstein	Density	n_D	m.p.	b.p.	Flash
d358	Dihydrolinalool	$(CH_3)_2C{=}CHCH_2CH_2{-}$ $C(OH)(CH_3)CH_2CH_3$	156.27		0.925^{25}	1.433^{20}	80	171^{11m}	178
d359	3,4-Dihydro-1(2H)-6-methoxynaphthalen-one		176.22	9^2, 889					16
d360	3,4-Dihydro-2-methoxy-2H-pyran		114.14			1.4425^{20}			62
d361	2,3-Dihydro-2-methylbenzofuran		134.18	17^1, 23	1.061	1.5308^{20}		198	62
d362	5,6-Dihydro-4-methyl-2H-pyran		98.15	17^3, 160	0.912	1.4495^{20}		118	21
d363	3,4-Dihydro-1(2H)-naphthalenone		146.19	7, 370	1.099	1.5685^{20}	5–6	116^{mmm}	>110
d364	3,4-Dihydro-2H-pyran		84.12		0.922^{15}	1.4410^{20}	−70	86	−15	s aq, alc

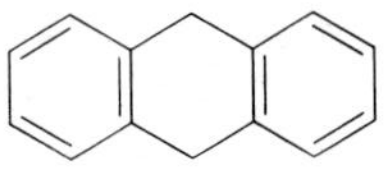

d352

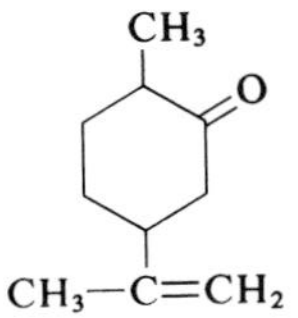

d353 (CH₃; $CH_3{-}C{=}CH_2$)

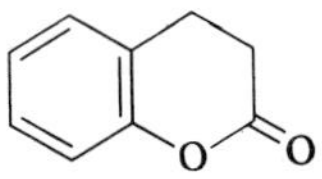

d354

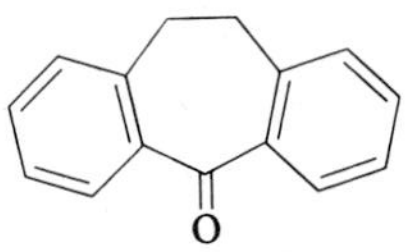

d355

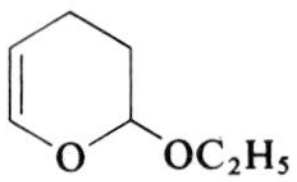

O

d356 (OC₂H₅)

d357

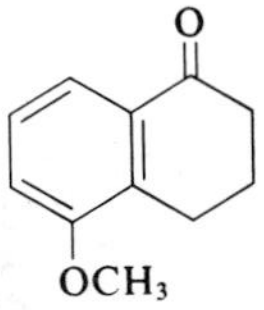

O

d358

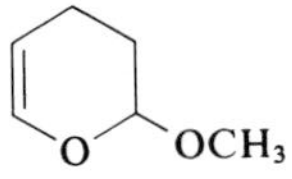

O

d359 (OCH₃)

d360 (OCH₃)

d361 (CH₃)

d362 (CH₃)

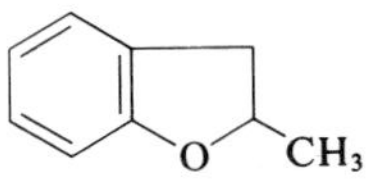

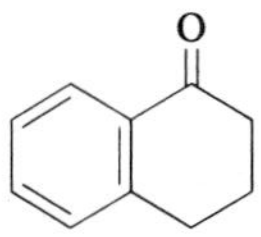

d363

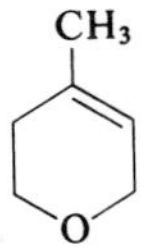

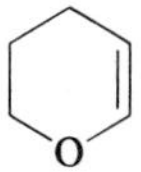

d364

TABLE 1.15 Physical Constants of Organic Compounds (*Continued*)

No.	Name	Formula	Formula weight	Beilstein reference	Density	Refractive index	Melting point	Boiling point	Flash point	Solubility in 100 parts solvent
d365	5,6-Dihydro-2*H*-pyran-3-carbaldehyde		112.13		1.100	1.4980^{20}		78^{12mm}	77	
d367	Dihydroterpineol		256.27		0.9075^{25}	1.4670^{20}			88	
d368	5,6-Dihydro-2,4,4,6-tetramethyl-4*H*-1,3-oxazine		141.21		0.886	1.4410^{20}		48^{17mm}		
d369	2,5-Dihydrothiophene-1,1-dioxide		118.15	17[3], 144				64–66	>110	s aq, alc, bz, chl, eth
d370	1,2-Dihydro-2,2,4-trimethylquinoline		173.26		0.934	1.5895^{20}		$90^{0.02mm}$	101	
d371	2',4'-Dihydroxyacetophenone	$(HO)_2C_6H_3C(=O)CH_3$	152.15	8, 266	1.180		147			s warm alc, HOAc, pyr; i bz, chl, eth
d372	1,2-Dihydroxyanthraquinone		240.21	8, 439			290	430		s alc, bz, chl, HOAc
d373	1,4-Dihydroxyanthraquinone		240.21	8, 450			196	450		s alc, alk, eth
d374	1,8-Dihydroxyanthraquinone		240.21	8, 458			193–197	subl		0.005 alc; 0.2 eth; s chl
d375	2,6-Dihydroxyanthraquinone		240.21	8, 463			360 d			sl s aq, alc
d376	2,4-Dihydroxybenzaldehyde	$(HO)_2C_6H_3CHO$	138.12	8, 241			135–136	226^{22mm}		v s aq, alc, chl, eth
d377	3,4-Dihydroxybenzaldehyde	$(HO)_2C_6H_3CHO$	138.12	8, 246			153			5 aq; 79 hot alc; v s eth
d378	1,2-Dihydroxybenzene	$C_6H_4(OH)_2$	110.11	6, 759	1.344^4		104–106	245.5	137	43 aq; s alc, bz, chl, eth; v s pyr, alk
d379	1,3-Dihydroxybenzene	$C_6H_4(OH)_2$	110.11	6[2], 802	1.272^{15}		109–110	276	171	110 aq; 110 alc; v s eth, glyc; sl s chl
d380 d381	1,4-Dihydroxybenzene 1,3-Dihydroxybenzene monoacetate	$C_6H_4(OH)_2$ $HOC_6H_4OOCCH_3$	110.11 152.15	6, 836 6, 816	1.332^{15}	1.5350^{20}	170–171	285–287 283	>110	7 aq; v s alc, eth
d382	2,5-Dihydroxy-*p*-benzenedisulfonic acid, K salt	$(HO)_2C_6H_2(SO_3^-K^+)_2$	346.43	11, 300			>300			v s aq

					mp		Solubility
d383	2,5-Dihydroxybenzenesulfonic acid, K salt	$(HO)_2C_6H_3SO_3K^+$	228.27	11, 300	251 d		v s aq
d384	2,4-Dihydroxybenzoic acid	$(HO)_2C_6H_3COOH$	154.12	10, 377	213		s hot aq, alc, eth
d385	2,5-Dihydroxybenzoic acid	$(HO)_2C_6H_3COOH$	154.12	10, 384	199–200		0.5 aq; s alc, eth
d386	3,5-Dihydroxybenzoic acid	$(HO)_2C_6H_3COOH$	154.12	10, 404	236 d		sl s aq; s alc, eth
d387	2,4-Dihydroxybenzophenone	$(HO)_2C_6H_3C(=O)C_6H_5$	214.22	8, 312	144–145		v s alc, eth, HOAc
d388	2,2'-Dihydroxybiphenyl	$HOC_6H_4C_6H_4OH$	186.21	6, 989	110	315	s alc, bz, eth; sl s aq
d389	4,6-Dihydroxy-2-mercaptopyrimidine		144.15	24, 476	236		

Dihydroresorcinol, c323
3,7-Dihydro-1,3,7-trimethyl-1H-purine-2,6-dione, c1
1,3-Dihydroxyacetone, d398

1,4-Dihydroxycyclohexane, c322a
2,2'-Dihydroxydiethylamine, d245

N,N-Di(hydroxyethyl)aminoacetic acid, b183
2,2-Dihydroxy-1,3-indandione, i16

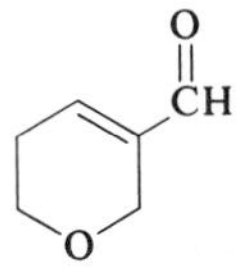

d365

d367

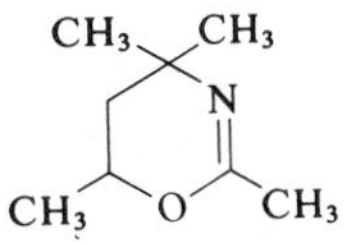

d368

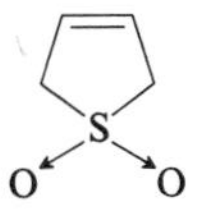

d369

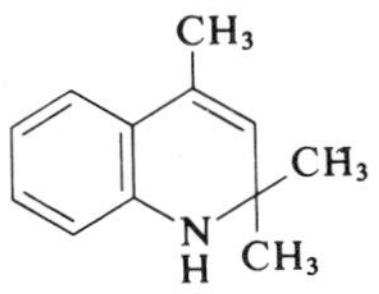

d370

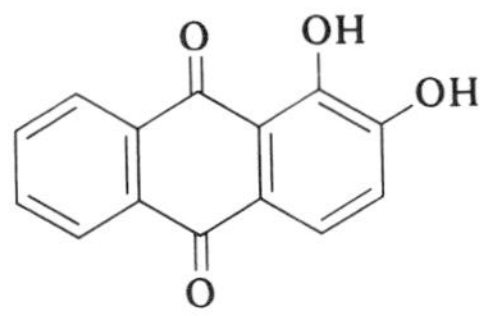

d372

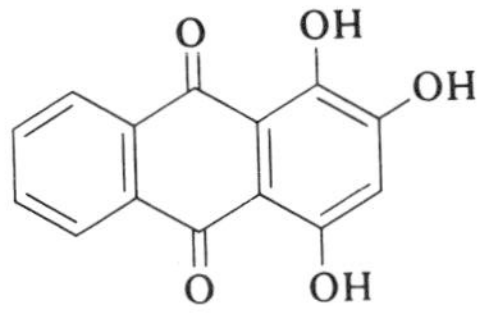

d373

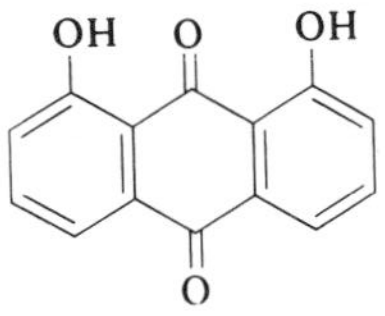

d374

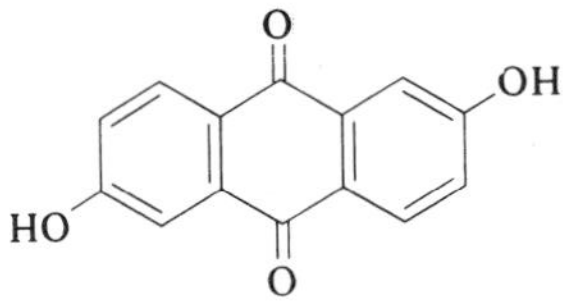

d375

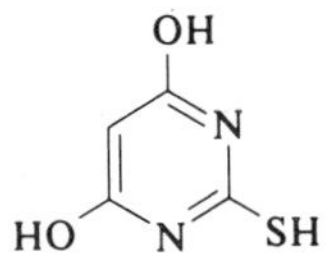

d389

TABLE 1.15 Physical Constants of Organic Compounds (*Continued*)

No.	Name	Formula	Formula weight	Beilstein reference	Density	Refractive index	Melting point	Boiling point	Flash point	Solubility in 100 parts solvent
d390	1,2-Dihydroxy-4-methylbenzene	$(HO)_2C_6H_3CH_3$	124.14	6, 878	1.1294^{74}	1.5425^{74}	67–69	251		v s aq, alc, eth
d391	1,3-Dihydroxy-2-methylbenzene	$(HO)_2C_6H_3CH_3$	124.14	6, 878			115–118	264		s aq, alc, bz, eth
d392	2,4-Dihydroxy-6-methylpyrimidine		126.12	24, 342			318 d			
d393	1,5-Dihydroxynaphthalene	$C_{10}H_6(OH)_2$	160.17	6, 980			259 d			sl s aq; s alc; v s eth
d394	1,7-Dihydroxynaphthalene	$C_{10}H_6(OH)_2$	160.17	6, 981			177–180			v s alc, eth
d395	2,3-Dihydroxynaphthalene	$C_{10}H_6(OH)_2$	160.17	6, 982			162–164			v s alc, eth
d396	2,7-Dihydroxynaphthalene	$C_{10}H_6(OH)_2$	160.17	6, 985			187 d			sl s aq; v s alc, eth
d398	1,3-Dihydroxy-2-propanone	$HOCH_2C(=O)CH_2OH$	90.08	1, 846	1.455^{18}_{18}		65–71			v s aq, alc, acet, eth
d399	2,3-Dihydroxypropionaldehyde	$HOCH_2CHOHCHO$	90.08	1, 845			145	$140^{0.8mm}$	>112	3 aq; i bz, PE
d400	7-(2,3-Dihydroxypropyl)theophylline		254.25				158			33 aq; 2 alc; 1 chl
d401	3,6-Dihydroxypyridazine		112.09	24, 312			d 260			sl s hot alc; s hot aq
d402	2,3-Dihydroxypyridine	$(HO)_2C_5H_3N$	111.10	$21^2, 107$			245 d			sl s alc; v s eth
d403	1,4-Diiodobenzene	$C_6H_4I_2$	329.91	5, 227			131–133	285		sl s aq; s alc, eth
d404	1,2-Diiodoethane	ICH_2CH_2I	281.86	1, 99	2.132^{10}		81	200		sl s aq; s alc, eth
d405	Diiodomethane	CH_2I_2	267.84	1, 71	3.3254^{20}_4	1.7411^{20}	5.6	181	>110	0.12 aq; misc alc, bz, eth, PE
d406	1,3-Diiodopropane	$ICH_2CH_2CH_2I$	295.88	1, 115	2.5755^{20}	1.6423^{20}	−13	222	>110	i aq; s chl, eth
d406a	Diisobutylaluminum chloride	$[(CH_3)_2CHCH_2]_2AlCl$	176.67	$4^4, 4403$	0.905	1.4506^{20}	−40	152^{10mm}		
d406b	Diisobutylaluminum hydride	$[(CH_3)_2CHCH_2]_2AlH$	142.22	$4^4, 4403$	0.798			118^{1mm}		
d407	Diisobutylamine	$[(CH_3)_2CHCH_2]_2NH$	129.25	4, 166	0.740	1.4081^{20}	−77	137–139	29	s alc, acet, eth, EtAc
d408	Diisobutyl ether	$[(CH_3)_2CHCH_2]_2O$	130.22		0.761^{15}			122–124	8	i aq; misc alc, eth

No.	Name	Formula	Formula wt.	Beilstein ref.	Density	n_D	m.p., °C	b.p., °C	Flash p., °C	Solubility
d409	Diisobutyl hexanedioate	$[(CH_3)_2CHCH_2OOCCH_2CH_2{-}]_2$	258.36		0.9502^{25}				160	
d410	Diisobutyl o-phthalate	$C_6H_4[COOCH_2CH(CH_3)_2]_2$	278.35	4[2], 711	1.0382^{25}				174	
d411	1,6-Diisocyanatohexane	$OCN(CH_2)_6NCO$	168.20		1.040	1.4525^{20}		255	140	
d411a	Diisooctyl nonanedioate	$C_8H_{17}O_2C(CH_2)_7CO_2C_8H_{17}$	412.66		0.905	1.4512^{20}		210^{2mm}	>110	11 aq
d411b	Diisopropenylbenzene	$C_6H_4[C(CH_3){=}CH_2]_2$	158.25	4, 154	0.925	1.5571^{20}		231	91	
d412	Diisopropylamine	$[(CH_3)_2CH]_2NH$	101.19	4[1], 430	0.7169^{20}	1.3924^{20}	−96.3	83.5	−6	
d413	2-(Diisopropylamino)ethanol	$[(CH_3)_2CH]_2NCH_2CH_2OH$	145.25		0.826	1.4417^{20}		187–192	57	
d414	2,6-Diisopropylaniline	$[(CH_3)_2CH]_2C_6H_3NH_2$	177.29	12, 168	0.940	1.5332^{20}	−45	257	123	misc alc, bz, eth, acet
d415	1,3-Diisopropylbenzene	$C_6H_4[CH(CH_3)_2]_2$	162.28	5, 447	0.8562^{20}_{4}	1.4890^{20}	−63	203	76	
d416	1,4-Diisopropylbenzene	$C_6H_4[CH(CH_3)_2]_2$	162.28	5[2], 339	0.8574^{20}_{4}	1.4889^{20}		203	76	misc alc, bz, eth, acet
d417	Diisopropylcyanamide	$[(CH_3)_2CH]_2NCN$	126.20	4[3], 279	0.839	1.4270^{20}		93^{25mm}	78	
d418	Diisopropyl ether	$[(CH_3)_2CH]_2O$	102.17	1, 362	0.7258^{20}_{4}	1.3689^{20}	−86.9	68.4	−12	1.2 aq; misc alc, bz, chl, eth
d419	N,N-Diisopropylethylamine	$[(CH_3)_2CH]_2NC_2H_5$	129.25		0.742	1.4133^{20}		127	10	
d420	2,6-Diisopropylphenol	$[(CH_3)_2CH]_2C_6H_3OH$	178.28	6[1], 272	0.962	1.5140^{20}	18	256	>110	misc alc, bz, eth, acet

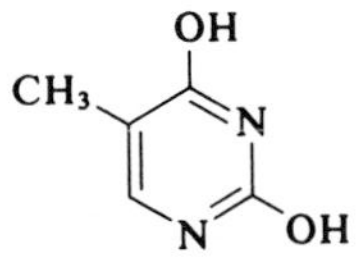

d392

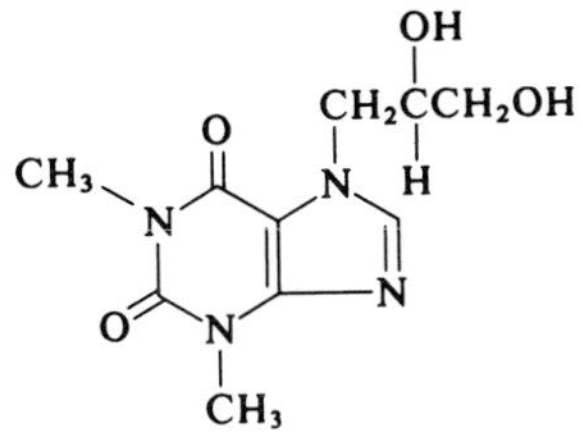

d400

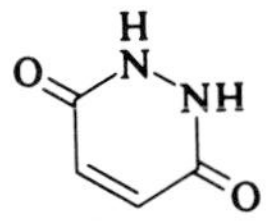

d401

TABLE 1.15 Physical Constants of Organic Compounds (*Continued*)

No.	Name	Formula	Formula weight	Beilstein reference	Density	Refractive index	Melting point	Boiling point	Flash point	Solubility in 100 parts solvent
d421	Diisopropyl phosphite	$[(CH_3)_2CHO]_2P(O)H$	166.16	1, 363	0.997	1.4070^{20}		75^{10mm}	>110	
d422	(+)-Diisopropyl L-tartrate	$[—CH(OH)COOCH(CH_3)_2]_2$	234.25	3, 517	1.114	1.4387^{20}		152^{12mm}	109	
d423	Diketene		84.07	17^3, 4297	1.073	1.4330^{20}		127	34	
d424	Dilauryl phosphite	$[CH_3(CH_2)_{11}O]_2P(O)H$	418.64		0.946	1.4520^{20}	42–43		>110	
d425	*threo*-1,4-Dimercapto-2,3-butanediol	$HSCH_2CH(OH)CH(OH)CH_2SH$	154.25							v s aq, alc, chl, eth
d426	2,3-Dimercapto-1-propanol	$HSCH_2CH(SH)CH_2OH$	124.22		1.2385^{25}_4	1.5720^{25}		120^{15mm}	>110	8 aq(dec); s alc, eth
d426a	2,5-Dimercapto-1,3,4-thiadiazole		150.24	27, 677			162 d			
d427	3′,4′-Dimethoxyacetophenone	$(CH_3O)_2C_6H_3COCH_3$	180.20	8^2, 298			49–51	286–288	>110	sl s aq, alc, eth
d428	2,4-Dimethoxyaniline	$(CH_3O)_2C_6H_3NH_2$	153.18	13, 784			34–37		>110	s alc, bz, eth
d429	2,5-Dimethoxyaniline	$(CH_3O)_2C_6H_3NH_2$	153.18	13, 788			80–82			s aq, alc
d430	3,4-Dimethoxyaniline	$(CH_3O)_2C_6H_3NH_2$	153.18	13, 780			88	270 sl dec 176^{22mm}		s hot eth
d431	3,4-Dimethoxybenzaldehyde	$(CH_3O)_2C_6H_3CHO$	166.18	8, 255			42–43	281	>110	v s alc, eth
d432	1,2-Dimethoxybenzene	$C_6H_4(OCH_3)_2$	138.17	6, 771	1.0819^{25}	1.5232^{25}	22.5	206.3	87	sl s aq; s alc, eth
d433	1,3-Dimethoxybenzene	$C_6H_4(OCH_3)_2$	138.17	6, 813	1.055	1.5240	−55	87^{7mm}	87	s alc, bz, eth; sl s aq
d434	1,4-Dimethoxybenzene	$C_6H_4(OCH_3)_2$	138.17	6, 843	1.0368^{88}_4		55–60	213		s alc; v s bz, eth
d435	3,4-Dimethoxybenzoic acid	$(CH_3O)_2C_6H_3COOH$	182.18	10^1, 188			180–181			0.047 aq; v s alc, eth
d435a	2,6-Dimethoxybenzoyl chloride	$(CH_3O)_2C_6H_3COCl$	200.62	10^3, 1402				64–66		
d435b	3,4-Dimethoxybenzyl alcohol	$(CH_3O)_2C_6H_3CH_2OH$	168.19	6, 1113	1.157	1.5520^{20}		297^{32mm}	>110	
d436	1,1-Dimethoxy-3-butanone	$(CH_3O)_2CHCH_2COCH_3$	132.16	10^3, 3097	0.993	1.4150^{20}		73^{20mm}	49	
d437	2,5-Dimethoxy-2,5-dihydrofuran		130.14		1.073	1.4339^{20}		160–162	47	

No.	Name	Formula	Mol. wt.	Ref.	Density	n_D	M.P.	B.P.	Fl.P.	Solubility
d438	Dimethoxydiphenyl-silane	$(C_6H_5)_2Si(OCH_3)_2$	244.4		1.0771^{20}	1.5447^{20}		161^{15mm}		s aq, alc, chl, eth
d439	1,1-Dimethoxyethane	$CH_3CH(OCH_3)_2$	90.12	1, 603	0.8502^{20}	1.3668^{20}	−113	64.5	−17	misc aq, alc; s PE
d440	1,2-Dimethoxyethane	$CH_3OCH_2CH_2OCH_3$	90.12	1, 467	0.8629^{20}	1.3796^{20}	−68	85.2	1	
d441	(2,2'-Dimethoxy)eth-ylamine	$H_2NCH_2CH(OCH_3)_2$	105.14	4^2, 758	0.965	1.4170^{20}		135^{95mm}	53	
d442	Dimethoxymethane	$CH_2(OCH_3)_2$	76.10	1, 574	0.8601^{20}	1.3534^{20}	−104.8	42.3	−17	32 aq
d443	1,1-Dimethoxy-2-methylaminoethane	$CH_3NHCH_2CH(OCH_3)_2$	119.16	4^2, 759	0.928	1.4115^{20}		140	29	
d444	Dimethoxymethyl-phenylsilane	$(CH_3O)_2Si(CH_3)C_6H_5$	182.30		0.993^{20}	1.469^{20}		199–200		v s alc, eth; s chl
d445	1,2-Dimethoxy-4-ni-trobenzene	$(CH_3O)_2C_6H_3NO_2$	183.16	6, 789	1.1888^{13}		95–98	230^{17mm}	>110	
d446	2,5-Dimethoxy-4'-ni-trostilbene	$(CH_3O)_2C_6H_3CH=CHC_6H_4NO_2$	285.30	6^2, 987			117–119			
d447	2,6-Dimethoxyphenol	$(CH_3O)_2C_6H_3OH$	154.17	6, 1081			53–56	261		s alc, alk; v s eth
d448	(3,4-Dimethoxy)phen-ylacetic acid	$(CH_3O)_2C_6H_3CH_2COOH$	196.20	10, 409			96–98			s aq; v s alc, eth
d449	(3,4-Dimethoxy)phen-ylacetonitrile	$(CH_3O)_2C_6H_3CH_2CN$	177.20	10^1, 198			62–63	178^{10mm}		
d450	2,2-Dimethoxy-2-phenylacetophenone	$C_6H_5C(O)C(OCH_3)_2C_6H_5$	256.20				67–70			
d451	1,1-Dimethoxy-2-phenylethane	$C_6H_5CH_2CH(OCH_3)_2$	166.22	7, 293	1.004	1.4950^{20}		221	83	
d452	2-(3,4-Dimethoxy-phenyl)ethylamine	$(CH_3O)_2C_6H_3CH_2CH_2NH_2$	181.24	13, 800	1.074	1.5464^{20}		188^{15mm}	>110	
d452a	1,2-Dimethoxypro-pane	$CH_3CH(OCH_3)CH_2OCH_3$	104.15	1^4, 2471	0.855	1.3835^{20}		96	0	
d453	2,2-Dimethoxypro-pane	$(CH_3)_2C(OCH_3)_2$	104.15	1, 648	0.847	1.3780^{20}		83	−11	

$CH_2=C-O$
$CH_2-C=O$ d423

Thiadiazole structure: $N-N$, HS, SH, S d426a

Furan structure: CH_3O ... OCH_3 d437

3,4-Dimethoxyphenethylamine, d452a

Dimedone, d510

TABLE 1.15 Physical Constants of Organic Compounds (*Continued*)

No.	Name	Formula	Formula weight	Beilstein reference	Density	Refractive index	Melting point	Boiling point	Flash point	Solubility in 100 parts solvent
d454	1,1-Dimethoxy-2-propanone	$CH_3C(O)CH(OCH_3)_2$	118.13	1[1], 395	0.976	1.3978[20]		143–147	37	
d455	3,3-Dimethoxy-1-propene	$(CH_3O)_2CHCH=CH_2$	102.13	1[1], 378	0.862	1.3954[20]		89–90	−2	
d456	1,2-Dimethoxy-4-propenylbenzene	$CH_3CH=CHC_6H_3(OCH_3)_2$	178.23	6, 956	1.055	1.5680[20]		262–264	>110	
d457	2,6-Dimethoxypyridine	$(CH_3O)_2C_5H_3N$	139.15		1.053	1.5029[20]		178–180	61	
d458	2,5-Dimethoxytetrahydrofuran	$(CH_3O)_2C_4H_6O$	132.16		1.020	1.4180[20]		145–147	35	
d459	N,N-Dimethylacetamide	$CH_3C(O)N(CH_3)_2$	87.12	4, 59	0.9366[25]	1.4356[25]	−20	165.5	70	misc aq, alc, bz, eth
d460	Dimethyl 1,3-acetonedicarboxylate	$[CH_3OOCCH_2]_2C=O$	174.15	3, 790	1.185	1.4434[20]		150[25mm]	>110	
d461	Dimethyl acetylenedicarboxylate	$CH_3OOCC\equiv CCOOCH_3$	142.11	2, 803	1.156	1.4470[20]		95–98[19mm]	86	
d461a	N,N-Dimethylacrylamide	$H_2C=CHCON(CH_3)_2$	99.13	4[3], 130	0.962	1.4730[20]		81[20mm]	71	
d461b	Dimethylaluminum chloride	$(CH_3)_2AlCl$	92.51	4[3], 1971	0.701					
d463	Dimethylamine	$(CH_3)_2NH$	45.09	4, 39	0.680[20]	1.4101[20]	−92.2	6.9		v s aq; s alc, eth
d463a	Dimethylaminoacetonitrile	$(CH_3)_2NCH_2CN$	84.12	4, 346	0.863			138	36	
d464	4-Dimethylaminobenzaldehyde	$(CH_3)_2NC_6H_4CHO$	149.19	14, 31			74	176[17mm]		s alc, chl, eth, HOAc
d465	p-(Dimethylamino)benzenesulfonic acid, Na salt	$(CH_3)_2NC_6H_4SO_3Na^+$	223.23	14[3], 2023			>300			
d466	4-Dimethylaminobenzoic acid	$(CH_3)_2NC_6H_4COOH$	165.19	14, 426			241 d			s alc; sl s eth
d467	2-(Dimethylamino)ethanol	$(CH_3)_2NCH_2CH_2OH$	89.14	4, 276	0.8876[20]	1.4294[20]		135	40	misc aq, alc, eth
d468	2-(Dimethylamino)ethyl benzoate	$C_6H_5COOCH_2CH_2N(CH_3)_2$	193.26		1.014	1.5077[20]		155[20mm]		

No.	Name	Formula	Mol. wt.	Beilstein	Density	n_D	M.p., °C	B.p., °C	Flash pt.	Solubility
d469	2-Dimethylaminoethyl chloride HCl	$(CH_3)_2NCH_2CH_2Cl \cdot HCl$	144.05	4, 133			205–208	182–192	70	v s alc, bz, eth, acet
d470	2-(Dimethylamino)ethyl methacrylate	$H_2C{=}C(CH_3)COOCH_2CH_2N(CH_3)_2$	157.22	4^3, 649	0.933	1.4391^{20}		73^{35mm}	38	
d471	4-Dimethylamino-3-methyl-2-butanone	$(CH_3)_2NCH_2CH(CH_3)COCH_3$	129.20	4^1, 452	0.841	1.4250^{20}				
d472	3-Dimethylaminophenol	$(CH_3)_2NC_6H_4OH$	137.18	13, 405	1.5895^{26}		82–84	265–268	105	s aq, alc, chl, eth
d473	3-(Dimethylamino)-1,2-propanediol	$(CH_3)_2NCH_2CH(OH)CH_2OH$	119.16	4, 302	1.004	1.4609^{20}		217	35	
d474	1-Dimethylamino-2-propanol	$CH_3CH(OH)CH_2N(CH_3)_2$	103.17		0.837	1.4193^{20}		121–127	36	
d475	3-Dimethylamino-1-propanol	$(CH_3)_2NCH_2CH_2CH_2OH$	103.17	4^1, 433	0.872	1.4360^{20}		164	62	
d476	3-(Dimethylamino)propionitrile	$(CH_3)_2NCH_2CH_2CN$	98.15	4^3, 1265	0.870	1.4258^{20}		171^{750mm}		
d477	3-Dimethylaminopropylamine	$(CH_3)_2N(CH_2)_3NH_2$	102.18	4^3, 554	0.812	1.4350	−43	133	15	v s aq, alc, bz, chl
d478	4-(Dimethylamino)pyridine	$(CH_3)_2N(C_5H_4N)$	122.17	22^2, 341			108–110			
d478a	Dimethyl 2-amino-1,4-phthalate	$H_2NC_6H_3(CO_2CH_3)_2$	209.20	14, 559			130–133			
d479	N,N-Dimethylaniline	$C_6H_5N(CH_3)_2$	121.18	12, 141	0.9559^{20}	1.5584^{20}	2.5	194.2	62 (CC)	v s alc, chl, eth
d480	2,3-Dimethylaniline	$(CH_3)_2C_6H_3NH_2$	121.18	12, 1101	0.9931^{20}	1.5685^{20}	2.5	222	96	sl s aq; s alc, eth
d481	2,4-Dimethylaniline	$(CH_3)_2C_6H_3NH_2$	121.18	12, 1111	0.9804	1.5586^{20}	11.5	218	90	s alc, bz, eth
d482	2,5-Dimethylaniline	$(CH_3)_2C_6H_3NH_2$	121.18	12, 1135	0.9790^1	1.5592^{20}	10–12	218	93	sl s aq; s alc, eth
d483	2,6-Dimethylaniline	$(CH_3)_2C_6H_3NH_2$	121.18	12, 1107	0.984^{20}	1.5601^{20}		216	91	sl s aq; s alc, eth
d484	3,4-Dimethylaniline	$(CH_3)_2C_6H_3NH_2$	121.18	12, 1103	1.076^{18}		49–51	226		sl s aq; s alc
d485	3,5-Dimethylaniline	$(CH_3)_2C_6H_3NH_2$	121.18	12, 1131	0.9724^{20}	1.5578^{20}		104^{14mm}	93	sl s aq; s alc
d486	Dimethylarsinic acid	$(CH_3)_2As(O)OH$	138.00	4, 610			195–196			v s alc; 200 aq; i eth

TABLE 1.15 Physical Constants of Organic Compounds (*Continued*)

No.	Name	Formula	Formula weight	Beilstein reference	Density	Refractive index	Melting point	Boiling point	Flash point	Solubility in 100 parts solvent
d486a	1,3-Dimethylbarbituric acid		156.14	24, 471			126			
d486b	N,N-Dimethylbenzamide	$C_6H_5CON(CH_3)_2$	149.19	9, 201			43–45	133^{15mm}	110	
d487	3,4-Dimethylbenzoic acid	$(CH_3)_2C_6H_3COOH$	150.18	9^2, 353			165–167	subl		s alc, bz
d488	2,5-Dimethylbenzonitrile	$(CH_3)_2C_6H_3CN$	131.18	9, 535	0.957	1.5284^{20}	13–14	223^{730mm}	92	
d489	N,N-Dimethylbenzylamine	$C_6H_5CH_2N(CH_3)_2$	135.21	12, 1019	0.900	1.5011^{20}	−75	183	54	
d490	2,3-Dimethyl-1,3-butadiene	$H_2C=C(CH_3)C(CH_3)=CH_2$	82.15	1^3, 991	0.7222^{25}_4	1.4362^{25}	−76.0	69.2	<1	v s hot aq, alc, eth
d491	2,2-Dimethylbutane	$CH_3CH_2C(CH_3)_3$	86.18	1, 150	0.6492^{20}_4	1.3688^{20}	−99.9	49.7	<−34	
d492	2,3-Dimethylbutane	$(CH_3)_2CHCH(CH_3)_2$	86.18	1, 151	0.6616^{20}_4	1.3750^{20}	−128.5	58.0	−33	
d493	2,3-Dimethyl-2,3-butanediol	$(CH_3)_2C(OH)C(OH)(CH_3)_2$	118.18	1, 487			41.1	174.4	77	
d493a	3,3-Dimethyl-1,2-butanediol	$(CH_3)_3CCH(OH)CH_2OH$	118.18	1, 487			37–39		98	
d494	2,2-Dimethyl-1-butanol	$CH_3CH_2C(CH_3)_2CH_2OH$	102.18	1^3, 1675	0.8286^{20}_4	1.4208^{20}	<−15	136.8		sl s aq; s alc, eth
d495	2,3-Dimethyl-1-butanol	$(CH_3)_2CHCH(CH_3)CH_2OH$	102.18	1^3, 1677	0.8300^{20}_4	1.4205^{20}		149		s alc, eth
d496	2,3-Dimethyl-2-butanol	$(CH_3)_2CHC(CH_3)_2OH$	102.18	1, 413	0.8236^{20}_4	1.4176^{20}	−10.6	118.7	29	s aq; misc alc, eth
d497	3,3-Dimethyl-1-butanol	$(CH_3)_3CCH_2CH_2OH$	102.18	1^3, 1677	0.8147^{20}	1.4120^{20}	−60	143	47	s alc, eth
d498	3,3-Dimethyl-2-butanol	$(CH_3)_3CCH(OH)CH_3$	102.18	1, 412	0.8185^{20}_4	1.4151^{20}	5.3	120.4	28	s alc; misc eth
d499	3,3-Dimethyl-2-butanone	$(CH_3)_3CCOCH_3$	100.16	1, 694	0.7250^{25}_5	1.3939^{25}	−52.5	106.2	23	2.5 aq; s alc, eth
d500	2,3-Dimethyl-2-butene	$(CH_3)_2C=C(CH_3)_2$	84.16	1, 218	0.7081^{20}_4	1.4124^{20}	−74.3	73.2	−16	s alc, eth
d501	3,3-Dimethyl-1-butene	$(CH_3)_3CCH=CH_2$	84.16	1, 217	0.6531^{20}_4	1.3762^{20}	−115.2	41.3	−28	
d501a	2,2-Dimethylbutyric acid	$C_2H_5C(CH_3)_2COOH$	116.16	2, 335	0.928	1.4154^{20}		96^{5mm}	79	

No.	Name	Formula	Mol. wt.	Beil. ref.	Density	n_D	mp, °C	bp, °C	Fl. p.	Solubility
d502	3,3-Dimethylbutyric acid	$(CH_3)_3CCH_2COOH$	116.16	2, 337	0.9124^{20}_{4}	1.4100^{20}	6–7	190	88	s alc, eth
d503	Dimethylcadmium	$(CH_3)_2Cd$	142.48		1.9846^{17}	1.5488	–4.5	105.5	>150 explodes	dec aq; s PE
d504	Dimethylcarbamyl chloride	$(CH_3)_2NCOCl$	107.54	4, 73	1.168	1.4540^{20}	–33	168	68	
d505	Dimethyl carbonate	$(CH_3O)_2C=O$	90.08	3, 4	1.0651^{17}	1.3682^{20}	0.5	90–91	18	i aq; misc alc, eth
d506	Dimethyl chlorothiophosphate	$(CH_3O)_2P(S)Cl$	160.56	11, 143	1.322	1.4819^{20}		67^{16mm}	105	
d507	Dimethylcyanamide	$(CH_3)_2NCN$	70.09	4, 74	0.867	1.4100^{20}		161–163	58	
d507a	Dimethyl N-cyanothioiminocarbonate	$(CH_3S)_2C=NCN$	146.23	3, 220			46–50	>110		
d508	cis-1,2-Dimethylcyclohexane	$(CH_3)_2C_6H_{10}$	112.22	5, 36	0.7692^{20}	1.4335^{20}	–49.9	129.7	12	i aq; s alc, bz
d509	trans-1,2-Dimethylcyclohexane	$(CH_3)_2C_6H_{10}$	112.22	5, 36	0.7772^{20}	1.4273^{20}	–88.2	123.4	6	i aq; s alc, eth
d509a	cis-1,3-Dimethylcyclohexane	$(CH_3)_2C_6H_{10}$	112.22	5, 36	0.784	1.4230^{20}		120	5	
d509b	cis-1,4-Dimethylcyclohexane	$(CH_3)_2C_6H_{10}$	112.22	5^2, 22	0.783	1.4297^{20}	–88	125	6	
d510	5,5-Dimethyl-1,3-cyclohexanedione	$(CH_3)_2C_6H_9OH$	140.18	7, 559			d 149			0.4 aq; s alc, bz
d511	2,3-Dimethylcyclohexanol	$(CH_3)_2C_6H_9OH$	128.22		0.934	1.4653^{20}			65	

2,4-Dimethyl-3-azapentane, d412
Dimethylbenzenes, x4, x5, x6
6,6-Dimethylbicyclo[3.1.1]hept-2-ene-2-ethanol, n105
Dimethyl (Z)-butenedioate, d546

Dimethyl 2-butynedioate, d461
Dimethyl cellosolve, d440
Dimethylchlorosilane, c93

(Z)-2-Dimethylcrotonic acid, m162
Dimethyl 1,4-cyclohexanedione-2,5-dicarboxylic acid, d516

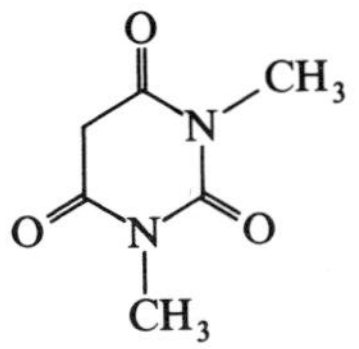

d486a

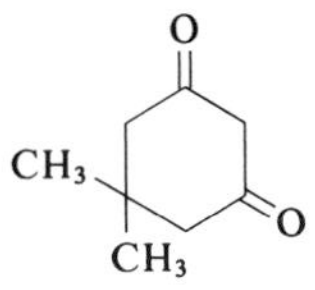

d510

TABLE 1.15 Physical Constants of Organic Compounds (*Continued*)

No.	Name	Formula	Formula weight	Beilstein reference	Density	Refractive index	Melting point	Boiling point	Flash point	Solubility in 100 parts solvent
d511a	2,5-Dimethylcyclohex-anol	$(CH_3)_2C_6H_9OH$	128.22	6, 19	0.898	1.4584^{20}		177	70	
d511b	2,6-Dimethylcyclohex-anol	$(CH_3)_2C_6H_9OH$	128.22	6, 18	0.944	1.4600^{20}		175	55	
d511c	3,4-Dimethylcyclohex-anol	$(CH_3)_2C_6H_9OH$	128.22	6, 17	0.949	1.4620^{20}		189	78	
d511d	3,5-Dimethylcyclohex-anol	$(CH_3)_2C_6H_9OH$	128.22	6, 18	0.892	1.4552	11–12	186	73	
d512	2,6-Dimethylcyclohex-anone	$(CH_3)_2C_6H_8(=O)$	126.20	7, 23	0.925	1.4460^{20}		175	51	i aq; s alc, eth
d512a	N,N-Dimethylcyclo-hexylamine	$C_6H_{11}N(CH_3)_2$	127.23		0.849	1.4535^{20}		159	42	
d513	2,3-Dimethylcyclohex-ylamine	$(CH_3)_2C_6H_9NH_2$	127.23		0.835	1.4595^{20}		160	51	
d513a	1,5-Dimethyl-1,5-cy-clooctadiene		136.24		0.867	1.4896^{20}		74^{16mm}	55	
d514	Dimethyl decanedioate	$CH_3OOC(CH_2)_8COOCH_3$	230.30	2, 719	0.983^{30}_{20}	1.4335^{28}	23	144^{5mm}	145	i aq; s alc, eth
d515	Dimethyl diethyl-malonate	$(C_2H_5)_2C(COOCH_3)_2$	188.23	2, 686	1.040	1.4275^{20}		97^{17mm}	80	
d515a	2,2-Dimethyl-1,3-diox-ane-4,6-dione		144.13				d 94			
d516	Dimethyl 2,5-dioxo-1,4-cyclohexanedi-carboxylate		228.20	10, 894			155–157			
d518	Dimethyl disulfide	CH_3SSCH_3	94.20	1, 291	1.046	1.5253^{20}	−84.7	109.8	24	i aq; misc alc, eth
d519	N,N-Dimethyldodec-ylamine	$CH_3(CH_2)_{11}N(CH_3)_2$	213.41	4^3, 409	0.775	1.4375^{20}	−20	112^{3mm}	110	
d520	Dimethyl ether	$(CH_3)_2O$	46.07	1, 281	0.661^{20}		−141.5	−24.9	−41	35% aq(5 atm); 15% bz; 11.8% acet
d522	N,N-Dimethylethyl-amine	$C_2H_5N(CH_3)_2$	73.14	4, 94	0.675	1.3720^{20}	−140	36–38	−36	
d523	N,N-Dimethylethyl-enediamine	$(CH_3)_2NCH_2CH_2NH_2$	88.15	4^2, 690	0.803	1.4260^{20}		106	23	
d524	N,N-Dimethylform-amide	$(CH_3)_2NCHO$	73.10	4, 58	0.9445^{25}	1.4282^{25}	−60.4	153.0	57	misc aq, alc, bz, eth

	Name	Formula								Solubility
d525	N,N-Dimethylformamide dimethyl acetal	$(CH_3)_2NCH(OCH_3)_2$	119.16		0.897	1.3972^{20}		103^{720mm}	7	sl s alc, eth
d526	Dimethyl fumarate	$CH_3OOCCH=CHCOOCH_3$	144.13	2,741	1.045^{106}	1.4414^{20}	105	193		i aq; misc alc, eth
d527	2,5-Dimethylfuran	$(CH_3)_2(C_4H_2O)$	96.13	17,41	0.9000^{20}		−62	93	−1	
d528	Dimethylglyoxime	$CH_3C(=NOH)C(=NOH)CH_3$	116.12	1,772			238–240			s alc, acet, eth, pyr
d529	2,4-Dimethyl-2,6-heptadienal	$H_2C=CHCH_2CH(CH_3)-CH=C(CH_3)CHO$	138.21		0.870	1.4664^{20}		47^{2mm}	64	
d530	2,4-Dimethyl-2,6-heptadien-1-ol	$H_2C=CHCH_2CH(CH_3)-CH=C(CH_3)CH_2OH$	140.23		1.351	1.4640^{20}		86^{10mm}	78	
d531	2,6-Dimethyl-2,5-heptadien-4-one	$(CH_3)_2C=CHC(=O)-CH=C(CH_3)_2$	138.21	1,751	0.8852^{4}	1.4968^{21}	28	199	79	sl s aq; s alc, eth
d532	Dimethyl heptanedioate	$CH_3OOC(CH_2)_5COOCH_3$	188.22	2^1, 281	1.0625^{20}_{4}	1.4314^{20}	−21	122^{11mm}	>110	s alc
d532a	2,6-Dimethyl-4-heptanol	$(CH_3)_2CHCH_2CH(OH)-CH_2CH(CH_3)_2$	144.26	1, 425	0.809	1.4236^{20}		178	66	
d533	2,6-Dimethyl-4-heptanone	$[(CH_3)_2CHCH_2]_2C=O$	142.24	1,710	0.806^{20}_{20}	1.4114^{20}	−41.5	168.1	48	0.06 aq; misc alc, bz, chl, eth
d533a	2,6-Dimethyl-5-heptenal	$(CH_3)_2C=CHCH_2-CH_2CH(CH_3)CHO$	140.23	1^2, 802	0.879	1.4441^{20}		124^{100mm}	60	
d534	2,5-Dimethyl-2,4-hexadiene	$(CH_3)_2C=CHCH=C(CH_3)_2$	110.20	1,259	0.7636^{20}	1.4741^{20}	12–14	132–134	29	i aq; s alc, eth
d536	Dimethyl hexanedioate	$CH_3OOC(CH_2)_4COOCH_3$	174.20	1,652	1.0600^{20}	1.4285^{20}	8	112^{10mm}	107	i aq; s alc, eth
d537	2,5-Dimethyl-2,5-hexanediol	$[(CH_3)_2C(OH)CH_2-]_2$	146.23	1,492			86–90	214–215	126	
d538	1,5-Dimethylhexylamine	$(CH_3)_2CH(CH_2)_3CH(NH_2)CH_3$	129.25		0.767	1.4209^{20}		154–156	48	i aq; s alc, eth

Dimethyl diphenyl sulfone 4,4'-dicarboxylate, s28
Dimethyleneimine, e134

Dimethylene oxide, e132
N,N-Dimethylethanolamine, d467

Dimethyl glutarate, d576
Dimethylglutaric acid, d577, d578

d512

d513a

d515a

d516

TABLE 1.15 Physical Constants of Organic Compounds (*Continued*)

No.	Name	Formula	Formula weight	Beilstein reference	Density	Refractive index	Melting point	Boiling point	Flash point	Solubility in 100 parts solvent
d539	2,5-Dimethyl-3-hexyne-2,5-diol	(CH$_3$)$_2$CC≡CC(CH$_3$)$_2$ with OH OH	142.20	1, 501			94–95	205–206		
d539a	(±)-3,5-Dimethyl-1-hexyn-3-ol	(CH$_3$)$_2$CHCH$_2$C(CH$_3$)-(OH)C≡CH	126.20	1^2, 507	0.859	1.4335^{20}		151	44	
d540	5,5-Dimethylhydantoin		128.13	24, 289			176–178			v s aq, alc, bz, chl, eth, acet
d541	1,1-Dimethylhydrazine	(CH$_3$)$_2$NNH$_2$	60.10	4, 547	0.7911^{22}	1.4075^{20}	−58	63.9	1	misc aq, alc, eth, PE
d542	1,2-Dimethylhydrazine	CH$_3$NHNHCH$_3$	60.10	4, 547	0.8274$^{20}_{4}$	1.4209^{20}		81	flammable	misc aq, alc, eth, PE
d543	Dimethyl hydrogen phosphonate	(CH$_3$O)$_2$P(=O)H	110.05	1, 285	1.200^{20}	1.4009^{20}		170–171	29	s aq(hyd); misc alc, acet, eth
d543a	2,2-Dimethyl-3-hydroxypropanal	HOCH$_2$C(CH$_3$)$_2$CHO	102.13	1^3, 3228				82^{4mm}	83	
d544	1,2-Dimethylimidazole		96.13	23, 66	1.084	1.4720^{20}	29–30	204	92	
d545	1,3-Dimethyl-2-imidazolidinone		114.15		1.044			108^{17mm}	80	
d546	Dimethyl maleate	CH$_3$OOCCH=CHCOOCH$_3$	144.13	2, 751	1.1513^{20}	1.4422^{20}	−17.5	200.4	91	8.7 aq
d547	Dimethyl malonate	CH$_3$OOCCH$_2$COOCH$_3$	132.12	2, 572	1.1542$^{20}_{4}$	1.4135^{20}	−62	180–181	90	sl s aq; misc alc, eth
d548	Dimethylmercury	(CH$_3$)$_2$Hg	230.66	4, 678	3.1874^{20}	1.5452^{20}		92^{740mm}		i aq; s alc, eth
d549	3,4-Dimethyl-1-methoxybenzene	(CH$_3$)$_2$C$_6$H$_3$OCH$_3$	136.19	6, 481	0.9744$^{14}_{4}$	1.5198^{14}		200		i aq; s alc, bz, eth
d550	3,5-Dimethyl-1-methoxybenzene	(CH$_3$)$_2$C$_6$H$_3$OCH$_3$	136.19	6, 493	0.9627$^{15}_{5}$	1.5107^{15}		193	65	i aq; s alc, bz, eth
d551	N,N-Dimethylmethyleneammonium iodide	H$_2$C=N(CH$_3$)$_2$I$^-$	185.01	4^4, 153			219 d			
d552	Dimethyl methylenesuccinate	CH$_3$OOCCH$_2$C(=CH$_2$)-COOCH$_3$	158.15	2, 762	1.1241$^{18}_{8}$	1.4442^{20}	38	208	100	s alc, eth
d553	Dimethyl methylphosphonate	(CH$_3$O)$_2$P(O)CH$_3$	124.08	4^1, 572	1.145	1.4130^{20}		181	43	
d554	Dimethyl methylsuccinate	CH$_3$OOCCH$_2$CH(CH$_3$)-COOCH$_3$	160.17	2^3, 1696	1.076	1.4200^{20}		196	83	
d555	2,6-Dimethylmorpholine		115.18		0.9346^{20}	1.4470^{20}	−85	147	48	misc aq, alc, bz

d558	1,2-Dimethyl-3-nitrobenzene	$(CH_3)_2C_6H_3NO_2$	151.17	5, 367	1.129	1.5434^{20}	7–9	245	107	i aq; s alc
d559	1,2-Dimethyl-4-nitrobenzene	$(CH_3)_2C_6H_3NO_2$	151.17	5, 368	1.139		29–31	143^{20mm}	>110	i aq; s alc
d560	1,3-Dimethyl-2-nitrobenzene	$(CH_3)_2C_6H_3NO_2$	151.17	5, 378	1.112	1.5220^{20}	14–16	225^{44mm}	87	i aq; s alc
d561	1,3-Dimethyl-4-nitrobenzene	$(CH_3)_2C_6H_3NO_2$	151.17	5, 378	1.117	1.5497^{20}	2	237–239	107	s alc, bz, chl, eth
d562	N,N-Dimethyl-4-nitrosoaniline	$(CH_3)_2NC_6H_4NO$	150.18	12, 677			86	flammable solid		i aq; s alc, eth
d563	Dimethyl 2-nitro-1,4-phthalate	$O_2NC_6H_3(COOCH_3)_2$	239.18	9, 826			72–75			
d564	cis-3,7-Dimethyl-2,6-octadienal		152.24		0.8888^{20}	1.4898^{20}		229	101	misc alc, eth, glyc
d565	trans-3,7-Dimethyl-2,6-octadienal		152.24		0.8869^{20}	1.4869^{20}		229	101	misc alc, eth, glyc
d566	3,7-Dimethyl-1-octanol	$(CH_3)_2CH(CH_2)_3CH(CH_3)CH_2CH_2OH$	158.29	1, 426	0.840	1.4355^{20}		96^{9mm}	95	
d566a	3,7-Dimethyl-3-octanol	$(CH_3)_2CH(CH_2)_3C(CH_3)(OH)C_2H_5$	158.29	1, 426	0.826	1.4336^{20}		73^{6mm}	76	
d566b	2,6-Dimethyl-2,4,6-octatriene	$CH_3CH=C(CH_3)CH=CHCH=C(CH_3)_2$	136.24	1^3, 1050	0.811	1.5429^{20}		75^{14mm}	68	

2,2-Dimethyl-3-hydroxypropionaldehyde, d543a
Dimethyl isophthalate, d591
1,4a-Dimethyl-7-isopropyl-1,2,3,4,4a,9,10,10a-octahydro-1-phenanthrenemethylamine, d20

Dimethyl itaconate, d552
2,2-Dimethyl-3-methoxyoxirane, m75

2,2-Dimethyl-3-methylenenorbornane, c2
6,6-Dimethyl-2-methylenenorpinene, p179

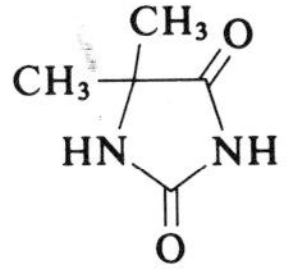

d540

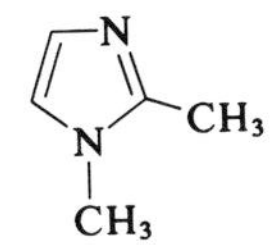

d544

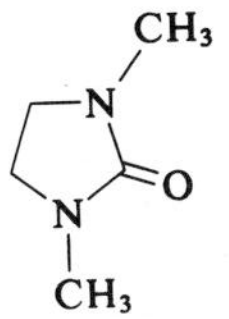

d545

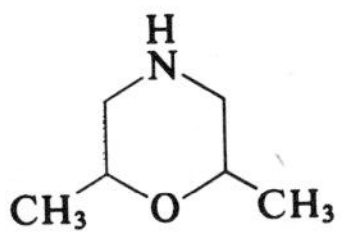

d555

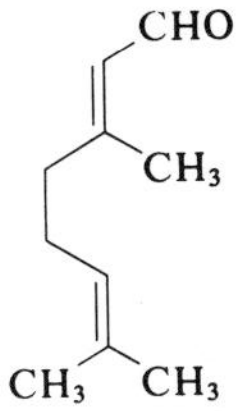

d564

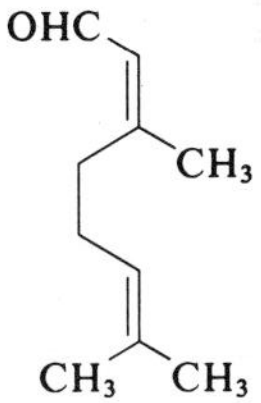

d565

TABLE 1.15 Physical Constants of Organic Compounds (*Continued*)

No.	Name	Formula	Formula weight	Beilstein reference	Density	Refractive index	Melting point	Boiling point	Flash point	Solubility in 100 parts solvent
d566c	N,N-Dimethyloctyl-amine	$CH_3(CH_2)_7N(CH_3)_2$	157.30	4[1], 386	0.765	1.4243^{20}	−57	195	65	
d566d	3,6-Dimethyl-4-oc-tyne-3,6-diol	$C_2H_5C(CH_3)(OH)\text{-}C{\equiv}CC(CH_3)(OH)C_2H_5$	170.35	1[1], 263			55	214^{680mm}	>110	i aq; s alc
d567	Dimethyl octanedioate	$CH_3OOC(CH_2)_6COOCH_3$	202.25	2, 693	1.0210^{20}	1.4325^{20}	−4.8	268	75	6 aq; s alc, eth
d568	Dimethyl oxalate	$CH_3OOCCOOCH_3$	118.09	2, 534	1.148^{54}	1.379^{80}	50–54	163.5		s aq, acids, alk
d569	N[1]-(4,5-Dimethyloxazol-2-yl)-sulfanilamide		267.31					193–194		
d569a	3,3-Dimethyloxetane	$CH_2{=}CHC({=}O)\text{-}NHC(CH_3)_2CH_2COCH_3$	86.13	17[2], 21	0.835	1.3990		81^{765mm}	−9	
d570	N-(1,1-Dimethyl-3-ox-obutyl)acrylamide		169.23				57–58	120^{8mm}		
d571	2,3-Dimethylpentanal	$CH_3CH_2CH(CH_3)CH(CH_3)\text{-}CHO$	114.19		0.832	1.4132^{20}			58	
d573	2,3-Dimethylpentane	$CH_3CH_2CH(CH_3)CH(CH_3)_2$	100.21	1[2], 120	0.6951^{20}_{4}	1.3920^{20}	glass	89.8	−6	i aq; s alc, eth
d576	Dimethyl pentane-dioate	$CH_3OOC(CH_2)_3COOCH_3$	160.17	2, 633	1.0934^{15}	1.4234^{20}		95^{13mm}	102	v s alc, eth
d577	2,2-Dimethylpentane-dioic acid	$HOOCC(CH_3)_2CH_2CH_2COOH$	160.17	2, 676			83–85			v s aq, alc, chl
d578	3,3-Dimethylpen-tanedioic acid	$(CH_3)_2C(CH_2COOH)_2$	160.17	2, 684			100–103			v s aq, alc, eth
d579	2,4-Dimethyl-3-pen-tanol	$(CH_3)_2CHCH(OH)CH(CH_3)_2$	116.20	1, 417	0.829^{20}_{4}	1.4254^{20}	<70	140	37	sl s aq; s alc, eth
d580	2,4-Dimethyl-3-pen-tanone	$(CH_3)_2CHC({=}O)CH(CH_3)_2$	114.19	1, 703	0.8062^{24}	1.3986^{20}	−80	124	15	misc alc, eth; s bz
d581	2,3-Dimethylphenol	$(CH_3)_2C_6H_3OH$	122.17	6, 480		1.5420^{20}	75	218		v s alc, bz, chl, eth
d582	2,4-Dimethylphenol	$(CH_3)_2C_6H_3OH$	122.17	6, 486		1.5390^{20}	27	212	>110	v s alc, bz, chl, eth
d583	2,5-Dimethylphenol	$(CH_3)_2C_6H_3OH$	122.17	6, 494			74.5	211.5		v s alc, bz, chl, eth
d584	2,6-Dimethylphenol	$(CH_3)_2C_6H_3OH$	122.17	6, 485	0.965^{80}		49.0	203	73	v s alc, bz, chl, eth
d585	3,4-Dimethylphenol	$(CH_3)_2C_6H_3OH$	122.17	6, 480	1.0642^{28}		62.5	225		v s alc, bz, chl, eth
d586	3,5-Dimethylphenol	$(CH_3)_2C_6H_3OH$	122.17	6, 492	1.0082^{28}		64–68	219.5		v s alc, bz, chl, eth
d587	N,N-Dimethyl-1,4-phenylenediamine	$(CH_3)_2NC_6H_4NH_2$	136.20	13, 72			36	262	90	
d588	4,5-Dimethyl-o-phenylenediamine	$(CH_3)_2C_6H_2(NH_2)_2$	136.20	13, 179			127–129			

1.188

No.	Name	Formula	Formula weight	Beilstein reference	Density	n_D	mp, °C	bp, °C	Flash p, °C	Solubility in 100 parts solvent
d589	2,2-Dimethyl-3-phenyl-1-propanol	$C_6H_5CH_2C(CH_3)_2CH_2OH$	164.25				35	126^{15mm}	109	
d590	Dimethyl 1,2-phthalate	$C_6H_4(COOCH_3)_2$	194.19	9, 797	1.1940^{20}_{20}	1.515^{21}	5.5	283.7	146 (CC)	0.4 aq; misc alc, chl, eth; i PE
d591	Dimethyl 1,3-phthalate	$C_6H_4(COOCH_3)_2$	194.19	9, 834	1.194^{20}_4	1.5168^{20}	67–68	282		i aq
d592	Dimethyl 1,4-phthalate	$C_6H_4(COOCH_3)_2$	194.19	9, 843			140–142	subl		0.3 hot aq; s hot alc; s eth
d593	2,6-Dimethylpiperidine		113.20	20, 108	0.840	1.4394^{20}		127	11	
d594	2,2-Dimethylpropane	$(CH_3)_4C$	72.15	4^3, 595	0.613^0	1.3476^6	−16.6	9.5		
d595	2,2-Dimethyl-1,3-propanediamine	$H_2NCH_2C(CH_3)_2CH_2NH_2$	102.18		0.851	1.4566^{20}	31	154	47	
d596	2,2-Dimethyl-1,3-propanediol	$(CH_3)_2C(CH_2OH)_2$	104.15	1, 483	1.11^{25}		127–128	208–210	107	180 aq; 12 bz; 60 acet; v s alc, eth
d597	2,2-Dimethyl-1-propanol	$(CH_3)_3CCH_2OH$	88.15	1, 406	0.812^{20}		52–54	113.1	36	3.6 aq; misc alc, eth
d598	2,2-Dimethylpropionaldehyde	$(CH_3)_3CCHO$	86.13		0.793	1.3794^{20}	6	74^{130mm}	<1	
d599	N,N-Dimethylpropionamide	$C_2H_5CON(CH_3)_2$	101.15	4^3, 126	0.920	1.4400^{20}	−45	175	62	
d600	2,2-Dimethylpropionic acid	$(CH_3)_3CCOOH$	102.13	2, 319	0.905^{50}	1.3931^{37}	35.5	163.8	63	2.5 aq; v s alc, eth
d601	2,2-Dimethylpropionic anhydride	$[(CH_3)_3CC(O)]_2O$	186.25	2, 320	0.918	1.4092^{20}		193	57	
d602	2,2-Dimethylpropionyl chloride	$(CH_3)_3CCOCl$	120.58	2, 320	0.979	1.4120^{20}		106	<1	dec aq, alc; v s eth

3,7-Dimethyl-6-octen-1-ol, c275
Dimethylolpropionic acid, b185
Dimethyl 3-oxoglutarate, d460
1,5-Dimethyl-2-phenyl-4-aminopyrazolone, z113

2,3-Dimethyl-1-phenyl-3-pyrazolin-5-one, a314
Dimethyl phosphite, d543
Dimethyl pimelate, d532

N,N-Dimethyl-1,3-propanediamine, d477
Dimethyl propanedioate, d547
1,1-Dimethylpropargylamine, d604

d569

d569a

d593

TABLE 1.15 Physical Constants of Organic Compounds (*Continued*)

No.	Name	Formula	Formula weight	Beilstein reference	Density	Refractive index	Melting point	Boiling point	Flash point	Solubility in 100 parts solvent
d603	1,1-Dimethylpropylamine	$CH_3CH_2C(CH_3)_2NH_2$	87.17	4, 179	0.7312^{25}_{4}	1.3996^{20}	-105	77	65	misc aq, alc, eth
d604	1,1-Dimethyl-2-propynylamine	$HC{\equiv}CC(CH_3)_2NH_2$	83.13		0.790	1.4235^{20}		79–80	2	
d605	3,5-Dimethylpyrazole		96.13	23, 74			108	218	50	s aq; v s bz, eth
d605a	2,3-Dimethylpyridine	$(CH_3)_2(C_5H_3N)$	107.16	20, 243	0.945	1.5080	-15	163	37	
d606	2,4-Dimethylpyridine	$(CH_3)_2(C_5H_3N)$	107.16	20, 244	0.9272^{25}	1.4991^{20}	<-60	158.3	33	17 aq; v s alc, bz, eth
d607	2,6-Dimethylpyridine	$(CH_3)_2(C_5H_3N)$	107.16	20, 244	0.9200^{25}	1.4956^{25}	-6.0	143–144	53	43 aq[45]; s alc, eth
d608	3,4-Dimethylpyridine	$(CH_3)_2(C_5H_3N)$	107.16	20, 246	0.9542^{5}	1.5100^{25}	-12	164	53	sl s aq; s alc, eth
d609	3,5-Dimethylpyridine	$(CH_3)_2(C_5H_3N)$	107.16	20, 246	0.9392^{5}	1.5033^{25}	-9	170	80	s aq, alc, eth
d609a	Dimethyl pyrocarbonate	$O(CO_2CH_3)_2$	134.09	3^4, 17	1.250	1.3933^{20}		46^{6} mm		
d610	Dimethyl succinate	$CH_3OOCCH_2CH_2COOCH_3$	146.14	2, 609	1.202^{8}_{4}	1.4190^{20}	19.5	195–200	85	0.83 aq; 2.9 alc
d611	Dimethylsulfamoyl chloride	$(CH_3)_2NSO_2Cl$	143.59	4, 84	1.337	1.4518^{20}		114^{75} mm	94	
d612	Dimethyl sulfate	$(CH_3O)_2SO_2$	126.13	1, 283	1.3322^{20}	1.3874^{20}	-31.8	188 d	83	2.8 aq(hyd); s acet, bz, diox, eth
d613	Dimethyl sulfide	$(CH_3)_2S$	62.13	1, 288	0.8462^{1}	1.4354^{20}	-98.3	37.3	-36	2 aq; s alc, eth
d614	Dimethyl sulfite	$(CH_3O)_2SO$	110.13	1, 282	1.294	1.4083^{20}		126–127	30	
d615	Dimethyl sulfone	$(CH_3)_2SO_2$	94.13	1, 289			109	238	143	v s aq, alc, acet
d616	Dimethyl sulfoxide	$(CH_3)_2SO$	78.13	1, 289	1.100^{20}	1.4783^{20}	18.5	189.0	95 (OC)	s alc, acet, bz, chl
d617	Dimethyl-d_6 sulfoxide	$(CD_3)_2SO$	84.18	1^4, 1279	1.190	1.4758^{20}		55^{5} mm	95	
d618	(+)-Dimethyl L-tartrate	$CH_3OOCCH(OH)CH(OH)COOCH_3$	178.14	3, 510	1.328^{20}_{4}		48–50	163^{23} mm	>110	s aq; 200 alc[15]; v s bz
d619	Dimethyltelluride	$(CH_3)_2Te$	157.68	1, 291				91–92		
d620	2,5-Dimethyltetrahydrofuran	$(CH_3)_2(C_4H_2O)$	100.16	17, 14	0.833	1.4041	-10	90–92	26	d aq; v s alc; i eth
d620a	1,3-Dimethyl-3,4,5,6-tetrahydro-2(1H)-pyrimidinone		128.18	24^3, 32	1.060	1.4880^{20}		146^{44} mm	>110	
d621	Dimethyl 3,3'-thiodipropionate	$(CH_3OOCCH_2CH_2)_2S$	206.26		1.198	1.4740^{20}		148^{18} mm	>110	
d622	N,N-Dimethylthioformamide	$(CH_3)_2NC(S)H$	89.16	4, 70	1.047	1.5757^{20}		58^{1} mm	99	

No.	Name	Formula	Mol wt	Beilstein ref.	Density	n_D	m.p., °C	b.p., °C	Flash	Solubility
d623	N,N'-Dimethylthiourea	(CH₃NH)₂C=S	104.18	4, 70			60–62			v s aq, alc, acet
d624	N,N-Dimethyl-p-toluidine	CH₃C₆H₄N(CH₃)₂	135.21	12, 902	0.937	1.5458^{20}		211	83	
d624a	N,N-Dimethyltrimethylsilylamine	(CH₃)₃SiN(CH₃)₂	117.27		0.732	1.3970^{20}		84	−19	v s aq, alc; i eth misc bz, PE; s eth
d625	1,3-Dimethylurea	(CH₃NH)₂C=O	88.11	4, 65			101–104	268–270		i aq; 0.75 alc
d626	Dimethylzinc	(CH₃)₂Zn	95.45		1.386^1		−40	46	ignites in air	
d627	2,4-Dinitroaniline	(O₂N)₂C₆H₃NH₂	183.12	12, 747	1.615^{14}		188			0.05 aq; 2.7 alc; v s bz, chl, EtAc
d628	1,3-Dinitrobenzene	C₆H₄(NO₂)₂	168.11	5, 258	1.575^8		89–90	303		s bz, HOAc; dec alc
d629	2,4-Dinitrobenzenesulfenyl chloride	(O₂N)₂C₆H₃SCl	234.62	6[2], 316			96			0.7 aq; v s alc, eth
d630	3,4-Dinitrobenzoic acid	(O₂N)₂C₆H₃COOH	212.12	9, 413			166	subl		1.9 hot aq; v s alc; sl s bz, eth
d631	3,5-Dinitrobenzoic acid	(O₂N)₂C₆H₃COOH	212.12	9, 413			207			dec aq, alc; s eth
d632	3,5-Dinitrobenzoyl chloride	(O₂N)₂C₆H₃COCl	230.56	9, 414			69.5	196^{11mm}		
d633	2,6-Dinitro-p-cresol	(O₂N)₂C₆H₂(OH)CH₃	198.13	6, 414			77–79 (anhyd) 87.5			v s alc, acet, eth, alk
d633a	4,6-Dinitro-o-cresol	(O₂N)₂C₆H₂(OH)CH₃	198.13	6, 368						

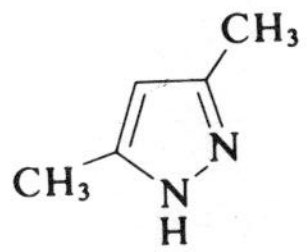

d605

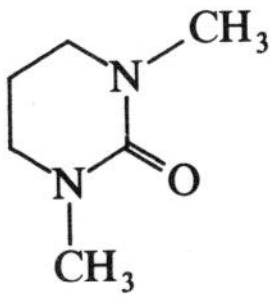

d620a

TABLE 1.15 Physical Constants of Organic Compounds (*Continued*)

No.	Name	Formula	Formula weight	Beilstein reference	Density	Refractive index	Melting point	Boiling point	Flash point	Solubility in 100 parts solvent
d634	2,4-Dinitrodiphenyl-amine	$(O_2N)_2C_6H_3NHC_6H_5$	259.22	12, 751			161			s bz, eth, glyc
d635	2,4-Dinitro-1-fluoro-benzene	$FC_6H_3(NO_2)_2$	186.10	5, 262		1.5690^{20}	26	178^{25mm}	>110	s bz; v s eth; sl s alc
d636	1,5-Dinitro-naphthalene	$C_{10}H_6(NO_2)_2$	218.17	5, 558			216–217	subl		s bz; v s eth; sl s alc
d637	2,4-Dinitrophenol	$(O_2N)_2C_6H_3OH$	184.11	6, 251	1.683		112–114			s alc, bz; 16 EtAc; 36 acet; 5 chl; 20 pyr
d638	2,4-Dinitrophenyl-acetic acid	$(O_2N)_2C_6H_3CH_2COOH$	226.15	9, 459			169–175			s alc, eth
d639	2,4-Dinitrophenyl-hydrazine	$(O_2N)_2C_6H_3NHNH_2$	198.14	15, 489			–200	flammable solid		sl s aq, alc; s acid
d640	3,5-Dinitrosalicylic acid	$(O_2N)_2C_6H_2(OH)COOH$	228.12	10, 122			169–172			s aq; v s alc, eth
d641	2,4-Dinitrotoluene	$CH_3C_6H_3(NO_2)_2$	182.14	5, 339	1.321^{71}	1.442	64–66	300 sl dec		1.2 alc; 9 eth
d642	2,6-Dinitrotoluene	$CH_3C_6H_3(NO_2)_2$	182.14	5, 341	1.2833^{111}	1.479	64–66			s alc
d643	3,4-Dinitrotoluene	$CH_3C_6H_3(NO_2)_2$	182.14	5, 341	1.2594^{111}		54–57		>110	i aq; s alc
d643a	Dinonyl hexanedioate	$C_9H_{19}OOC(CH_2)_4COOC_9H_{19}$	398.63		0.9172^{25}				218	
d644	Dioctadecyl phosphite	$(C_{18}H_{37}O)_2P(O)H$	586.97				57–59			
d645	Dioctylamine	$(C_8H_{17})_2NH$	241.46	4, 196	0.806	1.4318^{20}	14–16	298	>110	i aq; v s alc, eth
d645a	Dioctyl ether	$[CH_3(CH_2)_7]_2O$	242.45	1, 419	0.842	1.4610^{20}	–7.6	287	>110	
d646	Dioctyl sulfide	$(C_8H_{17})_2S$	258.51	1, 419	0.962	1.4609^{20}		180^{10mm}	>110	
d647	4,9-Dioxa-1,12-dodecanediamine	$H_2N(CH_2)_3O(CH_2)_4O(CH_2)_3NH_2$	204.32					$134–136^{4mm}$	>110	
d648	1,4-Dioxane		88.10	19, 3	1.0329^{20}	1.4224^{20}	11.7	101.2	12	misc aq, alc, bz, chl, eth, PE
d649	1,3-Dioxolane		74.08	19^2, 3	1.0602^{20}	1.4000^{20}	–95	74–75	1	misc aq; s alc, eth
d650	Dipentaerythritol	$(HOCH_2)_3CCH_2OCH_2-C(CH_2OH)_3$	254.28				215–218			
d651	Dipentene		136.24	5, 137	0.8402^{21}	1.4739^{20}		176	42	i aq; misc alc
d652	Dipentylamine	$(C_5H_{11})_2NH$	157.29	4^1, 378	0.777	1.4272		195–202	52	v s alc
d653	Dipentyl ether	$(C_5H_{11})_2O$	158.29	1^1, 193	0.7833^{20}	1.4120^{20}		186.8	57	misc alc, eth; s acet
d653a	N,N-Diphenylacet-amide	$CH_3CON(C_6H_5)_2$	211.26	12, 247			103	$130^{0.02mm}$		sl s aq; s alc, eth
d654	Diphenylacetic acid	$(C_6H_5)_2CHCOOH$	212.25	9, 673	1.2581^{15}		148	195^{5mm}		s hot aq, alc, chl, eth

No.	Name	Formula	Formula wt	Beilstein ref	Density	n_D	mp, °C	bp, °C		Solubility
d655	Diphenylacetonitrile	$(C_6H_5)_2CHCN$	193.25	9, 674	0.990		76	181^{12mm}		s alc, eth
d656	Diphenylacetylene	$C_6H_5C{\equiv}CC_6H_5$	178.23	5, 656			60–61	300		v s hot alc, eth
d657	Diphenylamine	$(C_6H_5)_2NH$	169.23	12, 174	1.160		53–54	302	152	45 alc; v s bz, eth
d658	*cis,cis*-1,4-Diphenyl-1,3-butadiene	$C_6H_5CH{=}CHCH{=}CHC_6H_5$	206.29	5, 676	0.9697^{4}	1.6347^{101} (He line)	70.5			s bz, chl, eth, PE
d659	*cis,trans*-1,4-Diphenyl-1,3-butadiene	$C_6H_5CH{=}CHCH{=}CHC_6H_5$	206.29	5, 676	0.9974^{22}	1.6053^{22}	88	$133^{0.1mm}$		s alc, bz, eth, chl
d660	1,3-Diphenyl-2-buten-1-one	$C_6H_5C(O)CH{=}C(C_6H_5)CH_3$	222.27	7², 433	1.1080^{15}	1.6343^{20}	−30 glass	246^{80mm}		i aq; s alc, eth
d661	Diphenylcarbamoyl chloride	$(C_6H_5)_2NCOCl$	231.68				82–84			
d662	1,5-Diphenylcarbohydrazide	$(C_6H_5NHNH)_2C{=}O$	242.28	15, 292			168–171			s hot alc, acet, HOAc
d663	Diphenyl carbonate	$(C_6H_5O)_2C{=}O$	214.22	6, 158	1.296		80–81	302–306		s hot alc, bz, eth
d664	Diphenyl chlorophosphate	$(C_6H_5O)_2P(O)Cl$	268.64	6, 179		1.5500^{20}		314^{272mm}	>110	
d665	Diphenyl diselenide	$C_6H_5SeSeC_6H_5$	312.13	6, 346	1.5574^{80}		61–64			s hot alc
d666	Diphenyl disulfide	$C_6H_5SSC_6H_5$	218.34	6, 323	1.3532^{20}		58–60	310		s alc, bz, eth; i aq
d667	Diphenylenimine		167.21	20, 433	1.104^{18}		246	355		0.8 bz; 3 eth; 16 pyr; 11 acet; i aq

d648 (1,4-dioxane ring structure)

d649 (1,3-dioxolane ring structure)

d651 (cyclohexene structure; CH₃, C=CH₂, CH₃)

d667 (carbazole structure; N–H)

TABLE 1.15 Physical Constants of Organic Compounds (*Continued*)

No.	Name	Formula	Formula weight	Beilstein reference	Density	Refractive index	Melting point	Boiling point	Flash point	Solubility in 100 parts solvent
d668	1,2-Diphenylethane	$C_6H_5CH_2CH_2C_6H_5$	182.27	5, 598	0.9952^{20}	1.5338	52.5	284	>110	s alc; v s chl, eth
d669	Diphenyl ether	$C_6H_5OC_6H_5$	170.21	6, 146	1.0661^{30}	1.5763^{30}	26.9	258.3	115	s alc, bz, eth, HOAc
d671	N,N'-Diphenylethyl-enediamine	$C_6H_5NHCH_2CH_2NHC_6H_5$	212.30	12, 543			67.5	228–230		v s alc, eth
d672	N,N'-Diphenyl-formamidine	$C_6H_5N{=}CHNHC_6H_5$	196.25	12, 236			138–141			s eth; v s chl
d673	1,3-Diphenylguanidine	$C_6H_5NHC({=}NH)NHC_6H_5$	211.27	12, 369	1.13		150	d 170		s alc, hot bz, chl
d674	5,5-Diphenyl-hydantoin		252.27	24, 410			295–298			i aq; 1.7 alc; 3.3 acet
d675	1,2-Diphenylhydrazine	$C_6H_5NHNHC_6H_5$	184.24	15, 123	1.1581^6		123–126			v s alc; sl s bz
d677	Diphenylmercury	$(C_6H_5)_2Hg$	354.81	16, 946	2.318^4		124–125	d > 306	>110	s chl; sl s hot alc
d678	Diphenylmethane	$C_6H_5CH_2C_6H_5$	168.24	5^2, 498	1.3421^{10}	1.5768	25.9	264.5		v s alc, bz, chl, eth
d679	Diphenylmethanol	$(C_6H_5)_2CHOH$	184.24	6, 678			66.7	298		0.05 aq; v s alc, chl, eth
d680	1,1-Diphenylmethyl-amine	$C_6H_5CH(NH_2)C_6H_5$	183.25	12, 1323	1.06352 super-cooled	1.5956^{99}	34	295	>112	sl s aq
d681	2,5-Diphenyloxazole	$(C_6H_5)_2C_6H_3OH$	221.26	27, 78			72–73	360		
d682	2,6-Diphenylphenol		246.31	6^3, 3631			100–102			
d682a	N,N'-Diphenyl-1,4-phenylenediamine	$C_6H_5NHC_6H_4NHC_6H_5$	260.34	13, 80	1.20		148	$225^{0.5mm}$		s bz, chl, EtAc, eth
d683	Diphenyl phosphite	$(C_6H_5O)_2P({=}O)H$	234.19	6^1, 94	1.223	1.5575^{20}	12	219^{26mm}	176	
d684	Diphenylphosphoryl azide	$(C_6H_5O)_2P({=}O)N_3$	275.20		1.277	1.5518^{20}		$157^{0.17mm}$	>110	
d684a	Diphenyl 1,2-phthalate	$C_6H_4(CO_2C_6H_5)_2$	318.33	9, 801			76			
d685	2,2-Diphenyl-1-picrylhydrazyl		394.32	16^2, 363			127 d			
d686	1,3-Diphenyl-2-propanone	$C_6H_5CH_2C({=}O)CH_2C_6H_5$	210.28	7, 445	1.2		32–34	330		i aq; v s alc, eth
d688	2,2-Diphenylpropionic acid	$CH_3C(C_6H_5)_2COOH$	226.28	9^2, 474			175–177	300		s alc; v s bz, eth
d689	Diphenylsilanediol	$(C_6H_5)_2Si(OH)_2$	216.31	16, 909	1.1181^5	1.6327^{20}	140 d	296	53	misc bz, eth, CS_2
d690	Diphenyl sulfide	$(C_6H_5)_2S$	186.28	6, 299			−40	379	>110	i aq; s hot alc, bz
d691	Diphenyl sulfone	$(C_6H_5)_2SO_2$	218.27	6, 300			128–129			
d692	Diphenyl sulfoxide	$(C_6H_5)_2S{=}O$	202.28	6, 300			69–71	207^{13mm}		

No.	Name	Formula	Formula wt	Beil. ref	Density	Index of refraction	Melting point	Boiling point	Flash pt	Solubility
d693	Diphenylthio-carbazone	C6H5N=NC(S)NNHNHC6H5	256.33	16, 26	1.32		168 d			i aq; v s chl, CCl4
d694	1,3-Diphenylthiourea	C6H5NHC(S)NHC6H5	228.32	12, 394	1.239		154	260 d		i aq; v s alc, eth
d695	1,3-Diphenylurea	C6H5NHC(O)NHC6H5	212.352	12, 352			238			0.015 aq; s eth, HOAc
d697	Dipiperidinomethane		182.31		0.915	1.4820^{20}		123^{15mm}	91	
d698	Dipropylamine	(C3H7)2NH	101.19	4, 138	0.7375^{20}_{4}	1.4043^{20}	−39.6	109.2	3	4 aq; v s alc, eth, PE
d699	Dipropylene glycol butyl ether	CH3CH(OH)CH2OCH2CH(OC4H9)CH3	190.3		0.917^{25}_{25}	1.425^{25}		229	113	
d700	Dipropylene glycol ethyl ether	CH3CH(OH)CH2OCH2CH(OC2H5)CH3	162.2		0.930^{25}_{25}	1.419^{25}		388	90	
d701	Dipropylene glycol iso-propyl ether	CH3CH(OH)CH2OCH2CH[OCH(CH3)2]CH3	176.2		0.878^{25}_{25}	1.421^{25}		80.1	90	
d702	Dipropylene glycol methyl ether	CH3CH(OH)CH2OCH2CH(OCH3)CH3	148.2		0.951^{20}_{20}	1.419^{20}	−117	188.3	85	
d703	Dipropyl ether	(C3H7)2O	102.18	1, 354	0.7466^{20}	1.3803^{20}	−123.2	89.6	4	0.4 aq
d704	Dipropyl hexanedioate	C3H7OOC(CH2)4COOC3H7	230.30	2^2, 574	0.9790^{20}	1.4314^{20}	−20	144^{10mm}	126	
d705	Dipropyl sulfate	(C3H7O)2SO2	182.24	1, 354	1.106^{20}		d 140	120^{20mm}		i aq; s alc, eth
d706	Dipropyl sulfone	(C3H7)2SO2	150.24	1, 359	1.028^{50}		28–30	270		v s PE

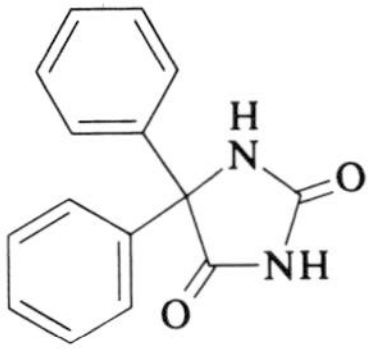

d674

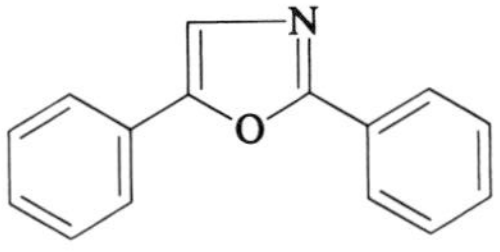

d681

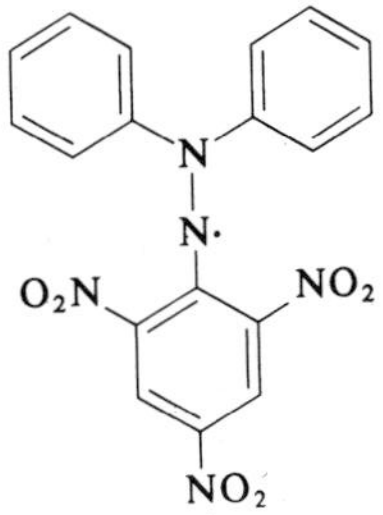

d685

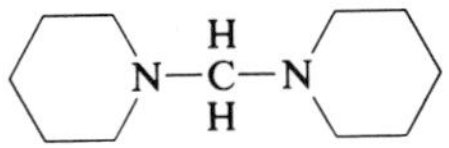

d697

TABLE 1.15 Physical Constants of Organic Compounds (*Continued*)

No.	Name	Formula	Formula weight	Beilstein reference	Density	Refractive index	Melting point	Boiling point	Flash point	Solubility in 100 parts solvent
d707	2,2'-Dipyridyl		156.19	23, 199			69.7	273		0.5 s aq; v s alc, bz, chl, eth, PE
d708	2,2'-Dipyridylamine		171.20	22^2, 630			89–90	222^{50mm}		
d709	1,3-Dithiane		120.24				53–55		90	
d710	4,4'-Dithiobutyric acid	$HOOC(CH_2)_3SS(CH_2)_3COOH$	238.32	3, 312			110			
d711	3,3'-Dithiopropionic acid	$HOOCCH_2CH_2SSCH_2CH_2COOH$	210.27				157–159			
d712	Dithiooxamide	$H_2NC(=S)C(=S)NH_2$	120.20	2, 565			170 d	subl		sl s aq; s alc; i eth
d712a	2,2'-Dithiosalicylic acid	$S_2(C_6H_4CO_2H)_2$	306.36	10, 129			290			
d713	1,3-Di-o-tolylguanidine	$(CH_3C_6H_4NH)_2C=NH$	239.32	12, 803	1.10^{20}_4		176–178			s hot alc, eth
d714	1,5-Di(vinyloxy)-3-oxapentane	$(CH_2=CHOCH_2CH_2)_2O$	158.20		0.975^{29}	1.445		81^{10mm}		
d715	1,3-Divinyltetra-methyldisiloxane	$[CH_2=CHSi(CH_3)_2]_2O$	186.39		0.811^{20}_4	1.412^{20}	−99.7	139		
d716	3,9-Divinyl-2,4,8,10-tetraoxaspiro[5.5]-undecane		212.25	19^3, 5679	1.251		40–45	120^{2mm}	110	
d717	Docosane	$CH_3(CH_2)_{20}CH_3$	310.61	1, 174	0.7782^{45}	1.4358^{45}	44.4	369	>110	i aq; sl s alc; v s eth
d719	1-Docosanol	$CH_3(CH_2)_{21}OH$	326.61	1, 431			65–72	$180^{0.22mm}$		sl s eth; s alc, chl
d721	Dodecane	$CH_3(CH_2)_{10}CH_3$	170.41	1, 171	0.7490^{26}_4	1.4216^{20}	−9.6	216.28	71	
d722	1,12-Dodecanediamine	$H_2N(CH_2)_{12}NH_2$	200.37	4, 273			71		155	
d723	Dodecanedioic acid	$HOOC(CH_2)_{10}COOH$	230.30	2, 729			128–130	245^{10mm}		
d724	1,2-Dodecanediol	$CH_3(CH_2)_9CH(OH)CH_2OH$	202.34	1^3, 2237			58–60			
d725	1,12-Dodecanediol	$HOCH_2(CH_2)_{10}CH_2OH$	202.34	1^2, 562			81–84	189^{12mm}		
d727	1-Dodecanethiol	$CH_3(CH_2)_{11}SH$	202.40	2, 359	0.8452^{20}_{20}	1.4587^{20}		266–283	87	i aq; s alc, eth
d728	Dodecanoic acid	$CH_3(CH_2)_{10}COOH$	200.32		0.869^{20}_4	1.4183^{82}	44	225^{100mm}	>110	s aq; 100 alc; v s bz, eth
d729	1-Dodecanol	$CH_3(CH_2)_{11}OH$	186.34	1, 428	0.83082^{25}	1.4413^{25}	24	259	>110	i aq; s alc, eth
d730	δ-Dodecanolactone		198.31		0.942	1.4602^{20}		126^{1mm}	>110	
d730a	Dodecanoyl chloride	$CH_3(CH_2)_{10}COCl$	218.77	2, 363	0.946	1.4459^{20}		137^{11mm}	>110	
d730b	Dodecanoyl peroxide	$[CH_3(CH_2)_{10}CO]_2O_2$	398.63	2^3, 893			55–57			s alc, eth, PE
d731	1-Dodecene	$CH_3(CH_2)_9CH=CH_2$	168.32	1, 225	0.7584^{20}_4	1.4294^{20}_4	−35.2	213.4	77	

No.	Name	Formula	Mol. wt.		Density	n_D	mp, °C	bp, °C	Flash	Solubility
d732	2-Dodecen-1-yl-succinic anhydride		266.38					180^{5mm}	177	
d732a	Dodecyl acetate	$CH_3CO_2(CH_2)_{11}CH_3$	228.38	2, 136	0.865	1.4318^{20}		150^{15mm}	110	misc alc, bz, chl, eth
d733	Dodecyl aldehyde	$CH_3(CH_2)_{10}CHO$	184.32	1, 714	0.835	1.4344^{20}		185^{100mm}	101	
d734	Dodecylamine	$CH_3(CH_2)_{11}NH_2$	185.36	4, 200			28–30	247–249	>110	
d735	Dodecyl methacrylate	$H_2C{=}C(CH_3)CO_2(CH_2)_{11}CH_3$	254.42	2^3, 1290	0.868	1.4460^{20}	−7	142^{4mm}	110	10 aq
d737	Dodecyl sulfate, Na salt	$CH_3(CH_2)_{11}SO_4^-Na^+$	288.38	1^3, 1786			204–207			
d738	Dodecyltrichlorosilane	$CH_3(CH_2)_{11}SiCl_3$	303.78	4^3, 1907	1.020	1.458^{20}		294	>110	sl s alc, bz, eth
d739	Dotriacontane	$CH_3(CH_2)_{30}CH_3$	450.88	1, 177	0.8124^{20}_{4}	1.4364^{70}	68–70	467		3.3 aq; sl s alc
d740	Dulcitol		182.17	1, 544	1.47^{20}		188–189	275^{1mm}		
e1	(−)-Ephedrine	$CH_3NHCH(CH_3)CH(OH)C_6H_5$	165.24	13, 636			39		85	s aq, alc, chl, eth

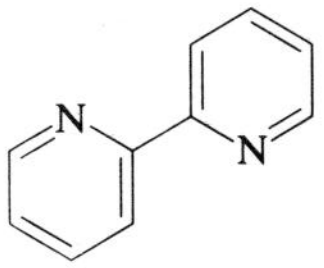

d707

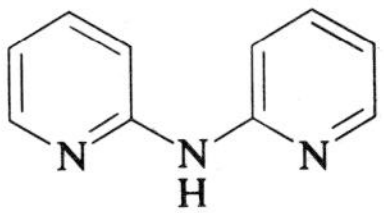

d708

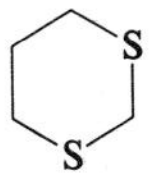

d709

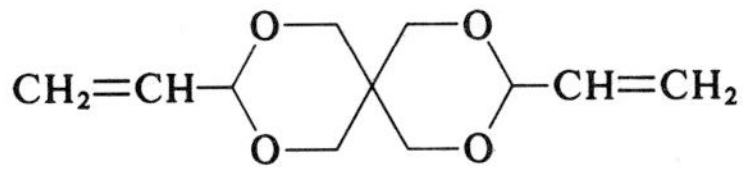

d716

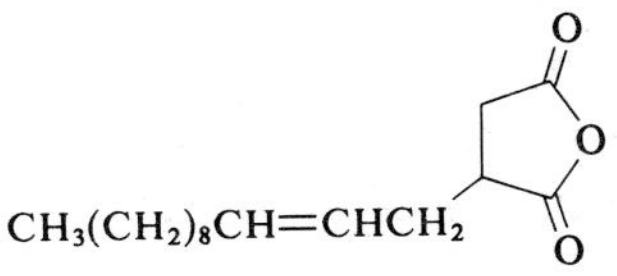

$CH_3(CH_2)_8CH{=}CHCH_2$

d732

$HOCH_2{-}C{-}C{-}C{-}C{-}CH_2OH$ (with OH, H substituents and CH_2OH)

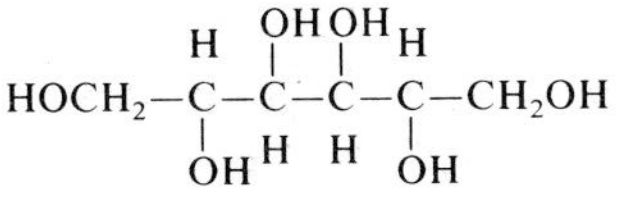

d740

TABLE 1.15 Physical Constants of Organic Compounds (*Continued*)

No.	Name	Formula	Formula weight	Beilstein reference	Density	Refractive index	Melting point	Boiling point	Flash point	Solubility in 100 parts solvent
e2	1,2-Epoxybutane	$CH_3CH_2CH{-}CH_2$ (O)	72.11	17[2], 17	0.8297[20]	1.3840[20]	−150	63.2	−12	6 aq; misc alc, bz, chl, eth
e3	1,2-Epoxy-5,9-cyclododecadiene		178.28		0.980	1.5045[20]		83[1mm]	>110	
e4	1,2-Epoxycyclododecane		182.31		0.939	1.4773[20]			>110	
e5	1,2-Epoxycyclopentane		84.12	17, 21	0.964	1.4336[20]		102	10	
e6	1,2-Epoxydecane	$CH_3(CH_2)_6CH_2CH{-}CH_2$ (O)	156.27	17, 18	0.840	1.4290[20]		94[15mm]	78	
e7	1,2-Epoxydodecane	$CH_3(CH_2)_8CH_2CH{-}CH_2$ (O)	184.32	17[3], 136	0.844	1.4355[20]		125[15mm]		
e8	1,2-Epoxyethylbenzene	$C_6H_5CH{-}CH_2$ (O)	120.15	17, 49	1.0523[16]	1.5338[20]	−37	194	79	i aq; s alc, eth
e9	1,2-Epoxyhexadecane	$CH_3(CH_2)_{12}CH_2{-}CH{-}CH_2$ (O)	240.43	17, 20	0.846	1.4452[20]	22	180[12mm]	93	
e9a	1,2-Epoxyhexane	$CH_3CH_2CH_2CH_2CH{-}CH_2$ (O)	100.16	17[4], 86	0.831	1.4056[20]		120	15	
e9b	1,2-Epoxy-5-hexene	$H_2C{=}CHCH_2CH_2{-}CH{-}CH_2$ (O)	98.15	17[3], 163	0.870	1.4252[20]		121	15	
e9c	1,2-Epoxy-7-octadecane	$CH_3(CH_2)_{14}CH_2CH{-}CH_2$ (O)	268.49	17[3], 140			35	137[0.5mm]	110	
e9d	1,2-Epoxyoctane	$CH_3(CH_2)_4CH_2CH{-}CH_2$ (O)	128.22	17, 17	0.839	1.4197		63[17mm]	37	
e9e	1,2-Epoxy-3-phenoxypropane	$C_6H_5OCH_2CH{-}CH_2$ (O)	150.18	17, 105	1.109	1.530[20]	3.5	245	>110	
e10	1,2-Epoxypropane	$CH_3{-}CH{-}CH_2$ (O)	58.08	17, 6	0.859[0]	1.3660[20]	−112.1	34.2	−37	41 aq; misc alc, eth
e11	2,3-Epoxy-1-propanol	$H_2C{-}CHCH_2OH$ (O)	74.08	17, 104	1.1143[25]	1.4315[20]		66[2.5mm]	81	misc aq

No.	Name	Formula	Formula wt.	Beil. ref.	Density	n_D	M.p., °C	B.p., °C	Flash pt.	Solubility
e11a	2,3-Epoxypropyl acrylate	$H_2C{=}CH{-}C({=}O)OCH_2{-}CH{-}CH_2$ (O)	128.13	17³, 1005	1.099	1.4495²⁰		115⁷⁸	76	
e11b	2,3-Epoxypropyl isopropyl ether	$(CH_3)_2CHOCH_2{-}CH{-}CH_2$ (O)	116.16	17³, 988	0.924	1.4103²⁰		132	33	
e12	2,3-Epoxypropyl methacrylate	$H_2C{=}C(CH_3)COOCH_2{-}CH{-}CH_2$ (O)	142.15		1.042	1.4494²⁰		189	76	
e13	1,2-Epoxy-3,3,3-trichloropropane	$Cl_3CCH{-}CH_2$ (O)	161.42	17², 14	1.495	1.4778²⁰		151⁷⁴⁵mm	66	
e13a	meso-Erythritol	$HOCH_2[CH(OH)]_2CH_2OH$	122.12	1, 525	1.356⁰ g/L		123	331		4.7 mL aq; 46 mL alc⁴
e14	Ethane	CH_3CH_3	30.07	1, 80			−172	−88		
e15	1,2-Ethanediamine	$H_2NCH_2CH_2NH_2$	60.10	4, 230	0.8977²⁰	1.4568²⁰	8.5	117.3	33	misc aq, alc; i bz
e16	1,2-Ethanediol	$HOCH_2CH_2OH$	62.07	1, 465	1.1135²⁰	1.4318²⁰	−12.6	197.3	110	misc aq, alc, glyc, pyr
e17	1,2-Ethanediol diacetate	$CH_3COOCH_2CH_2OOCCH_3$	146.14	2, 142	1.1043²⁰	1.4150²⁰	−31	190.2	82	misc alc, eth
e17a	1,2-Ethanediol dimethacrylate	$[H_2C{=}C(CH_3)CO_2CH_2{-}]_2$	198.22	2³, 1292	1.051	1.4549²⁰		100⁵mm	>110	
e18	1,2-Ethanedithiol	$HSCH_2CH_2SH$	94.20	1, 471	1.123²⁴	1.5580²⁰		146	50	v s alc, alk
e19	Ethanesulfonyl chloride	$CH_3CH_2SO_2Cl$	128.57	4², 526	1.357²²	1.4339²⁰		177	>110	d aq, alc; v s eth
e20	Ethanethiol	CH_3CH_2SH	62.13	1, 340	0.8315²⁵	1.420²⁵	−147.9	35.0	−17	0.7 aq; s alc, eth

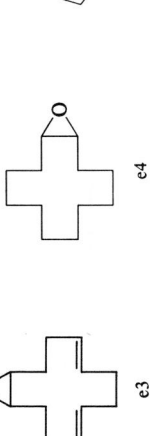

e3 e4 e5

TABLE 1.15 Physical Constants of Organic Compounds (*Continued*)

No.	Name	Formula	Formula weight	Beilstein reference	Density	Refractive index	Melting point	Boiling point	Flash point	Solubility in 100 parts solvent
e21	Ethanol	CH_3CH_2OH	46.07	1, 292	0.7894^{20}_4	1.3614^{20}	-114.5	78.3	8	misc aq, alc, eth, chl
e22	Ethanol-*d*	CH_3CH_2OD	47.08	1^3, 1287	0.801	1.3595^{20}		78.8	12	misc aq, alc, eth
e24	Ethoxyacetic acid	$CH_3CH_2OCH_2COOH$	104.11	3, 233	1.1021^{24}_4	1.4190^{20}		97^{11mm}	97	s aq, alc, eth
e24a	3-Ethoxyacrylonitrile	$C_2H_5OCH=CHCN$	97.12	3^3, 681	0.944	1.4545^{20}		91^{19mm}	81	
e24b	2-Ethoxyaniline	$CH_3CH_2OC_6H_4NH_2$	137.18	13, 359	1.051	1.5560^{20}	<-20	230	80	i aq; s alc
e25	4-Ethoxyaniline	$CH_3CH_2OC_6H_4NH_2$	137.18	13, 436	1.0652^{16}	1.5609^{20}	4	250	115	i aq; s alc
e26	2-Ethoxybenzaldehyde	$CH_3CH_2OC_6H_4CHO$	150.18	8, 43	1.074	1.5422^{20}	20	136^{24mm}	107	misc alc, eth
e27	4-Ethoxybenzaldehyde	$CH_3CH_2OC_6H_4CHO$	150.18	8, 73	1.080^{25}_{25}	1.5584^{20}	13–14	255	>110	v s alc, bz, eth
e28	2-Ethoxybenzamide	$CH_3CH_2OC_6H_4CONH_2$	165.19	10, 93			132–133			sl s aq; s alc, eth
e29	Ethoxybenzene	$CH_3CH_2OC_6H_5$	122.17	6, 140	0.9672^{20}_4	1.5074^{20}	-29.5	170.0	57	0.12 aq; misc alc, eth
e30	2-Ethoxybenzoic acid	$CH_3CH_2OC_6H_4COOH$	166.18	10, 64	1.105	1.5400^{20}	19.4	174^{15mm}	>110	sl s aq
e31	4-Ethoxybenzoic acid	$CH_3CH_2OC_6H_4COOH$	166.18	10, 156			197–199			sl s hot aq
e32	2-Ethoxybenzyl alcohol	$CH_3CH_2OC_6H_4CH_2OH$	152.19	6, 893	1.074	1.5321^{20}		265	>110	
e33	Ethoxycarbonyl isothiocyanate	$CH_3CH_2OC(=O)NCS$	131.15	3^3, 279	1.112	1.5000^{20}		56^{18mm}	50	
e34	Ethoxydimethylvinylsilane	$(CH_3)_2Si(OC_2H_5)CH=CH_2$	130.3		0.7902^{20}_4	1.398^{20}		99^{710mm}		
e35	2-Ethoxyethanol	$CH_3CH_2OCH_2CH_2OH$	90.12	1, 467	0.9295^{20}	1.4075^{20}	-90	134.8	44	misc aq, alc, eth, acet
e36	2-(2-Ethoxyethoxy)ethanol	$C_2H_5OCH_2CH_2OCH_2CH_2OH$	134.18	1^2, 520	0.9841^{25}_4	1.4254^{25}	-55	201.9	96	misc aq, alc, bz, chl, acet, pyr
e37	2-(2-Ethoxyethoxy)ethyl acetate	$C_2H_5OCH_2CH_2OCH_2-$ CH_2OOCCH_3	176.21		1.0096^{20}	1.4213^{20}		218.5	110	misc aq, alc, eth, oils
e38	2-Ethoxyethyl acetate	$CH_3COOCH_2CH_2OCH_2CH_3$	132.16	2^2, 155	0.9749^{20}_4	1.4023^{20}	-61.7	156.3	57	29 aq; misc alc, eth
e39	2-Ethoxyethylamine	$CH_3CH_2OCH_2CH_2NH_2$	89.14	4^2, 718	0.8512^{20}_4	1.4101^{20}		107	21	misc alc, eth
e39a	2-Ethoxyethyl methacrylate	$H_2C=C(CH_3)CO_2CH_2-$ $CH_2OC_2H_5$	158.20	2^3, 1291	0.964	1.4285^{20}		93^{35mm}	71	
e40	3-Ethoxy-4-hydroxybenzaldehyde	$C_2H_5OC_6H_3(OH)CHO$	166.18	8, 256			76–78			s eth, glycols; 50 alc
e41	3-Ethoxymethacrolein	$C_2H_5OCH=C(CH_3)CHO$	114.15	1^4, 4082	0.960	1.4792^{20}		$78-81^{14mm}$	35	
e42	3-Ethoxy-4-methoxybenzaldehyde	$C_2H_5OC_6H_3(OCH_3)CHO$	180.2	8, 256			53		>110	s alc, bz, chl, eth
e44	Ethoxymethylenemalononitrile	$CH_3CH_2OCH=C(CN)_2$	122.13	3^1, 162			64–66	160^{12mm}	>110	
e45	1-Ethoxynaphthalene	$C_{10}H_7OCH_2CH_3$	172.23	6, 606	1.060^{20}_4	1.6040^{20}	5.5	280	>110	i aq; v s alc, eth
e46	2-Ethoxyphenol	$C_2H_5OC_6H_4OH$	138.17	6, 771	1.090	1.5288^{20}	29	217	91	

No.	Name	Formula	Mol. wt.	Beilstein ref.	Density	n_D	M.p., °C	B.p., °C	Flash pt.	Solubility
e46a	3-Ethoxy-1,2-propanediol	$C_2H_5OCH_2CH(OH)CH_2OH$	120.15	1, 512	1.063	1.4407^{20}		222	110	
e47	*trans*-2-Ethoxy-5-(1-propenyl)phenol	$C_2H_5OC_6H_3\text{-}(CH{=}CHCH_3)OH$	178.23	6[2], 918			86–88			
e47a	3-Ethoxypropionitrile	$C_2H_5OCH_2CH_2CN$	99.14	3, 298	0.911	1.4065^{20}		172	63	
e48	3-Ethoxypropylamine	$C_2H_5OCH_2CH_2CH_2NH_2$	103.17	4[3], 739	0.861	1.4178^{20}		136–138	32	
e49	3-Ethoxysalicylaldehyde	$C_2H_5OC_6H_3(OH)CHO$	166.18	8[3], 267			68	264		
e50	2-Ethoxytetrahydrofuran	$C_2H_5O(C_4H_7O)$	116.16	17[4], 1020	0.908	1.4140^{20}		170–172	16	
e51	Ethoxytrimethylsilane	$(CH_3)_3SiOC_2H_5$	118.25	4[3], 1856	0.7573^{20}	1.3742^{20}		76	−18	9.7 aq; misc alc, acet, chl, eth
e52	Ethyl acetate	$CH_3COOC_2H_5$	88.11	2, 125	0.9006^{20}	1.3724^{20}	−84	77.1	−3	2.9 aq; misc alc, chl
e54	Ethyl acetoacetate	$CH_3COCH_2COOC_2H_5$	130.15	3, 632	1.0213^{25}	1.4194^{20}	−45	180.8	84 (CC)	1.5 aq; s alc, eth
e56	*p*-Ethylacetophenone	$C_2H_5C_6H_4COCH_3$	148.21	7[4], 1101	0.993	1.5293^{20}	−20.6	114^{11mm}	90	
e57	Ethyl acrylate	$CH_2{=}CHCOOCH_2CH_3$	100.12	2, 399	0.9405^{20}	1.4068^{20}	−71.2	99.5	15 (OC)	
e58	Ethylaluminum dichloride	$CH_3CH_2AlCl_2$	126.95	4[3], 1973	1.207^{50}		32	113^{50mm}		
e59	Ethylamine	$CH_3CH_2NH_2$	45.09	4, 87	0.6891^{5}		−81.0	16.6	−16	misc aq, alc, eth
e60	Ethyl 2-aminobenzoate	$H_2NC_6H_4COOCH_2CH_3$	165.19	14, 319	1.088^{15}		14	266–268	>110	i aq; s alc, eth
e61	Ethyl 4-aminobenzoate	$H_2NC_6H_4COOCH_2CH_3$	165.19	14, 422			88–90	310		0.04 aq; 20 alc; 50 chl; 25 eth; s dil acid
e62	Ethyl 3-aminocrotonate	$CH_3C(NH_2){=}CHCOOCH_2CH_3$	129.16	3, 654	1.021^{20}	1.4402^{20}	33–35	210–215	97	i aq; s alc, bz, eth
e63	2-(Ethylamino)ethanol	$CH_3CH_2NHCH_2CH_2OH$	89.14	4, 282	0.9142^{20}		−90	170	71	v s aq, alc, eth
e64	*N*-Ethylaniline	$C_6H_5NHCH_2CH_3$	121.18	12, 159	0.9558^{25}	1.5559^{20}	−63	204.5	85 (OC)	i aq; misc alc, eth
e65	2-Ethylaniline	$CH_3CH_2C_6H_4NH_2$	121.18	12[2], 584	0.9832^{2}	1.5590^{20}	−44	210	85 (OC)	sl s aq; v s alc, eth
e66	4-Ethylaniline	$CH_3CH_2C_6H_4NH_2$	121.18	12, 1090	0.9752^{2}	1.5542^{20}	−5	216	85	sl s aq; v s alc, eth

TABLE 1.15 Physical Constants of Organic Compounds (*Continued*)

No.	Name	Formula	Formula weight	Beilstein reference	Density	Refractive index	Melting point	Boiling point	Flash point	Solubility in 100 parts solvent
e67	2-Ethylanthraquinone		236.27	7¹, 425			108–111		31	
e68	Ethylbenzene-d_{10}	$C_6D_5CD_2CD_3$	116.25		0.867^{20}	1.4920^{20}		134.6	20	
e69	Ethylbenzene	$C_6H_5CH_2CH_3$	106.17	5², 274		1.4959^{20}	−95.0	136.2	84	0.01 aq; misc alc, bz, chl, eth
e70	Ethyl benzoate	$C_6H_5COOCH_2CH_3$	150.18	9, 110	1.050^{25}	1.5052^{20}	−34.7	212.4	140	0.05 aq; misc alc, chl, bz, eth, PE
e71	Ethyl benzoylacetate	$C_6H_5C(=O)CH_2COOCH_2CH_3$	192.21	10, 674	1.110	1.5338^{20}		265 d	>110	i aq; misc alc, eth
e71a	Ethyl 3-benzoylacrylate	$C_6H_5COCH=CHCO_2C_2H_5$	204.23	10², 501	1.112	1.5435^{20}		185^{25mm}	>110	
e72	Ethyl 2-benzylaceto-acetate	$CH_3COCH(CH_2C_6H_5)COOC_2H_5$	220.27	10, 710	1.036	1.4996^{20}		276		
e73	N-Ethylbenzylamine	$C_6H_5CH_2NHC_2H_5$	135.21	12, 1020	0.909	1.5117^{20}		194	66	
e74	Ethyl (2-benzyl)ben-zoylacetate	$C_6H_5COCH(CH_2C_6H_5)COOC_2H_5$	282.34	10, 764	1.110	1.5567^{20}		270^{80mm}	>110	
e75	Ethyl N-benzyl-N-cy-clopropylcarbamate	$C_6H_5CH_2N(C_3H_5)COOCH_2CH_3$	219.28		0.997	1.5104^{20}	<−20		>110	
e76	Ethyl bromoacetate	$BrCH_2COOCH_2CH_3$	167.01	2, 214	1.506^{20}_{20}	1.4510^{20}		159	47	i aq; misc alc, eth
e77	Ethyl 2-bromobutyrate	$CH_3CH_2CH(Br)COOCH_2CH_3$	195.06	2², 255	1.329^{20}_{20}	1.4470^{20}		177 d	58	i aq; misc alc, eth
e78	Ethyl 4-bromobutyrate	$BrCH_2CH_2CH_2COOCH_2CH_3$	195.06	2, 283	1.363	1.4559^{20}		82^{10mm}	90	
e78a	Ethyl 2-bromoheptan-oate	$CH_3(CH_2)_3CH(Br)CO_2C_2H_5$	237.14	2, 341	1.211	1.4524^{20}		109^{10mm}	104	
e79	Ethyl 2-bromoisobu-tyrate	$(CH_3)_2C(Br)COOCH_2CH_3$	195.06	2, 296	1.329^{20}	1.4446^{20}		67^{11mm}	60	i aq; misc alc, eth
e80	Ethyl 3-bromo-2-oxo-propionate	$BrCH_2C(=O)COOCH_2CH_3$	195.02	3², 409	1.554	1.4695^{20}		100^{10mm}	98	i aq; misc alc, eth
e81	Ethyl 2-bromopentan-oate	$CH_3(CH_2)_2CH(Br)COOCH_2CH_3$	209.09	2, 302	1.226	1.4486^{20}		190–192	77	i aq; misc alc, eth
e82	Ethyl 2-bromopro-pionate	$CH_3CH(Br)COOCH_2CH_3$	181.03	2, 255	1.447^{20}	1.4470^{20}		159–160	51	i aq; misc alc, eth
e83	Ethyl 3-bromopro-pionate	$BrCH_2CH_2COOCH_2CH_3$	181.03	2, 256	1.412^{20}_{18}	1.4569^{18}		136^{50mm}	79	i aq; misc alc, eth
e84	2-Ethyl-1-butanol	$(C_2H_5)_2CHCH_2OH$	102.18	1, 412	0.8330^{20}	1.4224^{20}	−114.4	146.5	58	0.63 aq
e85	2-Ethyl-1-butene	$(C_2H_5)_2C=CH_2$	84.16	1², 195	0.6696^{20}	1.3967^{20}	−131.5	64.7	−26	
e86	N-Ethylbutylamine	$CH_3(CH_2)_3NHCH_2CH_3$	101.19	4, 157	0.740^{20}	1.4050^{20}		108	18	
e87	2-Ethylbutylamine	$(C_2H_5)_2CHCH_2NH_2$	101.19	4, 192	0.776^{20}			121–125	21	s aq, alc, acet, eth
e88	2-Ethylbutyraldehyde	$(C_2H_5)_2CHCHO$	100.16	1, 693	0.816^{20}_{20}	1.4018^{20}	−89	116.7	21	0.31 aq

No.	Name	Formula	Mol. wt.	Beil. ref.	Density	n	m.p.	b.p.		Solubility
e89	Ethyl butyrate	$CH_3CH_2CH_2COOCH_2CH_3$	116.16	2, 270	0.8794^{20}	1.3928^{20}	−98.0	121.6	19	0.49 aq; misc alc, eth
e90	2-Ethylbutyric acid	$(C_2H_5)_2CHCOOCH_2CH_3$	116.16	2, 333	0.92225^{20}	1.4133^{20}	−15	194.2	87	0.22 aq
e91	Ethyl butyrylacetate	$CH_3(CH_2)_2C(O)CH_2COOC_2H_5$	158.20	3, 684	1.001	1.4295^{20}		104^{22mm}	78	200 aq; 125 alc; 111 chl; 67 eth
e92	Ethyl carbamate	$H_2NCOOCH_2CH_3$	89.09	3, 22	1.056		49–50	182–184	92	
e93	Ethyl carbazate	$H_2NNHCOOCH_2CH_3$	104.11	3, 98			44–47	110^{22mm}	86	
e94	N-Ethylcarbazole		195.27	20, 436			66–68			i aq; misc alc, eth
e95	Ethyl chloroacetate	$ClCH_2COOCH_2CH_3$	122.55	2, 197	1.1498^{20}_4	1.4227^{20}	−26	144–146	65	i aq; misc alc, eth
e96	Ethyl 2-chloroacetoacetate	$CH_3C(=O)CH(Cl)COOC_2H_5$	164.59	3, 662	1.190	1.4430^{20}		107^{14mm}	50	i aq; s alc, eth
e97	Ethyl 4-chloroacetoacetate	$ClCH_2C(=O)CH_2COOC_2H_5$	164.59	3, 663	1.218^{17}	1.4520^{20}		115^{14mm}	96	i aq; misc alc, eth
e98	Ethyl 4-chlorobutyrate	$ClCH_2CH_2CH_2COOC_2H_5$	150.61	2, 278	1.0754^{20}_4	1.4306^{20}		186	51	s alc, acet, eth
e99	Ethyl chloroformate	$ClCOOC_2H_5$	108.52	3, 10	1.1403^{20}	1.3941^{20}	−81	95	2	misc alc, bz, chl, eth
e100	Ethyl 2-chloropropionate	$CH_3CH(Cl)COOC_2H_5$	136.58	2, 248	1.0872^{20}_4	1.4185^{20}		147–148	38	i aq; misc alc, eth
e101	Ethyl 3-chloropropionate	$ClCH_2CH_2COOC_2H_5$	136.58	2, 250	1.1086^{20}_4	1.4249^{20}		163	54	misc alc, eth
e102	Ethyl chlorothioformate	$ClC(=O)SCH_2CH_3$	124.59	3, 134	1.195	1.4820^{20}		132	30	
e103	Ethyl chrysanthemumate		196.29	9^2, 45	0.906	1.4600^{20}		112^{10mm}		
e104	Ethyl *trans*-cinnamate	$C_6H_5CH=CHCOOCH_2CH_3$	176.22	9^2, 385	1.0495^{20}_4	1.5598^{20}	12	271.0	>110	misc alc, eth; i aq
e105	Ethyl crotonate	$CH_3CH=CHCOOCH_2CH_3$	114.14	2, 411	0.9175^{20}_4	1.4248^{20}		138	2	i aq; s alc, eth

e67

e94

e103

TABLE 1.15 Physical Constants of Organic Compounds (*Continued*)

No.	Name	Formula	Formula weight	Beilstein reference	Density	Refractive index	Melting point	Boiling point	Flash point	Solubility in 100 parts solvent
e106	Ethyl cyanoacetate	$NCCH_2COOCH_2CH_3$	113.12	2, 585	1.0564^{25}	1.4156^{20}	-22.5	206.0	110	i aq; misc alc, eth
e107	Ethyl cyanoformate	$NCCOOCH_2CH_3$	99.09	2, 547	1.0034^{20}	1.3820^{20}		116	24	
e108	Ethyl cyano(hydroxyimino)acetate	$NCC(=NOH)COOCH_2CH_3$	142.12	3, 775			130–132			
e109	Ethylcyclohexane	$C_2H_{11}CH_2CH_3$	112.22	5, 35	0.7879^{20}	1.4330^{20}	-111.3	131.8	18	
e110	cis-2-Ethylcyclohexanol	$CH_3CH_2C_6H_{10}OH$	128.22	6^2, 26	0.9272^{21}	1.4646^{20}		$74–79^{12mm}$	68	i aq
e111	4-Ethylcyclohexanol	$CH_3CH_2C_6H_{10}OH$	128.22	6^2, 26	0.889	1.4625^{20}		84^{10mm}	77	
e112	Ethyl cyclohexylacetate	$C_6H_{11}CH_2COOCH_2CH_3$	170.25	9, 14	0.948	1.4439^{20}		212	80	
e112a	N-Ethylcyclohexylamine	$C_6H_{11}NHC_2H_5$	127.23	12, 6	0.844	1.4525^{20}		165	43	
e113	Ethyl cyclopropanecarboxylate	$C_3H_5COOCH_2CH_3$	114.14	9, 4	0.960	1.4197^{20}		129–133	18	
e114	Ethyl decanoate	$CH_3(CH_2)_8COOCH_2CH_3$	200.32	2, 356	0.862^{20}	1.4248^{20}	-22	245	102	misc alc, chl, eth
e115	Ethyl diazoacetate	$N_2CHCOOCH_2CH_3$	114.10	3^1, 211	1.0852^{18}	1.4588^{18}	Explodes when heated	141^{720mm}	26	misc alc, bz, eth
e116	Ethyl 2,3-dibromopropionate	$BrCH_2CH(Br)COOCH_2CH_3$	259.94	2, 259	1.788_6	1.4986^{20}		214	91	s alc, eth
e117	Ethyl dichlorophosphate	$CH_3CH_2OP(O)Cl_2$	162.94	1, 332	1.373	1.4338^{20}		65^{10mm}	>110	
e118	Ethyl dichlorothiophosphate	$CH_3CH_2OP(S)Cl_2$	179.01	1, 333	1.353	1.5040^{20}		$55–68^{10mm}$	>110	
e119	N-Ethyldiethanolamine	$CH_3CH_2N(CH_2CH_2OH)_2$	133.19	4, 284	1.014	1.4665^{20}		246–252	123	
e120	Ethyl diethoxyphosphinylformate	$(C_2H_5O)_2P(O)COOC_2H_5$	210.17	3^2, 103	1.110	1.4230^{20}	-50	135^{13mm}	>110	
e121	Ethyl 3-(diethylamino)propionate	$(C_2H_5)_2NCH_2CH_2COOC_2H_5$	173.26	4, 404	0.881	1.4253^{20}		84^{12mm}	7	
e122	Ethyl 3,3-dimethylacrylate	$(CH_3)_2C{=}CHCOOC_2H_5$	128.17	2, 433	0.9247^{20}_4	1.4350^{20}		155	33	
e123	Ethyl 2-dimethylaminobenzoate	$(CH_3)_2NC_6H_4COOC_2H_5$	193.25		1.061	1.5425^{20}			98	

No.	Name	Formula								
e124	Ethyl 2,2-dimethyl-propionate	$(CH_3)_3CCOOCH_2CH_3$	130.19	2, 320	0.8584^{18}	1.3922^{18}		118.2	16	s alc, eth
e124a	Ethyl 3,5-dinitroben-zoate	$(O_2N)_2C_6H_3CO_2C_2H_5$	240.17	9, 414			95			
e124b	Ethylene	$H_2C{=}CH_2$	28.05	1, 180	1.260 g/L		-169.4	-104		11 mL aq^{25}, 200 alc^{25}; v s eth; s acet, bz
e125	Ethylene carbonate		88.06	19, 100	1.3208^{25}	1.4199^{40}	36.4	238	160	misc aq
e128	Ethylenediamine-N,N,N',N'-tetra-acetic acid	$(HOOCCH_2)_2NCH_2CH_2N{-}$ $(CH_2COOH)_2$	292.24	4^3, 1187			245 d			0.05 aq
e129	Ethylene glycol bis(mercaptoacetate)	$(HSCH_2CO_2CH_2{-})_2$	210.27		1.313	1.5211^{20}		139^{2mm}	>110	
e130	Ethylene glycol diglyci-dyl ether	$(H_2C{-}CH{-}CH_2OCH_2{-})_2$	174.20	1, 468	0.842	1.3923^{20}	-74	121	20	
e131	Ethylene glycol di-methacrylate	$[H_2C{=}C(CH_3)CO_2CH_2{-}]_2$	198.22	2^3, 1292	1.051	1.4549^{20}		100^{5mm}	>110	
e132	Ethylene oxide	$H_2C{-}CH_2$	44.05		0.891^0	1.35977	-112.44	10.6	-18	misc aq; s alc, eth

$$H_2C{-}O{-}C{=}O \quad H_2C{-}O$$

e125

TABLE 1.15 Physical Constants of Organic Compounds (*Continued*)

No.	Name	Formula	Formula weight	Beilstein reference	Density	Refractive index	Melting point	Boiling point	Flash point	Solubility in 100 parts solvent
e133	Ethylene sulfide	H_2C-CH_2 $\backslash S /$	60.12	17^2, 12	1.010	1.4935^{20}		55–56	10	sl s alc, eth
e134	Ethylenimine	H_2C-CH_2 $\backslash NH /$	43.07		0.8321^{25}	1.4123^{25}	−78.0	56	−24	misc aq; sl s alc
e134a	Ethyl 2-ethoxy-2-hydroxyacetate	$HOCH(OC_2H_5)CO_2C_2H_5$	148.16	3, 601	1.079	1.4200^{20}		137	49	
e135	Ethyl (ethoxymethylene)-cyanoacetate	$C_2H_5OCH=C(CN)COOC_2H_5$	169.18	3, 470			51–53	190^{30mm}	>110	
e136	Ethyl fluoroacetate	$FCH_2COOC_2H_5$	106.10	2, 193	1.0926^{21}	1.3755^{20}		119^{755mm}	30	s aq
e137	Ethyl fluorosulfonate	$FSO_2OC_2H_5$	128.12					$23-25^{12mm}$	32	
e138	Ethyl formate	$HCOOC_2H_5$	74.08	2, 19	0.9172^{20}	1.3599^{20}	−79.4	54.2	−19 (CC)	12 aq; misc alc, eth
e139	Ethyl 2-furoate		140.14	18, 275	1.117^{20}		33–36	196	70	i aq; s alc, eth
e140	Ethyl heptafluoro-butyrate	$CF_3CF_2CF_2COOC_2H_5$	242.09		1.394^{20}	1.3030^{20}		94–96	14	
e141	Ethyl heptanoate	$CH_3(CH_2)_5COOC_2H_5$	158.24	2^2, 295	0.8685^{20}	1.4144^{15}	−66	189	66	s alc, eth
e142	Ethyl hexadecanoate	$CH_3(CH_2)_{14}COOC_2H_5$	284.48	2^2, 336	0.85774^{25}	1.4347^{34}	22	191^{10mm}		s alc, eth
e142a	2-Ethylhexanal	$CH_3(CH_2)_3CH(C_2H_5)CHO$	128.22	1, 707	0.822	1.4155		$55^{13.5mm}$	42	sl s alc; s eth
e143	3-Ethylhexane	$CH_3CH_2CH_2CH(C_2H_5)_2$	114.24	1^3, 478	0.7136^{20}	1.4016^{20}		118.5		
e144	2-Ethyl-1,3-hexanediol	$C_3H_7CH(OH)CH(C_2H_5)$-CH_2OH	146.23		0.9325^{22}	1.4530^{22}	−40	244.2	129	0.6 aq; s alc
e145	Ethyl hexanoate	$CH_3(CH_2)_4COOC_2H_5$	144.21	2, 323	0.8714^{20}	1.4075^{20}	−67	166–168	49	i aq; misc alc, eth
e146	2-Ethylhexanoic acid	$CH_3(CH_2)_3CH(C_2H_5)COOH$	144.21	2, 349	0.9077^{20}	1.4241^{20}	−118.4	227.6	127	0.25 aq
e147	2-Ethyl-1-hexanol	$CH_3(CH_2)_3CH(C_2H_5)CH_2OH$	130.23		0.9344^{20}	1.4231^{20}	−76	184.3	77	0.07 aq; s alc, bz, chl
e148	2-Ethylhexanoyl chloride	$CH_3(CH_2)_3CH(C_2H_5)COCl$	162.66	2^2, 304	0.939	1.4335^{20}		68^{11mm}	69	
e149	2-Ethylhexyl acetate	$CH_3(CH_2)_3CH(C_2H_5)$-CH_2OOCCH_3	172.27		0.8718^{20}	1.4204^{20}	−93	198.6	82	0.03 aq; misc alc
e149a	2-Ethylhexyl acrylate	$H_2C=CHCO_2CH_2CH(C_2H_5)$-$CH_2CH_2CH_2CH_3$	184.28	2^3, 1229	0.885	1.4358		219	79	
e150	2-Ethylhexylamine	$CH_3(CH_2)_3CH(C_2H_5)CH_2NH_2$	129.31	4^3, 388	0.7922^{20}	1.4300^{20}	−76	165–169	52	i aq; s alc, acet, eth
e150a	2-Ethylhexyl chloroformate	$ClCO_2CH_2CH(C_2H_5)$-$CH_2CH_2CH_2CH_3$	192.69	3^4, 28	0.981	1.4312^{20}		107^{30mm}	81	

e150b	2-Ethylhexyl glycidyl ether	$CH_3(CH_2)_3CH(C_2H_5)CH_2\text{-}OCH_2\text{—}CH\text{—}CH_2$ (epoxide O)	186.30		0.891	1.4340^{20}		$61^{0.3mm}$	96	0.01 aq
e150c	2-Ethylhexyl methacrylate	$H_2C{=}C(CH_3)CO_2CH_2\text{-}CH(C_2H_5)(CH_2)_3CH_3$	198.31	2[3], 1289	0.885	1.4381^{20}		120^{18mm}	92	
e150d	2-Ethylhexyl nitrate	$CH_3(CH_2)_3CH(C_2H_5)\text{-}CH_2ONO_2$	175.23		0.963	1.4321^{20}	Explodes when heated		75	
e151	2-Ethylhexyl vinyl ether	$CH_3(CH_2)_3CH(C_2H_5)\text{-}CH_2OCH{=}CH_2$	156.26		0.8102^{20}_{20}	1.4273^{20}	<100 glass			
e152	Ethyl hydrogen hexanedioate	$HOOC(CH_2)_4COOC_2H_5$	174.20	2[1], 277		1.4387^{20}	28–29	180^{18mm}	177.7	v s alc, eth
e153	Ethyl hydroxyacetate	$HOCH_2COOC_2H_5$	104.11	3, 236	1.0871^{15}			160		0.07 aq; v s alc, eth
e154	Ethyl 4-hydroxybenzoate	$HOC_6H_4COOC_2H_5$	166.18	10, 159			116	297 d	>110	
e155	Ethyl 3-hydroxybutyrate	$CH_3CH(OH)CH_2COOC_2H_5$	132.16	3, 309	1.017^{20}	1.4205^{20}		170	64	s aq, alc
e156	Ethyl 2-hydroxyethyl sulfide	$HOCH_2CH_2SCH_2CH_3$	106.19	1[2], 525	1.020^{20}_{20}	1.4869^{20}		184.5	>110	s eth
e157	Ethyl 2-hydroxyisobutyrate	$(CH_3)_2C(OH)COOC_2H_5$	132.16	3, 315	0.965	1.4078^{20}		150	44	dec hot aq
e158	Ethyl 4-hydroxy-3-methoxyphenylacetate	$HOC_6H_4(OCH_3)CH_2\text{-}COOC_2H_5$	210.23	10[1], 198			44–47	185^{14mm}	>110	

Ethylenethiourea, i5
Ethylene trichloride, t231
Ethyleneurea, i7
N-Ethylethanamine, d267
Ethyl 2-ethoxyglycolate, e134a

Ethyl *N*-ethylcarbamate, e234
Ethyl fluoride, f17
2-Ethylhexyl alcohol, e147
2-Ethylhexyl bromide, b287a

Ethyl hexyl ketone, n99b
Ethyl homovanillate, e158
Ethyl hydrogen adipate, e152
N-Ethyl-*N*-(2-hydroxyethyl)-3-toluidine, e228

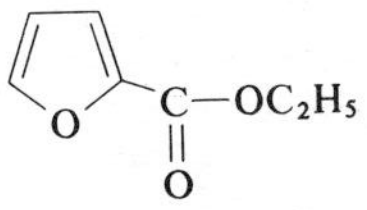

e139

TABLE 1.15 Physical Constants of Organic Compounds (*Continued*)

No.	Name	Formula	Formula weight	Beilstein reference	Density	Refractive index	Melting point	Boiling point	Flash point	Solubility in 100 parts solvent
e159	2-Ethyl-2-(hydroxymethyl)-1,3-propanediol	$CH_3CH_2C(CH_2OH)_3$	134.18	1[3], 2349			60–62	159–161[2mm]		
e159a	2-Ethyl-2-(hydroxymethyl)-1,3-propanediol triacrylate	$(H_2C=CHCO_2CH_2)_3CC_2H_5$	296.32		1.100	1.4736[20]			>110	
e159b	2-Ethyl-2-(hydroxymethyl)-1,3-propanediol trimethacrylate	$[H_2C=C(CH_3)CO_2CH_2]_3\text{-}CC_2H_5$	338.40		1.060	1.4724[20]			>110	
e160	N-Ethyl-3-hydroxypiperidine		129.20		0.970	1.4754[20]		93–95[15mm]	47	s alc, eth
e161	5-Ethylidene-2-norbornene		120.20		0.893	1.4895			38	
e162	2-Ethylimidazole		96.13	23, 78			86	268		i aq; misc alc, eth
e163	Ethyl isobutyrate	$(CH_3)_2CHCO_2C_2H_5$	116.16	2, 291	0.869	1.3869[20]		113	13	
e164	Ethyl isothiocyanate	CH_3CH_2NCS	87.14	4, 123	1.003[18]	1.5142[18]	−6	130–132	32	misc aq, alc, eth, esters, PE
e165	Ethyl (−)-lactate	$CH_3CH(OH)COOC_2H_5$	118.13	3, 264	1.0328[20]	1.4124[20]	−26	154.5	48	
e166	Ethyl DL-mandelate	$C_6H_5CH(OH)COOC_2H_5$	180.21	10, 202	1.0964[15]	1.4571[20]	37	253–255	>110	
e167	Ethyl 2-mercaptoacetate	$HSCH_2COOC_2H_5$	120.17	3, 255				54[12mm]	47	s alc, eth
e167a	Ethyl 3-mercaptopropionate	$HSCH_2CH_2CO_2C_2H_5$	134.20	3[3], 555	1.039	1.4570[20]		76[10mm]	72	
e168	Ethylmercuric chloride	CH_3CH_2HgCl	165.13		3.5		192	subl		
e169	Ethyl methacrylate	$H_2C=C(CH_3)COOC_2H_5$	114.14	2, 423	0.909[25]	1.4116[25]		118	15	0.78 eth; 2.6 chl
e170	Ethyl 4-methoxyphenylacetate	$CH_3OC_6H_4CH_2COOC_2H_5$	194.23	10[1], 83	1.097	1.5075[20]		138[7mm]	46	i aq; s alc, eth
e171	Ethyl 2-methylacetoacetate	$CH_3C(=O)CH(CH_3)\text{-}COOC_2H_5$	144.17	3, 679	1.019[20]	1.4182[20]		187	62	i aq; s alc, eth
e171a	N-Ethyl-2-methylallylamine	$H_2C=C(CH_3)CH_2NHC_2H_5$	99.18	4[4], 1104	0.753	1.4421[20]		105	7	
e171b	N-Ethylmethylamine	$CH_3CH_2NHCH_3$	59.11	4, 94	0.680	1.3740[20]		36	<−12	v s aq, alc

No.	Name	Formula	M.W.	Beilstein	Density	n_D	M.P.	B.P.		Solubility
e172	N-Ethyl-N-methyl-aniline	$C_6H_5N(CH_3)C_2H_5$	135.21	12, 162	0.9193^{55}	1.5474^{20}		203–205	74	i aq; misc alc, eth
e172a	Ethyl 2-methyl-benzoate	$CH_3C_6H_4CO_2C_2H_5$	164.21	9, 463	1.032	1.5070^{20}		221^{731mm}	91	
e172b	Ethyl 3-methyl-benzoate	$CH_3C_6H_4CO_2C_2H_5$	164.21	9, 476	1.030	1.5054^{20}		110^{20mm}	101	
e172c	Ethyl 4-methyl-benzoate	$CH_3C_6H_4CO_2C_2H_5$	164.21	9, 484	1.025	1.5085^{20}		235	99	
e172d	Ethyl 2-methylbutyrate	$CH_3CH_2CH(CH_3)CO_2C_2H_5$	130.19	2, 305	0.869	1.3969^{20}		133	26	0.2 aq; misc alc, bz
e173	Ethyl 3-methylbutyrate	$(CH_3)_2CHCH_2COOC_2H_5$	130.19	2^2, 275	0.868^{20}_{20}	1.3962^{20}	−99.3	134.7	26	s aq; misc alc, eth
e174	Ethyl methyl ether	$CH_3CH_2OCH_3$	60.09	1, 314	0.7250			10.8		
e175	2-Ethyl-4-methyl-imidazole		110.16	23^2, 72	0.975	1.4995^{20}		292–295	137	
e176	Ethyl 4-methyl-5-imidazolecar-boxylate		154.17	25^1, 534			204–206			
e177	3-Ethyl-2-methyl-pentane	$(C_2H_5)_2CHCH(CH_3)_2$	114.24	1^3, 489	0.7193^{20}	1.4040^{20}	−115.0	115.7		i aq; sl s alc; s eth
e178	3-Ethyl-3-methyl-pentane	$(C_2H_5)_3CCH_3$	114.24		0.7274^{20}_4	1.4078^{20}	−90.9	118.3		i aq; s eth

e160   e161   e162   e175   e176

TABLE 1.15 Physical Constants of Organic Compounds (*Continued*)

No.	Name	Formula	Formula weight	Beilstein reference	Density	Refractive index	Melting point	Boiling point	Flash point	Solubility in 100 parts solvent
e179	Ethyl 3-methyl-3-phenylglycidate		206.24		1.091^{15}	1.508^{20}				
e180	Ethyl 1-methyl-2-piperidinecarboxylate		171.24	22^1, 485	0.975	1.4519^{20}		$92{-}96^{11\text{mm}}$	73	
e181	Ethyl 1-methyl-3-piperidinecarboxylate		171.24		0.954	1.4510^{20}		$89^{11\text{mm}}$	68	
e182	2-Ethyl-2-methyl-1,3-propanediol	$HOCH_2C(C_2H_5)(CH_3)CH_2OH$	118.18	1, 487			41–44	226	>110	s alc, eth; sl s aq
e183	3-Ethyl-4-methylpyridine	$C_2H_5(CH_3)C_5H_3N$	121.18	20^2, 163	0.9286^{17}			198		
e184	5-Ethyl-2-methylpyridine	$C_2H_5(CH_3)C_5H_3N$	121.18	20, 248	0.9184^{23}	1.4974^{20}		$178^{747\text{mm}}$	66	s alc, bz, eth, acid
e185	Ethyl methyl sulfide	$CH_3CH_2SCH_3$	76.15	1, 343	0.8422^{20}	1.4403^{20}	−105.9	66.7	−15	i aq; misc alc, eth
e186	Ethyl (methylthio)acetate	$CH_3SCH_2COOC_2H_5$	134.20		1.043	1.4587^{20}			59	
e187	N-Ethylmorpholine		115.18	27^1, 203	0.9162^{20}_{20}	1.4410^{20}	−63	139	27	misc aq, alc, eth
e188	Ethyl nitrate	$CH_3CH_2ONO_2$	91.13	1, 329	1.1004^{25}	1.3849^{22}	−94.6	87.7 ex-plodes 85	10 (CC)	l aq; misc alc, eth
e189	Ethyl nitrite	CH_3CH_2ONO	75.07	1, 329	0.90^{15}_{15}			17 ex-plodes 90	−35	misc alc, eth
e190	4-Ethylnitrobenzene	$C_2H_5C_6H_4NO_2$	151.17	5, 358	1.118	1.5445^{20}	−32	246	>110	v s alc, eth
e190a	Ethyl nonanoate	$CH_3(CH_2)_7CO_2C_2H_5$	186.30	2, 353	0.8664^{18}	1.4219^{20}	−44	$119^{23\text{mm}}$	94	i aq; misc alc, eth
e191	Ethyl (Z,Z)-9,12-octadecadienoic acid	$H(CH_2)_5CH=CHCH_2$-$CH=CH(CH_2)_7COOC_2H_5$	308.51	2^2, 461	0.8846^{16}	1.4675^{20}		$193^{6\text{mm}}$	>110	misc DMF, oils
e192	Ethyl cis-9-octadecenoate	$CH_3(CH_2)_7CH=CH(CH_2)_7$-$COOC_2H_5$	310.52	2, 467	0.869^{20}_{4}	1.445^{25}	<−15	$216^{15\text{mm}}$		i aq; misc alc, eth
e193	Ethyl octanoate	$CH_3(CH_2)_6COOC_2H_5$	172.27	2, 348	0.878^{17}	1.4166^{20}	−47	206–208	75	i aq; misc alc, eth
e194	Ethyl oxalyl chloride	$CH_3CH_2OC(=O)COCl$	136.53	2, 541	1.2223^{20}	1.4164^{20}		135	41	d aq, alc; s bz, eth
e195	Ethyl oxamate	$CH_3CH_2OC(=O)CONH_2$	117.10	2, 544			114–116			s aq, eth; i bz
e196	2-Ethyl-2-oxazoline		99.13		0.982	1.4370^{20}	−62	128.4	29	

No.	Name	Formula	Mol. wt.	Beil. ref.	Density	n_D	mp, °C	bp, °C	Flash pt	Solubility
e197	Ethyl 2-oxocyclopentanecarboxylate	$(O{=})C_5H_7COOC_2H_5$	156.18	10, 597	1.054	1.4485^{20}		102^{11mm}	77	v s aq; misc alc
e198	Ethyl 4-oxopentanoate	$CH_3C({=}O)CH_2CH_2COOC_2H_5$	144.17	3, 675	1.0122^{20}	1.4222^{20}		206	45	sl s aq; misc alc, eth
e199	Ethyl 2-oxopropionate	$CH_3C({=}O)COOC_2H_5$	116.12	3, 616	1.060^{16}	1.408^{16}		144		i aq; s alc, eth
e200	3-Ethylpentane	$(C_2H_5)_3CH$	100.20	1^3, 441	0.6982^{20}	1.3934^{20}	−118.6	93.5		0.2 aq; misc alc, eth
e201	Ethyl pentanoate	$CH_3(CH_2)_3COOC_2H_5$	130.19	2, 301	0.8774^{20}	1.3732^{20}	−91.3	145.5	78	
e201a	2-Ethylphenol	$C_2H_5C_6H_4OH$	122.17	6, 470	1.037	1.5372^{20}	−18	197	94	
e201b	3-Ethylphenol	$C_2H_5C_6H_4OH$	122.17	6, 471	1.001	1.5330^{20}	−4	110^{15mm}	100	i aq; misc alc, eth
e202	4-Ethylphenol	$CH_3CH_2C_6H_4OH$	122.17	6, 472	1.0112^{25}	1.5239	47.0	219	77	i aq; misc alc, eth
e203	Ethyl phenylacetate	$C_6H_5CH_2COOC_2H_5$	164.20	9, 434	1.03333^{20}	1.4980^{20}		226	>110	
e204	Ethyl N-piperazinocarboxylate		158.20	23^2, 9	1.080	1.4765^{20}		273		s aq
e205	1-Ethylpiperidine		113.20	20, 17	0.82370^{20}	1.4440^{20}		131	18	
e206	2-Ethylpiperidine		113.20	20, 104	0.850	1.4510^{20}		143	31	
e208	Ethyl 3-piperidinecarboxylate		157.21		1.012	1.4601^{20}		104^{7mm}	90	

Ethyl N-methyl-N-phenylcarbamate, m367a
Ethyl 1-methylpipicolinate, e181
Ethyl nicotinate, e218

Ethyl nipecotate, e208
Ethyl oleate, e192
Ethyl pelargonate, e190a

Ethyl pentyl ketone, o35
Ethyl phenyl ether, e29
Ethyl picolinate, e217

Structure e179 — CH_3, $C{=}O$, $C{-}OC_2H_5$ (cyclopropane bearing phenyl)
Structure e180 — N-CH_3 piperidine, $C({=}O){-}OC_2H_5$
Structure e181 — N-CH_3 ring, $C{=}O$, O, OC_2H_5
Structure e187 — morpholine, N-C_2H_5
Structure e196 — C_2H_5 (oxazoline)
Structure e204 — H-N piperazine N-$C({=}O){-}OC_2H_5$
Structure e205 — piperidine N-C_2H_5
Structure e206 — piperidine N-H, C_2H_5
Structure e208 — piperidine N-H, $C({=}O){-}OC_2H_5$

TABLE 1.15 Physical Constants of Organic Compounds (*Continued*)

No.	Name	Formula	Formula weight	Beilstein reference	Density	Refractive index	Melting point	Boiling point	Flash point	Solubility in 100 parts solvent
e209	Ethyl 4-piperidine-carboxylate		157.21		1.010	1.4591²⁰		204	80	s aq, alc, bz, eth
e210	Ethyl N-piperidine-propionate		185.27	20, 62	0.927	1.4545²⁰		219	87	
e211	Ethyl propionate	$CH_3CH_2COOC_2H_5$	102.13	2, 240	0.891²⁰	1.3839²⁰	−73.9	99.1	12 (CC)	1.7 aq; misc alc, eth
e212	Ethyl propyl ether	$CH_3CH_2OCH_2CH_2CH_3$	88.15	1, 354	0.7392²⁰	1.3695²⁰	−79	63	32	sl s aq; misc alc, eth
e213	Ethyl propyl sulfide	$CH_3CH_2SCH_2CH_2CH_3$	104.21	1³, 1432	0.8270²⁰	1.4462²⁰	−117.0	118.5		s alc
e214	2-Ethylpyridine	$CH_3CH_2C_5H_4N$	107.16	20, 241	0.937	1.4964²⁰		149	29	sl s aq; s alc, eth
e215	3-Ethylpyridine	$CH_3CH_2C_5H_4N$	107.16	20, 242	0.954	1.5015²⁰		162−165	48	v s alc, eth; sl s aq
e216	4-Ethylpyridine	$CH_3CH_2C_5H_4N$	107.16	20, 243	0.9404²²	1.5009²⁰		168	47	sl s aq; s alc, eth
e217	Ethyl 2-pyridine-carboxylate		151.17	22, 35	1.1194²⁰	1.5088²⁰	2	241	107	misc aq, alc, eth
e218	Ethyl 3-pyridine-carboxylate		151.17	22, 39	1.1070²⁰	1.5040²⁰	8−9	224	93	v s aq, alc, eth; s bz
e219	Ethyl 4-pyridine-carboxylate		151.17	22², 37	1.009¹⁵	1.5009²⁰	23	220	87	i aq; s alc, bz, chl
e220	1-Ethyl-2-pyrrolidinone		113.16		0.992	1.4652²⁰		97²⁰mm	76	
e221	Ethyl salicylate	$C_6H_4(OH)COOC_2H_5$	166.18	10, 73	1.1312²⁰	1.5219²⁰	2−3	231−234	107	misc alc, eth; sl s aq
e222	Ethyl sorbate	$CH_3CH=CHCH=CHCOOC_2H_5$	140.18	2, 484	0.959	1.4942²⁰		195.5	69	
e223	S-Ethyl thioacetate	$CH_3C(=O)SCH_2CH_3$	104.16	2, 232	0.9762²⁸	1.4503²⁸		117	>110	i aq; v s alc, eth
e224	3-Ethylthio-1,2-propanediol	$C_2H_5SCH_2CH(OH)CH_2OH$	136.21		1.095	1.5065²⁰				
e225	Ethyl 4-toluene-sulfonate	$CH_3C_6H_4SO_2OC_2H_5$	200.26	11, 99	1.166¹⁵	1.5110²⁰	33	173¹⁵mm	157	i aq; s alc, eth
e226	N-Ethyl-m-toluidine	$CH_3C_6H_4NHC_2H_5$	135.21	12, 857	0.957	1.5451²⁰		221	89	
e227	6-Ethyl-o-toluidine	$CH_3CH_2C_6H_3(CH_3)NH_2$	135.21		0.968	1.5525²⁰	−33	231	89	
e228	2-(N-Ethyl-m-toluidino)ethanol	$CH_3C_6H_4N(C_2H_5)CH_2CH_2OH$	179.26		1.019	1.5540²⁰		115¹mm	>110	
e229	Ethyl trichloroacetate	$Cl_3CCOOC_2H_5$	191.44	2, 209	1.3834²⁰	1.4447²⁰		168	65	i aq; s alc, eth
e229a	Ethyltrichlorosilane	$C_2H_5SiCl_3$	163.51	4, 630	1.238	1.4252²⁰		99	−10	
e230	Ethyl trifluoroacetate	$F_3CCOOC_2H_5$	142.08	2², 186	1.194	1.3068²⁰	−106	60−62	−1	
e231	Ethyl (trimethylsilyl)-acetate	$(CH_3)_3SiCH_2COOC_2H_5$	160.29		0.876	1.4153²⁰		156−159	35	

No.	Name	Formula	Mol. wt.	Ref.	Density	n	m.p., °C	b.p., °C	Flash p., °C	Solubility
e232	Ethyl undecanoate	$CH_3(CH_2)_{10}COOC_2H_5$	214.35	2, 358	0.859	1.4280^{20}		105^{4mm}	>110	i aq; s org solv
e233	Ethylurea	$CH_3CH_2NHC(=O)NH_2$	88.11	4, 115	1.213^{18}		93–96			v s aq; 80 alc; i eth
e234	N-Ethylurethane	$CH_3CH_2NHCOOC_2H_5$	117.15	4, 114	0.9812^{4}	1.4211^{20}		85^{20mm}	75	63 aq
e235	Ethyl vinyl ether	$CH_3CH_2OCH=CH_2$	72.11	1, 433	0.7531	1.3754^{20}	-115.8^{20}	33	−45	0.9 aq
e236	N-Ethyl-2,3-xylidine	$(CH_3)_2C_6H_3NHC_2H_5$	149.24	12, 1101	0.917	1.5468^{20}		228	71	2.4 aq; misc alc, bz, acet, ketones, PE
e237	1-Ethynyl-1-cyclohexanol	$HOC_6H_{10}C{\equiv}CH$	124.18	62, 100	0.9672^{20}		30–31	180	62	
e238	1-Ethynylcyclohexylamine	$HC{\equiv}CC_6H_{10}NH_2$	123.20		0.913	1.4817^{20}		66^{20mm}	42	sl s alc; s bz, eth
f1	Fluoranthene		202.26	5, 685	1.252^{4}		107–110	384		v s HOAc; s bz, eth
f2	Fluorene		166.22	5, 625	1.203^{4}		114.8	295		s alc, bz; v s eth
f3	9-Fluorenone		180.21	7, 465	1.1300^{99}	1.6369^{99}	82–85	342		

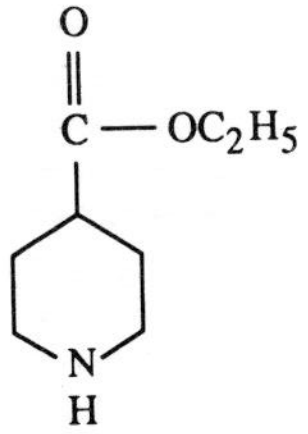

e209

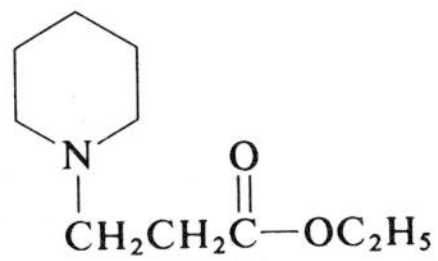

e210

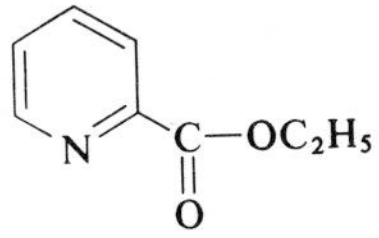

e217

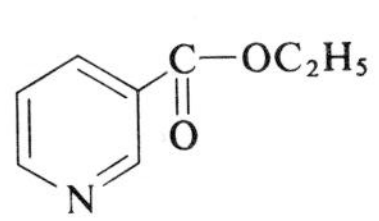

e218

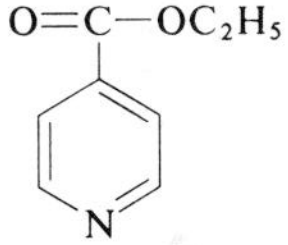

e219

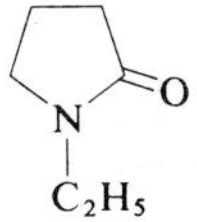

e220

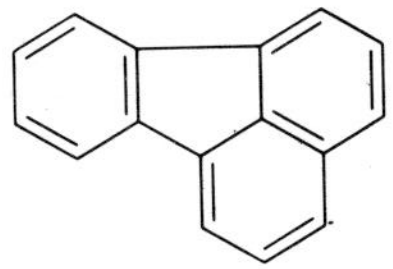

f1

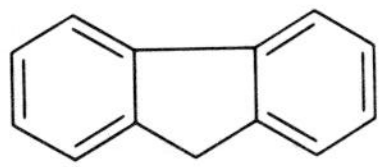

f2

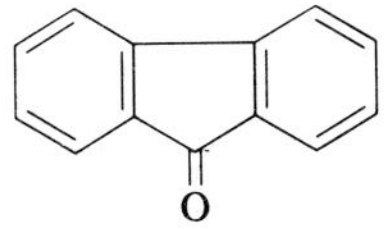

f3

TABLE 1.15 Physical Constants of Organic Compounds (*Continued*)

No.	Name	Formula	Formula weight	Beilstein reference	Density	Refractive index	Melting point	Boiling point	Flash point	Solubility in 100 parts solvent
f4	Fluorescein		332.31	19, 222			314 d			s hot alc, hot HOAc, alk; i bz, chl, eth
f5	Fluoroacetamide	$FCH_2C(O)NH_2$	77.06	2, 193			107 subl			v s aq; s acet
f6	Fluoroacetic acid	FCH_2COOH	78.04	2, 193			33	165	71	sl s aq, alc
f8	p-Fluoroacetophenone	$FC_6H_4COCH_3$	138.14		1.138	1.5110^{20}	-1.9	196	73	
f9	p-Fluoroaniline	$FC_6H_4NH_2$	111.12	12, 597	1.1725^{20}_4	1.5395^{20}	-44.5	187	55	sl s aq; s alc, eth
f10	o-Fluorobenzaldehyde	FC_6H_4CHO	124.11	7^1, 132	1.178	1.5220^{20}	-42.2	91^{46mm}	-12	
f11	Fluorobenzene	C_6H_5F	96.11	5, 198	1.0240^{20}_4	1.4657^{20}	123–125	84.7		0.15 aq; misc alc
f12	2-Fluorobenzoic acid	FC_6H_4COOH	140.11	9, 333	1.460^{25}		182.6			sl s aq; s alc, eth
f13	4-Fluorobenzoic acid	FC_6H_4COOH	140.11	9, 333	1.479^{25}		9			0.1 aq; s alc, eth
f14	4-Fluorobenzoyl chloride	FC_6H_4COCl	158.56	9^1, 137	1.342	1.5296^{20}		82^{20mm}	82	
f15	2-Fluorobenzyl alcohol	$FC_6H_4CH_2OH$	126.13	6^1, 222	1.173	1.5136^{20}			90	
f16	4-Fluorobenzyl chloride	$FC_6H_4CH_2Cl$	144.58		1.207	1.5130^{20}		82^{26mm}	60	
f17	Fluoroethane	CH_3CH_2F	48.06	1, 82	0.00220^0 g/L		-143.2	-37.7		198 mL aq; v s alc, eth
f18	Fluoromethane	CH_3F	34.04	1, 59	1.1951		-141.8	-78.4		166 mL aq; v s alc, eth
f19	4-Fluoro-1-methoxybenzene	$FC_6H_4OCH_3$	126.13	6^1, 98	1.114	1.4877^{20}	-45	157	43	s eth
f20	2-Fluoro-2-methyl-propane	$(CH_3)_3CF$	76.11	1^4, 286			-77	12.1	-12	
f20a	4-Fluoro-2-nitroaniline	$FC_6H_3(NO_2)NH_2$	156.12	12^1, 355			93		89	i aq; s alc, eth
f21	1-Fluoro-4-nitro-benzene	$FC_6H_4NO_2$	141.10	5, 241	1.3300^{20}_4	1.5312^{20}	21	205	83	
f22	4-Fluorophenol	FC_6H_4OH	112.10	6, 183	1.128		46–48	185	68	
f23	2-Fluoropyridine	FC_5H_4N	97.09	20^1, 80				126	28	
f24	2-Fluorotoluene	$FC_6H_4CH_3$	110.13	5, 290	1.0014^{17}	1.4680^{20}	-62.0	114.4	12	v s alc, eth
f25	3-Fluorotoluene	$FC_6H_4CH_3$	110.13	5, 290	0.9974^{20}	1.4716^{17}	-87.7	116.5	9	s alc, eth
f26	4-Fluorotoluene	$FC_6H_4CH_3$	110.13	5, 290	0.9975^{20}	1.4688^{20}	-56.7	116.6	40	s alc, eth
f27	Formaldehyde	$H_2C=O$	30.03	1, 558	0.815^{-20}_4	0.8153^{-20}	-92	-19.5	56	122 aq; s alc, eth
f28	Formamide	$HC(=O)NH_2$	45.04	2, 26	1.1334^{20}_4	1.4475^{20}	2.6	111^{20mm}	154 (COC)	misc aq, alc, acet
f29	Formamidine acetate	$HC(=NH)NH_2 \cdot HOOCCH_3$	104.11				158 d			

			108.12	31, 36			126 d			2.5 aq
f30	Formamidinesulfinic acid	$H_2NC(=NH)S(O)OH$	108.12	31, 36			126 d			2.5 aq
f31	Formanilide	C_6H_5NHCHO	121.14	12, 230	1.144	1.3714^{20}	47	271	>110	misc aq, alc, eth
f32	Formic acid	HCOOH	46.03	2, 8	1.220^{20}_4		8.5	100.8	68 (OC)	s aq; v s alc, eth
f33	2-Formylbenzoic acid	$C_6H_4(HCO)COOH$	150.13	10, 666	1.404		98		>110	v s alc, chl, eth; s bz
f34	Formylhydrazine	$HC(=O)NHNH_2$	60.06	2, 93		1.4848^{20}	54–56	237	>110	
f34a	4-Formylmorpholine		115.13	20, 45	1.145	1.4780^{20}		222	91	v s aq; 6.7 alc; s pyr
f35	N-Formylpiperidine		113.16	31, 321	1.019					0.6 aq; 9 alc; 0.7 eth
f36	D(−)-Fructose		180.16				287			dec aq, alc
f37	Fumaric acid	HOOCCH=CHCOOH	116.07	2, 737	1.635^{20}	1.4988^{20}	subl 200	161–164		8 aq; misc alc, eth
f38	Fumaroyl dichloride	ClC(=O)CH=CHC(=O)Cl	152.96	2, 743	1.408^{20}	1.5262^{20}		161.8	73	1 aq; misc alc, eth
f39	2-Furaldehyde		96.09	177, 305	1.15984^{20}	1.4214^{20}	−36.5	161.8	68 (CC)	
f40	Furan		68.07	17, 27	0.9371^{20}_4		−85.6	31.4	−35	

N-9H-(2-Fluorenyl)acetamide, a13
Fluorothane, b260
Fluorotrichloromethane, t236
Formaldehyde diethyl acetal, d255
Formic acid hydrazide, f34

Formylamide, f28
Formylphenols, h94, h95, h96
1-Formyl-piperazine, p183
Formylpyridines, p255, p256, p257
Freon-11, t236

Freon-12, d170
Freon-12B2, d75
Freon-21, d183
Freon-22, c85
Freon-114, d227

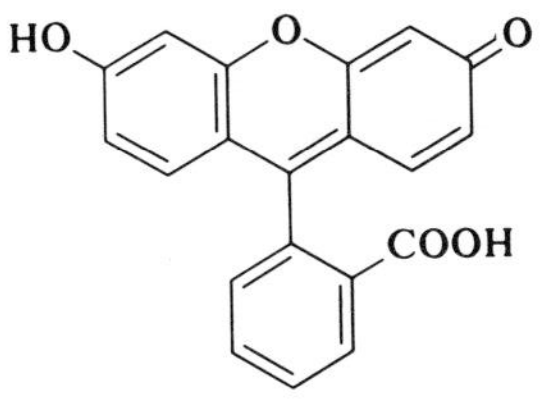

f4

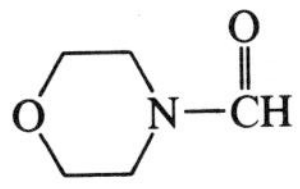

f34a

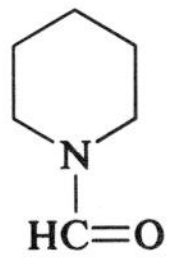

f35

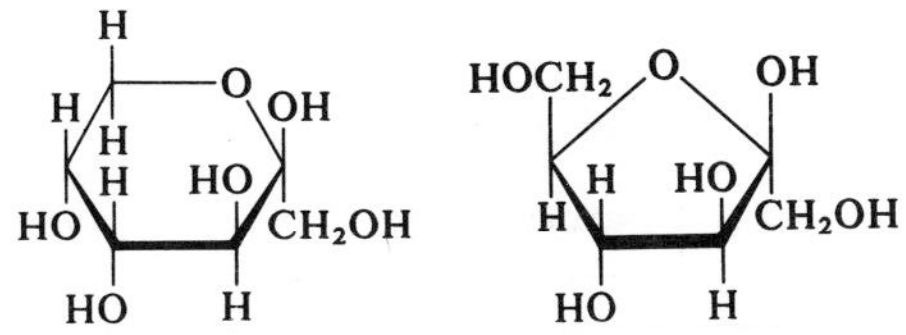

Pyranose form f36   Furanose form

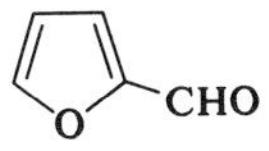

f39

f40

TABLE 1.15 Physical Constants of Organic Compounds (*Continued*)

No.	Name	Formula	Formula weight	Beilstein reference	Density	Refractive index	Melting point	Boiling point	Flash point	Solubility in 100 parts solvent
f41	2-Furanacrylic acid		138.12	18, 300			141	286		0.2 aq; 1.1 bz; s alc, eth, HOAc
f42	2-Furancarboxylic acid		112.08	18, 272			133–134	230–232		4 aq; s alc; v s eth
f43	2,5-Furandimethanol		128.13	17^1, 90			74–76			
f44	2-Furanmethanethiol		114.17	17^2, 116	1.132	1.5304^{20}		155	45	i aq; s alc, eth
f45	Furfuryl acetate		140.14	17^2, 115	1.1175^{20}_4	1.4618^{20}		175–177	65	misc aq(dec); v s alc
f46	Furfuryl alcohol		98.10	17, 112	1.1285^{20}_4	1.4868^{20}	−14.6	170.0	65	misc aq; s alc, eth
f47	Furfurylamine		97.12	18, 584	1.0995^{20}_4	1.4900^{20}	−70	145–146	45	dec aq, alc; s eth
f48	2-Furoyl chloride		130.53	18, 276	1.324	1.5310^{20}	−2	170	85	200 aq; s pyr, sl s alc
g1	D-(+)-Galactose		180.16	31, 295			167			i aq; misc alc, eth
g2	Geraniol	$(CH_3)_2C=CHCH_2CH_2$-$C(CH_3)=CHCH_2OH$	154.25	1, 457	0.8894^{20}_4	1.4760^{20}		230	76	v s aq; sl s alc; i eth
g4	D-Gluconic acid		196.16	3, 542			131			91 aq; 0.83 MeOH; s pyr
g6	α-D-(+)-Glucose		180.16	31, 83	1.5620^{18}		146			0.15 aq; 1.3 alc; 3 eth
g7	α-D-Glucose penta-acetate		390.34	31, 119			109–111			27 aq; 2.8 MeOH
g8	D-Glucurono-3,6-lactone		176.12				176–178			
g9	L-Glutamic acid		147.13	4, 488	1.538^{20}		d 247	subl 200		0.8 aq; i alc, eth
g10	L-Glutamine		146.15	4, 491			d 185			4 aq; 0.0035 MeOH; i bz, chl, eth, acet
g11	Glutaric acid	$HOOCCH_2CH_2CH_2COOH$	132.12	2, 631	1.429^{20}_4	1.4188^{106}	97.5	200^{20mm}		64 aq; v s alc, eth; s bz, chl; sl s PE
g12	Glutaric anhydride		114.10	17, 411			52–55	150^{10mm}	>110	misc aq, alc
g13	Glutaric dialdehyde	$OCHCH_2CH_2CH_2CHO$	100.12	1, 776	0.9888^{23}	1.3730^{20}	−6	187–189 d	none	s aq, alc, chl; i eth
g14	Glutaronitrile	$NCCH_2CH_2CH_2CN$	94.12	2, 635		1.4345^{20}	−29	286	>110	dec aq, alc; s eth
g15	Glutaryl dichloride	$ClC(=O)CH_2CH_2$-$CH_2C(=O)Cl$	169.01	2, 634	1.324	1.4720^{20}		216–218	106	
g16	Glycerol	$HOCH_2CH(OH)CH_2OH$	92.09	1, 502	1.2613^{20}_4	1.4746^{20}	18.18	182^{20mm}	160	misc aq, alc; 0.2 eth
g17	Glyceryl 1,2-diacetate	$HOCH_2CH(OOCCH_3)$-CH_2OOCCH_3	176.17	2, 147	$1.1846^{}_4$	1.1173^{15}	40	172^{40mm}		s aq, alc, bz, eth
g18	Glyceryl 1,3-diacetate	$CH_3COOCH_2CH(OH)$-CH_2OOCCH_3	176.17	2, 290	1.179^{15}	1.4395^{20}	42	172^{40mm}		s aq, alc, bz, chl
g19	Glyceryl tris(butyrate)		302.37	2, 273	1.032^{24}_4	1.4359^{20}	−75	305–310	173	i aq; v s alc, eth

2,5-Furandione, m2

2-Furanmethanol, f46

Furfural, f39

2-Furfuraldehyde, f39

Furfuryl mercaptan, f44

Furoic acid, f42

Furylacrylic acid, d41

2-Furyl methyl ketone, a44

Galactaric acid, t86

Galactitol, d740

Gallic acid, t311

Gallusic acid, t311

Gentisic acid, d385

Geranial, d564

D-Glucitol, s5

Glutaraldehyde, g13

Glyceraldehyde, d399

Glycerol dichlorohydrin, d220

Glycerol α-monochlorohydrin, c213

Glyceryl triacetate, p204

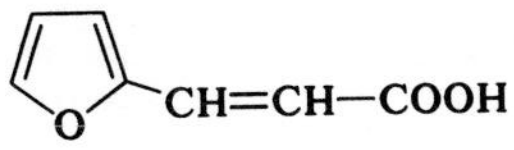

f41 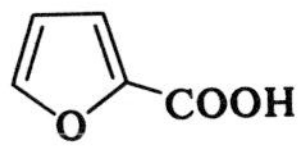   f42 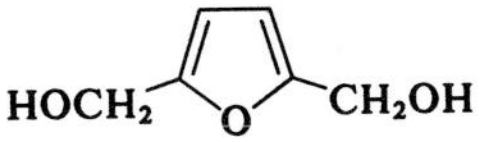   f43 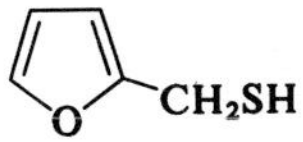   f44 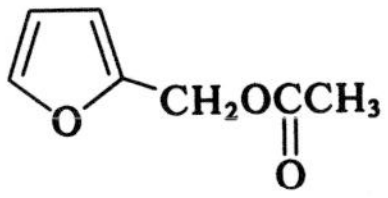   f45

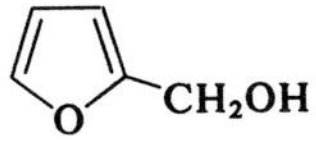

f46 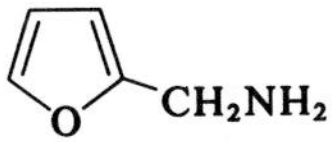   f47 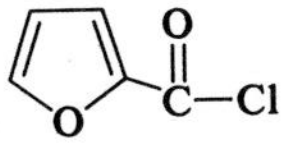   f48

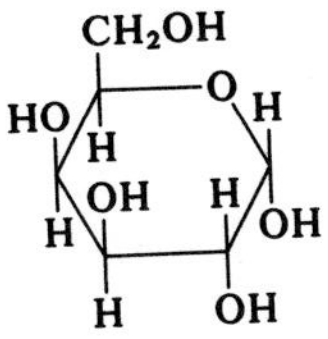

g1 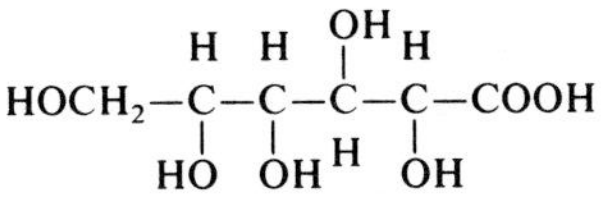   g4 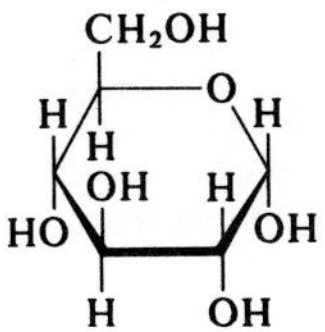   g6

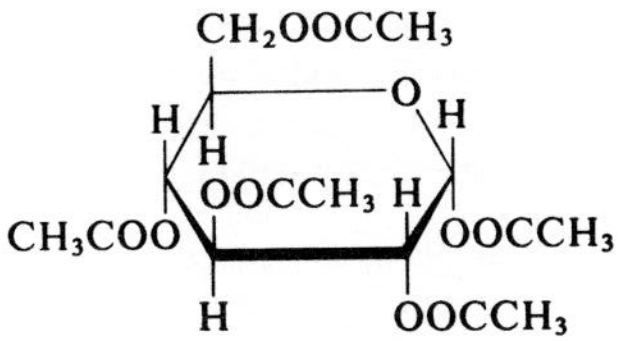

g7 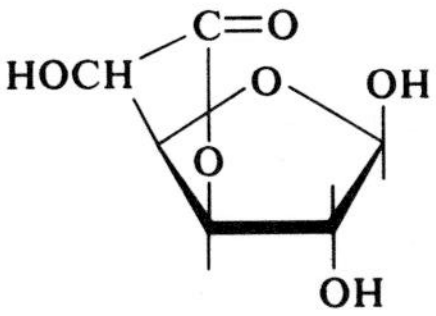   g8 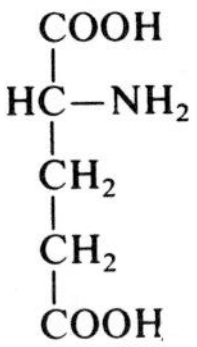   g9 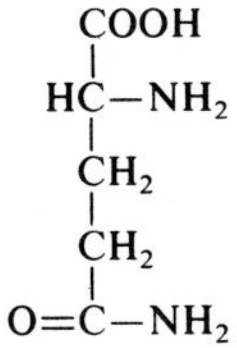   g10 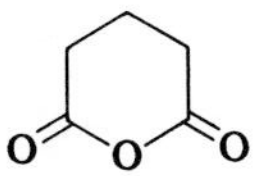   g12 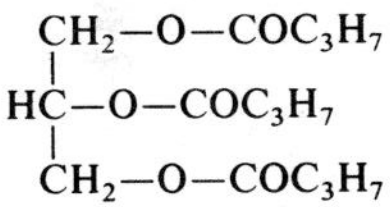   g19

1.217

TABLE 1.15 Physical Constants of Organic Compounds (*Continued*)

No.	Name	Formula	Formula weight	Beilstein reference	Density	Refractive index	Melting point	Boiling point	Flash point	Solubility in 100 parts solvent
g20	Glyceryl tris(dodecanoate)		639.02	2, 362	0.8946^{60}	1.4404^{60}	46			v s bz, eth; sl s alc
g21	Glyceryl tris(nitrate)	$O_2NOCH_2CH(ONO_2)CH_2ONO_2$	227.09	1, 516	1.5942^{4}_{4}	1.4786^{12}	13.3	160^{5mm}	explodes	0.18 aq; 54 alc; misc eth
g22	Glyceryl tris(oleate)		885.46	4, 468	0.9151^{5}	1.4621^{40}	−4 to −5	235^{15mm}	270	s chl, eth, CCl$_4$
g23	Glyceryl tris(palmitate)		807.35	2, 373	0.8663^{80}	1.4381^{80}	65–66	310–320		v s bz, chl, eth
g24	Glyceryl tris(tridecanoate)		723.18	2, 367	0.8854^{60}	1.4428^{60}	57			v s alc, bz, chl
g25	Glycine	H_2NCH_2COOH	75.07	4, 333	1.1607		d 233			25 aq; 0.6 pyr; i eth;
g26	N-Glycylglycine	$H_2NCH_2C(=O)NHCH_2COOH$	132.12	4, 371			d 262			s hot aq; sl s alc
g27	Glyoxal	$HC(=O)CHO$	58.04	1, 759	1.29^{20}_{4}	1.3826^{20}	15	51		viol rxn aq; s anhyd solv; mixtures with air may explode
g28	Glyoxylic acid	$HC(=O)COOH$	74.04	3, 594			98			v s aq; sl s alc, eth
g29	Guanidine	$H_2NC(=NH)NH_2$	59.07	3, 82			−60	d 160		v s aq, alc
h1	Heptadecane	$CH_3(CH_2)_{15}CH_3$	140.41	1, 173	0.7767^{22}	1.4360^{25}	22.0	302.2	148	s eth; sl s alc
h2	Heptafluorobutyric acid	CF_3CF_2COOH	214.04		1.645			120	none	
h3	Heptafluoro-2,3,3-trichlorobutane	$CF_3CCl_2CF(Cl)CF_3$	287.5		1.7484^{20}	1.3530^{20}	4	98		
h4	2,2,4,6,8,8-Heptamethylnonane	$(CH_3)_3CCH_2C(CH_3)_2CH_2CH(CH_3)CH_2C(CH_3)_2$	226.45		0.793	1.4391^{20}		240		
h5	Heptanal	$CH_3(CH_2)_5CHO$	114.19	1^2, 750	0.8216^{15}	1.4285^{20}	−43	153	35	misc alc, eth; sl s aq
h6	Heptane	$CH_3(CH_2)_5CH_3$	100.21	1, 154	0.6838^{20}_{4}	1.3877^{20}	−90.6	98.4	−4 (CC)	s alc, chl, eth
h7	1,7-Heptanediamine	$H_2N(CH_2)_7NH_2$	130.24	4, 271			27–29	147–149	87	5 aq; v s alc, eth
h8	Heptanedioic acid	$HOOC(CH_2)_5COOH$	160.17	2, 670	1.329^{15}		105.8	212^{10mm}		i aq
h9	1-Heptanethiol	$CH_3(CH_2)_6SH$	132.27	1, 415			−43.2	176.9	46	
h10	Heptanoic acid	$CH_3(CH_2)_5COOH$	130.19	2, 338	0.9181^{20}	1.4221^{20}	−7.5	223.0	>110	0.25 aq; s alc, eth
h11	Heptanoic anhydride	$[CH_3(CH_2)_5CO]_2O$	242.36	2, 340	0.9320^{20}	1.4332^{20}	−12.4	268	>110	i aq; s alc, eth
h12	1-Heptanol	$CH_3(CH_2)_6OH$	116.20	1, 414	0.8219^{20}	1.4242^{20}	−34.6	175.8	73	misc alc, eth
h13	2-Heptanol	$CH_3(CH_2)_4CH(OH)CH_3$	116.20	1, 415	0.8193^{20}	1.4210^{20}		160	41	0.35 aq; s alc, bz, eth
h14	3-Heptanol	$CH_3(CH_2)_3CH(OH)CH_2CH_3$	116.20	1^1, 205	0.818	1.4214^{20}		66^{20mm}	54	sl s aq
h15	2-Heptanone	$CH_3(CH_2)_4COCH_3$	114.19	1, 699	0.8197^{15}	1.4116^{15}	−35	151	47 (CC)	s alc, eth

No.	Name	Formula	Formula wt	Beilstein ref	Density	n_D	mp, °C	bp, °C	Flash pt	Solubility
h16	3-Heptanone	CH$_3$(CH$_2$)$_3$C(=O)CH$_2$CH$_3$	114.19	1, 699	0.8192^{20}	1.4085^{20}	−36.7	147.8	41 (CC)	0.43 aq; s alc, eth
h17	4-Heptanone	CH$_3$CH$_2$CH$_2$C(O)-CH$_2$CH$_2$CH$_3$	114.19	1, 699	0.8211^{15}	1.4068^{20}	−32.1	143.7	48 (CC)	0.53 aq; misc alc, eth
h18	Heptanoyl chloride	CH$_3$(CH$_2$)$_5$COCl	148.63	2, 340	0.960^{20}	1.4300^{20}		173	58	dec aq, alc; s eth
h19	1-Heptene	CH$_3$(CH$_2$)$_4$CH=CH$_2$	98.90	1, 219	0.6970^{20}	1.3999^{20}	−118.9	93.6	−1	0.1 aq; s alc, eth
h20	1-Heptylamine	CH$_3$(CH$_2$)$_6$NH$_2$	115.22	4, 193	0.777	1.4243^{20}	−23	154–156	35	s alc, acet, eth, PE
h22	1-Heptyne	CH$_3$(CH$_2$)$_4$C≡CH	96.17	1, 256	0.733	1.4075^{20}	−81	99–100	−2	sl s aq; s acet
h23	Hexachloroacetone	Cl$_3$CC(=O)CCl$_3$	264.75	1, 657	1.743	1.5112^{20}	−30	66^{6mm}	none	s bz, chl, eth
h24	Hexachlorobenzene	C$_6$Cl$_6$	284.78	5, 205	2.044^{24}		231	323–326	242	s alc, eth
h25	Hexachloro-1,3-butadiene	Cl$_2$C=CClCCl=CCl$_2$	260.76	1, 250	1.655	1.5550^{20}	−19	210–220	none	s bz, chl
h26	1,2,3,4,5,6-Hexachlorocyclohexane, isomer	C$_6$H$_6$Cl$_6$	290.83	5[1], 8	1.87^{20}		113		none	s bz, chl
h27	Hexachlorocyclo-1,3-pentadiene	(structure h27)	272.77		1.701^{25}	1.5644^{20}	−10	239	none	
h29	Hexachloroethane	Cl$_3$CCCl$_3$	236.74	1, 87	2.091^{20}	2.0914^{20}	187–188		none	s alc, bz, chl, eth

Structure h27 (hexachlorocyclopentadiene):

Cl — Cl (ring), positions bearing Cl; labeled **h27**

Glyceryl tris(laurate), g20
Glyceryl tris(myristate), g24
Glycidic acid, h87
Glycidol, e11
Glycidyl acrylate, e11a
Glycidyl methacrylate, e12
Glycidyl phenyl ether, e9e
Glycinonitrile, z106

Glycolaldehyde, h86
Glycolaldehyde diethyl acetal, d253
Glycol methacrylate, h121
Glyoxaline, i4
Guaiacol, m87
Heliotropin, m240
Heliotropyl alcohol, m243
Hemimellitene, t338

Hemimellitic acid, b28
Heptaldehyde, h5
sec-Heptyl alcohol, h13
Heptyl bromide, b293
Heptyl chloride, c129
Heptyl iodide, i37
Heptyl mercaptan, h9
Heptyl methyl ketone, n99a

g20
CH$_2$—O—COC$_{11}$H$_{23}$
HC—O—COC$_{11}$H$_{23}$
CH$_2$—O—COC$_{11}$H$_{23}$

g22
—CH$_2$O—CO(CH$_2$)$_7$CH=CH(CH$_2$)$_8$H
HCO—CO(CH$_2$)$_7$CH=CH(CH$_2$)$_8$H
—CH$_2$O—CO(CH$_2$)$_7$CH=CH(CH$_2$)$_8$H

g23
CH$_2$—O—COC$_{15}$H$_{31}$
HC—O—COC$_{15}$H$_{31}$
CH$_2$—O—COC$_{15}$H$_{31}$

g24
CH$_2$—O—COC$_{13}$H$_{27}$
HC—O—COC$_{13}$H$_{27}$
CH$_2$—O—COC$_{13}$H$_{27}$

1.219

TABLE 1.15 Physical Constants of Organic Compounds (*Continued*)

No.	Name	Formula	Formula weight	Beilstein reference	Density	Refractive index	Melting point	Boiling point	Flash point	Solubility in 100 parts solvent
h30	1,4,5,6,7,7-Hexachloro-5-norbornene-2,3-dicarboxylic anhydride		370.83				235–239			
h31	Hexachloropropene	$Cl_3CC(Cl)=CCl_2$	248.75	1, 200	1.765	1.5480^{20}		210	none	
h32	Hexadecane	$CH_3(CH_2)_{14}CH_3$	226.45	1, 172	0.7733^{20}	1.4345^{20}	18.2	286.8	135	misc eth
h33	1,2-Hexadecanediol	$CH_3(CH_2)_{13}CH(OH)CH_2OH$	258.45	1^3, 2244			72–74			
h34	1-Hexadecanethiol	$CH_3(CH_2)_{15}SH$	258.51	1, 430	0.840	1.4720^{20}	18–20	184^{mm}	101	sl s alc; s eth
h35	Hexadecanoic acid	$CH_3(CH_2)_{14}COOH$	256.43	2, 370	0.8524^{62}	1.4273^{80}	63–64	215^{15mm}	135	s hot, chl, eth
h36	1-Hexadecanol	$CH_3(CH_2)_{15}OH$	242.45	1, 429	0.8116^{60}	1.4355^{60}	49.3	344	132	s alc, chl, eth
h37	1-Hexadecene	$CH_3(CH_2)_{13}CH=CH_2$	224.43	1, 226	0.7834^{20}	1.4401	4.1	274	140	s alc, eth, PE
h38	1-Hexadecylamine	$CH_3(CH_2)_{15}NH_2$	241.46	4, 202			40–42	330	67	v s alc, eth; s bz, chl
h40	2,4-Hexadienal	$CH_3CH=CHCH=CHCHO$	96.13	1^2, 809	0.898^{20}	1.5386^{20}		76^{30mm}	–27	s alc, eth
h41	1,5-Hexadiene	$H_2C=CHCH_2CH_2CH=CH_2$	82.15	1, 253	0.6923^{24}	1.4042^{20}	–140.7	59.5	127	0.2 aq; 13 alc; 9 acet; 2.3 bz; 11 diox; 1 CCl$_4$
h42	2,4-Hexadienoic acid	$CH_3CH=CHCH=CHCOOH$	112.13	2, 483			134.5	119^{10mm}		
h43	Hexafluorobenzene	C_6F_6	186.05	5^3, 523	1.6182^{20}	1.3781^{20}	5.1	80.3	10	sl s alc, eth
h44	Hexafluoroethane	F_3CCF_3	138.01	1^3, 132	1.590^{-78}		–100.1	–78.3		s aq, bz, CCl$_4$
h45	1,1,1,3,3,3-Hexafluoro-2-propanol	$(CF_3)_2CHOH$	168.04		1.596^{25}	1.2750^{20}	–3	58.2	4	
h45a	Hexafluoropropene	$CF_3CF=CF_2$	150.02	1^3, 697			–153	–28		
h46	cis-Hexahydroindane		124.23	5, 82	0.876	1.4702	–53	167	23	s eth
h48	Hexamethylcyclotrisiloxane	$[-Si(CH_3)_2O-]_3$	222.48	4^3, 1884			64	133–135	32	
h49	1,1,1,3,3,3-Hexamethyldisilazane	$(CH_3)_3SiNHSi(CH_3)_3$	161.40		0.7744^{20}	1.4071^{20}		126	8	
h50	Hexamethyldisiloxane	$(CH_3)_3SiOSi(CH_3)_3$	162.38	20, 94	0.7644^{20}	1.3775^{20}	–67	101	–2	
h51	Hexamethyleneimine		99.18		0.880	1.4631^{20}		138^{749mm}	18	
h52	Hexamethylenetetramine		140.19	1, 583	1.331^{-5}		subl 263		250	67 aq; 8 alc; 10 chl
h53	Hexamethylphosphoramide	$[(CH_3)_2N]_3P(O)$	179.20		1.027^{20}	1.4588^{20}	7.2	233	105	misc aq

No.	Name	Formula								Solubility
h54	Hexanal	$CH_3(CH_2)_4CHO$	100.16	1^2, 745	0.8335^{20}	1.4035^{20}		131	32 (OC)	v s alc, eth; sl s aq
h55	Hexane	$CH_3(CH_2)_4CH_3$	86.18	1, 142	0.6594^{20}	1.3749^{20}	−95.4	68.7	−23	misc alc, chl, eth
h56	1,6-Hexanediamine	$H_2N(CH_2)_6NH_2$	116.21	4, 269			42	205	81	v s aq; sl s alc, bz
h57	1,6-Hexanedioic acid	$HOOC(CH_2)_4COOH$	146.14	2, 649	1.360^{25}		152	337.5	196 (CC)	1.4 aq; v s alc; s acet
h58	DL-1,2-Hexanediol	$CH_3(CH_2)_3CH(OH)CH_2OH$	118.18	1, 251	0.951	1.4425^{20}		224	>110	v s aq, alc
h59	1,6-Hexanediol	$HO(CH_2)_6OH$	118.18	1, 484	0.958	1.4579^{25}	42.8	243–250	101	s aq, alc, eth
h60	2,5-Hexanediol	$CH_3CH(OH)CH_2CH_2CH(OH)CH_3$	118.18	1, 485	0.9617^{16}	1.4465^{20}	−50 glass	220.8	101	
h60a	1,6-Hexanediol diacrylate	$[H_2C{=}CHCO_2(CH_2)_2{-}]_2$	226.28		1.010	1.4562^{20}				
h61	2,5-Hexanedione	$CH_3COCH_2CH_2COCH_3$	114.14	1, 788	0.973	1.4260^{20}	−6	191.4	70	misc aq, alc, eth
h62	Hexanedioyl dichloride	$ClC({=}O)(CH_2)_4COCl$	183.03	2, 653	1.259	1.4706^{20}		$105^{2\text{mm}}$	>110	i aq; s alc, eth
h63	Hexanenitrile	$CH_3(CH_2)_4CN$	97.16	2, 324	0.8052^{20}	1.4069^{20}	−80.3	163.6	43	i aq; v s alc, eth
h64	1-Hexanethiol	$CH_3(CH_2)_5SH$	118.24	1^3, 1659	0.8424^{20}	1.4496^{20}	−80.5	152.7	20	misc alc, acet; i bz
h65	1,2,6-Hexanetriol	$HOCH_2CH(OH)(CH_2)_3CH_2OH$	134.17	1^4, 2784	1.1063^{20}	1.4771	−32.8	178^{mm}	79	1.1 aq; v s alc, eth
h66	Hexanoic acid	$CH_3(CH_2)_4COOH$	116.16	2, 321	0.9265^{20}	1.4168^{20}	−4.0	205.7	104	s alc
h67	Hexanoic anhydride	$[CH_3(CH_2)_4C({=}O)]_2O$	214.31	2, 324	0.926	1.4280^{20}	−41	246–248	>110	8 aq; misc bz, eth; s alc
h68	1-Hexanol	$CH_3(CH_2)_5OH$	102.18	1, 407	0.8186^{20}	1.4182^{20}	−51.6	157.5	60	sl s aq; s alc, eth
h69	2-Hexanol	$CH_3(CH_2)_3CH(OH)CH_3$	102.18	1, 408	0.8108^{25}	1.4128^{25}	−47	139.9	41	sl s aq; s alc, eth

h30

h46

$(CH_2)_6NH$

h51

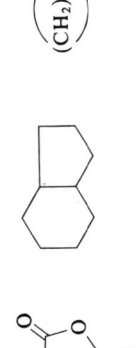

h52

TABLE 1.15 Physical Constants of Organic Compounds (*Continued*)

No.	Name	Formula	Formula weight	Beilstein reference	Density	Refractive index	Melting point	Boiling point	Flash point	Solubility in 100 parts solvent
h70	3-Hexanol	$CH_3CH_2CH_2CH(OH)CH_2CH_3$	102.18	1, 408	0.8193^{20}_4	1.4160^{20}		135	41	v s alc, eth
h71	6-Hexanolactone		114.14	17^2, 290	1.030	1.4630^{20}	−56.9	97^{15mm}	109	
h72	2-Hexanone	$CH_3(CH_2)_3COCH_3$	100.16	1, 689	0.8209^{20}_4	1.4024^{20}		127.2	35 (OC)	v s alc, eth
h72a	3-Hexanone	$CH_3CH_2CH_2COCH_2CH_3$	100.16	1, 690	0.815	1.4002^{20}		123	35	
h73	Hexanoyl chloride	$CH_3(CH_2)_4COCl$	134.61	2, 324	0.9754^{20}_4	1.4263^{20}	−87	153	50	d aq, alc; s eth
h75	1-Hexene	$CH_3(CH_2)_3CH{=}CH_2$	84.16	1, 215	0.6732^{20}	1.3879^{20}	−139.8	63.5	−26	0.005 aq
h76	trans-3-Hexenoic acid	$CH_3CH_2CH{=}CHCH_2COOH$	114.14	2, 435	0.963	1.4398^{20}	11—12	119^{22mm}	>110	
h77	trans-2-Hexen-1-ol	$CH_3CH_2CH_2CH{=}CHCH_2OH$	100.16	1^2, 486	0.849	1.4343^{20}		158—160	54	
h78	5-Hexen-2-one	$H_2C{=}CH_2CH_2CH_2COCH_3$	98.15	1, 734	0.847	1.4197^{20}		128—129	23	
h78a	trans-2-Hexenyl acetate	$CH_3CO_2CH_2CH{=}CHCH_2\text{-}CH_3$	142.20	2^2, 151	0.898	1.4275^{20}		166	58	
h79	Hexyl acetate	$CH_3(CH_2)_5OOCCH_3$	144.21	2, 132	0.8602^{20}_{20}	1.4090^{20}	−80	168—170	37	0.13 aq; v s alc, eth
h80	Hexylamine	$CH_3(CH_2)_5NH_2$	101.19	4, 188	0.7632^{25}	1.4180^{20}	−23	131—132	8	sl s aq; misc alc, eth
h81	4-Hexylaniline	$CH_3(CH_2)_5C_6H_4\text{-}NH_2$	177.29	12^3, 2759				148^{17mm}	110	
h81a	2-(2-Hexyloxyethoxy)-ethanol	$CH_3(CH_2)_5OCH_2CH_2OCH_2\text{-}CH_2OH$	190.29	1, 2100	0.935	1.4381	−40	260	140	
h82	1-Hexyne	$CH_3(CH_2)_3C{\equiv}CH$	82.14	1^3, 977	0.7152^{20}_4	1.3989^{20}	−131.9	71.3	−21	i aq; s alc, eth
h83	L-Histidine		155.16	25, 513			d 285			41 aq; v sl s alc
h84	Hydantoin		100.08	24, 242			220			s alc, alk; sl s eth
h86	Hydroxyacetaldehyde	$HOCH_2CHO$	60.05	1, 817	1.366^{100}		93—94	110^{12mm}		v s aq, alc; sl s eth
h87	Hydroxyacetic acid	$HOCH_2COOH$	76.05	3, 228			80			
h88	1'-Hydroxy-2'-aceto-naphthone	$C_{10}H_6(OH)COCH_3$	186.21	8, 149			98—100	325 sl d		i aq; v s bz; s HOAc
h89	Hydroxyacetone	$HOCH_2COCH_3$	74.08	1^1, 84	1.082	1.4315^{20}	−17	146	56	misc aq, alc, eth
h90	2'-Hydroxyaceto-phenone	$HOC_6H_4COCH_3$	136.15	8, 85	1.131^{21}_4	1.5584^{20}	4—6	213^{717mm}	>110	misc alc, eth; sl s aq
h91	3'-Hydroxyaceto-phenone	$HOC_6H_4COCH_3$	136.15	8, 86	1.100^{100}	1.535^{100}	87—89	296		s aq; v s alc, bz, eth
h92	4'-Hydroxyaceto-phenone	$HOC_6H_4COCH_3$	136.15	8, 87	1.109^{100}		106—107	147^{3mm}		v s alc, eth; sl s aq
h93	1-Hydroxyanthra-quinone		224.22	8, 338			196—198			
h94	2-Hydroxybenzalde-hyde	$C_6H_4(OH)CHO$	122.12	8, 31	1.1672^{20}_4	1.5718^{20}	−7	196.7	76	1.7 aq[86.]; s alc, eth

	Name	Formula	M.W.	Ref	Density	M.P.	B.P.	Solubility
h95	3-Hydroxybenzaldehyde	$C_6H_4(OH)CHO$	122.12	8, 58		100–102	191^{50mm}	s alc, bz, eth; sl s aq
h96	4-Hydroxybenzaldehyde	HOC_6H_4CHO	122.12	8, 64	1.1294^{30}	117–119	subl	1 aq; 70 acet; 4 bz[65]; v s alc, eth
h97	2-Hydroxybenzaldehyde oxime	$C_6H_4(OH)CH{=}NOH$	137.14	8, 49		57	dec	v s alc, bz, eth, acid
h98	2-Hydroxybenzamide	$C_6H_4(OH)CONH_2$	137.14	10, 87		140	d 270	0.2 aq; s alc, chl, eth
h98a	4-Hydroxybenzenesulfonic acid	$HOC_6H_4SO_3H$	174.18	11, 241			none	
h99	2-Hydroxybenzoic acid	$C_6H_4(OH)COOH$	138.12	10, 43	1.443^{20}_4	157–159	211^{20mm}	0.2 aq; 37 alc; 33 eth; 33 acet; 2 chl; 0.7 bz
h100	3-Hydroxybenzoic acid	$C_6H_4(OH)COOH$	138.12	10, 134	1.473	201–203		0.8 aq; 10 eth
h101	4-Hydroxybenzoic acid	HOC_6H_4COOH	138.12	10, 149	1.468^4	214–215		0.2 aq; v s alc; 23 eth
h102	4-Hydroxybenzophenone	$HOC_6H_4COC_6H_5$	198.22	82, 184		132–135		v s alc, eth; sl s aq
h103	1-Hydroxybenzotriazole		135.13	26, 41		155–158		

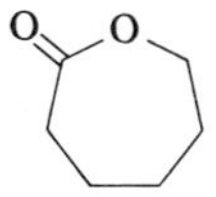

h71

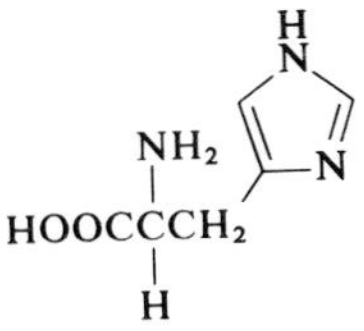

h83

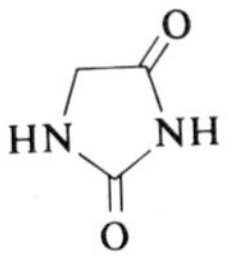

h84

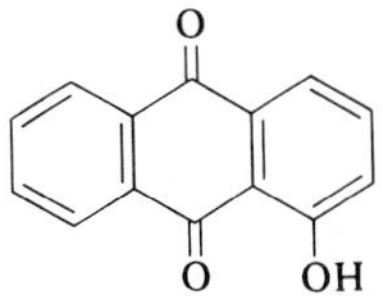

h93

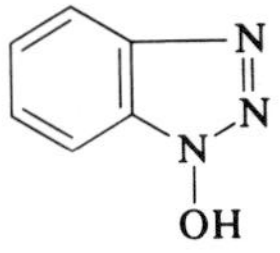

h103

TABLE 1.15 Physical Constants of Organic Compounds (*Continued*)

No.	Name	Formula	Formula weight	Beilstein reference	Density	Refractive index	Melting point	Boiling point	Flash point	Solubility in 100 parts solvent
h104	6-Hydroxy-1,3-benzoxathiol-2-one		168.17	19^4, 2508			158–160			
h105	2-Hydroxybenzyl alcohol	$HOC_6H_4CH_2OH$	124.13	6, 891	1.161^{25}		86–87	subl 100		6.6 aq; v s alc, chl; eth; s bz
h105a	1-Hydroxy-2-butanone	$CH_3CH_2COCH_2OH$	88.11	1, 826	1.026	1.4282^{20}		78^{60mm}	60	misc aq, alc; sl s eth
h106	3-Hydroxy-2-butanone	$CH_3COCH(OH)CH_3$	88.11	1, 827	0.9972^{17}	1.4171^{20}	15	148	50	s alc, eth; sl s aq
h107	4-Hydroxycinnamic acid	$HOC_6H_4CH{=}CHCOOH$	164.16	10, 297			210–213			
h108	4-Hydroxycoumarin		162.14	17, 488			213 d			s aq, alc, eth
h109	7-Hydroxycoumarin		162.14	18, 27			226–228	subl		v s alc, chl, alk, HOAc
h110	1-Hydroxy-1-cyclohexanecarbonitrile	$C_6H_{10}(OH)CN$	125.17	10, 5	1.031	1.4576^{20}	29		60	
h111	2-Hydroxy-3,5-diiodobenzoic acid	$I_2C_6H_2(OH)COOH$	389.91	10, 113			235 d			v s alc, eth; i bz, chl
h112	3-Hydroxydiphenylamine	$HOC_6H_4NHC_6H_5$	185.23	13, 410			82	340		
h113	2-Hydroxydiphenylmethane	$C_6H_5CH_2C_6H_4OH$	184.24	6, 675			54	312	>110	s alc, chl, eth, alk
h114	2-(2-Hydroxyethoxy)phenol	$HOCH_2CH_2OC_6H_4OH$	154.17	6^2, 782			100	$128^{0.7mm}$		
h115	N-(2-Hydroxyethyl)acetamide	$HOCH_2CH_2NHCOCH_3$	103.12	4^1, 430	1.1233^{20}_{20}	1.4575^{20}	63–65	155^{5mm}	176	misc aq; sl s bz
h116	2-Hydroxyethyl acetate	$CH_3COOCH_2CH_2OH$	104.11	2, 141	1.108^{15}	1.4201^{20}		188	88	misc aq, alc, chl, eth
h117	3-(1-Hydroxyethyl)aniline	$CH_3CH(OH)C_6H_4NH_2$	137.18	13^3, 1654			68–71			
h118	2-Hydroxyethyl disulfide	$HOCH_2CH_2SSCH_2CH_2OH$	154.25	1, 471	1.261	1.5655^{20}	25–27	$158^{3.5mm}$	>110	
h119	N-(2-Hydroxyethyl)ethylenediamine-N,N,N'-triacetic acid	$HOOCH_2N(CH_2CH_2OH)CH_2CH_2N(CH_2COOH)_2$	278.26	4^1, 562			212 d			
h120	2-Hydroxyethyl hydrazine	$HOCH_2CH_2NHNH_2$	76.10		1.119	1.4961^{20}	−70	220	73	misc aq; s alc
h121	2-Hydroxyethyl methacrylate	$HOCH_2CH_2OOCC(CH_3){=}CH_2$	130.14		1.034	1.4515^{20}		$67^{3.5mm}$	97	

1.224

No.	Name	Formula	Mol wt	mp	d	n_D	bp	fp	Solubility
h122	N-(2-Hydroxyethyl)morpholine		131.18	27.7	1.083	1.4760^{20}	227	99	misc aq
h123	1-(2-Hydroxyethyl)piperazine		130.19	23^2, 6	1.061	1.5065^{20}	246	>110	
h124	N-(2-Hydroxyethyl)piperazine-N′-ethanesulfonic acid		238.31				234 d		
h125	N-(2-Hydroxyethyl)piperidine		129.20	20, 25	1.0059^5	1.4804^{20}	199–202	68	v s aq, alc, chl
h126	N-(2-Hydroxyethyl)pyridine	$HOCH_2CH_2NC_5H_4$	123.16	21, 50	1.093	1.5368^{20}	$116^{4\mathrm{mm}}$	92	
h126a	N-(2-Hydroxyethyl)pyrrolidine	$HOCH_2CH_2NC_4H_8$	115.18	20^2, 5	0.985		$81^{13\mathrm{mm}}$	56	
h126b	2-Hydroxyethylsalicylate	$(HOC_6H_4CO_2CH_2CH_2OH)$	182.18	10, 81	1.224	1.4713^{20}	$166^{13\mathrm{mm}}$	>110	

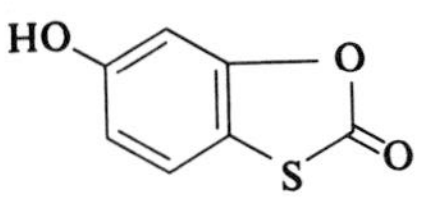

h104

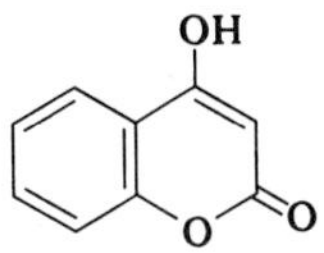

h108

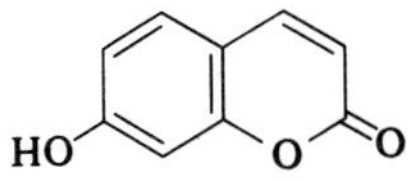

h109

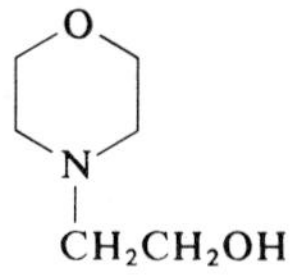

CH_2CH_2OH

h122

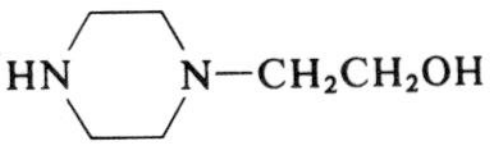

HN — N—CH_2CH_2OH

h123

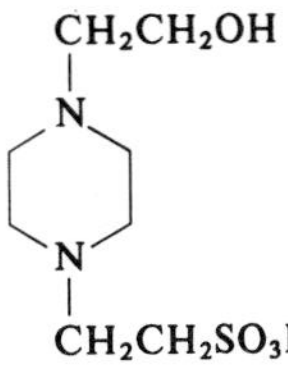

CH_2CH_2OH —N   N— $CH_2CH_2SO_3H$

h124

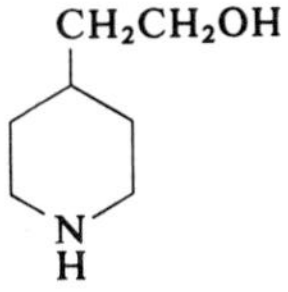

CH_2CH_2OH

h125

TABLE 1.15 Physical Constants of Organic Compounds (*Continued*)

No.	Name	Formula	Formula weight	Beilstein reference	Density	Refractive index	Melting point	Boiling point	Flash point	Solubility in 100 parts solvent
h126c	(2-Hydroxyethyl)-triphenylphospho-nium bromide	HOCH$_2$CH$_2$P(C$_6$H$_5$)$_3$Br	387.26	16, 761			219			
h126d	8-Hydroxy-7-iodo-5-quinolinesulfonic acid		351.12	22, 408			270 d			
h127	2-Hydroxyisobutyric acid	(CH$_3$)$_2$C(OH)COOH	104.11	3, 313			82	84$^{1.5mm}$		v s aq, alc, eth
h128	Hydroxylamine HCl	H$_2$NOH·HCl	69.49	24^3,	1.670		159 d			
h129	4-Hydroxy-2-mercapto-6-methyl-pyrimidine		142.18	1289			330 d			
h130	4-Hydroxy-2-mercapto-6-propyl-pyrimidine		170.23				219–221			0.1 aq; 1.7 alc; 1.7 acet; v s alk; i bz
h131	2-Hydroxy-3-methoxy-benzaldehyde	CH$_3$OC$_6$H$_3$(OH)CHO	152.15	8, 240			40–42	265–266		v s alc, eth; sl s aq
h132	4-Hydroxy-3-methoxy-benzaldehyde	CH$_3$OC$_6$H$_3$(OH)CHO	152.15	8, 247	1.056		80–81	285		1 aq; s alc, chl, pyr
h133	4-Hydroxy-3-methoxy-benzoic acid	CH$_3$OC$_6$H$_3$(OH)COOH	168.15	10, 392			210			0.12 aq; v s alc
h134	4-Hydroxy-3-methoxy-benzonitrile	CH$_3$OC$_6$H$_3$(OH)CN	149.15	10, 398			85–87			
h135	2-Hydroxy-4-methoxy-benzophenone	CH$_3$OC$_6$H$_3$(OH)COC$_6$H$_5$	228.25	8, 312			66	155^{5mm}		v s alc, chl, eth
h136	4-Hydroxy-3-methoxy-benzyl alcohol	CH$_3$OC$_6$H$_3$(OH)CH$_2$OH	154.17	6, 1113			113–115			
h137	N-(Hydroxymethyl)-acrylamide	H$_2$C=CHCONHCH$_2$OH	101.11	2^4, 1472	1.074				none	
h138	2-Hydroxy-3-methyl-benzoic acid	CH$_3$C$_6$H$_3$(OH)COOH	152.15	10, 220			165–166			s alc, chl, eth, alk
h139	2-Hydroxy-4-methyl-benzoic acid	CH$_3$C$_6$H$_3$(OH)COOH	152.15	10, 233			177			s alc, chl, eth, alk
h140	4-Hydroxy-3-methyl-2-butanone	HOCH$_2$CH(CH$_3$)COCH$_3$	102.13	1^1, 422	0.993	1.4340^{20}		92^{15mm}	81	

No.	Name	Formula	Mol. wt.	Beilstein ref.	Density	n	m.p., °C	b.p., °C	Flash	Solubility
h141	7-Hydroxy-4-methylcoumarin		176.17	18, 31			194–195			s alc, HOAc; sl s eth
h142	N-(Hydroxymethyl)-nicotinamide	$(C_5H_4N)CONHCH_2OH$	152.15	10, 4750			154			misc aq
h143	4-Hydroxy-4-methyl-2-pentanone	$(CH_3)_2C(OH)CH_2COCH_3$	116.16		0.9385^{20}	1.4235^{20}	−44	167.9	61	sl s aq, alc, bz
h144	N-(Hydroxymethyl)-phthalimide		177.16	21, 475			142–145			s aq, alc, chl, eth
h145	4-Hydroxy-N-methyl-piperidine		115.18	21¹, 188		1.4775^{20}	29–31	200		
h146	2-Hydroxy-2-methyl-propanenitrile	$(CH_3)_2C(OH)CN$	85.10	3, 316	0.9267^{25}	1.3992^{20}	−19	95	63	1.2 aq; v s hot aq; s alc, alk; sl s bz, eth
h147	3-Hydroxy-2-methyl-4-pyrone		126.11				161–162			
h148	2-Hydroxy-1-naphthal-dehyde	$C_{10}H_6(OH)CHO$	172.18	8, 143			82–85	192^{27mm}		

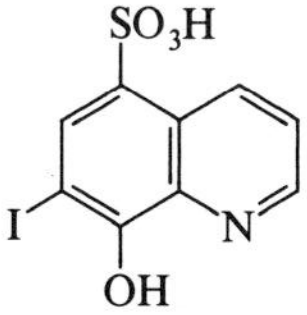

SO₃H, I, OH (quinoline) — h126d

H₃C, N, SH, OH (pyrimidine) — h129

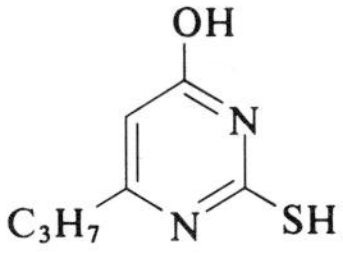

OH, N, SH, C₃H₇ (pyrimidine) — h130

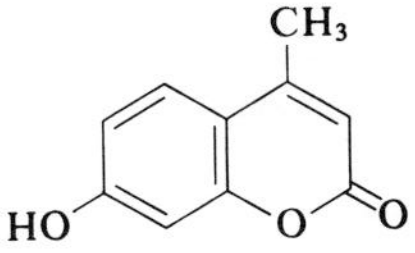

CH₃, HO (coumarin) — h141

N—CH₂OH (phthalimide) — h144

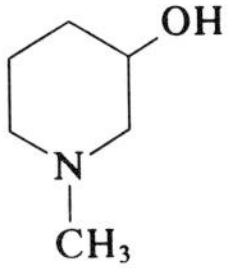

OH, N—CH₃ (piperidine) — h145

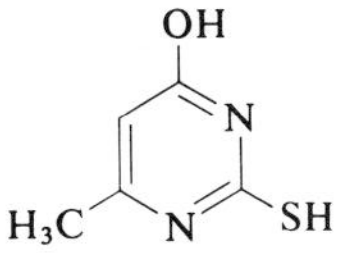

CH₃, O, OH (pyranone) — h147

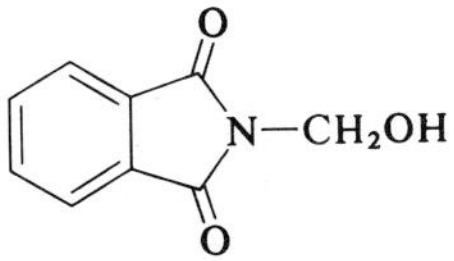

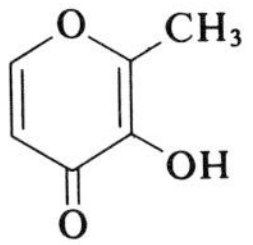

TABLE 1.15 Physical Constants of Organic Compounds (*Continued*)

No.	Name	Formula	Formula weight	Beilstein reference	Density	Refractive index	Melting point	Boiling point	Flash point	Solubility in 100 parts solvent
h149	1-Hydroxy-2-naphthalene-carboxylic acid	$C_{10}H_6(OH)COOH$	188.18	10, 331			191–192			v s alc, bz, eth, alk
h149a	2-Hydroxy-1-naphthalene-carboxylic acid	$C_{10}H_6(OH)COOH$	188.18	10, 328			167 d			
h150	3-Hydroxy-2-naphthalene-carboxylic acid	$C_{10}H_6(OH)COOH$	188.18	10, 333			222–223			v s alc, eth; s bz, chl
h151	2-Hydroxy-3,6-naphthalenedisulfonic acid, disodium salt	$C_{10}H_5(OH)(SO_3^-\,Na^+)_2$	348.25	11, 288						v s aq, alc; i eth
h152	4-Hydroxy-2,7-naphthalenedisulfonic acid, disodium salt	$C_{10}H_5(OH)(SO_3^-Na^+)_2$	348.25	11, 227			>300			
h153	2-Hydroxy-1,4-naphthoquinone		174.16	8, 300			d 185			s HOAc
h154	4-Hydroxy-3-nitro-benzenearsonic acid	$HOC_6H_3(NO_2)AsO(OH)_2$	263.04	161, 456			>300			v s alc, acet, HOAc, alk; sl s aq; i eth
h155	3-Hydroxy-4-nitroben-zoic acid	$HOC_6H_3(NO_2)COOH$	183.12	10, 146			229–231			
h156	2-Hydroxy-5-nitro-benzyl bromide	$HOC_6H_3(NO_2)CH_2Br$	232.04	6, 367			147–149			
h157	5-Hydroxy-1-pentanal	$HO(CH_2)_4CHO$	102.13		1.055	1.4530^{20}		115^{15mm}	>110	s aq
h158	5-Hydroxy-2-penta-none	$CH_3COCH_2CH_2CH_2OH$	102.13	1, 831	1.0072^{20}_4	1.4372^{20}		144^{100mm}	93	misc aq; s alc, eth
h161	4-Hydroxyphenylace-tic acid	$HOC_6H_4CH_2COOH$	152.15	10, 190			149–151			v s alc, eth; sl s aq
h163	4-(p-Hydroxyphenyl)-2-butanone	$HOC_6H_4CH_2CH_2COCH_3$	164.20				82–83	subl		
h164	4-Hydroxyphenyl-glycine	$HOC_6H_4CH(NH_2)COOH$	167.16	141, 659			240 d			sl s aq, alc, bz, acet

No.	Name	Formula	Mol. wt.	Beil. ref.	n_D	Density	M.p., °C	B.p., °C	Flash	Solubility
h165	N-(4-Hydroxyphenyl)-glycine	HOC$_6$H$_4$NHCH$_2$COOH	167.16	13, 488			220–248 d			s alk, acid; v sl s aq, alc, acet, bz, chl, eth
h165a	2'-Hydroxy-3-phenyl-propiophenone	HOC$_6$H$_4$COCH$_2$CH$_2$C$_6$H$_5$	226.28	8^2, 202	1.5968^{20}		37		>110	misc aq, alc; s eth
h166	1-(3-Hydroxyphenyl)-urea	HOC$_6$H$_4$NHCONH$_2$	152.15	13, 417			182–184			misc aq, alc, acet; 2.3 eth; i bz, PE
h167	N-Hydroxyphthali-mide		163.13	21, 500			233 d			v s alc, eth; sl s aq
h169	2-Hydroxypropio-nitrile	CH$_3$CH(OH)CN	71.08	3^2, 209	1.4027^{25}	0.9834^{25}	−40	103^{30mm}	76	v s alc, eth; sl s aq
h170	3-Hydroxypropioni-trile	HOCH$_2$CH$_2$CN	71.08	3, 298	1.4256^{20}	1.0404^{25}	−46	228	>110	misc aq, alc
h171	2'-Hydroxypropio-phenone	HOC$_6$H$_4$COCH$_2$CH$_3$	150.18	8, 102	1.5480^{20}	1.094		115^{15mm}	>110	
h172	4'-Hydroxypropio-phenone	HOC$_6$H$_4$COCH$_2$CH$_3$	150.18	8, 102			148			
h173	1-(2-Hydroxy-1-pro-poxy)-2-propanol	CH$_3$CH(OH)CH$_2$OCH$_2$-CH(OH)CH$_3$	134.18		1.4440^{20}	1.0252^{20}		231.8	138	
h173a	Hydroxypropyl methacrylate	H$_2$C=CH(CH$_3$)CO$_2$(CH$_2$)$_3$OH	144.17	2^4, 1532	1.4470^{20}	1.066		$57^{0.5mm}$	96	s aq, alc, bz; sl s eth
h174	2-Hydroxypyridine	HOC$_5$H$_4$N	95.10	21, 43			105–107	281		v s aq, alc; sl s eth
h175	3-Hydroxypyridine	HOC$_5$H$_4$N	95.10	21, 46			126–129	151^{3mm}		v s aq; i alc, bz, eth
h176	4-Hydroxypyridine	HOC$_5$H$_4$N	95.10	21, 48				230^{12mm}		

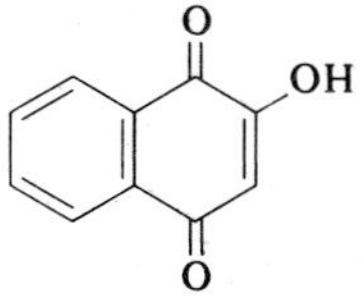

h153

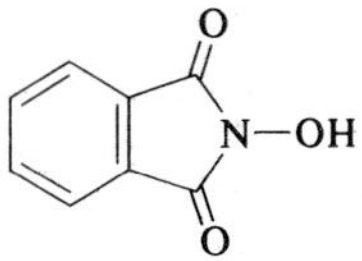

h167

TABLE 1.15 Physical Constants of Organic Compounds (*Continued*)

No.	Name	Formula	Formula weight	Beilstein reference	Density	Refractive index	Melting point	Boiling point	Flash point	Solubility in 100 parts solvent
h177	2-Hydroxypyridine-5-carboxylic acid	HO(C$_5$H$_3$N)COOH	139.11	22, 215			>300			sl s aq, alc, eth
h178	3-Hydroxypyridine-N-oxide	(HO)C$_5$H$_4$N=O	111.10				190–192			
h179	8-Hydroxyquinoline		145.16	21, 91			76	267		v s alc, acet, bz, chl
h180	8-Hydroxyquinoline-5-sulfonic acid		225.22	22, 407			213 d			v s aq; sl s alc, eth
h181	DL-Hydroxysuccinic acid	HOOCCH(OH)CH$_2$COOH	134.09	3, 435			131–133			56 aq; 45 EtOH; 18 acet 0.8 eth; 23 diox; i bz
h182	(−)-Hydroxysuccinic acid	HOOCCH(OH)CH$_2$COOH	134.09	3, 419			100			36 aq; 87 EtOH; 2.7 eth 61 acet; 75 diox
h183	N-Hydroxysuccinimide		115.09	21, 380			93–95			v s aq
h185	3-Hydroxy-3,7,11-trimethyl-1,6,10-dodecatriene	H$_2$C=CH(OH)(CH$_3$)CH$_2$CH$_2$-CH=C(CH$_3$)CH$_2$CH$_2$-CH=C(CH$_3$)$_2$	222.37		0.8760^{25}_4	1.4769^{25}		114^{1mm}	96	s abs alc
h186	3-Hydroxy-2,2,4-trimethyl-3-pentenoic acid β-lactone		140.18		0.947	1.4380^{20}	−18	170	62	
i1	1H,1H,11H-Icosafluoro-1-undecanol	HCF$_2$(CF$_2$)$_9$CH$_2$OH	531.1				95–97	181^{200mm}		
i2	Icosane	CH$_3$(CH$_2$)$_{18}$CH$_3$	282.56	1, 174	0.7777^{37}	1.4346^{40}	36.4	343.8	>112	v s aq, alc, chl, eth
i3	1-Icosene	CH$_3$(CH$_2$)$_{17}$CH=CH$_2$	280.54	1^3, 881			28.7	342.4		
i4	Imidazole		68.08	23, 45			90–91	257	145	2 aq; s alc, pyr; i bz, acet, chl, eth
i5	2-Imidazolidinethione		102.16	24, 4			203–204			
i7	2-Imidazolidone		86.09	24, 16			131			v s aq, hot alc
i8	2-(4-Imidazolyl)ethyl-amine		111.15	25, 315			83–84	209^{18mm}		v s aq, alc, hot chl
i9	3,3'-Iminobispropyl-amine	H$_2$NCH$_2$CH$_2$CH$_2$NHCH$_2$CH$_2$CH$_2$NH$_2$	131.22		0.938	1.4810^{20}	−14	151^{50mm}	118	2 aq; v sl s bz, eth
i10	Iminodiacetic acid	HOOCCH$_2$NHCH$_2$COOH	133.10	4, 365			243 d			s aq, alc; sl s eth
i11	Iminodiacetonitrile	NCCH$_2$NHCH$_2$CN	95.11	4, 367			77			
i12	Iminodibenzyl		195.27				105–108			

i13	Indan		118.18		0.9639^{20}_{4}	1.5360^{20}	−51.4	176.5	50	s alc, chl, eth; i aq
i14	5-Indanol		134.18	6, 575			51–53	255	>110	v s alc, eth; sl s aq
i15	1-Indanone		132.16	7, 360	1.1090^{45}_{4}	1.561^{45}	40–42	243–245	111	s alc, eth; sl s aq
i16	1,2,3-Indantrione hydrate		178.14				d 241			
i17	Indene		116.16	5, 515	0.9968^{20}_{4}	1.5762^{20}	−1.8	181.6	58	misc alc, bz, chl, eth
i18	Indole		117.15	20, 304	1.0643	1.609^{60}	52	253	>110	s hot aq, bz, eth
i19	Indole-3-acetic acid		175.19	22, 66			168–170			v s alc; s acet, eth
i20	Indole-3-carbaldehyde		145.16	21, 313			195–198			
i21	Indole-2,3-dione		147.13	21, 432			203.5 d			s hot aq, hot alc, alk

5-Hydroxyvaleraldehyde, h157
Imidodicarbonic diamide, b216

Indalone, b448

Indanamine, a201, a202

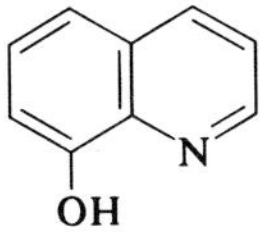
h179

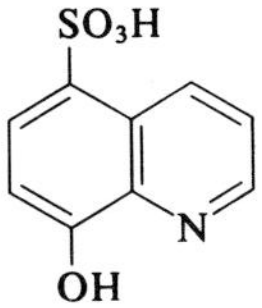
h180

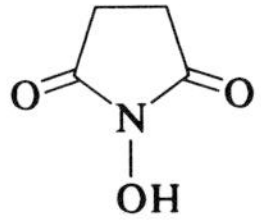
h183

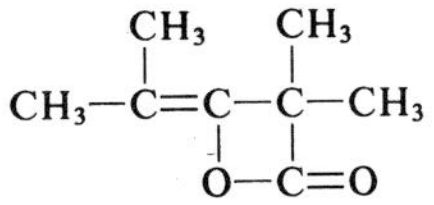
h186

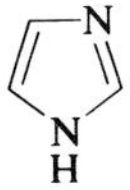
i4

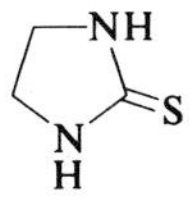
i5

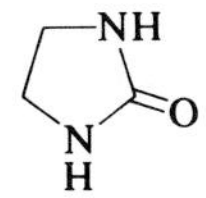
i7

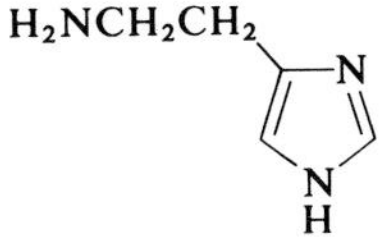
i8

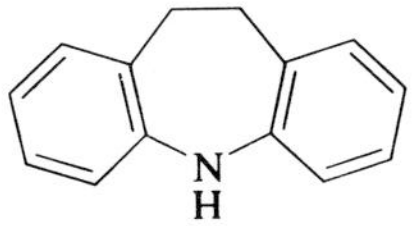
i12

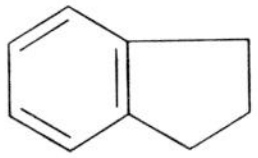
i13

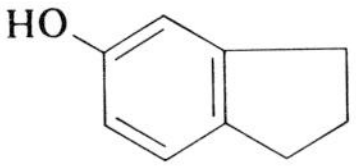
i14

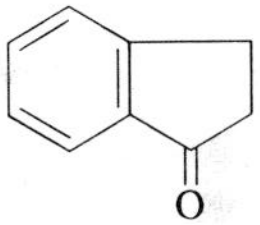
i15

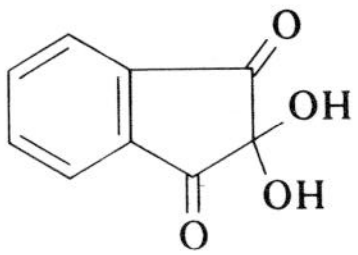
i16

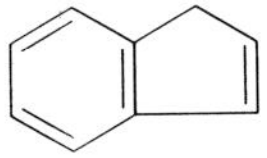
i17

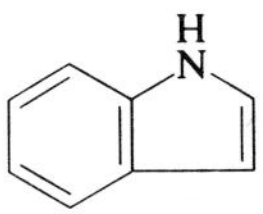
i18

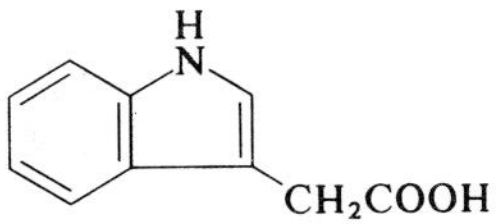
i19

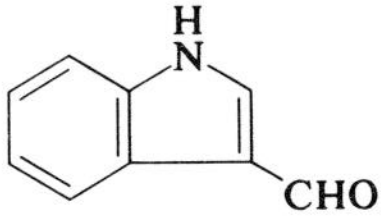
i20

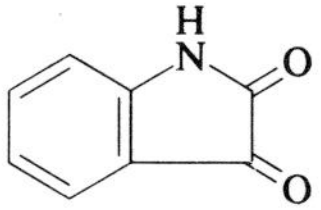
i21

TABLE 1.15 Physical Constants of Organic Compounds (*Continued*)

No.	Name	Formula	Formula weight	Beilstein reference	Density	Refractive index	Melting point	Boiling point	Flash point	Solubility in 100 parts solvent
i22	Indoline		119.17	20, 257	1.063	1.5906^{20}		221	92	sl s aq
i23	Inositol		180.16	6^2, 1157	1.752		225–227			14 aq; sl s aq; i eth
i24	Iodoacetamide	ICH_2CONH_2	184.96	2, 223			91–93			s hot aq
i25	Iodoacetic acid	ICH_2COOH	185.95	2, 222			82–83			s aq, alc,; v sl s eth
i26	3-Iodoaniline	$IC_6H_4NH_2$	219.03	12, 670	1.821	1.6820^{20}	25	146^{15mm}	>110	i aq; s alc, eth
i27	Iodobenzene	C_6H_5I	204.01	5, 215	1.83834^{25}	1.6211^8	−30	188.3	74	misc alc, chl, eth
i28	Iodobenzene diacetate	$C_6H_5I(OOCCH_3)_2$	322.10	5, 218			163–165			
i29	2-Iodobenzoic acid	IC_6H_4COOH	248.02	9, 363	2.249^{25}		162			s alc, eth; sl s aq
i30	1-Iodobutane	$CH_3CH_2CH_2CH_2I$	184.02	1, 123	1.6164^{20}	1.4999^{20}	−103.5	129–130	33	i aq; s alc, eth
i31	2-Iodobutane	$CH_3CH_2CH(I)CH_3$	184.02		1.5922^{20}	1.4991^{20}	−104.0	118–120	23	i aq; s alc, eth
i32	Iodocyclohexane	$C_6H_{11}I$	210.06	5^2, 13	1.6261^{15}	1.5472^{20}		180		i aq; s eth
i33	1-Iododecane	$CH_3(CH_2)_9I$	268.18	1, 168	1.2574^{20}	1.4827^{20}		132^{15mm}	>110	i aq; s alc, eth
i33a	1-Iodododecane	$CH_3(CH_2)_{11}I$	296.24	1^1, 67	1.201	1.4844^{20}	−3	160^{15mm}	>110	
i34	Iodoethane	CH_3CH_2I	155.97	1, 96	1.9358^{20}	1.5137	−110.9	72.4	none	0.4 aq; misc alc, bz, chl, eth
i35	2-Iodoethanol	ICH_2CH_2OH	171.97	1, 339	2.21970^4	1.5694^{20}		75^{5mm}	65	s aq; v s alc, eth
i36	Iodoform	CHI_3	393.73	1, 73	4.008		123		none	1.4 alc; 10 chl; 13 eth; v s bz, acet
i37	1-Iodoheptane	$CH_3(CH_2)_6I$	226.10	1, 155	1.3732^{20}	1.4900^{20}	−48.2	204	78	i aq; s alc, eth
i38	1-Iodohexadecane	$CH_3(CH_2)_{15}I$	352.35	1, 172	1.121	1.4806^{20}	23	206^{10mm}	>110	i aq
i39	1-Iodohexane	$CH_3(CH_2)_5I$	212.08	1, 146	1.4372^{20}	1.4926^{20}		179	61	i aq
i40	Iodomethane	CH_3I	141.94	1, 69	2.27894^4	1.5308^{20}	−66.5	42.4	none	1.4 aq; misc alc, eth
i41	4-Iodomethoxy-benzene	$IC_6H_4OCH_3$	234.04	6, 208			48–50	237^{726mm}	none	s hot alc, eth
i42	1-Iodo-3-methylbutane	$(CH_3)_2CHCH_2CH_2I$	198.06	1^3, 367	1.5094^{20}	1.4939^{20}		147.5	42	misc alc, eth; sl s aq
i43	1-Iodo-2-methyl-propane	$(CH_3)_2CHCH_2I$	184.02	1, 128	1.6034^{20}	1.4948^{20}	−93.5	119	12	i aq; misc alc, eth
i44	2-Iodo-2-methyl-propane	$(CH_3)_3CI$	184.02	1^3, 326	1.5710	1.4918^{20}	−38.2	100	7	d aq; misc alc, eth
i45	1-Iodo-3-nitrobenzene	$IC_6H_4NO_2$	249.01	5, 253	1.94775^0		36–38	280	71	i aq; s alc, eth
i45a	1-Iodo-4-nitrobenzene	$IC_6H_4NO_2$	249.01	5, 252			51	289	>110	i aq; s alc, eth
i45b	1-Iodononane	$CH_3(CH_2)_8I$	254.18	1, 166	1.288	1.4870^{20}		108^{8mm}	85	
i45c	1-Iodooctadecane	$CH_3(CH_2)_{17}I$	380.40	1, 173			35	197^{2mm}	110	
i46	1-Iodooctane	$CH_3(CH_2)_7I$	240.13	1, 160	1.3302^{20}	1.4889^{20}	−46	226	95	s alc, eth
i47	1-Iodopentane	$CH_3(CH_2)_4I$	198.06	1, 133	1.5124^{20}	1.4954^{20}	−85.6	155	51	sl s aq; s alc, eth

No.	Name and Formula	Formula wt	Beilstein ref	Density	n_D	mp, °C	bp, °C	Flash pt, °C	Solubility
i48	1-Iodopropane $CH_3CH_2CH_2I$	169.99	1, 113	1.7489^{20}	1.5058^{20}	−101	102.5	44	0.1 aq; misc alc, eth
i49	2-Iodopropane $(CH_3)_2CHI$	169.99	1, 114	1.7025^{20}	1.4992^{20}	−90	89	42	0.14 aq; misc alc, bz, chl, eth
i50	3-Iodo-1-propene $ICH_2CH{=}CH_2$	167.97	1, 202	1.845^{22}	1.5540^{21}	−99	103	18	misc alc, chl, eth
i51	5-Iodosalicylic acid $IC_6H_3(OH)COOH$	264.02	10, 112			189–191			v s alc; i bz, chl
i52	2-Iodothiophene	210.04	17, 34	1.902	1.6520^{20}	−40	73^{15mm}	71	v s eth
i53	2-Iodotoluene $IC_6H_4CH_3$	218.04	5, 310	1.713	1.6079^{20}		211	90	i aq; s alc, eth
i54	3-Iodotoluene $IC_6H_4CH_3$	218.04	5, 311	1.698	1.6040^{20}		80^{10mm}	82	i aq; misc alc, eth
i54a	4-Iodotoluene $IC_6H_4CH_3$	218.04	5, 312			36	211	90	
i55	Iodotrimethylsilane $(CH_3)_3SiI$	200.10		1.406^{4}	1.4710^{20}		106	−31	
i55a	1-Iodoundecane $CH_3(CH_2)_{10}I$	282.21	1^{1}, 66	1.220	1.4849^{20}		130^{5mm}	>110	
i56	α-Ionone	192.30	7, 168	0.932^{20}	1.4980^{20}		124^{11mm}	104	s alc, bz, chl, eth
i57	β-Ionone	192.30	7, 167	0.946^{17}	1.521^{17}		128^{12mm}	>110	s alc, bz, chl, eth
i58	Isatoic anhydride	163.13	27, 264			233 d		>110	sl s aq, hot alc, acet
i59	D-(−)-Isoascorbic acid	176.12				169 d			s aq, alc, acet, pyr

Indonaphthene, i17
4-Iodoanisole, i41
5-Iodoanthranilic acid, a205
Isatin, i21

Isoamyl acetate, i80
Isoamyl alcohol, m155
sec-Isoamyl alcohol, m156

Isoamyl bromide, b308
Isoamyl iodide, i42
Isoamyl nitrite, i81

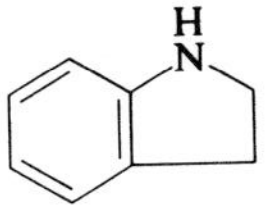

i22

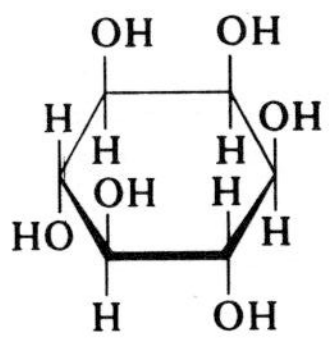

i23

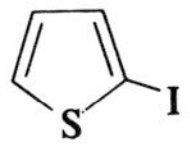

i52

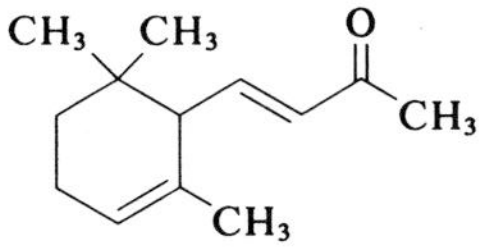

i56

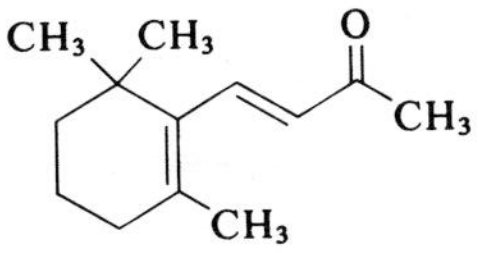

i57

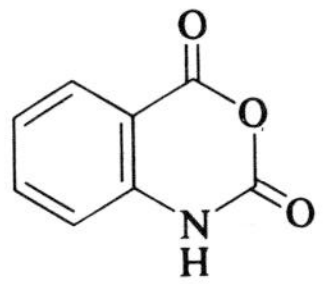

i58

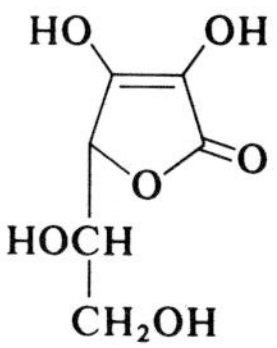

CH_2OH
HOCH
i59

TABLE 1.15 Physical Constants of Organic Compounds (*Continued*)

No.	Name	Formula	Formula weight	Beilstein reference	Density	Refractive index	Melting point	Boiling point	Flash point	Solubility in 100 parts solvent
i60	DL-Isoborneol		154.25	6^2, 80	1.022	1.5230^{20}	212	subl		v s alc, chl, eth
i61	2-Isobutoxy-1-isobutoxycarbonyl-1,2-dihydroquinoline		303.40					$140^{0.2mm}$	>110	
i62	Isobutyl acetate	$(CH_3)_2CHCH_2OOCCH_3$	116.16	2, 131	0.8745^{20}	1.3902^{20}	−98.9	118.0	21	0.7 aq; v s alc
i63	Isobutylamine	$(CH_3)_2CHCH_2NH_2$	73.14	4, 163	0.724^{20}	1.3972^{20}	−84.6	67.7	−20	misc aq, alc, acet, eth
i64	Isobutylbenzene	$C_6H_5CH_2CH(CH_3)_2$	134.22	5, 414	0.8673^{20}	1.4855^{20}	−51.5	172.8	55	misc alc, eth
i65	Isobutylchloroformate	$ClCOOCH_2CH(CH_3)_2$	136.58	3, 12	1.053	1.4070^{20}		128.8	27	misc bz, chl, eth
i66	Isobutyl formate	$HCOOCH_2CH(CH_3)_2$	102.13	2, 21	0.8884^{20}	1.3855^{20}	−94.5	98.4	10	1 aq; misc alc, eth
i67	Isobutyl isobutyrate	$(CH_3)_2CHCH_2OOCCH(CH_3)_2$	144.22	2, 291	0.8542^{20}	1.3999^{20}	−80.7	147.5	37	0.5 aq; misc alc
i68	Isobutyl lactate	$CH_3CH(OH)COOCH_2CH(CH_3)_2$	146.19	3^2, 188	0.971^{20}	1.4181^{25}		96^{40mm}	66	
i69	Isobutyl methacrylate	$H_2C{=}C(CH_3)COOCH_2CH(CH_3)_2$	142.19		0.882^{25}_{15}	1.4170^{25}		155	41	misc alc, eth
i70	Isobutyl nitrate	$(CH_3)_2CHCH_2ONO_2$	119.12	1, 377	1.015^{20}	1.4028^{20}		123	21	i aq; misc alc, eth
i71	Isobutyl nitrite	$(CH_3)_2CHCH_2ONO$	103.12	1, 377	0.870^{22}_4	1.3715^{22}		67	4	misc alc; sl s aq(dec)
i71a	Isobutyl propionate	$C_2H_5CO_2CH_2CH(CH_3)_2$	130.19	2, 241	0.888^{20}_4	1.3974^{20}	−71	137	26	i aq; misc alc
i71b	Isobutyl stearate	$CH_3(CH_2)_{16}CO_2CH_2CH(CH_3)_2$	340.57				−20			
i72	Isobutyl vinyl ether	$(CH_3)_2CHCH_2OCH{=}CH_2$	100.16	1^3, 1862	0.7702^{20}	1.3961^{20}	−112	83.4	−13	0.2 aq
i73	Isobutyraldehyde	$(CH_3)_2CHCHO$	72.11	1, 671	0.7988^{20}_4	1.3723^{20}	−65.9	64	−40 (CC)	11 aq; misc alc, bz, acet, chl, eth
i74	Isobutyramide	$(CH_3)_2CHCONH_2$	87.12	2, 293	1.013		127−129	216−220		
i75	Isobutyric acid	$(CH_3)_2CHCOOH$	88.11	2, 288	0.950^{20}_4	1.3925^{20}	−47	154	55 (TOC)	17 aq; misc alc, chl, eth
i75a	Isobutyric anhydride	$[(CH_3)_2CHCO]_2O$	158.20	2, 292	0.954	1.4052^{20}	−56	182	59	v s alc, eth; sl s aq
i76	Isobutyronitrile	$(CH_3)_2CHCN$	69.11	2, 294	0.7704^{20}	1.3734^{20}	−71.5	108	3	
i77	Isobutyrophenone	$C_6H_5COCH(CH_3)_2$	148.21	7, 316	0.988^{20}	1.5172		217	84	dec aq, dec alc; s eth
i78	Isobutyryl chloride	$(CH_3)_2CHCOCl$	106.55	2, 293	1.017	1.4073^{20}	−90	91−93	1	4 aq; sl s hot alc
i79	L-Isoleucine	$CH_3CH_2CH(CH_3)CH(NH_2)COOH$	131.18	4, 454			d 284	subl 168		
i80	Isopentyl acetate	$CH_3COOCH_2CH_2CH(CH_3)_2$	130.19	2, 132	0.876^{15}	1.4007^{20}	−78.5	142.0	80	0.25 aq; misc alc, eth
i81	Isopentyl nitrite	$(CH_3)_2CHCH_2CH_2ONO$	117.15	1, 402	0.872	1.3860^{20}		99	10	misc alc, eth; sl s aq
i82	Isophorone		138.21	7, 65	0.923	1.4759^{20}	−8.1	215.2	84	1.2 aq

No.	Name	Formula	M.W.	Beil.	Density	n_D	m.p.	b.p.	Flash pt.	Solubility
i82a	Isophorone diisocyanate		222.29		1.049	1.4841^{20}		159^{15mm}	110	
i83	DL-Isopinocampheol		154.25	6^3, 283	0.909	1.4005^{20}	39–41	217	93	
i84	Isopropenyl acetate	$CH_3COOC(CH_3)=CH_2$	100.12	2^2, 278	0.903	1.4104^{20}		94	18	
i84a	2-Isopropoxyethanol	$(CH_3)_2CHOCH_2CH_2OH$	104.15	1^2, 519	1.030	1.5157^{20}		44^{13mm}	45	
i85	2-Isopropoxyphenol	$(CH_3)_2CHOC_6H_4OH$	152.19	6^3, 4209	0.879^{25}	1.407^{25}		102^{11mm}	87	
i86	1-Isopropoxy-2-propanol	$CH_3CH(OH)CH_2OCH(CH_3)_2$	118.1					47.9	49	
i86a	3-Isopropoxypropylamine	$(CH_3)_2CHO(CH_2)_3NH_2$	117.19	4^3, 739	0.845	1.4195^{20}		79^{85mm}	39	
i87	Isopropyl acetate	$(CH_3)_2CHOOCCH_3$	102.13	2, 130	0.870^{20}_4	1.3770^{20}	−73	89	2 (CC)	3 aq; misc alc, eth

Isobutane, m378
Isobutene, m386
α-Isobutoxy-α-phenylacetophenone, b48
Isobutoxyacetylene, m355
Isobutyl alcohol, m384
Isobutyl bromide, b313
Isobutyl chloride, c162
Isobutyl chlorocarbonate, i65
Isobutyl 1,2-dihydro-2-isobutoxy-1-quinolinecarboxy-late, i61
Isobutyl ether, d408
Isobutyl heptyl ketone, t355
Isobutyl mercaptan, m382
Isobutyl octadecanoate, i71b
Isobutyraldehyde, m377

Isobutyramide, m391
Isobutyric acid, m393
Isocapronitrile, m342
Isocinchomeronic acid, p264
Isocrotonic acid, b404
5-Isocyanato-1-(isocyanatomethyl)-1,3,3-trimethylcy-clohexane, i82a
Isodurene, t100
Isoeugenol, m98
Isohexane, m339
Isoleucinol, a219
Isoniazid, p262
Isonicotinaldehyde, p257
Isonicotinic acid, p261
Isonicotinic acid hydrazide, p262

Isonicotinonitrile, c298
Isooctane, t362
Isopentane, m149
Isopentyl alcohol, m155
Isopentyl isovalerate, m171
Isophorone, t345
Isophthalic acid, b16
Isophthalonitrile, d237
Isophthaloyl dichloride, b14
Isoprene, m147
Isopropanolamine, a274
Isopropenyl acetate, p211
Isopropenylacetylene, m166
Isopropenyl methyl ether, m96

i60

i61

i82

i82a

i83

TABLE 1.15 Physical Constants of Organic Compounds (*Continued*)

No.	Name	Formula	Formula weight	Beilstein reference	Density	Refractive index	Melting point	Boiling point	Flash point	Solubility in 100 parts solvent
i88	Isopropylamine	$(CH_3)_2CHNH_2$	59.11	4, 152	0.6862^5	1.3711^{25}	−101	33–34	−26(OC)	misc aq, alc, eth
i89	2-Isopropylamino-ethanol	$(CH_3)_2CHNHCH_2CH_2OH$	103.17	4, 282	0.8970_4^{20}	1.4395^{20}		75^{11mm}		misc aq, alc, eth
i90	2-Isopropylaniline	$(CH_3)_2CHC_6H_4NH_2$	135.2	12, 1147	0.966	1.5477^{20}		222	95	
i90a	4-Isopropylbenzaldehyde	$(CH_3)_2CHC_6H_4CHO$	148.21	7, 318	0.977	1.5298^{20}		236	93	
i91	Isopropylbenzene	$C_6H_5CH(CH_3)_2$	120.20	5, 393	0.8644_4^{20}	1.4915^{20}	−96	152	39 (CC)	s alc, bz, eth
i92	4-Isopropylbenzyl alcohol	$(CH_3)_2CHC_6H_4CH_2OH$	150.22	6^3, 1911	0.982^{15}	1.5206^{20}	28	248.4	>110	misc alc, eth; i aq
i93	N-Isopropylbenzyl-amine	$C_6H_5CH_2NHCH(CH_3)_2$	149.24		0.892	1.5025^{20}		200	87	
i93a	Isopropyl butyrate	$CH_3CH_2CH_2CO_2CH(CH_3)_2$	130.19	2, 271	0.859	1.3932^{20}		131	30	
i94	Isopropylcyclohexane	$C_6H_{11}CH(CH_3)_2$	126.24	5, 41	0.8023_{20}^{20}	1.4399^{20}	−90	155	35	v s alc, eth
i95	N-Isopropylcyclo-hexamine	$C_6H_{11}NHCH(CH_3)_2$	141.26		0.859	1.4480^{20}		60^{12mm}	33	
i95a	Isopropyl hexadecanoate	$CH_3(CH_2)_{14}CO_2CH(CH_3)_2$	298.51	2^2, 336	0.852	1.4385^{20}			>110	
i96	4,4'-Isopropylidenebis-[2-(2,6-dibromo-phenoxy)ethanol]	$(CH_3)_2C[C_6H_2(Br)_2OCH_2\text{-}CH_2OH]_2$	632.01				107			
i97	4,4'-Isopropylidene-diphenol	$(CH_3)_2C[C_6H_4OH]_2$	228.29	6, 1011			159	220^{4mm}		
i98	Isopropyl isocyanate	$(CH_3)_2CHCNO$	85.11	4, 155	0.866	1.3825^{20}		74–75	−2	s aq, alc, eth
i99	Isopropyl S-(−)-lactate	$(CH_3)_2CHOOCCH(OH)CH_3$	132.16	3, 282	0.9983_4^{20}	1.4082^{25}		166–168	57	
i99a	2-Isopropyl-6-methyl-aniline	$(CH_3)_2CHC_6H_3(CH_3)NH_2$	149.24		0.957	1.5440^{20}			41	
i100	2-Isopropyl-1-methyl-benzene	$CH_3C_6H_4CH(CH_3)_2$	134.21	5, 419	0.8766_4^{20}	1.5006^{20}	−71.5	178.2		misc alc, eth
i101	3-Isopropyl-1-methyl-benzene	$CH_3C_6H_4CH(CH_3)_2$	134.21	5, 419	0.8610^{20}	1.4930^{20}	−63.75	175.1		misc alc, eth
i102	4-Isopropyl-1-methyl-benzene	$CH_3C_6H_4CH(CH_3)_2$	134.21	5, 420	0.8573_4^{20}	1.4909^{20}	−67.9	177.1	47	misc alc, eth
i102a	2-Isopropyl-5-methyl-phenol	$CH_3C_6H_3(OH)CH(CH_3)_2$	150.22	6, 532	0.925_4^{80}		49–51	232		i aq; v s alc, chl, eth

No.	Name	Formula	Mol. wt.	Ref.	Density	n_D	mp, °C	bp, °C	Flash	Solubility
i102b	5-Isopropyl-2-methyl-phenol	$(CH_3)_2CHC_6H_3(CH_3)OH$	150.22	6, 527	0.9764^{20}	1.5230^{20}	3	238	106	Volatile with steam; v s alc, eth; i aq
i102c	5-Isopropyl-3-methyl-phenol	$(CH_3)_2CHC_6H_3(CH_3)OH$	150.22	6, 526			51		110	
i103	N^1-Isopropyl-2-methyl-1,2-propane-diamine	$(CH_3)_2C(NH_2)CH_2NHCH(CH_3)_2$	130.24		0.822	1.4269^{20}		147–149	90	
i104	Isopropyl methyl sulfide	$(CH_3)_2CHSCH_3$	90.18	1, 367			−101.5	84.7		
i105	Isopropyl nitrate	$(CH_3)_2CHONO_2$	105.09	1, 363	1.0361^{9}	1.391^{20}		102.1	12	
i105a	Isopropyl nitrite	$(CH_3)_2CHONO$	89.09		0.8444^{25}	1.3520^{20}		39^{752mm}		
i106	2-Isopropylphenol	$(CH_3)_2CHC_6H_4OH$	136.19	6, 504	1.012^{20}	1.5259^{20}	15–16	212		misc alc, eth
i107	4-Isopropylphenol	$(CH_3)_2CHC_6H_4OH$	136.19	6, 505	0.990^{20}		59–61	212	88	316 alc; 350 eth
i107a	Isopropyl tetradecano-ate	$CH_3(CH_2)_{12}CO_2CH(CH_3)_2$	270.46	2^{3}, 923	0.850	1.4350^{20}		192^{20mm}	>110	
i108	Isopropyl vinyl ether	$(CH_3)_2CHOCH=CH_2$	86.13	6, 65	0.7532^{20}	1.3849^{20}	−140	5–6	78	v sl s aq
i109	Isopulegol		154.25	20, 380	0.911	1.4725^{20}		91^{12mm}		sl s aq; s acid
i110	Isoquinoline		129.16	1, 724	1.0910^{30}	1.6208^{30}	26.5	243.2	107	s acet, eth; d aq
k1	Ketene	$H_2C=C=O$	42.04				−150	−56		
L1	DL-Lactic acid	$CH_3CH(OH)COOH$	90.08	3, 268	1.249^{45}		16.8	122^{14mm}	>110	s aq, alc; i chl
L2	L-(+)-Lactic acid	$CH_3CH(OH)COOH$	90.08	3, 261	1.2060^{25}	1.4270^{20}	53	119^{12mm}	>110	v s aq, alc, eth

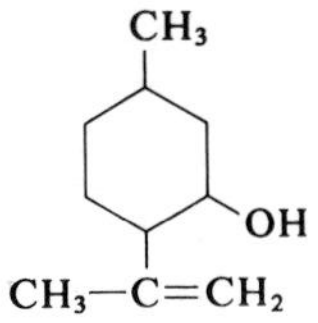

i109

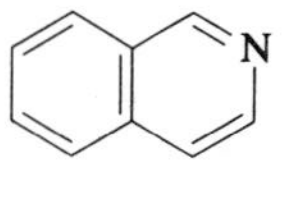

i110

TABLE 1.15 Physical Constants of Organic Compounds (*Continued*)

No.	Name	Formula	Formula weight	Beilstein reference	Density	Refractive index	Melting point	Boiling point	Flash point	Solubility in 100 parts solvent
L3	β-Lactose		342.30	31, 408	1.525^{20}		202	subl 293		20 aq; i alc, eth
L4	DL-Leucine	$(CH_3)_2CHCH_2CH(NH_2)COOH$	131.18	4, 447			d 332	subl 145		1 aq; 0.13 alc; i eth
L5	L-Leucine	$(CH_3)_2CHCH_2CH(NH_2)COOH$	131.18	4, 437	1.293^{18}		d 293			2.4 aq^{25}; 0.07 alc; 1 HOAc; i eth
L6	(+)-Limonene		136.24	5, 133	0.8411^{20}	1.4715	−96.5	176	48	misc alc, eth
L7	(−)-Limonene		136.24	5, 136	0.8412^{20}	1.4706^{20}	−96.5	176	48	misc alc, eth
L8	(+)-Limonene oxide		152.24	17, 44	0.929	1.4661^{20}		114^{50mm}	65	
L9	Linalool		154.25	1, 462	0.8655^{15}	1.4615^{20}		197^{720mm}	76	misc alc, eth
L10	Linalyl acetate		196.29	2, 141	0.8952^{20}	1.4460^{20}		220	90	misc alc, eth
L12	L-(+)-Lysine	$H_2N(CH_2)_4CH(NH_2)COOH$	146.19	4, 435			212 d			v s aq; sl s alc; i eth
m1	Maleic acid	HOOCCH=CHCOOH	116.07	2, 748	1.590		138–139			79 aq; 70 alc; 8 eth
m2	Maleic anhydride		98.06	17, 432	1.48		52.8	202.0	103	s aq (to acid), alc (to ester); 227 acet; 53 chl; 50 bz; 112 EtAc
m3	Malonic acid	$HOOCCH_2COOH$	104.06	2, 566	1.63		135 d			154 aq; 42 alc; 8 eth
m4	Malonodiamide	$H_2NCOCH_2CONH_2$	102.09	2, 582			168–170			9 aq; i alc, eth
m5	Malononitrile	$NCCH_2CN$	66.06	2, 589	1.1910^{20}	1.4146^{34}	32	218–219	112	13 aq; 40 alc; 20 eth
m6	Malonyl dichloride	$ClCOCH_2COCl$	140.95	2^1, 252	1.4486^{19}	1.4620^{20}		53^{19mm}	47	d hot alc; s eth
m7	D-(+)-Maltose hydrate		342.30	31, 386	1.540^{17}		102–103	d 130		v s aq; sl s alc; i eth
m8	DL-Mandelic acid	$C_6H_5CH(OH)COOH$	152.15	10, 197	1.300^{20}		119	dec		16 aq; 100 alc; s eth
m9	Mandelonitrile	$C_6H_5CH(OH)CN$	133.15	10, 193	1.117	1.5315^{20}	−10	d 170	97	v s alc, chl, eth; i aq
m10	Mannitol		182.17	1, 534	1.52^{20}		166–168	$290^{3.5mm}$		18 aq; 1.2 alc; i eth
m11	D-(+)-Mannose		180.16	31, 284	1.54^{20}		128–130			250 aq; 28 pyr; 0.8 alc
m12	(−)-Menthol		156.27	6, 28	0.890^{15}	1.4582^{5}	43–45	212	93	v s alc, chl, eth, PE
m13	(−)-Menthone		154.25	7, 38	0.8952^{20}	1.4510^{20}	−6	207	72	misc alc, eth; sl s aq
m14	Mercaptoacetic acid	$HSCH_2COOH$	92.12	3, 245	1.325	1.5030^{20}	−16.5	96^{5mm}	>110	misc aq, alc, bz, eth
m15	2-Mercaptobenzimidazole		150.20	24, 119			303–304			sl s aq; s alc
m16	2-Mercaptobenzoic acid	HSC_6H_4COOH	154.19	10, 125			164–165			v s alc, HOAc
m17	2-Mercaptobenzothiazole		167.25	27, 185	1.42^{20}		180–181	dec		2 alc; 1 eth; 10 acet; 1 bz; s alk; i aq
m18	2-Mercaptoethanol	$HSCH_2CH_2OH$	78.13	1, 470	1.1143^{20}	1.5006^{20}		156.9	73	misc aq, alc, bz, eth

Lactonitrile, h169
Lauraldehyde, d733
Lauric acid, d728
Lauryl alcohol, d729
Laurylamine, d734
Lauryl bromide, b277
Lauryl mercaptan, d727
Lauryl methacrylate, d735
Lauryl sulfate, d737
Lepidine, m411
Leucinol, a220

Levulinic acid, o58
Linoleic acid, o1
Linolenic acid, o7
Luminol, a153
2,6-Lupetidine, d593
β-Lutidine, e215
Lutidines, d605a, d606, d607, d608, d609
Maleic hydrazide, d401
Malic acid, h181, h182
Malonaldehyde bis(dimethyl acetal), t93
Malonamide nitrile, c287

Malonic acid diamide, m4
Malonylurea, b1
Melamine, t202
Mellitic acid, b19
MEM chloride, m67
Menadione, m313
1,8-Mentanediamine, d41
p-Mentha-1,8-diene, d651
p-Mentha-6,8-dien-2-one, c20
Mercaptobenzene, t162

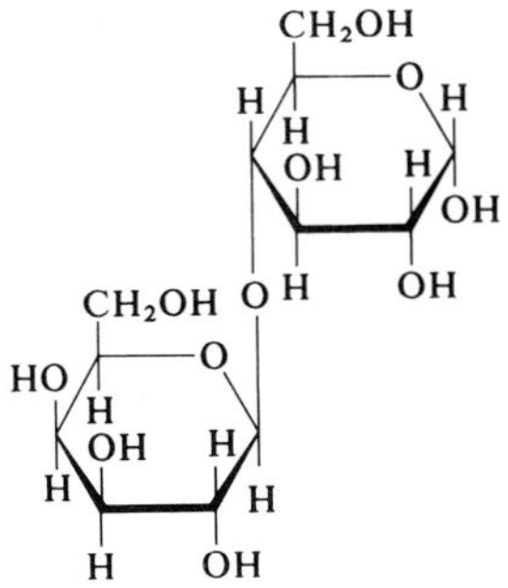

m2

L10

L9

m7

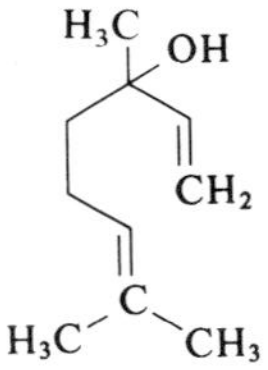

L8

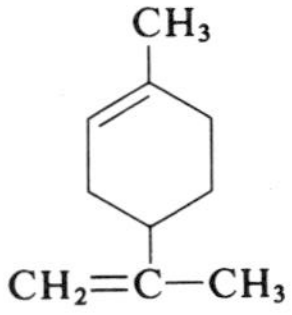

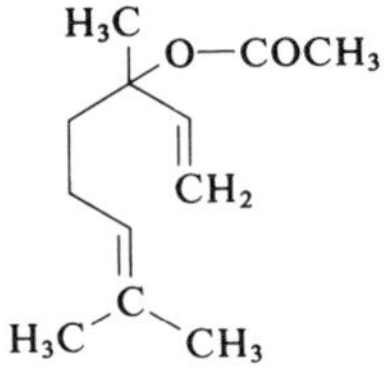

L6, L7

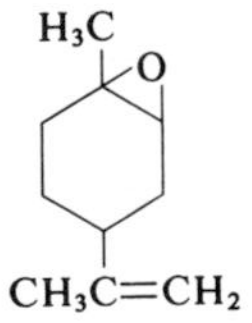

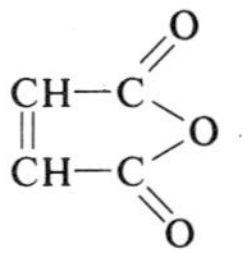

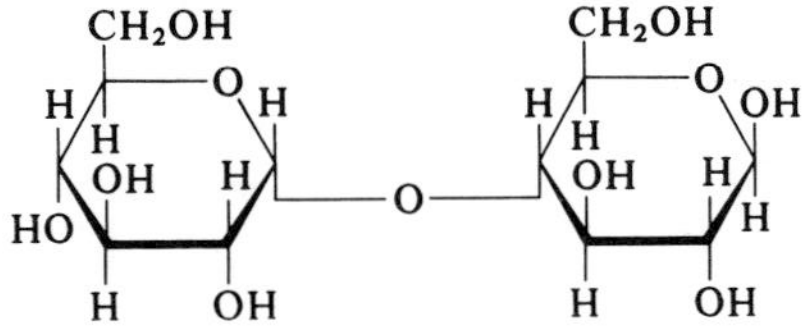

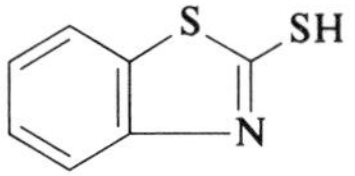

m17

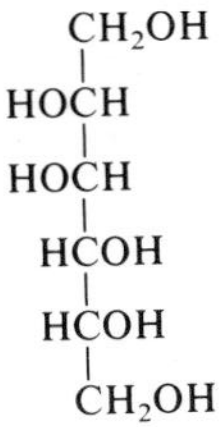

m15

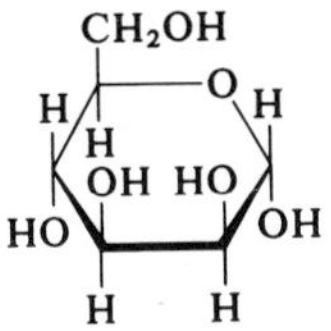

m13

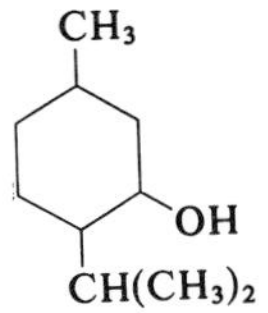

m12

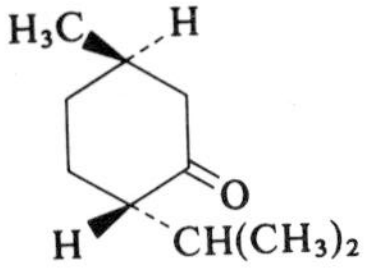

m11

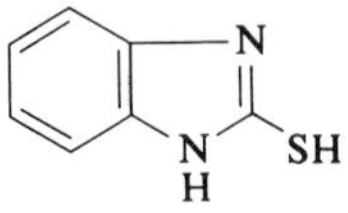

m10

L3

TABLE 1.15 Physical Constants of Organic Compounds (*Continued*)

No.	Name	Formula	Formula weight	Beilstein reference	Density	Refractive index	Melting point	Boiling point	Flash point	Solubility in 100 parts solvent
m19	2-Mercaptoethyltriethoxysilane	$HSCH_2CH_2Si(OC_2H_5)_3$	224.38		0.9884^{20}	1.432^{20}		210	104	misc alc; v s acet
m20	3-Mercapto-1,2-propanediol	$HSCH_2CH(OH)CH_2OH$	108.16	1, 519	1.2954^{14}	1.5243^{20}		118^{5mm}	>110	misc aq, alc, eth, acet
m21	2-Mercaptopropionic acid	$CH_3CH(SH)COOH$	106.14	3, 289	1.220^{15}	1.4809^{20}	10–14	102^{16mm}	87	misc aq, alc, eth
m21a	3-Mercaptopropionic acid	$HSCH_2CH_2CO_2H$	106.14	3, 299	1.218	1.4911^{20}	17–19	111^{15mm}	93	misc aq, alc, eth
m22	(3-Mercaptopropyl)trimethoxysilane	$HS(CH_2)_3Si(OCH_3)_3$	196.34		1.039^{20}	1.4416^{20}		93^{40mm}	48	
m23	Mercaptosuccinic acid	$HOOCCH_2CH(SH)COOH$	150.15	3, 439			152–154			50 aq; 50 alc; s eth
m24	Methacrylaldehyde	$H_2C=C(CH_3)CHO$	70.09	1, 731	0.8304^{20}	1.4160^{20}	−81	69	−15	6 aq; misc alc, eth
m25	Methacrylamide	$H_2C=C(CH_3)CONH$	85.11	2^2, 399			109–111			s alc; sl s eth
m26	Methacrylic acid	$H_2C=C(CH_3)COOH$	86.09	2, 421	1.0153^{20}	1.4314^{20}	16	163	76	9 aq; misc alc, eth
m26a	Methacrylic anhydride	$[H_2C=C(CH_3)CO]_2O$	154.17	2^3, 1293	1.035	1.4536^{20}		87^{13mm}	84	
m27	Methacrylonitrile	$H_2C=C(CH_3)CN$	67.91	2, 423	0.8001^{20}	1.4007^{20}	−35.8	90.3	13 (CC)	2.6 aq; misc acet, bz
m28	Methacryloyl chloride	$H_2C=C(CH_3)COCl$	104.54	2^2, 394	1.070	1.4447^{20}	−15	95–96	2	
m28a	Methallylidene diacetate	$(CH_3CO_2)_2CHC(CH_3)=CH_2$	172.18	2^4, 292	1.039	1.4245^{20}		191	83	
m29	Methane	CH_4	16.04	1, 56	0.4240^{bp} / 0.7168 g/L(g)		−182.5	−161.5		3.3 mL aq; 47 mL alc
m30	Methanesulfonic acid	CH_3SO_3H	96.10	4, 4	1.4812^{24}	1.4303^{20}	20	167^{10mm}	>110	1.5 bz; misc aq v s aq(dec)
m31	Methanesulfonic anhydride	$(CH_3SO_2)_2O$	174.19				71	138^{10mm}		
m32	Methanesulfonyl chloride	CH_3SO_2Cl	114.55	4, 5	1.4805^{18}	1.4518^{20}	−32	161	110	s alc, eth
m33	Methanethiol	CH_3SH	48.11	1, 288	0.8665^{20}		−123	6.0		2.3 aq; v s alc, eth
m34	Methanol	CH_3OH	32.04	1, 273	0.7913^{20}	1.3284^{20}	−97.7	64.7	12 (CC)	misc aq, alc, bz, chl, eth
m35	Methanol-*d*	CH_3OD	33.05	1^3, 1186	0.81277^{20}	1.3270^{20}	−110	65.5	11	misc aq, alc, eth
m36	Methanol-*d*₄	CD_3OD	36.07	1^3, 1187	0.888	1.3256^{20}		65.4	11	misc aq, alc, eth
m37	DL-Methionine	$CH_3SCH_2CH_2CH(NH_2)COOH$	149.21	4^2, 938	1.340		281 d			3 aq; i eth; v sl s alc
m38	Methoxyacetic acid	CH_3OCH_2COOH	90.08	3, 232	1.174	1.4158^{20}		202–204	>110	misc aq, alc, eth

No.	Name	Formula	Formula wt	Ref	Density	n_D	mp, °C	bp, °C	Flash	Solubility
m39	o-Methoxyacetophenone	$CH_3OC_6H_4COCH_3$	150.18	8, 85	1.090^{20}	1.5393^{20}		131^{18mm}	108	s aq
m40	m-Methoxyacetophenone	$CH_3OC_6H_4COCH_3$	150.18	8, 86	1.094	1.5410^{20}		239–241	110	v s alc, eth
m41	p-Methoxyacetophenone	$CH_3OC_6H_4COCH_3$	150.18	8, 87	1.0824^{1}	1.5335^{20}	36–38	154^{26mm}	>110	v s alc, eth
m42	2-Methoxyaniline	$CH_3OC_6H_4NH_2$	123.16	13, 358	1.098^{15}	1.5730^{20}	5	225	98	i aq; misc alc, eth
m43	3-Methoxyaniline	$CH_3OC_6H_4NH_2$	123.16	13, 404	1.096	1.5794^{20}	−10	251	>110	s alc, acid; sl s aq
m44	4-Methoxyaniline	$CH_3OC_6H_4NH_2$	123.16	13, 435	1.087		57	246	117	v s alc; sl s aq
m45	2-Methoxybenzaldehyde	$CH_3OC_6H_4CHO$	136.15	8, 43	1.127	1.560^{20}	35–36	236		sl s alc, bz; i eth
m45a	3-Methoxybenzaldehyde	$CH_3OC_6H_4CHO$	136.15	8, 59	1.119	1.5533^{20}		143^{50mm}	110	misc alc
m46	4-Methoxybenzaldehyde	$CH_3OC_6H_4CHO$	136.15	8, 67	1.119	1.5713^{20}	−1	248	108	s aq; v s alc; sl s eth
m47	4-Methoxybenzamide	$CH_3OC_6H_4CONH_2$	151.17	10^2, 100			164–167	295	108	1 aq; misc alc, eth
m48	Methoxybenzene	$C_6H_5OCH_3$	108.14	6, 138	0.9942^{20}	1.5170^{20}	−37.5	153.8	51	d aq; s alc, eth
m49	4-Methoxybenzenesulfonyl chloride	$CH_3OC_6H_4SO_2Cl$	206.65	11, 243			40–43		>110	
m50	2-Methoxybenzoic acid	$CH_3OC_6H_4COOH$	152.15	10, 64	1.180		100	200		0.5 aq; v s alc, eth
m51	3-Methoxybenzoic acid	$CH_3OC_6H_4COOH$	152.15	10, 137			104	172^{10mm}		s hot aq, alc, eth
m52	4-Methoxybenzoic acid	$CH_3OC_6H_4COOH$	152.15	10, 154	1.385^{4}		185	275–280		0.04 aq; v s alc, chl
m53	4-Methoxybenzoyl chloride	$CH_3OC_6H_4COCl$	170.60	10, 163		1.5810^{20}	22	145^{14mm}	87	i aq(dec); s alc(dec); s bz, acet
m54	4-Methoxybenzyl alcohol	$CH_3OC_6H_4CH_2OH$	138.17	6, 897	1.109^{25}	1.5442^{20}	23–25	259	>110	i aq; s alc, eth
m55	4-Methoxybenzylamine	$CH_3OC_6H_4CH_2NH_2$	137.18	13, 606	1.050^{15}	1.5462^{20}		236–237	>110	v s aq, alc, eth
m56	2-Methoxybiphenyl	$CH_3OC_6H_4C_6H_5$	184.24	6, 672	1.023	1.6105^{20}		274	>110	
m57	3-Methoxy-1-butanol	$CH_3OCH(CH_3)CH_2CH_2OH$	104.15		0.9229^{20}	1.4145^{20}	−85	161.1	46	misc aq
m58	4-Methoxy-3-buten-2-one	$CH_3OCH=CHCOCH_3$	100.12		0.982	1.4660^{20}		200	63	
m59	1-Methoxy-1-buten-3-yne	$CH_3OCH=CHC≡CH$	82.10		0.9064^{20}	1.4818^{20}		122–125	8	v s org solv

TABLE 1.15 Physical Constants of Organic Compounds (*Continued*)

No.	Name	Formula	Formula weight	Beilstein reference	Density	Refractive index	Melting point	Boiling point	Flash point	Solubility in 100 parts solvent
m60	2-Methoxycinnamaldehyde	$CH_3OC_6H_4CH{=}CHCHO$	162.19				44–48	$130^{0.6mm}$	>110	
m61	1-Methoxy-1,3-cyclohexadiene		110.16	6[3], 367	0.929	1.4885^{20}		40^{15mm}	26	
m62	1-Methoxy-1,4-cyclohexadiene		110.16	6[3], 367	0.940	1.4819^{20}		148–150	36	
m62a	2-Methoxydibenzofuran		198.22	17[3], 1590			42–45		110	
m63	7-Methoxy-3,7-dimethyloctanal	$(CH_3)_2C(OCH_3)CH_2CH_2CH(CH_3)CH_2CHO$	186.30		0.877	1.4374^{20}		$60^{0.45mm}$	98	misc aq
m64	2-Methoxy-1,3-dioxolane		104.11	19[4], 617	1.092	1.4091^{20}		129–130	31	misc aq, alc, bz, eth, ketones
m65	2-Methoxyethanol	$CH_3OCH_2CH_2OH$	76.10	1, 467	0.9646^{20}	1.4021^{20}	−85.1	124.6	46	misc aq
m66	2-(2-Methoxyethoxy)-ethanol	$CH_3OCH_2CH_2OCH_2CH_2OH$	120.15	6, 970	1.035^{20}_{4}	1.4264^{20}	−50	194.1	83	
m67	2-Methoxyethoxymethyl chloride	$CH_3OCH_2CH_2OCH_2Cl$	124.57		1.091	1.4270^{20}		50^{13mm}	>110	misc aq
m68	2-Methoxyethyl acetate	$CH_3COOCH_2CH_2OCH_3$	118.13	2, 141	1.0049^{20}	1.4022^{20}		144.5	43	
m68a	2-Methoxyethyl acetoacetate	$CH_3COCH_2CO_2CH_2CH_2OCH_3$	160.17		1.090	1.4339^{20}	−65.1	120^{20mm}	103	
m69	2-Methoxyethylamine	$CH_3OCH_2CH_2NH_2$	75.11	4[2], 718	0.864	1.4054^{20}		95	9	v s aq, alc
m70	1-Methoxy-2-indanol		164.20	6, 970				146^{11mm}	>110	
m71	2-Methoxy-5-methyl-aniline	$CH_3OC_6H_3(CH_3)NH_2$	137.18	13[2], 388		1.5482^{20}	52–54	235	>110	s aq; v s alc, bz, eth
m72	3-Methoxy-4-methyl-aniline	$CH_3OC_6H_3(CH_3)NH_2$	137.18	13, 574			51–54	250–252		
m73	4-Methoxy-2-methyl-aniline	$CH_3OC_6H_3(CH_3)NH_2$	137.18	13[2], 330	1.065	1.5647^{20}	13–14	248–249	>110	s alc
m74	2-Methoxy-4-methyl-phenol	$CH_3OC_6H_3(CH_3)OH$	138.17	6, 878	1.092	1.5372^{20}	5	222	99	
m75	1-Methoxy-2-methyl-propylene oxide	$(CH_3)_2C{-}CH(OCH_3)$ with O	102.13	17[3], 1035	0.904	1.3929^{20}		94	6	
m76	1-Methoxynaphthalene	$C_{10}H_7OCH_3$	158.20	6, 606	1.090	1.6220^{20}		135^{12mm}	>110	s bz, eth, CS_2
m77	2-Methoxynaphthalene	$C_{10}H_7OCH_3$	158.20	6, 640			72	272		

No.	Name	Formula	M.W.	Beilstein	Density	n_D	M.P.	B.P.	Flash	Solubility
m78	2-Methoxy-4-nitroaniline	$CH_3OC_6H_3(NO_2)NH_2$	168.15	13, 390			138–140			s alc, hot bz, HOAc
m79	2-Methoxy-5-nitroaniline	$CH_3OC_6H_3(NO_2)NH_2$	168.15	13, 389	1.207[156]		117–119			sl s aq; s alc, eth
m80	4-Methoxy-2-nitroaniline	$CH_3OC_6H_3(NO_2)NH_2$	168.15	13, 521			123–126			0.17 aq; s alc, eth
m81	2-Methoxynitrobenzene	$CH_3OC_6H_4NO_2$	153.14	6, 217	1.2527[20]	1.5619[20]	9.4	277	>110	i aq; v s alc, eth
m82	4-Methoxynitrobenzene	$CH_3OC_6H_4NO_2$	153.14	6, 230	1.233		54	260		
m83	4-Methoxy-3-nitrobenzoic acid	$CH_3OC_6H_3(NO_2)COOH$	197.15	10, 181			186–189			
m84	2-Methoxy-5-nitropyridine	$CH_3OC_5H_3N(NO_2)$	154.13	21[2], 33			108–109			
m85	4-Methoxy-2-nitrotoluene	$CH_3OC_6H_3(NO_2)CH_3$	167.16	6, 411	1.207	1.5525[20]	17	267	>110	
m86	p-Methoxyphenethylamine	$CH_3OC_6H_3CH_2CH_2NH_2$	151.21	13, 626	1.033	1.5379[20]		138[20mm]	>110	
m87	2-Methoxyphenol	$CH_3OC_6H_4OH$	124.14	6, 768	1.112(lg)	1.5429	28	205	82	1.5 aq; misc alc, eth
m88	3-Methoxyphenol	$CH_3OC_6H_4OH$	124.14	6, 813	1.131	1.5510[20]	<−17.5	115[5mm]	>110	misc alc, eth; sl s aq
m89	4-Methoxyphenol	$CH_3OC_6H_4OH$	124.14	6, 843			55–57	243	>110	v s bz; s alk
m90	3-(4-Methoxyphenoxy)-1,2-propanediol	$CH_3OC_6H_4OCH_2CH(OH)CH_2OH$	198.22	6[3], 4411			76–80			
m91	4-Methoxyphenylacetic acid	$CH_3OC_6H_4CH_2COOH$	166.18	10, 190			86–88	140[3mm]		i aq; v s alc; s eth

Methoxyethane, e174

2-Methoxyethoxychloromethane, m67

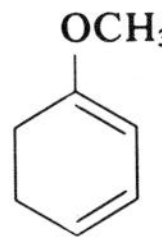

OCH_3 m61

OCH_3 m62

OCH_3 m62a

OCH_3 m64

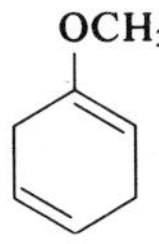

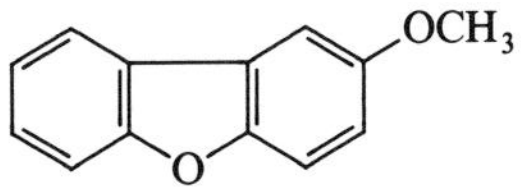

OH OCH_3 m70

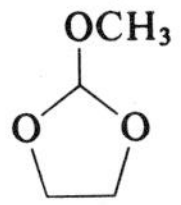

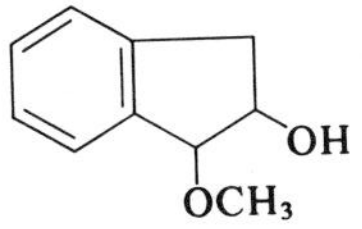

TABLE 1.15 Physical Constants of Organic Compounds (*Continued*)

No.	Name	Formula	Formula weight	Beilstein reference	Density	Refractive index	Melting point	Boiling point	Flash point	Solubility in 100 parts solvent
m92	o-Methoxyphenyl-acetone	$CH_3OC_6H_4CH_2COCH_3$	164.20	8³, 397	1.054	1.5250^{20}		130^{10mm}	>110	s alc, eth
m93	(o-Methoxyphenyl)-acetonitrile	$CH_3OC_6H_4CH_2CN$	147.18	10, 188			65–68	143^{15mm}		s hot bz
m93a	4-Methoxy-1,3-phenylenediamine	$CH_3OC_6H_3(NH_2)_2$	138.17	13¹, 204			66–68			
m94	2-Methoxy-p-phenylenediamine sulfate	$CH_3OC_6H_3(NH_2)_2 \cdot H_2SO_4$	236.26	13³, 1349			283 d			
m94a	3-Methoxy-1,2-propanediol	$CH_3OCH_2CH(OH)CH_2OH$	106.12	1, 512	1.114	1.4438^{20}		220	>110	
m95	1-Methoxy-2-propanol	$CH_3OCH_2CH(OH)CH_3$	90.1		0.9192^{20}_{20}	1.4021^{20}	−97	120.1	33	misc aq, acet, bz, eth
m96	2-Methoxypropene	$CH_3C(OCH_3){=}CH_2$	72.11	1, 435	0.753	1.3820^{20}		34–36	−29	
m97	trans-1-Methoxy-4-(1-propenyl)benzene	$CH_3OC_6H_4CH{=}CHCH_3$	148.21	6, 566	0.9883^{20}_{4}	1.5615^{20}	21.4	237	90	misc chl, eth; 50 alc; s bz, EtAc
m98	2-Methoxy-4-propenylphenol	$CH_3OC_6H_3(OH)CH{=}CHCH_3$	164.20	6, 955	1.087^{20}_{4}	1.5748^{20}	−10	266	>112	misc alc, eth; sl s aq
m99	2-Methoxy-4-(2-propenyl)phenol	$CH_3OC_6H_3(OH)CH_2CH{=}CH_2$	164.20	6, 961	1.0664^{20}	1.5408^{20}	−9.2	255	>112	misc alc, chl, eth; s HOAc, alk; i aq
m99a	3-Methoxypropionitrile	$CH_3OCH_2CH_2CN$	85.11	3¹, 113	0.937	1.4030^{20}		165	61	
m100	p-Methoxypropiophenone	$CH_3OC_6H_4COCH_2CH_3$	164.20	8, 103	1.071	1.5465^{20}	27–29	274	61	
m101	2-Methoxypyridine	$CH_3OC_5H_4N$	109.13	21, 44	1.038	1.5029^{29}		142	32	misc aq
m102	2-Methoxytetrahydrofuran		102.13	17⁴, 1019	0.972	1.4119^{20}		105–107	7	
m103	6-Methoxy-1,2,3,4-tetrahydronaphthalene		162.23	6², 537	1.033	1.5402^{20}		90^{1mm}	>110	
m104	6-Methoxy-1-tetralone		176.22	9², 889	0.9851^{15}	1.5161^{20}	77–79	171^{11mm}		i aq; v s alc, eth
m105	2-Methoxytoluene	$CH_3OC_6H_4CH_3$	122.17	6, 352	0.9697^{25}	1.5131^{20}		170–172	51	s alc, bz, eth; i aq
m106	3-Methoxytoluene	$CH_3OC_6H_4CH_3$	122.17	6, 376	0.969^{25}	1.5112^{20}		175–176	54	s alc, eth; i aq
m107	4-Methoxytoluene	$CH_3OC_6H_4CH_3$	122.17	6, 392				174	53	
m108	Methoxytrimethylsilane	$CH_3OSi(CH_3)_3$	104.23	4³, 1856	0.7560^{20}	1.3678^{20}		57–58	−30	

No.	Name	Formula	Mol wt	Beil. ref.	Density	n_D	mp, °C	bp, °C	Flash pt	Solubility
m109	Methoxytripropyl-silane	$CH_3OSi(C_3H_7)_3$	188.4		0.8222^{20}	1.428^{20}		83^{12mm}		s aq
m110	*N*-Methylacetamide	$CH_3CONHCH_3$	73.10	4, 58	0.9460^{35}	1.4253^{35}	30.6	206	108	24 aq; misc alc, eth
m111	Methyl acetate	CH_3COOCH_3	74.08	2, 224	0.9342^{20}	1.3619^{20}	−98	57	−10 (CC)	50 aq; misc alc
m112	Methyl acetoacetate	$CH_3COCH_2COOCH_3$	116.12	3, 632	1.0747^{20}	1.4186^{20}	−80	171.7	70	i aq; v s alc, eth
m113	*p*-Methylacetophenone	$CH_3C_6H_4COCH_3$	134.18	7, 307	1.0051	1.5328^{20}	22–24	226	92	6 aq; s alc, eth
m114	Methyl acrylate	$H_2C{=}CHCOOCH_3$	86.09	2, 399	0.9561^{20}	1.4117^{18}	−76.5	80.2	−3 (OC)	959 mL aq; 10.5 bz;
m115	Methylamine	CH_3NH_2	31.06	4, 32	0.699_4^{-11}		−93.5	−6.3	0 (CC)	s alc; misc eth
m115a	1-(Methylamino)-anthraquinone		237.26	14, 179			171			
m116	Methyl 2-amino-benzoate	$H_2NC_6H_4COOCH_3$	151.17	14, 317	1.168_4^{19}	1.5820^{20}	24	256	104	sl s aq; v s alc eth
m117	2-(*N*-Methylamino)-benzoic acid	$CH_3NHC_6H_4COOH$	151.17	14, 323			170–172 d			0.2 aq; s alc, eth
m118	Methyl 3-amino-crotonate	$CH_3C(NH_2){=}CHCOOCH_3$	115.13	3, 632			81–83			
m119	2-(Methylamino)-ethanol	$CH_3NHCH_2CH_2OH$	75.11	4, 276	0.937^{20}	1.4387^{20}		156	72	misc aq, alc, eth
m120	4-Methylaminophenol sulfate	$(CH_3NHC_6H_4OH)_2 \cdot H_2SO_4$	344.39	13, 441			260 d			4 aq; sl s alc; i eth
m121	Methyl 2-(amino-sulfonyl)benzoate	$H_2NSO_2C_6H_4CO_2CH_3$	215.23	11, 377			127			
m122	*N*-Methylaniline	$C_6H_5NHCH_3$	107.16	12, 135	0.989_4^{4}	1.5704^{20}	−57	196	78	sl s aq; s alc, eth

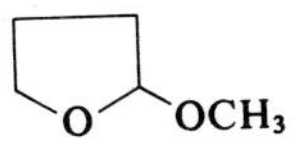

m102

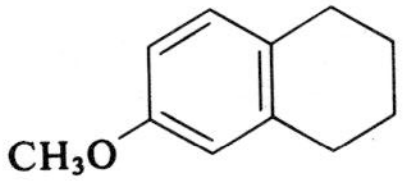

m103

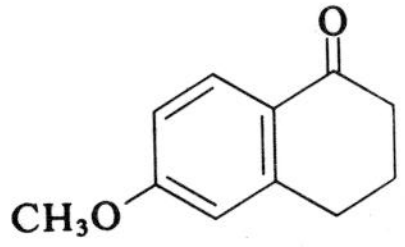

m104

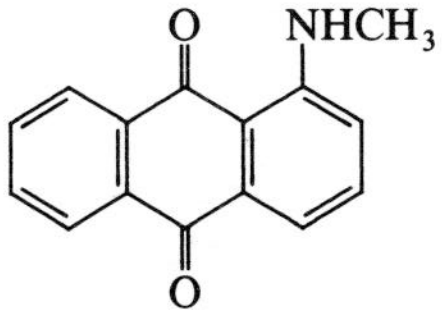

m115a

TABLE 1.15 Physical Constants of Organic Compounds (*Continued*)

No.	Name	Formula	Formula weight	Beilstein reference	Density	Refractive index	Melting point	Boiling point	Flash point	Solubility in 100 parts solvent
m123	N-Methylanilinium trifluoroacetate	$C_6H_5NHCH_3 \cdot HOOCCF_3$	221.18				65–66			
m124	2-Methylanthra-quinone		222.24	7, 809			177	subl		v s bz; s alc, eth
m125	Methylarsonic acid	$CH_3AsO(OH)_2$	139.96	4, 613			161			v s aq; s alc
m126	4-Methylbenzaldehyde	$CH_3C_6H_4CHO$	120.15	7, 297	1.0194^{17}	1.5447^{20}		205	80	misc alc, eth; sl s aq
m127	Methyl benzene-sulfonate	$C_6H_5SO_2OCH_3$	172.20	11^2, 20	1.2889^{4}	1.5151^{20}	−4	$154^{20\text{mm}}$		v s alc, chl, eth
m128	2-Methylbenzimid-azole		132.17	23, 145			176–177			s alk, hot aq; sl s alc
m129	Methyl benzoate	$C_6H_5COOCH_3$	136.15	9, 109	1.0933^{15}	1.5205^{15}	−12.1	199.5	82	0.2 aq; misc alc, eth
m130	2-Methylbenzoic acid	$CH_3C_6H_4COOH$	136.15	9, 462	1.062		107–108	259		sl s aq; v s alc
m131	3-Methylbenzoic acid	$CH_3C_6H_4COOH$	136.15	9, 475	1.054		111–113	263		0.09 aq; v s alc
m132	4-Methylbenzoic acid	$CH_3C_6H_4COOH$	136.15	9, 483			180–182	275		v s alc, eth
m133	2-Methylbenzo-phenone	$CH_3C_6H_4COC_6H_5$	196.25	7, 439	1.083	1.5958^{20}	< −18	309–311	>110	v s alc, org solv
m134	4-Methylbenzo-phenone	$CH_3C_6H_4COC_6H_5$	196.25	7, 440			59–60	326		v s bz, eth
m135	2-Methylbenzothiazole		149.22	27, 46	1.173	1.6170^{20}	12–14	238	102	s alc, HOAc; i aq
m136	2-Methylbenzotriazole		133.15	27, 46	1.121	1.5497^{20}	9–10	178	75	
m137	2-Methylbenzoxazole		133.15	27, 46	1.121	1.5497^{20}	8–10	178	75	
m137a	α-Methylbenzyl acetate	$CH_3CO_2CH(CH_3)C_6H_5$	164.20	6, 476	1.028	1.4945^{20}		$95^{12\text{mm}}$	91	v s alc; s bz, chl
m138	α-Methylbenzyl alcohol	$C_6H_5CH(CH_3)OH$	122.17	6, 475	1.0191^{13}	1.5265^{20}	20	$204^{745\text{mm}}$	85	
m139	3-Methylbenzyl alcohol	$CH_3C_6H_4CH_2OH$	122.17	6, 494	0.916^{17}	1.5334^{20}	< −20	217	105	5 aq; 5 alc, eth
m140	4-Methylbenzyl alcohol	$CH_3C_6H_4CH_2OH$	122.17	6, 498			59–61	217		s alc, eth; sl s aq
m141	(±)-α-Methylbenzyl-amine	$C_6H_5CH(CH_3)NH_2$	121.18	12, 1094	0.940	1.5254^{20}		185	79	
m142	4-Methylbenzylamine	$CH_3C_6H_4CH_2NH_2$	121.18	12, 1141	0.952	1.5340^{20}	12–13	195	75	
m143	Methyl bromoacetate	$BrCH_2COOCH_3$	152.98	2, 213	1.616	1.4586^{20}		$52^{15\text{mm}}$	62	
m144	(±)-Methyl 2-bromo-butyrate	$CH_3CH_2CH(Br)COOCH_3$	181.04	2, 282	1.573			$138^{50\text{mm}}$	68	s alc
m145	Methyl 4-bromo-crotonate	$BrCH_2CH{=}CHCOOCH_3$	179.02		1.522	1.4980^{20}		$85^{13\text{mm}}$	91	

No.	Name	Formula	MW	Beil. ref.	Density	n_D	mp	bp	Flash pt	Solubility
m146	Methyl 2-bromopropionate	$CH_3CH(Br)COOCH_3$	167.01	2, 253	1.497	1.5420^{20}		51^{19mm}	51	s alc
m147	2-Methyl-1,3-butadiene	$H_2C{=}C(CH_3)CH{=}CH_2$	68.12	1, 252	0.6812^{20}	1.4216^{20}	-146.0	34.1	-53	misc alc, eth
m148	3-Methyl-1,2-butadiene	$CH_3C(CH_3){=}C{=}CH_2$	68.12	1, 252	0.6944^{20}	1.4179^{20}	-113.6	40.9	-12	misc alc, eth
m149	2-Methylbutane	$CH_3CH_2CH(CH_3)_2$	72.15	1, 134	0.6197^{20}	1.3537^{20}	-159.9	27.9	-56	0.005 aq; misc alc
m150	2-Methyl-1-butanethiol	$CH_3CH_2CH(CH_3)CH_2SH$	104.22	1^2, 421	0.848	1.4465^{20}		119.0	19	s alc, eth; i aq
m151	2-Methyl-2-butanethiol	$CH_3CH_2C(CH_3)_2SH$	104.22	1^1, 196	0.842	1.4385^{20}	-103.9	99.1	-1	s alc, eth; i aq
m152	3-Methyl-1-butanethiol	$(CH_3)_2CHCH_2CH_2SH$	104.22	1, 405	0.835_4	1.4432^{20}	-133.5	118.4	18	misc alc, chl, eth
m153	2-Methyl-1-butanol	$CH_3CH_2CH(CH_3)CH_2OH$	88.15	1, 388	0.816_4	1.4100^{20}	< -70	128	50 (OC)	3 aq; misc alc, eth
m154	2-Methyl-2-butanol	$CH_3CH_2C(CH_3)_2OH$	88.15	1, 388	0.8090^{20}	1.4050^{20}	-9.0	102.0	21	11 aq; misc alc, bz, chl, eth
m155	3-Methyl-1-butanol	$(CH_3)_2CHCH_2CH_2OH$	88.15	1, 392	0.8129_4	1.4085^{15}	-117.2	132.0	45	2 aq; misc alc, bz, chl, eth, PE, HOAc
m156	3-Methyl-2-butanol	$(CH_3)_2CHCH(OH)CH_3$	88.15	1, 391	0.8179^{20}	1.4091^{20}		113–114	39 (CC)	2.8 aq; misc alc, eth
m157	3-Methyl-2-butanone	$(CH_3)_2CHCOCH_3$	86.13	1, 682	0.802^{20}	1.3890	-92	94–95	6	misc alc, eth
m157a	3-Methyl-2-butenal	$(CH_3)_2C{=}CHCHO$	84.12	1^3, 2990	0.872	1.4613^{20}		133–135	33	
m158	2-Methyl-1-butene	$CH_3CH_2C(CH_3){=}CH_2$	70.14	1, 211	0.6504^{20}	1.3777^{20}	-137.6	31.2	-34	misc alc, eth

Methyl o-anisate, m301b
Methyl p-anisate, m301c
2-Methyl-p-anisidine, m71
4-Methyl-m-anisidine, m72
5-Methyl-o-anisidine, m73
Methylanisoles, m105, m106, m107
Methyl anthranilate, m116

Methylanthranilic acid, a214, a215
N-Methylanthranilic acid, m117
Methylbenzene, t170
4-Methylbenzenesulfonic acid, t179
Methyl benzilate, m231a
α-Methylbenzyl alcohol, p114
N-Methylbenzylamine, b104

Methylbenzyl bromide, b371, b372
Methylbenzyl chlorides, c259, c260, c261
Methylbis(2-chloroethoxy)silane, b158
N-Methylbis(2-chloroethyl)amine, b160
Methyl bromide, b303
3-Methyl-1-buten-1-carboxylic acid, m352

m124

m128

m135

m136

m137

TABLE 1.15 Physical Constants of Organic Compounds (*Continued*)

No.	Name	Formula	Formula weight	Beilstein reference	Density	Refractive index	Melting point	Boiling point	Flash point	Solubility in 100 parts solvent
m159	2-Methyl-2-butene	CH$_3$CH=C(CH$_3$)$_2$	70.14	1, 211	0.6620^{20}	1.3878^{20}	−133.8	38.6	−45	misc alc, eth; i aq
m160	3-Methyl-1-butene	(CH$_3$)$_2$CHCH=CH$_2$	70.14	1, 213	0.6272^{20}	1.3638^{20}	−168.5	20.1	−56	misc alc, eth
m161	(E)-2-Methyl-2-butenoic acid	CH$_3$CH=C(CH$_3$)COOH	100.12	2, 430	0.969	1.4342^{81}	64	198.5		s alc, eth; v s hot aq
m162	(Z)-2-Methyl-2-butenoic acid	CH$_3$CH=C(CH$_3$)COOH	100.12	2, 428	0.9834^{7}	1.4437^{47}	45	185		s alc, eth; v s hot aq
m163	3-Methyl-2-butenoic acid	(CH$_3$)$_2$C=CHCOOH	100.12	2, 432	1.006^{24}		69	194–195		s aq, alc, eth
m164	2-Methyl-3-buten-2-ol	(CH$_3$)$_2$C(OH)CH=CH$_2$	86.13	1, 444	0.8672^{20}	1.4160^{20}	2.6	98–99	13	
m164a	3-Methyl-2-buten-1-ol	(CH$_3$)$_2$C=CHCH$_2$OH	86.13	1, 444	0.848	1.4412^{20}		140	43	
m165	3-Methyl-3-buten-1-ol	H$_2$C=C(CH$_3$)CH$_2$CH$_2$OH	86.13		0.853	1.4337^{20}			36	
m166	2-Methyl-1-buten-3-yne	H$_2$C=C(CH$_3$)C=CH	66.10	1^1, 126		1.4140^{20}	−113	32	−6	
m168	N-Methylbutylamine	CH$_3$CH$_2$CH$_2$CH$_2$NHCH$_3$	87.17	4, 157	0.736	1.3995^{20}	−75	91	1	misc aq, alc, eth
m169	1-Methylbutylamine	CH$_3$CH$_2$CH$_2$CH(CH$_3$)NH$_2$	87.17	4, 177	0.7384^{20}	1.4029^{20}		91	35	
m170	2-Methylbutylamine	CH$_3$CH$_2$CH(CH$_3$)CH$_2$NH$_2$	87.17	4^3, 342	0.738	1.4116^{20}		94–97	3	misc alc, eth
m171	3-Methylbutyl 3-methylbutyrate	(CH$_3$)$_2$CHCH$_2$CH$_2$OOCCH$_2$CH(CH$_3$)$_2$	172.27	2, 312	0.8541^{25}	1.4100^{25}		194.0		misc alc, eth
m172	3-Methyl-1-butyne	(CH$_3$)$_2$CHC=CH	68.12	1, 251	0.6662^{20}	1.3740^{20}	−89.8	26.4		misc alc, eth
m173	2-Methyl-3-butyne-2-ol	(CH$_3$)$_2$C(OH)C=CH	84.12	1^1, 235	0.8672^{20}	1.4209^{20}	2.6	104	25	misc aq, acet, bz
m174	2-Methylbutyraldehyde	CH$_3$CH$_2$CH(CH$_3$)CHO	86.13	1^1, 352	0.804	1.3919^{20}	−51	90–92	4	misc alc, eth; sl s aq
m175	3-Methylbutyraldehyde	(CH$_3$)$_2$CHCH$_2$CHO	86.13	1, 684	0.7852^{20}	1.3882^{20}		92–93	19	misc alc, eth; sl s aq
m176	Methyl butyrate	CH$_3$CH$_2$CH$_2$COOCH$_3$	102.13	2, 270	0.8982^{4}	1.3879^{20}	~ −95	102	14	1.4 aq; misc alc, eth
m177	2-Methylbutyric acid	CH$_3$CH$_2$CH(CH$_3$)COOH	102.13	2, 305	0.936	1.4055^{20}		176.5	73	4 aq; s alc, chl, eth
m178	3-Methylbutyric acid	(CH$_3$)$_2$CHCH$_2$COOH	102.13	2, 309	0.9308^{20}	1.4033^{20}	−30.0	176.5	70	misc alc, eth
m179	3-Methylbutyronitrile	(CH$_3$)$_2$CHCH$_2$CN	83.13	2^2, 278	0.7925^{19}	1.3927^{20}	−101	129		dec aq, alc; s eth
m180	3-Methylbutyryl chloride	(CH$_3$)$_2$CHCH$_2$COCl	120.58	2, 315	0.9834^{20}	1.4161^{20}		115–117	18	
m181	1-(3-Methylbutyryl)-pyrrolidine		155.24		0.938	1.4710^{20}			104	
m182	Methyl carbamate	H$_2$NCOOCH$_3$	75.07	3, 21	1.1364^{56}		56–58	177		220 aq; 73 alc; s eth
m183	Methyl chloroacetate	ClCH$_2$COOCH$_3$	108.52	2, 197	1.2382^{20}	1.4220^{20}	−33	130–132	51	i aq; misc alc, eth

No.	Name	Formula	M.W.	Solubility refs	Density	n_D	m.p.	b.p.	Flash	Solubility
m184	Methyl 2-chloroacetoacetate	$CH_3COCH(Cl)COOCH_3$	150.56		1.236	1.4465^{20}	−32.7	137	71	
m184a	Methyl 4-chloroacetoacetate	$ClCH_2COCH_2CO_2CH_3$	150.56	3^2, 426	1.305	1.4564^{20}		85^{4mm}	102	
m185	Methyl 3-chlorobenzoate	$ClC_6H_4COOCH_3$	170.60	9, 338	1.227	1.4923^{20}	21	101^{12mm}	104	s alc
m186	Methyl 4-chlorobenzoate	$ClC_6H_4COOCH_3$	170.60	9, 340	1.382^{20}	1.4321^{20}	44	176	106	v s eth; s alc, acet
m187	Methyl 4-chlorobutyrate	$ClCH_2CH_2CH_2COOCH_3$	136.58	2, 278	1.1268^{14}			176	59	misc alc, bz, chl, eth
m188	Methyl chloroformate	$ClCO_2CH_3$	94.50	3, 9	1.223^{20}	1.3865^{20}		71	17	
m189	Methyl 3-(chloroformyl)propionate	$CH_3OOCCH_2CH_2COCl$	150.56	2^2, 553	1.223	1.4402^{20}		65^{3mm}	73	
m190	Methyl 2-chloropropionate	$CH_3CH(Cl)COOCH_3$	122.55	2, 248	1.075	1.4193^{20}		132–133	38	s alc
m190a	Methyl 2-(chlorosulfonyl)benzoate	$ClSO_2C_6H_4CO_2CH_3$	234.66	11, 373	1.04071^{17}		59–64			
m191	2-Methylcinnamaldehyde	$C_6H_5CH=C(CH_3)CHO$	146.19	7, 369		1.6045^{20}		149^{27mm}	79	
m191a	(E)-Methyl cinnamate	$C_6H_5CH=CHCO_2CH_3$	162.19	9, 581			36–38	262	>110	
m192	6-Methylcoumarin		160.17	17, 337			75–76	303^{725mm}		
m193	Methyl crotonate	$CH_3CH=CHCOOCH_3$	100.12	2, 410	0.9444^{20}_{4}	1.4242^{20}		121	4	v s alc, eth; i aq
m194	Methyl cyanoacetate	$NCCH_2COOCH_3$	99.09	2, 584	1.1225^{25}	1.4166^{25}	−13.1	205.1	110	misc alc, eth

(Z)-2-Methyl-2-butenedioic acid, c271
Methyl 2-buten-1-oate, m193
3-Methylbutyl acetate, i80
2-Methylbutylamine, a254
Methyl tert-butyl ether, b463
2-Methylbutylisovalerate, m171
Methyl tert-butyl ketone, h72

Methyl caprate, m219
Methyl caproate, m266
Methyl caprylate, m332
Methyl carbazate, m272
Methyl carbitol, m66
4-Methylcatechol, d390

Methyl cellosolve, m65
Methyl cellosolve acetate, m68
β-Methylchalcone, d660
Methyl chlorocarbonate, m188
Methyl chloroform, t230
(E)-2-Methylcrotonic acid, m161

$$O=C\!-\!CH_2CH(CH_3)_2$$

m181

m192

TABLE 1.15 Physical Constants of Organic Compounds (*Continued*)

No.	Name	Formula	Formula weight	Beilstein reference	Density	Refractive index	Melting point	Boiling point	Flash point	Solubility in 100 parts solvent
m195	Methylcyclohexane	$C_6H_{11}CH_3$	98.19	5, 29	0.7694^{20}	1.4231^{20}	−126.6	100.9	−3	i aq; s alc, eth
m196	Methyl cyclohexanecarboxylate	$C_6H_{11}COOCH_3$	142.20	9^1, 5	0.9954_4^{20}	1.4445^{20}		183	60	misc alc, eth
m197	4-Methyl-1,2-cyclohexanedicarboxylic anhydride		168.19		1.162	1.4774^{20}				
m198	1-Methylcyclohexanol	$C_6H_{10}(CH_3)OH$	114.19	6, 11	0.9251^{25}	1.4587^{25}	26	168	67	i aq; s bz, chl
m199	(Z)-2-Methylcyclohexanol	$C_6H_{10}(CH_3)OH$	114.19	6^2, 17	0.9340_4^{20}	1.4654^{20}	7	165	58	misc alc, eth
m200	(E)-2-Methylcyclohexanol	$C_6H_{10}(CH_3)OH$	114.19	6, 11	0.9247_4^{20}	1.4616^{20}	−21	167.5	58	misc alc; s eth
m201	(Z)-3-Methylcyclohexanol	$C_6H_{10}(CH_3)OH$	114.19	6, 12	0.9155^{20}	1.4572^{20}	−6	94	62	misc alc, eth
m202	(E)-3-Methylcyclohexanol	$C_6H_{10}(CH_3)OH$	114.19	6, 12	0.9214^{20}	1.4580^{20}	−1	84	62	
m203	(Z)-4-methylcyclohexanol	$CH_3C_6H_{10}OH$	114.19	6, 14	0.9122_4^{20}	1.4614^{20}		173	70	misc alc, eth
m204	(E)-4-Methylcyclohexanol	$CH_3C_6H_{10}OH$	114.19	6, 14	0.9118_4^{21}	1.4559^{20}		175	70	misc alc; s eth
m205	2-Methylcyclohexanone	$CH_3C_6H_9(=O)$	112.17	7, 14	$0.9252_4^{}$	1.4478^{20}		162	46 (CC)	i aq; s alc, eth
m206	3-Methylcyclohexanone	$CH_3C_6H_9(=O)$	112.17	7, 15	0.9155_4^{20}	1.4460^{20}		169	51	i aq; s alc, eth
m207	4-Methylcyclohexanone	$CH_3C_6H_9(=O)$	112.17	7, 18	0.9162_4^{20}	1.4455^{20}		171	40	i aq; s alc, eth
m208	1-Methyl-1-cyclohexene		96.17	5, 66	0.8092^{20}	1.4502^{20}	−121	111	−3	i aq; s alc, eth
m209	4-Methyl-1-cyclohexene		96.17	5, 67	0.799	1.4412^{20}	−115.5	102	−1	i aq; s alc, eth
m210	N-Methylcyclohexylamine	$C_6H_{11}NHCH_3$	113.20	12, 6	0.868	1.4560^{20}		149	29	
m211	3-Methylcyclohexylamine	$C_6H_{10}(CH_3)NH_2$	113.20	12, 10	0.855	1.4525^{20}		150^{730mm}	22	
m212	4-Methylcyclohexylamine	$C_6H_{10}(CH_3)NH_2$	113.20	12, 12	0.855	1.4531^{20}		151–154	26	

No.	Name	Formula	Formula wt.	Beilstein ref.	Density	n_D	M.p., °C	B.p., °C		Solubility
m213	Methylcyclopentadiene dimer		160.26	5^4, 1435	0.941	1.4976^{20}	−51	200	26	0.013 aq
m214	Methylcyclopentane	$C_5H_9CH_3$	84.16	5, 27	0.7487^{20}	1.4097^{20}	−142.4	71.8	−23	s aq; v s alc, eth
m215	3-Methyl-1,2-cyclopentanedione		112.13	7, 310			105−107			
m216	2-Methylcyclopentanone		98.15	7^2, 13	0.9200^{20}	1.4347^{20}	−76	139	26	
m217	3-Methyl-2-cyclopenten-1-one		96.13	7^1, 46	0.971	1.4780^{20}		74^{15mm}	65	
m217a	Methyl cyclopropanecarboxylate	$C_3H_5CO_2CH_3$	100.12	9^1, 3	0.985	1.4181^{20}		119	17	
m217b	1-Methylcyclopropanecarboxylic acid	$CH_3C_3H_4CO_2H$	100.12	9^2, 5			35	184	84	
m218	Methylcyclopropane carboxylic acid		100.12	9, 6	1.027	1.4395^{20}		191^{745mm}	87	
m219	Methyl decanoate	$CH_3(CH_2)_8COOCH_3$	186.30	2, 356	0.873	1.4255^{25}	−18	223	94	i aq; misc alc, eth
m220	Methyl dichloroacetate	$Cl_2CHCOOCH_3$	142.97	2, 203	1.3808^{19}	1.4421^{20}	−52	143	80	i aq; s alc
m221	Methyl 2,2-dichloro-1-methylcyclopropanecarboxylate		183.03		1.245	1.4639^{20}		74^{8mm}	74	
m222	Methyl 2,3-dichloropropionate	$ClCH_2CH(Cl)COOCH_3$	157.00	2^1, 111	1.3282^{24}	1.4447^{20}		92^{50mm}	42	s alc

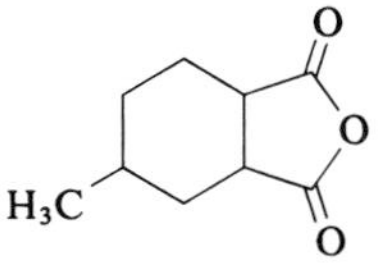

m197

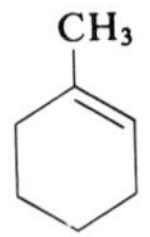

m208 (CH3)

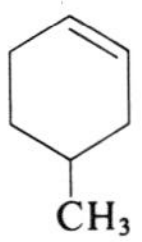

m209 (CH3)

m213 (CH3, CH3)

m215 (CH3, O)

m216 (CH3, O)

m217 (H_3C—, O)

m218 (H_3C—COOH)

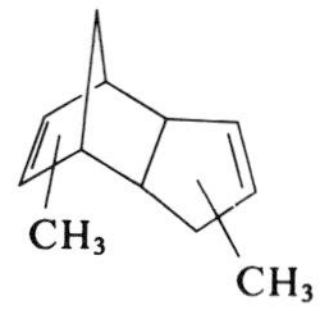

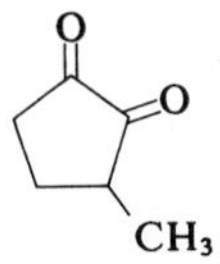

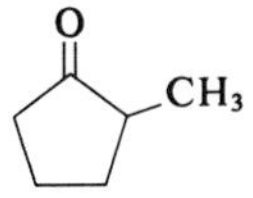

m221 (H_3C—COOCH3, Cl, Cl)

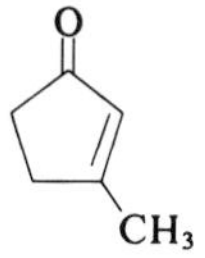

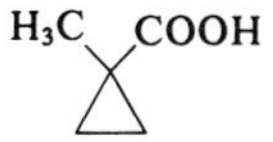

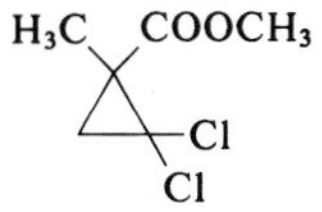

TABLE 1.15 Physical Constants of Organic Compounds (*Continued*)

No.	Name	Formula	Formula weight	Beilstein reference	Density	Refractive index	Melting point	Boiling point	Flash point	Solubility in 100 parts solvent
m224	N-Methyldiethanolamine	$CH_3N(CH_2CH_2OH)_2$	119.16	4, 284	1.0377^{20}	1.4685^{20}		248	126	misc aq alc
m225	O-Methyl-N,N'-diisopropylurea	$(CH_3)_2CHNHC(OCH_3){=}N{-}CH(CH_3)_2$	158.25		0.871	1.4358^{20}		$51^{0.1mm}$	35	
m226	Methyl 3,4-dimethoxybenzoate	$(CH_3O)_2C_6H_3CO_2CH_3$	196.20	10, 396			57–60	283		
m226a	Methyl 3,5-dimethoxybenzoate	$(CH_3O)_2C_6H_3CO_2CH_3$	196.20	10, 405			43	298	>110	
m227	Methyl 4,5-dimethoxy-2-nitrobenzoate	$(CH_3O)_2C_6H_2(NO_2)COOCH_3$	241.20	10, 403			141–144			
m227a	Methyl 3-(dimethylamino)propionate	$(CH_3)_2NCH_2CH_2CO_2CH_3$	131.18	4, 403	0.917	1.4184^{20}		154	51	
m227b	Methyl 2,5-dimethyl-3-furoate		154.17	18, 298	1.037	1.4750^{20}		198	80	
m228	Methyl 2,2-dimethyl-propionate	$(CH_3)_3CCOOCH_3$	116.16	21, 139	0.873	1.3880^{20}		101–103	−1	misc alc, eth; sl s aq
m229	2-Methyl-3,5-dinitrobenzoic acid	$CH_3C_6H_2(NO_2)_2COOH$	226.15	9, 474			205–207			
m230	N-Methyldioctylamine	$(C_8H_{17})_2NCH_3$	255.49	4, 381	1.066	1.4424^{20}	−30.1	165^{15mm}	>110	i aq; s alc, eth
m231	N-Methyldiphenylamine	$(C_6H_5)_2NCH_3$	183.26	12, 180	1.0484^{20}	1.6193^{20}	−7.6	135^{6mm}		
m231a	Methyl diphenylglycolate	$(C_6H_5)_2C(OH)CO_2CH_3$	242.27	10, 344			76	187^{13mm}		
m232	Methyleneaminoacetonitrile	$CH_2{=}NCH_2CN$	68.08				129			
m233	N,N'-Methylenebisacrylamide	$H_2C{=}CHCONHCH_2NH{-}COCH{=}CH_2$	154.17				>300			
m234	2,2'-Methylenebis(4-chlorophenol)	$CH_2[C_6H_3(Cl)OH]_2$	269.13				177			100 EtOH; 100 eth; s PE
m235	4,4'-Methylenebis(N,N-dimethylaniline)	$CH_2[C_6H_4N(CH_3)_2]_2$	254.38	13, 239			90			

m236	4,4'-Methylenebis(3-hydroxy-2-naphthoic) acid	CH₂[C₁₀H₅(OH)COOH]₂	388.38	10, 575			d>280			i aq, alc, eth, bz; sl s chl; s pyr
m237	1,1'-Methylenebis(3-methylpiperidine)	CH₂[CH₃C₅H₉N]₂	210.37		0.887	1.4734²⁰		160⁵⁰ᵐᵐ	110	
m238	Methylene blue		373.90	27, 393			190 d			4 aq; 1.3 alc; s chl 221
m239	4-4'Methylenedianiline	CH₂(C₆H₄NH₂)₂	198.26	13, 238			92			
m240	3,4-Methylenedioxybenzaldehyde		150.13	19, 115		1.5398	37	264	>110	0.2 aq; v s alc, eth
m241	1,2-Methylenedioxybenzene		122.12	19, 20	1.064			173		
m242	3,4-Methylenedioxybenzoic acid		166.13	19, 269			229	sub 210	55	sl s aq, chl, alc, eth
m243	3,4-Methylenedioxybenzyl alcohol		152.14	19, 67			53–55			

Methyl 4,6-dimethyl-2-oxo-2H-pyran-5-carboxylate, m289
Methyldinitrophenol, d633, d633a

Methyl enanthate, m262
Methylene bromide, d88
Methylene bromochloride, b258

Methylene chloride, d190
4,4'-Methylenedianiline, d35
Methylene dimethyl ether, d442

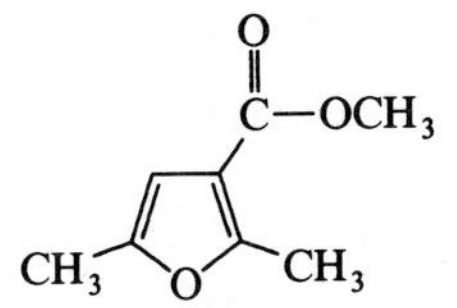

m227b

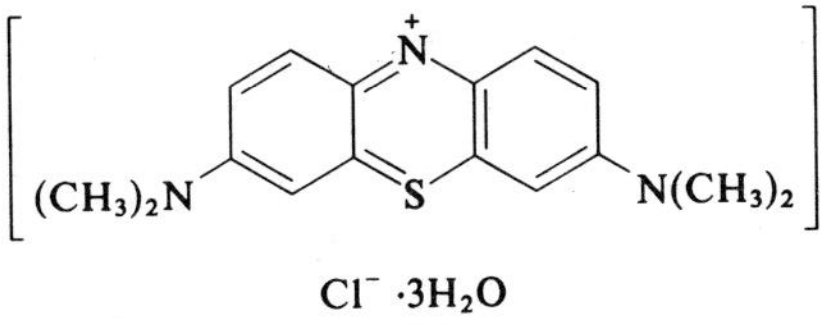

m238

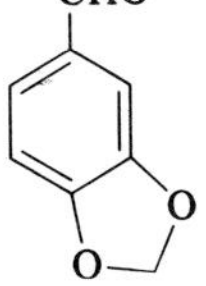

m240

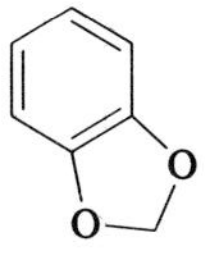

m241

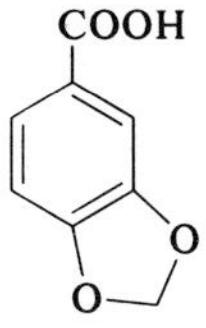

m242

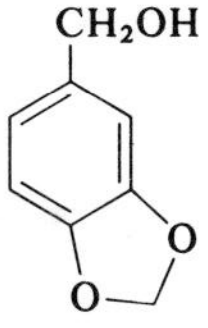

m243

TABLE 1.15 Physical Constants of Organic Compounds (*Continued*)

No.	Name	Formula	Formula weight	Beilstein reference	Density	Refractive index	Melting point	Boiling point	Flash point	Solubility in 100 parts solvent
m244	3,4-Methylenedioxy-6-propylbenzyldiethyleneglycol butyl ether		338.45	19^3, 779	1.059	1.498^{20}		$180^{1\text{mm}}$	171	misc alc, bz, geons
m245	5-Methylene-2-norbornene		106.17		0.981	1.4819^{20}			4	
m246	Methylenesuccinic acid	$H_2C{=}C(COOH)CH_2COOH$	130.10	2, 760	1.573		167			8.2 aq; 20 alc; v s bz, chl, eth, PE
m248	*N*-Methylethylenediamine	$CH_3NHCH_2CH_2NH_2$	74.13	4^1, 415	0.841	1.4395^{20}		114–116	41	misc aq
m249	*N*-Methylformamide	$HCONHCH_3$	59.07	4, 58	0.9988^{25}	1.4300^{25}	−40	185	98	23 aq; misc alc
m250	*N*-Methylformanilide	$C_6H_5N(CH_3)CHO$	135.17	12, 234	1.095	1.5593^{20}	8–13	244	126	s aq; v s alc; misc eth
m251	Methyl formate	$HCOOCH_3$	60.05	2, 18	0.9815^{15}	1.3451^{15}	−99	31.5	−19 (CC)	0.3 aq
m252	5-Methylfuraldehyde		110.11	17, 289	1.1072^{4}	1.5263^{20}		187	72	s alc, eth; sl s aq
m253	2-Methylfuran		82.10	17, 36	0.9152^{0}	1.4332^{20}	−88	63–66	−22	
m254	Methyl furoate		126.11	18, 274	1.179^{20}	1.4862^{20}		181	73	
m255	Methylgermanium tribromide	CH_3GeBr_3	327.35		2.6337^{4}	1.5770^{20}		168		
m256	*N*-Methyl-D-glucamine		195.22	31, 179			128–129			100 aq; 1.2 alc
m257	Methyl-α-D-glucopyranoside		194.19		1.46^{30}		168	$200^{0.2\text{mm}}$		63 aq; i alc, eth
m258	DL-2-Methylglutaronitrile	$NCCH_2CH(CH_3)CN$	108.14	2, 656	0.950			$125^{10\text{mm}}$	126	42 aq; sl s alc
m259	*N*-Methylglycine	CH_3NHCH_2COOH	89.09	4, 345			d 212			s aq; misc alc, eth
m260	Methyl glycolate	$HOCH_2COOCH_3$	90.08	3, 236	1.168^{18}		74	151		
m261	Methyl heptafluorobutyrate	$CF_3CF_2CF_2CO_2CH_3$	228.07	2^4, 812	1.472	1.2930^{20}		81	>110	
m262	Methyl heptanoate	$CH_3(CH_2)_5COOCH_3$	144.22	2, 339	0.8815^{20}	1.4115^{20}	−55.8	173.8	52	s alc, eth; sl s aq
m262a	6-Methyl-2-heptanol	$(CH_3)_2CH(CH_2)_3CH(OH)CH_3$	130.23	1, 421	0.803	1.4240^{20}		172	67	
m262b	5-Methyl-3-heptanone	$C_2H_5CH(CH_3)CH_2COC_2H_5$	128.22	1^1, 363	0.823	1.4142^{20}		157–162	43	
m263	6-Methyl-5-hepten-2-one	$(CH_3)_2C{=}CHCH_2CH_2COCH_3$	126.20	1^2, 797	0.8554^{16}	1.4392^{20}	−67	$73^{18\text{mm}}$	50	misc alc, eth
m264	Methyl hexadecanoate	$CH_3(CH_2)_{14}COOCH_3$	270.46	2, 372	0.852	1.4512^{20}	32–34	$196^{15\text{mm}}$	>110	s alc, chl, eth
m266	Methyl hexanoate	$CH_3(CH_2)_4COOCH_3$	130.19	2, 323	0.9038^{4}	1.4038^{23}	−71	151	45	v s alc, eth

m267	5-Methyl-2-hexanol	$(CH_3)_2CHCH_2CH_2CH(OH)CH_3$	116.20	1, 416	0.8144^{20}	1.4176^{20}		150	46	s alc, eth; i aq
m268	5-Methyl-2-hexanone	$(CH_3)_2CHCH_2CH_2COCH_3$	114.19	$1^2, 756$	0.8882^{20}	1.4062^{20}	-73.9	145	41	0.5 aq; misc alc, eth
m269	5-Methyl-3-hexen-2-one	$(CH_3)_2CHCH{=}CHCOCH_3$	112.17		0.850	1.4400^{20}			47	
m269a	1-Methylhexylamine	$H(CH_2)_5CH(NH_2)CH$	115.22	4, 194	0.7665^{18}	1.4175^{20}		144	54	sl s aq; s alc, eth
m270	1-Methylhydanotoin		114.10	24, 244			157	subl		s aq, alc; 3 eth
m271	Methylhydrazine	CH_3NHNH_2	46.07	$4^2, 957$	0.866	1.4235^{20}	-52.4	87.5	21	misc aq, alc; s PE
m272	Methyl hydrazinocarboxylate	$H_2NNHCOOCH_3$	90.08	$3^1, 46$			70–73	108^{12mm}		
m273	Methyl hydrogen glutarate	$HOOCCH_2CH_2COOCH_3$	146.14	$2^2, 565$	1.169	1.4381^{20}		151^{10mm}	>110	s alc
m274	Methyl hydrogen hexanedioate	$HOOC(CH_2)_4COOCH_3$	160.17	2, 652	1.081	1.4401^{20}	8–9	162^{10mm}	>110	

1,1'-Methylenedipiperidine, d697
Methylene iodide, d405
β-Methylene-β-propiolactone, d423
Methyl ethyl ketone, b396

Methyl fluoroform, t296
Methyl 2-furancarboxylate, m254
5-Methylfurfural, m252
Methyl gallate, m441a

α-Methyl-D-glucopyranoside, m257
N-Methyl guanidine acetic acid, c278
4-Methylhexahydrophthalic anhydride, m197

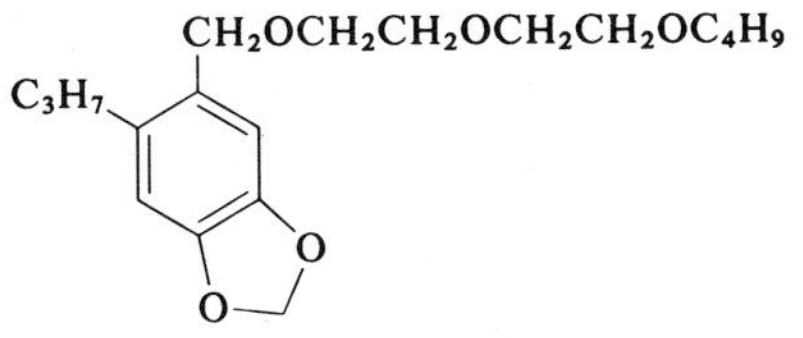

$CH_2OCH_2CH_2OCH_2CH_2OC_4H_9$

m244

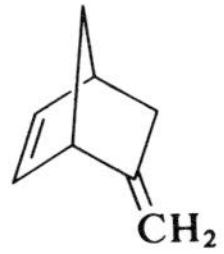

m245

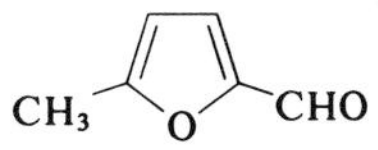

$CH_3NHCH_2{-}C{-}C{-}C{-}C{-}CH_2OH$

m256

m252

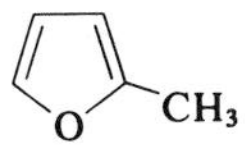

m253

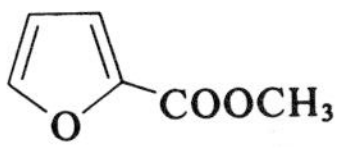

m254

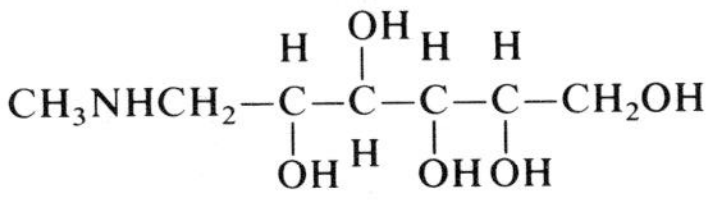

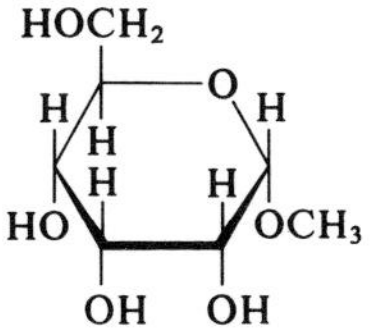

m257

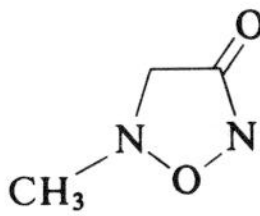

m270

TABLE 1.15 Physical Constants of Organic Compounds (*Continued*)

No.	Name	Formula	Formula weight	Beilstein reference	Density	Refractive index	Melting point	Boiling point	Flash point	Solubility in 100 parts solvent
m275	Methyl hydrogen succinate	$HOOCCH_2CH_2COOCH_3$	132.12	2, 608			56–59	151^{20mm}		v s aq, alc, eth
m276	Methyl hydroperoxide	CH_3OOH	48.04	1^2, 270	1.9971^{15}	1.3642^{15}		38^{65mm}		misc aq, alc, eth; s bz
m277	Methyl hydroquinone	$CH_3C_6H_3\text{-}1,4\text{-}(OH)_2$	124.14	6, 874			125–128			
m278	Methyl 4-hydroxybenzoate	$HOC_6H_4COOCH_3$	152.15	10, 158			128	270 d		v s alc, eth, acet
m279	Methyl 2-hydroxyisobutyrate	$(CH_3)_2C(OH)COOCH_3$	118.13	3^2, 223	1.023	1.4112^{20}		137	42	v s aq, alc
m280	Methyl 4-hydroxyphenylacetate	$HOC_6H_4CH_2COOCH_3$	166.18	10, 191	1.030	1.4970^{20}	57–60	$162\text{–}163^{5mm}$	92	
m281	1-Methylimidazole		82.11	23, 46			−60	198		misc aq
m282	2-Methylimidazole		82.11	23, 65			143	268		
m283	4-Methylimidazole		82.11	23, 69			48	263	>110	
m284	2-Methyl-1*H*-indole		131.18	20, 311	1.072^{20}	1.5681^{20}	58–60	273		v s alc, eth; s hot aq
m285	2-Methylindoline		133.19	20, 279	1.023			229	93	
m286	*N*-Methylisatoic anhydride		177.16	27, 265			165 d			
m287	Methyl isobutyrate	$(CH_3)_2CHCOOCH_3$	102.13	2, 290	0.891^{20}	1.3840^{20}	−84	90	3	misc alc, eth; sl s aq
m288	Methyl isocyanate	CH_3NCO	57.05	4, 77	0.967	1.3695^{20}	−17	39	−6	s aq
m289	Methyl isodehydroacetate		182.18	18, 410			60–63	167^{14mm}		
m290	Methyl isothiocyanate	CH_3NCS	73.12	4, 77	1.069	1.5258^{37}	35	119	32	v s alc, eth; sl s aq
m291	5-Methylisoxazole		83.09	27, 16	1.018	1.4386^{20}		122	30	
m292	Methyl lactate	$CH_3CH(OH)COOCH_3$	104.10	3, 280	1.0882^{20}	1.4131^{20}	−66	144	49	s aq(dec), alc, eth
m293	Methyl mandelate	$C_6H_5CH(OH)COOCH_3$	166.18	10, 202	1.1756^{20}		54	135^{12mm}	>110	s aq, alc, bz, chl
m294	Methyl mercaptoacetate	$HSCH_2COOCH_3$	106.14		1.187	1.4657^{20}		43^{10mm}	30	s alc, eth
m297	Methyl 3-mercaptopropionate	$HSCH_2CH_2COOCH_3$	120.17	3^2, 214	1.085	1.4640^{20}		55^{14mm}	60	
m298	Methylmercury chloride	CH_3HgCl	251.10		4.06^{25}		170			
m299	Methyl methacrylate	$H_2C{=}C(CH_3)COOCH_3$	100.12	2^2, 398	0.9433^{20}	1.4146^{20}	−48	100	10	1.6 aq; s ketones, esters, CCl_4
m300	Methyl methanesulfonate	$CH_3SO_2OCH_3$	110.13	4, 4	1.2943^{24}	1.4138^{20}		202–203	104	20 aq; 100 DMF
m301	Methyl methoxyacetate	$CH_3OCH_2COOCH_3$	104.11	3, 236	1.0511^{20}_{4}	1.3964^{20}		130	35	v s alc, eth; sl s aq

m301a	Methyl 4-methoxy-acetoacetate	$CH_3OCH_2COCH_2CO_2CH_3$	146.14	3[4], 1939	1.129		1.4316^{20}	89^{8mm}	89
m301b	Methyl 2-methoxy-benzoate	$CH_3OC_6H_4CO_2CH_3$	166.18	10, 71	1.157		1.5335^{20}	248	>110
m301c	Methyl 4-methoxyben-zoate	$CH_3OC_6H_4CO_2CH_3$	166.18	10, 159		51		245	>110
m302	Methyl 1-methoxy-bicyclo-[2.2.2]oct-5-ene-2-carboxylate		196.25		1.086		1.4886^{20}	105^{17mm}	103
m303	Methyl 4-methoxy-phenylacetate	$CH_3OC_6H_4CH_2COOCH_3$	180.20	10, 191	1.135		1.5165^{20}	158^{19mm}	36
m304	1-Methyl-4-(methyl-amino)piperidine		128.22		0.882		1.4672^{20}		55

Methyl hydroxyacetate, m260
Methyl 4-hydroxy-3-methoxybenzoate, m448
Methyl 2-hydroxypropionate, m292
2,2'-Methyliminodiethanol, m224
2,2'-Methyliminobis(acetaldehyde diethyl acetal), b170
Methyl iodide, i40

Methyl isoamyl ketone, m268
Methyl isobutenyl ketone, m353
Methyl isobutyl ketone, m349
Methyl isonicotinate, m405
Methyl isopentyl ketone, m268

2-Methyllactic acid, h127
Methyl mercaptan, m33
Methylmercaptoanilines, m425, m426
4-Methylmercaptobenzaldehyde, m427
Methylmercaptophenols, m431a, m432

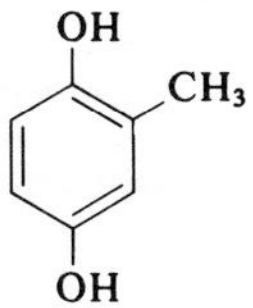

m277
m281 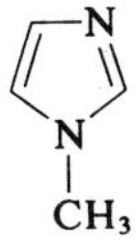
m282 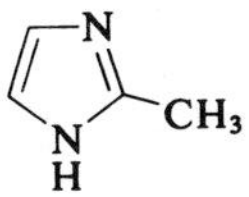
m283 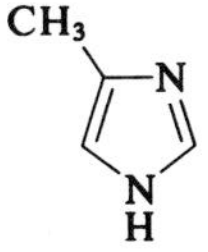
m284 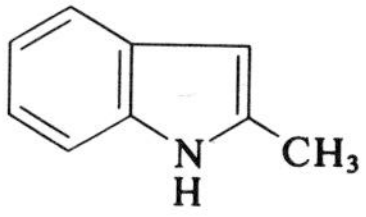
m285 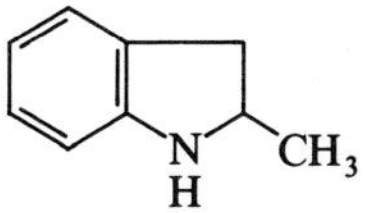
m286 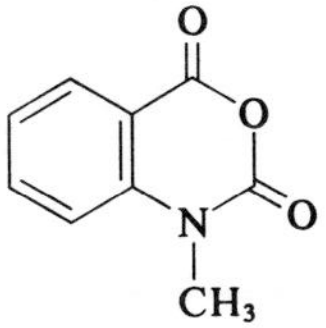
m289 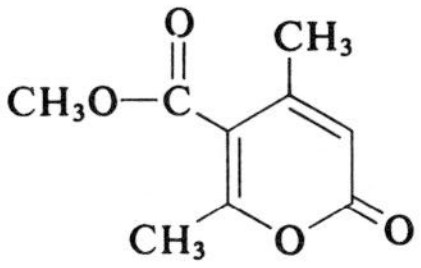
m291 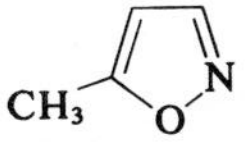
m302 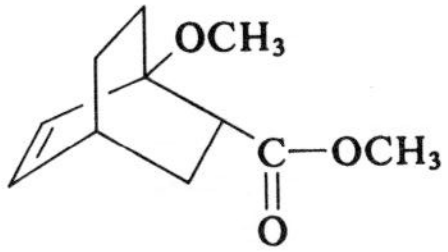
m304

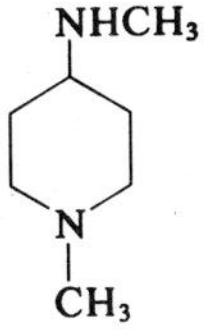

TABLE 1.15 Physical Constants of Organic Compounds (*Continued*)

No.	Name	Formula	Formula weight	Beilstein reference	Density	Refractive index	Melting point	Boiling point	Flash point	Solubility in 100 parts solvent
m304a	Methyl 2-methylbenzoate	$CH_3C_6H_4CO_2CH_3$	150.18	9, 463	1.073	1.5184[20]		208	82	
m304b	Methyl 3-methylbenzoate	$CH_3C_6H_4CO_2CH_3$	150.18	9, 475	1.063	1.5156[20]		113[27mm]	95	
m304c	Methyl 4-methylbenzoate	$CH_3C_6H_4CO_2CH_3$	150.18	9, 484			35–37	104[15mm]	95	
m305	Methyl 2-methylbutyrate	$C_2H_5CH(CH_3)CO_2CH_3$	116.16	2, 304	0.885	1.3931[20]		115	32	sl s aq; misc alc, eth
m306	2-Methyl-6-methylene-2-octanol	$C_2H_5C(=CH_2)(CH_2)_3$-$C(CH_3)_2OH$	156.27		0.784	1.4431[20]		84[10mm]	76	
m307	Methyl 2-methyl-3-furancarboxylate		140.14		1.116	1.4730[20]		75[20mm]	63	
m308	Methyl S-methylthiomethyl sulfoxide	$CH_3S(=O)CH_2SCH_3$	124.22		1.191	1.5487[20]		95[2.5mm]	>110	
m309	Methyl 3-(methylthio)propionate	$CH_3SCH_2CH_2COOCH_3$	134.20		1.077	1.4650[20]		75[13mm]	72	
m310	N-Methylmorpholine		101.15	27, 6	0.920	1.4349[20]	−66	116	23	s aq, alc, eth
m310a	4-Methylmorpholine N-oxide, 60 wt % in water		117.15		1.130	1.4374[20]	−20	118	none	
m311	1-Methylnaphthalene	$C_{10}H_7CH_3$	142.20	5, 566	1.0254[14]	1.6159[20]	−30.5	244.7	82	v s alc, eth
m312	1-Methylnaphthalene	$C_{10}H_7CH_3$	142.20	5, 567	1.029[20]	1.6026[40]	34.6	241.4	97	v s alc, eth
m312a	Methyl 1-naphthaleneacetate	$C_{10}H_7CH_2CO_2CH_3$	200.24	9³, 3206	1.142	1.5961[20]		162[5mm]	>110	
m313	2-Methyl-1,4-naphthoquinone		172.18	7², 656			105–107			1.4 alc; 10 bz; s chl
m314	Methyl 1-naphthyl ketone	$C_{10}H_7COCH_3$	170.21	7, 401	1.1336₄	1.6284[20]	11	302	>110	s alc; eth; i aq
m315	Methyl 2-naphthyl ketone	$C_{10}H_7COCH_3$	170.21	7, 402			53–55	301	>110	sl s alc; s CS$_2$
m316	Methyl nitrate	CH_3ONO_2	77.04	1, 284	1.20754[20]	1.3748[20]	−83	64 explodes		sl s aq; s alc, eth
m317	Methyl nitrite	CH_3ONO	61.04	1, 284	0.991(*lq*)			−17.35		
m317a	N-Methyl-4-nitroaniline	$O_2NC_6H_4NHCH_3$	152.15	12, 714			154			s alc, eth

No.	Name	Formula	M.W.	Beilstein	Density	n_D	B.p., °C		M.p., °C	Solubility
m317b	2-Methyl-3-nitroaniline	$CH_3C_6H_3(NO_2)NH_2$	152.15	12, 848					90	v s alc; s bz
m318	2-Methyl-4-nitroaniline	$CH_3C_6H_3(NO_2)NH_2$	152.15	12, 846	1.1586^{40}		305		133	s alc, acet, eth
m319	2-Methyl-5-nitroaniline	$CH_3C_6H_3(NO_2)NH_2$	152.15	12, 844					107	v s alc; s eth
m320	4-Methyl-2-nitroaniline	$CH_3C_6H_3(NO_2)NH_2$	152.15	12, 1000					116	s alc, eth
m321	Methyl 2-nitrobenzoate	$O_2NC_6H_4COOCH_3$	181.15	9, 372	1.280	1.5350^{20}	$106^{0.1mm}$	>110	−13	
m321a	Methyl 3-nitrobenzoate	$O_2NC_6H_4CO_2CH_3$	181.15	9, 378			279		80	
m321b	Methyl 4-nitrobenzoate	$O_2NH_6H_4CO_2CH_3$	181.15	9, 390					96	
m322	2-Methyl-3-nitrobenzoic acid	$CH_3C_6H_3(NO_2)COOH$	181.15	9, 471					182–184	
m322a	3-Methyl-4-nitrobenzoic acid	$CH_3C_6H_3(NO_2)COOH$	181.15	9, 481					218	
m323	4-Methyl-3-nitrobenzoic acid	$CH_3C_6H_3(NO_2)COOH$	181.15	9, 502					187–190	
m324	5-Methyl-2-nitrobenzoic acid	$CH_3C_6H_3(NO_2)COOH$	181.15	9, 482					134–136	
m325	2-Methyl-5-nitroimidazole		127.10	23^1, 23					252–254	
m326	3-Methyl-2-nitrophenol	$CH_3C_6H_3(NO_2)OH$	153.14	6, 385					35–39	

Methyl 2-methyllactate, m279
Methyl methyl-2-propenoate, m299
Methyl methylsulfinylmethyl sulfide, m308

Methyl myristate, m416
Methyl nicotinate, m404
4-Methyl-3-nitroanisole, m85

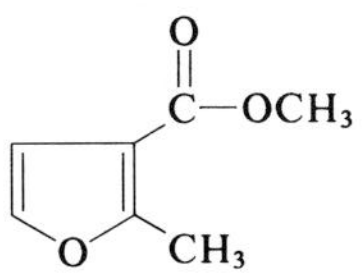

m307

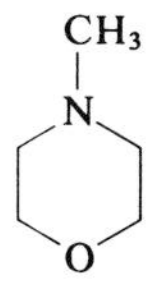

m310

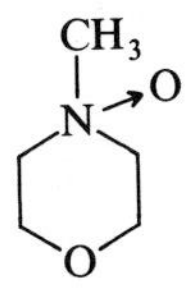

m310a

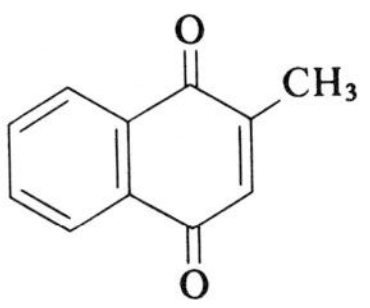

m313

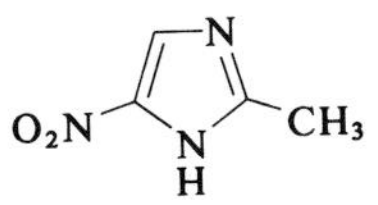

m325

7-Methyl-3-methylene-1,6-octadiene, m456
1-Methyl-4-(1-methylethenyl)cyclohexane, d651
5-Methyl-2-(1-methylethyl)cyclohexanol, m12
5-Methyl-2-(1-methylethyl)cyclohexanone, m13

TABLE 1.15 Physical Constants of Organic Compounds (*Continued*)

No.	Name	Formula	Formula weight	Beilstein reference	Density	Refractive index	Melting point	Boiling point	Flash point	Solubility in 100 parts solvent
m327	4-Methyl-2-nitrophenol	$CH_3C_6H_3(NO_2)OH$	153.14	6, 412	1.240^{40}_{4}	1.574^{40}	32–35	125^{22}mm	198	v s alc, eth
m327a	2-Methyl-2-nitro-1-propanol	$O_2NC(CH_3)_2CH_2OH$	119.12	1, 378			86–89	95^{10}mm		350 aq
m327b	N-Methyl-N-nitroso-4-toluenesulfonamide	$CH_3C_6H_4SO_2N(CH_3)NO$	214.24	11^1, 29			62			
m328	Methyl 2-nonynoate	$CH_3(CH_2)_5C{\equiv}CCO_2CH_3$	168.24	2, 490	0.915	1.4484^{20}		121^{20}mm	100	s alc, eth
m329	Methyl octadecanoate	$CH_3(CH_2)_{16}COOCH_3$	298.51	2, 379			38	215^{15}mm	>110	misc abs alc, eth
m330	Methyl cis-9-octadecenoate	$CH_3(CH_2)_7CH{=}CH(CH_2)_7COOCH_3$	296.50	2, 467	0.8791^{18}	1.4521^{20}	19.9	168^{2}mm		
m332	Methyl octanoate	$CH_3(CH_2)_6COOCH_3$	158.24	2, 348	0.8775^{20}	1.4160^{25}	−40	192.9	72	v s alc, eth; i aq
m333	Methyloctyldichlorosilane	$C_8H_{17}Si(CH_3)Cl_2$	227.3		0.9764^{20}	1.444^{20}		94^{6}mm		
m334	3-Methyl-2-oxazolidinone		101.11		1.170	1.4541^{20}	15	90^{1}mm	>110	
m335	2-Methyl-2-oxazoline		85.11	27, 13	1.005	1.4340^{20}		110	20	
m336	Methyl 2-oxocyclopentanecarboxylate	$(O{=})C_5H_7CO_2CH_3$	142.15	10, 597	1.145	1.4560^{20}		105^{19}mm	>110	misc alc, eth; sl s aq
m337	Methyl 2-oxopropionate	$CH_3C({=}O)COOCH_3$	102.09	3, 616	1.130	1.4065^{20}		134–137	39	
m338	Methyl 2-oxo-1-pyrrolidineacetate		157.17		1.131	1.4719^{20}			110	
m338a	2-Methyl-1,3-pentadiene	$CH_3CH{=}CHC(CH_3){=}CH_2$	82.15	1, 255	0.718	1.4469^{20}		76	−12	
m339	2-Methylpentane	$CH_3CH_2CH_2CH(CH_3)_2$	86.18	1, 148	0.6532^{20}	1.3725^{20}	−153.7	60.3	−23	
m340	3-Methylpentane	$(CH_3CH_2)_2CHCH_3$	86.18	1, 149	0.6643^{20}	1.3765^{20}		63.3	−6	
m340a	2-Methyl-1,5-pentanediamine	$H_2N(CH_2)_2CH(CH_3)CH_2NH_2$	116.21	4, 270	0.860	1.4590^{20}	80			
m341	2-Methyl-2,4-pentanediol	$(CH_3)_2C(OH)CH_2CH(OH)CH_3$	118.18	1, 486	0.9216^{20}_{4}	1.4270^{20}		198.3	93	misc aq
m342	4-Methylpentanenitrile	$(CH_3)_2CHCH_2CH_2CN$	97.16	2, 329	0.8035^{20}_{4}	1.4061^{20}	−51.1	153.5	45	s alc; misc eth
m343	Methyl pentanoate	$CH_3(CH_2)_3COOCH_3$	116.16	2, 301	0.875	1.3962^{20}		128	22	sl s aq; misc alc, eth
m344	2-Methylpentanoic acid	$CH_3CH_2CH_2CH(CH_3)COOH$	116.16	2^2, 288	0.9242^{20}	1.4135^{20}	−85 glass	196.4	107	1.3 aq
m346	2-Methyl-1-pentanol	$CH_3CH_2CH_2CH(CH_3)CH_2OH$	102.18	1, 409	0.8242^{20}	1.4190^{20}		148.0	50	s alc, eth

No.	Name	Formula	M.W.		Density	n	M.P.	B.P.		Solubility
m347	3-Methyl-3-pentanol	(CH₃CH₂)₂C(CH₃)OH	102.18	1, 411	0.8281^{20}	1.4186^{20}	<−38	122.4	46	misc alc, eth; sl s aq
m348	4-Methyl-2-pentanol	(CH₃)₂CHCH₂CH(OH)CH₃	102.18	1, 410	0.8080^{20}	1.4112^{20}	−90	131.7	41	1.6 aq
m349	4-Methyl-2-pentanone	(CH₃)₂CHCH₂COCH₃	100.16	1, 691	0.8006^{20}_{4}	1.3958^{20}	−83.5	115.7	13	1.7 aq; misc alc, bz, eth
m351	2-Methyl-2-pentenal	CH₃CH₂CH=C(CH₃)CHO	98.15	14, 3471	0.861	1.4503^{20}		138	31	s alc
m352	4-Methyl-2-pentenoic acid	(CH₃)₂CHCH=CHCOOH	114.14	2^{2}, 406	0.9529	1.4489	35	115^{20mm}	46	i aq; v s alc
m353	4-Methyl-3-penten-2-one	(CH₃)₂C=CHCOCH₃	98.15	1, 736	0.8548^{20}	1.4458^{20}	−42	129.5	30	3.1 aq
m354	1-Methylpentylamine	CH₃(CH₂)₃CH(NH₂)CH₃	101.19	4, 190	0.7672^{20}	1.4318^{20}	−19	116–118	13	s aq, alc, PE
m355	3-Methyl-1-pentyn-3-ol	CH₃CH₂C(CH₃)(OH)C≡CH	98.15	1^{2}, 506	0.86884^{20}		−30.6	122	26	13 aq; misc bz, acet, PE, EtAc; s eth
m356	4-Methylphenetole	CH₃C₆H₄OC₂H₅	136.19	6, 393	0.945	1.5044^{20}	150–153	191	70	s alc, EtAc, HOAc
m358	N-(4-Methylphenyl)-acetamide	CH₃C₆H₄NHCOCH₃	149.19	12, 920	1.212^{15}			307		
m359	Methyl phenylacetate	C₆H₅CH₂COOCH₃	150.18	9, 434	1.044	1.5075^{20}		215	90	i aq; misc alc, eth
m363	1-Methyl-1-phenylhydrazine	C₆H₅N(CH₃)NH₂	122.17	15, 117	1.0382^{22}	1.5834^{20}		118^{21mm}	96	misc alc, bz, chl, eth
m363a	2-Methyl-1-phenyl-2-propanol	C₆H₅CH₂C(CH₃)₂OH	150.22	6, 523	0.974	1.5145^{20}	81			
m364	1-Methyl-3-phenylpropyl acetate	C₆H₅CH₂CH₂CH(CH₃)OOCCH₃	192.26	6^{1}, 258	0.991			$74^{0.05mm}$	>110	
m364a	1-Methyl-3-phenylpropylamine	C₆H₅CH₂CH₂CH(CH₃)NH₂	149.24	12, 1165	0.922	1.5123^{20}		222	97	

m334

m335

m336

CH₂CO—OCH₃ m338

TABLE 1.15 Physical Constants of Organic Compounds (*Continued*)

No.	Name	Formula	Formula weight	Beilstein reference	Density	Refractive index	Melting point	Boiling point	Flash point	Solubility in 100 parts solvent
m365	3-Methyl-1-phenyl-2-pyrazolin-5-one		174.20	24, 20			130	287^{265mm}		
m367	Methyl phenyl sulfide	$C_6H_5SCH_3$	124.21	6, 297	1.058	1.5852^{20}	−15	188	57	i aq; s alc
m367a	N-Methyl-N-phenylurethane	$C_6H_5N(CH_3)CO_2C_2H_5$	179.22	12, 417	1.074	1.5149^{20}	110			
m368	N-Methylpiperazine		100.17	23, 17	0.903	1.4655^{20}		138	42	v s aq, alc, eth
m369	2-Methylpiperazine		100.17				65–67	155.6	22	78 aq; 37 acet; 32 bz
m370	4-Methyl-1-piperazine-propanol		158.25	23^3, 123		1.4835^{20}	28–30	121^{9mm}	>110	
m371	N-Methylpiperidine	$C_5H_{10}N{-}CH_3$	99.19	20, 19	0.816	1.4378^{20}		106	3	v s aq; misc alc, eth
m372	2-Methylpiperidine	$CH_3C_5H_9N$	99.19	20, 95	0.844	1.4459^{20}	−5	119	8	v s aq; misc alc, eth
m373	3-Methylpiperidine	$CH_3C_5H_9N$	99.19	20, 100	0.845	1.4470^{20}		126	17	v s aq
m374	4-Methylpiperidine	$CH_3C_5H_9N$	99.19	20, 101	0.838	1.4458^{20}		124	7	v s aq
m375	1-Methyl-3-piperidine-methanol	$CH_3C_5H_9N$	129.20	21^2, 8	1.013	1.4772^{20}		140–145	94	
m376	1-Methyl-4-piperidone		113.16	21^2, 215	0.920	1.4614^{20}	−65		60	9 aq; misc alc, bz, chl, eth
m377	2-Methylpropanal	$(CH_3)_2CHCHO$	72.11	1, 671	0.7891^{20}	1.3727^{20}		64.1	−40	13 mL aq; 1320 mL alc; 2890 mL eth
m378	2-Methylpropane	$(CH_3)_3CH$	58.12	1, 124	0.557^{20}		−159.6	−11.7		
m379	N-Methyl-1,3-propane-diamine	$H_2NCH_2CH_2CH_2NHCH_3$	88.15	4^1, 419	0.844	1.4468^{20}		139–141	35	
m380	2-Methyl-1,2-propane-diamine	$(CH_3)_2C(NH_2)CH_2NH_2$	88.15	4, 266	0.841	1.4410^{20}			23	
m381	1-Methyl-1-propane-thiol	$CH_3CH_2CH(SH)CH_3$	90.19	1, 373	0.8246^{25}_4	1.4338^{25}	−165	84–85	21	sl s aq; v s alc, eth
m382	2-Methyl-1-propane-thiol	$(CH_3)_2CHCH_2SH$	90.19	1, 378	0.8357^{25}_4	1.4396^{20}	−79	88.5	−9	v s alc, eth
m383	2-Methyl-2-propane-thiol	$(CH_3)_3CSH$	90.19	1, 383	0.7943^{25}_4	1.4198^{25}	1.1	64.2	−26	i aq
m384	2-Methyl-1-propanol	$(CH_3)_2CHCH_2OH$	74.12	1, 373	0.8016^{20}	1.3958^{20}	−108	107.9	37	10 aq; misc alc, eth
m385	2-Methyl-2-propanol	$(CH_3)_3COH$	74.12	1, 379	0.7858^{20}_4	1.3877^{20}	25.8	82.4	4	misc aq, alc, eth
m386	2-Methylpropene	$(CH_3)_2C{=}CH_2$	56.10	1, 207	0.6266^{-140}_4		−140.4	−6.9		v s alc, eth
m387	2-Methyl-2-propene-1-sulfonic acid, Na salt	$H_2C{=}C(CH_3)CH_2SO_3{}^-Na^+$	158.15				>300			

No.	Name	Formula	Formula weight	Beilstein reference	Density	Refractive index	Melting point, °C	Boiling point, °C	Flash point, °C	Solubility
m388	2-Methyl-2-propen-1-ol	$H_2C{=}C(CH_3)CH_2OH$	72.11	1, 443	0.857	1.4250^{20}		113–115	33	6 aq; misc alc, eth
m389	4-Methyl-2-(2-propenyl)phenol	$CH_3C_6H_3(CH_2CH{=}CH_2)OH$	148.21	6^1, 287	0.980	1.5385^{20}		238	101	
m390	6-Methyl-2-(2-propenyl)phenol	$CH_3C_6H_3(CH_2CH{=}CH_2)OH$	148.21	6^1, 287	0.992	1.5381^{20}		231–233	94	
m391	N-Methylpropionamide	$CH_3CH_2CONHCH_3$	87.12	4^3, 125	0.9305^{25}	1.4345^{25}	−43	146^{90mm}	105	
m392	Methyl propionate	$CH_3CH_2COOCH_3$	88.11	2, 239	0.9152^{20}	1.3770^{20}	−88	79.7	−2 (CC)	6 aq; misc alc, eth
m393	Methyl propionylacetate	$C_2H_5COCH_2CO_2CH_3$	130.15	3^3, 1212	1.037	1.4221^{20}		74^{5mm}	71	
m394	4'-Methylpropiophenone	$CH_3C_6H_4COCH_2CH_3$	148.21	7, 317	0.993	1.5280^{20}	7.2	239	96	
m396	Methyl propyl ether	$CH_3CH_2CH_2OCH_3$	74.12	1, 354	0.738^{20}			39.1		s aq
m397	2-Methyl-2-propyl-1,3-propanediol	$C_3H_7C(CH_3)(CH_2OH)_2$	132.20	1^1, 254			53–55	232	>110	sl s aq; misc alc, eth
m398	Methyl propyl sulfide	$CH_3SCH_2CH_2CH_3$	90.18	1^3, 1432	0.8424^{20}	1.4442^{20}	−113.0	95.5	−18	s aq
m399	Methyl 2-propynyl ether	$CH_3OCH_2C{\equiv}CH$	70.09	1, 454	0.830	1.3961^{20}		62		
m400	2-Methylpyrazine	(see structure)	94.12	23, 94	1.030	1.5042^{20}	−29	135	50	v s aq, alc, eth

Structure m365 (pyrazolone; labeled CH_3, N–N–phenyl, C=O)

Structure m368 (CH_3 on N-methylpiperazine, $H{-}N$)

Structure m369 (CH_3, 2-methylpiperazine, $HN{\cdots}NH$)

Structure m370: CH_3-N(piperazine)$N-CH_2CH_2CH_2OH$

Structure m375 (piperidine with CH_2OH and $N-CH_3$)

Structure m376 ($O{=}$ piperidinone with $N-CH_3$)

Structure m400 (2-methylpyrazine; CH_3, N, N)

TABLE 1.15 Physical Constants of Organic Compounds (*Continued*)

No.	Name	Formula	Formula weight	Beilstein reference	Density	Refractive index	Melting point	Boiling point	Flash point	Solubility in 100 parts solvent
m401	2-Methylpyridine	$CH_3C_5H_4N$	93.13	20, 234	0.9504^{15}	1.5010^{20}	−67	128–129	26	misc aq; s alc, eth
m402	3-Methylpyridine	$CH_3C_5H_4N$	93.13	20, 239	0.9611^{15}	1.5068^{20}	−18.3	143.3	36	misc aq, alc, eth
m403	4-Methylpyridine	$CH_3C_5H_4N$	93.13	20, 240	0.9574^{15}	1.5058^{20}	3.8	143–145	56	misc aq, alc, eth
m404	Methyl 3-pyridinecarboxylate	$(C_5H_4N)COOCH_3$	137.14	22, 39			39	209		s aq, alc, bz
m405	Methyl 4-pyridinecarboxylate	$(C_5H_4N)COOCH_3$	137.14	22, 46	1.001	1.5122^{20}	8.5	207–209	82	
m406	1-Methyl-2-pyridone		109.13	21, 268	1.112	1.5690^{20}	30–32	250^{740mm}	>110	i aq; misc alc, eth
m407	N-Methylpyrrole		81.12	20, 163	0.914	1.4875^{20}	−57	113	15	misc aq, eth
m408	N-Methylpyrrolidine		85.15	20, 4	0.8192^{20}	1.4247^{20}		80–81	−21	misc aq, alc, bz, eth
m409	N-Methyl-2-pyrrolidinone		99.13	21, 237	1.0279^{25}	1.4680^{25}	−24.4	202	86	misc aq, alc, bz, eth
m410	2-Methylquinoline		143.19	20, 387	1.058	1.6108^{20}	−2	248	79	i aq; s chl, eth
m411	4-Methylquinoline		143.19	20, 395	1.0826^{20}	1.6200^{20}	9–10	263	>110	misc alc, bz, eth
m412	2-Methylquinoxaline		144.18	23^1, 44	1.118	1.6156^{20}	180	246	107	misc aq
m413	Methyl salicylate	$HOC_6H_4COOCH_3$	152.15	10, 70	1.1831^{20}	1.5240^{20}	−8.6	223.0	>110	0.7 aq; s chl, eth; misc alc, HOAc
m414	α-Methylstyrene	$C_6H_5C(CH_3)=CH_2$	118.18	5, 484	0.909	1.5375^{20}	−23.2	165.5	45	
m414a	4-Methylstyrene	$CH_3C_6H_4CH=CH_2$	118.18	5, 485	0.897	1.5412^{20}		170–175	45	66 aq; v s alc, eth
m415	Methylsuccinic acid	$HOOCCH_2CH(CH_3)COOH$	132.12	2, 636	1.411	1.4303	110–112	dec		misc alc, bz, eth
m416	Methyl tetradecanoate	$CH_3(CH_2)_{12}COOCH_3$	242.40	2^2, 326	0.855	1.4362^{20}	18.4	323	>110	
m417	2-Methyl-3,3,4-tetrafluoro-2-butanol	$HCF_2CF_2C(CH_3)_2OH$	160.11		1.282	1.3524^{20}		117	73	
m418	2-Methyltetrahydrofuran		86.13	17, 12	0.860	1.4056^{20}		78–80	−11	
m418a	5-Methyltetrahydrofuran-2-methanol		116.16	17^3, 1141	1.008	1.4456^{20}		73^{14mm}	40	
m419	3-Methyltetrahydropyran		100.16	17^3, 77	0.863	1.4204^{20}		109^{733mm}	6	
m420	3-Methyltetrahydrothiophene-1,1-dioxide		134.20		1.191	1.4772^{20}		276	>110	
m420a	4-Methylthiazole		99.16	27, 16	1.090	1.5257^{20}		134	32	
m421	4-Methyl-5-thiazole-ethanol		143.21		1.196	1.5508^{20}		135^{7mm}	>110	

1.264

No.	Name	Formula	Mol. wt.		Density	n_D	m.p.	b.p.		Solubility
m422	2-Methyl-2-thiazoline		101.17	27, 13	1.067	1.5200^{20}		145	37	s alc, eth
m423	Methyl thioacetate	CH_3COSCH_3	90.14			1.4628	−101	98	10	
m424	(Methylthio)acetonitrile	CH_3SCH_2CN	87.14		1.039	1.4826^{20}		63^{15mm}	67	
m425	2-(Methylthio)aniline	$CH_3SC_6H_4NH_2$	139.22	13, 399	1.111	1.6239^{20}		234	>110	
m426	3-(Methylthio)aniline	$CH_3SC_6H_4NH_2$	139.22	13^1, 141	1.130	1.6423^{20}		165^{16mm}	>110	
m427	4-(Methylthio)benzaldehyde	$CH_3SC_6H_4CHO$	152.22	8^1, 533	1.144	1.6452^{20}		90^{1mm}		
m428	3-(Methylthio)-2-butanone	$CH_3CH(SCH_3)COCH_3$	118.20	1^4, 3993	0.975	1.4710^{20}		50–54^{20mm}	44	i aq; misc alc, eth
m429	Methyl thiocyanate	CH_3SCN	73.12	3, 175	1.068^{20}	1.4697^{20}	−51	133	38	
m429a	2-Methylthiophene		98.17	17, 37	1.014	1.5199^{20}	−63	113	7	

Methyl pyridyl ketones, a53, a54, a55
1-Methyl-2-(3-pyridyl)pyrrolidine, n20
Methyl pyruvate, m337
Methylresorcinol, d391

Methylsalicylic acid, h138, h139
Methyl stearate, m329
Methyl succinyl chloride, m189
Methylsulfonic acid, m30

Methyl theobromine, c1
3-Methyl-2-thiabutane, i104
Methyl thienyl ketone, a57
Methyl thioglycolate, m294

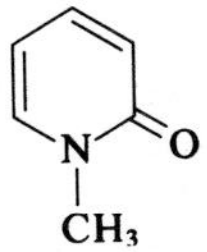

m406 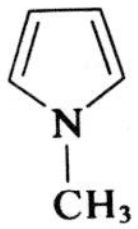 m407  m408 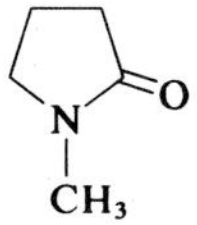 m409 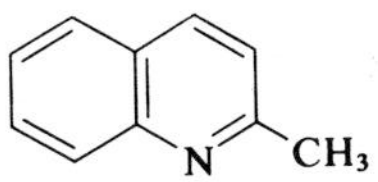 m410 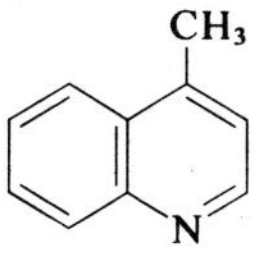 m411 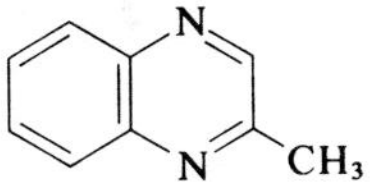 m412 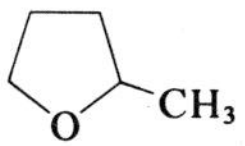 m418 m429a

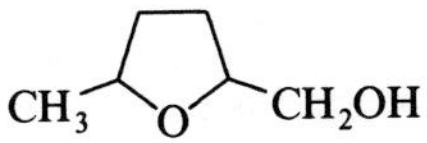

m418a 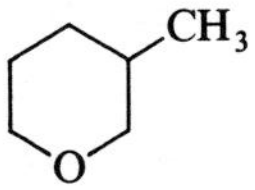 m419 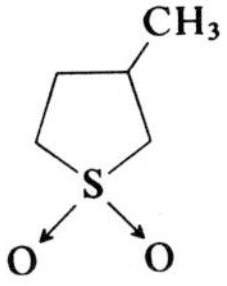 m420 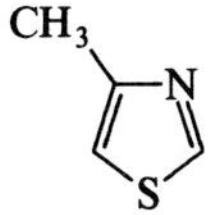 m420a 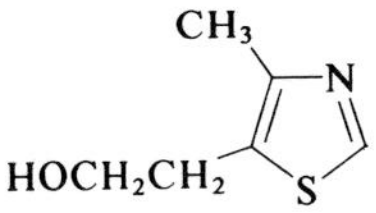 m421 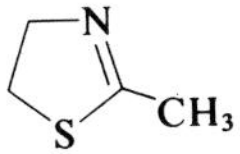 m422

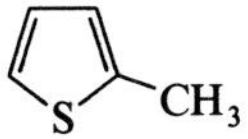

TABLE 1.15 Physical Constants of Organic Compounds (*Continued*)

No.	Name	Formula	Formula weight	Beilstein reference	Density	Refractive index	Melting point	Boiling point	Flash point	Solubility in 100 parts solvent
m430	3-Methylthiophene		98.17	17, 38	1.016	1.5180^{20}	−69.0	115.4	11	i aq; misc alc, eth
m431	5-Methyl-2-thiophene-carbaldehyde		126.18	17^1, 151	1.170	1.5825^{20}		114^{25mm}	87	
m432	4-(*S*-Methylthio)-phenol	$CH_3SC_6H_4OH$	140.20	6^1, 419			83–85	156^{20mm}		v s aq, alc
m434	*N*-Methylthiourea	$CH_3NHC(=S)NH_2$	90.15	4, 70			119–121			
m435	*N*-Methyl-*o*-toluamide	$CH_3C_6H_4CONHCH_3$	149.19	9, 465	1.168^{15}		69–71			
m436	*N*-Methyl-*p*-toluene-sulfonamide	$CH_3C_6H_4SO_2NHCH_3$	185.25	11, 105			76–79			
m437	Methyl *p*-toluenesulfonate	$CH_3C_6H_4SO_2OCH_3$	186.23	11, 99			27.5	145^{5mm}	>110	
m438	Methyltriacetoxysilane	$CH_3Si(OOCCH_3)_3$	220.26	4^3, 1896	1.175^{20}	1.408^{20}		88^{3mm}	85	
m439	Methyl trichloroacetate	$CCl_3CO_2CH_3$	177.42	2, 208	1.488	1.4558^{20}		153	72	
m440	Methyl trifluoroacetate	$CF_3CO_2CH_3$	128.05	2^3, 427	1.273	1.2907^{20}		43	−7	
m441	Methyl trifluoromethanesulfonate	$CF_3SO_2OCH_3$	164.10	3^4, 34	1.450	1.3244^{20}		94–99	38	
m441a	Methyl 3,4,5-trihydroxybenzoate	$(HO)_3C_6H_2CO_2CH_3$	184.15	10, 483			202			
m441b	Methyl 3,4,5-trimethoxybenzoate	$(CH_3O)_3C_6H_2CO_2CH_3$	226.23	10, 484			84	275		
m442	Methyltrimethoxysilane	$CH_3Si(OCH_3)_3$	136.22	4^4, 4203	0.955	1.3703^{20}		102	11	misc alc; eth; sl s aq
m442a	Methyl trimethylacetate	$(CH_3)_3CCO_2CH_3$	116.16	2, 320	0.873	1.3900^{20}		101	6	
m443	*N*-Methyl-*N*-(trimethylsilyl)trifluoroacetamide	$CF_3CON(CH_3)Si(CH_3)_3$	199.25		1.075	1.3802^{20}		132	25	
m445	(Methyl)triphenylphosphonium bromide	$[CH_3P(C_6H_5)_3]^+Br^-$	357.24	16, 760			233			
m446	2-Methylundecanal	$CH_3(CH_2)_8CH(CH_3)CHO$	184.32		0.830^{15}	1.4321^{20}		271	93	s alc, eth
m447	Methyl urea	$CH_3NHCONH_2$	74.08	4, 64	1.204		101–102	dec		v s aq, alc; i eth
m448	Methyl vanillate	$CH_3OC_6H_3(OH)COOCH_3$	182.18	10, 396			64–65	$285–287$		s hot alc, hot PE
m448a	*N*-Methyl-*N*-vinylacetamide	$CH_3CON(CH_3)CH=CH_2$	99.13	4^3, 442	0.959	1.4829^{20}		70^{25mm}	58	

m449	Methyl vinyl ether	$CH_3OCH{=}CH_2$	58.08	1^3, 1857	0.7511_4^{20}	1.3947	−123	5.5	−56	0.8 aq; v s alc
m451	Morpholine		87.12	27, 5	1.007_4^{20}	1.4542^{20}	−4.9	128.9	38 (OC)	misc aq, alc, bz, eth
m452	4-Morpholinecarbonitrile		112.12		1.109	1.4730^{20}		$73^{0.6mm}$	104	
m452a	4-Morpholinepropionitrile		140.19	27^3, 337	1.037	1.4715^{20}	21	121^{2mm}		
m453	*N*-Morpholino-1-cyclohexene		167.25		0.995	1.5128^{20}		117–122	68	
m455	3-(*N*-Morpholino)-1,2-propanediol		161.20		1.157		38^{30mm}	191	>110	
m456	Myrcene	$(CH_3)_2C{=}CHCH_2CH_2C({=}CH_2)CH{=}CH_2$	136.24	1, 264	0.794_4^{20}	1.4709^{20}		166–168	39	s alc, chl, eth, HOAc
n1	1-Naphthaldehyde	$C_{10}H_7CHO$	156.18	7, 400	1.150_4^{20}	1.6520^{20}	1–2	161^{15mm}	>110	s alc, eth
n2	Naphthalene	$C_{10}H_8$	128.17	5, 531	1.162_4^{20}	1.5821^{100}	80.2 subl above mp	217.7	78 (OC)	0.3 aq; 7 alc; 33 bz; 50 chl

4-Methyl-2-thiouracil, h120
Methyl *o*-toluate, m304a
Methyl *m*-toluate, m304b
Methyl *p*-toluate, m304c
Methyltrichlorosilane, t242
Methyltriethoxysilane, t266
Methyl trimethylacetate, m228
β-Methylumbelliferone, h141
Methyl undecyl ketone, t262a
Methyluracil, d392
Methyl urethane, m182

Methyl valerate, m343
4-Methylvaleronitrile, m342
Methyl veratrate, m226
Michler's ketone, b172
Monoethyl adipate, e152
Monoglyme, d440
Monomethyl adipate, m274
Monomethyl glutarate, m273
Monomethyl succinate, m275
Mordent violet 5, a60
4-Morpholinocarboxaldehyde, f34a

1-Morpholinocyclopentene, c362
4-Morpholinoethanol, h122
MSTFA, m443
Mucic acid, t86
Mucochloric acid, d208
Myristic acid, t40
Myristolyl chloride, t42
Myristyl alcohol, t41
Myristyl bromide, b35
Naphthacene, b7

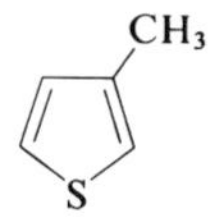

m430

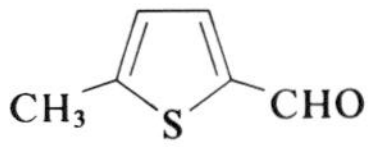

m431

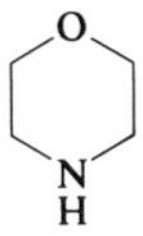

m451

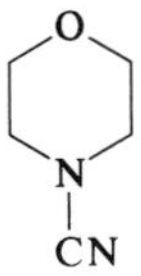

m452

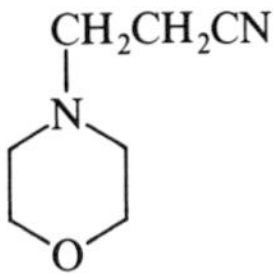

m452a

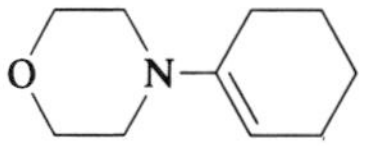

m453

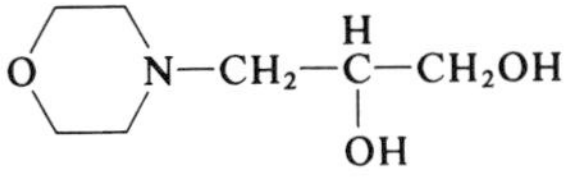

m455

TABLE 1.15 Physical Constants of Organic Compounds (*Continued*)

No.	Name	Formula	Formula weight	Beilstein reference	Density	Refractive index	Melting point	Boiling point	Flash point	Solubility in 100 parts solvent
n3	1-Naphthalenecarboxylic acid	$C_{10}H_7COOH$	172.18	9, 647			160–162	300		v s hot alc, eth
n4	1,5-Naphthalenediamine	$C_{10}H_6(NH_2)_2$	158.20	13, 203			185–187			s hot aq, hot alc
n5	1,8-Naphthalenediamine	$C_{10}H_6(NH_2)_2$	158.20	13, 204	1.1265^{99}_{4}	1.6828^{99}	66.5	205^{12mm}		sl s aq; s alc, eth
n6	1-Naphthalenesulfonic acid	$C_{10}H_7SO_3H$	208.24	11, 155			77–79			
n7	1,8-Naphthalic anhydride		198.18	17, 521			268			sl s HOAc
n8	1,8-Naphthalimide		197.19	21, 527			300			sl s alc; i bz, eth, aq
n9	1-Naphthol	$C_{10}H_7OH$	144.17	6, 596	1.0954^{99}_{4}	1.6206^{99}	96	288		v s alc, bz, chl, eth
n10	2-Naphthol	$C_{10}H_7OH$	144.17	6, 627	1.217^{4}		121–123	285–286	161	0.1 aq; 125 alc; 6 chl; 77 eth; s alk
n11	1,4-Naphthoquinone		158.16	7, 724	1.422		128	subl <100		s bz, chl, eth, alk
n12	2-Naphthoxyacetic acid	$C_{10}H_7OCH_2COOH$	202.21	6, 645			156			
n13	2-(1-Naphthyl)acetamide	$C_{10}H_7CH_2CONH_2$	185.23	9, 666			182			i aq; s bz, CS_2
n14	1-Naphthyl acetate	$C_{10}H_7OOCCH_3$	186.21	6, 608			43–46		>110	s alc, eth
n15	1-Naphthylacetic acid	$C_{10}H_7CH_2COOH$	186.21	9, 666			135	dec		3.3 alc; v s chl, eth
n16	1-Naphthylacetonitrile	$C_{10}H_7CH_2CN$	167.21	9, 667			33–35	194^{18mm}	>110	s alc
n17	1-Naphthylamine	$C_{10}H_7NH_2$	143.18	12, 1212	1.1235	1.6192^{20}	50	301	157	0.2 aq; v s alc, eth
n18	1-Naphthyl isocyanate	$C_{10}H_7NCO$	169.19	12, 1244	1.177	1.6703	4	267	110	0.6 aq; 2.4 acet; s alc
n19	1-(1-Naphthyl)-2-thiourea	$C_{10}H_7NHC(=S)NH_2$	202.28	12, 1241		1.6344^{20}	198			
n20	Nicotine		162.24	23, 117	1.0097^{20}_{4}	1.5282^{20}	−79	123^{17mm}	101	misc aq; v s alc, eth, eth, PE
n21	Nitrilotriacetic acid	$N(CH_2COOH)_3$	191.14	4, 369			246 d			0.1 aq; s hot alc
n22	m-Nitroacetophenone	$O_2NC_6H_4COCH_3$	165.15	7, 288			76–78	202		s alc, eth
n23	p-Nitroacetophenone	$O_2NC_6H_4COCH_3$	165.15	7, 288			78–80	202		s alc
n24	2-Nitroaniline	$O_2NC_6H_4NH_2$	138.13	12, 687	1.442^{15}		69–70	284		s hot aq, alc, chl
n25	3-Nitroaniline	$O_2NC_6H_4NH_2$	138.13	12, 698	1.43		114	306		0.1 aq; 5 alc; 6 eth
n26	4-Nitroaniline	$O_2NC_6H_4NH_2$	138.13	12, 711	1.437^{14}		146	260^{100mm}	165 (OC)	4 alc; 3.3 eth; s bz
n27	3-Nitrobenzaldehyde	$O_2NC_6H_4CHO$	151.12	7, 250	1.279^{20}_{4}		58	164^{23mm}		s alc, chl, eth

No.	Name	Formula	Formula wt	Density	n_D	mp, °C	bp, °C	Solubility
n28	4-Nitrobenzaldehyde	O₂NC₆H₄CHO	151.12	1.496		106–107	87 (CC)	s alc, bz, HOAc
n29	2-Nitrobenzamide	O₂NC₆H₄CONH₂	166.12	1.462^{32}		174–178	317	s hot aq, hot alc, eth
n30	Nitrobenzene	C₆H₅NO₂	123.11	1.205^{15}	1.5546^{15}	5.8	210.8	v s alc, bz, eth
n32	3-Nitrobenzene-1,2-dicarboxylic acid	O₂NC₆H₃(COOH)₂	211.13			216 d		2 aq; v s hot alc
n34	5-Nitrobenzene-1,3-dicarboxylic acid	O₂NC₆H₃(COOH)₂	211.13			260		0.15 aq; v s alc, eth
n35	2-Nitrobenzenesulfonyl chloride	O₂NC₆H₄SO₂Cl	221.62			65–67		s eth; d hot aq, alc
n36	6-Nitrobenzimidazole		163.14			207–209		s alc, acid
n37	2-Nitrobenzoic acid	O₂NC₆H₄COOH	167.12	1.58		146–148		0.7 aq; 33 alc; 22 eth
n38	3-Nitrobenzoic acid	O₂NC₆H₄COOH	167.12	1.494		142		0.3 aq; 33 alc; 40 acet
n39	4-Nitrobenzoic acid	O₂NC₆H₄COOH	167.12	1.58		242.8		9 alc; 2 eth; 5 acet
n40	4-Nitrobenzonitrile	O₂NC₆H₄CN	148.12			146–149		s HOAc; sl s aq, alc
n41	3-Nitrobenzoyl chloride	O₂NC₆H₄COCl	185.57			32–35	>110	d aq, alc; v s eth

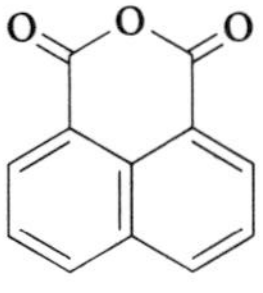

n7

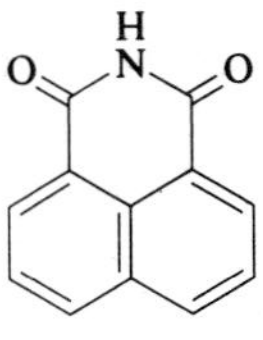

n8

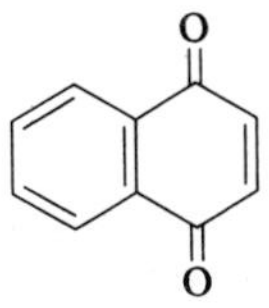

n11

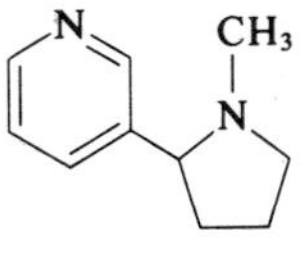

n20

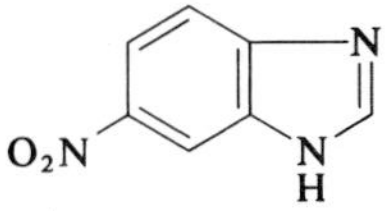

n36

TABLE 1.15 Physical Constants of Organic Compounds (*Continued*)

No.	Name	Formula	Formula weight	Beilstein reference	Density	Refractive index	Melting point	Boiling point	Flash point	Solubility in 100 parts solvent
n42	4-Nitrobenzoyl chloride	$O_2NC_6H_4COCl$	185.57	9, 394			75	205^{105mm}		d aq, alc; s eth
n43	N-(4-Nitrobenzoyl)glycine	$O_2NC_6H_4CONHCH_2COOH$	224.17	9, 395			131–133		>110	
n44	3-Nitrobenzyl alcohol	$O_2NC_6H_4CH_2OH$	153.14	6, 449			30–32	180^{3mm}		s aq, alc, eth
n45	4-Nitrobenzyl alcohol	$O_2NC_6H_4CH_2OH$	153.14	6, 450			92–94	185^{12mm}		v s alc, eth; sl s aq
n46	4-Nitrobenzyl bromide	$O_2NC_6H_4CH_2Br$	216.04	5, 334			98–100			2 alc; v s eth
n47	4-Nitrobenzyl chloride	$O_2NC_6H_4CH_2Cl$	171.58	5, 329			70–73			8 alc; s eth
n48	2-Nitrobiphenyl	$O_2NC_6H_4C_6H_5$	199.21	5, 582	1.442^{25}_4	1.613^{25}	36.7	325	179	s alc, acet, CCl_4
n49	4-Nitrobiphenyl	$O_2NC_6H_4C_6H_5$	199.21	5, 583			112–114	340		sl s alc; v s eth
n50	1-Nitrobutane	$CH_3CH_2CH_2CH_2NO_2$	103.18	1, 123	0.9752^{20}_{20}	1.4112	−81.3	152.8	47	sl s aq; misc alc, eth
n51	3-Nitro-2-butanol	$CH_3CH(NO_2)CH(OH)CH_3$	119.12	1, 373	1.12964^{25}_4	1.4414^{20}		92^{10mm}	91	i aq; s alc
n52	2-Nitrodiphenylamine	$O_2NC_6H_4NHC_6H_5$	214.22	12, 690			76–78			4.5 aq; misc alc, eth; s alk, chl
n53	Nitroethane	$CH_3CH_2NO_2$	75.07	1, 99	1.0528^{20}_{20}	1.3920^{20}	−90	114.1	30	
n54	1-Nitroguanidine	$O_2NNHC(=NH)NH_2$	104.07	3, 126			d 225			0.4 aq; sl s MeOH
n55	5-Nitro-1H-indazole		163.14	23, 129			207–209			s alc, bz, eth, acet
n56	Nitromethane	CH_3NO_2	61.04	1, 74	1.13322^{25}_4	1.3795^{25}	−28.4	101.2	35	11 aq; s alc, eth
n57	1-Nitronaphthalene	$C_{10}H_7NO_2$	173.17	5, 553	1.223		59–60	304		s alc; v s chl, eth
n58	2-Nitro-1-pentanol	$CH_3CH_2CH(NO_2)CH(OH)CH_3$	133.15	1, 385	1.08182^{25}	1.4430^{20}		100^{10mm}	90	
n60	2-Nitrophenol	$O_2NC_6H_4OH$	139.11	6, 213	1.495		44–45	214–216		s alc, bz, eth, alk
n61	4-Nitrophenol	$O_2NC_6H_4OH$	139.11	6, 226	1.270^{20}_4		113–114	279		s aq; v s alc, chl, eth
n62	4-Nitrophenyl acetate	$O_2NC_6H_4OOCCH_3$	181.15	6, 233			77–79			s aq; v s alc, bz, eth
n63	2-Nitrophenylacetic acid	$O_2NC_6H_4CH_2COOH$	181.15	9, 454			139–142			s hot aq, alc
n64	4-Nitrophenylacetic acid	$O_2NC_6H_4CH_2COOH$	181.15	9, 455			153			s alc, bz, eth
n65	4-Nitrophenylacetonitrile	$O_2NC_6H_4CH_2CN$	162.15	9, 456			117			s alc, eth
n66	4-Nitrophenyl chloroformate	$O_2NC_6H_4OOCCl$	201.57	6^1, 120			77–79	162^{19mm}		
n67	2-Nitro-1,4-phenylenediamine	$O_2NC_6H_3(NH_2)_2$	153.14	13, 120			137–140			
n68	4-Nitro-1,2-phenylenediamine	$O_2NC_6H_3(NH_2)_2$	153.14	13, 29			199–201			sl s aq; s HCl

No.	Name	Formula	Mol wt	Ref	Density	n_D	mp, °C	bp, °C	Flash pt	Solubility
n69	4-Nitrophenylhydrazine	$O_2NC_6H_4NHNH_2$	153.14	15, 468			156 d			s alc, chl, eth, hot bz
n70	2-Nitrophenyl phenyl ether	$O_2NC_6H_4OC_6H_5$	215.21	6^2, 222	1.2539^{22}	1.575^{20}	<-20	184^{8mm}		s alc, eth
n71	4-Nitrophenyl phenyl ether	$O_2NC_6H_4OC_6H_5$	215.21	6, 232			53–56	320	>110	s bz, eth
n71a	3-Nitro-1,2-phthalic acid	$O_2NC_6H_3(CO_2H)_2$	211.13	9, 823			213–216 d			
n71b	4-Nitro-1,2-phthalic acid	$O_2NC_6H_3(CO_2H)_2$	211.13	9, 828			170–172			
n72	3-Nitrophthalic anhydride		193.11	17, 486			163–165			sl s aq, bz
n73	1-Nitropropane	$CH_3CH_2CH_2NO_2$	89.09	1, 115	1.0009^{20}	1.4016^{20}	-108	131.6	34 (TCC)	1.4 aq; misc alc, eth
n74	2-Nitropropane	$(CH_3)_2CHNO_2$	89.09	1, 116	0.9876^{20}	1.3949^{20}	-93	120.3	24 (TCC)	1.7 aq; misc alc, eth
n75	2-Nitro-1-propanol	$CH_3CH(NO_2)CH_2OH$	105.09	1, 358	1.1841^{25}_{4}	1.4379^{20}		99^{10mm}	100	s aq, alc, eth
n76	4-Nitropyridine-N-oxide	$O_2NC_5H_4N(O)$	140.10				159–162			
n79	N-Nitrosodimethylamine	$(CH_3)_2NNO$	74.08	8, 84	1.0048^{20}	1.4368^{20}		151	61	v s aq, alc, eth
n80	4-Nitrosodiphenylamine	$C_6H_5NHC_6H_4NO$	198.22				144 d			v s alc, bz, chl, eth
n81	1-Nitroso-2-naphthol	$C_{10}H_6(NO)OH$	173.16	7, 712			109 d			3 alc; s bz, eth, alk
n82	1-Nitroso-2-naphthol-3,6-disulfonic acid, di-Na salt hydrate		377.26	112, 190			>300			2.5 aq; sl s alc

4-Nitrobenzyl cyanide, n65
Nitrocresols, m326, m327
Nitroglycerine, g21

n55

5-Nitroisophthalic acid, n34
3-Nitrophenyl disulfide, b195

n72

4-Nitrophenyl disulfide, b196
3-Nitro-o-phthalic acid, n32

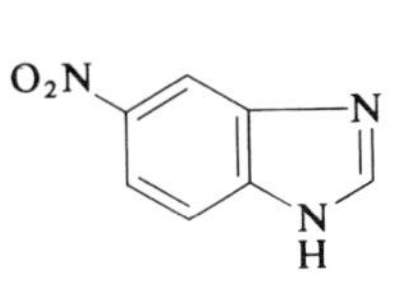

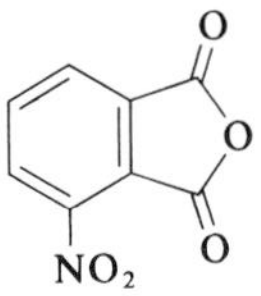

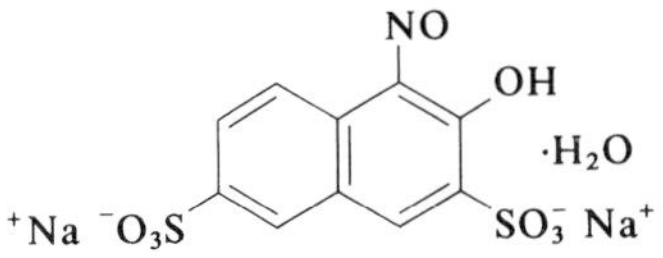

n82

TABLE 1.15 Physical Constants of Organic Compounds (*Continued*)

No.	Name	Formula	Formula weight	Beilstein reference	Density	Refractive index	Melting point	Boiling point	Flash point	Solubility in 100 parts solvent
n83	4-Nitrosophenol	ONC_6H_4OH	123.11	7, 622			132 d			s aq, v s alc, eth; explodes on contact with conc acid, alk or fire
n85	2-Nitrotoluene	$CH_2C_6H_4NO_2$	137.14	5, 318	1.1622^{19}	1.5472^{20}	−10	222	106	s alc, bz
n86	3-Nitrotoluene	$CH_3C_6H_4NO_2$	137.14	5, 321	1.1581^{20}_{4}	1.5459^{20}	15.5	231.9	101	misc alc, eth; s bz
n87	4-Nitrotoluene	$CH_3C_6H_4NO_2$	137.14	5, 323	1.392		53−54	238	106	s alc, bz, chl, eth
n88	2-Nitro-α,α,α-trifluorotoluene	$CF_3C_6H_4NO_2$	191.11	5^2, 251			31−32	105^{20mm}	95	v s alc, bz
n89	3-Nitro-α,α,α-trifluorotoluene	$CF_3C_6H_4NO_2$	191.11	5, 327	1.4364^{16}	1.4715^{20}	−2.4	200−205	87	s alc, eth
n89a	5-Nitrouracil		157.09	24, 320			>300			
n90	Nonadecane	$CH_3(CH_2)_{17}CH_3$	268.51	1, 174	0.7776^{32}_{4}	1.4335^{38}	31.9	330.6	168	s eth; sl s alc
n92	Nonane	$CH_3(CH_2)_7CH_3$	128.26	1, 165	0.7176^{20}_{4}	1.4054^{20}	−53.5	150.8	31 (CC)	s abs alc, eth
n93	1,9-Nonanediamine	$H_2N(CH_2)_9NH_2$	158.29	4, 272			37−38	258	>110	
n94	Nonanedinitrile	$NC(CH_2)_7CN$	150.23	2, 709	0.929	1.4460^{20}		176^{11mm}	>110	v s alc, bz, eth
n95	1,9-Nonanedioic acid	$HOOC(CH_2)_7COOH$	188.22	2, 707	1.029^{24}_{4}		106.5	286^{100mm}		0.24 aq; v s alc; 3 eth
n96	1,9-Nonanediol	$HO(CH_2)_9OH$	160.26	1, 493			47−49	177^{15mm}	>110	
n97	Nonanenitrile	$CH_3(CH_2)_7CN$	139.24	2, 354	0.8211^{15}	1.4260^{20}	−34.2	224.0	81	s alc, eth
n98	Nonanoic acid	$CH_3(CH_2)_7COOH$	158.24	2, 352	0.9062^{20}	1.4330^{20}	12.5	254	100	s alc, chl, eth
n98a	γ-Nonanoic lactone		156.23	17, 245	0.976	1.4475^{20}		122^{6mm}	>110	
n99	1-Nonanol	$CH_3(CH_2)_8OH$	144.26	1, 423	0.8274^{20}_{4}	1.4338^{20}	−5.5	213.1	75	0.6 aq; misc alc, eth
n99a	2-Nonanone	$CH_3(CH_2)_6COCH_3$	142.24	1, 709	0.832	1.4210^{20}	−21	192^{743mm}	64	misc alc, eth
n99b	3-Nonanone	$CH_3(CH_2)_5COC_2H_5$	142.24	1, 709	0.821	1.4204^{20}		188	67	
n100	5-Nonanone	$(C_4H_9)_2CO$	142.24	1, 710	0.8062^{20}	1.4190^{20}	−50	187	60	misc alc, eth
n101	Nonanoyl chloride	$CH_3(CH_2)_7COCl$	176.69	2, 353	0.9461^{15}	1.4377^{20}	−60.5	215.4	95	d aq, alc; s eth
n102	1-Nonene	$CH_3(CH_2)_6CH=CH_2$	126.24	1^2, 202	0.7292^{20}	1.4157^{20}	−81.4	146.9	46	
n102a	3-Nonen-2-one	$CH_3(CH_2)_4CH=CHCOCH_3$	140.23	1^3, 3017	0.848	1.4484^{20}		85^{12mm}	81	s eth
n103	Nonyl aldehyde	$CH_3(CH_2)_7CHO$	142.24	1, 708	0.827^{19}_{4}	1.4240^{20}		185	63	
n104	Nonylamine	$CH_3(CH_2)_8NH_2$	143.27	4, 198	0.782	1.4330^{20}		201	62	sl s aq; s alc, eth
n105	Nopol		166.26		0.973	1.4930^{20}		230−240	98	
n107	Norbornane		96.17	5^2, 45			82−84			s alc
n108	2-Norbornanone		110.16	7, 57			88−91	168−172	33	
n111	exo-2-Norbornyl formate		140.18		1.048	1.4622^{20}		67^{16mm}	53	

	Name	Formula	Mol. wt.	Beilstein ref.	Density	n_D	M.p. °C	B.p. °C	Solubility
n112	(+)-Norephedrine	$C_6H_5CH(OH)CH(CH_3)NH_2$	151.21	13[2], 371			51–54	>110	v s eth; 10 PE; s abs alc
o1	(Z,Z)-9,12-Octadecadienoic acid	$CH_3(CH_2)_4CH=CHCH_2$-$CH=CH(CH_2)_7COOH$	280.44	2, 496	0.9025[20]	1.4699[20]	–5	230[16mm]	s hot alc, hot eth
o2	Octadecanamide	$CH_3(CH_2)_{16}CONH_2$	283.50	2, 384			104	251[12mm]	s acet, eth; sl s alc
o3	Octadecane	$CH_3(CH_2)_{16}CH_3$	254.50	1, 173	0.77674[28]	1.4367[28]	28.2	316.7	s eth; sl s alc
o4	1-Octadecanethiol	$CH_3(CH_2)_{17}SH$	286.57			1.4648	29–31	360	4.9 alc; 20 bz; 50 chl; 3.9 acet
o5	Octadecanoic acid	$CH_3(CH_2)_{16}COOH$	284.48	2, 377	0.847[70]	1.4299[80]	70	383	s alc, eth
o6	1-Octadecanol	$CH_3(CH_2)_{17}OH$	270.50	1, 431	0.8123[58]	1.4388[20]	57.9	203[10mm]	s alc, eth
o7	9,12,15-Octadecatrienoic acid	$CH_3(CH_2CH=CH)_3CH_2$-$(CH_2)_6COOH$	278.44	2, 499	0.9144[18]	1.4800[20]		230[17mm]	s alc, bz, eth
o8	1-Octadecene	$CH_3(CH_2)_{15}CH=CH_2$	252.49	1, 226	0.7911[8]	1.4439[20]	17.7	314.9	s hot acet
o9	9-Octadecen-1-amine	$CH_3(CH_2)_7CH=CH$-$(CH_2)_8NH_2$	267.50		0.813	1.4578[20]			
o10	(Z)-9-Octadecenoic acid	$CH_3(CH_2)_7CH=CH$-$(CH_2)_7COOH$	282.47	2, 463	0.8906[20]	1.4571[20]	4	286[100mm]	misc alc, eth; s bz, chl
o11	(E)-9-Octadecenoic acid	$CH_3(CH_2)_7CH=CH$-$(CH_2)_7COOH$	282.47	2[2], 441	0.851[79]	1.4308[99]	44–45	288[100mm]	s bz, chl, eth
o12	(Z)-9-Octadecen-1-ol	$CH_3(CH_2)_7CH=CH(CH_2)_8OH$	268.49	1, 453	0.8492[4]	1.4610[20]	13–19	195[8mm]	s alc, eth

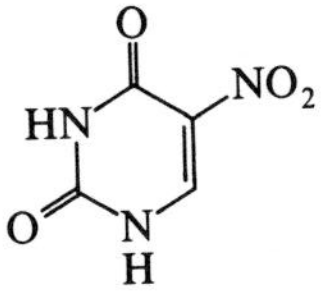

n89a

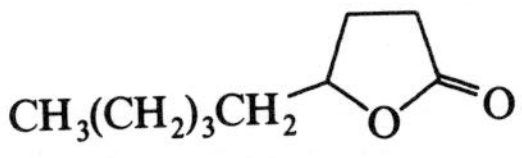

n98a

$CH_3(CH_2)_3CH_2$

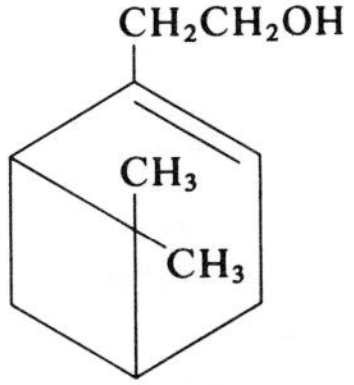

n105

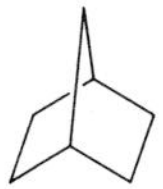

n107

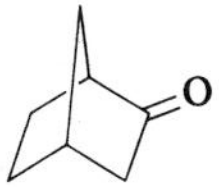

n108

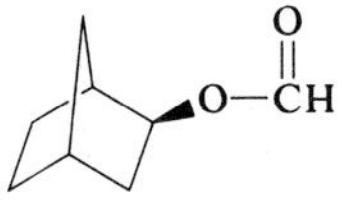

n111

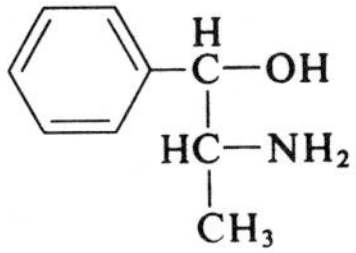

n112

TABLE 1.15 Physical Constants of Organic Compounds (*Continued*)

No.	Name	Formula	Formula weight	Beilstein reference	Density	Refractive index	Melting point	Boiling point	Flash point	Solubility in 100 parts solvent
o12a	9-Octadecenoyl chloride	$CH_3(CH_2)_7CH{=}CH(CH_2)_7COCl$	300.92	2, 469	0.912	1.4623^{20}		180^{3mm}	>110	s alc, bz, eth
o13	Octadecylamine	$CH_3(CH_2)_{17}NH_2$	269.52	4, 196	0.777^{27}		55–57	232^{32mm}	110	
o14	Octadecyl isocyanate	$CH_3(CH_2)_{17}NCO$	299.51	4^3, 439	0.847		15–16	173^{5mm}	148	
o15	Octadecyltrichlorosilane	$CH_3(CH_2)_{17}SiCl_3$	387.94		0.984	1.4602^{20}		223^{10mm}	89	
o16	Octadecyl vinyl ether	$CH_3(CH_2)_{17}OCH{=}CH_2$	296.54		0.821^{30}_{4}	1.4440^{30}	28	187^{5mm}	177	s bz, PE; sl s alc
o17	1,7-Octadiene	$H_2C{=}CH(CH_2)_4CH{=}CH_2$	110.20		0.746	1.4221^{20}		114–121	9	s eth; sl s alc
o18	1H,1H,5H-Octafluoro-1-pentanol	$HCF_2CF_2CF_2CF_2CH_2OH$	232.08		1.6647^{20}	1.3190^{20}		140–141	74	
o19	Octamethylcyclotetrasiloxane	$[-(CH_3)_2SiO-]_4$	296.6	4^3, 1885	0.956	1.3958^{20}	17–18	176	60	0.16 aq; 0.6 eth; s alc
o21	Octamethyltrisiloxane	$[(CH_3)_3SiO]_2Si(CH_3)_2$	236.5	4^3, 1879	0.8200^{20}	1.3838^{20}	−82	153	29	v s alc; sl s aq, eth
o22	Octane	$CH_3(CH_2)_6CH_3$	114.23	1, 159	0.7025^{20}_{4}	1.3974^{20}	−56.8	125.7	22 (OC)	s eth; sl s alc
o23	1,8-Octanediamine	$H_2N(CH_2)_8NH_2$	144.26	4, 271			50–52	225	165	s alc
o24	1,8-Octanedioic acid	$HOOC(CH_2)_6COOH$	174.20	2, 691			140–144	230^{15mm}		
o25	1,2-Octanediol	$CH_3(CH_2)_5CH(OH)CH_2OH$	146.23	1^3, 2217			36–38	132^{10mm}	>110	
o26	1,8-Octanediol	$HO(CH_2)_8OH$	146.23	1, 490			59–61	172^{20mm}		
o27	Octanenitrile	$CH_3(CH_2)_6CN$	125.22	2, 349	0.8135^{20}	1.4202^{20}	−45.6	205.2	73	
o27a	1,2,7,8-Octanetetrol	$[-CH_2CH_2CH(OH)CH_2OH]_2$	178.23				79–83	195^{1mm}		
o28	1-Octanethiol	$CH_3(CH_2)_7SH$	146.30	1^3, 1710	0.843	1.4525^{20}	−49.2	199.0	68	
o29	Octanoic acid	$CH_3(CH_2)_6COOH$	144.21	2, 347	0.9088^{20}_{4}	1.4279^{20}	16.6	239.3	110	0.07 aq; v s alc, chl, eth, PE
o30	1-Octanol	$CH_3(CH_2)_7OH$	130.23	1, 418	0.8258^{20}_{4}	1.4296^{20}	−15.0	195.2	81	0.06 aq; misc alc, chl, eth
o31	2-Octanol	$CH_3(CH_2)_5CH(OH)CH_3$	130.23	1, 419	0.8207^{20}_{4}	1.4202^{20}	−38.6	179–180	71	0.08 aq; misc alc, eth
o32	3-Octanol	$CH_3(CH_2)_4CH(OH)CH_2CH_3$	130.23	1, 208	0.8216^{20}	1.4262^{20}		174–176	65	
o33	4-Octanol	$CH_3(CH_2)_3CH(OH)CH_2CH_2CH_3$	130.23		0.8192^{20}	1.425^{20}		176.6	71	
o34	2-Octanone	$CH_3(CH_2)_5COCH_3$	128.22	1, 704	0.819^{20}	1.4150^{20}	−16	173	62	i aq; misc alc, eth
o35	3-Octanone	$CH_3(CH_2)_4COCH_2CH_3$	128.22	1, 706	0.8220^{20}_{4}	1.4150^{20}		167–168	46	i aq; misc alc, eth
o36	4-Octanone	$CH_3(CH_2)_3COCH_2CH_2CH_3$	128.22	1, 706	0.809	1.4139^{20}		164	45	
o37	Octanoyl chloride	$CH_3(CH_2)_6COCl$	162.66	2, 348	0.955^{15}_{5}	1.4350^{20}	<−70	195	80	d aq; alc; s eth
o38	Octaphenylcyclotetrasiloxane	$[-(C_6H_5)_2SiO-]_4$	793.2		1.185			340^{1mm}		s alc, bz, HOAc

No.	Name	Formula	Mol wt	Beil. ref.	Density	n_D	mp, °C	bp, °C	Flash pt	Solubility
o39	1-Octene	$CH_3(CH_2)_5CH{=}CH_2$	112.22	1, 221	0.7149^{20}	1.4087^{20}	−101.7	121.3	21	i aq; misc alc, eth
o39a	Octyl acetate	$CH_3CO_2(CH_2)_7CH_3$	172.27	2, 134	0.868	1.4184^{20}		211	88 (OC)	sl s aq; misc alc
o40	Octyl aldehyde	$CH_3(CH_2)_6CHO$	128.22	1, 704	0.821^{20}	1.4183^{20}	12–15	163.4	51	sl s aq; misc alc
o41	Octylamine	$CH_3(CH_2)_7NH_2$	129.25	4, 196	0.782	1.4290^{20}	−5 to −1	175–177	62	i aq; s alc, eth
o43	Octyltrichlorosilane	$CH_3(CH_2)_7SiCl_3$	247.67	4^3, 1907	1.070^{20}	1.4473^{20}		226^{730mm}	96	i aq; s alc, eth
o44	1-Octyne	$CH_3(CH_2)_5C{\equiv}CH$	110.19	1, 258	0.7457^{20}	1.4159^{20}	−79.3	126.2	17	i aq; s alc, eth
o45	1-Octyn-3-ol	$CH_3(CH_2)_4CH(OH)C{\equiv}CH$	126.20		0.864	1.4410^{20}			63	v s aq, alc; sl s eth
o46	L-(+)-Ornithine	$H_2N(CH_2)_3CH(NH_2)COOH$	132.16	4, 420			142			v s aq; alc; sl s eth
o47	Oxacycloheptane		100.16		0.890	1.440^{20}		122	10	8.3 aq^{20}, 24 alc; 1.3 eth
o48	Oxalic acid	$HOOCCOOH$	90.04	2, 502	1.90^{17}		189 d			14 aq; 40 alc; 1 eth
o49	Oxalic acid dihydrate	$HOOCCOOH \cdot 2H_2O$	126.07	2, 502	1.653^{19}		−2H$_2$O, 102			
o50	Oxalyl bromide	$BrCO{-}COBr$	215.84	2, 542		1.5220		103^{720mm}	none	s eth; viol d aq, alc
o51	Oxalyl chloride	$ClCO{-}COCl$	126.93	2, 559	1.488^{13}	1.4340^{13}	−12	64	none	s hot aq; sl s alc, eth
o52	Oxalyl dihydrazide	$H_2NNHCO{-}CONHNH_2$	118.10	2, 559			240 d			s alk; sl s aq; i eth
o53	Oxamic hydrazide	$H_2NCO{-}CONHNH_2$	103.08	2, 545			218 d			sl s hot aq, alc
o54	Oxamide	$H_2NCO{-}CONH_2$	88.07	27, 135	1.667^{20}		d 350			
o55	2-Oxazolidone		87.08				86–89	220^{48mm}		v s aq, alc; v sl s eth
o56	2-Oxobutyric acid	$CH_3CH_2C({=}O)COOH$	102.09	3, 629	1.200^{17}	1.3972^{20}	32–34	82^{16mm}	81	

o47

o55

TABLE 1.15 Physical Constants of Organic Compounds (*Continued*)

No.	Name	Formula	Formula weight	Beilstein reference	Density	Refractive index	Melting point	Boiling point	Flash point	Solubility in 100 parts solvent
o57	2-Oxohexamethylene-imine		113.16	21^2, 216	1.024^{25}	1.4935	69.2	180^{50mm}		84 aq
o57a	5-Oxohexanonitrile	$CH_3CO(CH_2)_3CN$	111.14	3^3, 1234	0.975	1.4328^{20}		240	107	v s aq, alc, bz, eth
o58	4-Oxopentanoic acid	$CH_3COCH_2CH_2COOH$	116.12	3, 671	1.1447^{25}	1.4396^{20}	33–35	245.8	137	s aq, alc
o59	2-Oxopropionaldehyde	CH_3COCHO	72.06	1, 762	1.0455^{24}	1.4209^{20}		72	none	misc aq, alc, eth
o60	2-Oxopropionic acid	$CH_3COCOOH$	88.06	3, 608	1.267^{4}	1.4315^{20}	11.8	165 d	82	v s aq, alc; sl s eth
o61	2,2'-Oxydiacetic acid	$HOOCCH_2OCH_2COOH$	134.09	3, 234			142–145	dec	218	
o62	4,4'-Oxydianiline	$H_2NC_6H_4OC_6H_4NH_2$	200.24	13, 441	1.043	1.4405^{20}	192		>110	
o63	3,3'-Oxydipropionitrile	$NCCH_2CH_2OCH_2CH_2CN$	124.14				156 d	$112^{0.5mm}$	71	
p1	Paraformaldehyde	$(CH_2O)_x$		1, 566						s(slow) aq; s alk; i alc, eth
p2	Paraldehyde	$[-C(CH_3)O-]_3$	132.16	19, 385	0.9984^{15}	1.4049^{20}	12	124		11 aq; misc alc, chl
p3	Parathion	$(C_2H_5O)_2P(=S)C_6H_4NO_2$	291.27	5, 357	1.265^{25}	1.5370^{25}	6	375		v s alc, bz, eth
p5	Pentabromoethylbenzene	$CH_3CH_2C_6Br_5$	500.67				137–139			
p6	Pentabromophenol	C_6Br_5OH	488.62	6, 206			223–226	subl		sl s alc, eth
p7	Pentachloroacetone	$Cl_2CHCOCCl_3$	230.31	1, 656	1.690	1.4967^{20}	21 anhyd	192	none	i aq; v s acet
p8	Pentachlorobenzene	C_6HCl_5	250.34	5, 205	1.8342^{16}		82–85	275–277		v s bz, chl, eth
p9	Pentachloroethane	Cl_2CHCCl_3	202.30	1, 87	1.6712^{25}	1.5030^{20}	−29.0	160.5	none	0.05 aq; misc alc, eth; s bz, chl
p10	Pentachloronitrobenzene	$C_6Cl_5NO_2$	295.34	5, 247	1.718^{5}		140–143			
p11	Pentachlorophenol	C_6Cl_5OH	266.34	6, 194	1.978^{22}		190–191	310 d		v s alc; s bz; 148 eth
p12	Pentachloropyridine	C_5Cl_5N	251.33	20, 232			124–126			
p13	Pentadecane	$CH_3(CH_2)_{13}CH_3$	212.42	1, 172	0.7684^{4}	1.4319^{20}	9.9	270.6	132	v s alc, eth
p14	8-Pentadecanone	$[CH_3(CH_2)_{12}]_2CO$	226.40	1, 717			41–43	178	110	s alc
p15	3-Pentadecylphenol	$C_{15}H_{31}C_6H_4OH$	304.52				50–53	195^{1mm}	>110	
p16	1,2-Pentadiene	$CH_3CH_2CH=C=CH_2$	68.12	1, 251	0.6926^{20}	1.4209^{20}	−137.3	44.9		
p17	(E)-1,3-Pentadiene	$CH_3CH=CHCH=CH_2$	68.12	1, 251	0.6760^{20}	1.4301^{20}	−87.5	42.0	−28	
p18	(Z)-1,3-Pentadiene	$CH_3CH=CHCH=CH_2$	68.12	1, 251	0.6910^{20}	1.4363^{20}	−140.8	44.1	−28	
p19	1,4-Pentadiene	$H_2C=CHCH_2CH=CH_2$	68.12	1, 251	0.6608^{22}	1.3888^{20}	−148.3	26.0	4	
p20	Pentaerythritol	$C(CH_2OH)_4$	136.15	1, 528	1.38^{25}	1.548	260	subl		6 aq; v sl s alc; i eth
p21	Pentaerythritol triacrylate	$(H_2C=CHCO_2CH_2)_3CCH_2OH$	298.30		1.180	1.4864^{20}			>110	

No.	Name	Formula	Mol wt	Beilstein ref.	Density	n_D	mp, °C	bp, °C	Flash pt, °C	Solubility
p22	Pentaerythrityl tetranitrate	$C(CH_2ONO_2)_4$	316.15	1², 602	1.7773^{20}_{4}	1.5096^{20}	140	explodes on percussion, shock	110	s acet; sl s eth, alc
p23	Pentaethylenehexamine	$H_2N(CH_2CH_2NH)_4CH_2NH_2$	232.38	4⁴, 1245	0.950	1.527^{20}			91	v s alc, bz
p24	Pentamethylbenzene	$C_6H(CH_3)_5$	148.25	5, 443	0.917^{20}_{4}	1.4733^{20}	54.4	231	44	
p25	1,2,3,4,5-Pentamethylcyclopentadiene		136.24		0.870			58^{13mm}		
p26	1,5-Pentamethylenetetrazole		138.17	26², 213			59–61	194^{12mm}		1.4 aq; misc alc, eth
p27	Pentanal	$CH_3CH_2CH_2CH_2CHO$	86.13	1, 676	0.8095^{4}	1.3942^{20}	−92	103	12	misc alc, eth
p28	Pentane	$CH_3CH_2CH_2CH_2CH_3$	72.15	1, 130	0.6262^{20}_{4}	1.3575^{20}	−129.7	36.1	−49	
p29	1,5-Pentanediamine	$H_2N(CH_2)_5NH_2$	102.18	4, 266	0.8734^{25}	1.4591^{20}		178–180	62	s aq, alc; sl s eth
p29a	1,2-Pentanediol	$CH_3CH_2CH_2CH(OH)CH_2OH$	104.15	1², 548	0.971	1.4397^{20}		206	104	s aq, alc; sl s eth
p30	1,5-Pentanediol	$HO(CH_2)_5OH$	104.15	1, 481	0.9941^{20}	1.4494^{20}	−15.6	242.5	129	
p31	2,3-Pentanedione	$CH_3CH_2COCOCH_3$	100.11	1, 776	0.957	1.4068^{20}	−52	110–112	19	
p32	2,4-Pentanedione	$CH_3COCH_2COCH_3$	100.11	1, 777	0.9721^{25}	1.4510^{20}	−23.1	140.6	40 (CC)	17 aq; misc alc, eth
p33	Pentanenitrile	$CH_3CH_2CH_2CH_2CN$	83.13	2, 301	0.80355^{4}	1.3991^{15}	−96.8	141.3	40	i aq; s alc, eth
p34	1-Pentanesulfonic acid, Na salt	$CH_3(CH_2)_4SO_3^-Na^+$	174.19	4³, 23			>300			4 aq

Structures: o57 (azepan-2-one; ring with N, H, O); p25 (pentamethylcyclopentadiene; five CH₃ groups — CH_3, CH_3, CH_3, CH_3, CH_3); p26 (pentamethylenetetrazole; fused N–N–N–N ring).

TABLE 1.15 Physical Constants of Organic Compounds (*Continued*)

No.	Name	Formula	Formula weight	Beilstein reference	Density	Refractive index	Melting point	Boiling point	Flash point	Solubility in 100 parts solvent
p35	1-Pentanethiol	$CH_3(CH_2)_4SH$	104.22	1, 384	0.840	1.4460^{20}	−75.7	126.6	18	i aq; misc alc, eth
p36	Pentanoic acid	$CH_3(CH_2)_3COOH$	102.13	2, 299	0.9390^4	1.4080^{20}	−33.7	185.5	88	2.4 aq; v s alc, eth
p37	1-Pentanol	$CH_3(CH_2)_4OH$	88.15	1, 383	0.8148^{20}	1.4100^{20}	−78.9	137.8	33 (CC)	2.7 aq; misc alc, eth
p38	2-Pentanol	$CH_3CH_2CH_2CH(OH)CH_3$	88.15	1, 384	0.8393^{20}	1.4064^{20}	glass	119.0	40 (CC)	16.6 aq; misc alc, eth
p39	3-Pentanol	$CH_3CH_2CH(OH)CH_2CH_3$	88.15	1, 385	0.8150^{25}	1.4079^{25}	−69	115.6	40	5.2 aq; s alc, eth
p41	2-Pentanone	$CH_3CH_2CH_2COCH_3$	86.13	1, 676	0.8095^{20}	1.3903	−77.8	101.7	7	misc acet, bz, eth, PE
p42	3-Pentanone	$CH_3CH_2COCH_2CH_3$	86.13	1, 679	0.8143^{20}	1.3923^{20}	−39.0	102.0	12	3.4 aq
p43	Pentanophenone	$C_6H_5CO(CH_2)_3CH_3$	162.23	7, 327	0.988	1.5143^{20}		107^{5mm}	102	s alc, eth
p44	Pentanoyl chloride	$CH_3CH_2CH_2CH_2COCl$	120.58	2, 301	1.016	1.4216^{20}		125–127	32	
p45	1,4,7,10,13-Pentaoxa-cyclopentadecane	$[—CH_2CH_2O—]_5$	220.27		1.109	1.4650^{20}		$135^{0.2mm}$	>110	
p47	1-Pentene	$CH_3CH_2CH_2CH=CH_2$	70.14	1, 210	0.6410^{20}	1.3714^{20}	−165.2	30.0	−28	misc alc, bz, eth
p48	(E)-2-Pentene	$CH_3CH_2CH=CHCH_3$	70.14	1, 210	0.6482^{20}	1.3793^{20}	−135	35.85	−45	misc alc, eth
p49	(Z)-2-Pentene	$CH_3CH_2CH=CHCH_3$	70.14	1, 210	0.6503^{20}	1.3830^{20}	−179	37.0	−27	misc alc, eth
p50	(Z)-2-Pentenenitrile	$CH_3CH_2CH=CHCN$	81.12	2^2, 400	0.820	1.4269^{20}		60^{72mm}	19	
p51	(E)-3-Pentenenitrile	$CH_3CH=CHCH_2CN$	81.12	2, 427	0.837	1.4221^{20}		144–147	40	
p52	Pentyl acetate	$CH_3(CH_2)_4OOCCH_3$	130.19	2, 131	0.8753^{20}	1.4028^{20}	−100	149.2	25 (CC)	0.17 aq; misc alc, eth
p53	Pentylamine	$CH_3(CH_2)_4NH_2$	87.17	4, 175	0.752	1.4110^{20}	−50	104	4	v s aq; misc alc, eth
p54	Pentylbenzene	$CH_3(CH_2)_4C_6H_5$	148.25	5, 434	0.8594^{20}	1.4885^{20}	−78.3	202.2	65	s alc; misc bz, eth
p55	2-Pentylcinnamalde-hyde	$C_6H_5CH=C[(CH_2)_4CH_3]CHO$	202.30	7^3, 310	0.970	1.5571^{20}		290	>110	s alc; misc bz, eth
p56	4-*tert*-Pentylphenol	$CH_3CH_2C(CH_3)_2C_6H_4OH$	164.25	6, 548	0.9624^{20}	1.3852^{20}	93	262.2		s alc, eth
p57	1-Pentyne	$CH_3CH_2CH_2C≡CH$	68.11	1, 250	0.6901^{20}	1.3876^{20}	−105.7	40.2	−34	v s alc; misc eth
p59	Peroxyacetic acid	$CH_3C(=O)OOH$	76.05	2, 169	1.226^{15}		0.1	105 explodes 110	41 (OC)	v s aq, alc, eth
p60	Petroleum ether	Principally pentanes and hex-anes			0.640			35–80	−40	misc bz, chl, eth, CCl4
p61	Phenanthrene		178.23	5, 667	1.179^{25}		100	340		1.6 alc; 50 bz; 30 eth s bz, eth, hot alc
p62	9,10-Phenanthrene-dione		208.22	7, 796	1.405^4		207	−360		
p63	1,10-Phenanthroline		180.21	23, 227			117			0.3 aq; 1.4 bz; s alc, acet
p64	Phenol	C_6H_5OH	94.11	6, 110	1.0576^{41}	1.5418^{41}	40.9	181.8	79 (CC)	6.7 aq; 8.2 bz; v s alc, chl, eth, alk

No.	Name	Formula	M.W.	Ref.	Density	n_D	M.P.	B.P.	Fl. P.	Solubility
p65	Phenolphthalein		318.33	18, 143	1.2994^{25}		258–262	371		8.2 alc; 1 eth
p66	Phenothiazine		199.28	27, 63			185.1	285 sl dec		v s bz; s eth; sl s alc
p68	Phenoxyacetic acid	$C_6H_5OCH_2COOH$	152.15	6, 161			98			1.3 aq; s alc, bz, HOAc, CS_2, eth
p68a	4'-Phenoxyacetophen-one	$C_6H_5OC_6H_4COCH_3$	212.25	8, 88	1.235	1.5340^{20}	52	200^{12mm}	>110	
p69	Phenoxyacetyl chloride	$C_6H_5OCH_2COCl$	170.60	6, 162				226	108	d aq, alc; s eth
p70	4-Phenoxyaniline	$C_6H_5OC_6H_4NH_2$	185.23	13, 438			84	189^{14mm}		s hot aq; v s alc, eth
p71	2-Phenoxybutyric acid	$CH_3CH_2CH(OC_6H_5)COOH$	180.20	6, 163	1.102^{24}	1.5370^{20}	79–83	258	110	sl s aq
p72	2-Phenoxyethanol	$C_6H_5OCH_2CH_2OH$	138.17	6, 146			14	245.2	110	s aq; v s alc, eth
p72a	3-Phenoxy-1,2-pro-panediol	$C_6H_5OCH_2CH(OH)CH_2OH$	168.19	6, 149			56	315		
p73	1-Phenoxy-2-propanol	$C_6H_5OCH_2CH(OH)CH_3$	152.19	61, 85	1.063^{25}	1.523^{20}	13–18	240	135	s alc; sl s aq
p74	Phenoxy-2-propanone	$C_6H_5OCH_2COCH_3$	150.18	6, 151	1.097	1.5210^{20}		230	85	
p75	DL-2-Phenoxypro-pionic	$CH_3CH(OC_6H_5)COOH$	166.18	6, 163			119	265		

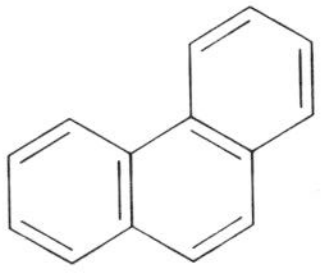

p61

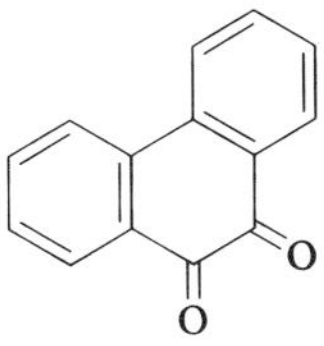

p62

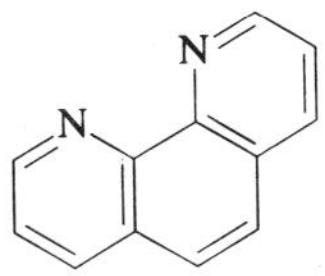

p63

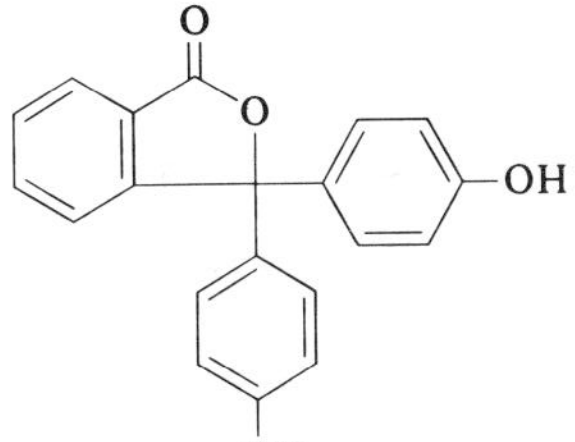

p65

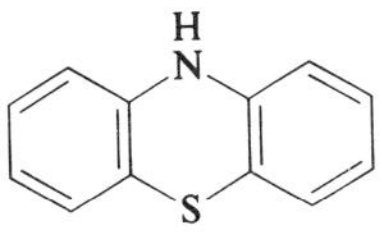

p66

TABLE 1.15 Physical Constants of Organic Compounds (*Continued*)

No.	Name	Formula	Formula weight	Beilstein reference	Density	Refractive index	Melting point	Boiling point	Flash point	Solubility in 100 parts solvent
p76	3-Phenoxytoluene	$C_6H_5OC_6H_4CH_3$	184.24	6, 377	1.051	1.5727^{20}		273	>110	sl s aq; s alc, eth
p76a	Phenylacetaldehyde	$C_6H_5CH_2CHO$	120.15	7, 292	1.0272^{25}	1.5273^{20}	33–34	195	86	
p77	Phenylacetaldehyde ethylene acetal		164.21	19^4, 220	1.100	1.5220^{20}		120^{12mm}	107	
p78	Phenyl acetate	$C_6H_5OOCCH_3$	136.15	6, 152	1.0734^{20}	1.5030^{20}		196	76	misc alc, eth, chl
p79	Phenylacetic acid	$C_6H_5CH_2COOH$	136.15	9, 431	1.0917^{77}		76.5	265.5		s hot aq, alc, eth
p80	Phenylacetonitrile	$C_6H_5CH_2CN$	117.15	9, 441	1.02141^{15}	1.5233^{20}	−23.8	233.5	101	i aq; misc alc, eth
p81	Phenylacetyl chloride	$C_6H_5CH_2COCl$	154.60	9, 436	1.169	1.5325^{20}		95^{12mm}	102	d aq, alc
p82	Phenylacetylene	$C_6H_5C{\equiv}CH$	102.14	5, 511	0.9300^{20}	1.5470^{20}	−44.9	142.4	31	misc alc, eth
p83	Phenylacetylurea	$C_6H_5CH_2CONHCONH_2$	178.19				212–216 d 283			sl s alc, bz, chl, eth
p84	L-3-Phenylalanine	$C_6H_5CH_2CH(NH_2)COOH$	165.19	14, 495			185 d			3 aq; s hot alc; i eth
p85	2-(Phenylamino)benzoic acid	$C_6H_5NHC_6H_4COOH$	213.24	14, 327						s hot alc
p86	Phenyl 4-aminosalicylate	$H_2NC_6H_3(OH)COOC_6H_5$	229.24				153			0.7 mg aq
p88	4-Phenylazoaniline	$C_6H_5N{=}NC_6H_4NH_2$	197.24	161, 310			128	>360		v s alc, bz, chl, eth
p89	Phenylazoformic acid 2-phenylhydrazide	$C_6H_5N{=}NCONHNHC_6H_5$	240.27	16, 24			156–159 d			
p90	4-Phenylazophenol	$C_6H_5N{=}NC_6H_4OH$	198.23	16, 96			155–157	230^{20mm}		v s alc, eth
p91	2-Phenylbenzimidazole		194.24	23, 230			291			s abs alc; sl s bz, chl
p92	Phenyl benzoate	$C_6H_5COOC_6H_5$	198.22	9, 116	1.235	1.5122^{20}	70	314	98	v s hot alc; sl s eth
p93	N-Phenylbenzylamine	$C_6H_5CH_2NHC_6H_5$	183.25	12, 1023	1.061	1.5836^{45}	37–38	307	>110	s alc, chl, eth
p94	1-Phenylbiguanide	$C_6H_5NHC(=NH)NHC(=NH)NH_2$	177.21				144–146			v s aq, alc
p97	4-Phenyl-2-butanone	$C_6H_5CH_2CH_2COCH_3$	148.21	7, 314	0.989	1.5112^{20}		235	98	s alc, eth
p98	(E)-4-Phenyl-3-buten-2-one	$C_6H_5CH{=}CHCOCH_3$	146.19	7, 364	1.00974^5	1.5196^{20}	41.5	261	65	v s alc, bz, chl, eth
p99	4-Phenylbutylamine	$C_6H_5CH_2CH_2CH_2CH_2NH_2$	149.24	12, 1165	0.944	1.5196^{20}		124^{17mm}	101	
p100	2-Phenyl-3-butyn-2-ol	$CH_3C(OH)(C_6H_5)C{\equiv}CH$	146.19	6^2, 559			51–52	217–218	96	0.8 aq; s alc, bz, acet
p101	2-Phenylbutyric acid	$CH_3CH_2CH(C_6H_5)COOH$	164.20	9^2, 356			42–44	270–272		s bz, eth
p103	DL-2-Phenylbutyronitrile	$CH_3CH_2CH(C_6H_5)CN$	145.21	9, 541	0.974	1.5086^{20}		114^{15mm}	105	
p104	Phenyl chloroformate	C_6H_5OOCCl	156.57	6, 159	1.248	1.5107^{20}	7.0	71^{9mm}	75	v s alc, eth
p106	Phenylcyclohexane	$C_6H_5C_6H_{11}$	160.26	5, 503	0.9427^{20}	1.5263^{20}		240.1	98	

No.	Name	Formula	Mol. wt.	Beilstein		Density	n	mp	bp	Flash	Solubility
p107	Phenyl dichlorophosphate	$C_6H_5OP(O)Cl_2$	210.98	6	179	1.412	1.5230^{20}		241–243	>110	5 aq; v s alc; 29 eth; 25 bz
p108	N-Phenyldiethanolamine	$C_6H_5N(CH_2CH_2OH)_2$	181.24	12	183	1.1206^{20}		56–58	350 sl d		v s alc, chl, eth
p109	1,2-Phenylenediamine	$C_6H_4(NH_2)_2$	108.14	13	6			103	258		s aq, alc, acet, chl
p110	1,3-Phenylenediamine	$C_6H_4(NH_2)_2$	108.14	3	33	1.1391^{5}		62–63	234–237		1 aq; s alc, chl, eth
p111	1,4-Phenylenediamine	$H_2NC_6H_4NH_2$	108.14	13	61			145	267	156	
p111a	1,4-Phenylene diisocyanate	$C_6H_4(NCO)_2$	160.13	13	105			97–101	112^{12mm}		
p112	1,2-Phenylene phosphorochloridite		174.52	27	809	1.466	1.5712^{20}	66–68	80^{20mm}	>110	v s aq, alc, bz, eth, chl, HOAc
p113	1-Phenyl-1,2-ethanediol	$C_6H_5CH(OH)CH_2OH$	138.17	6	907				272		2.3 aq
p114	1-Phenylethanol	$CH_3CH(C_6H_5)OH$	122.17	6	475	1.0150^{20}	1.5211^{20}	21.4	203.9	85	2 aq; misc alc, eth
p115	2-Phenylethanol	$C_6H_5CH_2CH_2OH$	122.17	6	478	1.0182^{25}	1.5317^{20}	−27	221	102	
p115a	2-Phenylethyl acetate	$CH_3CO_2CH_2CH_2C_6H_5$	164.20	9	510	0.984	1.4985^{20}		239	101	
p116	2-Phenylethylamine	$C_6H_5CH_2CH_2NH_2$	212.18	12	1096	0.9640^{25}	1.5332^{20}		195	90	s aq; v s alc, eth
p116a	1-Phenylethyl propionate	$C_2H_5CO_2CH(CH_3)C_6H_5$	178.23	6^3	1680	1.007	1.4895^{20}		92^{5mm}	94	
p117	(−)-2-Phenylglycine	$C_6H_5CH(NH_2)COOH$	151.17	14	460			302 d			

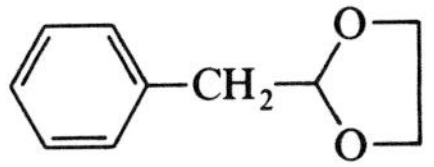

p77

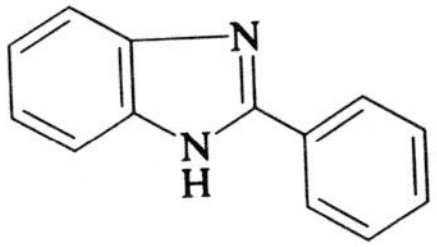

p91

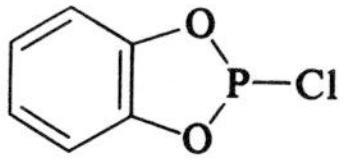

p112

TABLE 1.15 Physical Constants of Organic Compounds (*Continued*)

No.	Name	Formula	Formula weight	Beilstein reference	Density	Refractive index	Melting point	Boiling point	Flash point	Solubility in 100 parts solvent
p118	1-Phenylheptane	$C_6H_5(CH_2)_6CH_3$	176.30	5, 451	0.860	1.4842^{20}	−61	233	95	misc eth
p119	1-Phenylhexane	$C_6H_5(CH_2)_5CH_3$	162.28	5^2, 337	0.861	1.4865^{20}		226	83	misc alc, bz, chl, eth
p120	Phenylhydrazine	$C_6H_5NHNH_2$	108.14	15^2, 44	1.09782^4	1.6070^{20}	19.5	243.5 d	88	
p121	Phenyl 3-hydroxy-2-naphthoate	$C_{10}H_6(OH)COOC_6H_5$	264.28	10, 335			129–132	261^{160mm}		
p121a	2-Phenyl-2-imidazoline		146.19	23, 154			94–99			
p122	2-Phenylindole		193.25	20, 467			175	250^{10mm}		
p123	Phenyl isocyanate	C_6H_5NCO	119.12	12, 437	1.09562^{20}	1.5350^{20}	−30	163	55	d aq, alc; s eth
p124	Phenyl isothiocyanate	C_6H_5NCS	135.19	12, 453	1.12884^4	1.6497^{20}	−21	221	87	i aq; s alc, eth
p125	N-Phenylmaleimide		173.17	21, 400			89–90	163^{12mm}		s alc, chl, eth
p126	Phenylmalonic acid	$C_6H_5CH(COOH)_2$	180.16				155 d			0.17 aq; s alc, bz, acet
p127	Phenylmercury(II) acetate	$C_6H_5HgOOCCH_3$	336.74				149			
p128	Phenylmercury(II) chloride	C_6H_5HgCl	313.15	16, 952			250–252			s bz, eth, pyr
p129	Phenylmercury(II) hydroxide	C_6H_5HgOH	294.70				190 d			
p130	Phenylmethanethiol	$C_6H_5CH_2SH$	124.21	6, 453	1.058^{20}			194–195	70	1.0 aq; v s hot alc
p131	N-Phenylmorpholine		163.22	27, 6	1.058^{270}		57	268	>110	s alc, bz, chl, eth
p132	N-Phenyl-1-naphthylamine	$C_{10}H_7NHC_6H_5$	219.29	12, 1224			62	226^{15mm}		
p132a	N-Phenyl-2-naphthylamine	$C_{10}H_7NHC_6H_5$	219.29	12, 1275	1.213		109	395		
p133	2-Phenylphenol	$C_6H_5C_6H_4OH$	170.21	6^2, 623			57	282	123	s alc, chl, eth, alk
p134	4-Phenylphenol	$C_6H_5C_6H_4OH$	170.21	6, 674			164–165	305–308	165	s alc, chl, eth, alk
p135	N-Phenyl-1,4-phenylenediamine	$C_6H_5NHC_6H_4NH_2$	184.24	13, 76			73–75			
p137	Phenylphosphinic acid	$C_6H_5PH(O)OH$	142.09	16, 791			85–87			
p138	Phenylphosphonic acid	$C_6H_5P(O)(OH)_2$	158.09	16, 803			163–166			
p139	Phenylphosphonic dichloride	$C_6H_5P(O)Cl_2$	194.99	16, 804	1.375	1.5600^{20}	3	258	>110	
p140	Phenylphosphonothioic dichloride	$C_6H_5P(S)Cl_2$	211.05	16, 807	1.360	1.6244^{20}		205^{130mm}	>110	
p141	N-Phenylpiperazine		162.24	23^3, 49	1.06214^4	1.5875^{20}		286	>110	i aq; misc alc

No.	Name	Formula	Formula wt.	Beilstein ref.	Density	n_D	M.p., °C	B.p., °C	Flash p., °C	Solubility
p142	2-Phenyl-1,2-propanediol	$CH_3C(C_6H_5)(OH)CH_2OH$	152.19	6, 930			44–45	162^{26mm}	>110	misc alc, bz
p143	3-Phenyl-1-propanethiol	$C_6H_5CH_2CH_2CH_2SH$	152.26	6^1, 253	1.010	1.5494^{20}		109^{10mm}	90	s aq; misc alc, eth
p144	1-Phenyl-1-propanol	$C_6H_5CH(OH)CH_2CH_3$	136.19	6, 502	0.99154^{25}	1.5169^{23}	−18	219	109	v s alc, eth; misc bz
p145	3-Phenyl-1-propanol	$C_6H_5CH_2CH_2CH_2OH$	136.19	6, 503	1.008	1.5257^{20}	27	235	84	i aq; s alc
p146	1-Phenyl-2-propanone	$C_6H_5CH_2COCH_3$	134.18	7^2, 233	1.01572^{20}_4	1.5160^{20}		100^{13mm}	76	
p147	2-Phenylpropionaldehyde	$CH_3CH(C_6H_5)CHO$	134.18	7, 305	1.009^{20}	1.5175^{20}		202–205		
p148	3-Phenylpropionic acid	$C_6H_5CH_2CH_2COOH$	150.18	9, 508	1.047^{100}_4		47–48	280	>110	0.6 aq; s bz, alc, chl, eth, HOAc, PE
p150	1-Phenyl-3-pyrazolidinone		162.19	24, 2			121			10 hot aq; hot alc; s alk, acid
p151	2-Phenylpyridine	$C_6H_5C_5H_4N$	155.20	20, 424	1.086	1.6232^{20}		268–270	>110	s alc, eth
p152	2-Phenyl-4-quinoline-carboxylic acid		249.27	22, 103			214–215			0.8 alc; 1 eth; 0.3 chl
p153	Phenyl salicylate	$C_6H_5(OH)COOC_6H_5$	214.22	10, 76	1.25		41–43	173^{12mm}	>110	17 alc; 66 bz; s acet, chl, eth; 0.015 aq
p154	Phenylselenenyl chloride	C_6H_5SeCl	191.52	6^2, 319			63	120^{20mm}		
p155	Phenylsuccinic acid	$HOOCCH_2CH(C_6H_5)COOH$	194.19	9, 865			167–169	$-H_2O$, >160		s hot aq, alc, eth

p121 p122 p125 p131 p141 p150 p152

TABLE 1.15 Physical Constants of Organic Compounds (*Continued*)

No.	Name	Formula	Formula weight	Beilstein reference	Density	Refractive index	Melting point	Boiling point	Flash point	Solubility in 100 parts solvent
p156	S-Phenyl thioacetate	$C_6H_5SCOCH_3$	152.22	12, 388	1.3	1.5720^{20}		100^{6mm}	79	0.25 aq; s alc, alk
p157	1-Phenyl-2-thiourea	$C_6H_5NHC(S)NH_2$	152.22	16, 911			154			
p158	Phenyltrichlorosilane	$C_6H_5SiCl_3$	211.55	16, 911	1.329^{20}	1.5230^{20}		201	91	
p159	1-Phenyltridecane	$C_6H_5(CH_2)_{12}CH_3$	260.47		0.85554^{20}	1.4814^{20}	10	346	>110	
p160	Phenyltriethoxysilane	$C_6H_5Si(OC_2H_5)_3$	240.38		0.996	1.4604^{20}		113^{10mm}	42	
p161	Phenyltrimethoxysilane	$C_6H_5Si(OCH_3)_3$	198.3		1.064^{20}	1.4734^{20}		211		
p162	Phenyltrimethylammonium bromide	$[C_6H_5N(CH_3)_3]^+\ Br^-$	216.13	12, 158			210 d			v s aq; s hot alc
p163	Phenyltrimethylammonium chloride	$[C_6H_5N(CH_3)_3]^+\ Cl^-$	171.67	12, 158			237 subl			s aq; v s alc; sl s chl
p164	Phenyltrimethylammonium iodide	$[C_6H_5N(CH_3)_3]^+\ I^-$	263.12	12, 159			175			s aq, alc; sl s acet
p165	Phenyltrimethylammonium tribromide	$[C_6H_5N(CH_3)_3]^+\ Br_3^-$	375.95	12, 159			114–116			
p166	Phenyltrimethylsilane	$C_6H_5Si(CH_3)_3$	150.30	16^1, 525	0.873	1.4907^{20}		168–170	44	s hot aq, hot alc, eth
p167	Phenyltris(trimethylsiloxy)silane	$[(CH_3)_3SiO]_3SiC_6H_5$	372.8		0.9704^{25}	1.459^{25}		264–266	121	
p168	Phenylurea	$C_6H_5NHCONH_2$	136.15	12, 346	1.302		145–147	238		
p169	Phenylvinyldichlorosilane	$H_2C{=}CH(C_6H_5)SiCl_2$	203.2		1.1962^{25}	1.534^{25}		$87^{1.5mm}$		
p170	1,2-Phthalic acid	$C_6H_4(COOH)_2$	166.13	9, 791	1.593^{20}		−230 rapidly heated			0.6 aq; 10 alc; 0.5 eth; v sl s chl
p171	Phthalic anhydride		148.12	17, 469	1.53		130.8	295	151	0.6 aq(dec); s alc
p172	Phthalide		134.13	17, 310	1.164^{99}		72–74	290		s alc
p173	Phthalimide		147.13	21, 458			238	subl		v s alk; v sl s bz, PE dec by aq, alc; s eth
p174	1,2-Phthaloyl dichloride	$C_6H_4(COCl)_2$	203.02	9, 805	1.409^{20}	1.5684^{20}	15–16	280–282	>110	
p175	Phthalylsulfathioazole		403.44				272 d			s alk; sl s alc; i chl 1.3 aq; 8.2 alc; 10 bz; 2.9 chl; 1.6 eth
p176	Picric acid	$(O_2N)_3C_6H_2OH$	229.11	6, 265	1.763^{20}		122–123	explodes >300		
p177	(Z)-Pinane		138.3	5, 93	0.8392^{20}	1.4616^{20}	−50	167–168		
p178	(+)-α-Pinene		136.24	5, 146	0.8591^{20}	1.4660^{20}	−55	155–156	32	
p179	(−)-β-Pinene		136.24	5, 154	0.8590^{20}	1.4666^{20}	−61.5	166	32	misc alc, eth

No.	Name	Mol wt	Beil. ref	Density	n_D	bp	mp	FP	Solubility
p180	α-Pinene oxide	152.24	5, 152	0.964	1.4690^{20}	$103^{50\text{mm}}$		65	v s aq; 50 alc; i eth
p181	β-Pinene oxide	152.24	17², 44	0.976	1.4765^{20}	$100^{27\text{mm}}$		66	
p182	Piperazine	86.14	23, 4		1.446^{113}	146	108–110	109	
p183	1-Piperazinecarboxal-dehyde	114.15		1.107	1.5094^{20}	$97^{0.5\text{mm}}$		101	
p184	1,4-Piperazinedicar-bonitrile	136.16	23¹, 5				167–170		
p186	Piperidine	85.15	20, 6	0.8659^{15}	1.4525^{20}	106	−7	4	misc aq; s alc, bz, chl

Phorone, d531
Phthalaldehydic acid, f33
m-Phthalic acid, b16
p-Phthalic acid, b17
Phthalonitrile, d236
Picolinaldehyde, p255
Picolines, m401, m402, m403
Picolinic acid, p259, p261
Picolinonitrile, c296

Picolylamines, a226, a227, a228
Picramide, t388
Pimelic acid, h8
Pinacol, d493
Pinacolone, d499
Pinacolyl alcohol, d498
3-Pinanol, i83
Pipecolines, m372, m373, m374
1-Piperazinoethanol, h123

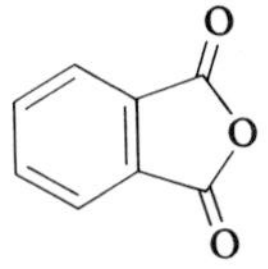

Phenyl sulfide, d690
Phenyl sulfone, d691
Phenylsulfonic acid, b22
Phenyl sulfoxide, d692
(Phenylthio)acetic acid, t163
Phenyl thiocarbamide, p157
α-Phenyl-o-toluic acid, b84
Phenyl m-tolyl ether, p76
Phloroglucinol, t310

p171
p172 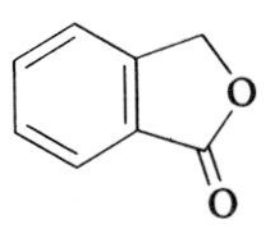
p173 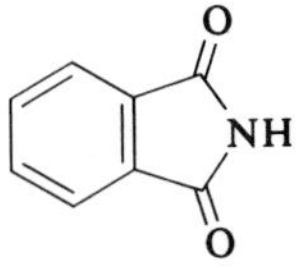
p175 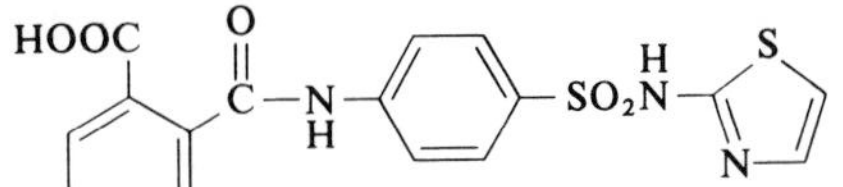
p177 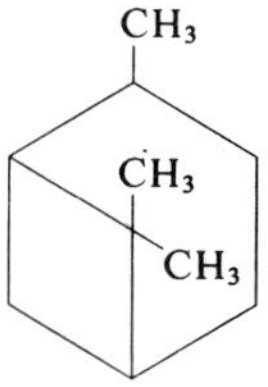
p178 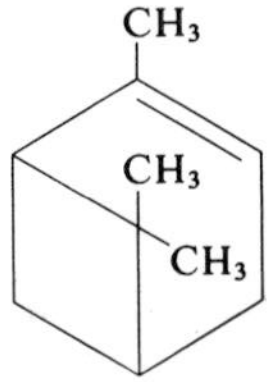
p179
p180 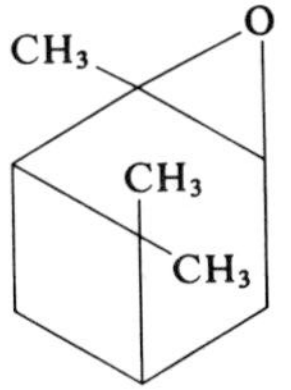
p181 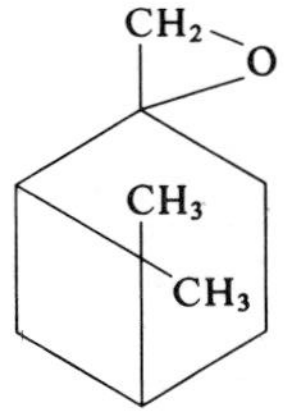
p182 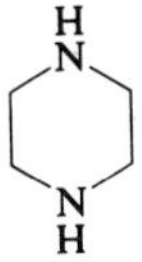
p183 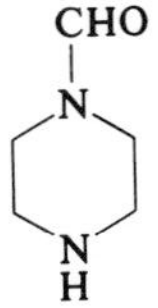
p184 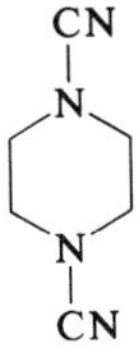
p186

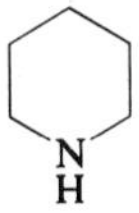

TABLE 1.15 Physical Constants of Organic Compounds (*Continued*)

No.	Name	Formula	Formula weight	Beilstein reference	Density	Refractive index	Melting point	Boiling point	Flash point	Solubility in 100 parts solvent
p187	1-Piperidinecarbonitrile		110.16	20, 56	0.951	1.4705^{20}		102^{10mm}	97	misc aq; s alc
p188	N-Piperidineethanol		129.20	20, 25	0.9732^{25}	1.4804^{20}		202	68	v s aq, alc, eth
p189	2-Piperidineethanol		129.20	21, 2	1.010^{17}		38–40	234	102	
p190	3-Piperidinemethanol		115.18	21^2, 8	1.026		66	$107^{3.5mm}$	>110	
p191	1-Piperidinepropionitrile		138.21		0.933	1.4695^{20}		111^{16mm}	102	
p191a	2-(2-Piperidinoethyl)-pyridine		190.29		0.985	1.5265^{20}		150^{17mm}	110	
p192	3-Piperidino-1,2-propenediol		159.23	20, 34			77–80			
p194	Propane	CH₃CH₂CH₃	44.10	1, 104	0.5842^{-42}	1.3397^{-42}	−187.7	−42.1	−104	6.5 mL aq; 790 mL alc; 926 mL eth; 1300 mL chl; 1450 mL bz
p195	1,2-Propanediamine	CH₃CH(NH₂)CH₂NH₂	74.13	4, 257	0.878^{15}	1.4460^{20}		119.7	33	misc aq, bz; s alc, eth
p196	1,3-Propanediamine	H₂NCH₂CH₂CH₂NH₂	74.13	4, 261	0.8842^{25}	1.4575^{20}	−12	140	48	misc alc, eth; s aq
p197	1,2-Propanediol	CH₃CH(OH)CH₂OH	76.10	1, 472	1.0364_4^{20}	1.4331^{20}	−60	188	107	misc aq, acet, chl; s alc, eth
p198	1,3-Propanediol	HOCH₂CH₂CH₂OH	76.10	1, 475	1.0597_4^{20}	1.4396^{20}	−26.7	214.4	79	misc aq, alc
p199	1,3-Propanedithiol	HSCH₂CH₂CH₂SH	108.23	1, 476	1.0772_4^{20}	1.5405^{20}	−79	169	58	misc alc, bz, eth, chl
p200	1-Propanesulfonyl chloride	CH₃CH₂CH₂SO₂Cl	142.60	4, 8	1.2864^{15}	1.4542^{20}		66^{8mm}	80	d hot aq, hot alc
p201	1,3-Propane sultone		122.14	19^3, 4	1.392	1.4380^{20}	30–33	180^{30mm}	>110	s alc, eth
p202	1-Propanethiol	CH₃CH₂CH₂SH	76.16	1, 359	0.8363_4^{5}	1.4255^{20}	−113.1	67.7	−20	misc alc, eth; sl s aq
p203	2-Propanethiol	CH₃CH(SH)CH₃	76.16	1, 367	0.8092_4^{5}	1.4302^{20}	−130.5	57–60	−34	7.2 aq; misc alc, bz, chl, eth
p204	1,2,3-Propanetriol tris(acetate)	H₃CCOOCH(CH₂OOCCH₃)₂	218.21	2, 147	1.1596^{20}		−78	258–260	148	
p205	1-Propanol	CH₃CH₂CH₂OH	60.10	1, 350	0.8037_4^{20}	1.3856^{20}	−126.2	97.2	15 (CC)	misc aq, alc, eth
p206	2-Propanol	(CH₃)₂CHOH	60.10	1, 360	0.7855_4^{20}	1.3772^{20}	−89.5	82.4	12 (CC)	misc aq, alc, chl, eth
p207	2-Propenal	H₂C=CHCHO	56.07	1, 725	0.841^{20}	1.4017^{20}	−81	49	−18 (CC)	21 aq; s alc, eth
p208	Propene	H₂C=CHCH₃	42.08	1, 196	0.6104_4^{-48}	1.3567^{-40}	−185.2	−47.7	−108	45 mL aq; 1200 mL alc; 500 mL acet
p209	2-Propene-1-thiol	H₂C=CHCH₂SH	74.15	1, 440	0.925^{23}	1.4765^{20}		67–68	21	misc alc, eth

No.	Name	Formula	Formula wt	Beilstein ref	Density	n_D	d 200	bp	Flash pt	Solubility
p210	1,2,3-Propenetricarboxylic acid		174.11	2, 849						50 aq[25]; 50 88% alc[12]; sl s eth
p211	1-Propen-2-yl acetate	$H_2C=C(OOCCH_3)CH_3$	100.12		0.909	1.4000^{20}		97	18	
p212	2-Propenylphenol	$CH_3CH=CHC_6H_4OH$	134.18	6^1, 279	1.044	1.5754^{20}		231	90	37 aq(hyd); misc alc (reacts), bz, eth, acet
p213	β-Propiolactone		72.06		1.1460^{20}_4	1.4131^{20}	−33.4	162.3	70	
p214	Propionaldehyde	CH_3CH_2CHO	58.08	1, 629	0.80712^{20}_4	1.3646^{19}	−81	48–50	−26	30 aq; misc alc, eth
p215	Propionamide	$CH_3CH_2CONH_2$	73.10	2, 243	0.95978^{80}_4	1.4160^{110}	79	222.2		v s aq, alc, chl, eth
p216	Propionic acid	CH_3CH_2COOH	74.09	2, 234	0.99344^{20}_4	1.3865^{20}	−21.5	141.1	58 (OC)	misc aq; s alc, chl, eth

p187

p188 (CH_2CH_2OH)

p189 (CH_2CH_2OH)

p190 (CH_2OH)

p191 (CH_2CH_2CN)

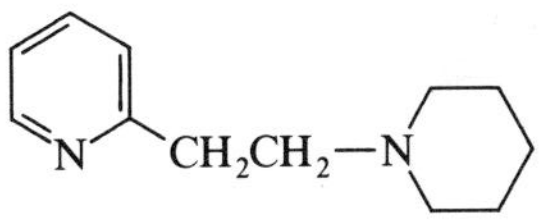

p191a

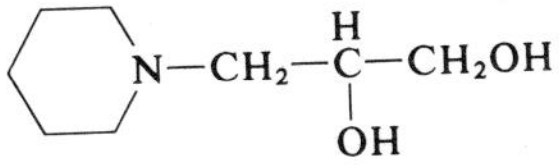

p192

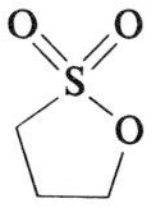

p201

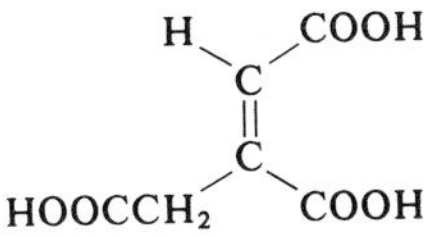

p210 ($HOOCCH_2$, COOH, COOH)

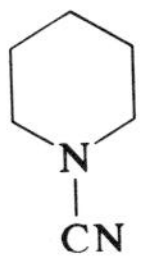

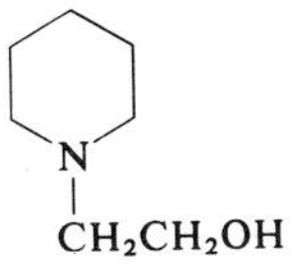

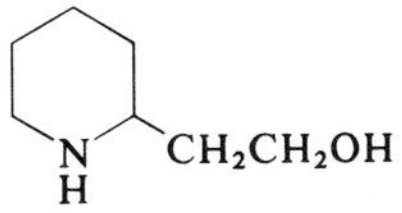

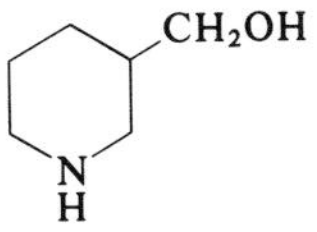

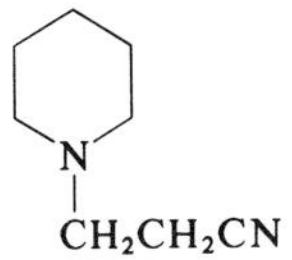

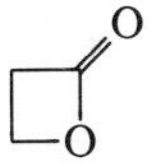

p213

TABLE 1.15 Physical Constants of Organic Compounds (*Continued*)

No.	Name	Formula	Formula weight	Beilstein reference	Density	Refractive index	Melting point	Boiling point	Flash point	Solubility in 100 parts solvent
p217	Propionic anhydride	[CH₃CH₂C(=O)]₂O	130.14	2, 242	1.0125^{20}_{4}	1.4047^{20}	−45	167	74 (OC)	d aq; s alc, chl, eth
p218	Propionitrile	CH₃CH₂CN	55.08	2, 245	0.7818^{20}_{4}	1.3658^{20}	−91.8	97.2	6	10 aq; misc alc, eth
p219	Propionyl chloride	CH₃CH₂COCl	92.53	2, 243	1.065^{20}	1.4051^{20}	−94	80	11	dec by aq, alc
p220	Propiophenone	C₆H₅COCH₂CH₃	134.18	7^2, 231	1.0105^{20}_{4}	1.5258^{20}	18.6	218.0	87	misc bz, eth, abs alc
p221	Propoxytrimethyl-silane	CH₃CH₂CH₂OSi(CH₃)₃	132.3		0.7684^{20}	1.384^{20}		$100^{735\text{mm}}$		
p222	Propyl acetate	CH₃CH₂CH₂OOCCH₃	102.13	2, 129	0.8362_{4}	1.3844^{20}	−92	101.6	14 (CC)	2.3 aq; misc alc, eth
p223	Propylamine	CH₃CH₂CH₂NH₂	59.11	4, 136	0.7173^{20}_{4}	1.3882^{20}	−83	48−49	−12 (CC)	misc aq, alc, eth
p224	2-(Propylamino)eth-anol	C₃H₇NHCH₂CH₂OH	103.17	4, 282	0.900	1.4415^{20}		$182^{746\text{mm}}$	78	
p225	Propylbenzene	CH₃CH₂CH₂C₆H₅	120.20	5, 390	0.8621^{20}_{4}	1.4912^{20}	−99.2	159.2	47	s alc, eth
p226	Propyl benzoate	C₆H₅COOCH₂CH₂CH₃	164.20	9, 112	1.0232^{20}	1.5003^{20}	−51.6	231.2	98	i aq; s alc, eth
p226a	Propyl butyrate	CH₃CH₂CH₂CO₂CH₂CH₂CH₃	130.19	2, 271	0.8791^{5}	1.4005^{20}	−95	143	38	sl s aq; misc alc, eth
p226b	Propyl chloroformate	ClCO₂CH₂CH₂CH₃	122.55	3, 11	1.090	1.4034^{20}		106	28	
p227	Propylcyclohexane	CH₃CH₂CH₂C₆H₁₁	126.24	5^2, 23 19^3, 1564	0.7929^{20}_{4}	1.4370^{20}	−94.9	156.7		s bz, eth
p228	Propylene carbonate		102.09		1.2041^{20}_{4}	1.4210^{20}	−55	240	132	v s aq, alc, bz, eth
p229	Propyleneimine	CH₃CH—CH₂ ＼NH／ (ring)	57.09	20, 3	0.8017^{25}	1.4084^{25}		66.0	−15	misc aq, alc, PE
p230	Propylene oxide	CH₃CH—CH₂ ＼O／ (ring)	58.08	17, 6	0.8287^{20}	1.3660^{20}	−112.1	34	−35 (CC)	41 aq; misc alc, eth
p231	Propylene sulfide	CH₃CH—CH₂ ＼S／ (ring)	74.15	17^2, 15	0.946	1.4760^{20}		72−75	10	
p232	Propyl formate	CH₃CH₂CH₂OOCH	88.10	2, 21	0.9006^{20}_{4}	1.3769^{20}	−92.9	80.9	−3 (CC)	2 aq; misc alc, eth
p233	Propyl 4-hydroxyben-zoate	HOC₆H₄COOCH₂CH₂CH₃	180.20	10, 160			95−98			0.05 aq; v s alc, eth
p234	Propyl isocyanate	CH₃CH₂CH₂NCO	85.11	4^1, 366	0.908	1.3970^{20}		83−84	26	s aq, alc, eth
p235	Propyl lactate	CH₃CH(OH)COOC₃H₇	132.16	3, 265	0.9962^{20}	1.4167^{25}		$86^{40\text{mm}}$		s alc, eth
p236	Propyl nitrate	CH₃CH₂CH₂ONO₂	105.09	1, 355	1.0538^{20}_{4}	1.3976^{20}	−100	110.1	23	s alc, eth
p237	2-Propylpentanoic acid	(CH₃CH₂CH₂)₂CHCOOH	144.21	2, 350	0.921	1.4250^{20}	May explode on heating	220	111	
p238	2-Propylphenol	CH₃CH₂CH₂C₆H₄OH	136.19	6, 499	1.015^{20}	1.5279^{20}		224−226	93	s alc, eth
p239	Propylphosphonic dichloride	CH₃CH₂CH₂P(O)Cl₂	160.97	4, 596	1.290	1.4643^{20}		$88\text{-}90^{50\text{mm}}$	>110	

No.	Name	Formula	Formula weight	Beilstein ref.	Density	n_D	mp, °C	bp, °C	Flash pt, °C	Solubility
p240	Propyltrichlorosilane	$CH_3CH_2CH_2SiCl_3$	177.53	4, 630	1.1851^{14}	1.429^{20}		123–124	2	0.5 aq; misc alc, eth
p241	1-Propyl-4-piperidone		141.22		0.936	1.4600^{20}		56^{1mm}	75	0.35 aq; 1 alc; 83 eth
p241a	Propyl propionate	$C_2H_5CO_2CH_2CH_2CH_3$	116.16	2, 240	0.883^{20}	1.3935^{20}	−76	124	19	
p242	Propyl 3,4,5-trihydroxybenzoate	$(HO)_3C_6H_2COOC_3H_7$	212.20				150			
p243	Propyne	$CH_3C{\equiv}CH$	40.06	1, 246	0.691^{-20}	1.3725^{-20}	−102.8	−23.2	58	v s alc; 3000 mL eth
p244	2-Propynoic acid	$HC{\equiv}CCOOH$	70.05	2, 477	1.1384^{20}	1.4320^{20}	9	102^{200mm}	33 (OC)	s aq, alc, eth
p245	2-Propyn-1-ol	$HC{\equiv}CCH_2OH$	56.06	1, 454	0.97152^{20}	1.4320^{20}	−52	114–115	82	misc aq, alc, bz, chl
p246	(+)-Pulegone		152.24	7, 81	0.93461^{15}	1.4850^{15}		224		misc alc, chl, eth
p247	Pyrazine		80.09	23, 91	1.0314^{61}	1.4953^{61}	53	115–118	55	v s aq, alc, eth
p248	Pyrazole		68.08	23, 39		1.4203	70	186–188		s aq, alc, bz, eth
p249	Pyrene		202.26	5, 693	1.271^{23}		156	404		s org solvents
p251	Pyridazine		80.09	23, 89	1.10354^{4}	1.5230^{23}	−8	208	85	misc aq, bz; v s alc, eth

p228 (propylene carbonate, ring with CH3)

$CH_3{-}CH{-}CH_2$ (N–H bridge) — p229

$CH_3CH{-}CH_2$ (O bridge) — p230

$CH_3{-}CH{-}CH_2$ (S bridge) — p231

p241 (1-propyl-4-piperidone; ring with $N{-}CH_2CH_2CH_3$ and C=O)

p246 (pulegone; ring with CH_3 and $C(CH_3)$)

p247 (pyrazine)

p248 (pyrazole, N–N–H)

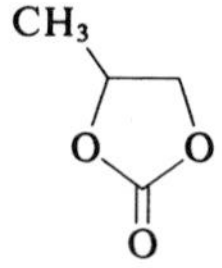

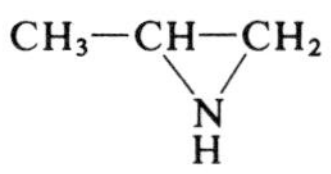

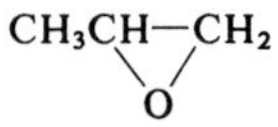

p249 (pyrene)

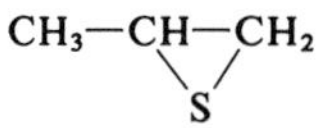

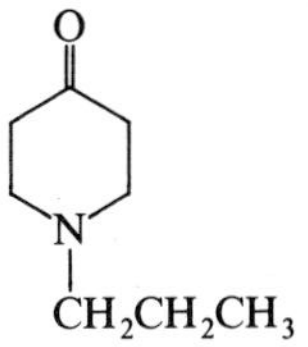

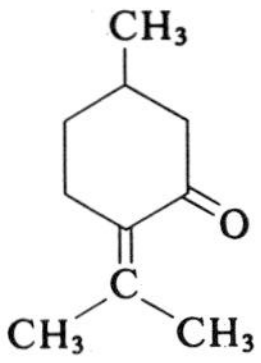

p251 (pyridazine, N–N)

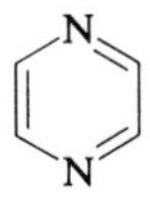

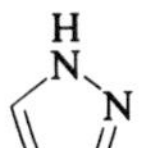

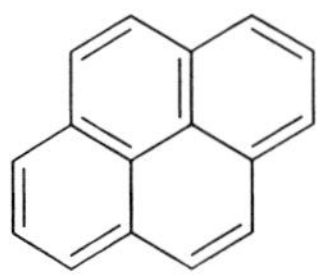

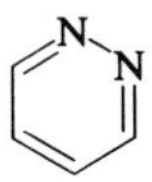

TABLE 1.15 Physical Constants of Organic Compounds (*Continued*)

No.	Name	Formula	Formula weight	Beilstein reference	Density	Refractive index	Melting point	Boiling point	Flash point	Solubility in 100 parts solvent
p252	Pyridine	C_5H_5N	79.10	20, 181	0.9827^{25}_{4}	1.5067^{25}	−41.6	115.2	20 (CC)	misc aq, alc, eth
p253	Pyridine-d_5	C_5D_5N	84.14	20^3, 2305	1.050	1.5079^{20}		114.4	20	
p254	2-Pyridinealdoxime	$(C_5H_4N)CH{=}NOH$	122.13	21^1, 288			110–112			
p254a	3-Pyridinecarbamide	$(C_5H_4N)CONH_2$	122.13	22, 40	1.400	1.466	130–133			100 aq; 66 alc
p255	2-Pyridinecarboxaldehyde	$(C_5H_4N)CHO$	107.11	21^1, 287	1.126	1.5370^{20}		181	54	
p256	3-Pyridinecarboxaldehyde	$(C_5H_4N)CHO$	107.11	21^1, 288	1.135	1.5493^{20}		97^{15mm}	60	s aq, eth
p257	4-Pyridinecarboxaldehyde	$(C_5H_4N)CHO$	107.11	21, 287	1.172	1.5440^{20}		78^{12mm}	54	s aq, alc, bz
p259	Pyridine-2-carboxylic acid	$(C_5H_4N)COOH$	123.11	22, 33			134–136	subl		1.4 aq; s alk
p260	Pyridine-3-carboxylic acid	$(C_5H_4N)COOH$	123.11	22, 38	1.473		236.6	subl		0.52 aq; i alc, bz, eth
p261	Pyridine-4-carboxylic acid	$(C_5H_4N)COOH$	123.11	22, 45			319	260^{15mm}		14 aq; 2 alc; 0.1 chl
p262	4-Pyridinecarboxylic hydrazide	$(C_5H_4N)CONHNH_2$	137.14	22^1, 504			171.4			0.56 aq; s alk
p263	2,3-Pyridinedicarboxylic acid	$(C_5H_3N)(COOH)_2$	167.12	22, 150			190 d			s hot acid
p264	2,5-Pyridinedicarboxylic acid	$(C_5H_3N)(COOH)_2$	167.12	22, 153			236–237	subl d		
p265	2,6-Pyridinedicarboxylic acid	$(C_5H_3N)(COOH)_2$	167.12	22, 154			250 d			sl s aq; v sl s alc
p266	Pyridine-N-oxide	$C_5H_5N(O)$	95.10	20^2, 131			66	270		v s aq
p267	3-Pyridinesulfonic acid	$(C_5H_4N)SO_3H$	159.16	22, 387			>300			v s aq, alc, eth
p268	2-Pyridylcarbinol	$(C_5H_4N)CH_2OH$	109.13	21^1, 203	1.131	1.5420^{20}		113^{16mm}	>110	v s aq, eth
p269	3-Pyridylcarbinol	$(C_5H_4N)CH_2OH$	109.13	21, 50	1.124	1.5445^{20}		154^{28mm}	>110	
p270	3-(3-Pyridyl)-1-propanol	$(C_5H_4N)CH_2CH_2CH_2OH$	137.18	21^3, 549	1.063	1.5305^{20}		133^{3mm}	110	
p271	Pyrimidine		80.09	23, 89	1.016	1.5035^{20}	20–22	123–124	31	misc aq; s alc, eth
p272	2,4(1H,3H)-Pyrimidinedione		112.09	24, 312			335			0.3 aq; s alk
p273	Pyrrole		67.09	20, 159	0.9691^{20}_{4}	1.5102^{20}	−23.4	129.8	39 (TCC)	4.5 aq; v s alc, eth

No.	Name	Mol. wt	Beilstein	Density	n_D	m.p.	b.p.	Flash pt	Solubility
p274	Pyrrolidine	71.12	20, 4	0.8520^{22}	1.4431^{20}	-58	89	2 (TCC)	misc aq; s alc, chl, eth
p274a	1-Pyrrolidinebutyronitrile	138.21		0.926	1.4605^{20}		115^{18mm}	99	
p275	1-Pyrrolidinecarbodithioic acid, ammonium salt	164.29				153–155		107	
p276	1-Pyrrolidinecarbonitrile	96.13		0.954	1.4690^{20}		$77^{1.8mm}$		
p278	1-Pyrrolidino-1-cyclohexene	151.25		0.940	1.5225^{20}		115^{15mm}	39	
p279	2-Pyrrolidinone	85.11	21, 236	1.1164^{25}	1.486^{25}	25	245	129 (OC)	misc aq, alc, bz, chl, eth, EtAc
p280	3-(N-Pyrrolidino)-1,2-propanediol	145.20	20', 4			46–48	158^{30mm}	>110	
q1	Quinhydrone	218.20	7, 617	1.401^{20}_{4}		171			s hot aq, alc, eth

Structures:

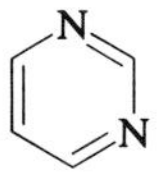

- p271
- p272 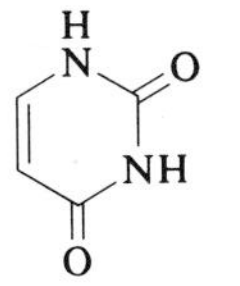
- p273 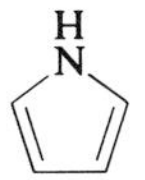
- p274 
- p274a $CH_2CH_2CH_2CN$ 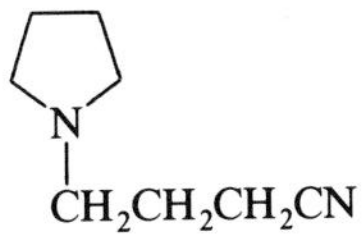
- p275 $S{=}C{-}S^-\ NH_4^+$
- p276 CN
- p278
- p279 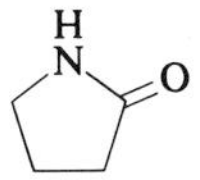
- p280 $CH_2CHOHCH_2OH$ 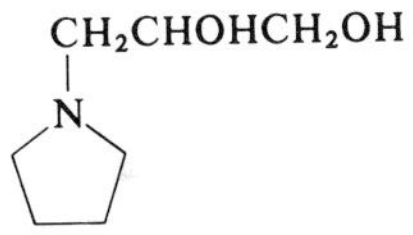
- q1

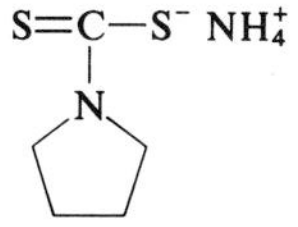

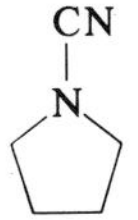

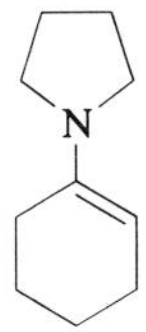

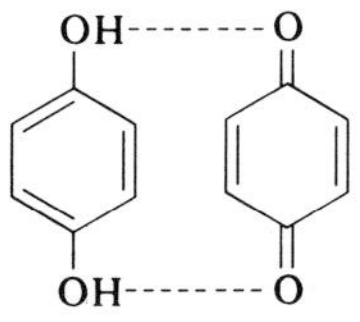

TABLE 1.15 Physical Constants of Organic Compounds (*Continued*)

No.	Name	Formula	Formula weight	Beilstein reference	Density	Refractive index	Melting point	Boiling point	Flash point	Solubility in 100 parts solvent
q2	Quinine		324.44	23, 511	1.0954^{20}	1.625	177d			125 alc; 1.2 bz; 83 chl
q3	Quinoline		129.16	20, 339		1.6273^{20}	−14.9	237	101	0.6 aq; misc alc, eth
q4	Quinoxaline		130.15	23, 176	1.13344^{48}	1.6231^{48}	29–30	223	98	v s aq, alc, bz, eth
q5	Quinuclidine		111.19	20, 144			156 sealed tube			v s aq, alc, eth
r1	D-Raffinose pentahydrate		594.52	31, 462			80	d 118		14 aq; 10 MeOH
r2	Rhodamine B		479.02	19, 345			165			v s aq, alc
r3	Rhodanine		133.19	27, 242	0.868		170 (may explode on rapid heating)			v s hot aq, alc, eth
r4	Riboflavin		376.37				d 278			v s alk(dec); i eth
r5	D-Ribose		150.13	1, 859			88–92			s aq; sl s alc
s1	Saccharin		183.19	27, 168			229–230			0.34 aq; 3 alc; 8 acet
s2	Safrole		162.19	19, 39	1.0954^{20}	1.5370^{20}	11.2	232–234	97	v s alc; misc chl, eth
s3	Semicarbazide	$H_2NNHCONH_2$	75.07	3, 98			96			v s aq, alc; i eth
s4	L-Serine	$HOCH_2CH(NH_2)COOH$	105.09	4, 505			222 d			s aq; v sl s alc, eth
s5	D-Sorbitol		182.17	1, 533	1.472^{-5}		110–112			83 aq; s hot alc, acet
s6	(−)-Sorbose		180.16	1, 927	1.65^{15}		165			55 aq; sl s alc

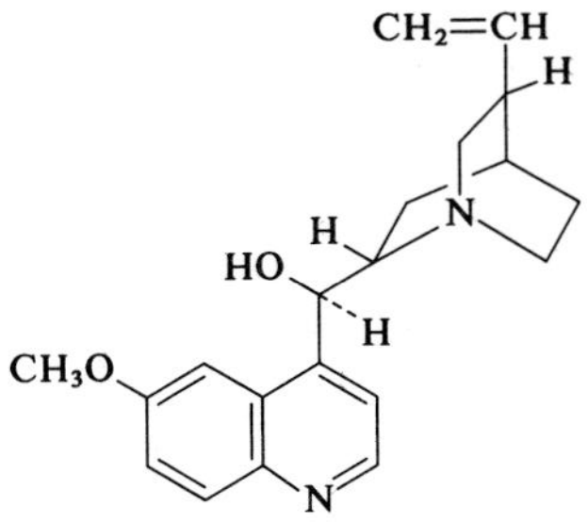

r1

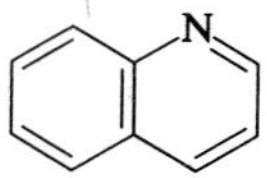

r5

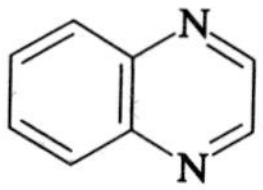

s1

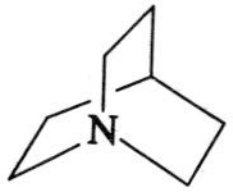

q5

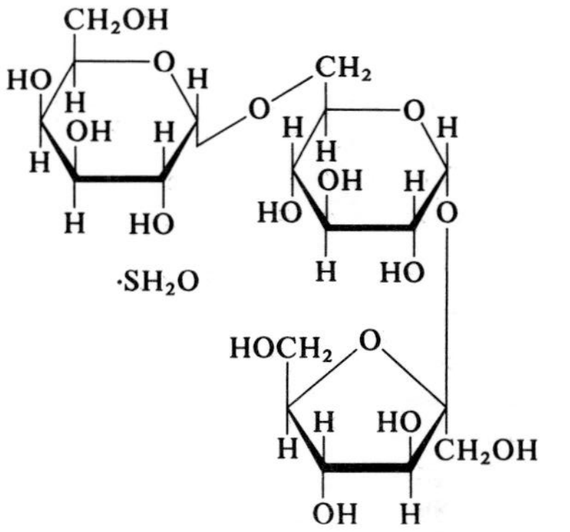

q4

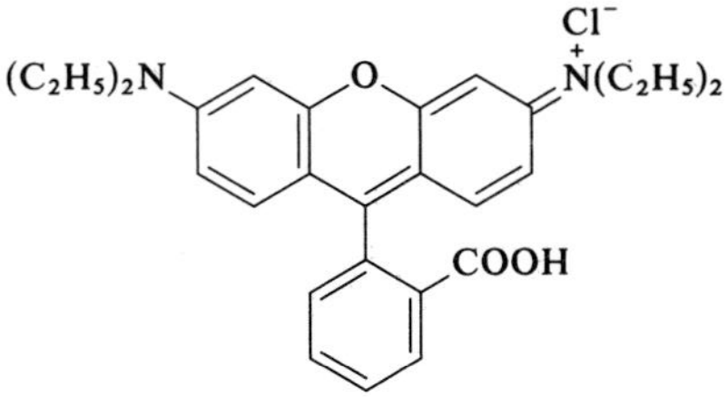

q3

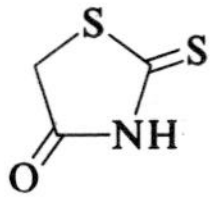

r3

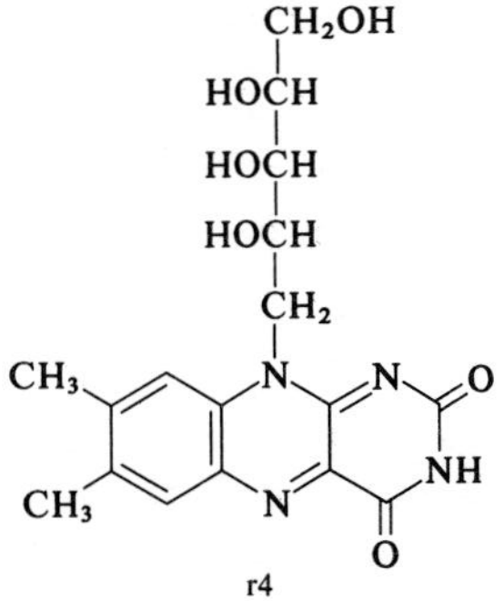

r4

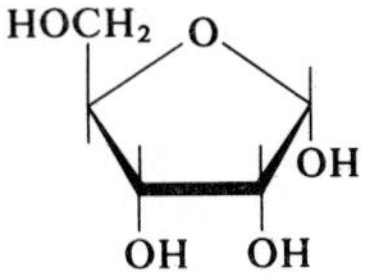

s5

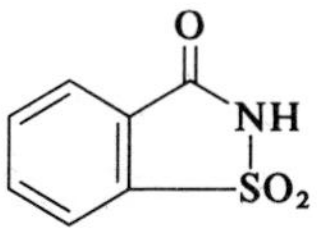

s6

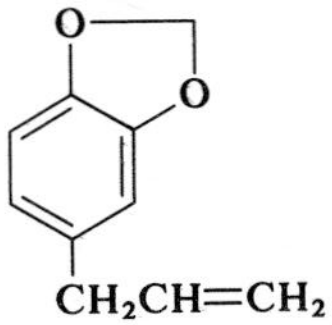

s2

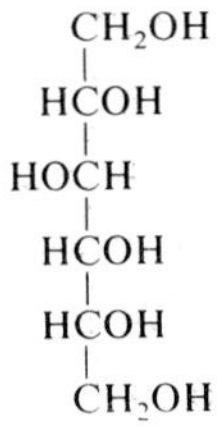

q2

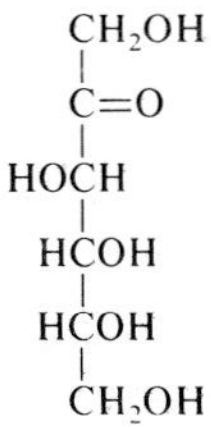

r2

TABLE 1.15 Physical Constants of Organic Compounds (*Continued*)

No.	Name	Formula	Formula weight	Beilstein reference	Density	Refractive index	Melting point	Boiling point	Flash point	Solubility in 100 parts solvent
s7	Squalane	$[(CH_3)_2CH(CH_2)_3CH(CH_3)(CH_2)_3CH(CH_3)CH_2CH_2-]_2$	422.80	1^1, 72	0.810	1.4530^{15}	−38	350	218	s bz, chl, eth, PE
s8	Squalene	$CH_3[C(CH_3)=CHCH_2CH_2]_5C(CH_3)=C(CH_3)_2$	410.73	1^1, 130	0.8584^{24}_4	1.4965^{20}	−75	285^{25mm}	200	v s eth, acet, PE
s9	*trans*-Stilbene	$C_6H_5CH=CHC_6H_5$	180.25	5, 630	0.970		124	307^{744mm}		v s bz, eth
s10	(−)-Strychnine		334.42	27^2, 723	1.36^{20}_4		284–286	270^{5mm}		6.2 alc; 20 chl; 0.55 bz; 15 mg aq
s11	Styrene	$C_6H_5CH=CH_2$	104.15	5, 474	0.9060^{20}	1.5468^{20}	−30.6	145.1	31 (CC)	s alc, acet, eth
s11a	Styrene oxide		120.15	17, 49	1.054	1.5338^{20}	−37	194	79	s aq; sl s alc; i eth
s13	Succinamic acid	$H_2NCOCH_2CH_2COOH$	117.10	2, 614			153–156			0.45 aq; i alc, eth
s14	Succinamide	$H_2NCOCH_2CH_2CONH_2$	116.12	2, 614			260 d	125 subl		7.7 aq; 5.4 alc; 2.8 acet; 0.88 eth; i bz
s15	Succinic acid	$HOOCCH_2CH_2COOH$	118.09	2, 601	1.552		187–190	235 d		
s17	Succinic anhydride		100.07	17, 407	1.41		119.6	261		s alc, chl; v sl s eth
s18	Succinimide		99.09	21, 369	0.985		125–127	287		33 aq; 4 alc; i eth
s19	Succinonitrile	$NCCH_2CH_2CN$	80.09	2, 615	1.395^{15}		46–48	265–267	>110	dec aq; s bz
s20	Succinyl chloride	$ClCOCH_2CH_2COCl$	154.98	2, 613		1.473^{15}	17	192	76	200 aq; 0.59 alc
s21	Sucrose		342.30	31, 424	1.587^{25}		d 160–186			
s22	Sulfamethazine		278.34				198–201			0.15 aq; s alk
s23	Sulfanilamide	$H_2NC_6H_4SO_2NH_2$	172.21	14, 698			164–166			0.76 aq; 2.7 alc; 20 acet; s acid, alk
s24	Sulfanilic acid	$4\text{-}(H_2N)C_6H_4SO_3H$	173.19	14, 695			d 288			1 aq; sl s hot MeOH
s25	Sulfoacetic acid	HO_3SCH_2COOH	140.11	4, 21			84–86	245 d		s aq, alc; i eth, chl
s26	2-Sulfobenzoic acid cyclic anhydride		184.17	19, 110				186^{18mm}		s bz, chl, eth; i aq
s27	4,4'-Sulfonylbis(2,6-dibromophenol)	$[HO(Br)_2C_6H_2]_2SO_2$	565.88	6, 865			289–292			
s28	4,4'-Sulfonylbis(methyl benzoate)	$(CH_3OOCC_6H_4)_2SO_2$	334.35	10^2, 109			196			
s29	4,4'-Sulfonyldiphenol	$(HOC_6H_4)_2SO_2$	250.27	6, 861	1.3663^{15}		245–247			s alc, eth, acet; i aq
s30	5-Sulfosalicylic acid	$HO_3SC_6H_3(OH)COOH$	254.21	11, 411			120			v s aq, alc; s eth
t1	(+)-Tartaric acid		150.09	3, 481	1.7598^{24}_4		168–170			139 aq; 33 alc; 0.4 eth

No.	Name	Formula	Mol. wt.	Beilstein ref.	Density	M.p., °C	B.p., °C	Solubility
t2	meso-Tartaric acid hydrate	$HOOCCH(OH)CH(OH)COOH \cdot xH_2O$	150.09	3, 528		140		125 aq
t2a	DL-Tartaric acid	$HO_2CCHOHCHOHCO_2H$	150.09	3, 522		206		5 alc[25]; 1 eth v s aq
t3	Tartrazine		534.37	25, 252				
t3a	m-Terphenyl	$C_6H_5C_6H_4C_6H_5$	230.31	5, 695	1.666^{20}_{4}	86–87	379	110
t4	o-Terphenyl	$C_6H_5C_6H_4C_6H_5$	230.31	5^2, 611		58–59	337	

Stearamide, o2
Stearic acid, o5
Stearyl bromide, b322
Styrene dibromide, d79
Styrene glycol, p113
Styrene oxide, e8, e9
Suberic acid, o24
Suberonitrile, d239
Succinic acid monoamide, s13

Succinonitrile, b383
Sulfanilic acid, z120
N-Sulfinylaniline, t155
3-Sulfoalanine, a293
Sulfolane, t108
3-Sulfolene, d369
Sulfonyldianiline, d36, d37
Sylvan, m253

Sylvic acid, a1
2,4,5-T, t246
TAPS, t434
Taurine, a161
Terephthaldehyde, b13
Terephthaldicarboxaldehyde, b13
Terephthalic acid, b17
Terephthaloyl chloride, b15

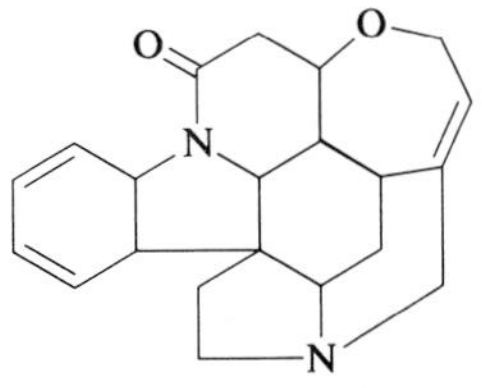

s10

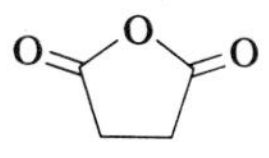

s17

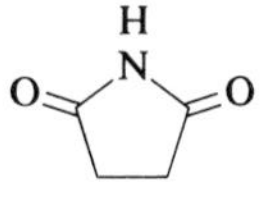

s18

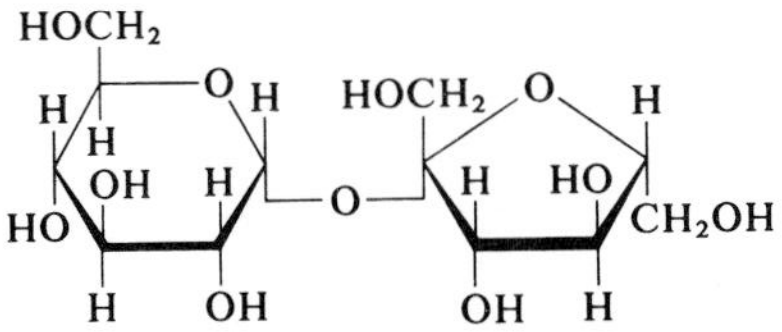

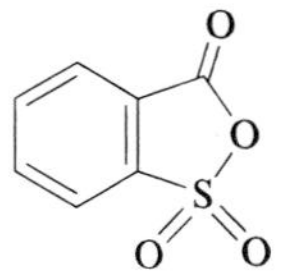

s26

s22

CH_2OH
$HOCH_2$ ... HO ... OH ... $HOCH_2$ ... OH ... HO
s21

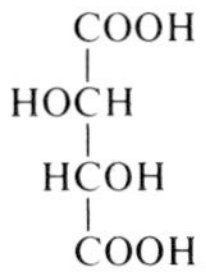

COOH
HOCH
HCOH
COOH
t1

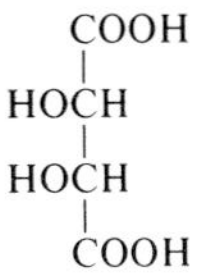

COOH
HOCH
HOCH
COOH
t2

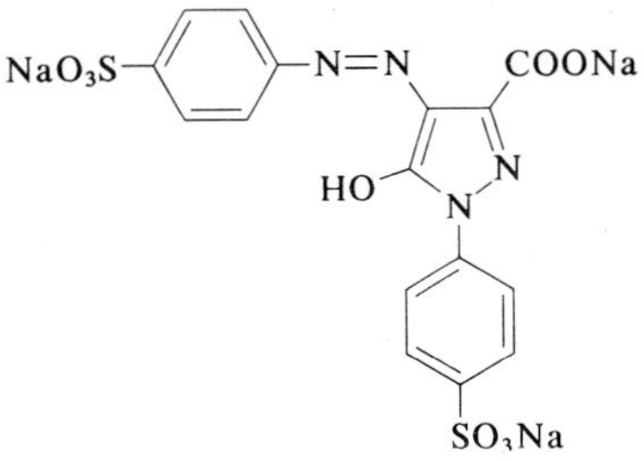

NaO_3S —N=N— ... COONa ... HO ... SO_3Na
t3

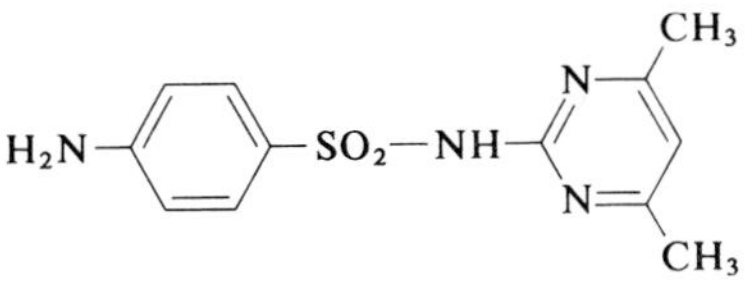

TABLE 1.15 Physical Constants of Organic Compounds (*Continued*)

No.	Name	Formula	Formula weight	Beilstein reference	Density	Refractive index	Melting point	Boiling point	Flash point	Solubility in 100 parts solvent
t5	α-Terpinene		136.24	5, 126	0.8752^{20}_{4}	1.4775^{20}		174	46	misc alc, eth
t6	γ-Terpinene		136.24	5, 128	0.8531^{15}	1.4754^{16}		183	51	v s alc, eth
t7	Terpinen-4-ol		154.25	6, 55	0.9338^{20}_{4}	1.4820^{20}		219	79	v s chl
t9	α-Terpineol		154.25	6, 57	0.933	1.4813^{20}	36.4	218	89	misc alc, chl, eth, HOAc
t10	1,1,2,2,-Tetrabromoethane	$Br_2CHCHBr_2$	345.67	1, 94	2.9638^{20}_{4}	1.638^{20}	0.0	243.5	none	sl s bz; i aq, alc
t11	Tetrabromophthalic anhydride		463.72	17, 485			274–276			v s chl
t12	α,α,α',α'-Tetrabromo-o-xylene	$C_6H_4(CHBr_2)_2$	421.77	5, 367			114–116			v s bz, chl
t13	α,α,α',α'-Tetrabromo-m-xylene	$C_6H_4(CHBr_2)_2$	421.77	5, 375			105–108			
t14	2,3,5,6-Tetrabromo-p-xylene	$C_6Br_4(CH_3)_2$	421.77	5, 386			254–256			
t15	Tetrabutylammonium bromide	$(C_4H_9)_4N^+Br^-$	322.38				103–104			
t16	Tetrabutylammonium chloride	$(C_4H_9)_4N^+Cl^-$	277.92	4[3], 292			83–86		>110	
t17	Tetrabutylammonium fluoride trihydrate	$(C_4H_9)_4N^+F^-\cdot3H_2O$	315.52	4[3], 292			62–63			sl s aq; s alc, eth
t18	Tetrabutylammonium hydrogen sulfate	$(C_4H_9)_4N^+HSO_4^-$	339.54				169–171			
t19	Tetrabutylammonium iodide	$(C_4H_9)_4N^+I^-$	369.38	4, 157			145–148			
t20	Tetrabutylammonium tetrafluoroborate	$(C_4H_9)_4N^+BF_4^-$	329.28	4[3], 293			160–162			
t20a	Tetrabutyl orthosilicate	$Si[O(CH_2)_3CH_3]_4$	320.55	1[2], 398	0.899^{20}_{4}	1.4131^{20}		275	78	
t20b	Tetrabutylphosphonium bromide	$[CH_3(CH_2)_3]_4PBr$	339.35				100–103			
t21	Tetrabutyltin	$(C_4H_9)_4Sn$	347.15	1, 656	1.057	1.4742^{20}		145^{10mm}	107	v s acet, chl
t22	1,1,3,3-Tetrachloroacetone	$Cl_2CHCOCHCl_2$	195.86		1.624^{15}_{4}	1.497^{18}	−97	182^{745mm}	none	
t23	1,2,3,4-Tetrachlorobenzene	$C_6H_2Cl_4$	215.89	5, 204			46–47	254	>110	v s eth; sl s alc

No.	Name	Formula	Formula wt.	Beilstein ref.	Density	n_D	mp, °C	bp, °C	Flash pt, °C	Solubility
t24	1,2,3,4-Tetrachloro-benzene	$C_6H_2Cl_4$	215.89	5, 205	1.858^{22}		138–140	240–246	>110	s bz, chl, eth
t25	Tetrachloro-1,2-benzo-quinone	$C_6Cl_4(=O)_2$	245.88	7, 602			127–129			s eth; sl s chl; i aq
t26	Tetrachloro-1,4-benzoquinone	$C_6Cl_4(=O)_2$	245.88	7, 636			290 d	subl		
t26a	1,2,3,4-Tetrachloro-butane	$ClCH_2CH(Cl)CH(Cl)CH_2Cl$	195.90	1, 119	1.440		70–74	140^{50mm}		
t27	Tetrachloro-1,2-difluoroethane	$Cl_2CFCFCl_2$	203.83		1.6447^{25}	1.4130^{25}	26.0	92.8		0.012 aq
t29	1,1,2,2-Tetrachloro-ethane	$Cl_2CHCHCl_2$	167.85	1, 86	1.5866^{25}_4	1.4910^{25}	−44	146.5	none	0.3 aq; misc alc, chl, eth, PE
t30	Tetrachloroethylene	$Cl_2C=CCl_2$	165.83	1, 187	1.6230^{20}_4	1.5057^{20}	−22.4	121.1	none	misc alc, chl, eth
t31	2,3,5,6-Tetrachloro-nitrobenzene	$HC_6Cl_4NO_2$	260.89	5, 247	1.744^{25}_4		98–101	304		s alc, bz, chl
t32	Tetrachlorophthalic anhydride		285.90	17, 484			254–258	371		d hot aq; sl s eth
t33	3,4,5,6-Tetrachloro-phthalimide		284.91	21, 505			>300		none	
t34	1,1,2,3-Tetrachloro-2-propene	$ClCH=C(Cl)CHCl_2$	179.86	1^1, 83	1.530	1.5163^{20}	165		none	

TES, t431
Tetrabromomethane, c13

Tetrabutoxysilane, t20a
Tetracene, b7

Tetrachloromethane, c14

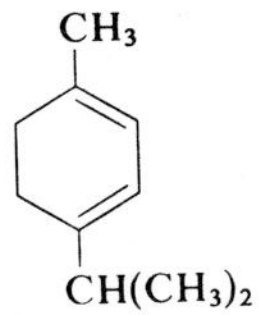

t5

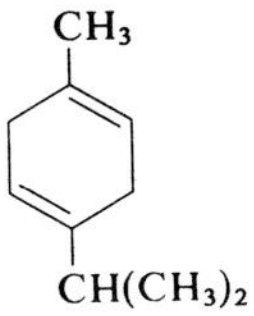

t6

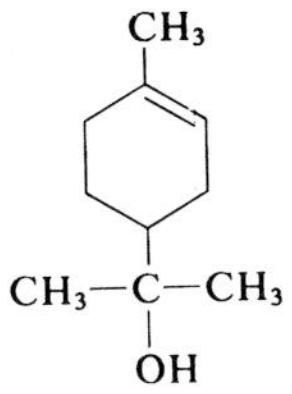

t7

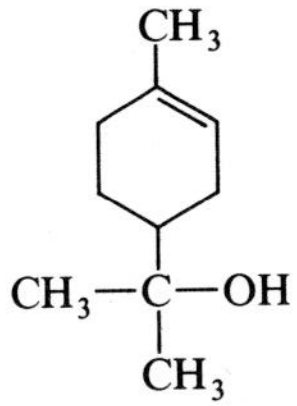

t9

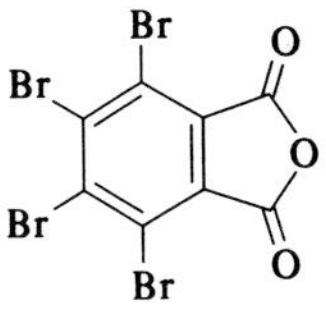

t11

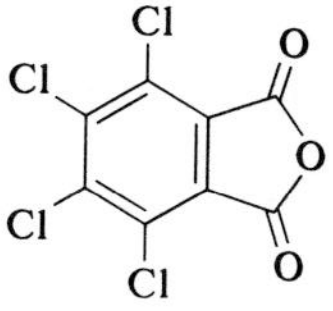

t32

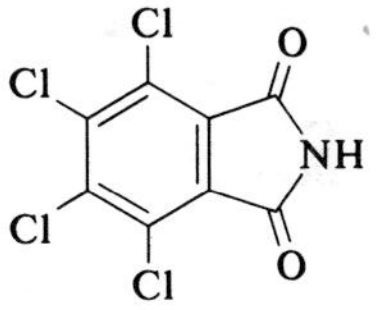

t33

1.297

TABLE 1.15 Physical Constants of Organic Compounds (*Continued*)

No.	Name	Formula	Formula weight	Beilstein reference	Density	Refractive index	Melting point	Boiling point	Flash point	Solubility in 100 parts solvent
t35	2,4,5,6-Tetrachloro-pyrmidine		217.87	23, 90			68–70			
t36	2,4,5,6-Tetrachloro-*m*-xylene	$C_6Cl_4(CH_3)_2$	243.95	5, 373			220–222			
t37	Tetracosane	$CH_3(CH_2)_{22}CH_3$	338.66	1, 175	0.7786^{51}	1.4283^{70}	51.1	391	>110	9.4 chl; s eth
t38	Tetracyanoethylene	$(NC)_2C{=}C(CN)_2$	128.09				200	subl 120		v s alc, eth
t39	Tetradecane	$CH_3(CH_2)_{12}CH_3$	198.40	1, 171	0.7627^{20}	1.4290^{20}	5.9	253.5	99	v s bz, chl, eth; s alc
t40	Tetradecanoic acid	$CH_3(CH_2)_{12}COOH$	228.38	2, 365	0.8528^{70}_4	1.4273^{70}	58.5	250^{100m}	>110	s eth; sl s alc
t41	1-Tetradecanol	$CH_3(CH_2)_{13}OH$	214.39	1, 428	0.8151^{50}	1.4358^{50}	37.8	264	115	d aq, alc; s eth
t42	Tetradecanoyl chloride	$CH_3(CH_2)_{12}COCl$	246.82	2, 368			−1	168^{15mm}		v s alc, eth
t43	1-Tetradecene	$CH_3(CH_2)_{11}CH{=}CH_2$	196.38	1, 226	0.7754^{15}	1.4351^{20}	−12.9	251.2	115	
t45	1-Tetradecylamine	$CH_3(CH_2)_{13}NH_2$	213.41	4, 201			40–42	162^{15mm}	>110	
t46	4-Tetradecylaniline	$CH_3(CH_2)_{13}C_6H_4NH_2$	213.41	$12^3,$ 2780			46–49			
t48	Tetraethoxysilane	$(CH_3CH_2O)_4Si$	208.33	1, 334	0.9344^{20}	1.383^{20}	−77	165.8	46	dec aq; s alc
t49	Tetraethylammonium bromide	$(CH_3CH_2)_4N^+Br^-$	210.16	4, 104	1.397^{24}_4		287 d			v s aq, alc, acet, chl
t50	Tetraethylammonium chloride	$(CH_3CH_2)_4N^+Cl^-$	165.71	4, 104	1.0801^{21}_4		37.5			141 aq; s alc; 8.2 chl
t51	Tetraethylammonium hydroxide	$(CH_3CH_2)_4N^+OH^-$	147.26	4, 103						misc aq
t52	Tetraethylene glycol	$(HOCH_2CH_2OCH_2CH_2)_2O$	194.23	1, 468	1.1253^{20}_{20}	1.4590^{20}	−6	307.8	176	misc aq, alc, bz, eth
t52a	Tetraethylene glycol diethyl ether	$C_6H_5(OCH_2CH_2)_4OC_2H_5$	250.34	$1^3,$ 2107	0.970	1.4324^{20}		159^{11mm}	>110	s aq
t53	Tetraethylene glycol dimethacrylate	$[H_2C{=}C(CH_3)COOCH_2{-}CH_2OCH_2CH_2]_2O$	330.37		1.08			220^{1mm}	62	
t53a	Tetraethylene glycol dimethyl ether	$CH_3(OCH_2CH_2)_4OCH_3$	222.28	$1^3,$ 2107	1.0087^{20}_4	1.4325^{20}	−27	276	140	s aq
t55	Tetraethylene glycol monomethyl ether	$CH_3O(CH_2CH_2O)_3{-}CH_2CH_2OH$	208.26		0.987	1.4453^{20}		166^{11mm}	>110	
t56	Tetraethylenepenta-mine	$(H_2NCH_2CH_2NHCH_2{-}CH_2)_2NH$	189.31		0.9992^{20}_{20}	1.5055^{20}	−40	340	185	misc aq, alc, eth
t57	*N,N,N′,N′*-Tetraethyl-ethylenediamine	$(C_2H_5)_2NCH_2CH_2N(C_2H_5)_2$	172.32	4, 251	0.808	1.4343^{20}		189–192	58	
t58	Tetraethylgermanium	$(C_2H_5)_4Ge$	188.84	4, 631	1.1989	1.4343^{20}	−90	165.5		s alc, eth; i aq

No.	Name	Formula	Mol wt	Beilstein ref	Density	n_D	mp, °C	bp, °C	Flash pt, °C	Solubility
t59	Tetraethyllead	$(C_2H_5)_4Pb$	323.45	4, 639	1.6532^{4}	1.5198^{20}	−136	$152^{291\text{mm}}$		s bz; misc eth
t60	Tetraethyl pyrophosphate	$[(C_2H_5O)_2P(O)]_2O$	290.20		1.1852^{4}	1.4196^{20}	d 170			d aq; misc alc, bz, chl
t61	Tetraethyl pyrophosphite	$[(C_2H_5O)_2P]_2O$	258.19		1.057	1.4341^{20}		$81^{11\text{mm}}$	>110	i aq
t62	Tetraethylsilane	$(C_2H_5)_4Si$	144.34	4, 625	0.7626^{20}	1.4246^{20}	−82	153	26	
t62a	N,N,N',N'-Tetraethylsulfamide	$(C_2H_5)_2NSO_2N(C_2H_5)_2$	208.33	4, 129	1.030	1.4485^{20}		251	>110	
t63	Tetraethylthiuram disulfide	$[(C_2H_5)_2NC(=S)S—]_2$	296.54	4, 122	1.30		70		53	3.8 alc; 7.1 eth; s bz; acet, chl; 0.02 aq
t64	Tetraethyltin	$(C_2H_5)_4Sn$	234.94	4, 632	1.1992^{20}		−112	181		i aq; s eth
t65	Tetrafluoroethylene	$F_2C=CF_2$	100.02	1[3], 638	1.1507^{-40}		−131.2	−75.6		i aq
t66	2,2,3,3-Tetrafluoro-1-propanol	$HCF_2CF_2CH_2OH$	132.06		1.48534^{20}	1.3197^{20}	−15	109	49	
t67	1,2,3,6-Tetrahydrobenzaldehyde	C_6H_9CHO	110.16	7[1], 48	0.940	1.4745^{20}		163	57	
t68	Tetrahydrofuran		72.11	17, 10	0.8892^{20}	1.4072^{20}	−108.5	66	−17	misc aq, alc, eth, PE
t69	2,5-Tetrahydrofurandimethanol		132.16	17[2], 106	1.1542^{25}	1.4766^{25}	<−50	265		misc aq, alc, bz, chl; s eth
t70	Tetrahydro-2-furanmethanol		102.13	18[2], 415	1.0524^{20}	1.4520^{20}	−80	178	83	misc aq, alc, bz, chl, eth, acet
t71	Tetrahydro-2-furanmethylamine		101.15		0.980	1.4560^{20}		$154^{744\text{mm}}$	45	
t72	2-(Tetrahydrofuryloxy)tetrahydropyran		186.25		1.030	1.4606^{20}			97	

t35

t68

t69 $HOCH_2$—(furan)—CH_2OH

t70   (furan)—CH_2OH

t71   (furan)—CH_2NH_2

t72

TABLE 1.15 Physical Constants of Organic Compounds (*Continued*)

No.	Name	Formula	Formula weight	Beilstein reference	Density	Refractive index	Melting point	Boiling point	Flash point	Solubility in 100 parts solvent
t73	1,2,3,4-Tetrahydroisoquinoline		133.19	20, 275	1.064	1.5668^{20}	-30	232–233	98	
t74	Tetrahydrolinalool	$(CH_3)_2CHCH_2CH_2CH_2C(CH_3)(OH)CH_2CH_3$	158.29	1, 426	0.925^{25}	1.4332^{20}	76			
t75	1,2,3,4-Tetrahydronaphthalene	$C_{10}H_{12}$	132.21	5, 491	0.9702^{20}_4	1.5414^{20}	-35.8	207.6	77	misc alc, bz, chl, eth, acet, PE
t76	cis-1,2,3,6-Tetrahydrophthalic anhydride		152.15	17, 461			72–76	102	157	
t77	cis-1,2,3,6-Tetrahydrophthalimide		151.17				134–138			
t78	Tetrahydropyran		86.14	17, 12	0.8814^{20}_4	1.4211^{20}	-49.2	88	-15	misc aq, alc, eth
t79	Tetrahydropyran-2-methanol		116.16		1.0254^{20}	1.4580^{20}	-70 glass	187	93	misc aq, alc, bz, eth
t80	1,2,3,6-Tetrahydropyridine		83.13	20^3, 1912	0.911	1.4800^{20}	-48	108	16	
t81	3,4,5,6-Tetrahydropyrimidinethiol		116.19	24, 5			210–212			
t82	1,2,3,4-Tetrahydroquinoline		133.19	20, 262	1.061	1.5924	15–16	249	100	s aq; misc alc, eth
t83	Tetrahydrothiophene		88.17	17^1, 5	0.9987^{20}	1.5048^{20}	-96.2	120.9	12	misc alc, eth; i aq
t84	1,4,9,10-Tetrahydroxyanthracene		242.23	8, 431			147–149			
t85	2,2',4,4'-Tetrahydroxybenzophenone	$[(HO)_2C_6H_3]_2C{=}O$	246.22	8, 496			200–203			
t86	Tetrahydroxyhexanedioic acid	$HOOC[CH(OH)]_4COOH$	210.14	3, 581			230 d			0.003 aq; s alk
t87	Tetrakis(dimethylamino)ethylene	$[(CH_3)_2N]_2C{=}C[N(CH_3)_2]_2$	200.23	4^4, 167	0.861	1.4800^{20}		$59^{0.9mm}$	53	
t88	Tetrakis(2-ethylhexoxy)silane	$[CH_3(CH_2)_3CH(C_2H_5)CH_2O]_4$-Si	549.95		0.880^{20}_4	1.4388^{20}		194^{1mm}	190	
t89	N,N,N',N'-Tetrakis(p-hydroxypropyl)ethylenediamine	$[[CH_3CH(OH)CH_2]_2NCH_2{-}]_2$	292.42	4^4, 1685	1.013	1.4812^{20}		$181^{0.8mm}$	>110	

No.	Name	Formula	Mol. wt.	Beilstein ref.	n_D	Density	m.p.	b.p.	Solubility
t90	Tetrakis(isopropoxy)-silane	$[(CH_3)_2CHO]_4Si$	264.4		1.385^{20}	0.877^{20}_{4}		64^{5mm}	
t91	Tetrakis(2-methoxy-ethoxy)silane	$(CH_3OCH_2CH_2O)_4Si$	328.4		1.422^{20}	1.079^{20}_{4}		182^{10mm}	
t92	Tetrakis(trimethylsi-loxy)titanium	$[(CH_3)_3SiO]_4Ti$	404.7		1.427^{20}	0.900^{20}_{4}		110^{10mm}	
t93	1,1,3,3-Tetramethoxy-propane	$[(CH_3O)_2CH]_2CH_2$	164.20		1.4081^{20}	0.997		183	54
t94	Tetramethoxysilane	$(CH_3O)_4Si$	152.22	1, 287	1.368^{20}	1.052^{20}_{4}		121–122	20; 55 aq
t95	Tetramethylammon-ium bromide	$(CH_3)_4N^+\ Br^-$	154.06	4, 51		1.56	d > 230	subl > 360	s aq, hot alc
t96	Tetramethylammon-ium chloride	$(CH_3)_4N^+\ Cl^-$	109.60	4, 51		1.169^{20}_{4}	d > 230	subl > 300	sl s aq; v s abs alc
t97	Tetramethylammon-ium iodide	$(CH_3)_4N^+\ I^-$	201.06			1.829	d 230		

Tetrahydrolinalool, d566a
Tetrahydro-2-methylfuran, m418
Tetrahydro-2-methylfuranmethanol, m418a
Tetrahydro-3-methylpyran, m419

6,7,8,9-Tetrahydro-5H-tetrazoloazepine, p26
Tetrahydrothiophene 1,1-dioxide, t108
Tetrahydrothiophene oxide, t109
Tetrahydroxyadipic acid, t86

Tetraiodomethane, c15a
Tetralin, t75
β-Tetralonehydantoin, b40

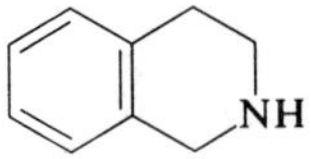

t73

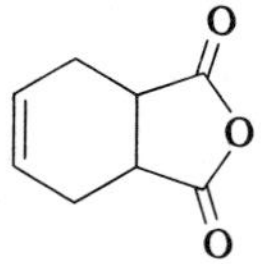

t76

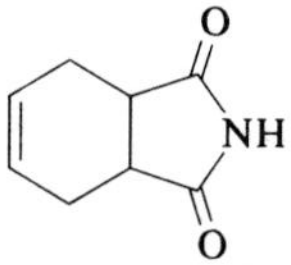

t77

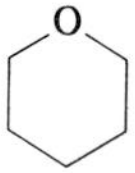

CH₂OH

t78

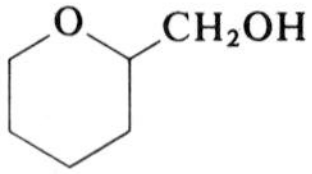

t79

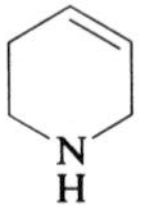

t80

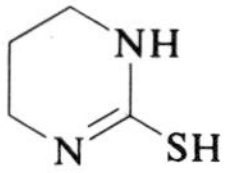

t81

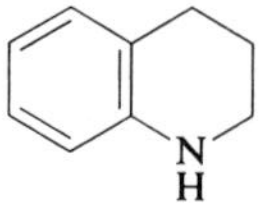

t82

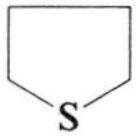

t83

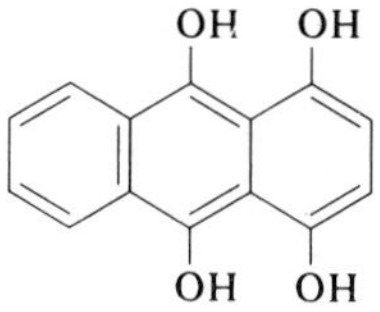

t84

TABLE 1.15 Physical Constants of Organic Compounds (*Continued*)

No.	Name	Formula	Formula weight	Beilstein reference	Density	Refractive index	Melting point	Boiling point	Flash point	Solubility in 100 parts solvent
t98	$N,N,3,5$-Tetramethylaniline	$(CH_3)_2C_6H_3N(CH_3)_2$	149.24	12, 1131	0.913	1.5443^{20}		226–228	90	misc alc, eth
t99	1,2,3,4-Tetramethylbenzene	$C_6H_2(CH_3)_4$	134.22	5, 430	0.905^{20}_4	1.5187^{20}	−6.2	205.0	68	s alc; v s eth
t100	1,2,3,5-Tetramethylbenzene	$C_6H_2(CH_3)_4$	134.22	5, 430	0.8906^{20}_4	1.5134^{20}	−23.7	198.0	63	v s alc, bz, eth
t101	1,2,4,5-Tetramethylbenzene	$C_6H_2(CH_3)_4$	134.22	5, 431	0.8838^1		79.2	196.8	73	
t102	2,2,3,3-Tetramethylbutane	$(CH_3)_3CC(CH_3)_3$	114.23	1, 165	0.656^{-120}		−120.7	106.5	4	
t103	N,N,N',N'-Tetramethyl-1,4-butanediamine	$(CH_3)_2N(CH_2)_4N(CH_3)_2$	144.26	4, 265	0.786^{20}	1.4280^{20}		169	46	s aq, alc, eth
t104	1,1,3,3-Tetramethylbutylamine	$(CH_3)_3CCH_2C(CH_3)_2NH_2$	129.25	4, 198	0.805	1.4240^{20}		137–143	32	s alc, eth, PE; i aq
t105	1,3,5,7-Tetramethylcyclotetrasiloxane	$[-SiH(CH_3)O-]_4$	240.5		0.9912^{20}_4	1.3870^{20}	−69	134–135		
t106	N,N,N',N'-Tetramethyldiaminomethane	$(CH_3)_2NCH_2N(CH_3)_2$	102.18	4, 54	0.749	1.4005^{20}		85	−12	
t107	Tetramethyldisiloxane	$[(CH_3)_2CH]_2O$	134.3	4^4, 3991	0.757^{20}_4	1.370^{20}		71^{731mm}	−10	
t108	Tetramethylene sulfone		120.71	17^1, 5	1.2614^{30}_0	1.4820^{30}	27.6	285	165	misc aq, acet, bz
t109	Tetramethylene sulfoxide		104.17		1.158	1.5200^{20}		122	>110	
t110	N,N,N',N'-Tetramethylethylenediamine	$(CH_3)_2NCH_2CH_2N(CH_3)_2$	116.21	4, 250	0.770	1.4179^{20}	−55		10	
t111	Tetramethylgermanium	$(CH_3)_4Ge$	132.73		1.006^0	1.3871^{20}	−88	43.4		
t112	1,1,3,3-Tetramethylguanidine	$[(CH_3)_2N]_2C=NH$	115.18	4^1, 335	0.918	1.4692^{20}		163	60	
t113	N,N,N',N'-Tetramethyl-1,6-hexanediamine	$[(CH_3)_2N(CH_2)_3-]_2$	172.32	4^1, 423	0.806	1.4359^{20}		210	73	

	Name	Formula	Mol wt	Beil. ref	Density	n_D	mp, °C	bp, °C	Flash	Solubility
t114	Tetramethyl lead	$(CH_3)_4Pb$	267.33	4, 639	1.9954^{20}	1.4005	−27.5	110	38	misc alc, eth
t115	N,N,N',N'-Tetramethylmethanediamine	$(CH_3)_2NCH_2N(CH_3)_2$	102.18	4, 54	0.7490^{20}			85	−12	
t117	2,6,10,14-Tetramethylpentadecane	$[(CH_3)_2CH(CH_2)_3CH(CH_3)-CH_2]_2CH_2$	268.53		0.7827^{20}	1.4379^{20}	−100	167^{11mm}	>110	s bz, chl, eth, PE
t118	2,3,5,6-Tetramethylphenol	$(CH_3)_4C_6HOH$	150.22	6, 547			108–110	250		
t119	2,2,6,6-Tetramethylpiperidino-N-oxy(free radical)		156.25				36–38		67	
t120	N,N,N',N'-Tetramethyl-1,3-propanediamine	$(CH_3)_2N(CH_2)_3N(CH_3)_2$	130.24	4, 262	0.779	1.4234^{20}		146	31	v s alc, eth
t121	Tetramethylpyrazine		136.20	23, 99			84–86	190	−27	
t122	Tetramethylsilane	$(CH_3)_4Si$	88.23	4, 625	0.6411^{20}	1.3585^{20}	−99.5	26.5	221	
t123	1,2,2,3-Tetramethyl-1,1,3,3-tetraphenyltrisiloxane	$[(C_6H_5)_2Si(CH_3)O]_2Si(CH_3)_2$	484.8		1.07^{20}	1.551^{25}		$235^{0.5mm}$		
t124	1,1,3,3-Tetramethyl-2-thiourea	$(CH_3)_2NC(=S)N(CH_3)_2$	132.23	4[1], 336		1.5201	75–77	245		
t125	Tetramethyltin	$(CH_3)_4Sn$	178.83	4, 631	1.3149^{25}		−54.8	78	−12	
t126	1,1,3,3-Tetramethylurea	$(CH_3)_2NC(=O)N(CH_3)_2$	116.16	4, 74	0.9687^{25}	1.4493^{25}	−1.2	176	65	misc aq, alc, chl, eth

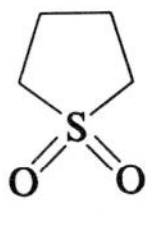

t108

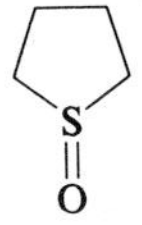

t109

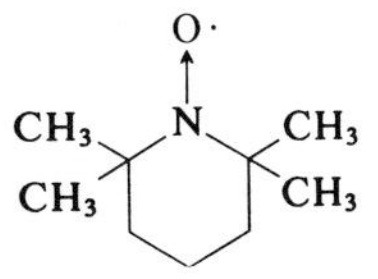

t119

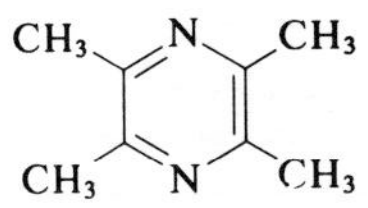

t121

TABLE 1.15 Physical Constants of Organic Compounds (*Continued*)

No.	Name	Formula	Formula weight	Beilstein reference	Density	Refractive index	Melting point	Boiling point	Flash point	Solubility in 100 parts solvent
t126a	Tetranitromethane	$C(NO_2)_4$	196.03	1, 80	1.6229^{25}_{4}	1.4358^{25}	13.5	126	>110	v s alc, eth, alk
t127	1,4,7,10-Tetraoxacyclododecane		176.21		1.089	1.4621^{20}	16		>110	
t128	2,4,8,10-Tetraoxaspiro[5.5]undecane		160.17	19, 436			52–55	$83^{1.5mm}$	108	
t129	Tetraphenoxysilane	$(C_6H_5O)_4Si$	400.5		1.141^{60}_{4}	1.554^{60}	48–49	237^{1mm}		v s aq, acet; s chl
t130	Tetraphenylboron sodium	$(C_6H_5)_4B^- Na^+$	342.23				>300			
t131	1,1,4,4-Tetraphenyl-1,3-butadiene	$(C_6H_5)_2C=CHCH=C(C_6H_5)_2$	358.49	5, 750			207–209			
t133	Tetraphenylethylene	$(C_6H_5)_2C=C(C_6H_5)_2$	332.45	5, 743			222–224	420		
t134	Tetraphenylsilane	$(C_6H_5)_4Si$	336.5		1.078^{20}		236–237	228^{3mm}	110	
t135	Tetraphenyltin	$(C_6H_5)_4Sn$	427.11		1.490^{0}		226	>420	95	
t136	Tetrapropoxysilane	$(C_3H_7O)_4Si$	264.44	1, 355	0.916^{20}_{4}	1.401^{20}		94^{5mm}		
t137	Tetrapropylammonium bromide	$(CH_3CH_2CH_2)_4N^+ Br^-$	266.27	4^1, 364			270 d			s aq
t138	1H-Tetrazole		70.06	26, 346			156–158	subl		s aq, alc, acet
t139	2-Thenoyltrifluoroacetone		222.18				40–44	98^{8mm}		
t140	Theobromine		180.17	26, 457			357	subl 290		0.05 aq; 0.045 alc; s alk; i bz, chl, eth
t141	Thiamine HCl		337.27				d 248			100 aq; 1 alc
t142	Thiazole		85.13	27, 15	1.200^{17}	1.5375^{20}		117–118	22	s alc, eth; sl s aq
t143	Thioacetamide	$CH_3C(=S)NH_2$	75.13	2, 232			112–114			16 aq; 16 alc; sl s eth
t144	Thioacetic acid	CH_3CO—SH	76.12	2, 230	1.065	1.4630^{20}	<−17	88–91	11	s aq; v s alc
t145	Thiobenzoic acid	C_6H_5CO—SH	138.19	9, 419	1.174	1.6020^{20}	15–18	dec	>110	misc eth; v s alc; i aq
t146	4,4'-Thiobis(2-tert-butyl-6-methylphenol)		358.54				127	316^{40mm}	240	
t148	Thiocarbanilide	$C_6H_5NHCSNHC_6H_5$	228.32	12, 394	1.32^{24}		154			v s alc, eth
t150	p-Thiocresol	$HSC_6H_4CH_3$	124.21	6, 416			42–44	195	68	s alc, eth; i aq
t151	2,2'-Thiodiacetic acid	$(HOOCCH_2)_2S$	150.15	3, 253			129			s aq, alc
t152	2,2'-Thiodiethanol	$(HOCH_2CH_2)_2S$	122.19	1, 470	1.1824^{20}	1.5203^{20}	−16	282	160 (OC)	
t153	4,4'-Thiodiphenol	$(HOC_6H_4)_2S$	218.27	6, 860			155			misc aq, alc; sl s eth

No.	Name, synonyms, and formula	M.W.	Beilstein ref.	Density	n_D	m.p., °C	b.p., °C	Solubility
t154	3,3′-Thiodipropionic acid; $(HOOCCH_2CH_2)_2S$	178.21				134		3,4 aq; v s alc, acet
t155	2-Thiohydantoin	116.14	24, 260			231 d	200	84 · sl s aq; i alc, eth
t156	N-Thionylaniline; $C_6H_5N{=}SO$	139.18	12, 578	1.236	1.6270^{20}	–38.2	84.2	–1 · misc alc, eth; i aq
t157	Thiophene; C_4H_4S	84.14	17, 29	1.05733^{25}	1.5257^{25}		160^{22m}	77
t158	2-Thiopheneacetic acid; $(C_4H_3S)CH_2COOH$	142.18	18, 293	1.200	1.5900^{20}	63–67	198	s eth
t159	2-Thiophenecarboxyaldehyde; $(C_4H_3S)CHO$	112.15	17, 285					73
t160	2-Thiophenecarboxylic	128.15	18, 289			128.5	260	s aq, chl; v s alc, eth
t161	2-Thiophenemethyl-amine; $(C_4H_3S)CH_2NH_2$	113.19	18^4, 7096	1.103	1.5569^{20}		99^{28mm}	

Structure labels (with visible atom labels):

t127

t128

t138 — N–N, N, H

t139 — CH_2–C(=O)–CF_3, C=O, S

t140 — CH_3, N, N, O, HN, O, N, CH_3

t141 — NH_2, N, CH_3, S, CH_2CH_2OH, CH_3, N^+–CH_2, Cl^-, ·HCl

t142 — S, N

t146 — CH_3, OH, $C(CH_3)_3$, S, $C(CH_3)_3$, CH_3, HO

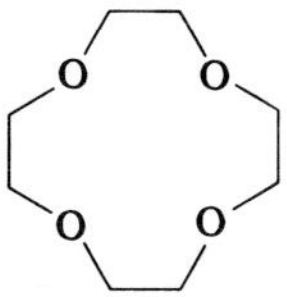

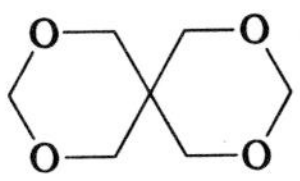

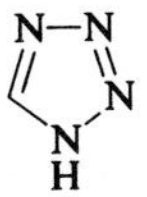

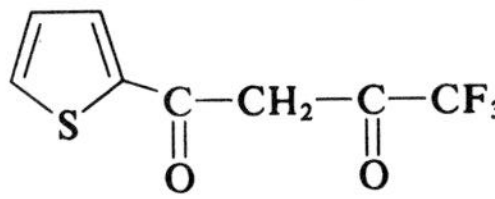

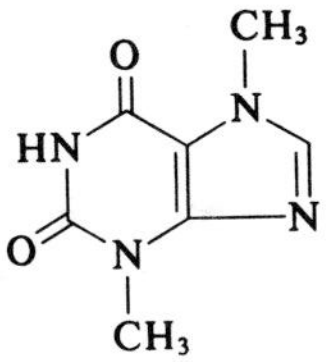

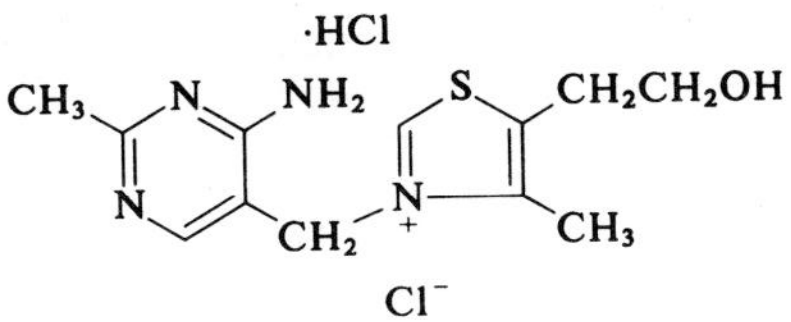

t155 — O, NH, HN, S

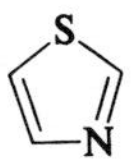

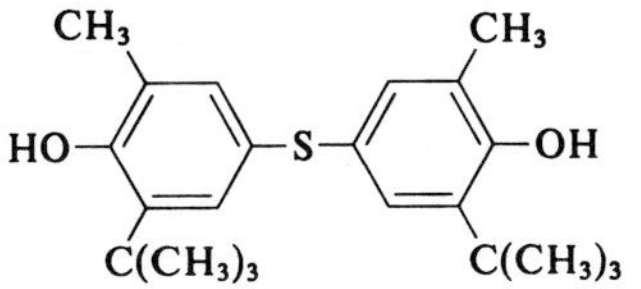

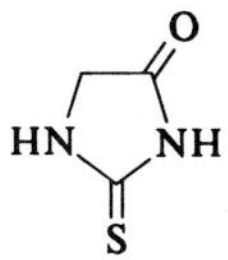

TABLE 1.15 Physical Constants of Organic Compounds (*Continued*)

No.	Name	Formula	Formula weight	Beilstein reference	Density	Refractive index	Melting point	Boiling point	Flash point	Solubility in 100 parts solvent
t162	Thiophenol	C_6H_5SH	110.18	6, 294	1.0766^{20}	1.5897^{20}	−14.9	169.1	50	v s alc; misc bz, eth
t163	Thiophenoxyacetic acid	$C_6H_5SCH_2COOH$	168.21	6, 313			64−66			
t164	Thiopropionic acid	$CH_3CH_2CO{-}SH$	90.14	2, 264	1.014	1.4640^{20}		108−110	11	s aq, alc
t165	3-Thiosemicarbazide	$H_2NC(=S)NHNH_2$	91.14	3, 195			182−184			9 aq; s alc; sl s eth
t166	Thiourea	$H_2NC(=S)NH_2$	76.12	3, 180	1.045		176−178			
t167	1,4-Thioxane		104.17	19, 3	1.114	1.5095^{20}		147	42	v s bz, chl, hot HOAc
t168	Thioxanthen-9-one		212.27	17, 357			211	273^{715mm}		
t168a	Titanium(IV) ethoxide	$Ti(OC_2H_5)_4$	228.15	1, 335	1.088	1.5043^{20}		152^{10mm}	28	
t169	Titanium(IV) isopropoxide	$Ti[OCH(CH_3)_2]_4$	284.26	1^2, 382	0.955	1.4654^{20}	18−20	218^{10mm}	22	
t169a	Titanium(IV) propoxide	$Ti(OCH_2CH_2CH_3)_4$	284.26	1^3, 1423	1.033	1.4986^{20}		170^{mm}	42	
t170	Toluene	$C_6H_5CH_3$	92.14	5, 280	0.8660^{20}_4	1.4969^{20}	−95	110.6	4 (CC)	misc alc, chl, eth, acet, HOAc; 0.067 aq
t171	2,4-Toluenediamine	$CH_3C_6H_3(NH_2)_2$	122.17	13, 124			97−99	285		s hot aq, alc, eth
t172	2,5-Toluenediamine	$CH_3C_6H_3(NH_2)_2$	122.17	13, 144			64	273−274		v s aq, alc, eth
t173	2,6-Toluenediamine	$CH_3C_6H_3(NH_2)_2$	122.17	13, 148			104−106			s aq, alc
t174	3,4-Toluenediamine	$CH_3C_6H_3(NH_2)_2$	122.17	13, 148			90−92			v s aq
t175	Toluene-2,4-diisocyanate	$CH_3C_6H_3(NCO)_2$	174.16	13, 138	1.2244^{20}_4	1.5689^{20}	20−21	251	132 (OC)	d aq, alc; misc bz, acet, eth
t176	p-Toluenesulfinic acid	$CH_3C_6H_4SO_2H$	172.20	11, 9			85			v s alc, eth; sl s aq
t177	p-Toluenesulfonamide	$CH_3C_6H_4SO_2NH_2$	171.22	11, 104			137−140			0.2 aq; 3.6 alc
t178	p-Toluenesulfonyl-hydrazide	$CH_3C_6H_4SO_2NHNH_2$	186.23	11^2, 66			110 d			
t179	p-Toluenesulfonic acid	$CH_3C_6H_4SO_3H$	172.20	11, 97			107 anhy	140^{20mm}		67 aq; s alc, eth
t180	p-Toluenesulfonyl chloride	$CH_3C_6H_4SO_2Cl$	190.65	11, 103			69−71	134^{10mm}		v s alc, bz, eth; i aq
t181	p-Toluenesulfonyl fluoride	$CH_3C_6H_4SO_2F$	174.19	11^2, 54			41−42	112^{16mm}		
t182	p-Toluenesulfonyl isocyanate	$CH_3C_6H_4SO_2NCO$	197.21			1.4355^{20}		144^{10mm}	>110	
t184	o-Toluidine	$CH_3C_6H_4NH_2$	107.16	12, 772	0.9984^{20}	1.5725^{20}	−16.1	200.4	85 (CC)	1.7 aq; s alc, eth
t185	m-Toluidine	$CH_3C_6H_4NH_2$	107.16	12, 853	0.989^{20}_4	1.5681^{20}	−30.4	203.4	85 (CC)	misc alc, eth

			Mol. wt.	Beil.	Density	n	M.p.	B.p.	Flash (CC)	Solubility
t186	p-Toluidine	CH₃C₆H₄NH₂	107.16	12, 880	1.046^{20}_4	1.5532^{29}	43.8	200.6	88 (CC)	7.4 aq; v s alc, eth
t187	1-(o-Toluidino)-1,3-butanedione	CH₃C₆H₄NHCOCH₂COCH₃	191.23	12, 823			104–106	143		
t188	o-Tolunitrile	CH₃C₆H₄CN	117.15	9, 466	0.9955^{20}_4	1.5279^{20}	−13	205.2	84	i aq; misc alc, eth
t189	m-Tolunitrile	CH₃C₆H₄CN	117.15	9, 477	0.976^{15}	1.5256^{20}	−23	210	86	0.09 aq; v s alc, eth
t190	p-Tolunitrile	CH₃C₆H₄CN	117.15	9, 489	0.9785^{30}_4		29.5	217.6	85	i aq; v s alc, eth
t191	o-Toluoyl chloride	CH₃C₆H₄COCl	154.60	9, 464	1.185	1.5549^{20}		90^{12mm}	76	
t192	m-Toluoyl chloride	CH₃C₆H₄COCl	154.60	9, 477	1.173	1.5485^{20}		86^{5mm}	76	
t193	p-Toluoyl chloride	CH₃C₆H₄COCl	154.60	9, 484	1.169	1.5535^{20}	−2	225–227	82	
t193a	p-Tolyl acetate	CH₃CO₂C₆H₄CH₃	150.18	6, 397	1.048	1.5013^{20}		211	90	
t194	m-Tolyl isocyanate	CH₃C₆H₄NCO	133.15	12, 864	1.033	1.5305^{20}		76^{12mm}	65	s alc, eth; i aq
t195	(p-Tolylsulfonyl)-methyl isocyanide	CH₃C₆H₄SO₂CH₂NC	195.24				114–115			
t197	1,2,4-Triacetoxy-benzene	C₆H₃(OOCCH₃)₃	252.22	6, 1089			98–100			
t198	Triacetoxyethylsilane	C₂H₅Si(OOCCH₃)₃	234.3		1.1428^{20}_4	1.4123^{20}		108^{8mm}	104	
t199	Triacetoxyvinylsilane	(CH₃COO)₃SiCH=CH₂	232.3		1.1674^{20}	1.423^{20}		113^{1mm}	17	
t200	1,3,5-Triacetylbenzene	C₆H₃(COCH₃)₃	204.23	7, 866			160–162	152^{4mm}	>110	
t201	Triallyl-s-triazine-2,4,6(1H,3H,5H)-trione		249.27		1.159	1.5129^{20}				

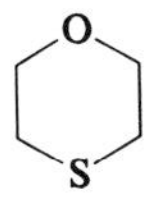

t167

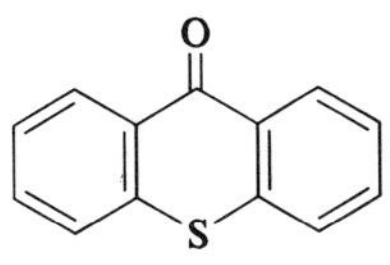

t168

t201

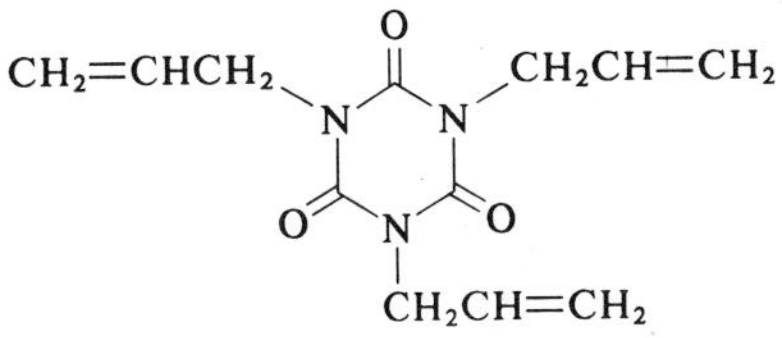

CH₂=CHCH₂ — CH₂CH=CH₂ — CH₂CH=CH₂

TABLE 1.15 Physical Constants of Organic Compounds (*Continued*)

No.	Name	Formula	Formula weight	Beilstein reference	Density	Refractive index	Melting point	Boiling point	Flash point	Solubility in 100 parts solvent
t202	2,4,6-Triamino-1,3,5-triazine		126.12	26, 245	1.573^{250}		>250	subl		sl s aq; i alc, eth
t203	1H-1,2,4-Triazole		69.07	26, 13			119–121	260 d		
t204	Tribenzylamine	$(C_6H_5CH_2)_3N$	287.41	12, 1038	0.9915		91–94		65	s aq, alc
t205	Tribromoacetaldehyde	Br_3CCHO	280.76	1, 626	2.665	1.5850^{20}		174	65	s hot alc, eth
t206	Tribromoacetic acid	Br_3CCOOH	296.76	2, 220			129–135	245 d		s aq, alc, chl, eth
t207	2,4,6-Tribromoaniline	$Br_3C_6H_2NH_2$	329.83	12, 663	2.35		120–122	300		s aq, alc, eth
t208	2,2,2-Tribromoethanol	Br_3CCH_2OH	282.77	1^2, 338			79–82	93^{10mm}		s hot alc, chl, eth
t209	1,1,2-Tribromoethylene	$BrCH=CBr_2$	264.74	1, 191	1.708^{21}	1.6247^{25}		162.5		2 aq; s alc, bz, eth
t210	Tribromomethane	$CHBr_3$	252.77	1, 68	2.9031^{15}	1.6005^{15}	8.1	149.6	none	0.3 aq; misc eth, MeOH
t211	2,4,6-Tribromophenol	$Br_3C_6H_2OH$	330.82	6, 203	2.55		94–96	244		s alc, chl, eth; i aq
t212	1,2,3-Tribromopropane	$BrCH_2CH(Br)CH_2Br$	280.78	1, 112	2.4114^{15}		16.5	220		s alc, eth
t213	Tributoxyborane	$(C_4H_9O)_3B$	230.16	1^2, 398	0.8580^{20}	1.4092^{20}	−70	230–235	93 (COC)	hyd aq
t214	Tributylamine	$(C_4H_9)_3N$	185.36	4, 157	0.7784^{20}	1.4283^{20}	−70	216	63 (CC)	v s alc, eth; s acet
t214a	Tributylborane	$(CH_3CH_2CH_2CH_2)_3B$	182.16	4^2, 1022	0.747			109^{20mm}	−36	i aq; s most org solv
t215	2,4,6-Tri-*tert*-butylphenol	$[(CH_3)_3C]_3C_6H_2OH$	262.44		0.8644^{27}		129–132	277		
t216	Tributyl phosphate	$(C_4H_9O)_3P(O)$	266.32	1^2, 397	0.972^{25}	1.4226^{25}	−79	178^{27mm}	146 (COC)	0.04 aq; misc org solv
t217	Tributylphosphine	$(C_4H_9)_3P$	202.32	4^2, 971	0.812	1.4619^{20}		150^{50mm}	37	
t218	Tributyl phosphite	$(C_4H_9O)_3P$	250.32	1^1, 187	0.925^{20}	1.4326^{20}		125^{7mm}	121	
t219	Tributyltin chloride	$(C_4H_9)_3SnCl$	325.49	4^3, 1926	1.200	1.4905^{20}		173^{25mm}	>110	misc alc, bz, eth, PE
t219a	Tributyltin ethoxide	$(C_4H_9)_3SnOC_2H_5$	335.10	4^4, 4312	1.098	1.4672^{20}		$92^{0.1mm}$	40	
t219b	Tributyltin hydride	$(C_4H_9)_3SnH$	291.05	4^4, 4331	1.082	1.4731^{20}		$80^{0.4mm}$	40	
t219c	Tributyltin methoxide	$(C_4H_9)_3SnOCH_3$	321.07		1.115	1.4723^{20}		$97^{0.06mm}$	98	
t219d	Trichloroacetamide	Cl_3CCONH_2	162.40	2, 211	1.510^{20}		141–143	240		v s aq (dec); s alc (dec); s eth
t219e	Trichloroacetaldehyde	Cl_3CCHO	147.40			1.4557^{20}	−57.5	97.8		120 aq; v s alc, eth
t220	Trichloroacetic acid	Cl_3CCOOH	163.39	2, 206	1.629^{61}	1.4838^{20}	57–58	196–197	none	
t220a	Trichloroacetic anhydride	$(CCl_3CO)_2O$	308.75	2, 210	1.690			141^{60mm}	none	
t221	Trichloroacetonitrile	Cl_3CCN	144.39	2, 212	1.4403^{25}	1.4409^{20}		85.7	none	

No.	Name	Formula	M.W.	Beil. ref.	Density	n_D	mp	bp	Flash p.	Solubility
t221a	2,2',4'-Trichloroacetophenone	$Cl_2C_6H_3COCH_2Cl$	223.49	7, 283			52–55		>110	
t222	Trichloroacetyl chloride	Cl_3CCOCl	181.83	2, 210	1.629	1.4689^{20}		114–116	none	s alc
t223	Trichloroacetyl isocyanate	$Cl_3CC(=O)NCO$	188.40		1.581	1.4809^{20}		85^{20mm}	65	s alc, eth
t224	2,4,5-Trichloroaniline	$Cl_3C_6H_2NH_2$	196.46	12, 627			93–95	270		v s bz, eth, CS_2
t225	2,4,6-Trichloroaniline	$Cl_3C_6H_2NH_2$	196.46	12, 627			73–75	262		
t226	1,2,3-Trichlorobenzene	$C_6H_3Cl_3$	181.45	5, 203	1.69	1.5776^{20}	52.6	221	113	misc bz, eth, PE
t227	1,2,4-Trichlorobenzene	$C_6H_3Cl_3$	181.45	5, 204	1.446^{25}	1.5707^{20}	17	213	>110	v s bz, eth, PE
t228	1,3,5-Trichlorobenzene	$C_6H_3Cl_3$	181.45	5, 204		1.5662^{19}	63.4	208.4	107	
t230	1,1,1-Trichloroethane	CH_3CCl_3	133.41	1, 85	1.3376^{20}_4	1.4384^{20}	−32.5	74.1	none	0.13 aq; s bz, eth
t231	1,1,2-Trichloroethane	$ClCH_2CHCl_2$	133.41	1, 85	1.4416^{20}_4	1.4711^{20}	−35	113–114	none	misc alc, eth
t232	2,2,2-Trichloroethanol	Cl_3CCH_2OH	149.40	1, 338	1.5572^{20}_4	1.4885^{20}	18	151–153	none	8 aq; misc alc, eth
t233	2,2,2-Trichloroethyl chloroformate	$ClCOOCH_2CCl_3$	211.86		1.539	1.4703^{20}		171–172	none	
t234	Trichloroethylene	$ClCH=CCl_2$	131.39	1, 187	1.4649^{20}_4	1.4775^{20}	−84.8	86.7	none	0.1 aq; misc alc, chl, eth
t235	Trichloroethylsilane	$C_2H_5SiCl_3$	163.51	4, 630	1.23733^{20}	1.4256^{20}	−106	99	−10	
t236	Trichlorofluoromethane	Cl_3CF	137.4		1.485^{21}	1.384^{20}	−111	23.8		
t237	α,α,2-Trichloro-6-fluorotoluene	$ClC_6H_3(F)CHCl_2$	213.47	5^3, 701	1.446	1.5506^{20}		228–230	>110	0.14 aq; s alc, eth
t238	Trichloroisocyanuric acid		232.41	25, 256			249–251			

1,3,5-Triazine-2,4,6-triol, c300
Tributyl borate, t213

Tributyrin, g19

β,β,β-Trichloroethoxycarbonyl chloride, t233

t202

t203

t238

TABLE 1.15 Physical Constants of Organic Compounds (*Continued*)

No.	Name	Formula	Formula weight	Beilstein reference	Density	Refractive index	Melting point	Boiling point	Flash point	Solubility in 100 parts solvent
t239	Trichloromethane-sulfenyl chloride	Cl_3CSCl	185.89	3, 135	1.700^{20}_{4}	1.5436^{20}		146–148	none	
t240	Trichloromethane-sulfonyl chloride	Cl_3CSO_2Cl	217.89	3, 19			139			s alc, eth
t241	1,1,1-Trichloro-2-methyl-2-propanol	$(CH_3)_2C(OH)CCl_3$	177.46	1, 382			99	167		s alc, bz, chl, eth
t242	Trichloromethylsilane	CH_3SiCl_3	149.48		1.275^{20}_{4}	1.4108^{20}	−90	66	−15	
t242a	1,2,4-Trichloro-5-nitrobenzene	$Cl_3C_6H_2NO_2$	226.45	5, 246	1.790^{20}		49–55	288	>110	v s bz, eth
t243	Trichloronitromethane	Cl_3CNO_2	164.38	1, 76	1.6558^{20}_{4}	1.4611^{20}	−64	112		misc alc, bz; s eth
t244	2,4,5-Trichlorophenol	$Cl_3C_6H_2OH$	197.45	6^2, 180			67	253		615 acet; 163 bz; 525 eth; s alc; i aq
t245	2,4,6-Trichlorophenol	$Cl_3C_6H_2OH$	197.45	6, 190	1.4901^{15}		69	246	none	525 acet; 113 bz; 354 eth; v s alc; i aq; s alc; v sl s aq
t246	(2,4,5-Trichloro-phenoxy)acetic acid	$Cl_3C_6H_2OCH_2COOH$	255.49	6^3, 702			153			
t247	2-(2,4,5-Trichloro-phenoxy)propionic acid	$Cl_3C_6H_2OCH(CH_3)COOH$	269.51				181.6			0.14 aq; 16 acet; 0.16 bz; 7.1 eth
t248	1,2,3-Trichloropropane	$ClCH_2CH(Cl)CH_2Cl$	147.43	1, 106	1.3880^{20}	1.4834^{20}	−14	156	82	misc alc, eth; i aq
t249	1,1,1-Trichloro-2-propanol	$CH_3CH(OH)CCl_3$	163.43	1, 365			50	162	82	2.9 aq; v s alc, eth
t250	2,4,6-Trichloropyrimidine		183.43	23, 90		1.5700^{20}	23–25	210–215	>110	
t251	Trichlorosilane	$HSiCl_3$	135.45		1.3417^{20}	1.4020^{20}	−127	31–32	−13	dec aq; s bz, chl
t252	α,α,α-Trichlorotoluene	$C_6H_5CCl_3$	195.48	5, 300	1.3756^{20}	1.5570^{20}	−7.5	219–223	97	s alc, bz, eth
t253	α,2,6-Trichlorotoluene	$Cl_2C_6H_3CH_2Cl$	195.48	5^4, 819	1.407	1.5761^{20}	−2.6	248	119	v s alc, eth
t253a	α,3,4-Trichlorotoluene	$Cl_2C_6H_3CH_2Cl$	195.48	5, 300	1.411	1.5766^{20}		124^{14mm}	110	
t254	2,4,6-Trichloro-1,3,5-triazine		184.41	26, 35			148	190		i aq; s alc
t255	1,1,2-Trichlorotri-fluoroethane	$Cl_2CFCClF_2$	187.38	1^3, 157	1.5635^{25}	1.35557^{25}	−36.4	47.6	none	0.017 aq
t256	Trichlorovinylsilane	$H_2C{=}CHSiCl_3$	161.49		1.243^{20}_{4}	1.4300^{20}	−95	90–93	−9	

No.	Name	Formula	Formula wt	Beilstein	Density	n_D	mp, °C	bp, °C	Flash	Solubility
t258	Tricyclo[5.2.1.0^{2,6}]decane		136.24	5, 164	1.063	1.5025^{20}	77–79	193	40	
t259	Tricyclo[5.2.1.0^{2,6}]decan-8-one		150.22	7?, 133			74–75	132^{30mm}		v s alc, eth
t260	1,3,5-Tricyclohexyl-hexahydro-s-triazine		333.57					97^{6mm}		v s alc, eth; i aq
t261	Tridecane	$CH_3(CH_2)_{11}CH_3$	184.37	1, 171	0.7563^{20}	1.4256^{20}	−5.4	235.4	79	
t262	Tridecanoic acid	$CH_3(CH_2)_{11}COOH$	214.35	2, 364			41–42	236^{100mm}	>110	s alc; v s eth
t262a	2-Tridecanone	$CH_3(CH_2)_{10}COCH_3$	198.35	1, 715	0.825	1.4350^{20}	29–32	134^{10mm}	110	
t262b	7-Tridecanone	$[CH_3(CH_2)_5]_2CO$	198.35	1, 715			30–32	264	>110	
t263	1-Tridecene	$CH_3(CH_2)_{10}CH{=}CH_2$	182.35	1, 225	0.7653^{20}	1.4334^{20}	−23.1	232.8	79	
t264	Triethanolamine	$(HOCH_2CH_2)_3N$	149.19	4, 285	1.1242^{20}	1.4835^{25}	21.6	335.4	185	misc aq, alc, acet; 4.5 bz; 1.6 eth; s chl
t264a	3,4,5-Triethoxybenzoic acid	$(C_2H_5O)_3C_6H_2COOH$	254.29	10, 481			110–112			
t265	Triethoxyborane	$(CH_3CH_2O)_3B$	145.99	1, 335	0.8642^{20}	1.3740^{20}		118	11	dec aq
t266	Triethoxymethylsilane	$CH_3Si(OC_2H_5)_3$	178.30	4, 629	0.8952^{20}	1.3845^{20}		143	23	s alc
t266a	Triethoxysilane	$(C_2H_5O)_3SiH$	164.28	1, 334	0.8752^{20}	1.3762		135	26	
t267	Triethoxyvinylsilane	$(C_2H_5O)_3SiCH{=}CH_2$	190.32		0.9032^{20}	1.3978^{20}		161	34	
t268	Triethylaluminum	$(C_2H_5)_3Al$	114.17	4, 643	0.832^{25}		−58	194		dec aq, air
t269	Triethylamine	$(C_2H_5)_3N$	101.19	4, 99	0.7326^{25}	1.3980^{25}	−114	89.6	−6 (CC)	5.5 aq; misc alc, eth; s acet, EtAc

t250   t254   t258   t259   t260

TABLE 1.15 Physical Constants of Organic Compounds (*Continued*)

No.	Name	Formula	Formula weight	Beilstein reference	Density	Refractive index	Melting point	Boiling point	Flash point	Solubility in 100 parts solvent
t270	Triethylantimony	$(C_2H_5)_3Sb$	208.94	4, 618	1.324^{16}	1.42	−29	159.5		i aq; misc alc, eth
t271	Triethylarsine	$(C_2H_5)_3As$	162.11	4, 602	1.150^{20}			140^{736mm}		i aq; v s alc, eth
t272	Triethylbismuthine	$(C_2H_5)_3Bi$	296.17	4, 622	1.82		Explodes when heated in air	107^{79mm}		
t273	Triethylborane	$(C_2H_5)_3B$	98.00	4, 641	0.6961^{23}	1.3971^{20}	−92.9	95		i aq; dec by air
t273a	Triethyl citrate	$HOC(CO_2C_2H_5)$-$(CH_2CO_2C_2H_5)_2$	276.29	3, 568	1.137	1.4426^{20}		127^{1mm}	110	
t274	Triethylenediamine		112.18				158	174		45 aq; 13 acet; 77 alc; 51 bz
t275	Triethylene glycol	$(HOCH_2CH_2OCH_2—)_2$	150.17	1, 468	1.1274^{15}	1.4578^{15}	−4.3	285		misc aq, alc, bz
t276	Triethylene glycol dibenzoate	$(C_6H_5COOCH_2CH_2OCH_2—)_2$	358.39		1.2715^{30}	1.5252^{50}	47		165	
t276a	Triethylene glycol dimethacrylate	$[H_2C=C(CH_3)$-$CO_2CH_2CH_2OCH_2—]_2$	286.33	2^4, 1531	1.092	1.4605^{20}		172^{5mm}	>110	
t276b	Triethylene glycol dimethyl ether	$[CH_3OCH_2CH_2OCH_2—]_2$	178.23		0.986	1.4224^{20}	−45	216	110	
t276c	Triethylene glycol monomethyl ether	$CH_3(OCH_2CH_2)_3OH$	164.20	1^3, 2105	1.026	1.4399^{20}		122^{10mm}	110	
t277	Triethylenetetramine	$(H_2NCH_2CH_2NHCH_2—)_2$	146.24	4, 255	0.982	1.4971	12	266	143	
t278	*N,N′*-Triethylethylenediamine	$(C_2H_5)_2NCH_2CH_2NHC_2H_5$	144.26	4^2, 691	0.804	1.4311^{20}		55^{13mm}	32	
t279	Triethylgallium	$(C_2H_5)_3Ga$	156.91	26, 2	1.0576^{30}	1.4595^{20}	−82.3	142.6		misc alc, chl, eth
t280	1,3,5-Triethylhexahydro-1,3,5-triazine		171.29		0.894			207	70	dec aq; s alc, eth
t281	Triethylindium	$(C_2H_5)_3In$	202.01		1.260^{20}	1.538^{20}	−32	144		v s alc, eth
t282	Triethyl orthoacetate	$CH_3C(OC_2H_5)_3$	162.23	2, 129	0.8847^{25}	1.3950^{25}		142	36	
t283	Triethyl orthoformate	$HC(OC_2H_5)_3$	148.20	2, 20	0.8912^{20}	1.3919^{20}	−76	146	30	
t284	Triethyl orthopropionate	$CH_3CH_2C(OC_2H_5)_3$	176.26	2, 240	0.876	1.3995^{20}		155–160	60	
t285	Triethyl phosphate	$(C_2H_5O)_3P(O)$	182.16	1, 332	1.0725^{19}	1.4045^{20}		215	115	s aq(dec); alc, eth
t286	Triethylphosphine	$(C_2H_5)_3P$	118.16	4, 582	0.800^5	1.4563^{20}	−88	129	−17	i aq; misc alc, eth pyrophoric
t287	Triethyl phosphite	$(C_2H_5O)_3P$	166.16	1, 330	0.9692^{20}	1.4131^{20}		65^{24mm}	55	i aq(hyd); misc alc, acet, bz, eth, PE
t288	Triethyl phosphonoacetate	$(CH_3CH_2O)_2P(O)CH_2$-$COOC_2H_5$	224.19	4^1, 573	1.130	1.4310^{20}		145^{9mm}	>110	

No.	Name	Formula	Formula wt	Beilstein ref	Density	n_D	m.p., °C	b.p., °C	Flash pt., °C	Solubility
t289	Triethylsilane	$(C_2H_5)_3SiH$	116.28	4, 625	0.7314^{20}	1.412^{20}		108	−3	i aq; misc alc, eth
t290	Triethyl thiophosphate	$(C_2H_5O)_3P(S)$	198.22	1, 333	1.082	1.4480^{20}		100^{16mm}	107	
t291	2,2,2-Trifluoroacet-amide	CF_3CONH_2	113.04	2^2, 186			75	162.5		
t292	Trifluoroacetic acid	CF_3COOH	114.02	2^2, 186	1.4890^{20}	1.2850^{20}	−15.3	71.8	none	misc aq
t293	Trifluoroacetic anhy-dride	$[CF_3C(O)]_2O$	210.03		1.487	>1.30	−65	39	none	
t294	1,1,1-Trifluoroacetone	CF_3COCH_3	112.05	1^2, 717	1.252	>1.30		22		
t295	α,α,α-Trifluoro-*m*-cresol	$CF_3C_6H_4OH$	162.11	6^1, 187	1.333	1.4588^{20}	−1.8	179	73	
t296	1,1,1-Trifluoroethane	CH_3CF_3	84.04	1^3, 1342			−111.3	−47.3		
t297	2,2,2-Trifluoroethanol	CF_3CH_2OH	100.04	2^4, 1462	1.3842^{20}	1.2907^{22}	−43.5	74.1	29	
t298	2,2,2-Trifluoroethyl acrylate	$CF_3CH_2OOCCH{=}CH_2$	154.0		2.142^{25}	1.3981^{25}		46^{125mm}	12	
t299	2,2,2-Trifluoroethyl trifluoroacetate	$CF_3CH_2OOCCF_3$	196.0	2^3, 427	1.4725^{18}	1.2812^{18}	−65.5	55	0	
t300	Trifluoromethane	HCF_3	70.01	1, 59	1.52^{-100}		−155.2	−82.2		75 mL aq; 500 mL alc
t301	Trifluoromethanesul-fonic acid	CF_3SO_3H	150.07	3^3, 34	1.695^{25}	1.3250^{25}	34	162	none	v s aq; misc eth
t302	Trifluoromethanesul-fonic anhydride	$(CF_3SO_2)_2O$	282.13	3^4, 35	1.677	1.3212^{20}		84	none	dec aq, alc
t303	3-(Trifluoromethyl)-benzonitrile	$CF_3C_6H_4CN$	171.12	9, 478	1.2813^{20}	1.4505^{20}	14.5	189	72	
t304	3-(Trifluoromethyl)-benzyl chloride	$CF_3C_6H_4CH_2Cl$	194.59		1.254	1.4605		70^{12mm}		

Triethyl borate, t265
Triethylenediamine, d45
Triethylene glycol, e130

Triethylene glycol dimethyl ether, b190
O,O-Triethyl phosphorothioate, t290
2,2,2-Trifluoroethyl mesylate, m441

2-(Trifluoromethyl)aniline, a129
3-(Trifluoromethyl)aniline, a130

t274

CH_3CH_2 N CH_2CH_3
CH_3CH_2 N CH_2CH_3

t280

TABLE 1.15 Physical Constants of Organic Compounds (*Continued*)

No.	Name	Formula	Formula weight	Beilstein reference	Density	Refractive index	Melting point	Boiling point	Flash point	Solubility in 100 parts solvent
t305	α,α,α-Trifluorotoluene	$C_6H_5CF_3$	146.11	5, 290	1.1886^{20}	1.4145^{20}	−29	102	12	v s alc, eth; i aq
t306	Trihexylamine	$[CH_3(CH_2)_5]_3N$	269.52	4, 188	0.794	1.4415^{20}		265	>110	
t307	Trihexylchlorosilane	$[CH_3(CH_2)_5]_3SiCl$	319.05	4^4, 4078	0.8712^0	1.456^{20}		155^{5mm}		
t308	Trihexylsilane	$[CH_3(CH_2)_5]_3SiH$	284.60	4^4, 3915	0.799	1.448^{20}		160	>110	
t309	1,2,3-Trihydroxybenzene	$C_6H_3(OH)_3$	126.11	6, 1071	1.45		133	309		59 aq; 77 alc; 62 eth
t310	1,3,5-Trihydroxybenzene	$C_6H_3(OH)_3$	126.11	6, 1092			218–221	subl d		1 aq; 10 alc; s eth
t311	3,4,5-Trihydroxybenzoic acid	$(HO)_3C_6H_2COOH$	170.12	10, 470			d 235			1.1 aq; 17 alc; 1 eth; 20 acet; i bz, chl, PE
t312	Triisobutylaluminum	$[(CH_3)_2CHCH_2]_3Al$	198.33	4, 643	0.781^{25}	1.4494^{20}	6	86^{10mm}	pyro-phoric	
t313	Triisodecyl phosphite	$[(CH_3)_2CH(CH_2)_7O]_3P$	502.80		0.8861^5	1.454^{25}	<0	$180^{0.1mm}$	235	
t314	Triisopropanolamine	$[CH_3CH(OH)CH_2]_3N$	191.27	4, 762	0.9996^{50}		46	305.4	152	v s aq
t315	Triisopropoxyborane	$[(CH_3)_2CHO]_3B$	188.08	1, 363	0.815	1.3764^{20}		139–141	17	
t316	Triisopropoxyvinylsilane	$[(CH_3)_2CHO]_3SiCH{=}CH_2$	232.4		0.8634^{25}	1.396^{25}		179–181		
t317	1,3,5-Triisopropylbenzene	$C_6H_3[CH((CH_3)_2]_3$	204.36	5, 458	0.845	1.4884^{20}		232–236	86	
t318	Triisopropyl phosphite	$[(CH_3)_2CHO]_3P$	208.24	1, 363	0.9124^{20}	1.4101^{20}		64^{11mm}	57	i aq(sl hyd)
t318a	Triisopropylsilane	$[(CH_3)_2CH]_3SiH$	158.36	4^3, 1851	0.773	1.4344^{20}		86^{35mm}	37	
t319	3,4,5-Trimethoxy-benzaldehyde	$(CH_3O)_3C_6H_2CHO$	196.20	8, 391			73–75	165^{10mm}		
t320	1,2,3-Trimethoxyben-zene	$C_6H_3(OCH_3)_3$	168.19	6, 1081	1.112		43–45	241	>110	
t321	3,4,5-Trimethoxyben-zoic acid	$(CH_3O)_3C_6H_2COOH$	212.20	10, 481			168–171	227^{10mm}		v s alc, eth; s chl
t322	3,4,5-Trimethoxyben-zoyl chloride	$(CH_3)_3C_6H_2COCl$	230.65	10, 487			79–81	185^{18mm}		
t323	3,4,5-Trimethoxyben-zyl alcohol	$(CH_3O)_3C_6H_2CH_2OH$	198.22	6, 1159	1.233	1.5439^{20}		228^{25mm}	>110	
t324	Trimethoxyborane	$(CH_3O)_3B$	103.91	1, 287	0.9202^{23}	1.3568^{20}	−34	67–68	−1	hyd aq; misc alc, eth
t325	Trimethoxyboroxine	$[{-}OB(OCH_3){-}]_3$	173.53		1.195	1.3996^{20}	10	130	10	

No.	Name	Formula	M			d	n	mp	bp	fp	Solubility
t326	1,3,3-Trimethoxybutane	$(CH_3O)_2C(CH_3)CH_2CH_2OCH_3$	148.20	1^3	3214	0.940	1.4096^{20}		63^{20mm}	45	
t326a	1,1,2-Trimethoxyethane	$CH_3OCH_2CH(OCH_3)_2$	120.15	1^3	3183	0.932	1.3921^{20}		59^{56mm}	23	
t327	1,1,3-Trimethoxypropane	$CH_3OCH_2CH_2CH(OCH_3)_2$	134.18	1	820	0.942	1.4004^{20}		46^{17mm}	40	
t327a	Trimethoxysilane	$(CH_3O)_3SiH$	122.20	1^2	274	0.960	1.3579^{20}	−115	81	−4	
t328	3-(Trimethoxysilyl)-propylamine	$H_2N(CH_2)_3Si(OCH_3)_3$	179.3			1.027	1.4260^{20}		92^{15mm}	83	
t329	N-[3-Trimethylsilyl)-propyl]ethylenediamine	$(CH_3O)_3Si(CH_2)_3NHCH_2CH_2NH_2$	222.4			1.010	1.4450^{20}		146^{15mm}	>110	
t331	3-(Trimethoxysilyl)-propyl methacrylate	$(CH_3O)_3SiCH_2CH_2CH_2OOCC(CH_3)=CH_2$	249.3			1.045^{20}	1.429^{25}		190	92	
t332	Trimethylaluminum	$(CH_3)_3Al$	72.09	4	643	0.752^{20}	1.432^{12}	15.4	20^{8mm}	pyrophoric	s alk; v sl s alc
t333	Trimethylamine	$(CH_3)_3N$	59.11	4	43	0.636	1.3443^{20}	−117.1	2.9	−6 (CC)	41 aq; misc alc; s bz, chl, eth
t334	Trimethylamine-N-oxide	$(CH_3)_3N(O)$	75.11					257			s aq, MeOH
t335	2,4,6-Trimethylaniline	$(CH_3)_3C_6H_2NH_2$	135.21	12	1160	0.963	1.5510^{20}		233	96	
t336	1,3,3-Trimethyl-6-aza-bicyclo[3.2.1]octane		153.27			0.902	1.4716^{20}		194	75	

t336

TABLE 1.15 Physical Constants of Organic Compounds (*Continued*)

No.	Name	Formula	Formula weight	Beilstein reference	Density	Refractive index	Melting point	Boiling point	Flash point	Solubility in 100 parts solvent
t337	3,3,5-Trimethyl-1-aza-cycloheptane		141.26		0.852	1.4563^{20}		180	67	i aq; s alc, eth
t338	1,2,3-Trimethylbenzene	C$_6$H$_3$(CH$_3$)$_3$	120.20	5, 399	0.8944^{20}_4	1.5139^{20}	-25.4	176.1	48	s alc, bz, eth
t339	1,2,4-Trimethylbenzene	C$_6$H$_3$(CH$_3$)$_3$	120.20	5, 400	0.87562^{20}	1.5048^{20}	-43.9	169.4	48	s alc, bz, eth
t340	1,3,5-Trimethylbenzene	C$_6$H$_3$(CH$_3$)$_3$	120.20	5, 406	0.86372^{20}	1.4994^{20}	-44.7	164.7	44	misc alc, bz, eth
t341	Trimethyl 1, 2, 4-benzenetricarboxylate	C$_6$H$_3$(COOCH$_3$)$_3$	252.22	9^1, 429	1.261	1.5214^{20}	38–40	194^{12mm}	>110	
t342	2,2,3-Trimethylbutane	(CH$_3$)$_2$CHC(CH$_3$)$_3$	100.20	1^2, 121	0.69014^{20}	1.3894^{20}	-24.9	80.9		s alc, eth
t343	2,3,3-Trimethyl-2-butanol	(CH$_3$)$_3$CC(CH$_3$)$_2$OH	116.20	1^2, 447	0.83804^{25}	1.4233^{22}	15–17	130.5		misc alc, eth
t345	3,5,5-Trimethylcyclohex-2-ene-1-one		138.2	7, 65	0.9252^{20}	1.478^{20}	-8.1	215	84	1.2 aq
t346	2,2,6-Trimethyl-1,3-dioxen-4-one		142.15	19^3, 1604	1.088	1.4622^{20}	12–13	67^{2mm}		
t347	4,4'-Trimethylene-bis(1-methylpiperidine)		238.42		0.896	1.4820^{20}	13	215^{50mm}	110	
t350	Trimethylene oxide		58.08	17, 6	0.8930^{25}	1.3895^{25}		50	-28	misc aq
t350a	Trimethylene sulfide		74.15	17^1, 3	1.025^{20}	1.5102^{20}	-73.3	95.0	<1	
t351	3,3,5-Trimethylhexahydroazepine		141.26		0.852	1.4563		180	67	
t351a	3,5,5-Trimethylhexanal	(CH$_3$)$_3$CCH$_2$CH(CH$_3$)CH$_2$CHO	142.24	1^3, 2894	0.817	1.4215^{20}		$68^{2.5mm}$	46	v s org solv
t352	2,2,5-Trimethylhexane	(CH$_3$)$_2$CHCH$_2$CH$_2$C(CH$_3$)$_3$	128.26	1^3, 516	0.7072^{20}	1.3997^{20}	-105.8	124.1		s alc, eth
t353	3,5,5-Trimethyl-1-hexanol	(CH$_3$)$_3$CCH$_2$CH(CH$_3$)CH$_2$-CH$_2$OH	144.25		0.8236^{20}	1.4300^{25}	<-70	194		
t354	Trimethylhydroquinone	(CH$_3$)$_3$C$_6$H(OH)$_2$	152.19	6, 931			172–174			s aq; v s alc, bz, eth
t355	2,6,8-Trimethyl-4-nonanone	(CH$_3$)$_2$CHCH$_2$CH(CH$_3$)C(O)-CH$_2$CH(CH$_3$)$_2$	184.31		0.8182^{20}		-75	218.4		
t356	1,3,3-Trimethyl-2-norbornanol		154.25	6, 70	0.9641^{20}		48	201	73	s alc, eth

t357	1,3,3-Trimethyl-2-nor-bornanone		152.24	7, 96	0.948^{18}	1.4635^{18}	5–6	192–194	52	v s alc, eth
t358	Trimethyl orthoacetate	$CH_3C(OCH_3)_3$	120.15	2[2], 128	0.9428^{25}	1.3859^{25}		105	16	v s alc, eth
t359	Trimethyl orthoformate	$HC(OCH_3)_3$	106.12	2, 19	0.9676^{20}	1.3790^{20}		100.6	15	
t360	2,4,4-Trimethyl-1-oxazoline		113.16		0.887	1.4213^{20}		113	12	s eth; sl s alc
t361	2,2,3-Trimethylpentane	$(CH_3)_3CCH(CH_3)CH_2CH_3$	114.23	1[1], 62	0.7160^{20}	1.4030^{20}	−112.3	109.8		
t362	2,2,4-Trimethylpentane	$(CH_3)_2CHCH_2C(CH_3)_3$	114.23	1[2], 127	0.6919^{20}	1.3915^{20}	−107.4	99.2	−7	s bz, chl, eth

Trimethylene chlorobromide, b259
Trimethylene chlorohydrin, c215
Trimethylenediamine, p196
Trimethylene dibromide, d93
Trimethylene glycol, p198
Trimethylethylene, m159

endo-1,7,7,-Trimethylbicyclo[2.2.1]heptan-2-ol, b217
Trimethyl borate, t324
Trimethylchlorosilane, c256
α,α,4-Trimethyl-3-cyclohexene-1-methanol, t7
3,5,5-Trimethylcyclohex-2-en-1-one, i82
1,2,2-Trimethyl-1,3-cyclopentanedicarboxylic acid, c5

Trimethylgermanium bromide, b366
3,3,5-Trimethylhexahydroazepine, t337
Trimethylolpropane, e159
Trimethylolpropane triacrylate, e159a
Trimethylolpropane trimethacrylate, e159b

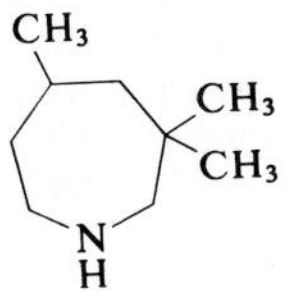

t337
t345 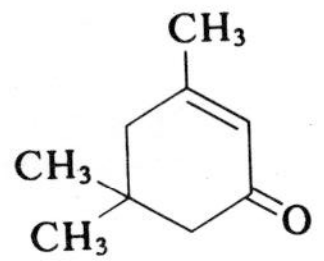
t346 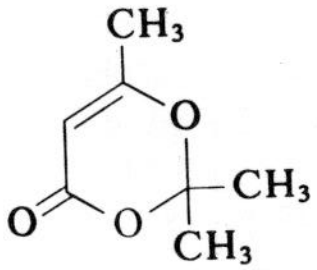
t347 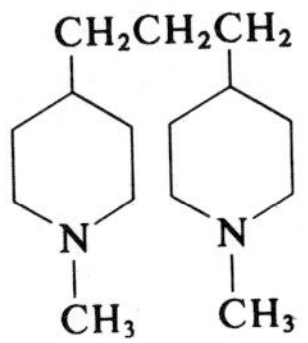
t349 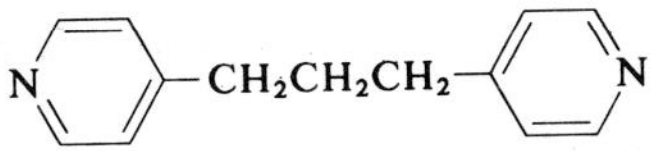
t350 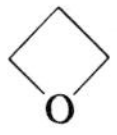
t350a  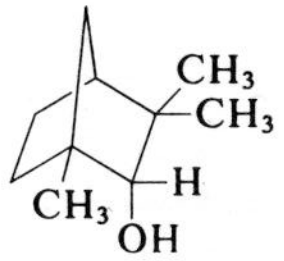
t356
α form
t357
t360

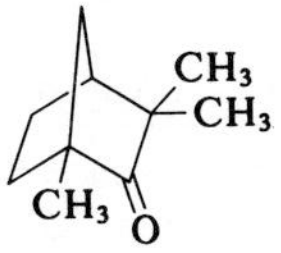

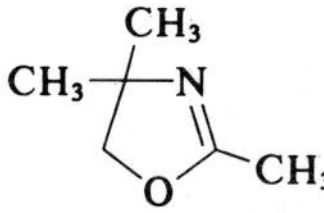

TABLE 1.15 Physical Constants of Organic Compounds (*Continued*)

No.	Name	Formula	Formula weight	Beilstein reference	Density	Refractive index	Melting point	Boiling point	Flash point	Solubility in 100 parts solvent
t363	2,3,4-Trimethylpentane	$(CH_3)_2CH[CH(CH_3)]_2CHCH_3$	114.24	1^3, 500	0.7190^{20}_4	1.4042^{20}	−109.2	113.5	5	a alc, org solv
t364	2,2,4-Trimethyl-1,3-pentanediol	$(CH_3)_2CHCH(OH)C(CH_3)_2CH_2OH$	146.22	1^3, 2225	0.928^{55}	1.4513^{15}	46	229	113	1.8 aq; 75 alc; 22 bz; 25 acet
t365	2,4,4-Trimethyl-1-pentene	$(CH_3)_3CCH_2C(CH_3){=}CH_2$	112.22	1^3, 849	0.7150^{20}_4	1.4112^{20}	−93	101.4	−6	
t365a	2,4,4-Trimethyl-2-pentene	$(CH_3)_3CCH{=}C(CH_3)_2$	112.22	1^3, 848	0.720	1.4159^{20}	−106	104	−1	
t366	2,3,5-Trimethylphenol	$(CH_3)_3C_6H_2OH$	136.19	6, 518			92–95	231		
t367	2,3,6-Trimethylphenol	$(CH_3)_3C_6H_2OH$	136.19				62–64			
t368	2,4,6-Trimethylphenol	$(CH_3)_3C_6H_2OH$	136.19	6, 518			68–71	220		
t369	Trimethyl phosphate	$(CH_3O)_3P(O)$	140.08	1, 286	1.197	1.3958^{20}	−46	197	none	100 aq; s alc
t370	Trimethyl phosphite	$(CH_3O)_3P$	124.08	1, 285	1.046^{20}_4	1.4080^{20}	< −78	111–112	40	d aq; misc alc, acet, bz, PE
t371	Trimethyl phosphonoacetate	$(CH_3O)_2P(O)CH_2COCH_3$	182.11		1.125	1.4370^{20}		$118^{0.85mm}$	>110	
t372	1,2,4-Trimethylpiperazine		128.22		0.851^{25}	1.4480^{25}	< −50	151^{746mm}		s aq, alc, acet, bz
t372a	1,2,2-Trimethylpropylamine	$(CH_3)_3CCH(NH_2)CH_3$	101.19	4, 193	0.755	1.4130^{20}	−20	103	1	
t373	2,4,6-Trimethylpyridine	$(C_5H_2N)(CH_3)_3$	121.18	20, 250	0.9166^{22}	1.4979^{20}	−43	170.5	57	3.5 aq; misc eth; s alc, bz, chl
t374	N-(Trimethylsilyl)acetamide	$CH_3CONHSi(CH_3)_3$	131.2				52–54	186	57	
t375	(Trimethylsilyl) acetate	$CH_3CO_2Si(CH_3)_3$	132.2	4^3, 1857	0.082	1.3880^2	−32	108	4	
t377	Trimethylsilyl bromoacetate	$BrCH_2COOSi(CH_3)_3$	211.14		1.284	1.4421^{20}		58^{9mm}	28	
t377a	Trimethylsilyl cyanide	$(CH_3)_3SiCN$	99.2	4, 3893	0.744	1.3924^{20}	11–12	119	1	
t378	N-(Trimethylsilyl)-diethylamine	$(CH_3)_3SiN(C_2H_5)_2$	145.3		0.767	1.4110^{20}		126	10	
t379	2-(Trimethylsilyl)ethanol	$(CH_3)_3SiCH_2CH_2OH$	118.25		0.825	1.4246^{20}		73^{35mm}	50	
t380	N-(Trimethylsilyl)imidazole		140.3		0.956	1.4751^{20}		94^{14mm}	5	

No.	Name	Formula	Formula wt	Beilstein ref	Density	n_D	Mp, °C	Bp, °C	Flash pt, °F	Solubility in 100 parts
t381	3-(Trimethylsilyloxy)-allene	$(CH_3)_3SiOCH_2CH=CH_2$	130.3		0.7830^{30}	1.4075^{25}		100–102		
t382	Trimethylsilyloxy-3-penten-2-one	$(CH_3)_3SiOC(CH_3)=CHCOCH_3$	172.3	4^4, 4003	0.912	1.4534^{20}		68^{4mm}	58	
t383	Trimethylsilyl trifluoroacetate	$(CH_3)_3SiOOCCF_3$	186.2		1.0772^{20}	1.3880^{20}		89–90	0	
t384	Trimethylsulfonium iodide	$[(CH_3)_3S]I$	204.07				215–220	subl		s hot acet; sl s alc
t385	Trimethylsulfoxonium iodide	$[(CH_3)_3S(O)]I$	220.07				169 d			
t387	Trimethylvinylsilane	$(CH_3)_3SiCH=CH_2$	100.24		0.6902^{20}	1.3920^{20}		55	<−34	
t388	2,4,6-Trinitroaniline	$(O_2N)_3C_6H_2NH_2$	228.12	12, 763	1.762^{14}		188–190	explodes		5.5 alc; 7.1 eth; i aq
t389	1,2,4-Trinitrobenzene	$C_6H_3(NO_2)_3$	213.11	5, 271	1.73^{16}		61–62	explodes		0.035 aq; 1.9 alc; 1.5 eth; 6.2 bz
t390	1,3,5-Trinitrobenzene	$C_6H_3(NO_2)_3$	213.11	5, 271	1.6884^{20}		122.5	explodes		v s bz, acet; sl s aq
t391	2,4,7-Trinitro-9-fluorenone		315.20	7^2, 410			175–176			
t392	Trinitromethane	$HC(NO_2)_3$	151.04	1, 79	1.5972^4		15	47^{22mm}		s aq, alk
t393	2,4,6-Trinitrotoluene	$(O_2N)_3C_6H_2CH_3$	227.13	5, 347	1.654^{20}		80.1	explodes		1.5 alc; 4 eth; s bz, acet; 0.01 aq
t394	Trioctylamine	$[CH_3(CH_2)_7]_3N$	353.68	4, 196	0.809	1.4485^{20}	64	367	>110	
t395	1,3,5-Trioxane		90.08	19, 381	1.170			115	45	17.2 aq; v s alc, bz, eth, EtAc
t396	Tripentaerythritol		372.41				245 d			

Trimethylsilyl cyanide, c299
(N-Trimethyl)silyldiethylamine, d343
(N-Trimethyl)silyldimethylamine, d624a

t372

t380

Trimethylsilyl iodide, i55
Trimethylsilylnitrile, c299
2,4,6-Trinitrophenol, p176

t391

Triolein, g22
Trioxymethylene, t395
Tripalmitin, g23

t395

$(HOCH_2)_3CCH_2OCH_2\overset{CH_2OH}{\underset{CH_2OH}{C}}CH_2OCH_2C(CH_2OH)_3$

t396

TABLE 1.15 Physical Constants of Organic Compounds (*Continued*)

No.	Name	Formula	Formula weight	Beilstein reference	Density	Refractive index	Melting point	Boiling point	Flash point	Solubility in 100 parts solvent
t397	2,4,6-Triphenoxy-s-triazine		357.37				232–234			
t398	Triphenoxyvinylsilane	$(C_6H_5O)_3SiCH{=}CH_2$	334.5		1.1302^{25}	1.562^{25}		210^{7mm}		s acet, eth; sl s alc
t399	Triphenylamine	$(C_6H_5)_3N$	245.33	12, 181	0.7740		125–127	348	>110	v s bz, eth; sl s alc
t400	Triphenylantimony	$(C_6H_5)_3Sb$	353.07	16, 891	1.4343^{25}		52–54	377		v s bz, eth; s alc
t401	Triphenylarsine	$(C_6H_5)_3As$	306.24	16, 828	1.2225^{48}	1.6139^{48}	60–62	233^{14mm}		v s bz; s abs alc, eth
t402	1,3,5-Triphenylbenzene	$(C_6H_5)_3C_6H_3$	306.41	5, 737	1.205		172–174	460		v s hot alc, eth; 49 chl; 7 bz; s PE; i aq
t404	Triphenylmethane	$(C_6H_5)_3CH$	244.34	5, 698	1.0134^{99}_{4}		93.4	360		v s alc, bz, eth; i aq
t405	Triphenylmethanol	$(C_6H_5)_3COH$	260.34	6, 713	1.199^{9}		164.2	360		misc alc; s bz, acet, chl, eth; i aq
t406	Triphenyl phosphate	$(C_6H_5O)_3P(O)$	326.29	6, 179			49–51	244^{10mm}	223	v eth; s bz, chl, HOAc; sl s alc; i aq
t407	Triphenylphosphine	$(C_6H_5)_3P$	262.29	16, 759	1.0754^{81}		80.5	377	181	
t408	Triphenylphosphine oxide	$(C_6H_5)_3P(O)$	278.29	16, 783			187–189			
t409	Triphenylphosphine sulfide	$(C_6H_5)_3P(S)$	294.36	16, 784			162–164			
t410	Triphenyl phosphite	$(C_6H_5O)_3P$	310.29	6, 177	1.1842^{5}	1.5903^{20}	22–24	360	218 (OC)	s alc, bz, chl, eth
t411	Triphenylsilane	$(C_6H_5)_3SiH$	260.41	16^2, 605			42–44	152^{2mm}		
t413	Tripropoxyborane	$(CH_3CH_2CH_2O)_3B$	188.08	1^2, 369	0.85762^{20}	1.3948^{20}		175	32	v s alc; misc eth
t414	Tripropylamine	$(CH_3CH_2CH_2)_3N$	143.27	4, 139	0.753	1.4160^{20}	−93	158	36	s aq, alc, eth
t415	Tripropylene glycol	$H(OCH_2CH_2CH_2)_3OH$	192.3		1.018	1.442^{25}		267.2	141	s aq
t416	Tripropylene glycol butyl ether	$HO(CH_2CH_2CH_2O)_3{-}(CH_2)_3CH_3$	248.4		0.934^{25}	1.430^{25}		276	135	
t417	Tripropylene glycol ethyl ether	$HO(CH_2CH_2CH_2O)_3CH_2CH_3$	220.3		0.948^{25}	1.427^{25}		486	132	
t418	Tripropylene glycol isopropyl ether	$HO(CH_2CH_2CH_2O)_3CH(CH_3)_2$	234.8		0.942^{25}	1.428^{25}		112.7	124	
t419	Tripropylene glycol methyl ether	$HO(CH_2CH_2CH_2O)_3CH_3$	206.29	1^4, 2475	0.9673^{25}	1.428^{25}	−42	242.4	127	misc aq, alc, eth
t420	Tripropyl orthoformate	$HC(OCH_2CH_2CH_3)_3$	190.28		0.8805^{20}_{4}	1.4072^{20}		108^{5mm}	>110	
t421	Tris(2-butoxyethyl) phosphate	$(C_4H_9OCH_2CH_2O)_3P(O)$	398.48		1.006	1.4359^{20}		228^{4mm}		

No.	Name	Formula	Formula wt	Beilstein	Density	n_D	m.p., °C	b.p., °C	Solubility
t422	Tris(2-chloroethoxy)silane	$(ClCH_2CH_2O)_3SiH$	267.6		1.2886^{20}_{4}	1.4577^{20}		118^{2mm}	misc alc, bz, eth
t423	Tris(2-chloroethyl) phosphate	$(ClCH_2CH_2O)_3P(O)$	285.49	1^2, 337	1.390	1.4721^{20}		232; 330	
t424	Tris(2-chloroethyl) phosphite	$(ClCH_2CH_2O)_3P$	269.49		1.3533^{20}_{4}	1.4863^{20}		190; 115^{2mm}	i aq
t426	Tris(dimethylamino)methane	$CH[N(CH_3)_2]_3$	145.25		1.4360^{20}		−15	43^{12mm}	
t427	Tris(dimethylamino)methylsilane	$[(CH_3)_2N]_3SiCH_3$	175.4		0.8502^{22}	1.432^{22}	−11	56^{17mm}	
t428	Tris(2-ethylhexyl) phosphate	$[C_4H_9CH(C_2H_5)CH_2O]_3P(O)$	434.65	1^3, 1734	0.924	1.4437^{20}		215^{4mm}	
t430	Tris(hydroxymethyl)aminomethane	$(HOCH_2)_3CNH_2$	121.14	4, 303			172	220^{10mm}	
t431	2-[Tris(hydroxymethyl)methyl]amino]-1-ethanesulfonic acid	$(HOCH_2)_3CNHCH_2CH_2SO_3H$	229.25				223−225		
t432	1,1,1-Tris(hydroxymethyl)ethane	$CH_3C(CH_2OH)_3$	120.15	1, 520					

Triphenylmethyl bromide, b369
Triphenylsilyl azide, a325
Triphenyltin chloride, c258

Tripropyl borate, t413
TRIS, t430
Tris(dimethylamino)silyl chloride, c90

Tris(3,6-dioxaheptyl)amine, t436
Tris(7-methylnonyl) phosphite, t313

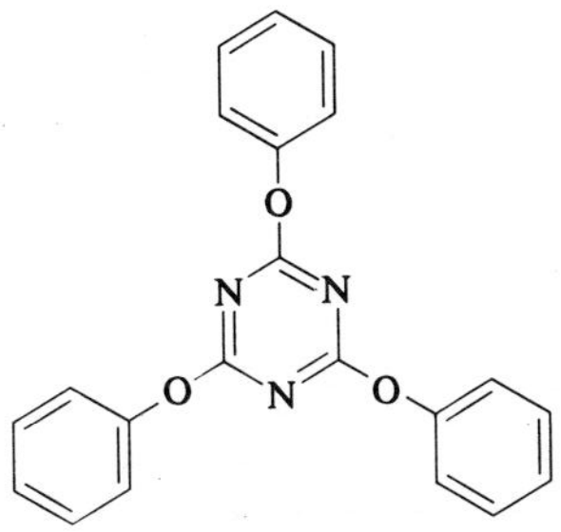

t397

TABLE 1.15 Physical Constants of Organic Compounds (*Continued*)

No.	Name	Formula	Formula weight	Beilstein reference	Density	Refractive index	Melting point	Boiling point	Flash point	Solubility in 100 parts solvent
t433	3-[N-Tris(hydroxymethyl)methylamino]-2-hydroxypropanesulfonic acid	$(HOCH_2)_3CNHCH_2CH(OH)CH_2SO_3H$	259.3				226			
t434	3-[Tris(hydroxymethyl)methylamino]-1-propanesulfonic acid	$(HOCH_2)_3CNHCH_2CH_2CH_2SO_3H$	243.28				240 d			
t435	N-[Tris(hydroxymethyl)methyl]-glycine	$(HOCH_2)_3CNHCH_2COOH$	179.17				184 d			
t435a	Tris(hydroxymethyl)nitromethane	$(HOCH_2)_3CNO_2$	151.12	1, 520			175 d			v s aq
t436	Tris[2-(2-methoxyethoxy)ethyl]amine	$[CH_3OCH_2CH_2OCH_2CH_2]_3N$	323.43		1.011	1.4486^{20}			>110	
t437	Tris(2-methoxyethoxy)vinylsilane	$H_2C=CHSi(OCH_2CH_2OCH_3)_3$	280.39	4[4], 4257	1.0342^5	1.427^{25}		284–286		
t438	Tris(2-methylallyl)amine	$[H_2C=C((CH_3)CH_2)]_3N$	173.91	4[3], 462	0.794	1.4575^{20}		85^{15mm}	53	
t439	Tris(2,2,2-trifluoroethyl) phosphite	$(CF_3CH_2O)_3P$	328.07	1[4], 1371	1.487	1.3245^{20}		131^{743mm}	>110	
t440	1,3,5-Trithiane	$(HS)_2C(S)$	138.27	19, 382	1.483^{20}	1.8225^{20}	216–218			s bz; sl s alc, eth
t441	Trithiocarbonic acid		110.21	3, 221	1.143	1.5553^{20}	−26.9	57.8		d aq, alc; sl s eth
t441a	Tritolyl phosphate	$(CH_3C_6H_4O)_3P(O)$	368.37					265^{10mm}	110	
t442	1,2,4-Trivinylcyclohexane	$(H_2C=CH)_3C_6H_9$	162.28		0.836	1.4780^{20}		88^{20mm}	68	
t443	L-(−)-Tryptophan		204.23	22, 546			280–285 d			1 aq; s hot alc, alk; i eth, chl
t444	L-Tyrosine	$(HOC_6H_4CH_2CH(NH_2)COOH$	181.19	14, 605			>300 d			0.03 aq; 0.01 alc; s alk; i eth
u1	Undecanal	$CH_3(CH_2)_9CHO$	170.30	1, 712	0.825	1.4322^{20}	−4	115^{5mm}	96	i aq; s alc, eth
u2	Undecane	$CH_3(CH_2)_9CH_3$	156.31	1, 170	0.7402^{20}	1.4173^{20}	−25.6	195.9	65 (OC)	i aq; misc alc, eth
u2a	Undecanenitrile	$CH_3(CH_2)_9CN$	167.30	2, 358	0.823	1.4330^{20}		253	110	
u3	Undecanoic acid	$CH_3(CH_2)_9COOH$	186.30	2, 358	0.8907	1.4294^{45}	28.5	228^{160mm}	>110	s alc, chl, eth; i aq
u4	1-Undecanol	$CH_3(CH_2)_{10}OH$	172.31	1, 427	0.8324^{20}	1.4402^{20}	15.9	242.8	>110	0.02 aq; s alc

	Name	Formula								Solubility
u5	2-Undecanone	$CH_3(CH_2)_8COCH_3$	170.30	1, 173	0.829	1.4280^{20}	11–12	231–232	88 (CC)	s alc, bz, chl, eth, acet; i aq
u5a	3-Undecanone	$CH_3(CH_2)_7COC_2H_5$	170.30	1, 713	0.827	1.4291^{20}	12–13	225–229	89	i aq; v s alc, eth
u6	6-Undecanone	$CH_3(CH_2)_4CO(CH_2)_4CH_3$	170.30	1, 174	0.831	1.4280^{20}	14.6	228	88	i aq; v s alc, eth
u7	10-Undecenal	$H_2C{=}CH(CH_2)_8CHO$	168.28	1, 225	0.810	1.4427^{20}			92	i aq; misc alc, eth
u8	1-Undecene	$CH_3(CH_2)_8CH{=}CH_2$	154.29		0.7632^{20}	1.4261^{20}	−49.2	192.7	62	s alc, chl, eth; i aq
u9	10-Undecenoic acid	$H_2C{=}CH(CH_2)_8COOH$	184.28	2, 458	0.9072^4	1.4493^{20}	24.5	137^{2mm}	148	
u10	10-Undecen-1-ol	$H_2C{=}CH(CH_2)_9OH$	170.30	1, 452	0.850^{15}	1.4500^{20}	−2	245	93	
u11	10-Undecenoyl chloride	$H_2C{=}CH(CH_2)_8COCl$	202.73	2, 459	0.944	1.4532^{20}		122^{10mm}	93	
u12	Urea	$(H_2N)_2CO$	60.06	3, 42	1.32^{18}		132.7	dec > mp		100 aq; 20 alc
u13	Uric acid		168.11	26, 513	1.893^{20}		>300	dec		s alk; i aq, alc, eth
u14	Uridine		244.20	31, 23			165	subl		s aq; hot alc, pyr
v1	L-Valine	$(CH_3)_2CHCH(NH_2)COOH$	117.15	4, 427	1.230		315			8.8 aq; v sl s alc, eth
v2	Vinyl acetate	$H_2C{=}CHOOCCH_3$	86.09	21, 63	0.9324^{20}	1.3954^{20}	−92.8	72.5	−8 (CC)	2 aq; misc alc, eth

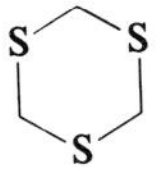

t440

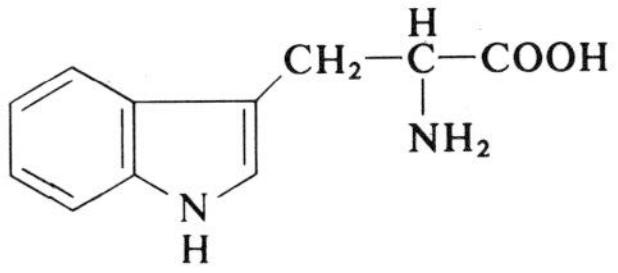

t443

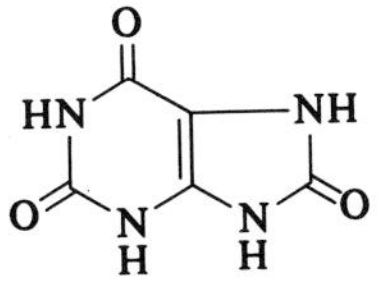

u13

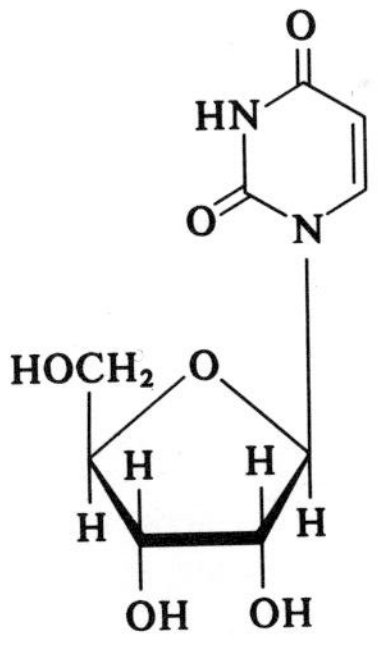

u14

TABLE 1.15 Physical Constants of Organic Compounds (*Continued*)

No.	Name	Formula	Formula weight	Beilstein reference	Density	Refractive index	Melting point	Boiling point	Flash point	Solubility in 100 parts solvent
v3	5-Vinylbicyclo-[2.2.1]-2-heptene		120.19		0.84	1.4802	−80	141		
v4	Vinyl crotonate	$CH_3CH{=}CHCOOCH{=}CH_2$	112.13	2[3], 1263	0.940	1.4488^{20}		134	27	
v5	Vinylcyclohexane	$C_6H_{11}CH{=}CH_2$	110.20	5[1], 35	0.805	1.4463^{20}		128	21	
v6	4-Vinyl-1-cyclohexene		108.18	5[1], 63	0.8302^{20}	1.4640^{20}	−101	127	20	
v7	1-Vinylimidazole		94.12	23[4], 569	1.039	1.5308^{20}		79^{13mm}	81	
v8	5-Vinyl-2-norbornene		120.20		0.841	1.4802^{20}	−80	141	27	
v9	2-Vinylpyridine	$(C_5H_4N)CH{=}CH_2$	105.14	20, 256	0.975	1.5490^{20}		158–159	46	v s alc, chl, eth
v10	4-Vinylpyridine	$(C_5H_4N)CH{=}CH_2$	105.14	20[2], 170	0.975	1.5500^{20}		65^{15mm}	51	sl s hot aq, hot alc
v11	N-Vinyl-2-pyrrolidin-one		111.14		0.980	1.5120^{20}		93^{13mm}	93	
x1	Xanthene		182.22	17, 73			101	310–312		s bz, eth; sl s alc, aq
x2	Xanthen-9-carboxylic acid		226.23	18[2], 279			217 d			s hot alc, eth
x3	9-Xanthenone		196.21	17, 354			174	350^{730mm}		0.5 alc; v s chl
x4	o-Xylene	$C_6H_4(CH_3)_2$	106.17	5, 362	0.8802^{20}	1.5054^{20}	−25.2	144.4	32	misc alc, eth; 0.017 aq
x5	m-Xylene	$C_6H_4(CH_3)_2$	106.17	5, 370	0.8684^{15}	1.4972^{20}	−47.9	139.1	25	misc alc, eth; 0.02 aq
x6	p-Xylene	$C_6H_4(CH_3)_2$	106.17	5, 382	0.8611^{20}	1.4958^{20}	13.3	138.4	27	v s eth; s alc; 0.02 aq
x7	Xylitol	$HOCH_2(CHOH)_3CH_2OH$	152.15	1, 531			95–97			s aq
x8	D-(+)-Xylose		150.13	31, 47	1.535^0		144–145			117 aq; s hot alc, pyr
x9	m-Xylylenediamine	$C_6H_4(CH_2NH_2)_2$	136.20	13, 186	1.032	1.5709^{20}		265^{745mm}	>110	

Vinylacetic acid, b406
Vinyl bromide, b286
Vinyl 2-butenoate, v4
Vinyl chloride, c110
Vinylidene chloride, d178
Vinyltrimethylsilane, t387

Vinyltris(2-methoxyethoxy)silane, t437
Vitamin B$_1$, t141
Vitamin B$_2$, r4
Vitamin C, a317
Xanthone, x3

Xylene-α,α′-diol, b18
Xylenols, d581, d582, d583, d584, d585, d586
o-Xylyl bromide, b371
Xylyl chlorides, c259, c260, c261
p-Xylylene glycol, b18

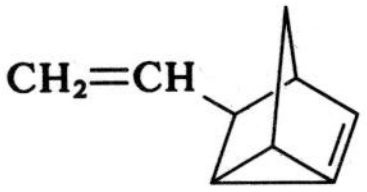

v3

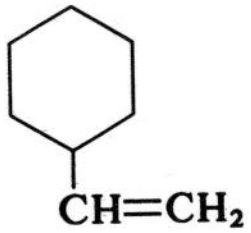

CH₂=CH

v6

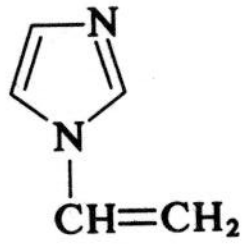

CH=CH₂

v7

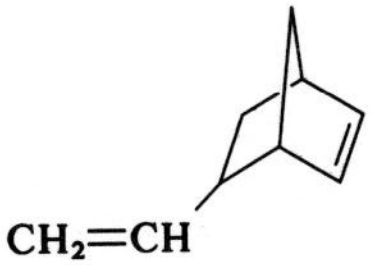

CH=CH₂

N

N

v8

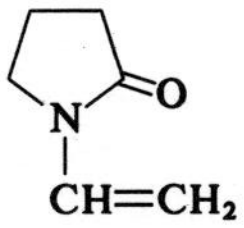

CH₂=CH

v11

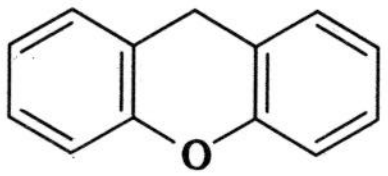

CH=CH₂

N

O

x1

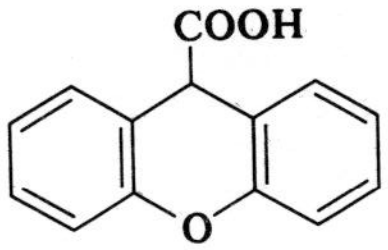

O

x2

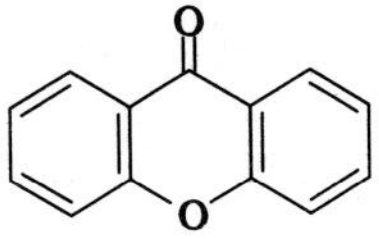

COOH

O

x3

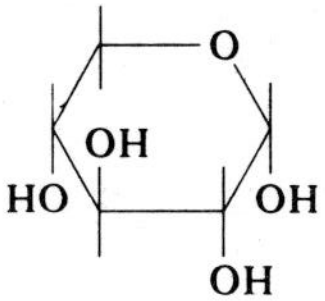

O

O

x8

OH

OH

OH

HO

SECTION 2

GENERAL INFORMATION, CONVERSION TABLES, AND MATHEMATICS

2.1 GENERAL INFORMATION AND CONVERSION TABLES

TABLE 2.1   Fundamental Physical Constants

E. R. Cohen and B. N. Taylor, CODATA Bull. *63:1–49 (1986).*

A. Defined values

Physical quantity	Name of SI unit	Symbol for SI unit	Definition
1. Base SI units			
Amount of substance	mole	mol	Amount of substance which contains as many specified entities as there are atoms of carbon-12 in exactly 0.012 kg of that nuclide. The elementary entities must be specified.
Electric current	ampere	A	Magnitude of the current that, when flowing through each of two straight parallel conductors of infinite length, of negligible cross section, separated by 1 meter in a vacuum, results in a force between the two wires of 2×10^{-7} newton per meter of length.
Length	meter	m	Distance light travels in a vacuum during 1/299 792 458 of a second.
Luminous intensity	candela	cd	Luminous intensity, in a given direction, of a source that emits monochromatic radiation of frequency 540×10^{12} hertz and that has a radiant intensity in that direction of $1/683$ watt per steradian.
Mass	kilogram	kg	Mass of a cylinder of platinum-iridium alloy kept at Paris.
Thermodynamic temperature	kelvin	K	Defined in the thermodynamic scale by assigning 273.16 K to the triple point of water (freezing point, 273.15 K = 0°C).
Time	second	s	Duration of 9 192 631 770 periods of the radiation corresponding to the transition between the two hyperfine levels of the ground state of the cesium-133 atom.
2. Supplementary SI units			
Plane angle	radian	rad	The plane angle between two radii of a circle which cut off on the circumference an arc equal in length to the radius.
Solid angle	steradian	sr	The solid angle which, having its vertex in the center of a sphere, cuts off an area of the surface of the sphere equal to that of a square with sides of length equal to the radius of the sphere.

TABLE 2.1 Fundamental Physical Constants (*Continued*)

B. Derived SI units

Physical quantity	Name of SI unit	Symbol for SI unit	Expression in terms of SI base units
Absorbed dose (of radiation)	gray	Gy	$J \cdot kg^{-1}$
Activity (radioactive)	becquerel	Bq	$s^{-1} = m^2 \cdot s^{-2}$
Capacitance (electric)	farad	F	$C \cdot V^{-1} = m^{-2} \cdot kg^{-1} \cdot s^4 \cdot A^2$
Charge (electric)	coulomb	C	$A \cdot s$
Conductance (electric)	siemens	S	$\Omega^{-1} = m^{-2} \cdot kg^{-1} \cdot s^3 \cdot A^2$
Energy, work, heat	joule	J	$N \cdot m = m^2 \cdot kg \cdot s^{-2}$
Force	newton	N	$m \cdot kg \cdot s^{-2}$
Frequency	hertz	Hz	s^{-1}
Illuminance	lux	lx	$cd \cdot sr \cdot m^{-2}$
Inductance	henry	H	$V \cdot A^{-1} \cdot s = m^2 \cdot kg \cdot s^{-2} \cdot A^{-2}$
Luminous flux	lumen	Lm	$cd \cdot sr$
Magnetic flux	weber	Wb	$V \cdot s = m^2 \cdot kg \cdot s^{-2} \cdot A^{-1}$
Magnetic flux density	tesla	T	$V \cdot s \cdot m^{-2} = kg \cdot s^{-2} \cdot A^{-1}$
Potential, electric (electromotive force)	volt	V	$J \cdot C^{-1} = m^2 \cdot kg \cdot s^{-3} \cdot A^{-1}$
Power, radiant flux	watt	W	$J \cdot s^{-1} = m^2 \cdot kg \cdot s^{-3}$
Pressure, stress	pascal	Pa	$N \cdot m^{-2} = m^{-1} \cdot kg \cdot s^{-2}$
Resistance, electric	ohm	Ω	$V \cdot A^{-1} = m^2 \cdot kg \cdot s^{-3} \cdot A^{-2}$
Temperature, Celsius	degree Celsius	°C	$\theta/°C = T/(K - 273.15)$

C. Recommended consistent values of constants

Quantity	Symbol	Value*
Anomalous electron moment correction	$\mu_e - 1$	0.001 159 615(15)
Atomic mass constant	$m_u = 1\ u$	$1.660\ 540\ 2(10) \times 10^{-27}$ kg
Avogadro constant	L, N_A	$6.022\ 136\ 7(36) \times 10^{23}$ mol^{-1}
Bohr magneton ($= eh/4\pi m_e$)	μ_B	$9.274\ 015\ 4(31) \times 10^{-24}$ J $\cdot$ T^{-1}
Bohr radius	a_0	$5.291\ 772\ 49(24) \times 10^{-11}$ m
Boltzmann constant	k	$1.380\ 658(12) \times 10^{-23}$ J $\cdot$ K^{-1}
Charge-to-mass ratio for electron	e/m_e	$1.758\ 805(5) \times 10^{-11}$ C $\cdot$ kg^{-1}
Compton wavelength of electron	λ_c	$2.426\ 309(4) \times 10^{-12}$ m
Compton wavelength of neutron	$\lambda_{c,n}$	$1.319\ 591(2) \times 10^{-15}$ m
Compton wavelength of proton	$\lambda_{c,p}$	$1.321\ 410(2) \times 10^{-15}$ m
Diamagnetic shielding factor, spherical water molecule	$1 + \sigma(H_2O)$	1.000 025 64(7)
Electron magnetic moment	μ_e	$9.284\ 770\ 1(31) \times 10^{-24}$ J $\cdot$ T^{-1}
Electron radius (classical)	r_e	$2.817\ 938(7) \times 10^{-15}$ m
Electron rest mass	m_e	$9.109\ 389\ 7(54) \times 10^{-31}$ kg
Elementary charge	e	$1.602\ 177\ 33(49) \times 10^{-19}$ C
Energy equivalents:		
1 electron mass		0.511 003 4(14) MeV
1 electronvolt	$1\ eV/k$	$1.160\ 450(36) \times 10^4$ K
	$1\ eV/hc$	$8.065\ 479(21) \times 10^3$ cm^{-1}
	$1\ eV/h$	$2.417\ 970(6) \times 10^{14}$ Hz
1 neutron mass		939.573 1(27) MeV
1 proton mass		938.279 6(27) MeV

* The digits in parentheses following a numerical value represent the standard deviation of that value in terms of the final listed digits.

TABLE 2.1   Fundamental Physical Constants (*Continued*)

C. Recommended consistent values of constants (*continued*)		
Quantity	Symbol	Value*
1 u		931.501 6(26) MeV
Faraday constant	F	96 485.309(29) C · mol^{-1}
Fine structure constant	α	0.007 297 353 08(33)
	α^{-1}	137.035 989 5(61)
First radiation constant	c_1	3.741 774 9(22) × 10^{-16} W · m^2
Gas constant	R	8.314 510(70) J · K^{-1} · mol^{-1}
g factor (Lande) for free electron	g_e	2.002 319 304 386(20)
Gravitational constant	G	6.672 59(85) × 10^{-11} m^3 · kg^{-1} · s^{-2}
Hartree energy	E_h	4.359 748 2(26) × 10^{-18} J
Josephson frequency-voltage ratio		4.835 939(13) × 10^{14} Hz · V^{-1}
Magnetic flux quantum	Φ_0	2.067 851(5) × 10^{-15} Wb
Magnetic moment of protons in water	μ_p/μ_B	1.520 993 129(17) × 10^{-3}
Molar volume, ideal gas, $p = 1$ bar, $\theta = 0°C$		22.711 08(19) L · mol^{-1}
Neutron rest mass	m_n	1.674 928(6(10)) × 10^{-27} kg
Nuclear magneton	μ_N	5.050 786 6(17) × 10^{-27} J · T^{-1}
Permeability of vacuum	μ_0	4π × 10^{-7} H · m^{-1} exactly
Permittivity of vacuum	ϵ_0	8.854 187 816 × 10^{-12} F · m^{-1}
	$\hbar = h/2\pi$	1.054 572 66(63) × 10^{-34} J · s
Planck constant	h	6.626 0.75 5(40) × 10^{-34} J · s
Proton magnetic moment	μ_p	1.410 607 61(47) × 10^{-26} J · T^{-1}
Proton magnetogyric ratio	γ_p	2.675 221 28(81) × 10^8 s^{-1} · T^{-1}
Proton resonance frequency per field in H$_2$O	$\gamma_p'/2\pi$	42.576 375(13) MHz · T^{-1}
Proton rest mass	m_p	1.672 623 1(10) × 10^{-27} kg
Quantum-charge ratio	h/e	4.135 701(11) × 10^{-15} J · Hz^{-1} · C^{-1}
Quantum of circulation	h/m_e	7.273 89(1) × 10^{-4} J · s · kg^{-1}
Ratio, electron-to-proton magnetic moments	μ_e/μ_p	6.582 106 88(7) × 10^2
Rydberg constant	R_∞	1.097 373 153 4(13) × 10^7 m^{-1}
Second radiation constant	c_2	1.438 769(12) × 10^{-2} m · K
Speed of light in vacuum	c_0	299 792 458 m · s^{-1} exactly
Standard acceleration of free fall	g_n	9.806 65 m · s^{-2} exactly
Standard atmosphere	atm	101 325 Pa exactly
Stefan-Boltzmann constant	σ	5.670 51(19) × 10^{-8} W · m^{-2} · K^{-4}
Thomson cross section	σ_e	6.652 448(33) × 10^{-29} m^2
Wien displacement constant	b	0.289 78(4) cm · K
Zeeman splitting constant	μ_B/hc	4.668 58(4) × 10^{-5} cm^{-1} · G^{-1}

* The digits in parentheses following a numerical value represent the standard deviation of that value in terms of the final listed digits.

TABLE 2.2 Physical and Chemical Symbols and Definitions

Symbols separated by commas represent equivalent recommendations. Symbols for physical and chemical quantities should be printed in *italic* type. Subscripts and superscripts which are themselves symbols for physical quantities should be italicized; all others should be in Roman type. Vectors and matrices should be printed in boldface italic type, e.g., $\boldsymbol{B}, \boldsymbol{b}$. Symbols for units should be printed in Roman type and should remain unaltered in the plural, and should not be followed by a full stop except at the end of a sentence. References: International Union of Pure and Applied Chemistry, *Quantities, Units and Symbols in Physical Chemistry,* Blackwell, Oxford, 1988; "Manual of Symbols and Terminology for Physico-chemical Quantities and Units," *Pure Applied Chem.* **31**:577–638 (1972), **37**:499–516 (1974), **46**:71–90 (1976), **51**:1–41, 1213–1218 (1979); **53**:753–771 (1981), **54**:1239–1250 (1982), **55**:931–941 (1983); IUPAP-SUN, "Symbols, Units and Nomenclature in Physics," *Physica* **93A**: 1–60 (1978).

A. Atoms and molecules

Name	Symbol	SI unit	Definition
Activity (radioactivity)	A	Bq	$A = -dN_B/dt$
Atomic mass constant	m_u	kg	$m_u = m_a(^{12}C)/12$
Bohr magneton	μ_B	$J \cdot T^{-1}$	$\mu_B = eh/4\pi m_e$
Bohr radius	a_0	m	$a_0 = 2\epsilon_0 h^2/m_e e^2$
Decay (rate) constant	λ	s^{-1}	$A = \lambda N_B$
Dissociation energy	D, E_d	J	
From ground state	D_0	J	
From the potential minimum	D_e	J	
Electric dipole moment of a molecule	$\boldsymbol{p}, \boldsymbol{\mu}$	$C \cdot m$	$E_p = -\boldsymbol{p} \cdot \boldsymbol{E}$
Electric field gradient	$\boldsymbol{q}$	$V \cdot m^{-2}$	$q_{\alpha\beta} = -\partial^2 V/\partial\alpha\partial\beta$
Electric polarizability of a molecule	α	$C \cdot m^2 \cdot V^{-1}$	$p(\text{induced}) = \alpha E$
Electron affinity	E_{ea}	J	
Electron rest mass	m_e	kg	
Elementary charge, proton charge	e	C	
Fine structure constant	α		$\alpha = e^2/2\epsilon_0 hc$
g factor	g		
Hartree energy	E_h	J	$E_h = h^2/4\pi^2 m_e a_0^2$
Ionization energy	E_i	J	
Larmor circular frequency	ω_L	s^{-1}	$\omega_L = (e/2m)B$
Larmor frequency	ν_L	Hz	$\nu_L = \omega_L/2\pi$
Longitudinal relaxation time	T_1	s	
Magnetogyric ratio	γ	$C \cdot kg^{-1}$	$\gamma = \mu/L$
Magnetic dipole moment of a molecule	$\boldsymbol{m}, \boldsymbol{\mu}$	$J \cdot T^{-1}$	$E_p = -\boldsymbol{m} \cdot \boldsymbol{B}$
Magnetizability of a molecule	ξ	$J \cdot T^{-2}$	$\boldsymbol{m} = \xi\boldsymbol{B}$
Mass of atom, atomic mass	m, m_a	kg	
Neutron number	N		$N = A - Z$
Nuclear magneton	μ_N	$J \cdot T^{-1}$	$\mu_N = (m_e/m_p)\mu_B$
Nucleon number, mass number	A		
Planck constant	h	$J \cdot s$	
Planck constant/2π	$\hbar$	$J \cdot s$	$\hbar = h/2\pi$
Principal quantum number (H atom)	n		$E = -hcR/n^2$
Proton number, atomic number	Z		
Quadrupole interaction	χ	J	$\chi_{\alpha\beta} = eQq_{\alpha\beta}$
Quadrupole moment of a molecule	$\boldsymbol{Q}; \boldsymbol{\Theta}$	$C \cdot m^2$	$E_p = 0.5\boldsymbol{Q}: V'' = \frac{1}{3}\boldsymbol{\Theta}V''$
Quadrupole moment	eQ	$C \cdot m^2$	$eQ = 2\langle\Theta_{zz}\rangle$
Rydberg constant	R_∞	m^{-1}	$R_\infty = E_h/2hc$
Transverse relaxation time	T_2	s	

TABLE 2.2 Physical and Chemical Symbols and Definitions (*Continued*)

B. Chemical reactions

Name	Symbol	SI unit	Definition
Amount (of substance)	n	mol	$n_B = N_B/L$
Atomic mass	m, m_a	kg	
Atomic mass constant[a]	m_u	kg	$m_u = m_a(^{12}C)/12$
Avogadro constant	L, N_A	mol^{-1}	
Concentration, amount (concentration)	c	$mol \cdot m^{-3}$	$c_B = n_B/V$
Degree of dissociation	α		
Density (mass)	ρ, γ	$kg \cdot m^{-3}$	$\rho = m_B/V$
Extent of reaction, advancement	ξ	mol	$\Delta\xi = \Delta n_B/\nu_{B_B}$
Mass (molecular or formula unit)	m, m_f	kg	
Mass fraction	w		$w_B = m_B/\Sigma m_i$
Molality (of a solute)	m	$mol \cdot kg^{-1}$	$m_B = n_B/m_A$
Molar mass	M	$kg \cdot mol^{-1}$	$M_B = m/n_B$
Molar volume	V_m	$m^3 \cdot mol^{-1}$	$V_{m,B} = V/n_B$
Molecular weight (relative molar mass)	M_r		$M_{r,B} = m_B/m_u$
Mole fraction[b], number fraction	x, y		$x_B = n_B/\Sigma n_i$
Number concentration	C, n	m^{-3}	$C_B = N_B/V$
Number of entities (e.g., molecules, atoms, ions, formula units)	N		
Pressure (partial)	p_B	Pa	$p_B = y_B p$
Pressure (total)	p, P	Pa	
Solubility	s	$mol \cdot m^{-3}$	$s_B = c_B$ (saturated solution)
Stoichiometric number	ν		
Surface concentration	Γ	$mol \cdot m^{-2}$	$\Gamma_B = n_B/A$
Volume fraction	ϕ		$\phi_B = V_B/\Sigma V_i$

Symbols for particles and nuclear reactions:

Alpha particle	α		Muon, positive	μ^+
Beta particle	β^-, β^+		Neutron	n, n^0
Deuteron	$d, {}^2H$		Photon	γ
Electron	e, e^-		Proton	p, p^+
Helion	h		Triton	$t, {}^3H$
Muon, negative	μ^-			

The meaning of the symbolic expression indicating a nuclear reaction:

$$\text{initial nuclide} \left(\begin{array}{c} \text{incoming particles} \\ \text{or quanta} \end{array} , \begin{array}{c} \text{outgoing particles} \\ \text{or quanta} \end{array} \right) \text{final nuclide}$$

Examples: $^{14}N(\alpha, p)^{17}O$, $^{23}Na(\gamma, 3n)^{20}Na$

States of aggregation:

am	amorphous solid	cd	condensed phase (solid or liquid)
aq	aqueous solution	cr	crystalline
aq, ∞	aqueous solution at infinite dilution	fl	fluid phase (gas or liquid)
		lc	liquid crystal
g	gas	vit	vitreous substance
l	liquid	mon	monomeric form
s	solid	pol	polymeric form
sln	solution	ads	species adsorbed on a substance

[a] In biochemistry this unit is called the dalton, with symbol Da.
[b] For condensed phases x is used, and for gaseous mixtures y may be used.

TABLE 2.2 Physical and Chemical Symbols and Definitions (*Continued*)

C. Chromatography

Name	Symbol	Definition
Adjusted retention time	t'_R	$t'_R = t_R - t_M$
Adjusted retention volume	V'_R	$V'_R = V_R - V_M$
Average linear gas velocity	μ	$\mu = L/t_M$
Band variance	σ^2	
Bed volume	V_g	
Capacity, volume	Q_v	
Capacity, weight	Q_w	
Column length	L	
Column temperature	θ	
Column volume	V_{col}	$V_{col} = \pi D d_c^2/4$
Concentration at peak maximum	C_{max}	
Concentration of solute in mobile phase	C_M	
Concentration of solute in stationary phase	C_S	
Density of liquid phase	ρ_L	
Diffusion coefficient, liquid film	D_f	
Diffusion coefficient, mobile phase	D_M	
Diffusion coefficient, stationary phase	D_S	
Distribution ratio	D_c	$= [A^+]_S/[A^+]_M$
		$= \dfrac{\text{amount of A per cm}^3 \text{ stationary phase}}{\text{amount of A per cm}^3 \text{ of mobile phase}}$
	D_g	$= \dfrac{\text{amount A per gram dry stationary phase}}{\text{amount A per cm}^3 \text{ of mobile phase}}$
	D_v	$= \dfrac{\text{amount A, stationary phase per cm}^3 \text{ bed volume}}{\text{amount A per cm}^3 \text{ of mobile phase}}$
	D_S	$= \dfrac{\text{amount of A per m}^2 \text{ of surface}}{\text{amount of A per cm}^3 \text{ of mobile phase}}$
Elution volume, exclusion chromatography	V_e	
Flow rate, column	F_c	$F_c = (\pi d_c^2/4)(\epsilon_{tot})(L/t_M)$
Gas/liquid volume ratio	β	
Inner column volume	V_i	
Interstitial (outer) volume	V_o	
Kovats retention indices	RI	
Matrix volume	V_g	
Net retention volume	V_N	$V_N = jV'_R$
Obstruction factor	γ	
Packing uniformity factor	λ	
Particle diameter	d_p	$d_p = L/Nh$
Partition coefficient	K	$K = C_S/C_M = (V_R - V_M)/V_S$
Partition ratio	k'	$k' = C_S V_S/C_M V_M = K(V_S/V_M)$
Peak asymmetry factor	AF	Ratio of peak half-widths at 10% peak height
Peak resolution	Rs	$Rs = (t_{R,2} - t_{R,1})/0.5(W_2 + W_1)$
Plate height	H	$H = L/N_{eff}$
Plate number	N_{eff}	$N_{eff} = L/H = 16(t'_R/W_b)^2 = 5.54(t'_R/W_{1/2})^2$
Porosity, column	ϵ	
Pressure, column inlet	p_i	
Pressure, column outlet	P_o	

TABLE 2.2 Physical and Chemical Symbols and Definitions (*Continued*)

C. Chromatography (*continued*)

Name	Symbol	Definition
Pressure drop	ΔP	
Pressure-gradient correction	j	$j = \dfrac{3[(p_i/p_o)^2 - 1]}{2[(p_i/p_o)^3 - 1]}$
Recovery factor	R_n	$R_n = 1 - (rD_c + 1)^{-n}; r = V_{\text{org}}/V_{\text{aq}}$
Reduced column length	λ	$\lambda = L/d_p$
Reduced plate height	h	$h = H/d_p$
Reduced velocity	v	$v = \mu d_p/D_M = Kd_p/t_M D_M$
Relative retention ratio	α	$\alpha = (k_2'/k_1')$
Retardation factor[c]	R_f	$R_f = d_{\text{solute}}/d_{\text{mobile phase}}$
Retention time	t_R	$t_R = t_M(1 + k') = L/\mu$
Retention volume	V_R	$V_R = t_R F_c$
Selectivity coefficient[d]	$k_{A,B}$	$k_{A,B} = [A^+]_r[B^+]/[B^+]_r[A^+]$
Separation factor	$\alpha_{A/B}$	$\alpha_{A/B} = (D_c)_A/(D_c)_B$
Specific retention volume	V_g°	$V_g^\circ = 273R/(p^\circ Mw_L)$
Thickness (effective) of stationary phase	d_f	
Total bed volume	V_{tot}	
Transit time of nonretained solute	t_M, t_0	
Vapor pressure	p	
Volume liquid phase in column	V_L	
Volume mobile phase in column	V_M	
Weight of liquid phase	w_L	
Zone width at baseline	W_b	$W_b = 4\sigma$
Zone width at ½ peak height	$W_{1/2}$	

D. Colloid and surface chemistry

Name	Symbol	SI unit	Definition
Adsorbed amount of B	n_B^s	mol	
Area per molecule	a, σ	m^2	$a_B = A/N_N^a$
Area per molecule in a filled monolayer	a_m	m^2	$a_{m,B} = A/N_{m,B}$
Average molar masses:			
Mass-average	M_m	kg · mol^{-1}	$M_m = \Sigma n_i M_i^2/\Sigma n_i M_i$
Number-average	M_n	kg · mol^{-1}	$M_n = \Sigma n_i M_i/\Sigma n_i$
Z-average	M_Z	kg · mol^{-1}	$M_Z = \Sigma n_i M_i^3/\Sigma n_i M_i^2$
Contact angle	θ	rad	
Film tension	Σ_f	N · m^{-1} ϕ	$\Sigma_f = 2\gamma_f$
Film thickness	t, h, δ	m	
Reciprocal thickness of the double layer	κ	m^{-1}	$\kappa = [2F^2 I_c/\epsilon RT]^{1/2}$
Retarded van der Waals constant	β, B	J	
Sedimentation coefficient[e]	s	s	$s = v/a$
Specific surface area	a, s, a_s	m^2/kg	$a = A/m$

[c] The distance d corresponds to the movement of solute and mobile phase from the starting (sample spotting) line.

[d] Subscript "r" represents an ion-exchange resin phase. Two immiscible liquid phases might be represented similarly using subscripts "1" and "2."

[e] v is the velocity of sedimentation and a is the acceleration of free fall or centrifugation.

TABLE 2.2 Physical and Chemical Symbols and Definitions (*Continued*)

D. Colloid and surface chemistry (*continued*)

Name	Symbol	SI unit	Definition
Surface coverage	θ		$\theta = N_B^a/N_B$
Surface excess of B	n_B^σ	mol	
Surface pressure	π^s, π	$N \cdot m^{-1}$	$\pi^s = \gamma^0 - \gamma$
Surface tension, interfacial tension	γ, σ	$J \cdot m^{-2}$	$\gamma = (\partial G/\partial A_s)_{T,p}$
Thickness of (surface or interfacial) layer	τ, δ, t	m	
Total surface excess concentration	Γ	$mol \cdot m^{-2}$	$\Gamma = \Sigma \Gamma_i$
van der Waals constant	λ	J	
van der Waals-Hamaker constant	A_H	J	

E. Electricity and magnetism

Name	Symbol	SI unit	Definition
Admittance	Y	S	$Y = 1/Z$
Capacitance	C	$F, C \cdot V^{-1}$	$C = Q/U$
Charge density	ρ	$C \cdot m^{-3}$	$\rho = Q/V$
Conductance	G	S	$G = 1/R$
Conductivity	κ	$S \cdot m^{-1}$	$\kappa = 1/\rho$
Dielectric polarization (dipole moment per volume)	$\boldsymbol{P}$	$C \cdot m^{-2}$	$\boldsymbol{P} = \boldsymbol{D} - \epsilon_0 \boldsymbol{E}$
Electrical resistance	R	Ω	$R = U/I = \Delta V/I$
Electric current	I	A	$I = dQ/dt$
Electric current density	j, J	$A \cdot m^{-2}$	$I = \int j \, dA$
Electric dipole moment	$\boldsymbol{p}, \boldsymbol{\mu}$	$C \cdot m$	$\boldsymbol{p} = Q\boldsymbol{r}$
Electric displacement	$\boldsymbol{D}$	$C \cdot m^{-2}$	$\boldsymbol{D} = \epsilon \boldsymbol{E}$
Electric field strength	$\boldsymbol{E}$	$V \cdot m^{-1}$	$\boldsymbol{E} = \boldsymbol{F}/Q = -\text{grad } V$
Electric flux	Ψ	C	$\Psi = \int \boldsymbol{D} \, dA$
Electric potential	V, ϕ	$V, J \cdot C^{-1}$	$V = dW/dQ$
Electric potential difference	$U, \Delta V$	V	$U = V_2 - V_1$
Electric susceptibility	χ_e		$\chi_e = \epsilon_r - 1$
Electromotive force	E	V	$E = \int (F/Q) \, ds$
Impedance	Z	Ω	$Z = R + iX$
Loss angle[f]	δ	rad	$\delta = (\pi/2) + \phi_I - \phi_U$
Magnetic dipole moment	$\boldsymbol{m}, \boldsymbol{\mu}$	$A \cdot m^2$	$E_p = -\boldsymbol{mB}$
Magnetic field strength	H	$A \cdot m^{-1}$	$\boldsymbol{B} = \mu \boldsymbol{H}$
Magnetic flux	Φ	Wb	$\Phi = \int \boldsymbol{B} \, dA$

[f] ϕ_I and ϕ_U are the phases of current and potential difference.

TABLE 2.2 Physical and Chemical Symbols and Definitions (*Continued*)

E. Electricity and magnetism (*continued*)

Name	Symbol	SI unit	Definition
Magnetization (magnetic dipole moment per volume)	M	$A \cdot m^{-1}$	$M = (B/\mu_0) - H$
Magnetic susceptibility	χ, κ		$\chi = \mu_r - 1$
Magnetic vector potential	A	$Wb \cdot m^{-1}$	$B = \nabla A$
Molar magnetic susceptibility	χ_m	n^3/mol	$\chi_m = V_m \chi$
Mutual inductance	M, L_{12}	H	$E_1 = L_{12}(dI_2/dt)$
Permeability	μ	$H \cdot m^{-1}$	$B = \mu H$
Permeability of vacuum	μ_0	$H \cdot m^{-1}$	
Permittivity	ϵ	$F \cdot m^{-1}$	$D = \epsilon E$
Permittivity of vacuum	ϵ_0	$F \cdot m^{-1}$	$\epsilon_0 = \mu_0^{-1} c_0^{-2}$
Poynting vector	S	$W \cdot m^{-2}$	$S = E \cdot H$
Quantity of electricity, electric charge	Q	C	
Reactance	X	Ω	$X = (U/I) \sin \delta$
Relative permeability	μ_r		$\mu_r = \mu/\mu_0$
Relative permittivity[g]	ϵ_r		$\epsilon_r = \epsilon/\epsilon_0$
Resistivity	ρ	$\Omega \cdot m$	$\rho = E/j$
Self-inductance	L	H	$E = -L(dI/dt)$
Susceptance	B	S	$Y = G + iB$

F. Electrochemistry

Name	Symbol	SI unit	Definition
Charge density (surface)	σ	$C \cdot n^{-2}$	$\sigma = Q/A$
Charge number of an ion	z		$z_B = Q_B/e$
Charge number of electrochemical cell reaction	$n, (z)$		
Conductivity (specific conductance)	κ	$S \cdot m^{-1}$	$\kappa = j/E$
Conductivity cell constant	K_{cell}	m^{-1}	$K_{cell} = \kappa R$
Current density (electric)	j	$A \cdot m^{-2}$	$j = I/A$
Diffusion rate constant, mass transfer coefficient	k_d	$m \cdot s^{-1}$	$k_{d,B} = |\nu_B| I_{1,B}/nFcA$
Electric current	I	A	$I = dQ/dt$
Electric mobility	μ	$m^2 \cdot V^{-1} \cdot s^{-1}$	$\mu_B = \nu_B/E$
Electric potential difference (of a galvanic cell)	$\Delta V, E, U$	V	$\Delta V = V_R - V_L$
Electrochemical potential	$\tilde{\mu}$	$J \cdot mol^{-1}$	$\tilde{\mu}_B^\alpha = (\partial G/\partial n_B^\alpha)$
Electrode reaction rate constant	k	(varies)	$k_{ox} = I_a/\left(nFA \prod_i c_i^{n_i}\right)$
Electrokinetic potential (zeta potential)	ζ	V	
Elementary charge (proton charge)	e	C	
emf, electromotive force	E	V	$E = \lim_{I \to 0} \Delta V$

[g] This quantity was formerly called the dielectric constant.

TABLE 2.2 Physical and Chemical Symbols and Definitions (*Continued*)

F. Electrochemistry (*continued*)

Name	Symbol	SI unit	Definition
emf of the cell	E	V	$E = E^0 - (RT/nF) \times \Sigma v_i \ln a_i$
Faraday constant	F	$C \cdot mol^{-1}$	$F = eL$
Galvani potential difference	$\Delta\phi$	V	$\Delta_\alpha^\beta \phi = \phi^\beta - \phi^\alpha$
Inner electrode potential	ϕ	V	$\nabla\phi = -E$
Ionic conductivity	λ	$S \cdot m^2 \cdot mol^{-1}$	$\lambda_B = |z_B| F u_B$
Ionic strength	I_c, I	$mol \cdot m^{-3}$	$I_c = \frac{1}{2}\Sigma c_i z_c^2$
Mean ionic activity	$a_\pm$		$a_\pm = m_\pm \gamma_\pm / m^0$
Mean ionic activity coefficient	$\gamma_\pm$		$\gamma_\pm^{(v_+ + v_-)} = (\gamma_\pm^{v_+})(\gamma_-^{v_-})$
Mean ionic mobility	$m_\pm$	$mol \cdot kg^{-1}$	$m_\pm^{(v_+ + v_-)} = (m_+^{v_+})(m_-^{v_-})$
Molar conductivity (of an electrolyte)	Λ	$S \cdot m^{-2} mol^{-1}$	$\Lambda_B = \kappa c_B$
pH	pH		$pH \approx -\log\left[\dfrac{c(H^+)}{mol \cdot dm^{-3}}\right]$
Outer electrode potential	ψ	V	$\psi = Q/4\pi\epsilon_0 r$
Overpotential	η	V	$\eta = E_I - E_{I=0} - IR_u$
Reciprocal radius of ionic atmosphere	κ	m^{-1}	$\kappa = (2F^2 I/\epsilon RT)^{1/2}$
Standard emf, standard potential of electrochemical cell reaction	E^0	V	$E^0 = -\Delta_r G^0/nF = (RT/nF) \ln K)$
Surface electric potential	χ	V	$\chi = \phi - \psi$
Thickness diffusion layer	δ	m	$\delta_B = D_B/k_{d,B}$
Transfer coefficient	α		$\alpha_c = \dfrac{-|v|RT}{nF}\dfrac{\partial \ln|I_c|}{\partial E}$
Transport number	t		$t_B = j_B/\Sigma j_i$
Volta potential difference	$\Delta\psi$	V	$\Delta_\alpha^\beta = \psi^\beta - \beta^\alpha$

G. Electromagnetic radiation

Name	Symbol	SI unit	Definition
Absorbance	α		$\alpha = \Phi_{abs}/\Phi_0$
Absorbance (decaidic)	A		$A = -\log(1 - \alpha_i)$
Absorbance (napierian)	B		$B = -\ln(1 - \alpha_i)$
Absorption coefficient:			
Linear (decaidic)	a, K	m^{-1}	$a = A/l$
Linear (napierian)	α	m^{-1}	$\alpha = B/l$
Molar (decaidic)	ϵ	$m^2 \cdot mol^{-1}$	$\epsilon = a/d = A/cl$
Molar (napierian)	κ	$m^2 \cdot mol^{-1}$	$\kappa = \alpha/c = B/cl$
Absorption index	k		$k = \alpha/4\pi\tilde{v}$
Angle of optical rotation	α	rad	
Circular frequency	ω	$s^{-1}, rad \cdot s^{-1}$	$\omega = 2\pi v$
Complex refractive index	$\hat{n}$		$\hat{n} = \eta + ik$
Concentration, amount of substance	c	$mol \cdot m^3$	
Concentration, mass	γ	$kg \cdot m^3$	
Einstein transition probabilities:			
Spontaneous emission	A_{nm}	s^{-1}	$dN_n/dt = -A_{nm}N_n$

TABLE 2.2 Physical and Chemical Symbols and Definitions (*Continued*)

G. Electromagnetic radiation (*continued*)

Name	Symbol	SI unit	Definition
Stimulated absorption	B_{mn}	$\text{s} \cdot \text{kg}^{-1}$	$dN_n/dt = \rho_{\tilde{v}}(\tilde{v}_{nm})B_{mn}N_m$
Stimulated emission	B_{nm}	$\text{s} \cdot \text{kg}^{-1}$	$dN_n/dt = \rho_{\tilde{v}}(\tilde{v}_{nm})B_{mn}N_m$
Emittance	ϵ		$\epsilon = M/M_{bb}$
By blackbody	M_{bb}		
First radiation constant	c_1	$\text{W} \cdot \text{m}^2$	$c_1 = 2\pi hc_0^2$
Frequency	v	Hz	$v = c/\lambda$
Irradiance (radiant flux received)	$E, (I)$	$\text{W} \cdot \text{m}^{-2}$	$E = d\Phi/dA$
Molar refraction	R, R_m	$\text{m}^3 \cdot \text{mol}^{-1}$	$R = \dfrac{(n^2 - 1)}{(n^2 + 2)} V_m$
Path length (absorbing)	l	m	
Optical rotatory power	$[\alpha]_\lambda^\theta$	rad	$[\alpha]_\lambda^\theta = \alpha/\gamma l$
Planck constant	h	$\text{J} \cdot \text{s}$	
Planck constant/2π	$\hbar$	$\text{J} \cdot \text{s}$	$\hbar = h/2\pi$
Radiant energy	Q, W	J	
Radiant energy density	ρ, w	$\text{J} \cdot \text{m}^{-3}$	$\rho = Q/V$
Radiant exitance, emitted radiant flux	M	$\text{W} \cdot \text{m}^{-2}$	$M = d\Phi/dA_{source}$
Radiant intensity	I	$\text{W} \cdot \text{sr}^{-1}$	$I = d\Phi/d\Omega$
Radiant power, radiant energy per time	Φ, P	W	$\Phi = dQ/dt$
Refractive index	n		$n = c_0/c$
Reflectance	ρ		$\rho = \Phi_{refl}/\Phi_0$
Second radiation constant	c_2	$\text{K} \cdot \text{m}$	$c_2 = hc_0/k$
Spectral radiant energy density:			
In terms of frequency	ρ_v, w_v	$\text{J} \cdot \text{m}^{-3} \cdot \text{Hz}^{-1}$	$\rho_v = d\rho/dv$
In terms of wavelength	ρ_λ, w_λ	$\text{J} \cdot \text{m}^{-4}$	$\rho_\lambda = d\rho/d\lambda$
In terms of wavenumber	$\rho_{\tilde{v}}, w_{\tilde{v}}$	$\text{J} \cdot \text{m}^{-2}$	$\rho_{\tilde{v}} = d\rho/d\tilde{v}$
Speed of light:			
In a medium	c	$\text{m} \cdot \text{s}^{-1}$	$c = c_0/n$
In vacuum	c_0	$\text{m} \cdot \text{s}^{-1}$	
Stefan-Boltzmann constant	σ	$\text{W} \cdot \text{m}^{-2} \cdot \text{K}^{-4}$	$M_{bb} = \sigma T^4$
Transmittance	τ, T		$\tau = \Phi_{tr}/\Phi_0$
Wavelength	λ	m	
Wavenumber:			
In a medium	σ	m^{-1}	$\sigma = 1/\lambda$
In vacuum	$\tilde{v}$	m^{-1}	$\tilde{v} = v/c_0 = 1/n\lambda$

H. Kinetics

Name	Symbol	SI unit	Definition
Activation energy	E_a, E	$\text{J} \cdot \text{mol}^{-1}$	$E_a = RT^2\, d \ln k/dT$
Boltzmann constant	k, k_B	$\text{J} \cdot \text{K}^{-1}$	
Collision cross section	σ	m^2	$\sigma_{AB} = \pi d_{AB}^2$
Collision diameter	d	m	$d_{AB} = r_A + r_B$
Collision frequency	Z_A	s^{-1}	
Collision frequency factor	z_{AB}, z_{AA}	$\text{m}^3 \cdot \text{mol}^{-1} \cdot \text{s}^{-1}$	$z_{AB} = Z_{AB}/Lc_A c_B$
Collision number	Z_{AB}, Z_{AA}	$\text{m}^{-3} \cdot \text{s}^{-1}$	

TABLE 2.2 Physical and Chemical Symbols and Definitions (*Continued*)

H. Kinetics (*continued*)

Name	Symbol	SI unit	Definition
Half-life	$t_{1/2}$	s	$c(t_{1/2}) = c_0/2$
Overall order of reaction	n		$n = \Sigma n_B$
Partial order of reaction	n_B		$v = k\Pi c_B^{n_B}$
Pre-exponential factor	A	$(\text{mol}^{-1} \cdot \text{m}^3)^{n-1} \cdot \text{s}^{-1}$	$k = A\exp(-E_a/RT)$
Quantum yield, photochemical yield	ϕ		
Rate of change of quantity X	$\dot{X}$	(varies)	$\dot{X} = dX/dt$
Rate of concentration change (chemical reaction)	r_B, v_B	$\text{mol} \cdot \text{m}^{-3} \cdot \text{s}^{-1}$	$r_B = dc_B/dt$
Rate constant, rate coefficient	k	$(\text{mol}^{-1} \cdot \text{m}^3)^{n-1} \cdot \text{s}^{-1}$	$v = k\,\Pi\,c_B^{n_B}$
Rate of conversion change due to chemical reaction	$\dot{\zeta}$	$\text{mol} \cdot \text{s}^{-1}$	$\dot{\zeta} = d\zeta/dt$
Rate of reaction (based on concentration)	v	$\text{mol} \cdot \text{m}^{-3} \cdot \text{s}^{-1}$	$v = \dot{\zeta}/V = v_B^{-1}\,dc_B/dt$
Relaxation time	τ	s	$\tau = 1/(k_1 + k_{-1})$
Standard enthalpy of activation	$\Delta H\ddagger$	$\text{J} \cdot \text{mol}^{-1}$	
Standard entropy of activation	$\Delta S_+^\ddagger$	$\text{J} \cdot \text{mol}^{-1} \cdot \text{K}^{-1}$	
Standard Gibbs energy of activation	$\Delta G_+^\ddagger$	$\text{J} \cdot \text{mol}^{-1}$	
Volume of activation	$\Delta_+^\ddagger V$	$\text{m}^3 \cdot \text{mol}^{-1}$	$\Delta_+^\ddagger V = -RT\,(\partial \ln k/\partial p)_T$

I. Mechanics

Name	Symbol	SI unit	Definition
Acoustic factors:			
Absorption	α_a		$\alpha_a = 1 - \rho$
Dissipation	δ		$\delta = \alpha_a - \tau$
Reflection	ρ		$\rho = P_r/P_0$
Transmission	τ		$\tau = P_{tr}/P_0$
Angular momentum	L	$\text{J} \cdot \text{s}$	$L = r \times p$
Bulk modulus, compression modulus	K	Pa	$K = -V_0(dp/dV)$
Density, mass density	ρ	$\text{kg} \cdot \text{m}^{-3}$	$\rho = m/V$
Energy	E	J	
Fluidity, kinematic viscosity	ϕ	$\text{m} \cdot \text{kg}^{-1} \cdot \text{s}$	$\phi = 1/\eta$
Force	F	N	$F = dp/dt = ma$
Friction coefficient	$\mu, (f)$		$F_{\text{frict}} = \mu F_{\text{norm}}$
Gravitational constant	G	$\text{N} \cdot \text{m}^2 \cdot \text{kg}^{-2}$	$F = Gm_1 m_2/r^2$
Hamilton function	H	J	$H(q, p) = T(q, p) + V(q)$
Kinematic viscosity	v	$\text{m}^2 \cdot \text{s}^{-1}$	$v = \eta/\rho$
Kinetic energy	E_k	J	$E_k = \tfrac{1}{2}mv^2$
Lagrange function	L	J	$L(q, \dot{q}) = T(q, \dot{q}) - V(q)$
Linear strain, relative elongation	ϵ, e		$\epsilon = \Delta l/l$
Mass	m	kg	

TABLE 2.2   Physical and Chemical Symbols and Definitions (*Continued*)

I. Mechanics (*continued*)

Name	Symbol	SI unit	Definition
Modulus of elasticity, Young's modulus	E	Pa	$E = \sigma/\epsilon$
Moment of inertia	I, J	$kg \cdot m^2$	$I = \Sigma m_i r_i^2$
Momentum	$\boldsymbol{p}$	$kg \cdot m \cdot s^{-1}$	$\boldsymbol{p} = m\boldsymbol{v}$
Normal stress	σ	Pa	$\sigma = F/A$
Potential energy	E_p	J	$E_p = \int -\boldsymbol{F} \cdot ds$
Power	P	W	$P = dW/dt$
Pressure	p, P	$Pa, N \cdot m^{-2}$	$p = F/A$
Reduced mass	μ	kg	$\mu = m_1 m_2/(m_1 + m_2)$
Relative density	d		$d = \rho/\pi^0$
Shear modulus	G	Pa	$G = \tau/\gamma$
Shear strain	γ		$\gamma = \Delta x/d$
Shear stress	τ	Pa	$\tau = F/A$
Sound energy flux	P, P_a	W	$P = dE/dt$
Specific volume	v	$m^3 \cdot kg^{-1}$	$v = V/\mu = 1/\rho$
Surface density	ρ_A, ρ_S	$kg \cdot m^{-2}$	$\rho_A = m/A$
Surface tension	γ, σ	$N \cdot m^{-1}, J \cdot m^{-2}$	$\gamma = dW/dA$
Torque, moment of a force	$\boldsymbol{T}, (\boldsymbol{M})$	$N \cdot m$	$\boldsymbol{T} = \boldsymbol{r} \times \boldsymbol{F}$
Viscosity (dynamic)	η, μ	$Pa \cdot s$	$\tau_{x,z} = \lambda(dv_x/dz)$
Volume (or bulk) strain	θ		$\theta = \Delta V/V_0$
Weight	$G, (W, P)$	N	$G = m \cdot g$
Work	W, w	J	$W = \int \boldsymbol{F} \cdot ds$

J. Solid state

Name	Symbol	SI unit	Definition
Acceptor ionization energy	E_a	J	
Bragg angle	θ	rad	$n\lambda = 2d \sin \theta$
Bloch function	$\boldsymbol{u}_k(\boldsymbol{r})$	$m^{-3/2}$	$\psi(\boldsymbol{r}) = \boldsymbol{u}_k(\boldsymbol{r}) \exp(i\boldsymbol{k} \cdot \boldsymbol{r})$
Burgers vector	$\boldsymbol{b}$	m	
Charge density of electrons	ρ	$C \cdot m^{-3}$	$\rho(\boldsymbol{r}) = -e\psi^*(\boldsymbol{r})\psi(\boldsymbol{r})$
Circular wave vector: For particles ($\boldsymbol{k}$) For phonons ($\boldsymbol{q}$)	$\boldsymbol{k}, \boldsymbol{q}$	m^{-1}	$\boldsymbol{k} = 2\pi/\lambda$
Conductivity tensor	σ_{ik}	$S \cdot m^{-1}$	$\sigma = \rho^{-1}$
Curie temperature	T_C	K	
Debye circular frequency	ω_D	s^{-1}	
Debye circular wavenumber	q_D	m^{-1}	
Debye-Waller factor	B, D		
Density of states	N_E	$J^{-1} \cdot m^{-3}$	$N_E = dN(E)/dE$
Density of vibrational modes (spectral)	N_ω, g	$s \cdot m^{-3}$	$N_\omega = dN(\omega)/d\omega$
Diffusion coefficient	D	$m^2 \cdot s^{-1}$	$dN/dt = -DA\, dn/dx$
Diffusion length	L	m	$L = (D\tau)^{1/2}$
Displacement vector of an ion	$\boldsymbol{u}$	m	$\boldsymbol{u} = \boldsymbol{R} - \boldsymbol{R}_0$
Donor ionization energy	E_d	J	

TABLE 2.2 Physical and Chemical Symbols and Definitions (*Continued*)

J. Solid state (*continued*)

Name	Symbol	SI unit	Definition
Effective mass	m^*	kg	
Equilibrium position vector of an ion	R_0	m	
Fermi energy	E_F	J	
Gap energy	E_g		
Grüneisen parameter	γ, Γ		$\gamma = \alpha V / \kappa C_V$
Hall coefficient	A_H, R_H	$m^3 \cdot C^{-1}$	$E = \rho \cdot j + R_H(B \times j)$
Lattice plane spacing	d	m	
Lattice vector	R, R_0	m	
Lorenz coefficient	L	$V^2 \cdot K^{-2}$	$L = \lambda / \sigma T$
Madelung constant	α		$E_{coul} = \dfrac{\alpha N_A z_+ z_- e^2}{4\pi \epsilon_0 R_0}$
Mobility	μ	$m^2 \cdot V^{-1} \cdot s^{-1}$	$\mu = v_{drift}/E$
Mobility ratio	b		$b = \mu_n / \mu_p$
Neel temperature	T_N	K	
Number density, number concentration	n	m^{-3}	
Order parameters:			
Long range	s		
Short range	σ		
Order of reflection	n		
Particle position vector:			
Electron	r	m	
Ion position	R_j	m	
Peltier coefficient	Π	V	
Reciprocal lattice vector (circular)	G	m^{-1}	$G \cdot R = 2\pi m$
Relaxation time	τ	s	$\tau = 1/v_F$
Residual resistivity	ρ_R	m	
Resistivity tensor	ρ	$\Omega \cdot m$	$E = \rho \cdot j$
Temperature	θ	K	
Thermal conductivity tensor	λ	$W \cdot m^{-1} \cdot K^{-1}$	$J_q = -\lambda \cdot \text{grad } T$
Thermoelectric force	E	V	
Thomson coefficient	μ	$V \cdot K^{-1}$	
Translation vectors for the reciprocal lattice (circular)	$b_1; b_2; b_3$ $a^*; b^*; c^*$	m^{-1}	$a_i \cdot b_k = 2\pi \delta_{ik}$
Translation vectors for crystal lattice	$a_1; a_2; a_3$ $a; b; c$	m	$R = n_1 a_1 + n_2 a_2 + n_3 a_3$
Work function	Φ	J	$\Phi = E_\infty - E_F$

K. Space and time

Name	Symbol	SI unit	Definition
Acceleration	$a, (g)$	$m \cdot s^{-2}$	$a = dv/dt$
Angular velocity	ω	$rad \cdot s^{-1}, s^{-1}$	$\omega = d\phi/dt$
Area	A, A_s, S	m^2	
Breadth	b	m	
Cartesian space coordinates	x, y, z	m	

TABLE 2.2 Physical and Chemical Symbols and Definitions (*Continued*)

K. Space and time (*continued*)

Name	Symbol	SI unit	Definition
Circular frequency, angular frequency	ω	$rad \cdot s^{-1}$, s^{-1}	$\omega = 2\pi\nu$
Diameter	d	m	
Distance	d	m	
Frequency	ν, f	Hz	$\nu = 1/T$
Generalized coordinate	q, q_i	(varies)	
Height	h	m	
Length	l	m	
Length of arc	s	m	
Path length	s	m	
Period	T	s	$T = t/N$
Plane angle	$\alpha, \beta, \gamma, \theta, \phi$	rad, l	$\alpha = s/r$
Position vector	r	m	$r = xi + yj + zk$
Radius	r	m	
Relaxation time, time constant	τ, T	s	$\tau = |dt/d \ln x|$
Solid angle	ω, Ω	sr, 1	$\Omega = A/r^2$
Speed	v, u, w, c	$m \cdot s^{-1}$	$v = |v|$
Spherical polar coordinates	r, θ, ϕ	m, l, 1	
Thickness	d, δ	m	
Time	t	s	
Velocity	v, u, w, c	$m \cdot s$	$v = dr/dt$
Volume	$V, (v)$	m^3	

L. Spectroscopy

Name	Symbol	SI unit	Definition
Asymmetry parameter	κ		$\kappa = \dfrac{2B - A - C}{A - C}$
Centrifugal distortion constants:			
A reduction	$\Delta_J \, \Delta_{JK} \, \Delta_K \, \delta_J \, \delta_K$	m^{-1}	
S reduction	$D_J \, D_{JK} \, D_K \, d_1 \, d_2$		
Degeneracy, statistical weight	g, d, β		
Electric dipole moment of a molecule	p, μ	$C \cdot m$	$E_p = -p \cdot E$
Electron spin resonance (ESR), electron paramagnetic resonance (EPR):			
Hyperfine coupling constant:			
In liquids	a, A	Hz	$\hat{H}_{hfs}/h = a\hat{S} \cdot I$
In solids	T	Hz	$\hat{H}_{hfs}/h = \hat{S} \cdot T \cdot I$
g factor	g		$h\nu = g\mu_B B$
Electronic term	T_e	m^{-1}	$T_e = E_e/hc$
Harmonic vibration wavenumber	$\omega_e; \omega_r$	m^{-1}	
Inertial defect	Δ	$kg \cdot m^2$	$\Delta = I_C - I_A - I_B$

TABLE 2.2 Physical and Chemical Symbols and Definitions (*Continued*)

	L. Spectroscopy (*continued*)		
Name	Symbol	SI unit	Definition
Interatomic distances:			
Equilibrium distance	r_e	m	
Ground state distance	r_0	m	
Substitution structure distance	r_s	m	
Zero-point average distance	r_z	m	
Longitudinal relaxation time	T_1	s	
Nuclear magnetic resonance (NMR), chemical shift, δ scale	δ		$\delta = 10^6(\nu - \nu_0)/\nu_0$
Coupling constant, direct (dipolar)	D_{AB}	Hz	
Magnetogyric ratio	γ	$C \cdot kg^{-1}$	$\gamma = 2\pi\mu/Ih$
Shielding constant	σ_A		$B_A = (1 - \sigma_A)B$
Spin-spin coupling constant	J_{AB}	Hz	$H/h = J_{AB}I_A \cdot I_B$
Principal moments of inertia	$I_A; I_B; I_C$	$kg \cdot m^2$	$I_A \le I_B \le I_C$
Rotational constants:			
In frequency	$A; B; C$	Hz	$A = h/8\pi^2 I_A$
In wavenumber	$\tilde{A}; \tilde{B}; \tilde{C}$	m^{-1}	$\tilde{A} = h/8\pi^2 I_A$
Rotational term	F	m^{-1}	$F = E_{rot}/hc$
Spin orbit coupling constant	A	m^{-1}	$T_{s.o.} = A \langle \tilde{L} \cdot \tilde{S} \rangle$
Total term	T	m^{-1}	$T = E_{tot}/hc$
Transition dipole moment of a molecule	$\boldsymbol{M}, \boldsymbol{R}$	$C \cdot m$	$\boldsymbol{M} = \int \psi' \boldsymbol{p} \psi'' \, d\tau$
Transition frequency	ν	Hz	$\nu = (E' - E'')/h$
Transition wavenumber	$\tilde{\nu}, (\nu)$	m^{-1}	$\tilde{\nu} = T' - T''$
Transverse relaxation time	T_2	s	
Vibrational anharmonicity constant	$\omega_e\chi_e; \chi_{rs}; g_{tt'}$	m^{-1}	
Vibrational coordinates:			
Internal coordinates	$R_i, r_i, \theta_j,$ etc.		
Normal coordinates, dimensionless	q_r		
Mass adjusted	Q_r		
Vibrational force constants:			
Diatomic	$f, (k)$	$J \cdot m^{-2}$	$f = \partial^2 V/\partial r^2$
Polyatomic			
Dimensionless normal coordinates	$\phi_{rst} \ldots, k_{rst} \ldots$	m^{-1}	
Internal coordinates	f_{ij}	(varies)	$f_{ij} = \partial^2 V/\partial r_i \partial r_j$
Symmetry coordinates	F_{ij}	(varies)	$F_{ij} = \partial^2 V/\partial S_i \partial S_j$
Vibrational quantum numbers	$v_r; l_t$		
Vibrational term	G	m^{-1}	$G = E_{vib}/hc$

TABLE 2.2 Physical and Chemical Symbols and Definitions (*Continued*)

L. Spectroscopy (*continued*)

Angular momentum types	Operator symbol	Quantum number symbol		
		Total	Z axis	z axis
Electron orbital	$\hat{L}$	L	M_L	Λ
One electron only	$\hat{l}$	l	m_l	λ
Electron orbital + spin	$\hat{L} + \hat{S}$			$\Omega = \Lambda + \Sigma$
Electron spin	$\hat{S}$	S	M_S	σ
One electron only	$\hat{s}$	s	m_s	Σ
Internal vibrational:				
Spherical top	$\hat{l}$	$l(l\zeta)$		K_l
Other	$\hat{j}, \hat{\pi}$			$l(l\zeta)$
Nuclear orbital (rotational)	$\hat{R}$	R		K_R, k_R
Nuclear spin	$\hat{I}$	I	M_I	
Sum of $J + I$	$\hat{F}$	F	M_F	
Sum of $N + S$	$\hat{J}$	J	M_J	K, k
Sum of $R + L(+j)$	$\hat{N}$	N		K, k

M. Thermodynamics

Name	Symbol	SI unit	Definition
Absolute activity	λ		$\lambda_B = \exp(\mu_B/RT)$
Activity (referenced to Henry's law):			
Concentration basis	a_c		$a_{c,B} = \exp\left[\dfrac{\mu_B - \mu_B^*}{RT}\right]$
Molality basis	a_m		$a_{m,B} = \exp\left[\dfrac{\mu_B - \mu_B^*}{RT}\right]$
Mole fraction basis	a_x		$a_{x,B} = \exp\left[\dfrac{\mu_B - \mu_B^*}{RT}\right]$
Activity (referenced to Raoult's law	a		$a_B = \exp\left[\dfrac{\mu_B - \mu_B^*}{RT}\right]$
Activity coefficient (referenced to Henry's law):			
Concentration basis	γ_c		$a_{c,B} = \gamma_{c,B} c_B/c^0$
Molality basis	γ_m		$a_{m,B} = \gamma_{m,B} m_B/m^0$
Mole fraction basis	γ_x		$a_{x,B} = \gamma_{x,B} x_B$
Activity coefficient (referenced to Raoult's law)	f		$f_B = a_B/x_B$
Affinity of reaction	A	$J \cdot mol^{-1}$	$A = -(\partial G/\partial \xi)_{p,T}$
Celsius temperature	θ, t	°C	$\theta/°C = T/K - 273.15$
Chemical potential	μ	$J \cdot mol^{-1}$	$\mu_B = (\partial G/\partial n_B)_{T,p,n}$
Compressibility:			
Isentropic	κ_S	Pa^{-1}	$\kappa_S = -(1/V)(\partial V/\partial p)_S$
Isothermal	κ_T	Pa^{-1}	$\kappa_T = -(1/V)(\partial V/\partial p)_T$
Compressibility factor	Z		$Z = pV_m/RT$
Cubic expansion coefficient	α, α_V, γ	K^{-1}	$\alpha = (1/V)(\partial V/\partial T)_p$
Enthalpy	H	J	$H = U + pV$

TABLE 2.2 Physical and Chemical Symbols and Definitions (*Continued*)

	M. Thermodynamics (*continued*)		
Name	Symbol	SI unit	Definition
Entropy	S	$J \cdot K^{-1}$	$dS \geq dq/T$
Equilibrium constant	K^0, K		$K° = \exp(-\Delta_r G°/RT)$
Equilibrium constant:			
Concentration basis	K_c	$(mol \cdot m^{-3})^{\Sigma\nu}$	$K_c = \Pi\, c^\nu$
Molality basis	K_m	$(mol \cdot kg^{-1})^{\Sigma\nu}$	$K_m = \Pi\, m^\nu$
Pressure basis	K_p	$Pa^{\Sigma\nu}$	$K_p = \Pi\, p^\nu$
Fugacity	f	Pa	$f_B = \lambda_B \lim\limits_{p \to 0} (p_B/\lambda_B)_T$
Fugacity coefficient	ϕ		$\phi_B = f_B/p_B$
Gibbs energy	G	J	$G = H - TS$
Heat	q, Q	J	
Heat capacity:			
At constant pressure	C_p	$J \cdot K^{-1}$	$C_p = (\partial H/\partial T)_p$
At constant volume	C_v	$J \cdot K^{-1}$	$C_v = (\partial U/\partial T)_v$
Helmholtz energy	A	J	$A = U - TS$
Internal energy	U		$\Delta U = q + w$
Ionic strength:			
Concentration basis	I_c, I	$mol \cdot kg^{-3}$	$I_c = \frac{1}{2}\Sigma m_B z_B^2$
Molality basis	I_m, I	$mol \cdot kg^{-1}$	$I_m = \frac{1}{2}\Sigma m_B z_B^2$
Joule-Thomson coefficient	μ, μ_{JT}	$K \cdot Pa^{-1}$	$\mu = (\partial T/\partial p)_H$
Linear expansion coefficient	α_l	K^{-1}	$\alpha_l = (1/l)(\partial l/T)$
Massieu function	J	$J \cdot K^{-1}$	$J = -A/T$
Molar quantity X	X_m	(varies)	$X_m = X/n$
Osmotic coefficient:			
Molality basis	ϕ_m		$\phi_m = (\mu_A^* - \mu_A)/ \\ (RTM_A \Sigma m_B)$
Mole fraction basis	ϕ_x		$\phi_x = (\mu_A - \mu_A^*)/(RT \ln x_A)$
Osmotic pressure (ideal dilute solution)	Π	Pa	$\Pi = c_B RT$
Partial molar quantity X	X_B	(varies)	$X = (\partial X/\partial n_B)_{T,p,n}$
Planck function	Y	$J \cdot K^{-1}$	$Y = -G/T$
Pressure coefficient	β	$Pa \cdot K^{-1}$	$\beta = (\partial P/\partial T)_v$
Ratio of heat capacities	γ		$\gamma = C_p/C_v$
Relative pressure coefficient	α_p	K^{-1}	$\alpha_p = (1/p)(\partial p/\partial T)_v$
Second virial coefficient	B	$m^3 \cdot mol^{-1}$	$pV_m = RT(1 + B/V_m + \cdots)$
Specific quantity X	x	(varies)	$x = X/m$
Standard chemical potential	μ^0	$J \cdot mol^{-1}$	
Standard partial molar enthalpy	H^0	$J \cdot mol^{-1}$	$H^0 = \mu^0 + TS$
Standard partial molar entropy	S^0	$J \cdot mol^{-1} \cdot K^{-1}$	$S^0 = -(\partial\mu^0/\partial T)$
Standard reaction enthalpy	$\Delta_r H^0$	$J \cdot mol^{-1}$	$\Delta_r H^0 = \Sigma\nu H^0$
Standard reaction entropy	$\Delta_r S^0$	$J \cdot mol^{-1} \cdot K^{-1}$	$\Delta_r S^0 = \Sigma\nu S^0$
Standard reaction Gibbs energy	$\Delta_r G^0$	$J \cdot mol^{-1}$	$\Delta_r G^0 = \Sigma\nu\mu^0$
Surface tension	γ, σ	$J \cdot m^{-2}, N \cdot m^{-1}$	$\gamma = (\partial G/\partial A_s)_{T,p}$
Thermodynamic temperature	T	K	
Work	w, W	J	

TABLE 2.2 Physical and Chemical Symbols and Definitions (*Continued*)

M. Thermodynamics (*continued*)

Symbols used as subscripts to denote a chemical reaction or process:

ads	adsorption	mix	mixing of fluids
at	atomization	r	reaction in general
c	combustion reaction	sol	solution of solute in solvent
dil	dilution of a solution	sub	sublimation (solid to gas)
f	formation reaction	trs	transition (two phases)
fus	melting, fusion (solid to liquid)		

Recommended superscripts:

‡	activated complex, transition state	∞	infinite solution
E	excess quantity	*	pure substance
id	ideal	°	standard

N. Transport properties

Name	Symbol	SI unit	Definition
Coefficient of heat transfer	$h, (k, K)$	$W \cdot m^{-2} \cdot K^{-1}$	$h = J_q/\Delta T$
Diffusion coefficient	D	$m^2 \cdot s^{-1}$	$D = J_n/(dc/dl)$
Flux (of a quantity X)	J_X, J	(varies)	$J_X = A^{-1}\, dX/dt$
Heat flow rate	ϕ	W	$\phi = dq/dt$
Heat flux	J_q	$W \cdot m^{-2}$	$J_q = \phi/A$
Mass flow rate	q_m, m	$kg \cdot s^{-1}$	$q_m = dm/dt$
Mass transfer coefficient	k_d	$m \cdot s^{-1}$	
Thermal conductance	G	$W \cdot K^{-1}$	$G = \phi/\Delta T$
Thermal conductivity	λ, k	$W \cdot m^{-1} \cdot K^{-1}$	$\lambda = J_q/(dT/dl)$
Thermal diffusivity	a	$m^2 \cdot s^{-1}$	$a = \lambda/\rho c_p$
Thermal resistance	R	$K \cdot W^{-1}$	$R = 1/G$
Volume flow rate	q_v, V	$m^3 \cdot s^{-1}$	$q_v = dV/dt$
Dimensionless quantities:			
Alfvén number	Al		$Al = v(\rho\mu)^{1/2}/B$
Cowling number	Co		$Co = B^2/\mu\rho v^2$
Euler number	Eu		$Eu = \Delta p/\rho v^2$
Fourier number	Fo		$Fo = at/l^2$
Fourier number for mass transfer in binary mixtures	Fo^*		$Fo^* = Dt/l^2$
Froude number	Fr		$Fr = v/(lg)^{1/2}$
Grashof number	Gr		$Gr = l^3\, go\, \Delta T\rho^2/\eta^2$
Grashof number for mass transfer in binary mixtures	Gr^*		$Gr^* = l^3 g\, (\partial\rho/\partial x)_{T,p}\, (\Delta x\pi/\eta)$
Hartmann number	Ha		$Ha = Bl(\kappa/\eta)^{1/2}$
Knudsen number	Kn		$Kn = \lambda/l$
Lewis number	Le		$Le = a/D$
Mach number	Ma		$Ma = v/c$
Magnetic Reynolds number	Rm, Re_m		$Rm = v\mu\kappa l$
Nusselt number	Nu		$Nu = hl/k$
Nusselt number for mass transfer in binary mixtures	Nu^*		$Nu^* = k_d l/D$

TABLE 2.2 Physical and Chemical Symbols and Definitions (*Continued*)

		N. Transport properties (*continued*)	
Name	Symbol	SI unit	Definition
Péclet number	Pe		$Pe = vl/a$
Péclet number for mass transfer in binary mixtures	Pe^*		$Pe^* = vl/D$
Prandtl number	Pr		$Pr = \eta/\rho a$
Rayleigh number	Ra		$Ra = l^3\, g\alpha\, \Delta T\rho/\eta a$
Reynolds number	Re		$Re = \rho vl/\eta$
Schmidt number	Sc		$Sc = \eta/\rho D$
Sherwood number	Sh		$Sh = k_d l/D$
Stanton number	St		$St = h/\rho v c_p$
Stanton number for mass transfer in binary mixtures	St^*		$St^* = k_d/v$
Strouhal number	Sr		$Sr = lf/v$
Weber number	We		$We = \rho v^2\, l/\gamma$

Symbols used in the definitions of dimensionless quantities:

Acceleration of free fall	g	Pressure	p
Area	A	Speed	v
Cubic expansion coefficient	α	Speed of sound	c
Density	ρ	Surface tension	γ
Frequency	f	Temperature	T
Length	l	Time	t
Mass	m	Viscosity	η
Mean free path	λ	Volume	V
Mole fraction	x		

TABLE 2.3 Mathematical Symbols and Abbreviations

Symbol or abbreviation	Meaning
$+$	Plus
$-$	Minus
$\pm$	Plus or minus
$\mp$	Minus or plus
$\equiv$	Identically equal to
$\times$, center dot	Multiplied by (ab, $a \times b$, $a \cdot b$)
$\div$	Divided by (a/b, ab^{-1})
$\neq$	Not equal to
$\approx$	approximately equal to
$\simeq$	Asymptotically equal to
$>$	Greater than
$<$	Less than
$\gg$	Much greater than
$\ll$	Much less than
$\geq$	Greater than or equal to
$\leq$	Less than or equal to
$\propto$, $\sim$	Proportional to
$\rightarrow$	Tends to, approaches

TABLE 2.3 Mathematical Symbols and Abbreviations (*Continued*)

Symbol or abbreviation	Meaning
∞	Infinity
$\lvert a \rvert$	Magnitude of a
a^n	nth power of a
$\sqrt[n]{a},\ a^{1/n}$	nth root of a
$\sqrt{a},\ a^{1/2}$	Square root of a
$\langle a \rangle,\ \bar{a}$	Mean value of a
$\displaystyle\prod_{i=1}^{n} a_i,\ \Pi a_i$	Product of a_i
$\log a$ or $\log_{10} a$	Common (or Briggsian) logarithm to the base 10 of a
$\log_a b$	Logarithm to the base a of b
$\ln b,\ \log_e b$	Natural (Napierian) logarithm (to the base e) of b
e	Base (2.718) of natural system of logarithms
π	Pi (3.1416)
i	Imaginary quantity, square root of minus one
$n!$	n factorial ($n! = 1 \cdot 2 \cdot 3 \cdots n$)
$\angle$	Angle
$\perp$	Perpendicular to
$\parallel$	Parallel to
a°	a degrees (angle)
a'	a minutes (angle); a prime
a''	a seconds (angle); a double prime
$\sin a$	sine of a
$\cos a$	cosine of a
$\tan a$	tangent of a
$\cot a$	cotangent of a
$\sec a$	secant of a
$\cos a$	cosecant of a
$\arcsin a,\ \sin^{-1} a$	Inverse sine of a (angle whose sine is a)
$\arccos a,\ \cos^{-1} a$	Inverse sine of a (angle whose cosine is a)
$\arctan a,\ \tan^{-1} a$	Inverse tangent of a (angle whose tangent is a)
$\sinh a$	Hyperbolic sine of a
$\cosh a$	Hyperbolic cosine of a
$\tanh a$	Hyperbolic tangent of a
$\coth a$	Hyperbolic cotangent of a
$P(x, y)$	Rectangular coordinate of point P
$P(r, \theta)$	Polar coordinate of point P
$f(x),\ F(x)$	Function of x
Δx	Increment of x
dy	Total differential of y
$\dfrac{dy}{dx}$ or $f'(x)$	Derivative of $y = f(x)$ with respect to x
$\dfrac{d^2 y}{dx^2}$ or $f''(x)$	Second derivative of $y = f(x)$ with respect to x
$\dfrac{\partial z}{\partial x}$	Partial derivative of z with respect to x
$\dfrac{\partial^2 z}{\partial x\,\partial y}$	Second partial derivative of z with respect to x and y
$\displaystyle\int$	Integral of
$\displaystyle\int_a^b$	Integral between the limits a and b

TABLE 2.3   Mathematical Symbols and Abbreviations (*Continued*)

Symbol or abbreviation	Meaning
$\lim\limits_{x \to a} f(x)$	limit of $f(x)$ as x tends to a
$\sum\limits_{i=1}^{n}$	Summation of a_i between the limits 1 and n

TABLE 2.4   SI Prefixes

Submultiple	Prefix	Symbol	Multiple	Prefix	Symbol
10^{-1}	deci	d	10	deka	da
10^{-2}	centi	c	10^2	hecto	h
10^{-3}	milli	m	10^3	kilo	k
10^{-6}	micro	μ	10^6	mega	M
10^{-9}	nano	n	10^9	giga	G
10^{-12}	pico	p	10^{12}	tera	T
10^{-15}	femto	f	10^{15}	peta	P
10^{-18}	atto	a	10^{18}	exa	E

Numerical (multiplying) prefixes

Number	Prefix	Number	Prefix	Number	Prefix
0.5	hemi	19	nonadeca	39	nonatriaconta
1	mono	20	icosa	40	tetraconta
1.5	sesqui	21	henicosa	41	hentetraconta
2	di (bis)*	22	docosa	42	dotetraconta
3	tri (tris)*	23	tricosa	43	tritetraconta
4	tetra (tetrakis)*	24	tetracosa	44	tetratetraconta
5	penta	25	pentacosa	45	pentatetraconta
6	hexa	26	hexacosa	46	hexatetraconta
7	hepta	27	heptacosa	47	heptatetraconta
8	octa	28	octacosa	48	octatetraconta
9	nona	29	nonacosa	49	nonatetraconta
10	deca	30	triaconta	50	pentaconta
11	undeca	31	hentriaconta	60	hexaconta
12	dodeca	32	dotriaconta	70	heptaconta
13	trideca	33	tritriaconta	80	octaconta
14	tetradeca	34	tetratriaconta	90	nonaconta
15	pentadeca	35	pentatriaconta	100	hecta
16	hexadeca	36	hexatriaconta	110	decahecta
17	heptadeca	37	heptatriaconta	120	icosahecta
18	octadeca	38	octatriaconta	130	triacontahecta

* In the case of complex entities such as organic ligands (particularly if they are substituted) the multiplying prefixes bis-, tris-, tetrakis-, pentakis-, . . . are used, i.e., -kis is added starting from tetra-. The modified entity is often placed within parentheses to avoid ambiguity.

TABLE 2.5 Greek Alphabet

Capital	Lower case	Name	Capital	Lower case	Name
A	α	Alpha	N	ν	Nu
B	β	Beta	Ξ	ξ	Xi
Γ	γ	Gamma	O	o	Omicron
Δ	δ	Delta	Π	π	Pi
E	ϵ	Epsilon	P	ρ	Rho
Z	ζ	Zeta	Σ	σ	Sigma
H	η	Eta	T	τ	Tau
Θ	θ	Theta	Υ	υ	Upsilon
I	ι	Iota	Φ	ϕ	Phi
K	κ	Kappa	X	χ	Chi
Λ	λ	Lambda	Ψ	ψ	Psi
M	μ	Mu	Ω	ω	Omega

TABLE 2.6 Abbreviations and Standard Letter Symbols

Abampere	abamp	Activity (radioactive)	A
Absolute	abs	Activity coefficient (referenced to Raoult's law)	f
Absolute activity	λ		
Absorbance (decaidic)	A		
Absorbance (napierian)	B	Activity coefficient (referenced to Henry's law):	
Absorptance	α		
Absorption coefficient, linear decaidic	a, K	Concentration basis	γ_c
		Molality basis	γ_m
Absorption coefficient, linear napierian	α	Mole fraction basis	γ_x
		Adjusted retention time	t'_R
Absorption coefficient, molar decaidic	μ, ϵ	Adjusted retention volume	V'_R
Absorption coefficient, molar napierian	κ	Admittance	Y
Absorption index	k	Affinity of reaction	A
Acceleration	a	Alcohol	alc
Acceleration due to gravity	g, g_n	Alfvén number	Al
		Alkaline	alk
Acetyl	Ac	Alpha particle	α
Acoustic absorption factor	α_a	Alternating current	ac
		Amorphous	am
Acoustic dissipation factor	δ	Amount concentration	c
		Amount of substance	n
Acoustic reflection factor	ρ	Ampere	A
Acoustic transmission factor	τ	Amplification factor	μ
		Angle of optical rotation	α
Activation energy	E_a	Angstrom	A, A
Activity (referenced to Raoult's law)	a	Angular dispersion	$d\theta/d\lambda$
		Angular momentum	π
Activity (referenced to Henry's law):		Angular momentum terms	j, J, l, L, N
Concentration basis	a_c	Angular velocity	ω
Molality basis	a_m	Anhydrous	anhyd
Mole fraction basis	a_x	Approximate (circa)	ca.

TABLE 2.6 Abbreviations and Standard Letter Symbols (*Continued*)

Aqueous solution	aq	Cartesian space coordinates	x, y, z
Aqueous solution at infinite dilution	aq, ∞	Celsius temperature	t, θ
Are, unit of area	a	Centimeter-gram-second system	cgs
Area	A, S	Centrifugal distortion constants:	
Area per molecule	a, σ	A reduction	Δ, δ
Astronomical unit	AU	S reduction	D, d
Asymmetry parameter	κ	Charge density of electrons	ρ
Atmosphere, unit of pressure	atm	Charge number of electrochemical reaction	n
Atomic mass	m_a		
Atomic mass constant	m_u	Charge number of an ion	z
Atomic mass unit	amu	Chemically pure	CP
Atomic number	Z	Chemical potential	μ
Atomic percent	at.%	Chemical shift	δ
Atomic weight	at. wt.	Circa (approximate)	ca.
Average	av	Circular frequency	ω
Average linear gas velocity	μ	Circular wave vector:	
Avogadro constant	L, N_A	For particles	$\boldsymbol{k}$
Axial angular momentum	λ, Λ, Ω	For phonons	$\boldsymbol{q}$
Axial spin angular momentum	σ, Σ	Circumference divided by the diameter	π
Bandwidth (10%) of a spectral filter	$\Delta\lambda_{0.1}$	Citrate	Cit
Band variance	σ^2	Coefficient of heat transfer	h
Bar, unit of pressure	bar	Collision cross section	σ
Barn, unit of area	b	Collision diameter	d
Barrel	bbl	Collision frequency	Z
Base of natural logarithms	e	Collision frequency factor	z
Becquerel	Bq	Collision number	Z
Bed volume	V_g	Column volume	V_{col}
Beta particle	β	Compare (confer)	cf.
Bloch function	$\boldsymbol{u_k(r)}$	Complex refractive index	$\hat{n}$
Body-centered cubic	bcc	Component of angular momentum	k, K, m, M
Bohr	b		
Bohr magneton	μ_B	Compressibility:	
Bohr radius	a_0	Isentropic	κ_S
Boiling point	bp	Isothermal	κ_T
Boltzman constant	k, k_B	Compression factor	Z
Bragg angle	θ	Compression modulus	K
Breadth	b	Compton wavelength of electron	λ_c
British thermal unit	Btu		
Bulk modulus	K	Compton wavelength of neutron	$\lambda_{c,n}$
Bulk strain	θ		
Burgers vector	$\boldsymbol{b}$	Compton wavelength of proton	$\lambda_{c,p}$
Butyl	Bu		
Calorie, unit of energy	cal	Concentration (amount of substance)	c
Calorie, international steam table	cal_{IT}		
		Concentration (mass)	γ
Candela	cd	Concentration at peak maximum	C_{max}
Capacitance	C		
Capacity, volume	Q_V	Concentration of solute in mobile phase	C_M
Capacity, weight	Q_w		

TABLE 2.6 Abbreviations and Standard Letter Symbols (*Continued*)

Concentration of solute in stationary phase	C_S		Diamagnetic shielding factor	$1 + \sigma$
Condensed phase (solid or liquid)	cd		Diameter	d
Conductance	G		Dielectric polarization	$\boldsymbol{P}$
Conductivity	γ, κ		Differential thermal analysis	DTA
Conductivity cell constant	K_{cell}		Diffusion coefficient	D
Conductivity tensor	σ_{ik}		Diffusion coefficient, liquid film	D_f
Contact angle	θ		Diffusion coefficient, mobile phase	D_M
Coordinate, position vector	r		Diffusion coefficient, stationary phase	D_S
Coulomb	C		Diffusion current	i_d
Counts per minute	cpm, c/m		Diffusion length	L
Coupling constant, direct dipolar	D_{AB}		Diffusion rate constant, mass transfer coefficient	k_d
Critical density	d_c			
Critical temperature	t_c		Dilute	dil
Cross section	σ		Dirac delta function	δ
Crystalline	cr, cryst		Direct current	dc
Cubic	cub		Direct dipolar coupling constant	D_{AB}
Cubic expansion coefficient	α, α_v, γ		Disintegration energy	Q
Curie	Ci		Disintegrations per minute	dpm
Cycles per second	Hz			
Curie temperature	T_c		Displacement vector of an ion	$\boldsymbol{u}$
Dalton (atomic mass unit)	Da		Dissociation energy	D, E_d
Day	d		From ground state	D_0
Debye, unit of electric dipole	D		From the potential minimum	D_e
Debye circular frequency	ω_D		Distribution ratio	D
Debye circular wavenumber	q_D		Donor ionization energy	E_d
Debye-Waller factor	D, B		Dropping mercury electrode	dme
Decay constant (radioactive)	λ		Dyne, unit of force	dyn
Decibel	dB		Einstein transition probabilities	A, B
Decompose	dec		Spontaneous emission	A_{nm}
Degeneracy, statistical weight	d, g, β		Stimulated absorption	B_{mn}
Degree of dissociation	α		Stimulated emission	B_{nm}
Degrees Baume	°Be		Electric charge	Q
Degrees Celsius	°C		Electric current	I
Degrees Fahrenheit	°F		Electric current density	j, J
Density (mass)	ρ, γ		Electric dipole moment of a molecule	$\boldsymbol{p}, \boldsymbol{\mu}$
Density, critical	d_c		Electric displacement	$\boldsymbol{D}$
Density, relative	d		Electric field gradient	q
Density of liquid phase	ρ_L		Electric field strength	$\boldsymbol{E}$
Density of states	N_E, ρ		Electric flux	Ψ
Density of vibrational modes (spectral)	N_ω		Electric mobility	u, μ
Detect, determine (d)	det(d)		Electric polarizability of a molecule	α
Determination	detn			
Deuteron	d		Electric potential	V, ϕ

TABLE 2.6 Abbreviations and Standard Letter Symbols (*Continued*)

Electric potential difference	$U, \Delta V$		Exponential	exp
Electric susceptibility	χ_e		Extent of reaction	ξ
Electrical conductivity	σ		Fano factor	F
Electrical conductance	G		Farad	F
Electrical resistance	R		Faraday constant	F
Electrochemical transfer coefficient	α		Fermi, unit of length	f
			Fermi energy	E_F
Electrokinetic potential	ζ		Film tension	Σ_f
Electromagnetic unit	emu		Film thickness	h, t
Electromotive force	E, emf		Fine structure constant	α
Electron	e^-, e		Finite change	Δ
Electron affinity	E_{ea}		First radiation constant	c_1
Electron magnetic moment	μ_e		Flow rate	q
			Flow rate, column chromatography	F_c
Electron paramagnetic resonance	EPR		Fluid phase (gas or liquid)	fl
			Fluidity	ϕ
Electron radius	r_e		Fluorescent efficiency	Φ_F
Electron rest mass	m_e		Fluorescent power	P_F
Electron spin resonance	ESR		Flux	F, J
Electronvolt	eV		Focal length	f
Electrostatic unit	esu		Foot	ft
Elementary charge	e		For example (exempli gratia)	e.g.
Elution volume, exclusion chromatography	V_e		Force	F
Emittance	ϵ		Force constant (vibrational levels)	k
By blackbody	M_{bb}			
Energy	E		Formal concentration	F
Energy density	w, ρ		Fourier number	Fo
Energy per electron hole pair of ion pair in detector	ϵ		Franklin, unit of electric charge	Fr
			Freezing point	fp
Enthalpy	H		Frequency	f, ν
Entropy	S		Friction coefficient	f, μ
Entropy unit	e.u.		Froude number	Fr
Equilibrium constant	K, K^0		Fugacity	f
Concentration basis	K_c		Fugacity coefficient	ϕ
Molality basis	K_m		Gallon	gal
Pressure basis	K_p		Galvani potential difference	$\Delta\phi$
Equilibrium position vector of an ion	$\boldsymbol{R}_0$			
			Gamma, unit of mass	γ
Equivalent weight	equiv wt		Gamma radiation	γ
Erg, unit of energy	erg		Gap energy (solid state)	E_g
Especially	esp.		Gas (physical state)	g
et alii (and others)	et al.		Gas constant	R
et cetera (and so forth)	etc.		Gauss	G
Ethyl	Et		g factor	g
Ethylenediamine	en		Gibbs energy	G
Ethylenediamine-N,N,N',N'-tetraacetic acid	EDTA		Grade	grad
			Grain, unit of mass	gr
			Gram	g
Euler number	Eu		Grand partition function	Ξ
Exempli gratia (for example)	e.g.		Grashof number	Gr
			Gravimetric	grav
Expansion coefficient	α		Gravitational constant	G

TABLE 2.6 Abbreviations and Standard Letter Symbols (*Continued*)

Gray	Gy		Zero-point average distance	r_z
Grüneisen parameter	γ, Γ		Internal energy	U
Half-life	$t_{1/2}$		Interstitial (outer) volume	V_o
Half-wave potential	$E_{1/2}$			
Hall coefficient	A_H, R_H		In the place cited (loco citato)	loc. cit.
Hamilton function	H			
Harmonic vibration wavenumber	ω		In the same place	ibid.
			In the work cited	op. cit.
Hartmann number	Ha		Ionic conductivity	λ, Λ
Hartree energy	E_h		Ionic strength	I
Heat	q, Q		Concentration basis	I_c
Heat capacity	C		Molality basis	I_m
At constant pressure	C_p		Ionization energy	E_i
At constant volume	C_v		Irradiance	E
Heat flow rate	ϕ		Joule	J
Heat flux	J		Joule-Thomson coefficient	μ, μ_{JT}
Hectare, unit of area	ha			
Height	h		Kelvin	K
Helion	h		Kilocalorie	kcal
Helmholtz energy	A		Kilogram	kg
Henry	H		Kilogram-force	kgf
Hertz	Hz		Kilowatt-hour	kWh
Hexagonal	hex		Kinematic viscosity	ν, ϕ
Horsepower	hp		Kinetic energy	K, T, E_k
Hour	h		Knudsen number	Kn
Hygroscopic	hygr		Kovats retention indices	RI
Hyperfine coupling constant	a, A		Lagrange function	L
			Lambda, unit of volume	λ
Hyperfine coupling tensor	$\boldsymbol{T}$		Landé g-factor	g, g_e
ibidem (in the same place)	ibid.		Larmor circular frequency	ω_L
id est (that is)	i.e.		Larmor frequency	ν_L
Ignition	ign		Lattice plane spacing	d
Impedance	Z		Lattice vector	$\boldsymbol{R}, \boldsymbol{R_0}$
Inch	in		Lattice vectors	$\boldsymbol{a}, \boldsymbol{b}, \boldsymbol{c}$
Indices of a family of crystallographic planes	hkl		Length	l, L
			Length of arc	s
Indirect spin-spin coupling constant	J_{AB}		Lewis number	Le
			Light year	l.y.
Inductance	L		Limit (mathematics)	lim
Inertial defect	Δ		Linear expansion coefficient	α_l
Infinitesimal change	δ			
Infrared	ir		Linear reciprocal dispersion	$D^{-1}, d\lambda/dx$
Inner column volume	V_i			
Inner electric potential	ϕ		Linear strain	e, ϵ
Inner electrode potential	ϕ		Liquid	l, lq
Inorganic	inorg		Liquid crystal	lc
Inside diameter	i.d.		Liter	L, l
Insoluble	insol		loco citato (in the place cited)	loc. cit.
Interatomic distances:				
Equilibrium distance	r_e		Logarithm, common	log
Ground-state distance	r_0		Logarithm, base e	ln
Substitution structure distance	r_s		Longitudinal relaxation time	T_1

TABLE 2.6   Abbreviations and Standard Letter Symbols (*Continued*)

Lorenz coefficient	L	Millimole	mM
Loss angle	δ	Minimum	min
Lumen	lm	Minute	m, min
Luminous intensity	I	Mixture	mixt
Lux	lx	Mobility	μ
Mach number	Ma	Mobility ratio	b
Madelung constant	α	Modulus of elasticity	E
Magnetic dipole moment of a molecule	$\boldsymbol{m, \mu}$	Molal	m
		Molality	b
Magnetic field strength	$\boldsymbol{H}$	Molar	M, м
Magnetic flux	Φ	Molar (decadic) absorption coefficient	ϵ
Magnetic flux density	B		
Magnetic moment of protons in water	μ_p/μ_B	Molar ionic conductivity	λ, Λ
		Molar magnetic susceptibility	χ_m
Magnetic quantum number	M_j		
		Molar mass	M
Magnetic Reynolds number	Rm	Molar quantity X	X_m
		Molar refraction	R, R_m
Magnetic susceptibility	κ, χ	Molar volume	V_m
Magnetic vector potential	A	Mole	mol
Magnetizability	ξ	Mole fraction, condensed phase	x
Magnetization	M		
Magnetogyric ratio	γ	Gaseous mixtures	y
Mass	m	Mole percent	mol %
Mass absorption coefficient	$\mu/\rho, \mu_m$	Molecular weight	mol wt
		Moment of inertia	I, J
Mass concentration	γ, ρ	Momentum	p
Mass density	ρ	Monoclinic	mn
Mass flow rate	q_m	Monomeric form	mon
Mass fraction	w	Muon, negative	μ^-
Massieu function	J	Muon, positive	μ^+
Mass number	A	Mutual inductance	M, L
Mass of atom	m, m_a	Napierian absorbance	B
Mass transfer coefficient	k_d	Napierian base	e
Matrix volume	V_g	Napierian molar absorption coefficient	κ
Maximum	max		
Maxwell, unit of magnetic flux	Mx	Neel temperature	T_N
		Net retention volume	V_N
Mean ionic activity	$a_\pm$	Neutrino	v_e
Mean ionic activity coefficient	$\gamma_\pm$	Neutron	n
		Neutron magnetic moment	μ_N
Mean ionic mobility	$W_\pm$		
Melting point	mp	Neutron number	N
Metallic	met	Neutron rest mass	m_n
Metastable	m	Newton	N
Metastable peaks	m^*	Normal concentration	N
Meter	m	Normal stress	σ
Methyl	Me	Nuclear magnetic resonance	NMR
Micrometer	μm		
Micron	μ	Nuclear magneton	μ_N
Mile	mi	Nuclear spin angular momentum	I
Miller indices	h, l, k		
Milliequivalent	meq	Nucleon number	A
Millimeters of mercury, unit of pressure	mmHg	Number concentration	C
		Number density	n

TABLE 2.6 Abbreviations and Standard Letter Symbols (*Continued*)

Number of entities	N	Parts per million, weight	μ/g
Numerical aperture	NA	Pascal	Pa
Nusselt number	Nu	Path length (absorbing)	l
Obstruction factor	γ	Peak asymmetry factor	AF
Oersted, unit of magnetic field	Oe	Peak resolution	Rs
		Péclet number	Pe
Ohm	Ω	Peltier coefficient	Π
opere citato (in the work cited)	op. cit.	Percent	%
		Period of time	T
Optical speed	f/number	Permeability	μ
Orbital angular momentum:		Permeability of vacuum	μ_0
		Permittivity	ϵ
Quantum number	$L = 0, 1, 2, 3, \ldots$	Permittivity of vacuum	ϵ_0
Series symbol	$S, P, D, F, \ldots$	pH, expressed in activity	paH
Orbital angular momentum (molecules):		Expressed in molarity	pH
		Phenyl	Ph, ϕ
Quantum number	$\Lambda = 0, 1, 2, \ldots$	Phosphorescent efficiency	Φ_P
Symbol	$\Sigma, \Pi, \Delta, \ldots$		
Orbital angular momenta of individual electrons	$l = 0, 1, 2, 3, \ldots$ $s, p, d, f, \ldots$	Phosphorescent power	P_P
		Photochemical yield	ϕ
		Photoluminescence power	P
Order of Bragg reflection	n		
Order of reaction	n	Photon	γ
Order of reflection	n	Pion	π
Order parameters (solid state), long range	s	Planck constant	h
		Planck constant/2π	$\hbar$
Short range	σ	Planck function	Y
Organic	org	Plane angle	$\alpha, \beta, \gamma, \theta, \phi$
Orthorhombic	o-rh	Plate height	H
Osmotic coefficient	ϕ	Plate number, effective	N_{eff}
Molality basis	ϕ_m	Poise	P
Mole fraction basis	ϕ_x	Polymeric form	pol
Osmotic pressure (ideal dilute solution)	Π	Porosity, column	ϵ
		Positron	β^+
Ounce	oz	Potential energy	V, Φ, E_p
Outer diameter	o.d.	Pound	lb
Outer electric potential	ψ	Pounds per square inch	psi
Overall order of reaction	n	Powder	pwd
Overpotential	η	Power	p
Oxalate	Ox	Poynting vector	S
Oxidant	ox	Prandtl number	Pr
Packing uniformity factor	λ	Pressure (partial)	p
Page(s)	p. (pp.)	Pressure (total)	p, P
Parsec, unit of length	pc	Pressure coefficient	β
Partial molar quantity	X	Pressure, column inlet	p_i
Partial order of reaction	n_B	Pressure, column outlet	p_o
Particle diameter	d_p	Pressure, critical	p_c
Particle position vector:		Pressure drop	ΔP
Electron	r	Pressure-gradient correction	j
Ion position	R_j		
Partition coefficient	K	Principal moments of inertia	$I_A; I_B; I_C$
Partition function	q, Q, z, Z, Ω		
Partition ratio	k'	Principal quantum number	n
Parts per billion, volume	ng/mL		
Parts per billion, weight	ng/g	Probability	P
Parts per million, volume	$\mu g/mL$	Probability density	P

TABLE 2.6 Abbreviations and Standard Letter Symbols (*Continued*)

Product sign	Π	Reference	ref
Propyl	Pr	Reflectance	ρ
Proton	p	Reflection plane	σ
Proton magnetic resonance	pmr	Refractive index	n
		Relative permeability	μ_r
Proton magnetrogyric ratio	γ_p	Relative permittivity (dielectric constant)	ϵ_r
Proton number	Z	Relative pressure coefficient	α_p
Proton rest mass	m_p		
Pyridine	py	Relative retention ratio	α
Quadrupole interaction energy tensor	χ	Relaxation time	τ
		Rem, unit of dose equivalent	rem
Quadrupole moment of a molecule	Q, Θ	Residual resistivity (solid state)	ρ_R
Quantity of electricity, electric charge	Q	Resistivity tensor	ρ
Quantum of energy	$h\nu$	Retardation factor	R_f
Quantum yield	ϕ	Retarded van der Waals constant	B, β
Rad, unit of radiation dose	rad	Retention time	t_R
Radian	rad	Retention volume	V_R
Radiant energy	Q, W	Revolutions per minute	rpm
Radiant energy density	ρ, w	Reynolds number	Re
Radiant energy flux	dQ/dt	Rhombic	rh
Radiant exitance	M	Rhombohedral	rh-hed
Radiant flux received	E	Roentgen	R
Radiant intensity	I	Root-mean-square	rms
Radiant intensity at time t after termination of excitation	$I(t)$	Rotational constants:	
		In frequency	A, B, C
		In wavenumber	$\tilde{A}; \tilde{B}; \tilde{C}$
Radiant power	Φ	Rotational term (spectroscopy)	F
Radiant power incident on sample	P_0	Rotation-reflection	S_n
Radiofrequency	rf	Rydberg, unit of energy	Ry
Radius	r	Rydberg constant	R, R_∞
Rate of concentration change	r	Saturated	satd
		Saturated calomel electrode	SCE
Rate constant	k		
Rate of reaction	v	Schmidt number	Sc
Ratio of heat capacities	γ	Second	s
Reactance	X	Second radiation constant	c_2
Reciprocal lattice	a^*, b^*, c^*		
Reciprocal lattice vector (circular)	G	Second virial coefficient	B
		Sedimentation coefficient	s
Vectors for	$a^*; b^*; c^*$		
Reciprocal radius of ionic atmosphere	κ	Selectivity coefficient	k
		Self-inductance	L
Reciprocal temperature parameter, $1/kT$	β	Separation factor	α
		Shear modulus	G
Reciprocal thickness of double layer	κ	Shear strain	γ
		Shear stress	τ
Reduced column length	λ	Sherwood number	Sh
Reduced mass	μ	Shielding constant (NMR)	σ
Reduced plate height	h		
Reduced velocity	v	Short-range order parameter	σ
Reductant	red		

TABLE 2.6 Abbreviations and Standard Letter Symbols (*Continued*)

Siemens	S	Spin orbit coupling constant	A
Sievert	Sv	Spin-spin coupling constant	J_{AB}
Signal-to-noise ratio	S/N	Spin-spin (or transverse) relaxation time	T_2
Slightly	sl		
Solid	c, s		
Solid angle	ω, Ω	Spin wavefunctions	α, β
Solid angle over which luminescence is measured (F, fluorescence; P, phosphorescence; DF, delayed fluorescence)	$\Omega_{F(P,DF)}$	Square	sq
		Standard	std
		Standard enthalpy of activation	$H\ddagger$
		Standard enthalpy of formation	ΔHf^0
Solid angle over which radiation is absorbed in cell	Ω_A	Standard entropy	S^0
		Standard entropy of activation	$\Delta S\ddagger$
Solubility	s	Standard Gibbs energy of activation	$\Delta G\ddagger$
Soluble	sol		
Solution	soln, sln	Standard Gibbs energy of formation	ΔGf^0
Solvent	solv		
Sound energy flux	P, P_a	Standard heat capacity	C_p
Spacing between crystal diffracting planes	d	Standard hydrogen electrode	SHE
Species adsorbed on a substance	ads	Standard partial molar enthalpy	H^0
Specific gravity	sp gr	Standard partial molar entropy	S^0
Specific retention volume	V_g^0		
Specific surface area	s	Standard potential of electrochemical cell reaction	E^0
Specific volume	v, υ		
Spectral bandwidth of emission monochromator	$\Delta\lambda_{em}$		
		Standard reaction enthalpy	$\Delta_r H^0$
Spectral bandwidth of excitation monochromator	$\Delta\lambda_{ex}$	Standard reaction entropy	$\Delta_r S^0$
		Standard reaction Gibbs energy	$\Delta_r G^0$
Spectral bandwidth of monochromator	$\Delta\lambda_m$		
Spectral radiant energy	$Q_\lambda, dQ/d\lambda$	Standard temperature and pressure	STP
Spectral radiant energy density:		Stanton number	St
In terms of frequency	ρ_v, w_v	Statistical weight	W, β, ω
In terms of wavelength	ρ_v, w_v	Statistical weight of atomic states	g
In terms of wavenumber	$\rho_{\tilde{v}} w_{\tilde{v}}$		
		Stefan-Boltzmann constant	σ
Spectral radiant energy flux	$d\phi/d\lambda$	Steradian	sr
Spectroscopic splitting factor	g	Stoichiometric number	ν
		Stokes	St
Speed	u, w	Summation sign	Σ
Speed of light:		Surface charge density	σ
In a medium	c	Surface concentration	Γ
In vacuum	c_0	Surface coverage	θ
Spherical polar coordinates	r, θ, ϕ	Surface density	ρ_A, ρ_S
		Surface electric potential	χ
Spin angular momentum	s, S	Surface pressure	π
Spin-lattice relaxation time	T_1	Surface tension	γ, σ
		Susceptance	B

TABLE 2.6 Abbreviations and Standard Letter Symbols (*Continued*)

Svedberg, unit of time	Sv	Ultrahigh frequency	uhf
Symmetrical	sym	Ultraviolet	uv
Symmetry coordinate	S	Unified atomic mass unit	u
Symmetry number	s, σ	United States Pharmaco-	USP
Tartrate	Tart	poeia	
Temperature	θ, Θ	Vacuum	vac
Temperature, thermody-	T	van der Waals constant	λ
namic		Vapor pressure	p, vp
Temperature at boiling	T_b	Velocity	u, w
point		Versus	vs
Term value spectroscopy	T	Vibrational anharmoni-	χ
Tesla	T	city constant	
Tetragonal	tetr	Vibrational coordinates:	
Thermal conductance	G	Internal coordinates	R_i, r_I, θ_j, etc.
Thermal conductivity	λ, k	Normal coordinates,	q_r
Thermal diffusivity	a	dimensionless	
Thermal resistance	R	Mass adjusted	Q_r
Thermoelectric force	E	Vibrational force con-	
Thickness of diffusion	δ	stants:	
layer		Diatomic	f
Thickness of layer	t	Polyatomic, dimen-	$\phi_{rst} \ldots; k_{rst} \ldots$
Thickness (effective) of	d_f	sionless normal	
stationary phase		coordinates	
Thickness of surface layer	τ	Internal coordinates	f_{ij}
Thickness of various	δ	Symmetry coordi-	F_{ij}
layers		nates	
Thomson coefficient	μ, τ	Vibrational quantum	v
Thomson cross section	σ_e	number	
Time	t	Vibrational term	G
Time interval, character-	T, τ	Viscosity	η, μ
istic		Vitreous substance	vit
Tonne	t, ton	Volt	V
Torr (mm of mercury)	Torr	Volt-ampere-reactive	var
Torque	$\boldsymbol{T}$	Volta potential difference	$\Delta\psi$
Total bed volume	V_{tot}	Volume	V, v
Total term (spectroscopy)	T	Volume flow rate	q_v
Transconductance	g_m	Volume fraction	ϕ
Transfer coefficient	α	Volume in space phase	Ω
Transit time of nonre-	t_M, t_0	Volume liquid phase in	V_L
tained solute		column	
Transition	tr	Volume mobile phase in	V_M
Transition dipole mo-	$\boldsymbol{M, R}$	column	
ment of a molecule		Volume of activation	$\Delta^{\ddagger}V$
Transition frequency	ν	Volume percent	vol %
Transition wavenumber	$\tilde{\nu}$	Volume per volume	v/v
Translation (circular)	$b_1; b_2; b_3$	Volume strain	θ
Translation vectors for	$a_1; a_2; a_3$	Watt	W
crystal lattice	$\boldsymbol{a}; \boldsymbol{b}; \boldsymbol{c}$	Wavefunction	ϕ, ψ, Ψ
Transmission factor	τ	Wavelength	λ
Transmittance	T, τ	Wavenumber (in a me-	σ
Transport number	t	dium)	
Transverse relaxation	T_2	Wavenumber in vacuum	$\tilde{\nu}$
time		Weber	Wb
Triclinic	tric	Weber number	We
Trigonal	trig	Weight	W
Triton (tritium nucleus)	t	Weight of liquid phase	w_L

TABLE 2.6 Abbreviations and Standard Letter Symbols (*Continued*)

Weight percent	wt %	Yard	yd
Weight per volume	w/v	Young's modulus	E
Wien displacement constant	b	Zeeman splitting constant	μ_B/hc
Work	w, W	Zone width at baseline	W_b
Work function	Φ	Zone width at one-half peak height	$W_{1/2}$
x unit	X		

TABLE 2.7 Conversion Factors

The data were compiled by L. P. Buseth for the Thirteenth Edition; some entries have been added or modified in view of recent data and SI units.

Relations which are exact are indicated by an asterisk (*). Factors in parentheses are also exact. Other factors are within ± 5 in the last significant figure.

To convert	Into	Multiply by
Abampere	ampere*	10
Abcoulomb	coulomb*	10
	statcoulomb	2.998×10^{10}
Abfarad	farad*	10^9
Abhenry	henry*	10^{-9}
Abmho	siemens*	10^9
Abvolt	volt	10^{-8}
Acre	hectare or square hectometer	0.404 685 64
	square chain (Gunter's)*	10
	square kilometer*	0.004 046 856
	square meter*	4046.856
	square mile*	(1/640)
	square rod*	160
	square yard*	4840
Acre (U.S. survey)	square meter	4046.872
Acre-foot	cubic foot*	4.3560×10^4
	cubic meter	1233.482
	gallon (U.S.)	3.259×10^5
Acre-inch	cubic foot*	3630
	cubic meter	102.7902
Amagat	centimeter per mole	0.022 414
Ampere per square centimeter	ampere per square inch*	6.4516
Ampere-hour	coulomb*	3600
	faraday	0.037 31
Ampere-turn	gilbert	1.256 637
Ampere-turn per centimeter	ampere-turn per inch	2.540
Angstrom	meter*	10^{-10}
	nanometer*	0.1
Apostilb	candela per square meter	0.318 309 9; $(1/\pi)$
	lambert*	10^{-4}
Are	acre	0.024 710 54
	square meter*	100
Assay ton	gram	29.1667
Astronomical unit	meter	$1.496\ 00 \times 10^{-11}$
	light year	$1.581\ 284 \times 10^{-5}$
Atmosphere	bar*	1.013 25

TABLE 2.7 Conversion Factors (*Continued*)

To convert	Into	Multiply by
Atmosphere (*continued*)	foot of water (at 4°C)	33.898 54
	inch of mercury (at 0°C)	29.921 26
	kilogram per square centimeter	1.033 227
	millimeter of mercury*	760
	millimeter of water (4°C)	$1.033\ 227 \times 10^4$
	newton per square meter*	$1.013\ 25 \times 10^5$
	pascal*	101 325
	pound per square inch	14.695 95
	ton per square inch	0.007 348
	torr*	760
Atomic mass unit	gram	1.6605×10^{-24}
Avogadro number	molecules per mole	$6.022\ 137 \times 10^{23}$
Bar	atmosphere	0.986 923
	dyne per square centimeter*	10^6
	kilogram per square centimeter	1.019 716
	millimeter of mercury	750.062
	millimeter of water (4°C)	$1.019\ 716 \times 10^4$
	newton per square meter	10^5
	pascal*	10^5
	pound per square inch	14.503 77
Barleycorn (British)	inch	(⅓)
Barn	square meter*	10^{-28}
Barrel (British)	gallon (British)*	36
	liter	163.659
Barrel (petroleum)	gallon (British)	34.9723
	gallon (U.S.)*	42
	liter	158.987
Barrel (U.S. dry)	bushel (U.S.)	3.281 22
	cubic foot	4.083 33
	liter	115.6271
	quart (U.S. dry)	104.9990
Barrel (U.S. liquid)	gallon (U.S.)	31.5 (variable)
	liter	119.2405
Barye	dyne per square centimeter*	1
Becquerel	curie*	2.7×10^{-11}
Biot	ampere*	10
Board foot	cubic foot	(¹⁄₁₂)
Bohr	meter	$5.291\ 77 \times 10^{-11}$
Bohr magneton	joule per tesla	$9.274\ 02 \times 10^{-24}$
Bolt (U.S. cloth)	foot*	120
	meter	36.576
Boltzmann constant	joule per degree	1.3806×10^{-23}
British thermal unit (Btu)	calorie	251.996
	cubic foot-atmosphere	0.367 717
	erg	1.0550×10^{10}
	foot-pound	778.169
	horsepower-hour (British)	$3.930\ 15 \times 10^{-4}$
	horsepower-hour (metric)	$3.984\ 66 \times 10^{-4}$
	joule	1055.06
	kilogram-calorie	0.2520
	kilogram-meter	107.5
	kilowatt-hour	$2.930\ 71 \times 10^{-4}$
	liter-atmosphere	10.4126

TABLE 2.7 Conversion Factors (*Continued*)

To convert	Into	Multiply by
Btu per cubic foot	kilocalorie per cubic meter	8.899 15
	joule per cubic meter	$3.725\ 895 \times 10^{-4}$
Btu per degree Fahrenheit	calorie per degree Celsius	453.592
	joule per degree Celsius	1899.10
Btu per (hour-square foot)	watt per square meter	3.154 59
Btu per minute	calorie per second	4.199 93
	foot-pound per second	12.9695
	horsepower (British)	0.023 580 9
	watt	17.5843
Btu per pound	calorie per gram	0.555 556
	joule per kilogram*	2.326
	watt-hour per pound	0.293 071
	watt-hour per kilogram	0.646 111
Btu per (pound-degree Fahrenheit)	calorie per (gram-degree Celsius)*	1
Btu per square foot	joule per square meter	$1.135\ 65 \times 10^4$
Bucket (British, dry)	gallon (British)*	4
Bushel (British)	bushel (U.S.)	1.032 057
	cubic foot	1.284 35
	gallon (British)*	8
	gallon (U.S.)	9.607 60
	liter	36.3687
Bushel (U.S.)	barrel (U.S., dry)	0.304 765
	bushel (British)	0.968 939
	cubic foot	1.244 456
	cubic meter	0.035 239 07
	gallon (British)	7.751 51
	gallon (U.S.)	9.309 18
	liter	35.239 07
	peck (U.S.)*	4
	pint (U.S., dry)*	64
	quart (U.S., dry)*	32
Butt	gallon (British)	108 or 126
Cable length (international)	foot	607.611 55
	meter*	185.2
	mile (nautical)*	0.1
Cable length (U.S. or British)	foot*	720
	meter	219.456
	mile (nautical)	0.118 407
	mile (statute)	0.136 364
Caliber	inch*	0.01
	millimeter*	0.254
Calorie	Btu	0.003 968 320
	foot-pound	3.088 03
	foot-poundal	99.3543
	horsepower-hour (British)	$1.559\ 61 \times 10^{-6}$
	joule*	4.184
	kilowatt-hour	1.163×10^{-6}
	liter-atmosphere	0.041 320 5
Calorie (15°C)	joule	4.1855
Calorie (international)	joule	4.1868
Calorie per minute	foot-pound per second	0.051 467 1
	horsepower (British)	$9.357\ 65 \times 10^{-5}$
	watt*	0.069 78

TABLE 2.7 Conversion Factors (*Continued*)

To convert	Into	Multiply by
Candela	Hefner unit	1.11
	lumen per steradian*	1
Candela per square centimeter	candela per square foot*	929.0304
	candela per square meter*	10^4
	lambert	3.141 593; (π)
Carat (metric)	gram*	0.2
Celsius (centigrade) temperature	Fahrenheit temperature	$(\%)°C + 32$
	kelvin	$°C - 273.15$
Cental	kilogram	45.359 237
	pound*	100
Centigrade heat unit or chu	Btu*	1.8
	calorie	453.592
	joule	1899.10
Centimeter	foot	0.032 808 4
	inch	0.393 700 8
	mil	393.7008
Centimeter per second	foot per minute	1.986 50
	kilometer per hour*	0.036
	knot	0.019 438 4
	mile per hour	0.022 369 4
Centimeter per second squared	foot per second squared	0.032 808 4
	meter per second squared*	0.01
Centimeter-dyne	erg*	1
	joule*	10^{-7}
	meter-kilogram	1.020×10^{-8}
	pound-foot	7.376×10^{-8}
Centimeter-gram	erg*	980.665
	joule*	$9.806\ 65 \times 10^{-5}$
Centipoise	kilogram per (meter-second)*	0.001
	pascal-second*	0.001
	pound per (foot-second)	0.006 72
Chain (Ramsden's)	foot*	100
	meter*	30.48
Chain (Gunter's)	foot*	66
	meter*	20.1168
Chaldron (British)	bushel (British)*	32
Circular inch	circular mil*	10^6
	square centimeter	5.067 075
	square inch	$(\pi/4)$
Circular millimeter	square millimeter	$(\pi/4)$
Circumference	degree*	360
	gon (grade)	400
	radian	(2π)
Clausius (cal_{th}/K)	joule per kelvin	4.184
Clo	(square meter $\times$ °C) per watt	0.155
Clove (British)	pound*	8
Coomb (British)	bushel (British)*	4
Cord	cord foot*	8
	cubic foot*	128
Coulomb	ampere-second*	1
Coulomb per square centimeter	coulomb per square inch*	6.4516
Cubic centimeter	cubic foot	$3.531\ 47 \times 10^{-5}$
	cubic inch	0.061 023 744
	drachm (British, fluid)	0.281 561

TABLE 2.7 Conversion Factors (*Continued*)

To convert	Into	Multiply by
Cubic centimeter (*continued*)	dram (U.S., fluid)	0.270 512 2
	gallon (British)	$2.199\ 69 \times 10^{-4}$
	gallon (U.S.)	$2.641\ 72 \times 10^{-4}$
	gill (British)	0.007 039 016
	gill (U.S.)	0.008 453 506
	liter*	0.001
	minim (British)	16.8936
	minim (U.S.)	16.230 73
	ounce (British, fluid)	0.035 195 1
	ounce (U.S., fluid)	0.033 814 02
	pint (British)	0.001 759 75
	pint (U.S., dry)	0.001 816 17
	pint (U.S., liquid)	0.002 113 376
Cubic centimeter-atmosphere	joule*	0.101 325
	watt-hour	$2.814\ 58 \times 10^{-5}$
Cubic centimeter per gram	cubic foot per pound	0.016 018 5
Cubic centimeter per second	cubic foot per minute	0.002 118 88
	liter per hour*	3.6
Cubic decimeter (dm³)	liter*	1
Cubic foot	acre-foot	$2.295\ 68 \times 10^{-5}$
	board foot*	12
	cord*	(¹⁄₁₂₈)
	cord foot*	(¹⁄₁₆)
	cubic inch*	1728
	cubic meter*	0.028 316 846 592
	cubic yard	(¹⁄₂₇)
	gallon (British)	6.228 835
	gallon (U.S.)	7.480 519
	liter	28.316 847
Cubic foot per hour	liter per minute	0.471 947
Cubic foot per pound	cubic meter per kilogram	0.062 428 0
Cubic foot × (pound per square inch)	Btu	0.185 050
	calorie	46.6317
	joule	195.238
	watt-hour	0.054 232 7
Cubic foot-atmosphere	Btu	2.719 48
	calorie	685.298
	joule	2869.205
	kilogram-meter	292.577
	liter-atmosphere	28.3168
	watt-hour	0.797 001
Cubic inch	cubic foot	(¹⁄₁₇₂₈)
	milliliter*	16.387 064
Cubic inch per minute	cubic centimeter per second	0.273 118
Cubic kilometer	cubic mile	0.239 913
Cubic meter per kilogram	cubic foot per pound	16.0185
Cubic yard	bushel (British)	21.0223
	bushel (U.S.)	21.6962
	cubic foot*	27
	cubic meter	0.764 554 86
	liter	764.555
Cubic yard per minute	cubic foot per second*	0.45
	gallon (British) per second	2.802 98

TABLE 2.7 Conversion Factors (*Continued*)

To convert	Into	Multiply by
Cubic yard per minute (*continued*)	gallon (U.S.) per second	3.366 23
	liter per second	12.742 58
Cubit	inch*	18
Cup (U.S.)	cubic centimeter	236.588
Cup (metric)	cubic centimeter*	200
Curie	becquerel*	3.7×10^{10}
Cycle per second	hertz*	1
Dalton	kilogram	$1.660\ 54 \times 10^{-27}$
	unified atomic mass*	1
Day (mean solar)	hour*	24
	minute*	1440
	second*	86 400
Debye	coulomb-meter	$3.335\ 64 \times 10^{-30}$
Decibel	neper	0.115 129 255
Degree (plane angle)	circumference	$(\frac{1}{360})$
	gon (grade)	1.111 11
	minute (angle)*	60
	quadrant	$(\frac{1}{90})$
	radian	$(\pi/180)$
	revolution	$(\frac{1}{360})$
	second (angle)*	3600
Degree (angle) per foot	radian per meter	0.057 261 5
Degree (angle) per second	radian per second	0.017 453 3
Degree Celsius	degree Fahrenheit*	1.8
	degree Rankine*	1.8
	kelvin*	1
Degree Fahrenheit	degree Celsius	$(\frac{5}{9})$
Degree Rankine	kelvin	$(\frac{5}{9})$
Denier	tex	$(\frac{1}{9})$
Dipole length (*e* cm)	coulomb-meter	$1.602\ 18 \times 10^{-21}$
Drachm (British)	dram (apothecaries or troy)*	1
Drachm (British, fluid)	cubic centimeter	3.551 633
	dram (U.S., fluid)	0.960 760
	minim (British)	60
	ounce (British, fluid)	$(\frac{1}{8})$
Dram (apothecaries or troy)	dram (weight)	2.194 285 7
	grain*	60
	gram*	3.887 934 6
	ounce (troy)*	$(\frac{1}{8})$
	pennyweight*	2.5
	pound (troy)*	$(\frac{1}{96})$
	scruple*	3
Dram (weight)	grain*	27.343 75
	gram	1.771 845 2
	ounce (weight)	$(\frac{1}{16})$
	pound (weight)	$(\frac{1}{256})$
Dram (U.S., fluid)	cubic centimeter	3.696 691 2
	gallon (U.S.)	$(\frac{1}{1024})$
	gill (U.S.)	$(\frac{1}{32})$
	milliliter	3.696 691 2
	minim (U.S.)*	60
	ounce (U.S., fluid)	$(\frac{1}{8})$
	pint (U.S., fluid)	$(\frac{1}{128})$

TABLE 2.7 Conversion Factors (*Continued*)

To convert	Into	Multiply by
Dyne	kilogram (force)	$1.019\ 716 \times 10^{-6}$
	newton*	10^{-5}
	pound (force)	$2.248\ 09 \times 10^{-6}$
Dyne per centimeter	newton per meter*	0.001
Dyne per square centimeter	bar*	10^{-6}
	kilogram per square centimeter	$1.019\ 716 \times 10^{-6}$
	millimeter of mercury	$7.500\ 617 \times 10^{-4}$
	millimeter of water	0.010 197 16
	newton per square meter*	0.1
	pascal*	0.1
	pound per square inch (psi)	$1.450\ 38 \times 10^{-5}$
Dyne-centimeter	erg*	1
	foot-pound (force)	$7.375\ 62 \times 10^{-8}$
	foot-poundal	$2.373\ 04 \times 10^{-6}$
	joule*	10^{-7}
	kilogram-meter (force)	$1.019\ 716 \times 10^{-8}$
	newton-meter*	10^{-7}
Dyne-second per square centimeter	poise*	1
	pascal-second*	0.1
Electron charge	coulomb	$1.602\ 18 \times 10^{-19}$
Electron charge-centimeter (e cm)	coulomb-meter	$1.602\ 18 \times 10^{-21}$
Electron charge-square centimeter	coulomb-meter squared	$1.602\ 18 \times 10^{-23}$
Electron mass	atomic mass unit	0.000 548 6
	gram	9.1096×10^{-28}
Electronvolt	erg	$1.602\ 18 \times 10^{-12}$
	joule	$1.602\ 18 \times 10^{-19}$
	kilojoule per mole	96.488
Ell	inch*	45
Em, pica	inch	0.167
	millimeter	4.217 52
Erg	dyne-centimeter*	1
	joule*	10^{-7}
	watt-hour	$2.777\ 78 \times 10^{-11}$
Erg per second	Btu	5.69×10^{-6}
	watt*	10^{-7}
Erg per ($cm^2 \times$ second)	watt per square meter*	0.001
Erg per gauss	ampere-centimeter squared*	10
	joule per tesla*	0.001
Fahrenheit scale	centigrade scale	(5⁄9)
Fahrenheit temperature (°F)	Celsius temperature (°C)	(°F − 32)(5⁄9)
Faraday constant	coulomb per equivalent	96 485.31
Fathom	foot*	6
Fermi	meter*	10^{-15}
Firkin (British, U.S.)	gallon (British, U.S.)	9
Foot	centimeter*	30.48
	inch*	12
	mile (nautical)	$1.645\ 788 \times 10^{-4}$
	mile (statute)	$1.893\ 939 \times 10^{-4}$
	yard	(1⁄3)
Foot of water (4°C)	atmosphere	0.029 499 8
	bar	0.029 499 8
	gram per square centimeter	30.48
	inch of mercury (0°C)	0.882 671
	pascal	2989.067

TABLE 2.7   Conversion Factors (*Continued*)

To convert	Into	Multiply by
Foot per minute	centimeter per second*	0.508
	knot	0.009 874 73
	mile per hour	0.011 363 6
Foot-candle	lumen per square foot*	1
	lumen per square meter	10.7639
	lux	10.7639
Foot-lambert	candela per square centimeter	$3.426\ 26 \times 10^{-4}$
	candela per square foot	$(1/\pi)$
	lambert	0.001 076 39
	meter-lambert	10.7639
Foot-pound	Btu	0.001 285 07
	calorie	0.323 832
	foot-poundal	32.1740
	horsepower (British)	$5.050\ 51 \times 10^{-7}$
	joule	1.355 818
	kilogram-meter	0.138 255
	liter-atmosphere	0.013 380 9
	newton-meter	1.355 818
	watt-hour	$3.766\ 161 \times 10^{-4}$
Foot-pound per minute	horsepower (British)	$3.030\ 30 \times 10^{-5}$
	horsepower (metric)	$3.072\ 33 \times 10^{-5}$
	watt	0.022 597 0
Foot-poundal	Btu	$3.994\ 11 \times 10^{-5}$
	calorie	0.010 064 99
	foot-pound	0.031 081 0
	joule	0.042 140 11
	kilogram-meter	0.004 297 10
	liter-atmosphere	$4.158\ 91 \times 10^{-4}$
	watt-hour	$1.170\ 56 \times 10^{-5}$
Franklin	coulomb	$3.335\ 64 \times 10^{-10}$
Franklin per cm³	coulomb per cubic meter	$3.335\ 64 \times 10^{-4}$
Franklin per cm²	coulomb per square meter	$3.335\ 64 \times 10^{-6}$
Furlong	chain (Gunter's)*	10
	foot*	600
	meter*	201.168
	mile	(⅛)
Gal, galileo	meter per second squared*	0.01
Gallon (British, imperial)	bushel (British)	(⅛)
	cubic decimeter, liter*	4.546 90
	cubic foot	0.160 544
	gallon (U.S., fluid)	1.200 95
	gill (British)*	32
	liter	4.546 09
	ounce (British)*	160
	quart (British)*	4
Gallon (U.S.)	barrel (petroleum)	(1/42)
	cubic decimeter, liter	3.785 41
	cubic foot	0.133 680 56
	gallon (British)	0.832 674
	liter	3.785 41
	ounce (U.S., fluid)*	128
	quart (U.S., fluid)*	4
Gallon (U.S.) per minute	cubic foot per hour	8.020 83
	cubic meter per hour	0.227 125

TABLE 2.7 Conversion Factors (*Continued*)

To convert	Into	Multiply by
Gallon (U.S.) per minute (*continued*)	liter per minute	3.785 412
Gamma	microgram*	1
Gas constant	calorie per mole-degree	1.987
	joule per mole-degree	8.3143
	liter-atmosphere per mole-degree	0.082 057
Gauss	tesla*	10^{-4}
	weber per square meter*	10^{-4}
Geepound	slug*	1
Gilbert	ampere-turn	0.795 775
Gill (British)	cubic centimeter, mL	142.065
	cubic inch	8.669 36
	gallon (British)	(1/32)
	gill (U.S.)	1.200 95
	ounce (British, fluid)*	5
	pint (British)	(1/4)
Gill (U.S.)	cubic centimeter, mL	118.2941
	gallon (U.S.)	(1/32)
	liter	0.118 294 1
	ounce (U.S., fluid)*	4
	quart (U.S.)	(1/8)
Gon (grade)	circumference	(1/400)
	minute (angle)*	54
	radian	($2\pi/400$)
Grade	radian	($2\pi/400$)
Grain	carat (metric)*	0.323 994 55
	milligram*	64.798 91
	ounce (weight)	0.002 285 714 3
	ounce (troy)	(1/480)
	pennyweight	(1/24)
	pound	(1/7000)
	scruple	(1/20)
Gram	carat (metric)*	5
	dram	0.564 383 39
	grain	15.432 358
	ounce (weight)	0.035 273 962
	ounce (troy)	0.032 150 747
	pennyweight	0.643 014 93
	pound	0.002 204 622 6
	ton (metric)*	10^{-6}
Gram per (centimeter-second)	poise*	1
Gram per cubic centimeter	kilogram per liter*	1
	pound per cubic foot	62.4280
	pound per gallon (U.S.)	8.345 40
Gram per square meter	ounce per square foot	0.327 706
Gram per ton (long)	gram per ton (metric)	0.984 207
	gram per ton (short)	0.892 857
Gram (force)	dyne*	980.665
	newton*	0.009 806 65
Gram per square centimeter	pascal*	98.0665
Gram-centimeter	joule*	$9.806\ 65 \times 10^{-5}$
Gram-square centimeter	pound-square foot	$2.373\ 04 \times 10^{-6}$
Gray	joule per kilogram*	1

TABLE 2.7 Conversion Factors (*Continued*)

To convert	Into	Multiply by
Hand	inch*	4
Hartree	joule	$2.179\ 87 \times 10^{-18}$
Hectare	acre	2.471 054
	are*	100
Hefner unit	candela	0.9
Hemisphere	sphere*	0.5
	spherical right angle*	4
	steradian	(2π)
Hertz	cycle per second*	1
Hogshead	gallon (U.S.)*	63
Horsepower (British)	Btu per hour	2544.43
	foot pound per hour*	1.98×10^6
	horsepower (metric)	1.013 87
	joule per second	745.700
	kilocalorie per hour	641.186
	kilogram-meter per second	76.0402
	watt	745.7
Horsepower-hour (British)	Btu	2544.43
	foot-pound*	1.98×10^6
	joule	$2.684\ 52 \times 10^6$
	kilocalorie	641.186
	kilogram-meter	$2.737\ 45 \times 10^5$
	watt-hour	745.7
Hour (mean solar)	day	$(\frac{1}{24})$
	minute*	60
	second*	3600
	week	$(\frac{1}{168})$
Hundredweight (long)	kilogram*	50.802 345 44
	pound*	112
	ton (long)	$(\frac{1}{20})$
	ton (metric)	0.050 802 345
	ton (short)*	0.056
Hundredweight (short)	hundredweight (long)	0.892 857
Inch	centimeter*	2.54
	foot	$(\frac{1}{12})$
	mil*	1000
Inch of mercury (0°C)	atmosphere	0.033 421 05
	inch of water (4°C)	13.5941
	millibar	33.863 88
	millimeter of water (4°C)	345.316
	pascal	3386.388
	pound per square inch (psi)	0.491 1541
Inch of water (4°C)	inch of mercury (0°C)	0.073 5559
	millibar	2.490 89
	millimeter of mercury (0°C)	1.868 32
	pascal	249.089
	pound per square inch (psi)	0.036 1273
Inch per minute	foot per hour*	5
	meter per hour*	1.524
	millimeter per second	0.423 333
Joule	Btu	$9.478\ 170 \times 10^{-4}$
	calorie*	0.2390
	centigrade heat unit (chu)	5.265 65

TABLE 2.7 Conversion Factors (*Continued*)

To convert	Into	Multiply by
Joule (*continued*)	centimeter-dyne*	10^7
	cubic foot-atmosphere	0.000 348 529
	cubic foot-(pound per square inch)	0.005 121 959
	erg*	10^7
	foot-pound	0.737 562
	foot-poundal	23.7304
	horsepower-hour (British)	$3.725\ 06 \times 10^{-7}$
	liter-atmosphere	0.009 869 233
	newton-meter*	1
	watt-second*	1
Joule per centimeter	kilogram (force)	10.197 16
	newton*	100
	pound (force)	22.4809
Joule per gram	Btu per pound	0.429 923
	kilocalorie per kilogram	0.238 846
	watt-hour per pound	0.125 998
Joule per second	watt*	1
Kilderkin (British)	gallon (British)	18
Kilogram (force)	dyne*	$9.806\ 65 \times 10^5$
	newton*	9.806 65
	pound (force)	2.204 62
	poundal	70.9316
Kilometer	astronomical unit	$6.684\ 59 \times 10^{-9}$
	mile (nautical)	0.539 956 80
	mile (statute)	0.621 371 192
Kilowatt	Btu per minute	56.8690
	foot-pound per second	737.562
	horsepower (British)	1.341 02
	horsepower (metric)	1.359 62
	joule per second*	1000
	kilocalorie per hour	859.845
Kilowatt-hour	Btu	3412.14
	horsepower-hour (British)	1.341 02
	joule*	3.6×10^6
	kilocalorie	859.845
Kip	pound (force)	1000
Knot	foot per minute	101.2686
	kilometer per hour*	1.852
	mile (nautical) per hour*	1
	mile (statute) per hour	1.150 78
Lambda	decimeter cubed*	10^{-6}
	microliter*	1
Lambert	candela per square centimeter	$(1/\pi)$
	candela per square inch	2.053 61
	foot-lambert	929.030
Langley	joule per square meter*	4.184×10^4
League (nautical)	mile (nautical)*	3
League (statute)	mile (statute)*	3
Light year	astronomical unit	$6.323\ 97 \times 10^4$
Line	maxwell*	1
Link	chain*	0.01
Liter	cubic decimeter (dm^3)*	1

TABLE 2.7 Conversion Factors (*Continued*)

To convert	Into	Multiply by
Liter (*continued*)	cubic foot	0.035 314 67
	gallon (British)	0.219 969
	gallon (U.S.)	0.264 172 1
	quart (British)	0.879 877
	quart (U.S.)	1.056 688
Liter per minute	cubic foot per hour	2.118 88
	gallon (British) per hour	13.198
	gallon (U.S.) per hour	15.8503
Liter-atmosphere	Btu	0.096 037 6
	calorie	24.2011
	cubic foot-atmosphere	0.035 314 7
	cubic foot-pound per square inch	0.518 983
	horsepower (British)	$3.774\,42 \times 10^{-5}$
	horsepower (metric)	$3.826\,77 \times 10^{-5}$
	joule*	101.325
	kilogram-meter	10.332 27
	watt-hour	0.028 145 8
Lumen per square centimeter	lux*	10^4
	phot*	1
Lumen per square meter	lumen per square foot	0.092 903 0
Lux	lumen per square meter*	1
Maxwell	weber*	10^{-8}
Meter	angstrom*	10^{10}
	fathom	0.546 807
	foot	3.280 839 895
	inch	39.370 078 740
	mile (nautical)	$5.399\,568 \times 10^{-4}$
	mile (statute)	$6.213\,712 \times 10^{-4}$
Meter per second	foot per minute	196.850
	kilometer per hour*	3.6
	knot	1.943 844
	mile per hour	2.236 936
Meter-candle	lux*	1
Meter-lambert	candela per square meter	$(1/\pi)$
	foot-lambert	0.092 903 0
	lambert*	10^{-4}
Mho (ohm^{-1})	siemen*	1
Micron	meter	10^{-6}
Mil	inch*	0.001
	micrometer*	25.4
Mile (nautical)	foot	6076.115 49
	kilometer*	1.852
	mile (statute)	1.150 78
Mile (statute)	chain (Gunter's)*	80
	chain (Ramsden's)*	52.8
	foot*	5280
	furlong*	8
	kilometer*	1.609 344
	light year	$1.701\,11 \times 10^{-11}$
	link (Gunter's)*	8000
	link (Ramsden's)*	5280
	mile (nautical)	0.868 976

TABLE 2.7 Conversion Factors (*Continued*)

To convert	Into	Multiply by
Mile (statute) (*continued*)	rod*	320
Mile per gallon (British)	kilometer per liter	0.354 006
Mile per gallon (U.S.)	kilometer per liter	0.425 144
Mile per hour	foot per minute	88
	kilometer per hour*	1.609 344
	knot	0.868 976
Milliliter	cubic centimeter*	1
Millimeter of mercury (0°C)	atmosphere	(1/760)
	dyne per square centimeter	1333.224
	millimeter of water (4°C)	13.5951
	pascal	133.322
	pound per square inch (psi)	0.019 336 8
	torr*	1
Millimeter of water (4°C)	atmosphere	0.009 678 41
	millibar*	0.098 066 5
	millimeter of mercury (0°C)	0.073 555 9
	pascal*	9.806 65
	pound per square inch	0.001 422 33
Minim (British)	milliliter	0.059 193 9
	minim (U.S.)	0.960 760
Minim (U.S.)	milliliter	0.061 611 5
Minute (plane angle)	circumference	$4.629\ 63 \times 10^{-5}$
	degree (angle)	(1/60)
	gon	(1/54)
	radian	($\pi/10\ 800$)
Minute	hour	(1/60)
	second	60
Month (mean of 4-year period)	day	30.4375
	hour	730.5
	week	4.348 21
Nail (British)	inch*	2.25
Nanometer	angstrom*	10
Neper	decibel	8.685 890
Neutron mass	atomic mass unit	1.008 66
	gram	1.6749×10^{-24}
Newton	dyne*	10^5
	kilogram (force)	0.101 971 6
	pound (force)	0.224 809
	poundal	7.233 01
Newton per square meter	*See* pascal	
Newton-meter	foot-pound	0.737 562
	joule*	1
	kilogram-meter	0.101 971 6
	watt-second*	1
Nit	candela per square meter*	1
Noggin (British)	gill (British)*	1
Nox	lux*	0.001
Nuclear magneton	joule per tesla	$5.050\ 79 \times 10^{-27}$
Oersted	ampere per meter (in practice)	($1000/4\pi$)
Ohm (mean international)	ohm	1.000 49
Ohm (U.S. international)	ohm	1.000 495
Ohm per foot	ohm per meter	3.280 84
Ounce (avoirdupois)	dram*	16
	grain*	437.5

TABLE 2.7 Conversion Factors (*Continued*)

To convert	Into	Multiply by
Ounce (avoirdupois) (*continued*)	gram*	28.3495
	ounce (troy)	0.911 458 33
	pound	(1/16)
Ounce (troy)	grain*	480
	gram*	31.1035
	ounce (avoirdupois)	1.097 142 9
	pennyweight*	20
	pound (avoirdupois)	0.068 571 429
	scruple*	24
Ounce (British, fluid)	cubic centimeter	28.413 06
	gallon (British)	(1/160)
	milliliter	28.413 06
	minim (British)	480
	ounce (U.S., fluid)	0.960 760
	pint (British)	(1/20)
	quart (British)	(1/40)
Ounce (U.S., fluid)	cubic centimeter	29.573 530
	gallon (U.S.)	(1/128)
	milliliter	29.573 530
	pint (U.S., fluid)	(1/16)
	quart (U.S., fluid)	(1/32)
Ounce (avoirdupois) per cubic foot	kilogram per cubic meter	1.001 154
Ounce (avoirdupois) per gallon (U.S.)	gram per liter	7.489 15
Ounce (avoirdupois) per ton (long)	gram per ton (metric)	27.9018
	milligram per kilogram	27.9018
Ounce (avoirdupois) per ton (short)	gram per ton (metric)*	31.25
	milligram per kilogram*	31.25
Pace	foot*	2.5
Palm (British)	inch*	3
Parsec	light year	3.261 636
Part per million	milligram per kilogram*	1
	milliliter per cubic meter*	1
Pascal	atmosphere	$9.869\ 233 \times 10^{-6}$
	bar*	10^{-5}
	dyne per square centimeter*	10
	inch of mercury	$2.953\ 00 \times 10^{-4}$
	millimeter of mercury	$7.500\ 62 \times 10^{-3}$
	millimeter of water	0.101 972
	newton per square meter*	1
	pound per square inch	$1.450\ 38 \times 10^{-4}$
	poundal per square foot	0.671 969
Pascal-second	poise*	10
Peck (British)	gallon (British)*	2
Peck (U.S.)	bushel (U.S.)*	0.25
Pennyweight	grain*	24
	gram*	1.555 173 84
	ounce (troy)	(1/20)
	pound	0.003 428 571 4
Perch	foot*	16.5
Phot	lux*	10^4
Pica (printer's)	inch	0.167
	point*	12
Pint (British)	gallon (British)	(1/8)
	liter	0.568 261

TABLE 2.7 Conversion Factors (*Continued*)

To convert	Into	Multiply by
Pint (British) (*continued*)	pint (U.S., fluid)	1.200 95
	quart (British)	0.5
Pint (U.S., dry)	bushel (U.S.)	(1/64)
	liter	0.550 610 5
	peck (U.S.)	(1/16)
	pint (British)	0.968 939
	quart (U.S., dry)	0.5
Pint (U.S., fluid)	gallon (U.S.)	(1/8)
	liter	0.473 176 5
	pint (British)	0.832 674
	quart (U.S., fluid)*	0.5
Planck's constant	joule-second	$6.626\ 08 \times 10^{-34}$
Point (printer's, Didot)	millimeter	0.376 065 03
Point (printer's, U.S.)	millimeter*	0.351 459 8
Poise	dyne-second per square centimeter*	1
	pascal-second*	0.1
Polarizability volume ($4\pi\epsilon_0$ cm³)	coulomb squared-(meter squared per joule)	$1.112\ 65 \times 10^{-16}$
Pole (British)	foot*	16.5
Pottle (British)	gallon (British)*	0.5
Pound	gram*	453.592 37
	ounce (weight)*	16
	ton (long)	$4.464\ 285\ 7 \times 10^{-4}$
	ton (short)	(1/2000)
Pound (troy)	grain	5760
	gram*	373.241 721 6
	ounce (troy)*	12
	pennyweight	240
	pound (weight)	0.822 857 14
	scruple*	288
Pound per cubic foot	kilogram per cubic meter	16.018 46
Pound per cubic inch	gram per cubic centimeter	27.679 905
	pound per cubic foot*	1728
Pound per foot	kilogram per meter	1.488 16
Pound per (foot-second)	pascal-second	1.488 16
Pound per gallon (U.S.)	gram per liter	119.8264
Pound per hour	kilogram per day	10.886 22
Pound per inch	kilogram per meter	17.857 97
Pound per minute	kilogram per hour	27.215 54
Pound per square foot	kilogram per square meter	4.882 43
Pound (force)	kilogram (force)	0.453 592
	newton	4.448 222
	poundal	32.1740
Pound per square inch	atmosphere	0.068 046 0
	bar	0.068 948 0
	inch of mercury (0°C)	2.036 02
	millimeter of mercury (0°C)	51.7149
	millimeter of water (4°C)	703.070
	pascal	6894.757
	pound per square foot	144
Pound-second per square inch	pascal-second	6894.76
Poundal	gram (force)	14.0981
	newton	0.138 255

TABLE 2.7 Conversion Factors (*Continued*)

To convert	Into	Multiply by
Poundal (*continued*)	pound (force)	0.031 081 0
Poundal per square foot	pascal	1.488 164
Poundal-foot	newton-meter	0.042 140 1
Poundal-second per square foot	pascal-second	1.488 164
Proof (U.S.)	percent alcohol by volume*	0.5
Proton mass	atomic mass unit	1.007 28
	gram	1.6726×10^{-24}
Puncheon (British)	gallon (British)	70
Quadrant	circumference*	0.25
	degree (angle)*	90
	gon (grade)*	100
	minute (angle)*	5400
	radian	$(\pi/2)$
Quadrupole area (*e* cm²)	coulomb meter squared	$1.602\ 18 \times 10^{-23}$
Quart (British)	gallon (British)*	0.25
	liter	1.136 523
	ounce (British, fluid)*	40
	pint (British)*	2
	quart (U.S., fluid)	1.200 95
Quart (U.S., dry)	bushel (U.S.)	(1/32)
	cubic foot	0.038 889 25
	liter	1.101 221
	peck (U.S.)	(1/8)
	pint (U.S., dry)*	2
Quart (U.S., fluid)	gallon (U.S.)*	0.25
	liter	0.946 353
	ounce (U.S., fluid)*	32
	pint (U.S., fluid)	2
	quart (British)	0.832 674
Quartern (British, fluid)	gill (British)*	0.5
Quintal (metric)	kilogram*	100
Rad	gray*	0.01
	joule per kilogram*	0.01
Radian	circumference	$(1/2\pi)$
	degree (angle)	57.295 780
	minute (angle)	3437.75
	quadrant	$(2/\pi)$
	revolution	$(1/2\pi)$
Radian per centimeter	degree per millimeter	5.729 58
	degree per inch	145.531
Radian per second	revolution per minute	9.549 30
Radian per second squared	revolution per minute squared	572.958
Rankin (degree)	kelvin	(5/9)
Ream	quire*	20
	sheet	480 or 500
Register ton	cubic foot*	100
	cubic meter	2.831 685
Rem	sievert*	0.01
Revolution	degree (angle)	360
	gon*	400
	quadrant*	4
	radian	(2π)
Revolution per minute	degree (angle) per second*	6
	radian per second	0.104 720
Revolution per minute squared	radian per second squared	0.001 745 33

TABLE 2.7   Conversion Factors (*Continued*)

To convert	Into	Multiply by
Revolution per second squared	radian per second squared	6.283 185
	revolution per minute squared	3600
Reyn	pascal-second	6894.76
	pound-second per square inch	1
Rhe	per pascal-second*	10
Right angle	degree*	90
	radian	$(\pi/2)$
Rod (British, volume)	cubic foot*	1000
Rod (surveyor's measure)	chain (Gunter's)*	0.25
	foot*	16.5
	link (Gunter's)*	25
	meter*	5.0292
Roentgen	coulomb per kilogram	2.58×10^{-4}
Rood (British)	acre*	0.25
	square meter	1011.7141
Rope (British)	foot*	20
Rydberg	joule	$2.179\ 87 \times 10^{-18}$
Sack (British)	bushel (British)	3
Scruple	dram (troy)	$(\frac{1}{3})$
	grain*	20
	gram*	1.295 978 2
	ounce (weight)	0.045 714 286
	ounce (troy)	$(\frac{1}{24})$
	pennyweight	$(\frac{10}{12})$
	pound	$(\frac{1}{350})$
Second (plane angle)	degree	$2.777\ 78 \times 10^{-4}$
	minute	$(\frac{1}{60})$
	radian	$(\pi/648\ 000)$
Section	square mile*	1
Shake	second*	10^{-8}
Siemens	mho (ohm^{-1})*	1
Sign	degree*	30
Slug	geepound*	1
	kilogram	14.593 90
	pound	32.1740
Span	inch*	9
Sphere	steradian	(4π)
Square centimeter	circular mil	$1.973\ 53 \times 10^5$
	circular millimeter	127.3240
	square inch	0.155 000 31
Square chain (Gunter's)	acre*	0.1
	square foot*	4356
	square meter	404.686
Square chain (Ramsden's)	square foot*	10^4
Square degree (angle)	steradian	$3.046\ 17 \times 10^{-4}$
Square foot	acre	$2.295\ 68 \times 10^{-5}$
	square centimeter	929.0304
	square meter	0.092 903 04
	square rod	0.003 673 09
Square inch	circular mil	$1.273\ 240 \times 10^6$
	circular millimeter	821.4432
	square centimeter	6.4516
Square kilometer	acre	247.1054

TABLE 2.7 Conversion Factors (*Continued*)

To convert	Into	Multiply by
Square kilometer (*continued*)	hectare*	100
	square mile	0.386 102 16
Square link (Gunter's)	square foot*	0.4356
Square link (Ramsden's)	square foot*	1
Square meter	are*	0.01
	square foot	10.763 91
	square mile	$3.861\ 01 \times 10^{-7}$
	square rod	0.039 536 9
	square yard	1.195 990
Square mile	acre*	640
	square kilometer	2.589 988 110
	township	(1/36)
Square rod	acre	(1/160)
	square foot	272.25
	square meter	25.292 853
Square yard	square foot*	9
	square inch*	1296
	square meter*	0.836 127 36
	square rod	0.033 057 85
Statampere	ampere	$3.335\ 64 \times 10^{-10}$
Statcoulomb	coulomb	$3.335\ 65 \times 10^{-10}$
Statfarad	farad	$1.112\ 650 \times 10^{-12}$
Stathenry	henry	$8.987\ 55 \times 10^{11}$
Statmho	siemens	$1.112\ 650 \times 10^{-12}$
Statohm	ohm	$8.987\ 55 \times 10^{11}$
Statvolt	volt	299.7925
Statweber	weber	299.7925
Steradian	sphere	$(1/4\pi)$
	spherical right angle	$(2/\pi)$
	square degree	3282.81
Stere	cubic meter*	1
Stilb	candela per square centimeter	1
Stokes	square meter per second*	10^{-4}
Stone (British)	pound*	14
Svedberg	second*	10^{-13}
Tablespoon (metric)	cubic centimeter*	15
Teaspoon (metric)	cubic centimeter*	5
Tesla	weber per square meter*	1
Tex	denier*	9
	gram per kilometer*	1
Therm	Btu*	10^5
Thou	mil*	1
Ton (assay)	gram	29.166 67
Ton (long)	hundredweight (long)*	20
	hundredweight (short)*	22.4
	kilogram	1016.046 908 8
	pound*	2240
	ton (metric)	1.016 046 9
	ton (short)	1.12
Ton (metric)	hundredweight (long)	19.684 131
	hundredweight (short)	22.046 226
	kilogram*	1000
	pound	2204.6226

TABLE 2.7   Conversion Factors (*Continued*)

To convert	Into	Multiply by
Ton (metric) (*continued*)	ton (long)	0.984 206 53
	ton (short)*	1.102 311 3
Ton (short)	kilogram	907.184 74
	pound*	2000
Ton (force, long)	newton	1186.553
Ton (force, metric)	newton	9806.65
Ton (force, short)	newton	8896.44
Ton (force, long) per square foot	bar	1.072 518
	pascal	$1.072\ 518 \times 10^5$
Ton (force, metric) per square meter	bar	0.098 066 5
	pascal	9806.65
Ton (force, short) per square foot	bar	0.957 605
	pascal	$9.576\ 05 \times 10^4$
Tonne (metric)	kilogram*	1000
Torr	atmosphere	($\frac{1}{760}$)
	millibar	1.333 224
	millimeter of mercury*	1
	pascal	133.322 ($^{101\ 325}/_{760}$)
Township (U.S.)	square kilometer	93.2396
	square mile*	36
Unified atomic mass unit	kilogram	$1.660\ 54 \times 10^{-27}$
Unit pole	weber	$1.256\ 637 \times 10^{-7}$
Volt (mean international)	volt	1.000 34
Volt (U.S. international)	volt	1.000 330
Volt-second	weber*	1
Watt	Btu per hour	3.412 14
	calorie per minute	14.3308
	erg per second*	10^7
	foot-pound per minute	44.2537
	horsepower (British)	0.001 341 02
	horsepower (metric)	0.001 359 62
	joule per second*	1
	kilogram-meter per second	0.101 972
Watt per square inch	watt per square meter	1550.003
Watt-hour	Btu	3.412 14
	calorie	859.845
	foot-pound	2655.22
	horsepower-hour (British)	0.001 341 02
	horsepower-hour (metric)	0.001 359 62
	joule*	3600
	liter-atmosphere	35.5292
Watt-second	joule*	1
Weber	maxwell*	10^8
Week	day*	7
	hour*	168
Wey (British, capacity)	bushel (British)	40 (variable)
Wey (British, mass)	pound	252 (variable)
X unit	meter	$1.002\ 02 \times 10^{-13}$
Yard	fathom*	0.5
	meter	0.9144
Year (mean of 4 years)	day	365.25
	week	52.178 87
Year (sidereal)	day (mean solar)	365.256 36

TABLE 2.8 Temperature Conversion Table

The column of figures in bold and which is headed "Reading in °F. or °C. to be converted" refers to the temperature either in degrees Fahrenheit or Celsius which it is desired to convert into the other scale. If converting from Fahrenheit degrees to Celsius degrees, the equivalent temperature will be found in the column headed "°C."; while if converting from degrees Celsius to degrees Fahrenheit, the equivalent temperature will be found in the column headed "°F." This arrangement is very similar to that of Sauveur and Boylston, copyrighted 1920, and is published with their permission.

°F.	Reading in °F. or °C. to be converted	°C.	°F.	Reading in °F. or °C. to be converted	°C.
....	−458	−272.22	....	−358	−216.67
....	−456	−271.11	....	−356	−215.56
....	−454	−270.00	....	−354	−214.44
....	−452	−268.89	....	−352	−213.33
....	−450	−267.78	....	−350	−212.22
....	−448	−266.67	....	−348	−211.11
....	−446	−265.56	....	−346	−210.00
....	−444	−264.44	....	−344	−208.89
....	−442	−263.33	....	−342	−207.78
....	−440	−262.22	....	−340	−206.67
....	−438	−261.11	....	−338	−205.56
....	−436	−260.00	....	−336	−204.44
....	−434	−258.89	....	−334	−203.33
....	−432	−257.78	....	−332	−202.22
....	−430	−256.67	....	−330	−201.11
....	−428	−255.56	....	−328	−200.00
....	−426	−254.44	....	−326	−198.89
....	−424	−253.33	......	−324	−197.78
....	−422	−252.22	....	−322	−196.67
....	−420	−251.11	....	−320	−195.56
....	−418	−250.00	....	−318	−194.44
....	−416	−248.89	....	−316	−193.33
....	−414	−247.78	....	−314	−192.22
....	−412	−246.67	....	−312	−191.11
....	−410	−245.56	....	−310	−190.00
....	−408	−244.44	....	−308	−188.89
....	−406	−243.33	....	−306	−187.78
....	−404	−242.22	....	−304	−186.67
....	−402	−241.11	....	−302	−185.56
....	−400	−240.00	....	−300	−184.44
....	−398	−238.89	....	−298	−183.33
....	−396	−237.78	....	−296	−182.22
....	−394	−236.67	....	−294	−181.11
....	−392	−235.56	....	−292	−180.00
....	−390	−234.44	....	−290	−178.89
....	−388	−233.33	....	−288	−177.78
....	−386	−232.22	....	−286	−176.67
....	−384	−231.11	....	−284	−175.56
....	−382	−230.00	....	−282	−174.44
....	−380	−228.89	....	−280	−173.33
....	−378	−227.78	....	−278	−172.22
....	−376	−226.67	....	−276	−171.11
....	−374	−225.56	....	−274	−170.00
....	−372	−224.44	−457.6	−272	−168.89
....	−370	−223.33	−454.0	−270	−167.78
....	−368	−222.22	−450.4	−268	−166.67
....	−366	−221.11	−446.8	−266	−165.56
....	−364	−220.00	−443.2	−264	−164.44
....	−362	−218.89	−439.6	−262	−163.33
....	−360	−217.78	−436.0	−260	−162.22

TABLE 2.8 Temperature Conversion Table (*Continued*)

°F.	Reading in °F. or °C. to be converted	°C.	°F.	Reading in °F. or °C. to be converted	°C.
−432.4	−258	−161.11	−216.4	−138	−94.44
−428.8	−256	−160.00	−212.8	−136	−93.33
−425.2	−254	−158.89	−209.2	−134	−92.22
−421.6	−252	−157.78	−205.6	−132	−91.11
−418.0	−250	−156.67	−202.0	−130	−90.00
−414.4	−248	−155.56	−198.4	−128	−88.89
−410.8	−246	−154.44	−194.8	−126	−87.78
−407.2	−244	−153.33	−191.2	−124	−86.67
−403.6	−242	−152.22	−187.6	−122	−85.56
−400.0	−240	−151.11	−184.0	−120	−84.44
−396.4	−238	−150.00	−180.4	−118	−83.33
−392.8	−236	−148.89	−176.8	−116	−82.22
−389.2	−234	−147.78	−173.2	−114	−81.11
−385.6	−232	−146.67	−169.6	−112	−80.00
−382.0	−230	−145.56	−166.0	−110	−78.89
−378.4	−228	−144.44	−162.4	−108	−77.78
−374.8	−226	−143.33	−158.8	−106	−76.67
−371.2	−224	−142.22	−155.2	−104	−75.56
−367.6	−222	−141.11	−151.6	−102	−74.44
−364.0	−220	−140.00	−148.0	−100	−73.33
−360.4	−218	−138.89	−144.4	−98	−72.22
−356.8	−216	−137.78	−140.8	−96	−71.11
−353.2	−214	−136.67	−137.2	−94	−70.00
−349.6	−212	−135.56	−133.6	−92	−68.89
−346.0	−210	−134.44	−130.0	−90	−67.78
−342.4	−208	−133.33	−126.4	−88	−66.67
−338.8	−206	−132.22	−122.8	−86	−65.56
−335.2	−204	−131.11	−119.2	−84	−64.44
−331.6	−202	−130.00	−115.6	−82	−63.33
−328.0	−200	−128.89	−112.0	−80	−62.22
−324.4	−198	−127.78	−108.4	−78	−61.11
−320.8	−196	−126.67	−104.8	−76	−60.00
−317.2	−194	−125.56	−101.2	−74	−58.89
−313.6	−192	−124.44	−97.6	−72	−57.78
−310.0	−190	−123.33	−94.0	−70	−56.67
−306.4	−188	−122.22	−90.4	−68	−55.56
−302.8	−186	−121.11	−86.8	−66	−54.44
−299.2	−184	−120.00	−83.2	−64	−53.33
−295.6	−182	−118.89	~79.6	−62	−52.22
~292.0	−180	−117.78	−76.0	−60	−51.11
−288.4	−178	−116.67	−72.4	−58	−50.00
−284.8	−176	−115.56	−68.8	−56	−48.89
−281.2	−174	−114.44	−65.2	−54	−47.78
−277.6	−172	−113.33	−61.6	−52	−46.67
−274.0	−170	−112.22	−58.0	−50	−45.56
−270.4	−168	−111.11	−54.4	−48	−44.44
−266.8	−166	−110.00	−50.8	−46	−43.33
−263.2	−164	−108.89	−47.2	−44	−42.22
−259.6	−162	−107.78	−43.6	−42	−41.11
−256.0	−160	−106.67	−40.0	−40	−40.00
−252.4	−158	−105.56	−36.4	−38	−38.89
−248.8	−156	−104.44	−32.8	−36	−37.78
−245.2	−154	−103.33	−29.2	−34	−36.67
−241.6	−152	−102.22	−25.6	−32	−35.56
−238.0	−150	−101.11	−22.0	−30	−34.44
−234.4	−148	−100.00	−18.4	−28	−33.33
−230.8	−146	−98.89	−14.8	−26	−32.22
−227.2	−144	−97.78	−11.2	−24	−31.11
−223.6	−142	−96.67	−7.6	−22	−30.00
−220.0	−140	−95.56	−4.0	−20	−28.89

TABLE 2.8 Temperature Conversion Table (*Continued*)

°F.	Reading in °F. or °C. to be converted	°C.	°F.	Reading in °F. or °C. to be converted	°C.
−0.4	−18	−27.78	+116.6	+47	+8.33
+3.2	−16	−26.67	+118.4	+48	+8.89
+6.8	−14	−25.56	+120.2	+49	+9.44
+10.4	−12	−24.44	+122.0	+50	+10.00
+14.0	−10	−23.33	+123.8	+51	+10.56
+17.6	−8	−22.22	+125.6	+52	+11.11
+19.4	−7	−21.67	+127.4	+53	+11.67
+21.2	−6	−21.11	+129.2	+54	+12.22
+23.0	−5	−20.56	+131.0	+55	+12.78
+24.8	−4	−20.00	+132.8	+56	+13.33
+26.6	−3	−19.44	+134.6	+57	+13.89
+28.4	−2	−18.89	+136.4	+58	+14.44
+30.2	−1	−18.33	+138.2	+59	+15.00
+32.0	±0	−17.78	+140.0	+60	+15.56
+33.8	+1	−17.22	+141.8	+61	+16.11
+35.6	+2	−16.67	+143.6	+62	+16.67
+37.4	+3	−16.11	+145.4	+63	+17.22
+39.2	+4	−15.56	+147.2	+64	+17.78
+41.0	+5	−15.00	+149.0	+65	+18.33
+42.8	+6	−14.44	+150.8	+66	+18.89
+44.6	+7	−13.89	+152.6	+67	+19.44
+46.4	+8	−13.33	+154.4	+68	+20.00
+48.2	+9	−12.78	+156.2	+69	+20.56
+50.0	+10	−12.22	+158.0	+70	+21.11
+51.8	+11	−11.67	+159.8	+71	+21.67
+53.6	+12	−11.11	+161.6	+72	+22.22
+55.4	+13	−10.56	+163.4	+73	+22.78
+57.2	+14	−10.00	+165.2	+74	+23.33
+59.0	+15	−9.44	+167.0	+75	+23.89
+60.8	+16	−8.89	+168.8	+76	+24.44
+62.6	+17	−8.33	+170.6	+77	+25.00
+64.4	+18	−7.78	+172.4	+78	+25.56
+66.2	+19	−7.22	+174.2	+79	+26.11
+68.0	+20	−6.67	+176.0	+80	+26.67
+69.8	+21	−6.11	+177.8	+81	+27.22
+71.6	+22	−5.56	+179.6	+82	+27.78
+73.4	+23	−5.00	+181.4	+83	+28.33
+75.2	+24	−4.44	+183.2	+84	+28.89
+77.0	+25	−3.89	+185.0	+85	+29.44
+78.8	+26	−3.33	+186.8	+86	+30.00
+80.6	+27	−2.78	+188.6	+87	+30.56
+82.4	+28	−2.22	+190.4	+88	+31.11
+84.2	+29	−1.67	+192.2	+89	+31.67
+86.0	+30	−1.11	+194.0	+90	+32.22
+87.8	+31	−0.56	+195.8	+91	+32.78
+89.6	+32	±0.00	+197.6	+92	+33.33
+91.4	+33	+0.56	+199.4	+93	+33.89
+93.2	+34	+1.11	+201.2	+94	+34.44
+95.0	+35	+1.67	+203.0	+95	+35.00
+96.8	+36	+2.22	+204.8	+96	+35.56
+98.6	+37	+2.78	+206.6	+97	+36.11
+100.4	+38	+3.33	+208.4	+98	+36.67
+102.2	+39	+3.89	+210.2	+99	+37.22
+104.0	+40	+4.44	+212.0	+100	+37.78
+105.8	+41	+5.00	+213.8	+101	+38.33
+107.6	+42	+5.56	+215.6	+102	+38.89
+109.4	+43	+6.11	+217.4	+103	+39.44
+111.2	+44	+6.67	+219.2	+104	+40.00
+113.0	+45	+7.22	+221.0	+105	+40.56
+114.8	+46	+7.78	+222.8	+106	+41.11

TABLE 2.8   Temperature Conversion Table (*Continued*)

°F.	Reading in °F. or °C. to be converted	°C.	°F.	Reading in °F. or °C. to be converted	°C.
+224.6	+107	+41.67	+332.6	+167	+75.00
+226.4	+108	+42.22	+334.4	+168	+75.56
+228.2	+109	+42.78	+336.2	+169	+76.11
+230.0	+110	+43.33	+338.0	+170	+76.67
+231.8	+111	+43.89	+339.8	+171	+77.22
+233.6	+112	+44.44	+341.6	+172	+77.78
+235.4	+113	+45.00	+343.4	+173	+78.33
+237.2	+114	+45.56	+345.2	+174	+78.89
+239.0	+115	+46.11	+347.0	+175	+79.44
+240.8	+116	+46.67	+348.8	+176	+80.00
+242.6	+117	+47.22	+350.6	+177	+80.56
+244.4	+118	+47.78	+352.4	+178	+81.11
+246.2	+119	+48.33	+354.2	+179	+81.67
+248.0	+120	+48.89	+356.0	+180	+82.22
+249.8	+121	+49.44	+357.8	+181	+82.78
+251.6	+122	+50.00	+359.6	+182	+83.33
+253.4	+123	+50.56	+361.4	+183	+83.89
+255.2	+124	+51.11	+363.2	+184	+84.44
+257.0	+125	+51.67	+365.0	+185	+85.00
+258.8	+126	+52.22	+366.8	+186	+85.56
+260.6	+127	+52.78	+368.6	+187	+86.11
+262.4	+128	+53.33	+370.4	+188	+86.67
+264.2	+129	+53.89	+372.2	+189	+87.22
+266.0	+130	+54.44	+374.0	+190	+87.78
+267.8	+131	+55.00	+375.8	+191	+88.33
+269.6	+132	+55.56	+377.6	+192	+88.89
+271.4	+133	+56.11	+379.4	+193	+89.44
+273.2	+134	+56.67	+381.2	+194	+90.00
+275.0	+135	+57.22	+383.0	+195	+90.56
+276.8	+136	+57.78	+384.8	+196	+91.11
+278.6	+137	+58.33	+386.6	+197	+91.67
+280.4	+138	+58.89	+388.4	+198	+92.22
+282.2	+139	+59.44	+390.2	+199	+92.78
+284.0	+140	+60.00	+392.0	+200	+93.33
+285.8	+141	+60.56	+393.8	+201	+93.89
+287.6	+142	+61.11	+395.6	+202	+94.44
+289.4	+143	+61.67	+397.4	+203	+95.00
+291.2	+144	+62.22	+399.2	+204	+95.56
+293.0	+145	+62.78	+401.0	+205	+96.11
+294.8	+146	+63.33	+402.8	+206	+96.67
+296.6	+147	+63.89	+404.6	+207	+97.22
+298.4	+148	+64.44	+406.4	+208	+97.78
+300.2	+149	+65.00	+408.2	+209	+98.33
+302.0	+150	+65.56	+410.0	+210	+98.89
+303.8	+151	+66.11	+411.8	+211	+99.44
+305.6	+152	+66.67	+413.6	+212	+100.00
+307.4	+153	+67.22	+415.4	+213	+100.56
+309.2	+154	+67.78	+417.2	+214	+101.11
+311.0	+155	+68.33	+419.0	+215	+101.67
+312.8	+156	+68.89	+420.8	+216	+102.22
+314.6	+157	+69.44	+422.6	+217	+102.78
+316.4	+158	+70.00	+424.4	+218	+103.33
+318.2	+159	+70.56	+426.2	+219	+103.89
+320.0	+160	+71.11	+428.0	+220	+104.44
+321.8	+161	+71.67	+431.6	+222	+105.56
+323.6	+162	+72.22	+435.2	+224	+106.67
+325.4	+163	+72.78	+438.8	+226	+107.78
+327.2	+164	+73.33	+442.4	+228	+108.89
+329.0	+165	+73.89	+446.0	+230	+110.00
+330.8	+166	+74.44	+449.6	+232	+111.11

TABLE 2.8 Temperature Conversion Table (*Continued*)

°F.	Reading in °F. or °C. to be converted	°C.	°F.	Reading in °F. or °C. to be converted	°C.
+453.2	+234	+112.22	+669.2	+354	+178.89
+456.8	+236	+113.33	+672.8	+356	+180.00
+460.4	+238	+114.44	+676.4	+358	+181.11
+464.0	+240	+115.56	+680.0	+360	+182.22
+467.6	+242	+116.67	+683.6	+362	+183.33
+471.2	+244	+117.78	+687.2	+364	+184.44
+474.8	+246	+118.89	+690.8	+366	+185.56
+478.4	+248	+120.00	+694.4	+368	+186.67
+482.0	+250	+121.11	+698.0	+370	+187.78
+485.6	+252	+122.22	+701.6	+372	+188.89
+489.2	+254	+123.33	+705.2	+374	+190.00
+492.8	+256	+124.44	+708.8	+376	+191.11
+496.4	+258	+125.56	+712.4	+378	+192.22
+500.0	+260	+126.67	+716.0	+380	+193.33
+503.6	+262	+127.78	+719.6	+382	+194.44
+507.2	+264	+128.89	+723.2	+384	+195.56
+510.8	+266	+130.00	+726.8	+386	+196.67
+514.4	+268	+131.11	+730.4	+388	+197.78
+518.0	+270	+132.22	+734.0	+390	+198.89
+521.6	+272	+133.33	+737.6	+392	+200.00
+525.2	+274	+134.44	+741.2	+394	+201.11
+528.8	+276	+135.56	+744.8	+396	+202.22
+532.4	+278	+136.67	+748.4	+398	+203.33
+536.0	+280	+137.78	+752.0	+400	+204.44
+539.6	+282	+138.89	+755.6	+402	+205.56
+543.2	+284	+140.00	+759.2	+404	+206.67
+546.8	+286	+141.11	+762.8	+406	+207.78
+550.4	+288	+142.22	+766.4	+408	+208.89
+554.0	+290	+143.33	+770.0	+410	+210.00
+557.6	+292	+144.44	+773.6	+412	+211.11
+561.2	+294	+145.56	+777.2	+414	+212.22
+564.8	+296	+146.67	+780.8	+416	+213.33
+568.4	+298	+147.78	+784.4	+418	+214.44
+572.0	+300	+148.89	+788.0	+420	+215.56
+575.6	+302	+150.00	+791.6	+422	+216.67
+579.2	+304	+151.11	+795.2	+424	+217.78
+582.8	+306	+152.22	+798.8	+426	+218.89
+586.4	+308	+153.33	+802.4	+428	+220.00
+590.0	+310	+154.44	+806.0	+430	+221.11
+593.6	+312	+155.56	+809.6	+432	+222.22
+597.2	+314	+156.67	+813.2	+434	+223.33
+600.8	+316	+157.78	+816.8	+436	+224.44
+604.4	+318	+158.89	+820.4	+438	+225.56
+608.0	+320	+160.00	+824.0	+440	+226.67
+611.6	+322	+161.11	+827.6	+442	+227.78
+615.2	+324	+162.22	+831.2	+444	+228.89
+618.8	+326	+163.33	+834.8	+446	+230.00
+622.4	+328	+164.44	+838.4	+448	+231.11
+626.0	+330	+165.56	+842.0	+450	+232.22
+629.6	+332	+166.67	+845.6	+452	+233.33
+633.2	+334	+167.78	+849.2	+454	+234.44
+636.8	+336	+168.89	+852.8	+456	+235.56
+640.4	+338	+170.00	+856.4	+458	+236.67
+644.0	+340	+171.11	+860.0	+460	+237.78
+647.6	+342	+172.22	+863.6	+462	+238.89
+651.2	+344	+173.33	+867.2	+464	+240.00
+654.8	+346	+174.44	+870.8	+466	+241.11
+658.4	+348	+175.56	+874.4	+468	+242.22
+662.0	+350	+176.67	+878.0	+470	+243.33
+665.6	+352	+177.78	+881.6	+472	+244.44

TABLE 2.8   Temperature Conversion Table (*Continued*)

°F.	Reading in °F. or °C. to be converted	°C.	°F.	Reading in °F. or °C. to be converted	°C.
+885.2	+474	+245.56	+1101.2	+594	+312.22
+888.8	+476	+246.67	+1104.8	+596	+313.33
+892.4	+478	+247.78	+1108.4	+598	+314.44
+896.0	+480	+248.89	+1112.0	+600	+315.56
+899.6	+482	+250.00	+1115.6	+602	+316.67
+903.2	+484	+251.11	+1119.2	+604	+317.78
+906.8	+486	+252.22	+1122.8	+606	+318.89
+910.4	+488	+253.33	+1126.4	+608	+320.00
+914.0	+490	+254.44	+1130.0	+610	+321.11
+917.6	+492	+255.56	+1133.6	+612	+322.22
+921.2	+494	+256.67	+1137.2	+614	+323.33
+924.8	+496	+257.78	+1140.8	+616	+324.44
+928.4	+498	+258.89	+1144.4	+618	+325.56
+932.0	+500	+260.00	+1148.0	+620	+326.67
+935.6	+502	+261.11	+1151.6	+622	+327.78
+939.2	+504	+262.22	+1155.2	+624	+328.89
+942.8	+506	+263.33	+1158.8	+626	+330.00
+946.4	+508	+264.44	+1162.4	+628	+331.11
+950.0	+510	+265.56	+1166.0	+630	+332.22
+953.6	+512	+266.67	+1169.6	+632	+333.33
+957.2	+514	+267.78	+1173.2	+634	+334.44
+960.8	+516	+268.89	+1176.8	+636	+335.56
+964.4	+518	+270.00	+1180.4	+638	+336.67
+968.0	+520	+271.11	+1184.0	+640	+337.78
+971.6	+522	+272.22	+1187.6	+642	+338.89
+975.2	+524	+273.33	+1191.2	+644	+340.00
+978.8	+526	+274.44	+1194.8	+646	+341.11
+982.4	+528	+275.56	+1198.4	+648	+342.22
+986.0	+530	+276.67	+1202.0	+650	+343.33
+989.6	+532	+277.78	+1205.6	+652	+344.44
+993.2	+534	+278.89	+1209.2	+654	+345.56
+996.8	+536	+280.00	+1212.8	+656	+346.67
+1000.4	+538	+281.11	+1216.4	+658	+347.78
+1004.0	+540	+282.22	+1220.0	+660	+348.89
+1007.6	+542	+283.33	+1223.6	+662	+350.00
+1011.2	+544	+284.44	+1227.2	+664	+351.11
+1014.8	+546	+285.56	+1230.8	+666	+352.22
+1018.4	+548	+286.67	+1234.4	+668	+353.33
+1022.0	+550	+287.78	+1238.0	+670	+354.44
+1025.6	+552	+288.89	+1241.6	+672	+355.56
+1029.2	+554	+290.00	+1245.2	+674	+356.67
+1032.8	+556	+291.11	+1248.8	+676	+357.78
+1036.4	+558	+292.22	+1252.4	+678	+358.89
+1040.0	+560	+293.33	+1256.0	+680	+360.00
+1043.6	+562	+294.44	+1259.6	+682	+361.11
+1047.2	+564	+295.56	+1263.2	+684	+362.22
+1050.8	+566	+296.67	+1266.8	+686	+363.33
+1054.4	+568	+297.78	+1270.4	+688	+364.44
+1058.0	+570	+298.89	+1274.0	+690	+365.56
+1061.6	+572	+300.00	+1277.6	+692	+366.67
+1065.2	+574	+301.11	+1281.2	+694	+367.78
+1068.8	+576	+302.22	+1284.8	+696	+368.89
+1072.4	+578	+303.33	+1288.4	+698	+370.00
+1076.0	+580	+304.44	+1292.0	+700	+371.11
+1079.6	+582	+305.56	+1295.6	+702	+372.22
+1083.2	+584	+306.67	+1299.2	+704	+373.33
+1086.8	+586	+307.78	+1302.8	+706	+374.44
+1090.4	+588	+308.89	+1306.4	+708	+375.56
+1094.0	+590	+310.00	+1310.0	+710	+376.67
+1097.6	+592	+311.11	+1313.6	+712	+377.78

TABLE 2.8   Temperature Conversion Table (*Continued*)

°F.	Reading in °F. or °C. to be converted	°C.	°F.	Reading in °F. or °C. to be converted	°C.
+1317.2	+714	+378.89	+1533.2	+834	+445.56
+1320.8	+716	+380.00	+1536.8	+836	+446.67
+1324.4	+718	+381.11	+1540.4	+838	+447.78
+1328.0	+720	+382.22	+1544.0	+840	+448.89
+1331.6	+722	+383.33	+1547.6	+842	+450.00
+1335.2	+724	+384.44	+1551.2	+844	+451.11
+1338.8	+726	+385.56	+1554.8	+846	+452.22
+1342.4	+728	+386.67	+1558.4	+848	+453.33
+1346.0	+730	+387.78	+1562.0	+850	+454.44
+1349.6	+732	+388.89	+1565.6	+852	+455.56
+1353.2	+734	+390.00	+1569.2	+854	+456.67
+1356.8	+736	+391.11	+1572.8	+856	+457.78
+1360.4	+738	+392.22	+1576.4	+858	+458.89
+1364.0	+740	+393.33	+1580.0	+860	+460.00
+1367.6	+742	+394.44	+1583.6	+862	+461.11
+1371.2	+744	+395.56	+1587.2	+864	+462.22
+1374.8	+746	+396.67	+1590.8	+866	+463.33
+1378.4	+748	+397.78	+1594.4	+868	+464.44
+1382.0	+750	+398.89	+1598.0	+870	+465.56
+1385.6	+752	+400.00	+1601.6	+872	+466.67
+1389.2	+754	+401.11	+1605.2	+874	+467.78
+1392.8	+756	+402.22	+1608.8	+876	+468.89
+1396.4	+758	+403.33	+1612.4	+878	+470.00
+1400.0	+760	+404.44	+1616.0	+880	+471.11
+1403.6	+762	+405.56	+1619.6	+882	+472.22
+1407.2	+764	+406.67	+1623.2	+884	+473.33
+1410.8	+766	+407.78	+1626.8	+886	+474.44
+1414.4	+768	+408.89	+1630.4	+888	+475.56
+1418.0	+770	+410.00	+1634.0	+890	+476.67
+1421.6	+772	+411.11	+1637.6	+892	+477.78
+1425.2	+774	+412.22	+1641.2	+894	+478.89
+1428.8	+776	+413.33	+1644.8	+896	+480.00
+1432.4	+778	+414.44	+1648.4	+898	+481.11
+1436.0	+780	+415.56	+1652.0	+900	+482.22
+1439.6	+782	+416.67	+1655.6	+902	+483.33
+1443.2	+784	+417.78	+1659.2	+904	+484.44
+1446.8	+786	+418.89	+1662.8	+906	+485.56
+1450.4	+788	+420.00	+1666.4	+908	+486.67
+1454.0	+790	+421.11	+1670.0	+910	+487.78
+1457.6	+792	+422.22	+1673.6	+912	+488.89
+1461.2	+794	+423.33	+1677.2	+914	+490.00
+1464.8	+796	+424.44	+1680.8	+916	+491.11
+1468.4	+798	+425.56	+1684.4	+918	+492.22
+1472.0	+800	+426.67	+1688.0	+920	+493.33
+1475.6	+802	+427.78	+1691.6	+922	+494.44
+1479.2	+804	+428.89	+1695.2	+924	+495.56
+1482.8	+806	+430.00	+1698.8	+926	+496.67
+1486.4	+808	+431.11	+1702.4	+928	+497.78
+1490.0	+810	+432.22	+1706.0	+930	+498.89
+1493.6	+812	+433.33	+1709.6	+932	+500.00
+1497.2	+814	+434.44	+1713.2	+934	+501.11
+1500.8	+816	+435.56	+1716.8	+936	+502.22
+1504.4	+818	+436.67	+1720.4	+938	+503.33
+1508.0	+820	+437.78	+1724.0	+940	+504.44
+1511.6	+822	+438.89	+1727.6	+942	+505.56
+1515.2	+824	+440.00	+1731.2	+944	+506.67
+1518.8	+826	+441.11	+1734.8	+946	+507.78
+1522.4	+828	+442.22	+1738.4	+948	+508.89
+1526.0	+830	+443.33	+1742.0	+950	+510.00
+1529.6	+832	+444.44	+1745.6	+952	+511.11

TABLE 2.8   Temperature Conversion Table (*Continued*)

°F.	Reading in °F. or °C. to be converted	°C.	°F.	Reading in °F. or C°. to be converted	°C.
+1749.2	+954	+512.22	+2498.0	+1370	+743.33
+1752.8	+956	+513.33	+2516.0	+1380	+748.89
+1756.4	+958	+514.44	+2534.0	+1390	+754.44
+1760.0	+960	+515.56	+2552.0	+1400	+760.00
+1763.6	+962	+516.67	+2570.0	+1410	+765.56
+1767.2	+964	+517.78	+2588.0	+1420	+771.11
+1770.8	+966	+518.89	+2606.0	+1430	+776.67
+1774.4	+968	+520.00	+2624.0	+1440	+782.22
+1778.0	+970	+521.11	+2642.0	+1450	+787.78
+1781.6	+972	+522.22	+2660.0	+1460	+793.33
+1785.2	+974	+523.33	+2678.0	+1470	+798.89
+1788.8	+976	+524.44	+2696.0	+1480	+804.44
+1792.4	+978	+525.56	+2714.0	+1490	+810.00
+1796.0	+980	+526.67	+2732.0	+1500	+815.56
+1799.6	+982	+527.78	+2750.0	+1510	+821.11
+1803.2	+984	+528.89	+2768.0	+1520	+826.67
+1806.8	+986	+530.00	+2786.0	+1530	+832.22
+1810.4	+988	+531.11	+2804.0	+1540	+837.78
+1814.0	+990	+532.22	+2822.0	+1550	+843.33
+1817.6	+992	+533.33	+2840.0	+1560	+848.89
+1821.2	+994	+534.44	+2858.0	+1570	+854.44
+1824.8	+996	+535.56	+2876.0	+1580	+860.00
+1828.4	+998	+536.67	+2894.0	+1590	+865.56
+1832.0	+1000	+537.78	+2912.0	+1600	+871.11
+1850.0	+1010	+543.33	+2930.0	+1610	+876.67
+1868.0	+1020	+548.89	+2948.0	+1620	+882.22
+1886.0	+1030	+554.44	+2966.0	+1630	+887.78
+1904.0	+1040	+560.00	+2984.0	+1640	+893.33
+1922.0	+1050	+565.56	+3002.0	+1650	+898.89
+1940.0	+1060	+571.11	+3020.0	+1660	+904.44
+1958.0	+1070	+576.67	+3038.0	+1670	+910.00
+1976.0	+1080	+582.22	+3056.0	+1680	+915.56
+1994.0	+1090	+587.78	+3074.0	+1690	+921.11
+2012.0	+1100	+593.33	+3092.0	+1700	+926.67
+2030.0	+1110	+598.89	+3110.0	+1710	+932.22
+2048.0	+1120	+604.44	+3128.0	+1720	+937.78
+2066.0	+1130	+610.00	+3146.0	+1730	+943.33
+2084.0	+1140	+615.56	+3164.0	+1740	+948.89
+2102.0	+1150	+621.11	+3182.0	+1750	+954.44
+2120.0	+1160	+626.67	+3200.0	+1760	+960.00
+2138.0	+1170	+632.22	+3218.0	+1770	+965.56
+2156.0	+1180	+637.78	+3236.0	+1780	+971.11
+2174.0	+1190	+643.33	+3254.0	+1790	+976.67
+2192.0	+1200	+648.89	+3272.0	+1800	+982.22
+2210.0	+1210	+654.44	+3290.0	+1810	+987.78
+2228.0	+1220	+660.00	+3308.0	+1820	+993.33
+2246.0	+1230	+665.56	+3326.0	+1830	+998.89
+2264.0	+1240	+671.11	+3344.0	+1840	+1004.4
+2282.0	+1250	+676.67	+3362.0	+1850	+1010.0
+2300.0	+1260	+682.22	+3380.0	+1860	+1015.6
+2318.0	+1270	+687.78	+3398.0	+1870	+1021.1
+2336.0	+1280	+693.33	+3416.0	+1880	+1026.7
+2354.0	+1290	+698.89	+3434.0	+1890	+1032.2
+2372.0	+1300	+704.44	+3452.0	+1900	+1037.8
+2390.0	+1310	+710.00	+3470.0	+1910	+1043.3
+2408.0	+1320	+715.56	+3488.0	+1920	+1048.9
+2426.0	+1330	+721.11	+3506.0	+1930	+1054.4
+2444.0	+1340	+726.67	+3524.0	+1940	+1060.0
+2462.0	+1350	+732.22	+3542.0	+1950	+1065.6
+2480.0	+1360	+737.78	+3560.0	+1960	+1071.1

TABLE 2.8   Temperature Conversion Table (*Continued*)

°F.	Reading in °F. or °C. to be converted	°C.	°F.	Reading in °F. or °C. to be converted	°C.
+3578.0	+1970	+1076.7	+4604.0	+2540	+1393.3
+3596.0	+1980	+1082.2	+4622.0	+2550	+1398.9
+3614.0	+1990	+1087.8	+4640.0	+2560	+1404.4
+3632.0	+2000	+1093.3	+4658.0	+2570	+1410.0
+3650.0	+2010	+1098.9	+4676.0	+2580	+1415.6
+3668.0	+2020	+1104.4	+4694.0	+2590	+1421.1
+3686.0	+2030	+1110.0	+4712.0	+2600	+1426.7
+3704.0	+2040	+1115.6	+4730.0	+2610	+1432.2
+3722.0	+2050	+1121.1	+4748.0	+2620	+1437.8
+3740.0	+2060	+1126.7	+4766.0	+2630	+1443.3
+3758.0	+2070	+1132.2	+4784.0	+2640	+1448.9
+3776.0	+2080	+1137.8	+4802.0	+2650	+1454.4
+3794.0	+2090	+1143.3	+4820.0	+2660	+1460.0
+3812.0	+2100	+1148.9	+4838.0	+2670	+1465.6
+3830.0	+2110	+1154.4	+4856.0	+2680	+1471.1
+3848.0	+2120	+1160.0	+4874.0	+2690	+1476.7
+3866.0	+2130	+1165.6	+4892.0	+2700	+1482.2
+3884.0	+2140	+1171.1	+4910.0	+2710	+1487.8
+3902.0	+2150	+1176.7	+4928.0	+2720	+1493.3
+3920.0	+2160	+1182.2	+4946.0	+2730	+1498.9
+3938.0	+2170	+1187.8	+4964.0	+2740	+1504.4
+3956.0	+2180	+1193.3	+4982.0	+2750	+1510.0
+3974.0	+2190	+1198.9	+5000.0	+2760	+1515.6
+3992.0	+2200	+1204.4	+5018.0	+2770	+1521.1
+4010.0	+2210	+1210.0	+5036.0	+2780	+1526.7
+4028.0	+2220	+1215.6	+5054.0	+2790	+1532.2
+4046.0	+2230	+1221.1	+5072.0	+2800	+1537.8
+4064.0	+2240	+1226.7	+5090.0	+2810	+1543.3
+4082.0	+2250	+1232.2	+5108.0	+2820	+1548.9
+4100.0	+2260	+1237.8	+5126.0	+2830	+1554.4
+4118.0	+2270	+1243.3	+5144.0	+2840	+1560.0
+4136.0	+2280	+1248.9	+5162.0	+2850	+1565.6
+4154.0	+2290	+1254.4	+5180.0	+2860	+1571.1
+4172.0	+2300	+1260.0	+5198.0	+2870	+1576.7
+4190.0	+2310	+1265.6	+5216.0	+2880	+1582.2
+4208.0	+2320	+1271.1	+5234.0	+2890	+1587.8
+4226.0	+2330	+1276.7	+5252.0	+2900	+1593.3
+4244.0	+2340	+1282.2	+5270.0	+2910	+1598.9
+4262.0	+2350	+1287.8	+5288.0	+2920	+1604.4
+4280.0	+2360	+1293.3	+5306.0	+2930	+1610.0
+4298.0	+2370	+1298.9	+5324.0	+2940	+1615.6
+4316.0	+2380	+1304.4	+5342.0	+2950	+1621.1
+4334.0	+2390	+1310.0	+5360.0	+2960	+1626.7
+4352.0	+2400	+1315.6	+5378.0	+2970	+1632.2
+4370.0	+2410	+1321.1	+5396.0	+2980	+1637.8
+4388.0	+2420	+1326.7	+5414.0	+2990	+1643.3
+4406.0	+2430	+1332.2	+5432.0	+3000	+1648.9
+4424.0	+2440	+1337.8	+5450.0	+3010	+1654.4
+4442.0	+2450	+1343.3	+5468.0	+3020	+1660.0
+4460.0	+2460	+1348.9	+5486.0	+3030	+1665.6
+4478.0	+2470	+1354.4	+5504.0	+3040	+1671.1
+4496.0	+2480	+1360.0	+5522.0	+3050	+1676.7
+4514.0	+2490	+1365.6	+5540.0	+3060	+1682.2
+4532.0	+2500	+1371.1	+5558.0	+3070	+1687.8
+4550.0	+2510	+1376.7	+5576.0	+3080	+1693.3
+4568.0	+2520	+1382.2	+5594.0	+3090	+1698.9
+4586.0	+2530	+1387.8	+5612.0	+3100	+1704.4

2.1.1 Conversion of Thermometer Scales

The following abbreviations are used: °F, degrees Fahrenheit; °C, degrees Celsius; K, degrees
Kelvin; °Ré, degrees Reaumur; °R, degrees Rankine; °Z, degrees on any scale; (fp)"Z", the
freezing point of water on the Z scale; and (bp)"Z", the boiling point of water on the Z scale.
Reference: Dodds, *Chemical and Metallurgical Engineering* **38**:476 (1931).

$$\frac{°F-32}{180}=\frac{°C}{100}=\frac{°Ré}{80}=\frac{K-273}{100}=\frac{°R-492}{180}=\frac{°Z-(fp)"Z"}{(bp)"Z"-(fp)"Z"}$$

Examples
(1)   To find the Fahrenheit temperature corresponding to −20°C:

$$\frac{°F-32}{180}=\frac{°C}{100}\quad\text{or}\quad\frac{°F-32}{180}=\frac{-20}{100}$$

$$°F-32=\frac{(-20)(180)}{100}=-36$$

$$°F=-4$$

(2)   To find the Reaumur temperature corresponding to 20°F:

$$\frac{°F-32}{180}=\frac{°Ré}{80}=\frac{20-32}{180}=\frac{°Ré}{80}$$

i.e., $20°F=-5.33°Ré$

(3)   To find the correct temperature on a thermometer reading 80°C and that shows a
reading of −0.30°C in a melting ice/water mixture and 99.0°C in steam at 760 mm pressure of
mercury:

$$\frac{°C}{100}=\frac{Z-(fp)"Z"}{(bp)"Z"-(fp)"Z"}=\frac{80-(-0.30)}{99.0-(-0.30)}$$

i.e., $°C=80.87$ (corrected)

2.1.2 Density and Specific Gravity

2.1.2.1 Hydrometers.   Various hydrometers and the relation between the various scales.

Alcoholometer.   This hydrometer is used in determining the density of aqueous ethyl
alcohol solutions; the reading in degrees is numerically the same as the percentage of
alcohol by volume. The scale known as Tralle gives the percentage by volume. Wine and
Must hydrometer relations are given below.

Ammoniameter.   This hydrometer, employed in finding the density of aqueous ammonia
solutions, has a scale graduated in equal divisions from 0° to 40°. To convert the reading to
specific gravity multiply by 3 and subtract the resulting number from 1000.

Balling Hydrometer.   See under Saccharometers.

Barkometer or Barktrometer.   This hydrometer, which is used in determining the density
of tanning liquors, has a scale from 0° to 80° Bk; the number to the right of the decimal
point of a specific gravity reading is the corresponding Bk degree; thus, a specific gravity of
1.015 is 15° Bk.

Baumé Hydrometers. For liquids heavier than water: This hydrometer was originally based on the density of a 10% sodium chloride solution, which was given the value of 10°, and the density of pure water, which was given the value of 0°; the interval between these two values was divided into 10 equal parts. Other reference points have been taken with the result that so much confusion exists that there are about 36 different scales in use, many of which are incorrect. In general a Baumé hydrometer should have inscribed on it the temperature at which it was calibrated and also the temperature of the water used in relating the density to a specific gravity. The following expression gives the relation between the specific gravity and several of the Baumé scales:

$$\text{Specific gravity} = \frac{m}{m - \text{Baumé}}$$

$m =$ 145 at 60°/60°F (15.56°C)   for the American Scale

$ =$ 144   for the old scale used in Holland

$ =$ 146.3 at 15°C   for the Gerlach Scale

$ =$ 144.3 at 15°C   for the Rational Scale generally used in Germany

For liquids lighter than water: Originally the density of a solution of 1 gram of sodium chloride in 9 grams of water at 12.5°C was given a value of 10° Bé. The scale between these points was divided into ten equal parts and these divisions were repeated throughout the scale giving a relation which could be expressed by the formula: Specific gravity = 145.88/ (135.88 + Bé), which is approximately equal to 146/(136 + Bé.) Other scales have since come into more general use such as that of the Bureau of Standards in which the specific gravity at 60°/60°F = 140/(130 + Bé.) and that of the American Petroleum Institute (A.P.I. Scale) in which the specific gravity at 60°/60°F = 141.5/(131.5 + API˚.).

See also special table for conversion to density and Twaddell scale.

Beck's Hydrometer. This hydrometer is graduated to show a reading of 0° in pure water and a reading of 30° in a solution with a specific gravity of 0.850, with equal scale divisions above and below these two points.

Brix Hydrometer. See under Saccharometers.

Cartier's Hydrometer. This hydrometer shows a reading of 22° when immersed in a solution having a density of 22° Baumé but the scale divisions are smaller than on the Baumé hydrometer in the ratio of 16 Cartier to 15 Baumé.

Fatty Oil Hydrometer. The graduations on this hydrometer are in specific gravity within the range 0.908 to 0.938. The letters on the scale correspond to the specific gravity of the various common oils as follows: *R*, rape; *O*, olive; *A*, almond; *S*, sesame; *HL*, hoof oil; *HP*, hemp; *C*, cotton seed; *L*, linseed. See also Oleometer below.

Lactometers. These hydrometers are used in determining the density of milk. The various scales in common use are the following:

New York Board of Health has a scale graduated into 120 equal parts, 0° being equal to the specific gravity of water and 100° being equal to a specific gravity of 1.029.

Quevenne lactometer is graduated from 15° to 40° corresponding to specific gravities from 1.015 to 1.040.

Soxhlet lactometer has a scale from 25° to 35° corresponding to specific gravities from 1.025 to 1.035 respectively.

Oleometer.   A hydrometer for determining the density of vegetable and sperm oils with a scale from 50° to 0° corresponding to specific gravities from 0.870 to 0.970. See also Fatty Oil Hydrometer above.

Saccharometers.   These hydrometers are used in determining the density of sugar solutions. Solutions of the same concentration but of different carbohydrates have very nearly the same specific gravity and in general a concentration of 10 grams of carbohydrate per 100 mL of solution shows a specific gravity of 1.0386. Thus, the wt. of sugar in 1000 mL soln. is (a) for conc. < 12g/100 mL: (wt. of 1000 mL soln. − 1000) ÷ 0.386; (b) for conc. > 12g/100 mL: (wt. of 1000 mL soln. − 1000) ÷ 0.385.

    Brix hydrometer is graduated so that the number of degrees is identical with the percentage by weight of cane sugar and is used at the temperature indicated on the hydrometer.

    Balling's saccharometer is used in Europe and is practically identical with the Brix hydrometer.

    Bates brewers' saccharometer which is used in determining the density of malt worts is graduated so that the divisions express pounds per barrel (32 gallons). The relation between degrees Bates (= b) and degrees Balling (= B) is shown by the following formula: B = 260b/ (360 + b).

    See also below under Wine and Must Hydrometer.

Salinometer.   This hydrometer, which is used in the pickling and meat packing plants, is graduated to show percentage of saturation of a sodium chloride solution. An aqueous solution is completely saturated when it contains 26.4% pure sodium chloride. The range from 0% to 26.4% is divided into 100 parts, each division therefore representing 1% of saturation. In another type of salinometer, the degrees correspond to percentages of sodium chloride expressed in grams of sodium chloride per 100 mL of water.

Sprayometer (Parrot and Stewart).   This hydrometer which is used in determining the density of *lime sulfur* solutions has two scales; one scale is graduated from 0° to 38° Baumé and the other scale is from 1.000 to 1.350 specific gravity.

Tralle Hydrometer.   See Alcoholometer above.

Twaddell Hydrometer.   This hydrometer, which is used only for liquids heavier than water, has a scale such that when the reading is multiplied by 5 and added to 1000 the resulting number is the specific gravity with reference to water as 1000. To convert specific gravity at 60°/60°F to Twaddell degrees, take the decimal portion of the specific gravity value and multiply it by 200; thus a specific gravity of 1.032 = 0.032 × 200 = 6.4° Tw. See also special table for conversion to density and Baumé scale.

Wine and Must Hydrometer.   This instrument has three scales. One scale shows readings of 0° to 15° Brix for sugar (see Brix Hydrometer above); another scale from 0° to 15° Tralle is used for sweet wines to indicate the percentage of alcohol by volume; and a third scale from 0° to 20° Tralle is used for tart wines to indicate the percentage of alcohol by volume.

2.1.2.2   *Conversion of Specific Gravity at 25°/25°C to Density at any Temperature from 0° to 40°C.* *   Liquids change volume with change in temperature, but the amount of this change, β (coefficient of cubical expansion), varies widely with different liquids, and to some extent for the same liquid at different temperatures.

    The table below, which is calculated from the relationship:

$$F_{\beta_t} = \frac{\text{density of water at 25°C} (= 0.99705)}{1 - \beta(25 - t)} \tag{1}$$

---

* Cf. Dreisbach, *Ind. Eng. Chem., Anal. Ed.* **12**:160 (1940).

may be used to find d^t, the density (weight of 1 mL) of a liquid at any temperature (t) between 0° and 40°C if the specific gravity at 25°/25°C (S) and the coefficient of cubical expansion (β) are known. Substitutions are made in the equations:

$$d^t = SF_{\beta_t} \tag{2}$$

$$S = \frac{d^t}{F_{\beta_t}} \tag{3}$$

Factors ($F\beta_t$)

Density $t°C = $ sp. gr. $25°/25° \times F_{\beta_t}$

°C. *$\beta \times 10^3$	0	5	10	15	20	25	30	35	40
1.3	1.0306	1.0237	1.0169	1.0102	1.0036	0.99705	0.99065	0.9843	0.9780
1.2	1.0279	1.0216	1.0154	1.0092	1.0031	0.99705	0.9911	0.9853	0.9794
1.1	1.0253	1.0195	1.0138	1.0082	1.0026	0.99705	0.9916	0.9963	0.9809
1.0	1.0227	1.0174	1.0123	1.0072	1.0021	0.99705	0.9921	0.9872	0.98234
0.9	1.0200	1.0153	1.0107	1.0060	1.0016	0.99705	0.99262	0.9882	0.9838
0.8	1.0174	1.0133	1.0092	1.0051	1.0011	0.99705	0.9931	0.98918	0.9851
0.7	1.0148	1.0113	1.0077	1.0041	1.0006	0.99705	0.9936	0.99015	0.98672
0.6	1.0122	1.0092	1.0061	1.0031	1.0001	0.99705	0.9941	0.9911	0.9882
0.5	1.0097	1.0072	1.0046	1.0021	0.99958	0.99705	0.9944	0.9921	0.9897
0.	1.0071	1.0051	1.0031	1.0011	0.99908	0.99705	0.9951	0.9931	0.9911

* β = coefficient of cubical expansion.

Examples. All examples are based upon an assumed coefficient of cubical expansion, β, of 1.3×10^{-3}.

Example 1. To find the density of a liquid at 20°C, d^{20}, which has a specific gravity (S) of $1.2500\frac{25}{25}$:

From the table above F_{β_t} at 20°C = 1.0036.

$$d^{20} = d^t = SF_{\beta_t} = 1.2500 \times 1.0036 = 1.2545$$

Example 2. To find the density at 20°C (d^{20}) of a liquid which has a specific gravity of $1.2500\frac{17}{4}$:

Since the density of water at 4°C is equal to 1, specific gravity at $17°/4° = d^{17} = 1.2500$. Substitution in Equation 3 with F_{β_t} at 17°C, by interpolation from the table, equal to 1.00756, gives

$$\text{Sp. gr. } 25°/25° = S = 1.2500 \div 1.00756$$

Substitution of this value for S in Equation 2 with F_{β_t} at 20°C, from the table, equal to 1.0036, gives

$$d^{20} = d^t = (1.2500 \div 1.00756) \times 1.0036 = 1.2451$$

Example 3. To find the specific gravity at 20°/4°C of a liquid which has a specific gravity of $1.2500\frac{25}{4}$:

Since the density of water at 4°C, is equal to 1, specific gravity 25°/4° = d^{25} = 1.2500; and, specific gravity 20°/4° = d^{20}.

Substitution in Equation 3, with d^t = 1.2500; and, with F_{β_t} at 25°C, from the table, equal to 0.99705, gives

$$\text{Sp. gr. } 25°/25° = S = 1.2500 \div 0.99705$$

Substitution of this value for S in Equation 2, with F_{β_t} at 20°C, from the table, equal to 1.0036, gives

$$\text{Sp. gr. } 20°/4° = d^{20} = (1.2500 \div 0.99705) \times 1.0036 = 1.2582$$

Example 4. To find the density at 25°C of a liquid which has a specific gravity of 1.2500$\frac{15}{15}$:

Since the density of water at 15° C = 0.99910,

$$d^{15} = \text{sp. gr. } 15°/15° \times 0.99910 = 1.2500 \times 0.99910$$

Substitution in Equation 3, with F_{β_t} at 15°C, from the table, equal to 1.0102, gives

$$\text{Sp. gr. } 25°/25° = S = (1.2500 \times 0.99910) \div 1.0102$$

Substitution of this value for S in Equation 2, with F_{β_t} at 25°, from the table, equal to 0.99705, gives

$$d^{26} = d^t = (1.2500 \times 0.99910 \div 1.0102) \times 0.99705 = 1.2326$$

2.1.3 Barometry and Barometric Corrections

In principle, the mercurial barometer balances a column of pure mercury against the weight of the atmosphere. The height of the column above the level of the mercury in the reservoir can be measured and serves as a direct index of atmospheric pressure. The space above the mercury in a barometer tube should be a Torricellian vacuum, perfect except for the practically negligible vapor pressure of mercury. The perfection of the vacuum is indicated by the sharpness of the click noted when the barometer tube is inclined. A barometer should be in a vertical position, suspended rather than fastened to a wall, and in a good light but not exposed to direct sunlight or too near a source of heat. The standard conditions for barometric measurements are 0°C and gravity as at 45° latitude and sea level. There are numerous sources of error, but corrections for most of these are readily applied. Some of the corrections are very small, and their application may be questionable in view of the probably larger errors. The degree of consistency to be expected in careful measurements is about 0.13 mm with a 6.4-mm tube, increasing to 0.04 mm with a tube 12.7 mm in diameter.

In reading a barometer of the Fortin type (the usual laboratory instrument for precision measurements), the procedure should be as follows: (1) Observe and record the temperature as indicated by the thermometer attached to the barometer. The temperature correction is very important and may be affected by heat from the observer's body. (2) Set the mercury in the reservoir at zero level, so that the point of the pin above the mercury just touches the surface, making a barely noticeable dimple therein. Tap the tube at the top and verify the zero setting. (3) Bring the vernier down until the view at the light background is cut off at the highest point of the meniscus. Record the reading.

The corrections to be made on the reading are as follows: (1) Temperature, to correct for the difference in thermal expansion of the mercury and the brass (or glass) to which the scale is attached. This correction converts the reading into the value of 0°C. The brass scale table is applicable to the Fortin barometer. See tables. (2a) Latitude-gravity correction, and (2b) altitude-gravity correction, to compensate for differences in gravity, which would affect the height of the mercury column by variation in mass. If local gravity is unknown, an approximate correction may be made from the tables. Local values of gravity are often subject to irregularities which lead to errors even when the corrections here provided are made. It is, therefore, advisable to determine the local value of gravity, from which the correction can be effected in the following manner:

$$Bt = Br + \left(\frac{g_1 - g_0}{g_0}\right) \times Br$$

in which Bt and Br are the true and the observed heights of the barometer, respectively. g_0 is standard gravity (980 665 cm · s^{-2}), and g_1 is the local gravity. It may be noted that for most localities, g_1 is smaller than g_0, which makes the correction negative. These corrections compensate the reading to gravity at 45° latitude and sea level. (3) Correction for capillary depression of the level of the meniscus. This varies with the tube diameter and actual height of the meniscus in a particular case. Some barometers are calibrated to allow for an average value of the latter and approximating the correction. See table. (4) Correction for vapor pressure of mercury. This correction is usually negligible, being only 0.001 mm at 20°C and 0.006 mm at 40°C. This correction is added. See table of vapor pressure of mercury.

The corrections above do not apply to aneroid barometers. These instruments should be calibrated at regular intervals by checking them against a corrected mercurial barometer.

For records on weather maps, meteorologists customarily correct barometer readings to sea level, and some barometers may be calibrated accordingly. Such instruments are not suitable for laboratory use where true pressure under standard conditions is required. Scale corrections should be specified in the maker's instructions with the instrument, and are also indicated by the lack of correspondence between a gauge mark usually placed exactly 76.2 cm from the zero point and the 76.2-cm scale graduation.

TABLE 2.9   Barometer Temperature Correction — Metric Units

The values in the table below are to be subtracted from the observed readings to correct for the difference in the expansion of the mercury and the glass scale at different temperatures.

A. Glass scale

Temp. °C.	Observed barometer height in millimeters						
	700	730	740	750	760	770	800
	mm.	mm.	mm.	mm.	mm.	mm.	mm.
0	0.00	0.00	0.00	0.00	0.00	0.00	0.00
1	0.12	0.13	0.13	0.13	0.13	0.13	0.14
2	0.24	0.25	0.26	0.26	0.26	0.27	0.27
3	0.36	0.38	0.38	0.39	0.40	0.40	0.42
4	0.49	0.51	0.51	0.52	0.53	0.53	0.55
5	0.61	0.63	0.64	0.65	0.66	0.67	0.69
6	0.73	0.76	0.77	0.78	0.79	0.80	0.83
7	0.85	0.89	0.90	0.91	0.92	0.93	0.97
8	0.97	1.01	1.03	1.04	1.05	1.07	1.11
9	1.09	1.14	1.15	1.17	1.18	1.20	1.25
10	1.21	1.26	1.28	1.30	1.32	1.33	1.39
11	1.33	1.39	1.41	1.43	1.45	1.47	1.52
12	1.45	1.52	1.54	1.56	1.58	1.60	1.66
13	1.58	1.64	1.67	1.69	1.71	1.73	1.80
14	1.70	1.77	1.79	1.82	1.84	1.87	1.94
15	1.82	1.90	1.92	1.95	1.97	2.00	2.08
16	1.94	2.02	2.05	2.08	2.10	2.13	2.21
17	2.06	2.15	2.18	2.21	2.23	2.26	2.35
18	2.18	2.27	2.30	2.33	2.37	2.40	2.49
19	2.30	2.40	2.43	2.46	2.50	2.53	2.63
20	2.42	2.52	2.56	2.59	2.63	2.66	2.77
21	2.54	2.65	2.69	2.72	2.76	2.79	2.90
22	2.66	2.78	2.81	2.85	2.89	2.93	3.04
23	2.78	2.90	2.94	2.98	3.02	3.06	3.18
24	2.90	3.03	3.07	3.11	3.15	3.19	3.32
25	3.02	3.15	3.20	3.24	3.28	3.32	3.45
26	3.14	3.28	3.32	3.37	3.41	3.46	3.59
27	3.26	3.40	3.45	3.50	3.54	3.59	3.73
28	3.38	3.53	3.58	3.63	3.67	3.72	3.87
29	3.50	3.65	3.70	3.75	3.80	3.85	4.00
30	3.62	3.78	3.83	3.88	3.93	3.99	4.14
31	3.74	3.90	3.96	4.01	4.06	4.12	4.28
32	3.86	4.03	4.08	4.14	4.20	4.25	4.42
33	3.98	4.15	4.21	4.27	4.33	4.38	4.55
34	4.10	4.28	4.34	4.40	4.46	4.51	4.69
35	4.22	4.40	4.47	4.53	4.59	4.65	4.83

TABLE 2.9 Barometer Temperature Correction — Metric Units (*Continued*)

The values in the table below are to be subtracted from the observed readings to correct for the difference in the expansion of the mercury and the glass scale at different temperatures.

B. Brass scale

Temp. °C.	Observed barometer height in millimeters						
	640	650	660	670	680	690	700
	mm.	mm.	mm.	mm.	mm.	mm.	mm.
0	0.00	0.00	0.00	0.00	0.00	0.00	0.00
1	0.10	0.11	0.11	0.11	0.11	0.11	0.11
2	0.21	0.21	0.22	0.22	0.22	0.23	0.23
3	0.31	0.32	0.32	0.33	0.33	0.34	0.34
4	0.42	0.42	0.43	0.44	0.44	0.45	0.46
5	0.52	0.53	0.54	0.55	0.55	0.56	0.57
6	0.63	0.64	0.65	0.66	0.66	0.67	0.68
7	0.73	0.74	0.75	0.76	0.78	0.79	0.80
8	0.84	0.85	0.86	0.87	0.89	0.90	0.91
9	0.94	0.95	0.97	0.98	1.00	1.01	1.03
10	1.04	1.06	1.07	1.09	1.11	1.12	1.14
11	1.15	1.16	1.18	1.20	1.22	1.24	1.25
12	1.25	1.27	1.29	1.31	1.33	1.35	1.37
13	1.35	1.38	1.40	1.42	1.44	1.46	1.48
14	1.46	1.48	1.50	1.53	1.55	1.57	1.59
15	1.56	1.59	1.61	1.64	1.66	1.68	1.71
16	1.67	1.69	1.72	1.74	1.77	1.80	1.82
17	1.77	1.80	1.82	1.85	1.88	1.91	1.94
18	1.87	1.90	1.93	1.96	1.99	2.02	2.05
19	1.98	2.01	2.04	2.07	2.10	2.13	2.16
20	2.08	2.11	2.15	2.18	2.21	2.24	2.28
21	2.18	2.22	2.25	2.29	2.32	2.35	2.39
22	2.29	2.32	2.36	2.40	2.43	2.47	2.50
23	2.39	2.43	2.47	2.50	2.54	2.58	2.62
24	2.49	2.53	2.57	2.61	2.65	2.69	2.73
25	2.60	2.64	2.68	2.72	2.76	2.80	2.84
26	2.70	2.74	2.79	2.83	2.87	2.91	2.96
27	2.81	2.85	2.89	2.94	2.98	3.02	3.07
28	2.91	2.95	3.00	3.05	3.09	3.14	3.18
29	3.01	3.06	3.11	3.15	3.20	3.25	3.29
30	3.12	3.16	3.21	3.26	3.31	3.36	3.41
31	3.22	3.27	3.32	3.37	3.42	3.47	3.52
32	3.32	3.37	3.43	3.48	3.53	3.58	3.63
33	3.42	3.48	3.53	3.59	3.64	3.69	3.75
34	3.53	3.58	3.64	3.69	3.75	3.80	3.86
35	3.63	3.69	3.74	3.80	3.86	3.91	3.97

TABLE 2.9   Barometer Temperature Correction — Metric Units (*Continued*)

B. Brass scale (*continued*)

Observed barometer height in millimeters								Temp. °C.
710	720	730	740	750	760	770	780	
mm.	mm.	mm.	mm.	mm.	mm.	mm.	mm.	
0.00	0.00	0.00	0.00	0.00	0.00	0.00	0.00	0
0.12	0.12	0.12	0.12	0.12	0.12	0.13	0.13	1
0.23	0.23	0.24	0.24	0.24	0.25	0.25	0.25	2
0.35	0.35	0.36	0.36	0.37	0.37	0.38	0.38	3
0.46	0.47	0.48	0.48	0.49	0.50	0.50	0.51	4
0.58	0.59	0.59	0.60	0.61	0.62	0.63	0.64	5
0.69	0.70	0.71	0.72	0.73	0.74	0.75	0.76	6
0.81	0.82	0.83	0.84	0.86	0.87	0.88	0.89	7
0.93	0.94	0.95	0.96	0.98	0.99	1.00	1.02	8
1.04	1.06	1.07	1.08	1.10	1.11	1.13	1.14	9
1.16	1.17	1.19	1.21	1.22	1.24	1.25	1.27	10
1.27	1.29	1.31	1.33	1.34	1.36	1.38	1.40	11
1.39	1.41	1.43	1.45	1.47	1.48	1.50	1.52	12
1.50	1.52	1.54	1.57	1.59	1.61	1.63	1.65	13
1.62	1.64	1.66	1.69	1.71	1.73	1.75	1.78	14
1.73	1.76	1.78	1.81	1.83	1.85	1.88	1.90	15
1.85	1.87	1.90	1.93	1.95	1.98	2.00	2.03	16
1.96	1.99	2.02	2.05	2.07	2.10	2.13	2.16	17
2.08	2.11	2.14	2.17	2.20	2.22	2.25	2.28	18
2.19	2.22	2.25	2.29	2.32	2.35	2.38	2.41	19
2.31	2.34	2.37	2.41	2.44	2.47	2.50	2.54	20
2.42	2.46	2.49	2.53	2.56	2.59	2.63	2.66	21
2.54	2.57	2.61	2.65	2.68	2.72	2.75	2.79	22
2.65	2.69	2.73	2.77	2.80	2.84	2.88	2.91	23
2.77	2.81	2.85	2.88	2.92	2.96	3.00	3.04	24
2.88	2.92	2.96	3.00	3.05	3.09	3.13	3.17	25
3.00	3.04	3.08	3.12	3.17	3.21	3.25	3.29	26
3.11	3.16	3.20	3.24	3.29	3.33	3.38	3.42	27
3.23	3.27	3.32	3.36	3.41	3.45	3.50	3.54	28
3.34	3.39	3.44	3.48	3.53	3.58	3.62	3.67	29
3.46	3.50	3.55	3.60	3.65	3.70	3.75	3.80	30
3.57	3.62	3.67	3.72	3.77	3.82	3.87	3.92	31
3.68	3.74	3.79	3.84	3.89	3.94	4.00	4.05	32
3.80	3.85	3.91	3.96	4.01	4.07	4.12	4.17	33
3.91	3.97	4.02	4.08	4.13	4.19	4.24	4.30	34
4.03	4.09	4.14	4.20	4.26	4.31	4.37	4.43	35

C. Correction of a barometer for capillarity (*Smithsonian Tables*)

Diameter of tube, millimeters	Height of meniscus in millimeters							
	0.4	0.6	0.8	1.0	1.2	1.4	1.6	1.8
	Correction to be added in millimeters							
4	0.83	1.22	1.54	1.98	2.37	....	....	....
5	0.47	0.65	0.86	1.19	1.45	1.80	....	....
6	0.27	0.41	0.56	0.78	0.98	1.21	1.43	....
7	0.18	0.28	0.40	0.53	0.67	0.82	0.97	1.13
8	....	0.20	0.29	0.38	0.46	0.56	0.65	0.77
9	....	0.15	0.21	0.28	0.33	0.40	0.46	0.52
10	....	....	0.15	0.20	0.25	0.29	0.33	0.37
11	....	....	0.10	0.14	0.18	0.21	0.24	0.27
12	....	....	0.07	0.10	0.13	0.15	0.18	0.19
13	....	....	0.04	0.07	0.10	0.12	0.13	0.14

TABLE 2.10   Barometric Latitude-Gravity Table—Metric Units

Smithsonian Tables.

The values in the table below are to be subtracted from the barometric reading for latitudes from 0 to 45° inclusive, and are to be added from 46 to 90°.

Deg. Lat.	Barometer readings, millimeters					
	680	700	720	740	760	780
	mm.	mm.	mm.	mm.	mm.	mm.
0	1.82	1.87	1.93	1.98	2.04	2.09
5	1.79	1.85	1.90	1.95	2.00	2.06
10	1.71	1.76	1.81	1.86	1.92	1.97
15	1.58	1.63	1.67	1.72	1.77	1.81
20	1.40	1.44	1.49	1.53	1.57	1.61
21	1.36	1.40	1.44	1.48	1.52	1.56
22	1.32	1.36	1.40	1.44	1.48	1.51
23	1.28	1.31	1.35	1.39	1.43	1.46
24	1.23	1.27	1.30	1.34	1.37	1.41
25	1.18	1.22	1.25	1.29	1.32	1.36
26	1.13	1.17	1.20	1.23	1.27	1.30
27	1.08	1.12	1.15	1.18	1.21	1.24
28	1.03	1.06	1.09	1.12	1.15	1.18
29	0.98	1.01	1.04	1.07	1.10	1.12
30	0.93	0.95	0.98	1.01	1.04	1.06
31	0.87	0.90	0.92	0.95	0.98	1.00
32	0.82	0.84	0.86	0.89	0.91	0.94
33	0.76	0.78	0.80	0.83	0.85	0.87
34	0.70	0.72	0.74	0.76	0.79	0.81
35	0.64	0.66	0.68	0.70	0.72	0.74
36	0.58	0.60	0.62	0.64	0.65	0.67
37	0.52	0.54	0.56	0.57	0.59	0.60
38	0.46	0.48	0.49	0.51	0.52	0.53
39	0.40	0.42	0.43	0.44	0.45	0.46
40	0.34	0.35	0.36	0.37	0.38	0.39
41	0.28	0.29	0.30	0.30	0.31	0.32
42	0.22	0.22	0.23	0.24	0.24	0.25
43	0.16	0.16	0.16	0.17	0.17	0.18
44	0.09	0.10	0.10	0.10	0.10	0.11
45	0.03	0.03	0.03	0.03	0.03	0.04
46	0.03	0.03	0.03	0.03	0.04	0.04
47	0.09	0.10	0.10	0.10	0.10	0.11
48	0.16	0.16	0.17	0.17	0.18	0.18
49	0.22	0.23	0.23	0.24	0.25	0.25
50	0.28	0.29	0.30	0.31	0.31	0.32
51	0.34	0.35	0.36	0.37	0.38	0.39
52	0.40	0.42	0.43	0.44	0.45	0.46
53	0.46	0.48	0.49	0.51	0.52	0.53
54	0.52	0.54	0.56	0.57	0.59	0.60
55	0.58	0.60	0.62	0.64	0.65	0.67
56	0.64	0.66	0.68	0.70	0.72	0.74
57	0.70	0.72	0.74	0.76	0.78	0.80
58	0.76	0.78	0.80	0.82	0.85	0.87
59	0.81	0.84	0.86	0.89	0.91	0.93
60	0.87	0.89	0.92	0.94	0.97	1.00
61	0.92	0.95	0.98	1.00	1.03	1.06
62	0.97	1.00	1.02	1.05	1.08	1.11
63	1.03	1.06	1.09	1.12	1.15	1.18
64	1.08	1.11	1.14	1.17	1.20	1.23
65	1.13	1.16	1.19	1.22	1.26	1.29
66	1.17	1.21	1.24	1.28	1.31	1.35
67	1.22	1.25	1.29	1.33	1.36	1.40
68	1.26	1.30	1.34	1.37	1.41	1.45
69	1.31	1.34	1.38	1.42	1.46	1.50
70	1.35	1.39	1.43	1.47	1.51	1.55
72	1.42	1.47	1.51	1.55	1.59	1.63
75	1.53	1.57	1.62	1.66	1.71	1.75
80	1.66	1.71	1.76	1.81	1.86	1.90
85	1.74	1.79	1.84	1.90	1.95	2.00
90	1.77	1.82	1.87	1.93	1.98	2.03

TABLE 2.11   Barometric Correction for Gravity — Metric Units

The values in the table below are to be subtracted from the readings taken on a mercurial barometer to correct for the decrease in gravity with increase in altitude.

Height above sea-level meters	Observed barometer height in millimeters								
	400 mm.	450 mm.	500 mm.	550 mm.	600 mm.	650 mm.	700 mm.	750 mm.	800 mm.
100	....	....	....	....	....	....	0.02	0.02	0.02
200	....	....	....	....	....	....	0.04	0.05	0.05
300	....	....	....	....	....	....	0.07	0.07	0.07
400	....	....	....	....	....	....	0.09	0.10	0.10
500	....	....	....	....	....	....	0.11	0.12	0.13
600	....	....	....	....	....	0.12	0.13	0.14	....
700	....	....	....	....	....	0.14	0.15	0.16	....
800	....	....	....	....	....	0.16	0.18	0.19	....
900	....	....	....	....	....	0.18	0.20	0.22	....
1000	....	....	....	0.18	0.19	0.20	0.22	0.24	....
1100	....	....	....	0.19	0.21	0.22	0.24	....	....
1200	....	....	....	0.21	0.23	0.24	0.26	....	....
1300	....	....	....	0.22	0.24	0.26	0.29	....	....
1400	....	....	....	0.24	0.26	0.28	0.31	....	....
1500	....	....	0.24	0.26	0.28	0.30	0.33	....	....
1600	....	....	0.25	0.28	0.30	0.32	....	....	....
1700	....	....	0.27	0.30	0.32	0.34	....	....	....
1800	....	....	0.28	0.31	0.34	0.36	....	....	....
1900	....	....	0.30	0.33	0.36	0.39	....	....	....
2000	....	0.28	0.31	0.34	0.38	0.41	....	....	....
2100	....	0.30	0.33	0.36	0.40	....	....	....	....
2200	....	0.31	0.35	0.38	0.41	....	....	....	....
2300	....	0.32	0.36	0.40	0.43	....	....	....	....
2400	....	0.34	0.38	0.42	0.45	....	....	....	....
2500	0.31	0.35	0.39	0.43	0.47	....	....	....	....
2600	0.33	0.37	0.41	....	....	....	....	....	....
2800	0.35	0.40	0.44	....	....	....	....	....	....
3000	0.38	0.42	0.47	....	....	....	....	....	....
3200	0.40	0.46	....	....	....	....	....	.....	....
3400	0.43	0.48	....	....	....	....	....	.....	....

TABLE 2.12   Reduction of the Barometer to Sea Level—Metric Units

A barometer located at an elevation above sea level will show a reading lower than a barometer at sea level by an amount approximately 2.5 mm (0.1 in) for each 30.5 m (100 ft) of elevation. A closer approximation can be made by reference to the following tables, which take into account (1) the effect of altitude of the station at which the barometer is read, (2) the mean temperature of the air column extending from the station down to sea level, (3) the latitude of the station at which the barometer is read, and (4) the reading of the barometer corrected for its temperature, a correction which is applied only to mercurial barometers since the aneroid barometers are compensated for temperature effects.

Example.   A barometer which has been corrected for its temperature reads 650 mm at a station whose altitude is 1350 m above sea level and at a latitude of 30°. The mean temperature (outdoor temperature) at the station is 20°C.

Table A (metric units) gives for these conditions a temperature-altitude factor
of 135.2
The Latitude Factor Table gives for 135.2 at 30°lat. a correction of . . . .   +0.17
Therefore, the corrected value of the temperature-altitude factor is 135.37

Entering Table B (metric units), with a temperature-altitude factor of 135.37
and a barometric reading of 650 mm (corrected for temperature), the correction is
found to be 109.6
Accordingly the barometric reading reduced to sea level is 650 + 109.6 = 759.6 mm.

Latitude Factor – English or Metric Units. For latitudes 0°–45° add the latitude factor, for 45°–90° subtract the latitude factor, from the values obtained in Table A.

Temp.—Alt. Factor From Table A	Latitude				
	0°	10°	20°	30°	45°
50	0.1	0.1	0.1	0.1	0.0
100	0.3	0.3	0.2	0.1	0.0
150	0.4	0.4	0.3	0.2	0.0
200	0.5	0.5	0.4	0.3	0.0
250	0.7	0.6	0.5	0.3	0.0
300	0.8	0.8	0.6	0.4	0.0
350	0.9	0.9	0.7	0.5	0.0
	90°	80°	70°	60°	45°

TABLE 2.12   Reduction of the Barometer to Sea Level—Metric Units (*Continued*)

*A. Values of the temperature-altitude factor for use in Table B.**

Altitude in Meters	Mean Temperature of Air Column in Centigrade Degrees										
	−16°	−8°	−4°	0°	6°	10°	14°	18°	20°	22°	26°
10	1.2	1.1	1.1	1.1	1.1	1.0	1.0	1.0	1.0	1.0	1.0
50	5.8	5.6	5.5	5.4	5.3	5.2	5.1	5.0	5.0	5.0	4.9
100	11.5	11.2	11.0	10.8	10.6	10.4	10.3	10.1	10.0	9.9	9.8
150	17.3	16.7	16.5	16.2	15.9	15.6	15.4	15.1	15.0	14.9	14.7
200	23.0	22.3	22.0	21.6	21.1	20.8	20.5	20.2	20.0	19.9	19.6
250	28.8	27.9	27.5	27.0	26.4	26.0	25.6	25.2	25.0	24.9	24.5
300	34.5	33.5	33.0	32.5	31.7	31.2	30.7	30.3	30.1	29.8	29.4
350	40.3	39.0	38.5	37.9	37.0	36.4	35.9	35.3	35.1	34.8	34.3
400	46.0	44.6	43.9	43.3	42.3	41.6	41.0	40.4	40.1	39.8	39.2
450	51.8	51.3	49.4	48.7	47.6	46.8	46.1	45.4	45.1	44.8	44.1
500	57.5	55.8	54.9	54.1	52.9	52.0	51.2	50.5	50.1	49.7	49.0
550	63.3	61.4	60.4	59.5	58.1	57.2	56.4	55.5	55.1	54.7	53.9
600	69.0	66.9	65.9	64.9	63.4	62.4	61.5	60.6	60.1	59.7	58.8
650	74.8	72.5	71.4	70.3	68.7	67.6	66.6	65.6	65.1	64.6	63.7
700	80.6	78.1	76.9	75.7	74.0	72.9	71.7	70.7	70.1	69.6	68.6
750	86.3	83.7	82.4	81.1	79.3	78.1	76.9	75.7	75.1	74.6	73.5
800	92.1	89.2	87.9	86.5	84.6	83.3	82.0	80.8	80.1	79.6	78.4
850	97.8	94.8	93.4	92.0	89.8	88.5	87.1	85.8	85.2	84.5	83.3
900	103.6	100.4	98.9	97.4	95.1	93.7	92.2	90.8	90.2	89.5	88.2
950	109.3	106.0	104.4	102.8	100.4	98.9	97.4	95.9	95.2	94.5	93.1
1000	115.1	111.5	109.8	108.2	105.7	104.1	102.5	100.9	100.2	99.4	98.0
1050	120.8	117.1	115.3	113.6	111.0	109.3	107.6	106.0	105.2	104.4	102.9
1100	126.6	122.7	120.8	119.0	116.3	114.5	112.7	111.0	110.2	109.4	107.8
1150	132.3	128.3	126.3	124.4	121.6	119.7	117.9	116.1	115.2	114.4	112.7
1200	138.1	133.8	131.8	129.8	126.8	124.9	123.0	121.1	120.2	119.3	117.6
1250	143.8	139.4	137.3	135.2	132.1	130.1	128.1	126 2	125.2	124.3	122.5
1300	149.6	145.0	142.8	140.6	137.4	135.3	133.2	131.2	130.2	129.3	127.4
1350	155.3	150.6	148.3	146.0	142.7	140.5	138.4	136.3	135.2	134.2	132.3
1400	161.1	156.2	153.8	151.4	148.0	145.7	143.5	141.3	140.2	139.2	137.2
1450	166.8	161.7	159.3	156.8	153.3	150.9	148.6	146.4	145.3	144.2	142.1
1500	172.6	167.3	164.8	162.3	158.5	156.1	153.7	151.4	150.3	149.1	147.0
1550	178.3	172.9	170.2	167.7	163.8	161.3	158.8	156.4	155.3	154.1	151.8
1600	184.1	178.5	175.7	173.1	169.1	166.5	164.0	161.5	160.3	159.1	156.7
1650	189.8	184.0	181.2	178.5	174.4	171.7	169.1	166.5	165.3	164.1	161.6
1700	195.6	189.6	186.7	183.9	179.7	176.9	174.2	171.6	170.3	169.0	166.5
1750	201.4	195.2	192.2	189.3	185.0	182.1	179.3	176.6	175.3	174.0	171.4
1800	207.1	200.8	197.7	194.7	190.2	187.3	184.5	181.7	180.3	179.0	176.3
1850	212.9	206.3	203.2	200.1	195.5	192.5	189.6	186.7	185.3	183.9	181.2
1900	218.6	211.9	208.7	205.5	200.8	197.7	194.7	191.8	190.3	188.9	186.1
1950	224.4	217.5	214.2	210.9	206.1	202.9	199.8	196.8	195.3	193.9	191.0
2000	230.1	223.0	219.7	216.3	211.4	208.1	204.9	201.9	200.3	198.8	195.9
2050	235.9	228.6	225.1	221.7	216.7	213.3	210.1	206.9	205.3	203.8	200.8
2100	241.6	234.2	230.6	227.1	221.9	218.5	215.2	211.9	210.4	208.8	205.7
2150	247.4	239.8	236.1	232.5	227.2	223.7	220.3	217.0	215.4	213.8	210.6
2200	253.1	245.4	241.6	237.9	232.5	228.9	225.4	222.0	220.4	218.7	215.5
2250	258.9	250.9	247.1	243.4	237.8	234.1	230.6	227.1	225.4	223.7	220.4
2300	264.6	256.5	252.6	248.8	243.1	239.3	235.7	232.1	230.4	228.7	225.3
2350	270.4	262.1	258.1	254.2	248.3	244.5	240.8	237.2	235.4	233.6	230.2
2400	276.1	267.7	263.6	259.6	253.6	249.7	245.9	242.2	240.4	238.6	235.1
2450	281.9	273.2	269.1	265.0	258.9	254.9	251.0	247.3	245.4	243.6	240.0
2500	287.6	278.8	274.5	270.4	264.2	260.1	256.2	252.3	250.4	248.5	244.9
2550	293.4	284.4	280.0	275.8	269.5	265.3	261.3	257.3	255.4	253.5	249.8
2600	299.1	290.0	285.5	281.2	274.8	270.5	266.4	262.4	260.4	258.5	254.7
2650	304.9	295.5	291.0	286.6	280.0	275.7	271.5	267.4	265.4	263.4	259.6
2700	310.6	301.1	296.5	292.0	285.3	280.9	276.6	272.5	270.4	268.4	264.5
2750	316.4	306.7	302.0	297.4	290.6	286.1	281.8	277.5	275.4	273.4	269.4
2800	322.1	312.3	307.5	302.8	295.9	291.3	286.9	282.6	280.4	278.3	274.3
2850	327.9	317.8	313.0	308.2	301.2	296.5	292.0	287.6	285.4	283.3	279.2
2900	333.6	323.4	318.4	313.6	306.4	301.7	297.1	292.6	290.4	288.3	284.1
2950	339.4	329.0	323.9	319.0	311.7	306.9	302.2	297.7	295.5	293.3	289.0
3000	345.1	334.5	329.4	324.4	317.0	312.1	307.4	302.7	300.5	298.2	293.8

* From *Smithsonian Meteorological Tables*, 3d ed., 1907.

TABLE 2.12   Reduction of the Barometer to Sea Level — Metric Units (*Continued*)

B. Values in millimeters to be added. *

Barometer Reading in Millimeters

Temp.—Alt. Factor	790	770	750	730	710	690	670
1	0.9	0.9	0.9	0.8	0.8	0.8	
5	4.6	4.4	4.3	4.2	4.1	4.0	
10	9.1	8.9	8.7	8.5	8.2	8.0	
15	13.8	13.4	13.1	12.7	12.4	12.0	
20	18.4	17.9	17.5	17.0	16.5	16.1	
25		22.5	21.9	21.3	20.7	20.1	
30		27.1	26.4	25.7	25.0	24.2	
35		31.7	30.8	30.0	29.2	28.4	
40		36.3	35.3	34.4	33.5	32.5	31.6
45			39.9	38.8	37.8	36.7	35.6

Temp.—Alt. Factor	750	730	710	690	670	650	630
50	44.4	43.3	42.1	40.9	39.7		
55	49.0	47.7	46.4	45.1	43.8		
60	53.6	52.2	50.8	49.3	47.9		
65	58.3	56.7	55.2	53.6	52.1		
70		61.3	59.6	57.9	56.2		
75		65.8	64.0	62.2	60.4		
80		70.4	68.5	66.6	64.6	62.7	60.8
85		75.0	73.0	70.9	68.9	66.8	64.8
90			77.5	75.3	73.1	71.0	68.8
95			82.1	79.7	77.4	75.1	72.8

Temp.—Alt. Factor	710	690	670	650	630	610
100	86.6	84.2	81.8	79.3	76.9	
105	91.2	88.7	86.1	83.5	81.0	
110	95.9	93.2	90.5	87.8	85.1	
115	100.5	97.7	94.8	92.0	89.2	
120		102.2	99.3	96.3	93.3	
125		106.8	103.7	100.6	97.5	94.4
130		111.4	108.2	104.9	101.7	98.5
135		116.0	112.7	109.3	105.9	102.6
140		120.7	117.2	113.7	110.2	106.7
145			121.7	118.1	114.5	110.8

Temp.—Alt. Factor	670	650	630	610	590	570
150	126.3	122.5	118.8	115.0		
155	130.9	127.0	123.1	119.2		
160	135.5	131.5	127.4	123.4		
165	140.2	136.0	131.8	127.6		
170		140.5	136.2	131.9	127.5	123.2
175		145.1	140.6	136.2	131.7	127.2
180		149.7	145.1	140.5	135.9	131.3
185		154.3	149.5	144.8	140.0	135.3
190		158.9	154.0	149.2	144.3	139.4
195			158.6	153.5	148.5	143.5

Temp.—Alt. Factor	630	610	590	570	550	530
200	163.1	157.9	152.8	147.6		
205	167.7	162.4	157.1	151.7		
210	172.3	166.8	161.4	155.9		
215	176.9	171.3	165.7	160.1	154.5	148.9
220		175.8	170.1	164.3	158.5	152.8
225		180.4	174.5	168.5	162.6	156.7
230		184.9	178.9	172.8	166.7	160.7
235		189.5	183.3	177.1	170.9	164.7
240		194.1	187.8	181.4	175.0	168.7
245		198.8	192.3	185.7	179.2	172.7

Temp.—Alt. Factor	590	570	550	530	510
250	196.8	190.1	183.4	176.8	
255	201.3	194.5	187.7	180.8	
260	205.9	198.9	191.9	185.0	178.0
265	210.5	203.3	196.2	189.1	181.9
270	215.1	207.8	200.5	193.2	185.9
275	219.8	212.3	204.9	197.4	190.0
280		216.8	209.2	201.6	194.0
285		221.4	213.6	205.8	198.1
290		225.9	218.0	210.1	202.1
295		230.5	222.4	214.3	206.3

Temp.—Alt. Factor	570	550	530	510	490
300	235.1	226.9	218.6	210.4	
305	239.8	231.4	223.0	214.6	206.1
310		235.9	227.3	218.7	210.1
315		240.4	231.7	222.9	214.2
320		245.0	236.1	227.2	218.3
325		249.6	240.5	231.4	222.4
330		254.2	244.9	235.7	226.5
335		258.8	249.4	240.0	230.6
340		263.5	253.9	244.4	234.8
345			258.4	248.7	238.9

* From *Smithsonian Meteorological Tables*, 3d ed., 1907.

TABLE 2.13 Viscosity Conversion Table

Centistokes to Saybolt, Redwood, and Engler units.

Poise = cgs unit of absolute viscosity   Centipoise = 0.01 poise

Stoke = cgs unit of kinematic viscosity   Centistoke = 0.01 stoke

Centipoises = centistokes × density (at temperature under consideration)

Reyn (1 lb · s per sq in) = 69 × 10⁵ centipoises

Cf. *Jour. Inst. Pet. Tech.,* Vol. 22, p. 21 (1936); *Reports of A. S. T. M. Committee D-2, 1936 and 1937.*

The values of Saybolt Universal Viscosity at 100°F and at 210°F are taken directly from the comprehensive *ASTM Viscosity Tables, Special Technical Publication No. 43A* (1953) by permission of the publishers, American Society for Testing Materials, 1916 Race St., Philadelphia 3, Pa.

Centistokes	Saybolt Universal Viscosity at			Redwood Seconds at			Engler Degrees at all Temps.
	100°F.	130°F.	210°F.	70°F.	140°F.	200°F.	
2.0	32.62	32.68	32.85	30.2	31.0	31.2	1.14
3.0	36.03	36.10	36.28	32.7	33.5	33.7	1.22
4.0	39.14	39.22	39.41	35.3	36.0	36.3	1.31
5.0	42.35	42.43	42.65	37.9	38.5	38.9	1.40
6.0	45.56	45.65	45.88	40.5	41.0	41.5	1.48
7.0	48.77	48.86	49.11	43.2	43.7	44.2	1.56
8.0	52.09	52.19	52.45	46.0	46.4	46.9	1.65
9.0	55.50	55.61	55.89	48.9	49.1	49.7	1.75
10.0	58.91	59.02	59.32	51.7	52.0	52.6	1.84
11.0	62.43	62.55	62.86	54.8	55.0	55.6	1.93
12.0	66.04	66.17	66.50	57.9	58.1	58.8	2.02
14.0	73.57	73.71	74.09	64.4	64.6	65.3	2.22
16.0	81.30	81.46	81.87	71.0	71.4	72.2	2.43
18.0	89.44	89.61	90.06	77.9	78.5	79.4	2.64
20.0	97.77	97.96	98.45	85.0	85.8	86.9	2.87
22.0	106.4	106.6	107.1	92.4	93.3	94.5	3.10
24.0	115.0	115.2	115.8	99.9	100.9	102.2	3.34
26.0	123.7	123.9	124.5	107.5	108.6	110.0	3.58
28.0	132.5	132.8	133.4	115.3	116.5	118.0	3.82
30.0	141.3	141.6	142.3	123.1	124.4	126.0	4.07
32.0	150.2	150.5	151.2	131.0	132.3	134.1	4.32
34.0	159.2	159.5	160.3	138.9	140.2	142.2	4.57
36.0	168.2	168.5	169.4	146.9	148.2	150.3	4.83
38.0	177.3	177.6	178.5	155.0	156.2	158.3	5.08
40.0	186.3	186.7	187.6	163.0	164.3	166.7	5.34
42.0	195.3	195.7	196.7	171.0	172.3	175.0	5.59
44.0	204.4	204.8	205.9	179.1	180.4	183.3	5.85
46.0	213.7	214.1	215.2	187.1	188.5	191.7	6.11
48.0	222.9	223.3	224.5	195.2	196.6	200.0	6.37
50.0	232.1	232.5	233.8	203.3	204.7	208.3	6.63
60.0	278.3	278.8	280.2	243.5	245.3	250.0	7.90
70.0	324.4	325.0	326.7	283.9	286.0	291.7	9.21
80.0	370.8	371.5	373.4	323.9	326.6	333.4	10.53
90.0	417.1	417.9	420.0	364.4	367.4	375.0	11.84
100.0*	463.5	464.4	466.7	404.9	408.2	416.7	13.16

* At higher values use the same ratio as above for 100 centistokes; *e.g.,* 102 centistokes = 102 × 4.635 Saybolt seconds at 100 F.
To obtain the Saybolt Universal viscosity equivalent to a kinematic viscosity determined at *t*°F., multiply the equivalent Saybolt Universal viscosity at 100°F. by 1 + (t − 100) 0.000064; *e.g.,* 10 centistokes at 210°F are equivalent to 58.91 × 1.0070, or 59.32 Saybolt Universal Viscosity at 210°F.

TABLE 2.14 Conversion of Weighings in Air to Weighings in Vacuo

If the mass of a substance in air is m_f, its density ρ_m, the density of weights used in making the weighing ρ_w, and the density* of air ρ_a, the true mass of the substance in vacuo, m_{vac}, is

$$m_{vac} = m_f + \rho_a m_f \left(\frac{1}{\rho_m} - \frac{1}{\rho_w} \right)$$

For most purposes it is sufficient to assume a density of 8.4 for brass weights, and a density of 0.0012 for air under ordinary conditions. The equation then becomes

$$m_{vac} = m_f + 0.0012 m_f \left(\frac{1}{\rho_m} - \frac{1}{8.4} \right)$$

The table which follows gives the values of k (buoyancy reduction factor), which is the correction necessary because of the buoyant effect of the air upon the object weighed; the table is computed for air with the density of 0.0012; m is the weight in grams of the object when weighed in air; weight of object reduced to "in vacuo" = $m + km/1000$.

Density of object weighed	Buoyancy reduction factor, k			
	Brass weights, density =8.4	Pt or Pt-Ir weights, density =21.5	Al or quartz weights, density =2.7	Gold weights, density =17
0.2	5.89	5.98	5.58	5.97
0.3	3.87	3.96	3.56	3.95
0.4	2.87	2.95	2.55	2.94
0.5	2.26	2.35	1.95	2.34
0.6	1.86	1.95	1.55	1.93
0.7	1.57	1.66	1.26	1.65
0.75	1.46	1.55	1.15	1.53
0.80	1.36	1.45	1.05	1.43
0.82	1.32	1.41	1.01	1.39
0.84	1.29	1.37	0.98	1.36
0.86	1.25	1.34	0.94	1.33
0.88	1.22	1.31	0.91	1.29
0.90	1.19	1.28	0.88	1.26
0.92	1.16	1.25	0.85	1.24
0.94	1.13	1.22	0.82	1.21
0.96	1.11	1.20	0.80	1.18
0.98	1.08	1.17	0.77	1.16
1.00	1.06	1.15	0.75	1.13
1.02	1.03	1.12	0.72	1.11
1.04	1.01	1.10	0.70	1.08
1.06	0.99	1.08	0.68	1.06
1.08	0.97	1.06	0.66	1.04
1.10	0.95	1.04	0.64	1.02
1.12	0.93	1.02	0.62	1.00
1.14	0.91	1.00	0.60	0.98
1.16	0.89	0.98	0.58	0.96
1.18	0.87	0.96	0.56	0.95
1.20	0.86	0.95	0.55	0.93
1.25	0.82	0.91	0.51	0.89
1.30	0.78	0.87	0.47	0.85
1.35	0.75	0.83	0.44	0.82
1.40	0.71	0.80	0.40	0.79
1.50	0.66	0.74	0.35	0.73
1.6	0.61	0.69	0.30	0.68
1.7	0.56	0.65	0.25	0.64
1.8	0.52	0.61	0.21	0.60
1.9	0.49	0.58	0.18	0.56
2.0	0.46	0.54	0.15	0.53
2.2	0.40	0.49	0.09	0.48
2.4	0.36	0.44	0.05	0.43

* See Table 5.15, Specific Gravity of Air at Various Temperatures.

TABLE 2.14 Conversion of Weighings in Air to Weighings in Vacuo (*Continued*)

Density of object weighed	Buoyancy reduction factor, k			
	Brass weights, density = 8.4	Pt or Pt-Ir weights, density = 21.5	Al or quartz weights density = 2.7	Gold weights, density = 17
2.6	0.32	0.41	0.01	0.39
2.8	0.29	0.37	−0.02	0.36
3.0	0.26	0.34	−0.05	0.33
3.5	0.20	0.29	−0.11	0.27
4	0.16	0.24	−0.15	0.23
5	0.10	0.18	−0.21	0.17
6	0.06	0.14	−0.25	0.13
7	0.03	0.12	−0.28	0.10
8	0.01	0.09	−0.30	0.08
9	−0.01	0.08	−0.32	0.06
10	−0.02	0.06	−0.33	0.05
12	−0.04	0.04	−0.35	0.03
14	−0.06	0.03	−0.37	0.02
16	−0.07	0.02	−0.38	0.00
18	−0.08	0.01	−0.39	0.00
20	−0.08	0.00	−0.39	−0.01
22	−0.09	0.00	−0.40	−0.02

TABLE 2.15 Hydrometer Conversion Table

This table gives the relation between density (c.g.s.) and degrees on the Baumé and Twaddell scales. The Twaddell scale is never used for densities less than unity. See also Sec. 2.1.2.1, Hydrometers.

Density	Degrees Baumé (NIST* scale)	Degrees Baumé (A.P.I.† scale)	Density	Degrees Baumé (NIST* scale)	Degrees Baumé (A.P.I.† scale)
0.600	103.33	104.33	0.895	26.42	26.60
0.605	101.40	102.38	0.900	25.56	25.72
0.610	99.51	100.47	0.905	24.70	24.85
0.615	97.64	98.58	0.910	23.85	23.99
0.620	95.81	96.73	0.915	23.01	23.14
0.625	94.00	94.90	0.920	22.17	22.30
0.630	92.22	93.10	0.925	21.35	21.47
0.635	90.47	91.33	0.930	20.54	20.65
0.640	88.75	89.59	0.935	19.73	19.84
0.645	87.05	87.88	0.940	18.94	19.03
0.650	85.38	86.19	0.945	18.15	18.24
0.655	83.74	84.53	0.950	17.37	17.45
0.660	82.12	82.89	0.955	16.60	16.67
0.665	80.52	81.28	0.960	15.83	15.90
0.670	78.95	79.69	0.965	15.08	15.13
0.675	77.41	78.13	0.970	14.33	14.38
0.680	75.88	76.59	0.975	13.59	13.63
0.685	74.38	75.07	0.980	12.86	12.89
0.690	72.90	73.57	0.985	12.13	12.15
0.695	71.43	72.10	0.990	11.41	11.43
0.700	70.00	70.64	0.995	10.70	10.71
0.705	68.57	69.21	1.000	10.00	10.00
0.710	67.18	67.80			
0.715	65.80	66.40			
0.720	64.44	65.03			

DENSITIES GREATER THAN UNITY

Density	Degrees Baumé (NIST* scale)	Degrees Twaddell
1.00	0.00	0
1.01	1.44	2
1.02	2.84	4
1.03	4.22	6
1.04	5.58	8
1.05	6.91	10
1.06	8.21	12
1.07	9.49	14
1.08	10.78	16
1.09	11.97	18
1.10	13.18	20
1.11	14.37	22
1.12	15.54	24
1.13	16.68	26
1.14	17.81	28
1.15	18.91	30
1.16	20.00	32
1.17	21.07	34
1.18	22.12	36
1.19	23.15	38
1.20	24.17	40
1.21	25.16	42
1.22	26.15	44
1.23	27.11	46
1.24	28.06	48
1.25	29.00	50
1.26	29.92	52
1.27	30.83	54
1.28	31.72	56
1.29	32.60	58
1.30	33.46	60
1.31	34.31	62

Density	Degrees Baumé (NIST* scale)	Degrees Baumé (A.P.I.† scale)
0.725	63.10	63.67
0.730	61.78	62.34
0.735	60.48	61.02
0.740	59.19	59.72
0.745	57.92	58.43
0.750	56.67	57.17
0.755	55.43	55.92
0.760	54.21	54.68
0.765	53.01	53.47
0.770	51.82	52.27
0.775	50.65	51.08
0.780	49.49	49.91
0.785	48.34	48.75
0.790	47.22	47.61
0.795	46.10	46.49
0.800	45.00	45.38
0.805	43.91	44.28
0.810	42.84	43.19
0.815	41.78	42.12
0.820	40.73	41.06
0.825	39.70	40.02
0.830	38.68	38.98
0.835	37.66	37.96
0.840	36.67	36.95
0.845	35.68	35.96
0.850	34.71	34.97
0.855	33.74	34.00
0.860	32.79	33.03
0.865	31.85	32.08
0.870	30.92	31.14
0.875	30.00	30.21
0.880	29.09	29.30
0.885	28.19	28.39
0.890	27.30	27.49

* NIST, National Institute for Science and Technology (formerly the National Bureau of Standards, U.S.).

† A.P.I. is the American Petroleum Institute.

TABLE 2.15 Hydrometer Conversion Table (*Continued*)

Density	Degrees Baumé (NIST* scale)	Degrees Twaddell	Density	Degrees Baumé (NIST* scale)	Degrees Twaddell
1.32	35.15	64	1.67	58.17	134
1.33	35.98	66	1.68	58.69	136
1.34	36.79	68	1.69	59.20	138
1.35	37.59	70	1.70	59.71	140
1.36	38.38	72	1.71	60.20	142
1.37	39.16	74	1.72	60.70	144
1.38	39.93	76	1.73	61.18	146
1.39	40.68	78	1.74	61.67	148
1.40	41.43	80	1.75	62.14	150
1.41	42.16	82	1.76	62.61	152
1.42	42.89	84	1.77	63.08	154
1.43	43.60	86	1.78	63.54	156
1.44	44.31	88	1.79	63.99	158
1.45	45.00	90	1.80	64.44	160
1.46	45.68	92	1.81	64.89	162
1.47	46.36	94	1.82	65.31	164
1.48	47.03	96	1.83	65.77	166
1.49	47.68	98	1.84	66.20	168
1.50	48.33	100	1.85	66.62	170
1.51	48.97	102	1.86	67.04	172
1.52	49.60	104	1.87	67.46	174
1.53	50.23	106	1.88	67.87	176
1.54	50.84	108	1.89	68.28	178
1.55	51.45	110	1.90	68.68	180
1.56	52.05	112	1.91	69.08	182
1.57	52.64	114	1.92	69.48	184
1.58	53.23	116	1.93	69.87	186
1.59	53.80	118	1.94	70.26	188
1.60	54.38	120	1.95	70.64	190
1.61	54.94	122	1.96	71.02	192
1.62	55.49	124	1.97	71.40	194
1.63	56.04	126	1.98	71.77	196
1.64	56.58	128	1.99	72.14	198
1.65	57.12	130	2.00	72.50	200
1.66	57.65	132			

* NIST, National Institute for Science and Technology (formerly the National Bureau of Standards, U.S.).

TABLE 2.16 Pressure Conversion Chart

psi	Inches H_2O at 4°C	Inches Hg at 0°C	mmH_2O at 4°C	mmHg at 0°C	atm	Pascals $(N \cdot m^{-2})$
0.01	0.2768	0.0204	7.031	0.517	0.0007	68.95
0.02	0.5536	0.0407	14.06	1.034	0.0014	137.90
0.03	0.8304	0.0611	21.09	1.551	0.0020	206.8
0.04	1.107	0.0814	28.12	2.068	0.0027	275.8
0.05	1.384	0.1018	35.15	2.586	0.0034	344.7
0.06	1.661	0.1222	42.18	3.103	0.0041	413.7
0.07	1.938	0.1425	49.22	3.620	0.0048	482.6
0.08	2.214	0.1629	56.25	4.137	0.0054	551.6
0.09	2.491	0.1832	63.28	4.654	0.0061	620.5
0.10	2.768	0.2036	70.31	5.171	0.0068	689.5
0.20	5.536	0.4072	140.6	10.34	0.0136	1 379.9
0.30	8.304	0.6108	210.9	15.51	0.0204	2 068.5
0.40	11.07	0.8144	281.2	20.68	0.0272	2 758
0.50	13.84	1.018	351.5	25.86	0.0340	3 447
0.60	16.61	1.222	421.8	31.03	0.0408	4 137
0.70	19.38	1.425	492.2	36.20	0.0476	4 826
0.80	22.14	1.629	562.5	41.37	0.0544	5 516
0.90	24.91	1.832	632.8	46.54	0.0612	6 205
1.00	27.68	2.036	703.1	51.71	0.0689	6 895
2.00	55.36	4.072	1 072	103.4	0.1361	13 790
3.00	83.04	6.108	2 109	155.1	0.2041	20 684
4.00	110.7	8.144	2 812	206.8	0.2722	27 579
5.00	138.4	10.18	3 515	258.6	0.3402	34 474
6.00	166.1	12.22	4 218	310.3	0.4083	41 369
7.00	193.8	14.25	4 922	362.0	0.4763	48 263
8.00	221.4	16.29	5 625	413.7	0.5444	55 158
9.00	249.1	18.32	6 328	465.4	0.6124	62 053
10.0	276.8	20.36	7 031	517.1	0.6805	68 948
14.7	406.9	29.93	10 332	760.0	1.000	101 325
15.0	415.2	30.54	10 550	775.7	1.021	103 421
20.0	553.6	40.72	14 060	1 034	1.361	137 895
25.0	692.0	50.90	17 580	1 293	1.701	172 369
30.0	830.4	61.08	21 090	1 551	2.041	206 843
40.0	1 107	81.44	28 120	2 068	2.722	275 790
50.0	1 384	101.8	35 150	2 586	3.402	344 738
60.0	1 661	122.2	42 180	3 103	4.083	413 685
70.0	1 938	142.5	49 220	3 620	4.763	482 633
80.0	2 214	162.9	56 250	4 137	5.444	551 581
90.0	2 491	183.2	63 280	4 654	6.124	620 528
100.0	2 768	203.6	70 307	5 171	6.805	689 476
150.0	4 152	305.4		7 757	10.21	1 034 214
200.0	5 536	407.2		10 343	13.61	1 378 951
250.0	6 920	509.0			17.01	1 723 689
300.0	8 304	610.8			20.41	2 068 427
400.0					27.22	2 757 903
500.0					34.02	3 447 379

1 bar = 10^{-5} pascal.

TABLE 2.17   Corrections to Be Added to Molar Values to Convert to Molal

Temperature, °C	Aqueous solution			
	$\Delta G°$ J·mol⁻¹	$\Delta H°$ J·mol⁻¹	$\Delta S°$ J·deg⁻¹·mol⁻¹	$\Delta C_p°$ J·deg⁻¹·mol⁻¹
0	0.4	−42.7	−0.17	55.2
10	0.8	58.1	0.21	45.6
20	4.2	148.1	0.50	38.9
30	10.9	230.5	0.79	35.1
40	20.1	313.4	1.09	33.0
50	32.2	397.9	1.34	32.6
60	46.8	482.4	1.59	32.2

TABLE 2.18   Molar Equivalent of 1 Liter of Gas at Various Temperatures and Pressures

The values in this table, which give the number of moles in 1 liter of gas, are based on the properties of an "ideal" gas and were calculated by use of the formula:

$$\text{Moles/liter} = \frac{P}{760} \times \frac{273}{T} \times \frac{1}{22.40}$$

where P is the pressure in millimeters of mercury and T is the temperature in kelvins ($= t\,°C + 273$).

To convert to moles per cubic foot multiply the values in the table by 28.316.

Pressure mm of mercury	Temperature °C					
	10°	12°	14°	16°	18°	20°
655	0.03712	0.03686	0.03660	0.03634	0.03610	0.03585
660	3731	3714	3688	3662	3637	3612
665	3768	3742	3716	3690	3665	3640
670	3796	3770	3744	3718	3692	3667
675	3825	3798	3772	3745	3720	3695
680	0.03853	0.03826	0.03800	0.03773	0.03747	0.03694
685	3881	3854	3827	3801	3775	3749
690	3910	3882	3855	3829	3802	3776
695	3938	3910	3883	3856	3830	3804
700	3967	3939	3911	3884	3858	3831
702	0.03978	0.03950	0.03922	0.03895	0.03869	0.03842
704	3989	3961	3934	3906	3880	3853
706	4000	3972	3945	3917	3891	3864
708	4012	3984	3956	3929	3902	3875
710	4023	3995	3967	3940	3913	3886
712	0.04035	0.04006	0.03978	0.03951	0.03924	0.03897
714	4046	4018	3989	3962	3935	3908
716	4057	4029	4001	3973	3946	3919
718	4068	4040	4012	3984	3957	3930
720	4080	4051	4023	3995	3968	3941
722	0.04091	0.04063	0.04034	0.04006	0.03979	0.03952
724	4103	4074	4045	4017	3990	3963
726	4114	4085	4057	4028	4001	3973
728	4125	4096	4068	4040	4012	3984
730	4136	4108	4079	4051	4023	3995
732	0.04148	0.04119	0.04090	0.04062	0.04034	0.04006
734	4159	4130	4101	4073	4045	4017
736	4171	4141	4112	4084	4056	4028
738	4182	4153	4124	4095	4067	4039
740	4193	4164	4135	4106	4078	4050
742	0.04204	0.04175	0.04146	0.04117	0.04089	0.04061
744	4216	4186	4157	4128	4100	4072
746	4227	4198	4168	4139	4111	4083
748	4239	4209	4179	4151	4122	4094
750	4250	4220	4191	4162	4133	4105
752	0.04261	0.04231	0.04202	0.04173	0.04144	0.04116
754	4273	4243	4213	4184	4155	4127
756	4284	4254	4224	4195	4166	4138
758	4295	4265	4235	4206	4177	4149
760	4307	4276	4247	4217	4188	4160
762	0.04318	0.04287	0.04258	0.04228	0.04199	0.04171
764	4329	4299	4269	4239	4210	4181
766	4341	4310	4280	4250	4221	4192
768	4352	4321	4291	4262	4232	4203
770	4363	4333	4302	4273	4243	4214
772	0.04375	0.04344	0.04314	0.04284	0.04254	0.04225
774	4386	4355	4325	4295	4265	4236
776	4397	4366	4336	4306	4276	4247
778	4409	4378	4347	4317	4287	4258
780	4420	4389	4358	4328	4298	4269

TABLE 2.18 Molar Equivalent of 1 Liter of Gas at Various Temperatures and Pressures (*Continued*)

Pressure mm of mercury	Temperature °C					
	22°	24°	26°	28°	30°	32°
655	0.03561	0.03537	0.03515	0.03490	0.03467	0.03444
660	3588	3564	3541	3516	3493	3470
665	3614	3591	3568	3543	3520	3496
670	3642	3618	3595	3569	3546	3523
675	3669	3645	3622	3596	3572	3549
680	0.03697	0.03672	0.03649	0.03623	0.03599	0.03575
685	3724	3699	3676	3649	3625	3602
690	3751	3726	3702	3676	3652	3628
695	3778	3753	3729	3703	3678	3654
700	3805	3780	3756	3729	3705	3680
702	0.03816	0.03790	0.03767	0.03740	0.03715	0.03691
704	3827	3801	3777	3750	3726	3701
706	3838	3812	3788	3761	3736	3712
708	3849	3823	3799	3772	3747	3722
710	3860	3834	3810	3783	3758	3733
712	0.03870	0.03844	0.03820	0.03793	0.03768	0.03744
714	3881	3855	3831	3804	3779	3754
716	3892	3866	3842	3815	3789	3765
718	3902	3877	3853	3825	3800	3775
720	3914	3888	3863	3836	3811	3786
722	0.03925	0.03898	0.03874	0.03847	0.03821	0.03796
724	3936	3909	3885	3857	3832	3807
726	3947	3920	3896	3868	3842	3817
728	3957	3931	3906	3878	3853	3828
730	3968	3941	3917	3889	3863	3838
732	0.03979	0.03952	0.03928	0.03900	0.03874	0.03849
734	3990	3963	3938	3910	3885	3859
736	4001	3974	3949	3921	3895	3870
738	4012	3985	3960	3932	3906	3880
740	4023	3995	3971	3942	3916	3891
742	0.04033	0.04006	0.03981	0.03953	0.03927	0.03901
744	4044	4017	3992	3964	3938	3912
746	4055	4028	4003	3974	3948	3922
748	4066	4039	4014	3985	3959	3933
750	4077	4049	4024	3996	3969	3943
752	0.04088	0.04060	0.04035	0.04006	0.03980	0.03954
754	4099	4071	4046	4017	3991	3964
756	4110	4082	4056	4028	4001	3975
758	4121	4093	4067	4038	4012	3985
760	4131	4103	4078	4049	4022	3996
762	0.04142	4114	4089	4060	4033	4006
764	4153	4125	4099	4070	4043	4017
766	4164	4136	4110	4081	4054	4027
768	4175	4147	4121	4092	4065	4038
770	4186	4158	4132	4102	4075	4048
772	0.04197	0.04168	0.04142	0.04113	0.04086	0.04059
774	4207	4179	4153	4124	4096	4070
776	4218	4190	4164	4134	4107	4080
778	4229	4201	4175	4145	4117	4091
780	4240	4211	4185	4155	4128	4101

TABLE 2.19 Factors for Reducing Gas Volumes to Normal (Standard) Temperature and Pressure (760 mmHg)

Examples: (*a*) 20 mL of dry gas at 22°C and 730 mm = 20 × 0.8888 = 17.78 mL at 0°C and 760 mm. (*b*) 20 mL of a gas over water at 22° and 730 mm = 20 × (factor corrected for aqueous tension; i.e., 730 − 19.8 or 710.2 mm) = 20 mL of dry gas at 22° and 710.2 mm = 20 × 0.86475 = 17.30 mL at 0°C and 760 mm. Mass in milligrams of 1 mL of gas at S.T.P.: acetylene, 1.173; carbon dioxide, 1.9769; hydrogen, 0.0899; nitric oxide (NO), 1.3402; nitrogen, 1.25057; oxygen, 1.42904.

Pressure mm of mercury	Temperature °C							
	10°	11°	12°	13°	14°	15°	16°	17°
670	0.8504	0.8474	0.8445	0.8415	0.8386	0.8357	0.8328	0.8299
672	0.8530	0.8500	0.8470	0.8440	0.8411	0.8382	0.8353	0.8324
674	0.8555	0.8525	0.8495	0.8465	0.8436	0.8407	0.8377	0.8349
676	0.8580	0.8550	0.8520	0.8490	0.8461	0.8431	0.8402	0.8373
678	0.8606	0.8576	0.8545	0.8516	0.8486	0.8456	0.8427	0.8398
680	0.8631	0.8601	0.8571	0.8541	0.8511	0.8481	0.8452	0.8423
682	0.8657	0.8626	0.8596	0.8566	0.8536	0.8506	0.8477	0.8448
684	0.8682	0.8651	0.8621	0.8591	0.8561	0.8531	0.8502	0.8472
686	0.8707	0.8677	0.8646	0.8616	0.8586	0.8556	0.8527	0.8497
688	0.8733	0.8702	0.8672	0.8641	0.8611	0.8581	0.8551	0.8522
690	0.8758	0.8727	0.8697	0.8666	0.8636	0.8606	0.8576	0.8547
692	0.8784	0.8753	0.8722	0.8691	0.8661	0.8631	0.8601	0.8572
694	0.8809	0.8778	0.8747	0.8717	0.8686	0.8656	0.8626	0.8596
696	0.8834	0.8803	0.8772	0.8742	0.8711	0.8681	0.8651	0.8621
698	0.8860	0.8828	0.8798	0.8767	0.8736	0.8706	0.8676	0.8646
700	0.8885	0.8854	0.8823	0.8792	0.8761	0.8731	0.8700	0.8671
702	0.8910	0.8879	0.8848	0.8817	0.8786	0.8756	0.8725	0.8695
704	0.8936	0.8904	0.8873	0.8842	0.8811	0.8781	0.8750	0.8720
706	0.8961	0.8930	0.8898	0.8867	0.8836	0.8806	0.8775	0.8745
708	0.8987	0.8955	0.8924	0.8892	0.8861	0.8831	0.8800	0.8770
710	0.9012	0.8980	0.8949	0.8917	0.8886	0.8856	0.8825	0.8794
712	0.9037	0.9006	0.8974	0.8943	0.8911	0.8880	0.8850	0.8819
714	0.9063	0.9031	0.8999	0.8968	0.8936	0.8905	0.8875	0.8844
716	0.9088	0.9056	0.9024	0.8993	0.8961	0.8930	0.8899	0.8869
718	0.9114	0.9081	0.9050	0.9018	0.8987	0.8955	0.8924	0.8894
720	0.9139	0.9107	0.9075	0.9043	0.9012	0.8980	0.8949	0.8918
722	0.9164	0.9132	0.9100	0.9068	0.9037	0.9005	0.8974	0.8943
724	0.9190	0.9157	0.9125	0.9093	0.9062	0.9030	0.8999	0.8968
726	0.9215	0.9183	0.9151	0.9118	0.9087	0.9055	0.9024	0.8993
728	0.9241	0.9208	0.9176	0.9144	0.9112	0.9080	0.9049	0.9017
730	0.9266	0.9233	0.9201	0.9169	0.9137	0.9105	0.9073	0.9042
732	0.9291	0.9259	0.9226	0.9194	0.9162	0.9130	0.9098	0.9067
734	0.9317	0.9284	0.9251	0.9219	0.9187	0.9155	0.9123	0.9092
736	0.9342	0.9309	0.9277	0.9244	0.9212	0.9180	0.9148	0.9117
738	0.9368	0.9334	0.9302	0.9269	0.9237	0.9205	0.9173	0.9141
740	0.9393	0.9360	0.9327	0.9294	0.9262	0.9230	0.9198	0.9166
742	0.9418	0.9385	0.9352	0.9319	0.9287	0.9255	0.9223	0.9191
744	0.9444	0.9410	0.9377	0.9345	0.9312	0.9280	0.9248	0.9216
746	0.9469	0.9436	0.9403	0.9370	0.9337	0.9305	0.9272	0.9240
748	0.9494	0.9461	0.9428	0.9395	0.9362	0.9329	0.9297	0.9265
750	0.9520	0.9486	0.9453	0.9420	0.9387	0.9354	0.9322	0.9290
752	0.9545	0.9511	0.9478	0.9445	0.9412	0.9379	0.9347	0.9315
754	0.9571	0.9537	0.9504	0.9470	0.9437	0.9404	0.9372	0.9339
756	0.9596	0.9562	0.9529	0.9495	0.9462	0.9429	0.9397	0.9364
758	0.9621	0.9587	0.9554	0.9520	0.9487	0.9454	0.9422	0.9389
760	0.9647	0.9613	0.9579	0.9546	0.9512	0.9479	0.9446	0.9414
762	0.9672	0.9638	0.9604	0.9571	0.9537	0.9504	0.9471	0.9439
764	0.9698	0.9663	0.9630	0.9596	0.9562	0.9529	0.9496	0.9463
766	0.9723	0.9689	0.9655	0.9620	0.9587	0.9554	0.9521	0.9488
768	0.9748	0.9714	0.9680	0.9646	0.9612	0.9579	0.9546	0.9513
770	0.9774	0.9739	0.9705	0.9671	0.9637	0.9604	0.9571	0.9538
772	0.9799	0.9764	0.9730	0.9696	0.9662	0.9629	0.9596	0.9562
774	0.9825	0.9790	0.9756	0.9721	0.9687	0.9654	0.9620	0.9587
776	0.9850	0.9815	0.9781	0.9746	0.9712	0.9679	0.9645	0.9612
778	0.9875	0.9840	0.9806	0.9772	0.9737	0.9704	0.9670	0.9637
780	0.9901	0.9866	0.9831	0.9797	0.9763	0.9729	0.9695	0.9662
782	0.9926	0.9891	0.9856	0.9822	0.9788	0.9754	0.9720	0.9686
784	0.9952	0.9916	0.9882	0.9847	0.9813	0.9778	0.9745	0.9711
786	0.9977	0.9942	0.9907	0.9872	0.9838	0.9803	0.9770	0.9736
788	1.0002	0.9967	0.9932	0.9897	0.9863	0.9828	0.9794	0.9761

TABLE 2.19 Factors for Reducing Gas Volumes to Normal (Standard) Temperature and Pressure (*Continued*)

Pressure mm of mercury	Temperature °C							
	18°	19°	20°	21°	22°	23°	24°	25°
670	0.8270	0.8242	0.8214	0.8186	0.8158	0.8131	0.8103	0.8076
672	0.8295	0.8267	0.8239	0.8211	0.8183	0.8155	0.8128	0.8100
674	0.8320	0.8291	0.8263	0.8235	0.8207	0.8179	0.8152	0.8124
676	0.8345	0.8316	0.8288	0.8259	0.8231	0.8204	0.8176	0.8149
678	0.8369	0.8341	0.8312	0.8284	0.8256	0.8228	0.8200	0.8173
680	0.8394	0.8365	0.8337	0.8308	0.8280	0.8252	0.8224	0.8197
682	0.8419	0.8390	0.8361	0.8333	0.8304	0.8276	0.8249	0.8221
684	0.8443	0.8414	0.8386	0.8357	0.8329	0.8301	0.8273	0.8245
686	0.8468	0.8439	0.8410	0.8382	0.8353	0.8325	0.8297	0.8269
688	0.8493	0.8464	0.8435	0.8406	0.8378	0.8349	0.8321	0.8293
690	0.8517	0.8488	0.8459	0.8430	0.8402	0.8373	0.8345	0.8317
692	0.8542	0.8513	0.8484	0.8455	0.8426	0.8398	0.8369	0.8341
694	0.8567	0.8537	0.8508	0.8479	0.8451	0.8422	0.8394	0.8366
696	0.8591	0.8562	0.8533	0.8504	0.8475	0.8446	0.8418	0.8390
698	0.8616	0.8587	0.8557	0.8528	0.8499	0.8471	0.8442	0.8414
700	0.8641	0.8611	0.8582	0.8553	0.8524	0.8495	0.8466	0.8438
702	0.8665	0.8636	0.8606	0.8577	0.8547	0.8519	0.8490	0.8462
704	0.8690	0.8660	0.8631	0.8602	0.8572	0.8543	0.8515	0.8486
706	0.8715	0.8685	0.8655	0.8626	0.8597	0.8568	0.8539	0.8510
708	0.8740	0.8710	0.8680	0.8650	0.8621	0.8592	0.8563	0.8534
710	0.8764	0.8734	0.8704	0.8675	0.8645	0.8616	0.8587	0.8558
712	0.8789	0.8759	0.8729	0.8699	0.8670	0.8640	0.8611	0.8582
714	0.8814	0.8783	p.8753	0.8724	0.8694	0.8665	0.8636	0.8607
716	0.8838	0.8808	0.8778	0.8748	0.8718	0.8689	0.8660	0.8631
718	0.8863	0.8833	0.8802	0.8773	0.8743	0.8713	0.8684	0.8655
720	0.8888	0.8857	0.8827	0.8797	0.8767	0.8738	0.8708	0.8679
722	0.8912	0.8882	0.8852	0.8821	0.8792	0.8762	0.8732	0.8703
724	0.8937	0.8906	0.8876	0.8846	0.8816	0.8786	0.8757	0.8727
726	0.8962	0.8931	0.8901	0.8870	0.8840	0.8810	0.8781	0.8751
728	0.8986	0.8956	0.8925	0.8895	0.8865	0.8835	0.8805	0.8775
730	0.9011	0.8980	0.8950	0.8919	0.8889	0.8859	0.8829	0.8799
732	0.9036	0.9005	0.8974	0.8944	0.8913	0.8883	0.8853	0.8824
734	0.9060	0.9029	0.8999	0.8968	0.8938	0.8907	0.8877	0.8848
736	0.9085	0.9054	0.9023	0.8992	0.8962	0.8932	0.8902	0.8872
738	0.9110	0.9079	0.9048	0.9017	0.8986	0.8956	0.8926	0.8896
740	0.9135	0.9103	0.9072	0.9041	0.9011	0.8980	0.8950	0.8920
742	0.9159	0.9128	0.9097	0.9066	0.9035	0.9005	0.8974	0.8944
744	0.9184	0.9153	0.9121	0.9090	0.9059	0.9029	0.8998	0.8968
746	0.9209	0.9177	0.9146	0.9115	0.9084	0.9053	0.9023	0.8992
748	0.9233	0.9202	0.9170	0.9139	0.9108	0.9077	0.9047	0.9016
750	0.9258	0.9226	0.9195	0.9164	0.9132	0.9102	0.9071	0.9041
752	0.9283	0.9251	0.9219	0.9188	0.9157	0.9126	0.9095	0.9065
754	0.9307	0.9276	0.9244	0.9212	0.9181	0.9150	0.9119	0.9089
756	0.9332	0.9300	0.9268	0.9237	0.9206	0.9174	0.9144	0.9113
758	0.9357	0.9325	0.9293	0.9261	0.9230	0.9199	0.9168	0.9137
760	0.9381	0.9349	0.9317	0.9286	0.9254	0.9223	0.9192	0.9161
762	0.9406	0.9374	0.9342	0.9310	0.9279	0.9247	0.9216	0.9185
764	0.9431	0.9399	0.9366	0.9335	0.9303	0.9272	0.9240	0.9209
766	0.9456	0.9423	0.9391	0.9359	0.9327	0.9296	0.9265	0.9233
768	0.9480	0.9448	0.9415	0.9383	0.9352	0.9320	0.9289	0.9258
770	0.9505	0.9472	0.9440	0.9408	0.9376	0.9344	0.9313	0.9282
772	0.9530	0.9497	0.9464	0.9432	0.9400	0.9369	0.9337	0.9306
774	0.9554	0.9522	0.9489	0.9457	0.9425	0.9393	0.9361	0.93?0
776	0.9579	0.9546	0.9514	0.9481	0.9449	0.9417	0.9385	0.9354
778	0.9604	0.9571	0.9538	0.9506	0.9473	0.9441	0.9410	0.9378
780	0.9628	0.9595	0.9563	0.9530	0.9498	0.9466	0.9434	0.9402
782	0.9653	0.9620	0.9587	0.9555	0.9522	0.9490	0.9458	0.9426
784	0.9678	0.9645	0.9612	0.9579	0.9546	0.9514	0.9482	0.9450
786	0.9702	0.9669	0.9636	0.9603	0.9571	0.9538	0.9506	0.9474
788	0.9727	0.9694	0.9661	0.9628	0.9595	0.9563	0.9531	0.9499

TABLE 2.19 Factors for Reducing Gas Volumes to Normal (Standard) Temperature and Pressure (*Continued*)

Pressure mm of mercury	Temperature °C							
	26°	27°	28°	29°	30°	31°	32°	33°
670	0.8049	0.8022	0.7996	0.7969	0.7943	0.7917	0.7891	0.7865
672	0.8073	0.8046	0.8020	0.7993	0.7967	0.7940	0.7914	0.7889
674	0.8097	0.8070	0.8043	0.8017	0.7990	0.7964	0.7938	0.7912
676	0.8121	0.8094	0.8067	0.8041	0.8014	0.7988	0.7962	0.7936
678	0.8145	0.8118	0.8091	0.8064	0.8038	0.8011	0.7985	0.7959
680	0.8169	0.8142	0.8115	0.8088	0.8061	0.8035	0.8009	0.7982
682	0.8193	0.8166	0.8139	0.8112	0.8085	p.8059	0.8032	0.8006
684	0.8217	0.8190	0.8163	0.8136	0.8109	0.8082	0.8056	0.8029
686	0.8241	0.8214	0.8187	0.8160	0.8133	0.8106	0.8079	0.8053
688	0.8265	0.8238	0.8211	0.8183	0.8156	0.8129	0.8103	0.8076
690	0.8289	0.8262	0.8234	0.8207	0.8180	0.8153	0.8126	0.8100
692	0.8313	0.8286	0.8258	0.8231	0.8204	0.8177	0.8150	0.8123
694	0.8338	0.8310	0.8282	0.8255	0.8227	0.8200	0.8174	0.8147
696	0.8362	0.8334	0.8306	0.8278	0.8251	0.8224	0.8197	0.8170
698	0.8386	0.8358	0.8330	0.8302	0.8275	0.8248	0.8221	0.8194
700	0.8410	0.8382	0.8354	0.8326	0.8299	0.8271	0.8244	0.8217
702	0.8434	0.8406	0.8378	0.8350	0.8322	0.8295	0.8268	0.8241
704	0.8458	0.8429	0.8401	0.8374	0.8346	0.8319	0.8291	0.8264
706	0.8482	0.8453	0.8425	0.8397	0.8370	0.8342	0.8315	0.8288
708	0.8506	0.8477	0.8449	0.8421	0.8393	0.8366	0.8338	0.8311
710	0.8530	0.8501	0.8473	0.8445	0.8417	0.8389	0.8362	0.8335
712	0.8554	0.8525	0.8497	0.8469	0.8441	0.8413	0.8386	0.8358
714	0.8578	0.8549	0.8521	0.8493	0.8465	0.8437	0.8409	0.8382
716	0.8602	0.8573	0.8545	0.8516	0.8488	0.8460	0.8433	0.8405
718	0.8626	0.8597	0.8569	0.8540	0.8512	0.8484	0.8456	0.8429
720	0.8650	0.8621	0.8592	0.8564	0.8536	0.8508	0.8480	0.8452
722	0.8674	0.8645	0.8616	0.8588	0.8559	0.8531	0.8503	0.8475
724	0.8698	0.8669	0.8640	0.8612	0.8583	0.8555	0.8527	0.8499
726	0.8722	0.8693	0.8664	0.8635	0.8607	0.8579	0.8550	0.8522
728	0.8746	0.8717	0.8688	0.8659	0.8631	0.8602	0.8574	0.8546
730	0.8770	0.8741	0.8712	0.8683	0.8654	0.8626	0.8598	0.8569
732	0.8794	0.8765	0.8736	0.8707	0.8678	0.8649	0.8621	0.8593
734	0.8818	0.8789	0.8759	0.8730	0.8702	0.8673	0.8645	0.8616
736	0.8842	0.8813	0.8783	0.8754	0.8725	0.8697	0.8668	0.8640
738	0.8866	0.8837	0.8807	0.8778	0.8749	0.8720	0.8692	0.8663
740	0.8890	0.8861	0.8831	0.8802	0.8773	0.8744	0.8715	0.8687
742	0.8914	0.8884	0.8855	0.8826	0.8796	0.8768	0.8739	0.8710
744	0.8938	0.8908	0.8879	0.8849	0.8820	0.8791	0.8762	0.8734
746	0.8962	0.8932	0.8903	0.8873	0.8844	0.8815	0.8786	0.8757
748	0.8986	0.8956	0.8927	0.8897	0.8868	0.8838	0.8809	0.8781
750	0.9010	0.8980	0.8950	0.8921	0.8891	0.8862	0.8833	0.8804
752	0.9034	0.9004	0.8974	0.8945	0.8915	0.8886	0.8857	0.8828
754	0.9058	0.9028	0.8998	0.8968	0.8939	0.8909	0.8880	0.8851
756	0.9082	0.9052	0.9022	0.8992	0.8962	0.8933	0.8904	0.8875
758	0.9106	0.9076	0.9046	0.9016	0.8986	0.8957	0.8927	0.8898
760	0.9130	0.9100	0.9070	0.9040	0.9010	0.8980	0.8951	0.8922
762	0.9154	0.9124	0.9094	0.9064	0.9034	0.9004	0.8974	0.8945
764	0.9178	0.9148	0.9118	0.9087	0.9057	0.9028	0.8998	0.8969
766	0.9202	0.9172	0.9141	0.9111	0.9081	0.9051	0.9021	0.8992
768	0.9227	0.9196	0.9165	0.9135	0.9105	0.9075	0.9045	0.9015
770	0.9251	0.9220	0.9189	0.9159	0.9128	0.9098	0.9069	0.9039
772	0.9275	0.9244	0.9213	0.9182	0.9152	0.9122	0.9092	0.9062
774	0.9299	0.9268	0.9237	0.9206	0.9176	0.9146	0.9116	0.9086
776	0.9323	0.9292	0.9261	0.9230	0.9200	0.9169	0.9139	0.9109
778	0.9347	0.9316	0.9285	0.9254	0.9223	0.9193	0.9163	0.9133
780	0.9371	0.9340	0.9308	0.9278	0.9247	0.9217	0.9186	0.9156
782	0.9395	0.9363	0.9332	0.9301	0.9271	0.9240	0.9210	0.9180
784	0.9419	0.9387	0.9356	0.9325	0.9294	0.9264	0.9233	0.9203
786	0.9443	0.9411	0.9380	0.9349	0.9318	0.9287	0.9257	0.9227
788	0.9467	0.9435	0.9404	0.9373	0.9342	0.9311	0.9281	0.9250

TABLE 2.19 Factors for Reducing Gas Volumes to Normal (Standard) Temperature and Pressure (*Continued*)

Pressure mm of mercury	Temperature °C.		
	34°	35°	36°
670	0.7839	0.7814	0.7789
672	0.7863	0.7837	0.7812
674	0.7886	0.7861	0.7835
676	0.7910	0.7884	0.7858
678	0.7933	0.7907	0.7882
680	0.7956	0.7931	0.7905
682	0.7980	0.7954	0.7928
684	0.8003	0.7977	0.7951
686	0.8027	0.8001	0.7975
688	0.8050	0.8024	0.7998
690	0.8073	0.8047	0.8021
692	0.8097	0.8071	0.8044
694	0.8120	0.8094	0.8068
696	0.8144	0.8117	0.8091
698	0.8167	0.8141	0.8114
700	0.8190	0.8164	0.8137
702	0.8214	0.8187	0.8161
704	0.8237	0.8211	0.8184
706	0.8261	0.8234	0.8207
708	0.8284	0.8257	0.8230
710	0.8307	0.8281	0.8254
712	0.8331	0.8304	0.8277
714	0.8354	0.8327	0.8300
716	0.8378	0.8350	0.8323
718	0.8401	0.8374	0.8347
720	0.8424	0.8397	0.8370
722	0.8448	0.8420	0.8393
724	0.8471	0.8444	0.8416
726	0.8495	0.8467	0.8440
728	0.8518	0.8490	0.8463
730	0.8541	0.8514	0.8486
732	0.8565	0.8537	0.8509
734	0.8588	0.8560	0.8533
736	0.8612	0.8584	0.8556
738	0.8635	0.8607	0.8579
740	0.8658	0.8630	0.8602
742	0.8682	0.8654	0.8626
744	0.8705	0.8677	0.8649
746	0.8729	0.8700	0.8672
748	0.8752	0.8724	0.8695
750	0.8775	0.8747	0.8719
752	0.8799	0.8770	0.8742
754	0.8822	0.8794	0.8765
756	0.8846	0.8817	0.8788
758	0.8869	0.8840	0.8812
760	0.8892	0.8864	0.8835
762	0.8916	0.8887	0.8858
764	0.8939	0.8910	0.8881
766	0.8963	0.8934	0.8905
768	0.8986	0.8957	0.8928
770	0.9009	0.8980	0.8951
772	0.9033	0.9004	0.8974
774	0.9056	0.9027	0.8998
776	0.9080	0.9050	0.9021
778	0.9103	0.9074	0.9044
780	0.9127	0.9097	0.9067
782	0.9150	0.9120	0.9091
784	0.9173	0.9144	0.9114
786	0.9197	0.9167	0.9137
788	0.9220	0.9190	0.9160

TABLE 2.20 Values of Absorbance for Percent Absorption

To convert percent absorption (% A) to absorbance, find the present absorption to the nearest whole digit in the left-hand column; read across to the column located under the tenth of a percent desired, and read the value of absorbance. The value of absorbance corresponding to 26.8% absorption is thus 0.1355.

% A	.0	.1	.2	.3	.4	.5	.6	.7	.8	.9
0.0	.0000	.0004	.0009	.0013	.0017	.0022	.0026	.0031	.0035	.0039
1.0	.0044	.0048	.0052	.0057	.0061	.0066	.0070	.0074	.0079	.0083
2.0	.0088	.0092	.0097	.0101	.0106	.0110	.0114	.0119	.0123	.0128
3.0	.0132	.0137	.0141	.0146	.0150	.0155	.0159	.0164	.0168	.0173
4.0	.0177	.0182	.0186	.0191	.0195	.0200	.0205	.0209	.0214	.0218
5.0	.0223	.0227	.0232	.0236	.0241	.0246	.0250	.0255	.0259	.0264
6.0	.0269	.0273	.0278	.0283	.0287	.0292	.0297	.0301	.0306	.0311
7.0	.0315	.0320	.0325	.0329	.0334	.0339	.0343	.0348	.0353	.0357
8.0	.0362	.0367	.0372	.0376	.0381	.0386	.0391	.0395	.0400	.0405
9.0	.0410	.0414	.0419	.0424	.0429	.0434	.0438	.0443	.0448	.0453
10.0	.0458	.0462	.0467	.0472	.0477	.0482	.0487	.0491	.0496	.0501
11.0	.0506	.0511	.0516	.0521	.0526	.0531	.0535	.0540	.0545	.0550
12.0	.0555	.0560	.0565	.0570	.0575	.0580	.0585	.0590	.0595	.0600
13.0	.0605	.0610	.0615	.0620	.0625	.0630	.0635	.0640	.0645	.0650
14.0	.0655	.0660	.0665	.0670	.0675	.0680	.0685	.0691	.0696	.0701
15.0	.0706	.0711	.0716	.0721	.0726	.0731	.0737	.0742	.0747	.0752
16.0	.0757	.0762	.0768	.0773	.0778	.0783	.0788	.0794	.0799	.0804
17.0	.0809	.0814	.0820	.0825	.0830	.0835	.0841	.0846	.0851	.0857
18.0	.0862	.0867	.0872	.0878	.0883	.0888	.0894	.0899	.0904	.0910
19.0	.0915	.0921	.0926	.0931	.0937	.0942	.0947	.0953	.0958	.0964
20.0	.0969	.0975	.0980	.0985	.0991	.0996	.1002	.1007	.1013	.1018
21.0	.1024	.1029	.1035	.1040	.1046	.1051	.1057	.1062	.1068	.1073
22.0	.1079	.1085	.1090	.1096	.1101	.1107	.1113	.1118	.1124	.1129
23.0	.1135	.1141	.1146	.1152	.1158	.1163	.1169	.1175	.1180	.1186
24.0	.1192	.1198	.1203	.1209	.1215	.1221	.1226	.1232	.1238	.1244
25.0	.1249	.1255	.1261	.1267	.1273	.1278	.1284	.1290	.1296	.1302
26.0	.1308	.1314	.1319	.1325	.1331	.1337	.1343	.1349	.1355	.1361
27.0	.1367	.1373	.1379	.1385	.1391	.1397	.1403	.1409	.1415	.1421
28.0	.1427	.1433	.1439	.1445	.1451	.1457	.1463	.1469	.1475	.1481
29.0	.1487	.1494	.1500	.1506	.1512	.1518	.1524	.1530	.1537	.1543
30.0	.1549	.1555	.1561	.1568	.1574	.1580	.1586	.1593	.1599	.1605
31.0	.1612	.1618	.1624	.1630	.1637	.1643	.1649	.1656	.1662	.1669
32.0	.1675	.1681	.1688	.1694	.1701	.1707	.1713	.1720	.1726	.1733
33.0	.1739	.1746	.1752	.1759	.1765	.1772	.1778	.1785	.1791	.1798
34.0	.1805	.1811	.1818	.1824	.1831	.1838	.1844	.1851	.1858	.1864
35.0	.1871	.1878	.1884	.1891	.1898	.1904	.1911	.1918	.1925	.1931
36.0	.1938	.1945	.1952	.1959	.1965	.1972	.1979	.1986	.1993	.2000
37.0	.2007	.2013	.2020	.2027	.2034	.2041	.2048	.2055	.2062	.2069
38.0	.2076	.2083	.2090	.2097	.2104	.2111	.2118	.2125	.2132	.2140
39.0	.2147	.2154	.2161	.2168	.2175	.2182	.2190	.2197	.2204	.2211
40.0	.2218	.2226	.2233	.2240	.2248	.2255	.2262	.2269	.2277	.2284
41.0	.2291	.2299	.2306	.2314	.2321	.2328	.2336	.2343	.2351	.2358

TABLE 2.20 Values of Absorbance for Percent Absorption (*Continued*)

% A	.0	.1	.2	.3	.4	.5	.6	.7	.8	.9
42.0	.2366	.2373	.2381	.2388	.2396	.2403	.2411	.2418	.2426	.2434
43.0	.2441	.2449	.2457	.2464	.2472	.2480	.2487	.2495	.2503	.2510
44.0	.2518	.2526	.2534	.2541	.2549	.2557	.2565	.2573	.2581	.2588
45.0	.2596	.2604	.2612	.2620	.2628	.2636	.2644	.2652	.2660	.2668
46.0	.2676	.2684	.2692	.2700	.2708	.2716	.2725	.2733	.2741	.2749
47.0	.2757	.2765	.2774	.2782	.2790	.2798	.2807	.2815	.2823	.2832
48.0	.2840	.2848	.2857	.2865	.2874	.2882	.2890	.2899	.2907	.2916
49.0	.2924	.2933	.2941	.2950	.2958	.2967	.2976	.2984	.2993	.3002
50.0	.3010	.3019	.3028	.3036	.3045	.3054	.3063	.3072	.3080	.3089
51.0	.3098	.3107	.3116	.3125	.3134	.3143	.3152	.3161	.3170	.3179
52.0	.3188	.3197	.3206	.3215	.3224	.3233	.3242	.3251	.3261	.3270
53.0	.3279	.3288	.3298	.3307	.3316	.3325	.3335	.3344	.3354	.3363
54.0	.3372	.3382	.3391	.3401	.3410	.3420	.3429	.3439	.3449	.3458
55.0	.3468	.3478	.3487	.3497	.3507	.3516	.3526	.3536	.3546	.3556
56.0	.3565	.3575	.3585	.3595	.3605	.3615	.3625	.3635	.3645	.3655
57.0	.3665	.3675	.3686	.3696	.3706	.3716	.3726	.3737	.3747	.3757
58.0	.3768	.3778	.3788	.3799	.3809	.3820	.3830	.3840	.3851	.3862
59.0	.3872	.3883	.3893	.3904	.3915	.3925	.3936	.3947	.3958	.3969
60.0	.3979	.3990	.4001	.4012	.4023	.4034	.4045	.4056	.4067	.4078
61.0	.4089	.4101	.4112	.4123	.4134	.4145	.4157	.4168	.4179	.4191
62.0	.4202	.4214	.4225	.4237	.4248	.4260	.4271	.4283	.4295	.4306
63.0	.4318	.4330	.4342	.4353	.4365	.4377	.4389	.4401	.4413	.4425
64.0	.4437	.4449	.4461	.4473	.4485	.4498	.4510	.4522	.4535	.4547
65.0	.4559	.4572	.4584	.4597	.4609	.4622	.4634	.4647	.4660	.4672
66.0	.4685	.4698	.4711	.4724	.4737	.4750	.4763	.4776	.4789	.4802
67.0	.4815	.4828	.4841	.4855	.4868	.4881	.4895	.4908	.4921	.4935
68.0	.4948	.4962	.4976	.4989	.5003	.5017	.5031	.5045	.5058	.5072
69.0	.5086	.5100	.5114	.5129	.5143	.5157	.5171	.5186	.5200	.5214
70.0	.5229	.5243	.5258	.5272	.5287	.5302	.5317	.5331	.5346	.5361
71.0	.5376	.5391	.5406	.5421	.5436	.5452	.5467	.5482	.5498	.5513
72.0	.5528	.5544	.5560	.5575	.5591	.5607	.5622	.5638	.5654	.5670
73.0	.5686	.5702	.5719	.5735	.5751	.5768	.5784	.5800	.5817	.5834
74.0	.5850	.5867	.5884	.5901	.5918	.5935	.5952	.5969	.5986	.6003
75.0	.6021	.6038	.6055	.6073	.6091	.6108	.6126	.6144	.6162	.6180
76.0	.6198	.6216	.6234	.6253	.6271	.6289	.6308	.6326	.6345	.6364
77.0	.6383	.6402	.6421	.6440	.6459	.6478	.6498	.6517	.6536	.6556
78.0	.6576	.6596	.6615	.6635	.6655	.6676	.6696	.6716	.6737	.6757
79.0	.6778	.6799	.6819	.6840	.6861	.6882	.6904	.6925	.6946	.6968
80.0	.6990	.7011	.7033	.7055	.7077	.7100	.7122	.7144	.7167	.7190
81.0	.7212	.7235	.7258	.7282	.7305	.7328	.7352	.7375	.7399	.7423
82.0	.7447	.7471	.7496	.7520	.7545	.7570	.7595	.7620	.7645	.7670
83.0	.7696	.7721	.7747	.7773	.7799	.7825	.7852	.7878	.7905	.7932
84.0	.7959	.7986	.8013	.8041	.8069	.8097	.8125	.8153	.8182	.8210
85.0	.8239	.8268	.8297	.8327	.8356	.8386	.8416	.8447	.8477	.8508
86.0	.8539	.8570	.8601	.8633	.8665	.8697	.8729	.8761	.8794	.8827
87.0	.8861	.8894	.8928	.8962	.8996	.9031	.9066	.9101	.9136	.9172
88.0	.9208	.9245	.9281	.9318	.9355	.9393	.9431	.9469	.9508	.9547
89.0	.9586	.9626	.9666	.9706	.9747	.9788	.9830	.9872	.9914	.9957

TABLE 2.21   Transmittance-Absorbance Conversion Table

From Meites, Handbook of Analytical Chemistry, *1963, McGraw-Hill Book Company; by permission.*

This table gives absorbance values to four significant figures corresponding to % transmittance values which are given to three significant figures. The values of % transmittance are given in the left-hand column and in the top row. For example, 8.4% transmittance corresponds to an absorbance of 1.076.

Interpolation is facilitated and accuracy is maximized if the % transmittance is between 1 and 10, by multiplying its value by 10, finding the absorbance corresponding to the result, and adding 1. For example, to find the absorbance corresponding to 8.45% transmittance, note that 84.5% transmittance corresponds to an absorbance of 0.0731, so that 8.45% transmittance corresponds to an absorbance of 1.0731. For % transmittance values between 0.1 and 1, multiply by 100, find the absorbance corresponding to the result, and add 2.

Conversely, to find the % transmittance corresponding to an absorbance between 1 and 2, subtract 1 from the absorbance, find the % transmittance corresponding to the result, and divide by 10. For example, an absorbance of 1.219 can best be converted to % transmittance by noting that an absorbance of 0.219 would correspond to 60.4% transmittance; dividing this by 10 gives the desired value, 6.04% transmittance. For absorbance values between 2 and 3, subtract 2 from the absorbance, find the % transmittance corresponding to the result, and divide by 100.

% Transmittance	0.0	0.1	0.2	0.3	0.4	0.5	0.6	0.7	0.8	0.9
0	. . . . .	3.000	2.699	2.523	2.398	2.301	2.222	2.155	2.097	2.046
1	2.000	1.959	1.921	1.886	1.854	1.824	1.796	1.770	1.745	1.721
2	1.699	1.678	1.658	1.638	1.620	1.602	1.585	1.569	1.553	1.538
3	1.523	1.509	1.495	1.481	1.469	1.456	1.444	1.432	1.420	1.409
4	1.398	1.387	1.377	1.367	1.357	1.347	1.337	1.328	1.319	1.310
5	1.301	1.292	1.284	1.276	1.268	1.260	1.252	1.244	1.237	1.229
6	1.222	1.215	1.208	1.201	1.194	1.187	1.180	1.174	1.167	1.161
7	1.155	1.149	1.143	1.137	1.131	1.125	1.119	1.114	1.108	1.102
8	1.097	1.092	1.086	1.081	1.076	1.071	1.066	1.060	1.056	1.051
9	1.046	1.041	1.036	1.032	1.027	1.022	1.018	1.013	1.009	1.004
10	1.000	0.9957	0.9914	0.9872	0.9830	0.9788	0.9747	0.9706	0.9666	0.9626
11	0.9586	0.9547	0.9508	0.9469	0.9431	0.9393	0.9355	0.9318	0.9281	0.9245
12	0.9208	0.9172	0.9136	0.9101	0.9066	0.9031	0.8996	0.8962	0.8928	0.8894
13	0.8861	0.8827	0.8794	0.8761	0.8729	0.8697	0.8665	0.8633	0.8601	0.8570
14	0.8539	0.8508	0.8477	0.8447	0.8416	0.8386	0.8356	0.8327	0.8297	0.8268
15	0.8239	0.8210	0.8182	0.8153	0.8125	0.8097	0.8069	0.8041	0.8013	0.7986
16	0.7959	0.7932	0.7905	0.7878	0.7852	0.7825	0.7799	0.7773	0.7747	0.7721
17	0.7696	0.7670	0.7645	0.7620	0.7595	0.7570	0.7545	0.7520	0.7496	0.7471
18	0.7447	0.7423	0.7399	0.7375	0.7352	0.7328	0.7305	0.7282	0.7258	0.7235
19	0.7212	0.7190	0.7167	0.7144	0.7122	0.7100	0.7077	0.7055	0.7033	0.7011
20	0.6990	0.6968	0.6946	0.6925	0.6904	0.6882	0.6861	0.6840	0.6819	0.6799

TABLE 2.21   Transmittance-Absorbance Conversion Table (*Continued*)

% Trans- mittance	0.0	0.1	0.2	0.3	0.4	0.5	0.6	0.7	0.8	0.9
21	0.6778	0.6757	0.6737	0.6716	0.6696	0.6676	0.6655	0.6635	0.6615	0.6596
22	0.6576	0.6556	0.6536	0.6517	0.6498	0.6478	0.6459	0.6440	0.6421	0.6402
23	0.6383	0.6364	0.6345	0.6326	0.6308	0.6289	0.6271	0.6253	0.6234	0.6216
24	0.6198	0.6180	0.6162	0.6144	0.6126	0.6108	0.6091	0.6073	0.6055	0.6038
25	0.6021	0.6003	0.5986	0.5969	0.5952	0.5935	0.5918	0.5901	0.5884	0.5867
26	0.5850	0.5834	0.5817	0.5800	0.5784	0.5766	0.5751	0.5735	0.5719	0.5702
27	0.5686	0.5670	0.5654	0.5638	0.5622	0.5607	0.5591	0.5575	0.5560	0.5544
28	0.5528	0.5513	0.5498	0.5482	0.5467	0.5452	0.5436	0.5421	0.5406	0.5391
29	0.5376	0.5361	0.5346	0.5331	0.5317	0.5302	0.5287	0.5272	0.5258	0.5243
30	0.5229	0.5214	0.5200	0.5186	0.5171	0.5157	0.5143	0.5129	0.5114	0.5100
31	0.5086	0.5072	0.5058	0.5045	0.5031	0.5017	0.5003	0.4989	0.4976	0.4962
32	0.4949	0.4935	0.4921	0.4908	0.4895	0.4881	0.4868	0.4855	0.4841	0.4828
33	0.4815	0.4802	0.4789	0.4776	0.4763	0.4750	0.4737	0.4724	0.4711	0.4698
34	0.4685	0.4672	0.4660	0.4647	0.4634	0.4622	0.4609	0.4597	0.4584	0.4572
35	0.4559	0.4547	0.4535	0.4522	0.4510	0.4498	0.4486	0.4473	0.4461	0.4449
36	0.4437	0.4425	0.4413	0.4401	0.4389	0.4377	0.4365	0.4353	0.4342	0.4330
37	0.4318	0.4306	0.4295	0.4283	0.4271	0.4260	0.4248	0.4237	0.4225	0.4214
38	0.4202	0.4191	0.4179	0.4168	0.4157	0.4145	0.4134	0.4123	0.4112	0.4101
39	0.4089	0.4078	0.4067	0.4056	0.4045	0.4034	0.4023	0.4012	0.4001	0.3989
40	0.3979	0.3969	0.3958	0.3947	0.3936	0.3925	0.3915	0.3904	0.3893	0.3883
41	0.3872	0.3862	0.3851	0.3840	0.3830	0.3820	0.3809	0.3799	0.3788	0.3778
42	0.3768	0.3757	0.3747	0.3737	0.3726	0.3716	0.3706	0.3696	0.3686	0.3675
43	0.3665	0.3655	0.3645	0.3635	0.3625	0.3615	0.3605	0.3595	0.3585	0.3575
44	0.3565	0.3556	0.3546	0.3536	0.3526	0.3516	0.3507	0.3497	0.3487	0.3478
45	0.3468	0.3458	0.3449	0.3439	0.3429	0.3420	0.3410	0.3401	0.3391	0.3382
46	0.3372	0.3363	0.3354	0.3344	0.3335	0.3325	0.3316	0.3307	0.3298	0.3288
47	0.3279	0.3270	0.3261	0.3251	0.3242	0.3233	0.3224	0.3215	0.3206	0.3197
48	0.3188	0.3179	0.3170	0.3161	0.3152	0.3143	0.3134	0.3125	0.3116	0.3107
49	0.3098	0.3089	0.3080	0.3072	0.3063	0.3054	0.3045	0.3036	0.3028	0.3019
50	0.3010	0.3002	0.2993	0.2984	0.2976	0.2967	0.2958	0.2950	0.2941	0.2933
51	0.2924	0.2916	0.2907	0.2899	0.2890	0.2882	0.2874	0.2865	0.2857	0.2848
52	0.2840	0.2832	0.2823	0.2815	0.2807	0.2798	0.2790	0.2782	0.2774	0.2765
53	0.2757	0.2749	0.2741	0.2733	0.2725	0.2716	0.2708	0.2700	0.2692	0.2684
54	0.2676	0.2668	0.2660	0.2652	0.2644	0.2636	0.2628	0.2620	0.2612	0.2604
55	0.2596	0.2588	0.2581	0.2573	0.2565	0.2557	0.2549	0.2541	0.2534	0.2526
56	0.2518	0.2510	0.2503	0.2495	0.2487	0.2480	0.2472	0.2464	0.2457	0.2449
57	0.2441	0.2434	0.2426	0.2418	0.2411	0.2403	0.2396	0.2388	0.2381	0.2373
58	0.2366	0.2358	0.2351	0.2343	0.2336	0.2328	0.2321	0.2314	0.2306	0.2299
59	0.2291	0.2284	0.2277	0.2269	0.2262	0.2255	0.2248	0.2240	0.2233	0.2226
60	0.2218	0.2211	0.2204	0.2197	0.2190	0.2182	0.2175	0.2168	0.2161	0.2154
61	0.2147	0.2140	0.2132	0.2125	0.2118	0.2111	0.2104	0.2097	0.2090	0.2083
62	0.2076	0.2069	0.2062	0.2055	0.2048	0.2041	0.2034	0.2027	0.2020	0.2013
63	0.2007	0.2000	0.1993	0.1986	0.1979	0.1972	0.1965	0.1959	0.1952	0.1945
64	0.1938	0.1931	0.1925	0.1918	0.1911	0.1904	0.1898	0.1891	0.1884	0.1878
65	0.1871	0.1864	0.1858	0.1851	0.1844	0.1838	0.1831	0.1824	0.1818	0.1811

TABLE 2.21 Transmittance-Absorbance Conversion Table (*Continued*)

% Trans- mittance	0.0	0.1	0.2	0.3	0.4	0.5	0.6	0.7	0.8	0.9
66	0.1805	0.1798	0.1791	0.1785	0.1778	0.1772	0.1765	0.1759	0.1752	0.1746
67	0.1739	0.1733	0.1726	0.1720	0.1713	0.1707	0.1701	0.1694	0.1688	0.1681
68	0.1675	0.1669	0.1662	0.1656	0.1649	0.1643	0.1637	0.1630	0.1624	0.1618
69	0.1612	0.1605	0.1599	0.1593	0.1586	0.1580	0.1574	0.1568	0.1561	0.1555
70	0.1549	0.1543	0.1537	0.1530	0.1524	0.1518	0.1512	0.1506	0.1500	0.1494
71	0.1487	0.1481	0.1475	0.1469	0.1463	0.1457	0.1451	0.1445	0.1439	0.1433
72	0.1427	0.1421	0.1415	0.1409	0.1403	0.1397	0.1391	0.1385	0.1379	0.1373
73	0.1367	0.1361	0.1355	0.1349	0.1343	0.1337	0.1331	0.1325	0.1319	0.1314
74	0.1308	0.1302	0.1296	0.1290	0.1284	0.1278	0.1273	0.1267	0.1261	0.1255
75	0.1249	0.1244	0.1238	0.1232	0.1226	0.1221	0.1215	0.1209	0.1203	0.1198
76	0.1192	0.1186	0.1180	0.1175	0.1169	0.1163	0.1158	0.1152	0.1146	0.1141
77	0.1135	0.1129	0.1124	0.1118	0.1113	0.1107	0.1101	0.1096	0.1090	0.1085
78	0.1079	0.1073	0.1068	0.1062	0.1057	0.1051	0.1046	0.1040	0.1035	0.1029
79	0.1024	0.1018	0.1013	0.1007	0.1002	0.0996	0.0991	0.0985	0.0980	0.0975
80	0.0969	0.0964	0.0958	0.0953	0.0947	0.0942	0.0937	0.0931	0.0926	0.0921
81	0.0915	0.0910	0.0904	0.0899	0.0894	0.0888	0.0883	0.0878	0.0872	0.0867
82	0.0862	0.0857	0.0851	0.0846	0.0841	0.0835	0.0830	0.0825	0.0820	0.0814
83	0.0809	0.0804	0.0799	0.0794	0.0788	0.0783	0.0778	0.0773	0.0768	0.0762
84	0.0757	0.0752	0.0747	0.0742	0.0737	0.0731	0.0726	0.0721	0.0716	0.0711
85	0.0706	0.0701	0.0696	0.0691	0.0685	0.0680	0.0675	0.0670	0.0665	0.0660
86	0.0655	0.0650	0.0645	0.0640	0.0635	0.0630	0.0625	0.0620	0.0615	0.0610
87	0.0605	0.0600	0.0595	0.0590	0.0585	0.0580	0.0575	0.0570	0.0565	0.0560
88	0.0555	0.0550	0.0545	0.0540	0.0535	0.0531	0.0526	0.0521	0.0516	0.0511
89	0.0506	0.0501	0.0496	0.0491	0.0487	0.0482	0.0477	0.0472	0.0467	0.0462
90	0.0458	0.0453	0.0448	0.0443	0.0438	0.0434	0.0429	0.0424	0.0419	0.0414
91	0.0410	0.0405	0.0400	0.0395	0.0391	0.0386	0.0381	0.0376	0.0372	0.0367
92	0.0362	0.0357	0.0353	0.0348	0.0343	0.0339	0.0334	0.0329	0.0325	0.0320
93	0.0315	0.0311	0.0306	0.0301	0.0297	0.0292	0.0287	0.0283	0.0278	0.0273
94	0.0269	0.0264	0.0259	0.0255	0.0250	0.0246	0.0241	0.0237	0.0232	0.0227
95	0.0223	0.0218	0.0214	0.0209	0.0205	0.0200	0.0195	0.0191	0.0186	0.0182
96	0.0177	0.0173	0.0168	0.0164	0.0159	0.0155	0.0150	0.0146	0.0141	0.0137
97	0.0132	0.0128	0.0123	0.0119	0.0114	0.0110	0.0106	0.0101	0.0097	0.0092
98	0.0088	0.0083	0.0079	0.0074	0.0070	0.0066	0.0061	0.0057	0.0052	0.0048
99	0.0044	0.0039	0.0035	0.0031	0.0026	0.0022	0.0017	0.0013	0.0009	0.0004

TABLE 2.22   Wavenumber/Wavelength Conversion Table

This table is based on the conversion: wavenumber (in cm^{-1}) = 10 000/wavelength (in μm). For example, 15.4 μm is equal to 649 cm^{-1}.

Wavelength (μm)	0	0.1	0.2	0.3	0.4	0.5	0.6	0.7	0.8	0.9 cm^{-1}
1.0	10000	9091	8333	7692	7143	6667	6250	5882	5556	5263
2.0	5000	4762	4545	4348	4167	4000	3846	3704	3571	3448
3.0	3333	3226	3125	3030	2941	2857	2778	2703	2632	2564
4.0	2500	2439	2381	2326	2273	2222	2174	2128	2083	2041
5.0	2000	1961	1923	1887	1852	1818	1786	1754	1724	1695
6.0	1667	1639	1613	1587	1563	1538	1515	1493	1471	1449
7.0	1429	1408	1389	1370	1351	1333	1316	1299	1282	1266
8.0	1250	1235	1220	1205	1190	1176	1163	1149	1136	1124
9.0	1111	1099	1087	1075	1064	1053	1042	1031	1020	1010
10.0	1000	990	980	971	962	952	943	935	926	917
11.0	909	901	893	885	877	870	862	855	847	840
12.0	833	826	820	813	806	800	794	787	781	775
13.0	769	763	758	752	746	741	735	730	725	719
14.0	714	709	704	699	694	690	685	680	676	671
15.0	667	662	658	654	649	645	641	637	633	629
16.0	625	621	617	613	610	606	602	599	595	592
17.0	588	585	581	578	575	571	568	565	562	559
18.0	556	552	549	546	543	541	538	535	532	529
19.0	526	524	521	518	515	513	510	508	505	503
20.0	500	498	495	493	490	488	485	483	481	478
21.0	476	474	472	469	467	465	463	461	459	457
22.0	455	452	450	448	446	444	442	441	439	437
23.0	435	433	431	429	427	426	424	422	420	418
24.0	417	415	413	412	410	408	407	405	403	402
25.0	400	398	397	395	394	392	391	389	388	386
26.0	385	383	382	380	379	377	376	375	373	372
27.0	370	369	368	366	365	364	362	361	360	358
28.0	357	356	355	353	352	351	350	348	347	346
29.0	345	344	342	341	340	339	338	337	336	334
30.0	333	332	331	330	329	328	327	326	325	324
31.0	323	322	321	319	318	317	316	315	314	313
32.0	313	312	311	310	309	308	307	306	305	304
33.0	303	302	301	300	299	299	298	297	296	295
34.0	294	293	292	292	291	290	289	288	287	287
35.0	286	285	284	283	282	282	281	280	279	279
36.0	278	277	276	275	275	274	273	272	272	271
37.0	270	270	269	268	267	267	266	265	265	264
38.0	263	262	262	261	260	260	259	258	258	257
39.0	256	256	255	254	254	253	253	252	251	251
40.0	250									

2.2 *MATHEMATICAL TABLES*

2.2.1 Logarithms

2.2.1.1 *Properties and Uses*

Definition of Logarithm. The *logarithm x* of the number N to the base b is the exponent of the power to which b must be raised to give N. That is,

$$\log_b N = x \quad \text{or} \quad b^x = N$$

The number N is positive and b may be any positive number except 1.

Properties of Logarithms

1. The logarithm of a product is equal to the sum of the logarithms of the factors; thus,

$$\log_b M \cdot N = \log_b M + \log_b N$$

2. The logarithm of a quotient is equal to the logarithm of the numerator minus the logarithm of the denominator; thus,

$$\log_b \frac{M}{N} = \log_b M - \log_b N$$

3. The logarithm of a power of a number is equal to the logarithm of the base multiplied by the exponent of the power; thus,

$$\log_b M^p = p \cdot \log_b M$$

4. The logarithm of a root of a number is equal to the logarithm of the number divided by the index of the root; thus

$$\log_b \sqrt[q]{M} = \frac{1}{q} \log_b M$$

Other properties of logarithms:

$$\log_b b = 1 \qquad \log_b \sqrt[q]{M^p} = \frac{p}{q} \log_b M$$

$$\log_b 1 = 0 \qquad \log_b N = \log_a N \cdot \log_b a = \frac{\log_a N}{\log_a b}$$

$$\log_b (b^N) = N \qquad b^{\log_b N} = N$$

Systems of Logarithms. There are two common systems of logarithms in use: (1) the *natural* (Napierian or hyperbolic) system which uses the base $e = 2.71828 \ldots$; (2) the *common* (Briggsian) system which uses the base 10.

We shall use the abbreviation $\log N \equiv \log_{10} N$ in this section.

Unless otherwise stated, tables of logarithms are always tables of common logarithms.

Characteristic of a Common Logarithm of a Number. Every real positive number has a real common logarithm such that if $a < b$, $\log a < \log b$. Neither zero nor any negative number has a real logarithm.

A common logarithm, in general, consists of an integer, which is called the *characteristic*, and a decimal (usually endless), which is called the *mantissa*. The characteristic of any number may be determined from the following rules:

Rule I.   The characteristic of any number greater than 1 is one less than the number of digits before the decimal point.

Rule II.*   The characteristic of a number less than 1 is found by subtracting from 9 the number of ciphers between the decimal point and the first significant digit, and writing -10 after the result.

Thus the characteristic of log 936 is 2; the characteristic of log 9.36 is 0; of log 0.936 is $9 - 10$; of log 0.00936 is $7 - 10$.

Mantissa of a Common Logarithm of a Number.   An important consequence of the use of base 10 is that the mantissa of a number is independent of the position of the decimal point. Thus 93 600, 93.600, 0.000 936, all have the same mantissa. Hence in Tables of Common Logarithms only mantissas are given. A five-place table gives the values of the mantissa correct to five places of decimals.

Since it is possible to obtain logarithms by using hand calculators, this Handbook contains no logarithm tables.

Helpful Hints

1. When connecting numbers to logarithms, use as many decimal places in the mantissa as there are significant digits in the number.

2. When finding the antilogarithm, keep as many significant digits as there are decimal places in the mantissa.

Examples: log 10.35 = 1.0149; antilog 0.065 = 1.16.

---

* Some writers use a dash over the characteristic to indicate a negative value; for example,

$$\log 0.004657 = 7.6681 - 10 = \overline{3}.6681$$

TABLE 2.23 Derivatives and Differentiation

Rules for differentiation.

From Baumeister and Marks, *Standard Handbook for Mechanical Engineers,* 7th ed., McGraw-Hill Book Company, New York (1967); by permission.

To find the derivative of a given function at a given point: (1) If the function is given only by a curve, measure graphically the slope of the tangent at the point in question; (2) if the function is given by a mathematical expression, use the following rules for differentiation. These rules give, directly, the differential, dy, in terms of dx; to find the derivative, dy/dx, divide through by dx.

Here $u, v, w, \ldots$ represents any functions of a variable x, or may themselves be independent variables. a is a constant which does not change in values in the same discussion; $e = 2.71828$.

---

1. $d(a + u) = du$ **2.** $d(au) = a\, du$

3. $d(u + v + w + \cdots) = du + dv + dw + \cdots$

4. $d(uv) = u\, dv + v\, du$

5. $d(uvw \ldots) = (uvw \ldots)\left(\dfrac{du}{u} + \dfrac{dv}{v} + \dfrac{dw}{w} + \cdots\right)$

6. $d\dfrac{u}{v} = \dfrac{v\, du - u\, dv}{v^2}$ **7.** $d(u^m) = mu^{m-1}\, du$

Thus, $d(u^2) = 2u\, du$; $d(u^3) = 3u^2\, du$; etc.

8. $d\sqrt{u} = \dfrac{du}{2\sqrt{u}}$ **9.** $d\left(\dfrac{1}{u}\right) = -\dfrac{du}{u^2}$

10. $d(e^u) = e^u\, du$ **11.** $d(a^u) = (\ln a)a^u\, du$

12. $d\ln u = \dfrac{du}{u}$ **13.** $d\log_{10} u = (\log_{10} e)\dfrac{du}{u} = (0.4343 \ldots)\dfrac{du}{u}$

14. $d\sin u = \cos u\, du$ **15.** $d\csc u = -\cot u\, \csc u\, du$

16. $d\cos u = -\sin u\, du$ **17.** $d\sec u = \tan u\, \sec u\, du$

18. $d\tan u = \sec^2 u\, du$ **19.** $d\cot u = -\csc^2 u\, du$

20. $d\sin^{-1} u = \dfrac{du}{\sqrt{1 - u^2}}$ **21.** $d\csc^{-1} u = -\dfrac{du}{u\sqrt{u^2 - 1}}$

22. $d\cos^{-1} u = -\dfrac{du}{\sqrt{1 - u^2}}$ **23.** $d\sec^{-1} u = \dfrac{du}{u\sqrt{u^2 - 1}}$

24. $d\tan^{-1} u = \dfrac{du}{1 + u^2}$ **25.** $d\cot^{-1} u = -\dfrac{du}{1 + u^2}$

26. $d\ln\sin u = \cot u\, du$ **27.** $d\ln\tan u = \dfrac{2\, du}{\sin 2u}$

28. $d\ln\cos u = -\tan u\, du$ **29.** $d\ln\cot u = -\dfrac{2\, du}{\sin 2u}$

30. $d\sinh u = \cosh u\, du$ **31.** $d\,\mathrm{csch}\, u = -\mathrm{csch}\, u\, \coth u\, du$

32. $d\cosh u = \sinh u\, du$ **33.** $d\,\mathrm{sech}\, u = -\mathrm{sech}\, u\, \tanh u\, du$

34. $d\tanh u = \mathrm{sech}^2 u\, du$ **35.** $d\coth u = -\mathrm{csch}^2 u\, du$

36. $d\sinh^{-1} u = \dfrac{du}{\sqrt{u^2 + 1}}$ **37.** $d\,\mathrm{csch}^{-1} u = -\dfrac{du}{u\sqrt{u^2 + 1}}$

38. $d\cosh^{-1} u = \dfrac{du}{\sqrt{u^2 - 1}}$ **39.** $d\,\mathrm{sech}^{-1} u = -\dfrac{du}{u\sqrt{1 - u^2}}$

40. $d\tanh^{-1} u = \dfrac{du}{1 - u^2}$ **41.** $d\coth^{-1} u = \dfrac{du}{1 - u^2}$

42. $d(u^v) = (u^{v-1})(u\ln u\, dv + v\, du)$

TABLE 2.24 Integrals

From Baumeister and Marks, Standard Handbook for Mechanical Engineers, *7th ed., McGraw-Hill Book Company, New York (1967); by permission.*

An integral of $f(x)\,dx$ is any function whose differential is $f(x)\,dx$, and is denoted by $\int f(x)\,dx$. All the integrals of $f(x)\,dx$ are included in the expression $\int f(x)\,dx + C$, where $\int f(x)\,dx$ is any particular integral, and C is an arbitrary constant. The process of finding (when possible) an integral of a given function consists in recognizing by inspection a function which, when differentiated, will produce the given function; or in transforming the given function into a form in which such recognition is easy. The most common integrable forms are collected in the following brief table; for a more extended list, see Peirce, *Table of Integrals,* Ginn, or Dwight, *Table of Integrals and other Mathematical Data,* Macmillan.

General formulas

1. $\displaystyle\int a\,du = a\int du = au + C$ 　　　　 **2.** $\displaystyle\int (u+v)\,dx = \int u\,dx + \int v\,dx$

3. $\displaystyle\int u\,dv = uv - \int v\,du$ 　　　　 **4.** $\displaystyle\int f(x)\,dx = \int f[F(y)]F'(y)\,dy,\ x = F(y)$

5. $\displaystyle\int dy \int f(x,y)\,dx = \int dx \int f(x,y)\,dy$

Fundamental integrals

6. $\displaystyle\int x^n\,dx = \frac{x^{n+1}}{n+1} + C,\ \text{when } n \neq -1$

7. $\displaystyle\int \frac{dx}{x} = \ln x + C = \ln cx$ 　　　　 **8.** $\displaystyle\int e^x\,dx = e^x + C$

9. $\displaystyle\int \sin x\,dx = -\cos x + C$ 　　　　 **10.** $\displaystyle\int \cos x\,dx = \sin x + C$

11. $\displaystyle\int \frac{dx}{\sin^2 x} = -\cot x + C$ 　　　　 **12.** $\displaystyle\int \frac{dx}{\cos^2 x} = \tan x + C$

13. $\displaystyle\int \frac{dx}{\sqrt{1-x^2}} = \sin^{-1} x + C = -\cos^{-1} x + c$

14. $\displaystyle\int \frac{dx}{1+x^2} = \tan^{-1} x + C = -\cot^{-1} x + c$

Rational functions

15. $\displaystyle\int (a+bx)^n\,dx = \frac{(a+bx)^{n+1}}{(n+1)b} + C$

16. $\displaystyle\int \frac{dx}{a+bx} = \frac{1}{b}\ln(a+bx) + C = \frac{1}{b}\ln c(a+bx)$

17. $\displaystyle\int \frac{1}{x^2}\,dx = -\frac{1}{x} + C$ 　　　　 **18.** $\displaystyle\int \frac{dx}{(a+bx)^2} = -\frac{1}{b(a+bx)} + C$

19. $\displaystyle\int \frac{dx}{1-x^2} = \tfrac{1}{2}\ln\frac{1+x}{1-x} + C = \tanh^{-1} x + C,\ \text{when } x < 1$

20. $\displaystyle\int \frac{dx}{x^2-1} = \tfrac{1}{2}\ln\frac{x-1}{x+1} + C = -\coth^{-1} x + C,\ \text{when } x > 1$

TABLE 2.24   Integrals (*Continued*)

Rational functions (*continued*)

21. $\int \dfrac{dx}{a + bx^2} = \dfrac{1}{\sqrt{ab}} \tan^{-1}\left(\sqrt{\dfrac{b}{a}}\, x\right) + C$

22. $\int \dfrac{dx}{a - bx^2} = \dfrac{1}{2\sqrt{ab}} \ln \dfrac{\sqrt{ab} + bx}{\sqrt{ab} - bx} + C$ $[a > 0,\, b > 0]$

$\qquad = \dfrac{1}{\sqrt{ab}} \tanh^{-1}\left(\sqrt{\dfrac{b}{a}}\, x\right) + C$

23. $\int \dfrac{dx}{a + 2bx + cx^2} = \dfrac{1}{\sqrt{ac - b^2}} \tan^{-1} \dfrac{b + cx}{\sqrt{ac - b^2}} + C$ $\{ac - b^2 > 0]$

$\qquad = \dfrac{1}{2\sqrt{b^2 - ac}} \ln \dfrac{\sqrt{b^2 - ac} - b - cx}{\sqrt{b^2 - ac} + b + cx} + C$

$\qquad = -\dfrac{1}{\sqrt{b^2 - ac}} \tanh^{-1} \dfrac{b + cx}{\sqrt{b^2 - ac}} + C$ $[b^2 - ac > 0]$

24. $\int \dfrac{dx}{a + 2bx + cx^2} = -\dfrac{1}{b + cx} + C$, when $b^2 = ac$

25. $\int \dfrac{(m + nx)\, dx}{a + 2bx + cx^2} = \dfrac{n}{2c} \ln (a + 2bx + cx^2) + \dfrac{mc - nb}{c} \int \dfrac{dx}{a + 2bx + cx^2}$

26. In $\int \dfrac{f(x)\, dx}{a + 2bx + cx^2}$, if $f(x)$ is a polynomial of higher than the first degree, divide by the denominator before integrating.

27. $\int \dfrac{dx}{(a + 2bx + cx^2)^p} = \dfrac{1}{2(ac - b^2)(p - 1)} \times \dfrac{b + cx}{(a + 2bx + cx^2)^{p-1}}$

$\qquad + \dfrac{(2p - 3)c}{2(ac - b^2)(p - 1)} \int \dfrac{dx}{(a + 2bx + cx^2)^{p-1}}$

28. $\int \dfrac{(m + nx)\, dx}{(a + 2bx + cx^2)^p} = -\dfrac{n}{2c(p - 1)} \times \dfrac{1}{(a + 2bx + cx^2)^{p-1}}$

$\qquad + \dfrac{mc - nb}{c} \int \dfrac{dx}{(a + 2bx + cx^2)^p}$

29. $\int x^{m-1}(a + bx)^n\, dx = \dfrac{x^{m-1}(a + bx)^{n+1}}{(m + n)b} - \dfrac{(m - 1)a}{(m + n)b} \int x^{m-2}(a + bx)^n\, dx$

$\qquad = \dfrac{x^m(a + bx)^n}{m + n} + \dfrac{na}{m + n} \int x^{m-1}(a + bx)^{n-1}\, dx$

Irrational functions

30. $\int \sqrt{a + bx}\, dx = \dfrac{2}{3b} \sqrt{(a + bx)^3} + C$ **31.** $\int \dfrac{dx}{\sqrt{a + bx}} = \dfrac{2}{b} \sqrt{a + bx} + C$

32. $\int \dfrac{(m + nx)\, dx}{\sqrt{a + bx}} = \dfrac{2}{3b^2} (3mb - 2an + nbx) \sqrt{a + bx} + C$

33. $\int \dfrac{dx}{(m + nx)\sqrt{a + bx}}$; substitute $y = \sqrt{a + bx}$, and use **21** and **22**

34. $\int \dfrac{f(x, \sqrt[n]{a + bx})}{F(x, \sqrt[n]{a + bx})}\, dx$; substitute $\sqrt[n]{a + bx} = y$

TABLE 2.24 Integrals (*Continued*)

Irrational functions (*continued*)

35. $\displaystyle\int \frac{dx}{\sqrt{a^2 - x^2}} = \sin^{-1}\frac{x}{a} + C = -\cos^{-1}\frac{x}{a} + c$

36. $\displaystyle\int \frac{dx}{\sqrt{a^2 + x^2}} = \ln(x + \sqrt{a^2 + x^2}) + C = \sinh^{-1}\frac{x}{a} + c$

37. $\displaystyle\int \frac{dx}{\sqrt{x^2 - a^2}} = \ln(x + \sqrt{x^2 - a^2}) + C = \cosh^{-1}\frac{x}{a} + c$

38. $\displaystyle\int \frac{dx}{\sqrt{a + 2bx + cx^2}} = \frac{1}{\sqrt{c}}\ln(b + cx + \sqrt{c}\sqrt{a + 2bx + cx^2}) + C,$ when $c > 0$

$\displaystyle\qquad = \frac{1}{\sqrt{c}}\sinh^{-1}\frac{b + cx}{\sqrt{ac - b^2}} + C,$ when $ac - b^2 > 0$

$\displaystyle\qquad = \frac{1}{\sqrt{c}}\cosh^{-1}\frac{b + cx}{\sqrt{b^2 - ac}} + C,$ when $b^2 - ac > 0$

$\displaystyle\qquad = \frac{-1}{\sqrt{-c}}\sin^{-1}\frac{b + cx}{\sqrt{b^2 - ac}} + C,$ when $c < 0$

39. $\displaystyle\int \frac{(m + nx)\,dx}{\sqrt{a + 2bx + cx^2}} = \frac{n}{c}\sqrt{a + 2bx + cx^2} + \frac{mc - nb}{c}\int \frac{dx}{\sqrt{a + 2bx + cx^2}}$

40. $\displaystyle\int \frac{x^m\,dx}{\sqrt{a + 2bx + cx^2}} = \frac{x^{m-1}X}{mc} - \frac{(m-1)a}{mc}\int \frac{x^{m-2}\,dx}{X} - \frac{(2m-1)b}{mc}\int \frac{x^{m-1}\,dx}{X},$

when $X = \sqrt{a + 2bx + cx^2}$

41. $\displaystyle\int \sqrt{a^2 + x^2}\,dx = \frac{x}{2}\sqrt{a^2 + x^2} + \frac{a^2}{2}\ln(x + \sqrt{a^2 + x^2}) + C$

$\displaystyle\qquad = \frac{x}{2}\sqrt{a^2 + x^2} + \frac{a^2}{2}\sinh^{-1}\frac{x}{a} + c$

42. $\displaystyle\int \sqrt{a^2 - x^2}\,dx = \frac{x}{2}\sqrt{a^2 - x^2} + \frac{a^2}{2}\sin^{-1}\frac{x}{a} + C$

43. $\displaystyle\int \sqrt{x^2 - a^2}\,dx = \frac{x}{2}\sqrt{x^2 - a^2} - \frac{a^2}{2}\ln(x + \sqrt{x^2 - a^2}) + C$

$\displaystyle\qquad = \frac{x}{2}\sqrt{x^2 - a^2} - \frac{a^2}{2}\cosh^{-1}\frac{x}{a} + c$

44. $\displaystyle\int \sqrt{a + 2bx + cx^2}\,dx = \frac{b + cx}{2c}\sqrt{a + 2bx + cx^2}$

$\displaystyle\qquad + \frac{ac - b^2}{2c}\int \frac{dx}{\sqrt{a + 2bx + cx^2}} + C$

Transcendental functions

45. $\displaystyle\int a^x\,dx = \frac{a^x}{\ln a} + C$

46. $\displaystyle\int x^n e^{ax}\,dx = \frac{x^n e^{ax}}{a}\left[1 - \frac{n}{ax} + \frac{n(n-1)}{a^2 x^2} - \cdots \pm \frac{n!}{a^n x^n}\right] + C$

47. $\displaystyle\int \ln x\,dx = x\ln x - x + C$

TABLE 2.24 Integrals (*Continued*)

Transcendental functions (*continued*)

48. $\int \dfrac{\ln x}{x^2} \, dx = -\dfrac{\ln x}{x} - \dfrac{1}{x} + C$

49. $\int \dfrac{(\ln x)^n}{x} \, dx = \dfrac{1}{n+1} (\ln x)^{n+1} + C$

50. $\int \sin^2 x \, dx = -\tfrac{1}{4} \sin 2x + \tfrac{1}{2}x + C = -\tfrac{1}{2} \sin x \cos x + \tfrac{1}{2}x + C$

51. $\int \cos^2 x \, dx = \tfrac{1}{4} \sin 2x + \tfrac{1}{2}x + C = \tfrac{1}{2} \sin x \cos x + \tfrac{1}{2}x + C$

52. $\int \sin mx \, dx = -\dfrac{\cos mx}{m} + C$ **53.** $\int \cos mx \, dx = \dfrac{\sin mx}{m} + C$

54. $\int \sin mx \cos nx \, dx = -\dfrac{\cos(m+n)x}{2(m+n)} - \dfrac{\cos(m-n)x}{2(m-n)} + C$

55. $\int \sin mx \sin nx \, dx = \dfrac{\sin(m-n)x}{2(m-n)} - \dfrac{\sin(m+n)x}{2(m+n)} + C$

56. $\int \cos mx \cos nx \, dx = \dfrac{\sin(m-n)x}{2(m-n)} + \dfrac{\sin(m+n)x}{2(m+n)} + C$

57. $\int \tan x \, dx = -\ln \cos x + C$ **58.** $\int \cot x \, dx = \ln \sin x + C$

59. $\int \dfrac{dx}{\sin x} = \ln \tan \dfrac{x}{2} + C$ **60.** $\int \dfrac{dx}{\cos x} = \ln \tan \left(\dfrac{\pi}{4} + \dfrac{x}{2} \right) + C$

61. $\int \dfrac{dx}{1 + \cos x} = \tan \dfrac{x}{2} + C$ **62.** $\int \dfrac{dx}{1 - \cos x} = -\cot \dfrac{x}{2} + C$

63. $\int \sin x \cos x \, dx = \tfrac{1}{2} \sin^2 x + C$ **64.** $\int \dfrac{dx}{\sin x \cos x} = \ln \tan x + C$

65.* $\int \sin^n x \, dx = -\dfrac{\cos x \sin^{n-1} x}{n} + \dfrac{n-1}{n} \int \sin^{n-2} x \, dx$

66.* $\int \cos^n x \, dx = \dfrac{\sin x \cos^{n-1} x}{n} + \dfrac{n-1}{n} \int \cos^{n-2} x \, dx$

67. $\int \tan^n x \, dx = \dfrac{\tan^{n-1} x}{n-1} - \int \tan^{n-2} x \, dx$

68. $\int \cot^n x \, dx = -\dfrac{\cot^{n-1} x}{n-1} - \int \cot^{n-2} x \, dx$

69. $\int \dfrac{dx}{\sin^n x} = -\dfrac{\cos x}{(n-1) \sin^{n-1} x} + \dfrac{n-2}{n-1} \int \dfrac{dx}{\sin^{n-2} x}$

70. $\int \dfrac{dx}{\cos^n x} = \dfrac{\sin x}{(n-1) \cos^{n-1} x} + \dfrac{n-2}{n-1} \int \dfrac{dx}{\cos^{n-2} x}$

71.* $\int \sin^p x \cos^q x \, dx = \dfrac{\sin^{p+1} x \cos^{q-1} x}{p+q} + \dfrac{q-1}{p+q} \int \sin^p x \cos^{q-2} x \, dx$

$\qquad = -\dfrac{\sin^{p-1} x \cos^{q+1} x}{p+q} + \dfrac{p-1}{p+q} \int \sin^{p-2} x \cos^q x \, dx$

72.* $\int \sin^{-p} x \cos^q x \, dx = -\dfrac{\sin^{-p+1} x \cos^{q+1} x}{p-1} + \dfrac{p-q-2}{p-1} \int \sin^{-p+2} x \cos^q x \, dx$

---

* If n, p, or q is an odd number, substitute $\cos x = z$ or $\sin x = z$

TABLE 2.24   Integrals (*Continued*)

Transcendental functions (*continued*)

73.* $\displaystyle\int \sin^p x \cos^{-q} x \, dx = \frac{\sin^{p+1} x \cos^{-q+1} x}{q-1} + \frac{q-p-2}{q-1}\int \sin^p x \cos^{-q+2} x \, dx$

74. $\displaystyle\int \frac{dx}{a+b\cos x} = \frac{2}{\sqrt{a^2-b^2}} \tan^{-1}\left(\sqrt{\frac{a-b}{a+b}}\tan \tfrac{1}{2}x\right) + C, \text{ when } a^2 > b^2$

$\qquad\qquad = \dfrac{1}{\sqrt{b^2-a^2}} \ln \dfrac{b+a\cos x + \sin x\sqrt{b^2-a^2}}{a+b\cos x} + C$

$\qquad\qquad = \dfrac{2}{\sqrt{b^2-a^2}} \tanh^{-1}\left(\sqrt{\dfrac{b-a}{b+a}}\tan \tfrac{1}{2}x\right) + C$ $\quad\Bigg\}\ [a^2 < b^2]$

75. $\displaystyle\int \frac{\cos x \, dx}{a+b\cos x} = \frac{x}{b} - \frac{a}{b}\int \frac{dx}{a+b\cos x} + C$

76. $\displaystyle\int \frac{\sin x \, dx}{a+b\cos x} = -\frac{1}{b} \ln (a+b\cos x) + C$

77. $\displaystyle\int \frac{A+B\cos x + C\sin x}{a+b\cos x + c\sin x}\, dx = A\int \frac{dy}{a+p\cos y}$

$\qquad + (B\cos u + C\sin u)\displaystyle\int \frac{\cos y \, dy}{a+p\cos y} - (B\sin u - C\cos u)\int \frac{\sin y \, dy}{a+p+\cos y},$

$\qquad$ where $b = p\cos u,\ c = p\sin u,$ and $x - u = y$

78. $\displaystyle\int e^{ax} \sin bx \, dx = \frac{a\sin bx - b\cos bx}{a^2+b^2}\, e^{ax} + C$

79. $\displaystyle\int e^{ax} \cos bx \, dx = \frac{a\cos bx + b\sin bx}{a^2+b^2}\, e^{ax} + C$

80. $\displaystyle\int \sin^{-1} x \, dx = x\sin^{-1} x + \sqrt{1-x^2} + C$

81. $\displaystyle\int \cos^{-1} x \, dx = x\cos^{-1} x - \sqrt{1-x^2} + C$

82. $\displaystyle\int \tan^{-1} x \, dx = x\tan^{-1} x - \tfrac{1}{2}\ln(1+x^2) + C$

83. $\displaystyle\int \cot^{-1} x \, dx = x\cot^{-1} x + \tfrac{1}{2}\ln(1+x^2) + C$

84. $\displaystyle\int \sinh x \, dx = \cosh x + C$ $\qquad\qquad$ **85.** $\displaystyle\int \tanh x \, dx = \ln \cosh x + C$

86. $\displaystyle\int \cosh x \, dx = \sinh x + C$ $\qquad\qquad$ **87.** $\displaystyle\int \coth x \, dx = \ln \sinh x + C$

88. $\displaystyle\int \operatorname{sech} x \, dx = 2\tan^{-1}(e^x) + C$ $\qquad$ **89.** $\displaystyle\int \operatorname{csch} x \, dx = \ln \tanh\left(\frac{x}{2}\right) + C$

90. $\displaystyle\int \sinh^2 x \, dx = \tfrac{1}{2}\sinh x \cosh x - \tfrac{1}{2}x + C$

91. $\displaystyle\int \cosh^2 x \, dx = \tfrac{1}{2}\sinh x \cosh + \tfrac{1}{2}x + C$

92. $\displaystyle\int \operatorname{sech}^2 x \, dx = \tanh x + C$ $\qquad\qquad$ **93.** $\displaystyle\int \operatorname{csch}^2 x \, dx = -\coth x + C$

---

* If n, p, or q is an odd number, substitute $\cos x = z$ or $\sin x = z$

2.2.2 Surface Areas and Volumes*

Let a, b, c, d, and s denote lengths, A denote areas, and V denote volumes.

Triangle. $A = bh/2$, where b denotes the base and h the altitude.

Rectangle. $A = ab$, where a and b denote the lengths of the sides.

Parallelogram (opposite sides parallel). $A = ah = ab \sin \theta$, where a and b denote the sides, h the altitude, and θ the angle between the sides.

Trapezoid (four sides, two parallel). $A = \frac{1}{2}h(a + b)$, where a and b are the sides and h the altitude.

Regular Polygon of n *Sides* (Fig. 2.1)

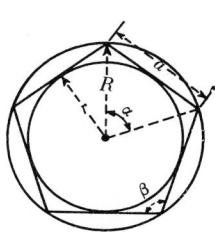

$$A = \frac{1}{4} na^2 \operatorname{ctn} \frac{180°}{n} \qquad \text{where } a \text{ is length of side}$$

$$R = \frac{a}{2} \csc \frac{180°}{n} \qquad \text{where } R \text{ is radius of circumscribed circle}$$

$$r = \frac{a}{2} \operatorname{ctn} \frac{180°}{n} \qquad \text{where } r \text{ is radius of inscribed circle}$$

$$\alpha = \frac{360°}{n} = \frac{2\pi}{n} \text{ radians}$$

FIGURE 2.1

$$\beta = \left(\frac{n-2}{n}\right) \cdot 180° = \left(\frac{n-2}{n}\right) \pi \text{ radians} \qquad \text{where } \alpha \text{ and } \beta \text{ are the angles indicated in Fig. 2.1}$$

$$a = 2r \tan \frac{\alpha}{2} = 2R \sin \frac{\alpha}{2}$$

Circle (Fig. 2.2). Let

C = circumference	S = length of arc subtended by θ
R = radius	l = chord subtended by arc S
D = diameter	h = rise
A = area	θ = central angle in radians

$$C = 2\pi R = \pi D \qquad \pi = 3.14159 \ldots$$

$$S = R\theta = \frac{1}{2}D\theta = D \cos^{-1} \frac{d}{R}$$

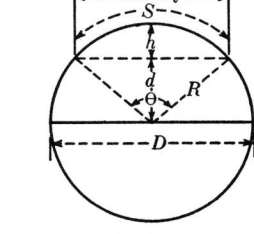

$$l = 2\sqrt{R^2 - d^2} = 2R \sin \frac{\theta}{2} = 2d \tan \frac{\theta}{2}$$

$$d = \frac{1}{2}\sqrt{4R^2 - l^2} = R \cos \frac{\theta}{2} = \frac{1}{2}l \operatorname{ctn} \frac{\theta}{2}$$

FIGURE 2.2

$$h = R - d$$

$$\theta = \frac{S}{R} = \frac{2S}{D} = 2 \cos^{-1} \frac{d}{R} = 2 \tan^{-1} \frac{l}{2d} = 2 \sin^{-1} \frac{l}{D}$$

---

* Adapted by permission from Burington, *Handbook of Mathematical Tables and Formulas,* 3d ed., McGraw-Hill Book Company, New York (1959).

$$A \text{ (circle)} = \pi R^2 = \frac{1}{4}\pi D^2$$

$$A \text{ (sector)} = \frac{1}{2}Rs = \frac{1}{2}R^2\theta$$

$$A \text{ (segment)} = A \text{ (sector)} - A \text{ (triangle)} = \frac{1}{2}R^2 (\theta - \sin \theta)$$

$$= R^2 \cos^{-1} \frac{R - h}{R} - (R - h) \sqrt{2Rh - h^2}$$

$$\text{Perimeter of } n\text{-sided regular polygon inscribed in a circle} = 2nR \sin \frac{\pi}{n}$$

$$\text{Area of inscribed polygon} = \frac{1}{2}nR^2 \sin \frac{2\pi}{n}$$

$$\text{Perimeter of } n\text{-sided regular polygon circumscribed about a circle} = 2nR \tan \frac{\pi}{n}$$

$$\text{Area of circumscribed polygon} = nR^2 \tan \frac{\pi}{n}$$

Radius of circle inscribed in a triangle of sides a, b, and c is

$$r = \sqrt{\frac{(s - a)(s - b)(s - c)}{s}} \qquad s = \frac{1}{2}(a + b + c)$$

Radius of circle circumscribed about a triangle is

$$R = \frac{abc}{4 \sqrt{s(s - a)(s - b)(s - c)}}$$

Ellipse (Fig. 2.3). $A = \pi ab$, where a and b are lengths of semimajor and semiminor axes, respectively.

Parabola (Fig. 2.4)

$$A = \frac{2ld}{3}$$

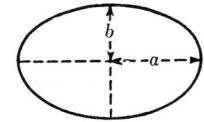

FIGURE 2.3

$$\text{Height of } d_1 = \frac{d}{l^2} (l^2 - l_1^2)$$

$$\text{Width of } l_1 = l \sqrt{\frac{d - d_1}{d}}$$

$$\text{Length of arc} = l\left[1 + \frac{2}{3} \left(\frac{2d}{l}\right)^2 - \frac{2}{5} \left(\frac{2d}{l}\right)^4 + \cdots \right]$$

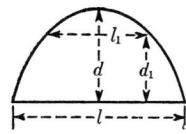

FIGURE 2.4

Area by Approximation (Fig. 2.5). If $y_0, y_1, y_2, \ldots, y_n$ are the length of a series of equally spaced parallel chords, and if h is their distance apart, the area enclosed by the boundary is given approximately by any one of the following formulae:

$$A_T = h[\frac{1}{2}(y_0 + y_n) + y_1 + y_2 + \cdots + y_{n-1}] \qquad \text{(Trapezoidal Rule)}$$

$$A_D = h[0.4(y_0 + y_n) + 1.1(y_1 + y_{n-1}) + y_2 + y_3 + \cdots + y_{n-2}] \qquad \text{(Durand's Rule)}$$

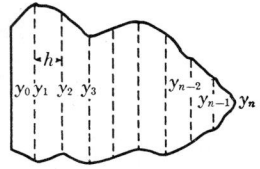

FIGURE 2.5

$$A_S = \tfrac{1}{3}h[(y_0 + y_n) + 4(y_1 + y_3 + \cdots + y_{n-1}) + 2(y_2 + y_4 + \cdots + y_{n-2})]$$

$(n$ even, Simpson's Rule)

In general, A_S gives the most accurate approximation.

The greater the value of n, the greater the accuracy of approximation.

Cube. $V = a^3$; $d = a\sqrt{2}$; total surface area $= 6a^2$, where a is length of side and d is length of diagonal.

Rectangular Parallelopiped. $V = abc$; $d = \sqrt{a^2 + b^2 + c^2}$; total surface area $= 2(ab + bc + ca)$, where a, b, and c are the lengths of the sides and d is length of diagonal.

Prism or Cylinder

$$V = \text{(area of base)} \cdot \text{(altitude)}$$

$$\text{Lateral area} = \text{(perimeter of right section)} \cdot \text{(lateral edge)}$$

Pyramid or Cone

$$V = \tfrac{1}{3}\text{(area of base)} \cdot \text{(altitude)}$$

$$\text{Lateral area of regular pyramid} = \tfrac{1}{2}\text{(perimeter of base)} \cdot \text{(slant height)}$$

Frustum of Pyramid or Cone. $V = \tfrac{1}{3}(A_1 + A_2 + \sqrt{A_1 \cdot A_2})h$, where h is the altitude and A_1 and A_2 are the areas of the bases.

$$\text{Lateral area of a regular figure} = \tfrac{1}{2}\text{(sum of perimeters of base)} \cdot \text{(slant height)}$$

Prismoid

$$V = \frac{h}{6}(A_1 + A_2 + 4A_3)$$

where $h = $ altitude, A_1 and A_2 are the areas of the bases, and A_3 is the area of the midsection parallel to bases.

Area of Surface and Volume of Regular Polyhedra of Edge l

Name	Type of surface	Area of surface	Volume
Tetrahedron	4 equilateral triangles	$1.73205l^2$	$0.11785l^3$
Hexahedron (cube)	6 squares	$6.00000l^2$	$1.00000l^2$
Octahedron	8 equilateral triangles	$3.46410l^2$	$0.47140l^3$
Dodecahedron	12 pentagons	$20.64578l^2$	$7.66312l^3$
Icosahedron	20 equilateral triangles	$8.66025l^2$	$2.18170l^3$

Sphere (Fig. 2.6)

$$A \text{ (sphere)} = 4\pi R^2 = \pi D^2$$

$$A \text{ (zone)} = 2\pi R h_1 = \pi D h_1$$

$$V \text{ (sphere)} = \tfrac{4}{3}\pi R^3 = \tfrac{1}{6}\pi D^3$$

$$V \text{ (spherical sector)} = \tfrac{2}{3}\pi R^2 h_1 = \tfrac{1}{6}\pi D^2 h_1$$

$$V \text{ (spherical segment of one base)} = \tfrac{1}{6}\pi h_3 (3 r_3^2 + h_3^2)$$

$$V \text{ (spherical segment of two bases)} = \tfrac{1}{6}\pi h_2 (3 r_3^2 + 3 r_2^2 + h_2^2)$$

$$A \text{ (lune)} = 2 R^2 \theta \qquad \text{where } \theta \text{ is angle in radians of lune}$$

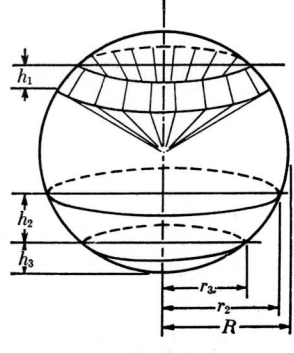

FIGURE 2.6

Ellipsoid. $V = \tfrac{4}{3}\pi abc$, where a, b, and c are the lengths of the semiaxes.

    Torus (Fig. 2.7)

$$V = 2\pi^2 R r^2$$

$$\text{Area of surface} = S = 4\pi^2 R r$$

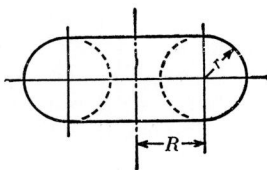

FIGURE 2.7

2.2.3   Trigonometric Functions of an Angle α

Let x be any angle whose initial side lies on the positive x axis and whose vertex is at the origin, and (x, y) be any point on the terminal side of the angle. (x is positive if measured along OX to the right, from the y axis; and negative, if measured along OX' to the left from the y axis. Likewise, y is positive if measured parallel to OY, and negative if measured parallel to OY'.) Let r be the positive distance from the origin to the point. The trigonometric functions of an angle are defined as follows:

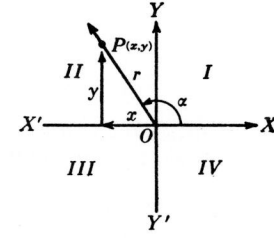

FIGURE 2.8

$$\text{sine } \alpha \quad = \sin \alpha \qquad = \frac{y}{r}$$

$$\text{cosine } \alpha \quad = \cos \alpha \qquad = \frac{x}{r}$$

$$\text{tangent } \alpha \quad = \tan \alpha \qquad = \frac{y}{x}$$

$$\text{cotangent } \alpha = \text{ctn } \alpha \quad = \cot \alpha = \frac{x}{y}$$

$$\text{secant } \alpha \quad = \sec \alpha \qquad = \frac{r}{x}$$

$$\text{cosecant } x \quad = \csc \alpha \qquad = \frac{r}{y}$$

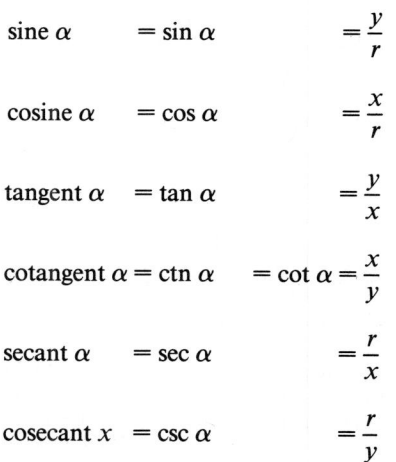

$$\text{exsecant } x = \text{exsec } \alpha \qquad = \sec \alpha - 1$$

$$\text{versine } \alpha = \text{vers } \alpha \qquad = 1 - \cos \alpha$$

$$\text{coversine } \alpha = \text{covers } \alpha \qquad = 1 - \sin \alpha$$

$$\text{haversine } \alpha = \text{hav } \alpha \qquad = \tfrac{1}{2} \text{ vers } \alpha$$

2.2.3.1 Signs of the Functions

Quadrant	sin	cos	tan	ctn	sec	csc
I	+	+	+	+	+	+
II	+	−	−	−	−	+
III	−	−	+	+	−	−
IV	−	+	−	−	+	−

2.2.3.2 Relations between the Functions of a Single Angle*

$$\sin^2 x + \cos^2 x = 1$$

$$\tan x = \frac{\sin x}{\cos x}$$

$$\cot x = \frac{1}{\tan x} = \frac{\cos x}{\sin x}$$

$$1 + \tan^2 x = \sec^2 x = \frac{1}{\cos^2 x}$$

$$1 + \cot^2 x = \csc^2 x = \frac{1}{\sin^2 x}$$

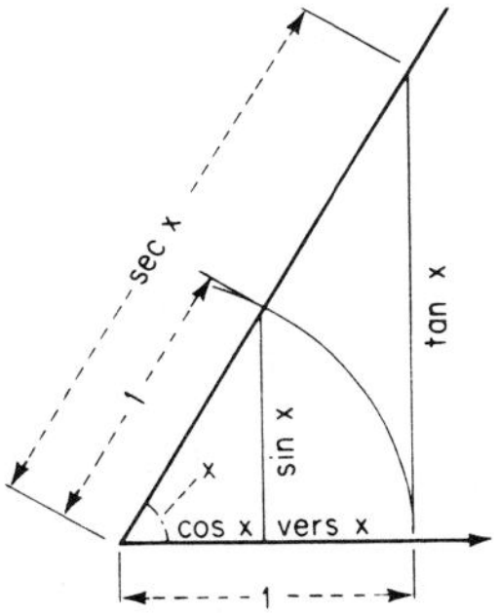

FIGURE 2.9

$$\sin x = \sqrt{1 - \cos^2 x} = \frac{\tan x}{\sqrt{1 + \tan^2 x}} = \frac{1}{\sqrt{1 + \cot^2 x}}$$

$$\cos x = \sqrt{1 - \sin^2 x} = \frac{1}{\sqrt{1 + \tan^2 x}} = \frac{\cot x}{\sqrt{1 + \cot^2 x}}$$

2.2.3.3 Functions of Negative Angles. $\sin(-x) = -\sin x$; $\cos(-x) = \cos x$; $\tan(-x) = -\tan x$.

2.2.3.4 Functions of the Sum and Difference of Two Angles

$$\sin(x + y) = \sin x \cos y + \cos x \sin y$$

$$\cos(x + y) = \cos x \cos y - \sin x \sin y$$

---

* From Baumeister and Marks, *Standard Handbook for Mechanical Engineers,* 7th ed., McGraw-Hill Book Company, New York (1967); by permission.

$$\tan(x + y) = \frac{\tan x + \tan y}{1 - \tan x \tan y}$$

$$\cot(x + y) = \frac{\cot x \cot y - 1}{\cot x + \cot y}$$

$$\sin(x - y) = \sin x \cos y - \cos x \sin y$$

$$\cos(x - y) = \cos x \cos y + \sin x \sin y$$

$$\tan(x - y) = \frac{\tan x - \tan y}{1 + \tan x \tan y}$$

$$\cot(x - y) = \frac{\cot x \cot y + 1}{\cot y - \cot x}$$

$$\sin x + \sin y = 2 \sin \tfrac{1}{2}(x + y) \cos \tfrac{1}{2}(x - y)$$

$$\sin x - \sin y = 2 \cos \tfrac{1}{2}(x + y) \sin \tfrac{1}{2}(x - y)$$

$$\cos x + \cos y = 2 \cos \tfrac{1}{2}(x + y) \cos \tfrac{1}{2}(x - y)$$

$$\cos x - \cos y = -2 \sin \tfrac{1}{2}(x + y) \sin \tfrac{1}{2}(x - y)$$

$$\tan x + \tan y = \frac{\sin(x + y)}{\cos x \cos y} \qquad \cot x + \cot y = \frac{\sin(x + y)}{\sin x \sin y}$$

$$\tan x - \tan y = \frac{\sin(x - y)}{\cos x \cos y} \qquad \cot x - \cot y = \frac{\sin(y - x)}{\sin x \sin y}$$

$$\sin^2 x - \sin^2 y = \cos^2 y - \cos^2 x = \sin(x + y) \sin(x - y)$$

$$\cos^2 x - \sin^2 y = \cos^2 y - \sin^2 x = \cos(x + y) \cos(x - y)$$

$$\sin(45° + x) = \cos(45° - x) \qquad \tan(45° + x) = \cot(45° - x)$$

$$\sin(45° - x) = \cos(45° + x) \qquad \tan(45° - x) = \cot(45° + x)$$

In the following transformations, a and b are supposed to be positive, $c = \sqrt{a^2 + b^2}, A =$ the positive acute angle for which $B = b/a$:

$$a \cos x + b \sin x = c \sin(A + x) = c \cos(B - x)$$

$$a \cos x - b \sin x = c \sin(A - x) = c \cos(B + x)$$

2.2.4 Expansion in Series*

The range of values of x for which each of the series is convergent is stated at the right of the series.

---

* From Baumeister and Marks, *Standard Handbook for Mechanical Engineers,* 7th ed., McGraw-Hill Book Company, New York (1967); by permission.

2.2.4.1 *Exponential and Logarithmic Series*

$$e^x = 1 + \frac{x}{1!} + \frac{x^2}{2!} + \frac{x^3}{3!} + \frac{x^4}{4!} + \cdots \qquad (-\infty < x < +\infty)$$

$$a^x = e^{mx} = 1 + \frac{m}{1!}x + \frac{m^2}{2!}x^2 + \frac{m^3}{3!}x^3 + \cdots \qquad (a > 0, -\infty < x < +\infty)$$

where $m = \ln a = 2.3026 \log_{10} a$.

$$\ln(1 + x) = x - \frac{x^2}{2} + \frac{x^3}{3} - \frac{x^4}{4} + \frac{x^5}{5} \cdots \qquad (-1 < x < +1)$$

$$\ln(1 - x) = -x - \frac{x^2}{2} - \frac{x^3}{3} - \frac{x^4}{4} - \frac{x^5}{5} - \cdots \qquad (-1 < x < +1)$$

$$\ln\left(\frac{1+x}{1-x}\right) = 2\left(x + \frac{x^3}{3} + \frac{x^5}{5} + \frac{x^7}{7} + \cdots\right) \qquad (-1 < x < +1)$$

$$\ln\left(\frac{x+1}{x-1}\right) = 2\left(\frac{1}{x} + \frac{1}{3x^3} + \frac{1}{5x^5} + \frac{1}{7x^7} + \cdots\right) \qquad (x < -1 \text{ or } +1 < x)$$

$$\ln x = 2\left[\frac{x-1}{x+1} + \frac{1}{3}\left(\frac{x-1}{x+1}\right)^3 + \frac{1}{5}\left(\frac{x-1}{x+1}\right)^5 + \cdots\right] \qquad (0 < x < \infty)$$

$$\ln(a + x) = \ln a + 2\left[\frac{x}{2a+x} + \frac{1}{3}\left(\frac{x}{2a+x}\right)^3 + \frac{1}{5}\left(\frac{x}{2a+x}\right)^5 + \cdots\right]$$
$$(0 < a < +\infty, -a < x < +\infty)$$

Series for the Trigonometric Functions. In the following formulas, *all angles must be expressed in radians.* If $D =$ the number of degrees in the angle, and $x =$ its radian measure, then $x = 0.017453D$.

$$\sin x = x - \frac{x^3}{3!} + \frac{x^5}{5!} - \frac{x^7}{7!} + \cdots \qquad (-\infty < x < +\infty)$$

$$\cos x = 1 - \frac{x^2}{2!} + \frac{x^4}{4!} - \frac{x^6}{6!} + \frac{x^8}{8!} - \cdots \qquad (-\infty < x < +\infty)$$

$$\tan x = x + \frac{x^3}{3} + \frac{2x^5}{15} + \frac{17x^7}{315} + \frac{62x^9}{2835} + \cdots \qquad \left(-\frac{\pi}{2} < x < +\frac{\pi}{2}\right)$$

$$\cot x = \frac{1}{x} - \frac{x}{3} - \frac{x^3}{45} - \frac{2x^5}{945} - \frac{x^7}{4725} - \cdots \qquad (-\pi < x < +\pi)$$

$$\sin^{-1} y = y + \frac{y^3}{6} + \frac{3y^5}{40} + \frac{5y^7}{112} + \cdots \qquad (-1 \leqq y \leqq +1)$$

$$\tan^{-1} y = y - \frac{y^3}{3} + \frac{y^5}{5} - \frac{y^7}{7} + \cdots \qquad (-1 \leqq y \leqq +1)$$

$$\cos^{-1} y = \tfrac{1}{2}\pi - \sin^{-1} y \qquad \cot^{-1} y = \tfrac{1}{2}\pi - \tan^{-1} y$$

Reversing a Series. If $y = x + bx^2 + cx^3 + dx^4 + ex^5 + \cdots$, then $x = y - by^2 + (2b^2 - c)y^3 - (5b^3 - 5bc + d)y^4 + (14b^4 - 21b^2c + 6bd + 3c^2 - e)y^5 + \cdots$, provided the latter series is convergent.

Fourier's Series. Let $f(x)$ be a function which is finite in the interval from $x = -c$ to $x = +c$ and whose graph has finite arc length in that interval.* Then, for any value of x between $-c$ and c,

$$f(x) = \tfrac{1}{2}a_0 + a_1 \cos \frac{\pi x}{c} + a_2 \cos \frac{2\pi c}{c} + a_3 \cos \frac{3\pi x}{c} + \cdots$$

$$+ b_1 \sin \frac{\pi x}{c} + b_2 \sin \frac{2\pi x}{c} + b_3 \sin \frac{3\pi x}{c} + \cdots$$

where the constant coefficients are determined as follows:

$$a_n = \frac{1}{c} \int_{-c}^{c} f(t) \cos \frac{n\pi t}{c}\, dt \qquad b_n = \frac{1}{c} \int_{-c}^{c} f(t) \sin \frac{n\pi t}{c}\, dt$$

In case the curve $y = f(x)$ is symmetrical with respect to the origin, the a's are all zero, and the series is a sine series. In case the curve is symmetrical with respect to the y axis, the b's are all zero, and a cosine series results. (In this case, the series will be valid not only for values of x between $-c$ and c, but also for $x = -c$ and $x = c$.) A Fourier series can always be integrated term by term; but the result of differentiating term by term may not be a convergent series.

TABLE 2.25 Some Constants

Constant	Number	Log$_{10}$ of Number
Pi (π)	3.14159 26535 89793 23846	0.49714 98726 94133 85435
Napierian Base (e)	2.71828 18284 59045 23536	0.43429 448
$M = \log_{10}e$	0.43429 44819 03251 82765	9.63778 43113 00536 78912 $-$ 10
$1 \div M = \log_e 10$	2.30258 50929 94045 68402	0.36221 569
$180 \div \pi =$ degrees in 1 radian	57.2957 795	1.75812 263
$\pi \div 180 =$ radians in $1°$	0.01745 329	8.24187 737 $-$ 10
$\pi \div 10800 =$ radians in $1'$	0.00029 08882	6.46372 612 $-$ 10
$\pi \div 648000 =$ radians in $1''$	0.00000 48481 36811 095	4.68557 487 $-$ 10

2.3 STATISTICAL TABLES

2.3.1 Statistics

Raw data are collected observations that have not been organized numerically. An array is an arrangement of raw numerical data in ascending or descending order of magnitude. An *average* is a value that is typical or representative of a set of data. Several averages can be defined, the most common being the arithmetic mean (or briefly the mean), the median, the mode, and the geometric mean.

---

* If $x = x_0$ is a point of discontinuity, $f(x_0)$ is to be defined as $\tfrac{1}{2}[f_1(x_0) + f_2(x_0)]$, where $f_1(x_0)$ is the limit of $f(x)$ when x approaches x_0 from below, and $f_2(x_0)$ is the limit of $f(x)$ when x approaches x_0 from above.

The *mean* of a set of N numbers, $x_1, x_2, x_3, \ldots, x_N$ is denoted by $\bar{x}$ and is defined as

$$\bar{x} = \frac{x_1 + x_2 + x_3 + \cdots + x_N}{N}$$

It is an estimation of the unknown true value μ of an infinite population.

The *median* of a set of numbers arranged in order of magnitude is the middle value or the arithmetic mean of the two middle values. The median allows inclusion of all data in a set without undue influence from outlying values; it is preferable to the mean for small sets of data.

The *mode* of a set of numbers is that value which occurs with the greatest frequency (the most common value). The mode may not exist, and even if it does exist it may not be unique. The empirical relation that exists between the mean, the mode, and the median for unimodal frequency curves which are moderately asymmetrical is

$$\text{Mean} - \text{mode} = 3(\text{mean} - \text{median})$$

The *geometric mean* of a set of N numbers is the Nth root of the product of the numbers:

$$\sqrt[N]{x_1 x_2 x_3 \ldots x_N}$$

2.3.1.1 *Distribution of Measurements.* When a large number of measurements are made, the individual measurements are not all identical and equal to the accepted value μ, which is the mean of an infinite population or universe of data, but are scattered about μ, owing to random error. If the magnitude of any single measurement is the abscissa and the relative frequencies (i.e., the probability) of occurrence of different-sized measurements are the ordinate, the smooth curve drawn through the points (Fig. 2.10) is the *normal distribution curve* (also called the *Gaussian distribution curve,* the *error curve,* or the *probability curve*). The term "error curve" arises when one considers the distribution of errors $(x - \mu)$ about the true value.

The breadth or spread of the curve indicates the precision of the measurements, and is determined by and related to the standard deviation, a relationship that is expressed in the equation for the normal curve,

$$Y = \frac{1}{\sigma\sqrt{2\pi}} e^{-1/2(x-\mu)^2/\sigma^2}$$

where σ is the standard deviation of the infinite population. The population mean μ expresses the magnitude of the quantity being measured; the standard deviation σ expresses the scatter. When $(x - \mu)/\sigma$ is replaced by the standardized variable z, then

$$Y = \frac{1}{\sqrt{2\pi}} e^{-1/2z^2}$$

The standardized variable measures the deviation from the population mean in units of standard deviation. Y is 0.399 for the most probable value, μ.

Table 2.26 lists the height of an ordinate (column Y) as a distance z from the mean, and the area under the normal curve (column A) at a distance z from the mean, expressed as fractions of the total area, 1.000.

Example 1. The true value of a quantity is 30.00, and σ for the method of measurement is 0.30. What is the probability that a single measurement will have a deviation from the mean

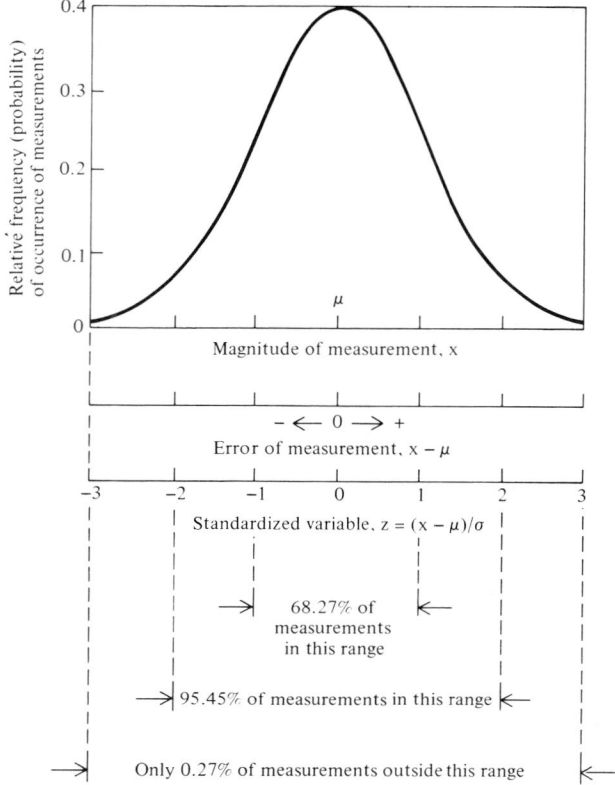

FIGURE 2.10   The normal distribution curve.

greater than 0.45; that is, what percentage of results will fall outside the range 30.00 ± 0.45?

$$z = \frac{x - \mu}{\sigma} = \frac{0.45}{0.30} = 1.5$$

From Table 2.26 the area under the normal curve from -1.5σ to $+1.5\sigma$ is 0.866, meaning that 86.6 percent of the measurements will fall within the range 30.00 ± 0.45 and 13.4 percent will lie outside this range. Half of these measurements, 6.8 percent, will be less than 29.55; and a similar percentage will exceed 30.45. In actuality the uncertainty in z is about 1 in 15; therefore, the value of z could lie between 1.4 and 1.6; the corresponding areas under the curve could lie between 84 and 89 percent.

Example 2. In the foregoing example, what expected number of samples will be 29.40 or less from a total of 400 samples?

$$z = \frac{29.40 - 30.00}{0.30} = 2.0$$

The area under the negative portion of the normal curve corresponds to $0.500 - 0.477 = 0.023$. Only $0.023(400) = 9.2$ or nine samples will be 29.40 or less.

TABLE 2.26 Ordinates and Areas between Abscissa Values $-z$ and $+z$ of the Normal Distribution Curve

From Perry, Chilton, and Kirkpatrick, Chemical Engineers' Handbook, 4th ed., McGraw-Hill Book Company, New York (1963); by permission.

z	X	Y	A	1 − A
0	μ	0.399	0.0000	1.0000
±0.05	μ ± 0.05σ	.398	.0399	0.9601
±.10	μ ± .10σ	.397	.0797	.9203
±.15	μ ± .15σ	.394	.1192	.8808
±.20	μ ± .20σ	.391	.1585	.8415
±.25	μ ± .25σ	.387	.1974	.8026
±.30	μ ± .30σ	.381	.2358	.7642
±.35	μ ± .35σ	.375	.2737	.7263
±.40	μ ± .40σ	.368	.3108	.6892
±.45	μ ± .45σ	.361	.3473	.6527
±.50	μ ± .50σ	.352	.3829	.6171
±.55	μ ± .55σ	.343	.4177	.5823
±.60	μ ± .60σ	.333	.4515	.5485
±.65	μ ± .65σ	.323	.4843	.5157
±.70	μ ± .70σ	.312	.5161	.4839
±.75	μ ± .75σ	.301	.5467	.4533
±.80	μ ± .80σ	.290	.5763	.4237
±.85	μ ± .85σ	.278	.6047	.3953
±.90	μ ± .90σ	.266	.6319	.3681
±.95	μ ± .95σ	.254	.6579	.3421
±1.00	μ ± 1.00σ	.242	.6827	.3173
±1.05	μ ± 1.05σ	.230	.7063	.2937
±1.10	μ ± 1.10σ	.218	.7287	.2713
±1.15	μ ± 1.15σ	.206	.7499	.2501
±1.20	μ ± 1.20σ	.194	.7699	.2301
±1.25	μ ± 1.25σ	.183	.7887	.2113
±1.30	μ ± 1.30σ	.171	.8064	.1936
±1.35	μ ± 1.35σ	.160	.8230	.1770
±1.40	μ ± 1.40σ	.150	.8385	.1615
±1.45	μ ± 1.45σ	.139	.8529	.1471
±1.50	μ ± 1.50σ	.130	.8664	.1336
±0.000	μ ± 0.126σ	0.3989	0.0000	1.0000
±.126	μ ± .253σ	.3958	.1000	.9000
±.253	μ ± .385σ	.3863	.2000	.8000
±.385	μ ± .524σ	.3704	.3000	.7000
±.524	μ ± .674σ	.3477	.4000	.6000
±.674	μ ± .842σ	.3178	.5000	.5000
±.842		.2890	.6000	.4000

z	X	Y	A	1 − A
±1.50	μ ± 1.50σ	0.1295	0.8664	0.1336
±1.55	μ ± 1.55σ	.1200	.8789	.1211
±1.60	μ ± 1.60σ	.1109	.8904	.1096
±1.65	μ ± 1.65σ	.1023	.9011	.0989
±1.70	μ ± 1.70σ	.0940	.9109	.0891
±1.75	μ ± 1.75σ	.0863	.9199	.0801
±1.80	μ ± 1.80σ	.0790	.9281	.0719
±1.85	μ ± 1.85σ	.0721	.9357	.0643
±1.90	μ ± 1.90σ	.0656	.9426	.0574
±1.95	μ ± 1.95σ	.0596	.9488	.0512
±2.00	μ ± 2.00σ	.0540	.9545	.0455
±2.05	μ ± 2.05σ	.0488	.9596	.0404
±2.10	μ ± 2.10σ	.0440	.9643	.0357
±2.15	μ ± 2.15σ	.0396	.9684	.0316
±2.20	μ ± 2.20σ	.0355	.9722	.0278
±2.25	μ ± 2.25σ	.0317	.9756	.0244
±2.30	μ ± 2.30σ	.0283	.9786	.0214
±2.35	μ ± 2.35σ	.0252	.9812	.0188
±2.40	μ ± 2.40σ	.0224	.9836	.0164
±2.45	μ ± 2.45σ	.0198	.9857	.0143
±2.50	μ ± 2.50σ	.0175	.9876	.0124
±2.55	μ ± 2.55σ	.0154	.9892	.0108
±2.60	μ ± 2.60σ	.0136	.9907	.0093
±2.65	μ ± 2.65σ	.0119	.9920	.0080
±2.70	μ ± 2.70σ	.0104	.9931	.0069
±2.75	μ ± 2.75σ	.0091	.9940	.0060
±2.80	μ ± 2.80σ	.0079	.9949	.0051
±2.85	μ ± 2.85σ	.0069	.9956	.0044
±2.90	μ ± 2.90σ	.0060	.9963	.0037
±2.95	μ ± 2.95σ	.0051	.9968	.0032
±3.00	μ ± 3.00σ	.0044	.9973	.0027
±4.00	μ ± 4.00σ	.0001	.99994	.00006
±5.00	μ ± 5.00σ	.000001	.999999994	.000000006
±1.036	μ ± 1.036σ	0.2331	0.7000	0.3000
±1.282	μ ± 1.282σ	.1755	.8000	.2000
±1.645	μ ± 1.645σ	.1031	.9000	.1000
±1.960	μ ± 1.960σ	.0584	.9500	.0500
±2.576	μ ± 2.576σ	.0145	.9900	.0100
±3.291	μ ± 3.291σ	.0018	.9990	.0010
±3.891	μ ± 3.891σ	.0002	.9999	.0001

Example 3. If the mean value of 500 determinations is 151 and $\sigma = 15$, how many results lie between 120 and 155 (actually any value between 119.5 and 155.5)?

$$z = \frac{119.5 - 151}{15} = -2.10 \qquad \text{Area:} \quad 0.482$$

$$z = \frac{155.5 - 151}{15} = \quad 0.30 \qquad \qquad \underline{0.118}$$

Total area: 0.600

$$500(0.600) = 300 \text{ results}$$

2.3.1.2 Measures of Dispersion. Several ways may be used to characterize the spread or dispersion in the original data. The *range* is the difference between the largest value and the smallest value in a set of observations. However, almost always the most efficient quantity for characterizing variability is the *standard deviation* (also called the *root mean square*).

The standard deviation is the square root of the average squared difference between the individual observations and the population mean:

$$\sigma = \sqrt{\frac{\sum\limits_{i=1}^{N}(x_i - \mu)^2}{N}}$$

The standard deviation may be estimated by calculating s drawn from a sample set as follows:

$$s = \sqrt{\frac{\sum\limits_{i=1}^{N}(x_i - \bar{x})^2}{N-1}} \qquad \text{or} \qquad s = \sqrt{\frac{x_1^2 + x_2^2 + \cdots - [(x_1 + x_2 + \cdots)^2]/N}{N-1}}$$

where $x_i - \bar{x}$ represents the deviation of each number in the array from the arithmetic mean. Since two pieces of information, namely s and $\bar{x}$, have been extracted from the data, we are left with $N - 1$ independent data available for measurement of precision. The divisor is termed the *degrees of freedom.* If a relatively large sample of data corresponding to $N > 30$ is available, its mean can be taken as a measure of μ, and $s \approx \sigma$.

So basic is the notion of a statistical *estimate* of a physical parameter that statisticians use Greek letters for the *parameters* and English letters for the estimates.

For many purposes, one uses the *variance,* which for the sample is s^2 and for the entire population is σ^2. The variance s^2 of a finite sample is an unbiased estimate of σ^2, whereas the standard deviation s is not an unbiased estimate of σ.

When a series of observations can be logically arranged into k subgroups, the variance is calculated by summing the squares of the deviations for each subgroup, and then adding all the k sums and dividing by $N - k$ because one degree of freedom is lost in each subgroup. For two groups of observations consisting of N_A and N_B members of standard deviations s_A and s_B, respectively, the variance is given by

$$s^2 = \frac{(N_A - 1)s_A^2 + (N_B - 1)s_B^2}{N_A + N_B - 2}$$

Another measure of dispersion is the *coefficient of variation,* which is merely the standard deviation expressed as a percentage of the arithmetic mean, viz.: $100s/\bar{x}$. It is useful mainly to show whether the relative or the absolute spread of values is constant as the values are changed.

2.3.1.3 Theoretical Distributions and Tests of Significance. If the data contained only random (or chance) errors, the cumulative estimates $\bar{x}$ and s would gradually approach the limits μ and σ. The distribution of results would be normally distributed with mean μ and standard deviation σ. Were the true mean of the infinite population known, it would also have some symmetrical type of distribution centered around μ. However, it would be expected that the dispersion or spread of this dispersion about the mean would depend on the sample size. The standard deviation of the distribution of means equals $\sigma/\sqrt{N}$. A distribution of this type is called the "Student's distribution"; and the corresponding test of significance, a measure of error between μ and $\bar{x}$, the t test. The Student t takes into account both the possible variation of the value of $\bar{x}$ from μ on the basis of the expected variance $\sigma^2/\sqrt{N}$ and the reliability of using s in place of σ. The distribution of the statistic is

$$\pm t = \frac{\bar{x} - \mu}{s/\sqrt{N}} \qquad \text{or} \qquad \mu = \bar{x} \pm \frac{ts}{\sqrt{N}}$$

This distribution is symmetrical about zero, and its dispersion is a function of the degrees of freedom $N - 1$. Its limits are called *confidence limits*. The percentage probability that μ lies within this interval is called the *confidence level*. The *level of significance* or *error probability* $(100 - \text{confidence level or } 100 - \alpha)$ is the percent probability that μ will lie outside the confidence interval, and represents the chances of being incorrect in stating that μ lies within the confidence interval. Values of t are in Table 2.27 for any desired degrees of freedom and various confidence levels.

Statistical methods are frequently used to give a yes or no answer to a particular question concerning the significance of data. The answer is qualified by a confidence level indicating the degree of certainty of the answer. A common procedure is to set up a *null hypothesis,* which states that there is no significant difference between two sets of data or that a variable exerts no significant effect. Generally confidence levels of 95 and 99 percent are chosen to express the probability that the answer is correct. These are also denoted as the 0.05 and 0.01 level of significance, respectively. When the hypothesis can be rejected at the 0.05 level of significance, but not at the 0.01 level, we can say that the sample results are probably significant. If, however, the hypothesis is also rejected at the 0.01 level, the results become highly significant. For a small number of samples we replace z, obtained from Table 2.26, by t from Table 2.27, and we replace σ by $[\sqrt{N/(N-1)}]s$.

Example 4. In the past a method gave $\mu = 0.050$ percent. A recent set of 10 results gave $\bar{x} = 0.053$ percent and $s = 0.003$ percent. Is everything satisfactory at a level of significance of 0.05? Of 0.01? We wish to decide between the hypotheses

$$H_0: \quad \mu = 0.050\% \qquad \text{and the method is working properly; and}$$
$$H_1: \quad \mu \neq 0.050\% \qquad \text{and the method is not working properly.}$$

A *two-tailed test* is required; that is, both tails on the distribution curve are involved:

$$t = \frac{0.053 - 0.050}{0.003} \sqrt{10 - 1} = -3.00$$

Enter Table 2.27 for nine degrees of freedom under the column headed $t_{.975}$ for the 0.05 level of significance, and the column $t_{.995}$ for the 0.01 level of significance. At the 0.05 level, accept H_0 if t lies inside the interval $-t_{.975}$ to $t_{.975}$, that is, within -2.26 and 2.26; reject otherwise. Since $t = -3.00$, we reject H_0. At the 0.01 level of significance, the corresponding interval is -3.25 to 3.25, which t lies within, indicating acceptance of H_0. Because we can reject H_0 at the 0.05

TABLE 2.27 Values of t

From Perry, Chilton, and Kirkpatrick, Chemical Engineers'
Handbook, *4th ed., McGraw-Hill Book Company, New York (1963);
by permission.*

df	$t_{.60}$	$t_{.70}$	$t_{.80}$	$t_{.90}$	$t_{.95}$	$t_{.975}$	$t_{.99}$	$t_{.995}$
1	0.325	0.727	1.376	3.078	6.314	12.706	31.821	63.657
2	.289	.617	1.061	1.886	2.920	4.303	6.965	9.925
3	.277	584	0.978	1.638	2.353	3.182	4.541	5.841
4	.271	.569	.941	1.533	2.132	2.776	3.747	4.604
5	.267	.559	.920	1.476	2.015	2.571	3.365	4.032
6	.265	.553	.906	1.440	1.943	2.447	3.143	3.707
7	.263	.549	.896	1.415	1.895	2.365	2.998	3.499
8	.262	.546	.889	1.397	1.860	2.306	2.896	3.355
9	.261	.543	.883	1.383	1.833	2.262	2.821	3.250
10	.260	.542	.879	1.372	1.812	2.228	2.764	3.169
11	.260	.540	.876	1.363	1.796	2.201	2.718	3.106
12	.259	.539	.873	1.356	1.782	2.179	2.681	3.055
13	.259	.538	.870	1.350	1.771	2.160	2.650	3.012
14	.258	.537	.868	1.345	1.761	2.145	2.624	2.977
15	.258	.536	.866	1.341	1.753	2.131	2.602	2.947
16	.258	.535	.865	1.337	1.746	2.120	2.583	2.921
17	.257	.534	.863	1.333	1.740	2.110	2.567	2.898
18	.257	.534	.862	1.330	1.734	2.101	2.552	2.878
19	.257	.533	.861	1.328	1.729	2.093	2.539	2.861
20	.257	.533	.860	1.325	1.725	2.086	2.528	2.845
21	.257	.532	.859	1.323	1.721	2.080	2.518	2.831
22	.256	.532	.858	1.321	1.717	2.074	2.508	2.819
23	.256	.532	.858	1.319	1.714	2.069	2.500	2.807
24	.256	.531	.857	1.318	1.711	2.064	2.492	2.797
25	.256	.531	.856	1.316	1.708	2.060	2.485	2.787
26	.256	.531	.856	1.315	1.706	2.056	2.479	2.779
27	.256	.531	.855	1.314	1.703	2.052	2.473	2.771
28	.256	.530	.855	1.313	1.701	2.048	2.467	2.763
29	.256	.530	.854	1.311	1.699	2.045	2.462	2.756
30	.256	.530	.854	1.310	1.697	2.042	2.457	2.750
40	.255	.529	.851	1.303	1.684	2.021	2.423	2.704
60	.254	.527	.848	1.296	1.671	2.000	2.390	2.660
120	.254	.526	.845	1.289	1.658	1.980	2.358	2.617
∞	.253	.524	.842	1.282	1.645	1.960	2.326	2.576
df	$-t_{.40}$	$-t_{.30}$	$-t_{.20}$	$-t_{.10}$	$-t_{.05}$	$-t_{.025}$	$-t_{.01}$	$-t_{.005}$

When the table is read from the foot, the tabled values are to be prefixed with
a negative sign. Interpolation should be performed using the reciprocals of
the degrees of freedom.

level but not at the 0.01 level of significance, we can say that the sample results are probably
significant and that the method is not working properly.

Example 5. Six samples from a bulk chemical shipment averaged 77.50 percent active
ingredient with $s = 1.45$ percent. The manufacturer claimed 80.00 percent. Can his claim be
supported? A one-tailed test is required.

$$t = \frac{77.50 - 80.00}{1.45} \sqrt{6 - 1} = -3.86$$

Since $t_{.95} = -2.01$, and $t_{.99} = -3.36$, the hypothesis is rejected at both the 0.05 and the 0.01
levels of significance. It is extremely unlikely that the claim is justified.

It is possible to compare the means of two relatively small sets of observations when the
variances within the sets can be regarded as the same, as indicated by the F test. Also the t test
can be applied to differences between pairs of observations. Perhaps only a single pair can be
performed at one time, or possibly one wishes to compare two methods using samples of

differing analytical content. It is still necessary that the two methods possess the same inherent standard deviation. An average difference $\bar{d}$ is calculated, and individual deviations from $\bar{d}$ are used to evaluate the variance of the differences.

Example 6. From the following data do the two methods actually give concordant results?

Sample	Method A	Method B	Difference
1	33.27	33.04	$d_1 = 0.23$
2	51.34	50.96	$d_2 = 0.38$
3	23.91	23.77	$d_3 = 0.14$
4	47.04	46.79	$d_4 = 0.25$
			$\bar{d} = 0.25$

$$s_d = \sqrt{\frac{\Sigma(d - \bar{d})^2}{N - 1}} = 0.099$$

$$t = \frac{0.25}{0.099}\sqrt{4 - 1} = 4.30$$

From Table 2.27, $t_{.975} = 3.18$ (at 95 percent probability) and $t_{.995} = 5.84$ (at 99 percent probability). The difference between the two methods is probably significant.

One can consider the distribution involving estimates of the true variance. With s_1^2 determined from a group of N_1 observations and s_2^2 from a second group of N_2 observations, the distribution of the ratio of the sample variances is given by the F test:

$$F = \frac{s_1^2}{s_2^2}$$

The larger variance is placed in the numerator. For example, the F test allows judgment regarding the existence of a significant difference in the precision between two sets of data or between two analysts. The hypothesis assumed is that both variances are indeed alike and a measure of the same σ.

Example 7. Suppose Analyst A made five observations and obtained a standard deviation of 0.06 whereas Analyst B with six observations obtained $s_B = 0.03$. The experimental variance ratio is

$$F = \frac{(0.06)^2}{(0.03)^2} = 4.00$$

From Table 2.28, with four degrees of freedom for A and five degrees of freedom for B, the value of F would exceed 5.19 five percent of the time. Therefore, the null hypothesis is valid, and comparable skills are exhibited by the two analysts. As applied, the F test was one-tailed. The F test may also be applied as a two-tailed test in which the alternative to the null hypothesis is $\sigma_1^2 \neq \sigma_2^2$. This doubles the probability that the null hypothesis is invalid and has the effect of changing the confidence level, in the above example, from 95 to 90 percent.

If improvement in precision is claimed for a set of measurements, the variance for the set against which comparison is being made should be placed in the numerator, regardless of magnitude. An experimental F smaller than unity indicates that the claim for improved precision cannot be supported.

TABLE 2.28 *F Distribution*

From Perry, Chilton, and Kirkpatrick, Chemical Engineers' Handbook, 4th ed., McGraw-Hill Book Company, New York (1963); by permission.

Upper 5% Points ($F_{.95}$)

	Degrees of freedom for numerator																		
Degrees of freedom for denominator	1	2	3	4	5	6	7	8	9	10	12	15	20	24	30	40	60	120	∞
1	161	200	216	225	230	234	237	239	241	242	244	246	248	249	250	251	252	253	254
2	18.5	19.0	19.2	19.2	19.3	19.3	19.4	19.4	19.4	19.4	19.4	19.4	19.4	19.5	19.5	19.5	19.5	19.5	19.5
3	10.1	9.55	9.28	9.12	9.01	8.94	8.89	8.85	8.81	8.79	8.74	8.70	8.66	8.64	8.62	8.59	8.57	8.55	8.53
4	7.71	6.94	6.59	6.39	6.26	6.16	6.09	6.04	6.00	5.96	5.91	5.86	5.80	5.77	5.75	5.72	5.69	5.66	5.63
5	6.61	5.79	5.41	5.19	5.05	4.95	4.88	4.82	4.77	4.74	4.68	4.62	4.56	4.53	4.50	4.46	4.43	4.40	4.37
6	5.99	5.14	4.76	4.53	4.39	4.28	4.21	4.15	4.10	4.06	4.00	3.94	3.87	3.84	3.81	3.77	3.74	3.70	3.67
7	5.59	4.74	4.35	4.12	3.97	3.87	3.79	3.73	3.68	3.64	3.57	3.51	3.44	3.41	3.38	3.34	3.30	3.27	3.23
8	5.32	4.46	4.07	3.84	3.69	3.58	3.50	3.44	3.39	3.35	3.28	3.22	3.15	3.12	3.08	3.04	3.01	2.97	2.93
9	5.12	4.26	3.86	3.63	3.48	3.37	3.29	3.23	3.18	3.14	3.07	3.01	2.94	2.90	2.86	2.83	2.79	2.75	2.71
10	4.96	4.10	3.71	3.48	3.33	3.22	3.14	3.07	3.02	2.98	2.91	2.85	2.77	2.74	2.70	2.66	2.62	2.58	2.54
11	4.84	3.98	3.59	3.36	3.20	3.09	3.01	2.95	2.90	2.85	2.79	2.72	2.65	2.61	2.57	2.53	2.49	2.45	2.40
12	4.75	3.89	3.49	3.26	3.11	3.00	2.91	2.85	2.80	2.75	2.69	2.62	2.54	2.51	2.47	2.43	2.38	2.34	2.30
13	4.67	3.81	3.41	3.18	3.03	2.92	2.83	2.77	2.71	2.67	2.60	2.53	2.46	2.42	2.38	2.34	2.30	2.25	2.21
14	4.60	3.74	3.34	3.11	2.96	2.85	2.76	2.70	2.65	2.60	2.53	2.46	2.39	2.35	2.31	2.27	2.22	2.18	2.13
15	4.54	3.68	3.29	3.06	2.90	2.79	2.71	2.64	2.59	2.54	2.48	2.40	2.33	2.29	2.25	2.20	2.16	2.11	2.07
16	4.49	3.63	3.24	3.01	2.85	2.74	2.66	2.59	2.54	2.49	2.42	2.35	2.28	2.24	2.19	2.15	2.11	2.06	2.01
17	4.45	3.59	3.20	2.96	2.81	2.70	2.61	2.55	2.49	2.45	2.38	2.31	2.23	2.19	2.15	2.10	2.06	2.01	1.96
18	4.41	3.55	3.16	2.93	2.77	2.66	2.58	2.51	2.46	2.41	2.34	2.27	2.19	2.15	2.11	2.06	2.02	1.97	1.92
19	4.38	3.52	3.13	2.90	2.74	2.63	2.54	2.48	2.42	2.38	2.31	2.23	2.16	2.11	2.07	2.03	1.98	1.93	1.88
20	4.35	3.49	3.10	2.87	2.71	2.60	2.51	2.45	2.39	2.35	2.28	2.20	2.12	2.08	2.04	1.99	1.95	1.90	1.84
21	4.32	3.47	3.07	2.84	2.68	2.57	2.49	2.42	2.37	2.32	2.25	2.18	2.10	2.05	2.01	1.96	1.92	1.87	1.81
22	4.30	3.44	3.05	2.82	2.66	2.55	2.46	2.40	2.34	2.30	2.23	2.15	2.07	2.03	1.98	1.94	1.89	1.84	1.78
23	4.28	3.42	3.03	2.80	2.64	2.53	2.44	2.37	2.32	2.27	2.20	2.13	2.05	2.01	1.96	1.91	1.86	1.81	1.76
24	4.26	3.40	3.01	2.78	2.62	2.51	2.42	2.36	2.30	2.25	2.18	2.11	2.03	1.98	1.94	1.89	1.84	1.79	1.73
25	4.24	3.39	2.99	2.76	2.60	2.49	2.40	2.34	2.28	2.24	2.16	2.09	2.01	1.96	1.92	1.87	1.82	1.77	1.71
30	4.17	3.32	2.92	2.69	2.53	2.42	2.33	2.27	2.21	2.16	2.09	2.01	1.93	1.89	1.84	1.79	1.74	1.68	1.62
40	4.08	3.23	2.84	2.61	2.45	2.34	2.25	2.18	2.12	2.08	2.00	1.92	1.84	1.79	1.74	1.69	1.64	1.58	1.51
60	4.00	3.15	2.76	2.53	2.37	2.25	2.17	2.10	2.04	1.99	1.92	1.84	1.75	1.70	1.65	1.59	1.53	1.47	1.39
120	3.92	3.07	2.68	2.45	2.29	2.18	2.09	2.02	1.96	1.91	1.83	1.75	1.66	1.61	1.55	1.50	1.43	1.35	1.25
∞	3.84	3.00	2.60	2.37	2.21	2.10	2.01	1.94	1.88	1.83	1.75	1.67	1.57	1.52	1.46	1.39	1.32	1.22	1.00

TABLE 2.28 *F* Distribution (*Continued*)

Upper 1% Points ($F_{.99}$)

Degrees of freedom for denominator	Degrees of freedom for numerator																		
	1	2	3	4	5	6	7	8	9	10	12	15	20	24	30	40	60	120	∞
1	4052	5000	5403	5625	5764	5859	5928	5982	6023	6056	6106	6157	6209	6235	6261	6287	6313	6339	6366
2	98.5	99.0	99.2	99.2	99.3	99.3	99.4	99.4	99.4	99.4	99.4	99.4	99.4	99.5	99.5	99.5	99.5	99.5	99.5
3	34.1	30.8	29.5	28.7	28.2	27.9	27.7	27.5	27.3	27.2	27.1	26.9	26.7	26.6	26.5	26.4	26.3	26.2	26.1
4	21.2	18.0	16.7	16.0	15.5	15.2	15.0	14.8	14.7	14.5	14.4	14.2	14.0	13.9	13.8	13.7	13.7	13.6	13.5
5	16.3	13.3	12.1	11.4	11.0	10.7	10.5	10.3	10.2	10.1	9.89	9.72	9.55	9.47	9.38	9.29	9.20	9.11	9.02
6	13.7	10.9	9.78	9.15	8.75	8.47	8.26	8.10	7.98	7.87	7.72	7.56	7.40	7.31	7.23	7.14	7.06	6.97	6.88
7	12.2	9.55	8.45	7.85	7.46	7.19	6.99	6.84	6.72	6.62	6.47	6.31	6.16	6.07	5.99	5.91	5.82	5.74	5.65
8	11.3	8.65	7.59	7.01	6.63	6.37	6.18	6.03	5.91	5.81	5.67	5.52	5.36	5.28	5.20	5.12	5.03	4.95	4.86
9	10.6	8.02	6.99	6.42	6.06	5.80	5.61	5.47	5.35	5.26	5.11	4.96	4.81	4.73	4.65	4.57	4.48	4.40	4.31
10	10.0	7.56	6.55	5.99	5.64	5.39	5.20	5.06	4.94	4.85	4.71	4.56	4.41	4.33	4.25	4.17	4.08	4.00	3.91
11	9.65	7.21	6.22	5.67	5.32	5.07	4.89	4.74	4.63	4.54	4.40	4.25	4.10	4.02	3.94	3.86	3.78	3.69	3.60
12	9.33	6.93	5.95	5.41	5.06	4.82	4.64	4.50	4.39	4.30	4.16	4.01	3.86	3.78	3.70	3.62	3.54	3.45	3.36
13	9.07	6.70	5.74	5.21	4.86	4.62	4.44	4.30	4.19	4.10	3.96	3.82	3.66	3.59	3.51	3.43	3.34	3.25	3.17
14	8.86	6.51	5.56	5.04	4.70	4.46	4.28	4.14	4.03	3.94	3.80	3.66	3.51	3.43	3.35	3.27	3.18	3.09	3.00
15	8.68	6.36	5.42	4.89	4.56	4.32	4.14	4.00	3.89	3.80	3.67	3.52	3.37	3.29	3.21	3.13	3.05	2.96	2.87
16	8.53	6.23	5.29	4.77	4.44	4.20	4.03	3.89	3.78	3.69	3.55	3.41	3.26	3.18	3.10	3.02	2.93	2.84	2.75
17	8.40	6.11	5.19	4.67	4.34	4.10	3.93	3.79	3.68	3.59	3.46	3.31	3.16	3.08	3.00	2.92	2.83	2.75	2.65
18	8.29	6.01	5.09	4.58	4.25	4.01	3.84	3.71	3.60	3.51	3.37	3.23	3.08	3.00	2.92	2.84	2.75	2.66	2.57
19	8.18	5.93	5.01	4.50	4.17	3.94	3.77	3.63	3.52	3.43	3.30	3.15	3.00	2.92	2.84	2.76	2.67	2.58	2.49
20	8.10	5.85	4.94	4.43	4.10	3.87	3.70	3.56	3.46	3.37	3.23	3.09	2.94	2.86	2.78	2.69	2.61	2.52	2.42
21	8.02	5.78	4.87	4.37	4.04	3.81	3.64	3.51	3.40	3.31	3.17	3.03	2.88	2.80	2.72	2.64	2.55	2.46	2.36
22	7.95	5.72	4.82	4.31	3.99	3.76	3.59	3.45	3.35	3.26	3.12	2.98	2.83	2.75	2.67	2.58	2.50	2.40	2.31
23	7.88	5.66	4.76	4.26	3.94	3.71	3.54	3.41	3.30	3.21	3.07	2.93	2.78	2.70	2.62	2.54	2.45	2.35	2.26
24	7.82	5.61	4.72	4.22	3.90	3.67	3.50	3.36	3.26	3.17	3.03	2.89	2.74	2.66	2.58	2.49	2.40	2.31	2.21
25	7.77	5.57	4.68	4.18	3.86	3.63	3.46	3.32	3.22	3.13	2.99	2.85	2.70	2.62	2.53	2.45	2.36	2.27	2.17
30	7.56	5.39	4.51	4.02	3.70	3.47	3.30	3.17	3.07	2.98	2.84	2.70	2.55	2.47	2.39	2.30	2.21	2.11	2.01
40	7.31	5.18	4.31	3.83	3.51	3.29	3.12	2.99	2.89	2.80	2.66	2.52	2.37	2.29	2.20	2.11	2.02	1.92	1.80
60	7.08	4.98	4.13	3.65	3.34	3.12	2.95	2.82	2.72	2.63	2.50	2.35	2.20	2.12	2.03	1.94	1.84	1.73	1.60
120	6.85	4.79	3.95	3.48	3.17	2.96	2.79	2.66	2.56	2.47	2.34	2.19	2.03	1.95	1.86	1.76	1.66	1.53	1.38
∞	6.63	4.61	3.78	3.32	3.02	2.80	2.64	2.51	2.41	2.32	2.18	2.04	1.88	1.79	1.70	1.59	1.47	1.32	1.00

Interpolation should be performed using reciprocals of the degrees of freedom.

TABLE 2.29 Percentiles of the χ^2 Distribution

From Perry, Chilton, and Kirkpatrick, Chemical Engineers' Handbook, 4th ed., McGraw-Hill Book Company, New York (1963); by permission.

df	\multicolumn{10}{c}{Per cent}									
	0.5	1	2.5	5	10	90	95	97.5	99	99.5
1	0.000039	0.00016	0.00098	0.0039	0.0158	2.71	3.84	5.02	6.63	7.88
2	.0100	.0201	.0506	.1026	.2107	4.61	5.99	7.38	9.21	10.60
3	.0717	.115	.216	.352	.584	6.25	7.81	9.35	11.34	12.84
4	.207	.297	.484	.711	1.064	7.78	9.49	11.14	13.28	14.86
5	.412	.554	.831	1.15	1.61	9.24	11.07	12.83	15.09	16.75
6	.676	.872	1.24	1.64	2.20	10.64	12.59	14.45	16.81	18.55
7	.989	1.24	1.69	2.17	2.83	12.02	14.07	16.01	18.48	20.28
8	1.34	1.65	2.18	2.73	3.49	13.36	15.51	17.53	20.09	21.96
9	1.73	2.09	2.70	3.33	4.17	14.68	16.92	19.02	21.67	23.59
10	2.16	2.56	3.25	3.94	4.87	15.99	18.31	20.48	23.21	25.19
11	2.60	3.05	3.82	4.57	5.58	17.28	19.68	21.92	24.73	26.76
12	3.07	3.57	4.40	5.23	6.30	18.55	21.03	23.34	26.22	28.30
13	3.57	4.11	5.01	5.89	7.04	19.81	22.36	24.74	27.69	29.82
14	4.07	4.66	5.63	6.57	7.79	21.06	23.68	26.12	29.14	31.32
15	4.60	5.23	6.26	7.26	8.55	22.31	25.00	27.49	30.58	32.80
16	5.14	5.81	6.91	7.96	9.31	23.54	26.30	28.85	32.00	34.27
18	6.26	7.01	8.23	9.39	10.86	25.99	28.87	31.53	34.81	37.16
20	7.43	8.26	9.59	10.85	12.44	28.41	31.41	34.17	37.57	40.00
24	9.89	10.86	12.40	13.85	15.66	33.20	36.42	39.36	42.98	45.56
30	13.79	14.95	16.79	18.49	20.60	40.26	43.77	46.98	50.89	53.67
40	20.71	22.16	24.43	26.51	29.05	51.81	55.76	59.34	63.69	66.77
60	35.53	37.48	40.48	43.19	46.46	74.40	79.08	83.30	88.38	91.95
120	83.85	86.92	91.58	95.70	100.62	140.23	146.57	152.21	158.95	163.64

For large values of degrees of freedom the approximate formula

$$x_\alpha{}^2 = n\left(1 - \frac{2}{9n} + z_\alpha \sqrt{\frac{2}{9n}}\,\right)^3$$

where z_α is the normal deviate and n is the number of degrees of freedom, may be used. For example, $X_{.99}{}^2 = 60[1 - 0.00370 + 2.326(0.06086)]^3 = 60(1.1379)^3 = 88.4$ for the 99th percentile for 60 degrees of freedom.

As we have seen, for each group of samples a standard deviation can be calculated. These estimates of σ possess a distribution called the chi-square (χ^2) distribution:

$$\chi^2 = \frac{s^2}{\sigma^2/df}$$

The upper and lower confidence limits for the standard deviation are obtained by dividing $(N-1)s^2$ by two numbers taken from Table 2.29.

Example 8. The variance obtained for 10 samples is $(0.65)^2$. How reliable is s^2 as an estimate of σ^2? σ^2 is known to be $(0.75)^2$.

$$\frac{s^2(N-1)}{\chi^2_{.975}} < \sigma^2 < \frac{s^2(N-1)}{\chi^2_{.025}}$$

$$\frac{(0.65)^2(10-1)}{19.02} < \sigma^2 < \frac{(0.65)^2(10-1)}{2.70}$$

$$0.20 < \sigma^2 < 1.43$$

Thus, only one time in 40 will $9s^2/\sigma^2$ be less than 2.70 by chance alone. Similarly, only one time in 40 will $9s^2/\sigma^2$ be greater than 19.02. Consequently, it is not unlikely that s^2 is a reliable estimate of σ^2. Stated differently:

Upper limit:
$$\sigma^2 = \frac{9s^2}{2.7} = 3.3s^2$$

Lower limit:
$$\sigma^2 = \frac{9s^2}{19.02} = 0.48s^2$$

Ten measurements give an estimate of σ^2 that may be as much as 3.3 times or only about one-half the true variance.

SECTION 3
INORGANIC CHEMISTRY

3.1 NOMENCLATURE OF INORGANIC COMPOUNDS

The following synopsis of rules for naming inorganic compounds and the examples given in explanation are not intended to cover all the possible cases. For a more comprehensive and detailed description, see G. J. Leigh (ed.), *Nomenclature of Inorganic Chemistry,* 3d ed., Blackwell Scientific Publications, Oxford, 1990. This 289-page publication contains the Recommendations 1990 of the Commission on Nomenclature of Inorganic Chemistry and was prepared under the auspices of the International Union of Pure and Applied Chemistry (IUPAC). In particular, the latest report should be consulted for coordination compounds, boron compounds, and crystalline phases of variable composition.

3.1.1 Writing Formulas

3.1.1.1 Mass Number, Atomic Number, Number of Atoms, and Ionic Charge. The mass number, atomic number, number of atoms, and ionic charge of an element are indicated by means of four indices placed around the symbol:

$$\begin{matrix} \text{mass number} \\ \text{atomic number} \end{matrix} \textbf{SYMBOL} \begin{matrix} \text{ionic charge} \\ \text{number of atoms} \end{matrix} \qquad {}^{15}_{7}\text{N}^{3-}_{2}$$

Ionic charge should be indicated by an Arabic superscript numeral preceding the plus or minus sign: Mg^{2+}, PO_4^{3-}.

3.1.1.2 Placement of Atoms in a Formula. The electropositive constituent (cation) is placed first in a formula. If the compound contains more than one electropositive or more than one electronegative constituent, the sequence within each class should be in alphabetical order of their symbols. The alphabetical order may be different in formulas and names; for example, $NaNH_4HPO_4$, ammonium sodium hydrogen phosphate.

Acids are treated as hydrogen salts. Hydrogen is cited last among the cations.

When there are several types of ligands, anionic ligands are cited before the neutral ligands.

3.1.1.3 Binary Compounds between Nonmetals. For binary compounds between nonmetals, that constituent should be placed first which appears earlier in the sequence:

Rn, Xe, Kr, Ar, Ne, He, B, Si, C, Sb, As, P, N, H, Te, Se, S, At, I, Br, Cl, O, F

Examples: $AsCl_3$, SbH_3, H_3Te, BrF_3, OF_2, and N_4S_4.

3.1.1.4 Chain Compounds. For chain compounds containing three or more elements, the sequence should be in accordance with the order in which the atoms are actually bound in the molecule or ion.

Examples: SCN^- (thiocyanate), HSCN (hydrogen thiocyanate or thiocyanic acid), HNCO (hydrogen isocyanate), HONC (hydrogen fulminate), and HPH_2O_2 (hydrogen phosphinate).

3.1.1.5 Use of Centered Period. A centered period is used to denote water of hydration, other solvates, and addition compounds; for example, $CuSO_4 \cdot 5H_2O$, copper(II) sulfate 5-water (or pentahydrate).

3.1.1.6 Free Radicals. In the formula of a polyatomic radical an unpaired electron(s) is(are) indicated by a dot placed as a right superscript to the parentheses (or square bracket for coordination compounds). In radical ions the dot precedes the charge. In structural formulas, the dot may be placed to indicate the location of the unpaired electron(s).

Examples: $(HO)^{\cdot}$ $(O_2)^{2\cdot}$ $(\overset{\cdot}{N}H_3^+)$

3.1.1.7 Enclosing Marks. Where it is necessary in an inorganic formula, enclosing marks (parentheses, braces, and brackets) are nested within square brackets as follows:

$$[\,(\,)\,], \quad [\,\{(\,)\}\,], \quad [\,\{[(\,)]\}\,], \quad [\,\{\{[(\,)]\}\}\,]$$

In an inorganic name the nesting order is different: $\{\,\{\,\{\,[\,(\,)\,]\,\}\,\}\,\}$, and so on.

3.1.1.8 Molecular Formula. For compounds consisting of discrete molecules, a formula in accordance with the correct molecular weight of the compound should be used.

Examples: S_2Cl_2, S_8, N_2O_4, and $H_4P_2O_6$; not SCl, S, NO_2, and H_2PO_3.

3.1.1.9 Structural Formula and Prefixes. In the structural formula the sequence and spatial arrangement of the atoms in a molecule are indicated.

Examples: $NaO(O{=}C)H$ (sodium formate), Cl—S—S—Cl (disulfur dichloride).

Structural prefixes should be italicized and connected with the chemical formula by a hyphen: *cis-, trans-, anti-, syn-, cyclo-, catena-, o-* or *ortho-, m-* or *meta-, p-* or *para-, sec-* (secondary), *tert-* (tertiary), *v-* (vicinal), *meso-, as-* for asymmetrical, and *s-* for symmetrical.

The sign of optical rotation is placed in parentheses, $(+)$ for dextrorotary, $(-)$ for levorotary, and $(\pm)$ for racemic, and placed before the formula. The wavelength (in nanometers is indicated by a right subscript; unless indicated otherwise, it refers to the sodium D-line.

The italicized symbols *d*- (for deuterium) and *t*- (for tritium) are placed after the formula and connected to it by a hyphen. The number of deuterium or tritium atoms is indicated by a subscript to the symbol.

Examples:	*cis*-[PtCl$_2$(NH$_3$)$_2$]	methan-*d$_3$*-ol
	di-*tert*-butyl sulfate	(+)$_{589}$[Co(en)$_3$]Cl$_2$
	methan-ol-*d*	

3.1.2 Naming Compounds

3.1.2.1 Names and Symbols for Elements. Names and symbols for the elements are given in Table 3.2. Wolfram is preferred to tungsten but the latter is used in the United States. In forming a complete name of a compound, the name of the electropositive constituent is left unmodified except when it is necessary to indicate the valency (see oxidation number and charge number, formerly the Stock and Ewens-Bassett systems). The order of citation follows the alphabetic listing of the names of the cations followed by the alphabetical listing of the anions and ligands. The alphabetical citation is maintained regardless of the number of each ligand.

Example: K[AuS(S$_2$)] is potassium (disulfido)thioaurate(1−).

3.1.2.2 Electronegative Constituents. The name of a monatomic electronegative constituent is obtained from the element name with its ending (-en, -ese, -ic, -ine, -ium, -ogen, -on, -orus, -um, -ur, -y, or -ygen) replaced by -ide. The elements bismuth, cobalt, nickel, zinc, and the noble gases are used unchanged with the ending -ide. Homopolyatomic ligands will carry the appropriate prefix. A few Latin names are used with affixes: cupr- (copper), aur- (gold), ferr- (iron), plumb- (lead), argent- (silver), and stann- (tin).

For binary compounds the name of the element standing later in the sequence in Sec. 3.1.1.3 is modified to end in -ide. Elements other than those in the sequence of Sec. 3.1.1.3 are taken in the reverse order of the following sequence, and the name of the element occurring last is modified to end in -ide; e.g., calcium stannide.

ELEMENT SEQUENCE

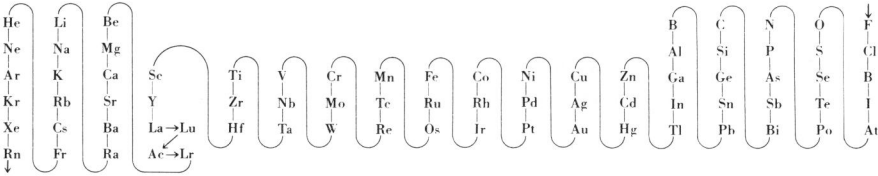

3.1.2.3 Stoichiometric Proportions. The stoichiometric proportions of the constituents in a formula may be denoted by Greek numerical prefixes: mono-, di-, tri-, tetra-, penta-, hexa-, hepta-, octa-, nona- (Latin), deca-, undeca- (Latin), dodeca-, . . . , icosa- (20), henicosa- (21), . . . , triconta- (30), tetraconta- (40), . . . , hecta- (100), and so on, preceding without a hyphen the names of the elements to which they refer. The prefix mono can usually be omitted; occasionally hemi- (½) and sesqui- (³⁄₂) are used. No elisions are made when using numerical prefixes except in the case of icosa- when the letter "i" is elided in docosa- and tricosa-. Beyond 10, prefixes may be replaced by Arabic numerals.

When it is required to indicate the number of entire groups of atoms, the multiplicative numerals bis-, tris-, tetrakis-, pentakis-, and so on, are used (i.e., -kis is added starting from tetra-). The entity to which they refer is placed in parentheses.

Examples: $Ca[PF_6]_2$, calcium bis(hexafluorophosphate); and $(C_{10}H_{21})_3PO_4$, tris(decyl) phosphate instead of tridecyl which is $(C_{13}H_{27}-)$.

Composite numeral prefixes are built up by citing units first, then tens, then hundreds, and so on. For example, 43 is written tritetraconta- (or tritetracontakis-).

In indexing it may be convenient to italicize a numerical prefix at the beginning of the name and connect it to the rest of the name with a hyphen; e.g., *di*-nitrogen pentaoxide (indexed under the letter "n").

3.1.2.4 *Oxidation and Charge Numbers.* The *oxidation number* (Stock system) of an element is indicated by a Roman numeral placed in parentheses immediately following the name of the element. For zero, the cipher 0 is used. When used in conjunction with symbols the Roman numeral may be placed above and to the right. The *charge number* of an ion (Ewens-Bassett system) rather than the oxidation state is indicated by an Arabic numeral followed by the sign of the charge cited and is placed in parentheses immediately following the name of the ion.

Examples: P_2O_5, diphosphorus pentaoxide or phosphorus(V) oxide; Hg_2^{2+}, mercury(I) ion or dimercury(2+) ion; $K_2[Fe(CN)_6]$, potassium hexacyanoferrate(II) or potassium hexacyanoferrate(4−); $Pb_2^{II}Pb^{IV}O_4$, dilead(II) lead(IV) oxide or trilead tetraoxide.

Where it is not feasible to define an oxidation state for each individual member of a group, the overall oxidation level of the group is defined by a formal ionic charge to avoid the use of fractional oxidation states; for example, O_2^-.

3.1.2.5 *Collective Names.* Collective names include:

Halogens (F, Cl, Br, I, At)

Chalcogens (O, S, Se, Te, Po)

Alkali metals (Li, Na, K, Rb, Cs, Fr)

Alkaline-earth metals (Ca, Sr, Ba, Ra)

Lanthanoids or lanthanides (La to Lu)

Rare-earth metals (Sc, Y, and La to Lu inclusive)

Actinoids or actinides (Ac to Lr, those whose $5f$ shell is being filled)

Noble gases (He to Rn)

A transition element is an element whose atom has an incomplete d subshell, or which gives rise to a cation or cations with an incomplete d subshell.

3.1.2.6 *Isotopically Labeled Compounds.* The hydrogen isotopes are given special names: ^1H (protium), ^2H or D (deuterium), and ^3H or T (tritium). The superscript designation is preferred because D and T disturb the alphabetical ordering in formulas.

Other isotopes are designated by mass numbers: ^{10}B (boron-10).

Isotopically labeled compounds may be described by inserting the italic symbol of the isotope in brackets into the name of the compound; for example, $H^{36}Cl$ is hydrogen chloride[*36Cl*] or hydrogen chloride-36, and $^2H^{38}Cl$ is hydrogen[*2H*] chloride[*38Cl*] or hydrogen-2 chloride-38.

3.1.2.7 Allotropes. Systematic names for gaseous and liquid modifications of elements are sometimes needed. Allotropic modifications of an element bear the name of the atom together with the descriptor to specify the modification. The following are a few common examples:

Symbol	Trivial name	Systematic name
H	Atomic hydrogen	Monohydrogen
O_2	(Common oxygen)	Dioxygen
O_3	Ozone	Trioxygen
P_4	White phosphorus	Tetraphosphorus
S_8	α-Sulfur, β-Sulfur	Octasulfur
S_n	μ-Sulfur (plastic sulfur)	Polysulfur

Trivial (customary) names are used for the amorphous modification of an element.

3.1.2.8 Heteroatomic and Other Anions. A few heteroatomic anions have names ending in -ide. These are

—OH, hydroxide ion (not hydroxyl)   —NH—, imide ion

—CN, cyanide ion   —NH—NH_2, hydrazide ion

—HF_2^-, hydrogen difluoride ion   —NHOH, hydroxylamide ion

—NH_2, amide ion   —HS^-, hydrogen sulfide ion

Added to these anions are

—I_3^-, triiodide ion   —O—O—, peroxide ion

—N_3, azide ion   —S—S—, disulfide ion

—O_3, ozonide ion

3.1.2.9 Binary Compounds of Hydrogen. Binary compounds of hydrogen with the more electropositive elements are designated hydrides (NaH, sodium hydride).

Volatile hydrides, except those of Periodic Group VII and of oxygen and nitrogen, are named by citing the root name of the element (penultimate consonant and Latin affixes, Sec. 3.1.2.2) followed by the suffix -ane. Exceptions are water, ammonia, hydrazine, phosphine, arsine, stibine, and bismuthine.

Examples: B_2H_6, diborane; $B_{10}H_{14}$, decaborane(14); $B_{10}H_{16}$, decaborane(16); P_2H_4, diphosphane; Sn_2H_6, distannane; H_2Se_2, diselane; H_2Te_2, ditellane; H_2S_5, pentasulfane; and PbH_4, plumbane.

3.1.2.10 Neutral Radicals. Certain neutral radicals have special names ending in -yl:

HO	hydroxyl	ClO_3	perchloryl*
CO	carbonyl	CrO_2	chromyl
ClO	chlorosyl*	NO	nitrosyl
ClO_2	chloryl*	NO_2	nitryl (nitroyl)

---

* Similarly for the other halogens.

PO	phosphoryl	SeO	seleninyl
SO	sulfinyl (thionyl)	SeO_2	selenonyl
SO_2	sulfonyl (sulfuryl)	UO_2	uranyl
S_2O_5	disulfuryl	NpO_2	neptunyl†

Radicals analogous to the above containing other chalcogens in place of oxygen are named by adding the prefixes thio-, seleno-, and so on; for example, PS, thiophosphoryl; CS, thiocarbonyl.

3.1.3 Cations

3.1.3.1 Monoatomic Cations. Monatomic cations are named as the corresponding element; for example, Fe^{2+}, iron(II) ion; Fe^{3+}, iron(III) ion.

This principle also applies to polyatomic cations corresponding to radicals with special names ending in -yl (Sec. 3.1.2.10); for example, PO^+, phosphoryl cation; NO^+, nitrosyl cation; NO_2^+, nitryl cation; O_2^+, oxygenyl cation.

Use of the oxidation number and charge number extends the range for radicals; for example, UO_2^{2+} uranyl(VI) or uranyl(2+) cation; UO_2^+, uranyl(V) or uranyl(1+) cation.

3.1.3.2 Polyatomic Cations. Polyatomic cations derived by addition of more protons than required to give a neutral unit to polyatomic anions are named by adding the ending -onium to the root of the name of the anion element; for example, PH_4^+, phosphonium ion; H_2I^+, iodonium ion; H_3O^+, oxonium ion; $CH_3OH_2^+$, methyl oxonium ion.

> *Exception:* The name ammonium is retained for the NH_4^+ ion; similarly for substituted ammonium ions; for example, NF_4^+, tetrafluoroammonium ion.

Substituted ammonium ions derived from nitrogen bases with names ending in -amine receive names formed by changing -amine into -ammonium. When known by a name not ending in -amine, the cation name is formed by adding the ending -ium to the name of the base (eliding the final vowel); e.g., anilinium, hydrazinium, imidazolium, acetonium, dioxanium.

Exceptions are the names uronium and thiouronium derived from urea and thiourea, respectively.

3.1.3.3 Multiple Ions from One Base. Where more than one ion is derived from one base, the ionic charges are indicated in their names: $N_2H_5^+$, hydrazinium(1+) ion; $N_2H_6^{2+}$, hydrazinium(2+) ion.

3.1.4 Anions

See Secs. 3.1.2.2 and 3.1.2.8 for naming monatomic and certain polyatomic anions. When an organic group occurs in an inorganic compound, organic nomenclature *(q.v.)* is followed to name the organic part.

3.1.4.1 Protonated Anions. Ions such as HSO_4^- are recommended to be named hydrogensulfate with the two words written as one following the usual practice for polyatomic anions. However, in the *Nomenclature of Organic Chemistry,* 1979 edition, hydrogen is used as a separate word; this practice is followed in this Handbook.

---

† Similarly for the other actinoid elements.

3.1.4.2 *Other Polyatomic Anions.* Names for other polyatomic anions consist of the root name of the central atom with the ending -ate and followed by the valence of the central atom expressed by its oxidation number. Atoms and groups attached to the central atom are treated as ligands in a complex.

Examples: $[Sb(OH)_6^-]$, hexahydroxoantimonate(V); $[Fe(CN_6]^{3-}$, hexacyanoferrate(III); $[Co(NO_2)_6]^{3-}$, hexanitritocobaltate(III); $[TiO(C_2O_4)_2(H_2O)_2]^{2-}$, oxobisoxalatodiaquatitanate(IV); $[PCl_6]^-$, hexachlorophosphate(V).

Exceptions to the use of the root name of the central atom are antimonate, bismuthate, carbonate, cobaltate, nickelate (or niccolate), nitrate, phosphate, tungstate (or wolframate), and zincate.

3.1.4.3 *Anions of Oxygen.* Oxygen is treated in the same manner as other ligands with the number of -oxo groups indicated by a suffix; for example, SO_3^{2-}, trioxosulfate.

The ending -ite, formerly used to denote a lower state of oxidation, may be retained in trivial names in these cases (note Sec. 3.1.5.3 also):

AsO_3^{3-}	arsenite	NOO_2^-	peroxonitrite
BrO^-	hypobromite	PO_3^{3-}	phosphite*
ClO^-	hypochlorite	SO_3^{2-}	sulfite
ClO_2^-	chlorite	$S_2O_5^{2-}$	disulfite
IO^-	hypoiodite	$S_2O_4^{2-}$	dithionite
NO_2^-	nitrite	$S_2O_2^{2-}$	thiosulfite
$N_2O_2^{2-}$	hyponitrite	SeO_3^{2-}	selenite

However, compounds known to be double oxides in the solid state are named as such; for example, Cr_2CuO_4 (actually $Cr_2O_3 \cdot CuO$) is chromium(III) copper(II) oxide (and not copper chromite).

3.1.4.4 *Isopolyanions.* Isopolyanions are named by indicating with numerical prefixes the number of atoms of the characteristic element. It is not necessary to give the number of oxygen atoms when the charge of the anion or the number of cations is indicated.

Examples: $Ca_3Mo_7O_{24}$, tricalcium 24-oxoheptamolybdate, may be shortened to tricalcium heptamolybdate; the anion, $Mo_7O_{24}^{6-}$, is heptamolybdate(6−); $S_2O_7^{2-}$, disulfate(2−); $P_2O_7^{4-}$, diphosphate(V)(4−).

When the characteristic element is partially or wholly present in a lower oxidation state than corresponds to its Periodic Group number, oxidation numbers are used; for example, $[O_2HP—O—PO_3H]^{2-}$, dihydrogendiphosphate(III,V)(2−).

A bridging group should be indicated by adding the Greek letter μ immediately before its names and separating this from the rest of the complex by a hyphen. The atom or atoms of the characteristic element to which the bridging atom is bonded, is indicated by numbers.

Examples: $[O_3P—S—PO_2—O—PO_3]^{5-}$, 1,2-$\mu$-thiotriphosphate(5−)
$[S_3P—O—PS_2—O—PS_3]^{5-}$, di-$\mu$-oxo-octathiotriphosphate(5−)

---

* Named for esters formed from the hypothetical acid $P(OH)_3$.

3.1.5   Acids

3.1.5.1   Acids and -ide Anions.   Acids giving rise to the -ide anions (Sec. 3.1.2.2) should be named as hydrogen . . . -ide; for example, HCl, hydrogen chloride; HN_3, hydrogen azide.

Names such as hydrobromic acid refer to an aqueous solution, and percentages such as 48% HBr denote the weight/volume of hydrogen bromide in the solution.

3.1.5.2   Acids and -ate Anions.   Acids giving rise to anions bearing names ending in -ate are treated as in Sec. 3.1.5.1; for example, H_2GeO_4, hydrogen germanate; $H_4[Fe(CN)_6]$, hydrogen hexacyanoferrate(II).

3.1.5.3   Trivial Names.   Acids given in Table 3.1 retain their trivial names due to long-established usage. Anions may be formed from these trivial names by changing -ous acid to -ite, and -ic acid to -ate. The prefix hypo- is used to denote a lower oxidation state and the prefix per- designates a higher oxidation state. The prefixes ortho- and meta- distinguish acids of differing water content; for example, H_4SiO_4 is orthosilicic acid and H_2SiO_3 is metasilicic acid. The anions would be named silicate(4−) and silicate(2−), respectively.

3.1.5.4   Peroxo- Group.   When used in conjunction with the trivial names of acids, the prefix peroxo- indicates substitution of —O— by —O—O—.

3.1.5.5   Replacement of Oxygen by Other Chalcogens.   Acids derived from oxoacids by replacement of oxygen by sulfur are called thioacids, and the number of replacements are given by prefixes di-, tri-, and so on. The affixes seleno- and telluro- are used analogously.

Examples: HOO—C=S, thiocarbonic acid; HSS—C=S, trithiocarbonic acid.

3.1.5.6   Ligands Other than Oxygen and Sulfur.   See Sec. 3.1.7, Coordination Compounds, for acids containing ligands other than oxygen and sulfur (selenium and tellurium).

3.1.5.7   Differences between Organic and Inorganic Nomenclature.   Organic nomenclature is largely built upon the scheme of substitution, that is, the replacement of hydrogen atoms by other atoms or groups. Although rare in inorganic nomenclature: NH_2Cl is called chloramine and $NHCl_2$ dichloroamine. Other substitutive names are fluorosulfonic acid and chlorosulfonic acid derived from HSO_3H. These and the names aminosulfonic acid (sulfamic acid), iminodisulfonic acid, and nitrilotrisulfonic acid should be replaced by the following based on the concept that these names are formed by adding hydroxyl, amide, imide, and so on, groups together with oxygen atoms to a sulfur atom:

HSO_3F	fluorosulfuric acid	$NH(SO_3H)_2$	imidobis(sulfuric) acid
HSO_3Cl	chlorosulfuric acid	$N(SO_3H)_3$	nitridotris(sulfuric) acid
NH_2SO_3H	amidosulfuric acid		

3.1.6   Salts and Functional Derivatives of Acids

3.1.6.1   Acid Halogenides.   For acid halogenides the name is formed from the corresponding acid radical if this has a special name (Sec. 3.1.2.10); for example, NOCl, nitrosyl chloride. In other cases these compounds are named as halogenide oxides with the ligands listed alphabetically; for example, BiClO, bismuth chloride oxide; VCl_2O, vanadium(IV) dichloride oxide.

TABLE 3.1   Trivial Names Retained for Acids

Alphabetically by characteristic element.

H_3AsO_4	arsenic acid	$H_4P_2O_7$	diphosphoric acid (or pyrophosphoric acid)
H_3AsO_3	arsenious acid		
H_3BO_3	orthoboric acid (or boric acid)	$H_4P_2O_8$	peroxodiphosphoric acid
HBO_2	metaboric acid	$(HO)_2OP$	diphosphoric(IV) acid or hypophosphoric acid
$HBrO_3$	bromic acid		
$HBrO_2$	bromous acid	$(HO)_2OP$	
$HBrO$	hypobromous acid	$(HO)_2P-O$	diphosphoric(III,V) acid
H_2CO_3	carbonic acid		
HOCN	cyanic acid	$(HO)_2P-O$	
HNCO	isocyanic acid	H_2PHO_3	phosphonic acid
HONC	fulminic acid	$H_2P_2H_2O_5$	diphosphonic acid
$HClO_4$	perchloric acid	HPH_2O_2	phosphinic acid (formerly hypophosphorous acid)
$HClO_3$	chloric acid		
$HClO_2$	chlorous acid	$HReO_4$	perrhenic acid
$HClO$	hypochlorous acid	H_2ReO_4	rhenic acid
H_2CrO_4	chromic acid	H_2SO_4	sulfuric acid
$H_2Cr_2O_7$	dichromic acid	$H_2S_2O_7$	disulfuric acid
H_5IO_6	orthoperiodic acid	H_2SO_5	peroxomonosulfuric acid
HIO_4	periodic acid	$H_2S_2O_3$	thiosulfuric acid
HIO_3	iodic acid	$H_2S_2S_6$	dithionic acid
HIO	hypoiodous acid	H_2SO_3	sulfurous acid
$HMnO_4$	permanganic acid	$H_2S_2O_5$	disulfurous acid
H_2MnO_4	manganic acid	$H_2S_2O_2$	thiosulfurous acid
HNO_4	peroxonitric acid	$H_2S_2O_4$	dithionous acid
HNO_3	nitric acid	$H_2S_xO_6$ $(x = 3, 4, \ldots)$	polythionic acid (tri-, tetra-, . . .)
HNO_2	nitrous acid		
H_2NO_2	nitroxylic acid	H_2SO_2	sulfoxylic acid
$H_2N_2O_2$	hyponitrous acid	$HSb(OH)_6$	hexahydrooxoantimonic acid
HOONO	peroxonitrous acid		
H_3PO_4	orthophosphoric acid (or phosphoric acid)	H_2SeO_4	selenic acid
		H_2SeO_3	selenious acid
HPO_3	metaphosphoric acid	H_4SiO_4	orthosilicic acid
H_3PO_5	peroxomonophosphoric acid	H_2SiO_3	metasilicic acid
		$HTcO_4$	pertechnetic acid
		H_2TeO_4	technetic acid
		H_6TeO_6	orthotelluric acid

3.1.6.2 Anhydrides. Anhydrides of inorganic acids are named as oxides; for example, N_2O_5, dinitrogen pentaoxide.

3.1.6.3 Esters. Esters of inorganic acids are named as the salts; for example, $(CH_3)_2SO_4$, dimethyl sulfate. However, if it is desired to specify the constitution of the compound, the nomenclature for coordination compounds should be used.

3.1.6.4 Amides. Names for amides are derived from the names of the acid radicals (or from the names of acids by replacing acid by amide); for example, $SO_2(NH_2)_2$, sulfonyl diamide (or sulfuric diamide); NH_2SO_3H, sulfamidic acid (or amidosulfuric acid).

3.1.6.5 Salts. Salts containing acid hydrogen are named by adding the word hydrogen before the name of the anion (however, see Sec. 3.1.4.1), for example, KH_2PO_4, potassium dihydrogen phosphate; $NaHCO_3$, sodium hydrogen carbonate (not bicarbonate); $NaHPHO_3$, sodium hydrogen phosphonate (only one acid hydrogen remaining).

Salts containing O^{2-} and HO^- anions are named oxide and hydroxide, respectively. Anions are cited in alphabetical order which may be different in formulas and names.

Examples: $FeO(OH)$, iron(III) hydroxide oxide; $VO(SO_4)$, vanadium(IV) oxide sulfate.

3.1.6.6 Multiplicative Prefixes. The multiplicative prefixes bis, tris, etc., are used with certain anions for indicating stoichiometric proportions when di, tri, etc., have been preempted to designate condensed anions; for example, $AlK(SO_4)_2 \cdot 12H_2O$, aluminum potassium bis(sulfate) 12-water (recall that disulfate refers to the anion $S_2O_7^{2-}$).

3.1.6.7 Crystal Structure. The structure type of crystals may be added in parentheses and in italics after the name; the latter should be in accordance with the structure. When the type-name is also the mineral name of the substance itself, italics are not used.

Examples: $MgTiO_3$, magnesium titanium trioxide (*ilmenite* type); $FeTiO_3$, iron(II) titanium trioxide (ilmenite).

3.1.7 Coordination Compounds

3.1.7.1 Naming a Coordination Compound. To name a coordination compound, the names of the ligands are attached directly in front of the name of the central atom. The ligands are listed in alphabetical order regardless of the number of each and with the name of a ligand treated as a unit. Thus "diammine" is listed under "a" and "dimethylamine" under "d." The oxidation number of the central atom is stated last by either the oxidation number or charge number.

3.1.7.2 Anionic Ligands. Whether inorganic or organic, the names for anionic ligands end in -o (eliding the final -e, if present, in the anion name). Enclosing marks are required for inorganic anionic ligands containing numerical prefixes, and for thio, seleno, and telluro analogs of oxo anions containing more than one atom.

If the coordination entity is negatively charged, the cations paired with the complex anion (with -ate ending) are listed first. If the entity is positively charged, the anions paired with the complex cation are listed immediately afterward.

The following anions do not follow the nomenclature rules:

F^-	fluoro	HO_2^-	hydrogen peroxo
Cl^-	chloro	S^{2-}	thio (only for single sulfur)
Br^-	bromo	S_2^{2-}	disulfido
I^-	iodo	HS^-	mercapto
O^{2-}	oxo	CN^-	cyano
H^-	hydrido (or hydro)	CH_2O^-	methoxo or methanolato
OH^-	hydroxo	CH_2S^-	methylthio or methanethiolato
O_2^{2-}	peroxo		

3.1.7.3 Neutral and Cationic Ligands. Neutral and cationic ligands are used without change in name and are set off with enclosing marks. Water and ammonia, as neutral ligands, are called "aqua" and "ammine," respectively. The groups NO and CO, when linked directly to a metal atom, are called nitrosyl and carbonyl, respectively.

3.1.7.4 Attachment Points of Ligands. The different points of attachment of a ligand are denoted by adding italicized symbol(s) for the atom or atoms through which the attachment occurs at the end of the name of the ligand; e.g., glycine-N or glycinato-O,N. If the same element is involved in different possible coordination sites, the position in the chain or ring to which the element is attached is indicated by numerical superscripts: e.g., tartrato(3−)-O^1,O^2, or tartrato(4−)-O^2,O^3 or tartrato(2−)-O^1,O^4.

3.1.7.5 Abbreviations for Ligand Names. Except for certain hydrocarbon radicals, for ligand (L) and metal (M), and a few with H, all abbreviations are in lowercase letters and do not involve hyphens. In formulas, the ligand abbreviation is set off with parentheses. Some common abbreviations are

Ac	acetyl	en	ethylenediamine
acac	acetylacetonato	Him	imidazole
Hacac	acetylacetone	H_2ida	iminodiacetic acid
Hba	benzoylacetone	Me	methyl
Bzl	benzyl	H_3nta	nitrilotriacetic acid
Hbg	biguanide	nbd	norbornadiene
bpy	2,2'-bipyridine	ox	oxalato(2−) from parent H_2ox
Bu	butyl	phen	1,10-phenanthroline
Cy	cyclohexyl	Ph	phenyl
D_2dea	diethanolamine	pip	piperidine
dien	diethylenetriamine	Pr	propyl
dmf	dimethylformamide	pn	propylenediamine
H_2dmg	dimethylglyoxime	Hpz	pyrazole
dmg	dimethylglyoximato(2−)	py	pyridine
Hdmg	dimethylglyoximato(1−)	thf	tetrahydrofuran
dmso	dimethylsulfoxide	tu	thiourea
Et	ethyl	H_3tea	triethanolamine
H_4edta	ethylenediaminetetraacetic acid	tren	2,2',2''-triaminotriethylamine
Hedta, edta	coordinated ions derived from H_4edta	trien	triethylenetetraamine
		tn	trimethylenediamine
Hea	ethanolamine	ur	urea

Examples: Li[B(NH$_2$)$_4$], lithium tetraamidoborate(1−) or lithium tetraamidoborate(III); [Co(NH$_3$)$_5$Cl]Cl$_3$, pentaamminechlorocobalt(III) chloride or pentaamminechlorocobalt(2+) chloride; K$_3$[Fe(CN)$_5$CO], potassium carbonylpentacyanoferrate(II) or potassium carbonyl-pentacyanoferrate(3−); [Mn{C$_6$H$_4$(O)(COO)}$_2$(H$_2$O)$_4$]$^-$, tetraaquabis[salicylato(2−)]manganate(III) ion; [Ni]C$_4$H$_7$N$_2$O$_2$)$_2$] or [Ni(dmg)] which can be named bis-(2,3-butanedione dioximato)nickel(II) or bis[dimethylglyoximato(2−)]nickel(II).

3.1.8 Addition Compounds

The names of addition compounds are formed by connecting the names of individual compounds by a dash (—) and indicating the numbers of molecules in the name by Arabic numerals separated by the solidus (diagonal slash). All molecules are cited in order of increasing number; those having the same number are cited in alphabetic order. However, boron compounds and water are always cited last and in that order.

Examples: $3CdSO_4 \cdot 8H_2O$, cadmium sulfate—water (3/8); $Al_2(SO_4)_3 \cdot K_2SO_4 \cdot 24H_2O$, aluminum sulfate—potassium sulfate—water (1/1/24); $AlCl_3 \cdot 4C_2H_5OH$, aluminum chloride—ethanol (1/4).

3.2 PHYSICAL PROPERTIES OF PURE SUBSTANCES

TABLE 3.2   Physical Constants of Inorganic Compounds

Names follow the IUPAC Nomenclature. Solvates are listed under the entry for the anhydrous salt. Acids are entered under Hydrogen and acid salts are entered as a subentry under hydrogen.

Formula weights are based upon the International Atomic Weights of 1988 and are computed to the nearest hundredth when justified. The actual significant figures are given in the atomic weights of the individual elements. When an atomic weight is given without an indicated uncertainty, it is stated to as many significant figures of decimal as possible but none more than would allow confidence that the last quoted significant decimal is better than one higher or lower by one in that figure. Figures in parentheses indicate uncertainties in the preceding digit when the uncertainty is other than one; for example, 107.8682(3) for silver implies an uncertainty of ± 0.0003 in the atomic weight.

Each element that has neither a stable isotope nor a characteristic natural isotopic composition is represented in this table by one of that element's commonly known radioisotopes identified by mass number and relative atomic mass.

Density values are given at room temperature unless otherwise indicated by the superscript figure; for example, 2.487^{15} indicates a density of 2.487 g/cm³ for the substance at 15°C. A superscript 20 over a subscript 4 indicates a density at 20°C relative to that of water at 4°C. For gases the values are given as grams per liter (g/L).

Melting point is recorded in a certain case as 250 d and in some other cases as d 250, the distinction being made in this manner to indicate that the former is a melting point with decomposition at 250°C while in the latter decomposition only occurs at 250°C and higher temperatures. Where a value such as $-6H_2O$, 150 is given it indicates a loss of 6 moles of water per formula weight of the compound at a temperature of 150°C. For hydrates the temperature stated represents the compound melting in its water of hydration.

Boiling point is given at atmospheric pressure (760 mm of mercury or 101 325 Pa) unless otherwise indicated; thus 82^{15mm} indicates that the boiling point is 82°C when the pressure is 15 mm of mercury. Also, subl 550 indicates that the compound sublimes at 550°C. Occasionally decomposition products are mentioned.

Solubility is given in parts by weight (of the formula weight) per 100 parts by weight of the solvent (i.e., percent by weight) and at room temperature. Another unit frequently used is grams per 100 mL of solvent (mL per 100 mL for liquids and gases). The symbols of the common mineral acids represent aqueous solutions of these acids.

TABLE 3.2   Physical Constants of Inorganic Compounds (*Continued*)

Abbreviations Used in the Table

a, acid	hex, hexagonal
abs, absolute	HOAc, acetic acid
abs alc, anhydrous ethanol	i, insoluble
acet, acetone	ign, ignites
alk, alkali (aq NaOH or KOH)	L, liter
anhyd, anhydrous	lq, liquid
aq, aqueous	MeOH, methanol
aq reg, aqua regia	min, mineral
atm, atmosphere	mL, milliliter
bz, benzene	org, organic
c, solid state	oxid, oxidizing
ca., approximately	PE, petroleum ether
chl, chloroform	pyr, pyridine
conc, concentrated	s, soluble
cub, cubic	satd, saturated
d, decomposes	sl, slightly
dil, dilute	soln, solution
disprop, disproportionates	solv, solvent(s)
EtAc, ethyl acetate	subl, sublimes
eth, diethyl ether	sulf, sulfides
EtOH, 95% ethanol	tart, tartrate
expl, explodes	THF, tetrahyrofuran
fcc, face-centered cubic	v, very
FP, flash point	vac, vacuum
fum, fuming	viol, violently
fus, fusion, fuses	volat, volatilizes
g, gas, gram	<, less than
glyc, glycerol	>, greater than
h, hot	

TABLE 3.2 Physical Constants of Inorganic Compounds (*Continued*)

Name	Formula	Formula weight	Density	Melting point, °C	Boiling point, °C	Solubility (in 100 parts solvent)
Actinum-227	Ac	227.0278	10.07	1050(50)	ca. 3300	d aq; s acids
bromide	AcBr$_3$	466.74	5.85	subl 800		s aq
Aluminum	Al	26.98154	2.70	660.30(1)	2518	s HCl, H$_2$SO$_4$, alk
acetylacetonate	Al(C$_5$H$_7$O$_2$)$_3$	324.31	1.27	190–193	315	i aq; v s alc; s bz, eth
ammonium bis(sulfate) 12-water	AlNH$_4$(SO$_4$)$_2$·12H$_2$O	453.33	1.65	anhyd >280		14.3 g/100 mL aq; s glyc; i alc
antimonide	AlSb	148.74		1060		v sl s aq, alc, eth
arsenide	AlAs	101.90		1740		i aq
bis(acetylsalicylate)	Al(OOCC$_6$H$_4$OCOCH$_3$)$_2$OH	402.30				
borate (2/1)	2Al$_2$O$_3$·B$_2$O$_3$	273.56		ca. 1050		
bromide	AlBr$_3$	266.71	3.205^{18}_{0}	97.5	subl 253	d (viol) aq; s alc, acet, bz, CS$_2$
butoxide, *sec-*	Al(C$_4$H$_9$O)$_3$	246.33	0.967		200–206^{30mm}	FP 27; v s org solv
butoxide, *tert-*	Al(C$_4$H$_9$O)$_3$	246.33	1.0252^{20}_{0}		subl 180	v s org solv
carbide (4/3)	Al$_4$C$_3$	143.96	2.360	2100	d >2200^{400mm}	d aq; fire hazard
chlorate	Al(ClO$_3$)$_3$	277.35				v s aq; s alc
chloride	AlCl$_3$	133.34	2.440^{25}	192.6	subl 181.1	g/100 mL: 70 aq (viol), 100^{12} abs alc; s CCl$_4$, eth; sl s bz
ethoxide	Al(C$_2$H$_5$O)$_3$	162.15	1.142^{20}	140	205^{14mm}	s hot aq (d); v sl s alc, eth
fluoride	AlF$_3$	83.98	2.882^{25}	1090	subl 1272	0.56^{25} aq; i a, alk, alc, acet
hydroxide	Al(OH)$_3$	78.01	2.42	to Al$_2$O$_3$, 300		i aq; s acids, alkalis
iodide	AlI$_3$	407.69	3.948^{17}	191.0(2)	382	d aq; s alc, eth, CS$_2$
isopropoxide	Al(C$_3$H$_7$O)$_3$	204.25	1.0346^{20}_{0}	118.5	135^{10mm}	d aq; s alc, bz, chl, PE
methoxide	Al(CH$_3$O)$_3$	72.07		0	130	
nitrate 9-water	Al(NO$_3$)$_3$·9H$_2$O	375.14		73	d 135	g/100 mL: 64^{25} aq, 100 alc; s acet
nitride	AlN	40.99	3.05	d 2517		d aq, acid, alkali
oxide (alpha-)	Al$_2$O$_3$	101.96	4.0^{20}	2054(6)	2980	i aq; v sl s a, alk
perchlorate 6-water	Al(ClO$_4$)$_3$·6H$_2$O	433.43	2.020	120.8	anhyd 178	133 g/100 mL^{20} aq
phenoxide	Al(C$_6$H$_5$O)$_3$	306.27	1.23	d 265		d aq; s alc, chl, eth
phosphide	AlP	57.95	2.85^{15}	2000		d aq
phosphinate (hypophosphite)	Al(H$_2$PO$_2$)$_3$	221.94		d to PH$_3$, 220		i aq; s HCl, warm alkali

Name	Formula					Solubility
potassium bis(sulfate) 12-water	$AlK(SO_4)_2 \cdot 12H_2O$	474.39	1.757^{20}	$-9H_2O, 92$	anhyd, 200	11.4 g/100 mL^{20} aq; v glyc; i alc
propoxide	$Al(C_3H_7O)_3$	204.25	1.0578^{20}	106	248^{14mm}	d aq; s alc
selenide	Al_2Se_3	290.84	3.437^{20}	947		d aq, acid
silicon oxide (1/1)	$Al_2O_3 \cdot SiO_2$	162.05	3.247			i aq; d HF; s fused alkali
sodium bis(sulfate) 12-water	$AlNa(SO_4)_2 \cdot 12H_2O$	458.28	1.675^{20}	61		110 g/100 mL^{15} aq; i alc
stearate	$Al(C_{18}H_{35}O_2)_3$	877.35	1.010	117–120		i aq; s alc, bz, alk
sulfate	$Al_2(SO_4)_3$	342.14	1.61	770 d		36.4 g/100 mL^{20} aq; sl s alc
sulfate 18-water	$Al_2(SO_4)_3 \cdot 18H_2O$	666.46	1.69^{17}	d 86.5		87 g/100 mL^{0} aq; i alc
sulfide	Al_2S_3	150.16	2.20^{13}	1097	subl 1500	hyd aq; s acid
tetrahydridoborate	$Al(BH_4)_3$	71.53		-64.5	44.5	d aq; ign air; expl in O_2, 20
Ammonia	NH_3	17.04	lq: 0.6818 at bp; g: 0.6175 at 15°, 7.2 atm	-77.75	-33.35	34^{20} aq; 13.2^{20} alc; s eth, organic solvents
Ammonium acetate	$NH_4C_2H_3O_2$	77.10	1.17^{20}	114	d	g/100 mL: 148^{4} aq, 7.9^{15} MeOH; s alc
amidosulfate	$NH_4SO_3NH_2$	114.13		131	d 160	v s aq; sl s alc
benzoate	$NH_4C_6H_5O_2$	139.02	1.260	198	subl 160	g/100 mL: 20^{15} aq, 2.8 alc; s glyc
bromide	NH_4Br	97.95	2.429^{25}	452 (subl under pressure)	d 397 vacuo	76 g/100 mL^{20} aq; v s acet, alc, eth
calcium arsenate 6-water	$NH_4CaAsO_4 \cdot 6H_2O$	305.13	1.905^{15}	d 140		0.02 aq; s NH_4Cl
carbamate	NH_4COONH_2	78.09		subl 60		v s aq; sl s alc; i eth
carbonate 1-water	$(NH_4)_2CO_3 \cdot H_2O$	114.10		volatilizes 60		v s aq; i alc
chloride	NH_4Cl	53.50	1.5274^{25}	237.8	520	g/100 mL: 26^{15} aq, 0.6^{19} abs alc; i acet, eth
chromate(VI)	$(NH_4)_2CrO_4$	152.09	1.91^{12}	d 180		34 g/100 mL^{20} aq; sl s MeOH, acet
chromium(III) bissulfate 12-water	$NH_4Cr(SO_4)_2 \cdot 12H_2O$	478.34	1.72	94		7.2 g/100 mL^{0} aq
copper(II) tetrachloride 2-water	$(NH_4)_2CuCl_4 \cdot 2H_2O$	277.46	1.993^{25}	anhyd, 110	d > 120	40.3 g/100 mL^{20} aq; s alc
cyanide	NH_4CN	44.06	1.002	d 36		v s aq, alc
dichromate(VI)	$(NH_4)_2Cr_2O_7$	252.10	2.1552^{25}	d 180 to Cr_2O_3		35.6 g/100 mL^{20} aq; s alc; flammable
dihydrogen phosphate	$NH_4H_2PO_4$	115.03	1.803^{19}	d 190		37 g/100 mL^{20} aq; sl s alc; i acet

TABLE 3.2 Physical Constants of Inorganic Compounds (*Continued*)

Name	Formula	Formula weight	Density	Melting point, °C	Boiling point, °C	Solubility (in 100 parts solvent)
Ammonium						
disulfatocobatate(II) 6-water	$(NH_4)_2[(Co(SO_4)_2]\cdot 6H_2O$	395.23	1.902			18 g/100 mL[20] aq; v sl s alc
disulfatoferrate(II) 6-water	$(NH_4)_2[Fe(SO_4)_2]\cdot 6H_2O$	392.14	1.864	d 100		36.4 g/100 mL[20] aq; i alc
disulfatoferrate(III) 12-water	$NH_4[Fe(SO_4)_2]\cdot 12H_2O$	482.19	1.71	39–41	d 230	124 g/100 mL aq
disulfatonickelate(II) 6-water	$(NH_4)_2[Ni(SO_4)_2]\cdot 6H_2O$	395.00	1.923	99 d		8.95 g/100 mL[20] aq
dithiocarbamate	$NH_4S(C{=}S)NH_2$	110.19	1.451[20]			v s aq; s alc; sl s eth
diuranate(VI)	$(NH_4)_2U_2O_7$	624.22				v sl s aq, alk; s acids
fluoride	NH_4F	37.04	1.009[25]	d to NH_3 + HF		100 g/100 mL[0] aq; s alc
formate	NH_4OOCH	63.06	1.27	116	d 180	143 g/100 mL[20] aq; s alc, eth
heptamolybdate(VI)(6−) 4-water	$(NH_4)_6Mo_7O_{24}\cdot 4H_2O$	1235.86	2.498	anhyd 90	d 190	43 g/100 mL aq; s acids; i alc
hexachloropalladate(IV)	$(NH_4)_2[PdCl_6]$	355.20	2.418	d		sl s aq
hexachloroplatinate(IV)	$(NH_4)_2[PtCl_6]$	443.89	3.065	d		0.5 aq
hexadecanoate	$NH_4OOC(CH_2)_{14}CH_3$	273.45		21–22		s aq; sl s bz; i alc, acet
hexafluoroaluminate(3−)	$(NH_4)_3[AlF_6]$	195.10	1.78	d >100		v s aq
hexafluorophosphate	$NH_4[PF_6]$	163.00	2.180^{18}_4	d		74.8 g/100 mL[20] aq; s alc, acet
hexafluorosilicate	$(NH_4)_2[SiF_6]$	178.14	2.011	d		18.6 g/100 mL[20] aq; i alc, acet
hexanitratocerate(IV)	$(NH_4)_2[Ce(NO_3)_6]$	548.26		d		135 g/100 mL[20] aq; s alc, HNO_3
hydrogen carbonate	NH_4HCO_3	79.06	1.586	107 (rapid heating)		g/100 mL: 17.4[20] aq, 10 glyc; i alc
hydrogen citrate	$(NH_4)_2HC_6H_5O_7$	226.19	1.48			100 g/100 mL aq; sl s alc
hydrogen difluoride	NH_4HF_2	57.05	1.51	124.6		v s aq; sl s alc
hydrogen oxalate hydrate	$NH_4HC_2O_4\cdot H_2O$	125.08	1.556	anhyd, 170		s aq; i bz, eth
hydrogen phosphate	$(NH_4)_2HPO_4$	132.08	1.619	d 155		69 g/100 mL[20] aq; i alc, acet
hydrogen sulfate	NH_4HSO_4	115.11	1.78	146.9		100 g/100 mL aq; i alc, acet
hydrogen sulfide	NH_4HS	51.11	1.17	d 25 to NH_3 + H_2S	d 350	128 g/100 mL[0] aq; s glyc; i alc, acet

Name	Formula	Formula wt	Density	m.p. °C	b.p. °C	Solubility
hydrogen sulfite	NH_4HSO_3	99.10	2.03	subl 150 in N_2		267 g/100 mL[10] aq
hydrogen (±)tartrate	$NH_4HC_4H_4O_6$	167.12	1.68	d 200		2.2[15] aq; i alc
hydroxide	NH_4OH	35.06		−77		49% dissolved in NH_3
iodate	NH_4IO_3	192.94	3.309[21]	d 150		2.6[15] aq
iodide	NH_4I	144.95	2.514[25]	subl 551	220 vacuo	167 g/100 mL[20] aq; v s alc, acet
lactate	$NH_4C_3H_5O_3$	107.11	1.2[15]	91–94		v s aq, alc, glyc; i acet, eth
magnesium arsenate 6-water	$NH_4MgAsO_4 \cdot 6H_2O$	289.36	1.932[15]	d		0.038[20] aq
molybdate(VI)(2−)	$(NH_4)_2MoO_4$	196.04	2.276[25]	d		s acids
nitrate	NH_4NO_3	80.04	1.725[25]	169.6	d 210	g/100 mL: 192[20] aq; 3.8[20] alc; 17[20] MeOH; s acet
octadecanoate	$NH_4OOC(CH_2)_{16}CH_3$	301.50		21–22		sl s aq; s alc; i acet
octanoate	$NH_4OOC(CH_2)_6CH_3$	161.24		d on standing		v s aq, alc, acet; sl s eth
oxalate hydrate	$(NH_4)_2C_2O_4 \cdot H_2O$	142.12	1.50	d 70		5.1[20] aq
oxodioxalatotitanate(IV)	$(NH_4)_2TiO(C_2O_4)_2$	276.02				v s aq
perchlorate	NH_4ClO_4	117.50	1.95	d 240		22 g/100 mL[20] aq; s MeOH; sl s acet
permanganate	NH_4MnO_4	136.97	2.208[10]	explodes, 110	expl 180	0.8[15] aq
peroxodisulfate	$(NH_4)_2S_2O_8$	228.20	1.982	d 120		58 g/100 mL[0] aq
phosphinate	$NH_4PH_2O_2$	83.04	1.634	200	d 240	g/100 mL: 100 aq, 5 alc; i acet
picrate	$NH_4C_6H_2N_3O_7$	246.14	1.719	d	expl 423	1.1[20] aq; sl s alc
selenate(VI)	$(NH_4)_2SeO_4$	179.03	2.193[20]	d		117 g/100 mL[7] aq; s HOAC; i alc
sulfate	$(NH_4)_2SO_4$	132.14	1.769[20]	d > 280		43.5 g/100 mL[25] aq; i alc, acet
sulfide	$(NH_4)_2S$	68.14	1.41	d		v s aq; s alc
Sulfite hydrate	$(NH_4)_2SO_3 \cdot H_2O$	134.15	1.601	d 60		75 g/100 mL[20] aq
(±)tartrate	$(NH_4)_2C_4H_4O_6$	184.15		d		58 g/100 mL[15] aq; sl s alc
tetraborate 4-water	$(NH_4)_2B_4O_7 \cdot 4H_2O$	263.44				s aq; i alc
tetrachloroaluminate	$NH_4[AlCl_4]$	186.84		304		s aq, eth
tetrachloropalladate(II)	$(NH_4)_2[PdCl_4]$	284.29	2.170	d		v s aq; i abs alc
tetrachloroplatinate(II)	$(NH_4)_2[PtCl_4]$	372.98	2.936	140 d		s aq
tetrachlorozincate	$(NH_4)_2[ZnCl_4]$	243.28	1.879	150 d	subl 341	v s aq
tetrafluoroborate	$NH_4[BF_4]$	104.84	1.871[15]	subl		25 g/100 mL[16] aq
thiocyanate	NH_4SCN	76.12	1.305	149.6	d 170	128 g/100 mL[0] aq; v s alc; s acet

3.17

TABLE 3.2 Physical Constants of Inorganic Compounds (*Continued*)

Name	Formula	Formula weight	Density	Melting point, °C	Boiling point, °C	Solubility (in 100 parts solvent)
Ammonium						
thiosulfate	$(NH_4)_2S_2O_3$	148.20	1.679	d 150		2.15^{15} aq; i alc, eth
vanadate(V)(1−)	NH_4VO_3	116.98	2.326	d 200		0.48^{20} aq
Antimony	Sb	121.75(3)	6.697^{25}	630.7	1635	s hot conc H_2SO_4, aqua regia
(III) chloride	$SbCl_3$	228.10	3.14^{20}_4	73.4	223.5	10 g/100 mL^{20} aq; s alc, bz, chl
(V) chloride	$SbCl_5$	299.01	2.336^{20}_4	3.5	79^{22mm}	d aq; s HCl, chl, CCl_4
(III) fluoride	SbF_3	178.75	4.379^{20}_4	292	376	444 g/100 mL^{20} aq
(V) fluoride	SbF_5	216.74	2.99^{23}	8.3	141	d viol aq; s HOAc; forms solids with alc, bz, CS_2, eth
hydride (stibine)	SbH_3	124.77	4.36^{15}	−91.5	−18.4	20 mL/100 mL^0 aq; s CS_2
(III) oxide	Sb_2O_3	291.50	5.2	655	1425	v sl s aq; s HCl, KOH
(V) oxide	Sb_2O_5	323.50	3.78	$−O_2, >300$		v sl s aq; sl s warm KOH, eth
(III) selenide	Sb_2Se_3	480.38	5.81	612		v sl s aq; s conc HCl
(III) sulfide	Sb_2S_3	339.68	4.64	546	565	0.002^{20} aq (d); s H_2SO_4
(V) sulfide	Sb_2S_5	403.82	4.120	d		i aq; s HCl (d), NaOH
(III) telluride	Sb_2Te_3	626.30	6.52	620		i aq; s HNO_3
triethyl	$Sb(C_2H_5)_3$	209.0	1.324^{14}	−29	159.5	i aq
trimethyl	$Sb(CH_3)_3$	166.9	1.523^{15}		80.6	sl s aq
Argon	Ar	39.948	1.7824 g/L^0	−189.38	−185.87	3.36 mL/100 mL^{20} aq
Arsenic	As	74.9216	5.727^{25}_4	$817^{36\,atm}$	subl 615	i aq; s HNO_3
(III) bromide	$AsBr_3$	314.65	3.397^{25}_4	31.1	220.0	hyd aq; s HCl, CS_2, PE
(III) chloride	$AsCl_3$	181.28	2.149^{25}_4	−16.2	130.2	misc chl, CCl_4, eth; s HCl
(di-) disulfide	As_2S_2	213.97	3.254^{19}	320	565	s alkali; v sl s bz
(III) fluoride	AsF_3	131.92	2.73^{15}	−5.95	57.8	s alc, bz, HF
(V) fluoride	AsF_5	169.91	7.71 g/L	−79.8	−53.2	hyd aq; s alc, bz, eth
(III) hydride (arsine)	AsH_3	77.95	2.695 g/L	−116.9	−62.5	28 mL/100 mL^{20} aq; s bz, chl
(III) oxide (dimer, monoclinic)	As_4O_6	395.68	4.15	313	465	1.8^{20} aq; s alc
(V) oxide	As_2O_5	229.82	4.32	d 800		66 g/100 mL^{20} aq; s alc
(III) selenide	As_2Se_3	386.72	4.75	260		s alkali; HNO_3
(III) sulfide	As_2S_3	246.04	3.46	300−325	707	i aq; s alk, slowly s hot hCl

Name	Formula		Density	mp	bp	Solubility
(V) sulfide	As_2S_5	310.16	6.25	subl 500		0.0003 aq; s alkali, HNO_3
(III) telluride	As_2Te_3	532.64		360		
Barium	Ba	137.33	3.51^{20}	726.9(3)	1845	d aq to $Ba(OH)_2$
acetate hydrate	$Ba(C_2H_3O_2)_2 \cdot H_2O$	273.46	2.19	anhyd 110	d 150	58.8 g/100 mL^0 aq; 0.014 alc
benzenesulfonate	$Ba(O_3SC_6H_5)_2$	451.70				s aq; sl s alc
bromate hydrate	$Ba(BrO_3)_2 \cdot H_2O$	411.17	3.99^{18}	d 260		0.96^{30} aq; s acet; i alc
bromide	$BaBr_2$	297.14	4.781	856(2)	1849	92 g/100 mL^0 aq; s MeOH, acet, diox
carbonate (rhombohedral)	$BaCO_3$	197.34	4.2865	d 1360		0.0024 aq; s acids
chlorate hydrate	$Ba(ClO_3)_2 \cdot H_2O$	322.26	3.179	anhyd 120	$-O_2$, 250	34 g/100 mL^{20} aq; sl s alc, acet
chloride	$BaCl_2$	208.24	3.856^{24}	962(1)	2029	36 g/100 mL^{20} aq; s MeOH; i acet, EtAc
chloride dihydrate	$BaCl_2 \cdot 2H_2O$	244.26	3.097	d 260		31.7 g/100 mL^0 aq
chromate(VI)	$BaCrO_4$	253.33	4.498^{20}	anhyd 113		0.001^{20} aq; s mineral acids
cyanide	$Ba(CN)_2$	189.38				80 g/100 mL^{14} aq; s alc
fluoride	BaF_2	175.33	4.89	1368	2272	0.161^{20} aq; s acids
hexafluorosilicate	$Ba[SiF_6]$	279.41	4.29^{21}	d 300		0.0235^{25} aq; s NH_4Cl soln; i alc
hydrogen phosphate	$BaHPO_4$	233.31	4.165^{15}	d 410		0.01 aq; s HCl, HNO_3
hydroxide 8-water	$Ba(OH)_2 \cdot 8H_2O$	315.48	2.18^{16}	78		3.9^{20} aq
iodate	$Ba(IO_3)_2$	487.14	5.23^{20}	d		0.033^{20} aq; s HCl
iodide	BaI_2	391.14	5.15	711	2027	169 g/100 mL^{20} aq; s alc, acet
manganate(VI)(2−)	$BaMnO_4$	256.27	4.85			disprop to $Ba(MnO_4)_2$ + MnO_2
molybdate	$BaMoO_4$	297.27	4.65	1480		0.0058^{23} aq
nitrate	$Ba(NO_3)_2$	261.35	3.24^{23}	592		5.0^0 aq; v sl s alc, acet
nitrite hydrate	$Ba(NO_2)_2 \cdot H_2O$	247.36	3.173^{20}	d 115	d	54.8 g/100 mL^0 aq; i alc
oxide	BaO	153.34	5.72	2013	3088	3.5^{20} aq
perchlorate 3-water	$Ba(ClO_4)_2 \cdot 3H_2O$	390.28	2.74	d 400		198 g/100 mL^{25} aq; s MeOH, sl s acet
permanganate	$Ba(MnO_4)_2$	375.20	3.77	d 200		v s aq
peroxide	BaO_2	169.34	4.96	450	$-O_2$, 800	1.5^0 aq
selenide	$BaSe$	216.30	5.02	1830		d aq
sulfate	$BaSO_4$	233.40	4.50^{15}	1580	d > 1600	0.000285^{30} aq
sulfide	BaS	169.40	4.25^{15}	>2227		7.9^{20} aq; dec in acids
sulfite	$BaSO_3$	217.40		d		0.02^0 aq; i alc

TABLE 3.2 Physical Constants of Inorganic Compounds (*Continued*)

Name	Formula	Formula weight	Density	Melting point, °C	Boiling point, °C	Solubility (in 100 parts solvent)
Barium tetracyanoplatinate(II) 4-water	Ba[Pt(CN)$_4$]·4H$_2$O	508.72	3.05			2.86 aq
thiocyanate 2-water	Ba(SCN)$_2$·2H$_2$O	289.53	2.286[18]	d 160		170 g/100 mL[25] aq; s alc, acet
thiosulfate hydrate	BaS$_2$O$_3$·H$_2$O	267.47	3.5[18]	d 220		0.21[20] aq; i alc, acet, eth, CS$_2$
titanate(IV)(2−) (tetragonal)	BaTiO$_3$	233.23	6.08	1625		
Berkelium-249	Bk	249.075	14.78	986(25)		
Beryllium	Be	9.012 18	1.847[20]	1287(5)	2467	i aq; s acid, alk
bromide	BeBr$_2$	168.83	3.465[25]	508(15)	521	v s aq; s alc; 18.6 pyr
carbide	Be$_2$C	30.04	1.90[15]	d >2127		d aq; s acids, alkali giving CH$_4$
chloride	BeCl$_2$	79.92	1.899[25]	415 (alpha)	482.3	42 g/100 mL aq; s alc, eth, pyr, CS$_2$
fluoride	BeF$_2$	47.01	1.986	555		v s aq (slowly)
hydride	BeH$_2$	11.03		−H$_2$, 220		d aq (slowly), acids (rapidly)
hydroxide	Be(OH)$_2$	43.03	1.909	93		s hot conc acids and alkali (viol)
iodide	BeI$_2$	262.82	4.2	480(15)	485(3)	hyd aq violently; s alc, eth, CS$_2$
nitrate 3-water	Be(NO$_3$)$_2$·3H$_2$O	177.07	1.557	60.5	d 125	166 g/100 mL[20] aq
nitride	Be$_3$N$_2$	55.05		2200(40)		d hot aq, alkali
oxide	BeO	25.01	3.025	2507 (alpha)	3787	s conc H$_2$SO$_4$
selenate 4-water	BeSeO$_4$·4H$_2$O	224.03	2.03	anhyd 300	d 560	49 g/100 mL[25] aq
silicate	Be$_2$SiO$_4$	110.11	3.0	1560		i aq
sulfate 4-water	BeSO$_4$·4H$_2$O	177.14	1.713[11]	anhyd 270	d 580	39 g/100 mL[20] aq; i alc
sulfide	BeS	41.08	2.36			i aq; s HNO$_3$
Bismuth	Bi	208.9804	9.78[20]	271.5	1504	i aq; s hot H$_2$SO$_4$
bromide oxide	BiBrO	304.89	8.082[15]	d		i aq; s acids
(III) chloride	BiCl$_3$	315.34	4.75	233.5	447	d aq; s HCl, alc, eth, acet
chloride oxide	BiClO	260.43	7.72[15]	d		i aq; s HCl
(III) fluoride	BiF$_3$	265.98	5.32[20]	727	900	i aq; s HF

Name	Formula	Formula weight	Density	Melting point	Boiling point	Solubility
(V) fluoride	BiF_5	303.98	5.40^{25}	154.4	subl 550	d (viol) aq giving O_3 + BiF_3
hydride	BiH_3	212.00			16.8	very unstable liquid
(III) hydroxide	$Bi(OH)_3$	260.00	4.962^{15}	−water, 100		d aq; s HCl
(III) iodide	BiI_3	589.69	5.778^{17}	408.6	subl 439	i aq; s HCl, alc
iodide oxide	$BiIO$	351.88	7.922	d red heat		i aq; s HCl
(III) nitrate 5-water	$Bi(NO_3)_3 \cdot 5H_2O$	485.07	2.83	anhyd 80		d aq; s HNO_3, acet, glyc
(III) oxide	Bi_2O_3	495.96	8.76	817	1890	i aq; s HCl, HNO_3
(V) oxide	Bi_2O_5	497.96	5.10	d 150		i aq; s KOH
(III) phosphate	$BiPO_4$	303.95	6.323^{15}	d		s conc HCl, HNO_3
(III) selenide	Bi_2Se_3	654.84	7.702^{20}	710		i aq; d aq reg
(III) sulfate	$Bi_2(SO_4)_3$	706.14	5.08	d 405	d	d aq, alc; s HCl
(III) sulfide	Bi_2S_3	514.15	6.78	850		i aq, EtAc; s HNO_3, HCl
(III) telluride	Bi_2Te_3	800.76	7.859	588.5		i aq
Boranes						
diborane(6)	B_2H_6	27.67	0.447^{-112}	−165	−92.5	FP −68; s NH_4OH, conc H_2SO_4
tetraborane(10)	B_4H_{10}	53.32	0.652	−120	18	sl s aq; s bz
pentaborane(9)	B_5H_9	63.13	0.760(c)	−46.81	60.0	hyd aq
pentaborane(11)	B_5H_{11}	65.14	0.745	−123	63	sl s aq; s bz, CS_2, eth
decaborane(14)	$B_{10}H_{14}$	122.22	0.948	−99.5	213	sl s aq (d)
Borazine	$B_3H_6N_3$	80.53	lq: 0.81^{bp}	−58	55	i aq
Boric acids, *see* under Hydrogen						
Boron	B	10.811(5)	2.46 alpha	2076(50)	3864	s fused alkalis
carbide	B_4C	55.26	$2.510(2)^{25}$	2470(20)	>3500	
tribromide	BBr_3	250.57	2.968^{0}	−46.0	91.3	d aq, alc
trichloride	BCl_3	117.19	1.351^{12}	−107	12.5	d aq, alc
trifluoride	BF_3	67.81	3.077 g/L STP	−127.1	−100.4	332 g/100 mL^{0} aq; s bz, chl, CCl_4
trifluoride 1-diethyl ether	$BF_3 \cdot O(C_2H_5)_2$	141.94	1.125	−60.4	125.7	d aq
trifluoride 1-methanol	$BF_3 \cdot HOCH_3$	131.89	1.203		594mm	
nitride	BN	24.82	2.25	d 2325(100)	subl <3000	sl s hot acids
oxide	B_2O_3	69.62	1.84^{25}	450.0(5)	2065	3.3 aq (slowly); s alc, glyc
Bromine	Br_2	159.808	3.10232^{25}	−7.25	59.47	3.4 g/100 mL^{20} aq; v s alc, chl, eth
fluoride, penta-	BrF_5	174.00	2.4716	−60.5	40.76	explodes with water; s HF
trifluoride	BrF_3	136.90	2.8030^{25}	8.77	125.74	d viol aq; d alk; smokes in air

TABLE 3.2 Physical Constants of Inorganic Compounds (*Continued*)

Name	Formula	Formula weight	Density	Melting point, °C	Boiling point, °C	Solubility (in 100 parts solvent)
Cadmium	Cd	112.41	8.65^{25}	321	767(2)	i aq, alk, s HNO_3, hot HCl
acetate	$Cd(C_2H_3O_2)_2$	230.50	2.341	256	d	v s aq
bromide	$CdBr_2$	272.22	5.192	566	963	99 g/100 mL^{20} aq; s acet; sl s eth
carbonate	$CdCO_3$	172.41	4.258^4	d 500		s acids, NH_4OH
chloride	$CdCl_2$	183.32	4.047^{25}	568	961	120 g/100 mL^{25} aq
cyanide	$Cd(CN)_2$	164.44	2.226	d 200		1.71 g/100 mL^{15} aq; sl s alc
fluoride	CdF_2	150.40	6.33	1049	1748	4.3 g/100 mL^{20} aq
hydroxide	$Cd(OH)_2$	146.11	4.79	d 200		0.00026^{20} aq; s acids
iodide	CdI_2	366.21	5.670^{30}	387	787	84.7 g/100 mL^{20} aq; s alc, acet, eth
nitrate 4-water	$Cd(NO_3)_2 \cdot 4H_2O$	308.47	2.455^{17}	59.4	132	167 g/100 mL^{25} aq; s alc, acet, EtAc
oxide	CdO	128.40	8.15 cubic	1540		i aq; s acids
selenide	CdSe	191.36	5.81^{15}	1264		i aq; d acids
sulfate-water (3/8)	$3CdSO_4 \cdot 8H_2O$	769.56	3.08	monohydrate, 80		94.4 g/100 mL^{25} aq; i alc, EtAc
sulfide	CdS	144.46	4.82 hex 4.50 cubic		subl 980	0.13^{18} aq; s acids
telluride	CdTe	240.00	6.20^{15}	1041		i aq; d HNO_3
tungstate(VI)	$CdWO_4$	360.33				i aq, dil acids; s alkali CN's
Calcium	Ca	40.078(4)	1.55^{20}	842(2)	1493(9)	d aq; s acids
acetate	$Ca(C_2H_3O_2)_2$	158.17	1.50	d > 160		37.4 g/100 mL^0 aq; i alc, bz, acet
arsenate	$Ca_3(AsO_4)_2$	398.08	3.620			0.013^{25} aq
bromide	$CaBr_2$	199.90	3.353^{25}	745	1810	143 g/100 mL^{20} aq; v s alc, acet
carbonate (calcite)	$CaCO_3$	100.09	2.711^{25}	d 825 to CaO		0.0013^{20}, s acids
chlorate 2-water	$Ca(ClO_3)_2 \cdot 2H_2O$	243.01	2.711	anhyd 100		167 g/100 mL^{20} aq; s alc
chloride	$CaCl_2$	110.99	2.16^{25}	772	ca. 1940	42 g/100 mL^{20} aq; s alc, acet
chloride 6-water	$CaCl_2 \cdot 6H_2O$	219.08	1.71^{25}	anhyd 200		74.5 g/100 mL^{20} aq; v s alc
chlorite	$Ca(ClO_2)_2$	174.99	2.71^{25}	100		167 g/100 mL aq; s alc
chromate(VI) 2-water	$CaCrO_4 \cdot 2H_2O$	192.09		anhyd 200		sl s aq; s dil acids

citrate 4-water	$CaC_6H_6O_7 \cdot 4H_2O$	570.51		anhyd 120		0.10 aq; i alc
cyanamide	$CaCN_2$	80.11	2.292^{20}_4	ca. 1340	subl 1175(25)	no known solv without de-composition
cyanide	$Ca(CN)_2$	92.12		s >350		s aq
dicarbide	CaC_2	64.10	2.222	ca. 2300		d aq giving C_2H_2
dichromate(VI)	$CaCr_2O_7$	256.10	2.370^{30}	d >100		v s aq; i eth; d alc
dihydrogen phosphate hydrate	$Ca(H_2PO_4)_2 \cdot H_2O$	252.07	2.220^{18}_4	anhyd 100	d 200	1.8^{30} aq
diphosphate (pyrophosphate)	$Ca_2P_2O_7$	254.10	3.09	1353		i aq; s HCl, HNO_3
fluoride	CaF_2	78.08	3.180	1418(5)	2533(30)	0.0015^{20} aq; s conc mineral acids
formate	$Ca(CHO_2)_2$	130.12	2.015	d		16.6 g/100 mL^{20} aq; i alc
(+)gluconate	$Ca[OOC(CHOH)_4CH_2OH]_2$	430.38				3.72^{20} aq
glycerophosphate	$Ca[C_3H_5(OH)_3]PO_4$	210.16		d >170		1.66^{20} aq; i alc
hexafluorosilicate	$Ca[SiF_6]$	182.17	$2.662^{17.5}$			i aq, acet
hydride	CaH_2	42.10	1.90	1.86		d aq, alc
hydroxide	$Ca(OH)_2$	74.09	2.343	$-H_2O$, 580		0.17^{10} aq; s acids
hypochlorite	$Ca(OCl)_2$	142.99	2.35	100 d		d aq evolving Cl_2; i alc
iodate	$Ca(IO_3)_2$	389.89	4.519^{15}_4	d >540		0.10^0 aq; i alc
iodide	CaI_2	293.89	3.956	779(2)	1755	68 g/100 mL^{20} aq; v s alc, acet; i eth
lactate 5-water	$Ca(C_3H_5O_3)_2 \cdot 5H_2O$	308.30		$-3H_2O$, 100	anhyd 120	5.4^{15} aq; v sl s alc
magnesium carbonate	$Ca[Mg(CO_3)_2]$	184.41	2.872	d 730		0.032^{18} aq; s HCl
molybdate(VI)(2−)	$CaMoO_4$	200.02	4.35			s conc mineral acids
nitrate	$Ca(NO_3)_2$	164.09	2.504^{18}	561		152 g/100 mL^{30} aq
nitride	Ca_3N_2	148.25	2.63^{17}	1195		d aq; s dilute acids (d)
nitrite 4-water	$Ca(NO_2)_2 \cdot 4H_2O$	204.15	1.674	d		84.5 g/100 mL^{18} aq; sl s alc
oleate	$Ca(C_{18}H_{33}O_2)_2$	603.01		83–84	d >400	0.04 aq; s chl, bz; v sl s alc, eth
oxalate hydrate	$CaC_2O_4 \cdot H_2O$	146.12	2.2	anhyd 200		0.0006 aq; s acids
oxide	CaO	56.08	3.3	2927(50)	3500	0.13^{25} aq; s acids
palmitate	$Ca(C_{16}H_{31}O_2)_2$	550.93		d >155		0.003 aq; sl s bz, chl, HOAc; i eth
(+)pantothenate (vitamin B₃)	$Ca[O_2CH_2CH_2NHOCH(OH)C(CH_3)_2CH_2OH]_2$	476.55		d 195–196		36 g/100 mL aq; sl s alc, acet
permanganate 5-water	$Ca(MnO_4)_2 \cdot 5H_2O$	368.03	2.4	d		338 g/100 mL aq
peroxide	CaO_2	72.08	2.92	explodes 275		sl s aq; s acids

TABLE 3.2 Physical Constants of Inorganic Compounds (*Continued*)

Name	Formula	Formula weight	Density	Melting point, °C	Boiling point, °C	Solubility (in 100 parts solvent)
Calcium						
phenoxide	Ca(OC$_6$H$_5$)$_2$	226.28	d in air			sl s aq, alc
phosphate	Ca$_3$(PO$_4$)$_2$	310.18	3.14	1670		0.03^{25} aq; s HCl, HNO$_3$; i alc
phosphide	Ca$_3$P$_2$	182.19	2.51	ca. 1600		d aq; s acids
phosphinate	Ca(PH$_2$O$_2$)$_2$	170.06		d > 300		15.4 g/100 mL aq; sl s glyc; i alc
propanoate	Ca(OOCC$_3$H$_5$)$_2$	186.22				s aq; sl s alc; i acet, bz
salicylate 2-water	Ca(C$_7$H$_5$O$_3$)$_2$·2H$_2$O	350.34		anhyd 200; d 240		2.8^{15} aq; 0.015^{16} EtOH
selenate 2-water	CaSeO$_4$·2H$_2$O	219.07	2.68^{20}	anhyd 200	d 698	9.2 g/100 mL^{25} aq
selenide	CaSe	119.04	3.82			i aq
silicate	Ca$_2$SiO$_4$	172.24	3.27	2130		0.004^{15} aq; s hot pyr; i acet, chl
stearate	Ca(C$_{18}$H$_{35}$O$_2$)$_2$	607.04		179–180		
succinate 3-water	CaC$_4$H$_4$O$_4$·3H$_2$O	212.22				1.28^{20} aq; s acids; i alc
sulfate	CaSO$_4$	136.14	2.960	1400		0.20 aq; s acids
sulfate hemihydrate	CaSO$_4$·0.5H$_2$O	145.15	2.32	anhyd 163		0.3^{20} aq; s acids, glyc
sulfate 2-water	CaSO$_4$·2H$_2$O	172.17	2.32	−1.5 H$_2$O, 128	anhyd 163	0.26^{20} aq; s acid, glyc
sulfide	CaS	72.14	2.59	2525		0.02 (d) aq; d acids
sulfite 2-water	CaSO$_3$·2H$_2$O	156.17		anhyd 100		0.004 aq; s acids d; sl s alc
(±)tartrate 4-water	CaC$_4$H$_4$O$_6$·4H$_2$O	260.21		anhyd 200		0.0045^{25} aq; s acids; sl s alc
telluride	CaTe	167.68	4.873			
tetraborate	CaB$_4$O$_7$	195.36		ign moist air		s dil acids
tetrahydridoaluminate	Ca[AlH$_4$]$_2$	102.10		d > 160		d viol aq, alc; i bz, eth
thiocyanate 3-water	Ca(SCN)$_2$·3H$_2$O	210.29		−H$_2$O, >95		150 g/100 mL aq; v s alc
thioglycollate 3-water	Ca(—OOCCH$_2$S—)·3H$_2$O	184.24		d > 43	d > 220	s aq; v sl s alc, chl; i bz, eth
thiosulfate 6-water	CaS$_2$O$_3$·6H$_2$O	260.30	1.872			92 g/100 mL^{25} aq; i alc
tungstate(VI)(2−)	CaWO$_4$	287.93	6.062^{20}			0.0032 aq; d hot acids
Californium-252	Cf	252.1				
chloride	CfCl$_3$	358.5	5.88	900(30)		
Carbon (diamond) (graphite)	C	12.011	3.5254^{20} 2.267	3500$^{63.5\ atm}$ subl 3915–4020	3930	i aq, alc
dioxide	CO$_2$	44.01	c: 1.56^{-79} g: 1.975 g/L^0	−78.44 subl		88 mL/100 mL^{20} aq

Name	Formula	Formula weight	Density	MP	BP	Solubility
diselenide	CSe_2	169.93	2.6626^{25}_4	−45.5	125.1	i aq; s acet, eth; misc CCl_4; d alc
disulfide	CS_2	76.14	1.2632^{20}_4	−111.6	46.56	FP −30; 0.29^{20} aq; s alc, eth s bz
hydride (methane)	CH_4	16.04	0.415^{-164}	−182.48	−161.49	2.3 mL/100 mL^{20} aq; 16 mL/100 mL^{20} alc; s HOAc, EtAc
monoxide	CO	28.01	lq: 0.814^{-195}; g: 1.250 g/L_4	−205.05	−191.49	d aq to malonic acid; sl s CS_2
suboxide	C_3O_2	68.03	1.114^2_4	−111.3	6.8	i aq; s alc, chl, eth
tetrabromide	CBr_4	331.65	3.42	90.1	190	0.05 mL/100 mL aq; s alc, chl, eth
tetrachloride	CCl_4	153.82	1.589^{25}_{25}	−22.9	76.7	sl s aq
tetrafluoride	CF_4	88.00	1.96^{-184}	−183.6	−127.8	slowly hyd aq; s bz, chl, eth
tetraiodide	CI_4	519.63	4.34^{20}	171	subl 130	hyd aq; s bz, HOAc
Carbonyl chloride (phosgene)	$COCl_2$	98.92	1.392^{19}	−118	8.2	hyd aq
fluoride	COF_2	66.01	lq: 1.139^{-114}	−114.0	−83.1	54 mL/100 mL^{20} aq; s alc, CS_2
sulfide	COS	60.07	1.073 g/L^0	−138.81	−50.23	
Cerium	Ce	140.12	6.773^{25}	795	3360(100)	i aq; s acids
(III) bromide	$CeBr_3$	379.84	5.18	733	1560	s aq, alc
(III) chloride	$CeCl_3$	246.48	3.97^{25}	822	1730	s aq, alc
(III) fluoride	CeF_3	197.12	6.157	1460	2327	i but slowly hyd aq; s H_2SO_4
(IV) fluoride	CeF_4	216.12	4.77	>650	d > 550	i aq
(III) iodide	CeI_3	520.83		752	1400	s aq
(III) nitrate 3-water	$Ce(NO_3)_3 \cdot 3H_2O$	380.17		anhyd 150	d 200	234 g/100 mL^{20} aq
(IV) oxide	CeO_2	172.13		>2600		i aq; s acids
(III) sulfate	$Ce_2(SO_4)_3$	568.42	3.912	d 1000		9.72 g/100 mL^{21} aq
(IV) sulfate	$Ce(SO_4)_2$	332.24	3.91	d 195		hyd aq; s dil H_2SO_4
Cesium	Cs	132.9054	1.8785^{15}	28.40(1)	668.2	d aq; s acids
bromide	$CsBr$	212.81	1.6984	635	1300	107 g/100 mL^{18} aq; s alc; i acet
carbonate	Cs_2CO_3	325.82		red heat	d 610	v s aq; 11 g/100 mL^{20} alc; s eth
chloride	$CsCl$	168.36	3.988	645	1303	g/100 mL: 187^{20} aq; 34^{25} MeOH; v s alc
fluoride	CsF	151.90	4.115	703	1231	322 g/100 mL^{18} aq
hydroxide	$CsOH$	149.91	3.675	315(1)	990	386 g/100 mL^{15} aq; s alc
iodate	$CsIO_3$	307.81	4.934^{20}	565		2.6^{25} aq

TABLE 3.2 Physical Constants of Inorganic Compounds (*Continued*)

Name	Formula	Formula weight	Density	Melting point, °C	Boiling point, °C	Solubility (in 100 parts solvent)
Cesium iodide	CsI	259.81	4.510	621	1280	76.5 g/100 mL[20] aq; s EtOH; i acet
nitrate	$CsNO_3$	194.91	3.685_4^{20}	414	d 849	23 g/100 mL[20] aq; s acet; v sl s alc
oxide	Cs_2O	281.81	4.25	272		v s aq
selenate	Cs_2SeO_4	408.77	4.4528_4^{20}	1003(5)		244 g/100 mL[12] aq
sulfate	Cs_2SO_4	361.87	4.243			179 g/100 mL[20] aq; i alc, acet, pyr
Chlorine	Cl_2	70.906	2.98^{20} g/L lq: 1.5649^{-35}	-100.99	-34.05	199 mL/100 mL[25] aq
dioxide	ClO_2	67.46	1.642^0	-59.6	10.9	11.2 g/100 mL[10] aq
fluoride	ClF	54.56	1.67^{-108}	-155.6	-100.1	d viol aq; organics burst into flame
heptoxide	Cl_2O_7	182.90	1.805^{25}	-91.5	82	hyd aq slowly; explodes on concussion or on contact with flame or I_2
monoxide (dichlorine oxide)	Cl_2O	86.91	3.02^2	-120.6	2.2	v s aq (forms HClO); s CCl_4
trifluoride	ClF_3	92.45	1.825^{bp}_0	-76.3	11.75	hyd viol aq; organic matter and glass wool burst into flame
trioxide (dimer)	$(ClO_3)_2$	166.90	1.92^{20}	3.5	d	d
Chromium	Cr	51.9961(6)	7.14^{20}	1857(20)	2679	s dil HCl
(II) acetate	$Cr(C_2H_3O_2)_2$	170.10	1.79			sl s aq, alc; i eth
(III) acetate	$Cr(C_2H_3O_2)_3$	229.13				s aq
(II) bromide	$CrBr_2$	211.81	4.236_4^{25}	842		s aq, alc
(III) bromide	$CrBr_3$	291.72	4.25			i aq; v s alc
(II) chloride	$CrCl_2$	122.90	2.751^{14}	814		v s aq
(III) chloride	$CrCl_3$	158.35	2.87^{25}	1152	subl 1300	s aq, alc (slow); i acet, eth
(II) fluoride	CrF_2	89.99	3.79	894	d > 1300	sl s aq; s hot HCl
(III) fluoride	CrF_3	108.99	3.8	1100	subl 1100(50)	i aq, alc; s HF, HCl
(III) formate 6-water	$Cr(CHO_2)_3 \cdot 6H_2O$	295.15		d > 300		s aq

Name	Formula	Mol. wt.	Density	mp, °C	bp, °C	Solubility
hexacarbonyl	Cr(CO)$_6$	220.06	1.77^{18}	d 130	explodes 210	i aq, alc; s eth, chl
(III) hydroxide	Cr(OH)$_3$	101.02		d		i aq; s acids
(III) nitrate 9-water	Cr(NO$_3$)$_3 \cdot$9H$_2$O	400.15		60	d > 100	208 g/100 mL^{15} aq; s alc
(III) oxide	Cr$_2$O$_3$	152.02	5.21	2330(15)	ca. 3000	i aq; sl s acids, alkalis
(IV) oxide	CrO$_2$	83.90	4.89	197	—O$_2$, 250	i aq; s HNO$_3$
(VI) oxide	CrO$_3$	99.99	2.70^{25}	198	d 250	61.7 g/100 mL^{20} aq; may ign organics
potassium bissulfate 12-water	CrK(SO$_4$)$_2 \cdot$12H$_2$O	499.41	1.826^{25}	89	anhyd 400	22 g/100 mL^{25} aq; i alc
(II) sulfate 7-water	CrSO$_4 \cdot$7H$_2$O	274.17		d 100		22.9 g/100 mL^{0} aq; sl s alc
(III) sulfate 18-water	Cr$_2$(SO$_4$)$_3 \cdot$18H$_2$O	716.45	1.7			220 g/100 mL^{20} aq
Chromyl chloride	CrO$_2$Cl$_2$	154.90	1.9145^{25}	−96.5	117	d aq; s bz, chl, eth, CCl$_4$, CS$_2$
fluoride	CrO$_2$F$_2$	121.99		31.6^{885mm}	subl 29.6	
Cobalt	Co	58.9332	8.90^{20}	1494	2897.1	i aq; s dil HNO$_3$
(II) acetate 4-water	Co(C$_2$H$_3$O$_2$)$_2 \cdot$4H$_2$O	249.08	1.705^{19}	anhyd 140		s aq; 2.1 g/100 mL^{15} MeOH
(III) acetate	Co(C$_2$H$_3$O$_2$)$_3$	236.07		d > 100		s aq, HOAc, alc
(II) bromide	CoBr$_2$	218.75	4.909^{25}	678 (in N$_2$)		112 g/100 mL^{20} aq; s alc, acet
(II) carbonate	CoCO$_3$	118.94	4.13	d		0.1^{15} aq; s hot acids
(II) chloride	CoCl$_2$	129.84	3.367^{75}	740(2)	1049	53 g/100 mL^{20} aq; s alc, acet, eth, glyc, pyr
(II) chloride 6-water	CoCl$_2 \cdot$6H$_2$O	237.93	1.924	anhyd 110		97 g/100 mL^{20} aq
(II) chromate	CoCrO$_4$	174.93		d		i aq; s acids
(II) cyanide	Co(CN)$_2$	110.99	1.872^{25}	d 300		0.0042^{18} aq; s KCN
(II) fluoride	CoF$_2$	96.93	4.43	1127(7)	1739	1.36^{20} aq; s warm mineral acids
(III) fluoride	CoF$_3$	115.93	3.88	926(200)		d aq
(II) formate 2-water	Co(CHO$_2$)$_2 \cdot$2H$_2$O	185.00	2.1294^{22}	anhyd 140	d 175	5.03 g/100 mL^{20} aq; i alc
(II) hydroxide	Co(OH)$_2$	92.95	3.597^{15}	168 (vacuo)		0.00018 aq; v s acids
(III) hydroxide	Co(OH)$_3$	109.96	4.46	—H$_2$O, 100	d	0.00032 aq; s acids
(II) iodide (alpha, black)	CoI$_2$	312.74	5.584^{25}	515 (vacuo)	570 (vacuo)	203 aq
(II) nitrate 6-water	Co(NO$_3$)$_2 \cdot$6H$_2$O	291.04	1.87	55		155 g/100 mL^{30} aq; v s alc
(II) oxalate	CoC$_2$O$_4$	146.95	3.021^{25}	d 250	d > 74	0.002^{18} aq
(II) oxide	CoO	74.93	5.7–6.7	1805(10)		i aq; s acids, alkalis
(II, III) oxide	Co$_3$O$_4$	240.80	6.07	d > 900		i aq; s acids, alkalis
(II) phosphate 8-water	Co$_3$(PO$_4$)$_2 \cdot$8H$_2$O	510.87	2.769	anhyd 200		v sl s aq; s mineral acids
(II) sulfate 7-water	CoSO$_4 \cdot$7H$_2$O	281.10	2.03^{25}	anhyd 420	d 1140	65 g/100 mL^{20} aq; sl s alc

TABLE 3.2 Physical Constants of Inorganic Compounds (*Continued*)

Name	Formula	Formula weight	Density	Melting point, °C	Boiling point, °C	Solubility (in 100 parts solvent)
Cobalt						
(II) sulfide	CoS	91.00	5.45^{18}	>1100		i aq; s acids
(II) thiocyanate 3-water	Co(SCN)$_2$·3H$_2$O	229.14		anhyd 105		7.8^{18} aq; s alc, eth
Copper	Cu	63.546(3)	8.92^{20}	1084.9	2571.5	i; s HNO$_3$, hot H$_2$SO$_4$
(II) acetate 1-water	Cu(C$_2$H$_3$O$_2$)$_2$·H$_2$O	199.65	1.882	115	d 240	8 g/100 mL aq; 0.48 MeOH; sl s eth
acetate metaarsenate (1/3)	Cu(C$_2$H$_3$O$_2$)$_2$·3Cu(AsO$_2$)$_2$	1013.77				unstable in acids, bases; s NH$_4$OH
(II) borate(1−)	Cu(BO$_2$)$_2$	149.16	3.859			s aq
(I) bromide	CuBr	143.45	4.98	504	1345	v sl s aq; s HCl, HBr, NH$_4$OH
(II) bromide	CuBr$_2$	223.35	4.72^{25}_{4}	498	900	126 g/100 mL aq; s alc, acet, pyr; i bz
(II) carbonate hydroxide (1/1) (malachite)	CuCO$_3$·Cu(OH)$_2$	221.11	4.0	d 200		i aq; s acids
(II) chlorate 6-water	Cu(ClO$_3$)$_2$·6H$_2$O	338.53	4.14^{25}_{4}	65	d 100	242 g/100 mL^{18} aq; v s alc; s acet
(I) chloride	CuCl	98.99	4.14^{25}_{4}	430	1212	0.024 aq; s conc HCl, conc NH$_4$OH
(II) chloride	CuCl$_2$	134.45	3.386^{25}_{4}	d > 300		73 g/100 mL^{20} aq; s alc, acet
(II) chloride 2-water	CuCl$_2$·2H$_2$O	170.47	2.51	anhyd 200	d > 300	76.4 g/100 mL^{25} aq; v s alc; s acet
(I) chromium(III) oxide (1/1)	Cr$_2$O$_3$·Cu$_2$O	295.07	5.24^{20}	d > 900		i aq; s HNO$_3$
(II) citrate 2.5-water	Cu$_2$C$_6$H$_4$O$_7$·2.5H$_2$O	360.21		anhyd 100	d	0.17 aq; s acids
(I) cyanide	CuCN	89.56	2.92	473 (in N$_2$)		i aq; s NH$_4$OH, KCN; d hot dil HCl
(II) fluoride	CuF$_2$	101.54	4.85	836(10)	1676	4.75 g/100 mL^{20} aq; s acids
(II) formate	Cu(CHO$_2$)$_2$	153.55	1.831			12.5 aq
(II) hexafluorosilicate 4-water	Cu[SiF$_6$]·4H$_2$O	277.70	2.56			124 g/100 mL^{20} aq
(II) hydroxide	Cu(OH)$_2$	97.55	3.368	d 160		i aq; s acids
(I) iodide	CuI	190.44	5.63^{25}_{4}	588–606	1290	i aq; s KCN, NH$_4$OH, KI

Name	Formula	Formula wt	Density	mp/°C	bp/°C	Solubility
(II) nitrate 3-water	$Cu(NO_3)_2 \cdot 3H_2O$	241.60	2.05	114.5	d 170	138 g/100 mL^0 aq; v s alc
(II) oleate	$Cu(OOCC_{17}H_{33})_2$	626.43				i aq; sl s alc; s eth
(II) oxalate hemihydrate	$CuC_2O_4 \cdot 0.5H_2O$	160.57			d 310	0.002 aq; s NH$_4$OH
(I) oxide	Cu_2O	143.08	6.0_4^{25}	1243.6	—O$_2$, 1800	i aq; s HCl
(II) oxide	CuO	79.54	6.315_4^{14}	d 1124		i aq, alc; s acids, KCN
(II) perchlorate	$Cu(ClO_4)_2$	262.43	2.225^{23}	anhydr >200	d > 130	146 g/100 mL^{30} aq; s eth, EtAc; i bz
(II) phosphate 3-water	$Cu_3(PO_4)_2 \cdot 3H_2O$	434.61		d		i aq; s acids
(II) salicylate 4-water	$Cu(C_7H_5O_3)_2 \cdot 4H_2O$	409.83		dehyd in air	d ca. 480	v s aq; s alc
(II) selenate 5-water	$CuSeO_4 \cdot 5H_2O$	296.57	2.559	anhyd 265		25 g/100 mL^{20} aq; v sl s acet; d HCl
(I) selenide	Cu_2Se	206.04	6.84_4^{21}	1113		s acids
(II) selenide	$CuSe$	142.50	6.0–6.6	d dull red heat		i aq, alc, eth; s hot bz, pyr
(II) stearate	$Cu(OOCC_{17}H_{35})_2$	630.46		ca. 250		14.3 g/100 mL^0 aq; i alc
(II) sulfate	$CuSO_4$	159.61	3.603	d > 560		32 g/100 mL^{20} aq; s MeOH, glyc
(II) sulfate 5-water	$CuSO_4 \cdot 5H_2O$	249.68	2.284_4^{16}	anhyd 200		i aq; d HNO$_3$, s KCN
(I) sulfide	Cu_2S	159.14	5.6_4^{20}	1130		i aq; s hot HNO$_3$, KCN
(II) sulfide	CuS	95.60	4.6	d		sl s aq; s HCl
(I) sulfite hydrate	$Cu_2SO_3 \cdot H_2O$	225.16	3.83^{15}			0.42^{20} aq; s acids, alkalis
(II) tartrate 3-water	$CuC_4H_4O_6 \cdot 3H_2O$	211.61				0.00044 aq; s NH$_4$OH, eth, alkali SCN
(I) thiocyanate	$CuSCN$	121.62	2.85	1084		0.1^{15} aq; d acids; s NH$_4$OH
(II) tungstate(VI)(2−) 2-water	$CuWO_4 \cdot 2H_2O$	347.42				
Curium-244	Cm	244.063	13.51^{25}	1340(40)	3110	s acids
Cyanogen	NC—CN	52.04	2.335 g/L	−27.84	−21.15	mL/100 mL: 450^{20} aq, 230 alc
azide	$NC-N_3$	68.04				s acetonitrile; pure azide detonates upon shock. Handle only in solvents
bromide	NCBr	105.93	2.015^{20}	51.4	61.35	v s aq, alc, eth
Deuterium	D_2 or $^2H_2$	4.03	0.169^{mp} (lq)	−252.89	−249.49	sl s aq
oxide	D_2O	20.03	1.1056^{20}	3.82	101.43	misc aq
Dysprosium	Dy	162.50(3)	8.540^{25}	1407	2567	s acids
bromide	$DyBr_3$	402.23	4.78	880	1480	s aq
chloride	$DyCl_3$	268.86	3.67	680	1530	s aq
fluoride	DyF_3	219.50	7.465	1154	2230	i aq

TABLE 3.2 Physical Constants of Inorganic Compounds (*Continued*)

Name	Formula	Formula weight	Density	Melting point, °C	Boiling point, °C	Solubility (in 100 parts solvent)
Dysprosium oxide	Dy_2O_3	373.00	7.81^{27}	2340		s aq
Erbium	Er	167.26(3)	9.045^{25}	1497	2868	s aq
chloride	$ErCl_3$	273.62		774	1500	s aq; sl s alc
oxide	Er_2O_3	382.52	8.640		d 630	0.0005^{25} aq; s acids
sulfate 8-water	$Er_2(SO_4)_3 \cdot 8H_2O$	766.87	3.205	anhyd 110		16.0 g/100 mL^{20} aq
Europium	Eu	151.96	5.244	826	1527	s acids
(III) chloride	$EuCl_3$	258.32	4.471^{35}	623 d		s acids
(III) oxide	Eu_2O_3	351.92	7.42	2050		
Fermium-257	Fm	257.0951				
Fluorine	F_2	37.996 806	1.513^{bp} (lq$_{25}$) 1.554 g/L	−219.61	−188.13	d aq viol; ignites organics and silicates
nitrate	$F(NO_3)$	81.01	1.507^{bp} (lq)	−175	−45.9	hyd aq; s acet; ignites alc, eth; liquid explodes on slight concussion
perchlorate	$F(ClO_4)$	102.45	4.85 g/L^{25}	−167.3	−15.9	d aq; explodes on slight concussion
Francium-233	Fr	223.0197				
Gadolinium	Gd	157.25(3)	7.898^{25}	1312	3250(25)	s acids
chloride	$GdCl_3$	263.61	4.52^{0}	609	1580	s aq
fluoride	GdF_3	214.25	7.047	1231	2277	i aq
nitrate 6-water	$Gd(NO_3)_3 \cdot 6H_2O$	451.36	2.322	91		s aq, alc
oxide	Gd_2O_3	362.50	7.407^{15}	2340		s acids
sulfate 8-water	$Gd_2(SO_4)_3 \cdot 8H_2O$	746.81	3.010^{15}	anhyd 400	d 500	4.08 aq
Gallium	Ga	69.723(4)	$5.904^{29.6}$ (c) $6.095^{29.8}$ (lq)	29.77(2)	2203	s acids
antimonide	GaSb	191.47	3.9	712		s HCl
arsenide	GaAs	144.64	5.31^{25}	1238		s HCl
chloride	$GaCl_3$	176.03	2.47	77.75	201.2	d aq; s bz, CCl$_4$, CS$_2$
fluoride	GaF_3	126.72	4.47	>1000	subl 950	0.004^{25} aq; s HF
nitrate x-water	$Ga(NO_3)_3 \cdot xH_2O$			d 110	to Ga$_2$O$_3$, 200	v s aq

name	formula	mol wt	density	mp	bp	solubility
phosphide	GaP	100.69		1465		
selenide	GaSe	148.68	5.03^{25}	667	d	
triethyl	$Ga(C_2H_5)_3$	146.90	1.058^{30}	-82.3	142.8	
trimethyl	$Ga(CH_3)_3$	114.84	1.151^{15}	-15.7	55.8	
Germanium	Ge	72.61(2)	5.323	937.4	2850	i aq; s hot H_2SO_4
(IV) bromide	$GeBr_4$	392.23	3.2329^{29}	26.1	186.5	hyd aq; s bz, eth
(IV) chloride	$GeCl_4$	214.42	1.879^{20}	-49.5	83.1	hyd aq; s bz, eth; sl s dil HCl
(IV) fluoride	GeF_4	148.61	2.161^{0}	-36.6	d > 1000	hyd aq; s dil HCl
hydride	GeH_4	76.64	1.523_4^{-142}	-164.8	-90	sl s hot HCl
(IV) oxide	GeO_2	104.61	4.228^{25}	1115	1200	0.43^{20} aq; s acids, alkalis
Gold	Au	196.9665	19.3 (c)	1064.76	2808	s aq reg, KCN, hot H_2SO_4
(I) chloride	AuCl	232.42	7.57	289		s HCl, HBr, KCN
(III) chloride	$AuCl_3$	303.33	3.9^{20}	254 d	subl 180	68 g/100 mL^{20} aq
(I) cyanide	AuCN	222.98	7.14^{20}	d		s aq reg, KCN, NH_4OH
(III) cyanide 3-water	$Au(CN)_3 \cdot 3H_2O$	329.07		d 50		v s aq; sl s alc
diantimonide	$AuSb_2$	440.47		460		
(III) fluoride	AuF_3	253.96		subl 300		
(III) oxide	Au_2O_3	441.93		d 110	d 500	s HCl, KCN
(I) sodium thiosulfate 2-water	$AuNa_3(S_2O_3)_2 \cdot 2H_2O$	526.24	3.09	anhyd 160		50 g/100 mL aq; i alc
stannide	AuSn	315.66		418		
(III) sulfide	Au_2S_3	490.13	8.754	d 197		i aq; s Na_2S
Hafnium	Hf	178.49	13.31^{20}	2227(20)	4450	s HF
chloride	$HfCl_4$	320.30		432	subl 250	hyd aq; s acet, MeOH
oxide	HfO_2	210.49	9.68^{20}	2714		i aq
Helium	He	4.00260	0.1785 g/L^0 (lq) 0.1249	-272.15^{25atm}	-268.935	0.861 mL/100 mL^{20} aq
Holmium	Ho	164.9304	8.78^{25}	1461	1720	s acids; oxidizes in moist air
bromide	$HoBr_3$	404.66	4.86	914	1470	s aq
chloride	$HoCl_3$	271.29		718	1510	s aq
Hydrazine	$H_2N—NH_2$	32.05	1.00362^{25}	2.0	113.5	FP 38; misc aq, alc
hydrate	$H_2N—NH_2 \cdot H_2O$	50.16	1.038^{21}	-51.7	119.4	misc aq, alc; i chl, eth
Hydrazinium						
(1+) chloride	$H_2N—NH_3Cl$	68.51		92.6	d 240	v s aq
(2+) chloride	$ClH_3N—NH_3Cl$	104.97	1.4226^{20}	198	d 200	v s aq; sl s alc
(1+) iodide	$H_2N—NH_3I$	159.98		127		s aq
(1) perchlorate	$H_2N—NH_3ClO_4$	132.51	1.939^{15}	137	d 145	d aq; s alc

TABLE 3.2 Physical Constants of Inorganic Compounds (*Continued*)

Name	Formula	Formula weight	Density	Melting point, °C	Boiling point, °C	Solubility (in 100 parts solvent)
Hydrazinium						
(2+) sulfate	$(H_3NNH_3)SO_4$	130.13	2.016^7	254	d	3.4^{20} aq; i alc
(1+) tartrate	$(H_2N-NH_3)_2C_4H_4O_6$	182.13		183		6.0 g/100 mL^0 aq
Hydrogen	H_2	2.0159	0.0899 g/L^0; 0.07099^{bp} (liq)	−259.76	−252.76	1.9 mL aq^0
amidosulfate (sulfamate)	H_2NSO_3H	97.09	2.126	205	d	14.7 g/100 mL aq; sl s alc, acet
azide	HN_3	43.03	1.126^0	−80	37	v s aq; (very explosive)
borate(1−) (cubic)	HBO_2	43.83	2.486	236(1)		v sl s aq
borate(3−) (ortho)	H_3BO_3	61.83	1.435^{15}	171.0(2)	d 357	5.56 g/100 mL^{30} aq
bromide	HBr	80.92	3.388 g/L^{20}	−86.87	−66.71	193 g/100 mL^{25} aq; misc alc
bromide (constant boiling)	48% HBr + H_2O		1.49	−11	126	v s aq
bromide-d	^2HBr	81.92	3.39 g/L^{20}	−87.46	−66.5	v s aq
bromosulfate	$HOSO_2Br$	240.90		−6 to −8	d	hyd aq
chlorate (40% solution)	$HClO_3$	84.46	1.282^{20}			
chloride	HCl	36.46	1.526^{20} g/L	−114.18	−85.05	72 g/100 mL^{20} aq
chloride (constant boiling)	20.24% HCl + H_2O		1.097		110	v s aq
chloride-d	^2HCl	37.47	1.49 g/L^{25}	−114.64	−84.72	v s aq
chlorosulfate	HSO_3Cl	116.52	1.753^{20}	−80	152	hyd viol aq giving HCl + H_2SO_4
cyanate	$HOCN$	43.03	1.140_4^{-20}	−86	23.5	s aq d; s bz, eth
cyanide	HCN	27.03	0.901 g/L	−13.24	25.70	v s aq
deuteride	$^1H^2H$ or HD	3.02		−256.56	−251.03	
diphosphate(IV)	$(HO)_2OP-PO(OH)_2$	162.01		d 100	d aq	
diphosphate(V)	$H_4P_2O_7$	177.98	70	61		709 g/100 mL^{23} aq
fluoride	HF	20.01	0.922 g/L^0	−83.57	19.52	v s aq, alc; 2.54 g/100 g^5 bz
fluoride (constant boiling)	35.35% HF + H_2O				120	v s aq
fluoride-d	^2HF	21.02		−83.6		s aq
fluoroborate	$H[BF_4]$	87.81		d 130	18.65	v s aq
fluorophosphate	H_2PO_3F	99.99	1.818	−80		
fluorosulfate	$HOSO_2F$	100.07	1.726_5^{25}	−87.3	165.5	s aq
hexafluorosilicate	$H_2[SiF_6]\cdot 2H_2O$	180.11	1.463	19		60–70% aq solution
iodate	HIO_3	175.91	4.629_4	d 110 (H_5IO_6)	d 220 (I_2O_5)	269 g/100 mL^{20} aq; s alc; i eth, chl

Name	Formula	Formula weight	Density	m.p., °C	b.p., °C	Solubility
iodide	HI	127.92	5.37 g/L^{20}	−50.79	−35.35	234 g/100 mL^{10} aq; misc alc
iodide (constant boiling)	57% HI + H_2O		1.70		127	v s aq
iodide-d	^2HI	128.92		−51.87	−35.7	v s aq
molybdate hydrate	$H_2MoO_4 \cdot H_2O$	179.97	3.124^{15}	−H_2O, 70		0.133^{18} aq; s alk
nitrate	HNO_3	63.02	1.5492 (lq)	−41.59	83	v s
nitrate (constant boiling)	69% HNO_3 + H_2O		1.41^{20}		120.5	misc aq
oxide (water)	H_2O	18.02	1.000^{4}	0.00	100.00	misc aq
oxide-d_2	D_2O or ^2H_2O	20.03	1.1044^{25}	3.81	101.42	
perchlorate 2-water	$HClO_4 \cdot 2H_2O$	136.49	1.67^{20}	−17.8	203	v s aq (commercial 72% acid)
periodate(1−) (meta)	HIO_4	191.91		subl 110	d 138	440 g/100 mL^{25} aq
periodate(5−)	H_5IO_6	227.94		122	d 130–140	misc aq; s alc
peroxide	H_2O_2	34.02	1.4639^{0}	−0.43	152	misc aq; s alc, eth
peroxodisulfate	$HO_3S-O-OSO_3H$	194.14	2.2–2.5	d 60		v s aq
phosphate(V)(1−) (meta)	HPO_3	79.98		subl	red heat	slowly s aq giving H_3PO_4; s alc
phosphate(V)(3−) (ortho)	H_3PO_4	98.00	1.868^{25}	42.35	d 213	v s aq
commercial 85% acid			1.685	anhyd 150	$H_4P_2O_7$, 200	to HPO_3, >300
phosphate(V)(3−)-d_3	$^2H_3PO_4$	101.03	1.908^{25}	46.0		v s aq
phosphide, *see* Phosphine						
phosphinate	HPH_2O_2	66.0	1.493^{19}	26.5	d 50	s aq
phosphonate (phosphorous acid)	H_2PHO_3	82.00	1.6512^{1}	ca. 73	d >180	v s aq, alc
selenate	H_2SeO_4	144.98	2.9508^{15}	58	260	v s aq (viol)
selenide	H_2Se	80.98	2.12^{-bp}	−65.73	−41.4	9.5 mL/100 mL^{20} aq; s CS_2
sulfate	H_2SO_4	98.08	1.8318^{20}	10.38	335.5	misc aq
sulfate-d_2	$^2H_2SO_4$ or D_2SO_4	100.09	1.8620	14.35		misc aq
sulfide	H_2S	34.08	1.5392 g/L^{0}	−85.52	−60.33	0.334 mL^{25} aq
tellurate(IV)	H_2TeO_3	177.63	3.0	d to TeO_2		0.0007 aq; s acid, alkali
tellurate(VI) (monoclinic)	H_6TeO_6	229.66	3.068	−2H_2O, 120	d 320 to TeO	30 g/100 mL^{18} aq
telluride	H_2Te	129.63	6.234 g/L	−49	−2	s aq d
trithiocarbonate	$(HS)_2CS$	110.21	1.483^{20}	−26.9	57.8	d aq, alc
tungstate(VI)(2−)	H_2WO_4	249.86	5.5	anhyd 100		i aq; s HF, alkalis
Hydroxylamine	$HONH_2$	33.03	1.2044^{40}	33.1	58^{22mm}	v s aq, MeOH; sl s bz, eth
Hydroxylammonium chloride	$HONH_3Cl$	69.49	1.680^{20}	150.5	d	g/100 mL: 83^{17} aq, 12.5^{20} MeOH, 5.1^{20} EtOH; s glyc
sulfate	$(HONH_3)_2SO_4$	164.14		170		69 g/100 mL^{20} aq

TABLE 3.2 Physical Constants of Inorganic Compounds (*Continued*)

Name	Formula	Formula weight	Density	Melting point, °C	Boiling point, °C	Solubility (in 100 parts solvent)
Indium	In	114.82	7.31^{20}	156.61	2080	s acids
antimonide	InSb	236.57	5.74^{mp} (c)	535		i aq
arsenide	InAs	189.74		943		
chloride	InCl$_3$	221.18	4.0	586	subl 500	212 g/100 mL^{25} aq
fluoride	InF$_3$	171.82	4.39^{25}	1170		0.040^{25} aq; s dilute acids
oxide	In$_2$O$_3$	277.64	7.179			s acids
phosphide	InP	145.79	10.8	1070		v sl s acids
telluride	In$_2$Te$_3$	612.44	5.798	669		
trimethyl	In(CH$_3$)$_3$	159.93	1.568	88.4	135.8	d aq; s acet, bz
Iodine	I$_2$	253.8090	4.660^{20}	113.60	185.24	g/100 mL^{25}: 0.029 aq, 14.1 bz, 16.5 CS$_2$, 21.4 EtOH, 25.2 eth, 2.6 CCl$_4$; s chl, HOAc
heptafluoride	IF$_7$	259.89	lq: 2.8^6	6.45	4.77 subl	s aq (d), s NaOH
monobromide	IBr	206.81	4.4157^0	40	116 d	s aq, alc, eth, CS$_2$
monochloride	ICl	162.36	3.10^{20}	27.38 alpha	97 d	d aq; s alc, eth
pentafluoride	IF$_5$	221.90	3.207^{25}	9.43	100.5	d aq
pentaoxide	I$_2$O$_5$	333.81	4.799^{25}	d 275		187 g/100 mL^{13} aq
trichloride	ICl$_3$	233.26	3.202^{-4}	33		d aq; s alc, bz, eth
Iridium	Ir	192.22(3)	22.65^{20}_4	2450(5)	4550(100)	s K$_2$SO$_4$ fusion, KOH + KNO$_3$ fusion
hexafluoride	IrF$_6$	306.19	4.82	44.4	53.6	d aq
(III) oxide	Ir$_2$O$_3$	432.40		d 1000		s boiling HCl
(IV) oxide	IrO$_2$	224.20	3.15	d 1100		0.0002^{20} aq; s HCl
trichloride	IrCl$_3$	298.56		d 763		i acids, alkalis
Iron	Fe	55.847(3)	7.874^{20}	1536(5)	2863.3	i sq; s acids
(III) arsenate 2-water	FeAsO$_4$·2H$_2$O	130.77	7.83	1020		v sl s aq
(II) bromide	FeBr$_2$	215.67	4.636^{25}	684	934	117 g/100 mL^{20} aq; v s alc
(III) bromide	FeBr$_3$	295.57				s aq, alc, eth, HOAc
(*tri-*) carbide	Fe$_3$C	179.55	7.694	1227		s acids
(II) carbonate	FeCO$_3$	115.85	3.8	d		0.072^{18} aq; s acids

Name	Formula	Formula wt	Density	mp, °C	bp, °C	Solubility
(II) chloride	FeCl$_2$	126.75	3.16^{25}	674	1024	62.5 g/100 mL^{20} aq; v s alc, acet
(III) chloride	FeCl$_3$	162.21	2.898^{25}	304	332	74 g/100 mL^0 aq; s alc, acet, eth
disulfide (pyrite)	FeS$_2$	119.98	5.0	d 602		s acids d
(II) fluoride	FeF$_2$	93.84	4.09	1100	1837	sl s aq; s dilute HF; i alc, bz, eth
(III) fluoride	FeF$_3$	112.84	3.87	subl 1000	726	0.091^{25} aq; s HF
(III) hexacyanoferrate(II)	Fe$_4$[Fe(CN)$_6$]$_3$	859.25	1.80	anhyd 250 d		i aq; s HCl
(II) hydroxide	Fe(OH)$_2$	89.86				0.006 aq; s acids
(III) hydroxide oxide	FeO(OH)	88.85	4.28	anhyd 136		i aq, alc; s HCl
(III) iodide	FeI$_2$	309.66	5.315	587(2)	1093	s aq
(III) nitrate 9-water	Fe(NO$_3$)$_3$·9H$_2$O	404.02	1.684^{21}	47	d 100	138 g/100 mL^{20} aq
(di-) nitride	Fe$_2$N	125.70	6.35	d 200		s HCl
(II) oxalate 2-water	FeC$_2$O$_4$·2H$_2$O	179.90	2.28	d 150–160		0.044^{18} aq; s mineral acids
(II) oxide	FeO	71.85	5.7	1360	d 3414	i aq; s acids
(II,III) oxide	Fe$_3$O$_4$	231.54	5.1	1597(2)		i aq; s acids
(III) oxide	Fe$_2$O$_3$	159.69	5.24	1462 d		i aq; s HCl
pentacarbonyl	Fe(CO)$_5$	195.00	1.46–1.52	−20.0(1)	103.9	FP −20; i aq; s alc, bz, eth
(II) phosphate 8-water	Fe$_3$(PO$_4$)$_2$·8H$_2$O	501.61	2.58			i aq; s acids
phosphide	FeP	142.67	6.85	1370		s hot mineral acids
(II) selenide	FeSe	134.81	6.78	d		s HCl
(II) silicate(2−)	FeSiO$_3$	131.93	3.5	1140		
(II) silicate(4−)	Fe$_2$SiO$_4$	203.78	4.34	1220		
(II) sulfate 7-water	FeSO$_4$·7H$_2$O	278.04	1.89	anhyd 300	d 671	d HCl
(III) sulfate	Fe$_2$(SO$_4$)$_3$	399.88	3.097^{18}	d 1178		48 g/100 mL^{20} aq; slowly s aq (hyd); sl s alc; i acet
(II) sulfide	FeS	87.92	4.82	1190(3)	d	0.0006^{18} aq; s acid
(III) thiocyanate	Fe(SCN)$_3$	230.09				v s aq
Krypton	Kr	83.80	3.7493 g/L^0	−157.2	−153.35	5.94 mL/100 mL^{20} aq
difluoride	KrF$_2$	122.80	3.24	subl −60		s anhyd HF
Lanthanum	La	138.90055(3)	6.166^{25}	920	3470	i aq; s HCl
chloride	LaCl$_3$	245.27	3.818	852	1812	v s aq
chloride 7-water	LaCl$_3$·7H$_2$O	371.38		anhyd 852 (in HCl atm)		v s aq; s alc
fluoride	LaF$_3$	195.91	4.49	1493	2327	
nitrate 6-water	La(NO$_3$)$_3$·6H$_2$O	433.02		40	d 126	181 g/100 mL^{20} aq; v s alc

TABLE 3.2 Physical Constants of Inorganic Compounds (*Continued*)

Name	Formula	Formula weight	Density	Melting point, °C	Boiling point, °C	Solubility (in 100 parts solvent)
Lanthanum						
oxide	La_2O_3	325.82	6.51	2320	4200	s acids
sulfate	$La_2(SO_4)_3$	566.00	3.60	d white heat		2.33 g/100 mL^{20} aq; i alc
sulfate 9-water	$La_2(SO_4)_3 \cdot 9H_2O$	728.14	2.821	anhyd 400		2.92 g/100 mL^{20} aq; i alc
Lead	Pb	207.2	11.34^{20} (fcc)	327.43	1740	s hot conc HNO_3, HCl, H_2SO_4
(II) acetate 3-water	$Pb(C_2H_3O_2)_2 \cdot 3H_2O$	379.3	2.55	75	d >200	g/100 mL: 63^{15} aq, 3.3 alc
(IV) acetate	$Pb(C_2H_3O_2)_4$	443.4	2.228^{17}	175–180		s hot HOAc, bz, chl, conc HX acids
(II) azide	$Pb(N_3)_2$	291.2		expl 350 or when shocked		0.023^{18} aq; v s HOAc
(II) borate(1−) hydrate	$Pb(BO_2)_2 \cdot H_2O$	310.8	5.598 anhyd	anhyd 160	mp 500	s acids
(II) bromide	$PbBr_2$	367.0	6.67	371(5)	912	0.450^{0} aq; s acids
(II) carbonate	$PbCO_3$	267.2	6.61	d 340 to PbO		i aq; s acids, alkalis
(II) chlorate	$Pb(ClO_3)_2$	374.1	3.89	d 230		140 g/100 mL^{18} aq; v s alc
(II) chloride	$PbCl_2$	278.1	5.85	501	950	0.99^{20} aq
(II) chromate(VI)(2−)	$PbCrO_4$	328.2	6.12^{15}	844	d	i aq; s dil HNO_3, alkalis
(II) fluoride	PbF_2	245.2	8.445	830	1297(5)	0.064^{20} aq
(IV) fluoride	PbF_4	283.2	6.7	ca. 600		hyd aq
(II) formate	$Pb(CHO_2)_2$	297.2	4.63	d 190		1.6 g/100 mL^{20} aq
(II) hydrogen arsenate	$PbHAsO_4$	347.1	5.79	d 280 to $Pb_2As_2O_7$		s HNO_3, alkalis
(II) hydroxide	$Pb(OH)_2$	241.2	7.41	d 145		0.016^{20} aq; s acids, alkalis
(II) iodide	PbI_2	461.1	6.16	410(2)	832	0.063^{20} aq; s KI, $Na_2S_2O_3$, alkalis
(II) molybdate(VI)(2−)	$PbMoO_4$	367.2	6.92	1065		s acids, alkalis
(II) nitrate	$Pb(NO_3)_2$	331.2	4.53^{20}	d 200		g/100 mL: 56^{20} aq, 1.3 MeOH
(II) oleate	$Pb(C_{18}H_{33}O_2)_2$	770.1				s alc, bz, eth
(II) oxalate	PbC_2O_4	295.2	5.28	d 300		s acids, alkalis
(II) oxide	PbO	223.2	9.53 (red)	886(5)	1472 d	0.0017^{20} aq; s HNO_3
(IV) oxide	PbO_2	239.2	9.375	d 290, Pb_3O_4	d 595, PbO	s HCl, dil HNO_3 + H_2O_2, $H_2C_2O_4$

	Formula	Formula wt.	Density	m.p., °C	b.p., °C	Solubility
(II,IV) oxide	Pb_2O_3	462.4			d 595 PbO	s HNO_3
(III,IV) oxide (red lead)	Pb_3O_4	685.6	8.32–9.16	d 370, Pb_3O_4 d 595, PbO		s HCl, dil HNO_3 + H_2O_2, $H_2C_2O_4$
(II) phosphate	$Pb_3(PO_4)_2$	811.6	6.9	1014		s HNO_3, alkalis
(II) selenide	PbSe	286.2	8.15	1070		s HNO_3
(II) silicate(2−)	$PbSiO_3$	283.3	6.5–6.65	764		s acids
(II) silicate(4−)	Pb_2SiO_4	506.5		743		
(II) stearate	$Pb(C_{18}H_{35}O_2)_2$	774.2		ca. 125		0.05^{35} aq; s hot alc
(II) sulfate	$PbSO_4$	303.3	6.2	d > 900		0.00425 aq; s NaOH
(II) sulfide	PbS	239.3	7.5	1113.4(15)	1300 subl	0.0006^{18} aq; s HNO_3, hot dil HCl
(II) telluride	PbTe	334.8	8.16_4^{20}	904	ca. 200	i acids and alkalis
tetraethyl	$Pb(C_2H_5)_4$	323.45	1.6532^{20}	−137 to −130	110	i aq; s bz, hydrocarbons
tetramethyl	$Pb(CH_3)_4$	267.35	1.9952^{20}	−30.2		s hydrocarbons
(II) thiocyanate	$Pb(SCN)_2$	323.4	3.82	d 190		0.44^{18} aq, s HNO_3, NaOH
Lithium	Li	6.941(2)	0.535^{20}	180.54	1341(5)	d aq to LiOH
acetate 2-water	$LiC_2H_3O_2 \cdot 2H_2O$	102.01		58	d	63 g/100 mL^{20} aq; v s alc
aluminate(1−)	$LiAlO_2$	65.92	2.554	1700(15)		
amide	$LiNH_2$	22.96	1.178^{18}	380–400	d 450 vacuo	d aq; i bz, eth
benzoate	$LiC_7H_5O_2$	128.05		>300		g/100 mL: 33 aq; 7.7 alc
borate(1−)	$LiBO_2$	49.75		844(1)	1719	2.7 g/100 mL^{20} aq; i alc
bromate	$LiBrO_3$	134.85	3.62			179 g/100 mL^{20} aq
bromide	LiBr	86.84	3.464	550	1289	164 g/100 mL aq; s alc, eth
carbonate	Li_2CO_3	73.89	$2.11^{17.5}$	720(1)	d	1.3 g/100 mL^{20} aq; i alc; s acids
chloride	LiCl	42.39	2.068^{25}	610(2)	1360	77 g/100 mL^{20} aq; s alc, acet
chromate(VI)(2−) 2-water	$Li_2CrO_4 \cdot 2H_2O$	165.00		anhyd 75		142 g/100 mL^{18} aq
citrate 4-water	$Li_3C_6H_5O_7 \cdot 4H_2O$	281.98		anhyd 105		61 g/100 mL^{15} aq; sl s alc
fluoride	LiF	25.94	2.640^{20}	848(1)	1717	0.13^{25} aq; s acids
hexafluoroaluminate(3−)	$Li_3[AlF_6]$	161.79		1012(3) (cryolite)		
hydride	LiH	7.95	0.76–0.77	688.7(3)	d 950	no solvent known; flammable
hydride-d	Li^2H or LiD	8.96	0.881	686		
hydroxide	LiOH	23.95	2.54	471.2	1626	12.4 g/100 mL^{20} aq; sl s alc
iodate	$LiIO_3$	181.84	4.502	450		66 g/100 mL aq; in alc
iodide	LiI	133.84	4.061	469	1174	165 g/100 mL^{20} aq & alc; v s acet

3.37

TABLE 3.2 Physical Constants of Inorganic Compounds (*Continued*)

Name	Formula	Formula weight	Density	Melting point, °C	Boiling point, °C	Solubility (in 100 parts solvent)
Lithium						
nitrate	LiNO$_3$	68.94	2.38	261		50 g/100 mL^{20} aq; s alc
nitride	Li$_3$N	34.82		813		d aq
oxide	Li$_2$O	29.88	2.013^{25}	1570(?)	2563	forms LiOH in aq
perchlorate	LiClO$_4$	106.40	2.43^{25}	236	d 400, LiCl	37.5 g/100 mL^{25} aq; v s organic solv
peroxide	Li$_2$O$_2$	45.88		d >195 to Li$_2$O		d dil HCl
silicate(2−)	Li$_2$SiO$_3$	89.97	2.52^{24}	1201(1)		34.5 g/100 mL^{20} aq; i alc
sulfate	Li$_2$SO$_4$	109.95	2.22	859(1)		sl s aq
tetraborate(2−)	Li$_2$B$_4$O$_7$	169.12		917(2)		d aq, alc; g/100 mL: 30 eth, 13 THF; flammable
tetrahydridoaluminate	LiAlH$_4$	37.95	0.917	d 137(10)		s aq pH >7; s eth, THF
tetrahydridoborate	LiBH$_4$	21.79	0.666	268	d 380	s acids
Lutetium	Lu	174.967	9.841^{25}	1652	3402	s aq
chloride	LuCl$_3$	281.33	3.98	892(2)	subl > 750	42.3 g/100 mL^{20} aq
sulfate 8-water	Lu$_2$(SO$_4$)$_3$·8H$_2$O	782.25				i aq; s dilute acids
Magnesium	Mg	24.305	1.738^{20}	650	1110(10)	53.4 g/100 mL^{20} aq; v s alc
acetate	Mg(C$_2$H$_3$O$_2$)$_2$	142.00	1.42	323 d		v sl s HCl
aluminate(2−)	MgAl$_2$O$_4$	142.27	3.6	2135		
amide	Mg(NH$_2$)$_2$	56.37	1.39^{25}	ign in air		d viol water giving NH$_3$
borate(1−) 8-water	Mg(BO$_2$)$_2$·8H$_2$O	254.06	2.30			sl s aq; s acids
bromide	MgBr$_2$	184.13	3.722	711(15)	1158	101 g/100 mL^{20} aq
carbonate	MgCO$_3$	84.32	2.958	d 402		0.01 aq; s acids
chloride	MgCl$_2$	95.23	2.41	714	1437	54.6 g/100 mL^{20} aq
fluoride	MgF$_2$	62.31	3.148	1248	2260	0.013^{25} aq; s HNO$_3$
(*di-*) germanide	Mg$_2$Ge	121.21		1115	ign in air	
hexafluorosilicate 6-water	Mg[SiF$_6$]·6H$_2$O	274.48	1.788	$-$SiF$_4$, 120		51 g/100 mL^{20} aq; i alc
hydride	MgH$_2$	26.34	1.45	d 287 vacuo	d 550	d aq and alc violently
hydrogen phosphate 3-water	MgHPO$_4$·3H$_2$O	174.34	2.123^{15}	anhyd 205		sl s aq; s acids
hydroxide	Mg(OH)$_2$	58.33	2.36	269.1 d		i aq; s acids
iodide	MgI$_2$	278.12	4.43	634(15)		140 g/100 mL^{20} aq; s alc
lactate 3-water	MgC$_6$H$_{10}$O$_6$·3H$_2$O	256.52				4 g/100 mL aq; sl s alc

Name	Formula					Solubility
mandelate	$MgC_{16}H_{14}O_6$	326.61				0.004^{100} aq; i alc
nitrate	$Mg(NO_3)_2 \cdot 6H_2O$	256.41	1.464	95	d 129	120 g/100 mL^{20} aq; v s alc
nitride	Mg_3N_2	100.95	2.712	d 800	subl 700 vac	d aq; s acids
oleate	$Mg(C_{18}H_{33}O_2)_2$	293.61				sl s alc, eth, PE
oxide	MgO	40.31	3.65–3.75	2832(30)	3260	i aq, alc; s acids
perchlorate	$Mg(ClO_4)_2$	223.23	2.21^{20}	d >251		49.6 g/100 mL aq
permanganate	$Mg(MnO_4)_2$	262.19				v s aq
peroxide	MgO_2	56.31				s acids
peroxoborate 7-water	$Mg(BO_3)_2 \cdot 7H_2O$	268.10				sl s aq d; s dilute acids
phosphate 5-water	$Mg_3(PO_4)_2 \cdot 5H_2O$	352.98	1.64^{15}	anhyd 400		0.02 aq; s acids
silicate(2−)	$MgSiO_3$	100.40	3.192^{25}	d 1557(2)		i aq; v sl s HF
silicate(4−)	Mg_2SiO_4	140.71	3.21	1898(11)		i aq; d hot HCl
(*di-*) silicide	Mg_2Si	76.73	2.0^{20}	1085		d aq, HCl
(*di-*) stannide	Mg_2Sn	167.34		778		
sulfate 7-water	$MgSO_4 \cdot 7H_2O$	246.49	1.67	anhyd 250		27.2 g/100 mL aq; sl s alc
sulfite 6-water	$MgSO_3 \cdot 6H_2O$	212.47	1.725	anhyd 200		0.66^{25} aq
tungstate(VI)(2−)	$MgWO_4$	272.18	5.66			i aq; d acids
Manganese	Mn	54.9380	7.47^{20}	1244	2095	d aq; s acids
acetate 4-water	$Mn(C_2H_3O_2)_2 \cdot 4H_2O$	245.08	1.589			38 g/100 mL^{50} aq; v s alc
bromide	$MnBr_2$	214.76	4.39	698	1027	147 g/100 mL^{20} aq; s alc
(*tri-*) carbide	Mn_3C	176.83	6.89^{17}	1520		d aq; s acid
carbonate	$MnCO_3$	114.94	3.125	d >200		0.0065^{25} aq; s acids
chloride	$MnCl_2$	125.84	2.977^{25}	650	1210(20)	74 g/100 mL^{20} aq; s alc, pyr; i eth
chloride 4-water	$MnCl_2 \cdot 4H_2O$	187.91	2.01	anhyd 198		143 g/100 mL aq; s alc; i eth
decacarbonyl	$Mn_2(CO)_{10}$	389.99	1.75^{25}	d 110		i aq; s organic solvents
diphosphate	$Mn_2P_2O_7$	283.82	3.707	1196		i aq; s acid
(II) fluoride	MnF_2	92.93	3.98	856	1820	0.66^{40} aq; s HF, conc HCl
(III) fluoride	MnF_3	111.93	3.54	d >600		hyd a; s acid
hydroxide	$Mn(OH)_2$	88.95	3.258	d		0.0021^{18} aq; s acids
iodide	MnI_2	308.75	5.0^1	638	1017	s aq
nitrate 6-water	$Mn(NO_3)_2 \cdot 6H_2O$	287.05	1.8	25.8		v s aq, alc
(II) oxide	MnO	70.94	5.37^{23}	1650		i aq; s acids
(III) oxide	Mn_2O_3	157.87	4.89^{25}	871–887 d		i aq; s HCl giving Cl_2
(IV) oxide	MnO_2	86.94	5.118^{25}	$-O_2$, 530		s HCl; i HNO_3, cold H_2SO_4
(II,IV) oxide	Mn_3O_4	228.81	4.7			i aq; s HCl
(VII) oxide	Mn_2O_7	221.87	2.396^{20}	ca. −20		v s aq
phosphinate hydrate	$Mn(PH_2O_2)_2 \cdot H_2O$	202.93		d to PH_3	ca. 25	15 g/100 mL aq; i alc

TABLE 3.2 Physical Constants of Inorganic Compounds (*Continued*)

Name	Formula	Formula weight	Density	Melting point, °C	Boiling point, °C	Solubility (in 100 parts solvent)
Manganese silicate(1−)	$MnSiO_3$	131.02	3.48	1270		i aq, HCl
sulfate	$MnSO_4$	151.00	3.25	700	d 850	52 g/100 mL aq; i alc
sulfate hydrate	$MnSO_4 \cdot H_2O$	169.01	2.95	anhyd 400–450		70 g/100 mL²⁰ aq
sulfate 7-water	$MnSO_4 \cdot 7H_2O$	277.11	2.09	anhyd 280		115 g/100 mL²⁰ aq
sulfide	MnS	87.00	3.99	1530		0.0006¹⁸ aq; s acids
titanate (IV)(2−)	Mn_2TiO_4	150.84	4.54	1360		
Mercury	Hg	200.59(3)	13.534(1)²⁵	−38.86	356.9	i aq; s HNO_3, hot conc H_2SO_4
(II) acetate	$Hg(C_2H_3O_2)_2$	318.69	3.280	178–180		g/100 mL: 40¹⁰ aq, 7.5¹⁵ MeOH
(II) benzoate	$Hg(C_7H_5O_2)_2$	4424.83		165		v s NaCl soln; sl s alc
(I) bromide	Hg_2Br_2	560.99	7.307	subl 393 d		i aq, alc, eth; d hot HCl
(II) bromide	$HgBr_2$	360.40	6.109²⁵	241	322 subl	g/100 mL: 0.56²⁰ aq, 20²⁵ alc; v s HCl, HBr
(I) chloride	Hg_2Cl	472.09	7.150	subl 382	d without melting	s aqua regia; i aq, alc, eth
(II) chloride	$HgCl_2$	271.52	5.44	277	304²⁰	g/100 mL²⁰: 7.15 aq, 26 alc, 4 eth 8.3 glyc, 0.5 bz; s HOAc, EtAc
(II) cyanide	$Hg(CN)_2$	252.65	3.996	d 320		g/100 mL²⁰: 9.3 aq, 25 MeOH, 8 EtOH
(I) fluoride	Hg_2F_2	439.22	8.73¹⁵	>570 d		hydrolyses in water
(II) fluoride	HgF_2	238.61	8.95¹⁵	645	d >650	hyd aq; s HF
(II) fulminate	$Hg(ONC)_2$	284.62	4.42	explodes		sl s aq; s alc; dangerously flammable
(I) iodide	Hg_2I_2	654.98	7.70	290 d	subl 140	i aq, alc, eth; s KI
(II) iodide	HgI_2	454.40	6.28	259	350 subl	g/100 mL: 0.006²⁵ aq, 0.8 alc, 0.8 eth, 1.7 acet
(I) nitrate 2-water	$Hg_2(NO_3)_2 \cdot 2H_2O$	561.22	4.79⁴	70 d		hyd aq; s HNO_3
(II) nitrate	$Hg(NO_3)_2$	324.63	4.39	79	d	v s aq; s acet
(I) oxide	Hg_2O	417.22	9.89	d 100		i aq; s HNO_3
(II) oxide	HgO	216.59	11.14	d 476		0.005²⁵ aq; s dil HCl, HNO_3, I^-, CN^-

Name	Formula	Formula wt	Density	mp, °C	bp, °C	Solubility
(I) sulfate	Hg_2SO_4	497.25	7.56	d	d	0.06^{25} aq; s HNO_3
(II) sulfate	$HgSO_4$	296.66	6.47	d	d	d aq; s acid
(II) sulfide (cinnabar)	HgS	232.66	8.10	subl 583		i aq; s aqua regia
(II) thiocyanate	$Hg(SCN)_2$	316.76		d 165		0.063^{25} aq; s HCl
Molybdenum	Mo	95.94	10.28^{20}	2623(8)	4683	s hot H_2SO_4, HNO_3, fused KNO_3
(III) bromide	$MoBr_3$	335.65		subl 977		d alkalis
(IV) chloride	$MoCl_4$	237.75		317	407	s conc acids
(V) chloride	$MoCl_5$	273.19	2.928	194	264	s conc acids, dry eth, dry alc
(VI) chloride	$MoCl_6$	198.85		subl 254		
(VI) fluoride	MoF_6	209.95	2.5341^{19}_{18}	17.61(2)	35.0	hyd aq; s alkalis; 31 g/100 g HF
hexacarbonyl	$Mo(CO)_6$	264.02	1.96	150 d	subl 102	s bz
(IV) oxide	MoO	127.94	6.47			i aq
(VI) oxide	MoO_3	143.94	4.6964^{26}	801(5)	1107(30)	0.05^{28} aq; s conc mineral acids, alk
(III) sulfide	Mo_2S_3	288.07	5.91^{15}	1807(20)	d 1867	d hot HNO_3, s aqua regia
(IV) sulfide	MoS_2	160.08	5.06^{15}	1750(50) [1 atm $S_2(g)$]	subl 450	s hot aq. acids
Neodymium	Nd	144.24(3)	7.004	1024	ca. 3100	
chloride	$NdCl_3$	250.60	4.134^{25}	760	1690	98 g/100 mL^{20} aq; s alc
oxide	Nd_2O_3	336.48	7.28	1900		s dilute acids
sulfate 8-water	$Nd_2(SO_4)_3 \cdot 8H_2O$	720.79	2.85	d 700–800		8.87 g/100 mL^{20} aq
Neon	Ne	20.179	0.8999 g/L^0	−248.67	−246.05	1.05 mL^{20} aq
Neptunium	Np	237.0482	20.45	637(2)	>3900	s HCl
Nickel	Ni	58.69	8.908^{20}	1455(4)	2884	i aq; s HNO_3
acetate 4-water	$Ni(C_2H_3O_2)_2 \cdot 4H_2O$	248.86	1.744	d		16 g/100 mL aq; s alc
acetylacetonate	$Ni(C_5H_7O_2)_2$	256.93	1.455^{17}	230	235^{11mm}	s aq, alc, bz, chl; i eth
bromide	$NiBr_2$	218.53	5.098	963	subl	100 g/100 mL^{20} aq
carbonate hydroxide (1/2)	$NiCO_3 \cdot 2Ni(OH)_2$	304.17	2.6			s dilute acids
chloride	$NiCl_2$	129.62	3.55	1030(4) 2.5 atm	subl 973	61 g/100 mL^{20} aq
chloride 6-water	$NiCl_2 \cdot 6H_2O$	237.70		anhyd 400 d		100 g/100 mL^{20} aq; s alc
cyanide 4-water	$Ni(CN)_2 \cdot 4H_2O$	182.81		subl 250		0.006^{18} aq; s KCN, NH_4OH
dimethylglyoxime	$Ni(HC_4H_6N_2O_2)_2$	288.91				i aq; s abs alc, dilute acids
(tri-) disulfide	Ni_3S_2	240.25	5.82	790(3)	d 2967(15)	s HNO_3
fluoride	NiF_2	96.71	4.72	1450	1740	4 g/100 mL^{20} aq; i alc, eth

TABLE 3.2 Physical Constants of Inorganic Compounds (*Continued*)

Name	Formula	Formula weight	Density	Melting point, °C	Boiling point, °C	Solubility (in 100 parts solvent)
Nickel						
formate 2-water	$Ni(CHO_2)_2 \cdot 2H_2O$	184.78	2.154^{20}	anhyd 130	d 180–200	s aq; i alc
nitrate 6-water	$Ni(NO_3)_2 \cdot 6H_2O$	290.81	2.05	56.7	136.7	150 g/100 mL^{20} aq
(II) oxide	NiO	74.71	7.45	2000(100)		s acids
Ni_2O_3	Ni_2O_3	165.42	4.83	$-O_2$, 600		s hot HCl, HNO_3, H_2SO_4
sulfate	$NiSO_4$	154.78	3.68	$-SO_3$, 840		29 g/100 mL^0 aq
sulfate 6-water	$NiSO_4 \cdot 6H_2O$	262.86	2.07	anhyd 280		40 g/100 mL^{20} aq
sulfide	NiS	90.77	5.3–5.6	976(3)	d 2047(200)	s HNO_3, KHS
tetracarbonyl	$Ni(CO)_4$	170.74	1.3185^{17}	$-19.3(1)$	42.3(2)	explodes 63; FP −4; s organic solvents
Niobium	Nb	92.9064	8.57^{20}	2468(10)	4860(100)	s fused alkai hydroxides
(V) chloride	$NbCl_5$	270.20	2.75	206(1)	247.0(5)	s HCl, CCl_4
(V) fluoride	NbF_5	187.91	2.696^{80}	80.0	234.9	hyd aq, alc; sl s CS_2, CCl_4
(V) oxide	Nb_2O_5	265.82	4.55	1512(30)		s HF, hot H_2SO_4
Nitrogen	N_2	28.0341	1.165 g/L^{20}	-210.01	-195.79	mL/100 mL: 1.6^{20} aq, 0.112 alc
	$^{15}N_2$	30.01	1.25 g/L^{20}	-209.952	-195.73	
(I) oxide	N_2O	44.02	1.843 g/L^{20}	-90.81	-88.46	130^0 mL aq; s alc, eth
(II) oxide	NO	30.01	1.249 g/L^{20}	-163.64	-151.76	4.6 mL/100 mL^{20} aq
(III) oxide	N_2O_3	76.02	1.447 g/L^2	-100.7	2	s eth
(IV) oxide dimer	N_2O_4	92.02	1.448_4^{20}	-9.3	21.15 d	s conc HNO_3, conc H_2SO_4, chl
(V) oxide	N_2O_5	108.01	2.05	30	47.0	v s chl; s CCl_4
selenide	N_4Se_4	371.87	4.2	explosive		sl s bz, CS_2
sulfide	N_4S_4	184.28	2.24^{18}	180	185	s organic solvents
trichloride	NCl_3	120.37	1.653^{20}	-27	71	i aq; s bz, CS_2, CCl_4
trifluoride	NF_3	70.01	2.96 g/L^{20}	-208.5	-129.06	
Nitrosyl chloride	NOCl	65.47	1.592^{-5}	-61.5	-5.5	hyd aq; s fuming H_2SO_4
fluoride	NOF	49.01	2.788 g/L^{20}	-132.5	-59.9	hyd aq
hydrogen sulfate	$NOHSO_4$	127.08		d 73.5		d aq; s H_2SO_4
tetrafluoroborate	$NO[BF_4]$	116.83	2.185^{25}	subl 250$^{0.01mm}$		d aq
Nitryl chloride	NO_2Cl	81.46	2.81 g/L^{100}	-145	-14.3	d aq
fluoride	NO_2F	65.00	2.7 g/L^{20}	-166.0	-72.4	d aq

Osmium	Os	190.2	22.61^{20}	3045(30)	5225(100)	s molten alkali or oxidizing fluxes
hexafluoride	OsF_6	304.2		32.1	45.9	hyd aq
tetrachloride	$OsCl_4$	332.0	4.38^{20}	subl 450		slow hyd aq
tetraoxide	OsO_4	254.20	4.91	40.6	130.0	g/100 mL: 7.24^{25} aq; 375^{25} CCl_4; s bz, eth, alc
Oxygen	O_2	31.9988(3)	1.331 g/L^{20}	−218.4	−182.96	mL/100 mL^{20}: 3.13 aq, 14.3 alc
difluoride	OF_2	54.00	2.26 g/L^{20}	−223.8	−145.3	6.8 mL/100 mL^{0} aq
(di-) difluoride	O_2F_2	70.00	1.45^{bp} (lq)	−154	d − 100	
Ozone	O_3	48.00	1.998 g/L^{20}	−192.5	−111.9	49.4 mL/100 mL^{0} aq
Palladium	Pd	106.42	12.023^{20}	1555	3167	s hot HNO_3, H_2SO_4
acetate	$Pd(C_2H_3O_2)_2$	224.49		205 d		i aq, alc; s acet, chl, eth
chloride	$PdCl_2$	177.30	4.0^{18}	680	d > 680	s alc, acet, HCl
nitrate	$Pd(NO_3)_2$	230.42		d		s dil HNO_3
oxide	PdO	122.40	8.70^{20}	879 d		s 48% HBr, sl s aqua regia
Perchloryl fluoride	ClO_3F	102.46	0.637 g/L	−147.74	−46.67	
Phosphorus (white)	P_4 molecules	123.8950	1.823^{25}	44.15(5)	280.3(5)	g/100 mL: 2.86 bz, 2.50 chl, 1.25 CS_2, 0.025 abs alc, 1.0 eth
(red)	P_4	123.8950	2.34	597	subl 416	i aq; ignites in air, 260
hydride, *see* Phosphine						
pentabromide	PBr_5	430.56	3.46^{20}	106 d	166 d	d aq; s CCl_4, CS_2
pentachloride	PCl_5	208.27	2.119^{20}	subl 100		hyd aq; s CCl_4, CS_2
pentafluoride	PF_5	125.98	5.805 g/L	−93.8	−84.6	hyd aq
pentaoxide (dimer)	P_4O_{10}	283.88	2.30	340	subl 360	d aq; s H_2SO_4
pentasulfide	P_2S_5	222.29	2.09	288(2)	514(1)	hyd aq; s alkali; 0.222^{17} CS_2
tribromide	PBr_3	270.73	2.85^{15}	−41.5	173.2	d aq, alc; s acet, CS_2
trichloride	PCl_3	137.35	1.5752^{4}	−93.6	76.1	d aq, alc; s bz, chl
trifluoride	PF_3	87.98	3.907 g/L	−151.30	−101.38	hyd aq
trioxide (dimer)	P_4O_6	219.90	2.1362^{4}	23.8	173 (N_2 atm)	hyd aq; s bz, CS_2
(*tetra*-) triselenide	P_4Se_3	360.80	1.31	245–246	360–400	flammable in air; s bz, acet, chl, CS_2
(*tetra*-) trisulfide	P_4S_3	220.09	2.03^{17}	167	407	100 g/100 mL^{17} CS_2; s tolune

TABLE 3.2 Physical Constants of Inorganic Compounds (*Continued*)

Name	Formula	Formula weight	Density	Melting point, °C	Boiling point, °C	Solubility (in 100 parts solvent)
Phosphine	PH_3	34.00	1.529 g/L	−133.81	−87.78	mL/100 mL[17]: 1025 CS_2, 726 bz, 319 HOAc, 26 aq; s alc, eth
Phosphonium iodide	PH_4I	161.91	2.86	18.5	subl 62.5	d aq
Phosphoryl chloride difluoride	$POClF_2$	120.43	1.656^0	−96.4	3.1	
dichloride fluoride	$POCl_2F$	136.89	1.5497^{20}	−80.1	52.90	s bz, CS_2, eth
tribromide	$POBr_3$	286.72	2.822	56	191.7 d	d aq, alc
trichloride	$POCl_3$	153.35	1.645^{25}	1.25	105	
Platinum	Pt	195.08(3)	21.09^{20}	1773	3824	s aqua regia, fused alkali
(II) chloride	$PtCl_2$	266.00	5.87	d 581		i aq, alc; s HCl, NH_4OH
(IV) chloride	$PtCl_4$	336.90	4.303^{25}	d 370		143 g/100 mL[25] aq
(VI) fluoride	PtF_6	309.08	3.826 (lq)	61.3	69.14	
(II) oxide	PtO	211.09	14.9^{15}	d 550		i aq; s HCl
(IV) oxide	PtO_2	227.09	10.2	450		i aqua regia
(II) sulfide	PtS_2	259.22	7.66	d 225		s HCl, HNO_3
Plutonium	Pu	239.052	19.816^{20}_4	639.5	3230	i aq; s acids
(III) bromide	$PuBr_3$	478.79	6.69	681	d > 1300	s aq
(III) chloride	$PuCl_3$	345.42	5.70	760	1767	i aq; v s acids
(III) fluoride	PuF_3	296.06	9.32	1425	d 2000	hyd aq
(IV) fluoride	PuF_4	315.05	7.00	1037 d		i aq
(VI) fluoride	PuF_6	353.05	4.86	51.59	62.16	
(II) hydride	PuH_2	241.08	10.40	ca. 727		
(III) hydride	PuH_3	242.08	9.61	ca. 327		
(II) oxide	PuO	255.05	13.9	1900		
(III) oxide	Pu_2O_3	526.12	10.2	2085 (in He)		
(IV) oxide	PuO_2	271.05	11.46	2390 (in He)	d 2800	
(III) sulfide	Pu_2S_3	574.30	9.95	1727		
Polonium	Po	208.9824	9.196 alpha / 9.398 beta	254	962	sl s aq; s acids
(IV) chloride	$PoCl_4$	350.79		300 (in Cl_2)	390 (in Cl_2)	sl hyd aq; v s HCl; s alc, acet
(IV) oxide	PoO_2	240.98		d 550		v s dilute HCl

Name	Formula	Formula weight	Density	mp, °C	bp, °C	Solubility
Potassium	K	39.0983	0.8562^{20}	63.2	765.5	d aq to KOH; s acids
acetate	$KC_2H_3O_2$	98.14	1.57^{25}	292		g/100 mL: 200 aq, 34 alc
arsenate	K_3AsO_4	256.23	2.8	1310		19 g/100 mL aq; slowly s glyc; s alc
borate(1−)	KBO_2	81.91		947(3)	1401	71 g/100 mL^{30} aq
bromate	$KBrO_3$	167.01	3.27^{17}	ca. 350	d 370	6.9 g/100 mL^{20} aq
bromide	KBr	119.01	2.75	734	1398	g/100 mL: 65^{20} aq, 22 glyc, 0.4 alc
carbonate	K_2CO_3	138.20	2.29	901(1)	d to K_2O	90 g/100 mL^{20} aq; i alc
chlorate	$KClO_3$	122.55	2.32^{20}	368	d > 400	g/100 mL: 7.3^{20} aq, 2 glyc
chloride	KCl	74.56	1.988	771	1437	g/100 mL: 34^{20} aq, 7 glyc, 0.4 alc
chromate(VI)	K_2CrO_4	194.20	2.732^{18}	975		64 g/100 mL^{20} aq; i alc
citrate hydrate	$K_3C_6H_5O_7 \cdot H_2O$	324.42	1.98	anhyd 180	d 230	167 g/100 mL^{15} aq
cyanate	KOCN	81.11	1.048	d 700–900		s aq; sl s alc
cyanide	KCN	65.12	1.52^{16}	622(2)	1625	g/100 mL: 50 aq, 50 glyc, 4 MeOH
dichromate(VI)	$K_2Cr_2O_7$	294.19	2.676^{25}	398	d 500	11.7 g/100 mL^{20} aq
dicyanoargentate(I)	$K[Ag(CN)_2]$	199.01	2.36			25 g/100 mL^{30} aq
dihydrogen arsenate	KH_2AsO_4	180.02	2.867	288		g/100 mL: 19^6 aq, 63 glyc; i alc
dihydrogen phosphate	KH_2PO_4	136.09	2.338	d 400 to KPO_3		22.6 g/100 mL^{20} aq; i alc
dioxide	KO_2	71.10	2.14	509(20)		v s aq with decomposition
diphosphate(V) 3-water	$K_4P_2O_7 \cdot 3H_2O$	384.40	2.33	anhyd 300	d	s aq; i alc
disulfate(IV)	$K_2S_2O_5$	222.32		flammable if ground		s aq
disulfate(VI) (pyrosulfate)	$K_2S_2O_7$	254.32	2.28	ca. 325		s aq
ethyldithiocarbonate	$KOCSSC_2H_5$	160.30	1.558^{22}	d 200		v s aq
fluoride	KF	58.10	2.481	859.9	1517	95 g/100 mL^{20} aq
formate	$KCHO_2$	84.10	1.91	167.5		250 g/100 mL aq
gluconate	$KC_6H_{11}O_7$	234.24		d 180	d > mp	v s aq; i alc, bz, chl
heptaiodobismuth-ate(III)(4−)	$K_4[BiI_7]$	1253.82				d aq; s alkali iodide solutions
hexachloroplatinate(IV)	$K_2[PtCl_6]$	486.03	3.499	d 250		0.48^{20} aq
hexacyanoferrate(II) 3-water	$K_4[Fe(CN)_6] \cdot 3H_2O$	422.41	1.853^{17}	anhyd 100	d	28 g/100 mL^{20} aq
hexacyanoferrate(III)	$K_3[Fe(CN)_6]$	329.26	1.89	d		40 g/100 mL^{20} aq (slow)
hexafluorosilicate	$K_2[SiF_6]$	220.25	2.27	d		sl s aq; i alc

3.45

TABLE 3.2 Physical Constants of Inorganic Compounds (*Continued*)

Name	Formula	Formula weight	Density	Melting point, °C	Boiling point, °C	Solubility (in 100 parts solvent)
Potassium						
hexafluorozirconate	$K_2[ZrF_6]$	283.41	3.58	d 200		2.7 g/100 mL^{20} aq
hexanitritocobaltate(III) 1.5-water	$K_3[Co(NO_2)_6] \cdot 1.5H_2O$	479.30				0.089^{17} aq; s HOAc; v sl s alc
hydride	KH	40.11	1.43	417 d		d aq
hydrogen carbonate	$KHCO_3$	100.11	2.17	d > 100		34 g/100 mL^{20} aq; i alc
hydrogen difluoride	KHF_2	78.11	2.37	238.80	d 477	39 g/100 mL^{20} aq; s alc
hydrogen phosphate	K_2HPO_4	174.18		d to $K_2P_2O_7$		150 g/100 mL aq
hydrogen phthalate	$KHC_8H_4O_4$	204.22	1.636^{25}	d		8.3 g/100 mL aq; sl s alc
hydrogen sulfate	$KHSO_4$	136.17	2.24	197	d to $K_2S_2O_7$	48 g/100 mL^{20} aq
hydrogen sulfide	KHS	72.17	1.70	455–510		s aq, alc
hydrogen tartrate	$KHC_4H_4O_6$	188.18	1.956			0.5^{20} aq; s acids; v sl s alc
hydroxide	KOH	56.11	2.044	406	1323	g/100 mL: 112^{20} aq, 33 alc, 40 glyc
iodate	KIO_3	214.02	3.89^{25}	560 d		8.1 g/100 mL^{20} aq; i alc
iodide	KI	166.02	3.12	681	1345	g/100 mL: 144^{20} aq, 4.5 alc, 50 glyc
manganate(VI)	K_2MnO_4	197.12		d 190	d 1400	s aq; stable in KOH
molybdate(VI)	K_2MoO_4	238.14	2.91^{18}	919	d 400	160 g/100 mL aq
nitrate	KNO_3	101.10	2.109^{16}	334.3		g/100 mL: 32^{20} aq, 0.16 alc, s glyc
nitrite	KNO_2	85.10	1.915	441	d 350	306 g/100 mL^{20} aq; sl s alc
oxalate hydrate	$K_2C_2O_4 \cdot H_2O$	184.24	2.127^{4}	anhyd 160	d to K_2CO_3	36 g/100 mL^{20} aq
oxide	K_2O	94.20	2.32^{20}	d 881^{600mm}		d aq to KOH
oxobisoxalatodiaquatitanate(IV)	$K_2[TiO(C_2O_4)_2(H_2O)_2]$	354.18				v s aq
perchlorate	$KClO_4$	138.56	2.5298^{25}	d 440		1.68^{20} aq; i alc
periodate	KIO_4	230.01	3.618^{5}	582		0.42^{20} aq
permanganate	$KMnO_4$	158.03	2.703	d 240		6.34 g/100 mL^{20} aq
peroxide	K_2O_2	110.20		490		d
peroxodicarbonate hydrate	$K_2C_2O_6 \cdot H_2O$	216.24				6.5 g/100 mL aq; d hot aq
peroxodisulfate	$K_2S_2O_8$	270.32	2.477	d 100		2.5 g/100 mL^{20} aq; i alc
perrhenate	$KReO_4$	289.30	4.38	555	1370	0.99^{20} aq
phenolsulfonate hydrate	$KC_6H_4(OH)SO_3 \cdot H_2O$	240.28	1.87			s aq, alc

Name	Formula	Formula wt	Density	mp, °C	bp, °C	Solubility
phosphate	K₃PO₄	212.28		1340		50.8 g/100 mL²⁵ aq; i alc
selenocyanate	KSeCN	144.08	2.564¹⁷	d 100		s aq
silicate(2−)	K₂SiO₃	154.29	1.633	976		s aq
sodium hexanitrito-cobaltate(III) hydrate	K₂Na[Co(NO₂)₆]·H₂O	454.18		d 135		0.07 aq
sodium tartrate 4-water	KNaC₄H₄O₆·4H₂O	282.23	1.790	70–80	anhyd 130–140	54 g/100 mL¹⁵ aq
sorbate	KC₆H₇O₂	150.22	1.363	d > 270		g/100 mL: 58.2²⁰ aq, 6.5 alc
stannate(IV) 3-water	K₂SnO₃·3H₂O	298.94	3.197	anhyd 140		100 g/100 mL²⁰ aq; i alc
stearate	KOOCC₁₇H₃₅	322.57				readily soluble hot aq or alc
sulfate	K₂SO₄	174.27	2.662	1069(1)	1670	g/100 mL: 11²⁰ aq, 1.3 glyc, i alc
sulfide	K₂S	110.27		948(10)		28.6 g/100 mL²⁰ aq
sulfite 2-water	K₂SO₃·2H₂O	194.30		d		138 g/100 mL²⁰ aq
tartrate hemihydrate	K₂C₄H₄O₆·0.5H₂O	235.28	1.98	anhyd 155	d 200	s aq
tellurate(IV)	K₂TeO₃	253.80				61.8 g/100 mL²⁰ aq
tetrachloroaurate(III)	K[AuCl₄]	377.88	3.75	d 357		
tetrafluoroborate	K[BF₄]	125.91	2.505²⁰	530		0.45²⁰ aq
tetrahydridoborate	K[BH₄]	53.95	1.11	d 497		g/100 mL: 21²⁵ aq, 3.5²⁰ MeOH
tetraiodocadmate 2-water	K₂[CdI₄]·2H₂O	698.21	3.3592¹			g/100 mL: 137¹⁵ aq, 71¹⁵ alc, 4 eth
tetraiodomercurate(II)	K₂[HgI₄]	786.48		173	d 500	v s aq; s alc, acet, eth
thiocyanate	KSCN	97.18	1.886¹⁴			g/100 mL: 217²⁰ aq, 200 acet, 8 alc
thiosulfate	K₂S₂O₃	190.33		d 400		155 g/100 mL²⁰ aq; i alc
trihydrogen bisoxalate 2-water	KH₃(C₂O₄)₂·2H₂O	254.20	1.836	d		1.8¹³ aq
trisoxalatoantimonate(III)	K₃[Sb(C₂O₄)₃]	503.12				a aq
trithiocarbonate	K₂CS₃	186.41		d		v s aq
uranyl(VI) acetate hydrate	K(UO₂)(C₂H₃O₂)₃·H₂O	504.28	3.296¹⁵	anhyd 275		s aq
Praseodymium	Pr	140.9077	6.775²⁵	935	3290(90)	s hot water and acids
chloride	PrCl₃	247.27	4.02	769–782	1710	104 g/100 mL¹³ aq; s alc
(III) oxide	Pr₂O₃	329.81	7.07	oxidizes to Pr₆O₁₁		i aq; s acids
(IV) oxide	PrO₂	172.91	6.82	tr 350 to Pr₆O₁₁		
Promethium-147	Pm	146.915	7.00²⁵	1080 or 1169	3000 est	
bromide	PmBr₃	386.7	5.38	727	1667	s aq
chloride	PmCl₃	153.4		737	1670	s aq

TABLE 3.2 Physical Constants of Inorganic Compounds (*Continued*)

Name	Formula	Formula weight	Density	Melting point, °C	Boiling point, °C	Solubility (in 100 parts solvent)
Protactinium	Pa	231.0359	15.37^{25}	1568(8)	4227	i aq; s HCl
(IV) chloride	$PaCl_4$	372.85	4.72	subl 400		hyd aq; s THF, CH_3CN
(V) chloride	$PaCl_5$	408.31	3.74	301	420	d aq; s acids
Radium	Ra	226.03	5.5^{20}	700.1	1737	s aq
bromide	$RaBr_2$	385.88	5.79	728	subl 900	s aq
chloride	$RaCl_2$	296.93	4.91	1000		
Radon	Rn	222.0	9.73 g/L	−71	−62	23 mL/100 mL^{20} aq; s org solv
Rhenium	Re	186.207(1)	21.02	3180	5678	s HNO_3
chloride trioxide	$ReClO_3$	269.65		4.5	128	hyd in water to $HReO_4$; s CCl_4
(IV) fluoride	ReF_4	262.19	5.38	124.5	795	hyd aq
(VI) fluoride	ReF_6	300.19	3.58	18.5	33.8	52.5 g/100 mL anhyd HF; s HNO_3
(VII) fluoride	ReF_7	319.19	3.65^{52}	48.3	73.7	hyd aq
(VI) oxide	ReO_3	234.20	6.9–7.4	disprop 400	750	s HNO_3
(VII) oxide	Re_2O_7	484.44	6.1	300.3	360.3	v s aq, org solv
(VII) sulfide	Re_2S_7	596.85	4.866	d 460		i aq; s HNO_3
(VI) tetrachloride oxide	$ReCl_4O$	344.01	3.309^{34}	34	225	hyd aq; s CCl_4
Rhodium	Rh	102.9055	12.41^{20}	1966	3727	s fused $KHSO_4$
(III) chloride	$RhCl_3$	209.28		d 450	subl 850	i aq; s KOH, KCN
(III) fluoride	RhF_3	159.90	5.38	subl 600		i acids, alkalis
(III) oxide	Rh_2O_3	253.81	8.20	d 1100		i aq reg, KOH
tetracarbonyldi-*μ*-chlorodi-chloride	$Rh_2(CO)_4Cl_2$	388.75		124–125		s org solv except hydrocarbons
Rubidium	Rb	85.4678(3)	1.532^{20}	39.0	691(3)	d aq to RbOH
acetate	$RbC_2H_3O_2$	144.52		246		86 g/100 mL^{45} aq
bromide	RbBr	165.39	3.35	681(1)	1346(6)	108 g/100 mL^{20} aq
carbonate	Rb_2CO_3	230.97		837	d 900	g/100 mL: 450^{20} aq, 0.74^{19} alc
chlorate	$RbClO_3$	168.94	3.184^{20}	342		5.4 g/100 mL^{20} aq

chloride	RbCl	120.94	2.76	715	1390	g/100 mL: 91^{20} aq, 1.1 MeOH
						s aq
dihydrogen phosphate	RbH$_2$PO$_4$	182.47		840		131 g/100 mL^{18} aq
fluoride	RbF	104.49	3.557	775	1390	0.028^{20} aq
hexachloroplatinate(IV)	Rb$_2$[PtCl$_6$]	578.75	3.94^{18}	d		d aq
hydroxide	RbOH	102.49	3.202^{11}	301		180 g/100 mL^{15} aq; s alc
iodide	RbI	212.37	3.55	642	1304	163 g/100 mL^{25} aq; s alc
nitrate	RbNO$_3$	147.47	3.11	310		19.5 g/100 mL^{20} aq
oxide	Rb$_2$O	186.94	3.72	477 d		s aq
sulfate	Rb$_2$SO$_4$	267.03	3.613^{20}	1070		48 g/100 mL^{20} aq
Ruthenium	Ru	101.07(2)	12.45^{20}_{4}	2450	4150	s fused alkali, oxidizing fluxes
(III) chloride (hexagonal)	RuCl$_3$	207.47	3.11	d>500		i HCl, alc
(V) fluoride	RuF$_5$	196.06	3.82	86.5	227	d aq
(IV) oxide	RuO$_2$	133.07	6.97	d		i aq; s fused alkali
Samarium	Sm	150.36(3)	7.536^{25}	1072	1803	s acids
(II) chloride	SmCl$_2$	221.26	3.687^{22}	848	2030	s aq dec; i alc
(III) chloride	SmCl$_3$	256.71	4.465	686	d	93.4 g/100 mL^{20} aq
(III) fluoride	SmF$_3$	207.35	6.643	1306	2427	i aq; s H$_2$SO$_4$
(III) oxide	Sm$_2$O$_3$	348.70	8.347	2350		s acids
(III) sulfate 8-water	Sm$_2$(SO$_4$)$_3\cdot$8H$_2$O	733.01	2.930^{18}	anhyd 450		2.7 g/100 mL^{20} aq
Scandium	Sc	44.95591	2.985 hex 3.19 cubic	1538	2730	d aq
chloride	ScCl$_3$	151.32	2.39	960	967	v s aq; i alc
oxide	Sc$_2$O$_3$	137.91	3.864	>2400		s hot or conc acids
sulfate 5-water	Sc$_2$(SO$_4$)$_3\cdot$5H$_2$O	468.17	2.519	anhyd 250	d 550	54.6 g/100 mL^{25} aq
Selenium (hexagonal)	Se	78.96(3)	4.819^{20}_{4}	217	685	s eth, KOH, KCN; i aq, alc
(IV) bromide	SeBr$_4$	398.62	4.029	d 70–80	subl 115	d aq; s HBr, chl, CS$_2$
(IV) chloride	SeCl$_4$	220.77	2.6	305	subl 196	d aq
(di-) dibromide	Se$_2$Br$_2$	317.75	3.6044^{15}		225–230	d aq; s chl, CS$_2$
dibromide oxide	SeBr$_2$O	254.79	3.38^{50}	46.1(1)	217 d	d aq
(di-) dichloride	Se$_2$Cl$_2$	228.83	2.789^{20}	−85	$127^{733\text{mm}}$ dec	d aq; s bz, chl, CS$_2$
dichloride oxide	SeCl$_2$O	165.87	2.44^{16}	ca. 5	177.2	d aq; misc bz, chl, CCl$_4$, CS$_2$
difluoride oxide	SeF$_2$O	132.97	2.801^{20}	15.0	126	hyd aq
(IV) fluoride	SeF$_4$	154.95	2.732^{20}	−9.5	106	hyd aq viol; misc alc, eth; s chl
(VI) fluoride	SeF$_6$	192.96	2.108^{-10}	−50.8	subl −63.8	i aq

TABLE 3.2 Physical Constants of Inorganic Compounds (*Continued*)

Name	Formula	Formula weight	Density	Melting point, °C	Boiling point, °C	Solubility (in 100 parts solvent)
Selenium						
(*di*-) hexasulfide	Se_2S_6	350.28	2.44	121.5		i aq; 1.2 g/100 mL[20] bz; s CS_2
(IV) oxide	SeO_2	110.96	3.954_4^{15}	340	subl 315	w/w %: 38[14] aq, 10[12] MeOH, 4.35[15] acet, 6.7[14] EtOH, 1.1[12] HOAc; s H_2SO_4
(*tetra*-) tetrasulfide	Se_4S_4	444.08	3.20	113 d		i aq; 0.04 g/100 mL[20] bz; s CS_2
Silane	SiH_4	32.09	0.68 g/L[−185]	−184.7	−111.9	d aq slowly; i alc, bz, chl, eth
chloro-	SiH_3Cl	66.56	3.033 g/L	−118.0	−30.4	
dichloro-	SiH_2Cl_2	101.01	4.60 g/L	−122.0	8.3	d aq
iodo-	SiH_3I	158.01	2.035^{15}	−57.0	45.5	d aq
trichloro-	$SiHCl_3$	135.45	1.331^{25}	−127(1)	31.8	d aq; s bz, chl
Silicon	Si	28.0855(3)	2.332^{25}	1412(3)	2680	s HF + HNO_3, fused alkali oxides
carbide (beta)	SiC	40.07	3.210^{20}	d 2986		s fused alkali oxides
dioxide (quartz)	SiO_2	60.08	2.65^0	1423(50)		i aq; s HF
dioxide-tungsten trioxide-water (1/12/26) (silicotungstic acid)	$SiO_2 \cdot 12WO_3 \cdot 26H_2O$	3310.66				v s aq, alc
disulfide	SiS_2	92.21	2.02	1090 & subl		s d aq, alc; i bz
tetrabromide	$SiBr_4$	347.72	2.81^{29}	5.4(10)	153.4(6)	hyd aq viol
tetrachloride	$SiCl_4$	169.89	1.48^{20}	−70	57.6	hyd aq; s bz, CCl_4, eth
tetrafluoride	SiF_4	104.06	4.69 g/L	$−90.3^{1318mm}$	subl −95.7	hyd aq; s HF
tetraiodide	SiI_4	535.70	4.198	121(3)	302.7	d aq; 2.2 g/100 mL[27] CS_2
tetraisothiocyanate	$Si(NCS)_4$	260.40	3.44	143.8	314.2	d aq
(*tri*-) tetranitride	Si_3N_4	140.28		d 1878		i aq; s HF
Silver	Ag	107.8682(3)	10.49^{15}	961.93	2164	s HNO_3
acetate	$AgC_2H_3O_2$	166.92	3.259^{15}	d		1.04^{20} aq; s dil HNO_3
antimonide	Ag_3Sb	445.35		559		
azide	AgN_3	149.89		252	297	i aq; s KCN, HNO_3 (explosive)
bromide	$AgBr$	187.80	6.473	430	1560	i aq; s KCN

3.50

Name	Formula	Mol wt	Density	m.p.	b.p.	Solubility
carbonate	Ag_2CO_3	275.77	6.077	d 220		0.003^{20} aq; s KCN, HNO_3, NH_4OH
chlorate	$AgClO_3$	191.34	4.430^{20}_4	231	d 270	10 g/100 mL^{15} aq
chloride	AgCl	143.34	5.56	455	1564	i aq; 7.7 g/100 mL NH_4OH, KCN, $Na_2S_2O_3$
chromate(VI)	Ag_2CrO_4	331.77	5.625^{25}			0.002^{20} aq; s HNO_3, NH_4OH
cyanide	AgCN	133.90	3.95	d 320		i aq; s KCN
fluoride	AgF	126.87^{16}	5.852	435	ca. 1150	182 g/100 mL^{20} aq; s HF, CH_3CN
(II) fluoride	AgF_2	145.87	4.57	690	d 700	hyd viol aq
iodate	$AgIO_3$	282.80	5.525^{20}	>200	d	0.053^{25} aq; 40 g/100 mL 10% NH_4OH
iodide (alpha)	AgI	234.80	5.683^{30}	558	1505	i aq; s KCN, KI, $(NH_4)_2CO_3$
nitrate	$AgNO_3$	169.89	4.352^{19}	210	d 440	g/100 mL: 216^{20} aq, 3.3 alc, 0.4 acet
nitrite	$AgNO_2$	153.89	4.453	d >140		0.33^{25} aq; d dilute acids
oxalate	$Ag_2C_2O_4$	303.78	5.029^4	explodes 140		0.004^{20} aq; s HNO_3, NH_4OH
oxide	Ag_2O	231.76	7.22^{25}	d 200 (d light)		0.002^{25} aq; v s dil HNO_3, NH_4OH
(II) oxide	AgO	123.88	7.483^{25}	d >100		i aq; s alkalis and acids with dec
perchlorate	$AgClO_4$	207.34	2.806^{25}	d 486		557 g/100 mL^{20} aq; s bz, glyc, pyr
permanganate	$AgMnO_4$	226.81	4.49	d by light		0.9 aq; d alc
phosphate	Ag_3PO_4	418.62	6.370	849		0.006 aq; v s dil HNO_3, KCN, $(NH_4)_2CO_3$
selenate(IV)	Ag_2SeO_3	342.72	5.9297	530	d >530	sl s aq; s HNO_3
sulfate	Ag_2SO_4	311.83	5.45^{30}	660	d 1085	0.80^{20} aq (slow); s HNO_3, NH_4OH, H_2SO_4
sulfide (agentite)	Ag_2S	247.83	7.234^{20}	845	d	i aq; s HNO_3, KCN
Sodium	Na	22.98977	0.968^{20}	97.83(2)	881.4	d aq to NaOH
acetate	$NaC_2H_3O_2$	82.04	1.528	324		75 g/100 mL^{20} aq
acetate 3-water	$NaC_2H_3O_2 \cdot 3H_2O$	136.08	1.45	anhyd 120	d >120	g/100 mL: 125^{20} aq, 5.1 alc
aluminate(1−)	$NaAlO_2$	81.97		1650		v s aq; i alc
amide	$NaNH_2$	39.02		210	subl 400	d aq viol
arsenate(III)(1−)	$NaAsO_2$	129.91	1.87			v s aq

3.51

TABLE 3.2 Physical Constants of Inorganic Compounds (*Continued*)

Name	Formula	Formula weight	Density	Melting point, °C	Boiling point, °C	Solubility (in 100 parts solvent)
Sodium						
ascorbate	$NaC_6H_7O_6$	198.12		d 218		62 g/100 mL^{25} aq
azide	NaN_3	65.01	1.846^{20}	d to Na + N_2		41 g/100 mL^{20} aq; 0.3 alc
benzoate	$NaO_2C_6H_5$	144.11				g/100 mL: 63^{25} aq; 1.3 alc
bismuthate(V)(1−)	$NaBiO_3$	280.00		d		i cold aq; dec by hot aq & acids
bismuthide	Na_3Bi	277.95	3.339^{17}	766		d aq
bromate	$NaBrO_3$	150.91	3.200^{20}	380 d		40 g/100 mL^{20} aq
bromide	$NaBr$	102.91		747	1447	g/100 mL: 90^{20} aq, 6 alc; 16 MeOH
carbonate	Na_2CO_3	106.00	2.533^{20}	850.0	d	29 g/100 mL^{20} aq; s glyc; i alc
carbonate hydrate	$Na_2CO_3 \cdot H_2O$	124.00	2.25	anhyd 100		g/100 mL: 33 aq, 14 glyc; i alc
carbonate 10-water	$Na_2CO_3 \cdot 10H_2O$	286.14	1.46	34		50 g/100 mL aq; s glyc
carbonate-hydrogen carbonate 2-water (trona)	$Na_2CO_3 \cdot NaHCO_3 \cdot 2H_2O$	226.04	2.112			13 g/100 mL^0 aq
chlorate(V)	$NaClO_3$	106.45	2.489^{15}	248	d > 300	g/100 mL: 96^{20} aq, 0.77 alc, 25 glyc
chloride	$NaCl$	58.45	2.164^{20}	800.8(10)	1465	g/100 mL: 36^{20} aq, 10 glyc
chlorite	$NaClO_2$	90.45	2.468	d 180−200		34 g/100 mL^{17} aq
chromate(VI)	Na_2CrO_4	161.97	2.723	792		84 g/100 mL^{20} aq
citrate 2-water	$Na_3C_6H_5O_7 \cdot 2H_2O$	294.10	1.893^{20}	anhyd 150		77 g/100 mL^{25} aq; i alc
cyanate	$NaOCN$	65.01		550		s aq d; 0.22^0 alc
cyanide	$NaCN$	49.02		562	1530	58.7 g/100 mL^{20} aq
cyanohydridoborate	$Na[BH_3CN]$	62.84	1.199^{28}	>242 d		g/100 mL: 212 aq, 37.2 THF; v s NaOH; i bz, eth
dichromate 2-water	$Na_2Cr_2O_7 \cdot 2H_2O$	298.00	2.348^{25}	anhyd 100; mp 356	d 400	73.1 g/100 mL^{20} aq
diethyldithiocarbamate	$NaS_2CN(C_2H_5)_2 \cdot 3H_2O$	225.31		anhyd 94−96		s aq, alc
dihydrogen arsenate(V) hydrate	$NaH_2AsO_4 \cdot H_2O$	181.94	2.53	anhyd 130	d 200	s aq
dihydrogen diphosphate(V)	$Na_2H_2P_2O_7$	221.97	1.862	d 220		4.5 g/100 mL^0 aq
dihydrogen phosphate(V) hydrate	$NaH_2PO_4 \cdot H_2O$	137.99	2.040	anhyd 100	d NaPO$_3$, 200	71 g/100 mL^0 aq; i alc

Name	Formula	Formula wt	Density	mp/°C	bp/°C	Solubility
dimethylarsonate 3-water (cacodylate)	NaO$_2$As(CH$_3$)$_2$	214.03		anhyd 120		g/100 mL: 200 aq, 40 alc
dioxide	NaO$_2$	54.99	2.45	552(10)		2.26^0 aq
diphosphate(V)	Na$_4$P$_2$O$_7$	265.00		988		13.4 g/100 mL^{20} aq; i alc
dithionate(V) 2-water	Na$_2$S$_2$O$_6\cdot$2H$_2$O	242.13	2.189	anhyd 110	d 267	22 g/100 mL^{20} aq
dithionate(III)	Na$_2$S$_2$O$_4$	174.13		d		i aq; s acids
diuranate(VI)	Na$_2$U$_2$O$_7$	634.04				
dodecylbenzenesulfonate	NaO$_3$SC$_6$H$_4$C$_{12}$H$_{25}$	348.49				10 g/100 mL aq
dodecylsulfate	NaO$_3$SOC$_{12}$H$_{25}$	288.38				d aq; s abs alc
ethoxide	NaOC$_2$H$_5$	68.06		>300		103 g/100 mL aq
ethylenebis(iminodiacetate) (EDTA)	(NaOOCCH$_2$)$_2$NC$_2$H$_4$-N(CH$_2$COONa)$_2$	380.20				
ethylsulfate	NaO$_3$SOC$_2$H$_5$	148.11				140 g/100 mL aq; s alc
fluoride	NaF	41.99	2.78	996(2)	1704	4 g/100 mL^{15} aq; i alc
formate	NaHCO$_2$	68.02	1.919	d >253		81 g/100 mL^{20} aq; s glyc; sl s alc
gluconate	NaC$_6$H$_{11}$O$_7$	218.13				59 g/100 mL^{25} aq; sl s alc; i eth
glycerophosphate	Na$_2$C$_3$H$_5$(OH)$_2$PO$_4$	216.03		d >130		67 g/100 mL aq; i alc
hexachloroaluminate	Na$_3$[AlCl$_6$]	308.70		507(20)		
hexachloroplatinate(IV) 6-water	Na$_2$[PtCl$_6$]$\cdot$6H$_2$O	561.88	2.50	−6H$_2$O, 110		v s aq; s alc
hexacyanoferrate(II) 10-water	Na$_4$[Fe(CN)$_6$]$\cdot$10H$_2$O	484.07	1.458	anhyd 82	d 435	28 g/100 mL^{20} aq
hexacyanoferrate(III) hydrate	Na$_3$[Fe(CN)$_6$]$\cdot$H$_2$O	298.93				18.9 g/100 mL^0 aq
hexafluoroaluminate (cryolite)	Na$_3$[AlF$_6$]	209.94		1012		
hexanitritocobaltate(III)	Na$_3$[Co(NO$_2$)$_6$]	403.98	1.396	d 425		v s aq; sl s alc
hydride	NaH	24.00				ign spontaneously moisture; d alc viol
hydrogen arsenate(V) 7-water	Na$_2$HAsO$_4\cdot$7H$_2$O	312.01	1.88	anhyd 130	d 150	61 g/100 mL^{15} aq; s glyc; sl s alc
hydrogen carbonate	NaHCO$_3$	84.01	2.20	to Na$_2$CO$_3$, 270		8 g/100 mL^{20} aq; i alc
hydrogen difluoride	NaHF$_2$	62.01		d		3.7 g/100 mL^{20} aq
hydrogen phosphate 7-water	Na$_2$HPO$_4\cdot$7H$_2$O	268.07	1.679	d	d	25 g/100 mL^{40} aq; v sl s alc
hydrogen sulfate	NaHSO$_4$	120.07	2.435	315		50 g/100 mL^{25} aq; d alc

TABLE 3.2 Physical Constants of Inorganic Compounds (*Continued*)

Name	Formula	Formula weight	Density	Melting point, °C	Boiling point, °C	Solubility (in 100 parts solvent)
Sodium						
hydrogen sulfide	NaHS	56.07	1.79	350		s aq, alc, eth
hydrogen sulfite	NaHSO$_3$	104.06	1.48	d		g/100 mL: 29 aq, 1.4 alc
hydroxide	NaOH	40.01	2.130^{20}	322	1557	g/100 mL: 108^{20} aq, 14 abs alc, 24 MeOH; s glyc
hydroxymethanesulfinate dihydrate	Na[HOCH$_2$SO$_2$]·2H$_2$O	154.12		63–64	d >64	v s aq; i abs alc, bz, eth
hypochlorite 5-water	NaClO·5H$_2$O	164.52	4.227^{20}	18		29 g/100 mL^0 aq
iodate	NaIO$_3$	197.90		d	d by CO$_2$	8.1 g/100 mL^{20} aq
iodide	NaI	149.92	3.667$^{25}_{4}$	660	1304	g/100 mL: 200^{20} aq, 100 glyc, 50 alc; s acet
lactate	NaOOCCHOHCH$_3$	112.07		d		misc aq, alc
methoxide	NaOCH$_3$	54.03		>300		d aq; s alc
molybdate(VI) 2-water	Na$_2$MoO$_4$·2H$_2$O	241.95	3.28	anhyd 100; mp 687		65 g/100 mL^{20} aq
nitrate	NaNO$_3$	85.01	2.257	308	d ca. 500	g/100 mL: 88^{20} aq, 0.8 alc
nitrite	NaNO$_2$	69.00	2.168^0	271	d >320	67 g/100 mL^{20} aq
oxalate	Na$_2$C$_2$O$_4$	134.01	2.27			3.4 g/100 mL^{20} aq; i alc
oxide	Na$_2$O	61.98	2.27	1132	d 1950	d aq to NaOH
pentacyanonitrosylferrate(III) 2-water (nitroprusside)	Na$_2$[Fe(CN)$_5$NO]·2H$_2$O	297.65	1.72			40 g/100 mL^{16} aq
perchlorate	NaClO$_4$	122.44	2.499	482		201 g/100 mL^{20} aq
periodate	KIO$_4$	213.91	3.865^6	d 300		10.3 g/100 mL^{20} aq
peroxide	Na$_2$O$_2$	77.99	2.805	675	d	v s aq (dec)
peroxoborate 4-water	NaBO$_3$·4H$_2$O	153.88		d 60		2.5 g/100 mL aq
peroxodisulfate(VI)	Na$_2$S$_2$O$_8$	238.13		d		55 g/100 mL aq; d by alc
perrhenate	NaReO$_4$	273.19	5.24	300		33 g/100 mL^{20} aq
phosphate	Na$_3$PO$_4$	163.94	2.537^{18}	1340		12.1 g/100 mL^{20} aq
phosphate 12-water	Na$_3$PO$_5$·12H$_2$O	380.12	1.62	73.4	−11H$_2$O, 100	28.3 g/100 mL^{15} aq; i alc
phosphinate hydrate	NaPH$_2$O$_2$·H$_2$O	105.99		anhyd 200	d to PH$_3$	100 g/100 mL^{20} aq; s glyc, alc; i eth

Name	Formula	Formula weight	Density	Melting point, °C	Boiling point, °C	Solubility
propanate	$NaOOCC_2H_5$	96.07				$g/100\ mL^{25}$: 100 aq, 4.1 alc
salicylate	$NaOOCC_6H_4OH$	160.11				$g/100\ mL$: 110^{20} aq, 11 alc, 25 glyc
selenate(VI)	Na_2SeO_4	188.94	3.098			27 $g/100\ mL^{20}$ aq
silicate(2−)	Na_2SiO_3	122.08	2.614	1089.0(5)		s aq; hyd by hot aq; i alc
silicate(2−) 5-water	$Na_2SiO_3 \cdot 5H_2O$	212.14	1.749	72.2	anhyd 100	v s aq
silicate(4−)	Na_4SiO_4	184.04		1018	d 2227	s aq
stannate(IV) 3-water	$Na_2SnO_3 \cdot 3H_2O$	266.71		d 140 (slow)		59 $g/100\ mL^{20}$ aq; i alc
stearate	$NaOOCC_{17}H_{35}$	306.47		d		sl s aq
sulfate	Na_2SO_4	142.06	2.664	884(1)		28 $g/100\ mL^{20}$ aq
sulfate 10-water	$Na_2SO_4 \cdot 10H_2O$	322.19	1.464	32.4	anhyd 100	67 $g/100\ mL^{25}$ aq; s glyc; i alc
sulfide	Na_2S	78.05	1.8564^{14}	1172(10) vacuo		18.6 $g/100\ mL^{20}$ aq; sl s alc
sulfide 9-water	$Na_2S \cdot 9H_2O$	240.18	1.4274^{16}	ca. 50		200 $g/100\ mL$ aq; sl s alc; i eth
sulfite	Na_2SO_3	126.06	2.6331^{15}	d	d	31 $g/100\ mL^{20}$ aq; s glyc; i alc
tartrate dihydrate	$Na_2C_4H_4O_6 \cdot 2H_2O$	230.08	1.818	anhyd 120		29 $g/100\ mL^{6}$ aq; i alc
tetraborate	$Na_2B_4O_7$	201.27	2.367	742.5		2.6^{20} aq
tetraborate 10-water (borax)	$Na_2B_4O_7 \cdot 10H_2O$	381.37	1.73	75	anhyd 320	$g/100\ mL$: 6.3 aq, 100 glyc; i alc
tetrachloroaluminate	$Na[AlCl_4]$	191.80		151	d	s aq
tetrachloroaurate	$Na[AuCl_4] \cdot 2H_2O$	397.80		d >100	d 315	166 $g/100\ mL^{20}$ aq; s alc, chl
tetrafluoroborate	$Na[BF_4]$	109.82	2.47^{20}	384	d >100	108 $g/100\ mL^{27}$ aq
tetrahydridoborate	$Na[BH_4]$	37.83	1.074	497		18^{25} DMF; 16.4^{20} MeOH (reacts)
thiocyanate	$NaSCN$	81.07		287		134 $g/100\ mL^{20}$ aq
thiosulfate	$Na_2S_2O_3$	158.11	2.345			s aq; i alc
thiosulfate 5-water	$Na_2S_2O_3 \cdot 5H_2O$	248.18	1.7502^{25}	anhyd 100		70 $g/100\ mL^{20}$ aq (dec slowly)
trimetaphosphate 6-water	$(NaPO_3)_3 \cdot 6H_2O$	414.04	1.786	53	anhyd 100	22 $g/100\ mL$ aq; i alc
tungstate(VI) dihydrate	$Na_2WO_4 \cdot 2H_2O$	329.86	3.245	anhyd 100; mp 695.6		88 $g/100\ mL^{0}$ aq; i alc
vanadate(V)	$NaVO_3$	121.93				s hot aq
Strontium	Sr	87.62	2.63^{20}	757	1366	d to $Sr(OH)_2$ in water
bromide	$SrBr_2$	247.43	4.216	643	2045(100)	100 $g/100\ mL^{20}$ aq
carbonate	$SrCO_3$	147.64	3.5	1100 d to SrO		i aq; s acids
chlorate	$Sr(ClO_3)_2$	254.54	3.152^{20}	d 120		167 $g/100\ mL^{20}$ aq
chloride	$SrCl_2$	158.52	3.052	874(1)	2058	52.9 $g/100\ mL^{20}$ aq
chromate(VI)	$SrCrO_4$	203.64	3.895^{15}			0.12^{20} aq; s HCl

TABLE 3.2 Physical Constants of Inorganic Compounds (*Continued*)

Name	Formula	Formula weight	Density	Melting point, °C	Boiling point, °C	Solubility (in 100 parts solvent)
Strontium						
fluoride	SrF_2	125.63	4.24	1477(2)	2460	0.011[20] aq; s hot HCl
hydrogen phosphate	$SrHPO_4$	183.60	3.544[15]			i aq; s acids
hydroxide	$Sr(OH)_2$	121.64	3.625	510(15)	$-H_2O$, 744	0.8[20] aq
iodate	$Sr(IO_3)_2$	437.43	5.045[15]			0.03[15] aq
iodide	SrI_2	341.42	4.42	538(10)	1773	178 g/100 mL[20] aq; s alc
lactate 3-water	$Sr(OOCCHOHCH_3)_2 \cdot 3H_2O$	319.81		anhyd 150		33 g/100 mL aq
nitrate	$Sr(NO_3)_2$	211.65	2.986	570	d 1100	69.5 g/100 mL[20] aq; sl s alc, acet
oxide	SrO	103.63	4.7	2430		0.69[20] aq
perchlorate	$Sr(ClO_4)_2$	286.52	3.00[25]			v s aq
peroxide	SrO_2	119.63	4.56	d 215		0.018[20] aq; d hot aq
sulfate	$SrSO_4$	183.70	3.96	1600		0.013[20] aq; sl s acid
sulfide	SrS	119.70	3.70[15]	>2000		sl s aq; s acid (dec)
Sulfinyl bromide	$SOBr_2$	207.88	2.67	−49.5	139.7	hyd aq (slow); misc bz, chl, CCl_4
chloride	$SOCl_2$	118.98	1.638[20]4	−104.5	75.8	hyd aq; misc bz, chl, CCl_4
fluoride	SOF_2	86.06	1.780[−100]	−110	−43.8	hyd aq; s bz, chl, eth
Sulfonyl chloride	SO_2Cl_2	134.98	1.6674[20]	−46	69.3	hyd aq; misc bz, eth, HOAc s aq
diamide	$SO_2(NH_2)_2$	96.11	1.807	91.5	d 250	mL gas/100 mL: 4 aq, 24 alc, 136 CCl_4, 210 toluene
fluoride	SO_2F_2	102.07	3.72 g/L	−135.8	−55.38	
Sulfur (gamma)	S	32.066(8)	1.92	106.8	444.72	i aq; 23 g/100 mL[0] CS_2; s alc, bz
(beta)	S_8	256.53	1.96[20]	115.21	444.72	23 g/100 mL[0] CS_2; s alc, bz
(alpha)	S_8	256.53	2.08[20]	115.21(10)	444.72	i aq; s organic solvents
(di-) decafluoride	S_2F_{10}	254.11	2.08[0]	−52.7	17	d fusion with KOH
(di-) dichloride	$ClSSCl$	135.03	1.688[15]	−77	138.1	hyd aq; s alc, bz, eth, CS_2, CCl_4
dichloride	SCl_2	102.97	1.622[15]	−78	59.5	hyd aq
dioxide	SO_2	64.07	2.716 g/L[20]	−75.47	−10.01	mL/100 mL: 3937[20] aq, 25 MeOH; s chl, eth
hexafluoride	SF_6	146.07	1.88[mp]	−50.8	subl 63.8	sl s aq; s alc, KOH

tetrafluoride	SF_4	108.07	1.919^{-73}	-121.0	-38	d aq viol; v s bz
trioxide (gamma)	SO_3	80.07	1.9225^{20}	16.86	43.4	v s aq (slow)
Sulfuryl, *see* Sulfinyl						
Tantalum	Ta	180.9479	16.65^{20}	2985(10)	5513(200)	s HF, fused alkali
(V) bromide	$TaBr_5$	580.49	4.989	265(15)	348.8	hyd aq; s abs alc, eth
carbide	TaC	192.96	13.9	4000(200)	d 5308	sl s HF
(*di-*) carbide	Ta_2C	373.91	13.9	3880		
(V) chloride	$TaCl_5$	358.24	3.68	218(2)	232.9(5)	hyd aq; s abs alc, inert org solvents
diboride	TaB_2	202.57	11.15	ca. 3100		s aq, eth, conc HNO_3
(V) fluoride	TaF_5	275.95	4.74^{20}	96.8	229.5	hyd aq; s eth
(V) iodide	TaI_5	815.47	5.80^{26}	496	543	sl s aqua regia
nitride	TaN	194.95	13.80	2800(50)		s HF; d fused $KHSO_4$ or KOH
(V) oxide	Ta_2O_5	441.90	8.2	1785(20)		
Technetium-98	Tc	97.9072	11.487	2250(50)	4567	s HNO_3, aq reg, conc H_2SO_4
(VI) fluoride	TcF_6	212.91		33.4	55.3	s HCl
(IV) oxide	TcO_2	130.91		subl 1000		s acid, alkali
(VII) oxide	Tc_2O_7	309.81	6.9	119.5	310.6	s aq
Tellurium	Te	127.60(3)	6.24^{20}	449.8	989.9	s HNO_3, KOH, conc H_2SO_4
(IV) bromide	$TeBr_4$	447.27	4.31^{15}	ca. 380	420 d	s HBr, eth, HOAc
(II) chloride	$TeCl_2$	198.51		208	328	disprop with eth, diox; s acid
(IV) chloride	$TeCl_4$	269.44	3.01	227.9	388	hyd aq; s HCl, abs alc, bz
(IV) fluoride	TeF_4	203.60		129.6	d > 194	d aq
(VI) fluoride	TeF_6	241.61	3.025^{-24}	-37.8	subl -38.9	hyd aq, KOH
(IV) iodide	TeI_4	635.29	5.05^{15}	280		hyd aq; s HI, alkali; sl s acet
(IV) oxide (tetrahedral)	TeO_2	159.60	5.76	732.6	subl 790	s HCl, HF, NaOH
Terbium	Tb	158.9254	8.253^{25}	1356	3230	s acids
chloride	$TbCl_3$	265.28	4.35^{0}	588	1550	v s aq
nitrate 6-water	$Tb(NO_3)_3 \cdot 6H_2O$	453.03		89.3		s aq
Thallium	Tl	204.383	11.85	303.5	1457	i aq; s HNO_3
(III) acetate sesquihydrate	$Tl(C_2H_3O_2)_3 \cdot 1.5H_2O$	408.53		182 d		
(I) bromide	TlBr	284.31	7.54	458(2)	820(5)	0.05^{20} aq; s alc
(I) carbonate	Tl_2CO_3	468.75	7.11	272		4.1 g/100 mL^{20} aq; i alc
(I) chloride	TlCl	239.85	7.004^{30}	430(1)	818(2)	0.33^{20} aq; i alc
(I) cyanide	TlCN	230.39	6.523	d		16.8 g/100 mL^{28} aq; s alc, acid

TABLE 3.2 Physical Constants of Inorganic Compounds (*Continued*)

Name	Formula	Formula weight	Density	Melting point, °C	Boiling point, °C	Solubility (in 100 parts solvent)
Thallium						
(I) ethoxide	TlOC$_2$H$_5$	249.43	3.493[20]	−3	d 130	s eth; sl s alc
(I) fluoride	TlF	223.39	8.23[4]	826	826	78.6%[15] aq
(III) fluoride	TlF$_3$	261.39	8.36[25]	550 d		d aq
(I) iodide (rhombic)	TlI	331.31	7.29	442	823	i aq, alc; s KI
(I) nitrate	TlNO$_3$	266.40	5.556	206	430; d 450	9.55 g/100 mL[20] aq; i alc
(I) oxide	Tl$_2$O	424.78	9.52[16]	ca. 300	1080	v s aq; s acid, alc
(III) oxide (hexagonal)	Tl$_2$O$_3$	456.78	10.19[22]	717	−O$_2$, 875	i aq; d by HCl, H$_2$SO$_4$
(I) selenate(VI)	Tl$_2$SeO$_4$	551.74	6.875	>400		2.8 g/100 mL[20] aq; i alc, eth
(I) selenide	Tl$_2$Se	487.74	9.05	339(1)		i aq, acid
(I) sulfate	Tl$_2$SO$_4$	504.85	6.77	632	d	4.87 g/100 mL[20] aq
(I) sulfide	Tl$_2$S	440.85	8.39	448.5	d	0.02[20] aq; s mineral acids
Thiocarbonyl chloride	S=CCl$_2$	114.98	1.509[15]		73.5	d aq; s eth
Thiocyanogen	(SCN)$_2$	116.16		ca. −2		d aq; s alc, CS$_2$, eth
Thionyl-, *see* Sulfinyl-						
Thiophosphoryl tribromide	PSBr$_3$	302.78	2.85[17]	38.0(2)	209(3) d	s aq, eth, CS$_2$
trichloride (alpha)	PSCl$_3$	169.41	1.635[25]	−40.8	125	hyd aq; s bz, chl, CS$_2$
trifluoride	PSF$_3$	120.03		−148.8	−52.2.	
Thiosulfinyl difluoride	S=SF$_2$	102.13		−165	−10.6	hyd aq
Thorium	Th	232.0381	11.724[25]	1842(20)	4820(35)	s acids
chloride	ThCl$_4$	373.88	4.60	770	922	s aq, alc
fluoride	ThF$_4$	308.03	5.71	1110		s acids
iodide	ThI$_4$	739.69	6.00	566	837	hyd aq
nitrate	Th(NO$_3$)$_4$	480.06		d > 216	d 630, ThO$_2$	191 g/100 mL[20] aq; v s alc
oxide	ThO$_2$	264.05	9.86	3390	4400	s hot H$_2$SO$_4$
sulfate 9-water	Th(SO$_4$)$_2$·9H$_2$O	586.31	2.77	anhyd 400		1.57 g/100 mL[25] aq
tetracyanoplatinate(II) 16-water	Th[Pt(CN)$_4$]$_2$·16H$_2$O	1118.6				sl s aq
Thulium	Tm	168.9342	9.318[25]	1545	1950	s acids
chloride	TmCl$_3$	275.29		824	1490	s aq, alc
fluoride	TmF$_3$	225.93	7.971	1158	2230	s H$_2$SO$_4$

Name	Formula					Solubility
Tin (tetragonal)	Sn	118.710(7)	7.31	231.89	2507	s conc HCl, hot H_2SO_4
(II) acetate	$Sn(C_2H_3O_2)_2$	236.79	2.31	182.5	240	d aq; s dilute HCl
(II) bromide	$SnBr_2$	278.53	5.12	215	623	85 g/100 mL^0 aq; s alc, eth
(IV) bromide	$SnBr_4$	438.36	3.340^{35}	30	202	v a (hyd) aq; s acet, alc
(II) chloride	$SnCl_2$	189.61	3.95^{25}	246.9(1)	635(12)	84 g/100 mL^0 aq; s acet, alc, eth
(IV) chloride	$SnCl_4$	260.53	2.234^{20}	−33.3	115	s aq (hyd), alc, acet, bz, eth
(IV) chromate(VI)	$Sn(CrO_4)_2$	350.72		d heat		s aq
(II) fluoride	SnF_2	156.70	4.57^{25}	213		30% aq
(IV) fluoride	SnF_4	194.70	4.78^{19}		subl 705	hyd aq
hexafluorozirconate	$Sn[ZrF_6]$	323.92	4.21			s aq
(II) iodide	SnI_2	372.54	5.285	320	340	0.98^{20} aq (d); s bz, chl, alk Cl^- or I^-
(IV) iodide	SnI_4	626.38	4.56^{20}	144.5	340	hyd aq; s alc, bz, chl, eth, CCl_4, CS_2
(II) oxalate	SnC_2O_4	206.72	3.56	to SnO_2, 300		s dilute HCl
(II) oxide	SnO	134.70	6.446^0	1630		s acids, conc KOH
(IV) oxide	SnO_2	150.70	6.95	861	subl 1900	s hot conc KOH (slow)
(II) selenide	SnSe	197.66	6.179^0	to SnO_2, 378		s aqua regia, alkali sulfides
(II) sulfate	$SnSO_4$	214.77				18.9 g/100 mL^{20} aq; s dilute H_2SO_4
(II) sulfide	SnS	150.77	5.08	881	1210	i aq; s conc HCl, hot conc H_2SO_4
(IV) sulfide	SnS_2	182.83	4.5	765		s aq reg, alkali hydroxides & sulfides
(II) telluride	SnTe	246.29	6.445	806	d	i aq
Titanium (hexagonal)	Ti	47.88(3)	4.507^{25}	1660(10)	3278	s hot acid, HF
(III) bromide	$TiBr_3$	387.63	4.24		subl 794	
(IV) bromide	$TiBr_4$	367.56	3.25^{20}	28.2	233.45	hyd aq; 187 g/100 mL abs alc, s abs eth
(II) chloride	$TiCl_2$	118.81	3.13	1035	subl 1026	d aq; s alc
(III) chloride	$TiCl_3$	154.27	2.71	subl 831 vac	d 5000	s aq (heat evolved), alc
(IV) chloride	$TiCl_4$	189.73	1.70^{20}	−24.10	136.4	s cold aq, alc
dihydride	TiH_2	49.92	3.752	d 450		
(IV) fluoride	TiF_4	123.90	2.798^{20}	>400	subl 285.5	s aq (slow hyd); s alc, pyr
(IV) iodide	TiI_4	555.50	4.40	155	378(1)	s dry nonpolar solvents
(IV) isopropoxide	$Ti[OCH(CH_3)_2]_4$	284.26	0.9711^{20}	ca. 20	220	
(II) oxide	TiO	63.90	4.888	1750(30)	3660	s H_2SO_4

TABLE 3.2 Physical Constants of Inorganic Compounds (*Continued*)

Name	Formula	Formula weight	Density	Melting point, °C	Boiling point, °C	Solubility (in 100 parts solvent)
Titanium						
(III) oxide	Ti_2O_3	143.80	4.486	1839		s H_2SO_4
(IV) oxide (rutile)	TiO_2	79.90	4.23	1857(20)		s HF, hot conc H_2SO_4
oxide sulfate	$TiOSO_4$	159.97				d aq
(III) sulfate	$Ti_2(SO_4)_3$	384.00				s dilute HCl, dilute H_2SO_4
Tungsten	W	183.85(3)	19.25^{25}	3407(20)	5900	s HNO_3 + HF, fusion NaOH + $NaNO_3$
(V) bromide	WBr_5	583.40		286(10)	360.4	hyd aq; s chl, eth
(VI) bromide	WBr_6	663.30	6.9	279	subl 327	hyd aq; s eth CS_2
(V) chloride	WCl_5	361.12	3.875	253	288.3	hyd aq
(VI) chloride	WCl_6	396.57	2.721^{mp} (lq)	282(2)	343.1	hyd aq; s CS_2, CCl_4
dichloride dioxide	WCl_2O_2	286.76		subl 260	d 369	hyd aq; s HCl
(VI) fluoride	WF_6	297.86	3.441^{15}	2.3	17.5	hyd aq; s anhyd HF
(IV) oxide	WO_2	215.85	12.11	1550	d 1724	s acids, KOH
(VI) oxide	WO_3	231.86	7.16	1472	1837	i aq; s hot alkali
(IV) sulfide	WS_2	247.98	7.5^{10}	d 1250		s HNO_3 + HF
tetrachloride oxide	WCl_4O	341.66	3.22^{mp} (lq)	211	220	hyd aq
tetrafluoride oxide	WF_4O	275.85		106	187	
Uranium	U	238.0289	19.05^{25}	1132.3	3927	s acid
(IV) bromide	UBr_4	557.67	5.55	519	777	v s aq
(III) chloride	UCl_3	344.44	5.51	842	1657	v s aq
(IV) chloride	UCl_4	379.90	4.725^{25}	590	790(1)	v s aq (d); s polar organic solvents
(V) chloride	UCl_5	415.30	3.56	327	527	d aq; s CS_2
(VI) chloride	UCl_6	450.75		179	392	hyd aq; s chl
(IV) fluoride	UF_4	314.07	6.70	1036	1457	s conc acids (d); alkali hydroxide (d)
(VI) fluoride	UF_6	352.07	$5.09^{20.7}$	64.8	subl 56.5	hyd aq; s chl, CCl_4
(III) hydride	UH_3	241.05	10.95			i aq
(IV) iodide	UI_4	745.65	5.6^{15}	506 (I_2 atm)	757	s aq
(IV) oxide (pitchblende)	UO_2	270.07	10.97	2865		s conc HNO_3
(VI) oxide	UO_3	286.07	7.29	d 1300		i aq; s HCl, HNO_3
peroxide 2-water	$UO_4 \cdot 2H_2O$	338.06		d 90–195 to U_2O_7 (slow)	d > 200 to UO_3	d by HCl

Name	Formula	Mol wt	Density	MP	BP	Solubility
(tri-) octaoxide	U_3O_8	842.09	8.30	d 1300 to UO_2		s HNO_3
Uranyl(VI) acetate 2-water	$UO_2(C_2H_3O_2)_2 \cdot 2H_2O$	424.15	2.893[15]	anhyd 110	d 275	7.7 g/100 mL[15] aq; sl s alc
chloride	UO_2Cl_2	340.98	5.43	577 d		320 g/100 mL[18] aq; s acet, alc
fluoride	UO_2F_2	308.03	6.37	d 300		v s aq
nitrate 6-water	$UO_2(NO_3)_2 \cdot 6H_2O$	502.13	2.807[13]	60.2	d 100	155 g/100 mL[20] aq; v s alc, eth
sulfate 3-water	$UO_2SO_4 \cdot 3H_2O$	420.14	3.28[16]	d 100		g/100 mL: 21 aq, 4 alc
Vanadium	V	50.9415	6.11[19]	1917(20)	3421(5)	s HF, HNO_3, hot H_2SO_4
(IV) chloride	VCl_4	192.75	1.82	−28(2)	148.5	hyd aq; s nonpolar solvents
dichloride oxide	VCl_2O	137.86	2.88[13]	disprop 384		hyd (slow) aq; s abs alc, HOAc
(III) fluoride	VF_3	107.95	3.363	ca. 1406	subl 800	i almost all organic solvents
(IV) fluoride	VF_4	126.95	3.15	subl 120 (vac) & disprop		s aq, acet, HOAc
(V) fluoride	VF_5	145.95	2.502[20]	19.5	47.9	hyd aq; v s anhyd HF, acet, alc, chl
(II) oxide	VO	66.94	5.55	1790(10)		s HCl
(III) oxide	V_2O_3	149.90	4.87	1940		sl s acids
(IV) oxide	VO_2	82.94	4.34	1360		s acids, alkalis
(IV) oxide, dimeric	V_2O_4	165.88		1545(15)		s acids
(V) oxide	V_2O_5	181.90	3.35	670(20)	d 1690	0.07 aq; s conc acids, alkalis
(IV) oxide sulfate	$VOSO_4$	163.00				s aq
(III) sulfate	$V_2(SO_4)_3$	390.10		d 410 (vac)		s (slow) aq, HNO_3
(III) sulfide	V_2S_3	198.10	4.72	d 600		s hot acids, alkali sulfides
Xenon	Xe	131.29(3)	5.897 g/L[0]	−111.8	−108.10	10.8 mL/100 mL[20] aq
difluoride	XeF_2	169.29	4.32	129.0	subl 114	2.5 g/100 mL[0] aq
hexafluoride	XeF_6	245.30	3.411[25]	49.5	75.6	hyd aq
tetrafluoride	XeF_4	207.30	4.04[25]	117.1	subl 116	hyd aq; s F_3CCOOH
trioxide	XeO_3	179.30	4.55	explodes 40		s aq giving xenic acid
Ytterbium	Yb	173.04(3)	6.973[25]	824	1196	s acids
(II) chloride	$YbCl_2$	243.95	5.08	702	1930	s aq
(III) chloride 6-water	$YbCl_3 \cdot 6H_2O$	387.49	2.575	anhyd 180; mp 865		v s aq
(III) fluoride	YbF_3	230.04	8.168	1157	2230	s H_2SO_4
(III) nitrate 4-water	$Yb(NO_3)_3 \cdot 4H_2O$	431.12	9.18			s aq
(III) oxide	Yb_2O_3	394.08				s dilute acids

TABLE 3.2 Physical Constants of Inorganic Compounds (*Continued*)

Name	Formula	Formula weight	Density	Melting point, °C	Boiling point, °C	Solubility (in 100 parts solvent)
Ytterbium (III) sulfate 8-water	$Yb_2(SO_4)_3 \cdot 8H_2O$	778.39	3.286			34.8 g/100 mL^{20} aq
Yttrium	Y	88.9059	4.472	1509	3300(40)	s hot water (d)
chloride	YCl_3	195.26	2.67	721	1510	79 g/100 mL^{20} aq; s alc
fluoride	YF_3	145.90	5.069	1152	2230	s conc acids (d)
nitrate 6-water	$Y(NO_3)_3 \cdot 6H_2O$	383.01	2.68	$-3H_2O$, 100		171 g/100 mL^{20} aq
oxide	Y_2O_3	225.81	5.03	2420	4300	s acids
sulfate 8-water	$Y_2(SO_4)_3 \cdot 8H_2O$	610.12	2.558	anhyd 400	d $>$ 1000	9.6 g/100 mL^{20} aq
Zinc	Zn	65.39(2)	7.14^{25}	419.58	910(2)	i aq; s acids, alkalis (slow)
acetate dihydrate	$Zn(C_2H_3O_2)_2 \cdot 2H_2O$	219.49	1.735	237		g/100 mL: 41.6^{20} aq, 3.3 alc
arsenate(III)(1−)	$Zn(AsO_2)_2$	279.23				s acids
bromide	$ZnBr_2$	225.20	4.22	394	697 d	g/100 mL: 471^{25} aq, 200 alc; s KOH, eth
carbonate	$ZnCO_3$	125.40	4.398	$-CO_2$, 300		0.02^{25} aq; s acids, KOH, NH$_4$ salts
chloride	$ZnCl_2$	136.30	2.907^{25}	ca. 290	732	g/100 ml: 395^{20} aq, 77 alc, 50 glyc; v s acet
chromate(VI)	$ZnCrO_4$	181.39	3.40			s acids
cyanide	$Zn(CN)_2$	117.43	1.852	d 800		0.058^{18} aq; s acids, KCN, KOH
fluoride	ZnF_2	103.39	5.00^{25}	872	1500	s HNO$_3$, HCl, NH$_4$OH
hexafluorosilicate 6-water	$Zn[SiF_6] \cdot 6H_2O$	315.56	2.104	d 100		v s aq
iodate	$Zn(IO_3)_2$	415.20	5.063^{25}	d		0.87^{20} aq; s HNO$_3$, KOH
iodide	ZnI_2	319.20	4.7364^{25}	ca. 446	ca. 625 d	g/100 mL: 332^{20} aq, 50 glyc; v s alc
nitrate 6-water	$Zn(NO_3)_2 \cdot 6H_2O$	297.49	2.065^{14}	36.4	$-6H_2O$, 131	146 g/100 mL^0 aq; v s alc
oxide	ZnO	81.39	5.607^{20}	d $>$ 150	subl 1800	i aq; s acids, KOH, NH$_4$OH
peroxide	ZnO_2	97.39	3.00	anhyd 120		d (slow) aq; s dilute acids (d)
1,4-phenolsulfonate 8-water	$Zn[C_6H_4(OH)SO_3]_2 \cdot 8H_2O$	555.84		900		g/100 mL: 63 aq, 56 alc
phosphate(V)	$Zn_3(PO_4)_2$	386.12	3.998^{15}	$>$ 420		s acids, NH$_4$OH
phosphide	Zn_3P_2	258.12	4.55		1100	d aq, HCl (viol); s bz, CS$_2$
propionate	$Zn(C_3H_5O_2)_2$	211.52				32%^{15} aq; 2.8%^{15} alc

				mp	bp	
selenide	ZnSe	144.35	5.42^{15}	1526		d dilute HNO_3
silicate(2−)	Zn_2SiO_4	222.87	4.103	1512		i aq or dilute acids
stearate	$Zn(C_{18}H_{35}O_2)_2$	632.33		ca. 120		d dil acids; s bz; i aq, alc, eth
sulfate	$ZnSO_4$	161.46	3.54	1200		$53.8\%^{20}$ aq
sulfate 7-water	$ZnSO_4 \cdot 7H_2O$	287.56	1.957	anhyd 280	d > 500	g/100 mL: 167 aq, 40 glyc; i alc
sulfide (wirzite)	ZnS	97.46	4.087	1722		i aq; s dilute mineral acid
telluride	ZnTe	192.99	6.34_4^{15}	1239		d (slow) aq or dilute HCl
thiocyanate	$Zn(SCN)_2$	181.56				0.14^{18} aq; s alc
Zirconium	Zr	91.224(2)	6.511^{25}	1852	3577	s aqua regia, HF, hot H_3PO_4, fusion with KOH, KNO_3
(IV) bromide	$ZrBr_4$	410.86		450(1)	subl 357	sl s conc H_2SO_4
carbide	ZrC	103.23	6.73	3532(125)	5100	d aq
(II) chloride	$ZrCl_2$	162.13	3.6^{18}	727	1292	hyd aq to $ZrCl_2O$; s alc, eth
(IV) chloride	$ZrCl_4$	233.05	2.803^{15}	437 (25 atm)	subl 334	v s aq, alc
diboride	ZrB_2	112.84	6.085	3050	d 4193	i aq
dichloride oxide 8-water	$ZrCl_2O \cdot 8H_2O$	322.25		anhyd 210		1.32 g/100 mL^{20} aq
dihydride	ZrH_2	93.24	5.61			s mineral acids
(IV) fluoride	ZrF_4	167.22	4.6^{16}	932	subl > 600	s aq (d), eth
(IV) hydroxide	$Zr(OH)_4$	159.25	3.25	to ZrO, 500		v s aq; s alc
(IV) iodide	ZrI_4	598.86		499 (sealed tube)	subl 432.5	s hot H_2SO_4, HF (slow)
(IV) nitrate 5-water	$Zr(NO_3)_4 \cdot 5H_2O$	429.32				very inert
(IV) oxide	ZrO_2	123.22	5.85	2678(2)	4300	
(IV) silicate(4−)	$ZrSiO_4$	183.31	4.56	d > 1538 to ZrO_2 + SiO_2		
sulfate 4-water	$Zr(SO_4)_2 \cdot 4H_2O$	355.41	3.22^{16}	anhyd 380		52.5 g/100 g aqueous solution

TABLE 3.3 Synonyms and Mineral Names

Acanthite, *see* Silver sulfide
Alabandite, *see* Manganese sulfide
Alamosite, *see* Lead(II) silicate(2−)
Altaite, *see* Lead telluride
Alumina, *see* Aluminum oxide
Alundum, *see* Aluminum oxide
Alunogenite, *see* Aluminum sulfate 18-water
Amphibole, *see* Magnesium silicate(2−)
Andalusite, *see* Aluminum silicon oxide (1/1)
Anglesite, *see* Lead sulfate
Anhydrite, *see* Calcium sulfate
Anhydrone, *see* Magnesium perchlorate
Aragonite, *see* Calcium carbonate
Arcanite, *see* Potassium sulfate
Argentite, *see* Silver sulfide
Argol, *see* Potassium hydrogen tartrate
Arkansite, *see* Titanium(IV) oxide
Arsenolite, *see* Arsenic(III) oxide dimer
Arsine, *see* Arsenic hydride
Auric and aurous, *see* under Gold
Azoimide, *see* Hydrogen azide
Azurite, *see* Copper(II) carbonate—dihydroxide (2/1)
Baddeleyite, *see* Zirconium(IV) oxide
Baking soda, *see* Sodium hydrogen carbonate
Barite (barytes), *see* Barium sulfate
Bieberite, *see* Cobalt sulfate 7-water
Bismuthine, *see* Bismuth hydride
Bismuthinite, *see* Bismuth sulfide
Bleaching powder, *see* Calcium hydrochlorite
Bleaching solution, *see* Sodium hydrochlorite
Blue copperas, *see* Copper(II) sulfate 7-water
Boracic acid, *see* Hydrogen borate
Borax, *see* Sodium tetraborate 10-water
Braunite, *see* Manganese(III) oxide
Brimstone, *see* Sulfur
Bromellite, *see* Beryllium oxide
Bromosulfonic acid, *see* Hydrogen bromosulfate
Bromyrite, *see* Silver bromide
Brookite, *see* Titanium(IV) oxide
Brucite, *see* Magnesium hydroxide
Bunsenite, *see* Nickel oxide
Cacodylate, *see* Sodium dimethylarsonate 3-water
Caesium, *see* under Cesium
Calamine, *see* Zinc carbonate
Calcia, *see* Calcium oxide
Calcite, *see* Calcium carbonate
Calomel, *see* Mercury(I) chloride

Caro's acid, *see* Hydrogen peroxosulfate
Cassiopeium, *see* Lutetium
Cassiterite, *see* Tin(IV) oxide
Caustic potash, *see* Potassium hydroxide
Caustic soda, *see* Sodium hydroxide
Celestite, *see* Strontium sulfate
Cementite, *see* *tri*-Iron carbide
Cerargyrite, *see* Silver chloride
Cerussite, *see* Lead carbonate
Chalcanthite, *see* Copper(II) sulfate 5-water
Chalcocite, *see* Copper(I) sulfide
Chalk, *see* Calcium carbonate
Chile nitre, *see* Sodium nitrate
Chile saltpeter, *see* Sodium nitrate
Chloromagnesite, *see* Magnesium chloride
Chlorosulfonic acid, *see* Hydrogen chlorosulfate
Cinnabar, *see* Mercury(II) sulfide
Claudetite, *see* Arsenic(III) oxide dimer
Clausthalite, *see* Lead selenide
Clinoenstatite, *see* Magnesium silicate(2−)
Columbium, *see* under Niobium
Corrosive sublimate, *see* Mercury(II) chloride
Corundum, *see* Aluminum oxide
Cotunite, *see* Lead chloride
Covellite, *see* Copper(II) sulfide
Cream of tartar, *see* Potassium hydrogen tartrate
Crocoite, *see* Lead chromate(VI)(2−)
Cryolite, *see* Sodium hexafluoroaluminate
Cryptohalite, *see* Ammonium hexafluorosilicate
Cupric and cuprous, *see* under Copper
Cuprite, *see* Copper(I) oxide
Dakin's solution, *see* Sodium hypochlorite
Dehydrite, *see* Magnesium perchlorate
Dental gas, *see* Nitrogen(I) oxide
Diamond, *see* Carbon
Dichlorodisulfane, *see* *di*-Sulfur dichloride
Diuretic salt, *see* Potassium acetate
Dolomite, *see* Calcium magnesium carbonate (1/1)
Dry ice, *see* Carbon dioxide (solid)
Enstatite, *see* Magnesium silicate(2−)
Epsom salts, *see* Magnesium sulfate 7-water
Epsomite, *see* Magnesium sulfate 7-water
Eriochalcite, *see* Copper(II) chloride
Fayalite, *see* Iron(II) silicate(4−)
Ferric and ferrous, *see* under Iron
Fluorine oxide, *see* Oxygen difluoride
Fluoristan, *see* Tin(II) fluoride

TABLE 3.3 Synonyms and Mineral Names (*Continued*)

Fluorite, *see* Calcium fluoride

Fluorosulfonic acid, *see* Hydrogen fluorosulfate

Fluorspar, *see* Calcium fluoride

Forsterite, *see* Magnesium silicate(4−)

Freezing salt, *see* Sodium chloride

Fulminating mercury, *see* Mercury fulminate

Galena, *see* Lead sulfite

Glauber's salt, *see* Sodium sulfate 10-water

Goethite, *see* Iron(II) hydroxide oxide

Goslarite, *see* Zinc sulfate 7-water

Graham's salt, *see* Sodium phosphate(1−)

Graphite, *see* Carbon

Greenockite, *see* Cadmium sulfide

Gruenerite, *see* Iron(II) silicate(2−)

Guanajuatite, *see* Bismuth selenide

Gypsum, *see* Calcium sulfate 2-water

Halite, *see* Sodium chloride

Hausmannite, *see* Manganese(II,IV) oxide

Heavy hydrogen, *see* Hydrogen[^2H] or name followed by -*d*

Heavy water, *see* Hydrogen[^2H] oxide

Heazlewoodite, *see tri*-Nickel disulfide

Hematite, *see* Iron(III) oxide

Hermannite, *see* Manganese silicate

Hessite, *see* Silver telluride

Hieratite, *see* Potassium hexafluorosilicate

Hydroazoic acid, *see* Hydrogen azide

Hydrophilite, *see* Calcium chloride

Hydrosulfite, *see* Sodium dithionate(III)

Hypo (photographic), *see* Sodium thiosulfate 5-water

Hypophosphite, *see* under Phosphinate

Ice, *see* Hydrogen oxide (solid)

Iceland spar, *see* Calcium carbonate

Iodyrite, *see* Silver iodide

Jeweler's borax, *see* Sodium tetraborate 10-water

Jeweler's rouge, *see* Iron(III) oxide

Kalinite, *see* Aluminum potassium bis(sulfate)

Kernite, *see* Sodium tetraborate

Kyanite, *see* Aluminum silicon oxide (1/1)

Laughing gas, *see* Nitrogen(I) oxide

Lautarite, *see* Calcium iodate

Lawrencite, *see* Iron(II) chloride

Lechatelierite, *see* Silicon dioxide

Lime, *see* Calcium oxide

Litharge, *see* Lead(II) oxide

Lithium aluminum hydride, *see* Lithium tetrahydridoaluminate

Lodestone, *see* Iron(II,III) oxide

Lunar caustic, *see* Silver nitrate

Lye, *see* Sodium hydroxide

Magnesia, *see* Magnesium oxide

Magnesite, *see* Magnesium carbonate

Magnetite, *see* Iron(II,III) oxide

Malachite, *see* Copper carbonate dihydroxide

Manganosite, *see* Manganese(II) oxide

Marcasite, *see* Iron disulfide

Marshite, *see* Copper(I) iodide

Mascagnite, *see* Ammonium sulfate

Massicotite, *see* Lead oxide

Mercuric and mercurous, *see* under Mercury

Metacinnabar, *see* Mercury(II) sulfide

Millerite, *see* Nickel sulfide

Mirabilite, *see* Sodium sulfate

Mohr's salt, *see* Ammonium iron(II) sulfate 6-water

Moissanite, *see* Silicon carbide

Molybdenite, see Molybdenum disulfide

Molybdite, *see* Molybdenum(VI) oxide

Molysite, *see* Iron(III) chloride

Montroydite, *see* Mercury(II) oxide

Morenosite, *see* Nickel sulfate 7-water

Mosaic gold, *see* Tin disulfide

Muriatic acid, *see* Hydrogen chloride, aqueous solutions

Nantokite, *see* Copper(I) chloride

Natron, *see* Sodium carbonate

Naumannite, *see* Silver selenide

Neutral verdigris, *see* Copper(II) acetate

Nitre (niter), *see* Potassium nitrate

Nitric oxide, *see* Nitrogen(II) oxide

Nitrobarite, *see* Barium nitrate

Nitromagnesite, *see* Magnesium nitrate 6-water

Nitroprusside, *see* Sodium pentacyanonitrosylferrate(II) 2-water

Oldhamite, *see* Calcium sulfide

Opal, *see* Silicon dioxide

Orpiment, *see* Arsenic trisulfide

Oxygen powder, *see* Sodium peroxide

Paris green, *see* Copper acetate arsenate(III) (1/3)

Pawellite, *see* Calcium molybdate(VI) (2−)

Pearl ash, *see* Potassium carbonate

Perborax, *see* Sodium peroxoborate

Periclase, *see* Magnesium oxide

Persulfate, *see* Peroxodisulfate

Phosgene, *see* Carbonyl chloride

TABLE 3.3 Synonyms and Mineral Names (*Continued*)

Phosphine, *see* Hydrogen phosphide

Pickling acid, *see* Hydrogen sulfate

Pitchblende, *see* Uranium(IV) oxide

Plaster of Paris, *see* Calcium sulfate hemihydrate

Plattnerite, *see* Lead(IV) oxide

Polianite, *see* Manganese(IV) oxide

Polishing powder, *see* Silicon dioxide

Potash, *see* Potassium carbonate

Potassium acid phthalate, *see* Potassium hydrogen phthalate

Prussic acid, *see* Hydrogen cyanide

Pyrite, *see* Iron disulfide

Pyrochroite, *see* Manganese(II) hydroxide

Pyrohytpophosphite, *see* diphosphate(IV)

Pyrolusite, *see* Manganese(IV) oxide

Pyrophanite, *see* Manganese titanate(IV)(2−)

Pyrophosphate, *see* Diphosphate(V)

Pyrosulfuric acid, *see* Hydrogen disulfate

Quartz, *see* Silicon dioxide

Quicksilver, *see* Mercury

Realgar, *see* di-Arsenic disulfide

Red lead, *see* Lead(II,IV) oxide

Rhodochrosite, *see* Manganese carbonate

Rhodonite, *see* Manganese silicate(1−)

Rochelle salt, *see* Potassium sodium tartrate 4-water

Rock crystal, *see* Silicon dioxide

Rutile, *see* Titanium(IV) oxide

Sal soda, *see* Sodium carbonate 10-water

Saltpeter, *see* Potassium nitrate

Scacchite, *see* Manganese chloride

Scheelite, *see* Calcium tungstate(VI)(2−)

Sellaite, *see* Magnesium fluoride

Senarmontite, *see* Antimony(III) oxide

Siderite, *see* Iron(II) carbonate

Siderotil, *see* Iron(II) sulfate 5-water

Silica, *see* Silicon dioxide

Silicotungstic acid, *see* Silicon oxide — tungsten oxide — water (1/12/26)

Sillimanite, *see* Aluminum silicon oxide (1/1)

Smithsonite, *see* Zinc carbonate

Soda ash, *see* Sodium carbonate

Spelter, *see* Zinc metal

Sphalerite, *see* Zinc sulfide

Spherocobaltite, *see* Cobalt(II) carbonate

Spinel, *see* Magnesium aluminate(2−)

Stannic and stannous, *see* under Tin

Stibine, *see* Antimony hydride

Stibnite, *see* Antimony(III) sulfide

Stolzite, *see* Lead tungstate(VI)(2−)

Strengite, *see* Iron(III) phosphate

Strontianite, *see* Strontium carbonate

Sugar of lead, *see* Lead acetate

Sulfamate, *see* Amidosulfate

Sulphate, *see* Sulfate

Sulfurated lime, *see* Calcium sulfide

Sulfuretted hydrogen, *see* Hydrogen sulfide

Sulphur, *see* Sulfur

Sulfuryl, *see* Sulfonyl

Sycoporite, *see* Cobalt sulfide

Sylvite, *see* Potassium chloride

Szmikite, *see* Manganese(II) sulfate hydrate

Tarapacaite, *see* Potassium chromate(VI)

Tellurite, *see* Tellurium dioxide

Tenorite, *see* Copper(II) oxide

Tephroite, *see* Manganese silicate(1−)

Thenardite, *see* Sodium sulfate

Thionyl, *see* Sulfinyl

Thorianite, *see* Thorium dioxide

Topaz, *see* Aluminum hexafluorosilicate

Tridymite, *see* Silicon dioxide

Troilite, *see* Iron(II) sulfide

Trona, *see* Sodium carbonate — hydrogen carbonate dihydrate

Tschermigite, *see* Aluminum ammonium bis(sulfate)

Tungstenite, *see* Tungsten disulfide

Tungstite, *see* Hydrogen tungstate

Uraninite, *see* Uranium(IV) oxide

Valentinite, *see* Antimony(III) oxide

Verdigris, *see* Copper acetate hydrate

Vermillion, *see* Mercury(II) sulfide

Villiaumite, *see* Sodium fluoride

Vitamin B_3, *see* Calcium (+)pantothenate

Washing soda, *see* Sodium carbonate 10-water

Whitlockite, *see* Calcium phosphate

Willemite, *see* Zinc silicate(4−)

Wolfram, *see* Tungsten

Wuestite, *see* Iron(II) oxide

Wulfenite, *see* Lead molybdate(VI)(2−)

Wurtzite, *see* Zinc sulfide

Zincite, *see* Zinc oxide

Zincosite, *see* Zinc sulfate

Zincspar, *see* Zinc carbonate

Zirconia, *see* Zirconium oxide

SECTION 4
PROPERTIES OF ATOMS, RADICALS, AND BONDS

4.1 ELEMENTS

The electronic configuration for an element's ground state (Table 4.1) is a shorthand representation giving the number of electrons (superscript) found in each of the allowed sublevels (s, p, d, f) above a noble gas core (indicated by brackets).

TABLE 4.1 Electronic Configuration and Properties of the Elements

Name	Symbol	At. no.	Electronic configuration	Thermal conductivity, $W \cdot (m \cdot K)^{-1}$ at 25°C	Electrical resistivity, $\mu\Omega \cdot cm$ at 20°C	Coefficient of linear thermal expansion, $m \cdot m^{-1} (\times 10^6)$
Actinium	Ac	89	$[Rn]6d^27s$	12		
Aluminum	Al	13	$[Ne]3s^23p$	234.3	2.6548 (20°C)	
Americium	Am	95	$[Rn]5f^77s^2$	10		
Antimony	Sb	51	$[Kr]4d^{10}5s^25p^3$	24.4	41.7 (20°C)	
Argon	Ar	18	$[Ne]3s^23p^6$	0.017 72		
Arsenic	As	33	$[Ar]3d^{10}4s^24p^3$	50.2	33.3 (20°C)	
Astatine	At	85	$[Xe]4f^{14}5d^{10}6s^26p^5$	1.7		
Barium	Ba	56	$[Xe]6s^2$	18.4	50.0 (20°C)	18
Berkelium	Bk	97	$[Rn]5f^86d7s^2$	10		
Beryllium	Be	4	$[He]2s^2$	190	4.46 (20°C)	16.5
Bismuth	Bi	83	$[Xe]4f^{14}5d^{10}6s^26p^3$	7.92	129 (20°C)	
Boron	B	5	$[He]2s^22p$	27.4	1.5×10^{12}	5–7
Bromine	Br	35	$[Ar]3d^{10}4s^24p^5$	0.122	7.8×10^{18}	
Cadmium	Cd	48	$[Kr]4d^{10}5s^2$	96.6	7.27 (22°C)	31.3
Calcium	Ca	20	$[Ar]4s^2$	201	3.5 (20°C)	
Californium	Cf	98	$[Rn]5f^{10}7s^2$			
Carbon	C	6	$[He]2s^22p^2$			
(amorphous)				1.59	0.8(1)	
(diamond)				900–2320	1375	
(graphite)				119–165	73	
Cerium	Ce	58	$[Xe]4f^55d6s^2$	11.3		
Cesium	Cs	55	$[Xe]6s$	35.9	19 (0°C)	
Chlorine	Cl	17	$[Ne]3s^23p^5$	0.008 9	$>10^9$	
Chromium	Cr	24	$[Ar]3d^54s$	93.9	12.7	62
Cobalt	Co	27	$[Ar]3d^74s^2$	100	6.24	
Copper	Cu	29	$[Ar]3d^{10}4s$	394	1.673	16.5
Curium	Cm	96	$[Rn]5f^76d7s^2$			
Dysprosium	Dy	66	$[Xe]4f^{10}6s^2$	10.7	89	
Einsteinium	Es	99	$[Rn]5f^{11}7s^2$			
Erbium	Er	68	$[Xe]4f^{12}6s^2$	14.5	86	
Europium	Eu	63	$[Xe]4f^76s^2$	13.9	81	
Fermium	Fm	100	$[Rn]5f^{12}7s^2$			

Element	Symbol	No.	Configuration			
Fluorine	F	9	[He]$2s^2 2p^5$	0.027 7		
Francium	Fr	87	[Rn]$7s$		126	120
Gadolinium	Gd	64	[Xe]$4f^7 5d 6s^2$	10.5		
Gallium	Ga	31	[Ar]$3d^{10} 4s^2 4p$	29.4(lq)	25.795(30°C)	
Germanium	Ge	32	[Ar]$3d^{10} 4s^2 4p^2$	60.2	53 000	6.0
Gold	Au	79	[Xe]$4f^{14} 5d^{10} 6s$	318	2.05 (0°C)	14.2
Hafnium	Hf	72	[Xe]$4f^{14} 5d^2 6s^2$	23.0	29.6 (0°C)	
Helium	He	2	$1s^2$	0.151 3		
Holmium	Ho	67	[Xe]$4f^{11} 6s^2$	16.2	94 (25°C)	
Hydrogen	H	1	$1s$	0.180 5		
Indium	In	49	[Kr]$4d^{10} 5s^2 5p$	81.8	8.37	
Iodine	I	53	[Kr]$4d^{10} 5s^2 5p^5$	449	1.3×10^{15} (0°C)	
Iridium	Ir	77	[Xe]$4f^{14} 5d^7 6s^2$	147	4.71	
Iron	Fe	26	[Ar]$3d^6 4s^2$	80.4	9.71	
Krypton	Kr	36	[Ar]$3d^{10} 4s^2 4p^6$	9.43		
Lanthanum	La	57	[Xe]$5d 6s^2$	13.4	54	
Lawrencium	Lr	103	[Rn]$4f^{14} 6d 7s^2$			
Lead	Pb	82	[Xe]$4f^{14} 5d^{10} 6s^2 6p^2$	35.3	20.65	29.1(20–100°C)
Lithium	Li	3	$1s^2 2s$	84.8	9.446	
Lutetium	Lu	71	[Xe]$4f^{14} 5d 6s^2$	16.4	54	
Magnesium	Mg	12	[Ne]$3s^2$	156	4.46	26.1(20–100°C)
Manganese	Mn	25	[Ar]$3d^5 4s^2$	7.81	185.0	22.8(0–100°C)
Mendelevium	Md	101	[Rn]$5f^{13} 7s^2$			
Mercury	Hg	80	[Xe]$4f^{14} 5d^{10} 6s^2$	8.30	95.8 lq; 21 c	0.23% (200°C)
Molybdenum	Mo	42	[Kr]$4d^5 5s$	138	5.0	
Neodymium	Nd	60	[Xe]$4f^4 6s^2$	16.5	64.0 (25°C)	
Neon	Ne	10	$1s^2 2s^2 2p^6$	0.049 1		
Neptunium	Np	93	[Rn]$5f^4 6d 7s^2$	6.3	122.0 (22°C)	
Nickel	Ni	28	[Ar]$3d^8 4s^2$	90.9	6.97	13.3(20–100°C)
Niobium	Nb	41	[Kr]$4d^4 5s$	53.7	13–16	7.1(20–100°C)
Nitrogen	N	7	$1s^2 2s^2 2p^3$	0.025 83		
Nobelium	No	102	[Rn]$5f^{14} 7s^2$			
Osmium	Os	76	[Xe]$4f^{14} 5d^6 6s^2$	87.6	8.12	
Oxygen	O	8	$1s^2 2s^2 2p^4$	0.026 58 g 0.149 lq		
Palladium	Pd	46	[Kr]$4d^{10}$	71.8	9.96	11.1(20°C)

TABLE 4.1 Electronic Configuration and Properties of the Elements (*Continued*)

Name	Symbol	At. no.	Electronic configuration	Thermal conductivity, W · (m · K)⁻¹ at 25°C	Electrical resistivity, $\mu\Omega \cdot$ cm at 20°C	Coefficient of linear thermal expansion, m · m⁻¹ ($\times 10^6$)
Phosphorus (white) (black)	P	15	$[Ne]3s^23p^3$	0.236 17	10	
Platinum	Pt	78	$[Xe]4f^{14}5d^96s$	12.1 71.6	10.6	9.1(20°C)
Plutonium	Pu	94	$[Rn]5f^67s^2$	6.70	146 (0°C) alpha	
Polonium	Po	84	$[Xe]4f^{14}5d^{10}6s^26p^4$	(0.2)	42 (0°C) alpha / 44 (0°C) beta	
Potassium	K	19	$[Ar]4s$	102.5	6.1	
Praseodymium	Pr	59	$[Xe]4f^36s^2$	12.5	68 (25°C)	
Promethium	Pm	61	$[Xe]4f^56s^2$	17.9	64 (25°C)	
Protactinium	Pa	91	$[Rn]5f^26d7s^2$	47	19.1 (22°C)	
Radium	Ra	88	$[Rn]7s^2$	18.6 (20)	100	
Radon	Rn	86	$[Xe]4f^{14}5d^{10}6s^26p^6$	0.003 61		
Rhenium	Re	75	$[Xe]5f^{14}5d^56s^2$	48.0	19.3	
Rhodium	Rh	45	$[Kr]4d^85s$	150	4.33	
Rubidium	Rb	37	$[Kr]5s$	58.2	11.0	
Ruthenium	Ru	44	$[Kr]4d^75s$	117	7.1 (0°C)	
Samarium	Sm	62	$[Xe]4f^66s^2$	13.3	91.4 (0°C)	
Scandium	Sc	21	$[Ar]3d4s^2$	15.8	50.5 (0°C)	
Selenium (amorphous)	Se	34	$[Ar]3d^{10}4s^24p^4$	0.519	1.2 (0°C)	37
Silicon	Si	14	$[Ne]3s^23p^2$	149	10^5	
Silver	Ag	47	$[Kr]4d^{10}5s$	429	1.59	10.68(20°C)
Sodium	Na	11	$[Ne]3s$	142	4.69	
Strontium	Sr	38	$[Kr]5s^2$	35.4	23	
Sulfur (amorphous)	S	16	$[Ne]3s^23p^4$	0.205	2×10^{23}	
Tantalum	Ta	73	$[Xe]4f^{14}5d^36s^2$	57.5	13.5	6.5(20°C)
Technetium	Tc	43	$[Kr]4d^55s^2$	50.6	22.6 (100°C)	
Tellurium	Te	52	$[Kr]4d^{10}5s^25p^4$	1.97–3.38	$(5.8–33) \times 10^3$	
Terbium	Tb	65	$[Xe]4f^96s^2$	11.1	116	
Thallium	Tl	81	$[Xe]4f^{14}5d^{10}6s^26p$	46.1	18	
Thorium	Th	90	$[Rn]6d^27s^2$	54.0	15.4 (22°C)	28
Thulium	Tm	69	$[Xe]4f^{13}6s^2$	16.9	90	

Element	Symbol	Z	Configuration		19.9(0°C)
Tin	Sn	50	$[\text{Kr}]4d^{10}5s^25p^2$	66.8	11
Titanium	Ti	22	$[\text{Ar}]3d^24s^2$	21.9	42.0
Tungsten	W	74	$[\text{Xe}]4f^{14}5d^46s^2$	173	4.9
Uranium	U	92	$[\text{Rn}]5f^36d^17s^2$	27.5	28
Vanadium	V	23	$[\text{Ar}]3d^34s^2$	30.7	24.8–26.0
Xenon	Xe	54	$[\text{Kr}]4d^{10}5s^25p^6$	0.005 65	
Ytterbium	Yb	70	$[\text{Xe}]4f^{14}6s^2$	34.9	28
Yttrium	Y	39	$[\text{Kr}]4d^15s^2$	17.2	57
Zinc	Zn	30	$[\text{Ar}]3d^{10}4s^2$	116	5.8
Zirconium	Zr	40	$[\text{Kr}]4d^25s^2$	22.6	40

4.2 IONIZATION ENERGY

TABLE 4.2 Ionization Energy of the Elements

The minimum amount of energy required to remove the least strongly bound electron from a gaseous atom (or ion) is called the ionization energy and is expressed in $MJ \cdot mol^{-1}$. Remember that 96.485 kJ = 1.000 eV = 23.0605 kcal. In Table 4.2 the successive stages of ionization are indicated by the heading of each column: I denotes first spectra arising from a neutral atom; viz.,

$$M(gas) \rightarrow M^+(gas) + e^-$$

II, second spectra from singly ionized atoms, and so on for successive stages of ionization. Italics denote values derived from ionization limits.

References: C. E. Moore, *National Standard Reference Data Series 34,* U.S. Government Printing Office, Washington, D.C., 1970; W. C. Martin et al., *J. Phys. Chem. Reference Data,* **3**:771 (1974) for the lanthanides and actinides.

At. no.	Element	Spectrum (in $MJ \cdot mol^{-1}$)					
		I	II	III	IV	V	VI
1	H	1.312					
2	He	2.372	5.250				
3	Li	0.520	7.298	11.815			
4	Be	0.899	1.757	14.848	21.006		
5	B	0.801	2.427	3.600	25.025	32.826	
6	C	1.086	2.353	4.620	6.223	37.830	47.276
7	N	1.402	2.856	4.578	7.475	9.445	53.265
8	O	1.314	3.388	5.300	7.469	10.989	13.326
9	F	1.681	3.374	6.147	8.408	11.022	15.164
10	Ne	2.081	3.952	6.122	9.370	12.177	15.238
11	Na	0.496	4.562	6.912	9.543	13.353	16.610
12	Mg	0.738	1.451	7.733	10.540	13.629	17.994
13	Al	0.578	1.817	2.745	11.577	14.831	18.377
14	Si	0.786	1.577	3.231	4.355	16.091	19.784
15	P	1.012	1.903	2.912	4.956	6.274	21.268
16	S	1.000	2.251	3.361	4.564	*7.013*	8.495
17	Cl	1.251	2.297	3.822	5.158	6.54	*9.362*
18	Ar	1.521	2.666	3.931	5.771	7.238	8.787
19	K	0.419	3.051	4.411	5.877	*7.975*	*9.649*
20	Ca	0.590	1.145	4.912	6.474	8.144	*10.496*
21	Sc	0.631	1.235	2.389	7.089	8.844	10.719
22	Ti	0.658	1.310	2.652	4.175	9.573	11.516
23	V	0.650	1.414	2.828	4.507	*6.294*	12.362
24	Cr	0.653	1.592	2.987	*4.74*	*6.69*	8.738
25	Mn	0.717	1.509	3.248	*4.94*	*6.99*	*9.2*
26	Fe	0.759	1.561	2.957	*5.29*	*7.24*	*9.6*
27	Co	0.758	1.646	3.232	*4.95*	*7.67*	*9.8*
28	Ni	0.737	1.753	3.393	*5.30*	*7.28*	*10.4*
29	Cu	0.745	1.958	3.554	*5.33*	*7.71*	*9.9*
30	Zn	0.906	1.733	3.833	*5.73*	*7.97*	*10.4*
31	Ga	0.579	1.979	2.963	6.2		
32	Ge	0.762	1.537	3.302	4.410	9.021	
33	As	0.947	1.798	2.735	4.837	6.043	12.31
34	Sc	0.941	2.045	2.974	4.143	6.59	7.883
35	Br	1.140	2.10	*3.5*	*4.56*	5.76	8.55
36	Kr	1.351	2.350	3.565	*5.07*	6.24	7.57
37	Rb	0.403	2.632	3.9	*5.08*	6.85	8.14
38	Sr	0.549	1.064	*4.21*	5.5	6.91	8.76

TABLE 4.2 Ionization Energy of the Elements (*Continued*)

At. no.	Element	Spectrum (in MJ·mol⁻¹)					
		I	II	III	IV	V	VI
39	Y	0.616	1.181	1.980	*5.96*	*7.43*	*8.97*
40	Zr	0.660	1.267	2.218	3.313	7.86	(9.5)
41	Nb	0.664	1.382	2.416	3.695	4.877	9.899
42	Mo	0.685	1.558	2.621	4.477	5.91	6.6
43	Tc	0.702	1.472	2.850			
44	Ru	0.711	1.617	2.747			
45	Rh	0.720	1.744	2.997			
46	Pd	0.805	1.875	3.177			
47	Ag	0.731	2.073	3.361			
48	Cd	0.868	1.631	3.616			
49	In	0.558	1.821	2.704	5.2		
50	Sn	0.709	1.412	2.943	3.930	6.974	
51	Sb	0.834	1.595	2.44	4.26	5.4	10.4
52	Te	0.869	1.795	2.698	3.610	5.668	6.82
53	I	1.008	1.846	*3.2*			
54	Xe	1.170	2.046	3.10			
55	Cs	0.376	2.42	(3.4)			
56	Ba	0.503	0.965	(3.6)			
57	La	0.538	1.067	1.850			
58	Ce	0.528	1.047	1.949	3.543		
59	Pr	0.523	1.018	2.086	3.758	5.543	
60	Nd	0.530	1.034				
61	Pm	0.535	1.052				
62	Sm	0.543	1.068				
63	Eu	0.547	1.085				
64	Gd	0.592	1.167				
65	Tb	0.564	1.112				
66	Dy	0.572	1.126				
67	Ho	0.581	1.139				
68	Er	0.589	1.151				
69	Tm	0.596	1.163	2.288			
70	Yb	0.603	1.174	2.43			
71	Lu	0.524	1.34				
72	Hf	0.68	1.44	2.25	3.21		
73	Ta	0.761	(1.5)				
74	W	0.770	(1.7)				
75	Re	0.760	1.26	2.51	3.64		
76	Os	0.84	(1.6)				
77	Ir	0.88	(1.6)				
78	Pt	0.87	1.791				
79	Au	0.890	1.98				
80	Hg	1.007	1.810	3.30			
81	Tl	0.589	1.971	2.878			
82	Pb	0.716	1.450	3.081	4.083	6.64	
83	Bi	0.703	1.610	2.466	4.371	5.40	8.52
84	Po	0.812					
85	At	(0.926)					
86	Rn	1.037					
87	Fr	(0.375)					
88	Ra	0.509	0.979				
89	Ac	0.67	1.17				

TABLE 4.2   Ionization Energy of the Elements (*Continued*)

At. no.	Element	Spectrum (in MJ·mol⁻¹)					
		I	II	III	IV	V	VI
90	Th	(0.671)	1.11	1.93	2.78		
91	Pa	0.568					
92	U	0.587	(1.42)				
93	Np	0.597					
94	Pu	0.585					
95	Am	0.579					
96	Cm	0.581					
97	Bk	0.601					
98	Cf	0.608					
99	Es	0.619					
100	Fm	0.627					
101	Md	0.636					
102	No	0.642					

TABLE 4.3 Ionization Energy of Molecular and Radical Species (in $MJ \cdot mol^{-1}$)

Species	Ionization energy	Species	Ionization energy
H_2	1.489	H_2CO	1.049
N_2	1.51	CH_3CHO	0.985
O_2	1.17	Acrolein	0.975
F_2	(1.59)	Acetone	0.935
Cl_2	1.11	HCO	(0.85)
Br_2	1.02	NH_2	1.09
I_2	0.90	HO_2	1.11
CH	1.07	CH_4	1.25
OH	(1.24)	C_2H_6	1.13
HCl	1.229	C_2H_4	1.01
HBr	1.121	C_2H_2	1.10
HI	1.002	NH_3	0.979
CO	1.35	C_2N_2	1.31
NO	0.892	HCN	1.34
NO_2	1.16	H_2O	1.215
N_2O	1.245	H_2S	1.009
CH_2	0.994	CH_3I	0.93
CH_3	0.96	CO	1.331
C_2H_5	0.84	SO_2	1.191
tert-Butyl	0.67	COS	1.079
Hexane	0.981	CS_2	0.973
Benzene	0.892	CH_3SH	0.911
Toluene	0.851	$(CH_3)_2S$	0.838
Xylene	0.815	CH_3SSCH_3	0.816
Styrene	0.817	Methanol	1.047
Phenol	0.820	Ethanol	1.011
Aniline	0.743	Formic acid	1.066
Pyridine	0.899	Acetic acid	1.001
1-Butene	0.924	Methylamine	0.865
Benzene	0.892	Acetonitrile	1.179
CH_3CO	0.76	Acrylonitrile	1.053
Chloromethane	1.078	1,2-Dichloroethane	1.073
Chloroform	1.102	Vinyl chloride	0.965
Carbon tetrachloride	1.107	1,1,2-Trichloroethene	0.912
1,1-Dichloroethane	0.948		

4.3 ELECTRON AFFINITY

TABLE 4.4 Electron Affinities of Elements, Molecules, and Radicals

The electron affinity of an atom (molecule or radicals) is defined as the energy released when an atom and an electron react to form a negative ion in the gas phase at 0 K.

$$A(g) + e^- = A^-(g)$$

Data are limited to those negative ions which, by virtue of their positive electron affinity, are stable. Uncertainty in the final data figures is given in parentheses. Calculated values are enclosed in parentheses.

A. Elements

Element	Electron affinity, kJ·mol^{-1}*	Element	Electron affinity, kJ·mol^{-1}*
Aluminum	44(3)	Molybdenum	97(19)
Antimony	101(5)	Neon	(−29)
Argon	(−34)	Nickel	111(10)
Arsenic	77(5)	Niobium	96(30)
Astatine	270(2)	Nitrogen	52.9(4)
Barium	(−52)	Osmium	106(30)
Beryllium	−241	Oxygen	141.1(3)
Bismuth	106(20)	Palladium	58(30)
Boron	27(1)	Phosphorus	71.7(10)
Bromine	324.7(4)	Platinum	205.3(2)
Cadmium	<0	Polonium	183(30)
Calcium	(−156)	Potassium	48.35(5)
Carbon	122.3(5)	Radon	(−41)
Cesium	45.49(5)	Rare earths	−48(estimate)
Chlorine	348.8(4)	Rhenium	14(10)
Chromium	64(5)	Rubidium	46.89(5)
Cobalt	67(20)	Ruthenium	106(30)
Copper	118.3(10)	Scandium	<0
Fluorine	332.6(3)	Selenium	194.96(3)
Francium	(44.0)	Silicon	133.6(5)
Gallium	29(15)	Silver	125.7(7)
Germanium	116(10)	Sodium	52.7(5)
Gold	222.75(7)	Strontium	(−167)
Hafnium	<0	Sulfur	200.42(5)
Helium	(−21)	Tantalum	77(29)
Hydrogen	72.770(3)	Technetium	68(30)
Indium	29(15)	Tellurium	190.15(3)
Iodine	2955(4)	Thallium	48(10)
Iridium	154(19)	Tin	121(10)
Iron	24(19)	Titanium	19(19)
Krypton	(−39)	Tungsten	58(40)
Lanthanum	48(30)	Vanadium	48(20)
Lead	106(20)	Xenon	(−40)
Lithium	59.8(7)	Yttrium	0.0(30)
Magnesium	(−232)	Zinc	~0
Manganese	<0	Zirconium	48(30)
Mercury	<0		

* 96.485 kJ · mol^{-1} = 1.00 eV = 23.06 kcal · mol^{-1}

TABLE 4.4 Electron Affinities of Elements, Molecules, and Radicals (*Continued*)

B. Molecules

Molecule	Electron affinity kJ · mol⁻¹*	Molecule	Electron affinity kJ · mol⁻¹*
BF_3	256	SF_6	138
1,4-Benzoquinone	129	2,3,5,6-Tetrachlorobenzo-	232
NO_2	377	quinone	
O_2	43	Tetracyanoethylene	278

C. Radicals

Radical	Electron affinity kJ · mol⁻¹*	Radical	Electron affinity kJ · mol⁻¹*
CH_3	104	OH	177
C_2H_5	86	CF_3O	130
C_6H_5	212	CH_3O	37
CCl_3	118	PH_2	154
CF_3	178	SH	211
CN	306	CH_3S	127
NH_2	108	SCN	209
C_6H_5NH	150	SeCN	255
$(C_6H_5)_2N$	115	SiF_3	323

* 96.485 kJ · mol⁻¹ = 1.00 eV = 23.06 kcal · mol⁻¹
Source: H. Hotop and W. C. Lineberger, *J. Phys. Chem. Reference Data* **4**:539 (1975); E. C. M. Chen and W. E. Wentworth, *J. Chem. Educ.* **52**:486 (1975).

4.4 ELECTRONEGATIVITY

Electronegativity χ is the relative attraction of an atom for the valence electrons in a covalent bond. It is proportional to the effective nuclear charge and inversely proportional to the covalent radius:

$$\chi = \frac{0.31(n + 1 \pm c)}{r} + 0.50$$

where n is the number of valence electrons, c is any formal valence charge on the atom and the sign before it corresponds to the sign of this charge, and r is the covalent radius. Originally the element fluorine, whose atoms have the greatest attraction for electrons, was given an arbitrary electronegativity of 4.0. A revision of Pauling's values based on newer data assigns 3.90 to fluorine. Values in Table 4.5 refer to the common oxidation states of the elements.

The greater the difference in electronegativity, the greater is the ionic character of the bond. The amount of ionic character I is given by

$$I = 0.46|\chi_A - \chi_B| + 0.035(\chi_A - \chi_B)^2$$

The bond is fully covalent when $(\chi_A - \chi_B) < 0.5$ (and $I < 6\%$).

TABLE 4.5 Electronegativities of the Elements

H																	
2.20																	
Li	Be											B	C	N	O	F
1.0	1.5											2.0	2.5	3.0	3.5	3.9
Na	Mg											Al	Si	P	S	Cl
0.9	1.2											1.5	1.8	2.1	2.4	2.8
K	Ca	Sc	Ti	V	Cr	Mn	Fe	Co	Ni	Cu	Zn	Ga	Ge	As	Se	Br
0.9	1.0	1.3	1.5	1.6	1.6	1.5	1.8	1.8	1.8	1.9	1.7	1.6	1.8	2.0	2.4	2.7
Rb	Sr	Y	Zr	Nb	Mo	Tc	Ru	Rh	Pd	Ag	Cd	In	Sn	Sb	Te	I
0.8	1.0	1.2	1.4	1.6	1.8	1.9	2.2	2.2	2.2	1.9	1.5	1.7	1.8	1.9	2.1	2.2
Cs	Ba	La–Lu	Hf	Ta	W	Re	Os	Ir	Pt	Au	Hg	Tl	Pb	Bi	Po	At
0.7	0.9	1.1	1.3	1.5	1.7	1.9	2.2	2.2	2.2	2.4	1.4	1.8	1.9	1.9	2.0	2.2
Fr	Ra	Ac	Th	Pa	U	Np–No										
0.7	0.9	1.1	1.5	1.5	1.7	1.3										

Source: L. Pauling, *The Chemical Bond,* Cornell University Press, Ithaca, New York, 1967.

4.5 BOND LENGTHS AND STRENGTHS

4.5.1 Atom Radius

The *atom radius* of an element is the shortest distance between like atoms. It is the distance of the centers of the atoms from one another in metallic crystals and for these materials the atom radius is often called the metal radius. Except for the lanthanides (CN = 6), CN = 12 for the elements. The atom radii listed in Table 4.6 are taken mostly from A. Kelly and G. W. Groves, *Crystallography and Crystal Defects,* Addison-Wesley, Reading, Mass., 1970.

4.5.2 Ionic Radii

One of the major factors in determining the structures of the substances that can be thought of as made up of cations and anions packed together is ionic size. It is obvious from the nature of wave functions that no ion has a precisely defined radius. However, with the insight afforded by electron density maps and with a large base of data, new efforts to establish tables of ionic radii have been made, the most successful being those of Shannon and Prewitt. Pertinent references: R. D. Shannon and C. T. Prewitt, *Acta Crystallographica* **B25**:925 (1969); **B26**:1046 (1970) and R. D. Shannon, *Acta Crystallographica* **A32**:751 (1976).

Shannon and Prewitt base their *effective ionic radii* on the assumption that the ionic radius of O^{2-} (CN 6) is 140 pm and that of F^- (CN 6) is 133 pm. Also taken into consideration is the coordination number (CN) and electronic spin state (HS and LS, high spin and low spin) of first-row transition metal ions. These radii are empirical and include effects of covalence in specific metal-oxygen or metal-fluorine bonds. Older "crystal ionic radii" were based on the radius of F^- (CN 6) equal to 119 pm; these radii are 14–18 percent larger than the effective ionic radii.

TABLE 4.6 Atom Radii and Effective Ionic Radii of Elements

Element	Atom radius, pm	Ion charge	Effective ionic radii, pm			
			Coordination number			
			4	6	8	12
Actinium	187.8	3+		111		
Aluminum	143.1	3+	39	53.5		
Americium	173	2+			126	
		3+		97.5	109	
		4+		89	95	
		5+		86		
		6+		80		
Antimony	145	3−		245		
		1+		89		
		3+	76	76		
		5+		60		
Arsenic	124.8	3−		222		
		3+		58		
		5+	33.5	46		
Astatine		1−		227		
		5+		57		
		7+		62		
Barium	217.3	2+		136	142	160
Berkelium		2+		118		
		3+		98		
		4+		87	93	
Beryllium	111.3	1−	195			
		2+	27	45		
Bismuth	154.7	3−		213		
		3+		103	111	
		5+		76		
Boron	86	1+	35			
		3+	11	27		
Bromine		1−		196		
		3+	59			
		5+	31*	47		
		7+		25		
Cadmium	148.9	2+	78	95	110	131
Calcium	197	2+		100	112	135
Californium	186(2)	2+		117		
		3+		95		
		4+		82.1		
Carbon		4−	260			
		4+	15	16		
Cerium	181.8	3+		102	114.3	134
		4+		87	97	114
Cesium	265	1+		167	174	188
Chlorine		1−		181		
		5+	34			
		7+	8	27		
Chromium	128	1+	81			
		2+		73 LS		
				80 HS		

* CN = 3

TABLE 4.6 Atom Radii and Effective Ionic Radii of Elements (*Continued*)

Element	Atom radius, pm	Ion charge	Effective ionic radii, pm			
			Coordination number			
			4	6	8	12
Chromium		3+		61.5		
(*continued*)		4+	41	55		
		5+	34.5	49	57	
		6+	26	44		
Cobalt	125	2+	38	65 LS	90	
				74.5 HS		
		3+		54.5 LS		
				61 HS		
		4+	40	53 HS		
Copper	128	1+	60	77		
		2+	57	73		
		3+		54 LS		
Curium	174	3+		97		
		4+		85	95	
Dysprosium	178.1	2+		107	119	
		3+		91.2	102.7	
Einsteinium	186(2)	3+		98		
Erbium	176.1	3+		89.0	100.4	
Europium	208.4	2+		117	125	135
		3+		94.7	106.6	
Fluorine	71.7	1−	131	133		
		7+		8		
Francium	270	1+		180		
Gadolinium	180.4	3+		93.8	105.3	
Gallium	135	2+		120		
		3+	47	62.0		
Germanium	128	2+		73		
		4+	39.0	53.0		
Gold	144	1+		137		
		3+	68	85		
Hafnium	159	4+	58	71	83	
Holmium	176.2	3+		90.1	101.5*	112
Hydrogen		1−		154		
Indium	167	1+		140		
		3+	62	80.0	92	
Iodine		1−		220		
		5+		95		
		7+	42	53		
Iridium	135.5	3+		68		
		4+		62.5		
		5+		57		
Iron	126	2+		61 LS		
			63 HS	78 HS	92 HS	
		3+		55 LS		
			49 HS	64.5 HS	78 HS	
		4+		58.5		
		6+	25			

* CN = 10

TABLE 4.6 Atom Radii and Effective Ionic Radii of Elements (*Continued*)

Element	Atom radius, pm	Ion charge	Effective ionic radii, pm			
			Coordination number			
			4	6	8	12
Lanthanum	183	3+		103.2	116.0	136
Lead	175	2+	98	119	129	149
		4+		78	94	
Lithium	152	1+	59	76		
Lutetium	173.8	3+		86.1	97.7	
Magnesium	160	2+	57	72.0	89	
Manganese	127	2+	66 HS	67 LS	96	
				83 HS		
		3+		58 LS		
				64.5 HS		
		4+	39	53		
		5+	33			
		6+	25.5			
		7+	25	46		
Mercury	151	1+	111*	119		
		2+	96	102	114	
Molybdenum	139	3+		69		
		4+		65.0		
		5+	46	61		
		6+	41	59	73†	
Neodymium	181.4	2+			129	
		3+		98.3	110.9	127
Neptunium	155	2+		110		
		3+		101		
		4+		87	98	
		5+		75		
		6+		72		
		7+		71		
Nickel	124	2+	55	69.0		
		3+		56 LS		
				60 HS		
		4+		48 LS		
Niobium	146	3+		72		
		4+		68	79	
		5+	48	64	74	
Nitrogen		3−	146			
		1+	25			
		3+		16		
		5+		13		
Nobelium		2+		110		
Osmium	135	4+		63.0		
		5+		57.5		
		6+		54.5		
		7+		52.5		
		8+	39			
Oxygen		2−	138	140	142	

* CN = 3
† CN = 7

TABLE 4.6　Atom Radii and Effective Ionic Radii of Elements (*Continued*)

Element	Atom radius, pm	Ion charge	Effective ionic radii, pm			
			Coordination number			
			4	6	8	12
Palladium	137	2+	64	86		
		3+		76		
		4+		61.5		
Phosphorus	108	3−		212		
		3+		44		
		5+	17	38		
Platinum	138.5	2+		80		
		4+		62.5		
		5+		57		
Plutonium	159	3+		100		
		4+		86	96	
		5+		74		
		6+		71		
Polonium	164	2−		(230)		
		4+		94	108	
		6+		67		
Potassium	232	1+	137	138	151	164
Praseodymium	182.4	3+		99	112.6	
		4+		85	96	
Promethium	183.4	3+		97	109.3	
Protoactinium	163	3+		104		
		4+		90	101	
		5+		78	91	
Radium	(220)	2+			148	170
Rhenium	137	4+		63		
		5+		58		
		6+		55		
		7+	38	53		
Rhodium	134	3+		66.5		
		4+		60		
		5+		55		
Rubidium	248	1+		152	161	172
Ruthenium	134	3+		68		
		4+		62.0		
		5+		56.5		
		7+	38			
		8+	36			
Samarium	180.4	2+			127	
		3+		95.8	107.9	124
Scandium	162	3+		74.5	87.0	
Selenium	116	2−		198		
		4+		50		
		6+		42		
Silicon	118	4+	26	40.0		
Silver	144	1+	100	115	130	
		2+	79	94		
		3+	67	75		
Sodium	186	1+	99	102	118	139

TABLE 4.6 Atom Radii and Effective Ionic Radii of Elements (*Continued*)

Element	Atom radius, pm	Ion charge	Effective ionic radii, pm			
			Coordination number			
			4	6	8	12
Strontium	215	2+		118	126	144
Sulfur	106	2−		184		
		4+		37		
		6+	12	29		
Tantalum	146	3+		72		
		4+		68		
		5+		64	74	
Technetium	136	4+		64.5		
		5+		60		
		7+	37	56		
Tellurium	142	2−		221		
		4+	66	97		
		6+	43	56		
Terbium	177.3	3+		92.3	104.0	
		4+		76	88	
Thallium	170	1+		150	159	170
		3+	75	88.5	98	
Thorium	179	4+		94	105	121
Thullium	175.9	2+		103		
		3+		88.0	99.4	105*
Tin	151	2+		118		
		4+	55	69.0	81	
Titanium	147	2+		86		
		3+		67.0		
		4+	42	60.5	74	
Tungsten	139	4+		66		
		5+		62		
		6+	42	60		
Uranium	156	3+		102.5		
		4+		89	100	117
		5+		76		
		6+	52	73	86	
Vanadium	134	2+		79		
		3+		64.0		
		4+		58	72	
		5+	35.5	54		
Xenon		8+	40	48		
Ytterbium	193.3	2+		102	114	
		3+		86.8	98.5	104*
Yttrium	180	3+		90.0	101.9	108*
Zinc	134	2+	60	74.0	90	
Zirconium	160	4+	59	72	84	89*

* CN = 11

4.5.3 Covalent Radii

Covalent radii (Table 4.7) are the distance between two kinds of atoms connected by a covalent bond of a given type (single, double, etc.).

TABLE 4.7   Covalent Radii for Atoms

Element	Single-bond radius, pm*	Double-bond radius, pm	Triple-bond radius, pm
Aluminum	126		
Antimony	141	131	
Arsenic	121	111	
Beryllium	106		
Boron	88		
Bromine	114	104	
Cadmium	148		
Carbon	77.2	66.7	60.3
Chlorine	99	89	
Copper	135		
Fluorine	64	54	
Gallium	126		
Germanium	122	112	
Hydrogen	30		
Indium	144		
Iodine	133	123	
Magnesium	140		
Mercury	148		
Nitrogen	70	60	55
Oxygen	66	55	
Phosphorus	110	100	93
Silicon	117	107	100
Selenium	117	107	
Silver	152		
Sulfur	104	94	87
Tellurium	137	127	
Tin	140	130	
Zinc	131		

* Single-bond radii are for a tetrahedral (CN = 4) structure.

TABLE 4.8   Octahedral Covalent Radii for CN = 6

Atom	Octahedral covalent radius, pm	Atom	Octahedral covalent radius, pm
Cobalt(II)	132	Nickel(III)	130
Cobalt(III)	122	Nickel(IV)	121
Gold(IV)	140	Osmium(II)	133
Iridium(III)	132	Palladium(IV)	131
Iron(II)	123	Platinum(IV)	131
Iron(IV)	120	Rhodium(III)	132
Nickel(II)	139	Ruthenium(II)	133

TABLE 4.9   Bond Lengths between Carbon and Other Elements

Bond type	Bond length, pm
Carbon-carbon	
Single bond	
Paraffinic: —C—C—	154.1(3)
In presence of —C=C— or of aromatic ring	153(1)
In presence of —C=O bond	151.6(5)
In presence of two carbon-oxygen bonds	149(1)
In presence of two carbon-carbon double bonds	142.6(5)
Aryl-C=O	147(2)
In presence of one carbon-carbon triple bond: —C—C≡C—	146.0(3)
In presence of one carbon-nitrogen triple bond: —C—C≡N	146.6(5)
In compounds with tendency to dipole formation, e.g., C=C—C=O	144(1)
In aromatic compounds	139.5(3)
In presence of carbon-carbon double and triple bonds: —C=C—C≡C—	142.6(5)
In presence of two carbon-carbon triple bonds: —C≡C—C≡C—	137.3(4)
Double bond	
Single: —C=C—	133.7(6)
Conjugated with a carbon-carbon double bond: —C=C—C=C—	133.6(5)
Conjugated with a carbon-oxygen double bond: —C=C—C=O	136(1)
Cumulative: —C=C=C— or —C=C=O	130.9(5)
Triple bond	
Simple: —C≡C—	120.4(2)
Conjugated: —C≡C—C=C—, —C≡C—C=O, or —C≡C—aryl	120.6(4)

Bond type	Bond length, pm			
Carbon-halogen				
	Fluorine	Chlorine	Bromine	Iodine
Paraffinic: R—X	137.9(5)	176.7(2)	193.8(5)	213.9(1)
Olenfinic: —C=C—X	133.3(5)	171.9(5)	189(1)	209.2(5)
Aromatic: Ar-X	132.8(5)	170(1)	185(1)	205(1)
Acetylenic: —C≡C—X	(127)	163.5(5)	179.5(10)	199(2)

Bond type	Bond length, pm
Carbon-hydrogen	
Paraffinic	
In methane (in CD_4, 109.2)	109.4
In monosubstituted carbon: H—C—Y	109.6(5)
In disubstituted carbon: H—C— (with X and Y substituents)	107.3(5)
In trisubstituted carbon: H—C—Y (with X and Z substituents)	107.0(7)

TABLE 4.9 Bond Lengths between Carbon and Other Elements (*Continued*)

Bond type	Bond length, pm
Carbon-hydrogen (continued)	
Olefinic	
Simple: H—C=C—	108.3(5)
Cumulative carbon-carbon double bonds: H—C=C=C—	107(1)
Cumulative carbon-carbon-oxygen double bonds: H—C—C=C=O	108(1)
Aromatic	108.4(5)
Acetylenic (in C_2H_2, 105.9)	105.5(5)
In small rings	108.1(5)
In presence of a carbon triple bond: H—C≡C—	111.5(4)
Carbon-nitrogen	
Single bond	
Paraffinic:	
3-covalent nitrogen: RNH_2, R_2NH, R_3N	147.2(5)
4-covalent nitrogen: RNH_3^+, R_3N-BX_3	147.9(5)
In —C—N=	147.5(10)
In aromatic compounds	143(1)
In conjugated heterocyclic systems (partial double bond)	135.3(5)
In —N—C=O (partial double bond)	132.2(5)
Double bond: —C=N—	132
Triple bond (in CN radical, 117.74): —C≡N	115.7(5)
Carbon-oxygen	
Single bond	
Paraffinic and saturated heterocyclic: —C—O—	142.6(5)
Strained, as in epoxides: —C—C— (with O below)	143.5(5)
In aromatic compounds, as Ar-OH	136(1)
Longer bond in carboxylic acids and esters (HCOOH, 131.2)	135.8(5)
In conjugated heterocyclics, as furan	137.1(16)
Double bond	
In CO^+	111.5
In CO	112.8
In CO_2^+	117.7
In HCO	119.8(8)
In carbonyls	114.5(10)
In aldehydes and ketones	121.5(5)
In acyl halides: R—CO—X	117.1(4)
Shorter bond in carboxylic acids and esters	123.3(5)
In zwitterion forms	126(1)
In O=C=	116.0(1)
In isocyanates: RN=C=O	117(1)
In conjugated systems, as in partial triple bond: O=C—C=C	121.5(5)
In 1,4-quinones	115(2)
In metal acetylacetonates	128(2)
In calcite: $CaCO_3$	129(1)

TABLE 4.9 Bond Lengths between Carbon and Other Elements (*Continued*)

Bond type	Bond length, pm
Carbon-selenium	
Single bond	
Paraffinic: —C—Se—	198(2)
In presence of fluorine, as in perfluorocompounds: —CF—Se—	195(2)
Double bond	
In Se=C=, as SeCS and SeCO	170.9(3)
In CSe radical	167
Carbon-silicon	
Alkyl substituent: H_3C—Si or H_2C—Si	187.0(5)
Aryl substituent: aryl—Si	184.3(5)
Electronegative substituent: R—Si—X	185.4(5)
Carbon-sulfur	
Single bond	
Paraffinic: —C—S—	181.7(5)
In presence of fluorine, as in perfluoro compounds: —CF—S—	183.5(1)
In heterocyclic systems: partial double bonds	171.8(5)
Double bonds	
In S=C: thiophene, $S=CR_2$	171(1)
In sulfoxides and sulfones	180(1)
In presence of second carbon-carbon double bond: S=C—C=C—	155.5(1)
In SC radical [in CS_2^+, 155.4(5)]	153.49(2)

Bond type	Bond length, pm	Bond type	Bond length, pm
Other elements and carbon			
C-Al	224(4)	C-In	216(4)
C-As	198(1)	C-Mo	208(4)
C-B	156(1)	C-Ni	210.7(5)
C-Be	193	C-Pb (alkyl)	230(1)
C-Bi	230	C-Pd	227(4)
C-Co	183(2)	C-Sb (paraffinic)	220.2(16)
C-Cr	192(4)	C-Sn	
C-Fe	184(2)	Alkyl	214.3(5)
C-Ge		Electronegative	218(2)
Alkyl	193(3)	substituent	
Aryl	194.5(5)	C-Te	190.4
C-Hg	207(1)	C-Tl	270.5(5)
in $Hg(CN)_2$	199(2)	C-W	206

TABLE 4.10   Bond Lengths between Elements Other than Carbon

Elements	Bond type	Bond length, pm	Elements	Bond type	Bond length, pm
	Boron			**Nitrogen** (*continued*)	
B-B	B_2H_6	177(1)	N=O	N_2O	118.6(2)
B-Br	BBr_3	187(2)		RNO_2	122(1)
B-Cl	BCl_3	172(1)		NO^+	106.19
B-F	BF_3, R_2BF	129(1)	N-Si	SiN	157.2
B-H	Boranes	121(2)			
	Bridge	139(2)		**Oxygen**	
B-N	Borazoles	142(1)			
B-O	$B(OH)_3$, $(RO)_3B$	136(5)	O-H	H_2O	95.8
				ROH	97(1)
	Hydrogen			OH^+	102.89
				HOOH	96.0(5)
H-Al	AlH	164.6		D_2O (^2H_2O)	95.75
H-As	AsH_3	151.9		OD	96.99
H-Be	BeH	134.3	O-O	HO—OH	148(1)
H-Br	HBr	140.8		O_2^+	122.7
H-Ca	CaH	200.2		O_2^-	126(2)
H-Cl	HCl	127.4		O_2^{2-}	149(2)
H-F	HF	91.7		O_3	127.8(5)
H-Ge	GeH_4	153	O-Al	AlO	161.8
H-I	HI	160.9	O-As	As_2O_6 bridges	179
H-K	KH	224.4	O-Ba	BaO	190.0
H-Li	LiH	159.5	O-Cl	ClO_2	148.4
H-Mg	MgH	173.1		OCl_2	168
H-Na	NaH	188.7	O-Mg	MgO	174.9
H-Sb	H_3Sb	170.7	O-Os	OsO_4	166
H-Se	H_2Se	146.0	O-Pb	PbO	193.4
H-Sn	SnH_4	170.1			
D-Br	DBr (^2HBr)	141.44		**Phosphorus**	
D-Cl	DCl	127.46			
D-I	DI	161.65	P-Br	PBr_3	223(1)
T-Br	TBr (^3HBr)	141.44	P-Cl	PCl_3	200(2)
T-Cl	TCl	127.40	P-F	$PFCl_2$	155(3)
			P-H	PH_3, PH_4^+	142.4(5)
	Nitrogen		P-I	PI_3	252(1)
			P-N	Single bond	149.1
N-Cl	NO_2Cl	179(2)	P-O	Single bond	144.7
N-F	NF_3	136(2)		p^3 bonding	167
N-H	NH_4^+	103.4(3)		sp^3 bonding	154(4)
	NH_3, RNH_2	101.2	P-S	p^3 bonding	212(5)
	H_2NNH_2	103.8		sp^3 bonding	208(2)
	R—CO—NH_2	99(3)		In rings	220(3)
	HN=C=S	101.3(5)	P-C	Single bond	156.2
N-D	ND (N^2H)	104.1		p^3 bonding	187(2)
N-N	HN_3	102(1)			
	R_2NNH_2	145.1(5)		**Silicon**	
	N_2O	112.6(2)			
	N_2^+	111.6	Si-Br	$SiBr_4$, R_3SiBr	216(1)
N-O	NO_2Cl	124(1)	Si-Cl	$SiCl_4$, R_3SiCl	201.9(5)
	RO—NO_2	136(2)			
	NO_2	118.8(5)			

TABLE 4.10 Bond Lengths between Elements Other than Carbon (*Continued*)

Elements	Bond type	Bond length, pm	Elements	Bond type	Bond length, pm
	Silicon (*continued*)			Sulfur	
Si-F	SiF_4, R_3SiF	156.1(3)	S-Br	$SOBr_2$	227(2)
	SiF_6	158	S-Cl	S_2Cl_2	158.5(5)
Si-H	SiH_4	148.0(5)	S-F	SOF_2	158.5(5)
	R_3SiH	147.6(5)	S-H	H_2S	133.3
Si-I	SiI_4	234		RSH	132.9(5)
	R_3SiI	246(2)		D_2S	134.5
Si-O	R_3SiOR	153.3(5)	S-O	SO_2	143.21
Si-Si	H_3SiSiH_3	230(2)		$SOCl_2$	145(2)
			S-S	RSSR	205(1)

TABLE 4.11 Bond Dissociation Energies

The bond dissociation energy (enthalpy change) for a bond A—B which is broken through the reaction

$$AB \rightarrow A + B$$

is defined as the standard-state enthalpy change for the reaction at a specified temperature, here at 298 K. That is,

$$\Delta Hf_{298} = \Delta Hf_{298}(A) + \Delta Hf_{298}(B) - \Delta Hf_{298}(AB)$$

All values refer to the gaseous state and are given at 298 K. Values of 0 K are obtained by subtracting $\frac{1}{2}RT$ from the value at 298 K.

To convert the tabulated values to kcal/mol, divide by 4.184.

Bond	ΔHf_{298}, kJ/mol	Bond	ΔHf_{298}, kJ/mol
	Aluminum		Aluminum (*continued*)
Al—Al	186(9)	Al—N	297(96)
Al—As	180	Al—O	512(4)
Al—Au	326(6)	AlCl—O	540(41)
Al—Br	439(8)	AlF—O	582
Al—C	255	Al—P	213(13)
Al—Cl	494(13)	Al—Pd	259(12)
AlCl—Cl	402(8)	Al—S	374(8)
AlCl_2—Cl	372(8)	Al—Se	334(10)
AlO—Cl	515(84)	Al—Si	251(3)
Al—Cu	216(10)	Al—Te	268(10)
Al—D	291	Al—U	326(29)
Al—F	664(6)		
AlF—F	546(42)		Antimony
AlF_2—F	544(46)		
AlO—F	761(42)	Sb—Sb	299(6)
Al—H	285(6)	Sb—Br	314(59)
Al—I	368(4)	Sb—Cl	360(50)
Al—Li	176(15)		

Source: T. L. Cottrell, *The Strengths of Chemical Bonds*, 2d ed., Butterworth, London, 1958; B. deB. Darwent, *National Standard Reference Data Series*, National Bureau of Standards, no. 31, Washington, 1970; S. W. Benson, *J. Chem. Educ.* **42**:502 (1965); and J. A. Kerr, *Chem. Rev.* **66**:465 (1966).

TABLE 4.11 Bond Dissociation Energies (*Continued*)

Bond	ΔHf_{298}, kJ/mol	Bond	ΔHf_{298}, kJ/mol
Antimony (*continued*)		Bismuth (*continued*)	
Sb—F	439(96)	Bi—Cl	305(8)
Sb—N	301(50)	Bi—D	284
Sb—O	372(84)	Bi—F	259(29)
Sb—P	357	Bi—Ga	159(17)
Sb—S	379	Bi—H	279
Sb—Te	277.4(38)	Bi—O	343(6)
		Bi—P	280(13)
Arsenic		Bi—Pb	142(15)
		Bi—S	316(5)
As—As	382(11)	Bi—Sb	251(4)
As—Cl	448	Bi—Se	280(6)
As—Ga	209.6(12)	Bi—Te	232(11)
As—H	272(12)	Bi—Tl	121(13)
As—N	582(126)		
As—O	481(8)	Boron	
As—P	534(13)		
As—S	(478)	B—B	297(21)
As—Se	96	H_3B—BH_3	146
As—Tl	198(15)	OB—BO	506(84)
		B—Br	435(21)
Astatine		B—C	448(29)
		B—Cl	536(29)
At—At	(115.9)	BO—Cl	460(42)
		B—D	341(6)
Barium		B—F	766(13)
		BF—F	523(63)
Ba—Br	370(8)	BF_2—F	557(84)
Ba—Cl	444(13)	B—H	330(4)
Ba—F	487(7)	B—I	381(21)
Ba—I	>431(4)	B—N	389(21)
Ba—O	563(42)	B—O	806(5)
Ba—OH	477(42)	BCl—O	715(41)
Ba—S	400(19)	B—P	347(17)
		B—S	581(9)
Beryllium		B—Se	462(15)
		B—Si	289(29)
Be—Be	59	B—Te	354(20)
Be—Br	381(84)		
Be—Cl	388(9)	Bromine	
BeCl—Cl	540(63)		
Be—F	577(42)	Br—Br	193.870(4)
Be—H	226(21)	Br—C	280(21)
Be—O	448(21)	Br—CH_3	284(8)
Be—S	372(59)	Br—CH_2Br	255(13)
		Br—$CHBr_2$	259(17)
Bismuth		Br—CBr_3	209(13)
		Br—CCl_3	218(13)
Bi—Bi	197(4)	Br—CF_3	285(13)
Bi—Br	267(4)	Br—CF_2CF_3	287.4(63)
		Br—$CF_2CF_2CF_3$	278.2(63)

TABLE 4.11 Bond Dissociation Energies (*Continued*)

Bond	ΔHf_{298}, kJ/mol	Bond	ΔHf_{298}, kJ/mol
Bromine (*continued*)		**Carbon** (*continued*)	
Br—CHF$_2$	289	CH$_3$—CN	506(21)
Br—Cl	218.8(4)	CH$_3$—CH$_2$CN	305(8)
Br—CN	381	CH$_3$—CH(CH$_3$)CN	331(8)
Br—CO—C$_6$H$_5$	268	CH$_3$—C(C$_6$H$_5$)CN(CH$_3$)	251
Br—F	233.8(2)	CH$_3$CH$_2$—CH$_2$CN	321.8(71)
Br—N	276(21)	NC—CN	603(21)
Br—NF$_2$	222	C$_6$H$_5$—C$_6$H$_5$	418
Br—NO	120.1(63)	CH$_3$—CF$_3$	423.4(46)
Br—O	235.1(4)	CH$_2$F—CH$_2$F	368(8)
		CF$_3$—CF$_3$	406(13)
Cadmium		CF$_2$=CF$_2$	318(13)
		CF$_3$—CN	501
Cd—Cd	11.3(8)	CH$_3$—CHO	314
Cd—Br	159(96)	CH$_3$—CO	342.7
Cd—Cl	206.7(34)	CH$_3$CO—CF$_3$	308.8
Cd—F	305(21)	CH$_3$CO—COCH$_3$	280(8)
Cd—H	69.0(4)	C$_6$H$_5$CO—COC$_6$H$_5$	277.8
Cd—I	138(21)	Aryl—CH$_2$COCH$_2$—aryl	273.6
Cd—In	138	C$_6$H$_5$CH$_2$—COOH	284.9
Cd—O	142(42)	(C$_6$H$_5$CH$_2$)$_2$CH—COOH	248.5
Cd—S	196	C—Cl	397(29)
Cd—Se	310	C—F	536(21)
		C—H	337.2(8)
Calcium		C—I	209(21)
		C—N	770(4)
Ca—Ca	14.98(46)	CF$_3$—NF$_2$	272(13)
Ca—Br	321(23)	CH$_3$—NH$_2$	331(13)
Ca—Cl	398(13)	C$_6$H$_5$CH$_2$—NH$_2$	301(4)
Ca—F	527(21)	CH$_3$—NHC$_6$H$_5$	285
Ca—H	167.8	CH$_3$—N(CH$_3$)C$_6$H$_5$	272
Ca—I	285(63)	C$_6$H$_5$CH$_2$—NHCH$_3$	289(4)
Ca—O	464(84)	C$_6$H$_5$CH$_2$—N(CH$_3$)$_2$	255(4)
Ca—S	314(19)	CH$_3$—(N=NCH$_3$)	219.7
		C$_2$H$_5$—(N=NC$_2$H$_5$)	209.2
Carbon		(CH$_3$)$_3$C—N=NC(CH$_3$)$_3$	182.0
		Aryl—CH$_2$N=NCH$_2$—aryl	157
C—C	607(21)	CF$_3$—(N=NCF$_3$)	231.0
H$_3$C—CH$_3$	368	H$_2$C=NH	644(21)
(CH$_3$)$_2$C—CH$_3$	335	HC≡N	937
(CH$_3$)$_2$C—C(CH$_3$)$_2$	282.4	CH$_3$—NO	174.9(38)
CH$_3$—C$_6$H$_5$	389	C$_2$H$_5$—NO	175.7(54)
CH$_3$—CH$_2$C$_6$H$_5$	301	C$_3$H$_7$—NO	167.8(75)
(CH$_3$)$_3$C—C(C$_6$H$_5$)$_3$	63	(CH$_3$)$_2$CH—NO	171.5(54)
CH$_3$—allyl	301	n-C$_4$H$_9$—NO	215.5(42)
CH$_3$—vinyl	121	C$_6$H$_5$—NO	215.5(42)
CH$_3$—C≡CH	490	Cl$_3$C—NO	134
CH$_2$=CH—CH=CH$_2$	418	F$_3$C—NO	130
HC≡C—C≡CH	628	C$_6$F$_5$—NO	211.3(42)
H$_2$C=CH$_2$	682	NC—NO	121(13)
HC≡CH	962	CH$_3$—NO$_2$	247(13)

TABLE 4.11   Bond Dissociation Energies (*Continued*)

Bond	ΔHf_{298}, kJ/mol	Bond	ΔHf_{298}, kJ/mol
Carbon (*continued*)		**Chlorine** (*continued*)	
C_2H_5—NO_2	259	Cl—CH_3	339(21)
C—O	1076.5(4)	Cl—CH_3^+	213
CH_3—OCH_3	335	Cl—$C(CH_3)_3$	328.4
CH_3—OC_6H_5	381	Cl—CH_2Cl	310(13)
CH_3—$OCH_2C_6H_5$	280	Cl—CCl_3	293(21)
C_2H_5—OC_6H_5	213	Cl—CF_3	360(33)
$C_6H_5CH_2$—$OCOCH_3$	285	Cl—CCl_2F	305(8)
$C_6H_5CH_2$—$OCOC_6H_5$	289	Cl—$CClF_2$	318(8)
CH_3CO—OCH_3	406	Cl—CF_2CF_2	346.0(71)
CH_3—$OSOCH_3$	280	Cl—CH=CH_2	351
CH_2=$CHCH_2$—$OSOCH_3$	209	Cl—CN	439
$C_6H_5CH_2$—$OSOCH_3$	222	Cl—COCl	328
C=O	749	Cl—$COCH_3$	349.4
H_2C=O	732	Cl—COC_6H_5	310(13)
OC=O	532.2(4)	Cl—Cl^+	393
SC=O	628	Cl—ClO	143.3(42)
C≡O	1075	O_3Cl—ClO_4	243
C—P	513(8)	Cl—F	250.54(8)
C—S	699(8)	O_3Cl—F	255
CH_3—SH	305(13)	Cl—N	389(50)
CH_3—SC_6H_5	285(8)	Cl—NCl	280
CH_3—$SCH_2C_6H_5$	247(8)	Cl—NCl_2	381
OC—S	310.4	Cl—NF_2	*ca.* 134
C—Se	582(96)	Cl—NH_2	251(25)
		Cl—NO	159(6)
Cerium		Cl—NO_2	142(4)
		Cl—O	272(4)
Ce—Ce	243(21)	OCl—O	243(13)
Ce—F	582(42)	O_2Cl—O	201(4)
Ce—N	519(21)	Cl—P	289(42)
Ce—O	795(13)	Cl—$SiCl_3$	464
Ce—S	573(13)		
Ce—Se	495(15)	**Chromium**	
Ce—Te	389(42)		
		Cr—Cr	155(21)
Cesium		Cr—Br	328(24)
		Cr—Cl	366(24)
Cs—Cs	41.75(93)	Cr—Cu	155(21)
Cs—Br	397.5(42)	Cr—F	437(20)
Cs—Cl	439(21)	Cr—Ge	170(29)
Cs—F	514(8)	Cr—H	280(50)
Cs—H	178.1(38)	Cr—I	287(24)
Cs—I	339(4)	Cr—N	378(19)
Cs—O	297(25)	Cr—O	427(29)
Cs—OH	385(13)	OCr—O	531(63)
		O_2Cr—O	477(84)
Chlorine		Cr—S	339(21)
Cl—Cl	242.580(16)		
Cl—C	338(42)		

TABLE 4.11 Bond Dissociation Energies (*Continued*)

Bond	ΔHf_{298}, kJ/mol	Bond	ΔHf_{298}, kJ/mol
Cobalt		**Europium** (*continued*)	
Co—Co	167(25)	Eu—O	557(13)
Co—Br	331(42)	Eu—S	364(15)
Co—Cl	398(8)	Eu—Se	301(15)
Co—Cu	162(17)	Eu—Te	243(15)
Co—F	435(63)		
Co—Ge	239(25)	**Fluorine**	
Co—I	235(81)		
Co—O	368(21)	F—F	156.9(96)
Co—S	343(21)	F—F$^+$	>251
		F—CH$_3$	452(21)
Copper		F—C(CH$_3$)$_3$	439
		F—C$_6$H$_5$	485
Cu—Cu	202(4)	F—CCl$_3$	444(21)
Cu—Br	331(25)	F—CCl$_2$F	460(25)
Cu—Cl	383(21)	F—CClF$_2$	490(25)
Cu—F	431(13)	F—CF$_3$	523(17)
Cu—Ga	216(15)	F—COCH$_3$	498
Cu—Ge	209(21)	F—FO	272(13)
Cu—H	280(8)	F—FO$_2$	81.0
Cu—I	197(21)	F—N	301(42)
Cu—Ni	206(17)	F—NF	318(25)
Cu—O	343(63)	F—NF$_2$	243(8)
Cu—S	285(17)	F—NO	235.6(42)
Cu—Se	293(38)	F—NO$_2$	197(25)
Cu—Sn	177(17)		
Cu—Te	176(38)	**Gadolinium**	
		Gd—F	590(27)
Curium		Gd—O	716(17)
		Gd—S	525(15)
Cm—O	736	Gd—Se	431(15)
Dysprosium		**Gallium**	
Dy—F	527(21)	Ga—Ga	138(21)
Dy—O	611(42)	Ga—Br	444(17)
Dy—Se	322(42)	(CH$_3$)$_3$Ga—CH$_3$	253
Dy—Te	234(42)	Ga—Cl	481(13)
		Ga—F	577(15)
Erbium		Ga—H	<274
		Ga—I	339(10)
Er—F	565(17)	Ga—O	285(63)
Er—O	611(13)	Ga—P	230(13)
Er—S	418(42)	Ga—Sb	209(13)
Er—Se	326(42)	Ga—Te	251(25)
Er—Te	239(42)		
		Germanium	
Europium		Ge—Ge	274(21)
		Ge—Br	255(29)
Eu—Eu	33.5(165)	Ge—Cl	431.8(4)
Eu—Cl	*ca.* 326		
Eu—F	528(18)		

TABLE 4.11 Bond Dissociation Energies (*Continued*)

Bond	ΔHf_{298}, kJ/mol	Bond	ΔHf_{298}, kJ/mol
Germanium (*continued*)		**Hydrogen** (*continued*)	
Ge—F	485(21)	H—CH	452(33)
Ge—H	321.3(8)	H—CH$_2$	473(4)
Ge—O	662(13)	H—CH$_3$	431(8)
Ge—S	551.0(25)	^{2}H—C^{2}H$_3$ or D—CD$_3$	442.75(25)
Ge—Se	490(21)	H—C≡CH	523(4)
Ge—Si	301(21)	H—CH=CH$_2$	427
Ge—Te	402(8)	H—CH$_2$CH$_3$	410(4)
		H—CH$_2$C≡CH	392.9(50)
Gold		H—CH$_2$CH=CH$_2$	356
		H—cyclopropyl	423(13)
Au—Au	221.3(21)	H—CH$_2$CH$_2$CH$_3$	410(8)
Au—B	368(11)	H—CH(CH$_3$)$_2$	395.4
Au—Be	285(8)	H—cyclobutyl	397(13)
Au—Bi	293(84)	H—CH$_2$CH(CH$_3$)$_2$	360
Au—Cl	343(10)	H—CH(CH$_3$)CH$_2$CH$_3$	397(4)
Au—Co	215(13)	H—C(CH$_3$)$_3$	381
Au—Cr	215(6)		
Au—Cu	232(9)	H—[cyclopentadienyl ring]	339(4)
Au—Fe	187(17)		
Au—Ga	294(15)	H—CH(CH=CH$_2$)(CH=CH$_2$)	335(4)
Au—Ge	277(15)		
Au—H	314(10)		
Au—La	80(5)	H—[cyclopentenyl ring]	343(4)
Au—Li	68.0(16)		
Au—Mg	243(42)		
Au—Mn	185(13)	H—CH$_2$C(CH$_3$)(CH$_3$)CH$_3$	414(4)
Au—Ni	247(21)		
Au—Pb	130(42)		
Au—Pd	143(21)	H—C(CH$_3$)$_2$CH=CH$_2$	331
Au—Rh	231(29)	H—cyclopentyl	395(42)
Au—S	418(25)	H—CH$_2$C(CH$_3$)$_3$	418(4)
Au—Si	312(12)	H—C$_6$H$_5$	431
Au—Sn	244(17)	H—CH$_2$C$_6$H$_5$	356(4)
Au—Te	247(67)	H—C(C$_6$H$_5$)$_3$	314
Au—U	318(29)		
		H—[cyclohexadienyl ring]	310
Hafnium			
		H—cyclohexyl	399.6(42)
Hf—C	548(63)	H—cycloheptyl	387.0(42)
Hf—N	534(29)	H—norbornyl	406(13)
Hf—O	791(8)	H—CH$_2$Br	410(25)
		H—CHBr$_2$	435
Hydrogen		H—CH$_2$Cl	423
		H—CHCl$_2$	414.2
H—H	436.002(4)	H—CCl$_3$	377(8)
H—^{2}H or H—D	439.446(4)	H—CBr$_3$	377(8)
^{2}H—^{2}H or D—D	443.546(4)		
H—Br	365.7(21)		
H—C	337.2(8)		

TABLE 4.11 Bond Dissociation Energies (*Continued*)

Bond	ΔHf_{298}, kJ/mol	Bond	ΔHf_{298}, kJ/mol
Hydrogen (*continued*)		Hydrogen (*continued*)	
H—CCl$_2$CHCl$_2$	393(8)	H—OOH	374(8)
H—CH$_2$F	423(8)	H—OOCCH$_3$	469(17)
H—CHF$_2$	423(8)	H—OOCCH$_2$CH$_3$	460(17)
H—CF$_3$	444(13)	H—OOCC$_3$H$_7$	431(17)
H—CF$_2$Cl	435(4)	H—P	343(29)
H—CH$_2$CF$_3$	446(45)	H—S	344(12)
H—CF$_2$CH$_3$	416(4)	H—SH	381(4)
H—CF$_2$CF$_3$	431(63)	H—SCH$_3$	*ca.* 368
H—CH$_2$I	431(8)	H—Se	305(2)
H—CHI$_2$	431(8)	H—Si	298.49(46)
H—CN	540(25)	H—SiH$_3$	393(13)
H—CH$_2$CN	*ca.* 389	H—Si(CH$_3$)$_3$	377(13)
H—CH(CH$_3$)CN	377(8)	H—Te	268(2)
H—C(CH$_3$)$_2$CN	364(8)		
H—CH$_2$NH$_2$	397(8)	Indium	
H—CH$_2$Si(CH$_3$)$_3$	414(4)		
H—CH$_2$COCH$_3$	393(75)	In—In	100(8)
H—Cl	431.8(4)	In—Br	418(21)
H—CO	126(8)	In—Cl	439(8)
H—CHO	364(4)	In—F	506(15)
H—COOH	377	In—O	360(21)
H—COCH$_3$	364(4)	In—P	197.9(85)
H—COCH$_2$CH$_3$	364(4)	In—S	289(17)
		In—Sb	152(11)
H—⟨furan ring⟩	385	In—Se	247(17)
		In—Te	218(17)
H—COC$_6$H$_5$	364(4)	Iodine	
H—COCF$_3$	381(8)		
H—F	568.6(13)	I—I	152.549(8)
H—I	298.7(8)	I—Br	179.1(4)
H—N	314(17)	I—CH$_3$	232(13)
H—NH	377(8)	I—C$_2$H$_5$	223.8
H—NH$_2$	435(8)	I—CH(CH$_3$)$_2$	222
H—NHCH$_3$	431(8)	I—C(CH$_3$)$_3$	207.1
H—N(CH$_3$)$_2$	397(8)	I—CH$_2$CF$_3$	234(4)
H—NHC$_6$H$_5$	335(13)	I—CF$_2$CH$_3$	216(4)
H—N(CH$_3$)C$_6$H$_5$	310(13)	I—C$_3$F$_7$	209(4)
HNF$_2$	318(13)	I—CH=CHCH$_3$	172
H—N$_3$	356	I—C$_6$H$_5$	268(4)
H—NO	<205	I—C$_6$F$_5$	276
H—O	428.0(21)	I—Cl	213.3(4)
H—OH	498.7(8)	I—COCH$_3$	219.7
H—OCH$_3$	436.8(42)	I—CN	305(4)
H—OCH$_2$CH$_3$	436.0	I—F	280(4)
H—OC(CH$_3$)$_3$	439(4)	I—N	159(17)
H—OC$_6$H$_5$	368(25)	I—NO	71(4)
H—ONO	327.6(25)	I—NO$_2$	75(4)
H—ONO$_2$	423.4(25)	I—O	184(21)

TABLE 4.11   Bond Dissociation Energies (*Continued*)

Bond	ΔHf_{298}, kJ/mol	Bond	ΔHf_{298}, kJ/mol
Iridium		**Lithium (*continued*)**	
Ir—O	352(21)	Li—O	341(6)
Ir—Si	463(21)	Li—OH	427(21)
Iron		**Lutetium**	
Fe—Fe	100(21)	Lu—Lu	142(34)
Fe—Br	247(96)	Lu—F	569(42)
Fe—Cl	*ca.* 352	Lu—O	695(13)
Fe—O	409(13)	Lu—S	507(15)
Fe—S	339(21)	Lu—Te	326(17)
Fe—Si	297(25)	**Magnesium**	
Krypton		Mg—Mg	8.522(4)
Kr—Kr	5.4(8)	Mg—Br	297(63)
Kr—F	54	Mg—Cl	318(13)
		Mg—F	462(21)
Lanthanum		MgF—F	569(42)
		Mg—H	197(50)
La—La	247(21)	Mg—I	*ca.* 285
La—C	506(63)	Mg—O	394(35)
La—F	598(42)	Mg—OH	238(21)
La—N	519(42)	Mg—S	310(75)
La—O	799(13)	**Manganese**	
La—S	577(25)	Mn—Mn	42(29)
Lead		Mn—Br	314(10)
		Mn—Cl	361(10)
Pb—Pb	339(25)	Mn—F	423(15)
Pb—Br	247(38)	Mn—I	283(10)
Pb(CH$_3$)$_3$—CH$_3$	207(42)	Mn—Cu	159(17)
Pb—Cl	301(29)	Mn—O	402(34)
Pb—F	356(8)	Mn—S	301(17)
Pb—H	176(21)	Mn—Se	201(13)
Pb—I	197(38)	**Mercury**	
Pb—O	378(4)	Hg—Hg	17.2(21)
Pb—S	346.0(17)	Hg—Br	72.8(42)
Pb—Se	303(4)	CH$_3$—HgCH$_3$	240.6
Pb—Te	251(13)	C$_2$H$_5$—HgC$_2$H$_5$	182.8(42)
Lithium		C$_3$H$_7$—HgC$_3$H$_7$	197.1
		Isopropyl—Hgisopropyl	170.3
Li—Li	106(4)	C$_6$H$_5$—HgC$_6$H$_5$	285
Li—Br	423(21)	Hg—Cl	100(8)
Li—Cl	469(13)	Hg—F	130(38)
Li—F	577(21)	Hg—H	39.8
Li—H	247		
Li—I	352(13)		
Li—Na	88		

TABLE 4.11 Bond Dissociation Energies (*Continued*)

Bond	ΔHf_{298}, kJ/mol	Bond	ΔHf_{298}, kJ/mol
Mercury (*continued*)		**Nitrogen**	
Hg—I	38	N—N	945.33(59)
Hg—K	8.24(21)	N—Br	276(21)
Hg—Na	>6.7	ON—Br	28.7(15)
Hg—S	213	N—Cl	389(50)
Hg—Se	(167)	ON—Cl	159(6)
Hg—Te	(142)	O_2N—Cl	142(4)
		N—F	301(42)
Molybdenum		FN—F	318(21)
		F_2F—N	243(8)
Mo—I	372	ON—F	236(4)
Mo—O	607(34)	O_2N—F	188(21)
MoO—O	678(84)	N—I	159(17)
MoO_2—O	565(84)	F_2N—NF_2	88(4)
		H_2N—NH_2	297(8)
Neodymium		H_2N—$NHCH_3$	271
		H_2N—$N(CH_3)_2$	264
Nd—F	545(13)	H_2N—NHC_6H_5	213
Nd—O	703(34)	HN—N_2	38
Nd—S	474(15)	ON—N	480.7(42)
Nd—Se	385(17)	ON—NO_2	39.8(8)
Nd—Te	305(17)	O_2N—NO_2	57.3(21)
		HN=NH	456(42)
Neon		N≡N	946
		N—O	630.57(13)
Ne—Ne	3.93	HN=O	481
		NN—O	167
Neptunium		ON—O	305
		N—P	617(21)
Np—O	720(29)	N—S	464(21)
Nickel		**Osmium**	
Ni—Ni	261.9(25)	O_3Os—O	301(21)
Ni—Br	360(13)		
Ni—Cl	372(21)	**Oxygen**	
Ni—F	435		
Ni—H	289(13)	O—O	498.34(20)
Ni—I	293(21)	O—Br	235.1(4)
Ni—O	391.6(38)	HO—CH_3	377(13)
Ni—S	360(21)	HO—CH=CH_2	364
Ni—Si	318(17)	HO—CH_2CH=CH_2	456
Niobium		HO—C_6H_5	431
		HO—$CH_2C_6H_5$	322
Nb—O	753(13)	HO—CHO	402(13)
		HO—$COCH_3$	452(21)
		HO—COC_2H_5	180
		O—Cl	272(4)
		HO—Cl	251(13)
		O—F	222(17)

TABLE 4.11 Bond Dissociation Energies (*Continued*)

Bond	ΔHf_{298}, kJ/mol	Bond	ΔHf_{298}, kJ/mol
Oxygen (*continued*)		Potassium (*continued*)	
O—FO	467	K—Cl	427(8)
FO—OF	261(84)	K—F	497.5(25)
O—I	184(21)	K—H	183(15)
HO—I	234(13)	K—I	331(13)
O—N	630.57(13)	K—Na	63.6(29)
HO—NCH$_3$	209	K—O	239(34)
HO—OC(CH$_3$)$_3$	192(8)	K—OH	343(8)
HO—OH	213.8(21)		
O—OH	268(4)	Praseodymium	
CF$_3$O—OCF$_3$	192		
CH$_3$O—OCH$_3$	157.3(8)	Pr—F	582(46)
C$_2$H$_5$O—OC$_2$H$_5$	159	Pr—O	753(17)
C$_3$H$_7$O—OC$_3$H$_7$	155	Pr—S	492.5(46)
		Pr—Se	446(23)
Palladium		Pr—Te	326(42)
Pd—O	234(29)	Promethium	
		Pm—F	540(42)
Phosphorus		Pm—O	674(63)
		Pm—S	423(63)
P—P	490(11)	Pm—Se	339(63)
P—Br	266.5	Pm—Te	255(63)
P—C	513(8)		
P—Cl	289(42)	Radium	
P—F	439(96)		
P—H	343(29)	Ra—Cl	343(75)
P—N	617(21)		
P—O	596.6	Rhodium	
Br$_3$P=O	498(21)		
Cl$_3$P=O	510(21)	Rh—Rh	285(21)
F$_3$P=O	544(21)	Rh—B	476(21)
P—S	346.0(17)	Rh—C	583.7(63)
P=S	347	Rh—O	377(63)
P—Se	363(10)	Rh—Si	395(18)
P—Te	298(10)	Rh—Ti	391(15)
Platinum		Rubidium	
Pt—B	478(17)	Rb—Rb	45.6(21)
Pt—H	352(38)	Rb—Br	389(13)
Pt—O	347(34)	Rb—Cl	448(21)
Pt—P	417(17)	Rb—F	494(21)
Pt—Si	501(18)	Rb—H	167(21)
		Rb—I	335(13)
Potassium		Rb—O	255(84)
		Rb—OH	351(8)
K—K	57.3(42)		
K—Br	383(8)		

TABLE 4.11 Bond Dissociation Energies (*Continued*)

Bond	ΔHf_{298}, kJ/mol	Bond	ΔHf_{298}, kJ/mol
Ruthenium		**Silicon** (*continued*)	
Ru—O	481(63)	Si—H	298.49(46)
O$_3$Ru—O	439	Si—I	339(84)
Ru—Si	397(21)	Si—N	439(38)
Ru—Th	592(42)	Si—O	798(8)
		Si—S	619(13)
Samarium		Si—Se	531(25)
		H$_3$Si—SiH$_3$	339(17)
Sm—Cl	423(13)	(CH$_3$)$_3$Si—Si(CH$_3$)$_3$	339
Sm—F	531(18)	(Aryl)$_3$Si—Si(aryl)$_3$	368(31)
Sm—O	619(13)	Si—Te	506(38)
Sm—S	389		
Sm—Se	331(15)	**Silver**	
Sm—Te	272(15)	Ag—Ag	163(8)
		Ag—Au	203(9)
Scandium		Ag—Bi	193(42)
		Ag—Br	293(29)
Sc—Sc	163(21)	Ag—Cl	341.4
Sc—Br	444(63)	Ag—Cu	176(8)
Sc—C	393(63)	Ag—F	354(16)
Sc—Cl	318	Ag—Ga	180(15)
Sc—F	589(13)	Ag—Ge	175(21)
Sc—N	469(84)	Ag—H	226(8)
Sc—O	674(13)	Ag—I	234(29)
Sc—S	478(13)	Ag—In	176(17)
Sc—Se	385(17)	Ag—O	213(84)
Sc—Te	289(17)	Ag—Sn	136(21)
		Ag—Te	293(96)
Selenium			
		Sodium	
Se—Se	332.6(4)	Na—Na	77.0
Se—Br	297(84)	Na—Br	370(13)
Se—C	582(96)	Na—Cl	410(8)
Se—Cl	322	Na—F	481(8)
Se—F	339(42)	Na—H	201(21)
Se—H	305(2)	Na—I	301(8)
Se—N	381(63)	Na—K	63.6(29)
Se—O	423(13)	Na—O	257(17)
Se—P	364(10)	Na—OH	381(13)
Se—S	381(21)	Na—Rb	59(4)
Se—Si	531(25)		
Se—Te	268(8)	**Strontium**	
		Sr—Br	332(19)
Silicon		Sr—Cl	406(13)
Si—Si	327(10)	Sr—F	542(7)
Si—Br	343(50)	Sr—H	163(8)
Si—C	435(21)	Sr—I	263(42)
Si—Cl	456(42)		
Si—F	540(13)		

TABLE 4.11   Bond Dissociation Energies (*Continued*)

Bond	ΔHf_{298}, kJ/mol	Bond	ΔHf_{298}, kJ/mol
Strontium (*continued*)		**Thorium**	
Sr—O	454(15)	Th—Th	289
Sr—OH	381(42)	Th—C	484(25)
Sr—S	314(21)	Th—N	577.4(21)
		Th—O	854(13)
Sulfur		Th—P	377
S—S	429(6)	**Thullium**	
S—Cl	255		
S—F	343(5)	Tm—F	569(42)
O_2S—F	71	Tm—O	557(13)
S—N	464(21)	Tm—S	368(42)
S—O	521.70(13)	Tm—Se	276(42)
OS—O	551.4(84)	Tm—Te	276(42)
O_2S—O	348.1(42)		
HS—SH	272(21)	**Tin**	
Tantalum		Sn—Sn	195(17)
		Sn—Br	339(4)
Ta—N	611(84)	BrSn—Br	326
Ta—O	805(13)	Br_3Sn—Br	272
		$(C_2H_5)_3Sn$—C_2H_5	*ca.* 238
Tellurium		Sn—Cl	406(13)
		Sn—F	467(13)
Te—B	354(20)	Sn—H	267(17)
Te—H	268(2)	Sn—I	234(42)
Te—I	193(42)	Sn—O	548(21)
Te—O	391(8)	Sn—S	464(3)
Te—P	298(10)	Sn—Se	401.3(59)
Te—S	339(21)	Sn—Te	319.2(8)
Te—Se	268(8)		
		Titanium	
Terbium		Ti—Ti	141(21)
		Ti—Br	439
Tb—F	561(42)	Ti—C	435(25)
Tb—O	707(13)	Ti—Cl	494
Tb—S	515(42)	Ti—F	569(34)
Tb—Te	339(42)	Ti—H	*ca.* 159
		Ti—I	310(42)
Thallium		Ti—N	464
		Ti—O	662(16)
Tl—Tl	63	Ti—S	426(8)
Tl—Br	333.9(17)	Ti—Se	381(42)
Tl—Cl	372.8(21)	Ti—Te	289(17)
Tl—F	445(19)		
Tl—H	188(8)	**Tungsten**	
Tl—I	272(8)		
		W—Cl	423(42)

TABLE 4.11 Bond Dissociation Energies (*Continued*)

Bond	ΔHf_{298}, kJ/mol	Bond	ΔHf_{298}, kJ/mol
Tungsten (*continued*)		**Ytterbium** (*continued*)	
W—F	548(63)	Yb—O	397.9(63)
W—O	653(25)	Yb—S	167
OW—O	632(84)		
O₂W—O	598(42)	**Yttrium**	
W—P	305(4)		
		Y—Y	159(21)
Uranium		Y—Br	485(84)
		Y—C	418(63)
U—O	761(17)	Y—Cl	527(42)
OU—O	678(59)	Y—F	605(21)
O₂U—O	644(88)	Y—N	481(63)
U—S	523(10)	Y—O	715.1(30)
		Y—S	528(11)
Vanadium		Y—Se	435(13)
		Y—Te	339(13)
V—V	242(21)		
V—Br	439(42)	**Zinc**	
V—C	469(63)		
V—Cl	477(63)	Zn—Zn	29
V—F	590(63)	Zn—Br	142(29)
V—N	477(8)	C₂H₅C—C₂H₅	*ca.* 201
V—O	644(21)	Zn—Cl	229(20)
V—S	490(16)	Zn—F	368(63)
V—Se	347(21)	Zn—H	85.8(21)
		Zn—I	138(29)
Xenon		Zn—O	284.1
		Zn—S	205(13)
Xe—Xe	6.53(30)	Zn—Se	136(13)
Xe—F	13.0(4)	Zn—Te	205
Xe—O	36.4		
		Zirconium	
Ytterbium			
		Zr—C	561(25)
Yb—Cl	322	Zr—F	623(63)
Yb—F	521(10)	Zr—N	565(25)
Yb—H	159(38)	Zr—O	760(8)
		Zr—S	575(17)

4.6 BOND AND GROUP DIPOLE MOMENTS

All bonds between equal atoms are given zero values. Because of their symmetry, methane and ethane molecules are nonpolar. The principle of bond moments thus requires that the CH_3 group moment equal one H—C moment. Hence the substitution of any aliphatic H by CH_3 does not alter the dipole moment, and all saturated hydrocarbons have zero moments as long as the tetrahedral angles are maintained.

The group moment always includes the C—X bond. When the group is attached to an aromatic system, the moment contains the contributions through resonance of those polar structures postulated as arising through charge shifts around the ring.

All values for bond and group dipole moments in Tables 4.12 and 4.13 were obtained in benzene solutions.

TABLE 4.12 Bond Dipole Moments

Bond	Moment, D*	Bond	Moment, D*
H—C		H—Cl	1.08
Aliphatic	0.3	H—F	1.94
Aromatic	0.0	C—Te	0.6
C—C	0.0	N—F	0.17
C≡C	0.0	P—I	0.3
C—O		P—Br	0.36
Ether, aliphatic	0.74	P—Cl	0.81
Alcohol, aliphatic	0.7	As—I	0.78
C=O		As—Br	1.27
Aliphatic	2.4	As—Cl	1.64
Aromatic	2.65	As—F	2.03
O—H	1.51	Sb—I	0.8
C—S	0.9	Sb—Br	1.9
C=S	2.0	Sb—Cl	2.6
S—H	0.65	S—Cl	0.7
S—O	(0.2)	Cl—O	0.7
S=O		I—Br	1.2
Aliphatic	2.8	I—Cl	1
Aromatic	3.3	Br—Cl	0.57
C—N, aliphatic	0.45	Br—F	1.3
C=N	1.4	Cl—F	0.88
C≡N (nitrile)	3.6	Li—C	1.4
NC (isonitrile)	3.0	K—Cl	10.6
N—H	1.31	K—F	7.3
N—O	0.3	Cs—Cl	10.5
N=O	2.0	Cs—F	7.9
N (lone pair on sp^3N)	1.0		
C—P, aliphatic	0.8	Dative (coordination) bonds	
P—O	(0.3)		
P=O	2.7	N → B	2.6
P—S	0.5	O → B	3.6
P=S	2.9	S → B	3.8
B—C, aliphatic	0.7	P → B	4.4
B—O	0.25	N → O	4.3
Se—C	0.7	P → O	2.9
Si—C	1.2	S → O	3.0
Si—H	1.0	As → O	4.2
Si—N	1.55	Se → O	3.1
H—Sb	−0.08	Te → O	2.3
H—As	−0.10	P → S	3.1
H—P	0.36	P → Se	3.2
H—I	0.38	Sb → S	4.5
H—Br	0.78		

* To convert debye units D into coulomb-meters, multiply by 3.33564×10^{-30}.

TABLE 4.13 Group Dipole Moments

	Moment, D*	
Group	Aromatic C—X	Aliphatic C—X
C—CH$_3$	0.37	0.0
C—C$_2$H$_5$	0.37	0.0
C—C(CH$_3$)$_3$	0.5	0.0
C—CH=CH$_2$	<0.4	0.6
C—C≡CH	0.7	0.9
C—F	1.47	1.79
C—Cl	1.59	1.87
C—Br	1.57	1.82
C—I	1.40	1.65
C—CH$_2$F	1.77	
C—CF$_3$	2.54	2.32
C—CH$_2$Cl	1.85	1.95
C—CHCl$_2$	2.04	1.94
C—CCl$_3$	2.11	1.57
C—CH$_2$Br	1.86	1.96
C—C≡N	4.05	3.4
C—NC	3.5	3.5
C—CH$_2$CN	1.86	2.0
C—C=O	2.65	2.4
C—CHO	2.96	2.49
C—COOH	1.64	1.63
C—CO—CH$_3$	2.96	2.49
C—CO—OCH$_3$	1.83	1.75
C—CO—OC$_2$H$_5$	1.9	1.8
C—OH	1.6	1.7
C—OCH$_3$	1.28	1.28
C—OCF$_3$	2.36	
C—OCOCH$_3$	1.69	
C—OC$_6$H$_5$	1.16	1.16
C—CH$_2$OH	1.58	1.68
C—NH$_2$	1.53	1.46
C—NHCH$_3$	1.71	
C—N(CH$_3$)$_2$	1.58	0.86
C—NHCOCH$_3$	3.69	
C—N(C$_6$H$_5$)$_2$	(0.3)	−0.3
C—NCO	2.32	2.8
C—N$_3$	1.44	
C—NO	3.09	
C—NO$_2$	4.01	2.70
C—CH$_2$NO$_2$	3.3	3.4
C—SH	1.22	1.55
C—SCH$_3$	1.34	1.40
C—SCF$_3$	2.50	
C—SCN	3.59	3.6
C—NCS	2.9	3.3
C—SC$_6$H$_5$	1.51	1.5
C—SF$_5$	3.4	
C—SOCF$_3$	3.88	
(C—)$_2$SO$_2$	5.05	4.53

* To convert debye units D into coulomb-meters, multiply by 3.33564×10^{-30}.

TABLE 4.13 Group Dipole Moments (*Continued*)

Group	Moment, D*	
	Aromatic C—X	Aliphatic C—X
$(C-)_2SO_2CH_3$	4.73	
$(C-)_2SO_2CF_3$	4.32	
C—SeH	1.08	
C—SeCH$_3$	1.31	1.32
C—Si(CH$_3$)$_3$	0.44	0.4

* To convert debye units D into coulomb-meters, multiply by 3.33564×10^{-30}.

4.7 MOLECULAR GEOMETRY

TABLE 4.14 Spatial Orientation of Common Hybrid Bonds

On the assumption that the pairs of electrons in the valency shell of a bonded atom in a molecule are arranged in a definite way which depends on the number of electron pairs (coordination number), the geometrical arrangement or shape of molecules may be predicted. A multiple bond is regarded as equivalent to a single bond as far as molecular shape is concerned.

Coordination Number	Orbitals Hybridized	Geometrical Arrangement	Minimum Radius Ratio
2	sp dp	Linear	
	p^2 ds d^2	Bent (angular)	
3	sp^2 ds^2	Trigonal planar	0.155
	p^3 d^2p	Trigonal pyramidal	
	sp^2d p^2d^2	Square planar	
4	sp^3 d^3s	Tetrahedral	0.225
	d^4	Tetragonal pyramidal	
5	sp^3d d^3sp	Trigonal bipyramidal	0.155
6	d^2sp^3	Octahedral	0.414
	d^4sp	Trigonal prism	
7		One atom above the face of an octahedron, which is distorted chiefly by separating the atoms at the corners of this face.	0.592
8	d^4sp^3	Square antiprism (dodecahedral)	0.645
		Cube	0.732
9		Formed by adding atoms beyond each of the vertical faces of a right triangular prism.	0.732
12		Cube-octahedron	1.000

TABLE 4.15 Crystal Structure

Unit cells of the different lattice types in each system are illustrated in Fig. 4.1.

System	Characteristics	Essential Symmetry	Axes in Unit Cell	Angles in Unit Cell
Cubic	Three axes equal and mutually perpendicular	Four threefold axes	$a = b = c$	$\alpha = \beta = \gamma = 90°$
Tetragonal	Two equal axes and one unequal axis mutually perpendicular	One fourfold axis	$a = b \neq c$	$\alpha = \beta = \gamma = 90°$
Orthorhombic (or rhombic)	Three unequal axes mutually perpendicular	Three mutually perpendicular twofold axes, or two planes intersecting in a twofold axis	$a \neq b \neq c$	$\alpha = \beta = \gamma = 90°$
Hexagonal or trigonal	Three equal axes inclined at 120° with a fourth axis unequal and perpendicular to the other three	One sixfold axis or one threefold axis	$a = b \neq c$ / $a = b = c$	$\alpha = \beta = 90°;$ $\gamma = 120°$ / $\alpha = \beta = \gamma \neq 90°$
Monoclinic	Two axes at an oblique angle with a third perpendicular to the other two	One twofold axis or one plane	$a \neq b \neq c$	$\alpha = \beta = 90°;$ $\gamma \neq 90°$
Triclinic	Three unequal axes intersecting obliquely	No planes or axes of symmetry	$a \neq b \neq c$	$\alpha \neq \beta \neq \gamma \neq 90°$
Rhombohedral	Two equal axes making equal angle with each other			

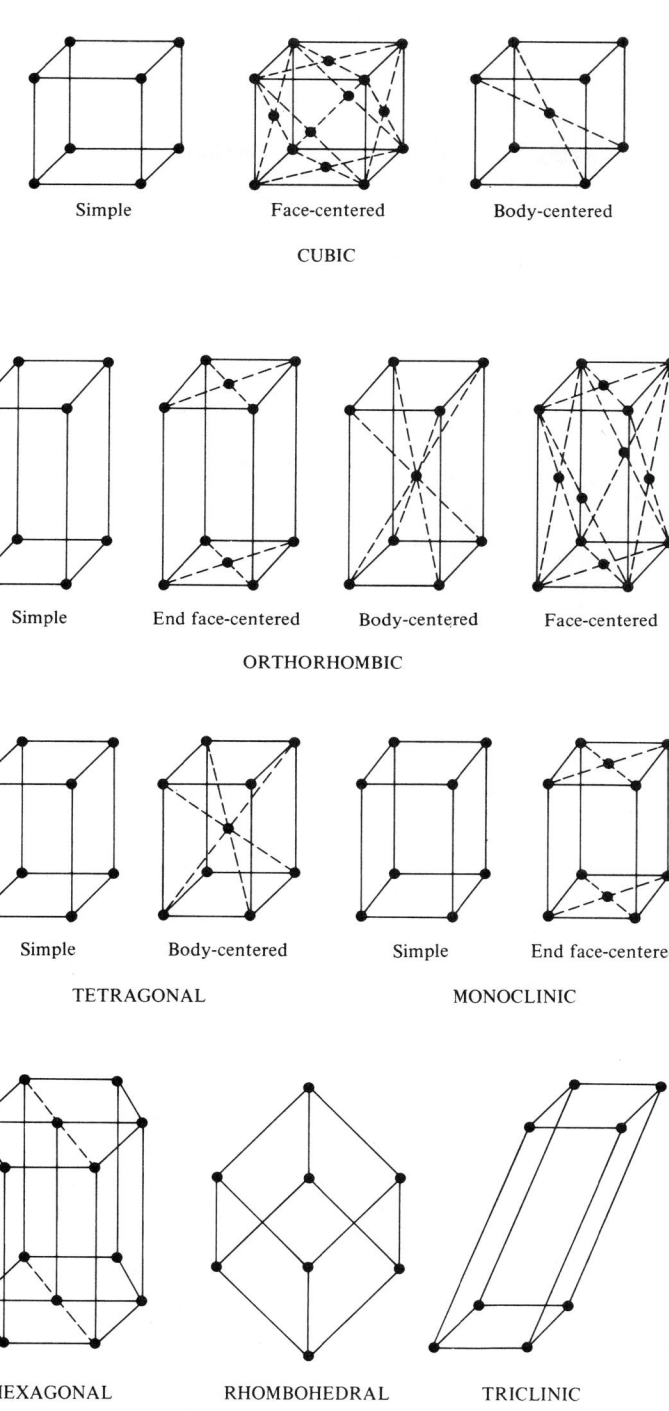

FIGURE 4.1 Crystal lattice types.

4.8 NUCLIDES

TABLE 4.16   Table of Nuclides

Explanation of Column Headings

Nuclide.   Each nuclide is identified by element name and the mass number A, equal to the sum of the numbers of protons Z and neutrons N in the nucleus. The m following the mass number (for example, ^{69m}Zn) indicates a metastable isotope. An asterisk preceding the mass number indicates that the radionuclide occurs in nature.

Half-life.   The following abbreviations for time units are employed: y = years, d = days, h = hours, min = minutes, and s = seconds.

Natural abundance.   The natural abundances listed are on an "atom percent" basis for the stable nuclides present in naturally occurring elements in the earth's crust.

Thermal neutron absorption cross section.   Simply designated "cross section," it represents the ease with which a given nuclide can absorb a thermal neutron (energy less than or equal to 0.025 eV) and become a different nuclide. The cross section is given here in units of barns (1 barn = 10^{-24} cm^2). If the mode of reaction is other than (n, γ), it is so indicated.

Major radiations.   In the last column are the principal modes of disintegration and energies of the radiations in million electronvolts (MeV). Symbols used to represent the various modes of decay are

α, alpha particle emission	K, electron capture
β^-, beta particle (negatron)	IT, isomeric transition
β^+, positron	e$^-$, internal conversion electron
γ, gamma radiation	x, X-rays of indicated element (e.g., O-x, oxygen X-rays)

For β^- and β^+, values of E_{max} are listed. Radiation types and energies of minor importance are omitted unless useful for identification purposes. For detailed decay schemes the literature should be consulted.

Element	A	Half-life	Natural abundance, %	Cross section, barns	Radiation (MeV)
Hydrogen	1		99.985	0.332	
	2		0.015	0.0005	
	3	12.33 y			β^-(0.0186)
Beryllium	7	53.29 d			K, γ(0.478)
	9		100	0.009	
	10	1.6 × 10^6 y			β^-(0.555)
Boron	10		19.9	3 837 (n, α)	
Carbon	11	20.39 min			β^+(0.961)
	14	5730 y			β^-(0.156)
Nitrogen	13	9.965 min			β^+(1.190)
	14		99.635	1.81 (n, p)	
Oxygen	19	29.1 s			β^-(4.60); γ (0.197, 1.37)
Fluorine	18	1.8295 h			β^+(0.635); K, O-x
	20	11.56 s			β^-(5.41), γ(1.63)
Sodium	22	2.602 y			β^+(0.545, 1.83); K Ne-x, γ(1.275)
	23		100	0.53	
	24	14.659 h			β^-(1.39); γ (2.75, 1.37)
Magnesium	25		10.00	0.18	

Source: C. M. Lederer and V. S. Shirley, eds., *Table of Isotopes,* 7th ed., Wiley-Interscience, New York, 1978, and V. S. Shirley, ed., *Table of Radioactive Isotopes,* 8th ed., Wiley-Interscience, New York, 1986.

TABLE 4.16 Table of Nuclides (*Continued*)

Element	A	Half-life	Natural abundance, %	Cross section, barns	Radiation (MeV)
Magnesium (*cont.*)	27	9.46 min			β^-(1.75), γ(0.84, 1.01)
	28	20.90 h			β^-(0.46); γ (1.34, 0.94, 0.40, 0.031)
Aluminum	26	7.2×10^5 y			β^+(1.16); K, Mg-x; γ(1.81, 1.12)
	27		100	0.235	
	28	2.31			β^-(2.85); γ(1.780)
Silicon	30		3.10	0.11	
	31	2.622 h			β^-(1.48); γ(1.27)
	32	104 y			β^-(0.213)
Phosphorus	31		100	0.19	
	32	14.282 d			β^-(1.71)
	33	25.34 d			β^-(0.25)
Sulfur	34		4.22	0.27	
	35	87.51 d			β^-(0.167)
	36	3.08×10^5 y		100	β^-(0.714)
	37	5.07 min			β^-(4.7, 1.6), γ(3.09)
	38	2.84 h			β^-(1.0, 3.0); γ(1.94)
Chlorine	35		75.77	44	
	36	3.01×10^5 y			β^-(0.71); K, S-x
	37		24.23	0.4	
	38	37.24 min			β^-(4.81, 1.11, 2.77); γ(2.17, 1.60)
	39	55.6			β^-(1.91, 2.18, 3.45); γ(1.27, 0.25, 1.52)
Argon	37	35.04 d			K, Cl-x
	41	1.827 h			β^-(1.20, 2.49); γ(1.29)
Potassium	*40	1.28×10^9 y	0.118	70	β^-(1.34); K, Ar-x; γ(1.46)
	41		6.730	1.2	
	42	12.360 h			β^-(3.52, 1.97); γ(1.46)
	43	22.3 h			β^-(0.83, 0.46, 1.22, 1.82); γ(0.618, 0.373, 0.39, 0.59, 0.22)
Calcium	44		2.086	0.7	
	45	163.8 d			β^-(0.255)
	47	4.536 d			β^-(1.98, 0.67); γ(1.30, 0.81, 0.49)
	49	8.8 min			β^-(1.95), γ(3.10, 4.1)
Scandium	46	83.83 d			β^-(0.357); γ(1.12, 0.889); Ti-x
	46*m*	19.5 s			γ(0.142)
Titanium	45	3.08 h			β^+(1.044); K, Sc-x
	50		5.4	0.14	
	51	5.79 min			β^-(2.14); γ(0.320, 0.605, 0.928)
Vanadium	48	16.0 d			β^+(0.696), γ(0.511, 0.945, 0.983, 1.312, 2.24)
	52	3.75 min			β^-(2.47); γ(1.434)

TABLE 4.16 Table of Nuclides (*Continued*)

Element	A	Half-life	Natural abundance, %	Cross section, barns	Radiation (MeV)
Chromium	51	27.70 d			K, V-x; γ(0.32)
	52		83.79	0.8	
Manganese	52	5.60 d			β^+(0.575), γ(0.511, 0.744, 0.935, 1.434)
	54	303 d			γ(0.835)
	55		100	13.3	
	56	2.5785 h			β^-(2.84, 1.03, 0.72); γ(0.847, 1.81, 2.11)
Iron	55	2.73 y			K, Mn-x
	59	44.496 d			β^-(0.273, 0.475); γ(1.10, 1.29, 0.192)
Cobalt	57	271.77 d			K, Fe-x; γ(0.136, 0.122)
	58	70.916 d			K, β^+(0.474), Fe-x; γ(0.811)
	59		100	19	
	60	5.271 y			β^-(0.318); γ(1.173, 1.332)
	60m	10.47 min			β^-(1.55)
Nickel	63	92 y			β^-(0.067)
	64		0.91	1.5	
	65	2.520 h			β^-(2.14, 0.65, 1.02); γ(1.48, 0.367, 1.12)
Copper	63		69.17	4.5	
	64	12.701 h			β^-(0.571); β^+(0.657); Ni-x
	66	5.10 min			β^-(2.63), γ(1.039)
Zinc	64		48.6	0.46	
	65	244.1 d			K, β^+(0.325), Cu-x; γ(1.12)
	68		18.8	1.0	
	69m	13.76 h			IT, Zn-x, γ(0.439)
	69	55.6 min			β^-(0.90)
Gallium	67	3.261 d			K, Zn-x; γ(0.093, 0.184, 0.300, 0.393)
	69		60.2	1.9	
	70	21.1 min			β^-(1.65); γ(0.173)
	71		39.8	5.0	
	72	14.10 h			β^-(0.64, 0.96, 1.51, 2.53, 3.17); γ(0.835, 0.63, 2.20, 2.50)
Germanium	71	11.4 d			Ga-x
	74		36.5	0.3	
	75	1.3797 h			β^-(1.19, 0.98, 0.92, 0.72); γ(0.265, 0.199)
	77	11.3 h			β^-(2.2), γ(0.21, 0.268, 0.368, 0.417, 0.568, 0.632, 0.73)
Arsenic	75		100	4.5	
	76	1.097 d			β^-(2.97, 2.41, 1.79); γ(0.559, 0.657)

TABLE 4.16 Table of Nuclides (*Continued*)

Element	A	Half-life	Natural abundance, %	Cross section, barns	Radiation (MeV)
Arsenic (*cont.*)	78	91 min			β^-(4.1); γ(0.614, 0.70, 0.83, 1.31)
Selenium	74		0.91	30	
	75	119.77 d			K, γ(0.265, 0.136, 0.280, 0.121, 0.401); As-x
	77m	17.5 s			γ(0.161)
	80		49.7	0.5	
	81	18.6 min			β^-(1.58); γ(0.56, 0.83)
Bromine	79		50.69	8.5	
	80	17.68 min			β^-(1.997, 1.38); K, β^+(0.85), Se-x; γ(0.616, 0.667)
	80m	4.42 h			IT, Br-x; γ(0.037, 0.049)
	81		49.31	3	
	82	1.4708 d			β^-(0.444); γ(0.554, 0.619, 0.698, 0.777, 0.818, 1.04, 1.32, 1.48)
Krypton	81m	13 s			IT, Kr-x; γ(0.19)
	84		56.90	0.10	
	85	10.72 y			β^-(0.67); γ(0.517)
Rubidium	85		72.16	0.9	
	86	18.66 d			β^-(1.78, 0.71); γ(1.08)
	87		27.84	0.12	
	88	17.8 min			β^-(5.080, 3.240, 2.350); γ(1.836, 0.898, 2.678)
Strontium	85	64.84 d			K, Rb-x; γ(0.514)
	87m	2.795 h			IT, γ(0.388)
	89	52.7 d			β^-(1.463); γ(0.91)
	90	28.5y			β^-(0.546)
Yttrium	90	2.671 d			β^-(2.288); γ(2.186)
	91	58.8 d			β^-(1.545); γ(1.21)
Zirconium	95	65.5 d			β^-(0.89, 0.396); γ(0.724, 0.756)
	97	16.90 h			β^-(1.91); γ(0.743)
Niobium	93		100	1	
	94m	6.29 min			γ(0.871)
	95	35.0 d			β^-(0.160); γ(0.765)
Molybdenum	98		24.13	0.51	
	99	2.7477 d			β^-(1.214); Tc-x; γ(0.181, 0.740, 0.780)
	101	14.6 min			β^-(2.23); γ(0.191, 0.51, 0.59, 0.70, 0.89)
Technetium	95	20.0 h			K, Mo-x, γ(0.766, 0.84)
	96	4.28 d			K, Mo-x, γ(0.778, 0.813, 0.850)
	97	2.6×10^6 y			K, Mo-x
	99	2.13×10^5 y			β^-(0.292)
	99m	6.006 h			IT, Tc-x; γ(0.141)

TABLE 4.16   Table of Nuclides (*Continued*)

Element	A	Half-life	Natural abundance, %	Cross section, barns	Radiation (MeV)
Ruthenium	97	2.88 d			K, Tc-x, λ(0.216, 0.324)
	102		31.6	1.4	
	103	39.254 d			β^-(0.12, 0.22); γ(0.497)
	106	1.020 y			β^-(0.0392)
Rhodium	103		100	144	
	103m	56.12 min			IT, Rh-x, γ(0.129)
	104	43 s			β^-(2.44), γ(0.56)
	104m	4.41 min			γ(0.051)
	105	1.4733 d			β^-(0.568, 0.249), γ(0.306, 0.319)
	106	29.80 s			β^-(3.53, 3.1, 2.4); γ(0.512, 0.622, 1.128)
Palladium	103	16.97 d			K, β^+(0.776, 0.488), Rh-x, γ(0.270, 0.296)
	109	13.7 h			β^-(1.028); Ag-x; γ(0.088, 0.311, 0.636)
Silver	105	41.29 d			K, β^+(0.325), Pd-x, γ(0.556, 0.768)
	108	2.42 min			β^-(1.64), β^+(0.90) γ(0.434, 0.511, 0.632)
	110m	249.76 d			β^-(0.087, 0.53); IT, γ(0.658, 0.885, 0.937)
	111	7.45 d			β^-(1.04, 0.69), γ(0.342)
Cadmium	111m	48.6 min			γ(0.159, 0.247)
	113		12.22	20 000	
	115	2.228 d			β^-(1.11, 0.58); In-x; γ(0.336, 0.528)
	115m	44.6 d			β^-(1.62), γ(0.934, 1.29, 0.485)
Indium	111	2.807 d			K, Cd-x; γ(0.172, 0.247)
	113m	1.658 h			IT, In-x; γ(0.393)
	114	1.1983 min			β^-(1.99, 0.67), K, β^+(0.40), Cd-x, γ(1.30)
	*115	4.41 × 10^{14} y			β^-(0.495)
	116m	54.0 min			β^-(1.00), γ(0.417, 0.819, 1.09, 1.293, 1.509, 2.111)
Tin	113	115.09 d			K, In-x, γ(0.392)
	125	9.64 d			β^-(2.35, 0.40); γ(0.823, 0.916, 1.067, 1.089)
	127	2.10 h			β^-(1.45), γ(1.114)
Antimony	121		57.25	6	
	122	2.70 d			β^-(1.414, 1.980, 0.723) γ(0.564); K, Sn-x
	123		42.75	3.3	
	124	60.20 d			β^-(2.301); γ(0.603, 1.69, 0.722)
Tellurium	127	9.35 h			β^-(0.70), I-x, γ(0.360, 0.418)

TABLE 4.16 Table of Nuclides (*Continued*)

Element	A	Half-life	Natural abundance, %	Cross section, barns	Radiation (MeV)
Tellurium (*cont.*)	129	1.160 h			β^-(1.453, 0.989, 0.69, 0.29), I-x, γ(0.460, 0.487)
	131m	1.25 d			β^-(2.46, 0.57, 0.42), IT, Te-x, I-x, γ(0.150, 0.774, 0.794)
	131	25.0 min			β^-(2.14, 1.69, 1.35); I-x; γ(0.150, 0.453)
Iodine	123	13.2 h			K, Te-x; γ(0.159)
	125	60.14 d			K, Te-x; λ(0.035)
	127		100	6.4	
	128	24.99 min			β^-(2.12); γ(0.441, 0.743, 0.969)
	131	8.040 d			β^-(0.607, 0.336); Xe-x γ(0.283, 0.364, 0.637, 0.723)
Xenon	133	5.245 d			β^-(0.346); Cs-x; γ(0.081)
Cesium	133		100	28	
	134	2.062 y			β^-(0.658, 0.089); γ(0.605, 0.796)
	134m	2.91 h			IT, β^-(0.55), Cs-x, γ(0.127)
	137	30.0 y			β^-(0.514, 1.18); Ba-x γ(0.662)
Barium	135		6.59	5	
	135m	1.196 d			IT, Ba-x; γ(0.268)
	137		11.23	4	
	137m	2.552 min			IT, Ba-x; γ(0.662)
	138		71.66	0.4	
	139	1.41 h			β^-(2.38, 2.23, 0.95); La-x; γ(0.166, 1.421)
	142	11 min			β^-(1.7), γ(0.080, 0.26, 0.89, 0.97, 1.08, 1.20)
Lanthanum	139		99.911	8.9	
	140	1.68 d			β^-(2.164, 1.680, 1.365, 1.150, 0.857, 0.510); γ(0.487, 1.596)
Cerium	140		88.48	0.6	
	141	32.50 d			β^-(0.444, 0.582); Pr-x; γ(0.145)
	142		11.07	1	
	143	1.38 d			β^-(1.40, 1.125, 0.74); Pr-x; γ(0.293)
	144	284.9 d			β^-(0.316), Pr-x, γ(0.080, 0.134)
Praseodymium	141		100	12	
	142	19.13 h			β^-(2.164); γ(1.576)
	143	13.58 d			β^-(0.932)
Neodymium	146		17.18	2	

TABLE 4.16 Table of Nuclides (*Continued*)

Element	A	Half-life	Natural abundance, %	Cross section, barns	Radiation (MeV)
Neodymium (*cont.*)	147	10.98 d			β^-(0.810, 0.369); γ(0.090, 0.531)
Promethium	143	265 d			K, Nd-x; λ(0.742)
	144	363 d			K, Nd-x; γ(0.477, 0.618, 0.696)
	146	5.53 y			K, β^-(0.795); Nd-x γ(0.453, 0.75)
	147	2.6234 y			β^-(0.224), γ(0.122)
	148m	41.29 d			β^-(0.69, 0.50, 0.40); IT, Pm-x, Sm-x; γ(0.550, 0.630)
	150	2.68 h			β^-(3.260); γ(0.344, 0.831, 0.88, 1.165, 0.133, 0.175)
Samarium	152		26.63	210	
	153	1.946 d			β^-(0.81); γ(0.103)
	155	23.5 min			β^-(1.53); γ(0.104, 0.246)
Europium	151		47.77	5 900	
	152	13.33 y			K, β^-(1.492, 0.690, 0.360), β^+(0.727, 0.479), Gd-x, Sm-x, γ(0.122, 0.344, 0.964, 1.086, 1.112, 1.408)
	152m	9.32 h			β^-(1.89); γ(0.122, 0.344, 0.841, 0.963)
	153		52.23	320	
	154	8.8 y			β^-(0.843); γ(0.123, 1.005, 1.274)
Gadolinium	158		24.9	3.4	
	159	18.56 h			β^-(0.95); Tb-x, γ(0.363)
	160		21.9	0.8	
	161	3.6 min			β^-(1.61); γ(0.102, 0.315, 0.361)
Terbium	159		100	46	
	160	72.3 d			β^-(1.76, 0.87); γ(0.299, 0.879, 0.966)
Dysprosium	164		28.18	2 000	
	165	2.334 d			β^-(1.305, 1.215); Ho-x; γ(0.095, 0.362)
	165m	1.26 min			β^-(1.04, 0.89), γ(0.108, 0.152, 0.362, 0.514)
Holmium	165		100	64	
	166	1.117 d			β^-(1.85, 1.78); Er-x; γ(0.081)
Erbium	170		14.88	9	
	171	7.52 h			β^-(1.49, 1.06); Tm-x; γ(0.112, 0.296, 0.308)
Thullium	169		100	125	
	170	128.6 d			β^-(0.968, 0.884); Yb-x γ(0.084)

TABLE 4.16   Table of Nuclides (*Continued*)

Element	A	Half-life	Natural abundance, %	Cross section, barns	Radiation (MeV)
Ytterbium	174		31.84	46	
	175	4.19 d			β^-(0.466); Lu-x; γ(0.114, 0.283, 0.396)
	176		12.73	7	
	177	1.9 h			β^-(1.40); Lu-x; γ(1.080)
Lutetium	175		97.40	18	
	176m	3.635 h			β^-(1.31); Hf-x; γ(0.0884)
	177	6.71 d			β^-(0.497, 0.385, 0.175), Hf-x, γ(0.113, 0.208)
Hafnium	179		13.629	65	
	179m	18.6 s			γ(0.217)
	180		35.100	10	
	180m	5.519 h			IT, Hf-x; γ(0.058, 0.215, 0.333, 0.444, 0.501)
	181	42.39 d			β^-(0.408); Ta-x; γ(0.133, 0.346, 0.482)
Tantalum	181		99.9877	21	
	182	115.0 d			β^-(1.713, 1.470); γ(0.068, 1.121, 1.189, 1.221, 1.231)
	182m	16.5 min			γ(0.147, 0.172, 0.184)
Tungsten	185	75.1 d			β^-(0.433), γ(0.125)
	186		28.4	40	
	187	23.9 h			β^-(1.312, 0.622); Re-x γ(0.072, 0.134, 0.480)
Rhenium	185		37.40	110	
	186	3.777 d			β^-(1.07, 0.933); K, W-x, Os-x; γ(0.137, 0.632, 0.768)
	187		62.60	70	
	188	16.98 h			β^-(2.12, 1.96); Os-x γ(0.155)
Osmium	190		26.4	8.6	
	190m	9.9 min			IT, Os-x; γ(0.187, 0.361, 0.502, 0.616)
	191	15.4 d			β^-(0.143); Os-x; γ(0.129)
	192		41.0	1.6	
	193	1.271 d			β^-(1.13); Ir-x; γ(0.139, 0.28, 0.38, 0.460, 0.558)
Iridium	191		37.3	750	
	192	73.831 d			β^-(0.672); K, Os-x, Pt-x, γ(0.296, 0.308, 0.316, 0.468, 0.604)
	193		62.7	110	
	194	19.15 h			β^-(2.24); γ(0.328)
Platinum	189	10.89 h			K, β^+(0.885, 0.479); Ir-x; γ(0.094, 0.114, 0.141, 0.187, 0.243, 0.31, 0.56, 0.61, 0.722)

TABLE 4.16   Table of Nuclides (*Continued*)

Element	A	Half-life	Natural abundance, %	Cross section, barns	Radiation (MeV)
Platinum (*cont.*)	195*m*	4.02 d			IT, Pt-x, γ(0.0311, 0.0991, 0.130)
	197*m*	1.573 h			IT, β^-(0.737), Pt-x, γ(0.279, 0.346)
	197	18.3 h			β^-(0.719, 0.642, 0.451), Au-x, γ(0.077, 0.191)
	199*m*	14.1 s			γ(0.393)
Gold	197		100	98.7	
	197*m*	7.8 s			IT, Au-x; γ(0.279)
	198	2.6935 d			β^-(1.371); γ(0.412, 0.676)
	199	3.139 d			β^-(0.296, 0.250, 0.462), Hg-x, γ(0.158, 0.208)
Mercury	196		0.146	880	
	197	2.6725 d			K, Au-x; γ(0.077, 0.191)
	197*m*	23.8 h			IT, K, Hg-x, γ(0.134)
	199		16.84	2 000	
	199*m*	43 min			γ(0.158, 0.375)
	202		29.80	4	
	203	46.60 d			β^-(0.214); γ(0.279)
Thallium	201	3.046 d			K, Hg-x; γ(0.135, 0.167)
	203		29.52	11	
	204	3.78 y			β^-(0.763); K, Hg-x
	205		70.48	0.11	
	206	4.20 min			β^-(1.53)
Lead	204*m*	1.120 h			IT, Pb-x, γ(0.375, 0.899, 0.912)
	209	3.253 h			β^-(0.645)
	210	22.3 y			α(3.72); γ(0.0465)
	212	10.64 h			β^-(0.569, 0.331); Bi-x; γ(0.239)
	214	26.8 min			β^-(0.59, 0.65, 1.03); γ(0.352)
Bismuth	209		100	0.019	
	210	5.013 d			β^-(1.16); α(4.69, 4.65)
	212	1.0092 h			β^-(2.25); γ(0.727) Tl-x; α(6.05, 6.09)
	214	19.9 min			β^-(3.26, 1.88, 1.51, 1.0, 0.4); γ(0.609) α(5.512, 5.448)
Polonium	208	2.898 y			α(5.11); K, Bi-x, γ(0.292, 0.571, 0.603, 0.862)
	209	102 y			α(4.88), I, Bi-x, γ(0.26)
	210	138.376 d			α(5.30); γ(0.803)
	212	298 ns			α(8.78)
	214	0.1637 ms			α(7.69); γ(0.799)
	216	150 ms			α(6.78)
	218	3.11 min			α(6.00)
Astatine	207	1.80 h			K, α(5.76), γ(0.588, 0.814)

TABLE 4.16   Table of Nuclides (*Continued*)

Element	A	Half-life	Natural abundance, %	Cross section, barns	Radiation (MeV)
Astatine (*cont.*)	208	1.63 h			K, α(5.641), Po-x, γ(0.177, 0.660, 0.685)
	209	5.41 h			K, α(5.65), Po-x, γ(0.545, 0.782, 0.790)
	210	8.1 h			K, α(5.52, 5.44, 5.36), Po-x, γ(0.245, 1.181, 1.483)
	211	7.214 h			K, α(5.87), Po-x, γ(0.67)
Radon	220	55.6 s			α(6.29); γ(0.550)
	222	2.825 d			α(5.49); γ(0.510)
Radium	*224	3.66 d			α(5.68, 5.45); Rn-x; γ(0.241)
	*226	1600 y			α(4.78, 4.60); Rn-x; γ(0.187)
	*228	5.75 y			γ(0.135)
Actinium	*227	21.77 y			β^-(0.046), α(4.95, 4.94), Th-x, γ(0.086, 0.100, 0.160)
	*228	6.13 h			β^-(2.18, 1.85); Th-x; γ(0.339, 0.911, 0.969)
Thorium	228	1.913 y			α(5.43, 5.34); Ra-x; γ(0.084, 0.132, 0.167, 0.214)
	*230	7.54×10^4 y			α(4.68, 4.62), Ra-x; γ(0.068)
	*232	1.405×10^{10} y			α(4.01, 3.95), γ(0.059)
	233	2.12 min			β^-(1.23), γ(0.029, 0.087, 0.171, 0.195, 0.453, 0.67, 0.895)
	*234	24.10 d			β^-(0.199, 0.104); Pa-x; γ(0.063, 0.093)
Protactinium	*231	3.276×10^4 y			α(5.06, 5.02, 5.01, 4.95, 4.73), Ac-x, γ(0.027, 0.284, 0.300, 0.303)
	233	27.0 d			β^-(0.257, 0.15, 0.568), U-x; γ(0.312)
	234*m*	1.17 min			β^-(2.29), IT, U-x; γ(0.765, 1.00)
Uranium	233	1.592×10^5 y			α(4.82, 4.78); Th-x; γ(0.029, 0.042, 0.055, 0.097, 0.119, 0.146, 0.164, 0.22, 0.291, 0.32)
	*234	2.454×10^5 y			α(4.77, 4.72); Th-x; γ(0.053, 0.121)
	*235	7.037×10^8 y			α(4.40, 4.37, 4.22); Th-x; γ(0.186)
	*238	4.468×10^9 y			α(4.20, 4.15); γ(0.050)
	239	23.47 min			β^-(1.21, 1.29); Np-x γ(0.044, 0.075)

TABLE 4.16 Table of Nuclides (*Continued*)

Element	A	Half-life	Natural abundance, %	Cross section, barns	Radiation (MeV)
Neptunium	236	1.55×10^5 y			K, β^-, $\gamma(0.160)$
	237	2.14×10^6 y			$\alpha(4.79, 4.77)$; Pa-x; $\gamma(0.029, 0.086)$
Plutonium	238	87.74 y			$\alpha(5.50, 5.46)$; U-x; $\gamma(0.0435, 0.998, 0.153)$
	239	2.41×10^4 y			$\alpha(5.16, 5.14, 5.11)$ U-x; $\gamma(0.052, 0.129)$
	242	3.763×10^5 y			$\alpha(4.90, 4.86)$, $\gamma(0.045, 0.103)$
	244	8.26×10^7 y			$\alpha(4.59, 4.55)$
Americium	241	432.7 y			$\alpha(5.49, 5.44)$; Np-x; $\gamma(0.060)$
	243	7380 y			$\alpha(5.28, 5.23)$; Np-x; $\gamma(0.075)$
Curium	242	162.94 d			$\alpha(6.12, 6.07)$; Pu-x; $\gamma(0.561, 0.605)$
	244	18.11 y			$\alpha(5.81, 5.77)$; $\gamma(0.043)$
Berkelium	249	320 d			$\alpha(5.42)$; $\beta^-(0.125)$ $\gamma(0.327)$
Californium	252	2.645 y			$\alpha(6.12, 6.08)$; Cm-x; $\gamma(0.043)$
Einsteinium	253	20.4 d			$\alpha(6.64)$; Bk-x; $\gamma(0.387, 0.389, 0.429)$
	254	275.7 d			$\alpha(6.44)$; Bk-x; $\gamma(0.034, 0.036, 0.043)$
Fermium	257	100.5 d			$\alpha(6.52)$; Cf-x; $\gamma(0.115, 0.241)$

4.9 WORK FUNCTION

TABLE 4.17 Work Functions of the Elements

The work function ϕ is the energy necessary to just remove an electron from the metal surface in thermoelectric or photoelectric emission. Values are dependent upon the experimental technique (vacua of 10^{-9} or 10^{-10} torr, clean surfaces, and surface conditions including the crystal face identification).

Element	ϕ, eV	Element	ϕ, eV
Ag	4.64	B	(4.75)
Al	4.19	Ba	2.35
As	(3.75)	Be	5.08
Au	5.32	Bi	4.36

Source: S. Trasatti, *J. Chem. Soc. Faraday Trans. I* **68**:229 (1972); N. D. Lang and W. Kohn, *Phys. Rev. B* **3**:1215 (1971).

TABLE 4.17 Work Functions of the Elements (*Continued*)

Element	ϕ, eV	Element	ϕ, eV
C	(5.0)	Pb	4.18
Ca	2.71	Pd	5.00
Cd	4.12	Po	4.6
Ce	2.80	Pr	2.7
Co	4.70	Pt	5.40
Cr	4.40	Rb	2.20
Cs	1.90	Re	4.95
Cu	4.70	Rh	4.98
Eu	2.5	Ru	4.80
Fe	4.65	Sb	4.56
Ga	4.25	Sc	3.5
Ge	5.0	Se	5.9
Gd	3.1	Si	4.85
Hf	3.65	Sm	2.95
Hg	4.50	Sn	4.35
In	4.08	Sr	2.76
Ir	5.6	Ta	4.22
K	2.30	Tb	3.0
La	3.40	Te	4.70
Li	3.10	Th	3.71
Mg	3.66	Ti	4.10
Mn	3.90	Tl	4.02
Mo	4.30	U	3.70
Na	2.70	V	4.44
Nb	4.20	W	4.55
Nd	3.1	Y	3.1
Ni	5.15	Zn	4.30
Os	4.83	Zr	4.00

4.10 RELATIVE ABUNDANCES OF NATURALLY OCCURRING ISOTOPES

TABLE 4.18 Relative Abundances of Naturally Occurring Isotopes

*The data were extracted from A. H. Wapstra and G. Audi, "The 1983 Atomic Mass Evaluation," Nucl. Phys. **A432:1–54** (1985).*

Element	Mass number	Percent	Element	Mass number	Percent
Aluminum	27	100		135	6.59(2)
Antimony	121	57.3(9)		136	7.85(4)
	123	42.7(9)		137	11.23(4)
Argon	36	0.337(3)		138	71.70(7)
	38	0.063(1)	Beryllium	9	100
	40	99.600(3)	Bismuth	209	100
Arsenic	75	100	Boron	10	19.9(2)
Barium	130	0.106(2)		11	80.1(2)
	132	0.101(2)	Bromine	79	50.69(5)
	134	2.42(3)		81	49.31(5)

TABLE 4.18   Relative Abundances of Naturally Occurring Isotopes (*Continued*)

Element	Mass number	Percent	Element	Mass number	Percent
Cadmium	106	1.25(3)	Gallium	69	60.1(2)
	108	0.89(1)		71	39.9(2)
	110	12.49(9)	Germanium	70	20.5(5)
	111	12.80(6)		72	27.4(4)
	112	24.13(11)		73	7.8(2)
	113	12.22(6)		74	36.5(7)
	114	28.7(2)		76	7.8(2)
	116	7.49(9)	Gold	197	100
Calcium	40	96.941(13)	Hafnium	174	0.162(2)
	42	0.647(3)		176	5.206(4)
	43	0.135(3)		177	18.606(3)
	44	2.086(5)		178	27.297(3)
	46	0.004(3)		179	13.629(5)
	48	0.187(3)		180	35.100(6)
Carbon	12	98.90(3)	Helium	4	100
	13	1.10(3)	Holmium	165	100
Cerium	136	0.19(1)	Hydrogen	1	99.985(1)
	138	0.25(1)		2	0.015(1)
	140	88.48(10)	Indium	113	4.3(2)
	142	11.08(10)		115	95.7(2)
Cesium	133	100	Iodine	127	100
Chlorine	35	75.77(5)	Iridium	191	37.3(5)
	37	24.23(5)		193	62.7(5)
Chromium	50	4.345(9)	Iron	54	5.8(1)
	52	83.789(12)		56	91.7(3)
	53	9.501(11)		57	2.2(1)
	54	2.365(5)		58	0.28(1)
Cobalt	59	100	Krypton	78	0.35(2)
Copper	63	69.17(2)		80	2.25(2)
	65	30.83(2)		82	11.6(1)
Dysprosium	156	0.06(1)		83	11.5(1)
	158	0.10(1)		84	57.0(3)
	160	2.34(5)		86	17.3(2)
	161	18.9(1)	Lanthanum	138	0.09(1)
	162	25.5(2)		139	99.91(1)
	163	24.9(2)	Lead	204	1.4(1)
	164	28.2(2)		206	24.1(1)
Erbium	162	0.14(1)		207	22.1(1)
	164	1.61(1)		208	52.4(1)
	166	33.6(2)	Lithium	6	7.5(2)
	167	22.95(1)		7	92.5(2)
	168	26.8(2)	Lutetium	175	97.41(2)
	170	14.9(1)		176	2.59(2)
Europium	151	47.81(5)	Magnesium	24	78.99(3)
	153	52.2(5)		25	10.00(1)
Fluorine	19	100		26	11.01(2)
Gadolinium	152	0.20(1)	Manganese	55	100
	154	2.18(3)	Mercury	196	0.14(10)
	155	14.80(5)		198	10.02(7)
	156	20.47(4)		199	16.84(11)
	157	15.65(3)		200	23.13(11)
	158	24.84(12)		201	13.22(11)
	160	21.86(4)		202	29.80(14)

TABLE 4.18 Relative Abundances of Naturally Occurring Isotopes (*Continued*)

Element	Mass number	Percent	Element	Mass number	Percent
	204	6.85(5)	Protoactinium	230	100
Molybdenum	92	15.84(4)	Rhenium	185	37.40(2)
	94	9.25(2)		187	62.60(2)
	95	15.92(4)	Rhodium	103	100
	96	16.68(4)	Rubidium	85	72.165(13)
	97	9.55(2)		87	27.835(13)
	98	24.13(6)	Ruthenium	96	5.52(5)
	100	9.63(2)		98	1.88(5)
Neodymium	142	27.13(10)		99	12.7(1)
	143	12.18(5)		100	12.6(1)
	144	23.80(10)		101	17.0(1)
	145	8.30(5)		102	31.6(2)
	146	17.19(8)		104	18.7(2)
	148	5.76(3)	Samarium	144	3.1(1)
	150	5.64(3)		147	15.0(2)
Neon	20	90.48(3)		148	11.3(1)
	21	0.27(1)		149	13.8(1)
	22	9.26(3)		150	7.4(1)
Nickel	58	68.27(1)		152	26.7(2)
	60	26.10(1)		154	22.7(2)
	61	1.13(1)	Scandium	45	100
	62	3.59(1)	Selenium	74	0.91(1)
	64	0.91(1)		76	9.0(2)
Niobium	93	100		77	7.6(2)
Nitrogen	14	99.634(9)		78	23.6(6)
	15	0.366(9)		80	49.7(7)
Osmium	184	0.02(1)		82	9.2(5)
	186	1.58(10)	Silicon	28	92.23(1)
	187	1.6(1)		29	4.67(1)
	188	13.3(2)		30	3.10(1)
	189	16.1(3)	Silver	107	51.839(5)
	190	26.4(4)		109	48.161(5)
	192	41.0(3)	Sodium	23	100
Oxygen	16	99.762(15)	Strontium	84	0.56(1)
	17	0.038(3)		86	9.86(1)
	18	0.200(12)		87	7.00(1)
Palladium	102	1.02(1)		88	82.58(1)
	104	11.14(8)	Sulfur	32	95.02(9)
	105	22.33(8)		33	0.75(1)
	106	27.33(5)		34	4.21(8)
	108	26.46(9)		36	0.02(1)
	110	11.72(9)	Tantalum	180	0.012(2)
Phosphorus	31	100		181	99.988(2)
Platinum	190	0.01(1)	Tellurium	120	0.096(2)
	192	0.79(5)		122	2.60(1)
	194	32.9(5)		123	0.908(3)
	195	33.8(5)		124	4.816(8)
	196	25.3(5)		125	7.14(1)
	198	7.2(2)		126	18.95(1)
Potassium	39	93.258(3)		128	31.69(2)
	40	0.0117(1)		130	33.80(2)
	41	6.730(3)	Terbium	159	100
Praseodymium	141	100	Thallium	203	29.524(9)

TABLE 4.18 Relative Abundances of Naturally Occurring Isotopes (*Continued*)

Element	Mass number	Percent	Element	Mass number	Percent
	205	70.476(9)	Xenon	124	0.10(1)
Thorium	228	100		126	0.09(1)
Thulium	169	100		128	1.91(3)
Tin	112	0.97(1)		129	26.4(6)
	114	0.65(1)		130	4.1(1)
	115	0.36(1)		131	21.2(4)
	116	14.53(11)		132	26.9(5)
	117	7.68(7)		134	10.4(2)
	118	24.22(11)		136	8.9(1)
	119	8.58(4)	Ytterbium	168	0.13(1)
	120	32.59(10)		170	3.05(5)
	122	4.63(3)		171	14.3(2)
	124	5.79(5)		172	21.9(3)
Titanium	46	8.0(1)		173	16.12(2)
	47	7.3(1)		174	31.8(4)
	48	73.8(1)		176	12.7(1)
	49	5.5(1)	Yttrium	89	100
	50	5.4(1)	Zinc	64	48.6(3)
Tungsten	180	0.13(3)		66	27.9(2)
	182	26.3(2)		67	4.1(1)
	183	14.3(1)		68	18.8(4)
	184	30.67(15)		70	0.6(1)
	186	28.6(2)	Zirconium	90	51.45(2)
Uranium	234	0.0055(5)		91	11.22(2)
	235	0.7200(12)		92	17.15(1)
	238	99.2745(15)		94	17.38(2)
Vanadium	50	0.250(2)		96	2.80(1)
	51	99.750(2)			

SECTION 5
PHYSICAL PROPERTIES

5.1 SOLUBILITIES

TABLE 5.1 Solubility of Gases in Water

The column (or line entry) headed "α" gives the volume of gas (in milliliters) measured at standard conditions (0°C and 760 mm or 101.325 kN·m^{-2}) dissolved in 1 mL of water at the temperature stated (in degrees Celsius) and when the pressure of the gas without that of the water vapor is 760 mm. The line entry "A" indicates the same quantity except that the gas itself is at the uniform pressure of 760 mm when in equilibrium with water.

The column headed "1" gives the volume of the gas (in milliliters) dissolved in 1 mL of water when the pressure of the gas plus that of the water vapor is 760 mm.

The column headed "q" gives the weight of gas (in grams) dissolved in 100 g of water when the pressure of the gas plus that of the water vapor is 760 mm.

Temp. °C	Acetylene		Air*		Ammonia		Bromine	
	α	q	$\alpha(\times 10^3)$	% oxygen in air	α	q	α	q
0	1.73	0.200	29.18	34.91	1130	89.5	60.5	42.9
1	1.68	0.194	28.42	34.87	——	——	——	——
2	1.63	0.188	27.69	34.82	——	——	54.1	38.3
3	1.58	0.182	26.99	34.78	——	——	——	——
4	1.53	0.176	26.32	34.74	1047	79.6	48.3	34.2
5	1.49	0.171	25.68	34.69	——	——	——	——
6	1.45	0.167	25.06	34.65	——	——	43.3	30.6
7	1.41	0.162	24.47	34.60	——	——	——	——
8	1.37	0.157	23.90	34.56	947	72.0	38.9	27.5
9	1.34	0.154	23.36	34.52	——	——	——	——
10	1.31	0.150	22.84	34.47	870	68.4	35.1	24.8
11	1.27	0.146	22.34	34.43	——	——	——	——
12	1.24	0.142	21.87	34.38	857	65.1	31.5	22.2
13	1.21	0.138	21.41	34.34	837	63.6	——	——
14	1.18	0.135	20.97	34.30	——	——	28.4	20.0
15	1.15	0.131	20.55	34.25	770	——	——	——
16	1.13	0.129	20.14	34.21	775	58.7	25.7	18.0
17	1.10	0.125	19.75	34.17	——	——	——	——
18	1.08	0.123	19.38	34.12	——	——	23.4	16.4
19	1.05	0.119	19.02	34.08	——	——	——	——
20	1.03	0.117	18.68	34.03	680	52.9	21.3	14.9
21	1.01	0.115	18.34	33.99	——	——	——	——
22	0.99	0.112	18.01	33.95	——	——	19.4	13.5
23	0.97	0.110	17.69	33.90	——	——	——	——
24	0.95	0.107	17.38	33.86	639	48.2	17.7	12.3
25	0.93	0.105	17.08	33.82	——	——	——	——
26	0.91	0.102	16.79	33.77	——	——	16.3	11.3
27	0.89	0.100	16.50	33.73	——	——	——	——
28	0.87	0.098	16.21	33.68	586	44.0	15.0	10.3
29	0.85	0.095	15.92	33.64	——	——	——	——
30	0.84	0.094	15.64	33.60	530	41.0	13.8	9.5
35	——	——	——	——	——	——	——	——
40	——	——	14.18	——	400	31.6	9.4	6.3
45	——	——	——	——	——	——	——	——
50	——	——	12.97	——	290	23.5	6.5	4.1
60	——	——	12.16	——	200	16.8	4.9	2.9
70	——	——	——	——	——	11.1	3.8	1.9
80	——	——	11.26	——	——	6.5	3.0	1.2
90	——	——	——	——	——	3.0	——	——
100	——	——	11.05	——	——	0.0	——	——

* Free from NH_3 and CO_2; total pressure of air + water vapor is 760 mm.

TABLE 5.1 Solubility of Gases in Water (*Continued*)

Temp. °C	Carbon dioxide		Carbon monoxide		Chlorine		Ethane		Ethylene		Hydrogen	
	α	q	α	q	l	q	α	q	α	q	α	q
0	1.713	0.3346	0.035 37	0.004 397	—	—	0.098 74	0.013 17	0.226	0.028 1	0.021 48	0.000 192 2
1	1.646	0.3213	0.034 55	0.004 293	—	—	0.094 76	0.012 63	0.219	0.027 2	0.021 26	0.000 190 1
2	1.584	0.3091	0.033 75	0.004 191	—	—	0.090 93	0.012 12	0.211	0.026 2	0.021 05	0.000 188 1
3	1.527	0.2978	0.032 97	0.004 092	—	—	0.087 25	0.011 62	0.204	0.025 3	0.020 84	0.000 186 2
4	1.473	0.2871	0.032 22	0.003 996	—	—	0.083 72	0.011 14	0.197	0.024 4	0.020 64	0.000 184 3
5	1.424	0.2774	0.031 49	0.003 903	—	—	0.080 33	0.010 69	0.191	0.023 7	0.020 44	0.000 182 4
6	1.377	0.268 1	0.030 78	0.003 813	—	—	0.077 09	0.010 25	0.184	0.022 8	0.020 25	0.000 180 6
7	1.331	0.258 9	0.030 09	0.003 725	—	—	0.074 00	0.009 83	0.178	0.022 0	0.020 07	0.000 178 9
8	1.282	0.249 2	0.029 42	0.003 640	—	—	0.071 06	0.009 43	0.173	0.021 4	0.019 89	0.000 177 2
9	1.237	0.240 3	0.028 78	0.003 559	—	—	0.068 26	0.009 06	0.167	0.020 7	0.019 72	0.000 175 6
10	1.194	0.231 8	0.028 16	0.003 479	3.148	0.997 2	0.065 61	0.008 70	0.162	0.020 0	0.019 55	0.000 174 0
11	1.154	0.223 9	0.027 57	0.003 405	3.047	0.965 4	0.063 28	0.008 38	0.157	0.019 4	0.019 40	0.000 172 5
12	1.117	0.216 5	0.027 01	0.003 332	2.950	0.934 6	0.061 06	0.008 08	0.152	0.018 8	0.019 25	0.000 171 0
13	1.083	0.209 8	0.026 46	0.003 261	2.856	0.905 0	0.058 94	0.007 80	0.148	0.018 3	0.019 11	0.000 169 6
14	1.050	0.203 2	0.025 93	0.003 194	2.767	0.876 8	0.056 94	0.007 53	0.143	0.017 6	0.018 97	0.000 168 2
15	1.019	0.197 0	0.025 43	0.003 130	2.680	0.849 5	0.055 04	0.007 27	0.139	0.017 1	0.018 83	0.000 166 8
16	0.985	0.190 3	0.024 94	0.003 066	2.597	0.823 2	0.053 26	0.007 03	0.136	0.016 7	0.018 69	0.000 165 4
17	0.956	0.184 5	0.024 48	0.003 007	2.517	0.797 9	0.051 59	0.006 80	0.132	0.016 2	0.018 56	0.000 164 1
18	0.928	0.178 9	0.024 02	0.002 947	2.440	0.773 8	0.050 03	0.006 59	0.129	0.015 8	0.018 44	0.000 162 8
19	0.902	0.173 7	0.023 60	0.002 891	2.368	0.751 0	0.048 58	0.006 39	0.125	0.015 3	0.018 31	0.000 161 6
20	0.878	0.168 8	0.023 19	0.002 838	2.299	0.729 3	0.047 24	0.006 20	0.122	0.014 9	0.018 19	0.000 160 3
21	0.854	0.164 0	0.022 81	0.002 789	2.238	0.710 0	0.045 89	0.006 02	0.119	0.014 6	0.018 05	0.000 158 8
22	0.829	0.159 0	0.022 44	0.002 739	2.180	0.691 8	0.044 59	0.005 84	0.116	0.014 2	0.017 92	0.000 157 5
23	0.804	0.154 0	0.022 08	0.002 691	2.123	0.673 9	0.044 35	0.005 67	0.114	0.013 9	0.017 79	0.000 156 1
24	0.781	0.149 3	0.021 74	0.002 646	2.070	0.657 2	0.042 17	0.005 51	0.111	0.013 5	0.017 66	0.000 154 8

TABLE 5.1 Solubility of Gases in Water (*Continued*)

Temp. °C	Carbon dioxide		Carbon monoxide		Chlorine		Ethane		Ethylene		Hydrogen	
	α	q	α	q	l	q	α	q	α	q	α	q
25	0.759	0.144 9	0.021 42	0.002 603	2.019	0.641 3	0.041 04	0.005 35	0.108	0.013 1	0.017 54	0.000 153 5
26	0.738	0.140 6	0.021 10	0.002 560	1.970	0.625 9	0.039 97	0.005 20	0.106	0.012 9	0.017 42	0.000 152 2
27	0.718	0.136 6	0.020 80	0.002 519	1.923	0.611 2	0.038 95	0.005 06	0.104	0.012 6	0.017 31	0.000 150 9
28	0.699	0.132 7	0.020 51	0.002 479	1.880	0.597 5	0.037 99	0.004 93	0.102	0.012 3	0.017 20	0.000 149 6
29	0.682	0.129 2	0.020 24	0.002 442	1.839	0.584 7	0.037 09	0.004 80	0.100	0.012 1	0.017 09	0.000 148 4
30	0.665	0.125 7	0.019 98	0.002 405	1.799	0.572 3	0.036 24	0.004 68	0.098	0.011 8	0.016 99	0.000 147 4
35	0.592	0.110 5	0.018 77	0.002 231	1.602	0.510 4	0.032 30	0.004 12	—	—	0.016 66	0.000 142 5
40	0.530	0.097 3	0.017 75	0.002 075	1.438	0.459 0	0.029 15	0.003 66	—	—	0.016 44	0.000 138 4
45	0.479	0.086 0	0.016 90	0.001 933	1.322	0.422 8	0.026 60	0.003 27	—	—	0.016 24	0.000 134 1
50	0.436	0.076 1	0.016 15	0.001 797	1.225	0.392 5	0.024 59	0.002 94	—	—	0.016 08	0.000 128 7
60	0.359	0.057 6	0.014 88	0.001 522	1.023	0.329 5	0.021 77	0.002 39	—	—	0.016 00	0.000 117 8
70	—	—	0.014 40	0.001 276	0.862	0.279 3	0.019 48	0.001 85	—	—	0.016 0	0.000 102
80	—	—	0.014 30	0.000 980	0.683	0.222 7	0.018 26	0.001 34	—	—	0.016 0	0.000 079
90	—	—	0.0142	0.000 57	0.39	0.127	0.017 6	0.000 8	—	—	0.016 0	0.000 046
100	—	—	0.0141	0.000 00	0.00	0.000	0.017 2	0.000 0	—	—	0.016 0	0.000 000

TABLE 5.1 Solubility of Gases in Water (*Continued*)

Temp. °C	Hydrogen sulfide α	q	Methane α	q	Nitric oxide α	q	Nitrogen* α	q	Oxygen α	q	Sulfur dioxide l	q
0	4.670	0.7066	0.055 63	0.003 959	0.073 81	0.009 833	0.023 54	0.002 942	0.048 89	0.006 945	79.789	22.83
1	4.522	0.6839	0.054 01	0.003 842	0.071 84	0.009 564	0.022 97	0.002 869	0.047 58	0.006 756	77.210	22.09
2	4.379	0.6619	0.052 44	0.003 728	0.069 93	0.009 305	0.022 41	0.002 798	0.046 33	0.006 574	74.691	21.37
3	4.241	0.6407	0.050 93	0.003 619	0.068 09	0.009 057	0.021 87	0.002 730	0.045 12	0.006 400	72.230	20.66
4	4.107	0.6201	0.049 46	0.003 513	0.066 32	0.008 816	0.021 35	0.002 663	0.043 97	0.006 232	69.828	19.98
5	3.977	0.6001	0.048 05	0.003 410	0.064 61	0.008 584	0.020 86	0.002 600	0.042 87	0.006 072	67.485	19.31
6	3.852	0.5809	0.046 69	0.003 312	0.062 98	0.008 361	0.020 37	0.002 537	0.041 80	0.005 918	65.200	18.65
7	3.732	0.5624	0.045 39	0.003 217	0.061 40	0.008 147	0.019 90	0.002 477	0.040 80	0.005 773	62.973	18.02
8	3.616	0.5446	0.044 13	0.003 127	0.059 90	0.007 943	0.019 45	0.002 419	0.039 83	0.005 632	60.805	17.40
9	3.505	0.5276	0.042 92	0.003 039	0.058 46	0.007 747	0.019 02	0.002 365	0.038 91	0.005 498	58.697	16.80
10	3.399	0.5112	0.041 77	0.002 955	0.057 09	0.007 560	0.018 61	0.002 312	0.038 02	0.005 368	56.647	16.21
11	3.300	0.4960	0.040 72	0.002 879	0.055 87	0.007 393	0.018 23	0.002 263	0.037 18	0.005 246	54.655	15.64
12	3.206	0.4814	0.039 70	0.002 805	0.054 70	0.007 233	0.017 86	0.002 216	0.036 37	0.005 128	52.723	15.09
13	3.115	0.4674	0.038 72	0.002 733	0.053 57	0.007 078	0.017 50	0.002 170	0.035 59	0.005 014	50.849	14.56
14	3.028	0.4540	0.037 79	0.002 665	0.052 50	0.006 930	0.017 17	0.002 126	0.034 86	0.004 906	49.033	14.04
15	2.945	0.4411	0.036 90	0.002 599	0.051 47	0.006 788	0.016 85	0.002 085	0.034 15	0.004 802	47.276	13.54
16	2.865	0.4287	0.036 06	0.002 538	0.050 49	0.006 652	0.016 54	0.002 045	0.033 48	0.004 703	45.578	13.05
17	2.789	0.4169	0.035 25	0.002 478	0.049 56	0.006 524	0.016 25	0.002 006	0.032 83	0.004 606	43.939	12.59
18	2.717	0.4056	0.034 48	0.002 422	0.048 68	0.006 400	0.015 97	0.001 970	0.032 20	0.004 514	42.360	12.14
19	2.647	0.3948	0.033 76	0.002 369	0.047 85	0.006 283	0.015 70	0.001 935	0.031 61	0.004 426	40.838	11.70

TABLE 5.1 Solubility of Gases in Water (*Continued*)

Temp. °C	Hydrogen sulfide		Methane		Nitric acid		Nitrogen*		Oxygen		Sulfur dioxide	
	α	q	α	q	α	q	α	q	α	q	l	q
20	2.582	0.384 6	0.033 08	0.002 319	0.047 06	0.006 173	0.015 45	0.001 901	0.031 02	0.004 339	39.374	11.28
21	2.517	0.374 5	0.032 43	0.002 270	0.046 25	0.006 059	0.015 22	0.001 869	0.030 44	0.004 252	37.970	10.88
22	2.456	0.364 8	0.031 80	0.002 222	0.045 45	0.005 947	0.014 98	0.001 838	0.029 88	0.004 169	36.617	10.50
23	2.396	0.355 4	0.031 19	0.002 177	0.044 69	0.005 838	0.014 75	0.001 809	0.029 34	0.004 087	35.302	10.12
24	2.338	0.346 3	0.030 61	0.002 133	0.043 95	0.005 733	0.014 54	0.001 780	0.028 81	0.004 007	34.026	9.76
25	2.282	0.337 5	0.030 06	0.002 091	0.043 23	0.005 630	0.014 34	0.001 751	0.028 31	0.003 931	32.786	9.41
26	2.229	0.329 0	0.029 52	0.002 050	0.042 54	0.005 530	0.014 13	0.001 724	0.027 83	0.003 857	31.584	9.06
27	2.177	0.320 8	0.029 01	0.002 011	0.041 88	0.005 435	0.013 94	0.001 698	0.027 36	0.003 787	30.422	8.73
28	2.128	0.313 0	0.028 52	0.001 974	0.041 24	0.005 342	0.013 76	0.001 672	0.026 91	0.003 718	29.314	8.42
29	2.081	0.305 5	0.028 06	0.001 938	0.040 63	0.005 252	0.013 58	0.001 647	0.026 49	0.003 651	28.210	8.10
30	2.037	0.298 3	0.027 62	0.001 904	0.040 04	0.005 165	0.013 42	0.001 624	0.026 08	0.003 588	27.161	7.80
35	1.831	0.264 8	0.025 46	0.001 733	0.037 34	0.004 757	0.012 56	0.001 501	0.024 40	0.003 315	22.489	6.47
40	1.660	0.236 1	0.023 69	0.001 586	0.035 07	0.004 394	0.011 84	0.001 391	0.023 06	0.003 082	18.766	5.41
45	1.516	0.211 0	0.022 38	0.001 466	0.033 11	0.004 059	0.011 30	0.001 300	0.021 87	0.002 858	—	—
50	1.392	0.188 3	0.021 34	0.001 359	0.031 52	0.003 758	0.010 88	0.001 216	0.020 90	0.002 657	—	—
60	1.190	0.148 0	0.019 54	0.001 144	0.029 54	0.003 237	0.010 23	0.001 052	0.019 46	0.002 274	—	—
70	1.022	0.110 1	0.018 25	0.000 926	0.028 10	0.002 668	0.009 77	0.000 851	0.018 33	0.001 856	—	—
80	0.917	0.076 5	0.017 70	0.000 695	0.027 00	0.001 984	0.009 58	0.000 660	0.017 61	0.001 381	—	—
90	0.84	0.041	0.017 35	0.000 40	0.026 5	0.00113	0.009 5	0.000 38	0.0172	0.000 79	—	—
100	0.81	0.000	0.017 0	0.000 00	0.026 3	0.000 00	0.009 5	0.000 00	0.017 0	0.000 00	—	—

* Atmospheric nitrogen containing 98.815% N_2 by volume + 1.185% inert gases.

TABLE 5.1 Solubility of Gases in Water (*Continued*)

Substance		0°	10°	20°	30°	40°	60°	80°
Argon	α	0.0528	0.0413	0.0337	0.0288	0.0251	0.0209	0.0184
Helium	A	0.0098	0.00911	0.0086	0.00839	0.00841	0.00902	$0.00942^{70°}$
Hydrogen bromide	l	612	582		$533^{25°}$	385	$469^{50°}$	$406^{75°}$
Hydrogen chloride	α	512	475	442	412		339	
Krypton	α	0.1105	0.0810	0.0626	0.0511	0.0433	0.0357	
Neon	A		$0.01179°$	0.0106	0.0100	$0.00948^{42°}$		$0.00984^{73°}$
Nitrous oxide	A		0.88	0.63				
Ozone	g · L^{-1}	0.0394	$0.02991^{2°}$	$0.02101^{9°}$	$0.01393^{27°}$	0.0042	0	
Radon	α	0.510	0.326	0.222	0.162	0.126	0.085	
Xenon	α	0.242	0.174	0.123	0.098	0.082		

TABLE 5.2 Solubilities of Inorganic Compounds and Metal Salts of Organic Acids in Water at Various Temperatures

Solubilities are expressed as the number of grams of substance of stated molecular formula which when dissolved in 100 g of water make a saturated solution at the temperature stated (°C).

Substance	Formula	0°	10°	20°	30°	40°	60°	80°	90°	100°
Aluminum chloride	AlCl₃	43.9	44.9	45.8	46.6	47.3	48.1	48.6		49.0
fluoride	AlF₃	0.56	0.56	0.67	0.78	0.91	1.1	1.32		1.72
nitrate	Al(NO₃)₃	60.0	66.7	73.9	81.8	88.7	106	132	153	160
perchlorate	Al(ClO₄)₃	122	128	133						182
sulfate	Al₂(SO₄)₃	31.2	33.5	36.4	40.4	45.8	59.2	73.0	80.8	89.0
thallium(I) sulfate	Al₂Tl₂(SO₄)₄	3.15	4.60	6.39	9.37	14.39	35.35			
Ammonium aluminum sulfate	NH₄Al(SO₄)₂	2.10	5.00	7.74	10.9	14.9	26.7			
azide	NH₄N₃	16.0		25.3		37.1				
bromide	NH₄Br	60.5	68.1	76.4	83.2	91.2	108	125	135	145
chloride	NH₄Cl	29.4	33.2	37.2	41.4	45.8	55.3	65.6	71.2	77.3
chloroiridate(IV)	(NH₄)₂IrCl₆	0.56	0.71	0.95	1.20	1.56	2.45	4.38		
chloroplatinate(IV)	(NH₄)₂PtCl₆	0.289	0.374	0.499	0.637	0.815	1.44	2.16	2.61	3.36
chromate	(NH₄)₂CrO₄	25.0	29.2	34.0	39.3	45.3	59.0	76.1		
chromium(III) sulfate	(NH₄)Cr(SO₄)₂	3.95			18.8	32.6				
cobalt(II) sulfate	(NH₄)₂Co(SO₄)₂	6.0	9.5	13.0	17.0	22.0	33.5	49.0	58.0	75.1
dichromate	(NH₄)₂Cr₂O₇	18.2	25.5	35.6	46.5	58.5	86.0	115		156
dihydrogen arsenate	NH₄H₂AsO₄	33.7		48.7		63.8	83.0	107	122	
dihydrogen phosphate	NH₄H₂PO₄	22.7	29.5	37.4	46.4	56.7	82.5	118		173
dithionate	(NH₄)₂S₂O₆	133	151	166	179					
formate	NH₄CHO₂	102		143		204	311	533		
hydrogen carbonate	NH₄HCO₃	11.9	16.1	21.7	28.4	36.6	59.2	109	170	354
hydrogen phosphate	(NH₄)₂HPO₄	42.9	62.9	68.9	75.1	81.8	97.2			
hydrogen tartrate	NH₄C₄H₅O₆	1.00	1.88	2.70						
iodide	NH₄I	155	163	172	182	191	209	229		250
iron(II) sulfate	(NH₄)₂Fe(SO₄)₂	12.5	17.2	26.4	33	46				

TABLE 5.2 Solubilities of Inorganic Compounds and Metal Salts of Organic Acids in Water at Various Temperatures (*Continued*)

Substance	Formula	0°	10°	20°	30°	40°	60°	80°	90°	100°
Ammonium magnesium										
sulfate	$(NH_4)_2Mg(SO_4)_2$	11.8	14.6	18.0	21.7	25.8	35.1	48.3		65.7
nickel sulfate	$(NH_4)_2Ni(SO_4)_2$	1.00	4.00	6.50	9.20	12.0	17.0			
nitrate	NH_4NO_3	118	150	192	242	297	421	580	740	871
oxalate	$(NH_4)_2C_2O_4$	2.2	3.21	4.45	6.09	8.18	14.0	22.4	27.9	34.7
perchlorate	NH_4ClO_4	12.0	16.4	21.7	27.7	34.6	49.9	68.9		
selenite	$(NH_4)_2SeO_3$	96	105	115	126	143	192			
sulfate	$(NH_4)_2SO_4$	70.6	73.0	75.4	78.0	81	88	95		103
sulfite	$(NH_4)_2SO_3$	47.9	54.0	60.8	68.8	78.4	104	144	150	153
tartrate	$(NH_4)_2C_4H_4O_6$	45.0	55.0	63.0	70.5	76.5	86.9			
thioantimonate(V)	$(NH_4)_3SbS_4$	71.2		91.2	120					
thiocyanate	NH_4SCN	120	144	170	208	234	346			
vanadate	NH_4VO_3			0.48	0.84	1.32	2.42			
zinc sulfate	$(NH_4)_2Zn(SO_4)_2$	7.0	9.5	12.5	16.0	20.0	30.0	46.6	58.0	72.4
Antimony(III) chloride	$SbCl_3$	602		910	1087	1368	[completely miscible at 72°]			
fluoride	SbF_3	385		444	562					
Arsenic hydride										
(760 mm), cc	AsH_3	42	30	28						
oxide (pent-)	As_2O_5	59.5	62.1	65.8	69.8	71.2	73.0	75.1		76.7
oxide (tri-)	As_2O_3	1.20	1.49	1.82	2.31	2.93	4.31	6.11		8.2
Barium acetate	$Ba(C_2H_3O_2)_2 \cdot 3H_2O$	58.8	62	72	75	78.5	75	74.0		74.8
azide	$Ba(N_3)_2$	12.5	16.1	17.41[7]						
bromate	$Ba(BrO_3)_2 \cdot H_2O$	0.29	0.44	0.65	0.95	1.31	2.27	3.52	4.26	5.39
bromide	$BaBr_2 \cdot 2H_2O$	98	101	104	109	114	123	135		149
n-butyrate	$Ba(C_4H_7O_2)_2$	37.0	36.1	35.4	34.9	35.2	37.2	41.7	45.5	48.1[95°]
caproate	$Ba(C_6H_{11}O_2)_2 \cdot 3.5H_2O$	11.71	8.38	6.89	5.87	5.79	8.39	14.71	19.28	
chlorate	$Ba(ClO_3)_2 \cdot H_2O$	20.3	26.9	33.9	41.6	49.7	66.7	84.8		105
chloride	$BaCl_2 \cdot 2H_2O$	31.2	33.5	35.8	38.1	40.8	46.2	52.5	55.8	59.4
chlorite	$Ba(ClO_2)_2$	43.9	44.6	45.4		47.9	53.8	66.6		80.8
fluoride	BaF_2		0.159	0.160	0.162					

Name	Formula									
formate	Ba(CHO$_2$)$_2$	26.2	28.0	29.9	31.9	34.0	38.6	44.2	47.6	51.3
hydroxide	Ba(OH)$_2$	1.67	2.48	3.89	5.59	8.22	20.94	101.4		
iodate	Ba(IO$_3$)$_2$			0.035	0.046	0.057				
iodide	BaI$_2 \cdot$2H$_2$O	182	201	223	250		264		291	301
nitrate	Ba(NO$_3$)$_2$	4.95	6.67	9.02	11.48	14.1	20.4	27.2		34.4
nitrite	Ba(NO$_2$)$_2 \cdot$H$_2$O	50.3	60	72.8		102	151	222	261	325
perchlorate	Ba(ClO$_4$)$_2 \cdot$3H$_2$O	239		336		416	495	575		653
propionate	Ba(C$_3$H$_5$O$_2$)$_2 \cdot$H$_2$O	57.2	56.8		57.5	59.0	62.0	67.8	73.0	82.7
isosuccinate	BaC$_4$H$_4$O$_4$	0.421	0.432	0.418	0.393	0.366	0.306	0.237		
sulfamate	Ba(SO$_3$NH$_2$)$_2$	18.3	22.3	26.8	32.5	38.5	49.6	61.5	67.34	73.5
sulfide	BaS	2.88	4.89	7.86	10.38	14.89	27.69	49.91		60.29
tartrate	Ba(C$_2$H$_2$O$_3$)$_2$	0.021	0.024	0.028	0.032	0.035	0.044	0.053		
Beryllium nitrate	Be(NO$_3$)$_2$	97	102	108	113	125	178			
sulfate	BeSO$_4$	37.0	37.6	39.1	41.4	45.8	53.1	67.2		82.8
Boric acid	H$_3$BO$_3$	2.67	3.73	5.04	6.72	8.72	14.81	23.62	30.38	40.25
Cadmium bromide	CdBr$_2$	56.3	75.4	98.8	129	152	153	156		160
chlorate	Cd(ClO$_3$)$_2$	299	308	322	348	376	455	713		
chloride	CdCl$_2 \cdot$2.5H$_2$O	90	100	113	132	135	136			
	CdCl$_2 \cdot$H$_2$O		135	135	135	135		140		147
formate	Cd(CHO$_2$)$_2$	8.3	11.1	14.4	18.6	25.3	59.5	80.5	85.2	94.6
iodide	CdI$_2$	78.7		84.7	87.9	92.1	100	111		125
nitrate	Cd(NO$_3$)$_2$	122	136	150	167	194	310			
perchlorate	Cd(ClO$_4$)$_2 \cdot$6H$_2$O		180	188	195	203	221	243		272
selenate	CdSeO$_4$	72.5	68.4	64.0	58.9	55.0	44.2	32.5	27.2	22.0
sulfate	CdSO$_4$	75.4	76.0	76.6		78.5	81.8	66.7	63.1	60.8
Calcium acetate	Ca(OAc)$_2 \cdot$2H$_2$O	37.4	36.0	34.7	33.8	33.2	32.7	33.5	31.1	29.7
benzoate	Ca(OBz)$_2 \cdot$3H$_2$O	2.32	2.45	2.72	3.02	3.42	4.71	6.87	8.55	8.70
bromide	CaBr$_2 \cdot$6H$_2$O	125	132	143	$185^{34°}$	213	278	295		$312^{105°}$
butyrate	Ca(C$_4$H$_7$O$_2$)$_2$	20.31	19.15	18.20	17.25	16.40	15.15	14.95		15.85
cacodylate	Ca(C$_2$H$_6$AsO$_2$)$_2 \cdot$9H$_2$O	48	52	59	71					
chloride	CaCl$_2 \cdot$6H$_2$O	59.5	64.7	74.5	100	128	137	147	154	159
chromate	CaCrO$_4$	4.5		2.25	1.83	1.49	0.83			
(mn)	CaCrO$_4 \cdot$2H$_2$O	17.3		16.6	16.1					
formate	Ca(CHO$_2$)$_2$	16.15		16.60		17.05	17.50	17.95		18.40
gluconate	Ca(C$_6$H$_{11}$O$_7$)$_2 \cdot$H$_2$O			3.72		5.29		12.11	36.80	$57.2^{96°}$
hydrogen carbonate	Ca(HCO$_3$)$_2$	16.15		16.60		17.05	17.50	17.95		18.40
hydroxide	Ca(OH)$_2$	0.189	0.182	0.173	0.160	0.141	0.121		0.086	0.076

TABLE 5.2 Solubilities of Inorganic Compounds and Metal Salts of Organic Acids in Water at Various Temperatures (*Continued*)

Substance	Formula	0°	10°	20°	30°	40°	60°	80°	90°	100°
Calcium iodate	$Ca(IO_3)_2 \cdot 6H_2O$	0.090		0.24	0.38	0.52	0.65	0.66	0.67	
iodide	CaI_2	64.6	66.0	67.6	69.0	70.8	74	78		81
lactate	$Ca(C_3H_5O_3)_2 \cdot 5H_2O$	3.1		$5.4^{15°}$	7.9					
levulinate	$Ca(C_{10}H_{14}O_6) \cdot 2H_2O$	38.1		$45.1^{16°}$	55.0	$70.3^{45°}$	$88.7^{55°}$			
malonate	$Ca(C_3H_2O_4)$	0.29	0.33	0.36	0.40	0.42	0.46	0.48		
nitrate	$Ca(NO_3)_2 \cdot 4H_2O$	102	115	129	152	191		358		363
nitrite	$Ca(NO_2)_2 \cdot 4H_2O$	63.9		$84.5^{18°}$	104		134	151	166	178
propionate	$Ca(C_3H_5O_2)_2 \cdot H_2O$	42.80		39.85			38.25	39.85	42.15	48.44
selenate	$CaSeO_4 \cdot 2H_2O$	9.73	9.77	9.22	8.79	7.14				
succinate	$Ca(C_3H_2O_2)_2 \cdot 3H_2O$	1.127	1.22	1.28		1.18	0.89	0.68		0.66
sulfamate	$Ca(SO_3NH_2)_2$	56.5	62.8	72.3	84.5	100.1	150.0	215.2	$242^{95°}$	
sulfate	$CaSO_4 \cdot {}^1/_2H_2O$			0.32	$0.29^{25°}$	$0.26^{35°}$	$0.21^{45°}$	$0.145^{65°}$	$0.12^{75°}$	0.071
	$CaSO_4 \cdot 2H_2O$	0.223	0.244	$0.255^{18°}$	0.264	0.265	$0.244^{65°}$	$0.234^{75°}$		0.205
tartrate	$CaC_4H_4O_6 \cdot 4H_2O$	0.026	0.029	0.034	0.046	0.063	0.091	0.130		
uranyl carbonate	$Ca_2UO_2(CO_3)_3 \cdot 10H_2O$	0.1		$0.4^{23°}$		0.8	$1.5^{55°}$			
valerate	$Ca(C_5H_9O_2)_2$	9.82	9.25	8.80	8.40	8.05	7.78	7.95	8.20	8.78
isovalerate	$Ca(C_5H_9O_2)_2 \cdot 3H_2O$	26.05	22.70	21.80	21.68	22.00	18.38	16.88	16.65	16.55
Carbon disulfide	CS_2	0.204	0.194	0.179	0.155	0.111				
oxide sulfide (STP) mL/100 mL	COS	133.3	83.6	56.1	40.3					
tetrafluoride (STP) mL/100 g	CF_4		0.595	0.490	0.415	0.366				
Cerium(III) ammonium nitrate	$Ce(NH_4)_2(NO_3)_5$		242	276	318	376	681			
(IV) ammonium nitrate	$Ce(NH_4)_2(NO_3)_6$			135	150	169	213			
(III) ammonium sulfate	$Ce(NH_4)(SO_4)_2$			5.53	4.49	3.48	2.02	1.33		
(III) selenate	$Ce_2(SeO_3)_3$	39.5	37.2	35.2	33.2	32.6	13.7	4.6	2.1	

(III) sulfate	$Ce_2(SO_4)_3 \cdot 9H_2O$	21.4		9.84	7.24	5.63	3.87	5.40	10.5	22.7
	$Ce_2(SO_4)_3 \cdot 8H_2O$	18.8		9.43	7.10	5.70	4.04			79.0
Cesium aluminum sulfate	$CsAl_2(SO_4)_4$	0.21	0.30	0.40	0.61	0.85	2.00			
bromate	$CsBrO_3$			$3.66^{25°}$	4.53	$5.30^{35°}$				
chlorate	$CsClO_3$	2.46	3.8	6.2	9.5	13.8	26.2	45.0	58.0	79.0
chloride	$CsCl$	161	175	187	197	208	230	250	260	271
chloroaurate(III)	$CsAuCl_4$		0.5	0.8	1.7	3.3	8.9	19.5	27.7	37.9
chloroplatinate(IV)	Cs_2PtCl_6	0.0047	0.0064	0.0087	0.0119	0.0158	0.0290	0.0525	0.0675	0.0914
formate	$CsCHO_2$	335	381	450	533	694				
iodide	CsI	44.1	58.5	76.5	96	$124^{45°}$	150	190	205	
nitrate	$CsNO_3$	9.33	14.9	23.0	33.9	47.2	83.8	134	163	197
perchlorate	$CsClO_4$	0.8	1.0	1.6	2.6	4.0	7.3	14.4	20.5	30.0
sulfate	Cs_2SO_4	167	173	179	184	190	200	210	215	220
Chlorine dioxide	ClO_2	2.76	6.00	$8.70^{15°}$						
Chromium(III) nitrate	$Cr(NO_3)_3$	$108^{5°}$	$124^{15°}$	$130^{25°}$	$152^{35°}$	172.5	183.9	191.6		206.8
(VI) oxide	CrO_3	164.8		167.2		163	227	241		257
(III) perchlorate	$Cr(ClO_4)_3$	104	123	130						
Cobalt(II) bromide	$CoBr_2$	91.9		112	128					
chlorate	$Co(ClO_3)_2$	135	162	180	195	214	316			
chloride	$CoCl_2$	43.5	47.7	52.9	59.7	69.5	93.8	97.6	101	106
iodate	$Co(IO_3)_2$			1.02	0.90	0.88	0.82	0.73		0.70
nitrate	$Co(NO_3)_2$	84.0	89.6	97.4	111	125	174	204	300	
nitrite	$Co(NO_2)_2$	0.076	0.24	0.40	0.61	0.85				
sulfate	$CoSO_4$	25.5	30.5	36.1	42.0	48.8	55.0	53.8	45.3	38.9
	$CoSO_4 \cdot 7H_2O$	44.8	56.3	65.4	73.0	88.1	101			
Copper(II) ammonium chloride	$CuCl_2 \cdot 2NH_4Cl$	28.2	$32.0^{12°}$	35.0	38.3	43.8	56.6	76.5	76.5	
ammonium sulfate	$CuSO_4 \cdot (NH_4)_2SO_4$	11.5	15.1	19.4	24.4	30.5	46.3	69.7	86.1	
bromide	$CuBr_2$	107	116	126	128	$131^{50°}$				
chloride	$CuCl_2$	68.6	70.9	73.0	77.3	87.6	96.5	104	108	107
fluorosilicate	$CuSiF_6$	73.5	76.5	81.6	$84.1^{25°}$	$91.2^{50°}$		$93.2^{75°}$		120
nitrate	$Cu(NO_3)_2$	83.5	100	125	156	163	182	208	222	247
potassium sulfate	$CuSO_4 \cdot K_2SO_4$	5.1	7.2	10.0	13.6	18.2				
selenate	$CuSeO_4$	12.04	14.53	17.51	21.04	25.22	36.50	53.68		
sulfate	$CuSO_4 \cdot 5H_2O$	23.1	27.5	32.0	37.8	44.6	61.8	83.8		
tartrate	$CuC_4H_4O_6 \cdot 3H_2O$		$0.020^{15°}$	0.042	0.089	0.142	0.197	0.144		
Gadolinium bromate	$Gd(BrO_3)_3 \cdot 9H_2O$	50.2	70.1	95.6	126	166				114
sulfate	$Gd_2(SO_4)_3$	3.98	3.30	2.60	2.32					

TABLE 5.2 Solubilities of Inorganic Compounds and Metal Salts of Organic Acids in Water at Various Temperatures (*Continued*)

Substance	Formula	0°	10°	20°	30°	40°	60°	80°	90°	100°
Germanium(IV) oxide	GeO_2		0.49	0.43	0.50	0.61				
Holmium sulfate	$Ho_2(SO_4)_3 \cdot 8H_2O$			8.18	6.7[25]	4.52				
Hydrazinium (1+) nitrate	$N_2H_5NO_3$		175	266	402	607	2127			
(2+) sulfate	$N_2H_6SO_4$			2.87	3.89	4.15	9.08	14.39		
(1+) sulfate	$(N_2H_5)_2SO_4$				221	300	554			
Hydrogen bromide	HBr	221.2	210.3	204.0[15]		171.5[50]		150.5[75]		130.0
chloride	HCl	82.3	77.2	72.1	67.3	63.3	56.1			
selenide, mL at STP	H_2Se	386	351	289						
Iodine	I_2	0.014	0.020	0.029	0.039	0.052	0.100	0.225	0.315	0.445
Iridium(IV) ammonium chloride	$(NH_4)_2IrCl_6$	0.556	0.706	0.77	1.21	1.57	2.46	4.38	dec	
sodium chloride	Na_2IrCl_6		34.46[15]		56.17	96.00	191.2	279.3		
Iron(II) ammonium sulfate	$FeSO_4 \cdot (NH_4)_2SO_4 \cdot 6H_2O$	17.23	31.0	36.47	45.0					
(II) bromide	$FeBr_2$	101	109	117	124	133	144	168	176	184
(II) chloride	$FeCl_2$	49.7	59.0	62.5	66.7	70.0	78.3	88.7	92.3	94.9
(III) chloride	$FeCl_3 \cdot 6H_2O$	74.4		91.8	106.8					
(II) fluoro-silicate	$FeSiF_6 \cdot 6H_2O$	72.1	74.4		77.0[25]		83.7[50]	88.1[75]		100.1[106]
(II) nitrate	$Fe(NO_3)_2 \cdot 6H_2O$	113	134				266			
(III) nitrate	$Fe(NO_3)_3 \cdot 9H_2O$	112.0		137.7		175.0				
(III) perchlorate	$Fe(ClO_4)_3$	289		368	422	478	772			
(II) sulfate	$FeSO_4 \cdot 7H_2O$	28.8	40.0	48.0	60.0	73.3	100.7	79.9	68.3	57.8
Lanthanum bromate	$La(BrO_3)_3$	98	120	149	200	168	247			
nitrate	$La(NO_3)_3$	100		136						
selenate	$La_2(SeO_3)_3$	50.5	45	45	45	45	18.5	5.4	2.2	
sulfate	$La_2(SO_4)_3$	3.00	2.72	2.33	1.90	1.67	1.26	0.91	0.79	0.68
Lead(II) acetate	$Pb(C_2H_3O_2)_2$	19.8	29.5	44.3	69.8	116				
bromide	$PbBr_2$	0.45	0.63	0.86	1.12	1.50	2.29	3.23	3.86	4.55
chloride	$PbCl_2$	0.67	0.82	1.00	1.20	1.42	1.94	2.54	2.88	3.20
fluorosilicate	$PbSiF_6$	190		222			403	428		463

Name	Formula									
iodide	PbI_2	0.044	0.056	0.069	0.090	0.124	0.193	0.294		0.42
nitrate	$Pb(NO_3)_2$	37.5	46.2	54.3	63.4	72.1	91.6	111		133
Lithium acetate	$LiC_2H_3O_2$	31.2	35.1	40.8	50.6	68.6				
ammonium sulfate	$LiNH_4SO_4$		55.2		55.9	56.1	56.5			
azide	LiN_3	61.3	64.2	67.2	71.2	75.4	86.6			100
benzoate	$LiC_7H_5O_2$	38.9	41.6	44.7	53.8					
borate (meta-)	$LiBO_2$	0.90	1.3	2.7	5.7	10.9				
bromate	$LiBrO_3$	154	166	179	198	221	269	308	329	355
bromide	$LiBr$	143	147	160	183	211	223	245		266
carbonate	Li_2CO_3	1.54	1.43	1.33	1.26	1.17	1.01	0.85		0.72
chlorate	$LiClO_3$	241	283	372	488	604	777			
chloride	$LiCl$	69.2	74.5	83.5	86.2	89.8	98.4	112	121	128
chloroaurate(III)	$LiAuCl_4$			136	167	206	324	599		
cyanoplatinate(II)	$Li_2Pt(CN)_4$	105	113	141	153	160	178	216	239	
formate	$LiCHO_2$	32.3	35.7	39.3	44.1	49.5	64.7	92.7	116	138
hydrogen phosphite	Li_2HPO_3	9.97			7.61	7.11	6.03			4.43
hydroxide	$LiOH$	11.91	12.11	12.35	12.70	13.22	14.63	16.56		19.12
iodide	LiI	151	157	165	171	179	202	435	440	481
molybdate	Li_2MoO_4	82.6		79.5	79.4	78.0				73.9
nitrate	$LiNO_3$	53.4	60.8	70.1	138	152	175	233	272	324
nitrite	$LiNO_2$	70.9	82.5	96.8	114	133	177			
perchlorate	$LiClO_4$	42.7	49.0	56.1	63.6	72.3	92.3	128	151	
phosphate (meta-)	$LiPO_3$	0.101	0.058	0.058[25]		0.048				
selenite	Li_2SeO_3	25.0	23.3	21.5	19.6	17.9	14.7	11.9	11.1	9.9
sulfate	Li_2SO_4	36.1	35.5	34.8	34.2	33.7	32.6	31.4	30.9	
tartrate (d-)	$Li_2C_4H_4O_6$	42.0	31.8	27.1	26.6	27.2	29.5			
thiocyanate	$LiSCN$			114	131	153				
vanadate	Li_3VO_4	2.50		4.82	6.28	4.38	2.67			
Magnesium acetate	$Mg(C_2H_3O_2)_2$	56.7	59.7	53.4	68.6	75.7	118			125
bromide	$MgBr_2$	98	99	101	104	106	112			
chlorate	$Mg(ClO_3)_2$	114	123	135	155	178	242		268	
chloride	$MgCl_2$	52.9	53.6	54.6	55.8	57.5	61.0	66.1	69.5	73.3
fluorosilicate	$MgSiF_6$	26.3		30.8		34.9	44.4			
formate	$Mg(CHO_2)_2$	14.0	14.2	14.4	14.9	15.9	17.9	20.5	22.2	23.9
iodate	$Mg(IO_3)_2$		7.2	8.6	10.0	11.7	15.2	15.5	15.6	
iodide	MgI_2	120		140		173		186		

TABLE 5.2 Solubilities of Inorganic Compounds and Metal Salts of Organic Acids in Water at Various Temperatures (*Continued*)

Substance	Formula	0°	10°	20°	30°	40°	60°	80°	90°	100°
Magnesium nitrate	$Mg(NO_3)_2$	62.1	66.0	69.5	73.6	78.9	78.9	91.6	106	
selenate	$MgSeO_4$	20.0	30.4	38.3	44.3	48.6	55.8	55.8		
sulfate	$MgSO_4$	22.0	28.2	33.7	38.9	44.5	54.6	55.8	52.9	50.4
sulfite	$MgSO_3$	0.339	0.446	0.573	0.751	0.959	0.779	0.642	0.622	
tartrate	$MgC_4H_4O_6$	0.54	0.78	1.06		1.02				
Manganese bromide	$MnBr_2$	127	136	147	157	169	197	225	226	228
chloride	$MnCl_2$	63.4	68.1	73.9	80.8	88.5	109	113	114	115
fluoride	MnF_2			1.06		0.67	0.44			0.48
nitrate	$Mn(NO_3)_2$	102	118	139	206					
oxalate	MnC_2O_4	0.020	0.024	0.028	0.033					
sulfate	$MnSO_4$	52.9	59.7	62.9	62.9	60.0	53.6	45.6	40.9	35.3
Mercury(II) bromide	$HgBr_2$	0.30	0.40	0.56	0.66	0.91	1.68	2.77		4.9
(II) chloride	$HgCl_2$	3.63	4.82	6.57	8.34	10.2	16.3	30.0		61.3
(I) perchlorate	$Hg_2(ClO_4)_2$	282	325	367	407	455	499	541		580
Molybdenum trioxide	MoO_3			0.134	0.285	0.454	1.08	1.74		
Neodymium bromate	$Nd(BrO_3)_3$	43.9	59.2	75.6	95.2	116				
chloride	$NdCl_3$		96.7	98.0	99.6	102	105			
nitrate	$Nd(NO_3)_3$	127	133	142	145	159	211			
selenate	$Nd_2(SeO_3)_3$	46.2	44.6	41.8	39.9	39.9	43.9	7.0	3.3	
sulfate	$Nd_2(SO_4)_3$	13.0	9.7	7.1	5.3	4.1	2.8	2.2	1.2	
Nickel bromide	$NiBr_2$	113	122	131	138	144	153	154		155
chlorate	$Ni(ClO_3)_2$	111	120	133	155	181	221	308		
chloride	$NiCl_2$	53.4	56.3	60.8	70.6	73.2	81.2	86.6		87.6
fluoride	NiF_2		2.55	2.56			2.56		2.59	
iodate	$Ni(IO_3)_2$				1.15		1.06		1.00	
	$Ni(IO_3)_2 \cdot 4H_2O$	0.74		1.09	1.43					
iodide	NiI_2	124	135	148	161	174	184	187	188	
nitrate	$Ni(NO_3)_2$	79.2		94.2	105	119	158	187	188	
perchlorate	$Ni(ClO_4)_2$	105	107	110	113	117				

Solubility table (values across eight unlabeled columns; temperature headings appear on the facing page). Colour notes for nickel sulfate: (pale blue) / (blue) / (green).

Substance	Formula								
Nickel sulfate	NiSO₄·6H₂O (pale blue / blue)	26.2	40.1	43.6	47.6	55.6	64.5	70.1	76.7
	NiSO₄·7H₂O (green)	32.4	37.7	43.4	50.4				
Osmium tetroxide	OsO₄	5.26	5.75	6.43					
Oxalic acid	H₂C₂O₄	3.54	6.08	9.52	14.23	21.52	44.32	84.5	120
Potassium acetate	KC₂H₃O₂	216	233	256	283	324	350	381	398
aluminum sulfate	KAl(SO₄)₂	3.00	3.99	5.90	8.39	11.7	24.8	71.0	109
azide	KN₃	41.4	46.2	50.8	55.8	61.0	70.7	82.1	106
benzoate	KC₇H₅O₂		65.8	70.7	76.7	82.1	94.9	99.2	104
bromate	KBrO₃	3.09	4.72	6.91	9.64	13.1	22.7	34.1	49.9
bromide	KBr	53.6	59.5	65.3	70.7	75.4	85.5	94.9	104
cadmium bromide	KCdBr₃	116	133	150	170	191	233	276	325
cadmium chloride	KCdCl₃	26.6	32.3	38.9	45.6	53.1	67.5	83.5	101
carbonate	K₂CO₃	105	108	111	114	117	127	140	156
chlorate	KClO₃	3.3	5.2	7.3	10.1	13.9	23.8	37.6	56.3
chloride	KCl	28.0	31.2	34.2	37.2	40.1	45.8	51.3	56.3
chloroaurate(III)	KAuCl₄								
chloroplatinate(IV)	K₂PtCl₆	0.48	0.60	0.78	1.00	1.36	2.45	3.71	5.03
chromate	K₂CrO₄	56.3	60.0	63.7	66.7	67.8	70.1	74.5	
citrate	K₃C₆H₅O₇	153	172	194					
cobalt(II) sulfate	K₂Co(SO₄)₂	8.5	11.7	15.5	19.3	23.3	32.5	47.7	
copper(II) sulfate	K₂Cu(SO₄)₂	5.1	7.2	10.0	13.6	18.2			
cyanoplatinate(II)	K₂Pt(CN)₄	11.6	19.8	33.9	52.0	78.3	139	177	194
dichromate	K₂Cr₂O₇	4.7	7.0	12.3	18.1	26.3	45.6	73.0	83.5
dihydrogen phosphate	KH₂PO₄	14.8	18.3	22.6	28.0	33.5	50.2	70.4	83.5
dithionate	K₂S₂O₆	2.6	4.2	6.6	9.3				
ferricyanide	K₃Fe(CN)₆	30.2	38	46	53	59.3	70		91
ferrocyanide	K₄Fe(CN)₆	14.3	21.1	28.2	35.1	41.4	54.8	66.9	74.2
fluoride	KF	44.7	53.5	94.9	108	138	142	150	
fluorogermanate(IV)	K₂GeF₆	0.25	0.36	0.50	0.66	0.96			
fluorosilicate	K₂SiF₆	0.077	0.102	0.151	0.202	0.253			
fluorotitanate(IV)	K₂TiF₆	0.55	0.91	1.28					
formate	KCHO₂	313	337	361	398	471	580	658	
hydrogen carbonate	KHCO₃	22.5	27.4	33.7	39.9	47.5	65.6		

TABLE 5.2 Solubilities of Inorganic Compounds and Metal Salts of Organic Acids in Water at Various Temperatures (*Continued*)

Substance	Formula	0°	10°	20°	30°	40°	60°	80°	90°	100°
Potassium hydrogen fluoride	KHF_2	24.5	30.1	39.2	46.8	56.5	78.8	114		
hydrogen selenite	$KH_3(SeO_3)_2$	115	162	215	300	408	900			122
hydrogen sulfate	$KHSO_4$	36.2		48.6	54.3	61.0	76.4	96.1		122
hydrogen tartrate	$KC_4H_5O_6$	0.231	0.358	0.523	0.762					
hydroxide	KOH	95.7	103	112	126	134	154			178
iodate	KIO_3	4.60	6.27	8.08	10.3	12.6	18.3	24.8		32.3
iodide	KI	128	136	144	153	162	176	192	198	206
iron(II) sulfate	$K_2Fe(SO_4)_2$	19.6	24.5	32.1	39.1	44.9	57.2			
magnesium sulfate	$K_2Mg(SO_4)_2$	14.0	19.5	25.0	30.4	36.6	50.2	63.4		
nickel sulfate	$K_2Ni(SO_4)_2$	3.37	4.50	5.94	7.72	9.85	15.4	23.0	27.8	33.4
nitrate	KNO_3	13.9	21.2	31.6	45.3	61.3	106	167	203	245
nitrite	KNO_2	279	292	306	320	329	348	376	390	410
oxalate	$K_2C_2O_4$	25.5	31.9	36.4	39.9	43.8	53.2	63.6	69.2	75.3
perchlorate	$KClO_4$	0.76	1.06	1.68	2.56	3.73	7.3	13.4	17.7	22.3
periodate	KIO_4	0.17	0.28	0.42	0.65	1.0	2.1	4.4	5.9	
permanganate	$KMnO_4$	2.83	4.31	6.34	9.03	12.6	22.1			
peroxodisulfate	$K_2S_2O_8$	1.65	2.67	4.70	7.75	11.0				
perrhenate	$KReO_4$	0.34	0.63	0.99	1.47	2.2	4.58	8.7		
phosphate	K_3PO_4		81.5	92.3	108	133				
salicylate	$KC_7H_5O_3$	21.2	32.4	47.1	61.3	78.6	116	156		
selenate	K_2SeO_4	107	109	111	113	115	119	121		122
selenite	K_2SeO_3	169	186	203	217	217	220			217
sulfate	K_2SO_4	7.4	9.3	11.1	13.0	14.8	18.2	21.4	22.9	24.1
sulfite	K_2SO_3	106		106	107	107	108			112
tellurate	K_2TeO_4	8.8		27.5	50.4					
thioantimonate(V)	K_3SbS_4	306	320		302	315		381		
thiocyanate	$KSCN$	177	198	224	255	289	372	492	571	675
thiosulfate	$K_2S_2O_3$	96		155	175	205	238	293	312	
zinc sulfate	$K_2Zn(SO_4)_2 \cdot 6H_2O$	13.0	18.9	25.9	35.0	44.9	72.1			

Solubility data (grams per 100 g water; columns correspond to successive temperatures, unlabeled on this page)

Compound	Formula									
Praseodymium bromate	$Pr(BrO_3)_3$	55.9	73.0	91.8	114	144				
nitrate	$Pr(NO_3)_3$			112	162	178				
selenate	$Pr_2(SeO_3)_3$	36.2			32.4	31.2	30.4	5.43	3.6	0.91
sulfate	$Pr_2(SO_4)_3$	19.8	15.6	12.6	9.89	2.56	5.04	3.5	1.1	
Rubidium aluminum sulfate	$Rb_2Al_2(SO_4)_4$	0.72	1.05	1.50	2.20	3.25	7.40	21.6	49	63
bromate	$RbBrO_3$				3.6	5.1				
bromide	$RbBr$	90	99	108	119	132	158			
chlorate	$RbClO_3$	2.1	3.4	5.4	8.0	11.6	22	38		
chloride	$RbCl$	77	84	91	98	104	115	127	133	143
chloroaurate(III)	$RbAuCl_4$	0.014	4.8	9.9	15.5	21.5	36.2	54.6	65.8	79.2
chloroplatinate(IV)	Rb_2PtCl_6		0.020	0.028	0.040	0.056	0.090	0.182	0.247	0.333
chromate	Rb_2CrO_4	62.0	67.5	73.6	78.9	85.6	95.7			
cobalt sulfate	$Rb_2Co(SO_4)_2$	5.10	7.47	10.8	14.5	18.2	30.2	44.9	55.0	70.1
dichromate (mn)	$Rb_2Cr_2O_7$			5.9	10.0	15.2	32.3			
(tric)				5.8	9.5	14.8	32.4			
formate	$RbCHO_2$		443	554	614	694	900			
iron(III) sulfate	$RbFe(SO_4)_2 \cdot 12H_2O$		8.0	20	35	52	200			
nitrate	$RbNO_3$	19.5	33.0	52.9	81.2	117		310	374	452
perchlorate	$RbClO_4$	1.09	1.19	1.55	2.20	3.26	6.27	11.0	15.5	22.0
salicylate	$RbC_7H_5O_3$		187	212	238	268	324			
sulfate	Rb_2SO_4	37.5	42.6	48.1	53.6	58.5	67.5	75.1	78.6	81.8
Samarium bromate	$Sm(BrO_3)_3$	34.2	47.6	62.5	79.0	98.5				
chloride	$SmCl_3$		92.4	93.4	94.6	96.9				
Selenic acid	H_2SeO_4	426	122.2	567	1328	344.4	383.1	383.1	385.4	
Selenious acid	H_2SeO_3	90.1	222	166.7	235.6		440			
Selenium dioxide	SeO_2			257	291	335				
Silver acetate	$AgC_2H_3O_2$	0.73	0.89	1.05	1.23	1.43	1.93	2.59	1.33	
bromate	$AgBrO_3$		0.11	0.16	0.23	0.32	0.57	0.94		
chlorate	$AgClO_3$		10.4	15.3	20.9	26.8				
fluoride	AgF	85.9	120	172	190	203	440	585		
nitrate	$AgNO_3$	122	167	216	265	311			652	733
nitrite	$AgNO_2$	0.16	0.22	0.34	0.51	0.73	1.39			
perchlorate	$AgClO_4$	455	484	525	594	635				793
sulfamate	$AgNH_2SO_3$	2.30	4.82	7.53	10.3	15.3	28.5			
sulfate	Ag_2SO_4	0.57	0.70	0.80	0.89	0.98	1.15	1.30	1.36	1.41

TABLE 5.2 Solubilities of Inorganic Compounds and Metal Salts of Organic Acids in Water at Various Temperatures (*Continued*)

Substance	Formula	0°	10°	20°	30°	40°	60°	80°	90°	100°
Sodium acetate	$NaC_2H_3O_2$	36.2	40.8	46.4	54.6	65.6	139	153	161	170
aluminum sulfate	$Na_2Al_2(SO_4)_4$	37.4	39.3	39.7	41.7	43.8				
azide	NaN_3	38.9	39.9	40.8						55.3
benzoate	$NaC_7H_5O_2$	62.6	62.8	62.8	62.9	63.1	64.5	68.6	70.6	73.3
borate (penta-)	$Na_2B_{10}O_{16}$	6.4	8.6	12.0	16.4	22.0	37.9	63.4	83.5	108
borate (tetra-)	$Na_2B_4O_7$	1.11	1.60	2.56	3.86	6.67	19.0	31.4	41.0	52.5
bromate	$NaBrO_3$	24.2	30.3	36.4	42.6	48.8	62.6	75.7		90.8
bromide	$NaBr$	80.2	85.2	90.8	98.4	107	118	120	121	121
carbonate	Na_2CO_3	7.00	12.5	21.5	39.7	49.0	46.0	43.9	43.9	
chlorate	$NaClO_3$	79.6	87.6	95.9	105	115	137	167	184	204
chloride	$NaCl$	35.7	35.8	35.9	36.1	36.4	37.1	38.0	38.5	39.2
chloroaurate(III)	$NaAuCl_4$		139	151	178	227	900	279		
chloroiridate(IV)	Na_2IrCl_6		31.6	39.3	56.2	96.1	192			
chromate	Na_2CrO_4	31.7	50.1	84.0	88.0	96.0	115	125		126
cyanide	$NaCN$	40.8	48.1	58.7	71.2					
dichromate	$Na_2Cr_2O_7$	163	172	183	198	215	269	376	405	415
diethyl barbiturate	$NaC_8H_{11}N_2O_3$		12.7	21.5	24.7				48.0	
dihydrogen phosphate (ortho-)	NaH_2PO_4	56.5	69.8	86.9	107	133	172	211	234	
dihydrogen phosphate (pyro-)	$Na_2H_2P_2O_7$	4.47	6.95	12.0	17.1	18.4				
dithionate	$Na_2S_2O_6$	6.3	11.1	15.1	19.6	24.7	36.1	49.3	56.3	64.7
dodecanesulfonate	$NaC_{12}H_{25}SO_3$			0.13	0.25	6.54				
dodecanoate	$NaC_{12}H_{23}O_2$				4.58	22.7	105	170		
EDTA (Y)*	$Na_2H_2Y \cdot 2H_2O$	10.6		11.1	12.8	14.2	17.0	22.2	24.3	27.0[98]
ferrocyanide	$Na_4Fe(CN)_6$	11.2	14.8	18.8	23.8	29.9	43.7	62.1		
fluoride	NaF	3.66		4.06	4.22	4.40	4.68	4.89		5.08
fluoroberyllate	Na_2BeF_4	1.33		1.44		1.92	2.24	2.62	2.73	
fluorogermanate	Na_2GeF_6	1.52	1.68		2.25	2.83		3.36		
fluorosilicate	Na_2SiF_6	4.35	5.7	7.2	8.6	10.3	14.3	18.7	21.5	24.5

Name	Formula									
formate	NaCHO₂	43.9	62.5	81.2	102	108	122	138	147	160
germanate	Na₂GeO₃	14.4	18.8	23.8	28.7	37.2	65.0	116	188	198
hydrogen arsenate	Na₂HAsO₄	5.9	13.0	33.9	49.3	69.5	144	186		104
hydrogen carbonate	NaHCO₃	7.0	8.1	9.6	11.1	12.7	16.0			
hydrogen phosphate	Na₂HPO₄	1.68	3.53	7.83	22.0	55.3	82.8	92.3	102	
hydrogen phosphite	Na₂HPO₃	418	424	429	566					
hydrogen succinate	NaC₄H₅O₄	17.5	25.3	34.8	47.7	61.6	74.5	90.1		
hydroxide	NaOH		98	109	119	129	174			
hydroxostannate(IV)	Na₂Sn(OH)₆	46.0		43.7	42.7	38.9				
hypochlorite	NaClO	29.4	36.4	53.4	100	110				33.0
iodate	NaIO₃	2.48	4.59	8.08	10.7	13.3	19.8	26.6	29.5	302
iodide	NaI	159	167	178	191	205	257	295		
molybdate	Na₂MoO₄	44.1	64.7	65.3	66.9	68.6	71.8	74.8		180
nitrate	NaNO₃	73.0	80.8	87.6	94.9	102	122	148		160
nitrite	NaNO₂	71.2	75.1	80.8	87.6	94.9	111	133		6.50
oxalate	Na₂C₂O₄	2.69	3.05	3.41	3.81	4.18	4.93	5.71		329
perchlorate	NaClO₄	167	183	201	222	245	288	306		
periodate	NaIO₄	1.83	5.6	10.3	19.9	30.4		60.0		77.0
phosphate	Na₃PO₄	4.5	8.2	12.1	16.3	20.2	29.9		68.1	
potassium tartrate	NaKC₄H₄O₆	31.9	46.6	67.8	102	117	130	144		
salicylate	NaC₇H₅O₃		44.7	95.3	111	117				72.7
selenate	Na₂SeO₄	13.3	25.2	26.9	77.0	81.8	78.6	74.8	73.0	82.5
selenite	Na₂SeO₃	78.6	81.2	86.2	94.2	96.5	91.6	86.6	84.5	42.5
sulfate	Na₂SO₄	4.9	9.1	19.5	40.8	48.8	45.3	43.7	42.7	
	Na₂SO₄ · 7H₂O	19.5	30.0	44.1						
sulfide	Na₂S	9.6	12.1	15.7	20.5	26.6	39.1	55.0	65.3	
sulfite	Na₂SO₃	14.4	19.5	26.3	35.5	37.2	32.6	29.4	27.9	
thioantimonate(V)	Na₃SbS₄	13.4	20.0	27.9	37.2	49.3	53.8	88.3		
thiocyanate	NaSCN	111	134	164	176	192	210	218		
thiosulfate	Na₂S₂O₃ · 5H₂O	50.2	59.7	70.1	83.2	104	90.8			97.2
tungstate	Na₂WO₄	71.5	73.0	73.0	77.0	77.6	33.0	40.8		
vanadate	NaVO₃		19.3	22.5	26.3	38.3	36.8			
Strontium acetate	Sr(C₂H₃O₂)₂	37.0	42.9	41.1	39.5	38.3	36.8	36.1	36.2	36.4
bromide	SrBr₂	85.2	93.4	102	112	123	150	182	218	223
chloride	SrCl₂	43.5	47.7	52.9	58.7	65.3	81.8	90.5		101
chromate	SrCrO₄	0.085	0.090					0.058		

* Properly called dihydrogen ethylenediaminetetraacetate (Na₂H₂EDTA · 2H₂O).

TABLE 5.2 Solubilities of Inorganic Compounds and Metal Salts of Organic Acids in Water at Various Temperatures (*Continued*)

Substance	Formula	0°	10°	20°	30°	40°	60°	80°	90°	100°
Strontium fluoride	SrF_2	0.0113		0.0117	0.0119					
formate	$Sr(CHO_2)_2$	9.1	10.6	12.7	15.2	17.8	25.0	31.9	32.9	34.4
hydroxide	$Sr(OH)_2$	0.91	1.25	1.77	2.64	3.95	8.42	20.2	44.5	91.2
iodide	SrI_2	165		178		192	218	270	365	383
nitrate	$Sr(NO_3)_2$	39.5	52.9	69.5	88.7	89.4	93.4	96.9	98.4	
nitrite	$Sr(NO_2)_2$			65	72	79	97	130	134	
oxide	SrO				1.03	1.05	3.40	9.15	13.13	12.15
sulfate	$SrSO_4$	0.0113	0.0129	0.0132	0.0138	0.0141	0.0131	0.0116	0.0115	
Sulfamic acid	H_2NSO_3H	14.7	18.6	21.3	26.1	29.5	37.1	47.1		
Telluric acid	H_2TeO_4	16.2	33.8	41.6	50.0	57.2	77.5	106		155
Terbium bromate	$Tb(BrO_3)_3 \cdot 9H_2O$	66.4	89.7	117	152	198				
Thallium(I) azide	TlN_3	0.171	0.236	0.364						
bromide	$TlBr$	0.022	0.032	0.048	0.068	0.097	0.177			
carbonate	Tl_2CO_3	2.00		5.3		12.7^{50}	12.2			27.2
chlorate	$TlClO_3$			3.92				36.6		57.3
chloride	$TlCl$	0.21	0.25	0.33	0.42	0.52	0.80	1.20		1.80
hydroxide	$TlOH$	25.4	29.6	35.0	40.4	49.4	73.3	106	126	150
iodide	TlI	0.002		0.006		0.015	0.035	0.070		0.120
nitrate	$TlNO_3$	3.90	6.22	9.55	14.3	21.0	46.1	110	200	414
nitrite	$TlNO_2$	17.9	28.9	40.3	53.2	83.6	216	1150	750	
perchlorate	$TlClO_4$	6.00	8.04	13.1	19.7	28.3	50.8	81.5		
picrate	$TlOC_6H_2(NO_2)_3$	0.135		0.40	0.57	0.83	1.73			
selenate	Tl_2SeO_4		2.17	2.80				8.50		10.8
sulfate	Tl_2SO_4	2.73	3.70	4.87	6.16	7.53	11.0	14.6	16.5	18.4
Thorium nitrate	$Th(NO_3)_4$	186	187	191						
sulfate	$Th(SO_4)_2 \cdot 4H_2O$	0.74	0.99	1.38	1.99	4.04	1.63			
	$Th(SO_4)_2 \cdot 9H_2O$			0.99	1.17	3.00				
Tin(II) iodide	SnI_2						2.11		3.58	4.20
Uranium(IV) sulfate	$U(SO_4)_2 \cdot 4H_2O$				10.1	9.0	7.7			
	$U(SO_4)_2 \cdot 8H_2O$			11.9	17.9	29.2	55.8			

Uranyl nitrate	UO₂(NO₃)₂	98	107	122	141	167	317	388	426	474
oxalate	UO₂C₂O₄		0.45	0.50	0.61	0.80	1.22	1.94		3.16
Ytterbium sulfate	Yb₂(SO₄)₃	44.2	37.5		22.2	17.2	10.4	6.4	5.8	4.7
Yttrium bromide	YBr₃	63.9		75.1	79.6	87.3	101	116	123	
chloride	YCl₃	77.3	78.1	78.8		80.8				
nitrate	Y(NO₃)₃	93.1	106	123	143	163	200			
sulfate	Y₂(SO₄)₃	8.05	7.67	7.30	6.78	6.09	4.44	2.89	2.2	
Zinc bromide	ZnBr₂	389		446	528	591	618	645		672
chlorate	Zn(ClO₃)₂	145	152	200	209	223				
chloride	ZnCl₂	342	363	395	437	452	488	541		614
formate	Zn(CHO₂)₂	3.70	4.30	5.20	6.10	7.40	11.8	21.2	28.8	38.0
iodide	ZnI₂	430		432		445	467	490		510
nitrate	Zn(NO₃)₂	98			138	211				
sulfate (rh)	ZnSO₄	41.6	47.2	53.8	61.3	70.5	75.4	71.1		
sulfate (mn)			54.4	60.0	65.5					60.5
tartrate	ZnC₄H₄O₆			0.022	0.041	0.060	0.104	0.059		

5.2 VAPOR PRESSURES

TABLE 5.3 Vapor Pressure of Mercury

Temp. °C	mm of Hg	Temp. °C	mm of Hg	Temp. °C	mm of Hg
0	0.000 185	78	0.078 89	158	3.873
2	0.000 228	80	0.088 80	160	4.189
4	0.000 276	82	0.100 0	162	4.528
6	0.000 335	84	0.112 4	164	4.890
8	0.000 406	86	0.126 1	166	5.277
10	0.000 490	88	0.1413	168	5.689
12	0.000 588	90	0.1582	170	6.128
14	0.000 706	92	0.1769	172	6.596
16	0.000 846	94	0.1976	174	7.095
18	0.001 009	96	0.2202	176	7.626
20	0.001 201	98	0.2453	178	8.193
22	0.001 426	100	0.2729	180	8.796
24	0.001 691	102	0.3032	182	9.436
26	0.002 000	104	0.3366	184	10.116
28	0.002 359	106	0.3731	186	10.839
30	0.002 777	108	0.4132	188	11.607
32	0.003 261	110	0.4572	190	12.423
34	0.003 823	112	0.5052	192	13.287
36	0.004 471	114	0.5576	194	14.203
38	0.005 219	116	0.6150	196	15.173
40	0.006 079	118	0.6776	198	16.200
42	0.007 067	120	0.7457	200	17.287
44	0.008 200	122	0.8198	202	18.437
46	0.009 497	124	0.9004	204	19.652
48	0.010 98	126	0.9882	206	20.936
50	0.012 67	128	1.084	208	22.292
52	0.014 59	130	1.186	210	23.723
54	0.016 77	132	1.298	212	25.233
56	0.019 25	134	1.419	214	26.826
58	0.022 06	136	1.551	216	28.504
60	0.025 24	138	1.692	218	30.271
62	0.028 83	140	1.845	220	32.133
64	0.032 87	142	2.010	222	34.092
66	0.037 40	144	2.188	224	36.153
68	0.042 51	146	2.379	226	38.318
70	0.048 25	148	2.585	228	40.595
72	0.054 69	150	2.807	230	42.989
74	0.061 89	152	3.046		
76	0.069 93	154	3.303		
		156	3.578		

TABLE 5.3 Vapor Pressure of Mercury (*Continued*)

Temp. °C	mm of Hg	Temp. °C	mm of Hg	Temp. °C	mm of Hg
232	45.503	302	257.78	372	994.34
234	48.141	304	269.17	374	1028.9
236	50.909	306	280.98	376	1064.4
238	53.812	308	293.21	378	1100.9
240	56.855	310	305.89	380	1138.4
242	60.044	312	319.02	382	1177.0
244	63.384	314	332.62	384	1216.6
246	66.882	316	346.70	386	1257.3
248	70.543	318	361.26	388	1299.1
250	74.375	320	376.33	390	1341.9
252	78.381	322	391.92	392	1386.1
254	82.568	324	408.04	394	1431.3
256	86.944	326	424.71	396	1477.7
258	91.518	328	441.94	398	1525.2
260	96.296	330	459.74	400	1574.1
262	101.28	332	478.13	430	2464
264	106.48	334	497.12	460	3715
266	111.91	336	516.74	490	5420
268	117.57	338	53 7.00	520	7691
270	123.47	340	557.90	550	10650
272	129.62	342	579.45	600	22.87 atm
274	136.02	344	601.69	650	35.49 atm
276	142.69	346	624.64	700	52.51 atm
278	149.64	348	648.30	750	74.86 atm
280	156.87	350	672.69	800	103.31 atm
282	164.39	352	697.83	850	138.42 atm
284	172.21	354	723.73	900*	180.92 atm
286	180.34	356	750.43	950	226.58 atm
288	188.79	358	777.92	1000	290.5 atm
290	197.57	360	806.23	1050	358.1 atm
292	206.70	362	835.38	1100	437.3 atm
294	216.17	364	865.36	1150	521.3 atm
296	226.00	366	896.23	1200	616.8 atm
298	236.21	368	928.02	1250	721.4 atm
300	246.80	370	960.66	1300	835.9 atm

* Critical point.

TABLE 5.4 Vapor Pressure of Ice in Millimeters of Mercury

For temperatures from −99 to 0°C.

The values in the table are for ice in contact with its own vapor. Where the ice is in contact with air at a temperature t °C, this correction must be added: Correction $= 20p/(100)(t + 273)$.

t, °C	p, mm Hg	t, °C	p, mm Hg	t, °C	p, mm Hg
−99	0.000 012	−49	0.033 4	−14.5	1.300
−98	0.000 015	−48	0.037 8	−14.0	1.361
−97	0.000 018	−47	0.042 6	−13.5	1.424
−96	0.000 022	−46	0.048 1	−13.0	1.490
−95	0.000 027	−45	0.054 1	−12.5	1.559
−94	0.000 033	−44	0.060 9	−12.0	1.632
−93	0.000 040	−43	0.068 4	−11.5	1.707
−92	0.000 048	−42	0.076 8	−11.0	1.785
−91	0.000 058	−41	0.086 2	−10.5	1.866
−90	0.000 070	−40	0.096 6	−10.0	1.950
−89	0.000 084	−39	0.108 1	−9.8	1.985
−88	0.000 10	−38	0.120 9	−9.6	2.021
−87	0.000 12	−37	0.135 1	−9.4	2.057
−86	0.000 14	−36	0.150 7	−9.2	2.093
−85	0.000 17	−35	0.168 1	−9.0	2.131
−84	0.000 20	−34	0.187 3	−8.8	2.168
−83	0.000 24	−33	0.208 4	−8.6	2.207
−82	0.000 29	−32	0.231 8	−8.4	2.246
−81	0.000 34	−31	0.257 5	−8.2	2.285
−80	0.000 40	−30.0	0.285 9	−8.0	2.326
−79	0.000 47	−29.5	0.301	−7.8	2.367
−78	0.000 56	−29.0	0.317	−7.6	2.408
−77	0.000 66	−28.5	0.334	−7.4	2.450
−76	0.000 77	−28.0	0.351	−7.2	2.493
−75	0.000 90	−27.5	0.370	−7.0	2.537
−74	0.001 05	−27.0	0.389	−6.8	2.581
−73	0.001 23	−26.5	0.409	−6.6	2.626
−72	0.001 43	−26.0	0.430	−6.4	2.672
−71	0.001 67	−25.5	0.453	−6.2	2.718
−70	0.001 94	−25.0	0.476	−6.0	2.765
−69	0.002 25	−24.5	0.500	−5.8	2.813
−68	0.002 61	−24.0	0.526	−5.6	2.862
−67	0.003 02	−23.5	0.552	−5.4	2.912
−66	0.003 49	−23.0	0.580	−5.2	2.962
−65	0.004 03	−22.5	0.609	−5.0	3.013
−64	0.004 64	−22.0	0.640	−4.8	3.065
−63	0.005 34	−21.5	0.672	−4.6	3.117
−62	0.006 14	−21.0	0.705	−4.4	3.171
−61	0.007 03	−20.5	0.740	−4.2	3.225
−60	0.008 08	−20.0	0.776	−4.0	3.280
−59	0.009 25	−19.5	0.814	−3.8	3.336
−58	0.010 6	−19.0	0.854	−3.6	3.393
−57	0.012 1	−18.5	0.895	−3.4	3.451
−56	0.013 8	−18.0	0.939	−3.2	3.509
−55	0.015 7	−17.5	0.984	−3.0	3.568
−54	0.017 8	−17.0	1.031	−2.8	3.360
−53	0.020 3	−16.5	1.080	−2.6	3.691
−52	0.023 0	−16.0	1.132	−2.4	3.753

TABLE 5.4 Vapor Pressure of Ice in Millimeters of Mercury (*Continued*)

t, °C	p, mm Hg	t, °C	p, mm Hg	t, °C	p, mm Hg
−51	0.026 1	−15.5	1.186	−2.2	3.816
−50	0.029 6	−15.0	0.241	−2.0	3.880
−1.8	3.946	−1.0	4.217	−0.2	4.504
−1.6	4.012	−0.8	4.287	0.0	4.579
−1.4	4.079	−0.6	4.359		
−1.2	4.147	−0.4	4.431		

TABLE 5.5 Vapor Pressure of Liquid Ammonia, NH_3

t°C.	p in atm	t°C.	p in atm	t°C.	p in atm
−78	0.0582	−6	3.3677	66	29.784
−76	0.0683	−4	3.6405	68	31.211
−74	0.0797	−2	3.9303	70	32.687
−72	0.0929	0	4.2380	72	34.227
−70	0.1078	+2	4.5640	74	35.813
−68	0.1246	4	4.9090	76	37.453
−66	0.1437	6	5.2750	78	39.149
−64	0.1651	8	5.6610	80	40.902
−62	0.1891	10	6.0685	82	42.712
−60	0.2161	12	6.4985	84	44.582
−58	0.2461	14	6.9520	86	46.511
−56	0.2796	16	7.4290	88	48.503
−54	0.3167	18	7.9310	90	50.558
−52	0.3578	20	8.4585	92	52.677
−50	0.4034	22	9.0125	94	54.860
−48	0.4536	24	9.5940	96	57.111
−46	0.5087	26	10.2040	98	59.429
−44	0.5693	28	10.8430	100	61.816
−42	0.6357	30	11.512	102	64.274
−40	0.7083	32	12.212	104	66.804
−38	0.7875	34	12.943	106	69.406
−36	0.8738	36	13.708	108	72.084
−34	0.9676	38	14.507	110	74.837
−32	1.0695	40	15.339	112	77.668
−30	1.1799	42	16.209	114	80.578
−28	1.2992	44	17.113	116	83.570
−26	1.4281	46	18.056	118	86.644
−24	1.5671	48	19.038	120	89.802
−22	1.7166	50	20.059	122	93.045
−20	1.8774	52	21.121	124	96.376
−18	2.0499	54	22.224	126	99.796
−16	2.2349	56	23.372	128	103.309
−14	2.4328	58	24.562	130	106.913
−12	2.6443	60	25.797	132	110.613
−10	2.8703	62	27.079	132.3	111.3(c.p.)
− 8	3.1112	64	28.407		

TABLE 5.6 Vapor Pressure of Water

For temperatures from − 10 to 120° C.

The values in the table are for water in contact with its own vapor. Where the water is in contact with air at a temperature t in degrees Celsius, the following correction must be added: Correction (when $t \leq 40°C$) $= p(0.775 - 0.000\ 313t)/100$; correction (when $t > 50°C$) $= p(0.0652 - 0.000\ 087\ 5\ t)/100$.

t, °C	p, mmHg	t, °C	p, mmHg	t, °C	p, mmHg	t, °C	p, mmHg
−10.0	2.149	11.5	10.176	22.2	20.070	30.8	33.312
−9.5	2.236	12.0	10.518	22.4	20.316	31.0	33.695
−9.0	2.326	12.5	10.870	22.6	20.565	31.2	34.082
−8.5	2.418	13.0	11.231	22.8	20.815	31.4	34.471
−8.0	2.514	13.5	11.604	23.0	21.068	31.6	34.864
−7.5	2.613	14.0	11.987	23.2	21.324	31.8	35.261
−7.0	2.715	14.5	12.382	23.4	21.583	32.0	35.663
−6.5	2.822	15.0	12.788	23.6	21.845	32.2	36.068
−6.0	2.931	15.2	12.953	23.8	22.110	32.4	36.477
−5.5	3.046	15.4	13.121	24.0	22.387	32.6	36.891
−5.0	3.163	15.6	13.290	24.2	22.648	32.8	37.308
−4.5	3.284	15.8	13.461	24.4	22.922	33.0	37.729
−4.0	3.410	16.0	13.634	24.6	23.198	33.2	38.155
−3.5	3.540	16.2	13.809	24.8	23.476	33.4	38.584
−3.0	3.673	16.4	13.987	25.0	23.756	33.6	39.018
−2.5	3.813	16.6	14.166	25.2	24.039	33.8	39.457
−2.0	3.956	16.8	13.347	25.4	24.326	34.0	39.898
−1.5	4.105	17.0	14.530	25.6	24.617	34.2	40.344
−1.0	4.258	17.2	14.715	25.8	24.912	34.4	40.796
−0.5	4.416	17.4	14.903	26.0	25.209	34.6	41.251
0.0	4.579	17.6	15.092	26.2	25.509	34.8	41.710
0.5	4.750	17.8	15.284	26.4	25.812	35.0	42.175
1.0	4.926	18.0	15.477	26.6	26.117	35.2	42.644
1.5	5.107	18.2	15.673	26.8	26.426	35.4	43.117
2.0	5.294	18.4	15.871	27.0	26.739	35.6	43.595
2.5	5.486	18.6	16.071	27.2	27.055	35.8	44.078
3.0	5.685	18.8	16.272	27.4	27.374	36.0	44.563
3.5	5.889	19.0	16.477	27.6	27.696	36.2	45.054
4.0	6.101	19.2	16.685	27.8	28.021	36.4	45.549
4.5	6.318	19.4	16.894	28.0	28.349	36.6	46.050
5.0	6.543	19.6	17.105	28.2	28.680	36.8	46.556
5.5	6.775	19.8	17.319	28.4	29.015	37.0	47.067
6.0	7.013	20.0	17.535	28.6	29.354	37.2	47.582
6.5	7.259	20.2	17.753	28.8	29.697	37.4	48.102
7.0	7.513	20.4	17.974	29.0	30.043	37.6	48.627
7.5	7.775	20.6	18.197	29.2	30.392	37.8	49.157
8.0	8.045	20.8	18.422	29.4	30.745	38.0	49.692
8.5	8.323	21.0	18.650	29.6	31.102	38.2	50.231
9.0	8.609	21.2	18.880	29.8	31.461	38.4	50.774
9.5	8.905	21.4	19.113	30.0	31.824	38.6	51.323
10.0	9.209	21.6	19.349	30.2	32.191	38.8	51.879
10.5	9.521	21.8	19.587	30.4	32.561	39.0	52.442
11.0	9.844	22.0	19.827	30.6	32.934	39.2	53.009

TABLE 5.6 Vapor Pressure of Water (*Continued*)

t, °C	p, mmHg	t, °C	p, mmHg	t, °C	p, mmHg	t, °C	p, mmHg
39.4	54.580	58.5	139.34	78.5	334.2	96.4	667.31
39.6	54.156	59.0	142.60	79.0	341.0	96.6	672.20
39.8	54.737	59.5	145.99	79.5	348.1	96.8	677.12
40.0	55.324	60.0	149.38	80.0	355.1	97.0	682.07
40.5	56.81	60.5	152.91	80.5	362.4	97.2	687.04
41.0	58.34	61.0	156.43	81.0	369.7	97.4	692.05
41.5	59.90	61.5	160.10	81.5	377.3	97.6	697.10
42.0	61.50	62.0	163.77	82.0	384.9	97.8	702.17
42.5	63.13	62.5	167.58	82.5	392.8	98.0	707.27
43.0	64.80	63.0	171.38	83.0	400.6	98.2	712.40
43.5	66.51	63.5	175.35	83.5	408.7	98.4	717.56
44.0	68.26	64.0	179.31	84.0	416.8	98.6	722.75
44.5	70.05	64.5	183.43	84.5	425.2	98.8	727.98
45.0	71.88	65.0	187.54	85.0	433.6	99.0	733.24
45.5	73.74	65.5	191.82	85.5	442.3	99.2	738.53
46.0	75.65	66.0	196.09	86.0	450.9	99.4	743.85
46.5	77.61	66.5	200.53	86.5	459.8	99.6	749.20
47.0	79.60	67.0	204.96	87.0	468.7	99.8	754.58
47.5	81.64	67.5	209.57	87.5	477.9	100.0	760.00
48.0	83.71	68.0	214.17	88.0	487.1	101.0	787.57
48.5	85.85	68.5	218.95	88.5	496.6	102.0	815.86
49.0	88.02	69.0	223.73	89.0	506.1	103.0	845.12
49.5	90.24	69.5	228.72	89.5	515.9	104.0	875.06
50.0	92.51	70.0	233.7	90.0	525.76	105.0	906.07
50.5	94.86	70.5	238.8	90.5	535.83	106.0	937.92
51.0	97.20	71.0	243.9	91.0	546.05	107.0	970.60
51.5	99.65	71.5	249.3	91.5	556.44	108.0	1004.42
52.0	102.09	72.0	254.6	92.0	566.99	109.0	1038.92
52.5	104.65	72.5	260.2	92.5	577.71	110.0	1074.56
53.0	107.20	73.0	265.7	93.0	588.60	111.0	1111.20
53.5	109.86	73.5	271.5	93.5	599.66	112.0	1148.74
54.0	112.51	74.0	277.2	94.0	610.90	113.0	1187.42
54.5	115.28	74.5	283.2	94.5	622.31	114.0	1227.25
55.0	118.04	75.0	289.1	95.0	633.90	115.0	1267.98
55.5	120.92	75.5	295.3	95.2	638.59	116.0	1309.94
56.0	123.80	76.0	301.4	95.4	643.30	117.0	1352.95
56.5	126.81	76.5	307.7	95.6	648.05	118.0	1397.18
57.0	129.82	77.0	314.1	95.8	652.82	119.0	1442.63
57.5	132.95	77.5	320.7	96.0	657.62	120.0	1489.14
58.0	136.08	78.0	327.3	96.2	662.45		

5.2.1 Vapor-Pressure Equations

Numerous mathematical formulas relating the temperature and pressure of the gas phase in equilibrium with the condensed phase have been proposed. The Antoine equation (Eq. 1) gives good correlation with experimental values. Equation 2 is simpler and is often suitable over restricted temperature ranges. In these equations, and the derived differential coefficients

TABLE 5.7 Vapor Pressure of Deuterium Oxide

t, °C	p, mmHg	t, °C	p, mmHg	t, °C	p, mmHg
0	3.65	20	15.2	80	331.6
1	3.93	30	28.0	90	495.5
2	4.29	40	49.3	100	722.2
3	4.65	50	83.6	101.43	760.0
3.8	5.05	60	136.6		
10	7.79	70	216.1		

for use in the Haggenmacher and Clausius-Clapeyron equations, the p term is the vapor pressure of the compound in millimeters of mercury (torr), the t term is the temperature in degrees Celsius, and the T term is the absolute temperature in kelvins ($t\,°C + 273.15$).

Eq.	Vapor-pressure equation	dp/dT	$-[d(\ln p)/d(1/T)]$
1	$\log p = A - \dfrac{B}{t+C}$	$\dfrac{2.303pB}{(t+C)^2}$	$\dfrac{2.303BT^2}{(t+C)^2}$
2	$\log p = A - \dfrac{B}{T}$	$\dfrac{2.303pB}{T^2}$	$2.303B$
3	$\log p = A - \dfrac{B}{T} - C \log T$	$p\left(\dfrac{2.303B}{T^2} - \dfrac{C}{T}\right)$	$2.303B - CT$

Equations 1 and 2 are easily rearranged to calculate the temperature of the normal boiling point:

$$t = \frac{B}{A - \log p} - C \tag{5.1}$$

$$T = \frac{B}{A - \log p} \tag{5.2}$$

The constants in the Antoine equation may be estimated by selecting three widely spaced data points and substituting in the following equations in sequence:

$$\left(\frac{y_3 - y_2}{y_2 - y_1}\right)\left(\frac{t_2 - t_1}{t_3 - t_2}\right) = 1 - \left(\frac{t_3 - t_1}{t_3 + C}\right)$$

$$B = \left(\frac{y_3 - y_1}{t_3 - t_1}\right)(t_1 + C)(t_3 + C)$$

$$A = y_2 + \left(\frac{B}{t_2 + C}\right)$$

In these equations, $y_i = \log p_i$.

TABLE 5.8 Vapor Pressures of Various Inorganic Compounds

Substance	State	Eq.	Range, °C	A	B	C
Aluminum						
$AlCl_3$		2	70–190	16.24	6 006	
Al_2O_3		2	1840–2000	14.22	28 200	
Ammonium						
NH_3	c*	1		9.963 82	1 617.907	272.55
	liq	1		7.360 50	926.132	240.17
NH_4Br	subl c	1		9.220 0	3 947	227.0
NH_4Cl	subl c	1		9.355 7	3 703.7	232.0
NH_4I	subl c	1		9.147 0	3 858	226.0
NH_4N_3	c	1		10.433 4	2 821.0	240.0
Antimony						
Sb	c	2	1070–1325	9.051	9 871	
$SbBr_3$		2	235–324	8.005	2 873	
$SbCl_3$		2	170–253	8.090	2 582.3	
SbI_3		2	330–445	7.831	3 350.55	
Sb_2Se_3	subl c	2		8.790 6	6 432.3	
Argon						
Ar	c	1		7.505 81	399.085	272.63
	liq	1		6.616 51	304.227	267.32
Arsenic						
As		2	440–815	10.800	6 947	
		2	800–860	6.692	2 460	
$AsCl_3$		2	50–100	7.953	2 042.7	
As_2O_3		2	100–310	12.127	5 815.81	
		2	315–490	6.513	2 722.2	
Barium						
Ba		2	930–1130	15.765	18 280	
BaH_2 [97% pure]		2	500–1000	6.86	4 000	
Bismuth						
Bi		2	1210–1420	8.876	10 446	
$BiCl_3$		2	91–213	2.681	685.519	
Boron						
BBr_3		2	−40 to 90	7.655	1 740.3	
BCl_3		1		6.188 11	756.89	214.0
$B(CH_3)_3$		2	−118 to −20	7.459 5	1 157.99	
B_2H_6	liq	1		6.366 38	521.490	241.98
B_5H_{11}	liq	2	−43 to 8.4	7.901	1 690.3	
Bromine						
Br_2	c	1		9.7209	2 041.3	260.1
	liq	1		6.877 80	1 119.68	221.38
BrF_3	liq	1		7.729 74	1 673.95	219.48
BrF_5	liq	1		7.273 68	1 219.28	236.40
BrO_2F	liq	1		7.436 51	1 195.8	260.1
Cadmium						
Cd		2	150–321	8.564	5 693	
		2	500–840	7.897	5 218	
CdI_2		2	385–450	9.269	6 383	
Calcium						
Ca		2	500–700	9.697	10 185	
		2	960–1100	16.240	19 325	
Carbon						
C [as C(g)]	liq	1		11.042 8	37 736	302.2
[as $C_2(g)$]	liq	1		12.583 2	43 281	318.3
[all species]	liq	1		9.381 3	27 240	264.0

* Crystalline solid.

TABLE 5.8 Vapor Pressures of Various Inorganic Compounds (*Continued*)

Substance	State	Eq.	Range, °C	A	B	C
Carbon						
CNBr	subl c	1		9.488 9	2 041.8	251.70
CNF		1	−76 to −47	6.778 9	697.61	224.95
CO	c l	1		7.414 8	342.50	269.0
	liq	1		6.694 22	291.743	267.99
CO_2	c	1		9.810 66	1 347.786	273.00
C_3O_2	liq	1	−71 to 7	7.188 99	1 100.94	249.15
$COCl_2$	liq	1		6.971 33	998.770	236.68
COF_2		1	−109 to −84	6.885 5	576.70	228.58
COS		1	−111 to −49	6.907 23	804.48	250.0
CS_2		1	3–80	6.942 79	1 169.11	241.59
CSe_2		1	0–50	6.776 73	1 353.20	219.95
CSeS		1	−16 to 84	6.699 6	1 161.97	219.59
Cesium						
Cs		2	200–350	6.949	3 833.7	
CsBr		2	978–1305	7.990	8 022.53	
CsCl		2	986–1295	8.340	8 523.94	
CsF		2	1033–1255	7.703	7 359.21	
CsH		2	245–378	11.79	5 900	
		2	340–440	9.25	4 410	
CsI		2	1052–1280	9.124	9 699.11	
Chlorine						
Cl_2	c	1		9.705 12	1 444.19	267.13
	liq	1		6.937 90	861.34	246.33
ClF	liq	1		6.989	682.1	256
ClF_3	liq	1		7.366 85	1 096.28	232.63
ClF_5		1		6.269 33	653.06	206.6
ClO_2	liq	1		6.036 11	590.09	176.15
Cl_2O	liq	1		7.132 68	1 021.56	238.16
$ClOClO_3$	liq	1		7.538 67	1 404.18	257.00
Cl_2O_7	liq	1		6.869 29	1 214.00	220.79
ClO_2F	liq	1		6.677 15	809.78	218.96
ClO_3F	liq	1		6.895 19	791.73	243.88
Copper						
CuBr		2	997–1351	5.460	4 173.2	
CuCl		2	878–1369	5.454	4 215.0	
CuI		2	991–1154	5.570	4 215.0	
Fluorine						
F_2	liq	1		6.765 88	304.35	266.54
FNO_3	liq	1		6.658 6	769.5	248.0
Germanium						
$GeCl_4$		2	10.4–86	7.340	2 010.9	
Helium						
^3He	liq	1	−271.13 to −270.86	4.272 7	5.594	273.840
	liq	1	−271.13 to −269.92	5.100 0	11.062	274.950
^4He		1	−271.4 to −270.1	4.558 7	8.1548	273.710
		1	−271.4 to −268.9	5.320 75	14.6515	274.950
		1	−271.4 to −268.1	6.004 60	24.0668	276.650
Hydrogen						
$^1H_2$ normal, 25% para	c	1		6.043 86	66.507	274.630
	liq	1		5.824 38	67.5078	275.700
equilibrium	c	1		6.042 07	65.961	274.60
	liq	1		5.814 64	66.7945	275.650
$^1H^2H$ (DH)	c	1		6.960 08	99.968	276.590
	liq	1		6.016 12	77.1349	275.620
$^2H_2$ (D_2) normal,	c	1		7.726 05	135.461	278.550
66.7% ortho	liq	1		6.128 25	83.5251	275.216

TABLE 5.8 Vapor Pressures of Various Inorganic Compounds (*Continued*)

Substance	State	Eq.	Range, °C	A	B	C
$^2H_2$ equilibrium,	c	1		7.751 10	135.58	278.50
97.8% ortho	liq	1		6.044 68	79.5888	274.680
$^3H_2$ (T$_2$) normal, 25%	c	1		6.184 03	76.7445	271.850
para	liq	1		6.089 21	81.8971	273.650
^1HBr	c	1		7.667 61	878.57	253.2
	liq	1		6.287 53	540.82	225.44
^2HBr (DBr)	c	1		7.500 93	820.68	247.3
	liq	1		6.162 38	505.68	220.6
^1HCl	c	1		8.134 73	941.57	268.06
	liq	1		7.170 00	745.80	258.88
^2HCl (DCl)	c	1		7.850 47	843.32	258.32
	liq	1		6.935 96	668.20	249.50
HCN	liq	1	−16 to 46	7.528 2	1329.5	260.4
^1HF	liq	1		7.680 98	1475.60	287.88
^2HF (DF)	liq	1		7.217 04	1268.37	273.87
^1HI	c	1		7.315 6	894.32	239.6
	liq	1		5.608 9	416.04	188.1
^2HI (DI)	c	1		7.314 9	889.52	238.8
	liq	1		5.601 8	413.98	187.8
HN$_3$	liq	1		6.857	1 066	232
HNO$_3$	liq	1		7.511 9	1 406	221.0
^1H_2O			[See Tables 5.4 and 5.6]			
^2H_2O (D$_2$O)			[See Table 5.7]			
H$_2$^{18}O		1	0–60	8.133 2	1 762.39	235.660
		1	60–120	7.972 08	1 668.84	227.700
H$_2$O$_2$	liq	1		7.969 17	1 886.76	220.6
HPO$_2$F	liq	1		6.735 3	1 342.9	232.0
H$_2$S	c	1		7.614 18	885.319	250.25
	liq	1		6.993 92	768.130	249.09
H$_2$S$_2$	liq	1		6.974	1 232	225
H$_2$S$_3$	liq	1		6.807	1 488	209
H$_2$S$_4$	liq	1		6.945	1 772	196
H$_2$S$_5$	liq	1		7.320	2 104	189
HSO$_3$Cl	liq	1		7.049	1 480	201
HSO$_3$F	liq	1		7.399 5	1 521	174.0
H$_2$Se	c	1		7.635 4	927.6	240.0
	liq	1		6.966 0	787.67	235.0
H$_2$Te	liq	1		7.000	935	229
Iodine						
I$_2$	c	1		9.810 9	2 901.0	256.00
	liq	1		7.018 1	1 610.9	205.0
ICl	liq	1		7.702 1	1 517.9	217.0
IF$_5$	c	1		10.964	2 538	245
	liq	1		7.464 8	1 460	216.0
IF$_7$	c	1		7.998	1 340	256
Iridium						
IrF$_6$	c	2	0.4–44	8.618	1 868	
	liq	2	44–54	7.952	1 657	
Iron						
FeCl$_2$	liq	2	708–834	9.794	7 455	
	liq	2	700–930	8.33	7 061	
FeCl$_3$	c	2	160–304	15.11	7 142	
FeI$_2$		2	517–577	13.183	10 778	
		2	601–686	9.674	7 716	

TABLE 5.8 Vapor Pressures of Various Inorganic Compounds (*Continued*)

Substance	State	Eq.	Range, °C	A	B	C
Krypton						
Kr	c	1		7.539 55	539.48	269.8
	liq	1		6.630 70	416.38	264.45
Lead						
Pb		2	525–1325	7.827	9 845.4	
PbBr$_2$		2	735–918	8.064	6 163.1	
PbCl$_2$		2	500–950	8.961	7 411.4	
PbF$_2$		2	1078–1289	8.391	8 623.2	
Lithium						
LiBr		2	1010–1265	8.068	7 975.5	
LiCl		2	1045–1325	7.939	8 142.7	
LiF		2	1398–1666	8.753	11 407	
LiH		2	500–650	11.227	9 600	
		2	700–800	9.926	8 204	
LiI		2	940–1140	8.011	7 500	
Magnesium						
Mg		2	900–1070	12.993	13 579.8	
MgH$_2$		2	337–415	9.78	3 857	
Mercury						
Hg				[See Table 5.3]		
HgBr$_2$		2	130–270	10.094	4 168.0	
HgCl$_2$		2	130–270	10.094	4 118.34	
		2	275–309	8.409	3 187.1	
Hg$_2$Cl$_2$		1		8.521 51	3 110.96	168.0
HgI$_2$		2	266–360	8.115	3 278.5	
Neon						
Ne	c	1		7.065 16	110.61	272.00
	liq	1		6.084 44	78.380	270.550
Neptunium						
NpF$_6$	liq	3	55.1–76.8	0.010 23	1 191.1	−2.582 5
Nickel						
Ni(CO)$_4$		2	2–40	7.780	1 556.5	
Niobium						
NbBr$_5$	liq	2		8.92	3 850	
NbCl$_5$	liq	2	210–254	8.37	2 827	
NbF$_5$	liq	2		8.439	2 824	
Nitrogen						
N$_2$ natural	c	1		7.345 12	322.222	269.980
	liq	1		6.494 57	255.680	266.550
^{15}N$_2$	c	1		7.363 96	323.17	269.88
	liq	1		6.494 14	255.535	266.451
NCl$_3$		1		6.956	1 190	221
NF$_3$	liq	1		6.779 66	501.913	257.79
NH$_3$				[See Table 5.5]		
N$_2$H$_4$	liq	1		7.801 9	1 679.07	227.7
NO natural	c	1		9.628 26	758.736	266.00
	liq	1		8.743 00	682.938	268.27
N$_2$O	c	1		9.437 00	1 174.020	268.22
	liq	1		7.003 94	654.260	247.16
N$_2$O$_4$ equilibrium	c	1		10.736 31	2 075.53	252.80
mixture	liq	1		8.917 12	1 798.54	276.80
N$_2$O$_5$	c	1		11.644 5	2 510	253.0
NOCl	c	1		8.540 8	1 397.3	261.0
	liq	1		7.361 54	1 094.73	249.70
N$_2$O$_3$		2	−25 to 0	10.30	2 057.9	
NOF	liq	1		6.443 5	556.13	216.0

TABLE 5.8 Vapor Pressures of Various Inorganic Compounds (*Continued*)

Substance	State	Eq.	Range, °C	A	B	C
Nitrogen						
NO_2Cl	liq	1		5.372 3	395.40	174.0
NO_2F	liq	1		6.833 4	654.55	238.0
Osmium						
OsF_5		2	75–180	9.75	3 429	
OsF_6		2	34–48	7.470	1 473	
OsF_8		2	38–47	7.650	1 525	
OsO_4		2	−38 to 40	10.710 0	2 951.00	
OsO_3F_2		2	59–105	7.994	1 911	
Oxygen						
O_2	liq	1		6.691 44	319.013	266.697
O_3	liq	1		6.837	552.5	251.0
OF_2	liq	1		7.236 19	545.05	269.91
O_2F_2	liq	1		6.779 02	756.39	250.16
O_3F_2		2	79–114	6.134 3	675.57	
Palladium						
$PdCl_2$		2	680–857	6.32	5 032	
Phosphorus						
P red, V	subl c	1		11.060	5 323	220
white	subl c	1		6.936 9	1 907.6	190.0
P_4 black, o-rh		1		12.405	6 671	247
PBr_3	liq	1	−40 to 173	6.915 5	1 590.5	221.0
PBr_5	liq	1	to 104	6.948	1 320	214
$PBrF_2$	liq	1	−133 to −16	6.904 2	885.12	236.0
PBr_2F	liq	1	−115 to 78	6.858 0	1 210.3	226.0
PCl_3	liq	1	−92 to 76	6.826 7	1 196	227.0
PCl_5	c	1	to 160	10.206 8	2 903.1	237.0
	liq	1		7.033	1 490	200.0
$PClF_2$	liq	1	−165 to −47	6.639 6	780.88	255.0
PCl_2F	liq	1	−144 to 14	6.796 56	982.332	237.00
$P(OCN)_3$	liq	2	−2 to 169	8.745 5	2 595	
PF_3	liq	1	−152 to −101	6.860 4	620.22	257.0
PF_5	liq	1	−93.8 to −84.5	6.914 4	647.21	245.0
PH_3	c	1		7.482 35	794.496	265.20
	liq	1		6.715 59	645.512	256.066
P_2H_4	liq	1		6.862 8	1 137	227.0
P_4O_6	liq	1	24–175	6.716 37	1 412.8	193.0
P_4O_{10}	c III	1		9.707 0	3 822	201.0
	c I	1		10.843 2	6 424	213
	liq	1		6.935 2	3 069	152
$POBr_3$	liq	1	51–192	7.007 8	1 609.2	198.0
$POBrCl_2$	liq	1	31–165	6.924	1 411	213
$POBrClF$	liq	1		6.914	1 214	222
$POBrF_2$	liq	1	−85 to 32	7.101 9	1 118.9	233.0
$POBr_2F$	liq	1	−117 to 110	6.721 2	1 328.9	236.0
$POCl_3$	liq	1	1.2–105	6.865 8	1 297.2	220.0
$POClF_2$	liq	1	−96 to 3	6.926 6	946.96	231.0
$POCl_2F$	liq	1	−80 to 53	7.084 65	1 201.86	233.00
POF_3	c	1		10.930 5	1 783	261.0
	liq	1		7.115 5	810.1	231.0
$PO(OCN)_3$		2	5–193	9.168 2	2 931	
$PO(SCN)_3$		2	14–300	8.533 0	3 240	
P_4S_{10}		2		9.17	4 940	
$PSBr_3$	c	2		10.105	3 196.2	
	liq	2		8.338 3	2 641.9	
$PS(OCN)_3$		2		10.032	3 492	

TABLE 5.8 Vapor Pressures of Various Inorganic Compounds (*Continued*)

Substance	State	Eq.	Range, °C	A	B	C
Platinum						
Pt		2	1425–1765	7.786	25 384	
PtF$_6$	liq	1	61.3–81.7	89.15	5 686	27.49
Polonium						
Po	liq	1		7.041 4	5 017.6	241.0
PoCl$_4$	liq	1		7.554	2 360	115
Potassium						
K		2	260–760	7.183	4 434.33	
KBr		2	1095–1375	7.936	8 555.3	
KCl		2	1116–1418	8.130	8 863.4	
KF		2	1278–1500	9.000	10 838	
KOH		2	1170–1327	7.330	7 103.3	
KI		2	1063–1333	7.949	8 132.2	
Protactinium	liq	2		17.27	7 377	
Radon						
Rn	c	1		7.495 5	884.41	255.0
	liq	1		6.701 5	718.25	250.0
Rhenium						
ReF$_5$	c	2		9.024	3 037	
ReF$_6$	c	3	−3.45 to 18.5	9.123 0	1 765.4	0.1790
	liq	3	18.5–48	18.208 1	1 956.7	3.599
ReF$_7$	c	3	−14.5 to 48.3	13.043 2	2 205.8	1.470 3
	liq	3	48.3–74.6	−21.583 5	244.28	−9.908 3
ReO$_2$	c	2	650–785	11.65	14 437	
	liq	2	480–660	5.345	4 742	
ReO$_3$	c	2	325–420	15.16	10 882	
	liq	2	300–480	7.745	4 966	
Re$_2$O$_7$	liq	2	230–360	8.98	3 868	
ReOF$_4$	liq	2	108–172	10.09	3 206	
ReOF$_5$	liq	2	41–73	7.727	1 679	
ReS$_2$	c	2	500–700	3.214	4 976	
Re$_2$S$_7$	c	2	260–410	8.86	4 800	
Rubidium						
Rb		2	250–370	6.976	3 969.5	
RbCl		2	1142–1395	9.111	10 373	
RbF		2	1142–1400	8.570	9 568.4	
Ruthenium						
RuOF$_4$		2	120–160	8.60	2 616	
Selenium						
Se	liq	1		7.631 6	4 213.0	202.0
SeCl$_4$	c	1		10.250 9	3 068.8	225.0
SeF$_4$	liq	1		7.888 7	1 603.0	215.0
SeF$_6$	c	1		8.385 4	1 121.4	250.0
SeO$_2$		1		6.577 81	1 879.81	179.0
SeOCl$_2$	liq	1		6.257 3	970.87	112.0
SeOF$_2$	liq	1		7.420	1 380	178
Silicon						
SiCl$_4$	liq	1	0–53	6.857 26	1 138.92	228.88
SiH$_4$		2	−160 to −112	6.881	645.9	
Si$_2$H$_6$		2	−115 to −14.6	7.258	1 133.4	
Si$_3$H$_8$		2	−70 to 52	7.676	1 559.1	
Silver						
AgCl		2	1255–1442	8.179	9 688.7	
Sodium						
Na		2	180–883	7.553	5 395.4	
NaCl		2	976–1155	8.329 7	9 417.07	

TABLE 5.8 Vapor Pressures of Various Inorganic Compounds (*Continued*)

Substance	State	Eq.	Range, °C	A	B	C
NaCl		2	1156–1430	8.548	9 704.3	
NaCN		2	800–1360	7.472	8 122.81	
NaF		2	1562–1701	8.640	11 396.6	
NaI		2	1063–1307	8.371	8 623.2	
NaOH		2	1010–1402	7.030	6 894	
Strontium						
Sr		2	940–1140	16.056	18 802.8	
Sulfur						
S equilibrium	liq	1		6.843 59	2 500.12	186.30
S_2Br_2	liq	1		7.177	1 660	185
SCl_2	liq	1		8.454	1 594	227
S_2Cl_2	liq	1		6.783 6	1 341	206.0
S_2F_2	liq	1		6.684	628	256
SF_4	liq	1		6.839 5	823.4	248.0
SF_6	c	1		8.416 0	1 096.5	262.0
S_2F_{10}	liq	1		7.067 6	1 100.6	234.0
SO_2	c	1		9.754 3	1 553.8	225.0
	liq	1		7.282 28	999.900	237.190
SO_3 "icelike"	c III	1		10.565 7	2 273.8	255.0
"woollike"	c II	1		11.590 1	2 665.6	264.0
	c I	1		14.255 9	3 692.1	273.0
	liq	1		9.050 85	1 735.31	236.50
$SOBr_2$	liq	1		7.056	1 445	206
$SOCl_2$	liq	1		7.287 45	1 446.7	252.7
SOClF	liq	1		7.173 1	1 100.1	244.00
SOF_2	liq	1		6.959 06	775.48	234.00
SOF_4	liq	1		7.071 8	840.3	249.0
$S_2O_2F_{10}$	liq	1		6.874	1 110	229
$S_2O_5Cl_2$	liq	1		7.019	1 460	202
S_2O_5ClF	liq	1		7.015 6	1 257.4	204.0
$S_2O_5F_2$	liq	1		6.881	1 120	229
$S_2O_5F_4$	liq	1		6.885	1 140	227
SO_2BrF	liq	1		7.142 8	1 155	231.0
SO_2Cl_2	liq	1		7.001 7	1 209	224.0
SO_2ClF	liq	1		6.521 5	793.73	210.70
SO_2F_2	liq	1		6.907 0	784.3	250
Tantalum						
$TaBr_5$	liq	2		8.11	3 260	
$TaCl_5$	liq	2	220–240	8.68	2 970	
TaF_5	liq	2		8.524	2 834	
TaI_5	liq	2		7.67	3 950	
Technetium						
TcF_6	liq	3	37.4–51.7	24.808 7	2 405	5.803 6
TcO_3F	liq	2	18.3–51.8	8.417	2 065	
Tc_2O_7	c	2		18.279	7 205	
	liq	2		8.999	3 571	
Tellurium						
Te	liq	1		7.301 0	5 370.6	221
$TeCl_4$	liq	1		7.558 6	2 355	115
TeF_6	liq	1		6.748 8	807.0	247.0
Te_2F_{10}	liq	1		6.901 8	1 150	227.0
TeO_2		2	450–733	12.328 4	13 222	
Thallium						
Tl		2	950–1200	6.1240	6 268	
TlF		2	282–298	12.52	5 484	

TABLE 5.8 Vapor Pressures of Various Inorganic Compounds (*Continued*)

Substance	State	Eq.	Range, °C	A	B	C
Thorium						
ThF_4	liq	2		10.821	15 270	
ThH_2		2	up to 883	9.50	7 650	
Tin						
$SnCl_4$		2	−52 to −38	9.824	2 441.23	
SnH_4		2	−148 to −49	7.400	999.68	
Titanium						
$TiCl_2$	subl c	2		9.30	8 500	
$TiCl_3$	subl c	2	455–550	10.401	8 296	
$TiCl_4$	liq	2	−23 to 136	7.683	1 964	
TiI_4	liq	2	160–360	7.577	3 054	
Tungsten						
W		2	2230–2770	9.920	46 850	
Uranium						
UF_6	liq	1	64–116	6.994 64	1 126.288	221.963
	liq	1	116–230	7.690 69	1 683.165	302.148
UH_3 dissociation		2	200–430	9.39	4 590	
U^2H_3 (UD_3)		2		9.43	4 500	
U^3H_3 (UT_3)		2		9.46	4 471	
Vanadium						
VBr_2	c	2	541–716	9.08	10 460	
	subl c	2	800–905	5.9	9 830	
VBr_3		2	314–427	11.12	7 470	
VCl_2	subl c	2	910–1100	5.725	9 721	
VCl_3		2	352–567	11.20	9 777	
VCl_4	liq	2	30–153	7.62	2 020	
VF_3	subl c	2	650–920	12.357	15 603	
VF_5	subl c	2	−20 to 19.5	8.168	2 608	
	liq	2	19.5–45.5	7.549	2 423	
VI_2	subl c	2	850–1016	2.56	5 600	
$VOCl_3$	liq	2	15.4–125	7.69	1 920	
Xenon						
Xe	c	1		7.484 5	714.896	264.0
	liq	1		6.642 89	566.282	258.660
XeF_2	subl c	1		10.019 47	2 683.96	261.68
XeF_4	subl c	1		10.913 87	3 095.06	269.56
Zinc						
Zn	c	2	250–419	9.200	6 946.6	

TABLE 5.9 Vapor Pressures of Various Organic Compounds

Substance	Eq.	Range, °C	A	B	C
Acenaphthene	1	147–187	7.728 19	2 534.234	245.576
	2	147–288	8.033	2 834.99	
Acetaldehyde	1	liq	8.005 52	1 600.017	291.809
Acetic acid	1	liq	7.387 82	1 533.313	222.309
Acetic anhydride	1	liq	7.149 48	1 444.718	199.817
Acetone	1	liq	7.117 14	1 210.595	229.664
Acetonitrile	1	liq	7.119 88	1 314.4	230
Acetophenone	2	30–100	9.135 2	2 878.8	
Acetyl bromide	1	liq	5.197 02	545.784	150.396
Acetyl chloride	1	liq	6.948 87	1 115.954	223.554
Acetylene	1	−130 to −83	9.140 2	1 232.6	280.9
	1	−82 to −72	7.099 9	711.0	253.4
Acetyl iodide	1	liq	4.181 44	355.452	108.160
Acrylic acid	1	20–70	5.652 04	648.629	154.683
Acrylonitrile	1	−20 to 140	7.038 55	1 232.53	222.47
Allyl isothiocyanate	1	10–50	5.126 58	791.434	154.019
m-Aminobenzotrifluoride	1	0–96	7.651 86	1 940.6	218.0
		96–300	7.170 30	1 650.21	193.58
p-Aminophenol	1	130–185	−3.357 50	699.157	−331.343
Aniline	1	102–185	7.320 10	1 731.515	206.049
Anthracene	2	100–160	8.91	3 761	
	1	176–380	7.674 01	2 819.63	247.02
9,10-Anthracenedione	2	224–286	12.305	5 747.9	
	2	285–370	8.002	3 341.94	
Benzene	1	−12 to 3	9.106 4	1 885.9	244.2
	1	8–103	6.905 65	1 211.033	220.790
Benzenethiol	1	52–198	6.990 19	1 529.454	203.048
Benzoic acid	2	60–110	9.033	3 333.3	
Benzonitrile	1	liq	6.746 31	1 436.72	181.0
Benzophenone	1	48–202	7.349 66	2 331.4	195.0
	1	200–306	7.162 94	2 051.855	173.074
Benzotrifluoride	1	−20 to 180	7.007 08	1 331.30	220.58
Benzoyl chloride	2	140–200	7.924 5	2 372.1	
Benzyl acetate	1	46–156	8.457 05	2 623.206	259.067
Benzyl alcohol	1	122–205	7.198 17	1 632.593	172.790
Biphenyl	1	69–271	7.245 41	1 998.725	202.733
2-(2-Biphenylyloxy)ethanol	1	240–300	8.005 87	2 776.761	206.914
Bromobenzene	1	56–154	6.860 64	1 438.817	205.441
2-Bromobenzyl cyanide	1	85–152	5.044 59	734.821	59.273
1-Bromobutane	1	−78 to 23	5.281 38	685.001	160.880
Bromochloromethane	1	16–68	6.496 06	942.267	192.587
Bromochlorodifluoromethane	1	−95 to 10	6.839 98	935.632	240.330
2-Bromo-2-chloro-1,1,1-tri-					
fluoroethane	1	−51 to 55	6.945 02	1 127.856	227.341
Bromocyclohexane	1	68–260	6.979 80	1 572.19	217.38
p-Bromodiphenyl ether	1	25–190	7.009 3	1 902.7	153.3
	1	190–400	6.681 43	1 683.84	132.90
Bromoethane	1	28–75	6.988 6	1 121.9	234.7
Bromoethene	1	−88 to 16	6.997 4	1 009.9	251.6
2-Bromoethylbenzene	1	127–217	7.800	2 235.4	238.7
4-Bromoethylbenzene	1	liq	6.982 09	1 632.60	193
2-Bromo-2-methylpropane	1	0–72.8	7.395 9	1 512.7	262.2
1-Bromonaphthalene	1	liq	7.003 50	1 927.05	186.0
o-Bromostyrene	1	liq	6.910 38	1 631.2	195

TABLE 5.9　Vapor Pressures of Various Organic Compounds (*Continued*)

Substance	Eq.	Range, °C	A	B	C
p-Bromostyrene	1		7.228 38	1 743.67	218.0
4-Bromotoluene	1	85–280	7.007 62	1 612.35	206.36
2-Bromovinylbenzene	1	110–129	0.564 97	82.913	−191.71
4-Bromovinylbenzene	1	119–147	12.504 2	7 349.00	559.02
1,2-Butadiene	1	−69 to −34	7.398 22	1 219.877	259.776
	1	−26 to 30	6.993 83	1 041.117	242.274
1,3-Butadiene	1	−80 to −62	7.035 55	998.106	245.233
	1	−58 to 15	6.849 99	930.546	238.854
n-Butane	1	−77 to 19	6.808 96	935.86	238.73
1-Butanethiol	1	−2 to 123	6.927 54	1 281.018	218.100
2-Butanethiol	1	−13 to 110	6.886 98	1 229.904	222.021
1-Butanol	1	15–131	7.476 80	1 362.39	178.77
2-Butanol	1	25–120	7.474 31	1 314.19	186.55
2-Butanone	1	43–88	7.063 56	1 261.34	221.97
1-Butene	1	−82 to 13	6.792 90	908.80	238.54
2-Butene　cis	1	−73 to 23	6.884 68	967.32	237.87
trans	1	−76 to 20	6.883 37	967.50	240.84
Butyl acetate	1	60–126	7.127 12	1 430.418	210.745
n-Butylamine trimethylboron	1	0–99	8.465 21	1 980.98	193.60
n-Butylbenzene	1	62–213	6.983 17	1 577.965	201.378
sec-Butylbenzene	1	87–174	6.942 19	1 533.95	204.39
t-Butylbenzene	1	84–170	6.922 55	1 505.987	203.490
n-Butyl borate	1	117–218	7.406 87	1 905.035	186.134
n-Butyl-t-butyl ether	1	83–124	6.955 56	1 348.702	206.303
Butyl carbitol	1	50–153	7.741 14	2 056.904	195.655
Butyl cellosolve	1	93–170	6.956 59	1 399.903	172.154
sec-Butyl chloroacetate	1	30–172	7.933 38	2 103.30	249.29
n-Butylcyclohexane	1	60–211	6.910 30	1 538.518	200.833
sec-Butylcyclohexane	1	91–180	6.890 96	1 530.70	202.373
t-Butylcyclohexane	1	84–173	6.856 80	1 501.724	206.108
n-Butylcyclopentane	1	41–185	6.899 35	1 457.08	205.99
n-Butyl formate	1	29–112	7.693 6	1 698.7	247.4
sec-Butyl formate	1	30–100	6.493	972.9	176.0
n-Butyl-α-hydroxyisobutyrate	1	112–185	8.421 7	2 617.32	287.09
1-n-Butylnaphthalene	1	25–170	7.434 47	2 227.7	202.2
	1	170–345	7.081 4	1 971.5	180
2-n-Butylnaphthalene	1	25–170	7.438 08	2 242.2	202.3
	1	170–345	7.084 8	1 984.3	180
n-Butyl nitrate	1	0–70	8.054 27	1 992.83	254.30
1-Butyl pentafluoropropionate	1	82–116	6.651 00	1 108.02	177.04
2-sec-Butylphenol	1	179–240	6.951 93	1 593.74	163.79
2-t-Butylphenol	1	135–225	7.217 56	1 822.81	196.23
4-t-Butylphenol	1	198–252	7.000 38	1 627.51	155.24
Butyl phenyl ether	1	119–210	7.299 7	1 882.70	215.82
n-Butyl propionate	1	32–93	9.484 89	2 852.58	296.98
n-Butyl trifluoroacetate	1	71–104	8.567 94	2 305.22	301.06
1-Butyl trimethylsilyl ether	1	71–124	7.763 00	1 884.68	261.31
1-Butyne	1	−68 to 27	6.981 98	988.75	233.01
2-Butyne	1	−51 to −34	7.037 91	896.91	199.06
	1	−31 to 47	7.073 38	1 101.71	235.81
n-Butyraldehyde	1	31–74	6.385 44	913.59	185.48
Butyric acid	1	90–163	7.739 9	1 764.7	199.9
Camphor	2	0–180	8.799	2 797.39	
	1	178–232	6.106	1 043.6	116.4
Capric acid	1	153–187	6.255 3	1 106.3	57.96

TABLE 5.9 Vapor Pressures of Various Organic Compounds (*Continued*)

Substance	Eq.	Range, °C	A	B	C
Caproic acid	1	98–179	6.924 9	1 340.8	126.6
Capronitrile	1	92–164	7.123 1	1 597.2	212.8
Caprylic acid	1	130–206	7.770 64	1 933.05	159.36
Carbazole	1	253–358	7.086 3	2 179.4	163.5
Carbitol	1	40–151	7.640 81	1 801.31	183.97
Chloroacetic acid	1	104–190	7.550 16	1 723.365	179.98
4-Chloroacetophenone	1	122–212	7.084 57	1 693.63	190.95
Chloroacetyl chloride	1	28–107	7.149 77	1 340.79	208.70
N-Chloroaniline	1	61–125	3.037 67	171.35	−14.99
2-Chloroaniline	1	20–108	7.562 65	1 998.6	220.0
	1	108–300	7.192 40	1 762.74	200.0
3-Chloroaniline	1	15–125	7.559 39	2 073.75	215
	1	125–310	7.236 03	1 857.75	196.64
o-Chloroanisole	1	115–186	7.121 36	1 655.80	188.77
Chlorobenzene	1	62–131.7	6.978 08	1 431.05	217.55
o-Chlorobenzotrichloride	1	30–150	7.504 30	2 228.07	220.0
	1	150–350	7.117 94	1 951.37	196.27
1-Chloro-4-bromobenzene	2	23–63	11.629	3 643.30	
1-Chlorobutane	1	−17 to 78.6	6.836 94	1 173.79	218.13
2-Chlorobutane	1	0–40	6.799 23	1 149.12	224.68
1-Chlorodecane	1	86–225.9	6.939 86	1 639.06	177.94
1-Chlorododecane	1	116–246	6.834 08	1 654.82	155.09
Chloroethane	1	−56 to 12.2	6.986 47	1 030.01	238.61
2-Chloroethylbenzene	1		6.981 69	1 556.0	201.0
3-Chloroethylbenzene	1		6.990 82	1 577.3	200
4-Chloroethylbenzene	1		6.983 09	1 577.0	200
Chloroethylene	1	−65 to −13	6.891 17	905.01	239.48
Chloroform	1	−35 to 61	6.493 4	929.44	196.03
1-Chloroheptane	1	34–160	6.916 70	1 453.96	199.83
1-Chlorohexadecane	1	166–327	7.282 03	2 152.61	162.73
1-Chlorohexane	1	15–136	7.051 36	1 461.72	215.57
Chlorohexylisocyanate	1	90–180	7.740 95	2 340.50	241.90
Chloromethane	1	−75 to −5	7.093 49	948.58	249.34
Chloromethoxytrichlorosilane	1	0–50	7.312 92	1 545.71	226.10
2-Chloro-2-methylpropane	1	22–47	4.896	334.99	114.0
1-Chlorononane	1	69–205	7.046 54	1 655.57	192.26
1-Chlorooctane	1	54–184	7.051 52	1 600.24	200.28
Chloropentafluorobenzene	1	36–140	7.068 83	1 389.19	213.75
p-Chlorophenetole	1	122–212	7.084 57	1 693.63	190.95
2-Chlorophenol	1	80–200	6.877 31	1 471.61	193.17
β-Chloro-β-phenylethyl alcohol	1	166–259	6.917 33	1 635.63	145.87
1-Chlorophenylisocyanate	1	50–160	12.265 9	6 532.55	499.59
m-Chlorophenylisocyanate	1	71–158	6.797 29	1 512.43	180.90
Chloroprene	1	20–60	6.161 50	783.45	179.7
1-Chloropropane	1	−25 to 47	6.926 48	1 110.19	227.94
2-Chloropropane	1	0–30	7.771	1 582	288
3-Chloro-1-propene	1	13–44	5.297 16	418.375	128.168
2-Chloropropionitrile	1	0–84	7.329 73	1 732.55	211.79
	1	84–240	7.200 85	1 657.25	205.3
γ-Chloropropyltrichlorosilane	1	87–179	7.156 4	1 679.07	210.38
1-Chlorotetradecane	1	142–296.8	7.200 7	2 018.9	170.6
o-Chlorotoluene	1	0–65	7.367 97	1 735.8	230.0
	1	65–220	6.947 63	1 497.2	209.0
1-Chloro-2,4,6-trinitrobenzene	1	200–270	3.080 9	184.93	−117.9
1-Chloroundecane	1	101–245	6.967 6	1 709.4	172.9

TABLE 5.9 Vapor Pressures of Various Organic Compounds (*Continued*)

Substance		Eq.	Range, °C	A	B	C
o-Chlorovinylbenzene		1	98–155	6.956 6	1 602.2	204.5
p-Chlorovinylbenzene		1	100–127	9.969 1	4 093.5	392.4
2-Chlorovinyldichloroarsine	cis	1	68–109	5.487 9	785.09	115.61
	trans	1	50–150	6.814 0	1 465.07	178.53
3-Chlorovinyldichloroarsine		1	66–110	2.810 5	97.17	−27.51
o-Cresol		1	120–191	6.911 7	1 435.50	165.16
m-Cresol		1	150–201	7.508 0	1 856.36	199.07
p-Cresol		1	128–202	7.035 08	1 511.08	161.85
Cyanic acid		1	−76 to −6	7.568 59	1 251.86	243.79
Cyclobutane		1	−60 to 12	6.916 31	1 054.54	241.37
Cyclobutanone		1	−24 to 25	6.116 68	933.95	183.19
Cyclobutene		1	−77 to 2	7.305 7	1 166.0	261.06
Cycloheptane		1	68–159	6.853 95	1 331.57	216.35
1,3,5-Cycloheptatriene		1	0–65	6.974 33	1 376.84	220.75
Cyclohexane		1	20–81	6.841 30	1 201.53	222.65
Cyclohexanethiol		1	84–203	6.886 73	1 476.70	209.83
Cyclohexanol		1	94–161	6.255 3	912.87	109.13
Cyclohexene		1		6.886 17	1 229.973	224.10
Cyclohexyl acetate		1	95–172	7.975 86	2 167.99	252.30
Cyclohexylamine		1	61–128	6.689 54	1 229.42	188.80
1-Cyclohexylamino-2-propanol		1	150–238	7.011 56	1 655.02	162.59
Cyclohexylpentafluoropropionate		1	82–155	7.725 5	1 844.73	224.89
Cyclohexyltrifluoroacetate		1	72–147	7.802 35	1 954.66	249.33
Cyclohexyltrimethylsilyl ether		1	91–168	8.090 52	2 276.62	267.94
Cyclooctane		1	97–194	6.861 87	1 437.79	210.02
1,3,5,7-Cyclooctatetraene		1	0–75	7.006 69	1 472.11	215.84
Cyclopentane		1	−40 to 72	6.886 76	1 124.162	231.36
Cyclopentanethiol		1	81–173	6.914 97	1 388.63	212.05
Cyclopentanone		1	0–26	2.902 47	162.90	63.22
Cyclopentene		1		6.920 66	1 121.818	223.45
Cyclopentyl-1-thiaethane		1	83–199	6.940 83	1 480.70	208.47
Cyclopropane		1	−90 to −32	6.887 88	856.01	246.50
o-Cymene		1	81–180	7.266 10	1 768.45	224.95
m-Cymene		1	79–176	7.123 74	1 644.95	212.76
p-Cymene		1	107–178	7.050 74	1 608.91	208.72
Decahydronaphthalene	cis	1	68–228	6.875 29	1 594.460	203.39
	trans	1	61–219	6.856 81	1 564.683	206.26
Decane		1	58–203	6.943 65	1 495.17	193.86
1-Decanethiol		1	109–271	6.998 1	1 713.6	177.0
1-Decanol		1	25–52	11.560	4 055	273.2
		1	103–230	6.922 44	1 472.01	133.98
1-Decene		1	54–199	6.934 77	1 484.98	195.707
Decylbenzene		1	203–298	7.035 96	1 903.98	160.33
Decylcyclohexane		1	197–298	7.019 37	1 899.33	161.35
Decylcyclopentane		1	182–279	6.999 12	1 822.05	163.05
Deuterodiborane		1	−155 to −94	6.480 83	545.20	244.73
Diacetone alcohol		1	28–115	8.502 42	2 400.56	263.79
1,3-Diacetylbenzene		1	50–145	0.056 24	64.188	−196.97
1,4-Diacetylbenzene		1	116–157	2.803 71	177.25	−46.43
Diacetylene		1	−78 to 0	4.990 79	356.36	143.22
Diallyl sulfide		1	10–40	4.829 30	643.18	142.34
4,4′-Diaminodiphenylmethane		1	198–272	3.172 31	210.49	−137.41
Diamyl ether		1	105–187	7.067 10	1 604.77	196.58
Dibenzyl ketone		2	285–325	8.257	3 244.42	

TABLE 5.9 Vapor Pressures of Various Organic Compounds (*Continued*)

Substance	Eq.	Range, °C	A	B	C
1,2-Dibromobenzene	1	20–117	7.501 28	2 093.7	230
	1	117–300	7.102 65	1 825.77	207.0
Dibromodichloroethane	1	25–130	5.197 53	763.44	110.81
Dibromodifluoromethane	1	−26 to 23	7.152 22	1 181.612	253.85
1,2-Dibromoethane	1	52–131	6.721 48	1 280.82	201.75
1,2-Dibromoethylene *cis*	1	26–78	7.038 74	1 349.84	209.26
trans	1	4–71	4.581 11	393.641	103.56
1,2-Dibromopropane	1	0–50	7.303 98	1 644.4	232.0
	1	50–250	6.891 05	1 419.60	212.0
1,3-Dibromopropane	1	0–71	7.549 84	1 890.56	240.0
	1	71–275	7.198 74	1 678.26	222.0
Di-*n*-butyl ether	1	89–140	6.796 3	1 297.29	191.03
Di-*t*-butyl ether	1	4–109	6.932 9	1 348.53	233.79
Di-*n*-butyl phthalate	1	126–202	6.639 80	1 744.20	113.69
Di-*n*-butyl sebacate	1	128–208	7.587 66	2 364.89	147.54
Di-*n*-butyl sulfide	1	10–40	6.769 3	1 208.80	217.51
1,2-Dichlorobenzene	1	131–181	7.143 78	1 704.49	219.42
1,3-Dichlorobenzene	1	91–173	7.040 1	1 607.05	213.38
1,4-Dichlorobenzene	1	95–174	7.020 8	1 590.9	210.2
Dichlorobenzotrichloride	1	20–167	7.439 54	2 190.0	200
	1	167–340	6.985 24	1 868.91	172.00
Dichlorobenzyl chloride	1	20–138	7.504 57	2 125.9	213.8
	1	138–350	7.147 35	1 881.38	192.93
1,1-Dichloroethane	1	−39 to 18	6.977 0	1 174.02	229.06
1,2-Dichloroethane	1	−31 to 99	7.025 3	1 271.3	222.9
1,1-Dichloroethylene	1	−28 to 32	6.972 2	1 099.4	237.2
1,2-Dichloroethylene *cis*	1	0–84	7.022 3	1 205.4	230.6
trans	1	−38 to 85	6.965 1	1 141.9	231.9
2,2′-Dichloroethyl sulfide	1	15–76	8.587 41	2 588.23	246.06
1,2-Dichloroethyltrichloro-silane	1	102–181	7.826	2 144.9	253.1
Dichloromethane	1	−40 to 40	7.409 2	1 325.9	252.6
2-(2,4-Dichlorophenoxy)-ethanol	1	212–286	7.240 09	2 004.31	157.25
3,4-Dichlorophenylisocyanate	1	60–190	8.679 3	3 312.3	333.9
1,2-Dichloropropane	1	45–96	6.980 7	1 308.1	222.8
3,4-Dichlorotoluene	1	0–105	7.343 94	1 882.5	215.0
	1	105–330	6.979 25	1 655.44	195.0
Diethanolamine	1	194–241	8.138 8	2 327.9	174.4
1,1-Diethoxyethane	1	0–70	6.757 63	1 191.60	203.12
Diethoxymethane	1	0–75	6.908 41	1 229.52	217.01
Diethylaluminum chloride	1	44–125	8.229 70	2 484.53	255.45
Diethylamine	1	31–61	5.801 6	583.30	144.1
N,N-Diethylaniline	1	50–218	7.466 0	1 993.57	218.5
1,2-Diethylbenzene	1	liq	6.987 80	1 576.940	200.51
1,3-Diethylbenzene	1	liq	7.003 60	1 575.310	200.96
1,4-Diethylbenzene	1	liq	6.998 20	1 588.310	201.97
Diethyldichlorosilane	1	48–128	6.862 9	1 346.3	207.7
Diethyl disulfide	1	15–61	7.349 89	1 695.00	227.29
	1	61–230	6.975 07	1 485.970	208.96
Diethylene glycol	1	130–243	7.636 7	1 939.4	162.7
Diethyl ether	1	−61 to 20	6.920 32	1 064.07	228.80
Diethyl ethylphosphate	1	76–134	4.101 6	315.17	15.50
N,N-Diethylformamide	1	30–90	6.395 4	1 203.8	165.6
Diethyl ketone	1		6.857 91	1 216.3	204
3,3-Diethylpentane	1	63–147	6.896 03	1 453.48	215.83

TABLE 5.9 Vapor Pressures of Various Organic Compounds (*Continued*)

Substance		Eq.	Range, °C	A	B	C
3,5-Diethylphenol		1	114–248	7.651 3	2 228	218.5
Diethylpropylphosphonate		1	87–134	4.558 1	446.50	26.17
Diethyl sulfide		1	0–150	6.928 36	1 257.83	218.66
1,2-*bis*-Difluoroamino-4-						
methylpentane		1	−20 to 20	8.009 11	1 944.92	245.44
Difluoromethane		1	−82 to −32	7.138 9	821.7	244.7
1,2-Dihydroxybenzene		1	118–246	7.577	2 054	187
1,3-Dihydroxybenzene		1	151–276	7.889	2 231	169
1,2-Diiodoethylene	*cis*	1	29–152	5.522	797.8	106.4
	trans	1	77–130	6.093 1	1 197.0	172.3
Diisoamyl sulfide		1	10–80	−1.959 8	390.61	−219.33
p-Diisopropylbenzene		1	120–211	6.993 3	1 663.88	194.41
Diisopropyl ether		1	23–67	6.849 5	1 139.34	218.7
2,4-Diisopropylphenol		1	122–255	6.714	1 506	138
1,2-Dimethoxyethane		1	0–60	6.718 9	1 050.5	209.2
N,N-Dimethylacetamide		1	30–90	9.720 9	3 273.8	334.5
Dimethylamine		1	−72 to 6.9	7.082 12	960.242	221.67
bis-Dimethylaminoborane		1	−25 to 62.5	5.584 52	774.371	170.64
N-Dimethylaminodiborane		1	−38 to 14	8.340 1	1 917.35	302.73
bis-Dimethylaminodifluorosilane		1	24–88	5.952	748.7	146.9
N,N-Dimethylaniline		1	71–197	7.367 7	1 857.08	220.36
Dimethyl beryllium		1	100–180	19.089 9	11 535.45	496.64
1,4-Dimethyl-bicyclo(2,2,1)-						
heptane		1	56–119	6.761 96	1 342.66	213.53
2,3-Dimethyl-bicyclo(2,2,1)-						
heptane	*trans*	1	72–138	6.868 15	1 420.32	212.94
2,3-Dimethyl-1,3-butadiene		1	0–68.5	7.119 7	1 299.69	238.09
2,2-Dimethylbutane		1	−42 to 73	6.754 83	1 081.176	229.34
2,3-Dimethylbutane		1	−35 to 81	6.809 83	1 127.187	228.90
2,3-Dimethyl-2-butanethiol		1	56–167	6.839 56	1 354.24	215.96
2,3-Dimethyl-1-butene		1	−36 to 78	6.862 36	1 134.675	229.37
2,3-Dimethyl-2-butene		1	−21 to 97	6.950 58	1 215.428	225.44
3,3-Dimethyl-1-butene		1	−47 to 64	6.677 51	1 010.516	224.91
Dimethyl cadmium		1	−2 to 23	6.490 55	1 126.36	201.07
1,1-Dimethylcyclohexane		1	10–147	6.798 21	1 321.705	217.85
1,2-Dimethylcyclohexane	*cis*	1	18–158	6.837 46	1 367.311	215.84
	trans	1	13–151	6.833 08	1 353.881	219.13
1,3-Dimethylcyclohexane	*cis*	1	11–147	6.838 83	1 338.473	218.07
	trans	1	15–152	6.834 55	1 343.687	215.39
1,4-Dimethylcyclohexane	*cis*	1	15–152	6.832 87	1 345.613	216.15
	trans	1	10–147	6.817 73	1 330.437	218.58
1,1-Dimethylcyclopentane		1	−12 to 113	6.817 24	1 219.474	221.95
1,2-Dimethylcyclopentane	*cis*	1	−3 to 125	6.850 08	1 269.140	220.21
	trans	1	−9 to 117	6.844 22	1 242.748	221.69
1,3-Dimethylcyclopentane	*cis*	1	−10 to 116	6.837 15	1 237.456	222.01
	trans	1	−9 to 117	6.838 17	1 240.023	221.62
Dimethyldichlorosilane		1	28–72	7.062 1	1 280.29	235.65
1,2-Dimethyldisilane		1	−46 to 0	4.024 3	255.4	129.2
Dimethyl ether		1	−71 to −25	6.976 03	889.264	241.96
N,N-Dimethylformamide		1	30–90	6.928 0	1 400.87	196.43
2,2-Dimethylhexane		1		6.837 15	1 273.59	215.07
2,3-Dimethylhexane		1		6.870 04	1 315.50	214.16
2,4-Dimethylhexane		1		6.853 05	1 287.88	214.79
2,5-Dimethylhexane		1		6.859 84	1 287.27	214.41
3,3-Dimethylhexane		1		6.851 21	1 307.88	217.44

TABLE 5.9 Vapor Pressures of Various Organic Compounds (*Continued*)

Substance	Eq.	Range, °C	A	B	C
3,4-Dimethylhexane	1		6.879 86	1 330.04	214.86
1,1-Dimethylhydrazine	1	−35 to 20	7.408 13	1 305.91	225.53
1,2-Dimethylhydrazine	1	1−25	5.611 9	633.59	143.17
N,N-Dimethylhydroxylamine	1	17−90	7.565 8	1 415.96	201.93
O,N-Dimethylhydroxylamine	1	−45 to 42.2	7.405 4	1 245.58	233.06
Dimethylmalononitrile	1	49−140	7.035 5	1 546.99	202.00
1,3-Dimethylnaphthalene	1	20−148	7.634 7	2 295.4	232.4
	1	148−310	7.269 8	2 076.0	210
1,4-Dimethylnaphthalene	1	20−148	7.634 7	2 345.8	232.6
(same for 1,6- and 1,7-)	1	148−310	7.269 8	2 076.0	210
1,8-Dimethylnaphthalene	1	25−150	7.407 89	2 123.2	201.2
	1	150−320	7.056 4	1 879	180
2,3-Dimethylnaphthalene	1	20−155	7.403 96	2 111.9	201.1
	1	155−315	7.052 7	1 869	180
2,6-Dimethylnaphthalene	1	20−150	7.396 8	2 080.3	200.8
	1	150−310	7.046 0	1 841	180
2,7-Dimethylnaphthalene	1	25−150	7.398 75	2 085.9	200.9
	1	150−310	7.047 8	1 846	180
2,2-Dimethylpentane	1	−19 to 103	6.814 80	1 190.033	223.30
2,3-Dimethylpentane	1	−10 to 115	6.853 82	1 238.017	221.82
2,4-Dimethylpentane	1	−17 to 105	6.826 21	1 192.04	225.32
3,3-Dimethylpentane	1	−14 to 112	6.826 67	1 228.663	225.32
2,4-Dimethyl-3-pentanone	1	48−125	6.968 53	1 382.84	213.06
Dimethyl-o-phthalate	1	82−151	4.522 32	700.31	51.42
2,2-Dimethylpropane	1	−14 to 29	6.604 27	883.42	227.78
2,2-Dimethyl-1-propanol	1	55−115	7.875 3	1 604.7	208.2
2,5-Dimethylpyrrole	1	100−199	7.203 06	1 509.60	181.76
2,4-Dimethylquinoline	1	185−269	7.025 4	1 830.29	174.44
2,6-Dimethylquinoline	1	188−267	6.931 12	1 748.73	166.37
Dimethyl sulfide	1	−22 to 20	7.150 9	1 195.58	242.68
3,3-Dimethyl-2-thiabutane	1	liq	6.847 09	1 259.648	218.69
2,2-Dimethyl-3-thiapentane	1	liq	6.850 86	1 323.24	212.89
2,4-Dimethyl-3-thiapentane	1	liq	6.871 18	1 327.12	212.55
2,3-Dimethylthiophene	1	50−205	6.924 9	1 430.0	212
2,4-Dimethylthiophene	1	50−205	6.993 9	1 450.7	212.0
2,5-Dimethylthiophene	1	47−200	6.961 1	1 427.7	213.2
3,4-Dimethylthiophene	1	54−205	6.996 1	1 467.1	211.5
1,3-Dinitrobenzene	1	252−292	4.337	229.2	−137
2,4-Dinitrotoluene	1	200−299	5.798	1 118	61.8
2,6-Dinitrotoluene	1	150−260	4.372	380	−43.6
3,5-Dinitrotoluene	1	220−270	1.556	30.59	−302
1,4-Dioxane	1	20−105	7.431 55	1 554.68	240.34
Dipentene	1	21−170	7.111 6	1 613.42	207.8
2,2'-Diphenol	1	171−325	8.193 5	3 067.6	253.1
Diphenyldichlorosilane	1	192−281	6.999 03	1 918.20	161.41
Diphenyl ether	1	204−271	7.011 04	1 799.71	177.74
Diphenylmethane	1	217−282	6.291	1 261	105
Di-n-propyl ether	1	26−89	6.947 6	1 256.5	219.0
Disilanyl chloride	1	−46 to 18	7.104 8	1 211.8	245.2
2,3-Dithiabutane	1	6−135	6.977 92	1 346.342	218.86
5,6-Dithiadecane	1	101−263	6.963 8	1 684.1	181.3
3,4-Dithiahexane	1	40−182	6.975 07	1 485.970	208.96
4,5-Dithiaoctane	1	72−226	6.975 29	1 603.793	195.85
Dodecane	1	91−247	6.997 95	1 639.27	181.84
1-Dodecanethiol	1		7.024 4	1 817.8	164.1

TABLE 5.9 Vapor Pressures of Various Organic Compounds (*Continued*)

Substance	Eq.	Range, °C	A	B	C
Dodecanoic acid	1	106–176	7.860 8	2 159.1	143.2
1-Dodecanol	1	138–214	7.539 86	2 003.29	168.13
1-Dodecene	1	89–244	6.976 07	1 621.11	182.45
Durenol	1	108–249	7.758	2 432	250
Eicosane	1	198–379	7.152 2	2 032.7	132.1
1-Eicosanethiol	1		7.114	2 125	119
1-Eicosene	1	liq	7.135 1	2 043.0	137.9
Ethane	1	−142 to −75	6.829 15	663.72	256.68
Ethanethiol	1	−49 to 56	6.952 06	1 084.531	231.39
Ethanol	1	−2 to 100	8.321 09	1 718.10	237.52
Ethanolamine	1	65–171	7.456 8	1 577.67	173.37
Ethyl acetate	1	15–76	7.101 79	1 244.95	217.88
m-Ethylacetophenone	1	19–143	3.767 2	708.05	182.6
p-Ethylacetophenone	1	21–94	4.274 6	629.34	120.9
Ethylamine	1	−20 to 90	7.054 13	987.31	220.0
N-Ethylaniline	1	50–207	7.422 8	1 903.4	214.3
Ethylbenzene	1	26–164	6.957 19	1 424.255	213.21
2-Ethyl-1-butene	1	−28 to 88	6.997 12	1 218.352	231.30
Ethyl butyl ether	1	38–92	6.944 4	1 256.4	216.9
Ethyl chloroacetate	1	25–146	6.967	1 355.9	188.2
p-Ethylchlorobenzene	1	109–184	6.951 1	1 557.1	198.1
Ethylcyclohexane	1	20–160	6.867 28	1 382.466	214.99
Ethylcyclopentane	1	−0.1 to 129	6.887 09	1 298.599	220.68
Ethylene	1	−153 to −91	6.744 19	594.99	256.16
Ethylene glycol	1	50–200	8.090 8	2 088.9	203.5
Ethylene glycol monoethyl ether	1	63–134	7.874 6	1 843.5	234.2
Ethylene glycol monomethyl ether	1	56–124	7.849 8	1 793.9	236.9
Ethylene oxide	1	−49 to 12	7.128 43	1 054.54	237.76
Ethyl formate	1	4–54	7.009 0	1 123.94	218.2
3-Ethylhexane	1		6.890 98	1 327.88	212.60
2-Ethyl-1-hexanol	1	74–184	6.914 7	1 339.7	147.8
2-Ethyl-2-hexenal	1	54–175	6.861 3	1 457.4	190.6
Ethyl iodoacetate	1	29–89	4.073 7	374.64	54.8
Ethyl isothiocyanate	1	10–50	7.106 0	1 567.5	234.2
Ethyl methyl ether	1	5–7.7	5.518	434.5	158
Ethyl methyl ketone	1		6.974 21	1 209.6	216
3-Ethyl-5-methylphenol	1	195–247	7.040 83	1 615.44	152.6
2-Ethyl-4-methyl-1-pentanol	1	70–176	6.582 6	1 134.6	129.2
Ethyl nitrate	1	0–60	7.163 7	1 338.8	224.9
3-Ethylpentane	1	−7 to 119	6.875 64	1 251.827	219.89
2-Ethylphenol	1	86–208	7.800 3	2 140.4	227
3-Ethylphenol	1	97–218	7.468	1 856	187
4-Ethylphenol	1	101–218	8.291	2 423	229
Ethyl phenyl ether	1	117–181	7.021 38	1 508.39	194.49
Ethyl *n*-propanoate	1	34–98	6.994 9	1 260.6	207.4
Ethyl *n*-propyl ether	1	20–63	6.985 1	1 188.5	226.4
Ethyl *n*-propyl ketone	1	75–133	7.000 82	1 365.79	208.01
m-Ethylstyrene	1		7.039 28	1 614.0	198
p-Ethylstyrene	1		6.900 71	1 570.9	198
Ethyl trichloroacetate	1	44–95	7.725 4	1 927.0	233.7
Ethyl trichlorosilane	1	28–96	6.606	1 118	201
Ethyl triexthoxysilane	1	64–153	6.886 8	1 377.9	183.0
Ethyl vinyldichlorosilane	1	45–122	6.859	1 331	210.8

TABLE 5.9 Vapor Pressures of Various Organic Compounds (*Continued*)

Substance	Eq.	Range, °C	A	B	C
Fenchyl alcohol	1	59–200	5.693	797.6	84.6
Fluoranthene	1	197–384	6.373	1 756	118
Fluorene	1	161–300	7.761 8	2 637.1	243.2
Fluorobenzene	1	−18 to 84	7.187 0	1 381.8	235.6
m-Fluorobenzotrifluoride	1	40–137	7.006 59	1 304.35	215.67
bis-(Fluorocarbonyl)-peroxide	1	−47 to −7	9.608 4	2 247.64	319.83
p-Fluorotoluene	1	68–155	6.994 26	1 374.055	217.40
Formaldehyde	1	−109 to −22	7.195 8	970.6	244.1
Formic acid	1	37–101	7.581 8	1 699.2	260.7
Formyl fluoride	1	−95 to −61	5.270	362	175
Furan	1	2–61	6.975 27	1 060.87	227.74
2-Furfuraldehyde	1	56–161	6.575 9	1 198.7	162.8
Glycerol	1	183–260	6.165	1 036	28
Glyceryl-1,3-diacetate	1	100–190	6.407 3	1 092.0	119.3
Guaiacol	1	82–205	6.161	1 051	116
Hemellitenol	1	123–248	6.972	1 563	134
Heptadecane	1	161–337	7.014 3	1 865.1	149.20
1-Heptadecene	1		7.008 67	1 868.9	152.50
Heptane	1	−2 to 124	6.896 77	1 264.90	216.54
1-Heptanethiol	1	58–206	6.952 49	1 525.311	197.70
Heptanoic acid	1	112–150	5.287 4	665.54	42.07
1-Heptanol	1	60–176	6.647 67	1 140.64	126.56
1-Heptene	1	−6 to 118	6.901 87	1 258.345	219.30
Hexadecane	1	149–321	7.028 67	1 830.51	154.45
1-Hexadecanethiol	1		7.075	1 990	140
1-Hexadecanol	1	50–103	7.281 7	1 909.7	128.1
	1	145–190	6.158 6	1 380.0	91
1-Hexadecene	1		7.040 11	1 840.52	157.57
1,5-Hexadiene	1	0–59	6.574 1	1 013.5	214.8
Hexafluoroacetone	1	−79 to −27	6.650 2	725.90	219.9
Hexafluorobenzene	1	5–114	7.032 95	1 227.98	215.49
Hexafluorodisiloxane	1	−39 to −23	7.471 2	1 169.3	278.1
Hexafluoroethane	1	−93 to −78	6.793 35	657.06	246.2
Hexahydroindane cis	1	77–168	6.868 22	1 497.33	207.67
trans	1	71–161	6.861 19	1 475.70	209.66
Hexamethyldisiloxane	1	36–138	6.773 79	1 202.03	208.25
Hexane	1	−25 to 92	6.876 01	1 171.17	224.41
1-Hexanethiol	1	40–181	6.946 64	1 454.004	204.95
1-Hexanol	1	35–157	7.860 45	1 761.26	196.66
2-Hexanol	1	25–142	7.261 0	1 371.7	173.2
3-Hexanol	1	25–138	7.689	1 670.0	211.8
1-Hexene	1	16–64	6.857 70	1 148.62	225.35
3-Hexyne	1	−20 to 24	5.895	863.3	194
Hydroquinone	1	159–286	8.137	2 461	183
3-Hydroxy-3-methyl-2-butanone	1	45–146	7.340 9	1 653.6	227.5
Iodobenzene	1	20–188	7.011 9	1 640.1	208.8
Iodoethane	1	30–60	6.959	1 232	229
Isoamyl acetate	1	41–95	7.436	1 606.6	216
Isobutylbenzene	1	86–174	6.935 56	1 530.05	204.59
Isobutyl borate	1	99–200	7.197	1 745.8	193
Isobutyl cellosolve	1	71–159	7.694 8	1 825.9	219.6
Isobutylcyclohexane	1	85–172	6.867 97	1 493.10	203.16
Isobutyl nitrate	1	0–70	8.164 3	2 022.7	262.4
Isobutyraldehyde	1	13–63	6.735 1	1 053.2	209.1
Isobutyric acid	1	58–152	4.894	382.6	38

TABLE 5.9 Vapor Pressures of Various Organic Compounds (*Continued*)

Substance	Eq.	Range, °C	A	B	C
Isocaproic acid	1	96–133	6.258	1 038.6	103
Isopropylbenzene	1	39–181	6.936 66	1 460.793	207.78
Isopropyl borate	1	65–139	8.070	2 120	269
o-Isopropylbromobenzene	1	132–210	6.717 8	1 462.7	170.9
Isopropyl caprate	1	90–178	9.959	4 013.9	326.5
Isopropyl caprylate	1	65–146	8.032 2	2 213.6	220.9
Isopropyl cellosolve	1	67–140	7.500 0	1 639.2	213.3
Isopropyl chloroacetate	1	35–153	8.382	2 328	275
Isopropylcyclohexane	1	71–155	6.873 14	1 453.20	209.44
Isopropylcyclopentane	1	47–127	6.887 36	1 380.12	218.05
Isopropyl laurate	1	117–196	8.532 6	2 951.6	240.7
Isopropyl myristate	1	140–193	10.418 0	4 866.48	314.17
Isopropyl nitrate	1	0–70	7.266 6	1 434.4	225.2
Isopropyl palmitate	1	160–197	10.916 4	5 572.0	364.8
o-Isopropylphenol	1	97–215	8.167	2 343	229
p-Isopropylphenol	1	108–228	8.666	2 810	258
Isopropyl phenyl ether	1	72–175	6.517 6	1 238.0	163.0
Isopropyl stearate	1	182–207	0.079 3	10.41	−221
Isopseudocumenol	1	106–233	5.602	768	49
Isoquinoline	1	167–244	6.912 2	1 723.4	184.3
Isovaleric acid	1	86–104	3.946 55	255.41	11.3
Ketene	1	−88 to −49	7.615	1 036	269
Lauric acid	1	106–176	7.860 8	2 159.1	143.2
Lepidine	1	199–266	7.271 2	1 946.14	177.64
2,3-Lutidine	1	155–162	7.447 8	1 832.6	240.1
2,4-Lutidine	1	150–160	7.339 0	1 733.4	230.4
2,5-Lutidine	1	85–157	7.081 0	1 539.6	209.6
2,6-Lutidine	1	79–144	7.056 7	1 470.2	208.0
3,4-Lutidine	1	172–180	7.362 0	1 840.1	231.5
3,5-Lutidine	1	163–173	7.333 1	1 783.6	228.7
Mesitol	1	94–221	6.659	1 392	148
Mesityl oxide	1	14–130	6.635 8	1 186.1	186.0
Methacrylonitrile	1		6.980 2	1 274.96	220.7
Methane c	1	−195 to −183	7.193 09	451.64	268.49
liq	1	−181 to −152	6.695 61	405.42	267.78
Methanol	1	−14 to 65	7.897 50	1 474.08	229.13
	1	64–110	7.973 28	1 515.14	232.85
Methoxybenzene	1	110–164	7.052 69	1 489.99	203.57
N-Methylacetamide	1	40–90	2.631 1	121.7	−9.3
Methyl acetate	1	1–56	7.065 2	1 157.63	219.73
Methylal	1	0–35	6.872 2	1 049.2	220.6
Methylamine	1	−83 to −6	7.336 9	1 011.5	233.3
N-Methylaniline	1	50–200	7.081 9	1 631.3	192.4
Methyl benzoate	1	111–199	7.273	1 847	221
Methyl borate	1	31–68	7.646 0	1 491.5	245.5
Methyl boric anhydride	1	0–55	8.004 1	1 726.1	257.9
2-Methyl-1,3-butadiene	1	−52 to −24	7.011 87	1 126.159	238.88
	1	−19 to 55	6.885 64	1 071.578	233.51
3-Methyl-1,2-butadiene	1	−45 to −20	7.151 95	1 194.537	239.47
	1	−20 to 62	6.943 50	1 103.901	230.89
2-Methylbutane	1	−57 to 49	6.833 15	1 040.73	235.45
2-Methyl-1-butanethiol	1	liq	6.913 85	1 347.317	215.07
3-Methyl-1-butanethiol	1	liq	6.914 91	1 342.509	214.45
2-Methyl-2-butanethiol	1	liq	6.828 37	1 254.885	218.76
2-Methyl-1-butanol	1	34–129	7.067 30	1 195.26	156.83

TABLE 5.9 Vapor Pressures of Various Organic Compounds (*Continued*)

Substance	Eq.	Range, °C	A	B	C
3-Methyl-1-butanol	1	25–153	7.258 21	1 314.36	169.36
2-Methyl-2-butanol	1	25–102	6.519 3	863.4	135.3
3-Methyl-2-butanol	1	25–111	6.942 1	1 090.9	157.2
2-Methyl-1-butene	1	−53 to 52	6.846 37	1 039.69	236.65
3-Methyl-1-butene	1	−63 to 41	6.824 55	1 012.37	236.65
2-Methyl-2-butene	1	−48 to 60	6.966 59	1 124.33	236.63
Methyl butyl ether	1	23–69	6.887 1	1 162.1	219.9
3-Methyl-1-butyne	1	−55 to 47	6.884 80	1 014.81	227.11
2-Methyl-3-butyn-2-ol	1	21–106	6.657 5	976.5	154.1
Methyl *n*-butyrate	1		6.972 11	1 272.73	208.5
Methyl caprate	1	107–188	7.190 0	1 783.8	181.6
Methyl caproate	1	44–105	7.409 3	1 672.74	218.98
Methyl caprylate	1	100–146	6.916 5	1 496.3	176.5
Methyl carbitol	1	112–193	7.424	1 751	192
Methyl cellosolve acetate	1	70–144	7.125 1	1 447.0	196.1
Methyl chloroacetate	1	45–130	7.004 4	1 306.3	187.3
Methylcyclohexane	1	−3 to 127	6.823 00	1 270.763	221.42
Methylcyclopentane	1	−24 to 96	6.862 83	1 186.059	226.04
Methyldichlorosilane	1	1–41	7.027 8	1 167.8	240.7
1-Methyl-2-ethylbenzene	1	48–194	7.003 14	1 535.374	207.30
1-Methyl-3-ethylbenzene	1	46–190	7.015 82	1 529.184	208.51
1-Methyl-4-ethylbenzene	1	46–191	6.998 02	1 527.113	208.92
1-Methyl-1-ethylcyclopentane	1	43–122	6.859 20	1 347.602	217.21
1-Methyl-2-ethylcyclo-					
pentane *cis*	1	49–129	6.905 88	1 388.412	216.89
2-Methyl-3-ethylpentane	1		6.867 31	1 318.12	215.31
3-Methyl-3-ethylpentane	1		6.867 31	1 347	219.68
3-Methyl-5-ethylphenol	1	111–233	7.958	2 236	208
2-Methyl-5-ethylpyridine	1	52–177	5.050	517	59
N-Methylformamide	1	96–200	7.497 4	1 849.4	201.1
Methyl formate	1	21–32	3.027	3.02	−11.9
2-Methylheptane	1	42–119	6.917 35	1 337.47	213.69
3-Methylheptane	1	43–120	6.899 44	1 331.53	212.41
4-Methylheptane	1		6.900 65	1 327.66	212.57
2-Methylhexane	1	−9 to 115	6.873 18	1 236.026	219.55
3-Methylhexane	1	−8 to 117	6.867 64	1 240.196	219.22
Methylhydrazine	1	2–25	6.576 2	1 007.5	181.4
N-Methylhydroxylamine	1	40–65	7.045 6	1 223.3	172.1
O-Methylhydroxylamine	1	−63 to 48	7.363 9	1 225.3	225.2
Methyl isobutyl ketone	1	22–116	6.672 7	1 168.4	191.9
1-Methyl-2-isopropylbenzene	1	liq	6.940 4	1 548.05	203.15
1-Methyl-3-isopropylbenzene	1	liq	6.940 5	1 539.05	203.93
1-Methyl-4-isopropylbenzene	1	liq	6.923 7	1 537.06	203.05
3-Methylisoquinoline	1	176–225	6.969 2	1 717.3	166.9
Methyl isothiocyanate	1	10–50	2.896 8	103.6	45.4
Methyl laurate	1	158–212	6.767 1	1 589.72	140.5
Methyl linolate	1	166–206	6.111 1	1 660.1	118.8
Methyl methacrylate	1	39–89	8.409 2	2 050.5	274.4
Methyl myristate	1	166–238	7.622 3	2 283.93	184.8
1-Methylnaphthalene	1	108–278	7.035 92	1 826.948	195.00
2-Methylnaphthalene	1	105–274	7.068 50	1 840.268	198.40
Methyl oleate	1	166–205	7.544 1	2 656.9	200.7
Methyl palmitate	1	148–202	9.594 4	4 146.43	297.76
2-Methylpentane	1	−32 to 83	6.839 10	1 135.410	226.57
3-Methylpentane	1	−30 to 87	6.848 87	1 152.368	227.13

TABLE 5.9 Vapor Pressures of Various Organic Compounds (*Continued*)

Substance		Eq.	Range, °C	A	B	C
2-Methyl-2-pentanethiol		1	56–165	6.858 5	1 343.79	212.8
2-Methyl-1-pentanol		1	25–150	7.520 1	1 564.7	189.2
2-Methyl-4-pentanol		1	25–133	8.467 1	2 174.9	257.8
2-Methyl-1-pentene		1	−30 to 85	6.850 30	1 138.516	224.70
3-Methyl-1-pentene		1	−38 to 77	6.755 23	1 086.316	226.20
4-Methyl-1-pentene		1	−38 to 77	6.835 29	1 121.302	229.687
2-Methyl-2-pentene		1	−26 to 90	6.923 67	1 183.837	225.51
3-Methyl-2-pentene	*cis*	1	−26 to 91	6.910 73	1 186.402	226.70
	trans	1	−23 to 94	6.926 34	1 194.527	224.83
4-Methyl-2-pentene	*cis*	1	−35 to 79	6.841 29	1 120.707	226.59
	trans	1	−33 to 81	6.880 30	1 142.874	227.14
Methyl phenyl ether		1	110–164	7.052 69	1 489.99	203.57
2-Methylpiperidine		1	51–158	6.818 59	1 274.61	205.40
2-Methylpropane		1	−87 to 7	6.910 48	946.35	246.68
2-Methyl-1-propanethiol		1	−10 to 113	6.887 46	1 237.282	220.31
2-Methyl-2-propanethiol		1	1–88	6.787 81	1 115.565	221.31
2-Methyl-1-propanol		1	20–115	7.327 05	1 248.48	172.92
2-Methyl-2-propanol		1	**26–83**	**9.170 6**	**2 206.4**	**267.9**
2-Methylpropene		1	−82 to 12	6.684 66	866.25	234.64
N-Methylpropionamide		1	30–90	−0.9103	119.4	−148.0
Methyl propionate		1	21–79	6.942 4	1 170.2	208.8
2-Methyl-2-propylamine		1	19–75	6.783 2	993.33	210.50
Methyl propyl ether		1	0–39	6.118 6	708.69	179.9
2-Methylpyridine		1	80–168	7.032 4	1 415.73	211.63
3-Methylpyridine		1	74–185	7.050 21	1 481.78	211.25
4-Methylpyridine		1	75–186	7.041 77	1 480.68	210.50
1-Methylpyrrole		1	49–149	7.085 0	1 368.66	212.80
6-Methylquinoline		1	187–266	6.927 2	1 746.08	166.46
7-Methylquinoline		1	238–258	7.597 7	2 229.4	214.9
Methyl salicylate		1	79–220	7.083 3	1 712.8	187.1
Methyl stearate		1	204–240	2.357 0	68.92	−156.5
o-Methylstyrene		1	32–112	7.212 9	1 664.08	214.59
		1	75–255	6.884 61	1 485.41	200.0
m-Methylstyrene		1	10–72	7.275 34	1 695.4	220.0
		1	72–250	6.879 28	1 471.44	200.0
p-Methylstyrene		1	68–170	7.011 2	1 535.1	200.7
α-Methylstyrene		1		6.923 66	1 486.88	202.4
β-Methylstyrene		1		6.923 39	1 499.80	201.0
Methyl sulfoxide		1	20–50	7.763 7	2 048.7	231.6
3-Methyl-2-thiabutane		1	−13 to 109	6.901 96	1 232.170	221.67
2-Methylthiacyclopentane		1	liq	6.944 12	1 409.503	214.41
3-Methylthiacyclopentane		1	67–179	6.949 1	1 431.8	213.6
2-Methyl-3-thiapentane		1	liq	6.891 30	1 293.05	215.04
Methyl-2-thiazole		1	80–128	7.042 1	1 407.05	209.33
2-Methylthiophene		1	9–138	6.938 97	1 326.48	214.31
3-Methylthiophene		1	11–141	6.986 11	1 363.83	216.78
Methyl trichlorosilane		1	13–64	7.088 2	1 289.2	239.9
2-Methyl-5-vinylpyridine		1	69–183	6.156	1 023	129
Morpholine		1	0–44	7.718 13	1 745.8	235.0
		1	44–170	7.160 30	1 447.70	210.0
Naphthalene	c	1	86–250	7.010 65	1 733.71	201.86
	liq	1	125–218	6.818 1	1 585.86	184.82
1-Naphthol		1	141–282	7.284 21	2 077.56	184.0
2-Naphthol		1	144–288	7.347 14	2 135.00	183.0
Nicotine		1	134–246	6.789	1 650	176

TABLE 5.9 Vapor Pressures of Various Organic Compounds (*Continued*)

Substance	Eq.	Range, °C	A	B	C
o-Nitroaniline	2	150–260	8.868 4	3 336.50	
m-Nitroaniline	2	170–260	8.818 8	3 440.9	
p-Nitroaniline	2	190–260	9.559 5	4 039.73	
Nitrobenzene	1	134–211	7.115 6	1 746.6	201.8
m-Nitrobenzotrifluoride	1	10–105	7.653 15	2 006.1	220.0
	1	104–280	7.180 25	1 710.60	195.12
Nitromethane	1	56–136	7.281 66	1 446.94	227.60
1-Nitropropane	1	59–131	7.114 6	1 467.45	215.23
o-Nitrotoluene	1	129–222	5.851	946	96
p-Nitrotoluene	1	148–233	6.994 8	1 720.39	184.9
Nonadecane	1	184–366	7.015 3	1 932.8	137.6
1-Nonadecene	1	liq	7.115 1	1 997.4	142.7
Nonafluorocyclopentane	1	17–75	6.945 3	1 051.7	220.1
Nonane	1	39–179	6.938 93	1 431.82	202.01
1-Nonanethiol	1	93–251	6.983 9	1 655.6	183.7
Nonanoic acid	1	137–177	3.235 9	143.97	−75.6
1-Nonanol	1	94–214	7.827 8	1 953.8	181.9
1-Nonene	1	35–175	6.954 30	1 436.20	205.69
Octadecane	1	172–352	7.002 2	1 894.3	143.30
1-Octadecanethiol	1	liq	7.096	2 061	129
1-Octadecanol	1	120–218	6.461 6	1 599	90
1-Octadecene	1		7.060 65	1 997.4	147.50
Octane	1	19–152	6.918 68	1 351.99	209.15
1-Octanethiol	1	76–229	6.969 09	1 593.0	190.61
1-Octanol	1	0–80	12.070 1	4 506.8	319.9
	1	70–195	6.837 90	1 310.62	136.05
2-Octanol	1	72–180	6.388 8	1 060.4	122.5
3-Octanol	1	76–176	5.221 5	560.3	64.7
4-Octanol	1	71–176	5.739 6	760.5	89.5
1-Octene	1	15–147	6.934 95	1 355.46	213.05
5-Oxyhydrindene	1	120–251	9.213 7	3 665.8	326.4
Pentachloroethane	1	25–162	6.740	1 378	197
Pentadecane	1	136–304	7.023 59	1 789.95	161.38
1-Pentadecene	1		7.022 91	1 788.58	163.347
1,2-Pentadiene	1	−42 to −26	7.259 90	1 250.293	241.96
	1	−21 to 67	6.918 20	1 104.991	228.85
1,3-Pentadiene cis	1	−43 to −22	7.193 87	1 223.602	240.62
	1	−18 to 66	6.910 89	1 101.923	229.37
trans	1	−45 to −20	7.102 12	1 185.389	239.41
	1	−18 to 64	6.913 17	1 103.840	231.72
1,4-Pentadiene	1	−57 to −37	7.174 01	1 155.378	244.30
	1	−33 to 47	6.835 43	1 017.995	231.46
2,3-Pentadiene	1	−39 to −18	7.202 53	1 231.768	237.56
	1	−14 to 70	6.962 16	1 126.837	227.84
Pentafluorobenzene	1	49–94	7.036 65	1 254.07	216.02
Pentafluorochloroacetone	1	−40 to 32	6.848 4	925.3	225.4
Pentafluorochloroethane	1	−95 to −39	6.833 34	802.97	242.27
Pentafluorophenol	1	105–155	7.066 0	1 379.15	183.91
2,2,3,3,3-Pentafluoropropanol	1	0–23	6.308 7	830.56	153.8
Pentafluorotoluene	1	39–138	7.084 78	1 392.20	213.67
bis-Pentamethyldisilanoxydi-silane	1	169–201	8.556 64	3 051.316	258.85
bis-Pentamethyldisilanyl ether	1	88–183	8.161 44	2 575.250	273.32
Pentane	1	−50 to 58	6.852 96	1 064.84	233.01
Pentanenitrile	1	69–141	7.104 9	1 519.4	218.4

TABLE 5.9 Vapor Pressures of Various Organic Compounds (*Continued*)

Substance		Eq.	Range, °C	A	B	C
1-Pentanethiol		1	19–153	6.933 11	1 369.479	211.31
Pentanoic acid		1	72–174	5.412	591	60
1-Pentanol		1	37–138	7.177 58	1 314.56	168.11
2-Pentanol		1	25–120	7.275 75	1 271.92	170.37
3-Pentanol		1	21–116	7.414 93	1 354.42	183.41
2-Pentanone		1	56–111	7.021 93	1 313.85	215.01
3-Pentanone		1	56–111	7.025 29	1 310.28	214.19
1-Pentene		1	−55 to 51	6.844 24	1 044.01	233.50
2-Pentene	*cis*	1	−49 to 58	6.843 08	1 052.44	228.69
	trans	1	−49 to 58	6.899 83	1 080.76	232.57
1-Pentyne		1	−44 to 61	6.967 34	1 092.52	227.18
2-Pentyne		1	−33 to 78	7.046 14	1 189.87	229.60
Perdeuterobenzene		1	10–82	6.892 35	1 198.39	219.43
Perdeuterocyclohexane		1	10–80	6.837 86	1 190.38	222.40
Perfluorobutane		1	−39 to −4	7.035 1	990.27	240.4
Perfluorobutene		1	−28 to 20	9.222	2 401.6	382
Perfluorocyclobutane		1	−32 to 0	6.815 29	862.49	225.19
Perfluorocyclohexane		1	19–65	6.04	597	136
Perfluorocyclopentane		1	17–56	7.039 6	1 069.3	234.6
Perfluoroheptane		1	−2 to 106	6.937 72	1 181.14	208.66
Perfluorohexane		1	30–57	6.875 2	1 080.8	213.4
Perfluoromethylcyclohexane		1	33–111	6.824 06	1 133.76	211.22
Perfluorooctane		1	37–105	5.902 5	1 225.93	198.99
Perfluoropentane		1	9–65	7.017 9	1 072.9	230.0
Perfluoropiperidine		1	29–81	6.853 4	1 059.95	217.2
Perfluoropropane		1	−79 to −36	6.919 4	825.8	241.2
Perfluoropropene		1	−41 to 20	7.355	1 012.1	257
Phenanthrene		1	176–379	7.260 82	2 379.04	203.76
Phenol		1	107–182	7.133 0	1 516.79	174.95
β-Phenylethyl acetate		1	149–233	6.834 3	1 555.2	160.8
α-Phenylethyl alcohol		1	82–190	1.508	91	−263
o-Phenylethylphenol		1	169–250	4.506 0	516.8	−32.1
p-Phenylethylphenol		1	174–251	4.304 1	459.3	−52.4
Phenylisocyanate		1	10–80	−0.708 0	106.4	−146.6
4-Phenylphenol		1	177–308	8.657 5	3 022.8	216.1
Phosgene		1	−68 to 68	6.842 97	941.25	230
Phthalic anhydride		2	160–285	8.022	2 868.5	
α-Pinene		1	19–156	6.852 5	1 446.4	208.0
β-Pinene		1	19–166	6.898 4	1 511.7	210.2
Piperidine		1	42–144	6.855 69	1 238.80	205.43
Propadiene		1	−99 to −16	5.713 7	458.06	196.07
Propane		1	−108 to −25	6.803 38	804.00	247.04
1-Propanethiol		1	−25 to 91	6.928 46	1 183.307	224.62
2-Propanethiol		1	−37 to 75	6.877 34	1 113.895	226.16
1-Propanol		1	2–120	7.847 67	1 499.21	204.64
2-Propanol		1	0–101	8.117 78	1 580.92	219.61
2-Propen-1-ol		1	21–97	11.187 0	4 068.5	392.7
Propionic acid		1	56–139.5	6.403	950.2	130.3
Propionic anhydride		1	67–167	5.819 5	810.3	108.7
Propionitrile		1	−84 to 22	5.278 2	665.52	159.10
Propiophenone		1	132–201	7.370	1 894	205
Propyl acetate		1	39–101	7.016 15	1 282.28	208.60
1-Propylamine		1	23–77	6.926 51	1 044.05	210.84
2-Propylamine		1	4–61	6.890 25	985.69	214.07
n-Propylbenzene		1	43–188	6.951 42	1 491.297	207.14

TABLE 5.9 Vapor Pressures of Various Organic Compounds (*Continued*)

Substance	Eq.	Range, °C	A	B	C
n-Propyl borate	1	85–179	7.399 8	1 741	206
n-Propyl caprate	1	97–186	8.701 22	2 945.99	253.63
n-Propyl caproate	1	43–120	8.667 1	2 556.0	262.9
n-Propyl caprylate	1	70–153	8.516 7	2 599.5	246.2
n-Propyl cellosolve	1	77–149	7.146 4	1 440.6	187.7
n-Propylcyclohexane	1	40–186	6.886 46	1 460.800	207.94
n-Propylcyclopentane	1	21–158	6.903 92	1 384.386	213.16
Propylene	1	−112 to −32	6.778 11	770.85	245.51
1,2-Propylene oxide	1	−35 to 130	7.064 92	1 113.6	232
n-Propyl formate	1	26–82	6.848	1 127	203
n-Propyl laurate	1	124–205	8.068 9	2 692.4	222.5
n-Propyl myristate	1	147–200	9.216 8	3 744.68	272.87
n-Propyl nitrate	1	0–70	6.954 9	1 294.4	206.7
n-Propyl palmitate	1	166–204	14.129 2	9 759.2	539.7
o-(n-Propyl)phenol	1	104–222	9.215	3 254	292
p-(n-Propyl)phenol	1	0–234	8.329 6	2 661	254
n-Propyl phenyl ether	1	101–190	7.734 3	2 146.2	252.3
Propyne	1	−90 to −6	6.784 85	803.73	229.08
Pseudocumenol	1	107–232	6.915	1 547	152
Pyrene	1	200–395	5.618 4	1 122.0	15.2
Pyridine	1	67–153	7.041 15	1 373.80	214.98
Pyrogallol	1	177–309	6.092	1 031	12
Pyrrole	1	66–166	7.294 70	1 501.56	210.42
Quinaldine	1	178–248	7.179 00	1 857.84	184.50
Quinoline	1	164–238	6.817 59	1 668.73	186.26
Spiropentane	1	3–71	6.917 00	1 090.08	231.10
Styrene	1	32–82	7.140 16	1 574.51	224.09
Terpenyl acetate	1	37–150	6.443 46	1 377.27	143.85
α-Terpineol	1	84–217	8.141 2	2 479.4	253.7
Terpinolene	1	40–179	7.169	1 706	211
Tetrabutyl tin	1	100–300	6.545	1 649	148
1,1,2,2-Tetrachloro-1,2-difluoroethane	1	10–91.5	10.995	4 437.1	455.2
1,1,1,2-Tetrachloroethane	1	59–130	6.898 75	1 365.88	209.74
1,1,2,2-Tetrachloroethane	1	25–130	6.631 7	1 228.1	179.9
Tetrachloroethylene	1	37–120	6.976 83	1 386.92	217.53
Tetrachloromethane	1		6.879 26	1 212.021	226.41
Tetradecane	1	122–286	7.013 00	1 740.88	167.72
1-Tetradecanethiol	1		7.048 5	1 909.2	151.9
1-Tetradecanol	1	130–264	6.674 1	1 204.5	54.0
1-Tetradecene	1	119–283	7.030 65	1 754.09	171.52
1,2,3,4-Tetrafluorobenzene	1	6–50	7.084 6	1 339.23	223.49
1,2,3,5-Tetrafluorobenzene	1	6–50	6.986 17	1 245.20	218.35
Tetrafluoroethylene	1	−131 to −65	6.896 59	683.84	245.93
Tetrafluoromethane	1		6.972 31	540.50	260.10
Tetrahydrofuran	1	23–100	6.995 15	1 202.29	226.25
Tetraiodothiophene	1	−65 to 24	5.585 44	871.25	175.59
Tetralin	1	94–206	7.070 55	1 741.30	208.26
1,2,3,4-Tetramethylbenzene	1	80–217	7.059 4	1 690.54	199.48
1,2,3,5-Tetramethylbenzene	1	75–228	7.077 9	1 675.43	201.14
1,2,4,5-Tetramethylbenzene	1	74–227	7.080 0	1 672.43	201.43
2,2,3,3-Tetramethylbutane	1	0–65	6.876 65	1 329.93	226.36
Tetramethyl lead	1	0–60	6.937 7	1 335.3	219.1
2,2,3,3-Tetramethylpentane	1	57–141	6.830 60	1 398.67	213.84

TABLE 5.9 Vapor Pressures of Various Organic Compounds (*Continued*)

Substance	Eq.	Range, °C	A	B	C
2,2,3,4-Tetramethylpentane	1	52–134	6.834 18	1 375.59	214.94
2,2,4,4-Tetramethylpentane	1	43–123	6.796 20	1 324.59	216.02
Tetramethylsilane	1	−64 to 21	6.822 39	1 033.72	235.62
2-Thiabutane	1	−26 to 90	6.938 49	1 182.562	224.78
Thiacyclobutane	1	−5 to 120	7.016 67	1 321.331	224.51
Thiacyclohexane	1	29–170	6.905 18	1 422.47	211.72
Thiacyclopentane	1	14–148	6.995 40	1 401.939	219.61
Thiacyclopropane	1	−35 to 77	7.037 25	1 194.37	232.42
3-Thiaheptane	1	33–172	6.941 02	1 421.32	205.81
4-Thiaheptane	1	32–170	6.935 77	1 413.44	205.73
2-Thiahexane	1	17–150	6.945 83	1 363.808	212.07
3-Thiahexane	1	14–144	6.933 80	1 341.57	212.51
2-Thiapentane	1	−4 to 120	6.955 45	1 284.32	219.66
3-Thiapentane	1	−13 to 109	6.928 36	1 257.833	218.66
2-Thiapropane	1	−47 to 58	6.948 79	1 090.755	230.80
Thiazole	1	63–118	7.142 01	1 425.35	216.26
Thiophene	1	−12 to 108	6.959 26	1 246.02	221.35
Toluene	1	6–137	6.954 64	1 344.800	219.48
o-Toluidine	1	118–200	7.082 03	1 627.72	187.13
m-Toluidine	1	122–203	7.093 67	1 631.43	183.91
p-Toluidine	1		7.260 22	1 758.55	201.0
m-Tolyl pentafluoro-propionate	1	98–174	7.427 20	1 707.59	201.70
p-Tolyl pentafluoro-propionate	1	99–176	8.078 6	2 223.8	252.1
m-Tolyl trifluoroacetate	1	91–166	7.681 0	1 874.84	223.48
p-Tolyl trifluoroacetate	1	92–169	7.913 8	2 055.41	238.99
Tribromomethane	1	30–101	6.821 8	1 376.7	201.0
1,2,3-Tribromopropane	1	128–205	7.037 2	1 735.32	195.42
Trichloroacetic acid	1	112–198	7.273 0	1 594.3	165.4
Trichloroacetonitrile	1	17–83	7.183 5	1 368.3	232.5
Trichloroacetyl chloride	1	32–119	6.990 75	1 390.47	220.11
1,1,1-Trichloroethane	1	−6 to 17	8.643 4	2 136.6	302.8
1,1,2-Trichloroethane	1	50–114	6.951 85	1 314.41	209.20
Trichloroethylene	1	18–86	6.518 3	1 018.6	192.7
Trichlorofluoromethane	1		6.884 28	1 043.004	236.88
Trichlorosilane	1	2–32	6.773 9	1 009.0	227.2
bis-Trichlorosilylethane	1	91–160	7.835 11	2 241.769	249.84
1,1,1-Trichloro-2,2,2-trifluoroethane	1	14–36	4.437 3	204.1	83.9
1,1,2-Trichloro-1,2,2-trifluoroethane	1	−25 to 83	6.880 3	1 099.9	227.5
Tridecane	1	107–267	7.007 56	1 690.67	174.22
1-Tridecene	1	105–264	6.981 02	1 672.00	174.95
Triethanolamine	1	252–305	10.067 5	4 542.78	297.76
Triethyl aluminum	1	57–126	11.646 1	4 466.59	322.87
Triethylamine	1	50–95	5.858 8	695.7	144.8
Triethyl borate	1	29–109	7.511 1	1 641.7	236.3
Triethylsilanol	1	24–140	7.793 7	1 756.1	202.4
Trifluoroacetic acid	1	12–72	6.147 76	1 228.60	216.09
Trifluoroacetic anhydride	1	−2 to 39	6.135 8	1 026.1	202.0
Trifluoroacetonitrile	1	−132 to −68	7.127 6	773.82	249.9
1,3,5-Trifluorobenzene	1	6–50	6.919 8	1 197.13	219.12
Trifluorochloroethylene	1	−67 to −11	6.896 16	848.33	239.64
1,1,1-Trifluoroethane	1	−110 to −48	6.903 78	788.20	243.23

TABLE 5.9 Vapor Pressures of Various Organic Compounds (*Continued*)

Substance	Eq.	Range, °C	A	B	C
2,2,2-Trifluoroethanol	1	−0.5 to 25	6.788 2	978.13	173.06
Trifluoromethane	1	−128 to −82	7.088 6	705.33	249.78
bis-(Trifluoromethyl)- acetoxyphosphine	1	0–40	7.391 31	1 426.254	220.37
2,2,2-Trifluoro-1-methyl- benzene	1	55–139	6.970 45	1 306.35	217.38
bis-(Trifluoromethyl)- chlorophosphine	1	−80 to 0	7.661 06	1 386.652	267.14
Trifluoromethylhypofluorite	1	145–189	6.950 6	650.1	−18.4
bis-(Trifluoromethyl)- iodophosphine	1	0–47	6.901 39	1 180.723	222.95
Triisobutylene	1	56–179	7.002 1	1 613.47	212.5
Trimethyl aluminum	1	64–127	7.570 29	1 734.72	242.78
Trimethylamine	1	−80 to 3	6.857 55	955.94	237.52
1,2,3-Trimethylbenzene	1	57–205	7.040 82	1 593.958	207.08
1,2,4-Trimethylbenzene	1	52–198	7.043 83	1 573.267	208.56
1,3,5-Trimethylbenzene	1	49–193	7.074 36	1 569.622	209.58
2,2,3-Trimethylbutane	1	−19 to 106	6.792 30	1 200.563	226.05
Trimethylchlorosilane	1	2–55	7.055 8	1 245.5	240.7
1,1,3-Trimethylcyclohexane	1	55–137	6.839 51	1 394.88	215.73
1,1,2-Trimethylcyclopentane	1	36–115	6.822 38	1 309.81	218.58
1,1,3-Trimethylcyclopentane	1	29–106	6.809 31	1 275.92	219.89
1,2,4-Trimethylcyclopentane					
cis, cis, trans	1	39–118	6.857 38	1 335.69	219.16
cis, trans, cis	1	33–110	6.851 3	1 307.10	219.92
1,3,5-Trimethyl-2-ethyl- benzene	1	88–210	6.790 8	1 505.8	174.7
1,4,5-Trimethyl-2-ethyl- benzene	1	87–132	3.029 3	116.4	−34.6
2,2,5-Trimethylhexane	1	46–125	6.837 75	1 325.54	210.91
2,4,4-Trimethylhexane	1	51–131	6.856 54	1 371.81	214.40
Trimethylhydrazine	1	−16 to 14	7.106 80	1 189.88	222.06
O,N,N-Trimethylhydroxylamine	1	−79 to 23	6.765 8	979.55	222.2
2,2,3-Trimethylpentane	1		6.825 46	1 294.88	218.42
2,2,4-Trimethylpentane	1	24–100	6.811 89	1 257.84	220.74
2,3,3-Trimethylpentane	1		6.843 53	1 328.05	220.38
2,3,4-Trimethylpentane	1	36–114	6.853 96	1 315.08	217.53
2,4,4-Trimethyl-1-pentene	1	−3 to 128	6.834 57	1 273.416	220.62
2,4,4-Trimethyl-2-pentene	1	2–131	6.859 22	1 272.717	214.99
2,3,5-Trimethylphenol	1	186–247	7.080 12	1 685.90	166.14
Trimethylsilanol	1	18–85	8.126 6	1 657.6	219.2
2,4,5-Trimethylstyrene	1	79–216	7.331 5	1 880.7	205.7
2,4,6-Trimethylstyrene	1	90–208	7.089 1	1 702.61	195.93
1,2,4-Trinitrobenzene	1	250–300	3.194	87	−199
1,3,5-Trinitrobenzene	1	202–312	5.534 5	993.6	11.2
2,4,6-Trinitrobenzene	1	249–342	9.621 1	4 987.9	329.9
2,4,6-Trinitrotoluene	1	230–250	7.671 52	2 669.4	205.6
α-Trioxane	1	56–114	7.818 6	1 783.3	247.1
Trivinylarsine	1	22–66	7.894 1	2 115.6	293.9
Trivinyl bismuth	1	20–74	7.237 2	1 667.0	215.1
Trivinylphosphine	1	16–61	7.928 4	2 102.0	301.3
Trivinylstibine	1	20–70	8.322 1	2 446.3	303.8
Undecane	1	75–226	6.972 20	1 569.57	187.70
1-Undecanethiol	1		7.012 2	1 767.4	170.4
1-Undecene	1	72–222	6.966 77	1 563.21	189.87

TABLE 5.9 Vapor Pressures of Various Organic Compounds (*Continued*)

Substance	Eq.	Range, °C	A	B	C
Urethane	1		7.421 64	1 758.21	205.0
Vinyl acetate	1	22–72	7.210 1	1 296.13	226.66
o-Xylene	1	32–172	6.998 91	1 474.679	213.69
m-Xylene	1	28–166	7.009 08	1 462.266	215.11
p-Xylene	1	27–166	6.990 52	1 453.430	215.31
2,3-Xylenol	1	149–218	7.053 97	1 617.57	170.74
2,4-Xylenol	1	144–212	7.055 39	1 587.46	169.34
2,5-Xylenol	1	144–212	7.051 56	1 592.70	170.74
2,6-Xylenol	1	145–204	7.070 70	1 628.32	187.60
3,4-Xylenol	1	172–229	7.079 19	1 621.45	159.26
3,5-Xylenol	1	155–223	7.130 76	1 639.86	164.16

5.3 BOILING POINTS

TABLE 5.10 Boiling Points of Water

A. Barometric Pressures at Various Temperatures

Temp. °C.	0.0° mm of Hg	0.2° mm of Hg	0.4° mm of Hg	0.6° mm of Hg	0.8° mm of Hg
80	355.40	358.28	361.19	364.11	367.06
81	370.03	373.01	376.02	379.05	382.09
82	385.16	388.25	391.36	394.49	397.64
83	400.81	404.00	407.22	410.45	413.71
84	416.99	420.29	423.61	426.95	430.32
85	433.71	437.12	440.55	444.01	447.49
86	450.99	454.51	458.06	461.63	465.22
87	468.84	472.48	476.14	479.83	483.54
88	487.28	491.04	494.82	498.63	502.46
89	506.32	510.20	514.11	518.04	521.99
90	525.97	529.98	534.01	538.07	542.15
91	546.26	550.40	554.56	558.75	562.96
92	567.20	571.47	575.76	580.08	584.43
93	588.80	593.20	597.63	602.09	606.57
94	611.08	615.62	620.19	624.79	629.41
95	634.06	638.74	643.45	648.19	652.96
96	657.75	662.58	667.43	672.32	677.23
97	682.18	687.15	692.15	697.19	702.25
98	707.35	712.47	717.63	722.81	728.03
99	733.28	738.56	743.87	749.22	754.59
100	760.00	765.44	770.91	776.42	781.95

B. Boiling Points of Water at Various Pressures

Pressure, atm.	Boiling Point, °C.	Pressure, atm.	Boiling Point, °C.	Pressure, atm.	Boiling Point, °C.	Pressure, atm.	Boiling Point, °C.
0.5	80.9	7	164.2	14	194.1	21	213.9
1	100.0	8	169.6	15	197.4	22	216.2
2	119.6	9	174.5	16	200.4	23	218.5
3	132.9	10	179.0	17	203.4	24	220.8
4	142.9	11	183.2	18	206.1	25	222.9
5	151.1	12	187.1	19	208.8	26	225.0
6	158.1	13	190.7	20	211.4	27	227.0

TABLE 5.11   Binary Azeotropic (Constant-Boiling) Mixtures

An azeotrope is a mixture that can be separated by distillation.

A. Binary azeotropes containing water

System	BP of azeotrope, °C	Composition, wt %	
		Water	Other component
Inorganic acids			
Hydrogen bromide	126	52.5	47.5
Hydrogen chloride	108.58	79.78	20.22
Hydrogen fluoride	111.35	64.4	35.6
Hydrogen iodide	127	43	57
Hydrogen peroxide	zeotrope		
Nitric acid	120.7	32.6	67.4
Perchloric acid	203	28.4	71.6
Organic acids			
Formic acid	107.2	22.6	77.4
Acetic acid	zeotrope		
Propionic acid	99.9	82.3	17.7
Isobutyric acid	99.3	79	21
Butyric acid	99.4	81.6	18.4
Pentanoic acid	99.8	89	11
Isopentanoic acid	99.5	81.6	18.4
Perfluorobutyric acid	97	71	29
Crotonic acid	99.9	97.8	2.2
Alcohols			
Ethanol	78.17	4	96
Allyl alcohol	88.9	27.7	72.3
1-Propanol	71.7	71.7	28.3
2-Propanol	80.3	12.6	87.4
1-Butanol	92.7	42.5	57.5
2-Butanol	87.0	26.8	73.2
2-Methyl-2-propanol	79.9	11.7	88.3
1-Pentanol	95.8	54.4	45.6
2-Pentanol	91.7	36.5	63.5
3-Pentanol	91.7	36.0	64.0
2,2-Dimethyl-2-propanol	87.35	27.5	72.5
1-Hexanol	97.8	67.2	32.8
1-Octanol	99.4	90	10
Cyclopentanol	96.25	58	42
1-Heptanol	98.7	83	17
Phenol	99.52	90.8	9.2
2-Methoxyphenol	99.5	87.5	12.5

TABLE 5.11 Binary Azeotropic (Constant-Boiling) Mixtures (*Continued*)

System	BP of azeotrope, °C	Composition, wt %	
		Water	Other component
Alcohols (*continued*)			
1-Phenylphenol	99.95	98.75	1.25
Benzyl alcohol	99.9	91	9
2,3-Dimethyl-2,3-butanediol	zeotrope		
Furfuryl alcohol	98.5	80	20
Aldehydes			
Propionaldehyde	47.5	2	98
Butyraldehyde	68	6	94
Pentanal	83	19	81
Paraldehyde	90	28.5	71.5
Furaldehyde	97.5	65	35
Amines			
N-Methylbutylamine	82.7	15	85
Furfurylamine	99	74	26
Piperidine	92.8	35	65
Pyridine	93.6	41.3	58.7
2-Methylpyridine	93.5	48	52
3-Methylpyridine	97	60	40
4-Methylpyridine	97.35	62.8	37.2
2,6-Dimethylpyridine	96.02	51.8	48.2
Dibutylamine	97	50.5	49.5
Dihexylamine	99.8	92.8	7.2
Triallylamine	95	38	62
Tributylamine	99.65	79.7	20.3
Aniline	98.6	80.8	19.2
N-Ethylaniline	99.2	83.9	16.1
1-Methyl-2-(2-pyridyl)pyrrolidine	99.85	97.5	2.5
Halogenated hydrocarbons			
Chloroform	56.1	2.8	97.2
Carbon tetrachloride	42.6	2.8	97.2
Trichloroethylene	73.4	17	83
Tetrachloroethylene	88.5	17.2	82.8
1,2-Dichloroethane	72	8.3	91.7
1-Chloropropane	44	2.2	97.8
1,2-Dichloropropane	78	12	88
Chlorobenzene	90.2	28.4	71.6

TABLE 5.11 Binary Azeotropic (Constant-Boiling) Mixtures (*Continued*)

System	BP of azeotrope, °C	Composition, wt %	
		Water	Other component
Esters			
Ethyl formate	52.6	5	95
Isopropyl formate	65.0	3	97
Propyl formate	71.6	2.3	97.7
Isobutyl formate	80.4	7.8	92.2
Butyl formate	83.8	14.5	85.5
Isopentyl formate	90.2	21	79
Pentyl formate	91.6	28.4	71.6
Benzyl formate	99.2	80	20
Ethyl acetate	70.38	8.47	91.53
Allyl acetate	83	14.7	85.3
Isopropyl acetate	76.6	10.6	89.4
Propyl acetate	82.4	14	86
Isobutyl acetate	87.4	16.5	83.5
Butyl acetate	90.2	28.7	71.3
Isopentyl acetate	93.8	36.3	63.7
Pentyl acetate	95.2	41	59
Hexyl acetate	97.4	61	39
Phenyl acetate	98.9	75.1	24.9
Benzyl acetate	99.6	87.5	12.5
Methyl propionate	71.4	3.9	96.1
Ethyl propionate	81.2	10	90
Isopropyl propionate	85.2	19.9	80.1
Propyl propionate	88.9	23	77
Isobutyl propionate	92.75	52.2	47.8
Isopentyl propionate	96.55	48.5	51.5
Methyl butyrate	82.7	11.5	88.5
Ethyl butyrate	87.9	21.5	78.5
Propyl butyrate	94.1	36.4	63.6
Isobutyl butyrate	96.3	46	54
Butyl butyrate	97.2	53	47
Isopentyl butyrate	98.05	63.5	36.5
Methyl isobutyrate	77.7	6.8	93.2
Ethyl isobutyrate	85.2	15.2	84.8
Propyl isobutyrate	92.2	30.8	69.2
Isobutyl isobutyrate	95.5	39.4	60.6
Isopentyl isobutyrate	97.4	56.0	44.0
Methyl isopentanoate	87.2	19.2	80.8
Ethyl isopentanoate	92.2	30.2	69.8
Propyl isopentanoate	96.2	45.2	54.8
Isobutyl isopentanoate	97.4	55.8	44.2
Isopentyl isopentanoate	98.8	74.1	25.9

TABLE 5.11 Binary Azeotropic (Constant-Boiling) Mixtures (*Continued*)

System	BP of azeotrope, °C	Composition, wt %	
		Water	Other component
Esters (continued)			
Ethyl pentanoate	94.5	40	60
Ethyl hexanoate	97.2	54	46
Methyl benzoate	99.08	79.2	20.8
Ethyl benzoate	99.4	84.0	16.0
Propyl benzoate	99.7	90.9	9.1
Butyl benzoate	99.9	94	6
Isopentyl benzoate	99.9	95.6	4.4
Ethyl phenylacetate	99.7	91.3	8.7
Methyl cinnamate	99.9	95.5	4.5
Methyl phthalate	99.95	97.5	2.5
Diethyl *o*-phthalate	99.98	98.0	2.0
Ethyl chloroacetate	95.2	45.1	54.9
Butyl chloroacetate	98.12	75.5	24.5
Methyl acrylate	71	7.2	92.8
Isobutyl carbonate	98.6	74	26
Ethyl crotonate	93.5	38	62
Methyl lactate	99	80	20
1,2-Ethanediol diacetate	99.7	84.6	15.4
Ethyl nitrate	74.35	22	78
Propyl nitrate	84.8	20	80
Isobutyl nitrate	89.0	25	75
Methyl sulfate	98.6	73	27
Ethers			
Ethyl vinyl ether	34.6	1.5	98.5
Diethyl ether	34.2	1.3	98.7
Ethyl propyl ether	59.5	4	96
Diisopropyl ether	62.2	4.5	95.5
Butyl ethyl ether	76.6	11.9	88.1
Diisobutyl ether	88.6	23	77
Dibutyl ether	92.9	33	67
Diisopentyl ether	97.4	54	46
1,1-Diethoxyethane	82.6	14.5	85.5
Diphenyl ether	99.33	96.75	3.25
Methoxybenzene	95.5	40.5	59.5
Hydrocarbons			
Pentane	34.6	1.4	98.6
Hexane	61.6	5.6	94.4
Heptane	79.2	12.9	87.1
2,2,4-Trimethylpentane	78.8	11.1	88.9

TABLE 5.11 Binary Azeotropic (Constant-Boiling) Mixtures (*Continued*)

System	BP of azeotrope, °C	Composition, wt %	
		Water	Other component
Hydrocarbons (continued)			
Nonane	94.8	82	18
Undecane	98.85	96.0	4.0
Dodecane	99.45	98	2
Acrolein	52.4	2.6	97.4
Cyclohexene	70.8	8.93	91.07
Cyclohexane	69.5	8.4	91.6
1-Octene	88.0	28.7	71.3
Benzene	69.25	8.83	91.17
Toluene	84.1	13.5	86.5
Ethylbenzene	92.0	33.0	67.0
m-Xylene	92	35.8	64.2
Isopropylbenzene	95	43.8	56.2
Naphthalene	98.8	84	16
Ketones			
Acetone	zeotrope		
2-Butanone	73.5	11	89
2-Pentanone	83.3	19.5	80.5
Cyclopentanone	94.6	42.4	57.6
4-Methyl-2-pentanone	87.9	24.3	75.7
2-Heptanone	95	48	52
3-Heptanone	94.6	42.2	57.8
4-Heptanone	94.3	40.5	59.5
4-Hydroxy-4-methyl-2-pentanone	98.8	87.3	12.7
4-Methyl-3-penten-2-one	91.8	34.8	65.2
Nitriles			
Acetonitrile	76.5	16.3	83.7
Isobutyronitrile	82.5	23	77
Butyronitrile	88.7	32.5	67.5
Acrylonitrile	70.6	14.3	85.7
Miscellaneous			
Hydrazine	120	32.3	67.7
Acetamide	zeotrope		
Nitromethane	83.59	23.6	76.4
Nitroethane	87.22	28.5	71.5
2,5-Dimethylfuran	77.0	11.7	88.3
Trioxane	91.4	30	70
Carbon disulfide	42.6	2.8	97.2

TABLE 5.11   Binary Azeotropic (Constant-Boiling) Mixtures (*Continued*)

B. Binary azeotropes containing organic acids

System	BP of azeotrope, °C	Composition, wt %	
		Acid	Other component
Formic acid			
2-Methylbutane	27.2	4	96
Pentane	34.2	20	80
Hexane	60.6	28	72
Methylcyclopentane	63.3	29	71
Cyclohexane	70.7	70	30
Methylcyclohexane	80.2	46.5	53.5
Heptane	78.2	56.5	43.5
Octane	90.5	63	37
Benzene	71.05	31	69
Toluene	85.8	50	50
o-Xylene	95.5	74	26
m-Xylene	92.8	71.8	28.2
Styrene	97.8	73	27
Iodomethane	42.1	6	94
Chloroform	59.15	15	85
Carbon tetrachloride	66.65	18.5	81.5
Trichloroethylene	74.1	25	75
Tetrachloroethylene	88.2	50	50
Bromoethane	38.2	3	97
1,2-Dibromoethane	94.7	51.5	48.5
1,2-Dichloroethane	77.4	14	86
1-Bromopropane	64.7	27	73
2-Bromopropane	56.0	14	86
1-Chloropropane	45.6	8	92
2-Chloropropane	34.7	1.5	98.5
1-Chloro-2-methylpropane	63.0	19	81
Bromobenzene	98.1	68	32
Chlorobenzene	93.7	59	41
Fluorobenzene	73.0	27	73
o-Chlorotoluene	100.2	83	17
Pyridine	127.43	61.4	38.6
2-Methylpyridine	158.0	25	75
2-Pentanone	105.3	32	68
3-Pentanone	105.4	33	67
Nitromethane	97.07	45.5	54.5
Diethyl sulfide	82.2	35	65
Diisopropyl sulfide	93.5	62	38
Dipropyl sulfide	98.0	83	17
Carbon disulfide	42.55	17	83

TABLE 5.11 Binary Azeotropic (Constant-Boiling) Mixtures (*Continued*)

System	BP of azeotrope, °C	Composition, w %	
		Acid	Other component
Acetic acid			
Hexane	68.3	6.0	94.0
Heptane	91.7	23	67
Octane	105.7	53.7	46.3
Nonane	112.9	69	31
Decane	116.75	79.5	20.5
Undecane	117.9	95	5
Cyclohexane	78.8	9.6	90.4
Methylcyclohexane	96.3	31	69
Benzene	80.05	2.0	98.0
Toluene	100.6	28.1	71.9
o-Xylene	116.6	78	22
m-Xylene	115.35	72.5	27.5
p-Xylene	115.25	72	28
Ethylbenzene	114.65	66	34
Styrene	116.8	85.7	14.3
Isopropylbenzene	116.0	84	16
Triethylamine	163	67	33
Nitromethane	101.2	96	4
Nitroethane	112.4	30	70
Pyridine	138.1	51.1	48.9
2-Methylpyridine	144.1	40.4	59.6
3-Methylpyridine	152.5	30.4	69.6
4-Methylpyridine	154.3	30.3	69.7
2,6-Dimethylpyridine	148.1	22.9	77.1
Carbon tetrachloride	76	98.46	1.54
Trichloroethylene	86.5	96.2	3.8
Tetrachloroethylene	107.4	61.5	38.5
1,2-Dibromoethane	114.4	55	45
2-Iodopropane	88.3	9	91
1-Bromobutane	97.6	18	82
1-Bromo-2-methylpropane	90.2	12	88
Chlorobenzene	114.7	58.5	41.5
Trichloronitromethane	107.65	80.5	19.5
1,4-Dioxane	119.5	77	23
Diisopropyl sulfide	111.5	48	52
Propionic acid			
Heptane	97.8	2	98
Octane	120.9	21.5	78.5

TABLE 5.11   Binary Azeotropic (Constant-Boiling) Mixtures (*Continued*)

		Composition, w %	
System	BP of azeotrope, °C	Acid	Other component
Propionic acid (continued)			
Nonane	134.3	54.0	46.0
Decane	139.8	80.5	19.5
o-Xylene	135.4	43	57
p-Xylene	132.5	34	66
1,3,5-Trimethylbenzene	139.3	77	23
Isopropylbenzene	139.0	65	35
Propylbenzene	139.5	75	25
Camphene	138.0	65	35
α-Pinene	136.4	58.5	41.5
Methoxybenzene	140.8	96	4
Pyridine	148.6	67.2	32.8
2-Methylpyridine	154.5	55.0	45.0
1,2-Dibromoethane	127.8	17.5	82.5
1-Iodo-2-methylpropane	119.5	9	91
Chlorobenzene	128.9	18	82
Dipropyl sulfide	136.5	45	55
Butyric acid			
Undecane	162.4	84.4	15.5
o-Xylene	143.0	10	90
m-Xylene	138.5	6	94
p-Xylene	137.8	5.5	94.5
Ethylbenzene	135.8	4	96
Styrene	143.5	15	85
1,2,4-Trimethylbenzene	159.5	45	55
1,3,5-Trimethylbenzene	158.0	38	62
Isopropylbenzene	149.5	20	80
Propylbenzene	154.5	28	72
Butylbenzene	162.5	75	25
Naphthalene	zeotrope		
Indene	163.7	84	16
Camphene	152.3	2.8	97.2
Methoxybenzene	152.9	12	88
Pyridine	163.2	92.0	8.0
2-Furaldehyde	159.4	42.5	57.5
1,2-Dibromoethane	131.1	3.5	96.5
1-Iodobutane	129.8	2.5	97.5
Chlorobenzene	131.75	2.8	97.2
1,4-Dichlorobenzene	162.0	57	43
o-Bromotoluene	163.0	72	28
m-Bromotoluene	163.6	79.5	20.5
p-Bromotoluene	161.5	75	25
α-Chlorotoluene	160.8	65	35
Ethyl bromoacetate	157.4	84	16
Propyl chloroacetate	160.5	40	60

TABLE 5.11 Binary Azeotropic (Constant-Boiling) Mixtures (*Continued*)

System	BP of azeotrope, °C	Composition, w % Acid	Composition, w % Other component
Isobutyric acid			
2,7-Dimethyloctane	148.6	48	52
o-Xylene	141.0	22	78
m-Xylene	139.9	15	85
p-Xylene	136.4	13	87
Styrene	142.0	27	73
1,2,4-Trimethylbenzene	152.3	63	37
Isopropylbenzene	146.8	35	65
Propylbenzene	149.3	49	51
Camphene	148.1	45	55
D-Limonene	152.5	78	22
Methoxybenzene	149.0	42	58
Ethyl bromoacetate	153.0	40	60
Ethyl 2-oxopropionate	153.0	60	40
1,2-Dibromoethane	130.5	6.5	93.5
1-Iodobutane	128.8	7	93
1-Bromohexane	148.0	35	65
Bromobenzene	148.6	35	65
Chlorobenzene	131.5	8	92
o-Bromotoluene	153.9	85	15
α-Chlorotoluene	153.5	80	20
Diisopentyl ether	154.2	93	7
Ethyl bromoacetate	153.0	40	60

C. Binary azeotropes containing alcohol

System	BP of azeotrope, °C	Composition, w % Alcohol	Composition, w % Other component
Methanol			
Pentane	30.9	7	93
Cyclopentane	38.8	14	86
Cyclohexane	53.9	36.4	63.6
Methylcyclohexane	59.2	54	46
Heptane	59.1	51.5	48.5
Octane	62.8	67.5	32.5
Nonane	64.1	83.4	16.6
Benzene	57.5	39.1	60.9
Fluorobenzene	59.7	32	68
Toluene	63.5	72.5	27.5
Bromomethane	3.55	99.55	0.45
Iodomethane	37.8	95.5	4.5

TABLE 5.11 Binary Azeotropic (Constant-Boiling) Mixtures (*Continued*)

System	BP of azeotrope, °C	Composition, w % Alcohol	Composition, w % Other component
		Alcohol	Other component
Methanol (*continued*)			
Bromodichloromethane	63.8	60	40
Chloroform	53.4	87.4	12.6
Carbon tetrachloride	55.7	79.44	20.56
Bromoethane	34.9	5.3	94.7
1,2-Dichloroethane	61.0	32	68
Trichloroethylene	59.3	38	62
1-Bromopropane	54.5	21	79
2-Bromopropane	48.6	15.0	85.0
1-Chloropropane	40.5	9.5	90.5
2-Chloropropane	33.4	6	94
2-Iodopropane	61.0	38	62
1-Chlorobutane	57.0	27	73
Isobutyl formate	64.6	95	5
Methyl acetate	53.5	19	81
Methyl acrylate	62.5	54	46
Methyl nitrate	52.5	73	27
Acetone	55.5	12.1	87.9
1,4-Dioxane	zeotrope		
Dipropyl ether	63.8	72	28
Methyl *tert*-butyl ether	51.3	14.3	85.7
Diethyl sulfide	61.2	62	38
Carbon disulfide	39.8	71	29
Thiophene	59.7	16.4	83.6
Nitromethane	64.4	9.1	90.9
Ethanol			
Pentane	34.3	5	95
Cyclopentane	44.7	7.5	92.5
Hexane	58.7	21	79
Cyclohexane	64.8	29.2	70.8
Heptane	70.9	49	51
Octane	77.0	78	22
Benzene	67.9	31.7	68.3
Fluorobenzene	70.0	75	25
Toluene	76.7	68	32
Bromodichloromethane	75.5	72	28
Iodomethane	41.2	96.8	3.2
Chloroform	59.3	93	7
Trichloronitromethane	77.5	34	66
Carbon tetrachloride	65.0	84.2	15.8
1,2-Dichloroethane	70.5	37	63
3-Chloro-1-propene	44	5	95

TABLE 5.11 Binary Azeotropic (Constant-Boiling) Mixtures (*Continued*)

System	BP of azeotrope, °C	Composition, w %	
		Alcohol	Other component
Ethanol (*continued*)			
1-Bromopropane	62.8	20.5	79.5
2-Bromopropane	55.6	10.5	89.5
1-Chloropropane	45.0	6	94
2-Chloropropane	35.6	2.8	97.2
1-Iodopropane	75.4	44	56
2-Iodopropane	71.5	27	73
1-Bromobutane	75.0	43	57
1-Chlorobutane	65.7	20.3	79.7
2-Butanone	74.8	40	60
1,1-Diethoxyethane	78.0	76	24
Dipropyl ether	74.5	44	56
Acetronitrile	72.5	44	56
Acrylonitrile	70.8	41	59
Nitromethane	76.1	29	71
Carbon disulfide	42.6	91	9
Diethyl sulfide	72.6	56	44
1-Propanol			
Hexane	65.7	4	96
Cyclohexane	74.7	18.5	81.5
Methylcyclohexane	87.0	34.7	65.3
Heptane	84.6	34.7	65.3
Octane	93.9	70	30
Benzene	77.1	16.9	83.1
Toluene	92.5	51.2	48.8
o-Xylene	zeotrope		
m-Xylene	97.1	94	6
p-Xylene	96.9	92.2	7.8
Styrene	97.0	8	92
Propyl formate	80.7	3	97
Butyl formate	95.5	64	36
Propyl acetate	94.7	51	49
Ethyl propionate	93.4	48	52
Methyl butyrate	94.4	49	51
Dipropyl ether	85.7	30	70
1,1-Diethoxyethane	92.4	37	63
1,4-Dioxane	95.3	55	45
Chloroform	zeotrope		
Carbon tetrachloride	73.4	92.1	7.9
Trichloronitromethane	94.1	58.5	41.5
Iodethane	70	93	7
1,2-Dichloroethane	80.7	19	81

TABLE 5.11   Binary Azeotropic (Constant-Boiling) Mixtures (*Continued*)

System	BP of azeotrope, °C	Composition, w %	
		Alcohol	Other component
1-Propanol (*continued*)			
Tetrachloroethylene	94.0	52	48
1-Bromopropane	69.7	9	91
1-Chlorobutane	74.8	18	82
Chlorobenzene	96.5	80	20
Flurobenzene	80.2	18	82
Nitromethane	89.1	48.4	51.6
1-Nitropropane	97.0	8.8	91.2
Carbon disulfide	45.7	94.5	5.5
2-Propanol			
Pentane	35.5	6	94
Hexane	62.7	23	77
Cyclohexane	69.4	32	68
Heptane	76.4	50.5	49.5
Octane	81.6	84	16
Benzene	71.7	33.7	66.3
Fluorobenzene	74.5	30	70
Toluene	80.6	69	31
Chloroform	60.8	4.2	95.8
Trichloronitromethane	81.9	35	65
Carbon tetrachloride	69.0	18	82
1,2-Dichloroethane	74.7	43.5	56.5
Iodoethane	67.1	15	85
3-Bromo-1-propene	66.5	20	80
1-Chloropropane	46.4	2.8	97.2
1-Bromopropane	66.8	20.5	79.5
2-Bromopropane	57.8	12	88
1-Iodopropane	79.8	42	58
2-Iodopropane	76.0	32	68
1-Chlorobutane	70.8	23	77
Ethyl acetate	75.3	25	75
Isopropyl acetate	81.3	60	40
Methyl propionate	76.4	37	63
Acrylonitrile	71.7	56	44
Butylamine	74.7	60	40
2-Butanone	77.5	32	68
1,1-Diethoxyethane	81.3	63	37
Ethyl propyl ether	62.0	10	90
Diisopropyl ether	66.2	14.1	85.9

TABLE 5.11 Binary Azeotropic (Constant-Boiling) Mixtures (*Continued*)

System	BP of azeotrope, °C	Composition, w % Alcohol	Other component
1-Butanol			
Cyclohexane	79.8	9.5	90.5
Cyclohexene	82.0	5	95
Hexane	68.2	3.2	96.8
Methylcyclohexane	95.3	20	80
Heptane	93.9	18	82
Octane	108.5	45.2	54.8
Nonane	115.9	71.5	28.5
Toluene	105.5	27.8	72.2
o-Xylene	116.8	75	25
m-Xylene	116.5	71.5	28.5
p-Xylene	115.7	68	32
Ethylbenzene	115.9	65.1	34.9
Butyl formate	105.8	23.6	76.4
Isopentyl formate	115.9	69	31
Butyl acetate	117.2	47	53
Isobutyl acetate	114.5	50	50
Ethyl butyrate	115.7	64	36
Ethyl isobutyrate	109.2	17	83
Methyl isopentanoate	113.5	40	60
Ethyl borate	113.0	52	48
Ethyl carbonate	116.5	63	37
Isobutyl nitrate	112.8	45	55
Dibutyl ether	117.8	82.5	17.5
Diisobutyl ether	113.5	48	52
1,1-Diethoxyethane	101.0	13	87
Carbon tetrachloride	76.6	97.6	2.4
Tetrachloroethylene	110.0	68	32
2-Bromo-2-methylpropane	90.2	7	93
2-Iodo-2-methylpropane	110.5	30	70
Chlorobenzene	115.3	56	44
Paraldehyde	115.8	52	48
Hexaldehyde	116.8	77.1	22.9
Ethylenediamine	124.7	35.7	64.3
Pyridine	118.6	69	31
1-Nitropropane	115.3	32.2	67.8
Butyronitrile	113.0	50	50
Diisopropyl sulfide	112.0	45	55
2-Methyl-2-propanol			
Cyclohexene	80.5	14.2	85.8
Cyclohexane	78.3	14	86

TABLE 5.11 Binary Azeotropic (Constant-Boiling) Mixtures (*Continued*)

System	BP of azeotrope, °C	Composition, w % Alcohol	Composition, w % Other component
2-Methyl-2-propanol (*continued*)			
Methylcyclopentane	71.0	5	95
Hexane	68.3	2.5	97.5
Methylcyclohexane	92.6	32	68
Heptane	90.8	27	73
2,5-Dimethylhexane	98.7	42	58
1,3-Dimethylcyclohexane	102.2	56	44
2,2,4-Trimethylpentane	92.0	27	73
Benzene	79.3	7.4	92.6
Chlorobenzene	107.1	63	37
Fluorobenzene	84.0	9	91
Toluene	101.2	45	55
Ethylbenzene	107.2	80	20
p-Xylene	107.1	88.6	11.4
Butyl formate	103.0	40	60
Isobutyl formate	97.4	12	88
Propyl acetate	101.0	17	83
Isobutyl acetate	107.6	92	8
Methyl butyrate	101.3	25	75
Ethyl isobutyrate	105.5	52	48
Methyl chloroacetate	107.6	12	88
Dipropyl ether	89.5	10	90
Isobutyl vinyl ether	82.7	6.2	93.8
1,1-Diethoxyethane	98.2	20	80
2-Pentanone	101.8	19	81
3-Pentanone	101.7	20	80
1,2-Dichloroethane	83.5	6.5	93.5
1-Bromobutane	95.0	21	79
1-Chlorobutane	77.7	4	96
2-Bromo-2-methylpropane	88.8	12	88
2-Iodo-2-methylpropane	104.0	36	64
1-Nitropropane	105.3	15.2	84.8
Isobutyl nitrate	105.6	36	64
Diisopropyl sulfide	105.8	73	27
3-Methyl-1-butanol			
Heptane	97.7	7	93
Octane	117.0	30	70
Toluene	109.7	10	90
Ethylbenzene	125.7	49	51
Isopropylbenzene	131.6	94	6
Camphene	130.9	24	76

TABLE 5.11 Binary Azeotropic (Constant-Boiling) Mixtures (*Continued*)

System	BP of azeotrope, °C	Composition, w %	
		Alcohol	Other component
3-Methyl-1-butanol (*continued*)			
Bromobenzene	131.7	85	15
o-Fluorotoluene	112.1	14.0	86.0
Butyl acetate	125.9	16.5	83.5
Paraldehyde	123.5	22.0	78.0
Dibutyl ether	129.8	65	35
Cyclohexanol			
o-Xylene	143.0	14	86
m-Xylene	138.9	5	95
Propylbenzene	153.8	40	60
Indene	160.0	75	25
Camphene	151.9	41	59
Cineole	160.6	92	8
Allyl alcohol			
Methylcyclohexane	85.0	42	58
Hexane	65.5	4.5	95.5
Cyclohexane	74.0	58	42
2,5-Dimethylhexane	89.3	50	50
Octane	93.4	68	32
Benzene	76.75	17.36	82.64
Toluene	92.4	50	50
Propyl acetate	94.2	53	47
Methyl butyrate	93.8	55	45
1,2-Dichloroethane	79.9	18	82
3-Iodo-1-propene	89.4	28	72
Chlorobenzene	96.2	85	15
Diethyl sulfide	85.1	45	55
Phenol			
2,7-Dimethyloctane	159.5	6	94
Decane	168.0	35	65
Tridecane	180.6	83.1	16.9
Butylbenzene	175.0	46	54
1,2,4-Trimethylbenzene	166.0	25	75
1,3,5-Trimethylbenzene	163.5	21	79
Indene	177.8	47	53
Camphene	156.1	22	78
Benzaldehyde	175.6	51.0	49.0

TABLE 5.11   Binary Azeotropic (Constant-Boiling) Mixtures (*Continued*)

System	BP of azeotrope, °C	Composition, w % Alcohol	Composition, w % Other component
Phenol (*continued*)			
1-Octanol	195.4	13	87
2-Octanol	184.5	50	50
Dipentyl ether	180.2	78	22
Diisopentyl ether	172.2	15	85
2-Methylpyridine	185.5	75.4	24.6
3-Methylpyridine	188.9	71.2	29.8
4-Methylpyridine	190.0	67.5	32.5
2,4-Dimethylpyridine	193.4	57.0	43.0
2,6-Dimethylpyridine	185.5	72.5	27.5
2,4,6-Trimethylpyridine	195.2	52.3	47.7
Aniline	185.8	41.9	58.1
Ethylene diacetate	195.5	39.2	60.8
Iodobenzene	177.7	53	47
Benzyl alcohol			
Naphthalene	204.1	60	40
D-Limonene	176.4	11	89
1,3,5-Triethylbenzene	203.2	57	43
o-Cresol	zeotrope		
m-Cresol	207.1	61	39
p-Cresol	206.8	62	38
N-Methylaniline	195.8	30	70
N,N-Dimethylaniline	193.9	6.5	93.5
N-Ethylaniline	202.8	50	50
N,N-Diethylaniline	204.2	72	28
Iodobenzene	187.8	12	88
Nitrobenzene	204.0	58	42
o-Bromotoluene	181.3	7	93
Borneol	205.1	85.8	14.2
2-Ethoxyethanol			
Methylcyclohexane	98.6	15	85
Heptane	96.5	14	86
Octane	116.0	38	62
Toluene	110.2	10.8	89.2
Ethylbenzene	127.8	48	52
p-Xylene	128.6	50	50
Styrene	130.0	55	45
Propylbenzene	134.6	80	20
Isopropylbenzene	133.2	67	33
Camphene	131.0	65	35
Propyl butyrate	133.5	72	28

TABLE 5.11 Binary Azeotropic (Constant-Boiling) Mixtures (*Continued*)

System	BP of azeotrope, °C	Composition, w %	
		Alcohol	Other component
2-Butoxyethanol			
Dipentene	164.0	53	47
1,3,5-Trimethylbenzene	162.0	32	68
Butylbenzene	169.6	73.4	26.6
Camphene	154.5	30	70
o-Cresol	191.6	15	85
Phenetole	167.1	52	48
Cineole	168.9	58.5	41.5
Benzaldehyde	171.0	91	9
Diisobutyl sulfide	163.8	42	58
1,2-Ethanediol			
Heptane	97.9	3	97
Decane	161.0	23	77
Tridecane	188.0	55	45
Toluene	110.1	2.3	97.7
Styrene	139.5	16.5	83.5
Stilbene	196.8	87	13
m-Xylene	135.1	6.55	93.45
p-Xylene	134.5	6.4	93.6
1,3,5-Trimethylbenzene	156	13	87
Propylbenzene	152	19	81
Isopropylbenzene	147.0	18	82
Naphthalene	183.9	51	49
1-Methylnaphthalene	190.3	60.0	40.0
2-Methylnaphthalene	189.1	57.2	42.8
Anthracene	197	98.3	1.7
Indene	168.4	26	74
Acenaphthene	194.65	74.2	25.8
Fluorene	196.0	82	18
Camphene	152.5	20	80
Camphor	186.2	40	60
Biphenyl	192.3	66.5	33.5
Diphenylmethane	193.3	68.5	31.5
Benzyl alcohol	193.1	56	44
2-Phenylethanol	194.4	69	31
o-Cresol	189.6	27	73
m-Cresol	195.2	60	40
3,4-Dimethylphenol	197.2	89	11
Menthol	188.6	51.5	48.5
Ethyl benzoate	186.1	46.5	53.5
o-Bromotoluene	166.8	25	75
Dibutyl ether	139.5	6.4	93.6

TABLE 5.11 Binary Azeotropic (Constant-Boiling) Mixtures (*Continued*)

System	BP of azeotrope, °C	Composition, w %	
		Alcohol	Other component
1,2-Ethanediol (*continued*)			
Methoxybenzene	150.5	10.5	89.5
Diphenyl ether	193.1	60	40
Benzyl phenyl ether	195.5	87	13
Acetophenone	185.7	52	48
2,4-Dimethylaniline	188.6	47	53
N,N-Dimethylaniline	175.9	33.5	66.5
m-Toluidine	188.6	42	58
2,4,6-Trimethylpyridine	170.5	9.7	90.3
Quinoline	196.4	79.5	20.5
Tetrachloroethylene	119.1	94	6
1,2-Dibromoethane	129.8	4	96
Chlorobenzene	130.1	94.4	5.6
α-Chlorotoluene	167.0	30	70
Nitrobenzene	185.9	59	41
o-Nitrotoluene	188.5	48.5	51.5
1,2-Ethanediol monoacetate			
Indene	180.0	20	80
1-Octanol	189.5	71	29
Phenol	197.5	65	35
o-Cresol	199.5	51	49
m-Cresol	206.5	31	69
p-Cresol	206.0	33	67
Dipentyl ether	180.8	42	58
Diisopentyl ether	170.2	28	72
m-Bromotoluene	182.0	32	68

D. Binary azeotropes containing ketones

System	BP of azeotrope, °C	Composition, w %	
		Ketone	Other component
Acetone			
Cyclopentane	41.0	36	64
Pentane	32.5	20	80
Cyclohexane	53.0	67.5	32.5
Hexane	49.8	59	41

TABLE 5.11 Binary Azeotropic (Constant-Boiling) Mixtures (*Continued*)

System	BP of azeotrope, °C	Composition, w %	
		Ketone	Other component
Acetone (*continued*)			
Heptane	55.9	89.5	10.5
Diethylamine	51.4	38.2	61.8
Methyl acetate	55.8	48.3	51.7
Diisopropyl ether	54.2	61	39
Chloroform	64.4	78.1	21.9
Carbon tetrachloride	56.1	11.5	88.5
Carbon disulfide	39.3	67	33
Ethylene sulfide	51.5	57	43
2-Butanone			
Cyclohexane	71.8	40	60
Hexane	64.2	28.6	71.4
Heptane	77.0	70	30
2,5-Dimethylhexane	79.0	95	5
Benzene	78.33	44	56
2-Methyl-2-propanol	78.7	69	31
Butylamine	74.0	35	65
Ethyl acetate	77.1	11.8	88.2
Methyl propionate	79.0	60	40
Butyl nitrite	76.7	30	70
1-Chlorobutane	77.0	38	62
Fluorobenzene	79.3	75	25

E. Miscellaneous binary azeotropes

System	BP of azeotrope, °C	Composition, w %	
		Solvent	Other component
Solvent: acetamide			
Dipentene	169.2	18	82
Biphenyl	213.0	50.5	49.5
Diphenylmethane	215.2	56.5	43.5
1,2-Diphenylethane	218.2	68	32
o-Xylene	142.6	11	89
m-Xylene	138.4	10	90
p-Xylene	137.8	8	92
Styrene	144	12	88
4-Isopropyl-1-methylbenzene	170.5	19	81

TABLE 5.11 Binary Azeotropic (Constant-Boiling) Mixtures (*Continued*)

System	BP of azeotrope, °C	Composition, w % Solvent	Composition, w % Other component

Solvent: acetamide (*continued*)

System	BP of azeotrope, °C	Solvent	Other component
Naphthalene	199.6	27	73
1-Methylnaphthalene	209.8	43.8	56.2
2-Methylnaphthalene	208.3	40	60
Indene	177.2	17.5	82.5
Acenaphthene	217.1	64.2	35.8
Camphene	155.5	12	88
Camphor	199.8	23	77
Benzaldehyde	178.6	6.5	93.5
3,4-Dimethylphenol	221.1	96	4
2-Methoxy-4-(2-propenyl)phenol	220.8	88	12
N-Methylaniline	193.8	14	86
N-Ethylaniline	199.0	18	82
N,N-Diethylaniline	198.1	24	76
Diphenyl ether	214.6	52	48
Safrole	208.8	32	68
Tetrachloroethylene	120.5	97.4	2.6

Solvent: aniline

System	BP of azeotrope, °C	Solvent	Other component
Nonane	149.2	13.5	86.5
Decane	167.3	36	64
Undecane	175.3	57.5	42.5
Dodecane	180.4	71.5	28.5
Tridecane	182.9	86.2	13.8
Tetradecane	183.9	95.2	4.8
Butylbenzene	177.8	46	54
1,2,4-Trimethylbenzene	168.6	13.5	86.5
1,3,5-Trimethylbenzene	164.3	12.0	88.0
Indene	179.8	41.5	58.5
1-Octanol	183.9	83	17
o-Cresol	191.3	8	92
Dipentyl ether	177.5	55	45
Diisopentyl ether	169.3	28	72
Hexachloroethane	176.8	66	34

Solvent: pyridine

System	BP of azeotrope, °C	Solvent	Other component
Heptane	95.6	25.3	74.7
Octane	109.5	56.1	43.9
Nonane	115.1	89.9	10.1
Toluene	110.1	22.2	77.8
Phenol	183.1	13.1	86.9
Piperidine	106.1	8	92

TABLE 5.11 Binary Azeotropic (Constant-Boiling) Mixtures (*Continued*)

System	BP of azeotrope, °C	Composition, w %	
		Solvent	Other component
Solvent: thiophene			
Methylcyclopentane	71.5	14	86
Cyclohexane	77.9	41.2	58.8
Hexane	68.5	11.2	88.8
Heptane	83.1	83.2	16.8
2,3-Dimethylpentane	80.9	64	36
2,4-Dimethylpentane	76.6	42.7	57.3
Solvent: benzene			
Methylcyclopentane	71.7	16	84
Cyclohexene	78.9	64.7	35.3
Cyclohexane	77.6	51.9	48.1
Hexane	68.5	4.7	95.3
Heptane	80.1	99.3	0.7
2,2-Dimethylpentane	75.9	46.3	53.7
2,3-Dimethylpentane	79.4	78.8	21.2
2,4-Dimethylpentane	75.2	48.3	51.7
2,2,4-Trimethylpentane	80.1	97.7	2.3
Solvent: bis(2-hydroxyethyl) ether			
Biphenyl	232.7	48	52
Diphenylmethane	236.0	52	48
1,3,5-Trimethylbenzene	210.0	22	78
Naphthalene	212.6	22	78
1-Methylnaphthalene	277.0	45	55
2-Methylnaphthalene	225.5	39	61
Acenaphthene	239.6	62	38
Fluorene	243.0	80	20
Benzyl acetate	214.9	7	93
Bornyl acetate	223.0	18	82
Ethyl fumarate	217.1	10	90
Dimethyl *o*-phthalate	245.4	96.3	3.7
Methyl salicylate	220.6	15	85
2-Hydroxy-1-isopropyl-4-methylbenzene	232.3	13	87
1,2-Dihydroxybenzene	259.5	46	54
Safrole	225.5	33	67
Isosafrole	233.5	46	54
Benzyl phenyl ether	241.5	80	20
Nitrobenzene	210.0	10	90
m-Nitrotoluene	224.2	25	75
o-Nitrophenol	216.0	10.5	89.5
Quinoline	233.6	29	71
p-Dibromobenzene	212.9	13	87

TABLE 5.12 Ternary Azeotropic Mixtures

A. Ternary azeotropes containing water and alcohols

System	BP of azeotrope, °C	Composition, wt %		
		Water	Alcohol	Other component
Methanol				
Chloroform	52.3	1.3	8.2	90.5
2-Methyl-1,3-butadiene	30.2	0.6	5.4	94.0
Methyl chloroacetate	67.9	6.3	81.2	13.5
Ethanol				
Acetonitrile	72.9	1	55	44
Acrylonitrile	69.5	8.7	20.3	71.0
Benzene	64.9	7.4	18.5	74.1
Butylamine	81.8	7.5	42.5	50.0
Butyl methyl ether	62	6.3	8.6	85.1
Carbon disulfide	41.3	1.6	5.0	93.4
Carbon tetrachloride	62	4.5	10.0	85.5
Chloroform	55.3	2.3	3.5	94.2
Crotonaldehyde	78.0	4.8	87.9	7.3
Cyclohexane	62.6	4.8	19.7	75.5
1,2-Dichloroethane	66.7	5	17	78
1,1-Diethoxyethane	77.8	11.4	27.6	61.0
Diethoxymethane	73.2	12.1	18.4	69.5
Ethyl acetate	70.2	9.0	8.4	82.6
Heptane	68.8	6.1	33.0	60.9
Hexane	56.0	3	12	85
Toluene	74.4	12	37	51
Trichloroethylene	67.0	5.5	16.1	78.4
Triethylamine	74.7	9	13	78
1-Propanol				
Benzene	67	7.6	10.1	82.3
Carbon tetrachloride	65.4	5	11	84
Cyclohexane	66.6	8.5	10.0	81.5
1,1-Dipropoxyethane	87.6	27.4	51.6	21.0
Dipropoxymethane	86.4	8.0	44.8	47.2
Dipropyl ether	74.8	11.7	20.2	68.1
3-Pentanone	81.2	20	20	60
Propyl acetate	82.5	17.0	10.0	73.0
Propyl formate	70.8	13	5	82
Tetrachloroethylene	81.2	12.5	20.7	66.8
2-Propanol				
Benzene	66.5	7.5	18.7	73.8
Butylamine	83	12.5	40.5	47.0

TABLE 5.12 Ternary Azeotropic Mixtures (*Continued*)

System	BP of azetrope, °C	Composition, wt %		
		Water	Alcohol	Other component
2-Propanol (*continued*)				
Cyclohexane	64.3	7.5	18.5	74.0
Toluene	76.3	13.1	38.2	48.7
Trichloroethylene	69.4	7	20	73
1-Butanol				
Butyl acetate	89.4	37.3	27.4	35.3
Butyl formate	83.6	21.3	10.0	68.7
Dibutyl ether	90.6	29.9	34.6	35.5
Heptane	78.1	41.4	7.6	51.0
Hexane	61.5	19.2	2.9	77.9
Nonane	90.0	69.9	18.3	11.8
Octane	86.1	60.0	14.6	25.4
2-Butanol				
Carbon tetrachloride	65	4.05	4.95	91.00
Cyclohexane	69.7	8.9	10.8	80.3
Isooctane	76.3	9	19	72
2-Methyl-1-propanol				
Isobutyl acetate	86.8	30.4	23.1	46.5
Isobutyl formate	80.2	17.3	6.7	76.0
Toluene	81.3	17.9	16.4	65.7
2-Methyl-2-propanol				
Benzene	67.3	8.1	21.4	70.5
Carbon tetrachloride	64.7	3.1	11.9	85.0
Cyclohexane	65.0	8	21	71
3-Methyl-1-butanol				
Isopentyl acetate	93.6	44.8	31.2	24.0
Isopentyl formate	89.8	32.4	19.6	48.0
Allyl alcohol				
Benzene	68.2	8.6	9.2	82.2
Carbon tetrachloride	65.2	5	11	84
Cyclohexane	66.2	8	11	81
Hexane	59.7	8.5	5.1	86.4

TABLE 5.12 Ternary Azeotropic Mixtures (*Continued*)

B. Other ternary azeotropes

System	BP of azeotrope, °C	Composition, wt %	System	BP of azeotrope, °C	Composition, wt %
Water	32.5	0.4	Water	71.4	7.9
Acetone		7.6	Nitromethane		29.7
2-Methyl-1,3-butadiene		92.0	Heptane		62.4
Water	66	8.2	Water	80.7	17.4
Acetonitrile		23.3	Nitromethane		58.3
Benzene		68.5	Nonane		24.3
Water	67	6.4	Water	77.4	12.4
Acetonitrile		20.5	Nitromethane		44.3
Trichloroethylene		73.1	Octane		43.3
Water	68.6	3.5	Water	33.1	2.1
Acetonitrile		9.6	Nitromethane		6.5
Triethylamine		86.9	Pentane		91.4
Water	63.6	5	Water	82.8	20.6
2-Butanone		35	Nitromethane		73.3
Cyclohexane		60	Undecane		6.1
Water	55.0	4	Water	93.5	40.5
Butyraldehyde		21	Pyridine		54.5
Hexane		75	Dodecane		5.0
Water	107.6	21.3	Water	93.1	38.5
Formic acid		76.3	Pyridine		51.0
Isopentanoic acid		2.4	Undecane		10.5
Water	107.0	15.5	Water	92.3	35.5
Formic acid		66.8	Pyridine		45.5
Isobutyric acid		17.7	Decane		19.0

Component 1	Component 2	Component 3	B.P. (°C)	% 1	% 2	% 3
Water	Formic acid	Butyric acid	107.6	19.5	75.9	4.6
Water	Formic acid	Propionic acid	107.2	18.6	71.9	9.5
Water	Hydrogen bromide	Chlorobenzene	105	11.0	10.4	78.6
Water	Hydrogen chloride	Chlorobenzene	96.9	20.2	5.3	74.5
Water	Hydrogen chloride	Phenol	107.3	64.8	15.8	19.4
Water	Hydrogen fluoride	Fluorosilic acid	116.1	54	10	36
Water	Nitroethane	Heptane	75.1	11.5	75.1	64.0
Water	Nitroethane	Hexane	59.5	8.4	9.3	82.3
Water	Nitromethane	Decane	82.4	19.1	68.1	12.8
Water	Nitromethane	Dodecane	83.1	21.5	75.3	3.2

Component 1	Component 2	Component 3	B.P. (°C)	% 1	% 2	% 3
Water	Pyridine	Nonane	90.5	30.5	37.0	32.5
Water	Pyridine	Octane	86.7	22.4	25.5	52.0
Water	Pyridine	Heptane	78.6	14.0	15.5	70.5
Acetic acid	Pyridine	Acetic anhydride	134.4	23	55	22
Acetic acid	Pyridine	Decane	134.1	31.4	38.2	30.4
Acetic acid	Pyridine	Ethylbenzene	129.1	13.5	25.2	61.3
Acetic acid	Pyridine	Heptane	98.5	3.4	10.6	86.0
Acetic acid	Pyridine	Nonane	128.0	20.7	29.4	49.9
Acetic acid	Pyridine	Octane	115.7	10.4	20.1	69.5
Acetic acid	Pyridine	o-Xylene	132.2	17.7	30.5	51.8

TABLE 5.12 Ternary Azeotropic Mixtures (*Continued*)

System	BP of azeotrope, °C	Composition, wt %	System	BP of azeotrope, °C	Composition, wt %
Acetic acid	129.2	10.2	Methanol	47.4	14.6
Pyridine		22.5	Methyl acetate		36.8
p-Xylene		67.3	Hexane		48.6
Acetic acid	163.0	75.0	Ethanol	63.2	10.4
2,6-Dimethylpyridine		13.8	Acetone		24.3
Undecane		11.2	Chloroform		65.3
Acetic acid	147.0	12.6	Ethanol	70.1	8
2,6-Dimethylpyridine		74.3	Acetonitrile		34
Decane		13.1	Triethylamine		58
Acetic acid	141.3	19.9	Ethanol	64.7	29.6
2-Methylpyridine		46.8	Benzene		12.8
Decane		33.3	Cyclohexane		57.6
Acetic acid	135.0	12.8	Ethanol	57.3	9.5
2-Methylpyridine		38.4	Chloroform		56.1
Nonane		48.8	Hexane		34.4
Acetic acid	121.3	3.6	1-Propanol	73.8	15.5
2-Methylpyridine		24.8	Benzene		30.4
Octane		71.6	Cyclohexane		54.2
Acetic acid	77.2	7.6	2-Propanol	69.1	31.1
Benzene		34.4	Benzene		15.0
Cyclohexane		58.0	Cyclohexane		53.9
Acetic acid	132	15	1-Butanol	77.4	4
2-Methyl-1-butanol		54	Benzene		48
Isopentyl acetate		31	Cyclohexane		48
Propionic acid	149.3	29.5	1-Butanol	108.7	11.9
2-Methylpyridine		32.0	Pyridine		20.7
Decane		38.5	Toluene		76.4

Component	Temp.	Comp.
Propionic acid	140.1	16.5
2-Methylpyridine		21.5
Nonane		42.0
Propionic acid	123.7	4.5
2-Methylpyridine		10.5
Octane		85.0
Propionic acid	153.4	43.0
2-Methylpyridine		40.0
Undecane		17.0
Propionic acid	147.1	55.5
Pyridine		26.4
Undecane		18.1
Methanol	57.5	23
Acetone		30
Chloroform		47
Methanol	47	14.6
Acetone		30.8
Hexane		59.6
Methanol	53.7	17.4
Acetone		5.8
Methyl acetate		76.8
Methanol	50.8	17.8
Methyl acetate		48.6
Cyclohexane		33.6
1,2-Ethanediol	185.0	8.7
Phenol		74.6
2,6-Dimethylpyridine		16.7
1,2-Ethanediol	185.1	5.9
Phenol		79.1
2-Methylpyridine		15.0
1,2-Ethanediol	186.4	15.9
Phenol		67.7
3-Methylpyridine		16.4
1,2-Ethanediol	188.6	29.5
Phenol		54.8
2,4,6-Trimethylpyridine		15.7
Acetone	60.8	3.6
Chloroform		68.8
Hexane		27.6
Acetone	49.7	51.1
Methyl acetate		5.6
Hexane		43.3
Chloroform	62.0	79.7
Ethyl formate		5.3
2-Bromopropane		15.7
1,4-Dioxane	101.8	44.3
2-Methyl-1-propanol		26.7
Toluene		29.0

5.4 FREEZING MIXTURES

TABLE 5.13 Compositions of Aqueous Antifreeze Solutions

FREEZING POINT OF ETHYL ALCOHOL-WATER MIXTURES*

Specific Gravity 20°/4°C. (68°F.)	% alcohol by weight	% alcohol by volume	Freezing Point °C.	°F.
0.99363	2.5	3.13	−1.0	30.2
0.98971	4.8	6.00	−2.0	28.4
0.98658	6.8	8.47	−3.0	26.6
0.98006	11.3	14.0	−5.0	23.0
0.97670	13.8	17.0	−6.1	21.0
0.97336	16.4	20.2	−7.5	18.5
0.97194	17.5	21.5	−8.7	16.3
0.97024	18.8	23.1	−9.4	15.1
0.96823	20.3	24.8	−10.6	12.9
0.96578	22.1	27.0	−12.2	10.0
0.96283	24.2	29.5	−14.0	6.8
0.95914	26.7	32.4	−16.0	3.2
0.95400	29.9	36.1	−18.9	−2.0
0.94715	33.8	40.5	−23.6	−10.5
0.93720	39.0	46.3	−28.7	−19.7
0.92193	46.3	53.8	−33.9	−29.0
0.90008	56.1	63.6	−41.0	−41.8
0.86311	71.9	78.2	−51.3	−60.3

FREEZING POINT OF METHYL (WOOD) ALCOHOL-WATER MIXTURES*

Specific Gravity 15.6°C. (60°F.)	% alcohol by weight	% alcohol by volume	Freezing Point °C.	°F.
0.993	3.9	5	−2.2	28
0.986	8.1	10	−5.0	23
0.980	12.2	15	−8.3	17
0.974	16.4	20	−11.7	11
0.968	20.6	25	−15.6	4
0.963	24.9	30	−20.0	−4
0.956	29.2	35	−25.0	−13
0.949	33.6	40	−30.0	−22
0.942	38.0	45	−35.6	−32

FREEZING POINT OF PRESTONE—WATER MIXTURES†

% Prestone By Weight	By Volume	Specific Gravity 15°/15C. (59°F.)	Freezing Point °C.	°F.
10	9.2	1.013	−3.6	25.6
15	13.8	1.019	−5.6	22.0
20	18.3	1.026	−7.9	17.8
25	23.0	1.033	−10.7	12.8
30	28.0	1.040	−14.0	6.8
40	37.8	1.053	−22.3	−8.2
50	47.8	1.067	−33.8	−28.8
60	58.1	1.079	−49.3	−56.7

*Values are for pure alcohol. Since some commercial antifreezes contain small amounts of water, slightly higher volume concentrations than those given in the table may be required. Antifreezes also contain corrosion inhibitors and other additives to make them function properly as cooling liquids. These affect freezing point slightly and specific gravity to a greater degree. If a protection table is furnished by the manufacturer it should be used in preference to the values given above for the pure substance.

†Eveready Prestone (manufactured by the National Carbon Co.), marketed for antifreeze purposes, is 97% ethylene glycol containing fractional percentages of soluble and insoluble ingredients to prevent foaming, creepage and water corrosion in automobile cooling systems.

TABLE 5.13 Compositions of Aqueous Antifreeze Solutions (*Continued*)

FREEZING POINT OF ETHYL ALCOHOL–WATER MIXTURES

Specific Gravity 15.6°C. (60°F.)	% alcohol by volume	Freezing Point	
		°C.	°F.
0.990	5	−1.7	29
0.984	10	−3.3	26
0.978	15	−6.1	21
0.972	20	−8.3	17
0.964	25	−11.1	12
0.955	30	−14.4	6
0.945	35	−17.8	0
0.933	40	−18.3	−1
0.922	45	−18.9	−2
0.910	50	−20.0	−4
0.899	55	−21.7	−7
0.887	60	−23.3	−10
0.875	65	−24.4	−12
0.864	70	−26.7	−16
0.852	75	−32.2	−26
0.840	80	−41.7	−43

FREEZING POINT OF PROPYLENE GLYCOL-WATER MIXTURES*

Specific Gravity 15.6°C. (60°F.)	% glycol by volume	Freezing Point	
		°C.	°F.
1.004	5	−1.1	30
1.006	10	−2.2	28
1.012	15	−3.9	25
1.017	20	−6.7	20
1.020	25	−8.9	16
1.024	30	−12.8	9
1.028	35	−16.1	3
1.032	40	−20.6	−5
1.037	45	−26.7	−16
1.040	50	−33.3	−28

FREEZING POINT OF GLYCEROL (GLYCERINE)—WATER MIXTURES†

% Glycerol by Weight	Specific Gravity 15°/15°C. (59°F.)	Specific Gravity 20°/20°C. (68°F.)	Freezing Point	
			°C.	°F.
10	1.02415	1.02395	−1.6	29.1
20	1.04935	1.04880	−4.8	23.4
30	1.07560	1.07470	−9.5	14.9
40	1.10255	1.10135	−15.5	4.3
50	1.12985	1.12845	−22.0	−7.4
60	1.15770	1.15605	−33.6	−28.5
70	1.18540	1.18355	−37.8	−36.0
80	1.21290	1.21090	−19.2	−2.3
90	1.23950	1.23755	−1.6	29.1
100	1.26557	1.26362	17.0	62.6

* See footnote on preceding page.
† The values are those reported by Bosart and Snoddy (*Jour. Ind. Eng. Chem.*, **19**, 506 (1927)), and Lane (*Jour. Ind. Eng. Chem.*, **17**, 924 (1925)) but modified by adding 2°F to all temperatures below 0°F in accordance with the suggestion of the Procter and Gamble Co.

TABLE 5.13   Compositions of Aqueous Antifreeze Solutions (*Continued*)

FREEZING POINT OF MAGNESIUM CHLORIDE BRINES

% MgCl$_2$ by weight	Spec. Grav. 15.6°C. (60°F.)	Freezing Point °C.	°F.	% MgCl$_2$ by weight	Spec. Grav. 15.6°C. (60°F.)	Freezing Point °C.	°F.
5	1.043	−3.11	26.4	18	1.161	−22.1	−7.7
6	1.051	−3.89	25.0	19	1.170	−25.6	−12.2
7	1.060	−4.72	23.5	20	1.180	−27.4	−17.3
8	1.069	−5.67	21.8	21	1.190	−30.6	−23.0
9	1.078	−6.67	20.0	22	1.200	−32.8	−27.0
10	1.086	−7.83	17.9	23	1.210	−28.9	−20.0
11	1.096	−9.05	15.7	24	1.220	−25.6	−14.0
12	1.105	−10.5	13.1	25	1.230	−23.3	−10.0
13	1.114	−12.1	10.3	26	1.241	−21.1	−6.0
14	1.123	−13.7	7.3	27	1.251	−19.4	−3.0
15	1.132	−15.6	4.0	28	1.262	−18.3	−1.0
16	1.142	−17.6	0.4	29	1.273	−17.2	+1.0
17	1.151	−19.7	−3.5	30	1.283	−16.7	2.0

FREEZING POINT OF SODIUM CHLORIDE BRINES

Compiled in collaboration with C. D. Looker, Ph.D., International Salt Co., Inc.

% NaCl by weight	Spec. Grav. 15°C. (59°F.)	Freezing Point °C.	°F.	% NaCl by weight	Spec. Grav. 15°C. (59°F.)	Freezing Point °C.	°F.
0	1.000	0.00	32.0	15	1.112	−10.88	12.4
1	1.007	−0.58	31.0	16	1.119	−11.90	10.6
2	1.014	−1.13	30.0	17	1.127	−12.93	8.7
3	1.021	−1.72	28.9	18	1.135	−14.03	6.7
4	1.028	−2.35	27.8	19	1.143	−15.21	4.6
5	1.036	−2.97	26.7	20	1.152	−16.46	2.4
6	1.043	−3.63	25.5	21	1.159	−17.78	+0.0
7	1.051	−4.32	24.2	22	1.168	−19.19	−2.5
8	1.059	−5.03	22.9	23	1.176	−20.69	−5.2
9	1.067	−5.77	21.6	23.3 (*E*)	1.179	−21.13	−6.0
10	1.074	−6.54	20.2	24	1.184	−17.0*	+1.4*
11	1.082	−7.34	18.8	25	1.193	−10.4*	13.3*
12	1.089	−8.17	17.3	26	1.201	−2.3*	27.9*
13	1.097	−9.03	15.7	26.3	1.203	0.0*	32.0*
14	1.104	−9.94	14.1				

*Saturation temperatures of sodium chloride dihydrate; at these temperatures NaCl·2H$_2$O separates leaving the brine of the eutectic composition (*E*).

5.4.1 Propylene Glycol–Glycerol

Propylene glycol, a satisfactory antifreeze with the advantage of being nontoxic, can be combined with glycerol, also an efficient nontoxic antifreeze, to give a mixture that can be tested for freezing point with an ethylene glycol (Prestone) hydrometer. A mixture of 70% propylene glycol and 30% glycerol (% by weight of water-free materials), when diluted, can be tested on the standard instrument used for ethylene glycol solutions.

5.5 *DENSITY AND SPECIFIC GRAVITY*

TABLE 5.14   Density of Mercury and Water

The density of mercury and pure air-free water under a pressure of 101 325 Pa (1 atm) is given in units of grams per cubic centimeter ($g \cdot cm^{-3}$). For mercury, the values are based on the density at 20°C being 13.545 884 $g \cdot cm^{-3}$. Water attains its maximum density of 0.999 973 $g \cdot cm^{-3}$ at 3.98°C. For water, the temperature (t_m, °C) of maximum density at different pressures (p) in atmospheres is given by

$$t_m = 3.98 - 0.0225(p - 1)$$

Density of Water	Temp., °C	Density of Mercury	Density of Water	Temp., °C	Density of Mercury
	−20	13.644 59	0.987 12	52	13.467 68
	−18	13.639 62	0.986 18	54	13.462 82
	−16	13.634 66	0.985 21	56	13.457 96
	−14	13.629 70	0.984 22	58	13.453 09
	−12	13.624 75	0.983 20	60	13.448 23
	−10	13.619 79	0.982 16	62	13.443 37
	−8	13.614 85	0.981 09	64	13.438 52
	−6	13.609 90	0.980 01	66	13.433 67
	−4	13.604 96	0.978 90	68	13.428 82
	−2	13.600 02	0.977 77	70	13.423 97
0.999 84	0	13.595 08	0.976 61	72	13.419 13
0.999 94	2	13.590 15	0.975 44	74	13.414 28
0.999 97	4	13.585 22	0.974 24	76	13.409 43
0.999 94	6	13.580 29	0.973 03	78	13.404 60
0.999 85	8	13.575 36	0.971 79	80	13.399 77
0.999 70	10	13.570 44	0.970 53	82	13.394 92
0.999 50	12	13.565 52	0.969 26	84	13.390 09
0.999 24	14	13.560 60	0.967 96	86	13.385 26
0.998 94	16	13.555 70	0.966 65	88	13.380 42
0.998 60	18	13.550 79	0.965 31	90	13.375 60
0.998 20	20	13.545 88	0.963 96	92	13.370 77
0.997 77	22	13.540 97	0.962 59	94	13.365 94
0.997 30	24	13.536 06	0.961 20	96	13.361 12
0.996 78	26	13.531 17	0.959 79	98	13.356 30
0.996 23	28	13.526 26	0.958 36	100	13.351 48
0.995 65	30	13.521 37		120	13.303 4
0.995 03	32	13.516 47		140	13.255 4
0.994 37	34	13.511 58		160	13.207 6
0.993 69	36	13.506 70		180	13.159 8
0.992 97	38	13.501 82		200	13.112 0
0.992 22	40	13.496 93		220	13.064 5
0.991 44	42	13.492 07		240	13.016 9
0.990 63	44	13.487 18		260	12.969 2
0.989 79	46	13.482 29		280	12.921 5
0.988 93	48	13.477 42		300	12.873 7
0.988 04	50	13.472 56			

TABLE 5.15 Specific Gravity of Air at Various Temperatures

The table below gives the weight in grams $\times$ 10^4 of 1 mL of air at 760 mm of mercury pressure and at the temperature indicated. Density in grams per milliliter is the same as the specific gravity referred to water at 4°C as unity. To convert to density referred to air at 70°F as unity, divide the values below by 12.00.

t°C.	Sp.Gr. $\times$10^4	t°C.	Sp.Gr. $\times$10^4	t°C.	Sp.Gr. $\times$10^4	t°C.	Sp.Gr. $\times$10^4
−25	14.240	15	12.255	60	10.596	140	8.541
−24	14.182	16	12.213	62	10.532	142	8.500
−23	14.125	17	12.170	64	10.470	144	8.459
−22	14.069	18	12.129	66	10.408	146	8.419
−21	14.013	19	12.087	68	10.347	148	8.379
−20	13.957	20	12.046	70	10.286	150	8.339
−19	13.902	21	12.004	72	10.227	155	8.242
−18	13.847	22	11.964	74	10.168	160	8.147
−17	13.793	23	11.923	76	10.109	165	8.054
−16	13.739	24	11.883	78	10.052	170	7.963
−15	13.685	25	11.843	80	9.995	175	7.874
−14	13.632	26	11.803	82	9.938	180	7.787
−13	13.580	27	11.764	84	9.882	185	7.702
−12	13.527	28	11.725	86	9.828	190	7.619
−11	13.476	29	11.686	88	9.773	195	7.537
−10	13.424	30	11.647	90	9.719	200	7.457
−9	13.373	31	11.609	92	9.666	205	7.379
−8	13.322	32	11.570	94	9.613	210	7.303
−7	13.272	33	11.533	96	9.561	215	7.228
−6	13.222	34	11.495	98	9.509	220	7.155
−5	13.173	35	11.458	100	9.458	230	7.013
−4	13.124	36	11.420	102	9.408	240	6.881
−3	13.075	37	11.383	104	9.358	250	6.753
−2	13.026	38	11.347	106	9.308	260	6.624
−1	12.978	39	11.310	108	9.259	270	6.504
0	12.931	40	11.274	110	9.211	280	6.389
+1	12.883	41	11.238	112	9.163	290	6.277
2	12.836	42	11.202	114	9.116	300	6.166
3	12.790	43	11.167	116	9.069	310	6.062
4	12.743	44	11.132	118	9.022	320	5.942
5	12.697	45	11.097	120	8.976	330	5.847
6	12.652	46	11.062	122	8.931	340	5.755
7	12.606	47	11.027	124	8.886	350	5.664
8	12.561	48	10.993	126	8.841	360	5.578
9	12.517	49	10.958	128	8.797	370	5.493
10	12.472	50	10.924	130	8.753	380	5.407
11	12.428	52	10.857	132	8.710	400	5.248
12	12.385	54	10.791	134	8.667	420	5.101
13	12.341	56	10.725	136	8.625	440	4.952
14	12.298	58	10.660	138	8.583	460	4.812

5.5.1 Density of Moist Air

The density of moist air depends upon the temperature, the humidity, and the barometric pressure. It is expressed by the equation

$$d_t = D_t \times \frac{P - 0.3783e}{760}$$

where d_t is the density of the moist air at the temperature t; D_t is the density of dry air at the temperature t (see Table 5.15, Specific Gravity of Air at Various Temperatures); P is the height of the barometer after correction and reduction to standard conditions, and is expressed in millimeters of mercury (see Sec. 2.1.3, Barometry and Barometric Corrections); e is the vapor pressure of water at the temperature of the dew point and is expressed in millimeters of mercury (see Table 5.6, Vapor Pressure of Water).

Example. To find the density of moist air at a temperature of 20°C, with a dew point of 10°C, and a corrected barometric pressure of 750 mm.

Reference to Table 5.15 shows that D at 20°C is equal to 0.001 204 6 g/mL. Reference to Table 5.6 shows that at 10°C (the temperature of the dew point) e is equal to 9.209 mm. Therefore,

$$d = 0.001\ 204\ 6 \times \frac{750 - (0.3783 \times 9.209)}{760}$$

$$= 0.001\ 183\ 2 \text{ g/mL} = 1.1832 \text{ g/L}$$

5.5.2 Specific Gravity Corrections for the Buoyant Effect of Air

5.5.2.1 *Determinations Made with a Pyknometer*

$$D_{\text{vac}} = \frac{W_2}{W_1} d - 0.0012 \left(\frac{W_2 d}{W_1} - 1 \right)$$

$$S_{\text{vac}} = \frac{W_2}{W_1} - 0.0012 \left(\frac{W_2}{W_1} - 1 \right)$$

where D_{vac} = density of the liquid in grams per milliliter at t °C corrected for the buoyant effect of air

W_1 = weight in air of the water required to fill the pyknometer at t °C

W_2 = weight in air of the liquid required to fill the pyknometer at t °C

d = density of water in grams per milliliter at t °C

S_{vac} = specific gravity of the liquid at t °C referred to water at t °C corrected for the buoyant effect of air

When the weight of the water is determined at a temperature of t °C, and that of the liquid at a different temperature t', the equations above are modified as follows:

$$D_{\text{vac}} = \frac{W_2}{W_1} d - 0.0012 \left(\frac{W_2}{W_1} d - 1 \right) + 0.000\ 026\ (t' - t°) \left(\frac{W_2}{W_1} d \right)$$

$$S_{\text{vac}} = \frac{W_2}{W_1} - 0.0012 \left(\frac{W_2}{W_1} - 1 \right) + 0.000\ 026\ (t' - t°) \left(\frac{W_2}{W_1} d \right)$$

5.5.2.2 Determinations Made with a Plummet or Sinker. The equations above may also be used when the density is determined with plummet or sinker, but in this case

W_1 = weight of the plummet in air minus its weight in water

W_2 = weight of the plummet in air minus its weight in the liquid

TABLE 5.16 Viscosity, Dielectric Constant, Dipole Moment, and Surface Tension of Selected Organic Substances

The temperature in degrees Celsius at which the viscosity, dielectric constant, dipole moment, and surface tension of a substance were measured is shown in parentheses after the value. In some cases, the dipole moment was determined with the substance dissolved in a solvent, and the solvent used is also shown in parentheses after the temperature.

For the majority of compounds the dependence of the surface tension γ on the temperature can be given as

$$\gamma = a - bt$$

where a and b are constants and t is the temperature in degrees Celsius.

Alternative names for entries are listed in Table 1.15 at the bottom of each double page.

List of Abbreviations

B, benzene	g, gas
C, CCl$_4$	Hx, hexane
cHx, cyclohexane	lq, liquid
D, 1,4-dioxane	

Substance	Viscosity, mN · s · m^{-2}	Dielectric constant ε	Dipole moment, D	Surface tension, dyn · cm^{-1}	
				a	b
Acetaldehyde	0.280 (0), 0.256 (10), 0.22 (20)	21.8 (10), 21.1 (21)	2.71 (g)	23.90	0.136 0
Acetaldoxime	1.415 (20)	3 (23)	0.830 (20, lq), 0.90 (25, B)	30.1 (35)	
Acetamide	1.32 (105), 1.06 (120)	59.2 (83), 60.6 (94)	3.90 (25, B), 2.44 (30, B)	47.66	0.102 1
Acetanilide	2.22 (120), 1.90 (130)		3.65 (25, B)	46.21	0.091 2
Acetic acid	1.232 (20), 0.796 (50)	6.15 (20), 6.29 (40)	1.76 (g), 1.92 (20, B)	29.58	0.009 4

TABLE 5.16 Viscosity, Dielectric Constant, Dipole Moment, and Surface Tension of Selected Organic Substances (*Continued*)

Substance	Viscosity, mN · s · m^{-2}	Dielectric constant ε	Dipole moment, D	Surface tension dyn · cm^{-1}	
				a	b
Acetic anhydride	0.907 (20), 0.699 (40)	23.3 (0), 21.2 (20)	2.8 (g), 3.15 (20, B)	35.52	0.143 6
Acetone					
(lq)	0.391 (0), 0.318 (20)	20.7 (25), 17.6 (56)	2.77 (22, B)	26.26	0.112
(g)	0.009 33 (100), 0.012 8 (225)	1.015 9 (100)	2.87		
Acetonitrile	0.397 (10), 0.329 (30)	37.5 (20), 26.6 (82)	3.97 (g), 3.47 (20, B)	29.58	0.117 8
Acetophenone	2.015 (15), 1.511 (30)	17.39 (25), 8.64 (202)	2.96 (30, B)	41.92	0.115 4
Acetyl bromide		16.2 (20)	2.45 (20, B)		
Acetyl chloride					
(lq)		16.9 (2), 15.8 (22)	2.47 (20, B)	26.7 (15)	
(g)		1.0217 (20)	2.71		
Acetylene (g)	0.010 2 (30), 0.012 6 (101)	1.001 34 (0)	0	3.42	0.193 5 (lq)
Acrylic acid	1.3 (20), 1.16 (25)			28.1 (30)	
Acrylonitrile	0.35 (20), 0.34 (25)	33.0 (20)	3.91 (g), 3.54 (25, B)	29.58	0.117 8
Allyl acetate	0.207 (30)			28.73	0.118 6
Allylamine	0.375 (25)			27.49	0.128 7
Allyl isothiocyanate		17.2 (18)	1.3 (25, B)	36.76	0.107 4
2-Aminoethanol	30.85 (15), 19.35 (25)	37.72 (25)	3.2 (20, B)	51.11	0.111 7
Aniline	5.30 (15), 4.40 (20), 3.18 (30)	6.89 (20), 5.93 (76)	2.59 (25, D)	44.83	0.108 5
Benzaldehyde	1.321 (25)	19.7 (0), 17.8 (20)	1.53 (20, B)	40.72	0.109 0
Benzaldehyde oxime (mp 30) (mp 128)		3.8 (20)	2.77 (20, lq) 1.2 (25, B) 1.5 (25, B)		
Benzene	0.649 (20), 0.566 (30), 0.395 (60)	2.292 (15), 2.274 (25), 1.002 8 (g)	0	28.88 (20)	27.56 (30)
Benzamide			3.42 (25, B)	47.26	0.070 5
Benzenesulfonyl chloride			4.50 (20, B)	45.48	0.111 7

5.92

Benzenethiol	1.239 (20), 1.144 (25)	4.38 (25)	1.13 (25, lq), 1.19 (20, B)	41.41	0.120 2
Benzonitrile	1.447 (15), 1.111 (30)	26.5 (20), 24.0 (40)	4.40 (g), 3.9 (20, B)	41.69	0.115 9
Benzophenone	4.79 (55), 1.38 (120)	14.60 (18), 11.4 (50)	3.09 (50, lq), 2.98 (25, B)	46.31	0.112 8
Benzoyl bromide	1.956 (20), 1.798 (25)	21.33 (20), 20.74 (25)	3.40 (20, B)	45.85	0.139 7
Benzoyl chloride		29 (0), 23 (20)	3.16 (25, B)	41.34	0.108 4
Benzyl acetate	1.399 (45)	5.1 (21)	1.80 (25, B)		
Benzyl alcohol	5.58 (20), 4.65 (30), 3.01 (45)	13.0 (20), 9.5 (70)	1.67 (25, B)	38.25	0.138 1
Benzylamine	1.59 (25)	5.5 (1), 4.6 (21)	1.15 (20, lq), 1.38 (25, B)	42.33	0.121 3
Benzyl benzoate	8.51 (25)	4.9 (20)	2.06 (30, B)	48.07	0.106 5
Benzyl butyl o-phthalate	65 (20)				
Benzyl chloride	1.400 (20), 1.290 (25)	7.0 (13)	1.83 (20, B)	39.92	0.122 7
Benzylethylamine		4.3 (20)			
Benzyl ethyl ether		3.9 (20)		32.83 (20)	29.97 (40)
Biphenyl		2.53 (75)	0	41.52	0.093 1
Bis(2-ethoxyethyl)ether			1.92 (25, B)	29.74	0.117 6
Bis(2-hydroxyethyl)ether	38.0 (20), 30.0 (25)	31.69 (20)	2.31 (20, B)	46.97	0.088 0
1,2-Bis(methoxyethoxy)-ethane	3.76 (20)				
Bis(2-methoxyethyl) ether	1.99 (20), 0.981 (25)		1.97 (25, B)	32.47	0.116 4
DL-Bornyl acetate		4.6 (21)	1.89 (22)		
3-Bromoaniline	6.81 (20), 3.70 (40)	13.0 (19)	2.67 (20, B)		
4-Bromoaniline	1.81 (80)	7.06 (30)	2.88 (25, B)		
Bromobenzene	1.196 (15), 0.985 (30)	5.40 (25)	1.70 (g), 1.50 (20, lq)	38.14	0.116 0
1-Bromobutane	0.633 (20), 0.597 (25)	7.88 (−10), 7.07 (20)	2.17 (g), 2.04 (20, lq)	28.71	0.112 6
DL-2-Bromobutane	1.434 (20)	8.64 (25)	2.22 (g), 2.14 (25, lq)	27.48	0.110 7
1-Bromo-2-chlorobenzene		6.80 (20)	2.15 (20, B)		
1-Bromo-3-chlorobenzene		4.58 (20)	1.52 (22, B)		
1-Bromo-4-chlorobenzene			0.1 (25, B)	40.03	0.100 2
Bromochloromethane	0.670 (20)	7.79	1.66 (25, B)	33.32 (20)	
Bromocyclohexane	2.0 (25)	11 (−65), 7.9 (25)	1.08 (25, lq), 2.3 (25, B)	36.13	0.111 7
1-Bromodecane		4.75 (1), 4.44 (25)	2.08 (20, lq), 1.90 (25, lq)	31.26	0.085 6

TABLE 5.16 Viscosity, Dielectric Constant, Dipole Moment, and Surface Tension of Selected Organic Substances (*Continued*)

Substance	Viscosity, mN · s · m^{-2}	Dielectric constant ε	Dipole moment, D	Surface tension dyn · cm^{-1}	
				a	b
Bromodichloromethane		4.07 (25)	1.31 (25, B)	35.11	0.129 4
1-Bromododecane		13.6 (−60), 9.39 (20)	2.01 (25, lq), 1.89 (25, B)	32.58	0.088 2
Bromoethane	0.397 (20), 0.348 (30)		2.03 (g), 2.04 (20, lq)	26.52	0.115 9
1-Bromo-2-ethoxyethane				31.98	0.112 9
1-Bromo-2-ethoxypentane		6.45 (25)	2.32 (25, B)		
2-Bromo-2-ethoxypentane		6.40 (25)	2.07 (25, B)		
3-Bromo-3-ethoxypentane		8.24 (25)	2.15 (25, B)		
Bromoethylene		4.78 (25)	1.42 (g)		
Bromoform	2.152 (15), 1.741 (30)	4.39 (20)	1.00 (g), 0.92 (25, lq)	48.14	0.130 8
1-Bromoheptane		5.33 (25), 4.48 (90)	2.17 (g), 2.02 (20, lq)	30.74	0.098 2
2-Bromoheptane		6.46 (22)	2.08 (20, B)		
3-Bromoheptane		6.93 (22)	2.06 (20, B)		
4-Bromoheptane		6.81 (22)	2.06 (20, B)		
1-Bromohexadecane		3.71 (25)	1.98 (20, lq), 1.96 (25, C)	33.37	0.086 1
1-Bromohexane		6.30 (1), 5.82 (25)	2.06 (20, lq)	29.81	0.096 7
Bromomethane		9.82 (0), 1.006 8 (100, g)	1.79 (g)	26.52	0.115 9
1-Bromo-3-methylbutane		8.04 (−56), 6.05 (20)	1.95 (20, B)	28.10	0.099 6
2-Bromo-3-methylbutyric acid		6.5 (20)			
1-Bromo-2-methylpropane	0.643 (20), 0.518 (40), 3.26 (90)	7.70 (0), 7.2 (25)	1.92 (25, lq), 1.99 (20, B)	26.96	0.105 9
1-Bromonaphthalene	5.99 (15), 3.20 (40)	5.83 (25), 5.12 (20)	1.29 (25, lq)	46.44	0.101 8
1-Bromononane		5.42 (−20), 4.74 (25)	1.95 (25, lq)	31.36	0.089 4
1-Bromooctane		6.35 (−50)	1.99 (20, lq), 1.88 (25, lq)	31.00	0.092 8
1-Bromopentadecane		3.9 (20)			

Compound						
1-Bromopentane			9.9 (−90), 6.32 (25)	2.21 (g), 2.09 (20, lq)	29.51	0.104 9
p-Bromophenol					48.88	0.107 0
1-Bromopropane	0.539 (15), 0.459 (30)		8.09 (25)	2.17 (g), 3.16 (20, lq)	28.30	0.121 8
2-Bromopropane	0.536 (15), 0.437 (30)		9.46 (25)	2.21 (g), 2.10 (25, lq)	26.21	0.118 3
1-Bromotetradecane			3.84 (25)	1.92 (20, lq), 1.83 (25, lq)	32.93	0.087 8
o-Bromotoluene	1.3 (25)		4.28 (58)	1.45 (20, B)	36.62	0.099 8
m-Bromotoluene			5.36 (58)	1.77 (20, B)		
p-Bromotoluene			5.49 (58)	1.95 (20, B)	36.40	0.099 7
Bromotrifluoromethane	0.15 (25)			0.65 (g)	4 (25)	
1-Bromoundecane			4.73 (−9)		31.94	0.086 1
Butane	0.007 39 (20, g), 0.008 39 (60, g)			0	14.87	0.120 6
1,3-Butanediol	130.3 (20), 89 (25)		28.8 (25)	3.93 (20, lq), 2.4 (15, D)	37.8 (25)	
1,4-Butanediol	65–70 (25)		33 (15), 30 (30)		36 (25)	0.097 7
2,3-Butanediol	121 (25)				37.33	0.114 2
Butanesulfonyl chloride				3.94 (25, D)	28.07	
1-Butanethiol	0.501 (20), 0.450 (30)		5.07 (25), 4.59 (50)	1.54 (25, lq or B)	27.18	0.089 8
1,2,4-Butanetriol	1227 (25)			1.66 (g; 20, B)	23.47 (20)	22.62 (30)
1-Butanol	2.948 (20), 1.782 (40)		17.8 (20), 8.2 (118)	1.66 (30, B)	26.77	0.112 2
DL-2-Butanol	3.907 (20), 0.527 (100)		16.6 (25)	3.2 (30, lq), 2.76 (25, B)	31.89	0.102 2
2-Butanone	0.428 (20), 0.349 (40)		18.5 (20), 15.3 (60)		15.19	0.132 3 (lq)
2-Butanone oxime			3.4 (20)	0.30	16.11	0.128 9
1-Butene (g)	0.007 6 (20), 0.010 0 (120)		1.003 2 (20)	0.33 (g, cis), 0 (g, trans)	31.40	0.108 5
2-Butene				4.53 (g)	28.18	0.081 6
3-Butenenitrile			28.1 (20)	2.08 (25, B)		
Butoxyethanol	3.15 (25), 1.51 (60)		9.30 (25)	2.05 (25, lq)	30.0 (25)	
Butoxyethyne			6.62 (25)		26.5 (25)	
2-(2-Butoxyethoxy)ethanol	4.76 (25)				27.55	0.106 8
1-Butoxy-2-propanol	2.55 (25)			1.86 (22, B)	23.33 (22)	21.24 (42)
Butyl acetate	0.734 (20), 0.688 (25)		6.85 (−73), 5.01 (20)	1.91 (25, B)	24.69	0.110 2
DL-sec-Butyl acetate						
tert-Butyl acetate						

TABLE 5.16 Viscosity, Dielectric Constant, Dipole Moment, and Surface Tension of Selected Organic Substances (*Continued*)

Substance	Viscosity, mN · s · m⁻²	Dielectric constant ε	Dipole moment, D	Surface tension dyn · cm⁻¹	
				a	*b*
Butylamine	0.681 (20)	4.88 (20)	1.00 (g), 1.22 (20, lq)	26.24	0.112 2
sec-Butylamine		4.4 (21)	1.28 (25, B)	23.75	0.105 7
tert-Butylamine			1.29 (25, B)	19.44	0.102 8
Butylbenzene	1.035 (20), 0.960 (25)	2.36 (20)	0.36 (20, lq)	31.28	0.102 5
sec-Butylbenzene		2.36 (20)	0.37 (20, lq)	30.48	0.097 9
tert-Butylbenzene		2.37 (20)	0.36 (20, lq)	30.10	0.098 5
Butyl butyrate	0.84 (25)			27.65	0.096 5
Butyl decyl o-phthalate	55 (20)				
N-Butyldiethanolamine	55 (25)				
4-tert-Butyl-2,5-dimethylphenol	8.30 (80)				
4-tert-Butyl-2,6-dimethylphenol	2.72 (80)				
6-tert-Butyl-2,4-dimethylphenol	2.10 (80)				
6-tert-butyl-3,4-dimethylphenol	3.50 (80)				
N-Butylethanolamine	17.4 (25)				
Butyl ethyl ether	0.421 (20), 0.397 (25)		1.24	22.75	0.104 9
Butyl formate	0.691 (20), 0.940 (0)	2.43 (80)	2.08 (26, lq), 2.03 (25, B)	27.08	0.102 6
Butyl methyl ether			1.25 (25, B)	22.17	0.105 7
2-tert-Butyl-4-methyphenol	2.55 (80)		1.31 (20, B)		
Butyl nitrate		13 (20)	2.99 (20, B)	30.35	0.112 6
2-(2-sec-Butylphenoxy)ethanol	65.1 (25)				
2-(4-tert-Butylphenoxy)ethanol	122.5 (25)				
Butyl propionate			1.79 (22, B)	27.37	0.099 3
4-tert-Butylpyridine	1.495 (20)		2.87 (25, C)	35.48	0.095 1
Butyl stearate	8.26 (25), 4.9 (50)	3.11 (30)	1.88 (24, B)	33.0 (25)	32.7 (30)
Butyl vinyl ether	0.5 (20)		1.25 (25, Hx)	21.99 (20)	

Butyraldehyde	0.455 (20), 0.367 (39)	13.4 (26)	2.45 (40, lq)	26.67	0.092 5
Butyric acid	1.540 (20), 0.980 (40)	2.97 (20)	1.65 (30, B)	28.35	0.092 0
Butyric anhydride	1.615 (20), 1.486 (25)	13 (20)		28.93 (20)	28.44 (25)
4-Butyrolactone	1.75 (25)	39.1 (20)			
Butyronitrile	0.624 (15), 0.515 (30)	20.3 (21)	4.12 (25, B)	29.51	0.103 7
Camphor		11.35 (20)	4.07 (g), 3.6 (20, B)	35.29	0.148 4
Carbon disulfide	0.363 (20)	3.0 (−112), 2.64 (20)	2.91 (20, B), 3.10 (25, B)	29.49	0.122 4
Carbon tetrachloride	0.965 (20), 0.793 (25)	2.24 (20), 2.23 (25)	0 (g), 0.12 (20, lq)	14 (−73)	0.092 0
Carbon tetrafluoride	0.020 (25)	1.000 6 (25, g)	0		
Carvone		11 (22)	0	36.54	0.111 7
Chloroacetic acid	3.15 (50), 1.92 (75)	20 (20), 12.3 (60)	2.8 (15, B)	43.27	0.090 4
o-Chloroaniline	0.925 (25)	13.4 (25)	2.31 (30, B)	43.41	
m-Chloroaniline		13.4 (19)	1.78 (20, B)		
p-Chloroaniline			2.68 (20, B)	48.69	
Chlorobenzene	0.799 (20), 0.631 (40)	5.71 (20), 4.2 (120)	2.99 (25, B)	35.97	0.109 9
1-Chlorobutane	0.469 (15)	9.07 (−30), 7.39 (20)	1.72 (g), 1.56 (20, lq)	25.97	0.119 1
2-Chlorobutane	0.439 (15)	7.09 (30)	2.13 (g), 2.0 (20, B)	24.40	0.111 7
Chlorocyclohexane		10.9 (−47), 7.6 (25)	2.14 (g), 2.1 (20, B)	33.90	0.111 8
Chlorodifluoromethane	0.23 (25), 0.013 (25, g)	6.11 (24)	2.2 (25, B)	8 (25)	0.110 1
1-Chlorododecane		4.2 (20)	1.4 (g)	31.56	0.090 4
1-Chloro-2-2,3-epoxypropane	1.03 (25)	25.6 (1), 22.6 (22)	2.11 (25, lq), 1.94 (20, B)	39.76	0.136 0
Chloroethane	0.279 (10)	1.013 (19, g)	1.8 (25, C)	21.18 (5)	20.58 (10)
2-Chloroethanol	3.913 (15)	25.8 (25), 13 (132)	2.0 (g), 1.96 (20, lq)	38.9 (20)	
Chloroform	0.596 (15), 0.514 (30)	4.81 (20), 4.31 (50)	1.77 (g), 1.90 (25, B)	29.91	0.129 5
1-Chloroheptane		4.48 (20)	1.1 (g), 1.1 (25, lq)	28.94	0.096 1
2-Chloroheptane		6.52 (22)	1.86 (22, B)		
3-Chloroheptane		6.70 (22)	2.05 (22, B)		
4-Chloroheptane		6.54 (22)	2.06 (22, B)		
1-Chlorohexane			2.06 (22, B)	28.32	0.103 8
Chloromethane			1.94 (20, B)		0.165 0
(g)	0.0106 (20), 0.012 9 (80)	1.006 9 (100)	1.87		
(lq)		12.6 (−20)	1.86 (20)	19.5	

TABLE 5.16 Viscosity, Dielectric Constant, Dipole Moment, and Surface Tension of Selected Organic Substances (*Continued*)

Substance	Viscosity, mN · s · m^{-2}	Dielectric constant ε	Dipole moment, D	Surface tension dyn · cm^{-1}	
				a	*b*
1-Chloro-3-methylbutane		7.63 (−70), 6.05 (20)	1.94 (20, B)	25.51	0.107 6
Chloromethyl methyl ether			1.88 (C)		
1-Chloro-2-methylpropane	0.462 (20), 0.373 (40)	7.87 (−38), 6.49 (14)	2.06 (g), 2.0 (25, B)	24.40	0.109 9
2-Chloro-2-methylpropane	0.543 (15)	10.95 (0), 9.96 (20)	2.11 (g), 2.13 (25, B)	20.06 (15)	18.35 (30)
1-Chloronaphthalene	2.940 (25)	5.04 (25)	1.33 (25, lq), 1.52 (25, B)	44.12	0.103 5
o-Chloronitrobenzene		38 (50), 32 (80)	4.62 (g), 6.22 (50, lq)	48.10	0.117 1
m-Chloronitrobenzene		21 (50), 18 (80)	3.72 (g), 3.30 (50, lq)	49.71	0.141 7
p-Chloronitrobenzene		8 (120)	2.81 (g), 2.83 (90, lq)	45.84	0.104 6
1-Chlorooctane		5.05 (25)	2.14 (25, lq)	29.64	0.096 1
Chloropentafluoroethane	0.26 (25), 0.013 (25, g)		0.5 (g)	5 (25)	
1-Chloropentane	0.580 (20)	6.6 (11)	2.14 (g), 1.94 (20, B)	27.09	0.107 6
o-Chlorophenol	2.250 (45), 4.11 (25)	6.31 (25)	2.19 (g), 1.46 (20, lq)	42.5	0.112 2
m-Chlorophenol	4.722 (45), 11.55 (25)		2.19 (25, B)	43.7	0.100 9
p-Chlorophenol	4.99 (50)		2.09 (20, B)	19.51	0.087 5
1-Chloropropane	0.372 (15), 0.318 (30)	7.7 (20)	2.05 (g), 1.96 (20, B)	24.41	0.124 6
2-Chloropropane	0.335 (15), 0.299 (30)	9.82 (20)	2.17 (g), 2.1 (20, B)	21.37	0.088 3
1-Chloro-2-propanone		30 (19)	2.22 (g), 2.37 (20, Hx)		
3-Chloro-1-propene	0.347 (15)	8.2 (20)	2.0 (g), 1.8 (20, B)	25.50	0.094 6
o-Chlorotoluene		4.45 (20), 4.2 (55)	1.57 (g), 1.41 (20, lq)		
m-Chlorotoluene		5.5 (20), 5.0 (60)	1.77 (20, lq), 1.8 (22, B)		
p-Chlorotoluene		6.08 (20), 5.6 (55)	2.21 (g), 1.90 (20, lq)	34.93	0.108 2
Chlorotrifluoromethane	0.016 (25)	1.001 3 (29, g)	0.50 (g)	14 (−73)	
Chlorotrimethylsilane			2.09 (20, B)	19.51	0.087 5
Cinnamaldehyde	3.506 (46)	17 (20), 16.9 (24)	3.74 (g), 3.30 (30, lq)		
o-Cresol		11.5 (25)	2.32 (25, lq), 1.45 (25, B)	39.43	0.101 1

m-Cresol	18.42 (20), 5.057 (45)	11.8 (25)	2.39 (20, lq), 1.61 (25, B)	38.00	0.092 4
p-Cresol	5.607 (45)	9.91 (58)	2.35 (20, lq), 1.54 (20, B)	38.58	0.096 2
Crotonic acid			2.13 (30 B)		
Cyanoacetic acid		33.4 (19)		35.02	0.092 3
Cycloheptanol					
1,3-Cyclohexadiene	0.980 (20), 0.534 (60)	2.6 (−89)	0.38 (20, B)	27.62	0.118 8
Cyclohexane		2.05 (15), 2.02 (25)	0		
Cyclohexanecarboxylic acid		2.6 (31)			
1,4-Cyclohexanedione		15.0 (25)	1.41 (g), 1.3 (30, B)		
Cyclohexanol	41.07 (30), 17.19 (45)	15.0 (25), 7.24 (100)	1.86 (25, C)	35.33	0.096 6
Cyclohexanone	2.453 (15), 1.803 (30)	20 (−40), 18.2 (20)	3.11 (20, B), 3.01 (25, B)	37.67	0.124 2
Cyclohexanone oxime		3.0 (89)	0.83 (25, B)		
Cyclohexene	0.650 (20)	2.6 (−105), 2.22 (25)	0.61 (g), 0.28 (20, lq)	29.23	0.122 3
Cyclohexylamine	1.662 (20), 1.16 (49)	4.73 (20)	1.22 (20, lq), 1.26 (20, B)	34.19	0.118 8
Cyclohexylbenzene	3.681 (0)		0.62 (20, B)		
Cyclohexylmethanol		9.7 (60), 8.1 (80)	1.68 (20, B)		
o-Cyclohexylphenol		3.97 (55)			
p-Cyclohexylphenol		4.42 (131)			
Cyclooctane			0	32.02	0.109 0
Cyclopentane	0.439 (20)	1.965 (20)		25.53	0.146 2
Cyclopentanol		25 (−20), 18 (20)	1.72 (25, C)	35.04	0.101 1
Cyclopentanone		16 (−51)	3.30 (g), 2.93 (25, B)	35.55	0.110 0
Cyclopentene			0.98 (25, Hx)	25.94	0.149 5
p-Cymene	3.402 (20)	2.243 (20)	0 (lq)	28.83	0.087 7
cis-Decahydronaphthalene	3.381 (20)	2.18 (20)	0	32.18 (20)	31.01 (30)
trans-Decahydronaphthalene	2.128 (20)	2.17 (20)	0	29.89 (20)	28.87 (30)
Decamethylcyclopentasiloxane		2.5 (20)		19.56	0.056 5
Decamethyltetrasiloxane	1.28 (20)	2.4 (20)		86.20 (25)	
Decane	0.928 (20), 0.775 (22)	1.991 (20), 1.844 (130)	0	25.67	0.092 0
1-Decanol		8.1 (20)	1.71 (20, B), 1.62 (25, B)	30.34	0.073 2
1-Decene	0.805 (20)		0.42 (20, B)	25.84	0.091 9
Diallyl sulfide		4.9 (20)	1.33 (25, B)		

TABLE 5.16 Viscosity, Dielectric Constant, Dipole Moment, and Surface Tension of Selected Organic Substances (*Continued*)

Substance	Viscosity, mN·s·m^{-2}	Dielectric constant ε	Dipole moment, D	Surface tension dyn·cm^{-1}	
				a	b
Dibenzofuran		3.0 (100)	0.88 (25, B)		
Dibenzylamine		3.6 (20)	0.97 (20, lq), 1.02 (20, B)	43.27	0.108 6
Dibenzyl decanedioate		4.6 (25)			
Dibenzyl ether	3.711 (25)		1.39 (21, B)	38.2 (35)	
o-Dibromobenzene		7.35 (20)	2.13 (20, B)		
m-Dibromobenzene		3.80 (20)	1.5 (20, B)		
p-Dibromobenzene		2.57 (95)	0	41.84	0.1007
1,4-Dibromobutane			2.16 (20, lq), 2.06 (20, B)	48.24	0.119 0
2,3-Dibromobutane		5.75 (25)	2.20 (g), 1.7 (25, lq)		
1,2-Dibromoethane	1.721 (20), 1.286 (40)	4.78 (25), 4.09 (131)	1.11 (g), 1.14 (20, lq)	35.43	0.142 8
cis-1,2-Dibromoethylene		7.7 (0), 7.08 (25)	1.35 (B)		
trans-1,2-Dibromoethylene		2.9 (0), 2.88 (25)	0		
1,2-Dibromoheptane		3.8 (25)	1.78 (25, D)		
2,3-Dibromoheptane		5.1 (25)	2.15 (25, B)		
3,4-Dibromoheptane		4.7 (25)	2.15 (25, B)		
Dibromomethane	1.5 (25)	7.77 (10), 6.7 (40)	1.43 (g), 1.85 (20, lq)	42.77	0.148 8
1,2-Dibromopropane	0.72 (25)	4.3 (20)	1.43 (25, B)	36.81	0.115 5
Dibromotetrafluoroethane	0.95 (20)	2.34 (25)	1.06 (20, lq), 1.05 (20, B)	18.9 (20)	18.1 (25)
Dibutylamine	9.03 (25)	2.978 (20)	2.64 (25, B)	26.50	0.095 2
Dibutyl decanedioate	0.602 (30)	4.54 (30)	1.18 (g), 1.19 (20, lq)	24.78	0.093 4
Dibutyl ether	5.62 (20), 4.76 (25)	3.06 (25)	2.70 (25, B)	32.46	0.086 5
Dibutyl maleate	3.47 (80)		1.68 (20, B)		
2,6-Di-tert-butyl-4-methylphenol	19.91 (20), 7.85 (45)		2.97 (20, lq), 2.85 (30, B)		
Dibutyl o-phthalate	3.23 (50), 1.92 (75)	6.436 (30), 5.99 (45)		33.40 (20)	0.0927
Dichloroacetic acid		8.2 (22), 7.8 (61)		37.8	

o-Dichlorobenzene	1.324 (25)	9.93 (25), 7.10 (90)	2.51 (g), 2.26 (24, B)	26.84 (20)	35.55 (30)
m-Dichlorobenzene	1.045 (23), 0.955 (33)	5.04 (25), 4.22 (90)	1.68 (g), 1.38 (24, B)	38.30	0.1147
p-Dichlorobenzene	0.839 (55), 0.668 (79)	2.41 (50)	0	34.66	0.0879
1,4-Dichlorobutane		8.9 (25)	2.22 (g), 2.13 (25, lq)	37.79	0.1174
Dichlorodifluoromethane	0.26 (25), 0.013 (25, g)	2.13 (29)	0.51 (g)	9 (25)	
1,1-Dichloroethane	0.505 (25), 0.430 (30)	10.1 (18), 10.86 (16)	2.06 (g), 2.00 (25, B)	27.03	0.1186
1,2-Dichloroethane	0.887 (15), 0.730 (30)	12.7 (−10), 10.65 (20)	1.48 (g), 1.7 (20, B)	35.43	0.1428
1,1-Dichloroethylene	0.442 (0), 0.358 (20)	4.67 (16)	1.30 (25, B)		
cis-1,2-Dichloroethylene	0.467 (20), 0.444 (25)	9.20 (25)	2.95 (g), 1.90 (25, B)	28 (20)	
trans-1,2-Dichloroethylene	0.423 (15), 0.404 (20)	2.14 (25)	0.70 (25, B)	25 (20)	
2,2'-Dichloroethyl ether	2.41 (20), 2.065 (25)	21.2 (20)	2.61 (20, B)	40.57	0.1306
Dichlorofluoromethane	0.34 (25), 0.011 (25, g)	5.34 (28)	1.3 (g)	18 (25)	0.1284
Dichloromethane	0.449 (15), 0.393 (30)	9.14 (20), 1.006 5 (100, g)	1.60 (g), 1.90 (20, B)	30.41	0.1221
2,4-Dichlorophenol			1.60 (25, B)	46.59	0.1240
1,2-Dichloropropane	0.865 (20), 0.700 (25)	8.925 (26), 7.90 (35)	1.87 (25, B)	31.42	0.1233
1,3-Dichloropropane			2.08 (g), 2.2 (25, B)	36.40	
2,2-Dichloropropane	0.769 (15), 0.619 (30)	11.37 (20)	2.62 (g), 2.20 (25, B)	23.60 (20)	22.53 (30)
1,1-Dichloro-2-propanone		14 (20)		12 (25)	
1,2-Dichlorotetrafluoroethane	0.38 (25), 0.011 (25, g)	2.26 (25)	0.53 (g)	41.26	0.1035
α,α-Dichlorotoluene		6.9 (20)	2.07 (20, B), 2.05 (25, B)		
Diethanolamine	368 (30), 196 (40)	2.81 (25)	2.84 (25, B)	23.46	0.1030
1,1-Diethoxyethane		3.80 (25)	1.08 (g)		
1,2-Diethoxyethane	0.65 (20)		1.99 (20, B), 1.65 (25, B)	23.87	
Diethoxymethane				22.71	
Diethylamine	0.388 (10), 0.273 (38)	3.6 (22)	0.92 (g), 1.11 (25, lq)	36.59	0.1291
N,N-Diethylaniline	1.15 (30), 0.750 (75)	5.5 (19)	1.40 (20, lq), 1.80 (20, B)	28.62	0.1143
Diethyl carbonate	0.868 (15), 0.748 (25)	2.82 (20)	1.07 (g), 0.91 (25, B)	34.68	0.1040
Diethyl decanedioate		5.0 (30)	2.38 (20, lq), 2.52 (20, B)	18.92	0.1100
Diethyl ether	0.247 (15), 0.245 (20)	4.335 (20), 3.97 (40)	1.15 (20, lq), 1.22 (16, lq)	30.63	0.0959
Diethyl ethyl phosphonate	1.627 (15), 0.969 (45)	11.00 (15), 9.86 (45)	2.95 (32, lq), 2.91 (20, C)		0.0908
Diethyl fumarate		6.5 (23)	2.40 (20, B)	34.34	0.0975
Diethyl glutarate		6.7 (30)	2.46 (30, lq)		0.1010

TABLE 5.16 Viscosity, Dielectric Constant, Dipole Moment, and Surface Tension of Selected Organic Substances (*Continued*)

Substance	Viscosity, mN · s · m^{-2}	Dielectric constant ε	Dipole moment, D	Surface tension dyn · cm^{-1}	
				a	b
Di(2-ethylhexyl)-2-ethylhexyl phosphonate	6.00 (45), 3.61 (65)	4.09 (45), 3.94 (65)			
Di(2-ethylhexyl) *o*-phthalate	33.67 (35), 21.40 (45)	4.91 (35), 4.77 (45)			
Diethyl maleate	3.57 (20), 3.14 (25)	8.58 (23)	2.56 (25, B)	34.67	0.103 9
Diethyl malonate	2.15 (20), 1.94 (25)	8.03 (25)	2.49 (20, lq), 2.54 (25, B)	33.91	0.104 2
Diethyl nonanedioate		5.13 (30)			
Diethyl oxalate	2.311 (15), 1.618 (30)	8.1 (21)	2.49 (20, D)	34.32	0.111 9
Diethyl *o*-phthalate	9.18 (35), 6.41 (45)	7.34 (35), 7.13 (45)	2.8 (25, B)	38.47	0.096 3
Diethyl succinate		6.64 (30)	2.3 (g), 2.37 (30, lq)	33.97	0.104 1
Diethyl sulfate		29 (20)	4.46 (25, D)	35.47	0.097 6
Diethyl sulfide	0.446 (20), 0.422 (25)	5.72 (25), 5.24 (50)	1.52 (g), 1.58 (20, B)	27.33	0.110 6
Diethyl sulfite		16 (20), 14 (50)			
Diethylzinc		2.5 (20)	0.62 (25, B)		
			2.30 (g)		
1,1-Difluoroethane	0.243 (21)				
1,2-Dihydroxybenzene		2.6 (−89)	2.60 (25, B)	47.6	0.084 9
1,3-Dihydroxybenzene		3.2 (18)	2.09 (44, B)	54.8	0.071 7
1,4-Dihydroxybenzene			1.4 (44, B)		
1,2-Diiodobenzene		5.7 (20)	1.70 (20, B)		
1,3-Diiodobenzene		4.3 (25)	1.22 (20, B)		
1,4-Diiodobenzene		2.9 (120)	0.19 (20, B)		
cis-1,2-Diiodoethylene		4.46 (83)	0.71 (B)		
trans-1,2-Diiodoethylene		2.19 (83)	0		
Diiodomethane	3.043 (15), 2.392 (30)	5.316 (25)	1.08 (25, B)	70.21	0.161 3
Diisobutylamine		2.7 (22)	1.10 (25, B)	24.00	0.091 2
Diisobutyl *o*-phthalate	30 (20)				

Diisopentylamine		2.5 (18)	1.48 (30, B)	26.04	0.085 8
Diisopentyl ether	1.40 (11), 1.012 (20)	2.82 (20)	0.98 (20, lq), 1.23 (25, B)	24.76	0.087 1
Diisopropylamine	0.40 (25)		1.26 (25, B)	21.83	0.107 7
Diisopropyl ether	0.379 (25)	3.88 (25)	1.13 (g), 1.26 (25, B)	19.89	0.104 8
1,2-Dimethoxybenzene	3.281 (25), 2.184 (40)	4.09 (25)	1.32 (25, B)	34.4	0.064 2
1,1-Dimethoxyethane				23.90	0.115 9
1,2-Dimethoxyethane	0.530 (10), 0.455 (25)	7.60 (10), 7.20 (25)	1.71 (25, B)	48.0 (25)	0.119 9
Dimethoxymethane	0.340 (15), 0.325 (20)	2.65 (20)	0.74 (g)	23.59	
N,N-Dimethylacetamide	2.141 (20), 0.838 (30)	37.78 (25)	3.80 (g), 4.60 (20, lq)	32.43 (30)	29.50 (50)
Dimethylamine	0.207 (15), 0.186 (25)	6.32 (0), 5.26 (25)	1.03 (g), 1.14 (25, lq)	29.50	0.126 5
N,N-Dimethylaniline	1.285 (25), 0.91 (50)	4.9 (20), 4.4 (70)	1.61 (g), 1.55 (25, B)	38.14	0.104 9
2,4-Dimethylaniline			1.40 (25, B)	39.34	0.099 6
2,2-Dimethylbutane	0.351 (25), 0.330 (30)	1.873 (25)	0	18.29	0.099 0
2,3-Dimethylbutane	0.361 (25), 0.342 (30)	1.890 (25)	0	19.38	0.100 0
2,3-Dimethyl-1-butanol				26.22	0.099 2
N,N-Dimethylbutyramide	1.271	2.00			
Dimethyl carbonate			0.90 (g), 0.96 (25, B)	31.94	0.134 3
1,1-Dimethylcyclopentane	11 (20)		0	23.78	0.101 6
2,2-Dimethyl-1,3-dioxolane-4-methanol					
Dimethyl ether	0.010 4 (60)	5.02 (25), 2.97 (110)	1.30 (g), 1.25 (25, B)	14.97	0.147 8
N,N-Dimethylformamide	0.845 (20), 0.598 (50)	38.3 (20), 36.71 (25)	3.86 (25, B)	36.76 (20)	34.40 (40)
2,4-Dimethylheptane		1.9 (20)	0	23.21	0.092 9
2,5-Dimethylheptane		1.9 (20)	0	23.21	0.092 9
2,6-Dimethylheptane		2 (20)	0	22.77	0.088 7
2,6-Dimethyl-4-heptanone	1.03 (20)		2.66 (25, C)	38.26	0.113 8
Dimethyl hexanedioate	14 (20)		2.28 (20, B)		
Dimethyl hydrogen phosphonate	1.08 (25)			40.73	0.122 0
Dimethyl maleate	3.54 (20), 3.21 (25)	10 (20)	2.48 (25, C)	39.72	0.120 8
Dimethyl malonate		1.91 (20)	2.41 (20, B)	19.94	0.095 7
2,2-Dimethylpentane		1.939 (20)	0	21.96	0.099 5
2,3-Dimethylpentane	0.406 (20)		0		

TABLE 5.16 Viscosity, Dielectric Constant, Dipole Moment, and Surface Tension of Selected Organic Substances (*Continued*)

Substance	Viscosity, mN · s · m^{-2}	Dielectric constant ε	Dipole moment, D	Surface tension dyn · cm^{-1}	
				a	*b*
2,4-Dimethylpentane	0.361 (20)	1.914 (20)	0	20.09	0.097 2
3,3-Dimethylpentane		1.94 (20)	0	21.59	0.099 6
2,4-Dimethylphenol			1.48 (20, B), 1.98 (60, B)	34.57	0.086 9
2,5-Dimethylphenol	1.55 (80)		1.43 (20, B), 1.52 (60, B)	36.72	0.085 0
3,4-Dimethylphenol	3.00 (80)	4.8 (17)	1.77 (20, B)	35.75	0.091 0
3,5-Dimethylphenol	2.42 (80)		1.76 (20, B)	34.09	0.080 7
Dimethyl *o*-phthalate	17.2 (25), 6.41 (45)	8.25 (25), 8.11 (45)	2.8 (25, B)		
2,2-Dimethylpropane	0.328 (0), 0.303 (5)	1.80 (20), 1.678 (98)	0	12.05 (20)	10.98 (30)
N,N-Dimethylpropionamide	0.935	33.1			
2,5-Dimethylpyrazine		2.43 (20)	0		
2,3-Dimethylquinoxaline		2.3 (25)	0		
Dimethyl succinate		5.1 (20)	2.09 (20, B)	39.00	0.119 1
Dimethyl sulfate		48.3 (20), 46.4 (20)	4.31 (25, D)	41.26	0.116 3
Dimethyl sulfide	0.289 (20), 0.265 (36)	6.2 (20)	1.45 (25, B)	26.07	0.080 5
Dimethyl sulfite	0.715 (30), 0.436 (80)	22.5 (23)	2.93 (20, B)	36.48	0.125 3
Dimethyl sulfoxide	2.47 (20), 1.192 (55)	48.9 (20), 41.9 (55)	3.9 (25, B)	43.54 (20)	42.41 (30)
2,4-Dimethyltetrahydrothiophene-1,1-dioxide	9.04	29.5			
N,N-Dimethyl-*o*-toluidine		3.4 (20)	0.88 (25, B)		
N,N-Dimethyl-*p*-toluidine		3.9 (20)	1.29 (25, B)		
Dinonyl hexanedioate	37 (20)		2.53 (25, B)		
Dinonyl *o*-phthalate		4.65 (35), 4.52 (45)			
Dioctyl decanedioate		4.0 (27)			
Dioctyl *o*-phthalate		5.1 (25)	3.06 (25, C)		
1,4-Dioxane	1.439 (15), 1.087 (30)	2.24 (20), 2.21 (25)	0	36.23	0.139 1

Dipentyl ether	1.188 (15), 0.922 (30)	2.77 (25)	0.98 (20, lq), 1.24 (25, B)	26.66	0.092 5
Dipentyl o-phthalate	17.03 (35), 11.51 (45)	5.79 (35), 5.62 (45)	2.71 (20, lq)	32.56	0.073 9
Dipentyl sulfide	4.66 (55), 1.04 (130)	3.83 (25)	1.59 (25, B)	29.55	0.087 6
Diphenylamine		3.3 (52)	1.31 (20, C), 1.01 (25, B)	45.36	0.101 7
1,2-Diphenylethane		2.4 (110)	0 (110, lq), 0.45 (25, B)		
Diphenyl ether	2.61 (40), 2.09 (50)	3.65 (30)	1.16	28.70	0.078 0
Diphenylmethane		2.7 (18), 2.57 (26)	0.26 (30, lq), 0.3 (25, B)		
1,1-Dipropoxyethane				25.03	0.097 2
Dipropoxymethane				25.17	0.095 3
Dipropylamine	0.534 (20), 0.427 (37)	3.07 (20)	1.01 (20, lq), 1.03 (20, B)	24.86	0.102 2
Dipropyl carbonate				28.94	0.101 5
Dipropylene glycol butyl ether	4.23 (25)			28.2 (25)	
Dipropylene glycol ethyl ether	3.11 (25)			27.7 (25)	
Dipropylene glycol isopropyl ether	386 (25)			25.9 (25)	
Dipropylene glycol methyl ether	3.1 (25)			28.8 (25)	
Dipropyl ether	0.448 (15), 0.376 (30)	3.39 (26)	1.21 (g), 1.17 (30, Hx)	22.60	0.104 7
Divinyl ether		3.9 (20)	1.07 (20, lq)	•	
Dodecamethylcyclohexasiloxane		2.6 (20)			
Dodecamethylpentasiloxane		2.5 (20)		17.08 (25)	
Dodecane	1.508 (20), 1.378 (25)	2.05 (−10), 2.01 (20)	0	27.12	0.088 4
1-Dodecanol		5.15 (20), 6.5 (25)	1.52 (20, B)	31.25	0.074 8
6-Dodecyne		2.17 (25)			
1,2-Epoxybutane	0.41 (20), 0.40 (25)		2.01 (20, B)	23.9 (20)	
Erythritol		28 (128)			
Ethane (g)	0.009 0 (20), 0.011 4 (100)	1.001 5 (0)	0	1.24	0.166 0 (lq)
1,2-Ethanediamine	1.54 (20), 1.226 (30)	16.8 (18), 14.2 (20)	1.96 (g), 1.92 (25, B)	44.77	0.139 8
1,2-Ethanediol	26.09 (15), 13.55 (30)	38.66 (20), 37.7 (25)	2.28 (g), 2.3 (25, D)	50.21	0.089 0
1,2-Ethanediol diacetate	3.13 (20)	13 (30)	2.34 (30, B)		

TABLE 5.16 Viscosity, Dielectric Constant, Dipole Moment, and Surface Tension of Selected Organic Substances (*Continued*)

Substance	Viscosity, mN · s · m^{-2}	Dielectric constant ε	Dipole moment, D	Surface tension dyn · cm^{-1}	
				a	*b*
Ethanesulfonic acid				45.74	0.082 4
Ethanesulfonyl chloride				43.43	0.117 7
Ethanethiol	0.003 16 (g)	6.9 (15)		25.06	0.079 3
Ethanol	1.209 (19), 0.991 (30)	25.00 (20), 20.21 (55)	3.89 (25, B)	24.05	0.083 2
Ethoxybenzene	1.364 (15), 1.040 (30)	4.22 (20)	1.57 (g), 1.40 (20, B)	35.17	0.110 4
2-Ethoxyethanol	2.04 (20), 1.85 (25)	29.6 (24)	1.69 (g), 1.71 (25, B)	30.59	0.089 7
2-(2-Ethoxyethoxy)ethanol	3.71 (25)		1.41 (g), 1.36 (25, CS$_2$)	31.8 (25)	27.2 (75)
2-Ethoxyethyl acetate	1.025 (25)	7.567 (30)	2.24 (30, B)	31.8 (25)	
1-Ethoxy-2-methylbutane		3.96 (20)			
1-Ethoxynaphthalene		3.3 (19)	2.25 (30, B)		
1-Ethoxypentane		3.6 (23)			
1-Ethoxy-2-propanol	1.68 (25)				
α-Ethoxytoluene		3.9 (20)		25.9 (25)	
Ethyl acetate	0.473 (15), 0.426 (25)	6.11 (20), 5.30 (77)	1.78 (g), 1.84 (25, lq)	26.29	0.116 1
Ethyl acetoacetate	1.419 (20), 1.508 (25)	15.7 (22)	3.22 (18, B, keto form), 2.04 (−80, CS$_2$, enol form)	34.42	0.101 5
Ethylamine	12.40 (25)	6.94 (10)	1.40 (25, B)	22.63	0.137 2
2-(Ethylamino)ethanol	2.04 (25), 1.08 (55)				
N-Ethylaniline	0.669 (20), 0.531 (40)	5.76 (20)		39.00	0.107 0
Ethylbenzene	2.407 (15), 1.751 (30)	2.41 (20)	0.37 (25, lq)	31.48	0.109 4
Ethyl benzoate		6.02 (20)	1.95 (g), 1.93 (25, B)	37.16	0.105 9
Ethyl α-bromobutyrate	8.021 (15), 5.892 (25)	8 (20)	2.40 (25, B)	25.06 (15)	24.32 (25)
2-Ethyl-1-butanol	0.771 (15), 0.613 (25)	6.19 (90)			
Ethyl butyrate		5.10 (18)	1.74 (22, B)	26.55	0.104 5

2-Ethylbutyric acid	3.3 (20)			26.3 (20)	0.117 7
Ethyl carbamate	0.916 (105), 0.715 (120)	14.2 (50)	2.59 (30, D)	34.18	0.108 4
Ethyl chloroacetate		11.4 (21)~	2.65 (25, B)	28.90	0.104 5
Ethyl chloroformate		11 (20)	2.56 (35, B)	39.99	0.106 6
Ethyl cinnamate	8.7 (20)	6.1 (18)	1.86 (20, B)	29.31	0.109 2
Ethyl crotonate		5.4 (20)	1.95 (24, B)	38.80	0.105 4
Ethyl cyanoacetate	3.256 (15), 2.148 (30)	26.9 (20)	4.04 (30, B)	27.78	0.115 8
Ethylcyclohexane	0.843 (20), 0.787 (25)	2.054 (20)	0 (g)	34.89	0.086 3
Ethyl dichloroacetate		12 (2), 10 (22)	2.63 (25, B)		
N-Ethyldiethanolamine	53 (25)			30.05	0.185 4
Ethyl dodecanoate		3.4 (20), 2.7 (143)	1.3 (20, lq)	−2.7	0.139 8
Ethylene	1.85 (40)	1.001 44 (0)	0 (g)	44.77	
Ethylene carbonate		89.6 (40), 69.4 (91)	4.87 (25, B)	49.1 (0)	
Ethylenediamine	1.540 (18)	16.0 (18), 14.2 (20)	1.98 (g), 1.92 (25, B)	47.33	
Ethylene dinitrate		28.3 (20)	3.58 (25, B)	50.21	46.7 (45)
2,2'-(Ethylenedioxy)diethanol	38 (20)	23.69 (20)	5.58 (lq)	27.66	0.088 0
Ethylene glycol	26.09 (15), 13.35 (30)	41.2 (20), 37.7 (25)	2.27 (g), 2.20 (15, lq)	7.9 (20)	0.089 0
Ethylene oxide	0.3 (0)	14 (−1)	1.88 (g), 1.92 (20, lq)	26.47	0.166 4
Ethyleneimine	0.418 (25)	18.3 (25)	1.89 (g), 1.77 (25, B)	33.90	
Ethyl formate	0.419 (15), 0.358 (30)	7.16 (25)	1.94 (g), 1.96 (25, lq)	32.86	0.131 5
Ethyl fumarate		6.5 (23)			0.105 6
Ethyl hexadecanoate	323 (20)	3.2 (20), 2.71 (104)	1.2 (lq)		0.085 9
2-Ethyl-1,3-hexanediol					
Ethyl hexanoate	7.7 (20)		1.80 (20, B)	27.73	0.096 0
2-Ethylhexanoic acid	9.8 (20)				
2-Ethyl-1-hexanol	1.5 (20)	4.41 (90)	1.74 (25, B)	30.0 (22)	0.104 6
2-Ethylhexyl acetate					
Ethyl isobutyrate		3.96 (20)		25.33	
Ethyl isopentyl ether			3.67 (20, B)		
Ethyl isothiocyanate		19.5 (21)	2.4 (20, B)	38.69	0.132 6
Ethyl lactate	2.44 (25)	13.1 (25)		30.72	0.098 3
Ethyl maleate		8.6 (23)			

TABLE 5.16 Viscosity, Dielectric Constant, Dipole Moment, and Surface Tension of Selected Organic Substances (*Continued*)

Substance	Viscosity, mN · s · m^{-2}	Dielectric constant ε	Dipole moment, D	Surface tension dyn · cm^{-1} a	b
Ethyl 3-methylbutyrate		4.71 (18)		25.79	0.100 6
Ethyl methyl ether			1.22 (g)	18.56	0.131 7
Ethyl methyl sulfide	0.373 (20), 0.354 (25)			27.63	0.128 6
Ethyl nitrate		19.4 (20)	2.93 (20, B)	30.81	0.134 5
Ethyl 9-octadecenoate		3.2 (25)	1.83 (20, lq)		
Ethyl 4-oxopentanoate		12 (21)			
3-Ethylpentane		1.94 (20)	0	22.52	0.103 2
Ethyl pentanoate	0.847 (20)	4.7 (18)	1.76 (28, B)	27.15	0.099 9
Ethyl pentyl ether		3.6 (23)	1.2 (20, B)	24.19	0.099 2
Ethyl phenylacetate		5.3 (21)	1.82 (30)		
Ethyl phenyl sulfide		5.65 (19)	4.08 (25, B)	39.30	0.113 1
Ethyl propionate	0.564 (15), 0.473 (30)		1.75 (22, B)	26.72	0.116 8
Ethyl propyl ether	0.323 (20), 0.225 (60)		1.16 (25, B)	21.92	0.105 4
Ethyl salicylate	1.772 (45)	7.99 (30)	2.85 (25, B)	31.00	0.109 1
Ethyl stearate		2.98 (40), 2.69 (100)	1.65 (40, lq)		
Ethyl thiocyanate		29.3 (21)	3.33 (20, B)	37.28	0.122 6
o-Ethyltoluene				32.33	0.106 0
p-Ethyltoluene		2.24 (25)	0	30.98	0.107 5
Ethyl trichloroacetate		7.8 (20)	2.56 (25, B)	32.97	0.107 3
Ethyl vinyl ether	0.2		1.26 (20, B)	19.00 (20)	
Ethynyl acetate				32.81 (20)	30.20 (40)
Fluorobenzene	0.620 (15), 0.517 (30)	5.42 (25), 4.7 (60)	1.61 (g)	29.67	0.120 4
1-Fluorohexane				23.41	0.100 1
2-Fluoro-2-methylbutane		5.89 (20)	1.92 (25, B)		

1-Fluoropentane	0.680 (20), 0.601 (30)	4.24 (20)	1.85 (25, B)	22.81	0.131 5
o-Fluorotoluene	0.608 (20), 0.534 (30)	4.22 (30), 3.9 (60)	1.35 (g), 1.26 (30, lq)	32.31	0.125 7
m-Fluorotoluene	0.622 (20), 0.522 (30)	5.42 (30), 4.9 (60)	1.86 (g), 1.66 (30, lq)	30.44	0.110 9
p-Fluorotoluene	4.320 (15), 2.296 (30)	5.86 (30), 5.3 (60)	2.00 (g), 1.76 (30, lq)	59.13	0.084 2
Formamide	1.65 (120)	111.0 (20), 103.5 (40)	3.73 (g)	44.30	0.087 5
Formanilide	1.966 (15), 1.219 (40)		3.37 (25, C)	39.87	0.109 8
Formic acid	2.475 (0), 1.494 (25)	58.5 (15), 57.0 (21)	1.35 (g), 1.20 (25, B)	46.41	0.132 7
2-Furaldehyde	0.380 (20), 0.361 (25)	41.9 (20), 34.9 (50)	2.13 (25, lq), 3.63 (25, B)	24.10 (20)	23.38 (25)
Furan	4.62 (25)	2.95 (25)	0.66 (g), 0.67 (20, B)	ca 38 (20)	
Furfuryl alcohol			1.92 (25, lq)		
Glycerol	945 (25), 134 (50)	42.5 (25)	2.68 (25, D)	63.14 (17)	62.5 (25)
Glycerol triacetate		7.2 (20)	2.73 (25, B)	37.88	0.081
Glycerol trinitrate	36.0 (20), 13.6 (40)	19 (20)	3.38 (25, B)	55.74	0.250 4
Glycerol trioleate		3.2 (26)	3.11 (23, B)	36.03	0.069 9
Glycerol tripalmitate		2.9 (65)	2.80 (23, B)	32.26	0.067 2
Glycerol tristearate		2.8 (70)	2.86 (23, B)	32.73	0.068 5
Heptanaldehyde	0.977 (15)	9.1 (20)	2.26 (40, lq), 2.58 (22, B)	28.64	0.092 0
Heptane	0.416 (20), 0.341 (40)	1.924 (20), 1.85 (70)	0	22.10	0.098 0
Heptanoic acid	3.40 (30)	2.6 (71)		29.88	0.084 8
1-Heptanol	7.014 (20), 8.53 (15)	12.1 (22)	1.73 (20, B)		
DL-2-Heptanol	5.06 (25)	9.21 (22)	1.73 (20, B)		
DL-3-Heptanol		6.9 (22)	1.73 (20, B)		
4-Heptanol		6.2 (22)	1.72 (20, B)		
2-Heptanone	0.854 (15), 0.686 (30)	11.95 (20), 8.27 (100)	2.61 (22, B)	28.76	0.105 6
3-Heptanone		12.9 (22)	2.81 (22, B)	28.24	0.101 5
4-Heptanone	0.736 (20)	12.60 (20), 9.46 (80)	2.74 (20, B)	28.11	0.106 0
1-Heptene	0.35 (20), 0.34 (25)	2.07 (20)	0.34 (20, lq)	22.28	0.099 1
Hexadecamethylcyclooctasiloxane		2.7 (20)			
Hexadecane	3.591 (22)		0	29.18	0.085 4
1-Hexadecanol		3.8 (50)	1.67 (25, B)		
1,5-Hexadiene	0.275 (20), 0.244 (36)				
2,4-Hexadiene		2.2 (25)	0.31 (25, B)		

TABLE 5.16 Viscosity, Dielectric Constant, Dipole Moment, and Surface Tension of Selected Organic Substances (*Continued*)

Substance	Viscosity, mN·s·m⁻²	Dielectric constant ε	Dipole moment, D	Surface tension dyn·cm⁻¹	
				a	b
Hexafluorobenzene		2.2 (20)	0	22.6 (20)	0.076 3
Hexamethyldisiloxane		30 (20)	0.37 (25, lq)	17.01	
Hexamethylphosphoramide	3.47 (20)		4.31 (25, lq)	33.8 (20)	0.102 2
Hexane	0.313 (20), 0.271 (40)	1.904 (15), 1.890 (20)	0	20.44	0.097 3
Hexanedinitrile	5.99	32.45	3.8 (25, B)	47.88	0.100 2
2,4-Hexanedione				32.22	0.090 7
Hexanenitrile		17.26 (25)		29.64	
Hexanoic acid	1.041 (15), 0.830 (30)	2.63 (71)	1.13 (25, lq)	28.05 (20)	27.55 (25)
1-Hexanol	3.525 (15), 2.511 (30)	13.3 (25), 8.5 (75)	1.55 (20, B)	27.81	0.080 1
2-Hexanone	6.203 (15), 3.872 (30)	14.6 (15)	2.68 (22, B)	28.18	0.109 2
1-Hexene	0.584 (25)	2.051 (20)	0.34 (20, lq)	20.47	0.102 7
Hexyl acetate	0.26 (20), 0.25 (25)			28.44	0.097 0
4-Hydroxy-4-methyl-2-pentanone	2.9 (25)	18.2 (25)	3.24 (20, B)	31.0 (20)	
Iodobenzene	1.774 (17), 0.488 (149)	4.62 (20)	1.71 (g), 1.3 (20, B)	41.52	0.112 3
1-Iodobutane		6.22 (20), 4.52 (130)	2.10 (g), 1.90 (20, B)	30.82	0.013 1
2-Iodobutane			2.06 (20, B)	30.32	0.105 6
1-Iodododecane		3.9 (20)	1.87 (20, C)		
Iodoethane	0.617 (15), 0.540 (30)	10.2 (−50), 7.82 (20)	1.91 (g), 1.69 (20, lq)	31.67	0.128 6
1-Iodoheptane		4.9 (22)	1.86 (22, B)	32.18	0.088 7
3-Iodoheptane		6.4 (22)	1.95 (22, B)		
1-Iodohexadecane		3.5 (20)		34.49	0.088 0
1-Iodohexane		5.37 (20)	1.94 (20, C)	31.63	0.084 5
Iodomethane	0.500 (20), 0.424 (40)	7.00 (20)	1.64 (g), 1.42 (20, B)	33.42	0.123 4
1-Iodo-3-methylbutane		5.6 (19)	1.85 (20, B)	30.37	0.091 5
2-Iodo-2-methylbutane		8.19 (20)	2.20 (20, B)		

1-Iodo-2-methylpropane	0.875 (20), 0.697 (40)		6.5 (20)	1.89 (20, B)	30.26	0.0172
1-Iodooctane			4.6 (25)	1.80 (25, lq), 1.90 (20, C)	32.51	0.0915
2-Iodooctane			5.8 (20)	2.07 (20, C)		
1-Iodopentane			5.81 (20)	1.90 (20, B)	31.41	0.1014
1-Iodopropane	0.837 (15), 0.670 (30)		7.00 (20)	2.03 (g), 1.86 (20, B)	31.64	0.1136
2-Iodopropane	0.732 (15), 0.620 (30)		7.87 (20)	2.01 (20, B)	29.35	0.1107
p-Iodotoluene			4.4 (35)	1.72 (22, B)	39.23	0.0965
α-Ionone			11 (18)		34.10	0.0949
β-Ionone			12 (20)		35.36	0.0950
Iron pentacarbonyl			2.6 (20)			
Isobutyl acetate			5.29 (20)	1.87 (22, B)	25.59	0.1013
Isobutylamine	0.553 (25)		4.43 (21)	1.27 (25, B)	24.48	0.1092
Isobutylbenzene			2.319 (20), 2.298 (30)	0.31 (20, lq)	29.39	0.0961
Isobutyl butyrate			4.1 (20)		24.47	0.0843
Isobutyl formate	0.680 (20)		6.41 (19)	1.89 (20, B)	26.14	0.1122
Isobutyl isobutyrate					30.92	0.1270
Isobutyl nitrate			2.7 (20)			
Isobutyl pentanoate			3.8 (19)			
Isobutyl propionate					28.97	0.1166
Isobutyric acid	1.44 (15)		2.7 (20)	1.09 (25, lq)	26.88	0.0920
Isobutyric anhydride			14 (20)			
Isobutyronitrile	0.551 (15), 0.456 (30)		20.4 (24)	3.61 (25, B)	24.93 (20)	23.84 (30)
Isopentyl acetate	0.872 (20), 0.790 (25)		4.81 (22, B), 4.63 (30)	1.84 (22, B), 1.76 (30, lq)	26.75	0.0989
Isopentyl butyrate			4.0 (20)		27.32	0.0918
Isopentyl pentanoate			3.6 (19)			
Isopentyl propionate			4.2 (20)	1.8 (28, B)		
Isopropyl acetate	0.559 (20)		5.45 (20)	1.86 (22, B)	24.44	0.1072
Isopropylamine	0.36 (25)		2.39 (20)	1.45 (25, B)	19.91	0.0972
Isopropylbenzene	0.791 (20), 0.739 (25)			0.65 (g), 0.39 (20, lq)	30.32	0.1054
Isopropyl formate	0.512 (20)				24.56	
1-Isopropyl-4-methylbenzene	3.402 (20), 1.600 (30)		2.24 (20)	0	29.44 (20)	0.1147
Isoquinoline	3.253 (30)		10.7 (20)	2.75 (g), 2.55 (25, B)		

TABLE 5.16 Viscosity, Dielectric Constant, Dipole Moment, and Surface Tension of Selected Organic Substances (*Continued*)

Substance	Viscosity, mN · s · m⁻²	Dielectric constant ε	Dipole moment, D	Surface tension dyn · cm⁻¹ a	b
Lactamide					
Lactic acid	40.33 (25)	22 (17)		38.31	0.096 0
Lactonitrile	2.01 (30)	38 (20)		29.50	0.092 9
D-Limonene		2.4 (20)	1.57 (25, B)	29.11	0.091 3
DL-Limonene		2.3 (20)	0.63 (25, B)		
DL-Mandelonitrile		17.8 (23)	1.55 (20, B)	45.90	0.098 8
Menthol	6.89 (35)				
2-Mercaptoethanol	3.4 (20)				
Methacrylic acid	1.32 (20)		1.65	26.5 (25)	
Methacrylonitrile	0.392 (20)		3.69 (g)	24.4 (20)	
Methane (g)	0.010 9 (20), 0.013 3 (100)	1.000 94 (0)	0	*	
Methanesulfonic acid				52.28	0.089 3
Methanethiol			1.26 (g)	28.09	0.169 6
Methanol	0.676 (10), 0.544 (25)	41.8 (−20), 33.62 (20)	1.69 (g), 1.68 (22, B)	24.00	0.077 3
o-Methoxybenzaldehyde			4.34 (20, B)	45.34	0.110 5
p-Methoxybenzaldehyde		22.3 (22), 10.4 (248)	3.26 (35, B)	44.69	0.104 7
Methoxybenzene		4.33 (25), 3.9 (70)	1.36 (g), 1.24 (20, B)	38.11	0.120 4
2-Methoxyethanol	1.152 (15), 0.789 (30)	16.93 (25), 16.0 (30)	2.04 (25, B)	33.30	0.098 4
2-(2-Methoxyethoxy)ethanol	1.72 (20), 1.60 (25)			34.8 (25)	29.9 (75)
2-Methoxyethyl acetate	3.48 (25), 1.61 (60)	8.25 (20)	2.13 (30, B)		
1-Methoxy-2-nitrobenzene				48.62	0.118 5
o-Methoxyphenol		12 (25)	4.83 (g)	41.2	0.094 3

* $38.618 - 0.1873\,T - 0.000356\,T^2$.

5.112

2-Methoxy-4-(2-propenyl)phenol	6.931 (25)				
o-Methoxytoluene		3.5 (20)			
m-Methoxytoluene		3.5 (20)			
p-Methoxytoluene		4.0 (20)	2.46 (25, B)	36.20	0.1071
N-Methylacetamide	3.88 (30), 2.54 (45)	178.9 (30), 138.6 (60)	4.39 (20, D)	33.67 (30)	30.62 (50)
Methyl acetate	0.388 (20), 0.320 (40)	7.03 (20), 6.68 (25)	1.70 (g), 1.75 (25, B)	27.95	0.1289
Methyl acetoacetate	1.704 (20)			34.98	0.0944
Methyl acrylate	1.398 (20)		1.77 (25, B)		
Methylamine	0.285 (15), 0.236 (0)	11.4 (−10), 10.0 (18)	1.29 (g)	22.87	0.1488
N-Methylaniline	2.02 (25), 1.084 (55)		1.67 (25, B)	39.32	0.0970
Methyl benzoate	2.298 (15), 1.673 (30)	6.59 (20)	1.86 (25, B)	40.10	0.1171
2-Methyl-1,2-butadiene	0.266 (0.3), 0.223 (20)	2.1 (25)	0.15 (g)		
2-Methylbutane	0.237 (15), 0.215 (25)	1.871 (0), 1.845 (20)	0.13 (g)	17.20	0.110 3
2-Methyl-1-butanol	5.50 (20), 1.44 (60)	14.7 (25)		21.5 (25)	
2-Methyl-2-butanol	5.48 (15), 2.81 (30)	5.82 (25)	1.72 (20, B)	24.18	0.0748
3-Methyl-1-butanol	4.81 (15), 2.96 (30)	14.7 (25), 5.82 (130)	1.82 (25, B)	25.76	0.0820
3-Methyl-2-butanol	3.51 (25)			23.0 (25)	
2-Methyl-1-butene		2.20 (20)	0.52 (20, lq)	18.81	0.1148
2-Methyl-2-butene			0.11 (25, lq), 0.34 (25, B)	19.70	0.127 1
3-Methyl-1-butene		1.002 8 (100, g)	0.25 (g)	16.42	0.103 1
2-Methylbutyl acetate	0.872 (20)	4.63 (30)	1.82 (22)	26.75	0.0989
Methyl butyrate	0.580 (20), 0.459 (40)	5.6 (20)	1.72 (22, B)	27.48	0.1145
3-Methylbutyric acid		2.64 (20)	0.63 (25)	27.28	0.0886
3-Methylbutyronitrile	2.731 (15), 2.411 (20)	18 (220)	3.62 (25, C)	27.58	0.0827
Methyl chloroacetate		12.9 (21)		37.90	0.1304
Methyl cyanoacetate	3.82 (50), 2.69 (65)	19.23 (50), 17.57 (65)		41.32	0.1074
Methylcyclohexane	0.734 (20), 0.685 (25)	2.02 (20), 2.07 (25)	0	26.11	0.113 0
cis-2-Methylcyclohexanol	18.08 (25), 13.60 (30)	13.3*	2.58 (30, lq),* 1.95 (25, B)*	32.45	0.077 0*
trans-2-Methylcyclohexanol	37.13 (25), 25.14 (30)				

* Mixed isomers.

5.113

TABLE 5.16 Viscosity, Dielectric Constant, Dipole Moment, and Surface Tension of Selected Organic Substances (*Continued*)

Substance	Viscosity, mN · s · m^{-2}	Dielectric constant ε	Dipole moment, D	Surface tension dyn · cm^{-1}	
				a	*b*
cis-3-Methylcyclohexanol	19.7 (25), 17.23 (30)	16.47 (20)	1.91	29.08	0.062 9*
trans-3-Methylcyclohexanol	25.52 (16), 15.60 (30)	8.05	1.75	28.80 (30)	
cis-4-Methylcyclohexanol	0.247 (25)	13.3*	2.70 (30, lq),* 1.9 (25, B)*	29.07	0.069 0*
trans-4-Methylcyclohexanol	0.385 (25)				
2-Methylcyclohexanone		16 (−15), 14 (20)	2.98 (25, B)	34.06	0.102 7
3-Methylcyclohexanone		18 (−80), 12 (20)	3.06 (25, B)	33.06	0.092 5
4-Methylcyclohexanone		15 (−41), 12 (20)	3.07 (25, B)	32.83	0.093 5
Methylcyclopentane	0.507 (20), 0.478 (25)	1.985 (20)	0	24.63	0.116 3
Methyl decanoate			1.65 (20, Hx)	30.33	0.091 2
Methyl dichloroacetate				37.00	0.121 9
Methyl dodecanoate			1.70 (20, Hx)	31.37	0.089 3
N-Methylformamide	1.99 (15), 1.65 (25)	200.1 (15), 182.4 (25)	3.86 (25, B)	37.96 (30)	35.02 (50)
Methyl formate	0.360 (15), 0.319 (29)	8.5 (20)	1.77 (g)	28.29	0.157 2
Methyl heptanoate				28.95	0.098 7
2-Methyl-2-heptanol		3.38 (−7), 2.46 (25)			
2-Methyl-3-heptanol		3.37 (20), 3.75 (60)			
2-Methyl-4-heptanol		3.30 (20), 3.65 (60)			
3-Methyl-3-heptanol		3.74 (20), 2.89 (60)	1.63 (20, B)		
3-Methyl-4-heptanol		9.1 (−20), 7.4 (20)			
4-Methyl-3-heptanol		5.25 (20), 4.62 (55)			
4-Methyl-4-heptanol		2.87 (20), 3.27 (60)			
Methyl hexadecanoate				31.50	0.077 5
2-Methylhexane	0.378 (20)	1.92 (20)	0	21.22	0.096 64

* Mixed isomers.

3-Methylhexane	0.372 (20)	1.93 (20)	0	21.73	0.097 0
Methyl hexanoate			1.70 (20, Hx)	28.47	0.104 5
Methyl isobutyrate	0.523 (20), 0.419 (40)		1.98 (20, B)	25.99	0.113 1
Methyl methacrylate	0.632 (20)	2.9 (20)	1.68 (25, B)	28–29 (30)	
Methyl o-methoxybenzoate		7.7 (21)			
Methyl p-methoxybenzoate		4.3 (33)			
1-Methylnaphthalene		2.7 (20)	0.23 (20, B)	39.96	0.0934
Methyl o-nitrobenzoate		28 (25)	3.67 (30, B)		
Methyl octadecanoate				32.20	0.77 5
2-Methyloctane		1.97 (20)	0	23.76	0.094 0
4-Methyloctane		1.97 (20)	0	24.22	0.094 0
Methyl octanoate				29.93	0.100 2
Methyl oleate	4.88 (20)	3.211 (20)	2.9 (0)	31.3 (25)	25.4 (100)
2-Methylpentane	0.310 (20), 0.295 (25)	1.88 (20)	0	19.37	0.099 7
3-Methylpentane	0.307 (25), 0.292 (30)	1.895 (20)		20.26	0.106 0
2-Methyl-2,4-pentanediol	34.4 (20)		2.9 (0)	33.1 (20)	
4-Methylpentanenitrile	0.980 (20), 9.843 (30)	15.5 (22)	3.53 (25, B)	28.89	0.0917
Methyl pentanoate	0.713 (20)	4.3 (19)	1.62 (22, B)	27.85	0.104 4
2-Methyl-1-pentanol				26.98	0.081 9
3-Methyl-1-pentanol				26.92	0.078 9
4-Methyl-1-pentanol				25.93	0.074 3
2-Methyl-2-pentanol				25.07	0.086 1
3-Methyl-2-pentanol				27.14	0.091 9
4-Methyl-2-pentanol	4.074 (25)			24.67	0.082 1
2-Methyl-3-pentanol				26.43	0.091 4
3-Methyl-3-pentanol				25.48	0.088 8
4-Methyl-2-pentanone	0.585 (20), 0.522 (30)	13.11 (20), 11.78 (40)	3.20 (25, B)	23.64 (20)	19.62 (60)
4-Methyl-3-penten-2-one	0.879 (25)	15.6 (0), 15.1 (20)	1.84 (15, B)		
1-Methyl-1-phenylhydrazine		7.3 (19)	1.38 (20, B)		
Methyl phenyl sulfide				42.81	0.123 8
2-Methylpropane	0.007 44 (20, g)		0	12.83	0.123 6
2-Methylpropanenitrile	0.551 (15), 0.456 (30)	20.2	4.07 (g), 3.60 (20, B)		

TABLE 5.16 Viscosity, Dielectric Constant, Dipole Moment, and Surface Tension of Selected Organic Substances (*Continued*)

Substance	Viscosity, $mN \cdot s \cdot m^{-2}$	Dielectric constant ε	Dipole moment, D	Surface tension $dyn \cdot cm^{-1}$ a	b
2-Methyl-1-propanol	4.70 (15), 2.876 (30)	26 (−34), 17.93 (25)	2.96 (30, lq), 1.78 (20, B)	24.53	0.079 5
2-Methyl-2-propanol	3.316 (20), 2.039 (40)	10.9 (30), 8.49 (50)	1.67 (22, B)	20.02 (15)	19.10 (30)
2-Methylpropene			0.50 (g)	14.84	0.131 9
N-Methylpropionamide	6.06 (20), 3.56 (40)	185 (20), 151 (40)	3.59 (g)	31.20 (20)	29.12 (50)
Methyl propionate	0.477 (15)	6.21	1.70 (22, B)	27.58	0.125 8
2-Methylpropionic acid	1.213 (25), 1.126 (30)	2.73 (40)	1.08 (25, lq)	25.55 (20)	25.13 (25)
1-Methylpropyl acetate				25.72	0.105 4
2-Methylpropyl acetate	0.702 (20), 0.366 (78)	5.29 (20)	1.87 (22, B)	25.59	0.101 3
2-Methylpropylamine	21.7 (25)	4.43 (21)	1.27 (27)	24.48	0.109 2
2-Methylpropyl formate	0.680 (20)	6.41 (19)	1.88 (22)	26.14	0.112 2
Methyl propyl ketoxime		3.3 (20)			
2-Methylpyridine	0.805 (20), 0.710 (30)	9.8 (20)	1.96 (25, B)	36.11	0.124 3
3-Methylpyridine			2.41 (25, B)	37.35	0.115 3
4-Methylpyridine			2.60 (25, B)	37.71	0.114 1
N-Methyl-2-pyrrolidinone	1.666 (25)	32.0 (25)	4.09 (30, B)		
Methyl salicylate		9.41 (30)	2.47 (25, B)	42.15	0.117 4
Methyl tetradecanoate			1.62 (25, B)	31.00	0.080 0
2-Methyltetrahydrofuran	0.601 (0), 0.536 (10)	6.92 (0), 6.63 (10)			
Methyl thiocyanate	64.3 (0)	4.3 (19)	3.34 (20, B)	40.66	0.130 5
Morpholine	2.53 (15), 1.79 (30)	7.33 (25)	1.75 (25, lq), 1.52 (25, B)	37.63 (20)	36.24 (30)
Naphthalene	0.780 (100), 0.967 (80)	2.54 (85)	0	42.84	0.110 7
1-Naphthonitrile		16 (70)			
2-Naphthonitrile		17 (70)			
o-Nitroaniline		34.5 (90)	4.28 (20, B)		

Compound					
p-Nitroaniline		56.3 (160)	6.3 (25, B)	60.62	0.092 3
o-Nitroanisole			4.83 (g)	48.62	0.118 5
Nitrobenzene	2.165 (15), 1.55 (35)	34.82 (25), 24.9 (90)	4.22 (g), 3.96 (25, B)	46.34	0.115 7
m-Nitrobenzyl alcohol		22 (20)			
2-Nitrobiphenyl	12 (45)		3.82 (20, B)		
Nitroethane	0.677 (20), 0.63 (35)	28.06 (30), 27.4 (35)	3.61 (g)	35.27	0.125 5
Nitromethane	0.692 (15), 0.596 (30)	35.87 (30), 35.1 (35)	3.46 (g)	40.72	0.167 8
1-Nitro-2-methoxybenzene			4.83 (g)	48.62	0.118 5
o-Nitrophenol	2.343 (45)	17 (50)	3.14 (25, B)	47.35	0.117 4
1-Nitropropane	0.798 (25), 0.70 (35)	23.24 (30), 22.7 (35)	3.60 (g)	32.62	0.100 9
2-Nitropropane	0.750 (25)	25.52 (30)	3.76 (g)	32.18	0.115 8
N-Nitrosodimethylamine		53 (20)	4.01 (20, B)		
o-Nitrotoluene	2.37 (20), 1.63 (40)	27.4 (20), 22.0 (58)	3.72 (20, B)	44.10	0.117 4
m-Nitrotoluene	2.33 (20), 1.60 (40)	24 (20), 22 (58)	4.20 (20, B)	43.54	0.111 8
p-Nitrotoluene	1.20 (60)	22 (52)	4.47 (25, B)	42.26	0.097 4
Nonane	0.713 (20), 0.666 (25)	1.972 (20), 1.85 (110)	0	24.72	0.093 5
1-Nonanol	14.3 (20)		1.72 (20, B)	29.79	0.078 9
1-Nonene	0.620 (20), 0.586 (25)		0.59 (20, B)	24.90	0.093 8
(Z,Z)-9,12-Octadecadienoic acid	2.20 (20)	2.70 (70), 2.60 (120)	1.40 (18, Hx)	20.19	0.081 1
Octamethylcyclotetrasiloxane	0.82 (20)	2.4 (20)	0.42 (25, lq), 0.67 (25, B)	67.56 (25)	0.095 1
Octamethyltrisiloxane	0.546 (20), 0.433 (40)	2.3 (20)	0.64 (25, lq)	23.52	0.080 2
Octane	1.811 (15), 1.356 (30)	1.95 (20), 1.83 (110)	0	29.61	
Octanenitrile	5.828 (20), 4.690 (25)	13.90 (25)		29.2 (20)	28.7 (25)
Octanoic acid	10.64 (15), 6.125 (30)	2.45 (20)	1.15 (25, lq)	29.09	0.079 5
1-Octanol		11.3 (10), 10.34 (20)	1.72 (20, B)	27.96	0.082 6
2-Octanol		8.20 (20), 6.52 (40)	1.65 (20, B)		
2-Octanone		10.39 (20), 7.42 (100)	2.72 (15, B)		
1-Octene	0.470 (20), 0.447 (25)	2.084 (20)	0.34 (20, lq)	23.68	0.095 8
Oleic acid	38.80 (20), 27.64 (25)	2.46 (20), 2.45 (60)	1.44 (25, lq)	32.80 (20)	27.94 (90)
Oxalyl chloride		3.5 (21)	0.93 (20, B)		
2-Oxohexamethyleneimine	9 (78)			41.69	
4-Oxopentanoic acid			3.88 (25, B)		0.076 3

TABLE 5.16 Viscosity, Dielectric Constant, Dipole Moment, and Surface Tension of Selected Organic Substances (*Continued*)

Substance	Viscosity, mN·s·m⁻²	Dielectric constant ε	Dipole moment, D	Surface tension dyn·cm⁻¹	
				a	b
Palmitic acid		2.3 (70)		28.28	0.106 2
Paraldehyde	15.30 (25)	13.9 (25)	1.91 (25, lq)	39.2 (25)	
Parathion			4.98 (25, B)		
Pentachloroethane	2.741 (15), 2.070 (30)	3.73 (20)	0.92 (g), 0.98 (25, lq)	37.09	0.117 8
Pentadecane	2.814 (22)		0	28.78	0.085 7
cis-1,3-Pentadiene		2.32 (25)	0.50 (25, B)		
Pentanaldehyde		10.1 (17)	2.59 (20, B)	27.96	0.101 0
Pentane	0.237 (15), 0.215 (25)	2.011 (−90), 1.84 (20)	0	18.25	0.112 1
1,5-Pentanediol	128 (20)		2.45 (20, D)	43.2 (20)	
2,4-Pentanedione	0.6 (20)	25.7 (20), 17.39 (25)	3.03 (g), 2.5 (20, B)	33.28	0.114 4
Pentanenitrile	0.779 (15), 0.637 (30)	17.4 (21)	3.57 (25, B)	27.44 (20)	26.33 (30)
1-Pentanethiol		4.55 (25), 4.23 (50)	1.54 (25, lq)		
Pentanoic acid	2.359 (15), 1.774 (30)	2.66 (20)	1.61 (20, D)	28.90	0.088 7
1-Pentanol	4.650 (15), 2.987 (20)	16.9 (20), 13.9 (25)	1.71 (20, B)	27.54	0.087 4
2-Pentanol	5.130 (15), 2.780 (30)	13.82 (22)	1.66 (22, B)	25.96	0.100 4
3-Pentanol	7.337 (15), 3.306 (30)	13.02 (22)	1.64 (22, B)	24.60 (20)	23.76 (30)
2-Pentanone	0.473 (25)	15.45 (20), 11.73 (80)	2.72 (22, B)	24.89	0.065 5
3-Pentanone	0.493 (15), 0.423 (30)	19.4 (−20), 17.00 (20)	2.72 (20, B)	27.36	0.104 7
1-Pentene	0.24 (0)	2.10 (20)	0.34 (20, lq)	18.20	0.109 9
cis-2-Pentene				19.73	0.117 2
trans-2-Pentene				18.90	0.099 7
Pentyl acetate	0.924 (20), 0.862 (25)	4.75 (20)	1.72 (g), 1.91 (25, B)	27.66	0.099 4
Pentylamine	1.018 (20)	4.5 (22)	1.55 (30, B)	24.4 (13)	
Pentyl formate		6.5 (30)		28.09	0.102 3
Pentyl nitrate		9 (18)			

Compound					
Phenanthrene	6.024 (35), 3.421 (50)	2.8 (20)	0	43.54	0.106 8
Phenol		9.78 (60)	1.53 (20, B)	46.26	0.078 8
Phenoxyacetaldehyde		4.8 (20)			
Phenoxyacetylene		4.8 (20)	1.42 (25, lq)		
2-Phenylacetamide	1.799 (45)				
Phenyl acetate	1.93 (25)	5.23 (20)	1.54 (22, B)	44.57	0.115 5
Phenylacetonitrile		19.0 (25), 8.5 (234)	3.47 (27, B)	42.88	0.103 8
Phenylacetylene		3.0 (20)	0.72 (20, B)		
1-Phenylethanol		13 (20), 7.6 (90)	1.51 (20, B)	48.14	0.129 2
Phenylhydrazine		7.2 (21)	1.67 (25, B)	42.73	0.108 6
Phenyl isocyanate		8.8 (20)			
Phenyl isothiocyanate		10 (20)		.	
1-Phenylpropene		2.7 (20)			
2-Phenylpropene		2.3 (20)			
3-Phenylpropene		2.6 (20)			
Phenyl propyl ether				34.27	0.105 6
Phenyl salicylate		6.3 (50)		45.20	0.097 6
Phosgene		4.7 (0), 4.3 (22)			
Phthalide		36 (75)	0.60 (25, B)	28.35	0.094 4
DL-α-Pinene	1.61 (25)	2.64 (25)		28.26	0.093 4
L-β-Pinene	1.70 (20), 1.41 (25)	2.76 (20)		31.79	0.115 3
Piperidine	1.679 (15), 1.224 (30)	5.8 (20)	1.19 (25, B)	9.22	0.087 4 (lq)
Propane (g)	0.008 1 (20), 0.0010 7 (125)	1.6 (0)	0		
1,2-Propanediamine	1.46	10.2	1.96 (25, B)	72.0 (25)	
1,3-Propanediamine	17.85	9.55	2.27 (25, D)		
1,2-Propanediol	56.0 (20), 18.0 (40)	32.0 (20)	2.52 (25, D)	47.43	0.090 3
1,3-Propanediol	56.0 (20), 18.0 (40)	35.0 (20)	1.55 (25, lq)	27.38	0.127 2
1-Propanethiol			1.64 (25, lq)	24.26	0.117 4
2-Propanethiol					
1-Propanol	2.522 (15), 1.722 (30)	22.2 (20), 20.33 (25)	1.67 (g), 1.75 (25, B)	25.26	0.077 7
2-Propanol	2.859 (15), 1.765 (30)	18.3 (25), 16.24 (40)	1.69 (g), 1.66 (30, B)	22.90	0.078 9
2-Propenaldehyde			3.04 (g), 2.90 (25, B)		

TABLE 5.16 Viscosity, Dielectric Constant, Dipole Moment, and Surface Tension of Selected Organic Substances (*Continued*)

Substance	Viscosity, mN · s · m⁻²	Dielectric constant ε	Dipole moment, D	Surface tension dyn · cm⁻¹ a	b
Propene (g)	0.008 43 (20), 0.009 33 (50)	1.88 (20), 1.44 (90)	0.35 (g)	9.99	0.142 7 (lq)
2-Propen-1-ol	1.363 (20), 0.914 (40)	21.6 (15)	1.63 (g)	27.53	0.090 2
Propionaldehyde	0.357 (15), 0.317 (27)	18.5 (17)	2.75 (g), 2.57 (20, B)		
Propionamide			3.4 (30, B)	39.05	0.090 9
Propionic acid	1.175 (15), 0.956 (30)	3.30 (10), 3.44 (40)	1.76 (g), 1.77 (25, D)	28.68	0.099 3
Propionic anhydride	1.144 (20), 1.061 (25)	18.3 (16)		30.30 (20)	29.70 (25)
Propionitrile	0.454 (15), 0.389 (30)	22.2 (20), 24.2 (50)	4.06 (g), 3.60 (20, B)	29.63	0.115 3
Propyl acetate	0.585 (20), 0.460 (40)	5.69 (19)	1.86 (25, B)	26.60	0.112 0
Propylamine	0.343 (25)	5.31 (20)	1.17 (g), 1.36 (20, B)	24.86	0.124 3
Propylbenzene		2.37 (20), 2.351 (30)	0.35 (25, lq)	31.13	0.107 5
Propyl benzoate				36.55	0.106 9
Propyl butyrate	0.831 (20)	4.3 (20)		27.06	0.100 0
Propyl chloroacetate				32.91	0.108 3
Propylene carbonate	2.53	64.4			
Propylene oxide	0.327 (20), 0.28 (25)		2.00 (g)		
Propyleneimine	0.491 (25)		1.77 (g, *cis*), 1.60 (g, *trans*)		
Propyl formate	0.574 (20), 0.417 (40)	7.72 (19)	1.91 (22, B)	26.77	0.111 9
Propyl isobutyrate	0.831 (20)			25.83	0.101 5
Propyl nitrate		14 (18)	3.01 (20, B)	29.67	0.123 7
Propyl pentoate	1.053 (20)	4 (19)		27.72	0.098 4
Propyl propionate	0.673 (20)	4.7 (20)	1.79 (22, B)	26.85	0.105 9
Propyne			0.75 (g)	14.51	0.148 2
2-Propyn-1-ol	1.68 (20)	24.5 (20)	1.78 (25, B)		
Pulegone		9.5 (20)	2.00 (25, B)	38.59	0.127 0

5.120

Pyradazine			3.97 (35, D)	50.55	0.103 6
Pyrazine	1.130 (10), 0.829 (30)	2.8 (54)	2.20 (20, B), 2.25 (g)	39.82	0.130 6
Pyridine		12.3 (25), 9.4 (116)	2.44 (35, D)	32.85	0.101 0
Pyrimidine	1.352 (20), 1.233 (25)	7.48 (18), 8.13 (25)	1.80 (25, B)	39.81	0.110 0
Pyrrole			1.58 (20, B)	31.48	0.090 0
Pyrrolidine	13.3 (25)		3.55 (25, B)		
2-Pyrrolidone					
Quinoline	4.354 (15), 3.37 (25)	9.00 (25)	2.18 (25, B)	45.25	0.106 3
Safrole	2.294 (25)	3.1 (21)			
Salicylaldehyde	2.90 (20), 1.67 (45)	13.9 (20)	3.1 (30, lq), 2.86 (20, B)	45.38	0.124 2
Squalane	6.08 (20)		0		
Squalene	12 (25)		0.68 (25, B)		
D-Sorbitol		33 (80)			
Stearic acid	11.6 (70)	2.29 (70), 2.26 (100)	1.76 (25, D)	32.0 (20)	30.98 (30)
Styrene	0.751 (20), 0.696 (25)	2.43 (25), 2.32 (75)	0.13 (25, lq)	53.26	0.107 9
Succinonitrile	2.591 (60), 2.008 (75)	56.5 (57), 54 (68)	3.68 (30, toluene)		
1,1,2,2-Tetrabromoethane	13.950 (11), 9.797 (20)	8.6 (3), 7.0 (22)	1.29 (20, Hx)	52.37	0.146 3
1,1,2,2-Tetrachlorodifluoroethane	1.21 (25), 1.208 (30)	2.52 (25)		26.13	0.113 3
1,1,2,2-Tetrachloroethane	1.844 (15), 1.456 (30)	8.20 (20)	1.29 (g), 1.45 (25, Hx)	38.75	0.126 8
Tetrachloroethylene	1.932 (15), 0.798 (30)	2.30 (25)	0	32.86 (15)	31.27 (30)
Tetradecamethylcyclohepta-siloxane		2.7 (20)			
Tetradecamethylhexasiloxane		2.5 (20)	1.58 (20, lq)	17.42 (25)	0.086 9
Tetradecane	2.131 (22)		0	28.30	0.093 2
Tetradecanoic acid			0.76 (25, B)	33.90	0.070 3
1-Tetradecanol		4.72 (38), 4.40 (48)	1.69 (25, C)	32.72	
Tetraethylene glycol	44.9 (25)		5.84 (20, lq)	45 (25)	
Tetraethyllead			0.3 (20, B)	30.50	0.096 9
Tetraethylsilane			0	25.22	0.107 9
Tetraethyl silicate		4.1 (20)	1.72 (32, B)	23.63	0.097 9
Tetrahydrofuran	0.55 (20), 0.460 (25)	11.6 (−70), 7.58 (25)	1.75 (25, B)	26.5 (25)	

TABLE 5.16 Viscosity, Dielectric Constant, Dipole Moment, and Surface Tension of Selected Organic Substances (*Continued*)

Substance	Viscosity, mN · s · m^{-2}	Dielectric constant ε	Dipole moment, D	Surface tension dyn · cm^{-1}	
				a	b
2,5-Tetrahydrofurandimethanol	225 (25)				
Tetrahydro-2-furanmethanol	6.24 (20)	13.61 (23)	2.12 (35, lq)	39.96	0.100 8
1,2,3,4-Tetrahydronaphthalene	2.202 (20), 2.003 (25)	2.76 (20)	0.60 (25, lq)	35.55	0.095 4
1,2,3,4-Tetrahydro-2-naphthol		11.7 (20), 6.7 (90)			
Tetrahydropyran	0.826 (20), 0.764 (25)	5.61 (25)	1.55 (25, B)	34.1 (25)	
Tetrahydropyran-2-methanol	11.0 (20)			35.5 (30)	
Tetrahydrothiophene-1,1-dioxide	9.87 (30)	43.3 (30)	4.81 (25, B)		
Tetrahydrothiophene oxide	52 (30), 19 (80)	42.5 (30)	3.47 (25, B)		
1,1,2,2-Tetramethylurea	1.76 (20)	23.06			
Tetranitromethane		2.32 (20)	0		
Tetrathiomethylmethane		2.82 (70)			
Thiacyclohexane	1.042 (20), 0.971 (25)		1.90 (25, B)	36.06 (20)	33.74 (40)
Thiacyclopentane				38.44	0.134 2
Thioacetic acid		12.8 (20)			
2,2'-Thiodiethanol	65.2 (20)			53.8 (20)	
Thiophene	0.662 (20), 0.353 (82)	2.76 (16), 2.57 (25)	0.55 (g), 0.52 (25, B)	34.00	0.132 8
Thymol			1.55 (25, B)	33.95	0.082 1
Toluene	0.623 (15), 0.523 (30)	2.385 (20), 2.364 (30)	0.45 (20, lq)	30.90	0.118 9
p-Toluenesulfonyl chloride				42.41	0.090 3
o-Toluidine	5.195 (15), 4.39 (20)	6.34 (18), 5.71 (58)	1.60 (25, B)	42.87	0.109 4
m-Toluidine	4.418 (15), 2.741 (30)	5.95 (18), 5.45 (58)	1.45 (25, B)	40.33	0.097 9
p-Toluidine	1.945 (45), 1.557 (60)	4.98 (54)	1.52 (25, B)	39.58	0.095 7
m-Tolunitrile			4.21 (22, B)	38.85	0.101 3
p-Tolunitrile			4.47 (20, B)	39.79	0.110 0
Tribenzylamine			0.65 (20, B)	42.41	0.095 3

Tributyl phosphite	1.9 (25)		1.92 (20, C)	27.57	0.086 5
2,2,2-Tribromoacetaldehyde		7.6 (20)	1.70 (20, B)		
Tribromoethane	2.152 (15), 1.741 (30)	4.39 (20)	0.99 (g)	48.14	0.130 8
1,2,3-Tribromopropane		6.45 (20)	1.59 (25, B)	47.99	0.126 7
Tributylamine	1.35 (25)		0.78 (25, B)	26.47	0.083 1
Tributyl borate	1.776 (20), 1.601 (25)		0.78 (25, C)	26.2 (20)	25.8 (25)
Tributyl phosphate	11.1 (15) 3.39 (25)	7.96 (30)	3.07 (25, B)	28.71	0.066 6
Trichloroacetaldehyde		7.6 (−40), 4.9 (20)	1.96 (25, B)	27.66	9.119 7
Trichloroacetic acid		4.6 (60)	1.1 (25, B, dimer)	35.4	0.089 5
Trichloroacetonitrile		7.85 (19)	1.93 (19, lq)		
1,1,1-Trichloroethane	0.903 (15), 0.725 (30)	7.1 (7), 7.52 (20)	1.79 (g), 1.6 (25, B)	28.28	0.124 2
1,1,2-Trichloroethane	0.119 (20), 0.110 (25)	8.78 (23)	1.45 (g)	37.40	0.135 1
Trichloroethylene	0.566 (20), 0.532 (25)	3.42 (16)	0.77 (30, lq), 0.95 (30, B)	29.5 (20)	28.8 (25)
Trichlorofluoromethane	0.42 (25), 0.011 (25, g)	2.28 (29)	0.45 (g), 0.49 (lq)	18 (25)	
Trichloromethylsilane	0.47 (20)		1.87 (25, B)	20.3 (20)	
2,4,6-Trichlorophenol		7.5 (20)	1.88 (25, D)	43.13	0.095 5
1,2,3-Trichloropropane			1.61 (g)	37.8 (20)	37.05 (25)
Trichlorosilane	0.332 (20), 0.316 (25)	6.9 (21)	0.86 (g), 0.98 (25, B)	20.43	0.107 6
α,α,α-Trichlorotoluene	3.07 (10), 2.55 (17)		2.17 (20, B)		
1,1,2-Trichloro-1,2,2-trifluoro-ethane	0.711 (20), 0.627 (30)	2.41 (25)		17.75 (20)	16.56 (30)
Tridecane	1.883 (20), 1.55 (23)		0	27.73	0.087 2
1-Tridecene				28.01	0.088 4
Triethanolamine	613.6 (25), 208.1 (40)	29.36 (25)	3.57 (25, B)		
Triethylaluminum		2.9 (20)			
Triethylamine	0.394 (15), 0.363 (30)	2.42 (25)	0.66 (g), 0.9 (25, B)	22.70	0.099 2
Triethylene glycol	49.0 (20), 8.5 (60)	23.7 (20)	5.58 (20, lq)	47.33	0.088 0
Triethyl phosphate	1.684 (40), 1.376 (55)	13.43 (15), 10.93 (65)	3.08 (25, B)	31.81	0.092 8
Triethyl phosphite	0.72 (25)	5.0	1.82 (25, D)	25.73	0.087 8
Trifluoroacetic acid	0.926 (20), 0.653 (40)	8.55 (20), 5.76 (50)	2.28 (g)	15.64	0.184 4
2,2,2-Trifluoroethanol	1.996 (20)		2.03 (25, cHx)	20.6 (33)	
α,α,α-Trifluorotoluene		9.2 (30), 8.1 (60)			

TABLE 5.16 Viscosity, Dielectric Constant, Dipole Moment, and Surface Tension of Selected Organic Substances (*Continued*)

Substance	Viscosity, mN · s · m^{-2}	Dielectric constant ε	Dipole moment, D	Surface tension dyn · cm^{-1}	
				a	b
Trimethylamine	0.321 (−33)	2.4 (25)		16.24	0.113 3
1,2,3-Trimethylbenzene	0.894 (15), 0.730 (30)	2.636 (20), 2.609 (30)	0.56 (20, lq)	30.91	0.104 0
1,2,4-Trimethylbenzene	1.154 (20)	2.38 (20), 2.36 (30)	0.30 (20, lq)	31.76	0.102 5
1,3,5-Trimethylbenzene		2.28	0	29.79	0.089 7
Trimethyl borate	0.579 (20)	8 (20)	0.82 (25, C)		
2,2,3-Trimethylbutane		1.93 (20)	0	20.70	0.097 3
cis,cis-1,3,5-Trimethyl-cyclohexane	0.632 (20), 0.558 (30)				
trans-1,3,5-Trimethyl-cyclohexane	0.714 (20), 0.624 (30)				
Trimethylene sulfide	0.638 (20), 0.607 (25)		1.78 (25, B)	36.3 (20)	35.0 (30)
3,5,5-Trimethyl-1-hexanol	11.06 (25)				
2,6,8-Trimethyl-4-nonanone	1.9 (20)				
1,3,5-Trimethyl-2-oxa bicyclo[2.2.2]octane		4.57 (24)	1.54 (25, C)	32.1 (20)	31.1 (25)
2,2,3-Trimethylpentane	0.598 (20)	1.962 (20)	0	22.46	0.089 5
2,2,4-Trimethylpentane	0.502 (20)	1.940 (20)	0	20.55	0.088 8
Trimethyl phosphite	0.61 (20)		1.83 (20, C)	27.18 (20)	24.88 (40)
2,4,6-Trimethylpyridine	1.498 (20), 1.496 (25)	6.6	1.95 (25, B)		
Triphenylamine				46.2	0.095 5
Triphenyl phosphite	25.18 (15), 6.95 (45)	3.67 (45), 3.57 (65)	2.04 (25, B)	24.58	0.087 8
Tripropylamine			0.58 (20, lq), 0.76 (20, B)	34 (25)	
Tripropylene glycol	56.1 (25)				
Tripropylene glycol butyl ether	6.58 (25)			28.8 (25)	
Tripropylene-glycol ethyl ether	5.17 (25)			28.2 (25)	

Tripropylene glycol isopropyl ether	7.7 (25)			27.4 (25)	
Tripropylene glycol methyl ether	5.96 (25)	30 (20)		30.0 (25)	
Tris(dimethylamino) phosphine oxide	3.34 (30)				
Tris(4-ethylphenyl) phosphite	30.22 (15), 9.047 (45)	3.74 (15), 3.61 (45)	2.08 (25, B)		
Tris(m-tolyl) phosphite	37.55 (15), 9.132 (45)	3.67 (15), 3.53 (45)	1.62 (25, B)		
Tris(p-tolyl) phosphite	35.52 (15), 8.794 (45)	3.88 (15), 3.74 (45)	1.77 (25, B)		
Tritolyl phosphate	38.8 (35), 16.8 (55)	6.92 (40)	2.84 (40, C)	40.9 (20)	
Undecane	1.186 (20), 0.761 (50)	2.00 (20), 1.84 (150)	0	26.26	0.090 1
2-Undecanone	1.61 (30)		2.71 (15, B)		
Urea			4.59 (25, D)		
Vinyl acetate	0.421 (20)		1.79 (25, B)	23.95 (20)	22.54 (30)
o-Xylene	0.809 (20), 0.627 (40)	2.57 (20), 2.54 (30)	0.62 (g), 0.52 (25, 1q)	32.51	0.110 1
m-Xylene	0.617 (20), 0.497 (40)	2.37 (20), 2.35 (30)	0.33 (20, 1q), 0.37 (20, B)	31.23	0.110 4
p-Xylene	0.644 (20), 0.513 (40)	2.26 (20), 2.22 (50)	0	30.69	0.107 4
Xylitol		40 (20)			

TABLE 5.17 Viscosity, Dielectric Constant, Dipole Moment, and Surface Tension of Selected Inorganic Substances

For the majority of compounds the dependence of the surface tension γ on the temperature can be given as

$$\gamma = a - bt$$

where a and b are constants and t is the temperature in degrees Celsius.

Substance	Viscosity, $mN \cdot s \cdot m^{-2}$	Dielectric constant ε	Dipole moment, D	Surface tension, $dyn \cdot cm^{-1}$	
				a	b
Air (20°C)	0.018 2	1.000 536 4			
AlBr$_3$		3.38[100]	5.2		
Ar					
(g, 20°C)	0.022 3	1.000 517 2			
(lq)		1.538[-191]	0	34.28	0.249 3
AsBr$_3$		8.83[35]	1.61	54.51	0.1043
AsCl$_3$		12.6[20]	1.59	41.67	0.097 81
AsH$_3$ (arsine)		2.05[20]	0.20		
BBr$_3$		2.58[0]	0	31.90	0.128 0
BCl$_3$			0		
BF$_3$			0	-2.92	0.203 0
B$_2$H$_6$ (diborane)		1.872[-92.5]	0	-3.13	0.178 5
B$_5$H$_9$			2.13		
B$_3$H$_6$N$_3$ (triborotriazine)			0		
Br$_2$					
(g, 20°C)		1.012 8			
(lq)	1.03[16]	3.09[20]	0	45.5	0.182 0
BrF$_3$	2.22[20]		1.1	38.30	0.099 9
BrF$_5$	0.62[24]	7.91[24.5]	1.51	25.24	0.109 8
Cl$_2$					
(g, 20°C)	0.013 2		0		
(lq)		1.91[14]			
ClF$_3$	0.48[12]	4.29[25]	0.554	26.9	0.166 0
ClO$_3$F (perchloryl fluoride)			0.023	12.24	0.157 6
CO					
(g)	0.017 5[20]	1.000 70[0]	0.112		
(lq)				-30.20	0.207 3
CO$_2$					
(g, 20°C)	0.014 7	1.000 922	0		
(lq)	0.071[20]	1.60[0°C, 50 atm]			
COCl$_2$		4.34[22]	1.17	22.59	0.145 6
COF$_2$			0.95		
COS			0.712	12.12	0.177 9
COSe		3.47[10]	0.73		
CS			1.98		
CS$_2$					
(g)		1.002 9[0]	0		
(lq)	0.375[20]	2.6[20]			

TABLE 5.17 Viscosity, Dielectric Constant, Dipole Moment, and Surface Tension of Selected Inorganic Substances (*Continued*)

Substance	Viscosity, $mN \cdot s \cdot m^{-2}$	Dielectric constant ε	Dipole moment, D	Surface tension, $dyn \cdot cm^{-1}$	
				a	b
CrO_2Cl_2 [chromyl(VI) chloride]		2.6^{20}	0.47		
D_2 (deuterium)		1.277^{-253}			
DH				6.537	0.188 3
D_2O	1.098^{25}	78.25^{25}	1.87	$(71.72^{20})*$	$(68.38^{40})*$
F_2		1.54^{-202}		-16.10	0.164 6
$GaCl_3$			0.85	35.0	0.100 0
$GeCl_4$		2.430^{25}	0	$(22.44^{30})*$	
H_2					
(g, 20°C)	0.008 8	1.000 253 8	0		
(lq)		$1.228^{20.4\,K}$			
HBr					
(g)		$1.003\ 13^0$	0.82		
(lq)	0.83^{-67}	3.82^{25}		13.10	0.207 9
HCl					
(g)		$1.004\ 6^0$	1.08		
(lq)	0.51^{-95}	4.60^{28}			
HCN	0.206^{18}	116^{20}	2.98	$(19.45^{10})*$	$(18.33^{20})*$
HCNO (isocyanate)			1.6		
HCNS (isothiocyanate)			1.7		
HF	0.256^0	83.6^0	1.82	10.41	0.078 67
HI					
(g)		$1.002\ 34^0$	0.44		
(lq)		2.90^{22}			
HN_3 (azide)			0.8		
H_2O (see Table 4-12)					
H_2O_2	1.25^{20}	84.2^0	2.2	78.97	0.154 9
HNO_3			2.17		
H_2S					
(g)		$1.00\ 4\ 0^0$	0.97		
(lq)	0.412^0	5.93^{10}		48.95	0.175 8
H_2Se			0.24	22.32	0.148 2
H_2SO_4	24.54^{25}	100^{25}			
HSO_3Cl (chlorosulfonic acid)	2.43^{20}	60^{20}			
HSO_2F (fluorosulfonic acid	1.56^{25}	$\sim120^{25}$			
H_2Te			<0.2	29.03	0.261 9
He					
(g, 20°C)	0.019 6	1.000 065 0	0		
Hg	1.552^{20}		0	490.6	0.204 9

* Actual values of surface tension.

TABLE 5.17 Viscosity, Dielectric Constant, Dipole Moment, and Surface Tension of Selected Inorganic Substances (*Continued*)

Substance	Viscosity, $mN \cdot s \cdot m^{-2}$	Dielectric constant ε	Dipole moment, D	Surface tension, $dyn \cdot 2m^{-1}$	
				a	b
I_2	1.98^{116}	11.1^{118}	0		
IF_5			2.18	33.16	0.131 8
Kr					
(g, 20°C)	0.025 0		<0.05		
(lq)				40.576	0.289 0
Ne (g, 20°C)	0.031 3	1.000 063 9	0		
N_2					
(g, 20°C)	0.017 6	1.000 548 0	0		
(lq)		1.454^{-203}		26.42	0.226 5
NH_3					
(g)		$1.007\ 2^0$	1.47		
(lq)	$0.254^{-33.5}$	$22.4^{-33.4}$		$(37.91^{-50})*$	$(35.38^{-40})*$
N_2H_4 (hydrazine)	0.97^{20}	52.9^{20}	1.75		
NO			0.153	−67.48	0.585 3
N_2O					
(g)	$0.014\ 6^{20}$	$1.001\ 13^0$	0.167		
(lq)		1.52^{15}		5.09	0.203 2
NO_2			0.316		
N_2O_4		2.56^{15}	0.5		
NOBr (nitrosyl bromide)		13.4^{15}	1.8		
NOCl		18.2^{12}	1.9	29.49	0.149 3
NOF			1.81	14.00	0.116 5
NO_2F (nitryl fluoride)			0.47	8.26	0.185 4
O_2					
(g, 20°C)	0.020 4	1.000 494 7	0		
(lq)		1.507^{-193}		−33.72	0.256 1
O_3			0.53	$(38.1^{-183})*$	
OF_2 (oxygen difluoride)			0.297		
OsO_4			0		
PBr_3		3.9^{20}	0.5	45.34	0.128 3
PCl_3		3.43^{25}	0.78	31.14	0.126 6
PCl_5		2.7^{165}	0.9		
PF_5			0		
PH_3		2.9^{15}	0.58		
PI_3		4.12^{65}	0	61.66	0.067 71
$POCl_3$	1.065^{25}	13.7^{25}	2.41	35.22	0.127 5
POF_3			1.76		
$PSCl_3$		5.8^{22}	1.42	37.00	0.127 2
$PbCl_4$		2.78^{20}			
S_2Cl_2 dimer		4.79^{15}	1.0	46.23	0.146 4

* Actual values of surface tension.

TABLE 5.17 Viscosity, Dielectric Constant, Dipole Moment, and Surface Tension of Selected Inorganic Substances (*Continued*)

Substance	Viscosity, $mN \cdot s \cdot m^{-2}$	Dielectric constant ε	Dipole moment, D	Surface tension, $dyn \cdot cm^{-1}$	
				a	b
S_2F_2					
FSSF isomer			1.45		
S=SF$_2$ isomer			1.03		
SF_4			0.632	12.87	0.173 4
SF_6			0	5.66	0.119 0
S_2F_{10}		2.020[20]	0		
SO_2					
(g)	0.012 6[29]	1.009 3[0]	1.63		
(lq)		15.0[0]		26.58	0.194 8
SO_3		3.11[18]	0		
$SOBr_2$ (thionyl bromide)		9.06[20]	9.11		
$SOCl_2$		9.25[20]	1.45	36.10	0.141 6
SO_2Cl_2 (sulfuryl chloride)		9.15[20]	1.81	32.10	0.132 8
$SbCl_3$		33.2[75]	3.93	47.87	0.123 8
$SbCl_5$		3.22[20]	0		
SbF_5				49.07	0.193 7
SbH_3			0.12		
SeF_4				38.61	0.127 4
SeF_6			0		
$SeOCl_2$		55[25]	2.64		
$SiCl_4$		2.40[16]	0	20.78	0.099 62
SiF_4			0		
SiH_4			0		
$SiHCl_3$			0.86	20.43	0.107 6
$SnBr_4$			0		
$SnCl_4$		2.89[20]	0	29.92	0.113 4
TeF_6			0		
$TiCl_4$		2.80[20]	0	(33.54[20])*	(31.06[40])*
UF_6					
(g)		1.002 92[67]	0		
(lq)		2.18[65]		25.5	0.124 0
VCl_4		3.05[25]	0		
$VOBr_3$		3.6[25]			
$VOCl_3$		3.4[25]	0.3	(36.36[20])*	(33.60[40])*
Xe (g, 20°C)	0.022 8	1.001 23	0		

* Actual values of surface tension.

TABLE 5.18 Refractive Index, Viscosity, Dielectric Constant, and Surface Tension of Water at Various Temperatures

Temp., °C	Refractive index n_D	Viscosity, $mn \cdot s \cdot m^{-2}$	Dielectric constant ε	Surface tension, $dyn \cdot cm^{-1}$
0	1.333 95	1.770 2	87.74	75.83
5	1.333 88	1.510 8	85.76	75.09
10	1.333 69	1.303 9	83.83	74.36
15	1.333 39	1.137 4	81.95	73.62
20	1.333 00	1.001 9	80.10	72.88
21	1.332 90	0.976 4	79.73	72.73
22	1.332 80	0.953 2	79.38	72.58
23	1.332 71	0.931 0	79.02	72.43
24	1.332 61	0.910 0	78.65	72.29
25	1.332 50	0.890 3	78.30	72.14
26	1.332 40	0.870 3	77.94	71.99
27	1.332 29	0.851 2	77.60	71.84
28	1.332 17	0.832 8	77.24	71.69
29	1.332 06	0.814 5	76.90	71.55
30	1.331 94	0.797 3	76.55	71.40
35	1.331 31	0.719 0	74.83	70.66
40	1.330 61	0.652 6	73.15	69.92
45	1.329 85	0.597 2	71.51	69.18
50	1.329 04	0.546 8	69.91	68.45
55	1.328 17	0.504 2	68.35	67.71
60	1.327 25	0.466 9	66.82	66.97
65	1.326 16	0.434 1	65.32	66.23
70	1.325 11	0.405 0	63.86	65.49
75	1.323 99	0.379 2	62.43	64.75
80		0.356 0	61.03	64.01
85		0.335 2	59.66	63.28
90		0.316 5	58.32	62.54
95		0.299 5	57.01	61.80
100		0.284 0	55.72	61.80

5.6.1 Refractive Index

The refractive index n is the ratio of the velocity of light in a particular substance to the velocity of light in vacuum. Values reported refer to the ratio of the velocity in air to that in the substance saturated with air. Usually the yellow sodium doublet lines are used; they have a weighted mean of 589.26 nm and are symbolized by D. When only a single refractive index is available, as in Tables 5.16 and 5.17, approximate values over a small temperature range may be calculated using a mean value of 0.000 45 per degree for dn/dt, and remembering that n_D decreases with an increase in temperature. If a transition point lies within the temperature range, extrapolation is not reliable.

The *specific refraction* r_D is given by the Lorentz and Lorenz equation,

$$R_D = \frac{n_D^2 - 1}{n_D^2 + 2} \cdot \frac{1}{\rho}$$

where ρ is the density at the same temperature as the refractive index, and is independent of temperature and pressure. The molar refraction is equal to the specific refraction multiplied by the molecular weight. It is a more or less additive property of the groups or elements comprising the compound. A set of atomic refractions is given in Table 5.19; an extensive discussion will be found in Bauer, Fajans, and Lewin, in *Physical Methods of Organic Chemistry,* 3d ed., A. Weissberger (ed.), vol. 1, part II, chap. 28, Wiley-Interscience, New York, 1960.

The empirical Eykman equation

$$\frac{n_D^2 - 1}{n_D + 0.4} \cdot \frac{1}{\rho} = \text{constant}$$

offers a more accurate means for checking the accuracy of experimental densities and refractive indices, and for calculating one from the other, than does the Lorentz and Lorenz equation.

The refractive index of moist air can be calculated from the expression

$$(n - 1) \times 10^6 = \frac{103.49}{T} p_1 + \frac{177.4}{T} p_2 + \frac{86.26}{T} \left(1 + \frac{5748}{T}\right) p_3$$

where p_1 is the partial pressure of dry air (in mmHg), p_2 is the partial pressure of carbon dioxide (in mmHg), p_3 is the partial pressure of water vapor (in mmHg), and T is the temperature (in kelvins).

TABLE 5.19 Atomic and Group Refractions

Group	Mr_D	Group	Mr_D
H	1.100	N (primary aliphatic amine)	2.322
C	2.418	N (sec-aliphatic amine)	2.499
Double bond (C=C)	1.733	N (tert-aliphatic amine)	2.840
Triple bond (C≡C)	2.398	N (primary aromatic amine)	3.21
Phenyl (C_6H_5)	25.463	N (sec-aromatic amine)	3.59
Naphthyl ($C_{10}H_7$)	43.00	N (tert-aromatic amine)	4.36
O (carbonyl) (C=O)	2.211	N (primary amide)	2.65
O (hydroxyl) (O—H)	1.525	N (sec amide)	2.27
O (ether, ester) (C—O—)	1.643	N (tert amide)	2.71
F (one fluoride)	0.95	N (imidine)	3.776
(polyfluorides)	1.1	N (oximido)	3.901
Cl	5.967	N (carbimido)	4.10
Br	8.865	N (hydrazone)	3.46
I	13.900	N (hydroxylamine)	2.48
S (thiocarbonyl) (C=S)	7.97	N (hydrazine)	2.47
S (thiol) (S—H)	7.69	N (aliphatic cyanide) (C≡N)	3.05
S (dithia) (—S—S—)	8.11	N (aromatic cyanide)	3.79
Se (alkyl selenides)	11.17	N (aliphatic oxime)	3.93
3-membered ring	0.71	NO (nitroso)	5.91
4-membered ring	0.48	NO (nitrosoamine)	5.37
		NO_2 (alkyl nitrate)	7.59
		(alkyl nitrite)	7.44
		(aliphatic nitro)	6.72
		(aromatic nitro)	7.30
		(nitramine)	7.51

Example: 1-Propynyl acetate has $n_D = 1.4187$ and density $= 0.9982$ at 20°C; the molecular weight is 98.102. From the Lorentz and Lorenz equation,

$$r_D = \frac{(1.4187)^2 + 1}{(1.4187)^2 + 2} \cdot \frac{1}{0.9982} = 0.2528$$

The molar refraction is

$$Mr_D = (98.102)(0.2528) = 24.80$$

From the atomic and group refractions in Table 5.19, the molar refraction is computed as follows:

6 H	6.600
5 C	12.090
1 C≡C	2.398
1 O(ether)	1.643
1 O(carbonyl)	2.211
	$Mr_D = 24.942$

5.6.2 Surface Tension

The surface tension of a liquid, γ, is the force per unit length on the surface that opposes the expansion of the surface area. In the literature the surface tensions are expressed in $dyn \cdot cm^{-1}$; $1 \; dyn \cdot cm^{-1} = 1 \; mN \cdot m^{-1}$ in the SI system. For the large majority of compounds the dependence of the surface tension on the temperature can be given as

$$\gamma = a - bt$$

where a and b are constants and t is the temperature in degrees Celsius. The values of a and b given in Tables 5.16 and 5.17 can be used to calculate the values of surface tension for the particular compound within its liquid range. For example, the least-squares constants for acetic anhydride (liquid from -73 to 140°C) are 35.52 and 0.1436, respectively. At 20°C, $\gamma = 35.52 - 0.1436(20) = 32.64 \; dyn \cdot cm^{-1}$.

A compilation of data of some 2200 pure liquid compounds has been prepared by Jasper, *J. Phys. Chem. Reference Data* **1**:841 (1972).

5.6.3 Dipole Moments

The permanent dipole moment of an isolated molecule depends on the magnitude of the charge and on the distance separating the positive and negative charges. It is defined as

$$\mu = \left(\sum_i q_i r_i \right)$$

where the summation extends over all charges (electrons and nuclei) in the molecule. The numerical values of the dipole moment, expressed in the c.g.s. system of units, are in debye units, D, where $1 \; D = 10^{-18}$ esu of charge $\times$ centimeters. The conversion factor to SI units is

$$1 \; D = 3.335 \; 64 \times 10^{-30} \; C \cdot m \quad \text{[coulomb-meter]}$$

Tables 5.16 and 5.17 contain a selected group of compounds for which the dipole moment is given. An extensive collection of dipole moments (approximately 7000 entries) is contained in A. L. McClellan, *Tables of Experimental Dipole Moments,* W. H. Freeman, San Francisco, 1963. A critical survey of 500 compounds in the gas phase is given by Nelson, Lide, and Maryott, NSRDS-NBS 10, Washington, D.C., 1967.

5.6.4 Dielectric Constants

If two oppositely charged plates exist in a vacuum, there is a certain force of attraction between them, as stated by Coulomb's law:

$$F = \frac{1}{4\pi\varepsilon_0} \cdot \frac{q_1 q_2}{\varepsilon r^2}$$

where F is the force, in newtons, acting on each of the charges q_1 and q_2, r is the distance between the charges, ε is the dielectric constant of the medium between the plates, and ε_0 is the permittivity of free space. q_1, q_2 are expressed in coulombs and r in meters. If another substance, such as a solvent, is in the space separating these charges (or ions in a solution), their attraction for each other is less. The dielectric constant is a measure of the relative effect a solvent has on the force with which two oppositely charged plates attract each other. The dielectric constant is a unitless number.

Dielectric constants for a selected group of inorganic and organic compounds are included in Tables 5.16 and 5.17. An extensive list has been compiled by Maryott and Smith, *National Bureau Standards Circular 514,* Washington, D.C., 1951.

For gases the values of the dielectric constant can be adjusted to somewhat different conditions of temperature and pressure by means of the equation

$$\frac{(\varepsilon - 1)_{t,p}}{(\varepsilon - 1)_{20^\circ,1\,\text{atm}}} = \frac{p}{760[1 + 0.003\,411(t - 20)]}$$

where p is the pressure (in mmHg) and t is the temperature (in °C). The errors associated with this equation probably do not exceed 0.02% for gases between 10 and 30°C and for pressures between 700 and 800 mm. The dielectric constants of selected gases will be found in Table 5.17.

5.6.5 Viscosity

The *dynamic viscosity,* or coefficient of viscosity, η of a Newtonian fluid is defined as the force per unit area necessary to maintain a unit velocity gradient at right angles to the direction of flow between two parallel planes a unit distance apart. The SI unit is pascal-second or newton-second per meter squared [N · s · m^{-2}]. The c.g.s. unit of viscosity is the poise [P]; 1 cP $\equiv$ 1 mN · s · m^{-2}. The dynamic viscosity decreases with the temperature approximately according to the equation: $\log \eta = A + B/T$. Values of A and B for a large number of liquids are given by Barrer, *Trans. Faraday Soc.* **39:**48 (1943).

Kinematic viscosity v is the ratio of the dynamic viscosity to the density of a fluid. The SI unit is meter squared per second [m^2 · s^{-1}]. The c.g.s. units are called stokes [cm^2 · s^{-1}]; poises = stokes × density.

Fluidity ϕ is the reciprocal of the dynamic viscosity.

The primary reference liquid for viscosity measurements is water. The absolute viscosity of water at 20°C is 1.0019 ($\pm$0.0003) mN · s · m^{-2} (or centipoise), as determined by Swindells, Coe, and Godfrey, *J. Research Natl. Bur. Standards* **48:**1 (1952). The relative viscosity of water, η/η_{20°, is 0.8885 at 25°C, 0.7960 at 30°C, and 0.6518 at 40°C. Values at temperatures between 15 and 60°C are best represented by Cragoe's equation:

$$\log \frac{\eta}{\eta_{20°}} = \frac{1.2348(20 - t) - 0.001\ 467(t - 20)^2}{t + 96}$$

The *Reynolds number* for flow in a tube is defined by $d\bar{v}\rho/\eta$, where d is the diameter of the tube, $\bar{v}$ is the average velocity of the fluid along the tube, ρ is the density of the fluid, and η is its dynamic viscosity. At flow velocities corresponding with values of the Reynolds number of greater than 2000, turbulence is encountered.

TABLE 5.20   Aqueous Glycerol Solutions

% Weight Glycerol	Grams per Liter	Relative Density 25°/25°C	Viscosity, mN · s · m^{-2}		
			20°C	25°C	30°C
100	1261	1.262 01	1 495	942	622
99	1246	1.259 45	1 194	772	509
98	1231	1.256 85	971	627	423
97	1216	1.254 25	802	521	353
96	1201	1.251 65	659	434	296
95	1186	1.249 10	543.5	365	248
80	966.8	1.209 25	61.8	45.72	34.81
50	563.2	1.127 20	6.032	5.024	4.233
25	265.0	1.061 15	2.089	1.805	1.586
10	102.2	1.023 70	1.307	1.149	1.021

TABLE 5.21   Aqueous Sucrose Solutions

% Weight Sucrose	Grams per Liter	Relative Density 20°/4°C	Viscosity, mN · s · m^{-2}		
			15°C	20°C	25°C
75	1034	1.379 0	4 039	2 328	1 405
70	943.0	1.347 2	746.9	481.6	321.6
65	855.6	1.316 3	211.3	147.2	105.4
60	771.9	1.286 5	79.49	58.49	40.03
50	614.8	1.229 6	19.53	15.43	12.40
40	470.6	1.176 4	7.463	6.167	5.164
30	338.1	1.127 0	3.757	3.187	2.735

5.7 COMBUSTIBLE MIXTURES

TABLE 5.22 Properties of Combustible Mixtures in Air

Additional compounds can be found in National Fire Protection Association, Fire Protection Handbook, *14th ed., 1976.*

Substance	Autoignition temperature, °C	Flammable (explosive) limits, percent by volume of fuel (25°C, 760 mm)	
		Lower	Upper
Acetaldehyde	175	4.0	57
Acetanilide	540		
Acetic acid, glacial	465	5.4	16.0
Acetic anhydride	390	2.9	10.3
Acetone	465	2.6	12.8
Acetonitrile	524	4.4	16.0
Acetophenone	571		
Acetylene	305	2.5	82
Acetyl chloride	390		
Acrolein	235*	2.8	31.0
Acrylonitrile	481	3.1	17
Allyl alcohol	378	2.5	18
Allylamine	374	2.2	22
Ammonia, anhydrous	651	16	25
Aniline	615	1.3	
Asphalt	485		
Benzaldehyde	376		
Benzene	562	1.4	8.0
Benzoyl peroxide	80		
Benzyl acetate	461		
Benzyl alcohol	436		
Benzyl benzoate	481		
Bis(2-aminoethyl)amine	399		
Bis(2-chloroethyl) ether	369		
Biscyclohexyl	245	0.7 (100°C)	5.1 (150°C)
Bis(2-hydroethyl) ether	229		
Bromobenzene	566		
1-Bromobutane	265	2.6 (100°C)	6.6 (100°C)
Bromoethane	511	6.7	11.3
Bromomethane	537	13.5	14.5
1-Bromopropane	490		
3-Bromopropene	295	4.4	7.3
1,3-Butadiene	420	2.0	11.5
Butanal (butyraldehyde)	230	2.5	12.5
Butane	405	1.9	8.5
1,3-Butanediol	394		
Butanoic acid	450	2.0	10.0
1-Butanol	365	1.4	11.2
2-Butanol	406	1.7 (100°C)	9.8 (100°C)
2-Butanone	516	1.8	11.5
(E)-2-Butenal (crotonaldehyde)	232	2.1	15.5

* Unstable

TABLE 5.22 Properties of Combustible Mixtures in Air (*Continued*)

Substance	Autoignition temperature, °C	Flammable (explosive) limits, percent by volume of fuel (25°C, 760 mm)	
		Lower	Upper
1-Butene	384	1.6	9.3
(E)-2-Butene	324	1.8	9.7
(Z)-2-Butene	324	1.7	9.0
1-Butene oxide		1.5	18.3
3-Buten-1-ol		4.7	34
2-(2-Butoxyethoxy)ethyl acetate	299		
Butyl acetate	425	1.4	7.5
sec-Butyl acetate		1.3	7.5
Butylamine	312	1.7	9.8
tert-Butylamine	380	1.7 (100°C)	8.9 (100°C)
Butylbenzene	410	0.8	5.8
sec-Butylbenzene	420	0.8	6.9
tert-Butylbenzene	450	0.7 (100°C)	5.7 (100°C)
Butyl formate	322	1.7	8.2
Butyl 2-methyl-2-propenoate	294	2	8
Butyl propanoate	427		
2-Butyne		1.4	
Carbon disulfide	125	1.3	50
Carbon monoxide	609	12.5	74.2
Carbonyl sulfide		12	28.5
Chlorobenzene	640	1.3	7.1
1-Chloro-1,3-butadiene		4.0	20.0
1-Chlorobutane		1.8	10.1
2-Chloro-2-butene		2.3	9.3
1-Chloro-1,1-difluoroethane		6.2	17.9
1-Chloro-2,4-dinitrobenzene		2.0	22
Chloroethane	519	3.8	15.4
2-Chloroethanol	425	4.9	15.9
Chloromethane	632	10.7	17.2
1-Chloro-2-methylpropane		2.0	8.7
3-Chloro-2-methyl-1-propene		2.3	9.3
1-Chloropentane	260	1.6	8.6
1-Chloropropane	520	2.6	11.1
1-Chloro-1-propene		4.5	16
2-Chloro-1-propene		4.5	16
3-Chloro-1-propene	485	2.9	11.2
Chlorotrifluorethylene		24	40.3
Cumene	424	0.9	6.5
Cyanogen		6.6	32
Cyclobutane		1.8	
Cyclohexane	245	1.3	8.4
Cyclohexanone	420	1.1 (100°C)	
Cyclohexene	310	1.2	
Cyclohexyl acetate	334		
Cyclohexylamine	293		
Cyclopentane	380		
Cyclopropane	500	2.4	10.4
Decahydronaphthalene	250	0.7 (100°C)	4.9 (100°C)

TABLE 5.22 Properties of Combustible Mixtures in Air (*Continued*)

Substance	Autoignition temperature, °C	Flammable (explosive) limits, percent by volume of fuel (25°C, 760 mm)	
		Lower	Upper
Decane	210	0.8	5.4
Diborane(6)	38–52	0.8	88
Dibutyl ether	194	1.5	7.6
1,2-Dichlorobenzene	648	2.2	9.2
1,1-Dichloroethane	456	5.6	
1,2-Dichloroethane	412	6.2	15.9
1,1-Dichloroethylene	570	7.3	16.0
(E)-1,2-Dichloroethylene	460	9.7	12.8
Dichloromethane	615	15.5	66.4
1,2-Dichloropropane	557	3.4	14.5
Diethylamine	312	1.8	10.1
Diethyl ether	160	1.85	36.0
Diethyl peroxide		2.3	15.9
Diethyl sulfate	436		
1,3-Dihydroxybenzene (resorcinol)	664		
1,4-Dihydroxybenzene	516		
Diisopropyl ether	443	1.4	7.9
Dimethoxymethane	237		
N,N-Dimethylacetamide	490	2.0	11.5
Dimethylamine (anhydrous)	400	2.8	14.4
N,N-Dimethylaniline	371		
2,2-Dimethylbutane	425	1.2	7.0
Dimethyl ether	350	3.4	27.0
N,N-Dimethylformamide	445	2.2 (100°C)	15.2 (100°C)
2,6-Dimethyl-4-heptanol		0.8 (100°C)	6.1 (100°C)
2,6-Dimethyl-4-heptanone		0.8 (100°C)	6.2 (100°C)
1,1-Dimethylhydrazine	249	2	95
2,3-Dimethylpentane	335	1.1	6.7
Dimethyl 1,2-phthalate	556		
2,2-Dimethylpropane	450	1.4	7.5
Dimethyl sulfate	188		
Dimethyl sulfide	206	2.2	19.7
Dimethyl sulfoxide	215	2.6	28.5
1,4-Dioxane	180	2.0	22.2
Diphenylamine	634		
Divinyl ether	360	1.7	27.0
1-Dodecanol	275		
Ethane	515	3.0	12.5
1,2-Ethanediamine	385		
1,2-Ethanediol	400	3.2	
Ethanethiol	299	2.8	18.2
Ethanol	423	3.3	19
2-Ethoxyethanol	235	1.8	14
2-Ethoxyethyl acetate	379	1.7	
1-Ethoxypropane		1.7	9.0
Ethyl acetate	427	2.2	11.0
Ethyl acrylate		1.8	

TABLE 5.22 Properties of Combustible Mixtures in Air (*Continued*)

Substance	Autoignition temperature, °C	Flammable (explosive) limits, percent by volume of fuel (25°C, 760 mm)	
		Lower	Upper
Ethylamine	385	3.5	14.0
Ethylbenzene	432	1.2	6.8
Ethyl benzoate	490		
2-Ethylbutanoic acid	463		
Ethyl chloroformate	500		
Ethylcyclobutane	210	1.2	7.7
Ethylene	490	2.7	36.0
Ethyleneimine	320	3.6	46
Ethylene oxide	429	3.0	100
Ethyl formate	455	2.7	13.5
2-Ethylhexanal	197		
Ethyl methyl ether		2.0	10.0
Ethyl nitrate	85 explodes	3.8	
Ethyl nitrite	90 explodes	3.0	50.0
Ethyl propanoate	440	1.9	11
Ethyl vinyl ether	202	1.7	28
Formaldehyde	430	7.0	73.0
Formic acid, 90%	434	18	57
2-Furaldehyde	316	2.1	19.3
Furan		2.3	14.3
Furfuryl alcohol	491	1.8	16.3
Gasoline, 50–100 octane	280–456	1.4	7.6
Heptane	223	1.05	6.7
1,1,2,3,4,4-Hexachlorobutadiene	610		
Hexane	225	1.2	7.5
1,6-Hexanedioic acid	420		
2-Hexanone	533	1.2	8.0
Hydrazine	(iron rust) 23–270 (glass)	4.7	100
Hydrogen	400	4.1	74.2
Hydrogen cyanide, 96%	538	5.6	40.0
Hydrogen sulfide	260	4	46
N-Hydroxyethyl-1,2-ethanediamine	368		
1-Hydroxy-2-methylbenzene	599	1.4 (149°C)	
1-Hydroxy-3-methylbenzene	559	1.1 (150°C)	
1-Hydroxy-4-methylbenzene (*p*-cresol)		1.1 (150°C)	
4-Hydroxy-4-methyl-2-pentanone	603	1.8	6.9
Isobutyl acetate	421	2.4	10.5
Isobutylamine	378		
Isobutylbenzene	430	0.8	6.0
Isopentane	420	1.4	7.6
Isopentyl acetate	360	1.0	7.5
Isoprene	220	2	9
Isopropyl acetate	460	1.7	7.8
Isopropyl alcohol	456	2.5	12
Isopropylamine	402	2.3	10.4
Isopropyl formate	485		
4-Isopropyl-1-methylbenzene	436		

TABLE 5.22 Properties of Combustible Mixtures in Air (*Continued*)

Substance	Autoignition temperature, °C	Flammable (explosive) limits, percent by volume of fuel (25°C, 760 mm)	
		Lower	Upper
Kerosene	210	0.7	5.0
Maleic anhydride	477	1.4	7.1
Methane	650	5.3	15.0
Methanethiol		3.9	21.8
Methanol	470	6.0	36.5
2-Methoxyethanol	285	2.5	14
2-Methoxyethyl acetate		1.7	8.2
Methyl acetate	502	3.1	16
Methyl acetylacetate	280		
Methyl acrylate		2.8	25
Methylamine	430	4.95	20.75
2-Methylbutane		1.4	7.6
2-Methyl-1-butanol		1.4	9.0
2-Methyl-2-butanol	437	1.2	9.0
3-Methyl-1-butanol	350	1.2	9.0
3-Methylbutyl acetate	360	1.0 (100°C)	7.5
2-Methyl-2-butene	275	1.6	8.7
3-Methyl-1-butene	365	1.5	9.1
2-Methyl-1-buten-3-one		1.8	9.0
Methyl chloroformate	504		
Methylcyclohexane	250	1.2	6.7
Methyl formate	465	5.9	20
Methylhydrazine	196	2.5	97 ± 2
2-Methyllactonitrile	688		
Methyl methacrylate		2.1	12.5
4-Methyl-2-pentanol		1.0	5.5
4-Methyl-2-pentanone	459	1.4	7.5
4-Methyl-3-penten-2-one	344		
2-Methylpropanal	223	1.6	10.6
2-Methylpropane	462		
Methyl propanoate	469	2.5	13
2-Methylpropanoic acid	501		
2-Methyl-1-propanol	427	1.2	10.9
2-Methyl-2-propanol	480	2.4	8.0
2-Methyl-1-propene	465	1.8	9.6
2-Methylpropyl acetate	423	2.4	10.5
2-Methylpyridine	538		
Methyl salicylate	454		
α-Methylstyrene	574	1.9	6.1
Methyl vinyl ether		2.6	39
Naphtha, coal tar	277		
Naphthalene	568	0.9	5.9
Neoprene		4.0	20
Nicotine	244	0.75	4.0
Nitrobenzene	482	1.8 (93°C)	
2-Nitrobiphenyl	179		
Nitroethane	414	4.0	
Nitroglycerine	270		
Nitromethane	418		

TABLE 5.22 Properties of Combustible Mixtures in Air (*Continued*)

Substance	Autoignition temperature, °C	Flammable (explosive) limits, percent by volume of fuel (25°C, 760 mm)	
		Lower	Upper
1-Nitropropane	420	2.2	
2-Nitropropane	428	2.6	
Nonane	190	0.8	2.9
Octadecanoic acid (stearic acid)	395		
(Z)-9-Octadecenoic acid	362		
Octane	220	1.0	4.7
Paraldehyde	238	1.3	
Pentaborane(9)		0.42	
Pentanamine		2.2	22
Pentane	309	1.5	7.8
1-Pentanol	300	1.2	10.0 (100°C)
2-Pentanol	343–385	1.2	9.0
3-Pentanol		1.2	9.0
2-Pentanone	505	1.5	8.2
3-Pentanone	450	1.6	
Pentyl acetate	360	1.1	7.5
Petroleum ether (solvent naphtha)	288	1.1	5.9
Propanamine	318	2.0	10.4
Phosphorus, red	260		
Phosphorus, white	30		
Phosphorus pentasulfide	142		
o-Phthalic anhydride	570	1.7	10.4
Picric acid	300 (explodes)		
2-Pinene	255		
1-Propanal	207	2.9	17.0
Propanamine	318	2.0	10.4
Propane	450	2.3	9.5
1,2-Propanediol	371	2.6	12.6
1,2,3-Propanetriol (glycerine)	370		
Propanoic acid	513		
1-Propanol	440	2.1	13.5
Propanonitrile		3.1	
Propene	460	2.4	10.1
Propyl acetate	450	2.0	8.0
Propylene oxide		2.8	37.0
Propyl nitrate	175	2	100
Propyne		1.7	
Pyridine	482	1.8	12.4
Quinoline	480		
Sodium	115 (dry air)		
Styrene	490	1.1	6.1
Sulfur (di-) dichloride	233		
Tetrabromoethylene	335		
Tetrahydrofuran	321	2	11.8
Tetrahydrofurfuryl alcohol	282	1.5	9.7
Tetrahydronaphthalene	385	0.8	5.0
Titanium, powder	250		

TABLE 5.22 Properties of Combustible Mixtures in Air (*Continued*)

Substance	Autoignition temperature, °C	Flammable (explosive) limits, percent by volume of fuel (25°C, 760 mm)	
		Lower	Upper
Toluene	480	1.3	7.0
Toluene diisocyanate		0.9	9.5
o-Toluidine (also *p*-)	482		
Trichloroethylene	420	12.5	90
Trichlorosilane	104		
1,1,2-Trichloro-1,2,2-trifluoroethane (Freon 113)	680		
Triethylamine		1.2	8.0
Triethylene glycol	371	0.9	9.2
Trimethylamine	190	2.0	11.6
1,2,4-Trimethylbenzene	516		
1,3,5-Trimethylbenzene	550		
1,1,3-Trimethyl-3-cyclohexen-5-one	462	0.8	3.8
2,2,4-Trimethylpentane	415	1.1	6.0
Trioxane	414	3.6	28.7
Tri-*o*-tolyl phosphate	385		
Turpentine		0.8	
Vinyl acetate	427	2.6	13.4
Vinyl butanoate		1.4	8.8
Vinyl chloride	472	4.0	22
4-Vinyl-1-cyclohexene	269		
Vinyl fluoride		2.6	21.7
Vinylidene	573	5.6	16.0
Xylene, *m*- and *p*-	530	1.1	7.0
o-Xylene	465	1.0	6.0

5.8 THERMAL CONDUCTIVITY

TABLE 5.23 Thermal Conductivities of Gases as a Function of Temperature

The coefficient k, expressed in $J \cdot s^{-1} \cdot cm^{-1} \cdot K^{-1}$, is the quantity of heat in joules, transmitted per second through a sample 1 centimeter in thickness and 1 cm^2 in area when the temperature difference between the two sides is 1 degree kelvin (or Celsius). The tabulated values are in microjoules. To convert to microcalories, divide values by 4.184.

Substance	\-40	\-20	0	20	40	60	80	100	120	140	160
						Temperature, °C					
Acetone		80	95	107	124	140	156	173	190	207	
Acetaldehyde				109	126	142	159	176	195		
Acetonitrile						112	124	137	151	166	
Acetylene	118[-75]		184	205	224	248	269	290			
Air			242	256	270	284	299	311	324	336	342[149]
Ammonia	164[-60]		218	238	259	280	301	321			
Argon			166	176	186	196	206				
Benzene						126	146	165	184	205	226
Bromine			42	45	50	54	59				
Bromomethane					82	94	104	117			
Butanamine			135[6.5]								
1-Butanamine								176[110]			
Butane			135	154	174	193	213	233			
Carbon dioxide			144	160	176	192	207	215			
Carbon disulfide			67	76	85						
Carbon monoxide			228	245	262	278					
Carbon tetrachloride					70	75	80	86		109[184]	
Chlorine	64	72	79	85	93	100					
Chlorodifluorimethane		103	110	116	122						
Chloroethane			90	105	120	134	151	167	186	204	
Chloroform					75	84	91	99	107	116	
Chloromethane			84		117	130	142	155			
Cyclohexane			77	99	120	141	163	184	206	230	256
Cyclopropane						192	218	243	270		
Deuterium	1150	1222	1297	1372	1448	1523					
Dibromomethane									74[110]		

5.142

Dichlorodifluoromethane		81	84	92	100			138			194[200]
1,1-Dichloroethane			69	81	93	105	117	129	144		
1,2-Dichloroethane								127	140		
Dichlorofluoromethane		91	94	97	100						
Dichloromethane			93	135	157			161			
Diethylamine			118	204	228	179	199	218	243	268	
Diethyl ether			113	141	155	178	200	222	244	269	351[213]
1,4-Dioxane								167	187	207	
Ethane	137	159				257	288	316			234
Ethanol			182					209	344		
Ethyl acetate			126		115	133	151	170	191	211	
Ethylamine			136	153	169	206					
Ethylene	137	158	178	220	241	262	282				
Ethylene oxide			79	100				193	256	279	
Ethyl formate						142	164	186	206	226	
Ethyl nitrate								159	178	197	
Fluorine	212	230	247	264	278	294	309	325			
Helium	1276	1343	1423	1481	1540	1598	1661	1720	1778		
Heptane			100	115	130			174			
Hexane			109				178	201	224	247	271
Hydrogen	1494	1607	1724	1828	1925	2025					
Hydrogen bromide	64	70	77	84	90	97	104				
Hydrogen chloride	107	117	128	138	148						
Hydrogen cyanide		99	110	121	132	143					
Hydrogen sulfide		116	129	143	156	169					
Iodomethane			46	53	60	68	75	82	89		
Krypton		79	85		95			110			
Methane	257	280	307	334	361	387	416	445			
Methanol						174	197	221	241	263	284
Methyl acetate			67			150[70]		177	195	215	237
2-Methylbutane			122					215			
2-Methylpropane			141	156	176	196		233[93]			
2-Methyl-2-propanol								225			
Neon	410	433	454	476	497	518	537	556			
Nitric oxide	205	221	238	254	269	285	301	317			
Nitrogen	211	226	241	256	270	282	295	307			
Nitromethane									320	333	
Nitrous oxide	121	137	152	168	184				139	155	

TABLE 5.23 Thermal Conductivities of Gases as a Function of Temperature (*Continued*)

Substance	Temperature, °C										
	−40	−20	0	20	40	60	80	100	120	140	160
Oxygen	211	228	245	261	278	294	311	328			
Pentane			130					218			
Propane	116	132	151	171	192	215	238	262	330	353	379
2-Propanol				151[31]						250[127]	
Sulfur dioxide			83					106			
Thiophene								152[110]			
Triethylamine								195	216	239	
Water		142	159	175	191	207	224	241	257		

TABLE 5.24   Liquid Thermal Conductivity of
Various Substances

All values of thermal conductivity, k, *in*
$mJ \cdot cm^{-1} \cdot s^{-1} \cdot K^{-1}$.

Substance	t, °C	k
Acetaldehyde	20	1.900
Acetic acid	20	1.565
Acetic anhydride	20	2.209
Acetone	−80	1.987
	0	1.711
	20	1.611
	40	1.510
Allyl alcohol	30	1.795
Aniline	17	1.774
Argon	−189	1.259
Benzene	20	1.477
	50	1.368
Bromobenzene	20	1.113
Bromoethane	20	1.029
1-Bromo-2-methylpropane	12	1.163
1-Bromopentane	20	0.983
Bromopropane	12	1.075
Butanoic acid	12	1.506
Butananol	20	1.536
	40	1.473
	60	1.439
Butyl acetate	20	1.368
Carbon disulfide	12	1.435
Carbon tetrachloride	−20	1.100
	0	1.071
	20	1.029
	50	0.974
Chlorobenzene	−40	1.406
	20	1.276
	80	1.113
Chloroform	20	1.029
1-Chloro-2-methylpropane	12	1.163
1-Chloropentane	12	1.184
Chloropropane	12	1.184
4-Chlorotuluene	20	1.297
m-Cresol	20	1.498
	80	1.452
Cyclohexane	20	1.243
Decane	30	1.401
	41	1.272
	76	1.188
1,2-Dichloroethane	20	1.264
Dichlorofluoromethane	0	0.134
Dichloromethane	−20	1.590
	0	1.561
	20	1.477
Diethyl ether	20	1.284
Diisopropyl ether	20	1.096
2,3-Dimethylbutane	32	1.038
	49	0.996
Ethanol	20	1.648

TABLE 5.24 Liquid Thermal Conductivity of
Various Substances (*Continued*)

Substance	t, °C	k
	40	1.519
	60	1.418
Ethoxybenzene	−20	1.497
Ethyl acetate	20	1.469
	60	1.410
Ethylbenzene	20	1.322
	80	1.176
Ethylene glycol	20	2.548
	77	2.238
Ethyl formate	12	1.581
Glycerol	20	2.941
Heptane	0	1.322
	20	1.259
	60	1.130
1-Heptanol	70	1.623
Hexane	20	1.218
	30	1.180
1-Hexanol	30	1.615
Hydrochloric acid, 38%	32	4.402
Hydrogen	−253	1.180
Iodobenzene	−20	1.063
	20	1.276
	80	0.937
Iodoethane	30	1.109
1-Iodo-2-methylpropane	12	0.870
1-Iodopentane	12	0.849
Iodopropane	12	0.920
Isopentyl acetate	20	1.297
Isopropylbenzene	20	1.247
4-Isopropyl-1-methylbenzene	30	1.347
Mercury	100	92.048
Methanol	20	2.021
Methyl acetate	12	1.611
Methyl butanoate	12	1.402
3-Methylbutanoic acid	12	1.305
3-Methyl-1-butanol	30	1.477
Methylcyclohexane	30	1.276
Methylcyclopentane	20	1.209
	38	1.151
2-Methylpentane	32	1.084
	49	1.033
Methyl pentanoate	12	1.318
4-Methylpentanoic acid	12	1.427
2-Methyl-1-propanol	12	1.423
2-Methyl-2-propanol	38	1.159
	77	1.067
Nitrobenzene	20	1.510
Nitromethane	30	2.151
Nonane	20	1.310
	80	1.151
1-Nonanol	30	1.678
Octane	20	1.289
	80	1.117

TABLE 5.24   Liquid Thermal Conductivity of
Various Substances (*Continued*)

Substance	t, °C	k
1-Octanol	20	1.657
	38	1.602
Palmitic acid	72	1.598
Pentachloroethane	20	1.251
Pentane	20	1.138
	30	1.109
Pentanoic acid	12	1.360
1-Pentanol	30	1.619
Pentyl acetate	20	1.289
Phenylhydrazine	25	1.724
Propanoic acid	12	1.728
2-Propanol	20	1.406
1,2-Propylene glycol	20	2.008
Propyl formate	12	1.494
Sodium	300	753.1
Sodium chloride (*aq*, satd)	20	5.732
Stearic acid	72	1.598
Sulfuric acid, 90%	32	3.540
1,1,2,2-Tetrachloroethane	20	1.138
Tetrachloroethylene	20	1.619
Toluene	−80	1.590
	20	1.347
	80	1.175
Trichloroethylene	−60	1.359
	20	1.160
Triethylamine	−80	1.464
	20	1.209
	44	1.113
1,3,5-Trimethylbenzene	20	1.360
2,2,4-Trimethylpentane	38	0.966
	77	0.841
Water	20	5.983
m-Xylene	25	1.577
o-Xylene	20	1.427
p-Xylene	20	1.360

TABLE 5.25   Thermal Conductivity of Various Solids

*All values of thermal conductivity, k, in mJ · cm⁻¹ · s⁻¹ · K⁻¹. For
values in millicalories, divide by 4.184.*

Substance	t, °C	k
Asphalt	20	7.447
Basalt	20	21.76
Bauxite	600	5.56
Boiler scale	66	13.1
Brick, common	20	6.3
Blotting paper	20	0.628
Cardboard	20	2.1
Cement, Portland	90	2.97
Chalk	20	9.2

TABLE 5.25 Thermal Conductivity of Various Solids (*Continued*)

Substance	t, °C	k
Chemical elements, *see* Table 4.1		
Coal	0	1.69
Concrete	20	9.2
Cork, sp. grav. = 0.2	30	0.54
Cork meal	100	0.556
Cotton, sp. grav. = 0.081	0	0.569
Diatomaceous earth	20	0.54
Ebonite	0	1.58
Eiderdown	20	0.046
Feathers (with air)	9	0.238
Feldspar	20	23.4
Felt (dark gray)	40	0.623
Fire brick	20	4.6
Flannel	60	0.148
Flint	20	10.0
Glass, crown	12.5	6.82
Flint	12.5	5.98
Jena	22	9.50
Quartz	0	13.89
	100	19.12
Soda	20	7.1
	100	7.5
Granite	20	34.2
Graphite, sp. grav. = 1.58	50	441.4
Graphite powder, sp. grav. = 0.7	40	11.92
Gypsum	0	13.0
Horse hair, sp. grav. = 0.172	20	0.510
Ice		23.8
Leather, cowhide	84	1.76
Linen	20	0.879
Magnesia brick	20	11.3
	1130	30.1
Marble, white		32.6
Mica	41	3.60
Naphthalene	0	3.77
Paper	20	1.3
Paraffin	0	2.88
Plaster of Paris	20	2.93
Porcelain	95	10.38
Quartz, parallel to axis	0	136.0
	100	90.0
Quartz, perpendicular to axis	0	72.43
	100	55.77
Plastics, *see* Sec. 10		
Roofing paper	0	1.90
Rubber, natural and synthetic, *see* Sec. 10		
Sand, dry	20	3.89
Sandstone, sp. grav. = 2.259	40	18.37
Silk, sp. grav. = 0.101	0	0.510
Slate	20	19.66
Soil, dry	20	1.38
Wax, bees	20	0.866

TABLE 5.25   Thermal Conductivity of Various Solids (*Continued*)

Substance	t, °C	k
Wood, maple, parallel to face	20	4.25
Perpendicular to face	50	1.82
Wood, oak, parallel to face	15	3.49
Perpendicular to face	15	2.09
Wood, pine, parallel to face	20	3.49
Perpendicular to face	15	1.51

5.9   MISCELLANY

TABLE 5.26   Compressibility of Water

In the table below are given the relative volumes of water at various temperatures and pressures. The volume at 0°C and one normal atmosphere (760 mm of Hg) is taken as unity.

P, atm	−10°C.	0°C.	10°C.	20°C.	40°C.	60°C.	80°C.
1	1.0017	1.0000	1.0001	1.0016	1.0076	1.0168	1.0287
500	0.9788	0.9767	0.9778	0.9804	0.9867	0.9967	1.0071
1000	0.9581	0.9566	0.9591	0.9619	0.9689	0.9780	0.9884
1500	0.9399	0.9394	0.9424	0.9456	0.9529	0.9617	0.9717
2000	0.9223	0.9241	0.9277	0.9312	0.9386	0.9472	0.9568
2500	0.9083	0.9112	0.9147	0.9183	0.9257	0.9343	0.9437
3000	0.8962	0.8993	0.9028	0.9065	0.9139	0.9225	0.9315
3500	0.8852	0.8884	0.8919	0.8956	0.9030	0.9115	0.9203
4000	0.8751	0.8783	0.8818	0.8855	0.8931	0.9012	0.9097
4500	0.8658	0.8692	0.8725	0.8762	0.8838	0.8919	0.9001
5000	0.8573	0.8606	0.8639	0.8675	0.8752	0.8832	0.8913
6000	......	0.8452	0.8481	0.8517	0.8595	0.8674	0.8752
7000	......	......	0.8340	0.8374	0.8456	0.8534	0.8610
8000	......	......	......	0.8244	0.8330	0.8408	0.8483
9000	......	......	......	0.8128	0.8219	0.8297	0.8371
10000	......	......	......	0.8027	0.8119	0.8196	0.8268
11000	......	......	......	......	0.8023	0.8101	0.8172
12000	......	......	......	......	0.7931	0.8009	0.8080

TABLE 5.27 Mass of Water Vapor in Saturated Air

The values in the table are grams of water contained in a cubic meter (m^3) of saturated air at a total pressure 101 325 Pa (1 atm).

°C	$g \cdot m^{-3}$	°C	$g \cdot m^{-3}$	°C	$g \cdot m^{-3}$
−30	0.341	12	10.65	53	95.56
−29	0.375	13	11.35	54	100.0
−28	0.413	14	12.05	55	104.5
−27	0.456	15	12.80	56	109.1
−26	0.504	16	13.60	57	114.1
−25	0.554	17	14.45	58	119.2
−24	0.607	18	15.35	59	124.7
−23	0.667	19	16.30	60	130.2
−22	0.733	20	17.30	61	136.0
−21	0.804	21	18.35	62	142.1
−20	0.883	22	19.40	63	148.4
−19	0.968	23	20.55	64	154.9
−18	1.063	24	21.75	65	161.3
−17	1.164	25	23.05	66	167.9
−16	1.273	26	24.35	67	175.1
−15	1.375	27	25.75	68	182.6
−14	1.510	28	27.20	69	190.3
−13	1.650	29	28.75	70	198.2
−12	1.800	30	30.35	71	206.5
−11	1.965	31	32.05	72	215.1
−10	2.140	32	33.80	73	223.7
−9	2.331	33	35.60	74	233.0
−8	2.539	34	37.55	75	242.0
−7	2.761	35	39.55	76	251.2
−6	3.003	36	41.65	77	261.1
−5	3.250	37	43.90	78	271.6
−4	3.512	38	46.20	79	282.3
−3	3.810	39	48.60	80	293.4
−2	4.131	40	51.21	81	304.8
−1	4.473	41	53.86	82	316.6
0	4.849	42	56.61	83	328.7
1	5.199	43	59.51	84	341.2
2	5.569	44	62.53	85	353.6
3	5.947	45	65.52	86	366.2
4	6.35	46	68.61	87	379.9
5	6.80	47	72.00	88	394.1
6	7.25	48	75.56	89	408.6
7	7.75	49	79.24	90	423.5
8	8.25	50	83.05	91	439.0
9	8.80	51	87.04	92	454.8
10	9.40	52	91.22	93	471.2
11	10.00				

TABLE 5.28 Van der Waals' Constants for Gases

$$\left(P + \frac{a}{V^2}\right)(V - b) = RT \qquad \text{for 1 mole}$$

To use the values of a and b in the table, P must be expressed in atmospheres and V in liters per mole; then $R = 0.082\ 057$ liter·atmospheres per mole per degree and T is degrees kelvin:

$$\left(P + \frac{n^2a}{V^2}\right)(V - nb) = nRT \qquad \text{for } n \text{ moles}$$

Substance	a, liter^2·atm·mole^{-2}	b, liter·mole^{-1}
Acetic acid	17.59	0.106 8
Acetic anhydride	19.90	0.126 3
Acetone	13.91	0.099 4
Acetonitrile	17.58	0.116 8
Acetylene	4.390	0.051 36
Ammonia	4.170	0.037 07
Aniline	26.50	0.136 9
Argon	1.345	0.032 19
Benzene	18.00	0.115 4
Benzonitrile	33.39	0.172 4
Bromobenzene	28.56	0.153 9
Butane	14.47	0.122 6
Butanonitrile	25.72	0.159 6
Carbon dioxide	3.592	0.042 67
Carbon disulfide	11.62	0.076 85
Carbon monoxide	1.485	0.039 85
Carbon tetrachloride	20.39	0.138 3
Carbonyl sulfide	3.933	0.058 17
Chlorine	6.493	0.056 22
Chlorobenzene	25.43	0.145 3
Chloroethane	10.91	0.086 51
Chloroform	15.17	0.102 2
Chloromethane	7.471	0.064 83
1-Chloropropane	15.91	0.114 1
m-Cresol	31.88	0.160 7
Cyanogen	7.667	0.069 01
Cyclohexane	22.81	0.142 4
Decane	48.55	0.290 5
1,2-Dibromoethane	13.98	0.086 64
1,1-Dichloroethane	15.50	0.107 3
1,2-Dichloroethane	16.91	0.108 5
Diethylamine	19.15	0.139 2
Diethyl ether	17.38	0.134 4
Diethyl sulfide	18.75	0.121 4
Diisopropyl	23.13	0.166 9
Dimethylamine	10.38	0.085 70
N,N-Dimethylaniline	37.49	0.197 0
Dimethyl ether	8.073	0.072 46
2,5-Dimethylhexane	34.97	0.229 6
Dimethyl sulfide	12.87	0.092 13
Diphenyl	52.79	0.248 0
Diphenylmethane	38.20	0.224 0
Dipropylamine	27.72	0.182 0
Ethane	5.489	0.063 80

TABLE 5.28   Van der Waals' Constants for Gases (*Continued*)

Substance	a, liter$^2 \cdot$ atm $\cdot$ mole^{-2}	b, liter $\cdot$ mole^{-1}
Ethanol	12.02	0.084 07
Ethyl acetate	20.45	0.141 2
Ethylamine	10.60	0.084 09
Ethylbenzene	28.60	0.166 7
Ethyl butanoate	30.07	0.191 9
Ethylene	4.471	0.057 14
Ethyl formate	14.80	0.105 6
Ethyl mercaptan	11.24	0.080 98
Ethyl methyl ether	11.95	0.097 75
Ethyl methyl sulfide	19.23	0.130 4
Ethyl phenyl ether	35.16	0.196 3
Ethyl phenyl oxide	35.16	0.196 3
Ethyl propanoate	24.39	0.161 5
Fluorobenzene	19.93	0.128 6
Fluoromethane	4.631	0.052 64
Germanium tetrachloride	22.60	0.148 5
Helium	0.034 12	0.023 70
Heptane	31.51	0.265 4
Hexane	24.39	0.173 5
Hydrogen	0.244 4	0.044 31
Hydrogen bromide	4.451	0.044 31
Hydrogen chloride	3.667	0.040 81
Hydrogen selenide	5.268	0.046 37
Hydrogen sulfide	4.431	0.042 87
Iodobenzene	33.08	0.165 6
Isopropylbenzene	35.64	0.202 5
Krypton	2.318	0.039 78
Mercury	8.093	0.016 96
Methane	2.253	0.042 78
Methanol	9.523	0.067 02
Methyl acetate	15.29	0.109 1
Methylamine	7.130	0.059 92
2-Methylbutane	18.05	0.141 7
Methyl butanoate	23.94	0.156 9
3-Methyl-1-butene	18.08	0.140 5
Methyl formate	10.84	0.080 68
Methyl pentanoate	28.96	0.184 5
Methyl 2-methylpropanoate	24.50	0.163 7
2-Methylpropane	12.87	0.114 2
Methyl propanoate	19.91	0.136 0
2-Methyl-1-propanol	17.03	0.114 3
2-Methylpropyl acetate	28.50	0.183 3
2-Methylpropylbenzene	38.59	0.214 4
2-Methylpropyl formate	22.54	0.147 6
Naphthalene	39.74	0.193 7
Neon	0.210 7	0.017 09
Nitric oxide	1.340	0.027 89
Nitrogen	1.390	0.039 13
Nitrogen dioxide	5.284	0.044 24
Nitrous oxide	3.782	0.044 15
Octane	37.32	0.236 8
Oxygen	1.360	0.038 03
Ozone	3.545	0.049 03

TABLE 5.28 Van der Waals' Constants for Gases (*Continued*)

Substance	a, liter$^2 \cdot$ atm $\cdot$ mole^{-2}	b, liter $\cdot$ mole^{-1}
Pentane	19.01	0.146 0
Pentanonitrile	34.16	0.198 4
1-Pentene	15.90	0.120 7
Pentyl formate	27.58	0.173 0
Phosphine	4.631	0.051 56
Phosphonium chloride	4.054	0.045 45
Phosphorus	52.94	0.156 6
Propane	8.664	0.084 45
Propanoic acid	20.11	0.118 7
1-Propanol	14.92	0.101 9
2-Propanol	13.78	0.098 04
Propanonitrile	16.44	0.106 4
Propene	8.379	0.082 72
Propyl acetate	24.63	0.161 9
Propylamine	14.99	0.109 0
Propylbenzene	35.85	0.202 8
Propyl formate	18.95	0.128 0
Silicon tetrafluoride	4.195	0.055 71
Silicon tetrahydride	4.320	0.057 86
Sulfur dioxide	6.714	0.056 36
1,2,4,5-Tetramethylbenzene	45.32	0.242 2
Thiophene	20.72	0.127 0
Tin(IV) chloride	26.91	0.164 2
Toluene	24.06	0.146 3
Triethylamine	27.17	0.183 1
Trimethylamine	13.02	0.108 4
1,2,4-Trimethylbenzene	36.61	0.202 1
1,3,5-Trimethylbenzene	34.32	0.197 9
Water	5.464	0.030 49
Xenon	4.194	0.051 05
m-Xylene	30.36	0.177 2
o-Xylene	29.98	0.175 5
p-Xylene	30.93	0.180 9

5.9.1 Some Physical Chemistry Equations for Gases

A number of physical chemistry relationships, not enumerated in other sections (*see* Index), will be discussed in this section.

Boyle's law states that the volume of a given quantity of a gas varies inversely as the pressure, the temperature remaining constant. That is,

$$V = \frac{\text{constant}}{P} \qquad \text{or} \qquad PV = \text{constant}$$

A convenient form of the law, true strictly for ideal gases, is

$$P_1 V_1 = P_2 V_2$$

TABLE 5.29 Triple Points of Various Materials

Substance	Triple point, K	Pressure, mmHg
Ammonia	195.46	45.58
Argon	83.78	516
Boron tribromide	226.67	
Bromine	280.4	44.1
Carbon dioxide	216.65	
Cyclopropane	145.59	
Deuterium oxide	276.97	
1-Hexene	133.39	
Hydrogen, normal	13.95	54
Hydrogen, para	13.81	
Hydrogen bromide	186.1	~232
Hydrogen chloride	158.8	
Iodine heptafluoride	279.6	
Krypton	115.95	548
Methane	90.67	87.60
Methane-d_1	90.40	84.52
Methane-d_2	90.14	81.80
Methane-d_3	89.94	80.12
Methane-d_4	89.79	79.13
Molybdenum oxide tetrafluoride	370.3	
Molybdenum pentafluoride	340	
Neon	24.55	324
Neptunium hexafluoride	328.25	758.0
Niobium pentabromide	540.6	
Niobium pentachloride	476.5	
Nitrogen	63.15	94
1-Octene	171.45	
Oxygen	54.34	
Phosphorus, white	863	32 760
Plutonium hexafluoride	324.74	533.0
Propene	103.95	
Radon	202	~500
Rhenium dioxide trifluoride	363	
Rhenium heptafluoride	321.4	
Rhenium oxide pentafluoride	313.9	
Rhenium pentafluoride	321	
Succinonitrile (NIST standard)	331.23	
Sulfur dioxide	197.68	1.256
Tantalum pentabromide	553	
Tantalum pentachloride	489.0	
Tungsten oxide tetrafluoride	377.8	
Uranium hexafluoride	337.20	1 139.6
Water	273.16	
Xenon	161.37	612

Charles' law, also known as *Gay-Lussac's law,* states that the volume of a given mass of gas varies directly as the absolute temperature if the pressure remains constant, that is,

$$\frac{V}{T} = \text{constant}$$

Combining the laws of Boyle and Charles into one expression gives

$$\frac{P_1 V_1}{T_1} = \frac{P_2 V_2}{T_2}$$

In terms of moles, *Avogadro's hypothesis* can be stated: The same volume is occupied by one mole of any gas at a given temperature and pressure. The number of molecules in one mole is known as the *Avogadro number constant* N_A.

The behavior of all gases that obey the laws of Boyle and Charles, and Avogadro's hypothesis, can be expressed by the ideal gas equation:

$$PV = nRT$$

where R is called the *gas constant* and n is the number of moles of gas. If pressure is written as force per unit area and the volume as area times length, then R has the dimensions of energy per degree per mole—8.314 $J \cdot K^{-1} \cdot mol^{-1}$ or 1.987 $cal \cdot K^{-1} \cdot mol^{-1}$.

Dalton's law of partial pressures states that the total pressure exerted by a mixture of gases is equal to the sum of the pressures which each component would exert if placed separately into the container:

$$P_{total} = p_1 + p_2 + p_3 + \cdots$$

There are two ways to express the fraction which one gaseous component contributes to the total mixture: (1) the pressure fraction, p_i/P_{total}, and (2) the mole fraction, n_i/n_{total}.

5.9.1.1 Equations of State (PVT Relations for Real Gases)

1. *Virial equation* represents the experimental compressibility of a gas by an empirical equation of state:

$$PV = A_p + B_p P + C_p P^2 + \cdots$$

or

$$PV = A_v + B_v V + \frac{C_v}{V^2} + \cdots$$

where $A, B, C, \ldots$ are called the virial coefficients and are a function of the nature of the gas and the temperature.

2. *Van der Waals' equation:*

$$\left(P + \frac{an^2}{V^2}\right)(V - nb) = nRT$$

where the term an^2/V^2 is the correction for intermolecular attraction among the gas molecules and the nb term is the correction for the volume occupied by the gas molecules. The constants a and b must be fitted for each gas from experimental data (Table 5.28); consequently the equation is semiempirical. The constants are related to the critical-point constants (Table 6.5) as follows:

$$a = 3P_c V^2$$

$$b = \frac{V_c}{3}$$

$$R = \frac{8P_c V_c}{3T_c}$$

Substitution into van der Waals' equation and rearrangement leads to only the terms P/P_c, V/V_c, and T/T_c, which are called the reduced variables P_R, V_R, and T_R. For 1 mole of gas,

$$\left(P_R + \frac{3}{V_R^2}\right)\left(V_R - \frac{1}{3}\right) = \frac{8}{3}T_R$$

3. *Berthelot's equation of state,* used by many thermodynamicists, is

$$PV = nRT\left[1 + \frac{9}{128}\frac{PT_c}{P_cT}\left(1 - 6\frac{T_c^2}{T^2}\right)\right]$$

This equation requires only knowledge of the critical temperature and pressure for its use and gives accurate results in the vicinity of room temperature for unassociated substances at moderate pressures.

5.9.1.2 Properties of Gas Molecules

Vapor Density. Substitution of the Antoine vapor-pressure equation for its equivalent log P in the ideal gas equation gives

$$\log \rho_{vap} = \log M - \log R - \log (t + 273.15) + A - \frac{B}{t + C}$$

where ρ_{vap} is the vapor density in $g \cdot mL^{-1}$ at $t°C$, M is the molecular weight, R is the gas constant, and A, B, and C are the constants of the Antoine equation for vapor pressure. Since this equation is based on the ideal gas law, it is accurate only at temperatures at which the vapor of any specific compound follows this law. This condition prevails at reduced temperatures (T_R) of about 0.5 K.

Velocities of Molecules. The mean square velocity of gas molecules is given by

$$\overline{u^2} = \frac{3kT}{m} = \frac{3RT}{M}$$

where k is Boltzmann's constant and m is the mass of the molecule.

The mean velocity is given by

$$\bar{u} = \left(\frac{8\overline{u^2}}{3\pi}\right)^{1/2}$$

Viscosity. On the assumption that molecules interact like hard spheres, the viscosity of a gas is

$$\eta = \left(\frac{5}{16\sigma^2}\right)\left(\frac{mkT}{\pi}\right)^{1/2}$$

where σ is the molecular diameter.

Mean Free Path. The mean free path of a gas molecule l and the mean time between collisions τ are given by

$$l = \frac{m}{\pi\rho\sigma^2\sqrt{2}}$$

$$\tau = \frac{l}{\bar{u}} = \frac{4\eta}{5P}$$

Graham's Law of Diffusion. The rates at which gases diffuse under the same conditions of temperature and pressure are inversely proportional to the square roots of their densities:

$$\frac{r_1}{r_2} = \left(\frac{\rho_2}{\rho_1}\right)^{1/2}$$

Since $\rho = MP/RT$ for an ideal gas, it follows that

$$\frac{r_1}{r_2} = \left(\frac{M_2}{M_1}\right)^{1/2}$$

Henry's Law. The solubility of a gas is directly proportional to the partial pressure exerted by the gas:

$$p_i = kx_i$$

Joule-Thompson Coefficient for Real Gases. This expresses the change in temperature with respect to change in pressure at constant enthalpy:

$$\mu_{JT} = \left(\frac{\partial T}{\partial P}\right)_H$$

SECTION 6
THERMODYNAMIC PROPERTIES

6.1 ENTHALPIES AND GIBBS ENERGIES OF FORMATION, ENTROPIES, AND HEAT CAPACITIES

The tables in this section contain values of the enthalpy and Gibbs energy of formation, entropy, and heat capacity at 298.15 K (25°C). No values are given in these tables for metal alloys or other solid solutions, for fused salts, or for substances of undefined chemical composition.

For a more complete listing of compounds see the tables in *Selected Values of Chemical Thermodynamic Properties,* by D. D. Wagman et al., National Bureau of Standards Technical Notes 270-3, 270-4, 270-5, 270-6, 270-7, and 270-8, Washington; *JANAF Thermochemical Tables,* 3d ed., by M. W. Chase et al., published by the American Chemical Society and the American Institute of Physics for the National Bureau of Standards, 1986; supplements to *JANAF* appearing in *J. Phys. Chem. Ref. Data;* D. R. Stull, E. F. Westrum, Jr., and G. C. Sinke, *The Chemical Thermodynamics of Organic Compounds,* Wiley-Interscience, New York, 1969; I. Barin and O. Knacke, *Thermochemical Properties of Inorganic Substances,* Springer-Verlag, Berlin, 1973; J. D. Cox and G. Pilcher, *Thermochemistry of Organic and Organometallic Compounds,* Academic Press, London, 1970; J. B. Pedley, R. D. Naylor, and S. P. Kirby, *Thermochemical Data of Organic Compounds,* 2d ed., Chapman and Hall, London, 1986; and V. Majer and V. Svoboda, *Enthalpies of Vaporization of Organic Compounds,* International Union of Pure and Applied Chemistry, Chemical Data Series No. 32, Blackwell, Oxford, 1985.

The physical state of each substance is indicated in the column headed "State" as crystalline solid (c), liquid (lq), or gaseous (g). Solutions in water are listed as aqueous (aq).

The values of the thermodynamic properties of the pure substances given in these tables are, for the substances in their standard states, defined as follows: For a pure solid or liquid, the standard state is the substance in the condensed phase under a pressure of 1 atm (101 325 Pa).

For a gas, the standard state is the hypothetical ideal gas at unit fugacity, in which state the enthalpy is that of the real gas at the same temperature and at zero pressure.

The values of $\Delta Hf°$ and $\Delta Gf°$ given in the tables represent the change in the appropriate thermodynamic quantity when one mole of the substance in its standard state is formed, isothermally at the indicated temperature, from the elements, each in its appropriate standard reference state. The standard reference state at 25°C for each element has been chosen to be the standard state that is thermodynamically stable at 25°C and 1 atm pressure. The standard reference states are indicated in the tables by the fact that the values of $\Delta Hf°$ and $\Delta Gf°$ are exactly zero.

The values of $S°$ represent the virtual or "thermal" entropy of the substance in the standard state at 298.15 K, omitting contributions from nuclear spins. Isotope mixing effects are also excluded except in the case of the ^{1}H-^{2}H system.

Solutions in water are designated as aqueous, and the concentration of the solution is expressed in terms of the number of moles of solvent associated with 1 mol of the solute. If no concentration is indicated, the solution is assumed to be dilute. The standard state for a solute in aqueous solution is taken as the hypothetical ideal solution of unit molality (indicated as std. state or ss). In this state the partial molal enthalpy and the heat capacity of the solute are the same as in the infinitely dilute real solution.

When available, the uncertainty of entries is indicated within parentheses immediately following the value; viz., an entry 34.5(4) implies 34.5 ± 0.4 and an entry 34.5(12) implies 34.5 ± 1.2.

6.1.1   Some Thermodynamic Relations

6.1.1.1   Enthalpy of Formation.   Once standard enthalpies are assigned to the elements, it is possible to determine standard enthalpies for compounds. For the reaction

$$C(\text{graphite}) + O_2(g) \rightarrow CO_2(g) \qquad \Delta H° = -393.51 \text{ kJ} \qquad (6.1)$$

Since the elements are in their standard states, the enthalpy change for the reaction is equal to the standard enthalpy of CO_2 less the standard enthalpies of C and O_2, which are zero in each instance. Thus,

$$\Delta Hf° = -393.51 - 0 - 0 = -393.51 \text{ kJ} \qquad (6.2)$$

Tables of enthalpies, such as Tables 6.1 and 6.3, can be used to determine the enthalpy for any reaction at 1 atm and 298.15 K involving the elements and any of the compounds appearing in the tables.

The solution of 1 mole of HCl gas in a large amount of water (infinitely dilute real solution) is represented by

$$HCl(g) + \inf H_2O \rightarrow H^+(aq) + Cl^-(aq) \qquad (6.3)$$

The heat evolved in the reaction is $\Delta H° = -74.84$ kJ. With the value of $\Delta Hf°$ from Table 6.3, one has for the reaction

$$\Delta Hf° = \Delta Hf°[H^+(aq)] + \Delta Hf°[Cl^-(aq)] - \Delta Hf°[HCl(g)] \qquad (6.4)$$

for the standard enthalpy of formation of the pair of ions H^+ and Cl^- in aqueous solution (standard state, $m = 1$). To obtain the $\Delta Hf°$ values for individual ions, the enthalpy of formation of $H^+(aq)$ is arbitrarily assigned the value zero at 298.15 K. Thus, from Eq. (6.4),

$$\Delta Hf°[Cl^-(aq)] = -74.84 + (-92.31) = -167.15 \text{ kJ}$$

With similar data from Tables 6.1 and 6.3, one can determine enthalpies of formation of other ions. Thus, for the $\Delta Hf°$[KCl(aq, std. state, $m = 1$ or aq, ss)] of -419.53 kJ and the foregoing value for $\Delta Hf°$[Cl$^-$(aq, ss)]

$$\Delta Hf°[\text{K}^+(\text{aq, ss})] = \Delta Hf°[\text{KCl(aq, ss)}] - \Delta Hf°[\text{Cl}^-(\text{aq, ss})]$$
$$= -419.53 - (-167.15) = -252.38 \text{ kJ} \tag{6.5}$$

6.1.1.2 Enthalpy of Vaporization (or Sublimation).

When the pressure of the vapor in equilibrium with a liquid reaches 1 atm, the liquid boils and is completely converted to vapor on absorption of the enthalpy of vaporization ΔHv at the normal boiling point T_b. A rough empirical relationship between the normal boiling point and the enthalpy of vaporization (*Trouton's rule*) is

$$\frac{\Delta Hv}{T_b} = 88 \text{ J} \cdot \text{mol}^{-1} \cdot \text{K}^{-1} \tag{6.6}$$

It is best applied to nonpolar liquids which form unassociated vapors.

To a first approximation, the enthalpy of sublimation ΔHs at constant temperature is

$$\Delta Hs = \Delta Hm + \Delta Hv \tag{6.7}$$

where ΔHm is the enthalpy of melting.

The *Clapeyron* equation expresses the dynamic equilibrium existing between the vapor and the condensed phase of a pure substance:

$$\frac{dP}{dT} = \frac{\Delta Hv}{T \Delta V} \tag{6.8}$$

where ΔV is the volume increment between the vapor phase and the condensed phase. If the condensed phase is solid, the enthalpy increment is that of sublimation.

Substitution of $V = RT/P$ into the foregoing equation and rearranging gives the *Clausius-Clapeyron* equation,

$$\frac{dP}{P \, dT} = \frac{\Delta Hv}{RT^2} \tag{6.9}$$

or

$$\Delta Hv = -R \frac{d(\ln P)}{1/T} \tag{6.10}$$

which may be used for calculating the enthalpy of vaporization of any compound provided its boiling point at any pressure is known. If an Antoine equation is available [such as Eq. (5.1), page 5.30], differentiation and insertion into the foregoing equation gives

$$\Delta Hv = \frac{4.5757 T^2 B}{(T + C - 273.15)^2} \tag{6.11}$$

Inclusion of a compressibility factor into the foregoing equation, as suggested by the *Haggenmacher* equation improves the estimate of Hv:

$$\Delta Hv = \frac{RT^2}{P} \left(\frac{dP}{dT}\right) \left(1 - \frac{T_c^3 P}{T^3 P_c}\right)^{1/2} \tag{6.12}$$

where T_c and P_c are critical constants (Table 6.5). Although critical constants may be unknown, the compressibility factor is very nearly constant for all compounds belonging to the same family, and an estimate can be deduced from a related compound whose critical constants are available.

6.1.1.3 Heat Capacity (or Specific Heat).

The temperature dependence of the heat capacity is complex. If the temperature range is restricted, the heat capacity of any phase may be represented adequately by an expression such as

$$C_P = a + bT + cT^2 \tag{6.13}$$

in which a, b, and c are empirical constants. These constants may be evaluated by taking three pieces of data: $(T_1, C_{p,1})$, $(T_2, C_{p,2})$, and $(T_3, C_{p,3})$, and substituting in the following expressions:

$$\frac{C_{p,1}}{(T_1 - T_2)(T_1 - T_3)} + \frac{C_{p,2}}{(T_2 - T_1)(T_2 - T_3)} + \frac{C_{p,3}}{(T_3 - T_2)(T_3 - T_1)} = c \tag{6.14}$$

$$\frac{C_{p,1} - C_{p,2}}{T_1 - T_2} - [(T_1 + T_2)c] = b \tag{6.15}$$

$$(C_{p,1} - bT_1) - cT_1^2 = a \tag{6.16}$$

Smoothed data presented at rounded temperatures, such as are available in Tables 6.2 and 6.4, plus the C_p° values at 298 K listed in Tables 6.1 and 6.3, are especially suitable for substitution in the foregoing parabolic equations. The use of such a parabolic fit is appropriate for interpolation, but data extrapolated outside the original temperature range should not be sought.

6.1.1.4 Enthalpy of a System.

The enthalpy increment of a system over the interval of temperature from T_1 to T_2, under constraint of constant pressure, is given by the expression

$$H_2 - H_1 = \int_{T_1}^{T_2} C_p \, dT \tag{6.17}$$

The enthalpy over a temperature range that includes phase transitions, melting, and vaporization is represented by

$$H_2 - H_1 = \int_{T_1}^{T_t} C_p(\text{s,II}) \, dT + \Delta Ht + \int_{T_1}^{T_m} C_p(\text{s,I}) \, dT + \Delta Hm$$
$$+ \int_{T_m}^{T_b} C_p(\text{lq}) \, dT + \Delta Hv + \int_{T_b}^{T_2} C_p(\text{g}) \, dT \tag{6.18}$$

Integration of heat capacities, as expressed by Eq. (6.13), leads to

$$\Delta H = a(T_2 - T_1) + \frac{b(T_2^2 - T_1^2)}{2} + \frac{c(T_2^3 - T_1^3)}{3} \tag{6.19}$$

6.1.1.5 Entropy.

In the physical change of state,

$$\Delta Sm = \frac{\Delta Hm}{T_m} \tag{6.20}$$

is the entropy of melting (or fusion),

$$\Delta Sv = \frac{\Delta Hv}{T_b} \tag{6.21}$$

is the entropy of vaporization, and

$$\Delta Ss = \frac{\Delta Hs}{Ts} \tag{6.22}$$

is the entropy of sublimation.

A general expression for the entropy of a system, involving any phase transitions, is

$$S_2 - S_1 = \int_{T_1}^{T_t} \frac{C_p(s,\text{II})\,dT}{T} + \frac{\Delta Ht}{T} + \int_{T_1}^{T_m} \frac{C_p(s,\text{I})\,dT}{T} + \frac{\Delta Hm}{T}$$

$$+ \int_{T_m}^{T_b} \frac{C_p(\text{lq})\,dT}{T} + \frac{\Delta Hv}{T} + \int_{T_b}^{T_2} \frac{C_p(\text{g})\,dT}{T} \tag{6.23}$$

If C_p is independent of temperature,

$$\Delta S = C_p(\ln T_2 - \ln T_1) = 2.303 C_p \log \frac{T_2}{T_1} \tag{6.24}$$

If the heat capacities change with temperature, an empirical equation like Eq. (6.13) may be inserted in Eq. (6.23) before integration. Usually the integration is performed graphically from a plot of either C_p/T versus T or C_p versus $\ln T$.

TABLE 6.1 Enthalpies and Gibbs Energies of Formation, Entropies, and Heat Capacities of Organic Compounds

Substance	State	$\Delta Hf°$, kJ·mol^{-1}	$\Delta Gf°$, kJ·mol^{-1}	$S°$, J·deg^{-1}·mol^{-1}	$C_p°$, J·deg^{-1}·mol^{-1}
Acenaphthene	c	70.3(26)			
Acenaphthylene	c	186.7(46)			
Acetaldehyde	lq	−191.8(5)	−128.3	160.4	274.7
	g	−166.1(5)	−133.4	246.4	54.7
Acetaldoxime	c	−77.9			
	lq	−81.6			
Acetamide	c	−317.0(8)			
Acetamidoguanidine nitrate	c	−494.0(46)			
1-Acetamido-2-nitroguanidine	c	−193.6(59)			
5-Acetamidotetrazole	c	−5.0			
Acetanilide	c	−210.6			
Acetic acid	lq	−484.4(3)	−390.2	159.9	124.4
Ionized; ss, $m = 1$	aq	−486.34	−369.65	86.7	−6.3
Acetic anhydride	lq	−624.4(9)	−489.14	268.8	
Acetone	lq	−242.1(17)	−155.8	200.6	126.5
Acetonitrile	lq	31.4(72)	99.2	149.7	91.5
Acetophenone	lq	−142.5(10)	−17.0	249.6	
Acetyl bromide	lq	−223.5(5)			
Acetyl chloride	lq	−272.9(7)	−208.2	201.0	117

TABLE 6.1 Enthalpies and Gibbs Energies of Formation, Entropies, and Heat Capacities
of Organic Compounds (*Continued*)

Substance	State	$\Delta Hf°$, kJ·mol^{-1}	$\Delta Gf°$, kJ·mol^{-1}	$S°$, J·deg^{-1}·mol^{-1}	$C_p°$, J·deg^{-1}·mol^{-1}
Acetylene	g	−226.7(8)	248.2	201.0	44.1
Acetylene-d_2	g	221.5	205.9	208.9	49.3
Acetylenedicarboxylic acid	c	−578.2			
Acetyl fluoride	g	−442.1(33)			
1-Acetylimidazole	c	−574.0			
Acetyl iodide	lq	−163.5(12)			
Acridine	c	200.9(67)			
Adamantane	c	−194.1(22)			
Adenine	c	96.0(9)	299.6	151.1	
(+)-Alanine	c	−561.2(6)	−369.4	132.3	
(−)-Alanine	c	−604.0(21)	−370.5	129.3	
(±)-Alanine	c	−563.6(6)	−372.3	132.3	
(±)-N-Alanylglycine	c	−777.8(9)	−489.9	213.5	
(−)-Alanylglycine	c	−827.0	−533.0	195.2	
Alloxan monohydrate	c	−1000.7(5)	−762.3	186.7	
Allyl *tert*-butyl sulfide	lq	−91.0(17)			
Allyl ethyl sulfone	lq	−406.0(10)			
Allyl methyl sulfone	lq	−385.1(11)			
Allyl trichloroacetate	lq	−395.3(84)			
Allyl (*see* Propene)					
3-Aminoacetophenone	c	−173.3(5)			
4-Aminoacetophenone	c	−182.1(5)			
2-Aminoacridine	c	166.4(13)			
9-Aminoacridine	c	159.2(71)			
2-Aminobenzoic acid	c	−400.9(9)			
3-Aminobenzoic acid	c	−411.6(23)			
4-Aminobenzoic acid	c	−412.9(10)			
2-Aminobiphenyl	c	112.2(63)			
4-Aminobiphenyl	c	81.2(63)			
4-Aminobutanoic acid	c	−577.9(9)			
2-Aminoethanesulfonic acid	c	−785.9	−562.3	154.1	140.7
Ionized; ss, $m = 1$	aq	−719.8	−509.8	200.1	
2-Aminohexanoic acid (norleucine)	c	−639.1(13)			
4-Aminohexanoic acid	c	−646.2(13)			
5-Aminohexanoic acid	c	−643.3(13)			
6-Aminohexanoic acid	c	−639.1(13)			
(−)-2-Amino-3-hydroxy-butanoic acid	c	−759.5			
5-Aminopentanoic acid	c	−604.1(9)			
5-Aminotetrazole	c	−207.8(23)			
3-Amino-1,2,4-triazole	c	76.8(39)			
Aniline	lq	31.3(10)	149.2	191.4	192.2
Anthracene	c	129.2(18)	286.0	207.6	208.1
9,10-Anthraquinone	c	−207.5(32)			
D-(−)-Arabinose [also (+)-]	c	−1057.9(16)			
(+)-Arginine	c	−623.5(13)	−240.5	250.8	
L-(+)-Ascorbic acid	c	−1164.6(10)			
L-(+)-Asparagine	c	−789.4(8)	−530.6	174.6	
L-(+)-Aspartic acid	c	−973.3(8)	−730.7	170.2	
(E)-Azobenzene	c	310.2(34)			

TABLE 6.1 Enthalpies and Gibbs Energies of Formation, Entropies, and Heat Capacities of Organic Compounds (*Continued*)

Substance	State	$\Delta Hf°$, kJ·mol^{-1}	$\Delta Gf°$, kJ·mol^{-1}	$S°$, J·deg^{-1}·mol^{-1}	$C_p°$, J·deg^{-1}·mol^{-1}
(Z)-Azobenzene	c	365.2(21)			
Azomethane	g	148.8	239.7	289.9	78.0
Azoisopropane	g	35.8(10)			
Azomethane-d_6	g	119.3	218.3	305.7	90.6
Azopropane	g	51.5(17)			
Azulene	g	289.1(34)	353.4	338.1	128.5
Barbituric acid	c	−637.2			
Benzaldehyde	lq	−87.0(21)	9.4		
Benzamide	c	−202.6(11)			
Benzanilide	c	−93.4			
1,2-Benzanthracene	c	170.9(24)			
2,3-Benzanthracene	c	160.4	359.2	215.5	
1,2-Benzanthracene-9,10-dione	c	−231.9(34)			
Benzene	lq	49.0(6)	124.4	173.4	81.7
	g	82.6(7)	129.7	269.2	
Benzeneboronic acid	c	−720.1			
1,2-Benzenediamine	c	−0.3(42)			
1,3-Benzenediamine	c	−7.8(42)			
1,4-Benzenediamine	c	3.1(7)			
Benzenethiol (thiophenol)	lq	63.7(8)	134.0	222.8	173.2
	g	112.4(9)	147.6	336.9	104.9
1,3-Benzenedicarboxylic acid	c	803.0			
1,4-Benzenedicarboxylic acid	c	816.1			
1,2,4,5-Benzenetetracarboxylic acid	c	1571.0			
1,2,3-Benzenetricarboxylic acid	c	−1160.0			
1,2,4-Benzenetricarboxylic acid	c	−1179.0			
1,3,5-Benzenetricarboxylic acid	c	−1190.0			
Benzoic acid	c	−385.2(5)	−245.3	167.6	146.3
Benzoic anhydride	c	−415.4(18)			
Benzonitrile	g	215.8(21)	260.8	321.0	109.1
Benzo[*def*]phenanthrene	c	125.5(11)	269.5	224.8	236.0
Benzophenone	c	−34.5(21)	140.2	245.2	
Benzo[*f*]quinoline	c	150.6(62)			
Benzo[*h*]quinoline	c	149.7(40)			
1,4-Benzoquinone	c	−185.7(11)	−83.6	162.8	
1,2,3-Benzotriazole	c	250.0(11)			
Benzotrifluoride	lq	−636.7(10)			
Benzoyl bromide	lq	−107.3(7)			
Benzoyl chloride	lq	−158.0(9)			
Benzoylformic acid	c	−482.4			
N-Benzoylglycine	c	−609.8(16)	−369.57	239.3	
Benzoyl iodide	lq	−53.5(15)			
3,4-Benzphenanthrene	c	184.9			
Benzylamine	lq	34.2(18)			
Benzyl alcohol	lq	−160.7(12)	−27.5	216.7	218.0
Benzyl bromide	lq	16.0(270)			
Benzyl chloride	lq	−32.6(26)			
N-Benzyldiphenylamine	c	184.7(100)			
Benzyl ethyl sulfide	lq	−4.9			
Benzyl iodide	lq	57.3(42)			

TABLE 6.1   Enthalpies and Gibbs Energies of Formation, Entropies, and Heat Capacities of Organic Compounds (*Continued*)

Substance	State	$\Delta Hf°$, kJ·mol⁻¹	$\Delta Gf°$, kJ·mol⁻¹	$S°$, J·deg⁻¹·mol⁻¹	$C_p°$, J·deg⁻¹·mol⁻¹
Benzyl methyl ketone	lq	−151.9(17)			
Benzyl methyl sulfide	lq	26.2			
Bicyclo[1.1.0]butane	g	217.1(8)			
Bicyclo[2.2.1]hepta-2,5-dione	lq	213.0(30)			
Bicylco[2.2.1]heptane	c	−95.1(46)			
Bicyclo[4.1.0]heptane	lq	−36.7(26)			
Bicylco[2.2.1]heptene	lq	90.0(32)	203.9		130.0
Bicyclo[3.1.0]hexane	g	38.6(7)			
Bicyclohexyl	lq	−273.7(15)			
Bicyclo[2.2.2]octane	c	−146.9(9)			
Bicyclo[4.2.0]octane	g	−26.2(27)			
Bicyclo[5.1.0]octane	g	−16.6(18)			
Bicyclo[2.2.2]oct-2-ene	g	−23.3(8)			
Bicyclopropyl	g	129.3(36)			
Biphenyl	c	99.4(18)	254.2	205.9	162.3
2-Biphenylcarboxylic acid	c	−349.0(63)			
(1,1′-Biphenyl)-4-4′-diamine	c	70.7(17)			
Biphenylene	c	334.0(33)			
Bis(dimethylthiocarbonyl) disulfide	c	41.6			
Bis(2-hydroxyethyl) ether	lq	−1621.0		441.0	135.1
	g	−571.1			
Bromobenzene	lq	60.9(41)	126.0	217.6	155.5
4-Bromobenzoic acid	c	−378.3(21)			
1-Bromobutane	g	−107.1(13)	−12.9	369.8	109.3
2-Bromobutane	lq	−154.8(12)	−19.25		
	g	−120.3(12)	−25.8	370.3	110.8
Bromochlorodifluoromethane	g	−471.5	−448.4	318.6	
Bromochlorofluoromethane	g	−295.0	−278.6	304.9	
Bromochloromethane	g	−50.2	−39.3	287.3	
1-Bromo-2-chloro-1,1,2-trifluoroethane	g	−700.9(208)			
Bromodichlorofluoromethane	g	−269.5	−246.8	330.0	
Bromodichloromethane	g	−58.6	−42.5	316.3	
Bromodifluoromethane	g	−424.9(10)	−447.3	295.0	
Bromoethane	lq	−90.1(28)	−27.8	198.7	100.8
Bromoethylene (vinyl bromide)	g	79.2(19)	80.75	275.4	55.4
Bromofluoromethane	g	−252.7	−241.5	276.0	
1-Bromoheptane	lq	−218.4(16)			
1-Bromohexane	lq	−194.2(16)			
Bromoiodomethane	g	50.2	39.2	307.5	
Bromomethane	g	−35.5(11)	−28.24	245.8	42.5
2-Bromo-2-methylpropane	lq	−163.8(19)			
	g	−132.4(18)	−28.2	332.0	116.5
1-Bromooctane	lq	−245.1(23)			
1-Bromopentane	lq	−170.2(15)			
	g	−129.0(15)	−5.7	408.8	132.2
1-Bromopropane	lq	−121.8(30)			
	g	−87.0(33)	−22.5	330.9	86.4
2-Bromopropane	lq	−130.5(26)			
	g	−99.4	−27.2	316.2	89.4

TABLE 6.1 Enthalpies and Gibbs Energies of Formation, Entropies, and Heat Capacities of Organic Compounds (*Continued*)

Substance	State	$\Delta Hf°$, kJ·mol^{-1}	$\Delta Gf°$, kJ·mol^{-1}	$S°$, J·deg^{-1}·mol^{-1}	$C_p°$, J·deg^{-1}·mol^{-1}
N-Bromosuccinimide	c	−336.18			
(α)-Bromotoluene	lq	23.4			
Bromotrichloromethane	g	−41.9(18)	−12.4	332.8	
Bromotrifluoromethane	g	−648.9(29)	−622.6	297.8(5)	69.3
Bromotrimethylsilane	lq	−325.9			
Brucine	c	−496.2			
1,2-Butadiene	g	162.3(6)	199.5	293.0	80.1
1,3-Butadiene	g	108.8(11)	150.7	278.7	79.5
1,3-Butadiyne	g	472.8	444.0	250.0	73.6
Butanal	g	−204.9(14)	−114.8	344.9	102.6
Butanamide	c	−364.4(7)			
Butane	g	−125.6(7)	−17.2	310.1	97.5
1,2-Butanediamine	lq	−120.2(8)			
1,2-Butanediol	lq	−523.6(25)			
1,3-Butanediol	lq	−501.0(21)			
1,4-Butanediol	lq	−503.3(20)			
2,3-Butanediol	lq	−541.5(25)			
Butanedinitrile	c	139.7(6)			
2,3-Butanedione	lq	−365.8(8)			
1,4-Butanedithiol	lq	−105.7(18)			
Butanenitrile	g	33.6(10)	108.7	325.4	97.0
1-Butanethiol	lq	−124.7(12)	4.1	276.0	
2-Butanethiol	lq	−131.0(9)	−0.17	271.4	
Butanoic acid	lq	−533.8(6)	−377.7	226.4	176.2
1-Butanol	lq	−327.3(4)	−162.5	226.4	177.0
	g	−275.0(4)	−150.8	362.8	122.6
(±)-2-Butanol	lq	−342.6(7)	−177.0	225.1	198.7
	g	−292.9(7)	−167.6	359.0	113.3
2-Butanone	lq	−273.3(9)	−151.4	238.8	158.9
Butanophenone	lq	−188.9(18)			
(E)-2-Butenal	lq	−138.7(16)			
(Z)-Butenedinitrile	c	268.2(17)			
1-Butene	g	0.1(10)	71.3	305.6	85.7
(E)-2-Butene	g	−11.4(10)	63.0	296.5	87.8
(Z)-2-Butene	g	−7.1(10)	65.9	300.8	78.9
(E)-2-Butenenitrile	lq	100.7(9)			
(Z)-2-Butenenitrile	lq	95.1(10)			
3-Butenenitrile	g	157.8(12)	193.4	298.4	82.1
(E)-2-Butenoic acid	c	−430.5			
(Z)-2-Butenoic acid	lq	−347.0			
(E)-2-Butenedioic acid	c	−811.1(18)			
(Z)-2-Butenedioic acid	c	−788.7(19)			
1-Buten-3-yne	g	304.6	306.0	279.4	73.2
N-Butylacetamide	lq	−380.8(16)			
Butyl acetate	lq	−529.2(7)			
Butylamine	g	−92.0(12)	49.2	363.3	118.6
sec-Butylamine	g	−109.8(10)	40.7	351.3	117.2
tert-Butylamine	g	−121.0(7)	28.9	337.9	120.0
Butylbenzene	g	−13.1(11)	144.7	439.5	416.3
sec-Butylbenzene	lq	−66.4(12)			
tert-Butylbenzene	lq	−70.7(12)			

TABLE 6.1   Enthalpies and Gibbs Energies of Formation, Entropies, and Heat Capacities of Organic Compounds (*Continued*)

Substance	State	$\Delta Hf°$, kJ·mol^{-1}	$\Delta Gf°$, kJ·mol^{-1}	$S°$, J·deg^{-1}·mol^{-1}	$C_p°$, J·deg^{-1}·mol^{-1}
sec-Butyl butanoate	lq	−492.6(38)			
Butyl chloroacetate	lq	−538.4(21)			
Butyl 2-chlorobutanoate	lq	−655.2			
Butyl 3-chlorobutanoate	lq	−610.9			
Butyl 4-chlorobutanoate	lq	−618.0			
Butyl 2-chloropropanoate	lq	−572.0			
Butyl 3-chloropropanoate	lq	−558.2			
Butyl crotonate	lq	−467.8			
Butylcyclohexane	g	−213.4(13)	56.4	458.5	207.1
Butylcyclopentane	g	−168.3	61.4	456.2	177.5
Butyl dichloroacetate	lq	−550.2(84)			
Butyl ethyl sulfide (3-thiaheptane)	g	−125.2	32.0	453.0	162.0
tert-Butyl ethyl sulfide	lq	−187.3(21)			
tert-Butyl hydroperoxide	lq	−293.6(50)			
Butyllithium	lq	−132.2			
Butyl methyl ether	lq	−290.6(12)			
Butyl methyl sulfide (2-thiahexane)	lq	−142.8(20)	17.1	307.5	
tert-Butyl methyl sulfide	lq	−156.9(8)			
Butyl methyl sulfone	lq	−535.8(8)			
tert-Butyl methyl sulfone	c	−556.0(28)			
tert-Butyl peroxide	lq	−380.9(9)			
Butyl trichloroacetate	lq	−545.8(84)			
1-Butyne	g	165.2(9)	202.1	290.8	81.4
2-Butyne	g	145.7(12)	185.4	283.3	78.0
2-Butynedinitrile	g	529.2(14)			
2-Butynedioic acid	c	−577.4(46)			
3-Butynoic acid	c	−241.8(10)			
9*H*-Carbazole	c	125.1(36)			
Chloroacetamide	c	−338.5			
Chloroacetic acid	c	−510.5(84)			
Chloroacetyl chloride	lq	−283.7(84)			
2-Chlorobenzaldehyde	lq	−118.4(84)			
3-Chlorobenzaldehyde	lq	−126.0(84)			
4-Chlorobenzaldehyde	c	−146.4(84)			
Chlorobenzene	lq	11.0(13)	89.2	209.2	150.2
2-Chlorobenzoic acid	c	−404.5(8)			
3-Chlorobenzoic acid	c	−424.3(16)			
4-Chlorobenzoic acid	c	−428.9(10)			
Chloro-1,4-benzoquinone	c	−220.6(84)			
1-Chlorobutane	g	−154.6(12)	−38.8	358.1	107.6
(±)-2-Chlorobutane	g	−161.2(84)	−53.5	359.6	108.5
2-Chlorobutanoic acid	lq	−575.5(84)			
3-Chlorobutanoic acid	lq	−556.3(84)			
4-Chlorobutanoic acid	lq	−566.3(84)			
Chlorocyclohexane	lq	−207.2(15)			
1-Chloro-2,2-difluoroethylene	g	−315.5	−289.1	302.9	72.1
2-Chloro-1,1-difluoroethylene	g	−331.4	−305.0	302.4	
Chlorodifluoromethane	g	−482.6(32)	−450.0	281.0(8)	57.1
2-Chloro-1,4-dihydroxybenzene	c	−382.81			

TABLE 6.1 Enthalpies and Gibbs Energies of Formation, Entropies, and Heat Capacities of Organic Compounds (*Continued*)

Substance	State	$\Delta Hf°$, kJ·mol^{-1}	$\Delta Gf°$, kJ·mol^{-1}	$S°$, J·deg^{-1}·mol^{-1}	$C_p°$, J·deg^{-1}·mol^{-1}
Chlorodimethylsilane	lq	−79.8			
1-Chloro-2,3-epoxypropane	lq	−148.5			
1-Chloroethane	g	−112.1(11)	−60.5	275.8	62.6
1-Chloro-2-ethylbenzene	lq	−54.1(16)			
1-Chloro-4-ethylbenzene	lq	−51.7(16)			
Chloroethylene (vinyl chloride)	g	37.3(13)	51.5	263.9	53.7
Chloroethyne	g	213.0	197.0	241.9	54.3
Chlorofluoroethane	g	−313.4(25)			
Chlorofluoromethane	g	−290.8	−265.5	264.3	47.0
Chlorohydroquinone	c	−382.8			
Chloroiodomethane	g	12.6	15.4	296.1	
Chloromethane	g	−81.9(5)	−58.5	234.2	40.8
1-Chloro-2-methylpropane	g	−159.4(84)	−49.7	355.0	108.5
2-Chloro-2-methylpropane	g	−182.2(23)	−64.1	322.2	114.2
1-Chloronaphthalene	lq	54.4(84)			
2-Chloronaphthalene	c	55.2(84)			
Chloropentafluoroacetone	g	−1121.0			
1-Chloropentane	g	−175.0(21)	−37.4	397.0	130.5
3-Chlorophenol	c	−206.4(84)			
4-Chlorophenol	c	−197.9(84)			
1-Chloropropane	g	−131.9(11)	−50.7	319.1	84.6
2-Chloropropane	g	−144.9(13)	−62.5	304.2	87.3
2-Chloro-1,3-propanediol	lq	−517.5(11)			
3-Chloro-1,2-propanediol	lq	−525.3(9)			
2-Chloropropanoic acid	lq	−522.5(84)			
3-Chloropropanoic acid	c	−549.3(84)			
2-Chloro-1-propene	g	−21.0(84)			
3-Chloro-1-propene (allyl chloride)	g	−0.63	43.6	306.7	75.4
N-Chlorosuccinimide	c	−358.1			
(*α*)-Chlorotoluene	lq	−32.6			
Chlorotrifluoromethane	g	−708.0(30)	−667.4	285.4(4)	66.9
Chlorotrimethylsilane	lq	−384.1			
Chlorotrinitromethane	lq	−27.1(18)			
Chrysene	c	145.3(21)			
(−)-Cinchonidine	c	29.7			
Cinchonine	c	31.0			
(E)-Cinnamic acid	c	−338.5(22)			
(Z)-Cinnamic acid	c	−315.0(28)			
Cinnamic anhydride	c	−347.7(84)			
Citric acid	c	−1543.9(46)	−1236.4	166.2	
Codeine monohydrate	c	−632.6			
Creatine	c	−537.2(9)	−264.9	189.5	
Creatinine	c	−238.5(5)	−29.2	167.8	
o-Cresol	c	−204.6(10)			
	g	−128.6(13)	37.1	357.6	130.3
m-Cresol	c	−194.0(7)			
	g	−132.3(13)	−40.5	356.8	122.5
p-Cresol	c	−199.3(8)			
	g	−125.4(16)	−30.9	347.6	124.5
Cuban	c	541.3			

TABLE 6.1 Enthalpies and Gibbs Energies of Formation, Entropies, and Heat Capacities of Organic Compounds (*Continued*)

Substance	State	$\Delta Hf°$, kJ·mol^{-1}	$\Delta Gf°$, kJ·mol^{-1}	$S°$, J·deg^{-1}·mol^{-1}	$C_p°$, J·deg^{-1}·mol^{-1}
Cyanamide	c	58.8(5)			
Cyanogen	g	306.7(8)	297.2	242.3	56.9
Cyclobutane	g	28.4(6)	110.0	265.4	72.2
Cyclobutanecarbonitrile	lq	103.0(12)			
Cyclobutene	g	156.7(15)	174.7	263.5	67.1
Cyclobutylamine	g	41.2(8)			
Cyclododecane	c	−306.6(14)			
1,3-Cycloheptadiene	g	94.3(11)			
Cycloheptane	lq	−156.6(9)	54.1	242.6	123.1
Cycloheptanone	lq	−299.4(13)			
1,3,5-Cycloheptatriene	lq	142.2(21)	243.1	214.6	162.8
Cycloheptene	g	−9.2(11)			
Cyclohexane	lq	−156.4(8)	26.7	204.4	106.3
	g	−123.4(8)	31.8	298.3	106.3
(E)-Cyclohexane-1,2-dicarboxylic acid	c	−970.7(75)			
(Z)-Cyclohexane-1,2-dicarboxylic acid	c	−961.1(75)			
Cyclohexanethiol	g	−96.1(9)			
Cyclohexanol	lq	−348.1(21)	−133.3	199.6	
Cyclohexanone	g	−225.1(21)	−90.8	322.2	109.7
Cyclohexene	lq	−38.5(6)	101.6	216.2	146.0
1-Cyclohexenylmethanol	lq	−382.4			
Cyclohexylamine	lq	−147.7(13)			
Cyclohexylcyclohexane	lq	−329.3			
Cyclooctane	lq	−167.7(10)			
Cyclooctanone	lq	−326.0(50)			
1,3,5,7-Cyclooctatetraene	lq	254.5(13)	358.6	220.3	184.0
Cyclooctene	lq	−74.0(20)			
Cyclopentadiene	g	133.8(18)	179.3	267.8	
Cyclopentane	lq	−105.1(13)	36.4	204.3	128.9
	g	−76.4(8)	38.6	292.9	83.0
(E)-Cyclopentane-1,2-diol	c	−489.9(29)			
(Z)-Cyclopentane-1,2-diol	c	−484.9(42)			
Cyclopentanethiol	lq	−89.5(8)	46.8	361.4	107.9
Cyclopentanol	lq	−300.1(16)	−127.8	205.8	
Cyclopentanone	lq	−235.7(18)			
Cyclopentene	lq	4.2(8)	108.5	201.3	122.3
	g	34.3(32)	110.8	291.8	75.1
1-Cyclopentenylmethanol	lq	34.3			
Cyclopentylamine	lq	−95.1(8)			
Cyclopropane	g	53.3(6)	104.4	237.4	55.9
Cyclopropene	g	277.1(25)	286.3	223.3	
Cyclopropylamine	g	77.0(7)			
Cyclopropylbenzene	lq	100.3(9)			
(−)-Cysteine	c	−515.5(21)			
(−)-Cystine	c	−1032.7(38)			
(E)-Decahydronaphthalene	lq	−230.6(10)	57.7	265.0	228.5
(Z)-Decahydronaphthalene	lq	−219.4(10)	68.9	265.0	232.0
Decanal	g	−330.9	−66.5	578.6	239.7
Decane	lq	−300.9(13)	17.5	425.5	234.5

TABLE 6.1 Enthalpies and Gibbs Energies of Formation, Entropies, and Heat Capacities of Organic Compounds (*Continued*)

Substance	State	$\Delta Hf°$, kJ·mol^{-1}	$\Delta Gf°$, kJ·mol^{-1}	$S°$, J·deg^{-1}·mol^{-1}	$C_p°$, J·deg^{-1}·mol^{-1}
Decanedioic acid	c	-1082.8			
1,10-Decanediol	c	$-693.5(26)$			
1-Decanenitrile	lq	$-158.4(13)$			
1-Decanethiol	g	$-211.5(15)$	61.4	610.1	255.6
Decanoic acid	c	$-713.7(10)$			
1-Decanol	lq	$-478.1(12)$	-132.2	430.5	
1-Decene	lq	$-173.8(19)$	105.0	425.0	
1-Decyne	g	41.2	252.2	524.5	219.7
Deoxybenzoin	c	$-71.0(29)$			
Diacetamide	c	-489.0			
Diacetyl peroxide	lq	$-535.3(92)$			
1,2-Diallyl phthalate	lq	-550.6			
2,6-Diaminopyridine	c	$-6.5(5)$			
Diazomethane	g	192.5	217.8	242.8	52.5
Dibenz[*de,kl*]anthracene	c	182.8			
1,2-Dibenzoylethane	c	$-255.6(15)$			
(E)-1,2-Dibenzoylethylene	c	$-114.7(24)$	109.8	319.2	
Dibenzoylmethane	c	$-223.5(30)$			
Dibenzoyl peroxide	c	$-369.6(46)$			
Dibenzyl	c	44.1	260.0	269.4	255.2
Dibenzyl sulfide	c	99.0(21)			
Dibenzyl sulfone	c	$-282.6(12)$			
1,2-Dibromobutane	g	$-91.5(17)$	-13.1	408.8	127.1
1,4-Dibromobutane	g	$-87.0(51)$			
2,3-Dibromobutane	g	$-102.0(13)$			
Dibromochlorofluoromethane	g	-231.8	-223.4	342.6	
Dibromochloromethane	g	-20.9	-18.8	327.7	
1,2-Dibromocycloheptane	lq	$-157.6(21)$			
1,2-Dibromocyclohexane	lq	$-162.8(23)$			
1,2-Dibromocyclooctane	lq	$-173.3(22)$			
Dibromodichloromethane	g	-29.3	-19.5	348.2	
Dibromodifluoromethane	g	-429.7	-419.1	324.9	
1,2-Dibromoethane	lq	$-79.2(13)$	-20.9	223.3	136.0
Dibromomethane	g	-14.8	-16.2	293.2	54.6
Dibromofluoromethane	g	-223.4	-221.1	316.7	
1,2-Dibromopropane	g	$-71.5(11)$	-17.7	376.1	102.8
Dibutoxymethane	lq	$-549.4(17)$			
Dibutylamine	lq	$-206.0(6)$			
Dibutyl disulfide	g	$-158.4(26)$	53.9	572.8	231.1
Di-*tert*-butyl disulfide	lq	-250.4			
Dibutyl ether	g	$-333.4(14)$	-88.5	500.4	204.0
Di-*sec*-butyl ether	g	$-360.9(17)$			
Di-*tert*-butyl ether	g	$-362.0(12)$			
Dibutylmercury	lq	-97.9			
Dibutyl peroxide	lq	-380.7			
Dibutyl 1,2-phthalate	c	$-842.6(126)$			
Dibutyl sulfate	lq	$-904.6(26)$			
Dibutyl sulfide	lq	$-220.7(15)$	32.2	405.1	
Di-*tert*-butyl sulfide	lq	-232.4			
Dibutyl sulfite	lq	$-693.1(40)$			
Dibutyl sulfone	c	$-610.2(20)$			

TABLE 6.1   Enthalpies and Gibbs Energies of Formation, Entropies, and Heat Capacities of Organic Compounds (*Continued*)

Substance	State	$\Delta Hf°$, kJ·mol^{-1}	$\Delta Gf°$, kJ·mol^{-1}	$S°$, J·deg^{-1}·mol^{-1}	$C_p°$, J·deg^{-1}·mol^{-1}
Dichloroacetic acid	lq	−496.3(84)			
Ionized	aq	−507.1(84)			
Dichloroacetyl chloride	lq	−280.8(84)			
1,2-Dichlorobenzene	g	30.2(21)	82.7	341.5	113.5
1,3-Dichlorobenzene	g	25.7(21)	78.6	343.5	113.8
1,4-Dichlorobenzene	g	22.5(15)	77.2	336.7	113.9
Dichlorodifluoromethane	g	−492.0(80)	−453.0	300.9(2)	72.4
Dichlorodimethylsilane	g	−461.1		335.4	101.1
Dichlorodiphenylsilane	lq	−278.2			
1,1-Dichloroethane	g	−127.7(12)	−73.3	305.1	76.4
1,2-Dichloroethane	g	−126.9(28)	−73.9	308.2	78.7
1,1-Dichloroethylene	g	2.6(13)	24.2	288.1	67.0
(E)-1,2-Dichloroethylene	g	5.0(85)	26.6	289.9	66.7
(Z)-1,2-Dichloroethylene	g	4.6(85)	24.4	289.5	65.1
Dichlorofluoromethane	g	−283.0(130)	−253.0	293.3(8)	61.0
Dichloromethane	g	−95.6(12)	−68.9	270.3	50.9
1,2-Dichloropropane	g	−162.8(12)	−83.1	354.8	98.2
1,3-Dichloropropane	g	−159.2(84)	−82.6	367.2	99.6
2,2-Dichloropropane	g	−173.2(85)	−84.6	326.0	105.9
1,3-Dichloro-2-propanol	lq	−385.4(21)			
Dicyanoacetylene	lq	500.4			
Dicyanobenzene	c	275.4			
1,4-Dicyano-2-butyne	c	366.5			
Dicyanodiamide	c	22.6	179.5	129.3	118.8
Dicyclopentadiene	c	116.7			
Dichloropentadienyliron	c	141.0			
Diethoxymethane	lq	−450.4(8)			
1,1-Diethoxyethane	lq	−491.4(23)			
1,2-Diethoxyethane	lq	−451.4(10)			
1,3-Diethoxypropane	lq	−482.1(14)			
2,2-Diethoxypropane	lq	−538.5(10)			
Diethylamine	g	−72.5(12)	72.1	352.2	115.7
Diethylbarbituric acid (veronal)	c	−747.7			
1,2-Diethylbenzene	g	−19.0	141.1	434.3	182.6
1,3-Diethylbenzene	g	−21.8	136.7	439.3	176.9
1,4-Diethylbenzene	g	−22.3	137.9	434.0	176.2
Diethyl carbonate	lq	−681.5(8)			
(E)-1,2-Diethylcyclopropane	lq	83.3(21)			
(Z)-1,2-Diethylcyclopropane	lq	−79.9(14)			
Diethyl disulfide	lq	−120.0	9.5	305	
	g	−74.6	22.3	414.5	141.3
Diethylenediamine	c	−13.4	240.2	85.8	
Diethylene glycol	lq	−628.5(24)			
	g	−571.1		441.0	135.1
Diethyl ether	lq	−279.3(8)	−116.7	253.1	170.7
	g	−252.1(8)	−122.3	342.7	112.5
Diethylmercury	lq	29.7			
Diethyl oxalate	lq	−805.5(84)			
Diethyl peroxide	lq	−223.3(13)			
Diethyl 1,2-phthalate	lq	−778.0			
Diethyl selenide	lq	−96.2			

TABLE 6.1   Enthalpies and Gibbs Energies of Formation, Entropies, and Heat Capacities of Organic Compounds (*Continued*)

Substance	State	$\Delta Hf°$, kJ·mol^{-1}	$\Delta Gf°$, kJ·mol^{-1}	$S°$, J·deg^{-1}·mol^{-1}	$C_p°$, J·deg^{-1}·mol^{-1}
Diethyl sulfate	lq	−813.2(11)			
Diethyl sulfide	g	−83.6(8)	17.8	368.0	117.0
Diethyl sulfite	lq	−600.7(9)			
Diethyl sulfone	c	−515.5(7)			
Diethyl sulfoxide	lq	−268.0(9)			
Diethylzinc	lq	16.7			
1,2-Difluorobenzene	g	−293.8(9)	−242.0	321.9	106.5
1,3-Difluorobenzene	g	−309.2(10)	−257.0	320.4	106.3
1,4-Difluorobenzene	g	−306.7(10)	−252.8	315.6	106.9
2,2′-Difluorobiphenyl	c	−295.9(21)			
4,4′-Difluorobiphenyl	c	−296.5(22)			
1,1-Difluoroethane	g	−497.0(84)	−443.0	282.4	68.0
1,1-Difluoroethylene	g	−335.0(45)	−321.5	265.2	59.2
Difluoromethane	g	−452.2(9)	−425.4	246.6	42.9
9,10-Dihydroanthracene	c	66.4(13)			
1,2-Dihydronaphthalene	lq	71.5(17)			
1,4-Dihydronaphthalene	lq	84.2(15)			
Dihydro-2H-pyran	lq	−157.4(13)			
5,12-Dihydrotetracene	c	106.4			
2,3-Dihydrothiophene	lq	52.9(12)			
2,5-Dihydrothiophene	lq	47.0(12)			
2,5-Dihydrothiophene-1,1-dioxide	c	318.9(13)			
2′,4-Dihydroxyacetophenone	c	−573.6			
1,2-Dihydroxybenzene	c	−361.1	−210.0	150.2	132.2
1,3-Dihydroxybenzene	c	−368.0(5)	−209.2	147.7	131.0
1,4-Dihydroxybenzene	c	−364.5(16)	−207.0	140.2	141.8
Dihydroxymalonic acid	c	−1216.3			
2,4-Dihydroxy-5-methyl-pyrimidine	c	−468.2			
2,4-Dihydroxy-6-methyl-pyrimidine	c	−456.9			
1,2-Diiodobenzene	c	172.4(42)			
1,3-Diiodobenzene	c	187.0(42)			
1,4-Diiodobenzene	c	160.7(42)			
1,2-Diiodoethane	g	66.8(17)	78.5	348.5	82.3
Diiodomethane	g	118.4	101.4	309.4	57.7
Diisobutylamine	lq	−218.5(6)			
Diisopropylamine	lq	−178.5(5)			
Diisopropyl ether	g	−319.2(16)	−121.9	390.2	158.3
Diisopropylmercury	lq	−13.0			
Diisopropyl sulfide	g	−142.1(15)	27.1	415.5	169.2
Diketene	lq	−233.1			
1,2-Dimethoxybenzene	lq	−290.4(21)			
1,1-Dimethoxybutane	lq	−468.1(15)			
2,2-Dimethoxybutane	lq	−485.1(10)			
1,1-Dimethoxyethane	lq	−420.2(8)			
1,2-Dimethoxyethane	lq	−377.7(9)			
Dimethoxymethane	lq	−450.4(8)			
1,1-Dimethoxypentane	lq	−494.6(24)			
2,2-Dimethoxypentane	lq	−509.2(12)			

TABLE 6.1 Enthalpies and Gibbs Energies of Formation, Entropies, and Heat Capacities of Organic Compounds (*Continued*)

Substance	State	$\Delta Hf°$, kJ·mol^{-1}	$\Delta Gf°$, kJ·mol^{-1}	$S°$, J·deg^{-1}·mol^{-1}	$C_p°$, J·deg^{-1}·mol^{-1}
1,1-Dimethoxypropane	lq	−443.3(11)			
2,2-Dimethoxypropane	lq	−459.0(19)			
1,1-Dimethoxy-2-methyl-propane	lq	−476.2(17)			
N,N-Dimethylacetamide	lq	−278.3(13)			
Dimethylamine	g	−18.6(8)	68.0	273.0	69.0
4-(Dimethylamino)-benzaldehyde	c	−137.6(9)			
Dimethylaminomethanol	lq	−253.6(12)			
N,N-Dimethylamino-trimethylsilane	lq	−279.5			
N,N-Dimethylaniline	lq	47.7(32)			
2,3-Dimethylbenzoic acid	c	−450.4(9)			
2,4-Dimethylbenzoic acid	c	−458.5(9)			
2,5-Dimethylbenzoic acid	c	−456.1(9)			
2,6-Dimethylbenzoic acid	c	−440.7(9)			
3,4-Dimethylbenzoic acid	c	−468.8(11)			
3,5-Dimethylbenzoic acid	c	−466.4(8)			
3,3′-Dimethylbiphenyl	lq	20.0			
2,2-Dimethylbutane	g	−186.1(10)	−9.2	358.2	141.9
2,3-Dimethylbutane	g	−178.3(10)	−4.1	365.8	140.5
3,3-Dimethyl-2-butanone	lq	−328.6(9)			
2,3-Dimethyl-1-butene	g	−62.6(13)	79.0	365.6	143.5
2,3-Dimethyl-2-butene	g	−68.2(11)	76.1	364.6	123.6
3,3-Dimethyl-1-butene	g	−60.5(13)	98.2	343.8	126.5
2,3-Dimethyl-2-butenoic acid	c	−455.6			
Dimethylcadmium	lq	63.6	139.3	201.9	
1,1-Dimethylcyclohexane	lq	−218.7(19)	26.5	267.2	
	g	−180.9(19)	35.2	365.0	154.4
(E)-1,2-Dimethylcyclohexane	g	−180.0(19)	34.5	370.9	159.0
(Z)-1,2-Dimethylcyclohexane	g	−172.1(18)	41.2	374.5	165.5
(E)-1,3-Dimethylcyclohexane	g	−176.5(17)	36.3	376.2	157.3
(Z)-1,3-Dimethylcyclohexane	g	−184.6(18)	29.8	370.5	157.3
(E)-1,4-Dimethylcyclohexane	g	−184.5(18)	31.7	364.8	157.7
(Z)-1,4-Dimethylcyclohexane	g	−176.6(18)	38.0	370.5	157.3
1,1-Dimethylcyclopentane	g	−138.2(11)	39.0	359.3	133.3
(E)-1,2-Dimethylcyclopentane	g	−136.6(12)	38.4	366.8	134.5
(Z)-1,2-Dimethylcyclopentane	g	−129.5(13)	45.7	366.1	134.14
(E)-1,3-Dimethylcyclopentane	g	−133.6(13)	41.5	366.8	134.5
(Z)-1,3-Dimethylcyclopentane	g	−135.9(11)	39.2	366.8	134.5
(Z)-2,4-Dimethyl-1,3-dioxane	lq	−465.2(42)			
4,5-Dimethyl-1,3-dioxane	lq	−451.6(21)			
5,5-Dimethyl-1,3-dioxane	lq	−461.3(23)			
4,4′-Dimethyldiphenylamine	c	−11.72			
Dimethyl disulfide	lq	−62.0	7.0	235.4	146.1
Dimethyl ether	g	−184.1(14)	−112.9	267.1	65.8
N,N-Dimethylformamide	lq	−239.3(16)			
Dimethyl fumarate	lq	−729.3			
Dimethylglyoxime	c	−177.86			
2,2-Dimethylhexane	lq	−261.9(13)	3.0	331.9	
2,3-Dimethylhexane	lq	−252.6(16)	9.1	342.7	

TABLE 6.1 Enthalpies and Gibbs Energies of Formation, Entropies, and Heat Capacities of Organic Compounds (*Continued*)

Substance	State	$\Delta Hf°$, kJ·mol^{-1}	$\Delta Gf°$, kJ·mol^{-1}	$S°$, J·deg^{-1}·mol^{-1}	$C_p°$, J·deg^{-1}·mol^{-1}
2,4-Dimethylhexane	lq	−257.0(13)	3.7	345.7	
2,5-Dimethylhexane	lq	−260.4(16)	2.5	338.7	
3,3-Dimethylhexane	lq	−257.5(13)	5.2	339.4	
3,4-Dimethylhexane	lq	−251.8(16)	8.5	347.2	
Dimethyl hexanedioate	lq	−886.6			
(E)-2,2-Dimethyl-3-hexene	lq	−144.9(16)			
(Z)-2,2-Dimethyl-3-hexene	lq	−126.4(28)			
(E)-2,5-Dimethyl-3-hexene	lq	−159.2(30)			
(Z)-2,5-Dimethyl-3-hexene	lq	−151.0(37)			
5,5-Dimethylhydantoin	c	−533.3(13)			
1,1-Dimethylhydrazine	lq	48.9(33)	206.7	200.0	164.1
1,2-Dimethylhydrazine	lq	52.7(42)	212.6	199.2	171.0
3,5-Dimethylisoxazole	lq	−63.2(29)			
Dimethyl maleate	lq	−703.8			
Dimethylmaleic anhydride	c	−581.6			
Dimethyl malonate	lq	−795.8			
Dimethylmercury	lq	58.6			
6,6-Dimethyl-2-methylene-bicyclo[3.1.1]heptane	lq	−7.7(30)			
Dimethyl oxalate	lq	−756.3(7)			
2,2-Dimethylpentane	g	−205.9(15)	0.1	392.9	166.0
2,3-Dimethylpentane	g	−198.9(15)	0.7	414.0	166.0
2,4-Dimethylpentane	g	−201.7(11)	3.1	396.6	166.0
3,3-Dimethylpentane	g	−201.2(11)	2.6	399.7	166.0
Dimethyl pentanedioate	lq	−205.9			
2,4-Dimethyl-3-pentanone	g	−311.5(38)			
2,4-Dimethyl-1-pentene	g	−83.8(14)			
4,4-Dimethyl-1-pentene	g	−81.6(20)			
2,4-Dimethyl-2-pentene	g	−88.7(11)			
(E)-4,4-Dimethyl-2-pentene	g	−88.8(11)			
(Z)-4,4-Dimethyl-2-pentene	g	−72.6(14)			
2,7-Dimethylphenanthrene	c	36.4(18)			
4,5-Dimethylphenanthrene	c	89.0(59)			
9,10-Dimethylphenanthrene	c	47.7(84)			
2,3-Dimethylphenol	c	−241.2(10)			
2,4-Dimethylphenol	lq	−228.7(10)			
2,5-Dimethylphenol	c	−246.6(9)			
2,6-Dimethylphenol	c	−237.4(11)			
3,4-Dimethylphenol	c	−242.3(12)			
3,5-Dimethylphenol	c	−244.4(12)			
Dimethyl-1,2-phthalate	lq	−678			
Dimethyl-1,3-phthalate	c	−715			
Dimethyl-1,4-phthalate	c	−711			
2,2-Dimethylpropane	g	−168.3(10)	−1.5	306.4	121.6
2,2-Dimethylpropanenitrile	lq	−39.8(8)			
2,2-Dimethylpropanoic acid	lq	−564.4			
2,2-Dimethylpropanoic anhydride	lq	−779.9			
2,3-Dimethylpyridine	lq	19.4(13)			
2,4-Dimethylpyridine	lq	16.2(8)			
2,5-Dimethylpyridine	lq	18.7(10)			

TABLE 6.1 Enthalpies and Gibbs Energies of Formation, Entropies, and Heat Capacities of Organic Compounds (*Continued*)

Substance	State	$\Delta Hf°$, kJ·mol^{-1}	$\Delta Gf°$, kJ·mol^{-1}	$S°$, J·deg^{-1}·mol^{-1}	$C_p°$, J·deg^{-1}·mol^{-1}
2,6-Dimethylpyridine	lq	12.7(15)			
3,4-Dimethylpyridine	lq	18.3(10)			
3,5-Dimethylpyridine	lq	22.5(9)			
Dimethyl succinate	lq	−835.1			
2,2-Dimethylsuccinic acid	c	−987.8(14)			
meso-2,3-Dimethylsuccinic acid	c	−977.5(15)			
Dimethyl sulfate	lq	−735.5(10)			
Dimethyl sulfide	g	−37.5(6)	7.0	285.9	74.1
Dimethyl sulfite	lq	−523.6(11)			
Dimethyl sulfone	c	−450.1(7)	−302.5	145.5	100.0
Dimethyl sulfoxide	lq	−204.2(7)	−99.2	188.3	147.3
1,5-Dimethyltetrazole	c	188.7			
2,2-Dimethylthiacyclopropane	lq	−24.2			
5,5-Dimethyl-4-thia-1-hexene	lq	−90.7			
Dimethylzinc	lq	25.1			
2,3-Dinitroaniline	c	−11.7(29)			
2,4-Dinitroaniline	c	−67.8(30)			
2,5-Dinitroaniline	c	−44.4(29)			
2,6-Dinitroaniline	c	−50.6(29)			
3,4-Dinitroaniline	c	−32.6(29)			
3,5-Dinitroaniline	c	−38.9(29)			
2,4-Dinitroanisole	c	−186.6			
2,6-Dinitroanisole	c	−189.1			
1,2-Dinitrobenzene	c	−1.8(7)	211.5	216.3	
1,3-Dinitrobenzene	c	−27.4(5)	184.6	220.9	
1,4-Dinitrobenzene	c	−38.7(5)			
1,1-Dinitroethane	lq	−148.2(9)			
1,2-Dinitroethane	lq	−165.2(15)			
Dinitromethane	lq	−104.9(8)			
1,5-Dinitronaphthalene	c	30.5(71)			
2,4-Dinitro-1-naphthol	c	−181.4(46)			
2,4-Dinitrophenol	c	−232.6(31)			
2,6-Dinitrophenol	c	−210.0(27)			
1,1-Dinitropropane	lq	−163.2(13)			
1,3-Dinitropropane	lq	−207.1(9)			
2,2-Dinitropropane	lq	−181.2(14)			
2,4-Dinitroresorcinol	c	−415.5			
2,4-Dinitrotoluene	c	−71.6			
2,6-Dinitrotoluene	c	−51.0			
1,3-Dioxane	lq	−379.7(42)			
1,4-Dioxane	lq	−353.9(8)	−188.1	195.3	
	g	−315.8(8)	−180.8	299.8	94.1
1,3-Dioxolane	g	−298.0(14)			
N,N-Diphenylacetamide	c	−43.1			
Diphenylacetylene	c	312.4			
Diphenylboron bromide	lq	−16.1			
(E),(E)-1,4-Diphenylbutadiene	c	178.8			
(Z),(Z)-1,4-Diphenylbutadiene	c	198.8			
Diphenylbutadiyne	c	518.4			
1,4-Diphenylbutane	c	−9.9(21)			
1,4-Diphenyl-1,4-butanedione	c	−256.2	7.8	324.7	

TABLE 6.1 Enthalpies and Gibbs Energies of Formation, Entropies, and Heat Capacities of Organic Compounds (*Continued*)

Substance	State	$\Delta Hf°$, kJ·mol^{-1}	$\Delta Gf°$, kJ·mol^{-1}	$S°$, J·deg^{-1}·mol^{-1}	$C_p°$, J·deg^{-1}·mol^{-1}
1,4-Diphenyl-2-butene-1,4-dione	c	−114.7(24)	111.5	319.2	
Diphenyl carbonate	c	−401.2(19)	−175.9	278.4	
Diphenyl disulfide	c	−148.5(28)			
Diphenyl disulfone	c	−643.2(16)			
Diphenyleneimine	c	126.8			
1,1-Diphenylethane	lq	48.7(21)	245.1	335.9	
1,2-Diphenylethane	lq	51.5(12)	267.2	270.3	
Diphenylethanedione	c	−154.0(28)			
Diphenyl ether	lq	−14.9(15)	144.2	291.3	268.6
1,1-Diphenylethylene	lq	172.4(13)			
Diphenylethyne	c	312.4(11)			
6,6-Diphenylfulvene	c	197.4(126)			
1,2-Diphenylhydrazine	c	221.3(13)			
Diphenylmercury	c	279.5			
Diphenylmethane	lq	89.7(14)	276.9	239.3	233.1
1,3-Diphenyl-2-propanone	c	−84.0(26)			
Diphenyl sulfide	lq	163.4(21)			
Diphenyl sulfone	c	−225.0(16)			
Diphenyl sulfoxide	c	9.7(10)			
1,3-Diphenylurea	c	−122.6(34)			
Dipropyl disulfide	lq	−171.3	19.1	373.6	
Dipropyl ether	g	−292.9(11)	−105.6	422.5	158.3
Dipropylmercury	lq	−20.9			
Dipropyl sulfate	lq	−859.0(13)			
Dipropyl sulfide	lq	−171.5(10)			
	g	−125.3(9)	33.2	448.4	161.2
Dipropyl sulfite	lq	−646.8(12)			
Dipropyl sulfone	lq	−548.2(8)			
Dipropyl sulfoxide	lq	−329.4(9)			
2,2′-Dipyridyl ketone	c	−19.7			
Divinyl ether	g	−39.8(13)			
Divinyl sulfone	lq	−207.4(35)			
Docosanoic acid	c	−983.0			
(E)-13-Docosenic acid	c	−960.7(28)			
(Z)-13-Docosenic acid	c	−866.0			
Dodecane	lq	−350.9(21)	28.1	490.6	376.0
	g	−289.7(21)	50.0	622.5	280.3
Dodecanedioic acid	c	−1130.0(29)			
Dodecanoic acid	c	−774.6(10)			
1-Dodecanol	lq	−528.5(9)			
1-Dodecene	g	−165.4(25)	137.9	618.3	269.6
1-Dodecyne	g	−0.04	268.6	602.4	265.4
Dulcitol	c	−1346.8			
Ergosterol	c	−789.9			
Ethane	g	−83.8(4)	−32.8	229.1	52.5
Ethane-d_6	g	−107.4	−47.3	244.5	64.6
1,2-Ethanediamine	lq	−63.0(5)		209.2	
1,2-Ethanediol (ethylene glycol)	lq	−455.3(7)	−323.2	166.9	149.8
	g	−387.5(18)	−304.5	323.6	97.1
1,2-Ethanedithiol	lq	−54.4(11)			

TABLE 6.1   Enthalpies and Gibbs Energies of Formation, Entropies, and Heat Capacities of Organic Compounds (*Continued*)

Substance	State	$\Delta H f°$, kJ·mol^{-1}	$\Delta G f°$, kJ·mol^{-1}	$S°$, J·deg^{-1}·mol^{-1}	$C_p°$, J·deg^{-1}·mol^{-1}
Ethanethiol	g	−46.3(6)	−4.7	296.1	72.7
Ethanol	lq	−277.0	−174.2	161.0	112.0
	g	−235.2(4)	−167.9	282.6	65.4
Ethene (*see* Ethylene)					
Ethoxybenzene	lq	−152.6(6)			
Ethyl acetate	lq	−473.3(5)	−332.7	259.4	169.9
	g	−444.1(6)	−327.4	362.8	113.6
Ethylamine	g	−47.4(7)	37.3	289.9	72.6
Ethyl 4-aminobenzoate	c	−418.0			
N-Ethylaniline	lq	4.0(42)	188.7	239.3	
Ethylbenzene	g	29.9(11)	130.6	360.5	128.4
2-Ethylbenzoic acid	c	−441.3(10)			
3-Ethylbenzoic acid	c	−445.8(8)			
4-Ethylbenzoic acid	c	−460.7(9)			
2-Ethyl-1-butene	g	−56.0(15)	80.0	376.6	133.6
Ethyl (E)-2-butenoate (ethyl crotonate)	lq	−420.1(21)			
Ethyl carbamate	c	−520.5			
Ethyl 4-chlorobutanoate	lq	−566.5(84)			
Ethyl chloroformate	lq	−505.1(18)			
Ethylcyclobutane	g	−26.3(11)			
Ethylcyclohexane	lq	−211.9(16)	29.1	280.9	209.2
	g	−171.7(16)	39.3	382.6	158.8
1-Ethylcyclohexene	lq	−106.7(1)			
Ethylcyclopentane	lq	−163.4(10)	37.3	280.3	185.8
1-Ethylcyclopentene	g	−19.7(9)			
Ethyl diethylcarbamate	lq	−592.3(22)			
Ethyl 2,2-dimethylpropanoate	g	−536.0(84)			
Ethylene	g	52.5(29)	68.4	219.3	42.9
Ethylene-d_4	g	38.2	59.2	230.5	51.9
Ethylene carbonate	c	−581.2			
Ethylenediaminetetraacetic acid	c	−1759.4			
Ethylenediammonium chloride	c	−513.4			
2,2′-(Ethylenedioxy)bisethanol	lq	−804.2			
Ethyleneimine	g	126.5(9)	178.0	250.6	52.6
Ethylene oxide	g	−52.6(6)	−13.1	242.4	48.3
2-Ethylhexanal	lq	−342.5(15)			
3-Ethylhexane	g	−210.7(13)			
Ethyl hydroperoxide	g	198.9(59)			
Ethylidenecyclohexane	lq	−103.5(7)			
Ethylidenecyclopentane	lq	−56.7(9)			
Ethyl isocyanide	lq	108.4(47)			
Ethyl isopropyl sulfide	lq	−156.1			
Ethyllithium	c	−58.6			
Ethylmercury bromide	c	−107.5			
Ethylmercury chloride	c	−141.1			
Ethylmercury iodide	c	−65.7			
1-Ethyl-2-methylbenzene	g	1.3(10)	131.1	399.2	157.9
2-Ethyl-3-methyl-1-butene	g	−79.5(14)			
Ethyl 2-methylbutanoate	lq	−566.8(84)			
Ethyl 3-methylbutanoate	lq	−570.9(84)			

TABLE 6.1   Enthalpies and Gibbs Energies of Formation, Entropies, and Heat Capacities
of Organic Compounds (*Continued*)

Substance	State	$\Delta Hf°$, kJ·mol^{-1}	$\Delta Gf°$, kJ·mol^{-1}	$S°$, J·deg^{-1}·mol^{-1}	$C_p°$, J·deg^{-1}·mol^{-1}
Ethyl methyl ether	g	−216.4(7)	−117.7	310.6	89.8
3-Ethyl-2-methylpentane	g	−211.0(14)	21.3	441.1	
3-Ethyl-3-methylpentane	g	−214.8(14)	19.9	433.0	
3-Ethyl-2-methyl-1-pentene	g	−100.3(13)			
Ethyl methyl sulfide	g	−59.6(12)	11.4	333.1	95.1
Ethyl nitrate	g	−154.1(10)	−36.9	348.3	97.4
Ethyl nitrite	g	−104.2		103.5	99.2
1-Ethyl-2-nitrobenzene	lq	−48.7(63)			
1-Ethyl-4-nitrobenzene	lq	−55.4(63)			
3-Ethylpentane	g	−189.6(12)	11.0	411.5	166.0
Ethyl pentanoate	lq	−553.0(26)			
2-Ethylphenol	c	−208.8(18)			
3-Ethylphenol	c	−214.3(16)			
4-Ethylphenol	c	−224.4(10)			
Ethylphosphonic acid	c	−1051.4			
Ethylphosphonic dichloride	lq	−613.4			
Ethyl propanoate	g	−463.3(8)	−323.7		
Ethyl propyl ether	g	−272.2(11)			
Ethyl propyl sulfide	g	−104.7(8)	23.6	414.1	139.3
2-Ethylpyridine	lq	7.4(42)			
S-Ethyl thioacetate	lq	−267.8			
2-Ethyltoluene	g	1.3(10)	131.1	399.2	157.9
3-Ethyltoluene	g	−1.8(11)	126.4	404.2	152.2
4-Ethyltoluene	g	−3.2(13)	85.3	398.9	151.5
Ethyl (β)-vinylacrylate	lq	−338.1			
Ethyl vinyl ether	g	−140.8(9)			
Ethynylbenzene	g	327.3	361.8	321.7	114.9
Fluoranthene	c	189.9(6)	345.6	230.5	
Fluoroacetamide	c	−496.6			
Fluoroacetic acid	c	−688.3			
Fluorobenzene	g	−116.0(14)	−69.0	302.6	94.4
2-Fluorobenzoic acid	c	−567.6(19)			
3-Fluorobenzoic acid	c	−582.0(14)			
4-Fluorobenzoic acid	c	−585.7(15)			
Fluoroethane	g	−263.2	−211.0	265.0	59.5
2-Fluoroethanol	lq	−465.7			
Fluoroethylene	g	−138.8(17)			
Fluoromethane	g	−237.8	−213.8	222.8	37.5
1-Fluoropropane	g	−285.9(23)	−200.3	304.2	82.6
2-Fluoropropane	g	−293.5(15)	−204.2	292.1	82.0
Fluorosyltrifluoromethane	g	−766.0	−707.0	322.4	79.4
4-Fluorotoluene	lq	−186.9(11)	−79.8	237.1	
Fluorotribromomethane	g	−190.4	−193.1	345.8	
Fluorotrinitromethane	lq	−220.9			
Formaldehyde	g	−108.6(5)	−109.9	218.8	35.4
Formamide	lq	−253.8			
	g	−186.2	−141.0	248.6	45.4
Formanilide	c	−151.5			
Formic acid	lq	−425.1(4)	−361.4	129.0	99.0
	g	−378.7(6)	−351.0	248.7	45.2
Formyl fluoride	g	−376.6	−368.1	246.5(8)	40.4

TABLE 6.1 Enthalpies and Gibbs Energies of Formation, Entropies, and Heat Capacities
of Organic Compounds (*Continued*)

Substance	State	$\Delta Hf°$, kJ·mol^{-1}	$\Delta Gf°$, kJ·mol^{-1}	$S°$, J·deg^{-1}·mol^{-1}	$C_p°$, J·deg^{-1}·mol^{-1}
D-(−)-Fructose	c	−1265.6(5)			
D-(+)-Fucose	c	−1099.1			
Fumaric acid	c	−811.7(8)	−655.6	166.1	
Fumaronitrile	c	268.2			
Furan	g	−34.9(7)	0.88	267.2	65.4
2-Furancarboxaldehyde	lq	−201.6(46)			
2-Furancarboxylic acid	c	−498.4(11)			
2-Furanmethanol	lq	−276.2(13)	−154.2	215.5	206.1
Furylacrylic acid	c	−459.0			
Furylethylene	lq	−10.5			
D-(+)-Galactose	c	−1286.3(5)	−918.8	205.4	
D-Gluconic acid	c	−1587.0			
D-(+)-Glucose	c	−1273.3(11)	−910.4	212.1	
D-(−)-Glutamic acid	c	−1009.7(8)	−727.5	191.2	
L-(+)-Glutamic acid	c	−1005.2(12)	−731.3	188.2	
L-Glutamine	c	−826.4(7)			
Glutaric acid	c	−960.0(12)			
Glyceraldehyde	lq	−598.0			
Glycerol	lq	−668.5(10)	−477.0	204.5	150.2
Glyceryl 1-acetate	lq	−909.1(37)			
Glyceryl 1-benzoate	c	−777.3(12)			
Glyceryl 2-benzoate	c	−772.8(12)			
Glyceryl 1,3-diacetate	lq	−1120.7(67)			
Glyceryl 1-dodecanoate	c	−1160.9(17)			
Glyceryl 2-dodecanoate	c	−1152.6(17)			
Glyceryl 1-hexadecanoate	c	−1281.5			
Glyceryl 1-hexanoate	c	−1109.0			
Glyceryl 2-hexanoate	c	−1095.8			
Glyceryl 1-octadecanoate	c	−1324.8			
Glyceryl 1-tetradecanoate	c	−1222.6(9)			
Glyceryl triacetate	lq	−1330.8(42)			
Glyceryl trinitrate	lq	−370.9(38)			
Glyceryl tris(dodecanoate)	c	−2046.0			
Glyceryl tris(tetradecanoate)	c	−2176.0			
Glycine	c	−528.5(5)	−368.6	103.5	99.2
Ionized; ss	aq	−469.8	−315.0	111.0	
$^+H_3NCH_2COOH$; ss	aq	−517.9	−384.2	190.2	
Glycylglycine	c	−747.7(6)	−490.6	190.0	
Glyoxal	g	−212.0(8)			
Glyoxime	c	−90.5			
Glyoxylic acid	c	−835.5			
Guanidine	c	−56.0(10)			
Guanidine carbonate	c	−971.9(6)	−557.4	295.4	258.9
Guanidine nitrate	c	−387.0			
Guanidine sulfate	c	−1205.0			
Guanine	c	−182.9	47.4	160.3	
Guanylurea nitrate	c	−427.2			
L-Gulonic acid-γ-lactone	c	−1219.6			
Heptadecane	g	−393.9	82.1	817.3	394.7
Heptadecanoic acid	c	−924.4(18)			
1-Heptadecene	g	−268.4	179.9	813.1	383.9

TABLE 6.1 Enthalpies and Gibbs Energies of Formation, Entropies, and Heat Capacities of Organic Compounds (*Continued*)

Substance	State	$\Delta Hf°$, kJ·mol^{-1}	$\Delta Gf°$, kJ·mol^{-1}	$S°$, J·deg^{-1}·mol^{-1}	$C_p°$, J·deg^{-1}·mol^{-1}
Heptanal	lq	−311.5(38)	−100.6	348.5	
	g	−264.0	−86.7	461.7	
Heptane	g	−187.7(13)	8.0	427.9	166.0
Heptanedioic acid	c	−1009.4(11)			
Heptanenitrile	lq	−82.8(9)			
1-Heptanethiol	g	−150.0(10)	36.2	493.3	186.9
Heptanoic acid	lq	−610.2(15)			
1-Heptanol	lq	−403.3(7)	−142.3	320.1	278.2
	g	−336.4(10)	−120.9	480.3	178.7
1-Heptene	g	−62.3(10)	95.8	423.6	155.2
(E)-2-Heptene	lq	−109.5(8)			
(Z)-2-Heptene	lq	−105.1(9)			
(E)-3-Heptene	lq	−109.3(10)			
(Z)-3-Heptene	lq	−104.3(8)			
1-Heptyne	g	103.0	226.7	407.7	151.1
Hexachlorobenzene	c	−127.6(42)	1.1	260.2	201.3
	g	−33.9	44.2	441.2	173.2
Hexachloroethane	g	−143.6(92)	−54.9	398.7	136.7
Hexadecafluoroethylcyclo-hexane	lq	−3420.0(36)			
Hexadecafluoroheptane	lq	−3420.8	−3093.0	561.8	
Hexadecane	g	−374.8(19)	83.7	778.3	371.8
Hexadecanoic acid	c	−891.5(17)	−316.1	452.4	
1-Hexadecanol	c	−686.7(19)	−98.7	451.9	438.5
	lq	−635.4	−96.6	606.7	
1-Hexadecene	g	−248.5(25)	171.5	774.1	361.0
2,4-Hexadienoic acid	c	−390.8			
Hexafluoroacetone	g	−1249.3			
Hexafluorobenzene	g	−955.4(15)	−79.4	383.2	156.6
Hexafluoroethane	g	−1344.2(55)	−1255.8	331.8	106.4
(E)-Hexahydroindane	g	−131.4			
(Z)-Hexahydroindane	g	−127.2			
Hexamethylbenzene	c	−161.5(5)	117.4	299.8	257.3
1,1,1,3,3,3-Hexamethyl-disilazane	lq	−518.0			
Hexamethyldisiloxane	lq	−814.6	−541.8	433.8	311.4
Hexamethylenetetramine	c	125.5	434.8	163.4	
Hexanal	g	−248.4	−100.1	422.9	148.2
Hexanamide	c	−423.0(5)			
Hexane	lq	−198.8	−3.8	296.1	189.1
	g	−167.1(8)	−0.25	388.4	143.1
1,6-Hexanedioic acid	lq	−985.4	−207.3		232.2
1,6-Hexanediol	c	−569.9(48)			
Hexanedinitrile	lq	85.1(5)			
1-Hexanethiol	g	−129.9(10)	27.8	454.3	164.1
Hexanoic acid	lq	−583.9(16)			
1-Hexanol	lq	−377.5(5)	−152.3	289.5	236.8
	g	−317.6	−135.6	441.4	155.6
2-Hexanol	lq	−392.9(36)			
3-Hexanol	lq	−392.4(17)			
2-Hexanone	lq	−322.0(10)			

TABLE 6.1 Enthalpies and Gibbs Energies of Formation, Entropies, and Heat Capacities of Organic Compounds (*Continued*)

Substance	State	$\Delta Hf°$, kJ·mol^{-1}	$\Delta Gf°$, kJ·mol^{-1}	$S°$, J·deg^{-1}·mol^{-1}	$C_p°$, J·deg^{-1}·mol^{-1}
3-Hexanone	lq	−320.2(9)			
1-Hexene	lq	−74.1(16)	83.6	295.1	183.3
	g	−43.5(16)	84.45	384.6	132.3
(E)-2-Hexene	g	−53.9(16)	76.4	380.6	132.4
(Z)-2-Hexene	g	−52.3(13)	76.2	386.5	125.7
(E)-3-Hexene	g	−54.4(13)	77.6	374.8	132.8
(Z)-3-Hexene	g	−47.6(13)	83.0	379.6	123.6
1-Hexyne	g	123.6	218.6	368.7	128.2
(−)-Histidine	c	−466.7(23)			
Hydantoin	c	−448.5			
Hydrazine	lq	50.6	149.2	121.2	98.9
Hydrazobenzene	c	221.3			
Hydroxyacetic acid	c	−663.6			
2′-Hydroxyacetophenone	c	−357.7			
3′-Hydroxyacetophenone	c	370.7			
4′-Hydroxyacetophenone	c	−364.4			
2-Hydroxybenzaldehyde	lq	−279.9			
2-Hydroxybenzaldoxime	c	−183.7(8)			
2-Hydroxybenzoic acid	c	−589.9(13)	−421.3	178.2	159.1
3-Hydroxybenzoic acid	c	−584.9	−417.3	177.0	157.3
4-Hydroxybenzoic acid	c	−584.5	−416.5	175.7	155.1
(±)-2-Hydroxybutanoic acid	lq	−679.1			
2-Hydroxy-2,4,6-cyclo-heptatrienone	c	−239.2			
2-Hydroxyisobutanoic acid	c	−744.3			
2-Hydroxy-1-isopropyl-4-methylbenzene	c	−309.6			
3-Hydroxy-4-methoxybenzal-dehyde	c	−453.6			
2-Hydroxymethyl-1,3-propanediol	c	−744.6			
3-Hydroxy-2-naphthalene-carboxylic acid	c	−547.7			
5-Hydroxy-1-pentanal	lq	−479.9			
(E)-(−)-4-Hydroxyproline	c	−661.1(18)			
(S)-2-Hydroxypropanoic acid	c	−694.0(10)			
2-Hydroxypropanonitrile	lq	−138.9	34.3		
2-Hydroxypyridine	c	−166.3(5)			
3-Hydroxypyridine	c	−132.0(4)			
4-Hydroxypyridine	c	−144.6(11)			
8-Hydroxyquinoline	c	−81.2(16)			
(−)-2-Hydroxysuccinic acid	c	−1103.7	−884.7		
(±)-2-Hydroxysuccinic acid	c	−1105.7			
Hypoxanthene	c	−110.8(8)	76.9	145.6	
Icosane	g	−455.8	117.3	934.1	463.3
Icosanoic acid	c	−1011.9(16)			
Icosene	g	−330.2	205.1	929.9	452.5
Imidazole	c	58.5(33)			
Indane	lq	11.5(16)	150.8	234.4	190.3
Indene	lq	110.6(16)	217.6	214.2	
1*H*-Indole	c	86.7(7)			

TABLE 6.1 Enthalpies and Gibbs Energies of Formation, Entropies, and Heat Capacities of Organic Compounds (*Continued*)

Substance	State	$\Delta Hf°$, kJ·mol^{-1}	$\Delta Gf°$, kJ·mol^{-1}	$S°$, J·deg^{-1}·mol^{-1}	$C_p°$, J·deg^{-1}·mol^{-1}
Indole-2,3-dione	c	−268.2(38)			
Indobenzene	g	164.9(59)	187.8	334.1	100.8
2-Iodobenzoic acid	c	−302.3(42)			
3-Iodobenzoic acid	c	−316.9(42)			
4-Iodobenzoic acid	c	−316.1(42)			
Iodocyclohexane	lq	−97.2(42)			
Iodoethane	g	−7.5(17)	21.3	296.3	65.9
Iodomethane	g	14.7(13)	15.6	253.7	44.1
2-Iodo-2-methylpropane	g	−72.0(33)	23.6	342.2	118.3
1-Iodonaphthalene	lq	161.5(63)			
2-Iodonaphthalene	c	144.3(63)			
2-Iodophenol	c	−95.8(42)			
3-Iodophenol	c	−94.5(42)			
4-Iodophenol	c	−95.4(42)			
1-Iodopropane	g	−30.3(39)			
2-Iodopropane	g	−40.3(38)	20.1	324.5	90.1
3-Iodopropanoic acid	c	−460.0(50)			
3-Iodo-1-propene	g	54.8			
(α)-Iodotoluene	lq	57.7			
3-Iodotoluene	lq	79.1(42)			
4-Iodotoluene	lq	67.4(42)			
Isobutanenitrile	g	25.4	103.6	313.3	96.4
Isobutylamine	lq	−132.6(5)			
Isobutylbenzene	lq	−69.8(13)			
Isobutyl trichloroacetate	lq	−554.0(84)			
(−)-Isoleucine	c	−637.9(9)	−347.2	208.0	188.3
(±)-Isoleucine	c	−635.3(18)			
Isopropenyl acetate	lq	−386.4(9)			
Isopropyl acetate	lq	−518.9(7)			
Isopropylamine	lq	−112.3(5)			
Isopropylbenzene	lq	−41.4(10)	124.3	279.8	
	g	4.0(10)	137.0	388.6	151.7
1-Isopropyl-2-methylbenzene	lq	−73.3(9)			
1-Isopropyl-3-methylbenzene	lq	−78.6(11)			
1-Isopropyl-4-methylbenzene	lq	−78.0(11)	119.1	306.6	
Isopropyl methyl ether	g	−252.0(10)	−120.9	332.3	111.1
2-Isopropyl-5-methylphenol	c	−309.7(96)			
Isopropyl methyl sulfide	g	−90.5(8)	13.4	359.3	117.2
Isopropyl nitrate	g	−101.0(14)	−40.7	373.2	120.7
Isopropyl thioacetate	lq	−298.2			
Isopropyl trichloroacetate	lq	−536.0(84)			
Isoquinoline	c	144.5(8)			
Ketene	g	−47.5(16)	−60.3	241.8	51.8
(+)-Lactic acid	c	−694.1	−522.9	142.3	
(±)-Lactic acid	lq	−674.5	−518.2	192.1	
(β)-Lactose	c	−2236.7(8)	−1567.0	386.2	
(+)-Leucine	c	−637.3(10)	−347.2	208.0	
(−)-Leucine	c	−637.4(9)	−346.3	211.8	201.0
(+)-Limonene	lq	−54.5(21)			
(±)-Lysine	c	−678.6(15)			
Malic acid	c	−789.4(8)	−625.1	159.4	135.4

TABLE 6.1   Enthalpies and Gibbs Energies of Formation, Entropies, and Heat Capacities of Organic Compounds (*Continued*)

Substance	State	$\Delta Hf°$, kJ·mol⁻¹	$\Delta Gf°$, kJ·mol⁻¹	$S°$, J·deg⁻¹·mol⁻¹	$C_p°$, J·deg⁻¹·mol⁻¹
Maleic anhydride	c	−469.8(8)			
(R)-Malic acid	c	−1105.7(6)			
(S)-Malic acid	c	−1103.6(42)			
Malonamide	c	−546.0			
Malonic acid	c	−891.0(4)			
Malonodiamide	c	−546.1			
Malononitrile	c	186.6			
D-(+)-Maltose	c	−2220.9	−1726.3		
(±)-Mandelic acid	c	−579.4(9)			
(+)-Mannitol	c	−1337.1(18)	−942.2	238.5	
D-(+)-Mannose	c	−125.2(34)			
2-Mercaptopropanoic acid	lq	−468.2(34)	−343.9	228.9	
Methane	g	−74.9(34)	−50.8	186.3(4)	35.6
Methane-d_4	g	−88.2	−59.5	198.9	40.3
Methanethiol	g	−22.9(7)	−9.9	255.1	50.3
Methanol	lq	−239.1(3)	−166.8	127.2	81.2
	g	−201.5(3)	−162.5	239.7	43.9
(−)-Methionine	c	−754.8	−505.8	231.5	
2-Methoxybenzaldehyde	c	−266.5(75)			
3-Methoxybenzaldehyde	lq	−276.1(75)			
4-Methoxybenzaldehyde	lq	−267.2(50)			
Methoxybenzene	g	−67.9(9)			
2-Methoxybenzoic acid	c	−538.5(7)			
3-Methoxybenzoic acid	c	−553.5(7)			
4-Methoxybenzoic acid	c	−561.7(8)			
2-Methoxytetrahydropyran	lq	−442.3(13)			
5-Methoxytetrazole	c	69.1(14)			
1-Methoxy-2,4,6-trinitro-benzene	c	−157.5(33)			
Methyl acetate	lq	−445.8(7)			155.6
Methyl acrylate	lq	−362.2(7)	−243.2		
	g	−333.0(8)	−237.6		
Methylamine	g	−23.0(5)	32.3	242.6	50.1
N-Methylaniline	lq	32.2			
Methyl benzoate	lq	−343.5(24)			
2-Methylbenzoic acid	c	−416.5(8)			
3-Methylbenzoic acid	c	−426.1(9)			
4-Methylbenzoic acid	c	−429.2(10)			
2-Methylbenzoic anhydride	c	−533.5(79)			
4-Methylbenzoic anhydride	c	−520.9(79)			
1-Methylbicyclo[4.1.0]heptane	lq	−59.9(15)			
1-Methylbicyclo[3.1.0]hexane	lq	−33.2(12)			
2-Methylbiphenyl	lq	108.0(71)			
3-Methylbiphenyl	lq	85.4(71)			
4-Methylbiphenyl	c	55.2(71)			
2-Methyl-1,3-butadiene	g	75.3(14)	145.9	315.6	104.6
3-Methyl-1,2-butadiene	g	129.7	198.6	319.7	105.4
2-Methylbutane	g	−154.0(10)	−14.8	343.6	118.8
2-Methyl-2-butanethiol	g	−127.1(10)	9.2	386.9	143.5
3-Methyl-1-butanethiol	g	−114.9(12)			
2-Methylbutanoic acid	lq	−554.4(59)			

TABLE 6.1 Enthalpies and Gibbs Energies of Formation, Entropies, and Heat Capacities of Organic Compounds (*Continued*)

Substance	State	$\Delta Hf°$, kJ·mol^{-1}	$\Delta Gf°$, kJ·mol^{-1}	$S°$, J·deg^{-1}·mol^{-1}	$C_p°$, J·deg^{-1}·mol^{-1}
3-Methylbutanoic acid	lq	−561.6(59)			
2-Methyl-1-butanol	lq	−354.6(6)			219.7
3-Methyl-1-butanol	lq	−356.4(6)			210.5
2-Methyl-2-butanol	lq	−379.5(5)	−175.3	229.3	247.7
3-Methyl-2-butanol	lq	−366.6(7)			232.2
3-Methyl-2-butanone	g	−262.5(9)			
2-Methyl-1-butene	g	−35.3(10)	65.6	339.5	110.0
3-Methyl-1-butene	g	−27.6(10)	74.8	333.5	118.6
2-Methyl-2-butene	g	−41.8(12)	59.7	338.6	105.0
(E)-2-Methyl-2-butenedioic acid [also (Z)]	c	−824.4(15)			
(E)-2-Methyl-2-butenoic acid	c	−490.8			
(Z)-2-Methyl-2-butenoic acid	c	−455.6			
3-Methyl-1-butyne	g	136.4	205.5	319.0	104.7
Methyl 2-butenoate	lq	−382.8(17)			
Methylcyclobutanecarboxylic acid	lq	−395.0(13)			
Methylcyclohexane	lq	−190.1(10)	20.3	247.9	184.9
	g	−154.7(10)	27.3	343.3	135.0
(E)-2-Methylcyclohexanol	lq	−415.8(88)			
(Z)-2-Methylcyclohexanol	lq	−390.4(50)			
(E)-3-Methylcyclohexanol	lq	−394.6(88)			
(Z)-3-Methylcyclohexanol	lq	−416.3(88)			
(E)-4-Methylcyclohexanol	lq	−433.5(105)			
(Z)-4-Methylcyclohexanol	lq	−413.4(50)			
2-Methylcyclohexene	lq	−81.2(8)			
Methylcyclopentane	lq	−138.0(8)	31.5	247.9	159.0
	g	−106.2(8)	35.8	339.9	109.8
1-Methylcyclopentanol	lq	−343.3			
2-Methylcyclopentanone	lq	−265.3(54)			
1-Methylcyclopentene	g	−3.8(7)	102.1	326.4	100.8
3-Methylcyclopentene	g	7.4(11)	115.0	330.5	100.0
4-Methylcyclopentene	g	14.6(20)	121.6	328.9	100.0
1-Methylcyclopropene	g	243.6(12)			
Methylenebutanedioic acid	c	−841.1(6)			
Methylenecyclohexane	lq	−61.3(38)			
Methylenecyclohexene	lq	−12.7(17)			
Methylenecyclopropane	g	200.5(18)			
Methyl decanoate	lq	−640.4			
Methyl 2,2-dimethylpropanoate	lq	−530.0(12)			
2-Methyl-1,3-dioxane	lq	−436.4(26)			
4-Methyl-1,3-dioxane	lq	−416.1(29)			
N-Methyldiphenylamine	lq	120.5(71)			
4-Methyldiphenylamine	c	49.0(71)			
Methyl dodecanoate	lq	−693.0(17)			
Methylenebutanedioic acid	c	−841.1(6)			
Methylenecyclohexane	lq	−61.3(38)			
2-Methylenecyclohexanol	lq	−277.6(33)			
3-Methylenecyclohexene	lq	−12.7(17)			
2-Methylenecyclopentanol	lq	46.9(38)			
Methylenecyclopropane	g	200.5(18)			

TABLE 6.1 Enthalpies and Gibbs Energies of Formation, Entropies, and Heat Capacities of Organic Compounds (*Continued*)

Substance	State	$\Delta Hf°$, kJ·mol^{-1}	$\Delta Gf°$, kJ·mol^{-1}	$S°$, J·deg^{-1}·mol^{-1}	$C_p°$, J·deg^{-1}·mol^{-1}
Methylenesuccinic acid	c	−841.2			
Methylene sulfate	c	−688.7			
Methyl formate	g	−355.5(8)	−297.2	301.3	66.5
Methyl 2-furancarboxylate	lq	−450.0(5)			
(α)-Methyl-(+)-glucoside	c	−1233.4			
N-Methylglycine	c	−513.3(3)			
Methylglyoxal	g	−27.1			
Methylglyoxime	c	−126.8			
2-Methylheptane	g	−215.4(15)	12.8	452.5	
3-Methylheptane	g	−212.5(13)	13.7	461.6	
4-Methylheptane	g	−212.0(13)	16.7	453.3	
Methyl heptanoate	lq	−567.1(9)			
2-Methylhexane	g	−194.6(10)	3.2	420.0	166.0
3-Methylhexane	g	−192.3(19)	4.6	424.1	166.0
Methyl hexanoate	lq	−540.2(10)			
5-Methyl-1-hexene	g	−65.7(10)			
(E)-3-Methyl-3-hexene	g	−76.8(11)			
(Z)-3-Methyl-3-hexene	g	−79.4(11)			
Methylhydrazine	lq	54.2(6)	179.9	165.9	134.9
	g	94.4	186.9	278.7	71.1
2-Methyl-1H-indole	c	60.7			
3-Methyl-1H-indole	c	68.2			
Methyl isocyanide	g	163.5(72)	165.7	246.8	52.9
Methyl isopropyl sulfide	g	−90.4	13.4	359.3	117.2
Methyl isothiocyanate	g	131.0	144.4	252.3	65.5
5-Methylisoxazole	lq	−5.6(8)			
Methylmercury bromide	c	−86.2			
Methylmercury chloride	c	−116.3			
Methylmercury iodide	c	−43.5			
Methyl 2-methylbutanoate	lq	−534.3(71)			
Methyl 3-methylbutanoate	lq	−538.9(71)			
7-Methyl-3-methylene-1,6-octadiene	lq	14.6			
1-Methylnaphthalene	lq	56.3(17)	189.4	254.8	224.4
2-Methylnaphthalene	c	44.9(15)	192.6	220.0	196.0
	g	116.1	216.2	380.0	159.8
Methyl nitrate	lq	−156.3(44)	−43.5	217.2	157.3
	g	−124.7	−39.3	301.9	76.5
Methyl nitrite	g	−66.4(10)	1.0	284.3	63.2
2-Methyl-5-nitroaniline	c	−91.3			
4-Methyl-3-nitroaniline	c	−71.7			
1-Methyl-2-nitrobenzene	lq	−9.7(38)			
1-Methyl-3-nitrobenzene	lq	−31.5(33)			
1-Methyl-4-nitrobenzene	c	−48.1(29)			
2-Methyl-2-nitropropane	c	−229.8(25)			
2-Methyl-2-nitro-1,3-propanediol	c	−575.3			
2-Methyl-2-nitro-1-propanol	c	−410.0(17)			
Methyl phenylcarbamate	c	−186.7(21)			
Methyl (Z)-9-octadecanoate	lq	−728.9			
Methyl octanoate	lq	−590.3(9)			

TABLE 6.1 Enthalpies and Gibbs Energies of Formation, Entropies, and Heat Capacities
of Organic Compounds (*Continued*)

Substance	State	$\Delta Hf°$, kJ·mol^{-1}	$\Delta Gf°$, kJ·mol^{-1}	$S°$, J·deg^{-1}·mol^{-1}	$C_p°$, J·deg^{-1}·mol^{-1}
2-Methylpentane	g	−174.8(10)	−5.0	380.5	144.2
3-Methylpentane	g	−172.1(10)	2.1	379.8	143.1
Methyl pentanoate	lq	−514.2(7)			
2-Methyl-3-pentanol	lq	−396.4(10)			
4-Methyl-2-pentanol	lq	−394.7(8)			
2-Methyl-3-pentanone	lq	−325.9(10)			
2-Methyl-1-pentene	g	−59.4(13)	77.6	382.2	135.6
2-Methyl-2-pentene	g	−66.9(15)	71.2	378.4	126.6
3-Methyl-1-pentene	g	−49.5(15)	86.4	376.8	142.4
(E)-3-Methyl-2-pentene	g	−63.1(13)	71.3	381.8	126.6
(Z)-3-Methyl-2-pentene	g	−62.3(15)	73.2	378.4	126.6
4-Methyl-1-pentene	g	−51.3(18)	90.0	367.7	126.5
(E)-4-Methyl-2-pentene	g	−61.5(14)	79.6	368.3	141.4
(Z)-4-Methyl-2-pentene	g	−57.5(12)	82.1	373.3	133.6
Methyl pentyl sulfide	g	122.9(24)	35.1	450.7	163.7
3-Methyl-1-phenyl-1-butanone	lq	−220.2(15)			
Methyl phenyl sulfide	lq	43.0(10)			
Methyl phenyl sulfone	c	−345.4(9)			
Methylphosphonic acid	c	−1054			
2-Methylpiperidine	lq	−124.9(10)			
2-Methylpropanal	g	−215.8(9)			
2-Methylpropane	g	−134.2(15)	−20.9	294.6	96.8
2-Methyl-1,2-propanediamine	lq	−133.9(7)			
2-Methyl-1,2-propanediol	lq	−539.7(25)			
2-Methylpropanenitrile	lq	−13.8(13)			
2-Methyl-1-propanethiol	g	−97.3(9)	5.6	362.9	118.3
2-Methyl-2-propanethiol	g	−109.6(19)	0.7	338.0	121.0
2-Methyl-1-propanol	g	−283.9(9)	−167.35	359.0	111.3
2-Methyl-2-propanol	g	−312.5(8)	−177.7	326.3	113.4
2-Methylpropene	g	−16.9(9)	58.1	293.6	89.1
1-Methyl-2-propylbenzene	lq	−72.5(10)			
1-Methyl-3-propylbenzene	lq	−76.2(12)			
1-Methyl-4-propylbenzene	lq	−75.1(10)			
Methyl propyl ether	g	−238.2(17)	−109.9	349.5	112.5
Methyl propyl sulfide	g	−82.3(10)	18.4	371.7	117.4
2-Methylpyridine	lq	56.7(8)	166.5	217.9	158.4
	g	99.2(8)	177.1	325.0	100.0
3-Methylpyridine	lq	61.9(6)	214.0	216.3	158.7
	g	106.4(6)	184.3	325.0	99.6
4-Methylpyridine	lq	59.2(9)			
1-Methyl-1H-pyrrole	lq	62.4(5)			
N-Methylpyrrolidone	lq	−262.2(5)			
2-Methylquinoline	c	164.4			
Methyl salicylate	lq	−531.8			
(α)-Methylstyrene	g	113.0	208.5	383.7	145.2
(E)-(β)-Methylstyrene	g	117.2	213.7	380.3	146.0
(Z)-(β)-Methylstyrene	g	121.3	216.9	383.7	145.2
Methylsuccinic acid	c	−958.2(11)			
Methylsuccinic anhydride	lq	−617.6(11)			
Methyl tetradecanoate	lq	−743.9			
2-Methylthiacyclopentane	g	−63.3			

TABLE 6.1   Enthalpies and Gibbs Energies of Formation, Entropies, and Heat Capacities of Organic Compounds (*Continued*)

Substance	State	$\Delta Hf°$, kJ·mol^{-1}	$\Delta Gf°$, kJ·mol^{-1}	$S°$, J·deg^{-1}·mol^{-1}	$C_p°$, J·deg^{-1}·mol^{-1}
4-Methylthiazole	lq	68.0(8)			
2-Methylthiophene	g	83.5(9)	122.9	320.6	95.4
3-Methylthiophene	g	82.6(9)	121.8	321.3	94.9
Methyl *p*-tolyl sulfone	c	−372.8			
5-Methyluracil	c	−462.8(9)			
Morphine monohydrate	c	−711.7			
Murexide	c	−1212.1			
Naphthalene	c	77.9(12)	201.0	166.9	
1-Naphthaleneacetic acid	c	−359.2(9)			
2-Naphthaleneacetic acid	c	−371.9(11)			
1-Naphthol	c	−121.0(10)			
2-Naphthol	c	−124.2(10)			
1,4-Naphthoquinone	c	−183.4(19)			
1-Naphthyl acetate	c	−288.2			
2-Naphthyl acetate	c	−304.3			
1-Naphthylamine	c	67.8(54)			
2-Naphthylamine	c	59.7(50)			
1-Naphthalenecarboxylic acid	c	−333.5(10)			
2-Naphthalenecarboxylic acid	c	−346.1(15)			
Nicotine	lq	39.3			
Nitrilotriacetic acid	c		−1307.5		
2-Nitroaniline	c	−26.1(5)	178.2	176.2	164.4
3-Nitroaniline	c	−38.3(15)	174.1	176.2	168.2
4-Nitroaniline	c	−42.0(8)	151.0	176.2	169.0
Nitrobenzene	lq	12.5(5)	146.2	224.3	185.8
2-Nitrobenzoic acid	c	−398.5(7)	−196.4	208.4	
3-Nitrobenzoic acid	c	−414.0(5)	−220.5	205.0	
4-Nitrobenzoic acid	c	−426.9(9)	−222.0	210.0	181.2
3-Nitrobiphenyl	c	65.1(63)			
4-Nitrobiphenyl	c	40.5(63)			
1-Nitrobutane	g	−143.9(15)	10.1	394.5	124.9
2-Nitrobutane	g	−163.6(16)	−6.2	383.3	123.5
3-Nitro-2-butanol	lq	−390.0(50)			
2-Nitrodiphenylamine	c	64.4			
Nitroethane	g	−102.3(6)	−4.9	315.4	78.2
2-Nitroethanol	lq	−350.7(23)			
1-Nitroguanidine	c	−92.5(42)			
Nitromethane	g	−74.3(5)	−7.0	275.0	57.3
(Nitromethyl)benzene	lq	−22.8(26)			
1-Nitronaphthalene	c	42.6(50)			
1-Nitroso-2-naphthol	c	−50.5(22)			
2-Nitroso-1-naphthol	c	−61.8(45)			
4-Nitroso-1-naphthol	c	−107.8(25)			
1-Nitropropane	g	−123.8(7)			
2-Nitropropane	g	−139.0(9)			
1-Nitro-2-propanone	c	−294.7(7)			
4-Nitrosodiphenylamine	c	213.0(32)			
(*β*)-Nitrostyrene	c	30.5			
Nonadecane	g	−435.1	108.9	895.2	440.4
1-Nonadecene	g	−309.6	196.7	891.0	429.7
1-Nonanal	g	−310.3	−74.9	539.6	216.8

TABLE 6.1 Enthalpies and Gibbs Energies of Formation, Entropies, and Heat Capacities of Organic Compounds (*Continued*)

Substance	State	$\Delta Hf°$, kJ·mol^{-1}	$\Delta Gf°$, kJ·mol^{-1}	$S°$, J·deg^{-1}·mol^{-1}	$C_p°$, J·deg^{-1}·mol^{-1}
Nonane	g	−228.2(7)	24.8	505.7	211.7
1-Nonanethiol	g	−190.8	53.0	571.2	232.7
Nonanoic acid	lq	−659.7(20)			
1-Nonanol	g	−376.3(14)	−110.5	558.6	224.3
2-Nonanone	lq	−397.2(13)			
1-Nonene	g	−103.5	112.7	501.5	201.0
Octadecane	g	−414.6(56)	100.5	856.2	417.6
Octadecanoic acid	c	−947.7(21)			
1-Octadecene	g	−289.0	188.3	852.0	406.8
(E)-9-Octadecenoic acid	c	−910.9			
(Z)-9-Octadecenoic acid	lq	−743.5			
1,7-Octadiyne	lq	334.4(52)			
Octafluorocyclobutane	g	−1542.6(107)	−1398.8	400.4	156.2
1-Octanal	g	−289.6	−83.3	500.7	194.0
Octanamide	c	−473.2(9)			
Octane	g	−208.6(14)	16.4	466.7	188.9
1,8-Octadecanoic acid	c	−1038.1(13)			
1-Octanenitrile	lq	−107.3(11)			
1-Octanethiol	g	−44.9	44.6	582.2	209.8
Octanoic acid	lq	−635.0(11)			
1-Octanol	lq	−426.5(7)	−143.1	377.4	325.1
2-Octanone	lq	−384.5	−140.3	373.8	273.3
1-Octene	g	−81.4(12)	104.2	462.5	178.1
(E)-2-Octene	lq	−135.7(42)			
(Z)-2-Octene	lq	−135.7(42)			
1-Octyne	g	82.4	235.4	494.6	174.0
(±)-Ornithine	c	−652.7			
Oxalic acid	c	−821.7(44)	−697.9	120.1	169.2
Oxalic acid dihydrate	c	−1492.0			
Oxaloyl dichloride	lq	−367.6(47)			
Oxaloyl dihydrazide	c	−295.2(6)			
Oxamic acid	c	−671.1			
Oxamide	c	−505.6(42)	−342.7	118.0	
Oxindole	c	−172.4			
2-Oxohexamethyleneimine	c	−328.6	−95.1	168.6	156.2
2-Oxo-1,5-pentanedioic acid	c	−1026.2(9)			
4-Oxopentanoic acid	c	−697.1			
2-Oxopropanoic acid	lq	−584.5	−463.4	179.5	
8-Oxypurine	c	−64.4			
Papaverine	c	−502.3			
Paraldehyde	lq	−687.0			
Pentachloroethane	g	−142.0(91)	−70.3	381.5	118.1
Pentachlorofluoroethane	g	−317.2	−234.0	391.8	
Pentachlorophenol	c	−292.4(30)	−144.1	251.9	202.0
Pentacyclo-[4.2.0.0^{2,5}.0^{3,8}.0^{4,7}]-octane	c	541.8(83)			
Pentadecane	g	−352.8	75.2	739.4	349.0
Pentadecanoic acid	c	−861.7(17)			
1-Pentadecene	g	−227.2	163.1	735.2	338.2
1-Pentadecyne	g	−61.8	293.9	719.3	33.41
1,2-Pentadiene	g	140.7(7)	210.4	333.5	105.4

TABLE 6.1   Enthalpies and Gibbs Energies of Formation, Entropies, and Heat Capacities of Organic Compounds (*Continued*)

Substance	State	$\Delta Hf°$, kJ·mol^{-1}	$\Delta Gf°$, kJ·mol^{-1}	$S°$, J·deg^{-1}·mol^{-1}	$C_p°$, J·deg^{-1}·mol^{-1}
(E)-1,3-Pentadiene	g	76.5(21)	146.73	319.7	103.3
(Z)-1,3-Pentadiene	g	81.5(23)	145.8	324.3	94.6
1,4-Pentadiene	g	105.7(16)	170.3	333.5	105.0
2,3-Pentadiene	g	133.1(8)	205.9	324.7	101.3
Pentaerythritol	c	−920.6(28)	−613.8	198.1	190.4
Pentaerythritol tetranitrate	c	−539.8(23)			
Pentafluorobenzoic acid	c	−1239.6(13)			
Pentafluoroethane	g	−1104.6	−1029.3	333.7	95.7
Pentafluorophenol	c	−1024.1(21)			
Pentamethylbenzene	c	−133.6(11)			
	g	−74.5	123.3	443.9	216.5
Pentamethylbenzoic acid	c	−536.1(14)			
Pentanal	g	−228.5(17)	−108.3	383.0	125.4
Pentanamide	c	−379.5(11)			
Pentane	lq	−173.5(10)	−9.3	262.7	
	g	−146.9(10)	−8.4	349.0	120.2
2,4-Pentanedione	lq	−424.0			
	g	−380.6(18)		397.9	120.1
1,5-Pentanedithiol	g	−71.0(4)			
Pentanenitrile	lq	−33.1(13)			
Pentanoic acid	g	−491.9(30)	−357.2	439.8	
1-Pentanol	g	−294.7(5)	−146.0	402.5	133.1
2-Pentanol	g	−312.7(17)			
3-Pentanol	g	−317.2(15)	−158.2	382.0	
2-Pentanone	g	−259.0(10)	−137.1	376.2	121.0
3-Pentanone	lq	−296.5(8)			
1-Pentene	g	−21.2(14)	79.1	345.8	109.6
(E)-2-Pentene	g	−31.9(13)	69.9	340.4	108.5
(Z)-2-Pentene	g	−28.1(15)	71.8	346.3	101.8
(E)-2-Pentenenitrile	lq	74.9(11)			
(Z)-2-Pentenenitrile	lq	71.8(12)			
(E)-3-Pentenenitrile	lq	80.9(13)			
2-Pentenoic acid	lq	−446.4(21)			
3-Pentenoic acid	lq	−434.8(21)			
4-Pentenoic acid	lq	−430.6(25)			
1-Pentyne	g	144.4	210.3	329.8	106.7
2-Pentyne	g	128.9	194.2	331.8	98.7
Perfluoropiperidine	lq	−2020.5	−1768.5	393.4	296.8
Perylene	c	182.8(8)			
(α)-Phellandrene	lq	41.3(43)			
Phenanthrene	c	116.2(13)	268.3	211.7	
9,10-Phenanthrenedione	c	−230.9(14)			
Phenazine	c	243.9(24)			
Phenol	c	−165.1(8)	−50.4	144.0	134.7
	g	−96.4	−32.9	315.6	103.6
Phenoxyacetic acid	c	−513.8			
Phenyl acetate	lq	−334.9(9)			
Phenylacetic acid	c	−398.7			
Phenylacetylene	g	327.3	363.5	321.7	114.9
(±)-3-Phenyl-2-alanine	c	−468.2(26)	−211.7	213.6	203.0
Phenyl benzoate	c	−241.0(26)			

TABLE 6.1 Enthalpies and Gibbs Energies of Formation, Entropies, and Heat Capacities of Organic Compounds (*Continued*)

Substance	State	$\Delta Hf°$, kJ·mol^{-1}	$\Delta Gf°$, kJ·mol^{-1}	$S°$, J·deg^{-1}·mol^{-1}	$C_p°$, J·deg^{-1}·mol^{-1}
Phenylboron dichloride	lq	−299.4			
1-Phenylcyclohexene	lq	−16.8(67)			
Phenylcyclopropane	lq	100.3(9)			
N-Phenyldiacetimide	c	−362.5			
1,4-Phenylenediamine	c	3.1			
Phenyl formate	lq	−268.7(28)			
N-Phenylglycine	c	−402.5			
(±)-2-Phenylglycine	c	−431.8			
Phenylhydrazine	lq	141.0(9)			
Phenyl 2-hydroxybenzoate	c	−436.6(46)			
Phenylmethanethiol	lq	43.5			
N-Phenyl-2-naphthylamine	c	159.8			
1-Phenyl-1-propanone	lq	−167.2(11)			
1-Phenyl-2-propanone	lq	−151.9(20)			
1-Phenylpyrrole	c	154.3(54)			
2-Phenylpyrrole	c	139.2(54)			
Phenylsuccinic acid	c	−841.0			
S-Phenyl thioacetate	lq	−122.0			
Phenyl vinyl ether	lq	−26.2(9)			
Phosgene	g	−220.9	−206.8	283.8	57.7
Phthalamide	c	−433.1(12)			
1,2-Phthalic acid	c	−782.0(9)	−591.6	207.9	188.3
1,3-Phthalic acid	c	−803.0(8)			
1,4-Phthalic acid	c	−816.1(7)			
Phthalic anhydride	c	−460.1(19)	−331.0	179.5	161.1
Phthalonitrile	c	280.6(8)			
Picric acid	c	−214.4			
Piperazine	c	−45.6(16)	240.2	85.8	
2,5-Piperazinedione	c	−446.5(13)			
Piperidine	lq	−86.4(6)			
2-Piperidone	c	−306.6(5)	−112.1	164.9	
Propadiene	g	190.5(12)	202.4	243.9	59.0
Propanal	g	−185.6(9)	−130.5	304.7	78.7
Propanamide	c	−338.2(5)			
Propane	g	−104.7(5)	−23.6	270.2	73.6
1,2-Propanediamine	lq	−97.8(4)			
1,2-Propanediol	lq	−485.7(23)			
1,3-Propanediol	lq	−464.9(23)			
1,2-Propanedione	lq	−309.1(42)			
Propanedinitrile	lq	186.4(13)			
1,2-Propanedithiol	lq	−79.4(13)			
1,3-Propanedithiol	lq	−78.9			
Propanenitrile	lq	15.5(6)	89.2	189.3	
1-Propanethiol	g	−67.9(7)	2.2	336.4	94.8
2-Propanethiol	g	−76.2(7)	−2.6	324.3	96.0
Propanoic acid	lq	−510.7(3)	−383.5		154.0
Propanoic anhydride	lq	−679.1(6)	−475.6		
1-Propanol	lq	−302.6(5)	−170.6	194.6	141.0
	g	−255.1(5)	−161.8	324.7	87.1
2-Propanol	lq	−318.1(5)	−180.3	180.6	150.9
	g	−272.8(5)	−173.4	309.9	88.7

TABLE 6.1 Enthalpies and Gibbs Energies of Formation, Entropies, and Heat Capacities of Organic Compounds (*Continued*)

Substance	State	$\Delta Hf°$, kJ·mol^{-1}	$\Delta Gf°$, kJ·mol^{-1}	$S°$, J·deg^{-1}·mol^{-1}	$C_p°$, J·deg^{-1}·mol^{-1}
2-Propenal	g	−85.8	−64.6		
Propene	g	20.0(8)	62.8	266.6	64.3
2-Propenenitrile	g	180.6(17)	195.4	274.1	63.8
(E)-1,2,3-Propenetricarboxylic acid	c	−1233.0			
(Z)-1,2,3-Propenetricarboxylic acid	c	−1224.7			
2-Propenoic acid	g	−336.5	−286.3	315.2	77.8
2-Propen-1-ol	g	−124.5(18)	−71.3	307.6	76.0
2-Propenyl acetate	lq	−386.2			
(E)-1-Propenylbenzene	g	117.2	213.7	380.3	146.0
(Z)-1-Propenylbenzene	g	121.3	216.9	383.7	145.2
2-Propenylbenzene	lq	88.0(12)			
Propylamine	g	−70.2(4)	39.8	324.2	85.8
Propylbenzene	g	7.9(8)	137.2	400.7	152.3
Propylcarbamate	c	−552.6(5)			
Propylchloroacetate	lq	−515.6(21)			
Propylcyclohexane	g	−192.5(10)	47.3	419.5	184.2
Propylcyclopentane	g	−147.1(10)	52.6	417.3	154.6
Propylene oxide	g	−92.8	−25.8	286.7	72.3
Propyl formate	lq	−500.3(420)			
Propyl nitrate	g	−173.9(14)	−27.3	385.4	121.3
Propyl thioacetate	lq	−294.1			
Propyl trichloroacetate	lq	−513.0(84)			
Propyl vinyl ether	lq	−190.9(9)			
Propyne	g	184.9(8)	194.4	248.1	60.7
Propynoic acid	lq	−193.2(29)			
2-Propynylamine	lq	205.7(9)			
Pyrazine	c	139.8(12)			
Pyrazole	c	116.0(46)			
Pyridazine	lq	224.8(10)			
Pyridine	lq	100.2(7)	181.3	177.9	132.7
	g	140.4(7)	190.2	282.8	78.1
3-Pyridinecarbonitrile	c	193.4(9)			
3-Pyridinecarboxylic acid	c	−344.9(7)			
Pyrimidine	lq	145.9(13)			
1*H*-Pyrrole	lq	63.1(4)			
Pyrrole-2-carbaldoxime	c	12.1(29)			
Pyrrole-2-carboxaldehyde	c	−106.4(25)			
Pyrrolidine	g	−3.4(8)	114.7	309.5	81.1
(±)-2-Pyrrolidinecarboxylic acid	c	−524.2(6)			
2-Pyrrolidone	c	−286.2(5)			
Quinhydrone	c	−82.8	−323.0	325.9	277.0
Quinidine	c	−160.3			
Quinine	c	−155.2			
Quinoline	lq	156.2	275.7	217.2	
Raffinose	c	−3184			
L-(+)-Rhamnose	c	−1073.2			
D-(−)-Ribose	c	−1051.1(17)			
Salicylaldehyde	lq	−279.9			
Salicylaldoxime	c	−183.7			

TABLE 6.1 Enthalpies and Gibbs Energies of Formation, Entropies, and Heat Capacities
of Organic Compounds (*Continued*)

Substance	State	$\Delta Hf°$, kJ·mol^{-1}	$\Delta Gf°$, kJ·mol^{-1}	$S°$, J·deg^{-1}·mol^{-1}	$C_p°$, J·deg^{-1}·mol^{-1}
Salicylic acid	c	−589.5(15)	−418.1	178.2	
Semicarbazide ss	aq	−166.9	−40.6	297.9	
(−)-Serine	c	−732.7(3)			
L-(−)-Sorbose	c	−1271.5(15)	−908.4	220.9	
5,5′-Spirobis(1,3-dioxane)	c	−702.1			
Spiro[2.2]pentane	g	185.2(8)	265.3	282.2	88.1
(E)-Stilbene	c	136.9(28)	317.6	251.0	
(Z)-Stilbene	lq	183.3(15)			
(−)-Strychnine	c	−171.5			
Styrene	lq	103.8(11)	202.4	237.6	182.6
	g	147.9(15)	213.8	345.1	122.1
Succinic acid	c	−940.5(4)	−747.4	175.7	149.8
Succinic acid monoamide	c	−581.2			
Succinic anhydride	c	−607.8(4)			
Succinimide	c	−459.0(3)			
(+)-Sucrose	c	−2226.1(30)	−1544.7	360.2	
(±)-Tartaric acid	c	−1290.8			
(−)-Tartaric acid	c	−1282.4			
meso-Tartaric acid	c	−1279.9			
(α)-Terpinene	g	−20.5(35)			
Tetrabromomethane	g	50.2	35.9	358.1	91.2
Tetrabutyltin	lq	−304.6			
Tetracene	c	158.8			
Tetrachloro-1,4-benzoquinone	c	−288.7			
1,1,1,2,-Tetrachloro-1,2-difluoroethane	g	−489.9	−407.1	382.8	123.4
1,1,1,2-Tetrachloroethane	g	−149.4	−80.3	355.9	103.2
1,1,2,2,-Tetrachloroethane	lq	−195.0(84)	−95.0	246.9	165.7
	g	−152.7	−85.6	362.7	100.8
Tetrachloroethylene	g	−10.8(84)	20.5	340.8	94.9
Tetrachloromethane	lq	−132.8	−62.6	216.2	
	g	−95.8	−53.6	309.9	83.4
1,1,2,2-Tetracyanocyclopropane	c	590			
Tetracyanoethylene	c	623.8(17)			
Tetracyanomethane	c	611.6(17)			
Tetradecane	g	−332.1	66.9	700.4	326.1
Tetradecanoic acid	c	−833.5(15)			
1-Tetradecanol	c	−629.6(8)			
1-Tetradecene	g	−206.5	154.8	696.2	315.3
Tetraethylene glycol	lq	−981.6(46)			
Tetraethylgermanium	lq	−210.5			
Tetraethyllead	lq	53.1	336.4	472.5	
Tetraethyltin	lq	−95.8			
1,1,1,2-Tetrafluoroethane	g	−895.8	−826.2	316.2	86.3
Tetrafluoroethylene	g	−658.9(49)	−623.7	300.0	80.5
Tetrafluoromethane	g	−933.6(14)	−888.3	261.3	61.0
Tetrahydrofuran	g	−184.2(8)			
Tetrahydro-2-furanmethanol	lq	−435.6			
1,2,3,4-Tetrahydronaphthalene	lq	−29.2(15)			218
5,6,7,8-Tetrahydro-1-naphthol	c	−285.3(54)			

TABLE 6.1 Enthalpies and Gibbs Energies of Formation, Entropies, and Heat Capacities of Organic Compounds (*Continued*)

Substance	State	$\Delta Hf°$, kJ·mol^{-1}	$\Delta Gf°$, kJ·mol^{-1}	$S°$, J·deg^{-1}·mol^{-1}	$C_p°$, J·deg^{-1}·mol^{-1}
Tetrahydro-2*H*-pyran	lq	−258.3(13)			
1,2,3,6-Tetrahydropyridine	lq	33.5(23)			
Tetraiodomethane	g	262.9	217.1	391.6	95.9
1,2,3,4-Tetramethylbenzene	lq	−90.2(12)	106.7	290.6	
1,2,3,5-Tetramethylbenzene	lq	−96.4(12)	98.7	416.5	240.7
1,2,4,5-Tetramethylbenzene	lq	−119.9(12)	101.3	300.5	215.9
2,3,5,6-Tetramethylbenzoic acid	c	−506.1			
2,2,3,3-Tetramethylbutane	g	−225.6(14)	22.0	389.4	192.5
1,1,2,2-Tetramethylcyclopropane	lq	−119.7(9)			
Tetramethyllead	lq	98.3	262.8	320.1	
	g	136.4	270.7	420.5	144.0
Tetramethylsilane	g	−286.6	−148.4	361.1	130.2
Tetramethylsuccinic acid	c	−1012.5(21)			
Tetramethylthiacyclopropane	c	−83.0			
Tetramethyltin	g	−19.3			
Tetranitromethane	lq	38.8(42)			
1,1,1,2-Tetraphenylethane	c	223.0(14)			
1,1,2,2-Tetraphenylethane	c	216.0(14)			
Tetraphenylethylene	c	311.5(14)			
Tetraphenylhydrazine	c	457.9(26)			
Tetraphenylmethane	c	247.1(26)	574.0		
Tetraphenyltin	c	412.1			
Tetrapropylgermanium	g	−229.7			
Tetrapropyltin	lq	−211.3			
1,2,3,4-(1*H*)-Tetrazole	c	237.0(9)			
Theobromine	c	−361.5			
2-Thiaadamantane	c	−143.5(11)			
Thiacyclobutane	g	61.1	107.5	285.2	69.3
Thiacycloheptane	g	−61.3	84.1	361.9	124.6
Thiacyclohexane	g	−63.3	53.1	323.3	108.2
Thiacyclopentane	g	−33.8	46.0	309.4	90.9
Thiacyclopropane	g	82.2	96.9	255.3	53.7
Thianthrene	c	−182.5			
Thioacetamide	c	−70.5(12)			
Thioacetic acid	g	−182.0	−154.0	313.2	80.9
1,2-Thiocresol	lq	44.2			
Thiohydantoic acid	c	−554.8			
Thiohydantoin	c	−249.0			
2-Thiolactic acid	lq	−466.9			
Thiophene	lq	80.6(12)	121.2	181.2	
	g	115.0(10)	126.8	278.9	72.9
Thiophenol	lq	64.1	134.0	222.8	173.2
	g	111.6	147.6	336.9	104.9
Thiosemicarbazide	c	25.1			
Thiourea	c	−93.1(21)	21.8	115.9	
(−)-Threonine	c	−807.2(9)			
(±)-Threonine	c	−758.8(5)			
Toluene	lq	12.4(6)	113.8	221.0	157.2
	g	50.4(16)	122.0	320.7	103.6

TABLE 6.1 Enthalpies and Gibbs Energies of Formation, Entropies, and Heat Capacities of Organic Compounds (*Continued*)

Substance	State	$\Delta Hf°$, kJ·mol^{-1}	$\Delta Gf°$, kJ·mol^{-1}	$S°$, J·deg^{-1}·mol^{-1}	$C_p°$, J·deg^{-1}·mol^{-1}
2,4,6-Triamino-1,3,5-triazine	c	−72.4	184.5	149.1	
2-Triazoethanol	lq	94.6			
Tribenzylamine	c	140.6(113)			
Tribromoacetaldehyde	lq	−130.3			
Tribromochloromethane	g	12.6	9.1	357.2	
Tribromofluoromethane	g	−190.0	−193.1	345.8	
Tribromomethane	g	16.7	7.5	330.6	71.0
Tributoxyborane	lq	−1199.6			
Tributylamine	lq	−281.6(12)			
Tributyl phosphate	lq	−1456			
Tributylphosphine oxide	c	−460			
Trichloroacetaldehyde	lq	−234.5(14)			
Trichloroacetamide	c	−358.2(84)			
Trichloroacetic acid	c	−503.3(84)			
ionized	aq	−517.6(84)			
Trichloroacetyl chloride	lq	−280.8(84)			
Trichlorobenzoquinone	c	−269.9			
1,1,1-Trichloroethane	g	−144.6(17)	−76.2	320.0	92.3
1,1,2-Trichloroethane	g	−151.2(29)	−77.5	337.1	89.8
Trichloroethylene	g	−8.0(88)	19.9	324.8	80.2
Trichlorofluoromethane	g	−288.7(63)	−249.3	309.7(2)	78.0
Trichloromethane	g	−103.6(13)	−70.1	295.5	65.4
1,2,2-Trichloropropane	g	−185.8	−97.8	382.9	112.2
1,2,3-Trichloropropane	g	−182.9(18)			
1,1,1-Tricyanoethane	c	351.0(50)			
Tricyanoethylene	c	439.3(25)			
Tridecane	g	−311.5	58.5	661.5	303.2
Tridecanoic acid	c	−806.6(14)			
1-Tridecene	g	−186.0	146.3	657.3	292.4
Triethoxyborane	lq	−1047.4			
Triethoxymethane	lq	−687.3(30)			
Triethylaluminum	lq	−236.8			
Triethylamine	g	−92.8(6)	110.3	405.4	160.9
Triethylaminoborane	lq	−198.6			
Triethyl arsenite	lq	−706.7			
Triethylarsine	lq	13.0			
Triethylbismuthine	lq	169.9			
Triethylborane	lq	−189.1			
Triethylenediamine	c	−14.2	239.7	157.6	
Triethylene glycol	lq	−804.2(36)			
Triethyl phosphate	lq	−1243			
Triethylphosphine	lq	−89.1			
Triethyl phosphite	lq	−861.5			
Triethylstibine	lq	5.0			
Triethylsuccinic acid	c	−1066.5			
Triethyl thiophosphate	lq	−972.8			
Trifluoroacetic acid	lq	−1069.9(15)			
Trifluoroacetonitrile	g	−495.4	−461.9	298.3	78.2
1,1,1-Trifluoroethane	g	−744.6(17)	−678.3	287.3	78.5
1,1,2-Trifluoroethane	g	−730.7(209)			
2,2,2-Trifluoroethanol	lq	−932.4(9)			

TABLE 6.1　Enthalpies and Gibbs Energies of Formation, Entropies, and Heat Capacities
of Organic Compounds (*Continued*)

Substance	State	$\Delta Hf°$, kJ·mol^{-1}	$\Delta Gf°$, kJ·mol^{-1}	$S°$, J·deg^{-1}·mol^{-1}	$C_p°$, J·deg^{-1}·mol^{-1}
Trifluoroethylene	g	−490.4(84)	−469.5	292.6	69.2
Trifluoroiodomethane	g	−589.9	−572.0	307.5	
Trifluoromethane	g	−695.3(27)	−658.9	259.6	51.1
(Trifluoromethyl)benzene	g	−599.1(10)	−511.3	372.6	130.4
Trihexylamine	lq	−433.0(16)			
(±)-Trihydroxyglutaric acid	c	−1490			
Triiodomethane	g	210.9	178.0	355.6	75.1
Triisopropyl phosphite	lq	−980.3			
Trimethoxyborane	g	−899.1			
Trimethoxyethane	lq	−612.0(11)			
Trimethoxymethane	lq	−570.0(24)			
Trimethylacetic acid	lq	−564.4			
Trimethylacetic anhydride	lq	−779.9			
2′,4′,5′-Trimethylacetophenone	lq	−252.3			
2′,4′,6′-Trimethylacetophenone	lq	−267.4			
Trimethylaluminum	lq	−151.0		209.4	155.6
Trimethylamine	g	−23.7(7)	98.9	288.8	91.8
std. state	aq	−76.0	93.0	133.5	
Trimethylamine-aluminum chloride adduct	c	−879.1			
Trimethylammonium ion, std. state	aq	−112.9	37.2	196.7	
Trimethyl arsenite	lq	−590.8			
Trimethylarsine	g	11.7			
1,2,3-Trimethylbenzene	lq	−58.5(12)	107.5	277.8	216.7
1,2,4-Trimethylbenzene	lq	−61.8(10)	102.3	284.2	215.9
1,3,5-Trimethylbenzene	lq	−63.4(13)	103.9	273.6	209.6
2,3,4-Trimethylbenzoic acid	c	−486.6(10)			
2,3,5-Trimethylbenzoic acid	c	−488.7(9)			
2,3,6-Trimethylbenzoic acid	c	−475.7(9)			
2,4,5-Trimethylbenzoic acid	c	−495.7(13)			
2,4,6-Trimethylbenzoic acid	c	−477.9(10)			
3,4,5-Trimethylbenzoic acid	c	−500.9(10)			
2,6,6-Trimethylbicyclo[3.1.1]-2-heptene	lq	16.4(21)			
Trimethylbismuthine	g	192.9			
2,2,3-Trimethylbutane	g	−204.5(13)	4.3	383.3	164.6
2,3,3-Trimethyl-1-butene	lq	−117.7(14)			
Trimethylchlorosilane	lq	−384.1			
(Z),(Z)-1,3,5-Trimethyl-cyclohexane	g	−215.4	33.9	390.4	179.6
1,1,2-Trimethylcyclopropane	lq	−96.2(8)			
Trimethylene oxide	g	−80.54	−9.8	273.9	
Trimethylgallium	g	−46.9			
Trimethylindium	g	170.7			
2,2,3-Trimethylpentane	lq	−256.9(16)	9.3	327.6	188.9
	g	−220.0(16)	17.1	425.2	
2,2,4-Trimethylpentane	lq	−259.2(15)	6.9	328.0	236.4
	g	−224.0(15)	13.7	423.2	
2,3,3-Trimethylpentane	lq	−253.5(15)	10.6	334.4	
	g	−216.3(15)	18.9	431.5	

TABLE 6.1 Enthalpies and Gibbs Energies of Formation, Entropies, and Heat Capacities of Organic Compounds (*Continued*)

Substance	State	$\Delta Hf°$, kJ·mol^{-1}	$\Delta Gf°$, kJ·mol^{-1}	$S°$, J·deg^{-1}·mol^{-1}	$C_p°$, J·deg^{-1}·mol^{-1}
2,3,4-Trimethylpentane	lq	−255.0(18)	10.7	329.3	
2,2,4-Trimethyl-3-pentanone	lq	−381.6(13)			
2,4,4-Trimethyl-1-pentene	lq	−145.9(14)	86.4	306.3	
2,4,4-Trimethyl-2-pentene	lq	−142.4(21)	88.0	311.7	
Trimethylphosphine	lq	−122.2			
Trimethylphosphine oxide	c	−477.8			
Trimethyl phosphite	lq	−741.0			
Trimethylsilanol	lq	−545.0			
Trimethylstibine	g	32.2			
Trimethylsuccinic acid	c	−1000.8(18)			
Trimethylsuccinic anhydride	c	−688.3			
Trimethylthiacyclopropane	lq	−60.5			
Trimethyltin bromide	lq	−185.4			
Trimethyltin chloride	lq	−213.0			
Trimethylurea	c	−330.5(42)			
2,4,6-Trinitroanisole	c	−157.3			
1,3,5-Trinitrobenzene	c	−37.2(5)			
1,1,1-Trinitroethane	lq	−96.9(11)			
Trinitromethane	lq	−32.8(16)			
2,4,6-Trinitrophenetole	c	−204.6			
2,4,6-Trinitrophenol	c	−214.3(14)			
2,4,6-Trinitrophenylhydrazine	c	36.8			
2,4,6-Trinitrotoluene	c	−65.5(31)			
2,4,6-Trinitro-1,3-xylene	c	−102.5(41)			
Trioctylamine	lq	−584.9(22)			
1,3,6-Trioxacyclooctane	lq	−515.9(10)			
1,3,5-Trioxane	c	−522.5(4)			
Triphenylamine	c	245.6	504.2		
Triphenylarsine	c	310.0			
Triphenylbismuthine	c	469.0			
Triphenylborane	c	48.5			
Triphenylene	c	151.8(13)	329.2	254.7	
1,1,1-Triphenylethane	c	157.2(21)			
1,1,2-Triphenylethane	c	130.2(21)			
Triphenylethylene	c	233.5(18)	514.6		
2,4,6-Triphenylimidazole	c	272			
Triphenylmethane	c	171.2(14)	412.5	312.1	295.0
Triphenylmethanol	c	−3.4	272.8	329.3	
Triphenyl phosphate	c	−757			
Triphenylphosphine	c	232.2			
Triphenylphosphine oxide	c	−60.3			
Triphenylstibine	c	329.3			
Tripropoxyborane	lq	−1127.2			
Tripropylamine	lq	−207.2			
Tripropynylamine	lq	814.2			
Tris(acetylacetonato)chromium	c	−1533.0			
Tris(diethylamino)phosphine	lq	−289.5			
1,1,1-Tris(hydroxymethyl)-ethane	c	−744.6			
Tris(hydroxymethyl)nitro-methane	c	−735.6			

TABLE 6.1   Enthalpies and Gibbs Energies of Formation, Entropies, and Heat Capacities of Organic Compounds (*Continued*)

Substance	State	$\Delta Hf°$, kJ·mol⁻¹	$\Delta Gf°$, kJ·mol⁻¹	$S°$, J·deg⁻¹·mol⁻¹	$C_p°$, J·deg⁻¹·mol⁻¹
Tris(isopropoxy)borane	lq	−293.3			
Tris(trimethylsilyl)amine	c	−725.1			
(−)-Tryptophane	c	−415.3(10)	−119.4	251.0	238.2
(−)-Tyrosine	c	−685.1(16)	−385.7	214.0	216.4
Undecane	lq	−327.2(26)	22.8	458.1	345.2
Undecanoic acid	c	−735.9(10)			
1-Undecanol	lq	−504.8(9)			
1-Undecene	g	−144.8	129.5	579.4	246.7
10-Undecenoic acid	c	−577			
Uracil	c	−429.4(9)			
Urea	c	−333.6(2)	−196.8	104.6	93.1
Urea nitrate	c	−564.0			
Urea oxalate	c	−1528.4			
5-Ureidohydantoin	c	−718.0	−434.0	195.1	
Uric acid	c	−618.1	−358.8	173.2	
(±)-Valine [also (−)-]	c	−617.9(6)	−359.0	178.9	168.8
Valylphenylalanine	c	−767.8			
Vinyl acetate	g	−314.9(6)			
Vinylbenzene	lq	103.8(11)			
Vinylcyclohexane	lq	−88.7(8)			
4-Vinylcyclohexene	lq	26.8(9)			
Vinylcyclopentane	lq	−34.8(11)			
Vinylcyclopropane	lq	122.5(42)			
2-Vinylpyridine	lq	157.1(38)			
Xanthine	c	−378.6	−165.9	161.1	
1,2-Xylene	lq	−24.4(11)	110.3	246.5	187.9
	g	19.1(11)	122.1	352.8	133.3
1,3-Xylene	lq	−25.4(8)	107.7	252.2	183.3
	g	17.3(8)	118.9	357.7	127.6
1,4-Xylene	lq	−24.4(10)	110.1	247.4	
	g	18.0(10)	121.1	352.4	126.9
Xylitol	c	−1118.5(7)			
D-(+)-Xylose	c	−1057.8(9)			

TABLE 6.2 Heats of Fusion, Vaporization, and Sublimation and Specific Heat at Various Temperatures of Organic Compounds

Abbreviations Used in the Table

ΔHm, enthalpy of melting (at the melting point) in $kJ \cdot mol^{-1}$

ΔHv, enthalpy of vaporization (at the boiling point) in $kJ \cdot mol^{-1}$

ΔHs, enthalpy of sublimation (at 298 K) in $kJ \cdot mol^{-1}$

C_p, specific heat (at temperature specified on the Kelvin scale) for the physical state in existence (or specified: c, lq, g) at that temperature in $J \cdot K^{-1} \cdot mol^{-1}$

ΔHt, enthalpy of transition (at temperature specified, superscript, measured in degrees Celsius) in $kJ \cdot mol^{-1}$

Substance	ΔHm	ΔHv	ΔHs	C_p 400 K	600 K	800 K	1000 K
Acenaphthene		54.73	86.2(8)				
Acenaphthylene			73.0(3)				
Acetaldehyde	3.2	25.8	25.7(1)	65.8	85.9	101.3	125.4
Acetamide	15.7	56.1	78.7(3)				
Acetanilide		64.7	80.8				
Acetic acid	11.7	23.7	51.6(15)	81.7	105.2	121.7	133.9
Acetic anhydride	10.5	41.2	48.3	129.1	174.1	204.6	226.4
Acetone	5.7	29.1	30.8(1)	92.1	122.8	146.2	163.8
Acetonitrile, $\Delta Ht = 0.22^{-56}$	8.2	29.8	32.9(1)	61.2	76.8	89.0	98.3
Acetophenone		38.8	55.9(13)				
Acetyl bromide			33.1(5)				
Acetyl chloride			30.1(4)	78.9	97.0	110.0	119.7
Acetylene	3.8	17.0	21.3	50.1	57.5	62.5	66.6
Acetylene-d_2				54.8	61.9	67.4	71.8
Acetylenedicarbonitrile			28.8	94.8	106.2	114.1	119.8
Acetyl fluoride			25.1(21)				
Acetyl iodide			38.5(33)				
Acrylic acid	11.1	44.1	54.3	96.0	123.4	142.0	155.3
Acrylonitrile	6.6	32.6	33.5(17)	76.8	96.7	110.6	120.8
Adamantane			59.7(8)				
Adenine			108.8(84)				
α-Alanine			138.1				
Allyl tert-butyl sulfide			44.4(13)				
Allyl ethyl sulfone			83.7(25)				
Allyl ethyl sulfoxide			71.6				
Allyl methyl sulfone			79.5(25)				
Allyl trichloroacetate			52.3(42)				
3-Aminoacetophenone	12.1(4)						
4-Aminoacetophenone	15.9(4)						
2-Aminobenzoic acid	20.5(4)		104.9(10)				
3-Aminobenzoic acid	21.8(4)		128.0(32)				
4-Aminobenzoic acid	20.9(4)		116.1(37)				
2-Aminoethanol	20.5	50.9					
Aniline	10.5	42.4	55.8(1)	143.0	192.8	225.1	230.9
Anthracene		56.5	101.5(20)				
9,10-Anthraquinone		88.5	112.1(59)				
(E)-Azobenzene	22.6(1)	93.8(1)					
(Z)-Azobenzene			92.9(1)				
Azomethane				93.9	123.1	145.7	162.6
Azomethane-d_6				110.7	142.8	165.2	180.6

TABLE 6.2 Heats of Fusion, Vaporization, and Sublimation and Specific Heat at Various Temperatures of Organic Compounds (*Continued*)

Substance	ΔHm	ΔHv	ΔHs	C_p			
				400 K	600 K	800 K	1000 K
Azoisopropane			36.0(21)				
Azopropane			39.9				
Azulene	12.1	55.5	76.8(2)	176.4	248.2	295.4	327.4
Benzaldehyde		42.7	49.8(8)				
1,2-Benzanthracene			123.0(30)				
2,3-Benzanthracene			126				
1,2-Benzanthracene-9,10-dione			82.8(42)				
Benzene	9.9	30.7	33.6(1)	111.4	157.9	188.5	209.9
1,3-Benzenedicarboxylic acid			106.7(21)				
1,4-Benzenedicarboxylic acid			98.3(25)				
Benzenethiol			48.7(2)				
Benzoic acid	18.1	50.6	91.1(30)	138.4	196.7	234.9	260.7
Benzoic anhydride	17.2(8)		96.4(42)				
Benzonitrile	10.9	46.0	52.5(17)	140.8	187.4	217.9	238.8
Benzo[*def*]phenanthrene			100.2(4)				
Benzophenone			94.1(10)				
1,4-Benzoquinone			62.8(33)				
Benzo[*f*]quinoline			83.1(36)				
Benzo[*h*]quinoline			80.8(26)				
Benzo[*b*]thiophene	11.8		58.6(63)				
$\Delta Ht = 3.0^{-11.6}$							
Benzotrifluoride			37.6(1)				
Benzoyl bromide			58.6(63)				
Benzoyl chloride			54.8(42)				
Benzoyl iodide			61.9(63)				
4-Benzphenanthrene			106.3				
Benzyl acetate		49.4					
Benzyl alcohol	17.1	50.5	60.3(4)				
Benzylamine			53.6(21)				
Benzyl benzoate			77.8				
Benzyl bromide			47.3(42)				
Benzyl chloride			51.5(17)				
Benzyl ethyl sulfide			56.9				
Benzyl iodide			47.3(42)				
Benzyl mercaptan			56.6(2)				
Benzyl methyl ketone			49.0(2)				
Benzyl methyl sulfide			53.6				
Bicyclo[1.1.0]butane			23.4(12)				
Bicyclo[2.2.1]hepta-2,5-dione		32.9(8)					
Bicyclo[2.2.1]heptane			40.2(8)				
Bicyclo[4.1.0]heptane			38.0(4)				
Bicyclo[2.2.1]-2-heptene			38.8(4)				
Bicyclo[3.1.0]hexane			32.8(4)				
Bicyclohexyl			58.0(3)				
Bicyclo[2.2.2]octane			48.0(4)				
Bicyclo[4.2.0]octane			42.0(10)				
Bicyclo[5.1.0]octane			43.5(6)				
Bicyclo[2.2.2]-2-octene			43.8(4)				
Bicyclopropyl			33.5(13)				
Biphenyl	16.7(5)	45.6	81.8(4)	221.0	307.7	363.7	401.7

TABLE 6.2 Heats of Fusion, Vaporization, and Sublimation and Specific Heat at Various Temperatures of Organic Compounds (*Continued*)

Substance	ΔHm	ΔHv	ΔHs	C_p 400 K	600 K	800 K	1000 K
Biphenylene			84.3(3)				
Bis(2-butoxyethyl) ether		55.9					
Bis(2-chloroethyl) ether	8.7	45.2					
Bis(2-ethoxyethyl) ether		49.0					
Bis(2-hydroxyethyl) ether		52.3	57.3				
Bis(2-methoxyethyl) ether		43.1					
Bromobenzene	10.6	37.9	44.6(1)	127.4	171.5	199.9	219.2
4-Bromobenzoic acid			87.9(42)				
1-Bromobutane	6.7	32.5	36.7(1)	136.6	180.0	211.2	234.4
2-Bromobutane			35.4(1)	138.1	214.7	238.2	
1-Bromo-2-chloroethane		33.7	38.2				
Bromochloromethane		30.0	32.8				
1-Bromo-3-chloropropane		37.6	44.1				
1-Bromo-2-chloro-1,1,2-trifluoroethane		28.3	30.1				
Bromochloro-2,2,2-trifluoroethane		28.1	29.8				
1-Bromododecane		74.8					
Bromoethane	5.9	27.0	27.6(13)	79.2	102.8	119.6	132.2
Bromoethylene			18.2	66.6	83.0	94.1	102.3
1-Bromoheptane			50.6(2)			74.8	
1-Bromohexadecane			94.4				
1-Bromohexane			45.9(3)				
Bromomethane, $\Delta Ht = 0.47^{-99.4}$	6.0	23.9	23.0(2)	50.0	62.7	72.2	79.5
1-Bromo-2-methylpropane		31.3	34.9				
2-Bromo-2-methylpropane	2.0	29.2	31.4(8)	146.1	190.7	220.3	241.6
$\Delta Ht = 5.7^{-64.5}$							
$\Delta Ht = 1.0^{-41.6}$							
1-Bromonaphthalene		52.5					
1-Bromooctane			55.4(3)				
1-Bromopentane	11.5	35.0	41.3(2)	165.6	219.0	257.5	286.0
1-Bromopropane	6.5	29.8	31.9(1)	107.5	140.8	164.9	182.8
2-Bromopropane		28.3	30.2(1)	110.2	144.0	167.7	185.2
Bromotrifluoromethane				79.3	91.3	97.5	100.9
Bromotrimethylsilane			32.6				
1,2-Butadiene	7.0	24.0	23.3(1)	98.4	128.5	150.7	167.4
1,3-Butadiene	8.0	22.5	22.0(1)	101.2	154.1	169.5	
1,3-Butadiyne				84.4	96.8	105.1	111.3
Butanal	11.1	33.7(4)	34.5(8)	126.4	165.7	195.0	216.3
Butanamide	17.6(8)		85.9(17)				
Butane, $\Delta Ht = 2.1^{-165.6}$	4.7	22.4	21.0(1)	123.9	168.6	201.8	226.9
1,2-butanediamine			46.3(2)				
Butanedinitrile	3.7	48.5	70.0(3)				
1,3-Butanediol		52.7	67.8(21)				
1,4-Butanediol			76.6(17)				
2,3-Butanediol			59.2(21)				
2,3-Butanedione			38.7(6)				
Butanenitrile	5.0	33.7	39.4	118.8	155.1	181.9	201.8
meso-1,2,3,4-Butanetetrol			135.1				
1,4-Butanethiol			49.7(1)				

TABLE 6.2 Heats of Fusion, Vaporization, and Sublimation and Specific Heat at Various Temperatures of Organic Compounds (*Continued*)

Substance	ΔHm	ΔHv	ΔHs	C_p 400 K	600 K	800 K	1000 K
1-Butanethiol	10.5	32.2	36.1(1)	146.2	194.7	233.0	263.4
2-Butanethiol	6.5	30.6	34.0(1)	148.0	194.2	227.2	251.1
1,2,4-Butanetriol		58.6					
Butanoic acid	10.5	41.8	58.0(41)				
Butanoic anhydride		50.0					
1-Butanol	9.4	43.3	52.3(1)	137.2	183.7	218.0	243.8
2-Butanol		40.8	49.7(1)	141.0	187.1	220.4	245.3
2-Butanone	8.4	31.3	34.7(2)	124.7	163.6	192.8	214.8
(E)-2-Butenal			34.5(1)				
1-Butene	3.9	22.1	20.6(10)	109.0	147.1	174.9	195.9
(E)-2-Butene	9.8	22.7	22.0	108.9	145.6	184.9	194.9
(Z)-2-Butene	7.3	23.3	22.7(10)	101.8	141.4	171.0	193.1
(Z)-2-Butenedinitrile			72.0(8)				
(E)-2-Butenedioic acid			136.3(63)				
(Z)-2-Butenedioic acid			110.0(25)				
(E)-2-Butene-1,4-diol		69.0					
(Z)-2-Butene-1,4-diol		66.1					
(E)-2-Butenenitrile			40.0(2)				
(Z)-2-Butenenitrile			38.9(2)				
3-Butenenitrile			40.0(1)				
(E)-2-Butenoic acid	13.0						
(Z)-2-Buten-1-ol		46.4					
1-Buten-3-yne				89.0	111.6	127.2	138.7
2-Butoxyethanol			56.6				
1-*tert*-Butoxy-2-ethoxyethane			50.9				
2-(2-Butoxyethoxy)ethanol		28.0					
2-Butoxyethyl acetate			59.5				
1-*tert*-Butoxy-2-methoxyethane		38.5	47.8				
N-Butylacetamide			76.1(13)				
Butyl acetate		36.3	43.6(2)				
Butylamine		31.8	35.7(2)	148.3	197.9	234.4	261.7
sec-Butylamine		29.9	32.6(2)	148.1	199.0	236.1	261.7
tert-Butylamine		28.3	29.7(2)	152.6	204.5	240.5	266.9
Butylbenzene	11.2	38.9	50.1(1)	229.1	314.6	373.9	416.3
sec-Butylbenzene	9.8	38.0	49.0(2)				
tert-Butylbenzene	8.4	37.6	48.1(2)				
sec-Butyl butanoate			47.3(13)				
Butyl chloroacetate			51.0(42)				
Butyl 2-chlorobutanoate			52.7				
Butyl 3-chlorobutanoate			53.1				
Butyl 4-chlorobutanoate			54.4				
Butyl 2-chloropropanoate			54.4				
Butyl 3-chloropropanoate			55.4				
Butyl crotonate			51.9				
sec-Butyl crotonate			49.4				
Butylcyclohexane	14.2	38.5	49.6(6)	276.1	289.5	469.9	525.9
Butylcyclopentane	11.3	36.2	45.9	241.7	336.3	407.3	480.3
N-Butyldiacetimide			64.4				
Butyl dichloroacetate			52.3(44)				
Butyl ethyl ether		31.6	36.4				

TABLE 6.2 Heats of Fusion, Vaporization, and Sublimation and Specific Heat at Various Temperatures of Organic Compounds (*Continued*)

Substance	ΔHm	ΔHv	ΔHs	C_p			
				400 K	600 K	800 K	1000 K
Butyl ethyl sulfide	12.4	37.1	44.5	202.4	271.8	325.3	367.2
tert-Butyl ethyl sulfide	7.1	33.5	39.3(13)				
Butyl formate		36.6	41.2				
tert-Butyl hydroperoxide			47.7(2)				
Butylisopropylamine		34.5	42.1				
Butyllithium			107.1				
Butyl methyl ether			32.4(2)				
Butyl methyl sulfide	12.5	34.5	40.7(1)	174.6	233.0	278.4	314.1
tert-Butyl methyl sulfide	8.4	31.5	35.8(1)				
Butyl methyl sulfone			76.2(21)				
tert-Butyl methyl sulfone			82.4(25)				
tert-Butyl peroxide			31.8(29)				
Butyl thiolacetate			48.1				
Butyl trichloroacetate			53.6(42)				
Butyl vinyl ether		31.6	36.2				
1-Butyne	6.0	24.5	23.3(1)	99.9	129.0	150.4	166.7
2-Butyne	9.2	26.5	26.6(1)	94.6	124.2	147.0	164.4
2-Butynedinitrile			28.8(6)				
4-Butyrolactone		52.2					
Butyrophenone			60.7(17)				
(+)-Camphor	6.8	59.5					
9*H*-Carbazole			84.5(8)				
Chloroacetic acid			75.3(42)				
Chloroacetyl chloride			38.9(21)				
2-Chloroaniline	11.9	44.4	56.8				
2-Chlorobenzaldehyde			53.1(25)				
Chlorobenzene	9.6	35.2	41.0(1)	128.1	172.2	200.4	219.6
2-Chlorobenzoic acid			79.5(33)				
3-Chlorobenzoic acid			82.0(33)				
4-Chlorobenzoic acid			87.9(33)				
Chloro-1,4-benzoquinone			69.0(84)				
1-Chlorobutane		30.4	33.5(1)	135.1	179.0	210.5	234.0
2-Chlorobutane		29.2	31.7(1)	136.1	180.7	212.7	236.8
Chlorocyclohexane			43.5(33)				
1-Chloro-1,1-difluoroethane		22.4					
Chlorodifluoromethane	4.1	20.2		67.1	80.4	88.3	93.2
2-Chloro-1,4-dihydroxybenzene			69.0				
Chlorodimethylsilane		26.2					
Chlorodiphenylsilane			69.5				
1-Chloro-2,3-epoxypropane		33.1	40.6				
Chloroethane	5.0	24.7		77.6	101.6	118.8	131.7
2-Chloroethanol		41.4					
1-Chloro-2-ethylbenzene			47.3(21)				
1-Chloro-4-ethylbenzene			48.1(21)				
Chloroethylene	4.7	21.3		65.0	82.1	93.5	101.9
2-Chloroethyl vinyl ether		38.2					
Chloroethyne				60.2	66.8	71.0	74.3
1-Chlorohexane		35.7	42.91				
Chlorohydroquinone			69.0				
Chloromethane	6.4	21.5		48.2	61.3	71.3	78.9

TABLE 6.2 Heats of Fusion, Vaporization, and Sublimation and Specific Heat at Various Temperatures of Organic Compounds (*Continued*)

Substance	ΔHm	ΔHv	ΔHs	C_p 400 K	600 K	800 K	1000 K
1-Chloro-2-methylpropane		29.2	31.7(1)	136.1	180.7	212.7	236.8
2-Chloro-2-methylpropane	2.0	27.6	29.0(1)	142.3	184.9	215.5	238.5
$\quad \Delta Ht = 1.7^{-90.1}$							
$\quad \Delta Ht = 5.8^{-53.6}$							
1-Chloronaphthalene		52.1	65.3(50)				
2-Chloronaphthalene			82.0(59)				
Chloropentafluoroacetone			25.3				
1-Chloropentane		33.2	38.2(1)	164.2	218.0	256.8	285.6
2-Chloropentane		31.8	36.1				
3-Chlorophenol			53.1(25)				
4-Chlorophenol			51.9(25)				
1-Chloropropane		27.2	28.5(2)	106.1	139.9	164.2	182.4
2-Chloropropane		26.3	27.2(8)	108.7	143.1	167.1	184.8
3-Chloro-1-propene		29.0	28.2	92.6	111.0	137.8	151.9
Chlorotrifluoroethylene	5.6	20.8					
Chlorotrifluoromethane				77.5	90.3	96.9	100.5
Chlorotrimethylsilane		27.6	30.1				
Chlorotrinitromethane			45.4(4)				
Chrysene			124.5(40)				
1,2-Cresol	15.8	42.7	76.0(8)	166.3	220.8	257.5	287.9
1,3-Cresol	10.7	61.7(10)	162.1	218.7	256.6	286.6	
1,4-Cresol	12.7	43.2	73.9(15)	161.7	218.0	255.7	286.5
Cubane			80.3				
Cyanamide	8.8	68.6					
Cyanogen	8.1	23.3	20.8(8)	61.9	68.2	72.9	76.4
Cyclobutane, $\Delta Ht = 5.8^{-126.8}$	1.1	24.2	24.6(3)	100.0	145.4	177.5	200.7
Cyclobutanenitrile			40.0(4)				
Cyclobutene				90.3	126.8	151.7	169.6
Cyclobutylamine			35.6(4)				
Cyclododecane			76.4(17)				
Cycloheptane	1.9	33.2	38.5(2)	175.0	261.2	322.3	365.7
$\quad \Delta Ht = 5.0^{-138.4}$							
$\quad \Delta Ht = 0.3^{-75.0}$							
$\quad \Delta Ht = 0.5^{-60.8}$							
Cycloheptanone			51.9(13)				
1,3,5-Cycloheptatriene,	1.2	38.7		155.4	209.5	245.1	270.2
$\quad \Delta Ht = 2.4^{-119.2}$							
Cyclohexane, $\Delta Ht = 6.7^{-87}$	2.7	30.0	33.0(8)	149.9	225.2	279.3	317.2
Cyclohexanethiol		37.1	44.6(2)				
Cyclohexanol, $\Delta Ht = 8.2^{-9.7}$	1.7	45.5	62.0(3)	172.1	248.1	302.0	339.5
Cyclohexanone		40.3	45.1(2)	150.6	221.3	272.0	305.4
Cyclohexene, $\Delta Ht = 4.3^{-134.4}$	3.3	30.5	33.5(3)	144.9	206.9	248.9	278.7
Cyclohexylamine		36.1	42.8(2)				
Cyclohexylcyclohexane		51.9	58.0				
(Z),(Z)-1,5-Cyclooctadiene			43.4				
Cyclooctane	2.4	35.9	43.3(2)	200.1	297.1	365.3	414.3
$\quad \Delta Ht = 6.3^{-106.7}$							
$\quad \Delta Ht = 0.5^{-89.4}$							
Cyclooctanone			54.4(21)				
1,3,5,7-Cyclooctatetraene	11.3	36.4	43.1(13)	160.9	220.8	260.4	288.2

TABLE 6.2 Heats of Fusion, Vaporization, and Sublimation and Specific Heat at Various Temperatures of Organic Compounds (*Continued*)

Substance	ΔHm	ΔHv	ΔHs	C_p 400 K	600 K	800 K	1000 K
Cyclooctene			47.0(12)				
Cyclopentadiene			28.4				
Cyclopentane	0.6	27.3	28.7(1)	118.7	178.1	220.1	250.4
$\quad \Delta Ht = 4.8^{-150.8}$							
$\quad \Delta Ht = 0.3^{-135.1}$							
Cyclopentanethiol	7.8	35.3	41.5(1)	144.5	203.6	245.2	275.5
Cyclopentanol			57.5(3)				
Cyclopentanone		36.4	42.7(2)				
Cyclopentene, $\Delta Ht = 0.5^{-186.1}$	3.4		28.1(3)	104.9	155.6	191.5	217.3
Cyclopentylamine			40.2(4)				
Cyclopropane	5.4	20.1	18.1	76.6	109.4	140.5	148.1
Cyclopropylamine			31.3(4)				
Cyclopropylbenzene			50.2(1)				
Decaborane(14)	32.5	48.5	76.7				
(E)-Decahydronaphthalene	14.5	40.2	43.5(21)	237.6	352.3	432.6	489.2
(Z)-Decahydronaphthalene,	9.5	41.0	50.2(21)	237.0	352.0	432.5	489.5
$\quad \Delta Ht = 2.1^{-57.1}$							
Decanal				300.4	400.4	472.8	525.9
Decane	28.7	38.8	51.4(1)	298.1	403.2	480.8	536.4
Decanedioic acid			160.7				
Decanenitrile			66.9(4)				
1-Decanethiol	31.0	46.4	65.0(5)	320.6	429.4	510.9	573.1
Decanoic acid	29.3		118.8(20)				
1-Decanol	37.7	49.8	81.5(8)	187.2	418.2	495.9	553.3
1-Decene, $\Delta Ht = 8.0^{-74.8}$	13.8	38.7	50.4(2)	283.6	381.9	453.0	505.9
1-Decyne				274.6	363.8	428.5	476.6
Deoxybenzoin			93.3(42)				
Dibenz[*de,kl*]anthracene			125.5				
Dibenzoyl peroxide	31.4(21)		102.5(84)				
Dibenzyl ether		20.2					
Dibenzyl sulfide			93.3(33)				
Dibenzyl sulfone			125.5(25)				
1,2-Dibromobutane			50.3(21)	153.9	195.4	224.3	244.8
1,4-Dibromobutane			53.1(1)				
2,3-Dibromobutane			37.7(21)				
1,2-Dibromochlorotrifluoro-		31.2	35.1				
$\quad$ethane							
1,2-Dibromocycloheptane			52.0(13)				
1,2-Dibromocyclohexane			50.5(13)				
1,2-Dibromocyclooctane			54.6(13)				
1,2-Dibromoethane	11.0	34.8	41.7(1)	99.7	122.3	137.8	149.8
1,2-Dibromoheptane			54.4				
Dibromomethane		32.9	37.0	63.0	74.8	82.5	88.0
1,2-Dibromopropane		35.6	41.7	124.4	157.4	179.5	195.6
1-3-Dibromopropane			47.5				
1,2-Dibromotetrafluoroethane		27.0	28.6				
1,2-Dibutoxyethane		47.8	58.8				
Dibutoxymethane			48.1(25)				
Dibutylamine		38.4	49.5(4)				
Dibutyl decanedioate		92.9					

TABLE 6.2 Heats of Fusion, Vaporization, and Sublimation and Specific Heat at Various
Temperatures of Organic Compounds (*Continued*)

Substance	ΔHm	ΔHv	ΔHs	C_p 400 K	600 K	800 K	1000 K
Dibutyl disulfide		46.9	64.5(17)	286.1	376.5	442.8	493.1
Di-*tert*-butyl disulfide			54.3				
Dibutyl ether		36.5	45.0(2)	254.3	340.1	403.8	451.3
Di-*sec*-butyl ether		34.1	40.6(13)				
Di-*tert*-butyl ether		32.2	37.6(2)				
Dibutylmercury			63.5				
Di-*tert*-butyl peroxide			31.8				
Dibutyl 1,2-phthalate		79.2	91.6(42)				
Dibutyl sulfate			75.9(17)				
Dibutyl sulfide	19.4		53.3(1)	259.8	348.6	420.8	475.8
Di-*tert*-butyl sulfide		33.3	43.8				
Dibutyl sulfite			67.8(17)				
Dibutyl sulfone			100.4(25)				
Dichloroacetyl chloride			39.3(21)				
1,2-Dichlorobenzene	13.4	40.6	47.7(17)	142.8	184.4	210.4	227.7
1,3-Dichlorobenzene		38.6	46.4(17)	143.0	184.5	210.4	227.7
1,4-Dichlorobenzene	18.2	38.8	64.9(8)	143.3	184.8	210.7	227.9
2,6-Dichlorobenzoquinone			69.9				
2.2′-Dichlorobiphenyl			96.2(42)				
4,4′-Dichlorobiphenyl			103.8(42)				
1,2-Dichlorobutane		33.9	39.6				
1,4-Dichlorobutane			46.4				
Dichlorodifluoromethane				82.4	93.6	99.1	100.0
Dichlorodimethylsilane			34.3				
Dichlorodiphenylsilane			69.5				
1,1-Dichloroethane	7.9	28.9	30.7(2)	91.4	113.7	128.8	139.8
1,2-Dichloroethane	8.8	32.0	35.4(1)	92.1	112.6	127.2	138.1
1,1-Dichloroethylene	6.5	26.1	26.5(2)	78.7	93.9	103.4	110.0
(E)-1,2-Dichloroethylene	12.0	28.9	29.3(13)	77.7	93.2	102.9	109.8
(Z)-1,2-Dichloroethylene	7.2	30.3	31.0(13)	77.0	93.0	102.9	109.8
2,2-Dichloroethyl ether		38.4					
Dichlorofluoromethane				70.2	82.4	89.6	94.2
1,2-Dichlorohexafluoropropane		26.3	27.3				
Dichloromethane	4.2	28.1	28.5(4)	59.6	72.4	80.8	86.8
1,2-Dichloropropane		31.8	36.0(17)	119.7	152.6	175.6	192.8
1,3-Dichloropropane		35.2	40.8(1)	120.0	151.5	173.9	190.4
2,2-Dichloropropane		29.3	32.6(13)	127.9	159.2	179.9	194.8
1,3-Dichloro-2-propanol			66.9(42)				
Dicyanoacetylene			28.8				
Dicyclopentadienyliron			73.6				
Diethanolamine	25.1	65.2					
1,1-Diethoxyethane		32.7	37.8(8)				
1,2-Diethoxyethane			43.2(1)				
Diethoxymethane		31.3	35.6(2)				
1,3-Diethoxypropane		37.2	45.9(2)				
2,2-Diethoxypropane			31.8(8)				
Diethylamine		29.1	31.2(4)	143.9	197.2	235.0	263.2
1,2-Diethylbenzene	16.8	39.4	52.8	234.4	316.6	374.6	416.3
1,3-Diethylbenzene	11.0	39.4	52.5	230.2	314.6	379.7	415.8
1,4-Diethylbenzene	10.6	39.4	52.5	228.8	313.1	372.5	414.9

TABLE 6.2 Heats of Fusion, Vaporization, and Sublimation and Specific Heat at Various Temperatures of Organic Compounds (*Continued*)

Substance	ΔHm	ΔHv	ΔHs	C_p			
				400 K	600 K	800 K	1000 K
Diethyl carbonate		36.2	43.6(2)				
Diethyl disulfide	9.4	37.6	45.2	171.1	218.6	251.8	276.0
Diethyl ether	7.3	26.5	27.2(2)	138.1	183.8	218.7	244.8
Diethyl oxalate		42.0	63.5(21)				
Diethyl peroxide			30.5(21)				
Diethylpentane		34.6	42.0				
Diethyl 1,2-phthalate			88.3				
Diethyl sulfide	11.9	31.8	35.9(1)	145.0	192.9	229.7	258.5
Diethyl sulfite			48.5(17)				
Diethyl sulfone			86.2(25)				
Diethyl sulfoxide			62.3(13)				
Diethylzinc			40.2				
1,2-Difluorobenzene	11.1	32.2	36.2(1)	137.1	181.3	209.7	229.0
1,3-Difluorobenzene		31.1	34.7(1)	137.0	180.5	207.8	225.6
1,4-Difluorobenzene		31.8	35.6(1)	137.4	180.1	207.8	225.7
2,2′-Difluorobiphenyl			95.0(42)				
4,4′-Difluorobiphenyl			91.2(42)				
1,1-Difluoroethane		21.4		83.4	107.5	124.3	136.3
1,1-Difluoroethylene				71.8	89.2	100.2	107.7
Difluoromethane				51.1	65.8	76.2	83.7
9,10-Dihydroanthracene			93.3(42)				
Dihydro-2*H*-pyran			32.2(8)				
5,12-Dihydrotetracene			115.9				
2,3-Dihydrothiophene		33.2	37.7(4)				
2,5-Dihydrothiophene		34.8	40.0(8)				
2,4-Dihydrothiophene-1,1-dioxide			62.8(25)				
1,4-Dihydroxybenzene			99.2				
1,2-Diisobenzene			64.9				
1,2-Diiodoethane			65.7(42)	96.0	116.8	131.3	141.6
Diiodomethane	12.6	41.8	51.0	65.9	76.9	83.9	89.1
Diisobutylamine			39.3(4)				
Diisobutyl ether		34.0	40.9				
Diisopropylamine		30.4	34.5(2)				
Diisopropyl ether	11.0	29.1	32.3(3)	196.2	262.0	311.3	348.0
Diisopropylmercury			53.6				
Diisopropyl sulfide	10.4	33.8	39.5(3)	211.9	277.1	322.7	356.6
Diketene		36.8	42.9				
1,2-Dimethoxybenzene	16.0	48.2	66.9(21)				
1,1-Dimethoxyethane			30.5				
1,2-Dimethoxyethane		32.4	36.5				
Dimethoxymethane	8.3		35.1(2)				
2,2-Dimethoxypropane			29.4				
N,N,-Dimethylacetamide	10.4	43.2	50.2				
Diethylamine	5.9	26.4	25.4(1)	87.4	118.9	142.0	159.8
Dimethylaminomethanol			50.2(42)				
N,N-Dimethylaminotrimethyl-silane			31.8				
1,4-Dimethylbicyclo[2.2.1]-heptane		33.3	38.9				

TABLE 6.2   Heats of Fusion, Vaporization, and Sublimation and Specific Heat at Various Temperatures of Organic Compounds (*Continued*)

Substance	ΔHm	ΔHv	ΔHs	C_p 400 K	600 K	800 K	1000 K
2,3-Dimethylbicyclo[2.2.1]-2-heptene		34.9	42.2				
N,N-Dimethylaniline			52.8(34)				
2,2-Dimethylbutane	0.6	26.3	27.7(10)	182.8	251.0	298.7	333.5
$\Delta Ht = 5.4^{-147.3}$							
$\Delta Ht = 0.3^{-132.3}$							
2,3-Dimethylbutane,	0.8	27.4	29.1(10)	181.2	247.7	314.6	331.0
$\Delta Ht = 6.5^{-137.1}$							
2,2-Dimethyl-1-butanol		42.6	56.1				
2,3-Dimethyl-1-butanol		47.3					
3,3-Dimethyl-1-butanol		46.4					
2,3-Dimethyl-2-butanol		40.4	51.0				
(±)-3,3-Dimethyl-2-butanol		43.9					
3,3-Dimethyl-2-butanone		33.4	37.9(4)				
2,3-Dimethyl-1-butene		27.4	30.7(15)	178.2	231.8	272.0	302.1
3.3-Dimethyl-1-butene,	1.1	25.7	27.1(14)	162.8	223.4	266.1	297.1
$\Delta Ht = 4.3^{-148.3}$							
2,3-Dimethyl-2-butene,	6.5	29.6	33.3(12)	156.8	216.7	262.7	297.7
$\Delta Ht = 3.5^{-76.3}$							
Di(3-methylbutyl) ether		35.2					
Dimethylcadmium			38.0				
1,1-Dimethylcyclohexane,	2.1	32.5	37.9(2)	212.1	310.0	379.5	427.6
$\Delta Ht = 6.0^{-120.0}$							
(E)-1,2-Dimethylcyclohexane	10.4	33.0	38.4(1)	217.2	312.1	378.7	425.5
(Z)-1,2-Dimethylcyclohexane,	1.6	32.5	39.7(1)	213.8	309.6	377.0	424.3
$\Delta Ht = 8.3^{-100.6}$							
(E)-1,3-Dimethylcyclohexane	9.9	33.4	89.2(1)	213.8	308.8	375.7	423.0
(Z)-1,3-Dimethylcyclohexane	10.8	32.9	38.2(1)	214.2	310.5	378.7	426.8
(E)-1,4-Dimethylcyclohexane	12.3	32.6	37.9(1)	215.9	312.1	378.9	425.7
(Z)-1,4-Dimethylcyclohexane	9.3	33.3	39.0(1)	213.8	308.8	375.7	423.0
1,1-Dimethylcyclopentane,	1.1	30.3	33.8(2)	182.2	262.6	318.7	359.1
$\Delta Ht = 6.5^{-126.4}$							
(E)-1,2-Dimethylcyclopentane	7.2	30.9	34.6(2)	182.9	262.2	317.3	357.4
(Z)-1,2-Dimethylcyclopentane,	1.7	31.7	35.7(2)	182.7	262.4	317.9	358.0
$\Delta Ht = 6.7^{-131.7}$							
(E)-1,3-Dimethylcyclopentane	7.3	30.8	34.5(2)	182.9	262.2	317.3	357.4
(Z)-1,3-Dimethylcyclopentane	7.4	30.4	34.2(2)	182.9	262.2	317.3	357.4
(Z)-2,4-Dimethyl-1,3-dioxane			39.9(13)				
4,5-Dimethyl-1,3-dioxane			42.5(13)				
5,5-Dimethyl-1,3-dioxane			41.3(13)				
Dimethyl disulfide	9.2	33.8	37.9	110.3	137.4	157.6	172.8
Dimethyl ether	4.9	21.5	19.3	79.6	105.3	125.7	141.4
N,N-Dimethylformamide	16.2	38.3	47.6(13)				
Dimethylglyoxime			97.1				
2,2-Dimethylhexane	6.8	32.1	37.3(1)				
2,3-Dimethylhexane		33.2	38.8(1)				
2,4-Dimethylhexane		32.5	37.8(1)				
2,5-Dimethylhexane	13.0	32.5	37.9(1)				
3,3-Dimethylhexane	7.1	32.3	37.5(1)				
3,4-Dimethylhexane		33.2	39.0(1)				

TABLE 6.2   Heats of Fusion, Vaporization, and Sublimation and Specific Heat at Various Temperatures of Organic Compounds (*Continued*)

Substance	ΔHm	ΔHv	ΔHs	C_p 400 K	600 K	800 K	1000 K
(E)-2,2-Dimethyl-3-hexene			37.3(2)				
(Z)-2,2-Dimethyl-3-hexene			37.2(2)				
1,1-Dimethylhydrazine	10.1	32.6	35.0(1)				
1,2-Dimethylhydrazine		35.2	39.3(1)				
3,5-Dimethylisoxazole			45.2(17)				
Dimethyl maleate	14.7		44.3				
Dimethylmercury			34.6				
6,6-Dimethyl-2-methylene-bicyclo[3.1.1]heptane		40.2	46.4(13)				
Dimethyl oxalate			47.4(5)				
2,2-Dimethylpentane	5.8	29.2	32.4(1)	211.0	285.9	340.7	381.6
2,3-Dimethylpentane		30.5	34.2(1)	211.0	285.9	340.7	381.6
2,4-Dimethylpentane	6.9	29.6	32.9(1)	211.0	285.9	340.7	381.6
3,3-Dimethylpentane	7.1	29.6	33.0(1)	211.0	285.9	340.7	381.6
2,2-Dimethyl-3-pentanone		36.1	42.4				
2,4-Dimethyl-3-pentanone		34.6	41.5(1)				
2,4-Dimethyl-1-pentene			33.2(14)				
4,4-Dimethyl-1-pentene			29.0(19)				
2,4-Dimethyl-2-pentene			34.4(11)				
(E)-4,4-Dimethyl-2-pentene			32.7(11)				
(Z)-4,4-Dimethyl-2-pentene			32.7(14)				
2,7-Dimethylphenanthrene			106.7(8)				
4,5-Dimethylphenanthrene			104.6(13)				
9,10-Dimethylphenanthrene			119.5(12)				
2,3-Dimethylphenol			84.0(10)				
2,4-Dimethylphenol			65.9(2)				
2,5-Dimethylphenol			85.0(3)				
2,6-Dimethylphenol			75.6(2)				
3,4-Dimethylphenol			85.7(2)				
3,5-Dimethylphenol			82.8(3)				
Dimethyl 1,2-phthalate	162.7						
2,2-Dimethylpropane, $\Delta Ht = 2.6^{-133.1}$	3.2	22.7	22.4(1)	157.1	218.5	254.3	283.7
2,2-Dimethylpropanenitrile		32.4	37.0(4)				
2,2-Dimethyl-1-propanol		9.6					
2,3-Dimethylpyridine		39.1	49.0(4)				
2,4-Dimethylpyridine		38.5	47.8(4)				
2,5-Dimethylpyridine			47.8(4)				
2,6-Dimethylpyridine		37.5	46.1(4)				
3,4-Dimethylpyridine		40.0	51.8(4)				
3,5-Dimethylpyridine		39.5	50.4(4)				
Dimethyl sulfate			48.5(17)				
Dimethyl sulfide	8.0	27.0	27.9(1)	88.4	113.0	132.2	147.2
Dimethyl sulfite			40.2(17)				
Dimethyl sulfone			77.0(29)				
Dimethyl sulfoxide	13.9	43.1	52.9(4)				
2,2-Dimethylthiacyclopropane			35.8				
Dimethylzinc			29.5				
Dinitromethane			46.0(42)				
2,4-Dinitrophenol			104.6(42)				

TABLE 6.2 Heats of Fusion, Vaporization, and Sublimation and Specific Heat at Various Temperatures of Organic Compounds (*Continued*)

Substance	ΔHm	ΔHv	ΔHs	C_p 400 K	600 K	800 K	1000 K
2,6-Dinitrophenol			112.1(42)				
1,1-Dinitropropane			62.5(6)				
1,3-Dioxane		34.4	39.1				
1,4-Dioxane, $\Delta Ht = 2.4^{-0.3}$	12.8	34.2	38.7(6)	126.5	181.8	218.2	243.3
1,3-Dioxolane			35.6(4)				
Diphenylamine			89.1(25)				
Diphenyl carbonate	23.4(13)		90.0(84)				
Diphenyl disulfide			95.0(29)				
Diphenyl disulfone			161.9(42)				
Diphenylenimine			84.5				
1,2-Diphenylethane		51.5	91.4(5)				
1,1-Diphenylethylene			73.2(42)				
Diphenyl ether	17.2	47.1	67.0				
6,6-Diphenylfulvene			104.6(84)				
Diphenylmercury			112.8				
Diphenylmethane	18.2(10)		67.5(3)				
1,3-Diphenyl-2-propanone			89.1(50)				
Diphenyl sulfide			67.8(21)				
Diphenyl sulfone			106.3(29)				
Diphenyl sulfoxide			97.1(29)				
Dipropylamine		33.5	40.0(1)				
Dipropyl disulfide	13.8	41.9	54.1(4)	186.2	298.3	350.2	390.0
Dipropyl ether		31.3	35.7(2)	196.2	262.0	311.3	348.0
Dipropylmercury			55.2				
Dipropyl sulfate			66.9(17)				
Dipropyl sulfide	12.1	36.6	44.6(2)	201.7	272.5	328.2	372.6
Dipropyl sulfite			58.6(17)				
Dipropyl sulfone			79.9(25)				
Dipropyl sulfoxide			74.5(13)				
Divinyl ether			26.2(8)				
Divinyl sulfone			56.5(8)				
Dodecane	36.8	43.7	61.3(3)	356.2	481.3	572.2	656.5
Dodecanedioic acid			153.1(29)				
Dodecanoic acid	36.7(15)		132.6(17)				
Dodecanol	31.4(63)	63.5	92.0				
1-Dodecene, $\Delta Ht = 4.6^{-60.2}$	20.0	43.0	60.8(21)	341.8	460.0	545.6	608.8
1,2-Epoxybutane		32.0					
1,2-Epoxypropane		21.6					
Ergosterol			118.4				
Ethane	2.9	14.7	9.8	65.5	89.3	108.0	122.6
Ethane-d_6				81.7	108.5	127.4	140.5
1,2-Ethanediamine	19.4	38.0	45.5(5)				
1,2-Ethanediol	11.6	49.6	67.8(17)	113.2	136.9	166.9	
1,2-Ethanedithiol		37.9	44.7(1)				
Ethanethiol	5.0	26.8	27.3(2)	88.2	113.9	133.2	148.0
Ethanol	5.0	38.6	42.3(1)	81.0	107.5	126.9	141.5
Ethoxybenzene		41.1	51.0(1)				
2-Ethoxyethanol		39.2	48.2				
2-(2-Ethoxyethoxy)ethanol		47.5					
2-(2-Ethoxyethoxy)ethyl acetate		91.2					

TABLE 6.2 Heats of Fusion, Vaporization, and Sublimation and Specific Heat at Various Temperatures of Organic Compounds (*Continued*)

Substance	ΔHm	ΔHv	ΔHs	C_p 400 K	600 K	800 K	1000 K
2-Ethoxyethyl acetate			52.7				
1-Ethoxy-2-methoxyethane		34.3	39.9				
Ethyl acetate	10.5	31.9	35.1(2)	137.4	182.6	213.4	234.5
Ethyl acrylate		34.6					
Ethylamine		28.0	26.6(4)	90.6	119.6	141.8	158.5
N-Ethylaniline			52.3(42)				
Ethylbenzene	9.2	35.6	42.3(1)	170.5	236.1	281.0	312.8
2-Ethylbenzoic acid			100.7(4)				
3-Ethylbenzoic acid			99.1(4)				
4-Ethylbenzoic acid			97.5(4)				
2-Ethyl-1-butanol		43.2	63.2				
Ethyl butanoate		35.5	42.7				
2-Ethylbutanoic acid		51.2					
2-Ethyl-1-butene		28.8	31.1(15)	170.3	228.0	269.5	300.8
Ethyl (E)-2-butenoate			44.4(13)				
Ethyl chloroacetate		40.4	49.5				
Ethyl 4-chlorobutanoate			52.7(42)				
Ethyl chloroformate			42.3(21)				
Ethyl cinnamate		58.6					
Ethyl crotonate			44.3				
Ethyl cyanoacetate		64.4					
Ethylcyclobutane		28.7	32.7(11)				
Ethylcyclohexane	8.3	34.0	40.2(4)	215.9	310.0	377.0	423.8
1-Ethylcyclohexene			43.3(2)				
Ethylcyclopentane	6.9	32.0	36.5(2)	183.6	258.2	314.7	356.3
1-Ethylcyclopentene		38.5(3)					
Ethyl 2,2-dimethylpropanoate			41.2(1)				
Ethylene	3.4	13.5		53.1	70.7	83.8	93.9
Ethylene-d_4				63.9	82.3	95.6	104.9
Ethylene carbonate	10.1	50.1	73.2				
2,2'-(Ethylenedioxy)bis(ethanol)		71.4	79.1				
Ethylene oxide	5.2	25.5	24.9(1)	62.6	86.3	102.9	114.9
Ethylenimine		30.3	34.6(6)	70.4	98.6	117.7	131.6
Ethyl formate	9.2	29.9	32.1				
2-Ethylhexanal			49.0(13)				
2-Ethylhexane		33.6	39.7(1)				
2-Ethylhexanoic acid		56.0					
2-Ethyl-1-hexanol	45.2						
2-Ethylhexyl acetate		43.5	48.1				
2-Ethyl hydroperoxide			43.1(42)				
Ethylidenecyclohexane			42.0(3)				
Ethylidenecyclopentane		18.1(9)					
Ethyl isocyanide			33.5(17)				
Ethyl isopentanoate	8.7	43.9					
Ethyl isopentyl ether		33.0	39.0				
Ethyl isopropyl ether		28.2	30.3				
Ethyl isopropyl sulfide	8.7	32.7	37.8				
Ethyl lactate		46.4	49.4				
Ethyllithium			116.7				
Ethylmercury bromide			76.6				

TABLE 6.2 Heats of Fusion, Vaporization, and Sublimation and Specific Heat at Various Temperatures of Organic Compounds (*Continued*)

Substance	ΔHm	ΔHv	ΔHs	C_p			
				400 K	600 K	800 K	1000 K
Ethylmercury chloride			76.1				
Ethylmercury iodide			79.5				
1-Ethyl-2-methylbenzene	10.0	38.9	47.7(2)	202.9	275.3	326.8	363.6
1-Ethyl-3-methylbenzene	7.6	38.5	46.9(2)	198.7	273.6	325.5	363.2
1-Ethyl-4-methylbenzene	13.4	38.4	46.6(2)	197.5	272.0	324.7	362.2
Ethyl 2-methylbutanoate			44.4(13)				
Ethyl 3-methylbutanoate		37.0	43.9(13)				
2-Ethyl-3-methyl-1-butene			34.5(2)				
Ethyl methyl ether		26.7		109.1	144.7	172.3	193.2
3-Ethyl-2-methylpentane	11.3	32.9	38.5(1)				
3-Ethyl-3-methylpentane	10.8	32.8	38.0(1)				
3-Ethyl-2-methyl-1-pentene			37.5(2)				
Ethyl methyl sulfide	9.8	29.5	32.0(1)	116.4	152.3	179.6	200.6
Ethyl nitrate	8.5	33.1	36.3(4)	120.2	155.1	178.7	195.4
1-Ethyl-2-nitrobenzene			59.8(21)				
1-Ethyl-4-nitrobenzene			62.8(21)				
3-Ethylpentane	9.6	31.1	35.2(1)	211.0	285.9	340.7	381.6
Ethyl pentanoate		37.0	46.1(14)				
2-Ethylphenol			63.6(10)				
3-Ethylphenol			68.2(5)				
4-Ethylphenol			80.3(5)				
Ethylphosphonic acid			50.6				
Ethylphosphonic dichloride			42.7				
Ethyl propanoate		33.9	39.1(1)				
Ethyl propyl ether		28.9	31.4(2)				
Ethyl propyl sulfide	10.6	34.2	40.0(6)	173.3	232.7	279.0	315.6
S-Ethyl thiolacetate	34.4	40.0					
Ethyl 2-vinylacrylate			48.5				
Ethyl vinyl ether		26.8	26.6(13)				
Fluoranthrene			99.2(8)				
Fluorobenzene	11.3	31.2	34.6(1)	125.5	171.0	200.1	220.0
4-Fluorobenzoic acid			91.2(13)				
Fluoroethane				74.1	98.6	116.4	129.7
Fluoromethane		16.7		44.2	57.9	68.8	77.2
1-Fluoropropane				102.7	137.3	162.7	181.5
2-Fluoropropane				103.5	138.7	163.8	182.2
2-Fluorotoluene		35.4					
4-Fluorotoluene	9.4	34.1	39.4(1)	152.4	207.9	245.2	271.3
Fluorotrichloromethane		25.0					
Fluorotrinitromethane			34.7				
Formaldehyde		23.3		39.3	48.2	55.9	62.0
Formamide	6.7	75.4	60.2				
Formic acid	12.7	22.7	46.3(5)	53.8	67.0	76.8	83.5
Formyl fluoride		21.7		46.4	56.2	63.1	67.9
Fumaric acid			136.0(63)				
Fumaronitrile			72.0				
2-Furaldehyde	14.4	43.2					
Furan, $\Delta Ht = 2.1^{-123.2}$	3.8	27.1	27.7(1)	88.7	122.6	164.9	158.5
2-Furancarboxaldehyde			50.6(4)				
2-Furancarboxylic acid			108.5(21)				

TABLE 6.2 Heats of Fusion, Vaporization, and Sublimation and Specific Heat at Various Temperatures of Organic Compounds (*Continued*)

Substance	ΔHm	ΔHv	ΔHs	C_p 400 K	600 K	800 K	1000 K
Furanmethanol	13.1	53.6	64.4(17)				
Glycerol	18.5	61.0	85.8(21)				
Glyceryl triacetate			82.0(8)				
Glyceryl tributanoate			107.1				
Glyceryl trinitrate			100.0(46)				
Heptadecane, $\Delta Ht = 11.0^{11.1}$	40.5	52.9	86.0	501.4	676.8	803.7	897.9
Heptadecanoic acid	58.8(23)						
1-Heptadecene	31.4	51.8	85.0	486.9	655.5	777.1	866.9
1-Heptanal	23.6		47.7(13)	213.4	283.3	333.9	371.1
Heptane	14.0	31.8	36.6(1)	211.0	285.9	340.7	381.6
1-Heptanenitrile			51.9(8)				
1-Heptanethiol	25.4(1)	39.8	50.6(2)	233.5	312.1	372.0	418.4
Heptanoic acid			74.0(20)				
1-Heptanol	13.2	48.1	66.8(2)	224.4	300.9	357.0	392.5
2-Heptanol		49.8					
2-Heptanone		39.8	47.2				
4-Heptanone		36.2					
1-Heptene, $\Delta Ht = 0.3^{-136}$	12.4	31.1	35.6(2)	196.5	264.6	314.1	351.0
Heptylamine			50.0				
Heptyl methyl ether			46.9				
Hexachlorobenzene	25.5		92.6(84)	201.2	233.4	250.9	260.8
Hexachloroethane, $\Delta Ht = 8.0^{71.3}$	9.8	45.9	59.0(17)	151.5	166.6	173.6	177.3
Hexadecafluoroethylcyclohexane			38.5				
Hexadecafluoroheptane			36.4(4)				
Hexadecane	51.8	51.2	81.4(4)	472.3	687.7	757.4	846.0
Hexadecanoic acid	53.6(20)		154.4(42)				
1-Hexadecanol, $\Delta Ht = 16.6^{34}$	35.4		169.5(21)	485.7	652.7	773.6	863.2
1-Hexadecene	30.2	50.4	80.2(4)	457.9	616.4	731.82	815.0
Hexadienoic acid	13.6						
Hexafluoroacetone		19.8	21.3				
Hexafluorobenzene	11.6	31.7	35.6(4)	183.6	219.9	241.1	253.7
Hexafluoroethane, $\Delta Ht = 3.7^{-169.2}$	2.7	16.2		125.6	149.0	160.7	166.8
(E)-Hexahydroindane			56.1				
(Z)-Hexahydroindane			57.5				
Hexamethylbenzene $\Delta Ht = 1.1^{-156.7}$ $\Delta Ht = 1.8^{110.7}$	20.6	48.2	74.7(21)	310.4	406.4	474.9	525.3
1,1,1,3,3,3-Hexamethyldisilazane			41.4				
Hexamethyldisiloxane			37.2				
Hexanal				184.2	243.9	287.4	319.7
Hexanamide	25.1(13)		98.7(17)				
Hexane	13.1	28.9	31.6(8)	181.9	246.8	294.4	330.1
1,6-Hexanedioic acid	16.7		129.3(25)				
1,6-Hexanediol	25.5(4)		83.3(17)				
Hexanenitrile		38.0	47.9				
1-Hexanethiol	18.0(1)	37.2	45.8(2)	204.5	273.1	325.1	366.7
Hexanoic acid	15.1	71.1	72.2(20)				

TABLE 6.2 Heats of Fusion, Vaporization, and Sublimation and Specific Heat at Various Temperatures of Organic Compounds (*Continued*)

Substance	ΔHm	ΔHv	ΔHs	C_p 400 K	600 K	800 K	1000 K
1-Hexanol	15.4	44.5	61.7(2)	195.3	261.8	310.7	346.9
2-Hexanol		41.0	58.5				
3-Hexanol	44.3	46.0					
2-Hexanone		36.4	42.3(4)				
3-Hexanone		35.4	41.9(2)				
1-Hexene	9.4	28.3	30.7(16)	167.5	225.5	267.9	299.3
(E)-2-Hexene		28.9	31.6(13)	166.1	223.4	266.1	297.9
(Z)-2-Hexene		29.1	31.6(13)	161.5	221.8	165.3	297.9
(E)-3-Hexene		28.9	31.7(13)	168.2	225.5	267.4	298.7
(Z)-3-Hexene		28.7	31.4(13)	161.1	222.6	265.7	297.9
Hexylamine		35.4	41.9(2)				
1-Hexyne				158.5	207.5	243.3	270.1
Hydrazine	12.7	45.3					
2-Hydroxybenzaldehyde		38.2					
2-Hydroxybenzoic acid			95.1(1)				
2-Hydroxy-2,4,6-cyclo-heptatrienone			83.7				
2-Hydroxy-1-isopropyl-4-methylbenzene			91.2				
4-Hydroxy-4-methyl-2-pentan-one		28.5	47.7				
3-Hydroxypropanonitrile		56.1					
2-Hydroxypyridine			86.6(13)				
3-Hydroxypyridine			88.3(13)				
4-Hydroxypyridine			103.8(17)				
8-Hydroxyquinoline			108.8(17)				
Icosane	69.9	57.5	100.8	588.5	794.0	942.6	1052.7
Icosanoic acid	72.0(28)		199.6(75)				
1-Icosene	34.3	55.9	99.8	574.0	772.7	916.0	1021.7
Indane		39.6	49.2(6)				
Indene			52.9(8)				
Indole			69.9(8)				
Iodobenzene	9.8	39.5	47.7(42)	130.1	173.3	201.1	220.1
4-Iodobenzoic acid			87.9(42)				
1-Iodobutane		34.7	40.7				
Iodocyclohexane			47.3(17)				
Iodoethane		29.4	31.9(1)	80.3	103.1	119.9	132.4
1-Iodohexane			49.8				
Iodomethane		27.3	28.0(13)	51.6	63.9	73.1	80.2
1-Iodo-2-methylpropane		33.5	38.9				
2-Iodo-2-methylpropane	14.5	31.4	35.4(1)	148.8	191.7	221.1	242.3
1-Iodonaphthalene			72.4(59)				
2-Iodonaphthalene			90.8(67)				
1-Iodopentane			45.3				
1-Iodopropane		32.1	36.2(1)	109.9	142.7	166.5	184.2
2-Iodopropane		30.7	34.1(1)	111.2	144.7	168.2	185.5
3-Iodopropane			38.1				
3-Iodo-1-propene			38.1				
2-Iodotoluene (also 3-, 4-)			54.4				
Isobutanonitrile		32.4	37.2	119.5	156.4	183.0	202.5
Isobutyl acetate		35.9					

TABLE 6.2 Heats of Fusion, Vaporization, and Sublimation and Specific Heat at Various Temperatures of Organic Compounds (*Continued*)

Substance	ΔHm	ΔHv	ΔHs	C_p 400 K	600 K	800 K	1000 K
Isobutylamine		30.6	33.9(2)				
Isobutylbenzene	12.5	37.8	48.3(1)				
Isobutylcyclohexane			47.6				
Isobutyl dichloroacetate			52.3				
Isobutyl formate		33.6					
Isobutyl isobutanoate		38.2	46.4				
Isobutyl isopropyl ether		31.6	36.6				
Isobutyl methyl ether		28.2	30.3				
Isobutyl propyl ether		28.3	30.3				
Isobutyl trichloroacetate			53.1(42)				
Isobutyl vinyl ether		30.7	34.6				
2-Isopropoxyethanol		40.4	50.1				
Isopropyl acetate		32.9	37.2(2)				
Isopropylamine		27.2	28.5(2)				
Isopropylbenzene	7.8	37.5	45.1(1)	200.8	277.0	328.9	365.3
Isopropylcyclopentane		34.7	41.1				
Isopropylmethylamine		28.7	30.9				
1-Isopropyl-2-methylbenzene	10.0	38.4	50.6				
1-Isopropyl-3-methylbenzene	13.7	38.1	50.0				
1-Isopropyl-4-methylbenzene	9.7	38.2	50.2				
Isopropyl methyl ether		26.1	26.8	138.0	184.8	220.4	247.2
2-Isopropyl-5-methylphenol			91.2(42)				
Isopropyl methyl sulfide	9.4	30.7	34.2(1)	145.1	192.5	229.9	260.6
Isopropyl nitrate		34.9	38.8(4)	150.5	195.9	226.5	247.9
Isopropyl trichloroacetate			51.9(42)				
Ketene			20.4(6)	59.5	70.7	78.7	86.4
(−)-Leucine			150.6(8)				
(+)-Limonene			48.1(21)				
Maleic acid			110.0(25)				
Maleic anhydride			71.5(50)				
Malononitrile			79.1				
D-Mannitol	22.6						
Methacrylonitrile		31.8					
Methane	0.94	8.2		40.5	52.2	62.9	71.8
Methane-d_4				48.6	63.4	74.8	83.0
Methanethiol, $\Delta Ht = 0.22^{-135.6}$	5.9	24.6	23.8(1)	58.7	73.5	85.0	94.1
Methanol, $\Delta Ht = 0.6^{-115.8}$	3.2	35.2	37.4(1)	51.4	67.0	79.7	89.5
4-Methoxybenzaldehyde		56.8	64.5(2)				
Methoxybenzene		39.0	46.8(4)				
2-Methoxybenzoic acid			104.7(3)				
3-Methoxybenzoic acid			107.4(4)				
4-Methoxybenzoic acid			109.8(6)				
3-Methoxy-1-butanol		50.8					
2-Methoxyethanol	37.5	45.2					
2-(2-Methoxyethoxy)ethanol		46.6					
2-Methoxyethyl acetate		41.1	50.3				
2-Methoxy-1-propoxyethane		36.3	43.7				
2-Methoxytetrahydropyran			42.7(13)				
1-Methoxy-2,4,6-trinitrobenzene			133.1(21)				
N-Methylacetamide	4.4	59.4					

TABLE 6.2 Heats of Fusion, Vaporization, and Sublimation and Specific Heat at Various Temperatures of Organic Compounds (*Continued*)

Substance	ΔHm	ΔHv	ΔHs	C_p			
				400 K	600 K	800 K	1000 K
Methyl acetate		30.3	34.2(16)				
Methyl acetoacetate		36.0					
Methyl acrylate		34.5	29.2(4)				
Methylamine	6.1	25.6	24.3(2)	60.2	78.9	93.9	105.7
Methyl benzoate		43.2	55.6(1)				
1-Methylbicyclo[4.1.0]heptane			39.2(4)				
1-Methylbicyclo[3.1.0]hexane		31.1	34.8(4)				
2-Methyl-1,3-butadiene	4.8	25.9	26.8(3)	133.1	173.2	200.8	221.3
3-Methyl-1,3-butadiene		27.2	28.0	129.7	168.6	197.5	219.2
2-Methylbutane	5.2	24.7	25.2(2)	152.7	208.7	249.8	280.8
3-Methylbutanenitrile		35.1	41.7				
2-Methylbutanethiol		33.8	39.5				
3-Methyl-1-butanethiol	7.5		39.4(2)				
2-Methyl-2-butanethiol, $\Delta Ht = 8.0^{-114.0}$	0.6	31.4	35.7(1)	179.0	236.7	279.4	308.8
Methyl butanoate		39.3	46.7				
3-Methylbutanoic acid	7.3	43.2	57.5(30)				
2-Methyl-1-butanol		43.9	54.6(13)				
3-Methyl-1-butanol		44.1	55.0(13)				
2-Methyl-2-butanol, $\Delta Ht = 2.0^{-127.2}$	4.9	39.0	48.7(13)				
3-Methyl-2-butanol		41.4	51.3(13)				
3-Methyl-2-butanone		32.4	36.9(2)				
2-Methyl-1-butene	7.9	25.5	25.9(3)	138.9	187.1	222.4	248.7
3-Methyl-1-butene	5.4	24.1	23.8(1)	147.5	192.1	225.3	250.3
2-Methyl-2-butene	7.6	26.3	27.1(1)	133.6	181.7	217.8	245.0
Methyl 2-butenoate			41.0(13)				
3-Methyl-1-butyne		26.2	25.8	130.1	169.9	198.3	219.2
2-Methylbutyl acetate		37.5					
Methyl chloroacetate		39.3	46.7				
Methyl cyanoacetate		48.2	61.7				
Methyl cyclobutanecarboxylate		37.1	39.7(4)				
Methylcyclohexane	6.8	31.3	35.4(3)	185.6	269.7	329.5	371.5
1-Methylcyclohexanol		35.2	80				
(E)-2-Methylcyclohexanol		53.1	63.2(21)				
(Z)-2-Methylcyclohexanol		48.5	63.2(21)				
(E)-3-Methylcyclohexanol			65.3(21)				
(Z)-3-Methylcyclohexanol			65.3(21)				
(E)-4-Methylcyclohexanol			66.1(21)				
(Z)-4-Methylcyclohexanol			65.7(21)				
1-Methylcyclohexene			37.9(9)				
Methylcyclopentane	6.9	29.1	31.7(8)	151.1	219.4	267.8	303.1
1-Methyl-1-cyclopentene			32.6(3)	136.0	195.8	238.5	269.0
3-Methyl-1-cyclopentene			31.0(12)	136.4	197.1	239.3	269.9
4-Methyl-1-cyclopentene			32.2(17)	136.4	196.7	238.4	269.5
Methyl cyclopropane-carboxyl-ate		35.3	41.3				
Methyl decanoate			66.7(10)				
Methyldichlorosilane			28.0				
Methyl 2,2-Dimethylpropanoate			35.7(4)				
2-Methyl-1,3-dioxane			38.6(13)				

TABLE 6.2 Heats of Fusion, Vaporization, and Sublimation and Specific Heat at Various Temperatures of Organic Compounds (*Continued*)

Substance	ΔHm	ΔHv	ΔHs	C_p 400 K	600 K	800 K	1000 K
4-Methyl-1,3-dioxane			39.2(13)				
N-Methylethanediamine		37.6	45.2				
1-Methylethyl acetate		32.9	37.3				
1-Methylethyl thiolacetate		35.7	42.3				
N-Methylformamide			56.2				
Methyl formate	7.5	27.9	30.6(4)	81.6	105.4	121.8	133.9
Methyl 2-furancarboxylate			45.2(8)				
Methylglyoxal			38.1				
2-Methylheptane	11.9	33.3	39.7(1)				
3-Methylheptane	11.6	33.7	39.8(1)				
4-Methylheptane	10.8	33.4	39.8(1)				
Methyl heptanoate			51.2(10)				
2-Methylhexane	9.2	30.6	34.9(2)	211.0	285.9	340.7	381.6
3-Methylhexane		30.9	35.1(2)	212.0	285.9	340.7	381.6
Methyl hexanoate		38.6	48.0(10)				
5-Methyl-1-hexene			34.3(10)				
(E)-3-Methyl-3-hexene			35.9(11)				
(Z)-3-Methyl-3-hexene			36.5(11)				
Methylhydrazine	10.4	40.4(1)					
Methyl isobutanoate		32.6	37.4				
Methyl isocyanide			30.8(10)				
3-Methylisoxazole			41.0				
5-Methylisoxazole			41.0(4)				
Methylmercury bromide			67.8				
Methylmercury chloride			64.4				
Methylmercury iodide			65.3				
Methyl methacrylate		36.0	60.7				
Methyl 2-methylbutanoate			41.8(13)				
Methyl-3-methylbutanoate			41.0(13)				
1-Methylnaphthalene, $\Delta Ht = 5.0^{-32.4}$	4.9	46.0		212.3	292.0	345.1	381.6
2-Methylnaphthalene, $\Delta Ht = 5.6^{15.4}$	11.8	46.0	61.7(8)	211.2	290.0	343.2	381.2
Methyl nitrate	8.2	31.6	32.1(22)	91.5	115.2	131.7	143.1
Methyl nitrite		20.9	22.6	76.3	97.7	112.8	123.5
1-Methyl-4-nitrobenzene			79.1(25)				
2-Methylnonane		38.2	49.6				
3-Methylnonane		38.3	49.7				
5-Methylnonane		58.1	49.4				
Methyl octanoate			56.4				
2-Methylpentane	6.3	27.8	29.8(10)	184.1	211.7	296.2	331.4
3-Methylpentane		28.1	30.3(10)	181.9	246.9	294.6	330.1
2-Methyl-2,4-pentanediol		51.5					
Methyl pentanoate		35.4	43.1(17)				
2-Methylpentanoic acid		52.1	57.5(30)				
2-Methyl-1-pentanol		46.2	55.7				
3-Methyl-1-pentanol		46.3	62.3				
4-Methyl-1-pentanol		44.5	60.5				
2-Methyl-2-pentanol		39.6	54.8				
3-Methyl-2-pentanol		43.4	56.9				
4-Methyl-2-pentanol		45.6	50.6				

TABLE 6.2 Heats of Fusion, Vaporization, and Sublimation and Specific Heat at Various Temperatures of Organic Compounds (*Continued*)

Substance	ΔHm	ΔHv	ΔHs	C_p 400 K	600 K	800 K	1000 K
2-Methyl-3-pentanol		41.8	54.4				
3-Methyl-3-pentanol		41.8					
3-Methyl-2-pentanone		34.2	40.6				
4-Methyl-2-pentanone		34.5	40.0(6)				
2-Methyl-1-pentene		28.1	30.6(16)	170.7	227.6	269.5	300.4
3-Methyl-1-pentene		26.9	28.7(15)	177.8	232.6	272.8	302.5
4-Methyl-1-pentene		27.1	28.7(18)	162.8	221.3	264.0	296.2
2-Methyl-2-pentene		29.0	31.6(15)	163.2	222.6	245.2	297.5
(E)-3-Methyl-2-pentene		29.3	31.5(13)	163.2	222.6	265.3	297.5
(Z)-3-Methyl-2-pentene		28.8	31.2(15)	163.2	222.6	265.3	297.5
(E)-4-Methyl-2-pentene		28.0	30.0(14)	171.1	229.3	269.9	300.4
(Z)-4-Methyl-2-pentene		27.6	29.5(12)	167.6	226.4	267.8	299.2
4-Methyl-3-penten-2-one			36.1	214.0			
Methyl pentyl ether		32.0	36.9				
Methyl pentyl sulfide		37.4	45.2(13)	203.6	272.2	324.6	366.0
3-Methyl-1-phenyl-1-butanone			59.5(17)				
2-Methyl-1-phenylpropane	12.5	37.8	49.5				
Methyl phenyl sulfide			54.3(2)				
Methyl phenyl sulfone			92.0(29)				
Methylphosphonic acid			48.1				
2-Methylpiperidine			40.5(2)				
2-Methylpropanal			31.5(13)				
2-Methylpropane	4.5	21.3	19.3(10)	124.6	169.5	202.9	227.6
2-Methylpropanenitrile			35.6(4)				
2-Methyl-1-propanethiol	5.0	31.0	34.7(1)	147.7	193.6	225.0	247.6
2-Methyl-2-propanethiol	2.5	28.5	30.9(2)	151.2	199.2	232.3	256.2
$\Delta Ht = 4.1^{-121.6}$							
$\Delta Ht = 0.7^{-116.2}$							
$\Delta Ht = 1.0^{-73.8}$							
Methyl propanoate		32.2	36.0				
2-Methylpropanoic acid	5.0	44.4	53.0				
2-Methyl-1-propanol	6.3	41.8	50.8(1)				
2-Methyl-2-propanol,	6.7	39.1	46.7(2)	142.9	189.8	222.9	247.5
$\Delta Ht = 0.8^{13}$							
2-Methylpropene	5.9	22.1	20.6(6)	111.2	147.7	175.1	196.0
Methyl propyl ether		26.8	27.8(2)	138.1	183.8	218.7	244.8
Methyl propyl sulfide	9.9	32.1	36.3(1)	144.9	191.9	227.8	255.8
2-Methylpyridine	9.7	36.1	42.5(1)	133.6	186.4	222.6	243.3
3-Methylpyridine	14.2	37.4	44.4(1)	133.1	186.1	222.3	247.8
4-Methylpyridine		37.5	44.9(4)				
1-Methyl-1*H*-pyrrole			40.8(2)				
Methyl salicylate		46.7					
α-Methylstyrene				187.4	254.0	300.4	333.9
(E)-β-Methylstyrene				189.1	256.1	301.3	334.7
(Z)-β-Methylstyrene				187.4	254.0	300.4	333.9
Methyl tetradecanoate			37.0				
2-Methylthiacyclopentane		36.4	41.8				
4-Methylthiazole		37.6	43.8(4)				
2-Methylthiophene	9.5	33.9	38.9(1)	123.1	165.6	194.3	214.6
3-Methylthiophene	10.5	34.2	39.5(2)	122.9	164.6	192.3	211.7

TABLE 6.2 Heats of Fusion, Vaporization, and Sublimation and Specific Heat at Various Temperatures of Organic Compounds (*Continued*)

Substance	ΔHm	ΔHv	ΔHs	C_p 400 K	600 K	800 K	1000 K
Methyl tridecanoate			82.7				
Methyl undecanoate			71.4				
5-Methyluracil			134.1(42)				
Morpholine		37.1	44.0				
Naphthalene	19.0	43.3	72.6(10)	179.2	249.7	296.1	327.9
1-Naphthalenecarboxylic acid			110.4				
2-Naphthalenecarboxylic acid			113.6				
1-Naphthol			91.2(4)				
2-Naphthol			94.2(5)				
1,4-Naphthoquinone			72.4(38)				
1-Naphthylamine			90.0(42)				
2-Naphthylamine			88.3(42)				
2-Nitroaniline	16.7(8)		90.0(42)				
3-Nitroaniline	23.8(8)		96.7(13)				
4-Nitroaniline	21.1		109				
Nitrobenzene	11.6	40.8	55.0(2)				
1-Nitrobutane		38.9	48.6(5)	157.5	210.1	247.0	273.6
2-Nitrobutane		36.8	43.8(4)	157.4	211.1	248.7	276.0
Nitroethane	9.9	38.0	41.6(4)	99.0	131.6	154.0	170.2
Nitromethane	9.7	34.0	38.4(1)	70.3	91.7	106.9	117.9
(Nitromethyl)benzene			53.6(13)				
1-Nitronaphthalene			107.1(21)				
1-Nitropropane		36.8	43.4(4)	128.5	171.0	200.7	222.0
2-Nitropropane		36.8	41.3(4)	129.2	172.3	201.8	222.8
2-Nitroso-1-naphthol			56.5(42)				
4-Nitroso-1-naphthol			87.4(42)				
1-Nitroso-2-naphthol			86.6(42)				
2-Nitrotoluene		16.5	47.2				
3-Nitrotoluene		15.0	49.9				
4-Nitrotoluene		15.5	50.2				
Nonadecane, $\Delta Ht = 13.8^{22.8}$	45.8	56.0	95.8	559.4	754.9	896.3	1000.8
1-Nonadecene	33.5	54.6	94.9	545.0	733.7	869.7	969.9
1-Nonal			72.3	271.1	361.5	426.4	474.5
Nonane, $\Delta Ht = 6.3^{-56.0}$	15.6	36.9	46.4(1)	269.0	364.1	433.3	484.9
1-Nonanethiol	33.5	44.4		291.6	390.3	464.6	521.5
Nonanoic acid			82.4(4)				
1-Nonanol		54.4	76.9(8)	282.4	379.1	449.6	501.7
2-Nonanone			56.0(4)				
1-Nonene	18.0	36.3	45.5	254.6	342.8	406.8	454.0
(E)-Octadecafluorodeca-hydronaphthalene		35.8	45.4				
(Z)-Octadecafluorodeca-hydronaphthalene		35.6	45.2				
Octadecafluoropropylcyclo-hexane		24.5	43.1				
Octadecafluorooctane		33.4	41.2				
Octadecane	62.0	54.5	152.8(50)	530.4	715.8	850.0	949.4
Octadecanoic acid	63.0(25)		166.5(42)				
Octadecanol			113.4				
1-Octadecene	32.6	53.3	90.0	516.0	694.5	823.4	918.4

TABLE 6.2   Heats of Fusion, Vaporization, and Sublimation and Specific Heat at Various
Temperatures of Organic Compounds (*Continued*)

Substance	ΔHm	ΔHv	ΔHs	C_p			
				400 K	600 K	800 K	1000 K
Octafluorocyclobutane	2.8	23.4		186.1	225.3	245.4	257.3
Octamethylcyclotetrasiloxane		45.6					
Octanal				242.3	322.2	380.3	422.6
Octanamide			110.5(29)				
Octane	20.7	34.4	41.5(1)	240.0	325.0	387.0	433.5
1,8-Octanedioic acid			143.1(38)				
Octanenitrile		41.3	56.8(3)				
1-Octanethiol	24.3	42.3		262.6	351.3	418.3	469.9
Octanoic acid	13.8	70.0	81.7(13)				
1-Octanol	42.3	46.9	71.1(4)	253.4	340.0	403.3	450.1
(±)-2-Octanol		44.4					
(±)-3-Octanol		36.5					
4-Octanol		40.5					
1-Octene	15.3	34.1	40.6(2)	225.6	303.7	360.5	402.5
1-Octyne		35.8	42.3	216.5	285.7	336.0	410.9
2-Octyne		37.3	44.5				
3-Octyne		36.9	44.0				
4-Octyne		36.0	42.8				
Oxalic acid			98.0(21)				
Oxaloyl chloride			31.8(42)				
Oxamide			113.0(21)				
2-Oxohexamethyleneimine	16.2	54.8	83.3				
2,2′-Oxybisethanol			57.3(59)				
Paraldehyde			41.4				
Pentachloroethane	11.3	37.2	45.6(21)	133.7	152.1	162.0	168.1
Pentachlorofluoroethane	1.9						
Pentachlorophenol			67.4(21)				
Pentacyclo[4.2.0.0^{2,5}.0^{3,8}.0^{4,7}]-octane			80.3(17)				
Pentadecane, $\Delta Ht = 9.2^{-2.25}$	34.8	49.5	76.1	443.3	598.6	711.1	794.5
Pentadecanoic acid	50.2(2)		162.7(42)				
1-Pentadecene	28.9	48.7	75.1	428.9	577.3	684.5	763.6
1,2-Pentadiene		27.6	28.7	131.4	170.7	199.6	220.9
(E)-1,3-Pentadiene		27.0	27.8	130.5	171.1	199.6	220.1
(Z)-1,3-Pentadiene		27.6	28.3	123.4	166.9	196.7	218.4
1,4-Pentadiene	6.1	25.2	25.7	131.0	170.2	220.5	
2,3-Pentadiene		28.2	29.5	125.1	164.9	195.0	217.6
Pentaerythritol		92	143.9(8)				
Pentaerythritol tetranitrate			151.9(21)				
Pentafluorobenzene	10.8(1)	32.2	35.3(20)				
Pentafluorobenzoic acid			91.6(42)				
Pentafluoroethane				113.8	137.8	151.1	158.9
Pentafluorophenol	16.4(1)		67.4(17)				
Pentamethylbenzene, $\Delta Ht = 2.0^{23.7}$	12.3	45.1	60.8	272.0	360.2	423.8	470.0
Pentanal			38.8(16)	155.2	205.0	241.4	267.8
Pentanamide			89.3(4)				
Pentane	8.4	25.8	26.7(2)	152.8	207.7	248.1	278.5
1,5-Pentanedithiol			59.3(4)				
2,4-Pentanedione		34.3	43.2(10)				

TABLE 6.2 Heats of Fusion, Vaporization, and Sublimation and Specific Heat at Various Temperatures of Organic Compounds (*Continued*)

Substance	ΔHm	ΔHv	ΔHs	C_p 400 K	600 K	800 K	1000 K
Pentanenitrile	4.7	36.1	43.6(2)				
1-Pentanethiol	17.5	34.9	41.1(1)	175.4	234.0	279.4	315.1
Pentanoic acid	14.2	44.1	62.4(30)				
1-Pentanol	9.8	44.4	56.9(2)	166.3	222.8	264.4	295.4
2-Pentanol		41.4	52.6(13)				
3-Pentanol		42.3	51.7(13)				
2-Pentanone		33.4	38.2(2)	152.4	202.2	239.0	266.1
3-Pentanone	11.6	33.5	38.6(1)				
1-Pentene	5.8	25.2	25.5(3)	138.5	186.4	221.5	247.7
(E)-2-Pentene	8.4	26.1	26.7(1)	136.7	184.2	219.5	246.1
(Z)-2-Pentene	7.1	26.1	26.8(1)	132.1	182.5	218.8	245.9
(E)-2-Pentenenitrile		37.8	44.9(2)				
(Z)-2-Pentenenitrile		36.4	43.2(2)				
(E)-2-Pentenenitrile		37.1	44.8(2)				
Pentyl acetate		41.0					
Pentylamine		34.0	40.2				
Pentyl propyl ether		35.0	42.8				
1-Pentyne		27.7	28.4	130.1	169.0	197.1	218.4
2-Pentyne		29.3	30.8	122.2	161.9	192.1	215.1
α-Phellandrene			50.6(2)				
Phenanthrene		55.7	91.8(45)				
9,10-Phenanthrenedione			91.6(46)				
Phenazine			99.9(25)				
Phenol	11.5	45.9	68.7(5)	135.8	182.2	211.8	232.2
Phenyl acetate			54.8(17)				
Phenylacetonitrile		52.9					
Phenylacetylene			41.8	150.4	200.9	233.4	255.9
(−)-3-Phenyl-1-alanine			155.2				
Phenyl benzoate			99.0(25)				
Phenylboron dichloride			33.9				
Phenylcyclopropane			50.2(1)				
N-Phenyldiacetimide			90.0				
Phenyl formate			52.9(10)				
Phenylhydrazine			61.7(8)				
1-Phenyl-1-propanone			58.5(17)				
1-Phenyl-2-propanone			49.0(2)				
Phenyl salicylate			92.1(42)				
Phenyl vinyl ether			49.9(17)				
Phthalamide			57.3(42)				
1,3-Phthalic acid			106.7				
1,4-Phthalic acid			98.3				
Phthalic anhydride			88.7(3)				
Phthalonitrile			86.9(15)				
Piperidine		31.7	39.3(2)				
Propadiene		18.6		72.0	92.1	106.4	117.2
Propanal		28.3	29.7(4)	96.6	126.4	148.3	164.0
Propanamide	17.6(8)		85.9(17)				
Propane	3.5	19.0	16.3	94.0	128.7	154.8	174.6
1,2-Propanediamine			44.2(2)				
1,3-Propanediamine		40.9	50.2				

TABLE 6.2 Heats of Fusion, Vaporization, and Sublimation and Specific Heat at Various
Temperatures of Organic Compounds (*Continued*)

Substance	ΔHm	ΔHv	ΔHs	C_p			
				400 K	600 K	800 K	1000 K
Propanedinitrile			79.1(8)				
1,2-Propanediol		54.1	58.0(41)				
1,3-Propanediol		57.9	37.1(20)				
1,2-Propanedione			38.1(21)				
1,2-Propanedithiol			49.7(1)				
Propanenitrile, $\Delta Ht = 1.7^{-96.2}$	5.0	31.8	36.0(4)	88.6	114.7	134.5	149.4
1-Propanethiol, $\Delta Ht = 4.0^{-131.1}$	5.5	29.5	32.0(1)	116.6	153.6	182.4	205.1
2-Propanethiol	5.7	27.9	29.6(1)	118.6	154.9	181.0	200.5
1,2,3-Propanetriol triacetate			82.0				
Propanoic acid	7.5	32.3	35.0(20)				
Propanoic anhydride		41.7	52.6(21)				
1-Propanol	5.2	41.4	47.4(1)	108.2	144.6	171.7	192.2
2-Propanol	5.4	39.9	45.4(2)	112.0	149.6	176.3	195.9
Propanolactone			47.0				
2-Propenal		28.3	31.3				
Propene	3.0	18.4		80.5	108.0	128.7	144.4
2-Propen-1-ol		40.0	47.3(13)	95.4	126.0	147.6	163.4
(Z)-1-Propenylbenzene				187.4	254.0	300.4	333.9
2-Propoxyethanol		41.4	52.1				
Propyl acetate		33.9	39.2				
1-Propylamine		29.6	31.3(2)	119.3	159.0	188.0	210.1
Propylbenzene	9.3	38.2	46.2(1)	200.1	275.6	327.6	364.7
Propyl benzoate		49.8	51.9				
Propyl carbamate			81.2(21)				
Propyl chloroacetate			48.5(42)				
Propylcyclohexane	10.4	36.1	44.7(4)	247.3	350.6	423.4	474.5
Propylcyclopentane	10.0	34.7	41.1(1)	212.7	297.2	361.0	407.9
Propylene oxide	6.5	27.4	28.3	92.7	125.8	149.3	166.5
Propyl formate		33.6	37.6				
Propyl nitrate		35.9	40.6(4)	149.8	194.5	225.4	247.2
Propyl propanoate		35.5	43.5				
Propyl trichloroacetate			53.1(42)				
Propyl vinyl ether			29.3(21)				
Propyne		22.1		72.5	91.2	105.2	115.9
2-Propyn-1-ol		42.1					
Pyrazine			56.3(5)				
Pyridazine			53.5(4)				
Pyridine	8.3	35.1	40.2(1)	106.4	149.5	177.8	197.4
Pyrimidine		43.1	50.0(4)				
1*H*-Pyrrole	7.9	38.8	45.2(1)				
Pyrrolidine, $\Delta Ht = 0.5^{-66}$	8.6	33.0	37.6(1)	114.4	168.7	206.5	233.6
Quinoline	10.8	49.7					
Salicylic acid			95.1(1)				
5,5'-Spirobis(1,3-dioxane)			72.8				
Spiro[2.2]pentane	6.4	26.8	27.5(2)	119.5	167.8	200.5	223.9
(E)-Stilbene			99.2(4)				
(Z)-Stilbene			69.0(13)				
Styrene	11.0	37.0(5)	43.9(4)	160.3	218.2	256.9	284.2
Succinic acid			117.5(33)				
1,1,3,3-Tetrabromoethane		48.7	70.0				

TABLE 6.2 Heats of Fusion, Vaporization, and Sublimation and Specific Heat at Various Temperatures of Organic Compounds (*Continued*)

Substance	ΔHm	ΔHv	ΔHs	C_p 400 K	600 K	800 K	1000 K
Tetrabromomethane		45.1	110	97.1	102.6	106.7	105.9
Tetrabutyltin			19.8				
Tetracene			125.5				
Tetrachloro-1,4-benzoquinone			98.7				
1,1,2,2-Tetrachloro-1,2-difluoroethane		35.0					
1,1,1,2-Tetrachloroethane				118.7	139.2	151.6	159.7
1,1,2,2-Tetrachloroethane		37.6	45.5(5)	116.7	137.7	150.0	158.0
Tetrachloroethylene	10.5	34.7	39.7(8)	105.0	116.6	122.6	125.8
Tetrachloromethane, $\Delta Ht = 4.6^{-47.9}$	2.5	29.8	32.4(1)	91.7	99.7	103.1	104.8
Tetracyanoethylene			81.2(49)				
Tetracyanomethane			61.1(88)				
Tetradecane	45.6	47.6	71.3	414.3	559.5	664.8	743.1
Tetradecanoic acid	44.7(17)		139.8(38)				
1-Tetradecanol	49.0(9)		102.2(23)				
1-Tetradecene	27.6	46.9	70.2	399.8	538.2	638.2	712.1
Tetraethylene glycol		62.6	98.7(100)				
Tetraethylgermanium			44.8				
Tetraethyllead			56.9				
Tetraethyltin			51.0				
1,1,1,2-Tetrafluoroethane				104.2	128.7	143.1	152.1
Tetrafluoroethylene	7.7	16.8		91.9	106.8	115.5	120.8
Tetrafluoromethane, $\Delta Ht = 1.5^{-196.9}$	0.7	12.6		72.4	86.8	94.5	98.8
Tetrahydrofuran		29.8	32.0(1)				
Tetrahydrofuran-2,5-dimeth-anol		63.6					
Tetrahydrofuran-2-methanol		45.3	51.6				
1,2,3,4-Tetrahydronaphthalene	12.5	43.9	55.2(13)				
Tetrahydropyran		31.2	34.9(8)				
Tetrahydropyran-2-methanol		44.4					
Tetraiodomethane				100.4	104.4	105.9	106.7
Tetramethoxysilane		194.6					
1,2,3,4-Tetramethylbenzene	11.2	45.0	57.2	237.7	316.7	374.1	416.2
1,2,3,5-Tetramethylbenzene	10.7	43.8	53.7	233.3	313.0	371.5	414.3
1,2,4,5-Tetramethylbenzene	21.0	45.5	53.4	232.2	311.2	369.9	413.0
2,2,3,3-Tetramethylbutane, $\Delta Ht = 2.0^{-120.7}$	7.5	31.4	43.4(2)				
Tetramethylene sulfone	1.4	61.5					
Tetramethyllead			38.1				
Tetramethyltin			33.1				
1,1,3,3-Tetramethylurea		45.6					
Tetranitromethane		40.7	43.5(21)				
Tetraphenylmethane			150.6(42)				
Tetraphenyltin			66.3				
Tetrapropylgermanium			61.5				
Tetrapropyltin			66.9				
1,2,3,4-(1H)-Tetrazole			97.5(42)				
Thiacycloheptane			47.3	175.7	272.0	330.5	368.2

TABLE 6.2　Heats of Fusion, Vaporization, and Sublimation and Specific Heat at Various Temperatures of Organic Compounds (*Continued*)

Substance	ΔHm	ΔHv	ΔHs	C_p 400 K	600 K	800 K	1000 K
Thiacyclohexane	2.5	36.0	41.6	149.4	219.1	267.8	302.7
$\quad \Delta Ht = 1.1^{-71.8}$							
$\quad \Delta Ht = 7.8^{-33.1}$							
Thiacyclopentane	7.4	34.7	39.5	121.1	167.5	199.4	222.3
Thiacyclopropane		29.2	30.3	69.2	92.0	107.2	118.0
Thioacetamide			83.3(3)				
Thioacetic acid			37.2	93.1	111.8	127.2	136.5
1,2-Thiocresol			51.5				
2,2'-Thiodiethanol		66.8					
Thiophene, $\Delta Ht = 0.6^{-101.6}$	5.1	31.5	34.7(1)	96.3	129.5	150.7	165.4
Thiophenol	11.5	39.9	47.6	137.1	184.6	215.9	237.6
Toluene	6.6	33.2	38.7(6)	140.1	197.5	236.9	264.9
1,2-Toluidine		44.6	56.7				
1,3-Toluidine	3.9	44.9	57.3				
1,4-Toluidine	17.3	41.2					
2,4,6-Triamino-1,3,5-triazine			124.3				
Tribromomethane		39.7	46.1	78.7	88.0	93.3	96.7
Tributoxyborane		56.2	52.3				
Tributyl phosphate		61.4	72.0				
Trichloroacetonitrile		34.1					
Trichloroacetyl chloride			41.0(2)				
Trichlorobenzoquinone			88.7				
1,1,1-Trichloroethane,	1.9	29.9	32.5(1)	107.6	128.4	141.1	149.8
$\quad \Delta Ht = 7.5^{-49.0}$							
1,1,2-Trichloroethane	11.4	34.8	40.3(1)	104.7	126.1	139.2	148.2
Trichloroethylene		31.4	35.5(17)	91.2	104.9	112.7	117.8
Trichloromethane	9.5	29.2	30.5(4)	74.3	85.3	91.5	95.5
1,2,3-Trichloropropane		8.9	11.2	31.7	38.9	43.8	47.3
1,1,1-Trichlorotrifluoroethane		26.9	28.3				
1,1,2-Trichlorotrifluoroethane		27.0	28.6				
1,1,1-Trichloro-3,3,3-trifluoropropane		32.2	36.8				
Tricyanoethylene			81.2(50)				
Tridecane, $\Delta Ht = 7.7^{-18.2}$	28.5	45.7	66.4	385.2	520.4	618.5	691.2
Tridecanoic acid	43.1(18)		146.4(21)				
1-Tridecene	25.9	45.0	65.3	370.8	499.1	592.0	660.2
Triethanolamine	27.2	67.5					
Triethoxyborane			43.9				
Triethoxymethane			46.0(8)				
Triethylaluminum			73.2				
Triethylamine		31.0	34.9(1)	203.8	276.6	328.7	367.4
Triethylaminoborane			60.7				
Triethylarsine			43.1				
Triethyl arsenite			50.6				
Triethylbismuthine			46.0				
Triethylborane			36.8				
Triethylenediamine,	6.1		61.9				
$\quad \Delta Ht = 9.6^{79.8}$							
Triethylene glycol		71.4	79.1(79)				
Triethylphosphine			39.8				

TABLE 6.2 Heats of Fusion, Vaporization, and Sublimation and Specific Heat at Various Temperatures of Organic Compounds (*Continued*)

Substance	ΔHm	ΔHv	ΔHs	C_p			
				400 K	600 K	800 K	1000 K
Triethyl phosphate			57.3				
Triethyl phosphite			41.8				
Triethylstibine			43.5				
Trifluoroacetic acid, ΔH(dimer dissoc) $= 58.8^{100}$		33.3	38.5(8)				
Trifluoroacetonitrile	5.0						
1,1,1-Trifluoroethane	6.2	19.2		95.2	118.7	133.8	144.1
2,2,2-Trifluoroethanol		40.0(5)					
Trifluoroethylene				81.1	97.5	107.5	113.9
Trifluoromethane	4.1	16.7		61.1	76.0	85.1	91.0
(Trifluoromethyl)benzene	13.8	32.6	37.6(1)	169.8	226.8	262.6	286.4
Triiodomethane	16.3		69.9	82.0	90.0	94.7	97.8
Triisopropylborane			41.8				
Triisopropyl phosphite			46.0				
Trimethoxyborane			34.7				
1,1,1-Trimethoxyethane			39.2(21)				
Trimethoxymethane			38.1(10)				
2′,4′,5′-Trimethylacetophenone			63.2				
2′,4′,6′-Trimethylacetophenone			62.3				
Trimethylaluminum			63.2				
Trimethylamine	6.5	22.9	22.0(1)	117.5	160.4	190.9	213.3
Trimethyl arsenite			42.3				
Trimethylarsine			28.9				
1,2,3-Trimethylbenzene	8.2	40.0	49.1(1)	196.2	267.8	320.9	359.4
$\Delta Ht = 0.7^{-54.5}$							
$\Delta Ht = 1.3^{-42.9}$							
1,2,4-Trimethylbenzene	13.2	39.3	47.9(1)	196.5	269.0	321.9	360.2
1,3,5-Trimethylbenzene	9.5	39.0	47.5(1)	194.2	268.1	321.5	360.1
2,6,6-Trimethylbicyclo[3.1.1]-2-heptene			44.8(13)				
Trimethylbismuthine			34.7				
Trimethylborane			20.2				
2,2,3-Trimethylbutane,	2.3	28.9	32.0(1)	212.7	291.3	346.1	386.3
$\Delta Ht = 2.5^{-151.8}$							
2,3,3-Trimethyl-1-butene			32.2(2)				
(Z),(Z)-1,3,5-Trimethylcyclohexane				242.9	351.2	427.6	482.0
Trimethylene sulfide,	8.3	32.3	36.0	91.6	127.4	152.3	170.2
$\Delta Ht = 0.7^{-96.5}$							
Trimethylgallium			38.1				
2,2,5-Trimethylhexane	6.2	33.7	40.2				
2,3,5-Trimethylhexane		34.4	41.5				
Trimethylindium			48.5				
2,4,7-Trimethyloctane		38.2	49.9				
2,2,3-Trimethylpentane	8.6	31.9	36.9(1)				
2,2,4-Trimethylpentane	9.2	30.8	35.1(1)				
2,3,3-Trimethylpentane,	0.9	32.1	37.2(1)				
$\Delta Ht = 7.7^{-109.0}$							
2,3,4-Trimethylpentane	9.3	32.4	37.7(1)				
2,2,4-Trimethyl-1,3-pentanediol	8.6	55.7					

TABLE 6.2 Heats of Fusion, Vaporization, and Sublimation and Specific Heat at Various Temperatures of Organic Compounds (*Continued*)

Substance	ΔHm	ΔHv	ΔHs	C_p 400 K	600 K	800 K	1000 K
2,2,4-Trimethyl-3-pentanone		35.6	43.3(1)				
2,4,4-Trimethyl-1-pentene		31.4	35.8(2)				
2,4,4-Trimethyl-2-pentene		32.6	37.5(2)				
Trimethylphosphine			28.0				
Trimethylphosphine oxide			50.2				
Trimethyl phosphate			36.8				
2,3,6-Trimethylpyridine		40.0	50.6				
2,4,6-Trimethylpyridine		39.9	50.3				
Trimethylsilanol			45.6				
Trimethylstibine			31.4				
Trimethylsuccinic anhydride			74.1				
Trimethylthiacyclopropane			39.3				
Trimethyltin bromide			47.3				
2,4,6-Trinitroanisole			133.1				
1,3,5-Trinitrobenzene	16.7(4)		99.6(21)				
Trinitromethane		32.6(4)	46.7(21)				
2,4,6-Trinitrophenetole			120.5				
2,4,6-Trinitrotoluene			104.7(17)				
1,3,6-Trioxacycloactane			48.8(2)				
1,3,5-Trioxane			56.6(2)				
Triphenylarsine			99.3				
Triphenylbismuthine			110.9				
Triphenylborane			81.6				
Triphenylene			118.0(42)				
Triphenylmethane			100.0(4)				
Triphenylphosphine			96				
Triphenylstibine			106.3				
Tripropoxyborane			49.4				
Tris(diethylamino)phosphine			60.7				
Tris(trimethylsilyl)amine			54.4				
Tropolone			83.7				
Undecane, $\Delta Ht = 6.9^{-36.6}$	22.1	41.5	56.3(2)	327.1	442.7	525.9	588.3
Undecanoic acid	25.9		121.3				
1-Undecene, $\Delta Ht = 9.2^{-55.8}$	17.0	40.9	55.4	312.7	421.1	499.3	557.3
Uracil			126.5(22)				
Urea	15.1	87.9					
(−)-Valine			162.8				
Vinyl acetate		34.4	34.8				
Vinyl benzene			39.6(2)				
Vinylcyclohexane			39.7(3)				
4-Vinyl-1-cyclohexene		33.5	38.3(8)				
1,2-Xylene	13.6	36.2	43.4(1)	171.7	234.2	278.8	311.1
1,3-Xylene	11.6	36.7	42.7(1)	167.5	232.2	277.9	310.6
1,4-Xylene	17.1	35.7	42.4(1)	166.1	230.8	276.7	309.7

TABLE 6.3 Enthalpies and Gibbs Energies of Formation, Entropies, and Heat Capacities of the Elements and Inorganic Compounds

Substance	State	$\Delta Hf°$, kJ·mol^{-1}	$\Delta Gf°$, kJ·mol^{-1}	$S°$, J·deg^{-1}·mol^{-1}	$C_p°$, J·deg^{-1}·mol^{-1}
Aluminum					
Al	c	0	0	28.28(8)	24.21
Al^{3+} std. state	aq	−531.37	−485.34	−321.75	
Al$_6$BeO$_{10}$	c	−5 624(5)	−5 317	175.6(3)	265.19
Al(BH$_4$)$_3$	lq	−16.3	144.8	289.1	194.6
AlBr$_3$	c	−511.8(8)	−488.5	180.2(11)	100.58
std. state	aq	−895	−799	−74.5	
Al$_4$C$_3$	c	−216(7)	−203	89(1)	
Al(CH$_3$)$_3$	lq	136.4	−10.0	209.4	155:6
Al(OAc)$_3$	c	−1 892.4			
AlCl$_3$	c	−705.63	−630.02	109.29	91.13
std. state	aq	−1 033	−878	−152	
AlCl$_3$·6H$_2$O	c	−2 692	−2 269	377	
AlF$_3$	c	−1 510.4(13)	−1 431.1	66.5(4)	75.13
AlF$_3$·3H$_2$O	c	−2 297	−2 052	209	
AlI$_3$	c	−303(2)	−301	196.1	98.9
std. state	aq	−699	−640	12.1	
AlK(SO$_4$)$_2$·12H$_2$O	c	−6 061.8	−5 141.7	687.4	651.0
AlN	c	−318.1(25)	−287.0	20.14(21)	30.10
Al(NO$_3$)$_3$ std. state	aq	−1 155	−820	117.6	
Al(NO$_3$)$_3$·6H$_2$O	c	−2 850.5	−2 203.9	467.8	433.0
Al(NO$_3$)$_3$·9H$_2$O	c	−3 757.1	−2 929.6	569	
AlO$_2^-$ std. state	aq	−918.8	−823.4	−21	
Al$_2$O$_3$ corundum	c	−1 676.0(12)	−1 582.3	50.95(8)	79.15
Al(OH)$_3$	c	−1 284	−1 306	71	93.1
Al(OH)$_4^-$ std. state	aq	−1 490.3	−1 297.9	117	
AlP	c	−166.5			
AlPO$_4$ berlinite	c	−1 692.0	−1 618.0	90.79	93.18
Al$_2$S$_3$	c	−651.0(38)	−640	116.85(21)	105.06
Al$_2$Se$_3$	c	−565			
Al$_2$SiO$_5$ andalusite	c	−2 592.0(20)	−2 444.8	93.2(4)	122.76
Al$_2$(SO$_4$)$_3$	c	−3 435	−3 507	239.3	259.4
std. state	aq	−3 790	−3 205	−583.3	
Al$_2$Te$_3$	c	−326			
Americium					
Am	c	0	0	62.7	
Am^{3+}	aq	−682.8	−671.5	−159.0	
Am^{4+}	aq	−511.7	−461.1	−372	
Am$_2$O$_3$	c	−1 757	−1 678	154.7	
AmO$_2$	c	−1 005.0	950.2	96.2	
Ammonium					
NH$_3$	g	−45.9(4)	−16.4	192.774(25)	35.65
undissoc; ss	aq	−80.29	−26.57	111.3	
ND$_3$	g	−58.6(4)	−26.0	203.9(50)	38.23
NH$_4^+$ std. state	aq	132.51	79.37	113.4	79.9
NH$_4$OH undissoc; ss	aq	−366.12	−263.76	181.2	
ionized; ss	aq	−362.50	−236.65	102.5	−68.6
NH$_4$ acetate	c	−616.14			
std. state	aq	−618.52	−448.78	200.0	73.6

TABLE 6.3 Enthalpies and Gibbs Energies of Formation, Entropies, and Heat Capacities of the Elements and Inorganic Compounds (*Continued*)

Substance	State	$\Delta Hf°$, kJ·mol^{-1}	$\Delta Gf°$, kJ·mol^{-1}	$S°$, J·deg^{-1}·mol^{-1}	$C_p°$, J·deg^{-1}·mol^{-1}
$NH_4Al(SO_4)_2$	c	−2 352.2	−2 038.4	216.3	226.44
std. state	aq	−2 481	−2 054	−168.2	
NH_4AsO_2 std. state	aq	−561.54	−429.41	154.8	
$NH_4H_2AsO_3$ std. state	c	−847.30	−666.60	223.8	
$NH_4H_2AsO_4$	c	−1 059.8	−833.0	172.05	151.17
std. state	aq	−1 042.07	−832.66	230.5	
$(NH_4)_2HAsO_4$ ss	aq	−1 171.1	−873.20	225.1	
$(NH_4)_3AsO_4$ ss	aq	−1 286.7	−886.63	177.4	
NH_4Br	c	−271.5	−176.1	112.84	88.66
std. state	aq	−254.05	−183.34	194.97	−61.9
NH_4BrO_3	aq	−199.58	−60.84	275.10	
NH_4 carbamate	c	−657.60	−448.07	133.5	
NH_4Cl	c	−314.5(8)	−203.1	95(2)	86.4
std. state	aq	−299.66	−210.62	169.9	−56.5
NH_4ClO_3 std. state	aq	−236.48	−87.40	275.7	
NH_4ClO_4	c	−295.8(2)	−88.6	184(1)	128.1
std. state	aq	261.84	−87.99	295.4	
$(NH_4)_2CO_3$ ss	aq	−942.15	−686.64	169.9	
NH_4CN std. state	aq	18.0	92.9	207.5	
NH_4CNO cyanate ss	aq	−278.7	−177.0	220.1	
$(NH_4)_2CrO_4$	c	−1 167.3			
std. state	aq	−1 144.3	−886.59	277.0	
NH_4 dithiocarbonate	c	−126.8			
NH_4F	c	−463.96	−348.78	71.97	65.27
std. state	aq	−465.14	−358.19	99.6	−26.8
NH_4 formate ss	aq	−558.06	−430.5	205	−8.0
NH_4I	c	−202.09	−111.97	113.0(63)	81.8
std. state	aq	−187.69	−130.96	224.7	−62.3
NH_4IO_3	c	−385.8			
std. state	aq	−354.0	−207.5	231.8	
NH_4N_3 azide	c	115.5	274.1	112.6	
	aq	142.7	268.6	221.3	
NH_4NO_3	c	−365.56	−184.01	151.08	139.3
std. state	aq	−339.87	−190.71	259.8	−6.7
$(NH_2)_2C_2O_4$ oxalate	c	−1 124.32			
$NH_4H_2PO_4$	c	−1 145.07	−1 210.56	151.96	142.26
std. state	aq	−1 428.79	−1 209.76	203.8	
$(NH_4)_2HPO_4$	c	−1 556.91			
std. state	aq	−1 557.16	−1 248.00	193.3	
$(NH_4)_3PO_4$	c	−1 671.9			
std. state	aq	−1 674.9	−1 256.9	117	
$(NH_4)_4P_2O_7$ ss	aq	−2 801.2	−2 236.8	335	
$(NH_4)_2PtCl_6$	c	−803.3			237.7
NH_4ReO_4	c	−945.6	−774.9	232.6	
NH_4HS	c	−156.9	−50.6	97.5	
$(NH_4)_2S$	aq	−231.8	−72.8	212.1	
NH_4HSO_4	c	−1 026.96			
std. state	aq	−1 019.85	−835.38	245.2	−3.8
$(NH_4)_2SO_4$	c	−1 180.25	−901.90	220.1	187.49
std. state	aq	−1 174.28	−903.37	246.9	−133.1

TABLE 6.3 Enthalpies and Gibbs Energies of Formation, Entropies, and Heat Capacities of the Elements and Inorganic Compounds (*Continued*)

Substance	State	$\Delta Hf°$, kJ·mol^{-1}	$\Delta Gf°$, kJ·mol^{-1}	$S°$, J·deg^{-1}·mol^{-1}	$C_p°$, J·deg^{-1}·mol^{-1}
$(NH_4)_2S_2O_8$	c	−1 648.08			
std. state	aq	−1 610.0	−1 273.6	471.1	
$(NH_4)_2SiF_6$	c	−2 681.69	−2 365.55	280.24	228.11
NH_4VO_3	c	−1 053.1	−888.3	140.6	129.33
Antimony					
Sb	c	0	0	45.22	25.02
$SbBr_3$	c	−259.8		179.9	108.53
$SbCl_3$	c	−382.0		187.03	106.86
$SbCl_5$	lq	−440.16	−350.2	301	
SbF_3	c	−915.5			
SbH_3	g	145.11	147.74	232.67	41.05
SbI_3	c	−96.2		215.5	97.57
Sb_2O_3	c	−708.8		123.01	101.25
Sb_2O_5	c	−971.9		125.1	117.61
Sb_2S_3	c	−174.9		182.0	117.74
Sb_2Te_3	c	−56.5	−55.2	234	
Argon	g	0	0	154.845(3)	20.79
Arsenic					
As	c	0	0	35.1	24.64
$AsBr_3$	g	−130	159	363.76	79.16
$AsCl_3$	lq	−305.0	−259.4	216.3	133.5
	g	−261.5	−245.0	327.06	75.73
AsH_3	g	66.44	88.91	222.67	38.07
AsI_3	c	−58.2	−59.4	213.05	105.77
As_2O_5	c	−924.87	−782.4	105.4	116.5
As_4O_6 octahedral	c	−1 313.94	−1 152.52	214.2	191.29
As_2S_3	c	−169.0	−168.6	163.6	116.3
Astatine					
At	c	0	0	121.3	
Barium					
Ba	c	0	0	62.48	28.10
Ba^{2+} std. state	aq	−537.64	−560.74	9.6	
Ba acetate	c	−1 484.5			
std. state	aq	−1 509.67	−1 299.55	182.8	
$BaBr_2$	c	−757.7(17)	−738.0	149(4)	77.0
std. state	aq	−780.73	−768.68	174.5	
$BaBr_2 \cdot 2H_2O$	c	−1 366.1	−1 230.5	226	
$Ba(BrO_3)_2$	c	−752.66	−577.4	243	
BaC_2O_4 oxalate	c	−1 368.6			
$BaCl_2$	c	−859(13)	−810	123.67(13)	75.14
$BaCl_2 \cdot 2H_2O$	c	−1 460.13	−1 296.45	202.9	161.96
$Ba(ClO_3)_2$	c	−762.7			
$Ba(ClO_3)_2 \cdot 3H_2O$	c	−1 691.6	−1 270.7	393	
$BaCO_3$ witherite	c	−1 216.3	−1 137.6	112.1	85.35
$BaCrO_4$	c	−1 446.0	−1 345.3	158.6	
BaF_2	c	−1 208.8(42)	−1 158.4	96.4(4)	72.20
std. state	aq	−1 202.90	−1 118.38	−17.0	
$Ba(HCO_3)_2$ std. state	aq	−1 921.63	−1 734.4	192.1	
$Ba(H_2PO_2)_2$	c	−1 762.3			

TABLE 6.3 Enthalpies and Gibbs Energies of Formation, Entropies, and Heat Capacities of the Elements and Inorganic Compounds (*Continued*)

Substance	State	$\Delta Hf°$, kJ·mol^{-1}	$\Delta Gf°$, kJ·mol^{-1}	$S°$, J·deg^{-1}· mol^{-1}	$C_p°$, J·deg^{-1}· mol^{-1}
BaI_2	c	−605.4(33)	−601.4	165.1(4)	77.49
std. state	aq	−648.02	−663.92	232.2	
$Ba(IO_3)_2$	c	−1 027.2	−864.8	249.4	187.4
std. state	aq	−980.3	−816.7	246.4	
$BaMnO_4$	c	−1 548	−1 439.7	138	140.6
$BaMoO_4$	c	−1 507.5	−1 439.7	144.3	114.7
$Ba(NO_2)_2$	c	−768.2			
$Ba(NO_3)_2$	c	−992.07	−796.72	213.8	151.38
std. state	aq	−952.36	−783.41	302.5	
BaO	c	−548.1(21)	−520.4	72.07(38)	47.28
BaO_2	c	−634.3			
$Ba(OH)_2$	c	−946.3	−859.5	107(8)	101.6
$Ba(OH)_2·8H_2O$	c	−3 342.2	−2 793.2	427	
BaS	c	−463.6(21)	−458.7	78.4(13)	49.37
$BaSe$	c	−372			
$BaSeO_3$	c	−1 040.6	−968.2	167	
$BaSiF_6$	c	−1 952.2	−2 794.1	163	
$BaSO_3$	c	−1 179.5			
$BaSO_4$	c	−1 473.19	−1 362.3	132.2	101.75
$BaTiO_3$	c	−1 659.8	−1 572.4	108.0	102.47
Beryllium					
Be	c	0	0	9.44(15)	16.38
Be^{2+} std. state	aq	−382.8	−379.7	−129.7	
$BeAl_2O_4$ chrysoberyl	c	−2 301.0(38)	−2 178.5	66.29(13)	105.38
$BeBr_2$	c	−356(13)	−337	100.4(42)	66.0
Be_2C	c	91(11)	−88	16.3(42)	43.2
$BeCl_2$ beta form	c	−496.2(33)	−449.5	75.81(21)	62.43
BeF_2 alpha form	c	−1 026.8(42)	−979.4	53.35(17)	51.82
BeI_2	c	−189(21)	−187	120.4(42)	69.0
Be_3N_2 cubic	c	−588.3(13)	−532.9	34.13	64.36
BeO alpha form	c	−608.4(33)	−579.1	13.77(21)	25.56
$3BeO·B_2O_3$	c	−3 105(8)	−2 939	100(13)	139.7
$Be(OH)_2$ beta form	c	−905.8(21)	−817.7	50.2(42)	65.7
BeS	c	−234.3(88)	−233.0	37.0(42)	34.0
$BeSeO_4$	c	−890.4			
std. state	aq	−982.0	−820.9	−75.7	
Be_2SiO_4	c	−2 117(29)	−2 003	64.19(40)	95.6
$BeSO_4$	c	−1 200.8(33)	−1 089.4	77.97	85.70
std. state	aq	−1 290.0	−1 124.3	−109.6	
$BeSO_4·4H_2O$	c	−2 423.75	−2 080.66	232.97	216.61
$BeWO_4$	c	−1 513(64)	−1 405	88.4(84)	97.3
Bismuth					
Bi	c	0	0	56.53	25.94
$BiBr_3$	c	264	234	226	109
$BiCl_3$	c	−379.1	−315.1	177.0	105
BiH_3	g	277.8			
BiI_3	c	−100.4	−175.3		
Bi_2O_3	c	−574.0		151.5	113.51
$BiOCl$	c	−366.9	−322.2	121	
Bi_2S_3	c	−176.6		155.6	128.53
Bi_2Te_3	c	−78.24		260.91	152.21

TABLE 6.3 Enthalpies and Gibbs Energies of Formation, Entropies, and Heat Capacities of the Elements and Inorganic Compounds (*Continued*)

Substance	State	$\Delta Hf°$, kJ·mol^{-1}	$\Delta Gf°$, kJ·mol^{-1}	$S°$, J·deg^{-1}·mol^{-1}	$C_p°$, J·deg^{-1}·mol^{-1}
Boron					
B	c	0	0	5.83	11.32
BBr$_3$	lq	−238.5(2)	−236.9	228.9	128.03
B$_4$C	c	−62.7(63)	−62.1	27.18(13)	53.76
BCl$_3$	g	−403.0(21)	−387.0	290.17	62.39
BF$_3$	g	−1 135.6(17)	−1 119.0	254.36(4)	50.45
BF$_4^-$ std. state	aq	−1 574.9	−1 487.0	179.9	
BH$_3$	g	107(10)	111	187.9	36.22
BH$_4^-$ std. state	aq	48.16	114.27	110.5	
B$_2$H$_6$ diborane(6)	g	41.0(167)	91.8	233.10	58.12
B$_5$H$_9$ pentaborane(9)	lq	42.8(67)	172	184.33	151.13
B$_{10}$H$_{14}$ decaborane(14)	c	−28.9(190)	216.1	353.0	221.2
BN	c	−250.9(15)	−225.0	14.80	19.72
B$_3$N$_3$H$_6$ borazine	g	−510	−389	288.61	96.94
BO$_2^-$ std. state	aq	−772.37	−678.94	−37.24	
B$_2$O$_3$	c	−1 271.9	−1 192.8	53.9(4)	62.59
B(OH)$_4$ std. state	aq	−1 344.03	−1 153.32	102.5	
B$_3$O$_3$H$_3$ boroxin	c	−1 262(42)	−11.56	167(42)	98.3
Bromine					
Br$^-$ std. state	aq	−121.55	−103.97	82.4	−141.8
Br$_2$	lq	0	0	152.21(4)	75.67
Br$_3^-$ std. state	aq	−130.42	−107.07	215.5	
BrCl	g	14.6	−0.96	239.91	34.98
BrF	g	−58.6	−73.6	288.86	32.97
BrF$_3$	lq	−301	−241		
BrF$_5$	lq	−458.6	−351.9		
	g	−428.9	−351.5	323.59	101.42
BrO$^-$ std. state	aq	−94.1	−33.5	42	
BrO$_3^-$ std. state	aq	−67.07	18.54	161.71	
Cadmium					
Cd	c	0	0	51.46	25.90
CdBr$_2$	c	−316.18	−296.31	137.2	
std. state	aq	−318.99	−285.52	91.6	
CdCl$_2$	c	−391.6		115.5	76.6
std. state	aq	−410.20	−340.12	39.8	
CdCl$_2$·½H$_2$O	c	−1 131.94	−944.08	227.2	
Cd(CN)$_2$	c	162.3			
std. state	aq	225.5	267.4	115.1	
CdCO$_3$	c	−750.6	−669.4	92.5	
Cd(C$_2$H$_3$O$_2$)$_2$ acetate	aq	−1 047.9	−816.4	100	
std. state					
CdF$_2$	c	−700.4	−647.7	77.4	
std. state	aq	−741.15	−635.21	−100.8	
CdI$_2$	c	−203.3	−201.4	161.1	80.0
std. state	aq	−186.3	−180.8	149.4	
CdI$_4^-$ std. state	aq	−341.8	−315.9	326	
Cd(NH$_3$)$_4^{2+}$ std. state	aq	−450.2	−226.4	336.4	
Cd(NO$_3$)$_2$	c	−456.3			
std. state	aq	−490.6	−300.2	219.7	
CdO	c	−255.6	−228.5	54.8	43.0
Cd(OH)$_2$	c	−560.7	−473.6	96	

TABLE 6.3 Enthalpies and Gibbs Energies of Formation, Entropies, and Heat Capacities of the Elements and Inorganic Compounds (*Continued*)

Substance	State	$\Delta Hf°$, kJ·mol^{-1}	$\Delta Gf°$, kJ·mol^{-1}	$S°$, J·deg^{-1}·mol^{-1}	$C_p°$, J·deg^{-1}·mol^{-1}
CdS	c	−149.4	−145.2	69.0	55.5
CdSO$_4$	c	−933.4	−822.4	123.0	100.4
std. state	aq	−985.2	−822.2	−53.1	
CdSO$_4$·⅔H$_2$O	c	−1 729.4	−1 465.3	229.6	213.3
CdSeO$_4$	c	−633.0	−531.8	164.4	
std. state	aq	−674.9	−518.8	−19.3	
CdTe	c	−92.5	−92.1	100	
Calcium					
Ca	c	0	0	43.07	25.95
Ca^{2+} std. state	aq	−542.83	−553.54	−53.1	
Ca(acetate)$_2$	c	−1 479.5			
std. state	aq	−1 514.73	−1 292.35	120.1	
Ca$_3$(AsO$_4$)$_2$	c	−3 298.7	−3 063.1	226	
Ca(BO$_2$)$_2$	c	−2 030.9	−1 924.1	104.85	103.98
CaB$_4$O$_7$	c	−3 360.3	−3 167.1	134.7	157.9
CaBr$_2$	c	−683.2(42)	−664.2	129.7(42)	75.04
std. state	aq	−785.9	−761.5	111.7	
CaC$_2$	c	−59(8)	−65	69.96	62.72
CaCN$_2$ cyanamide	c	−350.6			
Ca(CN)$_2$	c	−184.5			
CaCO$_3$	c	−1 206.92	−1 128.84	92.9	80.57
CaCl$_2$	c	−795.8(13)	−748.1	105(4)	
std. state	aq	−877.13	−816.05	59.8	
CaCl$_2$·2H$_2$O	c	−1 402.9			738
CaCrO$_4$	c	−1 379.1	−1 277.4	134	
CaF$_2$	c	−1 226(6)	−1 174	68.6(3)	68.6
Ca(formate)$_2$	c	−1 386.6			
CaH$_2$	c	−186.2	−147.3	42	
CaHPO$_4$·2H$_2$O	c	−2 403.58	−2 154.75	189.45	197.07
Ca(H$_2$PO$_2$)$_2$ hypophos-phite	c	−1 752.7			
Ca(H$_2$PO$_4$)$_2$ ss	aq	−3 135.41	−2 814.33	127.6	
Ca(H$_2$PO$_4$)$_2$·H$_2$O	c	−3 409.67	−3 058.42	259.8	258.82
CaI$_2$	c	−537(2)	−533	145.3(2)	77.16
std. state	aq	−653.2	−656.7	169.5	
Ca(IO$_3$)$_2$	c	−1 002.5	−839.3	230	
Ca[Mg(CO$_3$)$_2$] dolomite	c	−2 326.3	−2 163.6	155.18	157.53
CaMoO$_4$	c	−1 541.4	−1 434.7	122.6	114.3
Ca$_3$N$_2$	c	−439.3		105.0	113.0
Ca(NO$_2$)$_2$	c	−741.4			
Ca(NO$_3$)$_2$	c	−938.39	−743.20	193.3	149.37
std. state	aq	−957.55	−776.22	239.7	
CaO	c	−635.1(88)	−603.5	38.21(13)	42.12
Ca(OH)$_2$	c	−986(1)	−898	83.4(4)	87.5
Ca oxalate hydrate	c	−1 674.9	−1 514.0	156.5	152.8
Ca$_3$P$_2$	c	−506			
Ca$_3$(PO$_4$)$_2$	c	−4 120.8	−3 884.8	236.0	227.8
Ca$_2$P$_2$O$_7$	c	−3 338.8	−3 132.1	189.24	187.8
Ca$_{10}$(PO$_4$)$_6$F$_2$ fluoro-apatite	c	−13 744	−12 983	775.7	751.9

TABLE 6.3 Enthalpies and Gibbs Energies of Formation, Entropies, and Heat Capacities of the Elements and Inorganic Compounds (*Continued*)

Substance	State	$\Delta Hf°$, kJ·mol^{-1}	$\Delta Gf°$, kJ·mol^{-1}	$S°$, J·deg^{-1}·mol^{-1}	$C_p°$, J·deg^{-1}·mol^{-1}
CaS	c	−473(3)	−468	57.0(10)	47.4
CaSe	c	−368.2	−363.2	67	
CaSiO$_4$	c	−1 634.9	−1 549.7	81.92	85.27
Ca$_2$SiO$_4$	c	−2 307.5	−2 192.8	127.7	128.8
3CaO·SiO$_2$	c	−2 929.2	−2 784.0	168.6	171.9
CaSO$_3$·2H$_2$O	c	−1 752.7	−1 555.2	184	178.7
CaSO$_4$	c	−1 425.2	−1 309.1	108.4	99.0
CaSO$_4$·½H$_2$O	c	−1 576.7	−1 436.8	130.5	119.4
CaSO$_4$·2H$_2$O	c	−2 022.6	−1 797.5	194.1	186.0
Ca(VO$_3$)$_2$	c	−2 329.3	−2 169.7	179.1	166.8
CaWO$_4$	c	−1 645.15	−1 538.50	126.40	114.14
Carbon					
C graphite	c	0	0	5.74(21)	8.517
diamond	c	1.897	2.900	2.377	6.116
(CN)$_2$ cyanogen	g	308.9	297.2	242.3	56.9
CNN$_3$ cyanogen azide	c	387.4			
CNBr	c	136.0	165.1	248.36(4)	47.12
CNCl	g	137.95	131.02	236.33	45.27
CNI	c	160.3	169.4	128.9	
CO	g	−110.53(17)	−137.16	197.65(4)	29.14
CO$_2$	g	393.52(5)	1 394.39	213.80(12)	37.13
undissoc; ss	aq	−413.8	−386.0	117.6	
C$_3$O$_2$ suboxide	g	−93.7	−109.8	276.4	67.0
COCl$_2$ phosgene	g	−220.1(33)	−205.9	283.80	57.70
COF$_2$	g	−638.9(17)	−623.33	258.89(8)	47.26
COS carbonyl sulfide	g	−138.4(11)	−165.6	231.56	41.50
CS$_2$	g	117.1	66.9	237.8	45.48
Cerium					
Ce	c	0	0	72.0	26.9
Ce^{3+} std. state	aq	−696.2	−672.0	−205	
Ce^{4+} std. state	aq	−537.2	−503.8	−301	
CeCl$_3$	c	−1 053.5	−977.8	151	87.5
std. state	aq	−1 197.5	−1 065.7	38	
CeF$_3$	c	−1 635.9	−1 556	115.1	99.3
CeI$_3$	c	−649.8	−674	209	
Ce(NO$_3$)$_3$	c	−1 225.9			
CeO$_2$	c	−1 088.7	−1 024.7	62.30	61.63
Ce$_2$O$_3$	c	−1 796.2	−1 706.2	150.6	114.6
Ce$_2$(SO$_4$)$_3$	c	−3 954.3			
std. state	aq	−4 176.9	−3 652.6	−318	
Ce$_2$(SO$_4$)$_3$·8H$_2$O	c	−5 522.9	−5 607.4		
Cesium					
Cs	c	0	0	85.15(21)	32.20
	lq	2.087	0.025	92.1	32.4
Cs$^+$ std. state	aq	−258.3	−292.0	133.0	−10.5
CsBr	c	−405.68	−391.41	113.05	52.93
std. state	aq	−209.2	−241.0	267.3	37.07
Cs$_2$CO$_3$	c	−1 139.7	−1 054.4	204.5	123.9
std. state	aq	−1 193.7	−1 116.1	209.2	

TABLE 6.3 Enthalpies and Gibbs Energies of Formation, Entropies, and Heat Capacities of the Elements and Inorganic Compounds (*Continued*)

Substance	State	$\Delta Hf°$, kJ·mol^{-1}	$\Delta Gf°$, kJ·mol^{-1}	$S°$, J·deg^{-1}· mol^{-1}	$C_p°$, J·deg^{-1}· mol^{-1}
CsCl	c	−442.8(8)	414.4	101.18(2)	52.44
std. state	aq	−802.0	−423.3	−189.4	−146.9
CsF	c	−555(2)	−525	88(8)	52.0
std. state	aq	−590.9	−570.8	119.2	
CsI	c	−346.6	−340.6	123.1	52.8
std. state	aq	−313.5	−343.6	244.4	−152.7
CsIO$_3$	c	−525.9	−433.9		167
CsNO$_3$	c	−506.0	−406.6	155.2	
std. state	aq	−465.6	−403.3	279.5	−99.2
Cs$_2$O	c	−345.8	−308.2	146.9	76.0
CsOH	c	−416.7(8)	370.7	98.7(42)	67.9
std. state	aq	−488.3	−449.3	122.3	
Cs$_2$PtCl$_6$ std. state	aq	−1 184.9	−1 066.9	485.8	
Cs$_2$SO$_4$	c	−1 410.2	−1 306.4	264.8	108.7
std. state	aq	−1 425.8	−1 328.6	286.2	
Chlorine					
Cl$^-$ std. state	aq	−167.2	−131.3	56.4	−136.4
Cl$_2$	g	0	0	233.08(4)	33.95
ClF	g	−54.48	−55.94	217.82	32.08
ClF$_3$	g	−159(3)	−119	281.6	63.85
ClF$_5$	g	−239(63)	−147	310.74	97.17
ClO	g	101(2)	97	226.6	31.5
ClO$^-$ std. state	aq	−107.1	−36.8	41.8	
ClO$_2$	g	105(6)	122	257.2	42.00
ClO$_2^-$ std. state	aq	−66.5	17.2	101.3	
ClO$_3^-$ std. state	aq	−24.85	−1.92	38.8	
ClO$_3$F perchloryl fluoride	g	−21.4(29)	50.6	278.98(8)	64.92
ClO$_4^-$ std. state	aq	−129.33	−8.62	182.0	
Cl$_2$O	g	88(7)	105	267.9	47.81
Cl$_2$O$_7$	lq	238.1			
	g	1 138			
Chromium					
Cr	c	0	0	23.62(21)	23.43
Cr^{2+} std. state	aq	−143.5			
CrCl$_2$	c	−405.8		115.3	70.33
CrCl$_3$	c	−552.3		123.0	90.12
Cr(CO)$_6$ hexacarbonyl	c	−1 077.8		293.01	226.23
CrN	c	−117(8)	−93	38(2)	52.7
CrO$_4^{2-}$ std. state	aq	−881.15	−727.85	50.21	
HCrO$_4^-$ std. state	aq	−878.22	−764.84	184.1	
Cr$_2$O$_3$	c	−1 135(8)	−1 053	81.2(13)	120.4
Cr$_2$O$_7^{2-}$ std. state	aq	−1 490.3	−1 301.2	261.9	
Cr$_2$(SO$_4$)$_3$	c	−609.6		269.9	302.6
Cobalt					
Co	c	0	0	30.07(10)	24.80
Co^{2+} std. state	aq	−58.2	−54.4	−113	
Co^{3+} std. state	aq	92	134	−305	
CoBr$_2$	c	−220.9			
std. state	aq	−301.3	−262.3	50	

TABLE 6.3 Enthalpies and Gibbs Energies of Formation, Entropies, and Heat Capacities of the Elements and Inorganic Compounds (*Continued*)

Substance	State	$\Delta Hf°$, kJ·mol⁻¹	$\Delta Gf°$, kJ·mol⁻¹	$S°$, J·deg⁻¹· mol⁻¹	$C_p°$, J·deg⁻¹· mol⁻¹
$CoCl_2$	c	−312.5(17)	−269.6	109.3(2)	78.49
std. state	aq	−392.5	−316.7	0	
$CoCO_3$	c	−713.0			
CoF_2	c	−672(4)	−627	82.4(4)	68.9
CoF_3	c	−790(13)	−719	95(13)	92
$Co(NH_3)_6^{2+}$ std. state	aq	−584.9	−157.3	146	
$Co(NH_3)_6^{3+}$ std. state	aq		−189.5		
$Co(NO_3)_2$	c	−420.5			
std. state	aq	−472.8	−277.0	180	
CoO	c	−237.7(4)	−214.0	53.0(3)	55.3
Co_3O_4	c	−910(4)	−795	119(4)	123
$Co(OH)_2$	c	−539.7	−454.4	80	
$CoSO_4$	c	−888.3(13)	−782.4	117(4)	103
std. state	aq	−967.3	−799.1	−92	
$CoSO_4 \cdot 7H_2O$	c	−2 979.93	−2 473.83	406.06	390.49
Copper					
Cu	c	0	0	33.16(4)	24.44
Cu^+ std. state	aq	71.67	50.00	40.6	
Cu^{2+} std. state	aq	64.64	65.52	−99.6	
$Cu(acetate)_2$	c	−893.3			
std. state	aq	−907.25	−673.29	73.6	
$Cu_3(AsO_4)_2$ ss	aq	−1 581.97	−1 100.48	−804.2	
$CuBr$	c	−108.0	−100.8	96.2	54.77
$CuBr_2$	c	−141.84			
$CuCl$	c	−138(2)	−121	87(4)	48.5
$CuCl_2$	c	−206(6)	−162	108.09	71.88
$Cu(ClO_4)_2$ ss	aq	−193.89	48.28	264.4	
$CuCN$	c	95.0	108.4	90.00	61.04
$CuCNS$ std. state	aq	138.11	142.67	184.93	
$Cu(CNS)_2$ ss	aq	217.65	250.87	189.1	
CuF	c	−280(42)	−260	64.9(21)	51.9
CuF_2	c	−539(8)	−492	77.45(8)	65.55
$Cu(formate)_2$	aq	−786.34	−636.4	84	
CuI	c	67.8	−69.5	96.7	54.18
$Cu(NH_3)_4^{2+}$ ss	aq	−348.5	−111.3	273.6	
$Cu(NO_3)_2$	c	−302.9			
std. state	aq	−349.95	−157.15	193.3	
CuO	c	−156(2)	−128	42.6(4)	42.2
Cu_2O	c	−170.7(21)	−149.9	93.36(34)	52.54
$Cu(OH)_2$	c	−450(8)	−373	108.4	95.19
CuS	c	−48.5	−53.7	66.5	47.66
Cu_2S	c	−79.5	−86.2	120.9	81.59
Cu_2Se	c	−59.4		157.3	88.70
$CuSO_4$	c	−770.0(8)	−660.8	109.3(4)	98.87
std. state	aq	−844.50	−679.11	−79.5	
$CuSO_4 \cdot 5H_2O$	c	−2 279.65	−1 880.04	300.4	280
Dysprosium					
Dy	c	0	0	74.77	28.16
Dy^{3+} std. state	aq	−699	−665	−231.0	21
Dy_2O_3	c	−1 863.1	−1 771.5	149.8	116.27

TABLE 6.3 Enthalpies and Gibbs Energies of Formation, Entropies, and Heat Capacities of the Elements and Inorganic Compounds (*Continued*)

Substance	State	$\Delta Hf°$, kJ·mol^{-1}	$\Delta Gf°$, kJ·mol^{-1}	$S°$, J·deg^{-1}· mol^{-1}	$C_p°$, J·deg^{-1}· mol^{-1}
Erbium					
Er	c	0	0	73.18	28.12
Er^{3+} std. state	aq	−705.4	−669.0	−244.4	21
Er$_2$O$_3$	c	−1 897.9	−1 808.7	155.6	108.49
Europium					
Eu	c	0	0	77.78	27.66
Eu^{2+} std. state	aq	−523	540.2	4	
Eu^{3+}	aq	−605.0	−574.0	−222	8
Eu$_2$O$_3$ monoclinic	c	−1 651.4	−1 556.9	146	122.2
Fluorine					
F$_2$	g	0	0	154.6(6)	31.30
FNO$_3$	g	10.5(21)	73.7	292.9	65.22
Francium					
Fr	c	0	0	95.40	31.80
FrCl	c	−439		113.0	53.56
Fr$_2$O	c	−338	299.2	156.9	
Gadolinium					
Gd	c	0	0	68.07	37.03
Gd^{3+} std. state	aq	−686	−661	−205.9	0
GdCl$_3$	c	−1 008			
std. state	aq	−1 188	−1 059	−36.8	−410
Gd$_2$O$_3$ monoclinic	c	−1 819.6		150.6	106.7
Gallium					
Ga	c	0	0	40.8(2)	26.06
GaBr$_3$	c	−386.6	−359.8	180	
GaCl$_3$	c	−524.7	−454.8	142	
GaF$_3$	c	−1 163	−1 085.3	84	
Ga$_2$O$_3$ rhombic	c	−1 089.1	−998.3	84.98	80.67
GaSb	c	−41.8	−38.9	76.07	48.53
Germanium					
Ge	c	0	0	31.17	23.43
GeBr$_4$	lq	−347.7	−331.4	280.8	
GeCl$_4$	lq	−531.8	−462.8	245.6	
GeF$_4$	g	−1 189.9		302.8	81.84
GeH$_4$	g	90.8	113.4	217.02	45.02
GeI$_4$	c	−141.8	−144.4	271.1	
GeO$_2$	c	−554.8	−497.1	55.02	55.81
GeS	c	−69.0	−71.6	71	
Gold					
Au	c	0	0	47.32	25.36
AuCl	c	−34.7		92.9	48.74
AuCl$_3$	c	−115.06		148.1	94.81
AuCl$_4^-$ std. state	aq	−322.2	−237.32	266.9	
Au(CN)$_2^-$ std. state	aq	242.3	285.8	172	
AuF$_3$	c	−348.5		114.2	91.29
AuSb$_2$	c	−19.46		119.2	77.40
AuSn	c	−30.5		93.7	49.41

TABLE 6.3 Enthalpies and Gibbs Energies of Formation, Entropies, and Heat Capacities of the Elements and Inorganic Compounds (*Continued*)

Substance	State	$\Delta Hf°$, kJ·mol^{-1}	$\Delta Gf°$, kJ·mol^{-1}	$S°$, J·deg^{-1}· mol^{-1}	$C_p°$, J·deg^{-1}· mol^{-1}
Hafnium					
Hf hexagonal	c	0	0	43.56	25.69
HfC	c	-230.1		41.21	34.43
HfCl$_4$	c	-990.4		190.8	120.46
HfF$_4$ monoclinic	c	$-1\,930.5$	$-1\,830.5$	113	
HfO$_2$	c	$-1\,113.2$		59.37	60.25
Helium					
He	g	0	0	126.152(2)	20.786
Holmium					
Ho	c	0	0	75.3	27.15
Ho^{3+} std. state	aq	-705.0	-673.6	226.8	17
HoCl$_3$	c	$-1\,005.4$			88
std. state	aq	$-1\,206.7$	$-1\,067.3$	-57.7	-393
Ho$_2$O$_3$	c	$-1\,880.7$	$-1\,791.2$	158.2	115.0
Hydrogen					
H$^+$ std. state	aq	0	0	0	0
H$_2$	g	0	0	130.68(3)	28.84
H^{2}H	g	0.321(13)	-1.463	143.80(3)	29.20
^{2}H$_2$ deuterium	g	0	0	144.96(4)	29.19
HAsO$_2^-$ undissoc; ss	aq	-456.5	-402.71	125.9	
H$_2$AsO$_3^-$ undissoc; ss	aq	-714.79	-587.22	110.5	
H$_3$AsO$_3$ undissoc; ss	aq	-742.2	-639.90	195.0	
HAsO$_4^{2-}$ undissoc; ss	aq	-906.34	-714.70	-1.7	
H$_2$AsO$_4^-$ undissoc; ss	aq	-909.56	-753.29	117	
H$_3$AsO$_3$	c	-906.30			
undissoc; ss	aq	-902.5	-766.1	184	
HBO$_2$	c	$-802.8(8)$	-486.2	49.(4)	54.39
H$_3$BO$_3$	c	$-1\,094.0(8)$	-968.5	88.7(4)	81.34
HBr	g	-36.44	-53.51	198.61	29.16
std. state	aq	-121.55	-103.97	82.4	-141.8
HBrO undissoc; ss	aq	-113.0	-82.4	142	
HBrO$_3$ std. state	aq	-67.07	18.54	161.71	
HCl	g	-92.31	-95.30	186.8	29.12
std. state	aq	-167.15	-131.25	56.5	-136.4
^{2}HCl deuterium chloride	g	-93.35	-95.94	192.63	29.17
HClO	g	-92.1	-79.5	236.61	37.15
undissoc; std. state	aq	-120.9	-79.9	142	
HClO$_2$ undissoc; ss	aq	-51.9	5.9	188.3	
HClO$_3$ std. state	aq	-103.97	-8.03	162.3	
HClO$_4$	lq	-40.58			
std. state	aq	-129.33	-8.62	182.0	
HClO$_4$·H$_2$O	c	-302.21			
HClO$_4$·2H$_2$O	lq	-677.98			
HCN	g	134	126	201.71	35.86
	lq	108.87	124.93	112.84	70.63
HCN ionized; ss	aq	150.6	172.4	94.1	
undissoc; ss	aq	107.11	119.66	124.7	
HCNO ionized; ss	aq	-146.0	-97.5	106.7	
undissoc; ss	aq	-154.39	-117.2	144.8	

TABLE 6.3 Enthalpies and Gibbs Energies of Formation, Entropies, and Heat Capacities of the Elements and Inorganic Compounds (*Continued*)

Substance	State	$\Delta Hf°$, kJ·mol^{-1}	$\Delta Gf°$, kJ·mol^{-1}	$S°$, J·deg^{-1}·mol^{-1}	$C_p°$, J·deg^{-1}·mol^{-1}
HCNS ionized; ss	aq	76.44	92.68	144.4	−40.2
HCO$_3^-$ std. state	aq	−691.99	−586.85	91.2	
H$_2$CO$_3$ std. state	aq	−699.65	−623.16	187.4	
H$_2$CS$_3$ trithiocarbonic acid	lq	25.1	27.82	233.0	149.8
HF	g	−272.5(8)	−274.6	173.78(3)	29.14
	lq	−299.78		75.40	51.67
undissoc; ss	aq	−320.08	−296.86	88.7	
HF ionized; ss	aq	−332.63			
^{2}HF	g	−275.5(9)	−277.27(20)	179.70	29.14
HF$_2^-$ std. state	aq	−649.94	−578.15	92.5	
H$_2$F$_2$ dimer	g	−572.66	−544.51	238.(9)	44.89
H$_2$Fe(CN)$_6^{2-}$ ss	aq	455.6	658.44	218	
HFO	g	98.(4)	−86	226.8(2)	35.93
HI	g	26.36(21)	1.56	206.59	29.16
std. state	aq	−55.19	−51.59	111.3	−142.3
HIO undissoc; ss	aq	−138.1	−99.2	95.4	
H$_2$MoO$_4$	c	−1 046.0			
HN$_3$	lq	269.5	327.2		
^{2}H$_2$N$_2$ *cis*-diazine	g	207.(2)	241	224.09	39.02
HNCO isocyanic acid	g	−116.73	−107.36	238.11	44.85
HNCS isothiocyanic acid	g	127.61	112.88	248.03	46.40
HNO$_2$ *cis*	g	−76.7(13)	−41.9	249.4	45.35
trans	g	−78.8(13)	−43.9	249.3	46.05
HNO$_3$	g	−134.3(4)	−73.94	266.4	53.34
	lq	−173.22	−79.91	155.60	
std. state	aq	−207.36	−111.34	146.4	−86.6
H$_2$N$_2$O$_2$ hyponitrous	aq	−57.3	36.0	218	
HO hydroxyl	g	39.0(12)	34.27	183.64(4)	30.00
HO$^-$	aq	−229.99	−157.28	−10.75	−148.5
H$_2$O	c	−292.72			37.11
	lq	−285.83(4)	−237.14	69.95(8)	75.35
	g	−241.84	−228.61	188.74	33.60
^{1}H^{2}HO	g	−245.37(6)	−233.18	199.51	33.79
^{2}H$_2$O deuterium oxide	g	−249.20(7)	−234.54	198.33	34.25
H$_2$O$_2$ hydrogen peroxide	lq	−187.78	−120.42	109.6	89.41
	g	−136.11	−105.48	232.88	43.14
undissoc; ss	aq	−191.17	−134.10	143.9	
HO$_2^-$ std. state	aq	−160.33	67.4	23.9	
HOCN undissoc; ss	aq	−154.39	−117.2	144.8	
OCN$^-$ cyanate ss	aq	−146.02	−97.5	106.7	
HPO$_3$	c	−948.51			
HPO$_4^{2-}$ std. state	aq	−1 292.14	−1 089.26	−33.5	
H$_2$PO$_4^-$ std. state	aq	−1 296.29	−1 130.39	90.4	
H$_3$PO$_2$ hypophosphorous acid	c	−604.6			
H$_3$PO$_3$	c	−964.4			
H$_3$PO$_4$	c	−1 284.4(21)	−1 124.3	110.5(4)	106.1
	lq	−1 254.4	−1 111.7	150.6	145.06

TABLE 6.3 Enthalpies and Gibbs Energies of Formation, Entropies, and Heat Capacities of the Elements and Inorganic Compounds (*Continued*)

Substance	State	$\Delta Hf°$, kJ·mol^{-1}	$\Delta Gf°$, kJ·mol^{-1}	$S°$, J·deg^{-1}·mol^{-1}	$C_p°$, J·deg^{-1}·mol^{-1}
ionized; ss	aq	−1 277.4	−1 018.8	222	
undissoc; ss	aq	−1 288.34	−1 142.65	158.2	
$H_4P_2O_7$	c	−2 241.0			
undissoc; ss	aq	−2 268.6	−2 032.2	268	
HS$^-$ std. state	aq	−17.6	12.05	62.8	
H_2S	g	−20.5(8)	−33.3	205.7	34.19
undissoc; ss	aq	−39.8	−27.87	121	
^2H_2S	g	−23.9(8)	−35.3	215.3	35.76
$HSbO_2$ undissoc; ss	aq	−487.9	−407.5	46.6	
HSCN undissoc; ss	aq		97.53		
SCN$^-$ std. state	aq	76.44	92.68	144.5	−40.2
HSe$^-$ std. state	aq	15.9	43.9	80	
H_2Se	g	85.7			
HSeO$_3^-$ std. state	aq	−514.55	−411.54	135.1	
H_2SeO_3	c	−524.46			
undissoc; ss	aq	−507.48	−426.22	207.9	
HSeO$_4^-$ std. state	aq	−581.6	−452.3	149.4	
H_2SeO_4	c	−530.1			
H_2SiO_3	c	−1 188.67	−1 092.4	134	
undissoc; ss	aq	−1 182.8	−1 079.5	109	
H_4SiO_4	c	−1 481.1	−1 333.0	192	
undissoc; ss	aq	−1 468.6	−1 316.7	180	
HSO$_3^-$ std. state	aq	−626.22	−527.81	139.8	
HSO_3Cl	lq	−601.2			
HSO_3F	lq	−795.0			
	g	−753.(8)	−691	297.(3)	75.24
H_2SO_3 undissoc; ss	aq	−608.81	−537.90	232.2	
H_2SO_4	lq	−814.0(7)	−689.9	156.90(8)	138.58
std. state	aq	−909.27	−744.63	20.1	293
$H_2SO_4 \cdot H_2O$	lq	−1 127.6(8)	−950.3	211.5(2)	214.3
$H_2SO_4 \cdot 2H_2O$	lq	−1 427.1(8)	−1 199.6	276.4(4)	261.5
$H_2SO_4 \cdot 3H_2O$	lq	−1 720.4(8)	−1 443.9	345.4(4)	319.1
$H_2SO_4 \cdot 4H_2O$	lq	−2 011.2(8)	−1 685.8	414.5(4)	386.4
$H_2S_2O_7$	c	−1 273.6			
H_2Te	g	99.6		228.9	35.56
H_2WO_4	c	−1 131.8(17)	−1 003.9	145.(25)	113
Indium					
In	c	0	0	55.73	26.74
InAs	c	−58.6	−53.6	75.7	47.78
InH	g	215.5	190.3	207.53	29.58
In_2O_3	c	−925.27	−830.73	104.2	92
InP	c	−88.7	−77.0	59.8	45.44
InS	c	−138.1	−131.8	67	
In_2S_3	c	−427	−412.5	163.6	118.0
In_2Se_3	c	−343			
InSb	c	−30.5	−25.5	86.2	49.5
Iodine					
I$^-$ std. state	aq	−55.19	−51.59	111.3	−142.3
I_2	c	0	0	116.14(8)	54.44
	g	62.43	19.37	260.58	36.86
std. state	aq	22.6	16.40	137.2	

TABLE 6.3   Enthalpies and Gibbs Energies of Formation, Entropies, and Heat Capacities of the Elements and Inorganic Compounds (*Continued*)

Substance	State	$\Delta Hf°$, kJ·mol^{-1}	$\Delta Gf°$, kJ·mol^{-1}	$S°$, J·deg^{-1}· mol^{-1}	$C_p°$, J·deg^{-1}· mol^{-1}
I$_3^-$ std. state	aq	−51.5	−51.5	239.3	
IBr	c	−10.5			
ICl	c	−35.4(1)	−14.05	97.93	55.23
	lq	−23.93	−13.97	136.15	102.51
ICl$_3$	c	−89.5	−22.34	167.4	
IF	g	−94.8(38)	−117.7	236.3	33.56
IF$_5$	lq	−864.8			
	g	−840.1	−771.5	334.7	102.89
IF$_7$	g	−961.1(25)	−835.8	347.7(13)	134.5
IO$^-$ std. state	aq	−107.5	−38.5	118.4	
IO$_3^-$ std. state	aq	−221.3	−128.0	118.4	
IO$_4^-$ std. state	aq	−151.5	−58.6	222	
I$_2$O$_5$	c	−158.07			
Iridium					
Ir	c	0	0	35.48	25.06
IrCl$_3$	c	−245.6	180	113	
IrF$_6$	c	−579.65	−461.66	247.7	
IrO$_2$	c	−240.2		57.3	57.32
Iron					
Fe alpha	c	0	0	27.32(13)	25.09
Fe^{2+} std. state	aq	−89.1	−78.87	−137.7	
Fe^{3+} std. state	aq	−48.5	−4.6	−315.9	
FeBr$_2$	c	−248.9(21)	−237.4	140.7(13)	80.2
std. state	aq	−332.2	−286.81	27.2	
Fe$_3$C alpha-cementite	c	22.6	20.1	101.2	107.1
FeCl$_2$	c	−341.8(4)	−302.3	118.(3)	76.7
	aq	−423.4	−341.37	−24.7	
FeCl$_3$	c	−399.4(8)	−333.9	142.34	96.65
std. state	aq	−550.2	−398.3	−146.4	
Fe(CN)$_6^{3-}$ std. state	aq	561.9	729.3	270.3	
Fe(CN)$_6^{4-}$ std. state	aq	455.6	694.9	95.0	
FeCNS^{2+} std. state	aq	23.4	71.1	−130	
FeCO$_3$	c	−740.6	−666.7	92.9	82.1
Fe(CO)$_5$	lq	−766.(7)	−697	337.(4)	233.8
FeCr$_2$O$_4$	c	−1 446.0	−1 343.9	146.2	133.8
FeF$_2$	c	−706.(42)	−663	86.99(17)	68.12
std. state	aq	−754.4	−636.5	−165.3	
FeF$_3$	c	−1 042.(13)	−972	98.(8)	91.0
FeI$_2$	c	−104.6(84)	−111.7	167.4(84)	83.7
std. state	aq	−199.6	−182.1	84.9	
Fe$_2$N	c	−3.8		101.3	70.0
Fe(NO$_3$)$_3$ std. state	aq	−670.7	−338.5	123.4	
FeO	c	−272.0	−251.4	60.75	49.91
Fe$_2$O$_3$ hematite	c	−825.5(13)	−743.5	87.40	103.76
Fe$_3$O$_4$ magnetite	c	−1 120.9(8)	−1 051.6	145.27	147.24
FeOH$^+$ std. state	aq	−324.7	−277.4	−29	
Fe(OH)$^{2+}$ std. state	aq	−290.8	−229.4	−142	
Fe(OH)$_2$	c	−574.0(29)	−490.0	87.9(84)	97.1
Fe(OH)$_3$	c	−833.(13)	−705	104.6(84)	101.7

TABLE 6.3 Enthalpies and Gibbs Energies of Formation, Entropies, and Heat Capacities of the Elements and Inorganic Compounds (*Continued*)

Substance	State	$\Delta Hf°$, kJ·mol^{-1}	$\Delta Gf°$, kJ·mol^{-1}	$S°$, J·deg^{-1}· mol^{-1}	$C_p°$, J·deg^{-1}· mol^{-1}
FeS	c	$-101.7(8)$	-102.0	60.32(4)	50.52
FeS$_2$ marcasite	c	$-167.4(21)$	-156.1	53.87(13)	62.39
FeS$_2$ pyrite	c	$-171.5(21)$	-160.1	52.92(13)	62.12
FeSiO$_3$	c	$-1\ 155$		87.5	89.4
Fe$_2$SiO$_4$	c	$-1\ 479.9$		145.18	132.9
FeSO$_4$	c	$-929.(8)$	-825	121.0(13)	100.6
std. state	aq	-998.3	-823.4	-117.6	
Fe$_2$(SO$_4$)$_3$	c	$-2\ 583.0(17)$	$-2\ 262.7$	307.5(84)	264.8
std. state	aq	$-2\ 813.7$	$-2\ 243.0$	-570.7	
FeTiO$_3$	c	$-1\ 246.4$		105.9	99.5
Krypton					
Kr	g	0	0	164.084(3)	20.786
Lanthanum					
La	c	0	0	56.9	27.11
La^{3+}	aq	-707.1	683.7	-217.6	13
LaCl$_3$	c	$-1\ 071.1$		144.4	103.6
std. state	aq	$-1\ 208.8$	$-1\ 077.4$	-50	-423
LaCl$_3 \cdot 7H_2O$	c	$-3\ 178.6$	$-2\ 713.3$	462.8	431.0
La(NO$_3$)$_3$	c	$-1\ 254.4$			
std. state	aq	$-1\ 329.3$			
La$_2$O$_3$	c	$-1\ 793.7$	$-1\ 705.8$	127.32	108.78
La$_2$(SO$_4$)$_3$	c	$-3\ 941.3$		280	
La$_2$Te$_3$	c	-724	-714.6	231.63	132.13
Lead					
Pb	c	0	0	64.78(54)	26.84
Pb(acetate)$_2$	c	-964.4			
Pb(BO$_2$)$_2$	c	$-1\ 556.(6)$	$-1\ 450$	131.(12)	107.1
PbB$_4$O$_7$	c	$-2\ 858.(6)$	$-2\ 667$	167.(13)	168
PbBr$_2$	c	$-277.4(25)$	-260.7	161.1(21)	79.6
Pb(CH$_3$)$_4$	lq	97.9			
Pb(C$_2$H$_5$)$_4$	lq	52.7			
PbCl$_2$	c	$-359.4(8)$	-314.1	136.(2)	77.1
PbClF	c	-534.7	-488.3	121.8	
PbCO$_3$	c	-699.2	-625.5	131.0	87.40
PbC$_2$O$_4$	c	-851.4	-750.2	146.0	105.4
PbCrO$_4$	c	-930.9			
PbF$_2$ orthorhombic	c	$-677.(4)$	-631	$-113.(8)$	72.3
PbI$_2$	c	$-175.39(40)$	-173.58	174.84(21)	77.57
PbMoO$_4$	c	$-1\ 051.9$	-951.4	166.1	119.70
Pb(N$_3$)$_2$ monoclinic	c	478.2	624.7	148.1	
Pb(NO$_3$)$_2$	c	-451.9			
PbO red	c	$-219.4(8)$	-189.3	66.3(8)	45.8
PbO$_2$	c	$-274.5(29)$	-215.4	71.80(42)	61.17
Pb$_3$O$_4$	c	$-718.7(62)$	-601.6	212.0(67)	154.9
Pb$_3$(PO$_4$)$_2$	c	$-2\ 595.3$	$-2\ 432.6$	353.1	256.3
PbS	c	$-98.3(21)$	-96.7	91.3(17)	49.4
PbSe	c	-102.9	-101.7	102.5	50.2
PbSeO$_4$	c	-609.2	505.0	167.8	
PbSiO$_3$	c	$-1\ 145.2$	$-1\ 061.1$	110.1	90.04

TABLE 6.3 Enthalpies and Gibbs Energies of Formation, Entropies, and Heat Capacities of the Elements and Inorganic Compounds (*Continued*)

Substance	State	$\Delta Hf°$, kJ·mol^{-1}	$\Delta Gf°$, kJ·mol^{-1}	$S°$, J·deg^{-1}· mol^{-1}	$C_p°$, J·deg^{-1}· mol^{-1}
PbSiO$_4$	c	−2 023.8	−1 909.6	84.01	98.66
Pb$_2$SiO$_4$	c	−1 378.(15)	−1 267	188.8(21)	137.0
PbSO$_4$	c	−919.94	−813.20	148.57	103.22
PbSO$_4$·PbO	c	−1 182.0		225.06	150.16
PbTe	c	−70.7	−69.5	110.0	50.5
Lithium					
Li	c	0	0	29.085	24.62
Li$^+$ std. state	aq	−278.49	−293.30	13.4	68.6
Li$_3$AlF$_6$ cryolite	c	−3 317.(3)	−3 152	238.5(17)	215.7
LiAlH$_4$	c	−117.0(84)	−48.3	87.9(4)	86.4
LiAlO$_2$	c	−1 188.7(21)	−1 126.3	53.3	67.78
LiBeF$_3$	c	−1 651.8(63)	−1 576.3	89.2(42)	91.8
LiBH$_4$	c	−190.46(21)	−124.73	75.82	82.54
LiBH$_4$·tetrahydrofuran	c	−415.5	−220.5	289	
Li$_2$BeF$_4$	c	−2 274.(5)	−2 171	130.6(21)	135.3
LiBO$_2$	c	−1 019.2(8)	−963.1	51.7(2)	60.37
Li$_2$B$_4$O$_7$	c	−3 362.(6)	−3 170	156.(4)	183.0
LiBr	c	−351.20	−342.00	74.27	48.91
std. state	aq	−400.03	−397.27	95.81	−73.2
LiBrO$_3$	c	−346.98			
std. state	aq	−345.56	−274.89	174.9	
LiCl	c	−390.8	−372.2	78.43	48.03
LiClO$_4$	c	−381.(3)	−254	126.(4)	105
std. state	aq	−407.81	−302.1	195.4	−7.5
Li$_2$CO$_3$	c	−1 216.04(17)	−1 132.12	90.17(21)	96.23
LiF	c	−616.9(8)	−588.7	35.66(2)	41.82
std. state	aq	−611.12	−570.28	4.6	
LiH	c	−90.63(13)	−68.45	20.04(13)	27.96
LiI	c	−270.1(4)	−269.7	85.77	50.02
std. state	aq	−333.67	−344.8	124.7	−73.6
LiIO$_3$	c	−503.38			
std. state	aq	−499.82	−421.33	131.4	−55.2
Li$_3$N	c	−164.6(11)	−128.6	62.59(13)	75.27
LiNO$_3$	c	−481.6			
std. state	aq	−485.05	−404.30	160.7	
Li$_2$O	c	−598.7(21)	−562.1	37.89	54.09
Li$_2$O$_2$	c	−632.6	−578.9	56.5(42)	70.6
LiOH	c	−484.9(4)	−439	42.82(21)	49.59
std. state	aq	−508.40	−451.9	7.1	
Li$_2$SiO$_3$	c	−1 649.5(21)	−1 558.7	80.3(13)	100.6
Li$_2$Si$_2$O$_5$	c	−2 561.(4)	−2 417	125.5	138.1
Li$_2$SO$_4$	c	−1 436.4(17)	−1321.1	113.9(4)	117.6
std. state	aq	−1 464.44	−1 329.59	45.6	
Li$_2$TiO$_3$	c	−1 670.7	−1 579.8	91.8(4)	109.9
Lutetium					
Lu	c	0	0	50.96	26.86
Lu^{3+}	aq	−665	−628	−264	25
LuCl$_3$	c	−945.6			
std. state	aq	−1 167	−1 021	−96	−385
Lu$_2$O$_3$	c	−1 878.2	−1 789.1	109.96	101.75

TABLE 6.3 Enthalpies and Gibbs Energies of Formation, Entropies, and Heat Capacities of the Elements and Inorganic Compounds (*Continued*)

Substance	State	$\Delta Hf°$, kJ·mol^{-1}	$\Delta Gf°$, kJ·mol^{-1}	$S°$, J·deg^{-1}· mol^{-1}	$C_p°$, J·deg^{-1}· mol^{-1}
Magnesium					
Mg	c	0	0	32.67(10)	24.87
Mg^{2+} std. state	aq	−466.85	−454.8	−138.1	
MgAl$_2$O$_4$	c	−2 299.(1)	−2 177	89.0(42)	116.20
MgBr$_2$	c	−524.3(21)	−504.1	117.2(42)	73.16
std. state	aq	−709.94	−662.8	26.8	
MgBr$_2$·6H$_2$O	c	−2 410.0	−2 056.0	397	
MgCl$_2$	c	−641.6(5)	−592.1	89.63	71.38
std. state	aq	−801.15	−717.1	−25.1	
MgCl$_2$·6H$_2$O	c	−2 499.0	−2 115.0	315.1	
Mg(ClO$_4$)$_2$	c	−568.90			
std. state	aq	−725.51	−472.0	225.4	
Mg(ClO$_4$)$_2$·6H$_2$O	c	−2 445.6	−1 863.1	520.1	
MgCO$_3$	c	−1 111.(8)	−1 028	65.85	76.21
MgC$_2$O$_4$	c	−1 269.0			
std. state	aq	−1 292.0	−1 128.8	−92.5	
MgF$_2$	c	−1 124.0(10)	1 071	57.3(2)	61.5
Mg$_2$Ge	c	−108.8	−105.9	86.48	69.54
MgH$_2$	c	−76.2(92)	−36.7	31.1(8)	35.3
MgI$_2$	c	−367.0(60)	−361	130.0(40)	74.8
std. state	aq	−577.22	−552.2	84.5	
Mg$_3$N$_2$	c	−461.1(42)	400.9	87.9(84)	104.5
MgNH$_4$PO$_4$·6H$_2$O	c	−3 681.9			
Mg(NO$_3$)$_2$	c	−790.65	−589.5	164.0	141.9
std. state	aq	−881.6	−677.4	154.8	
Mg(NO$_3$)$_2$·6H$_2$O	c	−2 613.3	−2 080.7	452	
MgO microcrystal	c	−601.7(63)	−568.9	26.92(8)	37.11
Mg(OH)$_2$	c	−924.7(21)	−833.7	63.24	77.25
std. state	aq	−926.8	−769.4	−149.0	
Mg$_3$(PO$_4$)$_2$	c	−3 780.7	−3 538.8	189.20	213.47
MgSeO$_4$	c	−968.51			
std. state	aq	−1 066.1	−896.2	−84.1	
Mg$_2$Si	c	−77.8	−77.1	81.6	67.9
MgSiO$_3$ clinoenstatite	c	−1 548.9(42)	−1 462.0	67.8(8)	81.9
Mg$_2$SiO$_4$ forsterite	c	−2 176.9(42)	−2 057.9	95.1(8)	118.7
Mg$_3$Si$_4$O$_{10}$(OH)$_2$ talc	c	−5 922.5	−5 543.0	260.7	321.8
MgSO$_3$·3H$_2$O	c	−1 931.8			
MgSO$_3$·6H$_2$O	c	−2 817.5			
MgSO$_4$	c	−1 262.(21)	−1 147	91.4(8)	96.2
std. state	aq	−1 356.0	−1 212.3	−7.1	
MgSO$_4$·H$_2$O kieserite	c	−1 602.1	−1 428.8	126.4	
MgSO$_4$·7H$_2$O epsomite	c	−3 388.71	−2 871.9	372	
MgTiO$_3$	c	−1 497.6(63)	−1 420.1	111.08	91.88
Mg$_2$TiO$_4$	c	−2 164.0(60)	−2 048	115.0(60)	129
MgTi$_2$O$_5$	c	−2 509.(11)	−2 369	135.6(63)	146.9
Mg$_2$V$_2$O$_7$ triclinic	c	−2 835.9	−2 645.29	200.4	203.47
MgWO$_4$	c	−1 516.(34)	−1 404	101.2(8)	109.1
Manganese					
Mn	c	0	0	32.01(8)	26.30

TABLE 6.3 Enthalpies and Gibbs Energies of Formation, Entropies, and Heat Capacities of the Elements and Inorganic Compounds (*Continued*)

Substance	State	$\Delta Hf°$, kJ·mol^{-1}	$\Delta Gf°$, kJ·mol^{-1}	$S°$, J·deg^{-1}·mol^{-1}	$C_p°$, J·deg^{-1}·mol^{-1}
Mn^{2+} std. state	aq	−220.75	−228.0	−73.6	50
MnBr$_2$	c	−384.9	−372	138.1	75.31
std. state	aq	−464.0	−409.2		
Mn$_3$C	c	−4.6	5.4	98.7	93.51
MnCl$_2$	c	−482.0	−440.5	118.20	72.97
std. state	aq	−555.05	−490.8	38.9	−222
MnCO$_3$	c	−895.0	−816.7	85.8	80.3
Mn$_2$(CO)$_{10}$	c	−1 677.4			
MnF$_2$	c	−795.0	−749	92.26	67.99
MnI$_2$	c	−242.7		150.6	75.35
Mn(NO$_3$)$_2$	c	−576.26			
std. state	aq	−635.6	−451.0	218	−121
MnO	c	−384.9	−362.9	59.8	44.77
MnO$_2$	c	−520.1	−465.2	53.1	54.22
Mn$_2$O$_3$	c	−956.9	−881.2	110.5	98.7
MnO$_4^-$	aq	−541.4	−447.3	191.2	
MnO$_4^{2-}$	aq	−653	−500.8	59	
Mn$_3$O$_4$	c	−1 387.8	−1 283.2	155.6	139.7
Mn$_3$(PO$_4$)$_2$	c	−3 116.7			
MnS	c	−214.2	−218.4	78.2	50.0
MnSiO$_3$	c	−1 269.0	−1 240.6	89.1	86.4
MnSO$_4$	c	−1 065.3	−957.42	112.1	100.4
std. state	aq	−1 130.1	−972.8	−53.6	−243
MnTiO$_3$	c	−1 355.6		105.9	99.8
Mercury					
Hg	lq	0	0	76.0(2)	28.00
HgBr$_2$	c	−169.5(21)	−152.2	170.3(63)	75.3
Hg$_2$Br$_2$	c	−204.2(8)	−178.7	219.0(30)	104.6
Hg(CH$_3$)$_2$	lq	59.8	140.2	209	
Hg(C$_2$H$_5$)$_2$	lq	30.1			
HgCl$_2$	c	−230.0(40)	−184	144.0(60)	73.9
Hg$_2$Cl$_2$	c	−264.2(2)	−210.5	192.5	102.0
Hg(CN)$_2$	c	263.6			
HgF$_2$	c	−405	−362	134.3	74.86
Hg$_2$F$_2$	c	−485.(13)	−469	161.(8)	100.4
HgI$_2$	c	−105.4(17)	−102.2	181.3(63)	77.75
Hg$_2$I$_2$	c	−119.1(21)	−111.1	241.3(89)	105.9
Hg$_2$(N$_3$)$_2$	c	594.1	746.4	205	
HgO	c	−90.78(10)	−58.49	70.27(13)	44.06
HgS	c	−58.2	−50.6	81.6	46.4
HgSO$_4$	c	−707.5	−594		
Hg$_2$SO$_4$	c	−743.12	−625.88	200.66	131.96
Molybdenum					
Mo	c	0	0	28.61(5)	23.93
MoBr$_3$	c	−284.(25)	−259	175.(17)	105.4
MoCl$_4$	c	−477.(8)	−402	224.(13)	128
MoCl$_5$	c	−527.(8)	−423	238.(13)	155.6
MoCl$_6$	c	−523.(42)	−391	255.(17)	175
Mo(CO)$_6$	c	−984.5	−877.8	327.2	251.4
MoF$_6$	lq	−1 585.66(11)	−1 473.17	259.69(25)	169.8

TABLE 6.3 Enthalpies and Gibbs Energies of Formation, Entropies, and Heat Capacities of the Elements and Inorganic Compounds (*Continued*)

Substance	State	$\Delta Hf°$, kJ·mol^{-1}	$\Delta Gf°$, kJ·mol^{-1}	$S°$, J·deg^{-1}·mol^{-1}	$C_p°$, J·deg^{-1}·mol^{-1}
MoO_2	c	−587.9(29)	−532.0	46.46(42)	55.90
MoO_3	c	−745.2(4)	−668.1	77.8(13)	74.88
MoC_4^- std. state	aq	−997.9	−836.4	27.2	
MoS_2	c	−276.1(25)	−267.2	62.57(8)	63.56
Mo_2S_3	c	−270.3	−278.6	181.2	109.3
Neodymium					
Nd	c	0	0	71.6	27.5
Nd^{3+} std. state	aq	−696.2	−671.5	−206.7	−21
$NdCl_3$	c	−1 041.0			113
std. state	aq	−1 197.9	−1 065.7	−37.7	−431
$Nd(NO_3)_3$	c	−1 230.9			
Nd_2O_3	c	−1 807.9	−1 720.9	158.6	111.3
Neon					
Ne	g	0	0	146.327(3)	20.786
Neptunium					
Np	c	0	0		29.46
NpF_6	c	−1 937			
NpO_2	c	−1 029	−979	80.3	66.1
Nickel					
Ni	c	0	0	29.87(21)	25.98
Ni^{2+} std. state	aq	−54.0	−45.6	−128.9	
$Ni(acetate)_2$ ss	aq	−1 025.9	−784.5	44.4	
$NiCl_2$	c	−233.5	−203.4	151.8	71.66
std. state	aq	−388.3	−307.9	−15.1	
$Ni(CN)_4^{2-}$ std. state	aq	367.8	472.0	218	
$Ni(CO)_4$	lq	−632.0(80)	−589	320.(13)	203
NiC_2O_4	c	−856.9			
NiF_2	c	−651.5	−604.2	73.6	75.3
$Ni(NO_3)_2$	c	−415.1			
std. state	aq	−468.6	−268.6	164.0	
NiO	c	−240.6	−211.7	38.00	44.31
$Ni(OH)_2$	c	−529.7	−447.3	88	
NiS	c	−88.0(60)	−85	53.0(4)	47.1
Ni_3S_2	c	−216.0(50)	−210	133.9(4)	117.7
NiS_2	c	−131.4(17)	−124.7	72.0(83)	70.6
$NiSO_4$	c	−889.1	−759.8	97.1	138.3
std. state	aq	−2 682.8	−2 225.0	332.17	327.9
$NiSO_4 \cdot 7H_2O$	c	−2 976.3	−2 462.2	378.94	364.59
$NiWO_4$	c	−1 128.4		118.0	136.0
Niobium					
Nb	c	0	0	36.5(4)	24.7
$NbBr_5$	c	−556.(13)	−508	258.8(63)	147.9
NbC	c	−138.9	−136.8	34.98	36.23
$NbCl_5$	c	−797.5	−683.3	245.2	155.7
NbF_5	c	−1 813.8	−1 701.2	160.3	134.7
NbI_5	c	−268.6		343	155.6
NbN	c	−236.4	−205.9	34.5	39.0
NbO	c	−420.(13)	−392	46.0(84)	41.1
NbO_2	c	−710.9	−666.5	92.84	57.45

TABLE 6.3 Enthalpies and Gibbs Energies of Formation, Entropies, and Heat Capacities of the Elements and Inorganic Compounds (*Continued*)

Substance	State	$\Delta Hf°$, kJ·mol^{-1}	$\Delta Gf°$, kJ·mol^{-1}	$S°$, J·deg^{-1}·mol^{-1}	$C_p°$, J·deg^{-1}·mol^{-1}
Nb_2O_5	c	$-1\,899.5(42)$	$-1\,765.8$	137.3(13)	132.0
$NbOCl_3$	c	-879.5	-782	159	120.0
Nitrogen					
N_2	g	0	0	191.61(2)	29.124
NF_3	g	$-132.1(11)$	-90.6	260.8(2)	53.37
N_2F_2 *cis*	g	67	109	259.8	49.96
trans	g	81.2	120.5	262.6	53.47
N_2H_4 hydrazine	lq	50.6(11)	150.1	121.5(4)	98.84
$N_2^2H_4$ hydrazine-d_4	g	81.6	150.9	248.86	55.52
$N_2H_5^+$ std. state	aq	-7.5	82.4	151	70.3
N_2H_5Br	c	-155.6			
std. state	aq	-128.9	-21.8	233.1	-71.6
N_2H_5Cl	c	-197.1			
std. state	aq	-174.9	-49.0	207.1	-66.1
$N_2H_5Cl·HCl$	c	-367.4			
N_2H_5OH	lq	-242.7			
undissoc; ss	aq	-251.50	-109.2	207.9	73.2
$N_2H_5NO_3$	c	-251.58			
std. state	aq	-215.10	-28.91	297	
$(N_2H_5)_2SO_4$	c	-959.0			
std. state	aq	-924.7	-579.9	322	-151
NO	g	90.29(17)	86.60	210.76	29.85
NOBr	g	82.13	82.42	273.42	45.48
NOCl	g	51.71(42)	66.10	261.68(17)	44.59
NOF	g	$-65.7(17)$	-50.3	248.02	41.35
NOF_3	g	-163	-96	278.40	67.86
NO_2	g	33.1(8)	51.3	240.03(13)	36.97
NO_2Cl	g	12.1(17)	54	272.19	53.19
NO_2F	g	$-109.(21)$	-66	250.2	49.87
NO_3	g	69.41	114.35	252.5	46.9
N_2O	g	82.4(4)	104.2	220.0	38.62
N_2O_2	g	170.37	202.88	287.52	63.51
$N_2O_2^{2-}$ hyponitrite	aq	-17.2	138.9	27.6	
N_2O_3	g	82.8(8)	139.7	308.5(21)	65.62
N_2O_4	lq	-19.56	97.52	209.20	142.51
	g	9.08	97.79	304.38	77.26
N_2O_5	g	11.3(18)	118.0	346.5(42)	96.30
Osmium					
Os	c	0	0	32.6	24.7
$OsCl_3$	c	-190.4	-121	130	
$OsCl_4$	c	-254.8	-159	155	
OsO_4	c	-394.1	-305.0	143.9	
Oxygen					
O_2	g	0	0	205.147(35)	54.317
O_3	g	142.7(17)	163.2	238.9	39.24
OF_2	g	24.5(16)	41.8	247.5(4)	57.11
O_2F_2	g	19.79	61.42	268.11	54.06
Palladium					
Pd	c	0	0	37.91	25.94

TABLE 6.3 Enthalpies and Gibbs Energies of Formation, Entropies, and Heat Capacities of the Elements and Inorganic Compounds (*Continued*)

Substance	State	$\Delta Hf°$, kJ·mol^{-1}	$\Delta Gf°$, kJ·mol^{-1}	$S°$, J·deg^{-1}·mol^{-1}	$C_p°$, J·deg^{-1}·mol^{-1}
Pd^{2+} std. state	aq	149.0	176.6	−184	
PdBr$_2$	c	−104.2			
PdBr$_4^{2-}$ std. state	aq	−384.9	−318.0	247	
PdCl$_2$	c	−171.5	−125.1	105	
PdCl$_4^{2-}$ std. state	aq	−550.2	−416.7	167	
Pd$_2$H	c	−19.7	−5.0	91.6	
PdO	c	−90.8		56.1	31.5
PdS	c	−75	−67	46	
PdS$_2$	c	−81.2	−74.5	80	
Phosphorus					
P$_4$ white	c	0	0	41.08(8)	23.83
	g	58.9(21)	24.4	280.0(4)	67.16
red, V	c	−17.46	−12.46	22.85(8)	21.19
PBr$_3$	lq	−184.5	−175.5	240.2	
	g	−128.4	−157.3	348.15	76.02
PBr$_5$	c	−443.5			
PCl$_3$	lq	−319.7	−272.4	217.2	
	g	−271.1	−257.3	311.58	71.59
PCl$_5$	c	−443.5			
	g	−342.7	−278.2	364.0	111.9
PF$_3$	g	−958.(14)	−937	273.1(1)	58.69
PF$_5$	g	−1 594.4(29)	−1 520.7	300.8(21)	84.8
PH$_3$	g	22.9(17)	30.9	210.24	37.10
std. state	aq	−9.50	25.31	120.1	
PH$_4$Br	c	−127.6	−47.7	110.0	
PH$_4$Cl	c	−145.2			
PH$_4$I	c	−69.9	0.8	123.0	109.6
PH$_4$OH undissoc; ss	aq	−295.35	−211.88	190.0	
PI$_3$	c	−45.6			
PO$_3^-$	aq	−977.0			
PO$_4^{3-}$ std. state	aq	−1 277.4	−1 018.8	−222	
P$_2$O$_7^{4-}$ std. state	aq	−2 271.1	−1 919.2	−117	
(P$_2$O$_3$)$_2$ dimer	c	−1 640.1			
	g	−2 214.(33)	−2 085	346.(33)	143.9
P$_4$O$_{10}$	c	−3 009.9(89)	−2 723.3	228.78(42)	211.71
POBr$_3$	c	−458.6			
	g	−389.11	−390.91	−359.74	89.87
POCl$_3$	lq	−597.1	−520.9	222.46	138.62
	g	−542.3	−502.5	325.39	84.94
POClF$_2$	g	−970.7	−924.1	301.68	68.83
POCl$_2$F	g	−765.7	−721.6	320.38	79.32
POF$_3$	g	−1 254.0(80)	−1 206	285.4(3)	68.82
PSCl$_3$	g	−363.2	−347.7	337.23	89.83
PSF$_3$	g	−1 009.(63)	−985	298.1(2)	74.55
P$_4$S$_3$	c	−155	−159	201	146
Platinum					
Pt	c	0		41.63	25.86
PtBr$_2$	c	−82.0			
PtCl$_2$	c	−123.4		117	
PtCl$_3$	c	−182.0	−134	151	

TABLE 6.3 Enthalpies and Gibbs Energies of Formation, Entropies, and Heat Capacities of the Elements and Inorganic Compounds (*Continued*)

Substance	State	$\Delta Hf°$, kJ·mol^{-1}	$\Delta Gf°$, kJ·mol^{-1}	$S°$, J·deg^{-1}·mol^{-1}	$C_p°$, J·deg^{-1}·mol^{-1}
$PtCl_4$	c	−231.8	−172	176	
$PtCl_4^{2-}$ std. state	aq	−499.2	−361.5	155	
$PtCl_6^{2-}$ std. state	aq	−668.2	−482.8	220.1	
PtS	c	−81.6	−76.2	55.06	43.39
PtS_2	c	−108.8	−99.6	74.68	65.90
Plutonium					
Pu	c	0	0	51.5	35.5
Pu^{3+}	aq	−579.9	−587.9	−163	
Pu^{4+}	aq	−579.9	−1 490		
$PuBr_3$	c	−831.8	−804.6	192.88	107.86
$PuCl_3$	c	−961.5	−892.7	159.00	102.84
$PuCl_4$	c	−1 381			
PuF_3	c	−1 552	−1 478.8	112.97	96.82
PuF_4	c	−1 732	−1 644.7	161.9	120.8
PuF_6	c	25.48	27.2	222.59	167.36
PuH_2	c	−139.3	−101.7	59.8	39.0
PuH_3	c	−138	−82.4	64.9	43.2
PuI_3	c	−648.5	−643.9	214.2	111.8
PuO	c	−565	−538.9	70.7	51.3
PuO_2	c	−1 058.1	−1 005.8	82.4	68.6
Pu_2O_3 beta	c	−1 715.4	−1 632.3	152.3	131.0
$Pu(SO_4)_2$	c	−2 200.8	−1 969.5	163.18	181.96
PuS	c	−439.3	−436.7	78.24	53.97
Pu_2S_3	c	−989.5	−985.5	192.46	129.66
Polonium					
Po	c	0	0	62.8	26.4
PoO_2	c	−251	−197	71	61.5
Potassium					
K	c	0	0	64.67	29.50
	lq	2.284	0.264	71.46	32.72
K^+ std. state	aq	−252.38	−283.26	102.5	21.8
K acetate	c	−723.0			
	aq	−738.39	−652.66	189.1	15.5
$KAg(CN)_2$	aq	18.0	22.2	297	
$KAgCl_2$	aq	−497.4	−498.7	333.9	
K_2AgI_3	aq	−686.6	−720.5	458.1	
$KAlCl_4$	c	97.(10)	−1 094	197	156.4
K_3AlCl_6	c	−2 092.0(40)	−1 938	377.(8)	248.9
K_3AlF_6	c	3 290.7		284.5	221.1
$KAl(SO_4)_2$	c	−2 470.2	−2 240.1	204.47	192.92
K_3AsO_4 std. state	aq	−1 645.27	−1 498.29	144.8	
KBF_4	c	−1 887.(4)	−1 785	133.9(40)	114.48
std. state	aq	−1 827.2	−1 770.3	285	
KBH_4	c	−226.9(23)	−159.7	106.61	96.57
std. state	aq	−204.22	−168.99	212.97	
KBO_2	c	−995.0(80)	−936.6	79.98(12)	67.03
std. state	aq	−1 024.75	−962.19	65.3	
$K_2B_4O_7$	c	−3 334.2	−3 136.8	208.(6)	170.5
KBr	c	−393.8(4)	−380.4	95.94(4)	52.38

TABLE 6.3 Enthalpies and Gibbs Energies of Formation, Entropies, and Heat Capacities of the Elements and Inorganic Compounds (*Continued*)

Substance	State	$\Delta Hf°$, kJ·mol^{-1}	$\Delta Gf°$, kJ·mol^{-1}	$S°$, J·deg^{-1}· mol^{-1}	$C_p°$, J·deg^{-1}· mol^{-1}
std. state	aq	−373.92	−387.23	184.9	−120.1
KBrO$_3$	c	−360.24	−271.21	149.16	120.16
	aq	−319.45	−264.72	264.22	
KBrO$_4$	c	−287.86	−174.47	170.08	120.16
KCl	c	−436.68(25)	−408.76	82.55(17)	51.29
std. state	aq	−419.53	−414.51	159.0	−114.6
KClO std. state	aq	−359.4	−320.1	146	
KClO$_2$ std. state	aq	−318.8	−266.1	203.8	
KClO$_3$	c	−397.73	−296.31	143.1	100.3
std. state	aq	−356.35	−291.29	264.9	
KClO$_4$	c	−430.(4)	−300	151.0(8)	110.21
std. state	aq	−381.71	−291.88	284.5	
KCN	c	−104.15	−84.64	134.32	66.39
std. state	aq	−101.7	−110.9	196.7	
K$_2$CO$_3$	c	−1 150.2(21)	−1 064.5	155.5(4)	114.44
std. state	aq	−1 181.90	−1 094.41	148.1	
K$_2$C$_2$O$_4$	c	−1 346.8			
	aq	−1 329.72			
K$_2$CrO$_4$	c	−1 403.7	−1 295.8	200.12	145.98
std. state	aq	−1 385.91	−1 294.36	255.2	
K$_2$Cr$_2$O$_7$	c	−2 061.5	−1 882.0	291.2	219.2
K$_2$CuCl$_4$·2H$_2$O	c	−1 707.1	−1 492.9	355.43	253.22
KF	c	−568.6(4)	−538.9	66.5(2)	48.98
std. state	aq	−585.01	−562.08	88.7	−84.9
K$_3$Fe(CN)$_6$	c	−249.8	−129.7	426.06	
std. state	aq	−139.4	−120.5	577.8	
K$_4$Fe(CN)$_6$	c	−594.1	−453.1	418.8	322.2
std. state	aq	−554.0	−438.11	505.0	
K formate	c	−679.73			
std. state	aq	−677.93	−634.3	192	−66.1
K glycinate	aq	−722.16	−598.23	221.8	
KH	c	−57.82(5)	−53.01	50.21	37.91
K$_2$HAsO$_4$ std. state	aq	−1 411.10	−1 281.22	203.3	
KH$_2$AsO$_4$	c	−1 180.7	−1 036.0	155.02	126.73
std. state	aq	−1 161.94	−1 036.54	218	
KHCrO$_4$ std. state	aq	−1 130.5	−1 048.1	286.6	
KHCO$_3$	c	−963.2	−863.6	115.5	
std. state	aq	−944.33	−870.10	193.7	
KHC$_2$O$_4$ std. state	aq	−1 070.7	−981.7	251.9	
KHF	c	−931.2(13)	−863.1	104.3(4)	76.94
	aq	−902.32	−861.40	195.0	
KHgBr$_3$	c	−550.20			
std. state	aq	−545.6	−542.7	360	
K$_2$HgBr$_4$	c	−963.6			
std. state	aq	−935.5	−937.6	515	
KHgCl$_3$	c	−671.1			
std. state	aq	−641.0	−592.5	314	
K$_2$Hg(CN)$_4$	c	−32.2			
std. state	aq	21.8	51.9	510	
K$_2$HgI$_4$	c	−775.0			

TABLE 6.3 Enthalpies and Gibbs Energies of Formation, Entropies, and Heat Capacities of the Elements and Inorganic Compounds (*Continued*)

Substance	State	$\Delta Hf°$, kJ·mol^{-1}	$\Delta Gf°$, kJ·mol^{-1}	$S°$, J·deg^{-1}· mol^{-1}	$C_p°$, J·deg^{-1}· mol^{-1}
std. state	aq	−739.7	−778.2	565	
KH$_2$PO$_4$	c	−1 568.33	−1 415.95	134.85	116.57
std. state	aq	−1 548.67	−1 622.85	192.9	
K$_2$HPO$_4$ std. state	aq	−1 796.90	−1 655.78	171.5	
K$_2$H$_2$P$_2$O$_7$	c	−2 815.8			
	aq	−2 783.2	−2 576.9	368	
K$_3$HP$_2$O$_7$	aq	−3 032.1	−2 822.1	351	
KHS	c	−265.10			75.3
std. state	aq	−269.9	−271.21	165.3	
KHSO$_3$	aq	−878.60	−811.07	242.3	
KHSO$_4$	c	−1 160.6	−1 131.4	138.1	
std. state	aq	−1 139.72	−1 039.26	234.3	−63
KI	c	−327.9(4)	−323.0	106.39	52.78
	aq	−307.57	−334.85	213.8	−120.5
KIO$_3$	c	−510.03	−425.5	151.46	106.48
	aq	−473.6	−411.3	220.9	
KIO$_4$	c	−467.23	−361.41	176	
	aq	−403.8	−341.8	322	
KMnO$_4$	c	−837.2	−737.6	171.71	117.6
K$_2$MoO$_4$	c	−1 498.71			
std. state	aq	−1 502.5	−1 402.9	232.2	
KNH$_2$ amide	c	−128.9			
KNO$_2$	c	−369.82	−306.60	152.09	107.40
std. state	aq	−356.9	−315.5	225.5	
KNO$_3$	c	−494.63	−394.93	133.05	96.27
std. state	aq	−459.74	−394.59	249.0	−64.9
K$_2$Ni(CN)$_4$ std. state	aq	−136.8	−94.6	423	
K$_2$O	c	−363.2(21)	−322.1	94.1(63)	83.7
KO$_2$	c	−284.5(21)	−240.6	122.5(42)	77.53
K$_2$O$_2$	c	−495.8(42)	−429.8	113.0(63)	110
KOCN cyanate	c	−418.65			
std. state	aq	−398.3	−380.7	209.2	
KOH	c	−424.7(4)	−378.9	78.9(8)	64.9
std. state	aq	−482.37	−440.53	91.6	−126.8
K$_2$PdBr$_4$	c	−938.1			
std. state	aq	−889.5	−884.5	452	
K$_3$PO$_4$	c	−1 950.2			
std. state	aq	−2 034.7	−1 868.6	87.9	
K$_4$P$_2$O$_7$	aq	−3 280.7	−3 052.2	293	
K$_2$PtBr$_4$	c	−915.0			
std. state	aq	−872.8	−828.4	326.4	
K$_2$PtBr$_6$	c	−1 021.3			
std. state	aq	−975.3	−898.7	368	
K$_2$PtCl$_4$	c	−1 054.4			180.2
std. state	aq	−1 003.7	−928.0	360	
K$_2$PtCl$_6$	c	−1 229.3	−1 078.6	333.9	205.60
std. state	aq	−1 171.8	−1 049.4	424.7	
K$_2$ReCl$_6$	c	−1 310.4	−1 172.8	371.71	214.68
std. state	aq	−1 266.92	−1 156.0	460	
KReO$_4$	c	−1 097.0	−994.5	167.82	122.55
std. state	aq	−1 039.7	−977.8	303.8	8.4

TABLE 6.3 Enthalpies and Gibbs Energies of Formation, Entropies, and Heat Capacities of the Elements and Inorganic Compounds (*Continued*)

Substance	State	$\Delta Hf°$, kJ·mol^{-1}	$\Delta Gf°$, kJ·mol^{-1}	$S°$, J·deg^{-1}·mol^{-1}	$C_p°$, J·deg^{-1}·mol^{-1}
K$_2$S	c	−377.(13)	−363	115.(17)	74.7
std. state	aq	−471.5	−480.7	190.4	
K$_2$S$_2$	c	−432.2			
	aq	−474.5	−487.0	233.5	
KSCN	c	−200.16	−178.32	124.26	88.53
std. state	aq	−175.94	−190.58	246.9	−18.4
K$_2$SeO$_3$	c	−979.5			
std. state	aq	−1 013.8	−936.4	218.0	
K$_2$SeO$_4$	c	−1 110.02	−1 002.9	222	
std. state	aq	−1 103.7	−1 007.9	259.0	
K$_2$SiF$_6$	c	−2 956.0	−2 798.7	225.9	
std. state	aq	−2 893.7	−2 766.0	327.2	
K$_2$SiO$_3$	c	−1 548.1(84)	−1 455.7	146.1(8)	118.4
K$_2$SnBr$_6$	c	−1 218.0	−1 160.2	443.1	246.0
K$_2$SnCl$_6$	c	−1 477.0	−1 333.0	366.5	246.0
K$_2$SO$_3$	c	−1 125.5			
std. state	aq	−1 140.1	−1 053.1	176	
K$_2$SO$_4$	c	−1 437.7(4)	−1 319.6	175.5(8)	131.3
std. state	aq	−1 414.02	−1 311.14	225.1	−251
K$_2$S$_2$O$_3$	c	−1 173.6			
std. state	aq	−1 156.9	−1 089.1	272	
K$_2$S$_2$O$_4$	aq	−1 258.1	−1 166.9	297	
K$_2$S$_2$O$_7$	c	−1 986.6	−1 791.6	255	
K$_2$S$_2$O$_8$	c	−1 916.10	−1 697.41	278.7	213.2
std. state	aq	−1 849.3	−1 681.6	449.4	
K$_2$S$_4$O$_6$	c	−1 780.7	−1 613.43	309.66	230.79
std. state	aq	−1 728.8	−1 607.1	462.3	−24.3
KSO$_3$F	c	−1 159.0			
K$_2$UO$_4$	c	−1 921.3			
KVO$_4$	c	−1 154.8			
std. state	aq	−1 140.6	−1 066.9	155	
K$_2$Zn(CN)$_4$	c	−100.0			
std. state	aq	−162.3	−119.7	431	
Praseodymium					
Pr	c	0	0	73.2	27.20
Pr^{3+} std. state	aq	−704.6	−679.1	−209	−29
Pr(acetate)$_3$ std. state	aq	−2 147.52	−1 805.56	164.9	
PrCl$_3$	c	−1 056.9			100
std. state	aq	−1 206.3	−1 072.8	−42	−439
Pr(NO$_3$)$_3$	c	−1 229.3			
Pr$_2$O$_3$	c	−1 809.6			117.40
Promethium					
PmCl$_3$	c	−1 054.0			
Protactinium					
Pa	c	0	0	51.8	
Pa^{4+}	aq	−619.2			
PaBr$_4$	c	−824.3	−787.9	234.3	
PaBr$_5$	c	−862	−820	289	
PaCl$_4$	c	−1 043.1	−952.7	192.5	
PaCl$_5$	c	−1 144.7	−1 034.3	238	

TABLE 6.3 Enthalpies and Gibbs Energies of Formation, Entropies, and Heat Capacities of the Elements and Inorganic Compounds (*Continued*)

Substance	State	$\Delta Hf°$, kJ·mol^{-1}	$\Delta Gf°$, kJ·mol^{-1}	$S°$, J·deg^{-1}·mol^{-1}	$C_p°$, J·deg^{-1}·mol^{-1}
Radium					
Ra	c	0	0	71	
Ra^{2+}	aq	−527.6	−561.5	54	
RaCl$_2$ std. state	aq	−861.9	−823.8	167	
Ra(NO$_3$)$_2$	c	−992	−796.2	222	
std. state	aq	−942.2	−784.1	347	
RaSO$_4$	c	−1 441.1	−1 365.7	138	
std. state	aq	−1 436.8	−1 306.2	117	
Radon					
Rn	g	0	0	176.235	20.79
Rhenium					
Re	c	0	0	37.20	25.8
Re$^-$ std. state	aq	46	10.0	230	
ReCl$_3$	c	−264	−188	123.9	92.4
ReCl$_6^{2-}$ std. state	aq	−761	−590	251	
ReO$_2$	c	−423	−368	172	
ReO$_3$	c	−605.0	−531	257.3	
Re$_2$O$_7$	c	−1 240.1	−1 066.1	207.1	166.1
Rhodium					
Rh	c	0	0	31.51	24.98
Rh$_2$O$_3$	c	−285.8		110.9	104.0
Rubidium					
Rb	c	0	0	76.8(3)	31.06
Rb$^+$ std. state	aq	−251.17	−283.97	121.50	
RbBr	c	−394.59	−381.79	109.96	52.84
std. state	aq	−372.71	−387.94	203.93	
RbBrO$_3$	c	−367.27	−278.11	161.1	
Rb$_2$CO$_3$	c	−1 136.0	−1 051.0	181.33	117.61
std. state	aq	−1 179.5	−1 095.8	186.2	
RbCl	c	−435.35	−407.81	95.90	52.21
std. state	aq	−418.32	−415.22	178.0	
RbClO$_3$	c	−402.9	−300.4	151.9	103.2
std. state	aq	−355.14	−291.9	283.68	
RbClO$_4$	c	−437.19	−307.69	164.0	
std. state	aq	−380.49	−292.59	303.3	
RbF	c	−549.4		75.3	50.5
std. state	aq	−583.79	−562.79	107.53	
RbHCO$_3$	c	−963.2	−863.6	121.3	
std. state	aq	−943.16	−870.82	212.71	
RbHF$_2$	c	−922.6	−856.0	120.08	79.37
std. state	aq	−901.11	−862.11	213.8	
RbHSO$_4$	c	−1 159.0			
std. state	aq	−1 138.51	−1 039.98	253.1	
RbI	c	−333.80	−328.9	118.41	53.18
std. state	aq	−306.35	−335.56	232.6	
RbNO$_3$	c	−495.05	−395.85	147.3	102.1
std. state	aq	−458.52	−395.30	267.8	
Rb$_2$O	c	−339			
RbOH	c	−418.19			

TABLE 6.3 Enthalpies and Gibbs Energies of Formation, Entropies, and Heat Capacities of the Elements and Inorganic Compounds (*Continued*)

Substance	State	$\Delta Hf°$, kJ·mol^{-1}	$\Delta Gf°$, kJ·mol^{-1}	$S°$, J·deg^{-1}·mol^{-1}	$C_p°$, J·deg^{-1}·mol^{-1}
std. state	aq	−481.16	−441.24	110.75	
Rb$_2$PtCl$_6$	c	−1 245.6	−1 109.6	406	
std. state	aq	−1 170.7	−1 056.6	464	
RbReO$_4$	c	−1 102.9	−996.2	167	
std. state	aq	−1 038.5	−978.6	322.6	
Rb$_2$SeO$_4$	c	−1 114.2			
std. state	aq	−1 101.7	−1 009.2	297.1	
Rb$_2$SO$_4$	c	−1 435.61	−1 316.96	197.44	134.06
std. state	aq	−1 411.60	−1 312.56	263.2	
Ruthenium					
Ru	c	0	0	28.53	23.85
RuO$_2$	c	−305.0			
RuO$_4$	c	−239.3	152.3	146.4	
	1q	−228.5	−152.3	183.3	
Samarium					
Sm	c	0	0	69.58	29.54
Sm^{2+} std. state	aq	−691.6	−666.5	−211.7	−21
SmCl$_2$	c	−815.5			
SmCl$_3$	c	−1 025.9			
std. state	aq	−1 193.3	−1 060.2	−42.7	−431
SmF$_3$·½H$_2$O	c	−1 825.1			
SmI$_3$	c	−620.1			
Sm(IO$_3$)$_3$	c	−1 381			
Sm(NO$_3$)$_2$	c	−1 212.1			
Sm$_2$O$_3$	c	−1 823.0	−1 734.7	151.0	114.10
Sm$_2$(SO$_4$)$_3$	c	−3 899.1			
Scandium					
Sc	c	0	0	34.64	25.52
Sc^{3+} std. state	aq	−614.2	−586.6	−255	
ScCl$_3$	c	−899.6		121.3	93.64
ScF$_3$	c	−1 629.2	−1 555.6	92	
Sc$_2$O$_3$	c	−1 906.2	−1 819.41	76.99	103.97
Selenium					
Se	c	0	0	41.97	24.98
SeCl$_4$	c	−188.3			
SeF$_6$	g	−1 029.3		313.8	110.0
SeO$_2$	c	−238.5			
SeO$_3$	c	−166.9			
SeO$_3^{2-}$ std. state	aq	−509.2	−369.9	13	
SeO$_4^{2-}$	aq	−599.2	−441.4	54.0	
Silicon					
Si	c	0	0	44.46	20.00
SiBr$_4$	1q	−457.3(84)	−433.9	278.2(13)	146.4
SiC alpha	c	−71.5(63)	−69.1	16.49(13)	26.76
beta	c	−73.2(63)	−70.9	16.61(13)	26.84
SiCl$_4$	g	−662.8(13)	−622.8	330.9(2)	90.26
	1q	−686.93	−620.00	241.37	140.16
SiClF$_3$	g	−1 318.(63)	−1 280	309.(17)	79.4

TABLE 6.3   Enthalpies and Gibbs Energies of Formation, Entropies, and Heat Capacities of the Elements and Inorganic Compounds (*Continued*)

Substance	State	$\Delta Hf°$, kJ·mol^{-1}	$\Delta Gf°$, kJ·mol^{-1}	$S°$, J·deg^{-1}· mol^{-1}	$C_p°$, J·deg^{-1}· mol^{-1}
SiF_4	g	−1 614.9(8)	−1 572.7	282.8(4)	73.62
SiH_4	g	34.3(21)	56.8	204.65	42.83
SiH_3Cl	g	−142.(8)	−119	250.8(2)	51.10
SiH_2Cl_2	g	−320.5	−295.0	286.60	62.17
$SiHCl_3$	g	−496.(4)	−465	313.7(4)	75.45
SiH_3F	g	−377.(21)	−353	238.4(8)	47.20
Si_2H_6	g	80.3	127.2	272.55	80.79
SiI_4	c	−189.5	−191.6	258.1	108.0
	lq	−174.60	−187.49	294.30	159.79
Si_3N_4	c	−745.(29)	−647	113.(17)	99.5
SiO_2 quartz	c	−910.9(17)	−856.4	41.46(13)	44.59
high cristobalite	c	−905.5	−853.6	50.05	26.58
$SiOF_2$	g	−967.(25)	−951	271.3	53.69
SiS_2	c	−213.4	−212.6	80.3	77.5
Silver					
Ag	c	0	0	42.68	25.52
Ag^+ std. state	aq	105.58	77.12	72.68	21.8
Ag^{2+} in $4M$ $HClO_4$	aq	268.6	269.0	−88	
AgAt	c	−45.2		133.1	55.7
AgBr	c	−100.37	−96.90	107.11	52.38
$AgBrO_3$	c	−27.2	54.39	151.9	
AgCl	c	−127.07	−109.80	96.2	50.79
$AgClO_2$	c	8.79	75.7	134.56	87.32
$AgClO_3$	c	−30.3	64.4	142.3	
$AgClO_4$	c	−31.13		162.3	
std. state	aq	−23.77	68.49	254.8	
AgCN	c	146.0	156.9	107.19	66.73
$Ag(CN)_2^-$ std. state	aq	270.3	305.4	192	
Ag_2CrO_4	c	−731.74	−641.83	217.6	142.26
Ag_2CO_3	c	−505.9	−436.8	167.4	112.26
$Ag_2C_2O_4$	c	−673.2	−584.1	209	
AgF	c	−204.6		83.7	51.92
AgI	c	−61.84	−66.19	115.5	56.82
$AgIO_3$	c	−171.1	−93.7	149.4	102.93
AgN_3	c	308.8	376.1	104.2	
$Ag(NH_3)_2^+$ std. state	aq	−111.29	−17.24	245.2	
$AgNO_3$	c	−120.5	−33.47	140.92	93.05
std. state	aq	−101.80	−34.23	219.2	−64.9
AgO	c	−11.42	14.23	57.78	45.02
Ag_2O	c	−31.04	−11.21	121.3	65.86
Ag_2O_3	c	33.9	121.3	100	
Ag_2S orthorhombic	c	−32.59	−40.67	143.9	76.53
Ag_3Sb	c	−23.0		171.5	101.7
AgSCN	c	87.9	101.38	131.0	63
Ag_2Se	c	−38	−44.4	150.71	81.76
Ag_2SO_4	c	−715.90	−618.48	200.4	131.46
std. state	aq	−698.10	−590.36	165.7	−251
Ag_2Te	c	−37.2	−43.1	154.8	87.5
Sodium					
Na	c	0	0	51.455	28.15

TABLE 6.3 Enthalpies and Gibbs Energies of Formation, Entropies, and Heat Capacities of the Elements and Inorganic Compounds (*Continued*)

Substance	State	$\Delta Hf°$, kJ·mol^{-1}	$\Delta Gf°$, kJ·mol^{-1}	$S°$, J·deg^{-1}·mol^{-1}	$C_p°$, J·deg^{-1}·mol^{-1}
Na$^+$ std. state	aq	-240.12	-261.88	59.1	46.4
NaAg(CN)$_2$ std. state	aq	30.12	43.5	251	
Na acetate	c	-708.81	-607.27	123.0	79.9
std. state	aq	-726.13	-631.28	145.6	40.2
NaAlCl$_4$	c	$-1\,142.0(40)$	-996.4	188.3	154.98
Na$_3$AlCl$_6$	c	$-1\,979.0(40)$	$-1\,829$	347.0(80)	244.1
Na$_3$AlF$_6$	c	$-3\,361.2$	$-3\,136.7$	239.5	215.89
NaAlH$_4$	c	-115.5			
NaAlO$_2$	c	$-1\,137.3(7)$	$-1\,069.2$	70.40(4)	73.64
NaAl(SO$_4$)$_2$ std. state	aq	$-2\,590$	$-2\,238$	-222.6	
NaAlSiO$_4$	c	$-2\,092.8$	$-1\,978.2$	124.3	
NaAsO$_2$	c	-660.53			
std. state	aq	-669.15	-611.91	99.6	
Na$_3$AsO$_4$	c	$-1\,540$			
std. state	aq	$-1\,608.50$	$-1\,434.19$	14.2	
NaAu(CN)$_2$	aq	2.1	23.9	230	
NaBF$_4$	c	$-1\,844.7$	$-1\,750.17$	145.31	120.25
std. state	aq	$-1\,812.1$	$-1\,748.9$	243	
NaBH$_4$	c	$-191.84(30)$	-127.06	101.34	86.48
std. state	aq	-199.60	-147.61	169.5	
NaBO$_2$	c	$-975.7(21)$	-919.4	73.54(8)	65.94
std. state	aq	$-1\,012.49$	-940.81	21.8	
NaBO$_3$·4H$_2$O	c	$-2\,114.2$			
Na$_2$B$_4$O$_7$	c	$-3\,257$	$-3\,063$	194.6	187.4
std. state	aq	$-3\,271.1$	$-3\,076.9$	192.9	
Na$_2$B$_4$O$_7$·10H$_2$O	c	$-6\,298.6$	$-5\,516.6$	586	614.5
NaBr	c	-361.08	-349.00	86.82	51.38
std. state	aq	-361.66	-365.85	141.4	-95.4
NaBr$_3$ std. state	aq	-370.54	-368.95	274.5	
NaBrO std. state	aq	-384.3	-295.4	100	
NaBrO$_3$	c	-334.09	-242.84	128.9	
std. state	aq	-307.19	-243.34	220.9	
NaBrO$_4$ std. state	aq	-227.19	-143.93	-258.57	
Na$_2$[Cd(CN)$_4$]	aq	-52.3	-16.3	439	
NaCl	c	-385.9	-365.7	95.06	50.51
std. state	aq	-407.27	-393.17	115.5	-90.0
NaClO std. state	aq	-347.3	-298.7	100	
NaClO$_2$	c	-307.02		115.9	
std. state	aq	-306.7	-244.8	160.3	
NaClO$_3$	c	-366.77	-262.34	123.4	
std. state	aq	-344.09	-269.91	221.3	
NaClO$_4$	c	$-382.8(9)$	-254.2	142.3	111.3
std. state	aq	-369.45	-270.50	241.0	
NaCN	c	$-90.7(13)$	-80.4	118.0(20)	68.7
std. state	aq	-89.5	-89.5	153.1	
Na$_3$[Co(NO$_2$)$_6$]	c	$-1\,423.0$			
Na$_2$CO$_3$	c	$-1\,130.77(21)$	$-1\,048.01$	138.8(8)	111.00
Na$_2$CO$_3$·H$_2$O	c	$-1\,431.26$	$-1\,285.41$	168.11	145.60
Na$_2$CO$_3$·10H$_2$O	c	$-4\,081.32$	$-3\,428.20$	564.0	550.32

TABLE 6.3 Enthalpies and Gibbs Energies of Formation, Entropies, and Heat Capacities of the Elements and Inorganic Compounds (*Continued*)

Substance	State	$\Delta Hf°$, kJ·mol⁻¹	$\Delta Gf°$, kJ·mol⁻¹	$S°$, J·deg⁻¹· mol⁻¹	$C_p°$, J·deg⁻¹· mol⁻¹
$Na_2C_2O_4$	c	−1 318.0			142
std. state	aq	−1 305.4	−1 197.9	163.6	
Na_2CrO_4	c	−1 342.2	−1 235.0	176.61	142.13
std. state	aq	−1 361.39	−1 251.64	168.2	
$Na_2Cr_2O_7$	c	−1 978.6			
std. state	aq	−1 970.7	−1 825.1	379.9	
Na ethoxide	c	−413.80			
NaF	c	−575.4(8)	−545.1	51.21(8)	46.85
std. state	aq	−572.75	−540.70	45.2	−60.3
$Na_3[Fe(CN)_6]$ ss	aq	−158.6	−56.5	447.3	
$Na_4[Fe(CN)_6]$ ss	aq	−505.0	−352.63	231.0	
Na formate	c	−666.5	−600.00	103.76	82.68
std. state	aq	−666.67	−613.0	151	−41.4
NaH	c	−56.44(8)	−33.55	40.02	36.39
Na_2HAsO_4 std. state	aq	−1 386.58	−1 238.51	116.3	
NaH_2AsO_4 std. state	aq	−1 149.68	−1 015.16	176	
$NaHCO_3$	c	−950.81	−851.0	101.7	87.61
std. state	aq	−932.11	−848.72	150.2	
$NaHCrO_4$ std. state	aq	−1 118.4	−1 026.8	243.1	
$NaHF_2$	c	−920.27	−852.20	90.92	75.02
std. state	aq	−890.06	−840.02	151.5	
$Na_2H_2[Fe(CN)_6]$	aq	−24.7	134.64	335	
NaH_2PO_4	c	−1 536.8	−1 386.2	127.49	116.86
std. state	aq	−1 536.4	−1 392.27	149.4	
Na_2HPO_4		−1 748.1	−1 608.3	150.50	135.31
std. state	aq	−1 772.38	−1 613.06	84.5	
$Na_2H_2P_2O_7$	c	−2 764.8	−2 522.5	220.20	198.15
NaHS	c	−237.23			
std. state	aq	−257.73	−249.83	121.8	
$NaHSeO_3$	c	−759.23			
std. state	aq	−754.67	−673.41	194.1	
$NaHSeO_4$	c	−821.40			
std. state	aq	−821.74	−714.2	208.4	
$NaHSO_4$	c	−1 125.5	−992.9	113.0	
std. state	aq	−1 127.46	−1 017.88	190.8	−38
NaI	c	−287.9(8)	−284.6	98.50	52.23
std. state	aq	−295.31	−313.47	170.3	−95.8
NaI_3	aq	−291.6	−313.4	298.3	
$NaIO_3$	c	−481.79		135.1	92.1
std. state	aq	−461.50	−389.95	177.4	
$NaIO_4$	c	−429.28	−323.09	163.2	
std. state	aq	−391.62	−320.49	280	
Na methoxide	c	−367.8	−294.80	110.58	69.45
std. state	aq	−433.59	−332.46	17.6	
$NaMnO_4$ std. state	aq	−781.6	−709.2	250.2	
Na_2MnO_4	c	−1 156.0			
std. state	aq	−1 134	−1 024.7	176	
Na_2MoO_4	c	−1 468.12	−1 354.40	159.70	141.71
std. state	aq	−1 478.2	−1 360.2	145.2	
$Na_2Mo_2O_7$	c	−2 245.05	−2 058.19	250.6	217.15

TABLE 6.3　Enthalpies and Gibbs Energies of Formation, Entropies, and Heat Capacities of the Elements and Inorganic Compounds (*Continued*)

Substance	State	$\Delta Hf°$, kJ·mol^{-1}	$\Delta Gf°$, kJ·mol^{-1}	$S°$, J·deg^{-1}· mol^{-1}	$C_p°$, J·deg^{-1}· mol^{-1}
NaN_3	c	21.71	93.76	96.86	76.61
std. state	aq	35.02	86.2	166.9	
$NaNH_2$	c	−123.9	−64.0	76.90	66.15
$NaNbO_3$	c	−1 315.9	−1 233.0	117	
std. state	aq	−1 265.7	−1 194.1	155	
$NaNO_2$	c	−358.65	−284.60	103.8	
std. state	aq	−344.8	−294.1	182.0	−51.0
$NaNO_3$	c	−467.85	−367.06	116.52	92.88
std. state	aq	−447.48	−373.21	205.4	−40.2
$Na_2[Ni(CN)_4]$	aq	−112.6	−51.9	335	
NaO_2	c	−260.7	−218.7	115.9(13)	72.14
Na_2O	c	−418.0(42)	−379.1	75.04	69.10
Na_2O_2	c	−513.4(50)	−449.6	94.8(13)	89.3
NaOCN cyanate	c	−405.39	−358.2	96.7	86.6
std. state	aq	−386.2	−359.4	165.7	
NaOH	c	−425.9(4)	−379.4	64.4(8)	59.5
std. state	aq	−469.15	−419.20	48.1	−102.1
Na_3PO_4	c	−1 917.40	−1 788.87	173.80	153.47
std. state	aq	−1 997.9	−1 804.6	−46	
$Na_4P_2O_7$	c	−3 188	−2 969.4	270.29	241.12
std. state	aq	−3 231.7	−2 966.9	117	
$NaReO_4$	c	−1 057.09	−953.74	151.5	133.89
std. state	aq	−1 027.6	−956.5	260.2	
Na_2S	c	−366.(13)	−355	96.(17)	82.8
std. state	aq	−443.3	−438.1	103.3	
Na_2S_2	c	−397.0(80)	−392	151.(24)	
std. state	aq	−450.2	−444.3	146.4	
NaSCN	c	−170.50			
std. state	aq	−163.68	−169.20	203.84	6.3
Na_2Se	c	−341.4			
Na_2SeO_3	c	−958.6			
std. state	aq	−989.5	−893.7	130	
Na_2SeO_4	c	−1 069.0			
std. state	aq	−1 079.5	965.3	172.0	
Na_2SiF_6	c	−2 909.6	−2 747.6	187.0	
Na_2SiO_3	c	−1 561.4(42)	−1 467.3	113.8(13)	111.9
$Na_2Si_2O_5$	c	−2 470.1(42)	−2 324.1	164.1(42)	157.0
$NaSnBr_3$	aq	−615.1	−608.8	310	
$NaSnCl_3$	aq	−727.2	−692.0	318	
Na_2SO_3	c	−1 100.8	−1 012.5	145.94	120.25
std. state	aq	−1 115.87	−1 010.44	87.9	
Na_2SO_4	c	−331.64(15)	−303.50	35.89(10)	30.63
std. state	aq	−1 389.51	−1 268.40	138.1	−201
$Na_2SO_4 \cdot 10H_2O$	c	−4 327.26	−3 647.40	592.0	
$Na_2S_2O_3$	c	−1 123.0	−1 028.0	155	
std. state	aq	−1 132.40	−1 046.0	184.1	
$Na_2S_2O_3 \cdot 5H_2O$	c	−2 607.93	−2 230.1		
$Na_2S_2O_4$ dithionate, hydrosulfite	c	−1 232.2			
std. state	aq	−1 233.9	−1 124.2	209.2	

TABLE 6.3 Enthalpies and Gibbs Energies of Formation, Entropies, and Heat Capacities of the Elements and Inorganic Compounds (*Continued*)

Substance	State	$\Delta Hf°$, kJ·mol^{-1}	$\Delta Gf°$, kJ·mol^{-1}	$S°$, J·deg^{-1}·mol^{-1}	$C_p°$, J·deg^{-1}·mol^{-1}
$Na_2S_2O_7$	c	−1 925.1	−1 722.1	202.1	
$Na_2S_2O_8$	aq	−1 825.1	−1 638.9	362.3	
Na_2Te	c	−349.4			
Na_2TeO_4	c	−1 270.7			
Na_2TiO_3	c	−1 591.2	−1 496.2	121.67	125.65
Na_2UO_4 beta	c	−1 893.3	−1 777.78	166.02	146.65
Na_3UO_4	c	−2 025.1	−1 901.2	198.20	173.01
$NaVO_3$	c	−1 145.79	−1 064.12	113.68	97.57
std. state	aq	−1 128.4	−1 045.6	109	
Na_3VO_4	c	−1 757.87	−1 637.83	190.0	164.85
$Na_2V_2O_7$	c	−2 918.84	−2 712.52	318.4	269.74
Na_2WO_4	c	−1 544.7(84)	−1 429.8	160.3(21)	139.8
$Na_2[Zn(CN)_4]$	aq	−138.1	−77.0	343	
Strontium					
Sr	c	0	0	55.7(1)	26.79
Sr^{2+} std. state	aq	−545.80	−559.44	−32.6	
$Sr(acetate)_2$	c	−1 487.4			
$Sr_3(AsO_4)_2$	c	−3 317.1	−3 080.3	255	
$SrBr_2$	c	−718.0(17)	−698.8	143.4(42)	76.85
std. state	aq	−788.89	−767.39	132.2	
$SrCl_2$	c	−828.8(25)	−780.0	114.8(2)	75.59
std. state	aq	−880.10	−821.95	80.3	
$Sr(ClO_4)_2$	c	−762.69			
std. state	aq	−804.46	−576.68	331.4	
$SrCO_3$	c	−1 220.1	−1 140.1	97.1	81.42
SrC_2O_4	c	−1 370.7			
SrF_2	c	−1 217.(3)	−1 164	82.1(2)	70.0
$SrHPO_4$	c	−1 821.7	−1 688.7	121	
$Sr(H_2PO_4)_2$	c	−3 134.7			
SrI_2	c	−561.5(21)	−557.7	159.1(8)	77.95
std. state	aq	−656.18	−662.62	190.0	
$Sr(IO_3)_2$	c	−1 019.2	−855.2	234	
$SrMoO_4$	c	−1561.1		128.9	117.07
$Sr(NO_2)_2$	c	−762.3			
$Sr(NO_3)_2$	c	−978.22	−780.15	194.56	149.87
std. state	aq	−960.52	−782.12	260.2	
SrO	c	−592.0(33)	−561.4	55.5(4)	45.4
SrO_2	c	−654.4		54	79.45
$Sr(OH)_2$	c	−969.(9)	−881	97.(8)	74.9
$Sr_3(PO_4)_2$	c	−4 122.9			
SrS	c	−112.0(42)	−111	68.4(8)	48.7
SrSe	c	−385.8			
$SrSeO_3$	c	−1 047.7			
$SrSeO_4$	c	−1 142.7			
$SrSiO_3$	c	−1 633.9	−1 549.8	96.7	88.53
Sr_2SiO_4	c	−2 304.6	−2 191.2	153.1	134.26
$SrSO_3$	c	−1 177.0			
$SrSO_4$	c	−1 453.1	−1 341.0	117.6	107.78
Sr_2TiO_4	c	−2 287.4	−2 178.6	159.0	143.68

TABLE 6.3 Enthalpies and Gibbs Energies of Formation, Entropies, and Heat Capacities of the Elements and Inorganic Compounds (*Continued*)

Substance	State	$\Delta Hf°$, kJ·mol^{-1}	$\Delta Gf°$, kJ·mol^{-1}	$S°$, J·deg^{-1}·mol^{-1}	$C_p°$, J·deg^{-1}·mol^{-1}
Sulfur					
S	c	0	0	32.056(50)	22.70
monoclinic	c	0.360	−0.070	33.03(5)	23.23
S_8	g	101.25	49.16	430.20	156.06
SCl_2	lq	−50.0(20)	−28.5	184.(4)	91.0
S_2Cl_2	lq	−58.(2)	−39	224.(4)	124.3
SF_4	g	−763.(21)	−722	299.6(4)	77.60
SF_6	g	−1 220.5(8)	−1 116.5	291.5(4)	96.96
S_2F_{10}	g	−2 064.(29)	−1 861	397.(8)	176.7
SO_2	g	−296.84(21)	−300.13	248.12(8)	39.88
SO_3	g	−395.77(71)	−371.02	256.77	50.66
$SOCl_2$	g	−211.70	−196.52	306.39	66.44
SOF_2	g	−544.(105)	−502	276.52	57.11
SO_2Cl_2	g	−354.8	−310.5	311.00	77.11
SO_2ClF	g	−556.(21)	−513	303.(2)	71.6
SO_2F_2	g	−759.0(80)	−712	283.6(4)	65.83
Tantalum					
Ta	c	0	0	41.47(21)	25.30
TaB_2	c	−209.2		44.4	48.12
$TaBr_5$	c	−686		305.4	155.73
TaC	c	−144.1(38)	−142.7	42.37(84)	36.79
Ta_2C	c	−197.5		83.7	60.96
$TaCl_5$	c	−856.0(42)	−746	222.0(60)	148
TaF_5	c	−317.4		195.0	130.46
TaI_5	c	−490		343	155.6
TaN	c	−251		50.6	42.1
TaO_2	g	−201		280	44.0
Ta_2O_5	c	−2 046.0(42)	−1 911.0	143.1(13)	135.0
$TaOCl_3$	g	−780.7		361.5	98.53
Technetium					
Tc	c	0	0	33.47	24.27
Tc_2O_7	c	−1 113			
Tellurium					
Te	c	0	0	49.70	25.70
$TeCl_4$	c	−326.4		209	138.91
TeF_6	g	−1 369.0		335.77	116.90
TeO_2	c	−322.6	−270.3	79.5	63.89
Terbium					
Tb	c	0	0	73.22	28.91
Tb^{3+} std. state	aq	−682.8	−651.9	−226	
$TbCl_3$	c	−997.1			
std. state	aq	−1 184.1	−1 045.6	−59	−393
TbO_2	c	−971.5			
Tb_2O_3	c	−1 865.2			115.9
$Tb_2(SO_4)_3$ std. state	aq	−4 131.7	−3 597.4		
Thallium					
Tl	c	0	0	64.18	26.32
Tl^+ std. state	aq	5.36	−32.38	126	

TABLE 6.3 Enthalpies and Gibbs Energies of Formation, Entropies, and Heat Capacities of the Elements and Inorganic Compounds (*Continued*)

Substance	State	$\Delta Hf°$, kJ·mol^{-1}	$\Delta Gf°$, kJ·mol^{-1}	$S°$, J·deg^{-1}·mol^{-1}	$C_p°$, J·deg^{-1}·mol^{-1}
Tl^{3+} std. state	aq	197	214.6	−192	
TlBr	c	−173.2	−167.36	120.5	50.50
std. state	aq	−116.19	−136.36	207.9	
TlBrO$_3$	c	−136.4	−53.14	168.6	
std. state, $m = 1$	aq	−78.2	−30.5	288.7	
TlCl	c	−204.20	−184.93	111.20	50.92
std. state	aq	−161.80	−163.64	182.00	
TlCl$_3$	c	−315.1			
std. state	aq	−305.0	−179.1	−23.0	
TlClO$_3$	aq	−93.7	−35.6	287.9	
TlF	c	−324.6		83.3	54.77
std. state	aq	−327.27	−311.21	111.7	
TlI	c	−123.9	−125.39	127.6	52.51
std. state	aq	−49.83	−83.97	236.8	
TlNO$_3$	c	−243.93	−152.46	160.7	99.50
Tl$_2$O	c	−178.7	−147.3	126	
TlOH	c	−238.9	−195.8	88	
std. state	aq	−224.64	−189.66	114.6	
Tl$_2$SO$_4$	c	−931.8	−830.48	230.5	
std. state	aq	−898.56	−809.40	271.1	
Thorium					
Th	c	0	0	53.39	27.32
Th^{4+} std. state	aq	−769.0	−705.0	−422.6	
ThBr$_4$	c	−965.3	−927.2	230	
ThC$_{1.94}$	c	−146	−147.7	68.49	56.69
ThCl$_4$	c	−1 186.6	−1 094.5	190.4	120.75
ThF$_4$	c	−2 091.58	−1 997.02	142.05	110.54
undissoc; ss	aq	−2 115.0	−1 947.2	−105	
ThH$_2$	c	−139.8	−100.0	50.71	36.69
ThI$_4$	c	−664.8	−655.2	255	
ThN	c	−391.2	−363.6	56.07	45.2
Th$_3$N$_4$	c	−1 315.0	−1 212.9	201	155.90
Th(NO$_3$)$_4$	c	−1 441.4			
ThO$_2$	c	−1 226.41	−1 168.80	65.65	61.76
ThOCl$_2$	c	−1 232.2	−1 156.0	123.4	91.25
ThOF$_2$	c	−1 665.2	−1 589.5	105	
Th$_3$P$_4$	c	−1 140.2	−1 112.9	221.8	
ThS$_2$	c	−626.3	−620.1	96.2	
Th$_2$S$_3$	c	−1 083.7	−1 077.0	180	
Th(SO$_4$)$_2$	c	−2 542.6	−2 310.4	159.0	173.47
Thullium					
Tm	c	0	0	74.01	27.03
Tm^{3+} std. state	aq	−697.9	−661.9	−243	25
TmCl$_3$	c	−986.6			
std. state	aq	−1 199.1	−1 055.6	−75	−385
Tm$_2$O$_3$	c	−1 888.7	−1 794.5	139.8	116.7
Tin					
Sn white	c	0	0	51.55	26.99
gray	c	−2.09	0.13	44.14	25.77

TABLE 6.3 Enthalpies and Gibbs Energies of Formation, Entropies, and Heat Capacities of the Elements and Inorganic Compounds (*Continued*)

Substance	State	$\Delta Hf°$, kJ·mol^{-1}	$\Delta Gf°$, kJ·mol^{-1}	$S°$, J·deg^{-1}· mol^{-1}	$C_p°$, J·deg^{-1}· mol^{-1}
Sn^{2+} in aqueous HCl	aq	−8.9	−27.2	−16.7	
Sn^{4+} in aqueous HCl	aq	30.5	2.5	−117	
SnBr$_4$	c	−377.4	−350.2	264.4	136.44
SnCl$_2$	c	−325.1		130	79.33
std. state	aq	−329.7	−299.6	172	
SnCl$_4$	lq	−511.3	−440.2	258.6	165.3
SnH$_4$	g	162.8	188.3	227.57	48.95
SnO	c	−285.8	−256.9	56.5	44.31
SnO$_2$	c	−580.7	−519.7	52.3	52.59
SnS	c	−100	−98.3	77.0	49.25
SnS$_2$	c	−167.4		87.4	70.12
Titanium					
Ti	c	0	0	30.8(2)	25.24
TiB	c	−160.(38)	−160	35.(6)	29.7
TiB$_2$	c	−280.(17)	−275	28.5(4)	44.3
TiBr$_2$	c	−405.(21)	−383	108.(8)	78.7
TiBr$_3$	c	−550.2(84)	−525.6	176.4(33)	101.7
TiBr$_4$	c	−618.0(42)	−590.7	243.6(67)	131.5
TiC	c	−184.0(4)	−180	24.23(21)	33.81
TiCl$_2$	c	−515.(17)	−466	87.0(40)	69.8
TiCl$_3$	c	−722.0(40)	−654	140.0(10)	97.2
TiCl$_4$	c	−804.0(40)	−728	221.9(4)	145.2
TiF$_3$	c	−1 435.(42)	−1 362	88.(13)	92
TiF$_4$	c	−1 649.0(40)	−1 559	133.96	114.27
TiH$_2$	c	−144.4(84)	−105.1	29.71	30.09
TiI$_4$	c	−375.7(84)	−370.7	246.2(67)	125.6
TiN	c	−265.8	−243.8	52.73	37.08
TiO	c	−543.(13)	−513	34.8(21)	39.9
TiO$_2$	c	−938.7(21)	−883.3	49.9(3)	55.27
Ti$_2$O$_3$	c	−1 520.9(84)	−1 433.8	77.25(21)	95.81
Tungsten					
W	c	0	0	32.66(10)	24.30
WBr$_5$	c	−312.(13)	−270	272.(21)	155
WBr$_6$	c	−343.1	−290.8	314.(21)	181.4
W(CO)$_6$	c	−953.5		331.8	242.5
WCl$_4$	c	−443	−360	198.3	129.7
WCl$_5$	c	−515	−402	217.6	155.6
WCl$_6$	c	−594.(25)	−456	238.5	175.4
WF$_6$	g	−1 721.7(17)	−1 632.3	341.1	119.0
WO$_2$	c	−589.7(9)	−533.86	50.5(3)	55.74
WO$_3$	c	−842.9(8)	−764.1	75.9(13)	73.1
WOCl$_4$	c	−671.(8)	−549	173.(17)	146
WOF$_4$	c	−1 407.(63)	−1 298	176.0(4)	133.6
WO$_2$Cl$_2$	c	−780.0(60)	−703	200.8	104.4
Uranium					
U	c	0	0	50.21	27.66
UB$_2$	c	−161.6	−159.4	55.52	55.77
UBr$_3$	c	−699.2	−673.6	192	108.8
UBr$_4$	c	−802.5	−767.8	238.0	128.0

TABLE 6.3 Enthalpies and Gibbs Energies of Formation, Entropies, and Heat Capacities of the Elements and Inorganic Compounds (*Continued*)

Substance	State	$\Delta Hf°$, kJ·mol⁻¹	$\Delta Gf°$, kJ·mol⁻¹	$S°$, J·deg⁻¹·mol⁻¹	$C_p°$, J·deg⁻¹·mol⁻¹
UBr_5	c	−810.9	−769.9	293	160.7
UC	c	−98.3	−99.2	59.20	50.12
UCl_3	c	−866.5	−799	159.0	102.5
UCl_4	c	−1 019.2	−930.1	197.1	122.01
	aq	−1 259.8	−1 056.0	−184	
UCl_5	c	−1 058	−950	242.7	144.6
UCl_6	c	−1 092	−962	285.8	175.56
UF_3	c	−1 508.8	−1 431.8	123.43	95.10
UF_4	c	−1 920.9	−1 830.1	151.67	116.02
UF_5	c	−2 075.3	−1 958.6	199.6	132.3
UF_6	c	−2 197.0	−2 068.6	227.6	166.8
UH_3	c	−127.2	−72.8	63.68	49.29
UI_3	c	−460.7	−459.8	222	112.1
UI_4	c	−512.1	−506.7	264	134.3
UN	c	−290.8	−265.7	62.43	47.57
UO_2	c	−1 084.9	−1 031.8	77.03	63.60
UO_2^{2+} std. state	aq	−1 019.6	−953.5	−97.5	
UO_3	c	−1 223.8	−1 146.0	96.11	81.67
U_3O_8	c	−3 574.8	−3 369.8	282.59	238.36
$UOBr_2$	c	−973.6	−929.7	158.00	98.00
$UOCl_2$	c	−1 066.9	−996.2	138.32	95.06
UOF_2	c	−1 499.1	−1 428.8	119.2	
$UO_2(acetate)_2$	c	−1 963.55			
UO_2Br_2	c	−1 137.6	−1 066.5	169.5	
UO_2CO_3	c	−1 691.2	−1 562.7	138	
std. state	aq	−1 696.6	−1 481.6	−154.4	
$UO_2C_2O_4$	c	−1 796.94			
UO_2Cl_2	c	−1 243.9	−1 146.4	150.62	107.86
std. state	aq	−1 353.9	−1 215.9	15.5	
UO_2F_2	c	−1 648.1	−1 551.9	135.56	103.22
std. state	aq	−1 684.0	−1 551.3	−125.1	
$UO_2(NO_3)_2$	c	−1 349.3	−1 105.0	243	
std. state	aq	−1 434.3	−1 176.1	195.4	
$UO_2(OH)_2$ std. state	aq	−1 479.5	−1 267.8	−118.8	
UO_2SO_4	c	−1 845.1	−1 683.6	154.8	145.2
std. state	aq	−1 928.8	−1 698.3	−77.4	
US_2	c	−527	−526.4	110.42	74.64
US_3	c	−549.4	−547.3	138.49	95.60
Vanadium					
V	c	0	0	28.94(42)	24.90
VCl_2	c	−452	−406	97.1	72.22
VCl_3	c	−580.7	−511.3	131.0	93.18
VCl_4	lq	−569.9	−503.8	255	161.7
VF_5	lq	−1 480.3	−1 373.2	175.7	
	g	−1 433.9	−1 369.8	320.79	98.58
VN	c	−217.15	−191.08	37.28	38.00
VO	c	−431.8(63)	−404.2	39.0(8)	45.5
VO_2	c	−717.6		51.5	62.59
VO_2^+ std. state	aq	−649.8	−587.0	−42.3	
VO_2^{2+} std. state	aq	−486.6	−446.4	−133.9	

TABLE 6.3 Enthalpies and Gibbs Energies of Formation, Entropies, and Heat Capacities of the Elements and Inorganic Compounds (*Continued*)

Substance	State	$\Delta Hf°$, kJ·mol^{-1}	$\Delta Gf°$, kJ·mol^{-1}	$S°$, J·deg^{-1}·mol^{-1}	$C_p°$, J·deg^{-1}·mol^{-1}
VO_3^- std. state	aq	−888.3	−783.7	50.2	
V_2O_3	c	−1 218.8(63)	−1 139.0	98.1(13)	105.0
V_2O_4	c	−1 427.2(63)	−1 318.4	103.5(21)	115.4
V_2O_5	c	−1 550.6(63)	−1 419.3	130.5(21)	130.6
$VOCl_3$	lq	−734.7	−668.6	244.4	150.62
$VOSO_4$	c	−1 309.2	−1 169.9	108.8	
Xenon					
Xe	g	0	0	169.684(3)	20.786
XeF_2	c	−164.0			
XeF_4	c	−261.5	−123.0		
XeF_6	c	−360			
	g	−297			
XeO_3	c	402			
$XeOF_4$	lq	146			
Ytterbium					
Yb	c	0	0	59.87	26.74
Yb^{2+} std. state	aq		−527		
Yb^{3+} std. state	aq	−674.5	−643.9	238	25
$Yb(acetate)_3$ undissoc; std. state	aq	−2 105.0	−1 772.84	183.3	
$YbCl_2$	c	−799.6			
$YbCl_3$	c	−959.8			
std. state	aq	−1 176.1	−1 037.6	−71	−385
$Yb(NO_3)_3$ std. state	aq	−1 296.6			
Yb_2O_3	c	−1 814.6	−1 726.7	133.1	115.35
Yttrium					
Y	c	0	0	114.48	27.11
Y^{3+} std. state	aq	−723.4	−693.7	−251	
YCl_3	c	−975		136.8	97.95
YF_3	c	−1 718.8	−1 644.7	100	
Y_2O_3	c	−1 905.31	−1 816.65	99.08	102.51
Zinc					
Zn	c	0	−41.72(17)	25.40	
Zn^{2+} std. state	aq	−153.89	−147.03	−112.1	46
$ZnBr_2$	c	−328.65	−312.13	138.5	65.7
std. state	aq	−396.98	−354.97	52.72	−238
$ZnCl_2$	c	−415.05	−369.45	111.46	71.34
std. state	aq	−488.19	−409.53	0.84	−226
$Zn(CN)_4^{2-}$ std. state	aq	342.3	446.9	226	
$ZnCO_3$	c	−812.78	−731.57	82.4	79.71
ZnF_2	c	−764.4	−449.8	73.68	65.7
std. state	aq	−819.14	−704.67	−139.8	−167
ZnI_2	c	−208.03	−208.95	161.1	65.69
ZnO	c	−348.28	−318.32	43.64	40.25
$Zn(OH)_2$	c	−641.91	−553.59	81.2	
std. state	aq	−613.88	−461.62	−133.5	−251
ZnS	c	−205.98	−201.29	57.7	46.02
$ZnSO_4$	c	−980.14(10)	−868.77	110.5(8)	99.04
Zn_2SiO_4	c	−1 636.7	−1 523.2	131.42	123.3

TABLE 6.3 Enthalpies and Gibbs Energies of Formation, Entropies, and Heat Capacities of the Elements and Inorganic Compounds (*Continued*)

Substance	State	$\Delta Hf°$, kJ·mol^{-1}	$\Delta Gf°$, kJ·mol^{-1}	$S°$, J·deg^{-1}· mol^{-1}	$C_p°$, J·deg^{-1}· mol^{-1}
Zirconium					
Zr	c	0	0	38.87(20)	25.20
ZrB$_2$	c	−322.6(67)	−318.2	35.94(4)	48.24
ZrBr$_2$	c	−405.(42)	−382	116.(13)	86.7
ZrBr$_4$	c	−760.7(84)	−725.3	224.7(42)	124.8
ZrC	c	197.(13)	−193	33.32	37.90
ZrCl$_2$	c	−431.(42)	−386	110.(13)	72.6
ZrCl$_3$	c	−714.(63)	−646	146.(13)	96
ZrCl$_4$	c	−981.0(20)	−890	181.4(7)	119.8
ZrF$_2$	c	−962.(63)	−913	75.(8)	66
ZrF$_4$	c	−1 911.3(1)	−1 810.0	104.7(2)	103.6
ZrI$_2$	c	−259	−258	150.2	94.1
ZrI$_3$	c	−397.5	−394.9	204.6	103.8
ZrI$_4$	c	−488.7(63)	−485.4	260.3(42)	127.8
ZrN	c	−365.3(84)	−336.7	38.86(21)	40.44
ZrO$_2$	c	−1 097.5(17)	−1 039.7	50.36(33)	56.19
ZrSiO$_4$	c	−2 024.0(29)	−1 909.5	84.9(13)	98.7

TABLE 6.4 Heats of Melting, Vaporization, and Sublimation and Specific Heat at Various Temperatures of the Elements and Inorganic Compounds

Abbreviations Used in the Table

ΔHm, enthalpy of melting (at the melting point) in kJ·mol⁻¹

ΔHv, enthalpy of vaporization (at the boiling point) in kJ·mol⁻¹

ΔHs, enthalpy of sublimation (at 298 K) in kJ·mol⁻¹

C_p, specific heat (at temperature specified on the Kelvin scale) for the physical state in existence (or specified: c, lq, g) at that temperature in J·K⁻¹·mol⁻¹

ΔHt, enthalpy of transition (at temperature specified, superscript, measured in degrees Celsius) in kJ·mol⁻¹

Substance	ΔHm	ΔHv	ΔHs	C_p			
				400 K	600 K	800 K	1000 K
Aluminum							
Al	10.71	294.0	326.4	25.78	27.89	30.56	34.86(1q)
Al_6BeO_{10}	402			324.27	380.56	407.83	425.19
$AlBr_3$	11.3(8)	23.81		124.97	124.97	124.97	124.97
Al_4C_3				138.51	159.20	169.71	176.14
$AlCl_3$			116.(2)	100.08	117.65	135.23	152.80
AlF_3 $\Delta Ht = 0.563^{455}$	35.4(8)			86.29	97.31	98.52	100.83
AlI_3	98.(12)		112	108.5	121.3		
AlN	15.9(13)			36.69	43.54	46.79	48.55
Al_2O_3 corundum	111.(4)			96.09	112.55	120.14	124.77
AlOCl				64.31	72.63	76.86	79.25
Al_2SiO_5 andalusite				149.57	174.52	186.12	194.04
kyanite				148.28	176.23	188.28	196.21
sillimanite				147.48	173.03	185.00	193.54
$Al_6Si_2O_{13}$ mullite				390.74	459.82	494.13	513.38
Al_2S_3				114.95	124.14	129.70	134.05
Al_2TiO_5				162.04	182.84	192.87	200.04
Ammonium							
NH_3	5.652	23.35		38.72	45.29	51.24	56.49
N^2H_3, ammonia-d_3				42.92	51.47	58.59	64.26
NH_4Br $\Delta Ht = 3.22^{138}$							
NH_4Cl $\Delta Ht = 1.046^{-30.6}$				103			
$\Delta Ht = 3.950^{184.6}$							
NH_4ClO_4 $\Delta Ht = 9.6(8)^{240}$			168.5^{525}	148.7			
NH_4I $\Delta Ht = 2.93^{-13}$	20.9			89.0	103.3	117.7	

TABLE 6.4 Heats of Melting, Vaporization, and Sublimation and Specific Heat at Various Temperatures of the Elements and Inorganic Compounds (*Continued*)

Substance	ΔHm	ΔHv	ΔHs	C_p			
				400 K	600 K	800 K	1000 K
Antimony							
Sb	19.62	193.43		25.94	27.70	29.50	31.38
$SbBr_3$	14.6	69.9		125.52(1q)	81.63(g)	82.22	82.51
$SbCl_3$	12.69	43.5		123.4(1q)	81.63(g)	82.22	82.51
$SbCl_5$	10.0	48.4					
SbI_3	22.80	61.1		106.61(1q)	143.51(1q)	82.22(g)	82.51(g)
Sb_2O_3 $\Delta Ht = 7.1^{573}$	54.4	74.56		108.53	122.84	137.15	150.62
Sb_2S_3				123.34	134.39	145.44	
Argon							
Ar	1.183	6.469		20.79	20.79	20.79	20.79
Arsenic							
As	27.740			25.61	27.45	29.33	
$AsBr_3$	17.15	41.84					
$AsCl_3$		31.4		133.47(1q)	88.28(g)	88.28	
AsH_3				45.44	53.18	58.83	63.89
As_2O_3 $\Delta Ht = 2.80^{-33}$	18.41			116.36			
Barium							
Ba	8.0(6)	140.3		33.23	33.94(c)		39.07(1q)
$BaBr_2$	32.0(6)			79.22	83.54	87.86	92.19
$BaCl_2$ $\Delta Ht = 16.9(2)^{925}$	16.0(3)	246.4		77.33	80.43	84.30	89.47
$BaCO_3$ $\Delta Ht = 18.8^{806}$				99.0	113.0	124.2	134.6
BaF_2 $\Delta Ht = 2.67(84)^{1207}$	23.4(6)	285.4	405.1	75.94	80.33	84.94	94.56
BaI_2	26.5(5)	43.9	302.5	79.54	83.51	87.53(c)	112.97(1q)
$BaMoO_4$				129.5	143.5	152.2	159.3
BaO	59	330.6	424.3	49.90	53.22	55.41	57.13
$Ba(OH)_2$	16.7(13)			112.6	122.7(c)	141.0(1q)	
$BaSO_4$	40.58			119.37	131.63	135.90	137.90
$BaTiO_3$ $\Delta Ht = 0.067^5$				111.50	121.80	126.15	128.74
Beryllium							
Be	7.9(5)	297	291	19.97	23.34	25.46	27.27
$BeAl_2O_4$ chrysoberyl	170.0(38)			130.28	154.97	166.83	174.20

Compound	1	2	3	4	5	6	7
$BeBr_2$	9.8		515	70.6	77.6(c)	113.0(1q)	113.0
Be_2C	75.3	100.0		47.6	51.9	64.7	73.2
$BeCl_2$ $\Delta H_t = 6.8(6)^{403}$	15.3(6)	105	136.0	68.70	75.81(c)	121.42(1q)	121.42
BeF_2 $\Delta H_t = 0.92^{227}$	4.757	199.4		62.52	67.45	74.14(c)	85.64(1q)
BeI_2	20.9(126)	96	125	76.9	84.2	117.59	123.6
Be_3N_2	129.3			84.43	106.52	46.66	49.26
BeO $\Delta H_t = 6.7(17)^{2100}$	84.9(63)			33.76	42.38	55.1	57.9
BeS				42.6	51.0	166.0	174.1
Be_2SiO_4				120.8	149.2	149.79	174.43
$BeSO_4$ $\Delta H_t = 1.113^{590}$ $\Delta H_t = 19.55^{635}$				103.88	126.82		
$BeWO_4$				113.0	131.3	142.9	153.0
Bismuth							
Bi	10.88			26.99(c)	31.80(1q)	31.80	31.80
$BiBr_3$	21.71	75.40					
$BiCl_3$	10.88	72.59					
BiI_3		20.92					
Bi_2O_3 $\Delta H_t = 116.7^{717}$	28.5			116.90	123.60	130.29	136.98
Bi_2S_3				131.13	136.23	141.34	146.44
Bi_2Te_3	120.5			164.31	179.75	192.26	
Boron							
B	50.2(17)	506	552	15.69	20.78	23.36	24.98
BBr_3	105.(10)	30.54		72.63(g)	77.61	79.79	81.13
B_4C				76.36	98.37	107.65	114.30
BCl_3		23.85		68.42(g)	75.04	78.17	79.82
BF_3		19.33	57.54	67.07	72.57	75.80	
BH_3				38.93	45.45	52.34	58.40
B_2H_6	4.44	14.43		74.27	101.34	121.67	136.40
B_5H_9	6.134	28.41		130.2(g)	187.6	227.4	254.4
$B_{10}H_{14}$	21.97(4)	43.1		250.0(1q)	351.6(g)	417.21	460.4
BN			728	26.28	35.23	40.46	44.35
$B_3N_3H_6$ borazine		32.09		126.9	169.4	197.2	216.6
B_2O_3		390.4		77.94	98.12(c)	129.70(1q)	129.70
$B_3O_3H_3$ boroxin	24.1(4)			120.1	162.8	194.6	214.2
Bromine							
Br_2	10.571(8)	29.56		36.72(g)	37.30	37.59	37.79
$BrCl$	10.4	34.7					

TABLE 6.4 Heats of Melting, Vaporization, and Sublimation and Specific Heat at Various Temperatures of the Elements and Inorganic Compounds (*Continued*)

Substance	ΔHm	ΔHv	ΔHs	C_p 400 K	600 K	800 K	1000 K
BF_3	12.05	42.84		72.59	77.95	80.12	81.17
BrF_5				113.0	123.2	127.3	129.3
Cadmium							
Cd	6.4(2)	100.0(21)		27.15(c)	29.71(1q)	29.71	29.71
$CdBr_2$	20.92	112.97					
$CdCl_2$	30.1	124.93		79.83	86.27	92.72	104.60
CdF_2	22.6	217.6					
CdI_2	15.31	106.3					
$Cd(NO_3)_2 \cdot 4H_2O$	32.64						
CdO			225.1	43.8	45.6	47.3	49.1
CdS			209.6	55.5	56.2	57.0	57.7
$CdSO_4$				108.3	123.8	139.2	154.7
Calcium							
Ca $\Delta Ht = 0.93(8)^{43}$	8.54(10)	154.7		26.89	30.01	33.82	39.71
$Ca(BO_2)_2$	74.1			125.0	144.9	157.2	176.2
CaB_4O_7	113.4			202.0	243.0	267.7	287.8
$CaBr_2$	29.1(6)	200	298.3	77.99	80.50	83.47	88.62
CaC_2 carbide	32						
$CaCl_2$	28.5(8)	235		75.6	78.2	80.9	85.8
$CaCN_2$ cyanamide	0.432						
CaF_2 $\Delta Ht = 4.8(4)^{1151}$	29.7(4)	308.9	441(2)	73.9	78.5	83.9	90.1
CaI_2	41.8(8)	179.4	243(13)	79.16	83.14	87.07	91.00
$Ca[Mg(CO_3)_2]$ dolomite				143.30	163.34	176.77	188.28
$CaMoO_4$				131.3	144.9	153.5	160.6
Ca_3N_2				122.2	140.8		
$Ca(NO_3)_2$	21.42			173.68	210.50	243.38	
CaO	79.5			46.63	50.48	52.40	53.74
$Ca(OH)_2$ $\Delta Hdec$	99.2			98.4	107.4		
$Ca_3(PO_4)_2$ $\Delta Ht = 15.5^{1100}$				255.1	295.6	331.3	365.7
CaS				49.2	51.5	53.0	54.1
$CaSiO_3$ $\Delta Ht = 7.1^{1190}$	56.1			100.42	112.97	119.24	123.80

Ca_2SiO_4 $\Delta H_t = 4.44^{675}$; $\Delta H_t = 3.26^{1420}$				146.4	162.8	179.2	184.0
$3CaO \cdot SiO_2$	28.0	23.33		196.4	218.4	230.8	240.4
$CaSO_4$				109.7	129.5	149.2	169.0
$CaSO_4 \cdot \frac{1}{2}H_2O$				147.4	167.2	186.9	206.7
$CaSO_4 \cdot 2H_2O$				260.7	280.3	300.0	319.8
$CaTiO_3$ $\Delta H_t = 2.30^{1257}$				112.26	123.14	127.65	130.37
$Ca(VO_3)_2$				182.9	206.7	230.5	254.4
$CaWO_4$				127.57	140.16	147.28	152.80
Carbon							
C graphite	8.11			11.82	16.84	19.83	21.61
$(CN)_2$ cyanogen				61.9(g)	68.20	72.9	76.4
CNBr	11.38		45.40	50.14(g)	53.66	56.20	58.14
CNCl				48.72	52.83	55.81	57.75
CNI			59.4	50.84	53.75		57.40
CO $\Delta H_t = 0.632^{-211.62}$	0.837	6.042		29.34	30.44	31.90	33.18
CO_2	8.33	15.82	25.23	41.32	47.32	51.43	54.31
C_3O_2		$26.87^{-43.5}$		75.0	85.5	92.7	97.7
$COCl_2$	5.402	24.40		63.94	71.05	74.99	77.37
COF_2	5.738	16.15		54.78	64.88	70.81	74.42
COS	4.727	18.57		45.85	51.25	54.68	56.99
CS_2	4.40	27.03		49.7	54.6	57.5	59.3
Cerium							
Ce $\Delta H_t = 3.01^{730}$	5.2(12)	398	419	30.6	30.8	32.1	33.8
$CeCl_3$		170.1	326				
CeI_3	51.9						
CeO_2				66.9	69.0	71.1	73.2
Cesium							
Cs	2.087(4)	63.85	76.6(10)	31.52	31.00	30.93(1q)	20.79(g)
CsBr	23.64	151		52.93	55.02	57.20(c)	77.4(1q)
CsCl $\Delta H_t = 3.77^{470}$	15.9(42)	115.1		54.69	59.10	63.68(c)	77.40(1q)
CsF	21.7	115.5		53.8	57.4	60.9(c)	74.1(1q)
CsI	23.9	150.2		51.9	57.8(c)	65.5(1q)	67.8
$CsIO_3$	13.01						
CsOH $\Delta H_t = 1.30(13)^{137}$; $\Delta H_t = 6.1(6)^{220}$	4.56(40)	120		74.4(c)	81.6(1q)	81.6	81.6
Cs_2SO_4 $\Delta H_t = 4.3(8)^{667}$	35.7(8)		76.5(6)	112.1	132.2	163.2	194.2

TABLE 6.4 Heats of Melting, Vaporization, and Sublimation and Specific Heat at Various Temperatures of the Elements and Inorganic Compounds (*Continued*)

Substance	ΔH_m	ΔH_v	ΔH_s	C_p			
				400 K	600 K	800 K	1000 K
Chlorine							
Cl_2	6.406	20.41		35.30	36.55	37.11	37.44
ClF				33.77	35.61	36.53	37.03
ClF_3	7.61	27.53		70.61(g)	76.84	79.43	80.71
ClF_5		22.93		110.0	121.6	126.3	128.6
ClO				33.2	35.3	36.3	36.9
ClO_2		27.28		46.07	51.38	54.21	55.84
ClO_3F		19.33		75.95	89.20	96.09	99.92
Cl_2O	3.83	26.28		51.38	54.75	56.16	56.87
Cl_2O_7		34.69					
Chromium							
Cr $\Delta Ht = 0.0008$[38.4]	20.5(42)	339.5	397(4)	25.23	27.72	29.43	31.86
$CrCl_2$	32.2	196.7		72.59	77.03	81.46	85.90
$CrCl_3$			237.7	93.09	98.99	104.85	110.75
$Cr(CO)_6$			72.0	233.90			
CrN $\Delta Hdec = 112$			49.1	50.4	51.7	53.0	
CrO_2Cl_2	23.4	34.52					
CrO_2F_2	15.77	34.3					
CrO_3	129.7			63.85	72.55	76.65	78.83
Cr_2O_3				112.7	120.5	124.3	127.0
$Cr_2(SO_4)_3$				316.9	345.2	373.5	401.8
Cobalt							
Co $\Delta Ht = 0.452$[427.]	16.19(25)	377	424	26.52	29.67	32.43	36.99
$CoCl_2$	45.(6)	146	219.(6)	81.71	84.60	86.82	88.24
CoF_2	59.(13)	202	315.1(12)	75.7	80.8	82.9	84.2
CoF_3				97	100	102	104
CoO				52.9	54.3	54.8	56.0
Co_3O_4				143	163	185	210
$CoSO_4$ $\Delta Ht = 2.1(4)$[691]				119	141	152	158
Copper							
Cu	13.1(4)	300.4(20)	337.7(20)	25.32	26.48	27.49	28.66

6.112

$CuBr$ $\Delta H_t = 5.86^{380}$	9.6			56.48	59.79(c)	66.94(1q)	66.94
$\Delta H_t = 2.9^{465}$							
$CuCl$	10.2(21)	54.2(est)	241.8	56.9	61.5(c)	66.94(1q)	66.94
$CuCl_2$				75.06	78.79	73.14	78.03
$CuCN$			12.(4)		66.73		
CuF			268	55.5	59.6	87.03	90.37
CuF_2	55.(20)	156	261.(2)	72.38	81.92	60.21	66.94
CuI	10.9			55.40	57.82	53.3	55.3
CuO	11.80			46.8	50.9	77.67	81.26
Cu_2O	64.8(6)			67.69	73.45	53.18	55.40
CuS	10.9			48.79	50.96	85.02	85.02
Cu_2S $\Delta H_t = 3.85^{103}$				97.28	97.28		
$\Delta H_t = 0.84^{350}$							
Cu_2Se $\Delta H_t = 4.85^{110}$				90.88	91.71	92.55	93.39
$CuSO_4$				114.93	136.32	147.70	153.76
Dysprosium							
Dy	20	280	290.4				
Erbium							
Er	19.90	280	317.2				
Europium							
Eu	9.21	176	178				
Fluorine							
F_2 $\Delta H_t = 0.728^{-227.6}$	0.510	6.535					
FNO_3				75.11	87.76	94.78	98.86
Gadolinium							
Gd	10.05	301.3		36.57	35.52	34.52	33.51
Gd_2O_3				113.4	120.1	124.4	127.9
Gallium							
Ga	5.59(8)	270.3	286.2	27.15(1q)	26.69	26.57	26.57
Ga_2O_3				91.42	112.51	133.55	
Germanium							
Ge	36.8	328	283	24.94	26.44	27.41	28.20
$GeCl_4$		29.41					
GeH_4		15.10					
GeO_2	43.9			58.87	64.85	70.88	76.86

TABLE 6.4 Heats of Melting, Vaporization, and Sublimation and Specific Heat at Various Temperatures of the Elements and Inorganic Compounds (*Continued*)

Substance	ΔHm	ΔHv	ΔHs	C_p 400 K	600 K	800 K	1000 K
Gold							
Au	12.6(2)	343.(11)		25.82	26.78	27.78	28.83
AuSn	25.61			54.10	63.30(c)	60.57(1q)	
Hafnium							
Hf $\Delta Ht = 5.9(2)^{1750}$	25.9	571.(25)	618.4(60)	26.75	28.58	30.29	31.94
$HfCl_4$	75		99.6	125.44	105.77	106.69	107.15
HfO_2 $\Delta Ht = 10.5^{1700}$	104.6			67.66	73.93	77.32	79.87
Helium							
He	0.0138	0.0817		20.786	20.786	20.786	20.786
Holmium							
Ho	16.87	280	317				
Hydrogen							
H_2	0.117	0.904		29.18	29.33	29.62	30.21
$^1H^2H$				29.23	29.40	29.89	30.71
$^2H_2$				29.24	29.62	30.50	31.64
HBO_2				61.51(c)			
HBr			242.1	29.20	29.79	30.88	32.13
HCl $\Delta Ht = 1.188^{-174.77}$	2.406	17.615		29.16	29.58	30.50	31.63
^2HCl	1.992	16.142		29.41	30.61	32.14	33.47
HClO				39.96	43.97	46.57	48.49
HCN				39.41	44.18	47.91	50.96
HF	8.406	25.217		29.15	29.23	29.55	30.17
^2HF	3.93	7.5		29.16	29.56	30.46	31.58
H_2F_2 dimer				49.71	56.53	61.00	64.43
HFO				38.57	42.83	45.74	47.94
HI	2.870	19.77		29.33	30.35	31.80	33.13
HNCO isocyanic acid				50.58	58.28	63.55	67.49
HNCS isothiocyanic acid				53.18	60.96	65.86	69.33
HNO_2 *cis*				51.36	59.90	65.42	69.24
trans				52.07	60.35	65.63	69.28
HNO_3	10.47	39.46		63.19	76.77	85.04	90.43

Compound							
H₂O	6.009	40.66	50.92	34.26	36.33	38.72	41.27
¹H²HO				34.77	37.45	40.41	43.28
²H₂O	12.50	51.63	47.11	35.64	38.84	42.24	45.42
H₂O₂	12.68		52.43	48.45	55.69	59.83	66.65
²H₂O₂	9.67						
HPH₂O₂	12.84						
H₃PO₃	13.4(8)						
H₃PO₄				175.7	236.0	296.2	356.5
H₂S ΔHt = 1.531⁻¹⁶⁹·⁶¹	2.377	18.67	35.58	38.94	42.52	45.79	
ΔHt = 0.452⁻¹⁴⁶·⁹¹							
HSO₃F	10.71(13)			87.48	102.62	111.04	116.26
H₂SO₄	19.46(4)	50.2		158.2	197.0(1q)	125.9(g)	132.7
H₂SO₄·H₂O	18.24(17)			228.5			
H₂SO₄·2H₂O	24.0(2)			294.6			
H₂SO₄·3H₂O	30.64(25)			347.8			
H₂SO₄·4H₂O				410.3			
Indium							
In	3.26	231.84	243.1	28.52(c)	30.08(1q)	30.08	30.08
Iodine							
I₂	15.52(16)	41.95	62.42	80.67(1q)	37.61(g)	37.85	38.08
ICl	11.602		52.93	98.32(1q)	89.96	81.59	73.22
IF				35.08	36.58	37.30	37.74
IF₅				476.1(g)	516.7	533.0	541.4
IF₇				152.0(g)	167.6	173.9	177.0
Iridium							
Ir	3.26	231.84	243.1	28.53(c)	30.08(1q)	30.08	30.08
IrO₂				63.81	76.53	89.24	101.96
Iron							
Fe ΔHt = 0.90(42)⁹¹¹	13.8(4)	340.(13)	415.5(13)	27.39	32.05	37.95	54.43
ΔHt = 0.837(84)¹³⁹²							
FeBr₂ ΔHt = 0.418³⁷⁷	50.(13)		207.5	82.95	86.95	91.4	95.9
Fe₃C ΔHt = 0.75¹⁹⁰	51.46			115.7	114.7	117.2	119.8
FeCl₂	43.0(2)	26.3		79.7	83.1	85.5	102.2
FeCl₃	43.10	43.76		106.69(c)	133.9(1q)	82.34(g)	81.50
FeCO₃				93.5	115.9	138.3	
Fe(CO)₅	13.23(1)	33.72		189.0	209.8	223.1	232.2

TABLE 6.4 Heats of Melting, Vaporization, and Sublimation and Specific Heat at Various Temperatures of the Elements and Inorganic Compounds (*Continued*)

Substance	ΔH_m	ΔH_v	ΔH_s	C_p 400 K	600 K	800 K	1000 K
$FeCr_2O_4$				152.00	167.67	175.90	182.17
FeF_2	51.9	224.4	316.(8)	71.97	77.15	80.25	82.13
FeF_3			274	96.4	96.8	99.3	101.8
FeI_2 $\Delta H_t = 0.8^{377}$	44.77	104.6	192	83.9	84.4	110.9	113.0(lq)
Fe_2N				72.6	77.7	82.8	87.9
FeO	24.06			51.84	54.89	57.32	59.37
Fe_2O_3 $\Delta H_t = 0.67^{677}$	138.1(84)			120.12	141.17	158.22	150.62
Fe_3O_4			243.5(25)	171.13	212.55	252.88	
$Fe(OH)_2$				102.1	111.3	118.9	123.4
$Fe(OH)_3$				118.0	140.6	154.8	164.9
FeS $\Delta H_t = 0.40(20)^{138}$ $\quad\Delta H_t = 0.095(20)^{325}$	31.5(21)			89.20	62.03	58.56	59.00
FeS_2 marcasite				69.17	74.59	78.67	82.76
FeS_2 pyrite				68.85	74.31	78.35	82.45
$FeSiO_3$				100.8	114.3	124.5	133.9
Fe_2SiO_4	92			150.9	168.5	179.7	189.1
$FeSO_4$				116.7	138.0	149.4	
$Fe_2(SO_4)_3$				307.0	363.3	393.3	409.2
$FeTiO_3$ ilminite	90.8	111.4	122.0	128.1	132.8		
Krypton							
Kr	1.640	9.012					
Lanthanum							
La $\Delta H_t = 2.85^{868}$	8.49	402.1		28.49	29.83	31.17	32.51
$LaCl_3$	54.4	192.1		105.77	110.08	114.35	118.66
La_2O_3				117.32	124.68	128.91	132.26
Lead							
Pb	4.77(1)	178.0	195.2	27.72	29.40	30.02	29.40
$Pb(BO_2)_2$				129.7	162.3		
PbB_4O_7				207	265	305	330
$PbBr_2$	16.4(8)	118.1	173.0	81.3	88.8		
$Pb(CH_3)_4$	10.86					112.1(lq)	112.1

Pb(C₂H₅)₄ — Pb(C2H5)4	8.80			80.1	85.9	111.5(1q)	111.5

Compound							
Pb(C$_2$H$_5$)$_4$	8.80			80.1	85.9	111.5(1q)	111.5
PbCl$_2$	21.9(8)	126	185.3(3)	99.70	123.64	147.57	95.6
PbCO$_3$				76.1	82.5	89.1	
PbF$_2$ $\Delta H_t = 1.46^{310}$	14.7(13)	157		78.86	83.71(c)	108.57(1q)	108.57
PbI$_2$	23.4(8)	118.6	172.2	135.27	148.91	159.03	168.20
PbMoO$_4$		207		50.4	55.4	55.0	57.8
PbO $\Delta H_t = 0.17^{488}$	25.5(4)			67.61	190.8	199.2	
PbO$_2$				173.1			
Pb$_3$O$_4$				50.5	52.4	54.3	
PbS	18.8(63)	230		101.50	113.55	125.60	56.2
PbSiO$_3$	26.02			152.0	173.3	184.2	138.36
Pb$_2$SiO$_4$	51.0			108.74	128.57	152.38	189.1
PbSO$_4$ $\Delta H_t = 17.2^{866}$	40.2			157.32	182.51	211.67	177.32
PbSO$_4 \cdot$ PbO							241.96

Lithium

Compound							
Li	3.000(15)		159.3(10)	27.61(c)	29.54(1q)	28.94	28.84
Li$_3$AlF$_6$ $\Delta H_t = 9.5(17)^{562}$	110.(5)	147.1		236.4	262.8	290.8	318.6
LiAlO$_2$	87.(21)			81.55	92.69	98.17	102.00
LiBH$_4$	27.2			91.02	129.7(c)	159.0(1q)	159.0
LiBeF$_3$	44.0(21)			104.6	180.2(c)	232.1(1q)	232.1
Li$_2$BeF$_4$	33.8(5)	265		150.5	85.06	96.94	108.35
LiBO$_2$				81.15			
Li$_2$B$_4$O$_7$	121.(6)			197.6	241.1	274.4	300.2
LiBr	17.66	107.11		51.25	56.11	64.48(c)	65.27(1q)
LiCl	19.83(42)			50.97	67.69	65.80	161
LiClO$_4$	29.29			130.(c)	161.(1q)	161	
Li$_2$CO$_3$ $\Delta H_t = 0.561^{350}$ $\Delta H_t = 2.238^{410}$	44.8(4)			112.17	149.37	158.99	
LiF	27.09(2)	146.8	276.1	46.54	51.59	55.73	59.58
LiH	22.59(42)	97.5	231.3	34.79	46.38	57.30	63.18
LiI	14.64			53.43	58.66(c)	63.18(1q)	
LiIO$_3$ $\Delta H_t = 2.22^{260}$				87.10	106.43	124.38	140.95
Li$_3$N	58.6			64.03	73.82	80.56	86.21
Li$_2$O				82.7(c)	80.21(g)	81.42	82.05
Li$_2$O$_2$				58.03	68.21(c)	87.09(1q)	87.09
LiOH	20.88(21)	187.9	250.6		134.3	144.4	152.3
Li$_2$SiO$_3$	28.0(21)			118.8			

TABLE 6.4 Heats of Melting, Vaporization, and Sublimation and Specific Heat at Various Temperatures of the Elements and Inorganic Compounds (*Continued*)

Substance	ΔHm	ΔHv	ΔHs	C_p 400 K	600 K	800 K	1000 K
$Li_2Si_2O_5$ $\Delta Ht = 0.941$[936]	53.8			174.9	205.7	222.6	235.4
Li_2SO_4 $\Delta Ht = 28.5(17)$[575]	8.56(80)			139.2	168.5	196.1	223.4
Li_2TiO_3 $\Delta Ht = 11.51$[1212]	110.17			127.4	141.5	149.0	153.9
Lutetium							
Lu	78.03	414					
Magnesium							
Mg	8.48(42)	128	147	26.14	28.18	30.51	
$MgAl_2O_4$	192.(21)			137.99	157.89	169.47	178.66
$MgBr_2$	39.(4)	149	222	77.28	81.43	84.55	
$MgCl_2$	43.1(2)	156.2	249.2	75.71	79.87	82.55	
$MgCO_3$				89.86	109.04	122.26	131.80
MgF_2	58.7(4)	274.1	399.5	68.5	75.3	78.6	80.5
MgI_2	29.3(84)		206.8(80)	78.4	83.0	96.3(c)	100.4(1q)
Mg_3N_2 $\Delta Ht = 0.46$[550] $\Delta Ht = 0.92$[788]			107.6	113.8	119.9	123.8	
$Mg(NO_3)_2$				168.53	225.5		
MgO	79.(17)	(113)		42.56	47.43	49.74	51.21
$Mg(OH)_2$				91.71			
$Mg_3(PO_4)_2$	121			240.16	282.21	320.62	351.54
Mg_2Si	85.8			73.8	79.8	83.9	87.4
$MgSiO_3$ $\Delta Ht = 0.67$[630] $\Delta Ht = 1.63$[985]	(75)			94.2	107.0	115.8	120.3
Mg_2SiO_4	(71.0)			137.6	156.4	167.1	174.6
$MgSO_4$	14.6			110.0	127.6	140.5	151.7
$MgTiO_3$	(90.4)			105.2	118.5	125.4	129.9
Mg_2TiO_4	(130)			146	164	175	184
$MgTi_2O_5$	(146)			165.8	184.5	196.3	205.9
$Mg_2V_2O_7$				238.9	264.3	272.5	275.4
$MgWO_4$				123.4	137.0	146.1	154.8
Manganese							
Mn $\Delta Ht = 2.23(21)$[727]	12.1(10)	221(8)		28.53	31.90	34.92	37.55

$\Delta Ht = 2.12(33)$[101]
$\Delta Ht = 1.88(33)$[1137]

$MnBr_2$	33		113	77.82	82.80	87.74	127.40
Mn_3C $\Delta Ht = 14.94$[1037]				104.43	115.02	121.75	96.23(1q)
$MnCl_2$	37.7		149.0	77.19	81.84	85.14	
$Mn_2(CO)_{10}$			62.8			8075	
MnF_2	23.0			70.58	75.65	89.04	85.86
MnI_2	42			78.12	83.60	52.38	108.78
MnO	54.4			47.45	50.33	75.10	54.22
MnO_2				63.39	71.09	129.4	137.2
Mn_2O_3				109.0	120.8	53.7	55.2
MnS	26.4			50.7	52.2	179.7	189.3
Mn_3O_4 $\Delta Ht = 20.79$[1172]				157.3	169.5	119.5	124.2
$MnSiO_3$	66.9			100.9	113.1	147.7	
$MnSO_4$				119.0	136.7	125.7	
$MnTiO_3$				111.7	121.2		128.8
Mercury							
Hg	2.2953(4)	59.1(4)	61.38	27.41	27.14(1q)	20.79(g)	20.79
$HgBr_2$	17.9(33)	59.2		78.3	102.1(1q)	102.1	102.1
Hg_2Br_2	19.4(2)	58.9		109.6	115.6		
$HgCl_2$	23.0(40)	92.(17)		77.0(c)	102.9(1q)		
Hg_2Cl_2	18.97(21)	59.2(10)		106.0	112.1		
HgF_2	27.2(84)			77.0	81.2	85.4(c)	102.9(1q)
Hg_2F_2				104.7	111.7	116.9	
HgI_2 $\Delta Ht = 2.52(17)$[129]				82.01(c)	84.12(1q)	62.18(g)	62.24
Hg_2I_2				110.4(c)	136.4(1q)		
HgO				48.33	54.12	54.1	
HgS $\Delta Ht = 4.2$[386]				48.0	51.0		
Molybdenum							
Mo	36.0(54)	617.(21)	664(13)	25.08	26.46	27.44	28.37
$MoBr_3$				106.9	109.8	112.7	
$MoCl_4$	17.(8)	61.5		135.(c)	146.4(1q)	175.7	
$MoCl_5$	18.8(63)	45		167.4(c)	175.7(1q)	175.7	175.7
$Mo(CO)_6$	4.326(4)	27.24	69.9	133.1	145.3	150.4	153.0
MoF_6 $\Delta Ht = 8.171(8)$[-9.65]		27.99(12)		63.55	71.17	76.49	81.44
MoO_2				83.09	91.78	100.03	109.04
MoO_2	48.5(42)						

TABLE 6.4 Heats of Melting, Vaporization, and Sublimation and Specific Heat at Various Temperatures of the Elements and Inorganic Compounds (*Continued*)

Substance	ΔH_m	ΔH_v	ΔH_S	C_p 400 K	600 K	800 K	1000 K
MoS_2	130.(42)			68.91	73.60	76.23	78.24
Mo_2S_3				117.5	127.4	135.2	142.3
Neodymium							
Nd $\Delta H_t = 2.98^{862}$	7.14	289		28.2	32.1	36.9	42.0
Nd_2O_3				120.3	130.3	137.7	144.4
Neon							
Ne	0.335	1.741					
Neptunium							
Np $\Delta H_t = 8.37^{280}$	9.46	336		34.81			
Nickel							
Ni	17.15(40)	377.5		28.49	30.00	31.00	32.22
$NiCl_2$	77.2(8)		231.0(10)	76.29	79.88	80.90	
$Ni(CO)_4$	13.83	29.3		160.4(g)	173.2	182.1	188.6
NiF_2				76.4	78.5	82.6	
NiO				52.17	51.84	53.56	55.23
NiS $\Delta H_t = 6.4(4)^{379}$	30.1(29)			12.07	13.22	13.71	15.07
Ni_3S_2 $\Delta H_t = 56.24(4)^{556}$	19.7(4)			127.1	139.9	150.7	188.6
NiS_2	65.7(83)			72.8	79.0	81.0	85.2
$NiSO_4$				142.6	150.8	159.2	167.4
$NiWO_4$				138.9	144.6	150.3	155.9
Niobium							
Nb	26.9(8)	689.9	726.(20)	25.4	26.3	27.2	28.0
$NbBr_5$	24.0(63)	50.2	112.47	147.9(c)	147.9(1q)		
$NbCl_5$	34.0(10)	55.2		170.7(c)	127.9(g)	129.8	130.7
NbF_5	17.6	46.0		43.50(1q)			
NbI_5	37.7	58.6		182.0(c)			
NbN $\Delta H_t = 4.2^{1370}$	46.0			45.40	49.92	51.59	53.22
NbO	85.(21)	618.(21)		44.0	47.2	49.5	51.5
NbO_2 $\Delta H_t = 3.42^{817}$	92.(21)		598.0(50)	63.55	71.65	79.50	87.50
Nb_2O_5	104.0(20)			145.0	160.7	170.0	175.5

	ΔHm	ΔHv					
Nitrogen							
N_2 $\Delta Ht = 0.230^{-237.53}$	0.720	5.577		29.249	30.110	31.433	32.697
NF_3		11.59		61.86	71.43	75.97	78.36
N_2F_2 *cis*	15.36	91.6		58.16	68.28	73.64	76.61
trans	14.23	87.9		60.21	68.91	73.81	76.65
NH_3, *see* under Ammonium							
N_2H_4	12.66	40.59		61.70(g)	77.56	88.20	96.36
NO	2.301	13.774		29.94	31.24	32.77	33.99
$NOCl$		25.69		47.15	50.69	53.17	54.89
NOF		19.3		44.57	48.90	51.68	53.49
NOF_3				78.74	90.88	97.03	100.50
NO_2				40.17	45.82	49.71	52.17
NO_2Cl		20.92		59.64	68.14	73.15	76.15
NO_2F		18.0		57.01	66.42	71.93	75.26
NO_3				55.9	67.4	73.3	76.5
N_2O	6.540	16.552		42.68	48.39	52.24	54.87
N_2O_3		39.33		72.74	82.93	89.48	93.62
N_2O_4	14.653(13)	38.12		88.52	104.0	113.4	119.2
		(to $N_2O_4 + NO_2$)					
N_2O_5			62.3	110.9	128.4	137.0	141.4
Osmium							
Os	31.75	738		25.1	25.9	26.7	27.4
OsF_6	16.99	28.62					
OsO_4		39.54					
Oxygen							
O_2 $\Delta Ht = 0.092^{-249.49}$							
$\Delta Ht = 0.745^{-229.38}$	0.444	6.820	8.204	30.106	32.090	33.733	34.874
O_3		10.84		43.74	49.86	53.15	55.02
OF_2		11.09		64.28	72.37	76.38	78.57
O_2F_2		19.18					
Palladium							
Pd	17.6	362.(11)		26.53	27.66	28.83	29.96
$PdCl_2$	40.08						
PdO				37.6	49.5	61.3	
Phosphorus							
P_4 $\Delta Ht = 0.521(1)^{-77.8}$	0.659(2)	56.5(4)	58.9(8)	73.26(g)	78.39	80.39	81.36

TABLE 6.4 Heats of Melting, Vaporization, and Sublimation and Specific Heat at Various Temperatures of the Elements and Inorganic Compounds (*Continued*)

Substance	ΔH_m	ΔH_v	ΔH_s	C_p 400 K	600 K	800 K	1000 K
PBr_3		39.7		78.91	81.17	82.01	82.42
PCl_3	4.52	30.5		76.02(g)	79.75	81.17	81.88
PCl_5			64.9	120.1(g)	126.8	129.5	130.7
PF_3		14.6		66.27(g)	73.97	77.58	79.46
PF_5		17.20		99.2(g)	114.7	121.9	125.6
PH_3	1.130	14.60		41.78	50.91	58.51	64.30
$(P_2O_3)_2$	14.06	43.43		172.2	200.8	213.5	220.0
P_4O_{10}	34.3			260.25	336.0(c)		
$POCl_3$	34.3		106.0(3)	91.96(g)	99.12	102.51	108.53
$POClF_2$		35.2		79.33	91.58	97.74	101.09
$POCl_2F$		25.44		87.66	96.58	100.90	103.23
POF_3		30.96		79.09	91.23	97.44	100.85
$PSCl_3$		23.22	21.1	96.52	102.38	104.77	105.90
PSF_3	15.06	19.58		84.53	95.28	100.28	102.89
P_4S_3	9.2	59.8		184.1	184.1(1q)	155.0(g)	155.0
Platinum							
Pt	19.7	469.(25)	545.(21)	26.40	27.49	28.53	29.62
PtS				51.4	53.8	56.2	58.6
PtS_2				69.9	75.9	81.9	87.9
Plutonium							
Pu $\Delta Ht = 13.4$[122]	66.9	333.5		39.54	46.86	40.59	40.59
$\Delta Ht = 2.9$[206]							
$\Delta Ht = 3.3$[319]							
$\Delta Ht = 66.9$[480]							
$PuBr_3$	56.1	236.4	292.5				
$PuCl_3$	63.6	241.0	304.6				
PuF_3	54		374.9				
PuF_4	42.7		299.6				
PuF_6	18.64	29.9	48.5				
PuI_3	50.2						
PuO_2		559.8					

Polonium							
Po		102.91					
Potassium							
K	2.334	76.90	88.8(8)	31.50(1q)	30.14	29.75	30.36
$KAlCl_4$				165.5	183.2	196.6	202.1
K_3AlCl_6				259.2	279.5	295.8	
K_3AlF_6				244.5	269.4	286.8	302.0
KBF_4 $\Delta H_t = 14.06(16)^{283}$	17.66(16)			130.82	142.10	150.94	167.17
KBH_4				100.92	106.03	118.41	
KBO_2	31.4(42)			76.65	89.83	98.49	
$K_2B_4O_7$	104.5(42)	238.9		206.3	250.5	271.1	283.3
KBr	25.5	149.2		53.81	56.36	60.42	68.01
KCl	26.28	124.3		53.14	56.32	60.08	66.97
$KClO_4$ $\Delta H_t = 13.77^{299.6}$	14.64	157.1		138.49	165.27		66.51(1q)
KCN $\Delta H_t = 1.167^{-104.9}$	27.6(4)			66.32	66.39	66.47(c)	188.95
K_2CO_3	28.95			128.14	150.67	170.04	
K_2CrO_4	36.69						
$K_2Cr_2O_7$	27.20						
KF		141.8	241.8	51.04	54.27	57.40	61.17
KH				44.14	51.92		
KHF_2 $\Delta H_t = 11.22(4)^{196.7}$	6.619(13)			86.07(c)	104.6(1q)	62.64(c)	72.38(1q)
KI	24.02	190.9	202.4	53.93	57.30		
KNO_3 $\Delta H_t = 5.10^{128}$	9.62			108.41	120.50		
K_2O				91.5	97.4	105.4	113.5
KO_2 $\Delta H_t = 0.302^{-79.7}$ $\Delta H_t = 0.157^{-42.3}$				83.89	90.17		
K_2O_2				107	121		
KOH $\Delta H_t = 6.4(6)^{243}$	8.6(6)	142.7	192	71.67	78.66(c)	83.11(1q)	83.11
KPO_3	8.83						
K_3PO_4	37.24						
$K_4P_2O_7$	58.58						
$KReO_4$	85.4						
K_2S	16.15(17)	77.3	82.5	87.7			
K_2SiO_3	50.(13)			135.6	157.7	170.7	179.1
K_2SO_4 $\Delta H_t = 8.45(42)^{584}$	34.39			147.6	172.5	199.6	226.1
K_2WO_4	19.46						
K_2ZrCl_6	23.01						

TABLE 6.4 Heats of Melting, Vaporization, and Sublimation and Specific Heat at Various Temperatures of the Elements and Inorganic Compounds (*Continued*)

Substance	ΔHm	ΔHv	ΔHs	C_p 400 K	600 K	800 K	1000 K
Praseodymium							
Pr	11.3(21)	331	356				
Promethium							
Pm	7.13	289	328				
Protoactinium							
Pa	16.7	481					
$PaCl_5$	92.9	61.30					
Radium							
Ra	8.5	113					
Radon							
Rn	3.247	18.10					
Rhenium							
Re	34.(4)	704	779.(8)	26.02	26.94	28.00	29.08
ReF_5	4.63	58.07					
ReF_6		28.73					
ReF_7	7.53	38.30					
ReO_2			274.64				
ReO_3	21.76		208.4				
Re_2O_7	66.1	74.1					
$ReOCl_4$		45.6					
$ReOF_4$	13.51	61.04					
$ReOF_5$		32.30	37.40				
Rhodium							
Rh	21.6	494	556.(11)	25.98	28.00	30.00	32.01
Rh_2O_3				109.9	121.4	133.0	144.5
Rubidium							
Rb	2.19(9)	75.77		31.72	30.85	30.67	
RbBr	15.5	154.8		52.80	54.94	57.07(c)	66.94(1q)
RbCl	18.41	165.7		52.30	54.35	56.44(c)	64.02(1q)
$RbClO_4$ $\Delta Ht = 12.59^{284}$							

RbF	26.4	177.8		51.9	57.9	64.9	72.3
RbI	12.6	150.6			55.1	57.3(c)	66.9(1q)
RbNO$_3$	5.61						
RbOH	6.78						
Ruthenium							
Ru $\Delta Ht = 0.13^{1035}$ $\Delta Ht = 0.96^{1500}$	25.98	591.6		24.48	25.73	27.00	28.24
Samarium							
Sm $\Delta Ht = 3.11^{917}$	8.9(4)	165.(17)	207	33.30	39.08	44.27	49.29
Sm$_2$O$_3$ $\Delta Ht = 1.05^{922}$				125.19	135.31	141.38	146.27
Scandium							
Sc	15.77	332.71	376.(20)	96.69	102.68	108.66	114.64
ScCl$_3$							
Sc$_2$O$_3$				106.40	111.09	115.81	120.54
Selenium							
Se $\Delta Ht = 0.75^{150}$	5.41	95.90		28.07(c)	35.15(1q)	35.15	
SeF$_4$	8.4	46.9		127.9	141.3	147.1	150.7
SeF$_6$							
SeO$_2$	4.23	94.47	26.8				
SeOCl$_2$		42.7					
Silicon							
Si	50.2(4)	359	450.(12)	22.14	24.15	25.36	26.34
				104.9(g)	106.2	106.9	
SiBr$_4$		35.9	146.4(1q)	34.10	41.79	45.88	48.42
SiC beta		29.7(1)		96.79(g)	102.53	104.84	105.97
SiCl$_4$	7.74		25.73	83.14	94.09	99.43	102.29
SiF$_4$				88.3	97.5	101.7	103.8
SiF$_3$Cl		12.5		51.47	65.88	76.71	84.51
SiH$_4$	0.67	26.61		60.66	74.01	83.09	89.41
SiH$_3$Cl		25.2		71.50	82.93	89.96	94.64
SiH$_2$Cl$_2$		26.6		83.66	92.47	97.21	100.16
SiHCl$_3$				57.19	71.78	81.66	88.43
SiH$_3$F				164.0(1q)	106.0(g)	106.9	107.3
SiI$_4$	19.7(21)	64	79	110.7	129.7	145.8	158.2
Si$_3$N$_4$				53.40	64.42	73.70	68.95
SiO$_2$ quartz $\Delta Ht = 0.73(17)^{574}$ $\Delta Ht = 2.0(6)^{806}$	7.7(8)		600.(33)				

TABLE 6.4 Heats of Melting, Vaporization, and Sublimation and Specific Heat at Various Temperatures of the Elements and Inorganic Compounds

Substance	ΔH_m	ΔH_v	ΔH_s	C_p			
				400 K	600 K	800 K	1000 K
$SiOF_2$	20.9			61.32	70.37	75.03	77.62
SiS_2				78.6	81.7	83.4	85.4
Silver							
Ag	11.95	258.0(60)					
AgBr	9.2	192.1		25.65	26.82	28.37	30.00
AgCl	13.0	177.8		58.95	71.84(c)	62.34(1q)	62.34
Ag_2CO_3				56.86	54.39	54.39	54.39
AgF	16.7	179.1		54.10(c)	122.6		
AgI $\Delta H_t = 6.15$[147]	9.41	144.4		64.68	58.41		
$AgNO_3$ $\Delta H_t = 2.5$[160]	11.53			112.47	56.48	56.48	58.56(1q)
Ag_2O				73.01	128.03		
Ag_2S $\Delta H_t = 5.86$[176] $\Delta H_t = 5.86$[586]	11.3			86.57	90.54	90.54	90.54
Ag_3Sb				108.5	121.9	135.3	
Ag_2Se $\Delta H_t = 6.7$[133]				90.29			
Ag_2SO_4 $\Delta H_t = 15.69$[430]	18.0			143.34	166.69	190.04	205.02
Ag_2Te $\Delta H_t = 0.71$[137]				96.65	92.88	92.88	92.88
Sodium							
Na	2.602	97.42	107.5(7)	31.51(1q)	29.31	28.95	28.95
$NaAlCl_4$				164.81(c)			
Na_3AlCl_6				254.4	273.0		
Na_3AlF_6 $\Delta H_t = 8.37$[565] $\Delta H_t = 0.42$[880]	107.28			234.64	261.83	296.81	282.00
$NaAlO_2$ $\Delta H_t = 1.297$[467]				83.41	94.27	98.75	102.30
$NaBH_4$ $\Delta H_t = 0.999$–83.3				94.56	108.62		
$NaBO_2$	33.5(20)	239.7		75.40	88.58	97.24	103.22
$Na_2B_4O_7$	76.9(63)		322.2	221.7	268.6	444.9(1q)	
NaBr	26.11	160.75	217.48	53.47	56.07	58.58	61.09
NaCl	28.16(17)			52.35	55.48	59.31	72.51
$NaClO_3$	22.59						
$NaClO_4$ $\Delta H_t = 13.98$[308]				136.0(c)			

NaCN	8.79			68.7	68.8	69.0	179.83
Na_2CO_3 $\Delta Ht = 0.690^{450}$	29.64	148.1	172.8	125.10	163.30	153.34	59.51
NaF	33.35	176.1	284.9	49.60	52.68	55.71	
NaH				42.47	50.71		
NaI	23.60			53.81	56.23	58.49(c)	64.85(1q)
$NaIO_3$ $\Delta Ht = 35.10^{422}$							
$NaNO_3$	16.11			76.27	84.48	92.63	94.92
NaO_2 $\Delta Ht = 1.464^{-76.7}$							
$\Delta Ht = 1.548^{-49.9}$							
Na_2O $\Delta Ht = 1.76^{750.1}$	47.7			75.78	85.73	91.29	
$\Delta Ht = 11.92^{970.1}$							
Na_2O_2 $\Delta Ht = 5.73^{512}$				97.7	108.4	113.6	
NaOH $\Delta Ht = 72.(8)^{299.6}$	6.6(8)	175.3	228.2	64.90(c)	86.07(1q)	84.89	83.72
Na_2S	19.3(42)			20.15	20.85	21.55	22.00
Na_2S_2				104.3	115.4(c)	124.7(1q)	124.7
Na_2SiO_3	51.80			127.8	147.1	159.7	169.4
$Na_2Si_2O_5$ $\Delta Ht = 0.42^{678}$	35.6			183.4	217.6	235.2	292.9
Na_2SO_4 $\Delta Ht = 10.91(8)^{241}$	23.85(42)			145.1	175.3	187.3	200.3
Na_2TiO_3	70.3						
Na_2WO_4 $\Delta Ht = 30.85^{587.7}$	23.80			155.3	178.2	198.7	
$\Delta Ht = 4.113^{588.9}$							
Strontium							
Sr $\Delta Ht = 0.84(17)^{547}$	7.4(8)	136.9	164.0	27.78	29.81	31.94	34.14
$SrBr_2$ $\Delta Ht = 12.2(2)^{645}$	10.13(21)	194.1	310	79.04	82.68	87.57(c)	116.40(1q)
$SrCl_2$ $\Delta Ht = 6.0(13)^{727}$	16.2(6)	248.15	356(6)	78.89	83.66	90.79	105.83
$SrCO_3$ $\Delta Ht = 19.7^{924}$		320	451.0(12)	95.06	107.15	116.06	124.01
SrF_2 $\Delta Ht = 0.04^{1148}$	29.7(6)			74.7	79.8	81.0	85.8
$\Delta Ht = 0.04^{1211}$							
SrI_2	19.67(38)	189.7	286.6	80.75	86.27	91.76(c)	110.04(1q)
$SrMoO_4$				131.54	145.39	154.05	161.21
SrO	75			48.5	52.0	54.3	56.1
SrO_2				81.34	85.02		
$Sr(OH)_2$	21.0(20)			88.5	115.0(c)	157.78(1q)	157.78
SrS				50.2	53.2	54.9	56.2
$SrSO_4$				113.47	124.60	135.73	146.86
Sulfur							
S $\Delta Ht = 0.400(3)^{95.2}$	1.727(4)	45.35	64.22(17)	32.16	34.31(1q)	31.70(g)	

TABLE 6.4 Heats of Melting, Vaporization, and Sublimation and Specific Heat at Various Temperatures of the Elements and Inorganic Compounds (*Continued*)

Substance	ΔHm	ΔHv	ΔHs	C_p 400 K	600 K	800 K	1000 K
S_8				166.69	175.27	178.53	180.08
SCl_2		32.4(1)		53.6	56.0	56.9	57.4
S_2Cl_2		36.0(60)		124.3(1q)	80.83(g)	82.59	83.51
SF_4				87.50	97.34	101.65	103.84
SF_6		17.07	23.43	116.4	136.1	144.8	149.3
S_2F_{10}				211.4	246.4	261.8	269.2
SO_2	7.401	24.92		43.49	49.05	52.43	54.48
SO_3 gamma phase, III	1.97	41.8		57.67	67.26	72.76	75.97
II	10.33						
I	25.48						
$SOCl_2$		31.3		71.25	76.36	78.87	80.25
SOF_2		21.67		64.27	72.38	76.40	78.58
SO_2Cl_2		31.38		85.23	94.52	99.37	102.13
SO_2ClF				81.1	92.1	97.9	101.1
SO_2F_2		20.04		76.51	89.32	96.07	99.86
Tantalum							
Ta	36.6(42)	732.8	778.(6)	25.84	26.84	27.46	27.93
TaB_2	83.7			57.57	66.57	72.17	83.35
$TaBr_5$	37.7	58.6		168.20			
TaC	105.(21)			41.66	46.45	49.07	51.07
Ta_2C				66.65	72.43	76.23	79.50
$TaCl_5$	35.0(20)	56.9	94.1	148.(c)	129.(g)	131	132
TaF_5	18.8	46.0		182.0(1q)	182.0(c)		
TaI_5	41.8	64.9		164.6		120.0(g)	120.6
TaN	67			45.4	51.9	58.5	65.0
TaO_2				47.7	52.3	54.6	55.7
Ta_2O_5	120.(17)			147.5	164.4	175.2	182.8
Technetium							
Tc	23.81	585.2		25.10	26.78	28.45	30.12
TcF_6	4.72	31.07					
TcO_3F	22.50	39.55					

Tellurium							
Te	17.87	46.0		27.95	32.34(c)	37.66(1q)	37.66
$TeCl_4$	18.8	70.3		138.91(c)	222.59(1q)	108.78(g)	108.78
TeF_4		34.3					
TeF_6			28.2	132.21	143.76	148.66	151.67
Te_2F_{10}		39.50					
TeH_2		23.9					
TeO_2	29.08			67.86	72.55	76.07	79.24
Terbium							
Tb	10.79	293	389				
Thallium							
Tl $\Delta Ht = 0.38^{234}$	4.1	165	181	27.49(c)	30.12(1q)	30.12	30.12
TlBr	16.3	103.3		53.51	59.45(c)	75.49(1q)	67.82
TlCl	15.9	103.55		53.56	55.23(c)	59.41(1q)	59.41
TlF	13.81	115.9			66.78(1q)	67.28	
TlI	6.3	103.8		53.93	60.58(c)	71.96(1q)	71.96
$TlNO_3$	9.56						
Tl_2SO_4	23.01						
Thorium							
Th $\Delta Ht = 2.73^{1360}$	16.12	514		28.41	30.54	32.68	34.43
$ThCl_4$ $\Delta Ht = 5.0^{406}$	43.9	152.7		126.69	132.69	136.40	139.62
Th_3N_4				169.45	196.45	222.67	
ThO_2	1218.0			67.36	72.43	75.31	77.66
$ThOCl_2$				97.03	102.47	105.86	108.62
$Th(SO_4)_2$				196.98	243.17	289.37	
Thullium							
Tm	16.84	247	232.2	28.87	28.87(c)	28.66(1q)	28.66
Tin							
Sn $\Delta Ht = 2.09^{13}$	7.00	296.1	157.95(1q)	28.87	28.87(c)		
$SnBr_2$	7.20	136.0	83.26(c)	106.8(g)	107.3	107.5	
$SnBr_4$	11.92	41.0		92.05(1q)	92.05	92.05	
$SnCl_2$	12.76	81.6					
$SnCl_4$	9.2	33.9					
SnH_4		18.49					
SnO				45.81	48.74	51.67	54.60
SnO_2 $\Delta Ht = 1.88^{410}$				64.43	73.93	78.53	81.76

TABLE 6.4 Heats of Melting, Vaporization, and Sublimation and Specific Heat at Various Temperatures of the Elements and Inorganic Compounds (*Continued*)

Substance	ΔHm	ΔHv	ΔHs	C_p 400 K	600 K	800 K	1000 K
$\Delta Ht = 1.26^{540}$							
SnS $\Delta Ht = 0.67^{602}$				50.54	55.52	61.34	
SnS$_2$				71.92	75.44	78.95	82.47
Titanium							
Ti $\Delta Ht = 4.2(3)^{893}$	18.8	425.(11)	469.(4)	26.86	28.60	29.47	32.07
TiB				40.3	48.6	50.9	51.9
TiB$_2$	100.4			54.9	66.2	72.1	76.9
TiBr$_2$			206.2	79.9	82.1	84.4	86.7
TiBr$_3$			138.8	105.8	125.5	147.3	156.7
TiBr$_4$	12.89	45.19		151.88(1q)	106.1(g)	106.9	107.3
TiC	71			40.69	47.65	49.90	51.18
TiCl$_2$			212	73.4	78.4	82.2	85.9
TiCl$_3$			166.3	98.6	102.0	104.4	106.7
TiCl$_4$	9.966	41.09(8)		146.2(1q)	104.4(g)	106.0	106.7
TiF$_3$			222	93	98	103	109
TiF$_4$			97.9	126.65(c)	100.2(g)	103.27	104.92
TiH$_2$				39.33	53.81	63.14	68.53
TiI$_2$			217	87.0	88.4	89.9	91.3
TiI$_3$				117.5	119.0	120.4(c)	20.63(g)
TiI$_4$ $\Delta Ht = 9.9(6)^{106}$	19.8(6)	56.5		148.1(c)	156.5(1q)	25.70(g)	25.75
TiN	66.9			43.81	48.72	50.62	52.14
TiO $\Delta Ht = 4.2^{992}$	41.8			45.0	50.8	55.2	59.1
TiO$_2$ rutile	57.99		673	63.59	70.89	73.86	75.35
Ti$_2$O$_3$ $\Delta Ht = 1.138^{197}$	105			117.53	136.44	142.97	146.36
Tungsten							
W	35.(11)	806.7	851.(6)	24.93	25.89	26.67	27.56
WBr$_5$	17.15	81.50		166(c)	182(1q)	132.2(g)	132.5
WBr$_6$				192.5(c)	156.3(g)	157.0	157.4
WCl$_4$				135.3	146.2(c)	106.7(g)	107.2
WCl$_5$	20.5	68.06	100	167.4(c)	129.5(g)	131.0	131.8
WCl$_6$ $\Delta Ht = 4.1^{177}$	6.69	61.5	79.2	192.5(c)	200.8(1q)	155.8(g)	156.6

$W(CO)_6$			72.0				
WF_6 $\Delta H_t = 2.067^{-8.5}$	4.10	27.05	26.65(4)	132.4(g)	145.0	150.3	153.0
WO_2			666.3	63.43	71.30	75.47	78.17
WO_3 $\Delta H_t = 1.49^{777}$	73.4	76.6	550.2	82.2	93.1	98.2	101.7
$WOCl_4$	45.(13)	44.5	57.1	157.(c)	123.2(g)	127.0	129.1
WOF_4	5.0(42)	56.(8)		107.8	119.8	125.0	127.8
WO_2Cl_2				115.1	135.6(c)		
Uranium							
U $\Delta H_t = 2.93^{672}$	12.55	417.1	525	28.95	34.77	41.63	41.84
$\Delta H_t = 4.791^{772}$							
UBr_4	55.2	119.2		131.4	140.1(c)	163.2(1q)	163.2
UC				54.64	58.28	60.33	62.17
UCl_3	46.4	193.0		102.76	107.70	113.64	119.91
UCl_4	44.8	141.4		126.06	134.39	141.96	162.46
UCl_5	35.6	75.3		150.9	159.8(c)	186.7(1q)	134.52(g)
UCl_6	20.9	50.2		182.84	213.97	158.82	168.03
UF_3				98.99	104.89	111.04	117.24
UF_4	42.7	221.8		119.08	125.02	130.92	136.77
UF_5	33.5			136.4	143.1(c)	166.6(1q)	154.4
UF_6	19.196	28.899	48.20	140.5(g)	148.7	152.2	165.7
UH_3				50.88	57.36	66.14	
UI_4	70.7	130.6		140.6	149.5(c)	165.7(1q)	
UN				52.22	56.27	58.32	59.83
UO_2				72.68	79.79	83.18	85.49
UO_3				88.91	95.35	99.00	
U_3O_8				265.98	290.70	304.18	
$UOCl_2$				101.88	109.58	115.06	
UO_2Cl_2				118.07	126.23	130.00	
UO_2F_2				113.89	122.51	126.65	129.49
Vanadium							
V	22.8(63)	459.(29)	516	26.23	27.49	28.66	30.08
VCl_4	9.62	38.1					
VF_5	49.96	44.43	741	161.7(1q)	100.1(g)	102.6	104.7
VN $\Delta Hdec = 227.6^{2346}$				43.30	48.20	51.21	53.68
VO	54.4			49.6	53.5	57.1	60.5
VO_2 $\Delta H_t = 4.31^{72}$	56.90			67.20	74.35	77.78	80.17
V_2O_3 $\Delta H_t = 1.623^{-104.3}$	117.2			117.5	127.3	132.6	138.0

TABLE 6.4 Heats of Melting, Vaporization, and Sublimation and Specific Heat at Various Temperatures of the Elements and Inorganic Compounds (*Continued*)

Substance	ΔH_m	ΔH_v	ΔH_s	C_p 400 K	600 K	800 K	1000 K
V_2O_4 $\Delta H_t = 9.0(8)$[67]	112.1(25)			135.3	148.4	155.5	160.7
V_2O_5	64.5(31)	263.6		151.0	168.3	177.3	183.7
$VOCl_3$		33.5					
Xenon							
Xe	2.313	12.640		20.786(g)	20.786	20.786	20.786
Ytterbium							
Yb	7.66	159					
Yttrium							
Y $\Delta H_t = 4.97$[1485]	11.42	365	425.(8)	27.32	28.49	29.92	31.46
Y_2O_3 $\Delta H_t = 1.30$[1057]				113.34	121.29	124.73	126.86
Zinc							
Zn	7.32(11)	123.6(10)		26.35	28.59(c)	31.38(1q)	31.38
$ZnBr_2$	15.65	98.3		70.1(c)	78.8(1q)	113.8	61.5(g)
$ZnCl_2$	10.25	119.2		69.87(c)	100.83(1q)	100.83	100.83
ZnF_2				66.9	69.1	71.4	73.7
ZnO				45.35	49.54	51.63	53.18
ZnO $\Delta H_t = 13.4$[1020]				49.41	52.43	54.14	55.46
$ZnSO_4$ $\Delta H_t = 20.3(8)$[740]				116.02	137.44	139.70	142.01
Zn_2SiO_4				129.4	141.4	153.4	165.4
Zirconium							
Zr $\Delta H_t = 4.02(30)$[862]	20.9(63)	573.(5)	610.0(80)	25.94	27.28	28.97	31.13
ZrB_2	104.6			57.51	65.81	69.71	72.15
$ZrBr_2$	63.(42)	131.5	230	87.9	90.2	92.5	94.8
$ZrBr_4$				129.3	133.3(c)	107.24(g)	107.57
ZrC	79.5			43.59	49.38	52.27	53.39
$ZrCl_2$	27.(13)	45.0		76.0	80.0	83.1	85.9
$ZrCl_3$			190	101	106	109	112
$ZrCl_4$	50.(13)		110.5(5)	125.4	131.1(c)	106.5(g)	107.1
ZrF_2	38.(8)	289	404	70	76	81	84
ZrF_4	64.2(4)		237.7	113.5	124.0	129.4	134.1

ZrI_2	25.1	113		95.0	96.6	106.1	123.6
ZrI_3			176	105.9	106.7	107.1(c)	82.9(g)
ZrI_4			126.4(21)	131.0	134.6(c)	107.6(g)	107.6
ZrN		624		44.76	48.66	50.94	52.75
$ZrO_2, \Delta Ht = 5.02^{1205}$	67.4			63.85	70.24	73.45	75.75
$ZrSiO_4$	87.03			114.6	133.7	142.7	147.3

6.2 CRITICAL PHENOMENA

The *critical temperature* T_c of a gas is the temperature above which the gas cannot be liquefied no matter how high the pressure.

The *critical pressure* P_c is the lowest pressure which will liquefy the gas at its critical temperature.

The *critical volume* V_c is the volume of 1 mol at the critical temperature and the critical pressure. It can be computed from the critical density p_c as follows:

$$\frac{\text{Molecular weight (in g} \cdot \text{mol}^{-1})}{p_c(\text{in g} \cdot \text{cm}^{-3})} = V_c(\text{in cm}^3 \cdot \text{mol}^{-1})$$

The critical pressure, critical molar volume, and critical temperature are the values of the pressure, molar volume, and thermodynamic temperature at which the densities of coexisting liquid and gaseous phases just become identical. At this critical point, the *critical compressibility factor* Z_c is

$$Z_c = \frac{P_c V_c}{R T_c}$$

TABLE 6.5 Critical Properties

Substance	T_c, K	P_c, atm	V_c, cm$^3 \cdot$ mol^{-1}
Acetaldehyde	461	55	154
Acetic acid	594.4	57.1	171.3
Acetic anhydride	569	46.2	290
Acetone	508.1	46.4	209
Acetophenone	701	38	376
Acetonitrile	548	47.7	173
Acetyl chloride	508	58	204
Acetylene	308.3	60.6	113
Acrylic acid	615	56	210
Acrylonitrile	536	45	210
Air	132.5	37.2	92.7
Allene	393		
Allyl alcohol	545	56.4	203
Aluminum trichloride	629	26	261
Aminoethanol	614	44	196
Ammonia	405.6	111.3	72.5
Aniline	699	52.4	270
Anthracene	883		
Antimony tribromide	904.5	56	
Antimony trichloride	794		270
Argon	150.8	48.1	74.9
Arsine	373.0		
Benzaldehyde	695	21.5	
Benzene	562.1	48.3	259
Benzoic acid	752	45	341
Benzonitrile	699.4	41.6	
Benzyl alcohol	677	46	334
Biphenyl	789	38	502
Bismuth tribromide	1219		301

TABLE 6.5 Critical Properties (*Continued*)

Substance	T_c, K	P_c, atm	V_c, cm$^3 \cdot$ mol^{-1}
Bismuth trichloride	1179	118	261
Boron pentafluoride	470		
Boron tribromide	573		280
Boron trichloride	451.9	38.2	
Boron trifluoride	260.8	49.2	
Bromine	584	102	127
Bromobenzene	670	44.6	324
Bromochlorodifluoromethane	153.8	40.5	232
Bromoethane	503.8	61.5	215
Bromomethane	446.5	85	
Bromopentafluorobenzene	670	44.6	
1-Bromopropane	271.3		
2-Bromopropane	258.9		
Bromotrifluoromethane	340.2	39.2	200
1,2-Butadiene	443.7	44.4	219
1,3-Butadiene	425	42.7	221
Butanal	524	40	278
Butane	425.2	37.5	255
Butanenitrile	582.2	37.4	285
Butanoic acid	628	52.0	292
1-Butanol	562.9	43.6	274
2-Butanol	536.0	41.4	268
2-Butanone	535.5	41.0	267
1-Butene	419.6	39.7	240
(E)-2-Butene	428.6	40.5	238
(Z)-2-Butene	428.6	40.5	238
3-Butenenitrile	585	39	265
1-Buten-3-yne	455	49	202
Butyl acetate	579.0	31	400
1-Butylamine	524.0	41	288
sec-Butylamine	509.4		
N-Butylaniline	72	28	518
Butylbenzene	660.5	28.5	497
sec-Butylbenzene	664	29.1	510
tert-Butylbenzene	660	29.3	490
Butyl benzoate	723	26	561
Butyl butanoate	611		
Butylcyclohexane	667	31.1	
sec-Butylcyclohexane	669	26.4	
tert-Butylcyclohexane	659.0	26.3	
Butylcyclopentane	631.0		
Butyl ethyl ether	531.0	30	390
Butylisopropylamine	563.6		
tert-Butyl methyl sulfide	569.8		
1-Butyne	463.7	46.5	220
2-Butyne	488.6	502	221
4-Butyrolactone	709		
Carbon dioxide	304.2	72.8	94.0
Carbon disulfide	552	78.0	170
Carbon monoxide	132.9	34.5	93.1
Carbon tetrachloride	556.4	45.0	276
Carbon tetrafluoride	227.4	36.9	140
Carbonyl chloride	455	56	190

TABLE 6.5　Critical Properties (*Continued*)

Substance	T_c, K	P_c, atm	V_c, cm$^3 \cdot$ mol^{-1}
Carbonyl sulfide	375	58	140
Chlorine	417	76.1	124
Chlorine pentafluoride	415.7	51.9	230.9
Chlorine trifluoride	426.6		
Chlorobenzene	632.4	44.6	308
1-Chlorobutane	542.0	36.4	312
2-Chlorobutane	520.6	39	305
1-Chloro-1,1-difluoroethane	410.2	40.7	231
2-Chloro-1,1-difluoroethylene	400.6	44.0	197
Chlorodifluoromethane	369.2	49.1	165
1-Chloro-2,3-epoxypropane	624		
Chloroethane	460.4	52.0	199
Chloroform	536.4	54.0	239
1-Chlorohexane	594.6		
Chloromethane	416.3	65.9	139
2-Chloro-2-methylpropane	507	39	295
Chloropentafluoroacetone	410.7	28.4	
Chloropentafluorobenzene	571.0	31.8	
Chloropentafluoroethane	353.2	31.2	252
1-Chloropentane	568.5		
1-Chloropropane	503	45.2	254
2-Chloropropane	485	46.6	230
3-Chloropropane	514	47	234
Chlorotrifluoromethane	379	40	190
Chlorotrifluorosilane	308.5	34.2	
Chlorotrimethylsilane	497.8	31.6	
1,2-Cresol	697.6	49.4	282
1,3-Cresol	705.8	45.0	310
1,4-Cresol	704.6	50.8	277
Cyanogen	400	59	
Cyclobutane	459.9	49.2	210
Cycloheptane	589	36.7	390
Cyclohexane	553.4	40.2	308
(E)-Cyclohexanedimethanol	724	34.85	
Cyclohexanethiol	664.0		
Cyclohexanol	625	37	327
Cyclohexanone	629	38	312
Cyclohexene	560.4	42.9	292
Cyclohexylamine	614.6		
Cyclopentane	511.6	44.5	260
Cyclopentanethiol	633.5		
Cyclopentanone	626	53	268
Cyclopentene	506.0		
Cyclopropane	397.8	54.2	170
Cymene	658		
Decafluorobutane	386.4	22.9	378
(E)-Decahydronaphthalene	690.0	31	
(Z)-Decahydronaphthalene	702.2	31	
Decane	617.6	20.8	603
Decanenitrile	621.9	32.1	
1-Decanol	687.0	22	600
1-Decene	615	21.8	650
Decylcyclohexane	750	13.4	

TABLE 6.5 Critical Properties (*Continued*)

Substance	T_c, K	P_c, atm	V_c, cm³·mol⁻¹
2-Methylpropanoic acid	609	40	292
2-Methyl-1-propanol	547.7	42.4	273
2-Methyl-2-propanol	506.2	39.2	275
2-Methylpropene	417.9	39.5	239
Methyl propyl ether	476.3		
Methyl propyl sulfide	574.1		
2-Methylpyridine	621		
3-Methylpyridine	645		
4-Methylpyridine	646	44	311
1-Methylstyrene	654	33.6	397
2-Methyltetrahydrofuran	537	37.1	267
2-Methylthiophene	606.2		
3-Methylthiophene	610.8		
Methyl vinyl ether	436	47	205
Morpholine	618	54	253
Naphthalene	748.4	40.0	410
Neon	44.4	27.2	41.7
Niobium pentabromide	1010		469
Niobium pentachloride	807		400
Niobium pentafluoride	737	62	155
Nitric oxide	180	64	58
Nitrobenzene	732		
Nitroethane	557	37	
Nitrogen-14	126.2	33.5	89.5
Nitrogen-15	126.3	33.5	90.4
Nitrogen dioxide (equilibrium)	431.4	100	170
Nitrogen trifluoride	234.0	44.7	
Nitromethane	588.0	62.3	173
1-Nitropropane	675.1		
2-Nitropropane	617.9		
Nitrosyl chloride	440	90	139
Nitrous oxide	309.6	71.5	97.4
Nitryl fluoride	349.4		
Nonadecane	756	11.0	
Nonane	594.6	22.8	548
1-Nonanol	677		546
1-Nonene	592	23.1	580
Nonylcyclopentane	710.5	16.3	
Octadecafluorooctane	502	16.4	
Octadecane	745	11.9	
1-Octadecanol	747	14	
1-Octadecene	739	11.2	
Octafluorocyclobutane	388.5	27.5	325
Octafluoronaphthalene	673.0		
Octafluoropropane	345.0	26.5	299
Octane	568.8	24.5	492
1-Octanol	658	34	490
2-Octanol	637	27	494
1-Octene	566.6	25.9	464
(E)-2-Octene	580	27.3	
Octylcyclopentane	694	17.7	
Osmium tetroxide	405	170	
Oxygen	154.6	49.8	73.4

TABLE 6.5 Critical Properties (*Continued*)

Substance	T_c, K	P_c, atm	V_c, cm³·mol⁻¹
Oxygen difluoride	215.2	48.9	97.7
Ozone	161.3	53.8	88.9
Pentachloroethane	646.1		
Pentadecane	707.0	15	880
1-Pentadecene	704	14.4	
Pentadecylcyclopentane	780	10.1	
1,2-Pentadiene	503	40.2	276
(E)-1,3-Pentadiene	496	39.4	275
1,4-Pentadiene	478	37.4	276
Pentafluorobenzene	532.0	34.7	
1,1,2H-Pentafluoropropane	380.1	31.0	273
2,3,4,5,6-Pentafluorotoluene	548.6		
Pentanal	554	35	333
Pentane	469.6	33.3	304
Pentanethiol	597.7		
Pentanoic acid	651	46.1	290
1-Pentanol	588.2	38	326
2-Pentanol	551.6		
2-Pentanone	561.1	38.4	301
3-Pentanone	561.5	36.9	336
1-Pentene	464.7	35.0	300
(E)-2-Pentene	475.6	40.4	304
(Z)-2-Pentene	475.6	40.4	304
Pentyl acetate	605		
Pentyl formate	576		
1-Pentyne	493.4	40	278
Perchloryl fluoride	368.4	53.0	161
Phenanthrene	878		
Phenol	694.2	60.5	229
Phosgene	455	56	190
Phosphine	324.4	64.5	
Phosphonium chloride	322.2	72.7	
Phosphorus bromide difluoride	386		
Phosphorus chloride difluoride	362.3	44.6	
Phosphorus dibromide fluoride	527		
Phosphorus dichloride fluoride	463.0	49.3	
Phosphorus pentachloride	645		
Phosphorus trichloride	563		260
Phosphorus trifluoride	271.2	42.7	
Phosphoryl chloride difluoride	423.8	43.4	
Phosphoryl trichloride	602		
Phosphoryl trifluoride	346.5	41.8	
Phthalic anhydride	810	47	368
Piperidine	594.0	47	289
Propadiene	393	54.0	162
Propanal	496.2	47	223
Propane	369.8	41.9	203
1,2-Propanediol	625	60	237
1,3-Propanediol	658	59	241
Propanenitrile	564.4	41.3	230
1-Propanethiol	535.6		
2-Propanethiol	517.3		
Propanoic acid	612	53.0	230

TABLE 6.5 Critical Properties (*Continued*)

Substance	T_c, K	P_c, atm	V_c, cm$^3 \cdot$ mol^{-1}
1-Propanol	536.7	51.0	218.5
2-Propanol	508.3	47.0	220
2-Propenal	506	51	
Propene	365.0	45.6	181
2-Propen-1-ol	545.1		
Propyl acetate	549.4	32.9	345
Propylamine	497.0	46.8	233
Propylbenzene	638.3	31.6	440
Propyl butanoate	600		
Propylcyclohexane	609.0	27.7	
Propylcyclopentane	631.0	29.6	425
Propylene oxide	482.2	48.6	186
Propyl formate	538.0	40.1	285
Propyl 2-methylpropanoate	589		
Propyl 3-methylpropanoate	609		
Propyl propanoate	578		
Propyne	402.4	55.5	164
Pyridine	617.2	60.0	254
Pyrrole	639.8	56	
Pyrrolidine	568.6	55.4	249
Quinoline	794.4	57	
Radon	376.9	62	139
Rhenium(VII) oxide	942		334
Selenium	1766		
Silane	269.6	47.8	
Silicon chloride trifluoride	307.6	34.2	
Silicon tetrachloride	507	37	326
Silicon tetrafluoride	259.1	36.7	
Silicon trichloride fluoride	438.5	35.3	
Spiro[2.2]pentane	506.4		
Styrene	636.9	36.3	347
Sulfur	1314		
Sulfur dioxide	430.8	77.8	122
Sulfur hexafluoride	318.7	37.1	198
Sulfur tetrafluoride	364.0		
Sulfur trioxide	491.0	81	130
Tantalum pentabromide	974		461
Tantalum pentachloride	767		400
1,2-Terphenyl	891.0	38.5	769
1,3-Terphenyl	924.8	34.6	784
1,4-Terphenyl	926.0	32.8	779
1,1,2,2-Tetrachlorodifluoroethane	551	34	371
1,1,2,2-Tetrachloroethane	644.5		
Tetrachloroethylene	620.0	44.3	290
Tetradecafluoro-1-heptene	478.2		
Tetradecafluorohexane	447.6	18.8	
Tetradecafluoromethylcyclohexane	486.8	23	
Tetradecane	694.0	16	830
1-Tetradecene	689	15.4	
Tetradecylcyclopentane	772	11.1	
Tetrafluoroethylene	306.4	38.9	175
Tetrafluorohydrazine	309.4	37	
Tetrafluoromethane	227.6	36.9	140

TABLE 6.5   Critical Properties (*Continued*)

Substance	T_c, K	P_c, atm	V_c, cm³·mol⁻¹
Tetrahydrofuran	540.2	51.2	224
1,2,3,4-Tetrahydronaphthalene	710	34.7	428
Tetrahydrothiophene	631.9		
1,2,4,5-Tetramethylbenzene	675	29	480
2,2,3,3-Tetramethylbutane	567.8	28.3	461
2,2,3,3-Tetramethylhexane	623.1	24.8	
2,2,5,5-Tetramethylhexane	581.5	21.6	
2,2,3,3-Tetramethylpentane	607.6	27.0	
2,2,3,4-Tetramethylpentane	592.7	25.7	
2,2,4,4-Tetramethylpentane	574.7	24.5	
2,3,3,4-Tetramethylpentane	607.6	26.8	
Thiacyclopentane	631.9		
2-Thiapropane	503.1	54.6	201
Thiophene	579.4	56.2	219
Thiophenol	689.5		
Thymol	698		
Tin(IV) chloride	591.8	37.0	351
Titanium tetrachloride	638	46	340
Toluene	591.7	40.6	316
1,2-Toluidine	694	37	343
1,3-Toluidine	709	41	343
1,4-Toluidine	766	23.8	
Toluonitrile	723		
Tributoxyborane	743.1	19.6	863
Tributylamine	638.4	18	
1,1,1-Trichloroethane	548.4		
1,1,2-Trichloroethane	602	41	294
Trichloroethylene	571.0	48.5	256
Trichlorofluoromethane	471.2	43.5	248
Trichloromethane	536.4	54.9	
1,2,3-Trichloropropane	651	39	348
1,2,2-Trichlorotrifluoroethane	487.2	33.7	304
Tridecane	675.8	17.0	780
1-Tridecane	674	16.8	
Tridecylclopentane	761	11.9	
Triethanolamine	787.4	24.2	
Triethylamine	535.6	30	390
Trifluoroacetic acid	519	40	204
1,1,1-Trifluoroethane	346.2	37.1	221
Trifluoromethane	298.9	47.7	133
(Trifluoromethyl)benzene	559.9		
Trimethylamine	433.2	40.2	254
1,2,3-Trimethylbenzene	664.5	34.1	430
1,2,4-Trimethylbenzene	649.1	31.9	430
1,3,5-Trimethylbenzene	637.3	30.9	433
2,2,3-Trimethylbutane	531.1	29.2	398
2,2,3-Trimethyl-1-butene	533	28.6	400
Trimethylchlorosilane	497.7	31.6	
1,1,2-Trimethylcyclopentane	579.5	29.0	
1,1,3-Trimethylcyclopentane	569.5	27.9	
(Z),(E),(Z)-1,2,4-Trimethylcyclopentane	571	27.7	
(Z),(Z),(E)-1,2,4-Trimethylcyclopentane	579	28.4	
3,3,5-Trimethylheptane	609.6	22.9	

TABLE 6.5   Critical Properties (*Continued*)

Substance	T_c, K	P_c, atm	V_c, cm³·mol⁻¹
2,2,3-Trimethylhexane	588	24.6	
2,2,4-Trimethylhexane	573.7	23.4	
2,2,5-Trimethylhexane	567.9	23.0	519
2,4,7-Trimethyloctane	608.8		
2,2,3-Trimethylpentane	563.4	26.9	436
2,2,4-Trimethylpentane	543.9	25.3	468
2,3,3-Trimethylpentane	573.5	27.8	455
2,3,4-Trimethylpentane	566.3	26.9	461
2,2,4-Trimethyl-1,3-pentanediol	671	25.6	364.6
2,3,6-Trimethylpyridine	654.5		
2,4,6-Trimethylpyridine	653.0		
2,4,6-Trimethyl-1,3,5-trioxane	563		
1*H*-Undecafluoropentane	443.9		
Undecane	638.8	19.4	660
1-Undecene	637	19.7	
Uranium hexafluoride	505.8	45.5	250
Vinyl acetate	501.5	22.4	265
Vinyl chloride	429.7	55.3	169
Vinyl fluoride	327.8	51.7	114
Vinyl formate	475	57	210
Water	647.3	217.6	56.0
Xenon	289.7	57.6	118
1,2-Xylene	630.2	36.8	369
1,3-Xylene	617.0	35.0	376
1,4-Xylene	616.2	34.7	379
Zirconium tetrabromide	805		415
Zirconium tetrachloride	778	56.9	319
Zirconium tetraiodide	960		528

SECTION 7
SPECTROSCOPY

7.1 X-RAY METHODS

An X-ray tube operating at a voltage V (in keV) emits a continuous X-ray spectrum, the minimum wavelength of which is given by $\lambda_{min} = 12.398/V$ with the wavelength expressed in angstroms. For expressing the wavelength in kX units, divide by the factor 1.00202. Tables 7.1

TABLE 7.1 Wavelengths of X-Ray Emission Spectra in Angstroms

Atomic No.	Element	$K\alpha_2$	$K\alpha_1$	$K\beta_1$	$L\alpha_1$	$L\beta_1$
3	Li		240			
4	Be		113			
5	B		67			
6	C		44			
7	N		31.60			
8	O		23.71			
9	F		18.31			
10	Ne		14.616	14.464		
11	Na		11.909	11.617	407.6	
12	Mg		9.889	9.558	251.0	
13	Al	8.3392	8.3367	7.981	169.8	
14	Si	7.1277	7.1253	6.7681	123	
15	P		6.1549	5.8038		
16	S	5.3747	5.3720	5.0317		
17	Cl	4.7305	4.7276	4.4031		
18	Ar	4.1946	4.1916	3.8848		
19	K	3.7446	3.7412	3.4538	42.7	
20	Ca	3.3616	3.3583	3.0896	36.32	35.95
21	Sc	3.0345	3.0311	2.7795	31.33	31.01
22	Ti	2.75207	2.7484	2.5138	27.39	27.02
23	V	2.5073	2.5035	2.2843	24.26	23.85
24	Cr	2.29351	2.28962	2.08480	21.67	21.28
25	Mn	2.1057	2.1018	1.9102	19.45	19.12
26	Fe	1.93991	1.93597	1.75653	17.567	17.255
27	Co	1.79278	1.78892	1.62075	15.968	15.667
28	Ni	1.66169	1.65784	1.50010	14.566	14.279
29	Cu	1.54433	1.54051	1.39217	13.330	13.053
30	Zn	1.4389	1.4351	1.2952	12.257	11.985
31	Ga	1.3439	1.3400	1.20784	11.290	11.023
32	Ge	1.2580	1.2540	1.1289	10.435	10.174
33	As	1.1798	1.1758	1.0573	9.671	9.414
34	Se	1.1088	1.1047	0.9921	8.990	8.736
35	Br	1.0438	1.0397	0.9327	8.375	8.125
36	Kr	0.9841	0.9801	0.8785	7.822	7.574
37	Rb	0.9296	0.9255	0.8286	7.3181	7.076
38	Sr	0.8794	0.8752	0.7829	6.8625	6.6237
39	Y	0.8330	0.8279	0.7407	6.4485	6.2117
40	Zr	0.7901	0.7859	0.7017	6.0702	5.8358
41	Nb	0.7504	0.7462	0.6657	5.7240	5.4921
42	Mo	0.713543	0.70926	0.632253	5.4063	5.1768
43	Tc	0.6793	0.6749	0.6014	5.1126	4.8782
44	Ru	0.6474	0.6430	0.5725	4.8455	4.6204
45	Rh	0.6176	0.6132	0.5456	4.5973	4.3739
46	Pd	0.5898	0.5854	0.5205	4.3676	4.1460
47	Ag	0.563775	0.559363	0.49701	4.1541	3.9344

TABLE 7.1 Wavelengths of X-Ray Emission Spectra in Angstroms (*Continued*)

Atomic No.	Element	$K\alpha_2$	$K\alpha_1$	$K\beta_1$	$L\alpha_1$	$L\beta_1$
48	Cd	0.5394	0.5350	0.4751	3.9563	3.7381
49	In	0.5165	0.5121	0.4545	3.7719	3.5552
50	Sn	0.4950	0.4906	0.4352	3.5999	3.3848
51	Sb	0.4748	0.4703	0.4171	3.4392	3.2256
52	Te	0.4558	0.4513	0.4000	3.2891	3.0767
53	I	0.4378	0.4333	0.3839	3.1485	2.9373
54	Xe	0.4204	0.4160	0.3685	3.016	2.807
55	Cs	0.4048	0.4003	0.3543	2.9016	2.8920
56	Ba	0.3896	0.3851	0.3408	2.7752	2.5674
57	La	0.3753	0.3707	0.3280	2.6651	2.4583
58	Ce	0.3617	0.3571	0.3158	2.5612	2.3558
59	Pr	0.3487	0.3441	0.3042	2.4627	2.2584
60	Nd	0.3565	0.3318	0.2933	2.3701	2.1666
61	Pm	0.3249	0.3207	0.2821	2.282	2.0796
62	Sm	0.3137	0.3190	0.2731	2.1994	1.9976
63	Eu	0.3133	0.2985	0.2636	2.1206	1.9202
64	Gd	0.2932	0.2884	0.2544	2.0460	1.8462
65	Tb	0.2834	0.2788	0.2460	1.9755	1.7763
66	Dy	0.2743	0.2696	0.2376	1.9088	1.7100
67	Ho	0.2655	0.2608	0.2302	1.8447	1.6468
68	Er	0.2572	0.2525	0.2226	1.7843	1.5873
69	Tm	0.2491	0.2444	0.2153	1.7263	1.5299
70	Yb	0.2415	0.2368	0.2088	1.6719	1.4756
71	Lu	0.2341	0.2293	0.2021	1.6194	1.4235
72	Hf	0.2270	0.2222	0.1955	1.5696	1.3740
73	Ta	0.2203	0.2155	0.1901	1.5219	1.3270
74	W	0.213813	0.208992	0.184363	1.4764	1.2818
75	Re	0.2076	0.2028	0.1789	1.4329	1.2385
76	Os	0.2016	0.1968	0.1736	1.3911	1.1972
77	Ir	0.1959	0.1910	0.1685	1.3513	1.1578
78	Pt	0.1904	0.1855	0.1637	1.3130	1.1198
79	Au	0.1851	0.1802	0.1590	1.2764	1.0836
80	Hg	0.1799	0.1750	0.1544	1.2411	1.0486
81	Tl	0.1750	0.1701	0.1501	1.2074	1.0152
82	Pb	0.1703	0.1654	0.1460	1.1750	0.9822
83	Bi	0.1657	0.1608	0.1419	1.1439	0.9520
84	Po	0.1608	0.1559	0.1382	1.1138	0.9222
85	At	0.1570	0.1521	0.1343	1.0850	0.8936
86	Rn	0.1529	0.1479	0.1307	1.0572	0.8659
87	Fr	0.1489	0.1440	0.1272	1.030	0.840
88	Ra	0.1450	0.1401	0.1237	1.0047	0.8137
89	Ac	0.1414	0.1364	0.1205	0.9799	0.7890
90	Th	0.1378	0.1328	0.1174	0.9560	0.7652
91	Pa	0.1344	0.1294	0.1143	0.9328	0.7422
92	U	0.1310	0.1259	0.1114	0.9105	0.7200

TABLE 7.1 Wavelengths of X-Ray Emission Spectra in Angstroms (*Continued*)

Atomic No.	Element	$K\alpha_2$	$K\alpha_1$	$K\beta_1$	$L\alpha_1$	$L\beta_1$
93	Np	0.1278	0.1226	0.1085	0.8893	0.6984
94	Pu	0.1246	0.1195	0.1058	0.8682	0.6777
95	Am	0.1215	0.1165	0.1031	0.8481	0.6576
96	Cm	0.1186	0.1135	0.1005	0.8287	0.6388
97	Bk	0.1157	0.1107	0.0980	0.8098	0.6203
98	Cf	0.1130	0.1079	0.0956	0.7917	0.6023
99	Es	0.1103	0.1052	0.0933	0.7740	0.5850
100	Fm	0.1077	0.1026	0.0910	0.7570	0.5682

TABLE 7.2 Wavelengths of Absorption Edges in Angstroms

Atomic No.	Element	K	L_I	L_{II}	L_{III}
3	Li	226.5			
4	Be	110.68			
5	B	66.289			
6	C	43.68			
7	N	30.99			
8	O	23.32			
9	F	17.913			
10	Ne	14.183			
11	Na	11.478		400	
12	Mg	9.512	197.4	247.92	
13	Al	7.951	142.5	170	
14	Si	6.745	105.1	126.48	
15	P	5.787	81.0	96.84	
16	S	5.018	64.23	76.05	
17	Cl	4.397	52.08	61.37	62.93
18	Ar	3.871	43.19	50.39	50.60
19	K	3.436	36.35	42.02	42.17
20	Ca	3.070	31.07	35.20	35.49
21	Sc	2.757	26.83	30.16	30.53
22	Ti	2.497	23.39	26.83	27.37
23	V	2.269	20.52	23.70	24.26
24	Cr	2.07012	16.7	17.9	20.7
25	Mn	1.896	16.27	18.90	19.40
26	Fe	1.74334	14.60	17.17	17.53
27	Co	1.60811	13.34	15.53	15.93
28	Ni	1.48802	12.27	14.13	14.58
29	Cu	1.38043	11.27	13.01	13.29
30	Zn	1.283	10.33	11.86	12.13
31	Ga	1.195	9.54	10.61	11.15
32	Ge	1.116	8.73	9.97	10.23

TABLE 7.2   Wavelengths of Absorption Edges in Angstroms (*Continued*)

Atomic No.	Element	K	L_I	L_{II}	L_{III}
33	As	1.044	8.108	9.124	9.367
34	Se	0.9800	7.505	8.417	8.646
35	Br	0.9199	6.925	7.752	7.989
36	Kr	0.8655	6.456	7.165	7.395
37	Rb	0.8155	5.997	6.643	6.863
38	Sr	0.7697	5.582	6.172	6.387
39	Y	0.7276	5.233	5.756	5.962
40	Zr	0.6888	4.867	5.378	5.583
41	Nb	0.6529	4.581	5.025	5.223
42	Mo	0.61977	4.299	4.719	4.912
43	Tc	0.5888	4.064	4.427	4.629
44	Ru	0.5605	3.841	4.179	4.369
45	Rh	0.5338	3.626	3.942	4.130
46	Pd	0.5092	3.428	3.724	3.908
47	Ag	0.48582	3.254	3.514	3.698
48	Cd	0.4641	3.084	3.326	3.504
49	In	0.4439	2.926	3.147	3.324
50	Sn	0.4247	2.778	2.982	3.156
51	Sb	0.4066	2.639	2.830	3.000
52	Te	0.3897	2.510	2.687	2.855
53	I	0.3738	2.390	2.553	2.719
54	Xe	0.3585	2.274	2.429	2.592
55	Cs	0.3447	2.167	2.314	2.474
56	Ba	0.3314	2.068	2.204	2.363
57	La	0.3184	1.973	2.103	2.258
58	Ce	0.3065	1.891	2.009	2.164
59	Pr	0.2952	1.811	1.924	2.077
60	Nd	0.2845	1.735	1.843	1.995
61	Pm	0.2743	1.668	1.766	1.918
62	Sm	0.2646	1.598	1.702	1.845
63	Eu	0.2555	1.536	1.626	1.775
64	Gd	0.2468	1.477	1.561	1.709
65	Tb	0.2384	1.421	1.501	1.649
66	Dy	0.2305	1.365	1.438	1.579
67	Ho	0.2229	1.319	1.390	1.535
68	Er	0.2157	1.269	1.339	1.483
69	Tm	0.2089	1.222	1.288	1.433
70	Yb	0.2022	1.181	1.243	1.386
71	Lu	0.1958	1.140	1.198	1.341
72	Hf	0.1898	1.099	1.154	1.297
73	Ta	0.1839	1.061	1.113	1.255
74	W	0.17837	1.025	1.074	1.215
75	Re	0.1731	0.9901	1.036	1.177
76	Os	0.1678	0.9557	1.001	1.140
77	Ir	0.1629	0.9243	0.9670	1.106

TABLE 7.2 Wavelengths of Absorption Edges in Angstroms (*Continued*)

Atomic No.	Element	K	L_I	L_{II}	L_{III}
78	Pt	0.1582	0.8914	0.9348	1.072
79	Au	0.1534	0.8638	0.9028	1.040
80	Hg	0.1492	0.8353	0.8779	1.009
81	Tl	0.1447	0.8079	0.8436	0.9793
82	Pb	0.1408	0.7815	0.8155	0.9503
83	Bi	0.1371	0.7565	0.7891	0.9234
84	Po	0.1332	0.7322	0.7638	0.8970
85	At	0.1295	0.7092	0.7387	0.8720
86	Rn	0.1260	0.6868	0.7153	0.8479
87	Fr	0.1225	0.6654	0.6929	0.8248
88	Ra	0.1192	0.6446	0.6711	0.8027
89	Ac	0.1161	0.6248	0.6500	0.7813
90	Th	0.1129	0.6061	0.6301	0.7606
91	Pa	0.1101	0.5875	0.6106	0.7411
92	U	0.1068	0.5697	0.5919	0.7233
93	Np	0.1045	0.5531	0.5742	0.7042
94	Pu	0.1018	0.5366	0.5571	0.6867
95	Am	0.0992	0.5208	0.5404	0.6700
96	Cm	0.0967	0.5060	0.5246	0.6532
97	Bk	0.0943	0.4913	0.5093	0.6375
98	Cf	0.0920	0.4771	0.4945	0.6223
99	Es	0.0897	0.4636	0.4801	0.6076
100	Fm	0.0875	0.4506	0.4665	0.5935

TABLE 7.3 Critical X-Ray Absorption Energies in keV

Atomic No.	Element	K	L_I	L_{II}	L_{III}
1	H	0.0136			
2	He	0.0246			
3	Li	0.0547			
4	Be	0.112			
5	B	0.187			
6	C	0.284			
7	N	0.400			
8	O	0.532			
9	F	0.692			
10	Ne	0.874	0.048	0.022	
11	Na	1.08	0.055	0.034	
12	Mg	1.30	0.0628	0.0502	
13	Al	1.559	0.0870	0.0720	
14	Si	1.838	0.118	0.0977	
15	P	2.142	0.153	0.128	

TABLE 7.3 Critical X-Ray Absorption Energies in keV (*Continued*)

Atomic No.	Element	K	L_I	L_{II}	L_{III}
16	S	2.469	0.193	0.163	0.162
17	Cl	2.822	0.238	0.202	0.201
18	Ar	3.200	0.287	0.246	0.244
19	K	3.606	0.341	0.295	0.292
20	Ca	4.038	0.399	0.350	0.346
21	Sc	4.496	0.462	0.411	0.407
22	Ti	4.966	0.530	0.462	0.456
23	V	5.467	0.604	0.523	0.515
24	Cr	5.988	0.679	0.584	0.574
25	Mn	6.542	0.762	0.656	0.644
26	Fe	7.113	0.849	0.722	0.709
27	Co	7.713	0.929	0.798	0.783
28	Ni	8.337	1.02	0.877	0.858
29	Cu	8.982	1.10	0.954	0.935
30	Zn	9.662	1.20	1.05	1.02
31	Ga	10.39	1.30	1.17	1.14
32	Ge	11.10	1.42	1.24	1.21
33	As	11.87	1.529	1.358	1.32
34	Se	12.65	1.66	1.472	1.431
35	Br	13.48	1.791	1.599	1.552
36	Kr	14.32	1.92	1.729	1.674
37	Rb	15.197	2.064	1.863	1.803
38	Sr	16.101	2.212	2.004	1.937
39	Y	17.053	2.387	2.171	2.096
40	Zr	17.998	2.533	2.308	2.224
41	Nb	18.986	2.700	2.467	2.372
42	Mo	20.003	2.869	2.630	2.525
43	Tc	21.050	3.045	2.796	2.680
44	Ru	22.117	3.227	2.968	2.839
45	Rh	23.210	3.404	3.139	2.995
46	Pd	24.356	3.614	3.338	3.181
47	Ag	25.535	3.828	3.547	3.375
48	Cd	26.712	4.019	3.731	3.541
49	In	27.929	4.226	3.929	3.732
50	Sn	29.182	4.445	4.139	3.911
51	Sb	30.497	4.708	4.391	4.137
52	Te	31.817	4.953	4.621	4.347
53	I	33.164	5.187	4.855	4.559
54	Xe	34.551	5.448	5.103	4.783
55	Cs	35.974	5.706	5.360	5.014
56	Ba	37.432	5.995	5.629	5.250
57	La	38.923	6.264	5.902	5.490
58	Ce	40.43	6.556	6.169	5.728
59	Pr	41.99	6.837	6.446	5.968
60	Nd	43.57	7.134	6.728	6.215

TABLE 7.3 Critical X-Ray Absorption Energies in keV (*Continued*)

Atomic No.	Element	K	L_I	L_{II}	L_{III}
61	Pm	45.19	7.431	7.022	6.462
62	Sm	46.85	7.742	7.316	6.720
63	Eu	48.51	8.059	7.624	6.984
64	Gd	50.23	8.383	7.942	7.251
65	Tb	52.00	8.713	8.258	7.520
66	Dy	53.77	9.053	8.587	7.795
67	Ho	55.61	9.395	8.918	8.074
68	Er	57.47	9.754	9.270	8.362
69	Tm	59.38	10.12	9.622	8.656
70	Yb	61.31	10.49	9.985	8.949
71	Lu	63.32	10.87	10.35	9.248
72	Hf	65.37	11.28	10.75	9.567
73	Ta	67.46	11.68	11.14	9.883
74	W	69.51	12.09	11.54	10.20
75	Re	71.67	12.52	11.96	10.53
76	Os	73.87	12.97	12.38	10.86
77	Ir	76.11	13.41	12.82	11.21
78	Pt	78.35	13.865	13.26	11.55
79	Au	80.67	14.351	13.731	11.92
80	Hg	83.08	14.838	14.205	12.278
81	Tl	85.52	15.344	14.695	12.65
82	Pb	87.95	15.861	15.200	13.03
83	Bi	90.54	16.386	15.709	13.42
84	Po	93.16	16.925	16.233	13.81
85	At	95.73	17.481	16.777	14.21
86	Rn	98.45	18.054	17.331	14.61
87	Fa	101.1	18.628	17.893	15.02
88	Ra	103.9	19.228	18.473	15.44
89	Ac	107.7	19.829	19.071	15.86
90	Th	109.8	20.452	19.673	16.278
91	Pa	112.4	21.096	20.295	16.720
92	U	115.0	21.757	20.944	17.163
93	Np	118.2	22.411	21.585	17.606
94	Pu	121.2	23.117	22.250	18.062
95	Am	124.3	23.795	22.935	18.524
96	Cm	127.2	24.502	23.629	18.992
97	Bk	131.3	25.231	24.344	19.466
98	Cf	133.6	26.010	25.070	19.954
99	Es	138.1	26.729	25.824	20.422
100	Fm	141.5	27.503	26.584	20.912

TABLE 7.4 X-Ray Emission Energies in keV

Atomic No.	Element	$K\beta_1$	$K\alpha_1$	$L\beta_1$	$L\alpha_1$
3	Li		0.052		
4	Be		0.110		
5	B		0.185		
6	C		0.282		
7	N		0.392		
8	O		0.523		
9	F		0.677		
10	Ne		0.851		
11	Na	1.067	1.041		
12	Mg	1.297	1.254		
13	Al	1.553	1.487		
14	Si	1.832	1.740		
15	P	2.136	2.015		
16	S	2.464	2.308		
17	Cl	2.815	2.622		
18	Ar	3.192	2.957		
19	K	3.589	3.313		
20	Ca	4.012	3.691	0.344	0.341
21	Sc	4.460	4.090	0.399	0.395
22	Ti	4.931	4.510	0.458	0.452
23	V	5.427	4.952	0.519	0.512
24	Cr	5.946	5.414	0.581	0.571
25	Mn	6.490	5.898	0.647	0.636
26	Fe	7.057	6.403	0.717	0.704
27	Co	7.649	6.930	0.790	0.775
28	Ni	8.264	7.477	0.866	0.849
29	Cu	8.904	8.047	0.948	0.928
30	Zn	9.571	8.638	1.032	1.009
31	Ga	10.263	9.251	1.122	1.096
32	Ge	10.981	9.885	1.216	1.186
33	As	11.725	10.543	1.317	1.282
34	Se	12.495	11.221	1.419	1.379
35	Br	13.290	11.923	1.526	1.480
36	Kr	14.112	12.649	1.638	1.587
37	Rb	14.960	13.394	1.752	1.694
38	Sr	15.834	14.164	1.872	1.806
39	Y	16.736	14.957	1.996	1.922
40	Zr	17.666	15.774	2.124	2.042
41	Nb	18.621	16.614	2.257	2.166
42	Mo	19.607	17.478	2.395	2.293
43	Tc	20.612	18.370	2.538	2.424
44	Ru	21.655	19.278	2.683	2.558
45	Rh	22.721	20.214	2.834	2.696
46	Pd	23.816	21.175	2.990	2.838
47	Ag	24.942	22.162	3.151	2.984

TABLE 7.4 X-Ray Emission Energies in keV (*Continued*)

Atomic No.	Element	$K\beta_1$	$K\alpha_1$	$L\beta_1$	$L\alpha_1$
48	Cd	26.093	23.172	3.316	3.133
49	In	27.274	24.207	3.487	3.287
50	Sn	28.483	25.270	3.662	3.444
51	Sb	29.723	26.357	3.843	3.605
52	Te	30.993	27.471	4.029	3.769
53	I	32.292	28.610	4.220	3.937
54	Xe	33.644	29.779	4.422	4.111
55	Cs	34.984	30.970	4.620	4.286
56	Ba	36.376	32.191	4.828	4.467
57	La	37.799	33.440	5.043	4.651
58	Ce	39.255	34.717	5.262	4.840
59	Pr	40.746	36.023	5.489	5.034
60	Nd	42.269	37.359	5.722	5.230
61	Pm	43.811	38.726	5.956	5.431
62	Sm	45.400	40.124	6.206	5.636
63	Eu	47.027	41.529	6.456	5.846
64	Gd	48.718	42.983	6.714	6.059
65	Tb	50.391	44.470	6.979	6.275
66	Dy	52.178	45.985	7.249	6.495
67	Ho	53.934	47.528	7.528	6.720
68	Er	55.690	49.099	7.810	6.948
69	Tm	57.487	50.730	8.103	7.181
70	Yb	59.352	52.360	8.401	7.414
71	Lu	61.282	54.063	8.708	7.654
72	Hf	63.209	55.757	9.021	7.898
73	Ta	65.210	57.524	9.341	8.145
74	W	67.233	59.310	9.670	8.396
75	Re	69.298	61.131	10.008	8.651
76	Os	71.404	62.991	10.354	8.910
77	Ir	73.549	64.886	10.706	9.173
78	Pt	75.736	66.820	11.069	9.441
79	Au	77.968	68.794	11.439	9.711
80	Hg	80.258	70.821	11.823	9.987
81	Tl	82.558	72.860	12.210	10.266
82	Pb	84.922	74.957	12.611	10.549
83	Bi	87.335	77.097	13.021	10.836
84	Po	89.809	79.296	13.441	11.128
85	At	92.319	81.525	13.873	11.424
86	Rn	94.877	83.800	14.316	11.724
87	Fr	97.483	86.119	14.770	12.029
88	Ra	100.136	88.485	15.233	12.338
89	Ac	102.846	90.894	15.712	12.650
90	Th	105.592	93.334	16.200	12.966
91	Pa	108.408	95.851	16.700	13.291
92	U	111.289	98.428	17.218	13.613

TABLE 7.4   X-Ray Emission Energies in keV (*Continued*)

Atomic No.	Element	$K\beta_1$	$K\alpha_1$	$L\beta_1$	$L\alpha_1$
93	Np	114.181	101.005	17.740	13.945
94	Pu	117.146	103.653	18.278	14.279
95	Am	120.163	106.351	18.829	14.618
96	Cm	123.235	109.098	19.393	14.961
97	Bk	126.362	111.896	19.971	15.309
98	Cf	129.544	114.745	20.562	15.661
99	Es	132.781	117.646	21.166	16.018
100	Fm	136.075	120.598	21.785	16.379

and 7.2 are based on the K and L wavelength values as published by Y. Cauchois and H. Hulubei (*Tables de Constantes et Données Numériques*, I. *Longueurs d'Onde des Émissions X et des Discontinuités d'Absorption X,* Hermann, Paris, 1947) and by the International Union of Crystallography (*International Tables for X-Ray Crystallography,* Kynoch Press, Birmingham, England, 1962). Wavelength accuracy is only to about 1 in 25 000 except for the lines employed in X-ray diffraction work.

Use of energy-proportional detectors for X-rays creates a need for energy values of K and L absorption edges (Table 7.3) and emission series (Table 7.4). These values were obtained by a conversion to keV of tabulated experimental wavelength values and smoothed by a fit to Moseley's law. Although values are listed to 1 eV, chemical form may shift absorption edges and emission lines as much as 10 to 20 eV. S. Fine and C. F. Hendee [*Nucelonics,* **13**(3):36 (1955)] also give values for $K\beta_2$, $L\gamma_1$, and $L\beta_2$ lines.

The relative intensities of X-ray emission lines from targets varies for different elements. However, one can assume a ratio of $K\alpha_1/K\alpha_2 = 2$ for the commonly used targets. The ratio of $K\alpha_2/K\beta_1$ from these targets varies from 6 to 3.5. The intensities of $K\beta_2$ radiations amount to about 1 percent of that of the corresponding $K\alpha_1$ radiation. In practical applications these ratios have to be corrected for differential absorption in the window of the tube and air path, the ratio of scattering factors for and differential absorption in the crystal, and for sensitivity characteristics of the detector. Generalizing, the intensities of radiations from the K and L series are as follows:

Emission line	$K\alpha_1$	$K\alpha_2$	$K\beta_1$	$K\beta_2$	$L\alpha_1$	$L\alpha_2$	$L\beta_1$	$L\beta_2$	$L\gamma_1$
Relative intensity	500	250	80–150	5	100	10	30	60	40

For angles at which the $K\alpha_1$, $K\alpha_2$ doublet is not resolved, a mean wavelength $[\overline{K\alpha} = (2K\alpha_1 + K\alpha_2)/3]$ can be used.

Filters.   The K spectra of the light metals, often used as target material in the production of X-rays for diffraction studies, contain three strong lines, α_1, α_2 and β_1, of which the α lines form a doublet with a narrow wavelength separation. The $K\beta$ radiation can be eliminated by using a thin foil filter, usually of the element of next lower atomic number to that of the target

TABLE 7.5 β Filters for Common Target Elements

Target Element	$K\bar{\alpha}$, Å	Excitation Voltage, keV	$K\beta_1 K\alpha_1 = \frac{1}{100}$			% Loss $K\alpha_1$
			Absorber	Thickness, mm	g/cm²	
Ag	0.560834	25.52	Pd	0.062	0.074	60
Mo	0.71069	20.00	Zr	0.081	0.053	57
Cu	1.54178	8.981	Ni	0.015	0.013	45
Ni	1.65912	8.331	Co	0.013	0.011	42
Co	1.79021	7.709	Fe	0.012	0.009	39
Fe	1.93728	7.111	Mn	0.011	0.008	38
			MnO_2	0.026	0.013	45
Cr	2.29092	5.989	V	0.011	0.007	37
			V_2O_5	0.036	0.012	48
	$L\alpha_1$		$L\beta_1 L\alpha_1 = \frac{1}{100}$			% Loss $L\alpha_1$
W	1.4763	10.200	Cu	0.035		77

element: the $K\alpha$ lines are transmitted with a relatively small loss of intensity. Table 7.5, restricted to the K wavelengths of target elements in common use, lists the calculated thicknesses of β filters required to reduce the $K\beta_1/K\alpha_1$ integrated intensity ratio to $\frac{1}{100}$.

Interplanar Spacings. Diffractometer alignment procedures require the use of a well-prepared polycrystalline specimen. Two standard samples found to be suitable are silicon and α-quartz (including Novaculite). The 2θ values of several of the most intense reflections for these materials are listed in Table 7.6 (*Tables of Interplanar Spacings d vs. Diffraction Angle 2θ for Selected Targets,* Picker Nuclear, White Plains, N.Y., 1966). To convert to d for $K\bar{\alpha}$ or to d for $K\alpha_2$, multiply the tabulated d value (Table 7.6) for $K\alpha_1$ by the factor given below:

Element	$K\bar{\alpha}$	$K\alpha_2$
W	1.007 69	1.023 07
Ag	1.002 63	1.007 89
Mo	1.002 02	1.006 04
Cu	1.000 82	1.002 48
Ni	1.000 77	1.002 32
Co	1.000 72	1.002 16
Fe	1.000 67	1.002 04
Cr	1.000 57	1.001 70

Analyzing Crystals. The range of wavelengths usable with various analyzing crystals are governed by the d spacings of the crystal planes and by the geometric limits to which the goniometer can be rotated. The d value should be small enough to make the angle 2θ greater than approximately 10 or 15 deg, even at the shortest wavelength used: otherwise excessively long analyzing crystals would be needed to prevent the direct fluorescent beam from entering the detector. A small d value is also favorable for producing a large dispersion of the spectrum to give good separation of adjacent lines. On the other hand, a small d value imposes an upper

TABLE 7.6   Interplanar Spacings for $K\alpha_1$ Radiation, d versus 2θ

α-quartz (Including Novaculite)

hkl d (Å)	100 4.260	101 3.343	110 2.458	102 2.282	200 2.128	112 1.817	202 1.672	211 1.541	203 1.375	301 1.372
W $K\alpha_1$: 2θ	2.81	3.58	4.87	5.25	5.63	6.59	7.17	7.78	8.72	8.74
Ag $K\alpha_1$: 2θ	7.53	9.60	13.07	14.08	15.10	17.71	19.26	20.91	23.47	23.52
Mo $K\alpha_1$: 2θ	9.55	12.18	16.59	17.88	19.19	22.51	24.49	26.61	29.89	29.96
Cu $K\alpha_1$: 2θ	20.83	26.64	36.52	39.45	42.44	50.16	54.86	59.98	68.14	68.31
Ni $K\alpha_1$: 2θ	22.44	28.71	39.42	42.60	45.85	54.28	59.44	65.08	74.15	74.34
Co $K\alpha_1$: 2θ	24.24	31.04	42.68	46.15	49.71	58.98	64.68	70.96	81.16	81.38
Fe $K\alpha_1$: 2θ	26.27	33.66	46.38	50.20	54.11	64.38	70.75	77.83	89.50	89.74
Cr $K\alpha_1$: 2θ	31.18	40.05	55.52	60.22	65.09	78.11	86.42	95.96	112.73	113.11

Silicon

hkl d (Å)	111 3.1353	220 1.91997	311 1.63736	400 1.357630	331 1.24584	422 1.1085	511,333 1.0451	440 0.959986	531 0.917922	620 0.858637
W $K\alpha_1$: 2θ	3.82	6.24	7.32	8.83	9.62	10.82	11.48	12.50	13.07	13.98
Ag $K\alpha_1$: 2θ	10.24	16.75	19.67	23.78	25.95	29.23	31.04	33.88	35.48	38.02
Mo $K\alpha_1$: 2θ	12.99	21.29	25.02	30.28	33.08	37.32	39.67	43.36	45.45	48.79
Cu $K\alpha_1$: 2θ	28.44	47.30	56.12	69.13	76.38	88.03	94.96	106.71	114.10	127.55
Ni $K\alpha_1$: 2θ	30.66	51.16	60.83	75.26	83.42	96.80	104.96	119.42	129.12	149.76
Co $K\alpha_1$: 2θ	33.15	55.53	66.22	82.42	91.77	107.59	117.71	137.42	154.04	
Fe $K\alpha_1$: 2θ	35.97	60.55	72.48	90.96	101.97	121.67	135.70			
Cr $K\alpha_1$: 2θ	42.83	73.21	88.72	114.97	133.53					

TABLE 7.7   Analyzing Crystals for X-Ray Spectroscopy

Crystal	Reflecting Plane	2d Spacing, Å	Reflectivity
Quartz	50$5\bar{2}$	1.624	Low
Aluminum	111	2.338	High
Topaz	303	2.712	Medium
Quartz	20$\bar{2}$3	2.750	Low
Lithium fluoride	220	2.848	High
Silicon	111	3.135	High
Quartz	112	3.636	Medium
Lithium fluoride	200	4.028	High
Sodium chloride	200	5.639	High
Calcium fluoride	111	6.32	High
Quartz	10$\bar{1}$1	6.686	High
Quartz	10$\bar{1}$0	8.50	Mediun,
Pentaerythritol (PET)	002	8.742	High
Ethylenediamine tartrate (EDT)	020	8.808	Medium
Ammonium dihydrogen phosphate (ADP)	110	10.648	Low
Gypsum	020	15.185	Medium
Mica	002	19.92	Low
Potassium hydrogen phthalate (KAP)	10$\bar{1}$1	26.4	Medium
Lead palmitate		45.6	
Strontium behenate		61.3	
Lead stearate		100.4	Medium

limit to the range of wavelengths that can be analyzed. Actually the goniometer is limited mechanically to about 150 deg for a 2θ value. A final requirement is the reflection efficiency and minimization of higher-order reflections. Table 7.7 gives a list of crystals commonly used for X-ray spectroscopy.

The long-wavelength analyzers are prepared by dipping an optical flat into the film of the metal fatty acid about 50 times to produce a layer 180 molecules in thickness.

Lithium fluoride is the optimum crystal for all wavelengths less than 3 Å. Pentaerythritol (PET) and potassium hydrogen phthalate (KAP) are usually the crystals of choice for wavelengths from 3 to 20 Å. Two crystals suppress even-ordered reflections: silicon (111) and calcium fluoride (111).

Mass Absorption Coefficients. Radiation traversing a layer of substance is diminished in intensity by a constant fraction per centimeter thickness x of material. The emergent radiant power P, in terms of incident radiant power P_0, is given by

$$P = P_0 \exp(-\mu x)$$

which defines the total linear absorption coefficient μ. Since the reduction of intensity is determined by the quantity of matter traversed by the primary beam, the absorber thickness is best expressed on a mass basis, in g/cm^2. The mass absorption coefficient μ/ρ, expressed in units cm^2/g, where ρ is the density of the material, is approximately independent of the physical state of the material and, to a good approximation, is additive with respect to the elements composing a substance.

Table 7.8 contains values of μ/ρ for the common target elements employed in X-ray work. A more extensive set of mass absorption coefficients for K, L, and M emission lines within the wavelength range from 0.7 to 12 Å is contained in Heinrich's paper in T. D. McKinley, K. F. J.

TABLE 7.8 Mass Absorption Coefficients for $K\alpha_1$ Lines and W $L\alpha_1$ Line

Emitter Wavelength, Å	Ag $K\alpha_1$ 0.559	Mo $K\alpha_1$ 0.709	Cu $K\alpha_1$ 1.541	Ni $K\alpha_1$ 1.658	Co $K\alpha_1$ 1.789	Fe $K\alpha_1$ 1.936	Cr $K\alpha_1$ 2.290	W $L\alpha_1$ 1.476
Absorber								
1 H	0.37	0.38	0.43	0.4	0.4	0.5	0.5	0.4
2 He	0.16	0.18	0.37	0.4	0.4	0.5	0.7	0.3
3 Li	0.18	0.22	0.50	0.6	0.7	0.9	1.5	0.4
4 Be	0.22	0.30	1.2	1.5	1.9	2.3	3.7	1.1
5 B	0.30	0.45	2.5	3.1	3.9	4.9	7.9	2.2
6 C	0.42	0.50	4.6	5.7	7.1	8.8	14.2	4.1
7 N	0.60	0.83	7.5	9.3	11.5	14.4	23.1	6.7
8 O	0.80	1.45	12.9	15.8	19.5	24.5	39.4	11.4
9 F	1.00	1.9	16.5	20.3	25.2	31.4	50.3	14.6
10 Ne	1.41	2.6	22.8	27.9	34.6	43.1	69.0	20.1
11 Na	1.75	3.5	30.3	37.2	45.9	57.2	91.4	26.8
12 Mg	2.27	4.6	39.5	48.4	59.8	74.6	119.1	34.9
13 Al	2.74	5.8	49.6	60.7	75.0	93.4	149.0	43.9
14 Si	3.44	7.3	61.4	75.2	92.8	115.5	183.8	54.4
15 P	4.20	8.8	74.7	91.4	112.9	140.5	223.6	66.2
16 S	5.15	10.6	89.2	109.2	134.7	167.4	266.1	79.1
17 Cl	5.86	12.4	104.8	128.2	158.1	196.6	312.4	92.8

TABLE 7.8 Mass Absorption Coefficients for $K\alpha_1$ Lines and W $L\alpha_1$ Line (*Continued*)

Emitter	Ag $K\alpha_1$ 0.559	Mo $K\alpha_1$ 0.709	Cu $K\alpha_1$ 1.541	Ni $K\alpha_1$ 1.658	Co $K\alpha_1$ 1.789	Fe $K\alpha_1$ 1.936	Cr $K\alpha_1$ 2.290	W $L\alpha_1$ 1.476
Absorber								
18 Ar	6.40	14.5	121.4	148.5	183.0	227.3	360.7	107.6
19 K	8.0	16.7	139.8	171	211	262	415	124
20 Ca	9.7	18.9	158.6	194	239	296	469	141
21 Sc	10.5	21.8	180.5	221	272	337	534	160
22 Ti	11.8	25.3	203	247	304	378	597	180
23 V	13.3	27.7	228	278	342	424	77	202
24 Cr	15.7	31.0	254	311	382	474	88	226
25 Mn	17.4	34.5	282	344	423	63.5	101	250
26 Fe	19.9	38.1	311	380	57.6	71.4	113	276
27 Co	21.8	42.1	341	52.8	64.9	80.6	127	303
28 Ni	25.0	46.4	48.3	58.9	72.5	90.0	142	333
29 Cu	26.4	50.7	53.7	65.5	80.6	100.0	158	47.6
30 Zn	28.2	55.4	59.5	72.7	89.4	110.9	175	52.8
31 Ga	30.8	60.1	65.9	80.5	99.0	122.8	194	58.5
32 Ge	33.5	65.2	72.3	88.2	108.6	134.7	213	64.1
33 As	36.5	70.5	79.1	96.6	118.9	147	233	70.2
34 Se	38.5	76.0	86.1	105.1	129.4	161	254	76.4
35 Br	42.3	82.5	93.9	114.7	141.2	175	277	83.4
36 Kr	45.0	88.3	101.9	124.5	153.2	190	300	90.5
37 Rb	48	95	84	103	127	158	252	98
38 Sr	52	102	90	110	137	170	271	106
39 Y	56	109	97	119	147	183	292	114
40 Zr	61	17	104	128	158	197	314	122
41 Nb	66	18	112	138	170	212	338	132
42 Mo	71	19	119	146	180	225	358	140
43 Tc	76	20	128	157	194	241	384	150
44 Ru	12	22	137	168	207	258	410	160
45 Rh	13	23	146	179	221	275	438	171
46 Pd	14	24	155	190	235	292	466	182
47 Ag	15	26	165	202	249	310	493	193
48 Cd	15	28	174	213	263	327	520	204
49 In	16	30	185	227	280	347	553	217
50 Sn	17	32	195	239	295	367	583	229
51 Sb	19	34	206	252	310	386	612	241
52 Te	19	36	216	265	326	405	644	253
53 I	21	37	230	281	346	431	684	269
54 Xe	22	39	239	293	361	448	710	280
55 Cs	24	42	332	404	495	612	822	295
56 Ba	25	44	349	425	522	645	622	311
57 La	26	46	365	444	545	673	647	325
58 Ce	28	48	383	466	571	603	216	341
59 Pr	29	51	401	487	597	453	229	356
60 Nd	31	54	420	510	534	473	241	373

TABLE 7.8 Mass Absorption Coefficients for $K\alpha_1$ Lines and W $L\alpha_1$ Line (*Continued*)

Emitter Wavelength, Å Absorber	Ag $K\alpha_1$ 0.559	Mo $K\alpha_1$ 0.709	Cu $K\alpha_1$ 1.541	Ni $K\alpha_1$ 1.658	Co $K\alpha_1$ 1.789	Fe $K\alpha_1$ 1.936	Cr $K\alpha_1$ 2.290	W $L\alpha_1$ 1.476
61 Pm	32	56	440	535		164	254	392
62 Sm	33	59	L_I 456	473	417	173	268	406
63 Eu	35	61	405	354	148	182	282	423
64 Gd	36	64	L_{II} 424	370	156	191	296	
65 Tb	38	67	316	135	164	201	311	393 L_I
66 Dy	39	70	L_{III} 329	141	172	211	327	293 L_{II}
67 Ho	41	72	123	148	181	222	343	304
68 Er	43	75	129	156	189	233	360	316 L_{III}
69 Tm	45	79	135	163	199	244	377	120
70 Yb	46	82	141	171	208	256	395	126
71 Lu	48	84	148	179	218	267	414	132
72 Hf	51	88	155	187	228	280	433	138
73 Ta	52	91	162	196	238	293	453	144
74 W	55	95	169	204	249	306	473	151
75 Re	57	98	176	213	260	319	494	157
76 Os	59	102	184	223	271	333	515	164
77 Ir	61	106	192	232	283	347	538	171
78 Pt	64	109	200	242	295	362	560	179
79 Au	67	113	209	252	307	377	584	186
80 Hg	69	117	218	263	321	394	609	194
81 Tl	72	121	227	275	334	411	635	203
82 Pb	74	125	236	286	348	428	662	211
83 Bi	78	129	247	298	363	446	690	220
84 Po		131	258	311	380	466	721	230
85 At			269	325	397	487	753	240
86 Rn	85		281	340	414	509	787	251
87 Fr		89	294	356	433	532	823	262
88 Ra	91		307	372	453	556	861	274
89 Ac			322	389	474	582	900	287
90 Th	97		337	408	497	610	944	301
91 Pa			353	427	520	639	988	315
92 U	104		372	450	548	673	898	332
93 Np			392	474	578	709	945	350
94 Pu		54	418	505	615	755	835	373

Heinrich, and D. B. Wittry (eds.), *The Electron Microprobe,* Wiley, New York, 1966, pp. 351–377. This article should be consulted to ascertain the probable accuracy of the values and for a compilation of coefficients and exponents employed in the computations.

7.2 ULTRAVIOLET-VISIBLE SPECTROSCOPY

Molecules with two or more isolated chromophores (absorbing groups) absorb light of nearly the same wavelength as does a molecule containing only a single chromophore of a particular type. The intensity of the absorption is proportional to the number of that type of chromophore present in the molecule. Representative chromophores are given in Table 7.9.

The solvent chosen must dissolve the sample, yet be relatively transparent in the spectral region of interest. In order to avoid poor resolution and difficulties in spectrum interpretation, a solvent should not be employed for measurements that are near the wavelength of or are shorter than the wavelength of its ultraviolet cutoff, that is, the wavelength at which absorbance for the solvent alone approaches one absorbance unit. Ultraviolet cutoffs for solvents commonly used are given in Table 7.10.

TABLE 7.9   Electronic Absorption Bands for Representative Chromophores

Chromophore	System	λ_{max}	ε_{max}
Acetylide	$-C\equiv C-$	175–180	6 000
Aldehyde	$-CHO$	210	strong
		280–300	11–18
Amine	$-NH_2$	195	2 800
Azido	$>C=N-$	190	5 000
Azo	$-N=N-$	285–400	3–25
Bromide	$-Br$	208	300
Carbonyl	$>C=O$	195	1 000
		270–285	18–30
Carboxyl	$-COOH$	200–210	50–70
Disulfide	$-S-S-$	194	5 500
		255	400
Ester	$-COOR$	205	50
Ether	$-O-$	185	1 000
Ethylene	$-C=C-$	190	8 000
Iodide	$-I$	260	400
Nitrate	$-ONO_2$	270 (shoulder)	12
Nitrile	$-C\equiv N$	160	——
Nitrite	$-ONO$	220–230	1 000–2 000
		300–400	10
Nitro	$-NO_2$	210	strong
Nitroso	$-NO$	302	100
Oxime	$-NOH$	190	5 000
Sulfone	$-SO_2-$	180	——
Sulfoxide	$>S=O$	210	1 500
Thiocarbonyl	$>C=S$	205	strong
Thioether	$-S-$	194	4 600
		215	1 600
Thiol	$-SH$	195	1 400
	$-(C=C)_2-$ (acyclic)	210–230	21 000
	$-(C=C)_3-$	260	35 000
	$-(C=C)_4-$	300	52 000
	$-(C=C)_5-$	330	118 000

TABLE 7.9 Electronic Absorption Bands for Representative Chromophores (*Continued*)

Chromophore	System	λ_{max}	ε_{max}
	$-(C=C)_2-$ (alicyclic)	230–260	3 000–8 000
	$C=C-C\equiv C$	219	6 500
	$C=C-C=N$	220	23 000
	$C=C-C=O$	210–250	10 000–20 000
		300–350	weak
	$C=C-NO_2$	229	9 500
Benzene		184	46 700
		204	6 900
		255	170
Diphenyl		246	20 000
Naphthalene		222	112 000
		275	5 600
		312	175
Anthracene		252	199 000
		375	7 900
Phenanthrene		251	66 000
		292	14 000
Naphthacene		272	180 000
		473	12 500
Pentacene		310	300 000
		585	12 000
Pyridine		174	80 000
		195	6 000
		257	1 700
Quinoline		227	37 000
		270	3 600
		314	2 750
Isoquinoline		218	80 000
		266	4 000
		317	3 500

Appreciable interaction between chromophores does not occur unless they are linked directly to each other, or forced into close proximity as a result of molecular stereochemical configuration. Interposition of a single methylene group, or *meta* orientation about an aromatic ring, is sufficient to insulate chromophores almost completely from each other. Certain combinations of functional groups afford chromophoric systems which give rise to characteristic absorption bands.

Sets of empirical rules, often referred to as Woodward's rules or the Woodward-Fieser rules, enable the absorption maxima of dienes (Table 7.11) and enones and dienones (Table 7.12) to be predicted. To the respective base values (absorption wavelength of parent compound) are added the increments for the structural features or substituent groups present. When necessary, a solvent correction is also applied (Table 7.13).

Ring substitution on the benzene ring affords shifts to longer wavelengths (Table 7.14) and intensification of the spectrum. With electron-withdrawing substituents, practically no change in the maximum position is observed. The spectra of heteroaromatics are related to their isocyclic analogs, but only in the crudest way. As with benzene, the magnitude of substituent shifts can be estimated, but tautomeric possibilities may invalidate the empirical method.

When electronically complementary groups are situated *para* to each other in disubstituted benzenes, there is a more pronounced shift to a longer wavelength than would be expected from the additive effect due to the extension of the chromophore from the electron-donating group through the ring to the electron-withdrawing group. When the *para* groups are not complementary, or when the groups are situated *ortho* or *meta* to each other, disubstituted benzenes show a more or less additive effect of the two substituents on the wavelength maximum. Calculation of the principal band of selected substituted benzenes is illustrated in Table 7.15.

TABLE 7.10 Ultraviolet Cutoffs of Spectrograde Solvents

Absorbance of 1.00 in a 10.0 mm cell vs. distilled water.

Solvent	Wavelength, nm	Solvent	Wavelength, nm
Acetic acid	260	Hexadecane	200
Acetone	330	Hexane	210
Acetonitrile	190	Isobutyl alcohol	230
Benzene	280	Methanol	210
1-Butanol	210	2-Methoxyethanol	210
2-Butanol	260	Methylcyclohexane	210
Butyl acetate	254	Methylene chloride	235
Carbon disulfide	380	Methyl ethyl ketone	330
Carbon tetrachloride	265	Methyl isobutyl ketone	335
1-Chlorobutane	220	2-Methyl-1-propanol	230
Chloroform (stabilized		*N*-Methylpyrrolidone	285
with ethanol)	245	Nitromethane	380
Cyclohexane	210	Pentane	210
1,2-Dichloroethane	226	Pentyl acetate	212
Diethyl ether	218	1-Propanol	210
1,2-Dimethoxyethane	240	2-Propanol	210
N,N-Dimethylacetamide	268	Pyridine	330
N,N-Dimethylformamide	270	Tetrachloroethylene	
Dimethylsulfoxide	265	(stabilized with thymol)	290
1,4-Dioxane	215	Tetrahydrofuran	220
Ethanol	210	Toluene	286
2-Ethoxyethanol	210	1,1,2-Trichloro-1,2,2-	
Ethyl acetate	255	trifluoroethane	231
Ethylene chloride	228	2,2,4-Trimethylpentane	215
Glycerol	207	*o*-Xylene	290
Heptane	197	Water	191

TABLE 7.11   Absorption Wavelength of Dienes

Heteroannular and acyclic dienes usually display molar absorptivities in the 8000 to 20 000 range, whereas homoannular dienes are in the 5000 to 8000 range.
   Poor correlations are obtained for cross-conjugated polyene systems such as

$$-C=C\begin{array}{c}C=C-\\C=C-\end{array}$$

   The correlations presented here are sometimes referred to as Woodward's rules or the Woodward-Fieser rules.

Base value for heteroannular or open chain diene, nm	214
Base value for homoannular diene, nm	253
Increment (in nm) for	
double bond extending conjugation	30
Alkyl substituent or ring residue	5
Exocyclic double bond	5
Polar groupings:	
-O-acyl	0
-O-alkyl	6
-S-alkyl	30
-Cl, -Br	5
-N(alkyl)$_2$	60
Solvent correction (see Table 7.13)	———
Calculated wavelength =	total

TABLE 7.12   Absorption Wavelength of Enones and Dienones

$$O=C-\overset{\alpha}{C}=\overset{\beta}{C}\diagdown^{\beta} \qquad\qquad O=C-\overset{\alpha}{C}=\overset{\beta}{C}-\overset{\gamma}{C}=\overset{\delta}{C}\diagdown^{\delta}$$

Base values, nm	
Acyclic α,β-unsaturated ketones	215
Acyclic α,β-unsaturated aldehyde	210
Six-membered cyclic α,β-unsaturated ketones	215
Five-membered cyclic α,β-unsaturated ketones	214
α,β-Unsaturated carboxylic acids and esters	195
Increments (in nm) for	
Double bond extending conjugation:	
Heteroannular	30
Homoannular	69
Alkyl group or ring residue:	
α	10
β	12
γ,δ	18
Polar groups:	
—OH	
α	35
β	30
γ	50
—O—CO—CH_3 and —O—CO—C_6H_5: $\alpha,\beta,\gamma,\delta$	6
—OCH_3	
α	35
β	30
γ	17
δ	31
—S—alkyl, β	85
—Cl	
α	15
β	12
—Br	
α	25
β	30
—$N(alkyl)_2$, β	95
Exocyclic double bond	5
Solvent correction (see Table 7.13)	
Calculated wavelength =	total

TABLE 7.13 Solvent Correction for Ultraviolet-Visible Spectroscopy

Solvent	Correction, nm
Chloroform	+1
Cyclohexane	
Diethyl ether	+11
1,4-Dioxane	+5
Ethanol	0
Hexane	+11
Methanol	0
Water	−8

TABLE 7.14 Primary Bands of Substituted Benzene and Heteroaromatics

In methanol.

Base value: 203.5 nm.

Substituent	Wavelength shift, nm	Substituent	Wavelength shift, nm
—CH$_3$	3.0	—COOH	25.5
—CH=CH$_2$	44.5	—COO$^-$	20.5
—C≡CH	44	—CN	20.5
—C$_6$H$_5$	48	—NH$_2$	26.5
—F	0	—NH$_3^+$	−0.5
—Cl	6.0	—N(CH$_3$)$_2$	47.0
—Br	6.5	—NH—CO—CH$_3$	38.5
—I	3.5	—NO$_2$	57
—OH	7.0	—SH	32
—O$^-$	31.5	—SO—C$_6$H$_5$	28
—OCH$_3$	13.5	—SO$_2$CH$_3$	13
—OC$_6$H$_5$	51.5	—SO$_2$NH$_2$	14.0
—CHO	46.0	—CH=CH—C$_6$H$_5$	
—CO—CH$_3$	42.0	*cis*	79
—CO—C$_6$H$_5$	48	*trans*	92.0
		—CH=CH—COOH, *trans*	69.5

Heteroaromatic	Base value, nm	Heteroaromatic	Base value, nm
Furan	200	Pyridine	257
Pyrazine	257	Pyrimidine	ca 235
Pyrazole	214	Pyrrole	209
Pyridazine	ca 240	Thiophene	231

TABLE 7.15   Wavelength Calculation of the Principal Band of Substituted Benzene Derivatives

In ethanol.

Base value of parent chromophore, nm	
C_6H_5COOH or C_6H_5COO—alkyl	230
C_6H_5—CO—alkyl (or aryl)	246
C_6H_5CHO	250
Increment (in nm) for each substituent on phenyl ring	
—Alkyl or ring residue	
o-, m-	3
p-	10
—OH and —O— alkyl	
o-, m-	7
p-	25
—O^-	
o-	11
m-	20
p-	78*
—Cl	
o-, m-	0
p-	10
—Br	
o-, m-	2
p-	15
—NH_2	
o-, m-	13
p-	58
—NHCO—CH_3	
o-, m-	20
p-	45
—$NHCH_3$	
p-	73
—$N(CH_3)_2$	
o-, m-	20
p-	85

* Value may be decreased markedly by steric hindrance to coplanarity.

7.3 FLUORESCENCE

TABLE 7.16 Fluorescence Spectroscopy of Some Organic Compounds

Compound	Solvent	pH	Excitation wavelength, nm	Emission wavelength, nm
Acenaphthene	Pentane		291	341
Acridine	CF₃COOH		358	475
Adenine	Water	1	280	375
Adenosine	Water	1	285	395
Adenosine triphosphate	Water	1	285	395
Adrenalin			295	335
p-Aminobenzoic acid	Water	8	295	345
Aminopterin	Water	7	280, 370	460
1-Aminopyrene	CF₃COOH		330, 342	415
p-Aminosalicyclic acid	Water	11	300	405
Amobarbital	Water	14	265	410
Anilines	Water	7	280, 291	344, 361
Anthracene	Pentane		420	430
Anthranilic acid	Water	7	300	405
Azaindoles	Water	10	290, 299	310, 347
Benz[c]acridine	CF₃COOH		295, 380	480
Benz[a]anthracene	Pentane		284	382
1,2-Benzanthracene			280, 340	390, 410
Benzanthrone	CF₃COOH		370, 420	550
Benzo[b]chrysene	Pentane		283	398
11-H-Benzo[a]fluorene	Pentane		317	340
Benzoic acid	70% H₂SO₄		285	385
3,4-Benzopyrene	Benzene		365	390, 480
Benzo[e]pyrene	Pentane		329	389
Benzoquinoline	CF₃COOH		280	425
Benzoxanthane	Pentane		363	418
Bromolysergic acid diethyl amide	Water	1	315	460
Brucine	Water	7	305	500
Carbazole	N,N-Dimethyl formamide		291	359
Chlortetracycline			355	445
Chrysene	Pentane		250, 300, 310	260, 380
Cinchonine	Water	1	320	420
Coumarin	Ethanol		280	352
Dibenzo[a,c]anthracene	Pentane		280	381
Dibenzo[b,k]chrysene	Pentane		308	428
Dibenzo[a,e]pyrene	Pentane		370	401
3,4,8,9-Dibenzopyrene			370, 335, 390, 410	480, 510
5,12-Dihydronaphthacene	Pentane		282	340
1,4-Diphenylbutadiene	Pentane		328	370

TABLE 7.16 Fluorescence Spectroscopy of Some Organic Compounds (*Continued*)

Compound	Solvent	pH	Excitation wavelength, nm	Emission wavelength, nm
Epinephrine	Water	7	295	335
Ethacridine	Water	2	370, 425	515
Fluoranthrene	Pentane		354	464
Fluorene	Pentane		300	321
Fluorescein	Water	7–11	490	515
Folic acid	Water	7	365	450
Gentisic acid	Water	7	315	440
Griseofulvin	Water	7	295, 335	450
Guanine	Water	1	285	365
Harmine	Water	1	300, 365	400
Hippuric acid	70% H_2SO_4		270	370
Homovanillic acid	Water	7	270	315
m-Hydroxybenzoic acid	Water	12	314	430
p-Hydroxycinnamic acid	Water	7	350	440
7-Hydroxycoumarin	Ethanol		325	441
5-Hydroxyindole	Water	1	290	355
5-Hydroxyindoleacetic acid	Water	7	300	355
3-Hydroxykynurenine	Water	11	365	460
p-Hydroxymandelic acid	Water	7	300	380
p-Hydroxyphenylacetic acid	Water	7	280	310
p-Hydroxyphenylpyruvic acid	Water	7	290	345
p-Hydroxyphenylserine	Water	1	290	320
5-Hydroxytryptophan	Water	7	295	340
Imipramine	Water	14	295	415
Indoleacetic acid	Water	8	285	360
Indoles	Water	7	269, 315	355
Indomethacin	Water	13	300	410
Kynurenic acid	Water	7	325	405
		11	325	440
Lysergic acid diethylamide	Water	1	325	445
Menadione	Ethanol		335	480
9-Methylanthracene	Pentane		382	410
3-Methylcholanthrene	Pentane		297	392
7-Methyldibenzopyrene	Pentane		460	467
2-Methylphenanthrene	Pentane		257	357
3-Methylphenanthrene	Pentane		292	368
1-Methylpyrene	Pentane		336	394
4-Methylpyrene	Pentane		338	386
Naphthacene			290, 310	480, 515
1-Naphthol	0.1*M* NaOH 20% ethanol		365	480
2-Naphthol	0.1*M* NaOH 20% ethanol		365	426
Oxytetracycline			390	520
Phenanthrene	Pentane		252	362

TABLE 7.16 Fluorescence Spectroscopy of Some Organic Compounds (*Continued*)

Compound	Solvent	pH	Excitation wavelength, nm	Emission wavelength, nm
Phenylalanine	Water		215, 260	282
o-Phenylenepyrene	Pentane		360	506
Phenylephrine			270	305
Picene	Pentane		281	398
Procaine	Water	11	275	345
Pyrene	Pentane		330	382
Pyridoxal	Water	12	310	365
Quinacrine	Water	11	285	420
Quinidine	Water	1	350	450
Quinine	Water	1	250, 350	450
Reserpine	Water	1	300	375
Resorcinol	Water		265	315
Riboflavin	Water	7	270, 370, 445	520
Rutin	Water	1	430	520
Salicyclic acid	Water	11	310	435
Scoparone	Water	10	350, 365	430
Scopoletin	Water	10	365, 390	460
Serotonin	3M HCl		295	550
Skatole	Water		290	370
Streptomycin	Water	13	366	445
p-Terphenyl	Pentane		284	338
Thiopental			315	530
Thymol	Water	7	265	300
Tocopherol	Hexane-ethanol		295	340
Tribenzo[a,e,i]pyrene	Pentane		384	448
Triphenylene	Pentane		288	357
Tryptamine	Water	7	290	360
Tryptophan	Water	11	285	365
Tyramine	Water	1	275	310
Tyrosine	Water	7	275	310
Uric acid	Water	1	325	370
Vitamin A	1-Butanol		340	490
Vitamin B$_{12}$	Water	7	275	305
Warfarin	Methanol		290, 342	385
Xanthine	Water	1	315	435
2,6-Xylenol			275	305
3,4-Xylenol			280	310
Yohimbine	Water	1	270	360
Zoxazolamine	Water	11	280	320

TABLE 7.17 Fluorescence Quantum Yield Values

Compound	Solvent	Q_F value vs. Q_F standard	
	Q_F standard		
9-Aminoacridine	Water	0.99	
Anthracene	Ethanol	0.30	
POPOP*	Toluene	0.85	
Quinine sulfate dihydrate	$1 N$ H_2SO_4	0.55	
	Secondary standards		
Acridine orange hydrochloride	Ethanol	0.54	Quinine sulfate
		0.58	Anthracene
1,8-ANS† (free acid)	Ethanol	0.38	Anthracene
		0.39	POPOP
1,8-ANS (magnesium salt)	Ethanol	0.29	Anthracene
		0.31	POPOP
Fluorescein	$0.1 N$ NaOH	0.91	Quinine sulfate
		0.94	POPOP
Fluorescein, ethyl ester	$0.1 N$ NaOH	0.99	Quinine sulfate
		0.99	POPOP
Rhodamine B	Ethanol	0.69	Quinine sulfate
		0.70	Anthracene
2,6-TNS‡ (potassium salt)	Ethanol	0.48	Anthracene
		0.51	POPOP

* POPOP *p*-bis[2-(5-phenyloxazoyl)]benzene.
† ANS, anilino-8-naphthalene sulfonic acid.
‡ TNS, 2-*p*-toluidinylnaphthalene-6-sulfonate.

7.4 FLAME ATOMIC EMISSION, FLAME ATOMIC ABSORPTION, ELECTROTHERMAL (FURNACE) ATOMIC ABSORPTION, ARGON INDUCTION COUPLED PLASMA, AND PLASMA ATOMIC FLUORESCENCE

The tables of atomic emission and atomic absorption lines are presented in two parts. In Table 7.18 the data are arranged in alphabetic order by name of the element, whereas in Table 7.19 the sensitive lines of the elements are arranged in order of decreasing wavelengths. For additional lines and their relative intensities consult W. F. Meggers, C. H. Corliss, and B. F. Scribner, *Tables of Spectral-Line Intensities, Part I,* National Bureau of Standards Monograph 32, U.S. Government Printing Office, Washington, D.C., 1961.

The detection limits in the table correspond generally to the concentration of an element required to give a net signal equal to three times the standard deviation of the noise (background) in accordance with IUPAC recommendations. Detection limits can be confusing when steady-state techniques such as flame atomic emission or absorption, and plasma atomic

TABLE 7.18 Detection Limits in ng/mL

The detection limits in the table correspond generally to the concentration of analyte required to give a net signal equal to three times the standard deviation of the background in accordance with IUPAC recommendations.

Element	Wavelength, nm	Flame emission	Flame atomic absorption	Electrothermal atomic absorption	Argon ICP	Plasma atomic fluorescence
Aluminum	308.22		40		10	
	309.28		20	0.05	11	4
	394.40	3.6	45		36	
	396.15	7.5	30	0.01	20	5
Antimony	206.83				50	
	217.58		30		50	
	231.15	70			30	10
	259.81	200		0.08		0.1
Arsenic	189.04		160		35	
	193.76		120	1	50	
	197.20		240			
	228.81	455				
	234.90	250				10
Barium	455.36	3			0.9	
	493.41	4			1	
	553.55	1.5	9	0.04		2
Beryllium	234.86		1	0.05	0.4	
	313.04		2	0.003	1	
	313.11	100			1	0.2
Bismuth	223.06		18	0.35	30	
	227.66			2		
	306.77	60		0.5	30	2
Boron	182.59				8	
	249.77		700	15	3	60
(as BO₂)	518.0	50				
(as BO₂)	547.6	50				
Bromine	154.07				50	
Cadmium	214.44				1.0	
	226.50				0.6	
	228.80	6	1	0.008	228	
	326.11	3	0.5	0.014		0.001
Calcium	315.89				20	
	393.37				0.6	
	396.85				1.2	
	422.67	1.5	1	0.3		0.08
Carbon	193.09				44	
	247.86				1000	
Cerium	413.38				30	
	418.66				30	
	569.92	150				
Cesium	852.11	0.02	8	0.04		
	894.35	0.04	130			
Chlorine	134.72				50	
Chromium	267.72				3	
	283.58				20	
	284.98				30	
	357.87	6	2	0.05		0.4

TABLE 7.18 Detection Limits in ng/mL (*Continued*)

Element	Wavelength, nm	Flame emission	Flame atomic absorption	Electrothermal atomic absorption	Argon ICP	Plasma atomic fluorescence
Chromium	359.35	7				
(*cont.*)	360.53	13				
	425.44	3	6		66	
	427.48	4				
	428.97	5				
Cobalt	228.62				3	
	238.89				28	
	240.73	5	8	0.01	7	0.4
	345.35	30				
Copper	324.75	1.5	1	0.01	2	0.2
	327.40	3	2	0.02	4	
Dysprosium	353.17				3	
	340.78				6	
	404.60	30	50			300
	418.68		60			
	421.17		60			
Erbium	323.06				15	
	349.81				10	
	400.80	30	40	0.3		500
	408.77		40			
Europium	381.97				2	
	412.97				3	
	459.40	0.45	20	0.5		20
Gadolinium	335.05				10	
	368.41		4000			
	440.19	72	1000	8		800
Gallium	287.42		70			
	294.36		20		30	
	403.30	5	50			
	417.21	3	30	1	40	0.9
Germanium	209.43				50	
	219.87				100	
	265.12	400	40	7.5		50
Gold	242.80		10	0.5	5	
	267.60	500	8	0.5	10	0.3
Hafnium	263.87				10	
	277.34				10	
	307.29		2000			
Holmium	339.90				3	
	345.60				8	
	405.39	15	40	0.7		100
	410.38		30			
Indium	230.61				40	
	303.94	100	7	0.01		
	325.61	22	8			
	410.18	14	20			
	451.13	0.7	22		2	0.2
Iodine	178.38				20	
	183.0			3		
Iridium	208.88	400	500	0.5		
	212.68				20	
	224.27				20	

TABLE 7.18 Detection Limits in ng/mL (*Continued*)

Element	Wavelength, nm	Flame emission	Flame atomic absorption	Electrothermal atomic absorption	Argon ICP	Plasma atomic fluorescence
Iron	238.20				4	
	248.33		3	0.01		
	259.94				3	
	302.06	18	5			
	371.99	15	10			0.3
	385.99	12	21			
Lanthanum	379.48				15	
	392.76		8000			
	408.67				2	
	550.13	20				
	579.13	5	2000	0.5		
(as LaO)	441.82	100				
(as LaO)	560.25	300				
Lead	217.00		20	0.4		
	220.35				20	
	283.31	60	10	1		5
	368.35	30				
	405.78	20				
Lithium	460.29	0.06	30		50	
	610.36	0.001				
	670.78	0.003	0.3	1.5	5	0.4
Lutetium	261.54				1	
	307.76				6	
Magnesium	279.08				30	
	279.55				1.5	
	285.21	4.5	0.1	0.018	3.6	0.4
Manganese	256.37				2.7	
	257.61				0.5	
	259.37		60		3	
	260.57				6	
	279.48	1	1	0.05		0.4
	293.30				24	
	294.92				24	
	403.08	1.5	30			
Mercury	194.23				30	
	253.65	150	0.001	6	50	5
Molybdenum	202.03				5	
	203.84				8	
	281.62				1.2	
	313.26	220	30	0.06		12
	390.30	75	50			
Neodymium	292.45	200				
	401.23				10	
	430.36				30	
	492.45	150	600			2000
Nickel	231.60				6	
	232.00	8	4	0.5	10	
	341.48	15	2			
	352.45	8	2			2
Niobium	316.34				20	
	405.89	250	1000			1000
Osmium	225.58				20	

TABLE 7.18 Detection Limits in ng/mL (*Continued*)

Element	Wavelength, nm	Flame emission	Flame atomic absorption	Electrothermal atomic absorption	Argon ICP	Plasma atomic fluorescence
Osmium	228.23				40	
(*cont.*)	263.71	2000	80			
	290.91		110			
Palladium	244.8	20	20	0.5		40
	340.46	25	80		40	
	363.47	50			60	
Phosphorus	178.28				50	
	213.62				50	
(as HPO)	524.9	100				
Platinum	214.42				20	
	265.95	2000	100	0.2	40	300
Potassium	404.41	1.3	100			
	404.72	2.6				
	766.49	0.15	1	0.004	200	0.6
	769.90	0.3	2			
Praseodymium	390.84				20	
	414.31				30	
	493.97	300				1000
Rhenium	197.31				8	
	345.19	690				
	346.05	200	200	10		
	346.47	275				
Rhodium	343.49	10	2	0.1	20	100
	369.24	20			30	
Rubidium	780.02	0.0065	0.3		500	3
	794.76	0.013				
Ruthenium	240.27				50	
	349.89	80	70	10	150	500
Samarium	442.43				10	
	476.03	30	500		100	
Scandium	255.24				21	
	357.24				1	
	361.38				1.5	
	391.18	21	20	6	120	10
	402.04	30				
	402.34	30				
Selenium	196.03		90	2.5	6	10
Silicon	251.61		80	0.5	10	50
	283.16				15	
Silver	328.07	2	0.9	0.001	4.5	0.1
	338.29	4			3	
Sodium	330.23	125		0.7	15	
	330.30	250				
	589.00	0.01	0.2	0.004	20	0.2
	589.59	0.02				
Strontium	407.78				1	
	421.55				0.5	
	460.73	0.1	2	0.01		0.3
Sulfur	180.73		10		70	
(as S_2)	394.0	1600				
Tantalum	240.06				20	
	271.47		800			

TABLE 7.18 Detection Limits in ng/mL (*Continued*)

Element	Wavelength, nm	Flame emission	Flame atomic absorption	Electrothermal atomic absorption	Argon ICP	Plasma atomic fluorescence
Tellurium	214.27	150	15	0.5		2
	238.58				60	
Terbium	350.92				10	
	384.87				40	
	431.89	150	600			500
Thallium	190.86				50	
	276.78		9	0.15		
	351.92				150	
	377.57	3		0.5		4
	535.0	1.5				
Thorium	283.73				30	
	401.91				30	
Thulium	313.13				3	
	371.79	4	10			100
	384.80				7	
Tin	189.99				15	
	224.60		110	1	30	
	284.00	100	200			10
	286.33		160	1.5		
Titanium	334.19	400				
	334.94				6	
	337.28				8	
	364.27	210	60	2.5		30
	365.35	180				
	399.86	150				
Tungsten	207.91				30	
	209.48				50	
	400.87	450	1000			2000
Uranium	358.49	100		30		
	385.96				70	
	409.01				140	
Vanadium	292.40				7.8	
	310.23				10	
	318.34	18				
	318.54	25	50	1		30
	437.92	15				
Ytterbium	328.94				1	
	369.42				2	
	398.80	0.45	5	0.1		10
Yttrium	360.07				3	
	362.09	40	50	10		50
	371.03				1	
	410.24	30	50			
Zinc	202.55				4	
	213.86	1000	0.8	0.005	2	0.0003
Zirconium	339.20				5	
	343.82				7	
	349.62				45	
	360.12	1000	350			

TABLE 7.19 Sensitive Lines of the Elements

In this table the sensitive lines of the elements are arranged in order of decreasing wavelengths. A Roman numeral II following an element designation indicates a line classified as being emitted by the singly ionized atom. In the column headed Sensitivity, the most sensitive line of the non-ionized atom is indicated by U1, and other lines by U2, U3, and so on, in order of decreasing sensitivity. For the singly ionized atom the corresponding designations are V1, V2, V3, and so on.

Wavelength, nm	Element		Sensitivity	Wavelength, nm	Element		Sensitivity
894.35	Cs		U2	492.45	Nd		U1
852.11	Cs		U1	488.91	Re		U4
819.48	Na		U4	487.25	Sr		U3
818.33	Na		U3	483.21	Sr		U2
811.53	Ar		U2	482.59	Ra		U1
794.76	Rb		U2	481.95	Cl	II	V4
780.02	Rb		U1	481.67	Br	II	V3
769.90	K		U2	481.05	Zn		U3
766.49	K		U1	481.01	Cl	II	V3
750.04	Ar		U4	479.45	Cl	II	V2
706.72	Ar		U3	478.55	Br	II	V2
696.53	Ar		U3	476.03	Sm		U1
690.24	F		U3	470.09	Br	II	V1
685.60	F		U2	467.12	Xe		U2
670.78	Li		U1	462.43	Xe		U3
656.28	H		U2	460.73	Sr		U1
649.69	Ba	II	V4	460.29	Li		U4
624.99	La		U3	459.40	Eu		U1
614.17	Ba	II	V3	459.32	Cs		U4
610.36	Li		U2	455.54	Cs		U3
593.06	La		U4	455.40	Ba	II	V1
589.59	Na		U2	451.13	In		U1
589.00	Na		U1	450.10	Xe		U4
587.76	He		U3	445.48	Ca		U2
587.09	Kr		U2	442.43	Sm	II	V4
579.13	La		U1	440.85	V		U4
569.92	Ce		U1	440.19	Gd		U1
567.96	N	II	V2	439.00	V		U3
567.60	N	II	V4	437.49	Y	II	V4
566.66	N	II	V3	437.92	V		U1
557.02	Kr		U3	435.84	Hg		U3
553.55	Ba		U1	431.89	Tb		U1
550.13	La		U2	430.36	Nd	II	V2
546.55	Ag		U4	430.21	W		U1
546.07	Hg		U2	429.67	Sm		U1
545.52	La		U3	428.97	Cr		U3
535.84	Hg		U3	427.48	Cr		U2
535.05	Tl		U1	425.43	Cr		U1
521.82	Cu		U3	422.67	Ca		U1
520.91	Ag		U3	421.56	Rb		U4
520.84	Cr		U8	421.55	Sr	II	V1
520.60	Cr		U7	421.17	Dy		U2
515.32	Cu		U4	420.19	Rb		U3
498.18	Ti		U1	418.68	Dy		U2
496.23	Sr		U2	418.66	Ce	II	V1
493.97	Pr		U1	417.21	Ga		U1
493.41	Ba	II	V2	414.31	Pr	II	V2

TABLE 7.19   Sensitive Lines of the Elements (*Continued*)

Wavelength, nm	Element		Sensitivity	Wavelength, nm	Element		Sensitivity
414.29	Y		U4	386.41	Mo		U2
413.38	Ce	II	V1	385.99	Fe		U2
413.07	Ba	II	V5	385.96	U	II	V1
412.97	Eu	II	V2	384.87	Tb	II	V2
412.83	Y		U3	384.80	Tm	II	V2
412.38	Nb		U4	383.83	Mg		U2
412.32	La	II	V4	383.82	Mo		U2
411.00	N		U2	382.23	Mg		U3
410.38	Ho		U1	382.94	Mg		U4
410.24	Y		U1	381.97	Eu	II	V1
410.18	In		U2	379.94	Ru		U3
410.09	Nb		U3	379.63	Mo		U1
409.99	N		U3	379.48	La	II	V2
409.01	U	II	V2	379.08	La	II	V3
408.77	Er		U1	377.57	Tl		U3
408.67	La	II	V1	377.43	Y	II	V3
407.97	Nb		U2	374.83	Fe		U4
407.77	Sr	II	V2	373.49	Fe		U2
407.74	Y		U2	372.80	Ru		U1
407.74	La	II	V2	371.99	Fe		U1
407.43	W		U2	371.79	Tm		U1
405.89	Nb		U1	371.03	Y	II	V1
405.78	Pb		U1	369.42	Yb	II	V2
405.39	Ho		U2	369.24	Rh		U2
404.72	K		U4	368.41	Gd		U2
404.66	Hg		U5	368.35	Pb		U2
404.60	Dy		U1	365.48	Hg		U4
404.41	K		U3	365.35	Ti		U2
403.45	Mn		U3	365.01	Hg		U3
403.31	Mn		U2	364.28	Sc	II	V3
403.30	Ga		U2	364.27	Sn		U3
403.08	Mn		U1	363.47	Pd		U2
402.37	Sc		U3	363.07	Sc	II	V2
402.04	Sc		U3	362.09	Y		U2
401.91	Th	II	V1	361.38	Sc	II	V1
401.23	Nd	II	V1	360.96	Pd		U2
400.87	W		U1	360.12	Zr		U1
400.80	Er		U1	360.07	Y	II	V2
399.86	Cr		U1	360.05	Cr		U6
399.86	Ti		U1	359.62	Ru		U3
398.80	Yb		U1	359.34	Cr		U5
396.85	Ca	II	V2	359.26	Sm	II	V1
396.15	Al		U1	358.49	U		V1
394.91	La	II	V2	357.87	Cr		U4
394.40	Al		U2	357.25	Zr	II	V4
393.37	Ca	II	V1	357.24	Sc	II	V1
391.18	Sc		U1	356.83	Sn	II	V1
390.84	Pr	II	V1	355.31	Pd		U3
390.75	Sc		U2	354.77	Zr		U3
390.30	Mo		U1	353.17	Dy	II	V1
389.18	Ba		V4	352.98	Co		U3
388.86	He		U2	352.94	Tl		U4
388.63	Fe		U5	352.69	Co		U4

TABLE 7.19 Sensitive Lines of the Elements (*Continued*)

Wavelength, nm	Element		Sensitivity	Wavelength, nm	Element		Sensitivity
352.45	Ni		U2	324.75	Cu		U1
351.96	Zr		U3	324.27	Pd		U4
351.92	Tl		U2	323.45	Cr		V3
351.69	Pd		U3	323.26	Li		U3
351.36	Ir		U2	323.06	Er	II	V2
350.92	Tb	II	V1	322.08	Ir		U1
350.63	Co		U3	318.54	V		U3
350.23	Co		U2	318.40	V		U2
349.89	Ru		U2	317.93	Ca	II	V3
349.62	Zr	II	V3	316.34	Nb	II	V1
349.41	Er	II	V1	315.89	Ca	II	V4
348.11	Pd		U5	313.26	Mo		U2
347.40	Ni		U3	313.13	Tm	II	V1
346.47	Re		U2	313.11	Be		U1
346.05	Re		U1	313.04	Be		U2
345.60	Ho	II	V2	311.84	V	II	V4
345.58	Co		U5	311.07	V	II	V3
345.19	Re		U3	310.23	V	II	V2
345.14	B	II	V2	309.42	Nb	II	V1
344.36	Co		U2	309.31	V	II	V1
344.06	Fe		U2	309.27	Al		U3
343.82	Zr	II	V2	308.22	Al		U4
343.67	Ru		U2	307.76	Lu	II	V2
343.49	Rh		U1	307.29	Hf		U1
342.83	Ru		U4	306.77	Bi		U3
342.12	Pd		U3	306.47	Pt		U1
341.48	Ni		U3	303.94	In		U4
341.23	Co		U4	303.90	Ge		U2
340.78	Dy	II	V2	303.41	Sn		U3
340.51	Co		U2	302.06	Fe		U3
340.46	Pd		U2	300.91	Sn		U4
339.90	Ho	II	V1	294.91	Mn	II	V4
339.20	Zr	II	V1	294.44	W		U5
338.29	Ag		U2	294.36	Ga		U3
337.28	Ti	II	V3	294.02	Ta		U3
336.12	Ti	II	V2	293.30	Mn	II	V4
335.05	Gd	II	V1	292.98	Pt		U3
334.94	Ti	II	V1	292.45	Nd		U2
334.50	Zn		U2	292.40	V	II	V1
334.19	Ti		U4	290.91	Os		U2
332.11	Be		U3	289.80	Bi		U2
331.12	Ta		U3	289.10	Mo	II	V4
330.03	Na		U6	288.16	Si		U1
330.26	Zn		U3	287.42	Ga		U4
330.23	Na		U5	287.15	Mo	II	V3
328.94	Yb	II	V1	286.33	Sn		U2
328.23	Zn		U5	286.04	As		U2
328.07	Ag		U1	285.21	Mg		U1
327.40	Cu		U2	284.82	Mo	II	V2
326.95	Ge		U3	284.00	Sn		U1
326.23	Sn		U3	283.73	Th	II	V1
326.11	Cd		U1	283.58	Cr	II	V2
325.61	In		U3	283.31	Pb		U3

TABLE 7.19 Sensitive Lines of the Elements (*Continued*)

Wavelength, nm	Element		Sensitivity	Wavelength, nm	Element		Sensitivity
283.16	Si	II	V1	239.56	Fe	II	V2
283.03	Pt		U3	238.89	Co	II	V2
281.62	Al	II	V2	238.58	Te		U2
281.61	Mo	II	V1	238.32	Te		U3
280.27	Mg	II	V2	238.20	Fe	II	V1
280.20	Pb		U4	234.90	As		U4
279.83	Mn		U3	234.86	Be		U1
279.55	Mg	II	V1	232.00	Ni		U2
279.48	Mn		U3	231.60	Ni	II	V1
279.08	Mg	II	V2	231.15	Sb		U1
278.02	As		U1	230.61	In	II	V1
277.34	Hf	II	V1	228.81	As		U5
276.78	Tl		U4	228.80	Cd		U2
272.44	W		U4	228.71	Ni	II	V1
271.90	Fe		U5	228.62	Co	II	V1
271.47	Ta		U1	228.23	Os	II	V2
270.65	Sn		U4	227.66	Bi		U3
267.72	Cr	II	V1	227.02	Ni	II	V2
267.60	Au		U2	226.50	Cd	II	V2
266.92	Al	II	V1	226.45	Ni	II	V3
265.95	Pt		U1	225.58	Os	II	V1
265.12	Ge		U1	225.39	Ni	II	V4
265.05	Ba		U2	224.70	Cu	II	V3
264.75	Ta		U2	224.64	Ag	II	V3
263.87	Hf	II	V1	224.60	Sn		U1
263.71	Os		U1	224.27	Ir	II	V1
260.57	Mn	II	V3	223.06	Bi		U1
259.94	Fe	II	V1	220.35	Pb	II	V1
259.81	Sb		U2	219.87	Ge	II	V2
259.37	Mn		U2	219.23	Cu	II	V2
257.61	Mn	II	V1	217.58	Sb		U2
256.37	Mn	II	V2	217.00	Pb	II	V1
255.33	P		U3	214.44	Cd	II	V1
255.24	Sc	II	V3	214.42	Pt	II	V1
253.65	Hg		U1	214.27	Te		U1
253.57	P		U1	213.86	Zn		U1
252.85	Si		U2	213.62	P		U1
252.29	Fe		U3	213.60	Cu	II	V1
251.61	Si		U3	212.68	Ir	II	V1
250.69	Si		U4	209.48	W	II	V2
250.20	Zn	II	V4	209.43	Ge	II	V1
249.77	B		U1	208.88	Ir		U1
249.68	B		U2	207.91	W	II	V1
248.33	Fe		U3	207.48	Se		U4
247.86	C		U2	206.83	Sb		U1
245.65	As		U4	206.28	Se		U3
243.78	Ag	II	V2	206.19	Zn	II	V2
242.80	Au		U1	203.99	Se		U1
241.05	Fe	II	V4	203.84	Mo	II	V3
240.73	Co		U1	202.55	Zn	II	V1
240.49	Fe		V3	202.03	Mo	II	V2
240.27	Ru		V1	197.31	Re	II	V1
240.06	Ta	II	V1	197.20	As		U3

TABLE 7.19   Sensitive Lines of the Elements (*Continued*)

Wavelength, nm	Element		Sensitivity	Wavelength, nm	Element		Sensitivity
196.03	Se		U2	183.00	I		U2
194.23	Hg	II	V1	182.59	B	II	V2
193.76	As		U1	180.73	S		U1
193.09	C		U1	178.38	I		U1
190.86	Tl	II	V1	178.28	P		U1
189.99	Sn	II	V1	154.07	Br	II	V4
189.04	As		U2	134.72	Cl	II	V1

emission or fluorescence, which are steady-state techniques, are compared with the electro-thermal or furnace technique which uses the entire sample and detects an absolute amount of the analyte element. To compare the several methods on the basis of concentration, the furnace detection limits assume a 20-μL sample.

Data for the several flame methods assume an acetylene–nitrous oxide flame residing on a 5- or 10-cm slot burner. The sample is nebulized into a spray chamber placed immediately ahead of the burner. Detection limits are quite dependent on instrument and operating variables, particularly the detector, the fuel and oxidant gases, the slit width, and the method used for background correction and data smoothing.

7.4.1   Some Common Spectroscopic Relationships

7.4.1.1   Electromagnetic Radiation.   Electromagnetic radiation travels in straight lines in a uniform medium, has a velocity of 299 792 500 m · s^{-1} in a vacuum, and possesses properties of both a wave motion and a particle (photon). *Wavelength* λ is the distance from crest to crest; *frequency* v is the number of waves passing a fixed point in a unit length of time. Wavelength and frequency are related by the relation

$$c = \lambda v$$

where c is the velocity of light (in a vacuum). In any material medium the speed of propagation is smaller than this and is given by the product nc, where n is the refractive index of the medium.

Radiation is absorbed or emitted only in discrete packets called photons and quanta:

$$E = hv$$

where E is the energy of the quantum and h is Planck's constant.

The relation between energy and mass is given by the *Einstein equation*:

$$\Delta E = \Delta mc^2$$

where ΔE is the energy release and Δm is the loss of mass. Strictly, the mass of a particle depends on its velocity, but here the masses are equated to their rest masses (at zero velocity).

The *Wien displacement law* states that the wavelength of maximum emission, λ_m, of a blackbody varies inversely with absolute temperature; the product $\lambda_m T$ remains constant. When λ_m is expressed in micrometers, the law becomes

$$\lambda_m T = 2898$$

In terms of σ_m, the wavenumber of maximum emission:

$$\sigma_m = 3.48\,T$$

Another useful version is $h\nu_m = 5kT$, where k is the Boltzmann constant.

Stefan's law states that the total energy J radiated by a blackbody per unit time and area (power per unit area) varies as the fourth power of the absolute temperature:

$$J = aT^{-4}$$

where a is a constant whose value is 5.67×10^{-8} W $\cdot$ m^{-2} $\cdot$ K^{-4}.

The relationship between the voltage of an X-ray tube (or other energy source), in volts, and the wavelength is given by the *Duane-Hunt equation*:

$$\lambda = \frac{hc}{eV} = \frac{12\,398}{V}$$

where the wavelength is expressed in angstrom units.

7.4.1.2 Laws of Photometry. The time rate at which energy is transported in a beam of radiant energy is denoted by the symbol P_0 for the incident beam, and by P for the quantity remaining unabsorbed after passage through a sample or container. The ratio of radiant power transmitted by the sample to the radiant power incident on the sample is the *transmittance T*:

$$T = \frac{P}{P_0}$$

The logarithm (base 10) of the reciprocal of the transmittance is the *absorbance A*:

$$A = -\log T = \log\left(\frac{1}{T}\right)$$

When a beam of monochromatic light, previously rendered plane parallel, enters an absorbing medium at right angles to the plane-parallel surfaces of the medium, the rate of decrease in radiant power with the length of light path (cuvette interior) b, or with the concentration of absorbing material C (in grams per liter) will follow the exponential progression, often referred to as *Beer's law*:

$$T = 10^{-abC} \quad \text{or} \quad A = abC$$

where a is the absorptivity of the component of interest in the solution. When C is expressed in moles per liter,

$$T = 10^{-\epsilon bC} \quad \text{or} \quad A = \epsilon bC$$

where ϵ is the molar absorptivity.

The total fluorescence (or phosphorescence) intensity is proportional to the quanta of light absorbed, $P_0 - P$, and to the efficiency ϕ, which is the ratio of quanta absorbed to quanta emitted:

$$F = (P_0 - P)\phi = P_0\phi(1 - e^{-\epsilon bC})$$

When the terms ϵbC is not greater than 0.05 (or 0.01 in phosphorescence),

$$F = k\phi P_0 \epsilon bC$$

where the term k has been introduced to handle instrumental artifacts and the geometry factor because fluorescence (and phosphorescence) is emitted in all directions but is viewed only through a limited aperture.

The thickness of a transparent film or the path length of infrared absorption cells b, in centimeters, is given by

$$b = \frac{1}{2n_D} \left(\frac{n}{\bar{v}_1 - \bar{v}_2} \right)$$

where n is the number of fringes (peaks or troughs) between two wavenumbers $\bar{v}_1$ and $\bar{v}_2$, and n_D is the refractive index of the sample material (unity for the air path of an empty cuvette). If measurements are made in wavelength, as micrometers, the expression is

$$b = \frac{1}{2n_D} \left(\frac{n\lambda_1\lambda_2}{\lambda_2 - \lambda_1} \right)$$

7.4.1.3 Grating Equation. The light incident on each groove is diffracted or spread out over a range of angles, and in certain directions reinforcement or constructive interference occurs, as stated in the grating formula:

$$m\lambda = b(\sin i \pm \sin r)$$

where b is the distance between adjacent grooves, i is the angle of incidence, r is the angle of reflection (both angles relative to the grating normal), and m is the order number. A positive sign applies where incoming and emergent beams are on the same side of the grating normal.

The *blaze wavelength* is that wavelength for which the angle of reflectance from the groove face and the angle of reflection (usually the angle of incidence) from the grating are identical.

The *Bragg equation*

$$m\lambda = 2d \sin \theta$$

states the condition for reinforcement of reflection from a crystal lattice, where d is the distance between each set of atomic planes and θ is the angle of reflection.

7.4.1.4 Ionization of Metals in a Plasma. A loss in spectrochemical sensitivity results when a free metal atom is split into a positive ion and an electron:

$$M = M^+ + e^-$$

The degree of ionization, α_i, is defined as

$$\alpha_i = \frac{[M^+]}{[M^+] + [M]}$$

At equilibrium, when the ionization and recombination rates are balanced, the ionization constant K_i (in atm) is given by

$$K_i = \frac{[M^+][e^-]}{[M]} = \left(\frac{\alpha_i^2}{1 - \alpha_i^2} \right) p_{\Sigma M}$$

where $p_{\Sigma M}$ (in atm) is the total atom concentration of metal in all forms in the plasma.

The ionization constant can be calculated from the *Saha equation*:

$$\log K_i = -5040 \frac{E_i}{T} + \frac{5}{2} \log T - 6.49 + \log \frac{g_{M^+} g_{e^-}}{g_M}$$

where E_i is the ionization potential of the metal in eV (Table 4.2), T is the absolute temperature of the plasma (in kelvins), and the g terms are the statistical weights of the ionized atom, the electron, and the neutral atom. For the alkali metals the final term is zero; for the alkaline earth metals, it is 0.6.

To suppress the ionization of a metal, another easily ionized metal (denoted a *deionizer* or *radiation buffer*) is added to the sample. To ensure that ionization is suppressed for the test element, the product $(K_i)_M p_M$ of the deionizer must exceed the similar product for the test element one hundred-fold (for 1 percent residual ionization of the test element).

7.5 INFRARED SPECTROSCOPY

TABLE 7.20 Absorption Frequencies of Single Bonds to Hydrogen

Abbreviations Used in the Table

m, moderately strong   var, of variable strength
m–s, moderate to strong   w, weak
s, strong   w–m, weak to moderately strong

Group	Band, cm^{-1}	Remarks
Saturated C—H		
H | —C—H | H	2975–2950 (s) 2885–2865 (w) 1450–1260 (m)	Two or three bands usually; asymmetrical and symmetrical CH stretching, respectively. In presence of double bond adjacent to CH$_3$ group symmetrical band splits into two. Sensitive to adjacent negative substituents
H | —C— acyclic | H	ca 2930 (s) 2870–2840 (w) 1480–1440 (m) ca 720 (w)	Frequency increased in strained systems. Symmetrical band splits into two bands when double bond adjacent. Scissoring mode Rocking mode
Alkane residues attached to carbon		
Cyclopropane	ca 3050 (w) 540–500 470–460 (s)	CH stretching Aliphatic cyclopropanes
Cyclobutanes Cyclopentanes	580–490 (s) 595–490 (s)	Alkyl derivatives: 550–530 cm^{-1} Alkyl derivatives: 585–530 cm^{-1}
$\geq$C(CH$_3$)$_2$	ca 1380 (m) 1175–1165 (m) 1150–1130 (m)	A roughly symmetrical doublet If no H on central carbon, then one band at ca 1190 cm^{-1}
—C(CH$_3$)$_3$	1395–1385 (m) 1365 (s)	Split into two bands
Aryl-CH$_3$ Aryl-C$_2$H$_5$ Aryl-C$_3$H$_7$ (or C$_4$H$_9$)	390–260 (m) 565–540 (m–s) 585–565 (m)	 Two bands

TABLE 7.20 Absorption Frequencies of Single Bonds to Hydrogen (*Continued*)

Group	Band, cm⁻¹	Remarks

Group	Band, cm^{-1}	Remarks
Alkane residues attached to carbon (*continued*)		
$-(CH_2)_n-$		
$n = 1$	785–770 (w–m)	Rocking vibrations
$n = 2$	745–735 (w–m)	
$n = 3$	735–725 (w–m)	
$n \geq 4$	725–720 (w–m)	
Alkane residues attached to miscellaneous atoms		
Epoxide C—H	ca 3050 (m–s)	
$>C\overset{NH}{\underset{}{\diagup\diagdown}}CH_2$	ca 3050 (m–s)	
$-CH_2-$halogen	ca 3050 (m–s)	Halogens except fluorine
	1435–1385 (m)	
	1300–1240 (s)	
$-CHO$	2900–2800 (w)	
	2775–2700 (w)	
	1420–1370 (m)	
$-CO-CH_3$	3100–2900 (w)	
	1450–1400 (s)	
	1360–1355 (s)	
$-O-CH_3$ ethers	2835–2810 (s)	
	1470–1430 (m–s)	Two bands
	ca 1030 (w–m)	
$-O-C(CH_3)_3$	1200–1155 (s)	
$-O-CH_2-O-$	2790–2770 (m)	
$-O-CH_2-$ esters	1475–1460 (m–s)	
	1470–1435 (m–s)	Acyclic esters. Frequency increased ca 30 cm^{-1} for cyclic and small ring systems.
$-O-CO-CH_3$	1450–1400 (s)	Acetate esters
	1385–1365 (s)	The high intensity of these bands often dominates this region of the spectrum.
	1360–1355 (s)	
$-CH_2-\overset{\vert}{C}=C<$	1445–1430 (m)	
$-CH_2-SO_2-$	ca 1250 (m)	

TABLE 7.20 Absorption Frequencies of Single Bonds to Hydrogen (*Continued*)

Group	Band, cm^{-1}	Remarks
Alkane residues attached to miscellaneous atoms (*continued*)		
P—CH$_3$	1320–1280 (s)	
Se—CH$_3$	ca 1280 (m)	
B—CH$_3$	1460–1405 (m)	
	1320–1280 (m)	
Si—CH$_3$	1265–1250 (m-s)	
Sn—CH$_3$	1200–1180 (m)	
Pb—CH$_3$	1170–1155 (m)	
As—CH$_3$	1265–1240 (m)	
Ge—CH$_3$	1240–1230 (m)	
Sb—CH$_3$	1215–1195 (m)	
Bi—CH$_3$	1165–1145 (m)	
—CH$_2$—(Cd, Hg, Zn, Sn)	1430–1415 (m)	
N—CH$_3$ and N—CH$_2$—	2820–2780 (s)	
	1440–1390 (m)	Ethylenediamine complexes
N—CH$_2$—CH$_2$—N	1480–1450 (s)	Ethylenediamine complexes
N—CH$_3$		
Amine · HCl	1475–1395 (m)	
Amino acid · HCl	1490–1480 (m)	
Amides	1420–1405 (s)	
N—CH$_2$— amides	ca 1440 (m)	
S—CH$_3$	2990–2955 (m-s)	
	2900–2865 (m-s)	
	1440–1415 (m)	
	1325–1290 (m)	
	1030–960 (m)	
	710–685 (w-m)	
S—CH$_2$—	2950–2930 (m)	
	2880–2845 (m)	
	1440–1415 (m)	
	1270–1220 (s)	
—C≡CH	ca 3300 (s)	Sharp
	700–600	Bending
$\diagdown$C=C$\diagup$^{H}	3040–3010 (m)	
$\diagdown$C=C$\diagup$$^{H}_{H}$	3095–3075 (m)	CH stretching sometimes
	2985–2970 (m)	obscured by much stronger
		bands of saturated CH groups

TABLE 7.20 Absorption Frequencies of Single Bonds to Hydrogen (*Continued*)

Group	Band, cm^{-1}	Remarks
Alkane residues attached to miscellaneous atoms (*continued*)		
R, H / C=C / H, H	995–980 (s) 940–900 (s) ca 635 (s) 485–445 (m–s)	
R, H / C=C / R, H	895–885 (s) 560–530 (s) 470–435 (m)	
R, H / C=C / H, R	980–955 (s) 455–370 (m–s)	
H, H / C=C / R, R	730–655 (m) 670–455 (s)	
R, H / C=C / R, R	850–790 (m) 570–515 (s) 525–470 (s)	
$-O-CH=CH_2$	965–960 (s) 945–940 (m) 820–810 (s)	
$-S-CH=CH_2$	ca 965 (s) ca 860 (s)	
$-CO-CH=CH_2$	995–980 (s) 965–955 (m)	
$-CO-OCH=CH_2$	950–935 (s) 870–850 (s)	
$-CO-C=CH_2$	ca 930 (s)	
$-CO-OC=CH_2$	880–865	
$-O-CH=CH-$ *trans*	940–920 (s)	
$-CO-CH=CH-$ *trans*	ca 990 (s)	
Hydroxyl group O—H compounds		
Primary aliphatic alcohols	3640–3630 (s)	Only in very dilute solutions in nonpolar solvents
	1350–1260 (s)	OH bending
	1085–1030 (s)	Also broad band at 700–600 cm^{-1}

TABLE 7.20   Absorption Frequencies of Single Bonds to Hydrogen (*Continued*)

Group	Band, cm^{-1}	Remarks
Hydroxyl group O—H compounds (*continued*)		
Secondary aliphatic alcohols	3625–3620 (s)	See comments under primary aliphatic alcohols
	1350–1260 (s)	
	1125–1085 (s)	Also for α-unsaturated and cyclic tertiary aliphatic alcohols
Tertiary aliphatic alcohols	3620–3610 (s)	See comments under primary aliphatic alcohols
	1410–1310 (s)	
	1205–1125 (s)	
Aryl—OH	ca 3610 (s)	See comments under primary aliphatic alcohols
	1410–1310 (s)	
	1260–1180 (s)	
	1085–1030 (s)	Also for unsaturated secondary aliphatic alcohols
Carboxylic acids	3300–2500 (w–m)	Broad
	995–915 (s)	Broad diffuse band
Enol form of β-diketones	2700–2500 (var)	Broad
Free oximes	3600–3570 (w–m)	Shoulder
Free hydroperoxides	3560–3530 (m)	
Peroxy acids	ca 3280 (m)	
Phosphorus acids	2700–2560 (m)	Broad
Water in solution	3710	When solution is damp
Intermolecular H bond Dimeric	3600–3500	Rather sharp. Absorptions arising from H bond with polar solvents also appear in this region.
Polymeric	3400–3200 (s)	Broad

Intramolecular H bond
   Polyvalent alcohols
   Chelation
	3600–3500 (s)
	3200–2500
		Sharper than dimeric band above
		Broad and occasionally weak; the lower the frequency, the stronger the intramolecular bond

TABLE 7.20 Absorption Frequencies of Single Bonds to Hydrogen (*Continued*)

Group	Band, cm⁻¹	Remarks
Hydroxyl group O—H compounds (*continued*)		
Water of crystallation (solid state spectra)	3600–3100 (w)	Usually a weak band at 1640–1615 cm⁻¹ also. Water in trace amounts in KBr disks shows a broad band at 3450 cm⁻¹.
Amine, imine, ammonium, and amide N—H		
Primary amines Aliphatic	3550–3300 (m) 1650–1560 (m) 1090–1020 (w–m) 850–810 (w–m) 495–445 (m–s) ca 290 (s)	Two bands in this range With α-carbon branching at 795 cm⁻¹ and strong Broad Broad
Aromatic	1350–1260 (s) 445–345	Also for secondary aryl amines
Amino acids	3100–3030 (m) 2800–2400 (m) 1625–1560 (m) 1550–1550 (m)	Values for solid states; broad bands also (but not always) near 2500 and 200 cm⁻¹ Number of sharp bands; dilute solution
Amino salts	3550–3100 (m) ca 3380 ca 3280	Values for solid state Dilute solutions
Secondary amines	3550–3400 (w) 1580–1490 (w) 1190–1170 (m) 1145–1130 (m) 455–405 (w–m)	Only one band, whereas primary amines show two bands Often too weak to be noticed
Salts	ca 2500 ca 2400 1620–1560 (m–s)	Sharp; broad values for solid state Sharp; broad values for solid state
Tertiary amines ‘R₁R₂R₃NH⁺	2700–2250	Group of relatively sharp bands; broad bands in solid state
Ammonium ion	3300–3030 (s) 1430–1390 (s)	Group of bands

TABLE 7.20   Absorption Frequencies of Single Bonds to Hydrogen (*Continued*)

Group	Band, cm^{-1}	Remarks
	Amine, imine, ammonium, and amide N—H (*continued*)	
Imines =N=H	3350–3310 (w) 3490 (s) 3490 (s)	Aliphatic Aryl Pyrroles, indoles; band sharp
Imine salts	2700–2330 (m–s) 2200–1800 (m)	Dilute solutions One or more bands; useful to distinguish from protonated tertiary amines
Primary amide —CONH$_2$	ca 3500 (m) ca 3400 (m)	Lowered ca 150 cm^{-1} in the solid state and on H bonding; often several bands 3200–3050 cm^{-1}
Secondary amide —CONH—	3460–3400 (m) 3100–3070 (w)	Two bands; lowered on H bonding and in solid state. Only one band with lactams Extra band with bonded and solid-state samples
	Miscellaneous R—H	
—S—H	2600–2550 (w)	Weaker than OH and less affected by H bonding
P—H	2440–2350 (m)	Sharp
P $\overset{\displaystyle O}{\underset{\displaystyle OH}{}}$	2700–2560 (m)	Associated OH
R—D	100/137 times the corresponding RH frequency	Useful when assigning RH bands; deuteration leads to a known shift to lower frequency

TABLE 7.21   Absorption Frequencies of Triple Bonds

Abbreviations Used in the Table

m, moderately strong   var, of variable strength

m–s, moderate to strong   w–m, weak to moderately strong

s, strong

Group	Band, cm^{-1}	Remarks
Alkynes		
Terminal	3300 (s)	CH stretching
	2140–2100 (w–m)*	C≡C stretching
	1375–1225 (w–m)	
	695–575 (m–s)	Two bands if molecule has axial symmetry
	ca 630 (s)	Alkyl monosubstituted
Nonterminal	2260–2150 (var)*	Symmetrical or nearly symmetrical substitution makes the C≡C stretching frequency inactive. When more than one C≡C linkage is present, and sometimes when there is only one, there are frequently more absorption bands in this region than there are triple bonds to account for them.
R_1—C≡C—R_2	540–465 (m)	The longer the chain, the lower the frequency
Aryl—C≡C—	ca 550 (m)	
	ca 350 (var)	
—C≡C—halogen (Cl, Br, I)	185–160 (var)	
Nitriles —C≡N	2260–2200 (var)	Stronger and toward the lower end of the range when conjugated; occasionally very weak or absent
Aliphatic	580–555 (m–s)	
	560–525 (m–s)	
	390–350 (s)	
Aromatic	580–540 (s)	
	430–380 (m)	
Isonitriles R—N̄≡C̄	2175–2150 (s)	Very sensitive to changes in substituents
or R—N=C:	2150–2115 (s)	
	1595	Not found for nitriles
Cyanamides		
>N—C≡N ⇌ >N̄—C=N̄	2225–2210 (s)	

* Conjugation with olefinic or acetylenic groups lowers the frequency and raises the intensity. Conjugation with carbonyl groups usually has little effect on the position of absorption.

TABLE 7.21   Absorption Frequencies of Triple Bonds (*Continued*)

Group	Band, cm^{-1}	Remarks
Thiocyanates R—S—C≡N	2175–2140 (s)	Aryl thiocyanates at the upper end of the range, alkyl at the lower end
	404–400 (s) ca 600 (m–s)	Aliphatic derivatives
Nitrile *N*-oxides —C≡N→O	2305–2285 (s) 1395–1365 (s)	Aryl derivatives
Diazonium salts R—$\overset{+}{N}$≡N	2300–2230 (m–s)	
Selenocyanates R—Se—C≡N	ca 2160 (m–s) 545–520 ca 390 ca 350	

TABLE 7.22   Absorption Frequencies of Cumulated Double Bonds

Abbreviations Used in the Table

m–s, moderate to strong	vs, very strong
s, strong	w, weak

Group	Band, cm^{-1}	Remarks
Carbon dioxide O=C=O	2349 (s)	Appears in many spectra as a result of inequalities in path length
Isocyanates —N=C=O	2275–2250 (vs)	Position unaffected by conjugation
Isoselenocyanates —N=C=Se	2200–2000 (s) 675–605	Broad; usually two bands
Azides —N$_3$ or —N=$\overset{+}{N}$=$\overset{-}{N}$	2140–2030 (s) 1340–1180 (w)	Not observed for ionic azides
—N=C=N—	2155–2130 (s)	Split into unsymmetrical doublet by conjugation with aryl groups: 2145–2125 (vs) and 2115–2105 (vs)

TABLE 7.22 Absorption Frequencies of Cumulated Double Bonds (*Continued*)

Group	Band, cm^{-1}	Remarks
Isothiocyanates —N=C=S	2140–1990 (vs) 649–600 (m–s) 565–510 (m–s) 470–440 (m–s)	Broad; usually a doublet
Ketenes >C=C=O	ca 2150 (s)	
Ketenimines C=C=N—	2050–2000 (s)	
Allenes >C=C=C<	2000–1915 (m–s)	Two bands when terminal allene or when bonded to electron-attracting groups
Thionylamines —N=S=O	1300–1230 (s) 1180–1110 (s)	
Diazoalkanes $R_2C=\overset{+}{N}=\overset{-}{N}$ $—CH=\overset{+}{N}=\overset{-}{N}$	2030–2000 (s) 2050–2035 (s)	
Diazoketones $—CO—CH=\overset{+}{N}=\overset{-}{N}$	2100–2080 2075–2050	Monosubstituted Disubstituted

TABLE 7.23 Absorption Frequencies of Carbonyl Bands

All bands quoted are strong.

Groups	Band, cm^{-1}	Remarks
Acid anhydrides **—CO—O—CO—** Saturated	1850–1800 1790–1740	Two bands usually separated by about 60 cm^{-1}. The higher-frequency band is more intense in acyclic anhydrides, and the lower-frequency band is more intense in cyclic anhydrides.
Aryl and α,β-unsaturated	1830–1780 1790–1710	

TABLE 7.23   Absorption Frequencies of Carbonyl Bands (*Continued*)

Groups	Band, cm^{-1}	Remarks
Acid anhydrides (*continued*)		
—CO—O—CO—		
Saturated five-ring	1870–1820	
	1800–1750	
All classes	1300–1050	One or two strong bands due to CO stretching
Acid chlorides **—COCl**		
Saturated	1815–1790	Acid fluorides higher, bromides and iodides lower
Aryl and α,β-unsaturated	1790–1750	
Acid peroxide		
CO—O—O—CO—		
Saturated	1820–1810	
	1800–1780	
Aryl and α,β-unsaturated	1805–1780	
	1785–1755	
Esters and lactones		
—CO—O—		
Saturated	1750–1735	
Aryl and α,β-unsaturated	1730–1715	
Aryl and vinyl esters		
C=C—O—CO—alkyl	1800–1750	The C=C stretching band also shifts to higher frequency.
Esters with electronegative α substituents; e.g.,		
$>$CCl—CO—O—	1770–1745	
α-Keto esters	1755–1740	
Six-ring and larger lactones	Similar values to the corresponding open-chain esters	
Five-ring lactone	1780–1760	
α,β-Unsaturated five-ring lactone	1770–1740	When α-CH is present, there are two bands, the relative intensity depending on the solvent.
β,γ-Unsaturated five-ring lactone, vinyl ester type	ca 1800	
Four-ring lactone	ca 1820	
β-Keto ester in H bonding enol form	ca 1650	Keto from normal; chelate-type H bond causes shift to lower frequency than the normal ester. The C=C band is strong and is usually near 1630 cm^{-1}.
All classes	1300–1050	Usually two strong bands due to CO stretching

TABLE 7.23 Absorption Frequencies of Carbonyl Bands (*Continued*)

Groups	Band, cm^{-1}	Remarks
Aldehydes —CHO (See also Table 7.49 for C—H.) All values given below are lowered in liquid-film or solid-state spectra by about 10–20 cm^{-1}. Vapor-phase spectra have values raised about 20 cm^{-1}.		
Saturated	1740–1720	
Aryl	1715–1695	*o*-Hydroxy or amino groups shift this value to 1655–1625 cm^{-1} because of intramolecular H bonding.
α,β-Unsaturated	1705–1680	
$\alpha,\beta,\gamma,\delta$-Unsaturated	1680–1660	
β-Ketoaldehyde in enol form	1670–1645	Lowering caused by chelate-type H bonding
Ketones ⟩C=O All values given below are lowered in liquid-film or solid-state spectra by about 10–20 cm^{-1}. Vapor-phase spectra have values raised about 20 cm^{-1}.		
Saturated	1725–1705	
Aryl	1700–1680	
α,β-Unsaturated	1685–1665	
$\alpha,\beta,\alpha',\beta'$-Unsaturated and diaryl	1670–1660	
Cyclopropyl	1705–1685	
Six-ring ketones and larger	Similar values to the corresponding open-chain ketones	
Five-ring ketones	1750–1740	α,β Unsaturation, $\alpha,\beta,\alpha',\beta'$ unsaturation, etc., have a similar effect on these values as on those of open-chain ketones.
Four-ring ketones	ca 1780	
α-Halo ketones	1745–1725	Affected by conformation; highest values are obtained when both halogens are in the same plane as the C=O.
α,α'-Dihalo ketones	1765–1745	
1,2-Diketones, *syn-trans-* open chains	1730–1710	Antisymmetrical stretching frequency of both C=O's. The symmetrical stretching is inactive in the infrared but active in the Raman.

TABLE 7.23 Absorption Frequencies of Carbonyl Bands (*Continued*)

Groups	Band, cm^{-1}	Remarks
Ketones $>$C$=$O (*continued*)		
syn-cis-1,2-Diketones, six-ring	1760 and 1730	
syn-cis-1,2-Diketones, five ring	1775 and 1760	
o-Amino-aryl or *o*-hydroxy-aryl ketones	1655–1635	Low because of intramolecular H bonding. Other substituents and steric hindrance affect the position of the band.
Quinones	1690–1660	C$=$C band is strong and is usually near 1600 cm^{-1}.
Extended quinones	1655–1635	
Tropone	1650	Near 1600 cm^{-1} when lowered by H bonding as in tropolones
Carboxylic acids —CO$_2$H		
All types	3000–2500	OH stretching; a characteristic group of small bands due to combination bands
Saturated	1725–1700	The monomer is near 1760 cm^{-1}, but is rarely observed. Occasionally both bands, the free monomer, and the H-bonded dimer can be seen in solution spectra. Ether solvents give one band near 1730 cm^{-1}.
α,β-Unsaturated	1715–1690	
Aryl	1700–1680	
α-Halo-	1740–1720	
Carboxylate ions —CO$_2^-$		
Most types	1610–1550	Antisymmetrical and symmetrical stretching, respectively
	1420–1300	
Amides —CO—N$<$		
(See also Table 7.49 for NH stretching and bending.)		
Primary —CONH$_2$		
In solution	ca 1690	Amide I; C$=$O stretching
Solid state	ca 1650	
In solution	ca 1600	Amide II: mostly NH bending
Solid state	ca 1640	
		Amide I is generally more intense than amide II. (In the solid state, amides I and II may overlap.)
Secondary —CONH—		
In solution	1700–1670	Amide I
Solid state	1680–1630	
In solution	1550–1510	Amide II; found in open-chain amides only

TABLE 7.23   Absorption Frequencies of Carbonyl Bands (*Continued*)

Groups	Band, cm^{-1}	Remarks
Amides —CO—N< (*continued*)		
Solid state	1570–1515	Amide I is generally more intense than amide II.
Tertiary	1670–1630	Since H bonding is absent, solid and solution spectra are much the same.
Lactams		
Six-ring and larger rings	ca 1670	
Five-ring	ca 1700	Shifted to higher frequency when the N atom is in a bridged system
Four-ring	ca 1745	
R—CO—N—C=C		Shifted +15 cm^{-1} by the additional double bond
C=C—CO—N		Shifted by up to +15 cm^{-1} by the additional double bond. This is an unusual effect by α,β unsaturation. It is said to be due to the inductive effect of the C=C on the well-conjugated CO—N system, the usual conjugation effect being less important in such a system.
Imides —CO—N—CO—		
Cyclic six-ring	ca 1710 and ca 1700	Shift of +15 cm^{-1} with α,β unsaturation
Cyclic five-ring	ca 1770 and ca 1700	
Ureas N—CO—N		
RNHCONHR	ca 1660	
Six-ring	ca 1640	
Five-ring	ca 1720	
Urethanes R—O—CO—N	1740–1690	Also shows amide II band when nonsubstituted on N
Thioesters and Acids		
RCO—S—R′		
RCOSH	ca 1720	α,β-Unsaturated or aryl acid or ester shifted about −25 cm^{-1}
RCOS—alkyl	ca 1690	
RCOS—aryl	ca 1710	

7.5.1   Intensities of Carbonyl Bands

Acids generally absorb more strongly than esters, and esters more strongly than ketones or aldehydes. Amide absorption is usually similar in intensity to that of ketones but is subject to much greater variations.

7.5.2   Position of Carbonyl Absorption

The general trends of structural variation on the position of C=O stretching frequencies may be summarized as follows:

1. The more electronegative the group X in the system R—CO—X—, the higher is the frequency.
2. α, β Unsaturation causes a lowering of frequency of 15 to 40 cm^{-1}, except in amides, where little shift is observed and that usually to higher frequency.
3. Further conjugation has relatively little effect.
4. Ring strain in cyclic compounds causes a relatively large shift to higher frequency. This phenomenon provides a remarkably reliable test of ring size, distinguishing clearly between four-, five-, and larger-membered-ring ketones, lactones, and lactams. Six-ring and larger ketones, lactones, and lactams show the normal frequency found for the open-chain compounds.
5. Hydrogen bonding to a carbonyl group causes a shift to lower frequency of 40 to 60 cm^{-1}. Acids, amides, enolized β-keto carbonyl systems, and o-hydroxyphenol and o-aminophenyl carbonyl compounds show this effect. All carbonyl compounds tend to give slightly lower values for the carbonyl stretching frequency in the solid state compared with the value for dilute solutions.
6. Where more than one of the structural influences on a particular carbonyl group is operating, the net effect is usually close to additive.

TABLE 7.24   Absorption Frequencies of Other Double Bonds

Abbreviations Used in the Table

m, moderately strong	vs, very strong
m–s, moderate to strong	w, weak
var, of variable strength	

Group	Band, cm^{-1}	Remarks
	Alkenes $>$C=C$<$	
Nonconjugated	1680–1620 (w–m)	May be very weak if symmetrically substituted
Conjugated with aromatic ring	1640–1610 (m)	More intense than with unconjugated double bonds
Internal (ring)	3060–2995 (m)	Highest frequencies for smallest ring
Carbons: $n = 3$	ca 1665 (w–m)	
$n = 4$	ca 1565 (w–m)	
$n = 5$	ca 1610 (w–m)	
	1370–1340 (s)	Characteristic
$n \geq 6$	1650–1645 (w–m)	

TABLE 7.24 Absorption Frequencies of Other Double Bonds (*Continued*)

Group	Band, cm^{-1}	Remarks
Alkenes $>$C$=$C$<$ (*continued*)		
Exocyclic C$=$C(CH$_2$)$_n$ $n = 2$ $n = 3$ $n \geq 4$	1780–1730 (m) ca 1680 (m) 1655–1650 (m)	
Fulvene	1645–1630 (m) 1370–1340 (s) 790–765 (s)	
Dienes, trienes, etc.	1650 (s) and 1600 (s)	Lower-frequency band usually more intense and may hide or overlap the higher-frequency band
α, β-Unsaturated carbonyl compounds	1640–1590 (m)	Usually much weaker than the C$=$O band
Enol esters, enol ethers, and enamines	1700–1650 (s)	
Imines, oximes, and amidines $>$C$=$N$-$		

Imines and oximes
    Aliphatic
    α,β-Unsaturated and aromatic
    Conjugated cyclic systems
	1690–1640 (w)
	1650–1620 (m)
	1660–1480 (var)
	960–930 (s)
		NO stretching of oximes

Imino ethers $-$O$-$C$=$N$-$	1690–1640 (var)	Usually a strong doublet
Imino thioethers $-$S$-$C$=$N$=$	1640–1605 (var)	
Imine oxides $>$C$=\overset{+}{\text{N}}-\overset{-}{\text{O}}$	1620–1550 (s)	
Amidines $>$N$-$C$=$N$-$	1685–1580 (var)	
Benzamidines Aryl$-$C$=$N$-$N	1630–1590	
Guanidine $>$N$-$C$=$N$-$ $\underset{\text{N}}{\vert}$	1725–1625 (s)	
Azines $>$C$=$N$-$N$=$C$<$	1670–1600	
Hydrazoketones $-$CO$-$C$=$N$-$N	1600–1530 (vs)	

TABLE 7.24   Absorption Frequencies of Other Double Bonds (*Continued*)

Group	Band, cm^{-1}	Remarks
Azo compounds —N=N—		

Azo   —N=N—
      Aliphatic
      Aromatic
        cis
        trans
	
	 ca 1575 (var)
	
	 ca 1510 (w)
	 1440–1410 (w)
		Very weak or inactive

Azoxy $-\overset{+}{N}=N-$
          $\underset{O^-}{\vert}$
      Aliphatic
 	
      Aromatic
	
	 1590–1495 (m–s)
	 1345–1285 (m–s)
	 1480–1450 (m–s)
	 1340–1315 (m–s)	

Azothio $-N=\overset{+}{N}-\overset{-}{S}-$	1465–1445 (w) 1070–1055 (w)	
Nitro compounds N=O		
Nitro C—NO$_2$ Aliphatic	 ca 1560 (s) 1385–1350 (s)	The two bands are due to asymmetrical and symmetrical stretching of the N=O bond. Electron-withdrawing substituents adjacent to nitro group increase the frequency of the asymmetrical band and decrease that of the symmetrical frequency.
Aromatic	1570–1485 (s) 1380–1320 (s)	See above remark; also bulky orthosubstituents shift band to higher frequencies. Strong H bonding shifts frequency to lower end of range.
	865–835 (s)	Strong and sometimes at ca 750 cm^{-1}
α,β-Unsaturated Nitroalkenes	580–520 (var) 1530–1510 (s) 1360–1335 (s)	
Nitrates —O—NO$_2$	1650–1625 (vs) 1285–1275 (vs) 870–855 (vs) 760–755 (w–m) 710–695 (w–m)	
Nitramines >N—NO$_2$	1630–1550 (s) 1300–1250 (s)	

TABLE 7.24 Absorption Frequencies of Other Double Bonds (*Continued*)

Group	Band, cm^{-1}	Remarks
Nitro compounds N=O (*continued*)		
Nitrates —O—N=O	1680–1610 (vs) 815–750 (s) 850–810 (s) 690–615 (s)	Two bands *Trans* form *Cis* form
Thionitrites —S—N=O	730–685 (m–s)	
Nitroso ≥C—N=O	1600–1500 (s)	
N—N=O (aliphatic/aromatic) Aliphatic Aromatic	1530–1495 (m–s) 1480–1450 (m–s) 1335–1315 (m–s)	
Nitrogen oxides N → O Pyridine Pyrazine	1320–1230 (m–s) 1190–1150 (m–s) 1380–1280 (m–s) 1040–990 (m–s) ca 850 (m)	Affected by ring substituents

TABLE 7.25 Absorption Frequencies of Aromatic Bands

Abbreviations Used in the Table

m, moderately strong   var, of variable strength
m–s, moderate to strong   w–m, weak to moderately strong
s, strong

Group	Band, cm^{-1}	Remarks
Aromatic rings	ca 1600 (m) ca 1580 (m) ca 1470 (m) ca 1510 (m)	Stronger when ring is further conjugated When substituent on ring is electron acceptor When substituent on ring is electron donor
Five adjacent H	900–860 (w–m) 770–730 (s) 720–680 (s) 625–605 (w–m) ca 550 (w–m)	Substituents: C=C, C≡C, C≡N

TABLE 7.25 Absorption Frequencies of Aromatic Bands (*Continued*)

Group	Band, cm^{-1}	Remarks
1,2-Substitution	770–735 (s) 555–495 (w–m) 470–415 (m–s)	
1,3-Substitution	810–750 (s) 560–505 (m) 460–415 (m–s)	490–460 cm^{-1} when substituents are electron-accepting groups
1,4-Substitution	860–800 (s) 650–615 (w–m) 520–440 (m–s)	520–490 cm^{-1} when substituents are electron-donating groups
1,2,3-Trisubstitution	800–760 (s) 720–685 (s) 570–535 (s) ca 485	
1,2,4-Trisubstitution	900–885 (m) 780–760 (s) 475–425 (m–a)	
1,3,5-Trisubstitution	950–925 (var) 865–810 (s) 730–680 (m–s) 535–495 (s) 470–450 (w–m)	
Pentasubstitution	900–860 (m–s) 580–535 (s)	
Hexasubstitution	415–385 (m–s)	

TABLE 7.26 Absorption Frequencies of Miscellaneous Bands

Abbreviations Used in the Table

m, moderately strong vs, very strong
m–s, moderate to strong w, weak
s, strong w–m, weak to moderately strong
var, of variable strength

Group	Band, cm^{-1}	Remarks
Ethers		
Saturated aliphatic $\equiv$C—O—C$\equiv$	1150–1060 (vs)	Two peaks may be observed for branched chain, usually 1140–1110 cm^{-1}.
	1140–900 (s)	Usually 930–900 cm^{-1}; may be absent for symmetric ethers
Alkyl-aryl $=$C—O—C$\equiv$	1270–1230 (vs)	$=$CO stretching
	1120–1020 (s)	CO stretching
Vinyl	1225–1200 (s)	Usually about 1205 cm^{-1}
Diaryl $=$C—O—C$=$	1200–1120 (s)	
	1100–1050 (s)	
Cyclic	1270–1030 (s)	
Epoxides $>$C —— C$<$ over O	1260–1240 (m–s)	
	880–805 (m)	Monosubstituted
	950–860 (var)	*Trans* form
	865–785 (m)	*Cis* form
	770–750 (m)	Trisubstituted
Ketals and acetals	1190–1140 (s)	
	1195–1125 (s)	
	1100–1000 (s)	Strongest band
	1060–1035 (s)	Sometimes obscured
Phthalanes	915–895 (s)	
Aromatic methylenedioxy	1265–1235 (s)	
Peroxides		
—O—O—	900–830 (w)	
	1150–1030 (m–s)	Alkyl
	ca 1000 (m)	Aryl

TABLE 7.26 Absorption Frequencies of Miscellaneous Bands (*Continued*)

Group	Band, cm^{-1}	Remarks
	Sulfur compounds	
Thiols		
—S—H	2600–2450 (w)	
—CO—SH	840–830 (m)	
—CS—SH	ca 860 (s)	Broad
Thiocarbonyl		
>C=S	1200–1050 (s)	Behaves generally in manner similar to carbonyl band
>N—C=S	1570–1395	
	1420–1260	
	1140–940	
—S—C=S	ca 580 (s)	
Sulfoxides		
>S=O	1075–1040 (vs)	Halogen or oxygen atom bonded to sulfur increases the frequency.
	730–690 (var)	
	395–360 (var)	
Sulfones		
>SO$_2$	1360–1290 (vs)	Halogen or oxygen atom bonded to sulfur increases the frequency.
	1170–1120 (vs)	
	610–545 (m–s)	
	525–495 (m–s)	
Sulfonamides		
—SO$_2$—N<	1380–1330 (vs)	
	1170–1140 (vs)	
	950–860 (m)	
	715–700 (w–m)	
Sulfonates		
—SO$_2$—O—	1420–1330 (s)	May appear as doublet
	1200–1145 (s)	
Thiosulfonates		
—SO$_2$—S—	ca 1340 (vs)	
Sulfates —O—SO$_2$—O—	1415–1380 (s)	Electronegative substituents increase frequencies.
	1200–1185 (s)	
Primary alkyl salts	1315–1220 (s)	Strongly influenced by metal ion
	1140–1075 (m)	

TABLE 7.26 Absorption Frequencies of Miscellaneous Bands (*Continued*)

Group	Band, cm^{-1}	Remarks
	Sulfur compounds (*continued*)	
Sulfates —O—SO$_2$—O— (*continued*) Secondary alkyl salts	1270–1210 (vs) 1075–1050 (s)	Doublet; both bands strongly influenced by metal ion
Stretching frequencies of C—S and S—S bonds		
—S—CH$_3$	710–685 (w–m)	
—S—CH$_2$—	660–630 (w–m)	
—S—CH<	630–600 (w–m)	
—S—C≤	600–570 (w–m)	
—S—aryl	1110–1070 (m) 710–685 (w–m)	
R—S—S—R	705–570 (w) 520–500 (w)	
Aryl—S—S—aryl	500–430 (w–m)	
Polysulfides	500–470 (w–m)	
CH$_2$—S—CH$_2$—	695–655 (w–m)	CSC stretching
(R—S)$_2$C=O	880–825 (s) 570–560 (var)	
—CO—S—	1035–935 (s)	
—CS—S	ca 580 (s)	
=C with S— / S—	1050–900 (m–s) 980–850 (m–s) 900–800 (m–s)	Monoionic Ionic 1,1-dithiolates
	Phosphorus compounds	
P—H	2455–2265 (m) 1150–965 (w–m)	Sharp. Phosphines lie in the region 2285–2265 cm^{-1}.
—PH$_2$	1100–1085 (m) 1065–1040 (w–m) 940–910 (m)	
P—alkyl	795–650 (m–s)	
P—aryl	1130–1090 (s) 750–680 (s)	
P—O—alkyl	1050–970 (s)	Broad
P—O—aryl	1240–1190 (s)	
P—O—P	970–910	Broad

TABLE 7.26 Absorption Frequencies of Miscellaneous Bands (*Continued*)

Group	Band, cm^{-1}	Remarks
Phosphorus compounds (continued)		
P=O	1350–1150 (s)	May appear as doublet
P(=O)OH	2725–2520 (w–m)	H-bonded; broad
	2350–2080 (w–m)	Broad; may be doublet for aryl acids
	1740–1600 (w–m)	
	1335 (s)	P=O stretching
	1090–910 (s)	
	540–450 (w–m)	
P=S	865–655 (m–s)	
	595–530 (var)	
P(=S)OH	3100–3000 (w)	
	2360–2200 (w)	
	935–910 (s)	PO stretching
	810–750 (m–s)	P=S stretching
	655585 (var)	P=S stretching
Silicon compounds		
Si—H	2250–2100 (s)	
	985–800	SiH$_3$ has two bands.
Si—C≣	860–760	Accompanied by CH$_2$ rocking
Si—CH$_3$	1280–1250 (s)	Sharp
Si—C$_2$H$_5$	1250–1220 (m)	
	1020–1000 (m)	
	970–945 (m)	
Si—Aryl	1125–1090 (vs)	Splits into two bands when two aryl groups are attached to one silicon atom, but has only one band when three aryl groups attached
≣Si—OH	870–820	OH deformation band
≣Si—O—Si≣	1100–1000	
≣Si—N—Si≣	940–870 (s)	
≣Si—Cl	550–470 (s)	
	250–150	

TABLE 7.26 Absorption Frequencies of Miscellaneous Bands (*Continued*)

Group	Band, cm^{-1}	Remarks
Silicon compounds (continued)		
$>$SiCl$_2$	595–535 (s) 540–460 (m)	
—SiCl$_3$	625–570 (s) 535–450 (m)	
Boron compounds		
Boranes $>$BH or —BH$_2$	2640–2450 (m–s) 2640–2570 (m–s) 2535–2485 (m–s) 2380–2315 (s) 2285–2265 (s) 2140–2080 (w–m) 2580–2450 (m)	Free H in BH Free H in BH$_2$ plus second band In complexes; second band for BH$_2$ Bridged H Borazoles and borazines
BH$_4$$^-$	2310–2195 (s)	Two bands
B—N	1550–1330 750–635	Borazines and borazoles
B—O	1390–1310 (s) 1280–1200	BO stretching Metal orthoborates
B—Cl B—Br	1090–890 (s)	Plus other bands at lower frequencies for BX$_2$ and BX$_3$
B—F	1500–840 (var)	Isotope splitting present
XBF$_2$	1500–1410 (s) 1300–1200 (s)	
X$_2$BF	1360–1300 (s)	
BF$_3$ complexes	1260–1125 (s) 1030–800 (s)	Band splitting may be added to isotopic splittings.
BF$_4$$^-$	ca 1030 (vs)	

TABLE 7.26 Absorption Frequencies of Miscellaneous Bands (*Continued*)

Group	Band, cm^{-1}	Remarks
	Halogen compounds	
C—F		
Aliphatic, mono-F	1110–1000 (vs)	
	780–680 (s)	
Aliphatic, di-F	1250–1050 (vs)	Two bands
Aliphatic, poly-F	1360–1090 (vs)	Number of bands
Aromatic	1270–1100 (m)	
	680–520 (m–s)	
	420–375 (var)	
	340–240 (s)	
—CF$_3$		
Aliphatic	1350–1120 (vs)	
	780–680 (s)	
	680–590 (s)	
	600–540 (s)	
	555–505 (s)	
Aromatic	1330–1310 (m–s)	
	600–580 (s)	
C—Cl		
Primary alkanes	730–720 (s)	
	685–680 (s)	
	660–650 (s)	
Secondary alkanes	ca 760 (m)	
	675–655 (m–s)	
	615–605 (s)	
Tertiary alkanes	635–610 (m–s)	
	580–560 (m–s)	
Poly-Cl	800–700 (vs)	
Aryl:		
1,2-	1060–1035 (m)	
1,3-	1080–1075 (m)	
1,4-	1100–1090 (m)	
Chloroformates	ca 690 (s)	
	485–470 (s)	
Axial Cl	730–580 (s)	
Equatorial Cl	780–740 (s)	
C—Br		
Primary alkanes	645–635 (s)	
	565–555 (s)	
	440–430 (var)	
Secondary alkanes	620–605 (s)	
	590–575 (m–w)	
	540–530 (s)	

TABLE 7.26 Absorption Frequencies of Miscellaneous Bands (*Continued*)

Group	Band, cm^{-1}	Remarks
Halogen compounds (*continued*)		
C—Br (*continued*)		
Tertiary alkanes	600–595 (m–s)	
	525–505 (s)	
Axial	690–550 (s)	
Equatorial	750–685 (s)	
Aryl:		
1,2-	1045–1025 (m)	
1,3-; 1,4-	1075–1065 (m)	
Other bands	400–260 (s)	
	325–175 (m–s)	
	290–225 (m–s)	
C—I		
Primary alkanes	600–585 (s)	
	515–500 (s)	
Secondary alkanes	ca 575 (s)	
	550–520 (s)	
	490–480 (s)	
Tertiary alkanes	580–560 (s)	
	510–485 (m)	
	485–465 (s)	
Aromatic	1060–1055 (m–s)	
	310–160 (s)	
	265–185	
Axial	ca 640 (s)	
Equatorial	ca 655 (s)	
Inorganic ions		
Ammonium	3300–3030	Several bands, all strong
Cyanate	2220–2130 (s)	
Cyanide	2200–2000	
Carbonate	1450–1410	
Hydrogen sulfate	1190–1160 (s)	
	1180–1000 (s)	
	880–840 (m)	
Nitrate	1410–1350 (vs)	
	860–800 (m)	
Nitrite	1275–1230 (s)	
	835–800 (m)	Shoulder

TABLE 7.26 Absorption Frequencies of Miscellaneous Bands (*Continued*)

Group	Band, cm^{-1}	Remarks
Inorganic ions (*continued*)		
Phosphate	1100–1000	
Sulfate	1130–1080 (s)	
Thiocyanate	ca 2050 (s)	

TABLE 7.27 Absorption Frequencies in the Near Infrared

Values in parentheses are molar absorptivity.

Class	Band, cm^{-1}	Remarks
Acetylenes	9800–9430 6580–6400 (1.0)	Overtone of ≡CH stretching
Alcohols (nonhydrogen-bonded)	7140–7010 (2.0)	Overtone of OH stretching
Aldehydes Aliphatic	4640–4520 (0.5)	Combination of C=O and CH stretchings
Aromatic	ca 8000 ca 4525 ca 4445	
Formate	4775–4630 (1.0)	
Alkanes —CH$_3$	9000–8350 (0.02) 5850–5660 (0.1) 4510–4280 (0.3)	
—CH$_2$—	9170–8475 (0.02) 5830–6640 (0.1) 4420–4070 (0.25)	
≡CH	8550–8130 7000–6800 5650–5560	All bands very weak
Cyclopropane	6160–6060 4500–4400	

Alkenes
 	
  $\diagdown$C=C$\diagup$
  $\diagup$ $\diagdown$H
	6850–6370 (1.0)	

>C=CH$_2$ and —CH=CH$_2$	7580–7300 (0.02) 6140–5980 (0.2) 4760–4700 (1.2)	

TABLE 7.27 Absorption Frequencies in the Near Infrared (*Continued*)

Class	Band, cm^{-1}	Remarks
Alkenes (*continued*) H, H / C=C / (trans configuration)	4760–4660 (0.15)	*Trans* isomers have no unique bands.
—O—CH=CH$_2$ —CO—CH=CH$_2$	6250–6040 (0.3) 7580–7410 (0.02) 6190–5990 (0.3) 4820–4750 (0.2–0.5)	
Amides Primary	7400–6540 (0.7)	Two bands; overtone of NH stretch
	5160–5060 (3.0) 5040–4990 (0.5) 4960–4880 (0.5)	Second overtone of C=O stretch; second overtone of NH deformation; combination of C=O and NH
Secondary	7330–7140 (0.5) 5050–4960 (0.4)	Overtone of NH stretch Combination of NH stretch and NH bending
Amines, aliphatic Primary	9710–9350 6670–6450 (0.5) 5075–4900 (0.7)	Second overtone of NH stretch Two bands; overtone of NH stretch Two bands; combination of NH stretch and NH bending
Secondary	9800–9350 6580–6410 (0.5)	Second overtone of NH stretch Overtone of NH stretch
Amines, aromatic Primary	9950–9520 (0.4) 7040–6850 (0.2) 6760–6580 (1.4) 5140–5040 (1.5)	
Secondary	10 000–9710 6800–6580 (0.5)	
Aryl-H	7660–7330 (0.1) 6170–5880 (0.1)	Overtone of CH stretch
Carbonyl	5200–5100	
Carboxylic acids	7000–6800	
Epoxide (terminal)	6135–5960 (0.2) 4665–4520 (1.2)	Cyclopropane bands in same region

TABLE 7.27 Absorption Frequencies in the Near Infrared (*Continued*)

Class	Band, cm^{-1}	Remarks
Glycols	7140–7040	

Hydroperoxides
  Aliphatic
 	
  Aromatic
	6940–6750 (2.0)
	4960–4880 (0.8)
	7040–6760 (1.0)
	4950–4850 (1.3)
		Two bands

Imides	9900–9620 6540–6370	
Nitriles	5350–5200 (0.1)	
Oximes	7140–7050	
Phosphines	5350–5260 (0.2)	

Phenols
  Nonbonded
 	
  Intramolecularly bonded
	7140–6800 (3.0)
	5000–4950
	7000–6700	

Thiols	5100–4950 (0.05)	

TABLE 7.28 Infrared Transmitting Materials

Material	Wavelength range, μm	Wavenumber range, cm^{-1}	Refractive index at 2 μm
NaCl, rock salt	0.25–17	40 000–590	1.52
KBr, potassium bromide	0.25–25	40 000–400	1.53
KCl, potassium chloride	0.30–20	33 000–500	1.5
AgCl, silver chloride*	0.40–23	25 000–435	2.0
AgBr, silver bromide*	0.50–35	20 000–286	2.2
CaF$_2$, calcium fluoride (Irtran-3)	0.15–9	66 700–1 110	1.40
BaF$_2$, barium fluoride	0.20–11.5	50 000–870	1.46
MgO, magnesium oxide (Irtran-5)	0.39–9.4	25 600–1 060	1.71
CsBr, cesium bromide	1–37	10 000–270	1.67
CsI, cesium iodide	1–50	10 000–200	1.74
TlBr–TlI, thallium bromide-iodide (KRS-5)*	0.50–35	20 000–286	2.37
ZnS, zinc sulfide (Irtran-2)	0.57–14.7	17 500–680	2.26

* Useful for internal reflection work.

TABLE 7.28 Infrared Transmitting Materials (*Continued*)

Material	Wavelength range, μm	Wavenumber range, cm^{-1}	Refractive index at 2 μm
ZnSe, zinc selenide* (vacuum deposited) (Irtran-4)	1–18	10 000–556	2.45
CdTe, cadmium telluride (Irtran-6)	2–28	5 000–360	2.67
Al_2O_3, sapphire*	0.20–6.5	50 000–1 538	1.76
SiO_2, fused quartz	0.16–3.7	62 500–2 700	
Ge, germanium*	0.50–16.7	20 000–600	4.0
Si, silicon*	0.20–6.2	50 000–1 613	3.5
Polyethylene	16–300	625–33	1.54

* Useful for internal reflection work.

TABLE 7.29 Infrared Transmission Characteristics of Selected Solvents

Transmission below 80%, obtained with a 0.10-mm cell path, is shown as shaded area.

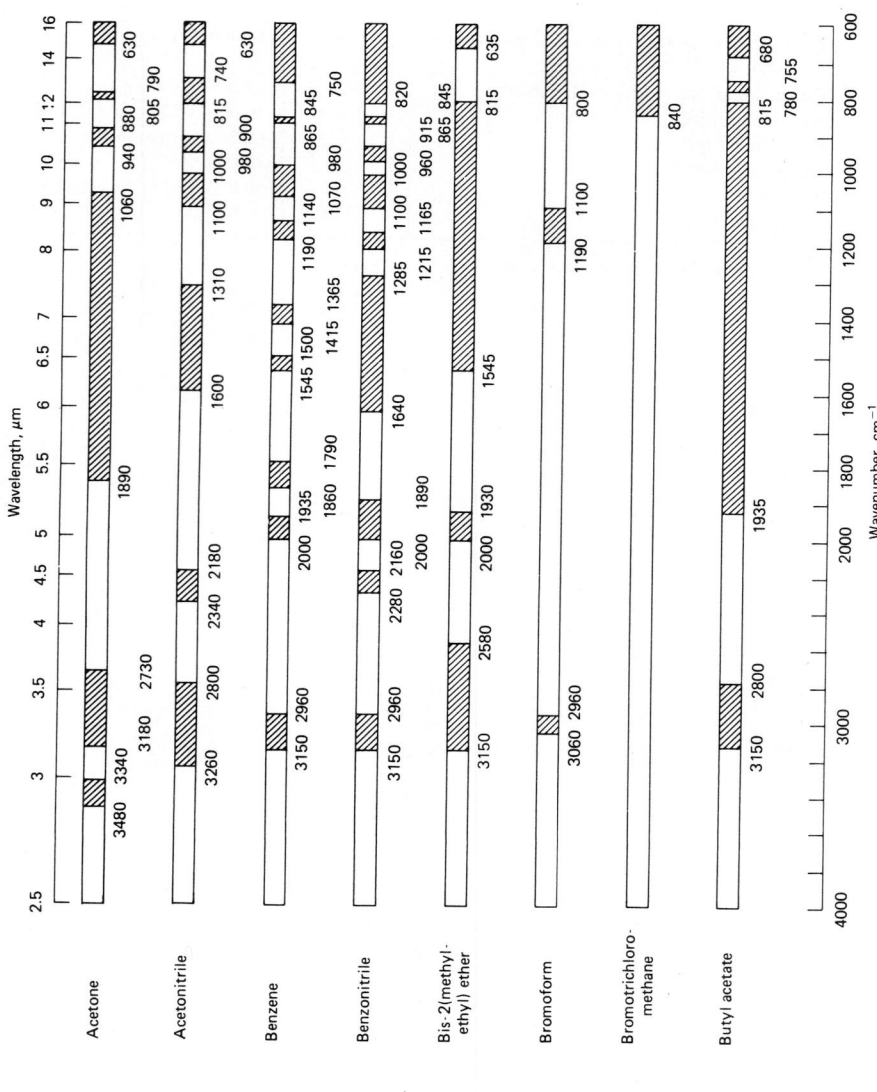

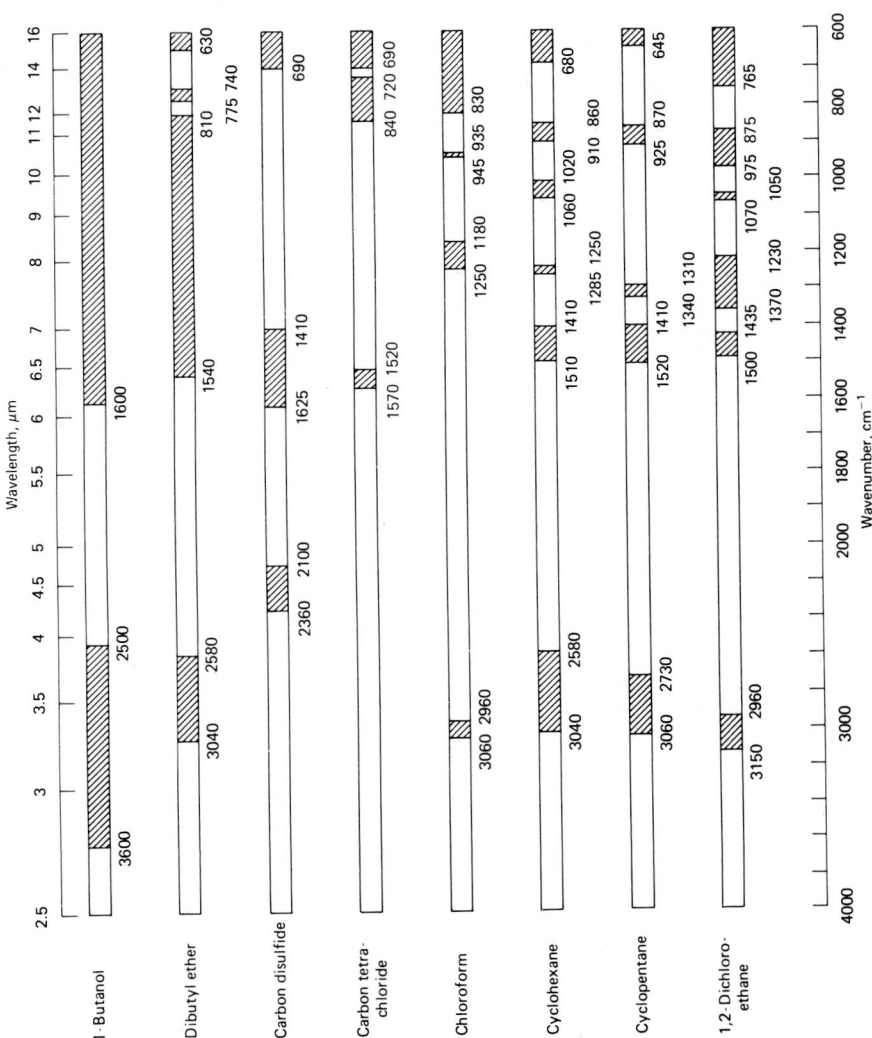

Wavelength, μm

| 2.5 | 3 | 3.5 | 4 | 4.5 | 5 | 5.5 | 6 | 6.5 | 7 | 8 | 9 | 10 | 11 12 | 14 | 16 |

1-Butanol
3600   2500   1600

Dibutyl ether
3040   2580   1540   810   775 740   630

Carbon disulfide
2360   2100   1625   1410   690

Carbon tetra-chloride
1570 1520   840 720 690

Chloroform
3060 2960   1250 1180   945 935 830

Cyclohexane
3040   2580   1510   1410   1285 1250   1060 1020   910 860   680

Cyclopentane
3060   2730   1520   1410   1340 1310   925 870   645

1,2-Dichloro-ethane
3150 2960   1500 1435   1370 1230   1070 975 875   765
1050

Wavenumber, cm⁻¹

4000   3000   2000   1800   1600   1400   1200   1000   800   600

7.73

TABLE 7.29 Infrared Transmission Characteristics of Selected Solvents (*Continued*)

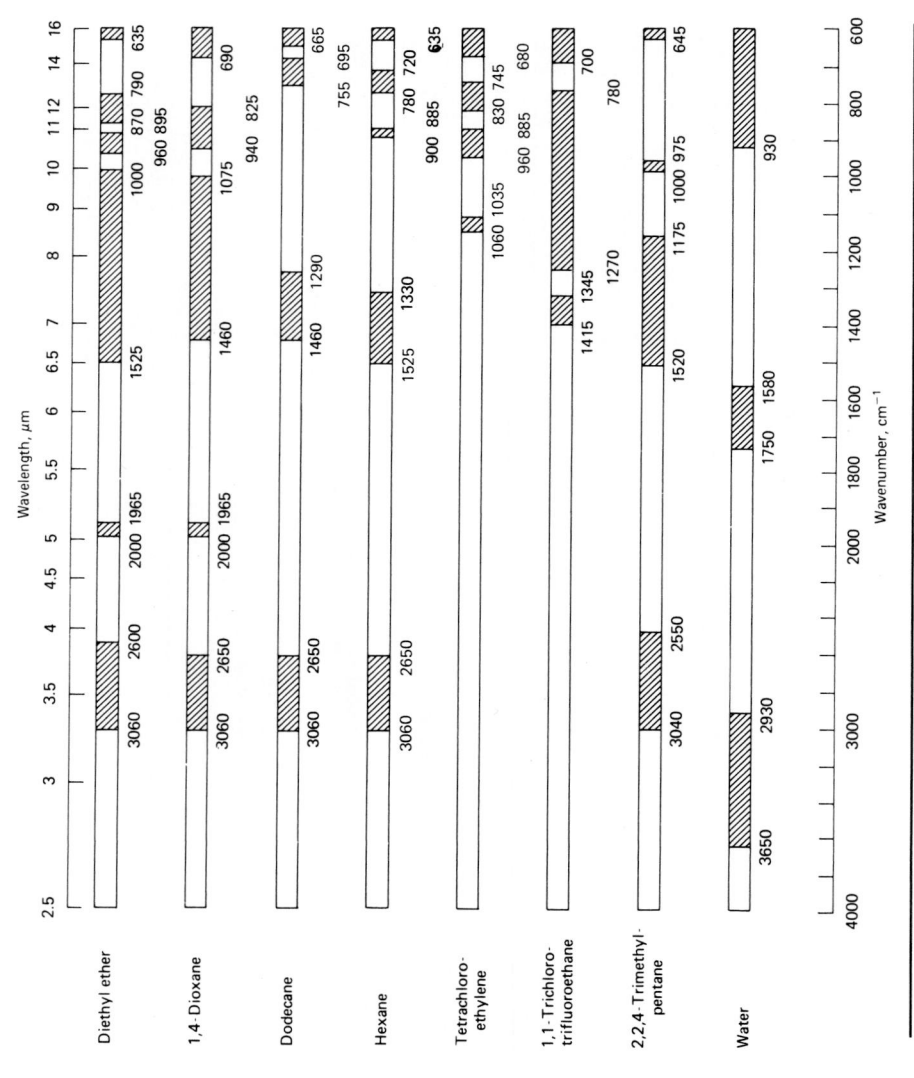

7.6 *RAMAN SPECTROSCOPY*

TABLE 7.30   Raman Frequencies of Single Bonds to Hydrogen and Carbon

Abbreviations Used in the Table

m, moderately strong	vw, very weak
m–s, moderate to strong	w, weak
m–vs, moderate to very strong	w–m, weak to moderately strong
s, strong	w–vs, weak to very strong
vs, very strong	

Group	Band, cm^{-1}	Remarks
	Saturated C—H and C—C	
—CH$_3$	2969–2967 (s)	
	2884–2883 (s)	
	ca 1205 (s)	In aryl compounds
	1150–1135	In unbranched alkyls
	1060–1056	In unbranched alkyls
	975–835 (s)	Terminal rocking of methyl group
	280–220	CH$_2$—CH$_3$ torsion
—CH$_2$—	2949–2912 (s)	
	2861–2849 (s)	
	1473–1443 (m–vs)	Intensity proportional to
	1305–1295 (s)	number of CH$_2$ groups
	1140–1070 (m)	Often two bands; see above
	888–837 (w)	
	425–150	
	500–490	Substituent on aromatic ring
—CH(CH$_3$)$_2$	1350–1330 (m)	
	835–750 (s)	If attached to C=C bond, 870–800 cm^{-1}. If attached to aryl ring, 740 cm^{-1}
—C(CH$_3$)$_3$	1265–1240 (m)	Not seen in *tert*-butyl bromide
	1220–1200 (m)	Not seen in *tert*-butyl bromide
	760–685 (vs)	If attached to C=C or aromatic ring, 760–720 cm^{-1}
Internal tertiary carbon atom	855–805 (w)	
	455–410	
Internal quaternary carbon atom	710–680 (vs)	
	490–470	

TABLE 7.30 Raman Frequencies of Single Bonds to Hydrogen and Carbon (*Continued*)

Group	Band, cm^{-1}	Remarks
Saturated C—H and C—C (*continued*)		
Two adjacent tertiary carbon atoms	730–920 770–725	Often a band at 530–524 cm^{-1} indicates presence of adjacent tertiary and quaternary carbon atoms.
Dialkyl substitution at α-carbon atom	800–700 (m–s) 680–650 (vs) 605–550	
Cyclopropane	3101–3090 3038–3019 1210–1180 (s)	Shifts to 1200 cm^{-1} for monoalkyl or 1,2-dialkyl substitution and to 1320 cm^{-1} for *gem*-1,1-dialkyl substitution
Cyclobutane	1001–960 (vs)	Shifts to 933 cm^{-1} for monoalkyl, to 887 cm^{-1} for *cis*-1,3-dialkyl, and to 891 cm^{-1} plus 855 cm^{-1} (doublet) for *trans*-1,3-dialkyl substition
Cyclopentane	900–800 (s)	
Cyclohexane	825–815 (vs) 810–795 (vs)	Boat configuration Chair configuration
Cycloheptane	ca 733	
Cyclooctane	ca 703	
$=C\underset{CH_3}{\overset{CH_3}{\diagdown}}$	1392–1377 450–400 (vw) 270–250 (m)	
$\underset{H}{\overset{CH_3}{\diagdown}}C=C\underset{CH_3}{\overset{H}{\diagup}}$	1380–1379 492–455 (vw) 220–200 (m)	

TABLE 7.30 Raman Frequencies of Single Bonds to Hydrogen and Carbon (*Continued*)

Group	Band, cm^{-1}	Remarks
Saturated C—H and C—C (*continued*)		
CH$_3$, CH$_3$ / C=C / H, H	1372–1368 970–952 (m) 592–545 (vw) 420–400 (m) 310–290 (m)	
CH$_3$, CH$_3$ / C=C / CH$_3$, H	1385–1375 522–488 (w)	
CH$_3$, CH$_3$ / C=C / CH$_3$, CH$_3$	1392–1386 690–678 (m–s) 510–485 (m) 424–388 (w)	
≡C—C—C≡ with ‖O	1170–1100 (w–m) 600–580 (m–s)	
≡C—C— with ‖O	1120–1090 (m–vs) 600–510 (w–m)	Tertiary or quaternary carbon adjacent to carbonyl group lowers the frequency 300 cm^{-1}.
—CH$_2$—CO—	1420–1410 (s)	
—CHO	2850–2810 (m) 2720–2695 (vs)	Often appears as a shoulder
Unsaturated C—H		
—C≡C—H	3340–3270 (w–m)	Alkyl substituents at higher frequencies; unsaturated or aryl substituents at lower frequencies
\C=C/ with H	3040–2995 (m)	
\C=C/ with H, H	3095–3050 (m) 2990–2983 (s)	Asymmetric =CH$_2$ stretch Symmetric =CH$_2$ stretch

TABLE 7.30 Raman Frequencies of Single Bonds to Hydrogen and Carbon (*Continued*)

Group	Band, cm^{-1}	Remarks
Unsaturated C—H (continued)		
H, R / C=C / H, H	1419–1415 (m) 1309–12888 (m)	Plus =CH and =CH stretching bands
H, R$_1$ / C=C / H, R$_2$	1413–1399 (m) 909–885 (m) 711–684 (w)	Plus =CH$_2$ stretching bands
R$_1$, R$_2$ / C=C / H, H	1270–1251 (m)	Plus =CH stretching band
R$_1$, H / C=C / H, R$_2$	1314–1290 (m)	Plus =CH stretching band
R$_1$, R$_3$ / C=C / R$_2$, H	1360–1322 (w) 830–800 (vw)	Plus =CH stretching band
Hydroxy O—H		
Free —OH Intermolecularly bonded Aromatic —OH	3650–3250 (w) 3400–3300 (w) ca 3160 (s)	
—OH	1460–1320 (w) 1276–1205 (w–m) 1260 (w–m)	Common to all OH substituents Primary Secondary
C—C—OH primary	1070–1050 (m–s) 1030–960 (m–s) 480–430 (w–m)	CCO stretching CCO deformation
C—C—OH Secondary Tertiary	1135–1120 (m–s) 825–815 (vs) 500–490 (w–m) 1210–1200 (m–s) 755–730 (vs) 360–350 (w–m)	
—CO—O—H	1305–1270	CO stretching

TABLE 7.30 Raman Frequencies of Single Bonds to Hydrogen and Carbon (*Continued*)

Group	Band, cm^{-1}	Remarks
N—H and C—N bonds		
Amine >N—H		
Associated	3400–3250 (s)	Primary amines show two bands.
Nonbonded	3550–3250 (s)	
Salts	2986–2974	Often obscured by intense CH stretching bands
—NH$_2$	1650–1590 (w–vs)	Bending
Amides		
Primary	3540–3500 (w)	Both bands lowered ca 150 cm^{-1} in solid state and H bonding
	3400–3380 (w)	
	1310–1250 (s)	Interaction of NH bending and CN stretching; lowered 50 cm^{-1} in nonbonded state
	1150–1095 (m)	Rocking of NH$_2$
Secondary	3491–3404 (m–s)	Two bands; lowered in frequency on H bonding and in solid sate
	1190–1130 (m)	
	931–865 (m–s)	
	430–395 (w–m)	
—CO—N	607–555 (m)	O=CN bending

C—N—C
   |
   C
	1070–1045 (m)	Stretching

≧C—N≦		
Primary carbon	1090–1060 (m)	CN stretching
Secondary α carbon	1140–1035 (m)	Two bands but often obscured. Strong band at 800 cm^{-1}
Tertiary α carbon	1240–1020 (m)	Two bands. Strong band also at 745 cm^{-1}

TABLE 7.31   Raman Frequencies of Triple Bonds

Abbreviations Used in the Table

m, moderately strong   s–vs, strong to very strong
m–s, moderate to strong   vs, very strong
s, strong

Group	Band, cm^{-1}	Remarks
R—C≡CH	2160–2100 (vs)	Monoalkyl substituted; C≡C stretch
	650–600 (m)	C≡CH deformation
	356–335 (s)	C≡C—C bending of monoalkyls
R$_1$—C≡C—R$_2$	2300–2190 (vs)	C≡C stretching of disubstituted alkyls; sometimes two bands
—C≡C—C≡C—	2264–2251 (vs)	
—C≡N	2260–2240 (vs)	Unsaturated nonaryl substituents lower the frequency and enhance the intensity.
	2234–2200 (vs)	Lowered ca 30 cm^{-1} with aryl and conjugated aliphatics
	840–800 (s–vs)	CCCN symmetrical stretching
	385–350 (m–s)	
	200–160 (vs)	Aliphatic nitriles
H—C≡N	2094 (vs)	
Azides —N̄—N̟≡N	2170–2080 (s)	Asymmetric NNN stretching
	1258–1206 (s)	Symmetric NNN stretching; HN$_3$ at 1300 cm^{-1}
Diazonium salts R—N̟≡N	2300–2240 (s)	
Isonitriles —N̟≡C̄	2146–2134	Stretching of aliphatics
	2124–2109	Stretching of aromatics
Thiocyanates —S—C≡N	2260–2240 (vs)	Stretching of C≡N
	650–600 (s)	Stretching of SC

TABLE 7.32 Raman Frequencies of Cumulated Double Bonds

Abbreviations Used in the Table

s, strong   vw, very weak
vs, very strong   w, weak

Group	Band, cm^{-1}	Remarks
Allenes C=C=C	2000–1960 (s) 1080–1060 (vs) 356	Pseudo-asymmetric stretching Symmetric stretching C=C=C bending

Carbodiimides (cyanamides)
—N=C=N—
	2140–2125 (s)
	
	2150–2100 (vs)
	
	1460
	
	1150–1140 (vs)
		Asymmetric stretching of
		  aliphatics
		Asymmetric stretching of
		  aromatics; two bands
		Symmetrical stretching of
		  aliphatics
		Symmetric stretching of aryls

Cumulenes (trienes) C=C=C=C	2080–2030 (vs) 878	
Isocyanates —N=C=O	2300–2250 (vw) 1450–1400 (s)	Asymmetric stretching Symmetric stretching
Isothiocyanates —N=C=S	2220–2100 690–650	Two bands Alkyl derivatives
Ketenes C=C=O	2060–2040 (vs) 1130 (s) 1374 (s) 1120 (s)	Pseudo-asymmetric stretching Pseudo-symmetric stretching Alkyl derivatives Aryl derivatives
Sulfinylamines R—N=S=O	1306–1214 (w) 1155–989 (s)	Asymmetric stretching Symmetric stretching

TABLE 7.33   Raman Frequencies of Carbonyl Bands

Abbreviations Used in the Table

m, moderately strong	s–vs, strong to very strong
m–s, moderate to strong	vs, very strong
s, strong	w, weak

Group	Band, cm^{-1}	Remarks
Acid anhydrides		
—CO—O—CO—		
Saturated	1850–1780 (m)	
	1771–1770 (m)	
Conjugated, noncyclic	1775	
	1720	
Acid fluorides —CO—F		
Alkyl	1840–1835	
Aryl	1812–1800	
Acid chlorides —CO—Cl		
Alkyl	1810–1770 (s)	
Aryl	1774	
	1731	
Acid bromides —CO—Br		
Alkyl	1812–1788	
Aryl	1775–1754	
Acid iodides —CO—I		
Alkyl	ca 1806	
Aryl	ca 1752	
Lactones	1850–1730 (s)	
Esters		
Saturated	1741–1725	Alkyl branching on carbon adjacent to C=O lowers frequency by 5–15 cm^{-1}.
Aryl and α,β-unsaturated	1727–1714	
Diesters		
Oxalates	1763–1761	
Phthalates	1738–1728	
C≡C—CO—O—	1716–1708	
Carbamates	1694–1688	
Aldehydes	1740–1720 (s–vs)	
Ketones		
Saturated	1725–1700 (vs)	
Aryl	1700–1650 (m)	

TABLE 7.33 Raman Frequencies of Carbonyl Bands (*Continued*)

Group	Band, cm^{-1}	Remarks
Ketones (*continued*)		
Alicyclic		
$n = 4$	1782 (m)	
$n = 5$	1744 (m)	
$n \geq 6$	1725–1699 (m)	
Carboxylic acids		
Mono-	1686–1625 (s)	These α-substituents increase the frequency: F, Cl, Br, OH.
Poly-	1782–1645	Solid state; often two bands
	1750–1710	In solution; very broad band
Amino acids	1743–1729	
Carboxylate ions	1690–1550 (w)	
	1440–1340 (vs)	
Amino acid anion	1743–1729	
	1600–1570 (w)	Often masked by water deformation band near 1630 cm^{-1}
Amides (see also Table 7.30)		
Primary		
Associated	1686–1576 (m–s)	
	1650–1620 (m)	
Nonbonded	1715–1675 (m)	
	1620–1585 (m)	
Secondary		
Associated	1680–1630 (w)	Both *cis* and *trans* forms
	1570–1510 (w)	*Trans* form
	1490–1440	*Cis* form
Nonbonded	1700–1650	Both *cis* and *trans* forms
	1550–1500	*Trans* form (no *cis* band)
Tertiary	1670–1630 (m)	
Lactams	1750–1700 (m)	

TABLE 7.34 Raman Frequencies of Other Double Bonds

Abbreviations Used in the Table

m, moderately strong

m–s, moderate to strong

s, strong

w–m, weak to moderately strong

vs, very strong

w, weak

s–vs, strong to very strong

Group	Band, cm^{-1}	Remarks
Alkenes $>C=C<$		
$>C=C<$	1680–1576 (m–s)	General range
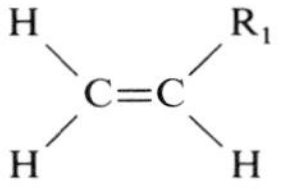 H, R$_1$ / C=C / H, H	1648–1638 (vs)	C=C stretching
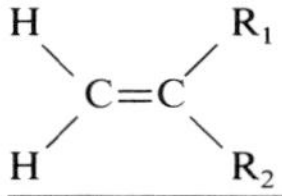 H, R$_1$ / C=C / H, R$_2$	ca 1650 (vs) 270–252 (w)	C=C stretching C=C—C skeletal deformation
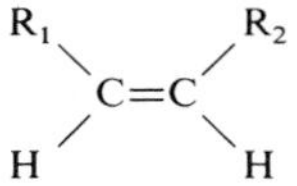 R$_1$, R$_2$ / C=C / H, H	ca 1660 (vs) 970–952 (w)	C=C stretching Asymmetric CC stretching
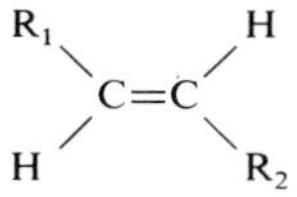 R$_1$, H / C=C / H, R$_2$	1676–1665 (s)	C—C stretching
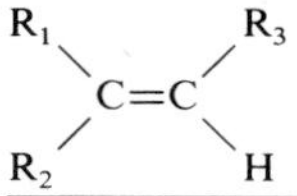 R$_1$, R$_3$ / C=C / R$_2$, H	1678–1664 (vs) 522–488 (w)	C=C stretching C=C—C skeletal deformation
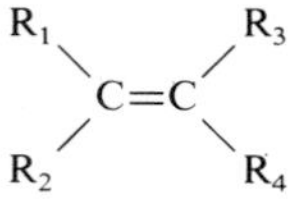 R$_1$, R$_3$ / C=C / R$_2$, R$_4$	1680–1665 (s) 690–678 (m–s) 510–485 (m) 424–388 (w)	C=C stretching Symmetrical CC stretching Skeletal deformation Skeletal deformation

Haloalkene	X = fluorine	X = chlorine	X = bromine	X-iodine
$>C=C<$ stretch of haloalkanes				
H$_2$C=CHX	1654	1603–1601	1596–1593	1581
HXC=CHX				
cis	1712	1590–1587	1587–1583	1543
trans	1694	1578–1576	1582–1581	1537
H$_2$C=CX$_2$	1728	1616–1611	1593	
X$_2$C=CHX	1792	1589–1582	1552	
X$_2$C=CX$_2$	1872	1577–1571	1547	1465 (solid)

TABLE 7.34 Raman Frequencies of Other Double Bonds (*Continued*)

Group	Band, cm^{-1}	Remarks
$>C=N-$ bonds		
Aldimines (azomethines) $R_1R_2C=N-R_2$ (H on R_1)	1673–1639 1405–1400 (s)	Dialkyl substituents at higher frequency; diaryl substituents at lower end of range
Aldoximines and Ketoximes $>C=N-OH$	1680–1617 (vs) 1335–1330 (w)	
Azines $>C=N-N=C<$	1625–1608 (s)	
Hydrazones $>C=N-N<$	1660–1610 (s–vs)	
Imido ethers O $>C=NH$	1658–1648	NH stretching at 3360–3327 cm^{-1}
Semicarbazones and thiosemicarbazones $>C=N-N<$ ... NH_2 C O (or S)	1665–1642 (vs) 1620–1610 (vs)	Aliphatic. Thiosemicarbazones fall in lower end of range. Aromatic derivatives
Azo compounds $-N=N-$		
$-N=N-$	1580–1570 (vs) 1442–1380 (vs) 1060–1030 (vs)	Nonconjugated Conjugated to aromatic ring CN stretching in aryl compounds
Nitro compounds $N=O$		
Alkyl nitrites	1660–1620 (s)	$N=O$ stretching
Alkyl nitrates	1635–1622 (w–m) 1285–1260 (vs) 610–562 (m)	Asymmetric NO_2 stretching Symmetric NO_2 stretching NO_2 deformation

TABLE 7.34 Raman Frequencies of Other Double Bonds (*Continued*)

Group	Band, cm^{-1}	Remarks
Nitro compounds N=O (continued)		
Nitroalkanes		
Primary	1560–1548 (m–s)	
	1395–1370 (s)	Sensitive to substituents attached to CNO$_2$ group
	915–898 (m–s)	
	894–873 (m–s)	
	618–609 (w)	
	640–615 (w)	Shoulder
	494–472 (w–m)	Broad; useful to distinguish from secondary nitroalkanes
Secondary	1553–1547 (m)	
	1375–1360 (s)	
	908–868 (m)	
	863–847 (s)	
	625–613 (m)	
	560–516 (s)	Sharp band
Tertiary	1543–1533 (m)	
	1355–1345 (s)	
Nitrogen oxides		
$\geqq\overset{+}{N}\rightarrow\bar{O}$	1612–1602 (s)	
	1252 (m)	
	1049–1017 (s)	
	835 (s)	
	541 (w)	
	469 (w)	

TABLE 7.35   Raman Frequencies of Aromatic Compounds

Abbreviations Used in the Table

m, moderately strong	var, of variable strength
m−s, moderate to strong	vs, very strong
m−vs, moderate to very strong	w, weak
s, strong	w−m, weak to moderately strong
s−vs, strong to very strong	

Group	Band, cm^{-1}	Remarks
Common features		
Aromatic compounds	3070–3020 (s) 1630–1570 (m-s)	CH stretching C—C stretching
Substitution patterns of the benzene ring		
Monosubstituted	1180–1170 (w–m) 1035–1015 (s) 1010–990 (vs) 630–605 (w)	Characteristic feature; found also with 1,3- and 1,3,5-substitutions
1,2-Disubstituted	1230–1215 (m) 1060–1020 (s) 740–715 (m)	Characteristic feature Lowered 60 cm^{-1} for halogen substituents
1,3-Disubstituted	1010–990 (vs) 750–640 (s)	Characteristic feature
1,4-Disubstituted	1230–1200 (s–vs) 1180–1150 (m) 830–750 (vs) 650–630 (m–w)	Lower frequency with Cl substituents
Isolated hydrogen	1379 (s–vs) 1290–1200 (s) 745–670 (m–vs) 580–480 (s)	Characteristic feature
1,2,3-Trisusbstituted	1100–1050 (m) 670–500 (vs) 490–430 (w)	The lighter the mass of the substituent, the higher the frequency.
1,2,4-Trisubstituted	750–650 (vs) 580–540 (var) 500–450 (var)	Lighter mass at higher frequencies

TABLE 7.35 Raman Frequencies of Aromatic Compounds (*Continued*)

Group	Band, cm^{-1}	Remarks
Substitution patterns of the benzene ring (*continued*)		
1,3,5-Trisubstituted	1010–990 (vs)	
Completeely substituted	1296 (s) 550 (vs) 450 (m) 361 (m)	
Other aromatic compounds		
Naphthalenes	1390–1370 1026–1012 767–762 535–512 519–512	Ring breathing α or β substituents β substituents α substituents β substituents
Disubstituted napthalenes	773–737 (s) 726–705 (s) 690–634 (s) 608 575–569 544–537	1,2-; 1,3-; 2,3-; 2,6-; 2,7- 1,3-; 1,4-(two bands); 1,6-; 1,7- (two bands) 1,2-; 1,4-(two bands); 1,5-; 1,8- (two bands) 1,3- 1,2-; 1,3-; 1,6- 1,2-; 1,7-; 1,8-
Anthracenes	1415–1385	Ring breathing

TABLE 7.36 Raman Frequencies of Sulfur Compounds

Abbreviations Used in the Table

m, moderately strong   s–vs, strong to very strong
m–s, moderate to strong   vs, very strong
s, strong   w–m, weak to moderately strong

Group	Band, cm^{-1}	Remarks
—S—H	2590–2560 (s)	SH stretching for both aliphatic and aromatic
$>$C$=$S	1065–1050 (m) 735–690 (vs)	Solid state
$>$S$=$O In $(RO_2)_2SO$ In $(R_2N)_2SO$	1209–1198 1108	One or two bands

TABLE 7.36 Raman Frequencies of Sulfur Compounds (*Continued*)

Group	Band, cm^{-1}	Remarks
$\geq$S=O (*continued*)		
In R$_2$SO	1070–1010 (w–m)	Broad
SOF$_2$	1308	
SOCl$_2$	1233	
SOBr$_2$	1121	
—SO$_2$—	1330–1260 (m–s)	Asymmetric SO$_2$ stretching
	1155–1110 (s)	Symmetric SO$_2$ stretching
	610–540 (m)	Scissoring mode of aryls
	512–485 (m)	Scissoring mode of alkyls
—SO$_2$—N<	ca 1322 (m)	Asymmetric SO$_2$ stretching
	1163–1138 (s)	Symmetric SO$_2$ stretching
	524–510 (s)	Scissoring mode
—SO$_2$—O	1363–1338 (w–m)	SO$_2$ stretching. Aryl substituents occur at higher range.
	1192–1165 (vs)	
	589–517 (w–m)	Scissoring (two bands). Aryl substituents occur at higher range of frequencies.
—SO$_2$—S—	1334–1305 (m–s)	
	1128–1126 (s)	
	559–553 (m–s)	
X—SO$_2$—X	1412–1361 (w–m)	
	(F) (Cl)	
	1263–1168 (s)	
	(F) (Cl)	
	596–531 (s)	
—O—SO$_2$—O—	1388–1372 (s)	
	1196–1188 (vs)	

—O—C—S—
  ‖
  S
	670–620 (vs)	C=S stretching

	480–450 (vs)	CS stretching
$\geqq$C—SH	920 (m)	C—SH deformation of aryls
	850–820 (m)	
$\equiv$C—S—	752 (vs), 731 (vs)	With vinyl group attached
	742–722 (m–s)	With CH$_3$ attached
	698 (w), 678 (s)	With allyl group attached
	693–639 (s)	Ethyl or longer alkyl chain
	651–610 (s–vs)	Isopropyl group attached
	589–585 (vs)	*tert*-Butyl group attached

TABLE 7.36 Raman Frequencies of Sulfur Compounds (*Continued*)

Group	Band, cm^{-1}	Remarks
≡C—S— (*continued*)		
(CH$_2$)$_n$ S		
$n=2$	1112	
$n=4$	688	
$n=5$	659	
≡C—(S—S)$_n$—C≡	715–620 (vs)	Two bands; CS stretching
	525–510 (vs)	Two bands; SS stretching
Didi-*n*-alkyl disulfides	576 (s)	CS stretching
Di-*tert*-butyl disulfide	543 (m)	SS stretching
Trisulfides	510–480 (s)	SS stretching

TABLE 7.37 Raman Frequencies of Ethers

Abbreviations Used in the Table

m, moderately strong   var, of variable strength
s, strong   vs, very strong

Group	Band, cm^{-1}	Remarks
≡C—O—C≡		
Aliphatic	1200–1070 (m)	Asymmetrical COC stretching. Symmetrical substitution gives higher frequencies
	930–830 (s)	Symmetrical COC stretching
	800–700 (s)	Braching at α carbon gives higher frequencies.
	550–400	
Aromatic	1310–1210 (m)	
	1050–1010 (m)	
≡C—O—C—O—C≡	1145–1129 (m)	
	900–800 (vs)	
	537–370 (s)	
	396–295	
>C —— C< (O)	1280–1240 (s)	Ring breathing
—O—O—	800–770 (var)	
(CH$_2$)$_n$ O $n=3$	1040–1010 (s)	
$n=4$	920–900 (s)	
$n=5$	820–800 (s)	

TABLE 7.38 Raman Frequencies of Halogen Compounds

Abbreviations Used in the Table

m–s, moderate to strong   var, of variable strength
s, strong   vs, very strong

Group	Band, cm^{-1}	Remarks
C—F	1400–870	Correlations of limited applicability because of vibrational coupling with stretching
C—Cl Primary Secondary Tertiary	350–290 (s) 660–650 (vs) 760–605 (s) 620–540 (var)	CCCl bending; general May be one to four bands May be one to three bands
=C—Cl	844–564 438–396 381–170	
=CCl$_2$	601–441 300–235	
C—Br	690–490 (s) 305–258 (m–s)	Often several bands; primary at higher range of frequencies. Tertiary has very strong band at ca 520 cm^{-1}.
=C—Br	745–565 356–318 240–115	
=CBr$_2$	467–265 185–145	
C—I	663–595 309 154–85	
=C—I	ca 180	Solid state
=CI$_2$	ca 265 ca 105	Solid state Solid state

TABLE 7.39 Raman Frequencies of Miscellaneous Compounds

Abbreviations Used in the Table

m, moderately strong   vs, very strong
s, strong   vvs, very very strong

Group	Band, cm^{-1}	Remarks
C—As	570–550 (vs)	CAs stretching
	240–220 (vs)	CAsC deformation
C—Pb	480–420 (s)	CPb stretching
C—Hg	570–510 (vvs)	CHg stretching
C—Si	1300–1200 (s)	CSi stretching
C—Sn	600–450 (s)	CSn stretching
P—H	2350–2240 (m)	PH stretching
Heterocyclic rings		
Trimethylene oxide	1029	
Trimethylene imine	1026	
Tetrahydrofuran	914	
Pyrrolidine	899	
1,3-Dioxolane	939	
1,4-Dioxane	834	
Piperidine	815	
Tetrahydropyran	818	
Morpholine	832	
Piperazine	836	
Furan	1515–1460	2-Substituted
	1140	
Pyrazole	1040–990	
Pyrrole	1420–1360 (vs)	
	1144	
Thiophene	1410 (s)	
	1365 (s)	
	1085 (vs)	
	1035 (s)	
	832 (vs)	
	610 (s)	
Pyridine	1030 (vs)	
	990 (vs)	

TABLE 7.40   Principal Argon-Ion Laser Plasma Lines

Wavelength, nm	Wavenumber, cm^{-1}	Relative intensity	Shift relative to 488.0 nm, cm^{-1}	Shift relative to 514.5 nm, cm^{-1}
487.9860	20 486.67	5000	0	
488.9033	20 448.23	200	38.4	
490.4753	20 382.70	130	104.0	
493.3206	20 265.13	970	221.5	
496.5073	20 135.07	960	351.6	
497.2157	20 106.39	330	380.3	
500.9334	19 957.16	1500	529.5	
501.7160	19 926.03	620	560.6	
506.2036	19 749.39	1400	737.3	
514.1790	19 443.06	360	1043.6	
514.5319	19 429.73	1000	1056.9	0
516.5774	19 352.79	38	1133.9	76.9
517.6233	19 313.69	41	1173.0	116.0
521.6816	19 163.44	20	1323.2	266.3
528.6895	18 909.43	150	1577.2	520.3
539.7522	18 521.87	18	1964.8	907.9
545.4307	18 329.04	19	2157.6	1100.7
555.8703	17 984.81	30	2501.9	1444.9
560.6734	17 830.75	48	2655.9	1599.0
565.0705	17 692.00	29	2794.7	1737.7
565.4450	17 680.28	27	2806.4	1749.4
569.1650	17 564.73	27	2921.9	1865.0
577.2326	17 319.24	69	3167.4	2110.5
581.2746	17 198.80	49	3287.9	2230.9
598.5920	16 701.24	23	3785.4	2728.5
610.3546	16 379.38	91	4107.3	3050.4
611.4929	16 348.90	1750	4137.8	3080.8
612.3368	16 326.36	100	4160.3	3103.4
613.8660	16 285.69	97	4201.0	3144.0
617.2290	16 196.96	1400	4289.7	3232.8
624.3125	16 013.19	590	4473.5	3416.5
639.9215	15 622.60	160	4864.1	3807.1
641.6308	15 580.98	50	4905.7	3848.8

7.7 NUCLEAR MAGNETIC RESONANCE

TABLE 7.41 Nuclear Properties of the Elements

In the following table the magnetic moment μ is in multiples of the nuclear magneton $\mu_N(eh/4\pi Mc)$ with diamagnetic correction. The spin I is in multiples of $h/2\pi$, and the electric quadrupole moment Q is in multiples of 10^{-28} square meters. Nuclei with spin ½ have no quadrupole moment. Sensitivity is for equal numbers of nuclei at constant field. NMR frequency at any magnetic field is the entry for column 5 multiplied by the value of the magnetic field in kilogauss. For example, in a magnetic field of 23.490 kG, protons will process at 4.2576×23.490 kG $= 100.0$ MHz. Radionuclides are denoted with an asterisk.

The data were extracted from A. H. Wapstra and G. Audi, "The 1983 Atomic Mass Evaluation," *Nucl. Phys.* **A432**:1–54 (1985), and M. Lederer and V. S. Shirley, *Table of Isotopes*, 7th ed., Wiley Interscience, New York, 1978.

Nuclide	Natural abundance, %	Spin I	Sensitivity at constant field relative to ^{1}H	NMR frequency for a 1-kG field, MHz	Magnetic moment μ/μ_N, J·T^{-1}	Electric quadrupole moment Q, 10^{-28} m^2
^{1}n		½	0.322	2.916 5	−1.913 0	
^{1}H	99.985(1)	½	1.000	4.257 6	2.792 846	
^{2}H	0.015(1)	1	0.009 64	0.653 6	0.857 438	0.002 875
^{3}H*		½	1.21	4.541 4	2.978 960	
^{3}He	1.38 × 10^{-4}	½	0.443	3.243 4	−2.127 624	
^{6}Li	7.5(2)	1	0.008 51	0.626 5	0.822 047	0.000 645
^{7}Li	92.5(2)	3/2	0.294	1.654 7	3.256 424	−0.036 6
^{9}Be	100	3/2	0.013 9	0.598 3	−1.177 9	0.053
^{10}B	19.9(2)	3	0.019 9	0.457 5	1.800 65	0.084 73
^{11}B	80.1(2)	3/2	0.165	1.366 0	2.688 637	0.040 65
^{13}C	1.10(3)	½	0.015 9	1.070 5	0.702 411	
^{14}N	99.634(9)	1	0.001 01	0.307 6	0.403 761	0.015 6
^{15}N	0.366(9)	½	0.001 04	0.431 5	−0.283 189	
^{17}O	0.003 8(3)	5/2	0.029 1	0.577 2	−1.893 80	−0.025 78
^{19}F	100	½	0.834	4.005 5	2.628 867	
^{21}Ne	0.27(1)	−3/2	0.027 2	0.336 11	−0.661 966	0.103 0
^{22}Na*		3	0.018 1	0.443 4	1.745	
^{23}Na	100	3/2	0.092 7	1.126 2	2.217 520	0.102
^{25}Mg	10.00(1)	5/2	0.026 8	0.260 6	−0.855 46	0.22
^{27}Al	100	5/2	0.207	1.109 4	3.641 504	0.140
^{29}Si	4.67(1)	½	0.078 5	0.846 0	−0.555 29	
^{31}P	100	½	0.066 4	1.723 5	1.131 60	
^{33}S	0.75(1)	3/2	0.022 6	0.326 6	0.643 821	−0.064
^{35}S*		3/2	0.008 50	0.508	1.00	0.045
^{35}Cl	75.77(5)	3/2	0.004 71	0.417 2	0.821 874	−0.082 49
^{36}Cl*		2	0.012 1	0.489 3	1.283 8	−0.016 8
^{37}Cl	24.23(5)	3/2	0.002 72	0.347 2	0.684 123	−0.064 93
^{39}K	93.258(3)	3/2	0.000 508	0.198 7	0.391 466	0.049
^{40}K*	0.011 9(1)	4	0.005 21	0.247 0	−1.298 099	−0.067
^{41}K	6.730(3)	3/2	0.000 083 9	0.109 2	0.214 870	0.060
^{43}Ca	0.135(3)	7/2	0.063 9	0.286 5	−1.317 27	
^{45}Sc	100	7/2	0.301	1.034 3	4.756 483	−0.22
^{47}Ti	7.3(1)	5/2	0.002 10	0.240 0	−0.788 48	0.29
^{49}Ti	5.5(1)	7/2	0.003 76	0.240 1	−1.104 17	0.24
^{50}V	0.250(2)	6	0.055 3	0.424 5	3.347 45	0.07
^{51}V	99.75(2)	7/2	0.383	1.119 3	5.151 4	−0.052

* Radioactive nuclide.

TABLE 7.41   Nuclear Properties of the Elements (*Continued*)

Nuclide	Natural abundance, %	Spin I	Sensitivity at constant field relative to ^{1}H	NMR frequency for a 1-kG field, MHz	Magnetic moment μ/μ_N, J·T^{-1}	Electric quadrupole moment Q, 10^{-28} m^2
^{53}Cr	9.50(1)	3/2	0.000 10	0.240 6	−0.473 45	0.022
^{55}Mn	100	5/2	0.178	1.053 3	3.453 2	0.40
^{59}Co	100	7/2	0.281	1.010 3	4.627	0.404
^{61}Ni	1.13(1)	3/2	0.003 50	0.380 48	−0.750 02	0.162
^{63}Cu	69.17(2)	3/2	0.093 8	1.128 5	2.223 3	−0.209
^{65}Cu	30.83(2)	3/2	0.116	1.209 0	2.381 7	−0.195
^{67}Zn	4.1(1)	5/2	0.002 86	0.263 5	0.875 479	0.150
^{69}Ga	60.1(2)	3/2	0.069 3	1.021 8	2.016 59	0.168 8
^{71}Ga	39.9(2)	3/2	0.142	1.298 4	2.562 27	0.106
^{73}Ge	7.8(2)	9/2	0.001 40	0.148 5	−0.879 467	−0.173
^{75}As	100	3/2	0.025 1	0.729 2	1.439 47	0.29
^{77}Se	7.6(2)	1/2	0.006 97	0.813 1	0.535 06	
^{79}Br	50.69(5)	3/2	0.078 6	1.066 7	2.106 399	0.293
^{81}Br	49.31(5)	3/2	0.098 4	1.149 8	2.270 560	0.27
^{83}Kr	11.5(1)	9/2	0.001 89	0.016 4	−0.970 669	0.270
^{85}Rb	72.16(1)	5/2	0.010 5	0.411 1	1.353 03	0.274
^{87}Rb	27.83(1)	3/2	0.177	1.393 2	2.751 24	0.132
^{87}Sr	7.00(1)	9/2	0.002 69	0.184 5	−1.092 83	0.16
^{89}Y	100	1/2	0.000 117	0.208 6	−0.137 415	
^{91}Zr	11.22(2)	5/2	0.009 4	0.40	−1.303 62	
^{93}Nb	100	9/2	0.482	1.040 7	6.170 5	−0.37
^{95}Mo	15.92(4)	5/2	0.003 22	0.277 4	−0.914 2	−0.019
^{97}Mo	9.55(2)	5/2	0.003 42	0.283 3	−0.933 5	−0.102
^{99}Ru	12.7(1)	5/2	0.000 195	0.144 3	−0.641 3	0.077
^{101}Ru	17.0(1)	5/2	0.001 4	0.210 4	−0.718 9	0.44
^{103}Rh	100	1/2	0.000 031 2	0.134 0	−0.088 40	
^{105}Pd	22.33(8)	5/2	0.000 779	0.174	−0.642	0.80
^{107}Ag	51.839(5)	1/2	0.000 066 9	0.172 2	−0.113 570	
^{109}Ag	48.161(5)	1/2	0.000 101	0.198 1	−0.130 691	
^{111}Cd	12.80(6)	1/2	0.009 54	0.902 8	−0.594 886	
^{113}Cd	12.22(6)	1/2	0.010 9	0.944 4	−0.622 300	
^{113}In	4.3(2)	9/2	0.345	0.931 0	5.528 9	0.846
^{115}In	95.7(2)	9/2	0.347	0.932 9	5.540 8	0.861
^{115}Sn	0.36(1)	1/2	0.035 0	1.322	−0.918 84	
^{117}Sn	7.68(7)	1/2	0.045 3	1.577	−1.001 05	
^{119}Sn	8.58(4)	1/2	0.051 8	1.587	−1.047 29	
^{121}Sb	57.3(9)	5/2	0.160	1.019	3.363 4	−0.20
^{123}Sb	42.7(9)	7/2	0.045 7	1.551 8	2.549 8	−0.26
^{123}Te	0.908(3)	1/2	0.018	1.115 7	−0.736 79	
^{125}Te	7.14(1)	1/2	0.031 6	1.345	−0.888 28	
^{127}I	100	5/2	0.093 5	0.851 9	2.813 28	−0.789
^{129}Xe	26.4(6)	1/2	0.021 2	1.178	−0.777 977	
^{131}Xe	21.2(4)	3/2	0.002 77	0.349 0	0.691 861	−0.12
^{133}Cs	100	7/2	0.047 4	0.558 5	2.582 024	−0.3
^{135}Ba	6.59(2)	3/2	0.004 99	0.425	0.837 943	0.18
^{137}Ba	11.23(4)	3/2	0.006 97	0.476	0.937 365	0.28
^{139}La	99.91(1)	7/2	0.059 2	0.601 4	2.783 2	0.22
^{141}Pr	100	5/2	0.234	1.13	4.136	−0.058 9
^{143}Nd	12.18(5)	7/2	0.002 81	0.22	−1.065	−0.484
^{145}Nd	8.30(5)	7/2	0.000 670	0.14	−0.656	−0.253

TABLE 7.41 Nuclear Properties of the Elements (*Continued*)

Nuclide	Natural abundance, %	Spin I	Sensitivity at constant field relative to ^1H	NMR frequency for a 1-kG field, MHz	Magnetic moment μ/μ_N, J·T^{-1}	Electric quadrupole moment Q, 10^{-28} m^2
^{147}Sm	15.0(2)	7/2	0.000 88	0.147	−0.814 9	−0.18
^{149}Sm	13.8(1)	7/2	0.000 47	0.119	−0.671 8	0.053
^{151}Eu	47.81(5)	5/2	0.168	1.0	3.471 8	1.16
^{153}Eu	52.2(5)	5/2	0.014 5	0.46	1.533 1	2.94
^{155}Gd	14.80(5)	3/2	0.000 279	0.162 59	−0.259 1	1.59
^{157}Gd	15.65(3)	3/2	0.000 544	0.203 26	−0.339 9	2.03
^{159}Tb	100	3/2	0.058 3	0.965 45	2.014	1.18
^{161}Dy	18.9(1)	5/2	0.000 417	0.140 26	−0.480 6	2.44
^{163}Dy	24.9(2)	5/2	0.001 12	0.195 12	0.672 6	2.57
^{165}Ho	100	7/2	0.181	0.873 27	4.173	2.74
^{167}Er	22.95(13)	7/2	0.000 507	0.123 05	−0.566 5	2.827
^{169}Tm	100	1/2	0.000 566	0.353 12	−0.231 6	
^{171}Yb	14.3(2)	1/2	0.004 19	0.69	0.493 67	
^{173}Yb	16.1(2)	5/2	0.001 19	0.198	−0.679 89	2.8
^{175}Lu	97.41(2)	7/2	0.049 4	0.57	2.232 7	5.69
^{177}Hf	18.606(3)	7/2	0.000 638	0.132 85	0.793 6	4.50
^{179}Hf	13.629(5)	9/2	0.000 216	0.079 61	−0.640 9	5.1
^{181}Ta	99.988(2)	7/2	0.026 0	0.46	2.371	3.9
^{183}W	14.3(1)	1/2	0.000 069 8	0.175	0.117 785	
^{185}Re	37.40(2)	5/2	0.133	0.958 6	3.187 1	2.36
^{187}Re	62.60(2)	5/2	0.137	0.986 4	3.219 7	2.24
^{189}Os	16.1(3)	3/2	0.002 24	0.330 7	0.659 933	0.91
^{191}Ir	37.3(5)	3/2	0.000 035	0.081	0.146 2	0.78
^{193}Ir	62.7(5)	3/2	0.000 042	0.086	0.159 2	0.70
^{195}Pt	33.9(5)	1/2	0.009 94	0.915 3	0.609 50	
^{197}Au	100	3/2	0.000 021 4	0.069 1	0.148 159	0.547
^{199}Hg	16.8(11)	1/2	0.005 72	0.761 2	0.505 885	
^{201}Hg	13.23(11)	3/2	0.001 90	0.308	−0.560 225	0.455
^{203}Tl	29.524(9)	1/2	0.187	2.433	1.622 257	
^{205}Tl	70.476(9)	1/2	0.192	2.457	1.638 214	
^{207}Pb	22.1(1)	1/2	0.009 13	0.889 9	0.582 19	
^{209}Bi	100	9/2	0.137	0.684 2	4.110 6	−0.46
^{235}U	0.720(1)	7/2	0.000 121	0.076 23	−0.35	4.55

TABLE 7.42 Proton Chemical Shifts

Values are given on the officially approved δ scale; $\tau = 10.00 - \delta$.

Abbreviations Used in the Table

R, alkyl group   Ar, aryl group

Substituent group	Methyl protons	Methylene protons	Methine proton
HC—C—CH$_2$	0.95	1.20	1.55
HC—C—NR$_2$	1.05	1.45	1.70
HC—C—C=C	1.00	1.35	1.70
HC—C—C=O	1.05	1.55	1.95
HC—C—NRAr	1.10	1.50	1.80

TABLE 7.42 Proton Chemical Shifts (*Continued*)

Substituent group	Methyl protons	Methylene protons	Methine proton
HC—C—H(C=O)R	1.10	1.50	1.90
HC—C—(C=O)NR$_2$	1.10	1.50	1.80
HC—C—(C=O)Ar	1.15	1.55	1.90
HC—C—(C=O)OR	1.15	1.70	1.90
HC—C—Ar	1.15	1.55	1.80
HC—C—OH	1.20	1.50	1.75
HC—C—OR	1.20	1.50	1.75
HC—C—C≡CR	1.20	1.50	1.80
HC—C—C≡N	1.25	1.65	2.00
HC—C—SR	1.25	1.60	1.90
HC—C—OAr	1.30	1.55	2.00
HC—C—O(C=O)R	1.30	1.60	1.80
HC—C—SH	1.30	1.60	1.65
HC—C—(S=O)R and HC—C—SO$_2$R	1.35	1.70	
HC—C—NR$_3^+$	1.40	1.75	2.05
HC—C—O—N=O	1.40		
HC—C—O(C=O)CF$_3$	1.40	1.65	
HC—C—Cl	1.55	1.80	1.95
HC—C—F	1.55	1.85	2.15
HC—C—NO$_2$	1.60	2.05	2.50
HC—C—O(C=O)Ar	1.65	1.75	1.85
HC—C—I	1.75	1.80	2.10
HC—C—Br	1.80	1.85	1.90
HC—CH$_2$	0.90	1.30	1.50
HC—C=C	1.60	2.05	
HC—C≡C	1.70	2.20	2.80
HC—(C=O)OR	2.00	2.25	2.50
HC—(C=O)NR$_2$	2.00	2.25	2.40
HC—SR	2.05	2.55	3.00
HC—O—O	2.10	2.30	2.55
HC—(C=O)R	2.10	2.35	2.65
HC—C≡N	2.15	2.45	2.90
HC—I	2.15	3.15	4.25
HC—CHO	2.20	2.40	
HC—Ar	2.25	2.45	2.85
HC—NR$_2$	2.25	2.40	2.80
HC—SSR	2.35	2.70	
HC—(C=O)Ar	2.40	2.70	3.40
HC—SAr	2.40		
HC—NRAr	2.60	3.10	3.60
HC—SO$_2$R and HC—(SO)R	2.60	3.05	
HC—Br	2.70	3.40	4.10
HC—NR$_3^+$	2.95	3.10	3.60
HC—NH(C=O)R	2.95	3.35	3.85
HC—SO$_3$R	2.95		
HC—Cl	3.05	3.45	4.05
HC—OH and HC—OR	3.20	3.40	3.60

TABLE 7.42 Proton Chemical Shifts (*Continued*)

Substituent group	Methyl protons	Methylene protons	Methine proton
HC−PAr$_3$	3.20	3.40	
HC−NH$_2$	3.50	3.75	4.05
HC−O(C=O)R	3.65	4.10	4.95
HC−OAr	3.80	4.00	4.60
HC−O(C=O)Ar	3.80	4.20	5.05
HC−O(C=O)CF$_1$	3.95	4.30	
HC−F	4.25	4.50	4.80
HC−NO$_2$	4.30	4.35	4.60
Cyclopropane		0.20	0.40
Cyclobutane		2.45	
Cyclopentane		1.65	
Cyclohexane		1.50	1.80
Cycloheptane		1.25	

Substituent group	Proton shift	Substituent group	Proton shift
HC≡CH	2.35	HO−C=O	10–12
HC≡CAr	2.90	HO−SO$_2$	11–12
HC≡C−C=C	2.75	HO−Ar	4.5–6.5
HAr	7.20	HO−R	0.5–4.5
HCO−O	8.1	HS−Ar	2.8–3.6
HCO−R	9.4–10.0	HS−R	1–2
HCO−Ar	9.7–10.5	HN−Ar	3–6
HO−N=C (oxime)	9–12	HN−R	0.5–5

Saturated heterocyclic ring systems

TABLE 7.42 Proton Chemical Shifts (*Continued*)

Substituent group	Methyl protons	Methylene protons	Methine proton

Unsaturated cyclic systems

TABLE 7.43 Estimation of Chemical Shift for Protons of —CH_2— and Methine Groups

$$\delta_{CH_2} = 0.23 + C_1 + C_2 \qquad \delta_{CH} = 0.23 + C_1 + C_2 + C_3$$

X*	C	X*	C	X*	C
—CH_3	0.5	—SR	1.6	—OR	2.4
—CF_3	1.1	—C≡C—Ar	1.7	—Cl	2.5
>C=C<	1.3	—CN	1.7	—OH	2.6
—C≡C—R	1.4	—CO—R	1.7	—N=C=S	2.9
—COOR	1.5	—I	1.8	—OCOR	3.1
—NR_2	1.6	—Ph	1.8	—OPh	3.2
—$CONR_2$	1.6	—Br	2.3		

* R, alkyl group; Ar, aryl group; Ph, phenyl group.

TABLE 7.44 Estimation of Chemical Shift for Proton Attached to a Double Bond

Positive Z values indicate a downfield shift, and an arrow indicates the point of attachment of the substituent group to the double bond.

$$\delta_{C=C\diagdown H} = 5.25 - Z_{gem} + Z_{cis} + Z_{trans}$$

R	Z_{gem}, ppm	Z_{cis}, ppm	Z_{trans}, ppm
→H	0	0	0
→alkyl	0.45	−0.22	−0.28
→alkyl—ring (5- or 6-member)	0.69	−0.25	−0.28
→CH$_2$O—	0.64	−0.01	−0.02
→CH$_2$S—	0.71	−0.13	−0.22
→CH$_2$X (X: F, Cl, Br)	0.70	0.11	−0.04
→CH$_2$N<	0.58	−0.10	−0.08
$\diagdown$C=C (isolated) $\diagup$	1.00	−0.09	−0.23
$\diagdown$C=C (conjugated) $\diagup$	1.24	0.02	−0.05
→C≡N	0.27	0.75	0.55
→C≡C—	0.47	0.38	0.12
$\diagdown$C=O (isolated) $\diagup$	1.10	1.12	0.87
$\diagdown$C=O (conjugated) $\diagup$	1.06	0.91	0.74
→COOH (isolated)	0.97	1.41	0.71
→COOH (conjugated)	0.80	0.98	0.32
→COOR (isolated)	0.80	1.18	0.55
→COOR (conjugated)	0.78	1.01	0.46
H | →C=O	1.02	0.95	1.17
$\diagdown$N$\diagup$ | →C=O	1.37	0.98	0.46
Cl | →C=O	1.11	1.46	1.01
→OR (R: aliphatic)	1.22	−1.07	−1.21
→OR (R: conjugated)	1.21	−0.60	−1.00
→OCOR	2.11	−0.35	−0.64
→CH$_2$—C=O; →CH$_2$—C≡N	0.69	−0.08	−0.06
→CH$_2$—aromatic ring	1.05	−0.29	−0.32
→F	1.54	−0.40	−1.02
→Cl	1.08	0.18	0.13
→Br	1.07	0.45	0.55
→I	1.14	0.81	0.88
→N—R (R: aliphatic)	0.80	−1.26	−1.21

TABLE 7.44 Estimation of Chemical Shift for Proton Attached to a Double Bond (*Continued*)

R	Z_{gem}, ppm	Z_{cis}, ppm	Z_{trans}, ppm
→N—R (R: conjugated)	1.17	−0.53	−0.99
→N—C=O	2.08	−0.57	−0.72
→aromatic	1.38	0.36	−0.07
→CF$_3$	0.66	0.61	0.32
→aromatic (*o*-substituted)	1.65	0.19	0.09
→SR	1.11	−0.29	−0.13
→SO$_2$	1.55	1.16	0.93

TABLE 7.45 Chemical Shifts in Monosubstituted Benzene

$$\delta = 7.27 + \Delta_i$$

Substituent	Δ_{ortho}	Δ_{meta}	Δ_{para}
NO$_2$	0.94	0.18	0.39
CHO	0.58	0.20	0.26
COOH	0.80	0.16	0.25
COOCH$_3$	0.71	0.08	0.20
COCl	0.82	0.21	0.35
CCl$_3$	0.8	0.2	0.2
COCH$_3$	0.62	0.10	0.25
CN	0.26	0.18	0.30
CONH$_2$	0.65	0.20	0.22
$\overset{+}{N}H_3$	0.4	0.2	0.2
CH$_2$X*	0.0–0.1	0.0–0.1	0.0–0.1
CH$_3$	−0.16	−0.09	−0.17
CH$_2$CH$_3$	−0.15	−0.06	−0.18
CH(CH$_3$)$_2$	−0.14	−0.09	−0.18
C(CH$_3$)$_2$	−0.09	0.05	−0.23
F	−0.30	−0.02	−0.23
Cl	0.01	−0.06	−0.08
Br	0.19	−0.12	−0.05
I	0.39	−0.25	−0.02
NH$_2$	−0.76	−0.25	−0.63
OCH$_3$	−0.46	−0.10	−0.41
OH	−0.49	−0.13	−0.2
OCOR	−0.2	0.1	−0.2
NHCH$_3$	−0.8	−0.3	−0.6
N(CH$_3$)$_2$	−0.60	−0.10	−0.62

* X = Cl, alkyl, OH, or NH$_2$.

TABLE 7.46   Proton Spin Coupling Constants

Structure	J, Hz	Structure	J, Hz
>C< with two H (geminal)	12–15	aziridine (N–H) cis	2
CH—CH (free rotation)	6–8	trans	6
>CH—OH (no exchange)	5	gem	4
>CH—NH	4–8	furan 2–3	1.8
CH—SH	6–8	3–4	3.5
CH—C=O (with H)	1–3	2–4	0–1
—N=C< (H, H)	8–16	2–5	1–2
C=C gem (H_t, H_g)	0–3	thiophene 2–3	5–6
cis	6–14	3–4	3.5–5.0
trans	11–18	2–4	1.5
H_c / CH cis	0.5–3	2–5	3.4
C=C trans	0.5–3	C6H5—F o	6–12
H_t / H_g gem	4–10	m	4–8
		p	1.5–2.5
>C=CH—CH=C<	10–13	C6H5—CH3 o	2.5
=CH—C=O (H)	6	—F m	1.5
—CH2—C≡C—CH<	0–3	p	0
>CH—C≡CH	0–3	cyclohexane a–a	8–10
C=C (ring) 3-member	0–2	a–e	2–3
4-member	2–4	e–e	2–3
5-member	5–7	Cyclopentane cis	4–6
6-member	6–9	trans	4–6
7-member	10–13	Cyclobutane cis	8
		trans	8
O (oxirane) cis	4–5	Cyclopropane cis	9–11
trans	3	trans	6–8
gem	5–6	gem	4–6
S (thiirane) cis	0	benzene (H) o	6–10
trans	7	m	1–3
gem	6	p	0–1
		naphthalene 1–2	8–9
		2–3	6
		pyridine 2–3	5–6
		3–4	7–9
		2–4	1–2
		3–5	1–2
		2–5	0–1
		2–6	0–1

TABLE 7.46 Proton Spin Coupling Constants (*Continued*)

Structure		J, Hz	Structure		J, Hz
pyrrole	1–2	2–3	H_t...H_g C=C / H$_c$...F	gem	72–90
	1–3	2–3		cis	−3 to 20
	2–3	2–3		trans	12–40
	3–4	3–4	F C=C / CH$_3$		2–4
	2–4	1–2			
	2–5	1–3			
			CF C=C / H		0–6
H C F		45–52	HC≡CF		21
			F_a F_e / H_e H_a	a–a	34
				a–e	12
CH—CF	gauche	0–12		e–e } e–e }	<5–8
	trans	10–45			

TABLE 7.47 Proton Chemical Shifts of Reference Compounds

Relative to tetramethylsilane.

Compound	δ, ppm	Solvent(s)
Sodium acetate	1.90	D_2O
1,2-Dibromoethane	3.63	$CDCl_3$
1,1,2,2-Tetrachloroethane	5.95	$CDCl_3$; CCl_4
1,4-Benzoquinone	6.78	$CDCl_3$; CCl_4
1,4-Dichlorobenzene	7.23	CCl_4
1,3,5-Trinitrobenzene	9.21	DMSO-d_6*
	9.55	$CHCl_3$

* DMSO, dimethyl sulfoxide.

TABLE 7.48 Solvent Positions of Residual Protons in Incompletely Deuterated Solvents

Relative to tetramethylsilane.

Solvent	Group	δ, ppm
Acetic-d_3 acid-d_1	Methyl	2.05
	Hydroxyl	11.5*
Acetone-d_6	Methyl	2.057
Acetonitrile-d_3	Methyl	1.95

* These values may vary greatly, depending upon the solute and its concentration.

TABLE 7.48 Solvent Positions of Residual Protons in Incompletely Deuterated Solvents (*Continued*)

Solvent	Group	δ, ppm
Benzene-d_6	Methine	6.78
tert-Butanol-d_1 (CH$_3$)$_3$COD	Methyl	1.28
Chloroform-d_1	Methine	7.25
Cyclohexane-d_{12}	Methylene	1.40
Deuterium oxide	Hydroxyl	4.7*
Dimethyl-d_6-formamide-d_1	Methyl	2.75; 2.95
	Formyl	8.05
Dimethyl-d_6 sulfoxide	Methyl	2.51
	Absorbed water	3.3*
1,4-Dioxane-d_8	Methylene	3.55
Hexamethyl-d_{18}-phosphoramide	Methyl	2.60
Methanol-d_4	Methyl	3.35
	Hydroxyl	4.8*
Dichloromethane-d_2	Methylene	5.35
Pyridine-d_5	C-2 Methine	8.5
	C-3 Methine	7.0
	C-4 Methine	7.35
Toluene-d_8	Methyl	2.3
	Methine	7.2
Trifluoroacetic acid-d_1	Hydroxyl	11.3*

* These values may vary greatly, depending upon the solute and its concentration.

TABLE 7.49 Carbon-13 Chemical Shifts

Values given in ppm on the δ scale, relative to tetramethylsilane.

Substituent group	Primary carbon	Secondary carbon	Tertiary carbon	Quaternary carbon
Alkanes				
C—C	5–30	25–45	23–58	28–50
C—O	45–60	42–71	62–78	73–86
C—N	13–45	44–58	50–70	60–75
C—S	10–30	22–42	55–67	53–62
C—halide (I to Cl)	3–25	3–40	34–58	35–75

Substituent group	δ, ppm	Substituent group	δ, ppm
Cyclopropane	−5–5	Alcohols R—OH	45–87
Cycloalkane C$_4$–C$_{10}$	5–25	Ethers R—O—R	57–87
Mercaptanes	5–70	Nitro R—NO$_2$	60–78
Amines:		Alkynes:	
R$_2$N—C	20–70	HC≡CR	63–73
Aryl—N	128–138	RC≡CR	72–95
Sulfoxides, sulfones	35–55	Acetals, ketals	88–112

TABLE 7.49 Carbon-13 Chemical Shifts (*Continued*)

Substituent group	δ, ppm	Substituent group	δ, ppm
Thiocyanates R—SCN	96–118	Esters:	
Alkenes:		Saturated	158–165
H$_2$C=	100–122	α,β-Unsaturated	165–176
R$_2$C=	110–150	Isocyanides R—NC	162–175
Heteroaromatics:		Carboxylic acids:	
C=N	100–152	Nonconjugated	162–165
C$_\alpha$	142–160	Conjugated	165–184
Cyanates R—OCN	105–120	Salts (anion)	175–195
Isocyanates R—NCO	115–135	Ketones:	
Isothiocyanates R—NCS	115–142	α-Halo	160–200
Nitriles, cyanides	117–124	Nonconjugated	192–202
Aromatics:		α,β-Unsaturated	202–220
Aryl-C	125–145	Imides	165–180
Aryl-P	119–128	Acyl chlorides R—CO—Cl	165–183
Aryl-N	128–138	Thioureas	165–185
Aryl-O	133–152	Aldehydes:	
Azomethines	145–162	α-Halo	170–190
Carbonates	159–162	Nonconjugated	182–192
Ureas	150–170	Conjugated	192–208
Anhydrides	150–175	Thioketones R—CS—R	190–202
Amides	154–178	Carbonyl M(CO)$_n$	190–218
Oximes	155–165	Allenes =C=	197–205

Saturated heterocyclic ring systems

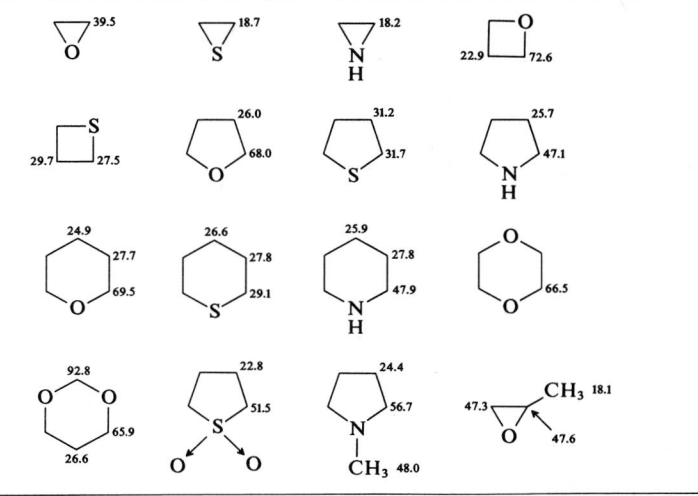

TABLE 7.49 Carbon-13 Chemical Shifts (*Continued*)

Unsaturated cyclic systems

TABLE 7.49 Carbon-13 Chemical Shifts (*Continued*)

Saturated alicyclic ring systems

TABLE 7.50 Estimation of Chemical Shifts of Alkane Carbons

Relative to tetramethylsilane.

Positive terms indicate a downfield shift.

$$\delta_C = -2.6 + 9.1n_\alpha + 9.4n_\beta - 2.5n_\gamma + 0.3n_\delta + 0.1n_\epsilon \quad \text{(plus any correction factors)}$$

where n_α is the number of carbons bonded directly to the ith carbon atom and n_β, n_γ, n_δ, and n_ϵ are the number of carbon atoms two, three, four, and five bonds removed. The constant is the chemical shift for methane.

Chain branching*	Correction factor	Chain branching*	Correction factor
1°(3°)	−1.1	4°(1°)	−1.5
1°(4°)	−3.4	2°(4°)	−7.2
2°(3°)	−2.5	3°(3°)	−9.5
3°(2°)	−3.7	4°(2°)	−8.4

* 1° signifies a CH_3— group; 2°, a —CH_2— group; 3°, a $\gtrsim CH$— group; and 4°, a $\gtrsim C \lesssim$ group. 1° (3°) signifies a methyl group bound to a $\gtrsim CH$— group, and so on.

Examples: For 3-methylpentane, CH_3—CH_2—$CH(CH_3)$—CH_2—CH_3,

$$\delta_{C-2} = -2.6 + 9.1(2) + 9.4(2) - 2.5 - 1(1)[2°(3°)] = 29.4$$

$$\delta_{C-3} = -2.6 + 9.1(3) + 9.4(2) + (2)[3°(2°)] = 36.2$$

TABLE 7.51   Effect of Substituent Groups on Alkyl Chemical Shifts

These increments are added to the shift value of the appropriate carbon atom as calculated from Table 7.50.

$$\text{Straight:}\quad Y-CH_2-CH_2-CH_3 \qquad \text{Branched:}\quad -CH_2-CH_2-\overset{\overset{\displaystyle Y}{|}}{CH}-CH_2-CH_2-$$
$$\qquad\qquad\qquad \alpha\quad\ \beta \qquad\qquad\qquad\qquad\quad \gamma\quad\ \beta\quad\ \alpha\quad\ \beta\quad\ \gamma$$

Substituent group Y*	α carbon		β carbon		γ carbon
	Straight	Branched	Straight	Branched	
—CO—OH	20.9	16	2.5	2	−2.2
—COO⁻ (anion)	24.4	20	4.1	3	−1.6
—CO—OR	20.5	17	2.5	2	−2
—CO—Cl	33	28		2	
—CO—NH₂	22	2.5			−0.5
—CHO	31		0		−2
—CO—R	30	24	1	1	−2
—OH	48.3	40.8	10.2	7.7	−5.8
—OR	58	51	8	5	−4
—O—CO—NH₂	51		8		
—O—CO—R	51	45	6	5	−3
—C—CO—Ar	53				
—F	68	63	9	6	−4
—Cl	31.2	32	10.5	10	−4.6
—Br	20.0	25	10.6	10	−3.1
—I	−8	4	11.3	12	−1.0
—NH₂	29.3	24	11.3	10	−4.6
—NH₃⁺	26	24	8	6	−5
—NHR	36.9	31	8.3	6	−3.5
—NR₂	42		6		−3
—NR₃⁺	31		5		−7
—NO₂	63	57	4	4	
—CN	4	1	3	3	−3
—SH	11	11	12	11	−6
—SR	20		7		−3
—CH=CH₂	20		6		−0.5
—C₆H₅	23	17	9	7	−2
—C≡CH	4.5		5.5		−3.5

* R, alkyl group; Ar, aryl group.

TABLE 7.52 Estimation of Chemical Shifts of Carbon Attached to a Double Bond

The olefinic carbon chemical shift is calculated from the equation

$$\delta_C = 123.3 + 10.6n_\alpha + 7.2n_\beta - 7.9n_{\alpha'} - 1.8n_{\beta'} \quad \text{(plus any steric correction terms)}$$

where n is the number of carbon atoms at the particular position, namely,

$$\begin{array}{cccc} \beta & \alpha & \alpha' & \beta' \\ \text{C} & \text{C} & \text{C} & \text{C} \end{array}$$
$$\text{C—C=C—C}$$

Substituents on both sides of the double bond are considered separately. Additional vinyl carbons are treated as if they were alkyl carbons. The method is applicable to alicyclic alkenes; in small rings carbons are counted twice, i.e., from both sides of the double bond where applicable. The constant in the equation is the chemical shift for ethylene. The effect of other substituent groups is tabulated below.

Substituent group	β	α	α'	β'
—OR	2	29	−39	−1
—OH	6			−1
—O—CO—CH$_3$	−3	18	−27	4
—CO—CH$_3$		15	6	
—CHO		13.6	13.2	
—CO—OH		5.2	9.1	
—CO—OR		6	7	
—CN		−15.4	14.3	
—F		24.9	−34.3	
—Cl	−1	3.3	−5.4	2
—Br	0	−7.2	−0.7	2
—I		−37.4	7.7	
—C$_6$H$_5$		12	−11	

Substituent pair		Steric correction term
α,α'	trans	0
α,α'	cis	−1.1
α,α	gem	−4.8
α',α'		+2.5
β,β		+2.3

TABLE 7.53 Carbon-13 Chemical Shifts in Substituted Benzenes

$$\delta_C = 128.5 + \Delta$$

Substituent group	Δ_{C-1}	Δ_{ortho}	Δ_{meta}	Δ_{para}
—CH$_3$	9.3	0.8	−0.1	−2.9
—CH$_2$CH$_3$	15.6	−0.4	0	−2.6
—CH(CH$_3$)$_2$	20.2	−2.5	0.1	−2.4
—C(CH$_3$)$_3$	22.4	−3.1	−0.1	−2.9
—CH$_2$O—CO—CH$_3$	7.7	0	0	0
—C$_6$H$_5$	13.1	−1.1	0.4	−1.2
—CH=CH$_2$	9.5	−2.0	0.2	−0.5
—C≡CH	−6.1	3.8	0.4	−0.2
—CH$_2$OH	12.3	−1.4	−1.4	−1.4
—CO—OH	2.1	1.5	0	5.1
—COO$^-$ (anion)	8	1	0	3
—CO—OCH$_3$	2.1	1.1	0.1	4.5
—CO—CH$_3$	9.1	0.1	0	4.2
—CHO	8.6	1.3	0.6	5.5
—CO—Cl	4.6	2.4	1	6.2
—CO—CF$_3$	−5.6	1.8	0.7	6.7
—CO—C$_6$H$_5$	9.4	1.7	−0.2	3.6
—CN	−15.4	3.6	0.6	3.9
—OH	26.9	−12.7	1.4	−7.3
—OCH$_3$	31.4	−14.0	1.0	−7.7
—OC$_6$H$_5$	29.2	−9.4	1.6	−5.1
—O—CO—CH$_3$	23.0	−6.4	1.3	−2.3
—NH$_2$	18.0	−13.3	0.9	−9.8
—N(CH$_3$)$_2$	22.4	−15.7	0.8	−11.5
—N(C$_6$H$_5$)$_2$	19	−4	1	−6
—NHC$_6$H$_5$	14.6	−10.7	0.7	−7.7
—NH—CO—CH$_3$	11.1	−9.9	0.2	−5.6
—NO$_2$	20.0	−4.8	0.9	5.8
—F	34.8	−12.9	1.4	−4.5
—Cl	6.2	0.4	1.3	−1.9
—Br	−5.5	3.4	1.7	−1.6
—I	−32.2	9.9	2.6	−1.4
—CF$_3$	−9.0	−2.2	0.3	3.2
—NCO	5.7	−3.6	1.2	−2.8
—SH	2.3	1.1	1.1	−3.1
—SCH$_3$	10.2	−1.8	0.4	−3.6
—SO$_2$—NH$_2$	15.3	−2.9	0.4	3.3
—Si(CH$_3$)$_3$	13.4	4.4	−1.1	−1.1

TABLE 7.54 Carbon-13 Chemical Shifts in Substituted Pyridines*

$$\delta_C(k) = C_k + \Delta_i$$

Substituent group	$C_2 = C_6 = 149.6$ Δ_{C-2} or Δ_{C-6}	Δ_{23}	Δ_{24}	Δ_{25}	Δ_{26}
—CH$_3$	9.1	−1.0	−0.1	−3.4	−0.1
—CH$_2$CH$_3$	14.0	−2.1	0.1	−3.1	0.2
—CO—CH$_3$	4.3	−2.8	0.7	3.0	−0.2
—CHO	3.5	−2.6	1.3	4.1	0.7
—OH	14.9	−17.2	0.4	−3.1	−6.8
—OCH$_3$	15.3	−13.1	2.1	−7.5	−2.2
—NH$_2$	11.3	−14.7	2.3	10.6	−0.9
—NO$_2$	8.0	−5.1	5.5	6.6	0.4
—CN	−15.8	5.0	−1.7	3.6	1.9
—F	14.4	−14.7	5.1	−2.7	−1.7
—Cl	2.3	0.7	3.3	−1.2	0.6
—Br	−6.7	4.8	3.3	−0.5	1.4

Substituent group	Δ_{32}	$C_3 = C_5 = 124.2$ Δ_{C-3} or Δ_{C-5}	Δ_{34}	Δ_{35}	
—CH$_3$	1.3	9.0	0.2	−0.8	−2.3
—CH$_2$CH$_3$	0.3	15.0	−1.5	−0.3	−1.8
—CO—CH$_3$	0.5	−0.3	−3.7	−2.7	4.2
—CH0	2.4	7.9	0	0.6	5.4
—OH	−10.7	31.4	−12.2	1.3	−8.6
—NH$_2$	−11.9	21.5	−14.2	0.9	−10.8
—CN	3.6	−13.7	4.4	0.6	4.2
—Cl	−0.3	8.2	−0.2	0.7	−1.4
—Br	2.1	−2.6	2.9	1.2	−0.9
—I	7.1	−28.4	9.1	2.4	0.3

Substituent group	$\Delta_{42} = \Delta_{46}$	$\Delta_{43} = \Delta_{45}$	$C_4 = 136.2$ Δ_{C-4}
—CH$_3$	0.5	0.8	10.8
—CH$_2$CH$_3$	0	−0.3	15.9
—CH=CH$_2$	0.3	−2.9	8.6
—CO—CH$_3$	1.6	−2.6	6.8
—CHO	1.7	−0.6	5.5
—NH$_2$	0.9	−13.8	19.6
—CN	2.1	2.2	−15.7
—Br	3.0	3.4	−3.0

* May be used for disubstituted, polyheterocyclic, and polynuclear systems if deviations due to steric and mesomeric effects are allowed for.

TABLE 7.55 Carbon-13 Chemical Shifts of Carbonyl Group

$$\underset{\underset{\displaystyle X-C-Y}{\|}}{O}$$

X	Y	δ_C	X	Y	δ_C
H—	—CH$_3$	199.7	CH$_3$—	—CH=CH$_2$	196.9
H—	—CCl$_3$	175.3	CH$_3$—	—C$_6$H$_5$	197.6
H—	—NH$_2$	165.5	CH$_3$—	—CH$_2$—CO—CH$_3$	201.9
					(keto)
H—	—N(CH$_3$)$_2$	162.4			191.4
					(enol)
H—	2-Furyl	153.3	CH$_3$—	—CH$_2$CHO	167.7
H—	2-Pyrrolyl	134.0	CH$_3$—	—C$_6$H$_5$—CH$_3$	196 (m, p)
H—	2-Thienyl	143.3			199 (o)
(CH$_3$)$_2$CH—	—OH	184.8	CH$_3$—	—2,6-(CH$_3$)$_2$C$_6$H$_5$	206
C$_6$H$_5$—	—OH	172.6	CH$_3$—	—OH	178
CF$_3$—	—OH	163.0	CH$_3$—	—O$^-$ (anion)	181.5
CCl$_3$—	—OH	168.0	CH$_3$—	—OCH$_3$	170.7
CH$_3$CH(NH$_2$)—	—OH	176.5	CH$_3$—	—O—CH=CH$_2$	167.7
CF$_3$—	—OCH$_2$CH$_3$	158.1	CH$_3$—	—O—CH(CH$_3$)$_2$	170.3
H$_2$N—	—OCH$_2$CH$_3$	157.8	CH$_3$—	—O—CO—CH$_3$	167.3
2-Furyl	—OCH$_3$	159.1	CH$_3$—	—NH$_2$	172.7
(CH$_3$)$_2$N—	—C$_6$H$_5$	170.8	CH$_3$—	—NHCH$_3$	172
CH$_2$=CHCH$_2$O— CO—	—OCH$_2$CH=CH$_2$	157.6	CH$_3$—	—N(CH$_3$)$_2$	169.5
CH$_3$CH$_2$—	—CH$_2$CH$_3$	211.4	CH$_3$—	—Cl	169.6
CH$_3$—CH$_2$—	—O—CO—CH$_2$CH$_3$	170.3	CH$_3$—	—Br	165.6
CH$_3$—	—CH$_3$	205.8	CH$_3$—	—I	158.9
CH$_3$—	—CH$_2$CH$_3$	207			

(CH$_2$)$_n$ C=O	
n	δ_C
3	207.9
4	218.2
5	211.3
6	211.4
7	216.0

TABLE 7.56 One-Bond Carbon-Hydrogen Spin Coupling Constants

Structure	J_{CH}, Hz	Structure	J_{CH}, Hz
$H-CH_3$	125.0	H_t H_g gem	177
$H-CH_2CH_3$	124.9	C=C cis	163
$CH_3-\underline{CH_2}-CH_3$	119.2	H_c CN trans	165
$H-C(CH_3)_2$	114.2		
$H-CH_2CH_2OH$	126.9	H OH cis	163
$H-CH_2CH=CH_2$	122.4	C=N trans	177
$H-CH_2C_6H_5$	129.4	CH_3	
$H-CH_2C\equiv CH$	132.0		
$H-CH_2CN$	136.1	$H-CH=O$; $CH_3-\underline{CH}=O$	172
$H-CH(CN)_2$	145.2	$H_2N-CH=O$	188.3
$H-CH_2-halogen$	149–152	$(CH_3)_2N-\underline{CH}=O$	191
$H-CHF_2$	184.5	$H-COOH$	222
$H-CHCl_2$	178.0	$H-COO^-$ (anion)	195
$H-CH_2NH_2$	133.0	$H-CO-OCH_3$	226
$H-CH_2NH_3^+$	145.0	$H-CO-F$	267
$H-CH_2OH$ (or $H-CH_2OR$)	140–141	$CH_3CH_2-O-\underline{CHO}$	225.6
$H-CH(OR)_2$	161–162	Cl_3-CHO	207
$H-C(OR)_3$	186	$H-C\equiv CH$	249
$H-C(OH)R_2$	143	$H-C\equiv CCH_3$	248
$H-CH_2NO_2$	146.0	$H-C\equiv CC_6H_5$	251
$H-CH(NO_2)_2$	169.4	$H-C\equiv CCH_2OH$	241
$H-CH_2COOH$	130.0	$H-CN$	269
$H-CH(COOH)_2$	132.0	Cyclopropane	161
$H-CH=CH_2$	156.2	Cyclobutane	136
$H-C(CH_3)=C(CH_3)_2$	148.4	Cyclopentane	131
$H-CH=C(tert-C_4H_9)_2$	152	Cyclohexane	123
$H-C(tert-C_4H_9)=$		Tetrahydrofuran 2,5	149
$C(tert-C_4H_9)_2$	143	3,4	133
Methylenecycloalkane		1,4-Dioxane	145
C_4-C_7	153–155	Benzene	159
$H-CH=C=CH_2$	168	Fluorobenzene 2,6	155
$H-C(C_6H_5)=CH(C_6H_5)$		3,5	163
cis	155	4	161
trans	151	Bromobenzene 2,6	171
Cyclopropene	220	3,5	164
		4	161
H_t H_g gem	200	Benzonitrile 2,6	173
C=C cis	159	3,6	166
H_c F trans	162	4	163
		Nitrobenzene 2,6	171
H_t H_g gem	195	3,5	167
C=C cis	163	4	163
H_c Cl trans	161	Mesitylene	154
H_t H_g gem	162	2,6	170
C=C cis	157	3,5	163
H_c CHO trans	162	4	152

TABLE 7.56 One-Bond Carbon-Hydrogen Spin Coupling Constants (*Continued*)

Structure	J_{CH}, Hz	Structure	J_{CH}, Hz
2,4,6-Trimethylpyridine	158	(imidazole) 2	208
(pyrrole) 2,5	183	4	199
3,4	170		
(furan) 2,5	201	(1,2,3-triazole)	205
3,4	175		
(thiophene) 2,5	185	(1,2,4-triazole)	216
3,4	167		
(pyrazole) 3,5	190		
4	178		

TABLE 7.57 Two-Bond Carbon-Hydrogen Spin Coupling Constants

Structure	$^2J_{CH}$, Hz	Structure	$^2J_{CH}$, Hz
CH_3-CH_2-H	-4.5	$(CH_2)_n$ $C=CH_2$ $n=4$	4.2
CCl_3-CH_2-H	5.9	$n=5$	5.2
$ClCH_2-CH_2Cl$	-3.4	$n=6$	5.5
$Cl_2CH-CHCl_2$	1.2	(HC=CH, Cl,Cl) *cis*	16.0
CH_3-CHO	26.7	*trans*	0.8
$CH_2=CH_2$	-2.4		
$(CH_3)_2C=O$	5.5	$HC\equiv CH$	49.3
$CH_2=CH-CH=O$	26.9	$C_6H_5O-C\equiv CH$	61.0
$(C_2H_5)_2CH-CHO$	26.9	$HC\equiv C-CHO$	33.2
$H_2NCH=CH-CHO$	6.0	$ClCH_2-CHO$	32.5
$H_2NCH-CH-CHO$	20.0	$Cl_2CH-CHO$	35.3
C_6H_6	1.0	Cl_3C-CHO	46.3
		$C_6H_5-C\equiv C-CH_3$	10.8

TABLE 7.58 Carbon-Carbon Spin Coupling Constants

Structure*	J_{CC}, Hz	Structure	J_{CC}, Hz
H_3C-CH_3	35	$H_3C-CH_2NH_2$	37
H_3C-CHR_2	37	$C-C=O$	38–40
H_3C-CH_2Ar	34	$C-C-C=O$	36
H_3C-CH_2CN	33	$C-C-Ar$	43
$H_3C-CH_2-CH_2OH$		$C-CO-O^-$ (anion)	52
C-1, C-2	38	$C-CO-N$	52
C-2, C-3	34	$C-CO-OH$	57

* R, alkyl group; Ar, aryl group.

TABLE 7.58 Carbon-Carbon Spin Coupling Constants (*Continued*)

Structure*	J_{CC}, Hz	Structure	J_{CC}, Hz
C—CO—OR	59	$C_6H_5NH_2$	
C—CN	52–57	1-2	61
C—C≡C $^2J_{CC} = 11.8$	67	2-3	58
$H_2C=CH_2$	68	3-4	57
$>C=C—CO—OH$	70–71	$^3J_{2-5}$	7.9
$>C=C—CN$	71	$C_6H_5CH_3$	44
$>C=C—Ar$	67–70	Pyridine	
C_6H_6	57	2-3	54
$C_6H_5NO_2$		3-4	56
1-2	55	$^3J_{2-5}$	14
2-3, 3-4	56	Furan	69
$^3J_{2-5}$	7.6	Pyrrole	69
C_6H_5I		Thiophene	64
1-2	60	$H_2C=C=C(CH_3)_2$	100
2-3	53	—C≡C—	170–176
3-4	58		
$^3J_{2-5}$	8.6	Structure	$^2J_{CC}$, Hz
$C_6H_5—OCH_3$		$\underset{.}{C}H_3—CO—\underset{.}{C}H_3$	16
2-3	58	$\underset{.}{C}H_3—C≡\underset{.}{C}H$	11.8
3-4	56	$\underset{.}{C}H_3CH_2—\underset{.}{C}N$	33

* R, alkyl group; Ar, aryl group.

TABLE 7.59 Carbon-Fluorine Spin Coupling Constants

Structure*	J_{CF}, Hz	Structure*	J_{CF}, Hz
F, H, H, H on C	−158	F, F, F, CH₃ on C	−271
F, H, F, H on C	−235	F, H, H, Ar on C	−165
F, F, F, H on C	−274	F—CH₂CH₂— or F—CR₃	−167
		p-F—C_6H_4—OR	−237
		p-F—C_6H_4—R	−241
		p-F—C_6H_4—CF₃	−252
F, F, F, F on C	−259	p-F—C_6H_4—CO—CH₃	−253
		p-F—C_6H_4—NO₂	−257
		F—C_6H_5 $^2J_{CF} = 21.0$ $^3J_{CF} = 7.7$ $^4J_{CF} = 3.4$	−244

* Ar, aryl group; R, alkyl group.

TABLE 7.59 Carbon-Fluorine Spin Coupling Constants (*Continued*)

Structure*	J_{CF}, Hz	Structure*	J_{CF}, Hz
F₂C=CH₂ (F\C(/F)=CH₂)	−287	CHF with F, H and CH₂OH (F\C(H)(/F)CH₂OH)	−241
F₂C=O	−308	CF₂(CH₂OH) (F\C(/F)CH₂OH)	−278
(F)(R)C=O	−353	CF₂(OCF₃) (F\C(/F)OCF₃)	−265
(F)(H)C=O	−369	CF₂(CO—CH₃) (F\C(/F)CO—CH₃)	−289

* Ar, aryl group; R, alkyl group.

TABLE 7.60 Carbon-13 Chemical Shifts of Deuterated Solvents
Relative to tetramethylsilane.

Solvent	Group	δ, ppm
Acetic-d_3 acid-d_1	Methyl	20.0
	Carbonyl	205.8
Acetone-d_6	Methyl	28.1
	Carbonyl	178.4
Acetonitrile-d_3	Methyl	1.3
	Carbonyl	117.7
Benzene-d_6		128.5
Carbon disulfide		193
Carbon tetrachloride		97
Chloroform-d_1		77
Cyclohexane-d_{12}		25.2
Dimethyl sulfoxide-d_6		39.5
1,4-Dioxane-d_6		67
Formic-d_1 acid-d_1	Carbonyl	165.5
Methanol-d_4		47–49
Methylene chloride-d_2		53.8
Nitromethane-d_3		57.3
Pyridine-d_5	C_3, C_5	123.5
	C_4	135.5
	C_2, C_6	149.9

TABLE 7.61 Carbon-13 Spin Coupling Constants with Various Nuclei

Nuclei	Structure	^1J, Hz	^2J, Hz	^3J, Hz	^4J, Hz
^{2}H	$CDCl_3$	32			
	$CD_3-CO-CD_3$	20			
	$(CD_3)_2SO$	22			
	C_6D_6	26			
^{7}Li	CH_3Li	15			
^{11}B	$(C_6H_5)_4B^-$	49		3	
^{14}N	$(CH_3)_4N^+$	10			
	CH_3NC	8			
^{29}Si	$(CH_3)_4Si$	52			
^{31}P	$(CH_3)_3P$	14			
	$(C_4H_9)_3P$	11	12	5	
	$(C_6H_5)_3P$	12	20	7	0
	$(CH_3)_4P^+$	56			
	$(C_4H_9)_4P^+$	48	4	15	
	$(C_6H_5)_4P^+$	88	11	13	3
	$R(RO)_2P=O$	142	5–7		
	$(C_4H_9O)_3P=O$		6	7	
^{77}Se	$(CH_3)_2Se$	62			
	$(CH_3)_3Se^+$	50			
^{113}Cd	$(CH_3)_2Cd$	513, 537			
^{119}Sn	$(CH_3)_4Sn$	340			
	$(CH_3)_3SnC_6H_5$	474	37	47	11
^{125}Te	$(CH_3)_2Te$	162			
^{199}Hg	$(CH_3)_2Hg$	687			
	$(C_6H_5)_2Hg$	1186	88	102	18
^{207}Pb	$(CH_3)_2Pb$	250			
	$(C_6H_5)_4Pb$	481	68	81	20

TABLE 7.62 Boron-11 Chemical Shifts

Values given in ppm on the δ scale, relative to $B(OCH_3)_3$.

Structure	δ, ppm	Structure	δ, ppm
R_3B	−67 to −68	$C_6H_5BCl_2$	−36
Ar_3B	−43	$C_6H_5B(OH)_2$	−14
BF_3	24	$C_6H_5B(OR)_2$	−10
BCl_3	−12	$M(BH_4)$	55–61
BBr_3	−6	$B(BF_4)$	19–20
BI_3	41		
$B(OH)_3$	36		
$B(OR)_3$	0–1	NH—BH / HB / NH / NH—BH	−12
$B(NR_2)_3$	−13		

TABLE 7.62 Boron-11 Chemical Shifts (*Continued*)

Structure	δ, ppm	Structure	δ, ppm
[structure: H₂B–N(R₂)–BH₂ bridged]	37	R_2O(or ROH)·BCl_3 R_2O(or ROH)·BBr_3 R_2O(or ROH)·BI_3	−7 to −8 23–24 74–82
[structure: HB(NR₂)₂BH bridged]	15	[pyridine]$N·BBr$,	24
$(CH_3)_2N-B(CH_3)_2$	62	**Boranes** B_2H_6 B_4H_{10}	1
Addition complexes $R_2O·BH_3$ $R_3N·BH_3$ $R_2NH·BH_3$	18–19 25 33	(BH_2) (BH)	25 60

			Base	Apex
[pyridine]$N·BH_3$	31			
		B_5H_9	31	70
		B_5H_{11}	−16	50
R_2O(or ROH)·BF_3	17–19	$B_{10}H_{14}$	7	54

TABLE 7.63 Nitrogen-15 (or Nitrogen-14) Chemical Shifts
Values given in ppm on the δ scale, relative to NH_3 liquid.

Substituent group	δ, ppm	Substituent group	δ, ppm
Aliphatic amines		Ureas	
Primary	1–59	Aliphatic	63–84
Secondary	7–81	Aryl	105–108
Tertiary	14–44	Sulfonamides	79–164
Cyclo, primary	29–44	Amides	
Aryl amines	40–100	HCO—NHR	
Aryl hydrazines	40–100	R = primary	100–115
Piperidines,		R = secondary	104–148
decahydroquinolines	30–82	R = tertiary	96–133
Amine cations		HCO—NH—Aryl	138–141
Primary	19–59	RCO—NHR or	103–130
Secondary	40–74	RCO—NR₂	
Tertiary	30–67	RCO—NH—Aryl	131–136
Quaternary	43–70	Aryl—CO—H—Aryl	ca 126
Enamines, tertiary type		Guanidines	
Alkyl	29–82	Amino	30–60
Cycloalkyl	55–104	Imino	166–207
Aminophosphines	59–100	Thioureas	85–111
Amine *N*-oxides	95–122	Thioamides	135–154

TABLE 7.63 Nitrogen-15 (or Nitrogen-14) Chemical Shifts (*Continued*)

Substituent group	δ, ppm	Substituent group	δ, ppm
Cyanamides		Diazo	
R$_2$N—	−12 to −38	Internal	226–303
—CN	175–200	Terminal	315–440
Carbodiimides	95–120	Nitrilium ions	123–150
Isocyanates		Azinium ions	185–220
Alkyl, primary	14–32	Azine *N*-oxides	230–300
Alkyl, secondary and tertiary	54–57	Nitrones	270–285
Aryl	ca 46	Imides	170–178
Isothiocyanates	90–107	Imimes	310–359
Azides	52–80	Oximes	340–380
	108–122	Nitramines	
	240–260	Amine	252–280
Lactams	113–122	—NO$_2$	328–355
Hydrazones		Nitrates	310–353
Amino	141–167	*gem*-Polynitroalkanes	310–353
Imino	319–327	Nitro	
Cyanates	155–182	Aryl	350–382
Nitrile *N*-oxides, fulminates	195–225	Alkyl	372–410
Isonitriles		Hetero, unsaturated	354–367
Alkyl, primary	162–178	Azoxy	330–356
Alkyl, secondary	191–199	Azo	504–570
Aryl	ca 180	Nitrosamines	222–250
Nitriles			525–550
Alkyl	235–241	Nitrites	555–582
Aryl	258–268	Thionitrites	720–790
Thiocyanates	265–280	Nitroso	
Diazonium		Aliphatic amines, NO	535–560
Internal	222–230	Aryl	804–913
Terminal	315–322		

TABLE 7.63 Nitrogen-15 (or Nitrogen-14) Chemical Shifts (*Continued*)

Substituent group	δ, ppm	Substituent group	δ, ppm
Saturated cyclic systems			

$(CH_2)_n$ N—H

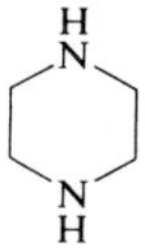

$n = 2$	−8.5
$n = 3$	25.3
$n = 4$	36.7
$n = 5$	37.7

Piperazine: 35.5

Quinuclidine: 7.5 (in C_6H_6), 18.0 (in H_2O)

Morpholine: 32.1

Decahydroquinoline:
cis	42.4
trans	52.9

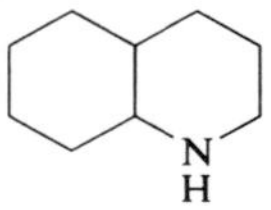

Unsaturated cyclic systems

Pyrrole: N 145 (H)

Pyrazole: N 245 (H)

Imidazole: N 205, N (H)

Triazole: N, N 240 (H)

Triazole: N 316, N 316, 244 N—H

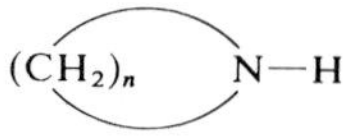

Oxazole: N 251

Isoxazole: N 383

Thiazole: N 323

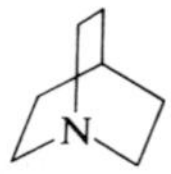

Isothiazole: N 298

270 N—N 351, N 207, 351 N—H

Pyridine: N 317

Pyridazine: N, N 396

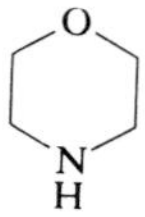

Pyrimidine: N 295

Pyrazine: 331

Triazine: N, N, N 383

Tetrazine: N, N, N, N 381

Indole: N 133 (H)

Benzimidazole: N 191, N (H)

Indazole: N 301, N 179 (H)

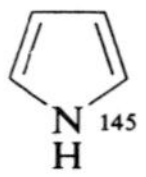

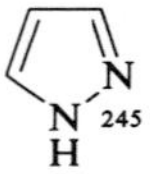

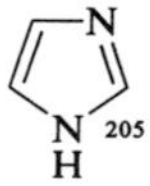

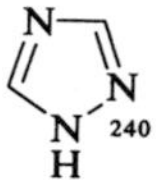

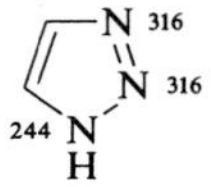

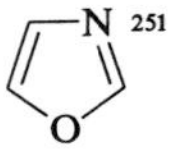

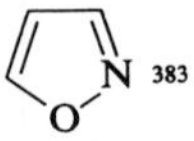

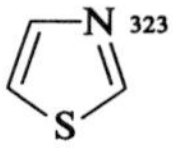

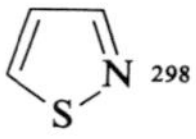

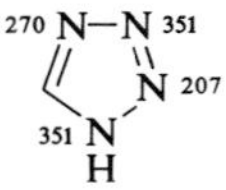

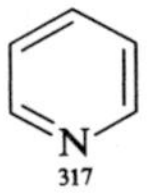

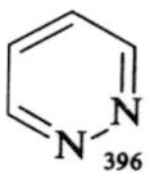

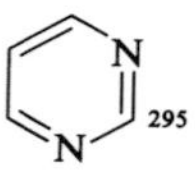

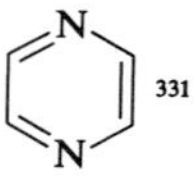

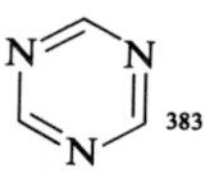

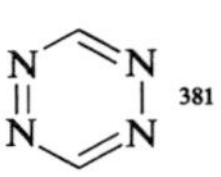

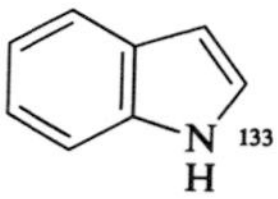

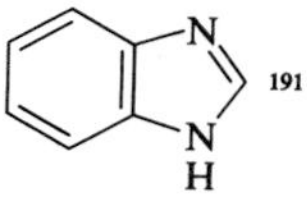

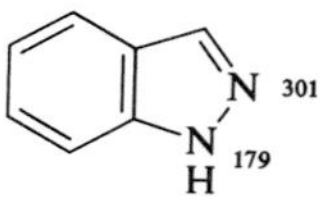

TABLE 7.63 Nitrogen-15 (or Nitrogen-14) Chemical Shifts (*Continued*)

Unsaturated cyclic systems (*continued*)

X	δ, ppm
O	517
S	331
Se	373

(Structures with chemical shift values, ppm)

291, 236, 316, 257, 516, 517, 331, 373, 316, 316, 293, 281, 330, 412, 361, 283, 114, 308, 280, 261, 301, 399, 351

TABLE 7.64 Nitrogen-15 Chemical Shifts in Monosubstituted Pyridine

$$\delta = 317.3 + \Delta_i$$

Substituent	$\Delta_{C\text{-}2}$	$\Delta_{C\text{-}3}$	$\Delta_{C\text{-}4}$
—CH$_3$	−0.4	0.3	−8.0
—CH$_2$CH$_3$	−1.8		−6.6
—CH(CH$_3$)$_2$	−5.1		−5.9
—C(CH$_3$)$_3$	−2.5		−5.8
—CN	−0.9	−0.8	10.6
—CHO	10	11	29
—CO—CH$_3$	−9	15	11
—CO—OCH$_2$CH$_3$	11.8		−5
—OCH$_3$	−49	0	−23
—OH	−126	−2	−118
—NO$_2$	−23	1	22
—NH$_2$	−45	10	−46
—F	−42	−18	
—Cl	−4	4	−6
—Br	2	8	7

TABLE 7.65 Nitrogen-15 Chemical Shifts for Standards

Values given in ppm, relative to NH_3 *liquid at 23°C.*

Substance	δ, ppm	Conditions
Nitromethane (neat)	380.2	For organic solvents and acidic aqueous solutions
Potassium (or sodium) nitrate (saturated aqueous solution)	376.5	For neutral and basic aqueous solutions
$C(NO_2)_4$	331	For nitro compounds
$(CH_3)_2$—CHO (neat)	103.8	For organic solvents and aqueous solutions
$(C_2H_5)_4N^+Cl^-$	64.4	Saturated aqueous solution
$(CH_3)_4N^+Cl^-$	43.5	Saturated aqueous solution
NH_4Cl	27.3	Saturated aqueous solution
NH_4NO_3	20.7	Saturated aqueous solution
NH_3	0.0	Liquid, 25°C
	−15.9	Vapor, 5 atm

TABLE 7.66 Nitrogen-15 to Hydrogen-1 Spin Coupling Constants

Structure	J, Hz	Structure	J, Hz
R—NH_2 and R_2NH	61–67	$\begin{array}{c} O \\ \diagup\diagdown \\ C-N \\ \diagup\quad\diagdown \\ R\qquad R \end{array}$ (with H)	88–92
Aryl—NH_2	78		
p-CH_3O—aryl—NH_2	79		
p-O_2N—aryl—NH_2	90–93		
Amine salts (alkyl and aryl)	73–76	Pyrrole	97
Aryl—NHOH	79	$HC\equiv NH^+$	133–136
Aryl—$NHCH_3$	87	$>$P—NH_2	82–90
Aryl—$NHCH_2F$	90	$(R_3Si)_2NH$	67
Aryl—$NHNH_2$	90	CF_3—S—NH_2	81
p-O_2N—aryl—$NHNH_2$	99	$(CF_3$—S$)_2NH$	99
Aryl—SO_2—NH_2	81	Pyridinium ion	90
Aryl—SO_2—NHR	86	Quinolinium ion	96
$\begin{array}{c} O \qquad H_{syn} \text{ (to —CO—)} \\ \diagdown\quad\diagup \\ C-N \\ \diagup\quad\diagdown \\ H \qquad H_{anti} \end{array}$	88 / 92–93		

TABLE 7.67 Nitrogen-15 to Carbon-13 Spin Coupling Constants

Structure	J, Hz	Structure	J, Hz
Alkyl amines	4–4.5	Alkyl—NO_2	11
Cyclic alkyl amines	2–2.5	R—CN	18
Alkyl amines protonated	4–5	CH_3—$\overset{+}{N}\equiv\bar{C}$	
Aryl amines	10–14	H_3C—N	10
Aryl amines protonated	9	—$N\equiv C$	9
CH_3CO—NH_2	14–15	Diaryl azoxy	
H_2N—CO—NH_2	20	*anti*	18
Aryl—NO_2	15	*syn*	13

TABLE 7.68 Nitrogen-15 to Fluorine-19 Spin Coupling Constants

Structure	J, Hz	Structure	J, Hz
NF_3	155	Pyridine	
F_4N_2	164	2-F	52
FNO_2	158	3-F	4
F_3NO	190	2,6-di-F	37
F_3C—O—NF_2	164–176	Pyridinium ion	
FCO—NF_2	221	2-F	23
$(NF_4)^+SbF_6^-$	323	3-F	3
$(NF_4)^+AsF_6^-$	328	Quinoline, 8-F	3
$(N_2F)^+AsF_6^-$	459	Aniline	
F_3C—NO_2	215	2-F	0
		3-F	0
$\underset{}{\overset{F}{\diagdown}}N{=}N\diagdown_{F}$ $(^2J = 10)$	190	4-F	1.5
		Anilinium ion	
		2-F	1.4
$\overset{F}{\diagdown}\underset{}{N}{=}N\overset{F}{\diagup}$ $(^2J = 52)$	203	3-F	0.2
		4-F	0

TABLE 7.69 Fluorine-19 Chemical Shifts

Values given in ppm on the δ scale, relative to CCl_3F.

Substituent group	δ, ppm	Substituent group	δ, ppm
—SO_2—F	−67 to −42 (aryl)(alkyl)	R—CF_2Cl	61–71
		$>$C—CF_3 and aryl—CF_3	56–73
—CO—F	−29 to −20	—CS—CF_3	70
$>$N—CO—F	−5	$>$CF—CF_3	71–73
Aryl—CF_2Cl	49	—S—CF_3	41
—CF_2I	56	—S—CF_2—S—	39
—CF_2Br	63	$>$P—CF_3	46–66

TABLE 7.69 Fluorine-19 Chemical Shifts (*Continued*)

Substituent group	δ, ppm	Substituent group	δ, ppm
$\geq$N$-$CF$_3$	40–58	Perfluorocycloalkane	131–138
$\geq$N$-$CF$_2$$-$C	85–127	$\geq$CF$-$CF$_3$	163–198
$-$O$-$CF$_2$$-$R	70–91	$\geq$CF(CF$_3$)$_2$	180–191
$-$O$-$CF$_2$$-CF_3$	70–91	$-$CFH$-$	198–231
$-$CH$_2$$-CF_3$	76–77	$-$CFH$_2$	235–244
HO$-$CO$-$CF$_3$	77	F$_2$C$=$CF$_2$	133
$-$CHF$-$CF$_3$	81		
$-$CF$_2$$-CF_3$	78–88		
$-$CS$-$F	81		
CF$_3$$-C-N\leq$	84–96		
$-$CO$-$CF$_2$$-CF_3$	83		
$-$CF$_2$$-$	86–126	*cis*	108
$-$CF$_2$Br	91	*trans*	92
$-$C$-$CF$_2$$-S-$	91–98	*gem*	192
$-$CF$=$	180–192		
$-$CF$_2$$-CF_3$	111		
$-$CO$-$CF$_2$$-$	116–131		
$-$C(halide)$-$CF$_2$$-$	119–128		
$-$CF$_2$$-CF_3$	121–125		
$-$CF$_2$$-CF_2$$-$	121–129		
$-$CF$_2$$-CH_2$$-$	122–133	F-1	126
$-$CF$_2$$-CHF_2$	128–132	F-2	155
$-$CF$_2$H	136–143	F-3	162
		ClCF$=$CH$-$CF$_3$	61
$\triangleright$F$_2$	151–156	Cycloalkenes	
		$=$CF$-$CF$_2$$-$	
		C(CF$_3$ or H)$-$	101–113
$\diamondsuit$F$_2$	147	$-$CF$_2$$-CF_2$$-$	
		C(CF$_3$ or CH$_3$)$=$	110–114
		$-$CF$_2$$-CF_2$$-CH=$	113–116
(cyclopentane)F$_2$	96–133	$-$CF$_2$$-CF_2$$-CF=$	119–122
		Aryl$-$F	113
		C$_{10}$H$_7$$-$F	
(cyclopentene)F	159	F-1	127
		F-2	114
Cyclohexane-F	210	C$_6$H$_5$$-C_6H_4$$-$F	
	(axial)	F-2	117
	to	F-3	113
	240	F-4	109
	(equatorial)	C$_6$F$_6$	163

TABLE 7.70 Fluorine-19 Chemical Shifts for Standards

Substance	Formula	δ, ppm
Trichlorofluoromethane	$CFCl_3$	0.0
α,α,α-Trifluorotoluene	$C_6H_5CF_3$	63.8
Trifluoroacetic acid	CF_3COOH	76.5
Carbon tetrafluoride	CF_4	76.7
Fluorobenzene	C_6H_5F	113.1
Perfluorocyclobutane	C_4F_8	138.0

TABLE 7.71 Fluorine-19 to Fluorine-19 Spin Coupling Constants

Structure	J_{FF}, Hz
F_2C cycloalkane	
gem	212–260
Unsaturated compounds $>C=C<$	
gem	30–90
trans	115–130
cis	9–58
Aromatic compounds, monocyclic	
ortho	18–22
meta	0–7
para	12–15
Alkanes	
$C\underline{F}Cl_2{-}C\underline{F}_2{-}CFCl_2$	6
$C\underline{F}Cl_2{-}C\underline{F}_2{-}CCl_3$	5
$C\underline{F}_2Cl{-}C\underline{F}_2{-}CF_2Cl$	1
$C\underline{F}_3{-}C\underline{F}_2{-}CF_2Cl$ (or $-CF_3$)	<1
$CF_3{-}C\underline{F}_2{-}C\underline{F}_2Cl$	2
$C\underline{F}_3{-}CF_2{-}C\underline{F}_2Cl$	9
$C\underline{F}_3{-}CF_2{-}C\underline{F}_3$	7

TABLE 7.72 Silicon-29 Chemical Shifts

Values given in ppm on the δ scale relative to tetramethylsilane.

Substituent group X in $(CH_3)_{4-n}SiX_n$	n			
	1	2	3	4
—F	35	9	−52	−109
—Cl	30	32	13	−19
—Br	26	20	−18	−94
—I	9	−34	−18	−346
—H	−19	−42	−65	−93
—C_2H_5	2	5	7	8

TABLE 7.72 Silicon-29 Chemical Shifts (*Continued*)

Substituent group X in $(CH_3)_{4-n}SiX_n$	n			
	1	2	3	4
—C_6H_5	−5	−9	−12	
—$CH=CH_2$	−7	−14	−21	−23
—Oalkyl	14–17	−3 to −6	−41 to −45	−79 to −83
—Oaryl	17	−6	−54	−101
—O—CO—alkyl	22	4	−43	−75
—$N(CH_3)_2$	6	−2	−18	−28

Structure	δ, ppm	Structure	δ, ppm
Hydrides		$CH_3\overset{\displaystyle O-}{\underset{\displaystyle O-}{Si}}$—O— (branching)	−65 to −66
H_3Si—	−39 to −60		
—H_2Si—	−5 to −37		
$HSi{\Large\lessgtr}$	−2 to −39		
Silicates		—O—$\overset{\displaystyle O-}{\underset{\displaystyle O-}{Si}}$—O— (cross-linked)	−105 to −110
Orthosilicate anions	−69 to −72		
Silicon in end position	−77 to −81		
Silicon in middle	−85 to −89	**Polysilanes**	
Branching silicons	−93 to −97	$F_3Si—SiF_3$	−74
Cross-linked silicons	−107 to −120	$Cl_3Si—SiCl_3$	−8
Methyl siloxanes		$(CH_3O)_3Si—Si(OCH_3)_3$	−53
$(CH_3)_2Si$—O— (end position)	6–8	$(CH_3)_3Si—Si(CH_3)_3$	−20
$(CH_3)_2Si\overset{\displaystyle O-}{\underset{\displaystyle O-}{\diagup}}$ (middle)	−18 to −23	$(CH_3)_2\underline{Si}[Si(CH_3)_3]_2$	−48
		$H\underline{Si}[Si(CH_3)_3]_3$	−117
$CH_3Si(H)\overset{\displaystyle O-}{\underset{\displaystyle O-}{\diagup}}$ (middle)	−35 to −36	$\underline{Si}[Si(CH_3)_3]_4$	−135

TABLE 7.73 Phosphorus-31 Chemical Shifts

Values given in ppm on the δ scale, relative to 85% H_3PO_4.

Structure	Identical atoms attached directly to phosphorus	Non-identically substituted phosphorus		
		R = CH_3	R = C_2H_5	R = C_6H_5
P_4	461			
PR_3		62	20	6
PHR_2		99	56	41
PH_2R		164	128	122
PH_3	241			
PF_3	−97			
PRF_2			−168	−207

TABLE 7.73 Phosphorus-31 Chemical Shifts (*Continued*)

Structure	Identical atoms attached directly to phosphorus	Non-identically substituted phosphorus		
		$R=CH_3$	$R=C_2H_5$	$R=C_6H_5$
PCl_3	−220			
$PRCl_2$		−192	−196	−162
PR_2Cl		−94	−119	−81
PBr_3	−227			
$PRBr_2$		−184	−194	−152
PR_2Br		−91	−116	−71
PI_3	−178			
$P(CN)_3$	136			
$P(SiR_3)_3$		251		
$P(OR)_3$		−141	−139	−127
$P(OR)_2Cl$		−169	−165	−157
$P(OR)Cl_2$		−114	−177	−173
$P(SR)_3$		−125	−115	−132
$P(SR)_2Cl$		−188	−186	−183
$P(SR)Cl_2$		−206	−211	−204
$P(SR)_2Br$				−184
$P(SR)Br_2$		−204		
$P(NR_2)_3$		−123	−118	
$P(NR_2)Cl_2$		−166	−162	−151
$PR(NR_2)_2$		−86	−100	−100
$PR_2(NR_2)$		−39	−62	
$F_2P—PF_2$	−226			
$Cl_2P—PCl_2$	−155			
$I_2P—PI_2$	−170			
$PH_2^- K^+$	255			
$P(CF_3)_3$	3			
P_4O_6	−113			

Structure	Identical atoms attached directly to phosphorus	Non-identically substituted phosphorus		
		$X = F$	$X = Cl$	$X = Br$
$P(NCO)_3$	−97			
$P(NCO)_2X$		−128	−128	−127
$P(NCO)X_2$		−131	−166	
$P(NCS)_3$	−86			
$P(NCS)_2X$			−114	−112
$P(NCS)X_2$			−155	−153

Structure	Identical atoms attached directly to phosphorus	Non-identically substituted phosphorus		
		$R = CH_3$	$R = C_2H_5$	$R = C_6H_5$
$O=PR_3$		−36	−48	−25
$O=PHR_2$		−63		−23
$O=PF_3$	36			

TABLE 7.73 Phosphorus-31 Chemical Shifts (*Continued*)

Structure	Identical atoms attached directly to phosphorus	Non-identically substituted phosphorus		
		$R = CH_3$	$R = C_2H_5$	$R = C_6H_5$
$O=PRF_2$		−27	−29	−11
$O=PCl_3$	−2			
$O=PRCl_2$		−45	−53	−34
$O=PR_2Cl$		−65	−77	−43
$O=P(OR)_3$		−1	1	18
$O=P(OR)_2Cl$		−6	−3	6
$O=P(OR)Cl_2$		−6	−6	−2
$O=PH(OR)_2$		−19	−15	
$O=PR_2(OC_2H_5)$		−50	−52	−31
$O=PR(OC_2H_5)_2$		−30	−33	−17
$O=P(NR_2)_3$		−23	−24	−2
$O=PR_2(NR_2)$		−44		−26
$O=P(OR)_2NH_2$		−15	−12	−3
$O=P(OR)_2(NCS)$			19	29
$O=P(SR)_3$		−66	−61	−55
$O=PBr_3$	103			
$O=P(NCO)_3$	41			
$O=P(NCS)_3$	62			
$O=P(NH_2)_3$	−22			

Structure	Identical atoms attached directly to phosphorus	Structure	Identical atoms attached directly to phosphorus
PF_5	35	O ‖ −O−P−O− | OR (middle group)	ca 18
PF_6^- H^+	144		
PBr_5	101		
$P(OC_2H_5)_5$	71		
PO_4^{3-}	−6		
$O=P[OSi(CH_3)_3]_3$	33		
$H_4P_2O_7$	11	O ‖ −O−P−O− | O | P (etc.) (branch group)	ca 30
Phosphonates	−24 to −2		
Phosphonium cations			
Alkyl	−43 to −32		
Aryl	−35 to −18		
$(O_3P-PO_3)^{4-}$	−9		
Polyphosphates			

$O=P-O-$
    |
    $(OR)_2$
      (end group)
	ca 6		

TABLE 7.73 Phosphorus-31 Chemical Shifts (*Continued*)

Structure	Identical atoms attached directly to phosphorus	Non-identically substituted phosphorus		
		R=CH$_3$	R=C$_2$H$_5$	R=C$_6$H$_5$
S=PR$_3$		−59	−55	−43
S=PCl$_3$	−29			
S=PRCl$_2$		−80	−94	−75
S=PR$_2$Cl		−87	−109	−80
S=PBr$_3$	112			
S=PRBr$_2$		−21	−42	−20
S=PR$_2$Br		−64	−98	
S=P(OR)$_3$		−73	−68	−53
S=P(OR)Cl$_2$		−59	−56	−54
S=P(OR)$_2$Cl		−73	−68	−59
S=PH(OR)$_2$		−74	−69	−59
S=P(SR)$_3$		−98	−92	−92
S=P(NH$_2$)$_3$	−60			
S=P(NR$_2$)$_3$		−82	−78	
Se=P(OR)$_3$		−78	−71	−58
Se=P(SR)$_3$		−82	−76	
P(OR)$_5$			71	86
PRF$_4$		30	30	42
PR$_2$F$_3$		−9	−6	

TABLE 7.74 Phosphorus-31 Spin Coupling Constants

Substituent group	J_{PH}, Hz	Substituent group	J_{PH}, Hz
>PH	180–225	>P—N—CH	8–25
—PH$_2^-$	134	>P—C—CH	0–4
RPH$_2$	160–210		
>P—CH$_3$	1–6		
>P—CH$_2$—	14		
H$_\alpha$ C=C H$_\beta$, >P, H$_\gamma$		>P—⟨phenyl⟩	
		ortho	7–10
		meta	2–4
α	12–22	O=PHR$_2$	210–500
β	30–40	O—PH(S)R	490–540
γ	14–20	O$_2$PHR	500–575
(Halogen)$_2$P—CH	16–20	O$_2$PH(N)	560–630
>P—NH	10–28	O$_2$PH(S or Se)	630–655
>P—O—CH$_3$	11–15	O$_3$PH	630–760
>P—O—CH$_2$—R	6–10	S(or Se)=P—H	490–650
>P—O—CHR$_2$	3–7		
>P—SCH	5–20	S(or Se)=PHR$_2$	420–454

TABLE 7.74 Phosphorus-31 Spin Coupling Constants (*Continued*)

Substituent group	J_{PH}, Hz	Substituent group	J_{PF}, Hz
$O=P-CH_3$	7–15	$P-F$ axial equatorial	600–860 800–1000
$O=P-CH=C$	15–30	$O=P-CF$	110–113
$O=P-CH-Aryl(or\ C=O)$	15–30	$O=P-F$	980–1190
$(Halogen)_2P-N-CH$	9–18	$\underline{P}-O-P-\underline{F}$	2
$S=P-CH$	11–15	**Substituent group**	J_{PB}, Hz
$\geqq P-CH_3{}^+$	12–17	H_3B-P-N	80
$\geqq P-H^+$	490–600		

Substituent group	J_{PP}, Hz	Substituent group	J_{PP}, Hz
$>P-F$	1320–1420 (1F) (3F)	$>P-P<$	220–400
RPF_2	1140–1290	$O=P-P=O$	330–500
R_2PF	1020–1110	$S=P-P=S$	15–500
$RP(N)F$	920–985 (alkyl) (aryl)	$P-C-P$	ca 70
$-O$ PF $-O$	1225–1305	$\geqq P-O-P\leqq$	20–40
		$\geqq P-S-P\leqq$	86–90
$(OCN)PF$	1310	$O=P-O-P=O$	15–25

$N-P$
    F
	1100–1200	$O=P-N-P=O$
		    H
			8–30

$>P-CF$	60–90	$P-N$ N P $P-N$	5–66
$>P-\bigcirc-F$		$P=N-P=N-$	5–65
ortho *meta* *para*	0–60 1–7 0–3		

7.8 MASS SPECTROMETRY

7.8.1 Correlation of Mass Spectra with Molecular Structure

7.8.1.1 Molecular Identification. In the identification of a compound, the most important information is the molecular weight. The mass spectrometer is able to provide this information, often to four decimal places. One assumes that no ions heavier than the molecular ion form when using electron-impact ionization. The chemical ionization spectrum will often show a cluster around the nominal molecular weight.

Several relationships aid in deducing the empirical formula of the parent ion (and also molecular fragments). From the empirical formula hypothetical molecular structures can be proposed, using the entries in the formula indices of Beilstein and *Chemical Abstracts*.

7.8.1.2 Natural Isotopic Abundances. The relative abundances of natural isotopes produce peaks one or more mass units larger than the parent ion (Table 7.75a). For a compound $C_wH_xO_zN_y$, a formula allows one to calculate the percent of the heavy isotope contributions from a monoisotopic peak, P_M, to the P_{M+1} peak:

$$100\frac{P_{M+1}}{P_M} = 0.015x + 1.11w + 0.37y + 0.037z$$

Tables of abundance factors have been calculated for all combinations of C, H, N, and O up to mass 500 (J. H. Beynon and A. E. Williams, *Mass and Abundance Tables for Use in Mass Spectrometry*, Elsevier, Amsterdam, 1963).

Compounds that contain chlorine, bromine, sulfur, or silicon are usually apparent from prominent peaks at masses 2, 4, 6, and so on, units larger than the nominal mass of the parent

TABLE 7.75 Isotopic Abundances and Masses of Selected Elements

(a) Abundances of some polyisotopic elements, %

Element	Abundance	Element	Abundance	Element	Abundance
^{1}H	99.985	^{16}O	99.76	^{33}S	0.76
^{2}H	0.015	^{17}O	0.037	^{34}S	4.22
^{12}C	98.892	^{18}O	0.204	^{35}Cl	75.53
^{13}C	1.108	^{28}Si	92.18	^{37}Cl	24.47
^{14}N	99.63	^{29}Si	4.71	^{79}Br	50.52
^{15}N	0.37	^{30}Si	3.12	^{81}Br	49.48

(b) Selected isotope masses

Element	Mass	Element	Mass
^{1}H	1.0078	^{31}P	30.9738
^{12}C	12.0000	^{32}S	31.9721
^{14}N	14.0031	^{35}Cl	34.9689
^{16}O	15.9949	^{56}Fe	55.9349
^{19}F	18.9984	^{79}Br	78.9184
^{28}Si	27.9769	^{127}I	126.9047

or fragment ion. For example, when one chlorine atom is present, the $P + 2$ mass peak will be about one-third the intensity of the parent peak. When one bromine atom is present, the $P + 2$ mass peak will be about the same intensity as the parent peak. The abundance of heavy isotopes is treated in terms of the binomial expansion $(a + b)^m$, where a is the relative abundance of the light isotope, b is the relative abundance of the heavy isotope, and m is the number of atoms of the particular element present in the molecule. If two bromine atoms are present, the binomial expansion is

$$(a + b)^2 = a^2 + 2ab + b^2$$

Now substituting the percent abundance of each isotope (^{79}Br and ^{81}Br) into the expansion,

$$(0.505)^2 + 2(0.505)(0.495) + (0.495)^2$$

gives
$$0.255 + 0.500 + 0.250$$

which are the proportions of $P : (P + 2) : (P + 4)$, a triplet that is slightly distorted from a $1 : 2 : 1$ pattern. When two elements with heavy isotopes are present, the binomial expansion $(a + b)^m(c + d)^n$ is used.

Sulfur-34 enhances the $P + 2$ peak by 4.2%; silicon-29 enhances the $P + 1$ peak by 4.7% and the $P + 2$ peak by 3.1%.

7.8.1.3 *Exact Mass Differences.*
If the exact mass of the parent or fragment ions are ascertained with a high-resolution mass spectrometer, this relationship is often useful for combinations of C, H, N, and O (Table 7.75b):

$$\frac{\text{Exact mass difference from nearest integral mass} + 0.0051z - 0.0031y}{0.0078} = \text{number of hydrogens}$$

One substitutes integral numbers (guesses) for z (oxygen) and y (nitrogen) until the divisor becomes an integral multiple of the numerator within 0.0002 mass unit.

For example, if the exact mass is 177.0426 for a compound containing only C, H, O, and N (note the odd mass which indicates an odd number of nitrogen atoms), then

$$\frac{0.0426 + 0.0051z - 0.0031y}{0.0078} = 7 \text{ hydrogen atoms}$$

when $z = 3$ and $y = 1$. The empirical formula is $C_9H_7NO_3$ since

$$\frac{177 - 7(1) - 1(14) - 3(16)}{12} = 9 \text{ carbon atoms}$$

7.8.1.4 *Number of Rings and Double Bonds.*
The total number of rings and double bonds can be determined from the empirical formula ($C_wH_xO_zN_y$) by the relationship

$$\frac{1}{2(2w - x + y + z)}$$

when covalent bonds comprise the molecular structure. Remember the total number for a benzene ring is four (one ring and three double bonds); a triple bond has two.

7.8.1.5   General Rules

1. If the nominal molecular weight of a compound containing only C, H, O, and N is even, so is the number of hydrogen atoms it contains.

2. If the nominal molecular weight is divisible by four, the number of hydrogen atoms is also divisible by four.

3. When the nominal molecular weight of a compound containing only C, H, O, and N is odd, the number of nitrogen atoms must be odd.

7.8.1.6   Metastable Peaks.

If the mass spectrometer has a field-free region between the exit of the ion source and the entrance to the mass analyzer, metastable peaks m^* may appear as a weak, diffuse (often humped-shape) peak, usually at a nonintegral mass. The one-step decomposition process takes the general form:

$$\text{Original ion} \rightarrow \text{daughter ion} + \text{neutral fragment}$$

The relationship between the original ion and daughter ion is given by

$$m^* = \frac{(\text{mass of daughter ion})^2}{\text{mass of original ion}}$$

For example, a metastable peak appeared at 147.9 mass units in a mass spectrum with prominent peaks at 65, 91, 92, 107, 108, 155, 172, and 200 mass units. Try all possible combinations in the above expression. The fit is given by

$$147.9 = \frac{(172)^2}{200}$$

which provides this information:

$$200^+ \rightarrow 172^+ + 28$$

The probable neutral fragment lost is either $CH_2{=}CH_2$ or CO.

7.8.2   Mass Spectra and Structure

The mass spectrum is a fingerprint for each compound because no two molecules are fragmented and ionized in exactly the same manner on electron-impact ionization. In reporting mass spectra the data are normalized by assigning the most intense peak (denoted as base peak) a value of 100. Other peaks are reported as percentages of the base peak.

A very good general survey for interpreting mass spectral data is given by R. M. Silverstein, G. C. Bassler, and T. C. Morrill, *Spectrometric Identification of Organic Compounds*, 4th ed., Wiley, New York, 1981.

7.8.2.1   Initial Steps in Elucidation of a Mass Spectrum

1. Tabulate the prominent ion peaks, starting with the highest mass.

2. Usually only one bond is cleaved. In succeeding fragmentations a new bond is formed for each additional bond that is broken.

3. When fragmentation is accompanied by the formation of a new bond as well as by the breaking of an existing bond, a rearrangement process is involved. These will be even mass

peaks when only C, H, and O are involved. The migrating atom is almost exclusively hydrogen; six-membered cyclic transition states are most important.

4. Tabulate the probable groups that (*a*) give rise to the prominent charged ion peaks and (*b*) list the neutral fragments.

7.8.2.2 General Rules for Fragmentation Patterns

1. Bond cleavage is more probable at branched carbon atoms: tertiary > secondary > primary. The positive charge tends to remain with the branched carbon.
2. Double bonds favor cleavage beta to the carbon (but see rule 6).
3. A strong parent peak often indicates a ring.
4. Saturated ring systems lose side chains at the alpha carbon. Upon fragmentation, two ring atoms are usually lost.
5. A heteroatom induces cleavage at the bond beta to it.
6. Compounds that contain a carbonyl group tend to break at this group; the positive charge remains with the carbonyl portion.
7. For linear alkanes, the initial fragment lost is an ethyl group (never a methyl group), followed by propyl, butyl, and so on. An intense peak at mass 43 suggests a chain longer than butane.
8. The presence of Cl, Br, S, and Si can be deduced from the unusual isotopic abundance patterns of these elements. These elements can be traced through the positively charged fragments until the pattern disappears or changes due to the loss of one of these atoms to a neutral fragment.
9. When unusual mass differences occur between some fragments ions, the presence of F (mass difference 19), I (mass difference 127), or P (mass difference 31) should be suspected.

7.8.2.3 Characteristic Low-Mass Fragment Ions

Mass 30   Primary amines

Masses 31, 45, 59   Alcohol or ether

Masses 19 and 33   Alcohol

Mass 66   Monobasic carboxylic acid

Masses 77 and 91   Benzene ring

7.8.2.4 Characteristic Low-Mass Neutral Fragments from the Molecular Ion

Mass 18 (H_2O)   From alcohols, aldehydes, ketones

Mass 19 (F) and 20 (HF)   Fluorides

Mass 27 (HCN)   Aromatic nitriles or nitrogen heterocycles

Mass 29   Indicates either CHO or C_2H_5

Mass 30   Indicates either CH_2O or NO

Mass 33 (HS) and 34 (H_2S)   Thiols

Mass 42 CH_2CO via rearrangement from methyl ketone or an aromatic acetate or an aryl-$NHCOCH_3$ group

Mass 43 C_3H_7 or CH_3CO

Mass 45   COOH or OC_2H_5

SECTION 8

ELECTROLYTES, ELECTROMOTIVE FORCE, AND CHEMICAL EQUILIBRIUM

8.1 ACTIVITY COEFFICIENTS

Although it is not possible to measure an individual ionic activity coefficient, f_i, it may be estimated from the following equation of the Debye-Hückel theory:

$$-\log f_i = \frac{A z_i^2 \sqrt{I}}{1 + B\mathring{a}\sqrt{I}}$$

where I is the ionic strength of the medium, and $\mathring{a}$ is the ion-size parameter—the effective ionic radius (Table 8.2). The values of A and B vary with the temperature and dielectric constant of the solvent; values from 0 to 100°C for aqueous medium ($\mathring{a}$ in angstrom units) are listed in Table 8.3. Corresponding values of A and B for unit weight of solvent (when employing molality) can be obtained by multiplying the corresponding values for unit volume (molarity units) by the square root of the density of water at the appropriate temperature.

The ionic strength can be estimated from the summation of the product molarity times ionic charge squared for all the ionic species present in the solution, i.e., $I = 0.5(c_1 z_1^2 + c_2 z_2^2 + \cdots + c_i z_i^2)$.

Values for the activity coefficients of ions in water at 25°C are given in Table 8.1 in terms of their effective ionic radii.

At moderate ionic strengths a considerable improvement is effected by subtracting a term bI from the Debye-Hückel expression; b is an adjustable parameter which is 0.2 for water at 25°C. Table 8.4 gives the values of the ionic activity coefficients (for z_i from 1 to 6) with $\mathring{a}$ taken to be 4.6Å.

In general, the mean ionic activity coefficient is given by

$$f_\pm = \sqrt[(x+y)]{f_+^x f_-^y}$$

TABLE 8.1 Individual Activity Coefficients of Ions in Water at 25°C

Effective Ionic Radii $\dot{a}$ (in Å)	f_i at Ionic Strength of				
	0.001	0.005	0.01	0.05	0.1
Univalent Ions					
9	0.967	0.933	0.914	0.86	0.83
8	0.966	0.931	0.912	0.85	0.82
7	0.965	0.930	0.909	0.845	0.81
6	0.965	0.929	0.907	0.835	0.80
5	0.964	0.928	0.904	0.83	0.79
4	0.964	0.928	0.902	0.82	0.775
3.5	0.964	0.926	0.900	0.81	0.76
3	0.964	0.925	0.899	0.805	0.755
2.5	0.964	0.924	0.898	0.80	0.75
Divalent Ions					
8	0.872	0.755	0.69	0.52	0.45
7	0.872	0.755	0.685	0.50	0.425
6	0.870	0.749	0.675	0.485	0.405
5	0.868	0.744	0.67	0.465	0.38
4.5	0.868	0.741	0.663	0.45	0.36
4	0.867	0.740	0.660	0.445	0.355
Trivalent Ions					
6	0.731	0.52	0.415	0.195	0.13
5	0.728	0.51	0.405	0.18	0.115
4	0.725	0.505	0.395	0.16	0.095
Tetravalent Ions					
11	0.588	0.35	0.255	0.10	0.065
5	0.57	0.31	0.20	0.048	0.021
Pentavalent Ions					
9	0.43	0.18	0.105	0.020	0.009

where f_+, f_- are the individual ionic activity coefficients, and x, y are the charge numbers (z_+, z_-) of the respective ions. In binary electrolyte solution.

$$f_\pm = \sqrt{f_+ f_-}$$

In ternary electrolytes, e.g., $BaCl_2$ or K_2SO_4,

$$f_\pm = \sqrt[3]{f_+ f_-^2} \quad \text{or} \quad f_\pm = \sqrt[3]{f_+^2 f_-}$$

In quaternary electrolytes, e.g., $LaCl_3$ or $K_3[Fe(CN)_6]$,

$$f_\pm = \sqrt[4]{f_+ f_-^3} \quad \text{or} \quad f_\pm = \sqrt[4]{f_+^3 f_-}$$

TABLE 8.2 Approximate Effective Ionic Radii in Aqueous Solutions at 25°C

$\mathring{a}$ (in Å)	Inorganic Ions	$\mathring{a}$ (in Å)	Organic Ions
2.5	Rb^+, Cs^+, NH_4^+, Tl^+, Ag^+	3.5	$HCOO^-$, H_2Cit^-, $CH_3NH_3^+$, $(CH_3)_2NH_2^+$
3	K^+, Cl^-, Br^-, I^-, CN^-, NO_2^-, NO_3^-	4	$H_3N^+CH_2COOH$, $(CH_3)_3NH^+$, $C_2H_5NH_3^+$
3.5	OH^-, F^-, SCN^-, OCN^-, HS^-, ClO_3^-, ClO_4^-, BrO_3^-, IO_4^-, MnO_4^-	4.5	CH_3COO^-, $ClCH_2COO^-$, $(CH_3)_4N^+$, $(C_2H_5)_2NH_2^+$, $H_2NCH_2COO^-$, $HCit^{2-}$
4	Na^+, $CdCl^+$, Hg_2^{2+}, ClO_2^-, IO_3^-, HCO_3^-, $H_2PO_4^-$, HSO_3^-, $H_2AsO_4^-$, SO_4^{2-}, $S_2O_3^{2-}$, $S_2O_8^{2-}$, SeO_4^{2-}, CrO_4^{2-}, HPO_4^{2-}, $S_2O_6^{2-}$, PO_4^{3-}, $Fe(CN)_6^{3-}$, $Cr(NH_3)_6^{3+}$, $Co(NH_3)_6^{3+}$, $Co(NH_3)_5H_2O^{3+}$	5	Cl_2CHCOO^-, Cl_3COO^-, $(C_2H_5)_3NH^+$, $C_3H_7NH_3^+$, Cit^{3-}, succinate^{2-}, malonate^{2-}, tartrate^{2-}
4.5	Pb^{2+}, CO_3^{2-}, SO_3^{2-}, MoO_4^{2-}, $Co(NH_3)_5Cl^{2+}$, $Fe(CN)_5NO^{2-}$	6	benzoate$^-$, hydroxybenzoate$^-$, chlorobenzoate$^-$, phenyl-acetate$^-$, vinylacetate$^-$, $(CH_3)_2C{=}CHCOO^-$, $(C_2H_5)_4N^+$, $(C_3H_7)_2NH_2^+$, phthalate^{2-}, glutarate^{2-}, adipate^{2-}
5	Sr^{2+}, Ba^{2+}, Ra^{2+}, Cd^{2+}, Hg^{2+}, S^{2-}, $S_2O_4^{2-}$, WO_4^{2-}, $Fe(CN)_6^{4-}$	7	trinitrophenolate$^-$, $(C_3H_7)_3NH^+$, methoxybenzoate$^-$, pimelate^{2-}, suberate^{2-}, Congo red anion^{2-}
6	Li^+, Ca^{2+}, Cu^{2+}, Zn^{2+}, Sn^{2+}, Mn^{2+}, Fe^{2+}, Ni^{2+}, Co^{2+}, $Co(en)_3^{3+}$, $Co(S_2O_3)(CN)_5^{4-}$	8	$(C_6H_5)_2CHCOO^-$, $(C_3H_7)_4N^+$
8	Mg^{2+}, Be^{2+}		
9	H^+, Al^{3+}, Fe^{3+}, Cr^{3+}, Sc^{3+}, Y^{3+}, La^{3+}, In^{3+}, Ce^{3+}, Pr^{3+}, Nd^{3+}, Sm^{3+}, $Co(SO_3)_2(CN)_4^{5-}$		
11	Th^{4+}, Zr^{4+}, Ce^{4+}, Sn^{4+}		

8.4

TABLE 8.3   Constants of the Debye-Hückel Equation from 0 to 100°C

$$-\log f_i = \frac{A z_i^2 \sqrt{I}}{1 + B \mathring{a} \sqrt{I}}$$

Temp., °C	Unit Volume of Solvent		Temp., °C	Unit Volume of Solvent	
	A	B		A	B
0	0.4918	0.3248	55	0.5432	0.3358
5	0.4952	0.3256	60	0.5494	0.3371
10	0.4989	0.3264	65	0.5558	0.3384
15	0.5028	0.3273	70	0.5625	0.3397
20	0.5070	0.3282	75	0.5695	0.3411
25	0.5115	0.3291	80	0.5767	0.3426
30	0.5161	0.3301	85	0.5842	0.3440
35	0.5211	0.3312	90	0.5920	0.3456
40	0.5262	0.3323	95	0.6001	0.3471
45	0.5317	0.3334	100	0.6086	0.3488
50	0.5373	0.3346			

The values for unit weight of solvent (molality scale) can be obtained by multiplying the corresponding values for unit volume by the square root of the density of water at the appropriate temperature.

TABLE 8.4   Individual Ionic Activity Coefficients at Higher Ionic Strengths at 25°C

The values were calculated from the modified Debye-Hückel equation utilizing the modifications proposed by Robinson and by Guggenheim and Bates:

$$-\frac{\log f_i}{z_i^2} = \frac{0.511I}{1 + 1.5I} - 0.2I$$

where I is the ionic strength and $\mathring{a}$ is assumed to be 4.6 Å.

I	$-\dfrac{\log_{10} f_i}{z_i^2}$	f_i for $z_i =$					
		1	2	3	4	5	6
0.05	0.0756	0.840	0.498	0.209	0.0617	0.0129	0.00190
0.1	0.0896	0.814	0.438	0.156	0.0369	0.00576	0.000595
0.2	0.0968	0.800	0.410	0.138	0.0283	0.00380	0.000328
0.3	0.0936	0.806	0.422	0.144	0.0318	0.00457	0.000427
0.4	0.0858	0.821	0.454	0.169	0.0424	0.00716	0.000815
0.5	0.0753	0.841	0.500	0.210	0.0624	0.0131	0.00195
0.6	0.0631	0.865	0.559	0.270$_5$	0.0978	0.0265	0.00535
0.7	0.0496	0.892	0.633	0.358	0.161	0.0575$_5$	0.0164
0.8	0.0352	0.922	0.723	0.482	0.273	0.132	0.0541
0.9	0.0201	0.955	0.831	0.659	0.477	0.314	0.189
1.0	0.0044	0.900	0.960	0.913	0.850	0.776	0.694

8.2 EQUILIBRIUM CONSTANTS

TABLE 8.5 Ionic Product Constant of Water

This table gives values of pKw on a molal scale, where Kw is the ionic activity product constant of water. Values are from W. L. Marshall and E. U. Franck, *J. Phys. Chem. Ref. Data,* **10:**295 (1981).

Temp., °C	pKw	Temp., °C	pKw	Temp., °C	pKw
0	14.938	45	13.405	95	12.345
5	14.727	50	13.275	100	12.264
10	14.528	55	13.152	125	11.911
15	14.340	60	13.034	150	11.637
18	14.233	65	12.921	175	11.431
20	14.163	70	12.814	200	11.288
25	13.995	75	12.711	225	11.207
30	13.836	80	12.613	250	11.192
35	13.685	85	12.520	275	11.251
40	13.542	90	12.431	300	11.406

TABLE 8.6 Solubility Products

The data refer to various temperatures between 18 and 25°C, and were primarily compiled from values cited by Bjerrum, Schwarzenbach, and Sillen, *Stability Constants of Metal Complexes*, part II, Chemical Society, London, 1958.

Substance	pK_{sp}	K_{sp}	Substance	pK_{sp}	K_{sp}
Actinium			$BaCrO_4$	9.93	1.2×10^{-10}
$Ac(OH)_3$	15	1×10^{-15}	$Ba_2[Fe(CN)_6] \cdot 6H_2O$	7.5	3.2×10^{-8}
Aluminum			BaF_2	5.98	1.0×10^{-6}
$AlAsO_4$	15.8	1.6×10^{-16}	$BaSiF_6$	6	1×10^{-6}
cupferrate, AIL_3	18.64	2.3×10^{-19}	$Ba(IO_3)_2 \cdot 2H_2O$	8.82	1.5×10^{-9}
$Al(OH)_3$ amorphous	32.9	1.3×10^{-33}	$Ba(OH)_2$	2.3	5×10^{-3}
$AlPO_4$	18.24	6.3×10^{-19}	$Ba(MnO_4)_2$	9.61	2.5×10^{-10}
8-quinolinolate, AIL_3	29.00	1.00×10^{-29}	$BaMoO_4$	7.40	4.0×10^{-8}
Al_2S_3	6.7	2×10^{-7}	$Ba(NbO_3)_2$	16.50	3.2×10^{-17}
Al_2Se_3	24.4	4×10^{-25}	$Ba(NO_3)_2$	2.35	4.5×10^{-3}
Americium			BaC_2O_4	6.79	1.6×10^{-7}
$Am(OH)_3$	19.57	2.7×10^{-20}	$BaC_2O_4 \cdot H_2O$	7.64	2.3×10^{-8}
$Am(OH)_4$	56	1×10^{-56}	$BaHPO_4$	6.5	3.2×10^{-7}
Ammonium			$Ba_3(PO_4)_2$	22.47	3.4×10^{-23}
$NH_4UO_2AsO_4$	23.77	1.7×10^{-24}	$Ba_2P_2O_7$	10.5	3.2×10^{-11}
Arsenic			$BaHPO_3 \cdot 0.5H_2O$	3	1×10^{-3}
$As_2S_3 + 4H_2O \rightarrow$	21.68	2.1×10^{-22}	8-quinolinolate, BaL_2	8.3	5.0×10^{-9}
$2HAsO_2 + 3H_2S$			$Ba(ReO_4)_2$	1.28	5.2×10^{-2}
Barium			$BaSeO_4$	7.46	3.5×10^{-8}
$Ba_3(AsO_4)_2$	50.11	8.0×10^{-51}	$BaSO_4$	9.96	1.1×10^{-10}
$Ba(BrO_3)_2$	5.50	3.2×10^{-6}	$BaSO_3$	6.1	8×10^{-7}
$BaCO_3$	8.29	5.1×10^{-9}	BaS_2O_3	4.79	1.6×10^{-5}
$BaCO_3 + CO_2 +$	4.35	4.5×10^{-5}	Beryllium		
$H_2O \rightarrow Ba^{2+} +$			$BeCO_3 \cdot 4H_2O$	3	1×10^{-3}
$2HCO_3^-$					

TABLE 8.6 Solubility Products (*Continued*)

Substance	pK_{sp}	K_{sp}	Substance	pK_{sp}	K_{sp}
Be(OH)$_2$ amorphous	21.8	1.6×10^{-22}	CaSeO$_3$	5.53	8.0×10^{-6}
Be(OH)$_2$ + OH$^-$ →	2.50	3.2×10^{-3}	CaSiO$_3$	7.60	2.5×10^{-8}
HBeO$_2^-$ + H$_2$O			CaSO$_4$	5.04	9.1×10^{-6}
BeMoO$_4$	1.5	3.2×10^{-2}	CaSO$_3$	7.17	6.8×10^{-8}
Be(NbO$_3$)$_2$	15.92	1.2×10^{-16}	tartrate dihydrate	6.11	7.7×10^{-7}
Bismuth			CaWO$_4$	8.06	8.7×10^{-9}
BiAsO$_4$	9.36	4.4×10^{-10}	Cerium		
cupferrate	27.22	6.0×10^{-28}	CeF$_3$	15.1	8×10^{-16}
Bi(OH)$_3$	30.4	4×10^{-31}	Ce(IO$_3$)$_3$	9.50	3.2×10^{-10}
BiI$_3$	18.09	8.1×10^{-19}	Ce(IO$_3$)$_4$	16.3	5×10^{-17}
BiPO$_4$	22.89	1.3×10^{-23}	Ce(OH)$_3$	19.8	1.6×10^{-20}
Bi$_2$S$_3$	97	1×10^{-97}	Ce(OH)$_4$	47.7	2×10^{-48}
BiOBr	6.52	3.0×10^{-7}	Ce$_2$(C$_2$O$_4$)$_3 \cdot$ 9H$_2$O	25.5	3.2×10^{-26}
BiOCl	30.75	1.8×10^{-31}	CePO$_4$	23	1×10^{-23}
BiOOH	9.4	4×10^{-10}	Ce$_2$(SeO$_3$)$_3$	24.43	3.7×10^{-25}
BiO(NO$_2$)	6.31	4.9×10^{-7}	Ce$_2$S$_3$	10.22	6.0×10^{-11}
BiO(NO$_3$)	2.55	2.82×10^{-3}	(III) tartrate	19.0	1×10^{-19}
BiOSCN	6.80	1.6×10^{-7}	Cesium		
Cadmium			CsBrO$_3$	1.7	5×10^{-2}
anthranilate, CdL$_2$	8.27	5.4×10^{-9}	CsClO$_3$	1.4	4×10^{-2}
Cd$_3$(AsO$_4$)$_2$	32.66	2.2×10^{-33}	Cs$_2$[PtCl$_6$]	7.5	3.2×10^{-8}
[Cd(NH$_3$)$_6$](BF$_4$)$_2$	5.7	2×10^{-6}	Cs$_3$[Co(NO$_2$)$_6$]	15.24	5.7×10^{-16}
benzoate $\cdot$ 2H$_2$O	2.7	2×10^{-3}	Cs[BF$_4$]	4.7	5×10^{-5}
Cd(BO$_2$)$_2$	8.64	2.3×10^{-9}	Cs$_2$[PtF$_6$]	5.62	2.4×10^{-6}
CdCO$_3$	11.28	5.2×10^{-12}	Cs$_2$[SiF$_6$]	4.90	1.3×10^{-5}
Cd(CN)$_2$	8.0	1.0×10^{-8}	CsClO$_4$	2.4	4×10^{-3}
Cd$_2$[Fe(CN)$_6$]	16.49	3.2×10^{-17}	CsIO$_4$	2.36	4.3×10^{-3}
Cd(OH)$_2$ fresh	13.6	2.5×10^{-14}	CsMnO$_4$	4.08	8.2×10^{-5}
CdC$_2$O$_4 \cdot$ 3H$_2$O	7.04	9.1×10^{-8}	CsReO$_4$	3.40	4.0×10^{-4}
Cd$_3$(PO$_4$)$_2$	32.6	2.5×10^{-33}	Chromium(II)		
quinaldate, CdL$_2$	12.3	5.0×10^{-13}	Cr(OH)$_2$	15.7	2×10^{-16}
CdS	26.1	8.0×10^{-27}	Chromium(III)		
CdWO$_4$	5.7	2×10^{-6}	CrAsO$_4$	20.11	7.7×10^{-21}
Calcium			CrF$_3$	10.18	6.6×10^{-11}
Ca$_3$(AsO$_4$)$_2$	18.17	6.8×10^{-19}	Cr(NH$_3$)$_6$(BF$_4$)$_3$	4.21	6.2×10^{-5}
acetate $\cdot$ 3H$_2$O	2.4	4×10^{-3}	Cr(OH)$_3$	30.2	6.3×10^{-31}
benzoate $\cdot$ 3H$_2$O	2.4	4×10^{-3}	Cr(NH$_3$)$_6$(ReO$_4$)$_3$	11.11	7.7×10^{-12}
CaCO$_3$	8.54	2.8×10^{-9}	CrPO$_4 \cdot$ 4H$_2$O green	22.62	2.4×10^{-23}
CaCO$_3$ calcite	8.35	4.5×10^{-9}	violet	17.00	1.0×10^{-17}
CaCO$_3$ aragonite	8.22	6.0×10^{-9}	Cobalt		
CaCrO$_4$	3.15	7.1×10^{-4}	anthranilate, CoL$_2$	9.68	2.1×10^{-10}
CaF$_2$	8.28	5.3×10^{-9}	Co$_3$(AsO$_4$)$_2$	28.12	7.6×10^{-29}
Ca[SiF$_6$]	3.09	8.1×10^{-4}	CoCO$_3$	12.84	1.4×10^{-13}
Ca(OH)$_2$	5.26	5.5×10^{-6}	Co$_2$[Fe(CN)$_6$]	14.74	1.8×10^{-15}
Ca(IO$_3$)$_2 \cdot$ 6H$_2$O	6.15	7.1×10^{-7}	Co(NH$_3$)$_6$(BF$_4$)$_2$	5.4	4×10^{-6}
Ca[Mg(CO$_3$)$_2$] dolomite	11	1×10^{-11}	Co(OH)$_2$ fresh	14.8	1.6×10^{-15}
CaMoO$_4$	7.38	4.2×10^{-8}	Co(OH)$_3$	43.8	1.6×10^{-44}
Ca(NbO$_3$)$_2$	17.06	8.7×10^{-18}	Co(IO$_3$)$_2$	4.0	1.0×10^{-4}
CaC$_2$O$_4 \cdot$ H$_2$O	8.4	4×10^{-9}	quinaldate, CoL$_2$	10.8	1.6×10^{-11}
CaHPO$_4$	7.0	1×10^{-7}	Co[Hg(SCN)$_4$]	5.82	1.5×10^{-6}
Ca$_3$(PO$_4$)$_2$	28.70	2.0×10^{-29}	α-CoS	20.4	4.0×10^{-21}
8-quinolinolate, CaL$_2$	11.12	7.6×10^{-12}	β-CoS	24.7	2.0×10^{-25}
CaSeO$_4$	3.09	8.1×10^{-4}	8-quinolinolate, CoL$_2$	24.8	1.6×10^{-25}

TABLE 8.6   Solubility Products (*Continued*)

Substance	pK_{sp}	K_{sp}	Substance	pK_{sp}	K_{sp}
$CoHPO_4$	6.7	2×10^{-7}	AuI_3	46	1×10^{-46}
$Co_3(PO_4)_2$	34.7	2×10^{-35}	$Au_2(C_2O_4)_3$	10	1×10^{-10}
$CoSeO_3$	6.8	1.6×10^{-7}	Hafnium		
Copper(I)			$Hf(OH)_3$	25.4	4.0×10^{-26}
CuN_3	8.31	4.9×10^{-9}	Holmium		
$Cu[B(C_6H_5)_4]$			$Ho(OH)_3$	22.3	5.0×10^{-23}
tetraphenylborate	8.0	1×10^{-8}	Indium		
$CuBr$	8.28	5.3×10^{-9}	$In_4[Fe(CN)_6]_3$	43.72	1.9×10^{-44}
$CuCl$	5.92	1.2×10^{-6}	$In(OH)_3$	33.2	6.3×10^{-34}
$CuCN$	19.49	3.2×10^{-20}	quinolinolate, InL_3	31.34	4.6×10^{-32}
CuI	11.96	1.1×10^{-12}	In_2S_3	73.24	5.7×10^{-74}
$CuOH$	14.0	1×10^{-14}	$In_2(SeO_3)_3$	32.6	4.0×10^{-33}
Cu_2S	47.6	2.5×10^{-48}	Iron(II)		
$CuSCN$	14.32	4.8×10^{-15}	$FeCO_3$	10.50	3.2×10^{-11}
Copper(II)			$Fe(OH)_2$	15.1	8.0×10^{-16}
anthranilate, CuL_2	13.22	6.0×10^{-14}	$FeC_2O_4 \cdot 2H_2O$	6.5	3.2×10^{-7}
$Cu_3(AsO_4)_2$	35.12	7.6×10^{-36}	FeS	17.2	6.3×10^{-18}
$Cu(N_3)_2$	9.2	6.3×10^{-10}	Iron(III)		
$CuCO_3$	9.86	1.4×10^{-10}	$FeAsO_4$	20.24	5.7×10^{-21}
$CuCrO_4$	5.44	3.6×10^{-6}	$Fe_4[Fe(CN)_6]_3$	40.52	3.3×10^{-41}
$Cu_2[Fe(CN)_6]$	15.89	1.3×10^{-16}	$Fe(OH)_3$	37.4	4×10^{-38}
$Cu(IO_3)_2$	7.13	7.4×10^{-8}	$FePO_4$	21.89	1.3×10^{-22}
$Cu(OH)_2$	19.66	2.2×10^{-20}	quinaldate, FeL_3	16.9	1.3×10^{-17}
CuC_2O_4	7.64	2.3×10^{-8}	$Fe_2(SeO_3)_3$	30.7	2.0×10^{-31}
$Cu_3(PO_4)_2$	36.9	1.3×10^{-37}	Lanthanum		
$Cu_2P_2O_7$	15.08	8.3×10^{-16}	$La(BrO_3)_3 \cdot 9H_2O$	2.5	3.2×10^{-3}
quinaldate, CuL_2	16.8	1.6×10^{-17}	LaF_3	16.2	7×10^{-17}
8-quinolinolate, CuL_2	29.7	2.0×10^{-30}	$La(OH)_3$	18.7	2.0×10^{-19}
rubeanate	15.12	7.67×10^{-16}	$La(IO_3)_3$	11.21	6.1×10^{-12}
CuS	35.2	6.3×10^{-36}	$La_2(MoO_4)_3$	20.4	4×10^{-21}
$CuSeO_3$	7.68	2.1×10^{-8}	$La_2(C_2O_4)_3 \cdot 9H_2O$	26.60	2.5×10^{-27}
Dysprosium			$LaPO_4$	22.43	3.7×10^{-23}
$Dy_2(CrO_4)_3 \cdot 10H_2O$	8	1×10^{-8}	La_2S_3	12.70	2.0×10^{-13}
$Dy(OH)_3$	21.85	1.4×10^{-22}	$La_2(WO_4)_3 \cdot 3H_2O$	3.90	1.3×10^{-4}
Erbium			Lead		
$Er(OH)_3$	23.39	4.1×10^{-24}	acetate	2.75	1.8×10^{-3}
Europium			anthranilate, PbL_2	9.81	1.6×10^{-10}
$Eu(OH)_3$	23.05	8.9×10^{-24}	$Pb_3(AsO_4)_2$	35.39	4.0×10^{-36}
Gadolinium			$Pb(N_3)_2$	8.59	2.5×10^{-9}
$Gd(HCO_3)_3$	1.7	2×10^{-2}	$Pb(BO_2)_2$	10.78	1.6×10^{-11}
$Gd(OH)_3$	22.74	1.8×10^{-23}	$PbBr_2$	4.41	4.0×10^{-5}
Gallium			$Pb(BrO_3)_2$	1.70	2.0×10^{-2}
$Ga_4[Fe(CN)_6]_3$	33.82	1.5×10^{-34}	$PbCO_3$	13.13	7.4×10^{-14}
$Ga(OH)_3$	35.15	7.0×10^{-36}	$PbCl_2$	4.79	1.6×10^{-5}
8-quinolinolate, GaL_3	40.8	1.6×10^{-41}	$PbClF$	8.62	2.4×10^{-9}
Germanium			$PbCrO_4$	12.55	2.8×10^{-13}
GeO_2	57.0	1.0×10^{-57}	$Pb(ClO_2)_2$	8.4	4×10^{-9}
Gold(I)			$Pb_2[Fe(CN)_6]$	14.46	3.5×10^{-15}
$AuCl$	12.7	2.0×10^{-13}	PbF_2	7.57	2.7×10^{-8}
AuI	22.8	1.6×10^{-23}	$PbFI$	8.07	8.5×10^{-9}
Gold(III)			$Pb(OH)_2$	14.93	1.2×10^{-15}
$AuCl_3$	24.5	3.2×10^{-25}	$PbOHBr$	14.70	2.0×10^{-15}
$Au(OH)_3$	45.26	5.5×10^{-46}	$PbOHCl$	13.7	2×10^{-14}

TABLE 8.6 Solubility Products (*Continued*)

Substance	pK_{sp}	K_{sp}	Substance	pK_{sp}	K_{sp}
$PbOHNO_3$	3.55	2.8×10^{-4}	Hg_2CO_3	16.05	8.9×10^{-17}
PbI_2	8.15	7.1×10^{-9}	$Hg_2(CN)_2$	39.3	5×10^{-40}
$Pb(IO_3)_2$	12.49	3.2×10^{-13}	Hg_2Cl_2	17.88	1.3×10^{-18}
$PbMoO_4$	13.0	1.0×10^{-13}	Hg_2CrO_4	8.70	2.0×10^{-9}
$Pb(NbO_3)_2$	16.62	2.4×10^{-17}	$(Hg_2)_3[Fe(CN)_6]_2$	20.07	8.5×10^{-21}
PbC_2O_4	9.32	4.8×10^{-10}	$Hg_2(OH)_2$	23.7	2.0×10^{-24}
$PbHPO_4$	9.90	1.3×10^{-10}	$Hg_2(IO_3)_2$	13.71	2.0×10^{-14}
$Pb_3(PO_4)_2$	42.10	8.0×10^{-43}	Hg_2I_2	28.35	4.5×10^{-29}
$PbHPO_3$	6.24	5.8×10^{-7}	$Hg_2C_2O_4$	12.7	2.0×10^{-13}
quinaldate, PbL_2	10.6	2.5×10^{-11}	Hg_2HPO_4	12.40	4.0×10^{-13}
$PbSeO_4$	6.84	1.4×10^{-7}	quinaldate, Hg_2L_2	17.9	1.3×10^{-18}
$PbSeO_3$	11.5	3.2×10^{-12}	Hg_2SeO_3	14.2	8.4×10^{-15}
$PbSO_4$	7.79	1.6×10^{-8}	Hg_2SO_4	6.13	7.4×10^{-7}
PbS	27.9	8.0×10^{-28}	Hg_2SO_3	27.0	1.0×10^{-27}
$Pb(SCN)_2$	4.70	2.0×10^{-5}	Hg_2S	47.0	1.0×10^{-47}
PbS_2O_3	6.40	4.0×10^{-7}	$Hg_2(SCN)_2$	19.7	2.0×10^{-20}
$PbWO_4$	6.35	4.5×10^{-7}	Hg_2WO_4	16.96	1.1×10^{-17}
Lead(IV)			Mercury(II)		
$Pb(OH)_4$	65.5	3.2×10^{-66}	$Hg(OH)_2$	25.52	3.0×10^{-26}
Lithium			$Hg(IO_3)_2$	12.5	3.2×10^{-13}
Li_2CO_3	1.60	2.5×10^{-2}	1,10-phenanthroline	24.70	2.0×10^{-25}
LiF	2.42	3.8×10^{-3}	quinaldate, HgL_2	16.8	1.6×10^{-17}
Li_3PO_4	8.5	3.2×10^{-9}	$HgSeO_3$	13.82	1.5×10^{-14}
$LiUO_2AsO_4$	18.82	1.5×10^{-19}	HgS red	52.4	4×10^{-53}
Lutetium			HgS black	51.8	1.6×10^{-52}
$Lu(OH)_3$	23.72	1.9×10^{-24}	Neodymium		
Magnesium			$Nd(OH)_3$	21.49	3.2×10^{-22}
$MgNH_4PO_4$	12.6	2.5×10^{-13}	Neptunium		
$Mg_3(AsO_4)_2$	19.68	2.1×10^{-20}	$NpO_2(OH)_2$	21.6	2.5×10^{-22}
$MgCO_3$	7.46	3.5×10^{-8}	Nickel		
$MgCO_3 \cdot 3H_2O$	4.67	2.1×10^{-5}	$[Ni(NH_3)_6][ReO_4]_2$	3.29	5.1×10^{-4}
MgF_2	8.19	6.5×10^{-9}	anthranilate, NiL_2	9.09	8.1×10^{-10}
$Mg(OH)_2$	10.74	1.8×10^{-11}	$Ni_3(AsO_4)_2$	25.51	3.1×10^{-26}
$Mg(IO_3)_2 \cdot 4H_2O$	2.5	3.2×10^{-3}	$NiCO_3$	8.18	6.6×10^{-9}
$Mg(NbO_3)_2$	16.64	2.3×10^{-17}	$Ni_2(CN)_4 \rightarrow Ni^{2+} +$	8.77	1.7×10^{-9}
$Mg_3(PO_4)_2$	23–27	10^{-23} to 10^{-27}	$Ni(CN)_4^{2-}$		
8-quinolinolate, MgL_2	15.4	4.0×10^{-16}	$Ni_2[Fe(CN)_6]$	14.89	1.3×10^{-15}
$MgSeO_3$	4.89	1.3×10^{-5}	$[Ni(N_2H_4)_3]SO_4$	13.15	7.1×10^{-14}
$MgSO_3$	2.5	3.2×10^{-3}	$Ni(OH)_2$ fresh	14.7	2.0×10^{-15}
Manganese			$Ni(IO_3)_2$	7.85	1.4×10^{-8}
anthranilate, MnL_2	6.75	1.8×10^{-7}	NiC_2O_4	9.4	4×10^{-10}
$Mn_3(AsO_4)_2$	28.72	1.9×10^{-29}	$Ni_3(PO_4)_2$	30.3	5×10^{-31}
$MnCO_3$	10.74	1.8×10^{-11}	$Ni_2P_2O_7$	12.77	1.7×10^{-13}
$Mn_2[Fe(CN)_6]$	12.10	8.0×10^{-13}	8-quinolinolate, NiL_2	26.1	8×10^{-27}
$Mn(OH)_2$	12.72	1.9×10^{-13}	quinaldate, NiL_2	10.1	8×10^{-11}
$MnC_2O_4 \cdot 2H_2O$	14.96	1.1×10^{-15}	$NiSeO_3$	5.0	1.0×10^{-5}
8-quinolinolate, MnL_2	21.7	2.0×10^{-22}	α-NiS	18.5	3.2×10^{-19}
$MgSeO_3$	6.9	1.3×10^{-7}	β-NiS	24.0	1.0×10^{-24}
MnS amorphous	9.6	2.5×10^{-10}	γ-NiS	25.7	2.0×10^{-26}
crystalline	12.6	2.5×10^{-13}	Palladium		
Mercury(I)			$Pd(OH)_2$	31.0	1.0×10^{-31}
$Hg_2(N_3)_2$	9.15	7.1×10^{-10}	$Pd(OH)_4$	70.2	6.3×10^{-71}
Hg_2Br_2	22.24	5.6×10^{-23}	quinaldate, PdL_2	12.9	1.3×10^{-13}

TABLE 8.6 Solubility Products (*Continued*)

Substance	pK_{sp}	K_{sp}	Substance	pK_{sp}	K_{sp}
Platinum			$AgBrO_3$	4.28	5.3×10^{-5}
$PtBr_4$	40.5	3.2×10^{-41}	$AgBr$	12.30	5.0×10^{-13}
$Pt(OH)_2$	35	1×10^{-35}	Ag_2CO_3	11.09	8.1×10^{-12}
Plutonium			$AgClO_2$	3.7	2.0×10^{-4}
PuO_2CO_3	12.77	1.7×10^{-13}	$AgCl$	9.75	1.8×10^{-10}
PuF_3	15.6	2.5×10^{-16}	Ag_2CrO_4	11.95	1.1×10^{-12}
PuF_4	19.2	6.3×10^{-20}	$Ag_3[Co(NO_2)_6]$	20.07	8.5×10^{-21}
$Pu(OH)_3$	19.7	2.0×10^{-20}	cyanamide, Ag_2CN_2	10.14	7.2×10^{-11}
$Pu(OH)_4$	55	1×10^{-55}	$AgOCN$	6.64	2.3×10^{-7}
$PuO_2(OH)$	9.3	5×10^{-10}	$AgCN$	15.92	1.2×10^{-16}
$PuO_2(OH)_2$	24.7	2×10^{-25}	$Ag_2Cr_2O_7$	6.70	2.0×10^{-7}
$Pu(IO_3)_4$	12.3	5×10^{-13}	dicyanimide, $AgN(CN)_2$	8.85	1.4×10^{-9}
$Pu(HPO_4)_2 \cdot xH_2O$	27.7	2×10^{-28}	$Ag_4[Fe(CN)_6]$	40.81	1.6×10^{-41}
Polonium			$AgOH$	7.71	2.0×10^{-8}
PoS	28.26	5.5×10^{-29}	$Ag_2N_2O_2$	18.89	1.3×10^{-19}
Potassium			$AgIO_3$	7.52	3.0×10^{-8}
$K_2[PdCl_6]$	5.22	6.0×10^{-6}	AgI	16.08	8.3×10^{-17}
$K_2[PtCl_6]$	4.96	1.1×10^{-5}	Ag_2MoO_4	11.55	2.8×10^{-12}
$K_2[PtBr_6]$	4.2	6.3×10^{-5}	$AgNO_2$	3.22	6.0×10^{-4}
$K_2[PtF_6]$	4.54	2.9×10^{-5}	$Ag_2C_2O_4$	10.46	3.4×10^{-11}
K_2SiF_6	6.06	8.7×10^{-7}	Ag_3PO_4	15.84	1.4×10^{-16}
K_2ZrF_6	3.3	5×10^{-4}	quinaldate, AgL	17.9	1.3×10^{-18}
KIO_4	3.08	8.3×10^{-4}	$AgReO_4$	4.10	8.0×10^{-5}
$K_2Na[Co(NO_2)_6] \cdot H_2O$	10.66	2.2×10^{-11}	Ag_2SeO_3	15.00	1.0×10^{-15}
$K[B(C_6H_5)_4]$	7.65	2.2×10^{-8}	Ag_2SeO_4	7.25	5.7×10^{-8}
KUO_2AsO_4	22.60	2.5×10^{-23}	$AgSeCN$	15.40	4.0×10^{-16}
$K_4[UO_2(CO_3)_3]$	4.2	6.3×10^{-5}	Ag_2SO_4	4.84	1.4×10^{-5}
Praseodymium			Ag_2SO_3	13.82	1.5×10^{-14}
$Pr(OH)_3$	21.17	6.8×10^{-22}	Ag_2S	49.2	6.3×10^{-50}
Promethium			$AgSCN$	12.00	1.0×10^{-12}
$Pm(OH)_3$	21	1×10^{-21}	$AgVO_3$	6.3	5×10^{-7}
Radium			Ag_2WO_4	11.26	5.5×10^{-12}
$Ra(IO_3)_2$	9.06	8.7×10^{-10}	Sodium		
$RaSO_4$	10.37	4.2×10^{-11}	$Na[Sb(OH)_6]$	7.4	4.0×10^{-8}
Rhodium			Na_3AlF_6	9.39	4.0×10^{-10}
$Rh(OH)_3$	23	1×10^{-23}	$NaK_2[Co(NO_2)_6]$	10.66	2.2×10^{-11}
Rubidium			$Na(NH_4)_2[Co(NO_2)_6]$	11.4	4×10^{-12}
$Rb_3[Co(NO_2)_6]$	14.83	1.5×10^{-15}	$NaUO_2AsO_4$	21.87	1.3×10^{-22}
$Rb_2[PtCl_6]$	7.2	6.3×10^{-8}	Strontium		
$Rb_2[PtF_6]$	6.12	7.7×10^{-7}	$Sr_3(AsO_4)_2$	18.09	8.1×10^{-19}
$Rb_2[SiF_6]$	6.3	5.0×10^{-7}	$SrCO_3$	9.96	1.1×10^{-10}
$RbClO_4$	2.60	2.5×10^{-3}	$SrCrO_4$	4.65	2.2×10^{-5}
$RbIO_4$	3.26	5.5×10^{-4}	SrF_2	8.61	2.5×10^{-9}
Ruthenium			$Sr(IO_3)_2$	6.48	3.3×10^{-7}
$Ru(OH)_3$	36	1×10^{-36}	$SrMoO_4$	6.7	2×10^{-7}
Samarium			$Sr(NbO_3)_2$	17.38	4.2×10^{-18}
$Sm(OH)_3$	22.08	8.3×10^{-23}	$SrC_2O_4 \cdot H_2O$	6.80	1.6×10^{-7}
Scandium			$Sr_3(PO_4)_2$	27.39	4.0×10^{-28}
ScF_3	17.37	4.2×10^{-18}	8-quinolinolate, SrL_2	9.3	5×10^{-10}
$Sc(OH)_3$	30.1	8.0×10^{-31}	$SrSeO_3$	5.74	1.8×10^{-6}
Silver			$SrSeO_4$	3.09	8.1×10^{-4}
AgN_3	8.54	2.8×10^{-9}	$SrSO_3$	7.4	4×10^{-8}
Ag_3AsO_4	22.0	1.0×10^{-22}	$SrSO_4$	6.49	3.2×10^{-7}

TABLE 8.6 Solubility Products (*Continued*)

Substance	pK_{sp}	K_{sp}	Substance	pK_{sp}	K_{sp}
$SrWO_4$	9.77	1.7×10^{-10}	Uranium		
Terbium			UO_2HAsO_4	10.50	3.2×10^{-11}
$Tb(OH)_3$	21.70	2.0×10^{-22}	UO_2CO_3	11.73	1.8×10^{-12}
Tellurium			$(UO_2)_2[Fe(CN)_6]$	13.15	7.1×10^{-14}
$Te(OH)_4$	53.52	3.0×10^{-54}	$UF_4 \cdot 2.5H_2O$	21.24	5.7×10^{-22}
Thallium(I)			$UO_2(OH)_2$	21.95	1.1×10^{-22}
TlN_3	3.66	2.2×10^{-4}	$UO_2(IO_3)_2 \cdot H_2O$	7.5	3.2×10^{-8}
$TlBr$	5.47	3.4×10^{-6}	$UO_2C_2O_4 \cdot 3H_2O$	3.7	2×10^{-4}
$TlBrO_3$	4.07	8.5×10^{-5}	$(UO_2)_3(PO_4)_2$	46.7	2.0×10^{-47}
$Tl_2[PtCl_6]$	11.4	4.0×10^{-12}	UO_2HPO_4	10.67	2.1×10^{-11}
$TlCl$	3.76	1.7×10^{-4}	UO_2SO_3	8.59	2.6×10^{-9}
Tl_2CrO_4	12.00	1.0×10^{-12}	$UO_2(SCN)_2$	3.4	4×10^{-4}
$Tl_4[Fe(CN)_6] \cdot 2H_2O$	9.3	5×10^{-10}	Vanadium		
$TlIO_3$	5.51	3.1×10^{-6}	$VO(OH)_2$	22.13	5.9×10^{-23}
TlI	7.19	6.5×10^{-8}	$(VO)_3PO_4$	24.1	8×10^{-25}
$Tl_2C_2O_4$	3.7	2×10^{-4}	Ytterbium		
Tl_2SeO_3	38.7	2×10^{-39}	$Yt(OH)_3$	23.6	2.5×10^{-24}
Tl_2SeO_4	4.00	1.0×10^{-4}	Yttrium		
Tl_2S	20.3	5.0×10^{-21}	YF_3	12.14	6.6×10^{-13}
$TlSCN$	3.77	1.7×10^{-4}	$Y(OH)_3$	22.1	8.0×10^{-23}
Thallium(III)			$Y_2(C_2O_4)_3$	28.28	5.3×10^{-29}
$Tl(OH)_3$	45.20	6.3×10^{-46}	Zinc		
8-quinolinolate, TlL_3	32.4	4.0×10^{-33}	anthranilate, ZnL_2	9.23	5.9×10^{-10}
Thorium			$Zn_3(AsO_4)_2$	27.89	1.3×10^{-28}
$ThF_4 \cdot 4H_2O + 2H^+ \rightarrow$	7.23	5.9×10^{-6}	$Zn(BO_2)_2 \cdot H_2O$	10.18	6.6×10^{-11}
$ThF_2^{2+} + 2HF + 4H_2O$			$ZnCO_3$	10.84	1.4×10^{-11}
$Th(OH)_4$	44.4	4.0×10^{-45}	$Zn_2[Fe(CN)_6]$	15.39	4.0×10^{-16}
$Th(C_2O_4)_2$	22	1×10^{-22}	$Zn(IO_3)_2$	7.7	2.0×10^{-8}
$Th_3(PO_4)_4$	78.6	2.5×10^{-79}	$Zn(OH)_2$	16.92	1.2×10^{-17}
$Th(HPO_4)_2$	20	1×10^{-20}	ZnC_2O_4	7.56	2.7×10^{-8}
$Th(IO_3)_4$	14.6	2.5×10^{-15}	$Zn_3(PO_4)_2$	32.04	9.0×10^{-33}
Thulium			quinaldate, ZnL_2	13.8	1.6×10^{-14}
$Tm(OH)_3$	23.48	3.3×10^{-24}	8-quinolinolate, ZnL_2	24.3	5.0×10^{-25}
Tin			$ZnSeO_3$	6.59	2.6×10^{-7}
$Sn(OH)_2$	27.85	1.4×10^{-28}	α-ZnS	23.8	1.6×10^{-24}
$Sn(OH)_4$	56	1×10^{-56}	β-ZnS	21.6	2.5×10^{-22}
SnS	25.0	1.0×10^{-25}	$Zn[Hg(SCN)_4]$	6.66	2.2×10^{-7}
Titanium			Zirconium		
$Ti(OH)_3$	40	1×10^{-40}	$ZrO(OH)_2$	48.2	6.3×10^{-49}
$TiO(OH)_2$	29	1×10^{-29}	$Zr_3(PO_4)_4$	132	1×10^{-132}

8.2.1   Proton-Transfer Reactions

The pK_a values listed in Tables 8.7 and 8.8 are the negative (decadic) logarithms of the acidic dissociation constant, i.e., $-\log_{10} K_a = pK_a$. For the general proton-transfer reaction

$$HB = H^+ + B$$

the acidic dissociation constant is formulated as follows:

$$K_a = \frac{[H^+][B]}{[HB]}$$

The most common charge types for the acid HB and its conjugate base B are

$$CH_3COOH = H^+ + CH_3COO^- \text{ (acetic acid, acetate ion)}$$

$$HSO_4^- = H^+ + SO_4^{2-} \text{ (hydrogen sulfate ion, sulfate ion)}$$

$$NH_4^+ = H^+ + NH_3 \text{ (ammonium ion, ammonia)}$$

Acids which have more than one acidic hydrogen ionize in steps, as shown for phosphoric acid:

$$H_3PO_4 = H^+ + H_2PO_4^- \qquad pK_1 = 2.148 \qquad K_1 = 7.11 \times 10^{-3}$$

$$H_2PO_4^- = H^+ + HPO_4^{2-} \qquad pK_2 = 7.198 \qquad K_2 = 6.34 \times 10^{-8}$$

$$HPO_4^{2-} = H^+ + PO_4^{3-} \qquad pK_3 = 11.90 \qquad K_3 = 1.26 \times 10^{-12}$$

If the basic dissociation constant K_b for the equilibrium such as

$$NH_3 + H_2O = NH_4^+ + OH^-$$

is required, pK_b may be calculated from the relationship

$$pK_b = pK_w - pK_a$$

If a desired organic acid is not entered in Table 8.8, a useful estimate of its pK_a value can sometimes be obtained by making a comparison with recognizably similar compounds for which pK_a values are known: (1) alkyl chains, alicyclic rings, or saturated carbocyclic rings fused to aromatic or heterocyclic rings can be replaced by methyl or ethyl groups; (2) acid-strengthening inductive and mesomeric effects of a nitro group attached to an aromatic ring are very similar to those of a nitrogen atom located at the same position in a heteroaromatic ring (e.g., 3-hydroxypyridine and 3-nitrophenol).

Hammett and Taft substituent constants and, in particular, Tables 9.1 through 9.4 may also prove useful for estimating pK_a values.

TABLE 8.7 Proton Transfer Reactions of Inorganic Materials in Water at 25°C

Substance	Formula or remarks	pK_1	pK_2
Aluminic acid	H_3AlO_3	11.2	
Aluminum ion (aquo)	Al^{3+} (aquo)	4.98(4)	
Americium(III) ion	Am^{3+} (aquo) $\mu = 0.1$	5.92	
Ammonium ion	NH_4^+	9.246(2)	
Ammonium-d_3	ND_3H^+	9.757	
Antimonic acid	$HSb(OH)_6 = Sb(OH)_6^- + H^+ \mu = 0.5$	2.55	
Antimony(III) ion	$SbO^+ + H_2O = Sb(OH)_3 + H^+ \mu = 1.0$	1.42	
Barium ion	pK_b of $Ba(OH)^+ \mu = 0.1$	0.64	
Berkelium(III) ion	pK for hydrolysis of $Bk^{3+} \mu = 0.1$	5.66	
Beryllium(II) ion	Be^{2+} (aquo) $= BeOH^+ + H^+ \mu = 1.0$	6.5	
Bismuth(III) ion	$Bi^{3+} = BiOH^{2+} + H^+ \mu = 3.0$	1.58	
Boric acid, tetra-	$H_2B_4O_7$	4	9
Bromine	$Br_2 + H_2O = HBrO + H^+ + Br^-$	7.92	
Cadmium ion	Cd^{2+} (aquo) hydrolysis	9.2(1)	
Calcium ion	Ca^{2+} (aquo) hydrolysis	12.67(3)	
Californium(III) ion	Cf^{3+} (aquo) hydrolysis $\mu = 0.1$	5.62	
Carbon dioxide	CO_2 (aquo)	6.352(1)	10.329
	CO_2 in D_2O	6.77	10.93
Cerium(III) ion	Ce^{3+} (aquo) hydrolysis	ca. 9.3	
Cerium(IV) ion	Hydrolysis to $Ce(OH)^{3+}$ and $Ce(OH)_2^{2+}$	−1.15	0.82
Chromium(III) ion	Cr^{3+} (aquo) hydrolysis	3.95	
Cobalt(II) ion	Co^{2+} (aquo) hydrolysis	8.9	
Cobalt(III) ion	Co^{3+} (aquo) hydrolysis $m = 1$	1.75	
Copper(II) ion	Cu^{2+} (aquo) hydrolysis	7.34	
Curium(III) ion	Cm^{3+} (aquo) hydrolysis $m = 0.1$	6.00(5)	
Deuterium oxide	D_2O (molal scale)	14.956(1)	
Dysprosium(III) ion	Dy^{3+} (aquo) hydrolysis	8.10	
Erbium(III) ion	Er^{3+} (aquo) hydrolysis $\mu = 3$	9.0	
Europium(III) ion	Eu^{3+} (aquo) hydrolysis	8.03	
Fermium(III) ion	Fm^{3+} hydrolysis $\mu = 0.1$	3.8	
Gadolinium(III) ion	Gd^{3+} hydrolysis	8.27	

Source: J. J. Christensen, L. D. Hansen, and R. M. Izatt, *Handbook of Proton Ionization Heats and Related Thermodynamic Quantities*, Wiley-Interscience, New York, 1976; D. D. Perrin, *Ionisation Constants of Inorganic Acids and Bases in Aqueous Solution*, 2d ed., Pergamon Press, 1982.

TABLE 8.7 Proton Transfer Reactions of Inorganic Materials in Water at 25°C (*Continued*)

Substance	Formula or remarks	pK_1	pK_2
Gallium(III) ion	Ga^{3+} (successive values for hydrolysis)	2.92 pK_3 4.75	3.77
Gold(III) hydroxide	H_3AuO_3	<11.7	13.36
Hafnium(IV) ion	Hf^{4+} hydrolysis $\mu = 1$	−0.12	0.23
Hexaminotriphosphazene	$N_3P_3(NH_2)_6$	<3.2	7.68(3)
Holmium(III) ion	Ho^{3+} hydrolysis $\mu = 0.3$	8.04	
Hydrazinium(2+) ion	$^+H_3N-NH_3^+$	0.27	7.94(3)
Hydrogen amidodisulfonate	$HNSO(OH)_2$	pK_3 8.50	
Hydrogen amidophosphate	$H_2NPO(OH)_2$ (26°C)	2.739	8.102
Hydrogen arsenate	H_3AsO_4	2.223	6.980
Hydrogen-d_3 arsenate	D_3AsO_4	2.596	
Hydrogen arsenite	$HAsO_2$	9.18(10)	
Hydrogen azide	HN_3	4.72	
Hydrogen-d azide	DN_3 (in D_2O)	5.115	
Hydrogen borate (3−)	H_3BO_3	9.236	
Hydrogen bromate	$HBrO_3$ (in formamide)	1.02	
Hydrogen bromide	HBr	−8.72(15)	
Hydrogen chlorate	$HClO_3$ (theoretical prediction)	−2.7	
Hydrogen chloride	HCl	−6.2(1)	
Hydrogen-d chloride	DCl (in dimethylformamide)	3.58	
Hydrogen chlorite	$HClO_2$	1.94	
Hydrogen chromate	H_2CrO_4	−0.98	6.488
Hydrogen cyanate	HOCN	3.46	
Hydrogen cyanide	HCN	9.21	
Hydrogen-d cyanide	DCN (in D_2O) $\mu = 0.11$	8.97	
Hydrogen diamidophosphate	$(NH_2)_2PO(OH)$ (30°C)	1.279(+1)	4.889
Hydrogen diamidothiophosphate	$(NH_2)_2PO(SH)$ (20°C)	2.0(+1)	4.3
Hydrogen diimidotriphosphate	$(HO)_2PO(NH)PO(OH)(NH)PO(OH)_2$ $\mu = 0.1$	~1 pK_3 3.03 pK_5 9.84	~2 pK_4 6.61
Hydrogen diphosphate	$H_4P_2O_7$	1.52 pK_3 6.60	2.36 pK_4 9.25
Hydrogen disulfate	$H_2S_2O_7$ (theoretical prediction)	−12	−8
Hydrogen dithionate	$H_2S_2O_6$	−3.4	−0.2

Name	Formula		
Hydrogen dithionite	$H_2S_2O_4$	0.35	2.45
Hydrogen fluoride	H_2F_2	3.20(4)	12.82
Hydrogen germanate	H_2GeO_4	8.73	1.92
Hydrogen hexafluorosilicate	H_2SiF_6		2.50
Hydrogen hydrosulfite	$H_2S_2O_4$	0.35	
Hydrogen hypobromite	$HBrO$	11.8	
Hydrogen hypochlorite	$HClO$	7.537	
Hydrogen hypoiodite	HIO	10.1(5)	
Hydrogen hyponitrite	$H_2N_2O_2$	7.21	11.45(10)
Hydrogen iodate	HIO_3	0.804	
Hydrogen-d iodate	DIO_3 (in D_2O)	1.15	
Hydrogen iodide	HI	-8.56	
Hydrogen manganate(VI)	H_2MnO_4 (35°C) $\mu = 0.1$		10.15
Hydrogen nitrate	HNO_3	$-1.37(7)$	
Hydrogen nitrite	HNO_2	3.14(1)	
Hydrogen perchlorate	$HClO_4$	-7 to -8	
Hydrogen peroxide	H_2O_2	11.64(2)	
Hydrogen peroxophosphate	H_3PO_5, $\mu = 0.2$	1.1	5.5
		pK_3 12.8	
Hydrogen peroxosulfate	H_2SO_5	1.0	9.86
Hydrogen perrhenate	$HReO_4$	-1.25	
Hydrogen pertechnetate	$HTcO_4$	0.3	
Hydrogen perthiocarbonate	H_2CS_3	3.54	7.24
Hydrogen perxenate	H_4XeO_6	pK_3 10.5	
Hydrogen phosphate(3−)	H_3PO_4	2.148(20)	7.198(10)
		pK_3 11.90(6)	
Hydrogen-d_2 phosphate	D_2PO_4 (in D_2O)	7.780	
Hydrogen phosphinate	H_2PHO_2	1.23	
Hydrogen phosphonate	H_2PHO_3	1.43	6.68(14)
Hydrogen selenate	H_2SeO_4		1.66
Hydrogen selenide	H_2Se $\mu = 0.03$	3.89	11.0
Hydrogen selenite	H_2SeO_3	2.62	8.30(15)
Hydrogen silicate(4−)	H_4SiO_4	9.60(10)	11.8(1)
Hydrogen sulfamate	H_2NSO_3H	0.99	
Hydrogen sulfate	H_2SO_4		1.99(1)
Hydrogen sulfide	H_2S	6.97	12.90
Hydrogen sulfite	H_2SO_3	1.89	7.205
	$SO_2 + H_2O = HSO_3^- = H^+$		

TABLE 8.7 Proton Transfer Reactions of Inorganic Materials in Water at 25°C (*Continued*)

Substance	Formula or remarks	pK_1	pK_2
Hydrogen tellurate	H$_6$TeO$_6$	7.65(5)	11.00(5)
Hydrogen telluride	H$_2$Te (18°C)	2.64	11–12
Hydrogen tellurite	H$_2$TeO$_3$ (20°C)	6.92	9.43
Hydrogen tetracyanonickelate	H$_2$Ni(CN)$_4$	4.69	6.59
Hydrogen tetraperoxochromate	H$_3$CrO$_8$ (30°C) $\mu = 3$	7.16	
Hydrogen tetrapolyphosphate	H$_4$P$_4$O$_{13}$ $\mu = 0.034$	1.99	2.64
		pK_3 6.62	pK_4 8.2
Hydrogen tetrathiophosphate	H$_3$PS$_4$	1.5	3.5
		pK_3 6.6	
Hydrogen thiocyanate	HSCN $\mu = 3$	−0.7	
Hydrogen thiophosphate	H$_3$PO$_3$S	1.788	5.427
		pK_3 10.08	
Hydrogen thiosulfate	H$_2$S$_2$O$_3$	0.6	1.74
Hydrogen tripolyphosphate	H$_3$P$_3$O$_9$	~1	1.7
		pK_3 2.00(10)	
		pK_4 5.83(7)	
		pK_5 8.51(6)	
Hydrogen triselenocarbonate	H$_2$CSe$_3$	1.16	7.70
Hydrogen trithiocarbonate	H$_2$CS$_3$ (20°C)	2.68	8.18
Hydrogen tungstate	H$_2$WO$_4$	2.20	3.70
Hydrogen vanadate(−1)	HVO$_3$	3.80	
Hydrogen vanadate(3−)	H$_3$VO$_4$	3.78	7.78(4)
Hydroxylamine-*N*,*N*-disulfonic acid	HON(SO$_3$H)$_2$ $\mu = 1.6$	11.85	
		pK_3 11.85	
Hydroxylamine *O*-sulfonate	$^+$H$_3$NOSO$_3^-$ $\mu = 1$	1.48	
Imidodiphosphoric acid	(HO)$_2$PO(NH)PO(OH)$_2$ $\mu = 0.2$	~2	2.85
		pK_3 7.08	pK_4 9.72
Indium(III) ion	In^{3+} hydrolysis	3.54	4.28
Iridium(III) ion	Ir^{3+} hydrolysis $\mu = 1$	4.37	5.20
Iron(II) ion	Fe^{2+} hydrolysis $\mu = 1$	6.8	
Iron(III) ion	Fe^{3+} hydrolysis	2.19	
Lanthanum(III) ion	La^{3+} hydrolysis	9.06	
Lead(II) ion	Pb^{2+} hydrolysis $\mu = 0.3$	7.8	
Lead(IV) ion	Pb^{4+} hydrolysis	1.8	3.2
Lithium(I) ion	Li$^+$	0.22(10)	

Species	Equilibrium		
Lutetium(III) ion	Lu³⁺ hydrolysis	7.94	
Magnesium(II) ion	Mg²⁺ hydrolysis	11.41	
Manganese(II) ion	Mn²⁺ hydrolysis	10.59	
Managanese(III) ion	Mn³⁺ hydrolysis	0.4	
Mercury(I) ion	Hg$_2^{2+}$ hydrolysis $\mu = 0.5$	5.0	2.65
Mercury(II) ion	Hg²⁺ hydrolysis $\mu = 0.5$	3.70	
Neodymium(III) ion	Nd³⁺ hydrolysis $\mu = 3$	9.0(5)	
Neptunium(III) ion	Np³⁺ hydrolysis $\mu = 0.3$	7.43	
Neptunium(IV) ion	Np⁴⁺ hydrolysis $\mu = 2$	2.30	
Neptunium(V) ion	NpO$_2^+$ hydrolysis	8.90(2)	
Nickel(II) ion	Ni²⁺ hydrolysis	9.86	
Osmium tetroxide	OsO₄ hydrolysis $\mu = 1$	12.1	12.8
Palladium(II) ion	Pd²⁺ (stepwise pK_b values)	13.0	
Pentacyanoaquoferrate(II) ion	Fe(CN)₅(H₂O)³⁻ $\mu = 0.1$	2.63	
Plutonium(III) ion	Pu³⁺ hydrolysis $\mu = 0.07$	7.2(2)	
Plutonium(IV) ion	Pu⁴⁺ hydrolysis $\mu = 2$	1.26	
Plutonium(V) ion	PuO$_2^+$ hydrolysis $\mu = 0.003$	9.7	
Plutonium(VI) ion	PuO$_2^{2+}$ hydrolysis	3.33	4.05
Polonium(IV) ion	Po⁴⁺ hydrolysis	0.48	2.74
		pK_3 5.58	
Praseodymium(III) ion	Pr³⁺ hydrolysis $\mu = 0.3$	8.55	
Protoactinium(IV) ion	Pa⁴⁺ hydrolysis $\mu = 3$	0.14	
Protoactinium(V) ion	Pa⁵⁺ hydrolysis $\mu = 3$	1.05	
Scandium(III) ion	Sc³⁺ hydrolysis $\mu = 0.05$	4.58(3)	0.38
Silver(I) ion	Ag⁺ hydrolysis	>11.1	
Sodium ion	Na⁺ (aquo)	14.67(10)	
Strontium ion	Sr²⁺ (aquo)	13.18	
Terbium(III) ion	Tb³⁺ hydrolysis $\mu = 0.3$	8.16	
Thallium(I) ion	Tl⁺	13.36(15)	
Thallium(III) ion	Tl³⁺ hydrolysis $\mu = 3$	1.14	
Thorium(IV) ion	Th⁴⁺ hydrolysis $\mu = 0.5$	3.89	4.20
Tin(II) ion	Sn²⁺ hydrolysis $\mu = 3$	3.81(10)	
Titanium(III) ion	Ti³⁺ hydrolysis $\mu = 3$	2.55	
Titanium(IV)	TiO²⁺ + H₂O = TiO(OH)⁺ + H⁺	1.3	
Tritium oxide	pK_w for T₂O = T⁺ + OH⁻	15.21	
Uranium(IV) ion	U⁴⁺ hydrolysis	0.68	
Uranyl(VI) ion	UO$_2^{2+}$ $\mu = 0.035$	5.82	

TABLE 8.7 Proton Transfer Reactions of Inorganic Materials in Water at 25°C (*Continued*)

Substance	Formula or remarks	pK_1	pK_2
Vanadium(II) ion	V^{2+} hydrolysis	6.85	
Vanadium(III) ion	V^{3+} hydrolysis	2.92	3.5
Vanadyl(IV) ion	VO^{2+} hydrolysis	6.86(10)	
Vanadyl(V) ion	VO_2^+ (20°C) $\mu = 0.1$	1.83	
Xenon trioxide	$XeO_3 + H_2O = HXeO_4^- + H^+$	10.5	
Ytterbium(III) ion	Yb^{3+} hydrolysis	7.99(6)	
Yttrium(III) ion	Y^{3+} hydrolysis $\mu = 0.3$	8.34	
Zinc ion	Zn^{2+} hydrolysis	8.96	
Zirconium(IV) ion	Zr^{4+} hydrolysis $\mu = 1$	-0.32	0.06
		pK_3 0.35	

TABLE 8.8 pK_a Values of Organic Materials in Water at 25°C

Ionic strength μ is zero unless otherwise indicated. Protonated cations are designated by (+1), (+2), etc., after the pK_a value; neutral species by (0), if not obvious; and negatively charged acids by (−1), (−2), etc.

Substance	pK_1	pK_2	pK_3	pK_4
Abietic acid	7.62			
Acetamide	−0.37(+1)			
Acetamidine	1.60(+1)			
N-(2-Acetamido)-2-aminoethanesulfonic acid (20°C)	6.88			
2-Acetamidobenzoic acid	3.63			
3-Acetamidobenzoic acid	4.07			
4-Acetamidobenzoic acid	4.28			
2-(Acetamido)butanoic acid	3.716			
N-(2-Acetamido)iminodiacetic acid (20°C)	6.62			
3-Acetamidopyridine	4.37(+1)			
Acetanilide	0.4(+1)	13.39(0)$^{40°C}$		
Acetic acid	4.756			
Acetic acid-d (in D_2O)	5.32			
Acetoacetic acid (18°C)	3.58			
Acetohydrazine	3.24(+1)			
Acetone oxime	12.2			
2-Acetoxybenzoic acid (acetylsalicyclic acid)	3.48			
3-Acetoxybenzoic acid	4.00			
4-Acetoxybenzoic acid	4.38			
Acetylacetic acid (18°C)	3.58			
N-Acetyl-α-alanine	3.715			
N-Acetyl-β-alanine	4.455			
2-Acetylaminobutanoic acid	3.72			
3-Acetylaminopropionic acid	4.445			
2-Acetylbenzoic acid	4.13			
3-Acetylbenzoic acid	3.83			
4-Acetylbenzoic acid	3.70			
2-Acetylcyclohexanone	14.1			
N-Acetylcysteine (30°C)	9.52			
Acetylenedicarboxylic acid	1.75	4.40		
N-Acetylglycine	3.670			
N-Acetylguanidine	8.23(+1)			
N-α-Acetyl-L-histidine	7.08			
Acetylhydroxamic acid (20°C)	9.40			
N-Acetyl-2-mercaptoethylamine	9.92(SH)			
4-Acetyl-β-mercaptoisoleucine (30°C)	10.30			
2-Acetyl-1-naphthol (30°C)	13.40			
N-Acetylpenicillamine (30°C)	9.90			

TABLE 8.8 pK$_a$ Values of Organic Materials in Water at 25°C (*Continued*)

Substance	pK$_1$	pK$_2$	pK$_3$	pK$_4$
2-Acetylphenol	9.19			
4-Acetylphenol	8.05			
2-Acetylpyridine	2.643(+1)			
3-Acetypyridine	3.256(+1)			
4-Acetylpyridine	3.505(+1)			
Aconitine	8.11(+1)			
Acridine	5.60(+1)			
Acrylic acid	4.26			
Adenine	4.17(+1)	9.75(0)		
Adeninedeoxyriboside-5′-phosphoric acid	——	4.4	6.4	
Adenine-*N*-oxide	2.69(+1)	8.49(0)		
Adenosine	3.5(+1)	12.34(0)		
Adenosine-5′-diphosphoric acid	——	4.2(−1)	7.20(−2)	
Adenosine-2′-phosphoric acid	3.81(+1)	6.17(0)		
Adenosine-3′-phosphoric acid	3.65(0)	5.88(−1)		
Adenosine-5′-phosphoric acid	3.74(0)	6.05(−1)	13.06(−2)	
Adenosine-5′-triphosphoric acid	——	4.00(−1)	6.48(−2)	
Adipamic acid (adipic acid monoamide)	4.629			
Adipic acid	4.418	5.412		
α-Alanine	2.34(+1)	9.87(0)		
β-Alanine	3.55(+1)	10.238(0)		
α-Alanine, methyl ester (μ=0.10)	7.743(+1)			
β-Alanine, methyl ester (μ=0.10)	9.170(+1)			
N-D-Alanyl-α-D-alanine (μ=0.1)	3.32(+1)	8.13(0)		
N-L-Alanyl-α-L-alanine (μ=0.1)	3.32(+1)	8.13(0)		
N-L-Alanyl-α-D-alanine	3.12(+1)	8.30(0)		
N-α-Alanylglycine	3.11(+1)	8.11(0)		
Alanylglycylglycine	3.190(+1)	8.15(0)		
β-Alanylhistidine	2.64	6.86	9.40	
Albumin (bovine serum (μ=0.15)	10–10.3			
2-Aldoxime pyridine	3.42(+1)	10.22(0)		
Alizarin Black SN	5.79	12.8		
Alizarin-3-sulfonic acid	5.54	11.01		
Allantoin	8.96			
Allothreonine	2.108(+1)	9.096(0)		
Alloxanic acid	6.64			
Allylacetic acid	4.68			
Allylamine	9.69(+1)			
5-Allylbarbituric acid	4.78(+1)			
5-Allyl-5-(-methylbutyl)barbituric acid	8.08			
2-Allylphenol	10.28			
1-Allylpiperidine	9.65(+1)			
2-Allylpropionic acid	4.72			
3-Amidotetrazoline	3.95(+1)			

TABLE 8.8 pK_a Values of Organic Materials in Water at 25°C (*Continued*)

Substance	pK_1	pK_2	pK_3	pK_4
2-Aminoacetamide	7.95(+1)			
Aminoacetonitrile	5.34(+1)			
9-Aminoacridine (20°C)	9.95(+1)			
4-Aminoantipyrine	4.94(+1)			
2-Aminobenzenesulfonic acid	2.459(0)			
3-Aminobenzenesulfonic acid	3.738(0)			
4-Aminobenzenesulfonic acid	3.227(0)			
2-Aminobenzoic acid	2.09(+1)	4.79(0)		
3-Aminobenzoic acid	3.07(+1)	4.79(0)		
4-Aminobenzoic acid	2.41(+1)	4.85(0)		
2-Aminobenzoic acid, methyl ester	2.36(+1)			
3-Aminobenzoic acid, methyl ester	3.58(+1)			
4-Aminobenzoic acid, methyl ester	2.45(+1)			
3-Aminobenzonitrile	2.75(+1)			
4-Aminobenzonitrile	1.74(+1)			
4-Aminobenzophenone	2.15(+1)			
2-Aminobenzothiazole (20°C)	4.48(+1)			
2-Aminobenzoylhydrazide	1.85	3.47	12.80	
2-Aminobiphenyl	3.78(+1)			
3-Aminobiphenyl	4.18(+1)			
4-Aminobiphenyl	4.27(+1)			
4-Amino-3-bromomethylpyridine	7.47(+1)			
4-Amino-3-bromopyridine (20°C)	7.04(+1)			
2-Aminobutanoic acid	2.286(+1)	9.830(0)		
3-Aminobutanoic acid	——	10.14(0)		
4-Aminobutanoic acid	4.031(+1)	10.556(0)		
2-Aminobutanoic acid, methyl ester ($\mu=0.1$)	7.640(+1)			
4-Aminobutanoic acid, methyl ester ($\mu=0.1$)	9.838(+1)			
D-(+)-2-Amino-1-butanol	9.52(+1)			
3-Amino-N-butyl-3-methyl-2-butanone oxime	9.09(+1)			
4-Aminobutylphosphonic acid	2.55	7.55	10.9	
2-Amino-N-carbamoylbutanoic acid	3.886(+1)			
4-Amino-N-carbamoylbutanoic acid	4.683(+1)			
2-Amino-N-carbamoyl-2-methylpropanoic acid	4.463			
1-Amino-1-cycloheptanecarboxylic acid	2.59(+1)	10.46(0)		
1-Amino-1-cyclohexanecarboxylic acid	2.65(+1)	10.03(0)		
2-Amino-1-cyclohexanecarboxylic acid	3.56(+1)	10.21(0)		
1-Aminocyclopentane	10.65(+1)			
1-Aminocyclopropane	9.10(+1)			
10-Aminodecylphosphonic acid	——	8.0	11.25	

TABLE 8.8   pK_a Values of Organic Materials in Water at 25°C (*Continued*)

Substance	pK_1	pK_2	pK_3	pK_4
10-Aminodecylsulfonic acid	2.65(+1)			
1-Amino-2-di(aminomethyl)butane	3.58(+3)	8.59(+2)	9.66(+1)	
2-Amino-N,N-dihydroxyethyl- 2-hydroxyl-1,3-propanediol	6.484(+1)			
2-Amino-N,N-dimethylbenzoic acid	1.63(+1)	8.42(0)		
4-Amino-2,5-dimethylphenol	5.28(+1)	10.40(0)		
4-Amino-3,5-dimethylpyridine (20°C)	9.54(+1)			
12-Aminododecanoic acid	4.648(+1)			
2-Aminoethane-1-phosphoric acid	5.838	10.64		
1-Aminoethanesulfonic acid	−0.33	9.06		
2-Aminoethanesulfonic acid	1.5	9.061		
2-Aminoethanethiol (cysteamine) ($\mu=0.01$)	8.23(+1)			
2-Aminoethanol (ethanolamine)	9.50(+1)			
2-[2-(2- Aminoethyl)aminoethyl]pyridine	3.50	6.59	9.51	
2-Amino-2-ethyl-1-butanol	9.82(+1)			
3-(2-Aminoethyl)indole	——	10.2		
3-Amino-N-ethyl-3-methyl-2- butanone oxime	9.23(+1)			
N-(2-Aminoethyl)morpholine	4.06(+2)	9.15(+1)		
p-(2-Aminoethyl)phenol	9.3	10.9		
2-Aminoethylphosphonic acid	2.45(+1)	7.0(0)	10.8(−1)	
N-(2-Aminoethyl)piperidine (30°C)	6.38	9.89		
2-(2-Aminoethyl)pyridine ($\mu=0.5$)	4.24(+2)	9.78(+1)		
4-Amino-3-ethylpyridine (20°C)	9.51(+1)			
N-(2-Aminoethyl)pyrrolidine (30°C)	6.56(+2)	9.74(+1)		
2-Aminofluorine	10.34(+1)			
2-Amino-D-β-glucose ($\mu=0.05$)	2.20(+1)	9.08(0)		
2-Amino-N-glycylbutanoic acid	3.155(+1)	8.331(0)		
7-Aminoheptanoic acid	4.502			
2-Aminohexanoic acid	2.335(+1)	9.834(0)		
6-Aminohexanoic acid	4.373(+1)	10.804(0)		
C-Amino-C- hydrazinocarbonylmethane	2.38(+2)	7.69(+1)		
2-Amino-3-hydroxybenzoic acid	2.5(+1)	5.192(0)	10.118(OH)	
L-2-Amino-3-hydroxybutanoic acid (threonine)	2.088(+1)	9.100(0)		
DL-2-Amino-4-hydroxybutanoic acid ($\mu=0.1$)	2.265(+1)	9.257(0)		
DL-4-Amino-3-hydroxybutanoic acid ($\mu=0.1$)	3.834(+1)	9.487(0)		
2-Amino-2'-hydroxydiethyl sulfide	9.27(+1)			
4-Amino-2-hydroxypyrimidine (cytosine)	4.58(+1)	12.15(0)		
3-Amino-N-isopropyl-3-methyl- 2-butanone oxime	9.09(+1)			

TABLE 8.8 pK_a Values of Organic Materials in Water at 25°C (*Continued*)

Substance	pK_1	pK_2	pK_3	pK_4
4-Amino-3-isopropylpyridine (20°C)	9.54(+1)			
1-Aminoisoquinoline (20°C, $\mu = 0.01$)	7.62(+1)			
3-Aminoisoquinoline				
(20°C, $\mu = 0.005$)	5.05(+1)			
4-Aminoisoxazolidine-3-one	7.4(+1)			
Aminomalonic acid	3.32(+1)	9.83(0)		
DL-2-Amino-4-mercaptobutanoic				
acid	2.22(+1)	8.87(0)	10.86(SH)	
2-Amino-3-mercapto-				
3-Methylbutanoic acid	1.8(+1)	7.9(0)	10.5(SH)	
2-Amino-6-methoxybenzothiazole	4.50(+1)			
3-Amino-4-methylbenzenesulfonic	3.633			
acid				
4-Amino-3-methylbenzenesulfonic	3.125			
acid				
2-Amino-4-methylbenzothiazole	4.7(+1)			
1-Amino-3-methylbutane	10.64(+1)			
3-Amino-3-methyl-2-butanone oxime	9.09(+1)			
3-Amino-N-methyl-3-methyl-2-				
butanone oxime	9.23(+1)			
2-Amino-3-methylpentanoic acid	2.320(+1)	9.758(0)		
3-Aminomethyl-6-methylpyridine	8.70(+1)			
(30°C)				
Aminomethylphosphonic acid	2.35	5.9	10.8	
2-Amino-2-methyl-1,3-propanediol	8.801			
2-Amino-2-methyl-1-propanol	9.694(+1)			
2-Amino-2-methylpropanoic acid	2.357(+1)	10.205(0)		
(2-Aminomethyl)pyridine ($\mu = 0.5$)	2.31(+2)	8.79(+1)		
2-Amino-3-methylpyridine	7.24(+1)			
4-Amino-3-methylpyridine	9.43(+1)			
2-Amino-4-methylpyridine	7.48(+1)			
2-Amino-5-methylpyridine	7.22(+1)			
2-Amino-6-methylpyridine	7.41(+1)			
2-Amino-4-methylpyrimidine (20°C)	4.11(+1)			
Aminomethylsulfonic acid	5.75(+1)			
N-Aminomorpholine	4.19(+1)			
4-Amino-1-naphthalenesulfonic acid	2.81			
1-Amino-2-naphthalenesulfonic acid	1.71			
1-Amino-3-naphthalenesulfonic acid	3.20			
1-Amino-5-naphthalenesulfonic acid	3.69			
1-Amino-6-naphthalenesulfonic acid	3.80			
1-Amino-7-naphthalenesulfonic acid	3.66			
1-Amino-8-naphthalenesulfonic acid	5.03			
2-Amino-1-naphthalenesulfonic acid	2.35			
2-Amino-4-naphthalenesulfonic acid	3.79			
2-Amino-6-naphthalenesulfonic acid	3.79	8.94		
2-Amino-8-naphthalenesulfonic acid	3.89			

TABLE 8.8 pK_a Values of Organic Materials in Water at 25°C (*Continued*)

Substance	pK_1	pK_2	pK_3	pK_4
3-Amino-1-naphthoic acid	2.61	4.39		
4-Amino-2-naphthoic acid	2.89	4.46		
8-Amino-2-naphthol	4.20(+1)			
DL-2-Aminopentanoic acid				
(DL-norvaline)	2.318(+1)	9.808		
3-Aminopentanoic acid	4.02(+1)	10.399(0)		
4-Aminopentanoic acid	3.97(+1)	10.46(0)		
5-Aminopentanoic acid	4.20(+1)	9.758(0)		
5-Aminopentanoic acid, ethyl ester	10.151			
2-Aminophenol	9.28	9.72		
3-Aminophenol	9.83	9.87		
4-Aminophenol	8.50	10.30		
4-Aminophenylacetic acid (20°C)	3.60	5.26		
2-Aminophenylarsonic acid	ca 2	3.77	8.66	
3-Aminophenylarsonic acid	ca 2	4.02	8.92	
4-Aminophenylarsonic acid	ca 2	4.02	8.62	
3-Aminophenylboric acid	4.46	8.81		
4-Aminophenylboric acid	3.71	9.17		
4-Aminophenyl				
(4-chlorophenyl) sulfone	1.38			
2-Aminophenylphosphonic acid	——	4.10	7.29	
3-Aminophenylphosphonic acid	——	——	7.16	
4-Aminophenylphosphonic acid	——	——	7.53	
1-Amino-1,2,3-propanetricarboxylic				
acid ($\mu=2.2$)	2.10(+1)	3.60(0)	4.60(−1)	9.82(−2)
3-Aminopropanoic acid	3.551(+1)	10.235(0)		
1-Amino-1-propanol	9.96(+1)			
DL-2-Amino-1-propanol	9.469(+1)			
3-Amino-1-propanol	9.96(+1)			
3-Aminopropene	9.691(+1)			
3-Amino-N-propyl-3-methyl-				
2-butanone oxime	9.09(+1)			
2-Aminopropylsulfonic acid	——	9.15		
2-Aminopyridine	6.71(+1)			
3-Aminopyridine	6.03(+1)			
4-Aminopyridine	9.114(+1)			
2-Aminopyridine-1-oxide	2.58(+1)			
3-Aminopyridine-1-oxide	1.47(+1)			
4-Aminopyridine-1-oxide	3.54(+1)			
8-Aminoquinaldine	4.86(+1)			
2-Aminoquinoline (20°C, $\mu=0.01$)	7.34(+1)			
3-Aminoquinoline (20°C, $\mu=0.01$)	4.95(+1)			
4-Aminoquinoline (20°C, $\mu=0.01$)	9.17(+1)			
5-Aminoquinoline (20°C, $\mu=0.01$)	5.46(+1)			
6-Aminoquinoline (20°C, $\mu=0.01$)	5.63(+1)			
8-Aminoquinoline (20°C, $\mu=0.01$)	3.99(+1)			
4-Aminosalicyclic acid	1.991(+1)	3.917(0)	13.74	

TABLE 8.8   pK_a Values of Organic Materials in Water at 25°C (*Continued*)

Substance	pK_1	pK_2	pK_3	pK_4
5-Aminosalicyclic acid	2.74(+1)	5.84(0)		
2-Amino-3-sulfopropanoic acid	1.89(+1)	8.70(0)		
4-Amino-2,3,5,6-tetramethylpyridine (20°C)	10.58(+1)			
5-Amino-1,2,3,4-tetrazole (20°C)	1.76	6.07		
2-Aminothiazole (20°C)	5.36(+1)			
1-Amino-3-thiobutane (30°C)	9.18(+1)			
5-Amino-3-thio-1-pentanol (30°C)	9.12(+1)			
2-Aminothiophenol	<2(+1)	7.90(0)		
2-Amino-4,4,4-trifluorobutanoic acid		8.171(0)		
3-Amino-4,4,4-trifluorobutanoic acid		5.831(0)		
3-Amino-2,4,6-trinitroluene		9.5(+1)		
Angiotensin II	10.37			
Anhydroplatynecine	9.40			
Aniline	4.60(+1)			
2-Anilinoethylsulfonic acid	3.80(+1)			
3-Anilinoethylsulfonic acid	4.85(+1)			
Anthracene-1-carboxylic acid	3.68			
Anthracene-2-carboxylic acid	4.18			
Anthracene-9-carboxylic acid	3.65			
Anthraquinone-1-carboxylic acid (20°C)	3.37			
Anthraquinone-2-carboxylic acid (20°C)	3.42			
9,10-Anthraquinone monoxime	9.78			
9,10-Anthraquinone-1-sulfonic acid	0.27			
9,10-Anthraquinone-2-sulfonic acid	0.38			
Antipyrine	1.45(+1)			
Apomorphine (15°C)		8.92		
D-(−)-Arabinose	12.34			
L-(+)-Arginine		8.994(+1)	12.47(−1)	
Arsenazo III [pK_5 10.5(−4); pK_6 12.0(−5)]		1.2	2.7	7.9(−3)
Arsenoacetic acid		4.67	7.68	
Arsenoacrylic acid		4.23	8.60	
Arsenobutanoic acid		4.92	7.64	
2-Arsenocrotonic acid		4.61	8.75	
3-Arsenocrotonic acid		4.03	8.81	
Arsenopentanoic acid		4.89	7.75	
L-(+)-Ascorbic acid (vitamin C)	4.17	11.57		
L-(+)-Asparagine		8.80(0)		
L-Asparaginylglycine		4.53	9.07	
D-Aspartic acid		3.87(0)	10.00(−)	
Aspartic diamide (μ=0.2)	7.00			
Aspartylaspartic acid		3.40	4.70	8.26
α-Aspartylhistidine (38°C, μ=0.1)		3.02	6.82	7.98
β-Aspartylhistidine (38°C, μ=0.1)		2.95	6.93	8.72

TABLE 8.8 pK_a Values of Organic Materials in Water at 25°C (*Continued*)

Substance	pK_1	pK_2	pK_3	pK_4
N-Aspartyl-*p*-tyrosine ($\mu=0.01$)		3.57	8.92	10.23(OH)
Aspidospermine	7.65			
Atropine (17°C)	4.35(+1)			
1-Azacycloheptane	11.11(+1)			
1-Azacyclooctane	11.1(+1)			
Azetidine	11.29(+1)			
Aziridine	8.04(+1)			
Barbituric acid		8.372(0)		
m-Benzbetaine	3.217(+1)			
p-Benzbetaine	3.245(+1)			
Benzenearsonic acid (22°C)		8.48(−1)		
Benzene-1-arsonic acid-4-carboxylic acid		4.22 (COOH)	5.59	
Benzeneboronic acid	13.7			
Benzene-1-carboxylic acid-2-phosphoric acid		3.78	9.17	
Benzene-1-carboxylic acid-3-phosphoric acid		4.03	7.03	
Benzene-1-carboxylic acid-4-phosphoric aic	1.50	3.95	6.89	
Benzenediazine	11.08(+1)			
1,3-Benzenedicarboxylic acid (isophthalic acid)	3.62(0)	4.60(−1)		
1,4-Benzenedicarboxylic acid (terephthalic acid)	3.54(0)	4.46(−1)		
1,3-Benzenedicarboxylic acid mononitrile	3.60(0)			
1,4-Benzenedicarboxylic acid mononitrile	3.55(0)			
Benzenehexarboxylic acid (pK_5 6.32; pK_6 7.49)	0.68	2.21	3.52	5.09
Benzenepentacarboxylic acid (pK_5 6.46)	1.80	2.73	3.96	5.25
Benzenesulfinic acid	1.50			
Benzenesulfonic acid	2.554			
1,2,3,4-Benzenetetracarboxylic acid	2.05	3.25	4.73	6.21
1,2,3,5-Benzenetetracarboxylic acid	2.38	3.51	4.44	5.81
1,2,4,5-Benzenetetracarboxylic acid	1.92	2.87	4.49	5.63
1,2,3-Benzenetricarboxylic acid	2.88	4.75	7.13	
1,2,4-Benzenetricarboxylic acid	2.52	3.84	5.20	
1,3,5-Benzenetricarboxylic acid	2.12	4.10	5.18	
Benzil-α-dioxime	12.0			
Benzilic acid	3.09			
Benzimidazole	5.53(+1)	12.3(0)		
Benzohydroxamic acid (20°C)	8.89(0)			

TABLE 8.8 pK_a Values of Organic Materials in Water at 25°C (*Continued*)

Substance	pK_1	pK_2	pK_3	pK_4
Benzoic acid	4.204			
5,6-Benzoquinoline (20°C)	5.00(+1)			
7,8-Benzoquinoline (20°C)	4.15(+1)			
1,4-Benzoquinone monoxime	6.20			
Benzosulfonic acid	0.70			
1,2,3-Benzotriazole	8.38(+1)			
1-Benzoylacetone	8.23			
Benzoylamine	9.34(+1)			
2-Benzoylbenzoic acid	3.54			
Benzoylglutamic acid	3.49	4.99		
N-Benzoyglycine (hippuric				
acid	3.65			
Benzoylhydrazine	3.03(+2)	12.45(+1)		
Benzoylpyruvic acid	6.40	12.10		
3-Benzoyl-1,1,1-trifluoroacetone	6.35			
Benzylamine	9.35(+1)			
Benzylamine-4-carboxylic acid	3.59	9.64		
2-Benzyl-2-phenylsuccinic acid				
(20°C)	3.69	6.47		
2-Benzylpyridine	5.13(+1)			
4-Benzylpyridine-1-oxide	−1.018(+)			
1-Benzylpyrrolidine	9.51(+1)			
2-Benzylpyrrolidine	10.31(+1)			
Benzylsuccinic acid (20°C)	4.11	5.65		
3-(Benzylthio)propanoic acid	4.463			
Berberine (18°C)	11.73(+1)			
Betaine	1.832(+1)			
Biguanide	2.96(+2)	11.51(+1)		
2,2'-Biimidazolyl ($\mu=0.3$)	5.01(+1)			
2-Biphenylcarboxylic acid	3.46			
(1,1'-Biphenyl)-4,4'-diamine	3.63(+2)	4.70(+1)		
Bis(2-aminoethyl) ether (30°C)	8.62(+2)	9.59(+1)		
N,N'-Bis(2-aminoethyl)-				
ethylenediamine (20°C)	3.32(+4)	6.67(+3)	9.20(+2)	9.92(+1)
N,N-Bis(2-hydroxyethyl)-2-				
aminoethane sulfonic acid				
(BES) (20°C)	7.15			
N,N-Bis(2-hydroxyethyl)glycine				
(bicine)				
(20°C)	8.35			
Bis(2-hydroxyethyl)iminotris				
(hydroxymethyl)-				
methane (bis-tris)	6.46(+1)			
1,3-Bis[tris(hydroxymethyl)				
methylamino]propane (20°C)	6.80(+1)			
Bromoactic acid	2.902			
2-Bromoaniline	2.53(+1)			

TABLE 8.8 pK_a Values of Organic Materials in Water at 25°C (*Continued*)

Substance	pK_1	pK_2	pK_3	pK_4
3-Bromoaniline	3.53(+1)			
4-Bromoaniline	3.88(+1)			
2-Bromobenzoic acid	2.85			
3-Bromobenzoic acid	3.810			
4-Bromobenzoic acid	3.99			
2-Bromobutanoic acid (35°C)	2.939			
erythro-2-Bromo-3-chlorosuccinic acid				
(19°C, $\mu=0.1$)	1.4	2.6		
threo-2-Bromo-chlorosuccinic acid				
(19°C, $\mu=0.1$)	1.5	2.8		
trans-2-Bromocinnamic acid	4.41			
3-Bromo-4-(dimethylamino)pyridine				
(20°C)	6.52(+1)			
2-Bromo-4,6-dinitroaniline	−6.94(+1)			
3-Bromo-2-hydroxymethylbenzoic				
acid (20°C)	3.28			
6-Bromo-2-hydroxymethylbenzoic				
acid (20°C)	2.25			
7-Bromo-8-hydroxyquinoline-				
5-sulfonic acid	2.51	6.70		
3-Bromomandelic acid	3.13			
3-Bromo-4-methylaminopyridine				
(20°C)	7.49(+1)			
(2-Bromomethyl)butanoic acid	3.92			
Bromomethylphosphonic acid	1.14	6.52		
2-Bromo-6-nitrobenzoic acid	1.37			
2-Bromophenol	8.452			
3-Bromophenol	9.031			
4-Bromophenol	9.34			
2-(2′-Bromophenoxy)acetic acid	3.12			
2-(3′-Bromophenoxy)acetic acid	3.09			
2-(4′-Bromophenoxy) acetic acid	3.13			
2-Bromo-2-phenylacetic acid	2.21			
2-(Bromophenyl)acetic acid	4.054			
4-(Bromophenyl)acetic acid	4.188			
4-Bromophenylarsonic acid	3.25	8.19		
4-Bromophenylphosphinic acid				
(17°C)	2.1			
2-Bromophenylphosphonic acid	1.64	7.00		
3-Bromophenylphosphonic acid	1.45	6.69		
4-Bromophenylphosphonic acid	1.60	6.83		
3-Bromophenylselenic acid	4.43			
4-Bromophenylselenic acid	4.50			
2-Bromopropanoic acid	2.971			
3-Bromopropanoic acid	3.992			
Bromopropynoic acid	1.855			

TABLE 8.8　pK_a Values of Organic Materials in Water at 25°C (*Continued*)

Substance	pK_1	pK_2	pK_3	pK_4
2-Bromopyridine	0.71(+1)			
3-Bromopyridine	2.85(+1)			
4-Bromopyridine	3.71(+1)			
3-Bromoquinoline	2.69(+1)			
Bromosuccinic acid	2.55	4.41		
2-Bromo-*p*-tolylphosphonic acid	1.81	7.15		
Brucine (15°C)	2.50(+2)	8.16(+1)		
2-Butanamine (*sec*-butylamine)	10.56(+1)			
1,2-Butanediamine	6.399(+2)	9.388(+1)		
1,4-Butanediamine	9.35(+2)	10.82(+1)		
2,3-Butanediamine	6.91(+2)	10.00(+1)		
1,2,3,4-Butanetetracarboxylic acid	3.43	4.58	5.85	7.16
cis-2-Butenoic acid (isocrotonic acid)	4.44			
trans-2-Butenoic acid (*trans*-crotonic acid) (35°C)	4.676			
3-Butenoic acid (vinylacetic acid)	4.68			
3-Butoxybenzoic acid (20°C)	4.25			
Butylamine	10.64(+1)			
tert-Butylamine	10.685(+1)			
4-*tert*-Butylaniline	3.78(+1)			
N-tert-Butylaniline	7.10(+1)			
Butylarsonic acid (18°C)	4.23	8.91		
2-*tert*-Butylbenzoic acid	3.57			
3-*tert*-Butylbenzoic acid	4.199			
4-*tert*-Butylbenzoic acid	4.389			
N-Butylethylenediamine	7.53(+2)	10.30(+1)		
N-Butylglycine	2.35(+1)	10.25(0)		
tert-Butylhydroperoxide	12.80			
1-(*tert*-Butyl)-2-hydroxybenzene	10.62			
1-(*tert*-Butyl)-3-hydroxybenzene	10.119			
1-(*tert*-Butyl)-4-hydroxybenzene	10.23			
Butylmethylamine	10.90(+1)			
2-Butyl-1-methyl-2-pyrroline	11.84(+1)			
4-*tert*-Butylphenylactic acid	4.417			
Butylphosphinic acid	3.41			
tert-Butylphosphinic acid	4.24			
tert-Butylphosphonic acid	2.79	8.88		
1-Butylpiperidine (μ=0.02)	10.43(+1)			
2-*tert*-Butylpyridine	5.76(+1)			
3-*tert*-Butylpyridine	5.82(+1)			
4-*tert*-Butylpyridine	5.99(+1)			
2-*tert*-Butylthiazole (μ=0.1)	3.00(+1)			
4-*tert*-Butylthiazole (μ=0.1)	3.04(+1)			
2-Butyn-1,4-dioic acid	1.75	4.40		
2-Butynoic acid (tetrolic acid)	2.620			
Butyric acid	4.817			
4-Butyrobetaine (20°C)	3.94(+1)			

TABLE 8.8   pK_a Values of Organic Materials in Water at 25°C (*Continued*)

Substance	pK_1	pK_2	pK_3	pK_4
Caffeine (40°C)	10.4			
Calcein (pK_5>12)	<4	5.4	9.0	10.5
Calmagite	8.14	12.35		
D-Camphoric acid	4.57	5.10		
Canaline	2.40	3.70	9.20	
Canavanine	2.50(+2)	6.60(+1)	9.25(0)	
N-Carbamoylacetic acid	3.64			
N-Carbamoyl-α-D-alanine	3.89(+1)			
N-Carbamoyl-β-alanine	4.99(+1)			
DL-N-Carbamoylalanine	3.892(+1)			
N-Carbamoylglycine	3.876			
2-Carbamoylpyridine (20°C)	2.10(+1)			
3-Carbamoylpyridine	3.328(+1)			
4-Carbamoylpyridine (20°C)	3.61(+1)			
β-Carboxymethylaminopropanoic				
acid	3.61(+1)	9.46(0)		
Chloroacetic acid	2.867			
N-(2′-Chloroacetyl)glycine	3.38(0)			
cis-3-Chloroacrylic acid				
(18°C, μ=0.1)	3.32			
trans-3-chloroacrylic acid				
(18°C, μ=0.1)	3.65			
2-Chloraniline	2.64(+1)			
3-Chloroaniline	3.52(+1)			
4-Chloroaniline	3.99(+1)			
2-Chlorobenzoic acid	2.877			
3-Chlorobenzoic acid	3.83			
4-Chlorobenzoic acid	3.986			
2-Chlorobutanoic acid	2.86			
3-Chlorobutanoic acid	4.05			
4-Chlorobutanoic acid	4.50			
2-Chloro-3-butenoic acid	2.54			
3-Chlorobutylarsonic acid (18°C)	3.95	8.85		
trans-2′-Chlorocinnamic acid	4.234			
trans-3′-Chlorocinnamic acid	4.294			
trans-4′-Chlorocinnamic acid	4.413			
2-Chlorocrotonic acid	3.14			
3-Chlorocrotonic acid	3.84			
Chlorodifluoroacetic acid	0.46			
1-Chloro-1,2-dihydroxybenzene	8.522			
1-Chloro-2,6-dimethyl-				
4-hydroxybenzene	9.549			
4-Chloro-2,6-dinitrophenol	2.97			
2-Chloroethylarsonic acid	3.68	8.37		
3-Chlorohexyl-1-arsonic acid (18°C)	3.51	8.31		
2-Chloro-3-hydroxybutanoic acid	2.59			
3-Chloro-2-(hydroxymethyl)benzoic				
acid (20°C)	3.27			

TABLE 8.8 pK_a Values of Organic Materials in Water at 25°C (*Continued*)

Substance	pK_1	pK_2	pK_3	pK_4
6-Chloro-2-(hydroxymethyl)benzoic acid (20°C)	2.26			
7-Chloro-8-hydroxyquinoline-5-sulfonic acid	2.92	6.80		
2-Chloroisocrotonic acid	2.80			
3-Chloroisocrotonic acid	4.02			
3-Chlorolactic acid	3.12			
3-Chloromandelic acid	3.237			
3-Chloro-4-methoxyphenyl-phosphonic acid	2.25	6.7		
3-Chloro-4-methylaniline	4.05(+1)			
4-Chloro-N-methylaniline	3.9(+1)			
4-Chloro-3-methylphenol	9.549			
Chloromethylphosphonic acid	1.40	6.30		
2-Chloro-2-methylpropanoic acid	2.975			
2-Chloro-6-nitroaniline	−2.41(+1)			
4-Chloro-2-nitroaniline	−1.10(+1)			
2-Chloro-3-nitrobenzoic acid	2.02			
2-Chloro-4-nitrobenzoic acid	1.96			
2-Chloro-5-nitrobenzoic acid	2.17			
2-Chloro-6-nitrobenzoic acid	1.342			
4-Chloro-2-nitrophenol	6.48			
2-Chlorophenol	8.55			
3-Chlorophenol	9.10			
4-Chlorophenol	9.43			
(4-Chloro-3-nitrophenoxy)acetic acid	2.959			
2-Chloro-4-nitrophenylphosphonic acid	1.12	6.14		
3-Chloropentyl-1-arsonic acid (18°C)	3.71	8.77		
2-Chlorophenoxyacetic acid	3.05			
3-Chlorophenoxyacetic acid	3.07			
4-Chlorophenoxyacetic acid	3.10			
4-Chlorophenoxy-2-methylacetic acid	3.26			
2-Chlorophenylacetic acid	4.066			
3-Chlorophenylacetic acid	4.140			
4-Chlorophenylacetic acid	4.190			
2-Chlorophenylalanine	2.23(+1)	8.94(0)		
3-Chlorophenylalanine	2.17(+1)	8.91(0)		
DL-4-Chlorophenylalanine	2.08(+1)	8.96(0)		
4-Chlorophenylarsonic acid	3.33	8.25		
2-Chlorophenylphosphonic acid	1.63	6.98		
3-Chlorophenylphosphonic acid	1.55	6.65		
4-Chlorophenylphosphonic acid	1.66	6.75		
3-(2′-Chlorophenyl)propanoic acid	4.577			
3-(3′-Chlorophenyl)propanoic acid	4.585			
3-(4′-Chlorophenyl)propanoic acid	4.607			

TABLE 8.8   pK_a Values of Organic Materials in Water at 25°C (*Continued*)

Substance	pK_1	pK_2	pK_3	pK_4
3-Chlorophenylselenic acid	4.47			
4-Chlorophenylselenic acid	4.48			
4-Chloro-1,2-phthalic acid	1.60			
2-Chloropropanoic acid	2.84			
3-Chloropropanoic acid	3.992			
2-Chloropropylarsonic acid (18°C)	3.76	8.39		
3-Chloropropylarsonic acid (18°C)	3.63	8.53		
Chloropropynoic acid	1.845			
2-Chloropyridine	0.49(+1)			
3-Chloropyridine	2.84(+1)			
4-Chloropyridine	3.83(+1)			
7-Chlorotetracycline	3.30(+1)	7.44	9.27	
4-Chloro-2-(2'-thiazolylazo)phenol	7.09			
4-Chlorothiophenol	5.9			
N-Chloro-p-toluenesulfonamide	4.54(+1)			
3-Chloro-o-toluidine	2.49(+1)			
4-Chloro-o-toluidine	3.385(+1)			
5-Chloro-o-toluidine	3.85(+1)			
6-Chloro-o-toludine	3.62(+1)			
Chrome Azurol S	2.45	4.86	11.47	
Chrome Dark Blue	7.56	9.3	12.4	
Cinchonine	5.85(+2)	9.92(+1)		
cis-Cinnamic acid	3.879			
trans-Cinnamic acid	4.438			
Citraconic acid	2.29(0)	6.15(−1)		
Citric acid	3.128	4.761	6.396	
L-(+)-Citrulline	2.43(+1)	9.41(0)		
Cocaine	8.41(+1)			
Codeine	7.95(+1)			
Colchicine	1.65(+1)			
Coniine ($\mu=0.5$)	11.24(+1)			
Creatine (40°C)	3.28(+1)			
Creatinine	3.57(+1)			
o-Cresol	10.26			
m-Cresol	10.00			
p-Cresol	10.26			
Cumene hydroperoxide	12.60			
Cupreine	7.63(+1)			
Cyanamide	10.27			
Cyanoacetic acid	2.460			
Cyanoacetohydrazide	2.34(+2)	11.17(+1)		
2-Cyanobenzoic acid	3.14			
3-Cyanobenzoic acid	3.60			
4-Cyanobenzoic acid	3.55			
4-Cyanobutanoic acid	4.44			
trans-1-Cyanocyclohexane-2-carboxylic acid	3.865			
4-Cyano-2,6-dimethylphenol	8.27			

TABLE 8.8   pK_a Values of Organic Materials in Water at 25°C (*Continued*)

Substance	pK$_1$	pK$_2$	pK$_3$	pK$_4$
4-Cyano-3,5-dimethylphenol	8.21			
2-Cyanoethylamine	7.7(+1)			
N-(2-Cyano)ethylnorcodeine	5.68(+1)			
Cyanomethylamine	5.34(+1)			
2-Cyano-2-methyl-2-phenylacetic				
acid	2.290			
1-Cyanomethylpiperidine	4.55(+1)			
2-Cyano-2-methylpropanoic acid	2.422			
3-Cyanophenol	8.61			
o-Cyanophenoxyacetic acid	2.98			
m-Cyanophenoxyacetic acid	3.03			
p-Cyanophenoxyacetic acid	2.93			
2-Cyanopropanoic acid	2.37			
3-Cyanopropanoic acid	3.99			
2-Cyanopyridine	−0.26(+1)			
3-Cyanopyridine	1.45(+1)			
4-Cyanopyridine	1.90(+1)			
Cyanuric acid	6.78			
Cyclobutanecarboxylic acid	4.785			
1,1-Cyclobutanedicarboxylic acid	3.13	5.88		
cis-1,2-Cyclobutanedicarboxylic acid	3.90	5.89		
trans-1,2-Cyclobutanedicarboxylic				
acid	3.79	5.61		
cis-1,3-Cyclobutanedicarboxylic acid	4.04	5.31		
trans-1,3-Cyclobutanedicarboxylic				
acid	3.81	5.28		
Cyclohexanecarboxylic acid	4.90			
1,1-Cyclohexanediacetic acid	3.49	6.96		
cis-1,2-Cyclohexanediacetic acid				
(20°C)	4.42	5.45		
trans-1,2-Cyclohexanediacetic acid				
(20°C)	4.38	5.42		
cis-1,2-Cyclohexanediamine	6.43(+2)	9.93(+1)		
trans-1,2-Cyclohexanediamine	6.34(+2)	9.74(+1)		
1,1-Cyclohexanedicarboxylic acid	3.45	4.11		
cis-1,2-Cyclohexanedicarboxylic acid				
(20°C)	4.34	6.76		
trans-1,2-Cyclohexanedicarboxylic				
acid (20°C)	4.18	5.93		
cis-1,3-Cyclohexanedicarboxylic acid				
(16°C)	4.10	5.46		
trans-1,3-Cyclohexanedicarboxylic				
acid (19°C)	4.31	5.73		
trans-1,4-Cyclohexanedicarboxylic				
acid (16°C)	4.18	5.42		
1,3-Cyclohexanedione	5.26			
cis,*cis*-1,3,5-Cyclohexanetriamine	6.9(+3)	8.7(+2)	10.4(+1)	

TABLE 8.8 pK_a Values of Organic Materials in Water at 25°C (*Continued*)

Substance	pK$_1$	pK$_2$	pK$_3$	pK$_4$
Cyclohexanonimine	9.15			
cis-4-Cyclohexene-1,2-dicarboxylic acid (20°C)	3.89	6.79		
trans-4-Cyclohexene-1,2-dicarboxylic acid (20°C)	3.95	5.81		
Cyclohexylacetic acid	4.51			
Cyclohexylamine	10.64(+1)			
2-(Cyclohexylamino)ethanesulfonic acid (CHES) (20°C)	9.55			
3-Cyclohexylamino-1-propanesulfonic acid (CAPS) (20°C)	10.40			
4-Cyclohexylbutanoic acid	4.95			
Cyclohexylcyanoacetic acid	2.367			
1,2-Cyclohexylenedinitriloacetic acid (μ=0.1)	2.4	3.5	6.16	12.35
3-Cyclohexylpropanoic acid	4.91			
2-Cyclohexylpyrrolidine	10.76(+1)			
2-Cyclohexyl-2-pyrroline	7.91(+1)			
Cyclohexylthioacetic acid	3.488			
Cyclopentanecarboxylic acid	4.905			
cis-Cyclopentane-1-carboxylic acid-2-acetic acid	4.40	5.79		
trans-Cyclopentane-1-carboxylic acid-2-acetic acid	4.39	5.67		
Cyclopentane-1,2-diamine-N,N′,N′-tetraacetic acid (μ=0.1)	——	——	——	10.20
Cyclopentane-1,1-dicarboxylic acid	3.23	4.08		
cis-Cyclopentane-1,2-dicarboxylic acid	4.43	6.67		
trans-Cyclopentane-1,2-dicarboxylic acid	3.96	5.85		
cis-Cyclopentane-1,3-dicarboxylic acid	4.26	5.51		
trans-Cyclopentane-1,3-dicarboxylic acid	4.32	5.42		
Cyclopentylamine	10.65(+1)			
1,1-Cyclopentyldiacetic acid	3.80	6.77		
cis-Cyclopentyl-1,2-diacetic acid	4.42	5.42		
trans-Cyclopentyl-1,2-diacetic acid	4.43	5.43		
Cyclopropanecarboxylic acid	4.827			
Cyclopropane-1,1-dicarboxylic acid	1.82	5.43		
cis-Cyclopropane-1,2-dicarboxylic acid	3.33	6.47		
trans-Cyclopropane-1,2-dicarboxylic acid	3.65	5.13		
Cyclopropylamine	9.10(+1)			

TABLE 8.8 pK_a Values of Organic Materials in Water at 25°C (*Continued*)

Substance	pK_1	pK_2	pK_3	pK_4
5-Cyclopropyl-1,2,3,4-tetrazole	4.90(+1)			
L-Cysteic acid (3-sulfo-L-alanine)	1.89(+1)	8.7(0)		
L-(+)-Cysteine	1.71(+1)	8.39(0)	10.70(SH)	
L-(+)-Cysteine, ethyl ester	6.69 (NH$_3$$^+$)	9.17(SH)		
L-(+)-Cysteine, methyl ester	6.56 (NH$_3$$^+$)	8.99(SH)		
L-Cysteinyl-L-asparagine	2.97	7.09	8.47	
L-Cystine (35°C)	1.6(+2)	2.1(+1)	8.02(0)	8.71(−1)
Cystinylglycylglycine (35°C)	3.12	3.21	6.01	6.87
Cytidine	4.08(+1)	12.24(0)		
Cytidine-2′-phosphoric acid	0.8(+1)	4.36(0)	6.17(−1)	
Cytidine-3′-phosphoric acid	0.80(+1)	4.31(0)	6.04(−1)	13.2(sugar)
Cytidine-5′-phosphoric acid	——	4.39(0)	6.62(−1)	
Cytosine	4.58(+1)	12.15(0)		
Decanedioic acid (sebacic acid)	4.59	5.59		
Dehydroascorbic acid (20°C)	3.21	7.92	10.3	
2′-Deoxyadenosine ($\mu=0.1$)	3.8(+1)			
Deoxycholic acid	6.58			
2-Deoxyglucose	12.52			
2-Deoxyguanosine ($\mu=0.1$)	2.5(+1)			
5-Desoxypyridoxal ($\mu=0$)	4.17(+1)	8.14(OH)		
1,1-Diacetic acid semicarbazide (30°C, $\mu=0.1$)	2.96	4.04		
Diacetylacetone	7.42			
Diallylamine ($\mu=0.02$)	9.29(+1)			
5,5-Diallylbarbituric acid	7.78(0)			
1,3-Diamino-2-aminomethylpropane	6.44(+3)	8.56(+2)	10.38(+1)	
3,5-Diaminobenzoic acid	5.30			
1,3-Diamino-N,N'-bis-(2-aminoethyl)propane ($\mu=0.5$)	6.01(+4)	7.26(+3)	9.49(+2)	10.23(+1)
2,4-Diaminobutanoic acid (20°C)	1.85(+2)	8.24(+1)	10.40(0)	
2,2′-Diaminodiethyl sulfide (30°C)	8.84(+2)	9.64(+1)		
1,8-Diamino-3,6-dithiooctane (30°C)	8.43(+2)	9.31(+1)		
2,7-Diaminooctanedioic acid (20°C, $\mu=0.1$)	1.84(+2)	2.64(+1)	9.23(0)	9.89(−1)
1,8-Diamino-3,6-octanedione (30°C)	8.60(+2)	9.57(+1)		
1,8-Diamino-3-oxa-6-thiooctane	8.54(+2)	9.46(+1)		
2,3-Diaminopropanoic acid ($\mu=0.1$)	1.33(+2)	6.674(+1)	9.623(0)	
2,3-Diaminopropanoic acid, methyl ester ($\mu=0.1$)	4.412(+1)	8.250(0)		
1,3-Diamino-2-propanol (20°C)	7.93(+2)	9.69(+1)		
2,5-Diaminopyridine (20°C)	2.13(+2)	6.48(+1)		
1,4-Diazabicyclo[2.2.2]octane	2.90(+2)	8.60(+1)		
Dibenzylamine	8.52(+1)			
Dibenzylsuccinic acid (20°C)	3.96	6.66		

TABLE 8.8   pK_a Values of Organic Materials in Water at 25°C (*Continued*)

Substance	pK$_1$	pK$_2$	pK$_3$	pK$_4$
Dibromoacetic acid	1.39			
3,5-Dibromoaniline	2.35(+1)			
3,5-Dibromophenol	8.056			
2,2-Dibromopropanoic acid	1.48			
2,3-Dibromopropanoic acid	2.33			
rac-2,3-Dibromosuccinic acid (20°C)	1.43	2.24		
meso-2,3-Dibromosuccinic acid (20°C)	1.51	2.71		
3,5-Dibromo-*p*-L-tyrosine	2.17(+1)	6.45(0)	7.60(−1)	
Dibutylamine	11.25(+1)			
Di-*sec*-butylamine	10.91(+1)			
2,6-Di-*tert*-butylpyridine	3.58(+1)			
rac-2,3-Di-*tert*-butylsuccinic acid (μ=0.1)	3.58	10.2		
1,12-Dicarboxydodecaborane	9.07	10.23		
Dichloroacetic acid	1.26			
Dichloroacetylacetic acid	2.11			
3,5-Dichloroaniline	2.37(+1)			
1,3-Dichloro-2,5-dihydroxybenzene (μ=0.65)	7.30	9.99		
2,5-Dichloro-3,6-dihydroxy-*p*-benzoquinone	1.09	2.42		
Dichloromethylphosphonic acid	1.14	5.61		
2,4-Dichloro-6-nitroaniline	−3.00(+1)			
2,5-Dichloro-4-nitroaniline	−1.74(+1)			
2,6-Dichloro-4-nitroaniline	−3.31(+1)			
2,3-Dichlorophenol	7.44			
2,4-Dichlorophenol	7.85			
2,6-Dichlorophenol	6.78			
3,4-Dichlorophenol	8.630			
3,5-Dichlorophenol	8.179			
2,4-Dichlorophenoxyacetic acid (2,4-D)	2.64			
4,6-Dichlorophenoxy-2-methylacetic acid	3.13			
3,6-Dichlorophthalic acid	1.46			
2,2-Dichloropropanoic acid	2.06			
2,3-Dichloropropanoic acid	2.85			
rac-2,3-Dichlorosuccinic acid (20°C)	1.43	2.81		
meso-2,3-Dichlorosuccinic acid	1.49	2.97		
3,5-Dichloro-*p*-tyrosine	2.12	6.47	7.62	
2-Dicyanoethylamine	5.14(+1)			
2,2-Dicyanopropanoic acid	−2.8			
Dicyclohexylamine	11.25(+1)			
Dicyclopentylamine	10.93(+1)			
Didodecylamine	10.99(+)			

TABLE 8.8 pK_a Values of Organic Materials in Water at 25°C (*Continued*)

Substance	pK_1	pK_2	pK_3	pK_4
Diethanolamine	8.88(+1)			
Di(ethoxyethyl)amine	8.47(+1)			
3,5-Diethoxyphenol	9.370			
3-(Diethoxyphosphinyl)benzoic acid	3.65			
4-(Diethoxyphosphinyl)benzoic acid	3.60			
3-(Diethoxyphosphinyl)phenol	8.66			
4-(Diethoxyphosphinyl)phenol	8.28			
Diethylamine	10.8(+1)			
2-(Diethylamino)ethyl-4-aminobenzoate	8.85(+1)			
α-(Diethylamino)toluene	9.44(+1)			
N,N-Diethylaniline	6.56(+1)			
5,5-Diethylbarbituric acid (veronal)	8.020(0)			
N,N-Diethylbenzylamine	9.48(+1)			
Diethylbiguanide (30°C)	2.53(+1)	11.68(0)		
Diethylenetriamine	4.42(+3)	9.21(+2)	10.02(+1)	
Diethylenetriaminepentaacetic acid (pK_5, 10.58)	1.80(0)	2.55(−1)	4.33(−2)	8.60(−3)
N,N-Diethylethylenediamine	7.70(+2)	10.46(+1)		
2,2-Diethylglutaric acid	3.62	7.12		
N,N-Diethylglycine	2.04(+1)	10.47(0)		
Diethylglycolic acid (18°C)	3.804			
Diethylmalonic acid	2.151	7.417		
Diethylmethylamine	10.43(+1)			
rac-2,3-Diethylsuccinic acid	3.63	6.46		
meso-2,3-Diethylsuccinic acid	3.54	6.59		
N,N-Diethyl-o-toluidine	7.18(+1)			
Difluoroacetic acid	1.33			
3,3-Difluoroacrylic acid	3.17			
Diglycolic acid	2.96			
Diguanidine	12.8			
Dihexylamine	11.0(+1)			
Dihydroarecaidine	9.70			
Dihydroarecaidine, methyl ester	8.39			
Dihydrocodcine	8.75(+1)			
Dihydroergonovine	7.38(+1)			
α-Dihydrolysergic acid	3.57	8.45		
γ-Dihydrolysergic acid	3.60	8.71		
α-Dihydrolysergol	8.30			
β-Dihydrolysergol	8.23			
Dihydromorphine	9.35			
3,4-Dihydroxyalanine	2.32(+1)	8.68(0)	9.87(−1)	
1,2-Dihydroxyanthraquinone-3-sulfonic acid (alizarin-3-sulfonic acid)	——	5.54(−1)	11.01(−2)	
3,4-Dihydroxybenzaldehyde	7.55			

TABLE 8.8 pK_a Values of Organic Materials in Water at 25°C (*Continued*)

Substance	pK$_1$	pK$_2$	pK$_3$	pK$_4$
1,2-Dihydroxybenzene				
(pyrocatechol) (μ=0.1)	9.356(0)	12.98(−1)		
1,3-Dihydroxybenzene (resorcinol)	9.44(0)	12.32(−1)		
1,4-Dihydroxybenzene				
(hydroquinone)	9.91(0)	12.04(−1)		
4,5-Dihydroxybenzene-1,3-disulfonic				
acid	——	——	7.66(−2)	12.6(−3)
2,3-Dihydroxybenzoic acid (30°C)	2.98	10.14		
2,4-Dihydroxybenzoic acid				
(β-resorcyclic acid)	3.29	8.98		
2,5-Dihydroxybenzoic acid	2.97	10.50		
2,6-Dihydroxybenzoic acid	1.30			
3,4-Dihydroxybenzoic acid	4.48	8.67	11.74	
3,5-Dihydroxybenzoic acid	4.04			
2,5-Dihydroxy-*p*-benzoquinone	2.71	5.18		
3,4-Dihydroxy-3-cyclobutene-				
1,2-dione	0.541	3.480		
2,3-Dihydroxy-2-cyclopenten-1-one				
(20°C)	4.72			
1,4-Dihydroxy-2,6-dinitrobenzene	4.42	9.14		
Di(2,2'-hydroxyethyl)amine	8.8(+1)			
N,N-Di(2-hydroxyethyl)glycine	8.333			
Dihydroxymaleic acid	1.10			
Dihydroxymalic acid	1.92			
1,3-Dihydroxy-2-methylbenzene				
(μ=0.65)	10.05	11.64		
2,2-Di(hydroxymethyl)-				
3-hydroxypropanoic acid	4.460			
2,4-Dihydroxy-5-methylpyrimidine	9.90			
2,4-Dihydroxy-6-methylpyrimidine	9.52			
1,4-Dihydroxynaphthalene				
(26°C, μ=0.65)	9.37	10.93		
1,2-Dihydroxy-3-nitrobenzene	6.68			
1,2-Dihydroxy-4-nitrobenzene				
(μ=0.1)	6.701			
2,4-Dihydroxy-1-phenylazobenzene				
(μ=0.1)	11.98			
2,4-Dihydroxyoxazolidine	6.11(+1)			
2,4-Dihydroxypteridine	<1.3	7.92		
2,6-Dihydroxypurine	7.53(0)	11.84(−1)		
2,4-Dihydroxypyridine (20°C)	1.37(+1)	6.45(0)	13(−1)	
Dihydroxytartaric acid	1.95	4.00		
1,4-Dihydroxy-2,3,5,6-				
tetramethylbenzene (μ=0.65)	11.25	12.70		
3,5-Diiodoaniline	2.37(+1)			
2,5-Diiodohistamine	2.31(+2)	8.20(+1)	10.11(0)	
2,5-Diiodohistidine (μ=0.1)	2.72	8.18	9.76	

TABLE 8.8 pK_a Values of Organic Materials in Water at 25°C (*Continued*)

Substance	pK_1	pK_2	pK_3	pK_4
3,5-Diiodophenol	8.103			
3,5-Diiodotyrosine	2.117(+1)	6.479(0)	7.821(−1)	
Diisopropylmalonic acid	2.124	8.848		
Dilactic acid	2.955			
threo-1,4-Dimercapto-2,3-butanediol	8.9			
meso-2,3-Dimercaptosuccinic acid	2.71	3.48	8.89(SH)	10.79(SH)
3,5-Dimethoxyaniline	3.86(+1)			
2,6-Dimethoxybenzoic acid	3.44			
1,10-Dimethoxy-3,8-dimethyl-4,7-phenanthroline	7.21			
Di(2-methoxyethyl)amine	9.51(+1)			
3,5-Dimethoxyphenol	9.345			
(3,4-Dimethoxy)phenylacetic acid	4.333			
Dimethylamine	10.77(+1)			
4-Dimethylaminobenzaldehyde	1.647(+1)			
N,*N*-Dimethylaminocyclohexane	10.72(+1)			
4-Dimethylamino-2,3-dimethyl-1-phenyl-3-pyrazolin-5-one	4.18(+1)			
4-Dimethylamino-3,5-dimethylpyridine (20°C)	8.15(+1)			
2-(Dimethylamino)ethanol	9.26(+1)			
2-[2-(Dimethylamino)ethyl]pyridine	3.46(+2)	8.75(+1)		
3-(Dimethylaminoethyl)pyridine	4.30(+2)	8.86(+1)		
4-(Dimethylaminoethyl)pyridine	4.66(+2)	8.70(+1)		
4-(Dimethylamino)-3-ethylpyridine (20°C)	8.66(+1)			
4-(Dimethylamino)-3-isopropylpyridine (20°C)	8.27(+1)			
2-(Dimethylaminomethyl)pyridine	2.58(+2)	8.12(+1)		
3-(Dimethylaminomethyl)pyridine	3.17(+2)	8.00(+1)		
4-(Dimethylaminomethyl)pyridine	3.39(+2)	7.66(+1)		
4-(Dimethylamino)-3-methylpyridine (20°C)	8.68(+1)			
4-(Dimethylaminophenyl)phosphonic acid	2.0(+1)	4.2	7.35	
3-(Dimethylaminopropanoic acid	9.85(+1)			
4-(Dimethylamino)pyridine (20°C)	6.09(+1)			
N,*N*-Dimethylaniline	5.15(+1)			
2,3-Dimethylaniline	4.70(+1)			
2,4-Dimethylaniline	4.89(+1)			
2,5-Dimethylaniline	4.53(+1)			
2,6-Dimethylaniline	3.95(+1)			
3,4-Dimethylaniline	5.17(+1)			
3,5-Dimethylaniline	4.765(+1)			
N,*N*-Dimethylaniline-4-phosphonic acid (17°C)	2.0(+1)	4.2	7.39	
Dimethylarsinic acid (cacodylic acid)	6.273			

TABLE 8.8 pK_a Values of Organic Materials in Water at 25°C (*Continued*)

Substance	pK_1	pK_2	pK_3	pK_4
1,3-Dimethylbarbituric acid	4.68(+1)			
2,3-Dimethylbenzoic acid	3.771			
2,4-Dimethylbenzoic acid	4.217			
2,5-Dimethylbenzoic acid	3.990			
2,6-Dimethylbenzoic acid	3.362			
3,4-Dimethylbenzoic	4.41			
3,5-Dimethylbenzoic acid	4.302			
N,N-Dimethylbenzylamine	9.02(+1)			
Dimethylbiguanide	2.77(+1)	11.52		
2,2-Dimethylbutanoic acid (18°C)	5.03			
Dimethylchlorotetracycline (μ=0.01)	3.30(+1)			
2,6-Dimethyl-4-cyanophenol	8.27			
3,5-Dimethyl-4-cyanophenol	8.21			
5,5-Dimethyl-1,3-cyclohexanedione	5.15			
cis-3,3-Dimethyl-1,2-cyclopropanedicarboxylic acid	2.34	8.31		
trans-3,3-Dimethyl-1,2-cyclopropanedicarboxylic acid	3.92	5.32		
3,5-Dimethyl-4-(dimethylamino)-pyridine (20°C)	8.12(+1)			
2,2-Dimethyl-1,3-dioxane-4,6-dione	5.1			
1,1-Dimethylethanethiol (μ=0.1)	11.22			
N,N-Dimethylethylenediamine-N,N-diacetic acid	6.63	9.53		
N,N'-Dimethylethylenediamine-N,N'-diacetic acid	7.40	10.16		
N,N-Dimethylethylenediamine-N,N'-diacetic acid	5.99	9.97		
N,N-Dimethylglycine	2.146(+1)	9.940(0)		
Dimethylglycolic acid (18°C)	4.04			
N,N-Dimethylglycylglycine	3.11(+1)	8.09(0)		
Dimethylglyoxime	10.60			
5,5-Dimethyl-2,4-hexanedione	10.01			
5,5-Dimethylhydantoin	9.19			
2,4-Dimethyl-8-hydroxyquinoline	6.20(+1)	10.60(0)		
3,4-Dimethyl-8-hydroxyquinoline	5.80(+1)	10.05(0)		
2,4-Dimethyl-8-hydroxyquinoline-7-sulfonic acid	3.20 (NH$^+$)	10.14(OH)		
Dimethylhydroxytetracycline	7.5	9.4		
2,4-Dimethylimidazole	8.38(+1)			
Dimethylmalic acid	3.17	6.06		
2,2-Dimethylmalonic acid	3.17	6.06		
3,5-Dimethyl-4-(methylamino)pyridine (20°C)	9.96(+1)			
2,3-Dimethylnaphthalene-1-carboxylic acid	3.33			

TABLE 8.8 pK_a Values of Organic Materials in Water at 25°C (*Continued*)

Substance	pK_1	pK_2	pK_3	pK_4
2,6-Dimethyl-4-nitrophenol	7.190			
3,5-Dimethyl-4-nitrophenol	8.245			
α,α-Dimethyloxaloacetic acid	1.77	4.62		
3,3-Dimethylpentanedioic acid	3.70	6.34		
2,2-Dimethylpentanoic acid	4.969			
4,4-Dimethylpentanoic acid (18°C)	4.79			
2,3-Dimethylphenol	10.50			
2,4-Dimethylphenol	10.58			
2,5-Dimethylphenol	10.22			
2,6-Dimethylphenol	10.59			
3,4-Dimethylphenol	10.32			
3,5-Dimethylphenol	10.15			
2,6-Dimethylphenoxyacetic acid	3.356			
Dimethylphenylsilylacetic acid	5.27			
N,N'-Dimethylpiperazine	4.630(+2)	8.539(+1)		
1,2-Dimethylpiperidine	10.22			
cis-2,6-Dimethylpiperidine	11.07(+1)			
2,2-Dimethylpropanoic acid (pivalic acid)	5.031			
2,2'-Dimethylpropylphosphonic acid	2.84	8.65		
2,4-Dimethylpyridine (2,4-lutidine)	6.74(+1)			
2,5-Dimethylpyridine (2,5-lutidine)	6.43(+1)			
2,6-Dimethylpyridine (2,6-lutidine)	6.71(+1)			
3,4-Dimethylpyridine (3,4-lutidine)	6.47(+1)			
3,5-Dimethylpyridine (3,5-lutidine)	6.09(+1)			
2,4-Dimethylpyridine-1-oxide	1.627(+1)			
2,5-Dimethylpyridine-1-oxide	1.208(+1)			
2,6-Dimethylpyridine-1-oxide	1.366(+1)			
3,4-Dimethylpyridine-1-oxide	1.493(+1)			
3,5-Dimethylpyridine-1-oxide	1.181(+1)			
2,3-Dimethylquinoline	4.94(+1)			
2,6-Dimethylquinoline	5.46(+1)			
meso-2,2-Dimethylsuccinic acid	3.77	5.936		
rac-2,2-Dimethylsuccinic acid	3.93	6.20		
D-2,3-Dimethylsuccinic acid	3.82	5.93		
meso-2,3-Dimethylsuccinic acid	3.67	5.30		
rac-2,3-Dimethylsuccinic acid	3.94	6.20		
2,4-Dimethylthiazole ($\mu=0.1$)	3.98			
2,5-Dimethylthiazole ($\mu=0.1$)	3.91			
4,5-Dimethylthiazole ($\mu=0.1$)	3.73			
N,N-Dimethyl-*o*-toluidine	5.86(+1)			
N,N-Dimethyl-*p*-toluidine	7.24(+1)			
2,4-Dinitroaniline	−4.25(+1)			
2,6-Dinitroaniline	−5.23(+1)			
3,5-Dinitroaniline	0.229(+1)			
2,3-Dinitrobenzoic acid	1.85			
2,4-Dinitrobenzoic acid	1.43			

TABLE 8.8   pK$_a$ Values of Organic Materials in Water at 25°C (*Continued*)

Substance	pK$_1$	pK$_2$	pK$_3$	pK$_4$
2,5-Dinitrobenzoic acid	1.62			
2,6-Dinitrobenzoic acid	1.14			
3,4-Dinitrobenzoic acid	2.82			
3,5-Dinitrobenzoic acid	2.85			
1,1-Dinitrobutane (20°C)	5.90			
1,1-Dinitrodecane	3.60			
1,1-Dinitroethane (20°C)	5.21			
Dinitromethane (20°C)	3.60			
1,1-Dinitropentane	5.337			
2,4-Dinitrophenol	4.08			
2,5-Dinitrophenol	5.216			
2,6-Dinitrophenol	3.713			
3,4-Dinitrophenol	5.424			
3,5-Dinitrophenol	6.732			
2,4-Dinitrophenylacetic acid	3.50			
1,1-Dinitropropane (20°C)	5.5			
2,6-Dioxo-1,2,3,6-tetrahydro- 4-pyrimidinecarboxylic acid (orotic acid)	1.8(+1)	9.55(0)		
Diphenylacetic acid	3.939			
Diphenylamine	0.9(+1)			
2,2-Diphenylglutaric acid (20°C)	3.91	5.38		
1,3-Diphenylguanidine	10.12			
2,2-Diphenylheptanedioic acid (20°C)	4.28	5.39		
2,2-Diphenylhexanedioic acid (20°C)	4.17	5.40		
3,3-Diphenylhexanedioic acid	4.22	5.19		
Diphenylhydroxyacetic acid (35°C)	3.05			
Diphenylketimine	6.82			
2,2-Diphenylnonanedioic acid (20°C)	4.33	5.38		
meso-2,2-Diphenylsuccinic acid	3.48			
rac-2,2-Diphenylsuccinic acid	3.58			
2,2-Diphenylsuccinic acid, 1-methyl ester (20°C)	4.47			
2,2-Diphenylsuccinic acid, 4-methyl ester (20°C)	3.900			
Diphenylthiocarbazone	4.50	15		
Dipropylamine	10.91(+1)			
Dipropylenetriamine	7.72(+3)	9.56(+2)	10.65(+1)	
2,2-Dipropylglutaric acid	3.688	7.31		
Dipropylmalonic acid	2.04	7.51		
2,2'-Dipyridyl	−0.52(+2)	4.352(+1)		
2,3'-Dipyridyl (20°C)	1.52(+2)	4.42(+1)		
2,4'-Dipyridyl (20°C)	1.19(+2)	4.77(+1)		
3,3'-Dipyridyl (20°C, μ=0.2)	3.0(+2)	4.60(+1)		
3,4'-Dipyridyl (20°C, μ=0.2)	3.0(+2)	4.85(+1)		
4,4'-Dipyridyl	3.17(+2)	4.82(+1)		

TABLE 8.8   pK_a Values of Organic Materials in Water at 25°C (*Continued*)

Substance	pK_1	pK_2	pK_3	pK_4
Dithiodiacetic acid (18°C)	3.075	4.201		
1,4-Dithioerythritol	9.5			
Dithiooxamide (rubeanic acid)	10.89			
Dulcitol	13.46			
Ecgonine	10.91			
Emetine	7.36(+1)	8.23(0)		
Epinephrine enantiomorph	9.39(+1)			
Epinephrine, pseudo	9.53(+1)			
Ergometrinine	7.32(+1)			
Ergonovine	6.73(+1)			
Eriochrome Black T	6.3	11.55		
1,2-Ethanediamine	6.85(+2)	9.92(+1)		
Ethane-1,2-diamino-N,N'-dimethyl-N,N'-diacetic acid (20°C)	6.047(0)	10.068(−1)		
1,2-Ethanedithiol	8.96	10.54		
Ethanethiol ($\mu=0.015$)	10.61			
Ethoxyacetic acid (18°C)	3.65			
2-Ethoxyaniline (*o*-phenetidine)	4.47(+1)			
3-Ethoxyaniline	4.17(+1)			
4-Ethoxyaniline	5.25(+1)			
2-Ethoxybenzoic acid (20°C)	4.21			
3-Ethoxybenzoic acid (20°C)	4.17			
4-Ethoxybenzoic acid (20°C)	4.80			
Ethoxycarbonylethylamine	9.13(+1)			
2-Ethoxyethanethiol	9.38			
2-Ethoxyethylamine	6.26(+1)			
2-Ethoxyphenol	10.109			
3-Ethoxyphenol	9.655			
(4-Ethoxyphenyl)phosphonic acid	2.06	7.28		
4-Ethoxypyridine	6.67(+1)			
Ethyl acetoacetate	10.68			
3-Ethylacrylic acid	4.695			
N-Ethylalanine	2.22(+1)	10.22(0)		
Ethylamine	10.63(+1)			
(3-Ethylamino)phenylphosphonic acid	1.1(+1)	4.90(0)	7.24(−1)	
N-Ethylaniline	5.11(+1)			
2-Ethylaniline	4.42(+1)			
3-Ethylaniline	4.70(+1)			
4-Ethylaniline	5.00(+1)			
Ethylarsonic acid (18°C)	3.89	8.35		
Ethylbarbituric acid	3.69(+1)			
2-Ethylbenzimidazole ($\mu=0.16$)	6.27(+1)			
2-Ethylbenzoic acid	3.79			
4-Ethylbenzoic acid	4.35			
Ethylbiguanide	2.09(+1)	11.47(0)		

TABLE 8.8   pK_a Values of Organic Materials in Water at 25°C (*Continued*)

Substance	pK_1	pK_2	pK_3	pK_4
2-Ethylbutanoic acid (20°C)	4.710			
S-Ethyl-L-cysteine (μ=0.1)	2.03(+1)	8.60(0)		
Ethylenebiguanide (30°C)	1.74	2.88	11.34	11.76
Ethylenebis(thioacetic acid) (18°C)	3.382(0)	4.352(−1)		
Ethylenediamine-N,N'-diacetic acid	6.42	9.46		
Ethylenediamine-N,N-dimethyl-N',N'-diacetic acid	6.047	10.068		
Ethylenediamine-N',N-dipropanoic acid (30°C)	6.87	9.60		
Ethylenediamine-N,N,N',N'-tetraacetic acid (μ=0.1)	1.99	2.67	6.16	10.26
Ethylenediamine-N,N,N',N'-tetrapropanoic acid (30°C)	3.00	3.43	6.77	9.60
Ethylene glycol	14.22			
Ethyleneimine	8.04(+1)			
cis-Ethylene oxide dicarboxylic acid	1.93	3.92		
trans-Ethylene oxide dicarboxylic acid	1.93	3.25		
N-Ethylethylenediamine	7.63(+2)	10.56(+1)		
N-Ethylglycine (μ=0.1)	2.34(+1)	10.23(0)		
3-Ethylglutaric acid	4.28	5.33		
Ethyl hydroperoxide	11.80			
Ethyl hydrogen malonate	3.55			
3-Ethyl-2-hydroxypyridine	5.00(+1)			
Ethylmalonic acid	2.90(0)	5.55(−1)		
N-Ethyl mercaptoacetamide	8.14(SH)			
Ethyl 2-mercaptoacetate	7.95(SH)			
Ethyl 3-mercaptopropanoate	9.48(SH)			
3-Ethyl-4-(methylamino)pyridine (20°C)	9.90(+1)			
5-Ethyl-5-(1-methylbutyl)barbituric acid	8.11(0)			
Ethyl methyl ketoxime	12.45			
Ethylmethylmalonic acid	2.86(0)	6.41(−1)		
1-Ethyl-2-methylpiperidine	10.66(+1)			
3-Ethyl-6-methylpyridine (20°C)	6.51(+1)			
3-Ethyl-4-methylpyridine-1-oxide	−1.534(+1)			
5-Ethyl-2-methylpyridine-1-oxide	−1.288(+1)			
1-Ethyl-2-methyl-2-pyrroline	11.84(+1)			
Ethylmorphine (15°C)	8.08			
Ethyl nitroacetate	5.85			
3-Ethylpentane-2,4-dione	11.34			
2-Ethylpentanoic acid (18°C)	4.71			
5-Ethyl-5-pentylbarbituric acid	7.960			
2-Ethylphenol	10.2			
3-Ethylphenol	10.07			
4-Ethylphenol	10.0			

TABLE 8.8 pK_a Values of Organic Materials in Water at 25°C (*Continued*)

Substance	pK_1	pK_2	pK_3	pK_4
4-Ethylphenylacetic acid	4.373			
5-Ethyl-5-phenylbarbituric acid	7.445			
Ethylphosphinic acid	3.29			
Ethylphosphonic acid	2.43	8.05		
1-Ethylpiperidine ($\mu = 0.01$)	10.45(+1)			
2,2-Ethylpropylglutaric acid	3.511			
Ethylpropylmalonic acid	3.14	7.43		
2-Ethylpyridine	5.89(+1)			
3-Ethylpyridine (20°C)	5.80(+1)			
4-Ethylpyridine	5.87(+1)			
Ethyl 3-pyridinecarboxylate	3.35(+1)			
Ethyl 4-pyridinecarboxylate	3.45(+1)			
2-Ethylpyridine-1-oxide	−1.19(+1)			
3-Ethylpyridine-1-oxide	−0.965(+1)			
Ethylpyrrolidine	10.43(+1)			
2-Ethyl-2-pyrroline	7.87(+1)			
Ethylsuccinic acid	4.08(0)			
S-Ethylthioacetic acid	5.06			
N-Ethyl-o-toluidine	4.92(+1)			
N-Ethylveratramine	7.40(+1)			
β-Eucaine	9.35(+1)			
Fluoroacetic acid	2.586			
2-Fluoroacrylic acid	2.55			
2-Fluoroaniline	3.20(+1)			
3-Fluoroaniline	3.58(+1)			
4-Fluoroaniline	4.65(+1)			
2-Fluorobenzoic acid	3.27			
3-Fluorobenzoic acid	3.865			
4-Fluorobenzoic acid	4.14			
Fluoromandelic acid	4.244			
2-Fluorophenol	8.73			
3-Fluorophenol	9.29			
4-Fluorophenol	9.89			
2-Fluorophenoxyacetic acid	3.08			
3-Fluorophenoxyacetic acid	3.08			
4-Fluorophenoxyacetic acid	3.13			
4-Fluorophenylacetic acid	4.25			
2′-Fluorophenylalanine	2.14(+1)	9.01(0)		
3′-Fluorophenylalanine	2.10(+1)	8.98(0)		
4-Fluorophenylalanine	2.13(+1)	9.05(0)		
2-Fluorophenylphosphonic acid	1.64	6.80		
3-Fluorophenylselenic acid	4.34			
4-Fluorophenylselenic acid	4.50			
2-Fluoropyridine	−0.44(+1)			
3-Fluoropyridine	2.97(+1)			
5-Fluorouracil	8.00(0)	ca 13(−1)		

TABLE 8.8 pK_a Values of Organic Materials in Water at 25°C (*Continued*)

Substance	pK_1	pK_2	pK_3	pK_4
Folic acid (pteroylglutamic acid)	8.26			
Formic acid	3.751			
N-Formylglycine	3.43			
2-Formyl-3-hydroxypyridine (20°C)	3.40(+1)	6.95(OH)		
4-Formyl-3-hydroxypyridine	4.05(+1)	6.77(OH)		
2-Formyl-3-methoxypyridine (20°C)	3.89(+1)	12.95		
Formyl-3-methoxypyridine (20°C)	4.45(+1)	11.7		
D-(−)-Fructose	12.03			
Fumaric acid	3.10	4.60		
2-Furancarboxylic acid (2-furoic acid)	3.164			
D-(+)-Galactose	12.35			
Galactose-1-phosphoric acid	1.00	6.17		
Glucoascorbic acid	4.26	11.58		
D-Gluconic acid	3.86			
α-D-(+)-Glucose	12.28			
α-D-Glucose-1-phosphate	1.11(0)	6.504(−1)		
trans-Glutaconic acid	3.77	5.08		
D-(−)-Glutamic acid	2.162(+1)	4.272(0)	9.358(−1)	
L-Glutamic acid	2.13(+1)	4.31(0)	9.76(−1)	
Glutamic acid, 1-ethyl ester	3.85(+1)	7.84(0)		
Glutamic acid, 5-ethyl ester	2.15(+1)	9.19(0)		
L-Glutamine (μ=0.2)	2.15(+1)	9.00(0)		
Glutaric acid	3.77	6.08		
Glutaric acid monoamide	4.600(0)			
Glutarimide	11.43			
Glutathione	2.12(+1)	3.53(0)	8.66	9.12
DL-Glyceric acid	3.64			
Glycerol	14.15			
Glyceryl-1-phosphoric acid	——	6.656(−1)		
Glyceryl-2-phosphoric acid	1.335(0)	6.650(−1)		
Glycine	2.351(+1)	9.70(0)		
Glycine amide	8.03(+1)			
Glycine, ethyl ester	7.66(+1)			
Glycine hydroxamic acid	7.10	9.10		
Glycine, methyl ester	7.59(+1)			
Glycine-O-phenylphosphorylserine	2.96	8.07		
Glycolic acid	3.831			
N-Glycyl-α-alanine	3.15 (+1)	8.33(0)		
Glycylalanylalanine	3.38(+1)	8.10(0)		
N-Glycylasparagine	2.942			
Glycyclaspartic acid	2.81(+1)	4.45(0)	8.60(−1)	
Glycyl-DL-glutamine (18°C)	2.88(+1)	8.33(0)		
N-Glycylglycine	3.126(+1)	8.252(0)		
Glycylglycylcysteine (35°C)	2.71	2.71	7.94	7.94
Glycylglycylglycine	3.225(+1)	8.090(0)		

TABLE 8.8 pK_a Values of Organic Materials in Water at 25°C (*Continued*)

Substance	pK_1	pK_2	pK_3	pK_4
Glycyl-L-histidine ($\mu=0.16$)	6.79	8.20		
Glycylisoleucine	8.00			
N-Glycyl-L-leucine	3.180(+1)	8.327(0)		
Glycyl-O-phosphorylserine	2.90	6.02	8.43	
L-Glycylproline ($\mu=0.1$)	2.81(+1)	8.65(0)		
N-Glycylsarcosine ($\mu=0.1$)	2.98(+1)	8.55(0)		
N-Glycylserine	2.98(+1)	8.38(0)		
Glycylserylglycine	3.32	7.99		
Glycyltyrosine	2.93	8.45	10.49	
Glycylvaline	3.15	8.18		
Glyoxaline	7.03(+1)			
Glyoxylic acid	3.30(0)			
Guanidineacetic acid	2.82(+1)			
Guanine	3.3(+1)	9.2	12.3	
Guanine deoxyriboside-3'-phosphoric acid	——	2.9	6.4	9.7
Guanosine	1.9(+1)	9.25(0)	12.33(OH)	
Guanosine-5'-diphosphoric acid ($\mu=0.1$; pK$_5$ 9.6)	——	——	2.9	6.3
Guanosine-3'-phosphoric acid	0.7	2.3	5.92	9.38
Guanosine-5'-phosphoric acid ($\mu=0.1$)	——	2.4	6.1	9.4
Guanosine-5'-triphosphoric acid [$\mu=0.1$; pK$_5$ 7.10(−3); pK$_6$ 9.3(−4)]	——	——	——	3.0(−2)
Guanylurea	1.80	8.20		
Harmine (20°C)	7.61(+1)			
Heptafluorobutanoic acid	0.17			
4,4,5,5,6,6,6-Heptafluorohexanoic acid	4.18			
4,4,5,5,6,6,6-Heptafluoro-2-hexenoic acid	3.23			
Heptanedioic acid (pimelic acid)	4.484	5.424		
2,4-Heptanedione	8.43(keto); 9.15(enol)			
Heptanoic acid	4.893			
Heroin	7.6(+1)			
2,4-Hexadienoic acid (sorbic acid)	4.77			
1,1,1,3,3,3-Hexafluoro-2,2-propanediol	8.801			
1,1,1,3,3,3-Hexafluoro-2-propanol	9.42			
Hexahydroazepine	11.07			
Hexamethyldisilazine	7.55			
1,2,3,8,9,10-Hexamethyl-4,7-phenanthroline (20°C)	7.26			
1,6-Hexanediamine	9.830(+2)	10.930(+1)		

TABLE 8.8   pK_a Values of Organic Materials in Water at 25°C (*Continued*)

Substance	pK$_1$	pK$_2$	pK$_3$	pK$_4$
1,6-Hexanedioic acid	4.418	5.412		
2,4-Hexanedione	8.49 (enol); 9.32 (keto)			
2,2′,4,4′,6,6′-Hexanitrodiphenylamine	5.42 (+1)			
Hexanoic acid (20°C)	4.849			
trans-2-Hexenoic acid	4.74			
trans-3-Hexenoic acid	4.72			
3-Hexen-4-oic acid	4.58			
4-Hexen-5-oic acid	4.74			
Hexylamine	10.64(+1)			
Hexylarsonic acid	4.16	9.19		
Hexylphosphonic acid	2.6	7.9		
DL-Histidine	1.82(+2)	6.00(+1)	9.16(0)	
Histidine amide (μ=0.2)	5.78(+2)	7.64(+1)		
Histidine, methyl ester (μ=0.1)	5.01(+2)	7.23(+1)		
Histidylglycine	2.40(+2)	5.80(+1)	7.82(0)	
Histidylhistidine (μ=0.16)	5.40(+2)	6.80(+1)	7.95(0)	
DL-Homatropine	9.7(+1)			
DL-Homocysteine	2.222(+1)	8.87	10.86	
Homocysteine (μ=0.1)	1.593(+2)	2.523(+1)	8.676(0)	9.413(−1)
Hydantoin	9.12			
Hydrastine	6.23(+1)			
Hydrazine-N,N-diacetic acid	<0.1	2.8	3.8	
Hydrazine-$N′,N′$-diacetic acid	2.40	3.12	7.32	
4-Hydrazinocarbonylpyridine (20°C)	1.82	3.52	10.79	
N-Hydroxyacetamide	9.40			
2′-Hydroxyacetophenone	9.90			
3′-Hydroxyacetophenone	9.19			
4′-Hydroxyacetophenone	8.05			
1-Hydroxyacridine (15°C)	5.72			
2-Hydroxyacridine (15°C)	5.62			
3-Hydroxyacridine (15°C)	5.30			
α-Hydroxyasparagine	2.28(+1)	7.20(0)		
β-Hydroxyasparagine	2.09(+1)	8.29(0)		
Hydroxyaspartic acid	1.91(+1)	3.51(0)	9.11(−1)	
2-Hydroxybenzaldehyde (salicylaldehyde)	8.34			
3-Hydroxybenzaldehyde	9.00			
4-Hydroxybenzaldehyde	7.620			
2-Hydroxybenzaldehyde oxime	1.37(+1)	9.18	12.11	
2-Hydroxybenzamide	8.36			
2-Hydroxybenzenemethanol (2-hydroxybenzyl alcohol)	9.92			
3-Hydroxybenzenemethanol	9.83			
4-Hydroxybenzenemethanol	9.82			

TABLE 8.8 pK_a Values of Organic Materials in Water at 25°C (*Continued*)

Substance	pK_1	pK_2	pK_3	pK_4
4-Hydroxybenzenesulfonic acid	——	9.055(−1)		
2-Hydroxybenzohydroxamic acid	5.19			
2-Hydroxybenzoic acid (salicyclic acid)	2.98	12.38		
3-Hydroxybenzoic acid	4.076	9.85		
4-Hydroxybenzoic acid	4.582	9.23		
4-Hydroxybenzonitrile	7.95			
2-Hydroxy-5-bromobenzoic acid	2.61			
2-Hydroxybutanoic acid (30°C)	3.65			
L-3-Hydroxybutanoic acid (30°C)	4.41			
4-Hydroxybutanoic acid (30°C)	4.71			
2-Hydroxy-5-chlorobenzoic acid	2.63			
trans-2'-Hydroxycinnamic acid	4.614			
trans-3'-Hydroxycinnamic acid	4.40			
10-Hydroxycodeine	7.12			
cis-2-Hydroxycyclohexane-1-carboxylic acid	4.796			
trans-2-Hydroxycyclohexane-1-carboxylic acid	4.682			
cis-3-Hydroxycyclohexane-1-carboxylic acid	4.602			
trans-3-Hydroxycyclohexane-1-carboxylic acid	4.815			
cis-4-Hydroxycyclohexane-1-carboxylic acid	4.836			
trans-4-Hydroxycyclohexane-1-carboxylic acid	4.687			
1-Hydroxy-2,4-dihydroxymethylbenzene	9.79			
N-(Hydroxyethyl)biguanide	2.8(+2)	11.53(+1)		
N-(2-Hydroxyethyl)ethylenediamine	7.21(+2)	10.12(+1)		
N'-(2-Hydroxyethyl)ethylene-diamine-N,N,N'-triacetic acid	2.39	5.37	9.93	
N-(2-Hydroxyethyl)iminodiacetic acid (μ=0.1)	2.2	8.65		
N-(2-Hydroxyethyl)piperazine-N'-ethansulfonic acid (20°C)	7.55			
4'-(2-Hydroxyethyl)-1'-piperazine-propanesulfonic acid (20°C)	8.00			
2-Hydroxyethyltrimethylamine	8.94(+1)			
L-β-Hydroxyglutamic acid	2.09	4.18	9.20	
1-Hydroxy-4-hydroxymethylbenzene	9.84			
5-Hydroxy-2-(hydroxymethyl)-4H-pyran-4-one	7.90	8.03		
3-Hydroxy-2-hydroxymethylpyridine (20°C, μ=0.2)	5.00(+1)	9.07(OH)		
3-Hydroxy-4-hydroxymethylpyridine (20°C, μ=0.2)	5.00(+1)	8.95(OH)		

TABLE 8.8 pK_a Values of Organic Materials in Water at 25°C (*Continued*)

Substance	pK_1	pK_2	pK_3	pK_4
8-Hydroxy-7-iodoquinoline- 5-sulfonic acid	2.51(0)	7.417(−1)		
Hydroxylysine (38°C, μ=0.1)	2.13(+2)	8.62(+1)	9.67(0)	
2-Hydroxy-3-methoxybenzaldehyde	7.912			
3-Hydroxy-4-methoxybenzaldehyde (isovanillin)	8.889			
4-Hydroxy-3-methoxybenzaldehyde (vanillin)	7.396			
4-Hydroxy-3-methoxybenzoic acid	4.355			
1-Hydroxy-2-methoxybenzylamine	8.70(+1)	10.52(0)		
2-Hydroxy-1-methoxybenzylamine	8.89(+1)	10.52(0)		
3-Hydroxy-2-methoxybenzylamine	8.94(+1)	10.42(0)		
2-Hydroxymethyl-2-benzeneacetic acid	4.12			
(2-Hydroxy-5-methylbenzene)- methanol	10.15			
2-Hydroxy-3-methylbenzoic acid	2.99			
2-Hydroxy-4-methylbenzoic acid	3.17			
2-Hydroxy-5-methylbenzoic acid	4.08			
2-Hydroxy-6-methylbenzoic acid	3.32			
2-Hydroxy-2-methylbutanoic acid (18°C)	3.991			
3-Hydroxy-2-methylbutanoic acid (18°C)	4.648			
4-Hydroxy-4-methylpentanoic acid (18°C)	4.873			
1-Hydroxymethylphenol	9.95			
Hydroxymethylphosphoric acid	1.91	7.15		
2-Hydroxy-2-methylpropanoic acid (μ=0.1)	3.717			
2-Hydroxy-4-methylpyridine	4.529(+1)			
8-Hydroxy-2-methylquinoline	5.55(+1)	10.31(0)		
8-Hydroxy-4-methylquinoline	5.56(+1)	10.00(0)		
8-Hydroxy-2-methylquinoline- 5-sulfonic acid	4.80(0)	9.30(−1)		
8-Hydroxy-4-methylquinoline- 7-sulfonic acid	4.78(0)	10.01(−1)		
8-Hydroxy-6-methylquinoline- 5-sulfonic acid	4.20(0)	8.7(−1)		
2-Hydroxy-1-naphthoic acid (20°C)	3.29	9.68		
2-Hydroxy-2-nitrobenzoic acid	2.23			
2-Hydroxy-3-nitrobenzoic acid	1.87			
2-Hydroxy-5-nitrobenzoic acid	2.12			
2-Hydroxy-6-nitrobenzoic acid	2.24			
2-Hydroxy-4-nitrophenylphosphonic acid	1.22	5.39		
8-Hydroxy-7-nitroquinoline- 5-sulfonic acid	1.94(0)	5.750(−1)		

TABLE 8.8   pK_a Values of Organic Materials in Water at 25°C (*Continued*)

Substance	pK_1	pK_2	pK_3	pK_4
3-Hydroxy-4-nitrotoluene (μ=0.1)	7.41			
4-Hydroxypentanoic acid (18°C)	4.686			
4-Hydroxy-3-pentenoic acid	4.30			
3-Hydroxyphenazine (15°C)	2.67			
4-Hydroxyphenylarsonic acid	3.89	8.37 (phenol)	10.05	
3-Hydroxyphenylboric acid	8.55	10.84		
2-Hydroxy-2-phenylpropanoic acid	3.532			
2-(2-Hydroxyphenyl)pyridine (20°C)	4.19(+1)	10.64		
trans-4-Hydroxyproline	1.818(+1)	9.662(0)		
Hydroxypropanedioic acid (tartronic acid)	2.37	4.74		
2-Hydroxypropanoic acid	3.858			
1-Hydroxy-2-propylbenzene	10.50			
4-Hydroxypteridine	1.3(+1)	7.89(0)		
2-Hydroxypyridine	1.25(+1)	11.62(0)		
3-Hydroxypyridine	4.80(+1)	8.72(0)		
4-Hydroxypyridine	3.23(+1)	11.09(0)		
2-Hydroxypyridine-*N*-oxide	−0.62(+1)	5.97(0)		
2-Hydroxypyrimidine	2.24(+1)	9.17(0)		
4-Hydroxypyrimidine	1.85(+1)	8.59(0)		
8-Hydroxyquinazoline	3.41(+1)	8.65(0)		
2-Hydroxyquinoline (20°C)	−0.31(+1)	11.74		
3-Hydroxyquinoline (20°C)	4.30(+1)	8.06(0)		
4-Hydroxyquinoline (20°C)	2.27(+1)	11.25(0)		
5-Hydroxyquinoline (20°C)	5.20(+1)	8.54(0)		
6-Hydroxyquinoline (20°C)	5.17(+1)	8.88(0)		
7-Hydroxyquinoline (20°C)	5.48(+1)	8.85(0)		
8-Hydroxyquinoline(20°C)	4.91(+1)	9.81(0)		
8-Hydroxyquinoline-5-sulfonic acid	4.092(+1)	8.776(0)		
DL-Hydroxysuccinic acid (malic acid)	3.458	5.097		
L-Hydroxysuccinic acid	3.40	5.05		
Hydroxytetracycline	3.27(+1)	7.32(0)	9.11(−1)	
5-Hydroxy-1,2,3,4-tetrazole	3.32			
4-Hydroxy-3-(2′-thiazolyazo)toluene	8.36			
2-Hydroxytoluene	10.33			
3-Hydroxytoluene	10.10			
4-Hydroxytoluene	10.276			
4-Hydroxy-α,α,α-trifluorotoluene	8.675			
1-Hydroxy-2,4,6-trihydroxymethylbenzene	9.56			
Hydroxyuracil	8.64			
Hydroxyvaline	2.55(+1)	9.77(0)		
Hyoscyamine	9.68(+1)			
Hypoxanthene	1.79(+1)	8.91(0)	12.07(−1)	
Hypoxanthine	5.3			

TABLE 8.8 pK_a Values of Organic Materials in Water at 25°C (*Continued*)

Substance	pK_1	pK_2	pK_3	pK_4
Imidazole	6.993(+1)	10.58(0)		
Imidazolidinetrione (parabanic acid)	6.10			
4-(4-Imidazolyl)butanoic acid				
(μ=0.1)	4.26(+1)	7.62(0)		
2-(4-Imidazolyl)ethylamine	5.784(+2)	9.756(+1)		
3-(4-Imidazolyl)propanoic acid				
(μ=0.16)	3.96(+1)	7.57(0)		
3,3'-Iminobispropanoic acid	4.11(0)	9.61(−1)		
3,3'-Iminobispropylamine (30°C)	8.02(+2)	9.70(+1)	10.70(0)	
2,2'-Iminodiacetic acid (diglycine)				
(30°C, μ=0.1)	2.54(0)	9.12(−1)		
4-Indanol	10.32			
Indole-3-acetic acid	4.75			
Inosine	ca 1.5(+1)	8.96(0)	12.36	
Inosine-5'-phosphoric acid	1.54(0)	6.66(−1)		
Inosine-5'-triphosphoric acid				
[pK_5 7.68(−4)]	——	——	2.2(−2)	6.92(−3)
Iodoacetic acid	3.175			
2-Iodoaniline	2.54(+1)			
3-Iodoaniline	3.58(+1)			
4-Iodoaniline	3.82(+1)			
2-Iodobenzoic acid	2.86			
3-Iodobenzoic acid	3.86			
4-Iodobenzoic acid	4.00			
5-Iodohistamine	4.06(+2)	9.20(+1)	11.88(0)	
	(imidazole)	(NH$_3^+$)	(imino)	
7-Iodo-8-hydroxyquinoline-5-sulfonic				
acid	2.514	7.417		
Iodomandelic acid	3.264			
Iodomethylphosphoric acid	1.30	6.72		
2-Iodophenol	8.464			
3-Iodophenol	8.879			
4-Iodophenol	9.200			
2-Iodophenoxyacetic acid	3.17			
3-Iodophenoxyacetic acid	3.13			
4-Iodophenoxyacetic acid	3.16			
2-Iodophenylacetic acid	4.038			
3-Iodophenylacetic acid	4.159			
4-Iodophenylacetic acid	4.178			
2-Iodophenylphosphoric acid	1.74	7.06		
2-Iodopropanoic acid	3.11			
3-Iodopropanoic acid	4.08			
2-Iodopyridine	1.82(+1)			
3-Iodopyridine	3.25(+1)			
4-Iodopyridine (20°C)	4.02(+1)			
Isoasparagine	2.97(+1)	8.02(0)		
Isobutylacetic acid (18°C)	4.79			

TABLE 8.8 pK_a Values of Organic Materials in Water at 25°C (*Continued*)

Substance	pK_1	pK_2	pK_3	pK_4
Isobutylamine	10.41(+1)			
Isochlorotetracycline	3.1(+1)	6.7(0)	8.3(−1)	
Isocreatine	2.84(+1)			
Isogluatamine	3.81(+1)	7.88(0)		
Isohistamine ($\mu = 0.1$)	6.036(+2)	9.274(+1)		
L-Isoleucine	2.318(+1)	9.758(0)		
Isolysergic acid	3.33(0)	8.46(NH)		
Isopilocarpine (15°C)	7.18(+1)			
2-(Isopropoxy)benzoic acid (20°C)	4.24			
3-(Isopropoxy)benzoic acid (20°C)	4.15			
4-(Isopropoxy)benzoic acid (20°C)	4.68			
Isopropylamine	10.64(+1)			
N-Isopropylaniline	5.50(+1)			
5-Isopropylbarbituric acid	4.907(+1)			
2-Isopropylbenzene acid	3.64			
4-Isopropylbenzoic acid	4.36			
N-Isopropylglycine ($\mu = 0.1$)	2.36(+1)	10.06(0)		
Isopropylmalonic acid	2.94	5.88		
Isopropylmalonic acid mononitrile	2.401			
3-Isopropyl-4-(methylamino)pyridine (20°C)	9.96(+1)			
3-Isopropylpentanedioic acid	4.30	5.51		
4-Isopropylphenylacetic acid	4.391			
Isopropylphosphinic acid	3.56			
Isopropylphosphonic acid	2.66	8.44		
2-Isopropylpyridine	5.83(+1)			
3-Isopropylpyridine (20°C)	5.72(+1)			
4-Isopropylpyridine	6.02(+1)			
DL-Isoproterenol	8.64(+1)			
Isoquinoline	5.40(+1)			
Isoretronecanol	10.83			
L-Isoserine ($\mu = 0.16$)	2.72(+1)	9.25(0)		
Isothiocyanatoacetic acid	6.62			
L-(+)-Lactic acid	3.858			
L-Leucine	2.328(+1)	9.744(0)		
Leucine amide	7.80(+1)			
Leucine, ethyl ester ($\mu = 0.1$)	7.57(+1)			
L-Leucyl-L-asparagine	3.00(+1)	8.12(0)		
L-Leucyl-L-glutamine	2.99(+1)	8.11(0)		
DL-Leucylglycine	3.25(+1)	8.28(0)		
Leucylisoserine (20°C)	3.188(+1)	8.207(0)		
D-Leucyl-L-tyrosine	3.12(+1)	8.38(0)	10.35(−1)	
L-Leucyl-L-tyrosine	3.46(+1)	7.84(0)	10.09(−1)	
Lysergic acid	3.44(+1)	7.68(0)		
L-(+)-Lysine	2.18(+2)	8.95(+1)	10.53(0)	
Lysine, methyl ester ($\mu = 0.1$)	6.965(+1)	10.251(0)		

TABLE 8.8 pK_a Values of Organic Materials in Water at 25°C (*Continued*)

Substance	pK_1	pK_2	pK_3	pK_4
L-Lysyl-L-alanine	3.22(+1)	7.62(0)	10.70(−1)	
L-Lysyl-D-alanine	3.00(+1)	7.74(0)	10.63(−1)	
Lysylglutamic acid	2.93(+2)	4.47(+1)	7.75(0)	10.50(+1)
L-Lysyl-L-lysine ($\mu = 0.1$)	3.01(+2)	7.53(+1)	10.05(0)	10.01(−1)
L-Lysyl-D-lysine ($\mu = 0.1$)	2.85(+2)	7.53(+1)	9.92(0)	10.89(−1)
L-Lysyl-L-lysyl-L-lysine ($\mu = 0.1$)	3.08(+2)	7.34(+1)	9.80(0)	10.54(−1)
L-Lysyl-D-lysyl-L-lysine ($\mu = 0.1$)	2.91(+2)	7.29(+1)	9.79(0)	10.54(−1)
L-Lysyl-D-lysyl-lysine ($\mu = 0.1$)	2.94(+2)	7.15(+1)	9.60(0)	10.38(−1)
α-D-Lyxose	12.11			
Maleic acid	1.910	6.33		
Malonamic acid	3.641(0)			
Malonic acid	2.826	5.696		
Malonitrile (cyanoacetic acid)	2.460			
Mandelic acid	3.411			
D-(+)-Mannose	12.08			
Mercaptoacetic acid (thioglycolic acid)	3.60(0)	10.56(SH)		
2-Mercaptobenzoic acid (20°C)	4.05(0)			
2-Mercaptobutanoic acid	3.53(0)			
Mercaptodiacetic acid	3.32	4.29		
2-Mercaptoethanesulfonic acid (20°C)		9.5(−1)		
2-Mercaptoethanol	9.88			
2-Mercaptoethylamine	8.27(+1)	10.53(0)		
2-Mercaptohistidine	1.84(+1)	8.47(0)	11.4(SH)	
Mercapto-S-phenylacetic acid ($\mu = 0.1$)	3.9			
2-Mercaptopropane ($\mu = 0.1$)	10.86			
3-Mercapto-1,2-propanediol ($\mu = 0.5$)	9.43			
2-Mercaptopropanoic acid	4.32(0)	10.20(SH)		
3-Mercaptopropanoic acid	——	10.84(SH)		
2-Mercaptopyridine (20°C)	−1.07(+1)	10.00(0)		
3-Mercaptopyridine (20°C)	2.26(+1)	7.03(0)		
4-Mercaptopyridine (20°C)	1.43(+1)	8.86(0)		
2-Mercaptoquinoline (20°C)	−1.44(+1)	10.21(0)		
3-Mercaptoquinoline (20°C)	2.33(+1)	6.13(0)		
4-Mercaptoquinoline (20°C)	0.77(+1)	8.83(0)		
Mercaptosuccinic acid	3.30(0)	4.94(−1)	10.94(SH)	
Mesitylenic acid	4.32			
Mesoxaldialdehyde	3.60			
Methacrylic acid	4.66			
Methanethiol	10.70			
DL-Methionine	2.13(+1)	9.28(0)		
2-(N-Methoxyacetamido)pyridine	2.01(+1)			
3-(N-Methoxyacetamido)pyridine	3.52(+1)			
4-(N-Methoxyacetamido)pyridine	4.62(+1)			
Methoxyacetic acid	3.570			

TABLE 8.8 pK_a Values of Organic Materials in Water at 25°C (*Continued*)

Substance	pK_1	pK_2	pK_3	pK_4
3-Methoxy-D-α-alanine	2.037(+1)	9.176(0)		
2-Methoxyaniline	4.53(+1)			
3-Methoxyaniline	4.20(+1)			
4-Methoxyaniline	5.36(+1)			
2-Methoxybenzoic acid	4.09			
3-Methoxybenzoic acid	4.08			
4-Methoxybenzoic acid	4.49			
N,*N*-Methoxybenzylamine	9.68(+1)			
2-Methoxycarbonylaniline	2.23(+1)			
3-Methoxycarbonylaniline	3.64(+1)			
4-Methoxycarbonylaniline	2.38(+1)			
Methoxycarbonylmethylamine	7.66(+1)			
2-Methoxycarbonylpyridine	2.21(+1)			
3-Methoxycarbonylpyridine	3.13(+1)			
4-Methoxycarbonylpyridine	3.26(+1)			
trans-2-Methoxycinnamic acid	4.462			
trans-3-Methoxycinnamic acid	4.376			
trans-4-Methoxycinnamic acid	4.539			
2-Methoxyethylamine	9.45(+1)			
2-Methoxy-4-nitrophenylphosphonic acid	1.53	6.96		
2-Methoxyphenol	9.99			
3-Methoxyphenol	9.652			
4-Methoxyphenol	10.20			
(2'-Methoxy)phenoxyacetic acid	3.231			
(3'-Methoxy)phenoxyacetic acid	3.141			
(4'-Methoxy)phenoxyacetic acid	3.213			
4'-Methoxyphenylacetic acid	4.358			
(4-Methoxyphenyl)phosphinic acid (17°C)	2.35			
(2-Methoxyphenyl)phosphonic acid	2.16	7.77		
(4-Methoxyphenyl)phosphonic acid (17°C)	2.4	7.15		
3-(2'-Methoxyphenyl)propanoic acid	4.804			
3-(3'-Methoxyphenyl)propanoic acid	4.654			
3-(4'-Methoxyphenyl)propanoic acid	4.689			
3-Methoxyphenylselenic acid	4.65			
4-Methoxyphenylselenic acid	5.05			
2-Methoxy-4-(2-propenyl)phenol	10.0			
2-Methoxypyridine	3.06(+1)			
3-Methoxypyridine	4.91(+1)			
4-Methoxypyridine	6.47(+1)			
4-Methoxy-2-(2'-thiazoylazo)phenol	7.83			
2-Methylacrylic acid (18°C)	4.66			
N-Methylalanine	2.22(+1)	10.19(0)		
O-Methylallothreonine ($\mu = 0.1$)	1.92(+1)	8.90(0)		
Methylamine	10.62(+1)			

TABLE 8.8 pK_a Values of Organic Materials in Water at 25°C (*Continued*)

Substance	pK_1	pK_2	pK_3	pK_4
2-(N-Methylamino)benzoic acid	1.93(+1)	5.34(0)		
3-(N-Methylamino)benzoic acid	——	5.10(0)		
4-(N-Methylamino)benzoic acid	——	5.05		
Methylaminodiacetic acid (20°C)	2.146	10.088		
2-(Methylamino)ethanol	9.88(+1)			
2-(2-Methylaminoethyl)pyridine (30°C)	3.58(+2)	9.65(+1)		
2-(Methylaminomethyl)-6-methyl-pyridine ($\mu = 0.5$)	3.03(+2)	9.15(+1)		
2-(Methylaminomethyl)pyridine (30°C)	2.92(+2)	8.82(+1)		
4-Methylamino-3-methylpyridine (20°C)	9.83(+1)			
(3-Methylamino)phenylphosphonic acid	1.1(+1)	4.72(+1)	7.30(−1)	
(4-Methylamino)phenylphosphonic acid	——	——	7.85(−1)	
3-(Methylamino)pyridine (30°C)	8.70(+1)			
4-(Methylamino)pyridine (20°C)	9.65(+1)			
4-(Methylamino)-2,3,5,6-tetramethyl-pyridine (20°C)	10.06(+1)			
N-Methylaniline	4.85(+1)			
Methylarsonic acid (18°C)	3.41	8.18		
1-Methylbarbituric acid	4.35(+1)			
5-Methylbarbituric acid	3.386(+1)			
2-(N-Methylbenzamido)pyridine	1.44(+1)			
3-(N-Methylbenzamido)pyridine	3.66(+1)			
4-(N-Methylbenzamido)pyridine	4.68(+1)			
2-Methylbenzimidazole ($\mu = 0.16$)	6.29(+1)			
2-Methylbenzoic acid (o-toluic acid)	3.90			
3-Methylbenzoic acid	4.269			
4-Methylbenzoic acid	4.362			
N-Methyl-1-benzoylecgonine	8.65			
Methylbiguanidine	3.00(+2)	11.44(+1)		
2-Methyl-2-butanethiol	11.35			
2-Methylbutanoic acid	4.761			
3-Methylbutanoic acid (20°C)	4.767			
(E)-2-Methyl-2-butendioic acid (mesaconic acid)	3.09	4.75		
3-Methyl-2-butenoic acid	5.12			
(E)-2-Methyl-2-butenoic acid (tiglic acid)	4.96			
(Z)-2-Methyl-2-butenoic acid (angelic acid)	4.30			
4-Methylcarboxylphenol	8.47			
(E)-2-Methylcinnamic acid	4.500			

TABLE 8.8 pK_a Values of Organic Materials in Water at 25°C (*Continued*)

Substance	pK_1	pK_2	pK_3	pK_4
(*E*)-3-Methylcinnamic acid	4.442			
(*E*)-4-Methylcinnamic acid	4.564			
1-Methylcyclohexane-1-carboxylic acid	5.13			
cis-2-Methylcyclohexane-1-carboxylic acid	5.03			
trans-2-methylcyclohexane-1-carboxylic acid	5.73			
cis-3-methylcyclohexane-1-carboxylic acid	4.88			
trans-3-Methylcyclohexane-1-carboxylic acid	5.02			
cis-4-Methylcyclohexane-1-carboxylic acid	5.04			
trans-4-Methylcyclohexane-1-carboxylic acid	4.89			
2-Methylcyclohexyl-1,1-diacetic acid	3.53	6.89		
3-Methylcyclohexyl-1,1-diacetic acid	3.49	6.08		
4-Methylcyclohexyl-1,1,1-diacetic acid	3.49	6.10		
3-Methylcyclopentyl-1,1-diacetic acid	3.79	6.74		
S-Methyl-L-cysteine	8.97			
N-Methylcytidine	3.88			
5-Methylcytidine	4.21			
N-Methyl-2′-deoxycytidine	3.97			
5-Methyl-2′-deoxycytidine	4.33			
2-Methyl-3,5-dinitrobenzoic acid	2.97			
5-Methyldipropylenetriamine (30°C)	6.32(+3)	9.19(+2)	10.33(+1)	
2,2′-Methylenebis(4-chlorophenol)	7.6	11.5		
2,2′-Methylenebis(4,6-dichloro-phenol)	5.6	10.56		
Methylenebis(thioacetic acid (18°C)	3.310	4.345		
3,3′-(Methylenedithio)dialanine	2.200(+1)	8.16(0)		
Methylenesuccinic acid	3.85	5.45		
N-Methylethylamine	4.23(+1)			
N-Methylethylenediamine	6.86(+1)	10.15(+1)		
α-Methylglucoside	13.71			
3-Methylglutaric acid	4.24	5.41		
N-Methylglycine (sarcosine)	2.12(+1)	10.20(0)		
5-Methyl-2,4-heptanedione	8.52(enol); 9.10(keto)			
5-Methyl-2,4-hexanedione	8.66(enol); 9.31(keto)			
5-Methyl-4-hexenoic acid	4.80			
3-Methylhistamine	5.80(+1)	9.90(0)		
1-Methylhistidine	1.69	6.48	8.85	
2-Methylhistidine (18°C)	1.7	7.2	9.5	

TABLE 8.8 pK_a Values of Organic Materials in Water at 25°C (*Continued*)

Substance	pK_1	pK_2	pK_3	pK_4
2-Methyl-8-hydroxyquinoline				
($\mu = 0.005$)	4.58(+1)	11.71(0)		
4-Methyl-8-hydroxyquinoline	4.67(+1)	11.62(0)		
1-Methylimidazole	7.06(+1)			
4-Methylimidazole	7.55(+1)			
N-Methyliminodiacetic acid	2.15	10.09		
S-Methylisothiourea	9.83(+1)			
O-Methylisourea	9.72(+1)			
Methylmalonic acid	3.07	5.87		
2-(N-Methylmethane-				
sulfonamido)pyridine	1.73(+1)			
3-(N-Methylmethane-				
sulfonamido)pyridine	3.94(+1)			
4-(N-Methylmethane-				
sulfonamido)pyridine	5.14(+1)			
2-Methyl-6-methyl-				
aminopyridine (20°C)	3.17(+1)	8.84(0)		
3-Methyl-4-methyl-				
aminopyridine (20°C)	——	9.84(0)		
4-Methyl-2,2'-				
(4-methylpyridyl)pyridine	5.32(+1)			
N-Methylmorpholine	7.13(+1)			
2-Methyl-1-naphthoic acid	3.11			
N-Methyl-1-naphthylamine	3.70(+1)			
2-Methyl-4-nitrobenzoic acid	1.86			
2-Methyl-6-nitrobenzoic acid	1.87			
1-Methyl-2-nitroterephthalic acid	3.11			
4-Methyl-2-nitroterephthalic acid	1.82			
3-Methylpentanedioic acid	4.25	5.41		
3-Methylpentane-2,4-dione	10.87			
2-Methylpentanoic acid	4.782			
3-Methylpentanoic acid	4.766			
4-Methylpentanoic acid	4.845			
cis-3-Methyl-2-pentenoic acid	5.15			
trans-3-Methyl-2-pentenoic acid	5.13			
4-Methyl-2-pentenoic acid	4.70			
4-Methyl-3-pentenoic acid	4.60			
6-Methyl-1,10-phenanthroline	5.11(+1)			
(2-Methylphenoxy)acetic acid	3.227			
(3-Methylphenoxy)acetic acid	3.203			
(4-Methylphenoxy)acetic acid	3.215			
(2-Methylphenyl)acetic acid (18°C)	4.35			
(4-Methylphenyl)acetic acid	4.370			
5-Methyl-5-phenylbarbituric acid	8.011(0)			
3-(2-Methylphenyl)propanoic acid	4.66			
3-(3-Methylphenyl)propanoic acid	4.677			
3-(4-Methylphenyl)propanoic acid	4.684			

TABLE 8.8   pK_a Values of Organic Materials in Water at 25°C (*Continued*)

Substance	pK_1	pK_2	pK_3	pK_4
1-Methyl-2-phenylpyrrolidine	8.80			
5-Methyl-1-phenyl-1,2,3-triazole- 4-carboxylic acid	3.73			
Methylphosphinic acid	3.08			
Methylphosphonic acid	2.38	7.74		
3-Methyl-*o*-phthalic acid	3.18			
4-Methyl-*o*-phthalic acid	3.89			
N-Methylpiperazine ($\mu=0.1$)	4.94(+2)	9.09(+1)		
2-Methylpiperazine	5.62(+2)	9.60(+1)		
N-Methylpiperidine	10.19(+1)			
2-Methylpiperidine	10.95(+1)			
3-Methylpiperidine	11.07(+1)			
4-Methylpiperidine ($\mu=0.5$)	11.23(+1)			
2-Methyl-1,2-propanediamine	6.178(+2)	9.420(+1)		
2-Methyl-2-propanethiol	11.2			
2-Methylpropanoic acid	4.853			
2-Methyl-2-propylamine	10.682(+1)			
2-Methyl-2-propylglutaric acid	3.626			
2-Methylpyridine	5.96(+1)			
3-Methylpyridine	5.68(+1)			
4-Methylpyridine	6.00(+1)			
Methyl 4-pyridinecarboxylate	3.26(+1)			
6-Methylpyridine-2-carboxylic acid	5.83			
2-Methylpyridine-1-oxide	1.029(+1)			
3-Methylpyridine-1-oxide	10.921(+1)			
4-Methylpyridine-1-oxide	1.258(+1)			
O-Methylpyridoxal ($\mu=0.16$)	4.74			
Methyl-2-pyridyl ketoxime	9.97			
1-Methyl-2-(3-pyridyl)pyrrolidine	3.41	7.94		
1-Methylpyrrolidine	10.46(+1)			
1-Methyl-3-pyrroline	9.88(+1)			
5-Methylquinoline	4.62(+1)			
Methylsuccinic acid	4.13	5.64		
Methylsulfonylacetic acid	2.36			
3-Methylsulfonylaniline	2.68(+1)			
4-Methylsulfonylaniline	1.48(+1)			
3-Methylsulfonylbenzoic acid	3.52			
4-Methylsulfonylbenzoic acid	3.64			
4-Methylsulfonyl-3,5-dimethylphenol	8.13			
3-Methylsulfonylphenol	9.33			
4-Methylsulfonylphenol	7.83			

1-Methyl-1,2,3,4-tetrahydro-
    3-pyridinecarboxylic acid
    (arecaidine; isoguvacine)
	9.07			

5-Methyl-1,2,3,4-tetrazole	3.32			
2-Methylthiazole ($\mu = 0.1$)	3.40(+1)			
4-Methylthiazole ($\mu = 0.1$)	3.16(+1)			

TABLE 8.8 pK_a Values of Organic Materials in Water at 25°C (*Continued*)

Substance	pK_1	pK_2	pK_3	pK_4
5-Methylthiazole ($\mu = 0.1$)	3.03(+1)			
Methylthioacetic acid	3.72			
4-Methylthioaniline	4.40(+1)			
2-Methylthioethylamine (30°C)	9.18(+1)			
Methylthioglycolic acid	7.68			
3-(*S*-Methylthio)phenol	9.53			
4-(*S*-Methylthio)phenol	9.53			
2-Methylthiopyridine (20°C)	3.59(+1)			
3-Methylthiopyridine (20°C)	4.42(+1)			
4-Methylthiopyridine (20°C)	5.94(+1)			
5-Methylthio-1,2,3,4-tetrazole	4.00(+1)			
O-Methylthreonine	2.02(+1)	9.00(0)		
O-Methyltyrosine	2.21(+1)	9.35(0)		
1-Methylxanthine	7.70	12.0		
3-Methylxanthine	8.10	11.3		
7-Methylxanthine	8.33	ca 13		
9-Methylxanthine	6.25			
Morphine (20°C)	7.87(+1)	9.85(0)		
Morpholine	8.492(+1)			
2-(*N*-Morpholino)ethanesulfonic acid (MES) (20°C)	6.15			
3-(*N*-Morpholino)-2-hydroxy-propanesulfonic acid (37°C)	6.75			
3-(*N*-Morpholino)propanesulfonic acid (20°C)	7.20			
Murexide	0.0	9.20	10.50	
Myosmine	5.26			
1-Naphthalenecarboxylic acid (1-naphthoic acid)	3.695			
2-Naphthalenecarboxylic acid	4.161			
1-Naphthol (20°C)	9.30			
2-Naphthol (20°C)	9.57			
Naphthoquinone monoxime	8.01			
1-Naphthylacetic acid	4.236			
2-Naphthylacetic acid	4.256			
1-Naphthylamine	3.92(+1)			
2-Naphthylamine	4.11(+1)			
1-Naphthylarsonic acid	3.66	8.66		
1-Naphthylsulfonic acid	0.57			
Narceine (15°C)	3.5(+1)	9.3		
Narcotine	6.18(+1)			
Nicotine	3.15(+1)	7.87(0)		
Nicotyrine	4.76(+1)			
Nitrilotriacetic acid (NTA) (20°C)	1.65	2.94	10.33	
Nitroacetic acid	1.68			
2-Nitroaniline	−0.28(+1)			

TABLE 8.8 pK_a Values of Organic Materials in Water at 25°C (*Continued*)

Substance	pK_1	pK_2	pK_3	pK_4
3-Nitroaniline	2.46(+1)			
4-Nitroaniline	1.01(+1)			
2-Nitrobenzene-1,4-dicarboxylic acid	1.73			
3-Nitrobenzene-1,2-dicarboxylic acid	1.88			
4-Nitrobenzene-1,2-dicarboxylic acid	2.11			
2-Nitrobenzoic acid	2.18			
3-Nitrobenzoic acid	3.46			
4-Nitrobenzoic acid	3.441			
trans-2-Nitrocinnamic acid	4.15			
trans-3-Nitrocinnamic acid	4.12			
trans-4-Nitrocinnamic acid	4.05			
Nitroethane	8.57			
2-Nitrohydroquinone	7.63	10.06		
N-Nitroiminodiacetic acid	2.21	3.33		
3-Nitromesitol	8.984			
Nitromethane	10.21			
1-Nitro-6,7-phenanthroline ($\mu = 0.2$)	3.23(+1)			
5-Nitro-1,10-phenanthroline	3.232(+1)			
6-Nitro-1,10-phenanthroline	3.23(+1)			
2-Nitrophenol	7.222			
3-Nitrophenol	8.360			
4-Nitrophenol	7.150			
(2-Nitrophenoxy)acetic acid	2.896			
(3-Nitrophenoxy)acetic acid	2.951			
(4-Nitrophenoxy)acetic acid	2.893			
2-Nitrophenylacetic acid	4.00			
3-Nitrophenylacetic acid	3.97			
4-Nitrophenylacetic acid	3.85			
2-Nitrophenylarsonic acid	3.37	8.54		
3-Nitrophenylarsonic acid	3.41	7.80		
4-Nitrophenylarsonic acid	2.90	7.80		
7-(4-Nitrophenylazo)-8-hydroxy-5-quinolinesulfonic acid	3.14(0)	7.495(−1)		
3-Nitrophenylphosphonic acid	1.30	6.27		
4-Nitrophenylphosphonic acid	1.24	6.23		
3-(2′-Nitrophenyl)propanoic acid	4.504			
3-(4′-Nitrophenyl)propanoic acid	4.473			
3-Nitrophenylselenic acid	4.07			
4-Nitrophenylselenic acid	4.00			
1-Nitropropane	8.98			
2-Nitropropane	7.675			
2-Nitropropanoic acid	3.79			
2-Nitropyridine ($\mu = 0.02$)	−2.06(+1)			
3-Nitropyridine ($\mu = 0.02$)	0.79(+1)			
4-Nitropyridine ($\mu = 0.02$)	1.23(+1)			
N-Nitrosoiminodiacetic acid	2.28	3.38		
4-Nitrosophenol	6.48			

TABLE 8.8 pK_a Values of Organic Materials in Water at 25°C (*Continued*)

Substance	pK_1	pK_2	pK_3	pK_4
Nitrourea	4.15(+1)			
1,9-Nonanedioic acid (azelaic acid)	4.53	5.40		
Nonanoic acid (pelargonic acid)	4.95			
DL-Norleucine	2.335(+1)	9.834(0)		
Novocaine	8.85(+1)			
2,2,3,3,4,4,5,5-Octafluoropentanoic acid	2.65			
1,8-Octanedioic acid (suberic acid)	4.512	5.404		
Octanoic acid (caprylic acid)	4.895			
Octopine-DD	1.35	2.30	8.68	11.25
Octopine-LD	1.40	2.30	8.72	11.34
Octylamine	10.65(+1)			
L-(+)-Ornithine	1.94(+2)	8.65(+1)	10.76(0)	
Oxalic acid	1.271	4.272		
3,6-Oxaoctanedioic acid ($\mu = 1.0$)	3.055	3.676		
Oxoacetic acid	3.46			
2-Oxabutanedioic acid (oxaloacetic acid)	2.56	4.37		
2-Oxobutanoic acid	2.50			
5-Oxohexanoic acid (5-ketohexanoic acid) (18°C)	4.662			
3-Oxo-1,5-pentanedioic acid	3.10			
4-Oxopentanoic acid (levulinic acid)	4.59			
2-Oxopropanoic acid (pyruvic acid)	2.49			
Oxytetracycline	3.10(+1)	7.26	9.11	
Papaverine	5.90(+1)			
Pentamethylenebis(thioacetic acid) (18°C)	3.485	4.413		
3,3-Pentamethylenepentanedioic acid	3.49	6.96		
1,5-Pentanediamine	10.05(+2)	10.916(+1)		
2,4-Pentanedione	8.24(enol); 8.95(keto)			
1-Pentanoic acid (valeric acid)	4.842			
2-Pentenoic acid	4.70			
3-Pentenoic acid	4.52			
4-Pentenoic acid	4.677			
Pentylarsonic acid	4.14	9.07		
N-Pentylveratramine	7.28(+1)			
Perhydrodiphenic acid (20°C)	4.96	6.68		
Perlolidine (18°C)	4.01	11.39		
Peroxyacetic acid	8.20			
1,7-Phenanthroline	4.30(+1)			
1,10-Phenanthroline	4.857(+1)			
6,7-Phenanthroline	4.857(+1)			
Phenazine	1.2(+1)			

TABLE 8.8   pK_a Values of Organic Materials in Water at 25°C (*Continued*)

Substance	pK_1	pK_2	pK_3	pK_4
Phenethylthioacetic acid	3.795			
Phenol	9.99			
Phenol-3-phosphoric acid	1.78	7.03	10.2	
Phenol-4-phosphoric acid	1.99	7.25	9.9	
Phenolphthalein	9.4			
3-Phenolsulfonic acid	——	9.05(−1)		
Phenosulsulfonephthalein	7.9			
Phenoxyactic acid	3.171			
2-Phenoxybenzoic acid	3.53			
3-Phenoxybenzoic acid	3.95			
4-Phenoxybenzoic acid	4.52			
5-Phenoxy-1,2,3,4-tetrazole	3.49(+1)			
Phenylacetic acid	4.312			
L-3-Phenyl-α-alanine	2.16(+1)	9.31(0)		
3-Phenyl-α-alanine, methyl ester	7.05(+1)			
Phenylalanylarginine ($\mu = 0.01$)	2.66(+1)	7.57(0)	12.40(−1)	
Phenylalanylglycine ($\mu = 0.01$)	3.10(+1)	7.71(0)		
7-Phenylazo-8-hydroxy-				
5-quinolinesulfonic acid	3.41(0)	7.850(−1)		
5-Phenylbarbituric acid	2.544(+1)			
2-Phenyl-2-benzylsuccinic acid	3.69	6.47		
1-Phenylbiguanide	2.13(+2)	10.76(+1)		
4-Phenylbutanoic acid	4.757			
Phenylbutazone	4.5(+1)			
2-Phenylenediamine	<2(+2)	4.47(+1)		
3-Phenylenediamine	2.65(+2)	4.88(+1)		
4-Phenylenediamine	3.29(+2)	6.08(+1)		
2-Phenylethylamine	9.83(+1)			
β-Phenylethylboronic acid	10.0			
DL-α-Phenylglycine	1.83(+1)	4.39(0)		
Phenylguanidine	10.77(+1)			
Phenylhydrazine	5.20(+1)			
2-Phenyl-3-hydroxypropanoic acid	3.53			
3-Phenyl-3-hydroxypropanoic acid	4.40			
Phenyliminodiacetic acid (20°C)	2.40	4.98		
Phenylmalonic acid	2.58	5.03		
Phenylmethanethiol	10.70			
2-Phenyl-2-phenethylsuccinic acid				
(20°C)	3.74	6.52		
2-Phenylphenol	9.55			
3-Phenylphenol	9.63			
4-Phenylphenol	9.55			
Phenylphosphinic acid (17°C)	2.1			
Phenylphosphonic acid	1.83	7.07		
O-Phenylphosphorylserine	2.13(+1)	8.79		
O-Phenylphosphorylserylglycine	3.18(+1)	6.95(0)		
O-Phenylphosphoryl-L-seryl-				
L-leucine	3.16(+1)	7.12(0)		

TABLE 8.8 pK_a Values of Organic Materials in Water at 25°C (*Continued*)

Substance	pK_1	pK_2	pK_3	pK_4
N-Phenylpiperazine ($\mu = 0.1$)	8.71(+1)			
2-Phenylpropanoic acid	4.38			
3-Phenylpropanoic acid (35°C)	4.664			
3-Phenyl-1-propylamine	10.39(+1)			
Phenylpropynoic acid (35°C)	2.269			
Phenylselenic acid	4.79			
Phenylselenoacetic acid ($\mu = 0.1$)	3.75			
β-Phenylserine ($\mu = 0.16$)	8.79(0)			
Phenylsuccinic acid (20°C)	3.78	5.55		
Phenylsulfenylacetic acid	2.66			
Phenylsulfonylacetic acid	2.44			
5-Phenyl-1,2,3,4-tetrazole	4.38(+1)			
1-Phenyl-1,2,3-triazole-4-carboxylic acid	2.88			
1-Phenyl-1,2,3-triazole-4,5-dicarboxylic acid	2.13	4.93		
Phosphoramidic acid	3.08	8.63		
O-Phosphorylethanolamine	5.838(+1)	10.638(0)		
O-Phosphorylserylglycine	3.13	5.41	8.01	
O-Phosphoryl-L-seryl-L-leucine	3.11	5.47	8.26	
Phosphoserine	2.08	5.65	9.74	
Phthalamide	3.79(0)			
Phthalazine	3.47(+1)			
o-Phthalic acid	2.950	5.408		
Phthalimide	9.90(0)			
Physostigmine	1.76(+1)	7.88(0)		
Picric acid (2,4,6-trinitrophenol) (18°C)	0.419			
Pilocarpine	1.3(+1)	6.85(0)		
Piperazine	5.333(+2)	9.781(+1)		
1,4-Piperazinebis(ethanesulfonic acid) (20°C)	6.80			
Piperazine-2-carboxylic acid	1.5	5.41	9.53	
Piperdine	11.123(+1)			
2-Piperidinecarboxylic acid	2.12(+1)	10.75(0)		
3-Piperidinecarboxylic acid	3.35(+1)	10.64(0)		
4-Piperidinecarboxylic acid	3.73(+1)	10.72(0)		
1-(2-Piperidinyl)-2-propanone (15°C)	9.45			
Piperine (15°C)	1.98(+1)			
Proline	1.952(+1)	10.640(0)		
1,2-Propanediamine	6.607(+2)	9.720(+1)		
1,3-Propanediamine	8.49(+2)	10.47(+1)		
1-Propanethiol	10.86			
1,2,3-Propanetriamine	3.72(+3)	7.95(+2)	9.59(+1)	
1,2,3-Propanetricarboxylic acid	3.67	4.87	6.38	
Propanoic acid	4.874			
Propenoic acid	4.247			
N-Propionyglycine	3.718(0)			

TABLE 8.8 pK_a Values of Organic Materials in Water at 25°C (*Continued*)

Substance	pK_1	pK_2	pK_3	pK_4
2-Propoxybenzoic acid (20°C)	4.24			
3-Propoxybenzoic acid (20°C)	4.20			
4-Propoxybenzoic acid (20°C)	4.78			
N-Propylalanine	2.21(+1)	10.19(0)		
Propylamine	10.568(+1)			
Propylarsonic acid (18°C)	4.21	9.09		
Propylenimine	8.18(+1)			
N-Propylglycine ($\mu = 0.1$)	2.38(+1)	10.03(0)		
L-Propylglycine	3.19(+1)	8.97(0)		
Propylmalonic acid	2.97	5.84		
Propylphosphinic acid	3.46			
Propylphosphonic acid	2.49	8.18		
2-Propylpyridine	6.30(+1)			
N-Propylveratramine	7.20(+1)			
2-Propynoic acid	1.887			
Pseudoecgonine	9.70			
Pseudoisocyanine ($\mu = 0.2$)	4.59(+2)			
Pseudotropine	9.86(+1)			
Pteroylglutamic acid	8.26			
Purine	2.52(+1)	8.92(0)		
Pyrazine	0.6(+1)			
Pyrazinecarboxamide	0.5(+1)			
Pyrazole	2.61(+1)			
Pyridazine	2.33(+1)			
Pyridine	5.17(+1)			
Pyridine-d_5	5.83(+1)			
2-Pyridinealdoxime	3.56(+1)	10.17(0)		
3-Pyridinealdoxime	4.07(+1)	10.39(0)		
4-Pyridinealdoxime	4.73(+1)	10.03(0)		
2-Pyridinecarbaldehyde	3.84(+1)			
3-Pyridinecarbaldehyde	3.80(+1)			
4-Pyridinecarbaldehyde	4.74(+1)			
3-Pyridinecarbamide (nicotinamide)	3.33(+1)			
3-Pyridinecarbonitrile	1.35(+1)			
Pyridine-2-carboxylic acid (picolinic acid)	1.01(+1)	5.29(0)		
Pyridine-3-carboxylic acid (nicotinic acid)	2.07(+1)	4.75(0)		
Pyridine-4-carboxylic acid (isonicotinic acid)	1.84(+1)	4.86(0)		
Pyridine-2,3-dicarboxylic acid	2.36(+1)	7.08(0)		
Pyridine-2,4-dicarboxylic acid	2.23(+1)	7.02(0)		
Pyridine-2,6-dicarboxylic acid	2.16(+1)	6.92(0)		
Pyridine-1-oxide	0.688(+1)			
Pyridoxal	4.20(+1)	8.66(ring OH)		
Pyridoxal-5-phosphate ($\mu = 0.15$)	<2.5	4.14	6.20	8.69

TABLE 8.8 pK_a Values of Organic Materials in Water at 25°C (*Continued*)

Substance	pK_1	pK_2	pK_3	pK_4
Pyridoxamine ($\mu = 0.1$)	3.37(+2)	8.01(+1)	10.13(ring OH)	
Pyridoxamine-5-phosphate ($\mu = 0.15$; pK$_5$ 10.92)	2.5	3.69	5.76	8.61
Pyridoxine (vitamin B$_6$) (18°C)	5.00(+1)	8.96(ring OH)		
3-(2'-Pyridyl)alanine	1.37(+2)	4.02(+1)	9.22(0)	
3-(3'-Pyridyl)alanine	1.77(+2)	4.64(+1)	9.10(0)	
2-(2'-Pyridyl)benzimidazole ($\mu = 0.16$)	5.58(+1)			
2-(2'-Pyridyl)imidazole ($\mu = 0.005$)	8.98(+1)			
4-(2'-Pyridyl)imidazole ($\mu = 0.1$)	5.49(+1)			
Pyrimidine	1.30(+1)			
2,4(1H,3H)-Pyrimidinedione (uracil)	0.6(+1)	9.46(0)		
2,4,5,6(1H,3H)-Pyrimidinetetrone-5-oxime	4.57(0)			
Pyrocatecholsulfonephthaleine	7.82	9.76	11.73	
Pyroxilidine	11.11(+1)			
Pyrrole-1-carboxylic acid	4.45			
Pyrrole-2-carboxylic acid	4.45			
Pyrrole-3-carboxylic acid	4.453			
Pyrrolidine	11.305(+1)			
Pyrrolidine-2-carboxylic acid (proline)	1.952(+1)	10.640(0)		
2-[2-(N-Pyrrolidinyl)ethyl]pyridine	3.60(+2)	9.39(+1)		
3-[2-(N-Pyrrolidinyl)ethyl]pyridine	4.28(+2)	9.28(+1)		
4-[2(N-Pyrrolidinyl)ethyl]pyridine	4.65(+2)	9.27(+1)		
2-(1-Pyrrolidinylmethyl)pyridine	2.54(+1)	8.56(+1)		
3-(1-Pyrrolidinylmethyl)pyridine	3.14(+2)	8.36(+1)		
4-(1-Pyrrolidinylmethyl)pyridine	3.38(+2)	8.16(+1)		
3-Pyrroline	−0.27(+1)			
Quinidine	4.0(+1)	8.54(0)		
Quinine	4.11(+1)	8.52(0)		
Quinoline	4.80(+1)			
Quinoxaline	0.72(+1)			
D-Raffinose	12.74			
Riboflavin (vitamin B$_2$) ($\mu = 0.01$)	ca −0.2	9.69		
α-D-Ribofuranose	12.11			
D-Ribose-5'-phosphonic acid	——	6.70(−1)	13.05(−2)	
D-Saccharic acid	5.00(0)			
Saccharin (*o*-benzoic sulfimide)	2.32			
Sarcosine	2.12(+1)	10.20(0)		

TABLE 8.8 pK_a Values of Organic Materials in Water at 25°C (*Continued*)

Substance	pK_1	pK_2	pK_3	pK_4
Sarcosine amide	8.35(+1)			
Sarcosine dimethylamide	8.86(+1)			
Sarcosine methylamide	8.28(+1)			
Sarcosylglycine ($\mu = 0.16$)	3.15(+1)	8.56(0)		
Sarcosylleucine	3.15(+1)	8.67(0)		
Sarcosylsarcosine	2.92(+1)	9.15(0)		
Sarcosylserine	3.17(+1)	8.63(0)		
3-Selenosemicarbazide ($\mu = 0.1$)	0.8(+1)			
Semicarbazide ($\mu = 0.1$)	3.53(+1)			
L-Serine	2.186(+1)	9.208(0)		
Serine, methyl ester ($\mu = 0.1$)	7.03(+1)			
Serylglycine ($\mu = 0.15$)	2.10(+1)	7.33(0)		
L-Seryl-L-leucine	3.08(+1)	7.45(0)		
Solanine	7.34(+1)			
D-Sorbitol (17.5°C)	13.60			
L-(−)-Sorbose (18°C)	11.55			
Sparteine	4.49(+1)	11.76(0)		
Spinaceamine ($\mu = 0.1$)	4.895(+2)	8.90(+1)		
Spinacine	1.649(+2)	4.936(+1)	8.663(0)	
L-Strychnine (15°C)	2.50	8.20		
Succinamic acid (succinic acid monoamide)	4.39(0)			
Succinic acid	4.207	5.635		
DL-Succinimide	9.623			
β-(4′-Sulfaminophenyl)alanine	1.99(+1)	8.64(0)	10.26(−1)	
3-Sulfamylbenzoic acid	3.54			
4-Sulfamylbenzoic acid	3.47			
4-Sulfamylphenylphosphoric acid	1.42	6.38	10.0	
Sulfanilamide	10.43(+1)			
Sulfoacetic acid	——	4.0		
3-Sulfobenzoic acid	——	3.78		
4-Sulfobenzoic acid	——	3.72		
3-Sulfophenol	0.39	9.07		
4-Sulfophenol	0.58	8.70		
2-Sulfopropanoic acid	1.99			
5-Sulfosalicyclic acid	2.49	12.00		
Sylvic acid	7.62			
D-Tartaric acid	3.036	4.366		
meso-Tartaric acid	3.22	4.81		
Tetracycline ($\mu = 0.005$)	3.30(+1)	7.68	9.69	
Tetradehydroyohimbine	10.59(+1)			
Tetraethylenepentamine [$\mu = 0.1$; pK_5 9.67(+1)]	2.98(+5)	4.72(+4)	8.08(+3)	9.10(+2)
1,4,5,6-Tetrahydro-1,2-dimethylpyridine	11.38(+1)			
1,4,5,6-Tetrahydro-2-methylpyridine	9.53(+1)			

TABLE 8.8 pK_a Values of Organic Materials in Water at 25°C (*Continued*)

Substance	pK_1	pK_2	pK_3	pK_4
cis-Tetrahydronaphthalene-				
2,3-dicarboxylic acid (20°C)	3.98	6.47		
trans-Tetrahydronaphthalene-				
2,3-dicarboxylic acid (20°C)	4.00	5.70		
5,6,7,8-Tetrahydro-1-naphthol	10.28			
5,6,7,8-Tetrahydro-2-naphthol	10.48			
Tetrahydroserpentine	10.55(+1)			
2,3,5,6-Tetramethylbenzoic acid	3.415			
Tetramethylenebis(thioacetic acid)				
(18°C)	3.463	4.423		
Tetramethylenediamine	9.22(+2)	10.75(+1)		
N,N,N′,N′-				
Tetramethylethylenediamine	2.20(+2)	6.35(+1)		
2,3,5,6-Tetramethyl-				
4-methylaminopyridine	0.07(+1)			
2,2,6,6-Tetramethylpiperidine				
($\mu = 0.5$)	1.24(+1)			
2,3,5,6-Tetramethylpyridine (20°C)	7.90(+1)			
Tetramethylsuccinic acid	3.50	7.28		
1,2,3,4-Tetrazole	4.90			
Thebaine	7.95(+1)			
2-Thenoyltrifluoroacetone	5.70(0)			
Theobromine	0.68(+1)	7.89		
Theophylline	<1(+1)	8.80		
Thiazoline	2.53(+1)			
Thioacetic acid	3.33			
o-Thiocresol	6.64			
m-Thiocresol	6.58			
p-Thiocresol	6.52			
Thiocyanatoacetic acid	2.58			
2,2′-Thiodiacetic acid	3.32	4.29		
4,4′-Thiodibutanoic acid (18°C)	4.351	5.275		
3,3′-Thiodipropanoic acid (18°C)	4.085	5.075		
3-Thio-*S*-methylcarbazide ($\mu = 0.1$)	7.563(+1)			
1-Thionylcarboxylic acid	3.53			
2-Thionylcarboxylic acid	4.10			
2-Thiophenecarboxylic acid (30°C)	3.529			
3-Thiophenecarboxylic acid				
(3-thenoic acid)	4.10			
Thiophenol	6.50			
3-Thiosemicarbazide ($\mu = 0.1$)	1.5(+1)			
3-Thiosemicarbazide-1,1-diacetic				
acid (30°C)	2.94	4.07		
Thiourea	2.03(+1)			
Thorin	3.7	8.3	11.8	
Thymidine	9.79	12.85		
p-Toluenesulfinic acid	1.7			

TABLE 8.8 pK_a Values of Organic Materials in Water at 25°C (*Continued*)

Substance	pK_1	pK_2	pK_3	pK_4
Toluhydroquinone	10.03	11.62		
o-Toluidine	4.45(+1)			
m-Toluidine	4.71(+1)			
p-Toluidine	5.08(+1)			
o-Tolylacetic acid (18°C)	4.36			
p-Tolyacetic acid (18°C)	4.36			
o-Tolylarsonic acid	3.82	8.85		
m-Tolylarsonic acid	3.82	8.60		
p-Tolylarsonic acid	3.70	8.68		
o-Tolylphosphonic acid	2.10	7.68		
m-Tolyphosphonic acid	1.88	7.44		
p-Tolyphosphonic acid	1.84	7.33		
3-Tolylselenic acid	4.80			
4-Tolylselenic acid	4.88			
Triacetylmethane	5.81			
Triallylamine	8.31(+1)			
1,3,5-Triazine-2,4,6-triol	7.20	11.10		
1H-1,2,3-Triazole	——	9.26		
1H-1,2,4-Triazole	2.386(+1)	9.972		
1,2,3-Triazole-4-carboxylic acid	3.22	8.73		
1,2,3-Triazole-4,5-dicarboxylic acid	1.86	5.90	9.30	
1,2,4-Triazolidine-3,5-dione (urazole)	5.80			
Tribromoacetic acid	−0.147			
2,4,6-Tribromobenzoic acid	1.41			
Trichloroacetic acid	0.52			
Trichloroacrylic acid	1.15			
3,3,3-Trichlorolactic acid	2.34			
Trichloromethylphosphonic acid	1.63	4.81		
2,4,5-Trichlorophenol	7.37			
3,4,5-Trichlorophenol	7.839			
Tricine (20°C)	8.15			
Triethanolamine	7.76(+1)			
Triethylamine	10.72(+1)			
Triethylenediamine	4.18(+2)	8.19(+1)		
Triethylenetetramine (20°C)	3.32(+4)	6.67(+3)	9.20(+2)	9.92(+1)
Triethylsuccinic acid	2.74			
Trifluoroacetic acid	0.50			
Trifluoroacrylic acid	1.79			
4,4,4-Trifluoro-2-aminobutanoic acid	1.600(+1)	8.169(0)		
4,4,4-Trifluoro-3-aminobutanoic acid	2.756(+1)	5.822(0)		
4,4,4-Trifluorobutanoic acid	4.16			
α,α,α-Trifluoro-m-cresol	8.950			
4,4,4-Trifluorocrotonic acid	3.15			
5,5,5-Trifluoroleucine	2.045(+1)	8.942(0)		
3-(Trifluoromethyl)aniline	3.5(+1)			
4-(Trifluoromethyl)aniline	2.6(+1)			
3-Trifluoromethylphenol	8.950			

TABLE 8.8 pK_a Values of Organic Materials in Water at 25°C (*Continued*)

Substance	pK_1	pK_2	pK_3	pK_4
.5-Trifluoromethyl-1,2,3,4-tetrazole	1.70			
6,6,6-Trifluoronorleucine	2.164(+1)	9.463(0)		
5,5,5-Trifluoronorvaline	2.042(+1)	8.916(0)		
5,5,5-Trifluoropentanoic acid	4.50			
3,3,3-Trifluoropropanoic acid	3.06			
4,4,4-Trifluorothreonine	1.554(+1)	7.822(0)		
4,4,4-Trifluorovaline	1.537(+1)	8.098(0)		
1,2,3-Trihydroxybenzene (pyrogallol)	9.03(0)	11.63(−1)		
1,3,5-Trihydroxybenzene (phloroglucinol)	8.45(0)	8.88(−1)		
2,4,6-Trihydroxybenzoic acid	1.68(0)			
3,4,5-Trihydroxybenzoic acid	4.19(0)	8.85(−1)		
3,4,5-Trihydroxycyclohex-1-ene-1-carboxylic acid [D-(−)-shikimic acid]	4.15			
2,4,6-Tri(hydroxymethyl)phenol	9.56			
Triisobutylamine	10.42(+1)			
Trimethylamine	9.80(+1)			
3-(Trimethylamino)phenol	8.06			
4-(Trimethylamino)phenol	8.35			
2,4,6-Trimethylaniline	4.38(+1)			
2,4,6-Trimethylbenzoic acid	3.448			
Trimethylenebis(thioacetic acid) (18°C)	3.435	5.383		
2,3,4-Trimethylphenol	10.59			
2,4,5-Trimethylphenol	10.57			
2,4,6-Trimethylphenol	10.88			
3,4,5-Trimethylphenol	10.25			
2,3,6-Trimethylpyridine ($\mu = 0.5$)	7.60(+1)			
2,4,6-Trimethylpyridine	7.43(+1)			
2,4,6-Trimethylpyridine-1-oxide	1.990(+1)			
3-(Trimethylsilyl)benzoic acid	4.089			
4-(Trimethylsilyl)benzoic acid	4.192			
2,4,5-Trimethylthiazole ($\mu = 0.1$)	4.55			
2,4,6-Trinitroaniline (picramide)	−10.23(+1)			
2,4,6-Trinitrobenzene acid	0.654			
2,2,2-Trinitroethanol	2.36			
Trinitromethane (20°C)	0.17			
Triphenylacetic acid	3.96			
Tripropylamine	10.66(+1)			
Tris(2-hydroxyethyl)amine	7.762(+1)			
Tris(hydroxymethyl)aminomethane (TRIS)	8.08(+1)			
2-[Tris(hydroxymethyl)methyl amino]-1-ethanesulfonic acid (TES)	7.50			

TABLE 8.8 pK_a Values of Organic Materials in Water at 25°C (*Continued*)

Substance	pK_1	pK_2	pK_3	pK_4
3-[Tris(hydroxymethyl)methyl amino]-1-propanesulfonic acid (TAPS) (20°C)	8.4			
N-[Tris(hydroxymethyl)methyl]-glycine (tricine)	2.023(+1)	8.135		
Tris(trimethylsilyl)amine	4.70(+1)			
Trithiocarbonic acid (20°C)	2.64			
Tropacocaine (15°C)	9.88(+1)			
3-Tropanol (tropine)	10.33(+1)			
Trypsin ($\mu = 0.1$)	6.25			
L-Tryptophan	2.38(+1)	9.39(0)		
DL-Tyrosine	2.18(+1)	9.21(0)	10.47(OH)	
Tyrosine amide	7.48	9.89		
Tyrosine, ethyl ester	7.33	9.80		
Tyrosylarginine ($\mu = 0.01$)	2.65(+1)	7.39(0)	9.36(−1)	11.62(−2)
Tyrosyltyrosine	3.52(+1)	7.68(0)	9.80(−1)	10.26(−2)
α-Ureidobutanoic acid	3.886(0)			
γ-Ureidobutanoic acid	4.683(0)			
β-Ureidopropanoic acid	4.487(0)			
Uric acid	5.40	5.53		
Uridine	9.30			
Uridine-5′-diphosphoric acid	7.16			
Uridine-5′-phosphoric acid (5′-uridylic acid)	6.63			
Uridine-5′-triphosphoric acid	7.58			
DL-Valine	2.286(+1)	9.719(0)		
L-Valine	2.296(+1)	9.79(0)		
Valine amide ($\mu = 0.2$)	8.00			
L-Valine, methyl ester	7.49(+1)			
L-Valylglycine	3.23(+1)	8.00(0)		
Vetramine	7.49(+1)			
Veratrine	8.85(+1)			
Vinylmethylamine	9.69(+1)			
2-Vinylpyridine	4.98(+1)			
4-Vinylpyridine	5.62(+1)			
Vitamin B$_{12}$	7.64(+1)			
Xanthine (40°C)	0.68(+1)			
Xanthosine	<2.5(+1)	5.67(0)	12.00(−1)	
Xylenol Orange [pK_5 10.46(−4); pK_6 12.28(−5)	——	2.58(−1)	3.23(−2)	6.37(−3)
D-(+)-Xylose	12.15(0)			
Zincon	——	4	7.85	15

TABLE 8.9 Selected Equilibrium Constants in Aqueous Solution at Various Temperatures

Abbreviations Used in the Table

(+1), protonated cation
(0), neutral molecule
(−1), singly ionized anion
(−2), doubly ionized anion
pK_{auto}, negative logarithm (base 10) of autoprotolysis constant
pK_{sp}, negative logarithm (base 10) of solubility product

Substance	Temperature, °C									
	0	5	10	15	20	25	30	35	40	50
Acetic acid (0)	4.780	4.770	4.762	4.758	4.757	4.756	4.757	4.762	4.769	4.787
DL-N-Acetylalanine (+1)		3.699	3.699	3.703	3.708	3.715	3.725	3.733	3.745	3.774
β-Acetylaminopropionic (+1)		4.479	4.465	4.465	4.449	4.445	4.444	4.443	4.445	4.457
N-Acetylglycine (+1)		3.682	3.676	3.673	3.667	3.670	3.673	3.678	3.685	3.706
α-Alanine (+1)	2.42		2.39		2.35	2.34	2.33	2.33	2.33	2.33
α-Alanine (0)	10.59		10.29		10.01	9.87	9.74	9.62	9.49	9.26
2-Aminobenzenesulfonic acid (0), pK_2	2.633	2.591	2.556	2.521	2.448	2.459	2.431	2.404	2.380	2.338
3-Aminobenzenesulfonic acid (0), pK_2										
4-Aminobenzenesulfonic acid (0), pK_2	4.075	4.002	3.932	3.865	3.799	3.738	3.679	3.622	3.567	3.464
3-Aminobenzoic acid (0)	3.521	3.457	3.398	3.338	3.283	3.227	3.176	3.126	3.079	2.989
4-Aminobenzoic acid (0)					4.90	4.79	4.75		4.68	4.60
					4.95	4.85	4.90		4.95	5.10
2-Aminobutyric acid (+1)			2.334			2.286		2.289 (37.5°C)		2.297
2-Aminobutyric acid (0)			10.530			9.380		9.518 (37.5°C)		9.234
4-Aminobutyric acid (+1)			4.057	4.046	4.038	4.031	4.027	4.025	4.027	4.032
4-Aminobutyric acid (0)			11.026	10.867	10.706	10.556	10.409	10.269	10.114	9.874

The following table lists ionization (pKa) values as a function of temperature. The temperature column headers are not printed on this page (they appear on the preceding page); the columns run from the lowest temperature (left, values annotated "1 °C") to the highest temperature (right). Superscript annotations on individual values indicate the actual measurement temperature.

2-Aminoethylsulfonic acid (0)	—	—	9.452	9.316	9.186	9.061	8.940	8.824	8.712	9.499
2-Amino-3-methylpentanoic acid (+1)	$2.365^{1\,°C}$	—	$2.338^{12.5\,°C}$	—	—	2.320	—	$2.317^{37.5\,°C}$	—	2.332
(0)	$10.460^{1\,°C}$	—	$10.100^{12.5\,°C}$	—	—	9.758	—	$9.439^{37.5\,°C}$	—	9.157
2-Amino-2-methyl-1,3-propanediol	9.612	9.433	9.266	9.104	8.951	8.801	8.659	8.519	8.385	8.132
2-Amino-2-methylpropionic acid (+1)	$2.419^{1\,°C}$	—	$2.380^{12.5\,°C}$	—	—	2.357	—	$2.351^{37.5\,°C}$	—	2.356
(0)	$10.960^{1\,°C}$	—	$10.580^{12.5\,°C}$	—	—	10.205	—	$9.872^{37.5\,°C}$	—	9.561
2-Aminopentanoic acid (+1)	$2.376^{1\,°C}$	—	2.347	—	—	2.318	—	—	2.309	2.313
(0)	$10.508^{1\,°C}$	—	—	$10.154^{12.5\,°C}$	—	9.808	—	$9.490^{37.5\,°C}$	—	9.198
3-Aminopropionic acid (+1)	3.656	3.627	—	3.583	—	3.551	—	3.524	3.517	—
(0)	11.000	10.830	—	10.526	—	10.235	—	9.963	9.842	—
4-Aminopyridine (+1)	9.873	9.704	9.549	9.398	9.252	9.114	8.978	8.846	8.717	8.477
Ammonium ion (+1)	10.081	9.904	9.731	9.564	9.400	9.245	9.093	8.947	8.805	8.539
Arginine (+1)	1.914	1.885	1.870	1.849	1.837	1.823	1.814	1.801	1.800	1.787
(0)	9.718	9.563	9.407	9.270	9.123	8.994	8.859	8.739	8.614	8.385
Barbituric acid (+1)	—	—	—	3.969	3.980	4.02	4.00	4.008	4.017	4.032
(0)	—	—	—	8.493	8.435	8.372	8.302	8.227	8.147	7.974
Benzoic acid (0)	—	4.231	4.220	4.215	4.206	4.204	4.203	4.207	4.219	4.223
Boric acid (0)	9.508	9.439	9.380	9.327	9.280	9.236	9.197	9.161	9.132	9.080
Bromoacetic acid (0)	—	—	—	2.875	2.887	2.902	2.918	2.936	—	—
3-Bromobenzoic acid (0)	—	—	—	3.818	3.813	3.810	3.808	3.810	3.813	—
4-Bromobenzoic acid (0)	—	—	—	4.011	4.005	3.99	4.001	4.001	4.003	—
Bromopropynoic acid (0)	—	—	1.786	1.814	1.839	1.855	1.879	1.900	1.919	—
3-tert-Butylbenzoic acid (0)	—	—	—	4.266	4.231	4.199	4.170	4.143	4.119	—
4-tert-Butylbenzoic acid (0)	—	—	—	4.463	4.425	4.389	4.354	4.320	4.287	—
2-Butynoic acid (0)	—	—	2.618	2.626	2.611	2.620	2.618	2.621	2.631	—

TABLE 8.9 Selected Equilibrium Constants in Aqueous Solution at Various Temperatures (*Continued*)

Substance	Temperature, °C									
	0	5	10	15	20	25	30	35	40	50
Butyric acid (0)	4.806	4.804	4.803	4.805	4.810	4.817	4.827	4.840	4.854	4.885
DL-N-Carbamoylalanine (+1)		3.898	3.894	3.891	3.890	3.892	3.896	3.902	3.908	3.931
N-Carbamoylglycine (+1)		3.911	3.900	3.889	3.879	3.876	3.874	3.873	3.875	3.888
Carbon dioxide+water										
(0)	6.577	6.517	6.465	6.429	6.382	6.352	6.327	6.309	6.296	6.285
(−1)	10.627	10.558	10.499	10.431	10.377	10.329	10.290	10.250	10.220	10.172
Chloroacetic acid (0)				2.845	2.856	2.867	2.883	2.900		
3-Chlorobenzoic acid (0)				3.838	3.831	3.83	3.825	3.826	3.829	
4-Chlorobenzoic acid (0)				4.000	3.991	3.986	3.981	3.980	3.981	
Chloropropynoic acid (0)			1.766	1.796	1.820	1.845	1.864	1.879	1.893	
Citric acid										
(0)	3.220	3.200	3.176	3.160	3.142	3.128	3.116	3.109	3.099	3.095
(−1)	4.837	4.813	4.797	4.782	4.769	4.761	4.755	4.751	4.750	4.757
(−2)	6.393	6.386	6.383	6.384	6.388	6.396	6.406	6.423	6.439	6.484
Cyanoacetic acid (0)		2.445	2.447	2.452	2.460	2.460	2.482	2.496	2.511	
2-Cyano-2-methylpropionic acid (0)		2.342	2.360	2.379	2.400	2.422	2.446	2.471	2.498	
5,5-Diethylbarbituric acid (0)	8.40	8.30	8.22	8.169	8.094	8.020	7.948	7.877	7.808	7.673
Diethylmalonic acid										
(0)			2.129	2.136	2.144	2.151	2.160	2.172	2.187	
(−1)			7.400	7.401	7.408	7.417	7.428	7.441	7.457	
2,3-Dimethylbenzoic acid (0)				3.663	3.687	3.771	3.726	3.762	3.788	
2,4-Dimethylbenzoic acid (0)				4.154	4.187	4.217	4.244	4.268	4.290	
2,5-Dimethylbenzoic acid (0)				3.911	3.954	3.990	4.020	4.045	4.065	
2,6-Dimethylbenzoic acid (0)				3.234	3.304	3.362	3.409	3.445	3.472	

Dissociation constants (pK values) of organic acids at various temperatures. (Temperature-column headings appear on the facing page and are not reproduced here; values are given left-to-right in order of increasing temperature.)

Compound (ionic charge)										
3,5-Dimethylbenzoic acid (0)	4.306	4.306	4.304	4.302	4.299	4.292			4.306	4.306
N,N'-Dimethylethyleneamine-N,N'-diacetic acid (0)				5.803	5.926	6.047	6.169		6.294	
(−1)				9.684	9.882	10.068	10.268		10.446	
N,N-Dimethylglycine (0)	10.34	10.14	9.94	9.76						
3,5-Dinitrobenzoic acid (0)	2.60	2.73	2.85	2.96	3.07					
2-Ethylbutyric acid (0)	4.623	4.664	4.710	4.751	4.758	4.812	4.869			
5-Ethyl-5-phenylbarbituric acid (0)		7.592	7.517	7.445	7.377	$7.311^{37.5°C}$	7.248	7.130		
Fluoroacetic acid (0)	2.60	2.555	2.571	2.586	2.604	2.624				
Formic acid (0)	3.786	3.772	3.762	3.757	3.753	3.751	3.752	3.758	3.766	3.782
2-Furancarboxylic acid (0)	3.239	3.216	3.200	3.164	3.159	3.126				
Glucose-1-phosphate (0)	6.506	6.500	6.499	6.500	6.504	6.510	6.519	6.531	6.561	
Glycerol-1-phosphoric acid (−1)	6.642	6.641	6.643	6.648	6.656	6.666	6.679	6.695	6.733	
Glycerol-2-phosphoric acid (0)	1.223	1.245	1.271	1.301	1.335	1.372	1.413	1.457	1.554	
(−1)	6.657	6.650	6.646	6.646	6.650	6.657	6.666	6.679	6.712	
Glycine (+1)	2.397	2.380	2.36	2.351	2.34	2.33	2.327	2.32		
(0)	10.34	10.193	10.044	9.91	9.780	9.65	9.53	9.412	9.19	
Glycolic acid (0)	3.875	$3.844^{12.5°C}$	3.831	$3.833^{37.5°C}$	3.849					
Glycylasparagine (+1)	2.968	2.958	2.952	2.943	2.942	2.944	2.947	2.959		
N-Glycylglycine (+1)	3.201			3.126	3.159					
	$8.594^{12.5°C}$	8.252		$7.948^{37.5°C}$	7.668					
Hexanoic acid (0)	4.840	$4.839^{12.5°C}$	4.849	4.865	4.890	4.920				
Hydrogen cyanide (0)		9.63	9.49	9.36	9.21	9.11	8.99	8.88		
Hydrogen peroxide (0)	12.23		11.86	11.75	11.65	11.55	11.45	11.21		
Hydrogen sulfide (0)	7.33	7.24	7.13	7.05	6.97	6.90	6.82	6.79	6.69	
(−1)	13.5		13.2	12.90	12.75	12.6				
4-Hydroxybenzoic acid (0)		4.596	4.586	4.582	4.577	4.576	4.578			

TABLE 8.9 Selected Equilibrium Constants in Aqueous Solution at Various Temperatures (*Continued*)

Substance	Temperature, °C									
	0	5	10	15	20	25	30	35	40	50
Hydroxylamine (0)				6.186	6.063	5.948		5.730		
2-Hydroxy-1-naphthoic acid (0)					3.29		3.24		3.19	3.26
(−1)					9.68		9.65		9.61	9.58
4-Hydroxyproline (+1)	$1.900^{1°C}$		$1.850^{12.5°C}$			1.818		$1.798^{37.5°C}$		1.796
(0)	$10.274^{1°C}$		$9.958^{12.5°C}$			9.662		$9.394^{37.5°C}$		9.138
2-Hydroxypropionic acid (0)	3.880		3.868	3.861	3.857	3.858	3.861	3.867	3.873	3.895
DL-2-Hydroxysuccinic acid (0)	3.537	3.520	3.494	3.482	3.472	3.458	3.452	3.446	3.444	3.445
(−1)	5.119	5.108	5.098	5.096	5.096	5.097	5.099	5.104	5.117	5.149
Hypobromous acid (0)				8.83		8.60		8.47	$8.37^{45°C}$	
Hypochlorous acid (0)	7.82	7.75	7.69	7.63	7.58	7.54	7.50	7.46		7.05
Imidazole (+1)	7.581	7.467	7.334	7.216	7.103	6.993	6.887	6.784	6.685	6.497
Iodoacetic acid (0)				3.143	3.158	3.175	3.193	3.213		
DL-Isoleucine (+1)	2.365		$2.338^{12.5°C}$			2.318		$2.317^{37.5°C}$		2.332
(0)	10.460		$10.100^{12.5°C}$			9.758		$9.439^{37.5°C}$		9.157
Isopropylmalonic acid, mononitrile (0)		2.299	2.320	2.343	2.365	2.401	2.427	2.452	2.481	2.332
Lactic acid (0)	3.880	3.873	3.868	3.862	3.857	3.858	3.861	3.867	3.873	3.895
Lead sulfate, pK_{sp}	8.01			7.87		7.80		7.73		7.63
DL-Leucine (+1)	$2.383^{1°C}$		$2.348^{12.5°C}$			2.328		$2.327^{37.5°C}$		2.333
(0)	$10.458^{1°C}$		$10.095^{1.5°C}$			9.744		$9.434^{37.5°C}$		9.142

Malonic acid (−1)	5.670	5.665	5.667	5.673	5.683	5.696	5.710	5.730	5.753	5.803
Mannose (0)			12.45			12.08			11.81	
Mercury(I) chloride, pK$_{sp}$			18.65	18.48	18.27	17.88		16.79		
Methanol (solvent), pK$_{auto}$		17.12		16.84		16.71	16.65	16.53		
Methylamine (+1)	11.496		11.130		10.787	10.62	10.466		10.161	9.876
Methylaminodiacetic acid (0)	2.138		2.142		2.146		2.150		2.154	
(−1)	10.474		10.287		10.088		9.920		9.763	
3-Methylbenzoic acid (0)				4.303	4.285	4.269	4.256	4.244	4.235	
4-Methylbenzoic acid (0)				4.390	4.376	4.362	4.349	4.336	4.322	
3-Methylbutyric acid (0)	4.726		4.742		4.767		4.794		4.831	4.871
4-Methylpentanoic acid (0)	4.827		4.827		4.837		4.853		4.879	4.908
5-Methyl-5-phenylbarbituric acid (0)				8.104	8.057	8.011	7.966	7.922	7.879	7.797
2-Methylpropionic acid (0)	4.825		4.827		4.840	4.853	4.886		4.918	4.955
2-Methyl-2-propylamine (+1)		11.439	11.240	11.048	10.862	10.682	10.511	10.341		
Nitric acid (0)	−1.65					−1.38				−1.20
Nitrilotriacetic acid (0)	1.69		1.65		1.65		1.66		1.67	
(−1)	2.95		2.95		2.94		2.96		2.98	
(−2)	10.59		10.45		10.33		10.23			
4-Nitrobenzoic acid (0)				3.448	3.444	3.441	3.441	3.442	3.445	
Nitrous acid (0)				3.244	3.177	3.138		3.100		
DL-Norleucine (+1)	2.394		2.356$^{12.5°C}$					2.324$^{37.5°C}$		2.328
(0)	10.564		10.190$^{12.5°C}$					9.513$^{37.5°C}$		9.224
Oxalic acid (−1)	4.210	4.216	4.227	4.240	4.254	4.272	4.295	4.318	4.349	4.409
2,4-Pentanedione (0)	9.07					8.95			8.90	
Pentanoic acid (0)	4.823		4.763		4.835	4.842	4.851		4.861	4.906
Phenylalanine (0)			9.75			9.31			8.96	
Phosphoric acid (0)	2.056	2.073	2.088	2.107	2.127	2.148	2.171	2.196	2.224	2.277
(−1)	7.313	7.282	7.254	7.231	7.213	7.198	7.189	7.185	7.181	7.183

TABLE 8.9 Selected Equilibrium Constants in Aqueous Solution at Various Temperatures (*Continued*)

Substance	Temperature, °C										
	0	5	10	15	20	25	30	35	40	50
o-Phthalic acid										
(0)	2.925	2.927	2.931	2.937	2.943	2.950	2.958	2.967	2.978	3.001
(−1)	5.432	5.418	5.410	5.405	5.405	5.408	5.416	5.427	5.442	5.485
Piperidine (+1)	11.963	11.786	11.613	11.443	11.280	11.123	10.974	10.818	10.670	10.384
Proline										
(+1)	2.011		1.964$^{12.5°C}$			1.952		1.950$^{37.5°C}$		1.958
(0)	11.296		10.972$^{12.5°C}$			10.640		10.342$^{37.5°C}$		10.064
Propenoic acid (0)				4.267	4.250	4.247	4.249	4.267	4.301	
N-Propionylglycine (+1)		3.728	3.723	3.718	3.716	3.718	3.721	3.725	3.731	3.750
Propynoic acid (0)			1.791	1.829	1.867	1.887	1.940	1.932	1.963	
Pyrrolidine (+1)	12.17	11.98	11.81	11.63	11.43	11.30	11.15	10.99	10.84	11.56
Serine										
(+1)	2.296$^{1°C}$		2.232$^{12.5°C}$			2.186		2.154$^{37.5°C}$		2.132
(0)	9.880$^{1°C}$		9.542$^{12.5°C}$			9.208		8.904$^{37.5°C}$		8.628
Silver bromide, pK_{sp}		13.33		12.83	12.57	12.30	12.07	11.83	11.61	11.19
Silver chloride, pK_{sp}		10.595		10.152		9.749		9.381	9.21	8.88
Succinic acid										
(0)	4.285	4.263	4.245	4.232	4.218	4.207	4.198	4.191	4.188	4.186
(−1)	5.674	5.660	5.649	5.642	5.639	5.635	6.541	5.647	5.654	5.680
Sulfuric acid (−1)	1.778	1.812$^{4.3°C}$	1.74			1.987	2.05	2.095	2.17	2.246
Sulfurous acid (0)	1.63					1.89		1.98		2.12
D-Tartaric acid										
(0)	3.118	3.095	3.075	3.057	3.044	3.036	3.025	3.019	3.018	3.021
(−1)	4.426	4.407	4.391	4.381	4.372	4.366	4.365	4.367	4.372	4.391
2,3,5,6-Tetramethylbenzoic (0)				3.310	3.367	3.415	3.453	3.483	3.505	

Compound	1 °C		12.5 °C					37.5 °C		
Threonine										
(+1)	$2.200^{1\,°C}$		$2.132^{12.5\,°C}$			2.088		$2.070^{37.5\,°C}$		2.055
(0)	$9.748^{1\,°C}$		$9.420^{12.5\,°C}$			9.100		$8.812^{37.5\,°C}$		8.548
o-Toluidine (0)				4.58	4.495	4.45	4.345	4.28	4.20	
1,2,4-Triazole										
(+1)				2.451	2.418	2.386	2.327			
(0)				10.205	10.083	9.972	9.768			
3,4,5-Trihydroxybenzoic acid (0)					4.19		4.30		4.38	4.53
Tris(2-hydroxyethyl)amine (+1)	8.290	8.173	8.067	7.963	7.861	7.762	7.666	7.570	7.477	7.299
2,4,6-Trimethylbenzoic (0)				3.325	3.391	3.448	3.498	3.541	3.577	
3-Trimethylsilylbenzene acid (0)				4.142	4.116	4.089	4.060	4.029	3.996	
4-Trimethylsilylbenzoic acid (0)				4.270	4.230	4.192	4.155	4.119	4.084	
β-Ureidopropionic acid (0)		4.514	4.505	4.497	4.490	4.487	4.486	4.486	4.488	4.500
DL-Valine										
(+1)	2.320		$2.297^{12.5\,°C}$			2.296		$2.292^{37.5\,°C}$		2.310
(0)	10.413		$10.064^{12.5\,°C}$			9.719		$9.405^{37.5\,°C}$		9.124

8.2.1.1 Calculation of the Approximate pH Value of Solutions

Strong acid: $pH = -\log [\text{acid}]$

Strong base: $pH = 14.00 + \log [\text{base}]$

Weak acid: $pH = \frac{1}{2}pK_a - \frac{1}{2}\log [\text{acid}]$

Weak base: $pH = 14.00 - \frac{1}{2}pK_b + \frac{1}{2}\log [\text{base}]$

Salt formed by a weak acid and a strong base:

$$pH = 7.00 + \frac{1}{2}pK_a + \frac{1}{2}\log [\text{salt}]$$

Acid salts of a dibasic acid:

$$pH = \frac{1}{2}pK_1 + \frac{1}{2}pK_2 - \frac{1}{2}\log [\text{salt}] + \frac{1}{2}\log (K_1 + [\text{salt}])$$

Buffer solution consisting of a mixture of a weak acid and its salt:

$$pH = pK_a + \log \left(\frac{[\text{salt}] + [H_3O^+] - [OH^-]}{[\text{acid}] - [H_3O^+] + [OH^-]} \right)$$

8.2.1.2 Calculation of Concentrations of Species Present at a Given pH

$$\alpha_0 = \frac{[H^+]^n}{[H^+]^n + K_1[H^+]^{n-1} + K_1K_2[H^+]^{n-2} + \cdots + K_1K_2 \cdots K_n} = \frac{[H_nA]}{C_{\text{acid}}}$$

$$\alpha_1 = \frac{K_1[H^+]^{n-1}}{[H^+]^n + K_1[H^+]^{n-1} + K_1K_2[H^+]^{n-2} + \cdots + K_1K_2 \cdots K_n} = \frac{[H_{n-1}A^-]}{C_{\text{acid}}}$$

$$\alpha_2 = \frac{K_1K_2[H^+]^{n-2}}{[H^+]^n + K_1[H^+]^{n-1} + K_1K_2[H^+]^{n-2} + \cdots + K_1K_2 \cdots K_n} = \frac{[H_{n-2}A^{2-}]}{C_{\text{acid}}}$$

$$\vdots$$

$$\alpha_n = \frac{K_1K_2 \cdots K_n}{[H^+]^n + K_1[H^+]^{n-1} + K_1K_2[H^+]^{n-2} + \cdots + K_1K_2 \cdots K_n} = \frac{[A^{n-}]}{C_{\text{acid}}}$$

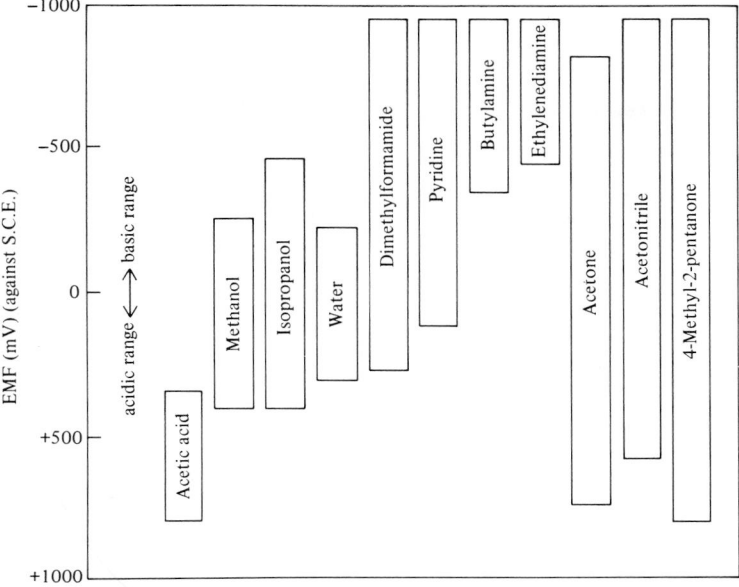

FIGURE 8.1 Approximate potential ranges in nonaqueous solvents.

TABLE 8.10 Properties of Common Acid-Base Solvents

Solvent	Potential Span, mV	$-\log K_s$	Dielectric Constant, 25°C
Acetic acid	400	14.5	6.1(20°)
Acetic anhydride	800	14.5	20.7(20°)
Acetone	1600		20.7
Acetonitrile	1600	26.5	37.5(20°)
Ammonia (at −50°C)		33	22(−33°)
n-Butanol	900		17.1
n-Butylamine	500		4.88(20°)
Chlorobenzene	1500		5.62
N,N-Dimethylformamide	1300	18.0	36.71
Dimethylsulfoxide		17.3	46.6
Ethanol	800	19.1	24.55
Ethanolamine		5.1	37.7
Ethyl acetate	1500		6.02
Ethylenediamine	500	15.3	14.2(20°)
Formic acid	200	6.2	58.5
Methanol	800	16.7	32.7
4-Methyl-2-pentanone (methyl *iso*butyl ketone)	1600	25.0	13.1(20°)
Nitromethane	1000		35.8(30°)
2-Propanol	900		19.92
Pyridine	1000		12.3
Sulfuric acid		3.85	101
Water	800	14.0	78.3

TABLE 8.11 pK_a Values for Proton-Transfer Reactions in Nonaqueous Solvents

Acid	Methanol	Ethanol	Other Solvents
Acetic acid	9.52	10.32	11.4[a], 9.75[d]
p-Aminobenzoic acid	10.25		
Ammonium ion	10.7		6.40[b]
Anilinium ion	6.0	5.70	
Benzoic acid		10.72	10.0[a]
Bromocresol purple	11.3	11.5	
Bromocresol green	9.8	10.65	
Bromophenol blue	8.9	9.5	
Bromothymol blue	12.4	13.2	
di-n-Butylammonium ion			10.3[a]
o-Chloroanilinium ion	3.4		
Cyanoacetic acid		7.49	
2,5-Dichloroanilinium ion			9.48[b]
Dimethylaminoazobenzene		5.2	6.32[b]
N,N'-Dimethylanilinium ion		4.37	
Formic acid		9.15	
Hydrobromic acid			5.5[c]
Hydrochloric acid			8.55[b], 8.9[c]
Methyl orange	3.8	3.4	
Methyl red (acid range)	4.1	3.55	
(alkaline range)	9.2	10.45	
Methyl yellow	3.4	3.55	
Neutral red	8.2	8.2	
o-Nitrobenzoic acid	7.6		
m-Nitrobenzoic acid	8.3		
p-Nitrobenzoic acid	8.4		
Perchloric acid			4.87[b]
Phenol	14.0		
Phenol red	12.8	13.4	
Phthalic acid, pK_2	11.65		11.5[d], 6.10[d](pK_1)
Picric acid	3.8	3.8	8.9[c]
Pyridinium ion			6.1[b]
Salicylic acid	8.7	7.9	
Stearic acid	10.0		
Succinic acid, pK_2	11.4		
Sulfuric acid, pK_1			7.24[b,c]
Tartaric acid, pK_2	9.9		
Thymol blue (alkaline range)	14.0	15.2	
(acid range)	4.7	5.35	
Thymolbenzein (acid range)	3.5		
(alkaline range)	13.1		
p-Toluenesulfonic acid			8.44[b]
p-Toluidinium ion		6.24	
Tribenzylammonium ion			5.40[b]
Tropeoline 00	2.2		
Urea (protonated cation)			6.96[b]
Veronal	12.6		

[a] Dimethylsulfoxide. [b] Glacial acetic acid. [c] Acetonitrile. [d] Acetone + 10% water.

8.2.2 Formation Constants of Metal Complexes

Each value listed in Tables 8.12 and 8.13 is the logarithm of the overall formation constant for the cumulative binding of a ligand L to the central metal cation M, viz.:

	Cumulative formation constant	Stepwise stability constants
$M + L = ML$	K_1	k_1
$M + 2L = ML_2$	K_2	$k_1 k_2$
. . . .		
$M + nL = ML_n$	K_n	$k_1 k_2 \cdots k_n$

As an example, the entries in Table 8.12 for the zinc ammine complexes represent these equilibria:

$$Zn^{2+} + NH_3 = Zn(NH_3)^{2+} \qquad K_1 = \frac{[Zn(NH_3)^{2+}]}{[Zn^{2+}][NH_3]}$$

$$Zn^{2+} + 2NH_3 = Zn(NH_3)_2^{2+} \qquad K_2 = \frac{[Zn(NH_3)_2^{2+}]}{[Zn^{2+}][NH_3]^2}$$

$$Zn^{2+} + 3NH_3 = Zn(NH_3)_3^{2+} \qquad K_3 = \frac{[Zn(NH_3)_3^{2+}]}{[Zn^{2+}][NH_3]^3}$$

$$Zn^{2+} + 4NH_3 = Zn(NH_3)_4^{2+} \qquad K_4 = \frac{[Zn(NH_3)_4^{2+}]}{[Zn^{2+}][NH_3]^4}$$

If the stepwise stability or formation constants of the reactions are desired, for the first step $\log K_1 = \log k_1 = 2.37$. For the second and succeeding steps the equilibria and corresponding constants are as follows:

$$Zn(NH_3)^{2+} + NH_3 = Zn(NH_3)_2^{2+} \qquad \log k_2 = \log K_2 - \log K_1 = 2.44$$

$$Zn(NH_3)_2^{2+} + NH_3 = Zn(NH_3)_3^{2+} \qquad \log k_3 = \log K_3 - \log K_2 = 3.50$$

$$Zn(NH_3)_3^{2+} + NH_3 = Zn(NH_3)_4^{2+} \qquad \log k_4 = \log K_4 - \log K_3 = 2.15$$

The reverse of the association or formation reactions would represent the dissociation or instability constant for the systems, i.e., $-\log K_f = \log K_{instab}$.

The data in the tables generally refer to temperatures of about 20 to 25°C. Most of the values in Table 8.12 refer to zero ionic strength, but those in Table 8.13 often refer to a finite ionic strength.

TABLE 8.12 Cumulative Formation Constants for Metal Complexes with Inorganic Ligands

	$\log K_1$	$\log K_2$	$\log K_3$	$\log K_4$	$\log K_5$	$\log K_6$
Ammonia						
Cadmium	2.65	4.75	6.19	7.12	6.80	5.14
Cobalt(II)	2.11	3.74	4.79	5.55	5.73	5.11
Cobalt(III)	6.7	14.0	20.1	25.7	30.8	35.2
Copper(I)	5.93	10.86				
Copper(II)	4.31	7.98	11.02	13.32	12.86	
Iron(II)	1.4	2.2				
Manganese(II)	0.8	1.3				
Mercury(II)	8.8	17.5	18.5	19.28		
Nickel	2.80	5.04	6.77	7.96	8.71	8.74
Platinum(II)						35.3
Silver(I)	3.24	7.05				
Zinc	2.37	4.81	7.31	9.46		
Bromide						
Astatine	2.51 [AtBr]					
Bismuth(III)	4.30	5.55	5.89	7.82		9.70
Bromine	1.24 [Br$_3^-$]					
Cadmium	1.75	2.34	3.32	3.70		
Cerium(III)	0.42					
Copper(I)		5.89				
Copper(II)	0.30					
Gold(I)		12.46				
Indium	1.30	1.88	2.48			
Iodine	2.64 [IBr]					
Iron(III)	−0.30	−0.50				
Lead	1.2	1.9		1.1		
Mercury(II)	9.05	17.32	19.74	21.00		
Palladium(II)				13.1		
Platinum(II)				20.5		
Rhodium(III)		14.3	16.3	17.6	18.4	17.2
Scandium	2.08	3.08				
Silver(I)	4.38	7.33	8.00	8.73		
Thallium(I)	0.93					
Thallium(III)	9.7	16.6	21.2	23.9	29.2	31.6
Tin(II)	1.11	1.81	1.46			
Uranium(IV)	0.18					
Yttrium	1.32					
Chloride						
Americium(III)	1.17					
Antimony(III)	2.26	3.49	4.18	4.72		
Bismuth(III)	2.44	4.7	5.0	5.6		
Cadmium	1.95	2.50	2.60	2.80		
Cerium(III)	0.48					
Copper(I)		5.5	5.7			
Copper(II)	0.1	−0.6				
Curium(III)	1.17					
Gold(III)		9.8				
Indium	1.42	2.23	3.23			
Iron(II)	0.36					
Iron(III)	1.48	2.13	1.99	0.01		
Lead	1.62	2.44	1.70	1.60		
Manganese(II)	0.96					

TABLE 8.12   Cumulative Formation Constants for Metal Complexes with Inorganic Ligands (*Continued*)

	$\log K_1$	$\log K_2$	$\log K_3$	$\log K_4$	$\log K_5$	$\log K_6$
Mercury(II)	6.74	13.22	14.07	15.07		
Palladium(II)	6.1	10.7	13.1	15.7		
Platinum(II)		11.5	14.5	16.0		
Plutonium(III)	1.17					
Silver(I)	3.04	5.04		5.30		
Thallium(I)	0.52					
Thallium(III)	8.14	13.60	15.78	18.00		
Thorium	1.38	0.38				
Tin(II)	1.51	2.24	2.03	1.48		
Tin(IV)						4
Uranium(IV)	0.8					
Uranium(VI)	0.22					
Zinc	0.43	0.61	0.53	0.20		
Zirconium	0.9	1.3	1.5	1.2		
Cyanide						
Cadmium	5.48	10.60	15.23	18.78		
Copper(I)		24.0	28.59	30.30		
Gold(I)		38.3				
Iron(II)						35
Iron(III)						42
Mercury(II)				41.4		
Nickel				31.3		
Silver(I)		21.1	21.7	20.6		
Zinc				16.7		
Fluoride						
Aluminum	6.10	11.15	15.00	17.75	19.37	19.84
Beryllium	5.1	8.8	12.6			
Cerium(III)	3.20					
Chromium(III)	4.41	7.81	10.29			
Gadolinium	3.46					
Gallium	5.08					
Indium	3.70	6.25	8.60	9.70		
Iron(III)	5.28	9.30	12.06			
Lanthanum	2.77					
Magnesium	1.30					
Manganese(II)	5.48					
Plutonium(III)	6.77					
Scandium						17.3
Thallium(I)	0.1					
Thallium(III) [TlO$^+$]	6.44					
Thorium	7.65	13.46	17.97			
Titanium(IV) [TiO^{2+}]	5.4	9.8	13.7	18.0		
Uranium(VI)	4.59	7.93	10.47	11.84		
Yttrium	4.81	8.54	12.14			
Zirconium	8.80	16.12	21.94			
Hydroxide						
Aluminum	9.27			33.03		
Antimony(III)		24.3	36.7	38.3		
Arsenic [as AsO$^+$]	14.33	18.73	20.60	21.20		
Beryllium	9.7	14.0	15.2			
Bismuth(III)	12.7	15.8		35.2		
Cadmium	4.17	8.33	9.02	8.62		

TABLE 8.12 Cumulative Formation Constants for Metal Complexes with Inorganic Ligands (*Continued*)

	$\log K_1$	$\log K_2$	$\log K_3$	$\log K_4$	$\log K_5$	$\log K_6$
Cerium(III)	14.6					
Cerium(IV)	13.28	26.46				
Chromium(III)	10.1	17.8		29.9		
Copper(II)	7.0	13.68	17.00	18.5		
Dysprosium	5.2					
Erbium(III)	5.4					
Gadolinium	4.6					
Gallium	11.0	21.7		34.3	38.0	40.3
Indium	9.9	19.8		28.7		
Iodine	9.49	11.24				
Iron(II)	5.56	9.77	9.67	8.58		
Iron(III)	11.87	21.17	29.67			
Lanthanum	3.3					
Lead(II)	7.82	10.85	14.58			61.0
Lutetium	6.6					
Magnesium	2.58					
Manganese(II)	3.90		8.3			
Neodymium	5.5					
Nickel	4.97	8.55	11.33			
Praseodymium	4.30					
Plutonium(III)	7.0					
Plutonium(IV)	12.39					
Plutonium [as PuO_2^{2+}]	8.3	16.6	20.9			
Samarium(III)	4.8					
Scandium	8.9					
Tellurium(IV)			41.6	53.0	64.8	72.0
Thallium(III)	12.86	25.37				
Titanium(III)	12.71					
Uranium(IV)	13.3				41.2	
Uranium(VI) [as UO_2^{2+}]	9.5	22.80		32.4		
Vanadium(III)	11.1	21.6				
Vanadium(IV) [as VO^{2+}]	8.6		[25.8 for $V_2O_4(OH)^-$]			
Vanadium(V) [as VO^{3+}]		25.2		46.2	58.5	
Yttrium	5.0					
Zinc	4.40	11.30	14.14	17.66		
Zirconium	14.3	28.3	41.9	55.3		
Iodide						
Bismuth	3.63			14.95	16.80	18.80
Cadmium	2.10	3.43	4.49	5.41		
Copper(I)		8.85				
Indium	1.00	2.26				
Iodine	2.89	5.79				
Iron(III)	1.88					
Lead	2.00	3.15	3.92	4.47		
Mercury(II)	12.87	23.82	27.60	29.83		
Silver	6.58	11.74	13.68			
Thallium(I)	0.72	0.90	1.08			
Thallium(III)	11.41	20.88	27.60	31.82		
Iodate						
Barium	1.05					
Calcium	0.89					
Magnesium	0.72					

TABLE 8.12 Cumulative Formation Constants for Metal Complexes with Inorganic Ligands (*Continued*)

	$\log K_1$	$\log K_2$	$\log K_3$	$\log K_4$	$\log K_5$	$\log K_6$
Strontium	1.00					
Thorium	2.88	4.79	7.15			
Nitrate						
Barium	0.92					
Beryllium	1.62					
Bismuth(III)	1.26					
Cadmium	0.40					
Calcium	0.28					
Cerium(III)	1.04	2.55				
Curium(III)	0.57					
Hafnium	0.92	2.43	4.32	6.40	8.48	10.29
Iron(III)	1.0					
Lanthanum	0.26	0.69	1.27			
Lead	1.18					
Mercury(II)	0.35					
Neodymium	0.52	1.18				
Neptunium(IV)	0.38					
Plutonium(III)	0.77	1.93	3.09			
Plutonium(IV)	0.54					
Strontium	0.82					
Thallium(I)	0.33					
Thallium(III)	0.92					
Thorium	0.78	1.89	2.89	3.63		
Uranium(IV)	0.20	0.37				
Uranium(VI)	0.34	0.45				
Ytterbium	0.45	1.30	2.42			
Zirconium [as ZrO^{2+}]		1.91		3.54		
Pyrophosphate						
Barium	4.6					
Calcium	4.6					
Cadmium	5.6					
Copper(II)	6.7	9.0				
Lead		5.3				
Magnesium	5.7					
Nickel	5.8	7.4				
Strontium	4.7					
Yttrium		9.7				
Zirconium		6.5				
Sulfate						
Cerium(III)	3.40					
Erbium	3.58					
Gadolinium	3.66					
Holmium	3.58					
Indium	1.78	1.88	2.36			
Iron(III)	2.03	2.98				
Lanthanum	3.64					
Neodymium	3.64					
Nickel	2.4					
Plutonium(IV)	3.66					
Praseodymium	3.62					
Samarium	3.66					
Thorium	3.32	5.50				

TABLE 8.12 Cumulative Formation Constants for Metal Complexes with Inorganic Ligands (*Continued*)

	$\log K_1$	$\log K_2$	$\log K_3$	$\log K_4$	$\log K_5$	$\log K_6$
Uranium(IV)	3.24	5.42				
Uranium(VI)	1.70	2.45	3.30			
Yttrium	3.47					
Ytterbium	3.58					
Zirconium	3.79	6.64	7.77			
Sulfite						
Copper(I)	7.5	8.5	9.2			
Mercury(II)		22.66				
Silver	5.30	7.35				
Thiocyanate						
Bismuth	1.15	2.26	3.41	4.23		
Cadmium	1.39	1.98	2.58	3.6		
Chromium(III)	1.87	2.98				
Cobalt(II)	−0.04	−0.70	0	3.00		
Copper(I)	12.11	5.18				
Gold(I)		23		42		
Indium	2.58	3.00	4.63			
Iron(III)	2.95	3.36				
Mercury(II)		17.47		21.23		
Nickel	1.18	1.64	1.81			
Ruthenium(III)	1.78					
Silver		7.57	9.08	10.08		
Thallium(I)	0.80					
Uranium(IV)	1.49	2.11				
Uranium(VI)	0.76	0.74	1.18			
Vanadium(III)	2.0					
Vanadium(IV)	0.92					
Zinc	1.62					
Thiosulfate						
Cadmium	3.92	6.44				
Copper(I)	10.27	12.22	13.84			
Iron(III)	2.10					
Lead		5.13	6.35			
Mercury(II)		29.44	31.90	33.24		
Silver	8.82	13.46				

TABLE 8.13 Cumulative Formation Constants for Metal Complexes with Organic Ligands

Temperature is 25 °C and ionic strengths are approaching zero unless indicated otherwise: (*a*) At 20 °C, (*b*) at 30 °C, (*c*) 0.1 *M* uni-univalent salt, (*d*) 1.0 *M* uni-univalent salt, (*e*) 2.0 *M* uni-univalent salt present.

	$\log K_1$	$\log K_2$	$\log K_3$	$\log K_4$
Acetate				
Ag(I)	0.73	0.64		
Ba(II)	0.41			
Ca(II)	0.6			
Cd(II)	1.5	2.3	2.4	
Ce(III)	1.68	2.69	3.13	3.18
Co(II)	1.5	1.9		
Cr(III)	1.80	4.72		
Cu(II) *a*	2.16	3.20		
Fe(II) *c*	3.2	6.1	8.3	
Fe(III) *a,d*	3.2			
In(III)	3.50	5.95	7.90	9.08
Hg(II)		8.43		
La(III) *a,e*	1.56	2.48	2.98	2.95
Mg(II)	0.8			
Mn(II)	9.84	2.06		
Ni(II)	1.12	1.81		
Pb(II)	2.52	4.0	6.4	8.5
Rare earths *a,e*	1.6–1.9	2.8–3.0	3.3–3.7	
Sr(II)	0.44			
Tl(III)				15.4
UO$_2$(II) *a,e*	2.38	4.36	6.34	
Y(III) *a,e*	1.53	2.65	3.38	
Zn(II)	1.5			
Acetylacetone				
Al(III) *b*	8.6	15.5		
Be(II)	7.8	14.5		
Cd(II)	3.84	6.66		
Ce(III)	5.30	9.27	12.65	
Cr(II)	5.9	11.7		
Co(II)	5.40	9.54		
Cu(II)	8.27	16.34		
Dy(III) *b*	6.03	10.70	14.04	
Er(III) *b*	5.99	10.67	14.09	
Eu(III) *b*	5.87	10.35	13.64	
Fe(II)	5.07	8.67		
Fe(III)	11.4	22.1	26.7	
Ga(III)	9.5	17.9	23.6	
Gd(III) *b*	5.90	10.38	13.79	
Hf(IV)	8.7	15.4	21.8	28.1
Ho(III)	6.05	10.73	14.13	
In(III)	8.0	15.1		
La(III) *b*	5.1	8.90	11.90	
Lu(III) *b*	6.23	11.00	13.63	
Mg(II)	3.65	6.27		
Mn(II)	4.24	7.35		
Mn(III)			3.86	
Nd(III)	5.6	9.9	13.1	
Ni(II) *a*	6.06	10.77	13.09	
Pd(II) *b*	16.2	27.1		
Pr(III) *b*	5.4	9.5	12.5	
Pu(IV) *c*	10.5	19.7	28.1	34.1
Sc(III) *b*	8.0	15.2		

TABLE 8.13 Cumulative Formation Constants for Metal Complexes with Organic Ligands (*Continued*)

	$\log K_1$	$\log K_2$	$\log K_3$	$\log K_4$
Sm(III) *b*	5.9	10.4		
Tb(III) *b*	6.02	10.63	14.04	
Th(*IV*)	8.8	16.2	22.5	26.7
Tm(IV) *b*	6.09	10.85	14.33	
U(IV) *a,c*	8.6	17.0	23.4	29.5
UO_2(II) *b*	7.74	14.19		
VO(II)	8.68	15.79		
V(II)	5.4	10.2	14.7	
Y(III) *b*	6.4	11.1	13.9	
Yb(III) *b*	6.18	11.04	13.64	
Zn(II) *b*	4.98	8.81		
Zr(IV)	8.4	16.0	23.2	30.1
Alizarin red				
Cr(VI)	4.7			
Cu(II)	4.1			
Hf(IV)		10.4		
Mo(VI)		9.6		
Pb(II)	6.0			
Th(IV)		8.24		
UO_2(II)	4.22			
V(V)		8.6		
W(VI)		7.8		
Arsenazo				
Hf(IV)	10.07			
Zr(IV)	12.95			
Aurintricarboxylic acid				
Be(II)	4.54			
Cu(II)	4.1	8.81		
Fe(III)	4.68			
Th(IV)	5.04			
UO_2(II)	4.77			
Benzoylacetone (75% dioxane)				
Ba(II)		9.4		
Be(II)	12.59	24.01		
Cd(II)	7.79	14.36		
Ce(III)	10.09	19.42	27.04	
Co(II)	9.42	17.83		
Cu(II)	12.05	23.01		
La(III)	6.33	11.66	16.78	
Mg(II)	7.69	14.09		
Mn(II)	8.66	15.78		
Ni(II)	9.58	18.00		
Pb(II)	8.84	16.35		
Pr(III)	7.02	13.62	18.74	
UO_2(II)	12.15	23.27		
Y(III)	8.24	14.98	20.57	
Zn(II)	9.62	17.90		
Calmagite				
Ca	6.05			
Mg	8.05			

TABLE 8.13 Cumulative Formation Constants for Metal Complexes with Organic Ligands (*Continued*)

	Complex of HL^{2-} Anion		Complex of L^{3-} Anion		Complex of H_2L^-
	$\log K_1$	$\log K_2$	$\log K_1$	$\log K_2$	$\log K_3$
Citric acid					
Ag	7.1				
Al	7.0		20.0		
Ba	2.98				
Be	4.52				
Ca	4.68				
Cd	3.98		11.3		
Ce(III)		6.18		9.65	3.2
Co(II)	4.8		12.5		
Cu(II)	4.35		14.2		
Eu(III)		6.46		9.80	
Fe(II)	3.08		15.5		
Fe(III)	12.5		25.0		
La		6.97		9.45	6.22
Mg	3.29				
Mn(II)	3.67				
Nd(III)		6.32		9.70	
Ni	5.11		14.3		
Pb	6.50				
Pr					3.4
Ra	2.36				
Sr	2.8				
Tl(I)	1.04				
UO_2	8.5	10.8			
Y					3.6
Yb				8	
Zn	4.71		11.4		

	$\log K_1$	$\log K_2$	$\log K_3$		
1,2-Diaminocyclohexane-*N,N,N',N'*-tetraacetic acid					
Al c	17.63				
Ba c	8.64				
Ca c	12.3				
Cd c	19.88				
Ce(III) c	16.76				
Co(II) c	19.57				
Cu(II) c	21.95				
Dy(III) c	19.69				
Er(III) c	20.20				
Eu(III) c	18.77				
Fe(III) c	27.48				
Ga c	22.91				
Gd c	18.80				
Hg(II) c	24.4				
Ho c	19.89				
La c	16.35				
Lu c	21.51				
Mg c	10.41				
Mn(II) c	17.43				
Nd c	17.69				
Ni c	19.4				
Pb c	20.33				
Pr c	17.23				

TABLE 8.13 Cumulative Formation Constants for Metal Complexes with Organic Ligands (*Continued*)

	$\log K_1$	$\log K_2$	$\log K_3$	$\log K_4$
Sm(III) *c*	18.63			
Sr *c*	8.92			
Tb *c*	19.30			
Tm *c*	20.46			
VO(II) *c*	19.40			
Y *c*	19.41			
Yb *c*	20.80			
Zn *c*	18.6			
Dibenzoylmethane (75% dioxane)				
Ba	6.10	11.50		
Be	13.62	26.03		
Ca	7.17	13.55		
Cd	8.67	16.63		
Ce(III)	10.99	21.53	30.38	
Co(II)	10.35	20.05		
Cu(II)	12.98	24.98		
Cs	3.42			
Fe(II)	11.15	21.50		
K	3.67			
Li	5.95			
Mg	8.54	16.21		
Mn(II)	9.32	17.79		
Na	4.18			
Ni	10.83	20.72		
Pb	9.75	18.79		
Rb	3.52			
Sr	6.40	12.10		
Zn	10.23	19.65		

	$\log K_1$	$\log K_2$	$\log K_3$	$\log K_f$ [MHL]
4,5-Dihydroxybenzene-1,3-disulfonic acid (Tiron)				
Al	19.02	31.10	33.5	
Ba	4.10			14.6
Ca	5.80			14.8
Cd *d*	7.69	13.29		
Ce(III)		3.75		
Co(II) *d*	8.19	14.41		15.7
Cu(II) *d*	12.76	23.73		18.1
Fe(III) *a,c*	20.7	35.9	46.9	22.6
La	12.9			18.6 [La(OH)L]
Mg *a,c*	6.86			14.6
Mn(II) *c*	8.6			
Ni *a,c*	8.56	14.90		15.6
Pb *d*	11.95	18.28		
Sr *c*	4.55			
UO$_2$(II) *c*	15.90			
VO(II)	15.88			
Zn *d*	9.00	16.91		15.9

	$\log K_1$	$\log K_2$	$\log K_f$ [M$_2$L$_3$]
2,3-Dimercaptopropan-1-ol (BAL)			
Fe(II)	15.8		
Fe(III)	30.6 [Fe(OH)L]		28
Mn(II)	5.23	10.43	
Ni		22.78	
Zn	13.48	23.3	40.6

TABLE 8.13 Cumulative Formation Constants for Metal Complexes with Organic Ligands (*Continued*)

	$\log K_1$	$\log K_2$	$\log K_3$	$\log K_4$
Dimethylglyoxime (50% dioxane)				
Cd	5.7	10.7		
Co(II)	9.80	18.94		
Cu(II)	12.00	33.44		
Fe(II)		7.25		
La	6.6	12.5		
Ni	11.16			
Pb	7.3			
Zn	7.7	13.9		
2,2′-Dipyridyl				
Ag	3.65	7.15		
Cd	4.26	7.81	10.47	
Co(II)	5.73	11.57	17.59	
Cr(II)	4.5	10.5	14.0	
Cu(I)		14.2		
Cu(II)	8.0	13.60	17.08	
Fe(II)	4.36	8.0	17.45	
Hg(II)	9.64	16.74	19.54	
Mg	0.5			
Mn(II) *d*	4.06	7.84	11.47	
Ni	6.80	13.26	18.46	
Pb	3.0			
Ti(III)			25.28	
V(II)	4.9	9.6	13.1	
Zn	5.30	9.83	13.63	
Eriochrome Black T				
Ca	5.4			
Mg	7.0			
Zn	13.5	20.6		
Ethanolamine				
Ag	3.29	6.92		
Cu(II)		6.68		16.48
Hg(II)	8.51	17.32		
Ethylenediamine				
Ag	4.70	7.70		
Cd *a*	5.47	10.09	12.09	
Co(II)	5.91	10.64	13.94	
Co(III)	18.7	34.9	48.69	
Cr(II)	5.15	9.19		
Cu(I)		10.8		
Cu(II)	10.67	20.00	21.0	
Fe(II)	4.34	7.65	9.70	
Hg(II)	14.3	23.3		
Mg	0.37			
Mn(II)	2.73	4.79	5.67	
Ni	7.52	13.84	18.33	
Pd(II)		26.90		
V(II)	4.6	7.5	8.8	
Zn	5.77	10.83	14.11	
Ethylenediamine-*N,N,N′,N′*-tetraacetic acid				
Ag	7.32			
Al	16.11			
Am(III)	18.18			
Ba	7.78			
Be	9.3			
Bi	22.8			
Ca	11.0			
Cd	16.4			
Ce(III)	16.80			

TABLE 8.13 Cumulative Formation Constants for Metal Complexes with Organic Ligands (*Continued*)

	$\log K_1$	$\log K_2$	$\log K_3$	$\log K_4$
Cf(III)	19.09			
Cm(III)	18.45			
Co(II)	16.31			
Co(III)	36			
Cr(II)	13.6			
Cr(III)	23			
Cu(II)	18.7			
Dy	18.0			
Er	18.15			
Eu(III)	17.99			
Fe(II)	14.33			
Fe(III)	24.23			
Ga	20.25			
Gd	17.2			
Hg(II)	21.80			
Ho	18.1			
In	24.95			
La	16.34			
Li	2.79			
Lu	19.83			
Mg	8.64			
Mn(II)	13.8			
Mo(V)	6.36			
Na	1.66			
Nd	16.6			
Ni	18.56			
Pb	18.3			
Pd(II)	18.5			
Pm(III)	17.45			
Pr	16.55			
Pu(III)	18.12			
Pu(IV)	17.66			
Pu(VI)	17.66			
Ra	7.4			
Sc	23.1			
Sm	16.43			
Sn(II)	22.1			
Sr	8.80			
Tb	17.6			
Th	23.2			
Ti(III)	21.3			
TiO(II)	17.3			
Tl(III)	22.5			
Tm	19.49			
U(IV)	17.50			
V(II)	12.70			
V(III)	25.9			
VO(II)	18.0			
V(V)	18.05			
Y	18.32			
Yb	18.70			
Zn	16.4			
Zr	19.40			
Glycine				
Ag	3.41	6.89		
Ba	0.77			
Be		4.95		

TABLE 8.13 Cumulative Formation Constants for Metal Complexes with Organic Ligands (*Continued*)

	$\log K_1$	$\log K_2$	$\log K_3$	$\log K_4$
Ca	1.38			
Cd	4.74	8.60		
Co(II)	5.23	9.25	10.76	
Cu(II)	8.60	15.54	16.27	
Dy		12.2		
Er		12.7		
Fe(II) *a*	4.3	7.8		
Fe(III) *a,d*	10.0			
Gd		11.9		
Hg(II)	10.3	19.2		
La		11.2		
Mg	3.44	6.46		
Mn(II)	3.6	6.6		
Ni	6.18	11.14	15	
Pb	5.47	8.92		
Pd(II)	9.12	17.55		
Pr		11.5		
Sm		11.7		
Sr	0.91			
Y		12.5		
Yb		13.0		
Zn	5.52	9.96		
***N'*-(2-Hydroxyethyl)ethylenediamine-*N,N,N'*-triacetic acid**				
Ba *c*	5.54			
Ca *c*	8.43			
Cd *c*	13.0			
Ce(III) *c*	14.11			
Co(II) *c*	14.4			
Cu(II) *c*	17.40			
Dy *c*	15.30			
Er *c*	15.42			
Eu(III) *c*	15.35			
Fe(II) *c*	11.6			
Fe(III) *c*	19.8			
Gd *c*	15.22			
Hg(II) *c*	20.1			
Ho *c*	15.32			
La *c*	13.46			
Lu *c*	15.88			
Mg *c*	5.78			
Mn(II) *c*	10.7			
Nd *c*	14.86			
Ni *c*	17.0			
Pb *c*	15.5			
Pr *c*	14.61			
Sm *c*	15.28			
Sr *c*	6.92			
Tb *c*	15.32			
Th *c*	18.5			
Tm *c*	15.59			
Y *c*	14.65			
Yb *c*	15.88			
Zn *c*	14.5			
8-Hydroxy-2-methylquinoline (50% dioxane)				
Cd	9.00	9.00	16.60	
Ce(III)	7.71			
Co(II)	9.63	18.50		

TABLE 8.13 Cumulative Formation Constants for Metal Complexes with Organic Ligands (*Continued*)

	$\log K_1$	$\log K_2$	$\log K_3$	$\log K_4$
Cu(II)	12.48	24.00		
Fe(II)	8.75	17.10		
Mg	5.24	9.64		
Mn(II)	7.44	13.99		
Ni	9.41	17.76		
Pb	10.30	18.50		
UO₂(II)	9.4	17		
Zn	9.82	18.72		
8-Hydroxyquinoline-5-sulfonic acid				
Ba	2.31			
Ca	3.52			
Cd	7.70	14.20		
Ce(III)	6.05	11.05	14.95	
Co(II)	8.11	15.05	20.41	
Cu(II)	11.92	21.87		
Er	7.16	13.34	18.56	
Fe(II)	8.4	15.7	21.75	
Fe(III)	11.6	22.8	35.65	
Gd	6.64	12.37	17.27	
La	5.63	10.13	13.83	
Mg	4.79	8.19		
Mn(II)	5.67	10.72		
Nd	6.3	11.6	16.0	
Ni	9.57	18.27	22.9	
Pb	8.53	16.13		
Pr	6.17	11.37	15.67	
Sm	6.58	12.28	17.04	
Sr	2.75			
Th	9.56	18.29	25.92	32.04
UO₂(II)	8.52	15.67		
Zn	8.65	16.15		
Lactic acid				
Ba	0.64			
Ca	1.42			
Cd	1.70			
Ce(III) *a,c*	2.76	4.73	5.96	
Co(II)	1.90			
Cu(II)	3.02	4.85		
Er	2.77	5.11	6.70	
Eu(III)	2.53	4.60	5.88	
Fe(III)	7.1			
Gd	2.53	4.63	5.91	
Ho	2.71	4.97	6.55	
La *a,c*	2.60	4.34	5.64	
Li	0.20			
Mg	1.37			
Mn(II)	1.43			
Nd	2.47	4.37	5.60	
Ni	2.22			
Pb	2.40	3.80		
Pr *a,c*	2.85	4.90	6.10	
Rare earths *a,c*	2.8–3.0	4.9–5.4	6.1–7.8	
Sm	2.56	4.58	5.90	
Sr	0.98			
Tb	2.61	4.73	6.01	
Y	2.53	4.70	6.12	
Yb	2.85	5.27	7.96	
Zn	2.20	3.75		

TABLE 8.13 Cumulative Formation Constants for Metal Complexes with Organic Ligands (*Continued*)

	$\log K_1$	$\log K_2$	$\log K_3$	$\log K_4$
Nitrilotriacetic acid				
Al	>10			
Ba *a*	5.88			
Ca	7.60	11.61		
Cd *c*	9.80	15.2		
Ce(III) *c*	10.83	18.67		
Co(II) *c*	10.38	14.5		
Cr(III)	>10			
Cu(II) *c*	13.10			
Dy *c*	11.74	21.15		
Er *c*	12.03	21.29		
Eu(III) *c*	11.52	20.70		
Fe(II) *c*	8.84			
Fe(III) *c*	15.87	24.32		
Gd *c*	11.54	20.80		
Hg(II)	12.7			
Ho *c*	11.90	21.25		
In	15			
La *c*	10.36	17.60		
Li *a*	3.28			
Lu *c*	12.49	21.91		
Mg *c*	5.36	10.2		
Mn(II)	8.60	11.1		
Na	2.15			
Nd *c*	11.26	19.73		
Ni	11.26	16.0		
Pb *a,c*	11.8			
Pr *c*	11.07	19.25		
Sm(III) *c*	11.53	20.53		
Sr	6.73			
Tb *c*	11.59	20.97		
Tl(I)	3.44			
Th *c*	12.4			
Tm *c*	12.22	21.45		
Y *c*	11.48	20.43		
Yb *c*	12.40	21.69		
Zn *c*	10.45	13.45		
Zr *c*	20.8			
1-Nitroso-2-naphthol (75% dioxane)				
Ag	7.74			
Cd	6.18	11.38		
Co(II)	10.67	22.81		
Cu(II)	12.52	23.37		
Mg	6.2	10.60		
Nd	9.5	17.7	25.6	
Ni	10.75	21.29	28.09	
Pb	9.73	17.31		
Pr	9.04	17.06	23.85	
Th *c*	8.50	16.13	24.03	30.29
Y	9.02	17.74	25.04	
Zn	9.32	17.02		
Zr	3.6			
Oxalate				
Ag	2.41			
Al	7.26	13.0	16.3	
Am(III)		9.8		[Am(HL)$_4^-$ 11.0]
Ba	2.31			
Be	4.90			

TABLE 8.13 Cumulative Formation Constants for Metal Complexes with Organic Ligands (*Continued*)

	$\log K_1$	$\log K_2$	$\log K_3$	$\log K_4$
Ca	3.0			
Cd	3.52	5.77		
Ce(III)	6.52	10.5	11.3	
Co(II)	4.79	6.7	9.7	
Co(III)			~20	
Cu(II)	6.16	8.5		
Er	4.82	8.21	10.03	
Fe(II)	2.9	4.52	5.22	
Fe(III)	9.4	16.2	20.2	
Gd	7.04			
Hg(II)		6.98		
Mg	3.43	4.38		
Mn(II)	3.97	5.80		
Mn(III) *e*	9.98	16.57	19.42	
Mo(III)	3.38			
Mo(VI)				$[MoO_3(L)^{2-}\ 13.0]$
Nd	7.21	11.5	>14	
Ni	5.3	7.64	~8.5	
NpO$_2$(II)	3.30	7.07		
Pb		6.54		
Pu(III)	9.31	18.70	28	
Pu(IV)	8.74	16.91	23.39	27.50
PuO$_2$(II)		11.4		
Sr	2.54			
Th				24.48
TiO(II)	2.67			
Tl(I)	2.03			
UO$_2$(II)		10.57		
VO(II)		9.80		
V(II)	~2.7			
Y	6.52	10.10	11.47	
Yb	7.30	11.7	>14	
Zn	4.89	7.60	8.15	
Zr	9.80	17.14	20.86	21.15
1,10-Phenanthroline				
Ag	5.02	12.07		
Ca	0.7			
Cd	5.93	10.53	14.31	
Co(II)	7.25	13.95	19.90	
Cu(II)	9.08	15.76	20.94	
Fe(II)	5.85	11.45	21.3	
Fe(III)	6.5	11.4	23.5	
Hg(II)		19.65	23.35	
Mg	1.2			
Mn(II)	3.88	7.04	10.11	
Ni	8.80	17.10	24.80	
Pb	4.65	7.5	9	
VO(II)	5.47	9.69		
Zn	6.55	12.35	17.55	
Phthalic acid				
Ba	2.33			
Ca	2.43			
Cd	2.5			
Co(II)	1.81	4.51		
Cu(II)	3.46	4.83		
La		7.74		
Ni	2.14			
Pb *d*	3.4			

TABLE 8.13 Cumulative Formation Constants for Metal Complexes with Organic Ligands (*Continued*)

	$\log K_1$	$\log K_2$	$\log K_3$	$\log K_4$
UO$_2$(II)	4.38			
Zn	2.2			
Piperidine				
Ag	3.30	6.48		
Hg(II)	8.70	17.44		
Pt(II)			$\log K_5$ 5.7	$\log K_6$ 8.2
Propylene-1,2-diamine				
Cd *b,c*		9.97	12.12	
Co(II) *d*	5.42	11.47	14.72	
Cu(II) *c*	6.41	20.06		
Hg(II) *c*	10.78	23.53	23.25	
Ni *d*	7.43	13.62	17.89	
Zn *b,c*	5.89	10.87	12.57	
Pyridine				
Ag	1.97	4.35		
Cd	1.40	1.95	2.27	2.50
Co(II)	1.14	1.54		
Cu(I)		3.34	4.51	5.44
				$\log K_6$ 6.89
Cu(II)	2.59	4.33	5.93	6.54
			$\log K_5$ 7.00	$\log K_6$ 10.2
Fe(II)	0.71			
Hg(II)	5.1	10.0	10.4	
Mn(II)	1.92	2.77	3.37	3.50
VO(II)	−1.70			
Zn	1.41	1.11	1.61	1.93
Pyridine-2,6-dicarboxylic acid				
Ba *a,d*	3.46			
Ca *a,d*	4.6	7.2		
Cd *a,d*	5.7	10.0		
Ce(III) *a,d*	8.34	14.42	18.80	
Co(II) *a,d*	7.0	12.5		
Cu(II) *a,d*	9.14	16.52		
Dy *a,d*	8.69	16.19	22.14	
Er *a,d*	8.77	16.39	22.14	
Eu(III) *a,d*	8.84	15.98	21.00	
Fe(II) *a,d*	5.71	10.36		
Fe(III) *a,d*	10.91	17.13		
Gd *a,d*	8.74	16.06	21.83	
Ho *a,d*	8.72	16.23	22.08	
La *a,d*	7.98	13.79	18.06	
Lu *a,d*	9.03	16.80	21.48	
Hg(II) *a,d*	20.28			
Mg *a,d*	2.7			
Mn(II) *a,d*	5.01	8.49		
Nd *a,d*	8.78	15.60	20.66	
Ni *a,d*	6.95	13.50		
Pb *a,d*	8.70	10.60		
Pr *a,d*	8.63	15.10	19.94	
Sm *a,d*	8.86	15.88	21.23	
Sr *a,d*	3.89			
Tb *a,d*	8.68	16.11	22.03	
Tm *a,d*	8.83	16.54	22.04	
Y *a,d*	8.46	15.73	21.34	
Yb *a,d*	8.85	16.61	21.83	
Zn *a,d*	6.35	11.88		

TABLE 8.13   Cumulative Formation Constants for Metal Complexes with Organic Ligands (*Continued*)

	$\log K_1$	$\log K_2$	$\log K_3$	$\log K_4$
1-(2-Pyridylazo)-2-naphthol (PAN)				
Co(II)	>12			
Cu(II)	16			
Mn(II)	8.5	16.4		
Ni	12.7	25.3		
Tl(III)	2.29			
Zn	11.2	21.7		

	$\log K_f$ [ML]	$\log K_f$ [MHL]	$\log K_f$ [M(HL)$_2$]
4-(2-Pyridylazo)resorcinal (PAR)			
Co(II)		>12	
Cu(II)	10.3		
Mn(II)		9.7	18.9
Ni		13.2	26.0
Sc	4.8		
Tl(III)	4.23		
Zn		12.4	23.5

	$\log K_f$ [ML]	$\log K_f$ [M$_2$L]	$\log K_f$ [MHL]
Pyrocatechol-3,5-disulfonate (Pyrocatechol Violet)			
Al	19.13	4.95	
Bi	27.07	5.25	
Cd	8.13		5.86
Co(II)	9.01		6.53
Cu(II)	16.47		11.18
Ga	22.18	4.65	
In	18.10	4.81	
Mg	4.42	4.6	3.66
Mn(II)	7.13		5.36
Ni	9.35	4.38	6.85
Pb	13.25		10.19
Th	23.36	4.42	
Zn	10.41	6.21	7.21
Zr	27.40	4.18	

	$\log K_1$	$\log K_2$	$\log K_3$	$\log K_4$
8-Quinolinol				
Ba	2.07			
Be	3.36			
Ca (75% dioxane)	7.3	13.2		
Cd	7.2	13.4		
Ce(III) (50% dioxane)	9.15	17.13		
Co(II)	9.1	17.2		
Cu(II)	12.2	23.4		
Fe(II)	8.58	16.93	22.23	
Fe(III)	12.3	23.6	33.9	
La	5.85	16.95		
Mg (50% dioxane)	6.38	11.81		
Mn(II) (50% dioxane)	8.28	15.45		
Ni (50% dioxane)	11.44	21.38		
Pb (50% dioxane)	10.61	18.70		

TABLE 8.13 Cumulative Formation Constants for Metal Complexes with Organic Ligands (*Continued*)

	$\log K_1$	$\log K_2$	$\log K_3$	$\log K_4$
Sm	6.84		19.50	
Sr	2.89	6.08		
Th	10.45	20.40	29.85	38.80
$UO_2(II)$ (50% dioxane)	11.25	20.89		
V(II)	12.8	23.6		
VO(II)	10.97	20.19		
Y	8.15	14.90	20.25	
Zn (50% dioxane)	9.96	18.86		

	$\log K_f$ [MHL$^+$]	$\log K_f$ [M(HL)$_2$]
Salicylaldoxime		
Ba	0.53	3.72
Be	<7	
Ca	0.92	3.72
Cd	<4.4	
Co(II)		8.13
Cu(II)		8.13
Mg	0.64	4.10
Ni		3.77
Sr		3.77
Zn	<5.2	

	$\log K_1$	$\log K_2$	$\log K_3$	$\log K_4$
Salicylic acid				
Al	14.11			
Be	17.4			
Cd	5.55			
Ce(III)	2.66			
Co(II)	6.72	11.42		
Cr(II)	8.4	15.3		
Cu(II)	10.60	18.45		
Fe(II)	6.55	11.25		
Fe(III) a,c	16.48	28.12	36.80	
La	2.64			
Mg (75% dioxane)	4.7			
Mn(II)	5.90	9.80		
Nd	2.70			
Ni	6.95	11.75		
Pr	2.68			
Th	4.25	7.60	10.05	11.60
TiO(II)	6.09			
$UO_2(II)$	13.4			
V(II)	6.3			
Zn	6.85			
Succinic acid				
Ba	2.08			
Be	3.08			
Ca	2.0			
Cd	2.2			
Co(II)	2.22			
Cu(II)	3.33			
Fe(III)	7.49			
Hg(II)		7.28		
La	3.96			

TABLE 8.13 Cumulative Formation Constants for Metal Complexes with Organic Ligands (*Continued*)

	$\log K_1$	$\log K_2$	$\log K_3$	$\log K_4$
Mg	1.20			
Mn(II)	2.26			
Nd	8.1			
Ni	2.36			
Pb	2.8			
Ra	1.0			
Sr	1.06			
Zn	1.6			
5-Sulfosalicylic acid				
Al *c*	13.20	22.83	28.89	
Be *c*	11.71	20.81		
Cd *c*	16.68	29.08		
Co(II) *c*	6.13	9.82		
Cr(II) *c*	7.1	12.9		
Cr(III) *c*	9.56			
Cu(II) *c*	9.52	16.45		
Fe(II) *c*	5.90			
Fe(III) *c*	14.64	25.18	32.12	
La *c*	9.11			
Mn(II) *c*	5.24	8.24		
NbO(III) *c*	4.0	7.7		
Ni *c*	6.42	10.24		
UO$_2$(II) *c*	11.14	19.20		
Zn *c*	6.05	10.65		
Tartaric acid				
Ba		1.62		
Bi			8.30	
Ca	2.98	9.01		
Cd	2.8			
Co(II)	2.1			
Cu(II)	3.2	5.11	4.78	6.51
				$\log K_I$ 19.14 [Cu(OH)$_2$L^{2-}]
Eu(III)	4.98	8.11		
Fe(III)	7.49			
La	3.06			
Mg		1.36		
Nd	9.0			
Pb	3.78		4.7	$\log K_I$ 14.1 [Pb(OH)$_2$L^{2-}]
Ra	1.24			
Sr	1.60			
Zn	2.68	8.32		
Thioglycolic acid				
Ce(III) *a,c*	1.99	3.03		
Co(II)	5.84	12.15		
Fe(II)		10.92		
Hg(II)		43.82		
La *a,c*	1.98	2.98		
Mn(II)	4.38	7.56		
Pb	8.5			
Ni	6.98	13.53		
Rare earths *a,c*	1.9–2.1	3.0–3.3		
Y *a,c*	1.91	3.19		
Zn	7.86	15.04		
Thiourea				
Ag	7.4	13.1		
Bi				$\log K_6$ 11.9
Cd	0.6	1.6	2.6	4.6

TABLE 8.13 Cumulative Formation Constants for Metal Complexes with Organic Ligands (*Continued*)

	$\log K_1$	$\log K_2$	$\log K_3$	$\log K_4$
Cu(I)			13	15.4
Hg(II)		22.1	24.7	26.8
Pb	1.4	3.1	4.7	8.3
Ru(III)	1.21		0.72	
Thoron				
Th		10.15		
Triethanolamine				
Ag	2.30	3.64		
Co(II)	1.73			
Cu(II)	4.30			
Hg(II)	6.90	13.08		
Ni	2.7			
Zn	2.00			
Triethylenetetramine (Trien)				
Ag	7.7			
Cd	10.75	13.9		
Co(II)	11.0			
Cu(II)	20.4			
Fe(II)	7.8			
Fe(III)	21.9			
Hg(II)	25.26			
Mn(II)	4.9			
Ni	14.0			
Pb	10.4			
Zn	11.9			
1,1,1-Trifluoro-3-2′-Thenoylacetone (TTA)				
Ba		10.6		
Cu(II)	6.55	13.0		
Fe(III)	6.9			
Ni	10.0			
Pr	9.53			
Pu(III)	9.53			
Pu(IV)	8.0			
Th	8.1			
U(IV)	7.2			
Zr	3.03 [as ZrL^{3+}]			
Xylenol orange				
Bi	5.52			
Fe(III)	5.70			
Hf	6.50			
Tl(III)	4.90			
Zn	6.15			
Zr	7.60			
Zincon				
Zn	13.1			

8.3 BUFFER SOLUTIONS

8.3.1 Standard Reference pH Buffer Solutions

The assigned values of pH_s, according to the Bates-Guggenheim convention [*Pure Applied Chem.* **1**:163 (1960)], for the primary standard solutions prepared from salts issued by the National Bureau of Standards (U.S.) are given in Table 8.14. These are smoothed values. The ionic strength of these reference solutions is 0.1 or less. Strictly speaking the NBS scale uses a molality concentration system; however, values are given in molarity units for convenience.

TABLE 8.14 National Bureau of Standards (U.S.) Reference pH Buffer Solutions

Temperature °C	Secondary standard 0.05 M K tetraoxalate	KH tartrate (saturated at 25°C)	0.05 M KH₂ citrate	0.05 M KH phthalate	0.025 M KH₂PO₄, 0.025 M Na₂HPO₄	0.0087 M KH₂PO₄, 0.0302 M Na₂HPO₄	0.01 M Na₂B₄O₇	0.025 M NaHCO₃, 0.025 M Na₂CO₃	Secondary standard Ca(OH)₂ (saturated at 25°C)
0	1.666		3.860	4.003	6.984	7.534	9.464	10.317	13.423
5	1.668		3.840	3.999	6.951	7.500	9.395	10.245	13.207
10	1.670		3.820	3.998	6.923	7.472	9.332	10.179	13.003
15	1.672		3.802	3.999	6.900	7.448	9.276	10.118	12.810
20	1.675		3.788	4.002	6.881	7.429	9.225	10.062	12.627
25	1.679	3.557	3.776	4.008	6.865	7.413	9.180	10.012	12.454
30	1.683	3.552	3.766	4.015	6.853	7.400	9.139	9.966	12.289
35	1.688	3.549	3.759	4.024	6.844	7.389	9.102	9.925	12.133
38	1.691	3.548		4.030	6.840	7.384	9.081		12.043
40	1.694	3.547	3.753	4.035	6.838	7.380	9.068	9.889	11.984
45	1.700	3.547		4.047	6.834	7.373	9.038		11.841
50	1.707	3.549	3.749	4.060	6.833	7.367	9.011	9.828	11.705
55	1.715	3.554		4.075	6.834		8.985		11.574
60	1.723	3.560		4.091	6.836		8.962		11.449
70	1.743	3.580		4.126	6.845		8.921		
80	1.766	3.609		4.164	6.859		8.885		
90	1.792	3.650		4.205	6.877		8.850		
95	1.806	3.674		4.227	6.886		8.833		
Dilution value ΔpH₁/₂	+0.186	+0.049	0.024	+0.052	+0.080	+0.070	+0.01	0.079	−0.28

Source: R. G. Bates, *J. Res. Natl. Bur. Stand. (U.S.)*, **66A**:179 (1962) and B. R. Staples and R. G. Bates, *J. Res. Natl. Bur. Stand. (U.S.)*, **73A**:37 (1969).

As a result of a variable liquid-junction potential, the measured pH may be expected to differ seriously from the pa_H determined from cells without a liquid junction in solutions of high acidity or high alkalinity. Merely to affirm the proper functioning of the glass electrode at the extreme ends of the pH scale, two secondary standards are included in Table 8.14. In addition, values for a 0.1 m solution of HCl are given to extend the pH scale up to 275°C [see R. S. Greeley, *Anal. Chem.* **32**:1717 (1960)]:

t, °C:	25	60	90	125	150	175	200	225–275
pH:	1.10	1.11	1.12	1.13	1.14	1.15	1.16	1.2

Uncertainties in the values are ±0.03 pH unit from 25 to 90°C, ±0.05 pH unit from 125 to 200°C, and ±0.1 pH unit from 225 to 275°C.

The buffer values for the NBS reference pH buffer solutions are given below:

Buffer solution	KH tartrate	0.05 M KH$_2$ citrate	0.05 M KH phthalate	0.025 M KH$_2$PO$_4$, 0.25 M Na$_2$HPO$_4$	0.0087 M KH$_2$PO$_4$, 0.0302 M Na$_2$HPO$_4$	0.01 M Na$_2$B$_4$O$_7$	0.025 M NaHCO$_3$, 0.025 M Na$_2$CO$_3$
Buffer value β	0.027	0.034	0.016	0.029	0.016	0.020	0.029

For the secondary pH reference standards, the buffer value is 0.070 for potassium tetroxalate and 0.09 for calcium hydroxide.

To prepare the standard pH buffer solutions recommended by the National Bureau of Standards (U.S.), the indicated weights of the pure materials in Table 8.15 should be dissolved in water of specific conductivity not greater than 5 micromhos. The tartrate, phthalate, and phosphates can be dried for 2 h at 110°C before use. Potassium tetroxalate and calcium hydroxide need not be dried. Fresh-looking crystals of borax should be used. Before use, excess solid potassium hydrogen tartrate and calcium hydroxide must be removed. Buffer solutions

TABLE 8.15 Compositions of Standard pH Buffer Solutions [National Bureau of Standards (U.S.)]

Air weight of material per liter of buffer solution.

Standard	Weight, g
KH$_3$(C$_2$O$_4$)$_2$ · 2H$_2$O, 0.05M	12.61
Potassium hydrogen tartrate, about 0.034M	Saturated at 25°C
Potassium hydrogen phthalate, 0.05M	10.12
Phosphate:	
KH$_2$PO$_4$, 0.025M	3.39
Na$_2$HPO$_4$, 0.025M	3.53
Phosphate:	
KH$_2$PO$_4$, 0.008665M	1.179
Na$_2$HPO$_4$, 0.03032M	4.30
Na$_2$B$_4$O$_7$ · 10H$_2$O, 0.01M	3.80
Carbonate:	
NaHCO$_3$, 0.025M	2.10
Na$_2$CO$_3$, 0.025M	2.65
Ca(OH)$_2$, about 0.0203M	Saturated at 25°C

TABLE 8.16 Composition and pH Values of Buffer Solutions

Values based on the conventional activity pH scale as defined by the National Bureau of Standards (U.S.) and pertain to a temperature of 25°C [Ref: Bower and Bates, *J. Research Natl. Bur. Standards (U.S.)*, **55**:197 (1955) and Bates and Bower, *Anal. Chem.*, **28**:1322 (1956)]. Buffer value is denoted by column headed β.

25 ml 0.2M KCl + x ml 0.2M HCl, Diluted to 100 ml			50 ml 0.1M KH Phthalate + x ml 0.1M HCl, Diluted to 100 ml			50 ml 0.1M KH Phthalate + x ml 0.1M NaOH, Diluted to 100 ml		
pH	x	β	pH	x	β	pH	x	β
1.00	67.0	0.31	2.20	49.5		4.20	3.0	0.017
1.20	42.5	0.34	2.40	42.2	0.036	4.40	6.6	0.020
1.40	26.6	0.19	2.60	35.4	0.033	4.60	11.1	0.025
1.60	16.2	0.077	2.80	28.9	0.032	4.80	16.5	0.029
1.80	10.2	0.049	3.00	22.3	0.030	5.00	22.6	0.031
2.00	6.5	0.030	3.20	15.7	0.026	5.20	28.8	0.030
2.20	3.9	0.022	3.40	10.4	0.023	5.40	34.1	0.025
			3.60	6.3	0.018	5.60	38.8	0.020
			3.80	2.9	0.015	5.80	42.3	0.015

50 ml 0.1M KH$_2$PO$_4$ + x ml 0.1M NaOH, Diluted to 100 ml			50 ml 0.1M Tris(hydroxymethyl)aminomethane + x ml of 0.1M HCl, Diluted to 100 ml ΔpH/$\Delta t \simeq -0.028$ $I = 0.001x$			50 ml of a Mixture 0.1M with Respect to Both KCl and H$_3$BO$_3$ + x ml 0.1M NaOH, Diluted to 100 ml		
pH	x	β	pH	x	β	pH	x	β
5.80	3.6		7.00	46.6		8.00	3.9	
6.00	5.6	0.010	7.20	44.7	0.012	8.20	6.0	0.011
6.20	8.1	0.015	7.40	42.0	0.015	8.40	8.6	0.015
6.40	11.6	0.021	7.60	38.5	0.018	8.60	11.8	0.018
6.60	16.4	0.027	7.80	34.5	0.023	8.80	15.8	0.022
6.80	22.4	0.033	8.00	29.2	0.029	9.00	20.8	0.027
7.00	29.1	0.031	8.20	22.9	0.031	9.20	26.4	0.029
7.20	34.7	0.025	8.40	17.2	0.026	9.40	32.1	0.027
7.40	39.1	0.020	8.60	12.4	0.022	9.60	36.9	0.022
7.60	42.4	0.013	8.80	8.5	0.016	9.80	40.6	0.016
7.80	44.5	0.009	9.00	5.7		10.00	43.7	0.014
8.00	46.1					10.20	46.2	

50 ml 0.025M Borax, + x ml 0.1M HCl, Diluted to 100 ml ΔpH/$\Delta t \simeq -0.008$ $I = 0.025$			50 ml 0.025M Borax + x ml 0.1M NaOH, Diluted to 100 ml ΔpH/$\Delta t \simeq -0.008$ $I = 0.001(25 + x)$			50 ml 0.05M NaHCO$_3$ + x ml 0.1M NaOH, Diluted to 100 ml ΔpH/$\Delta t \simeq -0.009$ $I = 0.001(25 + 2x)$		
pH	x	β	pH	x	β	pH	x	β
8.00	20.5		9.20	0.9		9.60	5.0	
8.20	19.7	0.010	9.40	3.6	0.026	9.80	6.2	0.014
8.40	16.6	0.012	9.60	11.1	0.022	10.00	10.7	0.016
8.60	13.5	0.018	9.80	15.0	0.018	10.20	13.8	0.015

TABLE 8.16 Composition and pH Values of Buffer Solutions (*Continued*)

50 ml 0.025M Borax, + x ml 0.1M HCl, Diluted to 100 ml $\Delta pH/\Delta t \simeq -0.008$ $I = 0.025$			50 ml 0.025M Borax + x ml 0.1M NaOH, Diluted to 100 ml $\Delta pH/\Delta t \simeq -0.008$ $I = 0.001(25 + x)$			50 ml 0.05M NaHCO$_3$ + x ml 0.1M NaOH, Diluted to 100 ml $\Delta pH/\Delta t \simeq -0.009$ $I = 0.001(25 + 2x)$		
pH	x	β	pH	x	β	pH	x	β
8.80	9.4	0.023	10.00	18.3	0.014	10.40	16.5	0.013
9.00	4.6	0.026	10.20	20.5	0.009	10.60	19.1	0.012
9.10	2.0		10.40	22.1	0.007	10.80	21.2	0.009
			10.60	23.3	0.005	11.00	22.7	

50 ml 0.05M Na$_2$HPO$_4$ + x ml 0.1M NaOH, Diluted to 100 ml $\Delta pH/\Delta t \simeq -0.025$ $I = 0.001(77 + 2x)$			25 ml 0.2M KCl + x ml 0.2M NaOH, Diluted to 100 ml $\Delta pH/\Delta t \simeq -0.033$ $I = 0.001(50 + 2x)$		
pH	x	β	pH	x	β
11.00	4.1	0.009	12.00	6.0	0.028
11.20	6.3	0.012	12.20	10.2	0.048
11.40	9.1	0.017	12.40	16.2	0.076
11.60	13.5	0.026	12.60	25.6	0.12
11.80	19.4	0.034	12.80	41.2	0.21
11.90	23.0	0.037	13.00	66.0	0.30

pH 6 or above should be stored in plastic containers and should be protected from carbon dioxide with soda-lime traps. The solutions should be replaced within 2 to 3 weeks, or sooner if formation of mold is noticed. A crystal of thymol may be added as a preservative.

8.3.2 Standards for pH Measurement of Blood and Biological Media

Blood is a well-buffered medium. In addition to the NBS phosphate standard of 0.025M (pH$_s$ = 6.480 at 38°C), another reference solution containing the same salts, but in the molal ratio 1:4, has an ionic strength of 0.13. It is prepared by dissolving 1.360 g of KH$_2$PO$_4$ and 5.677 g of Na$_2$HPO$_4$ (air weights) in carbon dioxide–free water to make 1 liter of solution. The pH$_s$ is 7.416 ± 0.004 at 37.5 and 38°C.

The compositions and pH$_s$ values of *tris*(hydroxymethyl)aminomethane, covering the pH range 7.0 to 8.9, are listed in Table 8.16.

The phosphate-succinate system gives the values of pH$_s$ shown below:

$\dfrac{\text{Molality}}{\text{KH}_2\text{PO}_4}=\dfrac{\text{Molality}}{\text{Na}_2\text{HC}_6\text{H}_5\text{O}_7}$	pH$_s$	$\Delta(pH_s/\Delta t)$
0.005	6.251	−0.000 86 deg^{-1}
0.010	6.197	−0.000 71
0.015	6.162	
0.020	6.131	
0.025	6.109	−0.000 4

TABLE 8.17 Standard Reference Values pH$^*_s$ for the Measurement of Acidity in 50 Weight Percent Methanol-Water

Temperature, °C	0.02m HOAc, 0.02m NaOAc, 0.02m NaCl	0.02m NaHSuc, 0.02m NaCl	0.02m KH$_2$PO$_4$, 0.02m Na$_2$HPO$_4$, 0.02m NaCl
10	5.560	5.806	7.937
15	5.549	5.786	7.916
20	5.543	5.770	7.898
25	5.540	5.757	7.884
30	5.540	5.748	7.872
35	5.543	5.743	7.863
40	5.550	5.741	7.858

OAc = acetate   Suc = succinate

Reference: R. G. Bates, *Anal. Chem.*, **40**(6):35A (1968).

TABLE 8.18 pH* Values for Buffer Solutions in Alcohol-Water Solvents at 25°C

Liquid-junction potential not included.

Solvent Composition (weight per cent alcohol)	0.01M H$_2$C$_2$O$_4$, 0.01M NH$_4$HC$_2$O$_4$	0.01M H$_2$Suc, 0.01M LiHSuc	0.01M HSal, 0.01M NaSal
Methanol-Water Solvents			
0	2.15	4.12	
10	2.19	4.30	
20	2.25	4.48	
30	2.30	4.67	
40	2.38	4.87	
50	2.47	5.07	
60	2.58	5.30	
70	2.76	5.57	
80	3.13	6.01	
90	3.73	6.73	
92	3.90	6.92	
94	4.10	7.13	
96	4.39	7.43	
98	4.84	7.89	
99	5.20	8.23	
100	5.79	8.75	7.53
Ethanol-Water Solvents			
0	2.15	4.12	
30	2.32	4.70	
50	2.51	5.07	
71.9	2.98	5.71	
100			8.32

Suc = succinate   Sal = salicylate

8.3.3 Buffer Solutions Other Than Standards

The range of the buffering effect of a single weak acid group is approximately one pH unit on either side of the pK_a. The ranges of some useful buffer systems are collected in Table 8.19. After all the components have been brought together, the pH of the resulting solution should be determined at the temperature to be employed with reference to standard reference solutions. Buffer components should be compatible with other components in the system under study; this is particularly significant for buffers employed in biological studies. Check tables of formation constants to ascertain whether metal-binding character exists.

When there are two or more acid groups per molecule, or a mixture is composed of several overlapping acids, the useful range is larger. Universal buffer solutions consist of a mixture of acid groups which overlap such that successive pK_a values differ by 2 pH units or less. The Prideaux-Ward mixture comprises phosphate, phenyl acetate, and borate plus HCl and covers the range from 2 to 12 pH units. The McIlvaine buffer is a mixture of citric acid and Na_2HPO_4 that covers the range from pH 2.2 to 8.0. The Britton-Robinson system consists of acetic acid, phosphoric acid, and boric acid plus NaOH and covers the range from pH 4.0 to 11.5. A mixture composed of Na_2CO_3, NaH_2PO_4, citric acid, and 2-amino-2-methyl-1,3-propanediol covers the range from pH 2.2 to 11.0.

TABLE 8.19 pH Values of Buffer Solutions for Control Purposes

Materials*	pH range
Glycine and HCl	1.0–3.7
Citrate and HCl	1.3–4.7
p-Toluenesulfonate and p-toluenesulfonic acid	1.1–3.3
Formate and HCl	2.8–4.6
Succinic acid and borax	3.0–5.8
Phenyl acetate and HCl	3.5–5.0
Acetate and acetic acid	3.7–5.6
Succinate and succinic acid	4.8–6.3
2-(N-Morpholino)ethanesulfonic acid and NaOH	5.2–7.1
2,2-Bis(hydroxymethyl)-2,2',2"-nitrilotriethanol and HCl	5.8–7.2
KH_2PO_4 and borax	5.8–9.2
N-Tris(hydroxymethyl)methyl-2-aminoethanesulfonic acid and NaOH	6.8–8.2
KH_2PO_4 and Na_2HPO_4	6.1–7.5
N-2-Hydroxyethylpiperazine-N'-2-ethanesulfonic acid and NaOH	6.9–8.3
Triethanolamine and HCl	6.9–8.5
Diethylbarbiturate (veronal) and HCl	7.0–8.5
Tris(hydroxymethyl)aminomethane and HCl	7.2–9.0
N-Tris(hydroxymethyl)methylglycine and HCl	
N,N-Bis(2-hydroxyethyl)glycine and HCl	
Borax and HCl	7.6–8.9
Glycine and NaOH	8.2–10.1
Ammonia (aqueous) and NH_4Cl	8.3–9.2
Ethanolamine and HCl	8.6–10.4
Borax and NaOH	9.4–11.1
Carbonate and hydrogen carbonate	9.2–11.0
Na_2HPO_4 and NaOH	11.0–12.0

* Bates, *Determination of pH, Theory and Practice,* Wiley, New York, 1964, pp. 121–122.

TABLE 8.19   pH Values of Buffer Solutions for Control Purposes (*Continued*)

*x* mL of 0.2*M* Sodium Acetate (27.199 g NaOAc · 3H$_2$O per liter) plus *y* mL of 0.2*M* Acetic Acid			*x* mL of 0.1*M* KH$_2$PO$_4$ (13.617 g · L^{-1}) plus *y* mL of 0.05*M* Borax Solution (19.404 g Na$_2$B$_4$O$_7$ · 10H$_2$O per Liter)					
pH	NaOAc, mL	Acetic Acid, mL	pH	KH$_2$PO$_4$, mL	Borax, mL	pH	KH$_2$PO$_4$, mL	Borax, mL
3.60	7.5	92.5	5.80	92.1	7.9	7.60	51.7	48.3
3.80	12.0	88.0	6.00	87.7	12.3	7.80	49.2	50.8
4.00	18.0	82.0	6.200	83.0	17.0	8.00	46.5	53.5
4.20	26.5	73.5	6.40	77.8	22.2	8.20	43.0	57.0
4.40	37.0	63.0	6.60	72.2	27.8	8.40	38.7	61.3
4.60	49.0	51.0	6.80	66.7	33.3	8.60	34.0	66.0
4.80	60.0	40.0	7.00	62.3	37.7	8.80	27.6	72.4
5.00	70.5	29.5	7.20	58.1	41.9	9.00	17.5	82.5
5.20	79.0	21.0	7.40	55.0	45.0	9.20	5.0	95.0
5.40	85.5	14.5						
5.60	90.5	9.5						

x mL of Veronal (20.6 g Na Diethylbarbiturate per Liter) plus *y* mL of 0.1*M* HCl			*x* mL of 0.2*M* Aqueous NH$_3$ Solution plus *y* mL of 0.2*M* NH$_4$Cl (10.699 g · L^{-1})			*x* mL of 0.1*M* Citrate (21.0 g Citric Acid Monohydrate + 200 mL 1*M* NaOH per Liter) plus *y* mL of 0.1*M* NaOH		
pH	Veronal, mL	HCl, mL	pH	Aq NH$_3$, mL	NH$_4$Cl, mL	pH	Citrate, mL	NaOH, mL
7.00	53.6	46.4	8.00	5.5	94.5	5.10	90.0	10.0
7.20	55.4	44.6	8.20	8.5	91.5	5.30	80.0	20.0
7.40	58.1	41.9	8.40	12.5	87.5	5.50	71.0	29.0
7.60	61.5	38.5	8.60	18.5	81.5	5.70	67.0	33.0
7.80	66.2	33.8	8.80	26.0	74.0	5.90	62.0	38.0
8.00	71.6	28.4	9.00	36.0	64.0			
8.20	76.9	23.1	9.25	50.0	50.0			
8.40	82.3	17.7	9.40	58.5	41.5			
8.60	87.1	12.9	9.60	69.0	31.0			
8.80	90.8	9.2	9.80	78.0	22.0			
9.00	93.6	6.4	10.00	85.0	15.0			

x mL of 0.2*M* NaOH Added to 100 mL of Stock Solution (0.04*M* Acetic Acid, 0.04*M* H$_3$PO$_4$, and 0.04*M* Boric Acid)

pH	NaOH, mL	pH	NaOH, mL	pH	NaOH, mL	pH	NaOH, mL
1.81	0.0	4.10	25.0	6.80	50.0	9.62	75.0
1.89	2.5	4.35	27.5	7.00	52.5	9.91	77.5
1.98	5.0	4.56	30.0	7.24	55.0	10.38	80.0
2.09	7.5	4.78	32.5	7.54	57.5	10.88	82.5
2.21	10.0	5.02	35.0	7.96	60.0	11.20	85.0
2.36	12.5	5.33	37.5	8.36	62.5	11.40	87.5
2.56	15.0	5.72	40.0	8.69	65.0	11.58	90.0
2.87	17.5	6.09	42.5	8.95	67.5	11.70	92.5
3.29	20.0	6.37	45.0	9.15	70.0	11.82	95.0
3.78	22.5	6.59	47.5	9.37	72.5	11.92	97.5

TABLE 8.19 pH Values of Buffer Solutions for Control Purposes (*Continued*)

x mL of 0.1M HCl plus y mL of 0.1M Glycine (7.505 g Glycine + 5.85 g NaCl per Liter)			x mL of 0.1M HCl plus y mL of 0.1M Citrate (21.008 g Citric Acid Monohydrate + 200 ml 1M NaOH per Liter)			x mL of 0.05M Succinic Acid (5.90 g · L⁻¹) plus y mL of Borax Solution (19.404 g $Na_2B_4O_7 \cdot 10H_2O$ per Liter)		
pH	HCl, mL	Glycine, mL	pH	HCl, mL	Citrate, mL	pH	Succinic Acid, mL	Borax, mL
1.20	84.0	16.0	3.50	52.8	47.2	3.60	90.5	9.5
1.40	71.0	29.0	3.60	51.3	48.7	3.80	86.3	13.7
1.60	61.8	38.2	3.80	48.6	51.4	4.00	82.2	17.8
1.80	55.2	44.8	4.00	43.8	56.2	4.20	77.8	22.2
2.00	49.1	50.9	4.20	38.6	61.4	4.40	73.8	26.2
2.20	42.7	57.3	4.40	34.6	65.4	4.60	70.0	30.0
2.40	36.5	63.5	4.60	24.3	75.7	4.80	66.5	33.5
2.60	30.3	69.7	4.80	11.0	89.0	5.00	63.2	36.8
2.80	24.0	76.0				5.20	60.5	39.5
3.00	17.8	82.2				5.40	57.9	42.1
3.30	10.8	89.2				5.60	55.7	44.3
3.60	6.0	94.0				5.80	54.0	46.0

x mL of 0.2M $Na_2HPO_4 \cdot 2H_2O$ (35.599 g · L⁻¹) plus y mL of 0.1M Citric Acid (19.213 g · L⁻¹)

pH	Na_2HPO_4, mL	Citric Acid, mL	pH	Na_2HPO_4, mL	Citric Acid, mL	pH	Na_2HPO_4, mL	Citric Acid, mL
2.20	2.00	98.00	4.20	41.40	58.60	6.20	66.10	33.90
2.40	6.20	93.80	4.40	44.10	55.90	6.40	69.25	30.75
2.60	10.90	89.10	4.60	46.75	53.25	6.60	72.75	27.25
2.80	15.85	84.15	4.80	49.30	50.70	6.80	77.25	22.75
3.00	20.55	79.45	5.00	51.50	48.50	7.00	82.35	17.65
3.20	24.70	75.30	5.20	53.60	46.40	7.20	86.95	13.05
3.40	28.50	71.50	5.40	55.75	44.25	7.40	90.85	9.15
3.60	32.20	67.80	5.60	58.00	42.00	7.60	93.65	6.35
3.80	35.50	64.50	5.80	60.45	39.55	7.80	95.75	4.25
4.00	38.55	61.45	6.00	63.15	36.85	8.00	97.25	2.75

General directions for the preparation of buffer solutions of varying pH but fixed ionic strength are given by Bates.* Preparation of McIlvaine buffered solutions at ionic strengths of 0.5 and 1.0 and Britton-Robinson solutions of constant ionic strength have been described by Elving et al.† and Frugoni,‡ respectively.

---

* Bates, *Determination of pH, Theory and Practice*, Wiley, New York, 1964, pp. 121–122.
† Elving, Markowitz, and Rosenthal, *Anal. Chem.*, **28**:1179 (1956).
‡ Frugoni, *Gazz. Chim. Ital.*, **87**:L403 (1957).

8.4 REFERENCE ELECTRODES

TABLE 8.20 Potentials of Reference Electrodes in Volts as a Function of Temperature
Liquid-junction potential included.

Temp., °C	0.1M KCl Calomel*	1.0M KCl Calomel*	3.5M KCl Calomel*	Satd. KCl Calomel*	1.0M KCl Ag/AgCl†	1.0M KBr Ag/AgBr‡	1.0M KI Ag/AgI§
0	0.3367	0.2883		0.25918	0.23655	0.08128	−0.14637
5					0.23413	0.07961	−0.14719
10	0.3362	0.2868	0.2556	0.25387	0.23142	0.07773	−0.14822
15	0.3361			0.2511	0.22857	0.07572	−0.14942
20	0.3358	0.2844	0.2520	0.24775	0.22557	0.07349	−0.15081
25	0.3356	0.2830	0.2501	0.24453	0.22234	0.07106	−0.15244
30	0.3354	0.2815	0.2481	0.24118	0.21904	0.06856	−0.15405
35	0.3351			0.2376	0.21565	0.06585	−0.15590
38	0.3350		0.2448	0.2355			
40	0.3345	0.2782	0.2439	0.23449	0.21208	0.06310	−0.15788
45					0.20835	0.06012	−0.15998
50	0.3315	0.2745		0.22737	0.20449	0.05704	−0.16219
55					0.20056		
60	0.3248	0.2702		0.2235	0.19649		
70					0.18782		
80				0.2083	0.1787		
90					0.1695	0.0251	

* Bates et al., *J. Research Natl. Bur. Standards*, **45**, 418 (1950).
† Bates and Bower, *J. Research Natl. Bur. Standards*, **53**, 283 (1954).
‡ Hetzer, Robinson and Bates, *J. Phys. Chem.*, **66**, 1423 (1962).
§ Hetzer, Robinson and Bates, *J. Phys. Chem.*, **68**, 1929 (1964).

Temp., °C	125	150	175	200	225	250	275
1.0M KCl Ag/AgCl*	0.1330	0.1032	0.0708	0.0348	−0.0051	−0.054	−0.090
1.0M KBr Ag/AgBr†	−0.0048	−0.0312	−0.0612	−0.0951			

* Greeley et al., *J. Phys. Chem.*, **64**, 652 (1960).
† Towns et al., *J. Phys. Chem.*, **64**, 1861 (1960).

TABLE 8.21 Potentials of Reference Electrodes (in Volts) at 25°C for Water–Organic Solvent Mixtures

Electrolyte solution of 1M HCl.

Solvent, wt %	Methanol, Ag/AgCl	Ethanol, Ag/AgCl	2-Propanol, Ag/AgCl	Acetone, Ag/AgCl	Dioxane, Ag/AgCl	Ethylene glycol, Ag/AgCl	Methanol, calomel	Dioxane, calomel
5	0.2153	0.2146	0.2180	0.2190		0.2190		
10	0.2090	0.2075	0.2138	0.2156		0.2160		
20		0.2003	0.2063	0.2079	0.2031	0.2101	0.255	0.2501
30	0.1968	0.1945				0.2036		
40		0.1859		0.1859	0.1635	0.1972	0.243	0.2104
45				0.158				
50	0.1818	0.173						
60		0.158			0.0659	0.1807	0.216	0.1126
70	0.1492	0.136						
80	0.1135							
82				−0.034	−0.0614			−0.0014
90	0.0841	0.196						
94.2		0.0215						
98							0.103	
99								
100	−0.0099	−0.0081		−0.53				

The values of several additional reference electrodes at 25°C are listed:

Ag/AgCl, satd. KCl	0.198
Ag/AgCl, 01M KCl	0.288
Hg/HgO, 1.0M NaOH	0.140
Hg/HgO, 0.1M NaOH	0.165
Hg/Hg$_2$SO$_4$, satd. K$_2$SO$_4$ (22°C)	0.658
Hg/Hg$_2$SO$_4$, satd. KCl	0.655

8.4.1 Electrometric Measurement of pH

The pH value is defined for an aqueous solution in an operational (arbitrary but reproducible) manner according to the Bates-Guggenheim convention:

$$pH_x = pH_s + \frac{E_x - E_s}{2.3026RT/F}$$

where R is the gas constant per mole, T is the temperature on the absolute scale, and F is the faraday. The pH$_x$ of the unknown medium is calculated from that of an accepted standard (pH$_s$) and the measured difference in the emf (E) of the electrode combination when the standard solution is removed from the cell and replaced by the unknown. The double vertical line marks a liquid junction. Electrodes as fabricated exhibit variations in the reproducibility of the reference electrode, in the liquid-junction potential, and, with glass electrodes, in the asymmetry potential. These differences are all eliminated in the standardizing procedure with standard reference pH buffers. (See R. G. Bates, *Determination of pH, Theory and Practice*, Wiley, New York, 1964.)

Electrode reversible to hydrogen ions	Standard reference buffer or unknown solution	Salt bridge (KCl. 3.5M or saturated)	Reference electrode

An electrometric pH-measurement system consists of (1) pH-responsive electrode, (2) reference electrode, and (3) potential-measuring device — some form of high-impedance electronic voltmeter for glass-electrode combinations and this or a potentiometer arrangement for other pH-responsive electrodes. Electronic pH meters are simply voltmeters with scale divisions in pH units which are equivalent to the values of 2.3026RT/F (in mV) per pH unit. Values of this function at several temperatures are given in Table 8.22. There is no compensation incorporated in the meter for the changes in pH of the test solution as a function of temperature. Reliability of an indicator–reference electrode combination must be ascertained by standardization of the pH meter with one standard buffer and checking the pH response by immersing the combination in a second and different reference buffer.

The temperature compensator on a pH meter varies the instrument definition of a pH unit from 54.20 mV at 0°C to perhaps 66.10 mV at 60°C. This permits one to measure the pH of the sample (and reference buffer standard) at its actual temperature and thus avoid error due to dissociation equilibria and to junction potentials which have significant temperature coefficients.

TABLE 8.22 Values of $2.3026RT/F$ at Several Temperatures

In millivolts.

$t\,°C$	Value	$t\,°C$	Value	$t\,°C$	Value	$t\,°C$	Value
0	54.197	25	59.157	50	64.118	80	70.070
5	55.189	30	60.149	55	65.110	85	71.062
10	56.181	35	61.141	60	66.102	90	72.054
15	57.173	38	61.737	65	67.094	95	73.046
18	57.767	40	62.133	70	68.086	100	74.038
20	58.165	45	63.126	75	69.078		

Report of the National Academy of Sciences: National Research Council Committee of Fundamental Constants, 1963.

8.5 INDICATORS

TABLE 8.23 Indicators for Aqueous Acid-Base Titrations

This table lists some selected indicators. The pH range or transition interval given in the third column may vary appreciably from one observer to another, and, in addition, it is affected by ionic strength, temperature, and illumination; consequently only approximate values can be given. They should be considered to refer to solutions having low ionic strengths and a temperature of about 25°C. In the fourth column the pK_a ($-\log K_a$) of the indicator as determined spectrophotometrically is listed. In the fifth column the wavelength of maximum absorption is given first for the acidic and then for the basic form of the indicator, and the same order is followed in giving the colors in the sixth column. The abbreviations used to describe the colors of the two forms of the indicator are as follows:

B, blue   G, green

V, violet   P, purple

Y, yellow   R, red

O, orange   O-Br, orange-brown

C, colorless

Indicator	Chemical name	pH range	pK_a	$\lambda_{\max}$, nm	Color change
Cresol red (acid range)	o-Cresolsulfone-phthalein	0.2–1.8			R–Y
Cresol purple (acid range)	m-Cresolsulfonephthalein	1.2–2.8	1.51	533, ——	R–Y
Thymol blue (acid range)	Thymolsulfonephthalein	1.2–2.8	1.65	544, 430	R–Y
Tropeolin OO	Diphenylamino-p-benzene sodium sulfonate	1.3–3.2	2.0	527, ——	R–Y
2,6-Dinitrophenol	2,6-Dinitrophenol	2.4–4.0	3.69		C–Y
2,4-Dinitrophenol	2,4-Dinitrophenol	2.5–4.3	3.90		C–Y
Methyl yellow	Dimethylaminoazobenzene	2.9–4.0	3.3	508, ——	R–Y
Methyl orange	Dimethylaminoazobenzene sodium sulfonate	3.1–4.4	3.40	522, 464	R–O
Bromophenol blue	Tetrabromophenolsulfone-phthalein	3.0–4.6	3.85	436, 592	Y–BV

TABLE 8.23 Indicators for Aqueous Acid-Base Titrations (*Continued*)

Indicator	Chemical name	pH range	pK_a	λ_{max}, nm	Color change
Bromocresol green	Tetrabromo-*m*-cresol-sulfonepthalein	4.0–5.6	4.68	444, 617	Y–B
Methyl red	*o*-Carboxybenzeneazo-dimethylaniline	4.4–6.2	4.95	530, 427	R–Y
Chlorophenol red	Dichlorophenolsulfone-phthalein	5.4–6.8	6.0	——, 573	Y–R
Bromocresol purple	Dibromo-*o*-cresolsulfone-phthalein	5.2–6.8	6.3	433, 591	Y–P
Bromophenol red	Dibromophenolsulfone-phthalein	5.2–6.8		——, 574	Y–R
p-Nitrophenol	*p*-Nitrophenol	5.3–7.6	7.15	320, 405	C–Y
Bromothymol blue	Dibromothymolsulfone-phthalein	6.2–7.6	7.1	433, 617	Y–B
Neutral red	Aminodimethylaminotolu-phenazonium chloride	6.8–8.0	7.4		R–Y
Phenol red	Phenolsulfonephthalein	6.4–8.0	7.9	433, 558	Y–R
m-Nitrophenol	*m*-Nitrophenol	6.4–8.8	8.3	——, 570	C–Y
Cresol red	*o*-Cresolsulfonephthalein	7.2–8.8	8.2	434, 572	Y–R
m-Cresol purple	*m*-Cresolsulfonephthalein	7.6–9.2	8.32	——, 580	Y–P
Thymol blue	Thymolsulfonephthalein	8.0–9.6	8.9	430, 596	Y–B
Phenolphthalein	Phenolphthalein	8.0–10.0	9.4	——, 553	C–R
α-Naphtholbenzein	α-Naphtholbenzein	9.0–11.0			Y–B
Thymolphthalein	Thymolphthalein	9.4–10.6	10.0	——, 598	C–B
Alizarin Yellow R	5-(*p*-Nitrophenylazo)-salicyclic acid, Na salt	10.0–12.0	11.16		Y–V
Tropeolin O	*p*-Sulfobenzenazo-resorcinol	11.0–13.0			Y–O-Br
Nitramine	2,4,6-Trinitrophenyl-methylnitroamine	10.8–13.0			C–O-Br

TABLE 8.24 Mixed Indicators

Mixed indicators give sharp color changes and are especially useful in titrating to a given titration exponent (p*I*).

The information given in this table is from the two-volume work *Volumetric Analysis* by Kolthoff and Stenger, published by Interscience Publishers, Inc., New York, 1942 and 1947, and reproduced with their permission.

Composition of Indicator Solution		pI	Color		Notes
			Acid	Alkaline	
1 part 0.1% methyl yellow in alc. 1 part 0.1% methylene blue in alc.	*	3.25	Blue-violet	Green	Still green at pH 3.4, blue-violet at 3.2†
1 part 0.14% xylene cyanol FF in alc. 1 part 0.1% methyl orange in aq.	*	3.8	Violet	Green	Color is gray at pH 3.8
1 part 0.1% methyl orange in aq. 1 part 0.25% indigo carmine in aq.	*	4.1	Violet	Green	Good indicator, especially in artificial light
1 part 0.1% methyl orange in aq. 1 part 0.1% aniline blue in aq.		4.3	Violet	Green	
1 part 0.1% bromcresol green sodium salt in aq. 1 part 0.02% methyl orange in aq.		4.3	Orange	Blue-green	Yellow at pH 3.5, greenish yellow at 4.0, weakly green at 4.3
3 parts 0.1% bromcresol green in alc. 1 part 0.2% methyl red in alc.		5.1	Wine-red	Green	Very sharp color change†
1 part 0.2% methyl red in alc. 1 part 0.1% methylene blue in alc.	*	5.4	Red-violet	Green	Color is red-violet at pH 5.2, a dirty blue at 5.4, and a dirty green at 5.6
1 part 0.1% chlorphenol red sodium salt in aq. 1 part 0.1% aniline blue in water		5.8	Green	Violet	Pale violet at pH 5.8

*Keep in a dark bottle.   † Excellent indicator.

TABLE 8.24 Mixed Indicators (*Continued*)

Composition of Indicator Solution	p*I*	Color Acid	Color Alkaline	Notes
1 part 0.1% bromcresol green sodium salt in aq. 1 part 0.1% chlorphenol red sodium salt in aq.	6.1	Yellow-green	Blue-violet	Blue-green at pH 5.4, blue at 5.8, blue with a touch of violet at 6.0, blue-violet at 6.2
1 part 0.1% bromcresol purple sodium salt in aq. 1 part 0.1% bromthymol blue sodium salt in aq.	6.7	Yellow	Violet-blue	Yellow-violet at pH 6.2, violet at 6.6, blue-violet at 6.8
2 parts 0.1% bromthymol blue sodium salt in aq. 1 part 0.1% azolitmin in aq.	6.9	Violet	Blue	
1 part 0.1% neutral red in alc. 1 part 0.1% methylene blue in alc. *	7.0	Violet-blue	Green	Violet blue at pH 7.0†
1 part 0.1% neutral red in alc. 1 part 0.1% bromthymol blue in alc.	7.2	Rose	Green	Dirty green at pH 7.4, pale rose at 7.2, clear rose at 7.0
2 parts 0.1% cyanine in 50% alc. 1 part 0.1% phenol red in 50% alc.	7.3	Yellow	Violet	Orange at pH 7.2, beautiful violet at 7.4, color fades on standing
1 part 0.1% bromthymol blue sodium salt in aq. 1 part 0.1% phenol red sodium salt in aq.	7.5	Yellow	Violet	Dirty green at pH 7.2, pale violet at 7.4, strong violet at 7.6†
1 part 0.1% cresol red sodium salt in aq. 3 parts 0.1% thymol blue sodium salt in aq.	8.3	Yellow	Violet	Rose at pH 8.2, distinctly violet at 8.4†
2 parts 0.1% α-naphtholphthalein in alc. 1 part 0.1% cresol red in alc.	8.3	Pale rose	Violet	Pale violet at pH 8.2, strong violet at 8.4

*Keep in a dark bottle.   † Excellent indicator.

TABLE 8.24 Mixed Indicators (*Continued*)

Composition of Indicator Solution	p*I*	Color		Notes
		Acid	Alkaline	
1 part 0.1% α–naphtholphthalein in alc. 3 parts 0.1% phenolphthalein in alc.	8.9	Pale rose	Violet	Pale green at pH 8.6, violet at 9.0
1 part 0.1% phenolphthalein in alc. * 2 parts 0.1% methyl green in alc.	8.9	Green	Violet	Pale blue at pH 8.8, violet at 9.0
1 part 0.1% thymol blue in 50% alc. 3 parts 0.1% phenolphthalein in 50% alc.	9.0	Yellow	Violet	From yellow thru green to violet†
1 part 0.1% phenolphthalein in alc. 1 part 0.1% thymolphthalein in alc.	9.9	Colorless	Violet	Rose at pH 9.6, violet at 10; sharp color change
1 part 0.1% phenolphthalein in alc. 2 parts 0.2% Nile blue in alc.	10.0	Blue	Red	Violet at pH 10†
2 parts 0.1% thymolphthalein in alc. 1 part 0.1% alizarin yellow in alc.	10.2	Yellow	Violet	Sharp color change
2 parts 0.2% Nile blue in aq. 1 part 0.1% alizarin yellow in alc.	10.8	Green	Red-brown	

*Keep in a dark bottle.   † Excellent indicator.

8.119

TABLE 8.25 Fluoresecent Indicators

Name	pH Range	Color Change Acid to Base	Indicator Solution
Benzoflavine	−0.3 to 1.7	Yellow to green	1
3,6-Dihydroxyphthalimide	0 to 2.4	Blue to green	1
	6.0 to 8.0	Green to yellow/green	
Eosin (tetrabromofluor-escein)	0 to 3.0	Non-fl to green	4, 1%
4-Ethoxyacridone	1.2 to 3.2	Green to blue	1
3,6-Tetramethyldiamino-xanthone	1.2 to 3.4	Green to blue	1
Esculin	1.5 to 2.0	Weak blue to strong blue	
Anthranilic acid	1.5 to 3.0	Non-fl to light blue	2 (50% ethanol)
	4.5 to 6.0	Light blue to dark blue	
	12.5 to 14	Dark blue to non-fl	
3-Amino-1-naphthoic acid	1.5 to 3.0	Non-fl to green	2 (as sulfate
	4.0 to 6.0	Green to blue	in 50% ethanol)
	11.6 to 13.0	Blue to non-fl	
1-Naphthylamino-6-sulfonamide	1.9 to 3.9	Non-fl to green	3
(also the 1-, 7-)	9.6 to 13.0	Green to non-fl	
2-Naphthylamino-6-sulfonamide (also	1.9 to 3.9	Non-fl to dark blue	3
the 2-, 8-)	9.6 to 13.0	Dark blue to non-fl	
1-Naphthylamino-5-sulfonamide	2.0 to 4.0	Non-fl to yellow/orange	3
	9.5 to 13.0	Yellow/orange to non-fl	
1-Naphthoic acid	2.5 to 3.5	Non-fl to blue	4
Salicylic acid	2.5 to 4.0	Non-fl to dark blue	4 (0.5%)
Phloxin BA extra (tetrachloro-tetrabromo-fluorescein)	2.5 to 4.0	Non-fl to dark blue	2
Erythrosin B (tetraiodo-fluorescein)	2.5 to 4.0	Non-fl to light green	4 (0.2%)
2-Naphthylamine	2.8 to 4.4	Non-fl to violet	1
Magdala red	3.0 to 4.0	Non-fl to purple	
p-Aminophenylbenzene-sulfonamide	3.0 to 4.0	Non-fl to light blue	3
2-Hydroxy-3-naphthoic acid	3.0 to 6.8	Blue to green	4 (0.1%)
Chromotropic acid	3.1 to 4.4	Non-fl to light blue	4 (5%)
1-Naphthionic acid	3 to 4	Non-fl to blue	4
	10 to 12	Blue to yellow-green	
1-Naphthylamine	3.4 to 4.8	Non-fl to blue	1
5-Aminosalicylic acid	3.1 to 4.4	Non-fl to light green	1 (0.2% fresh)
Quinine	3.0 to 5.0	Blue to weak violet	1 (0.1%)
	9.5 to 10.0	Weak violet to non-fl	
o-Methoxybenzaldehyde	3.1 to 4.4	Non-fl to green	4 (0.2%)
o-Phenylenediamine	3.1 to 4.4	Green to non-fl	5
p-Phenylenediamine	3.1 to 4.4	Non-fl to orange/yellow	5
Morin (2',4',3,5,7-penta-hydroxyflavone)	3.1 to 4.4	Non-fl to green	6 (0.2%)
	8 to 9.8	Green to yellow/green	
Thioflavine S	3.1 to 4.4	Dark blue to light blue	6 (0.2%)
Fluorescein	4.0 to 4.5	Pink/green to green	4 (1%)
Dichlorofluorescein	4.0 to 6.6	Blue green to green	1

Indicator solutions: 1, 1% solution in ethanol; 2, 0.1% solution in ethanol; 3, 0.05% solution in 90% ethanol; 4, sodium or potassium salt in distilled water; 5, 0.2% solution in 70% ethanol; 6, distilled water.

Reference: G. F. Kirkbright, "Fluorescent Indicators," Chap. 9 in *Indicators,* E. Bishop (ed.), Pergamon Press, Oxford, 1972.

TABLE 8.25 Fluorescent Indicators (*Continued*)

Name	pH Range	Color Change Acid to Base	Indicator Solution
β-Methylesculetin	4.0 to 6.2	Non-fl to blue	1
	9.0 to 10.0	Blue to light green	
Quininic acid	4.0 to 5.0	Yellow to blue	6 (satd)
β-Naphthoquinoline	4.4 to 6.3	Blue to non-fl	3
Resorufin (7-oxyphen-oxazone)	4.4 to 6.4	Yellow to orange	
Acridine	5.2 to 6.6	Green to violet	2
3,6-Dihydroxyxanthone	5.4 to 7.6	Non-fl to blue/violet	1
5,7-Dihydroxy-4-methyl-coumarin	5.5 to 5.8	Light blue to dark blue	
3,6-Dihydroxyphthalic acid dinitrile	5.8 to 8.2	Blue to green	1
1,4-Dihydroxybenzene-disulfonic acid	6 to 7	Non-fl to light blue	4 (0.1%)
Luminol	6 to 7	Non-fl to blue	
2-Naphthol-6-sulfonic acid	5-7 to 8-9	Non-fl to blue	4
Quinoline	6.2 to 7.2	Blue to non-fl	6 (satd)
1-Naphthol-5-sulfonic acid	6.5 to 7.5	Non-fl to green	6 (satd)
Umbelliferone	6.5 to 8.0	Non-fl to blue	
Magnesium-8-hydroxy-quinolinate	6.5 to 7.5	Non-fl to yellow	6 (0.1% in 0.01 M HCl)
Orcinaurine	6.5 to 8.0	Non-fl to green	6 (0.03%)
Diazo brilliant yellow	6.5 to 7.5	Non-fl to blue	
Coumaric acid	7.2 to 9.0	Non-fl to green	1
β-Methylumbelliferone	>7.0	Non-fl to blue	2 (0.3%)
Harmine	7.2 to 8.9	Blue to yellow	
2-Naphthol-6,8-disulfonic acid	7.5 to 9.1	Blue to light blue	4
Salicylaldehyde semi-carbazone	7.6 to 8.0	Yellow to blue	2
1-Naphthol-2-sulfonic acid	8.0 to 9.0	Dark blue to light blue	4
Salicylaldehyde acetyl-hydrazone	8.3	Non-fl to green/blue	2
Salicylaldehyde thiosemi-carbazone	8.4	Non-fl to blue/green	2
1-Naphthol-4-sulfonic acid	8.2	Dark blue to light blue	4
Naphthol AS	8.2 to 10.3	Non-fl to yellow/green	4
2-Naphthol	8.5 to 9.5	Non-fl to blue	2
Acridine orange	8.4 to 10.4	Non-fl to yellow/green	1
Orcinsulfonephthalein	8.6 to 10.0	Non-fl to yellow	
2-Naphthol-3,6-disulfonic acid	9.0 to 9.5	Dark blue to light blue	4
Ethoxyphenyl-naphtho-stilbazonium chloride	9 to 11	Green to non-fl	1
o-Hydroxyphenylbenzothiazole	9.3	Non-fl to blue green	2
o-Hydroxyphenylbenzoxazole	9.3	Non-fl to blue/violet	2
o-Hydroxyphenylbenzimidazole	9.9	Non-fl to blue/violet	2
Coumarin	9.5 to 10.5	Non-fl to light green	
6,7-Dimetnoxyisoquinoline-1-carboxylic acid	9.5 to 11.0	Yellow to blue	0.1% in glycerine/ethanol/water in 2:2:18 ratio
1-Naphthylamino-4-sulfonamide	9.5 to 13.0	Dark blue to white/blue	3

TABLE 8.26 Selected List of Oxidation-Reduction Indicators

Name	Reduction Potential (30°C) in Volts at		Suitable pH Range	Color Change Upon Oxidation
	pH = 0	pH = 7		
Bis(5-bromo-1,10-phenanthroline) ruthenium(II) dinitrate	1.41*			Red to faint blue
Tris(5-nitro-1,10-phenanthroline) iron(II) sulfate	1.25*			Red to faint blue
Iron(II)-2,2',2''-tripyridine sulfate	1.25*			Pink to faint blue
Tris(4,7-diphenyl-1,10-phenanthroline) iron(II) disulfate	1.13 (4.6 M H_2SO_4)* 0.87(1.0 M H_2SO_4)*			Red to faint blue
o,m'-Diphenylaminedicarboxylic acid	1.12			Colorless to blue-violet
Setopaline	1.06 (*trans*)†			Yellow to orange
p-Nitrodiphenylamine	1.06			Colorless to violet
Tris(1,10-phenanthroline)-iron(II) sulfate	1.06 (1.00 M H_2SO_4)* 1.00 (3.0 M H_2SO_4)* 0.89 (6.0 M H_2SO_4)*			Red to faint blue
Setoglaucine O	1.01 (*trans*)†			Yellow-green to yellow-red
Xylene cyanole FF	1.00 (*trans*)†			Yellow-green to pink
Erioglaucine A	1.00 (*trans*)†			Green-yellow to bluish red
Eriogreen	0.99 (*trans*)†			Green-yellow to orange
Tris(2,2'-bipyridine)-iron(II) hydrochloride	0.97*			Red to faint blue
2-Carboxydiphenylamine [N-phenyl-anthranilic acid]	0.94			Colorless to pink
Benzidine dihydrochloride	0.92			Colorless to blue
o-Toluidine	0.87			Colorless to blue
Bis(1,10-phenanthroline)-osmium(II) perchlorate	0.859 (0.1 M H_2SO_4)			Green to pink
Diphenylamine-4-sulfonate (Na salt)	0.85			Colorless to violet
3,3'-Dimethoxybenzidine dihydrochloride [o-dianisidine]	0.85			Colorless to red
Ferrocyphen	0.81			Yellow to violet

Indicator				Color change
4'-Ethoxy-2,4-diaminoazobenzene	0.76			Red to pale yellow
N,N-Diphenylbenzidine	0.76			Colorless to violet
Diphenylamine	0.76			Colorless to violet
N,N-Dimethyl-p-phenylenediamine	0.76			Colorless to red
Variamine blue B hydrochloride	0.712‡	0.310	1.5–6.3	Colorless to blue
N-Phenyl-1,2,4-benzenetriamine	0.70			Colorless to red
Bindschedler's green	0.680‡	0.224	2–9.5	Colorless to blue
2,6-Dichloroindophenol (Na salt)	0.668‡	0.217	6.3–11.4	Colorless to blue
2,6-Dibromophenolindophenol	0.668‡	0.216	7.0–12.3	Colorless to blue
Brilliant cresyl blue [3-amino-9-dimethyl-amino-10-methylphenoxazine chloride]	0.583	0.047	0–11	Colorless to blue
Iron(II)-tetrapyridine chloride	0.59			Red to faint blue
Thionine [Lauth's violet]	0.563‡	0.064	1–13	Colorless to violet
Starch (soluble potato, I$_3$ present)	0.54			Colorless to blue
Gallocyanine (25°C)		0.021		Colorless to violet-blue
Methylene blue	0.532‡	0.011	1–13	Colorless to blue
Nile blue A [aminonaphthodiethylamino-phenoxazine sulfate]	0.406‡	−0.119	1.4–12.3	Colorless to blue
Indigo-5,5',7,7'-tetrasulfonic acid (Na salt)	0.365‡	−0.046	<9	Colorless to blue
Indigo-5,5',7-trisulfonic acid (Na salt)	0.332‡	−0.081	<9	Colorless to blue
Indigo-5,5'-disulfonic acid (Na salt)	0.291‡	−0.125	<9	Colorless to blue
Phenosafranine	0.280‡	−0.252	1–11	Colorless to violet-blue
Indigo-5-monosulfonic acid (Na salt)	0.262‡	−0.157	<9	Colorless to blue
Safranine T	0.24‡	−0.289	1–12	Colorless to violet-blue
Bis(dimethylglyoximato)-iron(II) chloride	0.155		6–10	Red to colorless
Induline scarlet	0.047‡	−0.299	3–8.6	Colorless to red
Neutral red		−0.323	2–11	Colorless to red-violet

* Transition point is at higher potential than the tabulated formal potential because the molar absorptivity of the reduced form is very much greater than that of the oxidized form.

† *Trans* = first noticeable color transition; often 60 mV less than $E°$

‡ Values of $E°$ are obtained by extrapolation from measurements in weakly acid or weakly alkaline systems.

8.6 ELECTRODE POTENTIALS

TABLE 8.27 Potentials of the Elements and Their Compounds at 25°C

Standard potentials are tabulated except when a solution composition is stated; the latter are formal potentials and the concentrations are in mol/liter.

Half-reaction	Standard or formal potential	Solution composition
Actinium		
$Ac^{3+} + 3e^- = Ac$	-2.13	
Aluminum		
$Al^{3+} + 3e^- = Al$	-1.676	
$AlF_6^{3-} + 3e^- = Al + 6F^-$	-2.07	
$Al(OH)_4^- + 3e^- = Al + 4OH^-$	-2.310	
Americium		
$AmO_2^{2+} + 4H^+ + 2e^- = Am^{4+} + 2H_2O$	1.20	
$AmO_2^{2+} + e^- = AmO_2^+$	1.59	
$AmO_2^+ + 4H^+ + e^- = Am^{4+} + 2H_2O$	0.82	
$AmO_2^+ + 4H^+ + 2e^- = Am^{3+} + 2H_2O$	1.72	
$Am^{4+} + e^- = Am^{3+}$	2.62	
$Am^{4+} + 4e^- = Am$	-0.90	
$Am^{3+} + 3e^- = Am$	-2.07	
Antimony		
$Sb(OH)_4^- + 2e^- = SbO_2^- + 2OH^- + 2H_2O$	-0.465	1 NaOH
$SbO_2^- + 2H_2O + 3e^- = Sb + 4OH^-$	0.639	1 NaOH
$Sb + 3H_2O + 3e^- = SbH_3 + 3OH^-$	-1.338	1 NaOH
$Sb_2O_5 + 6H^+ + 4e^- = 2SbO^+ + 3H_2O$	0.605	
$Sb_2O_5 + 4H^+ + 4e^- = Sb_2O_3 + 2H_2O$	0.699	
$Sb_2O_5 + 2H^+ + 2e^- = Sb_2O_4 + H_2O$	1.055	
$Sb_2O_4 + 2H^+ + 2e^- = Sb_2O_3 + H_2$	0.342	
$SbO^+ + 2H^+ + 3e^- = Sb + H_2O$	0.204	
$Sb + 3H^+ + 3e^- = SbH_3$	-0.510	
Arsenic		
$H_3AsO_4 + 2H^+ + 2e^- = HAsO_2 + 2H_2O$	0.560	
$HAsO_2 + 3H^+ + 3e^- = As + 2H_2O$	0.240	
$As + 3H^+ + 3e^- = AsH_3$	-0.225	
$AsO_4^{3-} + 2H^+ + 2e^- = AsO_2^- + 4OH^-$	-0.67	
$AsO_2^- + 2H_2O + 3e^- = As + 4OH^-$	-0.68	
$As + 3H_2O + 3e^- = AsH_3 + 3OH^-$	-1.37	
Astatine		
$HAtO_3 + 4H^+ + 4e^- = HAtO + 2H_2$	*ca.* 1.4	
$2HAtO + 2H^+ + 2e^- = At_2 + 2H_2O$	*ca.* 0.7	
$At_2 + 2e^- = 2At^-$	0.20	
Barium		
$BaO_2 + 4H^+ + 2e^- = Ba^{2+} + 2H_2O$	2.365	
$Ba^{2+} + 2e^- = Ba$	-2.92	

Source: A. J. Bard, R. Parsons, and J. Jordan (eds.), *Standard Potentials in Aqueous Solution* (prepared under the auspices of the International Union of Pure and Applied Chemistry), Marcel Dekker, New York, 1985; G. Charlot et al., *Selected Constants: Oxidation-Reduction Potentials of Inorganic Substances in Aqueous Solution*, Butterworths, London, 1971.

TABLE 8.27 Potentials of the Elements and Their Compounds at 25°C (*Continued*)

Half-reaction	Standard or formal potential	Solution composition
Berkelium		
$Bk^{4+} + 4e^- = Bk$	-1.05	
$Bk^{4+} + e^- = Bk^{3+}$	1.67	
$Bk^{3+} + 3e^- = Bk$	-2.01	
Beryllium		
$Be^{2+} + 2e^- = Be$	-1.99	
Bismuth		
$Bi_2O_4 \text{ (bismuthate)} + 4H^+ + 2e^- = 2BiO^+ + 2H_2O$	1.59	
$Bi^{3+} + 3e^- = Bi$	0.317	
$Bi + 3H^+ + 3e^- = BiH_3$	-0.97	
$BiCl_4^- + 3e^- = Bi + 4Cl^-$	0.199	
$BiBr_4^- + 3e^- = Bi + 4Br^-$	0.168	
$BiOCl + 2H^+ + 3e^- = Bi + H_2O + Cl^-$	0.170	
Boron		
$B(OH)_3 + 3H^+ + 3e^- = B + 3H_2O$	-0.890	
$BO_2^- + 6H_2O + 8e^- = BH_4^- + 8OH^-$	-1.241	
$B(OH)_4^- + 3e^- = B + 4OH^-$	-1.811	
Bromine		
$BrO_4^- + 2H^+ + 2e^- = BrO_3^- + H_2O$	1.853	
$BrO_3^- + 6H^+ + 6e^- = Br^- + 3H_2O$	1.478	
$BrO_3^- + 5H^+ + 4e^- = HBrO + 2H_2O$	1.444	
$2BrO_3^- + 12H^+ + 10e^- = Br_2 + 6H_2O$	1.5	
$2HBrO + 2H^+ + 2e^- = Br_2 + 2H_2O$	1.604	
$HBrO + H^+ + 2e^- = Br^- + H_2O$	1.341	
$BrO^- + H_2O + 2e^- = Br^- + 2OH^-$	0.76	1 NaOH
$Br_3^- + 2e^- = 3Br^-$	1.050	
$Br_2(aq) + 2e^- = 2Br^-$	1.087	
Cadmium		
$Cd^{2+} + 2e^- = Cd$	-0.403	
$Cd^{2+} + Hg + 2e^- = Cd(Hg)$	-0.352	
$CdCl_4^{2-} + 2e^- = Cd + 4Cl^-$	-0.453	
$Cd(CN)_4^{2-} + 2e^- = Cd + 4CN^-$	-0.943	
$Cd(NH_3)_4^{2+} + 2e^- = Cd + 4NH_3$	-0.622	
$Cd(OH)_4^{2-} + 2e^- = Cd + 4OH^-$	-0.670	
Calcium		
$CaO_2 + 4H^+ + 2e^- = Ca^{2+} + H_2O$	2.224	
$Ca^{2+} + 2e^- = Ca$	-2.84	
$Ca + 2H^+ + 2e^- = CaH_2$	0.776	
Californium		
$Cf^{3+} + 3e^- = Cf$	-1.93	
$Cf^{3+} + e^- = Cf^{2+}$	-1.6	
$Cf^{2+} + 2e^- = Cf$	-2.1	
Carbon		
$CO_2 + 2H^+ + 2e^- = CO + H_2O$	-0.106	
$CO_2 + 2H^+ + 2e^- = HCOOH$	-0.20	
$2CO_2 + 2H^+ + 2e^- = H_2C_2O_4$	-0.481	
$C_2O_4^{2-} + 2H^+ + 2e^- = 2HCOO^-$	0.145	
$HCOOH + 2H^+ + 2e^- = HCHO + H_2O$	0.034	
$C_2N_2 + 2H^+ + 2e^- = 2HCN$	0.373	

TABLE 8.27 Potentials of the Elements and Their Compounds at 25°C (*Continued*)

Half-reaction	Standard or formal potential	Solution composition
$HCNO + 2H^+ + 2e^- = CO + H_2O$	0.330	
$HCHO + 2H^+ + 2e^- = CH_3OH$	0.2323	
$CNO^- + H_2O + 2e^- = CN^- + 2OH^-$	−0.97	
Cerium		
$Ce(IV) + e^- = Ce(III)$	1.70	1 $HClO_4$
	1.61	1 HNO_3
	1.44	0.5 H_2SO_4
	1.28	1 HCl
$Ce^{3+} + 3e^- = Ce$	−2.34	
Cesium		
$Cs^+ + e^- = Cs$	−2.923	
$Cs^+ + Hg + e^- = Cs(Hg)$	−1.78	
Chlorine		
$ClO_4^- + 2H^+ + 2e^- = ClO_3^- + H_2O$	1.201	
$2ClO_4^- + 16H^+ + 14e^- = Cl_2 + 8H_2O$	1.392	
$ClO_4^- + 8H^+ + 8e^- = Cl^- + 4H_2O$	1.388	
$ClO_3^- + 2H^+ + e^- = ClO_2(g) + H_2O$	1.175	
$ClO_3^- + 3H^+ + 2e^- = HClO_2 + H_2O$	1.181	
$2ClO_3^- + 12H^+ + 10e^- = Cl_2 + 6H_2O$	1.468	
$ClO_3^- + 6H^+ + 6e^- = Cl^- + 3H_2O$	1.45	
$ClO_2(g) + H^+ + e^- = HClO_2$	1.188	
$HClO_2 + 2H^+ + 2e^- = HClO + H_2O$	1.64	
$HClO_2 + 3H^+ + 4e^- = Cl^- + 2H_2O$	1.584	
$2HClO_2 + 6H^+ + 6e^- = Cl_2(g) + 4H_2O$	1.659	
$2ClO^- + 2H_2O + 2e^- = Cl_2(g) + 4OH^-$	0.421	1 NaOH
$ClO^- + H_2O + 2e^- = Cl^- + 2OH^-$	0.890	1 NaOH
$Cl_3^- + 2e^- = 3Cl^-$	1.415	
$Cl_2(aq) + 2e^- = 2Cl^-$	1.396	
Chromium		
$Cr_2O_7^{2-} + 14H^+ + 6e^- = 2Cr^{3+} + 7H_2O$	1.36	
	1.15	0.1 H_2SO_4
	1.03	1 $HClO_4$
$CrO_4^{2-} + 4H_2O + 3e^- = Cr(OH)_4^- + 4OH^-$	−0.13	1 NaOH
$Cr^{3+} + e^- = Cr^{2+}$	−0.424	
$Cr^{3+} + 3e^- = Cr$	−0.74	
$Cr^{2+} + 2e^- = Cr$	0.90	
Cobalt		
$CoO_2 + 4H^+ + e^- = Co^{3+} + 2H_2O$	1.416	
$Co(H_2O)_6^{3+} + e^- = Co(H_2O)_6^{2+}$	1.92	
$Co(NH_3)_6^{3+} + e^- = Co(NH_3)_6^{2+}$	0.058	7 NH_3
$Co(OH)_3 + e^- = Co(OH)_2 + OH^-$	0.17	
$Co(en)_3^{3+} + e^- = Co(en)_3^{2+}$ [en = ethylenediamine]	−0.2	0.1 en
$Co(CN)_6^{3-} + e^- = Co(CN)_5^{2-} + CN^-$	−0.8	0.8 KOH
$Co^{2+} + 2e^- = Co$	−0.277	
$Co(NH_3)_6^{2+} + 2e^- = Co + 6NH_3$	−0.422	
$[Co(CO)_4]_2 + 2e^- = 2Co(CO)_4^-$	−0.40	
Copper		
$Cu^{2+} + 2e^- = Cu$	0.340	
$Cu^{2+} + e^- = Cu^+$	0.159	

TABLE 8.27 Potentials of the Elements and Their Compounds at 25°C (*Continued*)

Half-reaction	Standard or formal potential	Solution composition
$Cu^+ + e^- = Cu$	0.520	
$Cu^{2+} + Cl^- + e^- = CuCl$	0.559	
$Cu^{2+} + 2Br^- + e^- = CuBr_2^-$	0.52	1 KBr
$Cu^{2+} + I^- + e^- + CuI$	0.86	
$Cu^{2+} + 2CN^- + e^- = Cu(CN)_2^-$	1.12	
$Cu(NH_3)_4^{2+} + e^- = Cu(NH_3)_2^+ + 2NH_3$	0.10	1 NH$_3$
$Cu(en)_2^{2+} + e^- = Cu(en)^+ + en$	−0.35	
$Cu(CN)_2^- + e^- = Cu + 2CN^-$	−0.44	
$CuCl_3^{2-} + e^- = Cu + 3Cl^-$	0.178	1 HCl
$Cu(NH_3)_2^+ + e^- = Cu + 2NH_3$	−0.100	
Curium		
$Cm^{4+} + e^- = Cm^{3+}$	3.2	1 HClO$_4$
$Cm^{3+} + 3e^- = Cm$	−2.06	
Dysprosium		
$Dy^{3+} + 3e^- = Dy$	−2.29	
$Dy^{3+} + e^- = Dy^{2+}$	−2.5	
$Dy^{2+} + 2e^- = Dy$	−2.2	
Einsteinium		
$Es^{3+} + 3e^- = Es$	−2.0	
$Es^{3+} + e^- = Es^{2+}$	−1.5	
$Es^{2+} + 2e^- = Es$	−2.2	
Erbium		
$Er^{3+} + 3e^- = Er$	−2.32	
Europium		
$Eu^{3+} + 3e^- = Eu$	−1.99	
$Eu^{3+} + e^- = Eu^{2+}$	−0.35	
$Eu^{2+} + 2e^- = Eu$	−2.80	
Fermium		
$Fm^{3+} + 3e^- = Fm$	−1.96	
$Fm^{3+} + e^- = Fm^{2+}$	−1.15	
$Fm^{2+} + 2e^- = Fm$	−2.37	
Fluorine		
$F_2 + 2H^+ + 2e^- = 2HF$	3.053	
$F_2 + H^+ + 2e^- = HF_2^-$	2.979	
$F_2 + 2e^- = 2F^-$	2.87	
$OF_2 + 3H^+ + 4e^- = HF_2^- + H_2O$	2.209	
Francium		
$Fr^+ + e^- = Fr$	*ca.* −2.9	
Gadolinium		
$Gd^{3+} + 3e^- = Gd$	−2.28	
Gallium		
$Ga^{3+} + 3e^- = Ga$	−0.529	
$Ga^{3+} + e^- = Ga^{2+}$	−0.65	
$Ga^{2+} + 2e^- = Ga$	−0.45	
Germanium		
$GeO_2(tetr) + 2H^+ + 2e^- = GeO(yellow) + H_2O$	−0.255	
$GeO_2(tetr) + 4H^+ + 2e^- = Ge^{2+} + 2H_2O$	−0.210	

TABLE 8.27 Potentials of the Elements and Their Compounds at 25°C (*Continued*)

Half-reaction	Standard or formal potential	Solution composition
$GeO_2(hex) + 4H^+ + 2e^- = Ge^{2+} + 2H_2O$	-0.132	
$H_2GeO_3 + 4H^+ + 4e^- = Ge + 3H_2O$	0.012	
$Ge^{4+} + 2e^- = Ge^{2+}$	0.0	
$Ge^{2+} + 2e^- = Ge$	0.247	
$GeO + 2H^+ + 2e^- = Ge + H_2O$	-0.255	
$Ge + 4H^+ + 4e^- = GeH_4$	-0.29	
Gold		
$Au^{3+} + 3e^- = Au$	1.52	
$Au^{3+} + 2e^- = Au^+$	1.36	
$Au^+ + e^- = Au$	1.83	
$AuCl_4^- + 2e^- = AuCl_2^- + 2Cl^-$	0.926	
$AuBr_4^- + 2e^- = AuBr_2^- + 2Br^-$	0.802	
$Au(SCN)_4^- + 2e^- = Au(SCN)_2^- + 2SCN^-$	0.623	
$AuBr_4^- + 3e^- = Au + 4Br^-$	0.854	
$AuCl_4^- + 3e^- = Au + 4Cl^-$	1.002	
$Au(SCN)_4^- + 3e^- = Au + 4SCN^-$	0.662	
$Au(OH)_3 + 3H^+ + 3e^- = Au + 3H_2O$	1.45	
$AuBr_2^- + e^- = Au + 2Br^-$	0.960	
$AuCl_2^- + e^- = Au + 2Cl^-$	1.15	
$AuI_2^- + e^- = Au + 2I^-$	0.576	
$Au(CN)_2^- + e^- = Au + 2CN^-$	-0.596	
$Au(SCN)_2 + e^- = Au + 2SCN^-$	0.69	
Hafnium		
$Hf^{4+} + 4e^- = Hf$	-1.70	
$HfO_2 + 4H^+ + 4e^- = Hf + 2H_2O$	-1.57	
Holmium		
$Ho^{3+} + 3e^- = Ho$	-2.23	
Hydrogen		
$2H^+ + 2e^- = H_2$	0.0000	
$2D^+ + 2e^- = D_2$	0.029	
$2H_2O + 2e^- = H_2 + 2OH^-$	-0.828	
Indium		
$In^{3+} + 3e^- = In$	-0.338	
$In^{3+} + 2e^- = In^+$	-0.444	
$In^+ + e^- = In$	-0.126	
Iodine		
$H_5IO_6 + H^+ + 2e^- = IO_3^- + 3H_2O$	1.603	
$IO_3^- + 5H^+ + 4e^- = HIO + 2H_2O$	1.14	
$HIO_3 + 5H^+ + 2Cl^- + 4e^- = ICl_2^- + 3H_2O$	1.214	
$2IO_3^- + 12H^+ + 10e^- = I_2(c) + 3H_2O$	1.195	
$IO_3^- + 3H_2O + 6e^- = I^- + 6OH^-$	0.257	
$2IBr_2^- + 2e^- = I_2Br^- + 3Br^-$	0.821	
$2IBr_2^- + 2e^- = I_2(c) + 4Br^-$	0.874	
$2IBr + 2e^- = I_2Br^- + Br^-$	0.973	
$2IBr + 2e^- = I_2 + 2Br^-$	1.02	
$2ICl + 2e^- = I_2(c) + 2Cl^-$	1.20	
$2ICl_2^- + 2e^- = I_2(c) + 4Cl^-$	1.07	
$2ICN + 2H^+ + 2e^- = I_2(c) + 2HCN$	0.695	
$2ICN + 2H^+ + 2e^- = I_2(aq) + 2HCN$	0.609	

TABLE 8.27 Potentials of the Elements and Their Compounds at 25°C (*Continued*)

Half-reaction	Standard or formal potential	Solution composition
$2HIO + 2H^+ + 2e^- = I_2 + 2H_2O$	1.45	
$HIO + H^+ + 2e^- = I^- + H_2O$	0.985	
$I_3^- + 2e^- = 3I^-$	0.536	
$I_2(aq) + 2e^- = 2I^-$	0.621	
$I_2(c) + 2e^- = 2I^-$	0.5355	
Iridium		
$IrBr_6^{2-} + e^- = IrBr_6^{3-}$	0.805	
$IrCl_6^{2-} + e^- = IrCl_6^{3-}$	0.867	
$IrI_6^{2-} + e^- = IrI_6^{3-}$	0.49	
$IrO_2 + 4H^+ + e^- = Ir^{3+} + 2H_2O$	0.223	
$IrO_2 + 4H^+ + 4e^- = Ir + 2H_2O$	0.935	1 H_2SO_4
$Ir^{3+} + 3e^- = Ir$	1.156	
$IrCl_6^{2-} + 4e^- = Ir + 6Cl^-$	0.835	
$IrCl_6^{3-} + 3e^- = Ir + 6Cl^-$	0.77	
Iron		
$FeO_4^{2-} + 8H^+ + 3e^- = Fe^{3+} + 4H_2O$	2.2	
$FeO_4^{2-} + 2H_2O + 3e^- = FeO_2^- + 4OH^-$	0.55	10 NaOH
$Fe^{3+} + e^- = Fe^{2+}$	0.771	
	0.70	1 HCl
	0.67	0.5 H_2SO_4
	0.44	0.3 H_3PO_4
$Fe(CN)_6^{3-} + e^- = Fe(CN)_6^{4-}$	0.361	
	0.71	1 HCl
$Fe(EDTA)^- + e^- = Fe(EDTA)^{2-}$	0.12	0.1 EDTA, pH 4–6
$Fe(OH)_4^- + e^- = Fe(OH)_4^{2-}$	−0.73	1 NaOH
$Fe^{2+} + 2e^- = Fe$	−0.44	
$[Fe(CO)_4]_3 + 6e^- = 3Fe(CO)_4^{2-}$	−0.70	
Lanthanum		
$La^{3+} + 3e^- = La$	−2.38	
Lawrencium		
$Lr^{3+} + 3e^- = Lr$	−2.0	
Lead		
$Pb^{4+} + 2e^- = Pb^{2+}$	1.65	
$PbO_2(alpha) + SO_4^{2-} + 4H^+ + 2e^- = PbSO_4 + 2H_2O$	1.690	
$PbO_2 + 4H^+ + 2e^- = Pb^{2+} + 2H_2O$	1.46	
$PbO_2 + 2H^+ + 2e^- = PbO + H_2O$	0.28	
$PbO^{2-} + H_2O + 2e^- = HPbO_2^- + 3OH^-$	0.3	2 NaOH
$Pb^{2+} + 2e^- = Pb$	−0.126	
$HPbO_2^- + H_2O + 2e^- = Pb + 3OH^-$	−054	
$PbHPO_4 + 2e^- = Pb + HPO_4^{2-}$	−0.465	
$PbSO_4 + 2e^- = Pb + SO_4^{2-}$	−0.356	
$PbF_2 + 2e^- = Pb + 2F^-$	−0.344	
$PbCl_2 + 2e^- = Pb + 2Cl^-$	−0.268	
$PbBr_2 + 2e^- = Pb + 2Br^-$	−0.280	
$PbI_2 + 2e^- = Pb + 2I^-$	−0.365	
$Pb + 2H^+ + 2e^- = PbH_2$	−1.507	
Lithium		
$Li^+ + e^- = Li$	−3.040	
$Li^+ + Hg + e^- = Li(Hg)$	−2.00	

TABLE 8.27 Potentials of the Elements and Their Compounds at 25°C (*Continued*)

Half-reaction	Standard or formal potential	Solution composition
Lutetium		
$Lu^{3+} + 3e^- = Lu$	-2.30	
Magnesium		
$Mg^{2+} + 2e^- = Mg$	-2.356	
$Mg(OH)_2 + 2e^- = Mg + 2OH^-$	-2.687	
Manganese		
$MnO_4^- + e^- = MnO_4^{2-}$	0.56	
$MnO_4^- + 4H^+ + 3e^- = MnO_2(beta) + 2H_2O$	1.70	
$MnO_4^- + 2H_2O + 3e^- = MnO_2 + 4OH^-$	0.60	
$MnO_4^- + 8H^+ + 5e^- = Mn^{2+} + 4H_2O$	1.51	
$MnO_4^{2-} + e^- = MnO_4^{3-}$	0.27	
$MnO_4^{2-} + 2H_2O + 2e^- = MnO_2 + 4OH^-$	0.62	
$MnO_4^{3-} + 2H_2O + e^- = MnO_2 + 4OH^-$	0.96	
$MnO_2 + 4H^+ + e^- = Mn^{3+} + 2H_2O$	0.95	
$MnO_2(beta) + 4H^+ + 2e^- = Mn^{2+} + 2H_2O$	1.23	
$Mn^{3+} + e^- = Mn^{2+}$	1.5	
$Mn(H_2P_2O_7)_3^{3-} + 2H^+ + e^- = Mn(H_2P_2O_7)_2^{2-} + H_4P_2O_7$	1.15	$0.4\ H_2P_2O_7^{2-}$
$Mn(CN)_6^{3-} + e^- = Mn(CN)_6^{4-}$	-0.24	$1.5\ NaCN$
$Mn^{2+} + 2e^- = Mn$	-1.17	
Mendelevium		
$Md^{3+} + 3e^- = Md$	-1.7	
$Md^{3+} + e^- = Md^{2+}$	-0.15	
$Md^{2+} + 2e^- = Md$	-2.4	
Mercury		
$2Hg^{2+} + 2e^- = Hg_2^{2+}$	0.911	
$2HgCl_2 + 2e^- = Hg_2Cl_2 + 2Cl^-$	0.63	
$Hg^{2+} + 2e^- = Hg(lq)$	0.8535	
$HgO(c,red) + 2H^+ + 2e^- = Hg + H_2O$	0.926	
$Hg_2^{2+} + 2e^- = 2Hg$	0.7960	
$Hg_2F_2 + 2e^- = 2Hg + 2F^-$	0.656	
$Hg_2Cl_2 + 2e^- = 2Hg + 2Cl^-$	0.2682	
$Hg_2Br_2 + 2e^- = 2Hg + 2Br^-$	0.1392	
$Hg_2I_2 + 2e^- = 2Hg + 2I^-$	-0.0405	
$Hg_2SO_4 + 2e^- = 2Hg + SO_4^{2-}$	0.614	
Molybdenum		
$MoO_4^{2-} + 4H_2O + 6e^- = Mo + 8OH^-$	-0.913	
$H_2MoO_4 + 6H^+ + 6e^- = Mo + 4H_2O$	0.114	
$H_2MoO_4 + 2H^+ + 2e^- = MoO_2 + 2H_2O$	0.646	
$MoO_2 + 4H^+ + 4e^- = Mo + 2H_2O$	-0.152	
$H_2MoO_4 + 6H^+ + 3e^- = Mo^{3+} + 4H_2O$	0.428	
$Mo(CN)_8^{3-} + e^- = Mo(CN)_8^{4-}$	0.725	
$Mo^{3+} + 3e^- = Mo$	-0.2	
Neodynium		
$Nd^{3+} + 3e^- = Nd$	-2.32	
$Nd^{3+} + e^- = Nd^{2+}$	-2.6	
$Nd^{2+} + 2e^- = Nd$	-2.2	
Neptunium		
$NpO_3^+ + 2H^+ + e^- = NpO_2^{2+} + H_2O$	2.04	
$NpO_2^{2+} + e^- = NpO_2^+$	1.34	

TABLE 8.27   Potentials of the Elements and Their Compounds at 25°C (*Continued*)

Half-reaction	Standard or formal potential	Solution composition
$NpO_2^{2+} + 4H^+ + 2e^- = Np^{4+} + 2H_2O$	0.95	
$Np^{4+} + e^- = Np^{3+}$	0.18	
$Np^{4+} + 4e^- = Np$	-1.30	
$Np^{3+} + 3e^- = Np$	-1.79	
Nickel		
$NiO_4^{2-} + 4H^+ + 2e^- = NiO_2 + 2H_2O$	1.8	
$NiO_2 + 4H^+ + 2e^- = Ni^{2+} + 2H_2O$	1.593	
$NiO_2 + 2H_2O + 2e^- = Ni(OH)_2 + 2OH^-$	0.490	
$Ni(CN)_4^{2-} + e^- = Ni(CN)_3^{2-} + CN^-$	-0.401	
$Ni^{2+} + 2e^- = Ni$	-0.257	
$Ni(OH)_2 + 2e^- = Ni + 2OH^-$	-0.72	
$Ni(NH_3)_6^{2+} + 2e^- = Ni + 6NH_3$	-0.49	
Niobium		
$Nb_2O_5 + 10H^+ + 4e^- = 2Nb^{3+} + 5H_2O$	-0.1	
$Nb_2O_5 + 10H^+ + 10e^- = 2Nb + 5H_2O$	-0.65	
$Nb^{3+} + 3e^- = Nb$	-1.1	
Nitrogen		
$2NO_3^- + 4H^+ + 2e^- = N_2O_4 + 2H_2O$	0.803	
$NO_3^- + 3H^+ + 2e^- = HNO_2 + H_2O$	0.94	
$N_2O_4 + 2H^+ + 2e^- = 2HNO_2$	1.07	
$HNO_2 + H^+ + e^- = NO + H_2O$	0.996	
$2HNO_2 + 4H^+ + 4e^- = N_2O(g) + 3H_2O$	1.297	
$2HNO_2 + 4H^+ + 4e^- = H_2N_2O_2 + 2H_2O$	0.86	
$2NO + 2H^+ + 2e^- = H_2N_2O_2$	0.71	
$2NO + 2H^+ + 2e^- = N_2O + H_2O$	1.59	
$H_2N_2O_2 + 6H^+ + 4e^- = 2HONH_3^+$	0.496	
$N_2O + 2H^+ + 2e^- = N_2 + H_2O$	1.77	
$N_2O + 6H^+ + H_2O + 4e^- = 2HONH_3^+$	-0.05	
$N_2 + 2H_2O + 4H^+ + 2e^- = 2HONH_3^+$	-1.87	
$N_2 + 5H^+ + 4e^- = N_2H_5^+$	-0.23	
$HONH_3^+ + 2H^+ + 2e^- = NH_4^+ + H_2O$	1.35	
$2HONH_3^+ + H^+ + 2e^- = N_2H_5^+ + 2H_2O$	1.41	
$N_2H_5^+ + 3H^+ + 2e^- = 2NH_4^+$	1.275	
$3N_2 + 2H^+ + 2e^- = 2HN_3$	-3.40	
Nobelium		
$No^{3+} + 3e^- = No$	-1.2	
$No^{3+} + e^- = No^{2+}$	1.4	
$No^{2+} + 2e^- = No$	-2.5	
Osmium		
$OsO_4(aq) + 4H^+ + 4e^- = OsO_2 \cdot 2H_2O + 2H_2O$	0.964	
$OsO_4(c, yellow) + 8H^+ + 8e^- = Os + 4H_2O$	0.85	
$OsO_2 + 4H^+ + 4e^- = Os + 2H_2O$	0.687	
$OsCl_6^{2-} + e^- = OsCl_6^{3-}$	0.45	
$OsBr_6^{2-} + e^- = OsBr_6^{3-}$	0.35	
Oxygen		
$O_3 + 2H^+ + 2e^- = O_2 + H_2O$	2.075	
$O_3 + H_2O + 2e^- = O_2 + 2OH^-$	1.240	1 NaOH
$O_2 + 4H^+ + 4e^- = 2H_2O$	1.229	
$O_2 + 2H^+ + 2e^- = H_2O$	0.695	

TABLE 8.27   Potentials of the Elements and Their Compounds at 25°C (*Continued*)

Half-reaction	Standard or formal potential	Solution composition
$O_2 + H_2O + 2e^- = HO_2^- + OH^-$	-0.076	
$H_2O_2 + 2H^+ + 2e^- = 2H_2O$	1.763	
$HO_2^- + H_2O + 2e^- = 3OH^-$	0.867	1 NaOH
$O_2 + 2H_2O + 4e^- = 4OH^-$	0.401	
Palladium		
$PdO_3 + 2H^+ + 2e^- = PdO_2 + H_2O$	2.030	
$PdCl_6^{2-} + 2e^- = PdCl_4^{2-} + 2Cl^-$	1.470	
$PdBr_6^{2-} + 2e^- = PdBr_4^{2-} + 2Br^-$	0.99	
$PdI_6^{2-} + 2e^- = PdI_4^{2-} + 2I^-$	0.48	
$Pd^{2+} + 2e^- = Pd$	0.915	
$PdCl_4^{2-} + 2e^- = Pd + 4Cl^-$	0.62	1 HCl
$PdBr_4^{2-} + 2e^- = Pd + 4Br^-$	0.49	
$Pd(NH_3)_4^{2+} + 2e^- = Pd + 4NH_3$	0.0	1 NH$_3$
$Pd(CN)_4^{2-} + 2e^- = Pd + 4CN^-$	-1.35	1 KCN
Phosphorus		
$H_3PO_4 + 2H^+ + 2e^- = H_3PO_3 + H_2O$	-0.276	
$2H_3PO_4 + 2H^+ + 2e^- = H_4P_2O_6 + 2H_2O$	-0.933	
$H_4P_2O_6 + 2H^+ + 2e^- = 2H_3PO_3$	0.380	
$H_3PO_3 + 2H^+ + 2e^- = HPH_2O_2 + H_2O$	-0.499	
$HPH_2O_2 + H^+ + e^- = P + 2H_2O$	-0.365	
$H_3PO_3 + 3H^+ + 3e^- = P + 3H_2O$	-0.502	
$2P(white) + 4H^+ + 4e^- = P_2H_4$	-0.100	
$P_2H_4 + 2H^+ + 2e^- = 2PH_3$	-0.006	
$P(white) + 3H^+ + 3e^- = PH_3$	-0.063	
Platinum		
$PtO_3 + 2H^+ + 2e^- = PtO_2 + H_2O$	2.0	
$PtO_2 + 2H^+ + 2e^- = PtO + H_2O$	1.045	
$PtCl_6^{2-} + 2e^- = PtCl_4^{2-} + 2Cl^-$	0.726	
$PtBr_6^{2-} + 2e^- = PtBr_4^{2-} + 2Br^-$	0.613	1 KBr
$PtI_6^{2-} + 2e^- = PtI_4^{2-} + 2I^-$	0.321	1 KI
$Pt^{2+} + 2e^- = Pt$	1.188	
$PtCl_4^{2-} + 2e^- = Pt + 4Cl^-$	0.758	
$PtBr_4^{2-} + 2e^- = Pt + 4Br^-$	0.698	
Plutonium		
$PuO_2^{2+} + e^- = PuO_2^+$	1.02	
$PuO_2^{2+} + 4H^+ + 2e^- = Pu^{4+} + 2H_2O$	1.04	
$Pu^{4+} + e^- = Pu^{3+}$	1.01	
	0.80	1 H$_3$PO$_4$
	0.50	1 HF
$Pu^{4+} + 4e^- = Pu$	-1.25	
$Pu^{3+} + 3e^- = Pu$	-2.00	
Polonium		
$PoO_2 + 4H^+ + 2e^- = Po^{2+} + 2H_2O$	1.1	
$Po^{4+} + 4e^- = Po$	0.73	
$Po^{2+} + 2e^- = Po$	0.37	
$Po + 2H^+ + 2e^- = H_2Po$	*ca.* -1.0	
Potassium		
$K^+ + e^- = K$	-2.924	
$K^+ + Hg + e^- = K(Hg)$	*ca.* -1.9	

TABLE 8.27 Potentials of the Elements and Their Compounds at 25°C (*Continued*)

Half-reaction	Standard or formal potential	Solution composition
Praseodymium		
$Pr^{4+} + e^- = Pr^{3+}$	3.2	
$Pr^{3+} + e^- = Pr$	-2.35	
Promethium		
$Pm^{3+} + 3e^- = Pm$	-2.42	
Protoactinium		
$PaOOH^{2+} + 3H^+ + e^- = Pa^{4+} + 2H_2O$	-0.10	
$PaOOH^{2+} + 3H^+ + 5e^- = Pa + 2H_2O$	-1.19	
$Pa^{4+} + 4e^- = Pa$	-1.46	
Radium		
$Ra^{2+} + 2e^- = Ra$	-2.916	
Rhenium		
$ReO_4^- + 2H^+ + e^- = ReO_3 + H_2O$	0.768	
$ReO_4^- + 4H^+ + 3e^- = ReO_2 + 2H_2O$	0.51	
$ReO_4^- + 2H_2O + 3e^- = ReO_2 + 4OH^-$	-0.594	
$ReO_4^- + 6Cl^- + 8H^+ + 3e^- = ReCl_6^{2-} + 4H_2O$	0.12	
$2ReO_4^- + 10H^+ + 8e^- = Re_2O_3 + 5H_2O$	-0.808	
$ReO_3 + 2H^+ + 2e^- = ReO_2 + H_2O$	0.63	
$ReO_2 + 4H^+ + 4e^- = Re + 2H_2O$	0.22	
$ReCl_6^{2-} + 4e^- = Re + 6Cl^-$	0.51	
$Re + e^- = Re^-$	-0.10	
Rhodium		
$RhO_2 + 4H^+ + e^- = Rh^{3+} + 2H_2O$	1.881	
$Rh^{3+} + 3e^- = Rh$	0.76	
$RhCl_6^{3-} + 3e^- = Rh + 6Cl^-$	0.5	
Rubidium		
$Rb^+ + e^- = Rb$	-2.924	
$Rb^+ + Hg + e^- = Rb(Hg)$	-1.81	
Ruthenium		
$RuO_4 + e^- = RuO_4^-$	0.89	
$RuO_4 + 4H^+ + 4e^- = RuO_2 + 2H_2O$	1.4	
$RuO_4 + 8H^+ + 8e^- = Ru + 4H_2O$	1.04	
$RuO_4^- + e^- = RuO_4^{2-}$	0.593	
$RuO_4^{2-} + 4H^+ + 2e^- = RuO_2 + 2H_2O$	2.0	
$RuO_2 + 4H^+ + 4e^- = Ru + 2H_2O$	0.68	
$Ru(H_2O)_6^{3+} + e^- = Ru(H_2O)_6^{2+}$	0.249	
$Ru(NH_3)_6^{3+} + e^- = Ru(NH_3)_6^{2+}$	0.10	
$Ru(CN)_6^{3-} + e^- = Ru(CN)_6^{4-}$	0.86	
$Ru^{3+} + e^- = Ru^{2+}$	0.249	
Samarium		
$Sm^{3+} + 3e^- = Sm$	-2.30	
$Sm^{3+} + e^- = Sm^{2+}$	-1.55	
$Sm^{2+} + 2e^- = Sm$	-2.67	
Scandium		
$Sc^{3+} + 3e^- = Sc$	-2.03	
Selenium		
$SeO_4^{2-} + 4H^+ + 2e^- = H_2SeO_3 + H_2O$	1.151	
$H_2SeO_3 + 4H^+ + 4e^- = Se + 3H_2O$	0.74	

TABLE 8.27   Potentials of the Elements and Their Compounds at 25°C (*Continued*)

Half-reaction	Standard or formal potential	Solution composition
$Se(c) + 2H^+ + 2e^- = H_2Se(aq)$	-0.115	
$Se + H^+ + 2e^- = HSe^-$	-0.227	
$Se + 2e^- = Se^{2-}$	-0.670	1 NaOH
Silicon		
$SiO_2(quartz) + 4H^+ + 4e^- = Si + 2H_2O$	-0.909	
$SiO_2 + 2H^+ + 2e^- = SiO + H_2O$	-0.967	
$SiO_2 + 8H^+ + 8e^- = SiH_4 + 2H_2O$	-0.516	
$SiF_6^{2-} + 4e^- = Si + 6F^-$	-1.37	
$SiO + 2H^+ + 2e^- = Si + H_2O$	-0.808	
$Si + 4H^+ + 4e^- = SiH_4(g)$	-0.143	
Silver		
$AgO^+ + 2H^+ + e^- = Ag^{2+} + H_2O$	1.360	
$Ag_2O_3 + 2H^+ + 2e^- = 2AgO + H_2O$	1.569	
$Ag_2O_3 + H_2O + 2e^- = 2AgO + 2OH^-$	0.739	1 NaOH
$Ag_2O_3 + 6H^+ + 4e^- = 2Ag^+ + 3H_2O$	1.670	
$Ag^{2+} + e^- = Ag^+$	1.980	
$AgO + 2H^+ + e^- = Ag^+ + H_2O$	1.772	
$Ag^+ + e^- = Ag$	0.7991	
$Ag_2SO_4 + 2e^- = 2Ag + SO_4^{2-}$	0.653	
$Ag_2C_2O_4 + 2e^- = 2Ag + C_2O_4^{2-}$	0.47	
$Ag_2CrO_4 + 2e^- = 2Ag + CrO_4^{2-}$	0.447	
$Ag(NH_3)_2^+ + e^- = Ag + 2NH_3$	0.373	
$AgCl + e^- = Ag + Cl^-$	0.2223	
$AgBr + e^- = Ag + Br^-$	0.071	
$AgCN + e^- = Ag + CN^-$	-0.017	
$AgI + e^- = Ag + I^-$	-0.152	
$Ag(CN) + e^- = Ag + 2CN^-$	-0.31	
$AgSCN + e^- = Ag + SCN^-$	0.09	
$Ag_2S + 2e^- = 2Ag + S^{2-}$	-0.71	
Sodium		
$Na^+ + e^- = Na$	-2.713	
$Na^+ + Hg + e^- = Na(Hg)$	-1.84	
Strontium		
$SrO_2 + 4H^+ + 2e^- = Sr^{2+}$	2.33	
$Sr^{2+} + 2e^- = Sr$	-2.89	
Sulfur		
$S_2O_8^{2-} + 2e^- = 2SO_4^{2-}$	1.96	
$S_2O_8^{2-} + 2H^+ + 2e^- = 2HSO_4^-$	2.08	
$2SO_4^{2-} + 4H^+ + 2e^- = S_2O_6^{2-} + 2H_2O$	-0.25	
$SO_4^{2-} + 4H^+ + 2e^- = SO_2(aq) + H_2O$	0.158	
$SO_4^{2-} + H_2O + 2e^- = SO_3^{2-} + 2OH^-$	-0.936	
$S_2O_6^{2-} + 4H^+ + 2e^- = 2H_2SO_3$	0.569	
$S_2O_6^{2-} + 2e^- = 2SO_3^{2-}$	0.037	
$2HSO_3^- + 2H^+ + 2e^- = S_2O_4^{2-} + 2H_2O$	0.099	
$2SO_3^{2-} + 2H_2O + 2e^- = S_2O_4^{2-} + 4OH^-$	-1.13	
$4H_2SO_3 + 4H^+ + 6e^- = S_4O_6^{2-} + 6H_2O$	0.507	
$4HSO_3^- + 8H^+ + 6e^- = S_4O_6^{2-} + 6H_2O$	0.577	
$2SO_2(aq) + 2H^+ + 4e^- = S_2O_3^{2-} + H_2O$	0.400	
$2SO_3^{2-} + 3H_2O + 4e^- = S_2O_3^{2-} + 6OH^-$	-0.576	1 NaOH

TABLE 8.27 Potentials of the Elements and Their Compounds at 25°C (*Continued*)

Half-reaction	Standard or formal potential	Solution composition
$SO_3^{2-} + 3H_2O + 4e^- = S + 6OH^-$	-0.59	1 NaOH
$S_4O_6^{2-} + 2e^- = 2S_2O_3^{2-}$	0.080	
$S_2O_3^{2-} + 6H^+ + 4e^- = 2S + 3H_2O$	0.5	
$SF_4(g) + 4e^- = S + 4F^-$	0.97	
$S_2Cl_2(g) + 2e^- = 2S + 2Cl^-$	1.19	
$S + H^+ + 2e^- = HS^-$	0.287	
$S + 2H^+ + 2e^- = H_2S(aq)$	0.144	
$S + 2H^+ + 2e^- = H_2S(g)$	0.174	
$S + 2e^- = S^{2-}$	-0.407	
Tantalum		
$Ta_2O_5 + 10H^+ + 10e^- = 2Ta + 5H_2O$	-0.81	
$TaF_7^{2-} + 5e^- = Ta + 7F^-$	-0.45	
Technetium		
$TcO_4^- + 4H^+ + 3e^- = TcO_2 + 2H_2O$	0.738	
$TcO_4^- + 2H^+ + e^- = TcO_3 + H_2O$	0.700	
$TcO_4^- + e^- = TcO_4^{2-}$	0.569	
$TcO_4^- + 8H^+ + 7e^- = Tc + 4H_2O$	0.472	
$TcO_4^{2-} + 4H^+ + 2e^- = TcO_2 + 2H_2O$	1.39	
$TcO_2 + 4H^+ + 4e^- = Tc + 2H_2O$	0.272	
$Tc + e^- = Tc^-$	*ca.* -0.5	
Tellurium		
$H_2TeO_4 + 6H^+ + 2e^- = Te^{4+} + 4H_2O$	0.929	
$H_2TeO_4 + 2H^+ + 2e^- = TeO_2(c) + 2H_2O$	1.02	
$TeO_4^{2-} + 2H^+ + 2e^- = TeO_3^{2-} + H_2O$	0.897	
$TeOOH^+ + 3H^+ + 4e^- = Te + 2H_2O$	0.559	
$H_2TeO_3 + 4H^+ + 4e^- = Te + 3H_2O$	0.589	
$TeO_3^{2-} + 6H^+ + 4e^- = Te + 3H_2O$	0.827	
$TeO_3^{2-} + 3H_2O + 4e^- = Te + 6OH^-$	-0.415	
$TeO_2(c) + 4H^+ + 4e^- = Te + 2H_2O$	0.521	
$Te + 2H^+ + 2e^- = H_2Te(aq)$	-0.740	
$Te + H^+ + 2e^- = HTe^-$	-0.817	
$Te^{2-} + 2H^+ + 2e^- = 2HTe^-$	-0.794	
Terbium		
$Tb^{3+} + 3e^- = Tb$	-2.31	
Thallium		
$Tl^{3+} + 2e^- = Tl^+$	1.25	1 HClO₄
	0.77	1 HCl
$Tl^{3+} + 3e^- = Tl$	0.72	
$Tl^+ + e^- = Tl$	-0.336	
$TlCl + e^- = Tl + Cl^-$	-0.557	
$TlBr + e^- = Tl + Br^-$	-0.658	
$TlI + e^- = Tl + I^-$	-0.752	
Thorium		
$Th^{4+} + 4e^- = Th$	-1.83	
Thullium		
$Tm^{3+} + 3e^- = Tm$	-2.32	
Tin		
$Sn^{4+} + 2e^- = Sn^{2+}$	0.154	
$SnCl_6^{2-} + 2e^- = SnCl_4^{2-} + 2Cl^-$	0.14	

TABLE 8.27 Potentials of the Elements and Their Compounds at 25°C (*Continued*)

Half-reaction	Standard or formal potential	Solution composition
$SnO_3^{2-} + 6H^+ + 2e^- = Sn^{2+} + 3H_2O$	0.849	
$SnF_6^{2-} + 4e^- = Sn + 6F^-$	−0.200	
$Sn^{2+} + 2e^- = Sn$	−0.1375	
$SnCl_4^{2-} + 2e^- = Sn + 4Cl^-$	−0.19	1 HCl
$HSnO_2^- + H_2O + 2e^- = Sn + 3OH^-$	−0.91	
$Sn + 4H^+ + 4e^- = SnH_4$	−1.07	
Titanium		
$TiO^{2+} + 2H^+ + e^- = Ti^{3+} + H_2O$	−0.10	
$TiO^{2+} + 2H^+ + 4e^- = Ti + H_2O$	−0.86	
$Ti^{3+} + e^- = Ti^{2+}$	−0.37	
$Ti^{3+} + 3e^- = Ti$	−1.21	
$Ti^{2+} + 2e^- = Ti$	−1.63	
Tungsten		
$2WO_3 + 2H^+ + 2e^- = W_2O_5 + H_2O$	−0.029	
$WO_3 + 6H^+ + 6e^- = W + 3H_2O$	−0.090	
$WO_4^{2-} + 4H_2O + 6e^- = W + 8OH^-$	−1.074	
$WO_4^{2-} + 2H_2O + 2e^- = WO_2 + 4OH^-$	−1.259	
$W_2O_5 + 2H^+ + 2e^- = 2WO_2 + H_2O$	−0.031	
$W(CN)_8^{3-} + e^- = W(CN)_8^{4-}$	0.457	
$WO_2 + 4H^+ + 4e^- = W + 2H_2O$	−0.119	
$WO_2 + 2H_2O + 4e^- = W + 4OH^-$	−0.982	
Uranium		
$UO_2^{2+} + e^- = UO_2^+$	0.16	
$UO_2^{2+} + 4H^+ + 2e^- = U^{4+} + 2H_2O$	0.27	
$UO_2^+ + 4H^+ + e^- = U^{4+} + 2H_2O$	0.38	
$U^{4+} + e^- = U^{3+}$	−0.52	
$U^{4+} + 4e^- = U$	−1.38	
$U^{3+} + 3e^- = U$	−1.66	
Vanadium		
$VO_2^+ + 2H^+ + e^- = VO^{2+} + H_2O$	1.000	
$VO_2^+ + 4H^+ + 2e^- = V^{3+} + 2H_2O$	0.668	
$VO_2^+ + 4H^+ + 3e^- = V^{2+} + 2H_2O$	0.361	
$VO_2^+ + 4H^+ + 5e^- = V + 4H_2O$	−0.236	
$VO^{2+} + 2H^+ + e^- = V^{3+} + H_2O$	0.337	
$V^{3+} + e^- = V^{2+}$	−0.255	
$V^{2+} + 2e^- = V$	−1.13	
Xenon		
$H_4XeO_6 + 2H^+ + 2e^- = XeO_3 + 3H_2O$	2.42	
$HXeO_6^{3-} + 2H_2O + e^- = HXeO_4 + 4OH^-$	0.9	
$XeO_3 + 6H^+ + 2F^- + 4e^- = XeF_2 + 3H_2O$	1.6	
$XeO_3 + 6H^+ + 6e^- = Xe(g) + 3H_2O$	2.10	
$XeF_2 + e^- = XeF + F^-$	0.9	
$XeF_2 + 2H^+ + 2e^- = Xe(g) + 2HF$	2.64	
$XeF + e^- = Xe(g) + F^-$	3.4	
Ytterbium		
$Yb^{3+} + e^- = Yb^{2+}$	−1.05	
$Yb^{2+} + 2e^- = Yb$	−2.8	
$Yb^{3+} + 3e^- = Yb$	−2.22	

TABLE 8.27 Potentials of the Elements and Their Compounds at 25°C (*Continued*)

Half-reaction	Standard or formal potential	Solution composition
Yttrium		
$Y^{3+} + 3e^- = Y$	-2.37	
Zinc		
$Zn^{2+} + 2e^- = Zn$	-0.7626	
$Zn(NH_3)_4^{2+} + 2e^- = Zn + 4NH_3$	-1.04	
$Zn(CN)_4^{2-} + 2e^- = Zn + 4CN^-$	-1.34	
$Zn(tartrate)_2^{6-} + 2e^- = Zn + 4(tartrate)^{2-}$	-1.15	
$Zn(OH)_4^{2-} + 2e^- = Zn + 4OH^-$	-1.285	
Zirconium		
$Zr^{4+} + 4e^- = Zr$	-1.55	
$ZrO_2 + 4H^+ + 4e^- = Zr + 2H_2O$	-1.45	

TABLE 8.28 Potentials of Selected Half-Reactions at 25°C

A summary of oxidation-reduction half-reactions arranged in order of decreasing oxidation strength and useful for selecting re-agent systems.

Half-reaction	$E°$, volts
$F_2(g) + 2H^+ + 2e^- = 2HF$	3.053
$O_3 + H_2O + 2e^- = O_2 + 2OH^-$	1.246
$O_3 + 2H^+ + 2e^- = O_2 + H_2O$	2.075
$Ag^{2+} + e^- = Ag^+$	1.980
$S_2O_8^{2-} + 2e^- = 2SO_4^{2-}$	1.96
$HN_3 + 3H^+ + 2e^- = NH_4^+ + N_2$	1.96
$H_2O_2 + 2H^+ + 2e^- = 2H_2O$	1.763
$Ce^{4+} + e^- = Ce^{3+}$	1.72
$MnO_4^- + 4H^+ + 3e^- = MnO_2(c) + 2H_2O$	1.70
$2HClO + 2H^+ + 2e^- = Cl_2 + H_2O$	1.630
$2HBrO + 2H^+ + 2e^- = Br_2 + H_2O$	1.604
$H_5IO_6 + H^+ + 2e^- = IO_3^- + 3H_2O$	1.603
$NiO_2 + 4H^+ + 2e^- = Ni^{2+} + 2H_2O$	1.593
$Bi_2O_4(bismuthate) + 4H^+ + 2e^- = 2BiO^+ + 2H_2O$	1.59
$MnO_4^- + 8H^+ + 5e^- = Mn^{2+} + 4H_2O$	1.51
$2BrO_3^- + 12H^+ + 10e^- = Br_2 + 6H_2O$	1.478
$PbO_2 + 4H^+ + 2e^- = Pb^{2+} + 2H_2O$	1.468
$Cr_2O_7^{2-} + 14H^+ + 6e^- = 2Cr^{3+} + 7H_2O$	1.36
$Cl_2 + 2e^- = 2Cl^-$	1.3583
$2HNO_2 + 4H^+ + 4e^- = N_2O + 3H_2O$	1.297
$N_2H_5^+ + 3H^+ + 2e^- = 2NH_4^+$	1.275
$MnO_2 + 4H^+ + 2e^- = Mn^{2+} + 2H_2O$	1.23
$O_2 + 4H^+ + 4e^- = 2H_2O$	1.229
$ClO_4^- + 2H^+ + 2e^- = ClO_3^- + H_2O$	1.201
$2IO_3^- + 12H^+ + 10e^- = I_2 + 3H_2O$	1.195
$N_2O_4 + 2H^+ + 2e^- = 2HNO_3$	1.07
$2ICl_2^- + 2e^- = 4Cl^- + I_2$	1.07
$Br_2(lq) + 2e^- = 2Br^-$	1.065
$N_2O_4 + 4H^+ + 4e^- = 2NO + 2H_2O$	1.039

TABLE 8.28   Potentials of Selected Half-Reactions at 25°C
(*Continued*)

Half-reaction	$E°$, volts
$HNO_2 + H^+ + e^- = NO + H_2O$	0.996
$NO_3^- + 4H^+ + 3e^- = NO + 2H_2O$	0.957
$NO_3^- + 3H^+ + 2e^- = HNO_2 + H_2O$	0.94
$2Hg^{2+} + 2e^- = Hg_2^{2+}$	0.911
$Cu^{2+} + I^- + e^- = CuI$	0.861
$OsO_4(c) + 8H^+ + 8e^- = Os + 4H_2O$	0.84
$Ag^+ + e^- = Ag$	0.7991
$Hg_2^{2+} + 2e^- = 2Hg$	0.7960
$Fe^{3+} + e^- = Fe^{2+}$	0.771
$H_2SeO_3 + 4H^+ + 4e^- = Se + 3H_2O$	0.739
$HN_3 + 11H^+ + 8e^- = 2NH_4^+$	0.695
$O_2 + 2H^+ + 2e^- = H_2O_2$	0.695
$Ag_2SO_4 + 2e^- = 2Ag + SO_4^{2-}$	0.654
$Cu^{2+} + Br^- + e^- = CuBr(c)$	0.654
$Au(SCN)_4^- + 3e^- = Au + 4SCN^-$	0.636
$2HgCl_2 + 2e^- = Hg_2Cl_2(c) + 2Cl^-$	0.63
$Sb_2O_5 + 6H^+ + 4e^- = 2SbO^+ + 3H_2O$	0.605
$H_3AsO_4 + 2H^+ + 2e^- = HAsO_2 + 2H_2O$	0.560
$TeOOH^+ + 3H^+ + 4e^- = Te + 2H_2O$	0.559
$Cu^{2+} + Cl^- + e^- = CuCl(c)$	0.559
$I_2^- + 2e^- = 3I^-$	0.536
$I_2 + 2e^- = 2I^-$	0.536
$Cu^+ + e^- = Cu$	0.53
$4H_2SO_3 + 4H^+ + 6e^- = S_4O_6^{2-} + 6H_2O$	0.507
$Ag_2CrO_4 + 2e^- = 2Ag + CrO_4^{2-}$	0.449
$2H_2SO_3 + 2H^+ + 4e^- = S_2O_3^{2-} + 3H_2O$	0.400
$UO_2^+ + 4H^+ + e^- = U^{4+} + 2H_2O$	0.38
$Fe(CN)_6^{3-} + e^- = Fe(CN)_6^{4-}$	0.361
$Cu^{2+} + 2e^- = Cu$	0.340
$VO^{2+} + 2H^+ + e^- = V^{3+} + H_2O$	0.337
$BiO^+ + 2H^+ + 3e^- = Bi + H_2O$	0.32
$UO_2^{2+} + 4H^+ + 2e^- = U^{4+} + 2H_2O$	0.27
$Hg_2Cl_2(c) + 2e^- = 2Hg + 2Cl^-$	0.2676
$AgCl + e^- = Ag + Cl^-$	0.2223
$SbO^+ + 2H^+ + 3e^- = Sb + H_2O$	0.212
$CuCl_3^{2-} + e^- = Cu + 3Cl^-$	0.178
$SO_4^{2-} + 4H^+ + 2e^- = H_2SO_3 + H_2O$	0.158
$Sn^{4+} + 2e^- = Sn^{2+}$	0.15
$S + 2H^+ + 2e^- = H_2S$	0.144
$Hg_2Br_2(c) + 2e^- = 2Hg + 2Br^-$	0.1392
$CuCl + e^- = Cu + Cl^-$	0.121
$TiO^{2+} + 2H^+ + e^- = Ti^{3+} + H_2O$	0.100
$S_4O_6^{2-} + 2e^- = 2S_2O_3^{2-}$	0.08
$AgBr + e^- = Ag + Br^-$	0.0711
$HCOOH + 2H^+ + 2e^- = HCHO + H_2O$	0.056
$CuBr + e^- = Cu + Br^-$	0.033
$2H^+ + 2e^- = H_2$	0.0000
$Hg_2I_2 + 2e^- = 2Hg + 2I^-$	-0.0405
$Pb^{2+} + 2e^- = Pb$	-0.125
$Sn^{2+} + 2e^- = Sn$	-0.136
$AgI + e^- = Ag + I^-$	-0.1522
$N_2 + 5H^+ + 4e^- = N_2H_5^+$	-0.225
$V^{3+} + e^- = V^{2+}$	-0.255

TABLE 8.28 Potentials of Selected Half-Reactions at 25°C (*Continued*)

Half-reaction	$E°$, volts
$Ni^{2+} + 2e^- = Ni$	-0.257
$Co^{2+} + 2e^- = Co$	-0.277
$Ag(CN)_2^- + e^- = Ag + 2CN^-$	-0.31
$PbSO_4 + 2e^- = Pb + SO_4^{2-}$	-0.3505
$Cd^{2+} + 2e^- = Cd$	-0.4025
$Cr^{3+} + e^- = Cr^{2+}$	-0.424
$Fe^{2+} + 2e^- = Fe$	-0.44
$H_3PO_3 + 2H^+ + 2e^- = HPH_2O_2 + H_2O$	-0.499
$2CO_2 + 2H^+ + 2e^- = H_2C_2O_4$	-0.49
$U^{4+} + e^- = U^{3+}$	-0.52
$Zn^{2+} + 2e^- = Zn$	-0.7626
$Mn^{2+} + 2e^- = Mn$	-1.18
$Al^{3+} + 3e^- = Al$	-1.67
$Mg^{2+} + 2e^- = Mg$	-2.356
$Na^+ + e^- = Na$	-2.714
$K^+ + e^- = K$	-2.925
$Li^+ + e^- = Li$	-3.045
$3N_2 + 2H^+ + 2e^- = 2HN_3$	-3.10

TABLE 8.29 Overpotentials for Common Electrode Reactions at 25°C

The overpotential is defined as the difference between the actual potential of an electrode at a given current density and the reversible electrode potential for the reaction.

Electrode	Current Density, A/cm²					
	0.001	0.01	0.1	0.5	1.0	5.0
	Overpotential, volts					
Liberation of H₂ from 1M H₂SO₄						
Ag	0.097	0.13	0.3		0.48	0.69
Al	0.3	0.83	1.00		1.29	
Au	0.017		0.1		0.24	0.33
Bi	0.39	0.4			0.78	0.98
Cd		1.13	1.22		1.25	
Co		0.2				
Cr		0.4				
Cu			0.35		0.48	0.55
Fe		0.56	0.82		1.29	
Graphite	0.002		0.32		0.60	0.73
Hg	0.8	0.93	1.03		1.07	
Ir	0.0026	0.2				
Ni	0.14	0.3			0.56	0.71
Pb	0.40	0.4			0.52	1.06
Pd	0	0.04				
Pt (smooth)	0.0000	0.16	0.29		0.68	
Pt (platinized)	0.0000	0.030	0.041		0.048	0.051
Sb		0.4				
Sn		0.5	1.2			
Ta		0.39	0.4			
Zn	0.48	0.75	1.06		1.23	
Liberation of O₂ from 1M KOH						
Ag	0.58	0.73	0.98		1.13	
Au	0.67	0.96	1.24		1.63	
Cu	0.42	0.58	0.66		0.79	
Graphite	0.53	0.90	1.09		1.24	
Ni	0.35	0.52	0.73		0.85	
Pt (smooth)	0.72	0.85	1.28		1.49	
Pt (platinized)	0.40	0.52	0.64		0.77	
Liberation of Cl₂ from saturated NaCl solution						
Graphite			0.25	0.42	0.53	
Platinized Pt	0.006		0.026	0.05		
Smooth Pt	0.008	0.03	0.054	0.161	0.236	
Liberation of Br₂ from saturated NaBr solution						
Graphite		0.002	0.027	0.16	0.33	
Platinized Pt		0.002	0.012	0.069	0.21	
Smooth Pt		0.002	0.006*	0.26	0.38†	
Liberation of I₂ from saturated NaI solution						
Graphite	0.002	0.014	0.097			
Platinized Pt		0.006	0.032		0.196	
Smooth Pt		0.003	0.03	0.12	0.22	

* At 0.23 A/cm².   † At 0.72 A/cm².

The overpotential required for the evolution of O_2 from dilute solutions of $HClO_4$, HNO_3, H_3PO_4 or H_2SO_4 onto smooth platinum electrodes is approximately 0.5 V.

TABLE 8.30 Half-Wave Potentials of Inorganic Materials

All values are in volts vs. the saturated calomel electrode.

Element	$E_{1/2}$, volts	Solvent system
Aluminum		
3+	-0.5	$0.2M$ acetate, pH 4.5–4.7, plus 0.07% azo dye Pontochrome Violet SW; reduction wave of complexed dye is 0.2 V more negative than that of the free dye.
Antimony		
3+ to 0	-0.15	$1M$ HCl
	$-0.31(1)$	$1M$ HNO$_3$ (or $0.5M$ H$_2$SO$_4$)
	-0.8	$0.5M$ tartrate, pH 4.5
	$-1.0; -1.2$	$0.5M$ tartrate, pH 9 (waves not distinct)
	-1.26	$1M$ NaOH; also anodic wave (3+ to 5+) at -0.45
	-1.32	$0.5M$ tartrate plus $0.1M$ NaOH
5+	$0.0; -0.257$	$6M$ HCl. First wave (5+ to 3+) starts at the oxidation potential of Hg; second wave is 3+ to 0.
5+ to 0	-0.35	$1M$ HCl plus $4M$ KBr
Arsenic		
3+ to 5+	-0.26	$0.5M$ KOH (anodic wave); only suitable wave
3+	$-0.8; -1.0$	$0.1M$ HCl; ill-defined waves
	$-0.7; -1.0$	$0.5M$ H$_2$SO$_4$ (or $1M$ HNO$_3$)
Barium		
2+ to 0	-1.94	$0.1M$ (C$_2$H$_5$)$_4$NI
Bismuth		
3+ to 0	$-0.025(15)$	$1M$ HNO$_3$ (or $0.5M$ H$_2$SO$_4$)
	-0.09	$1M$ HCl
	-0.29	$0.5M$ tartrate, pH 4.5
	-0.7	$0.5M$ tartrate (pH 9), wave not well-developed
	-1.0	$0.5M$ tartrate plus $0.1M$ NaOH, poor wave
Bromine		
5+ to 1−	-1.75	$0.1M$ alkali chlorides (or $0.1M$ NaOH)
	0.13	$0.05M$ H$_2$SO$_4$
0 to 1−	0.0	Wave (anodic) starts at zero; Hg$_2$Br$_2$ forms
Br−	0.1	Oxidation of Hg to form mercury(I) bromide
Cadmium		
2+ to 0	-0.60	$0.1M$ KCl, or $0.5M$ H$_2$SO$_4$, or $1M$ HNO$_3$
	-0.64	$0.5M$ tartrate at pH 4.5 or 9
	-0.81	$1M$ NH$_4$Cl plus $1M$ NH$_3$
Calcium		
2+ to 0	-2.22	$0.1M$ (C$_2$H$_5$)$_4$NCl
	-2.13	$0.1M$ (C$_2$H$_5$)$_4$NCl in 80% ethanol
Cerium		
3+ to 0	-1.97	$0.02M$ alkali sulfate
Cesium		
1+ to 0	-2.05	$0.1M$ (C$_2$H$_5$)$_4$NOH in 50% ethanol
Chlorine		
Cl−	0.25	Oxidation of Hg to form Hg$_2$Cl$_2$
Chromium		
6+ to 3+	-0.85	CrO$_4^{2-}$ to CrO$_2^-$ in 0.1 to $1M$ NaOH
3+ to 0	$-0.35; -1.70$	$1M$ NH$_4$Cl-NH$_3$ buffer (pH 8–9); 3+ to 2+ to 0
3+ to 2+	-0.95	$0.1M$ pyridine–$0.1M$ pyridinium chloride

TABLE 8.30 Half-Wave Potentials of Inorganic Materials (*Continued*)

Element	$E_{1/2}$, volts	Solvent system
2+ to 0	−1.54	$1M$ KCl
2+ to 3+	−0.40	$1M$ KCl (anodic wave)
Cobalt		
3+ to 0	−0.5; −1.3	$1M$ NH$_4$Cl plus $1M$ NH$_3$; 3+ to 2+ to 0
2+ to 0	−1.07	$0.1M$ pyridine plus pyridinium chloride
	−1.03	Neutral $1M$ potassium thiocyanate
	−1.4	Co(H$_2$O)$_6^{2+}$ in noncomplexing systems
3+ to 2+	0.0	$1M$ sodium oxalate in acetate buffer (pH 5); diffusion current measured between 0 and −0.1 V
Copper		
2+ to 0	0.04	$0.1M$ KNO$_3$, $0.1M$ NH$_4$ClO$_4$, or $1M$ Na$_2$SO$_4$
	−0.085	$0.1M$ Na$_4$P$_2$O$_7$ plus $0.2M$ Na acetate, pH 4.5
	−0.09	$0.5M$ Na tartrate, pH 4.5
	−0.20	$0.1M$ potassium oxalate, pH 5.7 to 10
	−0.22	$0.5M$ potassium citrate, pH 7.5
	−0.4	$0.5M$ Na tartrate plus $0.1M$ NaOH (pH 12)
	−0.568	$0.1M$ KNO$_3$ plus $1M$ ethylenediamine
2+	0.04; −0.22	$1M$ KCl; consecutive waves: 2+ to 1+ to 0
	−0.02; −0.39	$0.1M$ KSCN; consecutive waves: 2+ to 1+ to 0
	0.05; −0.25	$0.1M$ pyridine plus $0.1M$ pyridinium chloride; consecutive waves: 2+ to 1+ to 0
	−0.24; −0.50	$1M$ NH$_4$Cl plus $1M$ NH$_3$; consecutive waves
Gallium		
3+ to 0	−1.1	Not more than $0.001M$ HCl or wave masked by hydrogen wave which immediately follows
Germanium		
2+ to 0	−0.45	$6M$ HCl; prior reduction with HPH$_2$O$_2$ to 2+
Gold		
3+ to 1+	0	$1M$ KCN; wave starts at 0 V
1+ to 0	−1.4	Au(CN)$_2^-$ wave best for analytical purposes
Indium		
3+ to 0	−0.60	$1M$ KCl In Na acetate, pH 3.9 to 4.2
Iodine		
IO$_4^-$	0.36	First wave at pH 0 (shifts to −0.08 at pH 12); second wave corresponds to iodate reduction
IO$_3^-$	−0.075	$0.2M$ KNO$_3$ (shifts −0.13 V/pH unit increase)
	−0.305	$0.1M$ hydrogen phthalate, pH 3.2
	−0.500	$0.1M$ acetate plus $0.1M$ KCl, pH 4.9
	−0.650	$0.1M$ citrate, pH 5.95
	−1.050	$0.2M$ phosphate, pH 7.10
	−1.20	$0.05M$ borax + $0.1M$ KCl, pH 9.2; or NaOH plus $0.1M$ KCl, pH 13.0
0 to 1−	0.0	Wave starts from zero in acid media; Hg$_2$I$_2$ formed
1−	−0.1	Oxidation of Hg to form Hg$_2$I$_2$
Iron		
3+	−0.44; −1.52	$1M$ (NH$_4$)$_2$CO$_3$; two waves: 3+ to 2+ to 0
	−0.17; −1.50	$0.5M$ Na tartrate, pH 5.8; two waves: 3+ to 2+ to 0
	−0.9; −1.5	0.1 to $5M$ KOH plus 8% mannitol; 3+ to 2+ to 0
3+ to 2+	−0.13	$0.1M$ EDTA plus $2M$ Na acetate, pH 6–7

TABLE 8.30   Half-Wave Potentials of Inorganic Materials (*Continued*)

Element	$E_{1/2}$, volt	Solvent system
	−0.27	0.2M Na oxalate, pH 7.9 or less
	−0.28	0.5M Na citrate, pH 6.5
	−1.46(2)	1M NH$_4$ClO$_4$
	−1.36	0.1M KHF$_2$, pH 4 or less
2+ to 3+	−0.28	0.5M Na citrate, pH 6.5
	−0.27	0.2M Na oxalate, pH 7.9 or less
	−0.17	0.5M Na tartrate, pH 5.8
	−1.36	0.1M KHF$_2$, pH 4 or less
Lead		
2+ to 0	−0.405	1M HNO$_3$
	−0.435	1M KCl (or HCl)
	−0.49(1)	0.5M Na tartrate, pH 4.5 or 9
	−0.72	1M KCN
	−0.75	1M KOH or 0.5M Na tartrate plus 0.1M NaOH
Lithium		
1+ to 0	−2.31	0.1M (C$_2$H$_5$)$_4$NOH in 50% ethanol
Magnesium		
2+ to 0	−2.2	0.1M (C$_2$H$_5$)$_4$NCl (poorly defined wave)
Manganese		
2+ to 0	−1.65	1M NH$_4$Cl plus 1M NH$_3$
	−1.55	1M KCNS
	−1.33	1.5M KCN
Molybdenum		
6+	−0.26; −0.63	0.3M HCl, two waves: 6+ to 5+ to 3+
Nickel		
2+ to 0	−0.70	1M KSCN
	−0.78	1M KCl plus 0.5M pyridine
	−1.09	1M NH$_4$Cl plus 1M NH$_3$
	−1.1	Ni(H$_2$O)$_6^{2+}$ in NH$_4$ClO$_4$ or KNO$_3$
	−1.36	Ni(CN)$_4^{2-}$ in 1M KCN (alkaline media)
Niobium		
5+ to 3+	−0.80(4)	1M HNO$_3$
Nitrogen		
Nitrate	−1.45	0.017M LaCl$_3$ (reduced to hydroxylamine)
HNO$_2$	−0.77	0.1M HCl
C$_2$N$_2$	−1.2; −1.55	0.1M Na acetate, two waves
Oxamic acid	−1.55	0.1M Na acetate
Cyanide	−0.45	0.1M NaOH; anodic wave starts at −0.45
Thiocyanate	0.18	Anodic wave; neutral or weakly alkaline medium
Osmium		
OsO$_4$	0.0; −0.41; −1.16	Sat'd Ca(OH)$_2$. Three waves: first starts at 0; second wave is OsO$_4^{2-}$ to Os(V); and third wave is Os(V) to Os(III)
Oxygen		
O$_2$	−0.05; −0.9	Buffer solutions of pH 1 to 10. Two waves: O$_2$ to H$_2$O$_2$, and H$_2$O$_2$ to H$_2$O. Second wave extends from −0.5 to −1.3
H$_2$O$_2$	−0.9	Very extended wave (see above); sharper in presence of Aerosol OT
Palladium		
2+ to 0	−0.31	1M pyridine plus 1M KCl

TABLE 8.30 Half-Wave Potentials of Inorganic Materials (*Continued*)

Element	$E_{1/2}$, volts	Solvent system
	-0.64	$0.1M$ ethylenediamine plus $1M$ KCl
	-0.72	$1M$ NH$_4$Cl plus $1M$ NH$_3$
Potassium		
1+ to 0	-2.10	$0.1M$ (C$_2$H$_5$)$_4$NOH in 50% ethanol
Rhenium		
7+ to 4+	-0.44	$2M$ HCl or (better) $4M$ HClO$_4$
4+ to 3+	-0.51	ReCl$_6^{2-}$ ion in $1M$ HCl
Rhodium		
3+ to 2+	-0.41	$1M$ pyridine plus $1M$ KCl
Rubidium		
1+ to 0	-1.99	$0.1M$ (C$_2$H$_5$)$_4$NOH in 50% ethanol
Scandium		
3+ to 0	-1.80	$0.1M$ LiCl, KCl, or BaCl$_2$
Selenium		
4+ to 2−	-1.44	$1M$ NH$_4$Cl plus NH$_3$, pH 8.0
	-1.54	Same system adjusted to pH 9.5
2−	-0.49	Anodic wave at pH 0 due to HgSe
	-0.94	Anodic wave at pH 12 (0.01M NaOH)
Silver		
1+ to 0		Wave starts at oxidation potential of Hg
1+ to 0	-0.3	0.0014M KAg(CN)$_2$ without excess cyanide
Sodium		
1+ to 0	-2.07	$0.1M$ (C$_2$H$_5$)$_4$NOH in 50% ethanol
Strontium		
2+ to 0	-2.11	$0.1M$ (C$_2$H$_5$)$_4$NI, water or 80% ethanol
Sulfur		
SO$_2$	-0.38	$1M$ HNO$_3$ (or other strong acid); 4+ to 2+
S$_2$O$_4^{2-}$	-0.43	$0.5M$ (NH$_4$)$_2$HPO$_4$ plus $1M$ NH$_3$ (anodic wave)
S$_2$O$_3^{2-}$	-0.15	$1M$ strong acid; anodic mercury wave
0 to 2−	-0.50	90% methanol, 9.5% pyridine, 0.5% HCl (pH 6)
HS$^-$	-0.76	$0.1M$ NaOH (anodic mercury wave)
Tellurium		
4+ to 0	-0.4	Citrate buffer, pH 1.6 (second of two waves)
	-0.63	Ammoniacal buffer, pH 9.4
4+ to 2−	-1.22	$0.1M$ NaOH
2− to 0	-0.72	$1M$ HCl (true anodic reversible wave)
	-0.08	$1M$ NaOH (same as above; intermediate values at pH 1 to 13)
Thallium		
3+ to 0	-0.48	$1M$ KCl, KNO$_3$, K$_2$SO$_4$, KOH, or NH$_3$
Tin		
4+ to 2+	$-0.25; -0.52$	$4M$ NH$_4$Cl + $1M$ HCl; two waves: 4+ to 2+ to 0
2+ to 0	-0.59	$0.5M$ tartrate, pH 4.3
	-1.22	$1M$ NaOH (stannite ion to tin)
2+ to 4+	-0.28	$0.5M$ Na tartrate, pH 4.3 (anodic wave)
	-0.73	$1M$ NaOH (stannite ion to stannate ion)
Titanium		
4+ to 3+	-0.173	$0.1M$ K$_2$C$_2$O$_4$ plus $1M$ H$_2$SO$_4$
	-1.22	$0.4M$ tartrate, pH 6.5

TABLE 8.30 Half-Wave Potentials of Inorganic Materials (*Continued*)

Element	$E_{1/2}$, volts	Solvent system
Tungsten		
6+	0.0; -0.64	$6M$ HCl; two waves: first wave starts at zero and is W(VI) to W(V), the second wave is W(V) to W(III)
Uranium		
6+	-0.180; -0.92	UO_2^{2+} to UO_2^+, then U^{3+} in $0.02M$ HCl
Vanadium		
5+ to 4+ to 2+	-0.97; -1.26	$1M$ NH$_4$Cl plus $1M$ NH$_3$ and $0.08M$ Na$_2$SO$_3$
4+ to 2+	-0.98	$0.05M$ H$_2$SO$_4$
3+ to 2+	-0.55	$0.5M$ H$_2$SO$_4$
4+ to 5+	-0.32	$1M$ NH$_4$Cl, $1M$ NH$_3$, and $0.08M$ Na$_2$SO$_3$
4+ to 5+	0.76	$0.05M$ H$_2$SO$_4$; anodic wave starting from zero
2+ to 3+	-0.55	$0.5M$ H$_2$SO$_4$; anodic wave
Zinc		
2+ to 0	-0.995	$0.1M$ KCl
	-1.01	$0.1M$ KSCN
	-1.15	$0.5M$ tartrate, pH 9
	-1.23	$0.5M$ tartrate, pH 4.5
	-1.33	$1M$ NH$_4$Cl plus $1M$ NH$_3$
	-1.53	$1M$ NaOH

TABLE 8.31   Half-Wave Potentials (vs. Saturated Calomel Electrode) of Organic Compounds at 25°C

The solvent systems in this table are listed below:

    A, acetonitrile and a perchlorate salt such as $LiClO_4$ or a tetraalkyl ammonium salt

    B, acetic acid and an alkali acetate, often plus a tetraalkyl ammonium iodide

    C, 0.05 to 0.175M tetraalkyl ammonium halide and 75% 1,4-dioxane

    D, buffer plus 50% ethanol (EtOH)

Abbreviations Used in the Table

Bu, butyl	Me, methyl
Et, ethyl	MeOH, methanol
EtOH, ethanol	PrOH, propanol
M, molar	

Compound	Solvent system	$E_{1/2}$
Unsaturated aliphatic hydrocarbons		
Acrylonitrile	C but 30% EtOH	−1.94
Allene	C	−2.29
1,3-Butadiene	A	−2.03
	C	−2.59
1,3-Butadiyne	C	−1.89
1-Buten-2-yne	C	−2.40
1,4-Cyclohexadiene	A	−1.6
Cyclohexene	A	−1.89
1,3,5,7-Cyclooctatetraene	B	−1.42
	C	−1.51
Diethyl fumarate	B, pH 4.0	−0.84
Diethyl maleate	B, pH 4.0	−0.95
2,3-Dimethyl-1,3-butadiene	A	−1.83
Dimethylfulvene	C	−1.89
Diphenylacetylene	C	−2.20
1,1-Diphenylethylene	B	−1.52
	C	−2.19
Ethyl methacrylate	0.1 N LiCl+25% EtOH	−1.9
2-Methyl-1,3-butadiene	A	−1.84
2-Methyl-1-butene	A	−1.97
1-Piperidino-4-cyano-4-phenyl-1,3-butadiene	$LiClO_4$ in dimethylformamide	−0.16
trans-Stilbene	B	−1.51
Tetrakis(dimethylamino)ethylene	A	−0.75
Aromatic hydrocarbons		
Acenaphthene	A	−0.95
	B	−1.36
	C	−2.58
Anthracene	A	−0.84
	B	−1.20
	C	−1.94

TABLE 8.31 Half-Wave Potentials (vs. Saturated Calomel Electrode) of Organic Compounds at 25°C (*Continued*)

Compound	Solvent system	$E_{1/2}$
Aromatic hydrocarbons (*continued*)		
Azulene	A	−0.71
	C	−1.66, −2.26, −2.56
1,2-Benzanthracene	C	−2.03, −2.54
2,3-Benzanthracene	A	−0.54, −1.20
Benzene	A	−2.08
1,2-Benzo[*a*]pyrene	A	−0.76
Biphenyl	A	−1.48
	B	−1.91
	C	−2.70
Chrysene	A	−1.22
1,2,5,6-Dibenzanthracene	A	−1.00, −1.26
1,2-Dihydronaphthalene	C	−2.57
9,10-Dimethylanthracene	A	−0.65
2,3-Dimethylnaphthalene	A	−1.08, −1.34
9,10-Diphenylanthracene	A	−0.92
Fluorene	A	−1.25
	B	−1.65
	C	−2.65
Hexamethylbenzene	A	−1.16
	B	−1.52
Indan	A	−1.59, −2.02
Indene	A	−1.23
	C	−2.81
1-Methylnaphthalene	A	−1.24
	B	−1.53
	C	−2.46
2-Methylnaphthalene	A	−1.22
	B	−1.55
	C	−2.46
Naphthalene	A	−1.34
	B	−1.72
Pentamethylbenzene	A	−1.28
	B	−1.62
Phenanthrene	A	−1.23
	B	−1.68
	C	−2.46, −2.71
Phenylacetylene	C	−2.37
Pyrene	A	−1.06, −1.24
trans-Stilbene	B	−1.51
	C	−2.26
Styrene	C	−2.35

TABLE 8.31 Half-Wave Potentials (vs. Saturated Calomel Electrode) of Organic Compounds at 25°C (*Continued*)

Compound	Solvent system	$E_{1/2}$
Aromatic hydrocarbons (continued)		
1,2,3,5-Tetramethylbenzene	A	−1.50, −1.99
1,2,4,5-Tetramethylbenzene	A	−1:29
Tetraphenylethylene	C	−2.05
1,4,5,8-Tetraphenylnaphthalene	A	−1.39
Toluene	A	−1.98
1,2,3-Trimethylbenzene	A	−1.58
1,2,4-Trimethylbenzene	A	−1.41
1,3,5-Trimethylbenzene	A	−1.50
	B	−1.90
Triphenylene	A	−1.46, −1.55
Triphenylmethane	C	−1.01, −1.68, −1.96
o-Xylene	A	−1.58, −2.04
m-Xylene	A	−1.58
p-Xylene	A	−1.56
Aldehydes		
Acetaldehyde	B, pH 6.8–13	−1.89
Benzaldehyde	McIlvaine buffer, pH 2.2	−0.96, −1.32
Bromoacetaldehyde	pH 8.5	−0.40
	pH 9.8	−1.58, −1.82
Chloroacetaldehyde	Ammonia buffer, pH 8.4	−1.06, −1.66
Cinnamaldehyde	Buffer + EtOH, pH 6.0	−0.9, −1.5, −1.7
Crotonaldehyde	B, pH 1.3–2.0	−0.92
	Ammonia buffer, pH 8.0	−1.30
Dichloroacetaldehyde	Ammonia buffer, pH 8.4	−1.03, −1.67
3,7-Dimethyl-2,6-octadienal	0.1 M Et$_4$NI	−1.56, −2.22
Formaldehyde	0.05 M KOH+0.1 M KCl, pH 12.7	−1.59
2-Furaldehyde	pH 1–8	−0.86−0.07 pH
	pH 10	−1.43
Glucose	Phosphate buffer, pH 7	−1.55
Glyceraldehyde	Britton-Robinson buffer, pH 5.0	−1.47
	Britton-Robinson buffer, pH 8.0	−1.55
Glycolaldehyde	0.1 M KOH, pH 13	−1.70
Glyoxal	B, pH 3.4	−1.41
4-Hydroxybenzaldehyde	Britton-Robinson buffer, pH 1.8	−1.16
	Britton-Robinson buffer, pH 6.8	−1.45
4-Hydroxy-2-methoxybenzaldehyde	McIlvaine buffer, pH 2.2	−1.05
	McIlvaine buffer, pH 5.0	−1.16, −1.36
	McIlvaine buffer, pH 8.0	−1.47

TABLE 8.31　Half-Wave Potentials (vs. Saturated Calomel Electrode) of Organic Compounds at 25°C (*Continued*)

Compound	Solvent system	$E_{1/2}$
Aldehydes (*continued*)		
o-Methoxybenzaldehyde	Britton-Robinson buffer, pH 1.8	−1.02
	Britton-Robinson buffer, pH 6.8	−1.49
p-Methoxybenzaldehyde	Britton-Robinson buffer, pH 1.8	−1.17
	Britton-Robinson buffer, pH 6.8	−1.48
Methyl glyoxal	A, pH 4.5	−0.83
m-Nitrobenzaldehyde	Buffer + 10% EtOH, pH 2.0	−0.28, −1.20
Phthalaldehyde	Buffer, pH 3.1	−0.64, −1.07
	Buffer, pH 7.3	−0.89, −1.29
2-Propenal (acrolein)	pH 4.5	−1.36
	pH 9.0	−1.1
Propionaldehyde	0.1 M LiOH, pH 13	−1.93
Pyrrole-2-carbaldehyde	0.1 M HCl + 50% EtOH	−1.25
Salicylaldehyde	McIlvaine buffer, pH 2.2	−0.99, −1.23
	McIlvaine buffer, pH 5.0	−1.20, −1.30
	McIlvaine buffer, pH 8.0	−1.32
Trichloroacetaldehyde	Ammonia buffer, pH 8.4	−1.35, −1.66
	0.1 M KCl + 50% EtOH	−1.55
Ketones		
Acetone	B, pH 9.3	−1.52
	C	−2.46
Acetophenone	D + McIlvaine buffer, pH 4.9	−1.33
	D + McIlvaine buffer, pH 7.2	−1.58
	D + McIlvaine buffer, pH 1.3	−1.08
7H-Benz[de]anthracen-7-one	0.1 N H$_2$SO$_4$ + 75% MeOH	−0.96
Benzil	D + McIlvaine buffer, pH 1.3	−0.27
	D + McIlvaine buffer, pH 4.9	−0.50
Benzoin	D + McIlvaine buffer, pH 1.3	−0.90
	D + McIlvaine buffer, pH 8.6	−1.49
Benzophenone	D + McIlvaine buffer, pH 1.3	−0.94
	D + McIlvaine buffer, pH 8.6	−1.36
Benzoylacetone	Buffer, pH 2.6	−1.60
	Buffer, pH 5.3 and pH 7.6	−1.68
	Buffer, pH 9.7	−1.72
Bromoacetone	0.1 M LiCl	−0.29
2,3-Butanedione	0.1 M HCl	−0.84
3-Buten-2-one	0.1 M KCl	−1.42
Butyrophenone	0.1 M NH$_4$Cl + 50% EtOH	−1.55
D-Carvone	0.1 M Et$_4$NI + 80% EtOH	−1.71
Chloroacetone	0.1 M LiCl	−1.18
Coumarin	McIlvaine buffer, pH 2.0	−0.95
	McIlvaine buffer, pH 5.0	−1.11, −1.44

TABLE 8.31 Half-Wave Potentials (vs. Saturated Calomel Electrode) of Organic Compounds at 25°C (*Continued*)

Compound	Solvent system	$E_{1/2}$
	Ketones (*continued*)	
Cyclohexanone	C	−2.45
cis-Dibenzoylethylene	D, pH 1	−0.30
	D, pH 11	−0.62, −1.65
trans-Dibenzoylethylene	D, pH 1	−0.12
	D, pH 11	−0.57, −1.52
Dibenzoylmethane	D, pH 1.3	−0.59
	D, pH 11.3	−1.30, −1.62
9,10-Dihydro-9-oxoanthracene	D, pH 2.0	−0.93
1,5-Diphenyl-1,5-pentanedione	A	−2.10
1,5-Diphenylthiocarbazone	D, pH 7.0	−0.6
Flavanone	Acetate buffer + Me$_4$NOH +50% 2-PrOH, pH 6.1	−1.30
	Acetate buffer + Me$_4$NOH +50% 2-PrOH, pH 9.6	−1.51
Fluorescein	Acetate buffer, pH 2.0	−0.50
	Phthalate buffer, pH 5.0	−0.65
	Borate buffer, pH 10.1	−1.18, −1.44
Fructose	0.02 M LiCl	−1.76
Girard derivatives of aliphatic ketones	pH 8.2	−1.52
o-Hydroxyacetophenone	D, pH 5	−1.36
p-Hydroxyacetophenone	D, pH 5	−1.46
1,2,3-Indantrione (ninhydrin)	Britton-Robinson buffer, pH 2.5	−0.67, −0.83
	Britton-Robinson buffer, pH 4.5	−0.73, −1.01
	Britton-Robinson buffer, pH 6.8	−0.10, −0.90, −1.20
	Britton-Robinson buffer, pH 9.2	−1.35
α-Ionone	C	−1.59, −2.08
Isatin	Phosphate buffer + citrate buffer, pH 2.9	−0.3, −0.5
	Phosphate buffer + citrate buffer, pH 4.3	−0.3, −0.5, −0.8
	Phosphate buffer + citrate buffer, pH 5.4	−0.8
4-Methyl-3,5-heptadien-2-one	A	−0.64
4-Methyl-2,6-heptanedione	A	−1.28
4-Methyl-3-penten-2-one	D + McIlvaine buffer, pH 1.3	−1.01
	D + McIlvaine buffer, pH 11.3	−1.60
4-Phenyl-3-buten-2-one	D, pH 1.3	−0.72
	D, pH 8.6	−1.27
Phthalide	0.1 M Bu$_4$NI + 50% dioxane	−0.20

TABLE 8.31 Half-Wave Potentials (vs. Saturated Calomel Electrode) of Organic Compounds at 25°C (*Continued*)

Compound	Solvent system	$E_{1/2}$
	Ketones (*continued*)	
Phthalimide	pH 4.2	−1.1, −1.5
	pH 9.7	−1.2, −1.4
Pulegone	C	−1.74
Quinalizarin	Phosphate buffer + 1% EtOH, pH 8.0	−0.56
Testosterone	D + Britton-Robinson buffer, pH 2.6	−1.20
	D + Britton-Robinson buffer, pH 5.8	−1.40
	D + Britton-Robinson buffer, pH 8.8	−1.53, −1.79
	Quinones	
Anthraquinone	Acetate buffer + 40% dioxane, pH 5.6	−0.51
	Phosphate buffer + 40% dioxane, pH 7.9	−0.71
o-Benzoquinone	Britton-Robinson buffer, pH 7.0	+0.20
	Britton-Robinson buffer, pH 9.0	+0.08
2,3-Dimethylnaphthoquinone	D, pH 5.4	−0.22
1,2-Naphthoquinone	Phosphate buffer, pH 5.0	−0.03
	Phosphate buffer, pH 7.0	−0.13
1,4-Naphthoquinone	Britton-Robinson buffer, pH 7.0	−0.07
	Britton-Robinson buffer, pH 9.0	−0.19
	Acids	
Acetic acid	A	−2.3
Acrylic acid	pH 5.6	−0.85
Adenosine-5′-phosphoric acid	$HClO_4 + KClO_4$, pH 2.2	−1.13
4-Aminobenzenesulfonic acid	0.05 M Me_4NI	−1.58
3-Aminobenzoic acid	pH 5.6	−0.67
Anthranilic acid	pH 5.6	−0.67
Ascorbic acid	Birtton-Robinson buffer, pH 3.4	+0.17
	Britton-Robinson buffer, pH 7.0	−0.06
Barbituric acid	Borate buffer, pH 9.3	−0.04
Benzoic acid	A	−2.1
Benzoylformic acid	Britton-Robinson buffer, pH 2.2	−0.48
	Britton-Robinson buffer, pH 5.5	−0.85, −1.26
	Britton-Robinson buffer, pH 7.2	−0.98, −1.25
	Britton-Robinson buffer, pH 9.2	−1.25

TABLE 8.31 Half-Wave Potentials (vs. Saturated Calomel Electrode) of Organic Compounds at 25°C (*Continued*)

Compound	Solvent system	$E_{1/2}$
	Acids (*continued*)	
Bromoacetic acid	pH 1.1	−0.54
2-Bromopropionic acid	pH 2.0	−0.39
Crotonic acid	C	−1.94
Dibromoacetic acid	pH 1.1	−0.03, −0.59
Dichloroacetic acid	pH 8.2	−1.57
5,5-Diethylbarbituric acid	Borate buffer, pH 9.3	0.00
Flavanol	D, pH 5.6	−1.25
	D, pH 7.7	−1.40
Folic acid	Britton-Robinson buffer, pH 4.6	−0.73
Formic acid	0.1 M KCl	−1.66
Fumaric acid	HCl + KCl, pH 2.6	−0.83
	Acetate buffer, pH 4.0	−0.93
	Acetate buffer, pH 5.9	−1.20
2,4-Hexadienedioic acid	Acetate buffer, pH 4.5	−0.97
Iodoacetic acid	pH 1	−0.16
Maleic acid	Britton-Robinson buffer, pH 2.0	−0.70
	Britton-Robinson buffer, pH 4.0	−0.97
	Britton-Robinson buffer, pH 6.0	−1.11, −1.30
	Britton-Robinson buffer, pH 10.0	−1.51
Mercaptoacetic acid	B, pH 6.8	−0.38
Methacrylic acid	D + 0.1 M LiCl	−1.69
Nitrobenzoic acids	Buffer + 10% EtOH, pH 2.0	−0.2, −0.7
Oxalic acid	B, pH 5.4–6.1	−1.80
2-Oxo-1,5-pentanedioic acid	HCl + KCl, ph 1.8	−0.59
	Ammonia buffer, pH 8.2	−1.30
2-Oxopropionic acid	Britton-Robinson buffer, pH 5.6	−1.17
	Britton-Robinson buffer, pH 6.8	−1.22, −1.53
	Britton-Robinson buffer, pH 9.7	−1.51
Phenolphthalein	Phthalate buffer, pH 2.5	−0.67
	Phthalate buffer, pH 4.7	−0.80
	D, pH 9.6	−0.98, −1.35
Picric acid	pH 4.2	−0.34
	pH 11.7	−0.36, −0.56, −0.96
1,2,3-Propenetricarboxylic acid	pH 7.0	−2.1
Trichloroacetic acid	Ammonia buffer, pH 8.2	−0.84, −1.57
	Phosphate buffer, pH 10.4	−0.9, −1.6
3,4,5-Trihydroxybenzoic acid	Phosphate buffer, pH 2.9	+0.50
	Phosphate buffer, pH 8.8	+0.1
p-Aminophenol	Britton-Robinson buffer, pH 6.3	+0.14
	Britton-Robinson buffer, pH 8.6	−0.04
	Britton-Robinson buffer, pH 12.0	−0.16

TABLE 8.31 Half-Wave Potentials (vs. Saturated Calomel Electrode) of Organic Compounds at 25°C (*Continued*)

Compound	Solvent system	$E_{1/2}$
	Acids (*continued*)	
o-Chlorophenol	pH 5.6	−0.63
m-Chlorophenol	pH 5.6	−0.73
p-Chlorophenol	pH 5.6	−0.65
o-Cresol	pH 5.6	−0.56
m-Cresol	pH 5.6	−0.61
p-Cresol	pH 5.6	−0.54
1,2-Dihydroxybenzene	pH 5.6	−0.35
1,3-Dihydroxybenzene	pH 5.6	−0.61
1,4-Dihydroxybenzene	pH 5.6	−0.23
o-Methoxyphenol	pH 5.6	−0.46
m-Methoxyphenol	pH 5.6	−0.62
p-Methoxyphenol	pH 5.6	−0.41
1-Naphthol	A	−0.74
2-Naphthol	A	−0.82
1,2,3-Trihydroxybenzene	Britton-Robinson buffer, pH 3.1	+0.35
	Britton-Robinson buffer, pH 6.5	+0.10
	Britton-Robinson buffer, pH 9.5	−0.10
	Halogen compounds	
Bromobenzene	A	−1.98
	C	−2.32
1-Bromobutane	C	−2.27
Bromoethane	C	−2.08
Bromomethane	C	−1.63
1-Bromonaphthalene (also 2-bromonaphthalene)	A	−1.55, −1.60
3-Bromo-1-propene	C	−1.29
p-Bromotoluene	A	−1.72
Carbon tetrachloride	C	−0.78, −1.71
Chlorobenzene	A	−2.07
Chloroform	C	−1.63
Chloromethane	C	−2.23
3-Chloro-1-propene	C	−1.91
α-Chlorotoluene	C	−1.81
p-Chlorotoluene	A	−1.76
N-Chloro-p-toluenesulfonamide	0.5 M K_2SO_4	−0.13
9,10-Dibromoanthracene	A	−1.15, −1.47
p-Dibromobenzene	C	−2.10
1,2-Dibromobutane	D+1% Na_2SO_3	−1.45
Dibromoethane	C	−1.48
meso-2,3-Dibromosuccinic acid	Acetate buffer, pH 4.0	−0.23, −0.89
Dichlorobenzenes	C	−2.5

TABLE 8.31   Half-Wave Potentials (vs. Saturated Calomel Electrode) of Organic Compounds at 25°C (*Continued*)

Compound	Solvent system	$E_{1/2}$
Halogen compounds (continued)		
Dichloromethane	C	−1.60
Diiodomethane	C	−1.12, −1.53
Hexabromobenzene	C	−0.8, −1.5
Hexachlorobenzene	C	−1.4, −1.7
Iodobenzene	A	−1.72
Iodoethane	C	−1.67
Iodomethane	A	−2.12
	C	−1.63
Tetrabromomethane	C	−0.3, −0.75, −1.49
Tetraidomethane	C	−0.45, −1.05, −1.46
Tribromomethane	C	−0.64, −1.47
α,α,α-Trichlorotoluene	C	−0.68, −1.65, −2.00
Nitro and nitroso compounds		
1,2-Dinitrobenzene	Phthalate buffer, pH 2.5	−0.12, −0.32, −1.26
	Borate buffer, pH 9.2	−0.38, −0.74
1,3-Dinitrobenzene	Phthalate buffer, pH 2.5	−0.17, −0.29
	Borate buffer, pH 9.2	−0.46, −0.68
1,4-Dinitrobenzene	Phthalate buffer, pH 2.5	−0.12, −0.33
	Borate buffer, pH 9.2	−0.35, −0.80
Methyl nitrobenzoates	Buffer + 10% EtOH, pH 2.0	−0.20 to −0.25 −0.68 to −0.74
p-Nitroacetophenone	Britton-Robinson buffer, pH 2.2	−0.16, −0.61, −1.09
	Britton-Robinson buffer, pH 10.0	−0.51, −1.40, −1.73
o-Nitroaniline	0.03 *M* LiCl + 0.02 *M* benzoic acid in EtOH	−0.88
m-Nitroaniline	Britton-Robinson buffer, pH 4.3	−0.3, −0.8
	Briton-Robinson buffer, pH 7.2	−0.5
	Britton-Robinson buffer, pH 9.2	−0.7
p-Nitroaniline	pH 2.0	−0.36
	Acetate buffer, pH 4.6	−0.5
o-Nitroanisole	Buffer + 10% EtOH, pH 2.0	−0.29, −0.58

TABLE 8.31   Half-Wave Potentials (vs. Saturated Calomel Electrode) of Organic Compounds at 25°C (*Continued*)

Compound	Solvent system	$E_{1/2}$
\multicolumn{3}{c}{Nitro and nitroso compounds (*continued*)}		
p-Nitroanisole	Buffer + 10% EtOH, pH 2.0	−0.35, −0.64
1-Nitroanthraquinone	Britton-Robinson buffer, pH 7.0	−0.16
Nitrobenzene	HCl + KCl + 8% EtOH, pH 0.5	−0.16, −0.76
	Phthalate buffer, pH 2.5	−0.30
	Borate buffer, pH 9.2	−0.70
Nitrocresols	Britton-Robinson buffer, pH 2.2	−0.2 to −0.3
	Britton-Robinson buffer, pH 4.5	−0.4 to −0.5
	Britton-Robinson buffer, pH 8.0	−0.6
Nitroethane	Britton-Robinson buffer + 30% MeOH, pH 1.8	−0.7
	Britton-Robinson buffer + 30% MeOH, pH 4.6	−0.8
2-Nitrohydroquinone	Phosphate buffer + citrate buffer, pH 2.1	−0.2
	Phosphate buffer + citrate buffer, pH 5.2	−0.4
	Phosphate buffer + citrate buffer, pH 8.0	−0.5
Nitromethane	Britton-Robinson buffer + 30% MeOH, pH 1.8	−0.8
	Britton-Robinson buffer + 30% MeOH, pH 4.6	−0.85
o-Nitrophenol	Britton-Robinson buffer + 10% EtOH, pH 2.0	−0.23
	Britton-Robinson buffer + 10% EtOH, pH 4.0	−0.4
	Britton-Robinson buffer + 10% EtOH, pH 8.0	−0.65
	Britton-Robinson buffer + 10% EtOH, pH 10.0	−0.80
m-Nitrophenol	Britton-Robinson buffer + 10% EtOH, pH 2.0	−0.37
	Britton-Robinson buffer + 10% EtOH, pH 4.0	−0.40
	Britton-Robinson buffer + 10% EtOH, pH 8.0	−0.64
	Britton-Robinson buffer + 10% EtOH, pH 10.0	−0.76
p-Nitrophenol	Britton-Robinson buffer + 10% EtOH, pH 2.0	−0.35
	Britton-Robinson buffer + 10% EtOH, pH 4.0	−0.50
	Britton-Robinson buffer + 10% EtOH, pH 8.0	−0.82
1-Nitropropane	Britton-Robinson buffer + 30% MeOH, pH 1.8	−0.73

TABLE 8.31 Half-Wave Potentials (vs. Saturated Calomel Electrode) of Organic Compounds at 25°C (*Continued*)

Compound	Solvent system	$E_{1/2}$
Nitro and nitroso compounds (continued)		
1-Nitropropane (*continued*)	Britton-Robinson buffer + 30% MeOH, pH 8.6	−0.88
	Britton-Robinson buffer + 30% MeOH, pH 8.0	−0.95
2-Nitropropane	McIlvaine buffer, pH 2.1	−0.53
	McIlvaine buffer, pH 5.1	−0.81
Nitrosobenzene	McIlvaine buffer, pH 6.0	−0.03
	McIlvaine buffer, pH 8.0	−0.14
1-Nitroso-2-naphthol	D + buffer, pH 4.0	+0.02
	D + buffer, pH 7.0	−0.20
	D + buffer, pH 9.0	−0.31
N-Nitrosophenylhydroxylamine	pH 2.0	−0.84
o-Nitrotoluene	Phthalate buffer, pH 2.5	−0.35, −0.66
	Phthalate buffer, pH 7.4	−0.60, −1.06
m-Nitrotoluene (also *p*-nitrotoluene)	Phthalate buffer, pH 2.5	−0.30, −0.53
	Phthalate buffer, pH 7.4	−0.58, −1.06
Tetranitromethane	pH 12.0	−0.41
1,3,5-Trinitrobenzene	Phthalate buffer, pH 4.1	−0.20, −0.29, −0.34
	Borate buffer, pH 9.2	−0.34. −0.48, −0.65
Heterocyclic compounds containing nitrogen		
Acridine	D, pH 8.3	−0.80, −1.45
Cinchonine	B, pH 3	−0.90
2-Furanmethanol	Britton-Robinson buffer, pH 2.0	−0.96
	Britton-Robinson buffer, pH 5.8	−1.38, −1.70
2-Hydroxyphenazine	Britton-Robinson buffer, pH 4.0	−0.24
8-Hydroxyquinoline	B, pH 5.0	−1.12
	Phosphate buffer, pH 8.0	−1.18, −1.71
3-Methylpyridine	D + 0.1 *M* LiCl	−1.76
4-Methylpyridine	D + 0.1 *M* LiCl	−1.87
Phenazine	Phosphate buffer + citrate buffer, pH 7.0	−0.36
Pyridine	Phosphate buffer + citrate buffer, pH 7.0	−1.75
Pyridine-2-carboxylic acid	B, pH 4.1	−1.10
	B, pH 9.3	−1.48, −1.94
Pyridine-3-carboxylic acid	0.1 *M* HCl	−1.08
Pyridine-4-carboxylic acid	Britton-Robinson buffer, pH 6.1	−1.14
	pH 9.0	−1.39, −1.68

TABLE 8.31 Half-Wave Potentials (vs. Saturated Calomel Electrode) of Organic Compounds at 25°C (*Continued*)

Compound	Solvent system	$E_{1/2}$
Heterocyclic compounds containing nitrogen (continued)		
Pyrimidine	Citrate buffer, pH 3.6	−0.92, −1.24
	Ammonia buffer, pH 9.2	−1.54
Quinoline-8-carboxylic acid	pH 9	−1.11
Quinoxaline	Phosphate buffer + citrate buffer, pH 7.0	−0.66, −1.52
Azo, hydrazine, hydroxylamine, and oxime compounds		
Azobenzene	D, pH 4.0	−0.20
	D, pH 7.0	−0.50
Azoxybenzene	Buffer + 20% EtOH, pH 6.3	−0.30
Benzoin 1-oxime	Buffer, pH 2.0	−0.88
	Buffer, pH 5.6	−1.08
	Buffer, pH 8.2	−1.67
Benzoylhydrazine	0.13 M NaOH, pH 13.0	−0.30
Dimethylglyoxime	Ammonia buffer, pH 9.6	−1.63
Hydrazine	Britton-Robinson buffer, pH 9.3	−0.09
Hydroxylamine	Britton-Robinson buffer, pH 4.6	−1.42
	Britton-Robinson buffer, pH 9.2	−1.65
Oxamide	Acetate buffer	−1.55
Phenylhydrazine	McIlvaine buffer, pH 2	+0.19
	0.13 M NaOH, pH 13.0	−0.36
Phenylhydroxylamine	McIlvaine buffer + 10% EtOH, pH 2	−0.68
	McIlvaine buffer + 10 EtOH, pH 4–10	−0.33 0.061 pH
Salicylaldoxime	Phosphate buffer, pH 5.4	−1.02
Thiosemicarbazide	Borate buffer, pH 9.3	−0.26
Thiourea	0.1 M sulfuric acid	+0.02
Indicators and dyestuffs		
Brilliant Green	HCl + KCl, pH 2.0	−0.2, −0.5
Indigo carmine	pH 2.5	−0.24
Indigo disulfonate	pH 7.0	−0.37
Malachite Green G	HCl + KCl, pH 2.0	−0.2, −0.5
Metanil yellow	Phosphate buffer + 1% EtOH, pH 7.0	−0.51
Methylene blue	Britton-Robinson buffer, pH 4.9	−0.15
	Britton-Robinson buffer, pH 9.2	−0.30

TABLE 8.31 Half-Wave Potentials (vs. Saturated Calomel Electrode) of Organic Compounds at 25°C (*Continued*)

Compound	Solvent system	$E_{1/2}$
	Indicators and dyestuffs (*continued*)	
Methylene green	Phosphate buffer + 1% EtOH, pH 7.0	−0.12
Methyl orange	Phosphate buffer + 1% EtOH, pH 7.0	−0.51
Morin	D, pH 7.6	−1.7
Neutral red	Britton-Robinson buffer, pH 2.0	−0.21
	Britton-Robinson buffer, pH 7.0	−0.57
	Peroxide	
Ethyl peroxide	0.02 M HCl	−0.2

8.7 CONDUCTANCE

TABLE 8.32 Limiting Equivalent Ionic Conductances in Aqueous Solutions

In mho · cm² · equiv⁻¹.

Ion	Temperature, °C		
	0	18	25
Inorganic cations			
Ag^+	33	54.5	61.9
Al^{3+}	29		61
Ba^{2+}	33.6	54.3	63.9
Be^{2+}			45
Ca^{2+}	30.8	51	59.5
Cd^{2+}	28	45.1	54
Ce^{3+}			70
Co^{2+}	28	45	53
$Co(NH_3)_6^{3+}$			100
$Co(en)_3^{3+}$			74.7
Cr^{3+}			67
Cs^+	44	68	77.3
Cu^{2+}	28	45.3	56.6
D^+ (deuterium)		213.7	
Dy^{3+}			65.7
Er^{3+}			66.0
Eu^{3+}			67.9
Fe^{2+}	28	45.3	53.5
Fe^{3+}			69
Gd^{3+}			67.4

TABLE 8.32 Limiting Equivalent Ionic Conductances in Aqueous Solutions (*Continued*)

Ion	Temperature, °C		
	0	18	25
H^+	224.1	315.8	350.1
Hg_2^{2+}			68.7
Hg^{2+}			63.6
Ho^{3+}			66.3
K^+	40.3	64.6	73.50
La^{3+}	35.0	59.2	69.6
Li^+	19.1	33.4	38.69
Mg^{2+}	28.5	46	53.06
Mn^{2+}	27	44.5	53.5
NH_4^+	40.3	64	73.7
$N_2H_5^+$ (hydrazinium 1+)			59
Na^+	25.85	43.5	50.11
Nd^{3+}			69.6
Ni^{2+}	28	45	50
Pb^{2+}	37.5	60.5	71
Pr^{3+}			69.6
Ra^{2+}	33	56.6	66.8
Rb^+	43.5	67.5	77.8
Sc^{3+}			64.7
Sm^{3+}			65.8
Sr^{2+}	31	51	59.46
Tl^+	43.3	66	74.9
Tm^{3+}			65.5
UO_2^{2+}			32
Y^{3+}			62
Yb^{3+}			65.2
Zn^{2+}	28	45.0	53.5
Inorganic anions			
$Au(CN)_2^-$			50
$Au(CN)_4^-$			36
$B(C_6H_5)_4^-$			21
Br^-	43.1	67.6	78.4
Br_3^-			43
BrO_3^-	31.0	49.0	55.8
Cl^-	41.4	65.5	76.35
ClO_2^-			52
ClO_3^-	36	55.0	64.6
ClO_4^-	37.3	59.1	67.9
CN^-			78
CO_3^{2-}	36	60.5	72
$Co(CN)_6^{3-}$			98.9
CrO_4^{2-}	42	72	85
F^-		46.6	55.4
$Fe(CN)_6^{4-}$			111
$Fe(CN)_6^{3-}$			101
$H_2AsO_4^-$			34
HCO_3^-			44.5
HF_2^-			54.4
HPO_4^{2-}			57

TABLE 8.32 Limiting Equivalent Ionic Conductances in Aqueous Solutions (*Continued*)

Ion	Temperature, °C		
	0	18	25
$H_2PO_4^-$		28	36
HS^-	40	57	65
HSO_3^-	27		50
HSO_4^-			50
$H_2SbO_4^-$			31
I^-	42.0	66.5	76.9
IO_3^-	21.0	33.9	41.0
IO_4^-		49	54.5
MnO_4^-	36	53	62.8
MoO_4^{2-}			74.5
N_3^-			69.5
$N(CN)_2^-$			54.5
NO_2^-	44	59	72
NO_3^-	40.2	61.7	71.42
$NH_2SO_3^-$ (sulfamate)			48.6
OCN^- (cyanate)		54.8	64.6
OH^-	117.8	175.8	199.2
PF_6^-			56.9
PO_3F^{2-}			63.3
PO_4^{3-}			69.0
$P_2O_7^{4-}$			81.4
$P_3O_9^{3-}$			83.6
$P_3O_{10}^{5-}$			109
ReO_4^-		46.5	54.9
SCN^- (thiocyanate)	41.7	56.6	66.5
$SeCN^-$			64.7
SeO_4^{2-}		65	75.7
SO_3^{2-}			79.9
SO_4^{2-}	41	68.3	80.0
$S_2O_3^{2-}$			85.0
$S_2O_4^{2-}$	34		66.5
$S_2O_6^{2-}$			93
$S_2O_8^{2-}$			86
WO_4^{2-}	35	59	69.4
Organic cations			
Decylpyridinium			29.5
Diethylammonium			42.0
Dimethylammonium			51.5
Dipropylammonium			30.1
Dodecylammonium			23.8
Ethylammonium			47.2
Ethyltrimethylammonium			40.5
Isobutylammonium			38.0
Methylammonium			58.3
Piperidinium			37.2
Propylammonium			40.8
Tetrabutylammonium			19.1
Tetraethylammonium			33.0
Tetramethylammonium			45.3

TABLE 8.32 Limiting Equivalent Ionic Conductances in Aqueous Solutions (*Continued*)

	Temperature, °C		
Ion	0	18	25
Tetrapropylammonium			23.5
Trimethylammonium			34.3
Triethylsulfonium			36.1
Trimethylammonium			46.6
Trimethylsulfonium			51.4
Tripropylammonium			26.1
Organic anions			
Acetate	20	34	41
Benzoate			32.4
Bromobenzoate			30
Butanoate			32.6
Chloroacetate			39.7
Chlorobenzoate			33
Citrate(3−)			70.2
Cyanoacetate			41.8
Cyclohexanecarboxylate			28.7
Cyclopropane-1,1-dicarboxylate			53.4
Decylsulfonate			26
Dichloroacetate			38.3
Diethylbarbituate(2−)			26.3
Dihydrogen citrate			30
3,5-Dinitrobenzoate			28.3
Dodecylsulfonate			24
Ethylsulfonate			39.6
Fluorobenzoate			33
Formate		47	54.6
Hydrogen oxalate(1−)			40.2
Lactate			38.8
Methylsulfonate			48.8
Octylsulfonate			29
Phenylacetate			30.6
Propanoate			35.8
Propylsulfonate			37.1
Salicylate			36
Succinate(2−)			58.8
Tartrate(2−)		55	64
Trichloroacetate			36.6

TABLE 8.33 Standard Solutions for Calibrating Conductivity Vessels

The values of conductivity κ are corrected for the conductivity of the water used. The cell constant θ of a conductivity cell can be obtained from the equation

$$\theta = \frac{\kappa R R_{solv}}{R_{solv} - R}$$

where R is the resistance measured when the cell is filled with a solution of the composition stated in the table below, and R_{solv} is the resistance when the cell is filled with solvent at the same temperature.

Grams KCl per Kilogram Solution (in vacuo)	Conductivity in $ohm^{-1} \cdot cm^{-1}$ at		
	0°C	18°C	25°C
71.135 2	0.065 14$_4$	0.097 79$_0$	0.111 28$_7$
7.419 13	0.007 134$_4$	0.011 161$_2$	0.012 849$_7$
0.745 263*	0.000 773 2$_6$	0.001 219 9$_2$	0.001 408 0$_8$

* Virtually 0.0100 M.

From the data of Jones and Bradshaw, *J. Am. Chem. Soc.*, **55**, 1780 (1933). The original data have been converted from (int. ohm)$^{-1}$ cm^{-1}.

TABLE 8.34 Electrical Conductivity of Various Pure Liquids

Liquid	Temp. °C	mhos/cm or $ohm^{-1} \cdot cm^{-1}$	Liquid	Temp. °C	mhos/cm or $ohm^{-1} \cdot cm^{-1}$
Acetaldehyde	15	1.7×10^{-6}	Benzylamine	25	$<1.7 \times 10^{-8}$
Acetamide	100	$<4.3 \times 10^{-5}$	Benzyl benzoate	25	$<1 \times 10^{-9}$
Acetic acid	0	5×10^{-9}	Bromine	17.2	1.3×10^{-13}
	25	1.12×10^{-8}	Bromobenzene	25	$<2 \times 10^{-11}$
Acetic anhydride	0	1×10^{-6}	Bromoform	25	$<2 \times 10^{-8}$
	25	4.8×10^{-7}	iso-Butyl alcohol	25	8×10^{-8}
Acetone	18	2×10^{-8}			
	25	6×10^{-8}	Capronitrile	25	3.7×10^{-6}
Acetonitrile	20	7×10^{-6}	Carbon disulfide	1	7.8×10^{-18}
Acetophenone	25	6×10^{-9}	Carbon tetrachloride	18	4×10^{-18}
Acetyl bromide	25	2.4×10^{-6}	Chlorine	−70	$<1 \times 10^{-16}$
Acetyl chloride	25	4×10^{-7}	Chloroacetic acid	60	1.4×10^{-6}
Alizarin	233	$1.45 \times 10^{-6}(?)$	m-Chloroaniline	25	5×10^{-6}
Allyl alcohol	25	7×10^{-6}	Chloroform	25	$<2 \times 10^{-8}$
Ammonia	−79	1.3×10^{-7}	Chlorohydrin	25	5×10^{-7}
Aniline	25	2.4×10^{-8}	m-Cresol	25	$<1.7 \times 10^{-8}$
Anthracene	230	3×10^{-10}	Cyanogen	. . .	$<7 \times 10^{-9}$
Arsenic tribromide	35	1.5×10^{-6}	Cymene	25	$<2 \times 10^{-8}$
Arsenic trichloride	25	1.2×10^{-6}			
			Dichloroacetic acid	25	7×10^{-8}
			Dichlorohydrin	25	1.2×10^{-5}
Benzaldehyde	25	1.5×10^{-7}	Diethylamine	−33.5	2.2×10^{-9}
Benzene	. . .	7.6×10^{-8}	Diethyl carbonate	25	1.7×10^{-8}
Benzoic acid	125	3×10^{-9}	Diethyl oxalate	25	7.6×10^{-7}
Benzonitrile	25	5×10^{-8}	Diethyl sulfate	25	2.6×10^{-7}
Benzyl alcohol	25	1.8×10^{-6}	Dimethyl sulfate	0	1.6×10^{-7}

TABLE 8.34 Electrical Conductivity of Various Pure Liquids (*Continued*)

Liquid	Temp. °C	mhos/cm or ohm^{-1}·cm^{-1}	Liquid	Temp. °C	mhos/cm or ohm^{-1}·cm^{-1}
Epichlorohydrin	25	3.4×10^{-8}	Nitromethane	18	6×10^{-7}
Ethyl acetate	25	$<1 \times 10^{-9}$	o- or m-Nitrotoluene	25	$<2 \times 10^{-7}$
Ethyl acetoacetate	25	4×10^{-8}	Nonane	25	$<1.7 \times 10^{-8}$
Ethyl alcohol	25	1.35×10^{-9}	Oleic acid	15	$<2 \times 10^{-10}$
Ethylamine	0	4×10^{-7}			
Ethyl benzoate	25	$<1 \times 10^{-9}$	Pentane	19.5	$<2 \times 10^{-10}$
Ethyl bromide	25	$<2 \times 10^{-8}$	Petroleum	. . .	3×10^{-13}
Ethylene bromide	19	$<2 \times 10^{-10}$	Phenetole	25	$<1.7 \times 10^{-8}$
Ethylene chloride	25	3×10^{-8}	Phenol	25	$<1.7 \times 10^{-8}$
Ethyl ether	25	$<4 \times 10^{-13}$	Phenyl isothiocyanate	25	1.4×10^{-6}
Ethylidene chloride	25	$<1.7 \times 10^{-8}$	Phosgene	25	7×10^{-9}
Ethyl iodide	25	$<2 \times 10^{-8}$	Phosphorus	25	4×10^{-7}
Ethyl isothiocyanate	25	1.26×10^{-7}	Phosphorus oxychloride	25	2.2×10^{-6}
Ethyl nitrate	25	5.3×10^{-7}			
Ethyl thiocyanate	25	1.2×10^{-6}	Pinene	23	$<2 \times 10^{-10}$
Eugenol	25	$<1.7 \times 10^{-8}$	Piperidine	25	$<2 \times 10^{-7}$
			Propionaldehyde	25	8.5×10^{-7}
Formamide	25	4×10^{-6}	Propionic acid	25	$<1 \times 10^{-9}$
Formic acid	18	5.6×10^{-5}	Propionitrile	25	$<1 \times 10^{-7}$
	25	6.4×10^{-5}	n-Propyl alcohol	18	5×10^{-8}
Furfural	25	1.5×10^{-6}		25	2×10^{-8}
			iso-Propyl alcohol	25	3.5×10^{-6}
Gallium	30	36,800	n-Propyl bromide	25	$<2 \times 10^{-8}$
Glycerol	25	6.4×10^{-8}	Pyridine	18	5.3×10^{-8}
Glycol	25	3×10^{-7}			
Guaiacol	25	2.8×10^{-7}	Quinoline	25	2.2×10^{-8}
			Salicylaldehyde	25	1.6×10^{-7}
Heptane	. . .	$<1 \times 10^{-13}$	Stearic acid	80	$<4 \times 10^{-13}$
Hexane	18	$<1 \times 10^{-18}$	Sulfonyl chloride, SOCl$_2$	25	2×10^{-6}
Hydrogen bromide	−80	8×10^{-9}			
Hydrogen chloride	−96	1×10^{-8}	Sulfur	115	1×10^{-12}
Hydrogen cyanide	0	3.3×10^{-6}		130	5×10^{-11}
Hydrogen iodide	B.P.	2×10^{-7}		440	1.2×10^{-7}
Hydrogen sulfide	B.P.	1×10^{-11}	Sulfur dioxide	35	1.5×10^{-8}
			Sulfuric acid	25	1×10^{-2}
Iodine	110	1.3×10^{-10}	Sulfuryl chloride, SO$_2$Cl$_2$	25	3×10^{-8}
Kerosene	25	$<1.7 \times 10^{-8}$	Toluene	. . .	$<1 \times 10^{-14}$
			o-Toluidine	25	$<2 \times 10^{-6}$
Mercury	0	10,629.6	p-Toluidine	100	6.2×10^{-8}
Methyl acetate	25	3.4×10^{-6}	Trichloroacetic acid	25	3×10^{-9}
Methyl alcohol	18	4.4×10^{-7}	Trimethylamine	−33.5	2.2×10^{-10}
Methyl ethyl ketone	25	1×10^{-7}	Turpentine	. . .	2×10^{-13}
Methyl iodide	25	$<2 \times 10^{-8}$			
Methyl nitrate	25	4.5×10^{-6}	iso-Valeric acid	80	$<4 \times 10^{-13}$
Methyl thiocyanate	25	1.5×10^{-6}			
			Water	18	4×10^{-8}
Naphthalene	82	4×10^{-10}	Xylene	. . .	$<1 \times 10^{-15}$
Nitrobenzene	0	5×10^{-9}			

TABLE 8.35 Equivalent Conductivities of Electrolytes in Aqueous Solutions at 18°C

The unit of Λ in the table is $\Omega^{-1} \cdot cm^2 \cdot equiv^{-1}$. The entities to which the equivalent relates are given in the first column.

Electrolyte	Concentration, N										
	0.001	0.005	0.01	0.05	0.1	0.5	1.0	2.0	3.0	4.0	5.0
Acetic acid	41	20.0	14.3	6.48	4.60	2.01	1.32		0.54		0.29
$AgNO_3$	113.2	110.0	107.8	99.5	94.3	77.8	67.8	56.0	48.2	42.1	37.2
$\frac{1}{2}Ag_2SO_4$	116.3	108.4	102.9								
$\frac{1}{3}AlBr_3$ (25°)	132	124	119	103	97						
$\frac{1}{3}AlCl_3$	121.1	105.0	93.8			65.0	56.2	44.2	34.7	27.2	
$\frac{1}{3}AlI_3$ (25°)	131	124	119	108	88						
$\frac{1}{3}Al(NO_3)_3$ (25°)	123	115	110	94							
$\frac{1}{6}Al_2(SO_4)_3$ (25°)	107.2	76.8	60.6								
$\frac{1}{2}Ba(OAc)_2$	85.0	80.4	77.1	65.7	60.2	43.8	34.3				
$\frac{1}{2}Ba(BrO_3)_2$ (25°)	113.6	106.8	102.7								
$\frac{1}{2}BaCl_2$	115.6	112.3	106.7	96.0	90.8	77.3	70.1	60.3	52.3	23.4	
$\frac{1}{2}Ba(NO_3)_2$	111.7	105.3	101.0	86.8	78.9	56.6	48.4		29.8		
$\frac{1}{2}Ba(OH)_2$	216	213	207	191	180						
Butyric acid						1.66	0.98	0.46	0.26	0.18	0.11
$\frac{1}{2}Ca(OAc)_2$	79.6	75.0	71.9	60.3	54.0	36.3	26.3				
$\frac{1}{2}CaCl_2$	112.0	106.7	103.4	93.3	88.2	74.9	67.5	58.3	49.7	42.4	35.6
$\frac{1}{2}Ca(NO_3)_2$	108.5	103.0	99.5	88.4	82.5	65.7	55.9	43.5	35.5	26.0	21.5
$\frac{1}{2}Ca(OH)_2$		233	226								
$\frac{1}{2}CaSO_4$	104.3	86.3	77.4								
$\frac{1}{2}CdBr_2$		86.5	76.3	53.2	44.6	25.3	18.3	12.5	9.1	6.8	5.3
$\frac{1}{2}CdCl_2$		91	83	59	50	30.8	22.4	14.4	9.9	7.1	5.4
$\frac{1}{2}CdI_2$		76.7	65.6	40.1	31.0	18.3	15.4	12.3	9.7	8.0	
$\frac{1}{2}Cd(NO_3)_2$		100	96	86.4	80.8	63.9	54.5	41.0	31.4	23.7	17.6
$\frac{1}{2}CdSO_4$		79.7	70.3	49.6	42.2	28.7	23.6	17.7	14.0	11.0	8.35
$\frac{1}{3}CeCl_3$ (25°)	137.4		122.1		99.0						
$\frac{1}{6}Ce_2(C_2O_4)_3$ (25°)	85.5	54	45.8	29							

Compound	Values (left → right across the table)
Chloroacetic acid (25°)	88.4, 54, 42.5, 22.0, 42.9, 20.2, 13.6, 8.1, 5.6, 4.2, 3.3
Citric acid	99.3, 95.6, 82.3, 16.1, 7.3, 5.4, 40.3, 35.4, 30.5, 26.4
$\frac{1}{2}CoCl_2$	201, 195, 193, 191, 186, 51.5, 45.3, 44.8, 35.2, 24.5, 19.1
$\frac{1}{3}CrCl_3$	130.7, 127.5, 125.2, 113.5, 68.6, 56.8, 35.3, 27.8, 21.4
$\frac{1}{2}CrO_3(H_2CrO_4)$ (25°)	55.7, 50.6, 47.2, 34.9, 28.4, 104.3, 100.3, 95.7, 85.1
CsCl	104.3, 100.3, 95.7, 85.1
$\frac{1}{2}Cu(OAc)_2$ (25°)	107.9, 97.1, 93.7, 83.7, 78.2, 67.5, 56.8, 41.2, 31.5, 24.5, 19.1
$\frac{1}{2}CuCl_2$	98.5, 81.0, 71.7, 53.6, 43.8, 30.5, 25.6, 45.4, 35.3, 27.8, 21.4
$\frac{1}{2}Cu(NO_3)_2$ (15°)	119, 82, 44.6, 26.5, 16.3, 9.6
$\frac{1}{2}CuSO_4$	131, 125, 120, 103, 93, 119, 82, 44.6, 26.5, 16.3, 9.6
Dichloroacetic acid (25°)	82, 75, 70, 54, 44.5, 66.5, 52.9, 37.6, 28.1, 20.5, 15.9
$\frac{1}{2}FeCl_2$ (25°)	125.6, 230.0, 187.0, 103.4, 80.4, 30.8, 25.8, 19.5, 15.37, 2.39, 1.92
$\frac{1}{3}FeCl_3$	308.2, 5.18, 3.68, 2.93
$\frac{1}{2}FeSO_4$	13.5
Formic acid	401, 387, 373, 272, 356, 306, 282, 243, 214, 179
H_3AsO_4 (1 M) (25°)	377, 373, 370, 360, 156
H_3BO_3	413, 406, 402, 392, 351, 327, 301, 247, 215, 152.2
HBr	343, 358, 292, 207
HBrO$_3$ (25°)	386, 31.3, 25.7, 24.2, 24.0
HCl	343.3, 332.8, 323.9, 357, 347, 322, 297, 255, 215, 179
HClO$_3$	375, 371, 368, 253, 175, 141, 106, 87, 71
HClO$_4$ (25°)	318, 279, 255, 350, 324, 310, 220
HF	399, 394, 390, 377, 370, 205, 198, 166.8, 156
HI	361, 330, 308, 253, 225, 66, 53.1, 51.3
HIO$_3$	1.85, 1.23
HNO$_3$	98.3, 95.7, 94.0, 87.7, 53.9, 37.0, 28.7, 19.8, 14.4, 10.1
H_3PO_4 (1 M)	129.4, 126.4, 124.4, 117.8, 83.8, 71.6, 63.4, 50.0, 40.7, 31.4
HSCN (25°)	109.9, 106.9, 104.7, 97.3, 114.2, 105.4, 102.5, 98.0, 93.3, 87.9, 135.0
$\frac{1}{2}H_2SO_4$	109.9, 103, 87.8, 93.0, 102.5
$\frac{1}{2}HgCl_2$	80.8, 24.5
$\frac{1}{3}InBr_3$	127.3, 124.4, 122.4, 115.8, 112.0, 102.4, 98.3, 92.0, 88.9
KOAc	
KBr	
KBrO$_3$	
$\frac{1}{3}K_3$citrate	
KCl	127.3, 124.4, 122.4, 115.8, 112.0, 102.4, 98.3, 92.0, 88.9

TABLE 8.35 Equivalent Conductivities of Electrolytes in Aqueous Solutions at 18°C (*Continued*)

Electrolyte	Concentration, N										
	0.001	0.005	0.01	0.05	0.1	0.5	1.0	2.0	3.0	4.0	5.0
$KClO_3$	116.9	113.6	111.6	103.7	99.2	85.3					
$KClO_4$ (25°)	137.9	134.2	131.5	121.6	115.2						
KCN (15°)						104.2	99.7				
$\frac{1}{2}K_2CO_3$	133.0	121.6	115.5	100.7	94.1	77.8	70.7	65.0	55.6	49.2	42.9
$\frac{1}{2}K_2C_2O_4$	122.4	116.7	112.5	100.8	94.9	80.4	73.7	72.0	59.9		
$\frac{1}{2}K_2CrO_4$					100.5	86.4	79.5				
$\frac{1}{2}K_2Cr_2O_7$					98.2	85.4					
KF	108.9	106.2	104.3	97.7	94.0	82.6	76.0	63.4	56.5	51.7	46.5
$\frac{1}{3}K_3[Fe(CN)_6]$	163.1	150.7									
$\frac{1}{4}K_4[Fe(CN)_6]$	167.2	146.1	134.8	107.7	97.9						
$KHCO_3$ (25°)	115.3	112.2	110.1			86.5	78.9				
KH phthalate	119.3	103.7	99.9	89.3	83.8						
KHS						92.5	91.7	86.4	80.7		69.3
$KHSO_4$						21.0	18.4	15.2			
KH_2PO_4 (1 M) (25°)	107.1	100.8	98.0	90.7	85.6	60.0[18]	45.8[18]				
KI	128.2	125.3	123.4	117.3	114.0	106.2	103.6	101.3	96.4	89.0	81.2
KIO_3	96.0	93.2	91.2	84.1	79.7						
KIO_4 (25°)	124.9	121.2	118.5	106.7	98.1						
$KMnO_4$ (25°)	133.3		126.5		113						
KNO_3	123.6	120.5	118.2	109.9	104.8	89.2	80.5	69.4	61.3		
KOH	234	230	228	219	213	197	184		140.6		105.8
$KReO_4$ (25°)	125.1	121.3	118.5	106.4	97.4						
$\frac{1}{2}K_2S$							135.6	119.7	108.3	97.2	86.1
$KSCN$	118.6	115.8	113.9	107.7	104.3	95.7	91.6	86.8	74.6		
$\frac{1}{2}K_2SO_4$	126.9	120.3	115.8	101.9	94.9	78.5	71.6				
$\frac{1}{3}LaCl_3$ (25°)	137.0	127.5	121.8	106.2	99.1						
$\frac{1}{3}La(NO_3)_3$				86.1	72.1	65.4	54.0	39.1	28.5	19.9	
$\frac{1}{6}La_2(SO_4)_3$				25.7	21.5						
Lactic acid	108.9	53.5	39	18.1	13.2						

LiOAc					51.3	37.7	28.9	18.2	11.9	7.2	
LiBr				87.9	84.4	73.9	67.2	57.7		44.2	
LiCl	96.5	93.9	92.1	86.1	82.4	70.7	63.4	53.1	45.3		33.3
$LiClO_4$ (25°)	103.4	100.6	98.6	92.2	88.6						
$\frac{1}{2}Li_2CO_3$				64.2	59.1						
LiI						75.3	69.2	61.0			
$LiIO_3$	65.3	62.9	61.2	55.3	51.5	39.0	31.2	21.4	14.6		
$LiNO_3$	92.9	90.3	88.6	82.7	79.2	68.0	60.8	50.3	34.9	27.3	
LiOH						149.0	134.5	113.5	95.7		
$\frac{1}{2}Li_2SO_4$	96.4		86.9	74.7	68.2	50.5	41.3	30.7	23.3	18.1	13.9
$\frac{1}{2}MgCl_2$	106.4	101.3	98.1	88.5	83.4	69.6	61.5	52.3	43.3	35.0	28.0
$\frac{1}{2}Mg(NO_3)_2$	102.6	97.7	94.7	85.3	80.5	67.0	59.0	47.0	39.8		
$\frac{1}{2}MgSO_4$	99.8	84.5	76.2	56.9	49.7	35.4	28.9	23.0	17.3	12.9	9.3
$\frac{1}{2}MnCl_2$					86.0	68.5	61.0	48.5	38.8	30.2	23.0
$\frac{1}{2}MnSO_4$						27.6	24.4	18.3	14.0	10.5	7.3
NH_3(aq)	28.0	13.2	9.6	4.6	3.3	1.35	0.89		0.36		0.20
NH_4OAc		92.9	91.4	84.9		60.5	54.7	42.9	34.0	26.5	
NH_4Cl	127.3	124.3	122.1	115.2	110.7	101.4	97.0	92.1	88.2	85.0	80.7
NH_4F					90.1	74.5	65.7	55.3	47.9	42.2	
NH_4I				118.0	115.0	106.0	103.1	100.0		91.4	84.5
NH_4NO_3	124.5		118.0	110.0	106.6	94.5	88.8	85.1		71.9	47.6
NH_4SCN					104.3	94.0	89.9	84.7	79.2	74.0	
$\frac{1}{2}(NH_4)_2SO_4$		120.0	116.5		89.0	79.5	73.0	65.0		55.2	
NaOAc	75.2	72.4	70.2	64.2	61.1	49.4	41.2	29.8	21.5	15.3	10.5
NaBr				99.1	96.0	84.6	78.1	69.1		53.0	
$NaBrO_3$						61.8	54.5	44.1			
Na *n*-butyrate (25°)	80.3	77.6	75.8	69.3	65.3						
NaCl	106.5	103.8	102.0	95.7	92.0	80.9	74.3	64.8	56.5	49.4	42.7
$NaClO_4$	114.9^{25}	111.7^{25}	109.6^{25}	102.4^{25}	98.4^{25}	71.7	65.0	55.1	46.0	38.8	
$\frac{1}{2}Na_2CO_3$	112	102.5	96.2	80.3	72.9	54.5	45.5	34.5	27.2		
$\frac{1}{2}Na_2CrO_4$					82.5	66.4	57.7	46.6	38.3	31.1	
$\frac{1}{2}Na_2Cr_2O_7$ (25°)		103		98.3	94.9						
NaF	87.8	85.2	83.5	77.0	73.1	60.0	51.9				
$\frac{1}{4}Na_4[Fe(CN)_6]$ (25°)		129.6	120.0	97.0	88.2						
Na formate	88.6					61.4	53.7	43.1	34.8	28.2	

TABLE 8.35 Equivalent Conductivities of Electrolytes in Aqueous Solutions at 18°C (*Continued*)

Electrolyte	Concentration, N										
	0.001	0.005	0.01	0.05	0.1	0.5	1.0	2.0	3.0	4.0	5.0
$NaHCO_3$ (25°)	93.5	90.5	88.4	80.6	76.0						
$\frac{1}{3}Na_2HPO_4$	58.4		54.0		44.0	33.5	28.0				
NaH_2PO_4	67.9	65.8	64.4	57.8	54.1						
$\frac{1}{4}Na_2H_2P_2O_7$	41.1	39.4	38.2	34.6	32.5	25.4					
NaI	124.2	121.2	119.2	112.8	108.8	97.5	89.9	78.6	69.9	62.2	
$NaIO_3$	75.2	72.6	70.9	64.4	60.5						
$\frac{1}{2}Na_2MoO_4$	120.8	113	110								
NaN_3 (25°)	117.1	113.8	110.5	101.3	95.7		68.0				
$NaNO_2$ (25°)	102.9	100.1	98.2	91.4	87.2		75.9	63.1	53.6		39.7
$NaNO_3$						74.1	65.9	54.5	46.0	39.0	
$NaOH$	208	203	200	190	183	172	160		108.0		69.0
Na picrate (25°)	78.6	75.7	73.7	66.3	61.8						
$\frac{1}{3}Na_3PO_4$	125	122	119								
Na propionate (25°)	83.5	80.9	79.1								
$\frac{1}{2}Na_2S$						117.0	104.3	85.0	71.0	59.0	47.2
$NaSCN$	144	139	136	124	116	88	68.9	59.8	50.9	43.7	19
$\frac{1}{2}Na_2SiO_3$	106.7	100.8	96.8	83.9	78.4	74.3	72	51	38		
$\frac{1}{2}Na_2SO_4$	120	81.5	74.8	64.3	60.4	59.7	50.8	40.0	33.5	27	
(mono) Na tartrate	116.1	109.2	104.8	92.2	85.8						
$\frac{1}{2}Na_2WO_4$ (25°)											
$\frac{1}{2}NiSO_4$	96.3	79.5	70.8	51.0	43.8	30.4	25.1	19.3	15.1		
$\frac{1}{2}$Oxalic acid	180.7		158.2	132.9	116.9	75.9	59.4				
$\frac{1}{2}Pb(NO_3)_2$	116.1	108.6	103.5	86.3	77.3	53.2	42.0	31.0			
Propionic acid						1.57	1.00	0.54		0.20	
$RbCl$	130.3	127.4	125.3	117.8	113.9	204.8	101.9	97.1	92.7	87.2	
$RbOH$					220.6	216.8	192.0	170.0	148.3		
$\frac{1}{4}SnCl_4$	114.5	108.9	105.4	94.4	90.2	75.7	121.7	66.9	47.9	32.7	
$\frac{1}{2}SrCl_2$	108.3	102.7	99.0	87.3	80.9	62.7	68.5	58.7	49.9	42.2	
$\frac{1}{2}Sr(NO_3)_2$							52.1	38.0	29.3	29.3	16.4

							7.03	4.58	3.32	2.48	1.83
Tartaric acid (15°)	128.2	123.7	120.2			61.0	54.0	44.3	36.3	29.8	
$\frac{1}{4}$ThCl$_4$	113.3	108.2	105.4	97.4	92.6	78.8	71.5	62.7			
TlCl	124.7	121.1	118.4	107.9	101.2						
TlF	127.4	118.4	112.3	92.7	83.1						
TlNO$_3$											
$\frac{1}{2}$Tl$_2$SO$_4$											
Trichloroacetic acid (25°)						273	207	127	79	44	19
$\frac{1}{2}$UO$_2$F$_2$ (25°)	26.10	12.31	9.17	5.43	4.74	3.75	3.22				2.7
$\frac{1}{2}$UO$_2$SO$_4$ (25°)	106.5	63.2	49.2	27.6	22.2	14.4	11.6				
$\frac{1}{3}$YCl$_3$ (25°)	129	122	118	109							
$\frac{1}{2}$Zn(OAc)$_2$ (25°)	83	77	73	58	49						
$\frac{1}{2}$ZnCl$_2$	107	101	98	87	82	65	55	39.6	29.6	23.2	18.5
$\frac{1}{2}$Zn(NO$_3$)$_2$	120	114	111	100							
$\frac{1}{2}$ZnSO$_4$	98.4	82.1	73.2	53.0	45.6	32.3	26.6	20.0	15.9	12.0	9.0

TABLE 8.36 Conductivity of Very Pure Water at Various Temperatures and the Equivalent Conductances of Hydrogen and Hydroxyl Ions

Temp., °C	Conductivity, $\mu S \cdot cm^{-1}$	Resistivity, $M\Omega \cdot cm$	Equivalent conductance, $cm^2 \cdot ohm^{-1} \cdot equivalent^{-1}$	
			λ^0, H$^+$	λ^0, OH$^-$
0	0.011 61	86.14	224.1	117.8
5	0.016 61	60.21	250.0	133.6
10	0.023 15	43.21	275.6	149.6
15	0.031 53	31.71	300.9	165.9
18	0.037 54	26.64	315.8	491.6
20	0.042 05	23.78	325.7	182.5
25	0.055 08	18.15	350.1	199.2
30	0.070 96	14.09	374.0	216.1
35	0.090 05	11.10	397.4	233.0
40	0.112 7	8.88	420.0	267.2
45	0.139 3	7.18	442.0	267.2
50	0.170 2	5.88	463.3	284.3
55	0.205 5	4.86	483.8	301.4
60	0.245 7	4.06	503.4	318.5
65	0.291 2	3.43	522.0	335.4
70	0.341 6	2.93	539.7	352.2
75	0.397 8	2.51	556.4	368.8
80	0.459 3	2.18	572.0	385.2
85	0.525 8	1.90	586.4	401.4
90	0.597 7	1.67	599.6	417.3
95	0.675 3	1.48	611.6	432.8
100	0.756 9	1.32	622.2	448.1
150	1.84	0.543		
200	2.99	0.334	824	701
250	3.31	0.302		
300	2.42	0.413	894	821

Source: Data from T. S. Light and S. L. Licht, *Anal. Chem.*, **59**:2327–2330 (1987).

8.7.1 Common Conductance Relations*

Conductivity. The standard unit of conductance is electrolytic conductivity (formerly called specific conductance) κ, which is defined as the reciprocal of the resistance $[\Omega^{-1}]$ of a 1-m cube of liquid at a specified temperature $[\Omega^{-1} \cdot m^{-1}]$. See Table 8.33 and the definition of the cell constant.

In accurate work at low concentrations it is necessary to subtract the conductivity of the pure solvent (Table 8.34) from that of the solution to obtain the conductivity due to the electrolyte.

Resistivity (Specific Resistance)

$$\rho = \frac{1}{\kappa} \quad [\Omega \cdot m]$$

---

* SI units are in brackets.

Conductance of an Electrolyte Solution

$$\frac{1}{R} = \kappa \frac{S}{d} \quad [\Omega^{-1}]$$

where S is the surface area of the electrode, or the mean cross-sectional area of the solution [m^2], and d is the mean distance between the electrodes [m].

Equivalent Conductivity

$$\Lambda = \frac{\kappa}{C} \quad [\Omega^{-1} \cdot m^2 \cdot equiv^{-1}]$$

In the older literature, C is the concentration in equivalents per liter. The volume of the solution in cubic centimeters per equivalent is equal to $1000/C$, and $\Lambda = 1000 \, \kappa/C$, the units employed in Table 8.32 [$\Omega^{-1} \cdot cm^2 \cdot equiv^{-1}$]. The formula unit used in expressing the concentration must be specified; for example, NaCl, ½K$_2$SO$_4$, ⅓LaCl$_3$.

The equivalent conductivity of an electrolyte is the sum of contributions of the individual ions. At infinite dilution: $\Lambda^\circ = \lambda_c^\circ + \lambda_a^\circ$, where λ_c° and λ_a° are the ionic conductances of cations and anions, respectively, at infinite dilution (Table 8.35).

Ionic Mobility and Ionic Equivalent Conductivity

$$\lambda_c = Fu_c \quad \text{and} \quad \lambda_a = Fu_a \quad [\Omega^{-1} \cdot m^2 \cdot equiv^{-1}]$$

where F is the Faraday constant, and u_c, u_a are the ionic mobilities [m$^2 \cdot$ s$^{-1} \cdot$ V^{-1}].

$$\Lambda = \alpha F(u_c + u_a) = \alpha(\lambda_c + \lambda_a)$$

where α is the degree of electrolytic dissociation, Λ/Λ°. The electric mobility u of a species is the magnitude of the velocity in an electric field [m$\cdot$s^{-1}] divided by the magnitude of the strength of the electric field E [V$\cdot$m^{-1}].

Ostwald Dilution Law

$$K_d = \frac{\alpha^2 C}{1 - \alpha}$$

where K_d is the dissociation constant of the weak electrolyte. In general for an electrolyte which yields n ions:

$$K_d = \frac{C^{(n-1)} \Lambda^n}{\Lambda^{\circ (n-1)}(\Lambda^\circ - \Lambda)}$$

Transference Numbers or Hittorf Transport Numbers

$$T_c = \frac{\lambda_c}{\lambda_c + \lambda_a} \qquad T_a = \frac{\lambda_a}{\lambda_c + \lambda_a} \qquad T_c + T_a = 1$$

$$\frac{T_c}{T_a} = \frac{u_c}{u_a} = \frac{\lambda_c}{\lambda_a}$$

$$\lambda_c = T_c \Lambda \qquad \lambda_a = T_a \Lambda$$

SECTION 9
PHYSICOCHEMICAL RELATIONSHIPS

9.1 LINEAR FREE ENERGY RELATIONSHIPS

Many equilibrium and rate processes can be systematized when the influence of each substituent on the reactivity of substrates is assigned a characteristic constant σ and the reaction parameter ρ is known or can be calculated. The Hammett equation

$$\log \frac{K}{K^\circ} = \sigma\rho$$

describes the behavior of many *meta-* and *para*-substituted aromatic species. In this equation K° is the acid dissociation constant of the reference in aqueous solution at 25°C and K is the corresponding constant for the substituted acid. Separate sigma values are defined by this reaction for *meta* and *para* substituents and provide a measure of the total electronic influence (polar, inductive, and resonance effects) in the absence of conjugation effects. Sigma constants are not valid for substituents *ortho* to the reaction center because of anomalous (mainly steric) effects. The inductive effect is transmitted about equally to the *meta* and *para* positions. Consequently, σ_m is an approximate measure of the size of the inductive effect of a given substituent and $\sigma_p - \sigma_m$ is an approximate measure of a substituent's resonance effect. Values of Hammett sigma constants are listed in Table 9.1.

Taft sigma values σ^* perform a similar function with respect to aliphatic and alicyclic systems. Values of σ^* are listed in Table 9.1.

The reaction parameter ρ depends upon the reaction series but not upon the substituents employed. Values of the reaction parameter for some aromatic and aliphatic systems are given in Tables 9.2 and 9.3.

Since substituent effects in aliphatic systems and in *meta* positions in aromatic systems are essentially inductive in character, σ^* and σ_m values are related by the expression $\sigma_m = 0.217\sigma^* - 0.106$. Substituent effects fall off with increasing distance from the reaction center; generally a factor of 0.36 corresponds to the interposition of a $-CH_2-$ group, which enables σ^* values to be estimated for $R-CH_2-$ groups not otherwise available.

TABLE 9.1 Hammett and Taft Substituent Constants

Substituent	Hammett constants		Taft constant σ^*
	σ_m	σ_p	
$-AsO_3H^-$	−0.09	−0.02	0.06
$-B(OH)_2$	0.01	0.45	
$-Br$	0.39	0.23	2.84
$-CH_2Br$			1.00
m-BrC_6H_4-		0.09	
p-BrC_6H_4-		0.08	
$-CH_3$	−0.07	−0.17	0.0
$-CH_2CH_3$	−0.07	−0.15	−0.10
$-CH_2CH_2CH_3$	−0.05	−0.15	−0.12
$-CH(CH_3)_2$	−0.07	−0.15	−0.19

TABLE 9.1   Hammett and Taft Substituent Constants (*Continued*)

Substituent	Hammett constants		Taft constant
	σ_m	σ_p	σ^*
—$CH_2CH_2CH_2CH_3$	−0.07	−0.16	−0.13
—$CH_2CH(CH_3)_2$	−0.07	−0.12	−0.13
—$CH(CH_3)CH_2CH_3$		−0.12	−0.19
—$C(CH_3)_3$	−0.10	−0.20	−0.30
—$CH_2CH_2CH_2CH_2CH_3$			−0.25
—$CH_2CH_2CH(CH_3)_2$			−0.17
—$CH_2C(CH_3)_3$		−0.23	−0.12
—$CH_2CH_2CH_2CH_2CH_2CH_3$			−0.37
Cyclopropyl—	−0.07	−0.21	
Cyclohexyl—			−0.15
—3,4-$(CH_2)_2$ (fused)		−0.26	
—3,4-$(CH_2)_3$— (fused ring)		−0.48	
—3,4-$(CH)_4$— (fused ring)	0.06	0.04	
—$CH=CH_2$	0.02		0.56
—$CH=C(CH_3)_2$			0.19
—$CH=CHCH_3$, *trans*			0.36
—CH_2—$CH=CH_2$			0.0
—$CH=CHC_6H_5$	0.14	−0.05	0.41
—$C\equiv CH$	0.21	0.23	2.18
—$C\equiv CC_6H_5$	0.14	0.16	1.35
—CH_2—$C\equiv CH$			0.81
—C_6H_5	0.06	−0.01	0.60
p-$CH_3C_6H_4$—		−0.5	
Naphthyl— (both 1- and 2-)			0.75
—$CH_2C_6H_5$		0.46	0.22
—CH_2CH_2—C_6H_5			−0.06
—$CH(CH_3)C_6H_5$			0.37
—$CH(C_6H_5)_2$			0.41
—CH_2—$C_{10}H_7$			0.44
2-Furoyl—			0.25
3-Indolyl—			−0.06
2-Thienyl—			1.31
2-Thienylemethylene—			0.31
—CHO	0.36	0.22	
—$COCH_3$	0.38	0.50	1.65
—$COCH_2CH_3$		0.48	
—$COCH(CH_3)_2$		0.47	
—$COC(CH_3)_3$		0.32	
—$COCF_3$	0.65		3.7
—COC_6H_5	0.34	0.46	2.2
—$CONH_2$	0.28	0.36	1.68
—$CONHC_6H_5$			1.56
—CH_2COCH_3			0.60
—CH_2CONH_2			0.31
—$CH_2CH_2CONH_2$			0.19

TABLE 9.1   Hammett and Taft Substituent Constants (*Continued*)

Substituent	Hammett constants		Taft constant
	σ_m	σ_p	σ^*
$-CH_2CH_2CH_2CONH_2$			0.12
$-CH_2CONHC_6H_5$			0.0
$-COO^-$	-0.1	0.0	-1.06
$-COOH$	0.36	0.43	2.08
$-CO-OCH_3$	0.32	0.39	2.00
$-CO-OCH_2CH_3$	0.37	0.45	2.12
$-CH_2CO-OCH_3$			1.06
$-CH_2CO-OCH_2CH_3$			0.82
$-CH_2COO$			-0.06
$-CH_2CH_2COOH$	-0.03	-0.07	
$-Cl$	0.37	0.23	2.96
$-CCl_3$	0.47		2.65
$-CHCl_2$			1.94
$-CH_2Cl$	0.12	0.18	1.05
$-CH_2CH_2Cl$			0.38
$-CH_2CCl_3$			0.75
$-CH_2CH_2CCl_3$			0.25
$-CH=CCl_2$			1.00
$-CH_2CH=CCl_2$			0.19
$p\text{-}ClC_6H_4-$		0.08	
$-F$	0.34	0.06	3.21
$-CF_3$	0.43	0.54	2.61
$-CHF_2$			2.05
$-CH_2F$			1.10
$-CH_2CF_3$			0.90
$-CH_2CF_2CF_2CF_3$			0.87
$-C_6F_5$	-0.12	-0.03	
$-Ge(CH_3)_3$		0.0	
$-Ge(CH_2CH_3)_3$		0.0	
$-H$	0.00	0.00	0.49
$-I$	0.35	0.28	2.46
$-CH_2I$			0.85
$-IO_2$	0.70	0.76	
$-N_2^+$	1.76	1.91	
$-N_3$ (azide)	0.33	0.08	2.62
$-NH_2$	-0.16	-0.66	0.62
$-NH_3^+$	1.13	1.70	3.76
$-CH_2-NH_2$			0.50
$-CH_2-NH_3^+$			2.24
$-NH-CH_3$	-0.30	-0.84	
$-NH-C_2H_5$	-0.24	-0.61	
$-NH-C_4H_9$	-0.34	-0.51	
$-NH(CH_3)_2^+$			4.36
$-NH_2-CH_3^+$	0.96		3.74
$-NH_2-C_2H_5^+$	0.96		3.74

TABLE 9.1   Hammett and Taft Substituent Constants (*Continued*)

Substituent	Hammett constants		Taft constant σ^*
	σ_m	σ_p	
$-N(CH_3)_3^+$	0.88	0.82	4.55
$-N(CH_3)_2$	−0.2	−0.83	0.32
$-CH_2-N(CH_3)_3^+$			1.90
$-N(CF_3)_2$	0.45	0.53	
$p-H_2N-C_6H_6-$		−0.30	
$-NH-CO-CH_3$	0.21	0.00	1.40
$-NH-CO-C_2H_5$			1.56
$-NH-CO-C_6H_5$	0.22	0.08	1.68
$-NH-CHO$	0.25		1.62
$-NH-CO-NH_2$	0.18		1.31
$-NH-OH$	−0.04	−0.34	
$-NH-CO-OC_2H_5$	0.33		1.99
$-CH_2-NH-CO-CH_3$			0.43
$-NH-SO_2-C_6H_5$			1.99
$-NH-NH_2$	−0.02	−0.55	
$-CN$	0.56	0.66	3.30
$-CH_2-CN$	0.17	0.01	1.30
$-NO$		0.12	
$-NO_2$	0.71	0.78	4.0
$-CH_2-NO_2$			1.40
$-CH_2-CH_2-NO_2$			0.50
$-CH=CHNO_2$	0.33	0.26	
$m\text{-}O_2N-C_6H_4$		0.18	
$p\text{-}O_2N-C_6H_4$		0.24	
$(NO_2)_3C_6H_2-$ (picryl)	0.43	0.41	
$-N(CO-CH_3)(CO-C_6H_5)$			1.37
$-N(CO-CH_3)(naphthyl)$			1.65
$-O^-$	−0.71	−0.52	
$-OH$	0.12	−0.37	1.34
$-O-CH_3$	0.12	−0.27	1.81
$-O-C_2H_5$	0.10	−0.24	1.68
$-O-C_3H_7$	0.00	−0.25	1.68
$-O-CH(CH_3)_2$	0.05	−0.45	1.62
$-O-C_4H_9$	−0.05	−0.32	1.68
$-O-$cyclopentyl			1.62
$-O-$cyclohexyl	0.29		1.81
$-O-CH_2-$cyclohexyl	0.18		1.31
$-O-C_6H_5$	0.25	−0.32	2.43
$-O-CH_2-C_6H_5$		−0.42	
$-OCF_3$	0.40	0.35	
$3,4\text{-}O-CH_2-O-$		−0.27	
$3,4\text{-}O-(CH_2-)_2O-$		−0.12	
$-O-CO-CH_3$	0.39	0.31	
$-ONO_2$			3.86
$-O-N=C(CH_3)_2$			1.81

TABLE 9.1 Hammett and Taft Substituent Constants (*Continued*)

Substituent	Hammett constants		Taft constant σ^*
	σ_m	σ_p	
—ONH$_3^+$			2.92
—CH$_2$—O$^-$			0.27
—CH$_2$—OH	0.08	0.08	0.31
—CH$_2$—O—CH$_3$			0.52
—CH(OH)—CH$_3$			0.12
—CH(OH)—C$_6$H$_5$			0.50
p-HO—C$_6$H$_4$—		−0.24	
p-CH$_3$O—C$_6$H$_4$—		−0.10	
—CH$_2$—CH(OH)—CH$_3$			−0.06
—CH$_2$—C(OH)(CH$_3$)$_2$			−0.25
—P(CH$_3$)$_2$	0.1	0.05	
—P(CH$_3$)$_3^+$	0.8	0.9	
—P(CF$_3$)$_2$	0.6	0.7	
—PO$_3$H$^-$	0.2	0.26	
—PO(OC$_2$H$_5$)$_2$	0.55	0.60	
—SH	0.25	0.15	1.68
—SCH$_3$	0.15	0.00	1.56
—S(CH$_3$)$_2^+$	1.0	0.9	
—SCH$_2$CH$_3$	0.23	0.03	1.56
—SCH$_2$CH$_2$CH$_3$			1.49
—SCH$_2$CH$_2$CH$_2$CH$_3$			1.44
—S—cyclohexyl			1.93
—SC$_6$H$_5$	0.30		1.87
—SC(C$_6$H$_5$)$_3$			0.69
—SCH$_2$C$_6$H$_5$			1.56
—SCH$_2$CH$_2$C$_6$H$_5$			1.44
—CH$_2$SH	0.03		0.62
—CH$_2$SCH$_2$C$_6$H$_5$			0.37
—SCF$_3$	0.40	0.50	
—SCN	0.63	0.52	3.43
—S—CO—CH$_3$	0.39	0.44	
—S—CONH$_2$	0.34		2.07
—SO—CH$_3$	0.52	0.49	
—SO—C$_6$H$_5$			3.24
—CH$_2$—SO—CH$_3$			1.33
—SO$_2$—CH$_3$	0.60	0.68	3.68
—SO$_2$—CH$_2$CH$_3$			3.74
—SO$_2$—CH$_2$CH$_2$CH$_3$			3.68
—SO$_2$—C$_6$H$_5$	0.67		3.55
—SO$_2$—CF$_3$	0.79	0.93	
—SO$_2$—NH$_2$	0.46	0.57	
—CH$_2$—SO$_2$—CH$_3$			1.38
—SO$_3^-$	0.05	0.09	0.81
—SO$_3$H		0.50	
—SeCH$_3$	0.1	0.0	

TABLE 9.1 Hammett and Taft Substituent Constants (*Continued*)

Substituent	Hammett constants		Taft constant σ^*
	σ_m	σ_p	
—Se—cyclohexyl			2.37
—SeCN	0.67	0.66	3.61
—Si(CH$_3$)$_3$	−0.04	−0.07	−0.81
—Si(CH$_2$CH$_3$)$_3$		0.0	
—Si(CH$_3$)$_2$C$_6$H$_5$			−0.87
—Si(CH$_3$)$_2$—O—Si(CH$_3$)$_3$			−0.81
—CH$_2$Si(CH$_3$)$_3$	−0.16	−0.22	−0.25
—CH$_2$CH$_2$Si(CH$_3$)$_3$			−0.25
—Sn(CH$_3$)$_3$		0.0	
—Sn(CH$_2$CH$_3$)$_3$		0.0	

TABLE 9.2 pK_a° and Rho Values for Hammett Equation

Acid	pK_a°	ρ
Arenearsonic acids		
pK_1	3.54	1.05
pK_2	8.49	0.87
Areneboronic acids (in aqueous 25% ethanol)	9.70	2.15
Arenephosphonic acids		
pK_1	1.84	0.76
pK_2	6.97	0.95
α-Arylaldoximes	10.70	0.86
Benzeneseleninic acids	4.78	1.03
Benzenesulfonamides (20°C)	10.00	1.06
Benzenesulfonanilides (20°C)		
X—C$_6$H$_4$—SO$_2$—NH—C$_6$H$_5$	8.31	1.16
C$_6$H$_5$—SO$_2$—NH—C$_6$H$_4$—X	8.31	1.74
Benzoic acids	4.21	1.00
Cinnamic acids	4.45	0.47
Phenols	9.92	2.23
Phenylacetic acids	4.30	0.49
Phenylpropiolic acids (in aqueous 35% dioxane)	3.24	0.81
Phenylpropionic acids	4.45	0.21
Phenyltrifluoromethylcarbinols	11.90	1.01
Pyridine-1-oxides	0.94	2.09
2-Pyridones	11.65	4.28
4-Pyridones	11.12	4.28
Pyrroles	17.00	4.28
5-Substituted pyrrole-2-carboxylic acids	2.82	1.40
Thiobenzoic acids	2.61	1.0
Thiophenols	6.50	2.2
Trifluoroacetophenone hydrates	10.00	1.11
5-Substituted topolones	6.42	3.10

TABLE 9.2 pK_a° and Rho Values for Hammett Equation (*Continued*)

Acid	pK_a°	ρ
Protonated cations of		
Acetophenones	−6.0	2.6
Anilines	4.60	2.90
C-Aryl-N-dibutylamidines (in aqueous 50% ethanol)	11.14	1.41
N,N-Dimethylanilines	5.07	3.46
Isoquinolines	5.32	5.90
1-Naphthylamines	3.85	2.81
2-Naphthylamines	4.29	2.81
Pyridines	5.18	5.90
Quinolines	4.88	5.90

TABLE 9.3 pK_a° and Rho Values for Taft Equation

Acid	pK_a°	ρ
RCOOH	4.66	1.62
RCH$_2$COOH	4.76	0.67
RC≡C—COOH	2.39	1.89
H$_2$C=C(R)—COOH	4.39	0.64
(CH$_3$)$_2$C=C(R)—COOH	4.65	0.47
cis-C$_6$H$_5$—CH=C(R)—COOH	3.77	0.63
trans-C$_6$H$_5$—CH=C(R)—COOH	4.61	0.47
R—CO—CH$_2$—COOH	4.12	0.43
HON=C(R)—COOH	4.84	0.34
RCH$_2$OH	15.9	1.42
RCH(OH)$_2$	14.4	1.42
R$_1$CO—NHR$_2$	22.0	3.1*
CH$_3$CO—C(R)=C(OH)CH$_3$	9.25	1.78
CH$_3$CO—CH(R)—CO—OC$_2$H$_5$	12.59	3.44
R—CO—NHOH	9.48	0.98
R$_1$R$_2$C=NOH (R$_1$, R$_2$ not acyl groups)	12.35	1.18
(R)(CH$_3$CO)C=NOH	9.00	0.94
RC(NO$_2$)$_2$H	5.24	3.60
RSH	10.22	3.50
RCH$_2$SH	10.54	1.47
R—CO—SH	3.52	1.62
Protonated cations of		
RNH$_2$	10.15	3.14
R$_1$R$_2$NH	10.59	3.23
R$_1$R$_2$R$_3$N	9.61	3.30
R$_1$R$_2$PH	3.59	2.61
R$_1$R$_2$R$_3$P	7.85	2.67

* σ^* for R$_1$CO and R$_2$.

Two modified sigma constants have been formulated for situations in which the substituent enters into resonance with the reaction center in an electron-demanding transition state (σ^+) or for an electron-rich transition state (σ^-). σ^- constants give better correlations in reactions involving phenols, anilines, and pyridines and in nucleophilic substitutions. Values of some modified sigma constants are given in Table 9.4.

TABLE 9.4 Special Hammett Sigma Constants

Substituent	σ_m^+	σ_p^+	σ_p^-
—CH$_3$	−0.07	−0.31	−0.17
—C(CH$_3$)$_3$	−0.06	−0.26	
—C$_6$H$_5$	0.11	−0.18	
—CF$_3$	0.52	0.61	0.74
—F	0.35	−0.07	0.02
—Cl	0.40	0.11	0.23
—Br	0.41	0.15	0.26
—I	0.36	0.14	
—CN	0.56	0.66	0.88
—CHO			1.13
—CONH$_2$			0.63
—COCH$_3$			0.85
—COOH	0.32	0.42	0.73
—CO—OCH$_3$	0.37	0.49	0.66
—CO—OCH$_2$CH$_3$	0.37	0.48	0.68
—N$_2^+$			3.2
—NH$_2$	0.16	−1.3	−0.66
—N(CH$_3$)$_2$		−1.7	
—N(CH$_3$)$_3^+$	0.36	0.41	
—NH—CO—CH$_3$		−0.60	
—NO$_2$	0.67	0.79	1.25
—OH		−0.92	
—O$^-$			−0.81
—OCH$_3$	0.05	−0.78	−0.27
—SF$_5$			0.70
—SCF$_3$			0.57
—SO$_2$CH$_3$			1.05
—SO$_2$CF$_3$			1.36

SECTION 10
POLYMERS, RUBBERS, FATS, OILS, AND WAXES

10.1   POLYMERS

Polymers are mixtures of macromolecules with similar structures and molecular weights that exhibit some average characteristic properties. In some polymers long segments of linear polymer chains are oriented in a regular manner with respect to one another. Such polymers have many of the physical characteristics of crystals and are said to be *crystalline*. Polymers that have polar functional groups show a considerable tendency to be crystalline. Orientation is aided by alignment of dipoles on different chains. Van der Waals' interactions between long hydrocarbon chains may provide sufficient total attractive energy to account for a high degree of regularity within the polymers.

Irregularities such as branch points, comonomer units, and cross-links lead to *amorphous* polymers. They do not have true melting points but instead have glass transition temperatures at which the rigid and glasslike material becomes a viscous liquid as the temperature is raised.

Elastomers.   Elastomers is a generic name for polymers that exhibit rubberlike elasticity. Elastomers are soft yet sufficiently elastic that they can be stretched several hundred percent under tension. When the stretching force is removed, they retract rapidly and recover their original dimensions.

Polymers that soften or melt and then solidify and regain their original properties on cooling are called *thermoplastic*. A thermoplastic polymer is usually a single strand of linear polymer with few if any cross-links.

Thermosetting Polymers.   Polymers that soften or melt on warming and then become infusible solids are called *thermosetting*. The term implies that thermal decomposition has not taken place. Thermosetting plastics contain a cross-linked polymer network that extends through the finished article, making it stable to heat and insoluble in organic solvents. Many molded plastics are shaped while molten and are then heated further to become rigid solids of desired shapes.

Synthetic Rubbers.   Synthetic rubbers are polymers with rubberlike characteristics that are prepared from dienes or olefins. Rubbers with special properties can also be prepared from other polymers, such as polyacrylates, fluorinated hydrocarbons, and polyurethanes.

Structural Differences.   Polymers exhibit structural differences. A *linear* polymer consists of long segments of single strands that are oriented in a regular manner with respect to one another. *Branched* polymers have substituents attached to the repeating units that extend the polymer laterally. When these units participate in chain propagation and link together chains, a *cross-linked* polymer is formed. A *ladder* polymer results when repeating units have a tetravalent structure such that a polymer consists of two backbone chains regularly cross-linked at short intervals.

Generally polymers involve bonding of the most substituted carbon of one monomeric unit to the least substituted carbon atom of the adjacent unit in a *head-to-tail* arrangement. Substituents appear on alternate carbon atoms. *Tacticity* refers to the configuration of substituents relative to the backbone axis. In an *isotactic* arrangement, substituents are on the same plane of the backbone axis; that is, the configuration at each chiral center is identical.

$$\begin{array}{cccc} Y & Y & Y & Y \\ | & | & | & | \\ -C- & C- & C- & C- \end{array}$$

In a *syndiotactic* arrangement, the substituents are in an ordered alternating sequence, appearing alternately on one side and then on the other side of the chain, thus

$$\begin{array}{cccc} Y & & Y & \\ | & & | & \\ -C-&C-&C-&C- \\ & | & & | \\ & Y & & Y \end{array}$$

In an *atactic* arrangement, substituents are in an unordered sequence along the polymer chains.

Copolymerization.   Copolymerization occurs when a mixture of two or more monomer types polymerizes so that each kind of monomer enters the polymer chain. The fundamental structure resulting from copolymerization depends on the nature of the monomers and the relative rates of monomer reactions with the growing polymer chain. A tendency toward alternation of monomer units is common.

$$-X-Y-X-Y-X-Y-$$

Random copolymerization is rather unusual. Sometimes a monomer which does not easily form a homopolymer will readily add to a reactive group at the end of a growing polymer chain. In turn, that monomer tends to make the other monomer much more reactive.

In *graft copolymers* the chain backbone is composed of one kind of monomer and the branches are made up of another kind of monomer.

$$\begin{array}{cccccc} -X-&X-&X-&X-&X-&X- \\ | & & & | & & \\ Y & & & Y & & \\ | & & & | & & \\ Y & & & Y & & \end{array}$$

The structure of a *block copolymer* consists of a homopolymer attached to chains of another homopolymer.

$$-XXXX-YYY-XXXX-YYY-$$

Configurations around any double bond give rise to *cis* and *trans* stereoisomerism.

10.2   ADDITIVES TO POLYMERS

10.2.1   Antioxidants

Antioxidants markedly retard the rate of autoxidation throughout the useful life of the polymer. Chain-terminating antioxidants have a reactive —NH or —OH functional group and include compounds such as secondary aryl amines or hindered phenols. They function by transfer of hydrogen to free radicals, principally to peroxy radicals. Butylated hydroxytoluene is a widely used example.

Peroxide-decomposing antioxidants destroy hydroperoxides, the sources of free radicals in polymers. Phosphites and thioesters such as tris(nonylphenyl) phosphite, distearyl pentaerythritol diphosphite, and dialkyl thiodipropionates are examples of peroxide-decomposing antioxidants.

10.2.2   Antistatic Agents

External antistatic agents are usually quaternary ammonium salts of fatty acids and ethoxylated glycerol esters of fatty acids that are applied to the plastic surface. Internal antistatic agents are compounded into plastics during processing. Carbon blacks provide a conductive path through the bulk of the plastic. Other types of internal agents must bloom to the surface after compounding in order to be active. These latter materials are ethoxylated fatty amines and ethoxylated glycerol esters of fatty acids, which often must be individually selected to match chemically each plastic type.

Antistatic agents require ambient moisture to function. Consequently their effectiveness is dependent on the relative humidity. They provide a broad range of protection at 50% relative humidity. Much below 20% relative humidity, only materials which provide a conductive path through the bulk of the plastic to ground (such as carbon black) will reduce electrostatic charging.

10.2.3   Chain-Transfer Agents

Chain-transfer agents are used to regulate the molecular weight of polymers. These agents react with the developing polymer and interrupt the growth of a particular chain. The products, however, are free radicals that are capable of adding to monomers and initiating the formation of new chains. The overall effect is to reduce the average molecular weight of the polymer without reducing the rate of polymerization. Branching may occur as a result of chain transfer between a growing but rather short chain with another and longer polymer chain. Branching may also occur if the radical end of a growing chain abstracts a hydrogen from a carbon atom four or five carbons removed from the end. Thiols are commonly used as chain-transfer agents.

10.2.4 Coupling Agents

Coupling agents are molecular bridges between the interface of an inorganic surface (or filler) and an organic polymer matrix. Titanium-derived coupling agents interact with the free protons at the inorganic interface to form organic monomolecular layers on the inorganic surface. The titanate-coupling-agent molecule has six functions:

$$\overset{1}{(RO)_m}\text{—}Ti\text{—}(\overset{2}{O}\text{—}\overset{3}{Y}\text{—}\overset{4}{R^2}\text{—}\overset{5\ 6}{Z})_n$$

where

Type	m	n
Monoalkoxy	1	3
Coordinate	4	2
Chelate	1	2

Function 1 is the attachment of the hydrolyzable portion of the molecule to the surface of the inorganic (or proton-bearing) species.

Function 2 is the ability of the titanate molecule to transesterify.

Function 3 affects performance as determined by the chemistry of alkylate, carboxyl, sulfonyl, phenolic, phosphate, pyrophosphate, and phosphite groups.

Function 4 provides van der Waals' entanglement via long carbon chains.

Function 5 provides thermoset reactivity via functional groups such as methacrylates and amines.

Function 6 permits the presence of two or three pendent organic groups. This allows all functionality to be controlled to the first-, second-, or third-degree levels.

Silane coupling agents are represented by the formula

$$Z\text{—}R\text{—}SiY_3$$

where Y represents a hydrolyzable group (typically alkoxy); Z is a functional organic group, such as amino, methacryloxy, epoxy; and R typically is a small aliphatic linkage that serves to attach the functional organic group to silicon in a stable fashion. Bonding to surface hydroxy groups of inorganic compounds is accomplished by the $-SiY_3$ portion, either by direct bonding of this group or more commonly via its hydrolysis product $-Si(OH)_3$. Subsequent reaction of the functional organic group with the organic matrix completes the coupling reaction and establishes a covalent chemical bond from the organic phase through the silane coupling agent to the inorganic phase.

10.2.5 Flame Retardants

Flame retardants are thought to function via several mechanisms, dependent upon the class of flame retardant used. Halogenated flame retardants are thought to function principally in the vapor phase either as a diluent and heat sink or as a free-radical trap that stops or slows flame propagation. Phosphorus compounds are thought to function in the solid phase by forming a glaze or coating over the substrate that prevents the heat and mass transfer necessary for sustained combustion. With some additives, as the temperature is increased, the flame retardant acts as a solvent for the polymer, causing it to melt at lower temperatures and flow away from the ignition source.

Mineral hydrates, such as alumina trihydrate and magnesium sulfate heptahydrate, are used in highly filled thermoset resins.

10.2.6 Foaming Agents (Chemical Blowing Agents)

Foaming agents are added to polymers during processing to form minute gas cells throughout the product. Physical foaming agents include liquids and gases. Compressed nitrogen is often used in injection molding. Common liquid foaming agents are short-chain aliphatic hydrocarbons in the C_5 to C_7 range and their chlorinated or fluorinated analogs.

The chemical foaming agent used varies with the temperature employed during processing. At relatively low temperatures (15 to 200°C), the foaming agent is often 4,4'-oxybis(benzenesulfonylhydrazide) or p-toluenesulfonylhydrazide. In the midrange (160 to 232°C), either sodium hydrogen carbonate or 1,1'azobisformamide is used. For the high range (200 to 285°C), there are p-toluenesulfonyl semicarbazide, 5-phenyltetrazole and analogs, and trihydrazinotriazine.

10.2.7 Inhibitors

Inhibitors slow or stop polymerization by reacting with the initiator or the growing polymer chain. The free radical formed from an inhibitor must be sufficiently unreactive that it does not function as a chain-transfer agent and begin another growing chain. Benzoquinone is a typical free-radical chain inhibitor. The resonance-stabilized free radical usually dimerizes or disproportionates to produce inert products and end the chain process.

10.2.8 Lubricants

Materials such as fatty acids are added to reduce the surface tension and improve the handling qualities of plastic films.

TABLE 10.1 Plastic Families

Acetals	Alloys (*continued*)
Acrylics	Acrylonitrile-butadiene-styrene–
Poly(methyl methacrylate) (PMMA)	polycarbonate alloy (ABS-PC)
Poly(acrylonitrile)	Allyls
Alkyds	Allyl-diglycol-carbonate polymer
Alloys	Diallyl phthalate (DAP) polymer
Acrylic-poly(vinyl chloride) alloy	Cellulosics
Acrylonitrile-butadiene-styrene–	Cellulose acetate resin
poly(vinyl chloride) alloy (ABS-PVC)	Cellulose-acetate-propionate resin

TABLE 10.1   Plastic Families (*Continued*)

Cellulosics (*continued*)	Polyimide
Cellulose-acetate-butyrate resin	Poly(methylpentene)
Cellulose nitrate resin	Polyolefins (PO)
Ethyl cellulose resin	Low-density polyethylene (LDPE)
Rayon	High-density polyethylene (HDPE)
Chlorinated polyether	Ultrahigh-molecular-weight
Epoxy	polyethylene (UHMWPE)
Fluorocarbons	Polypropylene (PP)
Poly(tetrafluoroethylene) (PTFE)	Polybutylene (PB)
Poly(chlorotrifluoroethylene) (PCTFE)	Polyallomers
Perfluoroalkoxy (PFA) resin	Poly(phenylene oxide)
Fluorinated ethylene-propylene (FEP) resin	Poly(phenylene sulfide) (PPS)
Poly(vinylidene fluoride) (PVDF)	Polyurethanes
Ethylene-chlorotrifluoroethylene copolymer	Silicones
Ethylene-tetrafluoroethylene copolymer	Styrenics
Poly(vinyl fluoride) (PVF)	Polystyrene(PS)
Melamine formaldehyde	Acrylonitrile-butadiene-styrene (ABS) copolymer
Melamine phenolic	Sytrene-acrylonitrile (SAN) copolymer
Nitrile resins	Styrene-butadiene copolymer
Phenolics	Sulfones
Polyamides	Polysulfone (PSF)
Nylon 6	Poly(ether sulfone)
Nylon 6/6	Poly(phenyl sulfone)
Nylon 6/9	Thermoplastic elastomers
Nylon 6/12	Polyolefin
Nylon 11	Polyester
Nylon 12	Block copolymers
Aromatic nylons	Styrene-butadiene block copolymer
Poly(amide-imide)	Styrene-isoprene block copolymer
Poly(aryl ether)	Styrene-ethylene block copolymer
Polycarbonate (PC)	Styrene-butylene block copolymer
Polyesters	Urea formaldehyde
Poly(butylene terephthalate) (PBT)	Vinyls
[aso called polytetramethylene terephthalate (PTMT)]	Poly(vinyl chloride) (PVC)
Poly(ethylene terephthalate) (PET)	Poly(vinyl acetate) (PVAC)
Unsaturated polyesters (SMC, BMC)	Poly(vinylidene chloride)
Butadiene–maleic acid copolymer (BMC)	Poly(vinyl butyrate) (PVB)
Styrene–maleic acid copolymer (SMC)	Poly(vinyl formal)
	Poly(vinyl alcohol) (PVAL)

10.2.9   Plasticizers

Plasticizers are relatively nonvolatile liquids which are blended with polymers to alter their properties by intrusion between polymer chains. Diisooctyl phthalate is a common plasticizer. A plasticizer must be compatible with the polymer to avoid bleeding out over long periods of time. Products containing plasticizers tend to be more flexible and workable.

10.2.10 Ultraviolet Stabilizers

2-Hydroxybenzophenones represent the largest and most versatile class of ultraviolet stabilizers that are used to protect materials from the degradative effects of ultraviolet radiation. They function by absorbing ultraviolet radiation and by quenching electronically excited states.

Hindered amines, such as 4-(2,2,6,6-tetramethylpiperidinyl) decanedioate, serve as radical scavengers and will protect thin films under conditions in which ultraviolet absorbers are ineffective. Metal salts of nickel, such as dibutyldithiocarbamate, are used in polyolefins to quench singlet oxygen or electronically excited states of other species in the polymer. Zinc salts function as peroxide decomposers.

10.2.11 Vulcanization and Curing

Originally, vulcanization implied heating natural rubber with sulfur, but the term is now also employed for curing polymers. When sulfur is employed, sulfide and disulfide cross-links form between polymer chains. This provides sufficient rigidity to prevent *plastic flow.* Plastic flow is a process in which coiled polymers slip past each other under an external deforming force; when the force is released, the polymer chains do not completely return to their original positions.

Organic peroxides are used extensively for the curing of unsaturated polyester resins and the polymerization of monomers having vinyl unsaturation. The —O—O— bond is split into free radicals which can initiate polymerization or cross-linking of various monomers or polymers.

10.3 FORMULAS AND KEY PROPERTIES OF PLASTIC MATERIALS

10.3.1 Acetals

10.3.1.1 Homopolymer. Acetal homopolymers are prepared from formaldehyde and consist of high-molecular-weight linear polymers of formaldehyde.

$$
H-\underset{\underset{}{\overset{\overset{H}{|}}{C}}}{}=O \rightarrow \left[-\underset{\underset{H}{|}}{\overset{\overset{H}{|}}{C}}-O- \right]_n
$$

The good mechanical properties of this homopolymer result from the ability of the oxymethylene chains to pack together into a highly ordered crystalline configuration as the polymers change from the molten to the solid state.

Key properties include high melt point, strength and rigidity, good frictional properties, and resistance to fatigue. Higher molecular weight increases toughness but reduces melt flow.

10.3.1.2 Copolymer. Acetal copolymers are prepared by copolymerization of 1,3,5-trioxane with small amounts of a comonomer. Carbon-carbon bonds are distributed randomly in the polymer chain. These carbon-carbon bonds help to stabilize the polymer against thermal, oxidative, and acidic attack.

10.3.2 Acrylics

10.3.2.1 Poly(methyl Methacrylate). The monomer used for poly(methyl methacrylate), 2-hydroxy-2-methylpropanenitrile, is prepared by the following reaction:

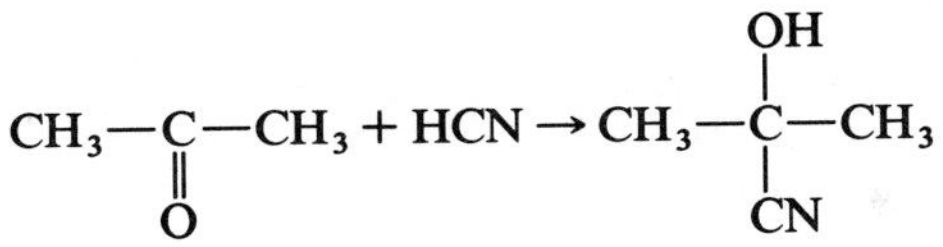

$$CH_3-\underset{\underset{O}{\|}}{C}-CH_3 + HCN \rightarrow CH_3-\underset{\underset{CN}{|}}{\overset{\overset{OH}{|}}{C}}-CH_3$$

2-Hydroxy-2-methylpropanenitrile is then reacted with methanol (or other alcohol) to yield methacrylate ester. Free-radical polymerization is initiated by peroxide or azo catalysts and produce poly(methyl methacrylate) resins having the following formula:

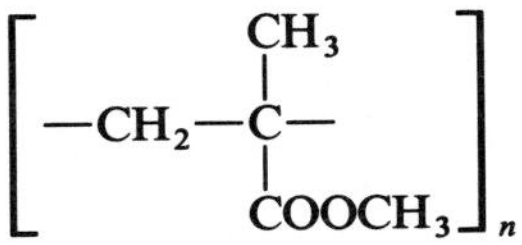

$$\left[-CH_2-\underset{\underset{COOCH_3}{|}}{\overset{\overset{CH_3}{|}}{C}}- \right]_n$$

Key properties are improved resistance to heat, light, and weathering. This polymer is unaffected by most detergents, cleaning agents, and solutions of inorganic acids, alkalies, and aliphatic hydrocarbons. Poly(methyl methacrylate) has light transmittance of 92% with a haze of 1 to 3% and its clarity is equal to glass.

10.3.2.2 Poly(methyl Acrylate). The monomer used for preparing poly(methyl acrylate) is produced by the oxidation of propylene. The resin is made by free-radical polymerization initiated by peroxide or azo catalysts and has the following formula:

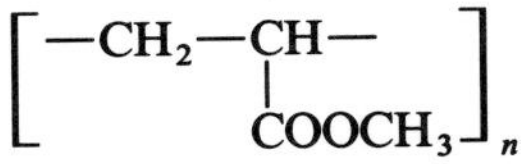

$$\left[-CH_2-\underset{\underset{COOCH_3}{|}}{CH}- \right]_n$$

Resins vary from soft, elastic, film-forming materials to hard plastics.

10.3.2.3 Poly(acrylic Acid) and Poly(methacrylic Acid). Glacial acrylic acid and glacial methacrylic acid can be polymerized to produce water-soluble polymers having the following structures:

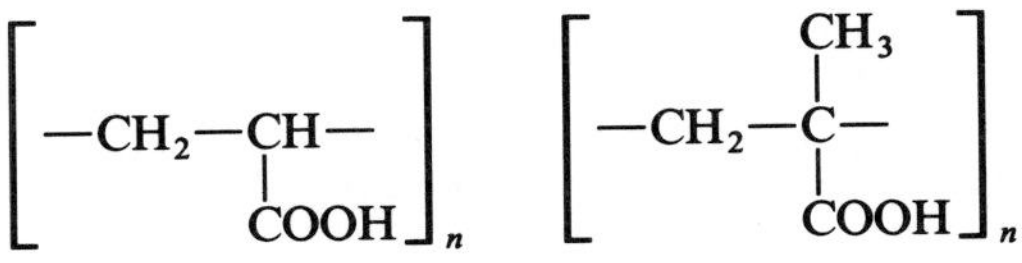

$$\left[-CH_2-\underset{\underset{COOH}{|}}{CH}- \right]_n \qquad \left[-CH_2-\underset{\underset{COOH}{|}}{\overset{\overset{CH_3}{|}}{C}}- \right]_n$$

These monomers provide a means for introducing carboxyl groups into copolymers. In copolymers these acids can improve adhesion properties, improve freeze-thaw and mechanical stability of polymer dispersions, provide stability in alkalies (including ammonia), increase resistance to attack by oils, and provide reactive centers for cross-linking by divalent metal ions, diamines, or epoxides.

10.3.2.4 Functional Group Methacrylate Monomers. Hydroxyethyl methacrylate and dimethylaminoethyl methacrylate produce polymers having the following formulas:

$$\left[-CH_2-\underset{\underset{COOCH_2CH_2OH}{|}}{\overset{\overset{CH_3}{|}}{C}}-\right]_n \qquad \left[-CH_2-\underset{\underset{COOCH_2CH_2N(CH_3)_2}{|}}{\overset{\overset{CH_3}{|}}{C}}-\right]_n$$

The use of hydroxyethyl (also hydroxypropyl) methacrylate as a monomer permits the introduction of reactive hydroxyl groups into the copolymers. This offers the possibility for subsequent cross-linking with an HO-reactive difunctional agent (diisocyanate, diepoxide, or melamine-formaldehyde resin). Hydroxyl groups promote adhesion to polar substrates.

Use of dimethylaminoethyl (also *tert*-butylaminoethyl) methacrylate as a monomer permits the introduction of pendent amino groups which can serve as sites for secondary cross-linking, provide a way to make the copolymer acid-soluble, and provide anchoring sites for dyes and pigments.

10.3.2.5 Poly(acrylonitrile). Poly(acrylonitrile) polymers have the following formula:

$$\left[-CH_2-\underset{\underset{CN}{|}}{CH}-\right]_n$$

10.3.3 Alkyds

Alkyds are formulated from polyester resins, cross-linking monomers, and fillers of mineral or glass. The unsaturated polyester resins used for thermosetting alkyds are the reaction products of polyfunctional organic alcohols (glycols) and dibasic organic acids.

Key properties of alkyds are dimensional stability, colorability, and arc track resistance. Chemical resistance is generally poor.

10.3.4 Alloys

Polymer alloys are physical mixtures of structurally different homopolymers or copolymers. The mixture is held together by secondary intermolecular forces such as dipole interaction, hydrogen bonding, or van der Waals' forces.

Homogeneous alloys have a single glass transition temperature which is determined by the ratio of the components. The physical properties of these alloys are averages based on the composition of the alloy.

Heterogeneous alloys can be formed when graft or block copolymers are combined with a compatible polymer. Alloys of incompatible polymers can be formed if an interfacial agent can be found.

10.3.5 Allyls

10.3.5.1 Diallyl Phthalate (and Diallyl 1,3-Phthalate). These allyl polymers are prepared from

$$\underset{CH_2-CH=CH_2}{\overset{CH_2-CH=CH_2}{\bigcirc}}$$

These resulting polymers are solid, linear, internally cyclized, thermoplastic structures containing unreacted allylic groups spaced at regular intervals along the polymer chain.

Molding compounds with mineral, glass, or synthetic fiber filling exhibit good electrical properties under high humidity and high temperature conditions, stable low-loss factors, high surface and volume resistivity, and high arc and track resistance.

10.3.6 Cellulosics

10.3.6.1 Cellulose Triacetate. Cellulose triacetate is prepared according to the following reaction:

$$C_6H_{10}O_5 + \begin{array}{c} CH_3-C \underset{O}{\overset{O}{\diagup}} \\ CH_3-C \underset{O}{\overset{O}{\diagdown}} \end{array} \longrightarrow \text{cellulose triester}$$

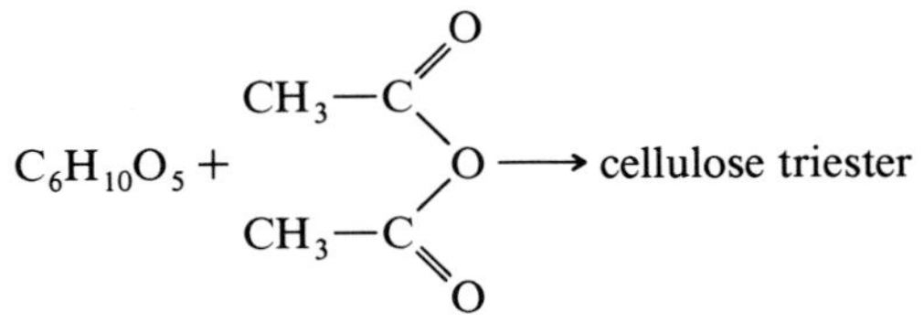

Because cellulose triacetate has a high softening temperature, it must be processed in solution. A mixture of dichloromethane and methanol is a common solvent.

Cellulose triacetate sheeting and film have good gauge uniformity and good optical clarity. Cellulose triacetate products have good dimensional stability and resistance to water and have good folding endurance and burst strength. It is highly resistant to solvents such as acetone. Cellulose triacetate products have good heat resistance and a high dielectric constant.

10.3.6.2 Cellulose Acetate, Propionate, and Butyrate. Cellulose acetate is prepared by hydrolyzing the triester to remove some of the acetyl groups; the plastic-grade resin contains 38 to 40% acetyl. The propionate and butyrate esters are made by substituting propionic acid and its anhydride (or butyric acid and its anhydride) for some of the acetic acid and acetic anhydride. Plastic grades of cellulose-acetate-propionate resin contain 39 to 47% propionyl and 2 to 9% acetyl; cellulose-acetate-butyrate resins contain 26 to 39% butyryl and 12 to 15% acetyl.

These cellulose esters form tough, strong, stiff, hard plastics with almost unlimited color possibilities. Articles made from these plastics have a high gloss and are suitable for use in contact with food.

10.3.6.3 Cellulose Nitrate. Cellulose nitrate is prepared according to the following reaction:

$$C_6H_{10}O_5 + HNO_3 \rightarrow [-C_6H_7O_2(OH)(ONO_2)_2-]_n$$

The nitrogen content for plastics is usually about 11%, for lacquers and cement base it is 12%, and for explosives it is 13%. The standard plasticizer added is camphor.

Key properties of cellulose nitrate are good dimensional stability, low water absorption, and toughness. Its disadvantages are its flammability and lack of stability to heat and sunlight.

10.3.6.4 Ethyl Cellulose. Ethyl cellulose is prepared by reacting cellulose with caustic to form caustic cellulose, which is then reacted with chloroethane to form ethyl cellulose. Plastic-grade material contains 44 to 48% ethoxyl.

Although not as resistant as cellulose esters to acids, it is much more resistant to bases. An outstanding feature is its toughness at low temperatures.

10.3.6.5 Rayon. Viscose rayon is obtained by reacting the hydroxy groups of cellulose with carbon disulfide in the presence of alkali to give xanthates. When this solution is poured (spun) into an acid medium, the reaction is reversed and the cellulose is regenerated (coagulated).

10.3.7 Epoxy

Epoxy resin is prepared by the following condensation reaction:

$$CH_2-CH-CH_2Cl \; + \; HO-\!\!\!\underset{CH_3}{\overset{CH_3}{\underset{|}{\overset{|}{C}}}}\!\!\!-OH \quad \xrightarrow{\text{aq NaOH}}$$

Bisphenol A

$$CH_2-CH-CH_2\!\!\left(\!O-\!\!\underset{CH_3}{\overset{CH_3}{\underset{|}{\overset{|}{C}}}}\!\!-O-CH_2-\underset{}{\overset{OH}{\overset{|}{CH}}}-CH_2\!\right)_n$$

The condensation leaves epoxy end groups that are then reacted in a separate step with nucleophilic compounds (alcohols, acids, or amines). For use as an adhesive, the epoxy resin and the curing resin (usually an aliphatic polyamine) are packaged separately and mixed together immediately before use.

Epoxy novolac resins are produced by glycidation of the low-molecular-weight reaction products of phenol (or cresol) with formaldehyde. Highly cross-linked systems are formed that have superior performance at elevated temperatures.

10.3.8 Fluorocarbon

10.3.8.1 Poly(tetrafluoroethylene). Poly(tetrafluoroethylene) is prepared from tetrafluoroethylene and consists of repeating units in a predominantly linear chain:

$$F_2C{=}CF_2 \rightarrow [-CF_2-CF_2-]_n$$

Tetrafluoroethylene polymer has the lowest coefficient of friction of any solid. It has remarkable chemical resistance and a very low brittleness temperature $(-100^\circ C)$. Its dielectric constant and loss factor are low and stable across a broad temperature and frequency range. Its impact strength is high.

10.3.8.2 Fluorinated Ethylene-Propylene Resin. Polymer molecules of fluorinated ethylene-propylene consist of predominantly linear chains with this structure:

$$\left[-CF_2-CF_2-CF_2-\underset{CF_3}{\overset{}{\underset{|}{CF}}}-\right]_n$$

Key properties are its flexibility, translucency, and resistance to all known chemicals except molten alkali metals, elemental fluorine and fluorine precursors at elevated temperatures, and concentrated perchloric acid. It withstands temperatures from -270° to $250^\circ C$ and may be sterilized repeatedly by all known chemical and thermal methods.

10.3.8.3 Perfluoroalkoxy Resin. Perfluoroalkoxy resin has the following formula:

$$\left[-CF_2-CF_2-\underset{\underset{R}{\overset{|}{O}}}{\overset{}{\underset{|}{CF}}}-CF_2-CF_2-\right]_n \qquad \text{where R is } -C_nF_{2n+1}$$

It resembles polytetrafluoroethylene and fluorinated ethylene propylene in its chemical resistance, electrical properties, and coefficient of friction. Its strength, hardness, and wear resistance are about equal to the former plastic and superior to that of the latter at temperatures above 150°C.

10.3.8.4 *Poly(vinylidene Fluoride)*.

Poly(vinylidene fluoride) consists of linear chains in which the predominant repeating unit is

$$[-CH_2-CF_2-]_n$$

It has good weathering resistance and does not support combustion. It is resistant to most chemicals and solvents and has greater strength, wear resistance, and creep resistance than the preceding three fluorocarbon resins.

10.3.8.5 *Poly(1-Chloro-1,2,2-Trifluoroethylene)*.

Poly(1-chloro-1,2,2-trifluoroethylene) consists of linear chains in which the predominant repeating unit is

$$\left[-CF_2-\underset{\underset{Cl}{|}}{CF}-\right]_n$$

It possesses outstanding barrier properties to gases, especially water vapor. It is surpassed only by the fully fluorinated polymers in chemical resistance. A few solvents dissolve it at temperatures above 100°C, and it is swollen by a number of solvents, especially chlorinated solvents. It is harder and stronger than perfluorinated polymers, and its impact strength is lower.

10.3.8.6 *Ethylene-Chlorotrifluoroethylene Copolymer*.

Ethylene-chlorotrifluoroethylene copolymer consists of linear chains in which the predominant 1 : 1 alternating copolymer is

$$\left[-CH_2-CH_2-CF_2-\underset{\underset{Cl}{|}}{CF}-\right]_n$$

This copolymer has useful properties from cryogenic temperatures to 180°C. Its dielectric constant is low and stable over a broad temperature and frequency range.

10.3.8.7 *Ethylene-Tetrafluoroethylene Copolymer*.

Ethylene-tetrafluoroethylene copolymer consists of linear chains in which the repeating unit is

$$[-CH_2-CH_2-CF_2-CF_2-]_n$$

Its properties resemble those of ethylene-chlorotrifluoroethylene copolymer.

10.3.8.8 *Poly(vinyl Fluoride)*.

Poly(vinyl fluoride) consists of linear chains in which the repeating unit is

$$[-CH_2-CHF-]_n$$

It is used only as a film, and it has good resistance to abrasion and resists staining. It also has outstanding weathering resistance and maintains useful properties from -100 to 150°C.

10.3.9 Nitrile Resins

The principal monomer of nitrile resins is acrylonitrile (see "Polyacrylonitrile"), which constitutes about 70% by weight of the polymer and provides the polymer with good gas barrier and chemical resistance properties. The remainder of the polymer is 20 to 30% methylacrylate (or styrene), with 0 to 10% butadiene to serve as an impact-modifying termonomer.

10.3.10 Melamine Formaldehyde

The monomer used for preparing melamine formaldehyde is formed as follows:

Hexamethylolmelamine

Hexamethylolmelamine can further condense in the presence of an acid catalyst; ether linkages can also form (see "Urea Formaldehyde"). A wide variety of resins can be obtained by careful selection of pH, reaction temperature, reactant ratio, amino monomer, and extent of condensation. Liquid coating resins are prepared by reacting methanol or butanol with the initial methylolated products. These can be used to produce hard, solvent-resistant coatings by heating with a variety of hydroxy, carboxyl, and amide functional polymers to produce a cross-linked film.

10.3.11 Phenolics

10.3.11.1 Phenol-Formaldehyde Resin. Phenol-formaldehyde resin is prepared as follows:

$$C_6H_5OH + H_2C{=}O \rightarrow [-C_6H_2(OH)CH_2-]_n$$

One-Stage Resins. The ratio of formaldehyde to phenol is high enough to allow the thermosetting process to take place without the addition of other sources of cross-links.

Two-Stage Resins. The ratio of formaldehyde to phenol is low enough to prevent the thermosetting reaction from occurring during manufacture of the resin. At this point the resin is termed *novolac* resin. Subsequently, hexamethylenetetramine is incorporated into the material to act as a source of chemical cross-links during the molding operation (and conversion to the thermoset or cured state).

10.3.12 Polyamides

10.3.12.1 Nylon 6, 11, and 12. This class of polymers is polymerized by addition reactions of ring compounds that contain both acid and amine groups on the monomer.

Nylon 6 is polymerized from 2-oxohexamethyleneimine (6 carbons); nylon 11 and 12 are made this way from 11- and 12-carbon rings, respectively.

10.3.12.2 Nylon 6/6, 6/9, and 6/12. As illustrated below, nylon 6/6 is polymerized from 1,6-hexanedioic acid (six carbons) and 1,6-hexanediamine (six carbons).

$$HOOC—(CH_2)_4—COOH + H_2N—CH_2—(CH_2)_4—CH_2—NH_2 →$$

1,6-Hexanedioic acid   1,6-Hexanediamine

$$\left[-NH—(CH_2)_6—NH—\underset{O}{\overset{||}{C}}—(CH_2)_4—\underset{O}{\overset{||}{C}}— \right]_n$$

Poly(hexamethylene 1,6-hexanediamide)

Other nylons are made this way from direct combinations of monomers to produce types 6/9, 6/10, and 6/12.

Nylon 6 and 6/6 possess the maximum stiffness, strength, and heat resistance of all the types of nylon. Type 6/6 has a higher melt temperature, whereas type 6 has a higher impact resistance and better processibility. At a sacrifice in stiffness and heat resistance, the higher analogs of nylon are useful primarily for improved chemical resistance in certain environments (acids, bases, and zinc chloride solutions) and for lower moisture absorption.

Aromatic nylons, $[—NH—C_6H_4—CO—]_n$ (also called aramids), have specialty uses because of their improved clarity.

10.3.13 Poly(amide-imide)

Poly(amide-imide) is the condensation polymer of 1,2,4-benzenetricarboxylic anhydride and various aromatic diamines and has the general structure:

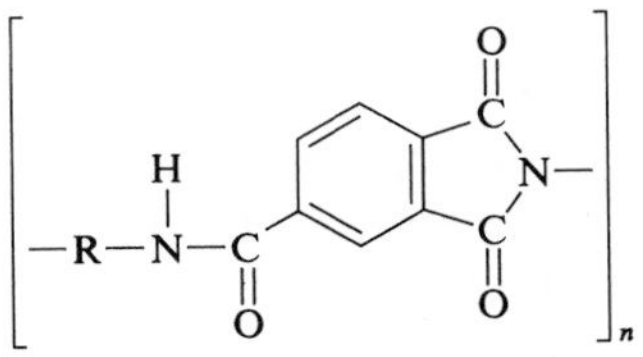

It is characterized by high strength and good impact resistance, and retains its physical properties at temperatures up to 260°C. Its radiation (gamma) resistance is good.

10.3.14 Polycarbonate

Polycarbonate is a polyester in which dihydric (or polyhydric) phenols are joined through carbonate linkages. The general-purpose type of polycarbonate is based on 2,2-bis(4'-hydroxybenzene)propane (bisphenol A) and has the general structure:

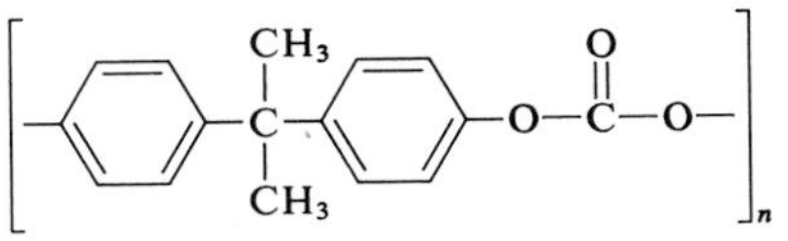

Polycarbonates are the toughest of all thermoplastics. They are window-clear, amazingly strong and rigid, autoclavable, and nontoxic. They have a brittleness temperature of -135°C.

10.3.15 Polyester

10.3.15.1 Poly(butylene Terephthalate). Poly(butylene terephthalate) is prepared in a condensation reaction between dimethyl terephthalate and 1,4-butanediol and its repeating unit has the general structure

This thermoplastic shows good tensile strength, toughness, low water absorption, and good frictional properties, plus good chemical resistance and electrical properties.

10.3.15.2 Poly(ethylene Terephthalate). Poly(ethylene terephthalate) is prepared by the reaction of either terephthalic acid or dimethyl terephthalate with ethylene glycol, and its repeating unit has the general structure

The resin has the ability to be oriented by a drawing process and crystallized to yield a high-strength product.

10.3.15.3 Unsaturated Polyesters. Unsaturated polyesters are produced by reaction between two types of dibasic acids, one of which is unsaturated, and an alcohol to produce an ester. Double bonds in the body of the unsaturated dibasic acid are obtained by using maleic anhydride or fumaric acid.

10.3.15.4 PCTA Copolyester. Poly(1,4-cyclohexanedimethylene terephthalic acid) (PCTA) copolyester is a polymer of cyclohexanedimethanol and terephthalic acid, with another acid substituted for a portion of the terephthalic acid otherwise required. It has the following formula:

10.3.15.5 Polyimides. Polyimides have the following formula:

They are used as high-temperature structural adhesives since they become rubbery rather than melt at about 300°C.

10.3.16   Poly(methylpentene)

Poly(methylpentene) is obtained by a Ziegler-type catalytic polymerization of 4-methyl-1-pentene.

Its key properties are its excellent transparency, rigidity, and chemical resistance, plus its resistance to impact and to high temperatures. It withstands repeated autoclaving, even at 150°C.

10.3.17   Polyolefins

10.3.17.1   Polyethylene.   Polymerization of ethylene results in an essentially straight-chain high-molecular-weight hydrocarbon.

$$CH_2{=}CH_2 \rightarrow [-CH_2-CH_2-]_n$$

Branching occurs to some extent and can be controlled. Minimum branching results in a "high-density" polyethylene because of its closely packed molecular chains. More branching gives a less compact solid known as "low-density" polyethylene.

A key property is its chemical inertness. Strong oxidizing agents eventually cause some oxidation, and some solvents cause softening or swelling, but there is no known solvent for polyethylene at room temperature. The brittleness temperature is $-100°C$ for both types. Polyethylene has good low-temperature toughness, low water absorption, and good flexibility at subzero temperatures.

10.3.17.2   Polypropylene.   The polymerization of propylene results in a polymer with the following structure:

$$CH_2{=}CH-CH_3 \rightarrow \left[\begin{array}{c} -CH_2-CH- \\ | \\ CH_3 \end{array} \right]_n$$

The desired form in homopolymers is the isotactic arrangement (at least 93% is required to give the desired properties). Copolymers have a random arrangement. In block copolymers a secondary reactor is used where active polymer chains can further polymerize to produce segments that use ethylene monomer.

Polypropylene is translucent and autoclavable and has no known solvent at room temperature. It is slightly more susceptible to strong oxidizing agents than polyethylene.

10.3.17.3   Polybutylene.   Polybutylene is composed of linear chains having an isotactic arrangement of ethyl side groups along the chain backbone.

$$CH_2{=}CH-CH_2-CH_3 \rightarrow \left[\begin{array}{c} -CH_2-CH- \\ | \\ CH_2 \\ | \\ CH_3 \end{array} \right]_n$$

It has a helical conformation in the stable crystalline form.

Polybutylene exhibits high tear, impact, and puncture resistance. It also has low creep, excellent chemical resistance, and abrasion resistance with coilability.

10.3.17.4   Ionomer. Ionomer is the generic name for polymers based on sodium or zinc salts of ethylene-methacrylic acid copolymers in which interchain ionic bonding, occurring randomly between the long-chain polymer molecules, produces solid-state properties.

The abrasion resistance of ionomers is outstanding, and ionomer films exhibit optical clarity. In composite structures ionomers serve as a heat-seal layer.

10.3.18   Poly(phenylene Sulfide)

Poly(phenylene sulfide) has the following formula:

The recurring *para*-substituted benzene rings and sulfur atoms form a symmetrical rigid backbone.

The high degree of crystallization and the thermal stability of the bond between the benzene ring and sulfur are the two properties responsible for the polymer's high melting point, thermal stability, inherent flame retardance, and good chemical resistance. There are no known solvents of poly(phenylene sulfide) that can function below 205°C.

10.3.19   Polyurethane

10.3.19.1   Foams. Polyurethane foams are prepared by the polymerization of polyols with isocyanates.

Commonly used isocyanates are toluene diisocyanate, methylene diphenyl isocyanate, and polymeric isocyanates. Polyols used are macroglycols based on either polyester or polyether. The former [poly(ethylene phthalate) or poly(ethylene 1,6-hexanedioate)] have hydroxyl groups that are free to react with the isocyanate. Most flexible foam is made from 80/20 toluene diisocyanate (which refers to the ratio of 2,4-toluene diisocyanate to 2,6-toluene diisocyanate). High-resilience foam contains about 80% 80/20 toluene diisocyanate and 20% poly(methylene diphenyl isocyanate), while semiflexible foam is almost always 100% poly(methylene diphenyl isocyanate). Much of the latter reacts by trimerization to form isocyanurate rings.

Flexible foams are used in mattresses, cushions, and safety applications. Rigid and semiflexible foams are used in structural applications and to encapsulate sensitive components to protect them against shock, vibration, and moisture. Foam coatings are tough, hard, flexible, and chemically resistant.

10.3.19.2   Elastromeric Fiber.   Elastomeric fibers are prepared by the polymerization of polymeric polyols with diisocyanates.

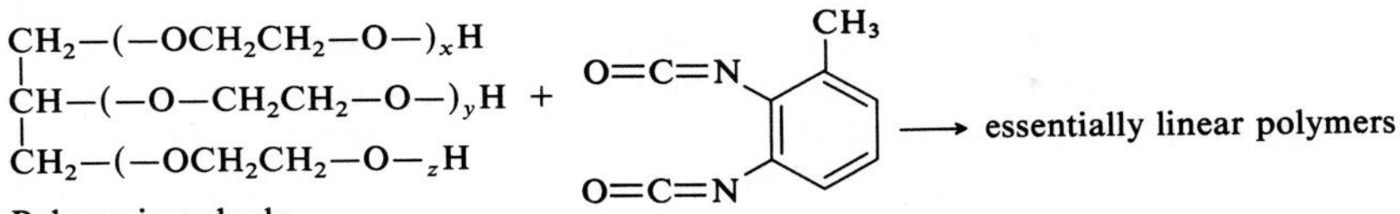

$$
\begin{array}{l}
CH_2-(-OCH_2CH_2-O-)_xH \\
CH-(-O-CH_2CH_2-O-)_yH \\
CH_2-(-OCH_2CH_2-O-_zH
\end{array}
+
$$

Polymeric polyols

Diisocyanate

→ essentially linear polymers

The structure of elastomeric fibers is similar to that illustrated for polyurethane foams.

10.3.20   Silicones

Silicones are formed in the following multistage reaction:

$$R_2SiCl_2 + 2H_2O \rightarrow R_2Si(OH)_2 + 2HCl$$
$$\downarrow$$
$$[-Si(R)_2-O-]_n$$

The silanols formed above are unstable and under dehydration. On polycondensation, they give polysiloxanes (or silicones) which are characterized by their three-dimensional branched-chain structure. Various organic groups introduced within the polysiloxane chain impart certain characteristics and properties to these resins.

Methyl groups impart water repellency, surface hardness, and noncombustibility.

Phenyl groups impart resistance to temperature variations, flexibility under heat, resistance to abrasion, and compatibility with organic products.

Vinyl groups strengthen the rigidity of the molecular structure by creating easier cross-linkage of molecules.

Methoxy and alkoxy groups facilitate cross-linking at low temperatures.

Oils and gums are nonhighly branched- or straight-chain polymers whose viscosity increases with the degree of polycondensation.

10.3.21   Styrenics

10.3.21.1   Polystytrene.   Polystyrene has the following formula:

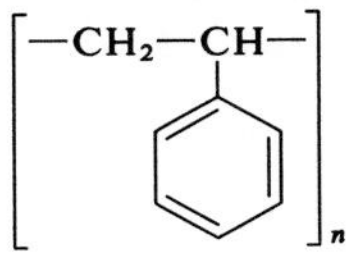

$$
\left[-CH_2-CH- \right]_n
$$

Polystyrene is rigid with excellent dimensional stability, has good chemical resistance to aqueous solutions, and is an extremely clear material.

Impact polystyrene contains polybutadiene added to reduce brittleness. The polybutadiene is usually dispersed as a discrete phase in a continuous polystyrene matrix. Polystyrene can be grafted onto rubber particles, which assures good adhesion between the phases.

10.3.21.2 Acrylonitrile-Butadiene-Styrene (ABS) Copolymers. This basic three-monomer system can be tailored to yield resins with a variety of properties. Acrylonitrile contributes heat resistance, high strength, and chemical resistance. Butadiene contributes impact strength, toughness, and retention of low-temperature properties. Styrene contributes gloss, processibility, and rigidity. ABS polymers are composed of discrete polybutadiene particles grafted with the styrene-acrylonitrile copolymer; these are dispersed in the continuous matrix of the copolymer.

10.3.21.3 Styrene-Acrylonitrile (SAN) Copolymers. SAN resins are random, amorphous copolymers whose properties vary with molecular weight and copolymer composition. An increase in molecular weight or in acrylonitrile content generally enhances the physical properties of the copolymer but at some loss in ease of processing and with a slight increase in polymer color.

SAN resins are rigid, hard, transparent thermoplastics which process easily and have good dimensional stability—a combination of properties unique in transparent polymers.

10.3.22 Sulfones

Below are the formulas for three polysulfones.

Polysulfone

Poly(ester sulfone)

Poly(phenyl sulfone)

The isopropylidene linkage imparts chemical resistance, the ether linkage imparts temperature resistance, and the sulfone linkage imparts impact strength. The brittleness temperature of polysulfones is $-100°C$. Polysulfones are clear, strong, nontoxic, and virtually unbreakable. They do not hydrolyze during autoclaving and are resistant to acids, bases, aqueous solutions, aliphatic hydrocarbons, and alcohols.

10.3.23 Thermoplastic Elastomers

10.3.23.1 Polyolefins. In these thermoplastic elastomers the hard component is a crystalline polyolefin, such as polyethylene or polypropylene, and the soft portion is composed of ethylene-propylene rubber. Attractive forces between the rubber and resin phases serve as labile cross-links. Some contain a chemically cross-linked rubber phase that imparts a higher degree of elasticity.

10.3.23.2 Styrene-Butadiene-Styrene Block Copolymers. Styrene blocks associate into domains that form hard regions. The midblock, which is normally butadiene, ethylene-butene, or isoprene blocks, forms the soft domains. Polystyrene domains serve as cross-links.

10.3.23.3 Polyurethanes. The hard portion of polyurethane consists of a chain extender and polyisocyanate. The soft component is composed of polyol segments.

10.3.23.4 Polyesters. The hard portion consists of copolyester, and the soft portion is composed of polyol segments.

10.3.24 Vinyl

10.3.24.1 Poly(vinyl Chloride) (PVC). Polymerization of vinyl chloride results in the formation of a polymer with the following formula:

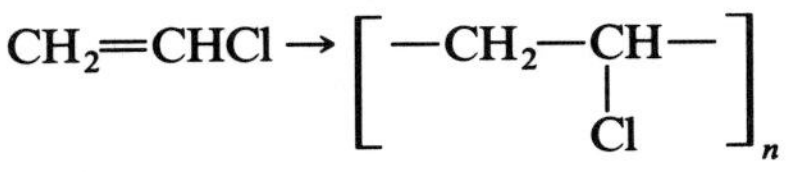

$$CH_2{=}CHCl \rightarrow \left[-CH_2-\underset{\underset{Cl}{|}}{CH}- \right]_n$$

When blended with phthalate ester plasticizers, PVC becomes soft and pliable.

Its key properties are good resistance to oils and a very low permeability to most gases.

10.3.24.2 Poly(vinyl Acetate). Poly(vinyl acetate) has the following formula:

$$\left[-CH_2-\underset{\underset{O-CO-CH_3}{|}}{CH}- \right]_n$$

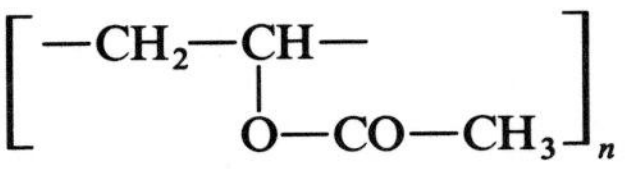

Poly(vinyl acetate) is used in latex water paints because of its weathering, quick-drying, recoatability, and self-priming properties. It is also used in hot-melt and solution adhesives.

10.3.24.3 Poly(vinyl Alcohol). Poly(vinyl alcohol) has the following formula:

$$\left[-CH_2-\underset{\underset{OH}{|}}{CH}- \right]_n$$

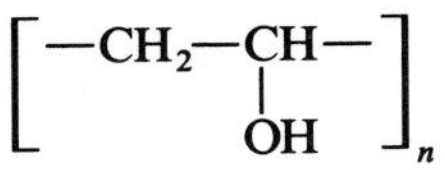

It is used in adhesives, paper coating and sizing, and textile warp size and finishing applications.

10.3.24.4 Poly(vinyl Butyral). Poly(vinyl butyral) is prepared according to the following reaction:

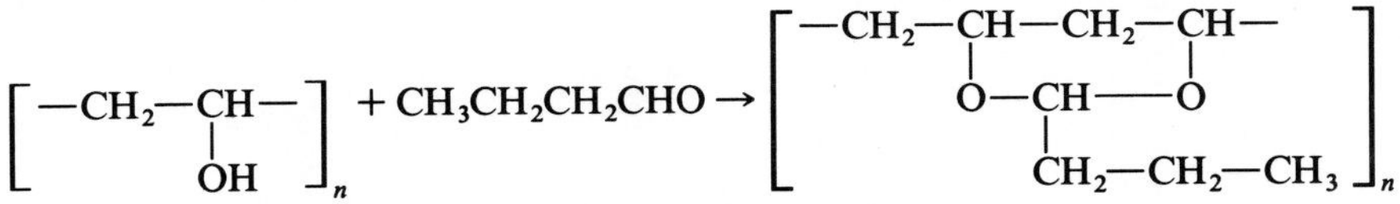

$$\left[-CH_2-\underset{\underset{OH}{|}}{CH}- \right]_n + CH_3CH_2CH_2CHO \rightarrow \left[\begin{array}{c} -CH_2-CH-CH_2-CH- \\ O-CH-\!\!-O \\ | \\ CH_2-CH_2-CH_3 \end{array} \right]_n$$

Its key characteristics are its excellent optical and adhesive properties. It is used as the interlayer film for safety glass.

TABLE 10.2 Properties of Commercial Plastics

Properties	Acetal				
	Homopolymer	Copolymer	20% glass-reinforced homopolymer	25% glass-reinforced copolymer	21% poly(tetrafluoroethylene)-filled homopolymer
Physical					
Melting temperature, °C					
Crystalline	175	175	181	175	181
Amorphous					
Specific gravity	1.42	1.41	1.56	1.61	1.54
Water absorption (24 h), %	0.25–0.40	0.22	0.25	0.29	0.20
Dielectric strength, KV $\cdot$ mm^{-1}	19.7	19.7	19.3	22.8	15.7
Electrical					
Volume (dc) resistivity, ohm-cm	10^{15}	10^{15}	5×10^{14}		3×10^{16}
Dielectric constant (60 Hz)	3.7	3.7	3.9		3.1
Dielectric constant (10^6 Hz)	3.7	3.7	3.9		3.1
Dissipation (power) factor (60 Hz)					
Dissipation factor (10^6 Hz)	0.005	0.005	0.005		0.005
Mechanical					
Compressive modulus, 10^3 lb $\cdot$ in^{-2}	670	450			

Compressive strength, rupture or 1% yield, 10^3 lb·in^{-2}	5.29	16 (10% yield)	18 (10% yield)	17 (10% yield)	13 (10% yield)
Elongation at break, %	25–75	40–75	7	3	15–22
Flexural modulus at 23°C, 10^3 lb·in^2	380–430	375	730	1100	340–350
Flexural strength, rupture or yield, 10^3 lb·in^{-2}	14	13	15	28	
Hardness, Rockwell (or Shore)	M94	M78	M90	M79	M78
Impact strength (Izod) at 23°C, J·m^{-1}	69–123	53–80	43	96	37–64
Tensile modulus, 10^3 lb·in^{-2}	520	410	1000	1250	
Tensile strength at break, 10^3 lb·in^{-2}	10	10	8.5	18.5	7.6
Tensile yield strength, 10^3 lb·in^{-2}	9.5–12	8.5			6.9–7.6
Thermal					
Burning rate, mm·min^{-1}	27.9				
Coefficient of linear thermal expansion, 10^{-6}°C	100	85	36–81		75
Deflection temperature under flexural load (264 lb·in^{-2}), °C	124	110	157	163	100
Maximum recommended service temperature, °C	84				
Specific heat, cal·g^{-1}	0.35				
Thermal conductivity, W·m^{-1}·K^{-1}	0.23	0.23			

TABLE 10.2 Properties of Commercial Plastics (*Continued*)

Properties	Acrylic				Alkyd, molded	Alloy	
	Poly(methyl methacrylate)	Cast sheet	Impact-modified	Heat-resistant		Acrylic poly(vinyl chloride) alloy	Acrylonitrile-butadiene-styrene-poly(vinyl chloride) alloy
Physical							
Melting temperature, °C							
Crystalline							
Amorphous	90–105	90–105	80–100	100–125		105	
Specific gravity	1.17–1.20	1.18–1.20	1.11–1.18	1.16–1.19	2.22–2.24		
Water absorption (24 h), %	0.1–0.4	0.2–0.4	0.2–0.8	0.2–0.3		0.06	
Dielectric strength, KV · mm^{-1}	15.7–19.9	17.7–21.7	15.0–19.9	15.7–19.9		>15.7	19.7
Electrical							
Volume (dc) resistivity, ohm-cm	>10^{14}	>10^{14}					
Dielectric constant (60 Hz)	3.3–4.5	3.5–4.5			3.8–5.0		
Dielectric constant (10^6 Hz)		3.0–3.5			3.6–4.7		
Dissipation (power) factor (60 Hz)		0.04–0.06			0.012–0.026		
Dissipation factor (10^6 Hz)		0.02–0.03			0.01–0.016		
Mechanical							
Compressive modulus, 10^4 lb · in^{-2}	370–460	390–475	240–370	350–460		330–400	

Compressive strength, rupture or 1% yield, 10^3 lb · in^{-2}	12–18	11–19	4–14	17	16–20	8.4	
Elongation at break, %	2–10	2–7	20–70	3–5		100	
Flexural modulus at 23°C, 10^3 lb · in^{-2}	420–460	390–475	200–380	460–500		330–400	340
Flexural strength, rupture or yield, 10^3 lb · in^{-2}	13–19	12–17	7–13	12–16		10.7	9.6
Hardness, Rockwell (or Shore)	M85–M105	M80–M100	R105–R120	M95–M105	E76	R99–R105	R100
Impact strength (Izod) at 23°C, J · m^{-1}	16–27	16–21	43–133	16–21	27–240	800	560
Tensile modulus, 10^3 lb · in^{-2}	380–450	350–450	200–400	350–460		330–335	330
Tensile strength at break, 10^3 lb · in^{-2}	7–11	8–11	5–9	10	4.5–6.5	6.5	5.8
Tensile yield strength, 10^3 lb · in^{-2}					10–13		
Thermal							
Burning rate, mm · min^{-1}		0.5–2.2			Self-extinguishing		
Coefficient of linear thermal expansion, 10^{-6}°C	50–90	50–90	50–80	50–60	40–55		46
Deflection temperature under flexural load (264 lb · in^{-2}), °C	74–99	71–102	74–95	83–104	177–204	71	
Maximum recommended service temperature, °C		60–71			220		
Specific heat, cal · g^{-1}	0.36	0.35					
Thermal conductivity, W · m^{-1} · K^{-1}	0.17–0.25	0.17–0.25	0.17–0.21	0.19			

TABLE 10.2 Properties of Commercial Plastics (*Continued*)

Properties	Alloy	Allyl			Cellulosic			
	Polycarbonate acrylonitrile-butadiene-styrene alloy	Allyl-diglycol-carbonate polymer	Diallyl phthalate molding		Cellulose acetate			Cellulose-acetate-butyrate resin
			Glass-filled	Mineral-filled	Sheet	Molding		Sheet
Physical								
Melting temperature, °C								
Crystalline								
Amorphous	150		Thermoset	Thermoset	230	230		140
		Thermoset						
Specific gravity	1.12–1.20	1.3–1.4	1.7–2.0	1.65–1.85	1.27–1.34	1.29–1.34		1.15–1.22
Water absorption (24 h), %	0.21–0.24	0.2	0.12–0.35	0.2–0.5	2–7	1.7–6.5		0.9–2.2
Dielectric strength, $kV \cdot mm^{-1}$	17.7	15.0	15.7–17.7	15.7–17.7	11–24	9–24		9–18
Electrical								
Volume (dc) resistivity, ohm-cm					10^{10}–10^{13}	10^{10}–10^{13}		10^{10}–10^{12}
Dielectric constant (60 Hz)					3.4–7.4	3.5–7.5		3.7–4.3
Dielectric constant (10^6 Hz)					3.2–7.0	3.2–7.0		3.3–3.8
Dissipation (power) factor (60 Hz)					0.01–0.06	0.01–0.06		0.01–0.04
Dissipation factor (10^6 Hz)					0.01–0.06	0.01–0.10		0.01–0.04
Mechanical								
Compressive modulus, 10^3 lb · in^{-2}		300						

Compressive strength, rupture or 1% yield, $10^3\ \mathrm{lb}\cdot\mathrm{in}^{-2}$	11	21–23	25–35	20–32	22–33	25–36	50–100
Elongation at break, %	10–15		3–5	3–5	17–40	6–40	
Flexural modulus at 23°C, $10^3\ \mathrm{lb}\cdot\mathrm{in}^{-2}$	300–400	250–330	1200–1500	1000–1400			740–1300
Flexural strength, rupture or yield, $10^3\ \mathrm{lb}\cdot\mathrm{in}^{-2}$	13.0–13.7	6–13	9–20	8.5–11	6–10	2–16	4–9
Hardness, Rockwell (or Shore)	R117	M95–M100	E80–E87	E61	R85–R120	R100–R123	R50–R95
Impact strength (Izod) at 23°C, $\mathrm{J}\cdot\mathrm{m}^{-1}$	560	11–21	21–800	16–43	107–454	53–214	133–288
Tensile modulus, $10^3\ \mathrm{lb}\cdot\mathrm{in}^{-2}$	370–380	300	1400–2200	1200–2200			200–250
Tensile strength at break, $10^3\ \mathrm{lb}\cdot\mathrm{in}^{-2}$	7.0–7.3	5–6	6–11	5–8	4.5–8.0	1.9–9.0	2.6–6.9
Tensile yield strength, $10^3\ \mathrm{lb}\cdot\mathrm{in}^{-2}$	8.5				2.2–7.4	4.1–7.6	
Thermal							
Burning rate, $\mathrm{mm}\cdot\mathrm{min}^{-1}$			0.68–2.4	2.8		1.3–3.8	1.3–3.8
Coefficient of linear thermal expansion, 10^{-6}°C	63–67	5.4–9.6			100–150	80–180	110–170
Deflection temperature under flexural load ($264\ \mathrm{lb}\cdot\mathrm{in}^{-2}$), °C	104–116	60–88	165–288+	160–288	44–91	51–98	49–58
Maximum recommended service temperature, °C							
Specific heat, $\mathrm{cal}\cdot\mathrm{g}^{-1}$					0.3–0.4	0.3–0.42	0.3–0.4
Thermal conductivity, $\mathrm{W}\cdot\mathrm{m}^{-1}\cdot\mathrm{K}^{-1}$	0.25–0.38	0.20–0.21	0.21–0.63	0.30–1.04	0.17–0.34	0.17–0.34	0.17–0.34

TABLE 10.2 Properties of Commercial Plastics (*Continued*)

Properties	Cellulosic				Chlorinated polyether	Epoxy	
						Bisphenol	
	Cellulose-acetate butyrate resin, molding	Cellulose-acetate-propionate resin, molding	Ethyl cellulose	Cellulose nitrate		Glass-fiber-reinforced	Mineral-filled
Physical							
Melting temperature, °C							
Crystalline							
Amorphous	140	190	135		125	Thermoset	Thermoset
Specific gravity	1.15–1.22	1.17–1.24	1.09–1.17	1.35–1.40	1.4	1.6–2.0	1.6–2.1
Water absorption (24 h), %	0.9–2.2	1.2–2.8	0.8–1.8			0.04–0.20	0.03–0.20
Dielectric strength, kV · mm^{-1}	9–13	12–17.7	13.8–19.7			9.8–15.7	9.8–15.7
Electrical							
Volume (dc) resistivity, ohm-cm	10^{10}–10^{12}			10^{10}			
Dielectric constant (60 Hz)	3.5–6.4			7.0–7.5			
Dielectric constant (10^6 Hz)	3.2–6.2		3.01	6.6			
Dissipation (power) factor (60 Hz)	0.01–0.04						
Dissipation factor (10^6 Hz)	0.01–0.04						
Mechanical							
Compressive modulus, 10^3 lb · in^{-2}						3000	

Compressive strength, rupture or 1% yield, 10^3 lb · in^{-2}	2.1–7.5	2.4–7.0		2.1–8.0	600–800	18,000–40,000 / 4	18,000–40,000
Elongation at break, %	40–88	29–100	5–40	40–45	5	2–4.5	
Flexural modulus at 23°C, 10^3 lb · in^{-2}	90–300	120–350					
Flexural strength, rupture or yield, 10^3 lb · in^{-2}	1.8–9.3	2.9–11.4	4–12	9–11		8–30	6–18
Hardness, Rockwell (or Shore)	R31–R116	R10–R122	R50–R115	R95–R115	R100	M100–M112	M100–M112
Impact strength (Izod) at 23°C, J · m^{-1}	53–582	27 to no break	21	267–374	21	16–533 / 3	16–22
Tensile modulus, 10^3 lb · in^{-2}	50–200	60–215		190–220			
Tensile strength at break, 10^3 lb · in^{-2}	2.6–6.9	2.0–7.8	2–8	7–8	1.5–1.8	5–20	4–10
Tensile yield strength, 10^3 lb · in^{-2}							
Thermal							
Burning rate, mm · min^{-1}	1.3–3.8				Self-extinguishing		
Coefficient of linear thermal expansion, 10^{-6}°C	110–170	110–170	100–200	80–120	6.6	11–50	20–60
Deflection temperature under flexural load (264 lb · in^{-2}), °C	44–94	44–109	45–88	60–71	185	107–260	107–260
Maximum recommended service temperature, °C					255		
Specific heat, cal · g^{-1}	0.3–0.4			0.31–0.41			
Thermal conductivity, W · m^{-1} · K^{-1}	0.17–0.30	0.17–0.30	0.16–0.30	0.23		0.17–0.42	0.17–1.48

TABLE 10.2 Properties of Commercial Plastics (*Continued*)

Properties	Epoxy			Fluorocarbon			
	Casting resin		Novolac resin	Poly(tetrafluoroethylene)		Poly(chloro-trifluoro-ethylene)	Perfluoroalkoxy
	Unfilled	Flexible	Mineral-filled	Granular	Glass-fiber-reinforced		
Physical							
Melting temperature, °C							
Crystalline	Thermoset	Thermoset	Thermoset	327	327	220	310
Amorphous							
Specific gravity	1.11–1.40	1.05–1.35	1.7–2.1	2.14–2.20	2.2–2.3	2.1–2.2	2.12–2.17
Water absorption (24 h), %	0.08–0.15	0.27–0.50	0.05–0.2	0.01		0.03	
Dielectric strength, kV · mm^{-1}	11.8–19.7	9.3–15.8	11.8–13.8	18.9	12.6	19.7–23	19.7
Electrical							
Volume (dc) resistivity, ohm-cm	10^{12}–10^{17}			10^{18}		10^{18}	
Dielectric constant (60 Hz)	3.5–5.0			2.1		2.3–2.7	
Dielectric constant (10^6 Hz)	3.5–5.0			2.1		2.3–2.5	
Dissipation (power) factor (60 Hz)				0.0002		0.001	
Dissipation factor (10^6 Hz)				0.0002		0.005	
Mechanical							
Compressive modulus, 10^3 lb · in^{-2}				60			

Compressive strength, rupture or 1% yield, 10^3 lb $\cdot$ in^{-2}	15–25	1–14	30	1.7		4.6–7.4	
Elongation at break, %	3–6	20–70	2–4	200–400	200–300	80–250	300
Flexural modulus at 23°C, 10^3 lb $\cdot$ in^{-2}			2000	80	235	120	
Flexural strength, rupture or yield, 10^{-3} lb $\cdot$ in^{-2}	13–21	1–13	16–20			7.4–9.3	
Hardness, Rockwell (or Shore)	M80–M110			(D50–D55)	(D60–D70)	R75–R95	(D64)
Impact strength (Izod) at 23°C, J $\cdot$ m^{-1}	10.7–53	187–267	21	160	144	133–160	No break
Tensile modulus, 10^3 lb $\cdot$ in^{-2}	350	1–350		58–80		150–300	
Tensile strength at break, 10^3 lb $\cdot$ in^{-2}	4–13	2–10	6–12	2–5	2–2.7	4.5–6	4–4.3
Tensile yield strength, 10^3 lb $\cdot$ in^{-2}			30				
Thermal							
Burning rate, mm $\cdot$ min^{-1}				Self-extinguishing	Self-extinguishing	Self-extinguishing	
Coefficient of linear thermal expansion, 10^{-6}°C	45–65	20–100	22–30	100	77–100	70	
Deflection temperature under flexural load (264 lb $\cdot$ in^{-2}), °C	46–288	23–121	149–260	121 (66 lb $\cdot$ in^{-2})		126 (66 lb $\cdot$ in^{-2})	74 (66 lb $\cdot$ in^{-2})
Maximum recommended service temperature, °C				260		200	
Specific heat, cal $\cdot$ g^{-1}				0.25		0.22	
Thermal conductivity, W $\cdot$ m^{-1} $\cdot$ K^{-1}	0.17–0.21			0.25	0.34–0.40	0.19–0.22	0.25

TABLE 10.2 Properties of Commercial Plastics (*Continued*)

Properties	Fluorinated ethylene-propylene resin	Poly(vinylidene fluoride)	Fluorocarbon		Ethylene-chlorotrifluoroethylene copolymer	Melamine formaldehyde	
			Ethylene-tetrafluoroethylene copolymer			Cellulose-filled	Glass-fiber-reinforced
			Unfilled	Glass-fiber-reinforced			
Physical							
Melting temperature, °C							
Crystalline	275	156	270	270	245	Thermoset	Thermoset
Amorphous							
Specific gravity	2.14–2.17	1.75–1.78	1.7	1.8	1.68	1.47–1.52	1.5–2.0
Water absorption (24 h), %	<0.01	0.04–0.06	0.03	0.02	0.01	0.1–0.8	0.09–1.3
Dielectric strength, kV · mm^{-1}	20–24	10	16	17	19	11–16	5–15
Electrical							
Volume (dc) resistivity, ohm-cm							
Dielectric constant (60 Hz)	2.1	8–9	2.6		2.6		
Dielectric constant (10^6 Hz)	2.1	8–9	2.6		2.6		
Dissipation (power) factor (60 Hz)		High					
Dissipation factor (10^6 Hz)		High					
Mechanical							
Compressive modulus, 10^3 lb · in^{-2}	120	120	120	1200	240		

Compressive strength, rupture or 1% yield, 10^3 lb · in^{-2}	2.2	8.7–10	7.1	10	200–300	33–45	20–35
Elongation at break, %	250–330	25–500	100–400	8	240	0.6–1.0	0.6
Flexural modulus at 23°C, 10^3 lb · in^{-2}	80–95	200	200	950		1100	
Flexural strength, rupture or yield, 10^3 lb · in^{-2}		8.6–11	5.5	10.7	7	9–16	14–23
Hardness, Rockwell (or Shore)	(D60–D65)	(D80)	R50 (D75)	R74	R95	M115–M125	M115
Impact strength (Izod) at 23°C, J · m^{-1}	No break	192–214	No break	480	No break	11–21	32–961
Tensile modulus, 10^3 lb · in^{-2}	50	120	120	1200	240	1.1–1.4	1.6–2.4
Tensile strength at break, 10^3 lb · in^{-2}	2.7–3.1	5.5–7.4	6.5	12	7	5–13	5–10.5
Tensile yield strength, 10^3 lb · in^{-2}							
Thermal							
Burning rate, mm · min^{-1}	Not combustible	Not combustible	Not combustible	Not combustible	Not combustible	Self-extinguishing	Self-extinguishing
Coefficient of linear thermal expansion, 10^{-6}°C	83–105	85	59	10–32	80	40–45	15–28
Deflection temperature under flexural load (264 lb · in^{-2}), °C	70 (66 lb · in^{-2})	80–90	71	210	77	177–199	190–204
Maximum recommended service temperature, °C	205	150				210	
Specific heat, cal · g^{-1}	0.28						
Thermal conductivity, W · m^{-1} · K^{-1}	0.25	0.19–0.24	0.24		0.16	0.27–0.41	0.41–0.49

TABLE 10.2 Properties of Commercial Plastics (*Continued*)

Properties	Melamine phenolic, woodflour- and cellulose-filled	Nitrile	Phenolic				
			Unfilled	Woodflour-filled	Glass-fiber-reinforced	Cellulose-filled	Mineral-filled
Physical							
Melting temperature, °C							
Crystalline							
Amorphous		95					
	Thermoset		Thermoset	Thermoset	Thermoset	Thermoset	Thermoset
Specific gravity	1.5–1.7	1.15	1.24–1.32	1.37–1.46	1.69–2.0	1.38–1.42	1.42–1.84
Water absorption (24 h), %	0.3–0.65	0.28	0.1–0.36	0.3–1.2	0.03–1.2	0.5–0.9	0.1–0.3
Dielectric strength, kV · mm^{-1}	8.7–12.8	8.7–9.5	9.8–15.8	10.2–15.8	5.5–15.8	11.8–15	7.9–13.8
Electrical							
Volume (dc) resistivity, ohm-cm		1.9×10^{15}	1×10^{12} to 7×10^{12}				
Dielectric constant (60 Hz)			6.5–7.5				
Dielectric constant (10^6 Hz)			4.0–5.5				
Dissipation (power) factor (60 Hz)			0.10–0.15				
Dissipation factor (10^6 Hz)			0.04–0.05				
Mechanical							
Compressive modulus, 10^3 lb · in^{-2}							

Compressive strength, rupture or 1% yield, 10^3 lb·in^{-2}	26–30	12	18–32	25–31	26–70	22–31	22.5–34.6
Elongation at break, %	0.4–0.8	3–4	1.5–2.0	0.4–0.8	0.2	1–2	0.1–0.5
Flexural modulus at 23°C, 10^3 lb·in^{-2}	1000–1200	500–590	700–1500	1000–1200	2000–33,000	900–1300	1000–2000
Flexural strength, rupture or yield, 10^3 lb·in^{-2}	8–10	14	11–17	7–14	15–60	5.5–11	11–14
Hardness, Rockwell (or Shore)	E95–E100	M72–M76	M93–M120	M100–M115	E54–E101	M95–M115	E88
Impact strength (Izod) at 23°C, J·m^{-1}	11–21	80–256	13–21	11–32	27–960	21–59	14–19
Tensile modulus, 10^3 lb·in^{-2}	800–1700	510–580	700–1500	800–1700	1900–3300		2400
Tensile strength at break, 10^3 lb·in^{-2}	6–8	9	6–9	5–9	7–18	3.5–6.5	6–9.7
Tensile yield strength, 10^3 lb·in^{-2}			12–15				
Thermal							
Burning rate, mm·min^{-1}			Self-extinguishing				
Coefficient of linear thermal expansion, 10^{-6}°C	10–40	66	68	30–45	8–21	20–31	19–26
Deflection temperature under flexural load (264 lb·in^{-2}), °C	140–154	73	74–80	149–188	177–316	149–177	320–246
Maximum recommended service temperature, °C							
Specific heat, cal·g^{-1}							
Thermal conductivity, W·m^{-1}·K^{-1}	0.17–0.30	0.26	0.15	0.17–0.34	0.34–0.59	0.25–0.38	0.42–0.57

TABLE 10.2 Properties of Commercial Plastics (*Continued*)

	Polyamide						
	Nylon 6				Nylon 6/6		
Properties	Molding and extrusion	30–35% glass-fiber-reinforced	High-impact copolymer	Molding	33% glass-fiber-reinforced	Molybdenum disulfide-filled	Nylon 6/6–nylon 6 copolymer
Physical							
Melting temperature, °C							
Crystalline	216	216	216	265	265	265	240
Amorphous							
Specific gravity	1.12–1.14	1.35–1.42	1.08–1.17	1.13–1.15	1.38	1.15–1.17	1.08–1.14
Water absorption (24 h), %	2.9	1.2	1.3–1.5	1.0–1.3	1.0	0.8–1.1	1.5–2.0
Dielectric strength, kV · mm^{-1}	15.8	15.8	22	24		14	15.8
Electrical							
Volume (dc) resistivity, ohm-cm	10^{12}			10^{12}–10^{15}			10^{10}
Dielectric constant (60 Hz)	9.8			4.0			16
Dielectric constant (10^6 Hz)	3.7			3.6			4
Dissipation (power) factor (60 Hz)	0.14			0.01–0.02			0.4
Dissipation factor (10^6 Hz)	0.12			0.02–0.03			0.1
Mechanical							
Compressive modulus, 10^3 lb · in^{-2}	250						

Compressive strength, rupture or 1% yield, 10^3 lb·in^{-2}	13–16	19		15 (yield)	24.9	12.5	40
Elongation at break, %	30–100	3–6	150–270	60	3	15	
Flexural modulus at 23°C, 10^3 lb·in^{-2}	390	1500	110–320	420	1300	450	150–410
Flexural strength, rupture or yield, 10^3 lb·in^{-2}	14	33	5–12	17	41	17	
Hardness, Rockwell (or Shore)	R119	M101	R81–R110	R120	M100	R119	R119
Impact strength (Izod) at 23°C, J·m^{-1}	32–53	160	96 to no break	43–53	117	240	37
Tensile modulus, 10^3 lb·in^{-2}	380	1450				550	150–410
Tensile strength at break, 10^3 lb·in^{-2}	11.8	25	7.5–11	12	28	13.7	7.4–12.4
Tensile yield strength, 10^3 lb·in^{-2}	8			8			
Thermal							
Burning rate, mm·min^{-1}	Self-extinguishing	Self-extinguishing	Self-extinguishing	Self-extinguishing	Self-extinguishing	Self-extinguishing	Self-extinguishing
Coefficient of linear thermal expansion, 10^{-6}°C	80–90	20–30	30–40	80	15–20	54	
Deflection temperature under flexural load (264 lb·in^{-2}), °C	68–85	210	45–54	75	249	127	77
Maximum recommended service temperature, °C	107			135			
Specific heat, cal·g^{-1}	0.4			0.4			
Thermal conductivity, W·m^{-1}·K^{-1}	0.24	0.24		0.24	0.22		

TABLE 10.2 Properties of Commercial Plastics (*Continued*)

Properties	Polyamide						Aromatic nylon (aramid), molded and unfilled	Poly(amide-imide), unfilled
	Nylon 6/9, molding and extrusion	Nylon 6/12		Nylon 11, molding and extrusion	Nylon 12, molding and extrusion			
		Molding	30–35% glass-fiber-reinforced					
Physical								
Melting temperature, °C								
Crystalline	205	217	217	194	179		275	
Amorphous								275
Specific gravity	1.08–1.10	1.06–1.08	1.31–1.38	1.03–1.05	1.01–1.02		1.30	1.40
Water absorption (24 h), %	0.5	0.4	0.2	0.3	0.25		0.6	0.28
Dielectric strength, kV · mm^{-1}	24	16	21	17	18		31	24
Electrical								
Volume (dc) resistivity, ohm-cm		10^{15}			10^{14}			
Dielectric constant (60 Hz)		4.0			3.8			
Dielectric constant (10^6 Hz)		3.5			3.0			
Dissipation (power) factor (60 Hz)		0.02			0.07			
Dissipation factor (10^6 Hz)		0.02			0.04			
Mechanical								
Compressive modulus, 10^3 lb · in^{-2}				180			290	413

Compressive strength, rupture or 1% yield, 10^3 lb·in^{-2}	1125	2.4			7.5	30	40
Elongation at break, %		150	4	300	300	5	12–18
Flexural modulus at 23°C, 10^3 lb·in^{-2}	290	290	1120	150	165	640	664
Flexural strength, rupture or yield, 10^3 lb·in^{-2}					1.5	25.8	30
Hardness, Rockwell (or Shore)	R111	R114	E40–E50	R108	R106–R109	E90	E78
Impact strength (Izod) at 23°C, J·m^{-1}	59	53	139	96	107–300	75	133
Tensile modulus, 10^3 lb·in^{-2}	275	290	1200	185	180		730
Tensile strength at break, 10^3 lb·in^{-2}	8.5	8.8	24	8	8–9	17.5	26.9
Tensile yield strength, 10^3 lb·in^{-2}		8.8					
Thermal							
Burning rate, mm·min^{-1}				Self-extinguishing			
Coefficient of linear thermal expansion, 10^{-6}°C	57–60	90	93–218	55–100	67–100	40	36
Deflection temperature under flexural load (264 lb·in^{-2}), °C		82		54	54	260	274
Maximum recommended service temperature, °C				100–120			260
Specific heat, cal·g^{-1}		0.4		0.58			
Thermal conductivity, W·m^{-1}·K^{-1}		0.22		0.34	0.22	0.22	0.25

TABLE 10.2 Properties of Commercial Plastics (*Continued*)

Properties	Poly(aryl ether), unfilled	Polycarbonate		Thermoplastic polyester			
				Poly(butylene terephthalate)		Poly(ethylene terephthalate)	
		Low viscosity	30% glass-fiber-reinforced	Unfilled	30% glass-fiber-reinforced	Unfilled	30% glass-fiber-reinforced
Physical							
Melting temperature, °C							
Crystalline	160	140	150	232–267	232–267	245	245
Amorphous							
Specific gravity	1.14	1.2	1.4	1.31–1.38	1.52	1.34–1.39	1.27
Water absorption (24 h), %	0.25	0.15	0.14	0.08–0.09	0.06–0.08	0.1–0.2	0.05
Dielectric strength, kV · mm^{-1}	17	15	19	16–22	18–22		22
Electrical							
Volume (dc) resistivity, ohm-cm		2×10^{16}	$>10^{16}$		10^{16}	10^{16}	
Dielectric constant (60 Hz)		3.17	3.35			3.25	
Dielectric constant (10^6 Hz)		2.96	3.31				
Dissipation (power) factor (60 Hz)		0.0009	0.011				
Dissipation factor (10^6 Hz)		0.010	0.007				
Mechanical							
Compressive modulus, 10^3 lb · in^{-2}		350	1300				

Compressive strength, rupture or 1% yield, 10^3 lb·in⁻²	80	12.5	18	8.6–14.5	18–23.5	11–15	25
Elongation at break, %	300	110	3–5	50–300	2–4	50–300	3
Flexural modulus at 23°C, 10^3 lb·in⁻²		340	1100	330–400	1100–1200	35–450	1440
Flexural strength, rupture or yield, 10^3 lb·in⁻²	11	13.5	23	12–16.7	26–29	14–18	33.5
Hardness, Rockwell (or Shore)	R117	M70	M92	M68–M78	M90	M94–M101	M100
Impact strength (Izod) at 23°C, J·m⁻¹	427	14	107	43–53	69–85	13–32	101
Tensile modulus, 10^3 lb·in⁻²	320	345	1250	280	1300	400–600	1440
Tensile strength at break, 10^3 lb·in⁻²	7.5	9.5	19	8.2	17–19	8.5–10.5	23
Tensile yield strength, 10^3 lb·in⁻²		9.0					
Thermal							
Burning rate, mm·min⁻¹		Self-extinguishing	Self-extinguishing				
Coefficient of linear thermal expansion, 10^{-6}°C	65	68	22	60–95	25	65	29
Deflection temperature under flexural load (264 lb·in⁻²), °C	149	138–145	146	50–85	220	38–41	224
Maximum recommended service temperature, °C		143					
Specific heat, cal·g⁻¹		0.3				0.27	
Thermal conductivity, W·m⁻¹·K⁻¹	0.30	0.20	0.22	0.18–0.30	0.30	0.15	

TABLE 10.2 Properties of Commercial Plastics (*Continued*)

| Properties | Thermoplastic polyester | | Thermosetting and alkyd polyester | | | | |
| | Aromatic polyester | | Unsaturated polyester | | Alkyd molding compounds | | Polyimide, unfilled |
	Extrusion-transparent	Injection molding	Styrene–maleic acid copolymer, low-shrink	Butadiene–maleic acid copolymer	Putty, mineral-filled	Glass-fiber-reinforced	
Physical							
Melting temperature, °C							
Crystalline							310–365
Amorphous	81						
Specific gravity		1.39					1.36–1.43
Water absorption (24 h), %		0.01					0.24
Dielectric strength, kV · mm^{-1}		14					22
Electrical							
Volume (dc) resistivity, ohm-cm							$>10^{16}$
Dielectric constant (60 Hz)							
Dielectric constant (10^6 Hz)							3–4
Dissipation (power) factor (60 Hz)							
Dissipation factor (10^6 Hz)							
Mechanical							
Compressive modulus, 10^3 lb · in^{-2}			Thermoset	Thermoset	2000–3000	Thermoset	

Compressive strength, rupture or 1% yield, 10^3 lb · in^{-2}	225	10	15–30	14–30	12–38	15–36	30–40
Elongation at break, %		7–10	3–5				8–10
Flexural modulus at 23°C, 10^3 lb · in^{-2}	290	700	1000–2500		2000	2000	450–500
Flexural strength, rupture or yield, 10^3 lb · in^{-2}	10.6	12	9–35	16–24	6–17	8.5–26	19–28.8
Hardness, Rockwell (or Shore)	R105		40–70 (Barcol)	50–60 (Barcol)	E98	E95	E52–E99
Impact strength (Izod) at 23°C, J · m^{-1}	101	300	133–800	214–694	16–27	27–854	80
Tensile modulus, 10^3 lb · in^{-2}			1000–2500	1500–2500	500–3000		300
Tensile strength at break, 10^3 lb · in^{-2}	6	11	4.5–20	5–10	3–9	4–9.5	10.5–17.1
Tensile yield strength, 10^3 lb · in^{-2}	7						12.5
Thermal							
Burning rate, mm · min^{-1}							
Coefficient of linear thermal expansion, 10^{-6}°C	63	29	6–30		20–50	15–33	45–56
Deflection temperature under flexural load (264 lb · in^{-2}), °C		282	190–260	160–177	177–260	204–260	277–360
Maximum recommended service temperature, °C							
Specific heat, cal · g^{-1}							0.27
Thermal conductivity, W · m^{-1} · K^{-1}	0.29	0.29		0.76–0.93	0.51–0.89	0.6–0.89	0.10–0.11

TABLE 10.2 Properties of Commercial Plastics (*Continued*)

Properties	Poly(methyl pentene), unfilled	Polyolefin — Polyethylene					Ethylene-vinyl acetate copolymer
		Low-density	Medium-density	High-density	Ultra high-molecular-weight	Glass-fiber-reinforced, high-density	
Physical							
Melting temperature, °C							
Crystalline	230–240	95–130	120–140	120–140	125–135	120–140	
Amorphous							65–90
Specific gravity	0.84	0.910–0.925	0.926–0.94	0.941–0.965	0.94	1.28	0.92–0.95
Water absorption (24 h), %	0.01	<0.01	<0.01	<0.01	<0.01	0.02	0.05–0.13
Dielectric strength, kV · mm^{-1}		18–39	18–39	18–39	28	20	24–30
Electrical							
Volume (dc) resistivity, ohm-cm		>10^{15}	>10^{15}	<10^{15}			
Dielectric constant (60 Hz)		2.3	2.3	2.3			
Dielectric constant (10^6 Hz)		2.3	2.3	2.3			
Dissipation (power) factor (60 Hz)		<0.0005	<0.0005	<0.0005			
Dissipation factor (10^6 Hz)		<0.0005	<0.0005	<0.0005			
Mechanical							
Compressive modulus, 10^3 lb · in^{-2}	114–171						

Compressive strength, rupture or 1% yield, 10^3 lb · in^{-2}	5–6.6	90–800	50–600	2.7–3.6	450–525	7	550–900
Elongation at break, %	10–50	8–60	60–115	20–130	130–140	1.5	1–20
Flexural modulus at 23°C, 10^3 lb · in^{-2}	110–260			100–260		800	
Flexural strength, rupture or yield, 10^3 lb · in^{-2}	4–6.5					11	
Hardness, Rockwell (or Shore)	L67–L74	(D40–D51)	(D50–D60)	R30–R50	R50	R75	No break
Impact strength (Izod) at 23°C, J · m^{-1}	16–64	No break	27–854	27–1068	No break	59	20–120
Tensile modulus, 10^3 lb · in^{-2}	160–280	14–38	25–55	60–180			
Tensile strength at break, 10^3 lb · in^{-2}	3.5–4	0.6–2.3	1.2–3.5	3.1–5.5	5.6	9	1.4–2.8
Tensile yield strength, 10^3 lb · in^{-2}		0.8–1.2	1.0–2.2	3–4	3.1–4.0		
Thermal							
Burning rate, mm · min^{-1}		1.0	1.0	1.0			
Coefficient of linear thermal expansion, 10^{-6}°C	117	100–220	140–160	110–130	130	48	160–200
Deflection temperature under flexural load (264 lb · in^{-2}), °C	41	32–41	41–49	43–54	43–49	121	34
Maximum recommended service temperature, °C	175	70	93	200			
Specific heat, cal · g^{-1}	0.55	0.55	0.55	0.46–0.55			
Thermal conductivity, W · m^{-1} · K^{-1}	0.17	0.34	0.34–0.42	0.46–0.51		0.46	

TABLE 10.2 Properties of Commercial Plastics (*Continued*)

Properties	Polybutylene extrusion	Polyolefin				Poly(phenylene sulfide)	
		Polypropylene			Polyallomer	Injection molding	40% glass-fiber-reinforced
		Homopolymer	Copolymer	Impact copolymer			
Physical							
Melting temperature, °C							
Crystalline	126	168	160–168		120–135	290	290
Amorphous							
Specific gravity	0.91–0.925	0.90–0.91	0.89–0.905	0.90	0.90	1.3	1.6
Water absorption (24 h), %	0.01–0.02	0.01–0.03	0.03	<0.03	<0.01	<0.02	0.05
Dielectric strength, kV · mm^{-1}	18	24	24	24	31	15	18
Electrical							
Volume (dc) resistivity, ohm-cm		10^{17}	10^{17}	10^{17}			
Dielectric constant (60 Hz)		2.2–2.6	2.3	2.3			
Dielectric constant (10^6 Hz)		2.2–2.6	2.3				
Dissipation (power) factor (60 Hz)		<0.0005	0.0001–0.0005				
Dissipation factor (10^6 Hz)		0.0005–0.002	0.0001–0.002	0.0003			
Mechanical							
Compressive modulus, 10^3 lb · in^{-2}	31	150–300					

Compressive strength, rupture or 1% yield, 10^3 lb·in⁻²	300–380	5.5–8.0	3.5–8.0	8–20	400–500	16	21
Elongation at break, %	45–50	100–600	200–700			1–2	1
Flexural modulus at 23°C, 10^3 lb·in⁻²		170–250	130–200	130–190	70–110	550	1700
Flexural strength, rupture or yield, 10^3 lb·in⁻²	2–2.3	6–8	5–7			14	29
Hardness, Rockwell (or Shore)		R80–R102	R50–R96	R40–R90	R50–R85	R123	R123
Impact strength (Izod) at 23°C, J·m⁻¹	No break	21–53	53–1068	80–900	91–203	<27	75
Tensile modulus, 10^3 lb·in⁻²	30–40	165–225	100–170			480	1100
Tensile strength at break, 10^3 lb·in⁻²	3.8–4.4	4.5–6	4–5.5	2.5–3.1	3–3.8	9.5	19.5
Tensile yield strength, 10^3 lb·in⁻²	1.7–2.5	4.5–5.4	3.5–4.3		3–3.4		
Thermal							
Burning rate, mm·min⁻¹							
Coefficient of linear thermal expansion, 10^{-6}°C	128–150	81–100	68–95	60–90	83–100	49	22
Deflection temperature under flexural load (264 lb·in⁻²), °C	54–60	48–57	45–57	90–105 (66 lb·in⁻²)	51–56	135	249
Maximum recommended service temperature, °C		160	240	140–160			
Specific heat, cal·g⁻¹		0.44–0.46	0.45–0.50	0.45–0.50			
Thermal conductivity, W·m⁻¹·K⁻¹	0.22	0.12	0.15–0.17	0.12–0.17	0.09–0.17	0.29	0.29

TABLE 10.2 Properties of Commercial Plastics (*Continued*)

Properties	Polyurethane			Silicone		Epoxy molding and encapsulating compound	Styrenic Polystyrene Crystal
	Casting resin		Thermoplastic elastomer	Cast resin, flexible	Mineral- and/or glass-filled		
	Liquid	Unsaturated					
Physical							
Melting temperature, °C							
Crystalline	Thermoset	Thermoset		Thermoset	Thermoset	Thermoset	
Amorphous			120–160				85–105
Specific gravity	1.1–1.5	1.05	1.05–1.25	0.99–1.5	1.8–1.94	1.84	1.04–1.05
Water absorption (24 h), %	0.02–1.5	0.1–0.2	0.7–0.9				0.03–0.10
Dielectric strength, kV·mm⁻¹	12–20		13–25	22	8–15	10	24
Electrical							
Volume (dc) resistivity, ohm-cm	10^{11}–10^{15}		10^{11}–10^{13}	10^{14}–10^{15}			$>10^{16}$
Dielectric constant (60 Hz)	4.0–7.5		5.4–7.6	2.7–4.2			2.5
Dielectric constant (10^6 Hz)							
Dissipation (power) factor (60 Hz)							
Dissipation factor (10^6 Hz)							
Mechanical							
Compressive modulus, 10^3 lb·in⁻²	10–100		4–9				

Compressive strength, rupture or 1% yield, 10^3 lb·in^{-2}	20		20		10–16	28	11.5–16
Elongation at break, %	100–1000	3–6	100–1100	100–700			1–2
Flexural modulus at 23°C, 10^3 lb·in^{-2}	10–100	610	10–350		1000–2500		380–450
Flexural strength, rupture or yield, 10^3 lb·in^{-2}	0.7–4.5	19	0.7–9		9–14	17	8–14
Hardness, Rockwell (or Shore)			(A65–D80)	(A15–A65)	M80–M90		M60–M75
Impact strength (Izod) at 23°C, J·m^{-1}	1334 to flexible	21	No break		13–427	16	13–21
Tensile modulus, 10^3 lb·in^{-2}	10–100	10–11	10–350				350–485
Tensile strength at break, 10^3 lb·in^{-2}	0.175–10		1.5–8.4	0.35–1.0	4–6.5	6–8	5.3–7.9
Tensile yield strength, 10^3 lb·in^{-2}							
Thermal							
Burning rate, mm·min^{-1}					0–78		
Coefficient of linear thermal expansion, 10^{-6}°C	100–200		100–200	300–800	20–50	30	70–80
Deflection temperature under flexural load (264 lb·in^{-2}), °C	Varies over wide range	87–93	Varies over wide range		260	74–100	
Maximum recommended service temperature, °C					371		93
Specific heat, cal·g^{-1}	0.43		0.43				0.3
Thermal conductivity, W·m^{-1}·K^{-1}	0.21		0.07–0.31	0.15–0.31	0.30	0.68	0.09–0.13

TABLE 10.2 Properties of Commercial Plastics (*Continued*)

		Styrenic						
	Polystyrene	Acrylonitrile-butadiene-styrene copolymer						
			Molding					
Properties	Heat-resistant	Extrusion	Heat-resistant	High-impact	Flame-retarded	Platable	20% glass-reinforced
Physical							
Melting temperature, °C							
Crystalline							
Amorphous	110–125	88–120	110–125	100–110	110–125	100–110	
Specific gravity	1.05–1.09	1.02–1.06	1.05–1.08	1.01–1.04	1.16–1.21	1.06–1.07	1.22
Water absorption (24 h), %	0.03–0.12	0.20–0.45	0.20–0.45	0.20–0.45	0.2–0.6		
Dielectric strength, kV · mm⁻¹	20	14–20	14–20	14–20	14–20	16–22	18
Electrical							
Volume (dc) resistivity, ohm-cm							
Dielectric constant (60 Hz)				2.4–5.0			
Dielectric constant (10⁶ Hz)				2.4–3.8			
Dissipation (power) factor (60 Hz)				0.003–0.008			
Dissipation factor (10⁶ Hz)				0.007–0.015			
Mechanical							
Compressive modulus, 10³ lb · in⁻²		150–390	190–440	140–300	130–310		

Compressive strength, rupture or 1% yield, 10^3 lb·in^{-2}	11.5-16	5.2-10	7.2-10	4.5-8	6.5-7.5		14
Elongation at break, %	2-60	20-100	3-20	5-70	5-25		
Flexural modulus at 23°C, 10^3 lb·in^{-2}	340-470	130-420	300-400	250-350	300-400	340-390	710
Flexural strength, rupture or yield, 10^3 lb·in^{-2}	8.9-14	4-14	10-13	8-11	9-14	10.5-11.5	15.5
Hardness, Rockwell (or Shore)	L80-L108	R75-R115	R100-R115	R85-R105	R100-R120	R103-R109	M85
Impact strength (Izod) at 23°C, J·m^{-1}	21-181	133-640	107-347	347-400	160-640	267-283	64
Tensile modulus, 10^3 lb·in^{-2}	320-460	130-380	300-350	230-330	320-400	330-380	740
Tensile strength at break, 10^3 lb·in^{-2}	5-7.8	2.5-8.0	6-7.5	4.8-6.3	5-8	6-6.4	11
Tensile yield strength, 10^3 lb·in^{-2}			5.5-7	4-5.5	4-6		
Thermal							
Burning rate, mm·min^{-1}		1.3		1.3			
Coefficient of linear thermal expansion, 10^{-6}°C	60-70	60-130	60-93	95-110	65-95	47-53	21
Deflection temperature under flexural load (264 lb·in^{-2}), °C	93-120 annealed	77-104 annealed	104-116 annealed	96-102 annealed	90-107 annealed	96-102 annealed	99
Maximum recommended service temperature, °C				110			
Specific heat, cal g^{-1}				0.3-0.4			
Thermal conductivity, W·m^{-1}·K^{-1}			0.19-0.34				

TABLE 10.2 Properties of Commercial Plastics (*Continued*)

	Styrenic			Sulfone			
	Styrene-acrylonitrile copolymer		Styrene-butadiene copolymer, high-impact	Polysulfone		Poly(ether sulfone)	Poly(phenyl sulfone)
Properties	Unfilled	20% glass-fiber-reinforced		Unfilled	20% glass-fiber-reinforced		
Physical							
Melting temperature, °C							
Crystalline							
Amorphous	115–125	115–125	90–110	200	200	230	220
Specific gravity	1.07–1.08	1.22	1.03–1.06	1.24	1.46	1.37	1.29
Water absorption (24 h), %	0.2–0.3	0.15–0.20	0.05–0.10	0.22	0.23	0.43	1.1–1.3 (saturated)
Dielectric strength, kV · mm^{-1}	16–20	20	18	17	17	17	16
Electrical							
Volume (dc) resistivity, ohm-cm				10^{15}			
Dielectric constant (60 Hz)				3.14	3.7		
Dielectric constant (10^6 Hz)				3.26	3.7		
Dissipation (power) factor (60 Hz)				0.004	0.002		
Dissipation factor (10^6 Hz)				0.008	0.009		
Mechanical							
Compressive modulus, 10^3 lb · in^{-2}	530			370			

Compressive strength, rupture of 1% yield, 10^3 lb·in^{-2}	14–17	19	4–9	13.9	22	30–80	60
Elongation at break, %	1–4	1–2	13–50	50–100	2		
Flexural modulus at 23°C, 10^3 lb·in^2	550	100–1100	280–450	390	1000	375	330
Flexural strength, rupture or yield, 10^3 lb·in^{-2}	14–17	20	5.3–9.4	15.4	23	18.7	12.4
Hardness, Rockwell (or Shore)	M80–M90	R122	M10–M68	M69, R120	R123	M88	
Impact strength (Izod) at 23°C, J·m^{-1}	19–27	53	32–192	64	59	85	640
Tensile modulus, 10^3 lb·in^{-2}	400–560	1150–1200	280–465	360	1200	350	310
Tensile strength at break, 10^3 lb·in^{-2}	9–12	15.8–18	3.2–4.9		17		
Tensile yield strength, 10^3 lb·in^{-2}			2.9–4.9	10.2		12.2	10.4
Thermal							
Burning rate, mm·min^{-1}							31
Coefficient of linear thermal expansion, 10^{-6}°C	36–38	38–40	70–101	52–56	25	55	
Deflection temperature under flexural load (264 lb·in^{-2}), °C	88–104	99	74–93	174	182	203	204
Maximum recommended service temperature, °C				149			
Specific heat, cal·g^{-1}							
Thermal conductivity, W·m^{-1}·K^{-1}	0.12	0.26–0.28	0.12–0.21	0.12	0.38	0.14–0.19	

TABLE 10.2 Properties of Commercial Plastics (*Continued*)

Properties	Thermoplastic elastomers					Vinyl	
						Poly(vinyl chloride) and poly(vinyl acetate)	
	Polyolefin	Polyester	Block copolymers of styrene and butadiene or styrene and isoprene	Block copolymers of styrene and ethylene or styrene and butylene	Urea formaldehyde, alpha-cellulose filled	Rigid	Flexible and unfilled
Physical							
Melting temperature, °C							
Crystalline		168–206			Thermoset	75–105	75–105
Amorphous							
Specific gravity	0.88–0.90	1.17–1.25	0.9–1.2	0.9–1.2	1.47–1.52	1.30–1.58	1.16–1.35
Water absorption (24 h), %	0.01		0.19–0.39		0.4–0.8	0.04–0.4	0.15–0.75
Dielectric strength, kV · mm^{-1}	24–26		16–21		12–16	14–20	12–16
Electrical							
Volume (dc) resistivity, ohm-cm					0.5–5.0	10^{12}–10^{15}	10^{11}–10^{14}
Dielectric constant (60 Hz)					7.7–9.5	3.2–4.0	5.0–9.0
Dielectric constant (10^6 Hz)					6.7–8.0	3.0–4.0	3.0–4.0
Dissipation (power) factor (60 Hz)					0.036–0.043	0.01–0.02	0.03–0.05
Dissipation factor (10^6 Hz)					0.025–0.035	0.006–0.02	0.06–0.1
Mechanical							
Compressive modulus, 10^3 lb · in^{-2}			3.6–120				

Compressive strength, rupture or 1% yield, 10^3 lb·in^{-2}	150–300	350–450	500–1350	600–800	25–45	8–13	0.9–1.7
Elongation at break, %	1.5–2.0	7–75	4–150	4–100	<1	40–80	200–450
Flexural modulus at 23°C, 10^3 lb·in^{-2}					1300–1600	300–500	
Flexural strength, rupture or yield, 10^3 lb·in^{-2}					10–18	10–16	
Hardness, Rockwell (or Shore)	(A65–A92)	(D40–D72)	(A40–A90)	(A50–A90)	M110–M120	(D65–D95)	(A50–A100)
Impact strength (Izod) at 23°C, J·m^{-1}	No break	208 to no break	No break	No break	13–21	21–1068	Varies over wide range
Tensile modulus, 10^3 lb·in^{-2}		1.1–2.5	0.8–50		1000–1500	350–600	
Tensile strength at break, 10^3 lb·in^{-2}	0.65–2.0	3.7–5.7	0.6–3.0	1–3	5.5–13	6–75	1.5–3.5
Tensile yield strength, 10^3 lb·in^{-2}							
Thermal							
Burning rate, mm·min^{-1}					Self-extinguishing	Self-extinguishing	Slow to self-extinguishing
Coefficient of linear thermal expansion, 10^{-6}°C	130–170		130–137		22–36	50–100	70–250
Deflection temperature under flexural load (264 lb·in^{-2}), °C			<0–49		127–143	60–77	
Maximum recommended service temperature, °C					77	70–74	80–105
Specific heat, cal·g^{-1}					0.6	0.2–0.28	0.36–0.5
Thermal conductivity, W·m^{-1}·K^{-1}	0.19–0.21	0.15			0.30–0.42	0.15–0.21	0.13–0.17

TABLE 10.2 Properties of Commercial Plastics (*Continued*)

Properties	Vinyl					
	Poly(vinyl chloride) and poly(vinyl acetate) Flexible and filled	Poly(vinyl chloride), 15% glass-fiber-reinforced	Poly(vinylidene chloride)	Poly(vinyl formal)	Chlorinated poly(vinyl chloride)	Poly(vinyl butyral), flexible
Physical						
Melting temperature, °C						
Crystalline			210			
Amorphous	75–105	75–105		105	110	49
Specific gravity	1.3–1.7	1.54	1.65–1.72	1.2–1.4	1.49–1.56	1.05
Water absorption (24 h), %	0.5–1.0	0.01	0.1	0.5–3.0	0.02–0.15	1.0–2.0
Dielectric strength, kV · mm^{-1}	9.8–12	24–31	16–24	19		14
Electrical						
Volume (dc) resistivity, ohm-cm			10^{14}–10^{16}			
Dielectric constant (60 Hz)			4.5–6.0			
Dielectric constant (10^6 Hz)						
Dissipation (power) factor (60 Hz)						
Dissipation factor (10^6 Hz)						
Mechanical						
Compressive modulus, 10^3 lb · in^{-2}					335–600	

Compressive strength, rupture or 1% yield, 10^3 lb · in^{-2}	1.0–1.8	9	2–2.7	5–20	9–22	150–450
Elongation at break, %	200–400	2–3	50–250		4–65	
Flexural modulus at 23°C, 10^3 lb · in^{-2}		750			380–450	
Flexural strength, rupture or yield, 10^3 lb · in^{-2}		13.5	4.2–6.2	17–18	14.5–17	
Hardness, Rockwell (or Shore)	(A50–A100)	R118	M50–M65	M85	R117–R122	A10–A100
Impact strength (Izod) at 23°C, J · m^{-1}	Varies over wide range	53	16–53	43–75	53–299	Varies over wide range
Tensile modulus, 10^3 lb · in^{-2}		870	50–80	350–600	360–475	
Tensile strength at break, 10^3 lb · in^{-2}	1–3.5	9.5	3–5	10–12	7.5–9	
Tensile yield strength, 10^3 lb · in^{-2}						0.5–3.0
Thermal						
Burning rate, mm · min^{-1}			Self-extinguishing			Slow
Coefficient of linear thermal expansion, 10^{-6}°C			190			
Deflection temperature under flexural load (264 lb · in^{-2}), °C		68	54–71	64	68–78	
Maximum recommended service temperature, °C			100	71–77	94–112	
Specific heat, cal · g^{-1}			0.32			
Thermal conductivity, W · m^{-1} · K^{-1}	0.13–0.17		0.13	0.16	0.14	

10.3.24.5 Poly(vinylidene Chloride). Poly(vinylidene chloride) is prepared according to the following reaction:

$$CH_2{=}CCl_2 + CH_2{=}CHCl \rightarrow [- CH_2{-}CCl_2{-}CH_2{-}CHCl{-}]_n$$

Random copolymer

10.3.25 Urea Formaldehyde

The reaction of urea with formaldehyde yields the following products, which are used as monomers in the preparation of urea formaldehyde resin.

$$H_2N{-}CO{-}NH_2 + H_2CO \rightarrow H_2N{-}CO{-}NH{-}CH_2OH$$
$$+ HOCH_2{-}NH{-}CO{-}NH{-}CH_2OH$$

The reaction conditions can be varied so that only one of those monomers is formed. 1-Hydroxymethylurea and 1,3-bis(hydroxymethyl)urea condense in the presence of an acid catalyst to produce urea formaldehyde resins. A wide variety of resins can be obtained by careful selection of the pH, reaction temperature, reactant ratio, amino monomer, and degree of polymerization. If the reaction is carried far enough, an infusible polymer network is produced.

Liquid coating resins are prepared by reacting methanol or butanol with the initial hydroxymethylureas. Ether exchange reactions between the amino resin and the reactive sites on the polymer produce a cross-linked film.

10.4 FORMULAS AND ADVANTAGES OF RUBBERS

10.4.1 Gutta Percha

Gutta percha is a natural polymer of isoprene (3-methyl-1,3-butadiene) in which the configuration around each double bond is *trans*. It is hard and horny and has the following formula:

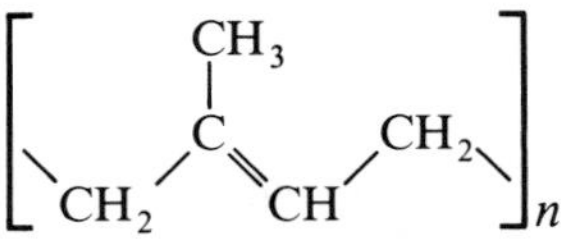

10.4.2 Natural Rubber

Natural rubber is a polymer of isoprene in which the configuration around each double bond is *cis* (or *Z*):

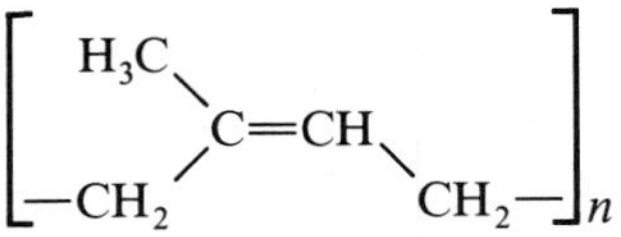

Its principal advantages are high resilience and good abrasion resistance.

10.4.3 Chlorosulfonated Polyethylene

Chlorosulfonated polyethylene is prepared as follows:

$$[-CH_2-CH_2-]_n + HSO_3Cl \rightarrow \left[\begin{array}{c} -CH_2-CH- \\ | \\ SO_3H \end{array}\right]_n + HCl$$

Cross-linking, which can occur as a result of side reactions, causes an appreciable gel content in the final product.

The polymer can be vulcanized to give a rubber with very good chemical (solvent) resistance, excellent resistance to aging and weathering, and good color retention in sunlight.

10.4.4 Epichlorohydrin

Epichlorohydrin is a product of covulcanization of epichlorohydrin (epoxy) polymers with rubbers, especially *cis*-polybutadiene.

Its advantages include impermeability to air, excellent adhesion to metal, and good resistance to oils, weathering, and low temperature.

10.4.5 Nitrile Rubber (NBR, GRN, Buna N)

Nitrile rubber can be prepared as follows:

$$CH_2= CH-CH=CH_2 + CH_2=CH-CN \rightarrow$$

$$\qquad \text{2 parts} \qquad\qquad\qquad \text{1 part}$$

$$\left[\begin{array}{c} -CH_2-CH=CH-CH_2-CH_2-CH-CH_2-CH=CH-CH_2- \\ | \\ CN \end{array}\right]_n$$

Nitrile rubber is also known as nitrile-butadiene rubber (NBR), government rubber nitrile (GRN), and Buna N.

It possesses resistance to oils up to 120°C and excellent abrasion resistance and adhesion to metal.

10.4.6 Polyacrylate

Polyacrylate has the following formula:

$$\left[\begin{array}{c} -CH_2-CH- \\ | \\ CN \end{array}\right]_n$$

It possesses oil and heat resistance to 175°C and excellent resistance to ozone.

10.4.7 *cis*-Polybutadiene Rubber (BR)

cis-Polybutadiene is prepared by polymerization of butadiene by mostly 1,4-addition.

$$CH_2=CH-CH=CH_2 \rightarrow [-CH_2-CH=CH-CH_2-]_n$$

The polybutadiene produced is in the Z (or *cis*) configuration.

cis-Polybutadiene has good abrasion resistance, is useful at low temperature, and has excellent adhesion to metal.

10.4.8 Polychloroprene (Neoprene)

Polychloroprene is prepared as follows:

$$CH_2{=}CH{-}\underset{\underset{Cl}{|}}{C}{=}CH_2 \rightarrow [-CH_2{-}CH{=}C(Cl){-}CH_2{-}]_n$$

It has very good weathering characteristics, is resistant to ozone and to oil, and is heat-resistant to 100°C.

10.4.9 Ethylene-Propylene-Diene Rubber (EPDM)

Ethylene-propylene-diene rubber is polymerized from 60 parts ethylene, 40 parts propylene, and a small amount of nonconjugated diene. The nonconjugated diene permits sulfur vulcanization of the polymer instead of using peroxide.

It is a very lightweight rubber and has very good weathering and electrical properties, excellent adhesion, and excellent ozone resistance.

10.4.10 Polyisobutylene (Butyl Rubber)

Polyisobutylene is prepared as follows:

$$H_3C{-}\underset{\underset{}{\overset{\overset{CH_3}{|}}{C}}}{}{=}CH_2 + CH_2{=}\underset{}{\overset{\overset{CH_3}{|}}{C}}{-}CH{=}CH_2 \rightarrow$$

98 parts   2 parts

$$\left[\left(-\underset{\underset{CH_3}{|}}{\overset{\overset{CH_3}{|}}{C}}{-}CH_2{-} \right)_n {-}CH_2{-}\underset{}{\overset{\overset{CH_3}{|}}{C}}{=}CH{-}CH_2{-} \right]$$

It possesses excellent ozone resistance, very good weathering and electrical properties, and good heat resistance.

10.4.11 (Z)-Polyisoprene (Synthetic Natural Rubber)

Polymerization of isoprene by 1,4-addition produces polyisoprene that has a *cis* (or *Z*) configuration.

$$\left[\underset{-CH_2}{\overset{H_3C}{}}{>}C{=}C{<}\underset{CH_2-}{\overset{H}{}} \right]_n$$

10.4.12 Polysulfide Rubbers

Polysulfide rubbers are prepared as follows:

$$\text{Cl}-\text{R}-\text{Cl} + \text{Na}-\text{S}-\text{S}-\text{S}-\text{S}-\text{Na} \rightarrow \text{HS}[-\text{R}-\text{S}-\text{S}-\text{S}-\text{S}-]_n\text{R}-\text{SH}$$

where R can be

$$-\text{CH}_2\text{CH}_2-, \quad -\text{CH}_2\text{CH}_2-\text{O}-\text{CH}_2\text{CH}_2-,$$

or
$$-\text{CH}_2\text{CH}_2-\text{O}-\text{CH}_2-\text{O}-\text{CH}_2\text{CH}_2-.$$

Polysulfide rubbers possess excellent resistance to weathering and oils and have very good electrical properties.

10.4.13 Poly(vinyl Chloride) (PVC)

Poly(vinyl chloride) as previously discussed in Sec. 10.3, Formulas and Key Properties of Plastic Materials, has the following structures:

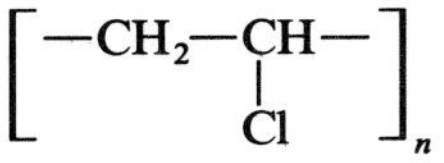

$$\left[\begin{array}{c} -\text{CH}_2-\text{CH}- \\ | \\ \text{Cl} \end{array}\right]_n$$

PVC polymer plus special plasticizers are used to produce flexible tubing which has good chemical resistance.

10.4.14 Silicone Rubbers

Silicone rubbers are prepared as follows:

$$\text{Cl}-\underset{\underset{\text{CH}_3}{|}}{\overset{\overset{\text{CH}_3}{|}}{\text{Si}}}-\text{Cl} \xrightarrow{\text{H}_2\text{O}} \text{HO}-\underset{\underset{\text{CH}_3}{|}}{\overset{\overset{\text{CH}_3}{|}}{\text{Si}}}-\text{OH} \xrightarrow{\text{polymerize}} \left[\begin{array}{c} \overset{\text{CH}_3}{|} \\ -\text{Si}-\text{O}- \\ | \\ \text{CH}_3 \end{array}\right]_n$$

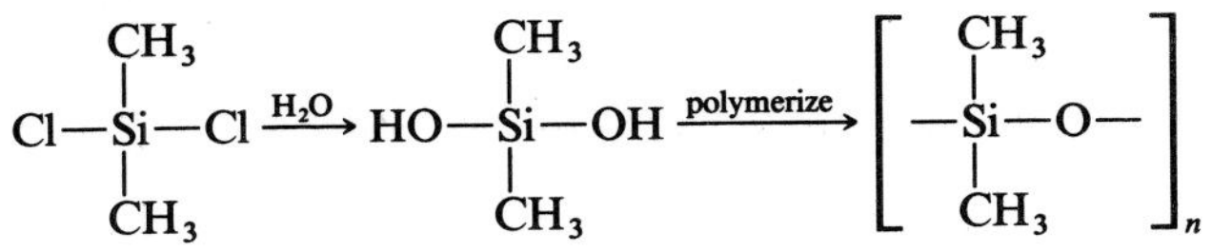

Other groups may replace the methyl groups.

Silicone rubbers have excellent ozone and weathering resistance, good electrical properties, and good adhesion to metal.

10.4.15 Styrene-Butadiene Rubber (GRS, SBR, Buna S)

Styrene-butadiene rubber is prepared from the free-radical copolymerization of one part by weight of styrene and three parts by weight of 1,3-butadiene. The butadiene is incorporated by both 1,4-addition (80%) and 1,2-addition (20%). The configuration around the double bond of the 1,4-adduct is about 80% *trans*. The product is a random copolymer with these general features:

trans-1,4-Adduct   1,2-Adduct   trans-1,4-Adduct   Styrene   cis-1,4-Adduct

Styrene-butadiene rubber (SBR) is also known as government rubber styrene (GRS) and Buna S.

10.4.16 Urethane

See Table 10.3.

TABLE 10.3 Properties of Natural and Synthetic Rubbers

Rubber	Specific gravity	Durometer hardness (or Shore)	Ultimate elongation % (23°C)	Tensile strength, lb·in^{-2} (23°C)	Service temperature, °C	
					Minimum	Maximum
Gutta percha (hard rubber)	1.2–1.95	(65–95)	3–8	4000–10,000	−56	104
Natural rubber (NR)	0.93	20–100	750–850	3000–4500		82
Chlorosulfonated polyethylene	1.10	50–95	100–500	500–3000	−54	121
Epichlorohydrin	1.27	60–90	100–400	1000–2500	−46	121
Fluoroelastomers	1.4–1.95	60–90	100–350	2000–3000	−40	232
Isobutene-isoprene rubber (IIR) [also known as government rubber I(GR-I)]	0.91	(40–70)	750–950	2300–3000		121
Nitrile rubber (butadiene-acrylonitrile rubber) (also known as Buna N and NBR)	1.00	30–100	100–600	500–4000	−54	121
Polyacrylate	1.10	40–100	100–400	1000–2200	−18	149
Polybutadiene rubber (BR)	0.93	30–100	100–700	2500–3000	−62	79–100
Polychloroprene (neoprene)	1.23	20–90	800–1000	2000–3500	−54	121
Poly(ethylene-propylene-diene) (EPDM)	0.85	30–100	100–300	1000–3000	−40	149
Polyisobutylene (butyl rubber)	0.92	30–100	100–700	1000–3000	−54	100
Polyisoprene	0.94	20–100	100–750	2000–3000	−54	79–82
Polysulfide (Thiokol ST)	1.34	20–80	100–400	700–1250	−54	82–100
Poly(vinyl chloride) (Koroseal)	1.32	(80–90)		2400–3000		71
Silicone, high-temperature				700–800		316
Silicone	0.98	20–95	50–800	500–1500	−84	232
Styrene-butadiene rubber (SBR) (also known as Buna S)	0.94	40–100	400–600	1600–3700	−60	107
Urethane	0.85	62–95	100–700	1000–8000	−54	100

10.5 CHEMICAL RESISTANCE

TABLE 10.4 Resistance of Selected Polymers and Rubbers to Various Chemicals at 20°C

The information in this table is intended to be used only as a general guide. The chemical resistance classifications are E = excellent (30 days of exposure causes no damage), G = good (some damage after 30 days), F = fair (exposure may cause crazing, softening, swelling, or loss of strength), N = not recommended (immediate damage may occur).

Polymers	Chemical												
	Acids, dilute or weak	Acids, strong and concentrated	Alcohols, aliphatic	Aldehydes	Alkalies, concentrated	Esters	Ethers	Glycols	Hydrocarbons, aliphatic	Hydrocarbons, aromatic	Hydrocarbons, halogenated	Ketones	Oxidizing agents, strong
Acetals	F	N	F	N	N	N	N	G	N	N	N	N	N
Acrylics: poly(methyl methacrylate)	G	N	E	—	N	N	E	E	G	N	N	N	N
Allyls: diallyl phthalate	G	—	—	—	N	—	—	—	E	G	G	N	—
Cellulosics: cellulose-acetate-butyrate and cellulose-acetate-propionate polymers	F	N	N	N	N	N	N	G	F	N	N	N	—
Fluorocarbons	E	E	E	E	E	E	E	E	E	E	E	E	E
Polyamides	N	N	G	E	E	G	—	G	G	F	F	G	N
Polycarbonates	G	N	G	F	N	N	N	G	N	G	N	N	N
Polyesters	G	G	G	—	N	N	F	G	F	G	F	N	F
Poly(methyl pentene)	E	E	N	G	E	G	N	E	G	G	N	F	F
Low-density polyethylene	E	E	E	E	E	G	N	E	F	F	N	G	F
High-density polyethylene	E	E	E	E	E	G	N	E	G	F	N	G	F
Polybutadiene	G	F	E	—	—	—	—	—	—	E	E	E	—

Material																
Polypropylene and polyallomer	E	E	E	E	E	N	G	N	E	E	E	E	G	F	N	G
Polystyrene	N	Z	E	–	N	–	N	Z	–	N	N	Z	Z	N	Z	N
Styrene-acrylonitrile copolymers	–	N	N	–	–	G	F	N	–	F	F	N	N	–	–	–
Styrene-acrylonitrile-butadiene copolymers	G	Z	G	–	F	Z	F	G	F	F	G	F	N	N	Z	N
Sulfones: polysulfone	E	N	E	F	G	E	E	G	G	F	F	E	G	F	Z	Z
Vinyls: poly(vinyl chloride)	E	G	E	G	G	G	F	G	F	F	G	G	N	Z	Z	G

Rubbers

Material																
Natural rubber	–	–	–	–	–	–	E	–	–	N	N	N	N	N	Z	–
Nitrile rubber	–	–	–	–	–	–	E	–	–	G	Z	E	E	E	Z	–
Polychloroprene	–	–	–	–	–	–	E	–	–	F	Z	F	F	F	Z	–
Polyisobutylene	–	–	–	–	–	–	E	–	–	F	F	E	E	F	Z	–
Polysulfide rubbers: Thiokol	–	–	–	–	–	–	E	–	–	E	E	E	E	E	Z	–
Styrene-butadiene rubber	–	–	–	–	–	–	E	–	–	N	Z	N	N	N	Z	–

TABLE 10.5 Gas Permeability Constants (10^{10} P) at 25°C for Polymers and Rubbers

The gas permeability constant P is defined as

$$P = \frac{\text{amount of permeant}}{(\text{area}) \times (\text{time}) \times (\text{driving forced across the film})}$$

The gas permeability constant is the amount of gas expressed in cubic centimeters passed in 1 s through a 1-cm² area of film when the pressure across a film thickness of 1 cm is 1 cmHg and the temperature is 25°C. All tabulated values are multiplied by 10^{10} and are in units of seconds⁻¹ (centimeters of Hg)⁻¹. Other temperatures are indicated by exponents and are expressed in degrees Celsius.

Polymer or rubber	Gas						
	He	N_2	H_2	O_2	CO_2	H_2O	Other
Cellulose (cellophane)	0.005^{20}	0.003 2	0.006 5	0.002 1	0.004 7	1 900	0.006^{45} (H_2S); 0.001 7 (SO_2)
Cellulose acetate	13.6^{20}	0.28^{30}	3.5^{20}	0.78^{30}	22.7^{30}	5 500	3.5^{30} (H_2S); 17^0 (ethylene oxide); 6.8^{60} (bromomethane)
Cellulose nitrate	6.9	0.12	2.0^{20}	1.95	2.12	6 290	57.1 (NH_3); 1.76 (SO_2)
Ethyl cellulose	400^{30}	8.4^{30}	87^{20}	26.5^{30}	41.0^{30}	$12\,000^{20}$	705 (NH_3); 204 (SO_2); 420^0 (ethylene oxide)
Gutta percha		2.17	14.4	6.16	35.4	510	15.7 (CO); 30.1 (CH_4);
Natural rubber	0.53^{20}	9.43	52.0	23.3	15.3	2 290	1.68 (C_3H_8); 98.9 (C_2H_2); 550 ($CH_3C{\equiv}CH$); 3.59 (SF_6)
Nylon 6		0.009 5³⁰		0.038^{30}	0.10^{30}	177	0.33^{30} (H_2S); 1.2^{20} (NH_3); 0.84^{60} (CH_3Br)
Nylon 11	1.95^{30}		1.78^{30}		1.00^{40}		0.344^{30} (Ne); 0.189^{40} (Ar); 13.6^{50} (propyne)
Poly(acrylonitrile)		0.000 2		0.000 2	0.000 8	300	

Polymer							Gas permeability values
Acrylonitrile-styrene copolymer (66:34)							
Poly(1,3-butadiene)	32.6	6.42		0.048	0.21	2 000	19.2 (Ne); 41.0 (Ar)
Poly(cis-1,4-butadiene)		19.2	41.9	19.0	138.0	5 070	
Butadiene-acrylonitrile copolymer (80:20)	12.2	1.06	15.9	3.85	30.8		24.8 (C_2H_2); 7.7 (propyne)
Butadiene-styrene copolymer (80:20)	13.4	1.71					5.01 (Ne); 4.49 (Ar)
Butadiene-styrene copolymer (92:8)	22.9	5.11					9.70 (Ne); 12.7 (Ar)
Polychloroprene	4.9	1.2	13.6^{30}	4.0	25.8		3.79 (Ar); 3.27 (CH_4)
Polyethylene, low-density		0.969	12.0^{30}	2.88	12.6	90	2.88 (CH_4); 6.81 (C_2H_6); 9.43 (C_3H_8); 1.48 (CO); 49^0 (ethylene oxide); 14.4 (propene); 42.2 (propyne); 0.170 (SF_6); 472^{60} (CH_3Br)
Polyethylene, high-density	1.14	0.143	3.0^{20}	0.403	0.36	12.0	0.388 (CH_4); 0.590 (C_2H_6); 0.537 (C_3H_8); 0.008 3 (SF_6); 1.69 (Ar); 4.01 (propene)
Poly(ethylene terephthalate) Crystalline	1.32	0.006 5		0.035	0.17		
Amorphous	3.28	0.013	3.70^{20}	0.059	0.30	130	0.003 2 (CH_4); 0.08^{60} (CH_3Br); 0.009 (CH_4)
Poly(ethyl methacrylate)	6.82	0.220		1.15	5.00	3 200	2.98 (Ne); 0.565 (Ar); 0.370 (Kr); 3.83 (H_2S); 0.000 001 65 (SF_6)
Isobutene-isoprene copolymer (98:2)	8.38	0.324	7.20	1.30	5.16	110^{38}	13.6^{50} (C_3H_8)
Isoprene-acrylonitrile copolymer (76:24)	7.77	0.181	7.41	0.852	4.32		

TABLE 10.5 Gas Permeability Constants ($10^{10}\ P$) at 25°C for Polymers and Rubbers (*Continued*)

Polymer or rubber	He	N$_2$	H$_2$	O$_2$	CO$_2$	H$_2$O	Other
					Gas		
Isoprene-methacrylonitrile copolymer (76:24)		0.596	13.6	2.34	14.1		
Methacrylonitrile-styrene-butadiene copolymer (88:7:5)				0.004 8	0.014	600	
Poly(methylpentene)	101	7.83	136	32.0	92.6		0.33^{20} (H$_2$S); 9.2^{20} (NH$_3$)
Polypropylene	38^{20}	0.44^{30}	41^{20}	2.3^{30}	9.2^{30}	51	191^{0} (Ne); 550^{0} (Ar);
Silicone rubber, 10% filler	233^{0}	227^{0}	464^{0}	489^{0}	3 240	43 000^{35}	1 020^{0} (Kr); 2 550^{0} (Xe); 19 000^{0} (butane)
Polystyrene	18.7	0.788	23.3	2.63	10.5	1 200	15.7 (NO$_2$); 37.5 (N$_2$O$_4$)
Poly(tetrafluoroethylene)	6.8^{20}	1.4	9.8	4.2	11.7		1.2^{0} (ethylene oxide); 4.6^{60} (CH$_3$Br)
Poly(trifluoroethylene)		0.003	0.94^{20}	0.025^{40}	0.048^{40}	0.29	2.64^{30} (Ne); 0.19^{30} (Ar); 0.078^{30} (Kr); 0.050^{30} (CH$_4$)
Poly(vinyl acetate)	12.6^{30}	<0.001^{14}	89^{30}	0.50^{30}			0.007 (H$_2$S); 0.002 (ethylene oxide)
Poly(vinyl alcohol)	0.001^{30}		0.009	0.008 9	0.001^{23}		
Poly(vinyl chloride)	2.05	0.011 8	1.70	0.045 3	0.157	275	3.92 (Ne); 0.011 5 (Ar); 0.028 6 (CH$_4$)
Poly(vinylidene chloride)	0.31^{34}	0.000 94^{30}		0.005 3^{30}	0.03^{30}	0.5	0.03^{30} (H$_2$S); 0.008^{60} (CH$_3$Br)

TABLE 10.6 Vapor Permeability Constants ($10^{10} P$) at 35°C for Polymers

All tabulated values are multiplied by 10^{10} and are in units of seconds^{-1} (centimeters of Hg)$^{-1}$.

Polymer	Vapor				
	Benzene	Hexane	Carbon tetrachloride	Ethanol	Ethyl acetate
Cellulose	1.4	0.912	0.836	85.8	13.4
Cellulose acetate	512	2.80	3.74	2 980	3 595
Poly(acrylonitrile)	2.61	1.59	1.47	0	1.34
Polyethylene, low-density	5 300	2 910	3 810	55.9	513
Polystyrene	10 600		6 820	0	soluble
Poly(vinyl alcohol)	3.58	2.34	1.61	32.7	2.53

10.7 FATS, OILS, AND WAXES

TABLE 10.7 Constants of Fats and Oils

Fat or oil	Solidification point, °C	Specific gravity (15°C/15°C)	Refractive index	Acid value	Saponification value	Iodine value
		Animal origin				
Butterfat	20–23	$0.91^{40°C}_{15°C}$	1.455	0.5–35	210–230	26–38
Chicken fat	21–27	0.924	$0.925^{25°C}$	1.2	193–205	66–72
Cod-liver oil	−3	0.92–0.93		5.6	171–189	137–166
Deer fat		0.96–0.97		0.8–5.3	195–200	26–36

TABLE 10.7 Constants of Fats and Oils (*Continued*)

Fat or oil	Solidification point, °C	Specific gravity (15°C/15°C)	Refractive index	Acid value	Saponification value	Iodine value
Animal origin (continued)						
Dolphin	−3 to +5	0.91–0.93		2–12	203 (body); 290 (jaw)	127 (body); 33 (jaw)
Goat butter	22–24	$0.91\text{–}0.94^{38°C/38°C}$			233–236	25–37
Goose fat		0.92–0.93		0.6	191–193	58–67
Herring oil	20–45	0.92–0.94	$0.900^{60°C}$	1.8–44	170–194	102–149
Horse fat	15	0.92–0.93	1.460	0–2.4	195–200	75–86
Human fat	−2 to +4	0.903	1.462		193–200	57–73
Lard oil	27–30	0.913–0.915	1.462	0.1–2.5	193–198	63–79
Lard oil, fatty tissue	−5	0.93–0.94		0.5–0.8	195–203	47–67
Menhaden oil	−2 to +10	0.92–0.93	$1.465^{60°C}$	3–12	189–193	148–185
Neat's-foot oil	−16	0.91–0.92	$1.464^{25°C}$	0.1–0.6	193–199	58–75
Porpoise, body oil		0.926		1.2	203	127
Rabbit fat	17–23	0.93–0.94		1.4–7.2	199–203	70–100
Sardine oil	20–22	0.92–0.93	$1.466^{60°C}$	4–25	188–196	130–152
Seal	3	0.915–0.926		1.9–40	188–196	130–152
Shark	15.5	0.916–0.919			157–164	115–139
Sperm oil		0.878–0.884		13	120–137	80–84
Tallow, beef	31–38	0.895		0.25	196–200	35–42
Tallow, mutton	32–41	0.937–0.953	$1.457^{40°C}$	2–14	195–196	48–61
Whale oil	−2 to 0	0.917–0.924	$1.460^{60°C}$	1.9	160–202	90–146
Plant origin						
Acorn	−10	0.916			199	100
Almond	−20 to −15	0.914–0.921	$1.443^{60°C}$	0.5–3.5	183–208	93–103
Babassu oil	22–26	$0.893^{60°C}$			247	16
Beechnut oil	−17	0.922			191–196	97–111

Castor oil	−18 to −17	0.960–0.967	1.477	0.1–0.8	175–183	84
Chaulmoogra oil, USP	<−25	$0.950^{25°C}$			196–213	98–110
Chinese vegetable tallow	24–34	0.918–0.922		2.4	179–206	23–41
Cocoa butter	21.5–23	0.964–0.974	$1.457^{40°C}$	1.1–1.9	193–195	33–42
Coconut oil	14–22	0.926	$1.449^{40°C}$	2.5–10	153–262	6–10
Corn (maize) oil	−20 to −10	0.921–0.928	$1.473^{40°C}$	1.4–2.0	187–193	111–128
Cottonseed oil	−13 to +12	$0.918^{25°C}$	$1.474^{40°C}$	0.6–0.9	194–196	103–111
Hazelnut oil	−18 to −17	0.917			191–197	87
Hemp-seed oil	−28 to −15	0.928–0.934	$1.478^{25°C}$	0.45	190–195	145–162
Linseed oil	−27 to −19	0.930–0.938	$1.475^{40°C}$	1–3.5	188–195	175–202
Mustard, black, oil	16	0.918–0.921	$1.462^{40°C}$	5.7–7.3	173–175	99–110
Neem oil	−3	0.917	$1.471^{40°C}$		195	71
Niger-seed oil		0.925			190	129
Oiticica oil		$0.974^{25°C}$				140–180
Olive oil	−6	0.914–0.918	$1.468^{40°C}$	0.3–1.0	185–196	79–88
Palm oil	35–42	0.915	$1.458^{40°C}$	10	200–205	49–59
Palm kernel oil	24	0.918–0.925	$1.457^{40°C}$	0.3–0.6	220–231	26–32
Peanut oil	3	0.917–0.926	$1.469^{40°C}$	0.8	186–194	88–98
Perilla oil		0.930–0.937	$1.481^{25°C}$		188–194	185–206
Pistachio-nut oil	−10 to −5	0.913–0.919			191	83–87
Poppy-seed oil	−18 to −16	0.924–0.926	$1.469^{40°C}$	2.5	193–195	128–141
Pumpkin-seed oil	−15	0.923–0.925			188–193	121–130
Rapeseed oil	−10	0.913–0.917	$1.471^{40°C}$	0.36–1.0	168–179	94–105
Safflower oil	−18 to −13	0.925–0.928	$1.462^{60°C}$	0.6	188–203	122–141
Sesame oil	−6 to −4	$0.919^{25°C}$	$1.465^{40°C}$	9.8	188–193	103–117
Soybean oil	−16 to −10	0.924–0.927	$1.473^{40°C}$	0.3–1.8	189–194	122–134
Sunflower-seed oil	−17	0.924–0.926	$1.469^{40°C}$	11.2	188–193	129–136
Tung oil	−2.5	0.94–0.95	$1.517^{25°C}$	2	190–197	163–171
White-mustard-seed oil	−16 to −8	0.912–0.916		5.4	171–174	94–98
Wheat-germ oil						125

TABLE 10.8 Constants of Waxes

Wax	Melting point, °C	Specific gravity (15°C/15°C)	Refractive index	Acid value	Saponification value	Iodine value
Bamboo leaf	79–80	$0.961^{25°C}$		14–15	43–44	7.8
Bayberry (myrtle)	47–49	0.99	$1.436^{80°C}$	3–4	205–212	4–9.5
Beeswax, ordinary	62–66	0.95–0.97	$1.44–1.48^{40°C}$	17–21	88–100	8–11
Beeswax, East Indian	61–67	0.95–0.97	$1.44^{40°C}$	5–10.5	87–117	4–10.5
Beeswax, white, USP	61–69	0.95–0.98	$1.45–1.47^{65°C}$	17–24	90–96	7–11
Candelilla	73–77	0.98–0.99	$1.45–1.46^{85°C}$	19–24	55–64	14–20
Cape berry	40–45	1.01	$1.45^{45°C}$	2.5–4.0	211–215	0.5–2.5
Caranda	80–85	0.99–1.00		5.0–9.5	64–79	8–9
Carnauba, No. 1 yellow	86–88	0.99–1.00		1.5–2.5	75–86	
Carnauba, No. 3, crude	86–90	0.99–1.01		3.0–8.5	75–89	
Carnauba, No. 3, refined	86–89	0.96–0.97	$1.47^{40°C}$	3.0–5.0	76–85	7–13.5
Castor oil, hydrogenated	83–88	$0.98–0.99^{20°C}$		1.0–5.0	177–181	2.5–8.5
Chinese insect	80–85	0.95–0.97	$1.46^{40°C}$	2–9	78–93	1.0–2.5
Cotton	68–71	0.96		32	71	25
Cranberry	207–218	0.97–0.98		42–59	131–134	44–53
Esparto	75–79	0.985–0.995		22–27	58–73	7–15
Flax	61–70	0.91–0.99		17–48	37–102	22–29
Japan	49–56	0.97–1.00		4–15	210–235	4–15
Jojoba	11–12	$0.86–0.90^{25°C}$	$1.465^{25°C}$	0.2–0.6	92–95	82–88
Microcrystalline, amber	64–91	0.91–0.94	$1.42–1.45^{80°C}$	0	0	0
Microcrystalline, white	71–89	0.93–0.94	$1.441^{80°C}$	0	0	0
Montan, crude	76–86	$1.01–1.02^{25°C}$		22–31	59–92	14–18
Montan, refined	77–84	1.02–1.04		23–45	72–115	10–14
Ouricury	86–89	0.99–1.01		12–19	88–96	6.9–7.8
Ozokerite	56–82	0.90–1.00		0	0	4–8
Palm	74–86	0.99–1.05		5–11	64–104	9–17

Wax	Melting point, °C	Specific gravity (15°C/15°C)	Refractive index	Acid value	Saponification value	Iodine value
Paraffin, American	49–63	0.896–0.925	$1.44-1.48^{80°C}$	0	0	0
Shellac	79–82	0.97–0.98		12–24	64–83	6–9
Sisal hemp	74–81	1.007–1.010		16–19	56–58	28–29
Spermaceti	41–49	0.905–0.960	$1.51^{25°C}$	0.5–3.0	121–135	2.5–8.5
Sugarcane, refined	76–82	0.96–0.98	$1.48^{40°C}$	8–23	55–70	13–29
Wool	38–40	0.97		6–22	82–130	15–47

SECTION 11
PRACTICAL LABORATORY INFORMATION

11.1

11.1 COOLING

TABLE 11.1 Cooling Mixtures

The table below gives the lowest temperature that can be obtained from a mixture of the inorganic salt with finely shaved dry ice. With the organic substances, dry ice ($-78°C$) in small lumps can be added to the solvent until a slight excess of dry ice remains or liquid nitrogen ($-196°C$) can be poured into the solvent until a slush is formed that consists of the solid-liquid mixture at its melting point.

Substance	Quantity of substance, g	Quantity of water, mL	Temperature, °C
Ammonium nitrate	100	94	-4.0
Sodium nitrate	75	100	-5.3
Sodium thiosulfate 5-water	110	100	-8.0
Sodium chloride	36	100	-10.0
Sodium nitrate	50	100	-17.8
Sodium bromide	66	100	-28
Magnesium chloride	85	100	-34
Calcium chloride 6-water	100	81	-40.3
	100	70	-55

Substance	Temperature, °C	Substance	Temperature, °C
Ethylene glycol	-13	Acetone	-77
1,2-Dichlorobenzene	-17	Ethyl acetate	-84
Carbon tetrachloride	-22.9	2-Butanone	-87
Bromobenzene	-31	Hexane	-95
Methoxybenzene	-37	Methanol	-98
Bis(2-ethoxyethyl) ether	-44	Carbon disulfide	-112
Chlorobenzene	-45	Bromoethane	-119
N-Methylaniline	-57	Pentane	-130
p-Cymene	-68	2-Methylbutane	-160

TABLE 11.2   Molecular Lowering of the Melting or Freezing Point
Cryoscopic constants.

The cryoscopic constant K_f gives the depression of the melting point ΔT (in degrees Celsius) produced when 1 mol of solute is dissolved in 1000 g of a solvent. It is applicable only to dilute solutions for which the number of moles of solute is negligible in comparison with the number of moles of solvent. It is often used for molecular weight determinations,

$$M_2 = \frac{1000 w_2 K_f}{w_1 \, \Delta T}$$

where w_1 is the weight of the solvent and w_2 is the weight of the solute whose molecular weight is M_2.

Compound	K_f	Compound	K_f
Acetamide	4.04	Diphenylamine	8.60
Acetic acid	3.90	Diphenyl ether	7.88
Acetone	2.40	1,2-Ethanediamine	2.43
Ammonia	0.957	Ethoxybenzene	7.15
Aniline	5.87	Formamide	3.85
Antimony(III) chloride	17.95	Formic acid	2.77
Benzene	5.12	Glycerol	3.3 to 3.7
Benzonitrile	5.34	Hexamethylphosphoramide	6.93
Benzophenone	9.8		
Bicyclohexane	14.52	*N*-Methylacetamide	6.65
Biphenyl	8.0	2-Methyl-2-butanol	10.4
Borneol	35.8	Methylcyclohexane	14.13
Bornylamine	40.6	Methyl *cis*-9-octadecenoate	3.4
Butanedinitrile	18.26	2-Methyl-2-propanol	8.37
Camphene	31.08	Naphthalene	6.94
Camphoquinone	45.7	Nitrobenzene	6.852
D-(+)-Camphor	39.7	Octadecanoic acid	4.50
Carbon tetrachloride	29.8	2-Oxohexamethyleneimine	7.30
o-Cresol	5.60	Phenol	7.40
p-Cresol	6.96	Pyridine	4.75
Cyclohexane	20.0	Quinoline	1.95
Cyclohexanol	39.3	Succinonitrile	18.26
Cyclohexylcyclohexane	14.52	Sulfuric acid	1.86
Cyclopentadecanone	21.3	1,1,2,2-Tetrabromoethane	21.7
cis-Decahydronaphthalene	19.47	1,1,2,2-Tetrachloro-	
trans-Decahydronaphthalene	20.81	1,2-difluoroethane	37.7
Dibenz[*de,kl*)anthracene	25.7	Tetramethylene sulfone	64.1
Dibenzyl ether	6.27	*p*-Toluidine	5.372
1,2-Dibromoethane	12.5	Tribromomethane	14.4
Diethyl ether	1.79	1,3,3-Trimethyl-2-oxabicyclo-	
1,2-Dimethoxybenzene	6.38	[2.2.2]octane	6.7
N,N-Dimethylacetamide	4.46	Triphenylmethane	12.45
2,2-Dimethyl-1-propanol	11.0	Water	1.86
Dimethyl sulfoxide	4.07	*p*-Xylene	4.3
1,4-Dioxane	4.63		

TABLE 11.3 Drying Agents

Drying Agent	Most Useful For	Residual Water, mg H_2O per Liter of Dry Air (25°C)	Grams Water Removed per Gram of Desiccant	Regeneration, °C
Al_2O_3	Hydrocarbons	0.002–0.005	0.2	175 (24 h)
$Ba(ClO_4)_2$[a]	Inert gas streams	0.6–0.8	0.17	140
BaO	Basic gases: hydrocarbons, aldehydes, alcohols	0.0007–0.003	0.12	1000
CaC_2[b]	Ethers		0.56	Impossible
$CaCl_2$[c]	Inert organics	0.1–0.2	0.15 (1 H_2O) 0.30 (2 H_2O)	250
CaH_2[d]	Hydrocarbons, ethers, amines, esters, higher alcohols	1×10^{-5}	0.85	Impossible
CaO	Ethers, esters, alcohols, amines	0.01–0.003	0.31	Difficult, 1000
$CaSO_4$	Most organic substances	0.005–0.07	0.07	225
Dow Desiccant 812[e]	Most materials	(5–200 ppm)		No
K_2CO_3	Most materials except acids and phenols		0.16	158
KOH	Amines	0.01–0.9	1.9	Impossible
$LiAlH_4$[f]	Hydrocarbons		0.24	Impossible
$Mg(ClO_4)_2$[a]	Gas streams	0.0005–0.002	0.45	250 (high vacuum)
MgO	All but acidic compounds	0.008	0.15–0.75	800
$MgSO_4$	Most organic compounds	1–12	0.18	Not feasible
Molecular sieves: 4X	Molecules with effective diameter > 4Å	0.001		250
5X	Molecules with effective diameter > 5Å	0.001	0.18	250
9.5% Na-Pb alloy[d]	Hydrocarbons, ethers	(For solvents only)	0.08	Impossible
Na_2SO_4	Ketones, acids, alkyl and aryl halides	12	1.25	150
P_2O_5	Gas streams; not suitable for alcohols, amines, ketones, or amines	2×10^{-5}	0.5	Not feasible
Silica gel	Most organic amines	0.002–0.07	0.2	200–350
Sulfuric acid	Air and inert gas streams	0.003–0.008	Indefinite	Not feasible

[a]May form explosive mixtures when contacting organic material. [b]Explosive C_2H_2 formed.
[c]Slow in drying action. [d]H_2 formed. [e]Used as column drying of organic liquids. [f]Strong reductant.

A saturated aqueous solution in contact with an excess of a definite solid phase at a given temperature will maintain constant humidity in an enclosed space. Table 11.4 gives a number of salts suitable for this purpose. The aqueous tension (vapor pressure, in millimeters of Hg) of a solution at a given temperature is found by multiplying the decimal fraction of the humidity by the aqueous tension at 100 percent humidity for the specific temperature. For example, the aqueous tension of a saturated solution of NaCl at 20°C is $0.757 \times 17.54 = 13.28$ mmHg and at 80°C it is $0.764 \times 355.1 = 271.3$ mmHg.

TABLE 11.4   Solutions for Maintaining Constant Humidity

Solid Phase	% Humidity at Specified Temperatures (°C)						
	10	20	25	30	40	60	80
$K_2Cr_2O_7$			98.0				
K_2SO_4	98	97	97	96	96	96	
KNO_3	95	93	92.5	91	88	82	
KCl	88	85.0	84.3	84	81.7	80.7	79.5
KBr		84	80.7		79.6	79.0	79.3
NaCl	76	75.7	75.3	74.9	74.7	74.9	76.4
$NaNO_3$			73.8	72.8	71.5	67.5	65.5
$NaNO_2$		66	65	63.0	61.5	59.3	58.9
$NaBr \cdot 2H_2O$		57.9	57.7		52.4	49.9	50.0
$Na_2Cr_2O_7 \cdot 2H_2O$	58	55	54		53.6	55.2	56.0
$Mg(NO_3)_2 \cdot 6H_2O$	57	55	52.9	52	49	43	
$K_2CO_3 \cdot 2H_2O$	47	44	42.8		42		
$MgCl_2 \cdot 6H_2O$	34	33	33.0	33	32	30	
$KF \cdot 2H_2O$				27.4	22.8	21.0	22.8
$KC_2H_3O_2 \cdot 1.5H_2O$	24	23	22.5	22	20		
$LiCl \cdot H_2O$	13	12	10.2	12	11	11	
KOH	13	9	8	7	6	5	
100% Humidity: Aqueous Tension (mm Hg)	9.21	17.54	23.76	31.82	55.32	149.4	355.1

TABLE 11.5 Concentrations of Solutions of H_2SO_4, NaOH, and $CaCl_2$ Giving Specified Vapor Pressures and Percent Humidities at 25°C

Percent humidity	Aqueous tension, mmHg	H_2SO_4		NaOH		$CaCl_2$	
		Molality	Weight %	Molality	Weight %	Molality	Weight %
100	23.76	0.00	0.00	0.00	0.00	0.00	0.00
95	22.57	1.263	11.02	1.465	5.54	0.927	9.33
90	21.38	2.224	17.91	2.726	9.83	1.584	14.95
85	20.19	3.025	22.88	3.840	13.32	2.118	19.03
80	19.00	3.730	26.79	4.798	16.10	2.579	22.25
75	17.82	4.398	30.14	5.710	18.60	2.995	24.95
70	16.63	5.042	33.09	6.565	20.80	3.400	27.40
65	15.44	5.686	35.80	7.384	22.80	3.796	29.64
60	14.25	6.341	38.35	8.183	24.66	4.188	31.73
55	13.07	7.013	40.75	8.974	26.42	4.581	33.71
50	11.88	7.722	43.10	9.792	28.15	4.990	35.64
45	10.69	8.482	45.41	10.64	29.86	5.431	37.61
40	9.50	9.304	47.71	11.54	31.58	5.912	39.62
35	8.31	10.21	50.04	12.53	33.38	6.478	41.83
30	7.13	11.25	52.45	13.63	35.29	7.183	44.36
25	5.94	12.47	55.01	14.96	37.45		
20	4.75	13.94	57.76	16.67	40.00		
15	3.56	15.81	60.80	19.10	43.32		
10	2.38	18.48	64.45	23.05	47.97		
5	1.19	23.17	69.44				

Concentrations are expressed in percentage of anhydrous solute by weight.
Source: Stokes and Robinson, *Ind. Eng. Chem.* **41**:2013 (1949).

TABLE 11.6   Relative Humidity from Wet and Dry Bulb Thermometer Readings

Dry bulb temperature, °C	Wet bulb depression, °C											
	0.5	1.0	1.5	2.0	2.5	3.0	3.5	4.0	4.5	5.0	5.5	6.0
	Relative humidity, %											
−10	83	67	51	35	19							
−5	88	76	64	52	41	29	18	7				
0	91	81	72	64	55	46	38	29	21	13	5	
2	91	84	76	68	60	52	44	37	29	22	14	7
4	92	85	78	71	63	57	49	43	36	29	22	16
6	93	86	79	73	66	60	54	48	41	35	29	24
8	93	87	81	75	69	63	57	51	46	40	35	29
10	94	88	82	77	71	66	60	55	50	44	39	34
12	94	89	83	78	73	68	63	58	53	48	43	39
14	95	90	85	79	75	70	65	60	56	51	47	42
16	95	90	85	81	76	71	67	63	58	54	50	46
18	95	91	86	82	77	73	69	65	61	57	53	49
20	96	91	87	83	78	74	70	66	63	59	55	51
22	96	92	87	83	80	76	72	68	64	61	57	54
24	96	92	88	84	80	77	73	69	66	62	59	56
26	96	92	88	85	81	78	74	71	67	64	61	58
28	96	93	89	85	82	78	75	72	69	65	62	59
30	96	93	89	86	83	79	76	73	70	67	64	61
35	97	94	90	87	84	81	78	75	72	69	67	64
40	97	94	91	88	85	82	80	77	74	72	69	67

Dry bulb temperature, °C	Wet bulb depression, °C											
	6.5	7.0	7.5	8.0	8.5	9.0	10.0	11.0	12.0	13.0	14.0	15.0
	Relative humidity, %											
4	9											
6	17	11	5									
8	24	19	14	8								
10	29	24	20	15	10	6						
12	34	29	25	21	16	12	5					
14	38	34	30	26	22	18	10					
16	42	38	34	30	26	23	15	8				
18	45	41	38	34	30	27	20	14	7			
20	48	44	41	37	34	31	24	18	12	6		
22	50	47	44	40	37	34	28	22	17	11	6	
24	53	49	46	43	40	37	31	26	20	15	10	5
26	54	51	49	46	43	40	34	29	24	19	14	10
28	56	53	51	48	45	42	37	32	27	22	18	13
30	58	55	52	50	47	44	39	35	30	25	21	17
32	60	57	54	51	49	46	41	37	32	28	24	20
34	61	58	56	53	51	48	43	39	35	30	26	23
36	62	59	57	54	52	50	45	41	37	33	29	25
38	63	61	58	56	54	51	47	43	39	35	31	27
40	64	62	59	57	54	53	48	44	40	36	33	29

TABLE 11.7   Relative Humidity from Dew Point Readings

Depression of dew point, °C	Dew point reading, °C				
	−10	0	10	20	30
	Relative humidity, %				
0.5	96	96	96	96	97
1.0	92	93	94	94	94
1.5	89	89	90	91	92
2.0	86	87	88	88	89
3.0	79	81	82	83	84
4.0	73	75	77	78	80
5.0	68	70	72	74	75
6.0	63	66	68	70	71
7.0	59	61	63	66	68
8.0	54	57	60	62	64
9.0	51	53	56	58	61
10.0	47	50	53	55	57
11.0	44	47	49	52	
12.0	41	44	47	49	
13.0	38	41	44	46	
14.0	35	38	41	44	
15.0	33	36	39	42	
16.0	31	34	37	39	
18.0	27	30	33	35	
20.0	24	26	29	32	
22.0	21	23	26		
24.0	18	21	23		
26.0	16	18	21		
28.0	14	16	19		
30.0	12	14	17		

11.3 BOILING POINTS AND HEATING BATHS

TABLE 11.8 Organic Solvents Arranged by Boiling Points

Name	BP, °C	Name	BP, °C
Ethylene oxide	10.6	2-Methyltetrahydrofuran	80.0
Chloroethane	12.3	Benzene	80.1
Furan	31.4	Cyclohexane	80.7
Methyl formate	31.5	Propyl formate	80.9
Diethyl ether	34.6	Acetonitrile	81.6
Propylene oxide	34.5	2-Propanol	82.4
Pentane	36.1	1,1-Dimethylethanol	82.4
Bromoethane	38.4	Cyclohexene	83.0
Dichloromethane	39.8	Diisopropylamine	83.5
Dimethoxymethane	42.3	1,2-Dichloroethane	83.7
Carbon disulfide	46.3	Thiophene	84.2
1-Isopropoxy-2-propanol	47.9	Trichloroethylene	87.2
Ethyl formate	54.2	Isopropyl acetate	88.2
Acetone	56.2	1-Bromo-2-methylpropane	91.5
Methyl acetate	56.3	2,5-Dimethylfuran	93–94
1,1-Dichloroethane	57.3	Ethyl chloroformate	94
Dichloroethylene	60.6	Allyl alcohol	96.6
Chloroform	61.2	1,2-Dichloropropane	96.8
Methanol	64.7	1-Propanol	97.2
Tetrahydrofuran	66.0	Heptane	98.4
Diisopropyl ether	68.0	1-Chloro-3-methylbutane	99
Hexane	68.7	Ethyl propionate	99.1
1-Chloro-2-methylpropane	68.9	2-Butanol	99.6
1,1,1-Trichloroethane	74.0	Formic acid	100.8
1,3-Dioxolane	74–75	Methylcyclohexane	100.9
Carbon tetrachloride	76.7	1,4-Dioxane	101.2
Ethyl acetate	77.1	Nitromethane	101.2
1-Chlorobutane	77.9	Propyl acetate	101.5
Ethanol	78.3	2-Pentanone	101.7
2-Butanone	79.6	3-Pentanone	102.0

TABLE 11.8 Organic Solvents Arranged by Boiling Points (*Continued*)

Name	BP, °C	Name	BP, °C
2-Methyl-2-butanol	102.0	*o*-Xylene	144.4
1,1-Diethoxyethane	102.7	2-Methoxyethyl acetate	144.5
Butyl formate	106.6	1,1,2,2-Tetrachloroethane	146.3
2-Methyl-1-propanol	107.9	3-Heptanone	147.8
Toluene	110.6	Tribromomethane	149.6
sec-Butyl acetate	112.3	Nonane	150.8
1,1,2-Trichloroethane	113.5	2-Heptanone	151
Nitroethane	114.1	Isopropylbenzene	152.4
Pyridine	115.2	*N,N*-Dimethylformamide	153.0
3-Pentanol	115.6	Methoxybenzene	153.8
4-Methyl-2-pentanone	115.7	Ethyl lactate	154.5
1-Chloro-2,3-epoxypropane	116.1	Cyclohexanone	155.7
1-Butanol	117.7	Bromobenzene	156.2
Acetic acid	117.9	1,2,3-Trichloropropane	156.9
Isobutyl acetate	118.0	1-Hexanol	157.5
2-Pentanol	119.3	Propylbenzene	159.2
1-Bromo-3-methylbutane	119.7	Cyclohexanol	161.1
1-Methoxy-2-propanol	120.1	Bis(2-methoxyethyl)ether	160
2-Nitropropane	120.3	Isopentyl propionate	160.2
Tetrachloroethylene	121.1	2-Heptanol	160.4
Ethyl butyrate	121.6	Pentachloroethane	160.5
3-Hexanone	123	2-Furaldehyde	161.8
2,4-Dimethyl-3-pentanone	124	2,6-Dimethyl-4-heptanone	168.1
2-Methoxyethanol	124.6	4-Hydroxy-4-methyl-	
Octane	125.7	2-pentanone	169.2
Butyl acetate	126.1	2-Furanmethanol	170.0
Diethyl carbonate	126.8	Ethoxybenzene	170
2-Hexanone	127.2	2-Butoxyethanol	170.2
1-Chloro-2-propanol	127.4	Diisopentyl ether	173.4
2-Chloroethanol	128.6	Decane	174.2
3-Methyl-1-penten-2-one	129.5	1,3-Dichloro-2-propanol	174.3
1-Nitropropane	131.2	Cyclohexyl acetate	174–175
Chlorobenzene	131.7	1-Heptanol	175.8
1,2-Dibromoethane	131.7	Furfuryl acetate	175–177
4-Methyl-2-pentanol	131.7	1,3,3-Trimethyl-	
3-Methyl-1-butanol	132.0	2-oxabicyclo[2.2.2]octane	177.4
Cyclohexylamine	134.8	4-Isopropyl-	
2-Ethoxyethanol	134.8	1-methylbenzene	177.1
Ethylbenzene	136.2	Isopentyl butyrate	178.6
1-Pentanol	138	Bis(2-chloroethyl) ether	178.8
p-Xylene	138.4	2-Octanol	179
m-Xylene	139.1	1,2-Dichlorobenzene	180.4
Acetic anhydride	140.0	Ethyl acetoacetate	180.8
2,4-Pentanedione	140.6	Phenol	181.8
Isopentyl acetate	142	2-Ethyl-1-hexanol	184.3
Dibutyl ether	142.4	Aniline	184.4
4-Heptanone	143.7	Benzyl ethyl ether	185.0

TABLE 11.8 Organic Solvents Arranged by Boiling Points (*Continued*)

Name	BP, °C	Name	BP, °C
Diethyl oxalate	185.4	2-(2-Ethoxyethoxy)ethyl	
1,2-Propanediol	188	acetate	218.5
Bis(2-ethoxyethyl) ether	188.4	Acetamide	221.2
Dimethyl sulfoxide	189.0	Methyl salicylate	223.0
1,2-Ethanediol diacetate	190.2	Diethyl maleate	225.3
Benzonitrile	191.0	1,4-Butanediol	230
2,5-Hexanedione	191.4	Propyl benzoate	231.2
2-(2-Methoxyethoxy)-		1-Decanol	230.2
ethanol	194.1	Phenylacetonitrile	233.5
N,N-Dimethylaniline	194.2	Quinoline	237
1-Octanol	195.2	Tributyl borate	238.5
1,2-Ethanediol	197.3	Propylene carbonate	240
Diethyl malonate	199.3	2-Phenoxyethanol	240
Methyl benzoate	199.5	Bis(2-hydroxyethyl) ether	245
o-Toluidine	200.4	Dibutyl oxalate	245.5
p-Toluidine	200.6	Butyl benzoate	250
2-(2-Ethoxyethoxy)ethanol	202	1,2,3-Propanetriol	
Acetophenone	202.1	triacetate	258–259
1,2-Dibutoxyethane	203.6	1-Chloronaphthalene	259.3
1-Phenylethanol	203.9	Isopentyl benzoate	262
m-Toluidine	203.4	*trans*-Ethyl cinnamate	271.0
Benzyl alcohol	205.5	Bis[2-(methoxyethoxy)-	
Camphor	207	ethyl]ether	275.3
1,3-Butanediol	207.5	1-Methoxy-2-nitrobenzene	277
1,2,3,4-Tetrahydro-		Isopentyl salicylate	277–278
naphthalene	207.6	1-Bromonaphthalene	281.1
γ-Valerolactone	207–208	Dimethyl *o*-phthalate	283.7
o-Chloroaniline	208.8	2,2'-(Ethylenedioxy)-	
Nitrobenzene	210.8	bisethanol	285
Ethyl benzoate	212.4	Glycerol	290
3,5,5-Trimethylcyclo-		Diethyl *o*-phthalate	295
hex-2-en-1-one	215.2	Benzyl benzoate	323.5
Naphthalene	217.7	Dibutyl *o*-phthalate	340.0
		Dibutyl decanedioate	344–345

TABLE 11.9 Molecular Elevation of the Boiling Point
Ebullioscopic constants.

Molecular weights can be determined with the relation

$$M = K_b \frac{1000 w_2}{w_1 \, \Delta T_b}$$

where ΔT_b is the elevation of the boiling point brought about by the addition of w_2 grams of solute to w_1 grams of solvent and K_b is the ebullioscopic constant. In the column headed "Barometric correction" is given the number of degrees for each millimeter of difference between the barometric reading and 760 mmHg to be subtracted from K_b if the pressure is lower, or added if higher, than 760 mm. In general, the effect is within experimental error if the pressure is within 10 mm of 760 mm.

Compound	Barometric correction	K_b
Acetic acid	0.000 8	3.07
Acetic anhydride		3.53
Acetone	0.000 4	1.71
Acetonitrile		1.30
Acetophenone		5.65
Aniline	0.000 9	3.52
Benzene	0.000 7	2.53
Benzonitrile		3.87
Bromobenzene	0.001 6	6.26
Bromoethane		2.53
2-Butanone		2.28
cis-2-Butene-1,4-diol		2.86
D-(+)-Camphor	0.001 5	5.611
Carbon disulfide	0.000 6	2.34
Carbon tetrachloride	0.001 3	5.03
Chlorobenzene	0.001 1	4.15
Chloroethane		1.95
Chloroform	0.000 9	3.63
Cyclohexane	0.000 7	2.79
1,2-Dibromoethane	0.001 6	6.608
1,1-Dichloroethane		3.13
1,2-Dichloroethane		3.44
Dichloromethane		2.60
Diethyl ether	0.000 5	2.02
Diethyl sulfide		3.23
Dimethoxymethane		2.125
N,N-Dimethylacetamide		3.22
Dimethyl sulfide		1.85
1,4-Dioxane		3.270
Ethanol	0.000 3	1.22
Ethoxybenzene		5.0
Ethyl acetate	0.000 7	2.77
Formic acid		2.4
Glycerol		6.52
Heptane	0.000 8	3.43
Hexane		2.75

TABLE 11.9 Molecular Elevation of the Boiling Point (*Continued*)

Compound	Barometric correction	K_b
2-Hydroxybenzaldehyde		4.96
Iodoethane		5.16
Iodomethane		4.19
4-Isopropyl-1-methylbenzene		5.52
Methanol	0.000 2	0.83
Methoxybenzene		4.502
Methyl acetate	0.000 5	2.15
2-Methyl-2-butanol		2.255
3-Methyl-1-butanol		2.65
3-Methylbutyl acetate		4.83
Methyl formate		1.649
2-Methyl-1-propanol		2.166
2-Methyl-2-propanol		1.745
Naphthalene	0.001 4	5.80
Nitrobenzene		5.24
Nitroethane		2.60
Nitromethane		1.86
Octane		4.02
Pentyl acetate		4.83
Phenol	0.000 9	3.60
Piperidine		2.84
1-Propanol		1.59
Propionic acid		3.51
Propionitrile		1.87
Pyridine		2.710
Quinoline		5.84
1,1,2,2-Tetrachloroethylene		5.50
1,2,3,4-Tetrahydronaphthalene		5.582
Toluene	0.000 8	3.33
p-Toluidine		4.14
Trichloroethylene		4.43
1,1,2-Trichloro-1,2,2-trifluoroethane		5.75
Triethylamine		3.45
Water	0.000 1	0.512

TABLE 11.10 Substances Which Can Be Used for Heating Baths

Medium	Melting point, °C	Boiling point, °C	Useful range, °C	Flash point, °C	Comments
Water	0	100	0–100	None	Ideal
Silicone oil	−50	—	30–250	315	Somewhat viscous at low temperature
Triethylene glycol	−7	285	0–250	165	Noncorrosive
Glycerol	18	290	−20 to 260	160	Water-soluble, nontoxic
Paraffin	50	—	60–300	199	Flammable
Dibutyl o-phthalate	−35	340	150–320	171	Generally used

TABLE 11.11 Solvents of Chromatographic Interest

| Solvent | Boiling point, °C | Solvent strength parameter | | Viscosity, mN · s · m⁻² (20°C) | Refractive index (20°C) | UV cutoff, nm |
		$e°$ (SiO$_2$)	$e°$ (Al$_2$O$_3$)			
Fluoroalkanes			−0.25		1.25	
Pentane	36	0.0	0.0	0.24$^{15°C}$	1.358	210
Hexane	69	0.0	0.0	0.31	1.375	210
2,2,4-Trimethylpentane	99		0.01	0.50	1.392	215
Decane	174		0.04	0.93	1.412	210
Cyclohexane	81	−0.05	0.04	0.98	1.426	210
Cyclopentane	49		0.05	0.44	1.407	210
Diisobutylene	101		0.06		1.411	
1-Pentene	30		0.08	0.24$^{0°C}$	1.371	
Carbon disulfide	46	0.14	0.15	0.36	1.626	380
Carbon tetrachloride	77	0.14	0.18	0.97	1.466	265
1-Chlorobutane	78		0.26	0.43	1.402	220
1-Chloropentane	98		0.26	0.58	1.412	225
o-Xylene	144		0.26	0.81	1.505	290
Diisopropyl ether	68		0.28	0.38$^{25°C}$	1.369	220
2-Chloropropane	35		0.29	0.33	1.378	225
Toluene	111		0.29	0.59	1.497	286
1-Chloropropane	47		0.30	0.35	1.389	225
Chlorobenzene	132		0.40	0.80	1.525	
Benzene	80	0.25	0.32	0.65	1.501	280
Bromoethane	38		0.37	0.40	1.424	

Diethyl ether	35	0.38	0.38	0.25	1.353	218
Diethyl sulfide	92		0.38	0.45	1.443	290
Chloroform	62	0.26	0.40	0.57	1.443	245
Dichloromethane	41		0.42	0.44	1.425	235
4-Methyl-2-pentanone	116		0.43	$0.42^{15°C}$	1.396	335
Tetrahydrofuran	66		0.45	0.55	1.407	220
1,2-Dichloroethane	84		0.49	0.80	1.445	228
2-Butanone	80		0.51	$0.42^{15°C}$	1.379	330
1-Nitropropane	131		0.53	$0.80^{25°C}$	1.402	380
Acetone	56	0.47	0.56	0.32	1.359	330
1,4-Dioxane	101	0.49	0.56	$1.44^{15°C}$	1.420	215
Ethyl acetate	77	0.38	0.58	0.45	1.372	255
Methyl acetate	56		0.60	$0.48^{15°C}$	1.362	260
1-Pentanol	138		0.61	4.1	1.410	210
Dimethyl sulfoxide	189		0.62	2.47	1.478	265
Aniline	184		0.62	4.40	1.586	
Diethylamine	56		0.63	0.33	1.386	275
Nitromethane	101		0.64	0.67	1.394	380
Acetonitrile	82	0.50	0.65	0.37	1.344	190
Pyridine	115		0.71	0.97	1.510	330
2-Butoxyethanol	170		0.74	$3.15^{25°C}$	1.420	220
1-Propanol	97		0.82	2.25	1.386	210
2-Propanol	82		0.82	2.50	1.377	210
Ethanol	78		0.88	1.20	1.361	210
Methanol	65		0.95	0.59	1.328	210
Ethylene glycol	198		1.11	21.8	1.432	210
Acetic acid	118		large	1.23	1.372	260
Water	100		large	1.00	1.333	191

TABLE 11.12 Solvents Having the Same Refractive Index and the Same Density at 25°C

Solvent 1	Solvent 2	Refractive index		Density, g/mL	
		1	2	1	2
Acetone	Ethanol	1.357	1.359	0.788	0.786
Ethyl formate	Methyl acetate	1.358	1.360	0.916	0.935
Ethanol	Propionitrile	1.359	1.363	0.786	0.777
2,2-Dimethylbutane	2-Methylpentane	1.366	1.369	0.644	0.649
2-Methylpentane	Hexane	1.369	1.372	0.649	0.655
Isopropyl acetate	2-Chloropropane	1.375	1.376	0.868	0.865
3-Butanone	Butyraldehyde	1.377	1.378	0.801	0.799
Butyraldehyde	Butyronitrile	1.378	1.382	0.799	0.786
Dipropyl ether	Butyl ethyl ether	1.379	1.380	0.753	0.746
Propyl acetate	Ethyl propionate	1.382	1.382	0.883	0.888
Propyl acetate	1-Chloropropane	1.382	1.386	0.883	0.890
Butyronitrile	2-Methyl-2-propanol	1.382	1.385	0.786	0.781
Ethyl propionate	1-Chloropropane	1.382	1.386	0.888	0.890
1-Propanol	2-Pentanone	1.383	1.387	0.806	0.804
Isobutyl formate	1-Chloropropane	1.383	1.386	0.881	0.890
1-Chloropropane	Butyl formate	1.386	1.387	0.890	0.888
Butyl formate	Methyl butyrate	1.387	1.391	0.888	0.875
Methyl butyrate	2-Chlorobutane	1.392	1.395	0.875	0.868
Butyl acetate	2-Chlorobutane	1.392	1.395	0.877	0.868
4-Methyl-2-pentanone	Pentanonitrile	1.394	1.395	0.797	0.795
4-Methyl-2-pentanone	1-Butanol	1.394	1.397	0.797	0.812
2-Methyl-1-propanol	Pentanonitrile	1.394	1.395	0.798	0.795
2-Methyl-1-propanol	2-Hexanone	1.394	1.395	0.798	0.810
2-Butanol	2,4-Dimethyl-3-pentanone	1.395	1.399	0.803	0.805
2-Hexanone	1-Butanol	1.395	1.397	0.810	0.812
Pentanonitrile	2,4-Dimethyl-3-pentanone	1.395	1.399	0.795	0.805
2-Chlorobutane	Isobutyl butyrate	1.395	1.399	0.868	0.860
Butyric acid	2-Methoxyethanol	1.396	1.400	0.955	0.960
1-Butanol	3-Methyl-2-pentanone	1.397	1.398	0.812	0.808
1-Chloro-2-methylpropane	Isobutyl butyrate	1.397	1.399	0.872	0.860
1-Chloro-2-methylpropane	Pentyl acetate	1.397	1.400	0.872	0.871
Methyl methacrylate	3-Methyl-2-pentanone	1.398	1.398	0.795	0.808
Triethylamine	2,2,3-Trimethylpentane	1.399	1.401	0.723	0.712
Butylamine	Dodecane	1.399	1.400	0.736	0.746
Isobutyl butyrate	1-Chlorobutane	1.399	1.401	0.860	0.875
1-Nitropropane	Propionic anhydride	1.399	1.400	0.995	1.007
Pentyl acetate	1-Chlorobutane	1.400	1.400	0.871	0.881
Pentyl acetate	Tetrahydrofuran	1.400	1.404	0.871	0.885
Dodecane	Dipropylamine	1.400	1.400	0.746	0.736
1-Chlorobutane	Tetrahydrofuran	1.401	1.404	0.871	0.885
Isopentanoic acid	2-Ethoxyethanol	1.402	1.405	0.923	0.926
Dipropylamine	Cyclopentane	1.403	1.404	0.736	0.740
2-Pentanol	4-Heptanone	1.404	1.405	0.804	0.813

TABLE 11.12   Solvents Having the Same Refractive Index and the Same Density at 25°C (*Continued*)

Solvent 1	Solvent 2	Refractive index 1	2	Density, g/mL 1	2
3-Methyl-1-butanol	Hexanonitrile	1.404	1.405	0.805	0.801
3-Methyl-1-butanol	4-Heptanone	1.404	1.405	0.805	0.813
Hexanonitrile	4-Heptanone	1.405	1.405	0.801	0.813
Hexanonitrile	1-Pentanol	1.405	1.408	0.801	0.810
Hexanonitrile	2-Methyl-1-butanol	1.405	1.409	0.801	0.815
4-Heptanone	1-Pentanol	1.405	1.408	0.813	0.810
2-Ethoxyethanol	Pentanoic acid	1.405	1.406	0.926	0.936
2-Heptanone	1-Pentanol	1.406	1.408	0.811	0.810
2-Heptanone	2-Methyl-1-butanol	1.406	1.409	0.811	0.815
2-Heptanone	Dipentyl ether	1.406	1.410	0.811	0.799
2-Pentanol	3-Isopropyl-2-pentanone	1.407	1.409	0.804	0.808
1-Pentanol	Dipentyl ether	1.408	1.410	0.810	0.799
2-Methyl-1-butanol	Dipentyl ether	1.409	1.410	0.815	0.799
Isopentyl isopentanoate	Allyl alcohol	1.410	1.411	0.853	0.847
Dipentyl ether	2-Octanone	1.410	1.414	0.799	0.814
2,4-Dimethyldioxane	3-Chloropentene	1.412	1.413	0.935	0.932
2,4-Dimethyldioxane	Hexanoic acid	1.412	1.415	0.935	0.923
Diethyl malonate	Ethyl cyanoacetate	1.412	1.415	1.051	1.056
3-Chloropentene	Octanoic acid	1.413	1.415	0.932	0.923
2-Octanone	1-Hexanol	1.414	1.416	0.814	0.814
2-Octanone	Octanonitrile	1.414	1.418	0.814	0.810
3-Octanone	3-Methyl-2-heptanone	1.414	1.416	0.830	0.818
3-Methyl-2-heptanone	1-Hexanol	1.415	1.416	0.818	0.814
3-Methyl-2-heptanone	Octanonitrile	1.415	1.418	0.818	0.810
1-Hexanol	Octanonitrile	1.416	1.418	0.814	0.810
Dibutylamine	Allylamine	1.416	1.419	0.756	0.758
Allylamine	Methylcyclohexane	1.419	1.421	0.758	0.765
Butyrolactone	1,3-Propanediol	1.434	1.438	1.051	1.049
Butyrolactone	Diethylmaleate	1.434	1.438	1.051	1.064
2-Chloromethyl-2-propanol	Diethyl maleate	1.436	1.438	1.059	1.064
N-Methylmorpholine	Dibutyl decanedioate	1.436	1.440	0.924	0.932
1,3-Propanediol	Diethyl maleate	1.438	1.438	1.049	1.064
Methyl salicylate	Diethyl sulfide	1.438	1.442	0.836	0.831
Methyl salicylate	1-Butanethiol	1.438	1.442	0.836	0.837
1-Chlorodecane	Mesityl oxide	1.441	1.442	0.862	0.850
Diethylene glycol	Formamide	1.445	1.446	1.128	1.129
Diethylene glycol	Ethylene glycol diglycidyl ether	1.445	1.447	1.128	1.134
Formamide	Ethylene glycol diglycidyl ether	1.446	1.447	1.129	1.134
2-Methylmorpholine	Cyclohexanone	1.446	1.448	0.951	0.943

TABLE 11.12 Solvents Having the Same Refractive Index and the Same Density at 25°C (*Continued*)

Solvent 1	Solvent 2	Refractive index		Density, g/mL	
		1	2	1	2
2-Methylmorpholine	1-Amino-2-propanol	1.446	1.448	0.951	0.961
Dipropylene glycol monoethyl ether	Tetrahydrofurfuryl alcohol	1.446	1.450	1.043	1.050
1-Amino-2-methyl-2-pentanol	2-Butylcyclohexanone	1.449	1.453	0.904	0.901
2-Propylcyclohexanone	4-Methylcyclohexanol	1.452	1.454	0.923	0.908
Carbon tetrachloride	4,5-Dichloro-1,3-dioxolan-2-one	1.459	1.461	1.584	1.591
N-Butyldiethanolamine	Cyclohexanol	1.461	1.465	0.965	0.968
D-α-Pinene	*trans*-Decahydronaphthalene	1.464	1.468	0.855	0.867
Propylbenzene	*p*-Xylene	1.490	1.493	0.858	0.857
Propylbenzene	Toluene	1.490	1.494	0.858	0.860
Phenyl 1-hydroxyphenyl ether	1,3-Dimorpholyl-2-propanol	1.491	1.493	1.081	1.094
Phenetole	Pyridine	1.505	1.507	0.961	0.978
2-Furanmethanol	Thiophene	1.524	1.526	1.057	1.059
m-Cresol	Benzaldehyde	1.542	1.544	1.037	1.041

TABLE 11.13 McReynolds' Constants for Stationary Phases in Gas Chromatography

The McReynolds' constants listed are differences in retention index units between the reference compound run on squalane and on the other phases listed. The last entry in the table shows the absolute retention indices for the reference compounds on squalane. Reference compounds are (1) benzene, (2) 1-butanol, (3) 2-pentanone, (4) 1-nitropropane, and (5) pyridine. [Note that Rohrschneider's constants are based on these reference compounds and may differ slightly from the McReynolds' constants. The reference compounds for Rohrschneider's constants are (1) benzene, (2) ethanol, (3) 2-butanone, (4) nitromethane, and (5) pyridine.] The minimum temperature is that at which normal gas-liquid chromatography (GLC) behavior is expected. Below that temperature, the phase will be a solid or an extremely viscous gum. The maximum temperature is that above which the bleed rate will be excessive.

Liquid phase	Chemical type	Similar liquid phases	Temperature, °C		Reference compounds					Sum
			Minimum	Maximum	1	2	3	4	5	
Squalane	(2,6,10,15,19,23-Hexamethyltetracosane)		20	150	0	0	0	0	0	0
Paraffin oil					9	5	2	6	11	33
Apolane-87	(24,24-Diethyl-19,29-dioctadecylhepta-tetracontane)		30	280	21	10	3	12	25	71
Apiezon L		SP-2100, SF 96	50	250	32	22	15	32	42	143
SE 30	Poly(dimethylsiloxane)	OV-1, DC 200, DC 410	50	350	15	53	44	64	41	217
OV-101			50	350	17	57	45	67	43	229
OV-73	Poly(diphenyldimethylsiloxane), 5% : 95%	SE 52	0	325	32	72	65	98	67	334
SE 54	Poly(diphenylvinyldimethyl-siloxane), 5% :1% :94%		50	300	33	72	66	99	67	337
OV-3	Poly(diphenyldimethylsiloxane), 10% :90%		0	350	44	86	81	124	88	423
Dexsil 300	Poly(carboranemethylsiloxane)		50	500	47	80	103	148	96	474
Kel F Wax				150	55	67	114	143	116	495
Apiezon H				300	59	86	81	151	129	506
Dexsil 400	Carborane and methylphenyl-silicone		50	500	72	108	118	166	123	587

TABLE 11.13 McReynolds' Constants for Stationary Phases in Gas Chromatography (*Continued*)

Liquid phase	Chemical type	Similar liquid phases	Temperature, °C		Reference compounds					Sum
			Minimum	Maximum	1	2	3	4	5	
OV-7	Poly(diphenyldimethylsiloxane), 20%:80%	DC 550	20	350	69	113	111	171	128	592
Di(2-ethylhexyl) sebacate			0	125	72	168	108	180	125	653
Diisodecyl adipate				175	71	171	113	185	128	668
Decyl octyl adipate					79	179	119	193	134	704
Bis(2-ethylhexyl)-tetrachlorophthalate			0	150	112	150	123	168	181	734
Diisodecyl phthalate	Poly(diphenyldimethylsiloxane), 35%:65%	DC 710	0	175	84	173	137	218	155	767
Dinonyl phthalate			20	150	83	183	147	231	159	803
OV-11			0	350	107	149	153	228	190	827
Dioctyl phthalate			20	125	92	186	150	236	167	831
Hallcomid M-18			40	150	79	268	130	222	146	845
OV-17	Poly(diphenyldimethylsiloxane), 50%:50%		0	325	119	158	162	243	202	884
Dexsil 410	Carborane and methylcyanoethylsilicone		50	500	72	286	174	249	171	952
UCON LB-550-X			0	200	118	271	158	243	206	996
Span 80			15	150	97	266	170	216	268	1017
OV-22	Poly(diphenyldimethylsiloxane), 65%:35%		0	350	160	188	191	283	253	1075
Polypropylene glycol			0	150	128	294	173	264	226	1085
Didecyl phthalate			10	175	136	255	213	320	235	1159
OV-25	Poly(diphenyldimethylsiloxane), 75%:25%		0	350	178	204	208	305	280	1175
Polyphenyl ether OS-138 (6 rings)			0	225	182	233	228	313	293	1249

11.22

Stationary phase	Synonyms / trade designation								
Neopentyl glycol sebacate	HI-EFF-3CP	50	225	172	327	225	344	326	1394
Squalene		0	100	152	341	328	329	344	1404
UCON 50-HB-280X		0	200	177	362	227	351	302	1419
Tricresyl phosphate		20	125	176	321	250	374	299	1420
Sucrose acetate isobutyrate		0	200	172	330	251	378	295	1426
QF-1	Poly(trifluoropropylsiloxane); SP-2401, FS 1265	0	250	144	233	355	463	305	1500
OV-210	Poly(trifluoropropylmethyl-siloxane)	0	275	146	238	358	468	310	1520
OV-215	XE 60	0	275	149	240	363	478	315	1545
UCON 50-HB-2000	Emulphor ON-870	0	200	202	394	253	392	341	1582
Triton X-100		0	200	203	399	268	402	362	1634
UCON 50-HB-5100		0	200	214	418	278	421	375	1706
Siponate DS-10		0	200	99	569	320	344	388	1720
Tween 80		0	150	227	430	283	438	396	1747
XE-60	Poly(cyanoethylphenyl-methylsiloxane)	0	250	204	381	340	493	367	1785
OV-225	Poly(cyanopropylphenyl-methylsiloxane)	0	265	228	369	338	492	386	1813
Neopentyl glycol adipate	HI-EFF-3AP	50	225	232	421	311	461	424	1849
UCON 75-H-90000	Igepal CO-880	100	200	255	452	299	470	406	1882
Triton X-305	HI-EFF-3BP	0	200	262	467	314	488	430	1961
Neopentyl glycol succinate		50	230	272	469	366	539	474	2120
Igepal CO 990	FFAP, SP-2300	100	200	298	508	345	540	475	2166
Carbowax 20M	Poly(ethylene glycol)	25	275	322	536	368	572	510	2308
Epon 1001		50	225	284	489	406	539	601	2319
Carbowax 4000		60	200	325	551	375	582	520	2353
Ethylene glycol isophthalate	HI-EFF-2EP	100	225	326	508	425	607	561	2427
Ethylene glycol adipate	HI-EFF-2AP	100	225	372	576	453	655	617	2673
Butane-1,4-diol succinate	HI-EFF-4BP	50	225	369	591	457	661	629	2707
Phenyldiethanolamine succinate	HI-EFF-10BP	0	200	386	555	472	674	654	2741

TABLE 11.13 McReynolds' Constants for Stationary Phases in Gas Chromatography (*Continued*)

Liquid phase	Chemical type	Similar liquid phases	Temperature, °C		Reference compounds					Sum
			Minimum	Maximum	1	2	3	4	5	
Diethylene glycol adipate		HI-EFF-1AP, LAC-1-R-296, SP-2330	25	275	378	603	460	665	658	2764
Carbowax 1540			50	175	371	639	453	666	641	2770
Hyprose SP-80			0	175	336	742	492	639	727	2936
SILAR-7CP			0	250	440	638	605	844	673	3200
ECNSS-M			30	200	421	690	581	803	732	3227
EGSS-X			90	200	484	710	585	831-	778	3388
Ethylene glycol phthalate		HI-EFF-2GP	100	200	453	697	602	816	872	3410
SILAR-9CP			0	250	489	725	631	910	778	3536
SILAR-10C		SP-2340	25	275	523	757	659	942	801	3682
Diethylene glycol succinate		HI-EFF-1BP, LAC-3-R-728	20	200	499	751	593	840	860	3543
Tetrahydroxyethylenediamine		THEED	0	150	463	942	626	801	893	3725
Tetracyanoethylated pentaerythritol			30	175	526	782	677	920	837	3742
Ethylene glycol succinate		HI-EFF-2BP	100	200	537	787	643	903	889	3759
1,2,3,4-Tetrakis-(2-cyanoethoxy)butane			110	200	617	860	773	1048	941	4239
1,2,3,4,5,6-Hexakis(2-cyanoethoxy)cyclohexane			125	150	567	825	713	978	901	3984
1,2,3-Tris-(2-cyanoethoxy)propane			0	175	593	857	752	1028	915	4145
N,N-Bis(2-cyanoethyl)-formamide			0	125	690	991	853	1110	1000	4644
OV-275	Dicyanoallylsilicone		25	250	781	1006	885	1177	1089	4938

Absolute retention index values on squalane for reference compounds | | | | | 653 | 590 | 627 | 652 | 699 | |

11.4.1 McReynolds' Constants

The *Kovats retention indices* (R.I.) indicate where compounds will appear on a chromatogram with respect to unbranched alkanes injected with the sample. By definition, the R.I. for pentane is 500, for hexane is 600, for heptane is 700, and so on, regardless of the column used or the operating conditions, although the exact conditions and column must be specified, such as liquid loading, particular support used, and any pretreatment. For example, suppose that on a 20% squalane column at 100°C, the retention times for hexane, benzene, and octane are found to be 15, 16, and 25 min, respectively. On a graph of ln t'_R (naperian logarithm of the adjusted retention time) of the alkanes versus their retention indices, a R.I. of 653 for benzene is read off the graph. The number 653 for benzene (see last line of Table 11.13 in the column headed "1" under "Reference compounds") means that it elutes halfway between hexane and heptane on a logarithmic time scale. If the experiment is repeated with a dinonyl phthalate column, the R.I. for benzene is found to be 736 (lying between heptane and octane), which implies that dinonyl phthalate will retard benzene slightly more than squalane will; that is, dinonyl phthalate is slightly more polar than squalane by $\Delta I = 83$ units (the entry in Table 11.13 for dinonyl phthalate in the column headed "1" under "Reference compounds"). The difference gives a measure of solute-solvent interaction due to all intermolecular forces other than London dispersion forces. The latter are the principal solute-solvent effects with squalane.

Now the overall effects due to hydrogen bonding, dipole moment, acid-base properties, and molecular configuration can be expressed as

$$\sum \Delta I = ax' + by' + cz' + du' + es'$$

where $x' = \Delta I$ for benzene (the column headed "1" in Table 11.13, intermolecular forces typical of aromatics and olefins), $y' = \Delta I$ for 1-butanol (the column headed "2" in Table 11.13, electron attraction typical of alcohols, nitriles, acids, and nitro and alkyl monochlorides, dichlorides and trichlorides), $z' = \Delta I$ for 2-pentanone (the column headed "3" in Table 11.13, electron repulsion typical of ketones, ethers, aldehydes, esters, epoxides, and dimethylamino derivatives), $u' = \Delta I$ for 1-nitropropane (the column headed "4" in Table 11.13, typical of nitro and nitrile derivatives), and $s' = \Delta I$ for pyridine (or dioxane) (the column headed "5" in Table 11.13).

TABLE 11.14 Characteristics of Selected Supercritical Fluids

Fluid	Critical temperature, K (°C)	Critical pressure, atm (psi)
Ammonia	406 (133)	111.3 (1636)
Argon	151 (−122)	48.1 (707)
Benzene	562 (289)	48.3 (710)
Butane	425 (125)	37.5 (551)
Carbon dioxide	304 (31)	72.8 (1070)
Carbon disulfide	552 (279)	78.0 (1147)
Chlorotrifluoromethane	379 (106)	40 (588)
2,2-Dimethylpropane	434 (161)	31.6 (464)
Ethane	305 (32)	48.2 (706)
Fluoromethane	318 (45)	58.0 (853)
Heptane	540 (267)	27.0 (397)
Hexane	507 (234)	29.3 (431)

TABLE 11.14 Characteristics of Selected Supercritical Fluids (*Continued*)

Fluid	Critical temperature, K (°C)	Critical pressure, atm (psi)
Hydrogen sulfide	373 (100)	88.2 (1296)
Krypton	209 (−64)	54.3 (798)
Methane	191 (−82)	45.4 (667)
Methanol	513 (240)	79.9 (1175)
2-Methylpropane	408 (65)	36.0 (529)
Nitrogen	126 (−147)	33.5 (492)
Nitrogen(I) oxide	310 (37)	71.5 (1051)
Pentane	470 (197)	33.3 (490)
Propane	470 (197)	41.9 (616)
Sulfur dioxide	431 (158)	77.8 (1144)
Sulfur hexafluoride	319 (46)	37.1 (545)
Trichloromethane	536 (263)	54.9 (807)
Trifluoromethane	299 (26)	47.7 (701)
Water	647 (374)	217.6 (3199)
Xenon	290 (17)	57.6 (847)

11.4.2 Chromatographic Behavior of Solutes

11.4.2.1 Retention Behavior. On a chromatogram the distance on the time axis from the point of sample injection to the peak of an eluted component is called the *uncorrected retention time* t_R. The corresponding retention volume is the product of retention time and flow rate, expressed as volume of mobile phase per unit time:

$$V_R = t_R F_c$$

The *average linear velocity u* of the mobile phase in terms of the column length L and the average linear velocity of eluent t_M (which is measured by the transit time of a nonretained solute) is

$$u = \frac{L}{t_M}$$

The *adjusted retention time* t'_R is given by

$$t'_R = t_R - t_M$$

When the mobile phase is a gas, a *compressibility factor j* must be applied to the adjusted retention volume to give the *net retention volume*:

$$V_N = j V'_R$$

The compressibility factor is expressed by

$$j = \frac{3[(P_i/P_o)^2 - 1]}{2[(P_i/P_o)^3 - 1]}$$

where P_i is the carrier gas pressure at the column inlet and P_o that at the outlet.

11.4.2.2 *Partition Ratio.* The partition ratio is the additional time a solute band takes to elute, as compared with an unretained solute (for which $k' = 0$), divided by the elution time of an unretained band:

$$k' = \frac{t_R - t_M}{t_M} = \frac{V_R - V_M}{V_M}$$

Retention time may be expressed as

$$t_R = t_M(1 + k') = \frac{L}{u}(1 + k')$$

11.4.2.3 *Relative Retention.* The relative retention α of two solutes, where solute 1 elutes before solute 2, is given variously by

$$\alpha = \frac{k'_2}{k'_1} = \frac{V'_{R,2}}{V'_{R,1}} = \frac{t'_{R,2}}{t'_{R,1}}$$

The relative retention is dependent on (1) the nature of the stationary and mobile phases and (2) the column operating temperature.

11.4.2.4 *Column Efficiency.* Under ideal conditions the profile of a solute band resembles that given by a Gaussian distribution curve (Fig. 11.1). The efficiency of a chromatographic system is expressed by the effective plate number N_{eff}, defined from the chromatogram of a single band,

$$N_{\text{eff}} = \frac{L}{H} = 16 \left(\frac{t'_R}{W_b} \right)^2 = 5.54 \left(\frac{t'_R}{W_{1/2}} \right)^2$$

where L is the column length, H is the plate height, t'_R is the adjusted time for elution of the band center, W_b is the width at the base of the peak ($W_b = 4\sigma$) as determined from the intersections of tangents to the inflection points with the baseline, and $W_{1/2}$ is the width at half the peak height. Column efficiency, when expressed as the number of theoretical plates N_{theor} uses the uncorrected retention time in the foregoing expression. The two column efficiencies are related by

$$N_{\text{eff}} = N_{\text{theor}} \left(\frac{k'}{k' + 1} \right)^2$$

11.4.2.5 *Band Asymmetry.* The peak asymmetry factor AF is often defined as the ratio of peak half-widths at 10% of peak height, that is, the ratio b/a, as shown in Fig. 11.2. When the asymmetry ratio lies outside the range 0.95–1.15 for a peak of $k' = 2$, the effective plate number should be calculated from the expression

$$N = \frac{41.7(t'_R/W_{0.1})}{(a/b) + 1.25}$$

11.4.2.6 *Resolution.* The degree of separation or resolution, Rs, of two adjacent peaks is defined as the distance between band peaks (or centers) divided by the average bandwidth using W_b, as shown in Fig. 11.3.

$$\text{Rs} = \frac{t_{R,2} - t_{R,1}}{0.5(W_2 + W_1)}$$

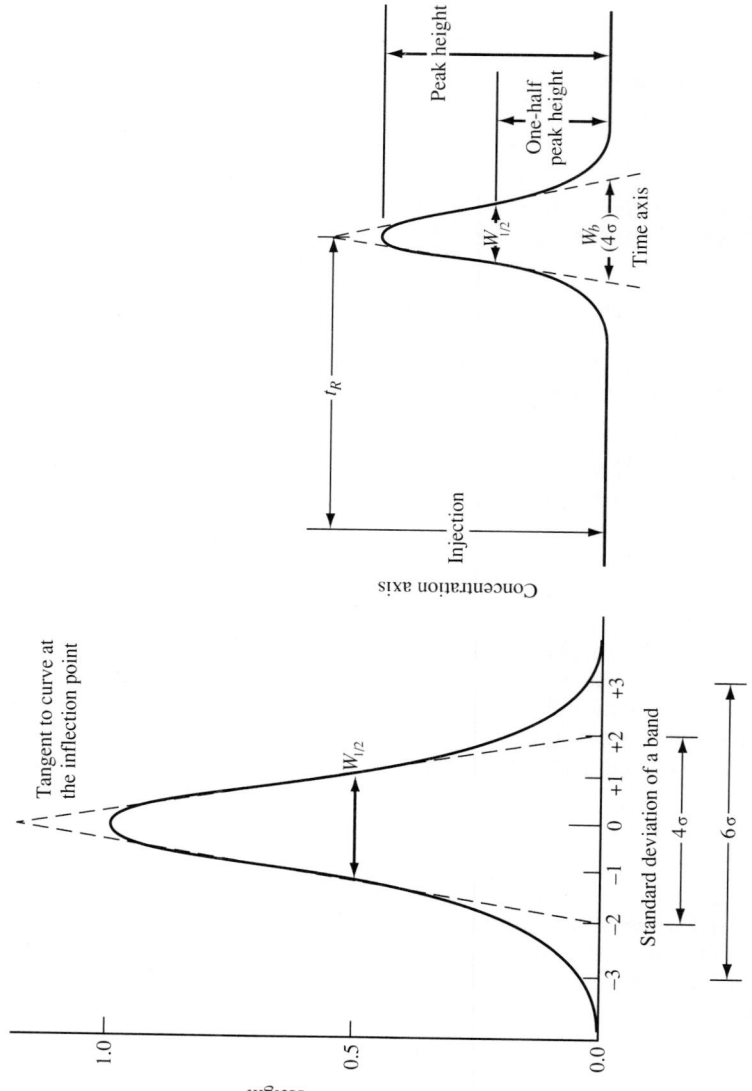

FIGURE 11.1 Profile of a solute band.

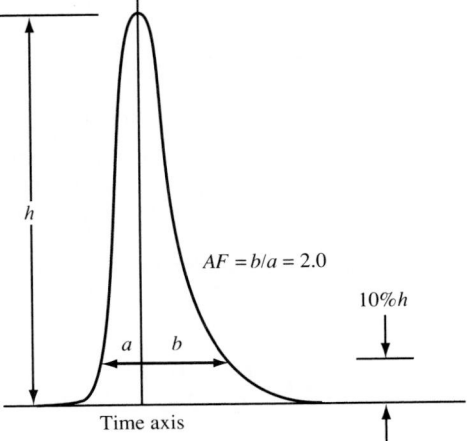

FIGURE 11.2 Band asymmetry.

For reasonable quantitative accuracy, peak maxima must be at least 4σ apart. If so, then $Rs = 1.0$, which corresponds approximately to a 3% overlap of peak areas. A value of $Rs = 1.5$ (for 6σ) represents essentially complete resolution with only 0.2% overlap of peak areas. These criteria pertain to roughly equal solute concentrations.

The fundamental resolution equation incorporates the terms involving the thermodynamics and kinetics of the chromatographic system:

$$Rs = \frac{1}{4}\left(\frac{\alpha - 1}{\alpha}\right)\left(\frac{k'}{1 + k'}\right)\left(\frac{L}{H}\right)^{1/2}$$

Three separate factors affect resolution: (1) a column selectivity factor that varies with α, (2) a capacity factor that varies with k' (taken usually as k_2), and (3) an efficiency factor that depends on the theoretical plate number.

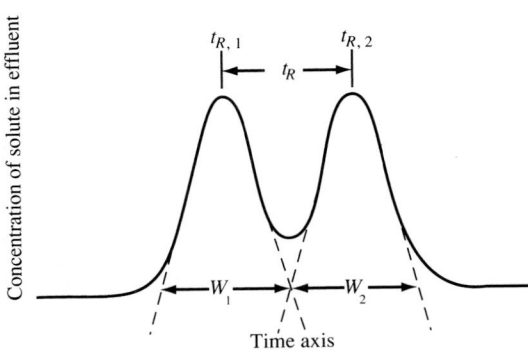

FIGURE 11.3 Definition of resolution.

11.4.2.7 Time of Analysis. The retention time required to perform a separation is given by

$$t_R = 16 \text{Rs}^2 \left(\frac{\alpha}{\alpha - 1} \right)^2 \left[\frac{(1 + k')^3}{(k')^2} \right] \left(\frac{H}{u} \right)$$

Now t_R is a minimum when $k' = 2$, that is, when $t_R = 3t_M$. There is little increase in analysis time when k' lies between 1 and 10. A twofold increase in the mobile-phase velocity roughly halves the analysis time (actually it is the ratio H/u which influences the analysis time). The ratio H/u can be obtained from the experimental plate height/velocity graph.

11.4.2.8 High-Performance Liquid Chromatography. Typical performances for various experimental conditions are given in Table 11.15. The data assume these reduced parameters: $h = 3$, $v = 4.5$. The *reduced plate height* is

$$h = \frac{H}{d_p} = \frac{L}{N d_p}$$

The *reduced velocity* of the eluent is

$$v = \frac{u d_p}{D_M} = \frac{L d_p}{t_M D_M}$$

In these expressions, d_p is the particle diameter of the stationary phase that constitutes one plate height. D_M is the diffusion coefficient of the solute in the mobile phase.

TABLE 11.15 Typical Performances in HPLC for Various Conditions

Performances		Column parameters		
N	t_M, s	L, cm	d_p, μm	P, atm (psi)
2 500	30	2.3	3	18.4 (270)
2 500	30	3.7	5	18.4 (270)
2 500	30	7.5	10	18.4 (270)
5 000	30	4.5	3	74 (1088)
5 000	30	7.5	5	74 (1088)
5 000	30	15.0	10	74 (1088)
10 000	30	9.0	3	300 (4410)
10 000	30	15.0	5	300 (4410)
10 000	30	30.0	10	300 (4410)
10 000	30	9.0	3	300 (4410)
10 000	60	9.0	3	150 (2200)
10 000	90	9.0	3	100 (1470)
15 000	90	2.3	3	223 (3275)
15 000	120	2.3	3	167 (2459)
11 100	30	10.0	3	369 (5420)
11 100	37	10.0	3	300 (4410)
11 100	101	10.0	3	100 (1470)
27 800	231	25.0	3	300 (4410)

Assumed reduced parameters: $h = 3$, $v = 4.5$. These are optimum values from a graph of reduced plate height versus reduced linear velocity of the mobile phase.

TABLE 11.16 Selectivity Coefficients for Ion Exchange Resins

Counterion	Selectivity coefficient for cation exchange AG 50W-X8 resin	Counterion	Selectivity coefficient for anion exchange AG 2-X8 resin
H^+	1.00	OH^-	1.00
Li^+	0.85	F^-	1.6
Na^+	1.5	Propanoate(1−)	2.6
NH_4^+	1.95	Acetate(1−)	3.2
K^+	2.5	Formate(1−)	4.6
Rb^+	2.6	$H_2PO_4^-$	5.0
Cs^+	2.7	IO_3^-	5.5
Cu^+	5.3	HCO_3^-	6.0
Ag^+	7.6	Cl^-	22
Mn^{2+}	2.35	NO_2^-	24
Mg^{2+}	2.5	BrO_3^-	27
Fe^{2+}	2.55	HSO_3^-	27
Zn^{2+}	2.7	CN^-	28
Co^{2+}	2.8	Br^-	50
Cu^{2+}	2.9	NO_3^-	65
Cd^{2+}	2.95	ClO_3^-	74
Ni^{2+}	3.0	HSO_4^-	85
Ca^{2+}	3.9	Phenate(1−)	110
Sr^{2+}	4.95	I^-	175
Hg^{2+}	7.2	Citrate(1−)	220
Pb^{2+}	7.5	Salicylate(1−)	450
Ba^{2+}	8.7	Benzenesulfonate(1−)	500

These are analytical grade resins with 8% cross-linking.

11.4.2.9 Ion-Exchange Equilibrium. For example, in the case of a cation-exchange resin, initially in the H-form, in contact with a solution containing K^+ ions, an equilibrium exists,

$$\text{Resin, } H^+ + K^+ = \text{resin, } K^+ + H^+$$

which is characterized by the *selectivity coefficient* $k_{K/H}$:

$$k_{K/H} = \frac{[K^+]_r [H^+]}{[H^+]_r [K^+]}$$

where the subscript *r* refers to the resin phase. Selectivity coefficients for selected cations and anions are listed in Table 11.16.

To accomplish any separation of two cations (or two anions), it is necessary that one of these cations be taken up by the resin in distinct preference to the other. This is expressed by the *separation factor* (or relative retention), $\alpha_{K/Na}$, using potassium and sodium ions as the example. Thus,

$$\alpha_{K/Na} = \frac{k_{K/H}}{k_{Na/H}} = k_{K/Na} = \frac{2.5}{1.5} = 1.7$$

11.5 GRAVIMETRIC ANALYSIS

TABLE 11.17 Gravimetric Factors

In the following table the elements are arranged in alphabetical order.

Example: To convert a given weight of Al_2O_3 to its equivalent of Al, multiply by the factor at the right, 0.52926; similarly to convert Al to Al_2O_3, multiply by the factor at the left, 1.8894.

Factor		Factor
	ALUMINUM	
	Al = 26.9815	
0.74971	$Al \leftrightarrow Al_4C_3$	1.3341
0.058728	$Al \leftrightarrow Al(C_9H_6ON)_3$ (oxinate)	17.027
0.65829	$Al \leftrightarrow AlN$	1.5191
1.8894	$Al_2O_3 \leftrightarrow Al$	0.52926
1.4165	$Al_2O_3 \leftrightarrow Al_4C_3$	0.70596
0.38233	$Al_2O_3 \leftrightarrow AlCl_3$	2.6155
0.41804	$Al_2O_3 \leftrightarrow AlPO_4$	2.3921
0.29800	$Al_2O_3 \leftrightarrow Al_2(SO_4)_3$	3.3557
0.15300	$Al_2O_3 \leftrightarrow Al_2(SO_4)_3 \cdot 18H_2O$	6.5361
0.10746	$Al_2O_3 \leftrightarrow K_2SO_4 \cdot Al_2(SO_4)_3 \cdot 24H_2O$	9.3055
0.11246	$Al_2O_3 \leftrightarrow (NH_4)_2SO_4 \cdot Al_2(SO_4)_3 \cdot 24H_2O$	8.8922
4.5197	$AlPO_4 \leftrightarrow Al$	0.22125
1.3946	$CaF_2 \leftrightarrow AlF_3$	0.71704
0.58196	$P_2O_5 \leftrightarrow AlPO_4$	1.7183
	AMMONIUM	
	NH$_4$ = 18.03858	
1.1013	$Ag \leftrightarrow NH_4Br$	0.90802
2.0166	$Ag \leftrightarrow NH_4Cl$	0.49590
0.74424	$Ag \leftrightarrow NH_4I$	1.3437
1.9171	$AgBr \leftrightarrow NH_4Br$	0.52161
2.6792	$AgCl \leftrightarrow NH_4Cl$	0.37323
1.6198	$AgI \leftrightarrow NH_4I$	0.61737
1.7663	$BaSO_4 \leftrightarrow (NH_4)_2SO_4$	0.56615
0.81583	$Br \leftrightarrow NH_4Br$	1.2257
1.9654	$Cl \leftrightarrow NH_4$	0.50881
0.66277	$Cl \leftrightarrow NH_4Cl$	1.5088
0.68162	$HCl \leftrightarrow NH_4Cl$	1.4671
0.87553	$I \leftrightarrow NH_4I$	1.1422
14.410	$MgNH_4PO_4 \cdot 6H_2O \leftrightarrow NH_3$	0.069398
13.604	$MgNH_4PO_4 \cdot 6H_2O \leftrightarrow NH_4$	0.073506
9.4249	$MgNH_4PO_4 \cdot 6H_2O \leftrightarrow (NH_4)_2O$	0.10610
0.82244	$N \leftrightarrow NH_3$	1.2159
0.77648	$N \leftrightarrow NH_4$	1.2879
0.26185	$N \leftrightarrow NH_4Cl$	3.8189
0.17499	$N \leftrightarrow NH_4NO_3$	5.7145
0.53793	$N \leftrightarrow (NH_4)_2O$	1.8590
0.21200	$N \leftrightarrow (NH_4)_2SO_4$	4.7169
0.94412	$NH_3 \leftrightarrow NH_4$	1.0592
0.35449	$NH_3 \leftrightarrow (NH_4)_2CO_3$	2.8210
0.21543	$NH_3 \leftrightarrow NH_4HCO_3$	4.6419
0.21277	$NH_3 \leftrightarrow NH_4NO_3$	4.6998

TABLE 11.17   Gravimetric Factors (*Continued*)

Factor		Factor
0.65407	$NH_3 \leftrightarrow (NH_4)_2O$	1.5289
0.48596	$NH_3 \leftrightarrow NH_4OH$	2.0578
0.25777	$NH_3 \leftrightarrow (NH_4)_2SO_4$	3.8794
3.1409	$NH_4Cl \leftrightarrow NH_3$	0.31838
2.9654	$NH_4Cl \leftrightarrow NH_4$	0.33723
2.0543	$NH_4Cl \leftrightarrow (NH_4)_2O$	0.48677
1.5263	$NH_4Cl \leftrightarrow NH_4OH$	0.65516
2.5020	$NH_4OH \leftrightarrow N$	0.39967
1.9428	$NH_4OH \leftrightarrow NH_4$	0.51472
13.032	$(NH_4)_2PtCl_6 \leftrightarrow NH_3$	0.076737
12.303	$(NH_4)_2PtCl_6 \leftrightarrow NH_4$	0.081279
4.1490	$(NH_4)_2PtCl_6 \leftrightarrow NH_4Cl$	0.24102
2.7728	$(NH_4)_2PtCl_6 \leftrightarrow NH_4NO_3$	0.36065
8.5235	$(NH_4)_2PtCl_6 \leftrightarrow (NH_4)_2O$	0.11732
6.3328	$(NH_4)_2PtCl_6 \leftrightarrow NH_4OH$	0.15791
3.3592	$(NH_4)_2PtCl_6 \leftrightarrow (NH_4)_2SO_4$	0.29769
1.3473	$(NH_4)_2SO_4 \leftrightarrow H_2SO_4$	0.74223
3.1710	$N_2O_5 \leftrightarrow NH_3$	0.31536
0.67470	$N_2O_5 \leftrightarrow NH_4NO_3$	1.4821
2.0740	$N_2O_5 \leftrightarrow (NH_4)_2O$	0.48215
5.7275	$Pt \leftrightarrow NH_3$	0.17460
5.4074	$Pt \leftrightarrow NH_4$	0.18493
1.8235	$Pt \leftrightarrow NH_4Cl$	0.54838
1.2187	$Pt \leftrightarrow NH_4NO_3$	0.82058
3.7462	$Pt \leftrightarrow (NH_4)_2O$	0.26694
2.7833	$Pt \leftrightarrow NH_4OH$	0.35928
1.4764	$Pt \leftrightarrow (NH_4)_2SO_4$	0.67733
2.3505	$SO_3 \leftrightarrow NH_3$	0.42545
0.60589	$SO_3 \leftrightarrow (NH_4)_2SO_4$	1.6505

ANTIMONY
Sb = 121.75

Factor		Factor
0.36460	$Sb \leftrightarrow KSbO \cdot C_4H_4O_6 \cdot \frac{1}{2}H_2O$	2.7428
0.83535	$Sb \leftrightarrow Sb_2O_4$	1.1971
0.75271	$Sb \leftrightarrow Sb_2O_5$	1.3285
0.43646	$Sb_2O_3 \leftrightarrow KSbO \cdot C_4H_4O_6 \cdot \frac{1}{2}H_2O$	2.2912
0.90106	$Sb_2O_3 \leftrightarrow Sb_2O_5$	1.1098
0.72184	$Sb_2O_3 \leftrightarrow Sb_2S_5$	1.3853
0.46042	$Sb_2O_4 \leftrightarrow KSbO \cdot C_4H_4O_6 \cdot \frac{1}{2}H_2O$	2.1719
1.2628	$Sb_2O_4 \leftrightarrow Sb$	0.79188
1.0549	$Sb_2O_4 \leftrightarrow Sb_2O_3$	0.94796
0.95053	$Sb_2O_4 \leftrightarrow Sb_2O_5$	1.0520
0.90523	$Sb_2O_4 \leftrightarrow Sb_2S_3$	1.1047
0.76147	$Sb_2O_4 \leftrightarrow Sb_2S_5$	1.3133
0.80110	$Sb_2O_5 \leftrightarrow Sb_2S_5$	1.2483
0.50862	$Sb_2S_3 \leftrightarrow KSbO \cdot C_4H_4O_6 \cdot \frac{1}{2}H_2O$	1.9661
1.3950	$Sb_2S_3 \leftrightarrow Sb$	0.71683
1.1653	$Sb_2S_3 \leftrightarrow Sb_2O_3$	0.85812
1.0500	$Sb_2S_3 \leftrightarrow Sb_2O_5$	0.95234
1.6584	$Sb_2S_5 \leftrightarrow Sb$	0.60299

TABLE 11.17 Gravimetric Factors (*Continued*)

Factor		Factor

ARSENIC
As = 74.9216

Factor		Factor
1.3203	$As_2O_3 \leftrightarrow As$	0.75738
0.86079	$As_2O_3 \leftrightarrow As_2O_5$	1.1617
1.5339	$As_2O_5 \leftrightarrow As$	0.65195
1.6420	$As_2S_3 \leftrightarrow As$	0.60903
1.2436	$As_2S_3 \leftrightarrow As_2O_3$	0.80413
1.0705	$As_2S_3 \leftrightarrow As_2O_5$	0.93418
0.79324	$As_2S_3 \leftrightarrow As_2S_5$	1.2606
2.0699	$As_2S_5 \leftrightarrow As$	0.48311
1.5678	$As_2S_5 \leftrightarrow As_2O_3$	0.63787
1.3495	$As_2S_5 \leftrightarrow As_2O_5$	0.74103
4.6729	$BaSO_4 \leftrightarrow As$	0.21400
3.5392	$BaSO_4 \leftrightarrow As_2O_3$	0.28255
3.0465	$BaSO_4 \leftrightarrow As_2O_6$	0.32825
2.8482	$BaSO_4 \leftrightarrow AsO_3$	0.35110
2.5202	$BaSO_4 \leftrightarrow AsO_4$	0.39680
2.0719	$Mg_2As_2O_7 \leftrightarrow As$	0.48265
1.5692	$Mg_2As_2O_7 \leftrightarrow As_2O_3$	0.63726
1.3509	$Mg_2As_2O_7 \leftrightarrow As_2O_5$	0.74032
1.2629	$Mg_2As_2O_7 \leftrightarrow AsO_2$	0.79186
1.1174	$Mg_2As_2O_7 \leftrightarrow AsO_4$	0.89493
1.2619	$Mg_2As_2O_7 \leftrightarrow As_2S_3$	0.79249
2.5397	$MgNH_4AsO_4 \cdot \frac{1}{2}H_2O \leftrightarrow As$	0.39374
1.9235	$MgNH_4AsO_4 \cdot \frac{1}{2}H_2O \leftrightarrow As_2O_3$	0.51988
1.6558	$MgNH_4AsO_4 \cdot \frac{1}{2}H_2O \leftrightarrow As_2O_5$	0.60395
1.5480	$MgNH_4AsO_4 \cdot \frac{1}{2}H_2O \leftrightarrow AsO_3$	0.64600
1.3697	$MgNH_4AsO_4 \cdot \frac{1}{2}H_2O \leftrightarrow AsO_4$	0.73008

BARIUM
Ba = 137.34

Factor		Factor
1.4369	$BaCO_3 \leftrightarrow Ba$	0.69592
0.94766	$BaCO_3 \leftrightarrow BaCl_2$	1.0552
0.76088	$BaCO_3 \leftrightarrow Ba(HCO_3)_2$	1.3143
1.2871	$BaCO_3 \leftrightarrow BaO$	0.77699
1.8446	$BaCrO_4 \leftrightarrow Ba$	0.54214
1.2165	$BaCrO_4 \leftrightarrow BaCl_2$	0.82205
1.2838	$BaCrO_4 \leftrightarrow BaCO_3$	0.77902
1.6521	$BaCrO_4 \leftrightarrow BaO$	0.60530
2.0345	$BaSiF_6 \leftrightarrow Ba$	0.49152
1.5936	$BaSiF_6 \leftrightarrow BaF_2$	0.62751
1.8222	$BaSiF_6 \leftrightarrow BaO$	0.54878
1.6994	$BaSO_4 \leftrightarrow Ba$	0.58843
1.1208	$BaSO_4 \leftrightarrow BaCl_2$	0.89224
0.95546	$BaSO_4 \leftrightarrow BaCl_2 \cdot 2H_2O$	1.0466
1.1827	$BaSO_4 \leftrightarrow BaCO_3$	0.84554
0.89308	$BaSO_4 \leftrightarrow Ba(NO_3)_2$	1.1197
1.5221	$BaSO_4 \leftrightarrow BaO$	0.65698
1.3783	$BaSO_4 \leftrightarrow BaO_2$	0.72554
1.3778	$BaSO_4 \leftrightarrow BaS$	0.72579

TABLE 11.17 Gravimetric Factors (*Continued*)

Factor		Factor
0.28701	$CO_2 \leftrightarrow BaO$	3.4842
0.22300	$CO_2 \leftrightarrow BaCO_3$	4.4842

BERYLLIUM
Be = 9.0122

8.8678	$BeCl_2 \leftrightarrow Be$	0.11277
2.7753	$BeO \leftrightarrow Be$	0.36033
0.31296	$BeO \leftrightarrow BeCl_2$	3.1953
0.14119	$BeO \leftrightarrow BeSO_4 \cdot 4H_2O$	7.0825

BISMUTH
Bi = 208.980

0.89699	$Bi \leftrightarrow Bi_2O_3$	1.1148
1.6648	$BiAsO_4 \leftrightarrow Bi$	0.60069
1.4933	$BiAsO_4 \leftrightarrow Bi_2O_4$	0.66968
0.48030	$Bi_2O_3 \leftrightarrow Bi(NO_3)_3 \cdot 5H_2O$	2.0820
0.81183	$Bi_2O_3 \leftrightarrow BiONO_3$	1.2318
1.2462	$BiOCl \leftrightarrow Bi$	0.80244
0.53689	$BiOCl \leftrightarrow Bi(NO_3)_3 \cdot 5H_2O$	1.8626
1.1178	$BiOCl \leftrightarrow Bi_2O_3$	0.89460
0.90748	$BiOCl \leftrightarrow BiONO_3$	1.1019
1.2301	$Bi_2S_3 \leftrightarrow Bi$	0.81291
1.1034	$Bi_2S_3 \leftrightarrow Bi_2O_3$	0.90627

BORON
B = 10.81

3.2199	$B_2O_3 \leftrightarrow B$	0.31057
0.81317	$B_2O_3 \leftrightarrow BO_2$	1.2298
0.59193	$B_2O_3 \leftrightarrow BO_3$	1.6894
0.89693	$B_2O_3 \leftrightarrow B_4O_7$	1.1149
0.56298	$B_2O_3 \leftrightarrow H_3BO_3$	1.7763
0.36510	$B_2O_3 \leftrightarrow Na_2B_4O_7 \cdot 10H_2O$	2.7389
6.4005	$B_6C \leftrightarrow C$	0.15624
11.646	$KBF_4 \leftrightarrow B$	0.085863
3.6171	$KBF_4 \leftrightarrow B_2O_3$	0.27647
2.0363	$KBF_4 \leftrightarrow H_3BO_3$	0.49108
1.3206	$KBF_4 \leftrightarrow Na_2B_4O_7 \cdot 10H_2O$	0.75723

BROMINE
Br = 79.90

1.3499	$Ag \leftrightarrow Br$	0.74079
0.84333	$Ag \leftrightarrow BrO_3$	1.1858
1.3331	$Ag \leftrightarrow HBr$	0.75013
2.3499	$AgBr \leftrightarrow Br$	0.42555
1.4681	$AgBr \leftrightarrow BrO_3$	0.68117
2.3206	$AgBr \leftrightarrow HBr$	0.43091
0.55756	$Br \leftrightarrow AgCl$	1.7935
9.9892	$Br \leftrightarrow O$	0.10010
1.1858	$BrO_3 \leftrightarrow Ag$	0.84333

TABLE 11.17   Gravimetric Factors (*Continued*)

Factor		Factor
	CADMIUM	
	Cd = 112.40	
0.61317	Cd ↔ CdCl$_2$	1.6309
0.47545	Cd ↔ Cd(NO$_3$)$_2$	2.1033
1.1423	CdO ↔ Cd	0.87539
0.70045	CdO ↔ CdCl$_2$	1.4276
0.54312	CdO ↔ Cd(NO$_3$)$_2$	1.8412
1.2852	CdS ↔ Cd	0.77807
0.78806	CdS ↔ CdCl$_2$	1.2689
0.61106	CdS ↔ Cd(NO$_3$)$_2$	1.6365
1.1251	CdS ↔ CdO	0.88883
0.69298	CdS ↔ CdSO$_4$	1.4430
1.8546	CdSO$_4$ ↔ Cd	0.53919
1.1372	CdSO$_4$ ↔ CdCl$_2$	0.87935
0.88177	CdSO$_4$ ↔ Cd(NO$_3$)$_2$	1.1341
1.6235	CdSO$_4$ ↔ CdO	0.61595
	CALCIUM	
	Ca = 40.08	
3.2352	BaSO$_4$ ↔ CaS	0.30910
1.7144	BaSO$_4$ ↔ CaSO$_4$	0.58329
1.3556	BaSO$_4$ ↔ CaSO$_4$·2H$_2$O	0.73766
0.36111	Ca ↔ CaCl$_2$	2.7692
0.51334	Ca ↔ CaF$_2$	1.9480
0.71471	Ca ↔ CaO	1.3992
2.4973	CaCO$_3$ ↔ Ca	0.40044
0.90179	CaCO$_3$ ↔ CaCl$_2$	1.1089
0.61742	CaCO$_3$ ↔ Ca(HCO$_3$)$_2$	1.6196
1.7848	CaCO ↔ CaO	0.56029
0.73520	CaCO$_3$ ↔ CaSO$_4$	1.3602
0.58134	CaCO$_3$ ↔ CaSO$_4$·2H$_2$O	1.7202
1.3726	CaCO$_3$ ↔ HCl	0.72856
0.50526	CaO ↔ CaCl$_2$	1.9792
0.71825	CaO ↔ CaF$_2$	1.3923
0.34593	CaO ↔ Ca(HCO$_3$)$_2$	2.8907
0.75685	CaO ↔ Ca(OH)$_2$	1.3213
0.41192	CaO ↔ CaSO$_4$	2.4276
0.32572	CaO ↔ CaSO$_4$·2H$_2$O	3.0701
2.5797	Ca$_3$(PO$_4$)$_2$ ↔ Ca	0.38765
1.8437	Ca$_3$(PO$_4$)$_2$ ↔ CaO	0.54239
0.75946	Ca$_3$(PO$_4$)$_2$ ↔ CaSO$_4$	1.3167
3.3967	CaSO$_4$ ↔ Ca	0.29440
1.2266	CaSO$_4$ ↔ CaCl$_2$	0.81526
1.3602	CaSO$_4$ ↔ CaCO$_3$	0.73520
1.7437	CaSO$_4$ ↔ CaF$_2$	0.57351
2.4276	CaSO$_4$ ↔ CaO	0.41192
1.7691	Cl ↔ Ca	0.56526
0.63885	Cl ↔ CaCl$_2$	1.5653
1.2644	Cl ↔ CaO	0.79089
0.78479	CO$_2$ ↔ CaO	1.2742
0.43970	CO$_2$ ↔ CaCO$_3$	2.2743
0.77989	Mg$_2$As$_2$O$_7$ ↔ Ca$_3$(AsO$_4$)$_2$	1.2822

TABLE 11.17 Gravimetric Factors (*Continued*)

Factor		Factor
0.71883	$MgO \leftrightarrow CaO$	1.3912
0.71755	$Mg_2P_2O_7 \leftrightarrow Ca_3(PO_4)_2$	1.3936
12.098	$(NH_4)_3PO_4 \cdot 12MoO_3 \leftrightarrow Ca_3(PO_4)_2$	0.082657
0.65824	$N_2O_5 \leftrightarrow Ca(NO_3)_2$	1.5192
0.45761	$P_2O_3 \leftrightarrow Ca_3(PO_4)_2$	2.1853
1.4277	$SO_3 \leftrightarrow CaO$	0.70044
0.58809	$SO_3 \leftrightarrow CaSO_4$	1.7004
0.46502	$SO_3 \leftrightarrow CaSO_4 \cdot 2H_2O$	2.1505
0.80523	$WO_3 \leftrightarrow CaWO_4$	1.2419

CARBON

C = 12.011

Factor		Factor
3.9913	$Ag \leftrightarrow HCN$	0.25054
1.6565	$Ag \leftrightarrow KCN$	0.60369
4.9541	$AgCN \leftrightarrow HCN$	0.20185
2.0561	$AgCN \leftrightarrow KCN$	0.48637
16.431	$BaCO_3 \leftrightarrow C$	0.060861
4.4842	$BaCO_3 \leftrightarrow CO_2$	0.22301
3.2887	$BaCO_3 \leftrightarrow CO_3$	0.30407
3.4842	$BaO \leftrightarrow CO_2$	0.28701
1.7421	$BaO \leftrightarrow CO_2$, bicarbonate	0.57402
0.19432	$CN \leftrightarrow AgCN$	5.1461
0.24120	$CN \leftrightarrow Ag$	4.1460
0.35000	$SCN \leftrightarrow AgSCN$	2.8572
0.47757	$SCN \leftrightarrow CuSCN$	2.0939
0.24885	$SCN \leftrightarrow BaSO_4$	4.0185
1.2742	$CaO \leftrightarrow CO_2$	0.78479
0.63712	$CaO \leftrightarrow CO_2$, bicarbonate	1.5696
0.33936	$CO_2 \leftrightarrow Ba(HCO_3)_2$	2.9467
3.6641	$CO_2 \leftrightarrow C$	0.27291
0.43970	$CO_2 \leftrightarrow CaCO_3$	2.2743
0.54297	$CO_2 \leftrightarrow Ca(HCO_3)_2$	1.8417
0.73341	$CO_2 \leftrightarrow CO_3$	1.3635
0.13507	$CO_2 \leftrightarrow Cs_2CO_3$	7.4033
0.22695	$CO_2 \leftrightarrow CsHCO_3$	4.4063
0.37986	$CO_2 \leftrightarrow FeCO_3$	2.6326
0.49483	$CO_2 \leftrightarrow Fe(HCO_3)_2$	2.0209
0.31843	$CO_2 \leftrightarrow K_2CO_3$	3.1404
0.43957	$CO_2 \leftrightarrow KHCO_3$	2.2749
0.46718	$CO_2 \leftrightarrow K_2O$	2.1405
0.59564	$CO_2 \leftrightarrow Li_2CO_3$	1.6789
0.64762	$CO_2 \leftrightarrow LiHCO_3$	1.5441
1.4730	$CO_2 \leftrightarrow Li_2O$	0.67887
0.52193	$CO_2 \leftrightarrow MgCO_3$	1.9159
0.60143	$CO_2 \leftrightarrow Mg(HCO_3)_2$	1.6627
1.0918	$CO_2 \leftrightarrow MgO$	0.91595
0.38286	$CO_2 \leftrightarrow MnCO_3$	2.6119
0.49737	$CO_2 \leftrightarrow Mn(HCO_3)_2$	2.0106
0.62041	$CO_2 \leftrightarrow MnO$	1.6118
0.41523	$CO_2 \leftrightarrow Na_2CO_3$	2.4083
0.52388	$CO_2 \leftrightarrow NaHCO_3$	1.9088
0.71008	$CO_2 \leftrightarrow Na_2O$	1.4083
0.45802	$CO_2 \leftrightarrow (NH_4)_2CO_3$	2.1833
0.55669	$CO_2 \leftrightarrow NH_4HCO_3$	1.7963
0.16471	$CO_2 \leftrightarrow PbCO_3$	6.0713

TABLE 11.17 Gravimetric Factors (*Continued*)

Factor		Factor
0.19055	$CO_2 \leftrightarrow Rb_2CO_3$	5.2477
0.30043	$CO_2 \leftrightarrow RbHCO_3$	3.3286
0.23542	$CO_2 \leftrightarrow Rb_2O$	4.2477
0.29811	$CO_2 \leftrightarrow SrCO_3$	3.3545
0.41984	$CO_2 \leftrightarrow Sr(HCO_3)_2$	2.3818
0.42474	$CO_2 \leftrightarrow SrO$	2.3545

CERIUM
Ce = 140.12

Factor		Factor
0.36100	$Ce \leftrightarrow Ce(NO_3)_4$	2.7701
0.24746	$Ce \leftrightarrow Ce(NO_3)_4 \cdot 2NH_4NO_3 \cdot H_2O$	4.0411
0.81408	$Ce \leftrightarrow CeO_2$	1.2284
0.85377	$Ce \leftrightarrow Ce_2O_3$	1.1713
0.49302	$Ce \leftrightarrow Ce_2(SO_4)_3$	2.0283
1.0527	$Ce_2(C_2O_4)_3 \cdot 3H_2O \leftrightarrow Ce_2(SO_4)_3$	0.94998
2.1351	$Ce_2(C_2O_4)_3 \cdot 3H_2O \leftrightarrow Ce$	0.46835
0.44345	$CeO_2 \leftrightarrow Ce(NO_3)_4$	2.2551
0.30397	$CeO_2 \leftrightarrow Ce(NO_3)_4 \cdot 2NH_4NO_3 \cdot H_2O$	3.2898
0.42284	$Ce_2O_3 \leftrightarrow Ce(NO_3)_4$	2.3650
0.28984	$Ce_2O_3 \leftrightarrow Ce(NO_3)_4 \cdot 2NH_4NO_3 \cdot H_2O$	3.4502
0.95352	$Ce_2O_3 \leftrightarrow CeO_2$	1.0487
0.57746	$Ce_2O_3 \leftrightarrow Ce_2(SO_4)_3$	1.7317

CESIUM
Cs = 137.905

Factor		Factor
0.85127	$AgCl \leftrightarrow CsCl$	1.1747
0.26675	$Cl \leftrightarrow Cs$	3.7489
0.21058	$Cl \leftrightarrow CsCl$	4.7488
0.78944	$Cs \leftrightarrow CsCl$	1.2667
0.57200	$Cs \leftrightarrow CsClO_4$	1.7483
0.81585	$Cs \leftrightarrow Cs_2CO_3$	1.2257
0.94326	$Cs \leftrightarrow Cs_2O$	1.0602
0.83693	$Cs_2O \leftrightarrow CsCl$	1.1948
0.77876	$Cs_2O \leftrightarrow Cs_2SO_4$	1.2841
2.5341	$Cs_2PtCl_6 \leftrightarrow Cs$	0.39461
2.0005	$Cs_2PtCl_6 \leftrightarrow CsCl$	0.49987
2.0675	$Cs_2PtCl_6 \leftrightarrow Cs_2CO_3$	0.48369
2.3903	$Cs_2PtCl_6 \leftrightarrow Cs_2O$	0.41835
1.3613	$Cs_2SO_4 \leftrightarrow Cs$	0.73457
1.0747	$Cs_2SO_4 \leftrightarrow CsCl$	0.93050
1.1106	$Cs_2SO_4 \leftrightarrow Cs_2CO_3$	0.90038
0.28410	$SO_3 \leftrightarrow Cs_2O$	3.5199

CHLORINE
Cl = 35.453

Factor		Factor
3.0426	$Ag \leftrightarrow Cl$	0.32866
2.9585	$Ag \leftrightarrow HCl$	0.33801
4.0425	$AgCl \leftrightarrow Cl$	0.24737
3.9308	$AgCl \leftrightarrow HCl$	0.25440
3.5728	$BaCrO_4 \leftrightarrow Cl$	0.27990
0.56526	$Ca \leftrightarrow Cl$	1.7691
0.97235	$Cl \leftrightarrow HCl$	1.0284
0.58227	$ClO_3 \leftrightarrow AgCl$	1.7174

TABLE 11.17 Gravimetric Factors (*Continued*)

Factor		Factor
1.1193	$ClO_3 \leftrightarrow KCl$	0.89340
1.4279	$ClO_3 \leftrightarrow NaCl$	0.70033
0.69391	$ClO_4 \leftrightarrow AgCl$	1.4411
1.3339	$ClO_4 \leftrightarrow KCl$	0.74967
1.7017	$ClO_4 \leftrightarrow NaCl$	0.58766
1.1029	$K \leftrightarrow Cl$	0.90668
2.1029	$KCl \leftrightarrow Cl$	0.47553
0.19572	$Li \leftrightarrow Cl$	5.1092
0.34288	$Mg \leftrightarrow Cl$	2.9165
1.3429	$MgCl_2 \leftrightarrow Cl$	0.74467
1.2261	$MnO_2 \leftrightarrow Cl$	0.81560
0.64846	$Na \leftrightarrow Cl$	1.5421
1.6485	$NaCl \leftrightarrow Cl$	0.60663
0.50881	$NH_4 \leftrightarrow Cl$	1.9654
1.4671	$NH_4Cl \leftrightarrow HCl$	0.68162
1.8121	$(NH_4)_2SO_4 \leftrightarrow HCl$	0.55185
4.5580	$PbCrO_4 \leftrightarrow Cl$	0.21939

CHROMIUM
Cr = 51.996

Factor		Factor
4.8721	$BaCrO_4 \leftrightarrow Cr$	0.20525
3.3335	$BaCrO_4 \leftrightarrow Cr_2O_3$	0.29998
2.5335	$BaCrO_4 \leftrightarrow CrO_3$	0.39472
2.1841	$BaCrO_4 \leftrightarrow CrO_4$	0.45786
0.70718	$BaCrO_4 \leftrightarrow Cr_2(SO_4)_3 \cdot 18H_2O$	1.4141
7.4935	$Cr_3C_2 \leftrightarrow C$	0.13345
1.9231	$CrO_3 \leftrightarrow Cr$	0.51999
1.4616	$Cr_2O_3 \leftrightarrow Cr$	0.68420
0.76000	$Cr_2O_3 \leftrightarrow CrO_3$	1.3158
0.65519	$Cr_2O_3 \leftrightarrow CrO_4$	1.5263
3.7349	$K_2CrO_4 \leftrightarrow Cr$	0.26774
1.9421	$K_2CrO_4 \leftrightarrow CrO_3$	0.51490
1.4710	$K_2Cr_2O_7 \leftrightarrow CrO_3$	0.67979
6.2155	$PbCrO_4 \leftrightarrow Cr$	0.16089
4.2527	$PbCrO_4 \leftrightarrow Cr_2O_3$	0.23515
3.2320	$PbCrO_4 \leftrightarrow CrO_3$	0.30941
2.7863	$PbCrO_4 \leftrightarrow CrO_4$	0.35890
0.90217	$PbCrO_4 \leftrightarrow Cr_2(SO_4)_3 \cdot 18H_2O$	1.1084
1.6642	$PbCrO_4 \leftrightarrow K_2CrO_4$	0.60090
2.1971	$PbCrO_4 \leftrightarrow K_2Cr_2O_7$	0.45515

COBALT
Co = 58.9332

Factor		Factor
0.20249	$Co \leftrightarrow Co(NO_3)_2 \cdot 6H_2O$	4.9385
0.78648	$Co \leftrightarrow CoO$	1.2715
0.20965	$Co \leftrightarrow CoSO_4 \cdot 7H_2O$	4.7698
7.6743	$K_3[Co(NO_2)_6] \leftrightarrow Co$	0.13030
6.0357	$K_3[Co(NO_2)_6] \leftrightarrow CoO$	0.16568
1.3620	$Co_3O_4 \leftrightarrow Co$	0.73422
1.0712	$Co_3O_4 \leftrightarrow CoO$	0.93355
2.4758	$Co_2P_2O_7 \leftrightarrow Co$	0.40391
1.9471	$Co_2P_2O_7 \leftrightarrow CoO$	0.51357

TABLE 11.17 Gravimetric Factors (*Continued*)

Factor		Factor
3.2233	CoNH$_4$PO$_4$·H$_2$O $\leftrightarrow$ Co	0.31024
2.5351	CoNH$_4$PO$_4$·H$_2$O $\leftrightarrow$ CoO	0.39447
2.6299	CoSO$_4$ $\leftrightarrow$ Co	0.38024
2.0684	CoSO$_4$ $\leftrightarrow$ CoO	0.48347
3.7514	CoSO$_4$·7H$_2$O $\leftrightarrow$ CoO	0.26657
7.0656	(CoSO$_4$)$_2$·(K$_2$SO$_4$)$_3$ $\leftrightarrow$ Co	0.14153
5.5569	(CoSO$_4$)$_2$·(K$_2$SO$_4$)$_3$ $\leftrightarrow$ CoO	0.17996

COPPER
Cu = 63.544

Factor		Factor
0.25071	Cu $\leftrightarrow$ Cu$_2$C$_2$H$_3$O$_2$·(AsO$_2$)$_3$	3.9887
0.79885	Cu $\leftrightarrow$ CuO	1.2518
0.25449	Cu $\leftrightarrow$ CuSO$_4$·5H$_2$O	3.9295
1.9141	CuSCN $\leftrightarrow$ Cu	0.52245
1.5291	CuSCN $\leftrightarrow$ CuO	0.65400
0.31856	CuO $\leftrightarrow$ CuSO$_4$·5H$_2$O	3.1391
1.1259	Cu$_2$O $\leftrightarrow$ Cu	0.88817
1.2523	Cu$_2$S $\leftrightarrow$ Cu	0.79854
1.0004	Cu$_2$S $\leftrightarrow$ CuO	0.99961
1.1122	Cu$_2$S $\leftrightarrow$ Cu$_2$O	0.89908
0.31869	Cu$_2$S $\leftrightarrow$ CuSO$_4$·5H$_2$O	3.1379
0.91872	Mg$_2$As$_2$O$_7$ $\leftrightarrow$ Cu$_2$C$_2$H$_3$O$_2$(AsO$_2$)$_3$	1.0885

ERBIUM
Er = 167.26

Factor		Factor
1.1435	Er$_2$O$_3$ $\leftrightarrow$ Er	0.87452

FLUORINE
F = 18.9984

Factor		Factor
1.5936	BaSiF$_6$ $\leftrightarrow$ BaF$_2$	0.62751
2.4513	BaSiF$_6$ $\leftrightarrow$ F	0.40795
2.3277	BaSiF$_6$ $\leftrightarrow$ 6HF	0.42960
1.9392	BaSiF$_6$ $\leftrightarrow$ H$_2$SiF$_6$	0.51568
2.6847	BaSiF$_6$ $\leftrightarrow$ SiF$_4$	0.37249
1.9666	BaSiF$_6$ $\leftrightarrow$ SiF$_6$	0.50848
1.6256	CaF$_2$ $\leftrightarrow$ H$_2$SiF$_6$	0.61516
1.6486	CaF$_2$ $\leftrightarrow$ SiF$_6$	0.60658
3.5829	CaSO$_4$ $\leftrightarrow$ F	0.27910
2.4024	CaSO$_4$ $\leftrightarrow$ HF	0.29391
0.48666	F $\leftrightarrow$ CaF$_2$	2.0548
0.51248	HF $\leftrightarrow$ CaF$_2$	1.9513
1.2641	H$_2$SiF$_6$ $\leftrightarrow$ F	0.79109
3.6011	H$_2$SiF$_6$ $\leftrightarrow$ 2HF	0.27769
1.2004	H$_2$SiF$_6$ $\leftrightarrow$ 6HF	0.83308
1.3844	H$_2$SiF$_6$ $\leftrightarrow$ SiF$_4$	0.72233
1.0141	H$_2$SiF$_6$ $\leftrightarrow$ SiF$_6$	0.98605
2.0556	KF·HF $\leftrightarrow$ 2F	0.48647
1.9520	KF·HF $\leftrightarrow$ 2HF	0.51228
0.67218	KF·HF $\leftrightarrow$ 2KF	1.4877
0.41489	KF·HF $\leftrightarrow$ 2(KF·2H$_2$O)	2.4103
1.9325	K$_2$SiF$_6$ $\leftrightarrow$ F	0.51748
1.8351	K$_2$SiF$_6$ $\leftrightarrow$ 6HF	0.54494
1.5288	K$_2$SiF$_6$ $\leftrightarrow$ H$_2$SiF$_6$	0.65412

TABLE 11.17 Gravimetric Factors (*Continued*)

Factor		Factor
1.8957	$K_2SiF_6 \leftrightarrow 2KF$	0.52751
1.5504	$K_2SiF_6 \leftrightarrow SiF_6$	0.64500
1.9495	$NH_4F \leftrightarrow F$	0.51295
1.5013	$NH_4F \cdot HF \leftrightarrow 2F$	0.66611
1.4256	$NH_4F \cdot HF \leftrightarrow 2HF$	0.70145
0.49090	$NH_4F \cdot HF \leftrightarrow 2KF$	2.0371
0.30300	$NH_4F \cdot HF \leftrightarrow 2(KF \cdot 2H_2O)$	3.3003
1.5629	$(NH_4)_2SiF_6 \leftrightarrow F$	0.63985
1.4841	$(NH_4)_2SiF_6 \leftrightarrow 6HF$	0.67381
1.2364	$(NH_4)_2SiF_6 \leftrightarrow H_2SiF_6$	0.80881
2.4050	$(NH_4)_2SiF_6 \leftrightarrow 2NH_4F$	0.41580
1.2539	$(NH_4)_2SiF_6 \leftrightarrow SiF_6$	0.79753
2.2101	$NaF \leftrightarrow F$	0.45246
1.6498	$Na_2SiF_6 \leftrightarrow F$	0.60614
1.5666	$Na_2SiF_6 \leftrightarrow 6HF$	0.63831
1.3052	$Na_2SiF_6 \leftrightarrow H_2SiF_6$	0.76619
2.2394	$Na_2SiF_6 \leftrightarrow 2NaF$	0.44654
1.3236	$Na_2SiF_6 \leftrightarrow SiF_6$	0.75550

GALLIUM
Ga = 69.72

Factor		Factor
1.3442	$Ga_2O_3 \leftrightarrow Ga$	0.74392
1.6898	$Ga_2S_3 \leftrightarrow Ga$	0.59178

GERMANIUM
Ge = 72.59

Factor		Factor
1.4408	$GeO_2 \leftrightarrow Ge$	0.69404
3.6476	$K_2GeF_6 \leftrightarrow Ge$	0.27415

GOLD
Au = 196.967

Factor		Factor
0.64936	$Au \leftrightarrow AuCl_3$	1.5400
0.47826	$Au \leftrightarrow HAuCl_4 \cdot 4H_2O$	2.0909
0.54995	$Au \leftrightarrow KAu(CN)_4 \cdot H_2O$	1.8183

HYDROGEN
H = 1.0079

Factor		Factor
8.9365	$H_2O \leftrightarrow H$	0.11190
7.9364	$O \leftrightarrow H$	0.12600
0.35607	$HSCN \leftrightarrow AgSCN$	2.8084
0.48586	$HSCN \leftrightarrow CuSCN$	2.0582
0.25317	$HSCN \leftrightarrow BaSO_1$	3.9499

INDIUM
In = 114.82

Factor		Factor
1.2090	$In_2O_3 \leftrightarrow In$	0.82711
1.4189	$In_2S_3 \leftrightarrow In$	0.70476

TABLE 11.17   Gravimetric Factors (*Continued*)

Factor		Factor
	IODINE	
	I = 126.904	
0.84333	Ag ↔ HI	1.1858
0.85004	Ag ↔ I	1.1764
1.1294	AgCl ↔ I	0.88543
1.8354	AgI ↔ HI	0.54483
1.8500	AgI ↔ I	0.54053
1.3423	AgI ↔ IO$_3$	0.74498
1.2298	AgI ↔ IO$_4$	0.81314
1.4066	AgI ↔ I$_2$O$_5$	0.71091
1.2836	AgI ↔ I$_2$O$_7$	0.77904
0.41592	Pd ↔ HI	2.4043
0.41921	Pd ↔ I	2.3854
1.4081	PdI$_2$ ↔ HI	0.71020
1.4192	PdI$_2$ ↔ I	0.70462
1.0297	PdI$_2$ ↔ IO$_3$	0.97113
0.94343	PdI$_2$ ↔ IO$_4$	1.0600
1.0791	PdI$_2$ ↔ I$_2$O$_5$	0.92671
0.98472	PdI$_2$ ↔ I$_2$O$_7$	1.0155
2.5899	TlI ↔ HI	0.38612
2.6105	TlI ↔ I	0.38307
1.8941	TlI ↔ IO$_3$	0.52797
1.7353	TlI ↔ IO$_4$	0.57627
1.9848	TlI ↔ I$_2$O$_5$	0.50383
1.8112	TlI ↔ I$_2$O$_7$	0.55211
	IRON	
	Fe = 55.847	
2.2598	Ag ↔ Fe$_7$(CN)$_{18}$ (Prussian blue)	0.44252
0.54503	CN ↔ Fe$_7$(CN)$_{18}$	1.8347
0.61256	CO$_2$ ↔ FeO	1.6325
0.37986	CO$_2$ ↔ FeCO$_3$	2.6326
0.49483	CO$_2$ ↔ Fe(HCO$_3$)$_2$	2.0209
0.31396	Fe ↔ Fe(HCO$_3$)$_2$	3.1851
0.44061	Fe ↔ FeCl$_2$	2.2696
0.77732	Fe ↔ FeO	1.2865
0.69944	Fe ↔ Fe$_2$O$_3$	1.4297
0.72359	Fe ↔ Fe$_3$O$_4$	1.3820
0.36763	Fe ↔ FeSO$_4$	2.7201
0.20087	Fe ↔ FeSO$_4$·7H$_2$O	4.9782
0.14242	Fe ↔ FeSO$_4$·(NH$_4$)$_2$SO$_4$·6H$_2$O	7.0217
0.62011	FeO ↔ FeCO$_3$	1.6126
0.40390	FeO ↔ Fe(HCO$_3$)$_2$	2.4759
0.89982	FeO ↔ Fe$_2$O$_3$	1.1113
0.49223	Fe$_2$O$_3$ ↔ FeCl$_2$	2.0316
0.68915	Fe$_2$O$_3$ ↔ FeCO$_3$	1.4511
0.44887	Fe$_2$O$_3$ ↔ Fe(HCO$_3$)$_2$	2.2278
0.33422	Fe$_2$O$_3$ ↔ Fe(HCO$_3$)$_3$	2.9920
1.1113	Fe$_2$O$_3$ ↔ FeO	0.89982
1.0345	Fe$_2$O$_3$ ↔ Fe$_3$O$_4$	0.96662
0.52941	Fe$_2$O$_3$ ↔ FePO$_4$	1.8889

TABLE 11.17 Gravimetric Factors (*Continued*)

Factor		Factor
0.52561	$Fe_2O_3 \leftrightarrow FeSO_4$	1.9026
0.28719	$Fe_2O_3 \leftrightarrow FeSO_4 \cdot 7H_2O$	3.4820
0.20361	$Fe_2O_3 \leftrightarrow FeSO_4 \cdot (NH_4)_2SO_4 \cdot 6H_2O$	4.9113
0.39934	$Fe_2O_3 \leftrightarrow Fe_2(SO_4)_3$	2.5041
2.7006	$FePO_4 \leftrightarrow Fe$	0.37029
2.0992	$FePO_4 \leftrightarrow FeO$	0.47637
1.5741	$FeS \leftrightarrow Fe$	0.63527
1.2236	$FeS \leftrightarrow FeO$	0.81726
1.1010	$FeS \leftrightarrow Fe_2O_3$	0.90825
0.79699	$Mg_2As_2O_7 \leftrightarrow FeAsO_4$	1.2547
1.1144	$SO_3 \leftrightarrow FeO$	0.89738
0.52704	$SO_3 \leftrightarrow FeSO_4$	1.8974

LANTHANUM
La = 138.91

Factor		Factor
1.1728	$La_2O_3 \leftrightarrow La$	0.85268

LEAD
Pb = 207.2

Factor		Factor
0.77541	$Pb \leftrightarrow PbCO_3$	1.2896
0.80141	$Pb \leftrightarrow (PbCO_3)_2 \cdot Pb(OH)_2$	1.2478
0.85901	$Pb \leftrightarrow Pb(OH)_2$	1.1641
0.92831	$Pb \leftrightarrow PbO$	1.0772
1.3422	$PbCl_2 \leftrightarrow Pb$	0.74502
1.2460	$PbCl_2 \leftrightarrow PbO$	0.80255
1.5598	$PbCrO_4 \leftrightarrow Pb$	0.64110
0.85198	$PbCrO_4 \leftrightarrow Pb(C_2H_3O_2)_2 \cdot 3H_2O$	1.1737
1.2501	$PbCrO_4 \leftrightarrow (PbCO_3)_2 \cdot Pb(OH)_2$	0.79997
1.4480	$PbCrO_4 \leftrightarrow PbO$	0.69061
1.4142	$PbCrO_4 \leftrightarrow Pb_3O_4$	0.70711
1.0657	$PbCrO_4 \leftrightarrow PbSO_4$	0.93833
0.83529	$PbO \leftrightarrow PbCO_3$	1.1972
0.67388	$PbO \leftrightarrow Pb(NO_3)_2$	1.4839
0.93311	$PbO \leftrightarrow PbO_2$	1.0717
1.1544	$PbO_2 \leftrightarrow Pb$	0.86622
0.72219	$PbO_2 \leftrightarrow Pb(NO_3)_2$	1.3847
1.1547	$PbS \leftrightarrow Pb$	0.86600
1.0720	$PbS \leftrightarrow PbO$	0.93287
0.78895	$PbS \leftrightarrow PbSO_4$	1.2675
1.2993	$PbSO_4 \leftrightarrow BaSO_4$	0.76966
1.4636	$PbSO_4 \leftrightarrow Pb$	0.68323
0.79944	$PbSO_4 \leftrightarrow Pb(C_2H_3O_2)_2 \cdot 3H_2O$	1.2509
1.1349	$PbSO_4 \leftrightarrow PbCO_3$	0.88112
1.1730	$PbSO_4 \leftrightarrow (PbCO_3)_2 \cdot Pb(OH)_2$	0.85254
0.91561	$PbSO_4 \leftrightarrow Pb(NO_3)_2$	1.0922
1.3587	$PbSO_4 \leftrightarrow PbO$	0.73599
1.2678	$PbSO_4 \leftrightarrow PbO_2$	0.78875
1.3270	$PbSO_4 \leftrightarrow Pb_3O_4$	0.75358

TABLE 11.17 Gravimetric Factors (*Continued*)

Factor		Factor
	LITHIUM	
	Li = 6.941	
0.59562	$CO_2 \leftrightarrow Li_2CO_3$	1.6789
0.64759	$CO_2 \leftrightarrow LiHCO_3$	1.5442
1.4729	$CO_2 \leftrightarrow Li_2O$	0.67894
6.1086	$LiCl \leftrightarrow Li$	0.16369
2.8378	$LiCl \leftrightarrow Li_2O$	0.35239
5.3228	$Li_2CO_3 \leftrightarrow Li$	0.18787
0.87147	$Li_2CO_3 \leftrightarrow LiCl$	1.1475
0.54364	$Li_2CO_3 \leftrightarrow LiHCO_3$	1.8395
2.4730	$Li_2CO_3 \leftrightarrow Li_2O$	0.40436
4.5491	$LiHCO_3 \leftrightarrow Li_2O$	0.21983
3.7371	$LiF \leftrightarrow Li$	0.26759
2.1525	$Li_2O \leftrightarrow Li$	0.46457
0.27176	$Li_2O \leftrightarrow Li_2SO_4$	3.6798
5.5609	$Li_2PO_4 \leftrightarrow Li$	0.17983
0.91047	$Li_3PO_4 \leftrightarrow LiCl$	1.0983
1.0447	$Li_3PO_4 \leftrightarrow Li_2CO_3$	0.95717
0.56797	$Li_3PO_4 \leftrightarrow LiHCO_3$	1.7607
2.5837	$Li_3PO_4 \leftrightarrow Li_2O$	0.38704
0.70214	$Li_3PO_4 \leftrightarrow Li_2SO_4$	1.4242
0.60331	$Li_3PO_4 \leftrightarrow Li_2SO_4 \cdot H_2O$	1.6575
7.9153	$Li_2SO_4 \leftrightarrow Li$	0.12634
1.2967	$Li_2SO_4 \leftrightarrow LiCl$	0.77118
2.6797	$SO_3 \leftrightarrow Li_2O$	0.37317
0.72823	$SO_3 \leftrightarrow Li_2SO_4$	1.3732
	MAGNESIUM	
	Mg = 24.305	
1.9390	$BaSO_4 \leftrightarrow MgSO_4$	0.51572
0.94693	$BaSO_4 \leftrightarrow MgSO_4 \cdot 7H_2O$	1.0560
6.5755	$Br \leftrightarrow Mg$	0.15208
0.86800	$Br \leftrightarrow MgBr_2$	1.1521
0.54691	$Br \leftrightarrow MgBr_2 \cdot 6H_2O$	1.8285
2.9173	$Cl \leftrightarrow Mg$	0.34278
0.74472	$Cl \leftrightarrow MgCl_2$	1.3429
0.25533	$Mg \leftrightarrow MgCl_2$	3.9165
0.28883	$Mg \leftrightarrow MgCO_3$	3.4683
10.4427	$I \leftrightarrow Mg$	0.095761
0.91261	$I \leftrightarrow MgI_2$	1.09576
0.34876	$Cl \leftrightarrow MgCl_2 \cdot 6H_2O$	2.8673
0.52193	$CO_2 \leftrightarrow MgCO_3$	1.9160
1.0918	$CO_2 \leftrightarrow MgO$	0.91595
0.57616	$MgCO_3 \leftrightarrow Mg(HCO_3)_2$	1.7356
10.094	$MgNH_4PO_4 \cdot 6H_2O \leftrightarrow Mg$	0.099067
6.0879	$MgNH_4PO_4 \cdot 6H_2O \leftrightarrow MgO$	0.16426
1.6581	$MgO \leftrightarrow Mg$	0.60311
0.47807	$MgO \leftrightarrow MgCO_3$	2.0918
0.27544	$MgO \leftrightarrow Mg(HCO_3)_2$	3.6305
0.33489	$MgO \leftrightarrow MgSO_4$ '	2.9860
4.5784	$Mg_2P_2O_7 \leftrightarrow Mg$	0.21841

TABLE 11.17 Gravimetric Factors (*Continued*)

Factor		Factor
1.1687	$Mg_2P_2O_7 \leftrightarrow MgCl_2$	0.85562
0.54737	$Mg_2P_2O_7 \leftrightarrow MgCl_2 \cdot 6H_2O$	1.8269
0.40049	$Mg_2P_2O_7 \leftrightarrow MgCl_2 \cdot KCl \cdot 6H_2O$	2.4969
1.3198	$Mg_2P_2O_7 \leftrightarrow MgCO_3$	0.75770
0.76040	$Mg_2P_2O_7 \leftrightarrow Mg(HCO_3)_2$	1.3151
2.7607	$Mg_2P_2O_7 \leftrightarrow MgO$	0.36223
0.92452	$Mg_2P_2O_7 \leftrightarrow MgSO_4$	1.0816
0.45150	$Mg_2P_2O_7 \leftrightarrow MgSO_4 \cdot 7H_2O$	2.2149
4.9523	$MgSO_4 \leftrightarrow Mg$	0.20193
1.9864	$SO_3 \leftrightarrow MgO$	0.50343
0.6651	$SO_3 \leftrightarrow MgSO_4$	1.5034
0.38482	$SO_3 \leftrightarrow MgSO_4 \cdot 7H_2O$	3.0786

MANGANESE
Mn = 54.9380

Factor		Factor
1.5457	$BaSO_4 \leftrightarrow MnSO_4$	0.64696
0.38286	$CO_2 \leftrightarrow MnCO_3$	2.6119
0.62041	$CO_2 \leftrightarrow MnO$	1.6118
0.47793	$Mn \leftrightarrow MnCO_3$	2.0924
0.77446	$Mn \leftrightarrow MnO$	1.2912
0.63193	$Mn \leftrightarrow MnO_2$	1.5825
0.69599	$Mn \leftrightarrow Mn_2O_3$	1.4368
0.76126	$MnCO_3 \leftrightarrow MnSO_4$	1.3136
1.5395	$Mn(HCO_3)_2 \leftrightarrow MnCO_3$	0.64955
0.61711	$MnO \leftrightarrow MnCO_3$	1.6205
0.40084	$MnO \leftrightarrow Mn(HCO_3)_2$	2.4947
0.89868	$MnO \leftrightarrow Mn_2O_3$	1.1127
0.46978	$MnO \leftrightarrow MnSO_4$	2.1286
1.3883	$Mn_3O_4 \leftrightarrow Mn$	0.72031
0.66351	$Mn_3O_4 \leftrightarrow MnCO_3$	1.5071
0.43098	$Mn_3O_4 \leftrightarrow Mn(HCO_3)_2$	2.3203
1.0752	$Mn_3O_4 \leftrightarrow MnO$	0.93008
0.96625	$Mn_3O_4 \leftrightarrow Mn_2O_3$	1.0349
0.87731	$Mn_3O_4 \leftrightarrow MnO_2$	1.1399
0.50510	$Mn_3O_4 \leftrightarrow MnSO_4$	1.9798
2.5831	$Mn_2P_2O_7 \leftrightarrow Mn$	0.38713
1.2345	$Mn_2P_2O_7 \leftrightarrow MnCO_3$	0.81002
2.0005	$Mn_2P_2O_7 \leftrightarrow MnO$	0.49987
1.6324	$Mn_2P_2O_7 \leftrightarrow MnO_2$	0.61261
0.93980	$Mn_2P_2O_7 \leftrightarrow MnSO_4$	1.0641
1.5836	$MnS \leftrightarrow Mn$	0.63146
0.75687	$MnS \leftrightarrow MnCO_3$	1.3212
1.2265	$MnS \leftrightarrow MnO$	0.81535
0.57617	$MnS \leftrightarrow MnSO_4$	1.7356
2.7486	$MnSO_4 \leftrightarrow Mn$	0.36383
1.1286	$SO_3 \leftrightarrow MnO$	0.88603
0.53021	$SO_3 \leftrightarrow MnSO_4$	1.8860

MERCURY
Hg = 200.59

Factor		Factor
0.73882	$Hg \leftrightarrow HgCl_2$	1.3535
0.92613	$Hg \leftrightarrow HgO$	1.0798

TABLE 11.17 Gravimetric Factors (*Continued*)

Factor		Factor
0.86220	Hg ↔ HgS	1.1598
1.1767	HgCl ↔ Hg	0.84981
0.86939	HgCl ↔ HgCl$_2$	1.1502
0.89889	HgCl ↔ HgNO$_3$	1.1125
1.1316	HgCl ↔ Hg$_2$O	0.88371
1.0898	HgCl ↔ HgO	0.91760
1.0146	HgCl ↔ HgS	0.98564
0.98564	HgS ↔ HgCl	1.0146
0.85691	HgS ↔ HgCl$_2$	1.1670
0.92091	HgS ↔ Hg(CN)$_2$	1.0859
0.88598	HgS ↔ HgNO$_3$	1.1287
0.71673	HgS ↔ Hg(NO$_3$)$_2$	1.3952
0.67903	HgS ↔ Hg(NO$_3$)$_2$·H$_2$O	1.4727
1.1153	HgS ↔ Hg$_2$O	0.89658
1.0741	HgS ↔ HgO	0.93097
0.78426	HgS ↔ HgSO$_4$	1.2751

MOLYBDENUM
Mo = 95.94

Factor		Factor
8.9876	MoC ↔ C	0.11126
1.5003	MoO$_3$ ↔ Mo	0.66653
0.73436	MoO$_3$ ↔ (NH$_4$)$_2$MoO$_4$	1.3617
2.0026	MoS$_3$ ↔ Mo	0.49935
1.3348	MoS$_4$ ↔ MoO$_3$	0.74918
0.98021	MoS$_3$ ↔ (NH$_4$)$_2$MoO$_4$	1.0202
1.0863	(NH$_4$)$_3$PO$_4$·12MoO$_3$ ↔ MoO$_3$	0.92058
0.79771	(NH$_4$)$_3$PO$_4$·12MoO$_3$ ↔ (NH$_4$)$_2$MoO$_4$	1.2536
3.8267	PbMoO$_4$ ↔ Mo	0.26132
2.5506	PbMoO$_4$ ↔ MoO$_3$	0.39207
1.8730	PbMoO$_4$ ↔ (NH$_4$)$_2$MoO$_4$	0.53390

NEODYMIUM
Nd = 144.24

Factor		Factor
1.1664	Nd$_2$O$_3$ ↔ Nd	0.85735

NICKEL
Ni = 58.71

Factor		Factor
0.20319	Ni ↔ Ni dimethylglyoxime	4.9215
0.20188	Ni ↔ Ni(NO$_3$)$_2$·6H$_2$O	4.9533
0.78585	Ni ↔ NiO	1.2725
0.20902	Ni ↔ NiSO$_4$·7H$_2$O	4.7842
3.8675	Ni dimethylglyoxime ↔ NiO	0.25856
0.25690	NiO ↔ Ni(NO$_3$)$_2$·6H$_2$O	3.8926
0.26598	NiO ↔ NiSO$_4$·7H$_2$O	3.7597
2.6362	NiSO$_4$ ↔ Ni	0.37934
0.53220	NiSO$_4$ ↔ Ni(NO$_3$)$_2$·6H$_2$O	1.8790
2.0716	NiSO$_4$ ↔ NiO	0.48271
0.55102	NiSO$_4$ ↔ NiSO$_4$·7H$_2$O	1.8148

TABLE 11.17   Gravimetric Factors (*Continued*)

Factor		Factor
	NIOBIUM	
	Nb = 92.906	
7.7351	Nb ↔ C	0.12928
8.7353	NbC ↔ C	0.11448
11.065	Nb$_2$O$_5$ ↔ 2C	0.090373
1.4305	Nb$_2$O$_5$ ↔ Nb	0.69904
	NITROGEN	
	N = 14.0067	
3.2731	AgNO$_2$ ↔ HNO$_2$	0.30552
4.0488	AgNO$_2$ ↔ N$_2$O$_3$	0.24698
1.8722	KNO$_3$ ↔ N$_2$O$_5$	0.53412
0.22229	N ↔ HNO$_3$	4.4987
0.30446	N ↔ NO$_2$	3.2845
0.36855	N ↔ N$_2$O$_3$	2.7134
0.22590	N ↔ NO$_3$	4.4268
0.25936	N ↔ N$_2$O$_5$	3.8556
6.0680	NaNO$_3$ ↔ N	0.16480
1.5738	NaNO$_3$ ↔ N$_2$O$_5$	0.63539
0.47619	NO ↔ HNO$_3$	2.1000
0.65222	NO ↔ NO$_2$	1.5332
0.78951	NO ↔ N$_2$O$_3$	1.2666
0.48393	NO ↔ NO$_3$	2.0664
0.55561	NO ↔ N$_2$O$_5$	1.7998
0.27028	NH$_3$ ↔ HNO$_3$	3.6999
1.2159	NH$_3$ ↔ N	0.82244
0.31536	NH$_3$ ↔ N$_2$O$_5$	3.1710
0.27467	NH$_3$ ↔ NO$_3$	3.6407
0.84890	NH$_4$Cl ↔ HNO$_3$	1.1780
0.86270	NH$_4$Cl ↔ NO$_3$	1.1591
0.99050	NH$_4$Cl ↔ N$_2$O$_5$	1.0096
3.8189	NH$_4$Cl ↔ N	0.26185
3.5221	(NH$_4$)$_2$PtCl$_6$ ↔ HNO$_3$	0.28393
15.845	(NH$_4$)$_2$PtCl$_6$ ↔ N	0.063112
4.1096	(NH$_4$)$_2$PtCl$_6$ ↔ N$_2$O$_6$	0.24333
3.5794	(NH$_4$)$_2$PtCl$_6$ ↔ NO$_3$	0.27938
4.7169	(NH$_4$)$_2$SO$_4$ ↔ N	0.21200
1.2234	(NH$_4$)$_2$SO$_4$ ↔ N$_2$O$_5$	0.81739
1.5480	Pt ↔ HNO$_3$	0.64599
6.9640	Pt ↔ N	0.14360
1.5732	Pt ↔ NO$_3$	0.63566
1.8062	Pt ↔ N$_2$O$_5$	0.55364
0.63528	SO$_3$ ↔ HNO$_3$	1.5741
2.8579	SO$_3$ ↔ N	0.34990
0.74125	SO$_3$ ↔ N$_2$O$_5$	1.3491
	OSMIUM	
	Os = 190.2	
1.3365	OsO$_4$ ↔ Os	0.74823

TABLE 11.17 Gravimetric Factors (*Continued*)

Factor		Factor
	PALLADIUM	
	Pd = 106.4	
0.49873	Pd ↔ $PdCl_2 \cdot 2H_2O$	2.0051
0.46179	Pd ↔ $Pd(NO_3)_2$	2.1655
3.3854	PdI_2 ↔ Pd	0.29538
3.7342	K_2PdCl_6 ↔ Pd	0.26779
1.8624	K_2PdCl_6 ↔ $PdCl_2 \cdot 2H_2O$	0.53695
	PHOSPHORUS	
	P = 30.9738	
13.514	Ag_3PO_4 ↔ P	0.073998
4.4075	Ag_3PO_4 ↔ PO_4	0.22689
5.8980	Ag_3PO_4 ↔ P_2O_5	0.16955
9.7730	$Ag_4P_2O_7$ ↔ P	0.10232
3.1874	$Ag_4P_2O_7$ ↔ PO_4	0.31374
4.2653	$Ag_4P_2O_7$ ↔ P_2O_5	0.23445
0.71833	Al_2O_3 ↔ P_2O_5	1.3921
1.2841	$AlPO_4$ ↔ PO_4	0.77877
1.7183	$AlPO_4$ ↔ P_2O_5	0.58196
2.1853	$Ca_3(PO_4)_2$ ↔ P_2O_5	0.45761
1.5881	$FePO_4$ ↔ PO_4	0.62970
2.1251	$FePO_4$ ↔ P_2O_5	0.47056
0.78392	$Mg_2P_2O_7$ ↔ Na_2HPO_4	1.2756
0.31073	$Mg_2P_2O_7$ ↔ $Na_2HPO_4 \cdot 12H_2O$	3.2182
0.53229	$Mg_2P_2O_7$ ↔ $NaNH_4HPO_4 \cdot 4H_2O$	1.8787
3.5929	$Mg_2P_2O_7$ ↔ P	0.27833
1.1718	$Mg_2P_2O_7$ ↔ PO_4	0.85340
1.5681	$Mg_2P_2O_7$ ↔ P_2O_5	0.63773
60.577	$(NH_4)_3PO_4 \cdot 12MoO_3$ ↔ P	0.016508
19.757	$(NH_4)_3PO_4 \cdot 12MoO_3$ ↔ PO_4	0.050616
26.438	$(NH_4)_3PO_4 \cdot 12MoO_3$ ↔ P_2O_5	0.037824
0.63773	P_2O_5 ↔ $Mg_2P_2O_7$	1.5681
0.49993	P_2O_5 ↔ Na_2HPO_4	2.0003
0.19816	P_2O_5 ↔ $Na_2HPO_4 \cdot 12H_2O$	5.0464
0.33946	P_2O_5 ↔ $NaNH_4HPO_4 \cdot 4H_2O$	2.9459
2.2913	P_2O_5 ↔ P	0.43644
58.057	$P_2O_5 \cdot 24MoO_3$ ↔ P	0.017225
18.935	$P_2O_5 \cdot 24MoO_3$ ↔ PO_4	0.052813
25.338	$P_2O_5 \cdot 24MoO_3$ ↔ P_2O_5	0.039466
11.526	$U_2P_2O_{11}$ ↔ P	0.086762
3.7590	$U_2P_2O_{11}$ ↔ PO_4	0.26603
5.0303	$U_2P_2O_{11}$ ↔ P_2O_5	0.19880
	PLATINUM	
	Pt = 195.09	
0.93839	K_2PtCl_6 ↔ $H_2PtCl_6 \cdot 6H_2O$	1.0657
2.4912	K_2PtCl_6 ↔ Pt	0.40141
1.4426	K_2PtCl_6 ↔ $PtCl_4$	0.69320
1.1383	K_2PtCl_6 ↔ $PtCl_4 \cdot 5H_2O$	0.87854
2.2753	$(NH_4)_2PtCl_6$ ↔ Pt	0.43950

TABLE 11.17 Gravimetric Factors (*Continued*)

Factor		Factor
1.3176	$(NH_4)_2PtCl_6 \leftrightarrow PtCl_4$	0.75897
1.0885	$(NH_4)_2PtCl_6 \leftrightarrow PtCl_6$	0.91872
0.37668	$Pt \leftrightarrow H_2PtCl_6 \cdot 6H_2O$	2.6548
0.57907	$Pt \leftrightarrow PtCl_4$	1.7269
0.45691	$Pt \leftrightarrow PtCl_4 \cdot 5H_2O$	2.1886

POTASSIUM
K = 39.098

Factor		Factor
0.90639	$Ag \leftrightarrow KBr$	1.1033
1.4469	$Ag \leftrightarrow KCl$	0.69116
0.88021	$Ag \leftrightarrow KClO_3$	1.1361
0.77856	$Ag \leftrightarrow KClO_4$	1.2844
1.6565	$Ag \leftrightarrow KCN$	0.60369
0.64978	$Ag \leftrightarrow KI$	1.5390
1.5779	$AgBr \leftrightarrow KBr$	0.63377
1.1244	$AgBr \leftrightarrow KBrO_3$	0.88939
1.9223	$AgCl \leftrightarrow KCl$	0.52020
1.1695	$AgCl \leftrightarrow KClO_3$	0.85508
1.0344	$AgCl \leftrightarrow KClO_4$	0.96672
2.0561	$AgCN \leftrightarrow KCN$	0.48637
1.4142	$AgI \leftrightarrow KI$	0.70712
1.0971	$AgI \leftrightarrow KIO_3$	0.91153
1.3045	$BaCrO_4 \leftrightarrow K_2CrO_4$	0.76659
1.7222	$BaCrO_4 \leftrightarrow K_2Cr_2O_7$	0.58065
1.7140	$BaSO_4 \leftrightarrow KHSO_4$	0.58342
2.1166	$BaSO_4 \leftrightarrow K_2S$	0.47245
1.3393	$BaSO_4 \leftrightarrow K_2SO_4$	0.74666
2.0436	$Br \leftrightarrow K$	0.48933
0.67145	$Br \leftrightarrow KBr$	1.4893
0.41473	$CaF_2 \leftrightarrow KF \cdot 2H_2O$	2.4112
0.72315	$CaSO_4 \leftrightarrow KF \cdot 2H_2O$	1.3828
0.90668	$Cl \leftrightarrow K$	1.1029
0.47553	$Cl \leftrightarrow KCl$	2.1029
0.28929	$Cl \leftrightarrow KClO_3$	3.4567
0.25589	$Cl \leftrightarrow KClO_4$	3.9080
0.75269	$Cl \leftrightarrow K_2O$	1.3286
0.46718	$CO_2 \leftrightarrow K_2O$	2.1405
0.31843	$CO_2 \leftrightarrow K_2CO_3$	3.1404
0.76441	$I \leftrightarrow KI$	1.3082
0.59299	$I \leftrightarrow KIO_3$	1.6864
0.31907	$K \leftrightarrow KClO_3$	3.1341
0.83016	$K \leftrightarrow K_2O$	1.2046
0.38673	$K \leftrightarrow KNO_3$	2.5858
3.0436	$KBr \leftrightarrow K$	0.32856
2.5267	$KBr \leftrightarrow K_2O$	0.39578
1.9067	$KCl \leftrightarrow K$	0.52447
1.0789	$KCl \leftrightarrow K_2CO_3$	0.92690
0.50685	$KCl \leftrightarrow K_2Cr_2O_7$	1.9730
0.74466	$KCl \leftrightarrow KHCO_3$	1.3429
0.73737	$KCl \leftrightarrow KNO_3$	1.3562
1.5829	$KCl \leftrightarrow K_2O$	0.63177
0.85563	$KCl \leftrightarrow K_2SO_4$	1.1687

TABLE 11.17 Gravimetric Factors (*Continued*)

Factor		Factor
1.6437	KClO$_3$ ↔ KCl	0.60836
3.5433	KClO$_4$ ↔ K	0.28222
1.8584	KClO$_4$ ↔ KCl	0.53811
2.9415	KClO$_4$ ↔ K$_2$O	0.33996
4.2456	KI ↔ K	0.23554
3.5245	KI ↔ K$_2$O	0.28373
0.38435	K$_2$O ↔ KClO$_3$	2.6018
0.68159	K$_2$O ↔ K$_2$CO$_3$	1.4672
0.32021	K$_2$O ↔ K$_2$Cr$_2$O$_7$	3.1229
0.47045	K$_2$O ↔ KHCO$_3$	2.1256
0.46584	K$_2$O ↔ KNO$_3$	2.1466
0.81194	KOH ↔ K$_2$CO$_3$	1.2316
1.1912	KOH ↔ K$_2$O	0.83946
6.2146	K$_2$PtCl$_6$ ↔ K	0.16091
3.5165	K$_2$PtCl$_6$ ↔ K$_2$CO$_3$	0.28438
3.2594	K$_2$PtCl$_6$ ↔ KCl	0.30680
2.4271	K$_2$PtCl$_6$ ↔ KHCO$_3$	0.41201
2.4034	K$_2$PtCl$_6$ ↔ KNO$_3$	0.41608
5.1592	K$_2$PtCl$_6$ ↔ K$_2$O	0.19383
2.7888	K$_2$PtCl$_6$ ↔ K$_2$SO$_4$	0.35857
0.51224	K$_2$PtCl$_6$ ↔ K$_2$SO$_4$·Al$_2$(SO$_4$)$_3$·24H$_2$O	1.9522
0.48659	K$_2$PtCl$_6$ ↔ K$_2$SO$_4$·Cr$_2$(SO$_4$)$_3$·24H$_2$O	2.0551
1.2609	K$_2$SO$_4$ ↔ K$_2$CO$_3$	0.79308
0.87031	K$_2$SO$_4$ ↔ KHCO$_3$	1.1490
0.63990	K$_2$SO$_4$ ↔ KHSO$_4$	1.5627
1.0238	K$_2$SO$_4$ ↔ KNO$_2$	0.97674
0.86179	K$_2$SO$_4$ ↔ KNO$_3$	1.1604
2.2285	K$_2$SO$_4$ ↔ K	0.44875
1.8499	K$_2$SO$_4$ ↔ K$_2$O	0.54056
1.5804	K$_2$SO$_4$ ↔ K$_2$S	0.63275
0.60582	Mg$_2$As$_2$O$_7$ ↔ K$_3$AsO$_4$	1.6506
0.71164	Mg$_2$As$_2$O$_7$ ↔ K$_2$HAsO$_4$	1.4052
0.40040	Mn$_2$O$_3$ ↔ K$_2$MnO$_4$	2.4975
0.49946	Mn$_2$O$_3$ ↔ KMnO$_4$	2.0022
0.44132	MnS ↔ K$_2$MnO$_4$	2.2659
0.55051	MnS ↔ KMnO$_4$	1.8165
0.13853	N ↔ KNO$_3$	7.2185
0.16844	NH$_3$ ↔ KNO$_3$	5.9368
0.29677	NO ↔ KNO$_3$	3.3697
0.44656	N$_2$O$_3$ ↔ KNO$_2$	2.2393
1.1466	N$_2$O$_5$ ↔ K$_2$O	0.87217
0.53412	N$_2$O$_5$ ↔ KNO$_3$	1.8722
2.4946	Pt ↔ K	0.40086
1.3084	Pt ↔ KCl	0.76431
2.0710	Pt ↔ K$_2$O	0.48287
0.38943	SiO$_2$ ↔ K$_2$SiO$_3$	2.5679
0.45941	SO$_3$ ↔ K$_2$SO$_4$	2.1767

PRASEODYMIUM
Pr = 140.908

1.1703	Pr$_2$O$_3$ ↔ Pr	0.85449

TABLE 11.17 Gravimetric Factors (*Continued*)

Factor		Factor
	RHODIUM	
	Rh = 102.905	
0.26758	Rh ↔ Na_3RhCl_6	3.7372
0.49178	Rh ↔ $RhCl_3$	2.0334
	RUBIDIUM	
	Rb = 85.468	
1.6768	AgCl ↔ Rb	0.59636
1.1852	AgCl ↔ RbCl	0.84371
0.41480	Cl ↔ Rb	2.4108
0.29319	Cl ↔ RbCl	3.4107
0.70683	Rb ↔ RbCl	1.4148
0.74016	Rb ↔ Rb_2CO_3	1.3511
0.91441	Rb ↔ Rb_2O	1.0936
0.64023	Rb ↔ Rb_2SO_4	1.5620
1.0472	RbCl ↔ Rb_2CO_3	0.95497
0.90577	RbCl ↔ Rb_2SO_4	1.1040
2.1636	$RbClO_4$ ↔ Rb	0.46220
0.78828	Rb_2CO_3 ↔ $RbHCO_3$	1.2686
0.77299	Rb_2O ↔ RbCl	1.2937
0.70015	Rb_2O ↔ Rb_2SO_4	1.4283
3.3857	Rb_2PtCl_6 ↔ Rb	0.29536
2.3931	Rb_2PtCl_6 ↔ RbCl	0.41787
2.5060	Rb_2PtCl_6 ↔ Rb_2CO_3	0.39905
1.9754	Rb_2PtCl_6 ↔ $RbHCO_3$	0.50623
3.0959	Rb_2PtCl_6 ↔ Rb_2O	0.32301
1.1561	Rb_2SO_4 ↔ Rb_2CO_3	0.86498
0.91133	Rb_2SO_4 ↔ $RbHCO_3$	1.0973
	SELENIUM	
	Se = 78.96	
0.61224	Se ↔ H_2SeO_3	1.6334
0.54466	Se ↔ H_2SeO_4	1.8360
0.71161	Se ↔ SeO_2	1.4053
0.62193	Se ↔ SeO_3	1.6079
	SILICON	
	Si = 28.086	
2.6847	$BaSiF_6$ ↔ SiF_4	0.37249
4.6504	$BaSiF_6$ ↔ SiO_2	0.21503
2.1163	K_2SiF_6 ↔ SiF_4	0.47249
3.6661	K_2SiF_6 ↔ SiO_2	0.27277
3.3384	SiC ↔ C	0.29954
0.91111	SiC ↔ CO_2	1.0976
0.76933	SiO_2 ↔ H_2SiO_3	1.2998
2.1393	SiO_2 ↔ Si	0.46744
0.57730	SiO_2 ↔ SiF_4	1.7322
0.78972	SiO_2 ↔ SiO_3	1.2663
0.65250	SiO_2 ↔ SiO_4	1.5326

TABLE 11.17 Gravimetric Factors (*Continued*)

Factor		Factor
1.6651	SiO$_2$ ↔ Si$_2$O	0.60057
0.62514	SiO$_2$ ↔ Si(OH)$_4$	1.5997

SILVER
Ag = 107.868

0.63501	Ag ↔ AgNO$_3$	1.5748
0.93096	Ag ↔ Ag$_2$O	1.0742
1.7408	AgBr ↔ Ag	0.57445
1.3286	AgCl ↔ Ag	0.75265
0.84371	AgCl ↔ AgNO$_3$	1.1852
1.2369	AgCl ↔ Ag$_2$O	0.80847
1.7935	AgCl ↔ Br	0.55756
1.2412	AgCN ↔ Ag	0.80566
2.1764	AgI ↔ Ag	0.45947
1.2935	Ag$_3$PO$_4$ ↔ Ag	0.77311
1.4031	Ag$_4$P$_2$O$_7$ ↔ Ag	0.71269
0.74079	Br ↔ Ag	1.3499
0.42555	Br ↔ AgBr	2.3499
0.32866	Cl ↔ Ag	3.0426
0.24737	Cl ↔ AgCl	4.0425
1.1764	I ↔ Ag	0.85004
0.54053	I ↔ AgI	1.8500

SODIUM
Na = 22.9898

1.0483	Ag ↔ NaBr	0.95393
1.8457	Ag ↔ NaCl	0.54179
0.71966	Ag ↔ NaI	1.3895
1.8249	AgBr ↔ NaBr	0.54798
2.4523	AgCl ↔ NaCl	0.40778
1.5663	AgI ↔ NaI	0.63845
1.9440	BaSO$_4$ ↔ NaHSO$_4$	0.51440
1.6905	BaSO$_4$ ↔ NaHSO$_4$·H$_2$O	0.59156
2.9906	BaSO$_4$ ↔ Na$_2$S	0.33438
1.8518	BaSO$_4$ ↔ Na$_2$SO$_3$	0.54002
0.92564	BaSO$_4$ ↔ Na$_2$SO$_3$·7H$_2$O	1.0803
1.6432	BaSO$_4$ ↔ Na$_2$SO$_4$	0.60857
0.72442	BaSO$_4$ ↔ Na$_2$SO$_4$·10H$_2$O	1.3804
0.69198	B$_2$O$_3$ ↔ Na$_2$B$_4$O$_7$	1.4451
0.36510	B$_2$O$_3$ ↔ Na$_2$B$_4$O$_7$·10H$_2$O	2.7389
3.4758	Br ↔ Na	0.28770
0.77657	Br ↔ NaBr	1.2877
2.5786	Br ↔ Na$_2$O	0.38781
0.94956	CaCl$_2$ ↔ NaCl	1.0531
0.94433	CaCO$_3$ ↔ Na$_2$CO$_3$	1.0590
0.92975	CaF$_2$ ↔ NaF	1.0756
0.52910	CaO ↔ Na$_2$CO$_3$	1.8900
1.2845	CaSO$_4$ ↔ Na$_2$CO$_3$	0.77854
1.5421	Cl ↔ Na	0.64846
0.60663	Cl ↔ NaCl	1.6485

TABLE 11.17 Gravimetric Factors (*Continued*)

Factor		Factor
1.1442	$Cl \leftrightarrow Na_2O$	0.87410
0.41520	$CO_2 \leftrightarrow Na_2CO_3$	2.4083
0.71008	$CO_2 \leftrightarrow Na_2O$	1.4083
1.2292	$H_3BO_3 \leftrightarrow Na_2B_4O_7$	0.81357
0.64853	$H_3BO_3 \leftrightarrow Na_2B_4O_7 \cdot 10H_2O$	1.5419
5.5198	$I \leftrightarrow Na$	0.18117
0.84662	$I \leftrightarrow NaI$	1.1812
4.0949	$I \leftrightarrow Na_2O$	0.24420
2.5029	$KBF_4 \leftrightarrow Na_2B_4O_7$	0.39954
1.3206	$KBF_4 \leftrightarrow Na_2B_4O_7 \cdot 10H_2O$	0.75724
0.91360	$Mg_2As_2O_7 \leftrightarrow Na_2HAsO_3$	1.0946
0.83497	$Mg_2As_2O_7 \leftrightarrow Na_2HAsO_4$	1.1976
0.81462	$MgCl_2 \leftrightarrow NaCl$	1.2276
0.67882	$Mg_2P_2O_7 \leftrightarrow Na_3PO_4$	1.4731
0.78392	$Mg_2P_2O_7 \leftrightarrow Na_2HPO_4$	1.2757
0.31073	$Mg_2P_2O_7 \leftrightarrow NaHPO_4 \cdot 12H_2O$	3.2182
0.53229	$Mg_2P_2O_7 \leftrightarrow NaNH_4 \cdot HPO_4 \cdot 4H_2O$	1.8787
0.49897	$Mg_2P_2O_7 \leftrightarrow Na_4P_2O_7 \cdot 10H_2O$	2.0041
4.4759	$NaBr \leftrightarrow Na$	0.22342
3.3205	$NaBr \leftrightarrow Na_2O$	0.30116
65.502	$NaOAc \cdot Mg(OAc)_2 \cdot UO_2(OAc)_2 \cdot 6\frac{1}{2}H_2O \leftrightarrow Na$	0.015267
14.635	Triple $MgOAc \leftrightarrow NaBr$	0.066331
28.416	Triple $MgOAc \leftrightarrow Na_2CO_3$	0.035192
25.768	Triple $MgOAc \leftrightarrow NaCl$	0.038809
17.926	Triple $MgOAc \leftrightarrow NaHCO_3$	0.055785
10.047	Triple $MgOAc \leftrightarrow NaI$	0.099535
37.650	Triple $MgOAc \leftrightarrow NaOH$	0.026560
48.594	Triple $MgOAc \leftrightarrow Na_2O$	0.020579
21.204	Triple $MgOAc \leftrightarrow Na_2SO_4$	0.047161
66.894	$NaOAc \cdot Zn(OAc)_2 \cdot UO_2(OAc)_2 \cdot 6H_2O \leftrightarrow Na$	0.014949
14.946	Triple $ZnOAc \leftrightarrow NaBr$	0.066909
29.020	Triple $ZnOAc \leftrightarrow Na_2CO_3$	0.034459
26.315	Triple $ZnOAc \leftrightarrow NaCl$	0.038002
18.307	Triple $ZnOAc \leftrightarrow NaHCO_3$	0.054624
10.260	Triple $ZnOAc \leftrightarrow NaI$	0.097464
38.451	Triple $ZnOAc \leftrightarrow NaOH$	0.026008
49.626	Triple $ZnOAc \leftrightarrow Na_2O$	0.020151
21.654	Triple $ZnOAc \leftrightarrow Na_2SO_4$	0.046180
2.5421	$NaCl \leftrightarrow Na$	0.39337
1.1028	$NaCl \leftrightarrow Na_2CO_3$	0.90678
0.69569	$NaCl \leftrightarrow NaHCO_3$	1.4374
0.82337	$NaCl \leftrightarrow Na_2HPO_4$	1.2145
1.8859	$NaCl \leftrightarrow Na_2O$	0.53025
0.82291	$NaCl \leftrightarrow Na_2SO_4$	1.2152
0.74267	$NaClO_3 \leftrightarrow AgCl$	1.3465
1.8213	$NaClO_3 \leftrightarrow NaCl$	0.54907
0.85432	$NaClO_4 \leftrightarrow AgCl$	1.1705
2.0950	$NaClO_4 \leftrightarrow NaCl$	0.47732
2.3051	$Na_2CO_3 \leftrightarrow Na$	0.43381
0.63084	$Na_2CO_3 \leftrightarrow NaHCO_3$	1.5852
1.7101	$Na_2CO_3 \leftrightarrow Na_2O$	0.58476
1.3250	$Na_2CO_3 \leftrightarrow NaOH$	0.75473
3.6541	$NaHCO_3 \leftrightarrow Na$	0.27367

TABLE 11.17 Gravimetric Factors (*Continued*)

Factor		Factor
2.7108	$NaHCO_3 \leftrightarrow Na_2O$	0.36889
6.5198	$NaI \leftrightarrow Na$	0.15338
4.8368	$NaI \leftrightarrow Na_2O$	0.20675
1.3480	$Na_2O \leftrightarrow Na$	0.74186
0.43659	$Na_2O \leftrightarrow Na_2HPO_4$	2.2905
0.36460	$Na_2O \leftrightarrow NaNO_3$	2.7427
0.77480	$Na_2O \leftrightarrow NaOH$	1.2907
0.93653	$Na_4P_2O_7 \leftrightarrow Na_2HPO_4$	1.0678
0.37122	$Na_4P_2O_7 \leftrightarrow Na_2HPO_4 \cdot 12H_2O$	2.6938
3.0892	$Na_2SO_4 \leftrightarrow Na$	0.32371
1.3401	$Na_2SO_4 \leftrightarrow Na_2CO_3$	0.74620
0.49640	$Na_2SO_4 \leftrightarrow Na_2CO_3 \cdot 10H_2O$	2.0145
2.2917	$Na_2SO_4 \leftrightarrow Na_2O$	0.43635
0.16480	$N \leftrightarrow NaNO_3$	6.0680
0.20038	$NH_3 \leftrightarrow NaNO_3$	4.9906
0.081461	$NH_3 \leftrightarrow NaNH_4HPO_4 \cdot 4H_2O$	12.276
0.35303	$NO \leftrightarrow NaNO_3$	2.8326
0.63539	$N_2O_5 \leftrightarrow NaNO_3$	1.5738
1.7427	$N_2O_5 \leftrightarrow Na_2O$	0.57383
0.49993	$P_2O_5 \leftrightarrow Na_2HPO_4$	2.0003
0.19816	$P_2O_5 \leftrightarrow Na_2HPO_4 \cdot 12H_2O$	5.0464
0.33946	$P_2O_5 \leftrightarrow NaNH_4HPO_4 \cdot H_2O$	2.9459
0.61564	$SO_2 \leftrightarrow NaHSO_3$	1.6243
0.50828	$SO_2 \leftrightarrow Na_2SO_3$	1.9674
0.25407	$SO_2 \leftrightarrow Na_2SO_3 \cdot 7H_2O$	3.9360
1.2918	$SO_2 \leftrightarrow Na_2O$	0.77414
0.56366	$SO_2 \leftrightarrow Na_2SO_4$	1.7741

STRONTIUM
Sr = 87.62

Factor		Factor
0.29811	$CO_2 \leftrightarrow SrCO_8$	3.3545
0.77265	$SO_3 \leftrightarrow SrO$	1.2942
0.43588	$SO_3 \leftrightarrow SrSO_4$	2.2942
0.41402	$Sr \leftrightarrow Sr(NO_3)_2$	2.4153
1.6849	$SrCO_3 \leftrightarrow Sr$	0.59351
0.93124	$SrCO_3 \leftrightarrow SrCl_2$	1.0738
0.70424	$SrCO_3 \leftrightarrow Sr(HCO_3)_2$	1.4200
0.69759	$SrCO_3 \leftrightarrow Sr(NO_3)_2$	1.4335
1.1826	$SrO \leftrightarrow Sr$	0.84559
0.65363	$SrO \leftrightarrow SrCl_2$	1.5299
0.70189	$SrO \leftrightarrow SrCO_3$	1.4247
0.49430	$SrO \leftrightarrow Sr(HCO_3)_2$	2.0231
0.48963	$SrO \leftrightarrow Sr(NO_3)_2$	2.0424
2.0963	$SrSO_4 \leftrightarrow Sr$	0.47703
1.1586	$SrSO_4 \leftrightarrow SrCl_2$	0.86308
1.2442	$SrSO_4 \leftrightarrow SrCO_3$	0.80373
0.86793	$SrSO_4 \leftrightarrow Sr(NO_3)_2$	1.1522
1.7726	$SrSO_4 \leftrightarrow SrO$	0.56413

TABLE 11.17   Gravimetric Factors (*Continued*)

Factor		Factor
	SULFUR	
	S = 32.06	
2.4064	$As_2S_3 \leftrightarrow H_2S$	0.41556
2.5577	$As_2S_3 \leftrightarrow S$	0.39097
3.8906	$BaSO_4 \leftrightarrow FeS_2$	0.25703
6.8486	$BaSO_4 \leftrightarrow H_2S$	0.14602
2.8436	$BaSO_4 \leftrightarrow H_2SO_3$	0.35166
2.3797	$BaSO_4 \leftrightarrow H_2SO_4$	0.42022
7.2792	$BaSO_4 \leftrightarrow S$	0.13738
3.6433	$BaSO_4 \leftrightarrow SO_2$	0.27448
2.9152	$BaSO_4 \leftrightarrow SO_3$	0.34302
2.4297	$BaSO_1 \leftrightarrow SO_4$	0.41158
4.2388	$CdS \leftrightarrow H_2S$	0.23591
4.5054	$CdS \leftrightarrow S$	0.22196
1.2250	$H_2SO_4 \leftrightarrow SO_3$	0.81631
1.6505	$(NH_4)_2SO_4 \leftrightarrow SO_3$	0.60589
1.3473	$(NH_4)_2SO_4 \leftrightarrow H_2SO_4$	0.74223
2.3492	$SO_3 \leftrightarrow H_2S$	0.42567
	TANTALUM	
	Ta = 180.948	
0.81898	$Ta \leftrightarrow Ta_2O_5$	1.2210
0.50515	$Ta \leftrightarrow TaCl_5$	1.9796
16.065	$TaC \leftrightarrow C$	0.062246
1.0664	$TaC \leftrightarrow Ta$	0.93776
0.61680	$Ta_2O_5 \leftrightarrow TaCl_5$	1.6213
1.0376	$Ta_2O_5 \leftrightarrow Ta_2O_4$	0.96379
	TELLURIUM	
	Te = 127.60	
0.65906	$Te \leftrightarrow H_2TeO_4$	1.5173
0.55565	$Te \leftrightarrow H_2TeO_4 \cdot 2H_2O$	1.7997
0.79950	$Te \leftrightarrow TeO_2$	1.2508
0.72665	$Te \leftrightarrow TeO_3$	1.3762
1.5645	$(TeO_2)_2SO_3 \leftrightarrow Te$	0.63918
	THALLIUM	
	Tl = 204.37	
0.87198	$Tl \leftrightarrow Tl_2CO_3$	1.1468
0.85218	$Tl \leftrightarrow TlCl$	1.1735
0.61693	$Tl \leftrightarrow TlI$	1.6209
0.76724	$Tl \leftrightarrow TlNO_3$	1.3034
0.96232	$Tl \leftrightarrow Tl_2O$	1.0391
1.2838	$Tl_2CrO_4 \leftrightarrow Tl$	0.77895
1.4750	$TlHSO_4 \leftrightarrow Tl$	0.67798
1.9977	$Tl_2PtCl_6 \leftrightarrow Tl$	0.50057
1.7024	$Tl_2PtCl_6 \leftrightarrow TlCl$	0.58740
1.7420	$Tl_2PtCl_6 \leftrightarrow Tl_2CO_3$	0.57406
1.2325	$Tl_2PtCl_6 \leftrightarrow TlI$	0.81139

TABLE 11.17   Gravimetric Factors (*Continued*)

Factor		Factor
1.5327	$Tl_2PtCl_6 \leftrightarrow TlNO_3$	0.65243
1.9225	$Tl_2PtCl_6 \leftrightarrow Tl_2O$	0.52017
1.6176	$Tl_2PtCl_6 \leftrightarrow Tl_2SO_4$	0.61821
1.2350	$Tl_2SO_4 \leftrightarrow Tl$	0.80971

THORIUM
Th = 232.038

Factor		Factor
1.1379	$ThO_2 \leftrightarrow Th$	0.87881
0.70627	$ThO_2 \leftrightarrow ThCl_4$	1.4159
0.44893	$ThO_2 \leftrightarrow Th(NO_3)_4 \cdot 6H_2O$	2.2275

TIN
Sn = 118.69

Factor		Factor
0.62600	$Sn \leftrightarrow SnCl_2$	1.5974
0.52604	$Sn \leftrightarrow SnCl_2 \cdot 2H_2O$	1.9010
0.45562	$Sn \leftrightarrow SnCl_4$	2.1948
0.32297	$Sn \leftrightarrow SnCl_4 \cdot (NH_4Cl)_2$	3.0962
0.88121	$Sn \leftrightarrow SnO$	1.1348
0.78764	$Sn \leftrightarrow SnO_2$	1.2696
0.79478	$SnO_2 \leftrightarrow SnCl_2$	1.2582
0.66786	$SnO_2 \leftrightarrow SnCl_2 \cdot 2H_2O$	1.4973
0.57846	$SnO_2 \leftrightarrow SnCl_4$	1.7287
0.41005	$SnO_2 \leftrightarrow SnCl_4 \cdot (NH_4Cl)_2$	2.4387
1.1188	$SnO_2 \leftrightarrow SnO$	0.89382

TITANIUM
Ti = 47.90

Factor		Factor
2.1063	$K_2TiF_6 \leftrightarrow F$	0.47477
3.0700	$K_2TiF_6 \leftrightarrow K$	0.32573
2.0662	$K_2TiF_6 \leftrightarrow 2KF$	0.48399
1.2753	$K_2TiF_6 \leftrightarrow 2(KF \cdot 2H_2O)$	0.78412
5.0123	$K_2TiF_6 \leftrightarrow Ti$	0.19951
3.0049	$K_2TiF_6 \leftrightarrow TiO_2$	0.33279
3.9880	$Ti \leftrightarrow C$	0.25075
4.9880	$TiC \leftrightarrow C$	0.20048
1.2508	$TiC \leftrightarrow Ti$	0.79952
1.6303	$TiF_4 \leftrightarrow F$	0.61338
1.6680	$TiO_2 \leftrightarrow Ti$	0.59951

TUNGSTEN
W = 183.85

Factor		Factor
3.9348	$FeWO_4 \leftrightarrow Fe_3O_4$	0.25414
1.3099	$FeWO_4 \leftrightarrow WO_3$	0.76344
6.7515	$MgWO_4 \leftrightarrow MgO$	0.14812
1.1739	$MgWO_4 \leftrightarrow WO_3$	0.85189
4.2684	$MnWO_4 \leftrightarrow MnO$	0.23428
1.3060	$MnWO_4 \leftrightarrow WO_3$	0.76571
2.0387	$PbWO_4 \leftrightarrow PbO$	0.49051

TABLE 11.17 Gravimetric Factors (*Continued*)

Factor		Factor
2.4751	$PbWO_4 \leftrightarrow W$	0.40403
1.9626	$PbWO_4 \leftrightarrow WO_3$	0.50952
15.307	$W \leftrightarrow C$	0.065330
0.96837	$W \leftrightarrow W_2C$	1.0327
0.93868	$W \leftrightarrow WC$	1.0653
31.614	$W_2C \leftrightarrow C$	0.031632
16.307	$WC \leftrightarrow C$	0.061324
1.1741	$WO_2 \leftrightarrow W$	0.85175
4.1515	$WO_3 \leftrightarrow Fe$	0.24088
1.2611	$WO_3 \leftrightarrow W$	0.79297

URANIUM
U = 238.03

Factor		Factor
1.1344	$UO_2 \leftrightarrow U$	0.88149
1.1792	$U_3O_8 \leftrightarrow U$	0.84800
1.0395	$U_3O_8 \leftrightarrow UO_2$	0.96200
0.55901	$U_3O_8 \leftrightarrow UO_2(NO_3)_2 \cdot 6H_2O$	1.7889
1.4998	$U_2P_2O_{11} \leftrightarrow U$	0.66675
1.3221	$U_2P_2O_{11} \leftrightarrow UO_2$	0.75639

VANADIUM
V = 50.941

Factor		Factor
5.2413	$VC \leftrightarrow C$	0.19079
1.7852	$V_2O_5 \leftrightarrow V$	0.56017
0.79120	$V_2O_5 \leftrightarrow VO_4$	1.2639

YTTERBIUM
Yb = 173.04

Factor		Factor
1.1387	$Yb_2O_3 \leftrightarrow Yb$	0.87820

ZINC
Zn = 65.38

Factor		Factor
2.3955	$BaSO_4 \leftrightarrow ZnS$	0.41745
0.81171	$BaSO_4 \leftrightarrow ZnSO_4 \cdot 7H_2O$	1.2320
0.80338	$Zn \leftrightarrow ZnO$	1.2447
2.7288	$ZnNH_4PO_4 \leftrightarrow Zn$	0.36646
2.1922	$ZnNH_4PO_4 \leftrightarrow ZnO$	0.45616
0.59707	$ZnO \leftrightarrow ZnCl_2$	1.6748
0.64898	$ZnO \leftrightarrow ZnCO_3$	1.5409
0.28298	$ZnO \leftrightarrow ZnSO_4 \cdot 7H_2O$	3.5338
2.3304	$Zn_2P_2O_7 \leftrightarrow Zn$	0.42911
1.8722	$Zn_2P_2O_7 \leftrightarrow ZnO$	0.53413
1.4905	$ZnS \leftrightarrow Zn$	0.67091
1.1974	$ZnS \leftrightarrow ZnO$	0.83512
0.33885	$ZnS \leftrightarrow ZnSO_4 \cdot 7H_2O$	2.9511

TABLE 11.17 Gravimetric Factors (*Continued*)

Factor		Factor
	ZIRCONIUM	
	Zr = 91.22	
2.4864	$K_2ZrF_6 \leftrightarrow F$	0.40219
2.4390	$K_2ZrF_6 \leftrightarrow 2KF$	0.41001
1.5054	$K_2ZrF_6 \leftrightarrow 2(KF \cdot 2H_2O)$	0.66427
3.1069	$K_2ZrF_6 \leftrightarrow Zr$	0.32187
2.3000	$K_2ZrF_6 \leftrightarrow ZrO_2$	0.43478
8.5946	$ZrC \leftrightarrow C$	0.11635
2.2004	$ZrF_4 \leftrightarrow F$	0.45447
1.3508	$ZrO_2 \leftrightarrow Zr$	0.74030
0.46470	$ZrO_2 \leftrightarrow ZrP_2O_7$	2.1519

TABLE 11.18   Tolerances for Analytical Weights

By Alan D. Westland with Fred E. Beamish.

This table gives the individual and group tolerances established by the National Bureau of Standards (Washington, D.C.) for classes M, S, S-1, and P weights. Individual tolerances are "acceptance tolerances" for new weights. Group tolerances are defined by the National Bureau of Standards as follows: "The corrections of individual weights shall be such that no combination of weights that is intended to be used in a weighing shall differ from the sum of the nominal values by more than the amount listed under the group tolerances."

For class S-1 weights, two-thirds of the weights in a set must be within one-half of the individual tolerances given below. No group tolerances have been specified for class P weights. See *Natl. Bur. Standards Circ.* 547, sec. 1 (1954).

Denomination	Class M		Class S		Class S-1, individual tolerance, mg	Class P, individual tolerance, mg
	Individual tolerance, mg	Group tolerance, mg	Individual tolerance, mg	Group tolerance, mg		
100 g	0.50		0.25	None	1.0	2.0
50 g	0.25	None	0.12	specified	0.60	1.2
30 g	0.15	specified	0.074		0.45	0.90
20 g	0.10		0.074	0.154	0.35	0.70
10 g	0.050		0.074		0.25	0.50
5 g	0.034		0.054		0.18	0.36
3 g	0.034	0.065	0.054	0.105	0.15	0.14
2 g	0.034		0.054		0.13	0.26
1 g	0.034		0.054		0.10	0.20
500 mg	0.0054		0.025		0.080	0.16
300 mg	0.0054	0.0105	0.025	0.055	0.070	0.14
200 mg	0.0054		0.025		0.060	0.12
100 mg	0.0054		0.025		0.050	0.10
50 mg	0.0054		0.014		0.042	0.085
30 mg	0.0054	0.0105	0.014	0.034	0.038	0.076
20 mg	0.0054		0.014		0.035	0.070
10 mg	0.0054		0.014		0.030	0.060
5 mg	0.0054		0.014		0.028	0.055
3 mg	0.0054		0.014		0.026	0.052
2 mg	0.0054	0.0105	0.014	0.034	0.025	0.050
1 mg	0.0054		0.014		0.025	0.050
½ mg	0.0054		0.014		0.025	.....

TABLE 11.19 Membrane Filters

Filter pore size, μm	Maximum rigid particle to penetrate, μm	Filter pore size, μm	Maximum rigid particle to penetrate, μm
14	17	0.65	0.68
10	12	0.60	0.65
8	9.4	0.45	0.47
7	9.0	0.30	0.32
5	6.2	0.22	0.24
3	3.9	0.20	0.25
2	2.5	0.10	0.108
1.2	1.5	0.05	0.053
1.0	1.1	0.025	0.028
0.8	0.95		

TABLE 11.20 Porosities of Fritted Glassware

Porosity	Nominal maximum pore size, μm	Principal uses
Extra coarse	170–220	Filtration of very coarse materials. Gas dispersion, gas washing, and extractor beds. Support of other filter materials.
Coarse	40–60	Filtration of coarse materials. Gas dispersion, gas washing, gas absorption. Mercury filtration. For extraction apparatus.
Medium	10–15	Filtration of crystalline precipitates. Removal of "floaters" from distilled water.
Fine	4–5.5	Filtration of fine precipitates. As a mercury valve. In extraction apparatus.
Very fine	2–2.5	General bacteria filtrations
Ultra fine	0.9–1.4	General bacteria filtrations.

TABLE 11.21 Cleaning Solutions for Fritted Glassware

Material	Cleaning solution
Fatty materials	Carbon tetrachloride.
Organic matter	Hot concentrated sulfuric acid plus a few drops of sodium or potassium nitrate solution.
Albumen	Hot aqueous ammonia or hot hydrochloric acid.
Glucose	Hot mixed acid (sulfuric plus nitric acids).
Copper or iron oxides	Hot hydrochloric acid plus potassium chlorate.
Mercury residue	Hot nitric acid.
Silver chloride	Aqueous ammonia or sodium thiosulfate.
Aluminous and siliceous residues	A 2% hydrofluoric acid solution followed by concentrated sulfuric acid; rinse immediately with distilled water followed by a few milliliters of acetone. Repeat rinsing until all trace of acid is removed.

TABLE 11.22 Common Fluxes

Flux	Melting point, °C	Types of crucible used for fusion	Type of substances decomposed
Na_2CO_3	851	Pt	For silicates, and silica-containing samples; alumina-containing samples; insoluble phosphates and sulfates
Na_2CO_3 plus an oxidizing agent such as KNO_3, $KClO_3$, or Na_2O_2		Pt (do not use with Na_2O_2) or Ni	For samples needing an oxidizing agent
NaOH or KOH	320–380	Au, Ag, Ni	For silicates, silicon carbide, certain minerals
Na_2O_2	Decomposes	Fe, Ni	For sulfides, acid-insoluble alloys of Fe, Ni, Cr, Mo, W, and Li; Pt alloys; Cr, Sn, Zn minerals
$K_2S_2O_7$	300	Pt or porcelain	Acid flux for insoluble oxides and oxide-containing samples
B_2O_3	577	Pt	For silicates and oxides when alkalis are to be determined
$CaCO_3$ plus NH_4Cl		Ni	For decomposing silicates in the determination of alkali element

11.6   VOLUMETRIC ANALYSIS

11.6.1   Equations and Equivalents for Volumetric Analysis

Aluminum
With EDTA:
$Al^{3+} + H_2Y^{2-}$ (minimum pH 4.2) $= AlY^- + 2H^+$

$$\frac{Na_2H_2Y}{1} \equiv \frac{Al}{1} \equiv \frac{Al_2O_3}{2}$$

Antimony
(1) $H_3SbO_4 + 2HI = H_3SbO_3 + H_2O + I_2$

$$I \equiv \frac{Sb}{2} \equiv \frac{H_3SbO_4}{2}$$

(2) $SbCl_5 + 2KI = SbCl_3 + 2KCl + I_2$

$$I \equiv \frac{Sb}{2} \equiv \frac{Sb_2O_5}{4}$$

(3) $Sb_2O_5 + 4HI = Sb_2O_3 + 2H_2O + 2I_2$

$$I \equiv \frac{Sb}{2} \equiv \frac{Sb_2O_5}{4}$$

(4) $Sb_2S_5 + 6HCl = 2SbCl_3 + 2S + 3H_2S$

$$3H_2S + 3I_2 = 6HI + 3S$$

$$I \equiv \frac{Sb}{3} \equiv \frac{Sb_2S_5}{6}$$

(5) $2H_3SbO_3 + KIO_3 + 2HCl = 2H_3SbO_4 + KCl + ICl + H_2O$

$$\frac{KIO_3}{6} \equiv \frac{Sb}{3} \equiv \frac{H_3SbO_3}{3}$$

(6) $2NaSbO_3 + 4HI = 2NaSbO_2 + 2H_2O + 2I_2$

$$I \equiv \frac{NaSbO_3}{2}$$

(7) $H_2SbO_3 + I_2 + H_3SbO_4 + 2HI$

$$I \equiv \frac{Sb}{2} \equiv \frac{H_3SbO_3}{2}$$

(8) $Sb_2O_3 + 2I_2 + 2H_2O = Sb_2O_5 + 4HI$

$$I \equiv \frac{Sb}{2} \equiv \frac{Sb_2O_3}{4}$$

(9) $SbCl_3 + I_2 + 2HCl = SbCl_5 + 2HI$

$$I \equiv \frac{Sb}{2} \equiv \frac{SbCl_3}{2}$$

(10) $Sb_2S_3 + 6HCl = 2SbCl_3 + 3H_2S$

$$3H_2S + 3I_2 = 6HI + 3S$$

$$I \equiv \frac{Sb}{3} \equiv \frac{Sb_2S_3}{6} \equiv \frac{H_2S}{2}$$

(11) $5Sb_2(SO_4)_3 + 4KMnO_4 + 24H_2O = 10H_3SbO_4 + 2K_2SO_4 + 4MnSO_4 + 9H_2SO_4$

$$\frac{KMnO_4}{5} \equiv \frac{Sb}{2} \equiv \frac{Sb_2O_3}{4} \equiv \frac{Sb_2(SO_4)_3}{4}$$

(12) $2NaSbO_2 + 2I_2 + 2H_2O = 2NaSbO_3 + 4HI$

$$I \equiv \frac{NaSbO_2}{2}$$

Arsenic

$H_3AsO_4 + 2HI$	Same as equation 1.
$AsCl_5 + 2KI$	Same as equation 2.
$As_2O_5 + 4HI$	Same as equation 3.
$As_2S_5 + 6HCl$	Same as equation 4.
$2H_3AsO_3 + KIO_3 + 2HCl$	Same as equation 5.
$H_3AsO_3 + I_2 + H_2O$	Same as equation 7.
$As_2O_3 + 2I_2 + 2H_2O$	Same as equation 8.
$AsCl_3 + I_2 + 2HCl$	Same as equation 9.
$As_2S_3 + 6HCl$	Same as equation 10.
$2NaAsO_2 + 2I_2 + 2H_2O$	Same as equation 12.

$2Na_3AsO_4 \equiv As_2O_5$

$$I \equiv \frac{Na_3AsO_4}{2} \equiv \frac{As_2O_5}{4}$$

$2Na_2HAsO_4 \equiv As_2O_5$

$$I \equiv \frac{Na_2HAsO_4}{2} \equiv \frac{As_2O_5}{4}$$

$2NaH_2AsO_4 \equiv As_2O_5$

$$I \equiv \frac{NaH_2AsO_4}{2} \equiv \frac{As_2O_5}{4}$$

$2Na_3AsO_3 \equiv As_2O_3$

$$I \equiv \frac{Na_3AsO_3}{2} \equiv \frac{As_2O_3}{4}$$

$2Na_2HAsO_3 \equiv As_2O_3$

$$I \equiv \frac{Na_2HAsO_3}{2} \equiv \frac{As_2O_3}{4}$$

$2NaH_2AsO_3 \equiv As_2O_3$

$$I \equiv \frac{NaH_2AsO_3}{2} \equiv \frac{As_2O_3}{4}$$

Boron
$Na_2B_4O_7 + 2HCl + 5H_2O + (glycerol) = 4H_3BO_3 + 2NaCl$

$H_3BO_3 + NaOH = NaBO_2 + 2H_2O$

$$NaOH \equiv B \equiv \frac{B_2O_3}{2} \equiv H_3BO_3 \equiv \frac{Na_2B_4O_7}{2}$$

$6HBO_2 + KIO_3 + 5KI + (mannitol) = 6KBO_2 + 3H_2O + 3I_2$

$$I \equiv HBO_2 \equiv H_3BO_3$$

Bromine
$Br_2 + 2KI = 2KBr + I_2$

$$I = Br$$

$KBrO_3 + 6KI + 6HBr = 7KBr + 3I_3 + 3H_2O$

$$I \equiv \frac{KBrO_3}{6}$$

$Ba(BrO_3)_2 + 6H_3AsO_3 = BaBr_2 + 6H_3AsO_4$

$$I \equiv \frac{Ba(BrO_3)_2}{12} \equiv \frac{H_3AsO_3}{2}$$

Cadmium
$CdS + I_2 + 2HCl = CdCl_2 + 2HI + S$

$$I \equiv \frac{Cd}{2} \equiv \frac{CdS}{2} \equiv \frac{S}{2}$$

Calcium
$5CaC_2O_4 + 2KMnO_4 + 8H_2SO_4 = 5CaSO_4 + K_2SO_4 + 2MnSO_4 + 10CO_2 + 8H_2O$

$$\frac{KMnO_4}{5} \equiv \frac{Ca}{2} \equiv \frac{CaC_2O_4}{2} \equiv \frac{CaCO_3}{2}$$

$CaCO_3 + 2HCl = CaCl_2 + CO_2 + H_2O$

$$HCl \equiv \frac{CaO}{2} \equiv \frac{CaCO_3}{2}$$

With methyl orange:
$Na_2CO_3 + 2HCl = 2NaCl + CO_2 + H_2O$

$$HCl \equiv \frac{Na_2CO_3}{2}$$

With methyl orange:
$Na_2CO_3 + H_2SO_4 = Na_2SO_4 + CO_2 + H_2O$

$$H_2SO_4 \equiv Na_2CO_3$$

With phenolphthalein:
$Na_2CO_3 + HCl = NaCl + NaHCO_3$

$$HCl \equiv Na_2CO_3$$

With methyl orange:
$NaHCO_3 + HCl = NaCl + CO_2 + H_2O$

$$HCl \equiv NaHCO_3$$

With methyl orange:
$2NaHCO_3 + H_2SO_4 = Na_2SO_4 + 2CO_2 + H_2O$

$$H_2SO_4 \equiv 2NaHCO_3$$

With EDTA:
$Ca^{2+} + H_2Y^{2-} \text{ (minimum pH 10)} = CaY^{2-} + 2H^+$

$$\frac{Na_2H_2Y}{1} \equiv \frac{Ca}{1} \equiv \frac{CaO}{1}$$

Carbon
$H_2C_2O_4 \cdot 2H_2O + 2NaOH = Na_2C_2O_4 + 4H_2O$

$$NaOH \equiv \frac{H_2C_2O_4 \cdot 2H_2O}{2}$$

$KHC_2O_4 + KOH = K_2C_2O_4 + H_2O$

$$KOH \equiv KHC_2O_4$$

$5Na_2C_2O_4 + 2KMnO_4 + 8H_2SO_4 = 5Na_2SO_4 + 2MnSO_4 + K_2SO_4 + 10CO_2 + 8H_2O$

$$\frac{KMnO_4}{5} \equiv \frac{Na_2C_2O_4}{2}$$

$KCN + I_2 = KI + ICN$

$$I \equiv \frac{KCN}{2}$$

$$5K_4Fe(CN)_6 + KMnO_4 + 4H_2SO_4 = 5K_3Fe(CN)_6 + MnSO_4 + 3K_2SO_4 + 4H_2O$$

$$\frac{KMnO_4}{5} \equiv K_4Fe(CN)_6 \equiv 6CN \equiv 6C \equiv FeO \equiv \frac{Fe_2O_3}{2} \equiv Fe$$

$$2K_3Fe(CN)_6 + 2KI = 2K_4Fe(CN)_6 = I_2$$

$$I \equiv K_3Fe(CN)_6 \equiv 6CN \equiv 6C \equiv FeO \equiv \frac{Fe_2O_3}{2} \equiv Fe$$

$$3KSCN + 4Al + 18HCl = 3KCl + 4AlCl_3 + 3NH_4Cl + 3C + 3H_2S$$

$$I \equiv \frac{KCNS}{2} \equiv \frac{H_2S}{2}$$

$$5HSCN + 6KMnO_4 + 4H_2SO_4 = 6MnSO_4 + 3K_2SO_4 + 5HCN + 4H_2O$$

$$\frac{KMnO_4}{5} \equiv \frac{HSCN}{6}$$

Cerium

$$2K_3Fe(CN)_6 + Ce_2O_3 + 2KOH = 2K_4Fe(CN)_6 + 2CeO_2 + H_2O$$

$$\frac{KMnO_4}{5} \equiv K_4Fe(CN)_6 \equiv Ce \equiv \frac{Ce_2O_3}{2}$$

$$2CeO_2 + 2KI + 8HCl = 2CeCl_3 + 2KCl + I_2 + 4H_2O$$

$$I \equiv Ce \equiv CeO_2$$

Chlorine

$$Cl_2 + 2KI = 2KCl + I_2$$

$$I_2 + 2Na_2S_2O_3 = 2NaI + Na_2S_4O_6$$

$$I \equiv Cl$$

$$CaOCl_2 + 2HCl = CaCl_2 + Cl_2 + H_2O$$

$$I \equiv Cl \equiv \frac{CaOCl_2}{2}$$

$$NaClO + 2KI + H_2O = NaCl + 2KOH + I_2$$

$$I \equiv \frac{NaClO}{2}$$

$$NaCl + AgNO_3 = AgCl + NaNO_3$$

$$NaCl \equiv Cl$$

$$KClO_3 + 6FeSO_4 + 3H_2SO_4 = KCl + 3Fe_2(SO_4)_3 + 3H_2O$$

$$\frac{KMnO_4}{5} \equiv \frac{KClO_3}{6}$$

$$KClO_3 + 6HI + H_2SO_4 = KHSO_4 + HCl + 3I_2 + 3H_2O$$

$$I \equiv \frac{KClO_3}{6}$$

$$KClO_4 = KCl + 2O_2$$

$$2O_2 + 8HI = 4I_2 + 4H_2O$$

$$I \equiv \frac{KClO_4}{8}$$

Chromium

$$Cr_2(SO_4)_3 + 5KI + KIO_3 + 3H_2O = 2Cr(OH)_3 + 3K_2SO_4 + 3I_2$$

$$I \equiv \frac{Cr}{3} \equiv \frac{Cr_2(SO_4)_3}{6}$$

$$K_2Cr_2O_7 + 6FeSO_4 + 8H_2SO_4 = Cr_2(SO_4)_3 + 3Fe_2(SO_4)_3 + 2KHSO_4 + 7H_2O$$

$$\frac{KMnO_4}{5} \equiv \frac{K_2Cr_2O_7}{6} \equiv \frac{Cr}{3} \equiv \frac{Cr_2O_3}{6}$$

$$K_2Cr_2O_7 + 6FeCl_2 + 14HCl = 2CrCl_3 + 6FeCl_3 + 2KCl + 7H_2O$$

$$Fe \equiv \frac{K_2Cr_2O_7}{6} \equiv \frac{Cr_2O_3}{6} \equiv \frac{Cr}{3}$$

$$K_2Cr_2O_7 + 6KI + 7H_2SO_4 = Cr_2(SO_4)_3 + 4K_2SO_4 + 3I_2 + 7H_2O$$

$$I \equiv \frac{Cr}{3} \equiv \frac{Cr_2O_3}{6} \equiv \frac{K_2Cr_2O_7}{6}$$

$$K_2Cr_2O_7 + 2KOH = 2K_2CrO_4 + H_2O$$

$$KOH \equiv \frac{K_2Cr_2O_7}{2}$$

$$2K_2CrO_4 + 6KI + 16HCl + 2CrCl_3 + 10KCl + 3I_2 + 8H_2O$$

$$I \equiv \frac{Cr}{3} \equiv \frac{CrO_3}{3} \equiv \frac{K_2CrO_4}{3}$$

$$2BaCrO_4 + 6FeSO_4 + 8H_2SO_4 = 2BaSO_4 + Cr_2(SO_4)_3 + 3Fe_2(SO_4)_3 + 8H_2O$$

$$\frac{KMnO_4}{5} \equiv FeSO_4 \equiv \frac{Cr}{3} \equiv \frac{CrO_3}{3} \equiv \frac{K_2CrO_4}{3} \equiv \frac{BaCrO_4}{3}$$

Cobalt

$$Co(NH_3)_6^{2+} + Fe(CN)_6^{3-} [Citrate-NH_3 \text{ buffer}] = Co(NH_3)_6^{3+} + Fe(CN)_6^{4-}$$

$$\frac{K_3Fe(CN)_6}{1} \equiv \frac{Co}{1}$$

Copper

$$2CuSO_4 + 4KI = 2CuI + 2K_2SO_4 + I_2$$

$$I \equiv Cu \equiv CuSO_4$$

$$2Cu(C_2H_3O_2)_2 + 4KI = 2CuI + 4KC_2H_3O_2 + I_2$$

$$I \equiv Cu \equiv Cu(C_2H_3O_2)_2$$

$$2CuSO_4 + 2KSCN + H_2SO_3 + H_2O = 2CuSCN + K_2SO_4 + 2H_2SO_4$$

$$CuSCN + KOH = KSCN + CuOH$$

$$5KSCN + 6KMnO_4 + 4H_2SO_4 + 3K_2SO_4 + 6MnSO_4 + 5KCN + 4H_2O$$

$$\frac{KMnO_4}{5} \equiv \frac{Cu}{6} \equiv \frac{CuSO_4}{6} \equiv \frac{KSCN}{6}$$

$$Cu_2Cl_2 + I_2 = CuCl_2 + CuI_2$$

$$I \equiv \frac{Cu_2Cl_2}{2} \equiv Cu$$

$$4CuSCN + 7KIO_3 + 10HCl = 4CuSO_4 + 4ICN + 3ICl + 7KCl + 5H_2O$$

$$\frac{KIO_3}{4} \equiv \frac{CuSCN}{7} \equiv \frac{Cu}{7}$$

With EDTA:

$$Na_2H_2Y + Cu^{2+} \text{ [minimum pH 4]} = CuY^{2-} + 2Na^+ + 2H^+$$

$$\frac{Na_2H_2Y}{1} \equiv \frac{Cu}{1}$$

Fluorine

$$4NaF + Th(NO_3)_4 = 4NaNO_3 + ThF_4$$

$$\frac{Th(NO_3)_4}{4} \equiv \frac{NaF}{1} \equiv \frac{H_2SiF_6}{6}$$

Gold

$$AuCl_3 + 2KI = AuCl + 2KCl + I_2$$

$$I \equiv \frac{Au}{2} \equiv \frac{AuCl_3}{2}$$

Iodine

$$I_2 + 2Na_2S_2O_3 = 2NaI + Na_2S_4O_6$$

$$I \equiv Na_2S_2O_3$$

$$10KI + 2KMnO_4 + 16HCl = 12KCl + 2MnCl_2 + 5I_2 + 8H_2O$$

$$\frac{KMnO_4}{5} \equiv I \equiv KI$$

$$KIO_3 + 5KI + 6HCl = 6KCl + 3I_2 + 3H_2O$$

$$I \equiv \frac{KIO_3}{6} \equiv \frac{5KI}{6}$$

$$Ca(IO_3)_2 + 10KI + 12HCl = CaCl_2 + 10KCl + 6I_2 + 6H_2O$$

$$I \equiv \frac{Ca(IO_3)_3}{12}$$

$$KIO_4 + 7KI + 8HCl = 8KCl + 4I_2 + 4H_2O$$

$$I \equiv \frac{KIO_4}{8}$$

Iron

$$6FeCl_2 + K_2Cr_2O_7 + 14HCl = 6FeCl_3 + 2CrCl_3 + 2KCl + 7H_2O$$

$$\frac{K_2Cr_2O_7}{6} \equiv Fe \equiv FeCl_2 \equiv \frac{Fe_2O_3}{2}$$

$$2FeSO_4 + Br_2 + H_2SO_4 = Fe_2(SO_4)_3 + 2HBr$$

$$Br \equiv FeSO_4 \equiv FeO \equiv \frac{Fe_2O_3}{2} \equiv Fe$$

$$2FeCl_3 + 2KI = 2KCl + 2FeCl_2 + I_2$$

$$I \equiv Fe \equiv FeCl_3 \equiv \frac{Fe_2O_3}{2}$$

$$Fe_2(SO_4)_3 + 2KI = 2FeSO_4 + K_2SO_4 + I_2$$

$$I \equiv Fe \equiv \frac{Fe_2(SO_4)_3}{2} \equiv \frac{Fe_2O_3}{2}$$

$$2FeCl_3 + SnCl_2 = 2FeCl_2 + SnCl_4$$

$$SnCl_2 + 2HgCl_2 = SnCl_4 + Hg_2Cl_2$$

$$2FeCl_2 + I_2 + 2HCl = 2FeCl_3 + 2HI$$

$$I \equiv Fe \equiv FeCl_3 \equiv \frac{Fe_2O_3}{2}$$

$$10FeSO_4 + 2KMnO_4 + 8H_2SO_4 = 5Fe_2(SO_4)_3 + 2MnSO_4 + K_2SO_4 + 8H_2O$$

$$\frac{KMnO_4}{5} \equiv Fe \equiv FeSO_4 \equiv \frac{Fe_2O_3}{2}$$

With EDTA:

$$Na_2H_2Y + FeCl_3 = FeY^- + 2NaCl + HCl + H^+$$

$$\frac{Na_2H_2Y}{1} \equiv \frac{FeCl_3}{1}$$

Lead

$2Pb(C_2H_3O_2)_2 + K_2Cr_2O_7 + H_2O = 2PbCrO_4 + 2KC_2H_3O_2 + 2HC_2H_3O_2$

$$Pb \equiv \frac{K_2Cr_2O_7}{2}$$

$2Pb(C_2H_3O_2)_2 + K_2Cr_2O_7 + H_2O = 2PbCrO_4 + 2KC_2H_3O_2 + 2HC_2H_3O_2$

$2PbCrO_4 + 6KI + 16HCI = 2PbCI_2 + 2CrCI_3 + 6KCI + 8H_2O + 3I_2$

$$\frac{K_2Cr_2O_7}{6} \equiv \frac{Pb}{3} \equiv \frac{PbCrO_4}{3} \equiv I$$

$PbO_2 + 4HI = PbI_2 + I_2 + 2H_2O$

$$I \equiv \frac{Pb}{2} \equiv \frac{PbO_2}{2}$$

$PbO_2 + H_2C_2O_4 = PbO + 2CO_2 + H_2O$

$5H_2C_2O_4 + 2KMnO_4 + 3H_2SO_4 = 10CO_2 + 2MnSO_4 + K_2SO_4 + 8H_2O$

$$\frac{KMnO_4}{5} \equiv \frac{H_2C_2O_4}{2} \equiv \frac{Pb}{2} \equiv \frac{PbO_2}{2}$$

$PbO_2 + H_2O_2 + 2HNO_3 = Pb(NO_3)_2 + 2H_2O + O_2$

$5H_2O_2 + 2KMnO_4 + 6HNO_3 = 2Mn(NO_3)_2 + 2KNO_3 + 8H_2O + 5O_2$

$$\frac{KMnO_4}{5} \equiv \frac{H_2O_2}{2} \equiv \frac{Pb}{2} \equiv \frac{PbO_2}{2}$$

$Pb_3O_4 + 4HC_2H_3O_2 = PbO_2 + 2Pb(C_2H_3O_2)_2 + 2H_2O$

$PbO_2 + 4KI = PbI_2 + 2K_2O + I_2$

$$I \equiv \frac{Pb}{2} \equiv \frac{Pb_3O_4}{2} \equiv \frac{PbO_2}{2}$$

With EDTA:

$Na_2H_2Y + Pb(NO_3)_2 = PbY^{2-} + 2NaNO_3 + 2H^+$

$$\frac{Na_2H_2Y}{1} \equiv \frac{Pb(NO_3)_2}{1}$$

Magnesium

$MgNH_4PO_4 + H_2SO_4 = MgSO_4 + NH_4H_2PO_4$

$$H_2SO_4 \equiv MgNH_4PO_4$$

With EDTA:

$Na_2H_2Y + Mg^{2+} [pH\ 10\ minimum) = MgY^{2-} + 2Na^+ + 2H^+$

$$\frac{Na_2H_2Y}{1} \equiv \frac{Mg}{1} \equiv \frac{MgCO_3}{1}$$

Manganese

$$MnO_2 + H_2C_2O_4 + H_2SO_4 = MnSO_4 + 2CO_2 + 2H_2O$$

$$\frac{KMnO_4}{5} \equiv \frac{H_2C_2O_4}{2} \equiv \frac{Mn}{2} \equiv \frac{MnO_2}{2}$$

$$MnO_2 + 4HCl = MnCl_2 + Cl_2 + 2H_2O$$

$$Cl_2 + 2NaI = 2NaCl + I_2$$

$$I \equiv Cl \equiv \frac{MnO_2}{2} \equiv \frac{Mn}{2}$$

$$MnO_2 + 2FeSO_4 + 2H_2SO_4 = MnSO_4 + Fe_2(SO_4)_3 + 2H_2O$$

$$FeSO_4 \equiv \frac{MnO_2}{2} \equiv \frac{Mn}{2}$$

Bismuthate Method:

$$2Mn(NO_3)_2 + 5NaBiO_3 + 16HNO_3 = 2HMnO_4 + 5Bi(NO_3)_3 + 5NaNO_3 + 7H_2O$$

$$2HMnO_4 + 5Na_3AsO_3 + 4HNO_3 = 2Mn(NO_3)_2 + 5Na_3AsO_4 + 3H_2O$$

$$\frac{HMnO_4}{5} \equiv \frac{Na_3AsO_3}{2} \equiv \frac{Mn}{5}$$

Ford Williams Method:

$$MnO_2 + 2FeSO_4 + 2H_2SO_4 = MnSO_4 + Fe_2(SO_4)_3 + 2H_2O$$

$$FeSO_4 \equiv \frac{KMnO_4}{5} \equiv \frac{MnO_2}{2} \equiv \frac{Mn}{2}$$

Lead Dioxide Method:

$$2MnSO_4 + 5PbO_2 + 6HNO_3 = 2HMnO_4 + 2PbSO_4 + 3Pb(NO_3)_2 + 2H_2O$$

$$2HMnO_4 + 5Na_3AsO_3 + 4HNO_3 = 2Mn(NO_3)_2 + 5Na_3AsO_4 + 3H_2O$$

$$\frac{Na_3AsO_3}{2} \equiv \frac{KMnO_4}{5} \equiv \frac{HMnO_4}{5} \equiv \frac{Mn}{5}$$

Persulfate Method:

$$2Mn(NO_3)_2 + 5(NH_4)_2S_2O_8 + 8H_2O = 2HMnO_4 + 5(NH_4)_2SO_4 + 5H_2SO_4 + 4HNO_3$$

$$2HMnO_4 + 10FeSO_4 + 7H_2SO_4 = 2MnSO_4 + 5Fe_2(SO_4)_3 + 8H_2O$$

$$FeSO_4 \equiv \frac{KMnO_4}{5} \equiv \frac{HMnO_4}{5} \equiv \frac{Mn}{5}$$

Volhard Method:

$$3MnSO_4 + 2KMnO_4 + 2H_2O = 5MnO_2 + K_2SO_4 + 2H_2SO_4$$

$$6MnSO_4 + 5ZnSO_4 + 4KMnO_4 + 14H_2O = 5[Zn(OH)_2 \cdot 2MnO_2] + 4KHSO_4 + 7H_2SO_4$$

$$\frac{KMnO_4}{5} \equiv \frac{3MnSO_4}{10} \equiv \frac{3Mn}{10}$$

Pyrophosphate method:
$$MnO_4^- + 4Mn^{2+} + 15H_2P_2O_7^{2-} + 8H^+ \text{ [pH range 4 to 7]} = 5Mn(H_2P_2O_7)_3^{3-} + 4H_2O$$

$$\frac{KMnO_4}{4} \equiv \frac{Mn}{1}$$

$$2KMnO_4 + 5H_2C_2O_4 + 3H_2SO_4 = 2MnSO_4 + K_2SO_4 + 10CO_2 + 8H_2O$$

$$\frac{KMnO_4}{5} \equiv \frac{H_2C_2O_4}{2}$$

$$K_2MnO_4 + 2H_2C_2O_4 + 2H_2SO_4 = MnSO_4 + K_2SO_4 + 4CO_2 + 4H_2O$$

$$\frac{KMnO_4}{5} \equiv \frac{H_2C_2O_4}{2} \equiv \frac{K_2MnO_4}{4}$$

Mercury
$$Hg_2Cl_2 + 6KI + I_2 = 2K_2HgI_4 + 2KCl$$

$$I \equiv \frac{Hg_2Cl_2}{2}$$

$$3HgCl_2 + 2Na_2S_2O_3 + 2H_2O = HgCl_2 \cdot 2HgS + 2Na_2SO_4 + 4HCl$$

$$I \equiv Na_2S_2O_3 \equiv \frac{3HgCl_2}{2} \equiv \frac{3Hg}{2}$$

$$2HgCl_2 + 2FeCl_2 = Hg_2Cl_2 + 2FeCl_3$$

$$\frac{KMnO_4}{5} \equiv FeCl_2 \equiv HgCl_2 \equiv Hg$$

$$Hg(CN)_2 + 2I_2 = HgI_2 + 2ICN$$

$$I \equiv \frac{Hg(CN)_2}{4} \equiv \frac{Hg}{4}$$

Molybdenum
$$2MoO_3 + 3Zn + 3H_2SO_4 = Mo_2O_3 + 3ZnSO_4 + 3H_2O$$

$$5Mo_2O_3 + 6KMnO_4 + 9H_2SO_4 = 10MoO_3 + 6MnSO_4 + 3K_2SO_4 + 9H_2O$$

$$\frac{KMnO_4}{5} \equiv \frac{Mo_2O_3}{6} \equiv \frac{MoO_3}{3} \equiv \frac{Mo}{3}$$

$$2MoO_3 + 4KI + 4HCl = 2MoO_2I + 4KCl + I_2 + 2H_2O$$

$$I \equiv MoO_3 \equiv Mo$$

Nickel
$$2NiSO_4 + 12NH_4OH = 2Ni(NH_3)_6SO_4 + 12H_2O$$

$$Ni(NH_3)_6SO_4 + 4KCN = K_2Ni(CN)_4 + K_2SO_4 + 6NH_3$$

$$\frac{KCN}{2} \equiv \frac{Ni(NH_3)_6SO_4}{8} \equiv \frac{Ni}{8}$$

$$3NiSO_4 + 5KI + KIO_3 + 3H_2O = 3Ni(OH)_2 + 3K_2SO_4 + 3I_2$$

$$I \equiv \frac{NiSO_4}{2} \equiv \frac{Ni}{2}$$

Nitrogen

$$2HNO_3 + 6HI = 2NO + 3I_2 + 4H_2O$$

$$I \equiv \frac{HNO_3}{3} \equiv \frac{N_2O_5}{6}$$

$$HNO_3 + 3HCl = NOCl + Cl_2 + 2H_2O$$

$$I \equiv Cl \equiv \frac{HNO_3}{2} \equiv \frac{N_2O_5}{4}$$

$$HNO_3 + 3MnCl_3 + 3HCl = NO + 3MnCl_4 + 2H_2O$$

$$3MnCl_4 = 3MnCl_2 + 3Cl_2$$

$$I \equiv Cl \equiv \frac{HNO_3}{3} \equiv \frac{N_2O_5}{6}$$

$$4FeSO_4 + 2HNO_3 + 2H_2SO_4 = 2Fe_2(SO_4)_3 + N_2O_3 + 3H_2O$$

$$\frac{K_2Cr_2O_7}{6} \equiv \frac{HNO_3}{2} \equiv \frac{N_2O_5}{4}$$

$$6FeSO_4 + 2HNO_3 + 3H_2SO_4 = 3Fe_2(SO_4)_3 + 2NO + 4H_2O$$

$$\frac{K_2Cr_2O_7}{6} \equiv \frac{HNO_3}{3} \equiv \frac{N_2O_5}{6}$$

$$2HNO_2 + 2HI = 2NO + I_2 + 2H_2O$$

$$I \equiv HNO_2 \equiv \frac{N_2O_3}{2}$$

$$2HNO_2 + MnCl_2 + 2HCl = 2NO + MnCl_4 + 2H_2O$$

$$MnCl_4 = MnCl_2 + Cl_2$$

$$I \equiv Cl \equiv HNO_2 \equiv \frac{N_2O_3}{2}$$

$$2KMnO_4 + 5HNO_2 + 3H_2SO_4 = 2MnSO_4 + 5HNO_3 + K_2SO_4 + 3H_2O$$

$$\frac{KMnO_4}{5} \equiv \frac{HNO_2}{2} \equiv \frac{N_2O_3}{4}$$

$$NH_3 + HCl = NH_4Cl$$

$$HCl \equiv NH_3 \equiv N \equiv \frac{N_2O_3}{2} \equiv \frac{N_2O_5}{2}$$

$$NH_3 + H_3BO_3 = NH_4H_2BO_3$$

$$NH_4H_2BO_3 + HCl = NH_4Cl + H_3BO_3$$

$$H_3BO_3 \equiv HCl \equiv NH_3 \equiv N \equiv \frac{N_2O_5}{2}$$

Oxygen
$$O_3 + 2KI + H_2O = O_2 + I_2 + 2KOH$$

$$I \equiv \frac{O_3}{2}$$

$$BaO_2 + 4HCl = BaCl_2 + 2H_2O + Cl_2$$

$$I \equiv Cl \equiv \frac{BaO_2}{2}$$

$$2KI + H_2O_2 = 2KOH + I_2$$

$$I \equiv \frac{H_2O_2}{2}$$

$$5H_2O_2 + 2KMnO_4 + 4H_2SO_4 = 2MnSO_4 + 2KHSO_4 + 5O_2 + 8H_2O$$

$$\frac{KMnO_4}{5} \equiv \frac{H_2O_2}{2}$$

Phosphorus
$$H_3PO_4 + 12(NH_4)_2MoO_4 + 21HNO_3 = (NH_4)_3[P(Mo_3O_{10})_4] + 21NH_4NO_3 + 12H_2O$$
$$2(NH_4)_3[P(Mo_3O_{10})_4] + 46NaOH + H_2O = 2(NH_4)_2HPO_4 + (NH_4)_2MoO_4$$
$$+ 23Na_2MoO_4 + 23H_2O$$

$$NaOH \equiv \frac{(NH_4)_3[P(Mo_3O_{10})_4]}{23} \equiv \frac{P_2O_5}{46} \equiv \frac{P}{23}$$

$$2(NH_4)_3[P(Mo_3O_{10})_4] + 46NH_4OH + H_2O = 24(NH_4)_2MoO_4 + 2(NH_4)_2HPO_4 + 23H_2O$$
$$(NH_4)_2MoO_4 + H_2SO_4 = (NH_4)_2SO_4 + H_2MoO_4$$
$$2MoO_3 + 3Zn + 3H_2SO_4 = Mo_2O_3 + 3ZnSO_4 + 3H_2O$$
$$5Mo_2O_3 + 6KMnO_4 + 9H_2SO_4 = 10MoO_3 + 6MnSO_4 + 3K_2SO_4 + 9H_2O$$

$$\frac{KMnO_4}{5} \equiv \frac{(NH_4)_3[P(Mo_3O_{10})_4]}{36} \equiv \frac{Mo_2O_3}{6} \equiv \frac{P_2O_5}{72} \equiv \frac{P}{36}$$

With methyl orange:
$$Na_2HPO_4 + HCl = NaH_2PO_4 + NaCl$$

$$HCl \equiv Na_2HPO_4$$

With methyl orange:
$$H_3PO_4 + NaOH = NaH_2PO_4 + H_2O$$

$$NaOH \equiv H_3PO_4$$

With phenolphthalein:
$$NaH_2PO_4 + NaOH = Na_2HPO_4 + H_2O$$

$$NaOH \equiv NaH_2PO_4$$

With phenolphthalein:
$$H_3PO_4 + 2NaOH = Na_2HPO_4 + 2H_2O$$

$$NaOH \equiv \frac{H_3PO_4}{2}$$

Potassium
$$10K_2NaCo(NO_2)_6 + 22KMnO_4 + 58H_2SO_4 = 21K_2SO_4 + 5Na_2SO_4 + 10CoSO_4$$
$$+ 22MnSO_4 + 60HNO_3 + 28H_2O$$

$$\frac{KMnO_4}{5} \equiv \frac{K_2O}{11}$$

Selenium
$$K_2SeO_4 + 4HCl = 2KCl + H_2SeO_3 + Cl_2 + H_2O$$

$$I \equiv Cl \equiv \frac{K_2SeO_4}{2} \equiv \frac{Se}{2} \equiv \frac{SeO_3}{2}$$

$$K_2SeO_3 + 4KI + 6HCl = 6KCl + Se + 2I_2 + 3H_2O$$

$$I \equiv \frac{K_2SeO_3}{4} \equiv \frac{Se}{4} \equiv \frac{SeO_2}{4}$$

$$4Na_2S_2O_3 + H_2SeO_3 + 4HCl = Na_2SeS_4O_6 + Na_2S_4O_6 + 4NaCl + 3H_2O$$

$$\frac{Na_2S_2O_3}{1} \equiv \frac{H_2SeO_3}{4}$$

Silicon
$$H_2SiF_6 + 6NaOH = 6NaF + Si(OH)_4 + 2H_2O$$

$$NaOH \equiv \frac{H_2SiF_6}{6}$$

Silver
$$AgNO_3 + KSCN = AgSCN + KNO_3$$

$$\frac{KSCN}{1} \equiv \frac{AgNO_3}{1}$$

$$AgNO_3 + KCl = AgCl + KNO_3$$

$$\frac{KCl}{1} \equiv \frac{AgNO_3}{1}$$

$$AgNO_3 + 2KCN = KAg(CN)_2 + KNO_3$$

$$2KCN \equiv AgNO_3$$

Sulfur

$H_2S + I_2 = 2HI + S$

$$I \equiv \frac{H_2S}{2} \equiv \frac{S}{2}$$

$Na_2SO_3 + I_2 + H_2O = Na_2SO_4 + 2HI$

$$I \equiv \frac{Na_2SO_3}{2} \equiv \frac{SO_2}{2} \equiv \frac{S}{2}$$

$2H_2SO_3 + KIO_3 + 2HCl = 2H_2SO_4 + KCl + ICl + H_2O$

$$\frac{KIO_3}{6} \equiv \frac{H_2SO_3}{3} \equiv \frac{SO_2}{3}$$

$2Na_2S_2O_3 + I_2 = Na_2S_4O_6 + 2NaI$

$$I \equiv Na_2S_2O_3$$

$Na_2S_2O_3 + 2KIO_3 + 2HCl = Na_2SO_4 + K_2SO_4 + 2ICl + H_2O$

$$\frac{KIO_3}{6} \equiv \frac{Na_2S_2O_3}{12}$$

$K_2S_2O_8 + H_2C_2O_4 = K_2SO_4 + H_2SO_4 + 2CO_2$

$$\frac{KMnO_4}{5} \equiv \frac{H_2C_2O_4}{2} \equiv \frac{K_2S_2O_8}{2}$$

$K_2S_2O_8 + 2FeSO_4 = Fe_2(SO_4) + K_2SO_4$

$$\frac{KMnO_4}{5} \equiv FeSO_4 \equiv \frac{K_2S_2O_8}{2}$$

$3Na_2S_2O_3 + 8KMnO_4 + H_2O = 3Na_2SO_4 + 3K_2SO_4 + 8MnO_2 + 2KOH$

$$\frac{KMnO_4}{5} \equiv \frac{3Na_2S_2O_3}{40}$$

$2BaS_2O_3 + I_2 = BaI_2 + BaS_4O_6$

$$I \equiv BaS_2O_3$$

$2Na_2S_2O_3 + I_2 = 2NaI + Na_2S_4O_6$

$$I \equiv Na_2S_2O_3$$

$H_2SO_4 + 2NaOH = Na_2SO_4 + H_2O$

$$NaOH \equiv \frac{H_2SO_4}{2} \equiv \frac{SO_4}{2} \equiv \frac{SO_3}{2} \equiv \frac{S}{2}$$

With phenolphthalein:

$H_2SO_3 + 2NaOH = Na_2SO_3 + 2H_2O$

$$NaOH \equiv \frac{H_2SO_3}{2} \equiv \frac{SO_2}{2}$$

With methyl orange:

$H_2SO_3 + NaOH = NaHSO_3 + H_2O$

$$NaOH \equiv H_2SO_3 \equiv SO_2$$

Tellurium

$Na_2TeO_3 + 4NaI + 6HCl = 6NaCl + Te + 3H_2O + 2I_2$

$$I \equiv \frac{Na_2TeO_3}{4} \equiv \frac{TeO_2}{4} \equiv \frac{Te}{4}$$

$Na_2TeO_4 + 4HCl = H_2TeO_3 + 2NaCl + Cl_2 + H_2O$

$$I \equiv Cl \equiv \frac{Na_2TeO_4}{2} \equiv \frac{TeO_3}{2} \equiv \frac{Te}{2}$$

$3H_2TeO_3 + Cr_2O_7^{2-} + 8H^+ = 3H_2TeO_4 + 2Cr^{3+} + 4H_2O$

$$\frac{K_2Cr_2O_7}{6} \equiv \frac{H_2TeO_3}{2}$$

Tin

$SnCl_2 + I_2 + 2HCl = SnCl_4 + 2HI$

$$I \equiv \frac{SnCl_2}{2} \equiv \frac{Sn}{2}$$

$SnCl_2 + 2K_3Fe(CN)_6 + 2KCl = SnCl_4 + 2K_4Fe(CN)_6$

$$\frac{KMnO_4}{5} \equiv K_4Fe(CN)_6 \equiv \frac{SnCl_2}{2} \equiv \frac{Sn}{2}$$

$SnCl_2 + 2FeCl_3 = SnCl_4 + 2FeCl_2$

$$\frac{KMnO_4}{5} \equiv \frac{SnCl_2}{2} \equiv \frac{Sn}{2}$$

Titanium

$Ti_2(SO_4)_3 + Fe_2(SO_4)_3 = 2Ti(SO_4)_2 + 2FeSO_4$

$$\frac{KMnO_4}{5} \equiv FeSO_4 \equiv \frac{Ti_2(SO_4)_3}{2} \equiv Ti$$

Tungsten

$WO_3 + 2NaOH = Na_2WO_4 + H_2O$

$$NaOH \equiv \frac{W}{2} \equiv \frac{WO_3}{2}$$

Uranium

$$U_3O_8 + 4H_2SO_4 = 2(UO_2)SO_4 + U(SO_4)_2 + 4H_2O$$

$$5U(SO_4)_2 + 2KMnO_4 + 2H_2O = 2KHSO_4 + 2MnSO_4 + 5(UO_2)SO_4 + 2MnSO_4$$

$$\frac{KMnO_4}{5} \equiv \frac{U}{2} \equiv \frac{UO_2}{2} \equiv \frac{U(SO_4)_2}{2}$$

$$U_3O_8 \equiv 2UO_3 + UO_2$$

$$\frac{KMnO_4}{5} \equiv \frac{3U}{2} \equiv \frac{U_3O_8}{2}$$

$$3UO_2(NO_3)_2 + 5KI + KIO_3 + 3H_2O = 3UO_2(OH)_2 + 6KNO_3 + 3I_2$$

$$I \equiv \frac{U}{2} \equiv \frac{UO_2}{2} \equiv \frac{UO_2(NO_3)_2}{2}$$

Vanadium

$$V_2O_5 + 2KI + 2HCl = V_2O_4 + 2KCl + I_2 + H_2O$$

$$I \equiv \frac{V_2O_5}{2} \equiv V$$

$$V_2O_5 + H_2C_2O_4 = V_2O_4 + 2CO_2 + H_2O$$

$$\frac{KMnO_4}{5} \equiv \frac{H_2C_2O_4}{2} \equiv \frac{V_2O_5}{2} \equiv V$$

$$V_2O_5 + 3Zn(+ H_3PO_4) + 3H_2SO_4 = V_2O_2 + 3ZnSO_4 + 3H_2O$$

$$5V_2O_2 + 6KMnO_4 + 9H_2SO_4 = 5V_2O_5 + 6MnSO_4 + 3K_2SO_4 + 9H_2O$$

$$\frac{KMnO_4}{5} \equiv \frac{V_2O_5}{6} \equiv \frac{V_2O_2}{6} \equiv \frac{V}{3}$$

$$V_2O_5 + SO_2 = V_2O_4 + SO_3$$

$$5V_2O_4 + 2KMnO_4 + 3H_2SO_4 = 5V_2O_5 + 2MnSO_4 + K_2SO_4 + 3H_2O$$

$$\frac{KMnO_4}{5} \equiv \frac{V_2O_5}{2} \equiv \frac{V_2O_4}{2} \equiv V$$

$$V_2O_4 + 2HI = V_2O_3 + I_2 + H_2O$$

$$I \equiv \frac{V_2O_4}{2} \equiv V$$

$$V_2O_4 + I_2 + H_2O = V_2O_5 + 2HI$$

$$I \equiv \frac{V_2O_4}{2} \equiv V$$

$$V_2O_4 + 2K_3Fe(CN)_6 + 2KOH = V_2O_5 + 2K_4Fe(CN)_6 + H_2O$$

$$\frac{KMnO_4}{5} \equiv K_4Fe(CN)_6 \equiv \frac{V_2O_4}{2} \equiv V$$

$$2H_3VO_4 + 2FeSO_4 + 3H_2SO_4 = V_2O_2(SO_4)_2 + Fe_2(SO_4)_3 + 6H_2O$$

$$2FeSO_4 + MnO_2 + 2H_2SO_4 = Fe_2(SO_4)_3 + MnSO_4 + 2H_2O$$

$$5V_2O_2(SO_4)_2 + 2KMnO_4 + 22H_2O = 10H_3VO_4 + 2MnSO_4 + K_2SO_4 + 7H_2SO_4$$

$$\frac{KMnO_4}{5} \equiv \frac{V_2O_2(SO_4)_2}{2} \equiv \frac{V_2O_5}{2} \equiv V$$

Zinc

$$Zn + Fe_2(SO_4)_3 = ZnSO_4 + 2FeSO_4$$

$$\frac{KMnO_4}{5} \equiv FeSO_4 \equiv \frac{Zn}{2}$$

$$15Zn + 30NaOH = 15Na_2ZnO_2 + 15H_2$$

$$5KIO_3 + 15H_2 = 5KI + 15H_2O$$

$$5KI + KIO_3 + 3H_2SO_4 = 3K_2SO_4 + I_2 + 3H_2O$$

$$\frac{KIO_3}{6} \equiv I \equiv \frac{15Zn}{6}$$

$$3Zn + K_2Cr_2O_7 + 7H_2SO_4 = 3ZnSO_4 + Cr_2(SO_4)_3 + K_2SO_4 + 7H_2O$$

$$K_2Cr_2O_7 + 6FeSO_4 + 8H_2SO_4 = Cr_2(SO_4)_3 + 3Fe_2(SO_4)_3 + 2KHSO_4 + 7H_2O$$

$$\frac{K_2Cr_2O_7}{6} \equiv \frac{Zn}{2}$$

$$3ZnCl_2 + 2K_4Fe(CN)_6 = K_2Zn_3[Fe(CN)_6]_2 + 6KCl$$

$$K_4Fe(CN)_6 \equiv \frac{3Zn}{2} \equiv \frac{3ZnCl_2}{2}$$

$$15ZnSO_4 + 20KI + 4KIO_3 + 12H_2O = 3Zn_5(OH)_8SO_4 + 12K_2SO_4 + 12I_2$$

$$I \equiv \frac{15ZnSO_4}{24} \equiv \frac{15Zn}{24}$$

$$Na_2ZnO_2 + 4HCl = ZnCl_2 + 2NaCl + 2H_2O$$

$$HCl \equiv \frac{Na_2ZnO_2}{4}$$

$$ZnS + 2FeCl_3 = ZnCl_2 + 2FeCl_2 + S$$

$$FeCl_3 \frac{ZnS}{2} \equiv \frac{ZnO}{2} \equiv \frac{Zn}{2}$$

With EDTA:

$$Na_2H_2Y + Zn^{2+}\ [pH\ 4.5] = ZnY^{2-} + 2Na^+ + 2H^+$$

$$\frac{Na_2H_2Y}{1} \equiv \frac{Zn}{1}$$

11.6.2 Standard Volumetric (Titrimetric) Solutions

Alkaline arsenite, $0.1 \, N \, As^{3+}$ to As^{5+}. Dissolve 4.9460 g of sublimed As_2O_3 in 40 mL of 30% NaOH solution. Dilute with 200 mL of water. Acidify the solution with 6 N HCl to the acid color of methyl red indicator. Add to this solution 40 g of $NaHCO_3$ and dilute to 1 liter. This solution may be used as a standard for iodine solutions employing starch as the indicator for the titration and for ceric ion solutions using *ortho*-phenanthroline indicator. One milliliter of the 0.1 N arsenite is equivalent to 0.012690 g of iodine.

Ammonium thiocyanate, 0.1 N. Dissolve 7.6120 g of NH_4SCN in water and dilute to a liter. Standardize against standard silver nitrate solution, using a ferric alum indicator.

Ceric sulfate, $0.1 \, N \, Ce^{4+}$ to Ce^{3+}. Dissolve 63.26 g of ceric ammonium sulfate, $Ce(SO_4)_2 \cdot 2(NH_4)_2SO_4 \cdot 2H_2O$, in 500 mL of 2 N sulfuric acid. Dilute the solution to a liter and standardize against the alkaline arsenite solution as follows: Measure, accurately, 30 to 40 mL of arsenite solution into an Erlenmeyer flask and dilute to 150 mL. Add slowly, to prevent excessive frothing, 20 mL of 4 N sulfuric acid, 2 drops of 0.01 M osmium tetroxide solution, and 4 drops of *ortho*-phenanthroline ferrous complex indicator. Titrate with the ceric sulfate solution to a faint blue endpoint. Compute the normality of the ceric solution from the normality of the arsenite solution.

Ferrous ammonium sulfate (Mohr's salt), $0.1 \, N \, Fe^{2+}$ to Fe^{3+}. Dissolve 39.2139 g of $FeSO_4 \cdot (NH_4)_2SO_4 \cdot 6H_2O$ in 500 mL of 1 N H_2SO_4 and dilute to a liter. Standardize against standard permanganate or standard dichromate solution.

Hydrochloric acid, 0.1 N. Add about 8 mL of concentrated HCl (37%, density 1.19) to approximately 1 liter of distilled water. Standardize against weighed portions of pure sodium carbonate using methyl orange indicator.

Iodine, $0.1 \, N \, I^0$ to I^-. Dissolve 12.690 g of resublimed iodine in 25 mL of a solution containing 15 g of KI which is free from iodate. Triturate the solution until all the iodine has dissolved and then dilute to a liter. This solution can be used as a primary standard if the weighing is carefully done, or it can be checked against a standard thiosulfate solution or an alkaline arsenite solution, using starch as the indicator.

Potassium cyanide, for nickel titration, $0.1 \, N = 0.4 \times KCN = 26.0479$ g per liter. To make up this solution dissolve 26.0479 g of KCN in water, add 10 g of KOH to prevent hydrolysis, and dilute to 1 liter. For silver titration, $0.1 \, N = 0.2 \times KCN$. In this case use 13.0240 g of KCN and 5 g of KOH.

Potassium dichromate, $0.1 \, N \, Cr^{6+}$ to Cr^{3+}. Weigh out 4.9030 g of $K_2Cr_2O_7$ that has been oven-dried and cooled in a desiccator, and dilute with water to 1 liter.

Potassium hydroxide, 0.1 N. This solution should contain 5.6109 g of KOH per liter, but it is best to dissolve about 7 g in a half-liter of water, add enough $BaCl_2$ to precipitate any carbonate that may be present, allow the $BaCO_3$ to settle, filter into a liter flask, and dilute to the mark with CO_2-free water. Standardize with potassium acid phthalate, succinic, sulfuric, or hydrochloric acid, using a phenolphthalein or methyl orange indicator. The solution made up in this manner is slightly more than 0.1 N, but it can easily be adjusted after an acid titration. Alkali solutions must be protected from air by having the solution siphoned from a bottle. The incoming air is filtered through a guard tube containing soda-lime or Ascarite. To standardize, titrate accurately weighed portions of potassium acid phthalate of approximately 0.8 g. These are dissolved in 50 mL of CO_2-free water and titrated using phenolphthalein indicator.

Potassium iodate, $0.1 \, N \, I^{5+}$ to I^-. Dry the KIO_3 in an oven at 120°C, cool in a desiccator, and weigh out exactly 3.5667 g of KIO_3 and about 15 g of KI, dissolve in distilled water, and dilute to 1 liter. This solution can be used as a standard if the iodate is pure, free from iodide, and is carefully weighed. The purity can be checked by dissolving a few crystals of the KIO_3 in 2 mL of 2 N sulfuric acid to which is added starch indicator. Any blue coloration is evidence of iodide contamination.

Potassium permanganate, 0.1 N Mn^{7+} to Mn^{2+}. A 0.1 N solution contains 3.1608 g $KMnO_4$ per liter. Dissolve a little more than the calculated amount (about 3.3 g) in a liter of distilled water. Allow this to stand for 2 or 3 days, then siphon it carefully through clean glass tubes or filter it through a Gooch crucible into the container in which it is to be kept, discarding the first 25 mL and allowing the last inch of liquid to remain in the bottle. In this way any dust or reducing substance in the water is oxidized, and the MnO_2 so formed is removed from the final solution. Permanganate solutions should never be allowed to touch rubber, filter paper, or any other organic matter. To standardize the $KMnO_4$, weigh accurately samples of about 0.3 g of pure, dry sodium oxalate into Erlenmeyer flasks, add 150 mL of distilled water and 4 mL of concentrated H_2SO_4, and heat to 70°C. When all the sodium oxalate has been dissolved, run in rapidly approximately three-fourths of the calculated volume of permanganate. Continue the addition of permanganate but add the last milliliter drop by drop until a faint, permanent pink color is obtained. The temperature should be kept around 70°C throughout the titration. Each milliliter of 0.1 N permanganate is equal to 0.00670 g of sodium oxalate. The equation is

$$2KMnO_4 + 5Na_2C_2O_4 + 8H_2SO_4 = K_2SO_4 + 2MnSO_4 + 5Na_2SO_4 + 10CO_2 + 8H_2O.$$

Silver nitrate, 0.1 N. Weigh out exactly 16.9875 g of silver nitrate (dried 1 hour at 110°C) and dissolve in exactly 1 liter of distilled water.

Sodium chloride, 0.1 N. This solution contains 5.8443 g of NaCl per liter. If the salt is dried, weighed, and diluted to a definite volume, it may be used as a primary standard.

Sodium EDTA, 0.01 M. Dry the dihydrate ($Na_2H_2Y \cdot 2H_2O$) at 80°C to remove superficial moisture. After cooling, weigh about 3.8 g into a 1-L flask and dilute with distilled water. Standardize against a Mg^{2+} solution of known strength.

Sodium hydroxide, 0.1 N. This solution contains 3.9997 g of NaOH per liter. It is prepared in the same manner as the potassium hydroxide solution, using about 4 g for the 0.1 N solution and about 42 g for the normal solution. Standardize as directed for potassium hydroxide.

Sodium thiosulfate, 0.1 N. This solution contains 24.8183 g of $Na_2S_2O_3 \cdot 5H_2O$ per liter. Weigh out 25 g of $Na_2S_2O_3 \cdot 5H_2O$ and dilute to 1 liter. Add 0.5 g of sodium carbonate, 0.5 mL of chloroform, or 1% potassium furoate as a preservative. Standardize against resublimed iodine in the following manner: Place approximately 5 mL of a saturated solution of potassium iodide in a weighing bottle. Weigh the stoppered bottle with its contents accurately. Add 0.5 g of pure iodine to the iodide solution, stopper the weighing bottle, and reweigh. After the weighing, transfer the stoppered bottle with its contents to a beaker containing 200 mL of water. Remove the stopper of the weighing bottle with a glass rod and titrate the solution with the thiosulfate using starch indicator. From the weight of the iodine and the volume of titrant calculate the normality of the thiosulfate. Each milliliter of 0.1 N thiosulfate is equivalent to 0.012690 g of iodine.

Sulfuric acid, 1 N. This solution contains 49.0388 g of H_2SO_4 per liter. The 0.1 N solution contains 4.9039 g per liter. Pour 30 mL (3 mL for the 0.1 N) of pure concentrated sulfuric acid of density 1.84 into 3 or 4 volumes of water. This must be done slowly and carefully. Cool the mixture and dilute to 1 liter. Standardize against a known KOH or NaOH solution using phenolphthalein or against weighed portions of pure sodium carbonate using a methyl orange indicator.

TABLE 11.23 Titrimetric (Volumetric) Factors

Acids

The following factors are the equivalent of 1 mL of *normal acid*. Where the normality of the solution being used is other than normal, multiply the factors given in the table below by the normality of the solution employed.

The equivalents of the esters are based on the results of saponification.

The indicators methyl orange and phenolphthalein are indicated by the abbreviations MO and PH, respectively.

Substance	Formula	Grams
Ammonia	NH_3	0.017031
Ammonium	NH_4	0.018039
Ammonium chloride	NH_4Cl	0.053492
Ammonium hydroxide	NH_4OH	0.035046
Ammonium oleate	$C_{17}H_{33}CO_2NH_4$	0.29950
Ammonium oxide	$(NH_4)_2O$	0.026038
Amyl acetate	$CH_3CO_2C_5H_{11}$	0.13019
Barium carbonate (MO)	$BaCO_3$	0.09867
Barium hydroxide	$Ba(OH)_2$	0.085677
Barium oxide	BaO	0.07667
Bornyl acetate	$CH_3CO_2C_{10}H_{17}$	0.19629
Calcium carbonate (MO)	$CaCO_3$	0.05004
Calcium hydroxide	$Ca(OH)_2$	0.037047
Calcium oleate	$(C_{17}H_{33}CO_2)_2Ca$	0.30150
Calcium oxide	CaO	0.02804
Calcium stearate	$(C_{17}H_{35}CO_2)_2Ca$	0.30352
Casein (N 6.38)	. . . .	0.089371
Ethyl acetate	$CH_3CO_2C_2H_5$	0.088107
Glue (N 5.60)	. . . .	0.078445
Hydrochloric acid	HCl	0.036461
Magnesium carbonate (MO)	$MgCO_3$	0.04216
Magnesium oxide	MgO	0.02016
Menthyl acetate	$CH_3CO_2C_{10}H_{19}$	0.19831
Methyl acetate	$CH_3CO_2CH_3$	0.074080
Nicotine	$C_{10}H_{14}N_2$	0.16224
Nitrogen	N	0.014007
Potassium carbonate (MO)	K_2CO_3	0.06911
Potassium carbonate, acid (MO)	$KHCO_3$	0.10012
Potassium nitrate	KNO_3	0.10111
Potassium oleate	$C_{17}H_{33}CO_2K$	0.32057
Potassium oxide	K_2O	0.04710
Potassium stearate	$C_{17}H_{35}CO_2K$	0.32258
Protein (N 5.70)	. . . .	0.079846
Protein (N 6.25)	. . . .	0.087550
Sodium acetate	CH_3CO_2Na	0.082035
Sodium acetate	$CH_3CO_2Na \cdot 3H_2O$	0.13608
Sodium borate, tetra- (MO)	$Na_2B_4O_7$	0.10061
Sodium borate, tetra- (MO)	$Na_2B_4O_7 \cdot 10H_2O$	0.19069
Sodium carbonate (MO)	Na_2CO_3	0.052994
Sodium carbonate (MO)	$Na_2CO_3 \cdot H_2O$	0.062002
Sodium carbonate (MO)	$Na_2CO_3 \cdot 10H_2O$	0.14307
Sodium carbonate, acid (MO)	$NaHCO_3$	0.084007
Sodium hydroxide	$NaOH$	0.39997
Sodium oleate	$C_{17}H_{33}CO_2Na$	0.30445
Sodium oxalate	$Na_2C_2O_4$	0.067000

TABLE 11.23   Titrimetric (Volumetric) Factors (*Continued*)

Acids (*continued*)

Substance	Formula	Grams
Sodium oxide	Na_2O	0.030990
Sodium phosphate (MO)	Na_2HPO_4	0.14196
Sodium phosphate (MO)	$Na_2HPO_4 \cdot 12H_2O$	0.35814
Sodium phosphate (MO)	Na_3PO_4	0.081970
Sodium phosphate (PH)	Na_3PO_4	0.16394
Sodium silicate	$Na_2Si_4O_9$	0.15111
Sodium stearate	$C_{17}H_{35}CO_2Na$	0.30647
Sodium sulfide (MO)	Na_2S	0.039022

Alkali

The following factors are the equivalent of the milliliter of *normal alkali*. Where the normality of the solution being used is other than normal, multiply the factors given in the table below by the normality of the solution employed.

The equivalents of the esters are based on the results of saponification.

The indicators methyl orange and phenolphthalein are indicated by the abbreviations MO and PH, respectively.

Substance	Formula	Grams
Abietic acid (PH)	$HC_{20}H_{29}O_2$	0.30246
Acetic acid (PH)	CH_3CO_2H	0.06005
Acetic anhydride (PH)	$(CH_3CO)_2O$	0.051045
Aluminum sulfate	$Al_2(SO_4)_3$	0.05702
Amyl acetate	$CH_3CO_2C_5H_{11}$	0.13019
Benzoic acid (PH)	$C_6H_5CO_2H$	0.12212
Borate tetra- (PH)	B_4O_7	0.03881
Boric acid (PH)	H_3BO_3	0.061833
Boric anhydride (PH)	B_2O_3	0.03486
Bornyl acetate	$CH_3CO_2C_{10}H_{17}$	0.19629
Butyric acid (PH)	$C_3H_7CO_2H$	0.088107
Calcium acetate	$(CH_3CO_2)_2Ca$	0.079085
Calcium oleate	$(C_{17}H_{33}CO_2)_2Ca$	0.30150
Calcium stearate	$(C_{17}H_{35}CO_2)_2Ca$	0.30352
Carbon dioxide (PH)	CO_2	0.022005
Chlorine	Cl	0.035453
Citric acid (PH)	$H_3C_6H_5O_7 \cdot H_2O$	0.070047
Ethyl acetate	$CH_3CO_2C_2H_5$	0.088107
Formaldehyde	$HCHO$	0.030026
Formic acid (PH)	HCO_2H	0.046026
Glycerol (sap. of acetyl)	$C_3H_5(OH)_3$	0.030698
Hydriodic acid	HI	0.12791
Hydrobromic acid	HBr	0.080917
Hydrochloric acid	HCl	0.036461
Lactic acid (PH)	$HC_3H_5O_3$	0.090079
Lead acetate	$(CH_3CO_2)_2Pb \cdot 3H_2O$	0.18966
Maleic acid (PH)	$(CHCO_2H)_2$	0.058037
Malic acid (PH)	$H_2C_4H_4O_5$	0.067045
Menthol (sap. of acetyl)	$C_{10}H_{19}OH$	0.15627
Menthyl acetate	$CH_3CO_2C_{10}H_{19}$	0.19831
Methyl acetate	$CH_3CO_2CH_3$	0.074080

TABLE 11.23 Titrimetric (Volumetric) Factors (*Continued*)

Alkali (*continued*)

Substance	Formula	Grams
Nitrate	NO_3	0.062005
Nitric acid	HNO_3	0.063013
Nitrogen	N	0.014007
Nitrogen pentoxide	N_2O_5	0.054005
Oleic acid (PH)	$C_{17}H_{33}CO_2H$	0.28247
Oxalic acid (PH)	$(CO_2H)_2$	0.045018
Oxalic acid (PH)	$(CO_2H)_2 \cdot 2H_2O$	0.063033
Phosphoric acid (MO)	H_3PO_4	0.097995
Phosphoric acid (PH)	H_3PO_4	0.048998
Potassium carbonate, acid (MO)	$KHCO_3$	0.10012
Potassium oleate	$C_{17}H_{33}CO_2K$	0.32056
Potassium oxalate, acid (PH)	KHC_2O_4	0.12813
Potassium phthalate, acid (PH)	$HC_8H_4O_4K$	0.20423
Potassium stearate	$C_{17}H_{35}CO_2K$	0.32258
Sodium benzoate	$C_6H_5CO_2Na$	0.14411
Sodium borate, tetra- (PH)	$Na_2B_4O_7$	0.050305
Sodium borate, tetra- (PH)	$Na_2B_4O_7 \cdot 10H_2O$	0.095343
Sodium carbonate, acid (MO)	$NaHCO_3$	0.084007
Sodium oleate	$C_{17}H_{33}CO_2Na$	0.30445
Sodium salicylate	$C_6H_5OCO_2Na$	0.16011
Stearic acid (PH)	$C_{17}H_{35}CO_2H$	0.28449
Succinic acid (PH)	$(CH_2CO_2H)_2$	0.059045
Sulfate	SO_4	0.048031
Sulfur dioxide (PH)	SO_2	0.032031
Sulfur trioxide	SO_3	0.040031
Sulfuric acid	H_2SO_4	0.049039
Sulfurous acid (PH)	H_2SO_3	0.041039
Tartaric acid (PH)	$H_2C_4H_4O_6$	0.975044
Tartaric acid (PH)	$H_2C_4H_4O_6 \cdot H_2O$	0.084052

Iodine

The following factors are the equivalent of 1 mL of *normal iodine*. Where the normality of the solution being used is other than normal, multiply the factors given in the table below by the normality of the solution employed.

Substance	Formula	Grams
Acetone	$(CH_3)_2CO$	0.0096801
Ammonium chromate	$(NH_4)_2CrO_4$	0.050690
Antimony	Sb	0.06088
Antimony trioxide	Sb_2O_3	0.07287
Arsenic	As	0.037461
Arsenic pentoxide	As_2O_5	0.057460
Arsenic trioxide	As_2O_3	0.049460
Arsenite	AsO_3	0.061460
Bleaching powder	$CaOCl_2$	0.063493
Bromine	Br	0.079909
Chlorine	Cl	0.035453
Chromic oxide	Cr_2O_3	0.02533
Chromium trioxide	CrO_3	0.033331
Copper	Cu	0.06354
Copper oxide	CuO	0.07954
Copper sulfate	$CuSO_4$	0.15960

TABLE 11.23 Titrimetric (Volumetric) Factors (*Continued*)

Iodine (*continued*)

Substance	Formula	Grams
Copper sulfate	$CuSO_4 \cdot 5H_2O$	0.24968
Ferric iron	Fe^{3+}	0.05585
Ferric oxide	Fe_2O_3	0.07985
Hydrogen sulfide	H_2S	0.017040
Iodine	I	0.126904
Lead chromate	$PbCrO_4$	0.10773
Lead dioxide	PbO_2	0.11959
Nitrous acid	HNO_2	0.023507
Oxygen	O	0.0079997
Potassium chlorate	$KClO_3$	0.020426
Potassium chromate	K_2CrO_4	0.064733
Potassium dichromate	$K_2Cr_2O_7$	0.049032
Potassium nitrite	KNO_2	0.042554
Potassium permanganate	$KMnO_4$	0.031608
Red lead	Pb_3O_4	0.34278
Sodium chromate	Na_2CrO_4	0.053991
Sodium dichromate	$Na_2Cr_2O_7$	0.043661
Sodium dichromate	$Na_2Cr_2O_7 \cdot 2H_2O$	0.049666
Sodium nitrite	$NaNO_2$	0.034498
Sodium sulfide	Na_2S	0.039022
Sodium sulfide	$Na_2S \cdot 9H_2O$	0.12009
Sodium sulfite	Na_2SO_3	0.063021
Sodium sulfite	$Na_2SO_3 \cdot 7H_2O$	0.12607
Sodium thiosulfate	$Na_2S_2O_3$	0.15811
Sulfur	S	0.016032
Sulfur dioxide	SO_2	0.032031
Sulfurous acid	H_2SO_3	0.041039
Tin	Sn	0.059345

Potassium dichromate

The following factors are the equivalent of 1 mL of *normal potassium dichromate*. Where the normality of the solution being used is other than normal, multiply the factors given in the table below by the normality of the solution employed.

Substance	Formula	Grams
Chromic oxide	Cr_2O_3	0.025332
Chromium trioxide	CrO_3	0.033331
Ferrous iron	Fe^{2+}	0.055847
Ferrous oxide	FeO	0.071846
Ferroso-ferric oxide	Fe_3O_4	0.077180
Ferrous sulfate	$FeSO_4$	0.15191
Ferrous sulfate	$FeSO_4 \cdot 7H_2O$	0.27802
Glycerol	$C_3H_5(OH)_3$	0.0065782
Lead chromate	$PbCrO_4$	0.10773
Zinc	Zn	0.032685

TABLE 11.23 Titrimetric (Volumetric) Factors (*Continued*)

Potassium permanganate

The following factors are the equivalent of 1 mL of *normal potassium permanganate*. Where the normality of the solution being used is other than normal, multiply the factors given in the table below by the normality of the solution employed.

Substance	Formula	Grams
Ammonium oxalate	$(NH_4)_2C_2O_4$	0.062049
Ammonium oxalate	$(NH_4)_2C_2O_4 \cdot H_2O$	0.071056
Ammonium peroxydisulfate	$(NH_4)_2S_2O_8$	0.11410
Antimony	Sb	0.060875
Barium peroxide	BaO_2	0.084669
Barium peroxide	$BaO_2 \cdot 8H_2O$	0.15673
Calcium carbonate	$CaCO_3$	0.050045
Calcium oxide	CaO	0.02804
Calcium peroxide	CaO_2	0.036039
Calcium sulfate	$CaSO_4$	0.068071
Calcium sulfate	$CaSO_4 \cdot 2H_2O$	0.086086
Ferric oxide	Fe_2O_3	0.079846
Ferroso-ferric oxide	Fe_3O_4	0.077180
Ferrous ammonium sulfate	$Fe(NH_4)_2(SO_4)_2 \cdot 6H_2O$	0.39214
Ferrous oxide	FeO	0.071846
Ferrous sulfate	$FeSO_4$	0.15191
Ferrous sulfate	$FeSO_4 \cdot 7H_2O$	0.27802
Formic acid	HCO_2H	0.023013
Hydrogen peroxide	H_2O_2	0.017007
Iodine	I	0.126904
Iron	Fe	0.055847
Manganese	Mn	0.010988
Manganese dioxide	MnO_2	0.043468
Manganous oxide (Volhard)	MnO	0.035469
Molybdenum trioxide titration from yellow ppt. after reduction	MoO_3	0.047979
Oxalic acid	$(CO_2H)_2$	0.045018
Oxalic acid	$(CO_2H)_2 \cdot 2H_2O$	0.063033
Phosphorus titration from yellow ppt. after reduction	P	0.0008604
Phosphorus pentoxide to titration from yellow ppt. after reduction	P_2O_5	0.0019715
Potassium dichromate	$K_2Cr_2O_7$	0.049032
Potassium nitrite	KNO_2	0.042552
Potassium persulfate	$K_2S_2O_8$	0.13516
Sodium nitrite	$NaNO_2$	0.034498
Sodium oxalate	$Na_2C_2O_4$	0.067000
Sodium persulfate	$Na_2S_2O_8$	0.11905
Tin	Sn	0.059345

TABLE 11.23　Titrimetric (Volumetric) Factors (*Continued*)

Silver nitrate

The following factors are the equivalent of 1 mL of *normal silver nitrate.* Where the normality of the solution being used is other than normal, multiply the factors given in the table below by the normality of the solution employed.

Substance	Formula	Grams
Ammonium bromide	NH_4Br	0.097948
Ammonium chloride	NH_4Cl	0.053492
Ammonium iodide	NH_4I	0.14494
Ammonium thiocyanate	NH_4SCN	0.076120
Barium chloride	$BaCl_2$	0.10412
Barium chloride	$BaCl_2 \cdot 2H_2O$	0.12214
Bromine	Br	0.079909
Cadmium chloride	$CdCl_2$	0.091653
Cadmium iodide	CdI_2	0.18310
Calcium chloride	$CaCl_2$	0.055493
Chlorine	Cl	0.035453
Ferric chloride	$FeCl_3$	0.054069
Ferrous chloride	$FeCl_2$	0.063377
Hydriodic acid	HI	0.12791
Hydrobromic acid	HBr	0.080917
Hydrochloric acid	HCl	0.036461
Iodine	I	0.126904
Lithium chloride	LiCl	0.042392
Lead chloride	$PbCl_2$	0.13905
Magnesium chloride	$MgCl_2$	0.047609
Magnesium chloride	$MgCl_2 \cdot 6H_2O$	0.10166
Potassium bromide	KBr	0.11901
Potassium chloride	KCl	0.074555
Potassium iodide	KI	0.16601
Potassium oxide	K_2O	0.047102
Potassium thiocyanate	KSCN	0.097184
Silver	Ag	0.10787
Silver iodide	AgI	0.23477
Silver nitrate	$AgNO_3$	0.16987
Sodium bromide	NaBr	0.10290
Sodium bromide	$NaBr \cdot 2H_2O$	0.13893
Sodium chloride	NaCl	0.058443
Sodium iodide	NaI	0.14989
Sodium iodide	$NaI \cdot 2H_2O$	0.18592
Sodium oxide	Na_2O	0.030990
Strontium chloride	$SrCl_2$	0.079263
Strontium chloride	$SrCl_2 \cdot 6H_2O$	0.13331
Zinc chloride	$ZnCl_2$	0.068138

TABLE 11.23 Titrimetric (Volumetric) Factors (*Continued*)

Sodium thiosulfate

The following factors are the equivalent of 1 mL of *normal sodium thiosulfate.* Where the normality of the solution being used is other than normal, multiply the factors given in the table below by the normality of the solution employed.

Substance	Formula	Grams
Acetone	$(CH_3)_2CO$	0.0096801
Ammonium chromate	$(NH_4)_2CrO_4$	0.050690
Antimony	Sb	0.06088
Antimony trioxide	Sb_2O_3	0.07287
Bleaching powder	$CaOCl_2$	0.063493
Bromine	Br	0.079909
Chlorine	Cl	0.035453
Chromic oxide	Cr_2O_3	0.02533
Chromium trioxide	CrO_3	0.033331
Copper	Cu	0.06354
Copper oxide	CuO	0.07954
Copper sulfate	$CuSO_4$	0.15960
Copper sulfate	$CuSO_4 \cdot 5H_2O$	0.24968
Iodine	I	0.126904
Lead chromate	$PbCrO_4$	0.10773
Lead dioxide	PbO_2	0.11959
Nitrous acid	HNO_2	0.023507
Potassium chromate	K_2CrO_4	0.064733
Potassium dichromate	$K_2Cr_2O_7$	0.049032
Red lead	Pb_3O_4	0.34278
Sodium chromate	Na_2CrO_4	0.053991
Sodium dichromate	$Na_2Cr_2O_7$	0.043661
Sodium dichromate	$Na_2Cr_2O_7 \cdot 2H_2O$	0.049666
Sodium nitrite	$NaNO_2$	0.034498
Sodium thiosulfate	$Na_2S_2O_3$	0.15811
Sodium thiosulfate	$Na_2S_2O_3 \cdot 5H_2O$	0.24818
Sulfur	S	0.016032
Sulfur dioxide	SO_2	0.032031
Tin	Sn	0.059345

TABLE 11.24 Tolerances of Volumetric Flasks

Capacity, mL	Tolerances,* ±mL		Capacity, mL	Tolerances,* ±mL	
	Class A	Class B		Class A	Class B
5	0.02	0.04	200	0.10	0.20
10	0.02	0.04	250	0.12	0.24
25	0.03	0.06	500	0.20	0.40
50	0.05	0.10	1000	0.30	0.60
100	0.08	0.16	2000	0.50	1.00

* Accuracy tolerances for volumetric flasks at 20°C are given by ASTM standard E288.

TABLE 11.25 Pipet Capacity Tolerances

Volumetric transfer pipets			Measuring and serological pipets	
	Tolerances,* ±mL			Tolerances,† ±mL
Capacity, mL	Class A	Class B	Capacity, mL	Class B
0.5	0.006	0.012	0.1	0.005
1	0.006	0.012	0.2	0.008
2	0.006	0.012	0.25	0.008
3	0.01	0.02	0.5	0.01
4	0.01	0.02	0.6	0.01
5	0.01	0.02	1	0.02
10	0.02	0.04	2	0.02
15	0.03	0.06	5	0.04
20	0.03	0.06	10	0.06
25	0.03	0.06	25	0.10
50	0.05	0.10		
100	0.08	0.16		

  * Accuracy tolerances for volumetric transfer pipets are given by ASTM standard E969 and Federal Specification NNN-P-395.
  † Accuracy tolerances for measuring pipets are given by Federal Specification NNN-P-350 and for serological pipets by Federal Specification NNN-P-375.

TABLE 11.26   Tolerances of Micropipets (Eppendorf)

Capacity, μL	Accuracy, %	Precision, %	Capacity, μL	Accuracy, %	Precision, %
10	1.2	0.4	100	0.5	0.2
40	0.6	0.2	250	0.5	0.15
50	0.5	0.2	500	0.5	0.15
60	0.5	0.2	600	0.5	0.15
70	0.5	0.2	900	0.5	0.15
80	0.5	0.2	1000	0.5	0.15

TABLE 11.27 Buret Accuracy Tolerances

		Accuracy, ±mL	
Capacity, mL	Subdivision, mL	Class A* and precision grade	Class B and standard grade
10	0.05	0.02	0.04
25	0.10	0.03	0.06
50	0.10	0.05	0.10
100	0.20	0.10	0.20

  * Class A conforms to specifications in ASTM E694 for standard taper stopcocks and to ASTM E287 for Teflon or polytetrafluoroethylene stopcock plugs. The 10-mL size meets the requirements for ASTM D664.

TABLE 11.28 Factors for Simplified Computation of Volume

The volume is determined by weighing the water, having a temperature of $t°C$, contained in or delivered by the apparatus at the same temperature. The weight of water, w grams, is obtained with brass weights in air having a density of 1.20 mg/mL.

For apparatus made of soft glass, the volume contained or delivered at 20°C is given by

$$v_{20} = wf_{20} \quad mL$$

where v_{20} is the volume at 20° and f_{20} is the factor (apparent specific volume) obtained from the table below for the temperature t at which the calibration is performed. The volume at any other temperature t' may then be obtained from

$$v' = v_{20}[1 + 0.00002(t' - 20)] \quad mL$$

For apparatus made of any other material, the volume contained or delivered at the temperature t is

$$v_t = wf_t \quad mL$$

where w is again the weight in air obtained with brass weights (in grams), and f_t is the factor given in the third column of the table for the temperature t. The volume at any temperature t' may then be obtained from

$$v'_t = v_t[1 + \beta(t' - t)] \quad mL$$

where β is the cubical coefficient of thermal expansion of the material from which the apparatus is made. Approximate values of β for some frequently encountered materials are given in Table 11.29.

t, °C	f_{20}	f_t	t, °C	f_{20}	f_t
0	1.001 62	1.001 22	20	1.002 86	1.002 86
1	54	16	21	1.003 05	1.003 07
2	48	12	22	26	30
3	43	09	23	47	53
4	41	09	24	69	77
5	1.001 39	1.001 09	25	1.003 93	1.004 03
6	40	12	26	1.004 17	29
7	42	16	27	42	56
8	45	21	28	68	84
9	50	28	29	95	1.005 13
10	1.001 56	1.001 36	30	1.005 23	1.005 43
11	63	45	31	1.005 52	1.005 74
12	72	56	32	1.005 82	1.006 06
13	82	68	33	1.006 13	1.006 39
14	93	81	34	1.006 44	1.006 72
15	1.002 06	1.001 96	35	1.006 77	1.007 07
16	20	1.002 12	36	1.007 10	1.007 42
17	35	29	37	1.007 44	1.007 78
18	51	47	38	1.007 79	1.008 15
19	68	66	39	1.008 15	1.008 53
			40	1.008 52	1.008 91

TABLE 11.29 Cubical Coefficients of Thermal Expansion

This table lists values of β, the cubical coefficient of thermal expansion, taken from "Essentials of Quantitative Analysis," by Benedetti-Pichler, and from various other sources. The value of β represents the relative increases in volume for a change in temperature of $1\,°C$ at temperatures in the vicinity of $25\,°C$, and is equal to 3α, where α is the linear coefficient of thermal expansion. Data are given for the types of glass from which volumetric apparatus is most commonly made, and also for some other materials which have been or may be used in the fabrication of apparatus employed in analytical work.

Material	β
Glasses	
Alkali-resistant, Corning 728	1.90×10^{-5}
Gerateglas, Schott G20	1.47
Kimble KG-33 (borosilicate)	0.96
N-51A ("Resistant")	1.47
R-6 (soft)	2.79
Pyrex, Corning 744	0.96
Vitreous silica	0.15
Vycor, Corning 790	0.24
Metals	
Brass	*ca.* 5.5
Copper	5.0
Gold	4.3
Monel metal	4.0
Platinum	2.7
Silver	5.7
Stainless steel	*ca.* 5.3
Tantalum	*ca.* 2.0
Tungsten	1.3
Plastics and other materials	
Hard rubber	24×10^{-5}
Polyethylene	45–90
Polystyrene	18–24
Porcelain	*ca.* 1.2
Teflon (polytetrafluoroethylene)	16.5

TABLE 11.30 Values for α_4 for EDTA in Solutions of Various pH

pH	α_4	$-\log \alpha_4$	pH	α_4	$-\log \alpha_4$
2.0	3.7×10^{-14}	13.44	8.0	5.4×10^{-3}	2.29
3.0	2.5×10^{-11}	10.60	9.0	0.052	1.29
4.0	3.6×10^{-9}	8.48	10.0	0.35	0.46
5.0	3.5×10^{-7}	6.45	11.0	0.85	0.07
6.0	2.2×10^{-5}	4.66	12.0	0.98	0.00
7.0	4.8×10^{-4}	3.33			

TABLE 11.31 Formation Constants of EDTA Complexes

Metal ion	$\log K_{MY}$	Metal ion	$\log K_{MY}$
Cr(III)	36	Ni(II)	18.56
Co(III)	36	Pb(II)	18.3
In(III)	25	Zn(II)	16.4
Fe(III)	24.23	Cd(II)	16.4
Th(IV)	23.2	La(III)	16.34
Bi(III)	22.8	Al(III)	16.11
Hg(II)	21.80	Fe(II)	14.33
Zr(IV)	19.40	Mn(II)	13.8
Cu(II)	18.7	Ca(II)	11.0
Co(II)	16.3	Mg(II)	8.64
		Sr(II)	8.80

TABLE 11.32 General Solubility Rules for Inorganic Compounds

Nitrates	All nitrates are soluble.
Acetates	All acetates are soluble; silver acetate is moderately soluble.
Chlorides	All chlorides are soluble except AgCl, $PbCl_2$, and Hg_2Cl_2. $PbCl_2$ is soluble in hot water, slightly soluble in cold water.
Sulfates	All sulfates are soluble except barium and lead. Silver, mercury(I), and calcium are only slightly soluble.
Hydrogen sulfates	The hydrogen sulfates are more soluble than the sulfates.
Carbonates, phosphates, chromates, silicates	All carbonates, phosphates, chromates, and silicates are insoluble, except those of sodium, potassium, and ammonium. An exception is $MgCrO_4$ which is soluble.
Hydroxides	All hydroxides (except lithium, sodium, potassium, cesium, rubidium, and ammonium) are insoluble; $Ba(OH)_2$ is moderately soluble; $Ca(OH)_2$ and $Sr(OH)_2$ are slightly soluble.
Sulfides	All sulfides (except alkali metals, ammonium, magnesium, calcium, and barium) are insoluble. Aluminum and chromium sulfides are hydrolyzed and precipitate as hydroxides.
Sodium, potassium, ammonium	All sodium, potassium, and ammonium salts are soluble. Exceptions: $Na_4Sb_2O_7$, $K_2NaCo(NO_2)_6$, K_2PtCl_6, $(NH_4)_2PtCl_6$, and $(NH_4)_2NaCo(NO_2)_6$.
Silver	All silver salts are insoluble. Exceptions: $AgNO_3$ and $AgClO_4$; $AgC_2H_3O_2$ and Ag_2SO_4 are moderately soluble.

11.7 LABORATORY SOLUTIONS

TABLE 11.33   Concentrations of Commonly Used Acids and Bases

Freshly opened bottles of these reagents are generally of the concentrations indicated in the table. This may not be true of bottles long opened and this is especially true of ammonium hydroxide, which rapidly loses its strength. In preparing volumetric solutions, it is well to be on the safe side and take a little more than the calculated volume of the concentrated reagent, since it is much easier to dilute a concentrated solution than to strengthen one that is too weak.

A concentrated C.P. reagent usually comes to the laboratory in a bottle having a label which states its molecular weight w, its density (or its specific gravity) d, and its percentage assay p. When such a reagent is used to prepare an aqueous solution of desired molarity M, a convenient formula to employ is

$$V = \frac{100wM}{pd}$$

where V is the number of milliliters of concentrated reagent required for 1 liter of the dilute solution.

Example: Sulfuric acid has the molecular weight 98.08. If the concentrated acid assays 95.5% and has the specific gravity 1.84, the volume required for 1 liter of a 0.1 molar solution is

$$V = \frac{100 \times 98.08 \times 0.1}{95.5 \times 1.84} = 5.58 \text{ mL}$$

Reagent	Formula Weight	Density, g · mL^{-1} (20°C)	Weight % (approx)	Molarity	V. mL*
Acetic acid	60.05	1.05	99.8	17.45	57.3
Ammonium hydroxide	35.05	0.90	56.6	14.53	60.0
(as NH$_3$)	17.03		28.0		
Ethylenediamine	60.10	0.899	100	15.0	66.7
Formic acid	46.03	1.20	90.5	23.6	42.5
Hydrazine	32.05	1.011	95	30.0	33.3
Hydriodic acid	127.91	1.70	57	7.6	132
Hydrobromic acid	80.92	1.49	48	8.84	113
Hydrochloric acid	36.46	1.19	37.2	12.1	82.5
Hydrofluoric acid	20.0	1.18	49.0	28.9	34.5
Nitric acid	63.01	1.42	70.4	15.9	63.0
Perchloric acid	100.47	1.67	70.5	11.7	85.5
Phosphoric acid	97.10	1.70	85.5	14.8	67.5
Pyridine	79.10	0.982	100	12.4	80.6
Potassium hydroxide (soln)	56.11	1.46	45	11.7	85.5
Sodium hydroxide (soln)	40.00	1.54	50.5	19.4	51.5
Sulfuric acid	98.08	1.84	96.0	18.0	55.8
Triethanolamine	149.19	1.124	100	7.53	132.7

*V, mL = volume in milliliters needed to prepare 1 liter of 1 molar solution.

TABLE 11.34 Standard Stock Solutions*

Element	Procedure
Aluminum	Dissolve 1.000 g Al wire in minimum amount of 2 M HCl; dilute to volume.
Antimony	Dissolve 1.000 g Sb in (1) 10 ml HNO_3 plus 5 ml HCl, and dilute to volume when dissolution is complete; or (2) 18 ml HBr plus 2 ml liquid Br_2; when dissolution is complete add 10 ml $HClO_4$, heat in a well-ventilated hood while swirling until white fumes appear and continue for several minutes to expel all HBr, then cool and dilute to volume.
Arsenic	Dissolve 1.3203 g of As_2O_3 in 3 ml 8 M HCl and dilute to volume; or treat the oxide with 2 g NaOH and 20 ml water; after dissolution dilute to 200 ml, neutralize with HCl (pH meter), and dilute to volume.
Barium	(1) Dissolve 1.7787 g $BaCl_2 \cdot 2H_2O$ (fresh crystals) in water and dilute to volume. (2) Dissolve 1.516 g $BaCl_2$ (dried at 250°C for 2 hr) in water and dilute to volume. (3) Treat 1.4367 g $BaCO_3$ with 300 ml water, slowly add 10 ml of HCl and, after the CO_2 is released by swirling, dilute to volume.
Beryllium	(1) Dissolve 19.655 g $BeSO_4 \cdot 4H_2O$ in water, add 5 ml HCl (or HNO_3), and dilute to volume. (2) Dissolve 1.000 g Be in 25 ml 2 M HCl, then dilute to volume.
Bismuth	Dissolve 1.000 g Bi in 8 ml of 10 M HNO_3, boil gently to expel brown fumes, and dilute to volume.
Boron	Dissolve 5.720 g fresh crystals of H_3BO_3 and dilute to volume.
Bromine	Dissolve 1.489 g KBr (or 1.288 g NaBr) in water and dilute to volume.
Cadmium	(1) Dissolve 1.000 g Cd in 10 ml of 2 M HCl; dilute to volume. (2) Dissolve 2.282 g $3CdSO_4 \cdot 8H_2O$ in water; dilute to volume.
Calcium	Place 2.4973 g $CaCO_3$ in volumetric flask with 300 ml water, carefully add 10 ml HCl; after CO_2 is released by swirling, dilute to volume.
Cerium	(1) Dissolve 4.515 g $(NH_4)_4Ce(SO_4)_4 \cdot 2H_2O$ in 500 ml water to which 30 ml H_2SO_4 had been added, cool, and dilute to volume. Advisable to standardize against As_2O_3. (2) Dissolve 3.913 g $(NH_4)_2Ce(NO_3)_6$ in 10 ml H_2SO_4, stir 2 min, cautiously introduce 15 ml water and again stir 2 min. Repeat addition of water and stirring until all the salt has dissolved, then dilute to volume.
Cesium	Dissolve 1.267 g CsCl and dilute to volume. Standardize: Pipette 25 ml of final solution to Pt dish, add 1 drop H_2SO_4, evaporate to dryness, and heat to constant weight at $>$ 800°C. Cs (in μg/ml) = (40)(0.734)(wt of residue)
Chlorine	Dissolve 1.648 g NaCl and dilute to volume.
Chromium	(1) Dissolve 2.829 g $K_2Cr_2O_7$ in water and dilute to volume. (2) Dissolve 1.000 g Cr in 10 ml HCl, and dilute to volume.
Cobalt	Dissolve 1.000 g Co in 10 ml of 2 M HCl, and dilute to volume.
Copper	(1) Dissolve 3.929 g fresh crystals of $CuSO_4 \cdot 5H_2O$, and dilute to volume. (2) Dissolve 1.000 g Cu in 10 ml HCl plus 5 ml water to which HNO_3(or 30%H_2O_2) is added dropwise until dissolution is complete. Boil to expel oxides of nitrogen and chlorine, then dilute to volume.
Dysprosium	Dissolve 1.1477 g Dy_2O_3 in 50 ml of 2 M HCl; dilute to volume.

* 1000 μg/mL as the element in a final volume of 1 liter unless stated otherwise.

From J. A. Dean and T. C. Rains, "Standard Solutions for Flame Spectrometry," in *Flame Emission and Atomic Absorption Spectrometry*, J. A. Dean and T. C. Rains (Eds.), Vol. 2, Chap. 13, Marcel Dekker, New York, 1971.

TABLE 11.34 Standard Stock Solutions (*Continued*)

Element	Procedure
Erbium	Dissolve 1.1436 g Er_2O_3 in 50 ml of 2 *M* HCl; dilute to volume.
Europium	Dissolve 1.1579 g Eu_2O_3 in 50 ml of 2 *M* HCl; dilute to volume.
Fluorine	Dissolve 2.210 g NaF in water and dilute to volume.
Gadolinium	Dissolve 1.152 g Gd_2O_3 in 50 ml of 2 *M* HCl; dilute to volume.
Gallium	Dissolve 1.000 g Ga in 50 ml of 2 *M* HCl; dilute to volume.
Germanium	Dissolve 1.4408 g GeO_2 with 50 g oxalic acid in 100 ml of water; dilute to volume.
Gold	Dissolve 1.000 g Au in 10 ml of hot HNO_3 by dropwise addition of HCl, boil to expel oxides of nitrogen and chlorine, and dilute to volume. Store in amber container away from light.
Hafnium	Transfer 1.000 g Hf to Pt dish, add 10 ml of 9 *M* H_2SO_4, and then slowly add HF dropwise until dissolution is complete. Dilute to volume with 10% H_2SO_4.
Holmium	Dissolve 1.1455 g Ho_2O_3 in 50 ml of 2 *M* HCl; dilute to volume.
Indium	Dissolve 1.000 g In in 50 ml of 2 *M* HCl; dilute to volume.
Iodine	Dissolve 1.308 g KI in water and dilute to volume.
Iridium	(1) Dissolve 2.465 g Na_3IrCl_6 in water and dilute to volume. (2) Transfer 1.000 g Ir sponge to a glass tube, add 20 ml of HCl and 1 ml of $HClO_4$. Seal the tube and place in an oven at 300°C for 24 hr. Cool, break open the tube, transfer the solution to a volumetric flask, and dilute to volume. Observe all safety precautions in opening the glass tube.
Iron	Dissolve 1.000 g Fe wire in 20 ml of 5 *M* HCl; dilute to volume.
Lanthanum	Dissolve 1.1717 g La_2O_3 (dried at 110°C) in 50 ml of 5 *M* HCl, and dilute to volume.
Lead	(1) Dissolve 1.5985 g $Pb(NO_3)_2$ in water plus 10 ml HNO_3, and dilute to volume. (2) Dissolve 1.000 g Pb in 10 ml HNO_3, and dilute to volume.
Lithium	Dissolve a slurry of 5.3228 g Li_2CO_3 in 300 ml of water by addition of 15 ml HCl; after release of CO_2 by swirling, dilute to volume.
Lutetium	Dissolve 1.6079 g $LuCl_3$ in water and dilute to volume.
Magnesium	Dissolve 1.000 g Mg in 50 ml of 1 *M* HCl and dilute to volume.
Manganese	(1) Dissolve 1.000 g Mn in 10 ml HCl plus 1 ml HNO_3, and dilute to volume. (2) Dissolve 3.0764 g $MnSO_4 \cdot H_2O$ (dried at 105°C for 4 hr) in water and dilute to volume. (3) Dissolve 1.5824 g MnO_2 in 10 HCl in a good hood, evaporate to gentle dryness, dissolve residue in water and dilute to volume.
Mercury	Dissolve 1.000 g Hg in 10 ml of 5 *M* HNO_3 and dilute to volume.
Molybdenum	(1) Dissolve 2.0425 g $(NH_4)_2MoO_4$ in water and dilute to volume. (2) Dissolve 1.5003 g MoO_3 in 100 ml of 2 *M* ammonia, and dilute to volume.
Neodymium	Dissolve 1.7373 g $NdCl_3$ in 100 ml 1 *M* HCl and dilute to volume.
Nickel	Dissolve 1.000 g Ni in 10 ml hot HNO_3, cool, and dilute to volume.
Niobium	Transfer 1.000 g Nb (or 1.4305 g Nb_2O_5) to Pt dish, add 20 ml HF, and heat gently to complete dissolution. Cool, add 40 ml H_2SO_4, and evaporate to fumes of SO_3. Cool and dilute to volume with 8 *M* H_2SO_4.
Osmium	Dissolve 1.3360 g OsO_4 in water and dilute to 100 ml. Prepare only as needed as solution loses strength on standing unless Os is reduced by SO_2 and water is replaced by 100 ml 0.1 *M* HCl.
Palladium	Dissolve 1.000 g Pd in 10 ml of HNO_3 by dropwise addition of HCl to hot solution; dilute to volume.
Phosphorus	Dissolve 4.260 g $(NH_4)_2HPO_4$ in water and dilute to volume.
Platinum	Dissolve 1.000 g Pt in 40 ml of hot aqua regia, evaporate to incipient dryness, add 10 ml HCl and again evaporate to moist residue. Add 10 ml HCl and dilute to volume.
Potassium	Dissolve 1.9067 g KCl (or 2.8415 g KNO_3) in water and dilute to volume.
Praseodymium	Dissolve 1.1703 g Pr_2O_3 in 50 ml of 2 *M* HCl; dilute to volume.
Rhenium	Dissolve 1.000 g Re in 10 ml of 8 *M* HNO_3 in an ice bath until initial reaction subsides, then dilute to volume.

TABLE 11.34 Standard Stock Solutions (*Continued*)

Element	Procedure
Rhodium	Dissolve 1.000 g Rh by the sealed-tube method described under iridium.
Rubidium	Dissolve 1.4148 g RbCl in water. Standardize as described under cesium. Rb (in $\mu g/ml$) = (40)(0.320)(wt of residue).
Ruthenium	Dissolve 1.317 g RuO_2 in 15 ml of HCl; dilute to volume.
Samarium	Dissolve 1.1596 g Sm_2O_3 in 50 ml of 2 M HCl; dilute to volume.
Scandium	Dissolve 1.5338 g Sc_2O_3 in 50 ml of 2 M HCl; dilute to volume.
Selenium	Dissolve 1.4050 g SeO_2 in water and dilute to volume or dissolve 1.000 g Se in 5 ml of HNO_3, then dilute to volume.
Silicon	Fuse 2.1393 g SiO_2 with 4.60 g Na_2CO_3, maintaining melt for 15 min in Pt crucible. Cool, dissolve in warm water, and dilute to volume. Solution contains also 2000 $\mu g/ml$ sodium.
Silver	(1) Dissolve 1.5748 g $AgNO_3$ in water and dilute to volume. (2) Dissolve 1.000 g Ag in 10 ml of HNO_3; dilute to volume. Store in amber glass container away from light.
Sodium	Dissolve 2.5421 g NaCl in water and dilute to volume.
Strontium	Dissolve a slurry of 1.6849 g $SrCO_3$ in 300 ml of water by careful addition of 10 ml of HCl; after release of CO_2 by swirling, dilute to volume.
Sulfur	Dissolve 4.122 g $(NH_4)_2SO_4$ in water and dilute to volume.
Tantalum	Transfer 1.000 g Ta (or 1.2210 g Ta_2O_5) to Pt dish, add 20 ml of HF, and heat gently to complete the dissolution. Cool, add 40 ml of H_2SO_4 and evaporate to heavy fumes of SO_3. Cool and dilute to volume with 50% H_2SO_4.
Tellurium	(1) Dissolve 1.2508 g TeO_2 in 10 ml of HCl; dilute to volume. (2) Dissolve 1.000 g Te in 10 ml of warm HCl with dropwise addition of HNO_3, then dilute to volume.
Terbium	Dissolve 1.6692 g of $TbCl_3$ in water, add 1 ml of HCl, and dilute to volume.
Thallium	Dissolve 1.3034 g $TlNO_3$ in water and dilute to volume.
Thorium	Dissolve 2.3794 g $Th(NO_3)_4 \cdot 4H_2O$ in water, add 5 ml HNO_3, and dilute to volume.
Thulium	Dissolve 1.142 g Tm_2O_3 in 50 ml of 2 M HCl; dilute to volume.
Tin	Dissolve 1.000 g Sn in 15 ml of warm HCl; dilute to volume.
Titanium	Dissolve 1.000 g Ti in 10 ml of H_2SO_4 with dropwise addition of HNO_3; dilute to volume with 5% H_2SO_4.
Tungsten	Dissolve 1.7941 g of $Na_2WO_4 \cdot 2H_2O$ in water and dilute to volume.
Uranium	Dissolve 2.1095 g $UO_2(NO_3)_2 \cdot 6H_2O$ (or 1.7734 g uranyl acetate dihydrate) in water and dilute to volume.
Vanadium	Dissolve 2.2963 g NH_4VO_3 in 100 ml of water plus 10 ml of HNO_3; dilute to volume.
Ytterbium	Dissolve 1.6147 g $YbCl_3$ in water and dilute to volume.
Yttrium	Dissolve 1.2692 g Y_2O_3 in 50 ml of 2 M HCl and dilute to volume.
Zinc	Dissolve 1.000 g Zn in 10 ml of HCl; dilute to volume.
Zirconium	Dissolve 3.533 g $ZrOCl_2 \cdot 8H_2O$ in 50 ml of 2 M HCl, and dilute to volume. Solution should be standardized.

11.7.1 General Reagents, Indicators, and Special Solutions

Unless otherwise stated, the term *g per liter* signifies grams of the formula indicated dissolved in water and made up to a liter of solution.

Acetic acid, $HC_2H_3O_2$—6N: 350 mL glacial acetic acid per liter.

Alcohol, amyl, $C_5H_{11}OH$: use as purchased.

Alcohol, ethyl, C_2H_5OH; 95% alcohol, as purchased.

Alizarin, dihydroxyanthraquinone (indicator): dissolve 0.1 g in 100 mL alcohol; pH range yellow 5.5–6.8 red.

Alizarin yellow R, sodium p-nitrobenzeneazosalicylate (indicator): dissolve 0.1 g in 100 mL water; pH range yellow 10.1–violet 12.1.

Alizarin yellow GG, salicyl yellow, sodium m-nitrobenzeneazosalicylate (indicator); dissolve 0.1 g in 100 mL 50% alcohol; pH range yellow 10.0–12.0 lilac.

Alizarin S, alizarin carmine, sodium alizarin sulfonate (indicator): dissolve 0.1 g in 100 mL water; pH range yellow 3.7–5.2 violet.

Aluminon (qualitative test for aluminum). The reagent consists of 0.1% solution of the ammonium salt of aurin tricarboxylic acid. A bright red precipitate, persisting in alkaline solution, indicates aluminum.

Aluminum chloride, $AlCl_3$ — $0.5N$: 22 g per liter.

Aluminum nitrate, $Al(NO_3)_3 \cdot 7.5H_2O$ — $0.5N$: 58 g per liter.

Aluminum sulfate, $Al_2(SO_4)_3 \cdot 18H_2O$ — $0.5N$: 55 g per liter.

Ammonium acetate, $NH_4C_2H_3O_2$ — $3N$: 231 g per liter.

Ammonium carbonate, $(NH_4)_2CO_3 \cdot H_2O$ — $3N$: 171 g per liter; for the anhydrous salt: 144 g per liter.

Ammonium chloride, NH_4Cl — $3N$: 161 g per liter.

Ammonium hydroxide, NH_4OH — $15N$: the concentrated solution which contains 28% NH_3; for $6N$: 400 mL per liter.

Ammonium molybdate, $(NH_4)_2MoO_4$ — N: dissolve 88.3 g of solid $(NH_4)_6Mo_7O_{24} \cdot 4H_2O$ in 100 mL $6N\,NH_4OH$. Add 240 g of solid NH_4NO_3 and dilute to 1 liter. Another method is to take 72 g of MoO_3, add 130 mL of water and 75 mL of $15N\,NH_4OH$; stir mechanically until nearly all has dissolved, then add it to a solution of 240 mL concentrated HNO_3 and 500 mL of water; stir continuously while solutions are being mixed; allow to stand 3 days, filter, and use the clear filtrate.

Ammonium nitrate, NH_4NO_3 — N: 80 g per liter.

Ammonium oxalate, $(NH_4)_2C_2O_4 \cdot H_2O$ — $0.5N$: 40 g per liter.

Ammonium polysulfide (yellow ammonium sulfide), $(NH_4)_2S_x$: allow the colorless $(NH_4)_2S$ to stand, or add sulfur.

Ammonium sulfate, $(NH_4)_2SO_4$ — $0.5\ N$: 33 g per liter; saturated: dissolve 780 g of $(NH_4)_2SO_4$ in water and make up to a liter.

Ammonium sulfide (colorless), $(NH_4)_2S$ — saturated: pass H_2S through 200 mL of concentrated NH_4OH in the cold until no more gas is dissolved, add 200 mL NH_4OH and dilute with water to a liter; the addition of 15 g of sulfur is sufficient to make the polysulfide.

Antimony pentachloride, $SbCl_5$ — $0.5N$: 39 g per liter.

Antimony trichloride, $SbCl_3$ — $0.5N$: 38 g per liter.

Aqua regia: mix 3 parts of concentrated HCl and 1 part of concentrated HNO_3 just before ready to use.

Arsenic acid, $H_3AsO_4 \cdot 0.5H_2O$ — $0.5N$ ($= \frac{1}{2}H_3AsO_4 \div 5$): 15 g per liter.

Arsenous oxide, As_2O_3 — $0.25N$: 8 g per liter for saturation.

Aurichloric acid, $HAuCl_4 \cdot 3H_2O$: dissolve in ten parts of water.

Aurin, *see* rosolic acid.

Azolitmin solution (indicator); make up a 1% solution of azolitmin by boiling in water for 5 minutes; it may be necessary to add a small amount of NaOH to make the solution neutral; pH range red 4.5–8.3 blue.

Bang's reagent (for glucose estimation): dissolve 100 g of K_2CO_3, 66 g of KCl, and 160 of $KHCO_3$ in the order given in about 700 mL of water at 30°C. Add 4.4 g of copper sulfate and dilute to 1 liter after the CO_2 is evolved. This solution should be shaken only in such a manner as not to allow the entry of air. After 24 hours 300 mL are diluted to a liter with saturated KCl solution, shaken gently and used after 24 hours; 50 mL $\equiv$ 10 mg glucose.

Barfoed's reagent (test for glucose): dissolve 66 g of cupric acetate and 10 mL of glacial acetic acid in water and dilute to 1 liter.

Barium chloride, $BaCl_2 \cdot 2H_2O$—0.5N: 61 g per liter.

Barium hydroxide, $Ba(OH)_2 \cdot 8H_2O$—0.2N: 32 g per liter for saturation.

Barium nitrate, $Ba(NO_3)_2$—0.5N: 65 g per liter.

Baudisch's reagent: *see* cupferron.

Benedict's qualitative reagent (for glucose): dissolve 173 g of sodium citrate and 100 g of anhydrous sodium carbonate in about 600 mL of water, and dilute to 850 mL; dissolve 17.3 g of $CuSO_4 \cdot 5H_2O$ in 100 mL of water and dilute to 150 mL; this solution is added to the citrate-carbonate solution with constant stirring. *See also* the quantitative reagent below.

Benedict's quantitative reagent (sugar in urine): This solution contains 18 g copper sulfate, 100 g of anhydrous sodium carbonate, 200 g of potassium citrate, 125 g of potassium thiocyanate, and 0.25 g of potassium ferrocyanide per liter; 1 mL of this solution $\equiv$ 0.002 sugar.

Benzidine hydrochloride solution (for sulfate determination): mix 6.7 g of benzidine $[C_{12}H_8(NH_2)_2]$ or 8.0 g of the hydrochloride $[C_{12}H_8(NH_2)_2 \cdot 2HCl]$ into a paste with 20 mL of water; add 20 mL of HCl (sp. gr. 1.12) and dilute the mixture to 1 liter with water; each mL of this solution is equivalent to 0.00357 g H_2SO_4.

Benzopurpurine 4B (indicator): dissolve 0.1 g in 100 mL water; pH range blue-violet 1.3–4.0 red.

Benzoyl auramine (indicator): dissolve 0.25 g in 100 mL methyl alcohol; pH range violet 5.0–5.6 pale yellow. Since this compound is not stable in aqueous solution, hydrolyzing slowly in neutral medium, more rapidly in alkaline, and still more rapidly in acid solution, the indicator should not be added until one is ready to titrate. The acid quinoid form of the compound is dichroic, showing a red-violet in thick layers and blue in thin. At a pH of 5.4 the indicator appears a neutral gray color by daylight or a pale red under tungsten light. The change to yellow is easily recognized in either case. Cf. Scanlan and Reid, *Ind. Eng. Chem., Anal. Ed.* **7**:125 (1935).

Bertrand's reagents (glucose estimation): (*a*) 40 g of copper sulfate diluted to 1 liter; (*b*) rochelle salt 200 g, NaOH 150 g, and sufficient water to make 1 liter; (*c*) ferric sulfate 50 g, H_2SO_4 200 g, and sufficient water to make 1 liter; (*d*) $KMnO_4$ 5 g and sufficient water to make 1 liter.

Bial's reagent (for pentoses): dissolve 1 g of orcinol in 500 mL of 30% HCl to which 30 drops of a 10% ferric chloride solution have been added.

Bismuth chloride, $BiCl_3$—0.5N: 52 g per liter, using 1:5 HCl in place of water.

Bismuth nitrate, $Bi(N_2O_3)_3 \cdot 5\ H_2O$—0.25$N$: 40 g per liter, using 1:5HNO_3 in place of water.

Bismuth standard solution (quantitative color test for Bi): dissolve 1 g of bismuth in a mixture of 3 mL of concentrated HNO_3 and 2.8 mL of H_2O and make up to 100 mL with glycerol. Also dissolve 5 g of KI in 5 mL of water and make up to 100 mL with glycerol. The two solutions are used together in the colorimetric estimation of Bi.

Boutron-Boudet solution: *see* soap solution.

Bromchlorophenol blue, dibromodichlorophenol-sulfonphthalein (indicator): dissolve 0.1 g in 8.6 mL 0.02 N NaOH and dilute with water to 250 mL; pH range yellow 3.2 – 4.8 blue.

Bromcresol green, tetrabromo-m-cresol-sulfonphthalein (indicator): dissolve 0.1 g in 7.15 mL 0.02 N NaOH and dilute with water to 250 mL; or, 0.1 g in 100 mL 20% alcohol; pH range yellow 4.0 – 5.6 blue.

Bromcresol purple, dibromo-o-cresol-sulfonphthalein (indicator): dissolve 0.1 g in 9.5 mL 0.02 N NaOH and dilute with water to 250 mL; or, 0.1 g in 100 mL 20% alcohol; pH range yellow 5.2 – 6.8 purple.

Bromine water, saturated solution: to 400 mL water add 20 mL of bromine; use a glass stopper coated with petrolatum.

Bromphenol blue, tetrabromophenol-sulfonphthalein (indicator): dissolve 0.1 g in 7.45 mL 0.02 N NaOH and dilute with water to 250 mL; or, 0.1 g in 100 mL 20% alcohol; pH range yellow 3.6 – 4.6 violet-blue.

Bromphenol red, dibromophenol-sulfonphthalein (indicator): dissolve 0.1 g in 9.75 mL 0.02 N NaOH and dilute with water to 250 mL; pH range yellow 5.2 – 7.0 red.

Bromthymol blue, dibromothymol-sulfonphthalein (indicator): dissolve 0.1 g in 8.0 mL 0.02 N NaOH and dilute with water to 250 mL; or, 0.1 g in 100 mL of 20% alcohol; pH range yellow 6.0 – 7.6 blue.

Brucke's reagent (protein precipitant): dissolve 50 g of KI in 500 mL of water, saturate with HgI_2 (about 120 g), and dilute to 1 liter.

Cadmium chloride, $CdCl_2$ — 0.5N: 46 g per liter.

Cadmium nitrate, $Cd(NO_3)_2 \cdot 4H_2O$ — 0.5N: 77 g per liter.

Cadmium sulfate, $CdSO_4 \cdot 4H_2O$ — 0.5N: 70 g per liter.

Calcium chloride, $CaCl_2 \cdot 6H_2O$ — 0.5N: 55 g per liter.

Calcium hydroxide, $Ca(OH)_2$ — 0.04N: 10 g per liter for saturation.

Calcium nitrate, $Ca(NO_3)_2 \cdot 4H_2O$ — 0.5N: 59 g per liter.

Calcium sulfate, $CaSO_4 \cdot 2H_2O$ — 0.03N: mechanically stir 10 g in a liter of water for 3 hours; decant and use the clear liquid.

Carbon disulfide, CS_2: commercial grade which is colorless.

Chloride reagent: dissolve 1.7 g of $AgNO_3$ and 25 g KNO_3 in water, add 17 mL of concentrated NH_4OH and make up to 1 liter with water.

Chlorine water, saturated solution: pass chlorine gas into small amounts of water as needed; solutions deteriorate on standing.

Chloroform, $CHCl_3$: commercial grade.

Chloroplatinic acid, $H_2PtCl_6 \cdot 6H_2O$ — 10% solution: dissolve 1 g in 9 mL of water; keep in a dropping bottle.

Chlorphenol red, dichlorophenol-sulfonphthalein (indicator): dissolve 0.1 g in 11.8 mL 0.02 N NaOH and dilute with water to 250 mL; or, 0.1 g in 100 mL 20% alcohol; pH range yellow 5.2 – 6.6 red.

Chromic chloride, $CrCl_3$—0.5N: 26 g per liter.

Chromic nitrate, $Cr(NO_3)_3$—0.5N: 40 g per liter.

Chromic sulfate, $Cr_2(SO_4)_3 \cdot 18H_2O$—0.5$N$: 60 g per liter.

Cobaltous nitrate, $Co(NO_3)_2 \cdot 6H_2O$—0.5N: 73 g per liter.

Cobaltous sulfate, $CoSO_4 \cdot 7H_2O$—0.5N: 70 g per liter.

Cochineal (indicator): triturate 1 g with 75 mL alcohol and 75 mL water, let stand for two days and filter; pH range red 4.8–6.2 violet.

Congo red, sodium tetrazodiphenyl-naphthionate (indicator): dissolve 0.1 g in 100 mL water; pH range blue 3.0–5.2 red.

Corallin (indicator): *see* rosolic acid.

Cresol red, *o*-cresol-sulfonphthalein (indicator): dissolve 0.1 g in 13.1 mL 0.02 N NaOH and dilute with water to 250 mL; or, 0.1 g in 100 mL 20% alcohol; pH range yellow 7.2–8.8 red.

***o*-Cresolphthalein** (indicator): dissolve 0.1 g in 250 mL alcohol; pH range colorless 8.2–10.4 red.

Cupferron (iron analysis): dissolve 6 g of ammonium nitrosophenyl-hydroxylamine (cupferron) in water and dilute to 100 mL. This solution is stable for about one week if protected from light.

Cupric chloride, $CuCl_2 \cdot 2H_2O$—0.5N: 43 g per liter.

Cupric nitrate, $Cu(NO_3)_2 \cdot 6H_2O$—0.5N: 74 g per liter.

Cupric sulfate, $CuSO_4 \cdot 5H_2O$—0.5N: 62 g per liter.

Cuprous chloride, $CuCl$—0.5N: 50 g per liter, using 1 : 5 HCl in place of water.

Cuprous chloride, acid (for gas analysis, absorption of CO): cover the bottom of a 2-liter bottle with a layer of copper oxide ⅜ inch deep, and place a bundle of copper wire an inch thick in the bottle so that it extends from the top to the bottom. Fill the bottle with HCl (sp. gr. 1.10). The bottle is shaken occasionally, and when the solution is colorless or nearly so, it is poured into half-liter bottles containing copper wire. The large bottle may be filled with hydrochloric acid, and by adding the oxide or wire when either is exhausted, a constant supply of the reagent is available.

Cuprous chloride, ammoniacal: this solution is used for the same purpose and is made in the same manner as the acid cuprous chloride above, except that the acid solution is treated with ammonia until a faint odor of ammonia is perceptible. Copper wire should be kept with the solution as in the acid reagent.

Curcumin (indicator): prepare a saturated aqueous solution; pH range yellow 6.0–8.0 brownish red.

Dibromophenol-tetrabromophenol-sulfonphthalein (indicator): dissolve 0.1 g in 1.21 mL 0.1N NaOH and dilute with water to 250 mL; pH range yellow 5.6–7.2 purple.

Dimethyl glyoxime, $(CH_3CNOH)_2$—0.01N: 6 g in 500 mL of 95% alcohol.

2,4-Dinitrophenol (indicator): dissolve 0.1 g in a few mL alcohol, then dilute with water to 100 mL; pH range colorless 2.6–4.0 yellow.

2,5-Dinitrophenol (indicator): dissolve 0.1 g in 20 mL alcohol, then dilute with water to 100 mL; pH range colorless 4–5.8 yellow.

2,6-Dinitrophenol (indicator): dissolve 0.1 g in a few mL alcohol, then dilute with water to 100 mL; pH range colorless 2.4–4.0 yellow.

Esbach's reagent (estimation of proteins): dissolve 10 g of picric acid and 20 g of citric acid in water and dilute to 1 liter.

Eschka's mixture (sulfur in coal): mix 2 parts of porous calcined MgO with 1 part of anhydrous Na_2CO_3; not a solution but a dry mixture.

Ether, $(C_2H_5)_2O$ — use commercial grade.

***p*-Ethoxychrysoidine,** *p*-ethoxybenzeneazo-*m*-phenylenediamine (indicator): dissolve 0.1 g of the base in 100 mL 90% alcohol; or, 0.1 g of the hydrochloride salt in 100 mL water; pH range red 3.5–5.5 yellow.

Ethyl bis-(2,4-dinitrophenyl) acetate (indicator): the stock solution is prepared by saturating a solution containing equal volumes of alcohol and acetone with the indicator; pH range colorless 7.4–9.1 deep blue. This compound is available commercially. The preparation of this compound is described by Fehnel and Amstutz, *Ind. Eng. Chem., Anal. Ed.* **16**:53 (1944), and by von Richter, *Ber.* **21**:2470 (1888), who recommended it for the titration of orange- and red-colored solutions or dark oils in which the endpoint of phenolphthalein is not easily visible. The indicator is an orange solid which after crystallization from benzene gives pale yellow crystals melting at 150–153.5°C, uncorrected.

Fehling's solution (sugar detection and estimation): (*a*) Copper sulfate solution: dissolve 34.639 g of $CuSO_4 \cdot 5H_2O$ in water and dilute to 500 mL. (*b*) Alkaline tartrate solution: dissolve 173 g of rochelle salts ($KNaC_4O_6 \cdot 4H_2O$) and 125 g of KOH in water and dilute to 500 mL. Equal volumes of the two solutions are mixed just prior to use. The Methods of the Assoc. of Official Agricultural Chemists give 50 g of NaOH in place of the 125 g KOH.

Ferric chloride, $FeCl_3$—0.5N: 27 g per liter.

Ferric nitrate, $Fe(NO_3)_3 \cdot 9H_2O$—0.5N: 67 g per liter.

Ferrous ammonium sulfate, Mohr's salt, $FeSO_4 \cdot (NH_4)_2SO_4 \cdot 6H_2O$—0.5$N$: 196 g per liter.

Ferrous sulfate, $FeSO_4 \cdot 7H_2O$—0.5N: 80 g per liter; add a few drops of H_2SO_4.

Folin's mixture (for uric acid): dissolve 500 g of ammonium sulfate, 5 g of uranium acetate, and 6 mL of glacial acetic acid, in 650 mL of water. The volume is about a liter.

Formal or Formalin: use the commercial 40% solution of formaldehyde.

Froehde's reagent (gives characteristic colorations with certain alkaloids and glycosides): dissolve 0.01 g of sodium molybdate in 1 mL of concentrated H_2SO_4; use only a freshly prepared solution.

Gallein (indicator): dissolve 0.1 g in 100 mL alcohol; pH range light brown-yellow 3.8–6.6 rose.

Glyoxylic acid solution (protein detection): cover 10 g of magnesium powder with water and slowly add 250 mL of a saturated oxalic solution, keeping the mixture cool; filter off the magnesium oxalate, acidify the filtrate with acetic acid and make up to a liter with water.

Guaiacum tincture: dissolve 1 g of guaiacum in 100 mL of alcohol.

Gunzberg's reagent (detection of HCl in gastric juice): dissolve 4 g of phloroglucinol and 2 g of vanillin in 100 mL of absolute alcohol; use only a freshly prepared solution.

Hager's reagent (for alkaloids): this reagent is a saturated solution of picric acid in water.

Hanus solution (for determination of iodine number): dissolve 13.2 g of iodine in a liter of glacial acetic acid that will not reduce chromic acid; add sufficient bromine to double the halogen content determined by titration (3 mL is about the right amount). The iodine may be dissolved with the aid of heat, but the solution must be cold when the bromine is added.

Hematoxylin (indicator): dissolve 0.5 g in 100 mL alcohol; pH range yellow 5.0–6.0.

Heptamethoxy red, 2,4,6,2′,4′,2″,4″-heptamethoxytriphenyl carbinol (indicator): dissolve 0.1 g in 100 mL alcohol; pH range red 5.0–7.0 colorless.

Hydriodic acid, HI—0.5N: 64 g per liter.

Hydrobromic acid, HBr—0.5N: 40 g per liter.

Hydrochloric acid, HCl—5N: 182 g per liter; sp. gr. 1.084.

Hydrofluoric acid, H_2F_2—48% solution: use as purchased, and keep in the special container.

Hydrogen peroxide, H_2O_2—3% solution: use as purchased.

Hydrogen sulfide, H_2S: prepare a saturated aqueous solution.

Indicator solutions: a number of indicator solutions are listed in this section under the names of the indicators; e.g., alizarin, aurin, azolitmin, et al., which follow alphabetically. *See also* various index entries.

Indigo carmine, sodium indigodisulfonate (indicator): dissolve 0.25 g in 100 mL 50% alcohol; pH range blue 11.6–14.0 yellow.

Indo-oxine, 5,8-quinolinequinone-8-hydroxy-5-quinoyl-5-imide (indicator): dissolve 0.05 g in 100 mL alcohol; pH range red 6.0–8.0 blue. Cf. Berg and Becker, *Z. Anal. Chem.* **119**:81 (1940).

Iodeosin, tetraiodofluorescein (indicator): dissolve 0.1 g in 100 mL ether saturated with water; pH range yellow 0–about 4 rose-red; *see also* under methyl orange.

Iodic acid, HIO_3—0.5N ($HIO_3/12$): 15 g per liter.

Iodine: *see* tincture of iodine.

Lacmoid (indicator): dissolve 0.5 g in 100 mL alcohol; pH range red 4.4–6.2 blue.

Lead acetate, $Pb(C_2H_3O_2)_2 \cdot 3H_2O$—0.5$N$: 95 g per liter.

Lead chloride, $PbCl_2$—saturated solution is 1/7N.

Lead nitrate, $Pb(NO_3)_2$—0.5N: 83 g per liter.

Lime water: *see* calcium hydroxide.

Litmus (indicator): powder the litmus and make up a 2% solution in water by boiling for 5 minutes; pH range red 4.5–8.3 blue.

Magnesia mixture: 100 g of $MgSO_4$, 200 g of NH_4Cl, 400 mL of NH_4Cl, 800 mL of water; each cc $\equiv$ 0.01 g phosphorus (P).

Magnesium chloride, $MgCl_2 \cdot 6H_2O$—0.5N: 50 g per liter.

Magnesium nitrate, $Mg(NO_3)_2 \cdot 6H_2O$—0.5N: 64 g per liter.

Magnesium sulfate, epsom salts, $MgSO_4 \cdot 7H_2O$—0.5N: 62 g per liter; saturated solution dissolve 600 g of the salt in water and dilute to 1 liter.

Manganous chloride, $MnCl_2 \cdot 4H_2O$—0.5N: 50 g per liter.

Manganous nitrate, $Mn(NO_3)_2 \cdot 6H_2O$—0.5N: 72 g per liter.

Manganous sulfate, $MnSO_4 \cdot 7H_2O$—0.5N: 69 g per liter.

Marme's reagent (gives yellowish-white precipitate with salts of alkaloids): saturate a boiling solution of 4 parts of KI in 12 parts of water with CdI_2; then add an equal volume of cold saturated KI solution.

Marquis reagent (gives a purple-red coloration, then violet, then blue with morphine, codeine, dionine, and heroine): mix 3 mL of concentrated H_2SO_4 with 3 drops of a 35% formaldehyde solution.

Mayer's reagent (gives white precipitate with most alkaloids in a slightly acid solution): dissolve 13.55 g of $HgCl_2$ and 50 g of KI in a liter of water.

Mercuric chloride, $HgCl_2$—0.5N: 68 g per liter.

Mercuric nitrate, $Hg(NO_3)_2$—0.5N: 81 g per liter.

Mercuric sulfate, $HgSO_4$—0.5N: 74 g per liter.

Mercurous nitrate, $HgNO_3$: mix 1 part of $HgNO_3$, 20 parts of H_2O, and 1 part of HNO_3.

Metacresol purple, *m*-cresol-sulfonphthalein (indicator): dissolve 0.1 g in 13.6 mL 0.02N NaOH and dilute with water to 250 mL; acid pH range red 0.5–2.5 yellow, alkaline pH range yellow 7.4–9.0 purple.

Metanil yellow, diphenylaminoazo-*m*-benzene sulfonic acid (indicator): dissolve 0.25 g in 100 mL alcohol; pH range red 1.2–2.3 yellow.

Methyl green, hexamethylpararosaniline hydroxymethylate (component of mixed indicator): dissolve 0.1 g in 100 mL alcohol; when used with equal parts of hexamethoxytriphenyl carbinol gives color change from violet to green at a titration exponent (pI) of 4.0.

Methyl orange, orange III, tropeolin D, sodium *p*-dimethylaminoazobenzenesulfonate (indicator): dissolve 0.1 g in 100 mL water; pH range red 3.0–4.4 orange-yellow. If during a titration where methyl yellow is being used a precipitate forms which tends to remove the indicator from the aqueous phase, methyl orange will be found to be a more suitable indicator. This occurs, for example, in titrations of soaps with acids. The fatty acids, liberated by the titration, extract the methyl yellow so that the endpoint cannot be perceived. Likewise methyl orange is more suitable for titrations in the presence of immiscible organic solvents such as carbon tetrachloride or ether used in the extraction of alkaloids for analysis. Iodeosin (*q.v.*) has also been proposed as an indicator for such cases. Cf. Mylius and Foerster, *Ber.* **24**:1482 (1891); *Z. Anal. Chem.* **31**:240 (1892).

Methyl red, *p*-dimethylaminoazobenzene-*o'*-carboxylic acid (indicator): dissolve 0.1 g in 18.6 mL of 0.02 N NaOH and dilute with water to 250 mL; or, 0.1 g in 60% alcohol; pH range red 4.4–6.2 yellow.

Methyl violet (indicator): dissolve 0.25 g in 100 mL water, pH range blue 1.5–3.2 violet.

Methyl yellow, *p*-dimethylaminoazobenzene, benzeneazodimethylaniline (indicator); dissolve 0.1 g in 200 mL alcohol; pH range red 2.9–4.0 yellow. The color change from yellow to orange can be perceived somewhat more sharply than the change of methyl orange from orange to rose, so that methyl yellow seems to deserve preference in many cases. *See also* under methyl orange.

Methylene blue, *N,N,N',N'*-tetramethylthionine (component of mixed indicator): dissolve 0.1 g in 100 mL alcohol; when used with equal part of methyl yellow gives color change from blue-violet to green at a titration exponent (pI) of 3.25; when used with equal part of 0.2% methyl red in alcohol gives color change from red-violet to green at a titration exponent (pI) of 5.4; when used with an equal part of neutral red gives color change from violet-blue to green at a titration exponent (pI) of 7.0.

Millon's reagent (gives a red precipitate with certain proteins and with various phenols): dissolve 1 part of mercury in 1 part of HNO_3 (sp. gr. 1.40) with gentle heating, then add 2 parts of water; a few crystals of KNO_3 help to maintain the strength of the reagent.

Mohr's salt: *see* ferrous ammonium sulfate.

α-Naphthol solution: dissolve 144 g of α-naphthol in enough alcohol to make a liter of solution.

α-Naphtholbenzein (indicator): dissolve 0.1 g in 100 mL 70% alcohol; pH range colorless 9.0–11.0 blue.

α-Naphtholphthalein (indicator): dissolve 0.1 g in 50 mL alcohol and dilute with water to 100 mL; pH range pale yellow-red 7.3–8.7 green.

Nessler's reagent (for free ammonia): dissolve 50 g of KI in the least possible amount of cold water; add a saturated solution of $HgCl_2$ until a very slight excess is indicated; add 400 mL of a 50% solution of KOH; allow to settle, make up to a liter with water, and decant.

Neutral red, toluylene red, dimethyldiaminophenazine chloride, aminodimethylaminotoluphenazine hydrochloride (indicator): dissolve 0.1 g in 60 mL alcohol and dilute with water to 100 mL; pH range red 6.8–8.0 yellow-orange.

Nickel chloride, $NiCl_2 \cdot 6H_2O$ — 0.5N: 59 g per liter.

Nickel nitrate, $Ni(NO_3)_2 \cdot 6H_2O$ – 0.5N: 73 g per liter.

Nickel sulfate, $NiSO_4 \cdot 6H_2O$ — 0.5N: 66 g per liter.

Nitramine, picrylmethylnitramine, 2,4,6-trinitrophenylmethyl nitramine (indicator): dissolve 0.1 g in 60 mL alcohol and dilute with water to 100 mL; pH range colorless 10.8–13.0 red-brown; the solution should be kept in the dark as nitramine is unstable; on boiling with alkali it decomposes quickly. Fresh solutions should be prepared every few months.

Nitric acid, HNO_3 — 5N: 315 g per liter; sp. gr. 1.165.

Nitrohydrochloric acid: *see* aqua regia.

p-Nitrophenol (indicator): dissolve 0.2 g in 100 mL water; pH range colorless at about 5–7 yellow.

Nitroso-β-naphthol, $HOC_{10}H_6NO$ — saturated solution: saturate 100 mL of 50% acetic acid with the solid.

Nylander's solution (detection of glucose): dissolve 40 g of rochelle salt and 20 g of bismuth subnitrate in 1000 mL of an 8% NaOH solution.

Obermayer's reagent (detection of indoxyl in urine): dissolve 4 g of $FeCl_3$ in a liter of concentrated HCl.

Orange III (indicator): *see* under methyl orange.

Oxalic acid, $H_2C_2O_4 \cdot 2H_2O$: dissolve in ten parts of water.

Pavy's solution (estimation of glucose): mix 120 mL of Fehling's solution and 300 mL of ammonium hydroxide (sp. gr. 0.88), and dilute to a liter with water.

Perchloric acid, $HClO_4$ – 60%: use as purchased.

Phenol red, phenol-sulfonphthalein (indicator): dissolve 0.1 g in 14.20 mL 0.02N NaOH and dilute with water to 250 mL; or, 0.1 g in 100 mL 20% alcohol; pH range yellow 6.8–8.0 red.

Phenol solution: dissolve 20 g of phenol (carbolic acid) in a liter of water.

Phenol sulfonic acid (determination of nitrogen as nitrate; water analysis for nitrate): dissolve 25 g pure, white phenol in 150 mL of pure concentrated H_2SO_4, add 75 mL of fuming H_2SO_4 (15% SO_3), stir well and heat for two hours at 100°C.

Phenolphthalein (indicator): dissolve 1 g in 60 mL of alcohol and dilute with water to 100 mL; pH range colorless 8.2–10.0 red.

Phosphoric acid, *ortho,* H_3PO_4 — 0.5N: 16 g per liter.

Poirrer blue C4B (indicator): dissolve 0.2 g in 100 mL water; pH range blue 11.0–13.0 red.

Potassium acid antimonate, KH_2SbO_4 — 0.1N: boil 23 g of the salt with 950 mL of water for 5 minutes, cool rapidly and add 35 mL of 6N KOH; allow to stand for one day, filter dilute filtrate to a liter.

Potassium arsenate, K_3AsO_4—$0.5N$ ($K_3AsO_4/10$): 26 g per liter.

Potassium arsenite, $KAsO_2$—$0.5N$ ($KAsO_2/6$): 24 g per liter.

Potassium bromate, $KBrO_3$—$0.5N$ ($KBrO_3/12$): 14 g per liter.

Potassium bromide, KBr—$0.5N$: 60 g per liter.

Potassium carbonate, K_2CO_3—$3N$: 207 g per liter.

Potassium chloride, KCl—$0.5N$: 37 g per liter.

Potassium chromate, K_2CrO_4—$0.5N$: 49 g per liter.

Potassium cyanide, KCN—$0.5N$: 33 g per liter.

Potassium dichromate, $K_2Cr_2O_7$—$0.5N$ ($K_2Cr_2O_7/8$): 38 g per liter.

Potassium ferricyanide, $K_3Fe(CN)_6$—$0.5N$: 55 g per liter.

Potassium ferrocyanide, $K_4Fe(CN)_6 \cdot 3H_2O$—$0.5N$: 53 g per liter.

Potassium hydroxide, KOH—$5N$: 312 g per liter.

Potassium iodate, KIO_3—$0.5N$ ($KIO_3/12$): 18 g per liter.

Potassium iodide, KI—$0.5N$: 83 g per liter.

Potassium nitrate, KNO_3—$0.5N$: 50 g per liter.

Potassium nitrate, KNO_2—$6N$: 510 g per liter.

Potassium permanganate, $KMnO_4$—$0.5N$ ($KMnO_4/10$): 16 g per liter.

Potassium pyrogallate (oxygen in gas analysis): weigh out 5 g of pyrogallol (pyrogallic acid), and pour upon it 100 mL of a KOH solution. If the gas contains less than 28% of oxygen, the KOH solution should be 500 g KOH in a liter of water; if there is more than 28% of oxygen in the gas, the KOH solution should be 120 g of KOH in 100 mL of water.

Potassium sulfate, K_2SO_4—$0.5N$: 44 g per liter.

Potassium thiocyanate, $KCNS$—$0.5N$: 49 g per liter.

Precipitating reagent (for group II, anions): dissolve 61 g of $BaCl_2 \cdot 2H_2O$ and 52 g of $CaCl_2 \cdot 6H_2O$ in water and dilute to 1 liter. If the solution becomes turbid, filter and use filtrate.

Quinaldine red (indicator): dissolve 0.1 g in 100 mL alcohol; pH range colorless 1.4–3.2 red.

Quinoline blue, cyanin (indicator): dissolve 1 g in 100 mL alcohol; pH range colorless 6.6–8.6 blue.

Rosolic acid, aurin, corallin, corallinphthalein, 4,4'-dihydroxy-fuchsone, 4,4'-dihydroxy-3-methyl-fuchsone (indicator): dissolve 0.5 g in 50 mL alcohol and dilute with water to 100 mL.

Salicyl yellow (indicator): *see* alizarin yellow GG.

Scheibler's reagent (precipitates alkaloids, albumoses and peptones): dissolve sodium tungstate in boiling water containing half its weight of phosphoric acid (sp. gr. 1.13); on evaporation of this solution, crystals of phosphotungstic acid are obtained. A 10% solution of phosphotungstic acid in water constitutes the reagent.

Schweitzer's reagent (dissolves cotton, linen, and silk, but not wool); add NH_4Cl and NaOH to a solution of copper sulfate. The blue precipitate is filtered off, washed, pressed, and dissolved in ammonia (sp. gr. 0.92).

Silver nitrate, $AgNO_3$—$0.25N$: 43 g per liter.

Silver sulfate, Ag_2SO_4—$N/13$ (saturated solution): stir mechanically 10 g of the salt in a liter of water for 3 hours; decant and use the clear liquid.

Soap solution (for hardness in water): (*a*) *Clark's or A.P.H.A. Stand. Methods*—prepare stock solution of 100 g of pure powdered castile soap in a liter of 80% ethyl alcohol; allow to stand over night and decant. Titrate against $CaCl_2$ solution (0.5 g $CaCO_3$ dissolved in a concentrated HCl, neutralized with NH_4OH to slight alkalinity using litmus as the indicator, make up to 500 mL; 1 mL of this solution is equivalent to 1 mg $CaCO_3$) and dilute with 80% alcohol until 1 mL of the resulting solution is equivalent to 1 mL of the standard $CaCl_2$ making due allowance for the lather factor (the lather factor is that amount of standard soap solution required to produce a permanent lather in a 50-mL portion of distilled water). One milliliter of this solution after subtracting the lather factor is equivalent to 1 mg of $CaCO_3$. (*b*) *Boutron-Boudet*—dissolve 100 g of pure castile soap in about 2500 mL of 56% ethyl alcohol and adjust so that 2.4 mL will give a permanent lather with 40 mL of a solution containing 0.59 g $Ba(NO_3)_2$ per liter of water; 2.4 mL of this solution is equivalent to 22 French degrees or 220 parts per million of hardness (as $CaCO_3$) on a 40-mL sample of water.

Sodium acetate, $NaC_2H_3O_2 \cdot 3H_2O$: dissolve 1 part of the salt in 10 parts of water.

Sodium acetate, acid: dissolve 100 g of sodium acetate and 30 mL of glacial acetic acid in water and dilute to 1 liter.

Sodium bismuthate (oxidation of manganese): heat 20 parts of NaOH nearly to redness in an iron or nickel crucible, and add slowly 10 parts of basic bismuth nitrate which has been previously dried. Add 2 parts of sodium peroxide, and pour the brownish-yellow fused mass on an iron plate to cool. When cold break up in a mortar, extract with water, and collect on an asbestos filter.

Sodium carbonate, Na_2CO_3—$3N$: 159 g per liter; one part Na_2CO_3, or 2.7 parts of the crystalline $Na_2CO_3 \cdot 10H_2O$ in 5 parts of water.

Sodium chloride, NaCl—$0.5N$: 29 g per liter.

Sodium chloroplatinite, Na_2PtCl_4: dissolve 1 part of the salt in 12 parts of water.

Sodium cobaltinitrite, $Na_2Co(NO_2)_6$—$0.3N$: dissolve 230 g of $NaNO_2$ in 500 mL of water, add 160 mL of $6N$ acetic acid and 35 g of $Co(NO_3)_2 \cdot 6H_2O$. Allow to stand one day, filter, and dilute the filtrate to a liter.

Sodium hydrogen phosphate, $Na_2HPO_4 \cdot 12H_2O$—$0.5N$: 60 g liter.

Sodium hydroxide, NaOH—$5N$: 220 g per liter.

Sodium hydroxide, alcoholic: dissolve 20 g of NaOH in alcohol and dilute to 1 liter with alcohol.

Sodium hypobromite: dissolve 100 g of NaOH in 250 mL of water and add 25 mL of bromine.

Sodium nitrate, $NaNO_3$—$0.5N$: 43 g per liter.

Sodium nitroprusside (for sulfur detection): dissolve about 1 g of sodium nitroprusside in 10 mL of water; as the solution deteriorates on standing, only freshly prepared solutions should be used. This compound is also called sodium nitroferricyanide and has the formula $Na_2Fe(NO)(CN)_5 \cdot 2H_2O$.

Sodium polysulfide, Na_2S_x: dissolve 480 g of $Na_2S \cdot 9H_2O$ in 500 mL of water, add 40 g of NaOH and 18 g of sulfur, stir mechanically and dilute to 1 liter with water.

Sodium sulfate, Na_2SO_4—$0.5N$: 35 g per liter.

Sodium sulfide, Na_2S: saturate NaOH solution with H_2S, then add as much NaOH as was used in the original solution.

Sodium sulfite, $Na_2SO_3 \cdot 7H_2O$—0.5N: 63 g per liter.

Sodium sulfite, acid (saturated): dissolve 600 g of $NaHSO_3$ in water and dilute to 1 liter; for the preparation of addition compounds with aldehydes and ketones: prepare a saturated solution of sodium carbonate in water and saturate with sulfur dioxide.

Sodium tartrate, acid, $NaHC_4H_4O_6$: dissolve 1 part of the salt in 10 parts of water.

Sodium thiosulfate, $Na_2S_2O_3 \cdot 5H_2O$: one part of the salt in 40 parts of water.

Sonnenschein's reagent (alkaloid detection): a nitric acid solution of ammonium molybdate is treated with phosphoric acid. The precipitate so produced is washed and boiled with aqua regia until the ammonium salt is decomposed. The solution is evaporated to dryness and the residue is dissolved in 10% HNO_3

Stannic chloride, $SnCl_4$—0.5N: 33 g per liter.

Stannous chloride, $SnCl_2 \cdot 2H_2O$—0.5N: 56 g per liter. The water should be acid with HCl and some metallic tin should be kept in the bottle.

Starch solution (iodine indicator): dissolve 5 g of soluble starch in cold water, pour the solution into 2 liters of water and boil for a few minutes. Keep in a glass-stoppered bottle.

Starch solution (other than soluble): make a thin paste of the starch with cold water, then stir in 200 times its weight of boiling water and boil for a few minutes. A few drops of chloroform added to the solution acts as a preservative.

Stoke's reagent: dissolve 30 g of ferrous sulfate and 20 g of tartaric acid in water and dilute to 1 liter. When required for use, add strong ammonia until the precipitate first formed is dissolved.

Strontium chloride, $SrCl_2 \cdot 6H_2O$—0.5N: 67 g per liter.

Strontium nitrate, $Sr(NO_3)_2$—0.5N: 53 g per liter.

Strontium sulfate, $SrSO_4$: prepare a saturated solution.

Sulfanilic acid (for detection of nitrites): dissolve 8 g of sulfanilic acid in 1 liter of acetic acid (sp. gr. 1.04).

Sulfuric acid, H_2SO_4—5N: 245 g per liter, sp. gr. 1.153.

Sulfurous acid, H_2SO_3: saturate water with sulfur dioxide.

Tannic acid: dissolve 1 g tannic acid in 1 mL alcohol and make up to 10 mL with water.

Tartaric acid, $H_2C_4H_4O_6$: dissolve one part of the acid in 3 parts of water; for a saturated solution dissolve 750 g of tartaric acid in water and dilute to 1 liter.

Tetrabromophenol blue, tetrabromophenol-tetrabromosulfonphthalein (indicator): dissolve 0.1 g in 5 mL 0.02N NaOH and dilute with water to 250 mL; pH range yellow 3.0–4.6 blue.

Thymol blue, thymol-sulfonphthalein (indicator): dissolve 0.1 g in 10.75 mL 0.02N NaOH and dilute with water to 250 mL; or dissolve 0.1 g in 20 mL warm alcohol and dilute with water to 100 mL; pH range (acid) red 1.2–2.8 yellow, and (alkaline) yellow 8.0–9.6 blue.

Thymolphthalein (indicator): dissolve 0.1 g in 100 mL alcohol; pH range colorless 9.3–10.5 blue.

Tincture of iodine (antiseptic): add 70 g of iodine and 50 g of KI to 50 mL of water; make up to 1 liter with alcohol.

o-**Tolidine solution** (for residual chlorine in water analysis): dissolve 1 g of pulverized *o*-tolidine, m.p. 129°C., in 1 liter of dilute hydrochloric acid (100 mL conc. HCl diluted to 1 liter).

Toluylene red (indicator): *see* neutral red.

Trichloroacetic acid: dissolve 100 g of the acid in water and dilute to 1 liter.

Trinitrobenzene, 1,3,5-trinitrobenzene (indicator): dissolve 0.1 g in 100 mL alcohol; pH range colorless 11.5 – 14.0 orange.

Trinitrobenzoic acid, 2,4,6-trinitrobenzoic acid (indicator): dissolve 0.1 g in 100 mL water; pH range colorless 12.0 – 13.4 orange-red.

Tropeolin D (indicator): *see* methyl orange.

Tropeolin O, sodium 2,4-dihydroxyazobenzene-4-sulfonate (indicator): dissolve 0.1 g in 100 mL water; pH range yellow 11.0 – 13.0 orange-brown.

Tropeolin OO, orange IV, sodium *p*-diphenylamino-azobenzene sulfonate, sodium 4'-anilino-azobenzene-4-sulfonate (indicator): dissolve 0.1 g in 100 mL water; pH range red 1.3 – 3.2 yellow.

Tropeolin OOO, sodium α-naptholazobenzene sulfonate (indicator): dissolve 0.1 g in 100 mL water; pH range yellow 7.6 – 8.9 red.

Turmeric paper (gives a rose-brown coloration with boric acid): wash the ground root of turmeric with water and discard the washings. Digest with alcohol and filter, using the clear filtrate to impregnate white, unsized paper, which is then dried.

Uffelmann's reagent (gives a yellow coloration in the presence of lactic acid): add a ferric chloride solution to a 2% phenol solution until the solution becomes violet in color.

Wagner's solution (phosphate rock analysis): dissolve 25 g citric acid and 1 g salicylic acid in water, and make up to 1 liter. Twenty-five to fifty milliliters of this reagent prevents precipitation of iron and aluminum.

Wijs solution (for iodine number): dissolve 13 g resublimed iodine in 1 liter of glacial acetic acid (99.5%), and pass in washed and dried (over or through H_2SO_4) chlorine gas until the original thio titration of the solution is not quite doubled. There should be only a slight excess of iodine and no excess of chlorine. Preserve the solution in amber colored bottles sealed with paraffin. Do not use the solution after it has been prepared for more than 30 days.

Xylene cyanole-methyl orange indicator, Schoepfle modification (for partially color blind operators): dissolve 0.75 g xylene cyanole FF (Eastman No. T 1579) and 1.50 g methyl orange in 1 liter of water.

p-**Xylenol blue,** 1,4-dimethyl-5-hydroxybenzene-sulfonphthalein (indicator): dissolve 0.1 g in 250 mL alcohol; pH range (acid) red 1.2 – 2.8 yellow, and (alkaline) yellow 8.0 – 9.6 blue.

Zinc chloride, $ZnCl_2$ — 0.5N: 34 g per liter.

Zinc nitrate, $Zn(NO_3)_2 \cdot 6H_2O$ – 0.5N: 74 g per liter.

Zinc sulfate, $ZnSO_4 \cdot 7H_2O$ — 0.5N: 72 g per liter.

TABLE 11.35　Some Common Reactive and Incompatible Chemicals

Chemical	Keep out of contact with
Acetic acid	Chromium(VI) oxide, chlorosulfonic acid, ethylene glycol, ethyleneimine, hydroxyl compounds, nitric acid, oleum, perchloric acid, peroxides, permanganates, potassium *tert*-butoxide, PCl$_3$
Acetylene	Bromine, chlorine, brass, copper and copper salts, fluorine, mercury and mercury salts, nitric acid, silver and silver salts, alkali hydrides, potassium metal
Alkali metals	Moisture, acetylene, metal halides, ammonium salts, oxygen and oxidizing agents, halogens, carbon tetrachloride, carbon, carbon dioxide, carbon disulfide, chloroform, chlorinated hydrocarbons, ethylene oxide, boric acid, sulfur, tellurium
Aluminum	Chlorinated hydrocarbons, halogens, steam
Ammonia, anhydrous	Mercury, halogens, hypochlorites, chlorites, chlorine(I) oxide, hydrofluoric acid (anhydrous), hydrogen peroxide, chromium(VI) oxide, nitrogen dioxide, chromyl(VI) chloride, sulfinyl chloride, magnesium perchlorate, peroxodisulfates, phosphorus pentoxide, acetaldehyde, ethylene oxide, acrolein, gold (III) chloride
Ammonium nitrate	Acids, metal powders, flammable liquids, chlorates, nitrites, sulfur, finely divided organic or combustible materials, perchlorates, urea
Ammonium perchlorate	Hot copper tubing, sugar, finely divided organic or combustible materials, potassium periodate and permanganate, powdered metals, carbon, sulfur
Aniline	Nitric acid, peroxides, oxidizing materials, acetic anhydride, chlorosulfonic acid, oleum, ozone
Benzoyl peroxide	Direct sunlight, sparks and open flames, shock and friction, acids, alcohols, amines, ethers, reducing agents, polymerization catalysts, metallic naphthenates
Bromine	Ammonia, carbides, dimethylformamide, fluorine, ozone, olefins, reducing materials including many metals, phosphine, silver azide
Calcium carbide	Moisture, selenium, silver nitrate, sodium peroxide, tin(II) chloride, potassium hydroxide plus chlorine, HCl gas, magnesium
Carbon, activated	Calcium hypochlorite, all oxidizing agents, unsaturated oils
Chlorates	Ammonium salts, acids, metal powders, sulfur, finely divided organic or combustible materials, cyanides, metal sulfides, manganese dioxide, sulfur dioxide, organic acids
Chlorine	Ammonia, acetylene, alcohols, alkanes, benzene, butadiene, carbon disulfide, dibutyl phthalate, ethers, fluorine, glycerol, hydrocarbons, hydrogen, sodium carbide, finely divided metals, metal acetylides and carbides, nitrogen compounds, nonmetals, nonmetal hydrides, phosphorus compounds, polychlorobiphenyl, silicones, steel, sulfides, synthetic rubber, turpentine
Chlorine dioxide	Ammonia, carbon monoxide, hydrogen, hydrogen sulfide, methane, mercury, nonmetals, phosphine, phosphorus pentachloride
Chlorites	Ammonia, organic matter, metals
Chloroform	Aluminum, magnesium, potassium, sodium, aluminum chloride, ethylene, powerful oxidants
Chlorosulfonic acid	Saturated and unsaturated acids, acid anhydrides, nitriles, acrolein, alcohols, ammonia, esters, HCl, HF, ketones, hydrogen peroxide, metal powders, nitric acid, organic materials, water
Chromic(VI) acid	Acetic acid, acetic anhydride, acetone, alcohols, alkali metals, ammonia, dimethylformamide, camphor, glycerol, hydrogen sulfide, phosphorus, pyridine, selenium, sulfur, turpentine, flammable liquids in general
Cobalt	Acetylene, hydrazinium nitrate, oxidants

TABLE 11.35 Some Common Reactive and Incompatible Chemicals (*Continued*)

Chemical	Keep out of contact with
Copper	Acetylene and alkynes, ammonium nitrate, azides, bromates, chlorates, iodates, chlorine, ethylene oxide, fluorine, peroxides, hydrogen sulfide, hydrazinium nitrate
Copper(II) sulfate	Hydroxylamine, magnesium
Cumene hydroperoxide	Acids (inorganic or organic)
Cyanides	Acids, water or steam, fluorine, magnesium, nitric acid and nitrates, nitrites
Cyclohexanol	Oxidants
Cyclohexanone	Hydrogen peroxide, nitric acid
Decaborane-14	Dimethyl sulfoxide, ethers, halocarbons
Diazomethane	Alkali metals, calcium sulfate
1,1-Dichloroethylene	Air, chlorotrifluoroethylene, ozone, perchloryl fluoride
Dimethylformamide	Halocarbons, inorganic and organic nitrates, bromine, chromium(VI) oxide, aluminum trimethyl, phosphorus trioxide
1,1-Dimethylhydrazine	Air, hydrogen peroxide, nitric acid, nitrous oxide
Dimethylsulfoxide	Acyl and aryl halides, boron compounds, bromomethane, nitrogen dioxide, magnesium perchlorate, periodic acid, silver difluoride, sodium hydride, sulfur trioxide
Dinitrobenzenes	Nitric acid
Dinitrotoluenes	Nitric acid
1,4-Dioxane	Silver perchlorate
Esters	Nitrates
Ethylamine	Cellulose, oxidizers
Ethers	Oxidizing materials, boron triiodide
Ethylene	Aluminum trichloride, carbon tetrachloride, chlorine, nitrogen oxides, tetrafluoroethylene
Ethylene oxide	Acids and bases, alcohols, air, 1,3-nitroaniline, aluminum chloride, aluminum oxide, ammonia, copper, iron chlorides and oxides, magnesium perchlorate, mercaptans, potassium, tin chlorides, alkane thiols
Ethyl ether	Liquid air, chlorine, chromium(VI) oxide, lithium aluminum hydride, ozone, perchloric acid, peroxides
Ethyl sulfate	Oxidizing materials, water
Flammable liquids	Ammonium nitrate, chromic acid, the halogens, hydrogen peroxide, nitric acid
Fluorine	Isolate from everything; only lead and nickel resist prolonged attack
Formamide	Iodine, pyridine, sulfur trioxide
Freon 113	Aluminum, barium, lithium, samarium, NaK alloy, titanium
Glycerol	Acetic anhydride, hypochlorites, chromium(VI) oxide, perchlorates, alkali peroxides, sodium hydride
Hydrazine	Alkali metals, ammonia, chlorine, chromates and dichromates, copper salts, fluorine, hydrogen peroxide, metallic oxides, nickel, nitric acid, liquid oxygen, zinc diethyl
Hydrides	Powerful oxidizing agents, moisture
Hydrocarbons	Halogens, chromium(VI) oxide, peroxides
Hydrogen	Halogens, lithium, oxidants, lead trifluoride
Hydrogen bromide	Fluorine, iron(III) oxide, ammonia, ozone
Hydrogen chloride	Acetic anhydride, aluminum, 2-aminoethanol, ammonia, chlorosulfonic acid, ethylenediamine, fluorine, metal acetylides and carbides, oleum, perchloric acid, potassium permanganate, sodium, sulfuric acid
Hydrogen fluoride	Acetic anhydride, 2-aminoethanol, ammonia, arsenic trioxide, chlorosulfonic acid, ethylenediamine, ethyleneimine, fluorine, HgO, oleum, phosphorus trioxide, propylene oxide, sodium, sulfuric acid, vinyl acetate

TABLE 11.35 Some Common Reactive and Incompatible Chemicals (*Continued*)

Chemical	Keep out of contact with
Hydrogen iodide	Fluorine, nitric acid, ozone, metals
Hydrogen peroxide	Copper, chromium, iron, most metals or their salts, alcohols, acetone, organic materials, flammable liquids, combustible materials
Hydrogen selenide	Hydrogen peroxide, nitric acid
Hydrogen sulfide	Fuming nitric acid, oxidizing gases, peroxides
Hydroquinone	Sodium hydroxide
Hydroxylamine	Barium oxide and peroxide, carbonyls, chlorine, copper(II) sulfate, dichromates, lead dioxide, phosphorus trichloride and pentachloride, permanganates, pyridine, sodium, zinc
Hypochlorites, salts of	Urea, amines, anthracene, carbon, carbon tetrachloride, ethanol, glycerol, mercaptans, organic sulfides, sulfur, thiols
Indium	Acetonitrile, nitrogen dioxide, mercury(II) bromide, sulfur
Iodine	Acetaldehyde, acetylene, aluminum, ammonia (aqueous or anhydrous), antimony, bromine pentafluoride, carbides, cesium oxide, chlorine, ethanol, fluorine, formamide, lithium, magnesium, phosphorus, pyridine, silver azide, sulfur trioxide
Iodine monochloride	Aluminum foil, organic matter, metal sulfides, phosphorus, potassium, rubber, sodium
Iodoform	Acetone, lithium, mercury(II) oxide, mercury(I) chloride, silver nitrate
Iodomethane	Silver chlorite, sodium
Iron disulfide	Water, powdered pyrites
Isothiourea	Acrylaldehyde, hydrogen peroxide, nitric acid
Ketones	Aldehydes, nitric acid, perchloric acid
Lactonitrile	Oxidizing materials
Lead	Ammonium nitrate, chlorine trifluoride, hydrogen peroxide, sodium azide and carbide, zirconium, oxidants
Lead(II) azide	Calcium stearate, copper, zinc, brass, carbon disulfide
Lead chromate	Iron hexacyanoferrate(4−)
Lead dioxide	Aluminum carbide, hydrogen peroxide, hydrogen sulfide, hydroxylamine, nitroalkanes, nitrogen compounds, nonmetal halides, peroxoformic acid, phosphorus, phosphorus trichloride, potassium, sulfur, sulfur dioxide, sulfides, tungsten, zirconium
Lead(II) oxide	Chlorinated rubber, chlorine, ethylene, fluorine, glycerol, metal acetylides, perchloric acid
Lead(II,IV) oxide	Same as for lead dioxide
Lithium hydride	Nitrous oxide, oxygen
Magnesium	Air, beryllium fluoride, ethylene oxide, halogens, halocarbons, HI, metal cyanides, metal oxides, metal oxosalts, methanol, oxidants, peroxides, sulfur, tellurium
Maleic anhydride	Alkali metals, amines, KOH, NaOH, pyridine
Manganese dioxide	Aluminum, hydrogen sulfide, oxidants, potassium azide, hydrogen peroxide, peroxosulfuric acid, sodium peroxide
Mercaptans	Powerful oxidizers
Mercury	Acetylenic compounds, chlorine, fulminic acid, ammonia, ethylene oxide, metals, methyl azide, oxidants, tetracarbonylnickel
Mercury(II) cyanide	Fluorine, hydrogen cyanide, magnesium, sodium nitrite
Mercury(I) nitrate	Phosphorus
Mercury(II) nitrate	Acetylene, aromatics, ethanol, hypophosphoric acid, phosphine, unsaturated organic compounds
Mercury(II) oxide	Chlorine, hydrazine hydrate, hydrogen peroxide, hypophosphorous acid, magnesium, phosphorus, sulfur, butadiene, hydrocarbons, methanethiol
Mesityl oxide	2-Aminoethanol, chlorosulfonic acid, nitric acid, ethylenediamine, sulfuric acid

TABLE 11.35 Some Common Reactive and Incompatible Chemicals (*Continued*)

Chemical	Keep out of contact with
Methanol	Beryllium dihydride, chloroform, oxidants, potassium *tert*-butoxide
Methylamine	Nitromethane
N-Methylformamide	Benzenesulfonyl chloride
Methyl isobutyl ketone	Potassium *tert*-butoxide
Methyl methacrylate	Air, benzoyl peroxide
4-Methylnitrobenzene	Sulfuric acid, tetranitromethane
2-Methylpyridine	Hydrogen peroxide, iron(II) sulfate, sulfuric acid
Methylsodium	4-Chloronitrobenzene
Molybdenum trioxide	Chlorine trifluoride, interhalogens, metals
Naphthalene	Chromium trioxide, dinitrogen pentaoxide
2-Naphthol	Antipyrine, camphor, phenol, iron(III) salts, menthol, oxidizing materials, permanganates, urethane
Neodymium	Phosphorus
Nickel	Aluminum, aluminum(III) chloride, ethylene, 1,4-dioxan, hydrogen, methanol, nonmetals, oxidants, sulfur compounds
Nickel carbonyl	Air, bromine, oxidizing materials
Niobium	Bromine trifluoride, chlorine, fluorine
Nitrates	Aluminum, BP, cyanides, esters, phosphorus, tin(II) chloride, sodium hypophosphite, thiocyanates
Nitric acid, fuming	Organic matter, nonmetals, most metals, ammonia, chlorosulfonic acid, chromium trioxide, cyanides, dichromates, hydrazines, hydrides, HCN, HI, hydrogen sulfide, sulfur dioxide, sulfur halides, sulfuric acid, flammable liquids and gases
Nitric oxide	Aluminum, BaO, boron, carbon disulfide, chromium, many chlorinated hydrocarbons, fluorine, hydrocarbons, ozone, phosphine, phosphorus, hydrazine, acetic anhydride, ammonia, chloroform, Fe, K, Mg, Mn, Na, sulfur
Nitrites	Organic nitrites in contact with ammonium salts, cyanides
Nitrobenzene	Nitric acid, nitrous oxide, silver perchlorate
Nitroethane	Hydroxides, hydrocarbons, metal oxides
Nitrogen trichloride	Ammonia, As, hydrogen sulfide, nitrogen dioxide, organic matter, ozone, phosphine, phosphorus, KCN, KOH, Se, dibutyl ether
Nitrogen dioxide	Cyclohexane, fluorine, formaldehyde, alcohols, nitrobenzene, petroleum, toluene
Nitrogen triiodide	Acids, bromine, chlorine, hydrogen sulfide, ozone
α-Nitroguanidine	Complex salts of mercury and silver
Nitromethane	Acids, alkylmetal halides, hydroxides, hydrocarbons, organic amines, formaldehyde, nitric acid, perchlorates
1-Nitropropane	*See* under Nitromethane; chlorosulfonic acid, oleum
Nitrosyl fluoride	Haloalkenes, metals, nonmetals
Nitrosyl perchlorate	Acetone, amines, diethyl ether, metal salts, organic materials
Nitrourea	Mercury(II) and silver salts
Nitrous acid	Phosphine, phosphorus trichloride, silver nitrate, semicarbazone
Nitryl chloride	Ammonia, sulfur trioxide, tin(IV) bromide and iodide
Oxalic acid	Furfuryl alcohol, silver, mercury, sodium chlorate, sodium chlorite, sodium hypochlorite
Oxygen	Acetaldehyde, acetone, alcohols, alkali metals, alkaline earth metals, Al-Ti alloys, ether, carbon disulfide, halocarbons, hydrocarbons, metal hydrides, 1,3,5-trioxane
Ozone	Alkenes, aromatic compounds, bromine, diethyl ether, ethylene, HBr, HI, nitric oxide, nitrogen dioxide, rubber, stibine
Palladium	Arsenic, carbon, ozonides, sulfur, sodium tetrahydridoborate
Paraformaldehyde	Liquid oxygen
Paraldehyde	Alkalies, HCN, iodides, nitric acid, oxidizers

TABLE 11.35 Some Common Reactive and Incompatible Chemicals (*Continued*)

Chemical	Keep out of contact with
Pentaborane-9	Dimethylsulfoxide
Pentacarbonyliron	Acetic acid, nitric oxide, transition metal halides, water, zinc
2-Pentanone	Bromine trifluoride
3-Pentanone	Hydrogen peroxide, nitric acid
Perchlorates	Carbonaceous materials, finely divided metals particularly magnesium and aluminum, sulfur, benzene, olefins, ethanol, sulfur, sulfuric acid
Perchloric acid	Acetic acid, acetic anhydride, alcohols, antimony compounds, azo pigments, bismuth and its alloys, methanol, carbonaceous materials, carbon tetrachloride, cellulose, dehydrating agents, diethyl ether, glycols and glycolethers, HCl, HI, hypophosphites, ketones, nitric acid, pyridine, steel, sulfoxides, sulfuric acid
Permanganates	All reducing agents, organic materials
Peroxides	Reducing agents, organic materials, thiocyanates
Peroxoacetic acid	Acetic anhydride, olefins, organic matter
Peroxobenzoic acid	Olefins, reducing materials
Peroxoformic acid	Metals and nonmetals, organic materials
Peroxosulfuric acid	Acetone, alcohols, aromatic compounds, catalysts
Phenol	Butadiene, peroxodisulfuric acid, peroxosulfuric acid, aluminum chloride plus nitrobenzene
Phenylhydrazine	Lead dioxide, oxidizers
Phosgene	Aluminum, alkali metals, 2-propanol
Phosphine	Air, boron trichloride, bromine, chlorine, nitric acid, nitrogen oxides, nitrous acid, oxygen, silver nitrate
Phosphorus pentachloride	Aluminum, chlorine, chlorine dioxide, chlorine trioxide, fluorine, magnesium oxide, nitrobenzene, diphosphorus trioxide, potassium, sodium, urea, water
Phosphorus pentafluoride	Water or steam
Phosphorus pentasulfide	Air, alcohols, water
Phosphorus pentoxide	Formic acid, HF, inorganic bases, metals, oxidants, water
Phosphorus, red	Organic materials
Phosphorus tribromide	Potassium, ruthenium tetroxide, sodium, water
Phosphorus trichloride	Acetic acid, aluminum, chromyl dichloride, dimethylsulfoxide, hydroxylamine, lead dioxide, nitric acid, nitrous acid, organic matter, potassium, sodium, water
Phosphorus, white	Air, oxidants of all types, halogens, metals
Phosphoryl chloride	Carbon disulfide, *N,N*-dimethylformamide, 2,5-dimethylpyrrole, 2,6-dimethylpyridine 1-oxide, dimethylsulfoxide, water, zinc
Phthalic acid	Nitric acid, sodium nitrite
Piperazine	Oxidizers
Platinum	Acetone, arsenic, hydrazine, lithium, peroxosulfuric acid, phosphorus, selenium, tellurium
Potassium	*See* under Alkali metals
Potassium *tert*-butoxide	Organic compounds, sulfuric acid
Potassium hydride	Air, chlorine, acetic acid, acrolein, acrylonitrile, maleic anhydride, nitroparaffins, *N*-nitrosomethylurea, tetrahydrofuran, water
Potassium perchlorate	Aluminum plus magnesium, carbon, nickel plus titanium, reducing agents, sulfur, sulfuric acid
Potassium permanganate	Organic or readily oxidizable materials
Potassium sodium alloy	Air, carbon dioxide, carbon disulfide, halocarbons, metal oxides
2-Propyn-l-ol	Alkali metals, mercury(II) sulfate, oxidizing materials, phosphorus pentoxide, sulfuric acid
Pyridine	Chlorosulfonic acid, chromium trioxide, formamide, maleic anhydride, nitric acid, oleum, perchromates, silver perchlorate, sulfuric acid
Pyrrolidine	Oxidizing materials

TABLE 11.35   Some Common Reactive and Incompatible Chemicals (*Continued*)

Chemical	Keep out of contact with
Quinoline	Dinitrogen tetroxide, linseed oil, maleic anhydride, thionyl chloride
Salicylic acid	Iodine, iron salts, lead acetate
Silicon	Alkali carbonates, calcium, chlorine, cobalt(II) fluoride, manganese trifluoride, oxidants, silver fluoride, sodium-potassium alloy
Silver	Acetylene, ammonium compounds, ethyleneimine, hydrogen peroxide, oxalic acid, sulfuric acid, tartaric acid
Sodium	*See* under Alkali metals
Sodium peroxide	Glacial acetic acid, acetic anhydride, aniline, benzene, benzaldehyde, carbon disulfide, diethyl ether, ethanol or methanol, ethylene glycol, ethyl acetate, furfural, glycerol, metals, methyl acetate, organic matter
Sulfides	Acids, powerful oxidizers, moisture
Sulfur	Oxidizing materials, halogens
Sulfur dioxide	Halogens, metal oxides, polymeric tubing, potassium chlorate, sodium hydride
Sulfuric acid	Chlorates, metals, HCl, organic materials, perchlorates, permanganates, water
Sulfuryl dichloride	Alkalis, diethyl ether, dimethylsulfoxide, dinitrogen tetroxide, lead dioxide, phosphorus
Tellurium	Halogens, metals
Tetrahydrofuran	Tetrahydridoaluminates, KOH, NaOH
Tetranitroaniline	Reducing materials
Tetranitromethane	Aluminum, cotton, aromatic nitro compounds, hydrocarbons, cotton, toluene
Thiocyanates	Chlorates, nitric acid, peroxides
Thionyl chloride	Ammonia, dimethylsulfoxide, linseed oil, quinoline, sodium
Thiophene	Nitric acid
Thymol	Acetanilide, antipyrine, camphor, chlorohydrate, menthol, quinine sulfate, urethene
Tin(II) chloride	Boron trifluoride, ethylene oxide, hydrazine hydrate, nitrates, Na, K, hydrogen peroxide
Tin(IV) chloride	Alkyl nitrates, ethylene oxide, K, Na, turpentine
Titanium	Aluminum, boron trifluoride, carbon dioxide, CuO, halocarbons, halogens, PbO, nitric acid, potassium chlorate, potassium nitrate, potassium permanganate, steam at high temperatures, water
Toluene	Sulfuric plus nitric acids, nitrogen dioxide, silver perchlorate, uranium hexafluoride
Toluidines	Nitric acid
2,4,6-Trinitrotoluene	Sodium dichromate, sulfuric acid
1,3,5-Trioxane	Oxidizing materials, acids
Urea	Sodium nitrite, phosphorus pentachloride
Vinylidene chloride	Chlorosulfonic acid, nitric acid, oleum

TABLE 11.36 Chemicals Recommended for Refrigerated Storage

A. Due to chemical decomposition or polymerization

Acetaldehyde	Isoprene
Acrolein	Lecithin
Adenosinetriphosphoric acid	Mercaptoacetic acid
Bromacetaldehyde, diethyl acetal	Methyl acrylate
Bromosuccinimide	2-Methyl-1-butene
3-Buten-2-one	Methylenedi-1,4-phenylene diisocyanate
tert-Butyl hydroperoxide	4-Methyl-1-pentene
2-Chlorocyclohexanone	α-Methylstyrene
Cupferron	1-Naphthyl isocyanate
1,3-Cyclohexadiene	1-Pentene
1,3-Dihydroxy-2-propanone	Isopentyl acetate
Divinylbenzene	Pyruvic acid
Ethyl methacrylate, monomer	Styrene, stabilized
Glutathione	Tetramethylsilane
Glycidol	Thioacetamide
Histamine, base	Veratraldehyde
Hydrocinnamaldehyde	Vitamin E (and the acetate)

B. Due to flammability and high volatility

Acetaldehyde	Iodomethane
Bromoethane	Isoprene
tert-Butylamine	Isopropylamine
Carbon disulfide	Methylal
1-Chloropropane	2-Methylbutane
3-Chloropropane	2-Methyl-2-butene
Cyclopentane	Methyl formate
Diethyl ether	Pentane
2,2-Dimethylbutane	Propylamine
Dimethyl sulfide	Propylene oxide
Furan	Trichlorosilane

TABLE 11.37 Chemicals Which Polymerize or Decompose on Extended Refrigeration

Formaldehyde	Sodium methoxide
Hydrogen peroxide	Sodium nitrate
Sodium chlorite [sodium chlorate (IV)]	Sodium peroxide
Sodium chromate(VI)	Strontium nitrate
Sodium dithionite	Urea
Sodium ethoxide	

11.8 SIEVES AND SCREENS

TABLE 11.38 U.S. Standard Sieve Series

Sieve no.	Sieve opening		Sieve no.	Sieve opening	
	mm	inch		mm	inch
	125	5.00	10	2.00	0.0787
	106	4.24	12	1.70	0.0661
	90	3.50	14	1.40	0.0555
	75	3.00	16	1.18	0.0469
	63	2.50	18	1.00	0.0394
	53	2.12	20	0.850	0.0331
	45	1.75	25	0.710	0.0278
	37.5	1.50	30	0.600	0.0234
	31.5	1.25	35	0.500	0.0197
	26.5	1.06	40	0.425	0.0165
	22.4	0.875	45	0.355	0.0139
	19.0	0.75	50	0.300	0.0117
	16.0	0.625	60	0.250	0.0098
	13.2	0.530	70	0.212	0.0083
	11.2	0.438	80	0.180	0.0070
	9.5	0.375	100	0.150	0.0059
	8.0	0.312	120	0.125	0.0049
	6.7	0.265	140	0.106	0.0041
3.5	5.60	0.223	170	0.090	0.0035
4	4.75	0.187	200	0.075	0.0029
5	4.00	0.157	230	0.063	0.0025
6	3.35	0.132	270	0.053	0.0021
7	2.80	0.111	325	0.045	0.0017
8	2.36	0.0937	400	0.038	0.0015

Specifications are from ASTM E.11-81/ISO 565. The sieve numbers are the approximate number of openings per linear inch.

11.9 THERMOMETRY

11.9.1 Temperature and Its Measurement

The new international temperature scale, known as ITS-90, was adopted in September 1989. However, neither the definition of thermodynamic temperature nor the definition of the kelvin or the Celsius temperature scales has changed; it is the way in which we are to realize these definitions that has changed. The changes concern the recommended thermometers to be used in different regions of the temperature scale and the list of secondary standard fixed points. The changes in temperature determined using ITS-90 from the previous IPTS-68 are always less than 0.4 K, and almost always less than 0.2 K, over the range 0–1300 K.

The ultimate definition of thermodynamic temperature is in terms of pV (pressure $\times$ volume) in a gas thermometer extrapolated to low pressure. The kelvin (K), the unit of thermodynamic temperature, is defined by specifying the temperature of one fixed point on the scale—the triple point of water which is defined to be 273.16 K. The Celsius temperature scale (°C) is defined by the equation

$$°C = K - 273.15$$

where the freezing point of water at 1 atm is 273.15 K.

The fixed points in the ITS-90 are given in Table 11.39. Platinum resistance thermometers are recommended for use between 14 K and 1235 K (the freezing point of silver), calibrated against the fixed points. Below 14 K either the vapor pressure of helium or a constant-volume gas thermometer is to be used. Above 1235 K radiometry is to be used in conjunction with the Planck radiation law,

$$L_\lambda = c_1 \lambda^{-5} (e^{c_2/\lambda T} - 1)^{-1}$$

where L_λ is the spectral radiance at wavelength λ. The first radiation constant, c_1, is $3.741\ 83 \times 10^{-16}\ \text{W} \cdot \text{m}^2$ and the second radiation constant, c_2, has a value of $0.014\ 388\ \text{m} \cdot \text{K}$.

TABLE 11.39 Fixed Points in the ITS-90

Fixed points	T, K	t, °C
Triple point of hydrogen	13.8033	−259.3467
Boiling point of hydrogen at 33 321.3 Pa	17.035	−256.115
Boiling point of hydrogen at 101 292 Pa	20.27	−252.88
Triple point of neon	24.5561	−248.5939
Triple point of oxygen	54.3584	−218.7916
Triple point of argon	83.8058	−189.3442
Triple point of mercury	234.3156	−38.8344
Triple point of water	273.16	0.01
Melting point of gallium	302.9146	29.7646
Freezing point of indium	429.7458	156.5985
Freezing point of tin	505.078	231.928
Freezing point of zinc	692.677	419.527
Freezing point of aluminum	933.473	660.323
Freezing point of silver	1234.93	961.78
Freezing point of gold	1337.33	1064.18
Freezing point of copper	1357.77	1084.62
Secondary reference points to extend the scale (IPTS-68):		
Freezing point of platinum	2042	1769
Freezing point of rhodium	2236	1963
Freezing point of iridium	2720	2447
Melting point of tungsten	3660	3387

11.10 THERMOCOUPLES

The thermocouple reference data in Tables 11.40 to 11.48 give the thermoelectric voltage in millivolts with the reference junction at 0°C. Note that the temperature for a given entry is obtained by adding the corresponding temperature in the top row to that in the left-hand column, regardless of whether the latter is positive or negative.

The noble metal thermocouples, Types B, R, and S, are all platinum or platinum-rhodium thermocouples and hence share many of the same characteristics. Metallic vapor diffusion at high temperatures can readily change the platinum wire calibration, hence platinum wires should only be used inside a nonmetallic sheath such as high-purity alumina.

Type B thermocouples (Table 11.41) offer distinct advantages of improved stability, increased mechanical strength, and higher possible operating temperatures. They have the unique advantage that the reference junction potential is almost immaterial, as long as it is between 0°C and 40°C. Type B is virtually useless below 50°C because it exhibits a double-value ambiguity from 0°C to 42°C.

Type E thermoelements (Table 11.42) are very useful down to about liquid hydrogen temperatures and may even be used down to liquid helium temperatures. They are the most useful of the commercially standardized thermocouple combinations for subzero temperature measurements because of their high Seebeck coefficient (58 μV/°C), low thermal conductivity, and corrosion resistance. They also have the largest Seebeck coefficient (voltage response per degree Celsius) above 0°C of any of the standardized thermocouples which makes them useful for detecting small temperature changes. They are recommended for use in the temperature range from −250 to 871°C in oxidizing or inert atmospheres. They should not be used in sulfurous, reducing, or alternately reducing and oxidizing atmospheres unless suitably protected with tubes. They should not be used in vacuum at high temperatures for extended periods of time.

Type J thermocouples (Table 11.43) are one of the most common types of industrial thermocouples because of the relatively high Seebeck coefficient and low cost. They are recommended for use in the temperature range from 0 to 760°C (but never above 760°C due to an abrupt magnetic transformation that can cause decalibration even when returned to lower temperatures). Use is permitted in vacuum and in oxidizing, reducing, or inert atmospheres, with the exception of sulfurous atmospheres above 500°C. For extended use above 500°C, heavy-gauge wires are recommended. They are not recommended for subzero temperatures. These thermocouples are subject to poor conformance characteristics because of impurities in the iron.

The Type K thermocouple (Table 11.44) is more resistant to oxidation at elevated temperatures than the Type E, J, or T thermocouple, and consequently finds wide application at temperatures above 500°C. It is recommended for continuous use at temperatures within the range −250 to 1260°C in inert or oxidizing atmospheres. It should not be used in sulfurous or reducing atmospheres, or in vacuum at high temperatures for extended times.

The Type N thermocouple (Table 11.45) is similar to Type K but it has been designed to minimize some of the instabilities in the conventional Chromel-Alumel combination. Changes in the alloy content have improved the order/disorder transformations occurring at 500°C and a higher silicon content of the positive element improves the oxidation resistance at elevated temperatures.

The Type R thermocouple (Table 11.46) was developed primarily to match a previous platinum – 10% rhodium British wire which was later found to have 0.34% iron impurity in the rhodium. Comments on Type S also apply to Type R.

The Type S thermocouple (Table 11.47) is so stable that it remains the standard for determining temperatures between the antimony point (630.74°C) and the gold point (1064.43°C). The other fixed point used is that of silver. The Type S thermocouple can be used from −50°C continuously up to about 1400°C, and intermittently at temperatures up to the freezing point of platinum (1769°C). The thermocouple is most reliable when used in a clean oxidizing atmosphere, but may also be used in inert gaseous atmospheres or in a vacuum for short periods of time. It should not be used in reducing atmospheres, nor in those containing

TABLE 11.40 Thermoelectric Values in Millivolts at Fixed Points for Various Thermocouples

Abbreviations Used in the Table

FP, freezing point
NBP, normal boiling point
BP, boiling point
TP, triple point

Fixed point	°C	Type B	Type E	Type J	Type K	Type N	Type R	Type S	Type T
Helium NPB	−268.934		−9.8331		−6.4569	−4.345			−6.2563
Hydrogen TP*	−259.347*		−9.7927		−6.4393	−4.334			−6.2292
Hydrogen NBP	−252.88*		−9.7447		−6.4167	−4.321			−6.1977
Neon TP	−248.594*		−9.7046		−6.3966	−4.271			−6.1714
Neon NBP	−246.048		−9.6776		−6.3827	−4.300			−6.1536
Oxygen TP	−218.792*		−9.2499		−6.1446	−4.153			−5.8730
Nitrogen TP	−210.001		−9.0629	−8.0957	−6.0346	−4.083			−5.7533
Nitrogen NBP	−195.802		−8.7168	−7.7963	−5.8257	−3.947			−5.5356
Oxygen NBP	−182.962		−8.3608	−7.4807	−5.6051	−3.802			−5.3147
Carbon dioxide SP	−78.474		−4.2275	−3.7187	−2.8696	−1.939			−2.7407
Mercury TP*	−38.834*		−2.1930	−1.4849		−0.985	−0.1830	−0.1895	−1.4349
Ice point	0.000	−0.000	0.000	0.000	0.000	0.000	0.000	0.000	0.000
Diphenyl ether TP	26.87	−0.0024	1.6091	1.3739	1.076	0.698	0.1517	0.1537	1.0679
Water BP	100.00	0.0332	6.3171	5.2677	4.0953	2.774	0.6472	0.6453	4.2773
Benzoic acid TP	122.37	0.0561	7.8468	6.4886	5.0160	3.446	0.8186	0.8129	5.3414
Indium FP	156.598*	0.1019	10.260	8.3743	6.0404	4.508	1.0956	1.0818	7.0364
Tin FP	231.928*	0.2474	15.809	12.552	9.4201	6.980	1.7561	1.7146	11.013
Bismuth FP	271.442	0.3477	18.821	14.743	11.029	8.336	2.1250	2.0640	13.219
Cadmium FP	321.108	0.4971	22.684	17.493	13.085	10.092	2.6072	2.5167	16.095
Lead FP	327.502	0.5182	23.186	17.846	13.351	10.322	2.6706	2.5759	16.473
Mercury BP	356.66	0.6197	25.489	19.456	14.571		2.9630	2.8483	18.218
Zinc FP	419.527*	0.8678	30.513	22.926	17.223		3.6113	3.4479	
Cu-Al eutectic FP	548.23	1.4951	40.901	30.109	22.696		5.0009	4.7140	
Antimony FP	630.74	1.9784	47.561	34.911	26.207		5.9331	5.5521	
Aluminum FP	660.37	2.1668	49.941	36.693	27.461		6.2759	5.8591	
Silver FP	961.93*	4.4908	73.495	55.669	39.779		10.003	9.1482	
Gold FP	1064.43*	5.4336		61.716	43.755		11.364	10.334	
Copper FP	1084.5	5.6263		62.880	44.520		11.635	10.570	
Nickel FP	1455	9.5766					16.811	15.034	
Cobalt FP	1494	10.025					17.360	15.504	
Palladium FP	1554	10.721					18.212	16.224	
Platinum FP	1772	13.262					21.103	18.694	

* Defining fixed points of the International Temperature Scale of 1990 (ITS-90). Except for the triple points, the assigned values of temperature are for equilibrium states at a pressure of one standard atmosphere (101 325 Pa).

TABLE 11.41 Type B Thermocouples: Platinum–30% Rhodium Alloy vs. Platinum–6% Rhodium Alloy

Thermoelectric voltage in millivolts; reference junction at 0°C.

°C	0	10	20	30	40	50	60	70	80	90
0	0.00	−0.0019	−0.0026	−0.0021	−0.0005	0.0023	0.0062	0.0112	0.0174	0.0248
100	0.0332	0.0427	0.0534	0.0652	0.0780	0.0920	0.1071	0.1232	0.1405	0.1588
200	0.1782	0.1987	0.2202	0.2428	0.2665	0.2912	0.3170	0.3438	0.3717	0.4006
300	0.4305	0.4615	0.4935	0.5266	0.5607	0.5958	0.6319	0.6690	0.7071	0.7462
400	0.7864	0.8275	0.8696	0.9127	0.9567	1.0018	1.0478	1.0948	1.1427	1.1916
500	1.2415	1.2923	1.3440	1.3967	1.4503	1.5048	1.5603	1.6166	1.6739	1.7321
600	1.7912	1.8512	1.9120	1.9738	2.0365	2.1000	2.1644	2.2296	2.2957	2.3627
700	2.4305	2.4991	2.5686	2.6390	2.7101	2.7821	2.8548	2.9284	3.0028	3.0780
800	3.1540	3.2308	3.3084	3.3867	3.4658	3.5457	3.6264	3.7078	3.7899	3.8729
900	3.9565	4.0409	4.1260	4.2119	4.2984	4.3857	4.4737	4.5624	4.6518	4.7419
1000	4.8326	4.9241	5.0162	5.1090	5.2025	5.2966	5.3914	5.4868	5.5829	5.6796
1100	5.7769	5.8749	5.9734	6.0726	6.1724	6.2728	6.3737	6.4753	6.5774	6.6801
1200	6.7833	6.8871	6.9914	7.0963	7.2017	7.3076	7.4140	7.5210	7.6284	7.7363
1300	7.8446	7.9534	8.0627	8.1724	8.2826	8.3932	8.5041	8.6155	8.7273	8.8394
1400	8.9519	9.0648	9.1780	9.2915	9.4053	9.5194	9.6338	9.7485	9.8634	9.9786
1500	10.0940	10.2097	10.3255	10.4415	10.5577	10.6740	10.7905	10.9071	11.0237	11.1405
1600	11.2574	11.3743	11.4913	11.6082	11.7252	11.8422	11.9591	12.0761	12.1929	12.3100
1700	12.4263	12.5429	12.6594	12.7757	12.8918	13.0078	13.1236	13.2391	13.3545	13.4696
1800	13.5845	13.6991	13.8135							

TABLE 11.42 Type E Thermocouples: Nickel-Chromium Alloy vs. Copper-Nickel Alloy

Thermoelectric voltage in millivolts; reference junction at 0°C.

°C	0	10	20	30	40	50	60	70	80	90
−200	−8.824	−9.063	−9.274	−9.455	−9.604	−9.719	−9.797	−9.835		
−100	−5.237	−5.680	−6.107	−6.516	−6.907	−7.279	−7.631	−7.963	−8.273	−8.561
−0	0.000	−0.581	−1.151	−1.709	−2.254	−2.787	−3.306	−3.811	−4.301	−4.777
0	0.000	0.591	1.192	1.801	2.419	3.047	3.683	4.394	4.983	5.646
100	6.317	6.996	7.683	8.377	9.078	9.787	10.501	11.222	11.949	12.681
200	13.419	14.161	14.909	15.661	16.417	17.178	17.942	18.710	19.481	20.256
300	21.033	21.814	22.597	23.383	24.171	24.961	25.754	26.549	27.345	28.143
400	28.943	29.744	30.546	31.350	32.155	32.960	33.767	34.574	35.382	36.190
500	36.999	37.808	38.617	39.426	40.236	41.045	41.853	42.662	43.470	44.278
600	45.085	45.891	46.697	47.502	48.306	49.109	49.911	50.713	51.513	52.312
700	53.110	53.907	54.703	55.498	56.291	57.083	57.873	58.663	59.451	60.237
800	61.022	61.806	62.588	63.368	64.147	64.924	65.700	66.473	67.245	68.015
900	68.783	69.549	70.313	71.075	71.835	72.593	73.350	74.104	74.857	75.608
1000	76.358									

TABLE 11.43 Type J Thermocouples: Iron vs. Copper-Nickel Alloy
Thermoelectric voltage in millivolts; reference junction at 0°C.

°C	0	10	20	30	40	50	60	70	80	90
−200	−7.890									
−100	−4.632	−8.096								
−0	0.000	−5.036	−5.426	−5.801	−6.159	−6.499	−6.821	−7.122	−7.402	−7.659
0	0.000	−0.501	−0.995	−1.481	−1.960	−2.431	−2.892	−3.344	−3.785	−4.215
100	5.268	0.507	1.019	1.536	2.058	2.585	3.115	3.649	4.186	4.725
200	10.777	5.812	6.359	6.907	7.457	8.008	8.560	9.113	9.667	10.222
300	16.325	11.332	11.887	12.442	12.998	13.553	14.108	14.663	15.217	15.771
400	21.846	16.879	17.432	17.984	18.537	19.089	19.640	20.192	20.743	21.295
500	27.388	22.397	22.949	23.501	24.054	24.607	25.161	25.716	26.272	26.829
600	33.096	27.949	28.511	29.075	29.642	30.210	30.782	31.356	31.933	32.513
700	39.130	33.683	34.273	34.867	35.464	36.066	36.671	37.280	37.893	38.510
		39.754	40.482	41.013	41.647	42.283	42.922			

TABLE 11.44 Type K Thermocouples: Nickel-Chromium Alloy vs. Nickel-Aluminum Alloy

Thermoelectric voltage in millivolts; reference junction at 0°C.

°C	0	10	20	30	40	50	60	70	80	90
−200	−5.891	−6.035	−6.158	−6.262	−6.344	−6.404	−6.441	−6.458		
−100	−3.553	−3.852	−4.138	−4.410	−4.669	−4.912	−5.141	−5.354	−5.550	−5.730
−0	0.000	−0.392	−0.777	−1.156	−1.517	−1.889	−2.243	−2.586	−2.920	−3.242
0	0.000	0.397	0.798	1.203	1.611	2.022	2.436	2.850	3.266	3.681
100	4.095	4.508	4.919	5.327	5.733	6.137	6.539	6.939	7.338	7.737
200	8.137	8.537	8.938	9.341	9.745	10.151	10.560	10.969	11.381	11.793
300	12.207	12.623	13.039	13.456	13.874	14.292	14.712	15.132	15.552	15.974
400	16.395	16.818	17.241	17.664	18.088	18.513	18.839	19.363	19.788	20.214
500	20.640	21.066	21.493	21.919	22.346	22.772	23.198	23.624	24.050	24.476
600	24.902	25.327	25.751	26.176	26.599	27.022	27.445	27.867	28.288	28.709
700	29.128	29.547	29.965	30.383	30.799	31.214	31.629	32.042	32.455	32.866
800	33.277	33.686	34.095	34.502	34.909	35.314	35.718	36.121	36.524	36.925
900	37.325	37.724	38.095	38.519	38.915	39.310	39.703	40.096	40.488	40.879
1000	41.269	41.657	42.045	42.432	42.817	43.202	43.585	43.968	44.349	44.729
1100	45.108	45.486	45.863	46.238	46.612	46.985	47.356	47.726	48.095	48.462
1200	48.828	49.192	49.555	49.916	50.276	50.633	50.990	51.344	51.697	52.049
1300	52.398	52.747	53.093	53.439	53.782	54.125	54.466	54.807		

TABLE 11.45 Type N Thermocouples: Nickel–14.2% Chromium–1.4% Silicon Alloy vs. Nickel–4.4% Silicon–0.1% Magnesium Alloy

Thermoelectric voltage in millivolts; reference junction at 0°C.

°C	0	10	20	30	40	50	60	70	80	90
−200	−3.990	−4.083	−4.162	−4.227	−4.277	−4.313	−4.336	−4.345		
−100	−2.407	−2.612	−2.807	−2.994	−3.170	−3.336	−3.491	−3.634	−3.766	−3.884
−0	0.000	−0.260	−0.518	−0.772	−1.023	−1.268	−1.509	−1.744	−1.972	−2.193
0	0.000	0.261	0.525	0.793	1.064	1.339	1.619	1.902	2.188	2.479
100	2.774	3.072	3.374	3.679	3.988	4.301	4.617	4.936	5.258	5.584
200	5.912	6.243	6.577	6.914	7.254	7.596	7.940	8.287	8.636	8.987
300	9.340	9.695	10.053	10.412	10.772	11.135	11.499	11.865	12.233	12.602
400	12.972	13.344	13.717	14.091	14.467	14.844	15.222	15.601	15.981	16.362
500	16.744	17.127	17.511	17.896	18.282	18.668	19.055	19.443	19.831	20.220
600	20.609	20.999	21.390	21.781	22.172	22.564	22.956	23.348	23.740	24.133
700	24.526	24.919	25.312	25.705	26.098	26.491	26.885	27.278	27.671	28.063
800	28.456	28.849	29.241	29.633	30.025	30.417	30.808	31.199	31.590	31.980
900	32.370	32.760	33.149	33.538	33.926	34.315	34.702	35.089	35.476	35.862
1000	36.248	36.633	37.018	37.402	37.786	38.169	38.552	38.934	39.315	39.696
1100	40.076	40.456	40.835	41.213	41.590	41.966	42.342	42.717	43.091	43.464
1200	43.836	44.207	44.577	44.947	45.315	45.682	46.048	46.413	46.777	47.140
1300	47.502									

metallic vapor (such as lead or zinc), nonmetallic vapors (such as arsenic, phosphorus, or sulfur), or easily reduced oxides, unless suitably protected with nonmetallic protecting tubes.

The Type T thermocouple (Table 11.48) is popular for the temperature region below 0°C (but see under Type E). It can be used in vacuum, or in oxidizing, reducing, or inert atmospheres.

TABLE 11.46 Type R Thermocouples: Platinum–13% Rhodium Alloy vs. Platinum

Thermoelectric voltage in millivolts; reference junction at 0°C.

°C	0	10	20	30	40	50	60	70	80	90
(Below zero)										
0	0.0000	−0.0515	−0.100	−0.1455	−0.1877	−0.2264				
0	0.0000	0.0543	0.1112	0.1706	0.2324	0.2965	0.3627	0.4310	0.5012	0.5733
100	0.6472	0.7228	0.8000	0.8788	0.9591	1.0407	1.1237	1.2080	1.2936	1.3803
200	1.4681	1.5571	1.6471	1.7381	1.8300	1.9229	2.0167	2.1113	2.2068	2.3030
300	2.4000	2.4978	2.5963	2.6954	2.7953	2.8957	2.9968	3.0985	3.2009	3.3037
400	3.4072	3.5112	3.6157	3.7208	3.8264	3.9325	4.0391	4.1463	4.2539	4.3620
500	4.4706	4.5796	4.6892	4.7992	4.9097	5.0206	5.1320	5.2439	5.3562	5.4690
600	5.5823	5.6960	5.8101	5.9246	6.0398	6.1554	6.2716	6.3883	6.5054	6.6230
700	6.7412	6.8598	6.9789	7.0984	7.2185	7.3390	7.4600	7.5815	7.7035	7.8259
800	7.9488	8.0722	8.1960	8.3203	8.4451	8.5703	8.6960	8.8222	8.9488	9.0758
900	9.2034	9.3313	9.4597	9.5886	9.7179	9.8477	9.9779	10.1086	10.2397	10.3712
1000	10.5032	10.6356	10.7684	10.9017	11.0354	11.1695	11.3041	11.4391	11.5745	11.7102
1100	11.8463	11.9827	12.1194	12.2565	12.3939	12.5315	12.6695	12.8077	12.9462	13.0849
1200	13.2239	13.3631	13.5025	13.6421	13.7818	13.9218	14.0619	14.2022	14.3426	14.4832
1300	14.6239	14.7647	14.9056	15.0465	15.1876	15.3287	15.4699	15.6110	15.7522	15.8935
1400	16.0347	16.1759	16.3172	16.4583	16.5995	16.7405	16.8816	17.0225	17.1634	17.3041
1500	17.4447	17.5852	17.7256	17.8659	18.0059	18.1458	18.2855	18.4251	18.5644	18.7035
1600	18.8424	18.9810	19.1194	19.2575	19.3953	19.5329	19.6702	19.8071	19.9437	20.0797
1700	20.2151	20.3497	20.4834	20.6161	20.7475	20.8777	21.0064			

TABLE 11.47 Type S Thermocouples: Platinum – 10% Rhodium Alloy vs. Platinum

Thermoelectric voltage in millivolts; reference junction at 0°C.

°C	0	10	20	30	40	50	60	70	80	90
(Below zero)										
0		-0.0527	-0.1028	-0.1501	-0.1944	-0.2357				
0	0.0000	0.0552	0.1128	0.1727	0.2347	0.2986	0.3646	0.4323	0.5017	0.5728
100	0.6453	0.7194	0.7948	0.8714	0.9495	1.0287	1.1089	1.1902	1.2726	1.3558
200	1.4400	1.5250	1.6109	1.6975	1.7849	1.8729	1.9617	2.0510	2.1410	2.2316
300	2.3227	2.4143	2.5065	2.5991	2.6922	2.7858	2.8798	2.9742	3.0690	3.1642
400	3.2597	3.3557	3.4519	3.5485	3.6455	3.7427	3.8403	3.9382	4.0364	4.1348
500	4.2336	4.3327	4.4320	4.5316	4.6316	4.7318	4.8323	4.9331	5.0342	5.1356
600	5.2373	5.3394	5.4417	5.5445	5.6477	5.7513	5.8553	5.9595	6.0641	6.1690
700	6.2743	6.3799	6.4858	6.5920	6.6986	6.8055	6.9127	7.0202	7.1281	7.2363
800	7.3449	7.4537	7.5629	7.6724	7.7823	7.8925	8.0030	8.1138	8.2250	8.3365
900	8.4483	8.5605	8.6730	8.7858	8.8989	9.0124	9.1262	9.2403	9.3548	9.4696
1000	9.5847	9.7002	9.8159	9.9320	10.0485	10.1652	10.2823	10.3997	10.5174	10.6354
1100	10.7536	10.8720	10.9907	11.1095	11.2286	11.3479	11.4674	11.5871	11.7069	11.8269
1200	11.9471	12.0674	12.1878	12.3084	12.4290	12.5498	12.6707	12.7917	12.9127	13.0338
1300	13.1550	13.2762	13.3975	13.5188	13.6401	13.7614	13.8828	14.0041	14.1254	14.2467
1400	14.3680	14.4892	14.6103	14.7314	14.8524	14.9734	15.0942	15.2150	15.3356	15.4561
1500	15.5765	15.6967	15.8168	15.9368	16.0566	16.1762	16.2956	16.4148	16.5338	16.6526
1600	16.6712	16.8895	17.0076	17.1255	17.2431	17.3604	17.4474	17.5942	17.7105	17.8264
1700	17.9417	18.0562	18.1698	18.2823	18.3937	18.5038	18.6124			

11.126

TABLE 11.48 Type T Thermocouples: Copper vs. Copper-Nickel Alloy

Thermoelectric voltage in millivolts; reference junction at 0°C.

°C	0	10	20	30	40	50	60	70	80	90
−200	−5.603	−5.753	−5.889	−6.007	−6.105	−6.181	−6.232	−6.258		
−100	−3.378	−3.656	−3.923	−4.177	−4.419	−4.648	−4.865	−5.069	−5.261	−5.439
−0	0.000	−0.383	−0.757	−1.121	−1.475	−1.819	−2.152	−2.475	−2.788	−3.089
0	0.000	0.391	0.789	1.196	1.611	2.035	2.467	2.908	3.357	3.813
100	4.277	4.749	5.227	5.712	6.204	6.702	7.207	7.718	8.235	8.757
200	9.286	9.820	10.360	10.905	11.456	12.011	12.572	13.137	13.707	14.281
300	14.860	15.443	16.030	16.621	17.217	17.816	18.420	19.027	19.638	20.252
400	20.869									

11.11 CORRECTION FOR EMERGENT STEM OF THERMOMETERS

When a thermometer which has been standardized for total immersion is used with a part of the liquid column at a temperature below that of the bulb, the reading is low and a correction must be applied. The stem correction, in degrees Celsius, is given by

$$KL(t_o - t_m) = \text{degrees Celsius}$$

where K = constant, characteristic of the particular kind of glass and temperature (see Table 11.49)

L = length of exposed thermometer, °C (that is, the length not in contact with vapor or liquid being measured)

t_o = observed temperature on thermometer

t_m = mean temperature of exposed column (obtained by placing an auxiliary thermometer alongside with its bulb midpoint)

For thermometers containing organic liquids, it is sufficient to use the approximate value, $K = 0.001$. In such thermometers the value of K is practically independent of the kind of glass.

TABLE 11.49 Values of K for Stem Correction of Thermometers

Temperature, °C	Soft glass	Heat-resistant glass
0–150	0.000 158	0.000 165
200	0.000 159	0.000 167
250	0.000 161	0.000 170
300	0.000 164	0.000 174
350		0.000 178
400		0.000 183
450		0.000 188

INDEX

ABOUT THE EDITOR

John A. Dean assumed editorship of *Lange's Handbook of Chemistry* in 1968 with the Eleventh Edition. He is currently Professor Emeritus of Chemistry at the University of Tennessee at Knoxville. The author of nine major chemistry reference books used throughout the world, John Dean's research interests, reflected in over 100 research papers and scholarly publications, include instrumental methods of analysis, flame emission and atomic absorption spectroscopy, chromatographic and solvent extraction methods, and polarography. He received his B.S., M.S., and Ph.D. in Chemistry from the University of Michigan. In 1991, he was awarded the Distinguished Service Award by the Society for Applied Spectroscopy.